1993 EDITION

WORLD RADIO TV Handbook®

WRTH

LISTEN TO THE WORLD...

hear the voices of people making history . . .

eavesdrop on local programs from around the globe . . .

discover music and cultures you never knew existed . . .

gain a new perspective on international events . . .

the World Radio TV Handbook (WRTH) is your guide to an exciting new world— the world of the International Listener!

This unique handbook is your personal 24-hour passport to:

- The world's broadcasters and their services listed by country.
- A special hour-by-hour guide to broadcasts in English directed to your area.
- Essential station information including frequencies, transmitter powers, operating times, languages, addresses, etc.
- Listings of stations in frequency order to help you identify them more easily.
- Maps of principal transmitter sites worldwide.
- Names and addresses of international radio listeners' clubs.
- Information on reception conditions, Time Signal Stations and other specialized subjects.
- Widely acclaimed annual test reports on receivers for the international listener.

The WRTH is easy to use. Begin your worldwide listening enjoyment by reading the User's Guide in your preferred language. (English, French, German or Spanish).

MESSAGE FROM THE EDITOR

Dear friends,

This edition of the WRTH will probably look unfamiliar to you. Not because we have made any drastic changes to the format of the book, but because within its pages you will see many references to countries and stations which did not exist when last year's edition of the WRTH was printed.

Keeping abreast of the all the changes, which have been happening at breathtaking speed, needs much time and dedication. Fortunately, the WRTH has never been short of collaborators with this sort of dedication - information has been pouring in by letter, fax and electronic mail throughout 1992.

At the same time, we have also been busy reorganizing the editorial assignments and procedures, an ongoing situation resulting from the progressive introduction of more and more automation. I would like to thank all members of our editorial team for their patience and cooperation.

A special word of appreciation, too, for one member of our team who has "retired from active service" after 21 years as a contributing editor. Tor-Henrik Ekblom has been loyal servant of WRTH, and his decision was only made following a well-deserved promotion at work.

Finally, it is now more than 20 years since a close working relationship was established between the BBC Monitoring Service, as it was then called, and the WRTH editorial office. In recognition of this, we presented a special WRTH award to BBC Monitoring last September. The fruits of the relationship may be found throughout the pages of WRTH. As always, I hope this edition helps you to get more enjoyment from your listening and viewing during 1993.

ANDREW G. SENNITT

VOLUME 47 · 1993

Editorial Office:
P.O. Box 9027
1006 AA Amsterdam
The Netherlands

Sales Offices:
BPI Communications
1515 Broadway, New York 10036, NY
USA

BPI Europe
23 Ridgmount Street,
London WC1E 7AH
United Kingdom

J. M. Frost
Søliljevej 44
DK-2650 Hvidovre
Denmark

CONTENTS

SHORTWAVE RECEPTION CONDITIONS

DX AND MEDIA PROGRAMMES 57

LONG & MEDIUMWAVE FREQUENCY TABLE

SHORTWAVE FREQUENCY TABLE 539

ORGANIZATIONS

INTERNATIONAL BROADCASTS IN ENGLISH 566

LISTEN TO THE WORLD 575

INFORMATION ABOUT THE WRTH

USER'S GUIDE TO THE WRTH

The World Radio TV Handbook (WRTH) is your practical guide to the World's radio and television services. Using the WRTH, you can tune into broadcasts of interest from all over the world, and identify stations you have not heard before. To get maximum benefit from the wealth of information between its covers, we recommend that you spend a few minutes reading about the following important sections, some of which are cross-referenced to make the WRTH as useful as possible in a variety of listening situations.

WORLD RADIO

This is a country-by-country compilation (subdivided by regions) of the essential facts on broadcasting stations and organisations, plus other useful data. A typical entry contains the following information:

L.T: Local time, indicated as the number of hours ahead of, or behind, Universal Time Coordinated (UTC). For an explanation of this relationship, see note a) below.

Pop: Population of the country.

Radio sets: Actual or estimated number in use.

Pr.L: Principal language(s) used in Home Service broadcasts.

E.C: Type, frequency and voltage of normal mains electricity supply.

ITU: The standard country code used by the International Telecommunication Union.

L.P: Leading personnel (these are senior administrative staff, not necessarily on-air personalities).

Stations: A list, in order of frequency, of all the transmitters operated by the broadcaster, with radiated power in kiloWatts (kW) and, if appropriate, the network or program carried by each transmitter.

D.Prgr: Details of operating hours, frequency schedules, languages and news bulletins of each separate network, with national services mentioned first, followed by regional and/or local services as appropriate.

Ann: Transcription of key identification announcements broadcast by the station.

Int-Sig (IS): In some cases, a description of regularly-used interval signals is included as an additional aid to identification.

V: Details of the station's policy concerning the written verification of listeners' reception reports. See note b) below.

External Services: These are listed at the end of the Home Service details, and in most cases include the following data:

Stations: A list of transmitting locations and facilities.

G.C: Exact location, in degrees and minutes, is given for the transmitter sites whenever this information is known.

kHz: A quick-reference list of scheduled frequencies in kiloHertz.

Areas: The official target areas designated by the station for its various transmissions. These vary widely from one station to another.

Beam: Where known, we indicate the azimuth of maximum radiation in degrees from true North of the transmitting site.

Ann, Int-Sig, V: As for Home Services (see above).

A few additional details, not mentioned above, are included wherever possible. If in doubt about their significance, the abbreviations list will help you.

Special notes:

a) **Universal Time Coordinated (UTC)** is used throughout the WRTH in accordance with international convention. Many broadcasters still refer to Greenwich Mean Time (GMT) which for practical listening purposes is the same thing. The **World Time Table** in this Handbook lists the differences between local time in each country and UTC. Daylight Saving Time, or Summer Time, is used in many countries as shown in the World Time Table. When this applies, Home Service broadcasts will be one hour **earlier** in relation to UTC. Many international broadcasters ignore this change and so their programs remain at the same UTC time all the year round. Others, however, change some or all of their times during Summer Time. Where the pattern is clear, this is indicated in the text.

b) **Verification of Reception Reports** is a complicated subject, and if you are interested in collecting such verifications (which are known as "QSL's") we advise you to study the guidelines issued by leading listeners' clubs and some radio stations (eg. Radio Nederland) before spending time and money on sending reception reports. Here we offer some very general hints.

Reports to **Home Service** broadcasters should use the language of the broadcast (or another language known to be widely understood), reception should be described in words, **not** in the SINPO code, and return postage **must** be included in the form of mint stamps of the country concerned or International Reply Coupons (IRC's). Some listeners' clubs produce report forms and/or vocabulary lists to help with reports in major languages.

Reports to **International Broadcasters** may be in English or in the language of the broadcast you heard. Return postage is appreciated but not always necessary.

Give some constructive comments on the program to prove that you really did listen to it. The SINPO code is a well-established and convenient way of indicating reception quality, namely:

S (Signal Strength)	I (Interference)	N (Noise)
5 excellent	5 nil	5 nil
4 good	4 slight	4 slight
3 fair	3 moderate	3 moderate
2 poor	2 severe	2 severe
1 barely audible	1 extreme	1 extreme

P Propagation)	O (Overall merit)
5 nil	5 excellent
4 slight	4 good
3 moderate	3 fair
2 severe	2 poor
1 extreme	1 unusable

The WRTH entries include an indication of whether the station verifies reception reports from listeners. Rp. means that return postage is necessary. The number of IRC's required is mentioned in brackets if stated by the broadcaster.

LONG & MEDIUM WAVE STATIONS

The stations operating between 150kHz and 1620kHz are listed, in frequency order, in five groups according to geographical location. Thus if you are in a particular area of the world, and want to identify a station heard on a certain frequency, you are most likely to find it amongst the stations listed for your region. If you are interested in hearing long distance ("DX") signals on these bands, you may need to refer to the other groups of stations. All these lists cross-refer to the World Radio Section, but may contain additional low power stations not separately indicated in the World Radio list (eg. USA stations).

SHORTWAVE STATIONS OF THE WORLD

This is a composite listing of all the stations between 2000 and 30,000kHz mentioned in the World Radio Section. As with the Long and Medium Wave lists, the purpose is to facilitate the identification of stations by cross-reference to the World Radio Section. However, see note c) below concerning frequency changes.

Special note:

c) **Frequency changes**

Some stations are forced to make changes without much advance notice due to interference, poor reception, changes in programming, technical problems etc. It is rare for all the scheduled frequencies of a station to be changed at once, therefore try all listed possibilities if you are unable to hear the broadcast you want at the first attempt.

PRACTICAL TIPS ON USING THE WRTH

What Can I Expect to Hear?

Long and Mediumwave Stations vary in power from a few Watts to as much as 2000 kiloWatts. When it is daylight at the transmitting station, the lowest powered stations will reach only a few kilometres, while the highest powered ones can cover a radius of several hundred kilometres.

At night, mediumwave signals are often reflected back to earth by the ionosphere (layers of gas around the earth), and stations can be received from great distances which are inaudible during daylight.

Shortwave Stations depend on variable solar activity for reception in their intended target area. The annual feature "High Frequency Reception Conditions" (see Table of Contents) describes in detail what may be expected during the currency of this edition.

FM (Frequency Modulation) Stations operate in the frequency ranges of 66-72MHz (Eastern Europe), 76-90MHz (Japan) and 87-108 MHz elsewhere. Reception is theoretically possible only on a 'line of sight' basis, but in practice signals do penetrate some distance beyond the optical horizon. The maximum range will of course depend on the effective radiated power (ERP) which is normally the figure mentioned in the WRTH listing. Under certain abnormal conditions, FM signals can be reflected in the same way as shortwave signals, and reception over many hundreds of kilometres may become possible, often with deep fading and with some distortion of the audio.

Finding a Particular Station

Let us suppose that you wish to hear a broadcast in your language from Radio X. The first thing to do, is to find the entry for Radio X in the World Radio section. If it is an International Broadcaster, you will find the available transmissions listed by language and/or target area. For a Home Service broadcast, find the transmission times and study the schedule to discover which frequencies carry the broadcast. Then select one which seems to have the best chance of being audible in your location (in the case of shortwave stations, refer to the Table of the Most Suitable Broadcasting Bands). If the broadcast is not audible at the time you want, you still may be able to hear the station at another time of the day. Do not assume that you are listening to the desired station until you have heard a positive identification. There is always a good chance that another station is occupying the frequency (see next section).

Identifying a Station

When you hear a signal which you cannot identify, note as accurately as possible what the frequency is. Listen carefully for clues as to its likely origin (language, news items, type of music etc.) and compare these clues with the list of stations in the Shortwave or Mediumwave tables likely to be using the frequency. When you have narrowed down the possibilities, check the listing for each possibility in the World Radio section to see if the schedule fits what you are hearing. Take into account the most likely reception possibilities (again referring to the Table of the Most Suitable Broadcasting Bands if appropriate), and wait for a positive identification before assuming that you have identified the station correctly.

If you are listening to an international broadcaster, you can expect a clear identification at the beginning and end of each transmission period, and at one or more points in between. Home Service stations may be more difficult to identify (particularly, for example, if a play or concert is in progress) but many will announce themselves on the hour or half hour. Note that stations in North America, Australia, New Zealand and a few other countries use callsigns of up to five letters and/or numbers, while in most other countries a station name is usually announced. In Latin America, the official name is frequently supplemented by a slogan.

GUIDE DE L'UTILISATEUR DU WRTH

Le World Radio TV Handbook (WRTH) est votre guide pratique pour les services de radiodiffusion et de télévision dans le monde. En utilisant le WRTH, vous pouvez vous syntoniser sur des émissions intéressantes du monde entier, et identifier les stations que vous n'aviez jamais entendues, auparavant. Mais, pour tirer le maximum de bénéfice de la richesse d'informations, d'un bout à l'autre de l'ouvrage, nous vous recommandons de consacrer quelques minutes à lire ces lignes d'explications.

Le WRTH est divisé en un certain nombre de sections, quelques-unes d'entre-elles se recoupant avec les autres, pour le rendre le plus utile possible dans une variété de situations d'écoute.

Si vous êtes un auditeur actif, vous trouverez vous-mêmes en vous référant fréquemment à plusieurs sections clé:

WORLD RADIO (Radio dans le monde)
C'est une compilation pays par pays (subdivisée par régions) des points essentiels sur les stations et organisations de radiodiffusion plus d'autres données utiles. Une présentation type contient les informations suivantes:

L.T: horaire local, indiqué en nombre d'heures en plus ou en moins par rapport au Temps Universel Coordonné (UTC). Pour une explication de ce rapport, voyez la note a), ci-dessous.

Pop: Nombre d'habitantes.

Radio sets: Nombre de recepteurs.

Pr.L: principales langues utilisées dans les émissions du Service Domestique.

E.C: type, fréquence et voltage de l'alimentation en courant électrique normal. Entre parenthèses, une indication de la stabilité du courant pour les horloges électriques, oui (yes) ou non (no).

ITU: symbole dé signant le pays.

L.P: personnalitiés dirigeantes (ce sont les reponsables administratifs, pas nécessairement celles que l'on entend sur les ondes).

Stations: une liste, par ordre de fréquences, de tous les émetteurs opérés par le radiodiffuseur, avec la puissance rayonnée en kilowatts (kW) et, si nécessaire, le réseau des programmes émis pour chaque émetteur.

D.Prgr: détails des heures de diffusion, grilles des fréquences, langues et bulletins de nouvelles de chaque réseau séparé, avec, au début, la mention des services nationaux, suivis des services régionaux et/ou locaux, quand cela est nécessaire.

Ann: transcription des annonces d'identification courantes diffusées par la station.

Int-Sig (IS): dans quelques cas, une description du signal d'intervalle d'identification régulièrement utilisé par la station est incluse comme aide supplémentaire à l'identification.

V: détails sur la politique de la station en ce qui concerne les vérifications écrites des rapports de réception des auditeurs. Voyez la note b), ci-dessous.

External Services (Services Extérieurs): ils sont listés à la fin des détails sur le Service Domestique et, dans la plupart des cas, incluert les données suivantes:

Stations: une liste des lieux d'émission et des facilités.

G.C. (Coordonnées Géographiques): une localisation exacte, en degrés et minutes, est donnée pour les sites d'émetteurs, toutes les fois que cette information est connue.

kHz: une liste de référence rapide des fréquences utilisées, en kiloHertz.

Areas (Régions): ce sont les régions ciblées officiellement désignées par la station pour ses diverses transmissions. Elles varient largement d'une station à une autre.

Beam (direction): quand il est connu, nous indiquons l'azimuth de la radiation maximum en degrés par rapport au Nord magnétique, à partir du site de transmission.

Ann, Int-Sig, V: comme pour les Services Domestiques (voir ci-dessus).

Quelques détails supplémentaires non mentionnés cidessus sont inclus, quand cela est possible. En cas de doute sur leur signification, la liste des abréviations vous aidera.

Notes Spéciales:

a) **Temps Universel Coordonné (UTC):** il est utilisé dans tout le WRTH en accord avec la convention internationale. De nombreuses stations de radiodiffusion se réfèrent encore à l'heure du méridien de Greenwich (GMT) qui, pour des buts pratiques d'écoute, est la même chose. La **World Time Table** (Table Mondiale des Horaires), dans cet ouvrage, liste la différence entre l'horaire local dans chaque pays et l'heure UTC. L'heure avancée, ou heure d'été, est utilisée dans de nombreux pays comme il est indiqué dans la World Time Table. Quand ce changement s'applique, les services domestiques de radiodiffusion seront avancés d'une heure par rapport au temps UTC. De nombreuses stations de radiodiffusion internationales ignorent ce changement, donc leurs programmes restent à la même heure UTC tout au long de l'année. Cependant, d'autres changent quelques-uns de leurs horaires pendant la période d'heure d'été. Quand la grille est claire, ceci est indiqué dans le texte.

b) La **Vérification des Rapports de Réception** est un sujet compliqué, et si vous êtes intéressés par la collection de telles vérifications (qui sont connues sous le nom de "carte QSL") nous vous conseillons de vous référer aux guides publiés par les principaux clubs d'auditeurs et quelques stations de radiodiffusion (par exemple Radio Nederland) avant de perdre du temps et d l'argent à envoyer des rapports de réception. Ici, nous vous proposons quelques conseils d'ordre général. Les rapports aux Stations Internationales doivent être rédigés en anglais ou dans la langue de l'émission que vous écoutez. Les frais de réponse sont appréciés, mais pas toujours exigés. Donnez quelques commentaires constructifs sur le programme, afin de prouver que vous l'avez réellement bien écouté. Le code SINPO est un moyen bien établi et pratique pour indiquer la qualité de la réception:

S (intensité du Signal)	**I** (Interférence)	**N** (Bruits)
5 excellent	5 nulle	5 nuls
4 bon	4 légère	4 légers
3 moyen	3 modérée	3 modérés
2 faible	2 sévère	2 sévères
1 médicre	1 extrème	1 extrèmes

P (propagation)	**O** (appréciation d'ensemble)
5 nulle	5 excellente
4 légère	4 bonne
3 modérée	3 moyenne
2 sévère	2 faible
1 extrème	1 inaudible

Les rapports aux stations de Service Domestique doivent être rédigés dans la langue de diffusion (ou une autre langue connue pour être largement comprise), les réceptions décrites en clair, et non en code SINPO, et les frais de réponse doivent être inclus sous la forme de timbres non oblitérés du pays concerné ou de coupons-réponse internationaux (IRC). Quelques clubs d'auditeurs fournissent des formules de rapports et/ou des listes de vocabulaire pour aider à rédiger des rapports dans les principales langues.

Les inscriptions du WRTH comprennent une indication sur comment la station reponde aux rapports de réception des auditeurs. "Rp." signifie si les frais de réponse sont nécessaires.

STATIONS EN ONDES LONGUES ET MOYENNES (Long & Medium Wave Stations)

Les stations diffusant entre 150 kHz et 1620 kHz sont listées, par ordre de fréquences, en cinq groupes d'après la situation géographique. Ainsi, si vous êtes dans une région particulière du monde et que vous vouliez identifier une station entendue sur une certaine fréquence, vous avez plus de chance de la trouver parmi les stations listées de votre région. Si vous êtes intéressés par l'écoute à longue distance ("DX") des signaux émis dans ces bandes, vous pouvez avoir besoin de vous référer à d'autres groupes de stations. Toutes ces listes se recoupent avec la "World Radio Section", mais peuvent contenir des stations supplémentaires de faible puissance que de sont pas indiquées dans la liste "World Radio" (par exemple, les stations des USA).

STATIONS ONDES COURTES DANS LE MONDE (Shortwave Stations of the World)

C'est une liste composée de toutes les stations entre 2000 et 30000 kHz mentionnées dans la "World Radio Section". Comme pour les listes en ondes longues et moyennes, le but est de faciliter l'identification des stations par recoupement avec la "World Radio Section". Cependant, lisez la note c), ci-dessous, concernant les changements de fréquences.

Note spéciale:

c) **changements de fréquences**

Les fréquences listées tout au long du WRTH sont celles programmées pour être utilisées pendant la période d'édition du livre.

Recherche d'une Station Particulière

Supposons que vous souhaitez écouter une émission dans votre langue diffusée par Radio X. La première chose à faire est de trouver les indications sur Radio X dans la "World Radio Section". Si c'est une station de radiodiffusion internationale, vous trouverez les transmissions effectuées listées par langue ou par zone visée.

Pour la diffusion d'une Service Domestique, recherchez les horaires de diffusion et étudiez la grille des programmes pour découvrir quelles fréquences diffusent le programme. Alors, sélectionnez celles qui semblent avoir le plus de chance d'être audibles dans vitre région (dans le cas des stations en ondes courtes, référez-vous a la table des bandes de radiodiffusion les plus favorables—Most Suitable Broadcasting Bands). Si l'émission n'est pas audible à l'heure que vous désirez, vous pouvez peut-être encore être en mesure d'entendre la station à une autre heure de la journée.

Ne déclarez pas que vous écoutez la bonne station jusqu'à ce que vous ayez entendu un identification positive. Il y a aussi de bonnes chances qu'une autre station occupe en même temps la fréquence (voir le châpitre suivant).

L'identification d'une station

Quand vous entendez un signal que vous ne pouvez pas identifier, notez le plus précisément possible qu'elle est la fréquence. Ecoutez soigneusement pour relever l'origine probable (langue, titres des nouvelles, musique, etc. . .) et comparez ces indices avec la liste des stations dans les tables ondes courtes ou ondes moyennes pour savoir qui utilise la fréquence. Quand vous avez sélectionné les possibilités, contrôlez la liste pour chaque possibilité dans la "World Radio Section" afin de vérifier si la grille correspond à ce que vous entendez. Prenez en compte les possibilités de réception les plus probables (en vous référant de nouveau à la table des bandes de radiodiffusion les plus propices, si nécessaire) et attendez une identification positive avant d'affirmer que vous avez identifé correctement la station.

Si vous écoutez une station de radiodiffusion internationale, vous pouvez espérer une identification précise au début et à la fine de chaque période d'émission et, une ou plusieurs fois, pendant. Les stations de service domestique peuvent être plus difficiles à identifier (particulièrement, par exemple, lors de la diffusion d'une pièce de théâtre ou d'un concert), mais beaucoup s'annonceront à l'heure et à la demi-heure. Notez que les stations en Amérique du Nord, Australie, Nouvelle-Zélande et de quelques autres pays utilisent des indicatifs d'appel allant jusqu'à cinq lettres et/ou chiffres, alors que dans la plupart des autres pays une station est couramment annoncée par son nom. En Amérique Latine, le nom officiel est fréquemment complété par un slogan.

HINWEISE FÜR BENUTZER DES WRTH

Das World Radio TV Handbook (WRTH) gibt Ihnen Auskunft über Rundfunk- und Fernsehstationen in aller Welt. Es hilft Ihnen, sich in Sie gerade interessierende Sendungen einzuschalten und noch unbekannte Stationen zu identifizieren. Nehmen Sie sich einige Minuten Zeit, um diese Hinweise zu lesen, dann werden Sie von der Fülle der gesammelten Information so viel wie möglich profitieren.

Das WRTH ist in verschiedene Rubriken eingeteilt, zwischen denen jedoch ab und zu Querverbindungen bestehen, so daß Sie in jeder nur denkbaren Situation den größten Nutzen daraus ziehen können.

Wer regelmäßig hört, wird wiederholt in folgenden Sektionen nachschlagen:

WORLD RADIO (RUNDFUNK AUS ALLER WELT)
Hier findet man eine nach Gebieten und -innerhalb der einzelnen Regionen- nach Ländern geordnete Aufstellung, die alle wichtigen und nützlichen Daten über Rundfunkstationen und -organisationen enthält. Eine typische Eintragung besteht aus folgenden Angaben:

L.T: Lokalzeit; diese bemißt sich danach, wieviele Stunden von der UTC (Weltzeit) abgezogen oder zugezählt werden müssen. Vgl. dazu unten die Anmerkung a).

Pop: Bevölkerung

Radio sets: Zahl der Empfänger.

Pr.L: Jene Sprache(n), in denen die Inlandsprogramme meist ausgestrahlt werden.

E.C: Art, Frequenz und Spannung des üblicherweise verwendeten Netzstromes. In Klammern steht, ob die Stromversorgung für den Betrieb elektrischer Uhren ausreicht (yes) oder nicht (no).

ITU: Die vorzugs Andeutung von Länder durch die International Telecommunication Union.

L.P.: Führungspersonal (dabei handelt es sich nicht unbedingt um populäre Programmgestalter, sondern meist um Verwaltungschefs).

Stationen: Die Liste enthält alle von der Station betriebenen Sender. Sie ist nach Frequenzen geordnet und nennt die Leistung in Kilowatt sowie das auf bestimmten Wellenlängen ausgestrahlte Programm, wenn eine Angabe darüber angebracht erscheint.

D.Prgr: Darunter fallen Angaben über Sendezeiten, Frequenzpläne, Sprachen und Nachrichtenprogramme jeder einzelnen Sendergruppe, wobei, je nach Bedarf, zuerst die landesweiten, danach die Regional- oder Lokaldienste genannt werden.

Ann: Niederschrift der gebräuchlichen Stationsansage.

Int-Sig (IS): Manchmal wird das regelmäßig verwendete Sendezeichen als zusätzliche Hilfe für die Identifizierung niedergeschrieben.

V: Angaben darüber, ob die Station Empfangsberichte der Hörer schriftlich bestätigt; siehe dazu unten die Anmerkung b)

Auslandsdienste: Sie sind stets nach den Inlandsdiensten genannt; eine Aufstellung besteht aus nachfolgenden Details:

Stationen: Eine Liste der Senderstandorte und ihrer Ausstattung.

G.C: Sofern der exakte Standort bekannt ist, wird er in Grad und Minuten genannt.

kHz: Alle benutzten Frequenzen sind für den eiligen Leser übersichtlich nach Kilohertz zusammengefaßt.

Bereiche: Jene Gebiete, für die bestimmte Sendungen gezielt ausgestrahlt werden. Hier gibt es erhebliche Abweichungen von Station zu Station.

Richtstrahlung: Sofern bekannt, wird der Azimut (Einfallswinkel) der Hauptsenderichtung in Grad angegeben.

Ann, Int-Sig, V: Die Aufstellung entspricht der bei den Inlandsdiensten (siehe oben).

Sofern möglich, werden darüberhinaus zusätzliche Angaben über die Station gemacht. Sollten Sie über deren Bedeutung im Unklaren sein, wird das Abkürzungsverzeichnis weiterhelfen.

Besondere Anmerkungen:

a) **Die Universal Time Coordinated (UTC)** = Weltzeit wird international angewendet und daher auch im WRTH zugrundegelegt. Viele Rundfunksender verwenden noch den Begriff Greenwich Mean Time (GMT), der, was das Hören angeht, mit UTC identisch ist. Mit der **Weltzeittabelle** in diesem Buch können Sie die Unterschiede zwischen der Lokalzeit jedes Landes und UTC ermitteln. Aus dieser Tabelle ersehen Sie auch, daß viele Länder die Sommerzeit eingeführt haben. In diesen Fällen beginnen die Inlandsdienste im Verhältnis zur UTC eine Stunde **früher.** Viele Auslandsdienste ignorieren diese Zeitverscheibung und strahlen ihre Programme während des gesamten Jahres zur gleichen UTC-Zeit aus, andere jedoch passen einige oder alle Sendezeiten entsprechend an. Wenn die Verfahrensweise der Station klar ist, wird dies im Text erwähnt.

b) Ein problematisches Thema ist die *Bestätigung von Empfangsberichten*. Wenn Sie gerne solche Bestätigungen (bekannt als 'QSL's') sammeln, lesen Sie die dazu erschienenen Empfehlungen führender Hörervereinigungen oder der Sendeanstalten (z.B. van Radio Nederland), bevor Sie Zeit und Geld mit Schreiben von Rapporten vergeuden. An dieser Stelle können wir nur sehr allgemeine Tips geben.

Berichte an *internationale Auslandsdienste* sollten in Englisch oder in der Sprache des gehörten Programms verfaßt werden. Rückporto ist häufig erwünscht, aber nicht stets erforderlich. Um zu beweisen, daß man die Station wirklich hörte, sind einige genaue Programmdetails anzugeben. Der SINPO-Code ist die gängigste und bequemste Möglichkeit, die Empfangsqualität darzustellen:

S (Signalstärke)	I (Interferenz)	N (Rauschen)
5 ausgezeichnet	5 keine	5 keines
4 stark	4 geringe	4 geringes
3 mittelstark	3 mittelstarke	3 mittleres
2 schwach	2 starke	2 starkes
1 kaum hörbar	1 sehr starke	1 sehr starkes

P (Schwund, Fading)	O (Gesamtbeurteilung)
5 keiner	5 hervorragend
4 leichter	4 gut
3 mittelstarker	3 zufriedenstellend
2 starker	2 mangelhaft
1 sehr starker	1 unbrauchbar

Berichte an *Inlandsdienste* sollten in der Sprache der gehörten Sendung oder einer anderen im Senderort weit verbreiteten Sprache geschrieben werden. Hier ist die Empfangsqualität mit Worten und *nicht* mittels des SINPO-Codes zu beschreiben. Rückporto *muß* in Form von Briefmarken des Landes oder Internationalen Antwortscheinen (IRCs) beigefügt werden. Einige Hörerklubs vertreiben Berichtsvordrucke und DX-Vokabulare für die verbreitetsten Sprachen.

Im WRTH wird angegeben, ob eine Station Hörberichte erwünscht. Ein (A) deutet an, daß sie nicht gerne gesehen sind, ein (B), daß man sich über Berichte freut. Der Hinweis Rp. zeigt, wenn Rückporto beigefügt werden muß.

LONG & MEDIUM WAVE STATIONS (LANG- UND MITTELWELLENSENDER)

In dieser Auflistung sind alle Stationen zwischen 150 und 1620 kHz, nach fünf geographischen Gebieten unterteilt, in Frequenzreihenfolge zusammengefaßt. Wenn Sie also an einem bestimmten Ort auf der Erde eine Station auf einer gewissen Frequenz identifizieren wollen, werden Sie sie sehr wahrscheinlich in der Tabelle dieser geographischen Region finden. Wer auf Lang- und Mittelwelle entfernte ("DX") Signale aufnimmt, wird in den Listen mehrerer Gebiete nachschlagen. Alle Aufstellungen decken sich mit den Angaben in der 'World Radio'-Rubrik; sie enthalten darüberhinaus aber auch Stationen mit sehr schwacher Sendeleistung (z.B. aus den USA), die unter 'World Radio' nicht genannt sind.

SHORTWAVE STATIONS OF THE WORLD (KURZWELLENSTATIONEN DER WELT)

Dabei handelt es sich um die Zusammenstellung aller Sender zwischen 2000 und 30000 kHz, die im World Radio-Abschnitt erwähnt werden. Wie bei den Lang- und Mittelwellensendern auch, erleichtert ein gleichzeitiges Nachschauen in beiden Rubriken die Identifizierung der Stationen. Beachten Sie aber den unten stehenden Hinweis c) (Frequenzwechsel).

Besondere Anmerkung:

c) **Frequenzwechsel**

Im WRTH sind die zur Zeit des Redaktionsschlusses bekannten Frequenzen genannt.

Was kann ich hören?

Lang- und Mittelwellensender haben unterschiedliche Leistungen von wenigen Watt bis hin zu 2000 Kilowatt. Ist es am Senderstandort hell, können die schwachen Stationen nur einen Radius von wenigen Kilometern bestreichen, die stärksten hingegen einen von mehreren Hundert Kilometern.

Zur Nachtzeit werden demgegenüber Mittelwellensignale von der Ionosphäre-der äußersten Gashülle der Erdatmosphäre-reflektiert, so daß dann weit entfernte Stationen, die man tagsüber nicht empfängt, hörbar werden.

Kurzwellenstationen wiederum sind von der Sonnenaktivität abhängig. Der alljährlich in diesem Buch erscheinende Beitrag "High Frequency Reception Conditions" (Empfangsbedingungen auf Kurzwelle, vgl. dazu das Inhaltsverzeichnis) stellt detailliert dar, mit welchen Bedingungen diesmal zu rechnen ist.

FM-(UKW-)Stationen nutzen in Osteuropa den Bereich 66-72 MHz, in Japan den Bereich 76-90MHz und überall sonst 87-108 MHz. Theoretisch ist ein Empfang nur bis zum optischen Horizont möglich, in der Praxis gehen die Möglichkeiten aber darüber hinaus.

Aufspüren einer bestimmten Station

Nehmen wir an, Sie wollen von Radio X eine Sendung in Ihrer Sprache hören. Zuerst müssen Sie dafür im Abschnitt 'World Radio' die entsprechende Eintragung finden. Sofern es sich um einen Auslandsdienst handelt, sind die verfügbaren Programme nach Sprache und/oder Zielgebiet aufgeführt. Bei einem Inlandsdienst gilt es, die Sendezeiten und die gerade benutzten Frequenzen zu suchen. Wählen Sie jene aus, die die größten Chancen versprechen, in Ihrem Land gehört zu werden (bei Kurzwellensendern bedienen Sie sich der Liste der am besten geeigneten Rundfunkbänder, die in diesem Buch enthalten ist). Ist die Station zu einer bestimmten Zeit nicht hörbar, probieren Sie es zu einer anderen Tageszeit. Nehmen Sie niemals an, tatsächlich den gewünschten Sender zu empfangen, bis Sie eine klare Stationsansage vernommen haben, denn es besteht immer die Möglichkeit, daß jetzt eine andere Station die Frequenz belegt (siehe dazu auch die Angaben im folgenden Abschnitt).

Identifizieren einer Station

Wenn Sie einen Sender aufnehmen, den Sie nicht identifizieren können, stellen Sie zuerst die Frequenz so exakt wie möglich fest. Versuchen Sie, aus dem Programm (Sprache, Nachrichtendetails, Art der Musik usw). Hinweise auf das Herkunftsland zu gewinnen und vergleichen Sie diese Anhaltspunkte mit den in der Kurz- oder Mittelwellenfrequenzliste genannten Stationen. Haben Sie auf diese Weise die Möglichkeiten eingegrenzt, prüfen Sie für jeden einzelnen Sender in der World Radio-Rubrik, ob die Eintragungen sich mit dem von Ihnen gehörten Programm decken. Berücksichtigen Sie stets die Empfangs- und Ausbreitungsbedingungen (dabei kann wiederum die oben genannte Tabelle helfen) und warten Sie auf eine eindeutige Identifikation, bevor Sie annehmen, die Station gehört zu haben. Bei Auslandsdiensten können Sie mit klaren Ansagen am Anfang und Ende jedes Programms, aber auch zwischendurch rechnen. Schwieriger ist es mit Inlandsdiensten (insbesondere wenn ein Hörspiel oder ein Konzert ausgestrahlt wird), doch Ansagen sind auch hier oft zur vollen oder halben Stunde zu erwarten. Während in den meisten Ländern der Stationsname genannt wird, sind in Nordamerika, Australien, Neuseeland und einigen anderen Staaten Rufzeichen gebräuchlich. Sie bestehen aus bis zu fünf Buchstaben und/oder Zahlen. In Lateinamerika wird der Stationsname häufig durch einen Slogan ergänzt.

40th Anniversary
Radio Korea 1953-1993
Keeping You Informed Around the Globe
11725 KHz
11740 KHz
11715 KHz
11810 KHz
9870 KHz
7550 KHz
9515 KHz
9525 KHz
13670 KHz
9750 KHz
7275 KHz
1170 KHz
5975 KHz
6135 KHz
9570 KHz
15170 KHz
9650 KHz
6480 KHz
891 KHz
6145 KHz
9640 KHz
15575 KHz
Radio Korea
Four Decades of Bringing the World Home to You
• English • Japanese • Korean • French • Russian • Chinese
• Spanish • Indonesian • Arabic • German • Portuguese • Italian
Korean Broadcasting System
18, Yoido-dong, Youngdungpo-gu, Seoul, Korea
Fax: (02) 781-3799 Tel: (02) 781-3710

GUÍA DEL USUARIO DEL MANUAL MUNDIAL DE RADIO Y TELEVISIÓN

El Manual Mundial de Radio y Televisión (WRTH) es su guía práctica de los servicios mundiales de radio y televisión. Utilizando el Manual, puede Vd. sintonizar emisiones de interés de todo el mundo, e identificar emisoras que no había escuchado antes. Pero para conseguir el máximo provecho del caudal de información contenido entre sus cubiertas, le recomendamos que dedique unos minutos a leer estas notas explicativas.

El Manual está dividido en varias secciones, algunas de las cuales se refieren unas a otras, para hacerlo lo más útil posible en un amplio abanico de situaciones de escucha.

Si es Vd. un escucha activo, se encontrará buscando referencias con frecuencia en varias secciones clave:

WORLD RADIO (LA RADIO EN EL MUNDO)

Es una relación país por país subdividida en regiones) de los datos esenciales sobre las emisoras y organizaciones radiodifusoras, además de otras informaciones de utilidad. Una nota típica contiene la siguiente información:

L.T: Hora local, indicada como un número de horas delante o detrás del Tiempo Universal Coordinado (UTC). Para una explicación más detallada de esta relación, ver nota a) más adelante.

Pop: población

Radio sets: Número de receptores

Pr.L: Idioma(s) principal(es) utilizado en las emisiones de ámbito local o nacional (interior del país).

E.C: Tipo, frecuencia y voltaje de las fuentes normales de electricidad. Entre paréntesis se indica si la fuente es lo suficientemente estable para hacer funcionar relojes eléctricos (yes) o no (no).

ITU: símbolo del país.

L.P: Personal directivo (compuesto por los principales directivos administrativos, no necesariamente las personas que salen en el aire).

Stations: (Emisoras). Una lista, por orden de frecuencia, de todos los transmisores operados por el emisor, con la potencia radiada en kilowatios (kW) y, en su caso, la red o el programa llevado a cabo por cada transmisor.

D.Prgr: Detalles de horarios de transmisión, lista de frecuencias, idiomas y nuevos boletines de cada red por separado, con los servicios nacionales mencionados en primer lugar, seguidos por los servicios regionales y/o locales si los hubiera.

Ann: Transcripción de los anuncios clave de identificación difundidos por la emisora.

Int-Sig (IS): (Sintonía). En algunos casos, se incluye una descripción de las sintonías que se usan normalmente como una ayuda adicional para la identificación.

V: Detalles de la política de la emisora en relación con la verificación escrita de los informes de recepción de los escuchas. Ver nota b) más abajo.

External Services: (Servicios exteriores)—Estos están relacionados al final de los detalles de los Servicios interiores, y en la mayoría de los casos incluyen los siguientes datos:

Stations: (emisoras)—Una lista de los lugares desde los que se emite y otros datos de utilidad.

G.C: Localización exacta, en grados y minutos, del lugar de transmisión, siempre que esta información sea conocida.

kHz: Una lista de rápida consulta de frecuencias en kilohercios.

Areas: Las zonas geográficas oficiales designadas por cada emisora para sus diferentes transmisiones. Estas varían ampliamente de una emisora a otra.

Beam: (Haz de emisión)—Se indica el azimut de irradiación máxima en grados desde el norte verdadero del lugar de transmisión, siempre que se conozcan estos datos.

Ann, Int-Sig, V: Igual que en el caso de los Servicios interiores (ver más arriba)

Algunos detalles adicionales, no mencionados más arriba, se incluyen siempre que es posible. En caso de duda acerca de su significado, le será de utilidad la lista de abreviaturas.

Notas Especiales:

a) **Tiempo Universal Coordinado (UTC).** Se utiliza a lo largo del Manual según las convenciones internacionales. Muchos radiodifusores utilizan como referencia todavía el Tiempo Meridiano Greenwich (GMT), lo cual a efectos prácticos de escucha es lo mismo. La **World Time Table** (Tabla Horaria Mundial) de este Manual contiene una lista de diferencias entre la hora local de cada país y el UTC. En muchos países, como se puede ver en la Tabla Horaria Mundial, se utiliza el horario de verano. Cuando esto ocurre, las emisiones del servicio interior serán **una hora antes** en relación con el UTC. Muchos radiodifusores internacionales ignoran este cambio y de este modo sus programas permanecen en la misma hora UTC durante todo el año. Sin embargo, otros cambian algunos o todos sus horarios durante el verano. Cuando el modelo está claro, está indicado en el texto.

b) **La verificación de informes de recepción** es un tema complicado, y si está interesado en coleccionar estas verificaciones (conocidas como tarjetas "QSLs") le recomendamos que estudie las instrucciones publicadas por los Clubs de Radioescuchas más importantes y algunas emisoras de radio (ej. Radio Nederland) antes de invertir su tiempo y dinero enviando informes de recepción. Aquí le ofrecemos unas instrucciones a nivel muy general.

Los informes a **Emisoras internacionales** podrán enviarse en inglés o en el idioma de la emisión que Vd. escuchó. Le agradecerán que envíe el franqueo postal para la respuesta, pero no siempre es necesario. Proporcione algunos comentarios constructivos sobre el programa para demostrar que realmente lo escuchó. El código SINPO es un método establecido y conveniente para indicar la calidad de recepción, y es como sigue:

S (Fuerza de la señal)	**I** (Interferencia)	**N** (Estática)
Signal Strength	Interference	Noise
5 excelente	5 ninguna	5 ninguna
4 fuerte	4 ligera	4 ligera
3 normal	3 moderada	3 moderada
2 pobre	2 fuerte	2 fuerte
1 casi inaudible	1 muy intensa	1 muy fuerte

P (Propagación)	**O** (Resúmen)
Propagation	Overall merit
5 ninguno	5 excelente
4 ligero	4 bueno
3 moderado	3 bastante bueno
2 notable	2 pobre
1 muy acentuado	1 inservible

En los informes para las emisoras del servicio interior, se deberá utilizar el idioma de la emisión (o cualquier otro idioma que se sepa que es conocido ampliamente), las condiciones de recepción deben estar descritas en palabras, **no** con el código SINPO, y **debe** enviarse el importe del franqueo postal para la respuesta, bien en sellos del país en cuestión, bien mediante los Cupones de Respuesta Internacional (IRC). Algunos clubs de radioescuchas disponen de formularios para usar en los informes, y/o de listas de vocabulario en los idiomas más importantes, para ayudar en la confección de informes.

En las diferentes notas del Manual se incluye una indicación de si la emisora quiere informes de recepción de los escuchas. Rp. quiere decir que es necesario enviar el importe del franqueo postal para la respuesta.

EMISORAS DE ONDA LARGA Y ONDA MEDIA

Las emisoras que emiten entre los 150 kHz y los 1620 kHz se relacionan, por orden de frecuencia, en cinco grupos, de acuerdo con la situación geográfica. De este modo, si Vd. se encuentra en una zona determinada del mundo y desea identificar una emisora escuchada con cierta frecuencia, lo más probable es que la encuentre entre las emisoras de la lista de su región. Si está interesado en escuchar señales de larga distancia ("DX") en estas bandas, podrá necesitar buscar la referencia en los otros grupos de emisoras. Todas estas listas se refieren a la Sección "World Radio" (la radio en el mundo), pero podrían incluír además emisoras de poca potencia que no estén indicadas por separado en la lista de esta sección (ej. las emisoras de los EE.UU.).

EMISORAS DE ONDA CORTA DEL MUNDO

Esta es una lista compuesta por todas las emisoras entre los 2000 y los 30000 kHz mencionadas en la sección "World Radio". Al igual que en las listas de Onda larga y media, el propósito es facilitar la identificación de las emisoras con referencias a la Sección "World Radio". Sin embargo, vea nota c) más abajo relativa al cambio de las frecuencias.

Nota especial:

c) **Cambios de frecuencia**

Las frecuencias que están listadas a lo largo del Manual son aquellas que estaban en vigor en el momento de cerrar esta edición.

CONSEJOS PRACTICOS PARA USAR EL MANUAL

Las emisoras de onda larga y media varían en potencia de unos pocos vatios a tanto como 2000 kilovatios. Cuando es de día en el lugar de emisión, las emisoras de menor potencia alcanzarán solamente unos pocos kilómetros, mientras que las más potentes pueden cubrir un radio de varios cientos de kilómetros.

Durante la noche, las señales de onda media se reflejan a menudo hacia la tierra por la ionosfera (capas de gas alrededor de la tierra), y se pueden recibir emisiones desde grandes distancias que son inaudibles durante el día.

Las emisoras de onda corta dependen de la variable actividad solar para su recepción en la zona pretendida. La característica anual "High Frequency Reception Conditions" ("Condiciones de recepción de las altas frecuencias", ver Indice General) describe en detalle lo que se puede esperar durante la vigencia de esta edición.

Para buscar una emisora determinada

Supongamos que Vd. sesea escuchar una emisión en su idioma de Radio X. Lo primero que debe hacer es encontrar la información de Radio X en la Sección "World Radio". Si es una emisora internacional, encontrará las emisiones disponibles por idiomas y/o por zonas. Para una emisión interior busque las horas de transmisión y estudie el horario para descubrir en qué frecuencias se emite. Después seleccione una que parezca ser la que más posibilidades tenga de ser audible en su zona (en el caso de emisoras de onda corta, busque en la "Table of the Most Suitable Broadcasting Bands" ("Tabla de Bandas de Emisión más adecuadas"). Si la emisión no es audible a la hora que desea, aún puede escuchar la emisora a otra hora del día. No dé por hecho que está escuchando la emisora correcta hasta que haya oído una identificación en sentido positivo. Siempre hay una amplia posibilidad de que otra emisora esté ocupando la frecuencia (ver la siguiente sección).

Identificar una emisora

Cuando escuche una señal que no puede identificar, anote con la mayor seguridad posible qué frecuencia es. Escuche con cuidado para encontrar pistas como su posible origen (idioma, temas informativos, tipo de música, etc.) y compare estos datos con la lista de emisoras en las tablas de onda corta o de onda media con la que con mayor probabilidad esté utilizando la frecuencia. Cuando haya reducido las posibilidades, compruebe las listas para cada posibilidad en la sección "World Radio" para ver si la tabla de frecuencias encaja con la que Vd. está oyendo. Tenga en cuenta las posibilidades de recepción más cercanas (consultando de nuevo la "Table of the Most Suitable Broadcasting Bands" ("Tabla de Bandas de Emisión más adecuadas"), en su caso), y espere a tener una identificación positiva antes de dar por hecho que ha identificado la emisora correctamente.

Si está escuchando una emisión internacional, puede esperarse una identificación clara al principio y al final de cada periodo de transmisión, y en uno o más puntos en medio. Las emisoras para servicios interiores podrían ser más difíciles de identificar (especialmente, por ejemplo, si se está en mitad de la transmisión de una obra de teatro o de un concierto), pero muchas se anunciarán a la hora o medio hora. Tenga en cuenta que las emisoras de Norteamérica, Australia, Nueva Zelanda y algunos otros países utilizan siglas de hasta cinco letras y/o números, mientras que en la mayor parte de los demás países se anuncia normalmente el nombre de la emisora. En Latinoamérica frecuentemente se añade al nombre oficial un slogan.

COLLABORATORS & MONITORS

In addition to the hundreds of individual readers who have sent us information during 1992, we wish to thank the following for special assistance in the compilation of this edition:

Argentina: Julian Anderson, Gabriel Iván Barrera, Daniel Camporini, Grupo Radioescucha Platense, Rubén G.Margenet, Emilio P.Povrzenic — **Australia:** DX Australia, Patrick McDonald, OZ DX — **Barbados:** Frankie P.Gittens —**Belgium:** Ludo Maes — **Brazil:** Rudolf W.Grimm, Claudio R. Moraes, Antonio Ribeiro da Motta — **Canada:** Alain Pepin — **Chile:** Pedro Morales C — **Colombia:** Yimber H.Gaviria, Rafael Rodríguez R.— **Czech Republic:** David Hertl — **Denmark:** Danish SW Clubs International — **Ecuador:** Rich McVicar — **Finland:** DXclusive, Timo Leponiemi — **France:** Roland Paget, Thierry Vignaud — **Germany:** Boris Böttcher, Uwe Bräutigam, Ingo Rose, Michael Schmitz — **India:** Manosij Guha, Dr.Supratik Sanatani — **Ireland:** Edward Dunne — **Italy:** Dario Monferini, Salvatore Placanica — **Japan:** Asian Broadcasting Institute, Tetsuya Hirahara, Takayuki Inoue Nozaki — **Lithuania:** Zigitas Zilionis, Rimantas Pleikys, Bernd Trutenau — **Netherlands:** Eric Beauchemin, Jan Willem Drexhage, Ehard Goddijn, Julius Hermans, Rocus de Joode, Jan Nieuwenhuis, Marcel Rommerts, Wian Stienstra, Herbert Visser, Liesbeth de Waard — **New Zealand:** Arthur Cushen, Paul Edwards, Paul Ormandy —**Nigeria:** Chijioke Anyanwu, Friday Okoloise — **Pakistan:** Svetomir Cukovic — **Panama:** Luis Angel Vargas — **Papua New Guinea:** Gordon Darling, Steve Lowe — **Peru:** Pedro F.Arrunátegui — **Spain:** Juan Franco Crespo —**Surinam:** Ravindranath G.Sewdien — **Sweden:** Stefan Bursell, Rikard Johansson, Shortwave Bulletin, Swedish DX Federation, Umeå Kortvågsklubb —**United Kingdom:** BBC Monitoring, British DX Club, Richard Hunt, David Thorpe — **Uruguay:** Gustavo Gil — **USA:** Kirk Baxter, Gary Bourgois, Jan Eric Österholm, Tom Sundström — **Venezuela:** Club DXistas de la Amistad, Manuel A.Rodríguez Lanza, Jairo Salazar Rojas.

WORLD BROADCAST RECEPTION CONDITIONS EXPECTED DURING 1993

Prepared by George Jacobs, MSEE Consulting Broadcast Engineer
P.O. Box 1714, Silver Spring, MD 20915, USA

Two events took place during 1992 that will influence considerably broadcast reception on the high frequency, or shortwave, bands during 1993 and for several years into the future. The first event was an act of nature, the second was an act of humans.

In September, 1986 nature created a new sunspot cycle, the 22nd since scientific telescopic observations were begun in the middle of the 18th century. Sunspot cycles wax and wane over an approximate 11-year period. Cycle 22 reached its peak intensity during 1989 with a count of 159. From then until the end of 1991, the cycle remained at a high plateau, with a slightly decreasing count ranging mainly in the 140's. But, in 1992 the cycle began a relatively steep decline which should carry it to its minimum level sometime during the latter half of this decade. The sunspot count during 1992 began at 124, and it is estimated to have dropped to 81 by the end of the year.

One may well ask why do spots on the face of the sun, about which science knows very little, be of interest to shortwave listeners?

An electrified region called the *ionosphere* exists in the upper atmosphere between approximately 100 and 600 kilometers above the earth's surface. The ionosphere is formed primarily by ultraviolet radiation from the sun. It is the ionosphere, acting much as a mirror does with light, that reflects high frequency or shortwave broadcasts over great distances.

Produced primarily by solar radiation, the strength of the ionosphere is dependent mainly on the relative position between the sun and the earth. The ionosphere is strongest near mid-day, and weakest during the hours of darkness. It is stronger at the equator and lower latitudes than at higher latitudes, and it is stronger on winter days than during summer days. The strength of the ionosphere is important for listeners to shortwave broadcasts because the stronger the ionosphere, the higher will be the MegaHertz band reflected, and the better will be shortwave reception conditions.

The ultraviolet radiation which produces the ionosphere is believed to originate mainly from violent volcano-like eruptions on the surface of the sun, called *sunspots.* The greater the number of sunspots, the stronger the ionosphere; the fewer the spots, the weaker the ionosphere.

The present cycle is expected to continue to decrease during 1993, dropping from an estimated level of 78 at the beginning of the year to approximately 56 by the end of 1993. The general state of the ionosphere is expected to be considerably weaker than it has been since 1987. Considering the natural relationship that exists between sunspots, solar radiation, the strength of the ionosphere and reception on the high frequency or shortwave bands, good reception should continue through 1993, but there will be some subtle changes noticed when compared to the super conditions of the past five years.

The second important event that occurred in 1992 was the convening of the **World Administrative Radio Conference (WARC-92).** The Conference, held under the auspices of the **International Telecommunication Union** (ITU) in Torremolinos, Spain from February 3 to March 3, was attended by more than 1400 delegates representing 127 countries. WARC-92 agreed to allocate radio spectrum for several future communication and broadcast systems, including satellite systems to provide world-wide cellular telephone service, High Definition Television (HDTV), and a Broadcasting Satellite Sound Service to provide compact disk digital audio quality. To readers of this column, however, the most important success of WARC-92 was the allocation of an additional 790 kiloHertz for shortwave broadcasting, including a new band between 18,900 and 19,020 kHz. While the new allocations are not officially to be available until the year 2007, and are presently reserved for single-side-band (SSB) broadcasting, they are expected to be used on a *de facto* basis by many broadcasting organizations during 1993.

The pre- WARC-92 Bands and the additions agreed to at WARC-92 appear in the following Table. While the 790 additional kiloHertz of spectrum allocated to h.f. Broadcasting fell considerably short of what many countries requested, the additional spectrum should permit clearer broadcasting in the future, with less interference and congestion than in the present bands.

HIGH FREQUENCY BROADCASTING BANDS

kiloHertz	MegaHertz Band		Meter Band
2,300-2,495	2		120*
3,200-3,400	3		90*
3,900-4,000	4		75**
4,750-5,060	5		60*
5,900-5,950	6	WARC-92 Allocation	49
5,950-6,200	6		49
7,100-7,300	7		41#
7,300-7,350	7	WARC-92 Allocation	41
9,400-9,500	9	WARC-92 Allocation	31
9,500-9,990	9		31
11,600-11,650	11	WARC-92 Allocation	25
11,650-12,050	11		25
12,050-12,100	11	WARC-92 Allocation	25
13,570-13,600	13	WARC-92 Allocation	22
13,600-13,800	13		22
13,800-13,870	13	WARC-92 Allocation	22
15,100-15,600	15		19
15,600-15,800	15	WARC-92 Allocation	19
17,480-17,550	17	WARC-92 Allocation	16
17,550-17,900	17		16
18,900-19,020	19	WARC-92 Allocation	15
21,450-21,750	21		13
25,600-26,100	26		11

*Tropical bands, for broadcasting use only in designated tropical areas.
**Regional band, used for broadcasting only in Europe and Asia.
#Not used for broadcasting in the Western Hemisphere.
All other bands are used on a world-wide basis.

The following is a general summary of reception conditions expected in each high frequency broadcasting band during 1993.

26 MHz Band (11 Meters): Conditions on this band are more susceptible to variation with changing solar activity than are any of the other bands. As solar activity decreases, fewer stations will make use of this band, but several broadcasting stations are planning to continue to use it during 1993. Long distance reception, up to several thousands of miles, should be best on this band during the daylight hours of the late fall, winter and early spring seasons. Signals will be strongest over north-south paths between the hemispheres during the equinoctial seasons, and strongest on east-west paths during the winter. Reception will be better in southern and tropical areas than at higher latitudes. During the summer months, very good reception should be possible over shorter distances, on the order of 800-1,200 miles, or 1,300-2,000 kilometers. Since only a small number of stations are expected to use this band during 1993, reception should be virtually free of interference.

21 MHz Band (13 Meters): This is another daytime band which will be suitable for long distance reception over several thousands of miles during 1993. Conditions should be best during the fall, winter and spring seasons. During the summer months, reception should be good on north-south paths, but considerably less so in easterly and westerly directions. Most major broadcasters are expected to make some use of this band during the equinoctial and winter seasons of 1993, and congestion and interference will at times be high. Because of decreasing solar activity, fewer stations plan to use this band during the summer months compared to the past several years of higher solar activity.

17 MHz Band (16 Meters): This is primarily a daytime band, with excellent reception expected throughout the year. Stations should be heard as close as 1,000 miles (1,750 km), to as far away as several thousand miles. A number of stations utilizing the 21 MHz band during the previous years of high solar activity are expected to move down to this already overcrowded band in 1993, and congestion and interference will often be very high. Relief may be sought by many stations in the new, but not yet official 19 MHz band.

15 MHz Band (19 Meters): During 1993, reception on this band is expected to be excellent for stations from several hundred to several thousands of miles away. This band is popular for both long distance reception of international broadcasts and for reception of shorter distance domestic broadcasts. Expect the strongest signals and reception of the most distant stations for an hour or two after local sunrise and again during the late afternoon and early evening hours. Medium and shorter distance reception should be possible throughout the daylight hours. During the summer months, excellent reception conditions may also extend well into the hours of darkness. Because of the popularity of this band, all major broadcasters are expected to use it, as well as a very large number of smaller broadcasters. For this reason, an exceptionally high level of congestion and interference is expected, particularly during the late afternoon and early evening.

13 MHz Band (22 Meters): This band was allocated to h.f. broadcasting by the ITU in 1979, but has not yet been officially authorized. Nevertheless, increasing use is being made of it, and every channel is expected to be occupied during 1993. The band should provide excellent reception during all seasons from sunrise to well past sundown. During the daylight hours, reception should be possible from stations located at distances between approximately 600 and 1800 miles (1,000 to 3,000 km), and considerably beyond this range during the winter season. During the later afternoon and early evening hours and again for several hours around the sunrise period, reception should favor longer distances, from approximately 1,000 miles (1,700 km) up to halfway around the world. During the summer months, excellent reception over long distances can also be expected during many hours of darkness. In time, this may well turn out to be one of the most popular bands for h.f. broadcast stations.

11 MHz Band (25 Meters): This is expected to be an excellent band for medium distance reception during the daylight hours, for stations located between approximately 500 and 2,000 miles away (800 to 3,500 km), during all seasons of the year. Longer distance reception, from stations up to several thousands of miles, should be possible for an hour or two after local sunrise, and again during the later afternoon and early evening. During the spring and summer months, medium and long distance reception should also be quite good throughout most of the hours of darkness. Heavy congestion and interference is expected during the late afternoon and early evening, with heavy use being made of this band by both international and domestic broadcasters.

9 MHz Band (31 Meters): This is a popular band during all seasons of the year for daytime broadcasting over medium distances, ranging between approximately 400 and 1,500 miles (700 to 2,500 km). During the spring, summer and fall months longer distance reception, up to distances of several thousands of miles, should also be possible during the hours of darkness, and until two to three hours after local sunrise. During the winter months, reception on this band is expected to

About the Author

George Jacobs was truly born into the age of shortwave, on the exact day in July, 1924 when Sir Edward Appleton, then with the BBC, discovered the ionosphere, that gaseous region in the earth's upper atmosphere upon which all shortwave radio depends.

This year marks George's 52nd in the field of broadcast engineering, and his 31st year with the **World Radio TV Handbook.**

During nearly forty years of Government Service, George was one of the principal architects of the world-wide broadcasting system of the **Voice of America** and the modernized technical facilities of **Radio Free Europe and Radio Liberty.** Retiring from Government Service in 1980 he continues to actively develop privately-owned international broadcasting systems. This past year he participated successfully in redesigning a Soviet jamming station into a 75 kW AM privately-owned broadcasting station providing the Moscow region with a 24-hour source of world and local news. Called the **OPEN RADIO,** it is fast becoming one of the most popular in the area.

become marginal by early evening, and remain so until daybreak. An exceptionally high level of interference is likely to occur, particularly during the late afternoon and early evening hours.

7 and 6 MHz Bands (41 and 49 Meters): These bands are expected to be very popular throughout the year for daytime domestic broadcasting, and for short to medium distance international broadcasting. Excellent reception conditions are expected for stations up to approximately 1,200 miles away (2,000 km). As nighttime approaches, reception of stations over increasingly longer distances should be possible. During the hours of darkness, reception of stations up to several thousands of miles away should improve. Higher static during the late spring and summer months may make long distance reception difficult at times. Almost half of the world's total high frequency broadcasting effort is expected to take place in these bands during 1993. This will result in an extremely high level of congestion and interference, peaking during the late afternoon and evening hours.

5, 4, 3 and 2 MHz Bands (60, 75, 90 and 120 Meters): All except the 4 MHz band are used exclusively in designated tropical areas for domestic broadcasting. The 4 MHz band is available throughout both Europe and Asia for domestic broadcasting. During the hours of daylight, reception in these bands should be possible over a range extending out to approximately 500 miles (800 km). After sunset, during the hours of darkness and the sunrise period, the range of reception should increase to a thousand miles or more, with the more powerful stations heard over greater distances. Very high static levels are likely to make reception difficult at times, particularly during the late spring, summer and early fall months.

DO-IT-YOURSELF PROPAGATION FORECAST TABLE FOR THE MOST SUITABLE MEGAHERTZ BROADCASTING BANDS 1993

Prepared by George Jacobs, MSEE, Consulting Broadcast Engineer
P.O. Box 1714, Silver Spring, MD 20915, USA

Thanks to computer programs, h.f. propagation forecasts, which at one time required a great deal of technical expertise, can be done on a do-it-yourself basis. The following **Table** is an example of the simplified nature of such forecasting.

The **Table** can be used to determine the MagaHertz band that should be best for reception in a given listener's area, for a specific radio transmitter location, month and time period. The **Table** should be as easy to use as reading a map or a transportation schedule.

Included in the **Table** are **Transmitter Station Locations** for each major geographical area of the world. Data appearing in the **Table** are based primarily on the **IONCAP** propagation program, which is an international standard. The influence that declining sunspot activity is expected to have upon the ionosphere during 1993 has been taken into account, as has been the planned operating schedules of a large number of World Band stations.

To use the **Table** properly, all that is required is a straight-edge. Begin by locating your geographical area in the left hand column marked **Listener's Area.** For the time period that you are interested in (given in the **Table** in both UTC and approximate local time), run your finger horizontally to the right until you reach the box containing the appropriate month within the block marked **Transmitting Station Location**. For the appropriate month within the this block continue to the right until your finger intersects the geographical area from which you want to listen to a broadcast. The best high frequency band for reception appears at this point of intersection. It may often also be possible to receive broadcasts from the desired geographical area on the next higher or lower MegaHertz bands. You may find it a bit more convenient to use a straight-edge rather than a finger to determine the best MegaHertz band.

Since many readers prefer the use of **Meter Bands** instead of **MegaHertz Bands** the following conversion table can be used.

MegaHertz Broadcasting Band	Meter Broadcasting Band
25	11
21	13
17	16
15	19
13*	22*
11	25
9	31
7	41
6	49
5/4**	60/75**

*This band does not appear in the **Table**. Reception conditions should be approximately the same as shown for the 11 MegaHertz Band (25 Meter Band).

These bands do not appear in the **Table. Reception conditions should be approximately the same as shown for the 6 MegaHertz Band (49 Meter Band).

A few words of caution. The months shown beneath the **Transmitter Station Location** are given as months and *not as seasons.* This is to make the **Table** usable in both the northern and southern hemispheres. For example, when the transmitter location is in the Northern Hemisphere, the summer period would be under the **May-August** column. But, if the transmitter location is in the Southern Hemisphere, the summer months would be under the **Jan./Feb. & Nov./Dec.** column. *To avoid this potential confusion, think in terms of the months shown, rather than of seasons, and the **Table** should produce a correct reading.*

Continued on page 36

DO-IT-YOURSELF PROPAGATION FORECAST TABLE FOR THE MOST SUITABLE MEGAHERTZ BROADCASTING BANDS 1993

Prepared by George Jacobs, MSEE, Consulting Broadcast Engineer
P.O. Box 1714, Silver Spring, Maryland, USA 20915

TRANSMITTING STATION LOCATION

LISTENER'S AREA	LOCAL TIME	APPROX. UTC TIME	JAN./FEB. & NOV./DEC.								MAR./APR. & SEPT./OCT.								MAY–AUGUST							
			EUR/NAF	N.AM(E)	N.AM(W)	C/S.AM	C/S.AF	ME/S.AS	E.AS	AUS/NZ	EUR/NAF	N.AM(E)	N.AM(W)	C/S.AM	C/S.AF	ME/S.AS	E.AS	AUS/NZ	EUR/NAF	N.AM(E)	N.AM(W)	C/S.AM	C/S.AF	ME/S.AS	E.AS	AUS/NZ
EUROPE AND NORTH AFRICA	0000-0400	2300-0300	6	6	6	9	9	9	9	9	6	9	9	11	9	9	9	9	9	11	11	11	9	9	9	11
	0400-0800	0300-0700	6	6	6	6	9	9	9	9	6	9	9	11	11	9	11	11	9	9	9	11	11	11	11	9
	0800-1200	0700-1100	11	6	9	9	17	15	15	15	11	9	9	11	15	15	17	15	11	9	9	11	15	17	17	17
	1200-1600	1100-1500	11	15	9	15	17	15	15	15	11	11	11	15	17	15	15	15	11	15	11	15	17	15	17	17
	1600-2000	1500-1900	9	17	15	17	15	11	9	9	9	17	17	17	17	15	11	11	11	15	15	15	15	15	15	9
	2000-2400	1900-2300	6	11	9	11	11	7	9	9	6	11	15	15	11	9	9	9	9	11	11	15	11	9	11	9
NORTH AMERICA (EAST)	10PM-2AM	0300-0700	6	6	9	9	9	6	11	11	9	6	11	9	9	9	11	11	9	9	11	9	9	9	11	15
	2AM-6AM	0700-1100	9	6	6	9	9	9	9	9	9	6	9	6	9	9	9	11	11	6	9	9	11	11	11	11
	6AM-10AM	1100-1500	11	9	9	9	17	15	9	9	15	15	9	11	17	15	11	9	15	15	9	11	15	15	11	9
	10AM-2PM	1500-1900	17	15	11	17	21	15	11	11	17	15	11	17	17	15	15	15	15	15	15	15	17	15	15	11
	2PM-6PM	1900-2300	15	11	15	15	15	15	9	11	11	15	15	15	15	15	15	11	15	15	15	15	15	15	15	15
	6PM-10PM	2300-0300	9	6	15	9	9	9	15	15	9	6	11	9	9	9	15	17	11	9	15	9	11	11	17	17
NORTH AMERICA (WEST)	12M-4AM	0800-1200	6	6	6	9	9	9	9	11	9	9	6	9	11	9	9	11	9	9	6	9	11	11	9	11
	4AM-8AM	1200-1600	9	6	9	9	11	11	9	11	11	11	6	11	15	11	9	11	11	11	6	11	11	11	11	11
	4AM-12N	1600-2000	11	15	9	15	17	15	11	11	17	15	9	15	17	15	11	11	15	15	9	15	17	15	15	11
	12N-4PM	2000-2400	11	15	11	15	15	11	15	17	15	11	15	15	15	11	15	17	15	15	15	15	15	15	15	15
	4PM-8PM	0000-0400	6	11	9	11	11	11	17	17	9	9	11	11	11	11	17	17	11	15	15	11	11	15	15	17
	8PM-12M	0400-0800	6	6	6	9	9	9	11	15	9	9	9	9	9	9	15	15	11	9	11	9	9	11	15	15
CENTRAL AND SOUTH AMERICA	12M-4AM	0400-0800	9	9	9	6	9	9	11	15	9	11	11	6	9	9	11	15	11	11	11	6	9	11	11	15
	4AM-8AM	0800-1200	9	6	9	9	15	11	9	11	11	9	9	9	11	11	11	11	11	9	9	9	11	11	11	15
	8AM-12N	1200-1600	17	15	11	11	17	15	11	11	17	17	11	11	17	17	11	11	17	15	11	11	15	17	11	11
	12N-4PM	1600-2000	17	17	15	15	15	15	11	11	17	17	15	17	15	15	15	11	17	17	15	15	17	15	15	11
	4PM-8PM	2000-2400	11	11	15	11	15	15	15	17	15	15	17	11	11	15	15	17	15	15	17	11	15	15	15	17
	8PM-12M	0000-0400	9	9	11	9	11	11	15	15	11	11	15	6	11	11	15	15	11	15	15	6	11	15	15	17
CENTRAL AND SOUTH AFRICA	0000-0400	2200-0200	9	11	15	11	7	9	9	15	11	11	15	11	9	9	11	15	11	15	15	11	9	11	11	11
	0400-0800	0200-0600	9	9	9	9	9	11	11	17	9	9	11	9	9	11	11	17	11	9	9	9	9	11	9	17
	0800-1200	0600-1000	15	9	11	15	11	15	15	15	17	11	11	11	11	15	15	15	15	9	11	15	11	15	15	15
	1200-1600	1000-1400	21	15	15	15	11	15	15	11	21	17	15	17	15	15	15	15	17	15	15	17	11	17	17	15
	1600-2000	1400-1800	17	17	15	17	11	15	15	9	17	17	15	17	11	15	15	9	17	17	17	17	11	15	15	11
	2000-2400	1800-2200	11	15	11	15	11	11	9	15	11	17	15	17	11	11	11	15	15	15	15	15	11	15	11	9
MIDDLE EAST AND SOUTH ASIA	0000-0400	2100-0100	9	9	9	9	11	9	9	9	9	9	11	11	11	9	9	11	11	11	11	11	11	9	9	11
	0400-0800	0100-0500	9	6	9	9	11	9	15	15	9	9	11	11	9	9	15	15	11	11	11	11	9	9	15	15
	0800-1200	0500-0900	11	6	9	11	15	11	15	17	11	9	9	15	15	15	17	17	15	11	11	11	15	11	17	17
	1200-1600	0900-1300	17	6	6	11	17	15	15	11	17	11	9	11	17	15	15	11	17	15	9	11	17	15	15	11
	1600-2000	1300-1700	17	15	11	15	17	11	11	9	17	17	11	17	21	11	11	9	17	15	11	15	17	11	11	9
	2000-2400	1700-2100	9	11	11	15	15	9	9	9	15	15	11	15	15	9	11	9	15	15	11	15	11	9	11	9
EAST ASIA AND FAR EAST	0000-0400	1600-2000	9	11	11	11	11	9	6	11	11	11	15	11	9	9	9	11	15	15	15	15	9	11	9	11
	0400-0800	2000-2400	9	9	11	11	11	11	7	15	9	15	15	15	11	11	9	15	11	15	15	15	11	11	9	15
	0800-1200	0000-0400	9	9	15	15	11	11	15	17	9	15	17	15	11	15	15	17	9	15	15	15	11	15	15	17
	1200-1600	0400-0800	11	9	11	11	15	17	15	17	11	11	15	11	15	17	15	17	15	11	15	11	15	17	15	17
	1600-2000	0800-1200	15	9	9	11	15	15	11	15	15	9	11	11	17	17	11	17	17	11	11	11	17	17	11	15
	2000-2400	1200-1600	11	11	9	11	11	11	9	11	15	11	9	11	15	15	9	11	15	11	9	15	15	15	9	9
AUSTRALIA AND NEW ZEALAND	0000-0400	1400-1800	15	15	9	11	11	9	11	9	15	15	9	11	9	9	11	9	11	11	11	11	11	9	11	9
	0400-0800	1800-2200	9	15	11	11	11	9	11	9	9	11	11	15	11	9	11	9	9	15	11	11	11	9	11	9
	0800-1200	2200-0200	11	15	17	17	11	15	15	11	9	17	21	15	11	15	15	15	9	17	17	17	11	11	15	15
	1200-1600	0200-0600	9	9	11	15	15	15	15	15	11	11	15	17	15	15	17	15	9	15	15	15	15	15	17	17
	1600-2000	0600-1000	9	9	11	15	15	15	15	11	11	11	11	15	17	15	15	11	15	11	11	15	17	15	15	15
	2000-2400	1000-1400	17	9	9	11	11	11	15	9	15	9	9	11	15	11	11	9	17	9	11	15	15	11	11	9

BAND SELECTIONS HAVE BEEN MADE TAKING INTO ACCOUNT BOTH PROPAGATION CONDITIONS AND STATION OPERATING SCHEDULES

WHO SAYS YOU CAN'T TAKE IT WITH YOU WHEN YOU GO.
ATS 818
LIGHT
POWER
SLEEP
TIME SET
DUAL TIME
FM
MW
LW
SW
MANUAL / AUTO
A WORLD OF LISTENING

Continued from page 30

Many of the larger international broadcasting organizations, such as the VOA, BBC, RFI and Deutsche Welle, maintain relay stations throughout the world. When using the **Table,** it is important that the *correct* transmitter location is used. For example, if a listener in France wanted to hear the VOA, the **Table** would produce one answer for a VOA transmitter at Greenville, which is located in **North America (East),** but very likely a different answer for a VOA transmitter located at Tangier, which is shown as **Europe and North Africa** in the **Table**.

Here are a few examples illustrating the proper use of the **Table**.

Assume that a listener in Durban, South Africa wants to listen to the **Monitor World Service** from WCSN, Maine at 6 am local time during December. Since Maine is located in the eastern part of the USA, first enter the column under the **Listener's Area** shown as **Central and South Africa.** Place a straight-edge horizontally at the **0400-0800** local time slot, since 6 am is equivalent to 0600 local time. Slide your finger horizontally along the straight-edge until you reach the **Jan./Feb. & Nov./Dec.** month column under **Transmitter Station Location.** Continue to run your finger along the straight-edge until you intersect the transmitter location **N.AM(E).** At this intersection read **9** MegaHertz. This would be the best band for reception of WCSN at 6 am in Durban during December. (The **Monitor World Service** is in fact transmitted by WCSN to southern Africa between 6 and 8 am local time on 9840 kHz.)

The **Table** can be used another way. Suppose that during March a listener in Vienna, Austria is free to tune his h.f. radio at 1700 local time, and he wants to have some idea of what he might hear on the **17** MegaHertz band. Place a straight-edge horizontally at the **1600-2000** local time slot under the Europe and North Africa **Listener's Area.** Run a finger along the straight-edge to the **March** month under the **Transmitting Station Location** column. Read along the straight-edge that the **17** MegaHertz band is expected to be optimum to **North America (East), North America (West), South and Central America** and **Central and South Africa** at 1700.

For more information about high frequency reception conditions expected during 1993, refer to **World Band Broadcast Reception Conditions Expected During 1993,** which can be found on Page 26.

SOLAR ACTIVITY 1993

by Mike Bird

That orbed continent,
the fire that severs day
from night.

Twelfth Night
William Shakespeare

Every day of our lives the Sun touches us in one way or another. The most obvious is the light and heat which not only produces night and day, but also the climate dynamo that allows life on Earth to exist.

The variation in the number of sunspots, the so-called solar cycle has fascinated man for a very long time. It is only in the past 60 years that we really started to understand the Sun's electromagnetic output. This was due in no small way to the discovery of the ionosphere and its ability to propagate shortwave signals. Since those early days we have learnt a lot about the solar terrestrial relationship and, in particular, not only how important the solar cycle is to shortwave broadcasting but also that it affects space exploration and satellite communications.

About six months ago, I was reading an article in a well-known radio magazine in which the author stated that "The present solar cycle, the 22nd observed since accurate records have been kept, appears to have stalled".

I stopped and thought about it, since solar activity affects radio propagation. What did the author really mean? Perhaps he was referring to the fact that the smoothed sunspot number had remained fairly static for many months, which in turn, had produced a graph which appeared constant. Most cycles are made up of three distinct phases, and they are the rise, maximum and decline. The current cycle started in September 1986 and rose very quickly, appearing to enter its maximum phase around December 1988. This lasted for over three years until about March 1992 when the monthly sunspot number started to drop quite rapidly and, we think, the cycle entered the declining phase. During the maximum phase the Sun's ultraviolet output was very high and produced an ionosphere which was capable of propagating frequencies well in excess of 30 MHz! Of course, with this increased output came an increase in solar activity such as solar flares and coronal holes.

One peculiar aspect of the solar cycle as seen in the graph of the smoothed sunspot number is the twin hump effect which looks somewhat like a camel's back. Certainly for the past few cycles the sunspot number has risen, peaked, dropped back slightly, and then peaked again at a lower value than the main one. Why this occurs is, as yet, unknown.

The Past Year (September 1991 - August 1992)

At the beginning of the year, September 1991, the smoothed sunspot number stood at 144, by January 1992 it was dropping rapidly and had reached 123. It is expected by December 1992 that it will have dropped even further to around 62.

Solar activity in terms of solar flares between September 1991 - August 1992 declined by around 40%. This is demonstrated by comparing solar flare activity for the past two years.

Solar Flare Activity

September 1990 - August 1991
M class = 513 X class = 31 Total 544
September 1991 - August 1992
M class = 321 X class = 18 Total 339

SRI on the Air
Swiss Radio International:
- worldwide on shortwave
- in Europe via satellite
Warning: Listening to SRI can be habit-forming!
For detailed information write to: Swiss Radio International, P.O.Box, CH-3000 Berne 15, Switzerland
Telephone: 031 43 92 22, Telefax: 031 43 95 44

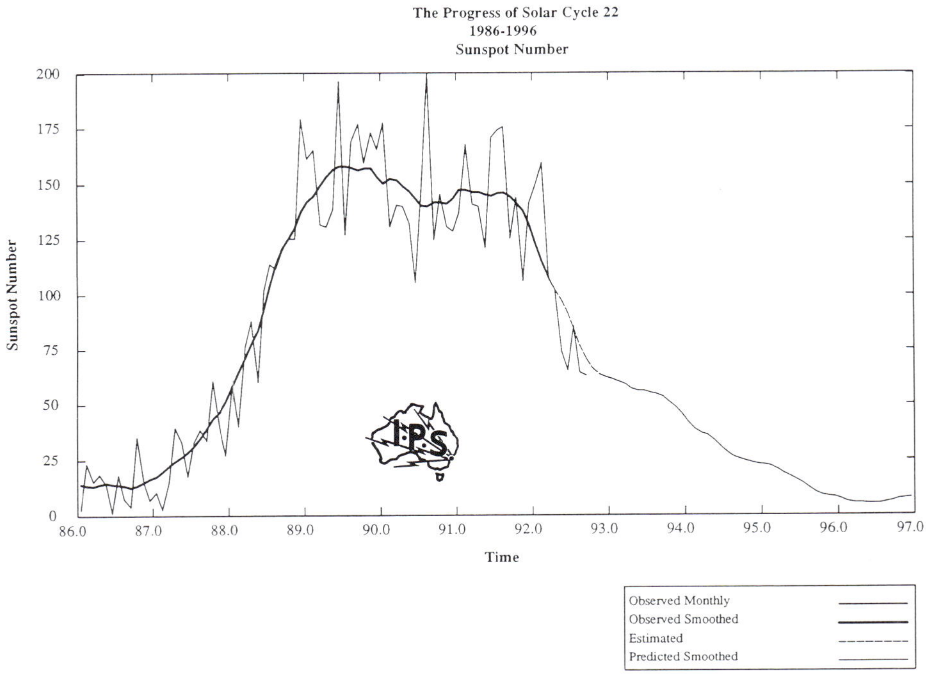
The Progress of Solar Cycle 22
1986-1996
Sunspot Number
Sunspot Number
200
175
150
125
100
75
50
25
0
86.0
87.0
88.0
89.0
90.0
91.0
92.0
93.0
94.0
95.0
96.0
97.0
Time
IPS
Observed Monthly
Observed Smoothed
Estimated
Predicted Smoothed

Fig. 1

SOLAR ACTIVITY 1993

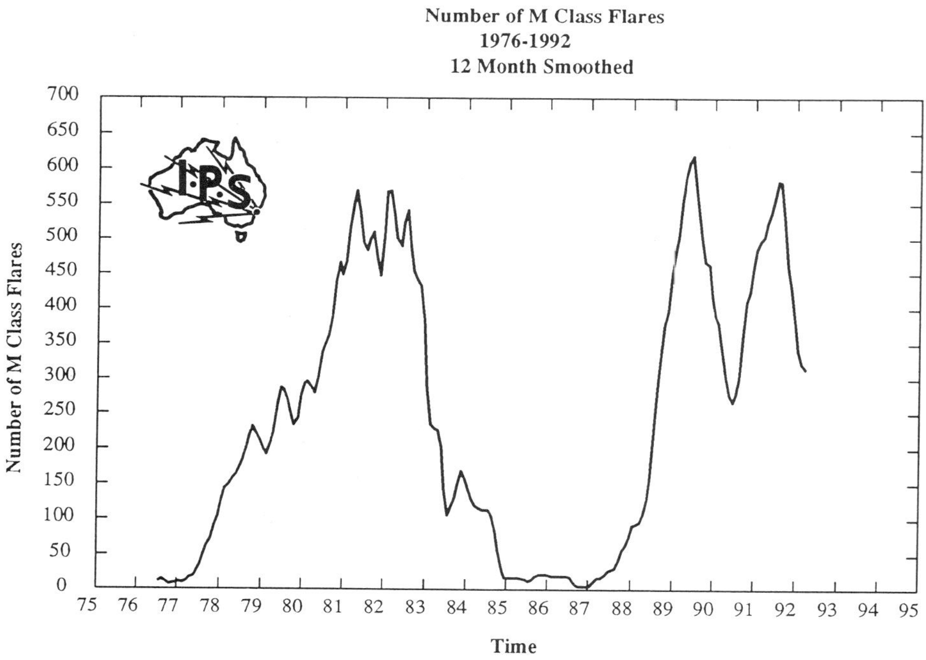

Fig. 2

Looking at the past year in more detail, solar activity was at relatively strong levels between September and November 1991. December 1991 saw a substantial increase in flare activity caused by four sunspot regions. By the end of the month 91 M class and 3 X class flares had been recorded. Activity returned to more moderate levels in January and February 1992, but between March and June 1992 activity declined to very low levels. Finally, between July and August 1992 we saw another slight increase.

The geomagnetic field whose state (or should I say disturbance) affects shortwave propagation appears to have suffered similar levels of disturbance this past year as the previous year. This is illustrated by the following table.

Geomagnetic Activity (days disturbed)

September 1990 - August 1991		
quiet 241	disturbed 77	stormy 47
September 1991 - August 1992		
quiet 228	disturbed 90	stormy 48

Geomagnetic disturbance indicator.

Planetary A index
0-14 = quiet 15-24 = disturbed 25+ = stormy

While there were no notable disturbances during September 1991, October saw a large storm take place on the 27th and 28th and the A index rose to 85 and 84. It appears that flare activity was the culprit. November saw more geomagnetic activity particularly on the 1st, 8th, 9th, 19th and 22nd. Solar flare activity appears to have been the main cause. December 1991 and January 1992 were magnetically quiet months, but February saw a return of more activity. The A index reached 58 on the 3rd, 48 on the 8th, and 53 on the 21st. Again all of these were flare induced.

March and April were quiet months, but this was short lived because in May a very large storm took place. On 8 May an M7 solar flare erupted, producing a shortwave fadeout between 1512 and 1610 UTC. The next day the A index reached 27, and on the 10th it rose to 116. Not only was this the largest A index for the year, shortwave propagation was all but non-existent! June, July and August 1992 while experiencing some geomagnetic storming were relatively quiet months. Even though the largest storms in 1991-1992 were still caused by large solar flares, coronal holes were becoming more and more the source of geomagnetic disturbances, particularly at the lower storm thresholds.

With the continuing downturn in solar cycle 22, the all important extreme ultraviolet (EUV) output from the Sun has also declined. EUV radiation from the Sun strips electrons from atoms in the upper atmosphere producing free electrons and positively charged ions, hence the name ionosphere. This allows shortwave signals to be reflected or more correctly, refracted, back to Earth.

At any given time the ionosphere can support a range of shortwave frequencies. The measure of the ionosphere's ability to do this is often referred to in terms of Maximum Usable Frequency (MUF). At solar maximum the range of frequencies spans the total shortwave spectrum and more, while at solar minimum the range of coverage is about half. During the past year the EUV has declined with the solar cycle. This has meant that the MUF has also declined by typically 20-30%.

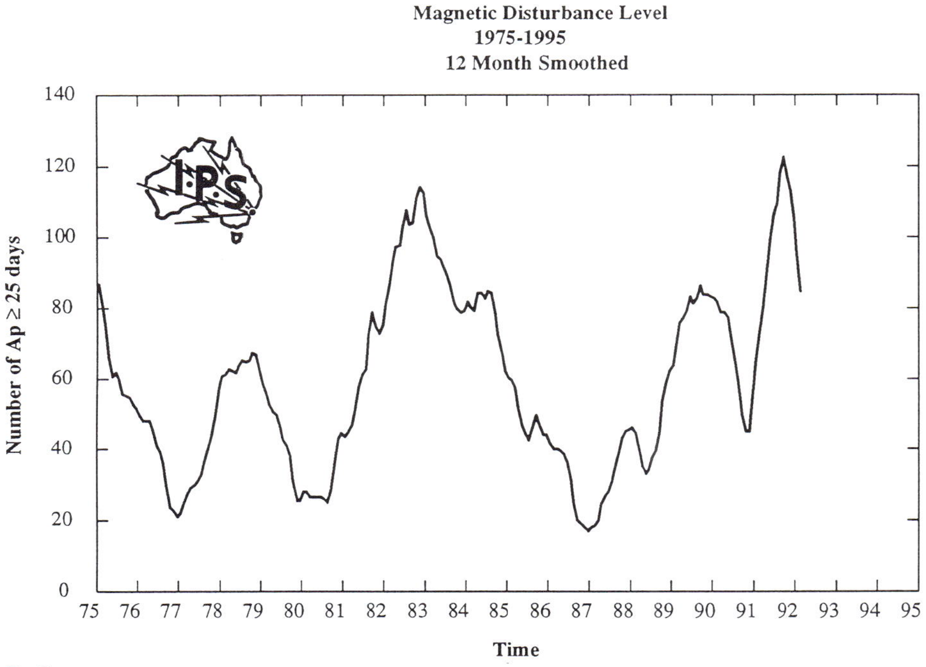
Magnetic Disturbance Level
1975-1995
12 Month Smoothed
I.P.S.
Number of Ap ≥ 25 days
0
20
40
60
80
100
120
140
75 76 77 78 79 80 81 82 83 84 85 86 87 88 89 90 91 92 93 94 95
Time

Fig. 3

SOLAR ACTIVITY 1993

Monthly Values

	10 CM FLUX Monthly Average	SUNSPOT Monthly Average	NUMBER Yearly Average	A INDEX Monthly Average	FLARES >M1.0
August 92	122.2	64.4		12.6	10
July 92	132.3	85.4		10.2	13
June 92	116.7	65.3		13.9	9
May 92	125.5	73.5		16.4	5
April 92	158.2	102.2		9.3	8
March 92	171.3	106.9		12.3	4
February 92	232.1	159.6	114.8	22.6	47
January 92	217.6	149.3	123.1	11.7	38
December 91	223.8	141.2	131.3	11.1	94
November 91	171.9	106.1	137.6	24.2	28
October 91	201.4	143.6	141.2	22.0	45
September 91	177.4	125.3	144.2	15.5	25

The All Powerful Sun

While solar physicists and broadcast engineers are well aware of what the Sun is capable of producing in terms of disruptions, it is often overlooked that the Sun has the ability to damage spacecraft and satellites as well as upset power utilities.

This was amply demonstrated during the previous solar maximum phase. In July 1979 after providing so much new solar related information it was ironic that the Sun was Skylab's source of destruction. As the solar cycle increased so did the heating of the Earth's atmosphere which in turn increased the drag on Skylab. Unfortunately the space shuttle project was badly behind schedule and didn't make its first flight until 1981. One of the shuttle's first missions would have been to boost Skylab into a higher orbit. Skylab tumbled prematurely from orbit and burnt up over Western Australia.

During 1989 and 1991 when solar activity was high United States astronauts were glad of the shielding which had been built into the space shuttle. Astronauts have to take particular care when solar flare activity is high, as the X-ray and proton emissions could be dangerous, if not fatal. Even when the astronauts retreated into the innermost shielded part of the shuttle, problems still occurred. On the Atlantis space shuttle mission during October 1989, a series of exceptionally large and longlasting flares took place emitting some very energetic high speed protons. Apparently these protons produced irritating "flashes" in the astronauts' eyes as the particles penetrated their optic nerves!

As we all know, geosynchronous and orbiting satellites are launched either by using rockets or the space shuttle. If the complexity of getting satellites to their point in the sky wasn't enough, the environment is also a hostile one and offers many engineering challenges.

For some time, the United States National Geophysical Data Center (NGDC) in Boulder, Colorado, has been collecting data on solar induced disturbances. During 1989 and 1991 when solar activity was high the NGDC received many reports of "hits", "losses" and damage.

The National Oceanic and Atmospheric Administration's Geosynchronous Orbiting Environmental Satellites (GOES) which monitor the Earth's weather, as well as the Sun and space environment, suffered significantly. Large solar flares at times poured out massive amounts of X-rays as well as high speed protons which bombarded these satellites, degrading their solar panels and shortening their operating lives.

On 29 September 1989 the GOES-7 satellite received a dose of high speed protons which degraded its solar panels as much as would normally have been expected in one year.

Proton events can also have disruptive effects on satellite circuitry. If a high energy proton particle penetrates a memory chip or RAM, it can upset the satellite's memory by changing the onboard software. This can usually be dealt with by re-programming but sometimes it cannot and the satellite's function becomes prematurely impaired.

Solar flare induced geomagnetic storms are also responsible for upsets to satellites. Storms create heat which causes the atmosphere to expand; this in turn creates drag on polar orbiting satellites. If the storm is big enough it can also make them descend rapidly to unexpected altitudes. Geosynchronous satellites can experience a static charge build-up on their external surfaces. When this charging is eventually released, it can physically damage the satellite and produce spurious commands.

Turning from satellites (and not wanting to appear alarmist), high flying aircraft can also be affected by the Sun. Again during the large solar flares which took place in September 1989 it was discovered that sensors on board Concorde supersonic jets recorded that passengers and crew had received radiation doses equivalent to a chest X-ray!

Finally and on terra firma, the Sun can also unleash great forces. During the great geomagnetic storms that took place in 1989 and 1991 power utilities situated within and close to the auroral regions experienced many interruptions to their services. The Hydro-Quebec Power Company experienced a massive failure that darkened most of Quebec Province for up to nine hours during March 1989. This was due to a geomagnetic storm which saturated transformers and tripped circuit breakers. Similar losses were also reported in Sweden. An engineer with Minnesota Electric has declared that since the big geomagnetic storms in 1989 many of their transformers have "aged" far quicker than normal.

Let's hope the large solar flares and geomagnetic storms are over for Cycle 22. While they are interesting to the scientific and lay community, they impose a heavy burden on the systems and services we depend on.

Forecast for 1993

Mankind has watched and marveled at the Sun for many thousands of years. While we've known about sunspots for many hundreds of years, it was only in the middle of the last century that the 11-year solar cycle was discovered. Like many aspects of the solar cycle, unpredictability is the dominant factor. Solar cycles on average last 11 years, but there is wide variation in the period of individual cycles. Some have lasted 7.5 years whereas others have lasted up to 16 years. Generally the rise of a cycle is faster than the decline. The average period from minimum to maximum is about 4.5 years, whereas the period between maximum and minimum is about 6.5 years.

The rise of this present cycle (22) was the fastest on record! The last cycle (21) only lasted 10 years. Will this happen this time? There is a growing opinion in the Solar Terrestrial scientific community that it will, and that the next minimum will take place in late 1995 or early 1996. If that is so, the decline will be more rapid than normal.

During 1993 it's believed that the solar cycle will continue on its way downwards. We should see a further drop in the number of solar flares and flare induced geomagnetic storms. But as history tells us, there is a period of up to four years after the sunspot cycle has peaked when solar flare-induced geomagnetic storming peaks. The two most likely periods of disturbances to the geomagnetic field and thus to the ionosphere should take place around the March and September equinox, when the coupling between the Sun and Earth by the solar wind is strongest.

IPS Predicted Sunspot Numbers Issued October 1992

Year	Jan	Feb	Mar	Apr	May	Jun	Jul	Aug	Sep	Oct	Nov	Dec
1992	123.1	114.8	107.6	102.2	97.2	91.4	84.3	76.4	70.2	66.2	63 .3	62.2
1993	61.2	60.0	58.8	56.8	56.0	55.8	55.2	54.4	53.2	51.0	48.4	45.6
1994	41.9	38.7	37.0	35.9	33.6	30.9	28.6	26.5	25.2	24.2	23.4	22.7
1995	22.4	21.8	20.4	18.6	17.1	15.8	14.2	12.1	10.1	8.8	8.3	7.8
1996	6.9	5.9	5.3	5.1	4.9	4.8	5.0	5.5	6.2	6.9	7.2	7.5
1997	8.3	9.8	11.3	12.7	14.6	16.9	16.8	24.9	22.3	27.4	30.7	34.0

In conclusion, it's thought that the all important MUF will continue to decline by as much as 20% in 1993. This means that shortwave broadcasters will have to seriously consider their optimum working frequencies and prepare for solar minimum.

Good listening

My thanks to I.P.S. Radio and Space Services, Sydney and to the United States National Geophysical Data Center, Boulder, Colorado for their help.

INTERGOVERNMENTAL ORGANIZATIONS

INTERNATIONAL TELECOMMUNICATION UNION

ADDR: Place des Nations. 1211 Geneva 20, Switzerland.
Te: +41 22 730 51 11. **Fax:** +41 22 733 72 56.
Telex: 421 000 UIT CH. **Teletex:** 228 468 151 00 = UIT.
Secretary Gen: Dr. P. Tarjanne (Finland). Dep. Secr. Gen: Jean Jipguep (Cameroon).

The International Telecommunication Union (ITU) was founded in 1865 and as such is the oldest intergovernmental organization. In 1947, it became a specialized agency of the United Nations and now has membership of 172 countries. It is the international organization responsible for the regulation and planning of telecommunications worldwide, for the establishment of equipment and systems operating standards, for the coordination and dissemination of information required for the planning and operation of telecommunications networks & services and for the promotion of and contribution to the development of telecommunications and the related infrastructures.

International Frequency Registration Board (IFRB)
Chmn: W.H. Bellchambers (UK). Vice-Chmn: M. Miura (Japan).
Members: G.C. Brooks (Canada), M. Harbi (Algeria), V. Kozlov (Russia).
Intern. Tel. & Tel. Consultative Committee (CCITT): Dir: T. Irmer.
Intern. Radio Consultative Committee (CCIR): Director: Richard C. Kirby; Study Group 10 (broadcasting), Chairman: C. Terzani (Italy); Study Group 11 (television), Chairman: M. I. Krivosheev (Russia).

Administrative Council: Americas: Argentina, Brazil, Canada, Colombia, Cuba, Jamaica, Mexico, USA. **We. Europe:** France, Germany (F.R.), Greece, Italy, Spain, Sweden, Switzerland. **Ea. Europe & No. Asia:** Bulgaria, Czechoslovakia, Russia, 1 vacant. **Africa:** Algeria, Benin, Burkina Faso, Cameroon, Cape Verde, Egypt, Kenya, Mali, Morocco, Nigeria, Senegal, Tanzania. **Asia & Australasia:** Australia, China (P.R.), India, Indonesia, Japan, Republic of Korea, Malaysia, Kuwait, Pakistan, Philippines, Saudi Arabia, Thailand.

UNITED NATIONS

ADDR: United Nations, New York, NY 10017, USA.
L.P: Secretary-General: Boutros Boutros-Ghali. Under-Sec-Gen for Public Inf.: E. Wyzner. Dir. Inf. Prod. Div: G. Leclère. Dir. Dissem. Div: V. Perfiliev. Dep. Dir Dissem. Div: K. Chhor. Electronic Magazine & Feature Sect./Inf. Prod. Div: S. Rowe.

UNESCO

(United Nations Educational, Scientific and Cultural Organization).
ADDR: 7 Place de Fontenoy, F-75700 Paris, France.
Te: +33 (1) 45 68 10 00.
L.P: Dir. Gen: Federico Mayor Zaragoza.

OFFICE OF PUBLIC INFORMATION (OPI)
Audio-Visual Production Division: Chief, Radio Production, English: Erin Faherty Mel a. Film enquiries and distribution: A. Fenton.

COMMUNICATION, INFORMATION & INFORMATICS SECTOR (CII): Asst. Dir. Gen.: Henrikas Iouchkiavitchious.
Communication Division (COM): Dir: Alain Modoux.
International Programme for the Development of Communication: (IPDC): Dir: Claude Ondobo.

ASIA-PACIFIC INSTITUTE FOR BROADCASTING DEVELOPMENT (AIBD)

Secretariat: Director and Secretary to Council: Dato Abdullah Mohamad. Deputy Dir: F. Hugh de Silva.
Address: P.O. Box 1137, Pantai, 59700 Kuala Lumpur, Malaysia.
Telex: MA 30083 APBRO. **Cable:** UNESBROAD Kuala Lumpur.
Te: +60 (3) 2821046/2824618/2823719/2827192. **Fax:** +60 (3) 2822761.

Nature: Intergovernmental institute for the development of broadcasting, operating since 1972 and established formally in 1977, serving countries of the United Nations Economic and Social Commission for Asia and the Pacific (UN ESCAP) region in the development and upgrading of their broadcasting personnel and systems.

Supreme Body: Governing Council of 11 voting member countries plus UNDP, UNESCO, ITU, ABU, CFTC and other non-voting representative private or public funding agencies invited to attend annual meeting.

Training Centre: At Malaysian Government National Broadcasting Training Centre (IPTAF), Angkasapuri, Kuala Lumpur, Malaysia.

Members: Afghanistan, Bangladesh, Brunei, China, Fiji, France, India, Indonesia, Iran, Republic of Korea, Lao People's Democratic Republic, Malaysia, Maldives, Nepal, Pakistan, Papua New Guinea, Philippines, Samoa, Singapore, Sri Lanka, Thailand and Vietnam.

INTERNATIONAL BROADCASTING ORGANIZATIONS

EUROPEAN BROADCASTING UNION (EBU)
UNION EUROPEENNE DE RADIODIFFUSION (UER)
(EUROVISION/EURORADIO)
ADDR: Ancienne Route, 17A, Case Postale 67, CH-1218 Grand-Saconnex (GE), Switzerland. **Cables:** Uniradio, Geneva. **Te:** +41 (22) 7172111. **Telex:** 415700 ebu ch.**Telefax:** +41 (22) 7985897.
Eurovision Operations: 32, Av. Albert Lancaster, B-1180 Brussels, Belgium. **Cables:** Uniradio Brussels. **Te:** +32 (2) 375 59 90. **Telex:** 21230, 21939. **Telefax:** +32 (2) 3755081. Receiving and Measuring Station: Jurbise, Belgium. Eurovision Control Centre: Brussels Belgium.
Administrative Council: President: A. Scharf (ARD, German Fed. Rep). Vice-Presidents: M. Checkland (BBC, UK), M. Fichera (RAI, Italy). Administrators: Representatives of broadcasting organisations in Algeria (ENTV/ENRS/ENTD), Denmark (DR), Egypt (ERTU), Finland (YLE), France (GRF), German Fed. Rep. (ARD/ZDF), Greece (ERT), Ireland (RTE), Italy (RAI), Netherlands (NOS), Spain (RTVE), Switzerland (SSR), Turkey (TRT), United Kingdom (BBC/UKIB), Yugoslavia (JRT).
Secretary Gen: Dr. J.B. Münch — **Dir. of Television Programme Dept:** H. Perez — **Dir. of Radio Programme Dept:** T. Alexanderson — **Dir. of Legal Affairs Dept:** Dr. W. Rumphorst — **Dir. of Technical Dept:** Dr. G. Waters.
PUBLICATIONS: EBU Technical Review, every two months in English and French—ESpace EBU monthly newsletter (ten editions)—Technical Documents—Technical Information Sheets—General, Legal and Technical Monographs—Lists of European Stations (long- and mediumwaves, VHF sound broadcasting, and television)—Reports of seminars and workshops for producers and directors of educational television and of programmes for children and young people.

Intergovernmental Organizations

Continued from page 42

ISLAMIC STATES BROADCASTING SERVICES ORGANIZATION (ISBO)

ADDR: P.O. Box 6351, Jeddah 21442, Saudi Arabia.
Telex: 601442 ISBO SJ. **Te:** +966 (2) 6721121/6722269.
Fax: +966 (2) 6722600.

Establishment: Founded as an affiliate of the Organization of the Islamic Conference (OIC) in 1975 in accordance with a resolution adopted by the Sixth Conference of Islamic Foreign Ministers.
Membership: Enjoyed by 45 Islamic states.

Objectives: a) Propagation of Islamic teaching — b) Protection of Islamic heritage — c) Strengthening fraternal relations among Muslim peoples — d) Promotion of Islamic solidarity — e) Defending Islamic causes — f) Furthering cooperative relations with regional and international information and cultural organizations.
Means: Production of radio and TV programmes — Exchange of programmes between broadcasting corporations of member states — Teaching Arabic language by audio-visual means — Training of personnel for broadcasting corporations in member states.

Governing Bodies:
General Assembly: Composed of all member states.
Executive Council: Composed of 15 members including the host state, Palestine and Representative of the OIC as permanent members.
Chairman: Dr. Foad Abdel-Salam Al-Farsi. (Host State).
General Secretariat: Responsible for implementation of resolutions adopted by the General Assembly and Executive Council and running the affairs of the Organisation.
Secretary General: Hussain Al-Askary.

International Relations: Concluded cooperation agreements with UNESCO and ALECSO. Attends Islamic Summits, Islamic Foreign & Information Ministers Conferences, meetings of Arab League and Arab Gulf Information Ministers and Committees.

Purpose: The EBU was founded on 12 February 1950 and is a professional association of broadcasting organisations whose activities include the world famous Eurovision news and programme exchanges and Euroradio programme exchanges. Its objectives are to promote cooperation among its members and with the broadcasting organizations of the entire world, and to represent the interests of its members in the programme, legal, technical and other fields.
NB: Since Jan 1st 1993, the EBU & the former OIRT have merged into a single organization.
Organization: Supreme authority is vested in the General Assembly composed of all Members. The Assembly elects a 15-man Administrative Council with executive powers.
The activities of the EBU are channelled through the Television Programme Committee, the Radio Programme Committee, the Legal Committee and the Technical Committee, with their Working Parties, Sub-Groups and ad hoc Groups. The Committees, although independent of each other, work closely together on all matters of common interest. The EBU has, in addition, a permanent staff of some 240 people, including programme, legal and technical officials.

ASSOCIATION OF COMMERCIAL TELEVISION (ACT)
L.P: Mme. Souné Wade (Mgr. & Gen. Secr), Dir. of Int. Affairs, TF1, 17 rue de l'Arrivée, F-75017 Paris, France. **Te:** +33 (1) 42 75 25 32. **Fax:** +33 (1) 42 75 26 88. **Telex:** 250 878 F.
Membership is open to all private broadcasting companies in Europe that approve ACT's charter and articles of association.

ASIA-PACIFIC BROADCASTING UNION (ABU)
Headquarters: Secretary-General's Office/Programme Services/Technical Centre: Asia-Pacific Broadcasting Union, 2nd Floor, Bangunan IPTAR, Angkasapuri, 50614 Kuala Lumpur, Malaysia.
Postal address: P.O. Box 1164, Pejabat Pos, Jalan Pantai Bahru, 59700 Kuala Lumpur, Malaysia. **Telephone:** +60 (3) 2823592, 2822480, 2823108. **Fax:** +60 (3)2825292. **Telex:** 32227 (ABU MA) Kuala Lumpur. **Cable:** ASBUN Kuala Lumpur.
ABU Tokyo Office: c/o NHK Broadcasting Centre, 150-01 Tokyo, Japan. **Te:** +81 (3) 3465-1749. **Fax:** +81 (3) 3465-1749.
Telex: 22377 (RADIONHK) Tokyo. **Cable:** ABUNI TOKYO.
Ag. President: Dato' Jaafar Kamin (DG, Radio Television Malaysia.)
Vice-Pres: Mr. Mikio Kawaguchi, Pres, NHK Japan.
Secretary-General: Mr. Hugh Leonard.
Administrative Council: Australian Broadcasting Corporation, Fiji Broadcasting Commission, Television Broadcasts Limited (Hong Kong), Radio Republik Indonesia/Televisi Republik Indonesia, All India Radio, Islamic Republic of Iran Broadcasting, Nippon Hoso Kyokai (Japan), Munhwa Broadc. Corp. (Korea), Radio Television Malaysia, Nepal Television Corporation, Pakistan Television Corporation Limited, Republic Broadc. System, Inc. (Philippines), Singapore Broadc. Corp.
Full Members (39): Radio-Television of Afghanistan (RTA-Afghanistan), Australian Broadcasting Corporation (ABC-Australia), Special Broadcasting Service Corporation (SBSA-Australia), National Broadcasting Authority of Bangladesh (NBAB), Radio Television Brunei (RTB-Brunei), Radio and Television of the People's Republic of China (RTPRC-China), Egyptian Radio and Television Union (ERTU-Egypt), Fiji Broadcasting Commission (FBC-Fiji), All India Radio (AIR-India), Doordarshan (DDI-India), Radio Republik Indonesia and Televisi Republik Indonesia (RRI/TVRI-Indonesia), Islamic Republic of Iran Broadcasting (IRIB-Iran), Iraqi Broadcasting and Television Establishment (IBTE-Iraq), Nippon Hoso Kyokai (NHK-Japan), Tokyo Broadcasting System Inc. (TBSJ-Japan), Jordan Radio and Television (JRTV-Jordan), Korean Broadcasting System (KBS-Korea), Lao National Radio and Television (RTNL-Laos), Radio Television Malaysia (RTM-Malaysia), Dept. of Inf. & Broadc. Maldives (DIB-Maldives), Nepal Television Corporation (NTV-Nepal), Radio Nepal (RNE-Nepal), Radio New Zealand Limited (RNZ-New Zealand), Television New Zealand (TVNZ-New Zealand), Pakistan Broadcasting Corporation (PBC-Pakistan), Pakistan Television Corporation (PTV-Pakistan), National Broadcasting Commission of Papua New Guinea (NBC/PNG-Papua New Guinea), Philippine Broadcasting Service/Peo-

ple's Television 4 (PBS/PTV 4-Philippines), Republic Broadcasting System, Inc., Philippines (GMA-Philippines), Saudi Arabian Radio and Television (SAB/SAR-Saudi Arabia), Singapore Broadcasting Corporation (SBC-Singapore), Solomon Islands Broadcasting Corporation (SIBC-Solomon Islands), Sri Lanka Broadcasting Corporation (SLBC-Sri Lanka), Sri Lanka Rupavahini Corporation (SLRC-Sri Lanka), National Broadcasting Services of Thailand (NBT-Thailand), TV Pool of Thailand (TPT-Thailand), Turkish Radio–Television Corporation (TRT-Turkey), Viet Nam Radio and Television (VNRT-Vietnam), Samoa Broadcasting Service (SBS-Western Samoa).
Associate Members (27): Federation of Australian Commercial Television Stations (FACTS-Australia), Nine Network Australia Limited (NNA-Australia), Canadian Broadcasting Corporation (CBC-Canada), Finnish Broadcasting Company Ltd. (YLE-Finland), Organismes Francais de Radiodiffusion et de Television (OFRT-France), Arbeitsgemeinschaft der offentlich-rechtlichen Rundfunkanstalten der Bundesrepublik Deutschland (ARD-Germany), Zweites Deutsches Fernsehen (ZDF-Germany), Radio Telefis Eireann (RTE-Ireland), National Association of Commercial Broadcasters in Japan (NAB-Japan), Radio Kiribati (BPA-Kiribati), Mauritius Broadcasting Corporation (M_ABC-Mauritius), Federated States of Micronesia Broadcasting Service (FSMBS-Micronesia), Radio Nederland (RN-Netherlands), Federal Radio Corporation of Nigeria (FRCN-Nigeria), Islands Broadcast Corporation (IBC-Philippines), Radiotelevisão Portuguesa, E.P. (RTP-Portugal), Swedish Broadcasting Corporation (SR-Sweden), Swiss Broadcasting Corporation (SwBC-Switzerland), British Broadcasting Corporation (BBC-United Kingdom), Independent Television Commission and Independent Television Association Ltd. (ITV-United Kingdom), American Broadcasting Companies, Inc. (ABC-USA), Columbia Broadcasting System Inc. (CBS-USA), Corporation for Public Broadcasting/Public Broadcasting Service/National Public Radio/American Public Radio Network (CPB/PBS/NPR/APRN-USA), National Association of Broadcasters (NAB-USA), National Broadcasting Co., Inc. (NBC-USA), Turner Broadcasting System, Inc. (TBS-USA), Voice of America (VOA-USA)
Additional Full Members (16): Asia Television Limited (ATV-Hong Kong), Hong Kong Commercial Broadcasting Co., Ltd. (HKCBC-Hong Kong), Metro Broadcast Corporation Limited (METRO-Hong Kong), Radio Television Hong Kong (RTHK-Hong Kong), Television Broadcasts Limited (TVB-Hong Kong), Pt. Rajawali Citra Televisi Indonesia/Pt. Surya Citra Televisi (RCTI/SCTV-Indonesia), Asahi National Broadcasting Co., Ltd., Japan (ANB-Japan), Christian Broadcasting System (CBS-Korea), Munhwa Broadcasting Corporation (MBC-Korea), The Radio and Television Broadcasting Committee of the Democratic People's Republic of Korea (KRT-Korea), Teledifusão de Macau, S.A.R.L. (TDM-Macau), Sistem Televisyen Malaysia Berhad (TV3-Malaysia), Mongol Radio and Television (MRTV-Mongolia), Norfolk Island Broadcasting Service (VL2NI-Norfolk Island), Radio Philippines Network, Inc. (RPN-Philippines), Independent Television Network Limited (ITN-Sri Lanka).
Aims: To foster and coordinate the development of broadcasting in the Asia-Pacific region, to develop ways and means of establishing closer collaboration and cooperation among broadcasting organisations, and to serve the professional needs of the broadcasters in Asia and the Pacific.
Publications: ABU News (bi-monthly), ABU Technical Review (bi-monthly). Both in English.

UNION OF NATIONAL RADIO AND TELEVISION ORGANISATIONS OF AFRICA (URTNA)

Permanent Services:
General Secretariat: 101 Rue Carnot, B.P. 3237 Dakar, Senegal. **Te:** +221 21-16-25, 21-59-70. **Fax:** +221 22-51-13. **Telex:** 51650 URTNA SG.
Technical Centre: B.P. 39, Bamako, Mali. **Te:** +223 222593. **Fax:** +223 224809. **Cable:** URTNA Bamako. **Tlx:** 2505.
Programme Exchange Centre: P.O. Box 50518, Nairobi, Kenya. **Te:** +254 333415. **Fax:** +221 33-68-56. **Telex:** 22675 Nairobi.
Inter-African Centre for Rural Radio Studies (CIERRO): B.P. 385, Ouagadougou, Burkina Faso. **Te:** +226 30-66-86. **Telex:** CIERRO 5349 BF.
Aims: URTNA has no commercial aim. Its objectives among others are:
a) to promote, maintain and develop cooperation among its members;
b) to support in every domain the interests of radio and television organisations which have accepted its statutes and to establish relations with other such organisations;
c) to promote and coordinate the study of all matters relating to radio and television and to ensure the exchange of information on all questions of general interest;
d) to use its best endeavours to ensure that all its members honour the provisions of international and intra-African agreements in all matters relating to radio and television;
e) to coordinate and collaborate closely with member organisations in the coverage of national and international events.
Governing Bodies: President, 1st Vice-President, 2nd Vice-President, Legal Adviser. The President and Vice-Presidents constitute the Bureau of the Union.
Administrative Council: Constituted by 15 active Member Organisations. The Administrative Council is the policy making body responsible to the General Assembly.
Specialized Commissions:
1) **Administrative and Legal Commission:** Constituted by 10 active Member Organizations.
2) **Technical Commission:** Constituted by 10 active Member Organizations.
3) **Programme Exchange and Cultural Affairs Commission:** Constituted by 10 active Member Organizations.
4) **TV News Exchange Commission:** Constituted by 10 active Member Organizations.
Active Members: Algeria (ENRS/ENTV), Angola (TPA), Benin (ORTB), Botswana (RB), Burkina Faso (RTB), Burundi (RTNB), Cameroon (CRTV), Cape Verde (RNC), Chad (RTN), Central African Rep. (RTC), Comoro, Congo (RTC), Côte d'Ivoire (RTI), Djibouti (RTD), Egypt (ERTU), Ethiopia (ETS), Gabon (RTG/Africa No. 1), Gambia (RG), Ghana (GBC), Guinea (RTG), Guinea-Bissau, Kenya (VOK/Kenya Television Network), Lesotho (LeBS), Liberia (LBS), Libya (LJB), Madagascar (RTM), Malawi (MBC), Mali (RTM), Mauritania (ORTM), Mauritius (MBC), Morocco (RTM), Mozambique (RM), Niger (ORTN), Nigeria (FRCN/NTA), Senegal (ORTS), Seychelles (RTS), Sierra Leone (SLBS), Somalia (SBS), Sudan (SNBC), Swaziland (STV), Tanzania (RTD), Togo (RTNM), Tunisia (RTT), Uganda (RU/UTV), Zaire (OZRT), Zambia (ZBS), Zimbabwe (ZBC).
Associate Members: Germany (ARD/ZDF), France (OFRT), Holland (RN), PLO, Switzerland (SSR).

All active and Associate Members constitute the General Assembly—the Supreme Body of the Union.
Publications: "URTNA Review" bi-annual/bi-lingual (English/French); studies, reports etc.

ARAB STATES BROADCASTING UNION (ASBU)

Nature: Intergovernmental union within the framework of the League of Arab States comprising the Arab Radio and TV organisations (established February 1969).
Supreme Body: General Assembly, meeting in ordinary sessions once a year. The Presidency of the Gen. Ass. held in rotation among the full members according to the alphabetical order of the names of the member States. The term of office: one year. The Vice-Presidency held by the member who shall assume the Presidency at the next session.
Administrative Council: Pres. + Vice Pres. + one full member according to the alphabetical order of the names of the member States.
General Secretariat: 22a Taha Hussein St, Zamalek, Cairo, Egypt. **Cable:** Asbunion, Cairo. **Te:** +20 (2) 805825 **Telex:** 347-ASBUUN
Technical Centre: P.O. Box 2231, Khartoum, Sudan.
Training Centre: P.O. Box 6333, Damascus, Syria.
Audience Research Centre: Iraqi Establishment for Radio and TV, Baghdad, Iraq.

NORTH AMERICAN NATIONAL BROADCASTERS ASSOCIATION (NANBA)

Headquarters:
Addr: 1500 Bronson Avenue, Ottawa, Ontario, Canada K1G 3J5. **Te:** +1 (613) 738-6553. **Telefax:** +1 (613) 738-6887. **Telex:** 06528046 NANBA Secretariat.
Secretary-General: Spencer Moore.
Board of Directors: President: Michael McEwen (Canadian Broadcasting Corporation); **Vice-Presidents:** Julius Barnathan (Capital Cities/ABC, Inc.), Jorge Kanahuati (Televisa s.a.) and one representative from each full member organization.

Advisory Council: comprises the Board of Directors, Secretary General, Committee Chairs, and representatives of NANBA member and associate member organizations.
Committees: Technical — Chmn. Carl Girod (PBS-USA), Legal — Chmn. Larry Loeb (Capital Cities/ABC Inc.–USA), News & Current Affairs — Chmn. Brian Knoblock (CBS-USA), Sports — Chmn. Gordon Craig (TSN-USA).
Members: CANADA — Canadian Broadcasting Corporation and CTV Television Network Ltd.; MEXICO — Televisa s.a.; UNITED STATES — Capital Cities/ABC Inc.; CBS Inc.; Corporation for Public Broadcasting (National Public Radio; Public Broadcasting Service; American Public Radio); National Broadcasting Company, Inc.; and Turner Broadcasting Systems, Inc. (CNN).
Associate Members: Home Box Office (HBO-USA), National Association of Broadcasters (NAB-USA), The Sports Network (TSN-Canada).
Meetings: The Advisory Council usually meets four times a year, including the annual meeting held in the month of January. The Board of Directors may meet separately. The Committees meet frequently as their work requires.
NANBA started as a loose ad hoc group in 1972. It became a formal organization in its present form in 1978. The Association is concerned with providing a framework for the identification, study and active solution of international questions affecting broadcasting and, working with other international Unions and Associations of broadcasters, to gain an effective voice in international forums on matters that affect broadcasting. Its objectives are not commercial in nature.
Publications: NANBA Bulletin published at intervals during the year. SNG/ENG Handbook — a new publication on satellite news gathering aimed at assisting crews abroad.

CARIBBEAN BROADCASTING UNION (CBU)

The CBU is an association of national broadcasting systems of the Commonwealth, Caribbean and other regional states formed for the purpose of promoting and facilitating interchange between member countries through the medium of broadcasting.

Secretariat: "Wilkins Lodge", Two Mile Hill, St. Michael, Barbados, W.I. **Te:** +1 (809) 429 9146. **Fax:** +1 (809) 429 2171. **Cable:** CARICAST. **Telex:** 2569 CARICAST.
Management Committee Members: Pres: Vic Fernandes, Gen Mgr: Barbados Rediffusion Sce Ltd.; Vice Pres (TV): Tino Barovier, Dep. Gen. Mgr: Jamaica Broadc. Corp; Vice Pres (Radio): Ms. Claudette Manchester, Ag. Gen. Mgr: ZIZ Radio and TV; Vice Pres. (Eng): Michael Thompson, Dir. of Eng: Broadc. Corp. of the Bahamas.

Members: Antigua: Antigua & Barbuda Broadc. Sce, CTV Entertainments Systems Ltd, Grenville Radio Ltd. **Aruba:** Tele Aruba. **Bahamas:** Broadc. Corp. of the Bahamas. **Barbados:** Barbados Rediffusion Sce. Ltd, Caribbean Broadc. Corp. **Belize:** Belize Broadc. Network. **Bermuda:** Bermuda Broadc. Co. Ltd. **Dominica:** Dominica Broadc. Corp, Marpin TV Co. Ltd. **Grenada:** Radio Grenada, Grenada Television Sces. Ltd. **Guyana:** Guyana Broadc. Corp. **Jamaica:** Jamaica Broadc. Corp, Island Broadc. Sces. Ltd, Radio Jamaica Ltd. **Montserrat:** Radio Montserrat. **Netherlands Antilles:** Tele Curacao, Foundation Broadc. Netw. (St. Maarten). **St. Kitts:** ZIZ Radio & TV. **St. Lucia:** St. Lucia Broadc. Corp, Helen Television System, Cablevision. **St. Vincent:** National Broadc. Corp, St. Vincent & Grenadines Broadc. Corp. **Surinam:** Surinaamse Televisie Stichting, ATV Telesur. **Trinidad & Tobago:** National Broadc. Sce, Trinidad Broadc. Co. Ltd, Trinidad & Tobago Television Co. Ltd. **Turks & Caicos:** Radio Turks & Caicos.

Associate Members: Belize: Great Belize Productions, Ltd. **Canada:** CBC, Radio Canada Int. **Germany:** Transtel GmbH. **Guyana:** Visual Production Centre, Guyana Public Communications Agency. **Jamaica:** Caribbean Institute of Mass Communication, Creative & Production Training Centre Ltd, Radio Education Unit, Independent Radio Co. Ltd **Trinidad:** Banyan Ltd. **United Kingdom:** BBC. **USA:** Cable News Network, Corp. for Public Broadc, National Assoc. of B'casters, USIA-Voice of America. **British Virgin Islands:** Dept. of Inf. & Pub. Rel.

ASOCIACION INTERNACIONAL DE RADIODIFUSION (AIR). INTERNATIONAL ASSOCIATION OF BROADCASTING (IAB). ASSOCIATION INTERNATIONALE DE RADIODIFUSION (AIR).
Addr: Cnel. Brandzen 1961, Apto-402, 11200 Montevideo, Uruguay. **Te:** +598 (2) 488121, +598 (2) 488129. **Fax:** +598 (2) 488121. **Cable:** AIRADIO. **Tlx:** 31173 **Mbx URUPAC:** 7482.31173. **Mbx EASYLINK:** 62809990.
L.P: Pres: Andrés García Lavín (Mexico); 1st Vice Pres: Dr. Luiz Eduardo Borgerth (Brazil); 2nd Vice Pres: Javier Gimeno de Priede (Spain); Vice Pres. Secr: Ricardo Londoño Londoño (Colombia); Vice Pres. Treas: Fernando Eleta Casanova (Panamá).
Titular Board Members: Dr. Carlos Alberto Mirson (Argentina); Dr. Juan Carlos Laborde (Argentina); Dr. Luiz Eduardo Borgerth (Brazil); Joaquim Mendonça (Brazil); Ricardo Londoño Londoño (Colombia); Luis Ajenjo Isasi (Chile); Edgar Yanez Villalobos (Ecuador); Ricardo Recinos (Salvador); Javier Gimeno de Priede (Spain); Augusto Delkader (Spain); John Abel (EE.UU.); Luciano Mecarozzi (Italy); Andrés García Lavín (México); Fernando Eleta Casanova (Panama); Genaro Delgado Parker (Peru); Oscar Imperio (Uruguay); Lic. Félix Cardona Moreno (Venezuela); Dr. Héctor Ponsdomenech (Venezuela).
Deputy Board Members: Alberto Guillermo Veiga (Argentina); Luis H. Tarsitano (Argentina); Hugo Romay Salvo (Uruguay); Emilio Amadei Beringhs Filho (Brazil); Ricardo Alarcón Gaviria (Colombia); Enrique González (Chile); Raúl Izurieta M. (Ecuador); Tommy Carrasquillo (Puerto Rico); Ramón de Rato Figaredo (Spain); Alfonso Ruiz de Assin (Spain); Walter Canals (U.S.A.); Eugenio Porta (Italy); Oscar Gutierrez (Mexico); José Gabriel Díaz (Panamá); Luis Vargas Hornes (Peru); Jorge Omar Sosa (Uruguay); Alfredo Vetancourt (Venezuela); Miguel Angel Contreras (Venezuela).
Director General: Dr. Héctor Oscar Amengual (Uruguay). **President of the Inter American Chapter:** Félix Cardona Moreno (Venezuela). **President of the European African Chapter:** Javier Gimeno de Priede (Spain). **President of the Legal Permanent Committee:** Dr. Rafael Peñaloza (Argentina). **President of the Technical Permanent Committee:** Luis Melide (Uruguay). **President of the Freedom of Expression Committee:** Dr. Héctor Ponsdomenech (Venezuela). **President of the Permanent Copyright Committee:** Ramón de Rato Figaredo (Spain).

COMMONWEALTH BROADCASTING ASSOCIATION
Secretariat Address: CBA Secretariat, BBC White City, 201 Wood Lane, London WI2 7TS. **Te:** +44 (81)-752-5022. **Fax:** +44 (81) 752 4137. **Telex:** 265781. **Secretary Gen:** Alva Clarke.
Members: Antigua & Barbuda Broadcasting Service, Australian Broadcasting Corporation, Broadcasting Corporation of the Bahamas, Bangladesh National Broadcasting Authority, Caribbean Broadcasting Corporation (Barbados), Broadc. Corp. of Belize Radio Botswana, British Broadcasting Corporation, Radio Television Brunei, Canadian Broadcasting Corporation, Radio Cayman, Cyprus Broadcasting Corporation, Dominica Broadcasting Corporation, Radio Gambia, Ghana Broadcasting Corporation, Gibraltar Broadcasting Corporation, Radio Grenada, Guyana Broadcasting Corporation, Radio Television Hong Kong, All India Radio, Doordarshan (India Television), Jamaica Broadcasting Corporation, Kenya Broadc. Corp, Radio Kiribati, Lesotho National Broadcasting Service, Malawi Broadcasting Corporation, Radio Television Malaysia, Malta Broadcasting Authority, Mauritius Broadcasting Corporation, Radio Montserrat, Namibian Broadcasting Corporation, Television New Zealand, Federal Radio Corporation of Nigeria, Nigerian Television Authority, Voice of Nigeria, Pakistan Broadcasting Corporation, Pakistan Television Corporation, Papua New Guinea National Broadcasting Commission, Radio and Television ZIZ (St. Kitts-Nevis), St. Lucia Broadcasting Corporation, St. Vincent & the Grenadines National Broadc. Corp., Seychelles Broadc. Corp., Sierra Leone Broadcasting Service, Singapore Broadcasting Corporation, Solomon Islands Broadcasting Corporation, Sri Lanka Broadcasting Corporation, Sri Lanka Rupavahini Corporation, Swaziland Broadcasting Service, Swaziland Television Authority, Radio Tanzania - Dar es Salaam, Tonga Broadcasting Commission, Trinidad and Tobago National Broadcasting Service (Radio), Trinidad and Tobago Television Co. Ltd., Radio Turks and Caicos, Radio Uganda and Uganda Television, Western Samoa Broadcasting Service, Zambia National Broadcasting Corporation, Zimbabwe Broadcasting Corporation.

The CBA was established in London in February 1945 and is an association of the national public service broadcasting organizations in 51 Commonwealth countries pledged to work for the professional improvement of broadcasting in member organisations through collective study and mutual assistance. There is a great deal of practical co-operation among member organisations of the CBA, which are drawn from all the continents of the world.

General Conferences of the Association are held every two years to discuss and exchange ideas, information and experience on all aspects of radio and television broadcasting—in programming, administration and engineering. The last General Conference was held in Botswana in 1992.

In between General Conferences, regional meetings are held in the main regions of the Association—e.g. the Caribbean, East, Central and Southern Africa, and West Africa—as part of the Association's policy of encouraging more effective co-operation among members in its main regions.

The Secretariat publishes a quarterly magazine COMBROAD, which has a wide international circulation.

INTERNATIONAL INSTITUTE OF COMMUNICATIONS
Headquarters: Tavistock House South, Tavistock Square, London, WC1H 9LF. **Te:** +44 (71) 388 0671. **Fax:** +44 (71) 380 0623. **Cable:** Widecast London. **Telex:** 24578 IICLDN G.
Officers: Pres: Brian Quinn (UK). Vice-Presidents: David Golden (Canada), Kiyoshi Hara (Japan). Chmn. of Exec. C'ttee: Henri Pigeat (France). Treasurer: Hans Kimmel, Germany (FR). Exec. C'ttee Members: John Berndt (USA), Laura Glässer-Weisser (Germany F.R.), Thomas P. Hardiman (Ireland), Lena Wennberg (Sweden).
Executive Director: Carol Joy. Asst. Dir: Gerry Jayasuriya.
The IIC provides an international, cross-disciplinary forum for the independent analysis of the social, economic, political, cultural, and legal issues related to communications, particularly electronic communications. It carries out research projects and consultancy, holds conferences and seminars, and publishes books and a regular bi-monthly journal "Inter Media". In this way it assists governments, industry and the public to formulate the necessary policies for handling the new information devices, systems and services. Current major activities include a series of activities on

the use of communications for development; an analysis of the relationship between (i) communications and (ii) regional development; a multinational study of the telecommunications structures; a forthcoming study of television and film production. Details of IIC activities and publications of the three categories of membership (individual, corporate and institutional) are available from the IIC Office.

MENTOR
International Institute for Communication & Management
Addr: P.O. 6723 St. Olavsplass, 0130 Oslo 1, Norway.
Te: +47 (2) 362989. **Fax:** +47 (2) 360304.
L.P: Dir: Thor Langbråthen. Training Projects: Geir Magnus Nyborg.
IMMI is a non profit agency supported by 25 Norwegian churches and church agencies.
Media consultancy: On request Mentor advises churches and organizations on the design and implementation of communication activities and projects.
Media education: Mentor trains people for radio and TV programming and engages in media-awareness training.
Communication research: Research and publications especially in the areas of communication and media strategy in relation to mission and development.

INTERNATIONAL BROADCASTING SOCIETY (IBS)
ADDR: c/o Korean Broadcasting System, #18 Yoido-dong, Youngdungpo-gu, Seoul 150, Korea: **Te:** +82 (2) 7803-636.
Telex: KBSKBK K24599. **Cable:** KBS SEOUL.
L.P: Pres: Park Hyon-Tae. Secr. Gen: Kim Doh-Jin.
Purpose: IBS was established in 1985, with the objective of helping to correct the imbalance in the information flow between the advanced and "third world" countries. It is involved in fostering cooperation and understanding between broadcasters of third world and developing countries.

BROADCASTING RESEARCH INSTITUTE
Addr: 39c Highbury Place, London N5 IQP, England. **Te:** +44 (1) 226 9903. **Fax:** +44 (1) 226 8666.
L.P: Chmn: Dr. Richard Hoggart. Dir: Dr. Timothy Leggatt.
Aims: The Broadcasting Research Institute is a plurally funded body established to undertake research into all aspects of broadcasting policy and practice.

RELIGIOUS BROADCASTING ORGANIZATIONS

AWR — Adventist World Radio
Worldwide HQ: 12501 Old Columbia Pike, Silver Spring, MD 20904-6600. **Te:** +1 (301) 680 6304. **Fax:** +1 (301) 680 6303.
Regional addr: AWR-Europe: C.P. 383, I-47100 Forli, Italy; AWR-Latin America: P.O. Box 1177, 4050 Alajuela, Costa Rica; AWR-Russia, Voice of Hope Media Center, 300000, Tula-Centre, Box 170, Russia.

CAM International (formerly Central American Mission)
Addr: 8625 La Prada Drive, Dallas, TX 75228, USA.
Operates TGNA Guatemala City, TGBA Barillas and TGMI San Sebastian, Guatemala (see World Radio section for details).

CEEH / MEBSH
Addr: c/o MFI Haiti, P.O. Box 15665, West Palm Beach, FL 33416, USA — CEEH Exec. Secr: Rigaud A. Antoine. MEBSH Pres: Raphael Lozama.
MEBSH is an association of Baptist Churches that operates the Radio Lumière network in Haiti ministering on behalf of the Evangelical churches of Haiti associated with CEEH, Concile des Eglises Evangéliques d'Haïti.

DX Radio Network
Addr: 3176 Elmwood Ave, Rochester, NY 14618-2096, USA. **Te:** +1 (716) 442 5070. **Fax:** +1 (716) 461 3169. **Telex:** 6731224 DXR—UW — Dir. Gen: Phillip M. Dampier.
Operates network of secular & religious programming material available to radio st's worldwide.

Evangeliums Rundfunk
Addr: Box 1444, D-6330 Wetzlar, Germany. **Te:** +49 (6441) 5050. **Fax:** +49 (6441) 53603 — Chmn: A. von Aufsess. Exec. Dir: Rev. Horst Marquardt.
ERF is the German-speaking partner of Trans World Radio.
Prgrs: German to Europe via TWR Monte Carlo, to So. America via TWR Bonaire, to the southern part of South America via RTM Uruguay, and to Ea. & So. Africa over TWR Swaziland (see World Radio section). Also via cable networks in Germany and via local and regional st's in Austria, Germany, Italy & Switzerland. Prgrs in other languages via TWR Monte Carlo.
New: prgrs via satellite — **V.** by QSL-card and letter.

Evangelische Radiomission "Christus Lebt"
Addr: D-4800 Bielefeld 14, PO Box 140 380, Federal Rep. of Germany. **Te:** +49 (521) 4486-24 — Exec: Rev. J. Eggebrecht.
A working group of Protestant Churches in Germany, concerned with prgrs in Estonian, Latvian, Polish & Czech via Trans World Radio Monte Carlo.

FEBC Radio International
Addr: Box 1, La Mirada, CA 90637, USA. **Te:** +1 (310) 947-4651. **Fax:** +1 (310) 943 0160. **MCI:** 335-0911 or FEBCOMLA. **Cable:** FEBCOM.
and
FEBA Radio
Addr: Ivy Arch Road, Worthing, West Sussex BN14 8BX, England. **Te:** +44 (903) 37281. **Cable:** FEBCOM WORTHING. **Telex:** 873835 FEBCOM G. **MCI:** 341-8572 or FEBA Radio.
An international group of private broadcasting facilities, broadcasting in over 140 languages for a total of approx. 9000 hours per month using 32 transmitters, and operating Christian radio stations in Korea, The Philippines, Saipan, The Seychelles and San Francisco, USA (see entries in World Radio section). Additional offices and/or studios in 18 countries.

Good News Broadc. Assoc. Inc. Back To The Bible Broadcast
Addr: Box 82808, Lincoln, NE 68501, USA. **Te:** +1 (402) 474-4567. **Fax:** +1 (402) 474 4519 — Dir: Woodrow Kroll. Exec. Dir: Brian Erickson.
Produces daily half hour prgrs of music & evangelical preaching heard on over 600 stations worldwide. Also produces prgrs in Hindi, Italian, Kannada, Konkani, Pidgin, Sinhala, Spanish, Tagalog, Tamil, Telugu, Urdu.
Pub: Free lists of st's carrying English prgrs.

IBRA Radio AB
Addr: S-105 36 Stockholm, Sweden. **Headquarters:** Gamelgårdsvägen 38-42, Store Essingen. **Te:** +46 (8) 130340. **Fax:** +46 (8) 579029. **Cable:** Ibraradio. **Telex:** 10888 DAGEN S.
Prgrs: The broadcasts of IBRA can be heard in over 100 countries in approx 60 languages. On shortwave, IBRA prgrs. are carried by High Adventure (Lebanon), FEBA Seychelles, FEBC Manila and local st's in Africa, Asia and Latin America.

W R T H
HF-225
MODE
16.195
VOLUME
TONE
PHONES
RECEIVER HF-150
VOLUME ON/OFF
PHONES
15.010
MEM
MODE
FAST

International Lutheran Laymen's League

Addr: 2185 Hampton Avenue, St. Louis, MO 63139-2983, USA. **Te:** +1 (314) 647 4900. **Fax:** +1 (314) 647 6923. **Telex:** 590083. **Cable:** LULALE — Exec. Dir: Larry E. Lumpe. Dir, Ministries: Kenneth M. Peterson. Dir. Communication: Ronald J. Schlegel.
Prgrs: The Lutheran Hour, Bringing Christ to the Nations weekly with 77 formats in 31 languages from 20 production centers.

Lutheran World Federation Department for Mission & Development

Head Office: 150 route de Ferney (P. O. Box 2100), CH-1211 Geneva 2, Switzerland. **Te:** +41 (22) 791 6111. **Fax:** +41 (22) 791 6401. **Cable:** LUTHWORLD GENEVA. **Telex:** 415 730 OIK CH — Chairperson of Program Committee: Rev. Karsten Nielsen (Denmark). Dir: Rev. Dr. Ishmael Noko (Zimbabwe). Secretary for Consultative Sces: Rev. Bernhardur Gudmundsson.

National Religious Broadcasters

Addr: 7839 Ashton Avenue, Manassas, VA 22110, USA. **Te:** +1 (703) 330 7000. **Fax:** +1 (703) 330 7100 — Exec. Dir: Dr. E. Brandt Gustavson. Pres: Dr. David Clark, KMC Media Inc, Dallas, TX, USA.
NRB is an association of over 1000 religious national and international broadcasting organizations in radio, TV and related services. Founded 1944.
Pub: "Religious Broadcasting" magazine (monthly), "Directory of Religious Broadcasting" (annual). **Conventions:** Sponsors annual convention in Washington DC and other cities alternating years, late Jan/Feb, drawing delegates from around the world.

Norea Radio (Nordic Radio Evangelistic Ass.)

Addr: Box 4807, N-4602 Kristiansand, Norway. **Te:** +47 (42) 93000 — Dir. Adm: S. Pettersen. Dir. Prgr: L. Taule. Tec. Dept: R. Gaustad.
Prgrs: In Norwegian and more than 20 other languages over TWR Monaco, TWR Cyprus, TWR Swaziland, TWR Sri Lanka, TWR Guam, FEBA Seychelles and st's in Japan, Korea, Taiwan, Bolivia & Peru — **V.** by QSL-card & letter.

Pontifical Council for Social Communications

Addr: Palazzo S. Carlo, 00120 Vatican City State. **Te:** +39 (6) 698 83197. **Fax:** +39 (6) 698 85373. **Telex:** 2019 PCCS VA — Pres: Archbishop John P. Foley. Secr: Msgr. Pierfranco Pastore. Under-Secr: Hans-Peter Röthlin. Tech. Consultant: Rev. Antonio Stefanizzi.

Radio Committee of the Evangelical Missionary Alliance

Addr: Whitefield House, 186 Kennington Park Rd, London SE11 4BT, U.K. (Correspondence to Committee Secretary). Brings together evangelical missionary societies in UK involved in all aspects of Christian radio; co-ordinates joint ventures informing supporters and recruiting missionary workers. Sponsors one week annual course in Christian radio production.

SIM International

operates Radio ELWA, P.O. Box 192, Monrovia, Liberia (see World Radio section). Also produces French prgrs from studios at 08 B. P. 886, Abidjan, Côte d'Ivoire.

Trans World Radio International Headquarters

Addr: P.O. Box 700, Cary, NC 27512-0700, USA. **Te:** +1 (919) 460 3700. **Fax:** +1 (919) 460 3702.
L.P: Founder & Pres: Paul E. Freed, Ph.D. Exec. VP: Thomas J. Lowell. Dir. of Pub. Rel: Rosemarie Jaszka.
TWR broadcasts the Gospel in over 90 languages from eight transmitting locations: Albania, Monaco, Netherlands Antilles, Swaziland, Cyprus, Sri Lanka, Guam and Uruguay.
European Headquarters: P. O. Box 2020, 1200 CA Hilversum, The Netherlands —Exec. Dir: William P. Mial.
Pub: TWR Magazine 4 times a year, TWReport 4 times a year.

UNDA — Association Catholique Internationale pour la Radio, la Télévision et l'Audiovisuel.

(International Catholic Association for Radio, Television & Audiovisuals).
UNDA General Secretariate: 12 rue de l'Orme, B-1040 Brussels, Belgium. **Te:** + 32 (2) 734 97 08. **Fax:** +32 (2) 734 70 18. **Cable:** Undabrussels. **Telex:** 051-91 8023 GEONET G. **E-Mail:** GEONET BOX:GEO2:UNDA.
Secretary General: Colm Murphy.
Membership: National Associations in 136 countries and territories, plus 23 International or internationally-oriented organisations.
Publications: Bimonthly newsletters in English & French, quarterly in English and French on Media Education, plus various newsletters published at continental and national levels.

World Association for Christian Communication (WACC)

Addr: 357 Kennington Lane, London SE11 5QY, U.K. **Te:** +44 (71) 582 9139. **Fax:** +44 (71) 735 0340. **Telex:** 8812669 WACC G. **Cables:** WACC London SE11 — Gen. Secr: Rev. Carlos A. Valle. Pres: Rev. Randy Naylor (Canada). Vice Pres: Ms. Manoushag Boyadjian (Lebanon). Secr: Ms. Murri Selle (Germany). Treasurer: Rev. Kevin Engel (Australia).
WACC is a fellowship of organisations and individuals committed to Christian communication & media development.
Pub: ACTION newsletter (10 times a year). Media Development (quarterly). Occasional conference and research reports.

World Christian Broadc. Corp.

Addr: 605 Bradley Court, Franklin, TN 37064, USA.
Operates SW-st. KNLS, Alaska. The broadcast format is diversified including informational and cultural prgrs. of general interest. (Further details in World Radio Section under "Alaska").

World Radio Bible Broadcasts

Addr: P.O. Box 2000, West Monroe, LA 71294-1000, USA.
A non-denominational Christian broadcast, heard throughout the world in 16 languages.

World Radio Missionary Fellowship, Inc.

Addr: PO Box 39800, Colorado Springs, CO 80949-9800, USA. **Te:** +1 (719) 590 9800. **Fax:** +1 (719) 590 9801. **MCI Mailbox:** 341-7450 — Pres: Ronald A. Cline. Exec. Dir: Scott Higgins. Operates HCJB, La Voz de los Andes, Quito, Ecuador. Affiliated with KVMV, KOIR, KBNJ, KBNL, KBNR, KRIO (USA).

CLUBS FOR DX-ERS AND INTERNATIONAL LISTENERS

Various clubs exist for shortwave listeners and DXers. Some are oriented to programme listening, and others to DX-ing, the reception of low power or distant stations. In most cases, bulletins are produced on a regular basis. Sample copies are generally available for return postage (3 or 4 IRCs). Unless otherwise stated, all clubs are international, publish bulletins in E and cover all aspects of the radio hobby.

EUROPE

European DX Council, P.O. Box 4, St. Ives, Huntingdon, PE17 4FE, England. Umbrella organisation of DX Clubs in Europe. Send 2 IRC's for more info.

AUSTRIA

Assoziation Junger DX-er in Österreich, Postfach 1000, 1081 Wien (G)

BELGIUM

DX-Antwerp, Postbus 16, 2710 Hoboken (Dutch)

BULGARIA

Bulgarian DX Club, Ivan Penev, P.O. Box 47, 1111 Sofia II.

CZECHOSLOVAKIA

Czechoslovak DX Club, c/o Václav Dosoudil, Horní 9, CS-76821 Kvasice. (Czech/Slovak/English).

DENMARK

The Danish Shortwave Clubs International, Tavleager 31, DK-2670 Greve.
Dansk DX Lytter Klub, P.O. Box 392, 8100 Aarhus (Danish)

FINLAND

Suomen DX Litto ry, P.O. Box 454, 00101 Helsinki (Umbrella organisation of Finnish-language DX Clubs)
Finlands Svenska DX-Förbund r.f, Postbox 402, SF-20101 Åbo, Finland (Umbrella organisation of various Swedish-language DX clubs in Finland).

FRANCE

Amitié Radio, B.P. 56, 94002 Creteil Cedex (F).
Club Européen de DX Radio TV, B.P. 114, 13652 Salon de Provence (F).
Club des Auditeurs de la Radio Mondiale, 79 Ave. Paul Vaillant Couturier, 94400 Vitry sur Seine (F).

GERMANY

Interessengemeinschaft Rundfunk fernempfang e.V., Postfach 10 13 11, 4650 Gelsenkirchen (G, exclusively for blind and disabled).
Arbeitsgemeinschaft DX, Postfach 1107, 8520 Erlangen (G, has 8 member clubs, details for Rp.).
Assoziation Deutschsprachiger DXer e.V., Postfach 130124, 4000 Düsseldorf 13 (G).
Shortwave Listeners Club Saar, Postfach 1170, 6688 Illingen (G)
East and West Radio Club, Eifelwall 46, 5000 Köln 1 (G)
Esperanto DX Club, Kafkastr. 48 5/M, 8000 München 83 (Esperanto).
Kurzwellenfreunde Bremen, Rockwinkler Heerstr. 59, 2800 Bremen 33 (G)
Kurzwellenfreunde Wuppertal, Postfach 22 03 42, 5600 Wuppertal 22 (G).
Kurzwellenclub Berlin e.V., Fregestr. 59, 1000 Berlin 41 (G).
Rhein-Main-Radio Club, Postfach M68, 6093 Flörsheim (G).
Vereinigung Berliner Empfangsamateure, Denkstrasse 2, 1000 Berlin 41 (G)
Radio Hör Club International, Gerhard Kriehn, Am Kieseberg 5, D/Ö 7420 Schmölln.

ITALY

Associazone la Radioascolto, C. P. 873, 34100 Trieste (I)
Coordinamento del Radioascolto, c/o Marco Cerutti, C. P. 146, 13100 Vercelli (I) Umbrella organization of various Italian DX Clubs.
Play-DX, c/o Dario Monferini, Via Davanzati 8, 20158 Milano (I/E, specialises in difficult DX)
Gruppo d'Ascolto della Marca Trevigiana, C. P. 3, Succursale 10, 31100 Treviso (I).
Gruppo di Ascolto di Torino, C. P. 108, 10024 Moncalieri (TO), (I)
Gruppo Ascolto Bergamasco, C. P. 8, 24020 Gorle (BG) (I)
Gruppo Ascolto Due Mari, c/o Carlo Pepe, via Cesare Battisti 27, 74100 Taranto (I)
Gruppo Ascolto Radio dello Stretto, c/o Giovanni Sergi, Via Crotone 33, 98010 Camaro Inferiore (ME) (I).
Gruppo Radioascolto Liguria, c/o Luca Botto Fiora, Salita Torre Menegotto 42/15, 16035 Rapallo (GE) (I).
World Radio Group BCL, c/o Alberto Lo Passo C.P. 172, 96100 Siracusa. (I)
FM Radio Club, c/o Gabriele Focosi, Via Pesciantina 711, 55010 Gragnano (LU) (I, specialises in FM DX)
Golf Romeo BCL Club, C.P.3, 27030 Palestro (PV). (I)
Gruppo BCL Veneziani Campagnoli, C.P. 343, 30100 Venezia. (I)
Radio Club Polistena, c/o Alessandro Pochì, Via Jerace 27, 89024 Polistena(RC) (I, specialises in FM DX)
Gruppo Utility Romagnolo, c/o Fabrizio Magrone, Via Marengo 33, 47100 Forlì. (I, specialises in Utility DX)

LITHUANIA

Lithuanian DX Club, P.O. Box 471, 235802 Klaipeda 2, Lithuania.

NETHERLANDS

Benelux DX Club, Postbus 150, 5270 AD St. Michielsgestel (Dutch/E)

NORWAY

DX Listeners' Club, P.O. Box 7080, Vestheiene, 4602 Kristiansand (Norwegian).

RUSSIA

Sankt-Peterburg DX Club, Aleksey Osipov, Kazanskaya 4-87, 195213 Sankt-Peterburg.

SPAIN

Asociacion DX Barcelona, Apartado 335, 08080 Barcelona (Sp)
Asociacion Española de Radioescucha (AER), P.O. Box 4031, 28080 Madrid (Sp)
Nuevo MAF, c/o Manuel Castro Vinagre, P.O. Box 21075, 08080 Barcelona (FM/TV, Sp)

SWEDEN

Sveriges DX Förbund, Box 3108, 103 62 Stockholm. (Sw, umbrella organisation of over 30 clubs)
Arctic Radio Club, Box 5050, 350 05 Växjö (MW, Sw/E)
Sveriges Radioklubb, Box 5083, 102 43 Stockholm (Sw language)

UNITED KINGDOM

British DX Club, 54 Birkhall Rd, Catford, London SE6 1TE.
DX Association of Great Britain, 44a Porter Terrace, Murton, Co. Durham SR7 9CT.
International Listeners Association, 1 Jersey Str, Hafod, Swansea SA1 2HF.
International Shortwave League, 6 Moorhead, Preston Upon The Weald Moors, Telford, Shropshire TF6 6DC (Also for amateur radio enthusiasts).

Mediumwave Circle, 137a Hampton Road, Southport PR8 5DY (LW/MW only)
World DX Club, 17 Motspur Drive, Northampton NN2 6LY.

AFRICA

COTE D'IVOIRE
DX-Ivoire, c/o Jibirila Liasu, 20 B.P. 197, Abidjan 20 (F).

TUNISIA
Club des Auditeurs et de l'Amitié, c/o De Riadh Sakka, Route de Gremda Merkez Sahnoun, 3012 Sfax (F).

NIGERIA
Africa DX Association, c/o Friday Okoloise, Radio/Carrier Room, Nitel, Ashaka, Bauchi State.

SOUTH AFRICA
South African DX Club, P.O. Box 72620, Lynnwood Ridge, Transvaal 0040.

TOGO
Club Inter Amitié Radio, CCF B.P. 2090, Lomé (F)
Groupe Endoc, B.P. 2667, Lomé (F)

ASIA

BANGLADESH
International Radio Listeners Club, Konabari, P.O. Nilnagor, Gazipur, Dhaka (E/Bengali).

INDIA
HAP-India, c/o R. Shreedar, 6/2 Nathamuni Street, T. Nagar, Madras 600 017.
Indian DX Club International, GPO Box 646, Calcutta 700 001.
Indo-Universal DXing Society, People's House, P.O. Bahudarada, Bhadrak, Orissa, India 756100.
North India DX-ers Association, c/o Mr. Kumar Radha Raman Kanth, P.O. Banmankhi, Dist-Purnea, Bihar 854202.
Southern Asia Radio Club, 63 K.D. Flats, Jamshedpur - 831 005.
Times R. Club, Creative Times, Times of India, Fraser Rd. Patna 800 001, Bihar.
Universal DX League, 408 Krishna Nagar, Ludhiana.

INDONESIA
Indonesian DX Club, P.O. Box 50, Kutoarjo 54201, Jawa Tengah (E/Indonesian).

JAPAN
Japanese Association of DX-ers, P.O. Box 1766, Tokyo 100-91 (J/E)
Asian Broadcasting Institute, C. P.O. Box 1334, Tokyo 100-91.
Japan BCL Federation, 4F, Matsuokadudan Bldg., 2-8 Kudanminami 2-chome, Chiyoda-ku, Tokyo 102 (J).
Japan Short Wave Club, P.O. Box 29, Sendai Central 980-91 (J/E).
Indonesian DX Circle, c/o Satoshi Wakisaka, 6-25-21 Senriyama Nishi, Suita City, Osaka 565.
Nagoya DXers Circle, c/o Shigenori Aoki, 2-51 Kasumori-cho, Nakamura-ku, Nagoya 453.
Radio Nuevo Mundo, c/o Tetsuyu Hirahara, 5-6-6 Nukuikita, Koganei-shi, Tokyo 184.

NEPAL
Radio Listeners' Centre of Nepal, Post Box 126, Biratnagar.

SRI LANKA
Union of Asian DXers, c/o Victor Goonetilleke, "Shangri-La" 298 Kolamunne, Piliyandala.

PACIFIC
South Pacific Association of Radio Clubs, 212 Earn Street, Invercargill, New Zealand (Umbrella organisation of most Australian and NZ DX Clubs).

AUSTRALIA

Australian Radio DX Club, Box 227, Box Hill, Vic. 3128 (DXing only, no programme listeners).
DX Australia, P.O. Box 285, Mt. Waverley, Vic. 3149.
Southern Cross DX Club Incorporated, G. P.O. Box 1487, South Adelaide, S.A. 5001.

NEW ZEALAND

New Zealand Radio DX League, P.O. Box 3011, Auckland.
New Zealand DX Radio Association, Long Beach RD1, Port Chalmers, New Zealand.

NORTH AMERICA

Association of North American Radio Clubs (ANARC), 2216 Burkey Drive, Wyomissing, PA 19610. **BBS:** +1 (913) 345 1978 24h 3/12/24 8-N-1.

CANADA

Canadian International DX Club, 79 Kipps Str, Greenfield Park, PQ J4V 3B1.
Ontario DX Association, P.O. Box 161, Station 'A', Willowdale, ON M2N 5S8.
Club Ondes Cortes du Quebec, C.P. 1001, Chicoutimi, PQ G7H 5G4 (F).

USA

All Ohio Scanner Club, 50 Villa Road, Springfield, OH 45503 (non-broadcast Public Service Bands only).
American Shortwave Listeners Club, 16182 Ballad Lane, Huntington Beach, CA 92649-2204.
Association of Clandestine Enthusiasts, P.O. Box 11201, Shawnee Mission, KS 66207-0201.
Association of DX Reporters, 7008 Plymouth Road, Baltimore, MD 21208.
Chicago Area DX Club, c/o Edward G.Stroh, 53 Arrowhead Drive, Thornton, IL 60476.
DX South Florida, 1673 Palace Drive, Clearwater, FL 34616-1833.
International Radio Club of America, 11300 Magnolia #43, Riverside, CA 92505 (mediumwave only)
Longwave Club of America, 45 Wildflower Road, Levittown, PA 19057.
Miami Valley DX Club (MVDXC), 4666 Larkhall Lane, Columbus, OH 43229.
Michigan Area Radio Enthusiasts, P.O. Box 311, Wixom, MI 48393-0311.
Minnesota DX Club, P.O. Box 3164, Burnsville, MN 55337.
National Radio Club, P.O. Box 164, Mannville, NY 13661-0164 (mediumwave only).
North American Shortwave Association, 45 Wildflower Road, Levittown, PA 19057 (SW only).
Radio Communications Monitoring Association, P.O. Box 542, Silverado, CA 92676 (utility stations)
Southern California Area DXers, 3809 Rose Avenue, Long Beach, CA 90807-4334.
SPEEDX, P.O. Box 196, DuBois, PA 15801-0196.
Washington Area DX Association, 606 Forest Glen Rd, Silver Spring, MD 20901.
Worldwide TV-FM DX Association, P.O. Box 514, Buffalo NY 14205-0514 (FM/TV/Satellites)

CENTRAL AMERICA & THE CARIBBEAN

MEXICO

Aguila DX, c/o Miguel Angel Rocha Gámez, Ap.Postal 24, 32000 Cd.Juàrez, Chih (Sp)
Asociación Diexista Becaleña, c/o José Acosta Gallareta, calle 33 #218, 29430 Bechal (Campeche).
Audio Pico DX Club, c/o César Granillo, Ap.Postal 309, 94301 Orizaba (Veracruz).

SOUTH AMERICA

ARGENTINA

Asociacón DX del Litoral, Casilla 10 Suc.5, 2000 Rosario (Santa Fé).
Grupo DX Suquia, Chalet "Monte Abedul", Villa Rivera Indarte km.14, 5149 Córdoba (Sp)
Grupo Radioescucha Platense, Casilla 465, 1900 La Plata (Sp)

BRAZIL

DX Club Paulista, C.P. 592, 13560 Sao Carlos SP (Portuguese)
Globo DX, C.P. 21.429, Sao Paulo 04698 (Portuguese)

CHILE

Amigos Radioescuchas de Santiago, Casilla 183, La Cisterna, Santiago 14 (Sp)

SURINAME

Suriname DX Club International, Bechaniestraat 58, Paramaribo.

URUGUAY

DXCU, P.O. Box 801, Montevideo (Sp).

Deadline for WRTH94:
October 15th 1993

DX AND MEDIA PROGRAMMES

ARGENTINA
RAE (DX Editor: Gabriel Ivan Barrera): **Arabic:** Tues 1820. **English:** Wed 1915, Thurs 0215. **French:** Thurs 2115, Fri 0315. **German:** Fri 2215. **Italian:** Fri 2030. **Japanese:** Thurs/Fri 1015. **Portuguese:** Thurs 1120, Fri 0120. **Spanish:** Tues 1230, 2330. Thurs 2330. Fri 0030, 1230.

AUSTRALIA
R. Australia: English: "Communicator": Tues 0930, 1130, 1530, 1730, 1930. Specialises in Pacific media news and technological developments. **French:** Sun 0000, 0300. **Japanese:** Sun 1040.

AUSTRIA
English: Austrian SW Panorama: Sun 1230, 1330, 1630, 2330. Mon 0330. Despite its name, covers a wide range of media topics. **French:** "Flash des Ondes" Sat 1730, 2130. Sun 0030. **German** (DX Editor: Wolf Harranth): Kurzwellen Panorama: Sat 0405, 0805, 1105, 1405, 1705, 2005. Sun 0205. DX-Telegramm (listening tips): Sun 0055, 0555, 0855, 1155, 1455, 1755, 2055.

BELGIUM
R. Vlaanderen Internationaal: English (DX Editor: Frans Vossen): Sun 0035. Mon 0735, 1005, 1405. Sat 0735, 1005, 1405, 1905, 2205. **French:** Mon 0810, 1040, 1440. Sat 0810, 1040, 1440, 1740, 2040, 2240. **German:** Wed 2000, Thurs 0930. **Spanish:** Sun 2150, Mon 0020.

BULGARIA
R. Sofia: English: Mon 1930. Thurs 0030, 0415. Fri 2115, 2330. Sat 0415. **Spanish:** Sun 2215. Mon 0145, 0545.

CHINA (Rep. of)
Voice of Free China: English: last Sun. of month 0240, 0340, 0740, 2240. **Spanish:** Sun 2335. Mon 0235, 0435.

COSTA RICA
Radio for Peace International: "World of Radio" (producer: Glenn Hauser): Sun 2300, Mon 0700, Tues 1900, Wed 0300 & 1100, Fri 2000, Sat 0400, 1200 & 1800, Sun 0200 & 1000. Subject to revision on the first day of Jan, Apr, July, Oct.

CUBA
R. Habana: English: Sun 0130, 0330, 0630. Tues 2130, 2230. Wed 0030, 0230, 0430, 0530. Sat 2130, 2235. **Spanish:** Sun 1105, 1335, 1910, 2100. Mon 0135.

CZECH & SLOVAK REP.
R. Czechoslovakia: English: Sun 0415. Wed 1815, 1945, 2115, 2215. Thurs 0015, 0315. **French:** Tues 1940. **Spanish:** Mon 1840, 2315.

COLOMBIA
R. Dif. Nacional de Colombia: Spanish: Sun 0115 on 11822/17865vkHz.

ECUADOR
HCJB: English: "DX Party Line" : Sat 0730, 1000, 1900. Sun 0030, 0230, 0500. **French** (produced in cooperation with Amitié Radio): Fri 0615, 0645, 2045, 0015. **German:** Sat 0600, 1830, 2100. **Japanese:** Tues/Fri 1130. **Swedish:** Fri 0500, 2000.

EGYPT
R. Cairo: French: Tues 2050. **Italian:** Wed 1845.

FRANCE
Radio France Internationale: French: R. Activité: Sat 0805. Courrier Technique: Sat 1223, 2144, Sun 0723. **Spanish:** To Latin America: Last Fri of month 2235, 2335, repeated Sat 0135, 0535. To Europe: Tues 2250.

GERMANY
Deutsche Welle: English: 1st Sun 0113, 0313, 0513. **German:** 2nd, 4th and 5th Sat of the month at 0915, 1315, 1715, 2115; repeated Sun 0115, 0515. **Japanese:** Tues following 2nd Sat at 1125. **Spanish:** 3rd Sun at 2325.

GUAM
KTWR: Japanese: Thurs 1135.
KSDA: English (DX Asiawaves, produced by Australian Radio DX Club): Sat 1615, 2315. Sun 0215, 1815. Aimed at the serious DXer.

HUNGARY
R. Budapest: English: Sun 0350, Tues 2250, Wed 0350 & 2250, Thurs 0350, Fri 2250, Sat 0350, 2250. **German:** Tues/Sat (1st prgr), Mon/Wed (2nd prgr).

INDIA
All India Radio: English: Mon 1040, 1435, 1840, 2130.

ISRAEL
Israel Radio: English (DX Editor: Ben Dalfen): Sun 2020. Mon 1425, 2255. Tues 2340.

ITALY
AWR Europe: Sun

JAPAN
R. Japan: Arabic: last Sat of month 0340, 1940. **English** (Media Roundup): Sun 0330, 0930, 1530, 2130, 2330. Covers SW and new media technology. **Korean:** Sat 1100, 1430. **Spanish:** Sun 0353, 0953, 2053. **Swahili:** last Sat of month 1725.
R. Tanpa: Japanese: "Sony World Network" Sat 1200.

KOREA (REP.)
Radio Korea: English: Sun 0835, 1035, 1135, 1250, 1435, 1635, 1835, 2105. Mon 0035, 0635. **Spanish:** Sat 1010, 1040, 1755, 1955, 2310. Sun 0240.

LITHUANIA
R. Vilnius: English: Mon 2253, Tues 0023. Useful source of news about the Lithuanian media scene.

MEXICO
R. México Int: Spanish: Sat 2200.

MONACO
Trans World Radio: German: Sat 1500.

NETHERLANDS
R. Netherlands: English: "Media Network" (producer: Jonathan Marks): Thurs 0150, 0750, 0950, 1150, 1350, 1350, 1550, 1750, 1950. Fri 0050, 0250, 0350. Award winning magazine covering all aspects of the electronic media. **French:** Thurs 1915, 2015. **Spanish:** "Radio Enlace" (producers: Alfonso Montealegre, Jaime Baguena): Fri 1248, 1948, 2303. Sat 0003, 0303, 0503. Excellent coverage of Latin American and Spanish language media.

NETHERLANDS ANTILLES
Trans World Radio: English: "Bonaire Wavelengths" (producer: Chuck Roswell): Sat 1130, Sun 0330. **Portuguese:** "Modulo DX" : Mon-Fri 2210, 0710. **Spanish:** "Monitor DX" : Sun 0035, Mon 0115.

NEW ZEALAND
R. NZ Int: English: Arthur Cushen's DX listening tips within "Mailbox" 1st & 3rd Mon 0430, repeated following Thurs 0830 and Fri 1930.

PHILIPPINES

FEBC: English: "DX Report": Sat 0100, 0900, Sun 1440. "DX-Spot": Wed 1300. "DX-Dial": Mon 0100, Thurs 1400.

POLAND

Polish R: English: Tues 1335, 1605, 1830, 1935. **Spanish:** Fri 0415, 2215 (Bi-weekly).

PORTUGAL

Radio Portugal: English: tri-weekly Fri 1645, 2015, 2115. Sat 0245. **French:** 2nd and 4th Fri. **Italian** (Producer: Giovanni Sergi): 1st Fri of each month. **Spanish:** Alt. Fri 2245, 0115, 0245.

ROMANIA

R. Romania Int: Sun 0245. Mon 1945, 2145. Wed 0245. Sat 1945.

RUSSIA

R. Moscow Int: Spanish: Sat 2240. Sun 0010, 0310, 0530, 2140. Mon 2040. Tues 1940. Wed 0045, 0510.

SEYCHELLES

FEBA Radio: English: Tues 1515.

SOUTH AFRICA

Channel Africa: French: Fri 0430, 0510, 1835.

SPAIN

R. Exterior: Arabic: Sun 1745, 2145. **English:** Sat 1930. Sun 0015, 0115, 0515. **French:** Sat 1845, 2045, 2345. **Spanish:** Sat 2210, Sun 0010.

SWEDEN

R. Sweden: English: "Media Scan" (formerly Sweden Calling DXers - producer: George Wood): 1st and 3rd Tues 1340, 1610, 2140, 2340. Repeated Wed 0110, 0210. Strong emphasis on satellite broadcasting news. **French:** Last Sun. of month at 1200, 1500, 1700, 2030, 0300. **German:** 1st & 3rd Tues.

SWITZERLAND

Swiss Radio International: English: "Swiss Shortwave Merry-Go-Round" (producer: Bob Zanotti with Bob Thomann): Sat 0618, 0918, 1118, 1318, 1518, 1718, 2018, 2218. Sun 0018, 0218, 0418. Mainly technical questions and answers. **French:** Sat/Sun 1115. **Italian:** 2nd & 4th Sat of the month in all transmissions.

TURKEY

Voice of Turkey: English/French/German/Greek: alt. Sat.

UKRAINE

R. Kiev: English: Sun 2200, Mon 0100.

UNITED KINGDOM

BBC World Service: English ("Waveguide"): Sat 1030, Wed 0415, Thurs 0130. Includes BBC frequency information and reception advice. **Spanish:** Mon 0115, Thurs 0400.

USA

Voice of America: English: "Communications World" (producer: Gene Reich): Sat 1210 (EAs), 1710 (SAs), 2110 (Eu). Sun 0110 (SAm/CAm). **Spanish:** Sun 0220.
World of Radio (producer: Glenn Hauser): On WWCR: Fri 2215 on 15690kHz, Sun 0405 & Mon 0000 on 7435kHz, Mon 1330 on 15690kHz, Tues 0730 on 7435kHz. On WRNO: Sat 2300, Sun 0300 & 2130. On WHUS 91.7 Storrs CT: Fri 1530. On WKPN 89.5 Bridgeport CT: Sat 1930. On WOI 640 Ames IA: Sat 1230. On WSUI 910 Iowa City IA: Sun 0430. On WNQM 1300 Nashville TN: Sat 0625. All times 1h earlier 1st Sun in April to last Sun in Oct. DX news & features of particular relevance to North American listeners.
WWCR: "Signals" Sun 0335 on 7435kHz. Crossband with Steve Coletti: Sun 0430 on 7435kHz.
Let's Talk Radio Satellite Network: "Signals" Fri 2330 and Sun 1600 on SpaceNet 3, 87W transponder 21, 6.2MHz wideband. "World of Radio" Fri 2330, Tues 0230 + other times.

UZBEKISTAN

R. Tashkent: English: 2nd Sat of month 1330, repeated Sun 1200, Mon 0100.

YUGOSLAVIA:

R. Yugoslavia: (bi-weekly) Fri 1850, 2120. Sat 0120, 0220.

STANDARD FREQUENCY AND TIME SIGNAL STATIONS

The SW-listener may notice a specialized technical prgr. of tones and time pulses usually received on carrier freqs of 2.5, 5, 10 or 15MHz and sometimes on 20 and 25MHz. These signals have many uses in science and technology. They are most useful to the non-specialist in obtaining accurate time, in checking the calibration of his radio receiver, and in gaining information on radio propagation conditions. The forecasts of the present and future state of the ionosphere can be quite useful in determining the possibilities of SW-reception. Such forecasts are given by JJY, Tokyo, Japan, for certain parts of the globe.

In recent years, a number of Standard Frequency and Time Signal stations have closed down due to increasing operational costs and the availability of such data from other sources. Regular users of any of the services listed below are urged to inform the stations that their broadcasts are useful.

WWV, WWVH, AND WWVB, U.S.A.

St. WWV, 2000 East County Rd. 58, Fort Collins, Colorado 80524, USA; St. WWVH, P.O. Box 417, Kekaha, Kauai, Hawaii 96752; WWVB, 2000 East County Rd. 58, Fort Collins, Colorado 80524, USA.
L.P: Time and Freq. Division: Roger Beehler, National Institute of Standards and Technology, 325 Broadway, Boulder, Colo. 80303. Eng. in Charge: James Maxton (WWV, WWVB). Eng. in Charge: Noboru Hironaka (WWVH).

Standard Radio Frequencies: WWV, 2500kHz 2.5kW, 5000kHz 10kW, 10000kHz 10kW, 15000kHz 10kW, 20000kHz 2.5kW. The broadc. are continuous night and day. WWVH 2500kHz 5kW, 5000kHz 10kW, 10000kHz 10kW, 15000kHz 10kW: Continuous — WWVB 60kHz 13kW: Continuous.
Programme: The broadcast formats of WWV and WWVH include voice time ann. every min. To eliminate confusion between the two stations, the WWV ann. are made by a male and the WWVH ann. by a female. At WWV, commencing about 7½ seconds before the min, a male voice ann. of UTC (Coordinated Universal Time) is given as follows: "At the tone . . . hours . . . minutes UTC (Coordinated Universal Time)". At WWVH, commencing about 15 seconds before the min. a female voice ann. of UTC is similarly given. (The terms of "GMT" and "Z" are accepted as the general equivalents of UTC in navigation and communication).
Standard audio tones of 500 and 600Hz of 45-second duration are broadc. on alternate min. from each station. The beginning of each h. is identified by a 1500Hz tone of 0.8 second duration.
Ann. of weather information are given in plain language. At WWV weather information is given between the 8th and 9th and between the 9th and 10th and between the 10th and 11th min. At WWVH weather information is given between the 48th and 49th and between the 49th and 50th and between the 50th and 51st and between the 51st and 52nd min. St. identification at WWV is given during the 1st and 31st min. after the h. St. identification at WWVH is given during the 30th and 60th min. after the h.
WWVB transmits a special BCD time code consisting of 10dB carrier reductions each second (format on request). Transmitted phase will be advanced 45° for 5 min. each h. between 10 min. and 15 min. after the hours. No other identification will be given — **V.** by QSL folder (B). **Pub:** Prgr. sched. on request.

OTHER STATIONS (in alphabetical order by country).

LOL, BUENOS AIRES, Argentina

Addr: Servicio de Hidográfica Naval, Observatorio Naval, Avenida España 2099, 1107 Buenos Aires, Argentina.
Standard Freq. and Time Signals: 5000kHz 2kW, 10000kHz 2kW, 15000kHz 2kW: 1100-1200, 1400-1500, 1700-1800, 2000-2100, 2300-2400. **Schedule:** 3 min. dash (1000 or 440Hz alternately) beginning at 00.00, 00.05, 00.10, etc. followed by 2 min. ann. at the mins 03/04, 08/09, 13/14, etc. . . . ("Observatorio Naval Argentina . . . Horas . . . minutos"). TS transmission at min. 55-58 in each h — **V.** by QSL-folder or letter. (B).

VNG, LLANDILO, Australia

Addr: c/o VNG Users Consortium, GPO Box 1090, Canberra, A.C.T. 2601.
Station: (G.C: 150.47E/33.42S): VNG 2500kHz 5000/8638kHz 10kW, 12984kHz 3kW, 16000kHz 5kW.
D. Prgr: 24h on 5000/8638/12984kHz, 2200-1000 on 16000kHz.
Ann: on 5000/16000kHz only during the 15th, 30th, 45th and 60th mins: "This is VNG, Llandilo, New South Wales, Australia on 2.5, 5, 8.638, 12.984 or 16MHz MegaHertz. VNG is an Australian standard frequency and time signal service. Inquiries may be directed to VNG Users Consortium, G.P.O. Box 1090, Canberra, ACT, Australia 2601." On 8638/12984kHz, "VNG" is transmitted in slow morse at a freq of approx 400Hz up to 6 times in the 15th, 30th, 45th and 60th min. **V.** by QSL-folder. Rp. Rec. acc.

OBSERVATORIO NACIONAL, Brazil

Responsible for Legal Time in Brazil, originates time signals, which are broadcast by stations:
PPE. Addr: Servico da Hora, Observatorio Nacional (CNPq), Rua General Bruce 586, 20921 Rio de Janeiro, ZC-08-RJ, Brazil. **Te:** +55 (21) 580-7313 **Fax:** +55(21)5806071 — Chief Eng: Paulo Mourilhe Silva.
Station: 8721kHz; 0025-0030, 1125-1130, 1325-1330, 1825-1830, 2025-2030, 2325-2330.
FM: 160.23, 171,13, 166.53MHz (20W). Continuous transmissions, omnidirectional.
F.Pl: New st. under construction in Brasilia, to operate on 2.5, 5, 10, 15, 20 & 25MHz with time signals & voice ann. (Power initially 1kW).
V. by QSL-card.

CHU, OTTAWA, Canada

Addr: R. Station CHU, National Research Council, Ottawa, Ontario, Canada, K1A OR6.
Stations: CHU 3330kHz 3kW; 7335kHz 10kW; 14670kHz 3kW (all antennas vertical).
TS: Continuous by voice and seconds pulses and baud FSK. Freq and time signal control by cesium clock.
Ann: English: "CHU Canada, Coordinated Universal Time . . .". French: "CHU Canada, UTC . . .". Identification of DUT1 = + UT1-UTC by split second code — **V.** by letter. (B). Re. welcomed — **Pub:** Station information free.

CBV, VALPARAISO, Chile

Addr: Instituto Hidrográfico de la Armada, Errázuriz 232, Casilla 324, Valparaíso. **Te:** +56 251056. **Cable:** Hidrovalparaíso. **Telex:** 230362 Hidro Cl.
L.P: Dir. Carlos Bari W.
Stations: 4298/8677kHz, 148.125MHz.
Transmissions: 4228/8677kHz: 1155-1200, 1555-1600, 1955-2000, 0055-0100 (one hour earlier 15 Oct-15 March); on 148.125MHz: hourly at :55-:00 — **V.** by letter. Rp.

TIME SIGNAL STATION BPM, China

Addr: Shaanxi Astronomical Observatory, Chinese Academy of Sciences, P.O. Box 18, Lintong, near Xian, China.
Stations: (10/20kW): 5000kHz (1400-2400), 10000kHz (24h), 15000kHz (0000-1400), 5430kHz (1000-1800 every 2h), 9351kHz (0600, then 1100-2300 every h).
D.Prgr: Time Pips. The Callsign BPM is sent in morse code 10 times at the 29th and 59th minute, followed by voice ann. in Standard Chinese read twice by a female speaker — **V.** by QSL-card.

BSF, Taiwan, R.O.C.

Addr: Telecommunication Laboratories, Ministry of Communications, P.O. Box 71, Chung-Li, Taiwan 32099, Republic of China. **Te:** +886 (3) 4244123. **Fax:** +886 (3) 4904464. **Telex:** 32194 TELELAB.
L.P: Dir: Mr. Yuan-Cheng Teng.
Station: BSF (G.C: 121.09E/24.56N): 5000/15000kHz 2kW 24h.
Schedule: From minutes 5-10, 15-20, 25-30, 45-50, 55-60: second pulses of 5ms duration (no modulation). From minutes 0-5, 10-15, 20-25, 30-35, 40-45, 50-55: second pulses of 5ms duration with 1kHz modulation. The modulation is interrupted 40ms before and after the pulses. Minute pulses are extended to 300ms. During the 9th, 19th, 29th, 49th and 59th minutes, a morse code and Chinese voice announcement of time—**V.** by QSL-card. (B)

STATION OMA, Czech Republic

Addr: Institute of Radio Engineering and Electronics of the Czechoslovak Academy of Sciences, Chaberská 57, 182 51 Praha 8, Czechoslovakia. **Te:** +42 (2) 840335. **Telex:** 122646. **Fax:** +42 (2) 840609.
L.P: Dir.: V. Trkal. **Stations:** Liblice OMA 50kHz, 7kW. Poděbrady OMA 50kHz 50W (reserve).
D.Prgr: Interruption of the carrier of 100 ms at the beginning of every second, of 500 ms at the beginning of every minute. Phase coded announcement of date, UT, local civil time change, leap second, day mark at 0300 of the local civil time, identification of the transmitter in operation. No DUT1 code.
Maintenance: from 0600 to 1200 on the 1st Wed of month. A reserve low power tr. at Poděbrady is used during this period.

HD2IOA, GUAYAQUIL, Ecuador

Addr: Instituto Oceanográfico de la Armada, Casilla 5940, Guayaquil, Ecuador. **Te:** +593 (4) 431105. **Fax:** +593 (4) 442151. **Telex:** 43572 INOCAR ED.
L.P: Dir: CPFG-EM Mario Pinto Ricaurte. Dir. Tec: CPBC-UN Fausto Lopez Villegas.
Stations: HD2IOA 1kW (G.C: 02.16S/79.54W) 1510kHz: 24h; 3810kHz: 0500-1700; 5000kHz: 1700-1800; 7600kHz: 1800-0500.
D.Prgr: On 3810 & 7600kHz: 01s-28s pulses of 100ms, 29s silence, 30s-50s pulses of 100ms, 51s-52s silence, 53s-58s time announcements (male voice), 59s silence, 60s minute pulse (of 300ms duration).
On 5000kHz: 1700-1715 600Hz tone, 1715-1730 440Hz tone, 1730-1800 unmodulated carrier.
Ann: between 59m 40s and 59m 50s: "Hora Oficial del Estado Ecuatoriano emitida por el Instituto Oceanográfico de la Armada, para el territorio continental del país." **V.** by QSL-card (B). Rp.

TIME SIGNAL STATIONS

DCF77 MAINFLINGEN, Germany

Addr: Physikalisch-Technische Bundesanstalt, Bundesallee 100, (Postfach 33 45), D-3300 Braunschweig, Fed. Rep. of Germany. **Te:** +49 531-5920. **Fax:** +49 531-5924006. **Telex:** 952 822 ptbd.
Station: DCF77 77.5kHz 150kW: 24h.
D.Prgr: standard frequency, second pulses of 0.1s or 0.2s duration by carrier reduction to 25% and pseudo-random phase-shift keying of the carrier, coded transmission of numbers of minute, hour, day, day of week, month and year in Legal Time of Fed. Rep. of Germany.
Pub: Prgr. schedule on request — **V.** by QSL-card.

HONG KONG TIME SIGNAL STATION

Addr: Royal Observatory, 134A Nathan Road, Kowloon, Hong Kong.
Te: +852 7329200. **Fax:** +852 7215034. **Cable:** Observahkg. **Telex:** 54777 GEOPH HX.

Stations	kHz	
VPS	500	every even hour
VPS8	4232.5	every odd hour 1100-2100
VPS35	8539	every odd hour
VPS60	13020.4	every odd hour 0100-1500
VPS80	17096	every odd hour 2100-1300
VPS22	22536	every odd hour 0100-0900

Type of transmission: a series of 6 dots, the 6th being on the hour. Hong Kong Aeradio 338, 6679, 8828, 13282kHz. At 15 and 45 minutes past each hour, a series of 6 dots, the 6th being exactly at 15 or 45 minutes past the hour. — **V.** by letter.

ATA, NEW DELHI, India

Addr: National Physical Laboratory, Hillside Road, New Delhi-110012.
Station: 5000kHz 8kW (1230-0330); 10000kHz 8kW (24h); 15000kHz 8kW (0330-1230).
D.Prgr: Seconds pulses, with or without tone modulation. Minute pulses are prolonged to 100ms. Voice ann. every 15 minutes.
V. by QSL-card (B). Rp.

VWC, CALCUTTA, India

Addr: Alipore Observatory, Calcutta.
Stations: VWC 434/4286/12745 kHz.
Schedule: 0825-0830 on 434/12745 kHz, 1625-1630 on 434/4286 kHz. Time signals preceded by callsign. — **V.** by letter. Rp.

INDONESIAN TIME SIGNAL STATIONS

Addr: Directorate General of Post and Telecommunications, Director of Frequency Management, Jl. Kebon Sirih 37, Jakarta.
Stations: 8542kHz 6kW(PKX), 11440kHz (PLC) 3kW.
Schedule: 0045-0055 callsign repeated in morse, 0055-0100 time signals. — **V.** by letter. Rp.

IAM, ROME, Italy

Addr: Istituto Superiore delle Poste e delle Telecommunicazioni, Ufficio 8° Rep. 2°, Viale Europa, I-00144, Rome.
Station: 5000kHz 1kW: 0730-0830, 1030-1130 (W), during summer one h earlier. Time is given in slow speed telegraphy at 0735, 0750, 0805, 0820, 1035, 1050, 1105, 1120 — **V.** by QSL-card. Rp.

IBF, TURIN, Italy

Addr: Istituto Elettrotecnico Nazionale, Corso Massimo d'Azeglio 42, I-10125 Torino. **Te:** +39 (11) 3488933. **Fax:** +39 (11) 6507611. **Telex:** 211553 IENGFI.
Stations: 5000kHz 5kW: W 0645-0700, 0845-0900, 0945-1000, 1045-1100, 1145-1200, 1245-1300, 1345-1400, 1445-1500, 1545-1600, 1645-1700, 1745-1800 (during summer one hour earlier).
Ann: "IBF, IBF, IBF, Emissioni di frequenze e tempo campioni dell'Istituto Elettrotecnico Nazionale, Torino". The voice ann. is repeated in French and English: "IBF, IBF, IBF, Signaux horaires et frequence etalon de l'Institut Electrotechnique Nationale, Turin, Italie" and "IBF, IBF, IBF, Standard Frequency and Time Signals from the National Electrotechnical Institute, Italy".
V. by QSL-card (B).

STATION JJY, TOKYO, Japan

Addr: Communications Research Laboratory, 2-1 Nukui-Kitamachi 4-chome, Koganei-shi, Tokyo 184, Japan. **Te:** +81 423 211211. **Fax:** +81 42327. **Telex:** 283 2611 DEMPA J.
L.P: Chief Standards & Measurements Div: A. Sugiura.
Stations: (all 2kW 24h): 2500kHz, 5000kHz, 8000kHz, 10000kHz, 15000kHz.
All freqs are interrupted at 35-39 mins past each h. Mod. seconds pulses of 8 cycles of 1600Hz and 1000Hz tone. International code of DUT1, consisting of 45 ms-long pulses of 1600Hz, will be tr. during 1-16 sec in each min.
ANN: Every ten mins in each h. identifying signals are tr. consisting of call-sign in Morse code (twice), time in JST once in Morse code, Call-sign twice in voice, time in JST once in voice and grade of propagation warning in Morse code — **V.** by QSL-card (B).

STATION NPO, Philippines

Addr: Commanding Officer, US Naval Communication Station, FPO San Francisco, CA 96656, USA.
Stations: NPO 4445/10440.5/12804kHz.
Schedule: 0555-0600, 1155-1200, 1755-1800, 2355-2400 (time signals).

STATION HLA, Korea

Addr: Time & Frequency Standards Laboratory, Korea Standards Research Institute, P.O. Box 3, Taedok Science Town, Taejeon 305-606, Republic of Korea.
Station: 5000kHz.
Schedule: 24h. Time Signals with station identification in voice on the h. and half h, and Korean Standard Time in voice between the 52nd and 59th second of each minute — **V.** by QSL-card.

OBC3, CALLAO, Peru

Addr: Callao Radio, Av. La Marina Cdra. 36, Callao-4, Peru.
Stations: OBC3 8650/12307kHz: 1555-1600, 2055-2100, 0155-0200 (time signals) — **V.** by QSL-card. Rp.

CAPETOWN RADIO, South Africa

Addr: Private Bag, Milnerton 7435, Cape Town, South Africa.
Stations: ZSC 418kHz 5kW, 4291/8461/12724/17018/22455kHz (all 10kW).
Schedule: 0755-0800, 1655-1700. Time signals (one pulse each second) — **V.** Reports addressed to Chief Technician, are verified by QSL-card. IRC's not valid. QSL-costs paid by staff fund.

EBC, SAN FERNANDO, Spain

Addr: Real Instituto, Observatorio de la Armada, San Fernando, Cádiz, Spain.
Station: EBC (1kW): 12008kHz 0959-1025, 6840kHz 1029-1055.
Schedule: Time signals at 1000-1025, 1030-1055. Tones of 1kHz of 0.1 sec duration, changed to 1250Hz and 0.5 sec duration at each complete minute. **V.** by letter. Rp.

4PB COLOMBO, Sri Lanka

Addr: Colombo Radio 4PB, Coastal Radio Station, Colombo 8, Sri Lanka.
Stations: 482, 8473kHz: 0553-0600, 1323-1330.
Schedule: Time signals every second of 0.1 sec duration, extended to 0.4 sec at each complete minute. During the first two minutes of each transmission the callsign is sent in slow morse — **V.** by letter Rp.

HBG, PRANGINS, Switzerland

Addr: Observatoire Cantonal de Neuchatel, 58, rue de l'Observatoire, CH-2000 Neuchatel, Switzerland. **Te:** +41 (38) 241861. **Fax:** +41 (38) 245941.
L.P: Dir: G. Busca.
Station: Prangins HBG 75kHz 20kW: 24h. Interruption of the carrier at the beginning of each second for 100ms. The minutes are identified by a double pulse, the hours by a triple pulse.
V. by letter. Rp.

ALPHABETICAL LIST OF COUNTRY CODES

* = provisional code not recognized by ITU

AFG = Afghanistan
AFS = South Africa
AGL = Angola
ALB = Albania
ALG = Algeria
ALS = Alaska
AND = Andorra
ANG = Anguilla
ANT = Antarctica
ARG = Argentina
ARM = Armenia
ARS = Saudi Arabia
ARU = Aruba
ASC = Ascension Is.
ATG = Antigua
ATN = Neth. Antilles
AUS = Australia
AUT = Austria
AZE = Azerbaidzhan
AZR = Azores
B = Brazil
BAH = Bahamas
BDI = Burundi
BEL = Belgium
BEN = Benin
BFA = Burkina Faso
BGD = Bangladesh
BHG* = Bosnia/Hercegovina
BHR = Bahrain
BHU = Bhutan
BLR = Belarus
BLZ = Belize
BOL = Bolivia
BOT = Botswana
BRB = Barbados
BRM = Myanmar (Burma)
BRU = Brunei
BUL = Bulgaria
CAF = Central Af. Rep.
CAN = Canada
CAR = Caroline Isl. (Palau)
CBG = Cambodia
CHL = Chile
CHN = China
CHR = Christmas Isl.
CKH = Cook Islands
CLM = Colombia
CLN = Sri Lanka
CME = Cameroon
CNR = Canary Is.
COG = Congo (Rep.)
COM = Comoro Rep.
CPV = Cape Verde
CTI = Cote d'Ivoire
CTR = Costa Rica
CUB = Cuba
CVA = Vatican City
CYM = Cayman Isl.
CYP = Cyprus
D = Germany
DJI = Djibouti
DMA = Dominica
DNK = Denmark
DOM = Dominican Rep.
E = Spain
EGY = Egypt
EQA = Ecuador
EST = Estonia
ETH = Ethiopia
F = France
FJI = Fiji
FLK = Falkland Isl.
FNL = Finland
FRI = Faroe Isl.
G = United Kingdom
GAB = Gabon
GDL = Guadeloupe
GEO = Georgia
GHA = Ghana
GIB = Gibraltar
GMB = Gambia
GNB = Guinea Bissau
GNE = Equatorial Guinea
GRC = Greece
GRD = Grenada
GRL = Greenland
GTM = Guatemala
GUF = French Guiana
GUI = Guinea (Rep.)
GUM = Guam
GUY = Guyana
HKG = Hong Kong
HND = Honduras
HNG = Hungary
HOL = Netherlands
HRV = Croatia
HTI = Haiti
HWA = Hawaii
I = Italy
ICO = Cocos Isl.
IND = India
INS = Indonesia
IRL = Ireland
IRN = Iran
IRQ = Iraq
ISL = Iceland
ISR = Israel
J = Japan
JMC = Jamaica
JOR = Jordan
KAZ = Kazakhstan
KEN = Kenya
KGZ = Kyrgyzstan
KIR = Kiribati
KOR = Korea (Rep.)
KRE = Korea (P.D.R.)
KWT = Kuwait
LAO = Laos
LBN = Lebanon
LBR = Liberia
LBY = Libya
LCA = St. Lucia
LSO = Lesotho
LTU = Lithuania
LUX = Luxembourg
LVA = Latvia
MAC = Macau
MAU = Mauritius
MCO = Monaco
MDA = Moldova
MDG = Madagascar
MDN* = Macedonia
MDR = Madeira
MEX = Mexico
MLA = Malaysia
MLD = Maldives
MLI = Mali
MLT = Malta
MNG = Mongolia
MOZ = Mozambique
MRA = No. Marianas
MRC = Morocco
MRT = Martinique
MSR = Montserrat
MTN = Mauritania
MWI = Malawi
MYT = Mayotte
NCG = Nicaragua
NCL = New Caledonia
NGR = Niger
NIG = Nigeria
NIU = Niue Isl.
NMB = Namibia
NOR = Norway
NPL = Nepal
NRU = Nauru
NZL = New Zealand
OCE = French Polynesia
OMA = Oman
PAK = Pakistan
PHL = Philippines
PNG = Papua New Guinea
PNR = Panama
PNZ = Panama Canal Zone
POL = Poland
POR = Portugal
PRG = Paraguay
PRU = Peru
PTR = Puerto Rico
QAT = Qatar
REU = Reunion
ROU = Romania
RRW = Rwanda
RUS = Russia
S = Sweden
SCN = St. Kitts & Nevis
SDN = Sudan
SEN = Senegal
SEY = Seychelles
SHN = St. Helena
SLM = Solomon Isl.
SLV = El Salvador
SMA = American Samoa
SMO = Western Samoa
SNG = Singapore
SOM = Somalia
SPM = St. Pierre & Miquelon
SRL = Sierra Leone
STP = Sao Tome e Pr.
SUI = Switzerland
SUR = Surinam
SVN = Slovenia
SWZ = Swaziland
SYR = Syria
TCA = Turks & Caicos
TCD = Chad
TCH = Czech & Slovak Rep.
TGO = Togo
THA = Thailand
TJK = Tadzhikistan
TKM = Turkmenistan
TON = Tonga
TRC = Tristan da Cunha
TRD = Trinidad & Tobago
TUN = Tunisia
TUR = Turkey
TUV = Tuvalu
TWN = Taiwan
TZA = Tanzania
UAE = United Arab Em.
UGA = Uganda
UKR = Ukraine
URG = Uruguay
USA = USA
UZB = Uzbekistan
VEN = Venezuela
VCT = St. Vincent & Grenadines
VIR = Virgin Isl. (Am.)
VRG = Virgin Isl. (Br.)
VTN = Vietnam
VUT = Vanuatu
WAL = Wallis & Futuna
YEM = Yemen
YUG = Yugoslavia
ZAI = Zaire
ZMB = Zambia
ZWE = Zimbabwe

Time Signal Stations *(Continued)*

MSF, RUGBY, U.K.

Addr: National Physical Laboratory, Div. of Electrical Science, Teddington, Middlesex TW11 OLW, Gt. Britain. **Te:** +44 (81) 9436493. **Cable:** Physics, Teddington. **Telex:** 262344. **Fax:** +44 (81) 9432155.
Station: 60kHz 27kW (ERP).
D.Prgr: Continuous but tr. interrupted for maintenance from 1000-1400 on the first Tuesday in each month.
Form of the time signals: Interruptions of the carrier of 100 ms for the second pulses, and of 500 ms for the minute pulses. The epoch is given by the beginning of the interruption. BCD NRZ code, 100 bits/s(month, day of month, hour, minute), during minute interruptions. BCD PWM code, 1 bit/s (year, month, day of month, day of week, hour, minute) from seconds 17 to 59 in each minute. DUT1:CCIR code by double pulses. — **V.** by QSL-card (B). Rp.

RCH, Uzbekistan

Station: Tashken 2500kHz 0500-0400, 5000kHz 1400-0400, 10000kHz 0500-1330 (all 1kW).

YVTO, CARACAS, Venezuela

Addr: Observatorio Cagigal, Apt. 6745, Armada 84-DHN, Caracas 103, Venezuela.
L.P: Dir CN: Fernando Aranda Griman. Dir. Tec: Jesús A. Escalona.
Station: YVTO (G.C: 66.55W/10.30N): 5000kHz 2kW: 24h. Second pulses of 1kHz modulation with 0.1 secs. duration. The min. is identified by a 800Hz tone and a 0.5 sec. duration. The h is identified by a 400Hz tone and a 0.5 sec. duration. Sec. 30 is omitted. Between secs 52 and 57 of each min, voice ann. of h, min & sec — **V.** by QSL-card. Rp. (B). Re. to "Técnico Encargado" YVTO.

WORLD TIME TABLE

DIFFERENCE BETWEEN LOCAL TIME AND COORDINATED UNIVERSAL TIME

The differences marked + indicate the number of hours ahead of UTC. Differences marked – indicate the number of hours behind UTC. Variations from standard time during part of the year (in some countries referred to as Summer Time) are decided annually and may vary from year to year. N = Normal Time; S = Summer Time.

	N	S
Afghanistan	+4½	+4½
Alaska	−9	−8
	−10	−9
Albania	+1	+2
Algeria	+1	+1
Andorra	+1	+1
Angola	+1	+1
Anguilla	−4	−4
Antigua	−4	−4
Argentina (Ea.)	−3	−2
Argentina (rest)	−3	−3
Armenia	+4	+5
Aruba	−4	−4
Ascension Isl.	UTC	UTC
Australia		
Victoria & NSW	+10	+11
Queensland	+10	+10
Tasmania	+10	+11
N. Territory	+9½	+9½
S. Australia	+9½	+10½
(we. part)	+9	+10
W. Australia	+8	+8
Austria	+1	+2
Azerbaidzhan	+4	+5
Azores	−1	UTC
Bahamas	−5	−4
Bahrain	+3	+3
Bangladesh	+6	+6
Barbados	−4	−4
Belarus	+2	+3
Belgium	+1	+2
Belize	−6	−6
Benin	+1	+1
Bermuda	−4	−3
Bhutan	+6	+6
Bolivia	−4	−4
Botswana	+2	+2
Boznia/Hercegovina	+1	+2
Brazil		
a) Oceanic Isl	−2	−2
b) Ea & Coastal	−3	−2
c) Manaos	−4	−3
d) Acre	−5	−4
Brunei	+8	+8
Bulgaria	+2	+3
Burkina Faso	UTC	UTC
Burundi	+2	+2
Cameroon	+1	+1
Canada		
a) NF, Labrador (So. Ea.)	−3½	−2½
b) Labrador (rest), NS, NB, PEI	−4	−3
c) ON, PQ	−5	−4
d) MB	−6	−5
e) AB, NWT	−7	−6
f) BC, YT	−8	−7
Cambodia	+7	+7
Canary Isl.	UTC	+1
Cape Verde Isl.	−1	−1
Cayman Isl.	−5	−4
Ce. African Rep.	+1	+1
Chad	+1	+1
Chile	−4	−3
China (P.R.)		
Beijing	+8	+9
Urumqi	+6	+7
Christmas Isl.	+7	+7
Cocos Isl.	+6½	+6½
Colombia	−5	−5
Comoro Rep.	+3	+3
Congo	+1	+1
Cook Isl.	−10	−9½
Costa Rica	−6	−5
Côte d'Ivoire	UTC	UTC
Croatia	+1	+2
Cuba	−5	−4
Cyprus	+2	+3
Czech & Slovak Rep.	+1	+2
Denmark	+1	+2
Diego Garcia	+5	+5
Djibouti	+3	+3
Dominica	−4	−4
Dom. Rep.	−4	−4
Easter Isl.	−6	−5
Ecuador	−5	−5
Egypt	+2	+2
El Salvador	−6	−6
Equatorial Guinea	+1	+1
Estonia	+2	+3
Ethiopia	+3	+3
Falkland Isl.	−4	−4
(Port Stanley)	−4	−3
Faroe Isl.	UTC	+1
Fiji	+12	+12
Finland	+2	+3
France	+1	+2
Gabon	+1	+1
Gambia	UTC	UTC
Georgia	+4	+5
Germany	+1	+2
Ghana	UTC	UTC
Gibraltar	+1	+2
Greece	+2	+3
Greenland		
Scoresbysund	−1	UTC
Thule area	−3	−3
Other areas	−3	−2
Grenada	−4	−4
Guadeloupe	−4	−4
Guam	+10	+10
Guatemala	−6	−5
Guiana (French)	−3	−3
Guinea (Rep.)	UTC	UTC
Guinea Bissau	UTC	UTC
Guyana (Rep.)	−3	−3
Haiti	−5	−4
Hawaii	−10	−10
Honduras (Rep.)	−6	−6
Hong Kong	+8	+8
Hungary	+1	+2
Iceland	UTC	UTC
India	+5½	+5½
Indonesia		
a) Java, Bali, Sumatra	+7	+7
b) Kalimantan, Sulawesi Timor	+8	+8
c) Moluccas, We. Irian	+9	+9
Iran	+3½	+3½
Iraq	+3	+4
Ireland	UTC	+1
Israel	+2	+3
Italy	+1	+2
Jamaica	−5	−4
Japan	+9	+9
Jordan	+2	+3
Kenya	+3	+3
Kazakhstan	+6	+7
Kiribati	−11	−11
Korea (Rep.)	+9	+10
Korea (D.P.R.)	+9	+9
Kuwait	+3	+3
Kyrgyzstan	+6	+7
Laos	+7	+7
Latvia	+2	+3
Lebanon	+2	+3
Lesotho	+2	+2
Liberia	UTC	UTC
Libya	+1	+2
Lithuania	+2	+3
Lord Howe Isl.	+10½	+11
Luxembourg	+1	+2
Macau	+8	+8
Macedonia	+1	+2
Madagascar	+3	+3
Madeira	UTC	+1
Malawi	+2	+2
Malaysia	+8	+8
Maldive Isl.	+5	+5
Mali	UTC	UTC
Malta	+1	+2
Marshall Isl.	+12	+12
Martinique	−4	−4
Mauritania	UTC	UTC
Mauritius	+4	+4
Mayotte	+3	+3
Mexico	−6	−6
(see country section for exceptions)		
Micronesia		
Truk, Yap	+10	+10
Pohnpei	+11	+11
Moldova	+2	+3
Monaco	+1	+2
Mongolia	+8	+9
Montserrat	−4	−4
Morocco	UTC	UTC
Mozambique	+2	+2
Myanmar	+6½	+6½
Nauru	+12	+12
Nepal	+5.45	+5.45
Netherlands	+1	+2
Neth. Antilles	−4	−4
New Caledonia	+11	+11
New Zealand	+12	+13
Nicaragua	−6	−6
Niger	+1	+1
Nigeria	+1	+1
Niue	−11	−11
Norfolk Isl.	+11½	+11½
N. Marianas	+10	+10
Norway	+1	+2
Oman	+4	+4
Pakistan	+5	+5
Palau	+9	+9
Panama	−5	−5
Papua N. Guinea	+10	+10
Paraguay	−4	−3
Peru	−5	−4
Philippines	+8	+8
Poland	+1	+2
Polynesia (Fr.)	−10	−10
Portugal	UTC	+1
Puerto Rico	−4	−4
Qatar	+3	+3
Reunion	+4	+4
Romania	+2	+3
Russia		
Moscow	+3	+4
Novosibirsk	+7	+8
Khabarovsk	+10	+11
Petropavlovsk	+12	+13
Rwanda	+2	+2
Samoa Isl.	−11	−11
S. Tomé	UTC	UTC
Saudi Arabia	+3	+3
Senegal	UTC	UTC
Seychelles	+4	+4
Sierra Leone	UTC	UTC
Singapore	+8	+8
Slovenia	+1	+2
Solomon Isl.	+11	+11
Somalia	+3	+3
S. Africa	+2	+2
Spain	+1	+2
Sri Lanka	+5½	+5½
St. Helena	UTC	UTC
St. Kitts-Nevis	−4	−4
St. Lucia	−4	−4
St. Pierre	−3	−2
St. Vincent	−4	−4
Sudan	+2	+2
Surinam	−3	−3
Swaziland	+2	+2
Sweden	+1	+2
Switzerland	+1	+2
Syria	+2	+3
Tadzhikistan	+6	+6
Taiwan	+8	+8
Tanzania	+3	+3
Thailand	+7	+7
Togo	UTC	UTC
Tonga	+13	+13
Transkei	+2	+2
Trinidad	−4	−4
Tristan da Cunha	UTC	UTC
Tunisia	+1	+2
Turks & Caicos	−4	−4
Turkey	+2	+3
Turkmenistan	+5	+5
Tuvalu	+12	+12
Uganda	+3	+3
Ukraine	+2	+3
United Arab Em.	+4	+4
United Kingdom	UTC	+1
Uruguay	−3	−2
USA		
a) Eastern*	−5	−4
*) Indiana	−5	−5
b) Central	−6	−5
c) Mountain*	−7	−6
*) Arizona	−7	−7
d) Pacific	−8	−7
Uzbekistan	+6	+6
Vanuatu	+11	+12
Venezuela	−4	−4
Vietnam	+7	+7
Virgin Isl.	−4	−4
Wake Isl.	+12	+12
Wallis & Futuna	+12	+12
Yemen	+3	+3
Yugoslavia	+1	+2
Zaire		
Kinshasa	+1	+1
Lubumbashi	+2	+2
Zambia	+2	+2
Zimbabwe	+2	+2

ABBREVIATIONS & SYMBOLS USED IN THIS HANDBOOK

☆ = station also relayed via satellite
a/c = alternating current
acc. = accepted
Addr. = Address
Admin. = Administrative
Ad = advertising
Af. = Africa
Ag. = Acting
alt. = alternate, alternative
AM = Amplitude Modulation
Am = America
Ann. = Announcement
Ant. = Antenna
approx. = approximately
Ap. = Apartado
As = Asia
Assoc. = Association
Asst. = Assistant
Aug. = August
Av., Ave = Avenida, Avenue
BBC = British Broadcasting Corporation
BBCWS = BBC World Service
B'casters = Broadcasters
Bldg. = Building
B.P. = Boite Postale
Broadc. = Broadcast(ing)
Cad. = Cadena
Carib = Caribbean
Cas. = Casilla
Cd. = Ciudad
Ce. = Central
Cel., Cnel. = Coronel
CE = Chief Engineer
cf. = refer to
Ch. = Channel
Chmn. = Chairman
Cl. = Club(e)
Co. = Company
Comm. = Commercial
Contr. = Controller
Corp. = Corporation
C.P. = Case/Caixa Postal
Cra = Carrera
Cult. = Cultura, Cultural
D = Daily
d = directional aerial
d/c = direct current
Dec. = December
Dem. = Democratic
Dep. = Deputy
Dept. = Department
Depto. = Departamento
Desp. = Despacho
Dif. = Difusora, Difusão
Diff = Diffusion
Dir. = Director
Div. = Division
DX = Long Distance
e = estimated (power)
Ea = East
E.C = Electric Current
Edif. = Edificio
Edo. = Estado
Educ. = Education, Educational, Educadora
Em. = Emis(s)ora
Eng. = Engineer, Engineering
ERP = Effective Radiated Power
Esq. = esquina
Eu = Europe
Exc. = except
Exec. = Executive
Ext. = External
FE = Far East
Feb. = February
Fed. = Federal
FM = Frequency modulation
For.Rel. = Foreign Relations
F.Pl. = Future plans
freq. = frequency
Fri. = Friday
Ft. = Fort
G.C. = Geographical coordinates given in degrees and minutes
Gen. = General
GM = General Manager
Gov. = Government, Governmental
Gte = Manager (Sp.)
H = Horizontal polarization
h = hour
HQ = Headquarters
ID = (Station) Identification
i.e. = that is
Inc. = Incorporated
incl. = including
Inf. = Information
Int. = International
Int.Sig. = Interval Signal
IRC = International Reply Coupon
irr. = irregular
Isl. = Island(s)
IS = Interval Signal
kHz = kiloHertz
kW = kiloWatts
Langs = Languages
L.L. = Language Lessons
L.P. = Leading Personnel
L.T. = Local Time
Ltd. = Limited
LW = Longwave
m. = metres
m.b. = metre band
Man. = Managing
max = maximum
ME = Middle East
MF = Mon-Fri
Mgr. = Manager
MHz = MegaHertz
Mil. = Military
Min. = Ministry, Ministerio
mins = minutes
Mon. = Monday
Mpal. = Municipal (Sp.)
Mpo. = Municipio
Mt. = Mount
MW = Mediumwave
N. = News, Network
Nal. = Nacional
Nat. = National
nd = nondirectional antenna
NE = North East
Netw. = Network
n.f. = nominal frequency
No. = North
Nom. = nominal
Nov. = November
Nte. = Norte
occ. = occasionally
Oce = Oceania
Oct. = October
Op's = Operations
Org. = Organization
Ote. = Oeste
Own = Owner
Pac. = Pacific
P = Programme
PD = Programme Director
Pol. = Polarization
Pop. = Population
P.O. = Post Office
PR = Public Relations
Pr. = Private
Pres. = President, Presidencia, Presidente
Prgr. = Programme
Pr.L. = Principal Language(s)
Prod. = Production
Prov. = Province, Provincial
Pte. = Presidente, Poniente, Puente
Pto. = Puerto
Pty = Party
Pub. = Publication(s)
QSL = confirmation of reception
R. = Radio
r = reported
Radiodiff. = Radiodiffusion
Re. = (reception) reports
Rec. = Recordings
Reg. = Regional
Rel. = Relay(ing), Relations
Rep. = Republic
Rlg. = Religious
Rp. = Return postage required
S. = San
Sat. = Saturday
Sce. = Service
Sched. = Schedule
Secr. = Secretary
Sen. = Senior
Sep. = September
SM = Station Manager
S/N, s/n = sin(sem) número
s/off = sign-off
s/on = sign-on
So. = South
Soc. = Sociedad(e)
Sp. = Spanish
SS = Sat/Sun
St. = Station, Saint
Sta. = Santa
Str. = Street
Su. = Summer
Sun. = Sunday
Superv. = Supervisor
Syst. = System
SW = Shortwave
TD = Technical Director
Te. = Telephone
Tech. = Technical
Thurs. = Thursday
Tlx. = Telex
Tr. = transmitter, transmission
Tues. = Tuesday
TV = Television
u.c. = under construction
UHF = Ultra High Frequency
Univ. = University, Universidad(e)
UTC = Coordinated Universal time
V. = Verification, Vertical polarization, Volts
VP = Vice-President
v. = varying
VHF = Very High Frequency
W. = Weekdays, Watt
We. = West
Wed. = Wednesday
Wi. = Winter
Wrp. = Weather report

WORLD RADIO

EUROPE

ALBANIA

L.T: UTC + 1h (Su: UTC+2h) — **Pop:** 3.395.000—**Radio sets:** 210.000 — **Pr.L:** Albanian — **ITU:** ALB.

RADIOTELEVISIONE SHQIPTAR (Gov.)

ADDR: Rruga Ismail Qemali, Tirana. **Te:** +355 422 3239. **Cable:** Albradio.
L.P: Dir. Gen: Skender Bucpapaj. Head of Foreign Rel. Dept: A. Papa Roko.
STATIONS: N = Nat. Prgr, 2 = 2nd Prgr, R = Reg. Prgr, E = Ext. Sce.

Mediumwaves

Station	kHz	kW	P		kHz	kW	P
Rrogozhina	648	15	R	Korgë	1206	10	
Pogradec	693	10	—	Lushnjë	1215	500	E
Kelcyra	864	1	R	Fier	1260	1	R
Korçë	909	15	R	Shkodër	1323	10	R
Puke	972	20	R	Tirana	1359	50	2
Kukës	990	15	R	Lushnjë	1395	1000	E
Durrës	1089	150	N/E	Lushnjë	1458	500	E

FM: Tirana 99.5MHz 100kW.
NAT. PRGR: 0400-2200. **N:** W 0500, 0700, 0900, 1100, 1300, 1430, 1700, 1900, 2100, 2230. Sun 0530, 0630, 1100, 1300, 1430, 1700, 1900, 2100.
2nd Prgr: 0930-1300 on 1359kHz.
Regional Prgrs: rel. Tirana & local prgrs (no details available).

EXTERNAL SERVICE (R. TIRANA)

Mediumwaves: Lushnjë (G.C: 19.40E/40.57N): 1215/1458kHz 500kW, 1395kHz 1000kW.
Shortwaves: Krujë (G.C: 19.37E/41.36N); Lushnjë (G.C: 19.40E/40.57N): 100/50kW.

kHz	kHz	kHz	kHz	kHz	kHz	kHz
5975	6145	7155	7260	9580	9730	11825
5985	7105	7170	7270	9630	9760	11835
6080	7110	7245	9530	9725	11745	11840
6120						

TO EUROPE

Albanian	kHz
0500-0700	5975, 7155
0900-1100	6145
1500-1700	7270, 5985, 1215
2000-2200	6170, 9725, 1458
English	
2200-2230	11825, 9760, 1395
French	
1800-1830	9630, 11835
1930-2000	7260, 9730
German	
0500-0530	7105, 9530, 1395
1330-1400	7105, 9530
1700-1730	7105, 9530
2030-2100	7105, 9530
Greek	
1800-1830	7270, 1215
Italian	
1300-1330	7110, 5985, 1215
1900-1930	6080, 7270, 1215
Serbian	
1430-1500	7110, 5985, 1215
1830-1900	7270, 1215
Spanish	
2130-2230	7245, 9530

TO MIDDLE EAST

Turkish	
2100-2130	7155, 9530

TO AFRICA

English	
1530-1600	7155, 9760
French	
1400-1430	7260, 9730

TO NORTH AMERICA

Albanian*	
0000-0200	6120, 9760, 11745
0200-0400	6120, 9760, 11745

*) Also to South America.

English	
0130-0200	9580, 11840
0230-0300	9580, 11840

TO CENTRAL AND SOUTH AMERICA

Spanish	
0200-0230	9580, 11840

ANN: Albanian: "Ju flet Tirana". E: "This is R. Tirana". **V.** by QSL-card.

TRANS WORLD RADIO (Rlg.)

ADDR: International HQ: P.O. Box 700, Cary, NC 27512-0700, USA.
STATION: Lushnjë 1395kHz 1000kW (cf. R. Tirana).
D.PRGR: 2030-2130 (**F.PL:** 2000-2300) in Ea. European langs.

ANDORRA

L.T: UTC + 1h (Su: UTC + 2h) — **Pop:** 54.450 — **Radio sets:** 10.000 (est.) — **Pr.L:** Catalan, French, Spanish — **ITU:** AND.

SERVEI DE TELECOMUNICACIONS D'ANDORRA (STA)

ADDR: Avgda. Merixtell 110, Andorra la Vella. **Te:** +33 628 23333. **Telex:** 299 MTG AND.
Station: Corroi 0.1kW relaying the following French & Spanish prgrs: 93.3MHz R. Valira, 102.0MHz France-Inter, 102.6MHz France-Musique, 10MHz France-Culture, 104.6MHz Catalunya R, 106.0MHz R4, 106.3MHz R. Barcelona FM, 106.8MHz RNE-1, 107.7MHz R3.

R. ANDORRA (Gov.)

ADDR: Avenida Merixtell 9, 6', Andorra La Vella. **Te:** +33 628 738 63777. **Fax:** +33 628 738 64999.
L.P: DG: Manuel Sansa. PD: Alberto Bartomeu.
Station: 94.2MHz 0.1kW.
D.Prgr: 0600-2300. **N:** every half h.

R. VALIRA (Comm.)

ADDR: Avgda. Merixtell 7, Andorra la Vella. **Te:** +33 628 9738/25461/27619 — **L.P:** St. Mgr: Gualberto Osorio.
Stations: 98.1MHz 0.3kW (st. located in Organya, Spain), Corroi 93.3MHz 0.2kW (repeater).
D.Prgr. in Catalan: W 0700-2000, Sun 0900-1300. **English:** 1330-1400. **French:** 1300-1330. Other times rel. Cadena 13 (R. 80 Serie de Oro), Spain — **V.** by QSL-card.

AUSTRIA

L.T: UTC + 1h (Su: UTC + 2h) — **Pop:** 7.680.000 — **Radio sets:** 4.700.000 — **Pr.L:** German — **E.C:** a/c 50, 220/380V (yes) — **ITU:** AUT.

ÖSTERREICHISCHER RUNDFUNK (ORF)

ADDR: 1136 Wien, Würzburggasse 30. **Te:** +43 (222) 878780. **Fax:** +43 (1) 50108250. **Blue Danube R:** +43 (1) 50101 8900. **Cable:** ORF Wien. **Telex:** 133601 orfz a.
L.P: Dir. Gen: Gerd Bacher. Adm. Dir: Dr. P. Radel. Tech. Dir: H. Doucha. MD (Radio): Rudolf Nagiller.

Mediumwaves	kHz	kW		kHz	kW
Wien	585	600**/240*	Linz-Kroustorf	1026	100
			Scharnitz	1026	0.05
Dornbirn-Lauterach	1026	25	Scharnitz	1314	0.05
			Wien	1476	600

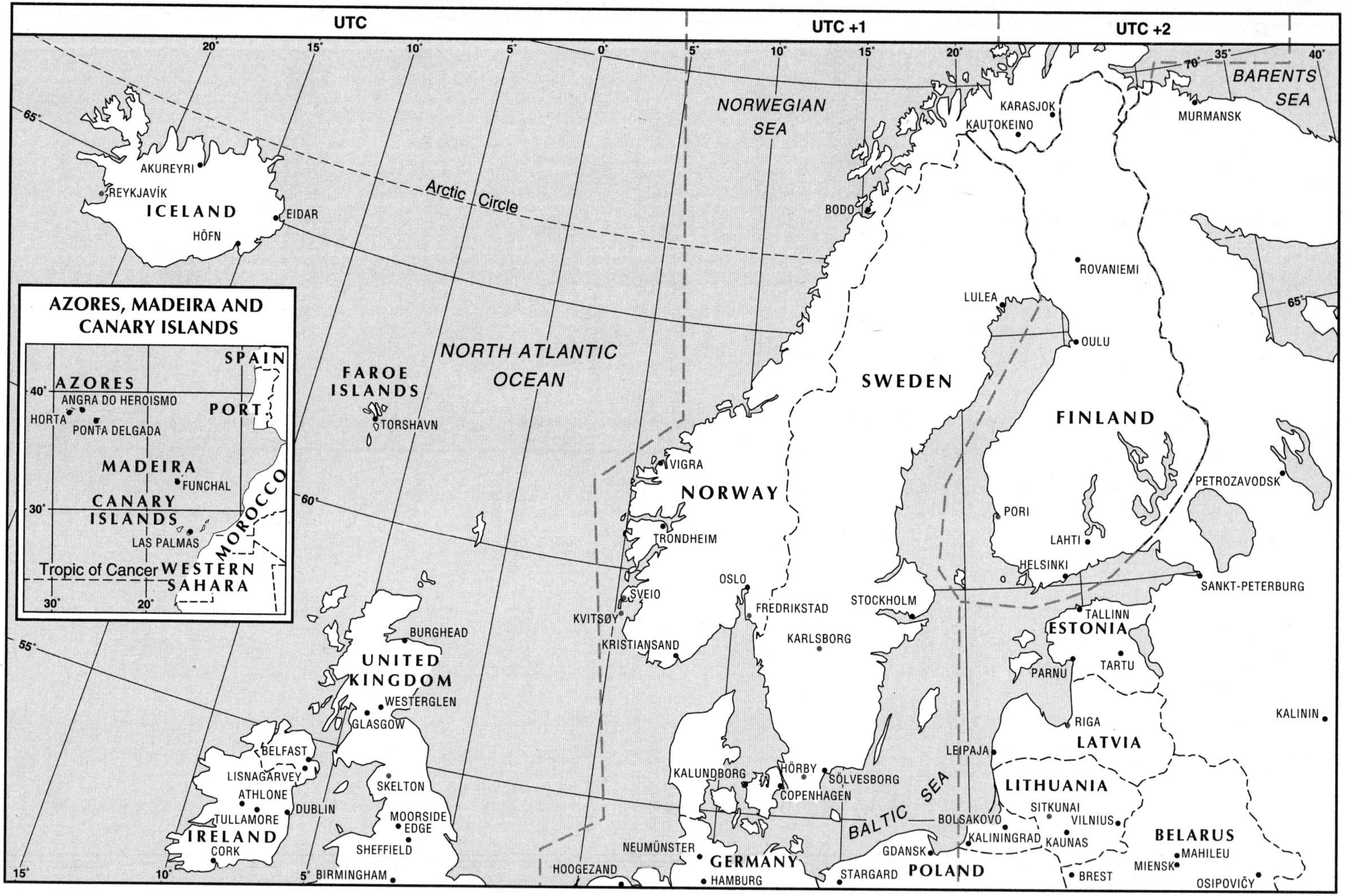
UTC
UTC +1
UTC +2
NORWEGIAN SEA
BARENTS SEA
NORTH ATLANTIC OCEAN
BALTIC SEA
Arctic Circle
ICELAND
AKUREYRI
REYKJAVÍK
EIDAR
HÖFN
FAROE ISLANDS
TORSHAVN
UNITED KINGDOM
BURGHEAD
WESTERGLEN
GLASGOW
SKELTON
MOORSIDE EDGE
SHEFFIELD
BIRMINGHAM
IRELAND
BELFAST
LISNAGARVEY
ATHLONE
DUBLIN
TULLAMORE
CORK
NORWAY
VIGRA
TRONDHEIM
OSLO
SVEIO
KVITSØY
KRISTIANSAND
FREDRIKSTAD
SWEDEN
BODO
LULEA
STOCKHOLM
KARLSBORG
HÖRBY
SÖLVESBORG
COPENHAGEN
KALUNDBORG
FINLAND
KARASJOK
KAUTOKEINO
ROVANIEMI
OULU
PORI
LAHTI
HELSINKI
MURMANSK
PETROZAVODSK
SANKT-PETERBURG
KALININ
ESTONIA
TALLINN
TARTU
PARNU
LATVIA
RIGA
LEIPAJA
LITHUANIA
SITKUNAI
VILNIUS
KAUNAS
BOLSAKOVO
KALININGRAD
BELARUS
MAHILEU
MIENSK
BREST
OSIPOVIČY
POLAND
GDANSK
STARGARD
GERMANY
NEUMÜNSTER
HAMBURG
HOOGEZAND
AZORES, MADEIRA AND CANARY ISLANDS
SPAIN
PORT.
AZORES
ANGRA DO HEROISMO
HORTA
PONTA DELGADA
MADEIRA
FUNCHAL
CANARY ISLANDS
LAS PALMAS
MOROCCO
WESTERN SAHARA
Tropic of Cancer

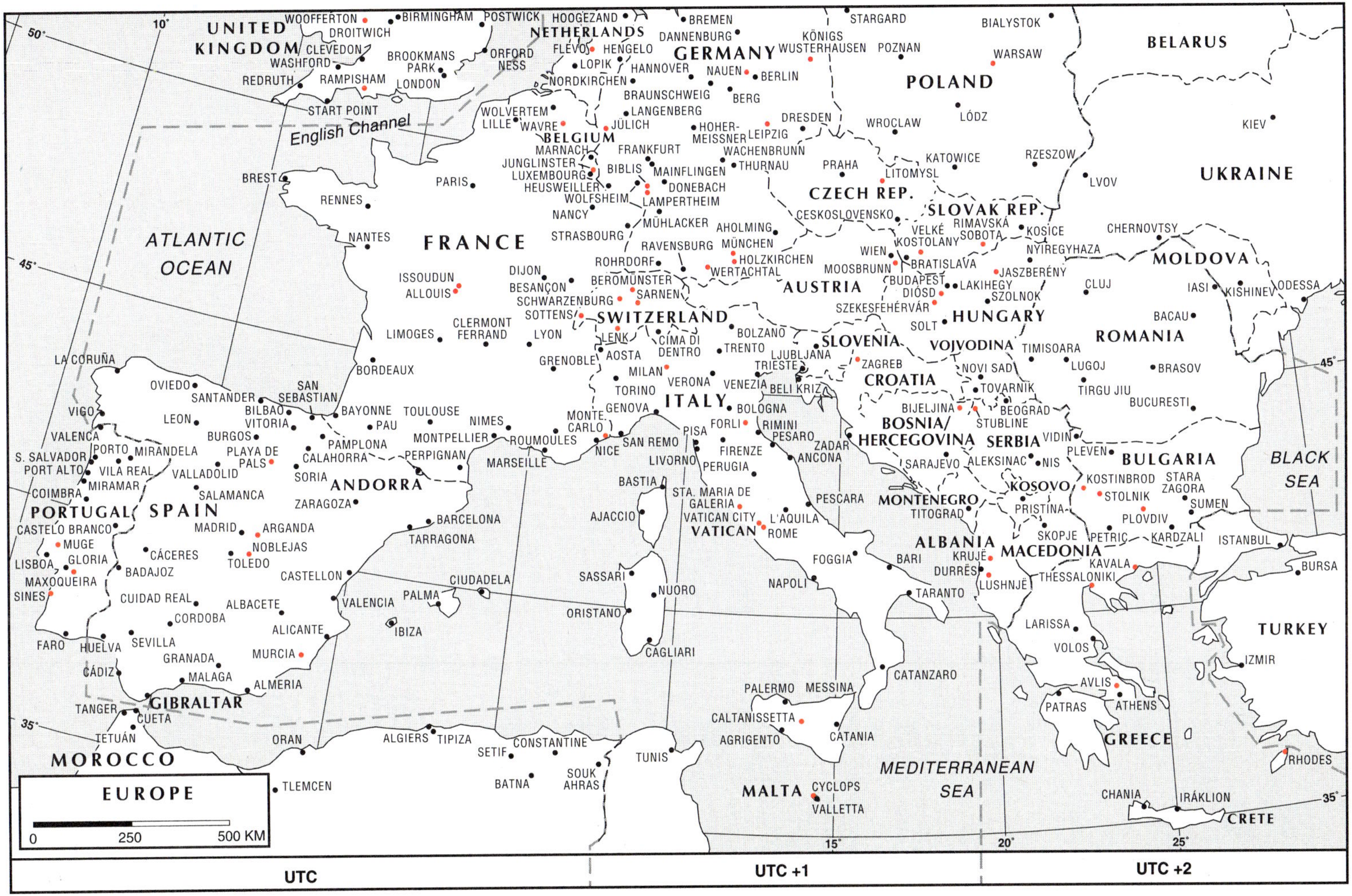
EUROPE
0
250
500 KM
UTC
UTC +1
UTC +2
10°
15°
20°
25°
35°
45°
50°
UNITED KINGDOM
NETHERLANDS
BELGIUM
GERMANY
POLAND
BELARUS
UKRAINE
CZECH REP.
SLOVAK REP.
MOLDOVA
FRANCE
SWITZERLAND
AUSTRIA
HUNGARY
ROMANIA
SLOVENIA
CROATIA
VOJVODINA
BOSNIA/ HERCEGOVINA
SERBIA
BULGARIA
MONTENEGRO
KOSOVO
ALBANIA
MACEDONIA
GREECE
TURKEY
ITALY
VATICAN
MALTA
CRETE
SPAIN
PORTUGAL
ANDORRA
GIBRALTAR
MOROCCO
ATLANTIC OCEAN
English Channel
BLACK SEA
MEDITERRANEAN SEA
WOOFFERTON
BIRMINGHAM
DROITWICH
CLEVEDON
WASHFORD
REDRUTH
RAMPISHAM
BROOKMANS PARK
LONDON
START POINT
POSTWICK
ORFORD NESS
HOOGEZAND
FLEVO
HENGELO
LOPIK
NORDKIRCHEN
WOLVERTEM
LILLE
WAVRE
JÜLICH
MARNACH
JUNGLINSTER
LUXEMBOURG
HEUSWEILLER
WOLFSHEIM
NANCY
STRASBOURG
BREMEN
DANNENBURG
HANNOVER
NAUEN
BERLIN
BRAUNSCHWEIG
BERG
LANGENBERG
HOHER-MEISSNER
LEIPZIG
DRESDEN
FRANKFURT
BIBLIS
MAINFLINGEN
WACHENBRUNN
THURNAU
DONEBACH
LAMPERTHEIM
MÜHLACKER
RAVENSBURG
AHOLMING
MÜNCHEN
HOLZKIRCHEN
WERTACHTAL
KÖNIGS WUSTERHAUSEN
STARGARD
POZNAN
BIALYSTOK
WARSAW
WROCLAW
LÓDZ
KATOWICE
RZESZOW
KIEV
LVOV
PRAHA
LITOMYSL
CESKOSLOVENSKO
VELKÉ KOSTOLANY
RIMAVSKÁ SOBOTA
KOSICE
NYIREGYHAZA
CHERNOVTSY
WIEN
MOOSBRUNN
BRATISLAVA
BUDAPEST
DIÓSD
LAKIHEGY
JASZBERÉNY
SZOLNOK
SZEKESFEHÉRVÁR
SOLT
CLUJ
IASI
KISHINEV
ODESSA
BACAU
TIMISOARA
LUGOJ
BRASOV
TIRGU JIU
BUCURESTI
BREST
RENNES
PARIS
NANTES
ISSOUDUN
ALLOUIS
DIJON
BESANÇON
SCHWARZENBURG
SOTTENS
ROHRDORF
BEROMÜNSTER
SARNEN
LENK
CIMA DI DENTRO
AOSTA
CLERMONT FERRAND
LIMOGES
LYON
GRENOBLE
BORDEAUX
BAYONNE
PAU
TOULOUSE
NIMES
MONTPELLIER
ROUMOULES
PERPIGNAN
MARSEILLE
MONTE CARLO
NICE
SAN REMO
GENOVA
TORINO
MILAN
VERONA
BOLZANO
TRENTO
VENEZIA
LJUBLJANA
TRIESTE
BELI KRIZ
ZAGREB
NOVI SAD
TOVARNIK
BIJELJINA
BEOGRAD
STUBLINE
SARAJEVO
ALEKSINAC
NIS
VIDIN
PLEVEN
KOSTINBROD
STOLNIK
STARA ZAGORA
SUMEN
PLOVDIV
KARDZALI
PETRIC
PRISTINA
SKOPJE
TITOGRAD
KRUJË
DURRËS
LUSHNJË
KAVALA
THESSALONIKI
ISTANBUL
BURSA
IZMIR
LARISSA
VOLOS
AVLIS
ATHENS
PATRAS
RHODES
CHANIA
IRÁKLION
BOLOGNA
FORLI
RIMINI
PESARO
ZADAR
ANCONA
FIRENZE
PISA
LIVORNO
PERUGIA
PESCARA
STA. MARIA DE GALERIA
VATICAN CITY
L'AQUILA
ROME
FOGGIA
BARI
TARANTO
NAPOLI
CATANZARO
PALERMO
MESSINA
CALTANISSETTA
AGRIGENTO
CATANIA
CYCLOPS
VALLETTA
BASTIA
AJACCIO
SASSARI
NUORO
ORISTANO
CAGLIARI
TUNIS
SOUK AHRAS
CONSTANTINE
SETIF
BATNA
ALGIERS
TIPIZA
ORAN
TLEMCEN
TANGER
CUETA
TETUÁN
LA CORUÑA
OVIEDO
SANTANDER
BILBAO
SAN SEBASTIAN
VITORIA
PAMPLONA
CALAHORRA
LEON
BURGOS
PLAYA DE PALS
SORIA
ZARAGOZA
BARCELONA
TARRAGONA
VALLADOLID
SALAMANCA
MADRID
ARGANDA
NOBLEJAS
TOLEDO
CÁCERES
BADAJOZ
CUIDAD REAL
ALBACETE
CORDOBA
SEVILLA
GRANADA
MALAGA
ALMERIA
MURCIA
ALICANTE
VALENCIA
CASTELLON
CIUDADELA
PALMA
IBIZA
HUELVA
FARO
CÁDIZ
VIGO
VALENCA
PORTO
MIRANDELA
VILA REAL
MIRAMAR
S. SALVADOR
PORT ALTO
COIMBRA
CASTELO BRANCO
MUGE
GLORIA
LISBOA
MAXOQUEIRA
SINES

FM (MHz)	Ö-1	Ö-Reg	Ö-3	BD	kW
Bad Gleichen.	—	94.9	—	—	6
Bludenz 1	87.6	96.0h	98.8	—	4
Bregenz 1	93.3	98.2h	89.6	102.1	100
Bruck/Mur 1	87.6	93.2f	98.7	102.1	20
Graz 1	91.2	95.4f	89.2	101.7	67
Innsbruck 1	92.5	96.4g	88.5	—	50
Klagenfurt 1	92.8	97.8b	90.4	102.9	100/60
Kufstein	97.5	95.4g	103.9	—	5
Lienz	89.3	93.8b	99.3	—	2.6
	—	95.9g	—	—	2.6
Linz 1	97.5	95.2d	88.8	104.0	100
	—	90.1c	—	—	10
Mattersburg	89.0	96.2a	100.3	—	3
Rechnitz	90.6	93.5a	87.9	—	3/6/3
	—	97.4f	—	—	3
Salzburg	90.9	94.8e	99.0	104.6	100/2×75
	—	101.2d	—	—	8
St. Pölten	97.0	91.5c	89.4	98.8	100
Schärding	—	99.5d	—	—	4
Schladming 1	94.3	96.3f	101.3	—	4
Semmering	90.3	95.8c	88.2	—	10.2
Spittal/Drau 1	91.6	100.4b	87.9	103.6	3
Weitra	92.7	95.7c	98.1	101.4	2
Wolfsberg 1	96.7	94.5b	99.5	102.3	1.5
Wien 1	92.0	97.9c	99.9	92.9	100
	—	94.7a	—	103.8	2.4/5

+ more than 500 low power trs — BD = Blue Danube R.

Österreich-1: 24h on 585/1026/1476kHz + FM. **N:** 0500, 0600, 0700, 0800, 0900 (W), 1100 (W), 1200, 1300 (Mon-Fri), 1500, 1600, 1700, 2100, 2200, 2300. **Newsreel:** 0600, 1100W, 1600 (Sun), 1700 (Mon-Fri), 2100 (Mon-Fri). **N. in English/French** (rel. Blue Danube Radio): 0805.
Ö-2: Key to regional sces (studios in brackets): a) Burgenland (Elsenstadt), b) Kärnten (Klagenfurt), c) Niederösterreich (Wien), d) Oberösterreich (Linz), e) Salzburg (Salzburg), f) Steiermark (Graz), g) Tirol (Innsbruck), h) Vorarlberg (Dornbirn).
Regional addr: a) Buchgraben 51,7001 Eisenstadt — b) Sponheimstr. 13, 9020 Klagenfurt — c) Argentinierstr. 30a, 1040 Wien — d) Franckstr. 2a, 4010 Linz — e) Nonntaler-Haupstr. 49a, 5010 Salzburg — f) Marburgerstr. 20, 8042 Graz — g) Rennweg 14, 6020 Innsbruck — h) Hoechster Str. 38, 6850 Dornbirn.
D.PRGR: 24h. **N:** W 0400, 0500, 0600, 0800, 0900, 1103, 1200, 1300, 1600, 1900, 2100, 2300 (Sat); Sun 0500, 0700, 1130, 1600, 1700, 1900, 2100. **Local Prgrs:** Mon-Fri 0505-0700, 0803-0900, 0903-1030, 1145-1650, 1705-1800, 1903-2100 (Thurs); Sat 0505-0745, 1003-1030, 1145-1800, 1820-1900; Sun 0605-0755, 1200-1500, 1607-1800, 1820-1900.
Ö-3: 24h. **N:** on the h.
Blue Danube R (BD): 0500-2400 (rel. Ö-1 0000-0500). Prgrs in English, French, German & Italian. **N:** on the half h. in English/German.
ANN: "Österreich 1" — "Ö2 . . . (Wien, Niederösterreich, Tirol . . .)" — "Ö3" — **INT-SIG:** Österreich 1 & Ö2: Composition by Austrian Composer Bert Breit — Ö3: Electronic Music.

FOREIGN SERVICE

ADDR: R. Austria International, A-1136 Vienna. **Te:** +43 (1) 87878-2130. **Fax:** +43 (1) 87878-4404. **Automatic Answerline:** +43 (1) 87878-3636. **Telex:** 133601 orfz a.
L.P: Dir: Prof. Paul Lendvai. Head of Prgrs: Dr. Edgar Sterbenz. Chief Eng's: Ing. Konrad Mayerhofer, Ing. Hans Willenbacher.
STATION: Moosbrunn (G.C: 16.28E/48.00N): 500/300/100kW — C = rel. via Sackville, Canada.

kHz	kHz	kHz	kHz	kHz	kHz
5945	6155	9880	13730	15450	21490
6015C	9875	11780	15410	17730	

Time	Area	kHz			
0000-0300	NAm	9875	1100-1400	FE	15450
0000-0400	NAm	13730			17730
0200-0400	CAm	9870	1400-1700	SAs/SEAs	11780
0400-0800	ME	15410		SWEu	21490
0400-1800	Eu	13730	1400-1700	WAf	21490
0400-2300	Eu	6155	1700-2000	ME	9880
0500-0700	NAm	6015C	1800-2200	SAf	13730
0500-0800	ME	21490	1800-2300	Eu	5945
0800-1100	Pac	15450	2000-2200	SWEu/WAf	9880
		21490	2200-0200	SAm	9870
1100-1300	NAm	13730	2200-2400	SAm	13730

N. in German: on the h. **Kurzwellen Panorama:** Sat 0405, 0805, 1105, 1405, 1705, 2005; Sun 0205. **DX-Telegramm:** Sun 0055, 0555, 0855, 1155, 1455, 1755, 2055.
English ("Report from Austria"): 0130 (NAm/LAm), 0330 (NAm/LAm), 0530 (Eu/ME/NAm), 0630 (NAm), 0830 (Eu/Pac), 1030 (Eu/Pac), 1230 (Eu/As/NAm), 1330 (As), 1530 (Eu/Af/As), 1630 (As), 1930 (Eu/Af/ME), 2330 (LAm). **Austrian SW Panorama:** Sun 1230, 1330, 1630, 2330; Mon 0330.
Esperanto: Sun 0405, 1305.
French: 0030 (NAm), 0730, (Eu/ME), 1130 (Eu/As/NAm), 1730 (Eu/ME), 2130 (Eu/AF). **Flash des Ondes:** Sat 1730, 2130; Sun 0030.
Spanish: 0030 (LAm), 0230 (NAm/LAm), 1330 (Eu), 2030, (Eu/Af), 2230 (Eu/LAm).
IS: First bars of the Blue Danube — **V.** by QSL-card. Rec. acc.

AZORES (Portuguese)

L.T: UTC – 1h (Su: UTC) — **Pop:** 312.000 — **Radio Sets:** 75.000 (est.) — **Pr.L:** Portuguese — **E.C:** a/c 50, 220/380V (no).

MEDIUMWAVES	kHz	kW		kHz	kW
4) Lajes	648	1	1) Faial	1259	1
1) Terceira Flores	693 / 828	10 / 1	2) Angra do Heroísmo	1394	1
1) Barrosa	837	10	5) Lajes	1503	0.1
2) Angra do Heroísmo	909	10	3) Vila do Porto	1566	10

FM: 1) Terceira 90.5MHz 22kW, Faial 88.9MHz 1kW, Barosa 97.9MHz 5kW, Terceira 90.5MHz 5kW + 4 st's 0.25kW 2) 88.8MHz 0.5kW — 3) 94.9MHz 1kW — 5) 96.1MHz 0.1kW.
Addresses & other information: St's 2) and 3) are comm.
1) **Radiodifusão Portuguesa, Centro Regional da R.D.P. — Açores,** R. Dr. Aristides Moreira da Mota, 33, 9500 Ponta Delgada, S. Miguel. **Te:** +351 (96) 22624. **Fax:** +351 (6) 27996 — Dir: José Alberto Rolão Bernardo. **D.Prgr:** 24h (Prgr. 1 on MW, Prgr. 2 on FM) — **Ann:** "Centro Regional dos Açores da RDP" — **Int-Sig:** Regional Hymn — **V.** by letter or QSL-card.
2) R. Club de Angra (CSB80), Ap. 12, 9701 Angra do Heroísmo Codex, Terceira. **Te:** +351 (96) 23101. **Cable:** Radioclub — Dir. Tec: António Santos Caiado — **D.Prgr:** 24h. **N:** 0730, 0815, 1100, 1300, 1800, 2000, 2345 — **Ann:** "Aqui estação CSB80 do R. Club do Angra (A Voz da Terceira)" — **Int-Sig:** Viola — **V.** by letter.
3) Estacão Emissora do Clube Asas do Atlântico (CSB81), Aeroporto de Santa Maria, 9580 Vila do Porto. **Te:** +351 (96) 86182. **Cable:** Asas do Atlântico — Pres & Dir. Tec: Manuel Pacheco de Carvalho — **D.Prgr:** 0545-0100. **N:** 0900, 1000, 1500, 1630, 2100 — **Ann:** "Estaçao Emissora do Clube Asas do Atlântico — **V.** by QSL-card or letter — **Int-Sig:** Opens and closes with Regional and National Hymn. **F.Pl:** 5kW FM-tr. at Barrosa.
4) R. Lajes (Portuguese Air Force), Comando Aereo dos Açores, 9760 Praia da Victoria. **Te:** +351 (96) 53300. **Telex:** 82297 — DG: Moises Couto Rocha. CE: Duarte Rosa e Carlos Martins. PD: Hernani Cardoso. Inf Dept: Manuel Sousa — **D. Prgr:** 0745-0130. **N:** on the h — **Ann:** "Sintoniza a Radio Lajes-Voz da Força Aerea Portuguesa." **Int-Sig:** Nat. Anthem at s/on & S/off — **V.** by letter. Rec. acc.
5) AFRTS (U.S. Air Force), Detachment 3, Air Force European Broadc. Squadron, APO New York, NY 09406-5000, USA — St. Mgr: William J. Kowalski. **D.Prgr. in English:** 24h. (FM st. carries separate prgrs of adult contemporary music).

OTHER FM STATIONS

Antena Nove, Rua S. Joao 28B, Apart. 118, 9900 Horta Codex — 91.3MHz 0.5kW.
R. Comercial dos Açores, Av. Inf. D. Henrique 13A, 1. Esq, 9500 Ponta Delgada — 99.4MHz 0.4kW.
Cooperativa de Radiodiff. do Pico, Rua D. Jaime García Goulart, 9950 Madalena do Pico — 100.2MHz 0.5kW.
R. Ciclone, Rua do Galo 13, 9700 Angra do Heroísmo — 104.4MHz 1kW.
Top Radio, Segunda Rua do Conde 40, 9500 Ponta Delgada — 105.4MHz 0.5kW.
Cooperativa Ecos del Norte, Rua Adolfo Medeiros, Conceicao, 9600 Ribeira Grande — 105.5MHz 0.5kW.
Musipelago, R. da Misericordia 52, Apart. 349, 9500 Ponta Delgada — 106.3MHz 1kW.

R. Ilha, Estrada 25 de Abril, Rua B, 9760 Praia da Vitoria — 106.4MHz 1kW.
R. Lumena, Bairro da Conceicao, 9800 Velas S. Jorge — 107.1MHz 0.5kW.
R. Insular, Alto das Covas 13, 9700 Angra do Heroísmo — 107.2MHz 1kW.
Cooperativa R. Graciosa, 9880 Santa Cruz de Graciosa — 107.9MHz 0.5kW.

BELARUS (Belorussia)

L.T: UTC + 2h (Su: UTC +3h) — **Pop:** 10.260.000 — **Pr.L:** Belorussian, Russian — **E.C:** a/c 50, 127/220 V — **ITU:** BLR.

BELARUSKAJE RADYJO (Gov.)

ADDR: vul. Makajonka 9, 220807 Miensk. **Te:** +7 (0172) 649286. **Fax:** +7 (0172) 648182. **Telex:** 252267 TV SU.
L.P: Chairman: A.G. Stoljarov.

Stations*	kHz	kW	H. of tr.	Prgr.
Osipovičy	171	1000	0200-2400	MR1
Osipovičy	+279e	500	0400-2300	BR1,Reg
Osipovičy	549	1000	0100-2400	RSM
Babrujsk	1008		0400-2300	BR2
Hrodna	1008		0400-2300	BR2
Slonim	1008	50	0400-2300	BR2
Smjataničy	1008		0400-2300	BR2
Ušačy	1008	50	0400-2300	BR2
Brest	1026	5	0400-2300	BR2
Heraneny	1026		0400-2300	BR2
Mahileŭ	1026	15	0400-2300	BR2
Mjadzel	1026		0400-2300	BR2
Pinsk	1026	5	0400-2300	BR2
Salihorsk	1026		0400-2300	BR2
Miensk	1125		0300-2200	RSY
Poseničy 1)	1125		0300-2200	RSY
Osipovičy	1170	150		
Baranavičy	1197		0400-2300	BR2
Braslaŭ	1197		0400-2300	BR2
Miensk	1197	50	0400-2300	BR2
Osipovičy	1278	150		
Baranavičy	1566		24h	RSM
Rakytnica 1)	1566		24h	RSM
Heraneny	1566		24h	RSM
Hrodna	1566		24h	RSM
Mjadzel	1566		24h	RSM
Pinsk	1566		24h	RSM
Mahileŭ	+5430d		0400-2300	BR1,Reg
?	6015		0400-1600	BR2
Miensk	+7210e	50	0400-2300	BR1,Reg,F

Relay via Balashikha, Russia: 6150kHz (0400-1700), 13645 (1700-2300) rel. BR1 + Miensk reg. prgr. 1) Brest oblast.
NB: Powers are tentative only.
*) Names transliterated from Belorussian using a traditional Latin system.

FM	BR1+	BR2$	MR1	RSM$
Babrujsk	66.02d	67.46	73.01	71.45
Brest		70.91		
Heraneny	71.54c	68.39	69.26	72.32
Homiel		66.20		
Hrodna	66.98c	68.90	67.75	66.20
Krascjukovičy	68.03d	66.47		
Mahileŭ	70.10d	71.96	71.18	72.74
Miensk	72.11e	+70.43e	72.99	71.33
Mjadzel		68.39		
Pinsk		67.10		
Slonim	69.44c	67.34	68.12	66.56
Viciebsk	70.67f	72.26	69.92	71.48

+) incl. regional prgrs, $ = stereo.
Regions: a = Brest, b = Homiel, c = Hrodna, d = Mahileŭ, e = Miensk, f = Viciebsk.
D.PRGR: BR1: 0400-2300 in Belorussian/Russian — **BR2 (Krynica):** 0700-1000 in Belorussian, at other times rel. RSM. Miensk 70.43MHz: 1400-2100 musical prgr — **MR1** = Moscow Ostankino R-1 — **RSY** = Moscow R. St. Yunost — **RSM** = Moscow R. St. Mayak — **F** = Foreign Sce, cf. below.
Regional Prgrs (within BR1): Mon-Fri 0430/0440-0500, 1600-1640.

Regional centres: vul. Kujbyšyva 64, 224030 **Brest** — vul. Puškina 6-a, 246000 **Homiel** — vul. Horkaha 85, 230015 **Hrodna** — zav. Kamunistyčny 1, 212000 **Mahileu** — vul. Kamunistyčnaja 8, 210026 **Viciebsk** — Radyjostancyja Stalica (Capital Radio), ul. Saharova 42, 220034 **Miensk.**

FOREIGN SERVICE: Radio Minsk

ADDR: ul. Krasnaja 4, 220807 Minsk.

Belorussian	**Days**	**Area**	**kHz**
1930-2000	1245	Eu	1215*, 5960, 6010, 7210, 7240, 7420, 9600
2230-2300	1234567	Eu	6160, 11960
0030-0100	1234567	NAm	11780, 13645, 15580
German			
1930-2000	36	Eu	1215*, 5960, 6010, 7210, 7240, 7420, 9600
1930-2000++	7	Eu	1215*, 5960, 6010, 7210, 7240, 7420, 9600

++) Rlg. prgr "Die Antwort" (**Addr:** Postfach 767, CH-1701 Frieburg, Switzerland) — *) via Kaliningrad.
ANN: Belorussian: "Havoryc Miensk"
V. by QSL-card. **PUB:** Belaruskaje telebačanne i radyjo (weekly for Home Sce.).

RADIO MOSCOW RELAY

Mediumwaves: Osipovičy 1170kHz 150 kW
Shortwaves: Osipovičy (G.C: 53.24N/28.32E) 1×240kW, 3×100kW.
Used for R. Moscow, and Ostankino RSY & RSM prgrs.

BELGIUM

L.T: UTC + 1h (Su: UTC + 2h) — **Pop:** 10m — **Radio sets:** 4.520.590 — **Pr.L:** Dutch, French — **E.C:** a/c 50, 220V — **ITU:** BEL.

BELGISCHE RADIO EN TELEVISIE (BRTN)
Dutch Language Network

Public Sce. grants by Flemish Council.
ADDR: Omroepcentrum, August Reyerslaan 52, B-1043 Brussels. **Te:** +32 (2) 741 31 11. **Cable:** BRT Brussel. **Telex:** 22486 BRTBIN B. **Fax:** +32 (2) 735 36 62.
L.P: Adm. Gen: C. Goossens. Dir. Gen (Radio): P. Van Roe. Dir. of Production Dept: C. Cleeren. Dir. of Prgr. Dept: H. Verheyden. Tech. Dept. Dir. Gen: H. van Roost. Dir. N. & Current Affairs: M. Gevaert. Dir. Educ. Prgr: Mrs L. Martel. Head of Int. Rel: Mrs. Jaklien Caenberghs.

MEDIUMWAVES (R = Regional Prgr)

Station	**kHz**	**Prgr.**	**kW**
Waver-Overijse	540	2	150
Wolvertem	927	1	150
Kuurne	1188	2	5

FM: (all freqs in MHz) all prgrs. in stereo.

STATIONS:	**I**	**II**	**III**	**IV**	**V**	**kW**
Brussegem	—	—	—	100.6	—	20
Brussel	—	—	—	102.1	101.5	1
Egem O.–Vl.	95.7	98.6	90.4	102.1	101.5	50
Egem W.–Vl.	—	100.1	—	—	—	50
Genk	99.9	97.9	89.9	101.4	102.0	10/20/20/50/40
Leuven	—	—	—	88.0	—	0.5
Schoten	94.2	97.5	—	100.9	89.0	10/10/25/10
Veltem	91.7	93.7	89.5	—	—	40

D.PRGR: Radio 1 on FM I (24h) **N:** Between 0500-0900 every half h. Between 0900-2400 every h on the h; **Radio 2** on FM II (24h) **N:** Between 0500-0800 every half h. Additional news at: 1000, 1300, 1800, 2400. Reg. prgrs. on weekdays from 0600-0800, 1200-1300, 1700-1800; **Radio 3** on FM III (24h) **N:** at 0700, 0800, 0900, 1300, 1800, 2300, 2400. Also relays own night program; **Studio Brussel** on FM IV **N:** at 0730, 1330, 1830, 2230, 2400; **Radio Donna** on FM V (24h) **N:** at 0600, 0730, 0830, 1330, 1830, 2400.
ANN: "Radio 1"; "Radio 2"; "Radio 3", "Studio Brussel"; "Radio Donna".

INTERNATIONAL PROGRAMMES
ADDR: Radio Vlaanderen Internationaal, P.O. 26, B-1000 Brussels. **Te:** +32 (2) 741 3807. **Fax:** +32 (2) 734 7804. **Cable:** BRT Brussels. **Telex:** 22486 BRTBIN B.
L.P: Dir: J. Vandersichel.
MEDIUMWAVE: Wolvertem 1512kHz 300kW (0500-2230 UTC).
SHORTWAVES: ORU2: 200kW; ORU4: 100kW; ORU8: 100kW.

kHz	kHz	kHz	kHz	kHz	kHz
5900	9930	11985	13675	15540	17550
9905	11695	13655	13710	17545	21810

Dutch	**Areas**	**kHz**
0500-0555	Af	13675
0600-0655	Eur	5900, 1512
0700-0725	AUS/Eur	11695, 5910, 9905, 1512
0830-0925 (Mon-Sat)	Eur	5900, 9905, 11695, 1512
1100-1225 (Mon-Sat)	Af/Eur	17545, 9905, 13675, 1512
0930-1225 (Sun)	Af/Eur	17545, 9905, 13675, 1512
1300-1355 (Mon-Sat)	NAm/SEAs	17545, 17550
1300-1325 (Sun)	NAm/SEAs	17545, 17550
1330-1725 (Sun)	Af/Eur	17545, 9905, 13710, 1512
1500-1725 (Mon-Sat)	Af/Eur	17545, 9905, 1512
1800-1825	Af/Eur	17545, 5900, 1512
1900-2055 (Wed/Sat)	Af	13720
2000-2125	Af/Eur	15540, 5900, 9905, 1512
2300-2355	NAm/SAm	9930, 13655
1230-1255 (Sun)	NAm/SEAs	17545, 17550
1400-1425 (Mon-Sat)	NAm/SEAs	17545, 17550
1900-1925	Af/Eur	15540, 5900, 1512
2200-2225	Eur	5900, 9905, 1512
0030-0055	NAm/SAm	9930, 13655
English		
0730-0755	AUS/Eur	11695, 5900, 9905, 1512
1000-1025 (Mon-Sat)	Af/Eur	17545, 5900, 9905, 1512
French		
0800-0825	Eur	5900, 9905, 11695, 1512
1030-1055 (Mon-Sat)	Eur	5900, 9905, 1512
1430-1455 (Mon-Sat)	NAm/SEAs	17545, 17550
1930-1955	Af/Eur	15540, 5900, 1512
2230-2255	NAm/SAm	9930, 13655
German		
0830-0855 (Sun)	Eur/Af	5900, 13675, 1512, 17545
0930-0955 (Mon-Sat)	Eur	5900, 9905, 1512
1830-1855	Af/Eur	17545, 5900, 1512
Spanish		
1230-1255 (Mon-Sat)	Eur	9905, 13675, 1512
2130-2155	Eur	5900, 9905
0000-0025	NAm/SAm	9930, 13655
Arabic		
1730-1755	Af/Eur	17550, 9855, 1512

Freqs subject to modifications. When summertime is in force all prgrs will be one hour earlier.
DX-Prgrs: (Editor: Frans Vossen): **English:** Sat/Mon; **French:** Sat/Mon; **German:** Mon/Fri; **Spanish:** Sun 2130; **Dutch:** Sat/Sun.
Mailbag: English: Tues/Wed/Fri/Sat. **French:** Tues/Wed/Sat. **German:** Sat/Sun. **Spanish:** Mon/Tues/Wed/Fri.
ANN: "Dit is Radio Vlaanderen International" (in all languages). E: "Brussels Calling"; F: "Ici Bruxelles"; D: "Hier ist Brüssel"; Sp: "Aqui Bruselas"; Ar: "Hoena Brussel".
INT-SIG: Starting: "Tussen Maas en Schelde".
V. by QSL-card — **PUB:** Detailed schedule (Free).

RADIO-TÉLÉVISION BELGE DE LA COMMUNAUTE FRANCAISE (R.T.B.F.)
French Language Network

Public sce. Grants by Parliament through French language Council.

ADDR: Cité de la Radio-Television, B-1044 Brussels. **Te:** +32 (2) 737 21 11. **Cable:** RTBF Brussels. **Telex:** 63132. **Fax:** +32 (2) 737 30 32.
L.P: Admin. Gen: R. Stéphane. Dir. Radio: Ph. Dasnoy. Tech. Dept: Chief Eng: F. Fantuzzi. Head Int. Rel: C. Vial.
Regional & Local Centres: Esplanade Anne Charlotte de Lorraine, 7000 Mons; Ave. Golenvaux 8, 5000 Namur; Passage de la Bourse, 6000 Charleroi; Rue de Verviers 203, 4821 Andrimont; Maison de la Culture, Parc des Expositions, 6700 Arlon; rue du Parc 17-21, 4020 Liège.

Mediumwaves	kHz	Prgr	kW
Wavre	621	1	300
La Louvière	1125	2	20
Liège	1233	2	5
Marche	1305	2	10

FM stereo (MHz)	I	II	III	IV	kW
Anderlues	—	92.3	96.6	99.1	50
Bruxelles	96.1	—	—	93.2	3
Léglise	87.6	91.5	94.1	96.4	10
Liège	96.4	90.5	99.5	95.6	1/3×50
Profondeville	102.7	92.8	98.3	90.8	10
Tourna	—	101.8	102.6	104.6	5
Verviers	—	103.0	—	—	2
Wavre	96.1	97.3	—	—	50/0.5

+ 9 low power tr's.
Radio 1: 0430 (Sat/Sun 0530)-2240. **N,Wrp:** 0430, 0500, 0530, 0600, 0630, 0700 and hourly to 1900, 2100, 2200.
Radio 2: 0430 (Sat/Sun 0530)-2240. **N,Wrp:** 0430, 0500, 0800, 0900, 1000, 1200, 1400, 1500, 1800, 2100. **R. Chouette** (computerized music prgr.): 2240-0430/0530 on FM only.
Reg. Prgrs: W 0530-0800, 1200-1300, 1600-1800; Fri 1800-2100 — R. Hainaut (Mons) on 1125kHz + 92.3/101.8MHz — R. Liège on 1233kHz + 90.5/103.0/94.6/89.1/89.4MHz — R. Namur on 1305kHz + 97.3/92.8/89.3/91.5/89.4/90.2MHz — R. Bruxelles on 93.2MHz.
Local Prgrs: R. Verviers (Radiolène) W 0000-1200 on 103.0/94.6/89.1/89.4MHz.
Radio 3: 0558-2245. **N,Wrp:** 0600, 0630, 0700, 0800, 1200, 1800-2240.
Radio 21: 0430 (Sat/Sun 0530)-2240, then R. Chouette (see R.2).
Bruxelles Capitale: Mon-Fri on 99.3MHz 1kW (rel. on 93.2MHz 0530-0300).
ANN: "Vous écoutez R. Une, R. Deux, R. Trois ou Radio 21" —
INT-SIG: Xylophone: Extract from melody: "Où peut on être mieux?" by André-Modeste Grétry.
V. by QSL-card. (Re. to Room 4U40). (Reg/Local centres: by letter).

BELGISCHES RUNDFUNK-UND FERNSEHZENTRUM DER DEUTSCHSPRACHIGEN GEMEINSCHAFT (BRF)

German Language Network (Public Sce.)
Grants by RDG-Rat der deutschsprachigen Gemeinschaft (German speaking community council).
ADDR: Herbesthaler Str. 82, B-4700 Eupen. **Te:** +32 (87) 59 4411/99. **Cable:** 49427 Brf eu b. (Regional addr: Blvd. Reyers 52, B-1040 Brussels. Malmedyer Str. 25, B-4780 St. Vith).
L.P: Pres. of Admin. Council: K. Klein. Dir. of BRF: H. Engels. Chief Editor: H. Engels; CE: J. Schifflers.

STATION (FM stereo)	MHz	kW		MHz	kW
Lüttich	88.5	50	Brussel	95.2	2
Auel	92.2	0.16	Eupen	100.5	1
Recht	94.9	5	Recht	104.1	20

D.PRGR: 0500-1900 (Sat/Sun 2000). **N,Wrp:** 0500, 0600, 0700, 0800, 0900, 1000, 1130*, 1300, 1500, 1600, 1700, 1730, 1900, 2000W **ANN:** "Hier ist der Belgischer Rundfunk, Sendungen in der deutschen Sprache" — **V.** by QSL-card.

COMMERCIAL NETWORKS:

RADIO CONTACT

ADDR: Avenue Télémaque 33, 1190 Brussels. **Te:** +32 (2) 347 4880. **Fax:** +32 (2) 347 5307 — Pres: Francis Lemaire.
Two networks serving Flanders and Wallonia respectively.

RADIO NOSTALGIE NETWORK (Dutch)

ADDR: Hof ter Lo 5-8, 2140 Antwerpen. **Te:** +32 (3) 235 7772. **Fax:** +32 (2) 235 6841 — PD: Max van Zanten.

RADIO NOSTALGIE NETWORK (French)

ADDR: Blvd. Ernest Melot 12, 5000 Namur. **Te:** +32 (81) 220 044. **Fax:** +32 (81) 241 254 — Dir: J. Fyon. PD: Pierre Joye.

PRIVATE FM STATIONS IN MAIN CITIES:
Brussels: R. Antipode 100.3 — Chérie FM 104.0 — R. Contact 102.2/103.1 — R. CRD 103.4 — R. Maranatha 103.7 — R. Nostalgie 106.1 (French) & 106.5 (Dutch) — R. Panik 105.4 — R. Plus 101.4MHz.
Antwerp: R. Antigoor 103.2 — R. Contact 104.2 — R. Expres 102.8 — R. Nostalgie 107.4/R. Totaal 104.8MHz.
Ghent: R. Contact 104.9 — R. Free 103.5 — R. Go 103.2 — R. Sis 102.8MHz.
Charleroi: Chérie FM 102.2 — R. Contact 104.0MHz.
Liège: Chérie FM 103.6 — R. Contact 100.9 — R. Nostalgie 107.5MHz.

AMERICAN FORCES NETWORK, SHAPE

ADDR: Box 7, 7010 SHAPE. (APO 09088). **Te:** +32 (65) 44 41 21.
L.P: Officer-in-charge: Cpt. G. Martel. Broadc. Superv: SFC D. Sweet. Chief Eng: René Libre.
STATIONS: Klein Brogel 106.2MHz 0.1kW, Brussels 101.7MHz 0.9kW, SHAPE 104.2/106.5MHz 4kW, Florennes 107.7MHz 0.1kW.
D.PRGR: 24h on 101.7/104.2/107.7MHz. Own prgrs Mon-Fri 0500-0800, 1400-1700; Sat 0800-1200. Other times rel. AFN Europe.
AFN-2: 24h easy listening stereo prgr. on 106.2/106.5MHz.
V. by QSL-card.

BRITISH FORCES BROADCASTING SERVICE

STATIONS: SHAPE 107.6MHz/Antwerp 107.4MHz 0.05kW.
D.PRGR: rel. BFBS Germany (see entry under "Germany").

VOA EUROPE: Brussels 104.5MHz.

BOSNIA/HERCEGOVINA

L.T: UTC + 1h (Su: UTC + 2h) — **Pop:** 4.408.000 — **Pr.L:** Serbian, Croatian — **E.C:** a/c 50, 220V.

RADIOTELEVIZIJA SARAJEVO (Gov.)

ADDR: VI Proloterske brigade 4, 71000 Sarajevo. **Te:** +38 (71) 652-333. **Fax:** +38 (71) 461-569. **Telex:** 41224—Dir.Gen: H. Ištuk.
NB: Due to the military activity in the country, the details below may not reflect the actual situation at press time.

Mediumwaves	kHz	kW		kHz	kW
Sarajevo 1	612	600	Mostar	1071	40 l
Livno	630	2	Brčko	1485	1
Trebinje	630	2	Drvar	1485	1
Mostar	756	10m	Jajce	1485	1
Bihac	774	2	Tuzla	1584	2
Sarajevo 2	945	100	Mostar	1584	1
Banja Luka	1071	20			

l = rel. Sarajevo 1 — m = rel. Sarajevo 2

FM (MHz)	I	II/III	L	kW
Banja Luka	+91.8			1
Bjelašnica	+98.1	+94.5		30
Hum	+95.7	+97.7		10
Kozara	+92.7	+95.3		30
Leotar	+92.8	+94.1		30
Majevica	+89.9	+98.9		30
Plješivica	+94.1	95.7		30
Sarajevo			93.1	0.5
Trovrh	+90.3	+99.5		30
Velež	+97.3	+91.7		30
Vlašić	+89.3	+97.0		100
Zvornik			91.1	1

L) local st's — +) stereo. +22 sts. less than 0.5kW.

R. Sarajevo 1: 0400-2400 (Sat/Sun 24h).
R. Sarajevo 2: 0700-1700.
R. Sarajevo 3: 1700-2300.

Radio "M" Sarajevo, Tršćanska 7, 71000 Sarajevo: 24h on 98.7MHz 1kW.

Local Stations	kHz	kW		kHz	kW
Prijedor	576	2	Bos. Brod	1206	1
Gračanica	639	1	Ilijaš	1233	1
Maglaj	729	1	Goražde	1251	1
Banovići	792	1	B. Krupa	1305	1
Čapljina	792	1	Prnjavor	1341	1
Sokolac	864	1	Sanski Most	1368	1
Foča	990	1	Zvornik	1467	1
Gazin	1017	1	Sarajevo 202	1485	1 *)
Bos Šamac	1035	1	Bos. Gradiška	1485	1
Zenica	1161	2	Zavidovići	1503	1
Bjeljina	1197	2	Derventa	1521	1
Bos. Novi	1197	1	Donji Vakuf	1530	1

*) **R. Sarajevo 202:** 0600-1800 also 93.1/90.3MHz + 20 lp st's not listed.
FM: See above.

BULGARIA

L.T: UTC + 2h (Su: UTC + 3h) — **Pop:** 8.892.000 — **Radio sets:** 3m (est.) — **Pr.L:** Bulgarian — **E.C:** a/c 50, 220V (yes) — **ITU:** BUL.

BULGARIAN NATIONAL RADIO (Gov.)

ADDR: 4 Dragan Tzankov Str, 1040 Sofia. **Te:** +359 (2) 652871. **Fax:** +359 (2) 662215. **Cable:** Bulgarradio. **Telex:** 22557/22558.
L.P: Head of Prgrs: Ivan Obreterov (Horizont), Eddi Emiyran (Christo Botev). Chief Eng: Gueorgui Trenev.
Regional addr: 32 Lavele Str, 1000 Sofia — RTV Center, 2 kn. Dondukov-Korsakov Str, 4000 Plovdiv — 22 blv. Primorski, 9000 Varna — 56 blv. A. Velitchkov, 2700 Blagoevgrad — 75 N. Nikolov Str, 6000 Stara Zagora — 7 D. Voynikov Str, 9700 Sumen.

LW/MW	P	kHz	kW		P	kHz	kW
Sofia	1	261	60	St.Zagora	2/R	873	50
Vidin	2	576	500	Sumen	1/R	963	50
Pleven	1	594	250	Sofia	1	963	40
Plovdiv	2/1/R	648	30	Pirin	1	963	40
Pirin	2	702	20	Kardzali	1	1017	50
Petrič	1/R	747	500	Vidin	1	1017	40
Sofia	1/4/R	774	40	Varna	1	1143	40
Varna	2/R	774	30	St.Zagora	1	1161	500
Sumen	2	828	500	Sofia	1	1161	40
Sofia	2	828	50	Vidin	1/F	1224	500
Blagoev-				Kardzali	2/1	1296	150
grad	2/1	864	50				

R) carries regional prgrs — F) carries Foreign Sce.

SW	P	kHz	kW	H. of tr.
Stolnik	1	7670	15	0400-2200
Plovdiv	1	11660	250	0330-1800

FM (MHz)	1	2	3/4	kW (ERP)
Belogradcik	67.68	66.52	67.1s	6
	102.3S			10
Botev Mt.	70.88	72.44	71.66s	10
	100.9s			
Burgas	67.58	66.02	66.8s	10
Dobric	71.42	65.36	68.6s	3
G.Deltchev	72.92R	72.14	71.36s	10
	100.3s			10
Kardzali	67.4	66.62	65.84s	10
	105.0s			10
Kjustendil	67.76R	66.2	66.98s	6
	102.1s			6
Petrohan	101.4s			10
Plovdiv		68.36R		3
Ruse	67.22	66.44	69.32s	3
	103.0			10
Silistra	71.78			3
Sliven	71.24	72.8	72.02s	10
Smolijan	67.19	66.41	67.97s	3
	101.6s			5
Sofia	68.06	66.50	69.26sR	10
		92.9s		10
	103.0			0.3
Stara Zagora			70.22sR	
Sumen	71.00	72.56	69.8s	10
	102.0s	87.6sR		10
Varna	67.31	68.09	69.41sR	3
	100.9s			5

s) stereo — R) carries regional prgrs (see below).
Prgr. 1 ("Horizont"): 24h music and news
Prgr. 2 ("Christo Botev"): 0400-2200.
Prgr. 3 ("Orphay"): 1405-2300 (Fri/Sat/Sun 2300).
Prgr. 4 ("Znanie"): W 0800-1000 & 1200-1400, Sun 0700-1000 (Sept-June only).
Regional Prgrs: Sofia: W 1100-1200 on 774kHz + 69.26MHz — **Blagoevgrad:** 0400-0800 on 747kHz + 72.92/1000.3MHz (Orelek) and 67.76/102.1MHz (Vidin), 1300-1600 on 747kHz — **Plovdiv:** 0400-0700 on 648kHz, 0900-1000 on 6836MHz, 1300-1600 on 648kHz — **Stara Zagora:** 0400-0700 on 873kHz + 70.22MHz, 1300-1600 on 873kHz — **Sumen:** 0400-0800 & 1300-1600 on 963kHz + 87.6MHz — **Varna:** 0400-0700 (Sat 0800) on 774kHz + 69.41MHz, 1300-1600 on 774kHz.
Foreign Language Prgr. from Varna: May-Sept 0700-1000 & 1100-1230 on 774kHz + 69.41MHz in English, French and German. Relayed on Prgr. 3 FM netw. July 1st-September 30th (Sun to 1000).

FOREIGN SERVICE

MEDIUMWAVE: Vidin (G.C: 22.40E/43.49N): 1224kHz 500kW.
SHORTWAVES: Stolnik (G.C: 23.35E/42.42N): 1×150kW (+ 1×15kW for Home Sce) — Kostinbrod (G.C: 23.13E/42.49N): 1×100kW, 4×50 kW — Plovdiv (G.C: 24.42E/42.10N): 2×500kW, 3×250kW.

kHz	kHz	kHz	kHz	kHz	kHz
5895	6235	7225	11630	11860	15160
5915	7115	7290	11635	11870	17780
5980	7120	9520	11660	13605	17825
6070	7150	9560	11700	13645	17830
6085	7170	9700	11720	13670	
6220	7185	9850	11765	15120	

Schedule to April 1993

Albanian	**Area**	**kHz**
1645-1730	Eu	1224, 5895
1945-2030	Eu	1224, 5895
Arabic		
1530-1615	ME	7150, 11870
1745-1830	ME	7150, 11870
2030-2130	ME	7150, 11870
Bulgarian		
0200-0300	NAm/LAm	9700, 11660, 11720, 13645
0400-0500	Eu/ME	1224, 5915, 5980+, 7115, 11735*
	NAm/SAm	7170, 9850
1300-1555	Eu/Pac	1224, 9700, 13670, 17780
1400-1655	Eu/NAm	11635, 11720, 11860
1900-1945	Eu	1224, 5895
2000-2030	ME	7150, 11870
English		
0100-0145	NAm	7225, 9700, 11720
0400-0540	Eu/NAm	7290, 9700, 11720
1130-1300	Eu/ME/NAm	11630, 11720, 13670, 17780, 17825
1830-2000	Eu/Am	6235, 9560, 9700, 11720
2045-2130	Eu/Am	6235, 9560
2245-0015	NAm/Eu	7225, 9700, 11720
French		
0015-0100	NAm	7225, 9700, 11720
0315-0400	NAm	9700, 11720
0715-0800	Eu/NAm	9700, 11720, 15160
0745-0830	Eu/NAf	6235, 9560
2000-2045	Eu/NAf	6235, 9560
2200-2245	Eu/NAm	11720
German		
0630-0715	Eu	9700, 11720, 15160
1100-1145	Eu	11765, 17830
1700-1745	Eu	6235, 7120, 9560, 9700, 11720
2015-2100	Eu	6070, 6085, 9700
Greek		
0500-0600Sun	Eu	5915, 5980, 7115+, 7170, 11735*
0600-0630W	Eu	1224, 5915, 5980, 7115+, 7170, 9520+, 11700*, 11735*
0900-1025Sun	Eu	9700, 11720, 11860
1000-1025W	Eu	7070, 9700, 11720, 11860
1730-1815	Eu	1224, 5895
2030-2115	Eu	1224, 5895, 6220

Italian

1745-1830	Eu	7120, 9700, 11720
2100-2145	Eu	6070, 6085, 9700
2200-2240	Eu	1224, 7225, 9700

Portuguese

0015-0100	Eu/Af	9850, 11660, 13645
2200-2245	SAm	9850, 11660, 13605

Serbo-Croat

0500-0600Sun	Eu	1224
0600-0700W	Eu	1224, 5915, 5980, 7115+, 7170, 9520+, 11700*, 11735*
0700-0855Sun	Eu	1224, 5915, 5980, 7115+, 7170, 9520+, 11700*, 11735*
1600-1645	Eu	1224, 5895, 6235, 9700
1800-1900	Eu	1224, 5895
2115-2155	Eu	1224, 6220

Spanish

0100-0200	Eu/SAm	9840, 11660, 13645
0500-0600	Eu/SAm	9520+, 9700, 11700*
2130-2230	Eu/SAm	9560, 15120

Turkish

0500-0600W	Eu/ME	1224, 5915, 5980, 7115+, 7170, 11735*
0600-0730Sun	Eu/ME	1224, 5915, 5980, 7115+, 7170, 9520+, 11700*, 11735*
1030-1130	Eu/ME	11720, 11870, 13670, 17825
1615-1745	ME	7150, 11870
1830-2000	ME	7150, 11870

+) to 7th March, *) from 8th March — all transmissions 1h earlier from March 28th.

ANN: Albanian: "Ju Flet Sofia". Arabic: "Huna Sofia". Bulgarian: "Govorry Sofia". E: "This is Sofia, Bulgaria calling". F: "Ici Sofia, Radiodiffusion Télévision Bulgarie". G: "Hier spricht R. Sofia, Bulgarien". Greek: "Milai Sofia". I: "Quie Sofie". Serbo-Croat: "Ovde Radio Sofia". Sp: "Habla Sofia". Turkish: "Burasi Sofia Radyosu". **IS:** Youth march played on organola — **V.** by QSL-card (Home Sce: R. Varna has own QSL-card). Rec. acc — **PUB:** "Radio i Televizija", weekly for Home Sce. Prgr. schedule for foreign sce.

Private Stations have been licensed incl. 6 in Sofia: FM+, R. Tangra, R. Express, R. Dal(?), R. 99 and Canal+. Freqs not available at press time.

FM relays of International Broadcasters

All tr's located in Sofia with 1kW ERP (stereo):
89.1MHz : Radio Free Europe in Bulgarian 1400-2200.
95.7MHz : Deutsche Welle 24h in German & Bulgarian.
97.6MHz : VOA Europe in English 24h.
98.9MHz : BBCWS 24h (English, Bulgarian & English By Radio).
105.0MHz : Radio France Internationale 24h in French.

CROATIA

L.T: UTC + 1h (Su: UTC + 2h) — **Pop:** 4.784.265 — **Radio sets:** 760.931 — **Pr.L:** Croatian — **E.C:** a/c 50, 220V — **ITU:** HRV.

HRVATSKA RADIO — TELEVIZIJA (HRT)

ADDR: Dežmanova 10, 41000 Zagreb, Croatia. **Te:** +38 (41) 456611. **Fax:** +38 (41) 424654. **Tlx:** 21326 HRT CRO.
L.P: GM: Antun Vrdoljak. **Te:** +38 (41) 272 372. MD. Transmitters & Links: Nikola Perčin. **Addr:** Radnička c. 22, 41000 Zagreb. **Te:** +38 (41) 615612.

HRVATSKI RADIO (HR)

ADDR: Jurišićeva 4, 41000 Zagreb. **Te:** +38 (41) 454333. **Fax:** +38 (41) 430633. **Tlx:** 21154 HR CRO.
L.P: MD: Tomislav Bakaric. **Te:** +38 (41) 450164.

Mediumwaves	**kHz**	**kW**	**Prgr.**
Osijek	594	20	1
Hvar	774	50	1
Deanovec	1125	100	1
Zadar+	1134	600/1200++	1
Tovarnik*	1143	300	1

+) currently limited operation — ++) 600kW ND, 1200kW bi-directional (beam 315°/135°) — *) inactive.

SHORTWAVES: Deanovec (G.C: 16.40E/45.70N).

kHz	**kW**	**Prgr.**	**Beam**
6210	10	1	125°/305°
9830	10	1	ND
13830	2	1	270°

Relay via WHRI, USA: 0000-0100 & 0500-0600 on 7315/15105kHz.

FM	**HR1**	**HR2**	**HR3**	**Local**	**kW**
Biokovo	89.7	98.9	102.0	—	80
Brač	99.8	88.8	104.5	—	3
Ivanščica	102.4	89.7	96.1	—	30/15/15
Kalnik	90.8	91.7	105.8	—	15/3/15
Labištica	91.3	96.1	101.0	105.5	3×30/5
Mircovica	91.3	93.3	102.7	—	30
Papuk	94.9	97.7	106.8	—	10
Psunj	97.3	99.7	102.0	—	80
Sljeme	92.1	98.5	101.0	88.1	3×120/5
Učka	99.3	96.9	104.7	—	80

NB: above list includes only those st's operating as of Oct 1992.
HR1: 24h. **N.** on the h. **N. in English:** 0705, 0905, 1305, 2205.
HR2: 0600-2300.
HR3: 1600-2300.
Local Prgrs: Sljeme-Zagreb on 88MHz, Split on 105.5MHz.
Ann: "Ovde Hrvatski Radio". E: "This is Croatian Radio".
IS: Tune to Dubrovnkik's poem "Beloved, Sweet, Dear Liberty" played on celeste, before news bulletins.
V. by letter. Re. in SINPO code to HRT Transmitters & Links, Radnička 22, Zagreb, Croatia or Fax: +38 (41) 615 559, Te: +38 (41) 619 222.

CZECH & SLOVAK REPUBLICS

L.T: UTC + 1h (Su: UTC + 2h) — **Pop:** 15.772.000 — **Radio sets:** 10.5m —**Pr.L:** Czech, Slovak — **E.C:** a/c 50 220V — **ITU:** TCH.

ČESKOSLOVENSKÝ ROZHLAS

ADDR: Czech Radio, 12099 Praha 2, Vinohradska 12.
Cable: Radiopraha. **Te:** +42 (2) 2115 111. **Fax:** +42 (2) 232 1020. **Telex:** 121100.
Slovak Radio, 81290 Bratislava, Mýtna 1. **Te:** +42 (7) 494462. **Fax:** +42 (7) 493626. **Cable:** Radiobratislava. **Telex:** 93353.
L.P: Dir. Gen: Peter Duhan. Dep. Dir. in charge of Central News Sce: Dr. V. Vrabec. Dep. Dir. Gen. & Dir. of Czech Radio: J. Mejstrik. Dep. Dir. Gen. & Dir. Foreign Sce: Ing. K. Lansky. Dep. Dir. Gen. & Dir. of Radio in Slovakia: Dr. V. Štefko. Dep. Dir. Gen. in charge of Technology & Management: Z. Škopán. Dir. Foreign Rel. & Prgr. Exchange: R. Sobková.
NB: The structure of domestic broadcasting as from Jan 1st 1993 may be revised. The above reflects the situation at press time.

CZECH REPUBLIC

Long & Mediumwaves — Reg = region (see regional addresses).

Reg	**Stations**	**kHz**	**kW**	**Prgr.**
	Uherské Hradiště	270	1500	Československo
5)	Broumov	558	2	Reg. H. Králové
2)	Lipno	558	1	Reg. Č. Budějovice
3)	Tachov	558	1	Reg. Plzeň
6)	Brno C. (Komárov)	603	25	Reg. Brno
	Praha (Liblice)	639	1500	Praha
5)	Hrádec Králové	774	20	Reg. H. Králové
1)	Praha-Zbraslav	792	40	Reg. Praha
7)	Ostrava-Svinov	846	30	Reg. Ostrava + Polish
7)	Olomouc	864	10	Reg. Ostrava + Polish
4)	Ústí nad Labem	864	10	Reg. Ústí nad Labem
3)	Plzeň	864	40	Reg. Plzeň
3)	Karlovy Vary	900	20	Reg. Plzeň
4)	Liberec	945	6	Reg. Ústí nad Labem
	Strakinice	954	40	Praha
	Brno	954	200	Praha
2)	České-Budějovice	981	30	Reg. Č. Budějovice
	Ostrava	981	2	Praha
1)	Mnich. Hradiště	1017	30	Reg. Praha
	České Budějovice	1071	7	Praha
	Karlovy Vary	1071	20	Praha
	Moravské B.-Domamil	1071	50	Praha
	Brno-Komárov	1233	50	Československo
	České Budějovice	1233	30	Československo

	Ostrava	1233	30	Československo
	Liberec	1233	2	Československo
	Plzeň Prestiče	1233	40	Československo
	Praha-Zbraslav	1233	40	Československo
	Litomyšl	1287	400	R. Free Europe
	Melník	1287	400	R. Free Europe
6)	Jihlava	1332	14/7	Reg. Brno
6)	Brno	1332	40	Reg. Brno
3)	Plzeň C.	1485	2	Reg. Plzeň
2)	Volary	1485	1	Reg. Č. Budějovice
	Hrádec Králové	1521	20	Československo
	Karlovy Vary	1521	40	Československo
	Mnichovo Hradiště	1521	50	Československo
	Strakonice	1521	7	Československo
	Ústí nad Labem	1521	10	Československo
	Lipno	1584	1	Československo
	Praha-Dejvice	1584	1	Re. Radio Rosii
	Tachov	1584	2	Československo
	Volary	1602	1	Československo

FM: Czech Republic

Reg	Location	Praha	Regina	Vltava	CS	kW
6)	Brno	102.0	76.65	69.86	71.87	30
2)	České Budějovice	91.1*	67.61	70.07	71.63	30/0.5
2)	Domažlice	98.0	103.8	94.0	90.3	2
5)	Hradec Králové		68.00	69.35	66.44	30
7)	Jeseník-Praděd	100.9	67.10	70.16	68.66	5
6)	Jihlava-Javořice	95.4	90.7	93.4	100.3	15
3)	Klatovy	88.6	98.4	99.8	102.4	2
4)	Liberec	103.9*	71.18	69.98	72.74	10/0.5
		71.96				10
7)	Ostrava	101.4*	67.88	69.08	66.32	30/0.5
3)	Plzeň		69.56	70.34	67.34	30
3)	Plzeň C.	101.3	99.2	93.3	90.0	2
1)	Praha	100.7*	70.85	68.96	66.83	60/10
1)	Praha C.	105.0	92.6	98.4	102.5	5
4)	Ústí nad Labem	102.0	69.26	70.85	72.20	30

+12 low power sites not mentioned — CS = Československo.

Praha: 0330 (Sun-0500)-2400 in Czech. **N:** 0330, 0400, 0500, 0530 (exc. Sat), 0600 (Sat), 0630, 0830 (Sat/Sun), 0930 (exc. Sat), 1130, 1430W, 1630 (Sun), 1730, 2030, 2255.
Vltava: 0555-0100 in Czech. **N:** Mon-Fri 1500, 2255; Sat 0900, 1400, 2255; Sun 0800, 1400, 2255.
Československo: 24h. in Czech/Slovak. **N:** 0500-2100 every half h, 2200-0400 on the h.
Green Wave (traffic information & music): Fri 1300-1700 & Sun 1600-2000 on 792kHz + 70.85/92.6MHz (Praha area), and 603kHz + 88.3MHz (Brno area).
R. Rossii: 24h.
Radio Free Europe: MF 0400-0700 & 1400-2300, Sun 1100-2300.

Regional Prgrs:
D.Prgr: 0730 (Sat/Sun 0800)-2030.
1) Regina Praha, Hybešova 10, 180 00 Praha 8.
2) Regina České Budějovice, U tri lvu 1, 37029 České Budějovice.
3) Regina Plzeň, Nám Miru 4, 32070 Plzeň.
4) Regina Ústí nad Labem, Na schodech 10, 400 91 Ústí nad Labem.
6) Regina Hradec Králové, Vrchlikého 985, 50182 Hradec Králové.
6) Regina Brno, Beethovenova 4, 65742 Brno.
7) Regina Ostrava, Dr. Smerala 2, 72991 Ostrava.
BBC World Sce. Relay: Praha 101.1MHz, Brno 92.6MHz + 4 low power st's.

PRIVATE STATIONS

	MHz	kW	Name	Location
1)	87.8	0.5	Radio R.I.O.	Praha/Petřín
2)	88.2	5.0	Europa 2	Praha/Žižkov
3)	89.5	1.0	Country Radio	Praha/Strahov
4)	89.6	—	Radio West	Plzeň/Krkavec
5)	90.3	1.0	Radio Golem	Praha/Rošických
6)	90.6	1.0	Radio Most	Most/Jedlová hora
7)	91.0	1.0	Radio Diana	Karlovy Vary/Tři kříže
8)	91.6	0.4	Radio Děčín	Děčín/Sněžnik
9)	91.9	1.0	Radio 1	Praha/Strahov
10)	93.1	0.83	Radio Hády	Brno/Hády
11)	93.2	0.07	Radio Egrensis	Cheb/Špitálský vrch
12)	93.6	0.2	Radio 21	České Budějovice/city
13)	93.7	1.0	RTL Praha	Praha/Strahov
14)	95.2	0.05	Radioclub	Ústí nad Labem/Klíše
15)	95.6	0.2	Radio Haná	Olomouc/Pohořany
16)	95.7	1.0	Radio Brno	Brno/Lesná
17)	96.6	5.0	Radio Plus	Praha/Žižkov
18)	96.9	1.0	Radio Profil	Pardubice/city
19)	98.1	0.35	Radio Agara	Chomutov/Jedlová hora
20)	98.1	0.05	VOA Europe	Praha/Petřín
32)	98.7	0.1	Radio Kobra	Praha/Nusle
21)	99.7	2.0	Radio Bonton	Praha/Michle
22)	99.7	2.0	Radio Faktor	České Budějovice/Včela
23)	99.7	1.0	Radio Dragon	Karlovy Vary/Tři kříže
24)	101.1	0.8	Star Radio Morava	Frýdek Místek/Čupek
25)	101.4	0.4	Radio Contact	Liberec/Ještěd
26)	101.5	1.0	Radio Vox	Praha/Petřín
27)	102.5	0.63	Radiohrad	Těrlicko/Babí hora
28)	103.6	1.2	Radio Orion	Ostrava/Hoštálkovice
29)	103.7	1.0	Radio Collegium	Praha/Nusle
30)	104.6	0.74	Radio Ekol S	Bystřice nad Olší
31)	106.1	1.0	Radio FM Plus	Plzeň/Krkavec

1) Kostelní 42, 17078 PRAHA 7; Mon-Fri 0500-2200, Sat 0600-2200, Sun 0600-2400.
2) Celetná 16, 11001 PRAHA 1; 24h.
3) Rudé armády 34, 18000 PRAHA 8; 0900-2100.
4) a part of "Evropa 2", see #2; 0000-2400.
5) Starochodovská 41, 14900 PRAHA 4; 0500-2300.
6) a part of "Evropa 2"; Průchova 1682, 43401 MOST; Mon-Sun 0000-2400.
7) AMBO ssro, Lidový dům, 36017 KRALOVY VARY — Stará Role; 24h.
8) Zelená 218/57, 40501 DĚČÍN 3; 0600-1800.
9) Lodecká 2, 11000 PRAHA 1; 0000-2400/ **English:** 0900, 1430.
10) Slovinská 18, 61200 BRNO; 24h.
11) Dyleňská 34, 35002 CHEB; 0500-2000.
12) Nemanická 14, 37168 ČESKÉ BUDĚJOVICE; 0300-2300.
13) 1.pluku 7, 18000 PRAHA 8 — Karlin; 24h. **English:** 2000-2400.
14) Mírové náměstí 25, 40000 ÚSTÍ NAD LABEM; 24h.
15) 24h.
16) Dusíkova 3, 63800 BRNO-LESNÁ; 24h.
17) Opletalova 5, 11144 PRAHA 1; Mon-Fri 0430-1200 and 1430-2200; Sat 0600-1245; Sun 1505-2300. At other times relays "Radio France Internationale".
18) Na Hrádku 105, 53002 PARDUBICE; Mon-Fri 0500-1900, Sat/Sun 0700-1500.
19) P.O. Box 64, 43001 CHOMUTOV; Mon-Fri 0400-2300, Sat/Sun 0600-0100.
20) 24h.
21) Zelený Pruh 99, 14050 PRAHA 4; 24h.
22) P.O. Box 118, 37004 ČESKÉ BUDĚJOVICE; Mon-Fri 0500-0400, Sat 0700-0400, Sun 0700-2200
23) Dubová 1, 36904 KARLOVY VARY; 24h.
24) c/o Hotel Centrum, 73801 FRÝDEK-MÍSTEK; 0630-2300.
25) Dr. M. Horákové 179/106, 46006 LIBEREC 6; 0000-2400.
26) c/o Palác kultury, 5.května 65, 14009 PRAHA 4; 24h.
27) Hradiště 245, 73542 TĚRLICKO; 24h.
28) Studentská 1770, 70032 OSTRAVA-PORUBA: 24h.
29) Novodvorská 994, 14221 PRAHA 4 — Braník; 1100-2300.
30) 73995 BYSTŘICE; 0530-1800.
31) P.O. Box 40, 32000 PLZEŇ; 0400-2215.
32) Sokolovská 16, 18000 PRAHA 8; 24h.

SLOVAK REPUBLIC

Long & Mediumwaves — Reg = region (see regional addresses)

Reg	Stations	kHz	kW	Prgr.
2)	Rimavská Sobota	567	14	Slovensko 3
2)	Žilina	567	14	Slovensko 3
2)	Dolny Kubín (Orava)	621	14	Slovensko 3
	Banská Bystrica	702	400	Slovensko 1
	Prešov	702	400	Slovensko 1
1)	Bratislava C.	792	6	Reg. Bratislava C.
2)	Čadcá	864	1	Slovensko 3
3)	Snina-Stackcín	864	1	Slovensko 3
3)	Poprad (Tatry)	900	14	Slovensko 3
1)	Trnava-V. Kostolany	927	40	Reg. Bratislava
1)	Bratislava C.	1017	6	Slovensko 3

3)	Košice C.	1017	2	Slovensko 3
1)	Nitra	1017	40	Slovensko 3
2)	Rimavská Sobota	1017	40	Slovensko 3
8)	Banská Bystrica	1035	7	Reg. B. Bystrica
3)	Prešov	1071	40	Slovensko 3
	Nitra-Járok	1098	1500	Slovensko 1
	Nitra	1233	40	Československo
	Rimavská Sobota	1233	14	Československo
	Trnava-V. Kostolany	1287	400	R. Free Europe
	Poprad (Tatry)	1287	14	R. Free Europe
	Prešov	1287	14	R. Free Europe
	Banská Bystrica	1521	14	Československo
	Bratislava C.	1521	10	Československo

FM

Location	A	B	C	D	E	kW
Banská Bystrica	70.94	72.50	102.0	69.68	101.5	30
Bardejov					101.7	1
Bratislava	67.76	71.12	99.3	68.84	96.6	60
Kosice	67.94	66.38	71.82	68.87	100.3	30
Kosice C.					101.9	1
Lucenec					103.6	1
Námestovo		72.02		100.4	102.4	2
Nitra					102.2	1
Poprad	68.06	67.28		69.20	104.3	30
Roznava					105.9	1
Ruzomberok		102.1		103.8	100.6	0.05
Sturovo					103.7	1
Trencin					101.2	2
Zilina	71.60	70.82	103.5	69.50	100.1	10

A = Československo, B = Slovensko 1, C = Slovensko 3, D = Slovensko 2, E = Rock FM Radio.

Československo: 24h. in Czech/Slovak. **N:** 0500-2100 every half h, 2200-0400 on the h.
Slovensko 1: 0300 (Sun 0400)-0100.
Slovensko 2: 0400-2400.
Slovensko 3: 24h. **N:** W 0330-0600 half hourly, 0930, 1130, 1600, 1730, 2030, 2255; Sun 0500, 0600, 0650, 0830, 1130, 1530, 1730, 2030, 2255.
Rock FM Radio: 24h.
Radio Free Europe: MF 0400-0700 & 1400-2300, Sat/Sun 1100-2300.

Regional Prgrs on Slovensko 3:
1) Regina Bratislava, Mýtna 1, 81290 Bratislava.
1) Regina Banská Bystrica, L. Sára 1, 97568 Banská Bystrica.
3) Regina Košice, Moyzesova 7/a, 04161 Košice.

BBC World Sce. Relay: Bratislava 93.8MHz 3kW + 2 low power relays.

Private Stations
Fun Radio Bratislava, Leškova 5, 80000 Bratislava — 24h on 94.3 MHz 15kW.
R. CD International, Mýtna 1, 84545 Bratislava — 0400-2400 on 101.8MHz 30kW.

FOREIGN SERVICE

CZECH & SLOVAK RADIO INTERNATIONAL

ADDR: Vinohradská 12, Praha 12099, Czech Republic. **Te:** +42 (2) 2360823. **Fax:** +42 (2) 2354760. **English Section:** Te: +42 (2) 2354918.
LP: Man. Dir: Dr. Richard Seemann. Head of Prgrs: Zdeněk Dohnal. Chief Eng: Oldrich Čip. Head of Pub. Rel: Miloš Šulch. Editor-in-Chief, English Section: Jan Valeška.
SHORTWAVES: Czech Republic: Litomyšl (G.C: 16.10E/49.48N): 2 × 100kW — **Slovak Republic:** Rimavská Sobota (G.C: 20.00E/48.23N): 2 × 250kW — Velké Kostolany (G.C: 17.44E/48.31N): 1 × 100kW.

kHz	kHz	kHz	kHz	kHz	kHz
5930	7345	9605	11985	15355	17725
5960	9505	9810	11990	15520	21705
6055	9580	11685	13715		

Czech/Slovak	Areas	kHz
0230-0257	NAm	9810, 7345, 5930
	SAm	11990
0500-1500*	Eu	5930
0800-0830	As/Pac	21705, 17725
1500-1600	Eu	7345, 6055, 5930
	NAf	11685
	ME	13715
	EAf	15520
1530-1730*	Eu	5930
1730-1757	Eu	7345, 6055, 5930
1800-1900*	Eu	5930
1900-1927	Eu	9605, 7345, 6055, 5960, 5930
1930-2115*	Eu	5930
2330-2357	NAm/SAm	11990, 9580, 7345, 5930

*) rel. Home Sce. "Ceskoslovensko" prgr.

English		
0000-0027	NAm	11990, 9580, 7345
0100-0130	NAm	9580, 7345, 5930
0300-0330	NAm	9810, 7345, 5930
0400-0430	NAm	9810, 7345, 5930
	ME/SAs	11985
	ME/EAf	13715
0700-0730	Eu	11990, 9505, 7345, 6055
0730-0800	As/Pac	21705, 17725
1130-1157	Eu	15355, 11990, 9505, 7345, 6055
1800-1827	Eu	9605, 7345, 6055, 5960
2100-2127	Eu	9605, 7345, 6055, 5960
2200-2225	Eu	9605, 7345, 6055, 5960
French		
0030-0100	NAm	9580, 7345, 5930
0600-0630	Eu	11990, 9505, 7345, 6055
0800-0830	Eu	11990, 9505, 6055
1200-1227	Eu	15355, 11990, 9505, 7345, 6055
1630-1657	Eu	7345, 6055
	NAf	11685
	NWAf	13715
	EAf	15520
1730-1800	Eu	9605, 5960
1930-2000	Eu	9605, 5960
2030-2057	Eu	9605, 7345, 6055, 5960
2230-2257	NAm	11990, 9580, 7345
German		
0630-0700	Eu	9505, 7345, 6055
1030-1100	Eu	11990, 9505, 7345, 6055
1300-1400	Eu	11990, 9505, 7345, 6055
1700-1730	Eu	9605, 7345, 6055
2000-2030	Eu	7345, 6055
Spanish		
0200-0227	LAm	11990, 9810, 7345, 5930
1230-1257	Eu	15355, 11990, 9505, 7345, 6055
1830-1900	Eu	9605, 7345, 6055, 5960
2000-2030	Eu	9605, 5960
2130-2157	Eu	9605, 7345, 6055, 5960
2300-2327	LAm	11990, 9580, 7345, 5960

INT-SIG: Fanfare from Dvořák's 9th Symphony ("From the New World"), played on French horn — **V.** by QSL-card. Rp. (31RC's for airmail reply). Rec. acc.

DENMARK

L.T: UTC + 1h (Su: UTC + 2h) — **Pop:** 5.233.000 — **Radio sets:** 2.235.000 — **Pr.L:** Danish — **EC:** a/c 50, 220/380V (yes) — **ITU:** DNK.

GENERAL DIRECTORATE OF POSTS AND TELEGRAPHS

ADDR: Tietgensgade 37, DK-1530 Copenhagen V.
Cable: Postgen — **L.P:** Dir. Gen: Helge Israelsen.

DANMARKS RADIO (Gov.)

ADDR: Radio House, Rosenørns Allé 22, 1999 Frederiksberg C. **Te:** + 45 (35) 204060. **Fax:** +45 (35) 202644. **Cable:** Danradio. **Telex:** 22695 DANRAD. **Listeners' Sce: Te:** +45 (35) 206200. **Fax:** +45 (35) 206205.
L.P: Dir. Gen: Hans Jørgen Jensen. Chief Adviser to the Dir. Gen: Bruno Grausen. Head of Inf. Dept: Flemming Hedemark. Dir. Tec: Jørgen Linthoe. Dir. of Finance: Frede Houkjær. Dir. of Personnel: Søren E. Jakobsen. Dir. Prgr. (radio): Hans Jørgen Skov. Head of Denmark Channel: Hans Chr. Andersen. Head of Channel 1: Finn Slumstrup. Head of P2 Music: Steen Frederiksen. Head of Channel 3: Palle Aarslev. Head of News and Actualities: Søren Elmquist. Head of Listener's Sce: John Plenov. Head of Regional Prgr:

Ole Damgaard. Head of Sport Dept. (radio): Henrik Brandt. Head of Foreign Rel: Bent Sørensen. For other personnel: see TV.
STATIONS: Long and Mediumwave K=Kanal (Prgr.).
Kalundborg 243kHz 300kW (1st Prgr.) — 1062kHz 250kW (2nd/3rd Prgr.).
FM (all stereo).

Station	I	II	III	kW/W
Bornholm	96.2	99.3	90.0	30
Copenhagen	90.8	96.5	93.9	60
Hammeren	91.6	93.7	88.4	35W
Hobro	94.9	96.7	92.0	160W
Funen	89.0	96.8	92.6	60
Odsherred	88.4	92.0	94.3	3
South Jutland	95.1	99.9	97.2	60
South Zealand	94.8	97.5	99.6	60
S.W. Jutland	88.7	99.0	92.3	10
		97.7		10
Thisted	91.4	95.6	99.2	2
Vejle	95.5	94.0	90.7	10
N. Jutland	91.0	94.4	96.6	10
Åalborg	93.3	98.1	89.7	30
Åarhus	88.1	95.9	91.7	30
West Jutland	90.2	98.5	92.9	60

1st Prgr: 0500-2310 MF, 0700-1830 (+ relay P2 1830-2400) Sat, 0700-2300 Sun. **N:** 0500 MF, 0600 MF, 0700, 0800, 1102-1130 (1120 SS), 1600-1645 MF, 1730-1745, 2200-2215. **Rlg:** 0725-0745 W, 1600-1700 Sat, 0855-1005 Sun. **Chimes** of the Copenhagen Town Hall: 1100. **N. English:** 0710-0715 MF. **Greenlandic:** Sat 1800-1830. **N. for Greenlanders** in Danish: 1725-1730 MF — Relays 3rd Prgr at other times (FM only).
2nd Prgr: Danmarkskanalen (The "Denmark Channel") 0500-1730 MF, 0700-1100 Sat: News, entertainment and regional prgr's. **N.** on the hour + 1730-1745. **The regions** may have separate prgr's 0500-0800, 1000-1100, 1130-1200, 1600-1730.
P-2 Music (classical music & jazz): 1745-2400 MF, 1120-2400 Sat, 0800-2400 Sun. Relays 3rd Prgr at other times.
3rd Prgr: Young music, news and sport: 24 hrs. **N.** on the hour + 0429, 0529, 0629, 0729, 1529, 1629, 1730.
Special Prgr's: on 243/1062 kHz: **Wrp:** 0450, 0750, 1050, 1650, 2150. **Navigational Warnings:** 1700. **Fish prices:** 1030-1045 MF. — On 1062 kHz: For Foreign Workers in Serbo-Croat, Arabic, Urdu and Turkish: 1545-1645. N. from Greenland in Greenlandic & Danish: 1720-1730 MF. Faoese: 1600-1630 Sat.

Radio Data System: RDS signals are broadcast on all tr's.
Regional Prgrs on the 2nd network:
Nordjyllands Radio, Fredrik Bajers vej 9, 9220 Ålborg Ø — **D.Prgr** on 94.4/96.7/98.1MHz: Mon-Fri 0510-0800, 1130-1200, 1630-1700; Sat 0730-1100.
Radio Midt- & Vestylland, Vestergade 1, 7500 Holstebro — **D.Prgr** on 95.6/97.7/98.5MHz: Mon-Fri 0510-0800, 1130-1200, 1600-1730; Sat 0710-1100.
Østjyllands Radio, Olof Palmes Alle 10-12, 8200 Århus N — **D.Prgr** on 95.9MHz: Mon-Fri 0510-0800, 1130-1200.
Kanal 94, Karl Bjarnhofs Vej 2, 7120 Vejle Ø — **D.Prgr** on 94.0MHz: Mon-Fri 0530-0800, 1130-1200, 1600-1700.
Radio Syd, H.P. Hanssensgade 11, 6220 Åbenrå — **D.Prgr** on 99.0/99.9MHz: Mon-Fri 0530-0800, 1030-1200, 1600-1730; Sat 0730-1100; Sun 0800-1000, 1600-1700.
Radio Fyn, Lille Tornbjergvej 10, 5220 Odense SØ — **D.Prgr** on 96.8MHz: Mon-Fri 0530-0800, 1130-1200, 1600-1700; Sat 0800-1100.
Regionalen, Vadestedet 1, 4700 Næstved —**D.Prgr** on 92.0/97.5MHz: Mon-Fri 0510-0800, 1130-1200, 1700-1800; Sat 0730-1100.
Københavns Radio, Landskronagade 68, 2100 Copenhagen Ø — **D.Prgr** on 96.5MHz: Mon-Fri 0425-0800, 1605-1730.
Bornholms Radio, Åkirkebyvej 52, 3700 Rønne — **D.Prgr** on 93.7/99.3MHz: Mon-Fri 0430-0800, 1130-1200, 1600-1730; Sat/Sun 0730-1000.
ANN: "Danmarks Radio Program et/to/tre" (1st, 2nd & 3rd prgr), "Radio Danmark".

FOREIGN SERVICE

ADDR: Radio Denmark, Radiohouse, DK-1999 Frederiksberg C. **Te:** +45 (35) 205785. **Telefax:** +45 (35) 205781 — **L.P:** Ag. Head of SW section: Flemming. Heltø.
STATION: Radio Denmark is broadcasting via the facilities of Radio Norway International. The main broadcast being at 1430. Please consult R. Norway listing for the complete schedule.

D.PRGR: All programs are in Danish. Letterbox: Saturdays in even weeks, hourly (24 times) from 1443. Sport: Mondays.
INT-SIG. and ANN: All trs preceded by repetition of the first bars of a piece of Danish music by Carl Nielsen, followed by announcements in English and Danish.
Message to Listeners: For the Danish speaking audience Radio Denmark operates an automatic telephone recorder for important reports/messages on +45 (35) 20 57 91. The current schedule is available in Danish on + 45 35 36 30 90 (Eastern Globe) and + 45 35 36 32 70 (Western Globe).
V: R. Denmark does **not** QSL. Foreign listeners must enclose 1 IRC for current schedule, available in Danish only.

Local Radio: Approx. 300 organizations are licensed to operate low-powered (max. 30W ERP at 40 m. height) FM-tr's. In many cases, several groups share a tr. A full listing of these st's, plus Radio Denmark and TV2, including all regional outlets, updated in June and December, called Radio-TV Håndbogen can be obtained from Hartvig Media Aps, P.O. Box 112, DK-8900 Randers, at a cost of 12 IRCs or equivalent.

ESTONIA

L.T: UTC + 2h (Su: UTC + 3h) — **Pop:** 1.600.000 — **Pr.L:** Estonian, Russian — **ITU:** EST.

TRANSPORDI-JA SIDEMINISTEERIUM (Transport and Telecommunications Ministry)

ADDR: Viru 9, EE-0101 Tallinn. **Te:** +7 (0142) 443852, 444404. **Fax:** +7 (0142) 449206.

EESTI RAADIO

ADDR: Gonsiori 21, EE-0100 Tallinn. **Te:** +7 (0142) 434115. **Fax:** +7 (0142) 434457. **Tlx:** 173271 Radio. **Note:** The Telephone country code will be changed to 372.
L.P: DG: Peeter Sookrus. MD: Hannes Valdma. Head of Prgrs: Margus Hunt. Dir.Tec: Avo-Rein Tereping. Head of Foreign Rel: Rein Liiver.

Mediumwaves	kHz	kW	Prgr
Tallinn (Laitse)	612	150	4
Tallinn	711	5	2
Kohtla-Nõmme	711	5	2
Kärdla	711	1	2, Reg
Pärnu	711	5	2
Tartu	711	5	2
Haapsalu	810	1	2, Reg
Võru	810	5	2, Reg
Tallinn (Laitse)	1035	500	1, Ext.
Orissaare	1215	30	1
Tartu	1215	50	1
Kohtla-Järve	1332	15	1
Pärnu	1332	30	1
Võru	1332	30	1
Tallinn	1512	30	3
Tartu	1566	2.5	4
Viljandi (Viiratsi)	1602	1	2

Shortwaves: Tallinn (Laitse, G.C: 59.11N/24.23E): 5925kHz 100kW (ND): Prgr 1 + External Sces 1600-2200, Sun also 0800-0930.

FM (MHz)	1	2	3	4
Kohtla-Nõmme	71.12	71.90	70.13	72.68
Orissaare	71.45	72.23	70.67	73.01
Pärnu	71.24	72.02	69.92	72.80
Tallinn	71.00	71.78	70.28	67.19
Valgjärve	71.36	72.14	70.58	72.92
Tartu		69.62		

D.PRGR: Prgr 1: in Estonian: 0330-2100.
Prgr 2 in Estonian: 0330 (Sat 0430)-2300 (Fri/Sat 2400) (Su: 30 min later sign off). Partly duplicates prgr 1. Rel. VOA Europe in E: 0000-0200.
Prgr 3 in Estonian: 0330 (Tallinn 0700)-2100. Includes stereo prgrs. Partly duplicates prgr 1. Rel. R. Liberty (cf. Germany): D 2000-2100 in Estonian.
Prgr 4 in Russian: 0400 (Sat/Sun 0600)-2300 (Fri/Sat 2400) (Su: 30 min later sign off). Prgrs from Tallinn: Mon-Fri 0400-0700, 1000-1200, 1600-1630; Sat 0600-0900, 1100-1600, incl. 1400-1500 in Ukrainian (produced in Kyyiv) and 1500-1600 in Belorus-

sian (produced in Miensk); Sun 0600-1500. Rel. R. Liberty: D 1700-1800 in Russian. At other times rel. R. St. Mayak from Moscow.

Local stations:
R. Tallinn: 69.32/101.6MHz in Estonian: Own prgrs and rel. prgr 1. 67.97MHz in Russian: Own prgrs and rebroadcast of R. Rossii from Moscow. Both st's 0330-2200.
Võru Raadio: Liiva 11, EE-2710 Voru. On 810kHz: Tue/Wed/Fri 1040-1100, Mon/Wed/Thurs/Fri 1610-1630, Thurs 1915-1930.
Mulgi Raadio: Riia 25, EE-2900 Viljandi. On 1602kHz: W 1040-1100, 1610-1630, Fri 1115-1200.
Lääne-Viru Raadio: On 71.90MHz: W 1115-1145.
Viru Raadio: On 71.90MHz W 1040-1100.
Saare Raadio (Leviraadio): Torni 1, EE-3300 Kuressaare. On 72.23MHz (also r. on 711/810kHz): W 1040-1100.

FOREIGN SERVICE
ADDR: R. Estonia, Gonsiori 21, EE-0100 Tallinn. **Te:** +7 (0142) 434115. **Fax/Tlx:** as for Home Sce.
STATION: Tallinn 1035/5925kHz, as Home Sce.
Prgrs: All transmissions on both freqs + R. Tallinn 69.32/101.6MHz.
Area: General. All prgrs 1h earlier during summer.
English: 1620-1630 Mon-Fri, 2130-2200 Mon/Thurs.
Esperanto: 2120-2130 Thurs, 0920-0930 Sun.
Estonian: 2130-2200 Tues/Wed/Fri.
Finnish: 0800-0900 Sun, 1000-1100 Sat, 1600-1610 Mon-Fri, 1600-1700 Sat, 2030-2100 Mon-Fri.
German: 1610-1620 Mon-Fri, 2000-2030 Tues/Fri.
Swedish: 0900-0930 Sun, 2100-2130 Mon-Fri.
DX-prgrs: English/Finnish: 1st Mon, **Swedish:** 2nd Tues.
V. by QSL-card. Rec. not acc. Rp. not req.

Private commercial stations:
Radio Tartu, Tartu. On 69.62MHz.
Radio Kuku, Tallinn. On 100.7MHz, 200W: 0400-1800.

FAROE ISLANDS (Danish)

L.T: UTC (Su: UTC + 1h) — **Pop:** 48.585 — **Radio sets:** 20.000 — **Pr.L:** Faroese — **E.C:** a/c 50, 220V — **ITU:** FRI.

ÚTVARP FØROYA

ADDR: P.O. Box 328, FR-110 Tórshavn. **Te:** +298 16566. **Fax:** +298 10471. **Telex:** 81226.
L.P: Mgr: Jógvan Jespersen.
MEDIUMWAVE: Akraberg 531kHz 200kW (usually operates at 100kW).
FM: Tórshavn 89.9MHz 3kW — Klaksvík 94.3MHz 2kW — Suduroy 97.5MHz 2kW + 10 st's below 1kW.
D.PRGR: Mon-Fri 0715-1400 & 1700-2100, Sat 0715-2400, Sun 1000-1900.
ANN: "Útvarp Føroya" — **INT-SIG:** Xylofon — **V.** by QSL-card.

FINLAND

L.T: UTC + 2h (Su: UTC + 3h) — **Pop:** 5m — **Radio sets:** 4.950.000 — **Pr.L:** Finnish, Swedish — **E.C:** a/c 50, 230/400V (yes) — **ITU:** FNL.

TELECOMMUNICATIONS ADMINISTRATION CENTRE

ADDR: P.O. Box 53, SF-00211 Helsinki. **Te:** +358 (0) 69661. **Telex:** 124545 thk sf. **Telefax:** +358 (0) 6966410. — CEO: Reijo Svensson.

OY YLEISRADIO AB (Non-comm. Public Broadc. Sce.)

ADDR: P.O. Box 10, SF-00241 Helsinki. **Te:** +358 (0) 14801. **Fax:** +358 (0) 1480 3390. **Telex:** 124735 radio sf. **L.P:** Dir. Gen: R. Paasilinna. Dep. Dir. Gen: J. Mykkänen. Dir. Radio 1 & 2: O. Alho. Dir. Radio 3: T. Siikala. Dir. of Swedish Radio & TV: B. Bergman. Dir. Eng: E. Larkka. Head of News: J. Heino. Head of Int. Rel: U-K. Haarma. Others: See TV.

Long and mediumwaves

	kHz	kW	N		kHz	kW	N
Lahti	252	200	1-4	Pori	963	600	1-4
Helsinki	558	50	1-4				

FM

Loc.	FM1	FM2	FM3	FM4	FM5	kW
Aavasaksa	87.9	89.8	94.7	—	—	3
Ahvenanmaa	—	—	100.3	93.1	91.3	10
Espoo	87.9	91.9	94.0	98.8	101.1	60
Eurajoki	87.7	92.0	94.8	103.3	—	30
Fiskars	90.9	93.1	97.0	102.5	99.7	3
Haapavesi	89.0	96.1	98.4	—	—	30
Iisalmi	87.7	92.8	96.5	—	—	2
Inari	88.4	92.8	98.8	—	—	50
Joutseno	88.0	90.9	98.5	—	—	30
Jyväskylä	89.9	92.5	87.6	103.5	—	30
Karigasniemi	89.5	93.4	96.8	—	—	2
Kerimäki	90.5	95.8	99.1	—	—	30
Kiihtelysvaara	88.4	94.9	97.2	—	—	5
Kilpisjärvi	88.0	90.9	96.0	—	—	0.1
Koli	90.2	93.4	99.6	102.4	—	30
Kruunupyy	91.4	94.0	97.6	99.7	102.7	60
Kuopio	91.6	93.9	98.1	100.2	—	50
Kuttanen	94.1	97.2	99.6	—	—	3
Lahti	93.2	95.5	97.9	100.6	—	50
Lammaskoski	88.5	91.4	98.7	—	—	5
Lapua	88.2	90.1	93.1	95.2	101.5	60
Mikkeli	88.9	92.1	94.6	—	—	30
Nuorgam	88.6	93.9	97.7	—	—	3
Nuvvus	88.1	90.2	94.4	—	—	0.05
Näätämö	87.6	89.9	97.4	—	—	1
Oulu	90.4	93.2	97.3	100.3	—	50
Pello	90.2	97.0	99.7	—	—	3
Pernaja	89.5	92.3	95.0	—	98.3	3
Pieksämäki	89.4	95.3	97.4	—	—	2
Pihtipudas	88.6	91.1	97.0	100.8	—	50
Posio	87.6	91.5	98.6	—	—	30
Pyhätunturi	91.0	97.6	99.9	—	—	50
Pyhävuori	88.9	91.0	94.2	98.6	102.6	30
Rovaniemi	88.2	94.0	96.7	—	—	30
Ruka	90.7	92.8	95.1	—	—	3
Anjalankoski	88.5	92.8	96.9	99.5	—	30
Taivalkoski	89.2	91.9	99.2	—	—	60
Tammela	89.2	91.2	96.0	—	—	5
Tampere	90.7	93.7	99.9	102.1	—	60
Tervola	88.6	92.6	95.6	—	—	30
Turku	89.8	92.6	94.3	98.2	101.4	50
Utsjoki	90.7	93.1	99.4	—	—	2
Vaasa	87.8	89.6	94.8	97.3	101.0	1
Vuokatti	92.3	94.3	98.9	—	—	60
Vuotso	87.8	90.1	94.3	—	—	3
Ylläs	92.2	95.3	98.1	—	—	60
Ähtäri	91.9	94.6	96.6	—	—	3

Additional transmitters

Loc.		MHz	kW	
Ahvenanmaa	P1	95.0	10	Swedish Radio
Ahvenanmaa	P2	97.1	10	Swedish Radio
Ahvenanmaa	P3	88.6	10	Swedish Radio
Espoo	FM6	97.5	3	FM1
Espoo	FM7	103.7	3	External programmes
Haapavesi	FM6	101.9	3	Regional programme
Inari	FM6	101.9	50	Regional programme
Jyväskylä	FM6	99.3	30	Regional programme
Karigasniemi	FM6	100.8	2	Regional programme
Kerimäki	FM6	97.7	6	Regional programme
Kerimäki	FM7	103.2	6	Regional programme
Kilpisjärvi	FM6	103.7	0.1	Regional programme
Kuttanen	FM6	102.2	3	Regional programme
Lammaskoski	FM6	101.2	5	Regional programme
Nuorgam	FM6	101.2	3	Regional programme
Nuvvus	FM6	101.7	0.05	Regional programme
Näätämö	FM6	103.3	1	Regional programme
Taivalkoski	FM6	103.6	60	Regional programme
Utsjoki	FM6	102.6	2	Regional programme
Vuotso	FM6	101.3	3	Regional programme
Ylläs	FM6	103.8	60	Regional programme

Network FM1: (national): Arts & music channel. Mon-Thurs 0330-2200, Fri/Sat 0330-2300, Sun 0400-2200.
Network FM2: (national): 24h. channel of popular culture.

Network FM3: (national): 24h. news and regional programme channel. N. every hour, also 1530. **N. in English:** 2155. **N. in German:** 2255.
Network FM4: Full-service Swedish language network covering southern and central Finland. Mon-Thurs 0330-2200, Fri/Sat 0330-2400. Sun 0400-2200.
Network FM5: Swedish language network for regional programmes.

REGIONAL AND LOCAL PROGRAMMES

The structure of YLE regional and local broadcasting is based on an administrative division of the country into ten areas, each having a regional centre and one or more editorial offices.

Programmes in Swedish on network FM5
Radio Mellannyland: 101.1MHz. **Addr:** Radiokatu 5, SF-002460 Helsinki.
Radio Östnyland: 98.3MHz. **Addr:** Brunnsgatan 40, SF-06100 Borgå.
Radio Västnyland: 99.7MHz. **Addr:** Långgatan 13, SF-10600 Ekenäs.
Radio Åboland: 101.4MHz. **Addr:** Hantverkaregatan 10, SF-20100 Åbo.
Radio Åland: 91.3MHz. **Addr:** Ålandsvägen 24, SF-22100 Mariehamn.
Radio Österbotten: 101.5 + 102.7 + 101.0 + 102.6MHz. **Addr:** Abborrvägen 20, SF-65200 Vasa.

Regional programme in Sámi on network FM6.
Sámi Radio: 101.9 + 101.3 + 103.8 + 102.2 + 101.2 + 103.7 + 100.8 + 101.7 + 102.6 + 101.2 + 103.3MHz. **Addr:** Tammukkatie, SF-99870 Inari.

FOREIGN SERVICE

ADDR: Box 10, SF-00241 Helsinki or Box 462, Windsor CT, 06095 USA.
Te: +358 (0) 1480 4320. Audience service in the US 1-203-688-5540 or (toll free) 1-800-221-9539. **Fax:** +358 (0) 1481169.
L.P: Head of External Broadcasting: Juhani Niinistö.
PUB: "Radio Finland" — twice a year, contains schedules. Free.
FM: In Helsinki the Capital FM 103.7MHz YLE Foreign Service programming (in English, French, German and Classical Latin), relays of VOA Europe (English), VOA Worldwide English (English), BBC World Service (English), Deutsche Welle (German), Radio France Internationale (French) and CSpan Radio (English). Mainly news and current events only.
MW & LW: 558kHz (Helsinki-Capital AM) and 252kHz (Lahti) foreign service programming for Europe and the Helsinki area in Finnish, Swedish, English, German, Russian, French and Classical Latin 0330-2330.
963kHz (Pori) simulcast of 558kHz and a relay of YLE FM3 2330-0330.

Shortwaves

kHz	kHz	kHz	kHz	kHz
6120	11735	11810	15330	17800
9560	11740	15120	15400	21550
9730	11755	15245	15440	

English	Area	kHz
0245-0310	NAm	15400, 11735
0745-0800*	Eu/ME/Af	11755, 9560, 6120
0900-0930	AUS/Pac	21550, 17800
0930-0955	EAs	17800, 15245
1230-1250MF	NAm	15400, 11735
1330-1355MF	NAm	21550, 15400
1400-1500SS	NAm	21550, 15400
1430-1500MF	NAm	21550, 15440
1503-1506*	Eu/ME/Af	21550, 15440, 11755, 9730, 6120
1930-2000*	Eu/ME/Af	11755, 9730, 6120
2215-2240*	Eu/ME/Af	11740, 9730
	EAs	11810
	SAm	11740
Finnish/Swedish		
0305-0425	NAm	11755, 9560
0400-0430*	Eu/ME/Af	6120
0450-0530*	Eu/ME/Af	15440, 11755, 6120
0550-0630*	Eu/ME/Af	15440, 11755, 6120
0645-0730*	Eu/ME/Af	11755, 9560, 6120
0800-0900	AUS/Pac	21550, 17800
0800-0930*	Eu/ME/Af	11755, 6120
0950-1503*	Eu/ME/Af	11755, 6120
0950-1100	EAs	17800, 15245
1100-1110	EAs	17800, 15245
1200-1230MF	NAm	15400, 11735
1300-1330MF	NAm	21550, 15400
1400-1430MF	NAm	21550, 15400
1400-1500SS	NAm	15400, 11735
1500-1600SS	NAm	21550, 15400
1506-1630*	Eu/ME/Af	21550, 15440, 11755, 9730, 6120
1700-1930*	Eu/ME/Af	11755, 9730, 6120
2000-2030*	Eu/ME/Af	11755
2130-2215*	Eu/ME/Af	11740, 9730
	EAs	11810
	SAm	11740
2240-2300*	Eu/ME/Af	11740, 9730
	EAs	11810
	SAm	11740
French		
0250-0305	NAm	11755, 9560
0730-0745*	Eu/ME/Af	11755, 9560, 6120
1110-1130	Eu/ME/Af	15330, 15120
2110-2130*	Eu/ME/Af	11755, 9730, 6120
German		
0630-0655*	Eu/ME/Af	11755, 9560, 6120
1130-1200	Eu/ME/Af	15330, 15120
1630-1700	Eu/ME/Af	11755, 9730, 6120
2030-2100*	Eu/ME/Af	11755, 9730, 6120
2300-2330	Eu	963
Latin		
0305-0310Sun	NAm	11755, 9560
0550-0555Sun*	Eu/ME/Af	15440, 11755, 6120
0920-0925	AUS/Pac	21550, 17800
1450-1455Sun	NAm	21550, 15400
1950-1955Sat*	Eu/ME/Af	11755, 9730, 6120
2235-2240Sat*	Eu/ME/Af	11740, 9730
	EAs	11810
	SAm	11740
Russian		
0430-0450*	RUS	6120
0930-0950*	RUS	11755, 6120
2000-2020*	RUS	9730, 6120

*) also on 963, 558 and 252kHz for No. Europe.

PRIVATE STATIONS

Approx. 60 st's are operating, mostly commercial. Sts in the major cities are as follows:
Helsinki: R. Ykkönen, PL 72, SF-00521 Helsinki (91.1MHz) — R. City, PL 291, SF-00181 Helsinki (96.2MHz) — Classic Radio, Bulevardi 17 D 20, SF-00120 Helsinki (92.9MHz) — Helsingin Lähiradio, Paasivuorenkatu 5 A, SF-00530 Helsinki (100.3MHz) — Radio Pasilan Asema PL 21, SF-00521 Helsinki (102.4MHz). **Tampere:** R 957, Kauppakatu 1, SF-33200 Tampere (95.7MHz) — Radio Sataplus, Hallituskatu 16, SF-33200 Tampere (100.9MHz) — Radio Moreeni, Kalevantie 4, SF-33100 Tampere (98.4MHz). **Turku:** R Auran Aalot, PL 320, SF-20101 Turku (90.5MHz) — Radio Sata, Aurakatu 8, SF-20100 Turku (100.1MHz) — Radio Robin Hood, Hämeenkatu 15, SF-20500 Turku (91.5MHz).

FRANCE

L.T: UTC + 1h (Su: UTC + 2h) — **Pop:** 5.682.000 — **Radio sets:** 49.000.000 — **Pr.L:** French — **E.C:** a/c 50, 220/240V — **ITU:** F.

CONSEIL SUPERIEUR DE L'AUDIOVISUEL (C.S.A.)

ADDR: 39 Quai André Citroen, F-75015 Paris.
Te: +33 (1) 40 58 38 00 — **L.P:** Pres: Jacques Boutet.
The CSA regulates press, TV and radio and issues broadcast licenses.

TÉLÉDIFFUSION DE FRANCE (TDF)

ADDR: 21/27 rue Barbès, 92542 Montrouge Cédex. **Te:** +33 (1) 49 65 10 00. **Telex:** 250738F.
L.P: Prés. Dir. Gen: Bruno Chetaille.
Regional Centres
Paris Centre Nord: 21/23 rue de la Vanne, 92120 Montrouge — **Est (East):** 43 Route de Mirecourt, 54042 Nancy Cédex — **Centre Est (Centre East):** 44 Bd. Vivier Merle, B.P. 68, 69398 Lyon Cédex 3 — **Sud Est (South East):** 14 Bd. Edouard Herriot, 13271

Marseille Cédex 2 — **Sud Ouest (South West):** La Cépière, 31081 Toulouse Cédex — **Ouest (West):** B.P. 79, Ave. de Belle Fontaine, 35510 Cesson Sévigné.
TDF operates the radio & TV transmitters.

RADIO TÉLÉVISION FRANÇAISE D'OUTRE-MER (RFO)

ADDR: 5 Avenue Recteur Poincaré, 75016 Paris. **Te:** +33 (1) 4524 7100 — **L.P:** Prés. Dir. Gen: François Giquel.
RFO controls broadcasting in the French overseas territories.

RADIO FRANCE

ADDR: 116, Av. du Président Kennedy, F-75786 Paris Cédex 16. **Te:** +33 (1) 42.30.22.22. **Fax:** +33 (1) 4230.14.88. **Telex:** 200 002 RANDFRAN F.
L.P: Prés. Dir. Gen: Jean Maheu. Dir. Gen: Jean Izard.

HOME SERVICE

Netw: A=France-Inter, B=Réseau Ondes Moyennes, S=Sorbonne R. France, E=rel. Ext. Sce (RFI).

Mediumwaves	Netw	kHz	kW				
Allouis	A	162	2000	Toulouse	A	1161	100
Paris	F	585	8	Bordeaux	B	1206	100
Lyon	B	603	300	Marseille	B	1242	150
Marseille	A	675	600	Strasbourg	B	1278	300
Rennes	B	711	300	Nancy	A	1350	100
Paris	E	738	4	Nice	A	1350	300
Limoges	B	792	300	Lille	B	1377	300
Nancy	B	837	200	Ajaccio	B	1404	20
Paris	B	864	300	Brest	B	1404	20
Toulouse	B	945	300	Dijon	B	1404	1
Paris	S	963	8	Grenoble	B	1404	1
Bastia	A	1071	20	Pau	B	1404	20
Brest	A	1071	20	Bastia	B	1494	20
Grenoble	A	1071	20	Bayonne	B	1494	4
Lille	A	1071	40	Besançon	B	1494	1
Montpellier	A	1071	10	Clermont-			
Ajaccio	A	1161	10	Ferrand	B	1494	20
Strasbourg	A	1161	200	Nice	B	1557	300

FM: C=France Inter (stereo), D=France-Culture (stereo), E=France-Musique (stereo).

Station	kW(ERP)	C	D	E
Abbeville	4	93.1	97.4	89.8
Ajaccio	5	92.4	97.6	88.0
Ajaccio La Punta	3.6	88.6	97.0	92.0
Alençon	10	93.0	88.0	91.0
Ales	1	87.6	96.1	98.6
Amiens	20	95.4	102.5	99.4
Angers	8	93.2	91.4	97.4
Argenton	5	101.9	89.8	97.2
Arnay le Duc	3	94.6	90.2	100.3
Augoulême	1.8	87.6	92.5	101.5
Aurillac	50	94.5	98.0	91.9
Autun	10	88.1	97.3	94.1
Auxerres	5	99.5	89.5	92.8
Avignon	4	97.4	90.7	93.2
Bar le Duc	10	90.9	88.4	92.7
Bastia	17	95.9	89.2	93.9
Bayonne	40	89.0	96.1	92.7
Bergerac	50	92.3	94.0	97.1
Besançon (Town)	10	98.7	89.3	95.0
	1	90.6	93.5	96.2
Besançon Lomont	22	90.0	97.7	92.9
Bordeaux	5	89.7	97.7	93.5
Boulogne	1	103.3	99.9	89.4
Bourges	100	94.9	88.5	91.8
Brest	50	95.4	97.8	89.4
Briançon	1	91.5	97.8	89.5
Brignoles	1.5	106.7	104.0	105.5
Caen	50	99.6	91.5	95.6
Carcassone	125	88.3	96.5	90.9
Chambery	8	93.5	90.5	98.6
Champagnole	1	88.5	91.7	98.3
Chartres	50	94.6	98.1	89.7
Chaumont	10	96.5	90.4	93.3
Cherbourg	1	94.1	89.2	92.3
Clermont-Ferrand	40	90.4	98.4	95.5
Corte	1	98.2	91.0	94.8
Dijon	10	95.9	93.7	99.2
Epinal	10	98.6	92.4	89.4
Evreux	1	87.7	98.9	102.0
Gap	5	98.3	88.5	95.3
Gex	20	94.4	96.7	89.6
Grenoble Chamrousse	2	99.4	88.2	91.8
Gueret	5	100.7	98.8	90.8
Hirson	5	94.4	99.7	97.2
Hyeres	1.5	91/6	97.5	94.5
Laval	5	95.1	88.3	92.1
Le Havre	1	88.9	93.3	98.5
Le Mans	100	92.6	89.0	97.0
Le Puy	10	99.3	89.3	92.8
Lesparre	1.5	92.4	90.3	95.1
Lille	150	103.7	98.0	88.7
Limoges	150	93.0	89.5	97.5
Longwy	5	98.1	88.3	91.0
Lyon Mont Pilat	50	99.8	88.8	92.4
Lyon (Town)	1	101.1	94.1	98.0
Mantes	5	95.0	92.4	97.1
Marseille	50	91.3	99.0	94.2
Marseille (Town)	1	91.7	98.6	94.7
Mende	10	90.1	96.9	93.7
Menton	4	97.0	89.6	91.7
Metz	50	99.8	94.5	89.7
Mezieres	10	95.8	90.1	93.5
Millau	2	94.9	91.2	88.9
Montargis	1	102.9	98.8	94.1
Montpellier	9	89.4	97.8	92.9
Mulhouse	50	95.7	88.6	91.6
Nancy	10	96.9	88.7	91.7
Nantes	200	90.6	94.2	98.9
Neufchatel-en-Bray	5	92.7	96.0	90.2
Nice	100	100.2	101.9	92.2
Nice (town)	1	—	—	94.4
Niort	200	99.4	96.4	91.1
Orleans	5	99.2	95.8	90.7
Paris (Tour Eiffel)	10	87.8	93.5	91.7
Paris (Romainville)	10	—	93.9	92.1
Parthenay	2	93.8	87.9	98.5
Perpignan	10	92.1	99.8	97.2
Porto Vecchio	1	96.8	90.8	98.9
Privas	1.5	89.8	96.5	94.7
Reims	150	96.8	98.8	89.2
Rennes	100	93.5	98.3	89.9
Rouen	100	96.5	94.0	92.0
St. Etienne	1	88.0	91.7	97.1
St. Maximin	1.5	106.7	104.0	105.5
Saint-Raphael	10	96.3	88.7	99.6
Sarrebourg	10	93.1	99.4	90.3
Sens	10	96.3	98.5	93.8
Strasbourg	50	97.3	87.7	95.0
Toulon	5	92.0	97.1	94.9
Toulouse (Pechbonnieu)	1	103.5	90.5	93.1
Toulouse (Pic du Midi)	56	87.9	95.7	91.5
Tours	10	99.9	97.8	92.2
Troyes	50	95.3	97.9	91.4
Ussel	10	96.0	88.2	99.7
Vannes	5	88.6	96.0	91.8
Verdun	5	92.1	99.3	97.4
Villers-Cotterets	13	91.1	89.6	92.9
Vittel	5	98.2	89.0	94.0
Voiron	1.25	91.5	89.2	95.7

+ sts under 1kW not mentioned.

D.PRGR. France-Inter: Network A on LW/MW, C on FM; Allouis 162kHz, Strasbourg 1161kHz, Nice 1350kHz & Marseille 675kHz: 24h exc. Tues 0005-0358. Other MW tr's: 0500-2100; FM tr's; 24h. **N:** Hourly, plus 0430, 0530, 0630, 0730, 1130.

Réseau B: 0530 (Sun 0558)-2200 on MW. **Prgrs. for immigrants:** W 0530-0600, 2100 (Sat 2030)-2200. **R. Bleue** (for older listeners): 0600-1800 — **Schools Prgrs:** W (exc. Wed) approx. 1400-1700. **University Broadcasts:** W. approx. 1700-1900 (term time only).

France-Culture: 24h on FM (Network D).
N: 0600, 0630, 0700, 0800, 1130, 1700, 1730, 2130, 2300.

France-Musique: (Network E in stereo): 24h. **N:** 0600, 0700, 0800, 1100, 1800, 2300.

Regional Stations ("F.I.P.")
D.PRGRS: Network F on MW: 0700-1500 (Su: 1600); On FM: 24h (stereo). Prgrs consist of music and N.
F.I.P. Bordeaux, 95 rue Judaïque, 33000 Bordeaux (96.7MHz 2.5kW).
F.I.P. Lille, 14 rue Léon Trulin, 59800 Lille (91.0MHz 0.5kW).
F.I.P. Metz, 3 Place St. Martin, 57000 Metz (Metz 98.5MHz 1kW, Forbach 98.8MHz 0.1kW).
F.I.P. Nantes, 1 rue de Alger, 44000 Nantes (Nantes 95.7MHz 20kW, Saint Nazaire 97.2MHz 1.5kW).
F.I.P. Strasbourg, Place de Bordeaux, 67011 Strasbourg Cédex (92.3MHz 0.05kW).
F.I.P. Paris, 116 avenue du Président Kennedy, 75786 Paris Cédex 16 (585kHz, 105.1MHz 10kW stereo).
Sorbonne R. France: Rel. University lectures during term at 0900-1100 & 1300-1600 (Mon-Fri) on 963kHz.

France-Info
24h news and information on FM: available in cities with up to 50,000 inhabitants using 140 tr's (principally 105.5MHz).

RADIO FRANCE INTERNATIONALE

Société Nationale de Radiodiffusion sonore pour l'étranger (RFI).
ADDR: B.P. 9516, F-75016 Paris. (Studios: 116, Av. du Président Kennedy, F-75016 Paris). **Te:** +33 (1) 42 30 22 22. **Fax:** +33 (1) 42 30 44 81. **Télex:** RFI 614171F.
L.P: Pres: André Larquié. Prgr. Dir: Philippe Sainteny. Head of Inf: Christian Bilmann. Tech. Dir: Jean Comiran.

HOME SERVICE

For Foreign Workers (on Mediumwave Network B): Mon-Fri 0435-0600, Sat 0500-0600 in Arabic, Cambodian, Lao, Portuguese, Serbo-Croat, Spanish, Turkish, Vietnamese & French for Africans. Schedules on request. Polish on Lille 1376 kHz: Mon-Fri 1800-1830.
Addr: R.F.I. Emissions pour les travailleurs étrangers immigrés, 116 Av. Prés. Kennedy, 75016 Paris.

FOREIGN SERVICE

STATIONS: Mediumwave: *Toulouse 945kHz 300kW, *Strasbourg 1278kHz 300kW — Cyprus 1233kHz 600kW — Kunming China 1296kHz 1000/500kW.
FM: Paris 89.0MHz 10kW (rel. World Sce./RFI plus Afrique in French 24h).
Shortwave: *Allouis (G.C: 02.00E/47.00N) & *Issoudun (G.C: 02.13E/47.10N); 12×100kW, 8×500kW, 1×4kW tr's. **Relay stations:** Moyabi, Gabon 500kW; *Montsinéry, French Guiana 500kW; Yamata, Japan 300kW; Beijing & X'ian, China 120kW; Hungary 250/100kW — *Operated by TDF.

kHz	kHz	kHz	kHz	kHz	kHz	kHz
4890	7280	11650	11890	15215	17620	21520
5910	7305	11660	11965	15285	17650	21530
5925	9550	11670	11975	15300	17690	21580
5945	9585	11680	11995	15315	17705	21620
5990	9605	11685	12005	15365	17710	21630
6040	9650	11695	12015	15405	17720	21635
6045	9715	11700	12025	15425	17775	21645
6150	9745	11705	13660	15435	17785	21645
6175	9790	11715	15135	15460	17795	21685
7120	9800	11790	15155	15485	17800	21685
7135	9805	11800	15180	15530	17845	21730
7145	9830	11845	15195	17560	17850	21765
7160	9845	11850	15200	17575	17860	25820
7175	9850					

Prgrs. in French

0000-0100	Am	9800
		11670
0100-0200	SAs	17710
0100-0300	Am	5920
0200-0300	Am	11670
0200-0500	Am	9800
0300-0400	Af	5945
		7175
0300-0500	Af	11705
	ME	7135
		7280
0300-0545	Eu	6045
		7280
0300-0600	Af	9805
	ME	9550
0300-0700	Af	7135
0300-0800	Af	9790
0400-0500	Am	11670
	ME	1233
0400-0545	Eu	5990
0400-0600	Af	11995
		4890
	ME	11695
	NAf	9845
0400-0700	Af	15155
1200-1400	Af	9790
1300-1345	SAs	21630
1300-1400	Eu	11670
		15155
		15195
1300-1500	Af/ME	21630
	Eu	9805
1300-1600	Am	21635
1330-1400	Am	17860
		21645
		21765
1500-1600	Am	17860
	As	12035
	Eu	15155
	Eu/ME	17650
1500-1700	Am	21645
1600-1700	Am	15365
		17720
	SAs	9790
	SEAs	1296
1600-1900	Af	15460
	Eu	11995
1600-2000	Af	21685
1700-0900	Eu	3965
1700-1800	Af	17795
	Eu	11670
1700-1900	NAf	11705
1700-2100	Af	11995
1700-2200	Af	17620
	Eu	6175
1800-1900	Af	12025
		15530
		17845
	Eu	9805
1800-2000	Eu	7135
1800-2200	Af	11705
		7160
1800-2300	Af	9790
1900-2000	ME	9790
	Af	11965
1900-2200	Eu	5910
		9830
2000-2200	Eu	6175
2100-2200	Eu	5990
	NAf	6175
2200-0300	Am	5945
		9715
		9790
		9800
		11670
		15435
2300-0030	As	11660
		17710
		7120
		9805
0500-0600	ME	15485
		7305
	NAf	11850
		6175
0500-0700	ME	15300
0500-0800	NAf	11700
0500-1230	Eu	9805
0500-1600	Af	17620
	Eu	6175
0500-2200	Af	15300
0600-0700	Af	15135
	Eu	5990
		7280
	NAf	13660
		15530
0600-0800	Af	11680
		17800
0600-0900	Eu	11790
0700-0800	Af	15155
	NAf	11850
		17690
0700-0900	Af	17850
	ME	17650
0700-1000	Eu	15180
		15425
0700-1230	Eu	11670
0700-1500	NAf	15315
0700-1700	Af	21620
0800-0900	Af	25820
	ME	21530
	NAf	17850
	SAs	21530
0800-1600	NAf	11845
0800-1700	Af	21580
1000-1100	ME	21530
	SAs	21530
1000-1200	Af	21520
1000-1230	Eu	15155
		15195
1000-1400	Eu/ME	17650
1000-1500	Af	17850
		21685
		25820
1030-1130	Am	11670
		15435
		17560
	Pac	15285
1030-1200	As	15215
	FE	9650
1100-1200	As	11890
1130-1230	Am	15365
		21645
1200-1300	Am	15435
		17575
		17860
		21765

Broadcasts in other languages

Arabic		
1600-1800	Af/ME	12025
		15315
1600-1700	Af	11705
		12015
		17620
		17850
		17795
	ME	15530
	Eu	6175
Chinese		
0930-1030	CHN	11715
	CHN	15215
1200-1300	CHN	9650
Creole		
1430-1500	CAm	17720
	CAm	21645
English		
1230-1300	Eu	9805
		11670
		15155
		15180
		15195
	NAm	15365
	CAm	21645
1400-1500	As	11910
		15405
	ME	17650
German		
1800-1900	WEu	1278
		6150
		7145
Persian		
1800-1900	ME	11975
Polish		
0545-0600	Eu	6045
		7280
1700-1800	Eu	7135
		9805
2200-2300	Eu	6175
		7135

At last—a COMPLETE and OBJECTIVE Buyers Guide to the current shortwave receiver market. Designed for the novice as well as the experienced listener to international radio, this essential guide explains how to make sense of the technical specifications and select the right radio for your listening needs.

EQUIPMENT REVIEWED INCLUDES:

- Inexpensive portables for listening on the move
- Specialized sets for picking up the hard-to-hear stations
- Top of the range communications receivers for the professional monitor and dedicated hobbyist
- Antennas—active and passive—why they matter
- Shortwave in the car
- Computer software and accessories
- All you need to know to listen to the world 24 hours a day
- and much more...

Written by the industry's most experienced review team—the technical staff of the *World Radio TV Handbook*

Available NOW from your bookseller or mail order vendor.

19.95 USA

ISBN 0-8230-5949-9

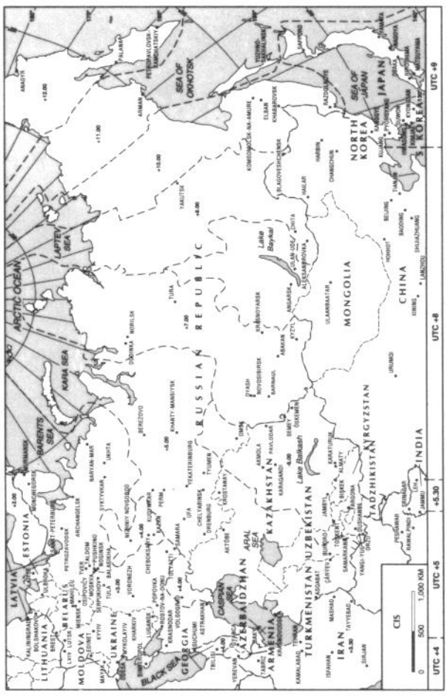

Get your Bearings Straight!

Available for the first time . . . the complete set of updated and accurate WRTH maps in a handy, pocket-sized, full-color edition. Pre-punched for inclusion in your pocket-sized agenda (3¾″x 6¾″page size; ⅜″ring binder). Laminated to withstand heavy usage.

World Time Zones (mercator projection) are shown on the front and detailed regional maps are shown on the reverse side. Principal transmitter sites are indicated in color.

Available NOW from your bookseller or mail order vendor.

Portuguese

0000-0100	SAm	11995
		15200
		17620
1700-1800	Af	12015
		15530
		17850
2200-2300	SAm	11965
		11995
		15200
		17620

Romanian

1600-1700	Eu	9805
		11670
2100-2200	Eu	7135
		9805

Russian

1400-1500	Eu	11670
		15195
		15425
1900-2000	Eu	9605
		9805

Serbian/Croatian

1500-1600	Eu	9805
		11670

Spanish

0100-0200	NAm	9800
	CAm	11670
		15200
		17620
	SAm	11965
		11995
0500-0600	CAm	9800
		11670
1000-1030	CAm	9790
		11670
	SAm	15435
1130-1200	CAm	11670
		15435
	SAm	17560
1300-1330	CAm	17860
		21645
	SAm	21765
2300-2400	Am	9800
	CAm	11670
		15200
		17620
	SAm	11965
		11995

Vietnamese

1430-1530	VTN	9650
1500-1600	VTN	1296

RFI Plus Afrique: alt. French prgrs. on most freq's beamed to Africa at 0400-0800, 1230-1300, 1600-1630, 1800-2000.
ANN: "Radio France Internationale" or "Ici Paris, Radio France Internationale".
INT-SIG: "La Marseillaise".
V. by QSL-card. (B).

SUD RADIO (Comm.)

ADDR: 4 place Alfonse Jourdain, 31071 Toulouse Cédéx. **Te:** +33 (61) 632020; 3, terrasse du 8 mai 1945, 33000 Bordeaux. (Addr. in Paris: 15-17, rue Saint Augustin, 75002 Paris).
MEDIUMWAVES: Toulouse 819kHz 20kW.
FM stereo: Albi 105.1/Bayonne 103.9/Arcachon 100.4/Bergerac 95.5/Bordeaux 106.0/Cahors 104.7/Carcassone 104.7/Cerdagne 103.4 & 106.0/Mende 101.9/Millau 103.2/Lesparre 103.2/Perigeux 104.9/Perpignan 103.2/Rodez 103.2/Sete 104.9/St. Afrique 105.1/Toulouse 102.0/Toulouse(ville) 105.1/Villeneuve/Lot 102.9Mhz.
D.PRGR: 24h (on FM, 0400-1930 on 819kHz).

EUROPE 1 (Priv/Comm.)

ADDR. & L.P: see Germany.
FM stereo tr's in France: 75 st's, mainly on 104.7MHz.

RADIO TELE LUXEMBOURG

ADDR. & L.P: see Luxembourg.
FM stereo tr's in France: 72 st's, mainly on 104.3MHz.

RADIO MONTE CARLO (Comm.)

ADDR. & L.P: see Monaco.
FM tr's in France: 71 st's, mainly between 104.3 & 105.1MHz.

OTHER STATIONS

As of Aug. 1st 1992, a total of 2501 licenses were allocated to private commercial and non-commercial FM st's. Approx. 920 st's are affiliated to one of the following private commercial networks. Prgrs are distributed via the Telecom 2B or ECS II f1 satellites. (no. of affiliates is approx).

Chérie FM, 39 Ave. d'léna, 75016 Paris. **Te:** +33 (1) 4720 53 48 — 60 st's affiliated.
Europe 2, 26 bis rue Francois 1er, 75008 Paris. **Te:** +33 (1) 47 23 10 63 — 150 st's affiliated.
Fun Radio, 143 ave. Charles de Gaulle, 92521 Neuilly sur Seine. **Te:** +33 (1) 47 47 11 72 — 120 st's affiliated.
NRJ, 39 ave. d'léna, 75016 Paris. **Te:** +33 (1) 47 20 06 06 — 170 st's affiliated.
M40, 12 Grand Galerie, Port St. Eustache, 75001 Paris. **Te:** +33 (1) 40 39 05 05 — 60 st's affiliated.
Radio Nostalgie, 54 rue Labrouste, 75015 Paris. **Te:** +33 (1) 45 32 02 02 — 140 st's affiliated.
RFM (La Radio FM), 15 Quai de Dion Bouton, 92816 Puteaux Cedex. **Te:** +33 (1) 40 99 15 15 — 80 st's affiliated.
Skyrock, 1 rue Pierre Lescot, 75001 Paris. **Te:** +33 (1) 42 36 89 39 — 85 st's affiliated.
R. Fourvière, 7 place St. Irénée, 69005 Lyon. **Te:** +33 (7) 72 38 20 22 — 50 st's affiliated.

GERMANY (Federal Republic)

L.T: UTC + 1h (Su: UTC + 2) — **Pop:** 70m — **Radio sets:** 30m — **Pr.L:** German — **E.C:** a/c 50, 220V.

NB: All towns in Germany will get new 5 digit postal codes in 1993.

BUNDESMINISTERIUM FÜR POST UND TELEKOMMUNIKATION (PTT)

ADDR: Heinrich-von-Stephan-Straße 1, 5300 Bonn 2. **Te:** +49 (228) 14-0 **Tlx:** 8861 101 bpm d **Fax:** +49 (228) 1488 72
L.P: Fed. Min. für Post und Telekommunikation: Christian Schwarz-Schilling

ARBEITSGEMEINSCHAFT DER ÖFFENTLICH-RECHTLICHEN RUNDFUNKANSTALTEN DER BUNDESREPUBLIK DEUTSCHLAND (ARD)

ADDR: Bertramstraße 8, 6000 Frankfurt 1. **Te:** +49 (69) 59 06 07 **Tlx:** 4111 27 **Fax:** +49 (69) 155 2075
NB: The ARD is the umbrella organisation which represents German public right (öffentlich-rechtlichen) broadcasters. The chairmanship rotates amongst the members. In 1993 ARD together with ZDF will start up a new public right body called **"Nationaler Hörfunk"**. It has been decided that **Deutschlandfunk** (ARD), **RIAS-1** and **Deutschlandsender Kultur** (ZDF) will become a part of this new organisation.

ARD members:

(Anstalten des öffentlichen Rechts)

Bayerischer Rundfunk (BR)
ADDR: Rundfunkplatz 1, 8000 Munich 2. **Te:** +49 (89) 5900 01 **Fax:** +49 (89) 5900 2375
L.P: Dir: Albert Scharf; Press: Rainer Tief

Deutschlandfunk (DLF)

ADDR: Raderberggürtel 40, 5000 Cologne 51. **Te:** +49 (221) 3451 **Fax:** +49 (221) 3807 66
L.P: Dir: vacant at time of editing; Press: Ralf Siepmann

Deutsche Welle (DW)

ADDR: Raderberggürtel 50, 5000 Cologne 51. **Te:** +49 (221) 3890 **Fax:** +49 (221) 3893 000
L.P: Dir Gen: Dieter Weirich; Press: Wilhelm Nöbel

Hessischer Rundfunk (HR)

ADDR: Bertramstraße 8, 6000 Frankfurt 1. **Te:** +49 (69) 1551 **Fax:** +49 (69) 1552 900
L.P: Dir: Hartwig Kelm; Press: Verena Metze-Mangold

Mitteldeutscher Rundfunk (MDR)

ADDR: Springerstraße 22-24, O-7022 Leipzig. **Te:** +49 (341) 5 66 30 **Fax:** +49 (341) 56 63 544
L.P: Dir: Udo Reiter; Press: Norbert John

Norddeutscher Rundfunk (NDR)

ADDR: Rothenbaumchaussee 132-134, 2000 Hamburg 13. **Te:** +49 (40) 4156 0 **Fax:** +49 (40) 4476 02
L.P: Dir: Jobst Plog; Press: Dagmar Reim

Ostdeutscher Rundfunk Brandenburg (ORB)

ADDR: August-Bebel-Sraße 26-53, O-1591 Potsdam-Babelsberg. **Te:** +49 (331) 72 36 80 **Fax:** +49 (331) 7 73 95
L.P: Dir: Hansjürgen Rosenbauer; Press: Manfred Füger

Radio Bremen (RB)

ADDR: Bürgermeister-Spitta-Allee 45, 2800 Bremen 33. **Te:** +49 (421) 2460 **Fax:** +49 (421) 2 46 1010/1020
L.P: Dir: Karl-Heinz Klostermeier; Press: Jochen Mangelsen

*RIAS Berlin

ADDR: Kufsteiner Straße 69, 1000 Berlin 62. **Te:** +49 (30) 8503 0 **Fax:** +49 (30) 8503 390
L.P: Dir: Helmut Drück; Press: Giso Deussen

Saarländischer Rundfunk (SR)

ADDR: Funkhaus Halberg, 6600 Saarbrücken 1. **Te:** +49 (681) 6020 **Fax:** +49 (681) 6023 874
L.P: Dir: Manfred Buchwald; Press: Rolf-Dieter Ganz

Sender Freies Berlin (SFB)

ADDR: Masurenallee 8-14, 1000 Berlin 19. **Te:** +49 (30) 3031 0 **Fax:** +49 (30) 3015 062
L.P: Dir: Günther von Lojewski; Press: Erich Nieswandt

Süddeutscher Rundfunk (SDR)

ADDR: Neckarstraße 230, 7000 Stuttgart 1. **Te:** +49 (711) 9290/1 **Fax:** +49 (711) 9292 600
L.P: Dir: Hermann Fünfgeld; Press: Hansjörg Bessler

Südwestfunk (SWF)

ADDR: Hans-Bredow-Straße, 7570 Baden-Baden 1. **Te:** +49 (7221) 920 **Fax:** +49 (7221) 92 20 10
L.P: Dir: Peter Voß; Press: Horst Walker

Westdeutscher Rundfunk (WDR)

ADDR: Appellhofplatz 1, 5000 Cologne 1. **Te:** +49 (221) 2201 **Fax:** +49 (221) 2 20 48 00
L.P: Dir: Friedrich Nowottny; Press: Stephan Piltz

***)** RIAS Berlin is not a full member of the ARD, but partakes in meetings.
D.PRGR: All ARD members (except DLF and DW) produce common programming, in turn, for nighttime broadcasts carried by most members. Foreign language programs are produced and carried by HR, NDR, RB, SFB and WDR.
Nachtexpress: (light music, news & traffic on the h.) 2130-0300 UTC; **Radiowecker:** 0300-0500 UTC; **Nachtkonzert:** (serious music) 2305-0500 UTC; **ARD-Popnacht:** 2305-0400 (Fri & Sat 0500) UTC

STATIONS:

BAYERISCHER RUNDFUNK

(Anstalt des öffentlichen Rechts)

Mediumwaves:	kHz	kW		kHz	kW
Hof	520	0.2	München	801	300/550
Würzburg	520	0.2	Dillberg/ Neumarkt	801	50

D.PRGR: 1700-2020 prgrs. for foreign workers (only on 801 kHz)
Shortwave: (G.C: 11.45E/48.15N) DMR24 6085 kHz 100kW ("**BR-Kurzwelle**")

FM Stereo MHz:

	I	II	III	IV	V	kW
Augsburg/ Heretsried	89.2	92.2	94.4	90.2	106.7	0.01/0.5
Bd Reichenhll	91.8	89.0OB	96.7	98.3	105.0	0.3
Bambg./Geisbg.	94.8	89.6F	99.8	102.9	97.4	25
Berchtesgaden/ Schönau	90.4	99.6OB	96.9	94.1	106.4	0.01/0.3
Brotjacklrgl.	92.1	96.5NO	94.4	100.9	106.9	100/50
Burgsinn	92.2	90.0MF	99.3	99.0	103.9*	0.01/0.2
Büttelberg/ Frankenhöhe	91.4	88.2F	99.3	95.5	104.0	25/10/25
Coburg/ Eckardtsberg	93.5	88.3F	99.2	97.7	101.8*	5/0.5/5
Dillberg/	88.9	92.3F/	97.9	87.6	102.0	25
Neumarkt	—	94.0	—	—	—	
G.-Partnkrchn.	93.5	89.2OB	97.7	95.9	104.9	0.5/0.5
Gelbelsee	101.6	90.5OB	97.6	88.0	106.1	25/10/25
Grünten/Allgäu	90.7	88.7S	95.8	101.0	106.9	100/50
Hohe Linie	95.0	93.0NO	99.6	97.0	105.0	25/5/25
Hohenpeißenbg.	92.8	94.2OB	99.2	100.4	106.3	25
Hoher Bogen/ Kötzting	96.8	91.6NO	94.7	88.3	104.4	50/5/25
Hühnerberg	91.9	96.1S	99.5	93.1	107.6	25/10/25
Kreuzberg/Rhön	98.3	93.1MF	96.3	107.9	105.3	100
Landshut-Altdorf	90.2	97.8NO	95.3	93.2	106.6	0.01/1
München	91.3	88.4M	97.3	103.2	90.0	25/5/25
Ochsenkopf/ Fichtelgeb.	90.7	96.0F	99.4	102.3	88.0	100/100/25
Passau/ Kühberg	87.7	93.2NO	90.4	95.6	107.4*	1
Pfaffenberg	95.6	88.4MF	93.4	98.0	106.4	25/1/25
Reit im Winkl	91.0	87.9OB	97.1	93.1	104.8	0.1
Tegernseer Tal	94.0	87.7OB	99.7	97.9	104.3*	0.5
Traunstein/ Hochberg	98.0	91.5OB	95.9	97.0	107.1	5/0.5/5
Wendelstein	93.7	89.5OB	98.5	102.3	105.7	100
Würzburg	90.9	90.0MF	97.6	89.0	106.9	5

+ 10 lps.
*) planned frequencies.
F = Mittel- and Oberfranken — NO = Niederbayern/Oberpfalz — M = München —MF = Mainfranken — S = Schwaben — OB = Oberbayern
D.PRGR: Bayern 1 on FM I, MW & SW (nights rel. ARD "Nachtexpress"); **Bayern 2 Wort** + regional prgrs. on FM II (nights rel. ARD "Nachtkonzert"); **Bayern 3** on FM III & SW (nights rel. SWF-3); **Bayern 4 Klassik** on FM IV (nights rel. ARD "Nachtkonzert"); **Bayern 5 Aktuell** on FM V, MW & SW in summer (nights rel. SWF 3). In addition nightly rel. of ORB and DS Kultur.
Ann: "Hier ist der Bayerische Rundfunk.."; **Bayern 3** ". . . Bayern Drei".
V: by QSL-card or letter.

DEUTSCHLANDFUNK

(Nationaler Hörfunk)

NB: Deutschlandfunk's European & foreign sce is broadcast via Deutsche Welle transmitters on SW and via satellite. For a detailed foreign language schedule see Deutsche Welle.

Long- and Mediumwaves:

	kHz	kW		kHz	kW
Donebach	153	500	Berlin	810	5
Aholming	207	500	Neumünster	1269*	600
Thurnau	549	200	Berlin	1359	250/100
Nordkirchen	549	100	Mainflingen	1539*	700
Ravensburg	756	100	Burg	1575	250
Braunschweig	756	800/200			

*) Foreign language prgrs. See ASTRA satellite.
D.PRGR: Czech & Slovak (2100-2200) on 1539; Danish (1900-1930/2130-2230) on 1269; Dutch (2000-2030) on 1269; English (1915-2000) on 1269; French (1900-1915/2030-2100) on 1539; Hungarian (2200-2230) on 1539; Italian (2230-2300) on 1539; Norwegian (2100-2130) on 1269; Polish (1930-2030W/1900-2030S) on 1539; Swedish (2030-2100) on 1269

FM Stereo:	MHz	kW			
Auerbach/ Schöneck	94.5	3	Hornisgrinde	106.3	0.2
Aurich	101.8	100	Höhbeck	102.2	100
Bonn	89.1	5	Leipzig	96.6	100
Brotjacklrgl.	100.1	100	Lingen	102.0	15
Chemnitz	97.0	100	Löbau	99.5	5
Dresden	97.3	100	Ochsenkopf	100.3	100
Eutin	101.9	100	Rhön	103.3	100
Flensburg	103.3*	16	Saarburg	104.6	50
Freiburg	105.1	0.2	Torfhaus	103.5	100
Hamburg	88.7	1	Wesel	102.8	100
Högl	100.3	15	Witthoh	100.6	40

*) prgrs. in Danish.
F.PL: further FM sts.
ANN: "Hier ist der Deutschlandfunk" — **V:** by QSL-card
F.PL: Will merge together with RIAS & DS Kultur under a new name **"Hörfunk Deutscher Länder"**.

DEUTSCHLANDSENDER KULTUR (DS KULTUR)

(Nationaler Hörfunk)

ADDR: Nalepastraße 10-50, O-1160 Berlin. **Te:** +49 (30) 6384 0
NB: At time of editing DS Kultur was still owned by ZDF.

Longwave:	kHz	kW			
Oranienburg	177	500	Königs-wusterhausen	177	100
FM Stereo:	**MHz**	**kW**			
Berlin	97.7	100			
Brocken	97.4	100	Frankfurt/O	92.7	5
Cottbus	90.8	30	Helpterberg	97.1	60
Dequede	96.9	10	Inselsberg	97.2	100

Marlow	96.7	40	Sonneberg	94.2	100
Schwerin	95.3	60	Suhl	98.8	0.1

D.PRGR: "DX-Aktuell" in German & English (Fri 2100-2200 & Mon 1000-1100 UTC)
V: by QSL-card
F.PL: Will merge together with RIAS & DLF under a new name **"Hörfunk Deutscher Länder"**.

HESSISCHER RUNDFUNK

(Anstalt des öffentlichen Rechts)

Mediumwaves:	kHz	kW
Frankfurt	594	1000/400*
H. Meissner	594	200

*) Directional, screened towards Sofia. 0400-2310 1000kW directional; 2310-0400 400kW n.d.

FM Stereo MHz:

	I	II	III	IV l.s.	kW
Biedenkopf	91.0	99.6	87.6	104.3L/	100
	—	—	—	102.3K	5
Frankfurt	—	101.4	—	*90.4RM/	1
	—	—	—	103.9RM	
Gr. Feldberg	94.4	96.7	89.3	102.5RM	100
Hardberg	90.6	95.3	92.7	101.6B	50
H. Meissner	99.0	95.5	89.5	101.7K	100
Rhön/Heidelst.	—	—	106.2	107.3F	50/50
Rimberg	91.3	95.0	97.7	91.9F	50/20
Würzberg	88.1	97.4	89.7	103.8B	5

+ 6 lps.
*) Freqs used by **Messeradio** during trade fairs — l.s.) local sts.
D.PRGR: HR1 on FM I & MW (nights rel. ARD "Nachtexpress"); **HR2** on FM II (nights rel. ARD "Nachtkonzert"); **HR3** on FM III (nights rel ARD "Popnacht"); **HR4** on FM IV (nights rel. ARD "Nachtexpress") — L = Lahn, K = R. Kurhessen, F = R. Fulda, RM = R. Frankfurt, B = R. Bensheim.
ANN: "Hier ist der Hessische Rundfunk" — **V:** by QSL-card.

MITTELDEUTSCHER RUNDFUNK

(Anstalt des öffentlichen Rechts)

NB: MDR is the new public broadcasting authority for Saxony, Saxony-Anhalt and Thuringia. The youth radio DT 64 (now operated by MDR) was forced to leave all FM freqs. and is now broadcasting to Saxony, Saxony-Anhalt and Thuringia only on MW Dresden (1044 kHz 250kW).
-Jugendradio DT 64, Nalepastraße 10-50, O-1160 Berlin. **Te:** +49 (30) 6384 4140. **F.PL:** audio subcarrier on the ASTRA satellite.

Mediumwaves:	kHz	kW	Prgr
Burg	783	250/20	MDR-Info
Dresden	1044	20	DT64 (24h)
Leipzig	531	50	MDR-Info
Reichenbach	1188	5	MDR-Info
Wachenbrunn	882	250/20	MDR-Info
Weida	1458	5	MDR1-Radio Thüringen

FM Stereo MHz:

	I	II	III	kW
Brocken	94.6SA	91.5	107.8	100
Chemnitz	92.8S	89.8	87.7	100
Dequede	94.9SA	98.9	89.4	7
Dresden	92.2S	90.1	95.4	80/100
Halle	100.8SA	—	—	3
Hoyerswerda	*100.4S	—	—	18
Inselsberg	92.5T	90.2	87.9	100/60
Leipzig	93.9S	98.5	96.2	100
	90.4SA	—	—	100
Löbau	98.2S	91.8	96.2	3/5
Oschatz	101.8S	103.7	—	2.5/10
Schöneck	88.7S	101.2	—	3
Sonneberg	91.7T	96.9	95.2	100/12/24
Suhl	93.7T	91.1	89.1	0.1
Weida	97.8T	—	—	5
Weimar	93.3T	—	—	3
Wittenberg	88.1SA	101.6	—	3

*) MDR1-Radio Sachsen together with Sorbischer Rundfunk.
D.PRGR: MDR-1 on FM I: 24h. (S = MDR1-Radio Sachsen, SA = MDR1-Radio Sachsen-Anhalt, T = MDR1-Radio Thüringen); **MDR-Life** on FM II; **MDR-Kultur** on FM III: 24 h.; **DT64** only on MW; **MDR-Info** only on MW: 24 h.
V: by QSL-card or letter.

NORDDEUTSCHER RUNDFUNK

(Anstalt des öffentlichen Rechts)

NB: The NDR is now also responsible for the Mecklenburg-Vorpommern region.

Mediumwaves:	kHz	kW		kHz	kW
Rostock	558	20	Flensburg	702	5/1.2
Schwerin	576	250	Putbus	729	12
Kiel	612	10	Lingen	792	5
Neubrandenb.	657	40	Hannover	828	100/5
Dannenberg	630*	80	Hamburg	972	300

*) only operates from 0500-1100 UTC.
NB: NDR prgr. 4 ("NDR Vier") with prgrs. for foreign workers (1800-2120 UTC) in Eastern Germany on 558 kHz, 576 kHz, 657 kHz & 729 kHz.
D.PRGR: At daytime rel. NDR 4 prgr on MW, at night NDR 2.

FM Stereo MHz:

	I	II	III	IV	kW
Aurich	95.8NS	98.1	90.0	96.4	25/10
Bremen(3)	—	—	—	95.0	1
Bremerhaven(3)	—	—	—	98.9	0.5
Bungsberg	97.8N	91.9	89.9	—	50/50/50
Cuxhaven	91.6NS	97.9	94.7	93.1	10/0.3/10
Dannenberg	91.2NS	96.4	93.3	—	15/5
Flensburg	89.6N	93.2	96.1	—	25
Göttingen	88.5NS	94.1	96.8	99.9	2/0.5/2
Hamburg	89.5N/	—	—	—	10
	90.3H	97.6	99.2	92.3	80/5
Hannover M.	*90.9NS	96.2	98.7	92.9	15/3/15/0.2
Harz/Torfhaus	98.0NS	92.1	89.9	99.5	100/50
Heide	90.5N	96.3	99.4	—	15
Helgoland	88.9N	—	—	—	0.01
Helpterberg	90.5MV	99.1	96.0	101.8	100
Holzminden	92.7NS	96.0	98.4	—	0.5
Kaltenkirchen	—	—	—	107.4	20
Kiel	91.3N	98.3	95.7	94.5	15/1/15
Lauenburg	94.2N	—	—	—	2
Lingen	92.8NS	97.8	90.2	—	15
Lübeck	93.1N	90.7	88.0	—	0.06/0.1
Marlow	91.0MV	93.5	88.2	102.8	80/40/80/24
Mölln	104.5N	—	—	—	1.5
Neumünster	101.1N	—	—	90.8	20
Osnabrück	92.4NS	89.2	98.8	87.6	8
Putbus	91.5MV	99.8	102.5	88.6	6/12/16
Röbel	88.5	107.0	94.7	100.4	10/50/10/4
Rosengarten	103.2NS	—	—	—	20
Schwerin	92.8MV	98.5	89.2	105.3	40/60/40/60
Stadthagen	100.8NS	102.6	104.4	—	25/0.5
Steinkimmen	91.1NS	99.8	94.4	—	100
Sylt	90.9N	98.7	94.3	—	1.8
Visselhövede	91.8NS	95.9	98.4	98.4	5

NS = Niedersachsen — N = Welle Nord — H = Hamburgwelle — MV = Mecklenburg-Vorpommern
(3) = Radio Bremen transmitter sites partially used by NDR
*) Freqs used by NDR-Messewelle only during Hannover trade fair.
-NDR-Messewelle, Rudolf-von-Bensingen-Ufer 22, 3000 Hannover. **Te:** +49 (511) 88621
D.PRGR: NDR 1, NDR "Welle Nord", NDR "Hamburgwelle", NDR "R. Niedersachsen" & Radio Mecklenburg-Vorpommern on FM I (NDR 1 Hamburgwelle on 90.3 at nights rel. NDR 2); **NDR 2** on FM II; **NDR 3** on FM III; **NDR 4** on FM IV. During the local summer **NDR-Ferienwelle** broadcasts a daily prgr for holidaymakers on the Baltic coast on FM: 95.5 (Putbus); 88.9 (Rostock); 99.5 (Schwerin), 97.4 (Röbel). **F.PL:** new MW frqs. for NDR Ferienwelle
ANN: "Hier ist NDR (1,2,3,4)" — **V:** by QSL-card.

OSTDEUTSCHER RUNDFUNK BRANDENBURG

(Anstalt des öffentlichen Rechts)

NB: ORB is the new public broadcasting authority for Brandenburg and represents 3 sts on FM and MW: Antenne Brandenburg, Radio Brandenburg & Rockradio Brandenburg.
-Antenne Brandenburg, Puschkinallee 4, O-1561 Potsdam. **Te:** +49 (331) 3200
-Radio Brandenburg & Rockradio Brandenburg, Nalepastraße 18-50, O-1160 Berlin. **Te:** +49 (30) 55 160

Mediumwave:	kHz	kW
Berlin-Uhlenhorst	693	40 (Antenne Brandenburg)

FM Stereo MHz:

	I	II	III	kW
Angermünde	91.1	—	100.1	4/4
Berlin-A'platz	99.7	95.8	102.6	100/100/15
Berlin-Kö.	93.1	—	—	30
Cottbus	98.6	95.1*	103.2	100/10/100
Dequede	91.7	—	—	2
Frankfurt/	87.6	89.1	101.5	5/7/9
Oder	—	96.8	—	4
Rhinow	99.9	—	103.1	3

*) Radio Brandenburg together with Sorbischer Rundfunk

D.PRGR: Antenne Brandenburg (entertainment); Radio Brandenburg (cultural); Rockradio Brandenburg (youth). **Antenne Brandenburg** on FM I; **Radio Brandenburg** on FM II; **Rockradio Brandenburg** on FM III. **V:** by QSL-card.

RADIO BREMEN

(Anstalt des öffenlichen Rechts)

Mediumwave:	**kHz**	**kW**
Bremen	936	100 (prgrs. for foreign workers)
Shortwave:	**kHz**	**kW**
Bremen	6190	50 (also carries SFB prgrs.)

FM Stereo MHz:

	I	II	III	IV	kW
Bremen	93.8	88.3	96.7	101.2	100/100/50/20
Bremerhaven	89.3	92.1	96.3	100.8	25

D.PRGR: RB-1 on FM I, MW & SW (nights rel. ARD "Nachtexpress"); **RB-2** on FM II (nights rel. ARD "Nachtexpress"); **RB-3** on FM III (nights rel. ARD "Nachtkonzert"); **RB-4** on FM IV (nights rel. ARD "Popnacht"). I = entertainment; II = cultural; III = classical; IV = youth prgrs.

ANN: "Hier ist Radio Bremen Eins-Hansawelle"; ". . . Radio Bremen Zwei-Kulturell"; ". . . Radio Bremen Zwei/Drei/Vier". **V:** by QSL-card

RIAS BERLIN

(Nationaler Hörfunk)

NB: RIAS Berlin now broadcasts only one instead of two prgrs. RIAS 1 identifies as "RIAS". The former RIAS 2 was privatized (taken over by Radio Information Audio Service Zwei GmbH) and operates under a new name **"R.S.2"** (Radio Sender 2). **R.S.2** broadcasts only in the greater Berlin area on FM 94.3 MHz. RIAS also broadcast on SW 6005 kHz.

Mediumwaves:	**kHz**	**kW**	**H. of tr.**
Hof	684	40	0700-1700
Berlin	855	—	
Berlin-Britz	990	300	24h.
Schwerin	999	20	

FM Stereo MHz:

	I	kW
Berlin	89.6	30
Hof	89.3	20

ANN: "Hier ist RIAS" — **V:** by QSL-card

F.PL: Will merge together with DLF & DS Kultur under a new name **"Hörfunk Deutscher Länder"**.

SAARLÄNDISCHER RUNDFUNK

(Anstalt des öffentlichen Rechts)

NB: Saarländischer Rundfunk started relaying "der Offene Kanal" in 1992.

-Offener Kanal, Eschberger Weg 65, W-6600 Saarbrücken. **Te:** +49 (681) 812026

Mediumwave:	**kHz**	**kW**
Heusweiler	1422	1200/600 (1800-0600 UTC)

FM Stereo MHz:

	I	II	III	IV	kW
Bliestal	92.3	98.0	89.1	105.0	5
Göttelborner Höhe	88.0	91.3	95.5	103.7	100
Merzig	—	—	—	105.1	100
Mettlach	—	—	—	106.1	100
Moseltal	91.9	88.6	96.1	106.2	5

D.PRGR: SR 1, Europawelle Saar on FM I & MW (nights rel. ARD "Popnacht"); **SR 2** on FM II and partially on FM IV (nights rel. ARD "Nachtkonzert"); **SR 3** on FM III (nights rel. ARD "Nachtexpress"); **SR 4 & "Offener Kanal"** (Tue 1400-1800, Sat 1100-1700), other times rel. **SR 2** on FM IV.

V: by QSL-card

SENDER FREIES BERLIN

(Anstalt des öffentlichen Rechts)

Mediumwaves:	**kHz**	**kW**
Berlin	567	100
Berlin	1449	100
Bremen	6190	100 (partially used by SFB)
Dannenberg	630	88 (only used by SFB 3 from 1100-1700 UTC)

FM Stereo MHz:

	I	II	III	IV	kW
Berlin	88.8	92.4	96.3	98.2	80/8

NB: SFB ceased transmission of its radio station for young people "Radio 4U" at the end of 1992.

SÜDDEUTSCHER RUNDFUNK

(Anstalt des öffentlichen Rechts)

NB: Süddeutscher Rundfunk is broadcasting only to northern Baden-Württemberg.

Mediumwaves:	**kHz**	**kW**		**kHz**	**kW**
Mühlacker	576	300	Werftheim	1413	0.2
Heidelberg	711	5	Heidenheim	1413	0.2
Heilbronn	711	5	Adelsheim	1485	0.2
Ulm	711	5	Buchen/		
Bopfingen	711	0.2	Walldürn	1485	0.1
Bd. Mergenth.	711	3			
Shortwave:	**kHz**	**kW**			
Mühlacker	6030	20			

FM Stereo MHz:

	I	II	III	IV	kW
Aalen	95.1	91.1	98.1	96.9S	50/5
Bd. Mergenth.	87.8	93.2	99.7	100.5F	10/0.005
Blaubeuren	89.6	91.5	98.9	—	0.002
Buchen/Walld.	91.9	—	—	100.6K	0.1
Creglingen	89.8	92.5	97.2	94.9F	0.01
Eberbach	97.2	87.8	94.2	—	0.01
Freudenberg	90.3	97.2	94.9	91.6F	0.01
Freudenberg/ Rauenberg	95.1	91.2	92.4	—	0.01
Geislingen	93.0	88.5	95.5	—	0.5
Heidelberg	97.8	88.8	99.9	104.1K	100
Heidenheim	87.6	99.1	97.6	89.8S	0.01
Heilbronn	—	—	—	99.5F	0.3
Ipf	—	—	—	94.1S	0.01
Karlsruhe	90.3	—	—	—	0.3
Mötzingen	—	—	97.2	87.6W	1
Mühlacker/L.	92.9	—	89.5	97.0B	5
Schw. Gmünd	—	—	—	100.9S	0.1
Sigmaringen	—	—	—	101.2W	0.1
Stuttgart/D	94.7	87.9	92.2	90.1S	1/100
Stuttgart-Fkh.	99.6	95.7	—	—	0.5
Ulm	92.6	89.2	97.4	94.5S	10
Vaihingen	—	104.7	—	—	0.1
Waldenburg	98.8	93.8	96.5	106.6F	100/50

+ 25 lps.

B = Badenradio; F = Frankenradio; K = Kurpfalz-Radio; S = Schwabenradio; W = Württemberg-Radio

D.PRGR: SDR 1 on FM I, MW & SW (nights rel. ARD "Nachtexpress"); **S2** on FM II (nights rel ARD "Nachtkonzert"); **SDR 3** on FM III (nights rel SWF 3); **S4** on FM IV

ANN: "Hier ist SDR 1/2/3" — **V:** by QSL-card

SÜDWESTFUNK

(Anstalt des öffentlichen Rechts)

NB: This publicly owned broadcasting organisation, based in Baden-Baden, is responsible for the Rhineland-Palatinate and Baden-Württemberg area. Südwestfunk also broadcasts S4 to southern Baden-Württemberg.

Mediumwaves:	**kHz**	**kW**		**kHz**	**kW**
Bodenseesender	666	300	Rheinsender	1017	800
Freiburg	828	40	Baden-Baden	1485	1
Shortwave:	**kHz**	**kW**			
Rohrdorf	7265	20			

FM Stereo MHz:

	I	II	III	IV	kW
Alf-Bullay	99.5	95.1	92.6	—	0.01
Baden-Baden	90.9	98.9	94.1	88.5	0.8/0.04
Bd. Marienberg	89.8	95.4	92.8	106.3	25
Blauen	89.2	92.6	97.0	—	8.4
Bleialf	88.3	94.6	98.9	—	0.1

Brandenkopf	95.4	—	99.7	97.6	0.5/0.1
Donautal	93.7	—	92.8	87.6	0.1
Donnersberg	99.1	92.0	101.1	—	60
Eifel	91.1	93.6	98.5	—	8
Enztal	92.5	98.0	94.3	—	0.01
Elzach	—	—	—	101.9RB	0.1
Eyachtal	—	—	—	99.5BW	0.1
Feldberg	89.8	97.9	93.8	—	5
Freiburg	107.0	94.7	99.2	91.1RB	0.5
Grenzach-Whln.	97.2	96.4	92.3	—	0.05
Grünten	98.7	—	103.0	—	30
Haardtkopf	97.7	93.0	90.0	107.1	50/25
Hohe Möhr	87.6	100.2	96.8	—	0.5
Hornisgrinde	93.5	96.2	98.4	106.3BW	80/0.2
Kettrichhof	100.8	—	—	—	5
Kirn	96.4	90.4	93.3	—	0.01
Koblenz	96.1	94.0	91.6	107.4	10/25
Lahntal	88.4	93.4	98.2	—	0.01
Laufenburg	94.0	91.4	90.1	—	0.05
Linz/Rhein	92.4	97.4	94.8	—	50
Mainz	87.7	103.2	93.7	91.4	1
Murgtal	89.1	—	99.6	—	0.01
Nagoldtal	97.4	90.4	99.5	—	0.01
Nahetal	88.5	95.1	98.1	—	0.01
Nierstein-O.	92.9R	—	87.9	—	0.1
Oberes Ahrtal	88.5	93.3	99.4	—	0.01
Potzberg	90.8	93.9	97.5	—	20
Raichberg	88.3	91.8	94.3	—	25
Rheinsender	—	—	—	94.9	0.1
Saarburg	99.2	93.8	90.6	101.2	5
Schramberg	89.4	98.8	91.2	—	0.1
St. Chriscona	87.9	92.0	89.5	98.3RB	5
Trier	94.9	89.4	98.2	—	0.1
Villingen-S.	—	—	93.1	—	0.1
Waldburg	—	94.9	—	91.2RB	25
Wannenberg	89.0	92.8	98.5	—	1
Weinbiet	95.9	—	89.9	—	25
Witthoh	92.4	90.4	97.1	100.6	40

RB = Radio Breisgau — BW = Baden-Württemberg — R = Rheinhessenwelle

D.PRGR: SWF 1 on FM I and MW; **S2** (in cooperation with SDR) on FM II (nights rel. ARD "Nachtkonzert"); **SWF 3** on FM III and SW; **SWF4** (in cooperation with SDR) on FM IV (SWF4 without loc. prgr. on 88.5/97.6/87.6/99.5/101.2/91.2 MHz; SWF4 with loc. prgr. **"Radio Breisgau"** on 91.1/98.3/101.9/102.2 MHz); **SWF 5** on FM 91.3 (0.1kW), 99.4 (0.2kW) prgrs. for foreign workers.

WESTDEUTSCHER RUNDFUNK

(Anstalt des öffentlichen Rechts)

Mediumwaves:	**kHz**	**kW**		**kHz**	**kW**
Herford	702	2	Bonn	702	3
Siegen	702	2	Langenberg	1593	400/800
Kleve	702	3			

FM Stereo MHz:

	I	**II**	**III**	**IV**	**V**	**kW**
Aachen	106.4	100.8	95.9	93.9	101.9A	20/5/0.5
Bärbelkreuz	—	101.0	96.3	104.4	89.6A	0.5/10
Bonn	102.4	100.4	93.1	90.7	88.0K	50/10/35
Ederkopf	—	101.8	—	100.7	95.8S	15/1
Kleve	103.7	93.3	97.3	101.7	99.7D	2
Köln	—	—	91.8	—	—	1
Langenberg	106.7	99.2	95.1	101.3	88.8D	80/100
	—	—	—	—	103.3E	35
Münster	107.9	94.1	89.7	100.0	92.0M	25
Nordhelle	102.7	93.5	98.1	103.8	90.3S	25/25
	—	—	—	—	104.7W	35
Olsberg	—	102.1	—	104.1	98.6S	10
Teutoburger W.	105.5	93.2	97.0	100.5	90.6B	100
Wittgenstein	—	92.3	88.7	—	—	15
Wuppertal	—	—	—	—	99.8W	1

+ 27 lps.

D.PRGR: WDR 1 on FM I & MW (nights rel. ARD "Popnacht"); **WDR 2** on FM II & MW (720 kHz) (nights rel ARD "Nachtexpress"); **WDR 3** on FM III & MW (774/1593 kHz); **WDR 4** on FM IV (nights rel. ARD "Nachtexpress"); **WDR 5** on FM V: prgrs. for foreign workers/rel. WDR 1 (WDR 5 regional prgrs: A = Aachen — B = Bielefeld — D = Düsseldorf — E = Essen — K = Köln — M = Münster — S = Siegen — W = Wuppertal)

ANN: "Hier ist WDR Eins/Zwei".. etc. — **V:** by QSL-card.

COMMERCIAL RADIO STATIONS

There are no nationwide commercial radio networks in Germany. The Federal Republic of Germany comprises 16 so called "Bundesländer" (states). Each state has its own authority, the so called "Landesmedienanstalt", which is responsible for the allocation of radio licenses to commercial radio stations. In general either one or two statewide high-power (up to 100kW) commercial radio stations, or a number of low power (± 0.1 kW) local commercial radio stations are allowed. Only in Bavaria both types of commercial radio exist. Statewide commercial radio stations operate in: Schleswig-Holstein — Hamburg — Lower Saxony — Saxony-Anhalt — Berlin — Saxony — Hesse — Rhineland-Palatinate — Saarland — Bavaria. Local commercial radio stations operate in: North Rhine-Westphalia — Baden-Württemberg — Bavaria. Further local and regional commercial radio stations are planned for: Bremen — Mecklenburg-Vorpommern — Brandenburg.

NB: Many sts. are also relayed via satellite. For detailed information see satellite section: KOPERNIKUS (23.5° East), ASTRA (19.2° East), TV-SAT (18.2° West).

SCHLESWIG-HOLSTEIN (State)

-Radio Schleswig-Holstein (RSH), Funkhaus Wittland, W-2300 Kiel 1. **Te:** +49 (431) 58 700/1 58 700.

-Radio 1 Nordfunk, Fleethörn 32, W-2300 Kiel 1

STATIONS:

Loc.	**R.Schleswig-H**	**R.1 Nordfunk**	**kW**
Eutin	100.2	105.6/	50
	—	104.1	50
Flensburg	101.4	—	20
Heide	103.8	100.4	15
Helgoland	100.0	103.5	0.05
Kaltenkirchen	102.9	—	20
Kiel	102.4	105.9	15
Lauenburg	102.5	105.6	1
Lübeck	101.5	107.9	20
Sylt	102.8	104.8	5

D.PRGR: Radio Schleswig-Holstein also rel. regional prgrs. from Flensburg, Kiel, Heide, Lübeck, Pinneberg and Hamburg.

HAMBURG (City)

-OK Radio, Spaldingstraße 218, W-2000 Hamburg 1. **Te:** +49 (40) 32 73 30.

-Offener Kanal (public), Stresemannstraße 375, Hamburg 50. **Te:** +49 (40) 89 81 51.

-Jazzwelle Plus, Rugenbarg 85, W-2000 Hamburg 53.

-Klassik Radio, Brandstwiete 4, W-2000 Hamburg 11. **Te:** +49 (40) 32 43 73 **Fax:** +49 (40) 32 63 05.

-Radio Hamburg, Speersort 10, P.O. Box 100123, W-2000 Hamburg 1. **Te:** +49 (40) 339 71 40.

-Alster Radio, Rödingsmarkt 29, W-2000 Hamburg 11.

STATIONS: OK Radio on 95.0 (0.1kW); Offener Kanal on 96.0 (0.1kW); Jazzwelle Plus on 97.1 (0.1kW); Klassik Radio on 98.1 (0.1kW); Radio Hamburg on 103.6 (80kW); Alster Radio on 106.8 (40kW)

D.PRGR: Radio Hamburg and Radio ffn (see Lower Saxony) have a common nighttime prgr. called **"Nordnacht"**.

LOWER SAXONY (State)

-Radio ffn (Press & PR), Martinistraße 43, W-2800 Bremen 1. **Te:** +49 (421) 32 09 50/32 81 20.

-Antenne Niedersachsen, Goseriede 9, W-3000 Hannover 1. **Te:** +49 (511) 91 180.

STATIONS:

Loc.	**Radio ffn**	**A. Nieders.**	**kW**
Aurich	103.1	104.9	25/2
Barsinghausen	101.9	103.8	25
Braunschweig	103.1	106.9	15/0.16
Cuxhaven	102.6	104.6	20/2
Dannenberg	102.7	106.1	25
Göttingen	102.8	106.0	5
Hamburg	—	105.1	?
Hann. Münden	100.7	106.7	0.5
Holzminden	102.2	105.7	0.5
Lingen	101.5	104.3	15
Lüneburg	—	95.5	1
Osnabrück	103.4	105.9	10
Rosengarten	100.6	105.1	20
Steinkimmen	102.3	105.7	100

Torfhaus	102.4	106.3	100
Visselhövede	101.7	104.2	10
Wolfsburg	—	95.1	0.1

D.PRGR: Radio ffn broadcasts loc. prgrs. from Hamburg, Lüneburg, Oldenburg, Bremen, Osnabrück, Hannover, Braunschweig, Göttingen.

BERLIN (City)

-Berliner Rundfunk, now privately owned
-R.S.2 (formerly RIAS 2), now privately owned
-Radio Hundert,6, Paulsborner Straße 44, W-1000 Berlin 33. **Te:** +49 (30) 89 69 40.
-Inforadio 101 Berlin, Kurfürstendamm 207-208, W-1000 Berlin 15. **Te:** +49 (30) 88 475-0/7738 759 (press) **Fax:** +49 (30) 8847 2514.
-Radio Energy 103.4, Potsdammer Straße 131, W-1000 Berlin 30. **Te:** +49 (30) 2169 393.
-104.6 RTL, Teplitzer Straße 33, W-1000 Berlin 33. **Te:** +49 (30) 825 70 17/8. **F.PL:** relay via satellite on subc. RTL Zwei.
STATIONS: Berliner Rundfunk on 91.4 (100kW); R.S. 2 on 94.3 (50kW); Radio Hundert, 6 on 100.6 (2kW); Inforadio 101 Berlin on 101.3 (2kW); Radio Energy 103.4 on 103.4 (10kW); RTL Berlin on 104.6 (2kW).

SAXONY-ANHALT (State)

-Radio Brocken, Brachwitzer Straße 16, O-4050 Halle/Saale. **Te:** +49 (345) 34 93 32.
-Radio SAW (Sachsen-Anhalt-Welle), Große-Driesdorfer-Straße 64, O-3031 Magdeburg.
NB: Radio SAW is the first private radio station in Saxony-Anhalt and began broadcasting on the 8th of September 1992.
STATIONS:

loc.	R. Brocken	R. SAW	kW
Brocken	89.0	101.4	100
Dequede	101.0	—	60
Halle	—	103.3	5
Leipzig	—	104.9	100
Magdeburg	—	100.1	20

SAXONY (State)

-Radio PSR, Delitzscher Straße 97, O-7021 Leipzig. **Te:** +49 (341) 59 18 72.
STATIONS: Leipzig 102.9 (25kW); Dresden 102.4 (60kW); Chemnitz 100.0 (100kW); Löbau 101.0 (5kW); Plauen 92.0 (?kW)

NORTH RHINE-WESTPHALIA (State)

NB: There are approx. 40 loc. radio stations occupying approx. 80 freqs.
Most popular sts. are:
-Radio Köln, Eupener Straße 161, W-5000 Köln 41. **Te:** +49 (221) 49 96 80.
-Antenne Düsseldorf, Speditions Straße 15, W-4000 Düsseldorf 1. **Te:** +49 (211) 39 10 21
-Radio DU(isburg), Güntherstraße 35, W-4100 Duisburg 1. **Te:** +49 (203) 92097/8.
-Radio Emscher-Lippe, Hochstraße 30, W-4650 Gelsenkirchen 2. **Te:** +49 (209) 37 51 39/360 880.
-Radio Bonn/Radio Rhein-Sieg, Kennedybrücke 2, W-5300 Bonn 3. **Te:** +49 (228) 400 710
-Radio FiV (Funk im Vest), Schaumburgstraße 14, W-4350 Recklinghausen 1. **Te:** +49 (2361) 12 024/5.
-Radio MK (Märkischer Kreis), Theodor-Heuss-Ring 24, P.O. Box 2248, W-5860 Iserlohn 1. **Te:** +49 (2371) 79 030
-Radio Bielefeld, Niedernstraße 23-27, W-4800 Bielefeld 1. **Te:** +49 (521) 555-0.
-Ruhrwelle Bochum, Grabenstraße 12, P.O. Box 102109, W-4630 Bochum 1. **Te:** +49 (234) 60 601/68 99 90.
-Radio Lippe, Ernst-Hilker-Straße 11, W-4930 Detmold. **Te:** +49 (5231) 21003
STATIONS:

Loc.	MHz	kW	St.
Altena	91.5	0.05	R. MK
Bielefeld	98.3	0.1	R. Bielefeld
Bochum	98.5	0.05	Ruhrwelle B
Bochum	105.0	0.1	Ruhrwelle B
Bonn	98.9	0.1	R. Bonn/Rhein-Sieg
Bottrop	104.5	0.1	R. Emscher-Lippe
Dorsten	105.2	0.1	Radio FiV
Duisburg	92.2	0.1	R. DU
Düsseldorf	104.2	0.2	A. Düsseldorf
Gelsenkirchen	96.1	0.1	R. Emscher-Lippe
Iserlohn	92.5	0.1	R. MK
Köln	98.6	0.4	R. Köln
Köln	107.1	0.5	R. Köln
Lemgo	106.6	0.4	R. Lippe
Lüdenscheid	100.2	0.1	R. MK
Meinerzhagen	88.3	0.05	R. MK
Plettenberg	99.5	0.05	R. MK
Recklinghausen	94.6	0.1	Radio FiV
Siegburg	91.2	0.2	R. Bonn/Rhein-Sieg
Werdohl	97.2	0.02	R. MK
Windeck	107.9	0.1	R. Bonn/Rhein-Sieg

THURINGIA (State)

-Antenne Thüringen, Berliner Straße 34, O-5300 Weimar. **Te:** +49 (3643) 2563.
STATIONS: 102.2 100kW (Inselsberg); 102.7 60kW (Sonneberg); 107.6 60kW (Saalfeld)

HESSE (State)

-Radio FFH, Graf-Vollrath-Weg 6, W-6000 Frankfurt 90. **Te:** +49 (69) 78 97 90.
STATIONS: Feldberg 105.9 (100kW); Hoher Meißner 105.1 (100kW); Rhön 100.9 (50kW); Rimberg 100.3 (50kW); Dillenburg 100.0 (30kW); Driedorf 106.8 (30kW); Kassel 103.7 (20kW); Südhessen 105.0 (20kW); Vogelsberg 104.8 (20); Frankfurt 100.2 (1kW); Korbach 107.7 (1kW); Dieburg 90.1 (0.8kW); Eschwege 104.6 (0.5kW); Wiesbaden 102.0 (0.1kW)

RHINELAND-PALATINATA (State)

-Radio RPR, Turmstraße 8, W-6700 Ludwigshafen. **Te:** +49 (621) 59 00 00.
STATIONS:

Loc.	Radio RPR	RPR 2	kW
Ahrweiler	103.5	104.9	30
Bad Dürkheim	98.1	—	0.05
Bad Marienberg	102.9	—	25
Eifel	102.1	—	20
Haardtkopf	100.1	—	50
Idar-Oberstein	100.3	—	1
Kalmit	103.6	106.7	25
Kettrichhof	104.7	—	5
Koblenz	101.5	104.0	40
Ober-Olm	—	104.5	20
Potzberg	103.1	—	25
Saarburg	102.6	—	20
Trier	102.9	106.4	0.1

NB: RPR-1 identifies as "Radio RPR", RPR-2 identifies as "RPR Zwei".
D.PRGR: RPR: pop music and reg. prgrs. from Koblenz, Mainz, Trier & Ludwigshafen; **RPR-2**: light music. Radio RPR and Radio Victoria (see Baden-Württemberg) have a common prgr. at nighttime called **"Starlight-Station"**.
-Radioropa Info, TechnicPark, W-5568 Daun. **Te:** +49 (6592) 2030
STATIONS: Burg 261 kHz (200kW) (see also under Radio Volga entry); Daun/Eifel 92.2 MHz (0.05kW); Mainz/Wiesbaden 105.2 MHz (0.1kW).
D.PRGR: Mon-Sat 0600-0800, 1100-1300, 1500-1700; Sun 1400-1700, 1900-2200, 1900-2200 on 261 kHz. On FM 24 hrs.

SAARLAND (State)

-Radio SALÜ, Richard-Wagner-Straße 58-60, W-6600 Saarbrücken. **Te:** +49 (681) 39 127.
STATIONS: Heusweiler II/Schoksberg 101.7 (100kW); Bliestal 100.0 (5kW); Perl 100.3 (5kW); Merzig 103.0 (0.1kW); Mettlach 104.2 (0.1kW)
D.PRGR: 24 h. hit radio.

BADEN-WÜRTTEMBERG (State)

NB: There are approx. 30 local & regional commercial radio stations on approx. 50 freqs. (0.1-75kW) in Baden-Württemberg. Some stations relay regional prgrs. from each other at certain times (e.g. Radio 7, Radio Regenbogen, RTL Radio & Stadtradio). At night most sts. relay RTL Radio.
Most popular sts. are:
-Radio 7, Schaffnerstraße 30, W-7900 Ulm 1. **Te** +49 (731) 14410

Regionalsender Ostalb, Bahnhofstraße 65, W-7080 Aalen 1. **Te:** +49 (7361) 57 27 0.
Regionalsender Witthoh, Königstraße 4, P.O. Box 7000, W-7200 Tuttlingen 1. **Te:** +49 (7461) 17000.
Regionalsender Göppingen, Rosenstraße 24, W-7320 Göppingen 1. **Te:** +49 (7161) 23 0 21.
Regionalsender Ulm, Kupfergasse 10, W-7900 Ulm 1. **Te:** +49 (731) 14 410
Regionalsender Biberach, Leipzigstraße 26, W-7950 Biberach 1. **Te:** +49 (7351) 2067/9.
Regionalsender Oberschwaben, R.-Roth-Straße 16, P.O. Box 7000, W-7970 Leutkirch 1. **Te:** +49 (7561) 8280 **Fax:** +49 (7561) 8282 8.
STATIONS:

Loc.	MHz	kW	St.
Aalen	103.7	50	Radio 7 Ostalb
Biberach	103.9	0.4	Radio 7 Biberach
Göppingen	100.3	5	Radio 7 Göppingen
Leutkirch/Höchsten	105.0	50	Radio 7 Oberschwaben
Ochsenhausen	88.0	0.1	repeater Radio 7 Biberach
Riedlingen	99.3	0.1	repeater Radio 7 Biberach
Ulm	101.8	10	Radio 7 Ulm
Witthoh	102.5	40	Radio 7 Witthoh

-Antenne 1 Stuttgart, P.O. Box 810150, W-7000 Stuttgart 10.
-Radio Ladies First, Frankreichstraße 4, W-7570 Baden-Baden 1. **Te:** +49 (7221) 700 70.
-Radio Regenbogen, Dudenstraße 12-16, W-6800 Mannheim 1. **Te:** +49 (621) 33 08 30.
-Radio Regional Heilbronn, Allee 2, W-7100 Heilbronn.
-Antenne RT 4, Burgplatz 5, W-7410 Reutlingen.
-Welle Fidelitas, Lammstraße 1-5, W-7500 Karlsruhe 1.
-Radio Victoria, Sophienstraße 3, W-7570 Baden-Baden 1. **Te:** +49 (7221) 22 631.
STATIONS:

Loc.	MHz	kW	St
Bad Mergentheim	101.2	0.45	R. Reg. Heilbronn
Blauen	101.1	8.4	Radio Regenbogen 101.1
Heidelberg	102.8	50	Radio Regenbogen
Heilbronn	103.2	25	R. Reg. Heilbronn
Hornisgrinde	100.4	80	R. Ladies First/R.Victoria
Karlsruhe	101.8	25	Welle Fidelitas
Langenburg	100.1	50	R. Reg. Heilbronn
Raichberg	103.4	100	Antenne RT 4
Stuttgart	101.3	75	Antenne 1 Stuttgart/A.RT4

-RTL Radio, Heusteigstraße 44, W-7000 Stuttgart 1.
STATIONS:

Loc.	MHz	kW	St.
Baden-Baden	100.9	0.8	Radio Merkur
Bühl	103.8	0.1	Radio Merkur
Esslingen	104.6	0.1	Radio ES(lingen)
Feldberg	104.0	5	Radio Freiburg FR 1
Freiburg	100.7	1	Radio Freiburg FR 1
Geislingen	100.3	5	RTL Radio 100.3
Göppingen	103.0	0.1	RTL Radio 103
Hohe Möhr	104.5	0.5	RTL Radio Hochrhein
Kirchheim	100.8	0.1	Radio ES(lingen)
Laufenburg	102.4	0.05	RTL Radio Hochrhein
Mühlacker	100.7	5	RTL Radio 100.7
Reutlingen	104.8	0.1	RTL Radio 104.8
Rheinfelden	103.4	0.1	RTL Radio Hochrhein
Waldshut	105.4	0.1	RTL Radio Hochrhein

-Radio T.O.N., Untere Wart 11-15, W-6990 Bad Mergentheim
STATIONS: Buchen-Walldürrn 102.1 (25kW); Bad Mergentheim 103.5 (20kW); Wertheim 104.7 (0.1kW); Mosbach 107.9 (0.1kW)
D.PRGR: 24h. 2300-0430 (Sat 1600-/Sun 1700-) relays **Star*Sat Radio**.

BAVARIA (State)

-Antenne Bayern, Münchener Straße 20, W-80043 Unterföhring. **Te:** +49 (89) 95 99 90
-Klassik Radio, Brandswiete 4, W-2000 Hamburg 11. **Te:** +49 (40) 3243 7375
STATIONS:

Loc.	A. Bayern	Klassik R.	kW
Augsburg	—	92.2	0.3
Bad Reichenhall	89.9	—	0.3
Bamberg	101.1	—	16
Berchtesgaden	107.9	—	0.1
Brotjacklriegl	103.5	—	100
Burgbernheim	101.5	—	25
Coburg	103.8	—	5
Dillberg	100.6	—	25
Gelbelsee	100.2	—	25
Grünten	104.4	—	10
Herzogstand	102.0	—	0.1
Hochries	107.7	—	10
Hohenpeienberg	103.8	—	25
Hoher Bogen	101.9	—	50
Landshut	104.1	—	1
Lindau	99.0	—	0.5
München	101.3	107.2	0.3
Nürnberg	—	107.4	0.9
Ochsenkopf	103.2	—	100
Passau	102.1	—	1
Pfaffenberg	103.0	—	25
Regensburg	103.0	91.1	25
Reit im Winkl	101.6	—	0.1
Rhön	101.9	—	100
Tegernseer Tal	101.1	—	0.5
Traunstein	103.7	—	5
Unterringingen	103.3	—	25
Würzburg	104.4	106.9*	0.16/5
Zugspitze	102.7	—	2

*) possible new user **Radio Charivari.**
NB: In addition approx. 50 loc. sts are broadcasting on approx. 100 freqs. (all lps).

MUNICH (City)

-Star*Sat Radio, Perlacher Straße 8, W-8022 Grünwald
-Radio Arabella & Charivari Radio, Paul-Heyse-Straße 2-4, W-8000 München 2
-Radio Xanadu, Pestalozzistraße 23, W-8000 München 5
-Radio Gong 2000, Nordendstraße 64, W-8000 München 40
STATIONS: Star*Sat R. on 89.0; R. Arabella on 92.4; R. Xanadu on 93.3; Charivari R. on 95.5; R. Gong 2000 on 96.3

NÜRNBERG (City)

-Radio N Eins, Bucherstraße 3, W-8500 Nürnberg 90
-Radio F, Marienplatz 3-5, W-8500 Nürnberg 1
-Radio Down-Town, Hauptstraße 106, W-8520 Erlangen 1
-Radio Gong, Innere-Cramer-Klett-Straße 6, W-8500 Nürnberg 1
-Radio Charivari, Königstraße 26, W-8500 Nürnberg
STATIONS: R. N Eins on 92.9; Radio F on 94.5; R. Down-Town on 95.8; Radio Gong on 97.1; R. Charivari on 98.6

AUGSBURG (City)

-Radio Kö/O.K., Konrad-Adenauer-Allee 11, W-8900 Augsburg
-Radio Fantasy, Maximilianstraße 41, W-8900 Augsburg
-Radio RT Eins, Curt-Frenzel-Straße 2, W-8900 Augsburg
STATIONS: R. Kö/O.K on 87.9; R. Fantasy on 93.4, R. RT Eins on 96.7

FOREIGN SERVICES

DEUTSCHE WELLE (DW)
(Anstalt des öffentlichen Rechts)

ADDR: P.O.Box 100 444, D-5000 Köln 1. **Te:** +49 (221) 3890 **Fax:** +49 (221) 389 3000 **Cable:** 888485 **Telex:** 221 386 DWK.
L.P: Dir. Gen: Dieter Weirich; Head of Prgrs: Joseph M. Gerwald; Chief Eng: Günter Roessler; Head of PR: Wilhelm Nöbel
STATIONS: Jülich (G.C: 06.33E/51.11N): 9 x 100kW — Wertachtal (G.C: 10.41E/48.05N): 9 x 500kW — Nauen (G.C: 12.54E/52.38N): 3 x 500kW, 1 x 100kW, 1 x 50kW — Königs Wusterhausen (G.C: 13.37E/52.18N): 4 x 100kW — Leipzig (G.C: 21.17E/51.11N): 1 x 100kW — **Relays:** Kigali/Rwanda (G.C: 30.07E/01.53S): 2 x 250kW — Sines/Portugal (G.C: 08.45W/37.57N): 2 x 250kW — Malta (G.C: 14.34E/35.50N): 3 x 250kW SW, 1 x 600kW MW — Antigua (G.C: 61.48W/17.06N): 2 x 250kW — Trincomalee/Sri Lanka (G.C: 81.10E/08.44N): 3 x 250kW SW, 1 x 400kW MW – additional relays via R.Canada Int, R.Bras (Brazil) and trs. in the CIS. **FM:** Sofia (BUL) on 95.7 MHz
Satellite: see satellite section (ECS II f1 13° East/ASTRA 19.2° East)

See *World Satellite Broadcasts* For details of German satellite radio and TV stations

kHz	kHz	kHz	kHz	kHz	kHz	kHz
1548	6120	7260	9670	11850	15205	17780
1557	6130	7265	9680	11865	15245	17800
3995	6140	7270	9690	11880	15270	17810
5960	6145	7275	9700	11905	15275	17820
5965	6160	7285	9705	11915	15350	17825
5980	6170	7315	9715	11945	15390	17830
5995	6180	7340	9730	11950	15400	17845
6000	6185	9510	9735	11960	15410	17860
6010	7105	9515	9750	11965	15415	17875
6015	7110	9525	9765	12045	15425	21465
6025	7120	9535	9770	12055	15435	21540
6035	7130	9545	9875	13610	15470	21560
6040	7140	9565	11705	13690	15545	21580
6045	7145	9570	11735	13760	15550	21600
6055	7150	9605	11740	13770	15595	21630
6065	7175	9610	11745	13780	17560	21640
6075	7185	9615	11765	13790	17675	21650
6085	7225	9625	11785	15105	17710	21680
6095	7235	9640	11795	15135	17715	25740
6100	7250	9650	11810	15185	17735	
6115	7255	9665	11835	15195	17765	

GERMAN LANGUAGE PROGRAMMES

A = Antigua, B = Brasilia, C = Canada, K = Kigali, M = Malta, R = CIS, S = Sines, T = Trincomalee.

H. of tr.	Areas	kHz
0600-0800	AUS/NZL	6075, 6115, 9615, 9745, 9690 9735, 11795, 11865, 11965A, 13780 17820, 17845, 17860, 21540, 21560 21600
0700-0800	AUS/NZL	21640T
0800-1000	AUS/NZL	6075, 6115, 9545, 9690A, 9735 11795, 11865, 13780, 15105, 15275 17845, 17860, 21540, 21560 21640T
1000-1200	J/EAs	1548T, 6075, 6115, 7340, 9545 13690, 13780, 15105, 15390T 17845, 21540, 21560, 21680
1100-1200	J/EAs	9545M, 25740
1200-1400	J/EAs	1548T, 1557M, 6075S, 6115, 7340 9545M, 9665, 11795, 13780, 15205 15245S, 15275A, 15390T, 17710 17715A, 17845, 21540, 21560 21680S, 25740
1225-1400	J/EAs	9545
1400-1600	SAs/NE	1557M, 6075S, 6115, 7315, 9525T 9545, 11795, 13780, 15135, 15275 15350, 17560, 17715A, 17830S 17845, 21560
1600-1700	SAs/NE/Eu	6115, 13780, 17715A
1600-1800	SAs/NE/Eu	1557M, 6075, 7315, 9525T, 9545M 9650, 11795, 15135, 15275, 15350 17560
1700-1800	SAs/NE/Eu	1548T, 21560T
1800-1900	Af/Eu	1557M, 9545M
1800-2000	Af/Eu	1548T, 6075, 7185, 9545, 9735 11795, 13610, 13690, 15275 17860K, 21560T
1900-2000	Af/Eu	6075S
2000-2120	Af/Eu	17810A
2000-2200	Af/Eu	3995, 6075S, 7185, 9545, 9735K 11765T, 11795, 13610, 13690 15275, 17810A, 17860K
2200-2400	SAm/CAm	1557M, 3995, 6010, 6075S, 6100 6115, 7110, 7275, 9545, 9715T 9730, 11785, 11915M, 15270K 15410A, 17860K
0000-0200	Am	3995, 6075S, 6100, 6115, 6180 7130, 7225, 9545, 9730, 11785 11795, 11915M, 15270K, 15410A
0100-0200	Am	6075B, 9640B, 11735T
0200-0250	Am	11795K, 15270K
0200-0300	Am	6075B, 9640B, 11735T
0200-0400	NAm/CAm	3995, 6075S, 6100, 6115, 6145 7105, 7140, 7250M, 9650, 9735M
0400-0600	NAm	3995, 6075M, 6075A, 6085S, 6100 6115, 6145, 9545S, 9605, 9735 11795, 13780

N: every hour on the hour (press review 0910, 1310, 1710, 2110, 0110, 0510).

FOREIGN LANGUAGE PROGRAMS

M = Malta, K = Kigali, A = Antigua, S = Sines, C = Canada, T = Trincomalee, B = Brasilia, R = CIS

Albanian		
2145-2215	ALB	5965, 7285S, 9615S
Amharic		
1400-1450	ETH	11810K, 15410K, 17765, 21600
1800-1850	WAf	11785, 13770, 13790, 15350, 15390 17810
Arabic		
0400-0550	NE/NAf	1557M, 6000, 6025M, 7235M, 9640
0700-0750	NAf	7270M
0735-0750	NAf	9750, 11905, 15435, 17715, 21580T 21650
1900-2120	NE/NAf	1557M, 6145, 7265M, 9515M, 9640 11850
Bengali		
0100-0150	SAs	1548T, 6035, 7285, 9615T, 9690 11945T, 15550R
Bulgarian		
1030-1120	BUL	7255, 9650, 11835, 11850, 11905 15545
1700-1750	BUL	7130, 9665, 11950S, 15470S
1930-2020	BUL	6015, 6035, 7130, 7225, 9605 9715
Chinese		
1200-1320	CHN	9535T, 9570, 11705, 11865, 13760 13790, 15185T, 15350R, 15595M 17825M
Croatian		
0830-0920	SEEu	5960, 6130, 7130, 7255, 9650
Czech/Slovak		
0530-0600	EEu	7270S, 9650S
1130-1220	EEu	5995, 6015, 7150, 9770
1930-2030	EEu	9615S, 11865S
Dari		
0800-0850	AFG	15185, 15435, 17675R, 17780T 17825, 17875, 21465, 21600T 21650, 21680
English		
0100-0150	NAm	6040A, 6055, 6085, 6145, 9515 9565M, 9610, 9700, 9770, 11865M
0200-0250	SAs	1548T, 6035, 7285, 9615T, 9690 11945T, 12055R
0300-0350	NAm	6045, 6055, 6085S, 6120, 9535 9545A, 9640, 9705, 9770M
0400-0450	SAf/EAf	6065, 6130, 7150, 7225K, 7275 9565K, 9665, 9765, 11765, 13770
0500-0550	NAm	5960, 6045, 6120, 6130, 9535 9670A, 9690
0600-0650	WAf	11765, 13610, 13790, 15185, 15435 17875
0900-0950	AUS/NZL	6160A, 11915T, 17780, 17820T 21465, 21650, 21680
0900-0950	ZAf/EAf	9665K, 15410K, 21600
1100-1150	WAf	15410K, 17765, 17800K, 17860 21465, 21600
1500-1550	ZAf/EAf	9735K, 11965K, 13610, 17735 17765, 21600
1600-1650	SAs	1548T, 6170T, 7225T, 9690, 9875R 11785M, 15105M, 15595
1900-1950	WAf	9765, 11765T, 11785, 11905, 13790 15350, 17810
2100-2150	AUS/NZL	6185, 9670T, 9690, 9765, 11785T
French		
0515-0550	SAf/EAf	9765, 11765, 15275
0700-0750	WAf	11765, 13610, 13790, 15185, 15205 17875
1200-1300	WAf	17765
1650-1750	SAf/EAf	13610, 17765, 21600
Greek		
1700-1750	GRC	6140, 9615
Hausa/French		
1200-1350	WAf	15410K
Hausa		
1300-1350	WAf	17800K, 17860, 21600
Hungarian		
1230-1300	HNG	6015, 7150, 9770
2030-2130	HNG	9615S, 11865S

Indonesian		
1330-1420	INS	13610, 15185, 15415T, 15545 17825M, 21465, 21650M
2230-2320	INS	5980, 7130, 9605, 11740, 12045R 15195T
Japanese		
1100-1150	J	9665, 9680, 11795, 11865M 11880T, 13790, 15185M, 15350R 15595
Macedonian		
1400-1430	SEEu	6130, 7175, 9770
Persian		
1000-1050	IRN	15275, 17715, 17735R, 17820T 21465, 21630T, 21650
1800-1850	IRN	9510T, 9615, 11905, 11950T 13780M, 15135M
Polish		
0600-0630	POL	7270S, 9650
1315-1350	POL	6015, 7150, 9770
1900-1930	POL	9615S, 11865S
Portuguese/French		
0500-0550	SAf/EAf	7225K
Portuguese		
0500-0515	SAf/EAf	9565K
2000-2050	Af	9765, 11810, 13790, 15105T, 15350 15425, 17765T
1000-1050	B	11795A, 11810A, 15205A, 17860A
2130-2300	B	6145, 9700, 11865M, 13790M, 15105A
2230-2320	POR	5965, 6130, 7235, 7260
Pushto		
0300-0350	AFG	7235, 9690, 11865, 12055R 15410T, 17800T, 21650T
Romanian		
0930-1020	ROU	7255, 9650, 11835, 11850, 11905 15545
1600-1650	ROU	7130, 11950S, 15470S
1800-1850	ROU	7130, 9665, 11865S, 15470S
Russian		
0400-0450	CIS	6055, 7120, 7285, 9650, 9690
1600-1920	CIS	5980, 6010, 6095, 7145, 9715 11915
2130-2220	CIS	5980, 6000, 6035, 7130, 7235 9535
Serbian		
1430-1515	SEEu	6130, 7175, 9770
Slovene		
1515-1550	SEEu	6130, 7175, 9770
Spanish		
0200-0250	SAm/CAm	6045, 6055, 6085, 9545, 9565 9700, 9770, 11810A, 11865M
1100-1150	SAm/CAm	11960A, 15205A
2030-2120	E	6000, 6130, 7130, 7235
2300-0050	SAm/CAm	6040, 6055, 6145, 9640, 9700
11810B, 11865M, 13790M, 15105		
Swahili		
0300-0400	SAf/EAf	6065, 7150, 7225K, 9565K, 9765 11765
1000-1050	SAf/EAf	9735K, 11785K, 17800, 21600
1550-1650	SAf/EAf	11965K
Swahili/French		
1550-1750	SAf/EAf	9735K
Turkish		
0600-0650	TUR	7235, 9615, 9625M, 9750, 11745 13690
1700-1750	TUR	6130M, 9640M, 11905, 13780, 15105T
Urdu/Hindi		
1430-1600	SAs	1548T, 6170T, 7225T, 9690, 9875R 11785M, 15105M, 15595

ANN: G: "Hier ist die Deutsche Welle"; E: "This is R.Deutsche Welle, the Voice of Germany"; F: "Ici La Voix de l'Allemagne"; Sp: "Habla la Voz de Alemania"; P: "Esta é Voz de Alemanha".
INT-SIG: Motif from "Fidelio" by Beethoven. "Es sucht der Bruder seine Brüder" played on celeste.
V. by QSL-card. **PUB:** Monthly prgr. schedule free.

RADIO FREE EUROPE/RADIO LIBERTY Inc.

NB: This is a non-profit, privately managed network broadcasting news and information in 23 languages to the peoples of Afghanistan, East Central and South Eastern Europe, the Baltic Countries and the successor states to the USSR. It is supported by U.S. government appropriations through the Board for International Broadcasting.

ADDR: Oettingenstrasse 67, 8000 Munich 22. **Te:** +49 (89) 2102 0. **Fax:** +49 (89) 2285188. **Telex:** 523228/523904. **Cable:** Radioferl.

L.P: Pres: Gene Pell; V.P. Eng: George Woodard; Head PR: Terry B. Shroeder

Washington Office: 1201 Connecticut Ave NW, Washington DC 2036, USA. **Te:** +1 (202) 457 6900.

New York Office: 1775 Broadway, New York, NY 10019, USA. **Te:** +1 (212) 397 5300.

STATIONS:

Biblis, Germany B1 to B10 (100kW); **Holzkirchen, Germany** H1 to H4 (250kW), HMWT (150kW); **Lampertheim, Germany** L1 to L9 (100kW); **Gloria, Portugal** G1 to G19 (250kW), G20 & G21 (50kW); **Playa de Pals, Spain** P1 to P6 (250kW); **Maxoqueira, Portugal** M1 to M6 (500kW).

NB: P1 & P2 or P3 & P4 can be combined to a single 500kW. P1 thru P4 can be combined to a single 1000kW.

Satellite: see satellite section Eutelsat II f1, 13° East.

kHz	kHz	kHz	kHz	kHz	kHz	kHz
720	6135	7245	9575	9750	11905	15370
3960	6170	7255	9595	11725	11925	15380
3970	7115	7275	9605	11770	11935	15445
5955	7130	7295	9625	11780	11970	17725
5970	7145	9505	9660	11815	15115	17735
5985	7155	9520	9680	11825	15130	17750
6050	7165	9530	9695	11855	15170	17760
6095	7180	9540	9705	11875	15215	17835
6105	7190	9555	9715	11885	15290	21510
6115	7220	9565	9725	11895	15340	21745

Tr. sites used on each freq. are also listed in "Shortwave Stations of the World"

Armenian

H. of tr.	site	kHz
0500-0530	G8	11725
0500-0530	G18	9680
0500-0530	L4	7180
0500-0530	L5	9595
0500-0530	L6	11875
1500-1530	G11	15170
1500-1530	L3	17750
1500-1530	L5	15340
1500-1530	L6	11875
1730-1800	G3	11825
1730-1800	G11	15170
1730-1800	L5	9540
1730-1800	L6	11875

Azerbaijani

H. of tr.	site	kHz
0430-0500	G8	11725
0430-0500	G18	9680
0430-0500	L4	7180
0430-0500	L5	9595
0430-0500	L6	11875
1700-1730	G3	11825
1700-1730	G11	15170
1700-1730	L5	9540
1700-1730	L6	11875
1900-1930	G9	11875
1900-1930	L3	7130
1900-1930	L5	9540
1900-1930	L7	9595

Belarus

H. of tr.	site	kHz
0400-0500	B2	3960
0400-0500	M5	5970
0400-0500	P6	7190
0400-0600	G10	9705
0400-0600	M4	7255
0500-0600	B5	5970
0500-0600	G4	11770
1500-1600	B2	7190
1500-1600	B5	5970
1500-1600	H2	7255
1500-1600	L7	9705
1500-1600	M6	11970
1800-1900	B2	3960
1800-1900	L4	5970
1800-1900	M6	11970
1800-1900	P4	9705

Bulgarian

H. of tr.	site	kHz
1400-1700	B7	9565
1400-1700	G12	17725
1400-1900	G16	15115
1400-2200	G4	11770
1700-2200	B3	6115
1700-2200	G18	9565
1900-2000	B4	7155
2000-2200	G17	7155

Estonian

H. of tr.	site	kHz
0500-0530	B2	3960
0500-0530	M3	9505
0500-0530	M5	6095
0500-0530	P6	7155
2000-2100	M3	5970
2000-2100	M6	7255
2200-2300	B2	3960
2200-2300	M6	7255
2200-2300	B10	6135
2200-2300	M3	5970

Georgian

H. of tr.	site	kHz
0330-0430	G8	11725
0330-0430	G18	9680
0330-0430	L4	7180
0330-0430	L5	9595
0330-0430	L6	11875
1600-1700	G3	11825
1600-1700	G11	15170
1600-1700	L5	15340
1600-1700	L6	11875
1800-1900	G9	11875
1800-1900	L3	7255
1800-1900	L5	9540
1800-1900	L7	9595

Hungarian

H. of tr.	site	kHz
1700-2300	G5	9695
1700-1800	B1	7115
1700-1800	L9	5985
1700-2000	G13	11895
1800-2000	B1	3970
1800-2300	G11	7115
1800-2300	M5	5985
2000-2300	G6	7145

Kazakh

H. of tr.	site	kHz
0000-0300	H1	6135
0000-0300	H2	7145
0000-0300	M3	9625
0000-0300	P5	9660
1200-1500	H2	11970
1200-1500	L3	17750
1200-1500	L6	15130
1200-1500	M1	15380

Kyrgyz

H. of tr.	site	kHz
0130-0200	L6	6050
0130-0200	L8	7295
0130-0200	M4	9680
0130-0200	P4	9505
0230-0300	L6	6050
0230-0300	L8	7295
0230-0300	M4	9680
0230-0300	P4	9505
1200-1300	B7	15445
1200-1300	B1	17760
1200-1300	G12	21745
1200-1300	G19	17835
1530-1600	H3	9595
1530-1600	H4	11905
1530-1600	M1	11895
1530-1600	M5	15370

Latvian

H. of tr.	site	kHz
0530-0600	B2	3960
0530-0600	M3	9505
0530-0600	M5	6095
0530-0600	P6	7155
1600-1700	H1	5970
1600-1700	H2	7255
1900-2000	G16	9705
1900-2000	G17	11970
1900-2000	M3	5970
1900-2000	M6	7255

Lithuanian

H. of tr.	site	kHz
0600-0630	B2	3960
0600-0630	M3	9505
0600-0630	M4	7255
0600-0630	M5	6095
1700-1800	H2	7255
1700-1800	M6	11970
1700-1800	P4	9705
1700-1800	H1	5970
2100-2200	M3	5970
2100-2200	M6	7255

Polish

H. of tr.	site	kHz
0500-0700	HMWT	720
0500-0700	M6	7190
1600-1800	B2	7190
1700-2300	HMWT	720
1800-2300	M4	7190

Pashto & Dari

H. of tr.	site	kHz
0230-0300	G17	7165
0230-0330	G8	11725
0230-0330	G10	9530
0230-0330	L5	9605
0300-0330	L4	7180
1300-1330	B7	15445
1300-1330	B1	17760
1300-1330	G12	17725
1300-1330	L5	15340
1530-1600	G11	15170
1530-1600	L3	17750
1530-1600	L5	15340
1530-1600	L6	11875

Romanian

H. of tr.	site	kHz
0500-0600	B1	3970
0500-0600	G17	7115
0500-0700	M1	7165
0500-0700	B3	5985
0500-0700	M2	9725
0600-0700	G8	11725
0600-0700*	B3	5985
0600-0700*	M1	7165
0600-0700*	M2	9725
0600-1000	B1	7115
0700-1000	B4	7165
0700-1000	B3	9725
0700-1000	M3	11725
0700-1000	M4	15215
0700-1000	B3	9725
0700-1800*	M4	15215
0700-1900*	B4	7165
0700-1900*	M3	11725
1000-1500*	B3	9725
1300-1900	B4	7165
1300-1900*	M3	11725
1500-1700	B1	9505
1500-1700	B3	9725
1500-1800	M4	15215
1500-1900	B4	7165
1500-1900	M3	11725
1700-1800	M5	9505
1700-2300	M2	9725
1800-2300	B5	5955
1800-2300	G14	9505
1900-2300	M1	7165
2000-2300	B1	3970

*) non-daily

Russian

H. of tr.	site	kHz
0000-0030	P4	7220
0000-0130	L8	7155
0000-0200	B9	7255
0000-0200	L3	7245
0000-0200	L9	6170
0000-0300	M5	6105
0000-0330	G19	9750
0000-0400	H4	7275
0000-0500	B5	5955
0000-0600	B7	9555
0000-0600	P1	9520
0030-0400	P6	7220
0130-0400	M6	7155
0200-0400	B10	5970
0200-0500	G6	7245
0300-0400	B8	6170
0300-0400	P5	9660
0300-0500	M3	9625
0300-0600	H1	6105
0330-0700	P4	11885
0400-0500	G4	11770
0400-0600	G12	11815
0400-0800	L2	11970
0400-1200	H2	7220
0500-0600	B4	9625
0500-0700	L3	15290
0500-1200	L8	7245
0530-0700	B10	7180
0530-0800	L5	9595
0600-1600	H1	9520
0600-1800	L4	6105
0700-1200	H4	11885
0700-1200	L6	15290
0700-1200	L7	9695
0700-1200	L9	11875
0700-1500	B2	7190

0700-1700	B8	6170
1200-1230	H4	11875
1200-1500	B10	7245
1200-1500	L7	9705
1200-1500	M5	15290
1200-1600	L2	11885
1200-1930	L8	7220
1300-1500	B9	9625
1300-1700	L9	5985
1330-1500	L5	15340
1500-1700	B9	7115
1500-1700	B10	7155
1500-1700	P5	9625
1500-1800	G6	15380
1500-1900	G17	15290
1600-1800	L2	11815
1600-1900	P3	11885
1600-2400	P1	9520
1700-1900	H3	9660
1700-1900	P6	7155
1700-2000	L9	6135
1700-2400	B7	9555
1800-2100	G12	11925
1800-2300	H2	7245
1800-2400	H1	6105
1900-2200	B2	3960
1900-2200	B10	6135
1900-2400	P5	9660
1930-2130	P3	7220
2000-2300	G9	9680
2100-2400	L9	6170
2130-2400	P4	7220
2200-2400	L8	7155
2300-2400	B5	5955
2300-2400	L3	7245

Tajik

H. of tr.	site	kHz
0200-0230	L6	6050
0200-0230	L8	7295
0200-0230	M4	9680
0200-0230	P4	9505
1230-1300	G10	21510
1230-1300	G11	17735
1230-1300	H3	15370
1230-1300	H4	11875
1500-1530	H3	9595
1500-1530	H4	11905
1500-1530	M1	11895
1500-1530	M5	15370

Tatar-Bashkir

H. of tr.	site	kHz
0400-0500	L3	7155
0400-0500	M2	9725

0400-0500	P3	11935
0400-0700	H3	9540
0400-0700	H4	11895
0500-0700	L7	9695
0500-0700	L9	11855
1600-1700	H3	9595
1600-1700	L3	7295
1600-1700	L7	9705
1600-1700	M1	11895
1600-1700	M6	11970
2000-2100	H4	7295
2000-2100	L6	9625
2000-2100	L7	9705
2000-2100	L9	6170

Turkmen

H. of tr.	site	kHz
0200-0300	B8	6170
0200-0300	B9	7255
0200-0300	G14	11875
0200-0300	G18	9715
0200-0300	L4	7180
1400-1500	G10	21510
1400-1500	G11	17735
1400-1500	H3	15370
1400-1500	H4	11875

Ukrainian

H. of tr.	site	kHz
0400-0500	M1	7165
0400-0700	B8	6170
0400-0700	B9	7145
0400-0700	P5	9660
0500-0700	P3	11935
1700-1900	M1	11935
1700-2000	B8	6170
1700-2000	B9	7295
1700-2000	P5	9625
1900-2000	G6	11935

Uzbek

H. of tr.	site	kHz
0200-0400	H3	9540
0200-0400	L3	7190
0200-0400	L9	5985
0200-0400	M2	9725
0200-0400	P3	11935
1300-1400	G10	21510
1300-1400	G11	17735
1300-1400	H3	15370
1300-1400	H4	11875
1700-1800	G12	11925
1700-1800	H3	9595
1700-1800	H4	11780
1700-1800	L7	9575

ANN: RFE, Sign on Sign off: Bulgarian: "Tuk e Radio Svobodna Evropa"; Hungarian: "Itt a Szabad Európa Rádio"; Polish: "Mowí Rozglosnia Polska Radio Wolna Evropa"; Romanian: "Aici e Radio Europa libera"; Estonian: "Siin Vaabaduseradio"; Latvian: "Te Runa Brivibas Raiditajs"; Lithuanian: "Kalba Lasives Radijas". **RL**, Azerbaijani: "Danyshyr Asadlyk Radiosu"; Armenian: "Chosume Azatutiun Rdiokayane"; Belarus: "Haworyts Radyio Svoboda"; Georgian: "Laparakobs Radio Thavisupleba"; Russian: "Govorit Radio Svoboda"; Tatar-Bashkir: "Monda Azatlyk Radiostansyese Suyili"; Ukranian: "Hovoryit Radio Svoboda"; Uzbek: "Azadiq Radiosidan Qapiramiz"; Turkmen: "Qepleyaer Azatik Radiosi"; Tajik: "Inja Radioi Azadi"; Kazakh: "Azzatiq Radioyosinan Sövlep Turmiz"; Kirghiz: "Azattiq Radiosunan Söylöbüz"; Dari: "Inja Radioi Afghanistan Azad Ast".
INT-SIG: RFE: Bulgarian: Part of an old patriotic Bulgarian folk song.; Polish: Polish patriotic song "Hail Glorious Dawn of May" (brass instruments); Romanian: "Romanian Rhapsody" by G. Enescu. **RL**: Russian sce: "Hymn to Freedom" by Gretchaninoff; Afghan sce: Afghan folklore song "Anar Anar".
V: by QSL-card sent to the Munich or New York office (B).

Others:

VOICE OF AMERICA (WERTACHTAL)

STATION: (G.C: 10.41E/48.05N): 500kW.

kHz	kHz	kHz	kHz	kHz	kHz	kHz
5965	6060	7170	9740	11825	15355	21570
5970	6090	7180	9770	11835	15445	
5975	6150	7210	11705	11840	17895	
5985	6160	9525	11710	15155	21455	
5995	7105	9530	11805	15195	21535	

D.PRGR: see USA.

VOICE OF AMERICA (MUNICH)

MEDIUMWAVE: 1197kHz 300kW

kHz	kHz	kHz	kHz	kHz	kHz
1197	3980	11895	11905	15195	17885

D.PRGR: see USA.

RADIO MOSCOW RELAY

STATION: Wachenbrunn 1323 kHz (1000kW 0400-1800), (150kW 1800-0400)
NB: Full schedule under Russia.

BBC BERLIN RELAY

ADDR: see UK
STATION: FM 90.2 MHz (50kW)
D.PRGR: Relays of BBC World Service, German Service and English service by radio. For more details see BBC schedules in the United Kingdom entry.

RADIO VOLGA (FOR RUSSIAN ARMED FORCES)

ADDR: Menselstraße 5, O-1560 Potsdam.
STATION: Burg 261 kHz (200kW)
D.PRGR: Radio Volga relays Radioropa and Russia's Radio. Mon-Sat: 0500 Radioropa Info, 0900 Radio Volga (Russian), 1000 Radioropa Info (German), 1200 Radio Volga (Russian; Tue/Wed/Fri 1300-1400 in German), 1400 Radioropa Info (German), 1800 Radio Volga (Russian), 2000-2100 Radioropa Info (German); Sun: 0500 Radio Volga (Russian), 1200 Radio Volga (German), 1300 Radioropa Info (German), 1900 Radio Volga (Russian), 2000-2100 Radioropa Info (German)
ANN: "Govorit Volga". **V:** by QSL-card against return postage.

AMERICAN FORCES NETWORK EUROPE (AFN)

ADDR: (local studios); APO 09757-4310, 6000 Frankfurt a/M.; APO 09742-4317, 1000 Berlin; APO 09094-5000, 6750 Kaiserslautern; APO 09696-4313 Nürnberg; APO 09108-4312, 8000 Munich; APO 09069-5000, 2850 Bremerhaven; APO 09154-4315, 7000 Stuttgart; APO 09036-4318, 8700 Würzburg.
STATIONS:

Mediumwaves	kHz	kW		kHz	kW
Frankfurt a/M	873	150	Stuttgart	1143	10
Berlin	1107	10	Heidelberg	1143	1
Grafenwöhr	1107	10	Karlsruhe	1143	1
Kaiserslautern	1107	10	Mönchengldb	1143	1
Nürnberg	1107	10	Hof	1143	1
München	1107	40	Bitburg	1143	1
Bremerhaven	1143	5	Augsburg	1485	1

+ 11 sts. 0.3kW on 1143 kHz (Bad Kissingen, Bamberg, Fulda, Giessen, Göppingen, Bad Hersfeld, Schweinfurt, Wertheim, Würzburg, Wildflecken) and 6 sts. on 1485 kHz (0.3kW) (Ansbach, Berchtesgaden, Grailsheim, Garmisch, Hohenfels, Regensburg).

FM:	MHz	kW		MHz	kW
Berlin*	89.7°	1	Stuttgart*	102.3°	100
Sögel	93.5	1	Ulm	102.6*	1
Frankfurt*	98.7	100	Rheinberg	105.1	1
Augsburg	100.0!	15	Bremerhaven	106.2	5
Kaiserslautern	100.2	5	Bonn	107.6	0.12

°) stereo — *) separate prgrs. — !) also rel. **Z. FM**
D.PRGR in English: 24h. — **N:** every h. on the h. (except 0400). major newscasts (30 min) 0500, 0600, 1700, 2100. — local prgrs: Mon-Fri 0405-0500, 0505-0600, 0705-0800, 0905-1055, 1205-1400, 1405-1700; Sat 0705-1100. FM separate prgrs. are mostly easy listening.
ANN: "This is the AFN, The American Forces Network Europe".
V: by QSL-card.

BRITISH FORCES BROADCASTING SERVICE

NB: A division of the Service Sound and Vision Corp.
ADDR: Wentworth Barracs, W-4900 Herford 1. **Te:** +49 (5221) 88 090

STATIONS:
BFBS 1st Network

FM	MHz	kW		MHz	kW
Braunschweig	93.0	80	Langenberg	96.5	50
Bonn	97.8	0.5	Visselhövede	97.6	30
Berlin	98.8	1.5	Bielefeld	103.0	70
Hameln	99.3	0.1	Jever	97.3	0.05

BFBS 2nd Network

FM	MHz	kW		MHz	kW
Bielefeld	101.6	0.16	Osnabrück	106.3	0.1
Münster	102.2	0.16	Lippstadt	105.0	0.16
Bergen-Hohne	104.8	0.16	Rheindahlen	104.3	0.3

+2 sts in Belgium & 2 sts. in The Netherlands listed separately.
D.PRGR: 24 h in English. **ANN:** "BFBS Germany".
V: by QSL-card

CANADIAN FORCES NETWORK (RESEAU DES FORCES CANADIENNES)

ADDR: Jammstraße 9, 7630 Lahr. **Te:** +49 (7821) 209 889 **Fax:** +49 (7821) 21 235
STATIONS: CFN Lahr on 90.5 (0.5kW). repeaters: Ramstein 101.9 (0.05kW), Heidelberg 105.4 (0.05kW); CFNS Söllingen 96.8 (0.5kW); RFC Lahr 94.0 (0.5kW). repeaters: Söllingen 90.5 (0.5kW) + 1st in The Netherlands.
ANN: "This is the Canadian Forces Network"; "Vous écoutez la radio des Forces Canadiennes". **V:** by QSL-card.

RADIO FORCES FRANÇAISES DE BERLIN (FFB)

ADDR: Kurt-Schumacher-Damm, Quartier Napoleon, 1000 Berlin-Reinickendorf
STATIONS: Berlin 93.6 (3kW), Goslar 92.7 (0.1kW) also rel. **Europe 1**.
D.PRGR: 1900-1800 in French (rel. Radio France); 1800-1900 in German (re. RFI).

EUROPE 1 (Independent Comm.)

ADDR: Europäische Rundfunk un Fernsehen AG, Europa Nr. 1, Postfach 111, 6600 Saarbrücken. **Te:** +49 (681) 30782
STATIONS: Felsberg 183kHz 2000kW.
D.PRGR: 0400-2200 (Sat-Sun 24h). **N:** hourly & 0530, 0630, 0730, 0830, 2230.
ANN: F: "Europe 1" **V:** by QSL-card (to: Europe 1, D-6630 Saarlouis, P.O. Box 1365, Germany.

GIBRALTAR (British Dependent Territory)

L.T: UTC + 1h (Su: UTC + 2h) — **Pop:** 30.000 — **Radio sets:** 17.000 — **Pr.L:** English, Spanish — **E.C:** a/c 50, 240V (yes) — **ITU:** GIB.

GIBRALTAR BROADC. CORPORATION (Partly Comm.)

ADDR: Broadcasting House, 18 South Barrack Rd, Gibraltar.
Te: +350 79760. **Telex:** 2229 GK. **Fax:** +350 78673.
L.P: Gen. Mgr: G. J. Valarino. Chief Eng: N. Celecia.
STATION: Wellington Front 1458kHz 0.5/2kW. **FM:** 92.6MHz 2kW, 91.3/100.5MHz 0.2kW (stereo).
D.PRGR. in English, Spanish: 0530-2300 (different prgrs on MW/VHF at certain times). **N. English:** 0600, 0700, 0830 (local), 0900, 1100, 1230 (local), 1600, 1700, 1800 (BBC), 2200, 2258. **Spanish:** 1300(W).
ANN: "This is Gibraltar" — **V.** by QSL-card. Re. in English or Spanish. Rec. acc. (B) from So. Europe.

BRITISH FORCES BROADC. SCE. GIBRALTAR

ADDR: BFBS, Gibraltar, BFPO 52. **Te:** +350 54211. **Fax:** +350 54443 — **L.P:** St. Mgr: Alan Clough. Sen. Eng: Steve Harris. Prgr. Dir: Kal Sutherland.
STATIONS: 89.4/99.5MHz (0.2kW), 93.5/97.8MHz (1kW).
D.PRGR: 24h (Prgr. 1 on 93.5/97.8MHz, Prgr. 2 on 89.4/99.5MHz).
N: 0000, 0700-1900 hourly, 2200, 2300.
ANN: "This is the British Forces Broadc. Sce. Gibraltar".

GREECE

L.T: UTC + 2h (Su: UTC + 3h) — **Pop:** 10.093.000 — **Radio sets:** 4.085.000 — **Pr.L:** Greek — **E.C:** a/c 50, 220V — **ITU:** GRC.

ELLINIKI RADIOPHONIA TELEORASSI S.A. (ERT S.A.)

ADDR: 432 Méssoghion Ave., 153 42 Aghia Paraskevi, Athens.
Te: +30 (1) 6395970-9. **Fax:** +30 (1) 6390652. **Tlx:** 216066.
L.P: Pres. & Man. Dir: Prof. Vassilis Silivridis.

ELLINIKI RADIOPHONIA (ERA)

Te: +30 (1) 6396700. **Fax:** +30 (1) 6380323.
L.P: Dir.Gen: George Tzavelas.
MEDIUMWAVES

Location	kHz	kW	Prgr.
Athens 3	666	15	3
Athens 1	729	150	1
Ioannina	765	10	R+1,2,4
Zakynthos	927	50	R+1,2,4
Larissa	945	5	R+1,2,4
Heraklion	954	10	R+1,2,4
Athens 4	981	200	4
Corfu Isl.	1008	50	R+1,2,4
Thessaloniki 1	1044	150	M1
Orestiada 2	1080	10	R+1,2,4
Thessaloniki 2	1179	50	M1,2,4
Florina	1278	10	R+1,2,4
Tripolis	1314	10	R+1,2,4
Pyrgos	1350	4	R+1,2,4
Athens 2	1386	50	2
Komotini	1404	50	R+1,2,4
Patra	1485	1	R+1,2,4
Volos	1485	1	R+1,2,4
Orestiada 1	1485	1	1
Rhodes Isl.	1494	50	R+1,2,4
Chania	1512	50	R+1,2,4
Serres	1584	1	R+1,2,4
Kavala	1602	1	R+1,2,4
Kozani	1602	1	R+1,2,4
Samos	1602	1	1

1 = ERA 1 — 2 = ERA 2 — 3 = ERA 3 — 4 = ERA 4 — R = Regional Station — M1 = Macedonia 1.

FM (MHz)

Location	FM 1	FM 2	FM 3	FM 4	kW (ERP)
Parnitha mt (Athens)	91.60+	93.60+	95.60+	—	100
Hymettos mt (Athens)	105.80+	103.70+	—	101.80+	35/35/100
Gerania mt (Corinth.)	97.90	99.90	—	—	3
Thessaloniki	88.00	90.00	92.00+	—	10
Alexandroupolis	89.80	91.8	—	—	3
Pilon mt (Volos)	92.80	94.80	96.80+	—	10
Heraklion	94.40	96.40	—	—	3
Thira isl. (Santorini)	96.90	98.90	—	93.30+*	10/10/35
Ioannina	97.80	99.80	—	—	3
Kavala	89.20	91.20	—	—	10
Kalamata	92.20	94.20*	—	—	3
Kastoria	88.60	90.60	—	—	10
Kerkyra isl. (Corfou)	91.80	93.80	—	—	3
Kephallonia isl.	96.90	98.90	—	—	10
Lesvos isl. (Mytilini)	92.30	94.30	—	104.40+*	3/3/35
Rhodos isl.	88.40	90.40	—	—	3
Tripolis	88.30	90.30	—	—	10
Chania	92.90	94.90	—	—	3

+) also carrying regional prgrs — *) stereo.

ERA regional programs on FM:
Macedonia 1: Thessaloniki 95.8MHz, 102.00MHz 5kW — **Aegaeon:** Thira Isl. 93.30MHz 35kW, Lesvos Isl. 104.40MHz 35kW (both stereo) — **Chania:** 104.00MHz 35kW — **Corfu:** 99.3MHz — **Florina:** 102.30MHz — **Heraklion:** 97.50MHz 3kW — **Ioannina:** 102.10MHz — **Kalamata:** 94.20MHz 3kW — **Kavala:** 96.30MHz — **Komotini:** 98.10MHz — **Kozani** 100.2MHz — **Larissa:**

98.30MHz — **Orestiada:** 103.50MHz — **Patra:** 92.50MHz — **Rhodes:** 92.70MHz — **Serres:** 101.50MHz — **Tripoli:** 95.30MHz — **Volos:** 101.20MHz 35kW — **Zakynthos:** 95.2MHz.

Regional Stations

Macedonia: Odos Yeorghikis Scholis 129, GR-546 39 Thessaloniki. **Ann:** "Elliniki Radiophonia, Radiophonikos Stathmos Makedonias".

Relays on SW: 1000 (Sat/Sun 0600)-2255 on 9935kHz (to the Middle East) and 1159kHz (to Europe).

FOREIGN SERVICE — "The Voice of Greece" (ERA 5th Prgr.)

ADDR: P.O. Box 60019, 15310 Aghia Paraskevi, Athens. **Te:** +30 (1) 6397375. **Fax:** +30 (1) 6009608 (Eng.)

L.P: MD: Filotas Giannotas. Head of Prgrs: George Apostolopoulous. CE: Demetri H. Vafeas. Head of PR: Gina Voyatzoglou.

MEDIUMWAVES: Kavala 792kHz 500kW (operated by the VOA).

SHORTWAVES: Avlis (G.C: 23.36E/38.23N): 2×100kW SW-tr's — Kavala (G.C: 24.50E/40.53N): 1×250kW SW-tr (operated by the VOA).

kHz	kHz	kHz	kHz	kHz	kHz
7450	9395	9420	9825	15630	17515
9375	9400	9425	11645	15650	17525

Times	Area	kHz
0000-0350	NAm	9395, 9420, 11645
0400-0450	ME/EAf	9425, 11645, 15630
0500-0615	CYP/TUR/ME	7450, 9395, 9425
0700-0750	Eu	7450, 9424, 11645
0800-0850	AUS	15650, 17525
0850-0950	AUS	17525
0900-0950	AZR	15650
1000-1050	J	15650, 17525
1100-1120	CYP/TUR/ME	9825, 11645
1130-1150	EGY/LBY	9825, 11645
1200-1250	NAm	15635, 15650, 17515
1300-1350	CAf	15650, 17515
1300-1350	UZB	11645
1400-1450	ME/EAf	11645, 15630, 17515
1500-1550	NAm/NEu	15630, 15650, 17525
1600-1710	SEEu	7450, 9395, 9425
1715-1755	CIS/POL	9425, 11645, 15630
1800-1850	SAf	15630, 17525
1800-2055	Eu	9425
1900-1950	Eu	7450, 9400
2000-2030	SEEu	792
2000-2050	SEEu	7450, 9400
2100-2150	AUS	7450, 9375, 9425
2150-2250	AUS	9425
2200-2300	CAm/SAm	7450, 9395
2300-2350	CAm/SAm	7450, 9395, 11645

N. Greek: 0000, 0200, 0400, 0500, 0700, 0800, 0900, 1000, 1200, 1300, 1400, 1500, 1600, 1715, 1800, 1900, 2100, 2200. **English:** 0130W, 0340W, 0940, 1040, 1235, 1530W, 1840W, 1900W, 2240+, 2335+. (+=except Sat). **French:** 1915. **German:** 1930. **Italian:** 1945. **Spanish:** 2320. **Portuguese:** 2305. **Turkish:** 0600, 1100. **Arabic:** 1130, 1430. **Albanian:** 1615, 2030. **Serbo Croat:** 1630, 2040. **Bulgarian:** 1645, 2000. **Romanian:** 1700, 2015. **Russian:** 1335, 1730. **Polish:** 1745. **Swedish:** 1540.

V. by QSL-card (re. in SINPO code). No rec.

Private stations in Athens

98.4FM Stereo 98.4MHz — Antenna FM Stereo 97.1MHz — Diavlos 10 92.8MHz — Ellada 88.5FM Stereo 88.5MHz — Galaxy 92 FM 92.0MHz — Klik FM 88.0MHz — Piraiki Ekklisia 91.2MHz — Pop FM 92.4 92.4MHz — R. Athina 99.1MHz — Seven-X 98.7MHz — Sky 100.4 FM Stereo 100.4MHz.

ARMED FORCES RADIO & TV SERVICE (United States Air Force)

ADDR: Det 4, Air Force European Broadc. Squadron (AFEBS), APO AE 09846, USA — SM: MSgt. Scott W. Kirby. **Station:** 1584kHz 0.25kW. 24h. **Ann:** "This is EBS Iraklion".

VOICE OF AMERICA RADIO STATION RHODES

MEDIUMWAVE: 1260kHz 500kW.

SHORTWAVE (G.C: 28.10E/36.15N): 2×50kW.

kHz	kHz	kHz	kHz	kHz	kHz
5965	7205	9545	11735	11905	11960

VOICE OF AMERICA RADIO STATION KAVALA

MEDIUMWAVE: 792kHz 500kW.

SHORTWAVE (G.C: 24.50E/40.53N): 10×250kW tr's.

kHz	kHz	kHz	kHz	kHz	kHz	kHz
5955	6160	7270	9650	11740	15145	15650
5965	7105	7280	9675	11805	15155	17780
5985	7130	7295	9680	11825	15160	17810
6010	7160	7450	9700	11835	15205	17875
6015	7170	9530	9705	11845	15225	21455
6060	7190	9535	9740	11875	15235	21520
6085	7205	9585	9745	11905	15400	21570
6140	7210	9635	11645	11945	15445	21720
6150						

HUNGARY

L.T: UTC + 1h (Apr-Oct UTC + 2h) — **Pop:** 10.550.000 — **Radio sets:** 6m — **Pr.L:** Hungarian — **E.C:** a/c 50 220V — **ITU:** HNG.

MAGYAR RADIO (Gov. Semi-Comm.)

ADDR: Bródy Sándor u. 5-7, H-1800 Budapest. **Cable:** Magyarradio. **Te:** +36 (1) 1-388-388. **Telex:** (22)5158.

L.P: Pres: Csaba Gombár. Man. Dir: Dr. Mihály Gálik. Dir. Tec.: Gábor Heckenast. Head of Foreign Rel. Dept.: László Óvári.

Mediumwaves	kHz	kW	Prgr
Solt	540	2000	1
Lakihegy	873	20	2+Reg.
Pécs	873	20	2+Reg.
Mosonmagyaróvár	1116	2.2	2+Reg.
Miskolc	1116	12	1+Reg.
Szolnok	1188	135	2
Szombathely	1188	25	2+Reg.
Nyíregyháza	1251	25	2+Reg.
Marcali	1251	500	2
Lakihegy	1341	300	2
Pecs	1350	10	2
Györ	1350	5	2+Reg.
Szolnok	1350	5	2+Reg.

Prgr: 1 = Kossuth (1st prgr); 2 = Petöfi (2nd prgr); 3 = Bartók (3rd prgr) stereo; Reg = Regional prgr.

Shortwave: Székesfehérvár (G.C: 18.24E/47.10N): 6025kHz 100kW (Prgr.1).

FM (MHz)	I	II	III	kW
Budapest	67.40	66.62	69.38	100
Györ	—	—	67.04	3
Kabhegy	72.98°	71.42	70.64	10
Kékes	71.21°	72.77	70.10	3
Komádi	66.14	66.92	68.24	3
Miskolc	66.80°	66.02	68.48	3
Nagykanizsa	71.03°	69.98	68.36	10
Pécs	71.81°	67.19	67.97	3/10/10
Sopron	72.86°	70.40	72.08	3
	102.0°			
Szeged	94.9°	—	—	1
Szentes	66.29°	67.85	68.72	3
Tokaj	71.33°	72.11	70.43	10

°) carries regional prgrs (see below).

HOME SERVICE

Prgr. I (Kossuth Radio): 0400 (Sun 0500)-2310. **Rel. BBC:** 2200-2245. **Rel. VOA:** Mon/Tues 1200-1300, Wed-Sun 1500-1600.

Prgr. II (Petöfi Radio): 0330 (Sun 0500)-2315. **N:** 0330 and half hourly to 0700, 0800, 1000, 1200, 1400, 1600, 1800, 2000, 2200.

Prgr. 3 (Bartók): 0500-0100. **N:** 0500, 0530, 0600, 0700, 1100, 1500, 1800, 2230.

Reg. Prgrs: Pécs on 873kHz + 67.69MHz, **Szolnok** on 1350kHz + 72.95MHz; **Nyiregyháza** on 1251kHz, **Györ** on 1116/1188/1350 kHz + 72.86/72.98MHz, **Miskolc** on 1116kHz + 66.80/71.21MHz, **Szeged** on 94.9 + 66.29MHz, **Debrecen** on 103.5MHz: W 0455-0700 & 1530-1730/1830, Sun 0800-1000 & 1500-1700/1800 (exact times vary).

R. Calypso: 0700-1800 on 873kHz (Lakihegy).

ANN: "Kossuth Radio, Budapest — Petöfi Radio, Budapest" — Bartók Radio, Budapest".

RADIO DANUBIUS (Comm.)

operated by Magyar Radio

ADDR: Benczúr u. 19, Budapest 1068. **Te:** +36 (1) 387678. **Fax:** +36 (1) 388027 — **L.P:** Dir.Gen: György Varga.

STATIONS: Kabhegy 100.5MHz 30kW — Sopron 102.0MHz 30kW — Budapest 103.3MHz 15kW — Szeged 94.9MHz 1kW — Debrecen 100.5MHz 30kW — **D.PRGR:** 0400-0100.

FOREIGN SERVICE

ADDR: R. Budapest, P.O. Box 1, H-1800 Budapest.
STATIONS: Jaszberény (G.C: 19.52E/47.35N), Diósd (G.C: 18.57E/47.25N): 250/100kW.

kHz	kHz	kHz	kHz	kHz	kHz	kHz
6025	6110	7220	9835	11910	15160	15220

TO EUROPE

English	**kHz**
2200-2300	6110, 9835, 11910
German	
1300-1400Sun	7220, 9835, 11910
1800-1900Sun	9835, 11910, 15160
1830-1900W	9835, 11910, 15160
2030-2100	9835, 11910, 15160
Hungarian	
1200-1300Sun	7220, 11910, 15220
1400-1500Sun	7220, 9835, 11910
1900-2030	9835, 11910, 15160
2300-0100Sun	9835, 11910, 15160

TO NORTH AMERICA

English	
0300-0400	6110, 9835, 11910
Hungarian	
0100-0300Sun	6025, 9835, 11910
0100-0230W	6025, 9835, 11910
2300-0100Sun	9835, 11910, 15160
2300-0030W	9835, 11910, 15160

ANN: E: "This is Radio Budapest, Hungary".
INT-SIG: Excerpts from the suite "1848" by T.K. Polgar (played on three trumpets 2 cornets) — **V.** by QSL-card. Rec. acc (B).
Relays of R. France Internationale: see schedule under "France".

OTHER STATIONS

R. Juventus Balabon: 0800-1600 on Siofok 810kHz 12kW — **R. Bridge,** P.O. Box 516, 1538 Budapest. **Te:** +36 (1) 2010298 — 24h on 102.1MHz 0.7kW. **N. in English:** on the h (rel. VOA Europe) — **R. Tizenegy** (R. 11): Budapest 71.63MHz — **Tilos R:** Budapest 95.5MHz.

ICELAND

L.T: UTC — **Pop:** 262.000 — **Radio sets:** 155.000 — **Pr.L:** Icelandic — **E.C:** a/c 50, 220V (yes) — **ITU:** ISL.

ICELANDIC NATIONAL BROADCASTING SERVICE RÍKISÚTVARPID (Gov.)

ADDR: Efstaleiti 1, 150 Reykjavik. **Te:** +354 (1) 693000. **Telex:** 2066 ISRADIO IS. **Fax:** +354 (1) 693010 — **Regional Addr:** Fjölnisgötu 3a, 600 Akureyri. **Te:** +354 (96) 27000. **Fax:** +354 (6) 27985 — Fagradalsbraut 9, 700 Egilsstaðir. **Te:** +354 (97) 12300. **Fax:** +354 (7) 11996 — Aðalstræti 22, 400 Isafjödur. **Te:** (94) 4404. **Fax:** +354 (4) 4638.
L.P: Dir. Gen: Heimir Steinsson. Chmn. Prgr. Council: Halldora J. Rafnar. Dir. Adm: Hördur Vilhjálmsson. Dir. Sound Radio: Elfa-Björk Gunnarsdóttir. Dir. N: Kari Jónasson. Head of Int. Rel: Dóra Ingvadóttic.

Stations	**kHz**	**kW**		**kHz**	**kW**
Reykjavik	207	100	Höfn	666	1
Eidar	207	20	Skjaldarvik	738	5

FM (MHz). All stereo.

Location	**I**	**II**	**kW**
Gagnheidi	99.8	87.7	8.8/11.8
Girdisholt	92.9		3.5
Háfell	93.8	98.7	14/34
Hegranes	90.6	98.8	3.1/5
Hnjúkar	89.1	95.5	6.0/6.2
Hvítabjarnarhóll	95.1	90.3	0.8/1
Höfn	90.3	104.8	1.2/1
Skálafell	92.4	99.9	24
Stykkishólmur	88.0	96.3	18/20
Thjódólfsholt		88.3	2.5
Vadlaheidi	91.6	96.5°	9.3
Vatnsendi	93.5	90.1°	2.5
Vidarfjall	88.1	96.1	3.3
Vestmannaeyar	97.1	88.1	17/24

+ 43 st's below 1kW. °) Carries regional prgrs (see below).

D.PRGR: Prgr. 1 on LW/MW + FM 1: 0645 (Sun 0800)-0110. **Main N:** 0800, 1220, 1600, 1900, 2200, 0000. **N. in English:** 0730.
Prgr II on FM II freqs: 24h.
Regional Prgrs on FM: Útvarp Nordurland (Akureyri) Mon-Fri 0810-0830 & 1835-1900; Útvarp Austurland (Egilsstadir) & Svaedisutvarp Vestfjarda (Ijafjore) Wed/Thurs/Fri 1835-1900.

SHORTWAVE SERVICE

STATIONS: Reykjavik (G.C: 21.50W/64.05N): 10kW.

kHz	kHz	kHz	kHz	kHz
11402	13835	13855	15770	15790

Times	**Area**	**kHz**
1215-1300	Eu	13835, 15770
1410-1440	NAm	13855, 15770
1855-1930	Eu	11402, 13855
1935-2010	NAm	13855, 15770
2300-2335	NAm	13855, 15770

All prgrs in Icelandic. **NB:** Frequencies change often.
ANN: "Utvarp Reykjavík" (P1), "Rás tvö" (P2).
V. by QSL-card.

NAVY BROADCASTING SCE. (U.S. Mil.)

ADDR: U.S. Navy Broadc. Sce, Detachment Keflavik, Box 25, Navairsta FPO New York 09571, USA (or Box 25, US Air Base, IS-235 Keflavikurflugvöllur).
L.P: St. Mgr: E. Barker. Chief Eng: S. Jonsson.
STATIONS: Keflavík 1485kHz 0.25kW — Höfn 1400kHz 0.1kW (relay).
D.PRGR. in English: 24h — Local Prgrs: Mon-Fri 0105-0400, 0605-0900, 1105-1200, 1505-1800, 1905-2100; Sat 0605-0900, 1105-1300; Sun 0705-0800, 1505-1700.
ANN: AM 1485, Keflavik, Iceland — **V.** by letter.

IRELAND

L.T: UTC (Su: UTC + 1h) — **Pop:** 3.564.000 — **Radio sets:** 2m — **Pr.L:** Irish, English — **E.C:** a/c 50, 220/380V (yes) — **ITU:** IRL.

RADIO TELEFÍS ÉIREANN (Statutory Corporation)

ADDR: Donnybrook, Dublin 4. **Te:** +353 (1) 643111. **Telex:** 93700.
Cable: Broadcasting. **Fax:** +353 (1) 643082.
L.P: Chmn: J.P. Sorohan. Dir. Gen: J. Barry. Dir. of Prgr's (radio): K. Healy. Dir. of Eng: M.J.C. Curley.
Networks: I = R. 1, II = 2FM, III = FM3, IV = R. Na Gaeltachta.

Mediumwaves	**kHz**	**kW**	**N**		**kHz**	**kW**	**N**
Conamara	540	2	IV	Corca Dhuibhne	828	1	IV
Tullamore	567	500	I	Tir Chonaill	963	10	IV
Athlone	612	100	II	Cork	1278	10	II
Cork	729	10	I/II	Dublin	1278	10	II

FM (MHz)	**I**	**II**	**III/IV**	**kW**
Achill	89.3	91.5	93.7	2
Athlone	90.2	—	—	0.1
Cahirciveen	89.5	91.7	93.9	2
Castletownbere	88.3	90.5	92.7	1
Clermont Carn	95.2	97.0	102.7	40
Cork City	89.2	91.4	93.6	1
Fanad	89.8	92.0	94.2	2
Holywell Hill	89.2	91.4	93.6	2
Kippure	89.1	91.3	93.5	40
Limerick City	89.3	91.5	—	0.4
Maghera	88.8	91.0	93.2	160
Moville	—	99.3	—	10
Mt. Leinster	89.6	91.8	94.0	100
Mullaghanish	90.0	92.2	94.4	160
Three Rock	88.5	90.7	92.9	10
Truskmore	88.2	90.4	92.6	80

HOME SCE: RTE Radio 1: 0630-0048 on 567/729*kHz + FMI. Also on FM III/IV at certain times (see below). *) exc. Mon-Fri 1203-1300 & 1700-1820.
N: English: 0630, 0700, 0730 (W), 0800, 0830 (W), 0900 (W), 1100 (W), 1140 (Sun), 1201 (W), 1300, 1600 (W), 1700 (W), 1801, 1830, 2200, 2300, 0000 (W), 0045.
ANN: "RTE Radio One" — **INT-SIG:** Musical Box playing 4 bars of Irish air "O'Donnell Abu."
Cork Local Radio: Mon-Fri 1100-1300 & 1430-1815 on 729kHz. + 89.2MHz. **N:** 1230, 1730.

2FM: 24h on 612/1278kHz and FMII. **N: English:** on the h. **ANN:** 2FM.
FM3: Opt-out sce. of classical music, literature & drama on FM III: Mon-Fri 0630-0800, 1930-2400; Sat 0730-1055, 1930-2400; Sun 0730-1055, 1930-2400. Other times rel. R1 or R. Na Gaeltachta.
V. by QSL-card. (B).

RADIÓ NA GAELTACHTA
(Nat. Sce. for Irish-speaking communities).
ADDR: Casla, Conamara, Co. na Gallimhe. **Te:** +353 (91) 72224. **Fax:** +353 (91) 72351.
L.P: MD: B. Feiritéar. PD: Sean O'Cuirreain. CE: M. Evans.
HOME SCE. in Irish on MW + FM III/IV: Mon-Fri 0800-1930, Sat/Sun 1100-1930.

ATLANTIC 252 (Comm.)
(Joint venture of CLT Luxembourg & Radio Telefis Eireann).
ADDR: Mornington House, Summerhill Rd, Trim, Co. Meath. **Te:** +353 (46) 36655. **Fax:** +353 (46) 36644. (Addr. in U.K.: P.O. Box 252, London W1E 2RA).
L.P: GM: Travis Baxter. Head of Music: Paul Kavanagh. CE: Niall MacCaughey.
STATION: 252kHz 500/100kW.
D.PRGR: 24h (100kW 1900-0600). Music prgrs aimed at the 18–34 age group.
V: by QSL-card. Re. to London addr.
F.PI: Broadcasting via Satellite.

COMMERCIAL STATIONS

INDEPENDENT RADIO & TV COMMISSION (IRTC)
ADDR: Clanwilliam Court, Marine House, Dublin 2. **Te:** +353 (1) 760966. **Fax:** +353 (1) 760948 — Chief Exec: Michael O'Keefe.

	MHz	kW	Name & location
24)	94.7		Tipperary North, Tipperary
10)	94.7	0.05	Highland R, Ballybofey
19)	94.8	2	Northern Sound, Cavan
1)	94.9	5	East Coast R, Bray
2)	95.0	5	R. Limerick 95FM, Limerick
16)	95.1	5	WLR FM, We. Waterford
10)	95.2	1	Highland R, We. Donegal
21)	95.2	0.25	R. Kerry, Dingle*
5)	95.4	0.02	MWR FM, Westport
14)	95.7	0.05	Shannonside 104FM
4)	95.8	5	R. LM/FM, Co. Louth
9)	95.8	0.1	Galway Bay FM, Galway
5)	96.1	5	MWR FM, Mayo
21)	96.2	3	R. Kerry, Cahirciveen
3)	96.2	0.025	South East Radio, Gorey
19)	96.3	5	Northern Sound, Monaghan
23)	96.3	5	North West R, Donegal
18)	96.4	2	Radio South, Cork
7)	96.4	5	Clare FM, Maghera
8)	96.6	5	R. Kilkenny, Kilkenny
9)	96.8	5	Galway Bay FM, Galway C.
21)	97.0	20	R. Kerry, Mullaghanish
11)	97.1	5	Suirland R, Woodruff
5)	97.1	1.5	MWR FM, Achill
5)	97.3	0.02	MWR FM, Ballina
12)	97.3	2.5	CKR, Carlow
16)	97.5	5	WLR FM, Ea. Waterford
19)	97.5	0.1	Northern Sound, Carrickmacross
21)	97.6	1	R. Kerry, No. Kerry
12)	97.6	1	CKR, Naas
13)	98.1	5	Classic Hits 98FM, Dublin
3)	99.2	2	South East R, Mt. Leinster
23)	102.5	2	North West R, Truskmore
1)	102.9	4	East Coast R, Ballyguile
10)	103.3	5	Highland R, No. Donegal
17)	103.3	5	County Sound, Cork C.
19)	103.4	0.05	Northern Sound, Kingscourt
6)	103.5	1.25	Radio 3, Tullamore
17)	103.7	2.5	County Sound
11)	103.9	1.6	Suirland R.
14)	104.1	5	Shannonside 104 FM, Roscommon
15)	104.4	5	FM104, Dublin
20)	104.8	5	Tipperary Mid-West Radio
4)	104.9	0.02	R. LM/FM, Dundalk

Addresses & other information:
H. of tr: 24h unless otherwise stated.
1) 9 Prince of Wales Terrace, Quinsboro Rd, Bray, Co. Wicklow — 0700-2400.
2) Radio House, Dooradoyle, Limerick.
3) Custom House Quay, Wexford Town, Co. Wexford.
4) Williamson Mall, Francis Str, Dundalk, Co. Louth — 0630 (Sat 0700, Sun 0800)-0200.
5) Abbey Str, Ballyhaunis, Co. Mayo — 0700-0100.
6) The Mall, William Str, Tullamore, Co. Offaly.
7) The Abbeyfield Centre, Francis Str, Ennis, Co. Clare — 0700-0100.
8) 56 Hebron Rd, Kilkenny — 0700-0200.
9) Unit 13, Sandy Rd, Galway — 0700-0200.
10) Pine Hill, Letterkenny, Co. Donegal — 0700 (Sat/Sun 0800)-0200.
11) Gashouse Bridge, Clonmel, Co. Tipperary — 0700-2400.
12) Lismard House, Tullow Str, Carlow — ACC House, Main Str, Naas, Co. Kildare — 0700-0200.
13) 7 Mount Str, Dublin 2.
14) Castle Str, Roscommon.
15) Ballast House, O'Connell Bridge, Dublin 2.
16) The Radio Centre, St. George's Str, Waterford.
17) Gouldshill, Mallow, Co. Cork.
18) Whites Cross, Cork.
19) 33 Glaslough Str, Monaghan, Co. Monaghan.
20) St. Michael's Str, Tipperary — 0900-2100.
21) Park View, Mary Str, Tralee, Co. Kerry — 0700-0100.
22) c/o ACC House, Abbey Str, Wicklow.
23) Market Yard, Sligo, Co. Sligo — 0700-0100.
24) Active Link Co-Op Society Ltd, c/o Ashbury Industrial Estate, Ashbury, Roscrea, Co. Tipperary.

ITALY

L.T: UTC + 1h (Su: UTC + 2h) — **Pop:** 57.880.000 — **Radio sets:** 15m — **Pr.L:** Italian — **E.C:** a/c 50, Voltages variable — **ITU:** I.

RAI-RADIOTELEVISIONE ITALIANA
(State Service by Public act to RAI).
ADDR: Viale Mazzini 14, 00195 Roma. **Te:** +39 (6) 36861. **Fax:** +39 (6) 3606070. **Cable:** Raidirro. **Telex:** 61142.
Technical Dept: Via Cernaia 33, 10121 Torino. **Fax:** +39 (11) 539429. **Telex:** 221035 RAI TOI.
L.P: Pres: Enrico Manca. Vice-Pres: Leo Birzoli. DG: Gianni Pascarelli. DG (Radio): Corrado Guerzoni.

HOME SERVICE
R = carries regional prgrs. Refer to corresponding number in address list below.
MEDIUMWAVES

R	Location	kHz	kW
20)	Caltanissetta	189	10
20)	Caltanissetta	567	20
9)	Bologna	567	20
21)	Sassari	567	10
17)	Salento	567	6
2)	Aosta	567	2
16)	Napoli	657	120
10)	Firenze	657	100
1)	Torino	657	50
5)	Bolzano	657	25
6)	Venezia	657	20
18)	Potenza	693	20
7)	Trieste	819	20
13)	Roma	846	540
3)	Milano	900	600
20)	Trapani	936	10
6)	Venezia	936	20
8)	Genova	936	10
7)	Trieste	981	10
18)	Potenza	990	10
1)	Torino	999	20
12)	Perugia	999	20
9)	Rimini	999	6
19)	Capo Vaticano	999	2
3)	Milano	1035	50
16)	Napoli	1035	20
7)	Trieste	1035	10
10)	Firenze	1035	2
14)	Pescara	1035	6
17)	Salento	1035	6
20)	Caltanissetta	1035	2
21)	Oristano	1035	2
21)	Cagliari	1062	25
17)	Squinzano	1062	25
20)	Catania	1062	2
7)	Udine	1062	2
6)	Verona	1062	2
10)	Pisa	1062	2
4)	Trento	1062	2
13)	Roma	1107	6
17)	Bari	1116	150
9)	Bologna	1116	60
10)	Pisa	1116	25
20)	Palermo	1116	10
2)	Aosta	1116	2
21)	Sassari	1143	10
20)	Messina	1143	6
8)	San Remo	1188	6
10)	Pisa	1305	2
11)	Ancona	1314	6
15)	Campobasso	1314	2
19)	Cantanzaro	1314	1

18)	Matera	1314	2
13)	Roma	1332	300
17)	Bari	1332	50
14)	Pescara	1332	25
20)	Palermo	1332	10
6)	Venezia	1368	20
16)	Napoli	1368	10
3)	Milano	1368	12
8)	Genova	1368	10
1)	Torino	1368	6
20)	Messina	1368	2
20)	Catania	1368	2
10)	Firenze	1368	2
21)	Sassari	1368	2
4)	Trento	1368	2
17)	Bari	1368	1
17)	Foggia	1431	2
11)	Pesaro	1431	2
17)	Taranto	1431	1
17)	Squinzano	1449	50
	Catania	1449	6
	10 st's	1449	2
	2 st's	1449	1
	8 st's	1449	0.15
	4 st's	1485	2
14)	L'Aquila	1485	1
	3 st's	1485	1
	4 st's	1485	0.15
20)	Palermo	1512	2
	Genova	1575	50
	Perugia	1575	20
	7 st's	1575	2
	5 st's	1575	1
	2 st's	1575	0.15
	7 st's	1602	2
	Calgiari	1602	1

*) only used for Regional Prgrs.

SHORTWAVES: Caltanissetta 6060/7175/9515kHz 50kW.

FM(MHz)	R1	R2	R3	(ERP)kW
Bologna Barbiano	89.5	91.7	93.9	12
Ca del Vento	92.1	96.5	98.5	20
Col Visentin	91.1	93.1	95.5	10
Firenze Morell	87.8	91.1	98.4	5
Genova-Portofino	89.5	91.9	95.1	25
M. Argentario	90.1	92.1	94.3	25
M. Beigua	91.5	94.6	98.9	6
M. Caccia	94.6	96.7	99.2	50
M. Cammarata	91.1	95.9	99.9	15
M. Cavo	87.6	91.2	98.4	12
M. Conero	88.3	90.3	92.3	6
M. Faito	94.1	96.1	98.1	20
M. Lauro	94.7	96.7	98.7	6
M. Limbara	88.9	95.3	99.3	6
M. Nerone	94.7	96.6	98.7	50
M. Peglia	95.7	97.7	99.7	20
M. Pellegrino	94.9	96.9	98.9	6
M. Penice	94.2	97.4	99.9	60
M. Sambuco	88.6	90.7	93.5	15
M. Scuro	88.5	90.5	92.5	6
M. Serpeddi	90.7	92.7	96.3	6
M. Serra	88.5	90.5	92.9	6
M. Venda	88.1	89.0	89.9	80
Martina Franca	89.1	91.1	93.1	50
Milano	90.6	93.7	99.4	15
P. Badde Urbara	91.3	93.3	97.3	6
Pescara	89.2	94.3	96.4	25
Pomarico	88.7	92.7	95.7	5
Roma M. Mario	89.7	91.7	93.7	20
S. Cerbone	95.3	97.3	99.3	6
Salento	90.7	95.5	97.5	10
Torino Eremo	92.1	95.6	98.2	25
Trieste M. Belv	91.5	93.6	95.8	10
Udine	94.9	97.2	99.8	25

+ over 1900 smaller sts not mentioned — M = Monte.

Additional stereo trs: Roma 100.3MHz 28kW, Torino 101.8MHz 40kW, Milano 102.2MHz 1.6kW, Napoli 103.9MHz 0.6kW. (These tr's relay classical music from cable radio 1442-2030 & Night Prgr 2300-0500.)

RADIO UNO (Prgr 1): 0500-2230. **N:** 0600 (W), 0700, 0900 (W), 0916 (Sun), 1100 (W), 1200, 1800, 2200. **Rel. on SW:** Continuous on 6060/9515kHz.

Separate Music Prgrs on FM ("StereoRAI") 1220-2300.

Night Prgrs: "Notturno Italiano": 2230-0500 on 846/900/6060kHz. **N. in Italian:** on the h. 2300-0400, 0445. **N. in English:** 0003, 0103, 0203, 0303, 0403. **N. in French:** 0006, 0106, 0206, 0306, 0406. **N. in German:** 0009, 0109, 0209, 0309, 0409.

"Stereonotte" on all R1/R2/R3 FM-tr's: 2300-0500.

Regional Prgrs: refer to MW list above, addresses below.

RADIO DUE (Prgr 2): 0500-2230. **N:** 0505, 0530, 0630, 0730, 0830, 1030, 1130, 1230, 1530, 1730, 1830, 2130. **Rel. on 7175KHz:** 0500-2230.

Separate Prgrs on FM (Music, N. & traffic reports "Radio Verde RAI"): 1150-2300.

Night Prgrs: see Radio Uno.

Regional Prgrs: refer to MW list above, addresses below.

RADIO TRE (Prgr 3): 0500, 2300, **N:** 0545, 0745, 1045, 1245, 1445, 1745, 1945, 2215.

Night Prgrs: see Radio Uno.

Regional Prgrs: refer to MW list above, addresses below.

REGIONAL PRGRS

H. of tr: W 0620-0630 on R1 freqs, 1110-1130 & 1315-1400 on R2 freqs. (Venezia 1368kHz also 1430-1530). **Regional Traffic. N:** W 1124, 1324, 1354. Freqs as indicated in MW list, plus additional low power MW-tr's not mentioned.

SPECIAL PRGRS

RAI 2 ISO FREQUENZA: 24h sce. for motorway users on 103.3MHz FM.

SENDER BOZEN (Bolzano): Prgrs in **German** on 1602kHz (Bolzano/Brunico/Merano 1kW, Bressanone 0.1kW) + FM (39 tr's of 1kW or less). MW-tr's relay Radiotre at other times.

D.Prgr: 0545 (Sun 0700)-2100. **N.** 0615 (W), 0845 (Sun), 0900 (W), 1100, 1200, 1600 (W), 1830.

Regional Prgr. in Slovene:

Trieste 981kHz 10kW + 103.9MHz 20kW (and 22 additional FM-tr's).

D.Prgr: 0600 (Sun 0700)-1830. **N:** W0600, 0700, 0900, 1200, 1300, 1600, 1800; Sun 0700, 1200, 1300, 1800. **N. in German:** 0900 (W).

N. in Arabic (for Mediterranean area): W 1430-1445 on Caltanissetta 567kHz + SW.

FOREIGN SERVICE

ADDR: Viale Mazzini 14, 00195 Roma. **Te:** +39 (6) 5919076.

STATIONS: Prato Smeralto (G.C: 12.31E/41.48N): 100kW.

kHz	kHz	kHz	kHz	kHz
5990	7290	11905	15385	21515
6010	9575	15245	17780	21535
7235	9710	15330	17795	21560
7275	11800	15340	17800	21690

EUROPEAN SERVICE

		kHz
Albanian	2000-2020	5990, 7235
Bulgarian	1540-1600	7235, 9710
Czech	1815-1830	7290, 9710
	2135-2155	6010, 7275
Danish	2000-2020a	7275, 9710, 11800
English	1935-1955	7275, 9710, 11800
Esperanto	2000-2020b	7275, 9710, 11800
French	1430-1455	7290, 9575, 11905
German	1535-1550	5990, 7290, 9575
	1750-1825	5990, 7275, 9575
Greek	1520-1540	7235, 9710
Hungarian	1935-1955	5990, 7235
Italian	1410-1730Sun	9710
	1430-1455	5990, 7235
	1555-1635	5990, 7290, 9575
Lithuanian	0445-0505	5990, 7275
Polish	1845-1905	7290, 9710
	2210-2225	6010, 7275
Portuguese	2110-2130	5990, 7275
Romanian	0510-0530	5990, 7275
	2135-2155	5990, 7235, 9710
Russian	0330-0345	5990, 7275 9710, 11800, 11905
	0535-0555	5990, 7275
	1605-1625	7235, 9710
Serbo-Croat	1500-1520	5990, 7290, 9575
	1910-1930	5990, 7275
Slovak	1830-1845	7290, 9710
	2155-2210	6010, 7275
Slovene	1520-1535	5990, 7290, 9575
Spanish	2050-2110	5990, 7275
Swedish	2000-2020c	7275, 9710, 11800
Turkish	1500-1520	7235, 9710
Ukrainian	0350-0410	5990, 7275

OVERSEAS SERVICE

AUS/NZL/FAR EAST

English	kHz
2200-2225	5990, 9710, 11800

TO AFRICA

Amharic	
0415-0435	15330, 17795, 21560

Arabic

1750-1810	7235, 9710
2025-2045	7275, 9710
French	
1630-1655	7235, 9710
Italian	
0435-0510	15330, 17795, 21560
1410-1730Sun	21515, 21690
1700-1745	7235, 9710, 15385, 17780, 21690
Somali	
0510-0530	15330, 17795, 21560
1640-1655	15385, 17780, 21690

TO NEAR EAST

Arabic

0535-0555	11800, 15330, 17795
1910-1930	9710, 11800, 15340
English	
2025-2045	7235, 9575, 11800

TO MEDITERRANEAN

Italian/English

0415-0440	5990, 7275

TO LATIN AMERICA

Italian

1410-1730Sun	21535
2230-0100	11905, 15245
0140-0305	9575, 11905, 15245
Portuguese	
0120-0135	11905, 15245
Spanish	
0100-0120	11905, 15245
0305-0325	9575, 11905, 15245

TO NO. AMERICA

English

0100-0120	9575, 11800
French	
0120-0135	9575, 11800
Italian	
1400-1425	17800, 21560
1410-1730Sun	17800
1830-1905	17780, 21560
2230-0100	9575, 11800
0140-0305	11800
Spanish	
0305-0325	11800

N. in English: 0100, 0425, 1935, 2025, 2200.
ANN: Home Sce: "Radiouno", "Radiodue", "Radiotre" as appropriate. Night Prgr: "RAI—Radiotelevisione Italiana stazioni a onda media di Roma kHz 846 pari a m.355, e di Milano kHz 900 pari a m.333, e stazione ad onda corta di kHz 6060 pari a m.49. 50 "Notturno Italiano" — Foreign Sce: I. "RAI—Radiotelevisione Italiana—stazioni ad onda corta di m..pari rispettivamente a..kHz—transmissione in lingua Italiana per. . ." E: "This is the Italian R. and Television Sce. Broadc: from Rome" . . .F: "Ici la R. Italienne". G: "Hier ist der Italienische Auslandssendedienst in Rom".
INT-SIG: Bird chirping; opening signal: Bells (composition by A. Parelli)—**V.** by QSL-card.
PUB: "Radiocorriere" (weekly for Home Sce). Prgr schedule 3 times a year in E, F, G, I. (Foreign Service).

RUNDFUNK ANSTALT SÜDTIROL (RAS)

ADDR: Eurpaallee 164/A, 39100 Bozen. **Te:** +39 (471) 932933.
L.P: Pres: Helmuth Hendrich. Man. Dir: Klaus Gruber. Dir. Tec: Helmuth Schäfer.
RAS is a public body of the autonomous region of Southern Tyrol whose purpose is to relay TV and radio from Germany, Austria and Switzerland to the German-speaking population.

Stations (MHz)	RAS 1	RAS 2	RAS 3	kW
Kronplatz	100.7	103.0	104.7	2
Meransen	101.3	103.9	107.3	1
Obervinschgau	100.5	103.0	106.1	0.6
Penegal	103.3	100.3	104.7	2
Perdonig	101.8	104.0	106.0	1
Plose	99.8	102.0	105.6	1
Vinschgau	101.1	102.9	105.0	2

+ low power st's.
D.Prgr: RAS 1: rel. OE-3 (Austria)
RAS-2: rel. OE-R (Austria)
RAS-3: rel. OE-1 (Austria)

RADIO TIROL GmbH

ADDR: Postfach 26/Aichweg 4, 39019 Dorf Tirol.
Te: +39 (473) 93656. **Telex:** 401406. **Fax:** +39 (473) 93663.
L.P: Gen. Mgr: Dr. Gerald Fleischmann.

Stations (FM Stereo)	MHz	kW		MHz	kW
Taufers/ Münstertal	99.3	8.5	Zirog-Alm	101.5, 105.3	67
Innichen	103.7	1.0			

+ low power relays.
D.PRGR. in German: 24h. **N:** W 0600, 0700, 0900, 1130, 1800;
Wrp: W 0535, 0615, 0705, 0905, 1215, 1815; Sun 0700, 1130.

PRIVATE STATIONS

This section was compiled by our collaborator Dario Monferini. Only st's r. with MW/SW broadcasts and FM networks are listed. A number of other st's are heard irr.

MEDIUMWAVES

	kHz	kW	Name, location and h. of tr.
1)	1318v	0.5	R. Campogalliano, Campogalliano: 0700-1700
2)	1503	2	Tele Radio Stereo, Roma: 0600-1700
3)	1584	0.5	R. San Luchino, Bologna: 24h
4)	1584	5	R. Verona, Verona: 0600-1800 irr.
5)	1586v	0.5	R. Studio X, Momigno: 0600-2300 irr.
6)	1620v	5	R. International, Verona: 1800-0600.

SHORTWAVES

2)	6012	2	Tele Radio Stereo, Roma: 24h
9)	6220	7	European Christian Radio, Ravenna: 0700-1900
10)	7140v	5	R. Italia Int., Spoleto: 0700-1500 irr.
11)	7297	0.5	R. Europe, Pioltello: 0800-1200
10)	11390	0.5	Marconi R. Int, Taranto: 0800-1200 irr.
13)	13667	5	The V. of Europe, Pordenone: 24h

Addresses and other information:
1) Via Madonna 22, I-41011 Campogalliano. **Te:** +39 (59) 526325 — St. Mgr. G. Zappelli. **V.** by letter. Re. to G. Zappelli, Via Roma 25, I-41011 Campogalliano.
2) Via Carlo Evangelisti 3, I-00136 Roma. **Te:** +39 (6) 3498301. **Fax:** +39 (6) 348300 — St. Mgr. D. Sbrigoli. **FM:** 92.7/94.5MHz 18kW (stereo). **V.** by QSL-card. Rp.
3) Via San Luca 35, I-40135 Bologna. **Te:** +39 (51) 434525. **Fax:** +39 (51) 435651 — St. Mgr. F. Piancastelli. **FM:** 100/100.3MHz 5kW. **V.** by letter. Rp.
4) Piazza Cittadella 26, I-37122 Verona. **Te:** +39 (45) 8000896. **Fax:** +39 (45) 8031500 — St. Mgr. S. Mischi. **FM:** 103MHz 5kW/ 103.3MHz 2kW (stereo). **V.** by letter. Rp.
5) Via Mamianese 687, I-51010 Momigno. **Te:** +39 (572) 619385 — St. Mgr: P.L. Belli. **FM:** 87.5MHz 1kW (stereo). **V.** by letter. Rp.
6) Corso Porta Palio 36, I-37122 Verona. **Te:** +39 (45) 8031100. **Fax:** +39 (45) 8031500 — — St. Mgr. F. Mischi. **FM:** 100.5/ 103.9MHz 1kW (stereo). **V.** by letter or QSL-card. Rp.
7) EFG, Postfach 500, A-2345 Brunn, Austria. **Te:** +39 (544) 34411 — **FM:** Ravenna 99.2MHz 5kW. **V.** by QSL-card. Rp.
8) Vicolo di Volusio, I-06049 Spoleto. **Te:** +39 (743) 223148. **Fax:** +39 (743) 223310 — St. Mgr. N. Mastoro. **FM:** 100.4/101MHz 2.5kW (stereo). **V.** by QSL-card (B) Rp.
9) c/o: PLAY-DX, Via Davanzati 8, I-20158 Milano — St. Mgr: A. Bertini. **V.** by QSL-card (B) Rp. note: **DX prgr:** Sun 1015-1045.
10) c/o Gruppo Ascolto Due Mari, c/o C. Pepe, via C. Battisti 27, I-74100 Taranto. **V.** by letter. Rp.
11) P.O. Box 26, I-33170 Pordenone. **Te:** +39 (434) 997667. **Fax:** +39 (434) 997687 — St. Mgr. L. Spadotto. **V.** by QSL-card Rp.

FM NETWORKS

1) 105 Classic (30 st's), Via Cenisio 32, I-20154 Milano. **Te:** +39 (2) 316583. **Fax:** +39 (2) 33106166.
2) Italia Radio (40 st's), Piazza del Gesù 47, I-00186 Roma. **Te:** +39 (6) 6791412, 6796539. **Fax:** +39 (6) 6781936 — St. Mgr. A. Seci.
3) Italia Radio Network (40 st's), P.O. Box 161, I-33100 Udine. **Te:** +39 (432) 530600, 530886. **Fax:** + 39 (432) 530644 — St. Mgr. M. Pinosa. **V.** by letter. Rp.
4) Latte e Miele l'Italiana (30 st's), c/o Tecnivest, Via Manzoni 4, I-40013 Castelmaggiore (BO). **Te:** +39 (51) 712601. **Fax:** +39 (51) 713919 — St. Mgr. F. Mignani. **V:** by letter. Rp.
5) R. Deejay Network (110 st's), Via Massena 2, I-20145 Milano. **Te:** +39 (2) 342522, 3311440. **Fax:** +39 (2) 3311569 — St. Mgr: U. La Bozzetta. **V.** by letter. Rp.
6) R. Dimensione Suono Network (55 st's), Viale Mazzini 119, I-00195 Roma. **Te:** +39 (6) 383466. **Fax:** +39 (6) 3725336 — St. Mgr. E. Montefusco. **V.** by letter. Rp.
7) R. Kiss Kiss Network (40 st's), Via Sgambati 61, I-80131 Napoli. **Te:** +39 (81) 5461212, 5466566. **Fax:** +39 (81) 5467789 — St. Mgr. L. Niespolo. **V.** by letter. Rp.
8) R. Italia solo Musica Italiana, (70 st's), Via F. Casati 2, I-20124 Milano. **Te:** +39 (2) 29401904. **Fax:** +39 (2) 29517463 — St. Mgr. M. Volanti. **V.** by letter. Rp.
9) R. Maria Netw. (160 st's), Via Marconi 80, I-22036 Arcellasco d'Erba (CO). **Te:** +39 (31) 610610. **Fax:** +39 (2) 610717 — St. Mgr: E. Ferrario. **V.** by letter. Rp.

10) R. Milano Int. Magic One-O-One Network (35 st's), Via Locatelli 6, I-20124 Milano. **Te:** +39 (2) 66982551, 66982147. **Fax:** +39 (2) 6704900 — St. Mgr. R. Borra. **V.** by QSL-card (B). Re. to D. Monferini, Via Davanzati 8, I-20158 Milano. **F.Pl:** reactivate MW 1391kHz + SW 7115kHz 5kW.
11) R. Montecarlo FM-Italia (30 st's), Via Turati 40, I-20121 Milano. **Te:** +39 (2) 6596116.
12) R. Radicale (70 st's), Via Principe Amedeo 2, I-00185 Roma. **Te:** +39 (6) 4880541. **Fax:** +39 (6) 4880196 — St. Mgr: P. Vigevano. **V.** by letter. Re. to R. Rodolico, Via Popoli Uniti 14, I-20125 Milano. **Fax:** +39 (2) 26111758.
13) Rete 105 The Radio (110 st's), P.O. Box 1448, I-20100 Milano. **Te:** +39 (2) 6596116. **Fax:** +39 (2) 6592272 — St. Mgr. J. Azan. **V.** by letter. Re. to F. Cavestro, Servizlo Tecnico, Via Turati 40, 7° piano, I-20121 Milano. Rp.
14) Rete Gamma Radio (35 st's), Radiant S.p.a., Strada 3a-Palazzo B-11, I-20094 Assago Milano Fiori (MI). **Te:** +39 (2) 89201630, 89201705. **Fax:** +39 (2) 89201852 — St. Mgr. C. Ticozzi. **V.** by letter Rp.
15) RTL Hit Radio s.r.l, (90 st's), Via Suardi 42, I-24040 Arcene (BG). **Te:** +39 (35) 879294-8. **Fax:** +39 (35) 878012. **V.** by letter. Rp.

ADVENTIST WORLD RADIO EUROPE (Rlg.)

ADDR: AWR-UICCA, Casella Postale 383, I-47100 Forli, Italy. **Te:** +39 (543) 766655.
L.P: Board Chmn: E. Ludescher/J. Paulsen. Mgr: G. Hodgson. Treasurer: E. Amelung.
MEDIUMWAVE: Sicily 1521kHz 0.25kW (F.Pl: 10kW).
STATIONS: F = Forli: 10kW — M = Moscow, Russia — S = Samara, Russia — Y = Yekaterinburg, Russia.
Shortwave:

kHz	kHz	kHz	kHz	kHz	kHz	kHz
7210	7310	11775	11820	11900	15125	17605
7230	9775	11785				

Arabic	**Area**	**kHz**
0230-0300	ME	11785M
0600-0630	ME	11775M
0800-0900	NAf	17605M
2200-2300	NAf	11820M
English		
0400-0430	NEu	15125S
0730-0800	NWEu	7210F
1030-1100	NWEu	7230F
1600-1630	NEu	15125S
French		
1230-1300SS	NEu	7230F
German		
0500-0530	CEu	15125S
0900-0930	CEu	7230F
1200-1230	CEu	7230F
1900-1930	CEu	7310Y
Italian		
0800-0900	CEu	7210F
1100-1200	CEu	7230F
1400-1500	SEu	9775M
Polish		
0530-0600	CEu	15125S
1930-2000	CEu	7310Y
Romanian		
0400-0430	SEu	11900Y
1530-1600	SEu	9775M
Russian		
0300-0330	RUS	11900Y
1700-1800	NEu	15125S
1800-1900	CEu	7310Y
Serbian/Croation		
0330-0400W	SEu	11900Y
1500-1530Su	SEu	9775M
Slovene		
0330-0400Su	SEu	11900Y
1500-1530Su	SEu	9775M
Spanish		
1000-1030SS	SWEu	7230F
1300-1330SS	SWEu	7230F
Swedish		
1630-1700	NEu	1512S
Ukrainian		
0600-0630Su	EEu	9605P

ANN: E: "This is Adventist World Radio, the Voice of Hope". F: "Ici la Radio Mondiale Adventiste, La Voix de l'Esperance". G: "Sie hören Adventist World Radio, die Stimme der Hoffnung". I: "Questa è la Radio Mondiale Avventista, la Voce della Speranza".
INT-SIG: Different arrangements of the melody: "Lift up the trumpet..."
V. by letter or QSL-card.

NEXUS — INTERNATIONAL BROADCASTING ASSOCIATION

ADDR: P.O. Box 10980, I-20110 Milano. **Te:** +39 (2) 266 6971. **Fax:** +39 (2) 706 38151. **E-mail:** 100020, 1013@compuserve. com.
CSI: 100020, 1013 — Pres: Alfredo Cotroneo.
FM: Globe Radio Milan FM 88.8MHz.
SW: Italian Radio Relay Sce. (IRRS): 7215kHz 10kW (USB with 6dB carrier reduction.

Times	**Days**	**Target Areas**
0300-0400	Daily	NAm
0630-0900	Mon-Fri	Eu/NAf/ME
0630-1000	Sat	Eu/NAf/ME
0630-1500	Sun	Eu/NAf/ME
1000-1500	Mon-Fri (irr.)	Eu/NAf/ME
1400-1600	Sat	Eu/NAf/ME
1700-2100	Mon-Thurs	Eu/NAf/ME
1700-2230	Fri/Sat/Sun	Eu/NAf/ME

Languages: English, French, Spanish, Russian, Polish. Includes prgrs. produced by Deutsche Welle, Swiss Radio International, UN Radio, UNESCO, European Gospel R. and own prgrs (write for current prgr. schedule).
V. by QSL-Card. Rp.

SOUTHERN EUROPEAN BROADC. SCE. (U.S. Military)

ADDR: APO New York, NY 09221, USA.
L.P: Commander Bernard L. Miles, MAJ, SC.
1st Prgr on 106.0 MHz. **Key Station:** Vicenza (1kW); other st's: Aviano/Livorno/Verona/San Vito (all 0.25kW), Napoli (1kW), Mt. Virgine/Rimini/Decimomannu/La Maddelena (all 0.05kW).
2nd Prgr on 107.0MHz (fully automated stereo easy listening sce). **Key Station:** Vicenza (1kW); other st's; Aviano/Livorno/Sigonella/La Maddelena (all 0.25kW), Napoli (1kW), San Vito (0.05kW).

LATVIA

L.T: UTC + 2h (Su: UTC + 3h) — **Pop:** 2.750.000 — **Radio sets:** n/a — **Pr.L:** Latvian, Russian — **E.C:** a/c 50, 220V (yes) — **ITU:** LVA.

LATVIJAS RADIO

ADDR: Doma laukumā 8, LV-6935 Rīgā. **Te:** +97 (2) 206788.
L.P: Chmn R. TV Committee: Zigmunds Skujins. Dir. Gen: Arnolds Klotiņš.
Mediumwave stations:

Location	kHz	kW	Prgr	Location	kHz	kW	Prgr
Rīga	576	500*	1	Liepāja	1422	7	1
Daugavpils	576	7	1	Rēzekne	1422	50	1
Rīga	945	150**	2	Valmiera	1422	50	1
Kuldīga	1071	30	2	Ventspils	1485	0.2	2
Rēzekne	1071	50	2	Cesvaine	1539	7	2
Valmiera	1071	50	2	Daugavpils	1539	7	2
Cesvaine	1350	50	1	Liepāja	1539	7	2
Kuldīga	1350	50	1	Ventspils	1602	0.2	1

*currently running at reduced powers.
**r. inactive.
Shortwaves: Rīga (Ulbroka, G.C: 24.16E/56.57N) 5935kHz 100kW, beams 250/300°: Prgr 1 + External sce's 1600-2300.

FM (MHz)	**1**	**2**	**3**	**R**	**kW (ERP)**
Cesvaine	67.85		66.29		12
Daugavpils	68.40		71.40		10
Dundaga	68.60		71.16		10
Kuldīga	66.45		69.475		10
Liepāja	68.325		71.375		10
Rēzekne	66.95	71.12	68.78	69.62	12
Rīga	68.15	71.72	69.71	72.50	30
Valmiera	68.27		66.68		12
Ventspils	70.91		72.90	71.84	10

D.PRGR: Prgr 1: 0400-2300 in Latvian. Stereo on FM — **Prgr 2:** 0400-2300 in Latvian, Russian; incl. minority languages. Stereo on FM — **Prgr 3:** 0400-2300 Mon-Fri only. 0400-1600 rel. prgr 2. Stereo on FM — **Prgr's in minority languages** (Estonian, Belorussian, Azeri, Lithuanian): Mon-Fri 1515-1600 on prgr 2 — **Prgr R:** Relay of R. Rossii (Moscow, Russia) 0400-2300 — R. Free Europe relay 1600-1700 on prgr 2 — Voice of America relay 1700-1730 on prgr 2.
ANN: "Raida Rīga" or "Runa Rīga", "Latvijas Radio pirma/otra/treša programma" (prgr 1/2/3).
PUB: "Rīgas viļņi" (weekly for Home Sce.).
INT-SIG: Melody from national Latvian song "Pūt, veijņi!"

FOREIGN SERVICE

Radio Rīga International — **Addr:** P.O. Box 266, Riga.
Stations: 5935kHz, as Home Sce.

English
0700-0730 (Sun) 5935kHz
1830-1900 (Sat) 5935kHz
2130-2135 (Mon-Fri) on prgr 1

German
2135-2140 (Mon-Fri) on prgr 1

Latvian
0530-0600 5935kHz
2030-2100 5935kHz

Swedish
1930-2000 (Sat/Sun) 5935kHz
2130-2200 (Sat/Sun) 5935kHz

ANN: E: "This is R. Rīga"; Swedish: "Här är Radio Rīga, Lettland". **V.** by QSL-card.
PUB: Free sched. on request.

Radio AA (comm.)

ADDR: P.O. Box 271, LV-6098 Rīga-Centrs. **Te:** +371 (2) 612285 — **STATION:** Rīga 102.7MHz.
D.PRGR: 0300-0200 own prgr's in Latvian, rel. R. ROKS in Russian and E. from Oslo (cf. Russia) and VOA Europe in night prgr.

LIEPĀJAS RADIO (municipal)

ADDR: Graudu ielā 45, LV-9700 Liepāja. **Te:** +371 (54) 23310.
STATION: Liepāja 71.375MHz.
D.PRGR: 0800-0900, 1100-1200 (exc. Mon, Thurs), 1730-1800 in Latvian — **F.PI:** Night prgr.

RADIO DEJAS (comm.)

ADDR: Terbatas ielā 6/8-81, LV-6050 Rīga. **Te:** +371 (2) 280698.
LP: Dir. Tec: Janis Kraukliņs.
STATION: Rīga 68.93MHz.
D.PRGR: 24h in Latvian, also rel. prgr's from R. Netherlands.
F.PI: High power FM tr.

RADIO SIGULDA (comm.)

ADDR: Siguldas Kulturas Nams, LV-9050 Sigulda. **Te:** +371 (49) 72678 — **STATION:** Sigulda 70.71MHz.
D.PRGR: 24h in Latvian.

LITHUANIA

L.T: UTC + 2h (Su: UTC + 3h) — **Pop:** 3.725.000 — **Radio sets:** 1.420.000 — **Pr.L:** Lithuanian, Russian, Polish, Byelorussian — **E.C:** a/c 50, 220V (yes) — **ITU:** LTU.

RYŠIŲ IR INFORMATIKOS MINISTERIJA (Ministry of Telecommunications and Informatics)

ADDR: Vilniaus 33, LT-2008 Vilnius. **Te:** +7012 (2) 620443. **Telex:** 261166 ptt lt. **Fax:** +7012 (2) 624402 or 225070 — **L.P:** Minister: Gintautas Zintelis. Dir. Foreign Relations Dept: Antanas Burkauskas.

LIETUVOS RADIJAS IR TELEVIZIJA (Gov.)

ADDR: Konarskio 49, LT-2674 Vilnius MTP. **Te:** +7012 (2) 633182. **Telex:** 261151 litvr. **Fax:** +7012 (2) 221571 — **L.P:** Dir. Gen: Skirmantas Valiulis. TD: Juozas Neverauskas.
Parent organization for Lietuvos Radijas, Lietuvos Televizija and LRT Kauno Programų Direkcija.

LIETUVOS RADIJAS (Gov.)

ADDR: Konarskio 49, LT-2674 Vilnius MTP. **Te:** +7012 (2) 634471. **Telex:** 261151 litvr. **Fax:** +7012 (2) 221571 — **L.P:** Dir: Virginijus Mičiulis. **Networks:** I = Prgr. 1/FS/Night Prgr/VOA, II = Prgr. 2/Prgr. 1/VOA/R. Free Europe, III = Radiocentras/R. st. M-1, IV = R. st. Majak (RUS).
MEDIUMWAVES: tr's operating on the same freq. are synchronized.

Location	kHz	kW	Prgr	Location	kHz	kW	Prgr
Vilnius	612	50	IV	Bubiai	1107	7	II
Sitkūnai	666	500	I	Visaginas	1107	1	II
Sitkūnai	1107	150	II	Pažagieniai	1134	1	II
Vilnius	1107	30	II	Vilnius	1557	25	I*
Viešintos	1107	25	II	Viešintos	1557	25	I*
Klaipėda	1107	25	II	Klaipėda	1557	25	I*
Druskininkai	1107	7	II	*) exc. Night Prgr.			

SHORTWAVES: Sitkūnai (G.C: 23.49E/55.02N): 9710kHz 50kW 259°. **Relay:** Balašiha, Russia (G.C: 37.54E/55.47N): 6010/9675kHz 260°.

FM (MHz)	I	II	III	IV	kW (ERP)
Bubiai	71.15s	70.07s	72.71	71.93	12
Juragiai	70.64s	72.98s	72.20	71.45	12
Klaipėda	67.13	66.35	69.11	67.91	12
Visaginas	69.50	67.01s	68.24	70.52s	12
Tauragė	66.89	68.51	—	—	9/8.1
Viešintos	68.03	66.47	—	—	12
Vilnius	69.80	72.59s	—	71.03s	12

s = stereo (polar modulation system).

HOME SERVICE

Prgr. 1: 0400-2200 (exc. for relays, cf. below) on 666/1557/9710kHz and FM I. **Rel.** via Balasiah: 0400-1700 on 6010kHz, 1700-2200 on 9675kHz. **N:** 0403, 0500, 0600, 0700(Sun), 0800, 0900, 1000, 1100, 1200, 1400, 1500, 1700, 1800, 1900, 2000. **Russian:** 1300-1400. **Polish:** 1600-1630. **Byelorussian:** 2005-2015 (N).
Prgr. 2: 0445-2100 on 1107/1134kHz and FM II. **N:** 0600, 0800, 1000, 1100MF, 1200(Sun), 1300W, 1400, 1500MF, 1600SS, 1830. **Russian:** 1700-1730. **Byelorussian:** Sat 0930-1000. **Yiddish:** 1st Sun 0930-1000. **Ukrainian:** 2nd & 4th Sat 1230-1300. **Tartar/Russian:** 1st & 3rd Sat 1230-1300.
Night Prgr: 2300-0400 on 666kHz & FM I. 2300-2400, 0030-0200, 0230-0400 on 9710kHz.
DX-PRGR: in Russian: Sun 1350-1400.
RELAYS: Voice of America in Lithuanian: 0500-0515 on Prgr. 2, 2030-2100 on Prgr. 1. **Radio Free Europe** in Lithuanian: 2100-2200 on Prgr. 2. **Blagovest** in Russian: 1930-2000, Sat also 0830-0900 on Prgr. 2. **Vilniaus Varpas** 1200-1300 (MF) on Prgr. 2.
ANN: "Kalba Vilnius", "Lietuvos Radijo pirmoji programa" (Prgr. 1), "Lietuvos Radijo antroji programa" (Prgr. 2), "Lietuvos Radijas, Nakties programa" (Night Prgr) — **PUB:** "Kalba Vilnius" and "RTV bangos" (weekly in Lithuanian), "Programmy TeleRadio" (weekly in Russian) — **INT-SIG:** 10-bar orchestral melody. **V:** by QSL-card or letter (B). Rec. acc. Rp (2 IRC). Re. to be addressed to the DX-editor S. Žilionis.

FOREIGN SERVICE (Radio Vilnius)

ADDR: see home sce — **LP:** Editor-in-chief: Audrius Braukyla.
MEDIUMWAVE: see home sce. **SHORTWAVES:** as home sce. B = rel. via Balašika (Russia). E = rel. via European site (Russia). F = rel. via Komsomolsk-na-Amure or Petropavlovsk-Kamchatskiy (Russia's Far East).

English		
2230-2258	Eu	666, 1557, 9675B, 9710 + FM I.
0000-0028	Eu & No. Am	7150E or 7400E, 9710, 17605F, 17690F.
Lithuanian		
2200-2228	Eu	as English 2230.
0200-0228	Eu & No. Am	9710, 17605F, 17690F.

DX-PRGR: in English: alt. Mon. 2253, rep. following Tues 0023.
ANN: E: "This is R. Vilnius". Lithuanian: "Vilniaus Radijas užsieniui". **INT-SIG:** Lithuanian: song "Lietuviais esame mes gimę" ("We Were Born Lithuanians"). English: melody played on kanklės (Lithuanian national instrument — **V:** by QSL-card (B). Rec. acc. IRCs appreciated. Re. to be addressed to the respective language dept.

LRT KAUNO PROGRAMŲ DIREKCIJA (Gov.)

ADDR: Daukanto 28a, LT-3000 Kaunas. **Te:** +7012 (7) 229461. **Fax:** +7012 (7) 205566 — **L.P:** Dir: Raimondas Šeštakauskas.
D.PRGR: Fri 0830-0900 on Lietuvos Radijas Prgr. 1. **ANN:** "Pusvalandis Kaune". **V:** does not verify.

MAJAK RELAY

ADDR: see Russia.
D.PRGR: Russian: 24h on 612kHz & FM IV.

PRIVATE COMMERCIAL STATIONS

RADIOCENTRAS

ADDR: P.d. 1792, LT-2019 Vilnius. **Te:** +7012 (2) 429463. **Telex:** 261143 losc. lt. **Fax:** +7012 (2) 220172. — **L.P:** Editor-in-chief: Rimantas Pleikys.
MEDIUMWAVE: Vilnius 1485kHz 3.5kW, Sitkūnai 1458kHz 12kW.
FM: Vilnius 71.81MHz 12kW (ERP) stereo (polar modulation), 101.0MHz 275W stereo (pilot tone).
SHORTWAVE: Sitkūnai (G.C: 23.49E/55.02N): 9710kHz 50kW 259°.
D.PRGR: Lithuanian: 24h on 101.0MHz, 0600-2200 on 1485kHz & 71.81MHz, 1100-1900 on 1458kHz & FM III. **N:** on the h exc. 9000, 1200, 1500, 1800 & 2200-0500.
On 9710kHz ("RC-Studio"): English: last Sat 0700-0735. **German:** last Sat 0735-0750. **Esperanto:** last Sat 0750-0800.
DX-PRGR: in English: last Sat 0700.
ANN: Lith.: "Kalba Radiocentras". **E:** "This is RC-Studio of Radiocentras, Lithuania". **G:** "Aus Vilnius, Hauptstadt der Republik Litauen, sendet RC-Studio von Radiocentras". **Esperanto:** "Parolas

RC-Studio de Radiocentras". **V:** by QSL-card or QSL-letter (B). Rec. acc. Rp (2 IRC). — **F.PI:** Kaunas 107.1MHz 1kW, Vilnius 105.6MHz 1kW. VOA Europe rel. in Vilnius on FM.

M-1
ADDR: A.d. 1747, LT-2019 Vilnius. **Te:** +7012 (2) 428370. **Fax:** +7012 (2) 421327 — **L.P:** Dir: Hubertas Grušnys.
STATIONS: Vilnius 73.34MHz 12kW (ERP) stereo (polar modulation), 106.4MHz 1kW (ERP) stereo (pilot tone), Kaunas 106.6MHz 1kW.
D.PRGR: Lithuanian: 24h on 106.4MHz, 0600-2200 on 73.34MHz, 0600-1100 & 1900-2200 on 106.6MHz & FM III. **N:** MF: 0600, 0700, 0800, 1500, 1600, 1700, 1800, 2000, 2100; SS: 1300, 1700, 2000. **English:** Tue/Thu 0800-1100, Mon/Wed/Fri 1200-1400, Thu 2100-2115, Sat 0845-0900 (15 min prgrs are prepared by the BBC).
ANN: "M-vienas" (jingle) — **V:** occasionally by letter (A) — **F.PI:** Own FM tr's in Klaipėda, Šiauliai, Panevežys and Anykščiai.

VILNIAUS VARPAS
ADDR: P.d. 2167, LT-2017 Vilnius. **Te:** +7012 (2) 222113, 222150. **Fax:** 223539. **L.P:** Dir: Kęstutis Sakalauskas.
STATION: Vilnius 69.05MHz 12kW (ERP).
D.PRGR: Lithuanian 0000-1200 on 69.05MHz, 1200-1300MF on 1107/1134kHz & FM II, exc. **Polish:** Mon/Wed/Fri 1030-1100. **Russian:** MF 0730-0800. **N:** on the h.
ANN: "Radiofonas Vilniaus Varpas — **V:** by letter (B). Rp. (1 IRC) — **F.PI:** several FM tr's in Lithuania.

ZNAD WILII
ADDR: Laisvės pr. 60, LT-2019 Vilnius. **Te:** +7012 (2) 429457. **Fax:** +7012 (2) 429465 — **L.P:** Dir: Walery Tankiewicz.
STATION: Vilnius 69.05MHz 12kW (ERP).
D.PRGR: Polish: Mon-Fri: 1200-1400, 1500-1600, 1700-2400. **N:** 1200-1230, 1300-1330, 1530MF, 1800, 1900, 1930. **Lithuanian:** 1400-1500 (MF), 1500-1600 (SS). **N:** 1400MF, 1430MF, 1500SS, 1530SS. **Russian:** 1600-1700. **N:** 1600-1630.
RELAYS BBC World Service in Polish: 1400-1410SS, 1500-1510MF, 1700-1730, 2100-2110, 2150-2200MF.
ANN: Polish: "Tu Radio Znad Wilii". Lith.: "Kalba radijo stotis Znad Wilii". Russian: "Govorit radiostancija Znad Wilii". **V:** by letter. Rp (1 IRC).

TITANIKA
ADDR: Radastų 14, LT-3000 Kaunas. **Te:** +7012 (2) 220213, 221431. **Fax:** +7012 (2) 706642 — **L.P:** Dir: Tomas Stasiulaitis.
STATION: Kaunas 73.73MHz 30W.
D.PRGR: Lithuanian 0600-0100 (SS -0300); partly in **English** Sat 1400-1430. **N:** on the h. exc. 1500, 1700, 1800, 1900.
ANN: Nepriklausoma radijo stotis iš Kauno — Titanika. **V:** by letter.

LUXEMBOURG

L.T: UTC + 1 h (Su: UTC + 2h) — **Pop:** 393.000 — **Radio sets:** 230.000 — **Pr.L:** Luxembourgian, French, German — **E.C:** a/c 50, 110/220V — **ITU:** LUX.

RADIO-TÉLÉ-LUXEMBOURG (Comm.)
ADDR: Head Office: RTL, 45 boulevard Pierre Frieden, L-1543 Luxembourg. **Te:** +352 421421. **Fax:** +352 4214 22756. **French Sce:** RTL, 22 Rue Bayard, F-75008 Paris, France.
L.P: Pres. Dir. Gen: G. Thorn. Dir. Gen. Asst: J. Felten. Dir. French Prgrs: Ph. Labro. Dir. German Prgrs: Arno Müller. Editor-in-Chief Luxembourg Sce: R. Rauchs.
STATIONS: Junglinster 234kHz 2000kW; Marnach 1440kHz 1200kW; Junglinster 6090kHz 500kW, 15350kHz 10kW.
FM: Marnach 88.9/97.0MHz 100kW, 92.5MHz 50kW — Dudelange 93.3MHz 100kW. **F.PI:** 100.7/107.7MHz 100kW — relay st's: see "France".
LOCAL SCE. on 92.5MHz: W 0525-0800, 1030-1930; Sun 0700-1930 — **English:** W 1300-1630 (Sat 1600) **German:** W 0800-1030, D 1930-2400. Other times in Luxembourgian
French: 24h on 234/15350kHz. + FM netw. in France.
German: 0400 (Sun 0600)-2400 on FM, s/on-1800 (Sun 1700) & 1900 (Sun 1815)-2400 on 1440kHz, 0600-2400 on 6090kHz.
Italian: W 1830-1845, Sun 1800-1815 on 1440kHz (prgrs produced by RAI, Italy).
Religious Prgrs: 1440kHz: W 1845-1900; **6090kHz:** W 0500-0515, 2100-2315; Sun 0500-0700, 2130-2250. **NB:** Times subject to variation.
ANN: Fr: "Ici R. Luxembourg". E: "This is R. Luxembourg" or "This is RTL International". Dutch: "Hier is R. Luxembourg". G: "Hier ist R. Luxembourg". Luxembourgian: "Hei ass R. Letzeburg" — **V.** by QSL-card.

MACEDONIA

L.T: UTC + 1h (Su: UTC + 2h) — **Pop:** 2m — **Pr.L:** Macedonian — **E.C:** a/c 50, 220V.

RADIOTELEVIZIJA MAKEDONIJE (Gov.)
ADDR: Goce Delčev bb, 91000 Skopje. **Te:** +38 (91) 227-711. **Fax:** +38 (91) 235-767. **Telex:** 51157 — Dir.Gen: T. Ilievski.

Mediumwaves	**kHz**	**kW**		**kHz**	**kW**
Strumica	567	10n	Bitola	1323	10n
Skopje 1	810	1000	Delčevo	1323	10n
Gevgelija	936	10n	Debar	1485	1j
Ohrid 1	1242	5j	Kičevo	1485	1j
Kriva Palanka	1197	1n	Berovo	1485	1n
Skopje 2/3	1314	100	Debar	1602	1k
Ohrid 2/3	1314	10k	Kičevo	1602	1k
Gostivar	1323	10k	Resen	1602	1n

j) rel. Skopje 1 — k) rel. Skopje 2/3 + local prgr — n) rel. Skopie 1 + local prgr.

FM (MHz)	**I**	**II/III**	**kW**
Belasica I	91.5	+97.8	10
Boskija	95.3	98.1	10
Crn Vrr	97.3	+94.1	100
Mali Vlaj	93.3	97.7	10
Pelister	92.3	+96.1	20
Turtel	93.3	+90.5	50

+13 st's less than 10kW — +) stereo.
R. Skopje 1: 24h.
R. Skopje 2: 0500-1900.
R. Skopje 3: 1900-2200.

Local Stations	**kHz**	**kW**		**kHz**	**kW**
Štip	639	1	Kratovo	1386	1
Struga	720	1	Probištip	1431	1
Tetovo	738	1	Kavadarci	1485	1
Kumanovo	945	2	Sveti Nikole	1530	1
Skopje	1026	5	Negotino	1548	1
Kočani	1287	1	Prilep	1584	1
Titov Veles	1341	1	Radovis	1584	1

+ 5 sts. less than 1kW.

MALTA (State of)

L.T: UTC + 1h (Su: UTC + 2h) — **Pop:** 360.000 — **Radio sets:** 90.000 — **Pr.L:** English, Maltese — **E.C:** a/c 50, 240V — **ITU:** MLT.

MALTA BROADCASTING AUTHORITY
An independent Authority to regulate and supervise sound and television broadc. in Malta with power to prescribe and produce prgr's and to select and appoint broadcasting licenses and contractors who will operate radio (at national and community levels) and television services throughout Malta.
ADDR: National Road, Blata-l-Bajda. **Te:** +356 247908. **Fax:** +356 240855. **Cable:** BROADCASTS MALTA.
L.P: Chief Exec: A.J. Ellul. Admin. Officers: H. Zammit Cordina (Secr), M. Camilleri.

PUBLIC BROADC. SCES. Ltd.
ADDR: P.O. Box 82, Valletta. **Te:** +356 225051. **Telex:** MW1443. **Fax:** +356 244601.
STATIONS: 999kHz 5kW — **FM:** 102.3MHz 4kW.
R. Malta 1: 0500-2200 on 999kHz. **N:** 0600, 0700, 0900, 1100, 1300, 1500, 1700, 2000.
R. Malta 2: 0500-2200 on 102.3MHz. **N:** 0600+, 0700, 0800+, 1100, 1500+, 1700, 2000+ — +) BBC.

COMMERCIAL STATIONS

	MHz	**Name**	**Location**
1)	92.7	Super One Radio	Marsa

2)	100.2	Radio One Live	Floriana
3)	101.0	Radio 101	San Gwann
4)	101.8	Island Sound Radio	Floriana
5)	102.6	Bay Radio	St. Julians
6)	103.0	RTK	Blata l-Bajda
7)	104.6	Smash Radio	Paola

OTHERS

VOICE OF THE MEDITERRANEAN (Gov.)

ADDR: P.O. Box 143, Valletta. **Te:** +356 248080, 240421. **Fax:** +356 241501.
L.P: Man. Dir: Richard Vella Laurenti. Dep. Man. Dir: Mohamed Sweidan. Head of N. & Prgrs: Atayalla Mezuchi.
STATION: Cyclops 1557kHz 600kW — 9765/11925kHz 250kW.

Arabic	kHz	English	kHz
0700-0800	9765, 1557	0600-0700	9765, 1557
1500-1600	11925	1400-1500	11925

VOM is a joint Maltese/Libyan operation — **F.Pl:** French sce.

DEUTSCHE WELLE RELAY MALTA

STATION: (G.C: 14.34E/35.50N): 1557kHz 600kW + 3 × 250kW SW-tr's.

kHz	kHz	kHz	kHz	kHz	kHz
6025	7250	9545	9735	11915	15135
6040	7265	9565	9770	13780	15185
6085	7270	9625	11785	13790	15595
7235	9515	9640	11865	15105	21650

D.PRGR: See Germany.

MOLDOVA (Moldavia)

L.T: UTC +2h (Su: UTC +3h) — **Pop:** 4.370.000 — **Pr.L:** Romanian, Russian, Ukrainian, Gagauz — **E.C:** a/c 50 127/220V — **ITU:** MDA

RADIOTELEVIZIUNEA NAŢIONALĂ (Gov.)

ADDR: Soseaua Hînceştilor 64, 277028 Chişinău. **Te:** +7042 (2) 721388. **Fax:** +7042 (2) 723537. **Tlx:** 163130 RADIO SU. Note: The telephone country code will be changed to 373.
L.P: Gen. Dir: Konstantin Pyrcak.

Stations	kHz	kW	H. of tr.	Prgr.
Grigoriopol	234	1000	0200-2400	MR1, RR
Grigoriopol	594	500	0000-1630	RSM
Chişinău	873	150		
Grigoriopol	999	300	0400-2200	RN1 (T, MR1)
Chişinău	1449	30	0400-2300	RN2 (RN1)
Grigoriopol	1467	500	0400-2300	RN2 (T, MR1)
Cahul	1494	30	0400-2300	RN2
Edineţ	1494	30	0400-2300	RN2
Chişinău	1593	5	0200-0100	RSM (RN2)

RN2 also rel. via Chernivtsi, Ukraine on 1449kHz.
NB: Powers are estimates only. Due to a take-over of the Grigoriopol facility by separatists, temporary changes as shown within brackets were in force at the time of editing. T = prgrs from separatist studios in Tiraspol. RN1 temporarily also rel. via Iasi, Romania, on 1053kHz 0400-0715, 1000-1400, 1900-2200.

FM:	RN1	RN2	MR1	RSM$
Bălţi	66.68			68.24
Cahul	67.37			69.14
Edineţ	67.46			70.035
Străşeni	72.02	66.80$	67.58	68.48
Ungheni	68.00			69.53

D.PRGR: RN1 (Programul 1): 0400-2300 in Romanian, exc. Russian: D 0700-0715 (Mon-Fri 0730), 1300-1320, 1500-1515, 1900-2045; Ukrainian: 1830-1900 — **RN2** (Luceafărul): 0400-2200 in Romanian — **MR1** = Moscow Ostankino R-1 — **RSM** = Moscow R. St. Mayak — $ = stereo.
ANN: Romanian: "Aici" + name of sce.
V. Rarely verifies. **PUB:** TeleRadio (weekly for Home Sce.).

FOREIGN SERVICE: Radio Moldova International

ADDR: as above
English: Planned service. Postponed due to the loss of control of the Grigoriopol facility.

RADIO MOSCOW RELAY

Mediumwaves: Grigoriopol (Mayak) 1548kHz 500kW (exact power not known).
Shortwaves: Grigoriopol (Mayak: G.C: 47.17N/29.24E) 5×500/1000kW.
Transmissions included in R. Moscow sched.

MONACO

L.T: UTC + 1h (Su: UTC + 2h) — **Pop:** 30.200 — **Radio sets:** 11.000 — **Pr.L:** French — **E.C:** a/c 50, 220V — **ITU:** MCO.

RADIO MONTE CARLO (Comm.)

ADDR: 16 Blvd. Princesse Charlotte, MC-98080 Monaco Cédex. **Te:** +33 (93) 151617, 151780. **Fax:** +33 (93) 151630. **Telex:** 469926.
L.P: Pres: C.C. Solamito. Man. Dir. (French): Jean Noel Tasser. Man. Dir. (Italian): Gerard Frederucci. Dir. Prgrs: Yves Mourousi (French), Novella Masaro (Italian), Thomas Weigt (German). Int. Rel. Dir: Lucien Allayena.
STATIONS: Monte Carlo (G.C: 7.25E/43.37N): 702kHz 300kW — Roumoules, France (G.C: 06.09E/43.47N): 216kHz 1400kW, 1467kHz 1000kW.
FM: 90.3/98.5/102.7/106.5MHz 50kW; 95.4/98.9/102.9/106.85MHz 1kW + 80 relay st's in France (see "France").
French: 24h on 98.5/98.9MHz + FM st's in France, 0400-0005 on 216kHz. **N:** 0500(W), 0530(W), 0600 (Mon-Fri), 0700, 0800, 0900, 1130, 1730, 2130. Summaries on the h. and half h.
Italian: 24h on 106.85MHz, 0530-1830 on 702kHz.
Arabic: 0500-1830 on 1467kHz.
R.M.C. Côte d'Azur: 90.3/93.8MHz.
R. Plus Monte Carlo: 24h in **German** on 95.4MHz.
R.M.C. Musique Classique: 102.7/102.9MHz.
V. by letter.

RIVIERA RADIO (Comm.)

ADDR: 16 Bd. Princess Charlotte, MC-98080 Monaco Cédex. **Te:** +33 (93) 254906. **Fax:** +33 (93) 304245.
L.P: MD & PD: Richard Younge.
STATIONS: FM 104.1/106.3/106.5MHz.
D.PRGR in English: 24h.

TRANS WORLD RADIO (Rlg.)

ADDR: B.P. 349, M-98007 Monte Carlo. **Te:** +33 (92) 16 5600. **Fax:** +33 (92) 16 5601. **European Office:** P.O. Box 2020, 1200 CA Hilversum, The Netherlands. **Te:** +31 (35) 217740. **Fax:** +31 (35) 234861.
MW: Romoules, France (G.C: 07.26E/43.44N): 1467kHz 1000kW — Monte Carlo 702kHz 300kW (cf. Radio Monte Carlo) — **SW** (G.C: 07.26E/43.44N): 2 × 100kW, 1 × 500kW tr's.
L.P: Pres: Dr. Paul Freed. MD: Bill Miall. Mgr. Broadc. Dept: Beate Kiebel. Freq. Coordinator: Bernhard Schraut. Head of PR: Felix Widmer.

kHz	kHz	kHz	kHz	kHz	kHz	kHz
5885	6230	7160	7385	9480	9795	12085
6200	7135	7355	9435	9650	11625	

Albanian	Days	kHz
2015-2030	1234567	1467
1715-1745	4	5885
1730-1745	7	5885
2115-2130	2	1467
Arabic		
0300-0315	1234567	1233
2000-2015	1234567	1467
2040-2145	1234567	1233
2145-2200	67	1467
2200-2300	1234567	1467
Armenian		
0315-0330	1234567	1233
1530-1545	67	12085
1530-1600	12345	12085
Berber		
2130-2145	17	1467
2130-2200	12345	1467
Bulgarian		
1917-1932	34567	7355
Croatian		
1332-1347	7	9490
1700-1715	6	5885
1700-1745	1235	5885
Czech		
0945-1000	46	9490
1129-1159	7	9490
1202-1217	6	9490
1202-1232	7	9490
1700-1745	1245	7385
1715-1745	3	7385
English		
0735-0920	12345	9480
0735-0935	6	9480
0735-0945	7	9480
2300-2320	6	1467
2300-2335	235	1467
2300-2350	14	1467
2300-2400	7	1467
Farsi		
0230-0300	1234567	1233
1600-1630	1234567	12085

French			**Lithuanian**		
0330-0400	1234567	216	1830-1845	23	7385
2000-2015	12345	1467	**Macedonian**		
German			1700-1730	7	5885
0445-0515	1234567	1467	**Norwegian**		
0830-0845	123456	9795	0800-0815	1234567	9650
		7160	**Polish**		
0830-0915	7	9795	1000-1015	124567	9490
		7160	1800-1830	123456	5885
1105-1120	123456	9795	2115-2130	7	1467
		7160	**Romanian**		
1105-1120G	1234567	6230	1845-1915	1234567	7355
1430-1500	1234567	9795	**Russian**		
		7160	1500-1630	7	11625
1830-1845G	1234567	6230	1500-1645	123456	11625
2030-2115	127	1467	1835-1920	36	7135
2030-2130	3456	1467	1835-1935	157	7135
G) for Guest Workers			1835-1950	24	7135
Greek			2000-2030	1234567	7135
2116-2130	27	1467	**Serbian**		
Hebrew			1700-1715	4	5885
2336-2350	5	1467	**Slovak**		
Hungarian			0945-1000	1235	9490
0925-0940	1234567	9490	1129-1159	6	9490
1800-1830	1234567	6200	1700-1715	367	7385
2016-2030	1	1467	**Slovene**		
Italian			1715-1730	6	5885
1830-1900	1234567	702	**Turkish**		
Kurdish			1630-1645	1234567	12085
0215-0230	1234567	1233	**Ukrainian**		
Latvian			1700-1730	67	9435
1640-1655	1567	9435	1700-1745	12345	9435

ANN: E: "This is Monte Carlo" or "You are listening to Monte Carlo. We invite our listeners in the British Isles to stay tuned to the following prgr's of Trans World Radio". F: "Ici Monte Carlo. Trans World Radio vous presente un programme en langue Francaise". G: "Hier ist Monte Carlo von Trans World Radio hören Sie ein Programm in Deutscher Sprache" — **INT-SIG:** Musical Box.

V. by QSL-card. (B.) **PUB:** MW & SW schedules available on request.

THE NETHERLANDS

L.T: UTC + 1h (Su: UTC + 2h) — **Pop:** 15m — **Radio Sets:** 12m — **Pr.L:** Dutch — **E.C:** a/c 50, 220V — **ITU:** HOL.

NEDERLANDSE OMROEPPROGRAMMA STICHTING (NOS)

ADDR: P.O. Box 444, 1200 JJ Hilversum. **Te:** +31 (35) 77 92 22 (TV/Radio) **Telex:** 43287 (TV/Radio) **Fax:** + (35) 77 26 49
L.P: Chmn: M. de Jong; Vice Chmn: A.H. van den Heuvel; Dir. Radio: W.J.M. van Beusekom; Press & PR: R. Reyners
Board Members: A. Grewel; M.J.E.M. Jager; G.W. van Montfrans-Hartman; W.M.C. de Vreij-Vringer; Drs. J.P. van der Reijden (VOO); Mr. A. Herstel (NCRV); A. van der Veer (EO); Mr. A.J. Heerma van Voss (VPRO); G.J.M. Braks (KRO); G.C. Wallis de Vries (AVRO); M. van Dam (VARA); *(TROS).
*) C. Wolzak left the TROS, therefore this organization was temporarily not represented at time of editing.
NB: The Dutch prgrs. are provided by the NOS and eight broadcasting organizations:
—Algemene Omroepvereniging AVRO, 's Gravelandseweg 52, 1217 ET Hilversum, Postbus 2 1200 JA Hilversum. **Te:** +31 (35) 717911 **Fax:** +31 (35) 717517
—Vereniging Evangelische Omroep EO, Oude Amersfoortseweg 79a, 1213 AC Hilversum, Postbus 565, 1200 AN Hilversum. **Te:** +31 (35) 882411 **Fax:** +31 (35) 882685
—Katholieke Radio Omroep KRO, Emmastraat 52, 1213 AL Hilversum, Postbus 9000, 1201 DH Hilversum. **Te:** +31 (35) 713911 **Fax:** +31 (35) 237345
—Nederlandse Christelijke Radio Vereniging NCRV, Bergweg 30, 1217 SC Hilversum, Postbus 121, 1200 JE Hilversum. **Te:** +31 (35) 719911 **Fax:** +31 (35) 719285
—TROS, Lage Naarderweg 45-47, 1217 GN Hilversum, Postbus 450, 1200 AL Hilversum. **Te:** +31 (35) 715715 **Fax:** +31 (35) 715236
—Omroepvereniging VARA, Heuvellaan 50, 1217 JN Hilversum, Postbus 175, 1200 AD Hilversum. **Te:** +31 (35) 711911 **Fax:** +31 (35) 711333
—Veronica Omroep Organisatie VOO, Laapersveld 75, 1213 VB Hilversum, Postbus 1234, 1200 BE Hilversum. **Te:** +31 (35) 716716 **Fax:** +31 (35) 249771
—Omroepvereniging VPRO, 's Gravelandseweg 63-73, 1217 EJ Hilversum, Postbus 11, 1200 JC Hilversum. **Te:** +31 (35) 712911 **Fax:** +31 (35) 712254

STICHTING ETHER RECLAME STER

ADDR: Laapersveld 70, 1213 VB Hilversum, Postbus 344, 1200 AH Hilversum. **Te:** +31 (35) 725500 **Fax:** +31 (35) 218940
NB: The STER is the only organization that provides commercials to the other public broadcasters, who are not allowed to carry any commercials themselves.

NEDERLANDSE OMROEP-ZENDERMAATSCHAPPIJ nv NOZEMA

ADDR: Bredewater 8, 2715 CA Zoetermeer, Postbus 104, 2700 AC Zoetermeer. **Te:** +31 (79) 713500 **Fax:** +31 (79) 512537
NB: The NOZEMA owns all Dutch radio- and tv transmitters. Amongst them are the transmitters of all national-, regional-, local- and commercial broadcasters as well as the transmitters of Radio Nederland Wereldomroep.
Mediumwaves:

Loc.	kHz	kW	Prgr.	Loc.	kHz	kW	Prgr.
Flevoland	747	400	Radio 1	Hulsberg	1251	10	Radio 1
Flevoland	1008	400	Radio 5	Lopik	675	120	Radio 3
Hulsberg	891	20	Radio 5				

FM:

Loc.	I	II	III	IV	kW
Arnhem	104.1	—	—	—	50
Goes	104.4	87.9	95.0	99.8	50/15
Haarlem	94.3	—	—	—	10
Hulsberg	105.3	92.1	103.9	98.7	10
Lelystad	103.0	—	—	—	50
Loon op Zand	98.2	—	—	—	40
Lopik	100.1	92.6	96.8	98.9	100
Markelo	104.6	91.4	96.2	98.4	100
Roermond	104.8	88.2	90.9	94.5	100
Rotterdam	90.2	—	—	—	10
Smilde	101.0	88.0	91.8	94.8	100
Wieringermeer	101.6	87.7	89.8	92.2	100

NB: The new Radio 1 network comes into effect as of January 1st 1993.
D.PRGR: Radio 1: 0600-1800 (rel. Radio 2 1800-0600) **N:** on the h.; Radio 2: 24h. **N:** on th h.; Radio 3: 0600(Sun 0700)-2300 **N:** on the h.; Radio 4: 0600(Sun 0700)-2300 **N:** on the h.; Radio 5: 0800-2130 (Sat 2200, Sun 2140, Mon 2100) **N:** various languages.
ANN: "Dit is de VARA.", "Dit is de VPRO.". etc. **V:** by QSL-card.

REGIONAL STATIONS

Omroep Brabant, Heggeranklaan 1, Postbus 108, 5600 AC Eindhoven. **Te:** +31 (40) 11 6262 — **D.Prgr:** 53 hrs, 22 min per week.
Omroep Flevoland, Meentweg 1, Postbus 567, 8200 AN Lelystad. **Te:** +31 (3200) 42221 — **D.Prgr:** 41 hrs, 46 min per week.
Omrop Fryslân, Groeneweg 1, Postbus 642, 8901 BK Leeuwarden. **Te:** +31 (58) 99 7799 — **D.Prgr:** 34 hrs, 56 min per week.
Omroep Gelderland, Rosendaalseweg 704, Postbus 747, 6800 AS Arnhem. **Te:** +31 (85) 71 3713 — **D.Prgr:** 61 hrs, 46 min per week.
Omroep Limburg, Bankastraat 3, Postbus 94, 6200 AB Maastricht. **Te:** +31 (43) 46 7777 — **D.Prgr:** 43 hrs, 43 min per week.
Omroep Zeeland, Kanaalstraat 64, Postbus 1090, 4388 ZH Oost-Souburg. **Te:** +31 (1184) 71820 — **D.Prgr:** 19 hrs, 28 min per hour.
Radio Drenthe, Beilerstraat 24, Postbus 999, 9400 AZ Assen. **Te:** +31 (5920) 100 48 — **D.Prgr:** 45 hrs, 17 min per week.
Radio Noord, Martinikerkhof 23, Postbus 30101, 9700 RP Groningen. **Te:** +31 (50) 18 3456 — **D.Prgr:** 57 hrs, 54 min per week.
Radio Noord-Holland, Lange Begijnestraat 11, 2011 HH Haarlem. **Te:** +31 (23) 15 5755 — **D.Prgr:** 48 hrs, 47 min.
Radio Oost, Beursstraat 2, Postbus 1000, 7550 BA Hengelo. **Te:** +31 (74) 45 6456 — **D.Prgr:** 65 hrs, 45 min per week.
Radio Rijnmond, Delftsestraat 21, Postbus 1515, 3000 BM Rotterdam. **Te:** +31 (10) 43 34433 — **D.Prgr:** 80 hrs per week.
Radio Utrecht, Laan van Beek en Rooyen 45, Postbus 666, 3700 AR Zeist. **Te:** +31 (3404) 35566 — **D.Prgr:** 59 hrs, 58 min per week.

Radio West, Van Vredenburghweg 71, Postbus 1220, 2280 CE Rijswijk. **Te:** +31 (70) 39 93131 — **D.Prgr:** 69 hrs, 23 min per week.

STATIONS:

Loc.	MHz	kW	Pol.	St.
Amsterdam	95.2	1	V	Radio Noord Holland
Arnhem	102.4	0.1	V	Omroep Gelderland
Den Haag	88.4	5	V	Radio West
Goes	101.9	50	V	Omroep Zeeland
Haarlem	97.6	1	V	Radio Noord Holland
Hilversum	93.1	0.1	V	Radio Noord Holland
Hoogezand	97.5	15	V	Radio Noord
Hulsberg	95.3	10	H	Omroep Limburg
Irnsum	88.6	15	V	Omrop Fryslân
Lelystad	102.1	5	V	Omroep Flevoland
Loon op Zand	91.9	2	V	Omroep Brabant
Lopik	100.7	10	H	Radio Utrecht
Losser	89.4	10	V	Radio Oost
Markelo	95.6	5	H	Radio Oost
Megen	89.1	5	V	Omroep Gelderland
Megen 2	95.8	5	V	Omroep Brabant
Middelburg	93.0	0.2	V	Omroep Zeeland
Mierlo	87.6	1	V	Omroep Brabant
Philippine	97.8	10	V	Omroep Zeeland
Roermond	100.3	100	H	Omroep Limburg
Roosendaal	95.4	1	V	Omroep Brabant
Rotterdam	93.4	10	V	Radio Rijnmond
Ruurlo	90.4	10	V	Omroep Gelderland
Smilde	90.8	4	H	Radio Drenthe
Ugchelen	103.5	20	V	Omroep Gelderland
Wieringermeer	93.9	10	H	Radio Noord Holland
Zwollerkerspel	99.4	25	V	Radio Oost

+ 264 local stations.

NB: Since June 1992 terrestrial frqs. have been allocated to a number of commercial satellite radio sts. Allocation of these freqs. after Jan 1st 1993 is subject to government approval.

STATIONS: Radio 10 Gold on 102.5 MHz 50kW (Wieringermeer), 100.4 MHz 50kW (Irnsum), 99.1MHz 1kW (Hoogezand); **Sky Radio** on 102.7 MHz 50kW (Rotterdam); **RTL Radio** on 101.2 MHz 10kW (Den Haag).

RADIO NEDERLAND WERELD OMROEP (RNW)

ADDR: P.O. Box 222, 1200 JG Hilversum. **Office:** Witte Kruislaan 55, 1217 AM Hilversum. **Te:** +31 (35) 724211 **Fax:** +31 (35) 724352 **Telex:** 43336 WOMR NL

L.P: Board of Management: M. Dijkstra (Chmn.) and C. Verheugd; Secr. of the Board: G.R. de Vries; Dir. of Prgr: F.H.J.M. Suasso de Lima de Prado; Dep. Dir. of Prgr: J. Heintz; Dir. Prgr. Export: H. de Wildt; Dep. Dir. Prgr. Export: A. van der Horst; Dir. Eng: G.J.M. Pappot; Gen. Mgr. Training Center: J.R. Swart; Dir. Pub. Rel: J.C. Veltcamp Helbach; Dir. Personnel: P. Aartsen Tuyn; Dir. Planning & Development: B.H. Steinkamp; Eng. Consultant: H. Bakhuizen; Freq. Mgr: J.W. Drexhage.

STATIONS: "Flevo"; 4x500kW (+ 1x 100kW reserve) (G.C: 05.27E/52.21N). "Bonaire RNW" (Neth. Antilles) 2x300kW (G.C: 68.19W/12.13N). "Talata/Volonondry" (Madagascar) 2x300kW (G.C: 37.37E/1845S)

kHz	kHz	kHz	kHz	kHz	kHz
5955	9630	11660	11895	15315	21500
6015	9715	11680	11900	15330	21515
6020	9720	11710	11935	15560	21590
6165	9775	11715	11950	15570	21640
7130	9855	11720	13700	17580	21745
9560	9860	11730	13770	17605	25970
9590	9895	11835	15120	17610	
9605	11655	11890	15150	21480	

Schedule until 28th March 1993

B = Bonaire, F = Flevo, M = Madagascar

Arabic	Area	kHz
1530-1625	ME/NEAf	F13700
	ME	F15560
1730-1925	NWAf	F9860
	NEAf	F11655
Dutch/Papiamento/Sarnami		
0900-0925	Car/SUR	B6020
2130-2225	SUR	B11950

Dutch		
0130-0225	cNAm	B6165
	wSAm	B15315
	eNAm	F6020
	eNAm	F9895
0330-0425	ME	F9855
	EAf	M9895
	ME	M13700
0530-0625	wNAm	B6165
	wNAm	B9715
	ME	F9895
	ME	F13700
0630-0725	NZL/AUS	B9630
	NZL/AUS	B15560
	SEEu	F9895
	SEEu	F11710
	CEu/SEEu	F7130
0630-0825	Eu	F5955
0730-0825	SWEu	F7130
	Eu	F9860
	SWEu	F11935
1030-1125	AUS	B9720
	AUS	B11895
	SEAs	M17580
	EAs/SEAs	M21480
1030-1125s	Eu	F9715
	EAs/SEAs	F21500
	SAs	F21640
	Af	F25970
1330-1425	SEAs	F21480
	WAUS	F21745
	SEAs	M13700
	SEAs	M17580
1430-1525	SAs	F17580
	SAs	F21480
1630-1725	Eu	F5955
	SEEu	F9860
	SWEu	F9860
	SEu	F13700
1630-1725s	SUR	B15120
	eNAm	B17605
	SAf	M6020
	EAf	M11655
1630-1725w	SAf	M6020
	SEAf	M11655
1730-1825	ME	F9895
	SEAf	F13700
2030-2125	WAf	B18605
	WAf	B21590
	SWEu	F6020
2130-2225	sSAm	B15315
	SWEu	F6015
	eNAm	F9895
	Car	F11730
	eSAm	F13700
2330-0025	SEAs	M9590
	SEAs	M11890
English		
0030-0125	eNAm	B11835
0030-0325	SAs	M9860
	SAs	M11655
0330-0424	wNAm	B9590
	wNAm	B11655
0730-0825	NZL/AUS	B9630
0730-1025	NZL/AUS	B11895
0930-1025	NZL/AUS	B9720
1130-1325	Eu	F5955
1330-1625	SAs	F13770
	SAs	F17610
1430-1625	SAs	M9895
	SAs/EAs	M15150
1730-1925	CAf	B21515
	SAf	M6020
	EAf	M9605
1730-2025	WAf	B21590
1930-2025	WAf	B17605
2330-0125	cNAm	B6165
	eNAm	F6020
French		
1830-2125	NWAf/WAf	F9895
	CAf/WAf	M9860
	CAf/WAf	M15570
Indonesian		
1130-1325	INS	F17610
	INS	F21480
	INS	M13700
	INS	M17580
2130-2325	INS	M9590
	INS	M11890
Portuguese		
0030-0125	B	F9895
	B	F11680
2230-2325	B	B15315
	B	B15560
Spanish		
0230-0325	nwSAm	B6165
	wSAm	B15315
	Car	F6020
	nSAm	F9895
0430-0525	CAm	B6165
	MEX	B9590
1130-1155	Car	B6020
	nwSAm	B11660
1200-1225	CAm	B9775
	CAm	B15330
1200-1325	SWEu	F11900
1930-2025	SWEu	F6020
2230-2325	Car	F6020
	eSAm	F9895
	eSAm	F11715
2330-0025	sSAm	B15315
	nSAm	F9895

w = weekdays — s = Sundays

Additional frequencies from Jan 1st 1993

From Jan 1st, additional frequencies will be used from tr's in the CIS. The tentative schedule is as follows:

Dutch	Area	kHz (D92)	kHz (M93)
0800-0925Sun*	CBG	15210C	17560C
1330-1425	EAs/SEAs	9810I	17655I
	FE/EAs	7260P	12065P
1430-1525	SAs	7115C	11695C
2330-0025	SEAs	9855T	15560T
English			
0030-0325	SAs	11675C	15595C
0930-1125	EAs/SEAs	9810I	17655I
	FE/EAs	7260P	12065P
Indonesian			
1130-1325	INS	9810I	17655I
	INS	17655I	21820I
2130-2325	INS	9855T	15560T

Tr. Sites: C = Chita, I = Irkutsk, P = Petropovlovsk-Kamchatskiy, T = Tashkent — *) for Dutch UN forces. D92 = Until 6th March, M93 = from 7th March.

Listener Return Line: 24h automatic answering sce. enables listeners to **English** programmes to phone in comments and questions. Dial country code 31 for Holland, then (35) 724222. **Computer line:** +31 (35) 45395 (300/1200/2400 Baud CCITT or BELL tones, 8-N-1 format).
ANN: Dutch: "Hier is R. Nederland Wereldomroep in Hilversum". English: "This is R. Netherlands, the Dutch International Service". French: "Ici R. Nederland, Hilversum, Hollande". Indonesian: "Inilah Radio Nederland di Hilversum dengan siaran dalam bahasa Indonesia." Spanish: "Transmite R. Nederland desde la ciudad de Hilversum en Holanda". Portuguese: "Esta e a R. Nederland. Emissora International da Holanda". Arabic: "Idha'atu 'l Holanda 'l Arabiya fi Hilversum".
INT-SIG: Traditional Dutch folk-song: "Merck toch hoe sterck" played on Breda carillon.
V. by QSL-card. Emphasis is on feedback to programme content. Reception reports appreciated when following guidelines given in free publication "Writing Useful Reception Reports". Taped reports discouraged.
PUB: Programme schedule covering all language services issued in March & September. Shortwave related pamphlets available in Arabic, French, and Spanish. Researched consumer advice on current SW receivers (English, French & Dutch) and list of radio related publications (English only). All free, ask for "Listeners' Services Catalogue".
TRANSCRIPTION SERVICE
A free sce to foreign st's, issues complete prgrs about The Netherlands, international cooperation, specific information, cultured and tailored to the requirements of the areas served.

AMERICAN FORCES NETWORK (SHAPE)

Stations	MHz	kW		MHz	kW
Schinnen	89.2	0.075	Volkel A.B.	93.6	0.05
Maastricht	93.4	0.025	Soesterberg	98.0	0.025

D.PRGR: 24h. Relays of AFN1-SHAPE & local prgrs.

BRITISH FORCES BROADCASTING SERVICE

STATIONS: Maastricht 87.7/Hoensbroek 90.2 MHz 0.05kW.
D.PRGR: Relays of BFBS Germany.

CANADIAN FORCES NETWORK—BRUNSSUM

STATION: Brunssum 91.5MHz 0.1kW (stereo).
D.PRGR: 0500-2310 rel. CFN or RFC Lahr (Germany), interrupted for local prgr's. **N:** on the h.

NORWAY

L.T: UTC + 1h (Su: UTC + 2h) — **Pop:** 4.230.000 — **Radio sets:** 3.300.000 — **Pr.L:** Norwegian — **EC:** a/c 50, 230V (yes) — **ITU:** NOR.

TELEDIREKTORATET

ADDR: Postboks 6701, St. Olavs Pl., N-0130 Oslo 1. **Cable:** Gentel. **Te:** +47 (2) 488990. **Telex:** 7 1203 gentl.

NORSK RIKSKRINGKASTING

A non-comm. enterprise operated by an indep. public foundation.
ADDR: Bjørnst. Bjørnsons Pl. 1, N-0340 Oslo 3. **Te:** +47 (2) 45 90 50. **Fax:** +47 (2) 45 7440.
Telex: 76820. **Cable:** Studio OSLO. **L.P:** Dir. Gen. (Radio and TV): Einar Førde. Dir. Adm. and Finance: Oddvar Svanevik. Ag. Tech. Dir: Dagfinn Bernstein. Head of Prgrs. (Radio): Tor Fuglevik. Head of Prgrs. (TV): Andreas Skartveit. Press and Inf. Officer: Mrs. H. Løchstøer.
LONG & MEDIUMWAVES (Prgr. 1):

	kHz	kW		kHz	kW
Tromsø	153	10	Bodø	675	10
Oslo	216	200	Vadsø	702	20
Vigra	630	100	Kvitsøy	1314	1200

FM (MHz)	P1	P2	kW		P1	P2	kW
Alta	89.7	94.6°	3.5	Gamlems-			
Bagn	91.7	95.3	30	veten	91.9	96.3	60
Bergen	89.1	94.8	60	Gausta	89.5	96.4	10
Bjerkreim	94.2	98.7	60		99.7	—	6
Bokn	93.5	97.3	100	Greipstad	88.8	92.4	30
Bremanger	93.6	98.1	30	Grong	91.9	96.6	60
Førde	92.8	88.7	15	Gulen	88.0	94.5	30

Hadsel	92.4	99.3	30	Mosvik	90.9	98.4	30
Halden	89.1	94.8	30	Narvik	88.8	98.9°	30
	94.1	—	75	Nordfjordeid	89.4	99.3	15
Hammer-				Nordhue	87.6	97.1	60
fest	96.6	87.7°	30		90.4	—	4
Hemnes	88.5	99.8	30	Nordkapp	89.2	95.4°	15
Hovdefjell	87.8	93.7	25	Oslo	88.7	100.0	100
Jetta	95.9	99.5	30	Reinsfjell	89.1	95.1	30
Karasjok	87.9	94.7°	3	Salten	93.3	95.5	30
Kautokeino	90.3	93.8°	35	Skien	88.2	92.3	30
Kistefjell	91.8	95.7°	30	Sogndal	91.5	95.1	30
Kongsberg	91.3	95.5	60	Steigen	90.3	97.8°	30
Kongsvinger	89.8	93.9	30	Stord	96.0	99.6	60
	96.1	—	11	Store Jaekkir	99.9	97.3°	1.15
Kopparen	88.3	94.5	30	Tana	92.5	97.0°	30
Lyngdal	97.6	88.3	30	Trolltind	88.2	94.0°	30
Lyngen	93.3	97.5°	4.2	Tron	98.3	88.6	30
Lønahorgi	93.3	88.3	60	Varanger	88.2	91.8°	30
Melhus	92.4	97.2	60	Vega	89.3	95.2	33

+ more than 1400 low power tr's.
°) carries Lappish sce. All tr's are stereo.
1st Prgr: 24h. **N:** 0500 (W), 0600, 0630 (W), 0700, 0800, 0900, 1130, 1400, 1530, 1730, 2100, 2300 — **Wrp:** 0500 (W), 0600 (Sun), 0700 (W), 0800 (Sun), 1400, 1730, 2050. **Rlg:** Sunday service 1000. Morning services W 0715.
Regional: See below.
Special prgrs on 1314kHz: 1st Prgr, with rel. of NRK Rogaland regional prgr. in mornings, also 0902 & 1404. **Wrp:** 1100-1115.
Rel. R. Norway Int: 1600-1700.
2nd Prgr: 24h. Rel. 1st Prgr 2300-0500 (Sun 0600). **N:** Mon-Fri on the h; Sat 0800, 0900; Sun 1130, 1400, 1530, 1730. **Rlg:** Sunday service 0830.
Oslo 93.0MHz 0.9kW: rel. Overseas Sce. plus special prgrs in Turkish (Sun 0900-0945, Wed 1545-1600), Urdu (Sat 1400-1445, Mon 1545-1600), & Lappish Prgr.
ANN: "Dette er Norsk Rikskringkasting". Lappish: "Datlae Sámeradio, Kárássjagás".

REGIONAL SERVICES

D.Prgr for all regional outlets (unless otherwise stated): Mon-Fri 0535-0600, 0603-0625, 0650-0700, 0902-0905, 1100-1115 (or 1130), 1404-1407, 1600-1700 (or 1730); Sat 0602-0625, 1100-1130.
NRK Finnmark, Postboks 190, 9801 Vadsø — 702kHz + 88.2/88.8/89.2/89.7/90.3/92.5/96.6MHz. — **V.** by letter.
NRK Troms, Postboks 220, 9001 Tromsø — 153kHz + 88.2/91.8MHz — **V.** by QSL-card.
NRK Nordland, Postboks 351,8001 Bodø — 675kHz + 88.5/88.8/89.3/90.3/92.4/93.3MHz. — **V.** by QSL-folder.
NRK Helgeland, Postboks 20, Mo i Rana — 88.5/89.3MHz — **V.** by QSL-folder.
NRK Nord-Trøndelag, Postboks 2,7701 Steinkjer — 90.9/91.9MHz — **V.** by letter.
NRK Sør-Trøndelag, Postboks 2250,7001 Trondheim — 88.3/92.4MHz — **V.** by QSL-card.
NRK Møre og Romsdal, Parktg.3 VI,6000 Ålesund – 630kHz + 89.1/91.9MHz —**V.** by QSL-card.
NRK Sogn og Fjordane, Postboks 100,6801 Førde — 88.0 89.4/91.5/92.8/93.6MHz. Mon-Fri 1100-1115 & 1600-1700, Sat 1100-1130 — **V.** by letter.
NRK Hordaland, Postboks 727,5001 Bergen — 89.1/93.3/96.0MHz. All times exc. Sat 0545-0600 — **V.** by letter.
NRK Rogaland, Postboks 17,4001 Stavanger — 93.5/94.2MHz — **V.** by letter.
NRK Sørlandet, Postboks 564, Kristiansand S — 87/8/88.8/97.6MHz — **V.** by QSL-card.
NRK Telemark, Borgeåsen, 3900 Porsgrunn — 88.2/89.5MHz. All time exc. Sat 0545-0600.
NRK Oppland, Postboks 340,2601 Lillehammer — 90.4/95.9/97.1MHz — **V.** by letter.
NRK Hedmark, Postboks 216,2401 Elverum — 87.6/89.8/98.3MHz. As listed, but s/on 0530 — **V.** by letter.
NRK Øsfold, Tøjhusgt.43/44,1600 Fredrikstad — 89.1 MHz. H. of tr. as above + 1430-1530MF — **V.** by QSL-card.
NRK Østlandssendingen, Staffelds Gt.2,0166 Oslo 1 — 88.7/87.8/91.9/96.1MHz. On 87.8/91.9MHz: Mon-Fri 0505-0800, 0902-0905, 1100-1115, 1400-1530, 1600-1730. On 88.7/96.1MHz: 0505-0600, then as indicated above — **V.** by QSL-card.
NRK Buskerud, Box 13, 3057 Solbergelva — 90.3/91.3/99.7MHz — **V.** by QSL-card.
NRK Vestfold, 3100 Tønsberg — 94.1MHz.

SPECIAL PRGRS IN LAPPISH (SÁMERADIO)
Addr: 9730 Karasjok. **Te:** +47 (84) 66209.
Stations: see main FM list.
D.Prgr: Mon-Fri 0700-0705, 1055-1100 (N), 1500-1545; Thurs 1710-1730; Fri 2200-2300; Sat 1055-1100 (N), 1500-1545; Sun 0830-0900.

V. Reception reports on long- and MW-trs must be on reg. transmissions to be verified.
F.Pl: 3rd Netw. in 1993.

FOREIGN SERVICE
ADDR: R. Norway International, N-0340 Oslo 3.
MEDIUMWAVE: Kvitsøy 1314kHz 1200kW.
SHORTWAVE: Fredrikstad (G.C: 10.58E/59.11N): 350kW — Kvitsøy (G.C: 05.27E/59.04N): 500kW — Sveio (G.C: 05.19E/59.37N): 500kW.

kHz	kHz	kHz	kHz	kHz	kHz	kHz
5965	9590	11775	11930	15220	17740	21595
7165	9615	11795	15165	15230	17785	21705
7210	9640	11850	15175	15330	17795	21705
7215	9645	11860	15180	15355	17860	21710
9560	9650	11865	15195	17730	17865	25730
9565	11735					

Schedule until March 27th 1993

Time	Area	kHz
0600-0700	Eu/Af	5965, 9590, 11735
		11735, 15175
0700-0800	Eu	9590, 11735
	WAf	11735
	FE/NZL	15175
0800-0900	As/Pac	17750, 25730
0900-1000	ME/EAf	21705
	As/AUS	25730
1000-1100	Eu/Af	21595, 21705
	AUS	21595
1100-1200	Atl/GRL	15165
	CAm/SAm	17785
1200-1300	NAm	11775, 15355
	As/AUS	**17860, 21705**
1300-1400	Eu	**9590, 25730**
	NAm	15355
	As/wAUS	17865
1400-1500	As/wAUS	15195
	NAm	17795, 21710
1500-1600	NAm	11870
	ME/As	15230
	ME/EAf	17860
1600-1700	ME	**11875**
	Af	**15230**
	NAm	15330
1700-1800	Eu	**9655**
	Af	9655, 15220
	NAm	15330
1800-1900	Eu/ME	9590
	Af	15220
1900-2000	Eu	9590
	Af	**15220,** 17730
	NZL/SAm	**17730**
2000-2100	Eu/NAm	9590
2100-2200	Atl/GRL	11930
	SAm/NZL	**15180**
2200-2300	SAm	11850, 11860
2300-2400	NAm	9645, **11795**
	CAm	**11795**
0000-0100	As	9640
	NAm	**9645**
	SAm	11850
0100-0200	NAm	**9565**
	CAm/SAm	9615
0200-0300	As	7165
	NAm	**9565,** 9615
0300-0400	ME/As	7165
	ME/EAf	9645
	NAm	9565, 9650
0400-0500	ME	7210, 9645
	EAf	9645
	NAm	**9560, 9650**
0500-0600	Eu	7215, 9645
	Af	7215, 11865
	ME/EAf	9645

Prgrs: Each programme lasts 30 mins. The second half of each transmission block is used by Radio Denmark.
English (Norway Today): Sat/Sun on freqs marked in **bold.**
ANN: "This is Radio Norway International".
SIG. TUNE: Each SW tr. begins with theme which originates from an ancient folk tune from Hallingdal.
V. by QSL-card. No Rp. (B).
PUB: SW prgr/freq guide. Free on request.

PRIVATE FM STATIONS

Approx. 350 st's are in operation. Advertising is permitted. However, many commercial st's have to share airtime with voluntary non-profit organizations.

AFRTS (U.S. Air Force)
ADDR: OL-A AFEBS, APO New York, NY 09085-5000, USA.
STATION: Oslo 105.5MHz 25W.
D.PRGR: 24h.

SVALBARD (SPITSBERGEN)

(Norwegian Territory)

L.T: UTC + 1h — **Pr.L:** Norwegian — **EC:** a/c 50, 230V (yes).

NRK SVALBARD

ADDR: 9170 Longyearbyen.
STATION: Longyearbyen (G.C: 25.24E/78.14N): 1485kHz 1kW.
FM: Svea 89.05MHz 15W — Longyearbyen 92.4MHz 15W.
D.PRGR: Rel. NRK Prgr. 1. During regional prgrs, rel. NRK Troms in morning & NRK Østlandssendinga at 0902 & 1404. Also rel. Wrp. at 1100-1115 (as 1314kHz) & R. Norway Int. at 1600-1700.
V. by letter.

POLAND

L.T: UTC + 1h (Su: UTC + 2h) — **Pop:** 38.598.000 — **Radio sets:** 10.400.000 — **Pr.L:** Polish — **E.C:** a/c 50, 220 — **ITU:** POL.

POLSKIE RADIO I TELEWIZJA

ADDR: Komitet do Spraw Radia i Telewizji "Polskie Radio i Telewizja", ul. J.P. Woronicza 17, 00-950 Warszawa (headquarters); ul. Myśliwiecka 3-5-7 (Radio 3,4); Al. Niepodleglości 75-89 (Radio 1,2), P-46, 00-950 Warszawa. **Te:** +48 (22) 478501, 445432 (Int. Rel.), 445714 (0800-2000UTC) Listeners Rel. Dept.
Telex: 825331.
L.P: Pres: Andrzej Drawicz. Dep. Pres: Jan Dworak (TV), Józef Kowalczyk (radio), Jerzy Mazurek (Tec). Int. Rel. Dir: Sergiusz Mikulicz.

HOME SERVICE

Long- and Mediumwaves:

	kHz	kW			
Warszawa III	198	200	Kraków	1368	60
Warszawa I	225	2000	Olsztyn	1485	2
Poznán	738	300	Gizycko	1485	1
Warszawa II	819	300	Kolobrzeg	1485	1
Katowice	1080	1500	Przemyśl	1485	1
Koszalin	1206	10	Szczecinek	1485	1
Lublin	1206	60	Zakopane	1485	1
Wroclaw	1206	200	Stargard	1503	300
Kielce	1206	10	Ostróda	1584	1
Starachowice	1260	2	Zielona Gora	1602	2
Szczecin	1260	10	Gorzów Wlkp	1602	1
Bialystok	1305	60	Mragowo	1602	1
Gdansk	1305	60	Lidzbark	1602	1
Lódz	1305	60	Radom	1602	1
Rzeszów	1305	100			

FM (MHz)	II°+L	III	IV	kW
Bialystok	70.01	72.02	72.80	74
Bydgoszcz	68.96	71.84	72.62	32
Częstochowa	68.96	66.23	67.79	10
Gdańsk	70.31	66.29	67.85	40
Jelenia Góra	71.72	68.78	—	0.1
Katowice	68.33	65.99	67.55	14
Kielce	70.49	72.71	71.15	60
Koszalin	69.92	66.95	67.73	54

Kraków	68.75	66.89°	67.67	120
Klodzko	67.64	69.74	72.44	1
Kudowa Zdrój	68.51	65.90	—	0.1
Lublin	69.92	71.81°	72.59	60
Lódź	68.51	72.23	73.01	80
Olsztyn	69.56	67.25	70.79	76
Opole	70.31	72.89	66.77	60
Pila	69.38	72.02	72.80	30
Plock (Sierpc)	68.72	70.97	72.53	10
Poznań	69.74	66.56°	67.40	38
Przemyśl	68.60	71.69	72.41	0.15
Rzeszów	68.24	65.90	67.46	7
Siedlce	70.22	66.41	68.03	25
Suwalki	68.60	71.12	72.68	30
Szczecin	68.78	66.74°	67.52	60/2×74
Warszawa	69.20	71.45°	67.94	100
Wroclaw	70.67	72.11	71.35	120
Zakopane	70.31	71.45	—	0.1
Zamość	69.38	66.68	67.61	30
Zgorzelec	68.24	69.56	67.46	52
Zielona Góra	69.14	71.72	72.50	30

°) stereo — L) local.

Radio 1: 24h on 225kHz. **Main N:** W 0600, 0700, 1500, 1615 (Vatican R.), 1800, 2200, 2348 (VOA). Sun 0600, 1105, 1500, 1800, 2200, 2348 (VOA).
Radio 2 (Dwójka): 0600-2400.
Radio 3 (Trójka): 0357 (Sun 0457)-0100.
Radio 4 (Czwórka): on FM + MW + LW 198 kHz (0750-1500): 0400 (Sun 0500)-2300. **N:** W 0405, 0500, 0600, 0700, 1130, 1700, 1830, 2255; Sun 0500, 1100, 1830, 2255. **BBC Prgr:** D 2100-2200.
ANN: "Tu Polski Radio Warszawa w programie pierwszym (1), drugim (2), trzecim (3), czwartym (4)" (or another station when local programme).

FOREIGN SERVICE

ADDR: P.O. Box 46, 00-950 Warsawa, Poland. **Te:** +48 (22) 459303. **Fax:** +48 (22) 444123. **Telex:** 814825.
L.P: MD: Jerzy Jagodiński. PD: Miroslaw Luboń. Man. Editor & Head of PR: Marek Lasota. CE: Jerzy Sliwinski.
MEDIUMWAVES: See Home Sce.
SHORTWAVES: Warszawa (G.C: 20.52E/52.04N): 100kW.

kHz	kHz	kHz	kHz
6095	7145	7285	9540
6135	7270	9525	11840

Schedule to May 1993 (all prgrs. beamed to Europe)

Byelorussian	kHz
0530-0555	7145, 6095, 1503
1900-1925	7285, 7145, 6095
2130-2155	7145, 5995
English	
1300-1355	11815, 9525, 7145, 6135, 1503
1600-1655	11840, 9525, 7285
1800-1855	9525, 7270
1930-2025	9525, 7270
2030-2125	9525, 7270, 7145, 6135, 6095, 1503
Esperanto	
1430-1455	9525, 7285, 7145, 6135, 6095, 1503
2130-2155	7285, 7270, 6135, 6095, 1503
German	
0630-0655	7145, 6095
1230-1255	11815, 9525, 7145, 6135, 1503
1400-1425	7285, 6135, 1503
1500-1525	9540, 7145, 6095, 1503
1600-1625	9540, 6135, 1503
1700-1725	7145
1730-1755	9525, 7285, 6135, 6095, 1503
1900-1925	7270, 6135, 1503
Lithuanian	
0600-0625	7145, 6095, 1503
1400-1425	9525, 7145, 6095
1800-1825	7145, 6095, 1503
Polish	
0630-0725	7270, 6135, 1503
0730-0825	7285, 6135, 1503
1500-1555	9525, 7285
1630-1725	9540, 6135, 1503
1930-2025	7145, 6135
2200-2255	7285, 6095, 1503
2302-2355	7285, 7145, 6135, 6095, 1503, 1305, 819
Russian	
1530-1555	11840, 9540, 6095
1700-1725	9525, 7285, 6095
1830-1855	11815, 7145, 6135, 6095, 1503
2000-2025	7285, 6095
Ukrainian	
0500-0252	7145, 6095, 1503
1600-1625	6095
1800-1825	7285, 6135
1930-1955	7285, 6095, 1503

ANN: Polish: "Tu Polskie Radio, Warszawa". E: "This is the Polish Radio, Warsaw".
IS: Revolutionary Etude (Chopin) — **V.** by QSL-card. Rec. acc.
F.PL: Czech and Slovak services to start during 1993. There are 'realistic prospects' of purchasing a 250kW SW-tr.

ROZGLOSINA HARCERSKA

(Operated by the Polish Pathfinders Union).
ADDR: Konopnickiej 6, 00-491 Warsaw.
STATION: 7205kHz 1kW.
D.PRGR: in Polish: Tues-Sat 1100-1700, Sun 0900-1700.
ANN: "Mowi Rozglosina Harcerska" (Polish Pathfinders Station speaking) —**V.** by QSL-card or letter. Rp. Re. in English.

RADIO MAZURY (Rlg.)

ADDR: P.O. Box 113, 14-100 Ostróda.
STATION: 1584kHz 2kW.
D.PRGR: details not available.

PORTUGAL

L.T: UTC (Su: UTC + 1h) — **Pop:** 10.420.000 — **Radio sets:** 2.200.000 — **Pr.L:** Portuguese — **E.C:** a/c 50, 220V — **ITU:** POR.

RADIODIFUSÃO PORTUGUESA (RDP)

ADDR: Central Sce. Av. Duarte Pacheco 5, 1000 Lisboa. **Te:** +351 (1) 3872041. **Cable:** RDPINS-P. **Telex:** 15649 (Int.Rel: 64774). **Fax:** +351 (1) 653468 (Int. Rel: 3870677).
L.P: Dir. Gen: Ing. Manuel Cardoso de Menzes. Administ: Dr. Fausto Correia, Drª. Isilda Branquinho, Jaime Fernandes. Prgr. Dir./Pub. Sce: Jorge Pego. N. Dir./Pub. Sce: Carlos Mendes. Comm. Dir: Jorge Dias. Comm. Prgr. Dir: Pedro Castelo. Comm. N. Dir: Jaime Marques de Almeida. Int. Sce. Dir: Dr. Garcia Marques de Freitas. Tech. Dir: Ing. Francisco Mascarenhas. Exploitation Dir: Ing. Celso de Albuquerque. Finance Dir: Dr. José Anibal Gomes Ferro de Carvalho. Ext. Rel. Dr. Duarte Guedes Vaz.
Reg. Addr: R. Castilho, 30-1106 Lisboa Codex; R. Cândido dos Reis, 74, 4099 Porto Codex; R. Dr. Alberto dos Reis, 3049 Coimbra Codex; Campo Senhora da Saúde, 8000 Faro.

HOME SERVICE

ANTENA 1:

Mediumwaves	**kHz**	**kW**
Guarda	558	10
Valença	567	10
Chaves	630	1
Miranda do Douro	630	1
Sta. Isabel	630	50
Lisboa	666	135
Bragança	666	1
Vila Real	666	10
Viseu	693	10
Norte (S. Salvador)	720	100
Mirandela	720	10
Faro	720	10
Castelo Branco	720	10
Elvas	720	10
Portalegre	1287	1
Canidelo (Porto)	1377	10
Covilhã	1562	1

FM	**MHz**	**kW**
Mendro	87.7	5
Lousã	87.9	50
Muro	88.3	5
Monchique (Fóia)	88.9	30
Bornes	92.8	5
Guarda	94.7	5
Minhéu	94.9	6
Marão	95.2	6
Lisboa I	95.7	100
Gardunha	96.4	6
Bragança	96.4	6
Porto (M. Virgem)	96.7	80
Faro (S. Miguel)	97.6	5
Portalegre	97.9	5
Valença	98.2	5
Montejunto	98.3	10
Santiago do Cacém	99.2	6
Lisboa II	99.4	8

D.Prgr: 24h. **N.** on the h. Main bulletins: 0700, 1200 (regional), 1300, 1900, 2330.

RDP is now also on satellite!
See *World Satellite Broadcasts*

Antena 2:

FM	MHz	kW		MHz	kW
Minheu	88.0	6	Braga	91.3	5
Guarda	88.4	5	Monchique (Fóia)	91.5	30
Montejunto	88.7	6	Porto (M. Virgem)	92.5	40
Lamego	88.7	5	Portalegre	92.9	5
Lousã	89.3	40	Faro (S. Miguel)	93.4	5
Santiago do Cacém	90.7	6	Gardunha	93.9	6
Mendro	91.1	5	Lisboa	94.4	100
Bornes	91.1	5	Valença	99.0	5

D.Prgr: 24h. **N:** 0800, 1330, 1830 (Mon-Fri), 2030, 0100.

Rádio Comercial:

Mediumwaves	kHz	kW		kHz	kW
Faro (1)	558	10	Guarda	828	1
Lisboa*	756	5	Covilhã	828	1
S. Isabel*	828	10	Lisboa	1035	135
Castelo Branco	828	1	Norte (S. Salvador)	1062	100
Coimbra	828	1	Vila Real	1170	10
Viseu	828	1	Canidelo (Porto)	1575	10

*) experimental.
D.Prgr: 24h — **Local prgrs:** Mon-Fri 0600-1000, 1700-2000; Sat: 0700-1000, 1300-1900; Sun: 1300-1900.

Rádio Alto Douro:

Mediumwaves	kHz	kW
Lamego	756	1

D.Prgr: 0700 (Sun 0800)-2000.

Rádio Comercial Norte

Mediumwaves	kHz	kW
Miramar	783	50

D.Prgr: 0600-2400.

Rádio Elvas: 1332kHz 1kW.
D.Prgr: 0700-1300, 1700-1900, 2100-2400.

Rádio Comercial-Estéreo

FM (stereo)	MHz	kW		MHz	kW
Monchique (Fóia)	88.1	15	Lisboa	97.4	100
Valença	89.6	6	Porto	97.7	40
Lousã	90.8	40	Gardunha	98.2	6
Bornes	91.9	5	Portalegre	98.9	6
Mendro	92.0	14	Braga	99.2	5
Santiago do Cacém	92.3	6	Marão	99.8	6
Faro	96.1	5	Montejunto	99.8	10
Guarda	96.1	5	Lisboa*	100.3	50

*) experimental, separate prgrs.
D.Prgr: 24h (same prgrs as 783kHz 0200-1000).
ANN: "Radiodifusão Portuguesa . . . Programa. . .".

Local Radio Stations

Location	MHz	kW		MHz	kW
Bragança	93.9	6h	Bornes	102.1	8h
Coimbra	94.9	4c	Lousã	102.2	14c
Santarém	98.8	0.4j	Mendro	102.4	0.5f
Porto	100.4	10d	Elvas	102.8	0.02f
Guarda	100.6	10g	Braga	103.0	0.5d
S. Miguel	100.7	2a	Elvas	103.8	8f
Covilhã	101.3	1e	Valença	104.0	0.4d
Lamego	101.5	1b	Bragança	104.2	0.01h
Viseu	101.8	1i	Minhéu	104.7	8b
Monchique	101.9	20a	Montejunto	105.2	1.5j
Muro	102.0	0.5d			

a) Algarve, b) R. Alto Douro, c) R. Coimbra, d) R. Comercial Norte, e) R. Covilhã, f) R. Elvas, g) R. Guarda, h) R. Nordeste, i) R. Viseu, j) R. Santarém.

RDP INTERNATIONAL — RADIO PORTUGAL

ADDR: Radiodifusão, Portuguesa, Rua S. Marçal 1, 1200 Lisboa. **Te:** +351 (1) 326185. **Telex:** 43247.
STATIONS: CEOC (S. Gabriel) (G.C: 08.40W/38.45N): 1×300kW, 7×100kW SW-tr's — G) Gloria (G.C: 08.40W/38.45N) 250kW (tr. operated by RFE/RL) — S) Sines (G.C: 08.45W/37.57N) 250kW.

kHz	kHz	kHz	kHz	kHz
6130	9630S	11760	15225	17745
9555	9635	11800G	15250	21515
9570	9705	11735	15295	21655
9600	9740	11840	17740	21700
9615S	11740			

English	Area	kHz			
			1000-1855SS	Eu	11800
1500-1530MF	IND/ME	21515	1000-1200MF	Timor	15225
2000-2030MF	Eu	11740		Af	17740
2100-2130MF	Af	15250			21700
0230-0300MF	USA	9570		GNB/	
		9705		CPV/B	15250
0230-0300MF	B	9600	1100-1300MF	Eu	11800
		11840	1300-2100SS	VEN	17745
French			1300-2100SS	USA	15295
2030-2100MF	Eu	11740	1400-2100SS	Af	15250
2130-2200MF	Af	15250	1500-1600W	IND	21515
Italian			1500-1900MF	Af	15250
2100-2130MF	Eu	11740	1800-2100MF	GNB/	
Portuguese				CPV/B	21655
0015-0115	B	9555	1800-2000MF	Af	11760
0130-0230MF	B	9600	1700-2000MF	Eu	9740
0130-0300SS	B	9600			11740
0600-0800MF	Eu	6130	1800-2000W	Eu	11800
0600-1100MF	Eu	11800	2200-0230MF	USA	9570
0645-0800MF	Eu	9630			9705
0800-1000SS	Timor	15225		B	11840
1000-1200MF	Timor	15225		VEN	9635
0800-0930SS	Eu	9615	2200-0300SS	USA	9570
0800-1430SS	Af	21700			9705
0900-2100SS	Eu	9740		B	11840
	GNB/	21655		VEN	9635
	CPV/B		2200-2400	B	9600

ANN: E: "This is R. Portugal broadcasting from Lisbon through the International Sce." F. "Ici la Radiodiffusion Portugaise, Service International, qui vous parle de Lisbonne". I: "Que le studie de Lisbona de la Radiodiffusione Portoghese".
INT-SIG: Opens and closes with National Anthem preceded by a time gong — **V.** by QSL-card. Rec. acc.

RÁDIO RENASCENÇA Lda — EMISSORA CATÓLICA PORTUGUESA (Rlg.).

ADDR: Rua Ivens 14, 1294 Lisboa Codex. **Te:** +351 (1) 3475270. **Telex:** 13030. **Fax:** +351 (1) 3422658 — Rua Sá da Bandeira 766, 4000 Porto. **Te:** +351 (2) 2005678. **Telex:** 25343.

Mediumwaves	kHz	kW		kHz	kW
Braga	576	10	Coimbra	981	10
Muge	594	100	Guarda	981	1
Vila Moura	891	10	Chaves	1251*	1
Evora	927*	1	Porto	1251*	10
Seixal	963*	10	Castelo Branco	1251	1
Bragança	981	1	Viseu	1251*	10
Vila Real	981	1			

*) includes local prgrs.

FM (MHz)	Ch.1	Ch.2	kW
Arrabida	105.8	89.9	10
Bornes	89.6	101.1	5/8
Bragança		99.5	1
Foia	98.6	104.9	28
Gardunha	103.4	99.5	8
Guarda		90.2	1
Lamego	98.6	106.2	10
Leiria/Fatima	95.0		10
Lisboa	103.4	93.2	100
Lousá	106.0	91.7	100
Marofa	94.2		6
Mendro	96.5	100.9	5/14
Minhéu	89.8	102.6	5/8
Montejunto	90.2	106.8	40
Muro	90.4	103.4	10
S. Mamede	95.3	101.1	8/4
S. Miguel	103.8	89.6	4/6
Valença	100.0	95.4	8
Valongo	93.7	104.1	100

D.PRGR: 24h. **N.** on the h. and half h.

EXTERNAL SERVICE

STATION: Muge (G.C: 08/41W/39.05N): 1×100kW tr.

kHz	kHz	kHz	kHz
6100	9575	9600	9680

D.PRGR. In Portuguese: 0500-0700 MF to Europe on 6100kHz, 1045-1200 (Sun) to Europe on 9575kHz. 1800 (Sat/Sun 1400)-2000 to Europe on 9680kHz. 0015-0115 to Brazil on 9600kHz — **V.** by letter.

RADIO ALTITUDE (Priv, Comm.)

ADDR: R. Batalha Reis, 6300 Guarda. **Te:** +351 (71) 212232. **Fax:** +351 (71) 221492.
L.P: Dir. Gen: Dr. H. Sequeira.
STATION: CSB21 1584kHz 1kW — **FM:** 90.9MHz 1kW (stereo).
D.PRGR: 24h. **N:** 0730, 0800, 0830, 1000, 1230, 1500, 1600, 1800, 1900, 2000, 2200, 2300.
V. by letter. Rec. acc.

OTHER STATIONS

Lisboa: R. Jornal 89.5MHz — R. Paris Lisboa 90.4MHz — R. Minuto 92.4MHz — Radiogeste 96.6MHz — Correlo da Manhá R. 104.8MHz — RCL Lisboa 102.4MHz.
More than 300 local FM-st's are operating throughout the country.

AMERICAN FORCES RADIO & TV SCE. (U.S. Mil.)

STATION: Costa da Caparica 88.4MHz 0.1kW — 24h.

SOCIEDADE DE RADIORETRANSMISSÃO Ld[a] (RARET)

ADDR: Av. Padre Manuel da Nóbrega 8-B, 3°, 1000 Lisboa. **Te:** +351 (1) 881131. RARET is responsible for the operation of the R. Free Europe/R. Liberty relay st's at Gloria and Maxoqueira (see RFE/RL listing under Germany F.R.).

RADIO TRANS EUROPE

ADDR: Rua Braamcamp 84, –6° Esq°, 1200 Lisboa
STATIONS: Sines (G.C: 08.45W/37.57N): 3×250kW SW-trs. (one reserve)
This station is used by various broadc. organisations (Radio Canada International, Deutsche Welle, IBRA Radio and R. Portugal). Full details of schedules and frequencies are included in the entries for these organisations.

VOICE OF AMERICA RELAY STATION

STATION: Gloria (G.C: 08.40W/38.45N): 250kW.
Transmitters operated by RFE/RL Inc. (see "Germany").

kHz	kHz	kHz	kHz	kHz	kHz	kHz
6095	7270	9575	9715	11710	11845	15225
6160	7285	9585	9735	11805	11855	17865
7200	9570	9690	9760	11825	11865	21625

D.PRGR: see USA.

ROMANIA

L.T: UTC + 2h (Su: UTC + 3h) — **Pop:** 23.515.000 — **Radio sets:** 3m (est.) — **Pr. L:** Romanian — **E.C:** a/c 50, 220V — **ITU:** ROU.

RADIO ROMÂNIA (Gov.)

ADDR: Str. General Berthelot 60-62 (P.O. Box 111), 70749 Bucureşti. **Te:** +40 (1) 6503055 or 6159350. **Fax:** +40 (1) 6121057. **Telex:** 11252.
L.P: Pres: Răzvan Teodorescu. Dir. Foreign Rel: Violeta Olteanu.
NETWORKS: 1=RTR-1, 2=RTR-2, 3=RTR-3, 4=Reg. + RTR-2

Mediumwaves	kHz	kW	Station	kHz	kW
Braşov 1	153	1200	Cluj 2	1152	950
Petroşani 2	531	14	Bacău 1	1179	20
Tg. Jiu 2	558	200	Bod 4	1197	14
Satu Mare 2	567	50	Constanţa 1	1314	14
Braşov 1	567	50	Craiova 1	1314	7
T. Severin 1	603	14	Timişoara 2	1314	30
Botoşani 1	603	50	T. Mureş 4	1323	7
Oradea 1	603	50	Galaţi 1	1332	50
Timişoara 1	630	400	Sighet 1	1404	50
Sighet 2	711	50	Baia Mare 1	1404	15
Predeal 1	720	14	Sibiu 1	1404	7
Isaccea 1	720	14	Pucioasa 2	1422	1
Borşa 1	720	7	Constanţa 2	1458	50
Lugoj 2+Ext	756	400	Mahmudia 1	1530	14
Bucureşti 2	855	1500	Suceava 1	1530	14
Reşita 1	909	50	Sibiu 4	1593	7
Cluj 4	909	50	Oradea 4	1593	7
M. Ciuc 2	945	14	Baneasa 1	1593	14
Iaşi 4	1053	1000			

FM-Stations (MHz)

Location	2	3	kW
Bîrlad	66.36	72.32	46/52
Sibiu	66.44	69.35	40/88
Baia Mare	66.56	68.12	35/37
Bacău	67.01	68.87	30/100
Băneceva	66.92	71.12	38
Constanţa	67.79	70.01	52
Cluj	68.36	66.76	16
Resita	68.52	72.36	40
Topolog	68.72	71.96	46
Varatec	69.11	67.34	15
Tulcea	69.65	70.76	46
T. Severin	69.68	72.71	38
Iasi	69.92	71.84	4/80
Focsani	70.22	67.88	80
Bucureşti	70.40	68.24	43
Deva	70.64	72.20	4
Oradea	71.00	69.86	60
P. Neant	71.06	66.17	46/40
Timisoara	71.72	69.65	60
Bistrita	72.44	69.74	15/50
Arad	72.56	70.79	46
Vilcea	72.68	67.25	12
Petrosani	71.06	72.80	40/50
Ploesti	72.92	71.30	46/90
Suceava	72.98	70.61	20/40
Novaci	70.04	71.90	46/300
Cimpulung	70.85	72.20	35/12
Zalau	67.67	65.96	15
Gheorgheni	65.96	68.60	180
Bihor		72.98	10

\+ low power relays.

HOME SERVICE

România Actualitati (RTR-1): 24h. **N:** on the h.
România Cultural (RTR-2): 0600-2200. **N:** 0500, 1000, 1300W, 1700Sun, 1900.
România Tineret (RTR-3): 0600-2300. **N:** 0600, 1000, 1300W, 1700Sun, 1900W.
Reg. Prgrs: Timisoara on 630kHz: 0400-2000 — **Cluj** on 909/1593 (Oreada)kHz: W 0600-1100 & 1400-1900, Sun 0600-1200 & 1400-1900 — **Tg. Mures** on 1197/1323/1593 (M. Ciuc)kHz: W 0600-1100 & 1400-1900 — **Iaşi** on 1053kHz: 0400-1100Sun, 1300-2100 — **Craiova** on 558kHz: W 0400-1100 & 1400-1700.
Rel. Moldovan R: 0300-0615 (SS 0500), 0900-1200 & 1800-2100 on 1053kHz (temporary sce, may be discontinued).

ANN: Opening 1st Prgr: "Aici Bucureşti, Radio România Actualităţi. Transmitem pe staţia de unde lungi, cu frecvenţa de...pe staţiile de unde medii cu frecvenţele de..." 2nd Prgr: "Aici Bucureşti, Radio România Cultural. Transmitem pe statiile de unde medii cu frecvenţele de... şi pe staţiile de unde ultrascurte cu frecvenţele de..." 3rd Prgr: "Aici Bucureşti, Radio România Tineret. Stimaţi ascultători, bună dimineaţa. Vă invitam să ne ascultati pe staţiile de unde ultrascurte cu frecvenţele de...".

SPECIAL PRGRS

ANTENA BUCUREŞTI

D.PRGR: 0900-1200 & 1400-1600 on 70.4MHz in English, French, German, Hungarian, Russian & Spanish.

RADIO VACANŢA (R. Holidays)

ADDR: Vila 27, 8741 Mamaia. **Te:** +40 (918) 31116.
STATIONS: Constanţa 1458 kHz 50kW + FM 100.1/101.0 MHz.
D.PRGR: May-October: 0700-1400, 1500-1700. Music and entertainment in Romanian, English, French, German and Russian for holidaymakers along the Black Sea coast — **V.** by letter.
R. Costinest: Oct-April 0600-1100 on 100.1MHz.
Antena Statelor: Oct-April 0300-0600 & 1500-1800 on 630/1458kHz + 101MHz.

FOREIGN SERVICE (RADIO ROMANIA INTERNATIONAL)

L.P: Dep. DG: Vasile Ionescu.
Mediumwave: Lugoj 756kHz 400kW.
Shortwaves: (G.C: 20.06E/44.25N): 120/250kW.

kHz	kHz	kHz	kHz	kHz	kHz	kHz
5955	7255	9625	11740	15250	15380	17790
5990	7290	9665	11775	15255	15405	17805
5995	9510	9690	11790	15335	15445	17815
6105	9530	9710	11810	15340	17720	17825
6155	9550	9750	11830	15365	17745	17850
6190	9570	11710	11940	15370	17775	21665
7195	9590	11720	11970			

TO EUROPE
Aromanian
1730-1800 1458, 756
2030-2100 756
English
0632-0643 11775, 9570, 9510, 7290
1300-1400 17850, 11720, 15365, 11940
1900-2000 9690, 7290, 7195, 6105
2100-2200 9690, 7290, 7195, 6105, 5955
French
0614-0623 11775, 9570, 9510, 7290
1130-1200 17850, 11970, 11940, 15445, 15365
1630-1700 15365, 11970, 11790
1900-1930 15365, 11790, 6190
2000-2100 9690, 9510, 7290, 7195, 5995
Greek
1800-1830 756
1930-2000 756
Italian
1600-1630 11810, 756
1900-1930 9530, 756
2000-2030 6105, 756
2100-2130 756
Portuguese
2130-2200 15250, 11970, 9750
Romanian
0800-0814 11775, 9750, 9710, 7255
1730-1800 11940, 11810, 9510
2000-2030 9625, 6190
2130-2230 756
Serbo-Croat
1700-1730 756
1830-1900 756
Spanish
1930-2000 15250, 11790, 9750
2030-2130 15250, 11790, 9750
Turkish
1400-1430 17775, 11830
1600-1630 11830, 9530
TO AFRICA
English
0530-0600 21665, 17745, 17790, 17720, 15380, 11840
1730-1800 17745, 17720, 15365, 15340
French
1130-1200 17790, 15380
1430-1500 17745, 15380, 15365
Arabic
1230-1300 17790, 17745, 15380, 15365, 17720
1400-1430 17745, 15365, 15380
1530-1630 17815, 15380
Portuguese
1800-1830 17720, 15340
TO NEAR EAST
Arabic
1230-1300 17720, 15405
1400-1430 17720, 15405
1530-1630 17720, 15405
Persian
1500-1530 17745, 15250, 11940
1700-1730 15250, 11940, 9690
TO ASIA
English
1430-1530 17720, 15335, 11810
Russian
1330-1400 15335, 11775
TO PACIFIC AREA
English
0645-0715 21665, 17805, 17720, 15335, 11940, 11810
Romanian
2300-2400 15365, 11810

TO NORTH AMERICA
English
0200-0300 11940, 11830, 9570, 9510, 6155, 5990
0400-0430 11940, 11830, 9570, 9510, 6155, 5990
Romanian
2300-2400 11940, 11830, 9570, 9510, 6155, 5990
0130-0200 11940, 11830, 9570, 9510, 6155, 5990
TO LATIN AMERICA
Portuguese
0100-0130 15380, 15255, 11940, 11830, 9570
Spanish
2200-2300 15380, 11940, 9570
0000-0100 15380, 11940, 11880, 9570, 9510, 6155
0300-0400 15255, 11940, 11830, 9570, 9510
ROMANIAN FOR SEAMEN
0715-0815Sun 17790, 17745, 17720, 15380, 15370, 15335
0815-0915Sun 17790, 17745, 17720, 15380, 15370, 15335
0915-1015Sun 11940, 11810, 11775, 9665, 9590, 9570

ANN: E: "This is Bucharest, Radio Romania International." F: "Ici Bucarest, Radio Roumanie Internationale." I: "Qui Bucarest, Radio Romania Internazionale!" S: "Aqui Bucarest, Radio Rumania Internacional." P: "Ouvem Radio Romania Internacional." G: "Hier spricht Radio Rumänien International." — **V.** by QSL-card.

PRIVATE STATIONS

FUN Radio, Hotel Dorbanti, Ap. 103, Calea Dorbanti, Bucuresti (67.81MHz) — **R. Contact Bucuresti,** Spl. Independentei 2 A, Bucuresti (96.1MHz) — **R. Delta,** P.O. Box 38-61, Bucuresti (98.5MHz 0.8kW). 0600-1000, 1500-1800 (other times rel. R. France Int.) — **R. Europa Nova,** Timisoara (98.2MHz) — **R. Nova 22,** Bucuresti (92.7MHz) — **R. Vox-T,** P.O. Box 2058, 6600 Iasi (72.1MHz) — **UNI-FAN,** Bucuresti (69.8MHz).

RUSSIA

L.T: Moscow: UTC + 3h (Su: UTC + 4h); See world time table for other zones — **Pop:** 148.100.000 — **Radio sets:** 90m (est.) — **Pr.L:** Russian, numerous minority languages — **E.C:** a/c 50 127/220 — **ITU:** RUS

MINISTERSTVO SVYAZI (Communications Ministry)

ADDR: ul. Delegatskaya 5, 103091 Moskva. **Te:** +7 (095) 2927070 — Minister: V.B. Bulgak.

MAIN CENTER FOR CONTROL OF BROADCASTING NETWORKS

ADDR: ul. Nikolavskaya 7, 103012 Moskva. **Fax:** +7 (095) 9211624. **Tlx:** 412912 — Chief: A.T. Titov.

ROSSIYSKAYA GOSUDARSTVENNAYA TELERADIOKOMPANIYA "OSTANKINO" (Gov.)

ADDR: ul. Akademika Koroleva 19, 127 427 Moskva (headquarters); ul. Pyatnitskaya 25, 113 326 Moskva (musical, drama, childrens prgr's); ul. Kachalova 24, 121812 Moskva (Radio Orfey). **Te:** +7 (095) 2337860 (Pyatnitskaya office). **L.P:** Gen. Dir: V.V. Lazutkin.
Ostankino" produces the radio prgr's Radio-1 (R1), R. St. Yunost (RSY), R. St. Mayak (RSM) and Radio Orfey (RO).

VSEROSSIYSKAYA GOSUDARSTVENNAYA TELERADIOKOMPANIYA (VGTRK, Gov.)

ADDR: Yamskogo polya 5-ya ul. 19/12, 125124 Moskva. **Te:** +7 (095) 2514050. **Fax:** +7 (095) 2500506.
L.P: Gen. Dir: Anatoliy Lysenko.
VGTRK produces Radio Rossii.

HOME SERVICES

Abbreviations: 1, 2, 3, 4 = "Ostankino" R1, RSY, RSM and RO prgrs, respectively — RR = R. Rossii — Ind = Independent stn or prgr — Reg = Regional prgrs — FLS = Foreign language service — F = Foreign sce's — NP = R. Novyy Peterburg — E = European part of Russia — S = Siberia — FE = Far East — v = variable times.

LONG & MEDIUMWAVES (Rg = Region)

Rg	Location	kHz	kW	H. of tr.	Prgr.
KH	Komsomolsk, FE	153	1000	2000-1800	RR,Reg
	incl. R. St. Vostok			0800-0930v	Ind
BA	Ufa, E	153	500	0000-2200	1
TM	Norilsk, S	162	150	2200-2100	RR,Reg
MO	Noginsk, E	171	500	0200-2400	1
KA	Bolshakovo, E	171	1200	0200-2400	1
KO	Syktyvkar, E	171	300	0200-2400	1
KD	Krasnodar, E	171	1000	0200-2400	1
NK	Oyash, S	171		2200-2000	RR,Reg
SA	Yakutsk, S	171	500	2000-1800	RR,Reg
CH	Atamanovka, S	180	150	1800-1700	3
KM	Petropavlovsk, FE	180	150	1800-1600	RR,Reg
AM	Belogorsk, S	189	1000	2000-1800	RR,Reg
MO	Noginsk, E	198	100	0000-2300	3
SP	Sankt-Peterburg, E	198	150	0000-2300	3
BA	Ufa, E	198	300	0000-2300	3
IR	Angarsk, S	198		24h	3
SL	Aleksandrovsk, FE	198	50		
AM	Tynda, S	207	30	2000-1900	3
KN	Krasnoyarsk, S	216	150	2200-2000	RR,Reg
YV	Birobidzhan, FE	216	30	2000-1800	RR,Reg
	incl. R. St. Vostok			0800-0930v	Ind
KY	Surgut, S	225	150	0000-2200	RR,Reg
SP	Sankt-Peterburg, E	234	1000	0200-2400	1
AR	Arkhangelsk, E	234	150	0200-2400	1
SM	Samara, E	234	1200	0200-2400	1
IR	Angarsk, S	234	250	2200-2000	RR,Reg
MA	Arman, FE	234	1000	1800-1600	RR,Reg
PM	Ussuriysk, FE	243	1000	2000-1800	1,RR,Reg
TS	Kazan, E	252	150	0200-2400	1,Reg
	Unknown, We. S	252		0000-2200	1
MO	Taldom, E	261	2000	0200-2400	RR
CH	Kruchina, S	261	1000	2000-1800	RR,Reg
OR	Orenburg, S	270	40	0000-2200	RR,Reg
NK	Novosibirsk, S	270	150	2200-2000	RR,Reg
KH	Khabarovsk, FE	270	150	2000-1800	1
YE	Yekaterinburg, S	279	50	0000-2200	RR,Reg
AT	Barnaul, S	279	50	2200-2000	RR,Reg
BU	Selenginsk, S	279	150	2158-2100	RR,Reg
SL	Yuzhno-Sakhalinsk, FE	279	150	1800-1600	RR,Reg
CV	Cheboksary, E	531	30	0145-2200	3,Reg
	Unknown, E	540		2300-2200	3
MO	Chkalovskaya, E	549	100	0100-2400	3
SP	Sankt-Peterburg, E	549	1000	0100-2400	3
KA	Kaliningrad, E	549	25	0100-2400	3
RO	Rostov-na-Donu	549		0100-2400	3
OM	Omsk, S	549	150	2000-1900	3
SA	Yakutsk, S	549	50	2000-1900	3
AM	Svobodnyy, S	549	150	2000-1900	3
MA	Magadan, FE	549	50	2000-1900	3
PM	Vladivostok, FE	549	150	2000-1900	3
BU	Unknown, S	558		0700-0600	3
VO	Volgograd, E	567	250	0200-2400	RR,Reg
TU	Kyzyl, S	567	50	2200-2000	RR,Reg
KB	Nalchik, E	576	25		
AS	Astrakhan, E	576	50	0100-2100	3
TY	Tyumen, S	576	100	0000-2200	RR,Reg
NK	Oyash, S	576	100	2200-2100	3
IR	Angarsk, S	576	200	2100-2000	3
KH	Khabarovsk, FE	576	150	1800-1700	3
KM	Petropavlovsk, FE	576	150	1700-1600	3
PR	Perm, E	585	40	0000-2200	1,RR,Reg
AM	Belogorsk, S	585	500	1900-1800	F
UD	Izhevsk, E	594	150	0200-2400	RR,Reg
SO	Vladikavkaz, E	594	22	0200-2200	3,Reg
KY	Surgut, S	594	500	2345-2300	3
KN	Krasnoyarsk, S	594	150	2200-2100	3
YN	Salekhard, S	603	50	0000-2200	RR,Reg
AM	Belogorsk, S	603		2100-1800	3
AM	Skovorodino, S	603		2100-1800	3
MO	R. Raduga, Moskva, E	612	50	1600-1700	Ind
MU	Murmansk, E	612	30	0300-0200	3
KT	Petrozavodsk, E	612	100	24h	3
	Kyzyl (?), S	612		2300-1800	3
KO	Syktyvkar, E	621	50	0200-2400	RR,Reg
DA	Makhachkala, E	621	50	0200-2400	RR,Reg
DA	Kochubey, E	621		0200-2400	RR,Reg
SA	Yakutsk, S	621		2100-1500	2
KH	Khabarovsk, FE	621	50	2000-1800	RR,Reg
	incl. R. St. Vostok			0800-0930v	Ind
SR	Saratov, E	630	50	0200-2400	3,Reg
KH	Komsomolsk, FE	630	500	1000-1500	F
	incl. R. Nadezhda			2000-1000	2
				0800-1000	Ind
OM	Omsk, S	639	50	0000-2200	RR,Reg
PM	Vladivostok, FE	648	1000	2000-1900	F
MU	Murmansk, E	657	150	0200-2400	RR,Reg
CI	Groznyy, E	657	50	0200-2400	3,Reg
SA	Verkhoyansk, S	657			
AD	Maykop, E	666	30	0200-2200	3
	Unknown, So. E	666		0300-2100	2
	Unknown, S	666		2000-1600	3
KH	Komsomolsk, FE	666	100	1700-1600	3
KO	Ukhta, E	675	50	0300-2400	2
NK	Oyash, S	675		0100-2000	2
SP	R. Ala, Sankt Peterburg, E	684	10	24h	Ind
BA	Ufa, E	693	30	0000-2200	RR,Reg
CK	Anadyr, FE	693	25	1800-1600	RR,Reg
	Unknown, E	702		0200-2400	1
IR	Bratsk, S	702		2200-1700	3
BU	Unknown, S	702		2200-1700	3
NE	Naryan-Mar, E	711	7	0200-2400	RR,Reg
KH	Khabarovsk, FE	711	50	1900-1500	3,Reg
SL	Yuzhno-Sakhalinsk, FE	720	500	0900-1600	F
				1800-0900	3
MO	Moskva, E	738	7.5	0300-2100	4
CB	Chelyabinsk, S	738	50	0000-2200	3,Reg
KK	Palana, FE	738	50	1800-1600	RR,Reg
SP	R. Baltika, Sankt-Peterburg, E	747	10	0400-1600	Ind
SP	Golos Nadezhdy, Sankt-Peterburg, E	747	10		Ind
KT	Petrozavodsk, E	765	150	0200-2400	RR,Reg
	Unknown, FE	765		1900-1800	3
VN	Voronezh, E	774	100	0200-0100	3,Reg
	Blagoveshchensk (?), S	783	1200	2000-1800	1
MO	Moskva, E	792	5	0200-0100	2
AS	Astrakhan, E	792	50	0200-2400	1,Reg
RK	Abakan, S	792	20	2200-2000	RR,Reg
SL	Aleksandrovsk, FE	792		1800-1600	RR,Reg
SP	Sankt-Peterburg, E	801	1000	0200-2400	1,Reg
CH	Kruchina, S	801	1000	2000-1800	1
SM	Samara, E	810	30	0100-2200	3
VO	Volgograd, E	810	150	0100-2200	3
UL	Ulyanovsk, E	810	50	0100-2200	3
YE	Yekaterinburg, S	810	30	24h	3
PM	Vladivostok, FE	810	50	1800-1600	1, 3,Reg
NN	Nizhniy Novgorod, E	828	50	0200-0100	3,Reg
DA	Unknown, E	828			?,Reg
TU	Kyzyl, S	828	50	1900-1800	3
	Unknown, FE	837		-1800	3
MO	Noginsk, E	846	50	2000-2400	RR
				0200-1900	RR,Reg
MO	R. St. Radonezh, Noginsk, E	846	50	1900-2000	Ind
KG	Elista, E	846	30	0200-2400	RR,Reg
PE	Kamenka, E	855		0200-2400	RR,Reg
AM	Blagoveshchensk, S	864		2100-1600	2
MO	Elektrostal, E	873	150	0200-2400	RR
SP	Sankt-Peterburg, E	873	150	0200-2400	RR
SM	Samara, E	873	150	0200-2400	RR
KA	Kaliningrad, E	873	150	0200-2400	RR
ST	Stavropol, E	882	50	0200-2200	2,Reg
	Unknown, E	891		0000-2200	1
ME	Yoshkar-Ola, E	900	50	0155-2200	3,Reg
	Unknown, So. E	909		0200-2200	3
YE	R. Sinitsa, Yekaterinburg, S	909		0400-2000v	Ind
	Unknown, FE	909			3
AR	Arkhangelsk, E	918	100	0200-2400	RR,Reg
MO	Otkrytoye R, Moskva, E	918	75	0300-2300	Ind
DA	Makhachkala, E	918	50	0200-0100	3,Reg
KU	Shadrinsk, S	918	5	0000-2000	3
KU	Shumikha, S	918	5	0000-2000	3
KH	Komsomolsk, FE	927	20	2000-1800	1
KN	Krasnoyarsk, S	936	5	2200-1800	2

OR	Matveyevka, E	936		0000-2300	RR,Reg
RO	Rostov-na-Donu, E	945	300	0200-2400	RR,Reg
MO	R. Nostalgie, Moskva, E	963	15	0200-	Ind
BU	Guzino-Ozersk, S	963		2158-2000	RR,Reg
AM	Blagoveshchensk, S	963		2000-1000	1
KH	Khabarovsk, FE	963		2100-1500	2
SL	Yuzhno-Sakhalinsk, FE	972	50	1800-1600	RR,Reg
YV	Birobidzhan, FE	999	7,5	2000-1500	3,Reg
	Unknown, So. E	1008		0200-2200	3
KM	Unknown, Kamchatka, FE	1008	50	1800-1400	RR,Reg
KH	Khabarovsk, FE	1008	20	2000-1600	2
	Unknown, No. E (3 st's)	1026	5	0200-2200	2
NK	Oyash, S	1026	200	2300-1700	3
SP	R. Polis, Sankt-Peterburg, E	1053	10	0430-1700	Ind
SP	R. Otkrytyy gorod, Sankt-Peterburg, E	1053	10	1700-2030	Ind
AM	Belogorsk, S	1053		2100-1600	2
MD	Saransk, E	1062	150	0200-2400	RR,Reg
AM	Zeya, S	1071		2100-1600	3
IR	Angarsk, S	1080	1000	2200-1000	1
				1030-1400	F
KD	Krasnodar, E	1089	300	0200-1600	RR,Reg
				1600-2400	3, F
KK	Tilichiki, FE	1089		1800-1600	RR
VG	Vologda + 1 st, E	1098	5	0200-2400	RR,Reg
VG	Chagoda, E	1098		0200-2400	RR,Reg
VG	Nokolsk, E	1098		0200-2400	RR,Reg
VG	Totma, E	1098		0200-2400	RR,Reg
VG	Ustyuzhna, E	1098		0200-2400	RR,Reg
	Unknown, E	1098		0130-2200	3
AR	Arkhangelsk, E	1107	5	0200-0100	2
KB	Nalchik, E	1107	25	0200-2400	RR,Reg
KB	Tegenekli, E	1107		0200-2400	RR,Reg
MO	R. Europa Plus, Moskva, E	1116	5	0300-0100	Ind
KA	Kaliningrad, E	1116	30	0200-2200	2,Reg
KD	Sochi, E	1116	30	0200-2400	RR
CH	Atamanovka, S	1116		2100-1500	2
YV	Birobidzhan, FE	1116	20	2100-1500	2
SP	Sankt-Peterburg, E	1125	150	0300-2100	4
	Unknown. No. E	1134		0300-2200	3
RO	Volgodonsk, E	1134		0300-2200	2
	Svedchiki, E	1134		0200-2200	3
KA	Bolshakovo, E	1143	150	0200-0100	3
SM	Samara, E	1143	100	0200-2200	3,Reg
BU	Unknown, S	1143		2200-1700	3
MO	Moskva, E	1152	50	0200-2400	1
KY	Khanty-Mansiysk, S	1152	20	0100-2100	3,Reg
KH	Komsomolsk, FE	1152	50	2000-1400	3,Reg
VO	Volgograd, E	1161		0300-2100	2,4
KD	Krasnodar, E	1170	500	0000-1630	3
				1630-2130	F
IR	Angarsk, S	1170	250	2200-1800	2
AT	Unk, S	1170		2300-1700	2
SP	Sakt-Peterburg, E	1188	10	DW,BBC	
KY	Berezovo, S	1197	5	0000-2200	RR,Reg
MO	Ekho Moskvy, Moskva, E	1206	10	24h	Ind
KA	Bolshakovo, E	1215	1200	0400-1400	3
				1400-2000	F
KA	R. Nadezhda, Bolshakovo, E	1215	1200	2000-2300	Ind
BU	Unknown, S	1224		2200-1700	3
KH	FEBC, Khabarovsk, FE	1224	100?	0900-1030v	Ind
	Unknown, No. E	1233		0200-2200	3
MO	R. 101, Moskva, E	1233		0300-2400	Ind
	Unknown, E	1251		0200-2400	1
KC	Urup, E	1251	5	0200-2400	RR,Reg
	Unknown, E	1251		0200-2200	2
PM	Vladivostok, FE	1251	1000	2000-2100	1
				2300-1800	1
PM	R. St. Vladivostok, FE	1251	1000	2105-2300	Ind
MO	R. St. SNC, Moskva, E	1260	10	0600-0400	Ind
SP	R. St. SNC, Sankt-Peterburg, E	1260	10	1000-	Ind
SP	Golos Nadezhdy, Sankt-Peterburg, E	1260	10	0300-0500	Ind

NN	Nizhniy Novgorod, E	1278	100	0200-2200	2
SR	Saratov, E	1278		0200-2200	2
SR	Balakovo, E	1278		0200-2200	2
SR	Balashov, E	1278		0200-2200	2
SR	Yershov, E	1278		0200-2200	2
UD	Balezino, E	1278		0200-2200	2
VO	Dubovka, E	1278		0200-2200	2
RO	Novocherkassk, E	1278	50	0200-2200	2
KD	Krasnodar, E	1278		0200-2200	2
BU	Severobaykalsk, S	1278		2220-2000	RR
BU	Bagdarin, S	1278		2220-2000	RR
SA	Unknown, S	1278			2
KL	Sukhinichi, E	1287		0200-2200	2
CI	Groznyy, E	1287		0200-2200	2
KO	Syktyvkar, E	1287		0200-2200	2
BA	Yazykovo (Ufa), E	1287		0200-2200	2
NK	Novosibirsk, S	1287		2300-1700	2
BU	Kyakhta, S	1287		2200-2000	RR
TU	Shagonar, S	1287			RR,Reg
TU	Shadan, S	1287			RR,Reg
	Unknown, FE	1287			3
MO	R. Vozrozhdeniye, Moskva	1305	10	0400-2200	Ind
	Unknown, We. S	1305		0000-2000	3
IR	R. St. Volna Baykala, Irkutsk, S	1305	0.2	0100-1300	Ind
	Unknown, Ural area	1323		0100-1900	2
	Unknown, So. E	1323		0100-2200	3
MO	Moskva, E	1332		0200-	FLS
IR	Inta R, Angarsk, S	1332	0,2	1000-1200v	Ind
IR	Angarsk, S	1332	0,2		3
AL	Onguday, S	1350		2200-2000	RR
AL	Ust-Kan, S	1350		2200-2000	RR
AL	Ust-Ulagan, S	1350		2200-2000	RR
MO	Chkalovskaya	1359	15	0300-1600	4
				1600-2200	NP
PR	Perm, E	1359	50	0100-2400	3
BU	Unkown, S	1359		2200-1700	3
YE	Yekaterinburg, S	1377		0100-2000	2
PM	Vladivostok, FE	1377	30	2000-1600	2
SL	Okha, FE	1377		1800-1600	RR
KA	Bolshakovo, E	1386	2500	1600-2100	F
KA	R. Ala, Bolshakovo, E	1386	2500	2100-2400	Ind
				0300-1600	Ind
IR	Inta R, Irkutsk, S	1386	0.4	0000-0300v	Ind
				0900-1200v	Ind
KT	Petrozavodsk, E	1395		0200-2200	2
OR	Bugurusian, S	1395		0000-2200	RR,Reg
YA	Volga, E	1395		0200-2200	2
YA	Lyubim, E	1395		0200-2200	2
YA	Dubki, E	1395		0200-2200	2
TS	Naberezhnyye Chelny, E	1395	5	0200-2200	3
BA	R Shark, Ufa, E	1395		0200-2000	Ind
BA	R. Rezonans, Ufa, E	1395			Ind
VO	Volgograd, E	1413		0200-2400	RR
SA	Sangar, S	1413		2000-1800	RR

CH	Kuanda, S	1422		0000-2000	2
AL	Ust-Koksa, S	1440		2200-2000	RR
MO	R. Rezonans, Moskva, E	1440	7.5	0700-1400	Ind
MO	R. Centre, Moskva, E	1440	7.5	0400-0500	Ind
MO	R. Nadezhda, Moskva, E	1440	7.5	1500-1800	Ind
				2000-2300	Ind
AL	Kosh-Agach, S	1440		2200-2000	RR
TV	Kushalino, E	1449	25	0200-2400	RR,Reg
	Unknown, No. E	1449		0245-2100	2
BR	Unecha, E	1449	5	0200-2200	3
	R. Ala, unknown, E (r.)	1449			Ind
BE	Valuyki, E	1458		0200-2400	2
KI	Pinyug, E	1458		0200-2200	3
KP	Kudymkar, E	1458	7	0100-2300	1,RR,Reg
OM	Tara, S	1458		0100-1900	2
MO	Moskva, E	1467	5	0200-2400	1
VO	R. Vedo, Volgograd, E	1467		0500-0900v	Ind
				1500-1930	Ind
TL	Aleksin, E	1476		0200-2400	1
PM	Vladivostok, FE	1476	150	0900-1600	F
				1700-0900	3
AL	Onguday, S	1476		2200-2000	RR
PR	Oktyabrskiy, E	1485	1	0000-2200	RR
	Unknown, Urals	1485			3
CB	Unknown, S	1485		0000-2200	RR,Reg
YN	Tazovskiy, S	1485		0000-2200	RR,Reg
YN	Krasnoselkup, S	1485		0000-2200	RR,Reg
SA	Tiksi, S	1485		2000-1800	RR
SA	Cherskiy, S	1485		2000-1800	RR
KK	Kamenskoye, FE	1485		1800-1600	RR
SP	R. Ala, Sankt-Peterburg	1494	1200	2100-1630	Ind
SP	Sankt-Peterburg, E	1494	1200	1700-2100	F
IR	R. St. Remira, Irkutsk, S	1500	0.2	0000-1000v	Ind
BE	Belgorod, E	1503		0200-2200	2
YN	Salekhard, S	1503		2300-2000	3
BE	Staryy Oskol, E	1512	30	0200-2200	2
KD	Sochi, E	1512		0200-2200	2
CV	Ibresi, E	1512	7	0200-2400	1,Reg
PR	Chaykovskiy, E	1512	5	0000-2200	1,RR,Reg
MU	Unknown, E	1521	5	0000-2300	3
	Unknown, Ce. E	1521		0100-2200	3
KN	Boguchany, S	1521		2200-2000	RR
IR	RFT, Shelekhov, S	1530	0.2	0400-1300v	Ind
AB	Aginskoye, S	1530			3,Reg
MO	R. St. Delovaya volna, Moskva, E	1539		0600-0900	Ind
				1100-1500	Ind
	Unknown, FE	1566		2000-1800	1
SL	Yuzhno-Sakhalinsk, FE	1575	5	1800-1600	RR,Reg
	Unknown, So. E	1584		0200-2400	RR
OR	Pleshanovo, E	1584		0000-2200	RR,Reg
DA	Khunzakh, E	1584		0200-2400	RR
BU	Taksino, S	1584		2200-2000	RR
KM	Klyuchi, FE	1584		1800-1600	RR
KK	Tigil, FE	1584		1800-1600	RR
	Unknown, No. E	1593			
SL	Neftegorsk, FE	1593		1800-1600	RR
IR	Angarsk, S	1593		2300-1700	2
DA	Botlikh, E	1602		0200-2400	RR
BE	Novyy Oskol, E	1602		0200-2400	2
KP	Gayny, E	1602	5	0000-2200	RR,Reg
TO	Kargasok, S	1602		2200-2000	RR,Reg
TO	Molchanovo, S	1602		2200-1800	RSM
BU	Novo-Ilinsk, S	1602		2200-2000	RR
BU	Ust-Barguzin, S	1602		2200-2000	RR
SL	Kurilsk, FE	1602		1800-1300	3
SL	Severo-Kurilsk, FE	1602		1800-1300	3

SHORTWAVES

Domestic Services on Shortwave (incl. 60mb Ext. Sces.). Powers quoted are estimates. *) low power. **) may be inactive. s = SSB transmission.

Rg	St. name, location	kHz	kW	H. of tr.	Prgr
NK	Kazakh R, Novosibirsk, S	3955		0000-2300	KR2
MO	R. Ala, Moskva, E	3995		0300-0200	Ind
TU	Kyzyl, S	3995	*	**	
KH	Khabarovsk, FE	3995	50	(Wi. only)	3
PM	Vladivostok, FE	4010	100	(Wi. only)	F
CK	Anadyr, FE	4030	*	**	RR,Reg
EV	Tura, S	4040	1.5	0030-	RR,Reg
PM	Vladivostok, FE	4040	50	0600-0500	3
SL	Yuzhno-Sakhalinsk, FE	4050s	50	1800-1700	RR,Reg
	Tver (?), E	4055	50	0200-2400	1
SA	Yakutsk, S	4395	*	2000-1600	3
BA	Ufa, E	4485	50	0000-2200	RR,Reg
KM	Petropavlovsk, FE	4485	100	1700-1600	RR,Reg
KK	Palana, FE	4520	5	2000-0800v	Reg
KY	Khanty-Mansiysk, S	4520	50	0000-2000	3
KH	Komsomolsk, FE	4610	15	2000-1800	RR,Reg
	incl. R. St. Vostok			0800-0930v	Ind
BU	Ulan-Ude, S	4795	*	2158-2100	RR,Reg
SA	Yakutsk, S	4800	50	variable	RR,Reg
PM	Vladivostok, FE	4810	100	0930-1500	F
	Unknown, S	4820		0000-2200	RR
KY	Khanty-Mansiysk, S	4820	50	0000-2200	RR,Reg
SA	Yakutsk, S	4825	50	(Wi. only)	RR,Reg
CH	Chita (Atamanovka), S	4860	100	2000-1800	RR,Reg
TY	Tyumen, S	4895	50	0000-2400	RR,Reg
SA	Yakutsk, S	4920	50	variable	RR,Reg
	Unknown, S	4930		0000-2300	3
SA	Yakutsk, S	4940	50	(Wi. only)	RR,Reg
AR	Arkhangelsk, E	5015	50	0200-2400	RR,Reg
PM	Vladivostok, FE	5015	50	1800-1700	RR,2,Reg
SP	R. Ala, Sankt-Peterburg, E	5040	200	1520-2400	Ind
				0100-0600	Ind
KY	R. Berezovo, Berezovo, S	5070		0515-0700v	Ind
PR	Perm, E	5290	20	0000-2200	RR,Reg
KN	Krasnoyarsk, S	5290	50	2200-2000	RR,Reg
BA	R. Shark, Ufa, E	5780	15	0200-2000	Ind
VO	R. Vedo, Volgograd, E	5915	20	0500-0900v	Ind
SA	R. St. Yakutsk, Yakutsk, S	5920	20	0600-1400	Ind
MU	Monchegorsk, E	5930	50	0200-2400	RR,Reg
MA	Arman, FE	5940	50	1700-1500	RR,Reg
KH	Khabarovsk, FE	6005	20	2300-1500	3,Reg
MO	Lithuanian R, Balashikha, E	6010	20	0400-1700	LTR1
KA	R. Ala, Bolshakovo, E	6015	100	1500-0600	Ind
SP	R. Ala, Sankt-Peterburg, E	6025	400	alt. to 7380kHz	
MU	R. St. Onega, Murmansk, E	6030	15	1000-1200v	Ind
SP	R. Polis, Sankt-Peterburg, E	6045	200	0430-1700	Ind
MU	Murmansk, E	6050	20	inactive	RR,Reg
YE	R. Dvizhenie, Yekaterinburg, S	6090		1300-1700	Ind
IR	R. St. ves Irkutsk, Irkutsk, S	6090	50	2330-1200v	Ind
SM	Tatar R, Tolyatti, E	6115	20	0330-1630	1,Reg
TS	Kazan, E	6120	20		1,Reg
YE	R. Pamyat, Yekaterinburg, S	6145	50	2300-0030	Ind
MO	Belorussian R, Balashikha, E	6150	20	0400-1700	BR1,Reg
KP	Kudymkar, E	6165	3	0000-2200	RR,Reg
YE	R. Dvizhenie, Yekaterinburg, S	6200			Ind
YE	R. N, Yekaterinburg, S	6200		0500-1000	Ind
VO	R. Vedo, Volgograd, E	7125		0500-0900v	Ind
SA	Yakutsk, S	7140	50	variable	RR,Reg
MO	Armenian R, Balashikha, E	7175	20	0300-1500	
PR	Tatar R, Perm, E	7185	3	0200-1745	RR,Reg
VO	R. Vedo, Volgograd, E	7185	20	0500-0900v	Ind
				1500-1930	Ind
SA	Yakutsk, S	7200	50	variable	RR,Reg
KH	Khabarovsk, FE	7210	50	2000-1900	RR,Reg
	incl. R. St. Vostok			0800-0930	Ind
YE	R. Pamyat, Yekaterinburg, S	7230	200	1430-1600 (Wi. only)	Ind

SA	Yakutsk, S	7265	50	variable	RR,Reg
MO	Azerbaidzhani R, Balashikha, E	7300	20	0300-1600	AZR2
MA	Arman, FE	7320	50	1500-1300	RR,Reg
SA	Yakutsk, S	7345	50	variable	RR,Reg
YE	R. Ala, Yekaterinburg, S	7370	400	1430-0600	Ind
SP	R. Ala, Sankt-Peterburg, E	7380	400	1300-1630	Ind
				1830-2300	Ind
SP	R. Ala, Sankt-Peterburg, E	7400	200	0630-1500	Ind
MA	Arman, FE	9530	50	1500-1300	RR,Reg
KH	FEBC, Khabarovsk, FE	9560	20	0900-1030v	Ind
MA	Arman, FE	9600	50	1700-1500	RR,Reg
VO	R. Vedo, Volgograd, E	9655	20	1500-1930	Ind
MO	Lithuanian R, Balashikha, E	9675	20	1700-2200	LTR1
MO	Kazakh R, Balashikha, E	9690	20		KR2
MO	R. Pamyat, Moskva, E	11650	20	1600-1900 (Su. only)	Ind
MO	R. Art, Moskva, E	11655		0515-0700	Ind
SP	R. Ala, Sankt-Peterburg, E	11685	400	0300-1400	Ind
VO	R. Vedo, Volgograd, E	11760	5	1500-1930	Ind
KP	Kudymkar, E	11770	3	1400-2200	RR,Reg
SL	Yuzhno-Sakhalinsk, FE	11840	50	1800-1700s	RR,Reg
SM	Tatar R, Tolyatti, E	11905		1630-1900	1,Reg
MO	Tatar R, Balashikha, E	11945	20	0330-1500	1,Reg
KA	R. Ala, Bolshakovo, E	11965	100	0630-1430	Ind
YE	R. Ala, Yekaterinburg, S	12030	200	0700-1400	Ind
YE	R Pamyat, Yekaterinburg, S	12030	200	1430-1600 (Su. only)	Ind
MO	R. Pamyat, Moskva, E	12060	20	1600-1900 (Wi. only)	Ind
MO	Belorussian R, Balashikha, E	13645	20	1700-2300	BR1
VO	R. Vedo, Volgograd, E	13710	20	1500-1930	Ind
PR	Tatar R, Perm, E	15200	3	1800-2100	RR,Reg
YE	R. Ala, Yekaterinburg, S	15255	200	0230-1400	Ind
MU	Murmansk, E	15350	20	inactive	RR,Reg
MO	Tatar R, Balashikha, E	17810	20	1505-	1,Reg
MO	Kazakh R, Balashikha, E	21490	20		

"OSTANKINO" PRGRS

Radio 1 (R-1): Aired in a "European" edition 0200-2400 for time zones UTC +2, +3, +4 and four time-shifted versions "Orbita-1" to "Orbita-4" opening at 1800, 2000, 2200, 0000, respectively. Each edition has a total transmission time of 22h every day of the week. Radio 1 is a general style "full service" prgr.

Radiostantsiya Yunost (RSY): The youth channel, with music and other prgrs for the younger generation. On the air continuously. European block: 0300-2200

Radiostantsiya Mayak (RSM): Light music, sports. On the air continuously. IS and news, weather or other information on the h and half h.

Radio Orfey RO: Classical music, drama. On the air to European Russia 0300-2100.

R-1, RSY and RSM are rebroadcast to a varying extent in other republics of the former USSR.

RADIO ROSSII PRGR

Radio Rossii is on the air continuously, but usually has daily breaks for transmitter maintenance at the same times as Radio 1. R. Rossii is a general style prgr with a heavy news emphasis. There are time shifted versions as with Ostankino R-1.

REGIONAL SERVICES (Gov.)

In most cases regional services produced by municipal companies are broadcast for a few hours a day on tr's shared with national services (R-1, RSY, RSM or R. Rossii). A considerable number of regions have no regional services on AM and use FM and/or wired networks only.

INDEPENDENT PRIVATE PRGR PRODUCERS

A number of companies produce prgrs using government owned studios and/or tr's. In some cases various prgr producers share transmitter time. Several of these production companies are joint ventures with the government owned radio.

INDEPENDENT STATIONS

Companies which possess their own production and transmitting equipment.

Addresses and other information:

a) Regional services

Resp. = respublika, obl. = oblast, ul. = ulitsa

AB) Aginskiy Buryatskiy avt. okrug: 674460 Aginskoye. **Reg.** prgrs on 1530kHz; no prgr details available.

AD) Resp. Adygeya: ul. Zhukovskogo 24, 352700 Maykop. **Reg:** On 1089kHz: Tues/Thurs/Sat 0500-0600, Sat 1200-1400, Mon-Fri 1300-1400.

AL) Resp. Gornyy Altay: pr. Kommunisticheskiy 37, 659700 Gorno-Altaysk. No recent freq/prgr details available. **Reg.** prgrs possibly via Barnaul 279kHz.

AM) Amurskaya obl: RTV Center, Internatsionalniy per. 15, 675000 Blagoveshchensk. **Reg:** On 189kHz: 2120-2140, 2230-2240W, 2230-2300 Sun, 2315-2345 Sat/Sun, 2325-2400 Tues-Fri, 2345-2400 Mon, 0215-0300 Tues-Fri, 0515-0600 Sat, 0515-0615 Sun, 0940-1010 Wed/Thurs, 0940-1030 Tues/Fri, 0940-1040 Mon, 1115-1200 Fri, 1215-1315 Mon/Wed/Sat. Prgr Zeya on 72.4MHz Sat/Sun 0100-0400.

AR) Arkhangelskaya obl: ul. Popova 2, 163000 Arkhangelsk. **Reg:** On 918kHz, 66.22MHz: Mon-Fri 0400-0500 & 1520-1600, Sat 0500-0600, 1100-1130 & 1300-1310.

AS) Astrakhanskaya obl: ul. Molodoy Gvardii 17, 414000 Astrakhan. **Reg:** On 792kHz: Mon-Fri 0330-0400, Sat/Sun 0515-0540, Tues-Fri 0525-0540, Sat/Sun 0830-0900, Sat 1030-1100, Sun 1215-1300, W 1500-1600.

AT) Altayskiy kray: Zmeinogorskiy trakt 27-A, 656020 Barnaul. **Reg:** On 279kHz: 0000-0100 Mon-Fri, 0115-0145 Sun, 0125-0200 & 0315-0600 Sat, 0430-0500 Sun, 0443-0500 Tues/Thurs/Fri, 1115-1200 Sun, 1145-1245 Mon-Sat. **German:** Sat 0425-0455.

BA) Resp. Bashkortostan: ul. Gafuri 9, 450076 Ufa. **Reg:** On 692/4485kHz, 66.88MHz in Russian/ Bashkirian: 0215-0300, Sat 0325-0545, Sat/Sun 0515-0700, 0800-0900, Mon-Fri 1300-1400 & 1415-1445 & 1500-1600 (irr.)

BE) Belgorodskaya obl: 308000 Belgorod. **Reg:** On FM only. Details not available.

BR) Bryanskaya obl: ul. Stanke Dimitrova 77, 241033 Bryansk. **Reg:** On FM only. Details not available.

BU) Resp. Buryatiya: RTV Center, Erbanova 7, 670000 Ulan-Ude. **Reg:** On 279/4795kHz in Russian/Buryat: D 2158-2200 & 2210-2300; Mon-Fri 2320-2400 & 1120-1200; Sun 0430-0500; Sat (varying) 0100-0200, 1000-1100.

CB) Chelyabinskaya obl: ul. Ordzhonikidze 54-B, 454000 Chelyabinsk. **Reg:** On 738/1485kHz, 69.65MHz: D 0054-0100, W 0130-0200 & 0215-0240, Sun 0315-0400, W 0325-0400, Sun 0600-0700, Mon 0615-0700, Tues-Thurs 0730-0800, Mon-Fri 1300-1340.

CH) Chitinskaya obl: RTV Center, ul. Kostushko-Grigorovicha 27, 672090 Chita. **Reg:** In Russian/Buryat on 261/4860kHz: 1959-2003, 2110-2200, Sat 0220-0300, Tues-Thurs 0310-0330, Mon-Fri 0910-1100, Sat/Sun 1110-1200. Prgr Dauriya: On 67.92MHz Mon-Sat 0200-0800, Sun 0300-0700.

CI) Resp. Checheno-Ingushiya: Dom Radio, ul. Dzerzhinskogo, 364021 Groznyy. **Reg:** On 657kHz in Russian, Chechen, Ingush: 0430-0500, 0515-0600, 0615-0700 Mon/Tues, 0815 (Sat 0715)-0900, 1215-1400 (var.), 1415-1600 Thurs/Fri, 1500-1540 /Mon-Wed, 1500-1600 Sat/Sun.

CK) Chukotskiy avt. okrug: ul. Lenina 18, 686710 Anadyr. **Reg:** On 692kHz (4030kHz inactive) in Russian/Chukchi/Eskimo: D 1825-1900, Mon-Fri 1940-2000, 2015-2100, Sat 2115-2145, Thurs 0115-0145, Mon-Fri 0700-0800, Mon/Wed/Fri 0815-0900 (also on 5940/9600kHz via Arman, MA). Also rel. R. Magadan, MA.

CV) Resp. Chuvashiya (Chavashen): ul. Nikolayeva 4, 428000 Cheboksary-20. **Reg:** On 531/1512kHz in Russian/Chuvash: W 0315-0400 & 0415-0430, Sun 0430-0500 & 0515-0600, Sat/Sun 0715-0800 & 1400/1415-1600, Mon-Fri 1500-1600 (Mon1540), Sun 1645-1745.

DA) Resp. Dagestan: ul. Magomed Gadzhieva 188, 367032 Makhachkala. **Reg:** 621/828/918/1584kHz (separate prgrs). Russian: D 0340-0400, Sat 0800-0900 & 1215-1300, Sun 0525-0600 & 0820-0900, Wed 0815-0900, 1500-1540/1600. At 0315 and other times in Avar, Azeri, Chechen, Dargin, Kumyk, Lak, Lezgin, Nogay, Tabasaran, Tat.

EV) Evenkiyskiy avt. okrug: ul. 50 let Oktyabrya 28, 663370 Tura. **Reg:** On 4040kHz in Russian, Evenkish: Mon-Fri 0030-0100 & 0515-0600, Mon/Wed/Fri/Sat 0115-0145, Tues/Thurs 0115-0200, Sat 0215-0300.

IR) Irkutskaya obl: RTV Center, ul. Gorkogo 15, 664000 Irkutsk. **Reg:** On 234kHz, 70.31MHz: Mon-Fri 2210-2300, 1100-1200, Tues-Thurs 2320-2400, Mon 0400-0500, Tues-Fri 0400-0410, Sat 0200-0300, Sun 0100-0130. Local stereo prgr on 72.02MHz 0815-1000 or 1100.

IV) Ivanovskaya obl: ul. Varentsovoy 24, 153647 Ivanovo. **Reg.** prgrs on FM only. Details not available.

KA) Kaliningradskaya obl: RTV Center, ul. Basseinaya 42, 236032 Kaliningrad; or (radio) ul. Komsomolskaya 61, 236000 Kaliningrad. **Reg:** On 1116kHz: Sat/Sun 0500-0545, Mon-Fri 0525-0545, W 0615-0700, Sat 0830-0900 & 1215-1300, Sun 1300-1400, Mon-Fri 1500-1540. On 1386kHz prgr Dlya tekh kto v more: Sun 0000-0030.

KB) Resp. Kabardino-Balkariya: ul. Nogmova 38, 360000 Nalchik. **Reg:** On 1107kHz: Russian: D 0455-0500, Tue-Sat 0525-0530, Thurs/Fri 0815-0900, Sun 1215-1300, W 1300-1400; Balkar: D 0515-0530, Tues-Sat 0520-0525, Wed 0815-0900, Sun 1300-1400, W 1400-1500; Kabardin: D 0430-0455, Tues-Sat 0515-0520, D 1500-1600.

KC) Karachayevo-Cherkesskaya SSR: Dom Radio, 357100 Cherkessk. **Reg:** On 1251kHz, 70.30/72.10 MHz in Russian, Abazin, Cherkess, Karachai, Nogai: 0330-0400, 0615-0700, 0715-0800 (Tues-Sun), 0815-0900 (Thurs/Fri), 1300-1400, 1415-1500 (Mon/Fri), 1645-1745 (Mon/Wed). Also rel. ST.

KD) Krasnodarskiy kray: RTV Center, ul. Radio 5, 350630 Krasnodar. **Reg:** On 1089kHz: Mon-Fri 0330-0400, Sat 0400-0500, Tues-Fri 0525-0545, Sun 0525-0600, Sat 0715-0900, Sat/Sun 0830-0900, Mon-Fri 1500-1600.

KE) Kemerovskaya obl: 650000 Kemorovo. **Reg:** On FM only. Details not available.

KG) Resp. Kalmykiya-Khalmg Tangch: M. Gorkogo 34, 358000 Elista. **Reg:** In Russian/Kalmyk on 846kHz: Tues-Sat 0430-0500, Sat 0515-0600, Sat/Sun 1100-1300, Fri 1215-1300, Mon/Wed 1500-1540, Thurs/Fri 1540-1600, Tues/Fri 1700-1800.

KH) Khabarovskiy kray: RTV Center, ul. Lenina 71, 680013 Khabarovsk, or: Dom Radio, pl. Slavy, 682632 Khabarovsk. **Reg: 1st prgr:** On 153/216/621/4610 (exc. for 2nd prgr)/7210kHz + FM (MHz): 66.32 (Birobidzhan), 68.70 (Komsomolsk), 70.16 (Chegdomyn), 70.28 (Khabarovsk), 70.70 (Bidzhan): 2010-2100, 2110-2200, Mon-Sat 2320-2400 & 0200-0300 & 0930-1100. R. St. Tikhiy Okean (cf. PM). Other times R. Rossii. **2nd prgr:** On 711/999/1152/4610(exc. 0800-0930)kHz: 0200-0800. At other times R. St. Mayak. Prgr. "Amur": 72.02MHz (Khabarovsk) 0300-0800. Komsomolsk local prgr: On 68.70MHz: Mon-Fri 0930-0945.

KI) Kirovskaya obl: ul. Uritskogo 34, 610002 Kirov. **Reg:** On FM only (Kirov 66.86MHz). Other details not available.

KK) Koryakskiy avt. okrug: 684620 Palana. **Reg:** On 738/4520kHz in Russian, Koryak: Mon-Sat 2000-2030, Tues-Thurs 2145-2200, Tues-Wed 0115-0145, Sat 0700-0800.

KL) Kaluzhskaya obl: 248000 Kaluga. **Reg:** On FM only. Details not available.

KM) Kamchatskaya obl: ul Sovetskaya 62 (or: Nikolskaya sopka), 683000 Petropavlovsk-Kamchatskiy. **Reg:** On 180/4485kHz, 69.68MHz: W1830-2000 (Sat 1900), Sat/Sun 2030-2100, Fri 0000-0100, Mon-Fri 0720-0800. Prgr for fishermen "Kamchatka rybackaya": Tues/Thurs/Sat 2000-2100 on 2 freq's in the 31, 25 or 21mb (200kW).

KN) Krasnoyarskiy kray: RTV Center, Sovetskaya 128, 660017 Krasnoyarsk or Radio, ul. Mechnikova 44a, 660028 Krasnoyarsk. **Reg:** On 216/5290kHz, 67.26MHz (rel. by TM 162kHz, RK 792kHz): Mon-Fri 2320-0100 & 1120-1200 & 1220-1300(irr), Sat 0000-0200 & 0410-0530 & 0900-1000(irr), Sun 0000-0020 & 0300-0400 & 1100-1200.

KO) Komi SSR: Oktyabrskiy pr. 164, 167610 Syktyvkar. **Reg:** On 621kHz in Russian, Komi: W 0315-0400, D 0525-0600, Mon-Fri 1055-1100 & 1500-1540, Sun 0430-0500 & 0820-0900.

KP) Komi-Permyatskiy avt. okrug: 617240 Kudymkar. **Reg:** On 1458/1602kHz & 6165 or 11770kHz: W 0250-0300 & 0315-0345 (Sat 0400), Mon-Fri 1415-1435/1500. Also rel. PR.

KR) Kurskaya obl: 305000 Kursk. **Reg:** On FM only. Details not available.

KS) Kostromskaya obl: ul. Nikitskaya 10, 156005 Kostroma. **Reg:** On FM only. Details not available.

KT) Resp. Kareliya (Karjalan tasavalta): ul. Frunze 20, 185630 Petrozavodsk. **Reg:** On 765kHz, 70.28MHz in Russian, Finnish, Karelian, Vepsian: Mon-Fri 0400-0600, Sat 0500-0700, Sun 0610-0700 & 0800-0900, Mon-Fri 1100-1200 (Wed/Thurs 1300), Sat 1200-1300, Mon-Fri 1520-1600, Thurs 1400-1500, Sat/Sun 1400-1500. Finnish: Wed/Fri 0445-0500, Mon-Fri 1100-1115 & 1520-1535, Wed/Fri 1115-1140, Thurs 1200-1300 & 1400-1500, Sat 1200-1300, Sun 0610-0630.

KU) Kurganskaya obl: ul. Sovetskaya 105, 640018 Kurgan. **Reg:** On 738kHz (Chelyabinsk) & FM (MHz) 66.89 (Shumikha), 68.48 (Makushino), 69.23 (Shadrinsk), 71.87 (Kurgan): W 0315-0330 & 0410-0440, Mon-Fri 1300-1400.

KY) Khanty-Mansiyskiy avt. okrug: ul. Mira 7 (or: RTV Center, ul. Lenin 21), 626200 Khanty-Mansiysk. **Reg:** On 225/1152/1197/4820/4520(irr.)kHz in Russian, Khanty, Mansi: Mon-Fri 0210-0230, 0800-0900, Sat 1000/1030-1100. Also rel. R. Tyumen, TY.

LI) Lipetskaya obl: 398000 Lipetsk. **Reg:** On FM only. Details not available.

MA) Magadanskaya obl: ul. Kommuny 8/12, 685000 Magadan-13. **Reg:** On 234/5930/7320/ 9530/9600kHz: D 0010-0100, Mon-Fri 2000-2100 & 0115-0200 & 0320-0400 & 0720-0800v, Sat/Sun 2020-2100, Sat 0111-0120(0200). 0815-0900 rel. R. St. Tikhiy Okean on 234kHz and available SW (D). Chukčhi/Eskimo prgr from R. Anadyr, CK on 5940/9600kHz (Mon/Wed/Fri) 0700-0800.

MD) Mordovskaya SSR: ul. Dokuchayeva 29, 430000 Saransk. **Reg:** On 1062kHz in Russian/ Mordovian: W 0310, Sun 0415, Mon-Fri 1300-1500.

ME) Mariyskaya SSR-Mari El: RTV Center, ul. Osipenko 50, 424014 Yoshkar-Ola. **Reg:** On 900kHz in Russian/Mari: W 0330-0400, D 0430-0500, D 0515-0545 (Sun 0600), Sat 1215-1300 & 1415-1500, Sun 1415-1600, Mon-Fri 1500-1540, Sat 1635-1730.

MO) Moskovskaya obl: ul. Akademika Koroleva 19, 127427 Moskva. **Reg:** R. Podmoskovskoye, ul. Pyatnitskaya 25, 113326 Moskva. On 846kHz/66.44MHz: D 0400-0500, 0600-0700, Sun-Thurs 1400-1600, Fri 1500-1600, Sat 1200-1300. For Moskva city on 846kHz, 66.44/69.80MHz: Mon-Fri 0315-0330, 0515-0545,; Sat 0315-0330, 0830-0900; Sun 0415-0500, 0515-0545. On 69.80MHz: W 1215-1230 (Mon/Fri/Sat 1300), Mon-Fri 1500-1600 (Mon/Wed 1540). **R-1** on 171/1152kHz, 66.44MHz. **RSY** on 792kHz, 66.84MHz. **R. St. Mayak** on 198/549kHz, 67.22MHz. **R. Orfey** on 738kHz, 1359kHz (to 1600), 72.14MHz. **R. Rossii** on 261/873kHz. **Prgr. Novyy Peterburg** on 1359kHz 1600-2200. **prgrs for foreigners** on 918/1116/1332kHz.

MU) Murmanskaya obl: RTV Center, sopka Varnichnaya, 183042 Murmansk. **Reg:** On 657/5930kHz: Mon-Fri 0400-0530, Sun 0600-0615, Sat 0720-0800, Sun 0820-0900, Mon-Fri 0910-1000, Sat 1000-1030, Sun 1030-1100 & 1120-1200 & 1300-1400, Mon-Fri 1500-1530. **R. Pasvik:** Wed 1500.

NE) Nenetskiy avt. okrug: ul. Smidovicha 19, 164700 Naryan-Mar. **Reg:** On 711kHz: 0330-0400.

NK) Novosibirskaya obl: POR-1, ul. Rimskogo-Korsakova 9, 630048 Novosibirsk or ul. Vertkovskaya 10, 630048 Novosibirsk. **Reg:** On 270kHz, 67.8MHz: Mon-Sat 2330-2400, Tues-Fri 0015-0100, Sat/Sun 0215-0300 & 0330-0400, Sat 0415-0500 & 0715-0800 & 0900-1000, Sun 0930-1000, Mon-Fri 1015-1100, Sun-Fri 1115-1145. Stereo prgr on 69.28MHz 1100-1500.

NN) Nizhnenovgorodskaya obl: ul. Belinskogo 9-A, 603600 Nizhniy Novgorod, GSP-228. **Reg:** On 828kHz, 67.72MHz: Mon-Fri 0230-0300 & 0330-0400 & 0415-0430 & 0710-0715, Sun 1000-1200, Mon-Fri 1400-1500, Fri 1500-1600.

NO) Novgorodskaya obl: ul. Sovorovskaya 16, 173610 Novgorod. **Reg:** On FM or wired netw. only. Mon-Fri 1300-1400, 1520-1600, Sat 0510, 1520, Sat/Sun 1800-1900 stereo prgrs on 72.95MHz.

OL) Orlovskaya obl: ul. 7 noyabrya, 302000 Orel. **Reg:** On FM only. Details not available.

OM) Omskaya obl: pr. Mira 2, 644008 Omsk. **Reg:** On 639kHz, 69.75MHz: 0059-0105, Sun-Fri 0115-0200, Sat/Sun 0215-0300, Thurs/Sat/Sun 0315-0400, Sat/Sun 0515-0600, Sun 0930-1000, Tues-Fri 1300-1340. German: Fri 0115-0200.

OR) Orenburgskaya obl: Televizionniy per. 3, 460024 Orenburg. **Reg:** On 270/1395/1584kHz Mon-Fri 0115-0200, Sat/Sun 0315-0400, Tues-Thurs 0615-0700, Sat 0600-0700, Sun 0930-1030, Mon-Fri 1415-1500. Tatar: Tues 1300-1340.

PE) Penzenskaya obl: RTV Center, ul. Lermontova 39, 440602 Penza. **Reg:** On 855kHz, 70.66MHz: W 0400-0500, Sat 0530-0600, Sat/Sun 0530-0600, Mon-Fri 1520-1600.

PM) Primorskiy kray: RTV Center, ul. Uborevitsa 20-A, 690000 Vladivostok-Central. **Reg:** 1st prgr on 234kHz, 5015kHz (part time), 71.84MHz: 1954-2000, Mon-Fri 2030-2200, Sat/Sun 2058-2315 & 0115-0200, Sat 0230-0400, Tues-Fri 0300-0400, 0815-0900, 0930-1010 (Mon-Fri 1030), Mon/Wed/Thurs 1045-1100, Fri 1100-1115. 2nd prgr on 810/5015 (from 0700) kHz: 0045-0100, 0500-0515, Wed/Thurs 0700-0715, Mon/Tues 0700-0815, Fri 0700-1400, Sat/Sun 0715-1400, other days 0930-1400. **Radiostantsiya Tikhiy Okean (R. St. Pacific Ocean):** 0815-0900, 1900-1945 on various LW/MW/SW freq's in eastern Russia, locally

also 0815 on 243kHz, 1215 on 810kHz. Segment in **English** Sat 0850, 1250, 1935.
PR) Permskaya obl: ul. Tekhnicheskaya 21, 614600 Perm GSP-128. **Reg**: On 585/1458/1512/1602/5290/6165/11770kHz: Mon-Fri 0057-0059 & 0130-0200, Thurs-Sat 0200-0245, Mon-Wed 0215-0245, Sun 0230-0300 & 0325-0400, Mon-Fri 0615-0700, Sat 1000-1100, Mon-Fri 1300-1340.
PS) Pskovskaya obl: Pskov. **Reg.** prgrs on FM only. Details not available.
RK) Resp. Khakasiya: ul. Rozy Luksemburg 94, 662000 Abakan. **Reg:** On 792kHz, 67MHz: in Russian, Khakass: Mon-Fri 0000-0100 & 1120-1200, Sat 0110-0200 & 0320-0400, Sun 0320-0400.
RO) Rostovskaya obl: ul. 1-a Barrikadnaya 18, 344101 Rostov-na-Donu. **Reg:** On 945kHz: Mon-Fri 0330-0400, Sat 0415-0500, Sun 0430-0500 & 0525-0545 & 0715-0800, W 1215-1300, Sat 1415-1500, Mon-Fri 1500-1540.
RY) Ryazanskaya obl: ul. 26 Bakinskikh komissarov 20, 390006 Ryazan. **Reg.** prgrs on FM only. Details not available.
SA) Resp. Sakha: Dom Radio, Ordzhonikidze 48, 677007 Yakutsk. **Reg:** In Russian/Yakut on 171kHz and various SW freq's (4 tr's): Mon-Fri 2145-2300 & 0900(0910)-1000 (Thurs 1035), Wed 0210-0230, Fri 0325-0500, Tues/Wed/Thurs 0400-0500, Sat 0210-0540, Sun 2210-2300 & 0115-0410.
SL) Sakhalinskaya obl: Dom Radio, Komsomolskaya 209, 693000 Yuzhno-Sakhalinsk. **Reg:** On 279/792/1575/4050/11840kHz: Mon-Sat 2000-2100, Sun 2025-2200, Mon-Fri 2135-2145 & 0000-0010, Mon-Sat 0115-0130, Wed 0930-1000. R. St. Tikhiy Okean (from Vladivostok): D 0815-0900. Korean on 279/1575kHz: Sun-Fri 0730-0800.
SM) Samarskaya obl: ul. Sovetskoy Armii 205, 443011 Samara. **Reg:** On 1143kHz: Mon-Fri 0400-0500, Sat/Sun 0700-0900, Mon-Fri 1500-1600. In Erzya: Sun 0715-0745.
SN) Smolenskaya obl: ul. Nakhimova 1, 214025 Smolensk. **Reg:** On 68.54MHz (Smolensk), 68.96MHz (Smogiri), 69.20MHz (Vyazma), 70.91MHz (Roslavl): Mon-Fri 0415-0500 & 1500-1540, Sat 0700-0800/0900 & 1215-1315, Sun 0700-0800/0900, Thurs 1700-1800.
SO) Severo-Osetinskaya SSR: Osetinskaya gorka 2, 362007 Vladikavkaz. **Reg:** In Osetian/ Russian on 594kHz: W 0430-0500 & 0525-0600 (Mon 0545), Sun 0800-0900 & 1000-1200, Sat 0945-1100, Wed 1300-1400, Mon-Fri 1500-1540/1600.
SP) Leningradskaya obl: ul. Italyanskaya 27, 191011 Sankt-Peterburg. **Reg:** On 801kHz 0315-0000. 0600, 0901, 1900 rel. R. Rossii, Mon-Fri 0700 rel. R. Baltika.
SR) Saratovskaya obl: 2-a Sadovaya 7, 410017 Saratov. **Reg:** On 630kHz: Mon-Fri 0330-0400, 0810-0815, 1500-1540; Tues-Sat 0525-0545; Sat 0435-0500, 1215-1300, 1500-1600; Sun 1645-1730.
ST) Stavropolskiy kray: Krayevoye radio, ul. Artema 35-A, 355000 Stavropol. **Reg:** On 882/ 1251kHz: W 0330-0400, Sat 0515-0540, Tues-Fri 0525-0600, Sun 0830-0900, Sat 1215-1300, Tues-Thurs 1415-1500, Mon-Fri 1500-1540, Fri/Sun 1500-1600.
TA) Tambovskaya obl: ul. Michurinskaya 8-a, 392720 Tambov. **Reg.** prgrs on FM only. Details not available.
TL) Tulskaya obl: 300000 Tula. **Reg.** prgrs on FM only. Details not available.
TM) Taymyrskiy (Dolgano-Nenetskiy) avt. okrug: ul. Sovetskaya 26, Dudinka (or Taymyrskiy teleradiokomitet, Gorkogo 32a, 663210 Dudinka). **Reg:** On 162kHz: Mon-Fri 0030-0100 & 0600-0700; Sat 0115-0145. In Dolgan, Nenets: D 0815-0900. Also rel. KN.
TO) Tomskaya obl: ul. Batenkova 3, 634032 Tomsk, or: RTV Center, ul Yakovleva 5, 634032 Tomsk or Pushkina 19, 634050 Tomsk. **Reg:** On 171/1602kHz: Mon-Fri 0000-0100 & 0115-0200 & 1120-1200, Mon 1210-1300, Sat 0030-0100 & 0115-0200, Sun 0600-0700.
TS) Resp. Tatarstan: ul. M. Gorkogo 15, 420015 Kazan. **Reg:** In Russian/Tatar on 252/6115/6120/7185/11905/11945/15200/17810kHz: Mon-Fri: 0330-0500 (Mon 0600), 1010-1100, 1500-1540; Mon 1645-1800, Wed 1600-1900, Thurs 1720-1800, Fri 1700-1730; Sat: 0400-0500, 0715-0800, 0830-0900, 0945-1100, 1215-1300. **ANN:** Tatar: "Kazan söyli".
TU) Resp. Tuva: ul. Gornaya 31, 667003 Kyzyl. **Reg:** On 567/1287kHz: W 0015-0100 & 0115-0200, Sun 0040-0100 & 0125-0200, Sat 0340-0400, Mon-Fri 1115-1200.
TV) Tverskaya obl: ul. Vagzhanova 9, 170000 Tver. **Reg:** Mon-Fri 0410-0500 & 1520-1600, Sat/Sun 0510-0600.
TY) Tyumenskaya obl: Telecenter, Permyakova 6, 625013 Tyumen. **Reg:** On 225/576/1152/1197/1485/4820/4895kHz + FM (MHz) 68.84 Surgut, 70.04 Tobolsk, 70.55 Omutinka, 71.00 Nizhne-Vartovsk, 71.66 Tyumen: Mon-Fri 0057-0100 & 0110-0300, Mon/Sat/Sun 0315-0400, Mon-Fri 1055-1100, Mon-Sat 1100-1200.
UB) Ust-Ordynskiy Buryatskiy avt. okrug: 666110 Ust-Ordynskiy. On 234kHz (Irkutsk) in Buryat: 2245-2250, 1145-1155.
UD) Resp. Udmurtiya: ul. Komunarov 216, 426000 Izhevsk. **Reg:** In Russian/Udmurt on 594kHz, 68.06MHz in Udmurt, Russian: Mon-Fri 0210-0300, 0330 (Sat, Sun 0320)-0400, Sat 0500-0610, Mon-Fri 0800-0900, Sun 0500-0700 & 0710-0800, Mon-Fri 1500 (Tues/Thurs 1410)-1610 (Mon 1700, Fri 1710).
UL) Ulyanovskaya obl: ul. Simbirskaya 5, 432700 Ulyanovsk. **Reg.** on FM only. Details not available.
VG) Vologodskaya obl: ul. Menzhinskogo 32, 160000 Vologda. **Reg.** on 1098kHz: Mon-Fri 0315-0400, Wed/Fri/Sat 0415-0500, Sun 0515-0600, Sat 0525-0600, Mon-Fri 1500-1540.
VL) Vladimirskaya obl: per. Vorontsovskiy 4, 600000 Vladimir. **Reg.** on FM only. Details not available.
VN) Voronezhskaya obl: ul. Karl Marksa 114, 394625 Voronezh. **Reg:** On 774kHz, 72.17MHz: Mon-Fri 0330-0400, D 0515-0545, Sun 1300-1400, Mon-Fri 1500-1540.
VO) Volgogradskaya obl: ul. Mira 9, 400066 Volgograd (radio); or: RTV center, Mamayev Kurgan, 400005 Volgograd. **Reg:** W 0330-0400, Sun 0510-0550, W 0625-0645, Sun 1215-1300, W 1500-1540, Wed 1630-1730.
YA) Yaroslavskaya obl: ul. Bogdanovicha 20, 150014 Yaroslavl. **Reg.** prgrs on FM only. Details not available.
YE) Sverdlovskaya obl: ul. Lunacharskogo 212, 620219 Yekaterinburg. **Reg:** On 279kHz, 71.66MHz: Tues-Fri 0115-0130, D 0230 (Mon 0215)-0300, Tues-Sun 0325-0345, Tues-Fri 0615-0700, Mon-Fri 0745-0800, Sun 0930-1030, D 1300-1340, Sun 1415-1500.
YN) Yamalo-Nenetskiy avt. okrug: ul. Lambinykh 3, 626600 Salekhard. **Reg:** On 603kHz: W 0200-0215, Mon-Fri 0800-0900 & 1000-1010, Mon-Thurs 1415-1445; Sat 1100-1200, 1300-1400.
YV) Yevreyskaya avt. oblast: 682200 Birobidzhan. **Reg:** On 216kHz, 66.32MHz in Russian, Yiddish: Mon 2115, Sun 2145, Sat 2315-2400, 0140-0300, Wed 0515, Fri 0950, and others. Also rel. KH.

b) INDEPENDENT PROGRAMMES AND STATIONS

Note: Changes among these st's are frequent. Stations using SW are subject to seasonal freq. changes. Most information as reported by monitors.
Avtoradio, Chelyabinsk: 71.96MHz 0200-2300.
Avtoradiostudiya Ton, Perm: 70.04MHz.
Echo Moskvy: ul. Nikolskaya, d. 7, 103012 Moskva (studio) or ul. Demyana Bednogo 24, 123308 Moskva (editorial). **Te:** +7 (095) 9251724. On 1206kHz 24h music with heavy news emphasis.
FEBC-Russia: P.O. Box 2128, 680020 Khabarovsk. On 1224/9560kHz 0900-1030v in Russian + irr. Korean.
Golos Nadezhdy: P.O. Box 170, 300000 Tula. On 1260kHz (Sankt-Peterburg).
Inta R: ul. Gorkogo 15, UAB 318, 664000 Irkutsk-Center. **Te:** +7 (3952) 241044. On 1386kHz Mon-Fri 0000-0300 & 0900-1200, Sat/Sun 0200-1000. DX prgr Mon, Thurs 1130-1145. On 1332kHz Mon-Fri 1000-1200.
Krishna R, Chelyabinsk: 72.56MHz.
M-Radio: ul. Akademika Koroleva 15, 127427 Moskva. 24h on 71.33MHz.
Otkrytoye R: ul. Pyatnitskaya 25, 113326 Moskva. **Te:** +7 (095) 2337224. On 918kHz in English 0400-1600 (BBC, VOA, R. Moscow English World Sce.), Russian 0300-0400, 1600-2100 (incl. rel. of R. Liberty (cf. Germany) 0300-0400, 1700-1900).
R. Ala: P.O. Box 159, 125047 Moskva. **Te:** +7 (095) 2337255/6. **Fax:** +7 (095) 2337842. **Tlx:** 411700 PTB SU Box 7362. 24h "music of the bards" on various MW/SW freq's via tr's in SP, YE, KA, MO.
R. Alef (rlg): P.O. Box 72, 123154 Moskva, or: ul. Pyatnitskaya 25, 113326 Moskva. Sun/Tues/Thurs 1700-1730 to Moskva on 1332kHz and Israel on 1170kHz + SW via R. Moscow.
R. St. Alpha & Omega (rlg): Izdatelstvo "Protestant", Mukomolnyy pr. 1/2, 123290 Moskva. **Te:** +7 (095) 2599397. 1500-1600 in Russian on SW via Yekaterinburg 200kW.
R. Art, Moskva. Te./Fax: +7 (095) 299 2449.
R. Aum Shinrikyo: Rlg. prgrs broadcast over the domestic (RSM) and R. Moscow nets. See Japan for details.
R. Baltika: Kamennoostrovskiy pr. 67, 197022 Sankt-Peterburg. **Te:** +7 (812) 2349848. **Fax:** +7 (812) 2356188. On 747kHz, 71.24MHz (stereo): 0400 (Sat/Sun 0500)-2100 (Sat 0000).

R. Berezovo, 626800 Berezovo. On 5070kHz Sun 0515-0600, 0625-0659.
R. Centre, Moskva. On 1440kHz 0400-0500.
R. St. Delovaya Volna: ul. Akademika Koroleva 19, 127427 Moskva. **Te:** +7 (095) 2179162, 2179320. **Fax:** +7 (095) 2178963. On 1539kHz 0500-0800, 1200-1700.
R. Dvizhenie, Yekaterinburg: 0200-1300 on 6200kHz, 1300-1700 on 6090kHz. Not recently confirmed.
R. Europa Plus: ul. Akademika Koroleva 19, 127427 Moskva. **Te:** +7 (095) 2159938. **Fax:** +7 (095) 2178986. On 1116kHz, 69.80MHz (15kW) 0300-0100. In Sankt-Peterburg on 72.68/100.5MHz (200W) 0400-0200. In Samara on 71.27MHz (1.5kW) 0400-0700, 1300-1600.
R. FM-Express, ul. Pobedy 53, Astrakhan. On 69.9MHz.
R. Galaktika (R. Galaxy): P.O. Box 7, 117418 Moskva. In English 1900-2200 (mostly pop music) on 1440kHz + one SW freq.
R. Irina/Voice of Freedom (Tieng Noi Tu Do): ul. Novinskiy 18/257, 121069 Moskva. prgrs in Vietnamese on SW (Irkutsk 500kW) to Vietnam 1400-1500.
R. Kray, Barnaul. On 67.64MHz 0600-1100.
R. St. Yakutsk: ul. Semena Dezhneva 75-2, Radiocenter, Yakutsk 677000. on 5920kHz: 0600-1400.
R. Kanal-3, Barnaul. **Te:** +7 (3852) 235619. On 71.57MHz W 0200-0600, 0800-1100, Sun ??-1800.
R. Maximum: ul. Tverskaya, d.16/2, Moskva. Te: +7 (095) 2001088. On 103.7 MHz (Moskva): 0300-0500 E, 1300-1900 r. In Perm: 846kHz 0130-0430 (W), 66.81MHz 0130-0300 (W) & 1200-1800 (D).
R. Modern: ul. Italyanskaya 27, 191011 Sankt-Peterburg. Te: +7 (812) 2199668. 0300-1500 on 69.47MHz.
R. Nadezhda: ul. Pyatnitskaya 25, 113326 Moskva. **Te:** +7 (095) 2336588. 0800-1100 on 630kHz (Komsomolsk) + SW to the Far East; 1100-1400 on SW to Siberia/Urals; 1500-1800 on 1440kHz (Moskva) + SW to ce. part of European Russia and Ce. Asia; 2000-2300 on 1215kHz (Bolshakovo), 1440kHz (Moskva) + SW to Europe. The station's purpose is to serve as a mouthpiece for women's public organisations.
R. N (Nezavisimoye R.), 620219 Yekaterinburg, GSP-808: On 6200kHz 0500-1000 exc Sun/Mon.
R. Nostalgie: ul. Pyatnitskaya 25, 113326 Moskva. On 963kHz 0500-1300 in F, Russian.
R. Onega: Murmansk. On 6030kHz 1000 or 1100 to 1100 or 1200, not daily.
R. Otkrytyy gorod, P.O. Box 250, Ligovskiy 52, 192007 Sankt-Peterburg: 1053kHz 1700-2030.
R. Pamyat: ul. Valovaya 32, kv. 4, 113054 Moskva. **Te:** +7 (095) 2373971. 2300-0030, 1430-1600, 1600-1900 on various SW frq's. At time of editing the 2300 & 1600 b'casts are reported to have been suspended.
R. Passazh: ul. Lenina 17, 420111 Kazan, Tatarstan. On 66.8MHz.
R. Polis: Ligovskiy pr. 253/2, 192007 Sankt-Peterburg (studio). **Te:** +7 (812) 1641439. On 1053/6045kHz 0430-1700.
R. St. Radonezh (the R. St. of the Russian Orthodox Church): Studio 158, Pyatnitskaya 25, 113326 Moskva. On 846kHz 1900-2000 to Moscow; on SW 1600-1700 to European part of CIS.
R. Raduga: 612kHz. Not recently confirmed.
R. St. Remira: ul. Rosa Luxembourg 285A, 664032 Irkutsk. Te: +7 (3952) 452958. On 1500kHz 0000-0200, 0300-0500, 0600-1000.
R. Rezonans: ul. Koroleva 19, 127427 Moskva. On 1440kHz 0700-1400.
R. ROKS: ul. Novyy Arbat (Kalinina) 21, 121019 Moskva. **Te:** +7 (095) 2919311. Moskva 102.8MHz, Sankt-Peterburg 102MHz, Kyyiv (Ukraine) 104/71.66MHz, Riga (Latvia) 102.7MHz. 0300-2200, incl. English 1400-1600. Most prgrs are produced in Oslo, Norway, and fed by satellite.
R. Rus: ul. Artema 35-A, 335000 Stavropol. On 71.6MHz Tues-Fri 0030-0100, 1200-1230, Sat/Sun 0800-0900.
R. Samara Mazimum: ul. Aerodromnaya 13, Samara. On 73.61MHz 1300-1900 Mom-Fri, 0500-1000 & 1200-2000 Sat/Sun.
R. Shark: pr. Oktyabrya 56/1, 450054 Ufa 54. Dir: Sergey Anatskiy. On 1395/5780kHz 0200-2000.
R. Sinitsa: P.O. Box 707, 620055 Yekaterinburg. On 909kHz, 69.92/71.84MHz Mon-Fri 0100-0400 & 1300-1900, Sat/Sun 0300-1900.
R. St. SNC: Krymskiy val. 9, 117049 Moskva. On 1260kHz 0500-0300 (Moskva), 1000-0100 (Sankt-Peterburg) pop/rock music. Note: At time of editing reported to be off the air.
R. Titan, Ufa. B'casting on freq's of R. Shark 1100-1200 (Sat 1300), including relays of Voice of America, R. Sweden.
Radiotrack: P.O. Box 932, 620063 Yekaterinburg. On 69.93MHz Mon-Fri 0100-0145, Sat/Sun 0300-0430.
R. Vedo: P.O. Box 1940, 400123 Volgograd. Te: +7 (844) 329391, 329609, 328002. On 1467/5915/7125/7185kHz Sat/Sun 0500-0900. On 1467/7185/9655/11760/13710kHz D 1500-1930.
R. St. Ves Irkutsk, ul. Lenina I, kab. 25, 664000 Irkutsk: On 6090kHz 2330-0030, 0500-0600, 1100-1200.
R. St. Vladivostok: P.O. Box 2234, Vladivostok 91. On 1251kHz 2105-2300.
R. St. Volna Baykala: ul. Sverdlova 37, 664011 Irkutsk. Te: +7 (3952) 288546. On 1305kHz 0100-1300.
R. St. Vostok: ul. Lenina 4, 680013 Khabarovsk. On 153/216/621/4610/7210kHz, 70.42MHz: Mon-Fri 0800-0930, Sun 0800-0900.
R. Vozrozhdeniye: ul. Koroleva 19, 127427 Moskva. **Te:** +7 (095) 2179611. On 1305kHz 0400-2200.
R. 7, Moskva: 73.385MHz 1kW (Moskva). **Te:** +7 (095) 9071521. 24h.
R. 101: ul. Pyatnitskaya 25, 113326 Moskva. 1233 kHz, 101.2MHz (stereo) 0300-0800, 1330-0000.
RFT (Radiokanal Firmy Telekomp): P.O. Box 35, 666020 Shelekhov-6 (Irkutsk oblast). On 1530kHz.
Rukhi Meroc (Moscow Islamic Center), Moskva: Fri 1600-1645 in Tatar on 1089kHz + SW. **Te:** +7 (095) 281 4904.
Vostochnaya Tribuna, Moskva: On 1539kHz 1700-1800 (sports).

FOREIGN SERVICES

Mediumwave facilities: Included in the Home Service LW/MW listing.

Shortwave facilities: As the Russian authorities have not made information about the actual facilities available all details are tentative.

High powered transmitters in Russia carrying Foreign and/or Home Sce prgrs (tr's only carrying regional Home Sce's and former jamming tr's now used for broadcasting are not included): Blagoveshchensk, S (G.C: 50.16N/127.32E): 1 x 50kW — Bolshakovo, E: 6 tr's (operated in 2 groups) — Chita (Aleksandrovka), S (G.C: 51.50N/113.43E): 3 x 500kW — Irkutsk (appr.15km west of Angarsk), S (G.C: 52.18N/104.18E): 3 x 50kW, 2 x 100kW, 2 x 250kW, 2 x 500kW — Khabarovsk, FE (G.C: 135.15E/48.33N): 9 x 50kW (3 used as feeders), 6 x 100kW, 1 x 200kW —Komsomolsk (Elban), FE: 4 x 50kW, 2 x 200kW — Krasnodar, E: 6 x 120kW, 3 x 250kW, 3 x 500kW — Moscow (Balashikha), E (G.C: 55.47N/37.54E): Several 20kW tr's — Moscow (central), E: 1 x 100kW, 1 x 500kW — Moscow (Noginsk), E: 6 x 120kW, 1 x 250kW — Moscow (Pushkino), E: 16 x 250kW (operated in 8 groups) —Moscow (Taldom), E (G.C: 56.44N/36.29E): 2 x 250kW, 1 x 500kW — Novosibirsk, S: 14 x 50/100kW — Novosibirsk (Oyash), S (G.C: 55.31N/83.45E): 2 x 250kW, 3 x 500kW — Petropavlovsk-Kamchatskiy, FE (G.C:52.59N/158.39E): 4 x 100/200kW —Samara, E (G.C: 53.20N/50.10E): 1 x 100/120kW, 4 x 200kW, 2 x 250kW — Sankt-Peterburg (Popovka), E (G.C: 59.39N/30.42E): 15 x 200kW, 2 x 100/200kW (operated in 10 groups) — Serpukhov, E (G.C: 54.54N/37.25E): 3 x 50kW, 1 x 100kW — Tver, E (G.C: 56.52N/35.35E): 2 x 50/100kW, 3 x 100kW, 1 x 250kW — Tula, E: (G.C: 54.12N/37.48E): 6 x 100kW — Ulan Ude, S: 2 x 50kW — Vladivostok (Razdolnoye), FE (G.C: 43.32N/131.57E): 6 x 50/100kW, 1 x 250kW, 2 x 500kW — Yekaterinburg, S (G.C: 56.50N/60.36E): Approximately 20 x 100kW operated in 9 groups. The exact locations of a few further tr's in the Far East remain unknown.

Foreign relay stations: Armenia, Azerbaidzhan, Belarus, Cuba, Germany, Kazakhstan, Moldova, Mongolia, Tadzhikistan, Ukraine, Uzbekistan. Relays on an exchange basis via China (People's Rep.).

Relay facilities in Russia are also leased to broadcasters from other countries.

SCHEDULE OF HF TRANSMISSIONS FROM THE CIS
The following is taken from the winter 92/93 coordinated transmission schedule supplied from Moscow. It includes transmissions from Russia and the other member states of the CIS, as well as relays via Bulgaria. Low power transmissions (20kW) are not included. Many of the listed frequencies carry programmes from

Radio Moscow International (see language schedule). However an increasing number of transmitters are carrying programmes produced by other organizations. The schedule is arranged in alphabetical order of the officially listed sites, however in some cases the actual site may be different. No information was supplied concerning transmissions between 7350 and 9500kHz, or in the 13MHz band.

kHz	kW	Times
Alma-Ata, Kazakhstan		
5915	100	1230-1700
		2330-0200
5960	100	0000-2300
5970	50	0000-2300
6135	100	1130-1700
		2230-0200
6180	100	0000-2000
7125	100	1200-1500
7180	100	1130-1500
7240	100	1130-1500
7255	240	2000-0100
7290	250	0000-0400
	1000	1530-2400
9505	100	0000-2300
9520	240	1230-1630
9575	1000	1530-2200
9790	500	2130-0300
9825	240	2100-2400
11655	100	0100-0930
11685	200	1800-0100
11760	100	1100-1700
11950	100	0000-2000
12000	500	1430-0600
12025	100	1000-1500
12040	500	2330-0800
12060	100	1530-2300
13615	250	0130-0900
13700	100	0900-1500
15105	240	0430-1600
15230	240	2100-1600
15445	1000	0530-0800
15460	240	1800-2200
15510	240	1030-1600
15515	240	1800-2400
15520	240	0130-1600
	500	1600-2200
15590	240	2000-0100
17645	1000	0900-1400
21525	240	0700-1700
Armavir, Ukraine		
5905	100	1430-2230
5940	250	2330-0700
5945	250	1530-1700
5980	100	1530-2300
6005	100	1530-2300
6105	240	0130-1000
6135	100	0300-1500
6145	250	2300-0400
6185	500	2130-0600
7105	240	1930-2130
7130	100	0100-2100
7135	240	0600-0930
7150	1000	1530-1100
7165	250	2330-0900
7190	100	1730-0500
7200	100	0330-0800
7205	250	1200-1800
7235	240	2230-0400
7310	240	1300-2200
7315	100	0200-0400
7320	1000	1530-2200
9530	500	0600-1100
	1000	2200-0600
9545	100	1539-2100
9565	250	1530-2100
9620	240	0530-0900
	240	1400-1600
9675	240	0130-0500
9705	100	1530-2300
9720	240	1800-0400
9735	100	1530-2030
9765	250	0700-1800
9820	250	0100-0800
	240	1200-1700
9830	500	1730-0300
11665	500	1130-1430
11670	100	0330-0500
		0600-1130
		2000-2230
11770	240	0800-1500
11775	500	1530-2230
11800	100	0500-0900
11850	250	2200-0400
11870	500	1530-2200
11900	500	0500-0900
	240	2230-0500
11920	100	1700-2200
11960	240	0900-1630
11975	100	1100-1800
11980	1000	2130-0500
12050	500	2200-0400
12055	120	0400-2200
13635	100	0330-0700
		1530-2230
13755	100	1130-1500
13790	240	0800-1800
15150	240	0700-1800
15175	240	0600-1700
15210	1000	0600-2100
15225	240	1000-1500
15240	240	1630-1800
15275	100	1130-2200
15305	1000	0400-1600
15395	500	1130-1200
		1230-0400
15440	100	0600-1600
15450	240	1330-2200
15470	240	0500-1530
15480	240	2130-0430
15500	100	1530-2200
15550	1000	2200-0900
15585	500	0700-1500
	1000	1630-0500
17560	100	0400-1100
17580	500	1630-2100
17625	100	0400-1600
17645	250	0900-1430
17655	240	1000-2200
17680	250	0430-0800
17690	100	0200-1500
17735	100	1800-2200
17755	100	1300-1500
17765	1000	0200-1500
17840	250	0800-1200
17855	100	0230-1200
17890	100	1200-1500
21450	100	0200-1700
21460	500	0700-1700
21530	240	1200-1600
21565	100	1130-1600
21610	100	0400-1100
21655	240	0430-1600
21665	100	0430-1200
21670	100	1700-2200
21680	100	0430-1500
21715	100	1000-1600
21785	1000	0600-1700
21825	100	0430-1600
21845	500	0430-1200
25780	100	0400-1700
Ashkhabad, Turkmenistan		
7120	1000	1430-2400
7265	100	2330-0330
15540	1000	1830-0200
17740	100	0330-0700
Biskek, Kyrgyzstan		
6080	50	0000-0200
	240	1230-1630
7105	100	2000-0130
7205	240	2000-2200
7325	250	1500-2100
9650	250	2200-0900
9695	240	2000-0300
11695	1000	1400-1800
	240	2100-2400
11820	240	0800-1800
11895	150	1330-2300
11905	100	0930-1600
12050	100	0930-1500
15265	250	0700-1000
15405	100	1330-2300
15420	500	2330-0900
15450	240	2000-0400
15490	1000	0900-1500
15535	240	0930-1700
17685	240	0300-1130
17745	100	0900-1500
17775	100	2300-1600
17815	240	0030-1200
17825	100	0900-1500
17840	250	0930-1530
21505	100	0330-1000
Blagoveschensk, Siberia		
5950	240	2000-2200
5970	100	1700-1600
7185	500	1230-2400
9580	240	0130-0900
		1430-1900
11790	240	1400-0700
11915	240	2100-0900
17645	100	2100-0900
Chita (Alexandrovka)		
5910	100	1000-1800
		2000-2400
6000	100	1030-2300
6045	250	1700-2100
6060	250	1100-1500
6195	50	0830-1800
		2000-0100
7195	1000	1930-2200
7220	1000	0830-0100
7265	250	1200-2100
7330	1000	1830-2400
11675	1000	0030-0900
11695	1000	1430-2000
11755	250	1030-2330
11925	240	1300-2030
12005	100	2330-0800
12060	50	0000-1200
15130	1000	1230-0100
15385	240	0000-0330
15530	1000	1700-2130
17870	1000	1930-1300
21585	500	0130-1000
Dushanbe, Tadzhikistan		
5985	50	2100-0300
7115	100	1300-1500
7320	1000	1530-2300
9535	100	0300-1100
9665	1000	1830-0400
9830	50	0930-1600
11670	1000	0300-0600
	1000	1430-0200
11705	100	2300-1130
11800	100	2300-1600
12015	200	0930-1600
12045	100	2200-0400
15140	100	0000-1400
15350	240	2230-1600
15545	1000	1330-1800
17580	500	0330-0900
		1000-1400
17700	500	0230-1000
21635	240	0130-1700
21820	1000	0430-0800
Grigoriopol, Moldova		
7195	500	0900-1400
		2230-0900
9620	500	1600-2300
9710	500	1930-0900
9750	1000	2130-0900
9765	500	1600-0600
9795	500	0530-0800
11745	1000	1400-1900
11750	500	2130-0600
11860	500	2130-0700
12045	1000	1730-2200
12070	500	1900-0600
13640	500	1900-2100
15265	1000	1930-0900
15510	500	1430-0800
15545	1000	0900-1800
17625	1000	0900-2100
17670	1000	0800-1900
17700	1000	1000-1900
17760	500	0800-2100
17850	1000	1000-1400
17880	500	0800-1800
Irkutsk, Siberia		
5990	100	1030-1800
		2100-2400
6060	250	1700-2400
6130	50	0830-1800
		2100-2400
7115	500	1430-2030
7135	250	0800-1900
7150	240	2000-2200
7160	100	1430-1700
7235	250	1630-1900
7245	240	1330-2300
7250	100	2000-2300
7305	240	1700-2400
7310	200	1230-1400
7315	500	1800-2230
9520	240	1630-2230
9565	240	0100-0300
	250	1800-2300
9575	100	2000-1800
9685	500	0900-1500
9735	1000	0900-1700
9820	100	2130-1800
9855	100	0930-2400
9875	250	1600-2230
11765	100	2200-0900
11775	500	2000-1500
12030	100	0030-1000
13675	250	2100-2400
13715	100	2100-2400
15170	50	0700-0900
15210	1000	0800-0400
15350	250	0730-1500
		2000-2200
15470	250	1800-2300
15475	100	2230-1100
15560	1000	0730-1400
	500	1800-0100
15580	1000	0830-2300
17695	1000	0930-1200
17825	1000	0030-0900
17880	250	0000-0600
17890	1000	2130-1200
21530	250	2230-0700
21565	500	0330-0900
21770	250	2330-0730
21790	250	2230-1300
Ivano-Frankovsk, Ukraine		
12030	250	1630-2100
13680	100	1730-2200
17745	240	0630-1700
Jigulevsk		
5980	150	1330-0500
6140	100	1500-2300

6175	100	1600-0800
6195	100	2200-0200
7115	100	1530-2100
7140	100	1300-2100
7235	250	1530-2200
7255	250	0000-0200
7260	100	2330-0600
7315	100	1400-1800
7325	100	2200-0200
9550	250	2200-0500
9600	250	1300-1600
9635	100	0900-1700
9755	100	0930-1800
		2100-0500
11800	250	1200-2200
11960	200	0500-0800
	240	2200-0500
12035	250	0200-0800
12045	100	1500-1930
13735	100	0600-2400
15185	240	0430-0800
	100	0800-1400
15200	100	0700-1700
	200	2100-0700
15220	240	0200-1500
15280	240	0000-1730
15470	250	1630-2100
15540	250	0430-2300
17560	100	0200-0330
17675	100	0000-1200
17780	250	0330-0800
17805	100	0630-1200
17890	150	1300-2300

Kalach

9640	250	2200-0430
9710	100	1730-2200
9790	100	0330-0430
11670	250	1130-2000
11700	100	1700-2130
	500	2200-0500
15360	100	0100-1600

Kaliningrad

5935	100	0200-2300
6015	100	0300-0600
		1430-0200
9540	240	1200-1530
9710	50	0300-1730
11965	100	0630-1500

Kazan

5905	100	1430-0300
5915	100	1730-2000
		2130-0700
5940	100	0000-0430
	240	1600-2400
5970	100	1730-2100
6025	100	0230-0330
		1030-1130
		1430-2000
6055	100	1430-2400
6085	150	1500-2300
7120	150	2300-0400
7145	100	0000-0530
7155	250	1400-1630
		2300-0300
7170	100	1400-2400
7210	240	1430-2300
7250	100	1400-2400
7280	200	1630-2300
7295	100	1200-0400
7300	100	1600-2100
7335	250	1130-1700
7340	100	0130-0700
9515	150	1730-2200
9560	240	1200-0500
9565	240	1530-1800
	200	2100-2300
9580	250	1400-0500
9605	100	2100-1530
9625	240	2200-0600
9630	160	1430-2300
9665	240	1730-1930
9695	250	1200-1500
	160	1530-2130
		2200-0400
9775	200	0730-1300
9780	100	1430-2400
9785	200	1430-1730
9815	100	2100-0400
9865	240	1400-0300
11675	100	1200-1600
11710	100	1130-2400
11750	100	0300-1530
11765	250	2300-0200
11780	100	2200-1600
11810	250	2200-0500
11965	250	2300-0300
11975	100	2200-0100
11990	100	0200-1400
12000	100	0430-1430
12015	100	2200-0100
12030	100	0330-0500
	250	1430-2200
12045	240	1500-1730
		2200-0500
13645	240	1000-1700
15350	150	1630-2330
15435	250	0000-1700
17735	240	0330-1800
21840	250	0530-1600

Kenga

5935	100	2200-2000
5995	50	2300-1800
6025	100	2200-2100
6100	100	1130-1600
6115	250	0230-0400
		2200-2330
7140	100	1100-2000
7165	500	1330-2300
7205	1000	1000-1400
7270	100	0930-1600
		2100-0300
7280	50	0930-1530
	100	1900-0100
7290	100	0830-1800
7315	1000	2230-2400
7335	100	2330-0100
9685	500	1800-2400
9700	100	2030-1100
9745	100	2100-0300
9845	1000	1000-1700
		2100-0200
9855	100	1130-2230
9865	100	2000-1700
9875	500	0500-1430
	1000	1430-2100
11665	100	0200-1100
11795	1000	2200-2400
11825	100	2300-1400
11835	100	2100-1500
11855	240	0300-1630
12020	200	1730-2100
13775	100	0100-1030
15375	240	0330-1600
15490	50	2330-1600
15595	500	0400-1200
		1530-2400
17590	1000	0130-1200

Khabarovsk, FE

5905	100	0830-1500
		2000-2200
5930	100	0930-2300
5985	100	2100-2400
6020	50	0930-1600
		1700-2130
6050	100	2130-0900
6065	100	0000-1300
6115	50	0730-1500
		1600-2200
6145	100	0900-2400
6165	100	0930-1300
		1400-2100
6175	50	0730-1600
		1800-2300
6180	100	1630-2100
7175	100	0630-0900
7210	100	2000-1900
7245	240	0930-1500
7250	100	1530-2100
7335	100	0930-1600
		1800-2200
9545	50	1900-1800
9610	200	1500-1700
9730	100	2000-1800
9805	100	1530-2000
		2230-1130
9810	100	1130-1500
9835	100	2230-0700
9855	100	0030-0830
9875	240	1030-1530
9885	120	2000-2200
11685	100	1800-2400
11770	200	2000-0900
11860	100	0700-0900
11870	240	0900-1700
11875	100	2230-0630
11970	100	1900-1600
12050	240	1400-0900
12060	200	1530-2100
12070	100	1530-2200
		2230-1000
15420	100	0930-1600
15550	240	1230-1430
17655	200	2030-1200
17700	100	2130-0500
17720	240	2230-0800
17805	200	0430-1130
17850	100	0830-0900
17895	500	1930-2200

Kharkiv, Ukraine

11950	500	2130-0900
15595	100	0700-1530

Kijiv, Ukraine

5960	200	1630-2200
6010	100	1630-2400
6020	100	0830-1500
	240	1700-0030
6050	240	1330-1630
6090	100	1600-2300
6145	240	0030-0330
7195	100	1630-2230
7230	240	0400-0800
		2230-0100
9800	100	2200-0430
9805	1000	1230-0800
11690	240	0400-1000
11705	100	0600-1700
11770	500	1800-0700
11960	500	1500-2100
12005	100	0100-0430
13720	240	1730-2000
15465	100	2200-2330
15525	100	1030-1700
		2200-0130
17680	240	0630-1800
5925	100	0330-0100
5945	150	0200-0600
7140	240	0530-1100
7300	240	1700-0600
9620	240	1630-2400
11930	500	0300-0900
11980	500	1400-2100
12015	200	0100-0800
12065	240	1330-2300
13635	200	1200-1800
13690	240	0100-1300
15465	500	0200-1730
17710	200	0730-1400
21515	240	1000-1900

Komsomolsk, FE

6080	100	0630-1600
		1900-2200
7135	1000	1930-2200
	100	2230-0900
7150	240	0930-1600
7170	250	0900-1400
	100	1400-2100
7195	100	0900-1700
7330	200	1000-1500
		1800-2200
9505	100	0330-0800
		1730-2100
9635	50	2100-0900
9670	100	2100-1300
9675	100	0930-1600
		1930-2030
9735	250	1730-2030
9795	100	0530-0900
		1430-2200
13645	100	1830-2200
15180	100	1930-0900
15535	100	0530-0900
17570	100	2100-1500
17690	240	2100-0900

Konevo

11800	500	2200-0500
11850	500	0400-0930

Krasnoyarsk, Siberia

6010	240	1300-1800
6010	240	2000-2400
7230	100	1130-1800
		2000-0100
11880	200	1500-2100
11995	250	0800-1600
12005	500	1230-1600
15315	150	0130-1400
15535	1000	0100-0200
	240	1630-0100
15545	240	2000-0900

Kursk

7335	100	2200-0400
9665	240	0400-0700
		1230-1400
		1430-1600
15330	240	1600-2400
15480	250	0430-0700
		1500-2230
17745	250	0400-1700
17860	100	0430-1500
	240	1700-2300

Lviv, Ukraine

7135	1000	2200-0800
7160	1000	0400-0700
9505	500	1730-2400
9560	240	0330-1630
9685	1000	1630-2200
9870	1000	1730-0700
11750	100	1000-1500
15495	240	0600-1500
17780	1000	1000-2100

Moscow (4 sites)

5945	100	1930-2200
5955	100	1330-1700
5970	100	0000-0400
6000	240	0000-0300
6020	100	1700-0030
6030	250	1400-0130
6045	240	1530-0400
6060	100	1530-0500
6110	100	2100-0430
6120	250	1600-2400
6130	500	1600-2200
6145	100	1400-2400
6165	250	1530-2130
	100	2130-0700
7140	250	1530-0500
7170	240	1530-0100
7180	200	1330-0500

7185	250	2200-0800
7200	240	1500-2200
7210	240	0100-0300
7275	200	1630-2100
7280	240	0530-1600
	250	2300-0300
7295	240	1430-0700
7310	1000	2230-0800
7320	240	2300-0900
7345	100	1530-2200
9515	250	1330-1500
	100	1530-0800
9530	250	1500-2300
	1000	2330-0600
9540	100	2000-1500
9550	240	0500-0700
		1200-1630
9585	150	0630-1600
	100	1630-1900
		2300-0400
9590	100	1530-2400
9600	100	1600-1830
	250	1830-0200
9610	200	0500-1200
9650	240	1500-2400
9655	500	0000-0500
9695	100	0030-0600
9720	500	1600-0600
9740	250	1530-0200
9785	240	0500-0800
	100	2330-0330
9800	50	0530-1500
9880	240	2100-0500
11690	240	1200-1700
	500	1700-0500
11700	100	1100-1430
11710	100	0300-0800
	240	0800-1030
11715	100	2100-2300
11745	240	0700-1900
		2300-0500
11755	100	0430-0600
		2200-2400
11775	240	0000-0500
11785	250	0200-0700
11820	240	0530-1400
		1600-2200
11830	240	0530-2100
	100	2100-2300
11850	200	1130-1500
11875	250	2130-2300
11880	240	2300-0800
11910	100	1130-1800
11920	240	0830-1800
11930	240	0100-0430
	100	0700-1330
	250	1330-1700
	200	1730-0100
11945	240	1400-1630
11980	150	0530-1400
11985	250	2130-0400
12010	100	0000-0400
12020	250	1930-2200
	500	2200-1930
12035	100	0200-0800
	500	1330-1700
	250	1830-0200
13635	100	1000-1100
13680	250	2230-0500
13705	250	2230-1700
13715	100	0330-1400
15135	100	1000-1500
	250	1530-2130
15150	240	0400-1400
15170	500	0000-0500
15180	500	1630-2400
15230	100	0930-1100
		1800-2300
15260	100	0700-1400
15330	240	1700-2200
15345	250	0830-1800
15370	200	0530-1500
15375	240	1330-1800
15460	250	0430-2100
15500	250	0500-1500
15510	100	0900-1500
15560	500	0430-1630
17590	240	1300-1630
17610	100	2300-1600
17660	240	0730-1600
17710	500	0430-1430
17720	250	0430-1300
17800	100	0730-1100
17835	240	0300-2000
17855	240	1700-1830
21450	240	0600-1200
21575	500	0900-1630
21600	250	0500-1700
21625	250	0930-1130
21830	500	0330-1500

Mykolayiv, Ukraine

5920	100	0930-2100
6090	100	0930-1700
		1800-2130
6155	100	0930-1500
	50	1330-1800
		1900-2200
7300	100	0830-1600
7310	100	1800-2130
9685	100	0530-0900
9780	100	0930-2300
	50	2300-0930
11980	100	2130-0900
15160	1000	2100-1600
15185	200	1530-2030
15295	120	1930-2200
15330	120	1000-1500
15415	240	1800-2100
17695	100	0800-0900

Nishnii Novgorod

11860	240	1230-2100
12040	250	1300-2300
17620	240	0330-1200

Novosibirsk, Siberia

6060	100	1430-1700
6070	100	1130-2000
		2200-0130
6120	100	2000-0300
6135	100	0830-2000
6150	100	1730-0100
6170	500	1430-2300
7120	100	2200-1800
7125	100	1130-0130
7170	500	2330-0300
7175	500	1530-0430
7185	240	1400-2200
7235	250	2100-0430
7250	100	2000-0100
7255	200	0200-1700
7305	1000	1230-1930
7340	1000	1000-1400
		2200-2400
9780	250	2300-0400
9790	100	1730-2130
11740	50	2200-1600
11940	240	2100-0300
12010	60	0400-1300
12025	100	2100-2300
12040	100	0030-1500
12045	1000	2230-2330
13625	100	2100-1600
13725	100	0130-0900
15330	1000	2130-2400
15385	240	0330-1600
	1000	1800-2300
15470	250	2300-0830
17620	500	0030-1000
21760	250	2200-1300
21780	500	2300-1300

Okhotsk

9530	100	1800-1700
9600	100	1800-1700

Omsk, Siberia

6190	100	0000-0330
		1330-2300
7160	100	1430-1800
9590	100	1130-1800
9625	100	0930-0400
11895	100	2330-0500
12030	50	0430-1430
21625	500	0000-1500

Orenburg, Siberia

6125	100	1400-2030
7105	240	0600-1430
11915	240	0230-1630

Osipovichy, Belarus

7255	150	1500-2200
7275	100	0000-0300
9570	240	2300-1430
11710	100	0000-0300
15470	240	0200-1100

Petropavlovsk/Kamchatskiy, Siberia

5905	100	0330-0900
5940	100	0930-1300
5950	100	0930-1400
5965	100	0330-0900
6165	100	0930-1300
7160	240	0930-1300
7185	100	0530-0900
7260	100	0230-1000
	240	0930-1300
		1400-2000
7270	250	0330-0900
7345	100	0630-0900
9540	250	1400-2100
9755	250	1400-2200
9825	100	0630-0900
		0930-2200
9895	100	0330-0600
		0930-1300
		1400-2100
11710	100	0400-0900
11910	100	0230-0500
11950	100	0930-1400
12010	100	0230-1000
		1400-2100
12030	200	1500-1700
12065	250	0830-1400
13665	100	0530-0900
15140	240	2100-0300
15240	100	2100-1000
15265	250	0930-1300
15395	250	0930-1300
15405	240	1430-1400
15425	100	1930-0900
15455	240	2130-0700
17605	240	1930-0500
17665	100	2130-0600
21480	100	2030-0500
21530	100	2330-0200

Petrozavodsk

6100	100	1800-0130
9735	250	0000-0500
11685	250	0300-1400

Plovdiv, Bulgaria

7115	500	2200-0400
9520	240	2200-0500
11735	500	2000-0400
17840	500	1100-1400
17860	500	1100-1400

Riazan (Moscow)

5905	240	2300-0500
5980	240	2300-0300
6035	100	0400-0600
		1500-2400
6185	240	1330-1900
		2100-2130
7245	100	1600-2100
9720	250	0430-1600
	240	2100-0500
9765	250	0400-1600
11850	240	1330-2300
11890	240	1030-1600
	500	1630-1030
12055	500	0000-0600
13740	240	1430-0400
15440	100	0800-2300
15540	240	1900-2230
17775	240	1800-2000
21630	200	0630-1900

Samara (Kuybyshev)

5950	240	1400-0400
7110	100	2200-0300
7160	100	0000-0530
7270	150	1730-2100
7285	100	2200-0100
9645	100	2200-2400
9720	240	1330-1630
9850	200	1330-1600
9885	100	2130-0330
11715	150	0430-1300
11720	150	1330-1700
11740	240	0930-1300
11925	240	0030-0700
13690	100	1400-1800
15560	100	0200-0400
17600	100	0330-1600
17685	250	1130-1430
21750	100	0430-1130

Sankt-Peterburg

5905	240	0400-0900
		1400-2400
5920	500	1230-0100
5975	500	1530-2000
		2030-2400
6025	250	1830-2300
6050	150	1700-0200
7105	500	0200-0400
	240	2130-2300
7130	500	1630-2200
7175	240	0100-0600
7265	500	1900-2300
7305	240	1830-2230
7315	500	1300-1530
7315	500	1730-2300
7330	500	1700-2200
7340	500	1430-2400
9510	500	0300-0200
9590	100	0000-0700
9800	100	0330-0600
		1500-2200
9820	500	1430-0100
9890	500	1330-0900
11745	100	0500-0700
11785	240	0000-1300
11950	200	0900-1600
12025	500	0000-0700
12045	200	0630-1500
12070	250	0830-1300
13625	200	1200-2200
13680	500	0730-1700
13775	240	0300-1600
15180	100	0630-1300
15480	500	0430-1900
15515	500	1030-1800
15530	500	0430-1100
15570	240	1430-2300
17655	500	1030-1600
21550	500	0430-1500
21750	240	1330-1700
21765	500	0830-1700
21775	500	0830-1700

Serpukhov

5910	100	1200-2300
6065	150	0100-0500
6095	100	0300-0600
		1630-2300

CIS
0
500
1,000 KM
UTC +4
UTC +5
+5.30
UTC +8
UTC +9
ARCTIC OCEAN
BARENTS SEA
KARA SEA
LAPTEV SEA
SEA OF OKHOTSK
SEA OF JAPAN
BLACK SEA
CASPIAN SEA
ARAL SEA
Lake Balkash
Lake Baykal
RUSSIAN REPUBLIC
LATVIA
ESTONIA
LITHUANIA
BELARUS
MOLDOVA
UKRAINE
GEORGIA
AZERBAIDZHAN
ARMENIA
TURKMENISTAN
UZBEKISTAN
KAZAKHSTAN
KYRGYZSTAN
TADZHIKISTAN
IRAN
INDIA
CHINA
MONGOLIA
NORTH KOREA
S. KOREA
JAPAN
+2.00
+3.00
+4.00
+5.00
-6.00
+7.00
+9.00
+10.00
+11.00
+12.00
+3.30
KALININGRAD
BOLSHAKOVO
LAITSE
ULBROKA
BREST
LVIV
LUTSK
MIENSK
MAHILEU
OSIPOVIČY
EDINET
MAYAK
KYYIV
ODESA
MYKOLAYIV
KHARKIV
SIMFEROPOL
LUGANSK
POPOVKA
ROSTOV-NA-DONU
KRASNODAR
VOLGOGRAD
SUCHUMI
ASTRAKHAN
TBILISI
YEREVAN
GYANCA
BAKY
TABRIZ
KRASNOVODSK
TEHRAN
KAMALABAD
ISFAHAN
MASHAD
TAYYEBAD
SIRJAN
ASQABAT
ČÄRYEV
BUHORO
TOSKENT
SAMARKAND
YANGI-YUL
ORZU
DUSHANBE
FARGONA
JAMBYL
BIŞKEK
ALMATY
KARATURUK
PESHAWAR
RAWALPINDI
SRINAGAR
JAMMU
LEH
MURMANSK
MONCHEGORSK
SANKT-PETERBURG
PETROZAVODSK
ARCHANGELSK
TVER
TALDOM
MOSKVA
PUSHKINO
NOGINSK
SERPUKHOV
TULA
BALASIKHA
VORONEZH
NARYAN-MAR
SYKTYVKAR
UKHTA
NIZHNIY NOVGOROD
CHEBOKSARY
KUDYMKAR
KAZAN
PERM
TOLYATTI
SAMARA
UFA
YEKATERINBURG
CHELYABINSK
ORENBURG
TYUMEN
AKTÖBE
KROSTANAY
BEREZOVO
KHANTY-MANSIYSK
OMSK
AKMOLA
PAVLODAR
KARAQANDI
SEMEY
ÖSKEMEN
DUDINKA
NORILSK
OYASH
NOVOSIBIRSK
BARNAUL
KRASNOYARSK
ABAKAN
KYZYL
TURA
ANGARSK
ULAN-UDE
CHITA
ALEKSANDROVKA
YAKUTSK
BLAGOVESHCHENSK
HAILAR
KOMSOMOLSK-NA-AMURE
ELBAN
KHABAROVSK
HARBIN
CHANGCHUN
RAZDOLNOYE
ULAANBAATAR
URUMQI
XINING
LANZHOU
HOHHOT
BEIJING
BAODING
SHIJIAZHUANG
TIANJIN
KUJANG
KANGGYE
PYONGYANG
SUWON
HWASONG
KIMJAE
KYONGSAN
ANADYR
PALANA
ARMAN
PETROPAVLOVSK-KAMCHATSKIY
YUZHNO-SAKHALINSK
SAPPORO
YAMATA
NAGOYA
OSAKA
HIROSHIMA
MATSUYAMA
60°
80°
180°
170°
160°
50°
150°
40°
140°
130°

6125	100	1900-2200
7105	500	0300-0900
	100	2300-0200
7130	250	2300-0300
7140	100	2300-0100
7165	50	0000-0500
	250	1500-2230
7205	100	2300-0400
7235	240	0100-0400
7250	100	0430-1600
7260	250	1500-2330
7325	50	0000-0500
7330	240	0830-1500
		1630-0700
7340	500	1730-0600
9525	150	0430-1400
9575	240	1000-1500
		2200-0600
9595	100	0300-0500
		0530-1700
9605	50	0000-0500
9630	250	0600-1100
		1330-2330
9725	240	1130-1330
9745	100	1130-0700
9795	100	0800-1000
	500	1530-0530
9865	240	0000-0800
9880	150	1100-0900
9895	200	1500-2400
11655	500	1400-2100
11725	240	1630-0300
11810	240	0500-1100
11870	240	0700-1500
11900	250	0900-1700
11995	140	0600-0800
12010	240	0630-0600
12070	250	1430-1900
	1000	2230-0800
13745	500	0030-0700
	240	0830-1200
15130	240	0200-0800
	240	1700-2330
15265	240	0200-1500
15295	100	1000-1100
		1400-1730
15320	250	0200-1200
15420	240	1200-1700
15475	240	2130-1600
15520	120	0330-1400
17615	50	0400-1700
17730	250	0400-1530
17795	500	0700-1600
17870	250	0830-1600

Starobelsk, Ukraine

6055	100	1700-0400
9620	100	0500-1530
9640	100	0100-2200

Tashkent, Uzbekistan

5930	100	0000-0330
5945	100	1100-1930
5955	100	0000-0330
5975	100	0230-0330
5975	100	1530-1830
5985	100	0600-1400
5995	100	0000-1900
6005	50	1430-2130
6045	100	1530-1800
6190	100	1530-2230
7145	100	1800-2400
7155	100	1530-1700
7190	240	0000-0330
7245	100	0000-1800
7275	100	1200-2400
7285	100	1500-2100
7325	100	1200-1930
7335	100	0100-0330
9510	100	0830-1230
		1800-0400
9540	100	1100-1930
9600	100	1200-1500
9630	1000	2200-0600
9640	1000	1430-2400
9675	500	0500-0600
	1000	1430-2100
		2130-0500
9705	240	1600-2130
9715	100	1630-0500
9840	100	1300-0330
9885	240	1230-1600
11660	100	0930-1130
11695	50	0000-1400
11730	240	0300-1600
11765	240	0330-2200
11785	100	0900-1430
11880	500	1000-1400
11975	100	2200-1100
11985	100	0230-0330
12055	240	1800-2400
13725	100	0830-1530
15295	240	0130-1700
15470	100	1100-1500
15570	200	1800-2400
17655	1000	0130-1000
17755	240	0130-1100
21615	100	0300-1700

Tbilisi, Georgia

5930	240	0200-0500
	100	0500-0900
		1500-1900
6085	100	1500-2300
6180	100	0200-0300
		1500-2230
7155	240	1630-2300
7160	100	1530-2330
7250	100	0000-0700
7310	240	1430-0200
7325	100	2330-0500
9825	240	1530-1700
11755	100	2330-0500
11795	100	0230-1800
11805	100	0500-0700
	200	1900-0200
11815	100	0530-1400
11845	100	0200-2000
11900	100	1900-2100
12050	250	0500-0600
13650	100	0330-0700
13650	100	1500-2300
15490	100	1530-2300
17560	1000	1100-1600
17590	240	1300-1600
17805	240	1200-1530
21725	100	0730-1530

Tula

5910	100	0200-2400
5935	240	2200-0500
5950	240	0400-0930
5985	240	1430-1900
5995	100	0200-0400
		1630-2130
		2230-0100
6005	150	0000-0500
6040	100	0000-0600
		1330-2000
6075	100	2130-0130
6125	100	1500-1900
	240	1930-2100
6190	100	0130-0430
7110	100	1600-2200
7120	100	0400-0530
7145	50	1430-2400
7215	240	1400-1000
7225	240	0330-0830
		1700-2300
7230	240	1230-1530
7240	240	0700-0900
		1200-1630
7305	100	2100-0430
9505	100	1530-0400
9555	100	1430-2400
9580	250	0500-0900
9715	240	1030-0500
9730	1000	1400-2300
9760	100	1330-2100
	240	2100-0700
9775	250	0300-0800
9790	120	1200-1900
9860	240	1630-2230
11675	250	1230-1430
11715	200	1430-1900
	100	1900-2300
11905	100	0430-0800
11925	100	0830-1600
12000	240	0600-0800
12030	500	2330-0700
12065	250	0300-0830
	240	1930-2230
13605	250	1130-1400
13615	250	1000-1100
		1400-2330
15190	240	0630-2000
15415	120	1130-1900
	200	2330-0500
15490	250	2330-0700
15535	500	0530-1500
15550	500	0300-1630
17635	500	0430-1500
17665	240	0400-2200
17840	250	0700-1100
		1400-2100
17880	250	1800-2200
21820	240	0800-1800

Tver (Kalinin)

6105	100	0100-0400
		1430-2100
		2200-0030
15175	500	0900-1600

Vinnitsa, Ukraine

7240	1000	1630-0700
9785	240	1730-1000
11840	100	0300-1630
12060	500	2330-0800
15490	1000	1730-2330
17565	1000	0900-1700
17595	250	0700-1800
17790	500	0700-1700

Vladivostok, Siberia

5960	100	0730-1700
		1930-2030
6035	100	0630-1600
		1800-2200
6130	240	2200-0800
7290	240	0430-0800
	100	0930-1500
	240	1930-2200
7300	200	0530-0800
9510	250	1500-1800
9530	100	2200-1000
9565	240	0930-1600
9705	1000	1030-1800
9745	100	0830-1500
9800	240	0000-1900
9845	100	1930-2030
11665	120	1530-0200
11675	240	0130-1200
	1000	1930-2230
	100	2000-2200
11755	100	0430-0800
15130	240	0730-1230
15505	100	2100-0700
21690	500	2230-1800
25715	100	2230-0700

Volgograd

5925	100	1530-0430
6110	200	0130-2100
7130	100	1930-2130
11730	250	1600-1900
12060	100	0500-1500
15475	240	0700-2100
15595	240	1100-1500
17620	240	0700-1800
17850	240	2100-1000

Yakutsk, Siberia

5940	50	1800-1700
7320	100	1400-1300
7345	50	2000-1800

Yekaterinburg (Sverdlovsk)

5920	100	1430-0200
5960	240	1400-1630
		2200-0400
5970	100	1400-1730
6000	240	0200-0730
		1500-0100
6030	240	2300-0200
6090	100	1200-1600
6115	240	2230-0200
6170	200	2300-0200
7140	200	1900-2200
7160	240	1500-2100
7185	240	1930-0200
7220	240	2200-2000
7230	200	1330-2200
7240	240	0230-0900
7300	200	1130-1700
9530	100	1700-2000
9595	50	2330-0500
9610	150	1400-2300
9640	100	0430-0800
9765	100	2300-1830
9775	250	1400-0300
9790	100	0230-1500
9795	240	1430-2300
9800	240	2300-0300
11745	240	1900-2300
11880	240	0230-1630
11940	100	0530-1300
12050	240	1530-2300
13720	240	1500-1730
15110	240	0230-1400
15130	240	1230-2000
15260	240	1400-1900
15340	240	0100-1300
15410	240	1400-2300

Yerevan, Armenia

6065	100	0230-2130
6080	100	1600-2400
9625	100	2330-0600
9775	100	1430-0900
9800	100	2330-0500
11720	100	0200-1530
11920	500	1630-0600
12015	100	1530-2230
12035	100	0200-0700
12060	1000	1700-1300
13625	200	0430-0700
13670	100	0300-0800
		1730-2400
15205	100	0730-1700
15455	1000	0430-1100
	250	1100-1530
15485	500	0630-1530
		2130-2300
15510	100	0800-0900
17570	1000	0600-2100
17605	100	0600-1800
17675	1000	1200-1730
21565	500	1630-2300
21740	100	1430-2200

Yevpatoria, Ukraine

5980	100	0300-0900
6050	100	2000-0400
6070	50	2130-0600
9610	500	1630-0800
9860	500	1700-0700
11780	100	0530-1730
11985	100	1100-1700
15195	1000	1530-0900
15485	500	1530-1800

15525	120	1730-0800	21800	1000	0900-1730
17810	1000	0700-2000	**Yuzhno-Sakhalinsk, FE**		
21755	240	0900-1900	11810	100	0900-1700
21765	500	0700-1900	11840	50	1900-1600

RADIO MOSCOW INTERNATIONAL

Foreign Language Schedule Winter 1992/93

*) Freq starts on the half hour

World Service in English

Times	kHz
0000-0100	21790, 21770, 21690, 21530, 21480, 17890, 17825*, 17775, 17655, 17620*, 17570, 15425, 15350, 12050, 12015, 11895*, 11775, 11655, 9885, 9870, 9755, 9750, 9735, 9725, 9715, 9695, 9685, 9655, 9625, 9530, 9480, 7390, 7355, 7305*, 7295, 7285, 7225, 7220, 7170, 7135, 7120, 7115, 7110, 6045, 6000, 5985, 4940, 648, 585
0100-0200	21790, 21770, 21690, 21585*, 21530, 21480, 17890, 17825, 17775, 17655, 17620, 17570, 15425, 15350, 15315, 12050, 12015, 11895, 11775, 11655, 9885, 9870, 9775*, 9755, 9735, 9725, 9715, 9695, 9685, 9655, 9625, 9565, 9530, 9525, 9480, 7390, 7340*, 7305, 7295, 7275, 7170, 7150, 7135, 7115, 7110, 6110, 6045, 6000, 5985, 4975, 4940, 648, 585
0200-0300	21790, 21770, 21690, 21585, 21530, 21480, 17890, 17825, 17775, 17655, 17620, 17610*, 17570, 15425, 15350, 15315, 12050, 12015, 11895, 11775, 11655, 9885, 9775*, 9755, 9735, 9725, 9715, 9695, 9685, 9655, 9565, 9530, 9525, 9480, 7390, 7340*, 7305, 7295, 7275, 7170, 7150, 7135, 7115, 7110, 6110, 6045, 6000, 5985, 4975, 4940, 648, 585
0300-0400	21830, 21790, 21770, 21690, 21585, 21530, 17890, 17825, 17775, 17655, 17620, 17610, 17570, 15545*, 15470*, 15425, 15385*, 15375*, 15350, 15315, 15280*, 12050, 12015, 11765*, 9775, 9755, 9725, 9715, 9655, 9530, 9480, 7350, 7340, 7305, 7295, 7270*, 7150, 7135, 7115, 6045, 6000, 5985, 4975, 4940, 648, 585
0400-0430	21845, 21830, 21790, 21770, 21690, 21615, 21585, 21565, 21530, 17890, 17825, 17775, 17685, 17655, 17625, 17620, 17610, 17590, 17580, 17570, 15545, 15470, 15385, 15375, 15350, 15315, 15280, 15230, 11975, 11765, 9870, 9775, 9665, 9580, 9535, 9530, 9490, 9480, 7340, 7295, 7150, 7135, 7115, 6165, 6110, 6000, 5950, 4975, 4940, 648, 585
0500-0600	21845, 21830, 21790, 21770, 21690, 21680, 21655, 21615, 21585, 21565, 21550, 21530, 17890, 17870, 17825, 17775, 17755, 17735, 17720, 17685, 17675, 17655, 17635, 17620, 17610, 17590, 17580, 17570, 15550, 15530, 15470, 15465*, 15350, 15280, 15230, 12055, 11975, 11765, 9890, 9870, 9840, 9665, 9535, 9530, 9490, 9450, 7390, 7340, 7330, 7295, 7180, 7165, 7150, 7135, 7130, 6165, 6110, 5950, 5905, 4975, 4940
0600-0700	21845, 21830, 21790, 21770, 21690, 21680, 21655, 21615, 21585, 21565, 21550, 21530, 17890, 17870, 17825, 17775, 17755, 17735, 17720, 17685, 17675, 17655, 17635, 17620, 17610, 17590, 17580, 17570, 15550, 15530, 15470, 15465, 15350, 15280, 15230, 15190*, 13650*, 12055, 11975, 11765, 9890, 9870, 9860, 9840, 9535, 9530, 9490, 9450, 7340, 7330, 7180, 7165, 7150, 7135, 7130, 6165, 6110, 5905, 4940, 1323
0700-0800	21845, 21830, 21790, 21785, 21690, 21680, 21655, 21615, 21585, 21565, 21550, 21450, 17890, 17870, 17825, 17805*, 17775, 17765, 17755, 17735, 17720, 17685, 17675, 17655, 17635, 17620, 17610, 17590, 17580, 17570, 15550, 15540, 15530, 15520*, 15470, 15465, 15350, 15345*, 15280, 15230, 15190, 13705*, 13650, 12055, 12010, 11975, 11765, 9890, 9535, 9530, 9490, 7380*, 7135*, 7130, 6110, 5960*, 5905, 4940, 1323
0800-0900	21845, 21830, 21825, 21790, 21785, 21690, 21680, 21655, 21630, 21615, 21550, 21450, 17870, 17860, 17805, 17775, 17765, 17755, 17735, 17720, 17685, 17655, 17635, 17625, 17610, 17580, 17570, 15550, 15540, 15530, 15520, 15475, 15470, 15465, 15440, 15435*, 15350, 15345, 15280, 15230, 15210, 15190, 15150*, 13705, 13650, 12070*, 12055, 12020, 12010, 11975, 11765, 9580, 9535, 7130, 6110, 5960, 4975, 4940, 1323
0900-1000	21845, 21825, 21785, 21755, 21690, 21680, 21655, 21630, 21625*, 21615, 21550, 21450, 17870, 17860, 17805, 17775, 17765, 17755, 17735, 17730, 17695*, 17685, 17655, 17635, 17625, 17610, 17600*, 17565, 15595, 15550, 15540, 15530, 15520, 15495*, 15475*, 15470, 15465, 15440, 15435, 15350, 15345, 15280, 15230, 15210, 15190, 15175, 15150, 13705, 13650, 12070, 12055, 12020, 12010, 11975, 11765, 11710*, 9855*, 9830*, 9780*, 9755*, 9685*, 9535, 7130, 6110, 4975, 4940, 1323
1000-1100	21845, 21825, 21785, 21755, 21680, 21630, 21625, 21615, 21550, 17840*, 17805, 17755, 17695, 17685, 17635, 17625, 17600, 17565, 15550, 15530, 15520, 15475, 15470, 15465, 15440, 15435, 15345, 15280, 15230, 15210, 15175, 13650, 12070, 12055, 12020, 12010, 11975, 11765, 11710, 11675, 9855, 9780, 9705*, 9535, 7130, 6110, 4975
1100-1200	21825, 21785, 21755, 21680, 21630, 21550, 17775, 17735, 17695, 17675, 17625, 17610, 17600, 17565, 15585*, 15550, 15530, 15520, 15510, 15475, 15470, 15465, 15440, 15435, 15350, 15345, 15280, 15210, 13650, 12070, 12055, 12020, 12010, 11710, 11675, 9780, 9745*, 9705, 7130, 6000, 4940
1200-1300	21825, 21785, 21755, 21680, 21630, 21615, 21550, 17860, 17840, 17735, 17625, 17600, 15585, 15540, 15520, 15510, 15480*, 15475, 15465, 15440, 15435, 15345, 15280, 15210, 13705, 13650, 12070, 12025, 12020, 11980, 11850*, 11775, 11765, 11710, 11675, 9895, 9855, 9780, 9745, 9705, 7380*, 7260, 7160, 7130, 6000, 5960, 5950, 5940, 1323, 720, 630
1300-1400	21785, 21755, 21680, 21615, 17860, 17840, 17775, 17610, 15550, 15540, 15440, 15345, 15210, 13705, 11980, 11890*, 11840*, 11765, 11715*, 11710, 11675, 9890*, 9855, 9840*, 9780, 9745, 9735*, 9725*, 9705, 7380, 7370*, 7330, 4940, 1323
1400-1500	21785, 21755, 21680, 21615, 17860, 17840, 17730, 15540, 15525, 15520, 15480, 15465, 15440, 15420, 15345, 15210, 12060*, 12050, 12025, 11930, 11840, 11765, 11715, 11675, 11655, 9895, 9890, 9855, 9840, 9755, 9745, 9725, 9705, 7380, 7370, 7335*, 7330, 7260, 7225, 7170, 7135, 7115*, 6055*, 5960, 5905, 4975*, 1323, 630
1500-1600	21755, 21680, 21615, 17730, 17610, 15550, 15540, 15480, 15465, 15420, 15210, 12065, 12060, 12050, 12045*, 11930, 11890, 11840, 9895, 9890, 9755, 9725, 9590*, 7390, 7370, 7360, 7335, 7330, 7290*, 7260, 7250*, 7185*, 7170*, 7155*, 7115, 6055, 5905, 4975, 4940, 1323
1600-1700	12060, 12045, 11840, 11685, 11655, 9895, 9890, 9830, 9755, 9590, 9540, 7420*, 7370, 7335, 7330, 7265, 7260, 7250, 7205*, 7185, 7170, 7155 (exc-.Fri), 7150, 7115, 6130*, 6055, 6000*, 5905, 1494, 1269, 1125
1700-1800	12060, 11840, 9895, 9890, 9860*, 9830, 9755, 9685, 9590, 9540, 9515*, 7420, 7340*, 7330, 7260, 7250, 7205, 7170
1800-1900	13670, 12050*, 11920, 11840, 11685, 9890, 9875, 9860, 9855, 9830, 9755, 9725, 9625, 9590, 9565, 7420, 7330, 7260, 7250, 7170, 7145, 6130*, 1215
1900-2000	17605*, 15425*, 13670, 12050, 11920, 11840, 11685, 9890, 9875, 9860, 9855, 9785, 9735, 9725, 9685, 9590, 9530, 7390, 7260, 7205, 7195*, 7170, 7145, 5950

2000-2030		17605, 15425, 13670, 12050, 11920, 11840, 11685, 9890, 9860, 9855, 9795, 9785, 9735, 9725, 9590, 9565, 9515, 9490, 9450, 7390, 7255, 7250, 7245, 7205, 7195, 7170, 7145, 5950, 4795
2100-2200		21480*, 17690*, 17665*, 17655, 17605, 17570*, 15425, 15130, 13670, 12050, 11840, 11655*, 9890, 9885*, 9870*, 9860, 9855, 9795, 9785, 9750*, 9735, 9725, 9695, 9565, 9530, 9490, 9480, 9450, 7390, 7370, 7255, 7245, 7220, 7205, 7195, 7180, 7170, 7150, 7145, 7115, 6055, 6030, 5950, 4795, 1170, 648, 585
2200-2300		21790*, 21770*, 21480, 17690, 17665, 17655, 17570, 15425, 15130, 12050, 11775*, 11655, 9890, 9885, 9870, 9860, 9750, 9735, 9725, 9695, 9685, 9530, 9520, 9480, 9450, 7390, 7300, 7295, 7285, 7255, 7245, 7220, 7180, 7150, 7145, 7135, 7115, 6060, 6055, 6045, 6030, 5985, 5980, 5920, 4860, 4825, 4795, 1170, 648, 585
2300-2400		21790, 21770, 21690*, 21530*, 21480, 17890*, 17665, 17655, 17570, 15425, 15350, 15130, 12050, 11775, 11655, 9885, 9870, 9755, 9735, 9725, 9715*, 9695, 9625, 9530, 9480, 9450, 7390, 7355*, 7315*, 7295, 7285, 7255, 7220, 7170*, 7145, 7135, 7120, 7115, 7110, 6045, 5985, 5895, 648, 585
Afrikaans	**Area**	**kHz**
1700-1800	Af	12045, 12030, 11670
Albanian		
1900-2000	Eu	9795, 7370, 7300, 7185, 5980, 5940, 1548
2100-2200	Eu	7230, 7185, 5940, 4825
Amharic		
1500-1600	Af	21785, 17570, 11695
Arabic		
1200-1400	ME/Af	17565, 15495, 12055, 11895
1400-1500	ME/Af	17565, 15495, 12065, 11895, 9675*
1500-1600	ME/Af	17565, 11895, 9675*, 9600*
1600-1700	ME/Af	17565, 11895, 9675, 9610*, 9600, 9530*, 7275*, 7200*, 1314, 1170, 1089
1700-1800	ME/Af	12015*, 11895, 9675, 8610, 9600, 9530, 9505*, 7275, 7200, 6135, 5905, 1314, 1296 (+ 5905, 1170, 1089 Mon-/Wed/Fri/Sat)
1800-1900	ME/Af	12015, 11895, 11765, 9675, 9610, 9600, 9505, 7210*, 6190, 6170, 6135, 6110, 6005, 1314, 1296
1900-2100	ME/Af	12015, 11895, 11765, 9675, 9610, 9600, 9505, 9480, 7210, 7200, 6170, 6135, 6110, 6005, 5905, 1314, 1296, 1170*
Bamar		
1100-1200	As	21625, 17840, 17560, 11775
Bengali		
1000-1100	As	21655, 17775, 15595, 15350
1200-1300	As	21655, 15595, 15550, 11775, 1269
1400-1500	As	17610, 17550, 15220, 11775
Bulgarian		
1700-1800	Eu	7350, 7185, 5940
1900-2000	Eu	7350, 7230, 6130, 6030, 4825
Cambodian		
1200-1300	As	17840, 17560, 15420*, 12050
Chinese		
0500-0800	As	15545, 15385, 15375, 15315, 9480, 648, 585
0800-0900	As	15545, 15375, 11905, 11775*, 11730, 1476, 585
0900-1000	As	15375, 13615, 11905, 11775, 11730, 9480, 9450, 7390, 7280*, 4010*, 1476, 648, 585
1000-1100	As	15375, 13615, 11950, 11775, 9845*, 9480, 7390, 7330, 7280, 1476, 1350, 801, 585
1100-1200	As	13615, 11905, 9845, 9450, 7390, 7330, 4010, 1476, 648, 585
1200-1300	As	13615, 11905, 9845, 7330, 1476, 585
1300-1400	As	13615, 11905, 9845, 9450, 7390*, 4010, 1476, 648, 585
1400-1500	As	15095, 11905, 9845, 6100, 4010, 1476, 1350, 648, 585
1500-1600	As	13615, 11905, 9845, 9480, 7280 (Exc-.Fri), 7135, 6100, 4010, 1476, 648, 585
1600-1700	As	11905, 9480, 7135, 6100, 4010, 1476, 585
Czech/Slovak		
1600-1700	Eu	7330, 5950, 5920, 1170
1800-2000	Eu	5950, 5920, 4860, 1170
Danish		
2000-2100	Eu	7150, 7115, 6055, 1494
Dari		
1230-1330	As	15395, 12015, 7335, 4975, 1125, 945
1400-1430	As	15395, 12015, 7335, 4975, 1125, 945
Dutch		
1800-1900	Eu	9480, 7290, 7205
English		
0400-0500	WNAm	17605, 12050, 9895, 9825*, 7270
0500-0600	WNAm	12050, 9895, 7270, 7175*, 5905*
0600-0700	WNAm	9825, 7345*, 7270, 7260*, 7175, 5905
0700-0800	WNAm	9825, 7345, 7270, 7175, 5905
Finnish		
1700-1800	Eu	7150, 7115, 6055, 1494
French		
0500-0600	Eu/Af	21825, 17860, 17780, 17625, 15475, 11900, 11890, 9880, 9630, 9610, 9580, 7310, 7135
0600-0700	Eu/Af	21825, 17860, 17780, 17625, 15475, 12020, 11900, 11890, 9880, 9630, 9610, 9580, 7310, 7135
0700-0800	Eu/Af	21825, 17860, 17780, 17625, 15475, 12020, 11890, 9880, 9630, 9610, 9580, 7310, 7135
1300-1400	Eu/Af	21825, 17625, 15585, 15510, 13650, 12020, 11850
1400-1500	Eu/Af	21825, 17625, 15585, 15510, 15475, 13650, 12055, 12020, 11850, 11725, 9880
1500-1600	Eu/Af	21825, 17625, 15585, 15510, 15475, 13650, 12055, 12020, 11690*, 9880. 7320*, 7215*, 7175*
1600-1700	Eu/Af	17625, 15510, 13650, 12070*, 12055, 12020, 11775*, 11690, 11670*, 9880, 9480, 7320, 7175, 6145
1700-1800	Eu/Af	15510, 12070, 11775, 11690, 11670*, 9880, 9720*, 9480, 7320, 7280*, 7215, 7175, 6145
1800-1900	Eu/Af	12055, 12045, 12030, 11775, 11670, 9880, 9745, 9720, 9630*, 7320, 7280, 7215, 6145, 1323
1900-2000	Eu/Af	12055, 11775, 11670, 9745, 9720, 9640, 9630, 7330, 7280, 7215, 1323
2000-2100	Eu/Af	12070, 12055, 11775, 11670, 9745, 9720, 9630, 7330, 7290, 7280, 7215, 7175, 1323
2100-2200	Eu/Af	12070, 9745, 9640*, 9630, 9540, 9505, 9480, 7330, 7290, 7280, 7215, 7175, 1323
2200-2300	Eu/Af	13670, 12070, 9745, 9640, 9630, 9590, 9505, 9490, 7215, 7205, 6185, 5980, 1323
2300-2400	Am	7150
0000-0100	Am	9860, 7180
German		
1000-1100	Eu	17860, 15540, 15495, 15190
1100-1200	Eu	15540, 15495, 15190, 13705, 11980, 1323
1600-1800	Eu	7390, 7360, 5905, 1386, 1323
1800-2000	Eu	7390, 7360, 5905, 1386
2000-2100	Eu	7190, 7360, 4795, 1386
Greek		
1800-1900	Eu	9795, 7350, 7185, 5940
2000-2100	Eu	7350, 7230, 7185, 5940
Hausa		
1330-1400	Af	15465, 11725
1800-2000	Af	12070, 12060, 9515, 7275
Hindi		
1100-1200	As	21655, 15450, 17870, 17765, 15595, 15175, 11880, 9830, 9755
1300-1400	As	17765, 15560, 15220*, 11985*, 11655*, 9830, 9755, 1269
1500-1600	As	11985, 11685, 11655, 9830, 7265, 1125

Hungarian		
1700-1800	Eu	5950, 5920, 4860, 1170
2000-2100	Eu	5950, 5920, 4860, 1170
Indonesian		
1000-1100	As	17730, 17610, 15150, 9685
1200-1300	As	17730, 17610, 15150, 9685
Italian		
1800-1900	Eu	11690, 7340, 7300*, 6030, 5980
2000-2100	Eu	7300, 6130, 6000, 1548
2100-2200	Eu	7300, 6130, 5980, 5920, 1548
Japanese		
0900-1000	As	7380, 7260*, 7160*, 7135, 5960, 5950*, 5940*, 1008, 720
1000-1100	As	7380, 7260, 7160, 5960, 5950, 5940, 720
1100-1200	As	7260, 7160, 7135, 5960, 5950, 5940, 720, 630
1300-1400	As	7135, 5960, 720, 630
Korean		
0800-0900	As	9480, 9450*, 7380, 7135, 648
1000-1100	As	9450, 7135, 4010, 648
1200-1300	As	9450, 7310*, 7135, 4010, 648
Mongolian		
1130-1400	As	9480, 7280, 6100, 1350, 1080, 801
1400-1500	As	9480, 7280, 1080, 801
Norwegian		
1800-1900	Eu	7150, 7115, 6055, 1494
Persian		
1330-1400	ME	15480, 12065, 11930
1530-1700	ME	11765, 6190, 6170
1700-1800	ME	11765, 6190, 6170, 6110
Polish		
1600-1700	Eu	9795, 7370, 7290, 7215, 7140*, 1215
1700-1800	Eu	7290, 7140, 4795, 1215
1900-2000	Eu	7370, 7290, 4795, 1215
Portuguese		
1900-2000	Eu/Af	12045, 12030, 9830, 9665*, 9575, 7340, 7275, 6145, 6000
2000-2100	Eu/Af	12045, 12030, 9830, 9665, 9575, 7340, 7275, 6145
2100-2200	Eu/Af	9720, 9675*, 9610, 9600, 7420, 7390, 7340, 7175, 6145
2200-2300	Eu/Af	11980, 11900*, 9720, 9675, 9665, 9610, 9600, 7470*, 7390, 7310*, 7175, 6145
2300-2400	Am	11900, 9675, 9665, 9610, 9470, 7310
0000-0200	Am	11980, 11900, 11690, 9675, 9665, 9610, 9470, 7310
Pushtu		
1130-1230	As	15395, 12015, 7335, 4975, 1125, 945
1330-1400	As	15395, 12015, 7335, 4975, 1125, 945
Romanian		
1700-1800	Eu	9795, 7370, 7230*, 6000, 1548
1800-1900	Eu	7370, 7230, 6000, 1548
Serbo-Croat		
1700-1800	Eu	7300, 6130, 6030, 5980*, 4825
2000-2100	Eu	7370, 6030, 5980, 4825
Sinhala		
1200-1300	As	17675, 11880, 11655
Slovene		
1800-1830	Eu	7300, 6130, 4825
Somali		
1600-1700	Af	17570, 11930, 11695
Spanish		
0500-0600	Eu/Am	11690, 9860, 7460, 7400, 7230, 6185
1900-2000	Eu	11690, 9880, 7320
2000-2100	Eu	11690, 7420, 7320
2100-2200	Eu	11690, 9665, 7400, 7320, 7290, 6000
2200-2300	Eu	11690, 9795, 7400, 7320, 7290, 6000
2300-2400	Eu/Am	12060, 11690, 9795, 9745, 9630, 9520, 9490, 7400, 7390, 7330, 7300, 7230, 7215, 7135, 6185, 6000
0000-0200	Am	12060, 9795, 9745, 9630, 9520, 9490, 9480, 7400, 7390, 7300, 7230, 7135, 6185
0200-0400	Am	12060, 11980, 11900, 11690, 9795, 9745, 9675, 9630, 9610, 9520, 9490, 9480, 9470, 7400, 7390, 7310, 7300, 7230, 7135, 6185
0400-0500	Am	11900, 11690, 9630, 9610, 9520, 9480, 9570, 7400, 7390, 7310, 7300, 7230, 7135, 6185
Swahili		
1700-1900	Af	17570, 11695
Swedish		
1900-2000	Eu	7150, 7115, 6055, 1494
Tamil		
1300-1400	As	17675, 15435, 11880
Thai		
1300-1400	As	17840, 17560, 15420, 12050
Turkish		
1500-1600	Eu/ME	11715, 9840 5905, 1170
1800-1900	Eu/ME	9530, 7275, 7175, 7140, 5905, 4825*, 4795, 1170
Urdu		
1000-1100	As	21450, 17870, 17765, 15490, 11880, 9830, 9755, 4940
1200-1300	As	17870, 17765, 15560*, 15490, 11880, 9830, 9755, 4940
1400-1500	As	17765, 15560, 15490, 11985, 9830, 9755, 7265*, 4940, 1125*
Vietnamese		
1100-1200	As	17730, 15150, 9685, 603
1300-1400	As	17730, 15535, 15520, 12025, 9685

ANN: E: "R.Moscow World Sce.". F: "R.Moscou Internationale".

RADIO AUM SHINRIKYO (Rlg.)

ADDR & L.P: see listing under Japan.

SLOVENIA

L.T: UTC + 1h (Su: UTC + 2h) — **Pop:** 2m — **Pr.L:** Slovene — **E.C:** a/c 50, 220V.

RADIOTELEVIZIJA SLOVENIA (Gov.)

Addr: Tavčarjeva, 61000 Ljubljana. **Te:** +38 (61) 311-922. **Fax:** +38 (61) 310-665. **Telex:** 32283 — Dir.Gen: J. Knez, a.i.

Mediumwaves	kHz	kW	Prgr		kHz	kW	Prgr
Beli Križ	549	20	1	Radlje	1485	1	1+L
Maribor	558	20	1+L	Trbovlje	1485	1	2+L
Cerkno	594	1	L	Šmarje	1566	2	2+L
M. Sobota	648	10	1+L	Brežice	1584	1	1+L
Ljubljana 1	918	600	1	Tržič	1584	1	2+L
Celje	963	2	2+L	Kobarid	1602	1	1
Črnomelj	1485	1	1	Ormož	1602	1	1+L
Ptuj	1485	1	1+L				

R. Ljubljana 1: 24h (Mon 2300-Tues 0300 on FM only). **N. in English:** Mon-Fri 2120. **N. in German:** Mon-Fri 2115.
R. Ljubljana 2: 0430 (Sun 0500)-2300. Pop + entertainment.
R. Ljubljana 3: 1500-2300. Serious music, educational.

Local stations	kHz	kW
B. Križ	1170	300*

*) **Studio Koper/Radio Capodistria,** P.O. Box 117, 66000 Koper - Capodistria — **D. Prgr. in Italian** on 89.3/97.7/101.0MHz (also rel. on 106.0MHz in Milano, Italy): 24h (on 1170kHz 0500-1900) — **V.** by QSL-card. IRC appreciated but not necessary.
FM: R. Student, Ljubljana 89.3/104.3MHz: 1200 (Sun 1100)-2300; R. Glas, Ljubljana: 102.4MHz Mon-Fri 0500-1900, Sat 0700-1800, Sun 1100-1800; Radio M.M.2 — Maribor: 102.8MHz 10kW 24h in German.

SPAIN

L.T: UTC + 1h (Su: UTC + 2h) — **Pop:** 39.500.000 — **Radio sets:** 12m — **Pr.L:** Spanish, Catalan, Valencian, Galician, Basque — **E.C:** a/c 50, 127/220V — **ITU:** E.

MINISTERIO DE TRANSPORTES, TURISMO Y COMUNICACIONES

Secretaría General de Comunicaciones.

ENTE PUBLICO RADIOTELEVISION ESPAÑOLA

Casa de la Radio, Prado del Rey, 28023 Madrid.

RADIO NACIONAL DE ESPAÑA

ADDR: Casa de la Radio, Prado del Rey, 28023 Madrid.
Te: +34 (1) 34610 00. **Telex:** 27366. **Fax:** 7112906. **Cable:** Radiogen.
L.P: Dir. RNE: Diego Carcedo. Dir. Prod: Gerardo Rodríguez. Dir. Inf. Sces: Antonio San José. Dir. R1: Alicia Fernández. Dir. R2: Miguel Alonso. Dir. R3: José Antonio Visuña. Dir. R5: Rosario Sacristán. Coordinador Em. Territoriales: José Antonio Pardellas. Dir. Gabinete: Victorino del Pozo. Dir. Tec: Juan Antonio Alberich Cid. Jefe Oficina de Prensa: Salvador López Ibinaga.

HOME SERVICE

(**Rg** = Region)

Radio 1	**kHz**	**kW**	**Rg**
Madrid	585	200	16)
Cadiz	585	5	1)
Vitoria	612	10	13)
Lérida-Lleida	612	10	8)
Jaén	621	10	1)
Avila	621	10	7)
Almería	639	20	1)
Zaragoza	639	20	2)
Albacete	639	10	6)
La Coruña	639	100	10)
Bilbao	639	20	13)
Palma de Mca	648	10	4)
Badajoz	648	10	9)
Sevilla	684	500	1)
Tortosa	693	2	8)
Zamora	702	10	7)
Málaga	729	20	1)
Oviedo	729	50	3)
Cuenca	729	5	6)
Valladolid	729	10	7)
Logroño	729	20	14)
Alicante	729	10	15)
Barcelona	738	500	8)
Granada	774	10	1)
León	774	10	7)
Soria	774	10	7)
Cácares	774	50	9)
Orense	774	20	10)
S Sebastián	774	50	13)
Valencia	774	50	15)
Castellón	801	10	15)
Cuidad Real	801	10	6)
Burgos	801	10	7)
Lugo	801	20	10)
Huelva	855	10	1)
Teruel	855	10	2)
Santander	855	20	5)
Ponferrada	855	10	7)
Salamanca	855	10	7)
Gerona	855	10	8)
Pontevedra	855	20	10)
Murcia	855	125	11)
Pamplona	855	10	12)
Marbella	855	5	1)
Socuellamos	864	2	6)
Monforte de Lemos	972	2	10)
Cabra	972	5	1)

FM

Rg.	**Location**	**R.1**	**R.2**	**R.3**	**kW**
1)	Jerez	—	94.5	96.7	5
	Pechina	—	92.4	94.9	5
	Córdoba	—	97.5	98.6	1
	Granada	—	96.4	94.4	1
	Sierra Lújar	—	90.4	94.2	1
	Parapanda	—	91.1	93.9	1
	Punta Umbria	—	92.6	99.0	5
	Sierra Almadén	—	90.0	96.0	1
	Mijas	106.6	98.1	98.8	10
	Guadalcanal	—	90.6	—	5
	Valencina	91.2	93.7	98.8	5
2)	Arguis	103.9	88.1	101.5	40
	Javalambre	—	90.0	93.9	1
	La Muela	104.4	90.9	96.3	10
3)	Gamonteiro	102.5	92.2	94.4	10
	El Naranco	—	96.0	90.3	1
4)	Alfabia	—	87.9	92.3	100
	Pollensa	—	95.4	97.4	1
5)	Liérganes	—	93.0	102.9	5
6)	Chinchilla	—	93.6	99.0	5
	La Mancha	88.4	89.8	94.5	10
	Cuenca	—	93.0	92.0	5
7)	Burgos	—	90.3	91.2	1
	Benavente	87.8	—	—	5
	Villadiego	—	102.3	103.3	1
	El Bierzo	—	93.0	99.9	1
	León	—	91.1	89.3	1
	Aranda de Duero	—	92.7	101.6	5
	Peña de Francia	—	92.4	95.4	1
	Villablino	—	89.0	106.0	1
	Villafranca	—	89.7	97.5	1
	Salamanca	—	88.1	91.4	6
	Santa Ana	—	91.5	94.3	5
	Valladolid	—	93.1	92.2	5
	El Viso	—	96.7	98.5	5
8)	Tibidado	88.3	93.0	98.7	100
	Rocacorba	93.3	91.1	95.9	5
	Alpicat	—	89.2	97.8	1
	Soriguera	99.9	—	—	1
	La Musara	—	91.5	94.5	1
	Montecaro	—	96.6	99.6	1
	Collsuspina	—	97.9	103.6	1
	Igualada	—	90.9	105.1	1
	S. Pere Ribes	—	95.2	97.5	1
9)	Montánchez	—	97.7	99.3	1
10)	Balladora	95.8	91.6	94.5	10
	Stgo de Compostela	—	98.1	99.0	5
	Páramo	—	88.2	99.6	70
	Meda	—	91.2	94.3	5
	Domayo	—	92.1	97.4	1
11)	Carrascoy	101.7	98.2	96.0	5
	Cartagena	—	94.5	97.5	20
12)	Monreal	—	93.0	97.5	5
	Tudela	—	91.3	—	5
13)	Beasain	—	98.4	94.9	25
	Eibar	—	98.7	95.9	25
	Jaizquibel	—	90.0	92.1	5
	Tolosa	—	98.8	96.0	10
	Archanda	100.7	90.6	99.2	5
	Sollube	—	93.9	95.4	1
	Vitoria	—	96.9	99.5	1
	Azcoitia	—	104.9	106.9	10
14)	Logroño	—	98.2	101.4	5
15)	Benicasim	—	90.3	92.8	10
	Torrente	89.8	92.0	95.1	10
	Altana	—	88.6	99.7	12
	Alcoy	—	92.3	91.1	1
16)	Madrid-Torrespaña	88.2	96.5	93.2	100
	Navacerrada	104.9	98.8	95.8	30

RADIO 1: 24 h. **N:** On the h. **Regional prgrs** (originating from key stn of each region): Mon-Fri 0555-0600, 0655-0700, 0905-0910, 1109-1200, 1505-1510; Sat-Sun 1109-1200.
RADIO 2: (Music & Cultural Prgrs): 24 h.
RADIO 3: (Young People's Music Prgr): 24 h. **N:** 0000, 0300, 0500, 0700, 0900, 1100, 1300, 1500, 1800, 2100.
RADIO 4: The R. 4 network, exc. Sevilla 88.3 and Barcelona 100.8, was closed down in July 1991. At the time of printing certain R. 4 outlets are still active relaying R. 5 or R. 1.
RADIO 4 & RADIO 5 (ex RCE)
RCE (Radiocadena Española) merged with Radio Nacional de España in 1989. RCE outlets operating on both MW and FM prior to the merger became Radio 5 and Radio 4, respectively. Stns that prior to the merger had been on FM only became Radio 5. Since transmitter and studio locations usually differ from those of Radio 1, 2 and 3, fq and address information is given under "Other stations".
ADDR. for regional outlets of Radio 1, 2 and 3:
1) **Andalucia:** Av. Rep. Argentina 31, 41001 Sevilla; Av. de la Aurora 49, 29002 Málaga; Real 24, 11300 La Línea; Martín Alonso Pinzón 1, 21003 Huelva.
2) **Aragón:** Albareda 1-3, 50004 Zaragoza.
3) **Asturiras:** Melquádes Alvarez 9, 33002 Oviedo.
4) **Baleares:** Aragón 26, 07012 Palma de Mallorca
5) **Cantabria:** Vargas 51, 39010 Santander.
6) **Castilla-La Mancha:** Núñez de Arve 10, 45003 Toledo; Camino de la Resiñeral 1, 16003 Cuenca.
7) **Castilla-León:** Angustias 1, 47003 Valladolid; Av. de Zequejo 21, 49003 Zamora; Gral. Yagüe 34, 09004 Burgos.
8) **Cataluna:** Paseo de Gracia 1, 08007 Barcelona.
9) **Extremadura:** Av. Alemania 36, 10001 Cáceres.
10) **Galicia:** Méndez Núñez s/n, 15006 La Coruña.
11) **Murcia:** Jaime el Conquistador 11-9°. 30008 Murcia.
12) **Navarra:** Arrieta 8, 31002 Pamplona.
13) **País Vasco:** San Prudencio 6, 01005 Vitoria; Ercilla 20, 48009 Bilbao; Andía 13, 20004 San Sebastián.
14) **La Rioja:** Av. Portugal 2, 26001 Logroño.
15) **Comunidad Valenciana:** Av. Colom 13, 46004 Valencia.
16) **Madrid:** Casa de la Radio, Prado del Rey, 28023 Madrid.

FOREIGN SERVICE

RADIO EXTERIOR DE ESPAÑA

ADDR: Apartado 156.202, 28080 Madrid. **Te:** +34 (1) 346 11 60. **Fax:** +34 (1) 261 63 88. **Tlx:** REXT 261 31 30. **Cable:** Radiogen.
L.P: Dir: Homero Valencia Benito. Asst. Dir: José Antonio Martin Morán. Head of Inf. Sce: Nuria Alonso Veiga. Foreign Language Prgr: Pennelope Eades. LF-Dept: Manuel Fernández Muñoz. Tech. Secr: Josefina E. Peña Márquez. Head of Freqs: Dr. Manuel Martínez Blanco.

STATIONS: Noblejas, Toledo (G.C: 03.31W/40.18N): 6 x 350kW — Arganda del Rey, Madrid (G.C: 03.26W/39.57N): 5 x 100kW — C = Cariari, Costa Rica: 3 x 100kW — B = Beijing, China (G.C:116.27E/39.57N): 120kW — X = Xi'an, China (G.C: 108.54E/34.12N): 120kW.

kHz	kHz	kHz	kHz	kHz	kHz	kHz
5970	7275	9685	9875	11910	15365	17845
6020	9530	9690	11790	11945	15375	17890
6055	9580	9745	11815	12035	15380	21460
6125	9620	9760	11890	15110	17715	21555
7145	9650					

Arabic	Area	kHz
1700-1900	ME	9580
2015-2200	ME/NAf	9580, 9875
English		
1900-2000	Af	15375
2100-2200	Eu	6125
0000-0200	NAm	9530
0500-0556	NAm	9530
French		
1800-1900	Eu	11890
2000-2100	Af	15375
2300-2400	NAm	9530
German		
1730-1800MF	Eu	6020
Portuguese		
1305-1310Sat	SAm	17715, 17845
Russian		
1700-1730MF	Eu	6020
Sefardi		
1945-2015Thurs	ME	9580
0235-0255Fri	SAm	9620
0415-0435Fri	NAm	9690
Spanish		
0100-0300	CAm	15110C
0100-0500	CAm	5970C
0200-0500	NAm	9530
0400-0700	Eu	7145
0500-0700	Eu	9685
	ME	15110
	AUS	9650, 9760
0600-0900	Eu	11790
0900-1700	Eu	12035
	ME	17890
0900-1900	SAm	17715
0900-2000x	CAf	21555
0900-2000Sun	SAm	15380
1000-1200	J	9620B
1200-1400	PHL	11910X
1200-1700	Eu	9875
1300-1900	SAm/CAm	17845
1500-2000	Eu	9620
1500-2000Sun	SAm	11815, 17845C
	CAm	9745C
	NAm	15365C
1700-2000	Eu	9685
1830-2100Sat	SAm	11815, 17845C
	CAm	9745C
	NAm	14365C
1900-2300	NAm	15110
2000-2300	Eu	7275
2200-2300	ME	9580
	NAf	9875
2200-0200	SAm	17845C
2200-0500	SAm	6125
2300-0100	CAm	21460C
2300-0200	SAm	11945
2300-0400	SAm	9620
2300-0500	NAm	6055

x) Thurs ends at 1935.

Spanish Minority Langs+		
0935-0955	Eu	12035
	SAm	15380, 17715
2235-2255	Eu	7275
	SAm	6125, 12035
	CAm	21460C
	NAm	15110

+) Catalan, Galican, Basque.

ANN: S: "Radio Nacional de España, Radio Exterior" or "Radio Exterior de España". E: "Spanish Foreign Radio:. A: "Idha'atu al-Watabia, Idhaa'atu Isbania al-Jariyia" — **V.** by QSL-card.

OTHER STATIONS
(Nationwide networks)

(SER) SOCIEDAD ESPAÑOLA DE RADIODIFUSION

ADDR: Gran Vía 32, 28013 Madrid. **Te:** +34 (1) 2328000. **Telex:** 27638 SER E.
L.P: Pres: Antonio Garrigues y Díaz Cañabate. Gen. Mgr: Eugenio Galdón Brugarolas. Dir. Tec: Rafael Bernardo Jiménez. Dir. Ext. Rel: Joaquín Peláez González.

(COPE) CADENA DE ONDAS POPULARES ESPAÑOLAS

ADDR: Valenzuela 1, 28004 Madrid. **Te:** +34 (1) 5216576. **Telex:** 41948 RAPOP.
L.P: Gen. Mgr: José Andrés Hernández Vicente. Dir. Tec: Agustín Ruiz de Aguirre.

(OCR) ONDA CERO RADIO

ADDR: Velásquez 54, 28001 Madrid. **Te:** +34 (1) 4316613.
L.P: Gen. Mgr: Ricardo Vaca Berdayes.

(A3) ANTENA 3 DE RADIO

ADDR: Oquendo 23, 28006 Madrid. **Te:** +34 (1) 411 70 11. **Telex:** 49369.
L.P: GM: Manuel Martín Ferrand. Dir. Tec: Juan Antonio Nieto.
Stations: Ch. 1 is called Antena 3, Ch. 2 is R. 80 Serie Oro.

(D) CADENA DIAL

ADDR: Gran Vía 32-B, 28013 Madrid. **Te:** 91-2319704, 2319705, 2215901.

(RM) RADIO MINUTO

ADDR: Miguel Yuste 40, 28037 Madrid. **Te:** +34 (1) 7544057.
L.P: Pres: Juan Tomás de Salas. Mgr: José María Baviano.

(Regional networks)

(CR) CORPORACIO CATALANA DE RADIO I TELEVISIO (Catalan Autonomous Govt.)

ADDR: Av. Diagonal 614-616, 08021 Barcelona. **Te:** +34 201 99 11. Ch. 1: Catalunya R.; Ch. 2: Catalunya Música; Ch. 3: R. Associaciò — **L.P:** Mgr: Jordi Daroca.

(N) CADENA NOVA (Pr.)

ADDR: Av. Diagonal 612, 08021 Barcelona. **Te:** +34 201 42 44. **Telex:** 98605 RAAV-E.
L.P: Gen. Mgr: Jordi Riba Torelló. Dir. Tec: Jaume Guillem.

(EI) EUSKO IRRATI TELEBISTA (Basque Autonomous Govt.)

ADDR: Ch. 1: Euskadi Irratia (in Basque). Andía 13, 20004 San Sebastián.
Ch. 2: R. Euskadi (in Sp.) Gran Vía 85, 48011 Bilbao.
Ch. 3: Euskadi Gaztea (bilingual youth prgr.) Pje. Postas 32, 01001 Vitoria.
L.P: Mgr: Mikel Lejarza Ortiz. Dir. Tec: Goio Torróntegui.

(RG) RADIOTELEVISION GALICIA (Galician Autonomous Govt.)

ADDR: Casa da Radio, San Caetano, 15701 Santiago de Compostela. **Te:** +34 (981) 562323. **Telex:** 86225 RTVG-E.
L.P: Mgr: Gerardo Rodríguez Rodríguez.
Stations: announce "R. Galicia — a nossa radio".

(CS) RADIO TELEVISION DE ANDALUCIA (Andalucian Autonomous Govt.)

ADDR: Ap. 132, 41920 San Juan Aznalfarache.
Ch. 1: Canal Sur 1
Ch. 2: Canal Sur 2 (Andalucian music)

(C9) RADIO TELEVISION VALENCIA (Valencian Autonomous Govt.)

ADDR: Blasco Ibáñez 134, 46022 Valencia.
Stations: Canal Novente.

(CSR) **CANAL SUR RADIO**

ADDR: Carretera San Juan de Aznalfarache, 41920 Sevilla. **Te:** 954-768759. **Fax:** 954-760126.

(M) **EMISORA MUNICIPAL.** Mostly low power FM tr's; about 150 stations operating in the various regions at present.

(L) **EMISORA LIBRE.** About 300 st's, mostly low power, on FM.

Call Signs (EAJ, EAK, EAM, EFJ, EFE, ECS etc), no longer to be heard on the air, and hardly ever to be seen in stn logos, have been omitted from this edition.

Spelling. Christian names and place names may be spelt in more than one way in regions where, in addition to Spanish, Catalan, Basque and Galician are spoken. Excepting certain regional networks, operating on FM mainly, the usual language in broadcasting is however still Spanish, even in these bilingual areas. In order to help those who may lack updated maps, certain place names are given in two languages, viz. Lleida-Lérida. Other address info is given as submitted by each network, which means that spelling usually is in Spanish.

Local progrs:

RNE R5: Mon-Fri: 0555-0600 (R1), 0625-0630 or 0630-0655, 0655-0700 (R1) 1000-1200, 1250-1500, 1505-1510 (R1), 1745-1800.

SER: Mon-Fri: 0520-0530, 0550-0600, 0630-0655, 1030-1230, 1245-1400, 1407-1700, 1720-1730, 1740-1800, 1830-1900. Sat: 0520-0530, 0550-0600, 0620-0655, 1000-1230, 1330-1500. Sun: 0620-0640, 0710-0900, 1000-1230, 1245-1400, 1830-2000.

COPE: Mon-Fri: 0525-0530 (reg), 0557-0600, 0625-0630 (reg/local). 1005-1200, 1210-1300, 1310-1400, 1820-1900 (reg/local). Sat: 0625-0730, 1005-1200, 1210-1400. Sun: 0630-1200, 1215-1330/1400.

OCR: Mon-Fri: 0530-0600, 0630-0700, 1000-1200, 1400-1600. Sat: 1000-1200, Sun: 1000-1100.

	kHz	kW	Net	Location	FM
43)	531	10	RNE-5	Pontevedra	
113)	531	5	RNE-5	Córdoba	92.2
2)	531	20	RNE-5	Oviedo	101.8+104.4
126)	531	10	RNE-5	Pamplona	95.7
87)	540	10	EI	R. Euskadi, Vitoria	
3)	558	10	RNE-5	La Coruña	95.8
4)	558	20	RNE-5	Valencia	88.2
88)	558	10	RNE-5	San Sebastián	93.3
44)	567	10	RNE-5	Socuéllamos	88.5
40)	567	50	RNE-5	Murcia	92.1
133)	567	5	RNE-5	Marbella	87.6
37)	576	50	RNE-5	Barcelona	88.3
5)	603	50	RNE-5	Sevilla	90.0
45)	603	10	RNE-5	Palencia	91.8
6)	621	50	OCR	Onda Cero R., Barcelona	
7)	657	20	RNE-5	Madrid	88.2+90.3
8)	666	10	COPE	R. Miramar, Barcelona	
66)	711	5	COPE	R. Pop. de Murcia	89.7
9)	747	10	RNE-5	Cádiz	103.5
87)	756	5	EI	R. Euskadi, Bilbao	
10)	792	20	SER	R. Sevilla	97.1
11)	810	20	SER	R. Madrid	93.9
87)	819	5	EI	R. Euskadi, San Sebastián	
12)	828	50	SER	R. Barcelona	96.9+93.9
13)	837	10	COPE	R. Pop. de Sevilla	99.6
14)	837	10	COPE	R. Pop. de Burgos	95.5
15)	837	1	COPE	R. Pop. de Ibiza	89.1
55)	837	2	COPE	R. Pop. de El Ferrol	88.7
16)	873	20	SER	R. Zaragoza	95.3
31)	882	2	COPE	R. Pop. de Gijón	103.6
17)	882	2	COPE	R. Pop. de Alicante	95.6
18)	882	2	COPE	R. Pop. de Valladolid	88.5
49)	882	5	COPE	R. Pop. de Málaga	89.4
19)	882	2	COPE	R. Sabadell	
46)	900	10	COPE	R. Pop. de Bilbao	97.8
67)	900	5	COPE	R. Pop. de Vigo	87.8
65)	900	5	COPE	R. Pop. de Granada	88.2
20)	909	10	RNE-5	Palma de Mallorca	90.1
21)	918	20	IND	R. Intercontinental, Madrid	95.1
1)	936	10	RNE-5	Bilbao	96.3
26)	936	10	RNE-5	Málaga	88.0+92.5
97)	936	10	RNE-5	Alicante	104.2
43)	936	5	RNE-5	Vigo	96.0
104)	936	10	RNE-5	Zaragoza	94.5
89)	936	20	RNE-5	Valladolid	97.3
23)	954	20	CI	R. España, Madrid	97.2
24)	990	10	SER	R. Bilbao	89.5
25)	999	50	COPE	R. Pop. de Madrid	99.5
84)	1088	5	SER	R. Extremadura, Badajoz	96.9
120)	1008	5	SER	R. Gerona	88.1+99.0
29)	1008	3	SER	R. Vigo	99.4
98)	1017	5	RNE-5	Granada	98.5
132)	1017	10	RNE-5	Burgos	93.6
123)	1026	2	SER	R. Salamanca	96.9
27)	1026	10	SER	R. Reus	101.4
28)	1026	3	SER	R. Alicante	91.0
30)	1026	2	SER	R. Jaén	97.1
144)	1026	2	SER	R. Jerez	97.8
134)	1026	5	SER	R. Asturias, Oviedo	97.5
82)	1044	10	SER	R. San Sebastián	97.2
48)	1053	10	COPE	R. Pop. de Zaragoza	97.9
87)	1071	5	EI	R. Euskadi Irratia, Bilbao	
32)	1080	5	SER	R. Mallorca	94.1
33)	1080	5	SER	R. Granada	92.8+95.4
34)	1080	2	SER	R. Coruña	91.0
35)	1080	3	SER	R. Huesca	96.9
36)	1080	5	OCR	Onda Cero R., Toledo	100.8
125)	1098	10	RNE-5	Lugo	92.8
127)	1098	5	RNE-5	Avila	87.6
110)	1098	10	RNE-5	Almería	100.9
99)	1098	5	RNE-5	Huelva	95.2
38)	1107	10	RNE-5	Santander	96.9
39)	1107	5	RNE-5	Cáceres	95.1
41)	1107	10	RNE-5	Ponferrada	
42)	1107	5	RNE-5	Teruel	95.6
155)	1107	25	RNE-5	Logroño	
135)	1116	5	SER	R. Pontevedra	89.1
138)	1116	5	SER	R. Albacete	89.6
108)	1125	10	RNE-5	Castellón	89.3
106)	1125	10	RNE-5	Vitoria	92.5
90)	1125	10	RNE-5	Soria	89.7
128)	1125	5	RNE-5	Badajoz	94.9
52)	1134	2	COPE	R. Pop. de Avilés	89.1
54)	1134	2	COPE	R. Pop. de Menorca	89.6
57)	1134	2	COPE	R. Pop. de Jaén	88.8
73)	1134	2	COPE	R. Pop. de Jerez	92.4
59)	1134	2	COPE	R. Pop. de Pamplona	87.9
60)	1134	2	COPE	R. Pop. de Salamanca	90.0
61)	1134	1	COPE	R. Pop. de Astorga	87.6
86)	1143	2	COPE	R. Pop. de Reus	89.7
58)	1143	2	COPE	R. Pop. de Orense	92.4
153)	1152	10	RNE-5	Zamora	101.8
93)	1152	10	RNE-5	Lérida-Lleida	87.9
107)	1152	10	RNE-5	Cartagena	103.5
112)	1152	10	RNE-5	Albacete	91.8
87)	1161	50	EI	R. Euskadi Irratia, San Sebastián	
63)	1179	10	SER	R. Galicia, Stgo de Compostela	
64)	1179	5	SER	R. Murcia	91.3
122)	1179	2	SER	R. Rioja, Logroño	91.7
87)	1197	5	EI	Euskadi Irratia, Vitoria	
70)	1215	2	COPE	R. Pop. de Castellón	91.7
62)	1215	1	COPE	R. Pop. de Lorca	93.5
78)	1215	2	COPE	R. Pop. de Santander	88.4
79)	1215	2	COPE	R. Pop. de Zamora	94.9
71)	1215	2	COPE	R. Pop. de Córdoba	87.6
68)	1224	2	COPE	R. Pop. de Albacete	95.4
69)	1224	2	COPE	R. Pop. de Cáceres	88.2
72)	1224	2	COPE	R. Pop. de Huelva	91.9
74)	1224	2	COPE	R. Pop. de Lugo	90.0
75)	1224	2	COPE	R. Pop. de Lérida-Lleida	96.0
76)	1224	2	COPE	R. Pop. de Mallorca	97.6
77)	1224	5	COPE	R. Pop. de San Sebastián	88.5
80)	1224	1	COPE	R. Pop. de Puertollano	97.5
51)	1224	2	COPE	R. Pop. de Almería	97.1
81)	1260	20	SER	R. Valencia	94.2
83)	1260	5	SER	R. Algeciras	95.7
85)	1269	5	COPE	R. Pop. de León	93.3
56)	1269	2	COPE	R. Pop. de Figueras	89.4
53)	1269	10	COPE	R. Pop. de Ciudad Real	93.6
50)	1269	3	COPE	R. Pop. de Badajoz	89.1
22)	1287	10	SER	R. Lérida-Lleida	92.6+96.6
47)	1296	10	COPE	R. Pop. de Valencia	99.0

114)	1305	10	RNE-5	Orense	95.1
102)	1305	10	RNE-5	Ciudad Real	95.7
100)	1305	10	RNE-5	León	97.1
89)	1314	10	RNE-5	Valladolid	97.3
91)	1314	4	RNE-5	Cabra	95.1
156)	1314	5	RNE-5	Cuenca	
95)	1341	2	OCR	Onda Cero R., Almería	93.8
96)	1341	5	SER	R. Terrassa-Tarrasa	95.5
101)	1395	2	RNE-5	La Rioja, Calahorra	88.8
103)	1395	2	RNE-5	Tortosa	90.7
105)	1413	10	RNE-5	Salamanca	94.5
109)	1413	5	RNE-5	Gerona	93.3
130)	1413	10	RNE-5	Jaén	
111)	1476	2	RNE-5	Toledo	102.0+99.9
115)	1485	2	SER	R. Alcoy	96.3
116)	1485	2	OCR	Onda Cero R., Antequera	96.9
117)	1485	5	SER	R. Cádiz	89.4
118)	1485	2	SER	R. Castellón	94.8
119)	1485	2	OCR	Onda Cero R., Ciudad Real	92.1
121)	1485	2	SER	R. León	88.2
124)	1485	2	SER	R. Santander	90.9
92)	1503	2	RNE-5	Monforte de Lemos	88.8
129)	1503	5	RNE-5	Tarragona	88.8
130)	1503	2	RNE-5	Baeza	105.4
131)	1503	2	RNE-5	Benavente	88.8
136)	1521	2	SER	R. Manresa	91.8
137)	1539	5	SER	R. Valladolid	90.9
139)	1575	5	SER	R. Córdoba	96.6
140)	1575	5	OCR	Onda Cero R., Villanueva	96.3
141)	1575	5	SER	R. Pamplona	92.2
142)	1584	2	SER	R. Elche-Elx	94.8
143)	1584	2	SER	R. Gandía	96.5
145)	1584	2	SER	R. Castilla, Burgos	89.1
146)	1584	2	SER	R. Orense	87.6+106.5
147)	1584	2	SER	R. Zamora	89.9
148)	1602	2	SER	R. Cartagena	102.3
149)	1602	5	SER	R. Linares	94.9
150)	1602	2	SER	R. Onteniente-Ontinyent	95.3
151)	1602	2	SER	R. Segovia	93.6
152)	1602	10	EI	R. Vitoria	104.1
154)	1602	2	SER	R. Lugo	91.8

STATIONS OPERATING ON FM ONLY

(St's less than 5 kW omitted)

	MHz	kW	Net	Location
157)	87.7	6	A3	Jerez de la Frontera
158)	88.2	5	SER	Vitória
159)	88.4	25	CR	Monte Caro
159)	88.6	20	CR	Soriguera
87)	88.9	20	EI	Bilbao
160)	89.0	21.4	A3	Madrid
161)	89.2	8	IND	Alicante
162)	89.3	6	IND	Pamplona
163)	89.4	10	RNE5	Talavera de la Reina
164)		5	RNE5	Torrelavega
165)	89.5	8	OCR	Palma de Mallorca
166)		6	OCR	Cáceres
167)	89.8	8	OCR	Barcelona
168)	89.9	20	RNE5	Gijón
169)	90.1	8	IND	Málaga
170)		5	OCR	Reinosa
171)	90.3	6	OCR	Jerez de la Frontera
37)	90.5	8	A3	Barcelona
172)		5	SER	Miranda de Ebro
173)	90.6	5	RM	Santiago de Compostela
174)	90.7	5		Torrejón de Ardoz
175)	90.8	5	RM	El Puerto de Santa Maria
87)	90.9	20	EI	Zaldiarán
176)	91.0	30	IND	Madrid
177)	91.1	6	A3	Oviedo
87)	91.2	20	EI	Oiz (Durango)
178)	91.4	10	SER	Placensia
179)		8	OCR	Córdoba
180)	91.5	8	OCR	San Sebastián
87)	91.7	20	EI	Bilbao
181)		10	D	Madrid
182)		5		Antequera
159)	91.9	80	CR	Alpicat
183)		6	OCR	Santander
184)	92.0	40	A3	Zaragoza
185)	92.1	5		Torrejón de Ardoz
186)	92.2	5		Badajoz
187)	92.3	44		Mas de las Matas
188)	92.4	13.5	A3	Madrid
189)	92.6	8	A3	La Coruña
190)	92.7	6	IND	Albacete
191)	93.1	30	IND	Vitoria
189)	93.4	8	A3	La Coruña (cfr 92.6)
192)	93.5	9.5	RM	Zaragoza
193)	94.0	20	CSR	Almonaster (Huelva)
194)		6	SER	Vigo
195)	94.1	20	RG	Xistral
87)	94.4	20	EI	Jaizquibel
87)	94.7	20	EI	Bilbao
196)	94.8	40	A3	Sevilla
197)	94.9	10	IND	Barcelona
87)	95.0	20	EI	Zaldiarán
198)	95.1	5	CSR	Baza
159)	95.4	20	CR	Soriguera
199)		8		Cádiz
200)	95.6	8		Córdoba
201)	95.9	40	OCR	Sevilla
202)	96.1	10	A3	Valencia
203)	96.2	6	OCR	Salamanca
204)	96.3	80	CSR	Cadiz
205)		5	OCR	Antequera
87)	96.5	20	EI	Jaizquibel
206)		8		Alicante
159)	96.7	25	CR	Rocacorba
207)		6	OCR	Cartegena
208)	96.9	40		Valencia
209)	97.0	6		Cáceres
210)	97.1	5	D	Zaragoza
159)	97.3	10	CR	Montserrat
203)	97.6	6	OCR	Salamanca (cfr 96.2)
211)	97.7	5	SER	Valencia
212)	97.9	60	CSR	Sierra de Almadén
213)	98.0	13	OCR	Madrid
214)	98.4	5	RM	Valencia
204)	98.5	20	CSR	San Roque
215)		6	OCR	Santander
216)		5	OCR	Talavera de la Reina
217)	98.6	40	A3	Zaragoza
218)	98.8	6		Cartagena
219)	99.2	6		Pamplona
220)	99.3	8		Murcia
221)	99.4	10	D	Barcelona
222)		8	OCR	Valladolid
223)	99.5	8		Granada
159)	99.7	5	CR	Collsuspina
224)	99.9	8	D	Cadiz
225)	100.0	8		Barcelona
226)	100.3	40	OCR	Sevilla
159)		25	CR	La Mussara
159)		10	CR	Massanet de Cabrenys
227)		8		Priego de Córdoba
228)	100.4	6	A3	Málaga
158)		5	SER	Vitoria
159)	100.5	5	CR	Collsuspina
159)	100.7	80	CR	Alpicat
195)	100.9	30		Xesteiras
229)	101.1	5	RM	Benalmádena
230)	101.2	20	OCR	Valencia
231)		5	OCR	Huelva
232)		5		Benidorm
233)	101.3	100		Torrespaña
234)		60	CSR	Lagar de la Cruz
235)	101.4	5	OCR	Inca
159)	101.5	80	CR	Barcelona
196)		40	A3	Sevilla
236)	102.0	10	COPE	Barcelona
159)		10	CR	La Seu d'Urgell
232)	102.2	100		Picayo
159)		70	CR	Rocacorba
237)		30	CSR	Alto Conquero
232)		10		Sagunto
195)	102.3	40		Domayo
159)	102.4	10	CR	Montserrat

238)	102.5	120	CSR	Pechina
159)		30	CR	Monte Caro
239)		8	IND	Granada
240)		5		La Carolina
241)	102.6	16	COPE	Bilbao
213)	102.7	100	OCR	Madrid
213)		8	OCR	La Coruña
159)	102.8	100	CR	Barcelona
242)		5		Eivissa
232)	103.0	120		Aitana
87)	103.2	20	EI	Olz
87)	103.5	20	EI	Jaizquibel
232)		10		Monduber
243)		5	IND	Barcelona
232)	103.7	90		Benicássim
244)	103.9	65	CSR	San Juan
195)		40		Bailadora
245)		8	IND	Murcia
191)	104.0	8	IND	Vitoria
246)	104.2	20	CSR	Jerez de la Frontera
37)		8	A3	Barcelona
247)		8	CSR	San Miguel
87)	104.4	20	EI	Oiz (Durango)
182)	104.6	60	CSR	Mijas
204)	104.8	80	CSR	San Cristóbal
195)		40		Monte Meda
198)	104.9	60	CSR	Paparanda
159)	105.0	5	CR	Barcelona
248)	105.2	8	OCR	Valladolid
181)	105.4	100	RM	Madrid
159)		40	CR	La Mussara
159)	105.5	10	CR	Massanet de Cabreny
249)		6	OCR	Burgos
250)	105.8	40	OCR	Zaragoza
182)		8	CSR	Malaga
233)	106.0	300		Puerto de Navacerrada
251)	106.1	8		Palma de Mallorca
252)		8	A3	Santiago de Compostela
253)	106.6	60		Barcelona
159)	106.8	5	CR	Alfábia
254)	106.9	40	IND	Sevilla
255)	107.2	5		Azuaga

Addresses and other information

1) Licenciado Pozas 55, 48013 Bilbao.
2) Asturias 9, 33004 Oviedo.
3) Paseo Méndez Nuñez s/n, 15006 La Coruña.
4) Milagro 2, 46003 Valencia.
5) Av. Rep. Argentina 24, 41001 Sevilla.
6) Ramblas 126, 08002 Barcelona.
7) Huertas 73, 28014 Madrid.
8) Diagonal 460, 08006 Barcelona.
9) Av. de Andalucía 3, 11006 Cádiz.
10) Rafael González Abreu 6, 41001 Sevilla.
11) Gran Vía 32, 28013 Madrid.
12) Caspe 6, 08010 Barcelona. **FM:** R. Trafic.
13) Vírgenes 24, 41004 Sevilla.
14) Av. del Cid 8, 09003 Burgos.
15) Felipe II 28, 07800 Ibiza.
16) Paseo de la Constitución 21, 50001 Zaragoza.
17) Rbl. Menden-Nuñez 45, 03002 Alicante.
18) Claudio Moyano 4, 47001 Valladolid.
19) Convento 22-24, 08202 Sabadell.
20) Aragón 26, 07006 Palma de Mallorca.
21) Modesto Lafuente 42, 28003 Madrid.
22) Marqués de Vila Antónia 5, 25007 Lérida-Lleida.
23) Manuel Silvela 9, 28010 Madrid. **FM:** Top 40.
24) Rodriguez Arias 6, 48008 Bilbao.
25) Valenzuela 1, 28014 Madrid.
26) Alameda de Colón 32, 29001 Málaga.
27) Arrabal de Santa Ana 54, 43201 Réus.
28) Calderón de la Barca 26, 03004 Alicante.
29) Policarpo Sanz 36, 36202 Vigo.
30) Obispo Aguilar 1, 23001 Jaén.
31) Crtra. de la Costa 87, 33205 Gijón.
32) Son Moix s/n, 07011 Palma de Mallorca.
33) Santa Paula 2, 18001 Granada.
34) Pl. de Orense 3, 15004 La Coruña.
35) Loreto 2, 22003 Huesca. **FM:** R. Aragón.
36) Pl. de la Merced 1, 45002 Toledo. **FM:** R. Tajo.
37) Paseo de Grácia 1, 08007 Barcelona.
38) Lealtad 14, 39002 Santander.
39) Rda. San Francisco s/n, 10003 Cáceres.
40) Callejonolma 27, 37005 Murcia.
41) Ave María 9, 24400 Ponferrada.
42) Pl. Catedral 9, 44001 Teruel.
43) García Barbón 32, 36201 Vigo.
44) Ramiro Ledesma 8, 13630 Socuéllamos (Cd. Real)
45) Becerro Bengoa 9, 34002 Palencia.
46) Alameda Mazarredo 47, 48009 Bilbao.
47) Pje. Dr. Serra 2, 46004 Valencia.
48) Sagasta 50, 50004 Zaragoza.
49) Santa María 2, 29006 Málaga.
50) Menacho 16, 06001 Badajoz.
51) Alcalde Muñoz 10, 04004 Almería.
52) La Cámara 71, 33400 Avilés.
53) Pje. San Isidro 3, 13001 Ciudad Real.
54) Av. Negrite 3, 07760 Ciudadela (Menorca)
55) Miramar s/n, 15401 El Ferrol.
56) San Lázaro 21, 17600 Figueres-Figueras.
57) Av. Madrid 68, 23008 Jaén.
58) Gral. Franco 101, 32003 Orense-Ourense.
59) Amaya 2-B, 31002 Pamplona.
60) Pl. Mayor 35, 37002 Salamanca.
61) Hermanos La Salle 2, 24700 Astorga.
62) Av. Juan Carlos I 63, 30800 Lorca (Murcia).
63) Ap. 469, 15701 Santiago de Compostela.
64) Radio Murcia 4, 30001 Murcia.
65) Gran Vía de Colón 28, 18001 Granada.
66) Arco de Santo Domingo s/n, 30001 Murcia.
67) Principe 57, 36202 Vigo.
68) Gallego 9, 02002 Albacete.
69) Av. España 15, Cáceres.
70) Av. Francisco Tárrega 69, 12540 Villareal (Castellón).
71) Pl. Cardenal Toledo 4, 14001 Córdoba.
72) Puerto 24, 21001 Huelva.
73) Ramón de Cala 2, 11401 Jerez de la Frontera.
74) Reina s/n, 27001 Lugo.
75) Academia 17, 25002 Lérida-Lleida.
76) Jaime III 18, 07002 Palma de Mallorca.
77) Garibay 19, 20004 San Sebastián.
78) Rualasal 5, 39001 Santander.
79) Pl. Fernández Duro 3, 49001 Zamora.
80) Alejandro Prieto 2, 13500 Puertollano (Cd. Real).
81) D. Juan de Austria 3, 46202 Valencia. **FM:** R. Mediterraneo.
82) Av. de la Libertad 27, 27004 San Sebastián.
83) Pl. Alta 2, 11201 Algeciras.
84) Santa Lucia 2, 06001 Badajoz.
85) Lope de Vega 1, 24002 León.
86) Arrabel de San Pedro 3, 43204 Reus.
87) Andía 13, 20004 San Sebastián; Postas 32, 01001 Vitoria; Gran Vía 85, 48011 Bilbao.
88) Easo 12, 20006 San Sebastián.
89) García Morato 27-29, 47007 Valladolid.
90) Campo 5, 42001 Soria.
91) Camina de la Resineral 1, 16003 Cuenca.
92) Pl. España 4, 27400 Monforte de Lemos (León).
93) Lluís Companys 1, 25003 Lérida-Lleida
94) Santa Ana 7, 25300 Tárrega.
95) Pl. de San Sebastián 5, 04003 Almería. **FM:** R. Luz.
96) Gutenberg 3-13, 08224 Terrassa-Tarrasa. **FM:** R. Club 25.
97) Angel Lorano 18, 03001 Alicante.
98) Severo Ochoa 13, 18001 Granada.
99) La Fuente 4, 21004 Huelva.
100) Ordoño II 28, 24001 León.
101) Bebricio 40, 26500 Calahorra (La Rioja).
102) Rambla del Carmen s/n, 13002 Ciudad Real.
103) Rep. Argentina 26, 43500 Tortosa.
104) Gran Vía 7, 50006 Zaragoza.
105) Pl. de Colón 4, 37001 Salamanca.
106) Simón Bolívar 13, 01003 Vitoria.
107) Av. Alfonso XII 51, 30203 Cartagena.
108) Paseo de Ribalta 5, 12001 Castellón.
109) Gran Vía Jaime I 60, 17001 Gerona-Girona.
110) Hermanos Machado 23, 04004 Almería.
111) Nuñez de Arce 12, 45003 Toledo.
112) Virgen de Medinaceli 1, 02002 Albacete.
113) Góngora 3, 14002 Córdoba.
114) Parque San Lázaro 14, 33008 Orense-Ourense.
115) Beato Nicolas Factor 2, 03800 Alcoy.

116) Rodaljarros 3, 29200 Antequera.
117) Po. Marítimo 1, Edif. Reina Victoria, 11010 Cádiz.
118) Moyano 5, 12002 Castellón.
119) Av. Alarcos 20, 13001 Ciudad Real. **FM:** R.Color.
120) Pompeu Fabra 11, 17002 Gerona-Girona.
121) Villafranca 6, 24001 León.
122) Av. Portugal 12, 26001 Logroño.
123) Arco 16, 37002 Salamanca.
124) Marcelino Sanz de Sautuola 4, 39003 Santander.
125) Orense 59-63, 27004 Lugo.
126) Aoiz 17, 31004 Pamplona.
127) Pl. de Santa Ana 7, 05001 Avila.
128) Av. de Colón 2, 06005 Badajoz.
129) Rambla Nova 23, 43003 Tarragona.
130) Paseo de la Estación 30, 23008 Jaén.
131) Portugal 35, 49600 Benavente (Zamora).
132) Caja de Ahorros Municipal 17, 09001 Burgos.
133) Av. Ricardo Soriano 11, 29600 Marbella.
134) Asturias 19, 33004 Oviedo.
135) Daniel de la Sota 5, 35001 Pontevedra.
136) Plana de l'Om 2, 08240 Manresa.
137) Montero Calvo 7, 47001 Valladolid.
138) Concepción 21, 02002 Albacete.
139) García Lovera 3, 14002 Córdoba.
140) Libertad 25, 08800 Villanueva y Geltrú-Vilanova i La Geltrú.
141) Yanguas y Miranda 17, 31002 Pamplona.
142) Dr. Caro 43, 03201 Elche-Elx.
143) Alcala del Olmo 15 y Loreto 38, 46700 Gandia.
144) Guadalete 12, 11403 Jerez de la Frontera.
145) Conde Jordana 1, 09004 Burgos.
146) José Antonio 14, 32003 Orense-Ourense.
147) Av. Ppe. de Asturias 1, 49002 Zamora.
148) Real 82, 30201 Cartagena.
149) Pérez Galdós 35, 23700 Linares (Jaén).
150) Ereta 2A, 46870 Onteniente-Ontinyent (Valencia).
151) San Agustín 3, 40001 Segovia.
152) Pasaje Postas 32, 01001 Vitoria.
153) Av. Requejo 21, 49003 Zamora.
154) Pl. Santo Domingo 3, 27001 Lugo.
155) Av. Portugal 2, 26001 Logroño.
156) Cervantes 11, 14940 Cabra (Córdoba).
157) Edificio Jerez 74, 11405 Jerez de la Frontera.
158) General Alava 10-6 Depto 9, 01005 Vitória.
159) Av. Diagonal 614-616, 08021 Barcelona.
160) Oquendo 23, 28006 Madrid.
161) **R. Pentagrama,** Alicante.
162) **R. Blanca,** Pamplona.
163) Ronda del Canillo 32, 45600 Talavera de la Reina.
164) Joaquin Hoyos 16, 39300 Torrelavega.
165) **R. Blanca,** Av. Alejandro Rosella 23-7, 07002 Palma de Mallorca.
166) Cáceres.
167) **Onda Rambla,** Barcelona.
168) Plaza de Instituto 3, 33201 Gijón.
169) **R. Sur 2000,** Malaga.
170) **Antena Norte,** Reinosa.
171) **R. Arenal,** Caballeros 2, 11402 Jerez de la Frontera.
172) Vitória 24, 09200 Miranda de Ebro.
173) Ramón Cabanillas 14, 15701 Santiago de Compostela.
174) Pl. Palmeras 13-3-D, Parque Liberty, 28850 Torrejón.
175) Misericordia 10, 11500 El Puerto de Santa María.
176) **R. 16,** Gran Vía 31, 38013 Madrid.
177) **Antena 3 de Asturias,** Marqués de la Vega de Anzo 1-2, 33007 Oviedo.
178) Santa Isabel 4, 10600 Placensia.
179) **R. Melodia,** Barroso 4-2, 14003 Córdoba.
180) **R. Blanca,** San Sebastián.
181) **Dial Madrid,** Gran Vía 32-B, 28013 Madrid.
182) **Canal Sur 1,** Carretera de Cádiz 307 km 232, 29006 Málaga.
183) Cuesta 4 entreplanta, 39002 Santander.
184) **Antena 3 de Zaragoza,** Paseo de la Constitución 13-7, 50001 Zaragoza.
185) **R. Sensación,** Torrejón de Ardoz.
186) **R. Guadiana,** Felipe Checa 15-2, 06001 Badajoz.
187) **R. Reo,** 44564 Mas de las Matas.
188) **Radiolé,** Quendo 23, 28006 Madrid.
189) **R. 80 Serie Oro,** Concepción Arenal 11-13, 15006 La Coruña.
190) **R. Arco Iris,** Albacete.
191) **La Voz de Euskadi,** Independencia 2, 01004 Vitoria-Gasteiz.
192) **R. Minuto,** Paseo de la Independencia 29 I, 50001 Zaragoza.
193) **Canal Sur 2,** Plaza de San Pedro 3 y 4, 21004 Huelva.
194) **R. Dial Vigo,** Vigo.
195) **R. TV Galicia,** Casa de la Radio, 15701 Santiago de Compostela.
196) **R. 80 Serie Oro,** Av. San Francisco Javier, 41005 Sevilla.
197) **R. Tiempo,** Teodor Roviralta 31, 08022 Barcelona.
198) **Canal Sur Radio,** Recogidas 24 al 32, 18002 Granada.
199) **R. y Tecnologia,** Cádiz.
200) **Radiosistemas,** Córdoba.
201) **R. Triana,** Miguel de Mañara 16 ático, 41004 Sevilla.
202) **R. 80 Serie Oro,** Barón de Cárcer 48-10, 46001 Valencia.
203) Paseo de Canelejas 68-72, 37001 Salamanca.
204) **Canal Sur 2,** Plaza de España 15-3, 11006 Cádiz.
205) **R. Color Málaga,** Edificio Galaxia, 29007 Málaga.
206) **Canal 9 R,** Blasco Ibañez 134, 46022 Valencia.
207) Puerta Murcia 11, 30201 Cartagena.
208) **R. Blanca,** Valencia.
209) **R. Estudio,** Cárceres.
210) **Dial Zaragoza,** Madre Rafols 2-8, 50004 Zaragoza.
211) **Intervalencia R.,** Av. Cid 2-4 C, 46018 Valencia.
212) **Canal Sur 1,** Av. del Ejército Español, 23007 Jaén.
213) **R. Cristal,** Valázquez 54-1, 28001 Madrid.
214) **R. Minuto Valencia,** Don Juan de Austria 3, 46002 Valencia.
215) Cuesta 4 entreplanta, 39002 Santander.
216) **R. Talavera,** Av. del Principe 25, 45600 Talavera de la Reina.
217) **R. 80 Serie Oro,** Coso 62-7, 50001 Zaragoza.
218) Cartagena.
219) **Cía Navarra de R. diff.,** Pamplona.
220) **Radio y Tecnologia,** Murcia.
221) **Dial Barcelona,** Teodor Roviralta 31, 08022 Barcelona.
222) Santiago 14-2-3, 47001 Valladolid.
223) Granada.
224) **Dial Bahia,** Paseo Maritimo 1, 11010 Cádiz.
225) **R. Nova Barcelona,** Av. Diagonal 612 entsol, 08021 Barcelona.
226) **R. Blanca,** Miguel de Mañara 16 atico, 41004 Sevilla.
227) **R. Priego,** Av. América B, Cana 22 D, 14800 Priego de Córdoba.
228) **Antena 3 de Málaga,** Compositor Lemberg Ruiz 62, 29007 Málaga.
229) **R. Minuto Málaga,** Palestina 3, 29007 Málaga.
230) **R. Color Valencia,** Cirilo Amoros 27-4, 46004 Valencia.
231) **R. Rocio,** Arqultecto Perez Carasa 14 y 16, 21001 Huelva.
232) **Canal 9 R,** Blasco Ibañez 134, 46022 Valencia.
233) **Onda Madrid,** García de Paredes 65, 28010 Madrid.
234) **Canal Sur 1,** Gran Capitán 2-4, 14008 Córdoba.
235) **R. Belear de Inca,** Llorenc Villalonga 25, 07300 Inca.
236) **R. Popular de Barcelona,** Av. Diagonal 441-1, 08037 Barcelona.
237) **Canal Sur 1,** Plaza de San Pedro 3 y 4, 21001 Huelva.
238) **Canal Sur 1,** Residencial Los Oliveros, 04004 Almería.
239) **R. Estilo,** Plaza del Carmen 2, Granada.
240) **R. Carolina,** 23200 La Coruña.
241) Fontecha Salazar 9-5, 48007 Bilbao.
242) **R. Diario de Ibiza,** Vía Púnica 10, 07800 Eivissa.
243) **R. Ambiente Musical,** Beethoven 11-1-1, 08021 Barcelona.
244) **Canal Sur 1,** San Juan Aznalfarache, 41920 Sevilla.
245) **R. 30,** Ronda Norte 9, Murcia.
246) **Canal Sur 1,** Hotel Capele, Corredera 58, 11402 Jerez de la Frontera.
247) **Canal Sur 1,** Recogidas 24 al 32, 18002 Granada.
248) **R. Valladolid,** Santiago 14-2-3, 47001 Valladolid.
249) **Onda Cero R.,** Burgos.
250) **R. Aragón Zaragoza,** Gran Vía 3, 50006 Zaragoza.
251) **R. 7,** Palma de Mallorca.
252) **Antena 3 de Compostela,** Rua de Villar 37, 15705 Santiago de Compostela.
253) Barcelona.
254) **R. América,** Placentines 2, 41004 Sevilla.
255) **R. Azuaga,** Plaza de la Merced 1, 06920 Azuaga.

AMERICAN FORCES RADIO & TV SCE. (Mil)

STATIONS (FM): 1) 101.1/101.5MHz; 2) Moron 92.1MHz 1kW (mono), Zaragoza 89.7MHz 1kW (stereo), Torrejon 100.2MHz 5kW (stereo); 3) 102.6MHz 5kW (stereo).
ADDR: 1) Det. 1, AFEBS, Zaragoza AB, Spain, APO New York, NY 09286-5000, USA; 2) Det 6. AFEBS, Torrejon, Spain, APO New York, NY 09286-6457, USA; 3) Navy Broadc. Sce, Det. 13, Naval Station Rota (Cadiz), Spain.
D.PRGR: All st's 24h.

SWEDEN

L.T: UTC + 1h (Su: UTC + 2h) — **Pop:** 8.600.000 — **Radio sets:** 7.271.556 — **Pr.L:** Swedish — **E.C:** a/c 220/V — **ITU:** S.

SVENSK RUNDRADIO AB (Teracom Swedish Broadcasting)

has the responsibility for the distribution of the prgrs. produced by Sveriges Radio AB and Community Radio.
ADDR: Medborgarplatsen 3, Stockholm (**Postal Addr:** Box 17666, S-118 92 Stockholm). **Te:** +46 (8) 671200. **Fax:** +46 (8) 6712001. **L.P:** Chmn. Board of Governors: Gösta Gunnarson. Dir. Gen: Valdemar Persson.

SVERIGES RADIO AB (Swedish Broadcasting Corporation) (Non-commercial broad. sce.)

ADDR: Radiohuset, Oxenstiernsgatan 20, Stockholm. **(Postal addr:** S-10510 Stockholm). **Te:** + 46 (8) 784 0000. **Fax:** + 46 (8) 662 6992.
L.P: Chmn. Board of Governors: Jan-Erik Wikström. MD: Ove Joanson. Controllers: Ewonne Winblad (Ch. 1) Christina Mattsson (Ch. 2), Lisa Soederberg (Ch. 3), Mie Jernbeck coord. (Ch 4.). Admin. Dir: Åke Levin. Dir. of Eng: Per-Erik Selemark. Legal Advisors: Håkan Fahlberg (corporate), Gunhild Frylén (programming). Head of Int. Rel: Carina Nilsson.

MEDIUMWAVE: Sölvesborg 1179kHz 600kW. (Carries Home Sce. 1st Prgr. from s/on until 1600, Foreign Sce. 1600-0030).

Shortwave: see Foreign Service.
FM (All freqs in MHz).

Station	Area°	I	II	III	IV	kW
Arvidsjaur	24)	89.4	94.2	97.1	100.6	60
Bollnäs	20)	88.4	91.7	96.0	103.8	60
Borlänge	19)	89.4	93.0	97.7	101.3	60
Borås	25)	88.5	94.6	97.9	102.9	10
Bäckefors	14)	92.7	96.8	99.1	102.2	60
Emmaboda	6)	93.0	96.7	99.7	101.8	60
	7)	—	—	—	95.6	60
Filipstad	16)	88.5	90.1	98.8	103.2	2.5
Finnveden	5)	90.6	94.2	99.9	103.4	30
Gällivare	24)	88.3	94.9	98.5	100.9	60
Gävle	20)	88.1	97.4	99.8	102.0	60
Göteborg	13)	89.3	96.3	99.4	101.9	60
Halmstad	12)	87.7	91.2	95.4	97.3	60
	10)	—	—	—	102.6	3
Helsingborg	11)	89.8	95.7	98.4	103.2	6
Hudiksvall/ Forsa	20)	87.6	90.2	93.8	100.7	60
Hörby	10)	88.8	92.4	97.0	101.4	60
	11)	—	—	—	89.5	5
Jönköping	5)	91.6	93.7	97.1	100.8	3
Kalix	24)	91.3	93.6	97.9	102.2	60
Karlshamn	9)	90.3	93.4	98.3	100.4	15
Karlskrona	9)	89.1	95.0	97.7	100.7	10
Karlstad	16)	90.5	94.2	96.5	103.5	15
Kiruna	24)	89.1	92.7	96.4	102.7	60
Kisa	4)	90.5	92.4	96.9	103.6	30
Lycksele	23)	92.9	95.4	98.7	103.3	60
Malmö	11)	87.9	93.3	98.0	102.0	6
Mora	19)	92.2	96.7	99.0	101.0	60
Motala	4)	91.1	94.0	98.2	101.2	20
Norrköping	3)	90.0	93.5	98.7	102.3	60
	4)	—	—	—	94.8	60
Nässjö	5)	89.6	92.1	99.0	102.1	60
Pajala	24)	90.8	93.0	95.9	100.2	60
Skellefteå	23)	93.8	96.3	100.0	103.9	10
Skövde	15)	88.9	95.1	97.5	100.3	60
Sollefteå	21)	89.3	93.5	98.1	101.2	60
Stockholm*	1)	92.4	96.2	99.3	103.3	60
	1)	—	—	—	93.8[1)]	0.9
Storuman	23)	87.6	91.2	99.0	102.5	60
Sundsvall	21)	92.7	96.9	99.2	102.8	60
Sunne	16)	90.9	94.5	98.5	101.8	60
Sveg	22)	90.6	94.9	97.9	102.2	60
Trollhättan	14)	91.9	95.7	99.8	103.7	3
Tåsjö	22)	89.9	94.7	97.5	100.8	60
	23)	—	—	—	88.2	60
Uddevalla	14)	89.9	93.1	97.2	103.3	10
Uppsala	2)	90.3	93.3	96.6	102.5	20
Varberg	12)	90.4	93.6	98.8	103.8	10
Visby	8)	87.6	94.1	97.2	100.2	60
Vislanda	6)	88.0	90.6	94.7	101.0	20
Väddö	1)	—	—	—	94.7	3
Vännäs	23)	88.5	92.1	95.8	103.6	60
Västervik	7)	88.3	91.8	96.0	102.7	60
Västerås	18)	90.7	95.8	98.0	100.5	60
Ånge	21)	93.2	95.6	99.6	103.1	60
	22)	—	—	—	94.5	60
Älvsbyn	24)	90.6	94.5	99.4	102.9	60
Örebro	17)	87.9	91.5	99.6	102.8	60
Örnsköldsvik	21)	90.8	94.4	97.8	100.1	60
Östersund	22)	87.9	91.5	94.0	100.4	60
Östhammar	2)	89.1	92.8	95.5	101.6	60
Överkalix	24)	88.9	91.7	99.0	103.2	3

\+ 350 low power tr's.
*) Additional tr on 89.6MHz, 0.7kW, relays Foreign Service.
°) Local radio area.
1) Immigrant languages

HOME SERVICE

First Prgr. (news & spoken word): Mon-Fri 0530-2310; Sat 6000-2350; Sun 600-2310. **N:** Mon-Fri 0430, 0500, 0530, 0600, 0700, 0800, 0900, 1000, 1130, 1300, 1400, 1500, 1554, 1645, 1600, 2000, 2100, 2200. **Wrp** (incl. forecast for Swedish waters): 0430, 0500, 0530, 0600, 0705, 1200, 1805, 2050 — **Time Signal:** D 1059-1100.
Relays of 1st programme on Shortwaves: See Foreign Sce. schedule.
Second Prgr: 24h serious music, Sami, Finnish and prgrs for immigrants.
Serious music: Mon-Fri 0000-0700, 1100-1500, 1700-2400, Sat 0000-0600, 0800-1200, 1400-1500, 1700-2300, Sun. 0000-0600.

0800-1500, 1700-2400. **Sami:** Mon-Fri 0700-0730, 1200-1230, 1530-1600, Sun. 1530-1600. **Finnish:** Mon-Fri 0730-0755. **N:** 1055-1100, 1500-1600. Sat. 1230-1400, 1500-1530, Sun 0600-0800, 1500-1530. **Prgrs for Immigrants:** Mon-Fri 1000-1055, 1600-1700. Sat 1600-1700. Sun 1600-1700.
Educational Prgrs: Mon-Fri 0800-1000. Sun 0600-0800.
Stockholm Classic, FM over Stockholm area: serious Music from Prgr. 0000-0600, 0700-2400 music.
Third Prgr: Prgrs for people up to 35/40 years of age: Light music, entertainment and actualities, 24h. **N:** short bulletins every hour on the hour. **Newsreel:** 0430, 0530, 0630.
Fourth Prgr: Progrs for people older than 35/40 years of age: Local progrs, news, light music, entertainment, and sports.
Ann: Nat. Prgr. "Sveriges Radio" and the service f. inst. "Sveriges Radio P1".

Frequencies for Local Programmes: Please refer to the FM list above, where the official area numbering is shown. These numbers are also shown below.
V.: All stations verify reception reports by QSL-card, exc. where otherwise is stated.
Channels:
1) R. Stockholm, Pipersgatan 45, 107 80 Stockholm — **V.** by letter.
2) R. Uppland, Box 1552, 751 45 Uppsala.
3) R. Sörmland, Box 641, 631 08 Eskilstuna.
4) R. Östergötland, Box 764, 601 17 Norrköping — **V.** by card or letter.
5) R. Jönköping, Box 1060, 55110 Jönköping.
6) R. Kronoberg, Box 62, 351 03 Växjö.
7) R. Kalmar, Box 805, 391 28 Kalmar.
8) R. Gotland, Box 1324, 621 24 Visby.
9) R. Blekinge, Box 305, 371 25 Karlskrona.
10) R. Kristianstad, Box 505, 291 25 Kristianstad.
11) R. Malmöhus, 203 55 Malmö.
12) R. Halland, Box 3007, 300 03 Halmstad.
13) R. Göteborg, Box 17400, 402 64 Göteborg.
14) R. Väst, Box 654, 451 24 Uddevalla.
15) R. Västmanland, Box 850, 721 22 Västerås — **V.** by form letter.
16) R. Skaraborg, Box 163, 541 24 Skövde.
17) R. Värmland, Box 98, 651 03 Karlstad.
18) R. Örebro, Box 1800, 701 18 Örebro.
19) R. Dalarna, Box 123, 791 23 Falun.
20) R. Gävleborg, Box 545, 801 07 Gävle.
21) R. Västernorrland, Box 110, 871 23 Härnösand.
22) R. Jämtland, Box 2031, 831 02 Östersund.
23) R. Västerbotten, V. Strandgatan 9, 901 20 Umeå.
24) R. Norrbotten, Box 951, 951 29 Luleå.
25) R. Sjuhärad, Box 27, 501 02 Borås.

FOREIGN SERVICE

Postal Addr: Radio Sweden, S-105 10 Stockholm. **Te:** + 46 (8) 7847281. **Fax:** +46 (8) 6676283. **Telex:** 10000 SRCENT S.
L.P: MD: Hans Wachholz. PD's: Jan Bergström, Anders Fridell. CE: Rolf Beckman. Head of PR: Ch. Adler. Marketing Officer: W. Matl.
STATIONS: Hörby (G.C: 13.44E/49N): 2×350kW SW-tr's, Karlsborg (G.C: 14/29E/58/29N); 1×350kW SW tr; Sölvesborg (G.C: 14.40E/55.59N): 1×600 MW-tr.

kHz	kHz	kHz	kHz	kHz	kHz	kHz
6000	9620	9695	11820	15240	15390	21625
6065	9655	11705	11995	15245	17870	
6185	9670	11710	15230	15270	21500	

Schedule to March 27th 1993

English	Areas	kHz
0100-0130	As/Pac	9695, 11820
0200-0230	NAm	9695, 11705
1330-1400	As/Pac	15240, 21625
1600-1630	Eu/Af/ME	15270
1600-1630	NAm	17870, 21500
2130-2230	As/Pac	11995
2130-2230	Eu/Af/ME	1179, 6065, 9655
2330-2400	Eu/Af/ME	1179, 6065
Estonian		
1730-1745	Eu	1179, 6000, 6065
1745-1800SS	Eu	1179, 6000, 6065
French		
0300-0330	NAm	9695
1200-1230	Eu/Af/ME	9620
1500-1530	Eu/Af/ME	15270
1500-1530	NAm	17870, 21500
1700-1730	Eu/Af/ME	6065
2030-2100	Eu/Af/ME	6065, 9655
German		
1000-1100SS	Eu	9620
1930-2030	Eu	1179, 6065, 9655
Latvian		
1745-1800	Eu	1179, 6000, 6065
1830-1900SS	Eu	1179, 6000, 6065
Russian		
1400-1430	As	15170, 17870
1630-7000	Eu	6000, 9670
1800-1830	Eu	1179, 6000
2030-2100	URS	1179, 6185
Spanish		
0000-0030	LAm	9695, 11705
0130-0200	NAm	9695, 11705
1230-1300	Eu/Af/ME	9620
1800-1830	Eu/Af/ME	6065
2230-2300	Eu/Af/ME	1179, 6065
	LAm	9695, 11705
Swedish		
0230-0300	NAm	9695
0030-0100	As/Pac	9695, 11820
0500-0715MF	Eu/Af/ME	6065, 15390
	NAm	9690
0700-0900Sat	Eu/Af/ME	6065, 9620, 15390
0800-1000Sun	Eu/Af/ME	6065, 9620, 15390
1100-1130SS	Eu/Af/ME	9620
	As/Pac	17710, 21625
1130-1200	Eu/Af/ME	9620, 15320, 15245
1300-1330	As/Pac	15240, 17870, 21625
1430-1500	NAm	17870, 21500
1530-1600	Eu/Af/ME	6065, 9670
	NAm	17870
1700-1715	Eu/Af/ME	1179, 9670, 15270
1830-1930MF	Eu/Af/ME	1179, 6065, 9655, 15270
1900-1930SS	Eu/Af/ME	1179, 6065, 9655, 15270
2100-2130	Eu/Af/ME	1179, 6065, 9655
2100-2130SS	As/Pac	11995
2300-2330	Eu/Af/Me	1179, 6065
	LAm	9695, 11705

ANN: E. "This is Radio Sweden". G: "Hier ist Radio Schweden". FR: "Ici Radio Suède". SP: "Aqui Radio Suécia, Estocolmo". SW: "Radio Sverige, Stockholm". RU: "Govorit Stokgolm".
INT-SIG: "To the Wide, Wide World" (Electronic music) composed by Ralph Lundsten.
V. by QSL-card, folder and/or letter. R.Sweden will appreciate all letters and reports. Prgr. schedules by writing to R.Sweden, S-105 10 Stockholm — **PUB:** twice a year. "Radio Sweden" in English, Estonian, German, French, Latvian, Spanish, Russian, Swedish.
F.Pl: 3×500kW SW-tr's at Hörby.

SVERIGES UTBILDNINGSRADIO AB (Swedish Educational Broadcasting Company) (Non-Comm Public Broadc. Sce).

ADDR: Post: S-115 80 Stockholm; Visitors: Tegeluddsvägen 23, Stockholm. **Te:** + 46 (8) 784 00 00 (Managing director, administration, pre-school, school programmes, adult education, University). Stockholmsvägen 30,, Stocksund. **Te:** + 46 (8) 850420 (Engineering and studios). **Cable:** Broadcast.
L.P: Chmn. Board of Governors: K. Mattson. Man. Dir: LK. Hansson. Int. Rel: Fred Fleicher.
Stations: FM: See Swedish Radio.

COMMUNITY RADIO

NÄRRADIONÄMNDEN

ADDR: Box 16334, S-103 26 Stockholm. **Te:** + 46 (8) 23 72 10.
STATIONS (FM): Independent broadcasting for local populations. Any non-commercial organization, whose main activity is other than broadcasting, may obtain a permit for community radio broadcasting. The transmitters are made available at a nominal fee. The number of tr's is 182, with powers between 5 and 300W ERP.

SWITZERLAND

L.T: UTC + 1h (Su: UTC + 2h) — **Pop:** 6.825.000 — **Radio sets:** 2.685.031 — **Pr.L:** German, Swiss German dialects, French, Italian, Romansch — **EC:** a/c 220V — **ITU:** SUI.

DIRECTION GENERALE DE L'ENTREPRISE DES PTT

ADDR: Viktoriastr. 21, CH-3030 Berne.**Te:** +41 (31) 621111. **Telex:** 32011 ptt ch. **Cable:** Gentel.
L.P: Pres. & Dir. Gen: Rudolf Trachsel; Dir. Peter Sollberger, Ing. (PTT responsible for the technical aspects of transmission).

SWISS BROADCASTING CORPORATION (SBC)

(A non-profit-making Comp. responsible for the radio & tv prgr. sce's).
ADDR: Schweizerische Radio-und Fernsehgesellschaft (SRG); Société suisse de radiodiffusion et télévision (SSR); Società Svizzera di Radiotelevisione (SSR); Giacomettistrasse 3, (P.O. Box 26), CH-3000 Berne 15. **Cable:** Radif. **Te:** +41 (31) 439111. **Fax:** +41 (31) 43 92 56. **Telex:** 911 590 ssr ch.
L.P: Pres. of the Board of Adm: Yann Richter. Dir. Gen: Antonio Riva. Dir. Human Resources: Raymond Zumsteg. Dir. Prgr. Sces (Radio and TV): Ulrich Kündig. Dir. Finance: François Landgraf. Tech. Dir: Daniel Kramer. Dir. Legal Dept: Beat Durrer. Prgr. Dept/Foreign Rel. Radio and TV: Félix Bollman. Press and Pub. Rel: Oswald Sigg. Documentation and Library: Dr. Emmanuel Belser.

GERMAN LANGUAGE NETWORK

"Radio und Fernsehen der deutschen und der raeto romanischen Schweiz" (DRS).
L.P: Radio Dir: Andreas Blum, Radiodirektion DRS Güterstrasse 91, 4024 Basel. **Te:** +41 (61) 343484. **Reg. Studios:** Postfach 9, 5001 Aarau; Inseliquai 8, 6002 Luzern; Rorschacherstr. 150, 9006 St. Fallen; Schwarztorstrasse 21, 3007 Berne; Novarastrasse 2, 4024 Basel; Brunnenhofstrasse 20, 8042 Zürich.
Mediumwaves: (Prgr. 1): Beromünster 531kHz 500kW (0500-2210), Sarnen 1566kHz 300kW (Su: 1800-2210, WI 1730-2210).

FM (MHz)	1	2	3	R	kW
Biel-Magglingen	90.2	92.2	107.3	—	1.0
Feschel	88.2	90.3	101.5	—	2.2
Froburg	96.0	98.7	91.3	—	1.2
Gebidem	89.4	93.9	103.9	—	1
Gotschnagrat	95.2	97.8	102.9	89.4	1.2
Grono	94.8	—	—	—	1.0
Lenzerheide	92.0	96.6	107.4	88.2	1.5
Monte Ceneri-Passo	96.9	—	—	—	1.1
Monte S. Salvatore	96.3	—	—	—	17.0
Niederhorn	93.6	97.2	105.8	—	35
Rigi	90.9	96.6	103.8	—	30
Ruschein	96.4	100.4	102.0	91.5	1.2
St. Chrischona	90.6	99.0	103.6	—	47
Säntis	99.9	95.4	105.6	—	58
Solothurn	89.7	98.0		—	1.5
Uetilberg	94.6	106.7	97.5	—	1.8

+ additional st's less than 1kW not mentioned — R = Romansch.

1st Prgr: 0400-2400 (Fri 0100). **Night Prgr:** 2300-0400. **N, Wrp, N:** on the h (exc. 1100SS, 1200, 1800, 2000. Also 0430, 0530, 0630, 1730).
Local Broadc. in German: MF 0553-0600, 1103-1112, 1603-1630, 1703-1725.
2nd Prgr: 0500-2300. **Night Prgr:** 2300-0500. **N:** 0500, 0600, 0700, 0800W, 0900W, 1000, 1130, 1300SS, 1500W, 1900W, 2100MF, 2200Sun, 2300. **N. in Romantsch:** 1820Sun, 1850W.
3rd Prgr: 0400-2300 (Fri 2400). **Night Prgr:** 2300-0400.
Romansch (R): MF 0900-1100 & 1600-1630, Mon/Thurs 0830-0900, Wed 1915-2100, Sat 1100-1130 & 1800-1830, Sun 1140-1200. Other times rel. Prgr. 2.
ANN: "Schweizer Radio DRS 1", "DRS 2" or "DRS 3".
V. by QSL-card.

FRENCH LANGUAGE NETWORK

"Radio-Télévision Suisse Romande".
L.P: Radio-TV Dir: Jean-Jacques Demartines, Radio-Télévision Suisse Romande, 6 ave de la gare, (P.O. Box 1275), CH-1001 Lausanne. **Te:** +41 (21) 205911. **Fax:** +41 (21) 23 75 06. Radio Prgr. RSR1: Gérald Sapey; RSR2: Esther Jouhet; RSR3: François Benedetti, Maison de la Radio, CH-1010 Lausanne. **Te:** +41 (21) 3181111. **Fax:** +41 (21) 652 37 19. TV Prgr. Dir: Guillaume Chenevière Télévision Suisse Romande, 20 quai Ernest Ansermet, (P.O. Box 234), CH-1211 Genève 8. **Te:** +41 (22) 292333. **Fax:** +41 (22) 781 52 18. **Radio Studios:** R. Genève, 66 Blvd. Carl-Vogt, 1205 Genève; R. Lausanne, Maison de la Radio, 1010 Lausanne.
Mediumwaves: Sottens 765kHz 600kW, (Prgr. 1 0500-2300, Prgr. 2 2300-0500), Savièse 1485kHz 1kW (Prgr 1.2).

FM (MHz)	I	II	III	kW
Bantiger	95.1	99.3		12
Biel-Magglingen	96.4	99.7	103.0	1
La Chaux-de-Fonds	92.3	96.3	103.4	1.5
La Dôle	94.8	91.2	100.1	45
Feschel	91.4	96.1	107.4	2.2
Gibloux	89.4	—	—	1.7
Grono	97.5	—	—	1.0
Haute-Nendaz	94.4	96.5	106.0	2.5
Les Ordons	94.2	99.6	104.8	10
Monte Ceneri-Passo	105.3	—	—	1.1
Mont-Pélerin	95.3	98.5	104.7	1
Monte S. Salvatore	104.0	—	—	17.0
Mont Salève (F)	105.6	102.5	100.7	1.0
Premier	96.2	97.8		1
Ravoire	94.0	95.7	100.5	1.2

+ st's less than 1kW not mentioned.
1st Prgr: 0500-2300. **N, Wrp, Comm, Current N. Items:** On the h. (exc. 2100, 2200). Also 0530, 0630, 1130, 2130.
2nd Prgr: 0500-2305. **N:** 0500, 0600, 0700, 0800, 1130, 1600, 1800, 2130, 2300. **N. in Romansch:** 2150.
3rd Prgr: "Couleur 3", 24h non-stop music and N.
ANN: "Radio Suisse Romande". Prgr. 1: "RSR-La Première". Prgr. 2: "RSR-Espace 2". Prgr. 3: "Couleur 3" — **V.** by QSL-card.

ITALIAN LANGUAGE NETWORK

Radiotelevisione della Svizzera Italiana.
L.P: Reg. Radio-TV Dir: Marco Blaser. Radio-TV Prgr. Dir: Dino Balestra, Radiotelevisione della Svizzera Italiana, Casella postale, CH-6903 Lugano-Besso. **Te:** +41 (91) 58 51 11. **Fax:** +41 (91) 58 91 50. **Radio Studio:** R. Lugano, 6903 Lugano-Besso. **Te:** +41 (91) 585111. **TV studio:** CH-6949 Comano. **Te:** +41 (91) 585111. **Fax:** +41 (91) 58 53 55.
Mediumwaves (Prgr. 1,2,3): Monte Ceneri-Cima 558kHz 300kW.

FM (MHz)	I	II	III	kW
Cardada	91.3	97.8	103.3	0.6
Grono	88.7	90.9	102.3	1
M. Ceneri-Passo	89.4	93.5	107.4	0.75
M. Morello	88.8	93.0	104.5	0.75
M. San Salvatore	88.1	91.5	106.0	2x15/1.5
Pizzo Matro	88.4	90.0	105.9	0.37

+ 22 low power st's not mentioned.
1st Prgr: 0500-2300. **Night Prgr:** 2300-0500.
N, Wrp. Comm, Current N. Items: W 0500-1000 hourly, 1130, 1300, 1700, 2000, 2200, 2400; Sun 0500-0900 hourly, 1130, 1300, 1700, 2100, 2200, 2400.
2nd Prgr: 0500-2300. **Night Prgr:** 2305-0500. **N:** W 0500, 0630, 0800, 1130, 1500; Sun 0500, 0700, 1130, 1325, 1700, 2200, 2300.
3rd Prgr: "Rete 3": non-stop music and N.
ANN: "Radio della Svizzera Italiana Rete 1", "Rete 2" or "Rete 3".

EXTERNAL SERVICE

SWISS RADIO INTERNATIONAL

ADDR: Giacomettistr. 1, 3000 Bern 15. **Te:** +41 (31) 439222. **Fax:** +41 (31) 439544. **Telex:** 911538 chri ch.
L.P: Dir: Roy Oppenheim; Head Prg. Dept: N.D. Lombard; Head Tec. Dept: P. Badertscher; Press Officer: W. Fankhauser.
STATIONS: Schwarzenburg (G.C. 07.24E/46.09N) 4×250kW, 2×100kW; Beromunster (G.C. 08.10E/47.12N) 2×250kW; Lenk (G.C. 07.26E/47.27N) 2×250kW; Sarnen (G.C. 08.15E/46.54N) 1×250kW; Sottens (G.C. 06.44E/46.39N) 1×500kW; **Relays:** Moyabi, Gabon (Africa No. 1) 1×500kW; Beijing, China 1×120kW; Brasilia 1×250kW.

EUROPEAN SCE: 0500-1100 & 1300-2045 on 3985/6165/9535kHz, 1100-1330 on 6165/12030kHz — **English:** 0500-0545, 0600-0615, 0700-0715, 1100-1145, 1700-1730 — **French:** 0630-0645, 0730-0745, 0800-1000*, 1000-1100, 1230-1300, 1830-1900

— **German:** 0615-0630, 0715-0730, 1200-1230, 1730-1815*, 1915-1930W, 1930-2030* — **Italian:** 0645-0700, 0745-0800, 1300-1330, 1400-1700*, 1900-1915(Sun 1930) — **Romansch:** 1330-1400 — **Personal Messages:** 0545-0600, 1145-1200, 1815-1830 — **Stock Market News:** Mon-Fri 2030-2045 — *) rel. of domestic sce. prgrs.

OVERSEAS SERVICE

Arabic	**Areas**	**kHz**
1900-2000	ME/NAf	9885, 13635, 15505
English		
0600-0630	Af	9860, 13635, 17535
0900-0930	Pac	9560, 13685, 17670, 21770
1100-1130	As/Pac	13635, 15505, 17670, 21770
1300-1330	SEAs/FE	7480, 11690, 13635, 15505, 17670, 21820
1500-1530	SEAs/CAs	13635, 15505, 17670, 21820
1700-1730	CAs/ME/Af	9885, 13635, 15430, 17635
2000-2030	Af	9885, 12035, 13635, 15505
2200-2230	SAm	6030, 9810, 9885, 12035
0000-0030	SAm/CAm	6135, 9650, 9885, 12035, 17730
0200-0230	CAm/NAm	6135, 9650, 9885, 12035
0400-0430	NAm	6135, 9860, 9885, 12035
French		
0630-0700	Af	9860, 13635, 17565
0700-0715*	Af	17565
0930-1000	Pac	9560, 13685, 17670, 21770
1000-1015*	Pac	13685, 21770
1130-1200	As/Pac	13635, 15505, 17670, 21770
1200-1215*	As/Pac	17670, 21770
1330-1400*	SEAs/FE	7480, 11690, 13635, 15505, 17670, 21820
1400-1415*	SEAs/FE	11690, 15505, 21820
1530-1600	SEAs/CAs	13635, 15505, 17670, 21820
1600-1615*	SEAs/CAs	15505, 21820
1730-1800	CAs/ME/Af	9885, 13635, 15430, 17635
1800-1815*	CAs/ME/Af	9885, 15430
2030-2100	Af	9885, 12035, 13635, 15505
2100-2115*	Af	9885, 12035
2230-2300	SAm	6030, 9810, 9885, 12035
2300-2315*	SAm	6030, 9810
0030-0100	SAm/CAm	6135, 9650, 9885, 12035, 17730
0100-0115*	SAm/CAm	9650, 12035
0230-0300	CAm/NAm	6135, 9650, 9885, 12035
0300-0315*	CAm/NAm	9650, 12035
0430-0500	NAm	6135, 9860, 9885, 12035
0500-0515*	NAm	6135, 9860
German		
0715-0730	Af	17565
0730-0745*	Af	9860, 13635, 17565
1015-1030	Pac	13685, 21770
1030-1045*	Pac	9560, 13685, 17670, 21770
1215-1230	As/Pac	17670, 21770
1230-1245*	As/Pac	13635, 15505, 17670, 21770
1415-1430	SEAs/FE	11690, 15505, 21820
1430-1445*	SEAs/FE	7480, 11690, 13635, 15505, 17670, 21820
1615-1630	SEAs/CAs	15505, 21820
1630-1645*	SEAs/CAs	13635, 15505, 17670, 21820
1815-1830	CAs/ME/Af	9885, 15430
1830-1845*	CAs/ME/Af	9885, 13635,15430, 17635
2115-2130	Af	9885, 12035
2130-2145*	Af	9885, 12035, 13635, 15505
2315-2330	SAm	6030, 9810
2330-2345*	SAm	6030, 9810, 9885, 12035
0115-0130	SAm/CAm	9650, 12035
0130-0145*	SAm/CAm	6135, 9650, 9885, 12035, 17730
0315-0330	CAm/NAm	9650, 12035
0330-0345*	CAm/NAm	6135, 9650, 9885, 12035
0515-0530	NAm	9885, 12035
0530-0545*	NAm	6135, 9860, 9885, 12035
Italian		
0700-0715	Af	9860, 13635
0715-0730*	Af	9860, 13635
1000-1015	Pac	9560, 17670
1015-1030*	Pac	9560, 17670
1200-1215	As/Pac	13635, 15505
1215-1230*	Pac	13635, 15505
1400-1415	SEAs/FE	7480, 13635, 17670
1415-1430*	SEAs/FE	7480, 13635, 17670
1600-1615	SEAs/CAs	13635, 17670
1615-1630*	SEAs/CAs	13635, 17670
1800-1815	CAs/ME/Af	13635, 17635
1815-1830*	CAs/ME/Af	13635, 17635
2100-2115	Af	13635, 15505
2115-2130*	Af	13635, 15505
2300-2315	SAm	9885, 12035
2315-2330*	SAm	9885, 12035
0100-0115	SAm/CAm	6135, 9885, 17730
0115-0130*	SAm/CAm	6135, 9885, 17730
0300-0315	CAm/NAm	6135, 9885
0315-0330*	CAm/NAm	6135, 9885
0500-0515	NAm	6135, 9860
0515-0530*	NAm	6135, 9860
Spanish		
2230-2300	LAm	9885, 12035
0030-0100	LAm	6135, 9885, 17730
0230-0300	LAm	6135, 9885

*) for Swiss citizens abroad.

INT-SIG: "Luegid vo Bärg und Tal" — Trad. Swiss-German folk song, played on a music box.
V. by QSL-card Re. to Swiss Radio International, CH-3000 Berne 15, Switzerland. Rp. — **PUB:** Programme schedule twice a year, in spring and autumn; report cards and report sheets free upon request.

BROADCASTS FROM THE INTERNATIONAL COMMITTEE OF THE RED CROSS, GENEVA

ADDR: Red Cross Broadcasting Service, 19 Ave de la Paix, 1202 Geneva. **Te:** +41 (22) 7346001. **Cable:** Intercroixrouge. **Telex:** 22269. **Fax:** +41 (22) 7348280.
L.P: Michèle Mercie, Ag. Head of Communication Dept; Carlos Bauverd, Head of Press Div.
To Europe: ICRC organises test transmissions, broadcasting Red Cross news, by courtesy of the Swiss postal and radio authorities on 7210kHz on the last Sun of each month as follows: 1100-1130 English, 1130-1200 French, 1200-1220 German, 1220-1240 Spanish. Prgrs repeated the following day at 1700-1840.
Worldwide Sce: within regular broadcasts of Swiss R. Int. as follows: **English:** last Thurs of month — **French:** 2nd & 4th Thurs of month — **Arabic:** last Sat of month — **Spanish:** last Thurs of month — **German:** first Mon of month — **Italian:** 3rd Sun of month.
ANN: E: "This is Geneva—Red Cross Broadcasting Service". F: "Ici Genève—Service de Radiodiffusion du Comité International de la Croix-Rouge". Sp: "Aqui Ginebra — Servicio de Radiodifusión del Comité Internacional de la Cruz Roja". G: "Hier ist der Rundfunkdienst des Internationalen Komitees vom Roten Kreuz in Genf". P: "Aqui Genebra, Servico de Radiodifusão do Comite Internacional da Cruz Vermelha".
V. by QSL-card. IRC appreciated. (B).

UNITED NATIONS BROADCASTING FROM GENEVA

ADDR: Nations Unies, Palais de Nations, CH-1211 Geneva 10. **Cable:** Omnipress. **Telex:** 41 29 62. **Te:** +41 (22) 7346011. Ext. 2566/4629. **Fax:** +41 (22) 7339879.
L.P: Dir. Inf: P. Méhu. Chief, Radio/TV: F. Séguillon.
FOREIGN SCE: Russian (Fri) on 7443kHz (4UZ USB) 1830-1840.
ANN: E. "This is United Nations R. in Geneva". F: "Ici la R des Nations Unies à Genève"

Private Stations in main cities:
Basel: R. Basilisk 94.5MHz.
Bern: R. Extra Bern 97.7MHz — R. Förderband Bern 96.7MHz.
Geneva: R. Cité 91.8MHz — R. Lac 97.6MHz — R. Tonic Genève 88.4MHz.
Zürich: Alternatives Lokal-Radio Zürich 104.5MHz — R. 24 102.8MHz — R. Z 100.9MHz — R. Zurisee 90.1MHz.

UKRAINE

L.T: UTC + 2h (Su: UTC + 3h) — **Pop:** 51.840.000 —**Pr.L:** Ukrainian, Russian - **E.C:** a/c 50 127/220V — **ITU:** UKR.

DERZHAVNA TELERADIOMOVNA KOMPANIYA UKRAYINY (Ukrainian Radio TV Corp.) (Gov.)

ADDR: vul. Khreshchatyk 26, 252001 Kyyiv.
L.P: Pres: Mykola Okhmakevych

UKRAYINSKO RADIO (Gov.)

ADDR: vul. Khreshchatyk 26, 252001 Kyyiv. **Te**: +7 (044) 2291285. **Fax:** +7 (044) 2991170, 2994557. **Telex:** 131163 TON.

Stations*	kHz	kW	H. of tr.	Prgr.
Lviv	171	500	0200-0000	MR1
Kyyiv	207	500	0400-2300	UR1
Bershad	549		0100-0000	RSM
Izmayil	549		0100-0000	RSM
Kerch	549		0100-0000	RSM
Kyyiv	549		0100-0000	RSM
Lviv	549		0100-0000	RSM
Mariupol	549		0100-0000	RSM
Mezhygirya	549		0100-0000	RSM
Sevastopol (1)	549		0100-0000	RSM
Tyachiv	549		0100-0000	RSM
Oktyabrskoye (1)	648	150	0400-2300	UR1/ MR1/Reg
Chernivtsi	657	25	0400-2200	UR3/F
Uzhgorod	675	50	0400-0300	UR3
Donetsk	711	150	0400-0300	UR1/RSM
Odesa	765	50	0400-2300	UR1/Reg
Kharkiv	810	6	24h	UR3
Lutsk	810			UR3
Tokmak	810			UR3
Kharkiv	837	150	24h	UR2
Kyyiv	873	50	0400-0700	RL
			1700-2000	RL
Dnipropetrovsk	873	10	as above	RL
Volochisk	873	25	as above	RL
Zarvantsi (Vinnytsya)	873	7	as above	RL
Simferopol (1)	873		inactive	
Uzhgorod	891	150	0400-2300	UR1/Reg
Lviv	936	300	0735-1730	UR1
			1800-0730	F
Mykolayiv	972	500	0200-0100	UR2
Dnipropetrovsk	1071	20	0400-2300	UR1
Kyyiv	1134		0400-0300	UR3
Lugansk (r.)	1134			UR3
Kyyiv	1242	150	0145-0030	UR2
Kholmy	1242		24h	UR2
Uzhgorod	1242	50	24h	UR2
Oktyabrskoye (1)	1242	50	24h	UR2
Donetsk	1242	30	24h	UR2
Lugansk	1242		24h	UR2
Odesa	1242	30	24h	UR2
Pervomaysk (Mykolayiv)	1242		24h	UR2
Starobilsk	1242		24h	UR2
Tyachiv	1242		24h	UR2
Volochisk	1242		24h	UR2
Kharkiv	1260	50	0400-0700	RL
			1700-2000	RL
Kherson (r.)	1278		0400-0700	RL
			1700-2000	RL
Petrivka (Odesa)	1278	150	as above	RL
Mykolayiv (r.)	1278		as above	RL
Simferopol (1)	1359		0400-0300	UR3
Chernivtsi	1377	50	24h	UR2
Lutsk	1377	50	24h	UR2
Tokmak	1377	25	24h	UR2
Tsybuliv	1377		24h	UR2
Zarvantsi (Vinnytsya)	1377	30	24h	UR2
Oktyabrskoye (1)	1395		0400-0300	UR3
Kovel	1404		24h	UR2
Lviv	1404	30	24h	UR2
Dnipropetrovsk	1404	30	24h	UR2
Izmayil	1404	25	24h	UR2
Kotovsk	1404		24h	UR2
Mezhygirya	1404		24h	UR2
Rovenki (Lugansk)	1404		24h	UR2
Strymba (Ivano-Frankovsk)	1404		24h	UR2
Mykolayiv	1431	400	0300-0200	UR3/F
Chernivtsi	1449		0400-2200	++
			2200-0200	UR3
Mariupol (r.)	1449			UR3
Yalta (1) (r.)	1467			UR3
Lviv	1476	120	inactive	UR3
Kharkiv	1485	1	0400-2300	UR1
Kyyiv	1485	2	0400-2300	UR1
			2330-0300	UR3
Simferopol (1)	1485		0400-2300	UR1
Trostyanets (Sumy)	1530	5	0345-0300	UR3
Zarvantsi (Vinnytsya)	1530		0400-2300	UR3
Izyum	1539		0400-2300	UR3
Odesa	1566	5	0400-0300	UR3
Lugansk	1566			RSM
Botanichnoye (1)	1575		inactive?	RSM
Simferopol	1584			UR2
Parkovoye (1)	1584		inactive?	RSM
Dnipropetrovsk	1593	5	0400-0300	UR3
Polonne (r.)	1593		0400-0300	UR3
Lugansk	1602		inactive	
Sevastopol (1)	1602		24h	UR2
Kyyiv	4940	50	0400-0300	UR3
Kyyiv	6020	100	variable	UR2
Unknown	7235	20	0400-1500	UR1
Lugansk	7245	20	0400-1730	UR1/Reg
Unknown	9735		0400-	UR1
Unknown (2 st's)	13795			UR2

Obs: Powers and locations are tentative only. (1) = Krym (Crimea). * in Ukrainian exc. for Krym in Russian. ++) rel. prgr. Luceaf rul from Moldavia.

FM: Reliable information not available. Numerous FM tr. centers are in operation. Most reg. sce's are carried on FM only.

D.PRGR: UR1 0400-2300 in Ukrainian: News & information. N. in Russian 1000-1010 — **UR2** 24h in Ukrainian and limited Russian: Music & entertainment — **UR3** ("R. MUZ") in Ukrainian: 0700-0200 cultural prgr, incl. Russian 0930 (Sat/Sun 0905)-1000. 2100-2200 rel. Voice of America. 0200-0700 rel. RSM —**MR1** = Moscow Ostankino R-1 — **RSM** = Moscow R. St. Mayak — **RL** = R. Liberty (cf. Germany) Addr. in Kiev: vul. Lenina 8 — **F** = Foreign Sce, cf. below.

Regional prgrs (within UR1): 0530-0600, 1600-1715.
Regional centres on MW/SW: vul. Demyokhina 25, 348000 Lugansk — vul. Vatutina 6, 290005 Lviv — vul. Troyitska 43-b, 270001 Odesa — Krymskoye Radio, ul. Studencheskaya 14, 333610 Simferopol — Kyyivska nab. 18, 294000 Uzhgorod.

FOREIGN SERVICE: Radio Ukraine International

ADDR: Kiev.

Stations: MW: Cf. Home Sce. **SW** (some shared with R. Moscow): Kharkiv 2x100kW; Crimea 4x250/1000kW — Kyyiv 8x100/200kW — Lviv 200kW, Mykolayiv 100kW. **Note:** Su. all prgr's 1h earlier.

English	Area	kHz
2200-2300	Eu	936, 5905, 5940, 6055, 6130, 7240, 9875, 9820
0100-0200	NAm	936, 5940, 7195, 7240, 9665, 9685, 11795, 12060, 15195, 17605, 17690
German		
0000-0100	Eu	936, 5940, 6130, 7250, 9785, 9820
1800-1900	Eu	936, 5940, 5960, 6020, 6130, 9600, 7240, 15455
2100-2200	Eu	936, 5905, 5940, 6020, 7250, 9875, 9820, 15455
Moldovan		
1800-1830	Eu	657, 891, 1431, 13795
2030-2100	Eu	1431, 13795
2200-2230	Eu	891, 1431
Ukrainian		
0200-1800	World Sce.	*
1900-2100	World Sce.	*
2300-2400	World Sce.	*

*) Up to 10 SW freq's used in parallel. Exact schedule not available.

Relays on SW of Home Sces:
UR1 on one tr. at Mykolayiv — UR2 24h on various freq's.

ANN: Ukrainian: "Hovorit Kyyiv".

V. by QSL-card. **PUB:** Govorit i pokazuye Ukrayina (weekly for Home Sce.).

RADIO MOSCOW RELAY

Shortwaves (shared with R. Ukraine): Kharkiv 3x100kW — Crimea ("Simferopol") 6x250/1000kW — Lviv 3x200kW, 1x500kW — Mykolayiv 2x100kW — Odesa 2x50. All powers are estimates. Used for R. Moscow, MR1 & RSM prgr's.

INDEPENDENT STATIONS

R.ROKS: Kyyiv 71.66, 104MHz. Cf. Russia.
Radio Vilne Slovo z Ukrayiny, P.O. Box 5656, 290016 Lviv. On 72 & 101MHz (100W) 0400-2300.
Radio Voskreseniye. Sat/Sun 1400, 1800 on 936kHz (Lviv), Tues-Fri 1905 on 936kHz + FM. 11735kHz also r.
Radio Nezaleznist, Lviv. r. on 1476kHz.
Further stations may be operating, but insufficient information available.

UNITED KINGDOM

L.T: UTC (Su: UTC + 1h) — **Pop:** 57.688.000 — **Radio sets:** 70m (est.) — **Pr.L:** English — **E.C:** a/c 50, 240V — **ITU:** G.

THE BRITISH BROADCASTING CORPORATION

The BBC is an independent body created by a Royal Charter and operated under a licence and agreement with the Minister exercising functions under the Wireless Telegraphy Acts.
ADDR: Broadc. House, London W1A 1AA. **Te:** +44 (71) 580-4468. **Fax:** +44 (71) 637-1630. **Telex:** 265781 BBC HQG. **Cable:** "Broadcasts".
Board of Management: Dir. Gen: M. Checkland. Dep. Dir. Gen: J. Birt; Man. Dir. Network TV: W. Wyatt; Man. Dir. Network Radio: D. Hatch; Man. Dir. World Sce: J. Tusa; Man. Dir. Reg. Broadc: R. Neil; Dir. Finance; I. Philips; Dir. Eng: W. Dennay; Dir. Corporate Affairs: H. James; Dir. Personnel: Margaret Salmon; Chief Exec. BBC Enterprises Ltd: J. Arnold-Baker..
Controllers: BBC-1 TV: J. Powell. BBC-2 TV: A. Yentob; Contr. Asst. Man. Dir. Netw. TV: Ms. J. Drabble; Planning & Prgr. Sces, TV: K.D. Anderson; Prod. Resources, TV: M.D. Lumley; Personnel, TV: R.S. Murdoch; Radio 1: J.W. Beerling; Radio 2: Miss. F.M. Line; Radio 3: N. Kenyon; Radio 4: M. Green. Radio 5: Miss P.A. Ewing; Contr. Educational Broadc: Dr. E.G. Jones; Dep. Man. Dir. World Sce: D.M.L. Witherow; Contr. English Sces: Mrs. E.J. Smith; Contr. European Sces: A. Taussig; Contr. Overseas Sces: S. Younger; Ag. Contr..Resources & Admin, World Sce: B. Whitehall. Contr. Inf. Sces. Int. Rel: D. Barlow; Contr. Reg. Broadc: Mark Byford; Contr. Inf. Sces: M. Bunce; Legal Adviser: G. Roscoe. Editorial Policy: J. Wilson. **Regional Contrs:** Wales: G. Talfan Davies; Scotland: J. McCormick; No. Irl: R. Walsh.

LONG- & MEDIUMWAVES

1FM	kHz	kW			
Barnstaple	1053	1	Start Point	1053	100
Barrow	1053	1	Brookmans Park	1089	150
Bexhill	1053	2	Fareham	1089	1
Brighton	1053	2	Lisnagarvey	1089	10
Burghead	1053	20	Moorside Edge	1089	150
Droitwich	1053	150	Redmoss	1089	2
Dundee	1053	1	Redruth	1089	2
Enniskillen	1053	1	Tywyn	1089	1
Folkestone	1053	1	Washford	1089	50
Hull	1053	1	Westerglen	1089	50
Londonderry	1053	1	Whitehaven	1089	1
Postwick	1053	10	Wallasey	1107	0.5
Stagshaw	1053	50	Bournemouth	1485	1

NB: Tr's will be taken out of sce. during 1993.

RADIO 4	kHz	kW			
Burghead	198	50	Londonderry	720	0.25
Droitwich	198	500	Redruth	756	2
Westerglen	198	50	Enniskillen	774	1
Newcastle	603	2	Plymouth	774	1
Lisnagarvey	720	10	Redmoss	1449	2
London (Lots Rd.)	720	0.75	Carlisle	1485	1

RADIO 5	kHz	kW			
Barrow	693	1	Redmoss	693	1
Bexhill	693	1	Stagshaw	693	50
Brighton	693	1	Start Point	693	50
Burghead	693	25	Bournemouth	909	0.25
Droitwich	693	150	Brookmans Park	909	150
Enniskillen	693	1	Clevedon	909	50
Folkestone	693	1	Exeter	909	1
Postwick	693	10	Fareham	909	1
Lisnagarvey	909	10	Westerglen	909	50
Londonderry	909	1	Whitehaven	909	1
Moorside Edge	909	200	Tywyn	990	1
Redruth	909	2			

FM (all stereo)

England, Isle of Man, Channel Isl.	1FM	R2	R3	R4	kW ERP
Barnstaple		88.5	90.7	92.9	1
Belmont	98.3	88.8	90.9	93.1	16
Bow Brickhill	98.2	88.6	90.8	93.0	10
Bristol		89.3	91.5	93.7	1.3
Chatton		90.1	92.3	94.5	5.6
Douglas	98.0	88.4	90.6	92.8	11
Guildford	97.7	88.1	90.3	92.5	3
Holme Moss	98.9	89.3	91.5	93.7	250
Keighley		88.9	91.1	93.3	1
Les Platons	97.1	89.6	91.1	94.8	16
Manningtree		88.1	90.3	92.5	5
Morecambe Bay	99.6	90.0	92.2	94.4	10
North Hessary Tor	97.7	88.1	90.3	92.5	160
Oxford	99.1	89.5	91.7	93.9	46
Pendle Forest		90.2	92.6	94.6	1
Peterborough	99.7	90.1	92.3	94.5	40
Pontop Pike	98.1	88.5	90.7	92.9	130
Redruth	99.3	89.7	91.9	94.1	25
Ridge Hill		88.6	90.8	93.0	10
Rowridge	98.2	88.5	90.7	92.9	250
Sandale	97.7	88.1	90.3	92.5	250
Stanton Moor		89.8	92.0	94.2	1.2
Sutton Coldfield	97.9	88.3	90.5	92.7	250
Swingate (Dover)	99.5	90.0	92.4	94.4	11
Tacolneston	99.3	89.7	91.9	94.1	250
Wenvoe	99.5	89.9	92.1	94.3	250
Winter Hill	98.2	88.6	90.8	93.0	4
Wrotham	98.8*	89.1	91.3	93.5	125*/250

+ low power tr's less than 1kW.

1FM: 24h. On MW: 0600-2400. **N:** Mon-Fri 0100-0500 hourly; 0530-1130, 1330-1630 & 1830-2330 on the half h; Sat 0100-0500 hourly; 0530-1330, 1530-2130 on the half h, 2330; Sun: as Mon-Fri.

RADIO 2: Light Music & Information Prgr. 24h. N: Mon-Fri on the h. Headlines 0530, 0630, 0730, 0830; Sat on the h exc. 2000 & 2100. Headlines 0630, 0730; Sun on the h exc. 2000. Headlines 0730.

RADIO 3: Serious Music & Cultural Prgr. 0655-0035 (Sat/Sun 0005). **N:** Mon-Fri 0730, 0830, 1300, 2330; Sat/Sun 0730, 0830, 1300, 0000.

RADIO 4 UK: Spoken Word, Current Affairs & Music Prgr. 0555-0040 (approx). **N:** Mon-Fri 0600-0900 every half h, 1000-1900 hourly (exc. Mon 1500), 2200, 0000; Sat 0600, 0700, 0730, 0800, 0830, 0900, 1000, 1100, 1300, 1400, 1600, 1800, 2200, 0000. Sun 0600, 0630, 0700, 0800, 0900, 1300, 1600, 1700, 1800, 2200, 0000. **Financial N:** Mon-Fri 2145.
Shipping forecast 0555, 1355, 1750, 0033.

RADIO 5: Sport, Education & Youth Prgr: 0600-0005. **N:** Mon-Fri 0600-0900 half-hourly, 1100-1700 hourly, 2200, 0000; Sat 0600-0900 half-hourly, 1000, 1100, 1200, 0000-Sun 0600, 0730, 0800, 0830, 0900-1300 hourly, 1800, 1900, 2200, 2300, 0000.
BBC World Sce. prgrs: Mon-Fri 1430-1635. **BFBS Worldwide:** Mon-Fri 1330-1430.

BBC Local Radio

FM: stereo exc. m = mono — **NB:** Some MW-tr's will be taken out of sce. during 1993 to accommodate new commercial radio sces.

	Station	kHz	kW	MHz	kW
34)	Stroud (r)	—	—	95.0	0.1
1)	R. Cornwall (Redruth)	630	2	103.9	17
31)	R. Bedfordshire (Luton)	630	0.2	103.8	0.5
1)	Bodmin (r)	657	0.5		
1)	Caradon Hill (r)			95.2	4.3
1)	Isles of Scilly (r)			96.0	0.06
2)	R. York	666	0.5	103.7	2
2)	Woolmoor (r)			104.3	0.5
33)	BBC Essex	729	0.2		

37)	BBC H&W, Worcester	738	0.37	94.7	2
37)	Kidderminster(r)			104.6	0.5
3)	R. Cumbria	756	1	95.6	15
32)	R. Shropshire	756	0.63	96.0	4.8
32)	Ludlow (r)			95.0	0.01
33)	Chelmsford (r)	765	0.5	103.5	12
4)	R. Leeds	774	0.5	92.4	5.6
4)	Wharfedale (r)			95.3	0.04
4)	Beercroft Hill (r)			103.9	0.1
4)	Keighley (r)			104.7	0.5
14)	Littlebourne (r)	774	0.7		
5)	Barnstaple (r)	801	2		
37)	BBC H&W, Hereford	819	0.37	104.0	2
6)	Wolverhampton (r)	828	0.2		
7)	R. Furness	837	1.5	96.1m	3.2
7)	Kendal (r)			95.2m	0.1
7)	Windermere (r)			104.2m	0.065
8)	R. Leicester	837	0.45	104.9	8
5)	R. Devon	855	1	103.4	10
5)	Okehampton (r)			96.0	0.07
9)	R. Lancashire	855	1.5	95.5	1.6
9)	Winter Hill (r)			103.9	1.8
10)	R. Norfolk	855	1.5	95.1	5.7
10)	We. Norfolk (r)	873	0.3	104.4m	4.2
5)	Exeter (r)	990	1	95.8	0.4
11)	R. Solent	999	1	96.1	10
12)	R. Cambridgeshire	1026	0.5	96.0	1
13)	R. Jersey	1026	1	88.8	3.8
14)	R. Kent	1035	0.5	96.7	8.7
14)	Swingate (r)			104.2	10
14)	Folkestone (r)			97.6	0.1
15)	R. Sheffield	1035	1	104.1	4.4
15)	Sheffield (r)			88.6	0.32
15)	Chesterfield (r)			94.7	0.2
16)	Geddington (r)			103.6	0.8
17)	R. Derby	1116	1.2	104.5	5.4
17)	Stanton Moor (r)			95.3	1.2
17)	Derby (r)			94.2	0.01
17)	Buxton (r)			96.0	0.15
18)	R. Guernsey	1116	0.5	93.2	1
31)	R. Bedfordshire (Bedford)	1161	0.1	95.5	1
	Milton Keynes			104.5	2.2
19)	Bexhill (r)	1161	1		
2)	Scarborough (r)	1260	0.5	95.5	0.25
29)	Taunton (r)	1323	0.63		
40)	Wilts Sound, Lacock	1332	0.3	104.3	0.6
11)	Bournemouth (r)	1359	0.85		
19)	Reigate (r)	1368	0.5	104.0	3.8
20)	R. Lincolnshire	1368	2	94.9	6
40)	Swindon	1368	0.1	103.6	0.5
40)	Salisbury (r)			103.5	0.02
12a)	R. Peterborough	1449	0.15	95.7	5.1
6)	R. WM	1458	5	95.6	11
3)	Whitehaven (r)	1458	0.5		
21)	R. Newcastle	1458	2	95.4	10
21)	Fenham (r)			104.4	0.042
21)	Chatton (r)			96.0	5.6
22)	GLR	1458	50	94.9	4
23)	Greater Manchester R.	1458	5	95.1	5.6
5)	Torbay (r)	1458	2	94.8	0.675
19)	R. Sussex	1485	1	95.3	1.2
19)	Heathfield (r)			104.5	10
19)	Horsham (r)			95.1	0.04
19)	Newhaven (r)			95.0	0.1
24)	R. Humberside	1485	1	95.9	9.6
26)	R. Merseyside	1485	1.2	95.8	7.5
27)	R. Stoke-on-Trent	1503	1	94.6	6.1
33)	Southend (r)	1530	0.15	95.3	1.2
29)	R. Bristol	1548	5	95.5	9
29)	Bath (r)			104.6	0.082
29)	Bristol (r)			94.9	0.95
9)	Lancaster (r)	1557	0.25	104.5	2.1
28)	Mansfield (r)	1584	1	95.5	2
32)	Woofferton (r)	1584	0.5		
14)	Rusthall (r)	1602	0.25		

STATIONS ON FM ONLY

	Station	MHz	kW
39)	R. Berkshire, Henley (r)	94.6	0.25
35)	BBC CWR, Meriden	94.8	2.2
30)	BBC Cleveland	95.0	5
34)	Stroud (r)	95.0	0.1
25)	BBC Oxford	95.2	5.8
35)	R. Berkshire, Windsor	95.4	0.5
38)	R. Suffolk, Lowestoft	95.5	0.05
34)	Cirencester (r)	95.8	0.08
30)	Whitby (r)	95.8	0.04
16)	Geddington (r)	103.6	0.8
35)	Lark Stoke	103.7	1.4
28)	BBC Nottingham	103.8	1
38)	R. Suffolk, Manningtree	103.9	5
35)	Nuneaton (r)	104.0	0.05
39)	R. Berkshire, Hannington	104.1	4
16)	BBC Northampton	104.2	4
39)	R. Berkshire, Reading	104.4	1
38)	R. Suffolk, Gt. Barton	104.6	2
41)	R. Surrey, Guildford	104.6	4
34)	R. Gloucestershire	104.7	2

Addresses & other information:

1) Phoenix Wharf, Truro, Cornwall TR1 1UA. 0700-1315, 1500-1800.

2) 20 Bootham Row, York YO3 7BR. 0600-1900.

3) Hilltop Heights, London Rd, Carlisle, Cumbria CA1 2NA. 0600 (Sat 0700)-1900.

4) Broadcasting House, Woodhouse Lane, Leeds LS2 9PN. 0600-1900.

5) P.O. Box 5, Plymouth PL1 1XT; P.O. Box 100, Exeter EX4 4DB. 0600-1800.

6) P.O. Box 206, Pebble Mill Rd, Birmingham B5 7SD. 0500-2400.

7) Broadcasting House, Hartington Str, Barrow-in-Furness, Cumbria LA14 5FH. 0733-0900. Other times rel. R. Cumbria.

8) Epic House, Charles Str, Leicester LE1 3SH. 0600-1900.

9) 20-26 Darwen Str, Blackburn, Lancs. BB2 2EA. 0600-1900.

10) Norfolk Tower, Surrey Str, Norwich NR1 3PA. 0500 (Sat/Sun 0600)-1900.

11) 10 Havelock Rd, Southampton S01 0XR. 0600-1800, 2200-2400.

12) Broadcasting House, 104 Hills Rd, Cambridge CB2 1LD. 0600-1900.

12a) Broadway Court, Peterborough PE1 1RP. 0600-1900 (partly rel. 12).

13) Broadcasting House, Rouge Bouillon, St. Helier, Jersey. Mon-Fri 0700-1800, Sat 0800-1330, Sun 0810-1300.

14) Sun Pier, Chatham, Kent ME4 4EZ; Graylaw House, 20/22 Watling Str, Canterbury. 0600 (Sat/Sun 0630)-2200 (Sun 2000).

15) Ashdell Grove, 60 Westbourne Rd, Sheffield S10 2QU. 0600 (Sun 0700)-1900 (Sat 1800).

16) P.O. Box 1107, Abingdon Str, Northampton NN1 2BE. 0600-1900.

17) P.O. Box 269, 56 St. Helens Str, Derby DE1 3HY. 0600-1900.

18) Commerce House, Les Banques, St. Peter Port, Guernsey. W0700-1800, Sun 0730-1400.

19) Marlborough Place, Brighton, Sussex BN1 1TU; 26 Gildredge Rd, Eastbourne BN21 4SA. 0530-1900.

20) P.O. Box 219, Newport, Lincoln LN1 3XY. 0600-1900.

21) Broadcasting Centre, Barrack Rd, Newcastle-upon-Tyne NE99 1RN. 0600 (Sat 0630, Sun 0800)-1900.

22) 35C Marylebone High Str, London W1A 4LG. Mon-Fri 0530-0200, Sat/Sun 0600-2400 (relays BBC World Sce. at night).

23) P.O. Box 90, New Broadcasting House, Oxford Rd, Manchester M60 1SJ. 0600-1900 (Thurs 2200).

24) 63 Jameson Str, Hull HU1 3NU. 0600 (Sat/Sun 0700)-1800.

25) 269 Banbury Rd, Oxford OX2 7DW. 0600-1900.

26) 55 Paradise Str, Liverpool L1 3BP. 0600-1900.

27) Conway House, Cheapside, Hanley, Stoke-on-Trent, Staffs ST1 1JJ. Mon-Thurs 0600-2100, Fri 0600-1900, Sat 0700-1800, Sun 0700-1630.

28) P.O. Box 222, York House, Mansfield Rd, Nottingham NG1 3HZ. Mon-Fri 0600-2400, Sat/Sun 0700-2100.

29) 3 Tyndalls Park Rd, Bristol BS8 1PP; 14-15 Paul Str, Taunton, Somerset TA1 3PF. 0530-1900.

30) P.O. Box 95FM, Broadc. House, Newport Rd, Middlesborough, Cleveland TS1 5DG. Mon-Fri 0600-1800, Sat 0700-1700, Sun 0730-1700.

31) P.O. Box 476, Hastings Str, Luton Beds LU1 5BA; P.O. Box 258, Lurke Str, Bedford MK40 3LZ. 0600-1900.
32) 2-4 Boscobel Drive, P.O. Box 397, Shrewsbury SY1 3TT. 0600-2100.
33) P.O. Box 765, 198 New London Rd, Chelmsford, Essex CM2 9AB. 0600-1900.
34) London Rd, Gloucester GL1 1SW.
35) 25 Warwick Rd, Coventry CV1 2WR.
37) Hylton Rd, Worcester WR2 5WW; 43 Broad Str, Hereford HR4 9HH. 0600-2100.
38) Broadc. House, St. Matthews Str, Ipswich IP1 3EP. 0530 (Sat/Sun 0700)-1900.
39) Thomas House, 42A Portman Rd, Reading RG3 1NB.
40) 56/58 Prospect Place, Swindon SN1 3RW. 0600-1800.
41) Broadc. House, Guildford GUI2 5SF.
D.PRGR: Local prgrs. as above st's generally carry regional prgrs until 2400, then R2 overnight.
V. by QSL-card or letter. Reports to Engineer-in-Charge.

BBC RADIO SCOTLAND

ADDR: Broadcasting House, Queen Margaret Drive, Glasgow G12 8DG. **Te:** +44 (41) 330 2345. **Fax:** +44 (41) 339 0614.
MEDIUMWAVES: 810kHz: Burghead 100kW, Westerglen 100kW, Redmoss 5kW — **R. Aberdeen:** 990kHz 1kW — **BBC Solway:** 585kHz 2kW.
FM: stereo exc. m = mono

Location	1FM	R2	R3	R4	RS/L	kW
Ashkirk	98.7	89.1	91.3	103.9	93.5g	50
Black Hill	99.5	89.9	92.1		94.3	250
Black Hill				95.8		200
Bressay		88.3	90.5		92.7abc	43
Darvel	99.1	89.5	91.7	104.3	93.9	10
Durris		89.4	91.6		93.8a	2.1
Eitshal	99.4	89.8	92.0	95.1	94.2de	2
					104.3m	2
Forfar	97.9	88.3	90.5	94.9	92.7	17
Fort William		89.3m	91.5m		93.7md	1.5
					98.9m	1.5
Glengorm		89.5m	91.7m		93.9md	4.6
					99.1m	4.6
Keelylang Hill		89.3	91.5		93.7ab	41
Meldrum	98.3	88.7	90.9	95.3	93.1a	150
Melvaig	98.7	89.1	91.3	95.7	93.5de	50
					103.9m	50
Oban		88.9m	91.1m		93.3md	3.6
					98.5m	3.6
Rosemarkie	99.2	89.6	91.8	103.6	94.0d	20
Rumster Forest		90.1	92.3		94.5d	10
Sandale	97.7	88.1	90.3	92.5	94.7f	250
Skriaig	98.1	88.5	90.7	94.8	92.9de	30
					104.7	30
So. Knapdale		89.3	91.5		93.7	1.1
					98.9m	1.1

+ low power tr's less than 1kW.
RS/L = R. Scotland + alternative & local prgrs — a) BBC Aberdeen — b) BBC Orkney — c) BBC Shetland — d) BBC Highland — e) R. nan Gaidheal — f) BBC Solway — g) BBC Tweed.

D.PRGR: 0600 (Sat/Sun 0630)-2400. Own prgrs on 810kHz + RS freqs: Mon-Fri 0610-1300, 1340-1402, 1635-1800, 1830-1920, 2200-2400; Sat 0740-1800, 1815-1935; Sun 0755-1705, 1750-1800. Other times relay R4 UK. **FM:** as 810kHz exc. separate prgrs on RS/L freqs: Mon-Fri 0600-0720, Sat 1330-1400, Sun 1530-1600. Rel. R4 UK Mon-Fri 2330-0023, Sun 1600-1800.
Local Services (Mon-Fri only). Freqs as above.
a) Beachgrove Terrace, Aberdeen AB9 2ZT. 0655-0700, 0730-0800, 1202-1300, 1655-1700, 1730-1755 — b) Commercial Assurance Bldg, Castle Str, Kirkwall, Orkney KW15 1DF. 0810-0845 — c) Brentham House, Lerwick, Shetland ZE1 0LR. 1730-1800, 1815-1930 — d) 7 Culduthel Rd, Inverness IV2 4AD. 0645-0700, 0730-0800, 1330-1400, 1700-1730 — e) Rosebank House, Church Str, Stornoway, Lewis PA87 2LS; Clydesdale Bank Bldg., Portree, Skye IV51 9EH. 0810-0900, 1200-1300, 1730-1800, 1815-1930, 2210-2400 — f) Elmbank, Lovers Walk, Dumfries DG1 1NZ. 0730-0800, 1200-1300 — g) Old Municipal Bldg, High Str, Selkirk TD7 4BU. 0730-0800, 1730-1800.

BBC RADIO WALES/RADIO CYMRU

ADDR: Broadcasting House, Llandaff, Cardiff CF5 2YQ. **Te:** +44 (222) 572 888. **Fax:** +44 (222) 555 286.
Mediumwaves: 882kHz: Penmon 10kW, Tywyn 5kW, Washford 100kW, Forden 1kW — Llandrindod Wells 1125kHz 1kW.

FM stereo	**R1**	**R2**	**R3**	**R4**	**R. Cymru**	**kW ERP**
Blaenplwyf	98.3	88.7	90.9	104.0	93.1	250
Carmel		88.4	90.6		92.8	3.2
Haverfordwest	98.9	89.3	91.5	104.9	93.7	20
Llanddona	99.4	89.8	92.0	103.6	94.2	21
Llandrindod Wells	98.7	89.1	91.3	103.8	93.5	2.8
Llangollen	98.5	88.9	91.1	93.3	104.3	15.6
Wenvoe	99.5	89.9	92.1	94.3	96.8	250

+ low power tr's less than 1kW.

D.PRGR: R. Wales on 882/1125kHz: Mon 0625-2202, Tues 0625-2300, Wed 0625-2230, Thurs 0625-2202, Fri 0625-2400, Sat 0659-1930, Sun 0730-1930. Other times rel. R.2 or R.4 prgrs.

R. Cymru on FM: Mon-Fri 0630-1830 & 2200-2330, Sat 0755-1800, Sun 0755-2330. Other times rel. R. Wales or R.4 prgrs.

BBC R. Clwyd

Addr: The Old School House, Glanrafon Rd, Mold, Clwyd CH7 1PA. **Station:** Wrexham 657kHz 2kW.
D.Prgr: Mon-Fri 0700-0730, 0800-0830, 0900-1030, 1655-1700; Mon also 1830-1900. Other times rel. R. Wales — **V.** by letter.

BBC RADIO ULSTER

ADDR: Broadcasting House, Ormeau Avenue, Belfast BT2 8HQ. **Te:** +44 (232) 244 400. **Fax:** +44 (232) 243 957.
Mediumwaves: Enniskillen 873kHz 1kW; Lisnagarvey 1341kHz 100kW.

FM stereo	**R1**	**R2**	**R3**	**R4**	**R. Ulster**	**kW ERP**
Brougher Mountain		89.4	91.6	—	93.8	9.8
Divis	99.7	90.1	92.3	94.5	96.0*	250/125*
Limavady		89.6	91.8	94.0	95.4	3.4
Londonderry	98.3	88.7	90.9	—	93.1h	31
				94.9		10

+ low power tr's less than 1kW — h) R. Foyle (see below).
D.PRGR: Own prgrs: Mon-Fri 0700-1930 & 2300-2400, Sat 0800-1800 & 2300-1400, Sun 0900-1900 & 2030-2400. Other times rel. R2 or R4.
V. by QSL-card.

BBC R. Foyle

Addr: 8 Northland Rd, Londonderry BT48 7NE.
Stations: Londonderry 792kHz 1kW, 93.1MHz 31kW.
D.Prgrs: As R. Ulster, exc. own prgrs: Mon-Fri 0830-0900.

V. Reports on national prgrs to Engineering Information Dept, White City, 201 Wood Lane, London W12 7TS. To all other sces, address reports to Engineer-in-Charge at local address.

BBC WORLD SERVICE

ADDR: Bush House, The Strand, London WC2B 4PH. **Te:** +44 (71) 240 3456. **Fax:** +44 (71) 240 8760. **Telex:** 265781.
L.P: Head of Prgrs. (English): Ernest Warburton. CE: Gordon Harold. Gead of PR: Richard McCarthy.

LONG & MEDIUMWAVES: Droitwich 198kHz 500kW — Orfordness 648/1296kHz 500kW.

SHORTWAVES: Rampisham (G.C: 02.38W/50.48N): 500kW — Skelton (G.C: 02.55W/54.44N): 250kW — Woofferton (G.C: 02.43W/52.19N): 300/250kW.

Overseas Relay St's: see listings under respective countries.

Rebroadcasters: BBC World Sce. prgrs in English and other languages are relayed by local st's in many countries. Full details can be found in BBC Worldwide magazine, available on subscription.

Frequencies used by tr's in the UK:

kHz	kHz	kHz	kHz	kHz	kHz	kHz
3955	6050	7200	9585	11680	13745	17695
3975	6110	7210	9600	11750	15070	17705
5875	6125	7260	9605	11760	15115	17715
5955	6150	7270	9635	11780	15135	17750
5970	6195	7285	9670	11815	15145	17815
5975	7105	7295	9750	11825	15165	17875
5990	7120	7320	9760	11835	15180	21640
6010	7125	7325	9770	12040	15235	21735
6025	7150	9410	9825	12095	15340	21745
6030	7175	9560	9915	13660	17640	

Winter Schedule 1992/1993

Relays: A = Ascension Isl, B = Bethany (USA), C = Cyprus, D = Delano (USA), E = Eastern Relay St. (Masirah Isl.), F = Far Eastern Relay (Singpore), G = Greenville (USA), H = Hong Kong, I = Indian Ocean Relay (Seychelles), L = Lesotho, R = Rangitaki (New Zealand), S = Sackville (Canada), T= Tashkent, W = West Indian Relay (Antigua) — sx) exc. Sat.

WORLD SERVICE IN ENGLISH

No. Eu/We. Eu

0300-0400	7120
0300-0445	1296
0300-0800	9410
0330-0500	6195
0400-0530	6180C
0430-0815	9410
0500-0530	6195
0500-0530	6195
0600-0730	15575C
0600-0730	6180C
0600-0730	6195
0600-0730	6195
0700-0730	1296
0700-0815	7325
0700-1030	12095
0700-1030	9760
0700-1129	648
0730-0900SS	9660C
0800-1500	17640
0800-1615	15070
0815-0915	7325
0900-0915	9410
0900-1030	9410
0900-1030	9750
0900-1515	9660C
1030-1200Sun	9410
1030-1200Sun	9750
1030-1200Sun	9760
1200-1400	648
1200-1615	9750
1200-1615	9760
1200-2030	9410
1400-1430sx	648
1430-1615	648
1500-2030	6195
1600-1615W	1296
1615-2200	9410
1700-2030	6180C
1700-2030	6195
2000-2030	6180C
2000-2200	7325
2055-2100sx	648
2100-0415	648
2100-2200	6180C
2100-2200	6180C
2100-2315	6195
2100-2315	6195
2200-2315	9410
2300-2315	1296
2345-0450	198

Ce./Ea. Europe

0030-0145	5965C
0145-0200SS	5965C
0400-0430	9410
0600-1030	15070
0800-0900	9410
1030-1200Sun	12095
1030-1200Sun	15070
1030-1200W	9750
1200-1745	12095
1200-2030	15070
1830-2200	9410
2200-0300	7325

Russia

0300-0330	6195C
0300-0530	3955
0330-0500	6195
0430-0815	9410
0500-0530	6195
0600-0730	3955
0600-0730	6195
0800-1500	17640
0800-1615	15070
0900-1030	9410
1030-1200Sun	9410
1200-2030	9410
1500-2030	6195
1700-2315	3955
2000-2030	6180C
2100-2200	6180C
2100-2315	6195

So. Eu/No. Africa

0400-0700	12095
0500-0530	3955
0600-0630	3955
0600-0730	15575C
0700-0815	15070
0700-2315	12095
0800-1515	17640
0815-2030	15070
0900-1615	17705
2000-2030	3955
2100-2315	3955

We. Africa

0400-0430	9610A
0500-0600	9600A
0600-0630MF	15400A
0600-0630MF	9600A
0700-0815	15400A
0800-0815	11860A
0815-0900SS	15400A
0900-1130	15400A
1300-1345	15105A
1500-1700	15400A
1615-1715	21470A
1745-1830	15400A
1900-1930MF	15400A
2030-2200	15400A
2200-2315	15400A

Ce. Africa

0300-0330	6190L
0300-0330	9600A
0400-0430	6190L
0400-0430	7105A
0400-0430	9600A
0500-0600	6190L
0500-0600	15400A
0600-0630MF	11940L
0600-0630MF	15400A
0700-0730	11940L
0700-0730	17790
0700-0815	15400A
0730-0815	17790A
0730-0815	21660A
0800-0815	11940L
0815-0900SS	11940L
0815-0900SS	17790A
0815-0900SS	21660A
0900-1500	17790A
0900-1700	11940L
0900-1745	21660A
1400-1700	17880A
1500-1515MF	17790A
1600-1700	17860A
1745-1830	11940L
1745-1830	15400A
1745-2030	17880A
1900-1930MF	6190L
1900-1930MF	15400A
2030-2200	15400A

Ea. Africa

0200-0330	11730I
0300-0330	9670C
0400-0430	15420I
0400-0430	17885I
0430-0730	21470C
0500-0545	15420I
0500-0545	17885I
0545-0600SS	15420I
0545-0600SS	17885I
0545-0630MF	15420I
0545-0630MF	17885I
0630-0730MF	15420C
0630-0730MF	17885C
0700-0900SS	15420I
0700-0900SS	17885I
0730-0900SS	21470C
0900-1130	15420I
0900-1400	17885I
0900-1615	21470C
1130-1215W	15420I
1245-1300W	15420I
1300-1400	15420I
1500-1515MF	15420I
1615-1700	15420I
1615-1700	9630I
1900-1930MF	6005I
1900-2030	9630I
2100-2200	6005I

So. Africa

0000-0300	6005A
0300-0330	3255L
0300-0330	6190L
0300-0330	9600A
0400-0430	90200L
0400-0430	1197L
0400-0430	3255L
0400-0430	6190L
0400-0430	9600A
0500-0545	3255L
0500-0545	6190L
0500-0700	1197L
0545-0600	3255L
0545-0600	6190L
0550-0600	15400A
0600-0630MF	11940L
0600-0630MF	15400A
0600-0630MF	6190L
0700-0730	11940L
0700-0815	15400A
0700-0815	6190L
0730-0815	21660A
0800-0815	11940L
0815-0900SS	11940L
0815-0900SS	21660A
0815-0900SS	6190L
0900-1700	11940L
0900-1700	6190L
0900-1745	21660A
1300-1700	1197L
1400-1700	17880A
1745-1830	11940L
1745-1830	15400A
1745-1830	1197L
1745-1830	6190L
1745-2030	17880A
1830-2030	3255L
1900-1930MF	1197L
1900-1930MF	6190L
1900-1930MF	15400A
2030-2200	15400A
2100-0330	1197L
2100-2200	3255L
2100-2200	6005I

Middle East

0030-0145	5965C
0030-0145	9580C
0145-0200SS	5965C
0145-0200SS	9580C
0200-0230	9580C
0200-0300	6195C
0200-0330	702E
0230-0430	11955C
0300-0330	639C
0300-0330	720C
0300-0330	6195C
0300-0400	1413E
0300-0730	1323C
0300-0815	11760E
0500-0530	6195
0600-0630	720C
0600-0815	639C
0600-0815	6195
0730-0900SS	1323C
0815-1030FSS	639C
0900-1400	11760E
0900-2315	1323C
1030-1130	639C
1130-1215W	639C
1215-1245	639C
1500-2030	6195
1745-1830	1413E
1900-2030	1413E
2100-2315	6195

We. Asia

0030-0145	5965C
0030-0145	9580C
0030-0145	15310C
0145-0200SS	5965C
0145-0200SS	9580C
0145-0200SS	15310C
0200-0230	9580C
0200-0300	15310C
0200-0300	6195C
0200-0330	7135C
0230-0430	11955C
0300-0330	6195C
0300-0400	1413E
0300-0815	11760E
0400-0730	15575C
0730-0900SS	15575C
0900-0930	15575C
0900-1400	11760E
0930-1500	15575C
1715-1830	7160E
1745-1830	1413E
1830-2030	9740C
1900-2030	1413E
1900-2030	7160E

So. We. Asia

0000-0030	1413E
0030-0145	11955E
0145-0200SS	11955E
0200-0230	11955E
0200-0230	15380F
0200-0230	1413E
0300-0330MF	15380F
0300-0815	15310E
0900-1515	15310E
1100-1615	11750F
1300-1400	1413E
1400-1405MF	1413E
1515-1700	7215E
1515-1830	9740F
1800-1830	11750E

So. Asia

0000-0030	5965E
0000-0030	9580E
0030-0145	11955E
0145-0200SS	11955E
0200-0230	11955E
0300-0815	15310E
0600-0815	17790E
0900-1030	11750F
0900-1100	17790E
0900-1400	9740F
0900-1515	15310E
1030-1100	11750F
1400-1500SS	9740F
1500-1515	9740F
1515-1830	15310E
1615-1700	5975F
1615-1830	9740F
1715-1830	5975F

So. Ea. Asia

0300-0815	15310E
0600-0815	17790E
0900-1030	11750F
0900-1100	17790E
0900-1400	9740F
0900-1515	15310E
0900-1615	6195F
1030-1100	11750F
1400-1500SS	9740F
1500-1515	9740F
1515-1830	15310E
1600-1745	3915F
1615-1700	5975F
1715-1830	5975F
2100-0030	6195F
2300-0030	11955F

Indonesia

0000-0330	15360F
0900-1030	11750F
0900-1030	9740F
0900-1615	6195F
1515-1615	9740F
2100-0030	6195F
2200-2400	9570F

China

0000-0030	9570F
0100-0300SS	15280H
0100-0330	17790F
0300-0915	15280H
0900-0915	11765H
1300-1400	7180H
1400-1500SS	7180H
1500-1615	7180H
2300-0030	11945H
2300-2400	9570F

Far East

0100-0300SS	21715H
0300-0900	21715H
0600-0915	15360F
0600-1030	17830F
0900-1030	21715F
1030-1400	9740F
1300-1400	11820H
1400-1500SS	9740F
1400-1500SS	11820H
1500-1515	9740F
2100-2200	7180H
2200-2300	11955F
2300-0030	11955F
2300-0030	15280H

Oceania

0600-0815	7150
0600-0915	11955F
0600-1030	17830F
1030-1400	9740F
1100-1130	9700R
1400-1500SS	9740F
1500-1515	9740F
1800-2200	11955F
2000-2315	15340F
2200-0030	11955F

No. America

2200-0430	7325
2200-0430	9915

No. America (Ea.)

1100-1400	15220W
1300-1830	9515S
1400-1615	17840W
2100-0030	9590S
2300-0330	6175S

No. America (We.)

1100-1400	5965S
2200-0430	7325
1400-1745	15260S
1745-1830	15999S

Ce. America

0030-0145	9590D
0145-0200SS	9590D
0200-0230	9590D
0230-0430	9915
0430-0730	5975W
0500-0815	9640W
2200-0230	9915

Caribbean

1100-1430	6195W
2000-0430	5975W

Ce./So. America

0500-0815	9640W
1100-1430	6195W
2000-0430	5975W
2200-0430	7325

So. America

0300-0330	6005A
0900-1100	15190A
1100-1400	15220W
1400-1615	17840W
2000-0330	15260A
2200-0330	9915
2200-0330	11750A

World N: 15 min: 0400, 1500. 9 min: 0300, 0600, 0800, 1200W, 1600, 2000, 2200. 5 min: 0900, 1400MF, 1700 exc. Sat, 1900MF, 2300, 0100 Tues-Sat. **Headlines:** 0100 (Sun/Mon), 1000, 1200Sun, 1400SS, 1700Sat, 1900SS. **Newshour:** 0500, 1300, 2100. **Newsdesk** (30 mins including correspondents' reports and British news & press review): 0000, 0200, 0700, 1100, 1800. **N. About Britain:** 0309, 0609, 1609, 2209. **The World Today:** Mon-Fri 1645, 2009; Tues-Sat 0615. **Worldbrief:** Fri 2315; Sat 0445, 0915. **World Business Report:** Mon-Fri 0905, 1705, 2305. Sat 0905. **World Business Review:** Sun 0905, 1705, 2305.

ALTERNATIVE ENGLISH PRGRS.

Europe	**kHz**
0430-0500	198, 648, 3955, 3975, 6010, 6180C, 6195
1930-2000	648, 3955, 6180C, 6195
Africa	
0330-0400	1197L, 3255L, 6005A, 6190L, 9600A, 11730I, 15420A
0415-0500	1197L, 3255L, 6005A, 6190L, 9600A, 15400A, 15420I
0600-0630SS	6190L, 9600A, 11940L, 15400A, 15420I, 17885I
0630-0700	6190L, 9600A, 11940L, 15400A, 17885I
0730-0800	11860A, 11940L, 15105A
1500-1515SS	11860I, 15420I, 17790A, 17895C, 21490A
1515-1530	11860I, 15420I, 17790A, 17895C, 21490A
1700-1745	1197L, 6190L, 9630I, 11940L, 15400A, 15420I, 17860A, 17880A
1830-1900	1197L, 6005A, 6190L, 9630I, 15400A
1900-1930SS	1197L, 6005A, 6190L, 15400A
1930-2030	1197L, 6005A, 6190L, 15400A
2030-2100	1197L, 3255L, 6005A
So. Asia	
0145-0200*	9580C, 11955E, 15310C
1700-1715	1413E, 5975F, 7215F, 9605F, 11750F
Ea. Asia	
1400-1415MF	7180H, 9740F, 11820H
Caribbean	
0145-0200*	9590D
2115-2130MF	6110W, 15390A, 17715G
Falkland Islands	
2130-2145	13660 (Tues/Fri only)

*) Tues-Sat

FOREIGN LANGUAGE SERVICES TO EUROPE

Bulgarian	**kHz**
0430-0445	5875, 6050C, 9770C
1645-1715	6085C, 9915, 11750
1715-1730SS	6085C, 9915, 11750
2015-2115	6050C, 7105, 9770
Croatian	
1200-1215	11780, 13660, 15145
1800-1830	6050C, 7105, 9770
2030-2045MF	6125C, 9915, 11835
Czech	
0515-0530MF	1296, 5875, 7260
0615-0630	1296, 7260, 15325C
1615-1630Sun	5875, 7210, 11680
1630-1700	5875, 7210, 11680
1900-1945	1296, 5875, 7210
2200-2215	6180C, 7230C, 9770C
Finnish	
0530-0545MF	6150, 11845
0900-10000Sun	11825, 15165
0930-0945MF	11680, 15340
1030-1045MF	11815, 15340
1600-1630	9915, 11780
1630-1645SS	9915, 11780
French	
0630-0700	648
1730-1800	648
German	
0445-0530	648, 1296(to 0500), 3975, 9750, 90.2MHz
1145-1200	648, 6125, 9600, 90.2MHz
1615-1630	1296, 6125, 9750, 90.2MHz
1630-1700	648, 1296, 6125, 9750, 90.2MHz
1745-1800	90.2MHz
1830-2000	648, 3975, 6125, 90.2MHz
Greek	
0600-0630MF	6085C, 7210, 9560C
1300-1315	9560C, 15145, 17875
1530-1600MF	6050C, 11760, 17695
2000-2015	6125C, 9915, 11835
2015-2045SS	6125C, 9915, 11835
2230-2300SS	6180C, 7105, 9635
Hungarian	
0530-0545MF	1296, 5875, 7260
0630-0645	1296, 5875, 7260, 15325C
1000-1130Sun	9635, 11680, 13745
1315-1400Sun	9635, 11680, 13745

1800-1900	1296, 5875, 7210
2200-2245	1296. 5875. 9735C
Polish	
0500-0515W	1296, 5875, 7260
0600-0615W	1296, 5875, 7260
0700-0730Sun	5875, 7260, 9825
1130-1230Sun	9635, 11680, 13745
1400-1500Sun	9635, 11680, 13745
1500-1530	1296, 9635, 11680, 13745
1700-1800	1296 (to 1730), 5875, 7210, 9750
2030-2200	1296, 3975, 5875
Portuguese	
2030-2115	5975, 7150, 9605
2330-2400	6030, 7175
Romanian	
0400-0430MF	6050C, 7210, 9750
0600-0630SS	6085C, 7210, 9560C
1600-1645	6050C, 9770, 11760
1900-2015	6050C, 7105, 9770
Russian	
0330-0430	5875, 6010, 7320, 9585, 11750C
0700-0715	11680, 13660, 15135, 15325C, 17750, 21745
0900-0915	11680, 13745, 15135C, 15340, 17750, 21745
1030-1100Sun	11815, 11970C, 15340, 17750, 21745
1100-1115	11815, 11970C, 15340, 17750, 21745
1300-1400	11845C, 15115, 15165C, 17695, 17815, 21735
1600-1700	9635C, 9825, 11845C, 12040, 13745
1700-1900	7120, 9635C, 9825, 11845C, 12040
1900-2030	5990, 6080, 7120, 9635C, 9825, 11845E
2030-2100	5990, 6010C, 7120, 9635C, 9750, 9825, 11845E
2100-2130	5990, 6010C, 7120, 9585E, 9635C, 9825, 11845E
Serbian	
1130-1145	11780, 13660, 15235
1700-1730	6050C, 7105, 9770
2015-2030MF	6125C, 9915, 11835
Slovak	
0545-0600MF	1296, 5875, 7260
0715-0730	6045, 7295, 9915
1530-1600sUN	1296, 5875, 7210, 11680
2000-2030	1296, 5875, 7210
Slovene	
1030-1100Sun	11780, 13660, 15235
1100-1115	11780, 13660, 15235
1730-1800	1296, 6050C, 7105, 9770
Turkish	
0515-0530W	6010, 6050C, 7270
0800-1000Sun	1296, 6015C, 9635, 9740C, 17695
1730-1800	6085C, 7270C, 9915
2045-2115	6125C, 7125, 9915
2115-2130W	6125C, 7125, 9915

TO AFRICA/MIDDLE EAST

Arabic	
0320-0330	9825, 11740C, 15235E, 15590C
0330-0445	639C, 702E, 720C, 7140C, 9825, 11740C, 15235E, 15590C
0445-0600	639C, 720C, 6110, 7140C, 7325, 9825, 11740C, 15235E, 15590C
1250-1615	639C, 702E, 720C, 7140C, 13660, 15590C, 17715, 17785C
1615-2100	639C, 702E, 720C (exc. 1830-1900SS), 6030E, 7140C, 9915 (to 2000), 11680 (from 1800), 11730C, 13660
French	
0430-0500	6155A, 7105A, 9610A, 17885I
0500-0530	5955, 7105A, 7285, 9610A, 9915, 11860A
0600-0630	6110, 7105A, 7295, 9610A, 9915, 11860A
0630-0700	7105A, 9610A, 11860A
1200-1245	15105A, 15180, 17715, 21640A
1215-1245	15420A
1815-1830	9630I, 11820A, 11850C, 15105A, 17830A
1830-1900	11820A, 11850C, 15105A, 17830A
1900-1915	11820A, 15105A, 17830A
Hausa	
0530-0600	7105A, 9610A, 11860A
1345-1415	15105A, 17810A, 21640
1915-1945	11880A, 15105A, 17830A
Portuguese	
0430-0500	6155A, 9610A, 11940L
0530-0600	6005A, 9610A, 15105A
1700-1730	6070, 7230, 9515
2030-2115	6190L, 9630I, 11820A, 15160A
Somali	
1430-1500	11860I, 15420I, 17895C, 21490A
1500-1515MF	11860I, 17895C, 21490A
1800-1830	6005I, 15360C
Swahili	
0315-0330	9615C, 11845C, 15420I
1530-1615	11860I, 15420I, 21490A
1745-1800	6005I, 7230, 9630I

TO ASIA

Bengali	
0030-0045	9600E, 11850E, 15380F
1330-1410	9605F, 11920F, 15245E
1630-1700	6065F, 7105F, 9605F
Burmese	
0000-0030	9600F, 11850F, 15380F
0245-0300SS	11850F, 15380F
1345-1430	6065F, 11765F, 15360F
Cantonese	
1300-1345	11765F, 15360F
Chinese (Mandarin)	
1000-1300	7180H, 11765H, 11955F, 15360F
2230-2300	5965F, 7180H, 9580H, 11750F
Hindi	
0045-0135	1413E, 7235C, 7315T, 9600E, 11850E, 15380F
1410-1500	1413E, 7215F, 9605F, 9725T, 11920F, 15245E
1715-1730	1413E, 7215F, 9605F, 11750F
Indonesian	
1100-1130	6065F, 9605F, 11920F, 21590
1300-1330	6065F, 9605F, 11920F, 15125E
2215-2245	6080F, 7160F, 11865F
Nepali	
1500-1520	6065F, 11920F, 15360F
Pashto	
0200-0230	7235C, 9600E, 11850E
0300-0315	7260C, 15590C, 11740E
0900-0915	11775C, 15590C, 17725C
1545-1615	1413E, 7160E, 11720E, 15245C
Persian	
0230-0300	720C, 1413E, 7235C, 9590E, 11740E
0915-0930	11775C, 15590C, 17725C
0930-1100Fri	11775C, 15590C
1615-1700	1413E, 7160E, 11720C
1700-1715	7160E, 11720C
1830-1900	720C, 1413E, 5975E, 7160E
Sinhala	
1520-1530	6065F, 7105F, 11920F
1530-1545SS	6065F, 7105F, 11920F
Tamil	
1545-1615	6065F, 7105F, 9605F
Thai	
1230-1245	6065F, 9605F, 11920F, 15590C
1615-1630	6065F, 7105F, 9605F
2330-2400	6080F, 7160F, 11865F
Urdu	
0135-0200	1413E, 7315T, 7235C, 9600E, 11850E, 15380F
1500-1545	1413E, 7160F, 9605F, 9725T, 11720C, 15245C
1730-1745	1413E, 7215F, 9605E, 11750F
Vietnamese	
1130-1200	6065F, 9605F, 11920F
1430-1500	6065F, 11765F, 15360F
2300-2330	6080F, 7160F, 11865F

TO LATIN AMERICA

Portuguese	
2130-2215	6110W, 9560A, 9825, 11765A, 15390A
Spanish	
1100-1130MF	9690D, 15190A, 21490A
1300-1330MF	6130D, 9690D, 15315B
0000-0115	5875, 6110W, 6110A, 11765A, 15390A
0115-0130*	5875, 6110W, 6110A, 11765A, 15390A
0300-0415	6110W, 6190, 9515D, 11965D, 15390A
0415-0430*	6110W, 6190, 9515D, 11965D, 15390A

*) Tues-Sat

BBC ENGLISH (Language Courses)

To Europe	
0530-0545	6010, 6050C, 7270
0645-0700	1296, 5875, 7260, 15325C

0730-0745	1296, 6010, 7295, 9825
1130-1145	90.2MHz (B)
1215-1230Sun	1296, 6125, 9600
1230-1300	1296, 6125, 9560, 9600, 11710C, 15115, 17875
1615-1630	648
1700-1730	6125, 7270
1715-1730MF	648
1730-1745	90.2MHz (B)
2130-2200	6125, 7125, 9770C
2245-2300	1296
To the CIS	
0715-0730	11680, 13660, 15135, 15325C, 17750, 21745
0915-0930	11680, 13745, 15135C, 15340, 17750, 21745
1115-1130	11815, 11970, 15125, 15340, 17750, 21745
1230-1300	11845C, 15115, 17695, 21735
To Africa	
0245-0300	11845I, 15420I
0630-0645	6110, 7295, 9915
0715-0730	11860A, 15105A
1400-1430	11860I, 15420I, 17895C, 21490A
2000-2030	1197L, 6005I, 6190L, 15400A
To Near East	
1215-1250FS	702E, 720C, 7140C, 13660, 15590C, 17785C
To Asia	
2200-2215	6080F, 7160F, 11865F
2200-2230	7180F, 9580H, 11750F, 11945H
2245-2300	6080F, 7160F, 11865F
2330-2345	6065F
0030-0045	6195F, 9570F, 11945H, 15280F, 17830H
0045-0100	17830H, 21715H
0230-0245	11850E, 15380F, 17780E
0945-1000	7180H, 11765H, 11955F, 15360F
1200-1215	6095, 9605F, 11920F
1415-1500MF	7180H, 9740F, 11820F
1530-1545MF	6065f, 7105F, 11920F
1745-1800	7215F, 9605F, 11750F
To South America	
2315-2330	6110W, 9560A, 9825, 11765A, 15390A
2345-2400	5875, 6110W, 6110A, 11765A, 15390A
0245-0300	5875, 6110w, 6110A, 9515D, 11965D, 15390A

INT-SIG: World Sce. in English: Bow Bells prior to transmission. Other Services: BBC in tonic scale.
PUB: "BBC Worldwide" (World Sce. Radio & TV in English) available on subscription. Full details from PO Box 76S, Bush House, Strand, London WC2B 4PH — BBC English Magazine (for those learning English). Details from PO Box 96, Cambridge — "Huna London" (Arabic Sce.) — Many other publications available from BBC World Sce. mail order, Bush House, Strand, London WC2B 4PH (ask for catalogue).
F.PL: Albanian sce.to start in 1993.

THE RADIO AUTHORITY (Regulatory Authority)

ADDR: Holbrook House, 14 Great Queen Str, London WC2B 5DG. **Te:** +44 (71) 430 2724. **Fax:** +44 (71) 405 7064 — Chmn: Lord Chalfont. Chief Exec: Peter Baldwin.

ASSOCIATION OF INDEPENDENT RADIO COMPANIES (AIRC)

ADDR: 46 Westbourne Grove, London W2 5SH. **Te:** +44 (71) 727 2646. **Fax:** +44 (71) 229 0352 — Dir: Brian West. (AIRC is the trade association of the independent radio companies).

INDEPENDENT RADIO NEWS (IRN)

ADDR: ITN House, 200 Grays Inn Road, London WC1 — MD: John Perkins (IRN provides national N. to most st's).

NETWORK NEWS

ADDR: Broadcast Centre, Crownhill, Milton Keynes MK8 0AB. **Te:** +44 (582) 666884. **Fax:** +44 (908) 564893 — Joint MD's: P. Chantler, J. Davies (national and international N. sce).

CLASSIC FM (Comm.)

ADDR: Academic House, 24/28 Oval Rd, London NW1 7DQ. **Te:** +44 (71) 284 3000. **Fax:** +44 (71) 284 2835.
L.P: Chmn: H. Meakin. Chief Exec: J. Spearman. PD: Michael Bukht.

FM	MHz	kW	FM	MHz	kW
Belmont	100.5	16 (FP)	Oxford	101.3	46
Black Hill	101.7	250	Peterborough	101.9	40
Darvel	101.3	10 (FP)	Pontop Pike	100.3	150
Divis	101.9	60	Rowridge	100.3	250
Dover	101.8	11 (FP)	Sandale	99.9	250
Forfar	100.1	17 (FP)	Sheffield	101.7	0.3 (FP)
Holme Moss	101.0	250	Sutton Coldfield	100.1	250
Llandonna	101.6	21 (FP)	Swansea	100.3	1 (FP)
Londonderry	100.5	31 (FP)	Tacolneston	101.5	250
Meldrum	100.5	150 (FP)	Wenvoe	101.7	250
No. Hessary Tor	100.0	160 (FP)	Wrotham	100.9	250

FP = Planned to open in 1993.
D.PRGR: 24h. **N:** on the h.

INDEPENDENT MUSIC RADIO (Comm.)

ADDR: Hawley Crescent, London NW1 8EF. **Te:** +44 (71) 288 4347.
L.P: Chmn: B. Gyngell. MD: J. Aumonier.
STATIONS: Synchronised MW netw. on 1215 & 1197kHz. Some tr's already testing.
F.PL: National pop and rock music sce. to start March 1993.

OTHER STATIONS (all commercial) — (sh) = shared transmitter, (sy) = synchronised.

	kHz	kW	Name or Slogan	Location
30)	558	0.8	Spectrum R. Int.	London
1a)	603	0.1	Invicta Supergold	Littlebourne
44)	603	0.1	CD 603	Gloucester
2)	666	0.34	Classic Gold	Exeter
47)	756		R. Maldwyn	Powys
4g)	774	0.14	SuperGold	Gloucester
4a)	792	0.2	SuperGold	Bedford
6)	819		Sunrise 819	Ludlow
20b)	828	0.12	Magic 828	Leeds
4b)	828	0.2	Supergold	Luton
7g)	828	0.27	2CR Classic Gold	Bournemouth
7)	936	0.18	"Classic Gold"	West Wilts
10a)	945	0.2	Gem AM	Derby
8)	954	0.16	R. Wyvern	Hereford
2a)	954	0.32	Classic Gold	Torquay
9)	990	0.09	WABC	Wolverhampton
21b)	990	0.25	Classic Gold	Doncaster
10)	999	0.25	Gem AM	Nottingham
20c)	999	0.8	Red Rose Gold	Preston
9)	1017	2	WABC	Shrewsbury
12)	1026	1.7	Downtown R.	Belfast
13a)	1035	0.78	NorthSound R.	Aberdeen
14)	1035	0.32	West Sound R.	Ayr
15)	1107	1.5	Moray Firth R.	Inverness
16)	1152	23.5	London Talkback R.	London
13)	1152	3.6	Clyde 2	Glasgow
10c)	1152	3	Xtra AM	Birmingham
19a)	1152	0.83	R. Broadland	Norwich
20a)	1152	1.5	Piccadilly Gold	Manchester
21a)	1152	1.8	Great North R.	Newcastle
7b)	1152	0.32	2CR Classic Gold	Plymouth
21c)	1161	0.35	Classic Gold	Hull
23a)	1161	1.4	R. Tay	Dundee
7c)	1161	0.16	Classic Gold	Swindon
25)	1170	0.58	Swansea Sound	Swansea
21a)	1170	0.32	Great North R.	Stockton
27)	1170	0.2	Signal R.	Stoke-on-Trent
1)	1170	0.12	South Coast R.	Portsmouth
19b)	1170	0.28	SGR	Ipswich
1a)	1242	0.32	Invicta Supergold	Maidstone
7f)	1242	0.5	Isle of Wight R.	Isle of Wight
19c)	1251	0.76	SGR	Bury St. Edmunds
24)	1260	0.29	Sunrise Midlands	Leicester
32)	1260	0.64	Marcher Gold	Wrexham
7a)	1260	1.6	Brunel Classic Gold	Bristol
21d)	1278	0.43	Classic Gold	Bradford
20e)	1305	0.2	Touch AM	Newport
21b)	1305	0.15	Classic Gold	Barnsley
1c)	1323	0.5	South Coast R.	Brighton
37)	1332	0.6	WGMS	Peterborough
10d)	1359	0.27	Xtra AM	Coventry
20e)	1359	0.2	Touch AM	Cardiff
4d)	1359	0.28	Breeze AM	Chelmsford
24)	1413	0.125	Sunrise R.	Southall

7e)	1431	0.14	210 Classic Gold	Reading
4d)	1431	0.35	Breeze AM	Southend
42a)	1476	1	County Sound R.	Guildford
42)	1521	0.64	County Sound R.	Reigate
21d)	1530	0.74	Classic Gold	Huddersfield
8)	1530	0.52	R. Wyvern	Worcester
35)	1530	0.04	KCBC R.	Kettering (sy)
35)	1530	0.01	KCBC R.	Corby (sy)
43)	1548	97.5	Capital Gold	London
21b)	1548	0.74	Classic Gold	Sheffield
45)	1548	4.4	City Talk	Liverpool
23)	1548	2.2	Max AM	Edinburgh
4c)	1557	0.76	Northants Supergold	Northampton
1)	1557	0.5	South Coast R.	Southampton
1b)	1557		Mellow 1557	Clacton
23a)	1584	0.21	R. Tay	Perth

	MHz	**ERP**	**Name or slogan**	**Location**
43)	95.8	4	Capital FM	London
1a)	95.9	0.27	Invicta FM	Thanet
1a)	96.1	0.25	Invicta FM	Ashford
21b)	96.1	0.05	Hallam FM	Rotherham
10)	96.2	1	Trent FM	Nottingham
17)	96.2	4	SIBC	Bressay
52)	96.2		Lantern 96.2FM	No. Devon
20b)	96.3	0.6	Aire FM	Leeds
7a)	96.3	2	GWR FM	Bristol
4d)	96.3	1.4	Essex R.	Southend
38)	96.3		Q96	Paisley
2a)	96.4	0.25	Devonair R.	Torquay
12)	96.4	2	Downtown R.	Limavady
10c)	96.4	9.8	BRMB FM	Birmingham
25)	96.4	1	Swansea Sound	Swansea
19c)	96.4	2	SGR-FM	Bury St. Edmunds
42a)	96.4	3	Premier R.	Guildford
23a)	96.4	0.35	R. Tay	Perth
27)	96.4	0.25	Signal Cheshire	Congleton
7)	96.5	0.06	GWR FM	Marlborough
51)	96.5	1	Orchard FM	Taunton
10)	96.5	0.02	Trent FM	Mansfield
50)	96.5	0.05	R. Wave	Blackpool
12)	96.6	10	Downtown R.	SW Ulster
4c)	96.6	4	Northants Radio	Northampton
7d)	96.6	0.04	R. in Tavistock	Tavistock
21e)	96.6	8.9	TFM	Stockton
53)	96.6		R. Ceredigion	?
14)	96.7	0.8	West Sound R.	Ayr
1)	96.7	0.5	Ocean Sound Classic Hits	Winchester
45)	96.7	8.2	City FM	Liverpool
46)	96.7	0.3	Classic Tracks BCR	Belfast
23b)	96.7	0.05	Central FM	Stirling
37c)	96.7	2.5	KLFM	King's Lynn
23c)	96.8	5	R. Borders	Selkirk
21c)	96.9	10	Viking FM	Hull
4b)	96.9	0.89	The Hot FM	Bedford
13a)	96.9	0.6	NorthSound R.	Aberdeen
1c)	96.9	0.1	Classic Hits	Newhaven
27)	96.9	0.1	Signal R.	Stafford
31)	96.9	0.1	Choice FM	So.London
1a)	97.0	0.5	Invicta FM	Dover
2)	97.0	1	Devonair R.	Exeter
7b)	97.0	1	Plymouth Sound	Plymouth
10d)	97.0	2	Mercia FM	Coventry
7e)	97.0	0.5	210 FM	Reading
21a)	97.1	10	Metro FM	Newcastle
19b)	97.1	4	SGR-FM	Ipswich
32)	97.1	0.2	MFM	Wrexham
42b)	97.1	0.25	Delta R.	Haslemere
51)	97.1	0.4	Orchard FM	Yeovil
9)	97.2	2	Beacon R.	Wolverhampton
7c)	97.2	0.72	GWR FM	Swindon
14a)	97.2	1	South West Sound	Dumfries
4e)	97.2	1	Galaxy R.	Bristol
16)	97.3	4	LBC Newstalk	London
23)	97.3	9.8	RFM	Edinburgh
20d)	97.4	1.9	Red Rose Rock FM	Preston
12)	97.4	2	Cool FM	Belfast
15)	97.4	2.8	Moray Firth R.	Inverness
20e)	97.4	0.5	Red Dragon FM	Newport
21b)	97.4	0.4	Hallam FM	Sheffield
49)	97.4	0.3	Fox FM	Banbury
48)	97.4	0.05	CN-FM	Newmarket
14)	97.5	0.15	West Sound	Girvan
1)	97.5	0.85	Classic Hits	Portsmouth
21d)	97.5	0.5	The Pulse	Bradford
42)	97.5	0.03	R. Mercury	Horsham
23c)	97.5	0.01	R. Borders	Berwick
54)	97.5		Heartland R.	Pitlochry
4a)	97.6	1	The Hot FM	Luton
8)	97.6	0.78	R. Wyvern	Hereford
23)	97.6	0.2	RFM	Bathgate
11)	100.0	2	Kiss 100 FM	London
5)	102.0	0.025	Sunset R.	Manchester
1c)	102.0	0.2	Classic Hits	Hastings
3)	102.0		Spire FM	Salisbury
7)	102.2	0.5	GWR FM	We.Wilts
33)	102.2	2	Jazz FM	London
17)	102.2	0.01	SIBC	Lerwick
36)	102.2		Lincs FM	Lincoln
26)	102.2		Pirate FM	Liskeard
7g)	102.3	1	2CR	Poole
1c)	102.4	8.2	Classic Hits	Eastbourne
4g)	102.4	2	Hot FM	Gloucester
12)	102.4	10	Downtown R.	Londonderry
19a)	102.4	3.3	R. Broadland	Norwich
41)	102.4	0.04	Buzz FM	Birmingham
13)	102.5	20	Clyde 1	Glasgow
21d)	102.5	0.63	The Pulse	Halifax
27)	102.6	4	Signal R.	Stoke-on-Trent
4d)	102.6	2	Essex R.	Chelmsford
49)	102.6	9	Fox FM	Oxford
51)	102.6	4	Orchard FM	Taunton
40)	102.6	0.08	R. Harmony	Coventry
37)	102.7	4	Hereward R.	Peterborough
42)	102.7	3.6	R. Mercury	Reigate
1a)	102.8	1	Invicta FM	Dunkirk
10a)	102.8	0.6	Trent FM	Derby
8)	102.8	1	R. Wyvern	Malvern
23a)	102.8	1.3	R. Tay	Dundee
26)	102.8		Pirate FM	Redruth
21b)	102.9	0.5	Hallam FM	Barnsley
7e)	102.9	4	210 FM	Basingstoke
10d)	102.9	0.05	Mercia FM	Leamington Spa
20a)	103.0	4	Key 103	Manchester
4g)	103.0	0.1	Severn Sound	Stroud
7)	103.0	0.08	GWR FM	Bath
37b)	103.0	1	CNFM	Cambridge
21a)	103.0	0.05	Metro FM	Fenham
2a)	103.0	2	South West 103	Ea.Devon
9)	103.1	1	Beacon R.	Shrewsbury
1a)	103.1	4	Invicta FM	Maidstone
23c)	103.1	0.03	R. Borders	Peebles
10b)	103.2	0.4	Leicester Sound FM	Leicester
20e)	103.2	2	Red Dragon R.	Cardiff
1)	103.2	2	Power FM	Southampton
24a)	103.2	0.5	Sunrise FM	Bradford
34)	103.3	0.025	London Greek R.	No.London (sh)
18)	103.3	0.025	WNK R.	No.London (sh)
53)	103.3		R. Ceredigion	Aberystwyth
4f)	103.3	2.2	Horizon R.	Milton Keynes
21b)	103.4	1.5	Hallam FM	Doncaster
32)	103.4	1.4	MFM	Wrexham
23c)	103.4	0.5	R. Borders	Eyemouth
29)	103.4		Wear FM	Sunderland
1c)	103.5	0.9	Classic Hits	Brighton
41)	102.7		Channel R.	Jersey
22)	103.8	0.16	RTM	So.Ea. London
39)	104.7		Minster FM	York
28)	104.7		Island FM	Guernsey
27a)	104.9	0.05	Signal Cheshire	Stockport
48)	104.9	2	Melody R.	London

Addresses & other information.

H. of tr: 24h unless otherwise stated. Stations belonging to groups generally have some shared programming, especially overnight, but local ID's may still be heard (eg. before N. on the h.).

1) Southern Radio, Whittle Ave, Segensworth West, Fareham, Hants PO15 5PA —1a) Radio House, John Wilson Business Park, Whitstable, Kent CT5 3TX — 1b) 21-23 Walton Rd, Frinton-on-Sea, Essex CO13 0AA — 1c) PO Box 2000, Brighton BN41 2SS.

2) West Country Broadc. Ltd, The Studio Centre, 35-37 St. David's Hill, Exeter EX4 4DA — 2a) The Studios, Harbour Point, Victoria Terrace, Torquay.
3) City Hall Studios, Malthouse Lane, Salisbury, Wilts.
4) Chiltern R. Netw: 4a) Chiltern Rd, Dunstable, Beds LU6 1HQ — 4b) 53 Goldington Rd, Bedford MK40 3LS — 4c) The Enterprise Park, Boughton Green Rd, Northampton NN2 7AH — 4d) P.O. Box 300, Southend-on-Sea, Essex SS1 1SY — 4e) 2nd Flr, 25 Portland Sq, Bristol BS2 8RZ — 4f) Broadcast Centre, Crown Hill, Milton Keynes MK8 0AB — 4g) Old Talbot House, 67 Southgate Str, Gloucester GL1 2DQ.
5) 23 New Mount Str, Manchester M4 4DE.
6) Highridge House, The Sheet, Ludlow, Shropshire SY8 4JT.
7) GWR Group: 7a) PO Box 2000, Bristol BS99 7SN — 7b) Earl's Acre, Alma Road, Plymouth PL3 4HX — 7c) PO Box 2000, Swindon SN4 7EX — 7d) 3a North Str, Tavistock, Devon PL19 4AN. W 0700-0900, Sun 0800-1200. Other times rel. 7b FM —7e) PO Box 210, Reading, Berks RG3 5RZ — 7f) Dodnor Park, Newport, Isle of Wight PO30 5XE — 7g) 5 Southcote Rd, Bournemouth, BH1 3LR.
8) 5/6 Barbourne Terrace, Worcester WR1 3JZ.
9) 267 Tettenhall Rd, Wolverhampton WV6 0DQ.
10) Midlands Radio Group, 29-31 Castle Gate, Nottingham NG1 7AP — 10a) Market Place, Derby DE1 3AA — 10b) Granville House, Granville Rd, Leicester LE1 7RW —10c) PO Box 555, Radio House, Aston Rd. North, Birmingham B6 4BX — 10d) Hertford Place, Coventry CV1 3TT.
11) Kiss House, 80 Holloway Rd, London N7 8JG.
12) Kiltonga Industrial Estate, Newtonards, Co. Down, BT23 4ES.
13) R. Clyde Group, South Ave, Clydebank Business Park, Clydebank, Glasgow G81 2RX — 13a) 45 King's Gate, Aberdeen, AB2 6BL. Relays 13 0100-0600 — 13b) The Spencers, 20 Augusta Str, Birmingham B18 6JA.
14) Radio House, 54 Holmston Road, Ayr KA7 3BE — 14a) Campbell House, Bankend Rd, Dumfries DG1 4TH.
15) PO Box 271, Scorguie, Inverness IV3 6SF.
16) Crown Communications, 72 Hammersmith Rd, London W14 8YE.
17) Market Str, Lerwick, Shetland ZE1 0JN.
18) 185b High Rd, Wood Green, London N22 6BA. 0400-0800, 1200-1600, 2000-2400. **Turkish:** 0000-0100.
19) East Anglian R: 19a) St. George's Plain, 47-49 Colegate, Norwich, NR3 1DB —19b) Electric House, Lloyd's Ave, Ipswich IP1 3HZ — 19c) Long Brackland, Bury St Edmunds IP33 1JY.
20) Trans World Communications, 130 Broadway, Piccadilly House, Salford, Lancs M5 2UW — 20a) 127-131 The Piazza, Piccadilly Plaza, Manchester, M1 4AW — 20b) P.O. Box 2000, 51 Burley Road, Leeds LS3 1LR — 20c) PO Box 999, St. Paul's Square, Preston PR1 1XR — 20d) PO Box 974, St. Paul's Square, Preston PR1 1XS —20e) Radio House, West Canal Wharf, Cardiff CF1 5XJ.
21) Metro Radio Group: 21a) Radio House, Long Rigg, Swalwell, Newcastle-upon-Tyne NE99 1BB — 21b) 900 Herries Rd, Hillsborough, Sheffield S6 1RH — 21c) Commercial Road, Hull HU1 2SG — 21d) PO Box 3000, Bradford BD1 5NE — 21e) Yale Crescent, Thornaby, Stockton-on-Tees, Cleveland TS17 6AA.
22) 19 Tavy Bridge, Thamesmead, London SE2 9UG.
23) R. Forth Group, Forth House, Forth Street, Edinburgh EH1 3LF — 23a) PO Box 123, 6 North Isla St, Dundee DD1 9UF — 23b) P.O. Box 967, Stirling FK7 7YJ. Relays st. 23 "Max AM" 2200-0600 — 23c) Tweedside Park, Galashiels, Selkirkshire TD1 3TD.
24) P.O. Box 212, Crosslances Rd, Hounslow, Middlesex TW3 2AD — 24a) 30 Chapel Str, Little Germany, Bradford BD1 5DN.
25) Victoria Road, Gowerton, Swansea, SA4 3AB.
26) Carn Brea Studios, Wilson Way, Redruth, Cornwall TR15 3XX.
27) Studio 257, 67-73 Stoke Rd, Stoke-on-Trent ST4 2SR — 27a) 1st Flr, Regent House, Heaton Lane, Stockport SK4 1BX.
28) 12 Westerbrook, St. Sampson, Guernsey, C.I.
29) Forster Bldg, Chester Rd, Sunderland SR1 3SD.
30) Suite 7, Endeavour House, Brent Cross, London NW2 1JT.
31) 16-18 Trinity Gardens, Brixton, London SW9 8DP.
32) The Studios, Mold Rd, Gwersyllt, Wrexham, Clwyd LL11 4AF.
33) 26/27 Castlereagh Str, London W1H 5YR.
34) Florentia Village, Vale Rd, London N4 1TD. 0100-0400, 0800-1200, 1600-2000.
35) P.O. Box 1530, Kettering, Northants NN16 8PU.
36) P.O. Box 102, Lincoln LN5 7JS.
37) Mid Anglia R: 37a) PO Box 225, Queensgate Centre, Peterborough PE1 1XJ —37b) P.O. Box 1000, Vision Park, Chivers Way, Histon, Cambridge CB4 4WW — 37c) P.O. Box 77, King's Lynn, Norfolk PE30 1NN.
38) 26 Lady Lane, Paisley PA1 2LG.
39) P.O. Box 123, Dunnington, York YO1 5ZX.
40) Ringway House, Hill Str, Coventry CV1 4AN.
41) Windsor House, St. Lawrence, Jersey, C.I.
42) Allied Radio Group, Broadfield House, Brighton Rd, Crawley RH11 9TT — 42a) The Friary, Guildford, Surrey GU1 4YX — 42b) 11 Lower Str, Haslemere, Surrey GU27 2NY.
43) Euston Tower, London, NW1 3DR.
44) P.O. Box 2, Cheltenham GL53 7YA.
45) P.O. Box 967, Liverpool L69 1TQ.
46) Russel Court Bldg, Claremont Str, Lisburn Rd, Belfast BT9 6JX.
47) c/o Davies Memorial Gallery, Newtown, Powys SY16 2NZ (12h daily).
48) 180 Brompton Rd, London SW3 1HF.
49) Brush House, Pony Rd, Cowley, Oxford OX4 2XR.
50) 965 Mowbray Drive, Blackpool, Lancs FY3 7JR.
51) Haygrove House, Shoreditch, Taunton, Somerset TA3 7BT.
52) Brigadoon, Durrant Lane, Northam, Bideford, Devon.
53) Unit 6E, Science Park, Cefnllan, Aberysytwyth. 0700-2400.
54) Lower Oakfield, Pitlochry, Perhshire PH16 2DS (limited hours).
V. Re. direct to individual st's.

Restricted Service Licences

Licences are granted for special event stations operating for up to 28 days on MW or FM with maximum power of 25 watts.

F.PL: Franchises will be advertised for MW-st's in Greater London (2), Birmingham, Manchester & Wolverhampton using ex-BBC freq's. Additional AM & FM st's planned for smaller communities.

MANX RADIO (Comm.)

Owned by the Isle of Man Government and operated by R. Manx Ltd.

ADDR: P.O. Box 1368, Broadcasting House, Douglas Head, Douglas, Isle of Man. **Te:** +44 (624) 661066. **Fax:** +44 (624) 661411. **Telex:** 629770 Manrad G.

L.P: GM: Stewart Watterson. PD: G. Ferguson. CE: E. Leeming.

STATIONS: Foxdale 1368kHz 20kW — Snaefell 89MHz 2kW — Carnane 97.2MHz 11.8kW — Jurby 103.7MHz 4kW.

D.PRGR: 0630-0100.

ANN: Broadcasting from Douglas in the Isle of Man on 1368kHz and stereo VHF, "This is Manx Radio" — **V.** by QSL-card. (B).

BRITISH FORCES BROADC. SCE. LONDON

(a division of the Services Sound & Vision Corp.)

ADDR: P.O. Box 1234, London W2. **Te:** +44 (71) 724-1234. **Fax:** +44 (71) 706 1582. **Telex:** 25704 BFBS G (UK). Studios: Bridge House, North Wharf Rd, London W2 1LA.

L.P: Dir. of Broadc: P. McDonagh. Prgr. Contr: Chris Russell. CE: Colin Rugg.

STATIONS: See under Germany, Cyprus, Gibraltar, Hong Kong, Belize, Brunei & Falkland Islands.

ANN: "This is the British Forces Broadc. Sce..." (Name of country or town) — **INT-SIG:** March "Force One", orchestra.

V. by QSL-card to individual station. Rp.

VOICE OF AMERICA RADIO STATION WOOFFERTON

STATION: (G.C: 02.43W/52.19N): 4 × 300kW, 6 × 250kW tr's operated by the BBC.

kHz	kHz	kHz	kHz	kHz	kHz	kHz
6015	6125	6160	7170	7325	9735	15235
6040	6130	6180	7180	9580	11760	15265
6060	6140	6195	7225	9585	11775	17855
6095	6150	7130	7295	9660	15185	

VATICAN CITY STATE

L.T: UTC + 1h (Su: UTC + 2h) — **Pop:** 1.000 — **ITU:** CVA.

VATICAN RADIO (Rlg.)

ADDR: Vatican Radio, I-00120 Vatican City. **Te:** +39 (6) 698 3551. (Int. Rel: +39 6 698 3945). **Fax:** +39 (6) 698 3237. **Telex:** 2023 VA. **Cable:** RADVAT.

L.P: DG: Rev. Pascal Borgomeo S.I. PD: Rev. Federico Lombardi S.I. CE: Pier Vincenzo Giudici. Head of Int. Rel: Mrs. Solange de Maillardoz.
STATIONS: MEDIUMWAVE: 526kHz 5kW — 1530kHz 300/600kW — 1611kHz 15kW.
FM: 93.3/96.3/103.8/105.0MHz 10kW (stereo).
SHORTWAVE: S.M. Galeria, Italy (G.C: 12.19E/42.03N): 100/250/500kW — +) Vatican City (G.C: 12.27E/41.54N): 80kW.

kHz	kHz	kHz	kHz	kHz	kHz	kHz
5882	7125	9500	9695	11640	15090	17865
6095	7250	9600	9755	11715	15120	21515
6150	7305	9605	9835	11740	15210	21670
6185	7310	9645	11620	11830	17525	21710
6245+	7365	9650	11625	11935	17730	21850

Winter Schedule
(Su: 1h earlier exc. transmissions marked * — H) Holy Days.
Local Sce. on 526kHz + 93.3/105.0MHz (stereo): music & prgrs in Italian, French, English & Spanish.

TO EUROPE

Albanian	Days	kHz
0330-0445	1234567	1530, 6245, FM93.3
1730-1745	1234567	1530, 6245, 7250, FM93.3
Arabic		
0745-0800	123456	1530, 6245, 7250, 9645, 15210, FM93.3
Armenian		
0310-0330	1234567	1611, 6185, 7365, 9755, FM96.3
1700-1720	1234567	1611, 6185, 7365, 9755, FM96.3
Bulgarian		
0530-0550	1234567	1611, 6185, 7365
1920-1940	1234567	1611, 6185, 7365
Byelorussian		
0420-0440	1234567	1611, 6185, 7365, FM96.3
1810-1830	1234567	1611, 6185, 7365, FM96.3
Czech		
0415-0430	1234567	1530, 6245, FM93.3
1830-1845	1234567	1530, 6245, 7250, FM93.3
English		
0600-0620	1234567	526, 1530, 6245, 7250, FM93.3, FM105.0
2050-2110	1234567	526, 1530, 5882, 7250, FM93.3, FM105.0
Esperanto		
0610-0630	4	1611, 6185, 7365, FM96.3
2020-2030	7+H	526, 1530, 6245, 7250, 11740, FM93.3, FM105.0
French		
0540-0600	1234567	526, 1530, 6245, 7250, FM93.3, FM105.0
2030-2050	1234567	526, 1530, 5882, 7250, FM93.3, FM105.0
German		
0520-0540	1234567	1530, 6245, FM93.3
1500-1515	1234567	1530, 6245, 7250, 9645, FM93.3
1920-1940	1234567	1530, 6245, 7250, FM93.3
Hungarian		
0445-0500	1234567	1530, 6245, FM93.3
1815-1830	1234567	1530, 6245, 7250, FM93.3
Italian		
0620-0630	1234567	526, 1530, 6245, 7250, FM93.3, FM105.0
1115-1200	7+H	526, 1530, 6245, 7250, FM93.3, FM105.0
1300-1330	1234567	526, 1530, 6245, 7250, 11740, 15210, 21670, FM93.3, FM105.0
1530-1600	5	526, 1530, 6245, 7250, FM93.3, FM105.0
1630-1700	1234567	526, 1530, 6245, 7250, 9645, FM93.3, FM105.0
2000-2030	1234567	526, 1530, 6245, 7250, 11740, FM93.3, FM105.0
2200-2310	1234567	526, 1530, 5882, FM93.3, FM105.0
Latvian		
0455-0510	1246	1611, 6185, 7365
1845-1900	1357	1611, 6185, 7365, FM96.3
Lithuanian		
0440-0455	1234567	1611, 6185, 7365
1830-1845	1234567	1611, 6185, 7365, FM96.3
Norwegian		
2000-2020	1	1611, 6185, 7365, FM96.3
Polish		
0500-0520	1234567	1530, 6245, FM93.3
1515-1530	1234567	1530, 6245, 7250, 9645, FM93.3
1900-1920	1234567	1530, 6245, 7250, FM93.3
Portuguese		
1415-1430	1234567	1530, 7250, 11740, 15210, FM93.3
2130-2145	1234567	1530, 5882, 7250, FM93.3
Romanian		
0510-0530	1234567	1611, 6185, 7365
1900-1920	1234567	1611, 6185, 7365
Russian		
0330-0400	1234567	1611, 6185, 7365, 9755, FM96.3
1330-1400	1234567	1530, 7250, 11740, 15210, FM93.3
1720-1750	1234567	1611, 6185, 7365, 9755, FM96.3
2100-2130	1234567	7365, 9500, FM96.3
Scandinavian Sce.		
0550-0610	1234567	1611, 6185, 7365, FM96.3
2000-2020	1234567	1611, 6185, 7365, FM96.3
Serbo-Croat		
0400-0415	1234567	1530, 6245, FM93.3
1800-1815	1234567	1530, 6245, 7250, FM93.3
Slovak		
0430-0445	1234567	1530, 6245, FM93.3
1845-1900	1234567	1530, 6245, 7250, FM93.3
Slovene		
0345-0400	1234567	1530, 6245, FM93.3
1745-1800	1234567	1530, 6245, 7250, FM93.3
Spanish		
1400-1415	1234567	1530, 7250, 11740, 15210, FM93.3
2110-2130	1234567	526, 1530, 5882, 7250, FM93.3, FM105.0
Ukrainian		
0400-0415	1234567	1611, 6185, 7365, 9755, FM96.3
1750-1810	1234567	1611, 6185, 7365, 9755, FM96.3
NEWS		
Italian/French/English		
0700-0745	123456	526, 1530, 6245, 7250, 9645, 11740, 15210, FM93.3, FM105.0
1100-1130	123456	526, 1530, 6245, 7250, 9645, 11740, 15210, FM93.3, FM105.0
French/English		
1700-1730	1234567	526, 1530, 6245, 7250, 9645, FM93.3, FM105.0
MUSIC		
1530-1600	12346	526, 1530, 6245, 7250, FM93.3, FM105.0
2310-0100	1234567	526, 1530, 5882, FM93.3, FM105.0
PRAYERS & LITURGY		
Angelus		
1100-1115	7+H	526, 1530, 6245, 7250, 9645, 11740, 15210, FM93.3, FM105.0
Holy Rosary		
1940-2000	1234567	526, 1530, 1611, 6185, 6245, 7250, 7365, 11740, FM93.3, FM96.3, FM105.0
Latin Mass		
0630-0700	1234567	526, 1530, 6245, 7250, 9645, 11740, 15210, FM93.3, FM105.0
Byzantine/Romanian		
0710-0810	7+H	1530, 7250, 11740, FM93.3
Italian Mass		
0830-0930	7+H	526, 1530, 7250, FM93.3, FM105.0
Lithuanian Mass		
0845-1000	7+H	1611, 9755, 15120, FM96.3
Oriental Mass		
0930-1100	7+H	1530, 7250, 9645, 11740, 15210, FM93.3
Byzantine/Ukrainian		
0715-0845	7+H	1611, 9755, 15120, FM96.3
Polish Mass		
1530-1630	7+H	1530, 6245, 7250, 9645, FM93.3
Vespers		
1600-1630	123456	526, 1520, 6245, 7250, 9645, FM93.3, FM105.0

TO AFRICA

Amharic/Tigre		
0400-0430*	1234567	9695, 11625, FM103.8
1630-1700	1234567	15090, 17730, FM103.8
Arabic		
0500-0525	1234567	9645, 11715, FM96.3
0745-0800	123456	1530, 6245, 7250, 9645, 15210, FM93.3

1910-1940	1234567	9645, 11740, FM96.3
2145-2200	1234567	1530, 5882, 7250, FM93.3
English		
0500-0530*	1234567	9695, 11625, 15090, FM103.8
0630-0700*	1234567	9695, 11625, 15090, FM103.8
1730-1800*	1234567	11625, 15090, 17730, FM103.8
2000-2030*	1234567	9645, 11625, 15090, FM103.8
French		
0430-0500*	1234567	9695, 11625, 15090, FM103.8
0600-0630*	1234567	9695, 11625, 15090, FM103.8
1700-1730*	1234567	11625, 15090, 17730, FM103.8
2030-2100*	1234567	9645, 11625, 15090, FM103.8
Portuguese		
0530-0600*	1234567	9695, 11625, 15090, FM103.8
1800-1830*	1234567	11625, 15090, 17730, FM103.8
Angelus		
1100-1115	7+H	1530, 21670, FM96.3
Holy Rosary		
1940-2000	1234567	9645, 11625, 15090, FM103.8
Mass in English		
1600-1630	6	15090, 17730, FM103.8

TO THE MIDDLE EAST

Arabic		
0500-0530	1234567	9645, 11715, FM96.3
1910-1940	1234567	9645, 11740, FM96.3

TO ASIA

Chinese		
1230-1315*	1234567	11640, 15090, 17525, FM103.8
2200-2245*	1234567	6150, 7310, 9600, FM103.8
English		
0145-0200*	1234567	7125, 9650, FM96.3
1345-1405*	1234567	15090, 17525, FM103.8
1545-1600*	1234567	15090, 17865, FM103.8
Hindi		
0100-0115*	1234567	7125, 9650, FM96.3
1500-1515*	1234567	15090, 17865, FM103.8
Japanese		
1315-1345*	1234567	9835, 11640, FM103.8
2130-2200*	1234567	6150, 7310, FM103.8
Malayalam		
0130-0145*	1234567	7125, 9650, FM96.3
1530-1545*	1234567	15090, 17865, FM103.8
Tamil		
0115-0130*	1234567	7125, 9650, FM96.3
1515-1530*	1234567	15090, 17865, FM103.8
Vietnamese		
1415-1500*	1234567	15090, 17525, FM103.8
2315-2400*	1234567	7310, 9600, FM103.8
Angelus		
1100-1115	7+H	1530, 21515, FM93.3
Mass in English		
1600-1630	6	15090, 17730, FM103.8

TO AUSTRALIA/NEW ZEALAND

English		
1345-1405*	1234567	15090, 17525, FM103.8
2245-2305*	1234567	9600, 11830, FM103.8

TO THE AMERICAS

English		
0250-0310*	1234567	6095, 7305, FM103.8
French		
0230-0250*	1234567	6095, 7305, FM103.8
Portuguese		
1100-1130*	1234567	21710, 21850, FM103.8
0030-0100*	1234567	7305, 9605, FM103.8
Spanish		
0100-0145*	1234567	7305, 9605, 11620, FM103.8
0145-0230*	1234567	7305, 9605, 11620, FM103.8
0315-0400*	1234567	6095, 7305, FM103.8
1130-1200*	1234567	21710, 21850, FM103.8

ANN: Before all transmissions: Latin: "Laudetur Jesus Christus" (Praised be Jesus Christ), repeated in the language of the broadcast, then station identification.
INT-SIG: "Christus Vincit" — **V.** by QSL-card (B).

YUGOSLAVIA (Federal Republic of)

L.T: UTC + 1h (Su: UTC + 2h) — **Pop:** 14.000.000 — **Radio sets:** 182.780 — **Pr.L:** Serbian — **E.C:** a/c 50, 220V — **ITU:** YUG.
NB: Information in this section reflects the situation at 01.01.1993.

UDRUŽENJE JUGOSLOVENSKIH RADIOTELEVIZIJA d.O.O. (JRT)

ADDR: Generala Ždanova 28, 11000 Beograd. **Te:** +38 (11) 330-194. **Fax:** +38 (11) 334-380. **Cable:** Jurate, Beograd. **Telex:** 71048 or 12158 yu jurate, Beograd.
L.P: Man. Dir: A. Todorović.
JRT comprises the following two broadcasting organizations plus the Ext. Sce (R. Jugoslavija):
RADIOTELEVIZIJA SRBIJE, Takovska 10, 11000 Beograd. **Te:** +38 (11) 342-00l. **Fax:** +38 (11) 543-178. **Telex:** 11884 — Dir. Gen: M. Vučelić. (R. Beograd: Hilendarska 2, 11000 Beograd. **Te:** +38 (11) 346-801).
RADIOTELEVIZIJA CRNE GORE, Cetinjski put bb, 81000 Podgorica. **Te:** +38 (81) 41-800. **Telex:** 61133—Dir. Gen: B. Bojanić.
R. Yugoslavia: see end of section.
Local Radio st's: These st's are part of the JRT, through their association with the main st's in the Republics.

SERBIA

Mediumwaves	kHz	kW		kHz	kW
Bosilegrad	675	10/5	Svetozarevo	1440	2x10b
Dimitrovgrad 1	675	10/5	Bujanovac	1467	1a
Beograd	684	2000	Crna Trava	1485	1d
Negotin	693	10/5d	Kladovo	1485	1a
Niš	711	100a	Pirot	1485	1a
Medvedja	765	1d	Trgovište	1485	1a
Kladovo	999	10/5d	Tutin	1485	1a
Beograd 2/3	1008	200	Zaječar	1485	1d
Aleksinac	1008	200b	Priboj	1485	1d
Novi Pazar	1062	1a	Leskovac	1602	1d
Vranje	1296	10d	Negotin	1602	1e

Shortwave: Beograd 7200kHz 50kW (rel. Beograd 1 0600-2305).
FM: Stereo

(MHz)	I	II	III	IV	L	kW
Avala	95.3	97.7	98.5	101.4	104.0	100
Obrenovac	88.9	—	—	—	—	10
Besna Kobila	91.7	95.3	—	105.6	—	15/60
Crni Vrh	87.7	99.3	—	101.0	—	25
Crveni Čot	94.5	96.5	—	101.8	103.2	75/50
Deli Jovan	87.7	94.9	—	—	—	25
Jastrebac	89.3	96.9	—	103.5	—	100
Kopaonik	102.1	93.7	—	90.9	—	100
Mokra Gora	—	101.1	—	103.2	—	50
Ovčar	88.1+	90.1	—	—	107.5	25
	101.6	—	—	—	—	—
Pirot	91.3	98.5	—	—	—	15
Tornik	90.6+	97.5	—	—	—	15
	100.2	—	—	—	—	—
Tupižnica	92.5+	96.1	—	100.4	—	25/60
	94.1	—	—	—	—	—
Cer	91.1	92.3	—	—	—	25
Rudnik	91.5	94.3	—	—	—	25
Maljen	93.4	96.3	—	—	—	25

+ additional low power st's not mentioned.
a) Beograd 1 + reg. prgr — b) rel. Beograd 2/3 — c) rel. Beograd 202 + reg. prgr — d) rel. Beograd 1 — e) rel. Beograd 2/3 + reg. prgr — L) local st's.
Music Radio: Beograd 101.4MHz 100kW.
Youth Radio: Beograd 92.5MHz 1kW.
Radio Politika: Makedonska 29, 11000 Beograd: 24h on 105.2MHz 10kW.
Ju Radio: Hilandarska 2, 11000 Beograd — 24h on 100.1MHz 1kW (operated by R. Jugoslavija).
Studio "B", Beograd, P.O. Box 10, 11001 Beograd, Masrikova 5. **Prgr. 1** on 1350kHz +99.1MHz: 24h — **Prgr. 2** on 900kHz +100.8 MHz+94.9MHz 0700-1700 — **Prgr. 3** on 94.9MHz: W 1700-2300: Only FM Stereo 106.4MHz + 107.2MHz.
Radio "Pingvin": Autoput 2, 11070 N. Beograd: 24h on 90.9MHz 1kW.
R. Beograd 1: 24h. **N:** W 0303, 0330, 0400, 0430, 0500, 0540, 0700, 0800, 0900, 1000, 1100, 1200, 1300, 1400, 1600, 1700, 1830, 2100, 2200, 2300; Sun 0430, 0500, 0530, 0600, 0700, 0800,

0900, 1000, 1100, 1200, 1400, 1600, 1830, 2000, 2200, 2300.
R. Beograd 2: W 0400-1900 (Sun 0600-1900). **N:** W 1130, 1230, 1330, 1500, 1600, 1850. Sun 0630, 0730, 0930, 1055, 1130, 1330, 1730, 1850.
R. Beograd 3: 1900-2300 (Serious prgr.)
R. Beograd 202: 0400-2400 on 1503kHz, 104.0MHz + FM IV (0000-0400 rel. R. Beograd 1).
Stereorama: Sat/Sun 0700-1700 on FM IV. At other times rel. Beograd 202.
Music Radio: 0700-1900 on 98.5/101.4/103.2MHz.
Youth Radio: 24h on 92.5MHz.

Loc. sts	**kHz**	**kW**	**MHz**				
				Loznica	1296	10	95.0
Užice	531	10	92.0	Zaječar	1341	10	98.1
Vrange	531	1		Studio "B" 1	1350	10	99.1
Soko Banja	639	1		Valjevo	1368	10	88.6
Lazarevac	648	1		Kladovo	1458	1	90.0
Šabac	702	10	99.4	Novi Pazar		1	90.0
Kruševac	738	10	92.2	Negotin		15	97.5
Arandjelovac	792	1	98.9	Priboj	1485	1	87.6
Studio "B" 2	900	2	100.8	Beograd 202	1503		98.5
			94.9	Smed. Palanka	1566	1	88.3
Smederevo	945	1	96.1	Prijepolje	1584	1	
Bor	981	10	91.7	Leskovac	1602	1	
Čačak	981	10	92.8	Obrenovac		1	102.3
Požarevac	990	1	90.1	Bosilegrad 2	999	10	
Kragujevac	1026	10	106.8	Sjenica	1602	1	
Svetozarevo	1062	10	97.3	Pirot		1	95.8
Vrnjačka Banja	1170	1	93.2	Bajina Bašta		1	87.6
Majdanpek	1206	1	96.7	Paraćin		1	97.8
Mladenovac	1215	1c	90.8	Gor.Milanovac		1	95.8
Kraljevo	1242	10	95.0	Knjaževac		10	94.5

VOJVODINA

RADIO NOVI SAD
ADDR: Žarka Zrenjanina 3, 21000 Novi Sad. **Te:** +38 (21) 611588. **Fax:** +38 (21) 26624. **Tlx:** 14127.

Mediumwaves	**kHz**	**kW**			
Sombor	666	10x	Sremska Mitrovica	1404	1y
Stara Pazova	783	1y	Beočin	1431	0.2y
Sombor	837	50°	Srbobran	1467	2y
Ruma	936	1y	Zrenjanin	1467	2y
Temerin	1044	1y	Kikinda	1485	1y
Novi Sad 2	1089	50	Novi Sad	1485	2x
Novi Sad 1	1107	150	Apatin	1494	0.2y
Kovin	1161	0.1y	Odžaci	1539	0.1y
Bačka Topola	1170	0.1y	Stara Pazova	1548	0.2y
Sombor	1269	750	Bač	1575	0.2y
Bački Petrovac	1224	0.1y	Pančevo	1584	1xy
Novi Sad	1269	750	Indjija	1593	0.2y
Šid	1323	1xy	Subotiva	1602	1yx
Vrbas-Kula	1359	2y	Vršac	1602	1xy
Kovačica	1359	0.2y			

x) rel. Novi Sad 1 + local prgr. — y) rel. Novi Sad 1 on second base.
o) Nat. Sce. 1 (0600-1800).

FM (MHz)	**I**	**II**	**M**	**L**	**kW**
Sombor	90.9	92.0	—	102.2	5/1
Novi Sad	89.5	—	100.0h	—	10
Temerin	—	—	—	93.5	0.5
Vršac	98.1	—	91.7	—	30
Subotica	99.3	92.5	—	91.5	3/5/10

L) local st's — M) prgr's for national minorities.
Panradio Novi Sad, Svetozara Miletića 28, 21000 Novi Sad: 0600-2000 on 100.6MHz 1kW.
R. Novi Sad 1485 kHz: 1300-1710 local prgr. in Serbian.

KOSOVO

RADIO PRIŠTINA
ADDR: Maršala Titabb, 3800 Priština. **Te:** +38 (38) 26255. **Fax:** +38 (38) 32073. **Tlx:** 18134.

Mediumwaves (kHz)	**I**	**II**	**L/R**	**kW**
Peć-Priština 2	—	549	—	100
Djakovica	936	936	936	10/2
Gnjilane	954	954	954	8
Mitrovica	1035	1035	1035	10
Prizren	1377	1377	1377	10
Priština 1	1413	—	—	1000
Peć	1539	1539	1539	2
Uroševac	—	—	1584	1

L/R) local and regional st's.

FM (MHz)	**I**	**II**	**L/R**	**kW**
Peć	—	—	98.4	10
Priština	91.9	100.6	—	10/50
Goleš	95.7	97.7	—	30
Cviljen	88.9	94.5	—	30/10
Prizren	—	—	94.0	30
Djakovica	—	—	91.8	10
Mitrovica	—	—	91.4	10
Gnjilane	—	—	96.4	10

R. Priština 1: 0400-2305 (Thurs/Fri/Sat 24h) in Albanian.
R. Priština 2: 0800-1700 in Albanian and Turkish.
Local/Regional Prgrs: 0400-2305 in Serbian.

MONTENEGRO

Mediumwaves	**kHz**	**kW**			
Pljevlja Grad	567	10L	Cetinje	1485	1L
Nikšić	594	10L	Plav	1485	1a
Bar	738	5a	Plužine	1485	1a
Podgorica	882	600/ 300a	Žabljak	1485	1a
			Ulcinj	1503	10a
Bijelo Polje	882	10a	Dragaš	1548	2/1a
Herceg Novi	882	10a	Dragaš	1548	2/1b
Pljevlja	882	10a	Bar	1548	1b
Rožaje	882	10a	Rožaje	1584	1a
Ulcinj	882	1a	Budva	1602	1a
Bar	1026	1L	Ivangrad	1602	1a
Ulcinj Grad	1161	1L	Pljevlja	1602	1a
Herceg Novi	1305	1L	Tivat	1602	1a
Nikšić	1458	1b	Šavnik	1602	1a

Radio Podgorica 1: 24h (a) — L: local prgr 0800-1500.
Radio Podgorica 2: 24h (b).

FM (MHz)	**I**	**II**	**III**	**L**	**kW**
Lovćen	94.9	98.0	101.0	89.8	60/1
Bjelasica	92.1	92.3	—	—	60
Sjenica	89.3	91.5	—	—	30/5
Velji Grad	89.6	99.7	—	—	10
Možura	93.4	97.3	—	—	10
Durmitor	91.3	96.1	—	—	60
Tvrdaš	96.8	98.3	—	—	10
Obrov	89.0	94.6	—	—	1/10
Sudjina Glava	88.0	98.9	—	—	10
Plav	89.9	97.2	—	—	10

FOREIGN SERVICE — RADIO JUGOSLAVIJA

ADDR: Hilendarska 2, 11000 Beograd. **Te:** +38 (11) 346 884. **Fax:** +38 (11) 332014. **Listener response: Te:** +38 (11) 344 455. **Fax:** +38 (11) 332 014. **Telex:** 12432.
L.P: DG: Dr Dragan Marković. CE: Rodoljub Medan. PD: D. Marković. Head of PR: Aleksandar Popović. Dep. Editor-in-Chief: Rade Tati.
MEDIUMWAVES: Niš 711 kHz 100kW, Podgorica 882 kHz 300 kW, Priština 1413 kHz 1000kW — **FM:** Ju Radio, Beograd 100.4 MHz 1 kW (0500-1700 in Serbian, 1700-0500 in English).
SHORTWAVES: Bijeljina (G.C:19.10E/44.42N):500/250 kW.

kHz	**kHz**	**kHz**	**kHz**	**kHz**	**kHz**
6005	7220	9580	11805	15175	17740
6100	7250	9620	11835	15180	21605
7165	9505	9720	15140		

Arabic	**Area**	**kHz**			
			German		
1830-1900	NEAf/ME	7250	1730-1800	WEu	9620
English			2030-2100	NWEu	6100
0100-0130	ENAm	9580	**Russian**		
0200-0230	WNAm	9580	1600-1630	RUS	9505
1230-1300	ENAm	17740		RUS	11805
	AUS	21605	1900-1930	RUS	6005
1930-2000	WEu	6100		RUS	7165
	SAf	15140	**Serbian**		
2200-2245	WEu	6100	0030-0100	ENAm	9580
		9505	0200-0230	ENAm	9580
French			2100-2130	WEu	6100
1700-1730	WEu	11835		AUS	9720
	NWAf	15175	**Spanish**		
2130-2200	WEu	6100	0000-0030	SAm	9720
	SWAf	15180			11835
			2000-2030	SWEu	7220

ANN: "This is R. Yugoslavia", "Ici R. Yugoslavie", "Hier Spricht R. Jugoslavien" — **V.** by QSL-card. Rec. acc.

AFRICA

ALGERIA

L.T: UTC+1h — **Pop:** 26.727.000 — **Radio sets:** 3.500.000 — **Pr.L:** French, Arabic, Berber dialects — **E.C:** a/c 50, 127/220 and 220/380V —**ITU:** ALG.

ÉTABLISSEMENT PUBLIC DE RADIODIFFUSION SONORE (RADIODIFFUSION ALGERIENNE)

ADDR: 21 Bld. des Martyrs, Alger. **Te:** +213 (2) 590700. **Fax:** +213 (2) 605814. **Telex:** 65265.
L.P: DG: Lamine Bechichi. Dir. Tec: Sadek Laskri. Dir. Coop: Abdelkader Lalmi.
Channels: 1 = Arabic, 2 = Berber, 3 = French, 4 = Int. Sce.

MW	kHz	kW	Ch.
Bechar	153	1000	1
Ouargla	198	1000	1
Tipaza	252	1500/750*	3/1
Ain-El-Beida	531	600/300*	1
Les Trembles	549	600/300*	1
Bechare	576	300	1
Tindouf	666	5	1
Reggane	693	5	1
Ain-El-Hammam	693	5	2
Ain-Amenas	738	5	1
Djanet	783	5	1
Ben-Abbas	837	5	1
Ghardaia	873	5	1
Alger	891	600/300*	1
Tamanrasset	909	5	1
Timimoun	927	5	1
Alger	981	600/300*	2
Hassi-Messaoud	1026	5	1
Adrar	1089	5	1
Ouargla	1098	5	1
Ain-Salah	1161	5	1
El-Golea	1287	5	1
Constantine	1305	20	3
Alger	1422	50	3

*) 1800-0600
FM: 94.2/94.7MHz (Ch. 3).
SW: B = Bouchaoui (G.C: 02.56E/36.44N), F = Ouled Fayet (G.C: 02.57E/36.43N).

kHz	kW	Ch.	Area	H. of tr.*
6145F	50	1	NAf	1700-2400
6150F	50	1	NAf	1800-2400
7145B	100	1/4	ME	1500-2400
7245F	50	1/3	NAf	1500-2400
9509F	50	2/4	NAf	0800-1900
9535B	100	2/4	ME	1800-2400
9640B	50	2/4	SAm	1900-0100
9685F	100	2	F	1800-2400
11715B	50	1/4	CEu	0600-2100
15160B	100	2	CEu	0700-1800
15205B	100	1/4	ME	0500-2200
15215B	100	1/2/4	WAf/SAm	0500-0400
17745B	100	2/4	EAf	0800-1500

***)** H. of tr. on all freq's varies widely.
Ch. 1 in Arabic: 24h. **N:** on the h.
Ch. 2 in Berber dialects: 0600-0100. **N:** on the h.
Ch. 3 in French: 0530-0100. **N:** on the h.
Ch. 4 International Sce:

Arabic	Areas	KHz
1900-2000	Eu/ME	15205, 11715, 7145
0100-0200	Eu/ME	252, 198, 153
English		
1800-1900	Eu/ME	9685, 9535, 9509, 981
2100-2200	Eu/ME	9685, 9535, 981
French		
2200-2400	NWAf	15215, 9640
Spanish		
1900-2000	CHL	15215, 981
0000-0100	Eu/ME	9640

ANN: A: "Huna El Djazair, idha'atu-El Djoumhouriya El Djazairia". F: "Alger chaine 3, Radiodiffusion Algerienne". E: "This is the Int. Sce. of R. Algiers".
IS: Oriental Lute (Ud) — **V.** by QSL-card.

ANGOLA (People's Republic)

L.T: UTC+1h — **Pop:** 8.902.325 — **Radio sets:** 450.000 — **Pr.L:** Portuguese + 10 ethnic — **ITU:** AGL.

RADIO NACIONAL DE ANGOLA (Gov.)

ADDR: C.P. 1329, Luanda. **Te:** +244 321558. **Fax:** +244 391234 or 320807. **Telex:** 3066 or 3121 EMISSORA AN.
L.P: DG: Guilherme Y. Mogas. PD: Luzia R. Fançony. TD: Wenceslau M. Machado.

Location	kHz	kW	Prgr.	H. of tr.
Cazenga	944	25	2	0500-1800
			E	1800-2200
Mulenvos	1088	25/100	1	24h
Cazenga	3355	10	2	0500-1800
			E	1800-2200
			1	2200-0500
Mulenvos	3377	10	1	2200-0500
Mulenvos	4960	10	1	1800-0500
	5500		1	1900-0250
Cazenga	7215	10	2	0500-1800
Mulenvos	7245	10/100	1	24h
Mulenvos	9535	100	2	0500-1800
			E	1800-2200
Mulenvos	9720	100	1	0500-1800
Mulenvos	11955	100	1	0500-1800

FM: Sala 93.5MHz (1:0500-2400), 101.4MHz (2:0500-1800, E: 1800-2200).
Main Prgr. (1) in Portuguese : 24h. **N:** on the h.
RNA Antenna 2 in Portuguese/Ethnic: 0500-1800.
Ext. Sce. (E): 1800-1900 **French**, 1900-2000 **Spanish**, 2000-2100 **English** (**N:** 2030), 2100-2200 **Portuguese.**
ANN: "De Luanda, capital da República Popular de Angola transmite a Rádio Nacional de Angola".
IS: Vibraphone — **V.** by QSL-card.

PROVINCIAL STATIONS (EP = Emissora Provincial).

Mediumwave	kHz	kW	H. of tr.
1) EP do Lobito	1000	1	24h
2) EP do Huambo	1010	1	0345-2300
3) EP do Zaire	1112	5	0500-1300
			1700-2200
4) EP do Kuanza-Sul	1110	1	alt. to 1115
4) EP do Kuanza-Sul	1115	10	24h
5) EP do Kuando-Kubango	1148r	10	0630-1300
			1700-2300
2) EP do Huambo	1160	10/1	0400-2300
6) EP do Malanje	1188	10/1	0455-2300
7) EP do Benguela	1200	1	24h
8) EP da Huila	1228	10	irr.
2) EP do Huambo	1232	1	0345-2300
9) EP da Lunda-Sul	1241	5	1930-2330
10) EP do Kuanza Norte	1259	10/1	0500-1400
			1700-2300
11) EP do Uige	1295v	1	0400-2300
8) EP da Huila	1313	1	0400-2300
12) EP do Namibe	1331	1	0455-2300
13) EP do Bié (alt. 1386)	1404	10	0455-2300
14) EP da Lunda-Norte	1484	1	
7) EP de Benguela	1502	10	0350-2300
15) EP da Cabinda	1570	10	
16) EP do Moxico	1586	10	0455-2300
Shortwave (all freqs variable)			
5) EP do Kuando-Kubango	4780	5	0500-0800
			1700-2300
8) EP da Huila	4820	3	0400-2300
9) EP da Lunda-Sul	4860		1930-2330
3) EP do Zaire	4885	5	*
6) EP do Malanje	4942	1	*
16) EP da Cabinda	4970	5	0455-2100
12) EP do Namibe	5015	1	*
7) EP de Benguela	5041	1	0350-2300

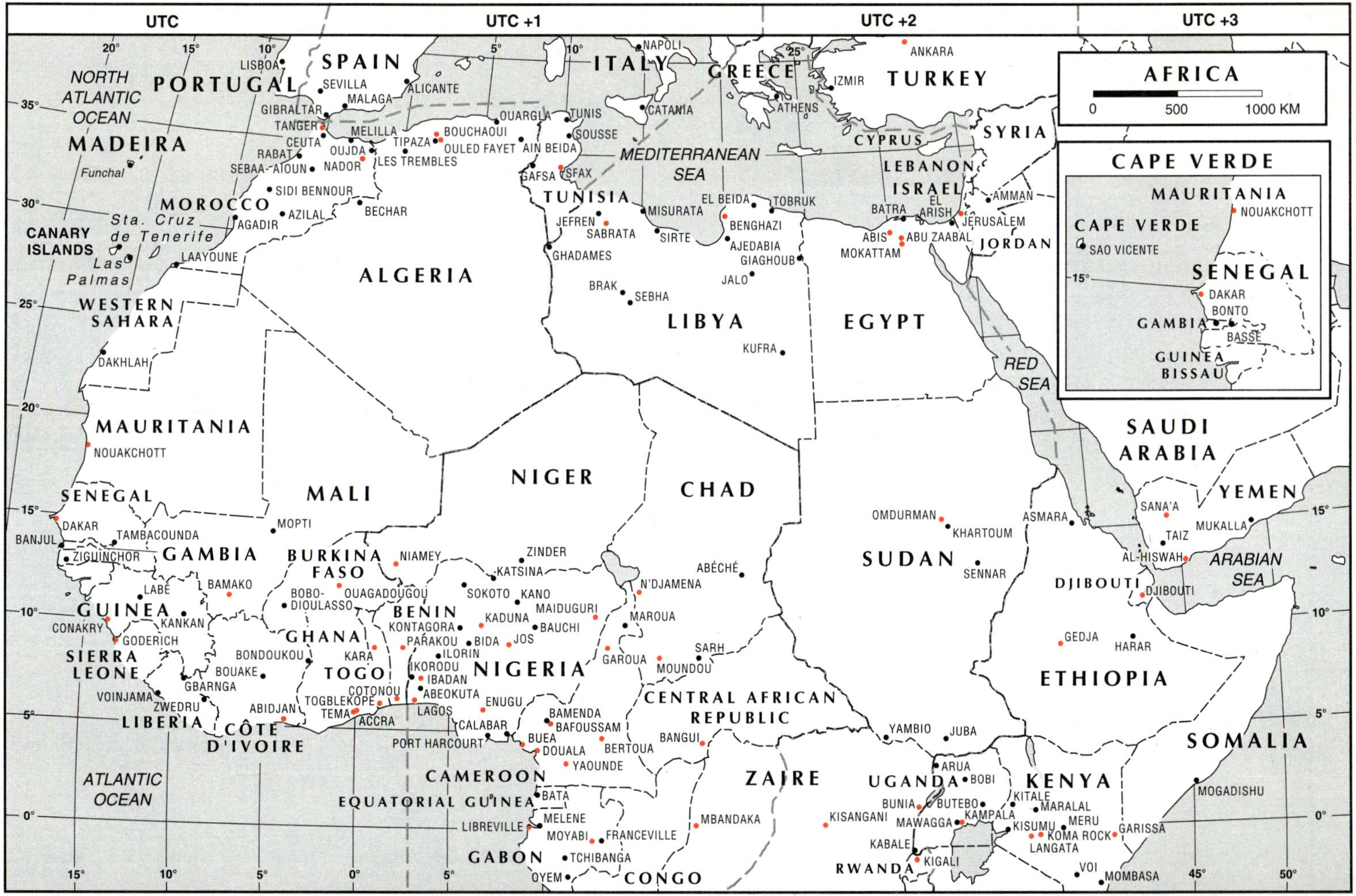
UTC
UTC +1
UTC +2
UTC +3
AFRICA
0
500
1000 KM
CAPE VERDE
MAURITANIA
NOUAKCHOTT
CAPE VERDE
SAO VICENTE
SENEGAL
DAKAR
BONTO
GAMBIA
BASSE
GUINEA BISSAU
NORTH ATLANTIC OCEAN
PORTUGAL
LISBOA
SPAIN
SEVILLA
MALAGA
ALICANTE
GIBRALTAR
TANGER
CEUTA
MELILLA
RABAT
OUJDA
SEBAA-AIOUN
NADOR
MADEIRA
Funchal
SIDI BENNOUR
MOROCCO
AZILAL
AGADIR
BECHAR
Sta. Cruz de Tenerife
CANARY ISLANDS
Las Palmas
LAAYOUNE
WESTERN SAHARA
DAKHLAH
ALGERIA
OUARGLA
TIPAZA
BOUCHAOUI
OULED FAYET
LES TREMBLES
AIN BEIDA
ITALY
NAPOLI
CATANIA
TUNIS
SOUSSE
GAFSA
SFAX
TUNISIA
MEDITERRANEAN SEA
GREECE
ATHENS
IZMIR
TURKEY
ANKARA
CYPRUS
SYRIA
LEBANON
ISRAEL
EL ARISH
JERUSALEM
AMMAN
JORDAN
BATRA
ABIS
ABU ZAABAL
MOKATTAM
EGYPT
JEFREN
SABRATA
MISURATA
SIRTE
EL BEIDA
BENGHAZI
TOBRUK
AJEDABIA
GIAGHOUB
JALO
GHADAMES
BRAK
SEBHA
LIBYA
KUFRA
RED SEA
SAUDI ARABIA
YEMEN
SANA'A
MUKALLA
TAIZ
AL-HISWAH
ARABIAN SEA
DJIBOUTI
MAURITANIA
NOUAKCHOTT
MALI
MOPTI
NIGER
CHAD
SENEGAL
DAKAR
BANJUL
TAMBACOUNDA
ZIGUINCHOR
GAMBIA
BURKINA FASO
NIAMEY
ZINDER
KATSINA
ABÉCHÉ
N'DJAMENA
BAMAKO
OUAGADOUGOU
BOBO-DIOULASSO
SOKOTO
KANO
MAIDUGURI
LABÉ
GUINEA
KANKAN
CONAKRY
GODERICH
SIERRA LEONE
BENIN
KONTAGORA
KADUNA
BAUCHI
MAROUA
GHANA
PARAKOU
BIDA
JOS
ILORIN
KARA
TOGO
BONDOUKOU
BOUAKE
GBARNGA
VOINJAMA
ZWEDRU
LIBERIA
ABIDJAN
CÔTE D'IVOIRE
TOGBLEKOPE
TEMA
ACCRA
COTONOU
IKORODU
IBADAN
ABEOKUTA
LAGOS
NIGERIA
ENUGU
CALABAR
PORT HARCOURT
GAROUA
MOUNDOU
SARH
CENTRAL AFRICAN REPUBLIC
BANGUI
BAMENDA
BAFOUSSAM
BUEA
DOUALA
BERTOUA
YAOUNDE
CAMEROON
EQUATORIAL GUINEA
BATA
MELENE
LIBREVILLE
MOYABI
FRANCEVILLE
GABON
TCHIBANGA
OYEM
CONGO
ZAIRE
MBANDAKA
ATLANTIC OCEAN
SUDAN
OMDURMAN
KHARTOUM
SENNAR
ASMARA
ETHIOPIA
GEDJA
HARAR
YAMBIO
JUBA
ARUA
UGANDA
BOBI
BUNIA
BUTEBO
KISANGANI
MAWAGGA
KAMPALA
KABALE
RWANDA
KIGALI
KENYA
KITALE
MARALAL
MERU
KISUMU
KOMA ROCK
LANGATA
GARISSA
VOI
MOMBASA
SOMALIA
MOGADISHU
20°
15°
10°
5°
0°
25°
35°
30°
45°
50°

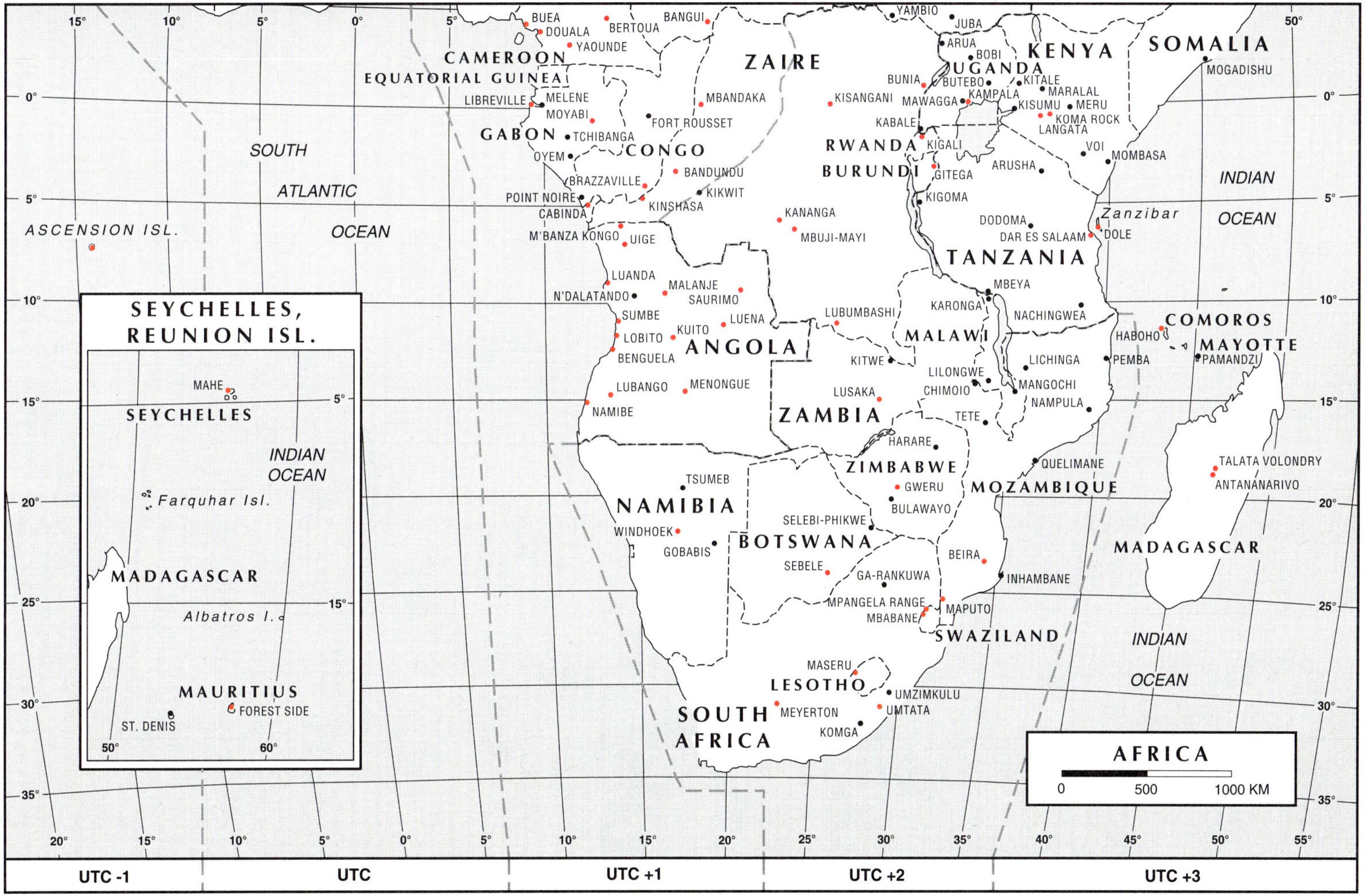
AFRICA
0
500
1000 KM
UTC -1
UTC
UTC +1
UTC +2
UTC +3
SOUTH
ATLANTIC
OCEAN
INDIAN
OCEAN
ASCENSION ISL.
SEYCHELLES,
REUNION ISL.
MAHE
SEYCHELLES
INDIAN
OCEAN
Farquhar Isl.
MADAGASCAR
Albatros I.
MAURITIUS
FOREST SIDE
ST. DENIS
CAMEROON
EQUATORIAL GUINEA
GABON
CONGO
ZAIRE
UGANDA
KENYA
SOMALIA
RWANDA
BURUNDI
TANZANIA
ANGOLA
ZAMBIA
MALAWI
MOZAMBIQUE
ZIMBABWE
NAMIBIA
BOTSWANA
SOUTH
AFRICA
LESOTHO
SWAZILAND
COMOROS
MAYOTTE
MADAGASCAR
Zanzibar
BUEA
DOUALA
YAOUNDE
BERTOUA
BANGUI
LIBREVILLE
MELENE
MOYABI
TCHIBANGA
OYEM
BRAZZAVILLE
POINT NOIRE
CABINDA
M'BANZA KONGO
UIGE
KINSHASA
KIKWIT
BANDUNDU
FORT ROUSSET
MBANDAKA
KISANGANI
KANANGA
MBUJI-MAYI
LUANDA
N'DALATANDO
MALANJE
SAURIMO
SUMBE
LOBITO
BENGUELA
KUITO
LUENA
LUBANGO
MENONGUE
NAMIBE
YAMBIO
JUBA
ARUA
BOBI
BUNIA
BUTEBO
MAWAGGA
KAMPALA
KABALE
KIGALI
GITEGA
KIGOMA
KITALE
MARALAL
KISUMU
MERU
KOMA ROCK
LANGATA
VOI
MOMBASA
ARUSHA
DODOMA
DAR ES SALAAM
DOLE
MOGADISHU
MBEYA
KARONGA
NACHINGWEA
LUBUMBASHI
KITWE
LUSAKA
LILONGWE
CHIMOIO
LICHINGA
MANGOCHI
NAMPULA
TETE
HABOHO
PEMBA
PAMANDZI
HARARE
GWERU
BULAWAYO
QUELIMANE
TALATA VOLONDRY
ANTANANARIVO
TSUMEB
WINDHOEK
GOBABIS
SELEBI-PHIKWE
SEBELE
GA-RANKUWA
BEIRA
INHAMBANE
MPANGELA RANGE
MBABANE
MAPUTO
MASERU
MEYERTON
UMZIMKULU
UMTATA
KOMGA

16) EP do Moxico	5191	5	0455-2300
7) EP de Benguela	6154	1	0350-2300
1) EP do Lobito	7151	1	0400-1700
8) EP da Huila	7350		
1) EP do Lobito	11815		

*) inactive as of late 1992.
FM: 2) 92 & 95MHz — 4) 104.5MHz.
Addresses & other information:
All st's rel N. from Luanda at 0500, 1200, 1900, 2200 and/or 2300. Many st's also rel. headlines.
1) C.P. 56, Lobito — Dir: Ribeiro dos Santos Tadeu. CE: António Mendes Filipe. **V.** by QSL-card. Rec. acc.
2) C.P. 125, Huambo — Dir: Nelson de Jesus Ferreira dos Santos. PD: José Manuel Lucas. TD: Francisco António Monteiro. Dir. Admin: Maria Isabel Faria Serrao. (Umbundo). **F.PL:** 50/10kW.
3) M'Banza Kongo.
4) C.P. 10, Sumbe — Dir: S. Daniel Neto. Tec. Asst: António Mendes Filipe.
5) C.P. 36, Menongue.
6) C.P. 83, Malanje.
7) C.P. 19, Benguela — Dir: José Cabral Sande. **V.** by letter.
8) C.P. 111, Lubango.
9) C.P. 116, Saurimo.
10) C.P. 258, N'dalatando — Dir: Agostino Mateus Augusto.
11) C.P. 140, Uige.
12) C.P. 174, Namibe — **Spanish:** irr. 2000/2030-2150v.
13) C.P. 33, Kuito — Dir: Cordeiro Chimo.
14) C.P. 1247, Luachimo.
15) Cabinda — **N. in French:** 2100. **V.** by letter.
16) C.P. 74, Luena.

ASCENSION ISLAND (British)

L.T: UTC — **Pop:** 1.800 — **Radio sets:** 1.000 (est.) — **Pr.L:** English — **E.C:** a/c 50, 220V — **ITU:** ASC.

VOLCANO RADIO (USAF)

Operated by the Ascension AAF personnel for the benefit of residents.
ADDR: Volcano Radio Station, Ascension Island P.O. Box 4235, Patrick AFB, Fla. 32925-0235, USA.
MEDIUMWAVE: ZD8VR 1602kHz 1kW authorized.
FM: 95.0MHz 40W and 98.8MHz 0.4kW (stereo).
D.PRGRS. in English: 24h Prgr. 1 on 1602kHz + 95.0MHz, Prgr. 2 on 98.8MHz (stereo).

RADIO ASCENSION (BBC operated)

Operated by the BBC Atlantic Relay for the benefit of residents.
ADDR: c/o BBC Atlantic Relay Station, English Bay, Ascension Island, South Atlantic. **Te:** 4457/4458.
STATION (English Bay): 1485kHz 0.5kW—**FM:** 93.2MHz 0.015kW.
D.PRGRS. in English: 24h (relay BBC World Sce. exc. for occ. local prgrs).

BBC ATLANTIC RELAY

ADDR: see R. Ascension. **Te:** 4458/4533. **Fax:** 6117. **Telex:** 3211 AV.
STATION: (G.C: 14.23W/7.54S): 6 × 250kW tr's.

kHz	kHz	kHz	kHz	kHz	kHz	kHz
6005	9560	11765	15105	15390	17830	21640
6110	9600	11820	15160	15400	17880	21660
6155	9610	11860	15190	17790	21470	
7105	11750	11880	15260	17810	21490	

VOICE OF AMERICA RELAY STATION

STATION: 2 × 250kW tr's operated by the BBC.

kHz	kHz	kHz	kHz
6155	15160	17810	17880
11820	15255	17830	21490

D.PRGR: see USA.

BENIN

L.T: UTC+1h — **Pop:** 5.000.000 — **Radio sets:** 350.000 — **Pr.L:** French, English + 18 ethnic — **E.C:** a/c 50, 220/380V — **ITU:** BEN.

RADIODIFFUSION DU BENIN (Gov.)

ADDR: B.P.366, Cotonou. **Te:** +229 301096/300481. **Telex:** ORTB 5132 CTNOU - 5208 A.B.P.
L.P: Dir: Emile Désiré Ologoudou. Chief Tech Sces: Anastase Adjoko.
MW: Cotonou 1476kHz 50/20kW — **SW:** Cotonou (G.C: 02.35E/06.21N): 4870kHz 30kW — Parakou (G.C: 02.38E/09.20N): 5025kHz 10kW — **FM:** Cotonou 88.4/94.7MHz 0.05kW.
Cotonou: MF 0458-0900, 1055-1400, 1600-2300. SS 0555-2300.
N. in French: 0615MF, 0800SS, 1200SS, 1230MF, 1930, 2115.
Parakou: MF 0455-0800, 1100-1400, 1700-2100. SS 0700-2300.
ANN: "Ici R. Bénin, émettant de Cotonou". "Ici Parakou, Radiodiffusion du Bénin, station regionale" — **IS:** Bénin Tam-Tam.
V. by QSL-card. Rp. Rec. Acc. (Parakou by letter).

BOTSWANA

L.T: UTC + 2h — **Pop:** 1.300.000 — **Radio sets:** 1.1m — **Pr.L:** English, SeTswana — **E.C:** a/c 50, 220V — **ITU:** BOT.

RADIO BOTSWANA (Gov.)

ADDR: Private Bag 0060, Gaborone. **Te:** +267 (31) 352541. **Fax:** +267 (31) 371588. **Cable:** Broadcasts Gaborone. **Telex:** 2409 RBINF BD — **L.P:** Dir: T. P. Makgekenene. Dep. Dirs: B. Tafa (Prgrs), H. Sosome (Tech). Asst. Dir : K. Reetsang (studios).

Tr's (50kW)	kHz	H.of tr.		kHz	H.of tr.
Selebi-Phikwe	621	0430-1700	Gaborone	4830	0300-2200
Orapa	648	0300-2200	Gaborone	7255	0300-1800
Gaborone	972	0300-2200	Gaborone	9600	0600-1530
Gaborone	3356	0300-0600 1530-2200			

FM (0300-2200): Gaborone 89.9MHz 5kW + 13 additional st's.
N. in English: 0510, 0800, 1110, 1400, 1610, 1910.
ANN: E: "This is R. Botswana broadcasting from Gaborone" or "You're listening to R. Botswana, the st. at the heart of the nation" or "This is the Gen. Sce. of R. Botswana". SeTswana: "Ke Seroma mowa sa Botswana mo Gaborone".
INT-SIG: Cow Bells. Imitation of farm animals at s/on. National Anthem at opening and closing — **V.** by letter.

VOICE OF AMERICA RELAY STATION

MW: Selebi-Phikwe 621kHz 50kW (cf. Radio Botswana).
SW: 4 x 100kW tr's.

kHz	kHz	kHz	kHz	kHz	kHz
6090	9885	13710	15375	15600	17705
7265	11940	15165	15445	15625	17870
7280	12080	15370	15495	17650	

D. PRGR: see VOA listing under "USA".

BURKINA FASO

L.T: UTC — **Pop:** 9.650.000 — **Radio sets:** 200.000 (est.) — **Pr.L:** French + 16 ethnic — **E.C:** a/c 50, 220/380V — **ITU:** BFA.

RADIODIFFUSION-TÉLÉVISION BURKINA (Gov.)

ADDR: B.P.7029, Ouagadougou. **Te:** +226 336805. **Telex:** MININFO 5132UV — Dir: Bayer Balawo.
STATIONS (G.C: 01.31W/12.22N):

kHz	kW	kHz	kW	kHz	kW	kHz	kW
747	100	1341*	1	4815	50	7230	50

*) Standby tr — **FM:** 92.0MHz 0.02kW.
D.PRGR. on MW/FM: 0530-0700MF, 0700-0900D, 0900-1200SS+Thurs, 1200-2400D. On 4815kHz: s/on-0800 & 1700-2400. Other times on 7230kHz. **N. in French:** 0600MF, 0630MF, 1000SS+Thurs, 1300, 1900, 2200.
ANN: "R. Burkina" — **IS:** Balafon — **V.** by letter. Rp.

RADIO BOBO (Reg. St.)

ADDR: B.P.392, Bobo-Dioulasso. **Te:** +226 991158. **Cable:** Radio-Bobo — **L.P:** St. Mgr: Daba Sereme — **MW:** 1008kHz 10kW. **FM:** 92.0MHz 0.02kW.
D.PRGR. in French & ethnic languages: MF 0600-0800, 1200-1400, 1600-2400. SS 0800-2400.

RADIO HORIZON FM STEREO — FRÉQUENCE MAGIQUE (Comm.)

ADDR: P.O. Box 2710, Ouagadougou. **Te:** +226 312858/308547. **Fax:** +226 312858 — Bobo-Dioulasso. **Te:** +226 972727/970666.
L.P: D.G: Moustapha Laabli Thiombano.
STATIONS (FM): Ouagadougou 104.4MHz — Bobo-Dioulasso 102.7MHz.
D.PRGR: 24h in French, English and ethnic languages.
NB: The first private radio station in Francophone We. Africa.

BURUNDI

L.T: UTC+2h — **Pop:** 5.856.000 — **Radio sets:** 500.000 (est.) — **Pr.L:** Kirundi, Swahili, French, English — **E.C:** a/c 50, 220/380V — **ITU:** BDI.

LA RADIODIFFUSION ET TÉLÉVISION NATIONALE DU BURUNDI (Gov.)

ADDR: B.P.1900, Bujumbura. **Te:** +257 23742. **Cable:** Radiodiffusion. **Tlx:** BDI 5119 — DG: Donatien Nahimana.
SW: Gitega (G.C: 29.21E/03.23S): 6140kHz 100kW.
FM: 88.5/91.6/95.6/98.2/99.2MHz.
D.PRGR: W 0300-0700 & 0900-2100, Sun 0300-2100. **N. in French:** 0420, 1100, 1800, 2000. **N. in English:** 1145, 1530.
FM (Ch. 1 in Kirundi, Ch. 2 in French/Swahili/English): as MW.
ANN: "Ici Bujumbura, Radiodiff. Nationale de la République du Burundi".
IS: Drums — **V.** by QSL-folder. Rec. acc.

CAMEROON (Republic of)

L.T: UTC + 1h — **Pop:** 11.698.000 — **Radio sets:** 2m — **Pr.L:** French, English, Vernaculars — **E.C:** a/c 50, 220/230V — **ITU:** CME.

CAMEROON RADIO TELEVISION (C.R.T.V.) (Gov.)

ADDR: B.P. 1634, Yaoundé. **Te:** +237 23088. **Cable:** Radiocam. **Telex:** 8888 KN.
L.P: Chmn: Henn Bandolo. Gen. Mgr: Prof. Mendoze.
HOME SCE: N = National Sce; P = Provincial Sce; U = Urban Sce.

kHz	kW	Pr	H. of Tr.	kHz	kW	Pr	H. of Tr.
899	20	N	All times	1286	1	U	All times
999	10	P	All times	4850	100	N	0430-2400

FM: 88.8MHz 10kW (N), 101.9MHz 10kW (P), 94.0MHz 0.1kW (U).
National Sce: 0400-0100. **N: in French:** 0530MF, 1200MF, 1900. **N. in English:** 0500MF, 0600Sun, 1400, 1800.
Provincial Sce. in French/English/Vernaculars: 0430-2400.
Urban Sce. in French: 0430-0600, 1700-1830. Relay Nat. Sce: 0600-1700.
ANN: N. "Ici Yaoundé, poste nationale de la CRTV". "This is Yaoundé, the National st. of CRTV". P: "Ici Yaoundé, station provinciale de CRTV du Centre". "This is Yaoundé, the provincial st. of CRTV"— **INT-SIG:** Drums or tam tam or flute or xylophone. **V.** by QSL-card. Rp. Re. in English, French.

PROVINCIAL STATIONS

Mediumwaves	kHz	kW		kHz	kW
5) Bertoua	972	20	1) Bamenda	1152	300/1
8) Maroua	1098	30	4) Garoua	1260	20
2) Douala	1106	5	6) Bafoussam	1448	20
Shortwaves	**kHz**	**kW**		**kHz**	**kW**
3) Buea	3970v	4	4) Garoua	5010	100
6) Bafoussam	4000	20	2) Douala	7150	100
5) Bertoua	4750	20	4) Garoua	7240	100
2) Douala	4795	100			

FM: 1) 94MHz, 2) 88.4/92.3/104.8MHz, 3) 101MHz, 5) 101MHz, 6) 88.3MHz.

Addresses & other information:
NB: All st's relay the N. in French & English from Yaoundé.
1) B.P. 4049, Bamenda, Province du Nord-ouest. **Te:** +237 361394. **D. Prgr:** 0430-0900, 1100-2300. **Local N. English:** 0515, 1705, 2130. **French:** 0640, 1930.
2) B.P. 986, Douala. **Te:** +237 429440. **Cable:** Radio Douala. **D. Prgr.** 0430-2300. On 7150kHz: 0650-1630. — **INT-SIG:** Tam-Tam — **ANN:** "Ici Douala, C.R.T.V., station provinciale de litoral". "This is Douala, the Literal provincial st. of CRTV" — **V.** by QSL-card. Rp.
3) P.M.B. Buea. **Te:** +237 322525. **Cable:** Radio Buea. **D.Prgr:** 0430-0800 (0645SS) & 1100-2305. **Local N: French/English:** 0545, 1715, 1920, 2130 — **ANN:** E: "This is Buea, the So.We. Provincial Station of CRTV". F: "Ici Buea, Radiodiffusion de la République du Cameroun, Station Provinciale du Sud-Ouest" — **V.** by QSL-card(B).
4) B.P. 103, Garoua. **Te:** +237 221179. **D. Prgr:** 0430-2300. On 7240kHz: 0650-1630. **Local N. in French:** 0615MF — **ANN:** "Ici Garoua, Station Provinciale de la C.R.T.V. du Nord". E: "This is Garoua, Provincial St. of CRTV" — **INT-SIG:** Flute and Tamtam — **V.** by Letter. Rp.
5) P.O. Box 230, Bertoua, Eastern Province. **D. Prgr:** 0430-2300. **Local N. in English:** 2100 — **ANN:** E: "This is Bertoua, the Eastern Provincial Station of R. Cameroon". F: "Ici Bertoua, Radiodiff. du Cameroun, Station Provinciale de l'est" — **INT-SIG:** Drums — **V.** by letter (2 IRC's req.)
6) B.P. 970, Bafoussam, Province de l'Ouest. **Te:** +237 441184. **D. Prgr:** 0430-0800 & 1630-2305 (0430-2305SS). **Local N. in French/English:** 0700, 0800, 1700, 2200 — **ANN:** "Ici Bafoussam, Station Provinciale de R. Cameroun". "This is Bafoussam, the Western Provincial Station of R. Cameroon." **INT-SIG:** Xylophone. All st's open and close with the National Anthem, and spoken id's in English and French. — **V.** by QSL-card.

CANARY ISLANDS

L.T: UTC (Su: UTC + 1h) — **Pop:** 1.650.000 — **Radio sets:** 600.000 (est.) — **Pr.L:** Spanish — **E.C:** a/c 50, 110/220V — **ITU:** CNR.
Abbreviations: GC — Gran Canaria; GCF — Fuerteventura; GCL — Lanzarote; TF — Tenerife; TFP — Isla de la Palma; TFG = Isla de la Gomera. (For network abbreviations please refer to Spain).

STATIONS: Mediumwaves

	kHz	kW	Net	Location	Island
1)	621	100	RNE-1	Sta. Cruz	TF
1)	720	20	RNE-5	Sta. Cruz	TF
1)	747	20	RNE-5	Las Palmas	GC
4)	837	10	COPE	Las Palmas	GC
5)	882	20	COPE	La Laguna	TF
6)	1008	10	SER	Las Palmas	GC
7)	1098	2	RNE-5	Sta. Cruz	TFP
9)	1179	20	SER	Sta. Cruz	TF
8)	1269	20	ECCA	Las Palmas	GC

FM:	MHz	kW	Net	Location	Island
7)	89.6	2	RNE-5	Sta. Cruz	TFP
28)	89.7	4	A3	Arrecife	GCL
1)	90.0	60	RNE-3	Izaña	TF
8)	90.4	2	ECCA	Las Palmas	GC
12)	90.7	4	SER	Arrecife	GCL
13)	91.1	8	A3	Sta. Cruz	TF
33)	91.6	2	D	Los Realejos	TF
4)	91.8	1	COPE	Las Palmas	GC
1)	92.3	60	RNE-1	Izaña	TF
1)	92.8	50	RNE-1	La Isleta	GC
8)	93.0	1	ECCA	Pto del Rosario	GCF
9)	93.2	25	SER	La Laguna	TF
9)	93.3	2.5	SER	La Gomera	TCF
1)	93.9	10	RNE-5	Finca España	TF
14)	94.4	8	R.80	Las Palmas	GC
32)	94.4	2	OCR	Tacoronte	TF
1)	95.1	50	RNE-2	La Isleta	GC
5)	95.1	2	COPE	Sta. Cruz	TFP
15)	95.3	4	SER	Maspalomas	GC
6)	95.8	8	SER	Las Palmas	GC
1)	96.2	60	RNE-2	Izaña	TF
16)	96.8	5	M	Arrecife	GCL
5)	97.1	2	COPE	La Laguna	TF
17)	97.5		L	Guía	GC
1)	97.7	—	RNE-5	La Isleta	GC
7)	98.4	5	RNE-5	El Paso	TFP
1)	98.5	50	RNE-3	La Isleta	GC
35)	98.6	2	A3	Llanos de Aridane	TFP

18)	99.1	2	RM	Pto. la Cruz	TF
8)	99.5	2	ECCA	Sta. Cruz	TFP
19)	99.8	2	SER	La Orotava	TF
31)	100.0	1	OCR	Adeje	TF
20)	100.0	5	IND	Arrecife	GCL
13)	100.1	8	R.80	Sta. Cruz	TF
11)	100.3	2	A3	Telde	GC
21)	101.1	2	RM	La Laguna	TF
23)	101.2		IND	Fuerteventura	GCF
5)	101.4	—	COPE	Isla Baja	TF
22)	101.4	2	RM	Las Palmas	GC
1)	102.1	10	RNE-2	Las Mesas	TF
23)	102.2		IND	Pto. del Rosario	GCF
10)	102.4	1	A3	Playa del Inglés	GC
1)	102.7	5	RNE-1	El Paso	TFP
24)	103.0	3.7	A3	Las Palmas	GC
25)	104.0	4	L	Arrecife	GCL
26)	104.1		M	Arucas	GC
1)	104.5	5	RNE-2	El Paso	TFP
29)	105.0		IND	Fuerteventura	GCF
27)	105.7		L	Las Palmas	GC
1)	105.7	10	RNE-3	Las Mesas	TF
1)	106.1	1	RNE-3	El Paso	TFP
30)	106.8	8	OCR	Las Palmas	GL
34)	107.0	—	L	Tegueste	TF

Addresses and other information

1) **R. Nacional de España,** Aptdo. 1233, Santa Cruz de Tenerife 38001. **Te:** +34 (22) 288162. **D.Prgr: R.1:** 24h on 621kHz. **N:** On the h. **R.2:** (classical music) 24h. **R.3:** 24h. **R.5:** 24h.

4) **R. Popular de Las Palmas,** Av. Escaleritas 60, Las Palmas 35011. **Te:** +34 (28) 255840. Dir: José Antonio Baeza Betancor. **D.Prgr:** 0500-0100 (occ. 24h). **N.** summaries in English, German & Swedish between 1330 & 1830, produced by **Canary Tourist R,** Paseo de Chil 117, 35014 Las Palmas. Dir: Xavier Palin — **FM:** 91.8 R. Insular.

5) **R. Popular de Tenerife,** Obispo Rey Redondo 49, La Laguna 38201. **Te:** +34 (22) 261211. Dir: José Silverio Pérez. **D.Prgr:** 0700-2400.

6) **R. Las Palmas,** Av. Rafael Cabrera 10, Las Palmas 35002. **Te:** 215063. Dir: Andrés Roda Díaz. **D.Prgr:** 0815-0100. FM: **R. Gran Canaria.**

7) **R. Nacional de España,** Plaza de la Constitución 4, Santa Cruz de la Palma 38700. **Te:** +34 (22) 288400. Dir: Julio Marante Díaz. **D.Prgr:** 0800-2300.

8) **R. ECCA,** Av. Mesa y López 38 (or: Aptdo. 994), Las Palmas 35007. **Te:** +34 (28) 275454. Dir: Luis Espina Cepeda. **D.Prgr:** 0655-2300 (Sun 0655-2100). FM 93.0: Plaza de España, Puerto del Rosario, Fuerteventura; FM 99.5: Av. Francisco de Abreu 6, Santa Cruz de la Palma 38700.

9) **R. Club Tenerife,** Av. de Anaga 35, Santa Cruz de Tenerife 38001. **Te:** +34 (22) 207400. Dir: Francisco Padrón García. **D.Prgr:** 24h.

10) **Antena 3, R. Canarias Sur,** Av. Tirajana 1, Playa del Inglés 35290. **Te:** 764863. **D. Prgr.** in Spanish 24h. exc: W. 0910-1000, 1700-1800, 1910-2100 **German,** 1000-1100 **Dutch,** 1800-1900 **Swedish.**

11) **Antena 3, R. Canarias,** Cruz de Ayala 16, Telde 35214. **Te:** +34 (22) 690294.

12) **R. Lanzarote,** Edif. Méjico 1, Av. Mancomunidad, Arrecife 35580. **Te:** +34 (28) 811517. Dir: Agustín Cruz. **D. Prgr.** in Spanish 0700-2400 exc: Mon-Fri 1530-1630 **German,** 1830-2000 **English.**

13) **Antena 3/R. 80 Serie Oro Tenerife,** Av. Villalba Hervás 1, Santa Cruz de Tenerife 38002. **Te:** +34 (22) 244222. Dir: Gabriel Mesa Barrera. **D. Prgr:** 24h.

14) **R. 80 Serie Oro,** José Franchy Roca 5, Las Palmas 35007. **Te:** +34 (28) 279911. Dir: Gabriel Mesa Barrera. **D.Prgr:** 24h. **DX Prgrs** produced by José Luis Martín: Sat 1300-1330 "El Mundo de las Ondas", Mon 2030-2100 "Radioaficionados", Mon 1205-1220 "Historia de la Radio".

15) **R. Maspalomas,** Morro Besudo s/n, Maspalomas 35100. **Te:** +34 (22) 765853. Dir: Juan José Roda. **D.Prgr.** in Spanish 24h. exc: W. 0830-0900, 1705-1800 **German,** 0905-0930 **English,** 0930-1000 **Swedish.**

16) **R. Onda Insular,** García Escámez 159, Arrecife 35580. **Te:** +34 (28) 815750. Dir: Armando de León Expósito. **D.Prgr:** 0700-2400.

17) **R. Guía,** 35450 Sta. María de Guía.

18) **R. Minuto,** Pl. de los Reyes Católicos 3, Puerto la Cruz 38400. **Te:** +34 (22) 380012. Dir: Ignacio Baute España.

19) **R. Club Sur,** Av. Marítima s/n, Playa de Américas 38660. **D.Prgr:** 0800-0100 (Fri 0300, Sat 0200).

20) **R. Tourist Lanzarote,** Av. Fred Olsen 14 (or: Aptdo. 260), Arrecife 35580. Prgr. Dir: Barbara Graf. **D.Prgr:** 0655-2300. **N.** in English on the h., in German on the halfhour. **German** 0740-0930, 1530-1630, **Scandinavian** 1630-1730, **English** 0930-1100, 1800-2000.

21) **R. Minuto.** Addr. as st. 9).

22) **R. Minuto,** Paseo de Chil 1013, Las Palmas 35010. **Te:** +34 (28) 268900.

23) **R. Fuerteventura,** Cervantes 122, 35600 Puerto del Rosario (Fuerteventura).

24) **Antena 3 Las Palmas,** c/ Mendoza, 6 Urb. El Cebadal, Las Palmas 35008. **Te:** 260812. Dir: Alfonso Oshanaham.

25) **R. 106,** Aptdo. 28, Arrecife 35580 (Lanzarote).

26) **R. Myal.,** Ayuntamto 35400 Arucas.

27) **Guiniguada R.,** Ap. 67, 35080 Las Palmas, Las Palmas de Gran Canaria.

28) **R. Volcán,** Luis Morote 20, 35500 Arrecife (Lanzarote).

29) **R. Archipielago,** La Juventud 31, 35600 Fuerteventura.

30) **Onda Cero Color,** León y Castillo 41, 35005 Las Palmas.

31) **Onda Cero R.,** C.C. San Eugenio Adeje, 38660 Playa Las Américas.

32) **Onda Cero, R. 21,** C.C. Tacoronte, 38350 Tacoronte.

CAPE VERDE (Rep. of)

L.T: UTC − 1h — **Pop:** 398.000 — **Radio sets:** 50.000 (est.) — **Pr.L:** Portuguese, Crioulo — **ITU:** CPV.

RADIO NACIONAL DE CABO VERDE (Gov.)

ADDR: C.P. 26, Praia, Ilha de Santiago. **Te:** +238 613729 (studio); C.P. 29, Mindelo, Ilha S. Vicente. **Te:** +238 312453/311731; C.P. 40, Espargos, Ilha do Sal. **Te:** +238 411444/411333.
L.P: Dir: Carlos Gonçalves.
STATIONS: 92.1/96.2/98.1MHz + 8 relay st's.
D.PRGR: 0730-0100.

CENTRAL AFRICAN REPUBLIC

L.T: UTC + 1h - **Pop:** 3.030.000 — **Radio sets:** 550.000 — **Pr.L:** French, Sango, ethnic — **E.C:** a/c 50 — **ITU:** CAF.

RADIODIFF.-TÉLÉVISION CENTRAFRICAINE (Gov.)

ADDR: B.P.940, Bangui. **Te:** +236 612588 or 611650.
L.P: TD: Michael Bata.
MW: 1440kHz 20/50kW — **SW** (G.C: 18.35E/04.21N): 5034v/7220kHz 20/50/100kW — **FM:** 87.78/90.2MHz.
D.PRGR: 0430-2300 on 1440kHz + FM, 0700-1630 on 7220kHz, other times on 5034kHz. **N. in French:** 0500, 0600, 0700, 0800, 1300, 1800, 2100.
ANN: F: "Ici Bangui, Station Nationale de la Radiodiffusion Télévision Centrafricaine" — **IS:** Repeated piano chord. Opens and closes with National Anthem — **V.** by QSL-card.

CHAD

L.T: UTC + 1h — **Pop:** 5.230.000 — **Radio sets:** 1.260.000 (est.) —**Pr.L:** French, Arabic, and 8 other languages — **E.C:** a/c 50, 220V — **ITU:** TCD.

RADIODIFFUSION NATIONALE TCHADIENNE (Gov.)

ADDR: B.P.892, N'Djamena. **Te:** +235 516071. **Cable:** Radio Tchad.
L.P: Dir.RNT: Khamis Togoï. S/Dir.of Tech.Sces: Raphael Mbaissané — **R. Rurale** (prgrs for rural listeners) : S/Dir: Amadou Batouré. **Te:** +235 515238.
STATIONS: 840kHz 20kW — 4904.5/7120kHz 100kW — **FM:** 94.05MHz 0.1kW.
D.PRGR. on MW/FM: 0430-0730 & 1025-2200MF, 0430-2200SS. On 7120kHz: 1025(0730SS)-1600. Other times on 4904.5kHz. **N. in French:** 0530, 1300, 1330, 1830, 1900, 2100. **N. in colloquial Arabic:** 1700.

ANN: "Ici N'Djamena, Radiodiffusion Nationale Tchadienne".
IS: Balafon — **V.** by QSL-card.

RADIO MOUNDOU (Gov.)

ADDR: Station Regionale de la RNT, B.P.122, Moundou. **Te:** +235 691322 — **L.P:** St. Dir: Derong Guela Ketchoua.
STATION: 5286.1kHz 5kW.
D.PRGR: 0500-0800, 1400-1800. **Arabic:** 0700-0730, 1515-1545. **Sara:** 0600-0700, 1430-1515. Other times in French. **N:** 0545, 1700.
ANN: "Ici Moundou, Station Régionale de la Radiodiffusion Nationale Tchadienne, émettant de Moundou" — **V.** by letter.

RADIO SARH (Gov.)

ADDR: Station Régionale de la RNT, B.P.270, Sarh. **Te:** +235 681422 — **L.P:** Dir: Dimanangar Djainta.
STATION: 850kHz 1kW — **D.PRGR. in French/Sara:** 1500-1800.

RADIO ABÉCHÉ (Gov.)

ADDR: Station Régionale de la RNT, B.P.105, Abéché. **Te:** +235 698149 — **L.P:** Dir: Sanoussi Saïd
STATION: 5900kHz 1kW.
D.PRGR. in French/Arabic: 0500-0600, 1630-1800.
ANN: "Ici Radio Abéché, station regionale de la Radiodiffusion Nationale Tchadienne émettant d'Abéché".

COMOROS (Fed. Islamic Rep.)

L.T: UTC + 3h — **Pr.L:** French, Comorian — **Pop:** 493.362 — **Radio sets:** 50.000 (est.) — **E.C:** a/c 50, 220V — **ITU:** COM.

RADIO COMORO (Gov.)

ADDR: B.P. 250, Moroni, Grand Comoro. **Te:** +269 2261.
L.P: Tech. Dir: Kombo Soulaimana.
STATIONS: Haboho (G.C: 43.15E/11.42S).

kHz	kW	kHz	kW	kHz	kW
1089	40	3330	60	7260	60

FM: 6×1kW, 3×0.5kW tr's.
D.PRGR: 0300-2000/2100. On 7260kHz: 0900-1500. Other times on 3330kHz. **N. in French:** 1000,1700.
ANN: "Ici R. Comoro" — **V.** by letter. Rp. Rec. acc.

CONGO (Republic of the)

L.T: UTC+1h — **Pop:** 2.382.000 — **Radio sets:** 250.000 (est.) — **Pr.L:** French, Lingala, Kikongo — **E.C:** a/c 50, 220V — **ITU:** COG.

RADIODIFF.-TÉLÉVISION CONGOLAISE (Gov.)

ADDR: B.P.2241, Brazzaville. **Te:** +242 812473. **Cable:** Radio-Congo. **Reg.Addr:** B.P.1063. Pt. Noire.
MW: Brazzaville 1476kHz 20kW, Ft. Rousset 863kHz 10kW, Pt. Noire 1520kHz 0.25kW — **FM:** 98.95/96.4MHz.
SW: Brazzaville (G.C: 15.18E/04.15S) — *) Pt. Noire (G.C: 11.52E/04.47S).

kHz	kW	H.of tr.	kHz	kW	H.of tr.
3265	25	Alt.to 5985	7105	25	0700-1100
4765	100	0400-0700	9610	50	0700-1600
		1700-2400	11710	100	1100-1700
5985	25	0400-0700	15190	50	1100-1700
		1700-2400			

National Netw. on MW/FM: 0400-2400. on SW: as above. **N. in French:** 0500, 0700, 1200, 1330, 1500, 1600, 1830, 2330.
Regional Netw. on 1520kHz: 0445-0700, 1030-1330, 1445-2000.
ANN: "Radiodiff. Nationale Congolaise", "R. Congo" or "RNC". "Ici Point Noire, station regionale de la Radiodiff. Nationale Congolaise" — **IS:** Zansi solo. Opens and closes with National Anthem — **V.** by letter. Rp. Rec. Acc.

COTE D'IVOIRE

L.T: UTC — **Pop:** 13.484.000 — **Radio sets:** 1.500.000 — **Pr.L:** French, Diola + 12 ethnic — **E.C:** a/c 50, 220/380V (yes) — **ITU:** CTI.

RADIODIFFUSION TELEVISION IVOIRIENNE (Gov.)

ADDR: B.P. V191, Abidjan 01. **Te:** +225 324800. **Cable:** Diraci 22635 — **L.P:** Dir. Gen: Mamadou Berté. Dir. Tec: Adje Kadjo.

MEDIUMWAVES: Abidjan 1493kHz 20kW, Bondoukou 1240kHz 1kW, Bouake 1578kHz 0.1kW.
SHORTWAVES: Abidjan (G.C: 03.57W/05.21N).

kHz	kW	kHz	kW	kHz	kW	kHz	kW
4940	25	6015	500	7215	20	11920	500

FM: Abidjan 88.9/92.0MHz + 9 relays.
Ch. 1: 0500-2400 on 1493/1578/4940/6015/7215kHz + 88.9MHz. **Main N:** 0530, 0615, 0700, 0745, 1245, 1900, 2200. **Headlines:** 0500, 0600, 0900, 1000, 1100, 1200, 1400, 1500, 2355. **Ethnic langs:** 1502-1830.
Ch. 2: 24h. on 1240/119.20kHz + 92.0MHz. **Main N:** 1330. **Headlines:** 0000-0400, 0830-1330 & 1530-1830 hourly, 2100, 2300. **Rel. Ch. 1:** 0455-0800. **English:** 1883-1930. **N:** 1845.
ANN: F: "R. Côte d, Ivoire Chaîne Un/Deux".
INT-SIG: Signs on with Clock Chimes and National Anthem.
V. by QSL-card. Rp.

DJIBOUTI (Rep.)

L.T: UTC + 3h — **Pop:** 355.315 — **Radio sets:** 30.000 (est.) — **Pr.L:** French, Somali, Afar, Arabic. — **E.C:** a/c 50, 220V — **ITU:** DJI.

RADIODIFFUSION-TÉLÉVISION DE DJIBOUTI (RTD-Gov.)

ADDR: B.P. 97, Djibouti. **Te:** +253 352294. **Telex:** 5863 DJ.
L.P: Dir: Mohamed Farah Moussa. Dir. Adj. Tech: Arreh Houssein. Tech. Chief: Moussed Yahya.
MEDIUMWAVES: 1170/1539kHz 20kW.
SHORTWAVES (G.C: 43.00E/11.30N): 4780kHz 20kW.
FM: 91.0/92.0MHz.
International Netw. in French on 1170kHz: Sat-Thurs 0300-0430, 1100-1300, 1500-1730. Fri 0500-0700, 0900-1100, 1500-1730. **N:** 0330, 1200, 1600. **Arabic:** Fri 0700-0900. **N:** 0700. Other times in French.
National Netw. on 1539/4780kHz+FM: 0300-0700 & 0900-2000 (Fri 0300-2000). **Somali:** 0300-0400, 0700-0900 (Fri), 1100-1300, 1600-1730 (Fri 1700). **Arabic:** 0600-0700 (exc. Fri), 1300-1600. **Afar:** 0430-0600 (Fri 0500-0700), 0900-1100, 1730-1900.
ANN: F: "Ici Djibouti"; Afar: "Ah Djibouti Radio"; Arabic: "Huna Djibouti"; Somali: "Hal Kani wa R. Djibouti".
V. by QSL-card or letter. Rp. Rec. acc.

EGYPT (Arab Republic of)

L.T: UTC+2h (Su: UTC+3h) — **Pop:** 57.702.000 — **Radio sets:** 14m —**Pr.L:** Arabic — **E.C:** a/c 50, 220V — **ITU:** EGY.

EGYPTIAN RADIO & TV UNION (Gov.)

ADDR: P.O. Box 1186, Cairo, Egypt. **Te:** +20 (2) 747120, 757155, 770355. **Cable:** CIBROTEV. **Tlx:** 92152.
L.P: Pres. ERTU: Amin Bassiouni. Chmn. Eng. Sector: Mahmoud M. Kishk. Chmn. Sound Prgr. Sector: Helmi El Bolok. Chmn. of TV Prgr. Sector: Abd. El Salam El Nadi.

Mediumwaves	kHz	kW	H. of tr. (Prgr.)
Cairo	558	40	0500-0700 (E), 0700-1000 Fri/Sun (E), 1000-2200 (E)
Batrah	621	1000	24h (6)
Tanta	621	25	24h (6)
El Kharga	702	10	0400-0600 (L), 0600-1000 (2), 1200-1600 (2), 1600-1900 (L), 2000-2200 (4)
Cairo	711	100	0600-1000 (2), 1200-2000 (2)
Qena	756	10	0400-0600 (4), 0600-1000 (2), 1200-2000 (2), 2000-2200 (4)
El Fayoum	756	1	0600-0800 (L), 0800-1100Fri (L), 1600-2000 (L)
Abis	774	500	0400-0630 (5), 1100-0100 (5)
Batra	819	1000	24h (1)
Cairo	819	50	24h (1)
Santah	864	500	0200-2200 (4)
Matruh	882	10	1100-0700 (1)

Hurghada	918	10	as 756kHz
Bawiti	918	10	24h (1)
Salum	936	10	24h (1)
Assiut	981	10	0400-0600 (4), 0600-0800 (L), 0800-1000 (2), 1200-1600 (2), 1600-2000 (L), 2000-2200 (4)
Baris	981	1	24h (1)
Abu Simbil	981	1	24h (1)
El Arish	1008	100	0400-0600 (7), 1200-1600 (7) 1600-0400 (5), 0600-1200 (5)
El Farafra	1008	1	24h (1)
Beni Suef	1008	1	as El Fayoum 756kHz
Shalatin	1044	7	24h (1)
Cairo	1071	60	0400-0700 (L), 1200-1500 (L), 1800-2200 (L)
El Minya	1080	10	24h (1)
Luxor	1080	10	24h (2)
Batrah	1107	600	0300-0500 (5), 0500-0600 (W), 0600-0800 (P), 1200-1500 (P) 1500-1800 (W), 2300-0300 (5)
Sohag	1143	10	24h (1)
Tanta	1161	60	0400-0600 (Fri 1100) (L) 1400-1900 (Thurs 2100) (L)
Quena	1179	10	24h (1)
Ras Gharib	1188	10	0400-2200 (1)
Alexandria	1197	25	0400-0700 (L), 0700-1200Fri (L), 1200-2200 (L)
Asswan	1278	10	24h (1)
Assiut	1305	10	24h (1)
Hurghada	1314	10	0400-2400 (1)
Nag Hamadi	1314	1	0400-2400 (1)
Abu Simbil	1314	1	as 756kHz
Cairo	1341	100	0400-1000 (3), 1300-0100 (3)
Bawiti	1341	10	as 756kHz
Idfu	1341	10	24h (1)
Siwa	1341	10	24h (1)
Quseir	1350	10	0400-2200 (1)
El Kharga	1368	10	0400-2200 (1)
El Farafra	1368	1	as 756kHz
El Dakhla	1386	10	0400-2200 (1)
Luxor	1386	10	as 756kHz
Ras Gharib	1422	10	as 756kHz
Salum	1422	10	as 756kHz
Sohag	1476	10	0400-0600 (4), 0800-1000 (2), 1200-1600 (2), 2000-2200 (4)
El Minya	1476	10	as Assiut 981kHz
El Tur	1485	1	0400-0700 (Fri 1100) (L) 0700-1000 exc.Fri (2), 1200-1600 (2), 1600-2000 (L), 2000-2200 (4)
El Arish	1503	25	0400-0700 (Fri 1100) (L), 0700-1000 exc.Fri (2), 1200-1600 (2), 1600-2000 (L)
Quseir	1575	10	0600-1000 (2), 1200-2000 (2)
Idfu	1584	10	0600-1000 (2), 1200-2000 (2)
Baris	1584	1	as 702kHz
Port Said	1584	1	0500-0900 (Fri 1100) (L) 1600-2200 (Thurs 2300) (L)
Matruh	1593	10	as 756kHz
El Dakhla	1602	10	as 702kHz
Nag Hamadi	1602	10	as 756kHz
Siwa	1602	10	as 756kHz

FM: As the FM tr's have limited coverage, details are not given here.

Prgrs: 1 = General Arabic Prgr, 2 = Youth Prgr (0600-1000, 1200-1700) & Education Prgr (1700-2000), 3 = Musical & Cultural Prgr, 4 = Holy Quran Prgr, 5 = Middle East Comm. Prgr, 6 = Voice of the Arabs, 7 = Hebrew Prgr, E = European Prgr. in English, French, German, Greek, Italian & Armenian, L = Local Prgr, P = Palestine Prgr, W = Wady El Nile Prgr.

SHORTWAVES: Z = Abu Zaabal (G.C: 31.22E/30.16N): 100/250kW, A = Abis (G.C: 30.05E/31.10N): 250kW, M = Mokattam (G.C: 31.15E/30.03N): 50/100kW.

kHz	kHz	kHz	kHz	kHz	kHz	kHz
6230	9755	9990	11875	15210	15335	17670
9475	9770	11560	11975	15220	15375	17690
9620	9800	11665	11980	15255	15420	17745
9670	9850	11715	12050	15285	15435	17770
9700	9900	11785	15115	15300	17595	17800
9740	9950	11865	15155			

RELAYS OF HOME SCES.

General Prgr.

kHz	Area	Times
9620Z	NAf/SEu	0350-0700
9670A	Eu	1800-2400
9770Z	NAf/SEu	0350-0700
9800Z	Gulf	0350-2400
9850A	NAf/SEu	1100-2400
11665Z	ME	0350-1800
11785	NAf/SEu	0700-1500
12050A	NAm/Eu	0200-0600
		0700-2400
	EAf	0700-1530
15115Z	WAf	0700-1100
17670Z	NAf/SEu	1300-1900

Voice of the Arabs

kHz	Area	Times
9700Z	NAf	1800-0030
9850A	NAf/SEu	0300-0600
11665Z	EAf/CAf	1900-0030
11980A	NAf/SEu	0600-1400
15285Z	Gulf	0300-0030

Holy Quran Prgr.

kHz	Area	Times
9755M	NAf/ME	0200-2200

FOREIGN SERVICE

	Area	kHz
Afar		
1530-1630	EAf/CAf	15155Z
Albanian		
1700-1800	Eu	9950A
Amharic		
1730-1900	EAf/CAf	15155Z
Arabic		
0030-0330	ENAm	9900A
0330-0430	WNAm	9900A
0900-1000	AUS	15435A
1100-1130	CAf/SAf	17800A
1115-1215	ME/AFG	17745A
1415-1615	WAf	15220Z
2345-0045	SAm	17770A
	CAm	15220Z
Bambara		
1930-2030	WAf	15375Z
Bengali		
1330-1430	SAs	17595A
English		
0200-0330	NAm	11865A
		9475A
1215-1330	SAs	17595A
1630-1830	CAf/SAf	15255A
2030-2200	WAf	15375Z
2115-2245	Eu	9900A
Fulani		
1930-2030	WAf	15335A
French		
2000-2115	Eu	9900A
2030-2230	WAf	15335A
German		
1900-2000	Eu	9900A
Hausa		
1900-2100	WAf	15210A
Hindi		
1500-1600	SAs	17690A
Indonesian		
1320-1450	SEAs	17770A
Italian		
1800-1900	Eu	9900A
Lingala		
1830-1915	CAf/SAf	15255A
Malay		
1215-1315	SEAs	17770A
Oulof (Wolof)		
1830-1930	WAf	15375Z
Persian		
1745-1945	ME	11560A
Portuguese		
2215-2330	SAm	15420Z
Pushtu		
1500-1600	AFG	15175A
Shona		
1645-1730	CAf/SAf	11875A
Ndebele		
1730-1815	CAf/SAf	11875A
Somali		
1630-1730	EAf/CAf	15155Z
Spanish		
0045-0200	NAm	9475A
	CAm	9740A
	SAm	11715A
Swahili		
1530-1730	EAf/CAf	11975Z
Thai		
1115-1215	SEAs	17770A
Turkish		
1615-1645	TUR	6230A
Urdu		
1600-1730	SAs	17690A
Yoruba		
2100-2200	WAf	15210A
Zulu		
1600-1645	CAf/SAf	11875A

ANN: General Prgr: "Idha'atu jumhuriya misr al'arabbiya min al-qahira". Voice of the Arabs: "Saut al-'arab, min al-qahira". Quran prgr: "Idha'atu-l-Quran min al-qahira". English to Africa: "This is the Voice of Africa from Cairo".

V. by QSL-card. Re. to Box 1186, Cairo. Schedules available on request.

EQUATORIAL GUINEA

L.T: UTC + 1h — **Pop:** 390.000 — **Radio sets:** 100.000 — **Pr.L:** Spanish, ethnic — **E.C:** a/c 50, 220/380V — **ITU:** GNE.

RADIO NACIONAL DE GUINEA ECUATORIAL

ADDR: Ap.749, Bata. **Te:** 382 — Ap.195, Malabo. **Te:** 2260/2097.
L.P: PD: Roman Manuel Mañe.
SW: Bata 4926v/5004vkHz 100kW — Malabo 6250kHz 10kW.
D.PRGR: Bata 0430-2200v, Malabo 0600-0800 and 1100-2300v. (Bata and Malabo carry common prgr. 0600-0800). **N:** 2100.
IS: Bata: opens with tribal music and nat. anthem. Malabo: opens and closes with nat. anthem.
V. Bata by QSL-card, Malabo by letter.

RADIO AFRICA/RADIO EAST AFRICA (Comm.)

ADDR: Box 851 Malabo — **SW:** 7200v/9585kHz 100kW.
R.Africa on 7200kHz: 1700-2200. **R.Ea.Africa** 9585kHz: 0500-1600SS (mostly religious prgrs. in **English**) — **V.** by QSL-card.

AFRICA 2000 (Cult.)
ADDR: Cooperación española, Embajada de España, Malabo — PD: Concha Chamorro.
SW: 6910kHz 10kW — **FM:** 90.9MHz 10kW.
D.PRGR: 1300(1045SS)-2130. **N:** 1400, 1415* — *) rel. R. Nacional.

ETHIOPIA

L.T: UTC + 3h — **Pop:** 54.840.000 — **Radio sets:** 3.300.000 — **Pr.L:** Amharic, English, French, Arabic, Somali, Afar, Tigre, Oromo — **E.C:** a/c 50, 127/220V — **ITU:** ETH.

VOICE OF ETHIOPIA (Gov.)
ADDR: P.O. Box 1020, Addis Ababa. **Te:** +251 (1) 121011, 121063. **Cable:** Ethiocast.

Mediumwaves	kHz	kW		kHz	kW
Harar	855	100	Addis Ababa	990	1
Addis Ababa	873	100	3 st's	1485	1

Shortwaves: Gedja (G.C: 38.38E/08.47N): 100kW.

kHz	kHz	kHz	kHz	kHz
5990	7110	7165	9560	9706.4

NATIONAL SCE. on 855/873/5990/7110/9706.4kHz: Mon-Fri 0330-0600, 0900-1000, 1500-2000; Sat 0330-0600, 1000-1400, 1500-2000; Sun 0400-0900, 1000-1400, 1500-2000.
ANN: Amharic: "Yeh Ye-Ethiopia Demts Beherawi Naw".
INT-SIG: Electronic keyboard followed by time signal.
EXT. SCE. on 990/7165/9560kHz: 1200-1300 Somali, 1300-1400 Afar, 1400-1500 Arabic, 1500-1600 **English,** 1600-1700 Amharic, 1700-1800 French.
ANN: "This is the Ext. Sce. of the V. of Ethiopia broadcasting in English" — **INT-SIG:** short tune on electronic keyboard — **V.** by letter (IRC's acc.).

VOICE OF THE BROAD MASSES OF ERITREA (operated by Provisional Government of Eritrea)
MW: Asmara 945kHz 50kW — **SW:** 3940/7020/7380kHz.
D.PRGR: 0300-0700, 0900-1100, 1400-1700.

GABON

L.T: UTC + 1h — **Pop:** 1.095.000 — **Radio sets:** 250.000 — **Pr.L:** French, Fang, Bopounou — **E.C:** a/c 50, 220V (no) — **ITU:** GAB.

RADIODIFFUSION TÉLÉVISION GABONAISE (Gov.)
ADDR: B.P. 10150, Libreville. **Te:** +241 732784 (Châine 1: +241 732025). **Telex:** RTG 5342 GO. **Regional addr:** B.P. 270, Franceville; B.P. 320, Port Gentil. **Te:** +241 752511 — B.P. 459, Oyem. **Te:** +241 986004 — B.P. 280, Makokou. **Te:** +241 503019.
L.P: Dir. Gen: William Oyonne. Dir. Gen. Adj. Radio: Jacques Litona Loumbi. Dir. Gen. Adj. TV: Henri Joseph Koumba. Dir. Gen. Adj. Tech: Jules Legnongo.

Mediumwaves	kHz	kW	N		kHz	kW	N
Oyem	549	20	P	Melene	1554	20	N
Epila	990	20	P	Tchibanga	1575	15	P
Oyem	1206	1	N	Franceville	1602	1	N
Port Gentil	1520	1	N				

Shortwaves: Libreville 4777/7270kHz 100kW.
FM: Franceville 87.86/Oyem 87.94/Port Gentil 88.03/Tchibanga 91.04/Libreville 96.54MHz.
National Sce. (N) in French: 0455-2300. **Shortwave:** 0800-1600 on 7270kHz, other times on 4777kHz. **N:** 0500, 0530, 0600, 0700, 0800, 0900, 1000, 1130, 1200, 1400, 1500, 1700, 1830, 2100, 2200, 2230.
Second Prgr: 0500-2305 on FM only.
Provincial Netw. (P) in French & ethnic languages:
Franceville on 990kHz: 0430-0700, 1030-1430 & 1600-2300.
Oyem on 549kHz: 0430-0630, 1030-1430, 1600-2300.
Tchibanga on 1575kHz: 0430-0630, 1020-1400, 1600-2130/2145.
ANN: "Ici Libreville, vouz écoutez Radio Gabon, châine 1" (or "Ici . . . (town) . . . station provinciale") de la Radiodiff.-Télévision Gabonaise".
IS: Indigenous instruments. Opens and closes with Nat. anthem — **V.** by QSL-folder.

AFRICA No. 1 (Comm.)
ADDR: B.P. 1, Libreville. **Te:** +241 760001. **Telex:** 5558 GO. (Addr. in France: Havas Media International, 78 ave. Raymond Poincaré, F-75116 Paris. **Telex:** 649454F).
L.P: Dir's: M. Koumbangoye, G. Marquet. Dir. Tec: Marguerite Bayimbi. Dir. Prgrs. & Adv: A. Letamba. Dir. N: J. J. Mbourou.
STATION: Moyabi-Moanda (G.C: 13.31E/01.40S): 5 × 500kW tr's.

kHz	kHz	kHz
9580	15475	17630

FM: Libreville 94.5MHz.
D.PRGR. in French: to CAf (beam 330°): 0455-2300 on 9580kHz. To WAf (beam 307°): 0700-1600 on 17630kHz, 1600-2000 on 15475kHz.
N: 0530, 0630, 0730, 1115, 1230, 1800, 1830, 2100.
N. Headlines in English: 1258.
Adventist World Radio: W 1700-1800 in French to WAf on 9625kHz, Sun 1200-1300 in English on 17890kHz.
Relays of other broadcasters: See under Radio France Internationale, Swiss Radio International, Radio Japan.
ANN: "Ici Africa Numero Un, émettant de Libreville et de Moyabi au Gabon" — **V.** by letter (for own prgrs only).

GAMBIA

L.T: UTC — **Pop:** 900.000 — **Radio sets:** 180.000 — **Pr.L:** English, Wollof, Mandinka, Fula + local languages — **E.C:** a/c 50, 230V — **ITU:** GMB.

RADIO GAMBIA (Gov.)
ADDR: Mile 7, Banjul. **Te:** +220 25060. **Fax:** +220 27230/95102.
L.P: Ag. Dir. of Inf. & Broadc: E. Sagnia. Asst. Dir. of Broadc: E. Cole. Principal Eng: Y.A. Toure. Contr. Prgrs: (1) Mrs. S. Johnson, (2) Mrs. L. Hydara.
MW: Bonto 648kHz 10kW, Basse 747kHz 1kW.
FM: 92.0/104.0MHz.
D.PRGR: Mon-Thurs 0625-0800, 0955-1400, 1745-2400. Fri-Sun 0625-1500, 1645-2400. **N. in English:** 0700, 1300, 2200.
ANN: "You are tuned to R. Gambia from Banjul".
IS: Cora (harp) — **V.** by letter (B).

RADIO SYD (Comm.)
ADDR: P.O.Box 279/280, Banjul. **Te:** +220 21870. **Fax:** +220 26490. **Cable:** SYD — GM: Miss Constance Wadner Enhörning.
MW: Banjul 909kHz 2.5kW.
D.PRGR: 0600-0200. **N:** rel. R. Gambia. **Swedish** (Nov 15-April 15): 0800-0830, 2000-2030 — **V.** by letter.

GHANA

L.T: UTC — **Pop:** 15.870.000 — **Radio sets:** 3.593.920 — **Pr.L:** English, Hausa, French + 12 Ghanaian languages — **E.C:** a/c 50, 220-240V — **ITU:** GHA.

GHANA BROADCASTING CORPORATION (Autonomous Statutory Corp.)
ADDR: P.O. Box 1633, Accra. **Cable:** Broadcast Accra. **Te:** +233 221161. **Telex:** 2114.
L.P: Dir. Gen: George Aryee. Dep. Dir. Gen: David Anaglate. Dir. Radio: K. Amoako. Ag. Dir. Eng: O.A. Kwawukume. Dir. of Admin: Seth Sosu. Dir. of Finance: G. Ac-Djangmah. Dir. of TV: Prof. Mark Duodo. Dep. Dir. Public Affairs: Mrs. Maud Blankson-Mills. Head of Tech. Sces: A.M. Attiogbe. Comm. Mgr: A. Torto-Laryea.
Shortwaves (G.C: 00.10E/05.31N): 3366/4915/6130/7295kHz 50/10kW. **FM:** Accra 95.7MHz 0.5kW, Bolgatanga 89.67MHz 5kW, Han 93.97MHz 5kW + community st's (10W) at Apam, Dormaa Ahenkro, Swedru.
Radio 1 in English/Ghanaian languages on 4915kHz: 0525-0905 & 1225-2305 (Sat/Sun 0526-2305). **N: in English:** 0600, 0700, 0800, 1100SS, 1300, 1400, 1800, 2000, 2100, 2245.
Schools Prgr: Mon-Fri 0905-1110.
Radio 2 in English: 0525-0905 (Sat/Sun 1225) & 1705-2305 on 3366kHz, 1226-1705 on 7295kHz. **Schools Prgr:** Mon-Fri 0909-1110 on 7295kHz. **N:** on the h. & rel. Radio 1.

EXTERNAL SCE. on 6130kHz: English: 0645-0800 & 1845-2000. **French:** 0800-0900 & 2000-2100. **Hausa:** 1745-1845.
ANN: "This is Radio 1/2".
INT-SIG: Sign on with drum beat — **V.** by QSL-card (B). Re. on technical matters to Ag. Dir. Eng.

GUINEA (Republic)

L.T: UTC — **Pop:** 7.640.000 — **Radio sets:** 130.000 — **Pr.L:** French, Fulah, Maninké, Soussou — **E.C:** a/c 50, 220/380V (no) — **ITU:** GUI.

RADIODIFFUSION TÉLÉVISION GUINÉENNE (Gov.)

ADDR: B.P.391, Conakry. **Te:** +224 44 22 05. **Telex:** 22341 RTG.
L.P: Dir.National: Justin Morel Jr.

Mediumwaves	kHz	kW		kHz	kW
Kankan	603	25	Labé	1386	50
Conakry	1295	1	Conakry*	1404	200

*)r.inactive
Shortwaves: Conakry 4910kHz 18kW (also irr. on 6155, 7125, 9650 and 15310kHz).
FM: Conakry 88.55MHz 2.5kW + 7 regional st's.
D.PRGR. in French/Nat. languages: W 0600-0805 & 1215-2400, Sun 0800-2400. **N. in French:** 0645W, 0800W, 0915Sun, 1245W, 1615, 1800, 1900, 1945, 2200, 2350. **N. in English:** 1845 (irr.).
ANN: F: "R.Conakry" or "R.Guinea" — **IS:** Guitar.
V. rarely by letter. Re. to Dir. Tec.

GUINEA-BISSAU (Rep.)

L.T: UTC — **Pop:** 1.048.000 — **Radio sets:** 35.000 — **Pr.L:** Portuguese — **E.C:** a/c 50, 220/380V — **ITU:** GNB.

RADIODIFUSÃO NACIONAL (Gov.)

ADDR: C.P. 191, Bissau.
STATION: (G.C: 15.35W/11.51N): 1034kHz 5kW + FM 92.9MHz.
HOME SCE. in Portuguese: 0600-0800 & 1200-2400.
N: 0700, 2215.
ANN: "Escutar a Radiodifusão Nacional de Republica de Guiné-Bissau" — **V.** by QSL-card. Rp.

KENYA

L.T: UTC + 3h — **Pop:** 26.150.000 — **Radio sets:** 4.200.000 — **Pr.L:** English, Swahili, Hindustani, Vernaculars — **E.C:** a/c 50, 240 (yes) + some d/c — **ITU:** KEN.

KENYA BROADC. CORP (KBC)

ADDR: Box 30456, Nairobi. **Te:** +254 (2) 334567. **Fax:** +254 (2) 332797. **Cable:** Broadcasts Nairobi. **Telex:** 25361 KBC KE.
Local Addr: Box 799, Nyeri — Box 1585, Nakuru — Box 1327, Meru — Box 3287, Kibirichia — Box 90584, Mombasa — Box 844, Kisumu.
L.P: Chm: J. Kiano; Man. Dir: D. K. ole Nasieku; Prod. Mgr: S. Anabwani; Public Rel: C. Opiyo.

MEDIUMWAVES

Location	kHz	kW	Sce.	Location	kHz	kW	Sce.
Voi	540	100	G	Voi	981	100	N
Kisumu	557	20	W	Nyeri	1017	20	N
Nairobi	612	100	N	Mombasa	1080	20	G
Meru	702	100	N	Nakuru	1206	20	G
Garissa	720	10	N	Garissa	1224	10	N
Nairobi	746	100	G	Nyeri	1268	20	C
Mombasa	792	20	N	Garissa	1314	10	?
Kisumi	846	100	N	Maralal	1359	5	N
Meru	900	100	G	Nyeri	1368	20	G
Nakuru	918	20	N	Kitale	1494	5	G
Kisumu	954	100	N				

SHORTWAVES: Nairobi (G.C: 36.41E/01.22S): 5/10/100 kW — Koma Rock: (G.C: 37.09E/01.16S): 2 × 250 kW.

kHz	Sce.	kHz	Sce.	kHz	Sce.
4885	E/N	6045	G	7120	E
4915	C	6075	N	7140	N
4935	G	6150	N	7270	G

FM: Nairobi 90.0 MHz (N), 95.6 (N) 10kW, 89.7 (G) 1kW, 104.4 (G) — Timboroa 88.63 (N), 91.5 (G) 10 kW — Mombasa 100.8 (G), 101.4 (N).
NATIONAL SCE (N) in Swahili: 0200 (Sun 0230)-2010 (Sat/Sun-2110) on MW + FM and **SW:** s/on-0625 on 6075kHz; 0625-1325 on 7140kHz; 1325-s/off on 6150 kHz, 1715-2010 on 4885kHz.
GENERAL SCE (G) in English: 0200 (Sat/Sun 0230)-2010 (Sat/Sun 2110) on MW + FM and **SW:** s/on-0630 on 6045/4935kHz; 0900 (Sat/Sun 0620)-1315 on 7270 kHz; 1300-s/off on 4935kHz.
N: 0215 (W), 0400, 0600, 1000, 1300, 1600, 1800, 2000 (W), 2100 (Sat/Sun).
CENTRAL SCE (C) in Kikiyu/Kikamba/Kimeru/Hindi: 0230 (Sun 0700)-1930 (Sun 1700) on 1269/4915kHz + FM.
WESTERN SCE (W) in Kuria/Kisi/Kalenjin/Teso/Luyha/Luo: 0300-0530, 0900-1100, 1300-1945 on 557 kHz.
NORTH-EASTERN AND COASTAL SCE (E) in Somali/Turkana/Rendille/Boran on FM and **SW:** 0200-0700 (W) on 4885kHz; 0600-1330(Sun 1400) on 7120kHz, 1330 (Sun 1400)-1745 (Sun 1700) on 4885kHz.

ANN: "This is the General Sce. of the Kenya Broadc. Corp., Nairobi" or "KBC Nairobi". Swahili: Nairobi "Hii ni Sauti ya Kenya"; Mombasa: "Hii ni Mombasa" — **INT-SIG:** Flute & drum melody. Nat. Anthem at s/on and s/off — **V.** by QSL-Aerogramme.

LESOTHO

L.T: UTC+2h — **Pop:** 1.848.000 — **Radio sets:** 420.000 — **Pr.L:** Sesotho, English — **E.C:** a/c 50, 240V — **ITU:** LSO.

LESOTHO NATIONAL BROADCASTING SERVICE (Gov. Comm.)

ADDR: P.O. Box 552, Maseru 100. **Te:** +266 323561. **Fax:** +266 310003. **Telex:** 4340 LENA CLO. **Cable:** Media, Maseru.
L.P: Dir: T. Ntsane. CE: F.N. Letele.
STATIONS: Maseru: 891kHz 100kW — 4800kHz 100kW.
D.PRGR: 0300-2200. **N. in English:** 0430MF, 0500, 1130, 1600.
ANN: "This is R. Lesotho" of "This is the Lesotho National Broadcasting Sce, Maseru". Sesotho: "Se-ea-le-moea sa Lesotho, Maseru" — **INT-SIG:** Nat. Anthem at opening and closing.
V. by QSL-card. Rec. acc.

BBC LESOTHO RELAY STATION

STATION: (G.C: 27.32E/29.19S): 2 × 100kW SW-tr's. on 3255/6190/11940kHz — **D.Prgr:** see United Kingdom.

LIBERIA (Rep.)

L.T: UTC — **Pop:** 2.823.000 — **Radio sets:** 600.000 — **Pr.L:** English and 15 ethnic — **E.C:** a/c 60, 120V — **ITU:** LBR.

ELBC (in support of ECOMOG)

ADDR: P.O. Box 594, 1000 Monrovia 10. **Te:** + 231 271146.
Cable: Broadcasts.
L.P: Ag.Dep.Dir of Broadc: J. Rufus Kaine.
FM: 98.9MHz 2.5kW — 7275kHz 10kW.
D.PRGR: 0655-1000, 1600-1900. **N. in English:** 0900, 1800.
ANN: "R.ELBC, the V. of peace, harmony & reconciliation from Monrovia." — **V.** by letter.

ELBC (National Patriotic Front of Liberia)

ADDR: Box 242, Danane, Côte d'Ivoire.
MW: Gbarnga 558kHz 10kW — **SW:** 3255/6090kHz.
D.PRGR: 0600-2400 on 558kHz, 0600-0800 & 1745-2000 on 3255kHz, 0800-1745 on 3255kHz.

ELCM (Catholic, Rlg.)

ADDR: P.O. Box 3569, Monrovia. **Te:** + 231 221658.
STATION: Monrovia 97.8MHz 2.5kW.
D.PRGR: 0500-0900, 1800-2300.

RADIO ELWA
(Cultural Missionary Broadc. Sce. of the Sudan Interior Mission)

ADDR: P.O. Box 192, Monrovia.
FM: Monrovia (freq unknown) — **SW:** 2 × 10kW tr's are being installed.
F.PL. Re-start Ext. Sce. in 25 We. African langs. (previously assigned freqs: 3230, 4760, 6070, 9550, 11830, 11955kHz).

LIBYAN ARAB JAMAHIRIYA

L.T: UTC + 1h (Su: UTC + 2h) — **Pop:** 4.500.000 — **Radio sets:** 1m (est.) — **Pr.L:** Arabic — **E.C:** a/c 50, 127/130V — **ITU:** LBY.

LIBYAN JAMAHIRIYA BROADCASTING (Gov.)

ADDR: P.O. Box 9933, Soug Al Jama, Tripoli. **Te:** + 218 (21) 603191/5. **Cable:** Ida Shaabia Tripoli. **Telex:** 20010.

Mediumwaves	kHz	kW		kHz	kW
Tobruk	648	300	Tripoli	1053	50
Benghazi	675	100	Ajedabia	1080	40
Ghadames	711	50	Jalo	1080	20
Jefren	711	50	El Beida	1125	500
Sebha	711	50	Tripoli	1251	500
Sirte	792	20	Tripoli	1404	20
Sebha	828	300	Misurata	1449	20
Giaghboub	909	20	Brak	1485	1
Kufra	909	10			

FM: Tripoli 94.0MHz 1kW; El Beida 87.9MHz 1kW; Tobruk 98.4MHz 2kW.
Shortwaves: Tripoli/Sabrata(G.C: 13.11E/32.54N) 500kW exc. +)100kW. —°) Benghazi (G.C: 20.04E/32.08N): 100kW.

kHz	kHz	kHz	kHz	kHz	kHz	kHz
6155	6185+	9600°	11815	15235	15415	15450

HOME SCE. in Arabic: 0500-2350 on 675/792/909/1053/1080/1449/6185/9600kHz + FM; 1315-1745 on 15235/15415/15450kHz. **N:** 0600, 0900, 1200, 1330, 1600, 1700, 1900, 2130, 2330.
FOREIGN LANGUAGE PRGR on 1485kHz: French: 1400-1600. **N:** 1530. **English:** 1600-1800. **N:** 1730.
HOLY QURAN PRGR. on 1404kHz: 0500-1900 (Su: 0300-1930). On 1485kHz: 0500-1400. Exact times vary acc. to sunrise/sunset.

EXTERNAL SERVICE

ADDR: P.O. Box 4677 or 4396, Tripoli — Dir: Abdulla Elmegri.
"Voice of the Great Homeland" in Arabic: 1115-1315 & 1745-0430 on 648/711/828/1125/1251/1449/15235/15415/15450kHz. **N:** 1200, 1230, 1300, 1815, 2115, 2300, 0015, 0215, 0400.
East. European Sce in German, Romanian, Hungarian, Polish, Bulgarian, Czech, Slovak: 1600-1715 on 15415kHz.
ANN: Home Sce: "Idha'at al-ğamahiriyah al-'arabbiya al-libiyya ash-sha'abiyya al-ishtirakiyya". Voice of the Great Homeland: "Idha'at saut al-uatan al-'arabiy, saut ath-thaurati ash-sha'abiyya".
INT-SIG: Prgrs open and close with Nat. Anthem — **V.** by QSL-card.

MADAGASCAR (Dem. Rep.)

L.T: UTC + 3h — **Pop:** 12.575.000 — **Radio sets:** 1.500.000 — **Pr.L:** Malagasy, French — **E.C:** a/c 50, 110/220V — **ITU:** MDG.

RADIO-TELEVISION MALAGASY (Gov.)

ADDR: B.P. 1202, Antananarivo. **Te:** +261 223 81 or +261 217 45 — **L.P:** Dir. Gen: Simon Andriamialison.

Mediumwaves	kHz	kW	Mediumwaves	kHz	kW
Antananarivo	630	2x75	Antananarivo	1502	4
Mahajanga	1260	1	Tolagnaro	1554	1
Fianarantsoa	1275	1	Toamasina	1562	1
Toliary	1322	1	Antsinarnana	1602	1
Antananarivo	1394	4			

+12 st's 0.25kW.
FM: Antananarivo 99/100MHz, Toamasina 99MHz, Ambositra 99MHz.
Shortwaves (G.C: 47.35E/18.50S)

kHz	kW	Prgr.	H.of tr.
2495	100	1	0800-1400
3232v	10	2	1600 (Sat 1400, Sun 0900)-2000
3288v	100	1	0200 (Sat/Sun 0300)-0500, 1500-1900 (Sat/Sun 2100)
5005v	100	1	as 3288kHz
6135	30	1	0500-1500
7155	100	1	0300-1500
9690	10	1	0500-1300

1st. Prgr. in Malagasy: 0200 (Sat/Sun 0300)-1900 (Sat 2100) on MW (exc. 1502kHz). Rel. on SW as above. **N:** 0330MF, 0430SS, 0900, 1600, 1830.
Frequence 2 in French: 1600 (Sat 1400, Sun 0900)-2000 on 1502kHz + FM and SW as above. **N:** 1000 Sun, 1600MF. **N.in English:** 1745.
ANN: Malagasy: "Radio Madagasikara ity". F: "Radio-Télévision Malagasy" — **V.** by QSL-card. Re. in French. Rp.

RADIO NEDERLAND RELAY STATION

STATION: Talata/Volondry (G.C: 43.37E/18.43S): 2 × 300kW SW-tr's.

kHz	kHz	kHz	kHz
6020	9895	11890	15570
9590	9860	13700	17580
9605	11655	15150	21480

MADEIRA (Portuguese)

L.T: UTC (Su: UTC + 1h) — **Pop:** 300.000 — **Radio sets:** 200.000 (est.) —**Pr.L:** Portuguese — **E.C:** a/c 50, 220/380V — **ITU:** MDR.

CENTRO REGIONAL RDP/MADEIRA (Gov.)

ADDR: Rua dos Netos 27, 9000 Funchal. **Te:** +351 (91) 29155. **Fax:** +351 (91) 27740. **Telex:** 72111 RDPMAD.
L.P: Dir: Afonso de Almeida.
MEDIUMWAVES: Porto Santo 531kHz 10kW, Pico de Arieiro 603kHz 10kW, Funchal 1332kHz 1kW.
FM: Funchal 89.8MHz + 4 st's on 90.8/93.3/94.1/96.5MHz.
D.PRGR: 24h (Canal 1 on MW, Canal 2 on FM).
ANN: Canal 1: "RDP Madeira, Rádio Provincial".
V. by QSL-card via RDP Lisboa.

POSTO EMISSOR DE RADIODIF. DO FUNCHAL (Pr,Comm.)

ADDR: Rua Ponte de S. Lázaro 3, 9000 Funchal. **Te:** +351 (91) 30393. **Fax:** +351 (91) 21797. **Telex:** 72519 PEF.
MEDIUMWAVE: 1530kHz 10kW — **FM:** 92.3MHz 2kW (stereo).
Relay st: Santana 1017kHz 1kW.
D.PRGR: 0600-2400 (Sat 0100) on MW, 24h on FM (different prgrs) — **ANN:** MW: "Posto Emissor de Radiodif. do Funchal".

ESTACAO RADIO DA MADEIRA (Pr.Comm.)

ADDR: Caixa Postal 450, Pico dos Barcelos, 9006, Funchal. **Te:** +351 (91) 64883. **Telex:** 72165 Esrama.
MEDIUMWAVE: 1485kHz 1kW — **FM:** 96.0MHz (stereo).
D.PRGR. on MW: 0600-0100 (Sat 0200). **Rádio Turista** (**Te:** +351 (91) 29563): W 0830-0900 Danish/Norwegian/Swedish, 0900-0930 German, 1745-1830 English, 1830-1900 French. **Canal + 96.0** on FM: 0700-2400 (Sat 0200).
ANN: MW: "Rádio Madeira".

OTHER STATIONS

1) **R. Jornal de Madeira,** Rua Dr. F. de Ornelas 35, 9000 Funchal — 88.8MHz 1kW.
2) **R. Clube,** Rua 31 Janeiro, 12F, 9000 Funchal — 106.8MHz 0.4kW.
3) **R. Girão,** Centro Com. de Estr, 9325 Estreito Camara de Lobos — 98.8MHz.
4) **R. Zarco,** Rua Gral. A. Texeira de Aguiar, 92000 Machico — 89.6MHz 1kW.
4) **R. Palmeira,** Sta. Cruz — 96.1MHz 0.5kW.
4) **R. Brava,** Ribeiro Brava — 98.4MHz 0.5kW.
4) **R. Sol,** Ponta do Sol — 103.7MHz 0.5kW.

MALAWI

L.T: UTC + 2h — **Pop:** 9.608.000 — **Radio sets:** 1.060.000 — **Pr.L:** Chichewa, English — **E.C:** a/c 50, 230V — **ITU:** MWI.

MALAWI BROADCASTING CORPORATION (Statutory Body)
ADDR: P.O. Box 30133, Chichiri, Blantyre 3. **Te:** +265 671222. **Fax:** +265 671257, +265 671353. **Cable:** Radomalawi, Blantyre. **Telex:** 0904 44425 Wailesi MI.
L.P: GM: H.R. Chirwa. Head of Production: T.J. Sineta. CE: P. Chinseu.

STATIONS	kHz	kW		kHz	kW
Mangochi	540	10	Nkhota Kota	1107	1
Karonga	558	10	Bangula	1278	1
Lilongwe	594	50	Kasungu	1278	1
Lusangazi	675	50	Blantyre	3380.6	100
Blantyre	756	50	Blantyre	5995	20/100

FM: Blantyre 92.2MHz 0.5kW, Lilongwe 92.7MHz 0.5kW + 15 st's 0.1kW.
D.PRGR: 0255-2210 on MW, 0255-0710 & 1400-2210 on 3380.6kHz, 0710-1400 on 5995kHz. **N. in English:** 0300, 0500 (W), 0600 (Sun), 0700 (W), 0800 (Sun), 0900 (W), 1000 (Sun), 1030 (Sat/Sun 1100), 1200 (MF), 1400 (MF), 1600, 1800, 2000 (W), 2100, 2200.
INTERNATIONAL SCE: 1600-1800 on all freqs.
ANN: E: "This is the MBC" or "This is the International Sce. of the MBC". Chichewa: "Kuno ndi ku Interanashonolo Sevesi ya MBC".
INT-SIG: 0255 Cock Crow and rapid drum beat. Pikisfon just before 1600 and 1700 — **V.** by letter.

MALI (Rep. of)

L.T: UTC — **Pop:** 8.538.000 — **Radio sets:** 150.000 (est.) — **Pr.L:** French + 9 national languages — **E.C:** a/c 50, 220V (yes) — **ITU:** MLI.

RADIODIFFUSION TÉLÉVISION DU MALI (Gov.)
ADDR: B.P. 171, Bamako. **Te:** +223 222019/222474. **Cable:** Radio-Mali — DG: Checkna Diarra. DG. Adj: Seydina O. Dicko.
STATIONS: Bamako (G.C: 07.58W/12.38N):

kHz	kW	kHz	kW	kHz	kW	kHz	kW
540	100	4835v	100	7285	100	11960v	50
4783v	100	5995	50	9635	50		

FM: 87.6/91.6MHz.
D.PRGR: Mon-Thurs/Sat 0555-1000 & 1155-2400, Fri 0555-2400, Sun 0655-2400. **On SW:** 0555-0800 & 1800-2400 on 4783/4835/5995kHz, 0555-2400 on 7285kHz, 0800-1800 on 11960/9635kHz. **N. in French:** 0700, 1300, 1500, 2000, 2230. **N. Magazine in English:** Sun 1845.
IS: Malian music on local guitar — **V.** by QSL-card. Rec. acc. Rp.(B).

RADIO BAMAKAN (Community st.)
ADDR: c/o Cooperative Culturelle Jamana, B.P. 2034, Bamako H — 12h daily on FM (exact times & freq. unknown).

CHINA RADIO INT. RELAY STATION
STATION (G.C: 08.01W/12.39N): 2×50kW tr's used for relays of China R. Int. Freqs and times listed under China (People's Rep.).

MAURITANIA (Islamic Rep.)

L.T: UTC — **Pop:** 2.150.000 — **Radio sets:** 300.000 — **Pr.L:** Arabic, French, Poular, Soninké, Ouolof — **E.C:** a/c 50, 220/380V — **ITU:** MTN.

RADIO MAURITANIE (Gov.)
ADDR: B.P. 2000, Nouakchott. **Te:** 52164/52679/52101. **Fax:** 52164. **Telex:** 515 MTN — **L.P:** DG: Mohamed Yahya Ould Haye.
MW: Nouakchott 1349kHz 20kW — **SW:** 4845/9610kHz 100kW — **FM:** 98.01MHz.
D.PRGR: 0630 (Fri 0800)-2400 on MW/FM, 0630-0800 & 1800-2400 on 4845kHz, 1200 (Fri 0800)-1700 on 9610kHz.
N: Arabic: 0700, 1200, 1500, 1700, 2030, 2230. **French:** 0730, 0930 (Fri), 1430, 1930.
ANN: F: "R. Mauritanie émettant de Nouakchott". A: "Huna Nuahshut iza'at gumhuriyati l-Islamiyya al mauritaniyya".
IS: Mauritanian guitar — **V.** by QSL-card or letter.

MAURITIUS

L.T: UTC + 4h — **Pop:** 1.162.000 — **Radio sets:** 250.000 — **Pr.L:** English, French, 6 Indian languages, Chinese — **E.C:** a/c 50,240V — **ITU:** MAU.

MAURITIUS BROADCASTING CORPORATION
ADDR: Broadcasting House, Louis Pasteur Str, Forest Side. **Te:** +230 675 5001. **Fax:** +230 675 7332. **Cable:** Broadcast, Mauritius. **Telex:** 4230 MAUBROD.
L.P: DG: Nando Bodha. Dep. DG: Pierre Ah Fat. Chief Exec. (Radio): Jacques Maunick. Adm. Mgr: Hootesh Ramburn. Dep. CE: Amoordalingum Pather. Pub. Rel. Officer: C.B. Yan Jankoo.
STATIONS (G.C: 57.13E/20.19S): 684/819kHz 10kW, 1575kHz 1kW (during cyclone emergencies Nov-April) — **FM:** 97.3/91.7MHz—Rodrigues 1206kHz 1kW.
D.PRGR: Ch.1: 0100-1900 on 684kHz. **English/French:** 0300-1030 (Sun 0900, Tues/Thurs 0930), 1430-1900. **N: French:** 0330, 0600, 0800 SS, 0830W, 1500, 1855. **English:** 0400 (local), 1700 (rel. BBC), 1800 (Sat rel. BBC). **Ch. 2:** 1300 (Sat 1430, Sun 0300)-1900 on 819 kHz. **Chinese:** 1400-1430. Other times in Indian languages.
ANN: E: "This is Mauritius Broadc. Corp." F: "Ici la M.B.C; office de Radiodiffusion — Télévision de Maurice". Opens & closes with Nat. Anthem. — **V.** by QSL-card. Rec. acc. (exc. BBC relays).

MAYOTTE (French Territory)

L.T: UTC+3h — **Pop:** 78.000 — **Radio sets:** 30.000 (est.) — **Pr.L:** French — **E.C:** a/c 50, 220V — **ITU:** MYT.

SOCIÉTÉ NATIONALE DE RADIO-TÉLÉVISION FRANÇAISE D'OUTRE-MER (RFO-MAYOTTE)
ADDR: B.P.103, F-97620 Dzaoudzi, Ile de Mayotte (par la Réunion). **Te:** +269 601017. **Fax:** +269 601852. **Telex:** 915822.
L.P: St. Dir: Bernard Regis. Dir. Tec: Joseph Blasco.
MEDIUMWAVE: Pamandzi 1458kHz 100kW — **FM:** Dzaoudzi 91.2MHz 50W, Lima Combani 88.8MHz 500W, Mamoudzou 89.5MHz 100W.
D.PRGR. in French & Mahorian: 24h. **N. in French:** 0330*, 0430, 0600, 0900, 0915*, 1500*, 1600*. **Rel. France-Inter:** 1930-0300 (Sat/Sun 0400). **ANN:** "Vous êtes à l'écoute de RFO-Mayotte".
INT-SIG: Melody on guitar — **V.** by letter (B).

MOROCCO

L.T: UTC (Su: UTC+1h) — **Pop:** 27.000.000 — **Radio sets:** 4.5m — **Pr.L:** Arabic, French, Spanish, English, Berber, Hassania — **E.C:** a/c 50, 127/220V (yes) — **ITU:** MRC.

RADIODIFFUSION TELEVISION MAROCAINE (Gov.)
ADDR: 1, Rue El Brihi (or B.P.1042), Rabat. **Te:** +212 (7) 764871. **Fax:** +212 (7) 762010. **Cable:** RTM Rabat. **Telex:** Radiomar 31010.
L.P: Dir. Gen: Mohamed Tricha. Dir. Radio: Abderrahman Achour. Dir. TV: Mohamed Issari. Dir. Tec: Jamal Eddine Tanane. Dir. Ext. Rel: Ali M'Barek. Dir. Fin. & Adm: Mehdi Bouzekri.

LONG & MEDIUMWAVES (N = Network)

Location	kHz	kW	N	Location	kHz	kW	N
Azilal	207	800	A	Ad-Dakhla*	999	10	A
Sidi Bennour*	540	600	A	Rabat	1026	1	B
Oujda	594	100	A+B	Sebaa-Aioun*	1044	300	A
Sebaa-Aioun	702	140	C	Tanger*	1053	600	A
Laayoune*	711	600	A	Casablanca	1080	1	A
Agadir	774	50	C	Ouarzazate	116	15	A
Rabat	819	25	A	Casablanca	1188	1	B
Oujda	828	100	B	Agadir	1197	20	B
Errachidia	864	15	A	Tanger	1233	200	R
Safi	909	5	B	Safi	1325	5	C
Agadir*	936	600	A	Casablanca	1485	1	C

*)carries regional prgrs.

SHORTWAVES: Tanger (G.C:05.50W/35.45N):

kHz	kW	H. of Tr.	kHz	kW	H. of Tr.
11920	50	1900-0100	15360	50	1045-1700
15105	50	2100-0100	17595	50	1400-1700
15330	50	1700-2100	17815	50	1045-1400
15335	100	0945-0100			1700-1900

FM	Prgr. A	Prgr. B	kW	FM	Prgr. A	Prgr. B	kW
Boukhouali		93.4	10	Oukaimeden	91.7	98.8	31
Cap-Spartel	88.7	91.8	20	Rabat		87.9	39
Casablanca		90.0	39	Zerhoun	98.4	95.1	10
Meghrez	89.8	99.4	12				

Network A in Arabic: 24h. **N:** on the h.
Regional Prgrs (exc. Sun): **Tetuan:** 0805-0900 & 1000-1200 on 1053kHz. **Tangier:** 1500-1600 & 1620-1800 on 1053kHz. **Oujda:** 1200-1300 & 1330-1500 on 594kHz. **Marrakech:** 1500-1600 & 1620-1800 on 540kHz. **Casablanca:** 1500-1600 & 1620-1800 on 1080kHz. **Agadir:** 1500-1600 & 1620-1800 on 936kHz. **Laayoune** on 711kHz & **Dhakla** on 999kHz: times not known.
Network B in French/English/Spanish: 0600-0100. **English:** Mon-Fri 1400-1500v on 17595kHz, Sat 1700-1800 on 17815kHz, Sun 1900-2000 on 11920kHz.
Spanish: 0900-1000. **N:** 0915, 0930.
Network C in Berber/Arabic: 0600-0100 (incl. relays of Network A).
ANN: Arabic: "Huna Ribat, idha'atu-l-mamlaka al Maghribiyya". French: "Ici Rabat, Radiodiff. Télévision Marocaine". Berber: "Dahab Rbad Lidaa Attalfaza Li mamlaka Lmaghrib".
V. by QSL-card. Rec. acc. Rp.

RADIO MEDITERRANÉE INTERNATIONALE (Gov./Comm.)

ADDR: 3 et 5, Rue Emsallah (B.P. 2055), Tanger. **Te:** +212 (9) 936363. **Telex:** RADMED 33711M. (Addr. in France: 78 av. Raymond Poincaré, 75116 Paris).
LW: Nador 171kHz 2000kW — **MW:** Sebaa Aioun 612kHz 300kW.
SW: Nador (G.C: 02.55W/34.58N): 9575kHz 250kW.
FM: Marrakech 105.3MHz — Rabat 97.5MHz — Casablanca 99.6MHz — Tanger 95.6MHz — Agadir 104.6MHz — Oujda 102.9MHz (all 1kW).
D.PRGR. in French/Arabic: 0500-0100. **N. in French:** 0730, 0830, 1230, 1930, 2200. **N. in Arabic:** 0700, 0800, 1200, 2030, 2330.
ANN: "Radio Mediterranée Internationale" or "Medi 1".
V. by letter.

VOICE OF AMERICA RELAY STATION TANGIER

SW: Tanger (G.C: 05.56W/35.41N): 100/50/35kW.

kHz	kHz	kHz	kHz	kHz	kHz	kHz
6005	9615	9740	11760	11920	15195	15320
6150	9650	9760	11780	15105	15205	15335
7210	9670	11720	11850	15125	15245	17595
9530	9715	11750	11895	15145	15290	17865

D.PRGR: see VOA entry under "USA".

Ceuta (Spanish)

R. NACIONAL DE ESPAÑA, S.A.

ADDR: Beatriz de Silva 12, E-11701 Ceuta.
FM: 105.1MHz 1kW (rel. RNE R1).

RADIO CEUTA (Comm.)

ADDR: Ap.180, Real 90, E-11701 Ceuta.
MW: 990kHz 5kW 24h (rel. SER Madrid 2200-0500)
FM: 88.5MHz (music), 96.2MHz (**R. Sol**).

R. Perla, Grupo Alfau 4, 3° izquierda, E-11701 Ceuta.
FM: 104MHz 24h.

Melilla (Spanish)

R. NACIONAL DE ESPAÑA, S.A.

ADDR: Ap. de Correos 222, E-29801 Melilla. **Te:** +952 681907. **Fax:** +952 683108 — **L.P:** Dir: Pedro A. Medina Barrenechea.
MW: 1359kHz 5kW (R1) — **FM:** 107.6MHz (R2), 105.3MHz (R3).

R. MELILLA (Comm.)

ADDR: Edif. No. 8, Zona del Puerto, E-29805 Melilla. **Te:** +952 681708 — Dir: Gaspar Diaz Cerdá.
MW: EAJ21 1539kHz 4kW — **FM:** 96.3MHz 0.4kW — **D.PRGR:** 24h.

ANTENA 3 DE RADIO S.A.

ADDR: Edif. Melilla-Urbaniz Rusadir. **Te:** +952 688840.
FM: 101.2MHz.

MOZAMBIQUE (People's Republic of)

L.T: UTC + 2h — **Pop:** 15.808.500 — **Radio sets:** 500.000 — **Pr.L:** Portuguese, Vernaculars — **E.C:** a/c 50, 220V (yes) — **ITU:** MOZ.

RADIO MOÇAMBIQUE (Gov/Comm.)

ADDR: C.P. 2000, Maputo. **Te:** +258 420844/6/7. **Cable:** Moradio. **Telex:** 6317 & 6340 RADIOMO.
L.P: Dir: Manuel Tomé. Dir Tec: Rufino de Matos. Comm. Dir: António A. Fonseca.
STATIONS: Maputo (G.C: 32.22E/25.59S).

kHz	kW	Netw.	kHz	kW	Netw.	kHz	kW	Netw.
737	50	N	4858v	20	I/E	7242	100	N
1008	50	I	4863.4	25	N	9525	10	I
1079	5	E	4910v	7.5	I	9618	120/25	N/E
3210.5v	100	N	6112	10	N	11818	120	N/E
3265	25	E	7113v	10	I	15295v	25	N
3338.2v	10	I						

FM: 98.01MHz 5kW.
EMISSÃO NACIONAL(N) in Portuguese: 0250-2210 on 737kHz, 0250-0500 & 1700-2210 on 3210kHz, 0250-0700 & 1600-2210 on 6112kHz, 0400-2210 on 7242/9618kHz, 0250-0500 & 1600-2210 on 4863.4kHz, 0500-1700 on 11818kHz, 0700-1600 on 15295kHz. **N:** 0300-0900 hourly, 1030, 1200-1600 hourly, 1730, 1900, 2100, 2200.
EMISSÃO INTERPROVINCIAL DE MAPUTO E GAZA(I) in Tsonga/Portuguese: 0255-2210. **N. in Portuguese:** 0530, 1700.
EXTERNAL SCE(E) to Southern Africa in English: 1100-1130 on 1079/9525/11818kHz+FM, 1800-1900 on 1079/3265/4858/9618kHz+FM. **N:** 1100, 1800.
ANN: "Dos seus estúdios em Maputo escutam a Rádio Moçambique, Emissão Nacional/Emissão C", "R. Moçambique, dos seus estúdios em Maputo transmite a Emissão Interprovincial de Maputo e Gaza". E: "This is the External Sce. of R. Mozambique" or "This is R. Maputo, the External Sce. of R. Mozambique".
INT-SIG: "Mbira" (indigenous xylophone) and spoken ID in P/E/F & Swahili. Opens and closes with the National Anthem. 4-note gong between prgr's — **V.** by QSL-card.

PROVINCIAL STATIONS

MEDIUMWAVES	kHz	kW		kHz	kW
4) Inhambane	557	5	8) Tete	1160	5
5) Lichinga	782	5	2) Nampula	1223	5
3) Chimoio	836	5	6) Quelimane	1295	10
1) Beira 1	872	50	1) Beira 1	1430	1
1) Beira (F.Pl.)	935	1	7) Pemba	1493	5
SHORTWAVES	**kHz**	**kW**		**kHz**	**kW**
1) Beira 2	3281v	100	1) Beira 1	6025v	10
1) Beira 1	3370.2	10	1) Beira 2	9637	100

1) E. PROVINCIAL DE SOFALA
ADDR: C.P. 1781, Beira. (G.C. (Dondo) : 34.52E/18.49S).
"A." Prgr. in Portuguese/Chissena/Ndau: 0255-2205. **N. in Swahili:** 1930.
"B." Prgr. In Nyanja/Ajaua/Makonde/Swahili: 0255-2255. **N. in Swahili:** 1930.
2) E. PROVINCIAL DE NAMPULA, C.P. 93, Nampula.
D. PRGR. in Portuguese/Macua: 0255-2215 on 1223kHz +FM.
3) E. PROVINCIAL DA MANICA, C.P. 390, Chimoio. **Te:** +258 22422 — Dir: Luisa Meneses.
D. PRGR. in Portuguese/Chimanhika/Chiutee: Mon-Fri 0255-0515, 0955-1215, 1455-2010. Sat 0255-0515, 0955-2210. Sun 0255-2010.
4) E. PROVINCIAL DO INHAMBANE, C.P. 196, Inhambane.
D. PRGR. in Portuguese/Vernaculars: Mon-Fri 0400-0515, 1000-1215, 1600-2000; Sat 0400-0515, 1000-2000; Sun 0400-0700, 1000-2000.
5) E. PROVINCIAL DO NIASSA, C.P. 171, Lichinga.
D. PRGR. in Portuguese/Vernaculars: 0255-2000.
6) E. PROVINCIAL DA ZAMBÉZIA, C.P. 333, Quelimane. **Te:** +258 3171 — Dir: Antonio Germane Barros.
D. PRGR. in Portuguese/Chuabo: Mon-Fri 0255-0515, 1000-2115; Sat 0255-0515 & 1000-2215; Sun 0255-2155.

7) E. PROVINCIAL DE CABO DELGADO, C.P. 45, Pemba.
D. PRGR. in Portuguese/Macua do Medo: Mon-Fri 0400-0515, 1000-1215, 1600-2000; Sat 0400-0515 & 1000-2000; Sun 0400-0700, 1000-2000.
8) E. PROVINCIAL DE TETE, C.P. 384, Tete.
D. PRGR. in Portuguese/Nyanja/Nhungwe: Mon-Fri 0255-0515 & 0955-2000; Sat 0255-0515 & 0955-2215; Sun 0255-2000.
ANN: "R. Moçambique dos seus estúdios . . . (town) . . . a sua Emissor Provincial . . . (Province) or "de . . . (town) . . . transmite EP de R. Moçambique do . . . (Province)" — **INT-SIG:** As Maputo with local ID in P/E/F/Swahili.
V. Reports to most provincial st's should be sent to Maputo. St. 2) verifies direct.

NAMIBIA

L.T: UTC + 2h — **Pop:** 1.575.242 — **Pr.L:** English (official), Afrikaans, German, Oshiwambo, Otjiherero, Tswana, Damara, Nama, Kwangali, Mbukushu, Gcirku, Lozi — **ITU:** NMB.

NAMIBIAN BROADCASTING CORP.

ADDR: P.O. Box 321, Windhoek 9000. **Te:** +264 (61) 291 911/215 811. **Fax:** + 264 (61) 217760. **Telex:** 622.
L.P: Chmn: Dr. Z. Ngavirue. Dir. Gen: N. Gorelick. Sen. Controller Radio & TV: E. Katjevena. TD: M. Venter.

MEDIUMWAVES	kHz	kW		kHz	kW
Tsumeb	594a	100	Gobabis	747b	100

a) Oshiwambo Sce — b) Otjiherero Sce/Tirelo ya Setswana.
SHORTWAVES: Windhoek (G.C: 17.13E/22.33S): 2 × 100kW SW-tr's.

kHz	Prgr.	H. of tr.	kHz	Prgr.	H. of tr.
3270	1	1600-2000	6060	1	0800-1600
3290	2	1600-0600	6175	2	0600-1600

Prgr. 1 = Otjiherero Sce, Damara/Nama Sce, Prgr. 2 = Afrikaans, German. Exact times of relays for each sce. are given below.
FM: 31 medium & low power st's.
National Sce: 0400-2300 on 21 FM-tr's. On all freq's MF 0400-0800 & 1900-2300, and on all Windhoek-based sces MF 1500-1600.
All Night Sce: 2300-0400 on all freq's.
Oshiwambo Sce: 0800(Sat/Sun 0400)-1900 on 594kHz + 10 FM-tr's. **Addr:** Box 123, Oshakati. **IS:** Call of the Fish Eagle and wooden pestles.
Otjiherero Sce. & Tirelo ya Setswana: MF 0800-1500 & 1600-1900, SS 0400-1900 on 747kHz + 7 FM tr's.
Damara/Nama Sce: MF 0800-1500 & 1600-1900, SS 0400-1900 on 18 FM-tr's. **Damara/Nama & Otjiherero:** 0800-1600 on 6060kHz, 1600-2000 on 3270kHz.
German Sce: MF 0800-1500 & 1600-1900, SS 0400-1900 on 14 FM-tr's.
Afrikaans Sce: MF 0800-1500 & 1600-1900, SS 0400-1900 on 22 FM-tr's.
German/Afrikaans: 0600-1600 on 6175kHz, 0400-0600 & 1600-2300 on 3290kHz.
Rukavango Sce: 0800(Sat/Sun 0400)-1900 on 7 FM-tr's. **Addr:** Box 1000, Rundu.
Caprivi Sce: 0800(Sat/Sun 0330)-1900 on 1 FM-tr. **Addr:** Box 1066, Katima Mulilo.
ANN: All Night Sce: "Here is the National Sce. of the NBC Nationwide". G: "Hier ist das Deutsche Hörfunkprogramm der NBC". Dam/Nam: "Nes ge Damara/Nama Gowab loabas NBC's disa". A: "Dit is die Afrikaanse diens van die NBC". Otjiher: "Indji oradio ja Namibia morupa rueraka Otjiherero".
IS: at s/on: Nat. Anthem with choir/orchestra.
V. by QSL-card. Re. to NBC Shortwave Section, P.O. Box 321, Windhoek, Namibia.

NIGER

L.T: UTC + 2h — **Pop:** 8 m — **Radio sets:** 400.000 (est.) — **Pr.L:** French + ethnic — **E.C:** a/c 50, 220/380V — **ITU:** NGR.

LA VOIX DU SAHEL (Gov.)

ADDR: B.P. 361, Niamey. **Te:** +227 722272. **Telex:** 5229 NI.
L.P: DG: M. Diallo. Dir. R: A. Khamed. TD: Y.A. Tidjani. Dir. Admin: Y. Ibrahim.
MEDIUMWAVES: Niamey 1125kHz 20kW, 1575kHz 1kW — Zinder 1511kHz 1kW + relay st's on 1041/1215/1584kHz.
FM: Niamey 90.2/93.6MHz + 7 relay st's.
SHORTWAVES: Niamey (G.C: 02.06E/13.30N):

kHz	kW	H. of tr.
3260	4	0430-0700, 1700(Sun 1630)-2200(Sun 2300)
5020	20/100	0430-0700, 1700(Sun 1630)-2200(Sun 2300)
6060	4	0700-1630
7155	20	0430-1400(Sun 1630), 1700-2200(Sun 2300)
9705	100	1100-1400(Sun 1630)

D. PRGR: W 0430-1400 & 1700-2200, Sun 0430-2300. **N. in French:** 0545, 1200, 1705, 1900, 2199, 2150. **N. in Arabic:** 1715.
N.B: splits into two local language networks at various times incl. 1745-1845 — **ANN:** "Ici la Voix du Sahel".

NIGERIA (Federal Republic of)

L.T: UTC + 1h — **Pop:** 122.840.000 — **Radio sets:** 10m — **Pr.L:** English, Yoruba, Hausa, Igbo — **E.C:** a/c 50, 230V — **ITU:** NIG.

NIGERIAN TELECOMMUNICATIONS LIMITED

ADDR: Dir. of Telecommunications, International Relations Division, Nigerian Telecommunications Limited, Private Mail Bag 12557, Lagos. **Cable:** Dintrel.

Mediumwaves	kHz	kW		kHz	kW
2) Akure	531	50	15) Ogbomosho	891	10
3) Sokoto	540	50	18) Abeakuta	900	25
14) Kano	549	50	19) Abuja	909	50
23) Calabar	558	50	20) Ikeja	918	50
13) Owerri	567	50	10) Makurdi	918	50
4) Ibadan	576	50	28) Awka	954	20
6) Ugwuaji	585	50	25) Katsina	972	
5) Kaduna	593	100	35) Lokoja	972	25
18) Ibese	603	30	20) Ikeja	990	20
16) Ilorin	612	50	21) Bauchi	990	50
9) Maiduguri	630	50	12) Kontagora	1008	50
22) Port Harcourt	630	50	15) Ibadan	1008	10
8) Kaduna	639	25	11) Yola	1017	10
7) Warri	639	5	6) Onitsha	1062	10
4) Ibadan	657	100	1) Lagos	1089	20
7) Benin	666	50	11) Yola	1089	50
15) Aha	675	10	5) Kaduna	1107	25
12) Bida	702	10	23) Ogoja	1134	10
31) Wukari	702	25	12) Bida	1143	50
11) Bali	722	20	18) Abeokuta	1170	40
33) Dutshe	729	25	11) Hong	1206	10
9) Damagun	756	25	24) Jos	1224	50
15) Ibadan	756	10	31) Jalingo	1260	10
12) Minna	756	50	9) Maiduguri	1296	20
11) Bali	774	10	15) Iwo	1359	10
35) Anyagba	783	25	23) Abak	1395	10
15) Idominasi	792	10	26) Uyo	1395	20
30) Damaturu	801	20	21) Gombe	1404	10
17) Enugu	828	25	1) Lagos	1458	5
21) Azare	846	10			

Shortwaves	kHz	kW		kHz	kW
1) Lagos	3326	50	4) Ibadan	6050	50
17) Enugu	3970+	?	5) Kaduna	v6090	250/50
5) Kaduna	4770	50	9) Maiduguri	6100	10
1) Lagos	4990	50	5) Kaduna	7275	100
24) Jos	5965	10	1) Lagos	7285	50
17) Enugu	6025	10	5) Kaduna	9570	50

+) r. testing Oct. 1992.
FM: 24) 88.63MHz, 7) 91.8MHz, 1) 92.93/97.66MHz, 4) 93.54MHz, 6) 96.1MHz, 7) 88.6/95.8MHz, 9) 94.5MHz, 10) 98.4MHz, 11) 90.61/95.77MHz, 13) 94.4MHz, 14) 96.9/89.0/89.3MHz, 15)98.5/95.8MHz, 16) 91.9MHz, 17) 92.85MHz, 18) 90.35MHz, 21) 94.5MHz, 22) 99.15MHz (stereo), 24) 90.5MHz.

Addresses and other information:
1) **FEDERAL RADIO CORPORATION OF NIGERIA** (Statutory Corporation).
ADDR: P.M.B. 12504, Ikoyi, Lagos. **Te:** +234 (1) 603010-4. **Cable:** Broadcasts, Lagos. **Telex:** 21484.
L.P: DG: B. Balogun. Dir.Corp.Affairs: S.O. Ajani. MD (Lagos): I. Ndaguba. Asst.Dir. (SD): M.A. Yakubu. Asst.Dir.N: P. Obazele.

HOME SCES: Channel 1 in English on 1089/4990kHz + FM 92.93MHz: 0430-2300; On 3326kHz: 0430-1000, 1700-2300. On 7285kHz: 1000-1600. **N:** On the h.
NB: Nigerian N. at 0600, 1500 & 2100 is relayed by all FRCN stations.
Channel 2 in English on 1458kHz/97.66MHz (stereo): 0500-2300. **N:** on the half h.
Channel 3 in Pidgin English/Yoruba/Hausa/Igbo on 92.43MHz: 0430-2300.
INTERNATIONAL SCE: See below.
ANN: Ch 1: "This is R. Nigeria, Lagos". Ch 2: "The Sunshine Station" — **INT-SIG:** Talking Drum. (Gangan). Opens and closes with the Nat. Anthem and spoken pledge.
2) Ondo State Radiovision, Akure, Ondo State — GM: S. Ade-Ayeni. 0400-2300 in English/Yoruba. **Ann:** "This is OSRC Akure".
3) Rima R, Sokoto State Broadc. Corp, Private Mail Bag 2156, Sokoto, Sokoto State — GM: S. Isa — 0430-2315 in English/Hausa. **N:** 1100, 1430, 1900. **Ann:** "This is Rima R, Sokoto". **V.** by letter.
4) R. Nigeria, P.M.B. 5003, Ibadan, Oyo State — MD: J.O. Oyinsan. 0430-2305 in English/Yoruba/Edo/Igala/Urhobo. **Ann:** "This is R. Nigeria Ibadan". **V.** by letter. **F.Pl:** more powerful tr's.
5) R. Nigeria, P.O. Box 250, Kaduna, Kaduna State — Man. Dir: Muhammad Suleiman. Ch 1 in Hausa: 0400-2305 on 594/6090 kHz. Ch 2 in English/Hausa/Fulfulde/Kanuri/Nupe: 0430-2305 on 1107/4770kHz. **N.** on the h. **English N.** (both ch's): 1600, 1700. **Ann:** "This is Ch 2, the English Sce. of R. Nigeria, Kaduna". **V.** by letter.
6) Enugu Broadc. Corp., P.M.B. 01600, Enugu, Enugu State — MD: A. Okafor — **D.Prgr:** in English/Igbo on 585kHz: 0430-2300. **N:** on the h — **Prgr. 2** on 1062kHz: 0500-1800 (Sat/Sun 2400). **N. in English:** 0530, 1430, 2000 (rel. 1st Prgr). Relay R. Nigeria N. on all freq's: 0600, 1500. **Prgr. 3** on FM: 0500-2100.
7) Edo State Broadc. Sce, P.M.B. 1012, Benin City, Edo State — GM: N.O. Ilube. 0400-2305 in English + 12 local languages. **N:** 0530, 0600, 1200, 1500, 2100. **Ann:** "BBS Radio".
8) Kaduna State Broadc. Corp., P.M.B. 2013, Kaduna, Kaduna State — GM: S. Para. 0430-2305 in English/Hausa **Ann:** "This is the Kaduna State Broadc. Corp, Kaduna".
9) Borno Radio & TV Corp., P.M.B. 1020, Maiduguri, Borno State — GM: A. U. Bukar. 0400-2305 in English/Hausa/Kanuri. **Ann:** "Radio Sce. of BRTV broadcasting from Maiduguri".
10) R. Benue, P.M.B. 102202, Makurdi, Benue State — GM: I. Edime — **Prgr. 1:** 0430-2305 in English/Hausa + 7 local languages. **Prgr. 2 on FM:** 0500-2105. **N:** 0800, 1100, 1600. **Ann:** "This is R. Benue, Makurdi".
11) Adamawa State Broadc. Corp, P.M.B. 2123, Yola, Adamawa State — GM: A.B. Jijiwa. 0430-2300 in English/Hausa + 6 Nigerian languages. **N:** on the h + 0530, 1445. **Ann:** "This is GBC Yola, your No. 1 Radio Station".
12) R. Niger, Minna, Niger State — GM: A. Goro. 0430-2330 on 702kHz, 0530-1115 & 1330-1900 on 1008kHz in English/Hausa.
13) Imo Broadc. Corp, P.O. Box 329, Owerri, Imo State — Dir: A. Ude. 0425-2305 on MW, 0445-2305 on FM. **English:** 0430-0630, 1100-1830, 2100-2300 (Sat/Sun 0100). **N.** 0500, 0530, 1100, 1300, 1400, 1600, 1700, 2100, 2200, 2300. **Distance Education Broadcast:** Mon-Fri 1630-2200, Sat 1930-2130. **Rel. R. Nigeria:** 0600, 1200, 1500. Other times in Igbo. **Ann:** "This is Imo Broadc. Corp, Owerri".
14) Kano State Broadc. Corp, P.M.B. 3014, Kano, Kano State — GM: A. A. Gwadabe. 0430-2320 in English/Hausa. **Prgr 2:** on FM: 0550-2320. — **Ann:** "This is KBC, the Kano State Broadc. Corp".
15) Broadc. Corp. of Oyo State, PMB 1, Ile Akede, Basorun, Ibadan, Oyo State — GM: J. Fademi. **Prgr. 1:** 0400-2200 in English/Yoruba. **Prgr 2:** on FM: 0700-2100. **Ann:** "R. Oyo Ibadan".
16) Kwara State Broadc. Corp., P.M.B. 1345, Ilorin, Kwara State — GM: R. Omotosho. 0425-2305 in English/Hausa/Baruba/Kabba/Igbira. **N. in English:** 0500, 0600, 0900, 1300, 1400 (State news), 1600, 1700, 1900, 2200, 2300. **Ann:** "This is R. Kwara, Ilorin". **V.** by letter.
17) R. Nigeria Broadcasting House, P.M.B. 1051, Enugu, Enugu State — MD: D.A. Bassey. 0430-2300 on MW in English/Igbo; 0430-2300 on 6025kHz (separate prgrs) in English/Igbo/Tiv/Efik and Izon. **V.** by letter. **F.Pl:** 3 new SW-tr's.
18) Ogun State Broadc. Corp, P.M.B. 2084, Abeokuta, Ogun State — GM: Chief T. Elegbede. CE: A.A. Majiyagbe. Ch. 1 on MW, Ch. 2 on FM: 0430-2300 in English/Yoruba/Egun. **Ann:** "Nation's Model Station", "Ogun-Radio", "Voice from the Rocks".
19) R. Nigeria P.M.B. 71, Abuja, Federal Capital Territory — Asst.Dir: S.D. Ibrahim. 0430-2200 in English/Hausa/Igbo/Yoruba and local languages. **Ann:** "The Capital R, the Voice of Unity".
20) R. Lagos, P.M.B. 21035, Ikeja, Lagos State — Gen.Mgr: O. B. Olemoh. 0430-2330 in English/Yoruba/Ogu. **Ann:** "This is R. Lagos, Ikeja".
21) Bauchi Radio Corp, P.M.B. 0133, Bauchi, Bauchi State — GM: A. Wunti. 0430-2300 on MW, 0500-1700 in Fulfulde (separate prgrs) on 94.5MHz. **F.Pl:** 24h.
22) Rivers State Broadc. Corp P.M.B. 5170, Port Harcourt, Rivers State — GM: E. Iyaye. **Prgr. 1:** 0450-2310 on 630kHz, in English/Ikwerre/Kalabari/Khana/Kolokoma/Ekpeye/Okrika/Gokana/Nembe/Pidgin. **Prgr. 2:** 0450-2310 on FM. **Ann:** "This is R. Rivers".
23) Cross River State Broadc. Corp, P.M.B. 1035, Calabar, Cross Rivers State. GM: B.A. Bassey. Prgrs in English/Efik/Ejagham/Bekwara: 0430-2315 on 558kHz, 0500-1200 & 1400-2030 on 1134kHz. **N:** on the h. **Ann:** "This is Cross River Radio, Calabar". **V.** by letter.
24) Plateau Radio Television Corp, Jos, Plateau State — GM: T. Yale. Ch.1 on MW: 0430-2305 in English/Idoma/Hausa/Tiv/Alago/Eggon/Birom/Angas/Efik/Fulfulde. Ch.2 on FM: 0500-2200. **Ann:** "This is Radio Plateau, Jos", "This is PRTV 2, Jos". **V.** by letter.
25) Radio Katsina, Broadc. House, Katsina, Katsina State — GM: I. Hamisu. 0500-2300 in English/Hausa. **Ann:** "This is Katsina State R."
26) Akwa-Ibom Radio, Broadc. House, Uyo, Akwa-Ibom State — GM: Rev. F. Mbaba. 0500-2300. **Ann:** "This is Akwa-Ibom R, Uyo".
27) Kebbi State Radio, Broadc. House, Birnin Kebbi, Kebbi State — 0500-2300 (freq's not yet known).
28) Anambra State Broadc. Corp, Awka, Anambra State — 0500-2300.
29) Delta State Broadc. Sce, Broadc. House, Asaba, Delta State — 0500-2300 (freq's not yet known).
30) Yobe Broadc. Sce, Broadc. House, Damaturu, Yobe State — 0500-2300.
31) Taraba Broadc. Corp, Broadc. House, Jalingo, Taraba State — 0500-2300.
32) Abia State Radio, Broadc. House, Umuahia, Abia State — 0500-2300 (freq's not yet known).
33) Jigawa State Broadc. Corp, Broadc. House, Dutse, Jigawa State — 0500-2300.
34) Osun State Radio, Broadc. House, Oshogbo, Osun State — 0500-2300 (freq's not yet known).
35) Kogi State Broadc. Sce, Broadc. House, Lokoja, Kogi State — 0500-2300.
ANN: Exc. where stated, FRCN st's identify as "R. Nigeria" followed by location.

INTERNATIONAL SCE

VOICE OF NIGERIA (Statutory Corp.)

ADDR: V. Of Nigeria, Broadc. House, PMB 40003, Falomo, Ikoyi, Lagos.
L.P: DG: Y. Abukabar. PD: Y. Saulawa. Dir.N: T. Oseni. Dir.Eng: G. Ugwu.
STATION: Ikorodu (G.C: 03.56E/07.23N):5×500kW, 1×300kW, 2×100kW tr's (**NB:** only 1×500kW tr. using 250kW on the air at time of editing).
We. African Sce. on 7255kHz: 0455-0700 **English,** 0700-0800 French, 0800-0900 Hausa, 0900-1100 **English,** 1100-1200 French, 1200-1400 Hausa, 1400-1500 Swahili, 1500-1700 **English**, 1700-1800 Arabic, 1800-1900 French, 1900-2100 **English,** 2100-2200 French, 2200-2300 Hausa. **N. in English:** 0530, 0630, 1030, 1045 (Nigerian N.), 1530, 2000.
ANN: "This is the Voice of Nigeria".
INT-SIG: As Home Sce. Also bells playing the first bars of the Nat. Anthem 15 mins. before the commencement of each block period.
V. by QSL-card — **F.Pl:** Reactivation of additional tr's.

REUNION

L.T: UTC + 4h — **Pop:** 600.000 — **Radio sets:** 150.000 (est.) — **Pr.L:** French — **E.C:** a/c 50, 220V — **ITU:** REU.

SOCIÉTÉ NATIONALE DE RADIO-TÉLÉVISION FRANÇAISE D'OUTRE MER (RFO)

ADDR: RFO-Réunion, B.P. 309, St. Denis. **Te:** +262 213456.
L.P: Regional Dir: J.P. Lannes. Dir. of N: M. Grimaud. Prgr. Dir: F. Dussaugey.
STATIONS: St. Pierre 666kHz 20kW — St. André 729kHz 20kW — Le Port 603kHz 4kW + 13 low power FM-st's.
D.PRGR: 0200-1900 (Sat 2000). **N:** W 0230, 0300, 0400, 0600, 0830, 0915, 1100°, 1505, 1800°; Sun 0400, 0830, 1100°, 1505, 1800° — °) relay France-Inter.
ANN: "Société Nationale de Radio-Télévision Française d'Outre-Mer, Station de la Réunion" — **INT-SIG:** "Madina" (Réunion Folklore) played on guitar — **V.** by QSL-card. Rp.

PRIVATE STATIONS: Over 40 st's broadcast locally on FM.

RWANDA (Republic of)

L.T: UTC + 2h — **Pop:** 8.200.000 — **Radio sets:** 630.000 — **Pr.L:** Kinyarwanda, Swahili, French, English — **E.C:** a/c 50, 220V — **ITU:** RRW.

RADIODIFF. DE LA RÉPUBLIQUE RWANDAISE (Gov.)

ADDR: B. P. 83, Kigali. **Te:** + 250 (7) 5665. **Fax:** +250 (7) 6185. **Cable:** Radiorwanda. **Telex:** ORINFOR 90922557 — Dir. of Inf: Ferdinand Nahimana.
SHORTWAVES (G.C: 30.04E/01.57S): 3330vkHz 20kW, 6055kHz 50kW — **FM:** 87.6MHz 10kW + 6 st's.
HOME SCE: W 0255-0600 & 0900-2100, Sun 0255-2100.
N. in French: 0430, 1028, 1115, 1500, 1630, 1800, 1930.
N. in English: 1430.
ANN: F: "Ici Kigali, Radiodiff. Nationale de la République Rwandaise" or "Ici R. Rwanda". Swahili: "Hii ni idhaayataifa ya R. Rwanda". E: "This is R. Rwanda broadcasting from Kigali, Rwanda." — **IS:** Inanga (local harps) — **V.** by QSL-card or letter.

DEUTSCHE WELLE RELAY STATION KIGALI

STATION: Kigali (G.C: 30.04E/01.58S): 250kW.

kHz	kHz	kHz	kHz	kHz	kHz
7225	9735	11795	11965	15410	17860
9565	11785	11810	15270	17800	

ANN: At opening and close down in F: "Ici le Deutsche Welle station relaise Kigali, Rwanda".
For prgr. details see D. Welle, Germany.

SAO TOME E PRINCIPE

L.T: UTC — **Pop:** 132.000 — **Radio sets:** 31.000 — **Pr.L:** Portuguese — **E.C:** a/c 50, 220V — **ITU:** STP.

RÁDIO NACIONAL DE S. TOMÉ E PRINCÍPE (Gov.)

ADDR: Caixa Postal 44, São Tomé. **Te:** +239 21342. **Cable:** Radio Nacional — **L.P:** Juvenal Vaz Rodrigues.
STATIONS: 945kHz 20kW — **FM:** 90.5/95.43/99.5MHz.
HOME SCE. in Portuguese: 0530-2300.
N: 0600, 0700, 1000, 1200, 1400, 1600, 1745, 1930, 2200.
ANN: "Aqui São Tomé transmite R. Nacional da República Democratica de S. Tomé e Princípe".
INT-SIG: 1 note Gong, Guitar — **V.** by letter. Rp.

VOICE OF AMERICA RELAY STATION

F.Pl: 1 × 600kW MW-tr, 4 × 500kW SW-tr's are being installed. A temporary 100kW MW-tr. will operate from March 1993 (freq. not announced at press time).

SENEGAL

L.T: UTC — **Pop:** 8.200.000 — **Radio sets:** 850.000 — **Pr.L:** French, ethnic — **E.C:** a/c 50, 220V — ITU: SEN.

OFFICE DE RADIODIFFUSION TÉLÉVISION SENEGALAISE — RTS (Gov.)

ADDR: Triangle Sud (B.P. 1765), Dakar. **Te:** +221 237286. **Cable:** Radio Sénégal Dakar. **Telex:** 21818.
L.P: DG: M. Guila Thiam. Dir. Radio: Ibrahim Sane. Dir. Tech. Sces: Joseph Nesseim.

STATIONS	kHz	kW	Sce.	H. of tr.
1) Dakar	765	300/200	N	0558(Sun 0700)-0100
Podor	810	1	N	0558(Sun 0700)-0100
Matam	963	1	N	0558(Sun 0700)-0100
4) Ziguinchor	1224	20/1	R	0600-0800, 1200-1400, 1800-2400
5) Kaolack	1287	4	R	0600-0800, 1200-1400, 1800-2230
1) Dakar	1305	20	I	As below
Linguére	1323	1	N	0558(Sun 0700)-0100
2) Saint-Louis	1368	20	R	0600(Sun 0700)-0100
3) Tambacounda	1503	20	R	0600-2200
1) Dakar	1539	10	N	0558(Sun 0700)-0100
1) Dakar	4890v	100	N	0558(Sun 0700)-0815, 1515-0100
1) Dakar	7170	100	N	0815-1515

FM: Dakar 94.5MHz (N), 92.0MHz (FM92).
Chaine Nationale: as above. **N. in French:** 0600MF, 0630MF, 0700SS, 0800-1100 hourly, 1300-1500 hourly, 1800-2200 hourly, 2355.
Chaine Inter (I): MF 0558-1405 & 1758-0005, SS 0658-0005. **N. in French:** on the h. **N. in Portuguese/Arabic/English:** MF 2005-2040.
FM92: 24h (rel. **Radio France Internationale** plus local prgrs).
ANN: N: "Radiodiffusion Télévision du Sénégal émettant de Dakar". I: "Senegal-Inter".
IS: Melody on "Cora" (local harp) — **V.** by QSL-card. Rp.
Regional Centres:
2) B.P. 375, Saint-Louis. **ANN:** "La Voix du Nord".
3) Tambacounda.
4) B.P.173, Ziguinchor.
5) B.P.321, Kaolack. **Te:** +221 412265.

SEYCHELLES

L.T: UTC + 4h — **Pop:** 72.000 — **Radio sets:** 30.000 — **Pr.L:** Creole, English, French — **E.C:** a/c 50, 240V — **ITU:** SEY.

SEYCHELLES BROADC. CORP. (Comm.)

ADDR: Box 321, Hermitage, Mahé. **Te:** +248 24161. **Fax:** +248 24914. **Tlx:** 2315 INFORTV SZ.
L.P: MD: I. Afic. Dir. Broadcasting: Bryant Marriott. Dir. Admin: Mrs. Fauzia Rose. Prgr. Mgr. (Radio): Miss M. Hermitte. Prgr. Mgr. (TV): Jean-Claude Matombe. CE: J. Rene.
STATION: ZCQ3 1368kHz 50kW.
D.PRGR: Mon-Fri 0200-0905 & 1300-1800, Sat/Sun 0200-1800.
N: English: 0300, 0600, 0900, 1500. **French:** 0330, 0700, 0830, 1700. **Creole:** 0230, 0500, 0800, 1600.
ANN: E: "This is RTS — Radio Television Seychelles". F: "Ici RTS — Radio Télévision Seychelloise". C: "Isi RTS — Radyo Televizyon Sesel".
IS: Instrumental music — **V.** by QSL-card (B).

FAR EAST BROADCASTING ASSOCIATION (FEBA) (Missionary)

ADDR: Box 234, Mahe. **Te:** +248 24449. **Cable:** FEBCOM. **Fax:** +248 25171.
L.P: Dir: B.T. Rome. Chief Eng: P.W. Williams. Freq. Mgr: R.W. Whittington.
STATION (G.C: 55.28E/04.36S): 100kW.

kHz	kHz	kHz	kHz	kHz	kHz
7270	9810	11790	11840	15145	15430
9565	11665	11805	11930	15325	15445
9660	11675	11810	15125	15330	17760
9770	11710				

Days: 1 = Mon, 2 = Tues, 3 = Wed, 4 = Thurs, 5 = Fri, 6 = Sat, 7 = Sun.

Afar	Days	Area	kHz
1130-1200	1234567	ETH	15430
Amharic			
0330-0400	167	ETH	15145
1630-1700	34567	ETH	11840
1730-1800	1234567	ETH	11840

Arabic			
0345-0430	1234567	ME	15325
0430-0445	457	ME	15325
0445-0500	5	ME	15325
1100-1145	1234567	ME	15125
1903-1948	1234567	ME	11840
1948-2003	57	ME	11840
Azeri			
1739-1759	1234567	ME	11710
Bengali			
0000-0030	1234567	SAs	7270
Bhili			
1345-1400	1	SAs	15330
Bhojpuri			
1245-1300	567	SAs	15445
Borana			
0330-0400	23	ETH	15145
Chattisgarhi			
1345-1400	567	SAs	15330
Dari			
0245-0300	1234567	SAs	11665, 15445
English			
1500-1600	1234567	SAs	11710
1500-1600	23456	SAs	9810, 15330
Farsi			
0300-0330	1234567	IRN	15445
1800-1830	1234567	IRN	11710
French			
0915-0930	12345	EAf	15430
0915-1000	67	EAf	15430
1833-1848	1234567	WAf	9565
1848-1903	14567	WAf	9565
Gujurati			
1330-1345	1234567	SAs	15330
Hindi			
0030-0100	1234567	SAs	7270
1415-1458	1234567	SAs	11675, 15330
Hindko			
0214-0229	56	As	11665
Kannada			
1330-1345	1234567	SAs	11675
1345-1400	23456	SAs	11675
Konkani			
1330-1345	4	SAs	11675
Malagasy			
1000-1015	67	MDG	15430
1628-1658	1234567	MDG	11930
Malayalam			
1300-1330	1234567	SAs	11675
1330-1345	23	SAs	11675
Mundari			
1245-1300	23	SAs	15445
Nagpuria			
1245-1300	4	SAs	15445
Nepali			
1230-1245	1234567	NPL	15445
Oriya			
1245-1300	1	SAs	15445
Oromo			
0330-0400	45	ETH	15145
0400-0430	34567	ETH	15145
1700-1730	1234567	ETH	11840
Portuguese			
1830-1845	1234567	EAf	9660
1845-1900	67	EAf	9660
Punjabi			
0214-0229	12	SAs	11665
1345-1415	34	SAs	15330
Pushtu			
0230-0245	1234567	SAs	11665, 15445
Sindhi			
0214-0229	1234567	SAs	9770
Sinhala			
1330-1345	1234567	SAs	9810
1345-1400	267	SAs	9810
Siraiki			
0214-0229	34	As	11665
Somali			
1700-1730	1234567	SOM	9770 11785
Swahili			
0315-0400	1234567	EAf	11810
1613-1643	1234567	EAf	11790
1643-1658	123567	EAf	11790
Tamil			
0130-0215	1234567	SAs	7270
1259-1330	1234567	SAs	15445
Telugu			
0100-0130	1234567	SAs	7270
Tibetan			
1215-1230	1234567	SAs	15445
Tigrigna			
1800-1830	1234567	ETH	11840
Tulu			
1330-1345	56	SAs	11675
Turkish			
0415-0430	1234567	TUR	17760
Urdu			
0159-0213	1234567	SAs	11665, 9770
0303-0328	5	SAs	15145
1345-1400	2	SAs	15330
1400-1415	12567	SAs	15330
1442-1457	6	SAs	9810
Yao			
1733-1748	1234567	EAf	11805

N. in English: 1530.
ANN: "This is FEBA Radio" — **IS:** "What a friend we have in Jesus" (xylophone) and spoken ID — **V.** by QSL-card. IRC's appreciated. 3 IRC's req. for airmail reply. Rec. Acc.(B) — **PUB:** "Programme Guide" (half yearly).

BBC INDIAN OCEAN RELAY STATION

ADDR: Grande Anse, Mahe, Seychelles. **Te:** +248 78496/78269. **Fax:** +248 78500. **Telex:** 2273 BBC SZ.
STATION (G.C: 55.28E/04.36S): 2×250kW tr's.

kHz	kHz	kHz	kHz	kHz	kHz
6005	9630	11730	11860	15420	17885

D.PRGR: see United Kingdom.
V. by letter. Re. to Peter Lee, Resident Engineer.

SIERRA LEONE

L.T: UTC — **Pop:** 4.385.000 — **Radio sets:** 900.000 — **Pr.L:** English, Mende, Temne, Krio, Limba — **E.C:** a/c 50, 230V(yes) — **ITU:** SRL.

SIERRA LEONE BROADCASTING SERVICE (Gov.)

ADDR: New England, Freetown. **Te:** + 232 40123. **Cable:** Broadcasts, Freetown. **Telex:** 3334 RADTEX SL.
L.P: Dir. Gen: M.J. Tunis. Ag Dir. Prgr. Sces: J. D. Ade Campbell. Dir. Eng: A. C. Ellis.
MEDIUMWAVES: Goderich 1206kHz 50/10kW.
SHORTWAVE: Goderich (G.C: 13.14W/08.30N): 3316kHz 10kW.
D.PRGR: 0558-2230. **N. in English:** 0700, 0800, 1900, 2000, 2100, 2200.
ANN: "This is the SLBS in Freetown" or "You are tuned to Freetown" — **INT-SIG:** 5 notes chime followed by "The Rain is Coming" played by Military Band — **V.** by letter. Rp(B).

SOMALI DEMOCRATIC REP.

L.T: UTC + 3h — **Pop:** 6.930.000 — **Radio sets:** 400.000 — **Pr.L:** Somali — **E.C:** a/c 50, 230V (no) — **ITU:** SOM.

R. MOGADISHU

ADDR: Private Postbag, Mogadishu. **Te:** + 249 20947. **Telex:** 621 — **STATION:** 962kHz 150kW — **FM:** 95.4MHz.
D.PRGR: 0300-0500(Fri 0600) & 1000-1200 on FM only, 1500-1900 on MW/FM. **N. in English:** 1830.
ANN: "Codka Jamhuriyada Somalida"; Arabic: "Huna Mogadishu idha'atu-l-ğumhuriya as-sumaliyya".
INT-SIG: Opens and closes with Nat. Anthem.
V. by QSL-card or letter. Rec. acc. Rp. (3 IRC's).

NORTHERN SOMALIA

RADIO HARGEISA
ADDR: P.O. Box 14, Hargeisa.
STATION: 7120kHz 1kW (mobile tr.).
D.PRGR: 1000-1230, 1500-1700.

SOUTH AFRICA (Republic)

L.T: UTC + 2h — **Pop:** 41.700.000 (incl. Homelands) — **Radio sets:** 10.000.000 (est.) — **Pr.L:** English, Afrikaans, N. & S. Sotho, Zulu, Tswana, Xhosa, Venda, Tsonga — **E.C:** a/c 50, 220/380V + d/c — **ITU:** AFS.

SOUTH AFRICAN BROADCASTING CORPORATION (SABC)
ADDR (Head Office): Broadcasting Centre, Auckland Park, Johannesburg 2092/Private Bag XI, Auckland Park 2086. **Te:** +27 (11) 714-9111. **Fax:** +27 (11) 714-5055. **Cable:** Broadcast. **Telex:** (095) 4 24 116 SA (Admin), (095) 4 26 259. SA (Tech). **Fax:** +27 (11) 714-3106. — (SW Freq. Division): P.O. Box 6, Honeydew 2040.

Reg. Addresses: P.O. Box 4559, Johannesburg 2000; Radio 5, P.O. Box 4301, Johannesburg 2000; P.O. Box 2551, Cape Town 8000; P.O. Box 1588, Durban 4000; P.O. Box 563, Bloemfontein 9300; P.O. Box 962, Pretoria 0001; P.O. Box 7054, Port Elizabeth 6000; P.O. Box 395, Pietersburg 0700.
L.P: Chmn (Board): Prof. H.C. Viljoen. Dir. Gen: W.J.J. Harmse. Dep. Dir. Gen (Prgrs): T. van Heerden. Dep. Dir. Gen. (Technology): N. Smuts. Dep. Dir. Gen. (Finance & Admin): Q. B. Green.

HOME SERVICES (Comm.)
Mediumwaves a) R. Metro (stereo), b) R. Xhosa, c) R. Swazi/R. Zulu, d) R. Ndebele/R. Zulu.

Location	kHz	kW/St.	Location	kHz	kW/St.
Bloemendal	576	100a	Welgedacht*	1287	2c
Komga	846	100b	Welgedacht*	1404	2d

*Johannesburg East.

FM: (MHz) Location	R. Suid-Afrika	R. South Africa	R. 2000	R.5
Cape Town	102.1	105.7	98.6	89.0s
Johannesburg	101.5	105.1	99.7	98.0s
Durban	100.8	104.4	97.3	89.9s
Pretoria	101.0	104.6	97.5	
Pt. Elizabeth	102.3	105.9	98.8	89.2s
Ea. London	101.6	105.2	98.1	88.5s

+ additional tr's at 119 locations — s) stereo.

NATIONAL SHORTWAVE SCES.
STATION: H.F. Verwoerd Station (Meyerton)
(G.C: 28.08E/28.35S): 4×100kW tr's + 1×100kW standby tr.

kHz	Sce.	H. of Tr.	kHz	Sce.	H. of Tr.
3215	R.Oranje	0300-0510	7285	R.S. Afrika	0440-0655
		1620-2200	9630	R.Oranje	0515-1615
3320	R. Orion	2300-0300	11770	R.S.Afrika	0540-1640
	R.S. Afrika	0300-0500			
		1530-2300			

R. So. Africa in English: 0300-1930 on FM only. **N:** 0330 (W), 0400 (W), 0500, 0600, 0900, 1100, 1400 & 1500 (Mon-Fri), 1600, 1700 (W), 1900.
R. Allegro: 1900-2300 on R. So. Africa freqs (serious music). **N:** 2100.
R. Suid Afrika in Afrikaans: 0300-1930 on FM. **SW:** as above.
Radio 5 in English: 0300-2200 on MW/FM. **N:** 0400 (W), 0500, 0600, 1100, 1500, 1630 (Mon-Fri). **Headlines** (Mon-Fri): 0430, 0530, 0630, 1430, 1530, 1630.
R. Orion in English/Afrikaans: 2300-0300 on R. Suid Afrika FM freqs, 2200-0300 on R5 FM & MW freqs (easy listening music). **N:** 2100, 2230. **SW:** as above.
R. 2000: 0300-2200.

REGIONAL SCES in English/Afrikaans
Highveld Stereo (Johannesburg): 94.0/94.7/95.2MHz + 2 relays.
R. Jacaranda (Pretoria): 94.2MHz 5kW + 21 relays.
R. Oranje (Bloemfontein): 96.2MHz 10kW + 22 relays.
R. Port Natal (Durban): 94.0MHz 10kW + 15 relays.
R. Lotus (Durban): 87.7MHz 10kW + 7 relays.
R. Good Hope (Cape Town): 95.3 MHz 3kW (stereo) + 30 relays.
Radio Algoa (Port Elizabeth): 95.5MHz 3kW + 18 relays.
D.PRGR. in English/Afrikaans (exc. R. Lotus): 0300-1930. After 1930, FM tr's simulcast TV4 sound until close, then rel. R. Orion.
N. every h. on the half h. 0430-1830, also 1100 & 1500.
Bilingual (Highveld Stereo): 0530, 1130, 1530 (Mon-Fri).
R. Lotus: in English: 0300-1930. **N:** 0445, 0900, 1100, 1500, 1900.

VERNACULAR SERVICES
R. Metro: Bloemendal 576kHz 100kW stereo. **FM:** Johannesburg 96.4MHz, Durban 93.0MHz, Durban No. 107.9MHz, Cape Town 93.0MHz, Port Elizabeth 87.9/100.5MHz, East London 107.7MHz.
Radio Sesotho: Bloemfontein 89.9/Johannesburg 88.4/Kroonstad 90.3MHz 10kW + 12 relays.
Radio Tswana: Bloemfontein 93.0MHz 10kW + 16 relays.
Radio Venda: 87.9/90.7/99.1/103.1/107.8MHz 3kW.
Radio Xhosa: King Williams Town 93.0MHz 3kW + 25 relays.
Radio Zulu: Johannesburg 91.5MHz 10kW + 19 relays.
Radio Swazi: 90.8/92.5/92.8/93.0MHz + 1 relay.
Radio Ndebele: Middelburg 91.8MHz 3kW + 2 relays.

Addresses & other information
R. Metro, P.O. Box 93116, Johannesburg 2000 — 24h in **English** on 576kHz. **N:** 0400 & 0500 (Mon-Fri), 0600, 1100, 1400 (Mon-Fri), 1600. Rel. on all FM/MW st's: 2100-0300 (exc. R. Xhosa 2200-0200) — **Ann:** "This is R. Metro 5-7-6 stereo", the Action st.", "The Sound of the City".
R. SeSotho, P.O. Box 1970, Johannesburg 2000 — 0300-2100 on **Ann:** "Ena ke R. Sotho".
R. Tswana, P.O. Box 28151, Sunnyside 0132 — 0300-2100 on **Ann:** "Eno ke R. Tswana".
R. Lebowa, P.O. Box 1869, Pietersburg 0700 — 0300-2100. **Ann:** "Ye ke R. Lebowa".
R. Tsonga (Pietersburg): 0300-0630 & 1400-1930 (Sat/Sun 0300-1930) at other times rel. R. Lebowa **Ann:** "Leyi i R. Tsonga".
R. Venda (Pietersburg): as R. Tsonga. **Ann:** "Heydi ndi R. Venda".
R. Xhosa, P.O. Box 36, King William's Town 5600 — 0200-2200 **Ann:** "Le yi R. Xhosa".
R. Zulu, P.O. Box 4559, Johannesburg 2000; P.O. Box 1588, Durban 4000 — 0300-2100. **Ann:** "Lona Umsakazo R. Zulu".
R. Swazi, P.O. Box 962, Pretoria 0001 — 0300-0630 & 1400-1930 (Sat/Sun 0300-1930). Other times rel. R. Zulu.
R. Ndebele: as R. Swazi.

CHANNEL AFRICA
ADDR: P.O. Box 91313, Auckland Park 2006.
Station: Meyerton (G.C: 28.08E/26.35E): 250/500kW.

kHz	kHz	kHz	kHz	kHz	kHz	kHz
3995	7270	11745	11815	11900	15430	17780
5960	9555	11800	11860	15145	15440	17815
7230	9585					

Chi-Chewa	Areas	kHz	Lozi	Areas	kHz
1500-1800	EAF	7270	1500-1800	ZMB	9555
English			**Portuguese**		
0200-0400	EAf/CAf	11745	0400-0500	MOZ	7230
0300-0500	SAf	7270	0500-0600	AGL	15145
0400-0700	WAf	15430	1100-1130MF	MOZ	11860
1000-1100	EAf	17780	1130-1200MF	AGL	15145
1100-1200	CAf/SAf	11900	1900-2000	MOZ	3995
1600-1800	Af	5960	2000-2100	AGL	11900
1600-1800	WAf	15430	**Swahili**		
French			1400-1700	EAf/CAf	11900
0300-0400	Ind Oc	11815	**Tsonga**		
0300-0600	CAf/WAf	11800	0900-1000SS	MOZ	9585
0500-0700	WAf	17815			
1800-2000	Af	11900			
		15440			

IS: Call of the Bokmaikerie accompanied by first bars of "Ver in de Wereld Kitty" on guitar — **V.** by QSL-card.

BOPHUTHATSWANA (Republic)
L.T: UTC + 2h — **Pop:** 2.500.000 — **Pr.L:** SeTswana, English, Afrikaans.

BOPHUTHATSWANA BROADCASTING (Comm.)
ADDR: P/Bag X2150, Mmabatho 8681. **Te:** +27 (140) 89-7111/7180. **Fax:** +27 (140) 89-7156. **Telex:** (0937) 3049 BP.

L.P: Dir. Gen: J.E. Procter. Dir. TV: R.R. Minton. Dir. Radio Sces: W. Tlaletse. Dir. Tech. Sces: D.G. Stewart. Head of Prgrs: R.S. Magau.

STATIONS: Mediumwave: Ga-Rankuwa 540kHz 100kW, 1098kHz 100kW (both stereo) — **FM:** Mmabatho 91.8MHz 1kW (Ch. 1), 88.7MHz 1kW (Ch. 2) — Ga-Rankuwa 100.4MHz stereo (Ch. 1), 103.9MHz (Ch. 2) — Madibogo 94.9MHz 1kW (Ch. 2) + 9 relay st's. at 17 locations.
Ch. 1 (R. Bop) in English on 540kHz + FM: 24h. **N:** 0400*, 0500W, 0600 (exc. Sat), 0900*, 1100, 1400*, 1600* — *) Mon-Fri. Headlines every h. 0400-2000.
Ch. 2 (R. Mmabatho) in SeTswana: 0400-2000 (Fri/Sat 2200). Other times rel. R. Bop.
Ann: E "This is R. Bop MW 540 Stereo". S: "ke R. Mmabatho, lentswe la Setshaba".
INT-SIG: R. Bop: instrumental music. Nat. Anthem at s/off.
PUB: Prgr. schedules on request — **V.** by QSL-card (re. to technical section).

RADIO 702 (Comm.)

ADDR:(Head Office): P.O. Box 5572, Rivonia 2128, South Africa; (studio): P.O. Box 203, Ga-Rankuwa 0208, Bophuthatswana. **Te:** +27 (11) 884 8400. **Fax:** +27 (11) 883 1982.
L.P: Man. Dir: I. Kirsh. Gen. Mgr: S. Katz. St. Mgr: Rina Broomberg. Sales Mgr: T. Williams. Op's Mgr: R. Vickers. Chief Eng: B. Richardson.
STATION: Ga Rankuwa 702kHz 100kW.
D.PRGR. in English (to So.Africa): 24h. **N:** on the half h.
ANN: "We do it for you" — **V.** by letter (re. to studio addr). — **F.PL:** FM stereo.

CISKEI (Republic)

L.T: UTC + 2h — **Pop:** 1.088.000 — **Pr.L:** Xhosa, English, Afrikaans.

CISKEIAN BROADC. CORP. (Gov.)

ADDR: P.O. Box 104, Bisho. **Telex:** (095) 25-0786 CX. **Cable:** Sasaza.
L.P: MD: F. E. Lubisi. PD: M. Xengana. CE: C. N. Stansfield. Sales Dir: S. F. Soga.
STATION: Mt. Kemp (Bisho) 89.9MHz 30kW, Queenstown 97.8MHz 2kW, Bisho 97.0MHz 0.05kW, 104.1MHz 0.05kW.
D.PRGR. in Xhosa/English: 24h.
N. in English: 0500, 1100, 1500, 1700.
ANN: "Le yi Radio Ciskei, umbomo yeCiskei". E: "The st. that blows your blues away".
INT-SIG: Call of the Blue Crane — **V.** by letter.

TRANSKEI (Republic)

L.T: UTC + 2h — **Pop:** 4.740.000 — **Pr.L:** Xhosa, English, Sesotho.

DEPT. OF POSTS & TELECOMMUNICATIONS

ADDR: P/Bag X5001, Umtata 5100. **Te:** +27 (471) 23311. **Cable:** Postgen. **Telex:** 75-440.

TRANSKEI BROADC. CORP. (Gov,Comm.)

ADDR: P.O. Box 794, Umtata 5100. **Te:** +27 (471) 312611. **Fax:** +27 (471) 24297. **Telex:** (0968) 709TT. **Cable:** Usasazo.
L.P: MD: L. Xinwa. Corp. Sec: H. Gxaweni. Tech. Man: K. Herold.
STATIONS: FM stereo (3kW) — +) RDS.

Location	MHz	Location	MHz
Butterworth	91.1+	Queenstown	107.6
Cala	93.4+	Umtata	92.0+
Matatiele	91.5	Umzimkulu	91.9+
Port St. Johns	93.7+	Sterkspruit	107.9
Mt. Ayliff	93.2		

D.PRGR. in Xhosa, English & S. Sotho: 24h. **N:** 0430, 0530, 0730, 0830, 0930, 1130, 1530, 1630, 1830, 2000.
ANN: 'Le yi R. Transkei", "This is R. Transkei (or R. TKI), the St. with a difference broadcasting from Umtata".
INT-SIG: Nat. Anthem & clock chimes at 2300 — **V.** by letter.

SABC Relays by agreement with TBC

FM	R.Suid Afrika	R.South Africa	R.2000	R.SeSotho
Butterworth	101.1+	104.7+	—	—
Cala	103.4°+	107.0°+	—	90.3+
Mt. Ayliff	103.2+	106.8+	99.7	—
Port St. Johns	103.7°+	107.3°+	100.2°+	—
Umtata	102.0°+	105.6°+	98.5°	—

Power: °) 0.3kW, others 3kW. — +) RDS.

CAPITAL RADIO (TRANSKEI) PTY. LTD. (Comm.)

ADDR: Head Office: Box 806, Umtata 5100. **Te:** +27 (471) 23134. **Fax:** +27 (471) 311607. **Telex:** (0968) 823 TT. **Studios:** Mermaid Lido Bldg., So. Beach, Durban 4001, So. Africa (Postal addr: P.O. Box 604, Durban 4000). **Te:** +27 (31) 3681604. **Fax:** +27 (31) 371604.
L.P: Chmn: G.S.K. Nota. Man. Dir: H. Jikela. Gen. Mgr: P. Chamberlain. Prgr. Mgr: A. Duke. Mgr. Tech. Sces: A. Cherry.

Stations	kHz	kW		kW	kW
Umtata	558	100	Umtata	3930v	10
Umzimkulu	603	100	Umtata	7150v	10

FM: (3kW): Umtata 95.2MHz, Mt. Ayliff 96.4MHz, Butterworth 94.3MHz, Cala 96.6MHz, Queenstown 104.2MHz, Port St. John's 96.9MHz.
D.PRGR. in English: 0230-2300 on MW/FM, 0230-0445 & 1530-2300 on 3930kHz, 0445-1530 on 7150kHz.
N. 0400-1600 on the h. **Headlines:** Mon-Fri 0430, 0530, 0630, 1430, 1530, 1630 and 20/20 format.
ANN: "This is Capital, on the Beach".
V. by QSL-letter. Re. to Mgr. Tec. Sces at Umtata addr.

VENDA (Republic)

L.T: UTC + 2h — **Pop:** 718.000 — **Pr.L:** LuVenda, English.

RADIO THOHOYANDOU (Gov,Comm.)

ADDR: P/bag 5011, Posts & Communications, Thohoyandou. **Te:** (0159) 21121/5. **Fax:** (0159) 22998.
L.P: Postmaster General: J.M. Marema. Dep. DG: T.J. Tshiololi. Prgr. Mgr: T.I. Dagada.
MEDIUMWAVE: Tshisahulu 1035kHz 100kW.
FM: Sibasa 106.9MHz 0.3kW, Mulima 93.0MHz 0.3kW. Dzamba 93.3MHz 0.3kW, Kutama 107.9Hz 0.02kW, Gaba 88.2MHz 0.05kW.
D.PRGR. in LuVenda/English: 24h (separate prgrs. on MW/FM).
N: on the h.
ANN: "Hei ndi R. Thohoyandou Ipfi la Riphabuliki ya Venda".
IS: Drums & trumpeting elephant — **V.** by letter — **F.PL:** SW-tr.

ST. HELENA (British)

L.T: UTC — **Pop:** 5.550 — **Radio sets:** 2.500 — **Only L:** English — **E.C:** a/c 50, 240V — **ITU:** SHN.

ST. HELENA GOVT. BROADCASTING SCE.

ADDR: Broadway House, Jamestown, Island of St. Helena, So. Atlantic Ocean. **Te:** +290 4669. **Fax:** +290 2598. **Tlx:** 4202.
L.P: Broadc. Officer: Tony Leo.
MEDIUMWAVE (G.C: 15.55S/05.32W): 1548kHz 1/0/1kW.
SHORTWAVE: uses SSB tr. of Cable & Wireless one day a year for special DX test. The last transmission, on Oct. 23rd 1992, used 11092kHz at 2000-2100 and 2220-2300 UTC.
D.PRGR. on MW: 0900-1800MF rel. BBC prgrs, 1800-2230D local prgrs incl. BBC N. 2000 & 2200.
ANN: "R. St. Helena" — **IS:** Trumpet call 11 seconds repeated at 5 second intervals for 64 seconds followed by "Life on the Ocean Waves" — **V.** by QSL-card. Rp.

SUDAN (Democratic Republic)

L.T: UTC + 2h — **Pop:** 28m — **Radio sets:** 6m (est.) — **Pr.L:** Arabic — **E.C:** a/c 50, 240V (yes) — **ITU:** SDN.

SUDAN NATIONAL BROADCASTING CORP.

ADDR: P.O. Box 572, Omdurman. **Te:** + 249 53151. **Cable:** IZZA. **Regional addr:** Box 126, Juba.
L.P: DG: Abbas al-Tawm. DG: Eng. & Tech. Affairs: Abbas Sidig.

Mediumwaves	kHz	kW		kHz	kW
Nyala	540	50	Atbara	783	5
Soba	576	100	Dongola	819	5
Arda	576	20	Port Sudan	860	7
Obeid	639	5	Medani	873	5
Kassala	666	5	Kadugli	945	100
Juba	693	100	Soba*	963	100
Arda	765	50	Rebia	1296	1500

*) Carries Special Prgrs, others General Prgr.

SHORTWAVES: (G.C: 32.28E/15.30N): 7200/9190 (alt: 9170) kHz. 100kW.

General Prgr: MW: 0300-2200 (on 1296kHz: 0300-0500 & 1700-2200). **SW:** 0300-2200 on 7200kHz, 0300-0600, 0800-1100 & 1900-2200 (times variable) on 9190 or 9170kHz. **N:** 0315, 0430, 0530 (exc. Fri), 0800, 1000 (exc. Fri), 1300, 1445 (local), 1500, 1700, 1800 (local), 1930, 2145. **English:** 1800-1900.

R. Juba (regional st.): 1400-1500 on 693/9540 (alt: 9550)kHz. **N. in English:** 1400.

ANN: "Huna Omdurman idha'atu-l-ğumhuriya as-sudan ad-dimuqratiyya".

INT-SIG: Omdurman: Interlude of Sudanese music. Opens and closes with National Anthem — **V.** by letter.

NATIONAL UNITY RADIO (Mil.)

ADDR: Omdurman.

SHORTWAVE: 9190 (alt: 9170) kHz 100kW.

D.PRGR: 1300-1700. **N. in English:** 1500.

SWAZILAND

L.T: UTC+2h — **Pop:** 882.000 — **Radio sets:** 65.000 — **Pr.L:** English, SiSwati — **E.C:** a/c 50, 220V — **ITU:** SWZ.

SWAZILAND BROADC. SCE. (Gov., Comm.)

ADDR: P.O. Box 338, Mbabane. **Te:** +268 47263/5. **Fax:** +268 46953. **Cable:** Broadcast, Mbabane. **Tlx:** 0964 2035 WD.

L.P: Dir: Abner T. Tempe. Dep. Dir: E.N. Khumalo.

MW: Sidvokodvo 954kHz 50/30kW (stereo) — **FM:** 93.02/95.6MHz 10kW + 4 low power relays (all stereo).

D.PRGR: 0330-1930. **English:** 0330-0700, 1600-1930. **N:** 1600, 1800. Rel BBC African Sce. 1615-1645MF.

ANN: "This is the English sce. of R. Swaziland 954 AM stereo". "Lona ngu Mawakato waka Ngwane".

IS: at s/on, Cilongo (Swazi instrument). English Sce: cock crow, fanfare, spoken ID, instrumental theme.

SWAZILAND COMMERCIAL RADIO (Pty) Ltd.

ADDR: P.O. Box 5569, Rivonia 2128, Transvaal, So. Africa. **Te:** +27 (11) 884 8400. **Fax:** +27 (11) 883 1982.

L.P: MD: Issie Kirsh. PD: Rob Vickers. CE: Guy Doult.

MW: Sandlane 1377kHz 50kW.

SW (G.C: 30.48E/26.38S): 6155kHz 100kW.

D.PRGR. in English to So. Africa: Mon-Fri 1700-2030, Sat 1700-2000, Sun 0500-0600 & 1700-2045.

ANN: "This is Swazi Radio" — **V.** by letter. Re. to Chief Eng.

COMMUNITIES BROADCASTING SCES. cc (Comm.)

ADDR: P.O. Box 15535, Doornfontein 2028, So. Africa. **Te:** + 27 (11) 434 1431/2. **Fax:** +27 (11) 434 1867 — PD: D. Bettancourt.

SW (G.C: 30.48E/26.38S): 6155kHz 100kW — **FM:** 99/100MHz.

Portuguese ("R. Cidade"): Mon-Fri 0700-1000, Sat 0900-1200, Sun 1200-1500.

Portuguese/English ("R. Cidade Jovem): Sat 1200-1600.

Italian ("Transmissione Italiana"): Sat 0600-0900.

Greek ("Hellenic Heart Beat"): Sun 1100-1200.

ANN: P: "Radio Cidade . . . ques ouve, gosta".

TRANS WORLD RADIO (Rlg.)

ADDR: P.O. Box 64, Manzini. **Te:** +268 52781-5. **Fax:** +268 55333. **Cable:** VOTAN.

L.P: Pres: Dr. Paul E. Freed. St. Dir: Peter Prediger. Dir. of Prgrs: Rev. D.A. MacDonald. Chief Eng: James Burnett. Head of PR: Lionel Wright. Regional Dir (Africa): Rev. Stephen Boakye-Yiadom.

MEDIUMWAVE: 1170kHz 50kW.

SHORTWAVE: Mpangela Ranch (G.C: 31.59E/26.34S): 4 × 25kW and 1 × 100kW tr's.

kHz	kHz	kHz	kHz	kHz
3200	4790	7180	9520	11740
3240	5055	7200	9640	11805
3365	5965	9500	9650	15380
4760	6020	9515	9655	

Schedule until March 6th 1993

Chewa		
0400-0430	MWI	9655
0530-0615SS	MWI	9655
0615-0645Sat	MWI	9655
1600-1700	MWI	9520
English		
0430-0500	SAf	5055
0430-0530	CAf	9655
0430-0700	SAf	5965
0430-0820	Cape	11740
0600-0820	SAf	7200
0820-0835SS	Cape	11740
0820-0835SS	SAf	7200
1600-1700	EAf	9500
1700-1730	CAf	9520
1730-2115	SAf	1170
1800-1845	EAf	9500
1900-2115	SAf	3240
French		
1515-1545	MDG	9650
1930-2000	ZAI	9520
2000-2015SS	ZAI	9520
German		
0400-0430	SAf	5055
	NMB	5965
Lingala		
1900-1930	ZAI	9520
Lomwe		
0400-0415	MOZ	6020
1530-1545	MOZ	9515
Malagasy		
1500-1515	MDG	9650
Ndebele		
0300-0330	ZWE	3200
1800-1830	ZWE	3240
Pedi/Tswana		
1730-1745SS	SAf	4760
Portuguese		
0300-0305MF	SAf	3365
0305-0345	SAf	3365
1440-1445MF	SAf	9515
1445-1530	MOZ	9515
1900-1945	AGL	7180
1945-1950MF	AGL	7180
Shona		
0300-0345	ZWE	3240
1745-1830	ZWE	4790
Tshwa		
1600-1645	MOZ	4760
Tswana		
1700-1730	SAf	4760
Umbundu		
1845-1900	AGL	7180
Urdu		
1530-1545	PAK	15380
Xhosa		
1830-1900	SAf	3240
Yao		
1445-1500	MWI/MOZ	11805
Zulu		
1700-1730	SAf	1170
1700-1800	SAf	3240

ANN: "This is Trans World Radio — Swaziland" or "from the beautiful Kingdom of Swaziland, this is Trans World Radio".

INT-SIG: Last bar of "We've a story to tell the nations" on Hand Bells.

V. by QSL-folder. Rec. Acc. preferably from target areas. IRC's appreciated (3 IRC's for airmail reply).

PUB: Programme guides available in English free on request.

TANZANIA

L.T: UTC + 3h — **Pop:** 27.782.000 — **Radio sets:** 4m (est.) — **Pr.L:** Swahili, English — **E.C:** a/c 50, 230V — **ITU:** TZA.

RADIO TANZANIA (Gov. Comm.)

ADDR: P.O. Box 9191, Dar es Salaam. **Te:** +255 (51) 38011-6. **Cable:** Sauti, Dar Es Salaam. **Telex:** 41201 Sauti.

L.P: Dir. of Broadc: Nkwabi Ng'wanakilala. Chief Eng: P.I. Mhumbira. Ag. Contr. of Prgrs: A. Ngarawa.

Networks: C=Commercial Sce, N=National Sce, R=Regional Sces, E=External Sce.

Mediumwaves	kHz	kW	Netw.
Dar es Salaam	531	10	N
Dodoma	603	100/10	N/R
Mbeya	621	50/10	N/R
Nachingwea	648	100/10	N
Dar es Salaam	657	50	N
Kigoma	711	100/10	N/R
Mwanza	720	5/10	N/R
Dar es Salaam	837	1	N/E
Arusha	1215	50/10	N/R

Shortwaves: Dar es Salaam (G.C: 39.14E/06.50S):

kHz	kW	Netw.	kHz	kW	Netw.
4785	50	N	7165	10	N
5050.1	10	N/E	7280	30	N
6105	100	N			

FM: (N) Dar es Salaam 89.9/92.3MHz, Dodoma 87.7MHz, Kigoma 88.4MHz, Masasi 92.3/105.9MHz.

National Sce. in Swahili: 0200-2100. **N:** 0400, 0700, 1000, 1300, 1600 (rel. Zanzibar), 1700, 1900, 2100. **Schools Prgrs in Swahili/**

English: Mon-Fri 0530-0830.
Regional Sces: No recent information received.
External Sce. in English to Ea, Ce. & So. Africa: 0300-0430 and 1530-1915. To Ea. & Ce. Africa: 0900-1030 (Sat/Sun 1530). **N:** 0400, 1000, 1300 (Sat/Sun), 1600, 1800, 1900.
ANN: National Sce: "Hii Nii Idhaa ya Taifa ya Radio Tanzania Dar es Salaam". Ext. Sce: "This is the Ext. Sce. of R. Tanzania Dar es Salaam".
INT-SIG: Celeste (Nat./Ext. Sces), Piano (Comm. Sce.).
V. by QSL-card or letter.

ZANZIBAR & PEMBA

SAUTI YA TANZANIA ZANZIBAR (Gov.)

ADDR: P.O. Box 1178, Zanzibar, Tanzania. **Te:** +255 31088/9. **Telex:** 57207 STZ.
L.P: Dir. of Broadc: J.C. Asama. Chief Eng: Abdulrahaman M. Said.
Mediumwave: Chumbum 585kHz 10kW.
Shortwave: Dole (G.C: 39.14E/06.05S): 6015/11734.4kHz 50kW.
D.PRGR: 0300-0500 on 585kHz, 1100-2000 on 585/6105kHz, 1500-1815 (irr. 1830) on 585/11734.4kHz. **N. in Swahili:** 0900, 1300, 1500, 1645. **Rel. Dar es Salaam:** 0400, 1700, 1900.
ANN: "Hii ni Sauti ya Tanzania Zanzibar".
INT-SIG: Indigenous xylophone — **V:** rarely by letter.
F.PL: Expand h. of tr. to 18h daily. Ext. Sce. in English.

TOGO

L.T: UTC — **Pop:** 3.950.000 — **Radio sets:** 700.000 — **Pr.L:** French, Ewe, Kabye, Mina, Dagombo — **E.C:** a/c 50, 127/220V — **ITU:** TGO.

RADIODIFFUSION TOGOLAISE (Gov.)

ADDR: B.P. 434, Lomé. **Te:** +228 21-24-93 — B.P. 21, Kara. **Te:** +228 606060. **Cable:** Radio Togo.
L.P: Dir. (Lomé): Pitang Tchalla. Chief of H.F. Division: K.O. Akpaki. Dir: (Kara): M'Ba Kpenougou. Chief of HF Div (Kara): D. Soares.
STATIONS: Lomé (G.C: 01.12E/06.16N), Kara (G.C: 01.09E/09.35N).

Station	kHz	kW	Times
Lomé	1395	20/1	0525-0005
Kara	1503	10	W 0525-0900, 1200-2305
Kara	3222v	10	0525-0900, 1200-2305
Lomé	5047v	100	0525-0903, 1603-0005
Lomé	7265v	50/100	0903-1603

FM: No details available.
D.Prgr: Lomé: as above. **Main N. in French:** 0600, 1230, 1900, 2200. **Headlines:** 0535, hourly 0700-1200 & 1400-1800, 2300, 2359 — **N. in English:** 0940, 1945.
Kara: as above. **N. in French:** 0535, 0600, 0700, 0800, 0900, 1300, 1425, 1930, 2145, 2300.
ANN: "Radiodiffusion Togolaise - Chaine Inter", "R. Togo", "R. Lomé", "R. Kara".
INT-SIG: Soft tempo chime (Lomé). At opening and close down: Hymne Togolaise — **V.** by QSL-card (R. Kara has own QSL-card) Re. in F, E, Rec. acc. (B) — **PUB:** Prgr. sched. (weekly).

TRISTAN DA CUNHA (British)

L.T: UTC — **Pop:** 311 — **Pr.L:** English — **E.C:** a/c 50, 220V.

TRISTAN BROADCASTING SCE. (Gov.)

ADDR: The Administrator, Tristan da Cunha, South Atlantic Ocean, via Cape Town, South Africa.
L.P: Dir. of Broadc: A. Swain.
STATION: Edinburgh (G.C: 37.00S/12.00W): 3290kHz 0.04kW.
D. PRGR: Mon-Fri 0709-0815, 1230-1300, 1600-1700; Mon/Wed/Fri 1945-2200; Sun 1000-1200. **N:** 2000 (BBC), 2009 (local).
ANN: "This is Tristan Radio, the Broadc. Sce. of Tristan da Cunha" — **V.** by letter.

TUNISIA

L.T: UTC + 1h (Su: UTC + 2h) — **Pop:** 8.450.000 — **Radio sets:** 1.700.000 — **Pr.L:** Arabic — **E.C:** a/c 50, 115/220 and 220/380V — **ITU:** TUN.

ÉTABLISSEMENT DE LA RADIODIFFUSION TÉLÉVISION TUNISIENNE (ERTT)

ADDR: 71, ave. de la Liberté, Tunis. **Te:** +216 (1) 287.300. **Telex:** 12365 TN.
L.P: DG: Abdelhafidh Herguem. Dir. R. Monastir: Salah Chebil, Dir. Ext. Rel: Mokhtar Rassaâ.

Mediumwaves	kHz	kW		kHz	kW
Gafsa	585	350	Remada	882	1
Sousse	603	10	Tunis	963	200
Tunis	630	600	Sfax	1566	1200
Sfax	720	100			

Shortwaves: Sfax (G.C: 10.35E/34.58N): 100kW.

kHz	kHz	kHz	kHz	kHz	kHz	kHz
7475	9675	11550	12005	15450	17500	21535

FM (MHz)	Nat. Netw.	Int. Netw.	kW
Ain Draham	90.3	96.6	5.6
Biadha	95.0	101.8	40
Gorrâa	89.1	95.4	3.8
Kasserine	89.6	99.2	49
Sfax	93.0	99.5	60
Tunis	88.6	98.2	1.2
Zaghouan	94.0	96.5	20
		92.0	
Zarzis	93.9	97.2	72

Nat. Netw. in Arabic: 24h on 630kHz + FM, 0425-2330 on 585kHz, 1600-2330 on 1566kHz, 2200-2330 on 963kHz.
N. on the h.
Rel. on SW: 0430-2300 on 12005kHz, 0430-0700 & 1800-2300 on 7475/9675/11550kHz, 0700-1800 on 11550/17500/21535kHz.
International Netw: in French: 0500-2200 on 963kHz + FM.
Italian: 1330-1430.
N: 0500, 0800, 0900, 1000, 1100, 1245, 1500, 1630, 1830, 2130.
Sfax Regional Prgr: 0430-2330 in Arabic on 720kHz.
Monastir Regional Prgr: 0550-1715 on 603kHz.
ANN: F: "Ici Radio Tunisia Internationale"; Arabic: "Idha'atu-l-ğumhuriya at-tunisiyya" — **V.** by letter (Re. in E, F to Dir. Gen. des Telecommunications. Rec. acc.).

UGANDA

L.T: UTC + 3h — **Pop:** 19.400.000 — **Radio sets:** 3.500.000 — **Pr.L:** English, Swahili, ethnic — **E.C:** a/c 50, 240/415V (yes) — **ITU:** UGA.

RADIO UGANDA (Gov.)

ADDR: P.O. Box 7142, Kampala. **Te:** +256 (41) 254461. **Fax:** +256 (41) 256888. **Telex:** 61084 UG. **Cable:** Knowledge Kampala.
L.P: Ag. Dir. of Broadc: J.C. Sserwadda. Ag. Dep. Dir. of Broadc: M.A.P. Mworoga. Ag. Chief Broadc. Eng: R. Adelia.
STATIONS (G.C: Kampala 32.36E/00.20N) — B=Blue Netw; R=Red Netw.

Location	kHz	kW	Net	Location	kHz	kW	Net
Mawagga	576	100	B	Arua	1485t	10	B
Kampala	639	50	B	Kampala	4976	50	B
Butebo	729	100	R	Kampala	5026v	20	B
Bobi	810	100	R	Kampala	7110	20/10	B
Kampala	909	20	R	Kampala	7195v	20/10	B
Kabale	999	100	B				

t) freq. tentative.
HOME SCE: NATIONAL PRGR: (Blue Netw.) in English, Swahili, French, Vernaculars: Mon-Fri 0300-0600, 1300-2100; Sat/Sun 0345-0600, 1400-2100. **On 7110/7195kHz:** 0600-1130. **N. in English:** 0400, 0700, 1400, 1800, 1900. **N. in French:** 1545 (Sat 1600).
REGIONAL PRGR: (Red Netw.) in English, Swahili and Vernaculars: Mon-Fri 0330-0545, 0600-1130, 1300-2100; Sat 0330-1345, 1400-2100; Sun 0330-1415, 1430-2100.
N. in English: 0400, 0700, 1000, 1400, 1700, 1900. **N. in Swahili:** 0430, 0945, 1830. **N. in Arabic:** 1530.
Reg. St's: 8h daily in regional languages. **N:** rel. Kampala.

ANN: E: "This is Radio Uganda, Kampala".
INT-SIG: Local Xylophone — **V.** by QSL-card. Rec. acc.

ZAIRE

L.T: Kinshasha/Mbandaka UTC+1h; Lubumbashi/Kisangani/Bukavu/Mbuji-Mayi/Kananga: UTC+2h — **Pop:** 39.100.000 — **Radio sets:** 3.400.000 — **Pr.L:** French, ethnic — **E.C:** a/c 50, 220V — **ITU:** ZAI.

LA VOIX DU ZAIRE (Gov.)

ADDR: Station Nationale, B.P. 3164 or B.P. 3171, Kinshasa-Gombe. **Te:** +243 (12) 23171-5. **Telex:** 21583 VOZAKIN ZR.
L.P: President-Delegue Gen: Dongo Badjanga: Dir. Inf. Rad: M.M. ma Mongo; Dir. Prg. Rad: A. Mwan'ango; Dir. Tec: L. Saye.

MEDIUMWAVES	**kHz**	**kW**			
7) Matadi	1160	50	1) Kinshasa	1448	2
5) Mbandaka	1268	2			
SHORTWAVES (all tr's irr.)					
3) Bukavu	4862v	3	2) Lubumbashi	7202	10
4) Mbandaka	5995	10	5) Mbuji-Mayi	7295	10
6) Kananga	6125*	10	6) Kananga	7865*	10
1) Kinshasa	7100	5/10	1) Kinshasa	9650	—
*Inactive			8) Kisangani	r11455	

FM: 1) 100MHz 10kW, 92.0MHz 3kW, 3) 88.04MHz 1.5kW, 4) 93.3MHz 1.5kW/90MHz 0.05kW, 6) 89MHz 50kW/92Mhz 1kW.
Addresses and other information:
1) B.P. 3171, Kinshasa-Gombe. 24h in French/Swahili/Lingala/Tshiluba/Kikongo. **N. in French:** 0100, 0300, 0500, 0600, 0700, 0900, 1000, 1130, 1300, 1400, 1500, 1800, 2000, 2200, 2300 — also re. by other sts. **V.** by letter.
2) B.P. 7296, Lubumbashi. W0300-0700, 1000-1245, 1500-1830v; Sun 0300-1830v. **V.** by letter.
3) B.P. 475 Bukavu — Dir: Baruti Lusorgela. W0300-0700, 1000-1245, 1500-1830; Sun 0300-1830. **V.** by letter.
4) B.P. 1061 Mbandaka. W0500-0800, 1000-1200, 1600-2100; Sat 0500-0800, 1000-2200; Sun 0500-2200. **N.** 0430 1030 1630. **V.** by letter. Rec. Acc.
5) B.P. 1232 Mbuji-Mayi.
6) B.P. 708, Kananga, Kasai Occidental — Dir: Hery Mambo Djugudjambe.
7) B.P. 704, Matadi.
8) B.P. 1745, Kisangani — 0400-
ANN. "Ici la Voix du Zaire, Station Nationale du Radiodiffusion." **V.** See above.

RADIO CANDIP-ISP (Educ.)

ADDR: Centre d'Animation et de Diffusion Pedagogique, Institut Supérieur Pédagogique de Bunia, B.P. 373, Bunia.
STATIONS (1kW)
3390.2 kHz: 0330-0600, 1530-1845 (Sun 0330-1845).
5066.4 kHz: 0330-0600, 1300 (Sat 1000)-1845.
7150 kHz: 1330 (Sat 1000, Sun 0600)-1530.
FM: 91.4/95/98MHz 0.02kW.
D.PRGR: In French/Swahili/Lingala & 6 dialects as above.
INT-SIG: Trad. music — **ANN:** "Vous êtes a l' écoute de la radio scolaire et development CANDIP Centre d'Animation et de Diffusion Pedagogique, Bunia en République du Zaire."

ZAMBIA (Republic)

L.T: UTC + 2h — **Pop:** 8.741.000 — **Radio sets:** 1.660.360 — **Pr.L:** English, ethnic — **E.C:** a/c 50, 220V — **ITU:** ZMB.

ZAMBIA NATIONAL BROADCASTING CORP. (Independent Statutory Body, Comm.)

ADDR: Broadcasting House, P.O. 50015, Lusaka. **Te:** +260 (1) 229648. **Fax:** +260 (1) 254013. **Cable:** Broadcasts Ridgeway. **Telex:** (0902) 41221 ZA — P.O. Box 20748, Kitwe. **Te:** +260 (2) 213555. **Cable:** Broadcasts Kitwe.
L.P: Ag. DG: E. Chayi. Dir. of Eng: E.C. Mulenga. Dir. of Finance: R.S. Nambi. Corp. Secr: F.M. Hamakando. Contr. Personnel & Training: C.A.J. Njobvu. Contr. Radio: F. Siliya. Ag. Contr. TV: H. Mapulanga. Reg. Contr. (Kitwe): P. Tembo.

Mediumwaves	**kHz**	**kW**	**Sce.**				
Kitwe	549	100	R2	Kasama	882	10	R1
Kasama	567	10	R1	Solwezi	909	10	R2
Mongu	603	10	R1	Mansa	918	10	R1
Lusaka	630	500	R2	Livingsone	927	10	R2
Chipata	666	10	R1	Kabwe	936	1	R1
Livingstone	729	10	R1	Kitwe	1071	100	R1
Lusaka	819	200	R1	Solweizi	1125	10	R2
Mongu	828	10	R2	Mansa	1269	10	R2
Chipata	855	10	R2	Kabwe	1458	1	R1

FM: R2: Lusaka 88.5MHz. R4: Lusaka/Kitwe/Kabwe 92.2, Ndola 94.5, Livingstone 95.5MHz.
Shortwaves: Lusaka (G.C:28.15E/15.30S):

kHz	**kW**	**Sce.**	**kHz**	**kW**	**Sce.**
4910v	50	R1	7220	50	R1
6165	50	R2	7235v	50	R2

Radio One in Zambian languages: 0245-2205. **N. in English** (rel. R2): 0500, 1100, 1800.
Radio Two in English: 0240-2210. **N:** 0400, 0500, 0600W, 0800W, 1000W, 1100, 1400MF, 1600, 1800, 2000, 2100.
Radio Four (R. Mulungushi): 24h in stereo.
ANN: R1 Bemba: "Kuni ni ku Radio One wa ZNBC". Kaonde: "Kunoni ko ku Radio One was ZNBC". Lozi: "Ye ki Radio One ya ZNBC". Lunda: "Ku fuma kunu Radio One ya ZNBC". Luvale: "Kuno minakwivwilila kukiko ku Radio ya ZNBC". Nyanja: "Kuno ndi ku Radio One wa ZNBC". Tonga: "Ukunu nkoku Radio One ya ZNBC". **R2:** "This is Radio Two of ZNBC Radio and Television Network" — **INT-SIG:** "Call of the Fish Eagle" — **V.** by QSL-card. Rp.

ZIMBABWE

L.T: UTC + 2h — **Pop:** 11.031.000 — **Radio sets:** 522.000 — **Pr.L:** English, Shona, Ndebele, Chewa — **E.C:** a/c 50, 220V — **ITU:** ZWE.

ZIMBABWE BROADCASTING CORPORATION (Indep. Statutory Body) (Comm.)

ADDR: P.O. Box HG 444, Highlands, Harare. **Te:** +263 (4) 707222. **Fax:** +263 (4) 795698. **Cable:** Broadcasts. **Telex:** 4175 ZBCHOV ZW.
Addr: Radio 2 Sce. (Mbare): P.O. Box 9048, Harare; Radio 2 Sce. (Bulawayo): P.O. Box 2279, Bulawayo. **Te:** +263 (9) 71811. **Cable:** Broadcasts Bulawayo. **Telex:** 3345 ZW.
L.P: Dir. Gen: T.J. Kangai. Dep. Dir. Gen: O.O. Chekeche. Asst. Dir. Gen. (Eng. & Tech. Sces): D. Woodward. Asst. Dir. Gen. (Prgrs, N. & Current Affairs): G. Tichatonga. Asst. Dir. Gen. (Admin. & Personnel): D. Mandizha. Dir. Eng. (Tr's & P.I.D.): I. McPhee. Dir. Eng. (Studios): R. Mhende. Dir. N: T. Mandigora. Dir. (Radio Prgrs): V. Mhizha-Murira. Dir (TV Prgrs): J. Johnhera. Dir. Finance: M. Gora. Dir. Admin. & Personnel (Bulawayo): G. Makoni.

FM (MHz)	**R1**	**R2**	**R3**	**R4**	**kW**
Harare	92.8	96.0	99.3	102.8	10
Mutare	102.2	89.1	95.4	98.7	3
Bulawayo	90.0	96.3	99.6	103.1	10
Gweru	90.7	93.9	97.2	100.7	5
Kadoma	88.5	94.8	98.1	101.6	10

+ relay st's at 17 locations.
R1 (mainly English): 0300-2200. **N:** 0400, 0500, 0600, 0800MF, 1000W, 1100, 1400MF, 1600, 1800, 2000, 2100.
R2 in Shona/Ndebele/English: 0300-2200.
R3 in English: 0300-2200. **N:** on the h.
R4 (Educational): Mon-Fri 0800-2000. **N. in English:** 1130, 1800.
ANN: E: "This is the Zimbabwe Broadcasting Corporation" (or "Voice of Zimbabwe") R1, 2, 3 or 4". R3 also "Zimbabwe 3". R4 also "You're tuned to R4, the educational st. of the Zimbabwe Broadc. Corp."

Clandestine broadcasting in Africa:
See *Listen to the World* section.

NEAR & MIDDLE EAST

AFGHANISTAN

L.T: UTC + 4½h — **Pop:** 17.300.000 — **Radio sets:** 1.5m (est.) — **Pr.L:** Pushtu, Dari — **E.C:** a/c 50, 220V (yes) — **ITU:** AFG.

RADIO AFGHANISTAN (Gov.)

ADDR: P.O. Box 544, Kabul. **Te:** 25241. **Cable:** Radioaf.

Mediumwaves	kHz	kW		kHz	kW
Taluqan	648	10	Kabul	1107	1000
Kabul	657	25	Mazar i Sharif	1206	500
Baghlan	936	10	Kabul	1278	100
Farah	945	10			

Provincial MW Stations: no recent information available.
Shortwaves: Kabul 7.5/50/100kW.

kHz	kHz	kHz
6100	7200	9635

FM: Kabul 96.55 MHz (stereo).
1st Prgr: 0130-1930 on 648/657/945/1107/1278/7200kHz.
2nd Prgr: 1030-1100, 1100-1130 Nuristani, 1130-1230 Turkmen on 6100/9635kHz, 1230-1330 Uzbek on 9635kHz.
Kabul City Sce: 0130-0230, 1230-1530 & 1730-2030 on 1206/6100kHz.
Afghan Voice: W 0230-0500 & 1430-1700 on 936kHz.
Voice of Asmaye: 1530-1930 on FM 96.55MHz.

EXTERNAL SERVICE on 9635kHz:
Arabic: 1630-1700
English: 1830-1900
French: 1900-1930
German: 1800-1830
Pushtu/Dari: 1700-1800
Russian: 1600-1630
ANN: Pushtu: "Da R. Afghanistan Kabul Dai". Dari: "Injá Radyu Afghanistan Kabul ast".
IS: A record on a national flute — **V.** by QSL-card. Rp.

ARMENIA

L.T: UTC + 3h (Su: UTC + 4h) — **Pop:** 3.500.000 — **Pr.L:** Armenian, Russian, Kurdish — **E.C:** a/c 50 127/220V — **ITU:** ARM.

ARMENIAN RADIO (Gov.)

ADDR: Alek Manukyan Street 5, 375025 Yerevan. **Te:** +7 (8852) 558010. **Fax:** +7 (8852) 551513. **Telex:** 243260 HORZIB SU.

Stations	kHz	kW	H. of tr.	Prgr.
Kamo	234	100	0200-0000	MR1
Yerevan	252	150	0300-0000	AR1
Yerevan	702		2300-2200	RSM
Yerevan	864	150	0300-2200	AR1, F
Yerevan	1314	250		F
Sisian	1377		0300-	AR1
Yerevan	1395		0300-2200	AR½
Kamo	1521	5	0300-2200	RSM
Yerevan	1566	5		RSM
Yerevan	4040	50	0200-2000	AR1, RSM
Yerevan	4810	50	0300-2200	AR1, F

Rel. via Balashikha, Russia: 7175kHz 20kW 0300-1500: AR1.
NB: Powers and locations are tentative only.
FM: AR1: 66.22MHz; AR2/RSM: 69.77/70.12MHz; AR3: 72.55MHz; no other details available.
D.PRGR: AR1 in Armenian, exc. Russian W0620-0700, Kurdish 1330-1400 & relay of MR1 2000-2400 — **AR2** (Tsiatsan): 1300-2000 in Armenian. At other times rel. AR1 — **AR3**: relay of AR1 & F — **MR1** = Moscow Ostankino R-1 — **RSM** = Moscow R. St. Mayak — **F** = Foreign Sce, cf. below.

FOREIGN SERVICE: Radio Yerevan

	Area	kHz
Armenian/ English		
2100-2200	Eu	11920, 12050*, 17660*
2200-2300	Eu	9450*, 11920
0330-0400	NAm	11675*, 11790*, 13645
Armenian/ French		
0900-1000	Eu	17625**, 17680**
Armenian/ Spanish		
2300-2400	Eu/LAm	9450*, 11920, 12050*
Armenian		
1730-1830	ME	864, 4810, 4990, 6065, 7390
Arabic		
1830-1930	ME	864, 4810, 4990, 6065, 7390
Kurdish		
1600-1730	ME	864, 4810, 4990, 6065, 7390
Russian		
1500-1600	Eu	864, 4810, 4990, 6065
Various +		
1930-2200	ME	864, 4810, 4990, 6065, 7390

+ Armenian, **English**, Turkish, Farsi.
N: English: 2140, 2245, 0345. **French:** 0951 (Sun). **Spanish:** 2330 —*) Rel.via Russia, **) Rel.via Ukraine.
ANN: Home Sce: "Khosum-e Yerevana". Foreign Sce: "Yerevan-e khosum" —**V.** by QSL-card.
PUB: Yetherum-e Yerevane (weekly for Home Sce.).

RADIO MOSCOW RELAY STATION

Mediumwaves: Yerevan 1314 kHz, 250 kW
Shortwaves: 4x120, 2x500 kW
Transmissions included in R. Moscow & R. Yerevan schedules.

VOICE OF AMERICA RELAY

Mediumwaves: Yerevan 1314 kHz, 250 (1000?) kW
Arabic: 1900-2000.

AZERBAIDZHAN

L.T: UTC +3h (Su: UTC +4h) — **Pop:** 7.130.000 — **Pr.L:** Azeri, Russian — **E.C:** a/c 50 127/220V — **ITU:** AZE.

AZERBAIDZHANI RADIO (Azärbaycan Radio) (Gov.)

ADDR: Mehdi Huseyn köçesi 1, 370011 Baky. **Te:** +7 (8922) 398585.
Fax: +7 (8922) 395452. **Telex:** 142214 TEMBR.
NB: The telephone country code will be changed to 994.
L.P: Head of Radio & TV: Mammad Ismaylov.

Stations*	kHz	kW	H. of tr.	Prgr.
Gyanca	218	500	0200-2100	AZR1
Gyanca	549	65	0200-0100	RSM
Baky	612	25	0300-0200	RSM
Baky	801	150	0200-2100	AZR1
Pirsagat	801		0200-2100	AZR1
Baky	891	30	0200-0000	MR1
Pirsagat	1296	150	0200-2300	RSM,F
Baky	1476	25	0300-2100	AZR2
Gyanca	1476		0300-2100	AZR2
Pirsagat	1476		0300-2100	AZR2
Baky	1530	5	2300-2200	RSM
Gyanca	4785	50	0200-2100	AZR1
Baky	4957.5	50	0300-2100	AZR2
Gyanca	6110	200	0030-2100	RSM,F
Gyanca	6135	100	0300-2100	AZR2,F
Baky	6195	50	0300-0200	RSM

Rel. via Balashikha, Russia: 7300kHz 20kW: AZR2 (daytime).
*names in Azeri. **NB:** Powers are estimates only.
FM: 67.75MHz; no other details available.
D.PRGR: AZR1 in Azeri/Russian, incl. relays of MR1 — **AZR2** (Araz) in Azeri — **MR1** — Moscow Ostankino R-1 — **RSM** = Moscow R. St. Mayak — **F** = Foreign Sce, cf. below.
Regional Service: Artsach R., Stepanakert (Nagorno-Karabach) r. on 1377kHz in Armenian & Russian.

FOREIGN SERVICE: Radio Baku

ADDR: M. Guzein Street, 370011 Baku, or as for Home Sce.

	Area	kHz
Arabic		
1700-1800	ME	1296, 6110, 6135
Azeri		
1400-1500	ME	1296, 6110, 6135

NEAR EAST & SOUTH ASIA
0
500
1000 KM
UTC +2
UTC +3
UTC +4
UTC +5
+6.00
+5.30
+4.30
+3.30
30°
60°
15°
TURKEY
ARMENIA
AZER.
TURKMENISTAN
UZBEK.
TADZHIKISTAN
KYRGYSTAN
CYPRUS
SYRIA
LEBANON
ISRAEL
JORDAN
IRAQ
IRAN
AFGHANISTAN
PAKISTAN
INDIA
KUWAIT
BAHRAIN
QATAR
UNITED ARAB EMIRATES
SAUDI ARABIA
OMAN
YEMEN
EGYPT
SUDAN
MEDITERRANEAN SEA
CASPIAN SEA
RED SEA
GULF OF OMAN
ARABIAN SEA
INDIAN OCEAN
Masirah I.
ANKARA
IZMIR
DENIZLI
ANTALYA
TARSUS
MALATYA
DIYARBAKIR
GAZIANTEP
ERZURUM
AGRI
MAZAR-E
HAKKARI
TABRIZ
BONAB
KRASNOVODSK
RASHT
SARI
GORGAN
BANDAR-E TORKMAN
AŞGABAT
ÇÄRYEV
BUHORO
SAMARKAND
DUSHANBE
YANGI-YUL
ORZU
TALUGAN
BAGHLAN
MAZAR-E SHARIF
MASHAD
TAYYEBAD
GHESLAGH
TEHRAN
KAMALABAD
CAPE GRECO
YENI ISKELE
LIMASSOL
TARTUS
HOMS
DAMASCUS
ADRA
HAIFA
JERUSALEM
TEL AVIV
AMMAN
AL KARANAH
BATRA
ABIS
ABU ZAABAL
MOKATTAM
EILAT
DUBA
NINEVEH
KIRKUK
QASR- E-SHIRIN
SALAH EL DEEN
BAGHDAD
ABUGHRAIB
BABYLON
ILAM
RUTBAH
AN NASIRIYAH
BASRAH
ABADAN
AHWAZ
ISFAHAN
SHIRAZ
BUSHEHR
SIRJAN
ZAHEDAN
ZABOL
FARAH
KANDAHAR
KABUL
PESHAWAR
ISLAMABAD
RAWALPINDI
SRINAGAR
JAMMU
LEH
SHIMLA
DELHI
ALIGARH
QUETTA
KARACHI
BANDAR-E ABBAS
CHAH BAHAR
AL-DAMMAN
ABU HAYAN
ALARISH
AL KHAISAH
DUBAI
ABU DHABI
MUSCAT
SEEB
RIYADH
DIRRIYA
JIDDAH
MECCA
THUMRAIT
SALALAH
MUKALLA
SANA'A
TAIZ
MOCHA
AL-HISWAH
ASMARA
OMDURMAN
KHARTOUM
BHOPAL
BOMBAY
HYDERABAD
PANAJI
BANGALORE

Persian		
0430-0515	ME	1296, 6110, 6135
1200-1245	ME	1296, 6110, 6135
1300-1357	ME	1296, 6110, 6135
Turkish		
1500-1557	ME	1296, 6110, 6135
1600-1656	ME	1296, 6110, 6135

ANN: Azeri: "Danyşyr Baky"
V. by QSL-card (irr.) — **PUB:** Azärbaycan televiziyasy vä radiosunun programlary (weekly for Home Sce.).

RADIO MOSCOW RELAYS

Mediumwaves: Pirsagat 1296kHz 150 kW
Shortwaves: Gyanca (Gyandzha) 6110/6135kHz 200/100kW
Transmissions included in R. Moscow schedule.

BAHRAIN

L.T: UTC + 3h — **Pop:** 541.000 — **Radio sets:** 320.000 — **Pr.L:** Arabic — **E.C:** a/c 50, 230V — **ITU:** BHR.

BAHRAIN BROADCASTING (Gov.)

ADDR: P.O. Box 702, Manama. **Arabic Sce: Te:** +973 781888/629009. **Fax:** +973 681544 — **English Sce (R. Bahrain): Te:** +973 629062/629085. **Fax:** +973 780911.
L.P: Dir. of Broadc: Abdulrahman Abdulla. Head of R. Bahrain: Ahmed Suliman.
MW: 612/801kHz 100kW, 1584kHz 1kW — **FM:** 90.9/93.3/96.5/101.0MHz stereo.
SW: Abu Hayan (G.C: 50.37E/26.02N) 6010/9746kHz 60kW.
D.PRGR: Arabic: 24h on 612/801kHz + FM 90.9MHz, 0300-2115 on 9746kHz. **N:** on the h.
R. Bahrain in English: 24h on 1584kHz + 93.3/96.5/101.0MHz, 0300-2106 on 6010kHz. **N:** on the h.
ANN: A: "Huna al Bahrain". E: "Rrrrradio Bahrrrrain".
IS: Local composition on guitar and violin — **V.** by letter.

CYPRUS

L.T: UTC + 2h (Su: UTC + 3h) — **Pop:** 716.000 — **Radio sets:** 270.000 — **Pr.L:** Greek, English, Turkish, Armenian, Arabic — **E.C:** a/c 50, 240V—**ITU:** CYP.

CYPRUS BROADCASTING CORPORATION

ADDR: P.O. Box 4824, Nicosia. **Te:** + 357 (2) 422231. **Fax:** +357 (2) 314050. **Cable:** Broadcasts. **Telex:** 2333 CYBC.
L.P: Dir. Gen: D. Kyprianou. Dir. Prgr: Ch. Papadopoulos. Head of Radio Prgr: Chr. Karageorghis. Head of TV Prgr: P. Ioannides. Dir. of Tech. Sces: O.C. Angelides. Dir. Admin. & Finance: G. Potamites. Head of Pub. & Int. Rel: Mrs. N. Roussou.

STATIONS	kHz	kW		kHz	kW
Paphos	558	10	Nicosia	963	100
Nicosia	603	100	Limassol	1044	10

FM: Mt. Olympos 91.1/94.8/97.2MHz — Larnace 90.2/96.0MHz.

Ch. 1 in Greek: 24h on 963/558/1044kHz + 90.2/97.2MHz.
Ch. 2: 0400-2200 on 91.1/92.4MHz. **Turkish:** 0400-0800, 1000-1130, 1400, 1800 — **English:** 1130-1300 (Sun 1230). **N: English/Arabic:** 1800-2200. Other times in Greek/Turkish/English. **N. in English:** 0800, 1800 — **Armenian:** 1300 (Sun 1230)-1400.
Ch. 3 in Greek: 24h on 603kHz + 94.8/96.0MHz.
SHORTWAVE SCE. in Greek (to the U.K.): 2215-2245 (Fri/Sat/Sun) via BBC relay st. on 6180/7125/9770kHz.
ANN: Greek: "Radiofonikon Idryma Kyprou". Turkish: "Burasi Kibris Radyo Yayin Korporasyonu" — **INT-SIG:** Avkoritssa played on Guitar — **V.** by QSL-card. Rec. acc. (B).
PRIVATE STATIONS
A large number of local st's are operating, using low power FM-tr's.

R. MONTE CARLO MIDDLE EAST (Comm.)

ADDR: B.P. 2026, Nicosia. **Telex:** 2162 SOMERA C.
L.P: St. Mgr: M. Pavlides.
STATION: Cape Greco: 1233kHz 600kW.
D.PRGR. in French, Arabic: 0500-2100. **N. in Arabic:** On the h, plus 0530, 1730. **N. in French:** 1000, 1830. Arabic prgr. for No. America via R. Canada Int: 0400-0415 on 5960/9755kHz.
TRANS WORLD RADIO: Arabic: 0300-0315, 2040-2145. **Armenian:** 0315-0330. **Farsi:** 0230-0300.
RADIO FRANCE INTERNATIONALE: 0400-0500 in French (see schedule under "France").
ANN: French: "R. Monte Carlo", Arabic: "Huna Monte Carlo".
V. Reports to B.P. 128, Monte Carlo; Listeners' Requests to B.P. 2026, Nicosia.

BRITISH FORCES BROADC. SCE. CYPRUS

ADDR: BFBS Akrotiri, BFPO 57. **Te:** +357 5278518. **Fax:** +357 5278580.
L.P: Gen. Mgr: D. Raven. Eng. Mgr: M.E. Townley.

Location	Ch. 1 MHz	Ch. 2 MHz	kW
Dhekelia	99.6	95.3	25
Akrotiri	92.1	89.9	25
Nicosia	99.6	95.3	1.5

D.PRGR: 24h. (Ch. 2 is satellite-delivered from U.K.).
ANN: "This is the British Forces Broadc. Sce. Cyprus."
INT-SIG: "Force One" (Raymonde), light orchestra.
V. by QSL-folder. Rp. (B) but only from outside Europe.

BBC EAST MEDITERRANEAN RELAY

ADDR: P.O. Box 219, Limassol.
MEDIUMWAVE: 639/720kHz 500kW — 1323kHz 200kW.
SHORTWAVES: 2 × 20, 4 × 100, 4 × 250kW tr's.

kHz	kHz	kHz	kHz	kHz	kHz	kHz
5965	6180	7260	9660	11720	15135	15575
6010	6195	7270	9670	11730	15165	15590
6015	7125	7285	9735	11740	15235	17750
6050	7135	9560	9740	11750	15245	17785
6080	7140	9580	9750	11845	15310	17885
6085	7180	9590	9770	11850	15325	17895
6125	7235	9635	11710	11955	15420	21470
6150						

D.PRGR: See BBC (United Kingdom).

NORTHERN CYPRUS

BAYRAK RADIO & TELEVISION CORPORATION

ADDR: Atatürk Square, Lefkosa via Mersin 10, Turkey.
Te: +90 5 20 71281. **Fax:** +90 5 20 81 991. **Telex:** 57 264 brtk.
L.P: DG: Erdal Onurhan. Head of Tr's: A. Ziya Dincer. Head of Admin: Süleyman Türem. Head of Prgrs: Hüseyin Cobanoglu. Head of Sales: Mehmet Kircalilar. Head of Int. N: Huriye Dimililer.
Mediumwave: Yeni Iskele 1098kHz 100kW, 1494kHz 10kW.
Shortwave: Yeni Iskele (G.C: 33.55E/35.17N): 6150vkHz 7.5kW.

FM (stereo)	Prgr.1	Prgr.2	Prgr.3	kW
Sinan Dagi	87.8	98.1	90.6	32
Selvili Tepe	92.0	102.0	105.0	0.7

Ch. 1 in Turkish on 1098kHz + FM: 0400-2200. **N:** 0500, 0600, 0800, 1000, 1330, 1500, 1730, 2130.
Ch. 2 in Greek/English/Arabic/German on 1494/6150kHz + FM: 0400-2200. **N. in English:** 1215, 1730.
Ch. 3: 0330-2100. **F.PL:** 24h.
ANN: "This is R. Bayrak, the Voice of the Turkish Republic of Northern Cyprus" — **V.** by QSL-folder (B).

GEORGIA

L.T: UTC + 3h (Su: UTC + 4h) — **Pop:** 5.500.000 — **Radio sets:** — **Pr.L:** Georgian, Russian, Abkhazian, Ossetian — **E.C:** a/c 50 127/220 — **ITU:** GEO.

GEORGIAN RADIO (Gov.)

ADDR: 68 M. Kostava Street, 380071 Tbilisi. **Te:** +7 (8832) 368331. **Fax:** +7 (8832) 362319 (manual). **Telex:** 212106 RADIO SU.

Stations	kHz	kW	H. of tr.	Prgr.
Dusheti	189	500	0200-0000	GR1,MR1
Batumi	693*		0300-2100	GR1,Reg
Tbilisi	810	150		
Tbilisi	1044	100	0330-2300	GR2,RSM

Tskhinvali	1323	30	0200-2200	RSM,Reg
Sukhumi	1350	50	0300-2100	GR1,Reg
Batumi	1422	30	0300-2100	GR1,Reg
Tbilisi	4760	100		inactive
Tbilisi	4875	100	0400-1900	GR2,RSM (irr.)
Tbilisi	5040	100	0300-2100	GR1

* not confirmed. **NB:** Powers and locations are tentative only.

FM: Location	GR1	GR2	MR1	RSM
Batumi	+67.46	65.90		
Gori	67.04	68.69		
Kutaisi	72.50	70.94		
Sukhumi	+68.45	66.05		
Tbilisi	71.60	69.59	70.82	72.38

+) carries reg. prgr's at times (see below).
D.PRGR: GR1: 0300-2100 in Georgian/Russian, incl. relays of MR1. **N. in Russian:** 1530-1545. 2100-0000 rel. MR1 — **GR2:** 0330-2300, own prgrs and rel. GR1 — **MR1** = Moscow Ostankino R-1 — **RSM** = Moscow R. St. Mayak.
Regional services:
Sukhumi: Abkhaz R, 384900 Sukhumi, Abkhaz Rep. — 1350kHz, 68.45MHz: 0500-0600, W 0900-1000, Mon-Fri 1600-1800, Sat 1500-1600, Sun 1100-1330.
Batumi: Adzharian R, pr. Stalina 57, 384500 Batumi —693/ 1422kHz, 67.46MHz: Mon-Fri 0600-0700, 1310-1500, Sat/ Sun 0730-0930.
Tskhinvali: South Ossetian R: 1323kHz; no further details available.

FOREIGN SERVICE: Radio Georgia

ADDR: As above.

English	Area	kHz
0300-0400*	ME	9830
0630-0700*	Eu/ME	12050, 11805
1730-1800*	Eu	12050
2000-2200*	Eu	9830
2130-2200	Eu	11760
2230-2300	Eu	11760
Russian		
0600-0630*	Eu	12050, 11805
2100-2130	Eu	11760
2200-2230	Eu	11760

*) Times / freqs based on anns. Unconfirmed from monitoring.
ANN: Georgian: "Laparakobs Tbilisi". E: "This is Georgia". R: "Govorit Radiostantsiysa Gruziya".
V. by QSL-card. **PUB:** Etershia Sakartvelo (weekly for Home Sce.).

IRAN

L.T: UTC + 3.30h — **Pop:** 61.770.000 — **Radio sets:** 12m — **Pr.L:** Farsi (Persian) — **E.C:** a/c 50,220V — **ITU:** IRN.

ISLAMIC REPUBLIC OF IRAN BROADCASTING (IRIB) (Gov.)

ADDR: P.O. Box 19395-333, Tehran. (Int. Tech. Affairs: P.O. Box 15875-4344, Tehran). **Te:** +98 (21) 21961. **Fax:** +98 (21) 295056/291051. **Cable:** IRIB.
L.P: Pres: Mohammad Hashemi. Ag. Pres: Mehdi Tabeshian. Vice-Presidents: Mohammad Davai (Admin. & Tech. Affairs), A. Hemmati (Political Affairs), M. Ahmadi Aliabadi (Tech. Development & Planning), Hassan Tabatabai (Tech. Op. & Maintenance), S.H. Shahidi Moadab (Ext. Prgr. Affairs), Shakrollah Heidari (Training Affairs), M. Zoragh (Sound Programming Affairs).DG (Int. Rel): Ali Reza Kamezi. Dir. Int. Tech. Affairs & Tech. Liaison Officer: Reza Moazzami. Dir. Bureau of Broadc. Union & Int. Rel: J. Mottaghi.

Mediumwaves	kHz	kW
Iranshahr	531	20
Mashhad	540	200
Gheslagh	558*	1000
Abadan	576	600
Tehran area	582	100
Sirjan	594	10
Zahedan	594	100
Birjand	621	20
Bonab	639*	400
Mashhad	684*	100
Rasht	702*	800
Tayebad	720*	400
Mashhad	738	10
Bandar-e-Torkaman	747a	800
Chahbahar	765*	1200
Arak	774	10
Iranshahr	783	20
Sari	819	20
Shiraz	837	400
Damghan	846	2
Mahabad	882	10
Yasuj	891	20
Tehran	900	50
Jiroft	918	20
Urumiyeh	936	10
Shahr-e-kord	936	10
Marivan	945	20
Birjand	963	20
Ilam	972	10
Shiraz	v990*	400
Semnan	1008	20
Chalus	1026	20
Tabriz	1026	100
Yazd	1035	20
Khorramabad	1053	20
Kerman	1062	20
Abadan	1080*	20
Zabol	1098	200
Bandar Lengeh	1116	125
Bojnurd	1134	20
Yasuj	1143	20
Tabriz	1152	100
Abadan	1160	10
Damghan	1170	2
Tehran	1188*	100
Moghan	1197	20
Kerman	1242	10
Sari	1251	20
Kermanshah	1278	200
Bushehr	1305	20
Ardabil	1314	20
Jolfa	1323	20
Tehran	1332	100
Bam	1341	1
Bandar-Abbas	1350	10
Ahwaz	1386	400
Rasht	1404*	800
Bakhtaran	1422	100
Esfahan	1431	200
Bandar-e-Torkaman	1449b*	800
Isfahan	1467b	200
Sanandaj	1476	200
Abadan	1485	10
Sabzevar	1485	1
Izeh	1485	1
Bushehr	1503	100
Kiashahr	1521	600
Yazd	1530	20
Marivan	1548	20
Ardabil	1551	10
Bandar-Abbas	1566	100
Isfahan	1593	100
Zanjan	1602	2

*) also carries Ext.Sce — a) daytime, b) nighttime.

FM: Tehran 91.3/97.8/100.7/104.7MHz 193kW, Khosrow Abad 93.5/100.1/106.3MHz 180kW, Tehran 93.9/97.7MHz 101.3MHz 96kW, Tehran 91.3/97.7MHz 96kW, Ziba Kenar 92.5/98.7/ 105.1MHz 80kW, Mashhad 88.1/94.5/100.9MHz 13kW, Maharlu 88.3/94.7/101.1MHz 9kW, Tabriz 89.9/98.1MHz 12.5kW +low power st's less than 5kW ERP.

HOME SERVICES
1st Prgr: 24h in Farsi on MW + FM 93.9MHz. Relayed in part by regional st's. Also relayed on SW (see below).
N: on the half h. Main N. at 0230, 0430, 1030, 1630, 2030.
2nd Prgr. on 1188kHz + FM 91.3MHz: no details received.
Quran Prgr: 0200-0500 & 1330-1630 on 1332kHz + FM97.8 MHz. Additional Quran Prgrs from Ahvaz on 1080kHz & Bushehr on 1305kHz.
Foreign Language Prgr. on 100.7MHz: rel. Ext. Sce. (see below).
Regional Stations: Exact details not received. There are studios in 39 centres producing prgrs in Farsi and local languages, including some Ext. Sce. prgrs. These st's also relay some prgrs from Tehran. **Regional N.** usually at 1330, 1430 or 1500.

EXTERNAL SERVICE

ADDR: P.O. Box 3333, Tehran.
L.P: Dep. MD for foreign Broadcasts: Shaban Shahidi.
MEDIUMWAVES: See Home Sce. list.
SHORTWAVES: Kamalabad (G.C: 51.27E/35.46N):100/250/ 350/500kW — **F.Pl:** Ahwaz (G.C: 48.40E/31.20N): 500kW — Mashhad (G.C: 59.33E/36.15N): 500kW. Zahedan (G.C: 60.53E/ 29.28N): 500kW.

kHz	kHz	kHz	kHz	kHz
5995	7190	9720	11790	15084
6005M	9022	9745	11895	15260
6175	9525	11715	11910	15315
7115	9610	11780	11930	15365
7155	9640M			

M) Mashhad, others Kamalabad.
Relays of Home Sce. 1st Prgr:

Times	Areas	kHz
0000-2400	Eu	15084
0000-2400	ME	5995
0000-2400	IRN	15365
1630-?	As	6005M, 9640M
Arabic		
0230-0330	Am	15260, 9022
0330-0730	ME	11930, 11780, 7155, 6175
0930-1130	ME	15260, 11715, 9525, 1224
1715-2130	ME	11930, 9610, 1386, 1224, 765
	Af	7190, 7155

Armenian		
1500-1530	WAs	1404
Assyrian		
1830-1930	ME	936
Azeri		
1230-1400	WAs	11895, 702
1630-1830	WAs	9745, 404, 1323, 1026
Baluchi		
1430-1530	EAs	594, 531
Bengali		
1230-1330	ME	11715
	EAs	11930, 11910, 11790
1530-1630	EAs	11790
Dari		
1430-1530	AFG	720
English		
0030-0130	NAm	15315, 15260, 9022
1130-1230	As	11930, 11910, 11790
	ME	11715, 9525, 1224
1400-1500	ME	702
1930-2030	Eu	15260, 9022
French		
0630-0730	Af	15260, 9720, 9022
1830-1930	Eu	15260, 9022
2230-2330	NAm	15315, 15260, 9022
German		
1730-1830	Eu	15260, 9022
Kurdish		
1330-1830	ME	1422, 639
Persian		
1430-1630	As	15084
1830-2100	Eu	15084, 4985
2230-0130	Eu/NAm	15084
Pushtu		
0200-0230	AFG	1098, 765, 594
1330-1430	AFG	11930, 11910, 11790
	ME	11715, 1224
1430-1530	AFG	11930, 594
1530-1630	AFG	720
Russian		
1400-1500	EEu	15260, 11895
1530-1630	EEu	15260, 9022
	NWAs	11790, 1449, 1404
1830-1930	EAs	11790
1930-2030	NWAs	11790, 1404
Spanish		
0130-0230	Am	15260, 9765, 9022
0530-0630	Eu	15260, 9745, 9022
2030-2130	Eu	9022, 6140
Tadzhik		
1530-1600	As	6005M, 9640M
Turkish		
0430-0530	TUR	15260, 11895, 9022
1630-1730	TUR	15260, 11895, 9022
1830-1930	TUR	1404, 639
Turkmen		
1630-1800	WAs	1449
Urdu		
0100-0200	EAs	1098, 765, 594
1230-1330	EAs	11930, 11910, 11790
	ME	11715, 1224
1430-1530	EAs	11930, 765
1530-1630	EAs	11930, 11910, 1098, 765
Uzbek		
1600-1630	As	6005M, 9640M

ANN: E: "This is Tehran the Voice of the Islamic Republic of Iran"; Persian: "Injā Tehrān ast, sedā-ye jomhūri-ye eslāmi-ye Irān"; F: "Ici Tehran la Voix de la République Islamique de l'Iran"; R: "Govorit Tegeran, Golos Islamskoy Respubliki Iran"; Turk: "Burasi Tehran, Iran Islam Gumhuriyetinin Sesi Radyosu"; Arabic: "Huna Tehran saut al-ǧumhuriyati-l-islamiyya fi l-Iran".
F.Pl: Kazakh & Kyrgyz sces.
V. By QSL-card. Rec. Acc. (B).

PUB: "Soroush"—weekly magazine in Farsi. Monthly editions in Arabic, English and French. Books in Farsi, Arabic, English, French, German, Kurdish, Pushtu, Russian, Spanish, Turkish & Urdu.

IRAQ

L.T: UTC + 3h (Su: UTC + 4h) — **Pop:** 19.000.000 — **Radio sets:** 3.500.000 — **Pr.L:** Arabic, Kurdish, Turkuman, Assyrian — **E.C:** a/c 50, 220V — **ITU:** IRQ.

BROADCASTING SERVICE OF THE REPUBLIC OF IRAQ

ADDR: Iraqi Broadcasting and TV Establishment, Salihiya, Baghdad. **Te:** +964 (1) 31151-9. **Cable:** Ida'ah. **Telex:** 2446 Ida'ah IK.
L.P: Dir. Gen: Hamid Said. Dir. Tec: M.F. Rashid. Dir. of FM Sce: Ghassan al-Mani'.

MW:	**kHz**	**kW**	**Prgr**				
Rutba	558*	300	GS	Babylon	1035*	1000	2
Nineva	603	300	GS	Rutba	1116	300	GS
Kirkuk	630*	20	+	Nineva	1197	300	K
Tanaf	684*	1000	GS	Nasiriya	1224	300	FS
Basrah	692	600	2	Kirkuk	1360	120	K
Baghdad	756*	300	GS	Tanaf	1377	1000	2
Nasiriya	846	300	GS	Missan	1431*	1000	GS
Baghdad	909	300	2	Arbat	1539*	1000	FS

*) not heard recently
FM: Baghdad 91.1/91.9/92.5/93.3MHz.
GS = General Sce, 2 = 2nd Prgr, FS = Foreign Sce, K = Kurdish, + = Turkmen/Assyrian.
Shortwaves: Abu Ghraib (G.C: 44.15E/33.19N); Babylon (G.C: 44.30E/32.30N); Salah El Deen (G.C: 44.10E/33.58N); Salman Pak (G.C: 44.35E/33.09N).
Freqs: R. Baghdad's SW freq usage is highly irr. Reported freqs incl. 4750, 6560, 7350, 9725, 11945, 15180, 17740.
General Sce in Arabic: 0225-2300. **Main N:** 0400, 1130, 1500, 1700, 1900, 2100, 2245.
2nd Prgr. in Arabic: 0355-1405 on 909kHz + 7350kHz. **N. summaries:** 0700, 0800, 0900, 1300, 1400.
Kurdish: 0400-1930 on 1197kHz + 6560kHz.
Persian: 1930-2130 on 1197kHz + 6560kHz.
FM Programme: 6h daily (Fri also 0700-1000) incl. N. in French, English, Russian & German.

FOREIGN SCE: R.IRAQ INTERNATIONAL

ADDR: P.O. Box 8145, Baghdad.

Times	**Area**	**kHz**
0000-0300	SAm	17740, 15150
0200-0500	NAm	15180
0700-0900	ME	15320, 13680
1200-1400	ARS	11880v
1400-1700	Eu/ME	15250v
1700-1900	IRN	9650
1900-2400	Eu	15210

D.PRGR: Mostly Arabic. **N. in English** (irr): 0030, 0330, 2000.
N. in French/German/Spanish: 2030.
ANN: HS: "Idha'at Jumhuriyat al-'Iraq min Baghdad". FS: A: "Huna Baghdad, idha'at al-' Iraq al-Duwaliyah. E: "This is Baghdad, R. Iraq International."

ISRAEL

L.T: UTC + 2h (Su: UTC + 3h) — **Pop:** 4.550.000 — **Radio sets:** 2.250.000 — **Pr.L:** Hebrew, Arabic — **E.C:** a/c 50, 230/400V (yes) — **ITU:** ISR.

ISRAEL BROADCASTING AUTHORITY

ADDR: P.O. Box 6387, Jerusalem. **Te:** +972 (2) 291888. **Fax:** +972 (2) 242944. **Telex:** 26488.

KOL ISRAEL — THE VOICE OF ISRAEL

Main Studios: Heleni Hamalka 21, P.O. Box 1082, Jerusalem 91010 **Te:** +972 (2) 302222. **Fax:** +972 (2) 248392. **Telex:** 25263.
News & Actuality Dept: Studios: Torah M'Zion 15, Jerusalem — **Te:** +972 (2) 383150/383173. **Cable:** Kolisrael.
L.P: MD: Amnon Nadav. Dir. Radio: Amnon Nadav. Dir. TV: Joseph Bar'el. Dir. Eng. Radio: Meir Khameides. Dir. Israel Radio

International: Shmuel Ben-Zvi. Legal Advisor: Nathan Cohen. Dir. Foreign Relations & Transcription Sce. (Radio): Hannah Lieblein.

STATIONS: Mediumwaves

	kHz	kW	Prgr.		kHz	kW	Prgr.
Jerusalem	531	100	C	Haifa	927	50	R
Tel Aviv	576	200	A	Jerusalem	954	100	R
Tel Aviv	657	200	B	Tel Aviv	1026	50	D
Jerusalem	675	50	D	Jerusalem	1080	5	B
Tel Aviv	738	1200	D	Haifa	1206	50	B
Safad	846	5	B	Jerusalem	1458	10	A
Beit Hillel	882	1	B	Eilat	1458	10	A
Eilat	927	1	B	Haifa	1575	5	X

FM: MHz	Prgr. B	Prgr. C	Prgr. M	kW
Beersheva	103.8	101.4	88.8	100
Eitanim	95.5	97.8	91.3	40
Haifa	103.2	93.3	89.4	100

+17 tr's 1kW or less.

Additional FM-tr's: Jerusalem 90.3MHz 1kW (D), 88.2MHz 1kW (E).

HOME SCE: PRGR. "A": 0400 (Sat 0500)-2210. Multilingual prgrs for immigrants incl. **English:** 0500-0515, 1800-1815. **French:** 0515-0530, 1815-1830. **Other languages:** 1600-1800 & 1830-1900.
Prgr. "B" (Commercial): 24h (Fri 0500-2310). **N:** On the hour.
Prgr. "C" (Commercial): Pop and light music 0400 (Sat 0500)-2310. **N:** on the h (rel. Prgr B).
Prgr. "D" in **Arabic:** 0400-2215. **N:** 0430, 0530, 0730, 0830, 0930 (Fri 0915), 1030, 1130 (exc. Fri), 1230, 1330 (local), 1430, 1530, 1630, 1730, 1830, 2030, 2200. **Russian** on 738kHz (rel. Ext. Sce.): 0300-0355.
Prgr. "X" (Traffic Prgr.): 0500-1700. **N:** on the h. (rel. Prgr B).
Prgr "R": 0500-1700 in **Russian,** 1700-1900 in **Amharic.**
Voice of Music (M): 0400 (Sat & holidays 0500)-2210 in stereo.
ANN: Prgr. "A": "Kolyisrael Mi Yreushalayim b'Reshet Aleph". Prgr. "B": Kolyisrael b'Reshet Beth". Prgr. "C": "Kolyisrael b'Reshet Gimel". Prgr. "D": "Sout el Israil-Min Urshalem-El Kuds". Voice of Music: "Kol Hamusica". Prgr. X: "Boker Tov Hayom Yom".
Shortwaves: 2×20, 1×50, 2×100, 4×300, 1×500kW tr's.

kHz	kHz	kHz	kHz	kHz
5900	9815	11603	15480	15650
5915	9845	11675	15590	17545
7465	11585	13753	15615	17575
9388	11587	15095	15640	17590
9435	11588			

	Areas	kHz
Bukharian		
1430-1445	CIS	11603, 15640, 17575
English		
0500-0515	Eu/NAm	9435
1100-1130	Eu/NAm	17545
1430-1425	Eu/NAm	11587, 11603, 15640, 17575
	As/Pac	15650
1800-1815	Eu/NAm	7465, 11587, 11675
	SAf/CAf	17575
2000-2030	Eu/NAm	7465, 9435, 11587, 11675
	SAf/CAf	17575
	CAm/SAm	11603
2230-2300	Eu/NAm	7465, 11587, 11603, 11675
	LAm	9435, 17575
French		
0515-0530	As 0500 English	
1130-1200	As 1100 English	
1815-1830	As 1800 English	
2030-2055	As 2000 English	
2200-2230	As 2230 English	
Georgian		
1445-1500	As	11603, 15640, 17575
Hebrew		
1725-1745	Eu/NAm	11587, 11675
Hungarian		
1745-1800	Eu/RUS/NAm	9435, 11603, 11587, 11675, 15640, 17575
Ladino		
1745-1800	Eu/NAm	11587, 11675
Persian		
1500-0555+	ME	11603, 15640, 17575
Romanian		
1830-1845	Eu	11587, 11675, 15640, 17575
Russian		
1400-1430	RUS	11587, 11603, 15640, 15650, 17575, 17590
1500-1555	RUS	11587, 11603, 15640, 15650, 17575, 17590
1600-1700	RUS	9435, 11587, 11603, 11675, 15640, 17575
1830-1900	Eu/NAm	9435, 11603
1900-1955	RUS	9435, 9845, 11587, 11603, 11675, 15640
2100-2125	RUS	7465, 9435, 11587, 11603, 11675, 15590
Spanish		
2200-2230	LAm	9435, 11587, 11603, 11675, 15640, 17575
Tat		
1500-1515	RUS	11587, 15640, 17590 (Sun/Wed)
Yiddish		
1700-1725	Eu/NAm	9435, 11587, 11603, 15640, 15650, 17575
1830-1900	Eu/NAm	9435, 11603
Yemenite		
1330-1355	YEM	15640 (Mon/Tues/Wed/Thurs)
Relays of Home Sce. "Reshet Bet:"		
0400-2300	Eu/NAm	9388, 13753, 15615
0400-0455	Eu/NAm	9435
0530-0610	Eu/NAm	9435
0615-1050	Eu/NAm	17545
1205-1355	Eu/NAm	17545
Home Sce. Prgr D in Arabic		
0400-2215	ME/Af/As	5900, 5915, 9815, 15095, 15480

+) exc. Fri/Sat + Jewish Holy Days.
DX-Prgr in English: Sun 2000, 2230.

ANN: Yiddish: "Hert zu der yisrael odizie im yiddish fyn yerushalayion". Ladino: "Emission de yisrael en a langua Judeo-espagnol". Sp: "Esta es Kol Israel", E: "This is Kol Israel broadcasting from Jerusalem". Fr: "Ici Kol Israel, Radiodiffusion Israelienne" — **INT-SIG:** Foreign Sce (Hatiqua) Nat. Anthem.
V. by QSL-card. Re. to IBA, Overseas Sces, P.O. Box 1082, Jerusalem 91010. (Correspondence for tech. operation: BEZEQ, The Israel Telecommunication Corp. Ltd., P.O. Box 29555, Tel Aviv 61290). — **PUB:** Overs. Prgr. schedule free of charge.

GALEI ZAHAL (Israel Defence Forces Station)

ADDR: Military Post Office Box 01005, Israel. **Te:** +972 (3) 5694276. **Fax:** +972 (3) 814697.
L.P: Dir: M. Shlonsky. Dir. Eng: S. Kasif.

STATIONS	kHz	kW	STATIONS	kHz	kW
Beersheba	1224	20	Shivta	1368	20
Tel Aviv	1287	100	Rosh-pina	1368	50
Haifa	1305	20	Jerusalem	1404	10
Eilat	1305	5			

FM (stereo)	MHz	kW(ERP)	FM (stereo)	MHz	kW(ERP)
Tel Aviv	92.4	0.3	Abugosh	96.6	20
Bnei-Brak	91.9	3.5	Beith-Shean	99.8	3.5
Jerusalem	93.9	0.3	Beersheba	99.8	20
N. Gallilee	93.9	20	Haifa	102.3	40
Eilat	93.9	0.3	N. Negev	106.4	1

D.PRGRS: "Galei Zahal 1" on MW + 92.4/93.9 (Jerusalem) MHz: 24h. **N:** on the h. (own bulletins or rel. IBA Prgr B).
"Galei Zahal 2": 1200-1500 & 1600-2200 on FM stereo network.
ANN: "Galei Zahal, Shidure Tsva Hagana Le'Yisrael".
V. by QSL-card.

JORDAN

L.T: UTC + 2h(Su: UTC + 3h) — **Pop:** 3.555.000 — **Radio sets:** 700.000 — **Pr.L:** Arabic, English — **E.C:** a/c 50, 220V — **ITU:** JOR.

JORDAN RADIO & TELEVISION CORP. (JRTV) (Gov.)

ADDR: P.O. Box 1041, Amman. **Te:** +962 (6) 773111. **Fax:** +962 (6) 788115. **Telex:** 24213 RTVENG 10.
L.P: DG: R. Alkhas. Dir. Radio: I. Shakzaleh. Dir. Eng: O. Asfoura. Prgr. Dir's: J. Hajjat (Arabic), J. Zada (English). Dir. Admin: A. Zabi.

MEDIUMWAVES	kHz	kW		kHz	kW
Al Karanah	207	600	Amman	855	10
Amman	612	200	Amman	1449	10
Ajlun	801	2000	Aqaba	1485	1
Amman	801	200	Al Karanah	1494	1000

SHORTWAVES: Al Karanah (G.C: 36.26E/31.44N): 3×500kW tr's. A=rel. Arabic Sce, E=rel. English Sce. F = Foreign Sce. in Arabic.

kHz	Sce.	H. of tr.	kHz	Sce.	H. of tr.
9560	E/A/F	1420-0030	11940	A	1800-2300
9835	A	1600-2330	11955	A	0330-0830
11810	A/F	0330-2330	13655	F/E	1100-1415
11820	A/F	0815-1615	15435	A	2130-2330
11940	A	0330-0630			

FM: Amman 93.1MHz 5kW, 96.3MHz 0.6kW, 99MHz 120kW, 103.0MHz 3kW, 105MHz 3kW; Aqaba 98.7MHz 3kW; Ajlun 90.9MHz 5kW; Irpit 95.4MHz 0.5kW.
HOME SCE. in Arabic: 0330-2330 on 207/612/801kHz+95.4/96.3/105MHz, 1600-2330 on 1449kHz. **N:** 0400, 0500, 0600, 0800, 1000 (exc. Fri), 1100, 1200, 1400, 1500, 1600, 1900, 2100, 2200, 2300.
English: 0500-2200 on 855kHz+90.9/98.7/99.0/103.0MHz. **On SW:** 1100-1200 on 13655kHz, 1400-1630 on 9560kHz. **N:** 0530, 0800, 1000 (exc. Fri), 1100 (Mon-Fri), 1200, 1400, 1600, 1700, 1900, 2000, 2100.
Quran Prgr: 0500-2200 on 93.1MHz.
Foreign Sce. in Arabic: 0900-1000 to AUS/FE on 11810kHz, 1000-1100 to SAm on 11820kHz, 1100-1200 to NAm on 13655kHz, 2130-2230 to ME/AUS/FE on 11810kHz, 2330-0030 to NAF/CAm on 9560kHz.
ANN: Arabic: "Huna 'amman, Idha'atu-l-mamlaka al-urduniyya al-hashimiyya". E: "This is R. Jordan, broadc. from Amman".
V. by QSL-card. Rec. acc. (Re. to P.O. Box 909, Amman).

KUWAIT

L.T: UTC+3h — **Pop:** 1.3m (approx) — **Radio sets:** 1m — **E.C:** a/c 50Hz, 240V — **ITU:** KWT.

MINISTRY OF INFORMATION

ADDR: P.O. Box 193, 13002 Safat. **Te:** +965 2415301. **Fax:** +965 2434511. **Cable:** ALIRSHAD. **Telex:** 490 MI 46151 (or 46152) KT.
L.P: Minister: Dr. Bader Jassim Al-Yacoub. Under-Secr: Hamed Yousef Al-Roumy. Asst. Under-Secr. (Eng. Affairs): Dr. Abdulaziz Al-Mansour. Asst. Under-Secr. (Int. Media): Mrs. Amal Al-Hamad. Dir. of Eng. (Radio): Ahmed Abdullah. Dir. Eng. (Communications): Abdulaziz Al-Baghli.

RADIO KUWAIT (Gov.)

ADDR: P.O. Box 397, 13004 Safat. **Te:** +965 2423774. **Fax:** +965 2456660 or 2415946. **Cable:** ALIRSHAD. **Telex:** 490 MI 46285 KT.

MW (kHz)	kW (Now)	kW (F.PL.)	MW (kHz)	kW (Now)	kW (F.PL.)
540	10	600(1600)	1269		50
963		50	1341	10	200
1134	10	600(1600)			

SW: Kabd (G.C: 47.55E/29.16N): 6 x 500kW and 3 x 250kW tr's (some under installation at time of editing).

kHz	H. of tr.	P	kHz	H. of tr.	P
6055	0400-1305	A	13620	1800-2100	E
9750	1000-1510	A	15345	0400-0700	A
9840	0230-0600	A	15495	0715-1305	A
	1530-1800	A	15505	1800-2300	A
9880	1600-2115	A	17850	0400-0600	A
11665	0400-0800	A	17895	0400-0730	A
11990	1315-1745	A	21675	1315-1800	A

FM: 98.9MHz 5kW, 92.5MHz 5kW, 97.5MHz 2kW.
Main Arabic Prgr. (A): 0200-2300 on 540/1134kHz + 87.9/98.9MHz. Also on 1341kHz (0500-1000, 1600-1800, 2100-2300), 97.5MHz (0500-1000) and on SW as above. **N:** 0300, 0400, 0500, 0730, 1000, 1300, 1500, 1700, 1900, 2100, 2200.
Second Arabic Prgr: 1000-1600 on 1341kHz + 97.5MHz.
Holy Quran Prgr: 0200-0500 on 1341kHz + 97.5MHz, 1600-2100 on 97.5MHz.
English (E): 0300-2100 on 97.5MHz, 1800-2100 on 1341/13620kHz. **N:** 1830.
ANN: A: "Huna Al-Kuwait". E: "This is R. Kuwait" — **IS:** old Kuwaiti tune on clarinet — **V.** by QSL-folder. Rec. acc.

LEBANON

L.T: UTC+2h (Su: UTC+3h) — **Pop:** 3.432.000 — **Radio sets:** 2.150.000 — **Pr.L:** Arabic, French, English, Armenian — **E.C:** a/c 50, 110/190V — **ITU:** LBN.

RADIO LEBANON (Gov.)

ADDR: Ministry of Information, Lebanese Broadcasting Station, Beirut. **Te:** +961 346880. **Cable:** Radio Liban. **Telex:** Minform 29786.
L.P: Dir: Fouad Hamdan. Chief, Tech. Dept: Micheal Samaha. Chief, Prgr. Dept: Waheed Jalal. Chief, Pub. Rel: Faouzi Fehmy.
MW: Beirut 837/999kHz 10kW — **FM:** Beirut 96.2MHz 120kW.
1st Prgr. in Arabic: 0330-2330 on MW (on 837kHz: 0900-1105 & 1800-2005, other times on 989kHz). **N:** 0530, 0900, 1130, 1400, 1630, 1900, 2000, 2100.
2nd Prgr. on FM: French: 0445-0900, 1100-1430, 1700-2200. **N:** 0530, 0700, 0800, 1130, 1200, 1800. **English:** 0900-1000, 1430-1630 (Sun 1700). **N:** 1505. **Armenian:** W 1630-1700.
ANN: A: "Iza'at Loubnan min Beirut". F: "Ici Radio Liban émettant de Beyrouth" — **IS:** Opening notes from the Lebanese National anthem played on guitar — **V.** by QSL-card.

OTHER STATIONS

NB: There are dozens of AM and almost 150 FM-st's operating in Lebanon. Many of these are unlicensed, therefore available details are incomplete, and many st's change freq's often.

	kHz	kW	Name & status
1)	756		Voice of the South, Marjayoun
2)	873	80	R. Voice of Lebanon (Phalangist/commercial)
3)	963		Voice of the People (Communist)
4)	945	12	Voice of Hope (Christian/commercial)
5)	953	100	R. Free Lebanon (Christian Militia)
6)	1017		R. of Islam (Hezbollah)
7)	1050		Voice of the National Resistance
8)	1080	50	Voice of the Mountain (Druze/Progressive Socialist)
9)	1134		Voice of the Homeland (Conservative Sunni Moslem/commercial)
3)	1188		Voice of the People (Communist)
3)	1269		Voice of the People (Communist)
10)	1310		Voice of Free & Unified Lebanon, Zgharta
11)	1368		The Lebanese Radio
12)	1377		Voice of Arab Lebanon, Murabitun
1)	1485	5	Voice of the South, Marjayoun
	MHz	**kW**	**Name & status**
13)	88.6	100	R. Orient, Beirut (commercial)
14)	90.0		R. Paradise (commercial)
15)	91.1		R. Byblos, Beirut
16)	91.6		R. Melody, Beirut
17)	92.0		Voice of Light, Beirut (Hezbollah)
1)	92.4	0.5	Voice of the South, Marjayoun
2)	93.2	2.5	R. Voice of Lebanon (Phalangist/commercial)
18)	93.5	0.01	R. Scorpion, Tibnine (UN Forces)
19)	95.1		R. One, Broumana
20)	95.5		Midway Radio, Beirut
31)	95.9		R. Flash, Beirut (commercial)
32)	96.2		Green Lebanon Radio (Dem. Socialist)
21)	98.0	0.05	R. Norbatt (UN Forces)
10)	98.3		Voice of Free & Unified Lebanon, Zgharta
22)	98.4		R. Mirage, Beirut
23)	99.0		R. Mont Liban, Beirut (commercial)
24)	99.5		R. New Lebanon, Beirut
1)	99.9	0.05	Voice of the South, Marjayoun
23)	101.0		R. Mont Liban, Beirut (commercial)
25)	101.6		BBS, Beirut
26)	102.0		Magic 102, Beirut (commercial)
5)	102.8		R. Free Lebanon (Christian Militia)
3)	103.6		Voice of the People (Communist)
27)	104.0		Fox Radio, Beirut (commercial)
28)	104.2		R. Dutchbatt (UN Forces)
4)	104.5		Voice of Hope (Christian/commercial)

29)	104.8		Voice of Truth, Antelias
4)	105.1	100	Voice of Hope (Christian/commercial)
19)	105.5		R. One, Broumana
30)	107.0		R. Delta, Beirut
8)	108.0		Voice of the Mountain (Druze/Progressive Socialist)

	SW kHz	kW	Name & H. of tr.
4)	6280	10	King of Hope, So. Lebanon: 0500-2200
2)	6550		R. Voice of Lebanon: 24h
4)	11530	25	Wings of Hope, So. Lebanon: 0500-2200

Addresses & other information

1) Marjayoun — 0500-1900. **N. in Arabic:** 0530, 0630, 0830, 1030, 1230, 1430, 1630, 1830. **ANN:** "Saout al Djenoub".
2) P.O. Box 165271, Ashrafieh, Beirut. **Te:** +961 323458. **Tlx:** VDL 23203 LE — GM: Simon El-Khazen. CE's: William Merheb, Maroun Salameh — **D.PRGR:** 24h. **N: Arabic:** 0015-1215 hourly, 1415, 1515, 1715, 1915-2315 hourly. **French:** 0800, 1300, 1615. **English:** 0900, 1315, 1815 — **ANN:** A: "Huna Saout Loubnan, saout al-Hurriyah wa al-Karamah". F: "Ici La Voix du Liban". E: "This is the Voice of Lebanon" — **IS:** Opens with River Kwai March and Lebanese National Anthem. Closes with River Kwai March and Katäeb Party Anthem — **V.** by letter.
3) Rumaylah (near Sidon) — 0400-2300. **N. in Arabic:** on the half h. **ANN:** A: "Saout al Shab".
4) P.O. Box 3379, Limassol, Cyprus (USA Addr: High Adventure Missionaries, Box 7466, Van Nuys, CA 91409) — Pres: George Otis. VP: Charles Pollak. GM: Don Otis. Int. PD: John Tayloe. VP (Eng): Paul Hunter — **D.PRGR:** 0300-0100 on MW/FM, 0300-2200 on SW. **Arabic:** 0500-0700, 1100-1400. **English:** 0700-1100, 1400-1700. **Russian & Ea. European languages:** 1700-2000 — **ANN:** A: "Saout al Amal". E: "Voice of Hope" (on MW), "King of Hope" (on 6280kHz), "Wings of Hope" (on 11530kHz) — **V.** by QSL-card or letter.
5) Zuq Musbih (near Juniyah) — 0400-2300. **N. in Arabic:** hourly at :45 (exc. 1745). **N. in English:** 0830, 1230, 1530, 1630. **N. in French:** 0930, 1330, 1730 — **ANN:** A: "Idha'at Loubnan al-Hurr".
6) Ba'labakk — 0415-1800. **N:** 0515, 0530-1630 hourly, 1700, 1730.
7) We. Beirut. **Te:** +961 831606-8 — 0400-2200. **N. in Arabic:** on the half h. (exc. 0630, 1630). **N. in French:** 1630 — **ANN:** "Saout al Muqawama al-Wataniyyah".
8) Alayh — 0400-2200. **N. in Arabic:** 0430, 0500, 0530, 0700, 0800-1200 hourly, 1230, 1400-1700 hourly, 1730, 1900-2200 hourly. **N. in French:** 1105, 1300, 1505 — **ANN:** "Saout al Jabal".
9) Beirut — 0355-2330. **N:** on the h. **N. in French:** 1125. **ANN:** "Saout al-Watan".
10) Zgharta, No. Lebanon — 0500-1925. **N:** 0530, 0600, 0700, 0830, 0930, 1030, 1130, 1330, 1430, 1530, 1730, 1830, 1900.
11) ?-2200. **ANN:** "al-Idha'ah al-Lunaniyah".
13) Avenue Foch 50, F-75116, Paris, France. **Te:** +33 (1) 45 00 78 78.
15) 0700-0100. **N:** Rel. R. Lebanon & Lebanese TV.
16) Beirut — CD music prgrs.
20) P.O. Box 11-1462, Rawché, Beirut. **Te:** +961 800451/2—Rock music prgrs. in English (non-rlg, non-political).
22) Beirut — **ANN:** "Idha'at at Miraj".
23) B.P. 166600, Beirut — 24h pop music prgrs in French/English (different prgrs on each freq.).
24) **ANN:** "Idha'at at Loubnan al-Jadid".
26) 24h — pop music prgrs in French/English.
27) 24h — pop music prgrs in French/English.
29) B.P. 373, Antelias. **Te:** +961 460570 — Mgr: Simon Karam. 0400-2200. **N:** 0600, 1200, 1700. **N. in English:** 1600. **N. in French:** 1400 — **ANN:** A: "Saout al-Haqq".
31) Beirut — Owner: Murun Khalil.

OMAN (Sultanate of)

L.T: UTC + 4h — **Pop:** 1.587.000 — **Radio sets:** 900.000 — **Pr.L:** Arabic — **E.C:** a/c 50, 220V — **ITU:** OMA.

RADIO OMAN (Gov.)

ADDR: Ministry of Information, P.O. Box 600, Muscat. **Te:** +968 603222. **Fax:** +968 602831. **Telex:** 5265 INFORM ON. **Cable:** Information.
L.P: Dir. Gen: Ali Bin Abdallah Al Mujeni. Dir. Gen. (Tech. Affairs): H.Y. Al-Kindy. Head of Maintenance: Rashid Haroon Al Jabry. Dir. Prgrs: Rajab Al Kathiry. Ag. Dir. Foreign Sce: Fathiyah Al Hinai. Head of Studios: Aboud Al-Sawafi.

Mediumwaves	kHz	kW		kHz	kW
Salalah	1035	100	Muscat	1242	200

SHORTWAVES: S=Seeb (G.C: 58.10E/23.40N), T=Thumrait (G.C: 53.56E/17.38N) 50/100kW.

kHz	H. of tr.	kHz	H. of tr.
6085T	0200-0400	9735T	0400-0800
6085S	1600-2200		1400-2100
7230S	0200-0400	11890S	0600-1600
	1200-2200	15375T	0800-1400
7270S	0200-0600		

FM: Muscat 90.4MHz, 94.3MHz 1kW (stereo) — Khasab Bukka area 99.9MHz + 4 low power relays.
D.PRGR. in Arabic: 0200-2145 on 1035/1242kHz + 99.9MHz. Shortwaves as above. **N:** 0300, 0700 (exc. Fri), 0830 (Fri), 1300, 1600, 1700, 1900, 2000.
English: 0300-1800 on 90.4/94.3MHz. **N:** 0430, 0700, 0900, 1030, 1400.
ANN: "Idha'atu-l-'umaniya min Musqat". E: "This is the English Sce. of R. Oman" — **V.** by QSL-folder (B).

BBC EASTERN RELAY ST. (Masirah Island)

ADDR: P.O. Box 6898, Ruwi PO, Muscat, Oman.
MW: 702/1413kHz 750kW — **SW** (G.C: 58.53E/20.36N): 4 x 100kW tr's.

kHz	kHz	kHz	kHz	kHz	kHz	kHz
5965	7235	9600	11740	11845	15125	15310
5975	9580	9750	11750	11850	15235	17780
6030	9585	11720	11760	11955	15245	17790
7160	9590					

D.PRGR: see United Kingdom — **V:** by letter. Re. to Tr. Mgr.

QATAR

L.T: UTC + 3h — **Pop:** 546.000 — **Radio sets:** 250.000 — **Pr.L:** Arabic — **E.C:** a/c 50, 240V — **ITU:** QAT.

MINISTRY OF INFORMATION & CULTURE

ADDR: P.O. Box 1836, Doha. **Fax:** + 974 831447. **Cable:** "INFORMATION", Doha.
Telex: 4229 INFOR DH.
L.P: Min. of Inf: H.E. Sheikh Hamad Bin Suhaim Al-Thani. Undersecretary Min. of Inf: Mana Abdul Hadi Al-Hajiri. Asst. Undersecretary, Min. of Inf: S.A.R. Al-Kuwari. Contr. of Tech. Office & Ag. Head of TPC: Hassan Al-Mass.

QATAR BROADCASTING SERVICE (Gov.)

ADDR: QBS, P.O. Box 3939, Doha. **Cable:** Broadcasting. **Telex:** 4597 Q Radio DH.
L.P: Dir. of Broadc: A.S. Al-Madhadi. Head of Eng: Q.A. Qassim.
MEDIUMWAVES: Al Khaisah — +) Al Arish — *) Doha.

kHz	kW	P	H. of tr.	kHz	kW	P	H. of tr.
675*	100	A	0245-2130				
954+	1500	A	0245-2130	1233	100	E	1400-1830
999	50	S	1200-1500			F	1100-1400
		U	1600-1730	1485	1	S	1200-1500
1233	100	E	0300-1100	1602*	1	S	1200-1500
						U	1600-1730

SHORTWAVES: Al Khaisah (G.C: 51.25E/25.25N).

kHz	kW	H. of tr.	kHz	kW	H. of tr.
9665	250	0245-0700	17800	250	1300-1700
		1700-2130	21555	250	0700-1300
11820	100	0245-0700		100	1300-1700
		1700-2130			

FM: Doha 97.5MHz (E) — Al-Jumailiyah 90.8MHz (A), 100.8MHz (F), 102.6MHz (E).
Arabic Prgr (A): as above. **N:** 0330, 0430, 0530, 0700, 0900, 1030, 1200, 1530, 1700, 1800, 1900 (exc. Fri/Sat), 2000, 2100.
English Prgr (E): 0300-2130 on FM; MW as above. **N:** 0330, 0430, 0530, 1000, 1500, 1700.
French Prgr (F): 1100-1400 on 1233kHz.
Shaabi (Folklore) Prgr (S): 1200-1500 on 999/1602kHz.
Urdu Prgr (U): 1600-1730 on 1602kHz.
ANN: "Idha'at at Qatar Min al-Doha". E: "This is the Qatar Broadcasting St. Doha"—**INT-SIG:** Arabic melody on string instrument (23 notes) — **V.** by QSL-card.

SAUDI ARABIA

L.T: UTC + 3h — **Pop:** 18.620.000 — **Radio sets:** 5m — **Pr.L:** Arabic — **E.C:** a/c 50 or 60, 127 or 220V (variable) — **ITU:** ARS.

MINISTRY OF P.T.T., Riyadh
Cable: Gentel. **L.P:** Minister of P.T.T: H.E. Dr. Alawi Kayal.

BROADCASTING SERVICE OF THE KINGDOM OF SAUDI ARABIA (Gov.)
ADDR: Ministry of Information, Riyadh. **Cable:** Information Ministry. **Te:** +966 (1) 4041692 or 4014440. **Telex:** 401040 SJ.
L.P: Minister of Inf: His Excellency Ali Hasan al-Shaer. Dep. Minister for Adm. Affairs: Ibrahim Al-Gadhi. Dep. Minister for Inf. Affairs: Dr. Fouad al-Farsy. Asst. Dep. Minister for Eng. Affairs: Fouad A. Taher (Addr: P.O. Box 61718, Riyadh 11575).

Mediumwaves	kHz	kW	H. of Tr. & Prgr.
Qurayyat	549	2000	0300-1500 (A), 1500-1700 (CI)
Gizan	549	1	0300-1500 (A), 1500-1800 (CI) 1800-2300 (A)
Rafha	549	20	0300-1500 (A), 1500-1700 (CI) 1700-2300 (A)
Jeddah	558	1	as Dammam 1098kHz
Abha	567	5	0300-2100 (H)
Afif	567	15	0300-2100 (H)
Gizan	576	1	0300-2100 (H)
Riyadh	585	1200	0300-1500 (A), 1500-1700 (CI) 1700-2300 (A)
Duba	594	2000	0300-1500 (A), 1500-1700 (CI)
Hail	612	5	0300-2100 (H)
Khamasin	612	15	0300-1500 (A), 1500-1700 (CI) 1700-2300 (A)
Gizan	630	20	0300-2100 (H)
Najran	630	10	0300-2100 (H)
Jeddah	648	2000	0300-1500 (A), 1500-1700 (CI) 1700-2300 (A)
Rafha	657	20	0300-2100 (H)
Abha	675	5	0300-1500 (A), 1500-2100 (A2)
Afif	675	20	0300-1500 (A), 1500-1700 (CI) 1700-2300 (A)
Riyadh	684	10	0300-2100 (A2)
Jeddah	684	50	0300-2100 (A2)
Duba	702	40	0300-2100 (A2)
Najran	747	1	as 585kHz
Qurayyat	765	10	0300-2100 (H)
Khamasin	765	20	0300-2100 (H)
Dammam	783	100	0300-2100 (A2)
Jeddah	792	50	0300-2100 (H)
Medinah	828	20	as 585kHz
Dammam	864	100	0300-2100 (H)
Qassim	873	10	0300-1500 (A), 1500-2100 (H)
Dammam	882	100	0100-0300 (CI), 0300-2100 (H)
Qurayyat	900	1000	as 585kHz
Riyadh	936	5	0100-0300 (CI), 0300-2100 (H)
Hail	945	5	as 585kHz
Madinah	981	20	0100-0300 (CI), 0300-2100 (H)
Tabuk	999	20	0300-2100 (H)
Qurayyat	1098	10	0300-2100 (A2)
Dammam	1098	5	0500-0800 (M), 0800-1300 (F) 1300-1400 (M), 1400-2100 (F)
Makkah	1287	5	0300-2100 (H)
Taif	1305	1	0300-2100 (A2)
Riyadh	1422	20	as Dammam 1098kHz
Dammam	1440	1600	as 585Hz
Jeddah	1485	1	1500-2100 (F)
Jeddah	1512	1000	0100-0300 (CI), 1300-2100 (H)
Duba	1521	2000	as 585kHz

A = General Arabic Prgr (from Riyadh); A2 = Second Arabic Prgr (from Jeddah); H = Holy Quran (from Riyadh); CI = Call of Islam (from Jeddah); F = Foreign Language Prgr; M = Music Prgr.

FM: Riyadh 91.2MHz (A/CI), 97.7MHz (M/F), 100.0MHz (H) — Jeddah 93.2MHz (A/CI), 98.0MHz (M/F) + 20 st's.
D. PRGRS. in Arabic: as above. **N: General Prgr:** 0400, 0600, 0800, 1030, 1130, 1300 (local), 1430, 1630, 1800, 2000, 2230. **Second Prgr:** 0330, 0500, 0700, 0900, 1000, 1130 (rel. 1st Prgr), 1400, 1500, 1700, 1900, 2030. **Call of Islam:** 0200, 1630.

Foreign Language Prgrs: English: 1000-1300, 1600-2100. **N:** 1830, 2045. **French:** 0800-1000, 1400-1600.

SHORTWAVES: Dirriya (G.C: 46.37E/24.39N); Jeddah (G.C: 39.10E/21.32N); Riyadh (G.C: 46.23E/24.30N): 50/350/500kW.

Relays of Home Sce. Prgrs. in Arabic:

kHz	Area	Times	kHz	Area	Times
6000	ME	1000-1700	9720	EEu	0300-0830
6020	ME	1700-2100*			2130-2300
7150	WEU	1700-2000	9870	WEu	1700-2130
7210	WEU	2000-2300	9885	WEu	2130-2300
7220	ME	1000-1500	11780	NAf	1500-1700
7280	ME	0300-0830	15140	EEu	0500-1100
9570	ME	0300-1700*	15435	WEu	0500-1100
9705	WEu	2130-2300	21505	NAf	1100-1700

*) rel-A2, others rel. A

Holy Quran Prgr.

kHz	Area	Times	kHz	Area	Times
7250	CAf	1500-2100	15240	CAs	0600-0800
7275	CAf	1900-2100	21510	CAs	0600-0800
11730	CAs	1400-1600	21510	SEAs	0800-1000
11935	NAf	1900-2100	21670	SEAs	0800-1000
15170	CAf	1000-1400			

EXTERNAL SERVICE

ADDR: P.O. Box 570, Riyadh 1116.

kHz	kHz	kHz	kHz	kHz	kHz	kHz
7275	9720	11685	11825	15345	17760	17895
9565	9730	11730	15060	15430	17845	21670
9705						

Bambara	Area	kHz	Somali		
1800-1900	CAf	7275, 15245	0400-0500	EAf	15430, 17845
Bengali			**Swahili**		
1600-1700	SEAs	11825, 15345	0500-0600	EAf	17760
English			**Turkestani**		
1600-2100	WEu	9705	1700-1800	CAs	9730
Indonesian			**Turkish**		
1000-1200	SEAs	17895, 21670	0400-0600	TUR	11685, 15060
Persian			**Urdu**		
1400-1600	IRN	9565, 11730	1200-1400	SAs	11825, 15345

ANN: Arabic: "Ithaa t'ill Mamlakah t'ill Arabiyah al-Saudiyah". E: "This is the Broadcasting Sce. of the Kingdom of Saudi Arabia".
INT-SIG: 'Ud' (oriental lute). Opens and closes with National Anthem — **V.** by QSL-card. (Re. to Freq. Mgr.)

ARAMCO RADIO (Priv.)
Serving staff of Saudi Aramco.
ADDR: Gen. Mgr. Public Affairs, Bldg. 3030, LIP. Dhahran 31311.
L.P: Coordinator, Media Broadc. Div: M. Al-Atani.
STATIONS: (FM): Pop/Country & Western music: Dhahran 91.4MHz, Safaniyah 103.8MHz, Udhailiyah 88.8MHz; Classical/ Easy Listening music: Dhahran 101.4MHz, Safaniya 107.9MHz, Udhailiyah 91.9MHz.
D.PRGR: 24h. **N:** 0945, 1030, 1530. **Time Signal:** on the h.

SYRIAN ARAB REPUBLIC

L.T: UTC+2h (March-Nov UTC+3h) — **Pop:** 13.450.000 — **Radio sets:** 2.850.000 — **Pr.L:** Arabic — **E.C:** a/c 50, 115/200V — **ITU:** SYR.

SYRIAN BROADCASTING AND TELEVISION ORGANISATION (Gov.)
ADDR: Ommayad Square, Damascus. **Te:** +963 (11) 720700. **Telex:** SY 411138 GD.
L.P: Dir. Gen: Khudr Omran. Dir. Eng: Dr. Eng. M. Bara. Dir. Public Rel: Mrs. Awafet Haffar. Dir. Admin & Finance: Zuheir Breidi.

Mediumwaves	kHz	kW		kHz	kW
Damascus-Adra	567	1000	Al-Hassake	918	200
Homs	617v	300	Deir el Zawr	954	60
Damascus-Saboura	666	600	Aleppo	1044	20
Sarakeb	747	100	Tartus	1071	60
Tartus	783	600	Al-Hassake	1125	200
Deir el Zawr	828	100	Aleppo	1314	10
Damascus-Kharabo	877v	10	Homs	1485	7

FM: Damascus 92.1/99.0MHz (Gen.Sce.), 97.1MHz (2nd Prgr.)
General Sce: 0325-0010 on 567/666/747/828/918/1485kHz + FM; 0325-1430 & 1630-0010 on 783kHz. **N:** 0415, 0515, 0615, 0715, 0815, 0915, 1015, 1115 (exc. Fri), 1215, 1315 (Fri), 1415 (Fri), 1515, 1615, 1815, 1915, 2115, 2215, 2315. **Rel. on SW:** see Ext. Sce.
2nd Prgr. on 617/877/954/1071/1125/1314kHz + FM: 0500-1500.
Rel. Foreign Sce: 1800-0100 on 877/1071/1314kHz.

EXTERNAL SERVICE

MEDIUMWAVES: See Home Sce.
SHORTWAVES: Adra (G.C: 36.30E/33.27N): 500kW.

kHz	kHz
12085	15095

Arabic	Area	kHz
0300-0400	IRQ	828, 1125, 12085
0500-1730*	ME	12085
1100-1630*	ME	15095
1700-1800	IRQ	828, 1125, 12085
2300-2400	SAm	17710, 15095

*) rel. Home Sce.

English		
2005-2105	Eu/NAm	15095, 12085
2110-2210	AUS/NAm	15095, 12085
French		
1905-2005	Eu/NAm	15095, 12085
German		
1805-1835	Eu/NAm	15095, 12085
Hebrew		
0400-0530	ME	783
Portuguese		
0000-0015	SAm	15095, 12085
Russian		
1835-1905	URS	15095
Spanish		
0015-0100	SAm	12085, 15095
Turkish		
1300-1400	ME	15095, 12085

ANN: Arabic: "Idha'atu-l-gumhuriyati-l-'arabiyya as-suriyya min dimashq". E: "This is the Syrian Republic Arab Broadc. Sce. from Damascus". F: "Ici Damas". Hebrew: "Kol Damesek".
INT-SIG: Guitar — **V.** by QSL-card.

TURKEY

L.T: UTC + 2h (Su: UTC + 3h) — **Pop:** 59.870.000 — **Radio sets:** 7.100.000 — **Pr.L:** Turkish — **E.C:** a/c 50, 220/380V (yes) — **ITU:** TUR.

TURKISH RADIO-TELEVISION CORPORATION

ADDR: TRT Türkiye Radyo Televizyon Kurumu, Nevzat Tandoğan Caddesi 2, Kavaklidere, Ankara. **Cable:** TRT.
Te: +90 (41) 28 22 30. **Telex:** 42374. (**Regional Addr:** R. Çukurova, 33130 Mersin).
L.P: Dir. Gen: Kerim Aydin Erdem. Dep. Dir. Gen (Tech): Coşkun Arslan. Legal Adviser: Akin Beşiroğlu. Secr. Gen: Ziya Arikan. Dep. Gen. Secr: Halid Ertuğrul. Dep. Dir. Int. Rel: Güleser Güler. Dir. Foreign Sales Dept: Sedat Örsel.
Regional Addr: Atatürk Bulvari, Sihhiye, Ankara; Harbiye, Istanbul; Kültür Parki, Izmir.
STATIONS: — *) Ankara — **)Tarsus — ***)Istanbul — °) Çukurova.

Location	N	kHz	kW			kHz	kW
Agri	2	162d	1000	Gaziantep	1	765	600
Polatli*	2	180	1200	Antalya	1	891	600
Etimesgut*	1	198	120	Izmir	1	927	200
Erzurum	1	245	200	Trabzon	1	954	300
Denizli	1	558	600	Mundanya***	1	1017	1200
Malatya	1	594	600	Hakkâri	P	1035	2
Tarsus°	1	630d	300	Diyarbakir	1	1062	300
Catalca***	2	702	150	Van	P	1179	2

N = Network 1, 2, & 3; P = Provincial.

FM (MHz)	TRT-2	TRT-3°	TRT-4	kW
Istanbul	—	88.2	91.4	60/30
Izmir	94.7	88.0	91.2	30/60/30
Eskişehir	—	94.4	96.8	30
Ankara	—	91.2	88.0	60/30
Edirne	—	91.0	89.0	30
Bursa	—	97.6	95.0	60/30
Adana	—	89.2	92.5	30
Gaziantep	—	95.2	97.6	30
Denizli	—	90.0	93.2	30
Konya	—	90.8	98.6	30
Muğla	88.8	96.6	92.0	30
Antalya	—	91.6	—	30
Çanakkale	89.5	97.0	93.0	30
Sivas	—	90.4	98.3	30
Isparta	—	99.2	89.6	60
Zonguldak	—	99.2	—	30
Ordu-Giresun	—	97.6	—	30
Trabzon	—	92.0	—	30
Kayseri	—	99.2	97.2	30
Izmit	—	93.6	96.0	30
Amasya	—	99.6	—	30
Samsun	—	90.8	—	30
Van	—	92.8	89.3	30
Diyarbakir	—	88.4	95.5	60
Malatya	—	89.6	—	30
Ağri	—	95.2	92.2	30
Erzincan	—	91.2	93.2	30
Erzurum	—	96.8	98.8	30
Kars	—	91.3	89.5	30
Silifke	—	94.4	—	30
Kastamonu	—	90.0	—	30
Kirşehir	—	88.8	92.0	30
Akşehir	—	94.4	95.0	30
Elbistan	—	94.0	96.4	30
Muş	—	90.2	97.2	30
Hatay	—	88.0	91.2	5
Artvin	—	88.4	94.4	5
Bolu	—	94.8	—	5
Bitlis	—	90.6	94.2	5
Edremit	—	90.4	95.4	5
Marmaris	—	90.9	95.0	5
Kuşadasi	—	90.2	—	5
Geşme	—	99.1	—	5
Elaziğ	—	—	92.8	30
Cizre	—	—	97.7	30

TRT-1: 24h (exc. Mon 2305-0255). **N:** 0300, 0400MF, 0500MF, 0530, 0700, 0800, 0900SS, 1000MF, 1100, 1300Sun, 1400SS, 1500MF, 1600MF, 1700, 1900Wed, 2100, 2200, 2255. **Regional N:** 0960MF, 1300W, 1400MF, 1900, (exc. Wed).
TRT-2: 0500-2300. **N:** 0500, 0530, 0700, 0900, 1100, 1300, 1500, 1700, 2100, 2255.
TRT-3: 0500-2300. **N:** 0500, 0700, 1000, 1200, 1500, 1700, 2000, 2255. **N:** in English/French/German (3 min's each): 0703, 1003, 1203, 1503, 1703, 2003.
Tourist Prgrs: English: Sat/Sun 1715. **French:** Sat/Sun 1515. **German:** Sat/Sun 2015.
TRT-4: 0500-1700. **N:** 0500, 0700, 0800, 0900, 1000, 1200, 1400, 1500.
Provincial st's: Van-Hakkari: W0355-0800, 0955-1200, 1355-2115. Sun 0355-1200, 1355-2115. **Erzurum:** 0255-2300. **Diyarbakir:** 0255-2300.
Tourist Station: FM-stereo prgrs in English for tourists at 0530-1045 & 1630-2000: Izmir 100.5, Antalya 100.6, Marmaris 101.0, Pamukkale 101.0, Kusadasi 101.9, Idis 103.2MHz (all 5kW). **N.** in English, French & German at 0630, 0830, 1030, 1630 & 1930.
ANN: "Burasi TRT-1 (or Radio Bir), Burasi TRT-2 (or Radio Iki), Burasi TRT-3 (or Radio Üç), Burasi TRT-4 (or Radio Dört)."

FOREIGN SERVICE

ADDR: P.O. Box 333, 06443 Ankara. **Te:** +90 (41) 189453. **Fax:** +90 (41) 353816. **Telex:** 42832 DISY FR.
L.P: MD: Savas Kiratli. CE: Erdogan Gönüllü. Chief English Dept: Semra Eren.
STATION: Ankara (G.C: 30.42E/39.54N): 2 × 500kW, 3 × 250kW tr's.

kHz	kHz	kHz	kHz	kHz	kHz	kHz
5980	7185	9555	9710	11875	11955	15325
6140	9445	9575	9795	11895	15160	15350
7110	9460	9685	11775	11925	15220	15405
7180						

Albanian	Area	kHz	Hungarian		
1200-1230	ALB	11875	1030-1100	HNG	11775
Arabic			**Persian**		
0900-1100	ME	11955	1400-1500	IRN	9675
1000-1100	NAf	15160	1930-2030	IRN	9675
1600-1700	ME/NAf	11955	**Romanian**		
		15160	1100-1130	ROU	11875
1700-1730	ME	11955	**Russian**		
Azerbaijani			1500-1530	URS	11735
0800-0900	SWAs	9590	**Serbo-Croat**		
1600-1700	NWAs	11925	1130-1200	YUG	11875
Bulgarian			**Turkish**		
1700-1800	BUL	5980	0000-0400	Eu	11895
2200-2300	BUL	5980	0000-0400	NAm	9445
Chinese			0400-0600	SEEu	6140
1230-1300	FE	9710	0500-1000	Eu	15220
English					9665
1330-1400	As	9675		As	15405
2100-2200	Eu	9445			11925
2300-2400	Eu/ME/	11985	1000-1700	Eu	15325
	NAm	9455			15350
		7185	1100-1600	ME	11955
0400-0500	NAm	9445	1100-1600	NAf	15160
French			1700-2300	NAf	11775
2200-2300	Eu	9445		Eu	9685
German			1700-0200	Eu	9460
1730-1830	Eu	9795	1800-2200	SEEu	5980
2030-2100	Eu	9445	**Urdu**		
Greek			1300-1330	PAK	9675
1530-1600	SEEu	7180			
1830-1930	SEEu	7110			

ANN: E: "This is the voice of Turkey". G: "Hier ist die Stimme der Turkei". Turkish: "Burasi Turkiyenin Sesi Radyosu" — **INT-SIG:** As for Home Sce — **V.** by QSL-card — **PUB:** "The Voice of Turkey Prgr sch" in English, French, German; "Turkiyenin Sesi Radyosu Program Bülteni" (Turkish every 6 months, free).

OTHER STATIONS

	kHz	kW	Times
2)	6900	5	0458-1000, 1200-1600 (Sun 1700)
1)	7370	10?	0458-1600
FM:	**MHz**	**kW**	**Times**
3)	94.1	—	0600-1700 (stereo)
4)	94.5	0.05	1600-2200
1)	95.5	—	as 7370kHz
5)	99.6	0.25	—

Addresses and other information

1) Türkiye Polis Radyosu, T.C. Içişleri Bakanliği, Emniyet Genel Müdürlüğü, Ankara. **Prgrs:** Turkish music & short ann — **Ann:** "Burasi Türkiye Polis Radyosu." **V.** by QSL-card.

2) Turkish State Meteorological Sce, P.O. Box 401, Ankara. **Te:** 141100. **Cable:** Meteor — Dir: Faysal Geyik — **Prgrs:** Turkish music, weather repts on the h. **Ann:** Burasi "Meteorolojinin Sesi Radyosu". **V.** by letter. Rp. Rec. acc.

3) Istanbul Polis Radyosu, Gamlica, Istanbul — **Ann:** "Burasi Istanbul Polis Radyosu".

4) Ankara Polis Radyosu. (stereo).

5) Istanbul Technical University R.

UNITED ARAB EMIRATES

L.T: UTC + 4h — **Pop:** 2.526.000 — **Radio sets:** 420.000 — **Pr.L:** Arabic — **E.C:** a/c 50, 220V — **ITU:** UAE.

MINISTRY OF INFORMATION & CULTURE

ADDR: P.O. Box 17, Abu Dhabi. **Te:** +971 (2) 453000. **Fax:** +971 (2) 452504. **Telex:** 22283 INFOR EM.

L.P: Minister of Inf. & Culture: Khalfan M. al Roumi. Under-Secr: Awad Mohammed al Otaiba. Dir. Gen. Tech. Dept: Ibrahim Rashid al Zaabi. Tech. Advisor: Fawzi Saleh al Ali.

RADIO OF THE UNITED ARAB EMIRATES (Gov.)

ADDR: P.O. Box 63, Abu Dhabi. **Te:** +971 (2) 451000. **Fax:** +971 (2) 451155.

L.P: Dir: Abdul Wahab al Radhwan. Contr. of Prgrs: Abdul Hadi al Mobarak. Chief Eng: Fawzi Saleh al Ali. Head of Transmission: Rushdi Hattab. Contr. of Foreign Prgrs: Ahmed A. Shouly.

Mediumwaves	kHz	kW			
Sadiyat	657	2 x 50	Fujairah	972	2 x 50
Sadiyat	729	2 x 750	Dabiya	1314	2 x 2000
Maqtaa	810	50	Sadiyat	1539	2 x 50
al Ain	828	1	Sharjah	1575	50

FM: 93.5/98.0MHz 2kW.

D.PRGR. in Arabic: 0200-2200 on 729/1314kHz, 0200-1800 on 657kHz, 0200-1400 on 972kHz. **N:** 0630, 0930, 1230, 1400, 1430.

Local Prgr: 1130-1500 on 1539kHz.

Holy Quran Sce: 1200-1500 on 98.0MHz.

English Comm. Sce. (Capital Radio): 0300-2000 on 93.5MHz.

Foreign Language Sce on 810kHz: 0800-1100 English, 1100-1300 French, 1300-1600 Urdu, 1600-1800 Pilipino.

Rel. Voice of the Arabs from Cairo: 24h on 1575kHz.

Shortwaves (G.C: 54.17E/24.23N): 500kW.

Arabic	Area	kHz			
0000-0200	NAm	9605	1000-1400	WAs	11815
		11710	1100-1300	AUS	25690
		11815		FE	15315
0200-0400	WAs	6180	1200-1700	ME	15265
0200-0600	ME	9695	1300-1400	AUS	17855
0300-0400	ME	9770	1300-1500	AUS	21735
		7215	1300-1600	Eu	13605
0300-0700	ME	15265	1400-1500	AUS	11750
0400-0500	NWAs	9760	1500-1600	ME	7215
0400-0700	ME	17820	1600-2000	ME	9695
0500-1000	WAs	15315	1600-2200	Eu	9770
0600-0800	Eu	15315		ME	6180
0600-0700	Eu	15380	**English**		
0700-0900	Eu	17855	2200-2400	NAm	9505
0800-1300	Eu	21735			9605
0900-1000	FE	25690			11710
0900-1100	FE	21735			11815
		17855			

ANN: A: "Idha'atu-l-imarat al-arabiyya al-muttahida min Abu Zabiy" — **V.** by QSL-folder. R. to Contr. of Foreign Prgrs.

U.A.E. RADIO & TELEVISION - DUBAI (Gov.)

ADDR: P.O. Box 1695, Dubai. **Te:** +971 (4) 370255. **Telex:** 45605 BRCAST EM.

L.P: Dir. of Inf: Sheikh Hasher al Maktoum. Ag. Dir. Gen: Ahmed Saeed al Gaoud. Contr. Radio: Hassan Ahmed. Contr. of Eng: Ahmed Najeeb.

Mediumwave

kHz	kW	kHz	kW	kHz	kW
1107	10	1251	600	1481	1500

FM: 92.0MHz 10kW.

D.PRGR. in Arabic: 0300-2100. **N:** 0315, 0500, 1000, 1300, 1700, 1900, 2030.

English: 24h on 92.0MHz. **N:** 0330, 0530, 1030, 1330, 1630.

Shortwaves (G.C: 55.16E/25.14N): 1 x 500kW, 3 x 300kW tr's.

Times	Area	KHz			
0230-0400	NAm	13675	0615-1445	Eu	21605
		11945	0615-1645	Eu	15435
		15435			15400
		9595	0615-2100	Eu	13675
0415-0600	FE	17830		NAf	15320
		21700	1700-2100	Eu	11795
		15435			9595

D.PRGR: Rel. Home Sce. in Arabic exc. **English:** 0330-0400, 0530-0600, 1030-1100, 1330-1400, 1630-1700.

ANN: A: "Idha'at al imarat al Arabiyyah al Mutahhida min Dubai" — E:
"This is United Arab Emirates Radio in Dubai" — **V.** by QSL-card (B).

RAS AL KHAIMAH BROADC. ST. (Gov.)

ADDR: P.O. Box 141, Ras al Khaimah. **Te:** +971 (7) 851151.

STATIONS: 1152kHz 50kW, 1175kHz 10kW — **FM:** 95.3MHz.

D.PRGR: 0200-1900. **Urdu:** 0830-1000. **Malayalam:** 1000-1100.

ANN: "Idha'at al imarat al Arabiyyah al Mutahhida min Ras al Khaima" — **V.** by letter.

UMM AL QIWAIN BROADC. ST. (Gov.)

ADDR: P.O. Box 444, Umm al Qiwain. **Te:** +971 (6) 666044. **Fax:** +971 (6) 666055. **Telex:** 69711 EM.

L.P: Dir. Gen: Ali Jassim Ahmed. Chief Eng: Saeed el Naqueeb.

STATION: 846kHz 20kW.

D.PRGR: 0200-1900. **Sinhalese:** 0800-0900. **Urdu:** 0900-1055.

F.Pl: New 20kW station, prgrs. in Hindi, Malayalam, Filipino.

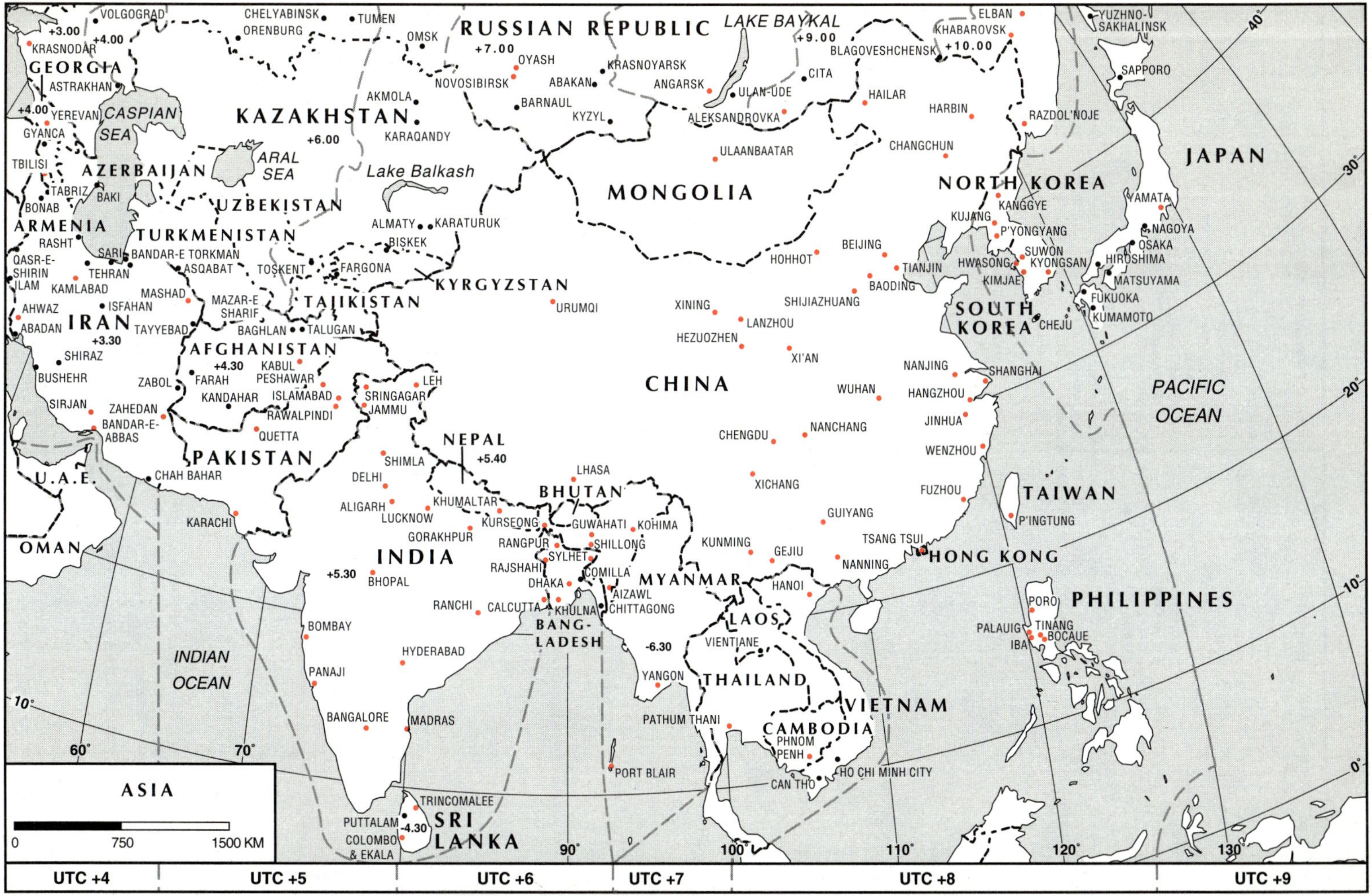
ASIA
0
750
1500 KM
UTC +4
UTC +5
UTC +6
UTC +7
UTC +8
UTC +9
RUSSIAN REPUBLIC
KAZAKHSTAN
MONGOLIA
CHINA
JAPAN
NORTH KOREA
SOUTH KOREA
TAIWAN
HONG KONG
PHILIPPINES
VIETNAM
CAMBODIA
THAILAND
LAOS
MYANMAR
BANG-LADESH
BHUTAN
NEPAL
INDIA
SRI LANKA
PAKISTAN
AFGHANISTAN
IRAN
TAJIKISTAN
KYRGYZSTAN
UZBEKISTAN
TURKMENISTAN
AZERBAIJAN
ARMENIA
GEORGIA
U.A.E.
OMAN
CASPIAN SEA
ARAL SEA
Lake Balkash
LAKE BAYKAL
PACIFIC OCEAN
INDIAN OCEAN
+3.00
+4.00
+4.00
+6.00
+7.00
+9.00
+10.00
+3.30
+4.30
+5.30
+5.40
-6.30
-4.30
VOLGOGRAD
CHELYABINSK
ORENBURG
TUMEN
OMSK
NOVOSIBIRSK
OYASH
BARNAUL
ABAKAN
KRASNOYARSK
KYZYL
ANGARSK
ULAN-UDE
ALEKSANDROVKA
CITA
BLAGOVESHCHENSK
KHABAROVSK
ELBAN
YUZHNO-SAKHALINSK
SAPPORO
HAILAR
HARBIN
CHANGCHUN
RAZDOL'NOJE
ULAANBAATAR
KRASNODAR
ASTRAKHAN
YEREVAN
GYANCA
TBILISI
TABRIZ
BAKI
BONAB
RASHT
SARI
BANDAR-E TORKMAN
ASQABAT
QASR-E-SHIRIN
ILAM
TEHRAN
KAMLABAD
MASHAD
ISFAHAN
AHWAZ
ABADAN
TAYYEBAD
SHIRAZ
BUSHEHR
SIRJAN
ZAHEDAN
BANDAR-E-ABBAS
ZABOL
CHAH BAHAR
AKMOLA
KARAQANDY
ALMATY
KARATURUK
BISKEK
FARGONA
TOSKENT
MAZAR-E SHARIF
BAGHLAN
TALUGAN
KABUL
PESHAWAR
FARAH
KANDAHAR
ISLAMABAD
RAWALPINDI
QUETTA
KARACHI
SRINGAGAR
JAMMU
LEH
SHIMLA
DELHI
ALIGARH
LUCKNOW
KHUMALTAR
GORAKHPUR
KURSEONG
BHOPAL
RANCHI
CALCUTTA
BOMBAY
PANAJI
HYDERABAD
BANGALORE
MADRAS
TRINCOMALEE
PUTTALAM
COLOMBO & EKALA
LHASA
GUWAHATI
KOHIMA
SHILLONG
RANGPUR
SYLHET
RAJSHAHI
COMILLA
DHAKA
AIZAWL
KHULNA
CHITTAGONG
YANGON
PORT BLAIR
PATHUM THANI
VIENTIANE
HANOI
PHNOM PENH
HO CHI MINH CITY
CAN THO
URUMQI
XINING
LANZHOU
HEZUOZHEN
XI'AN
SHIJIAZHUANG
HOHHOT
BEIJING
TIANJIN
BAODING
CHENGDU
NANCHANG
XICHANG
KUNMING
GEJIU
GUIYANG
NANNING
TSANG TSUI
WUHAN
NANJING
HANGZHOU
SHANGHAI
JINHUA
WENZHOU
FUZHOU
KUJANG
KANGGYE
P'YONGYANG
HWASONG
SUWON
KYONGSAN
KIMJAE
CHEJU
YAMATA
NAGOYA
OSAKA
HIROSHIMA
MATSUYAMA
FUKUOKA
KUMAMOTO
P'INGTUNG
PORO
PALAUIG
TINANG
BOCAUE
IBA
10°
20°
30°
40°
0°
10°
60°
70°
90°
100°
110°
120°
130°

YEMEN (Republic of)

L.T: UTC + 3h — **Pop:** 10.385.000 — **Radio sets:** 325.000 — **Pr.L:** Arabic — **E.C:** a/c 50, 230V — **ITU:** YEM.

YEMEN RADIO & TV CORP. (Gov.)

ADDR: PO Box 2182, Sana'a. **Te:** +969 (2) 231181. **Fax:** +969 (2) 230761. **Tlx:** 2645 YARTV.
Networks: S = Sana'a, A = Aden.

Mediumwaves	kHz	kW	N.	H. of tr.
Mocha	711	200	S	1400-2115
Al-Hiswah	792	750	A	0300-0600, 1100-2100+
Sana'a	837	30	S	
Sana'a	1008	600	S	
Taiz	1112	30	S	1100-2115
Al-Hiswah	1188	400	A	0300-0600, 1100-2100+
Shortwaves	**kHz**	**kW**	**N.**	**H. of tr.**
Sana'a	4853v	20	S	0300-0705, 1000-1100
Sana'a	5950	300	S	0300-0705, 1000-2115+
Al-Hiswah	5970	100	A	0300-0600, 1100-2100+
Sana'a	6135	300	S	0300-0705, 1000-1600+
Al-Hiswah	7190	100	A	0300-0600, 1100-2115
Sana'a	9779.4	50	S	0300-0705, 1000-2115+

+) no transmission break on Fri — **FM:** Sana'a 92.6MHz.
N: in Arabic: Sana'a: 0430, 0600, 0700, 1030, 1200, 1400, 1500, 1700, 1830, 2000, 2100. **Aden:** 0330, 0430, 0555, 0700Fri, 0800Fri, 1100Fri, 1130 (exc. Fri), 1230, 1400, 1500, 1630, 1800, 1900, 2030.
English: 1600-1630 on all Aden freq's. **N:** 1600.
ANN: "Idha'atu-l-jumhuriya al-yamanniya min Sana'a/Aden".
V. QSL-card or letter. Rp.

ASIA

BANGLADESH (People's Rep. of)

L.T: UTC + 6h — **Pop:** 120.000.000 — **Radio sets:** 4.500.000 — **Pr.L:** Bengali — **E.C:** a/c 50, 220/440V — **ITU:** BGD.

RADIO BANGLADESH (Gov.)

ADDR: National Broadc. Authority, NBA House, 121 Kazi Nazrul Islam Ave, Dhaka-1000. **Te:** +880 865294. **Cable:** Broadcasts. **Telex:** 642228 NBA BJ.
L.P: DG: Dr. Sa'dat Hossain. CE: Shamsuzzaman Chowhury. Dep. Dir. Gen (Prgr): M.N. Mustafa. Dep. Dir. Gen (News): A.K.M. A.R. Sharif. Sen. Engr. Research & Receiving Centre: Md. Abdur Rashid.

HOME SERVICE

STATIONS	kHz	kW	H. of tr.
Khulna	558	100	0000-0400 0600-0930, 1030-1730
Dhaka-A	693	1000	0000-0430, 1030-1730
Dhaka-B	819	100	0000-0230, 0430-1730
Rajshahi	846	100	0000-0400, 0600-0930, 1030-1730
Chittagong	873	10	0000-0400, 0600-0930, 1030-1730
Sylhet	963	20	0000-0400, 1030-1730
Thakurgaon	999	10	1200-1600
Rangpur	1053	20	0000-0430, 1030-1430
Rajshahi	1080	10	0000 as 846kHz
Comilla	1413	10	1200-1600
Dhaka	4879v	100	0000-0430, 1230-1600
Dhaka	6195	50	0430-0905
Dhaka	7080	7.5	0900-1235
Dhaka	15520	50	1230-1730

FM: Dhaka 100.0MHz 2kW, 104.0MHz 1kW; Sylhet 101.5/Chittagong 102.5/Rajshahi 103.0/Rangpur 105.5/Khulna 105.5MHz (all 2kW) + 6 st's less than 1kW.
N. in English: 0110, 0305, 0700, 1255, 1530, 1705.

EXTERNAL SERVICE

ADDR: P.O. Box 2204, Dhaka.
LP: Dir: Mobarak Hossain Khan.
STATION: Dhaka (G.C: 90.26E/23.43N): 2×250kW tr's.

kHz	kHz	kHz	kHz	kHz
7115	11705	13620	15255	17750
9570	12030	15200	15520	17825

All freqs variable.

	kHz		kHz
Arabic		**Hindi**	
1600-1630	11705, 13620	1515-1545	11705, 15255
Bengali		**Nepali**	
1630-1730	11705, 13620	1315-1345	17115, 11705
1915-2000	9570, 12030	**Urdu**	
English		1400-1430	15255, 17825
1230-1300	15200, 17750	**Voice of Islam**	
1815-1900	9570, 12030	1745-1815	9570, 12030

ANN: Bengali & English: "R. Bangladesh"; for Ext. Sce. "This is the General Overseas Sce. of R. Bangladesh" — **INT-SIG:** Local composition of violin and tanpura — **V.** by letter. Re. to Senior Eng. (Research Wing), NBA House, 121 Kazi Nazrul Islam Ave, Dhaka-1000 — **PUB:** "Betar Bangla" in Bengali quarterly for HS (Annual subs. TK 24,-), R. Bangladesh in English quarterly (free).

BHUTAN

L.T: UTC + 6h — **Pop:** 1.630.000 — **Radio sets:** 28.000 (est.) — **Pr. L:** Dzongkha, Sharchopkha, Lhotsam, English.

BHUTAN BROADCASTING SERVICE—BBS (Gov.)

ADDR: P.O. Box 101, Thimphu. **Te:** +975 23070/1/2. **Fax:** +975 23073. **Telex:** 212 Infobro TPU BT.
L.P: Dir. of Inf: Tashi Phuntshog. Joint Dir: Phub Tshering. Prgr. Officer: Tashi Dhendup. St. Eng: Sonam Tobgay.
STATION: 5025/6035/9615kHz 50/10kW. **FM:** 96/98.
D.PRGR: W 1100-1500 (**English** 1415-1500) on 5025kHz. Sun 0400-1000 (**English** 0900-1000) on 6035 or 9615kHz. **N. in English:** W 1415.
V. by QSL-card. 15 mins prgr. details req. 2 IRC's appreciated.

BRITISH INDIAN OCEAN TERRITORY

L.T: UTC + 5h — **Pr.L:** English — **Pop:** 3.000 — **E.C:** a/c 60, 110/220V — **ITU:** BIO.

ARMED FORCES RADIO AND TELEVISION SERVICE

ADDR: U.S. Navy Broadcasting Service Detachment Diego Garcia, Box 14, U.S. Navy Support Facility, F.P.O. San Francisco, CA 96685-2902, USA.
L.P: Officer-in-Charge: Bill W. Love. Chief Eng: M.S. Patio.
STATION: Diego Garcia (G.C: 72.22E/07.16S): 1485kHz 0.25kW.
FM: 99.1/101.9MHz 0.25kW — **D.PRGR:** 24h.
ANN: "Armed Forces Radio and Television Service, 1485kHz on your dial, Diego Garcia, British Indian Ocean Territory" — **V.** by letter.

BRUNEI DARUSSALAM

L.T: UTC + 8h — **Pop:** 422.837 — **Radio sets:** 100.000 (est.) — **Pr.L:** Malay, English, Chinese, Gurkha — **E.C:** a/c 50, 220V (no) — **ITU:** BRU.

RADIO TELEVISYEN BRUNEI (Gov.)

ADDR: Jalan Elizabeth II, Bandar Seri Begawan 2042, Negara Brunei Darussalam. **Te:** +673 (2) 243111. **Fax:** +673 (2) 222190/227204/243008. **Cable:** Broadcast. **Regional Studio:** Kuala Belait.
L.P: Dir: Hj Alidin Hj Othman. Dep. Dir: Sapawi Bolhassan. Supt. Eng: Hj Md Yusof Abd Rahman. Ag. Head of Prgr: Pg Hjh Normah Pg Hj Daud. Hj Mohamed. Chief N. Editor: Zainal Tinggal.

MEDIUMWAVES	kHz	kW		kHz	kW
Tutong	594	200d	S. Hanching	711	20
Serasa	675	200	Kuala Belait	972	1

FM: (stereo): 92.3/93.8/95.9/96.9MHz (all 5kW).
Malay Sce: 2030-1430 on 594/675kHz + 92.3/93.8MHz.
English Sce: 2230-0030, 0300-0800, 1200-1400 on 711/972kHz + 95.9/96.9MHz. **N:** 2245, 2330, 0430, 1315.
Chinese Sce: 0030-0300, 0800-1100 on English fqs.
Gurkha Sce: 0530-0630 (Sat) & 1100-1200 on English fqs.
ANN: E: "This is Radio Brunei". Malay: "Inilah Radio Brunei".
INT-SIG: Synthesized chimes, TS — **PUB:** "Pelita" in Malay, "Brunei Darussalam" in English — **V.** by QSL-card.

BRITISH FORCES. BROADC. SCE. (Mil.)

ADDR: B.F.P.O. II — **FM:** 92.0MHz 1kW, 89.5MHz 0.05kW. **D.PRGR:** 24h.

CAMBODIA (State of)

L.T: UTC + 7h — **Pop:** 7.300.000 — **Radio sets:** 800.000 (est.) — **Pr.L:** Cambodian (Khmere) — **E.C:** a/c 50, 120/208 or 220V — **ITU:** CBG.

VOICE OF THE PEOPLE OF CAMBODIA (Gov.)

ADDR: 28 Av. Sandech Choun Nath, Phnom-Penh (via Hanoi, Vietnam).
MEDIUMWAVES: Phnom Penh 740kHz 120kW—1360kHz 20kW. Battambang 1453kHz (r.)
SHORTWAVES: (G.C: 104.51E/11.34N): 4910/6090/11938kHz (all freqs variable).
HOME SCE. in Cambodian on 740/4910/6090kHz: 2300v-0130 (Sun 0215), 0600-0730 (Sun 0645), 1100-1430 (Sun 1445).
R.V. of the People of Battambang Province: r. 1200-1400 on 1453kHz. Additional broadcasts ann. at 2230 & 0500.
EXT. SCE on 1360/11938kHz: **English:** 0000-0015, 1200-1215; **French:** 0015-0030, 1215-1230; **Laotian:** 0045-0100, 1245-1300. **Thai:** 0030-0045, 1230-1245; **Vietnamese:** 0100-0115, 1300-1315.
ANN: "Thini Phnom-Penh". E: "R. Phnom Penh" or "Radio the Voice of the People of Cambodia."—**V.** by letter.

CHINA (People's Rep. of)

L.T: UTC + 8h (exc. Xinjiang Uighur Aut. Reg: UTC + 6h) — **Pop:** 1.148.177.000 — **Radio sets:** 20m (est.) — **Pr.L:** Chinese, Amoy, Cantonese, Chaozhou, Hakka, Kazakh, Korean, Mongolian, Tibetan, Uighur, Zhuang, a.o. — **E.C:** a/c 50, 220V — **ITU:** CHN.

DIRECTORATE GENERAL OF TELECOMMUNICATIONS

Ministry of Posts and Telecommunications, Xi Chang'an Jie, Beijing 100031. **Cable:** Gentel.

MINISTRY OF RADIO, FILM AND TELEVISION (Gov.)

ADDR: 2 Fuxingmenwai Dajie, Beijing 100866 or P.O. Box 4501, Beijing — **L.P:** Minister: Ai Zhisheng.

Official P.R.C Abbreviations: The 30 regions of People's Republic of China, With their abbreviations and name in Pinyin (Chinese Phonetic Alphabet) version followed by the old spelling in brackets: AH: Anhui (Anhwei) — BJ: Beijing M. (Peking) — FJ: Fujian (Fukien) — GD: Guangdong (Kwangtung) — GS: Gansu (Kansu) — GX: Guangxi Zhuang A.R. (Kwangsi) — GZ: Guizhou (Kweichow) — HAN: Hainan (Hainan) — HB: Hubei (Hupeh) — HEB: Hebei (Hopeh) — HEN: Henan (Honan) — HL: Heilongjiang (Heilungkiang) — HN: Hunan (Hunan) — JL: Jilin (Kirin) — JS: Jiangsu (Kiangsu) — JX: Jiangxi (Kiangsi) — LN: Liaoning (Liaoning) — NM: Nei Menggu A.R. (Inner Mongolia) — NX: Ningxia Hui A.R. (Ningsia) — QH: Qinghai (Tsinghai) — SC: Sichuan (Szechwan) — SD: Shandong (Shantung) — SH: Shanghai M. (Shanghai) — SN: Shaanxi (Shensi) — SX: Shanxi (Shansi) — TJ: Tianjin M. (Tientsin) — XJ: Xinjiang Uighur A.R. (Sinkiang) — XZ: Xizang A.R. (Tibet) — YN: Yunnan (Yunnan) — ZJ: Zhejiang (Chekiang).
Regional Services: Add "Renmin Guangbo Diantai" (People's Broadcasting Station) to the station name shown in the table below to obtain the full name in Standard Chinese. **Abbreviations:** 1 = 1st prgr, 2 = 2nd prgr, 3 = 3rd prgr, E = Educational prgr, EC = Economic prgr, G = General prgr, L = Literary prgr, M = Music prgr, N = News prgr.
Languages: Standard Chinese (Putonghua), based on the Beijing dialect, is used in broadcasts throughout China. Various dialects and minority languages are included in the relevant regional services and in Central broadcasts to Taiwan.
Abbreviations: Standard Chinese = Ch, Kazakh = Kz, Korean = Ko, Mongolian = Mo, Tibetan = Tb, Uighur = Ug.
Mediumwaves:

	kHz	kw	Station	Location
116)	531	10	Zhejiang	ZJ
1)	540		CPBS 1	
1)	549	100	CPBS Taiwan 1	FJ
40)	549	10	Chifeng-Ch	NM
47)	549	1	Alashan	NM, Bayanhot
170)	549	10	Zhengzhou	HEN
142)	558	50	Fujian 1	FJ
285)	558	20	Xinjiang-Ug	XJ, Urumqi
38)	558	10	Baotou	NM
1)	567		CPBS 1	
116)	576v		Zhejiang	ZJ
254)	576	50	Yunnan G	YN
82)	576		Yanji	JL
102A)	585	10	Jiangsu EBS	JS, Nanjing
23)	585v		Chengde Diqu	HEB, Chengde
150)	585	1	Xinyu	JX
1)	594	200	CPBS Taiwan 2	FJ
154)	594	100	Shandong 1	SD, Jinan
264)	594	300	Xizang-Tb	XZ, Lhasa
249)	600		Fuling	SC
300)	603		R. Beijing DS	SN, Xi'an
212)	603		Guangdong-Eg	GD, Guangzhou
43)	603		Hulun Bei'er-Ch	NM, Hailar
292)	603		Yining-Ch	XJ
52)	612	10	Liaoning	LN
142)	612	100	Fujian 1	FJ
236)	612	10	Sichuan 1	SC
86)	621	100	Heilongjiang 1	HL, Harbin
279)	621	20	Haixi	QH, Da Qaidam
1)	630		CPBS 2	
1)	639		CPBS 1	
101)	648		Shanghai TS	SH
212)	648	150	Guangdong N	GD, Guangzhou
130)	648	1	Huainan	HEN
168)	657	300	Henan	HEN, Zhengzhou
75)	657		Hunjiang	JL
122)	657		Jiaxing	ZJ
276)	666	200	Qinghai 1	QH, Xining
60)	666	2	Jinzhou	LN
72)	666		Siping	JL
95)	666	10	Jiamusi	HL
125)	666		Jinhua	ZJ
	(Nom 675kHz)			
129)	666	1	Hefei	AH
301)	666		China Huayi BC	FJ, Fuzhou
	(Nom 668kHz)			
163)	666	Jining SD		
36)	675	150	Nei Menggu-Ch	NM, Hohhot
285)	675		Xinjiang-Ch	XJ
1)	684		CPBS Taiwan 2	FJ
285)	684		Xinjiang-Ch	XJ
16)	684	10	Tangshan	HEB
57)	684	10	Fushun	LN
94)	684	10	Mudanjiang	HL
127)	684	7.5	Zhoushan	ZJ
138)	684		Suxian Diqu	AH, Suzhou
1)	693		CPBS Taiwan 1	FJ
266)	693	100	Shaanxi 1	SN, Xi'an
66)	700		Chaoyang	LN, Lingyuan
102)	702	150	Jiangsu 1	JS, Nanjing
285)	702		Xinjiang-Ch	XJ
257)	702		Honghe	YN, Gejiu
276)	711	10	Qinghai 1	QH, Golmud
126)	711		Quzhou 1	ZJ
139)	711		Lu'an	AH
242)	711		Panzhihua	SC
245)	711		Mianyang	SC
253)	711		Liupanshui	GZ
1)	720		CPBS 2	
244)	720	1	Deyang	SC

146)	729	150	Jiangxi 1	JX, Nanchang
181A)	729		Shangqiu	HEN
69)	738	100	Jilin	JL, Changchun
202)	738	150	Hunan 1	HN, Changsha
285)	738	10	Xinjiang-Ch	XJ, Urumqi
124)	738		Shaoxing	ZJ
236)	747		Sichuan 1	SC
39)	747	1	Wuhai	NM
250)	747		Nanchong	SC
258)	747		Xishuangbanna	YN, Jinghong
66)	750		Chaoyang	LN, Jianping
233)	750		Beihai	GX
1)	756		CPBS 1	
1)	765v		CPBS Taiwan 1	FJ
36)	765	10	Nei Menggu-Ch	NM
251)	765	10	Guizhou 1	GZ
218)	765		Shaoguan	GD
184)	774	100	Hubei 1	HB, Wuhan
9)	783	100	Hebei 1	HEB, Shijiazhuang
1)	792		CPBS Taiwan 1	FJ
101)	792	50	Shanghai 2	SH
228)	792	100	Guangxi 1	GX, Nanning
45)	792		Erdox BS-Ch	NM, Otog
54)	792	5	Shenyang	LN
238)	792		Chengdu 1	SC
300)	801		R. Beijing DS	TJ, Tianjin
213)	801	50	Zhujiang EBS	GD, Maoming
107)	801		Huaiyin	JS
177)	801		Xinxiang	HEN
281)	801		Yinchuan	NX
117)	810	200	Zhejiang	ZJ, Hangzhou
285)	810		Xinjiang-Ug	XJ
66)	810	1	Chaoyang	LN
73)	810	1	Liaoyuan	JL
267)	810	10	Xi'an	SN
286)	810		Urumqi-Ch	XJ
28)	819	200	Shanxi 1	SX, Taiyuan
291)	819		Kuitun-Kz	XJ
7)	828	50	Beijing 1	BJ
168)	828	10	Henan	HEN
212)	828	50	Guangdong N	GD, Heyuan
182)	828v		Zhoukou	HEN
1)	837		CPBS 1	
87)	837		Harbin 1	HL
28)	846	20	Shanxi 1	SX
228)	846		Guangxi 2	GX, Nanning
264)	846		Xizang-Ch	XZ, Lhasa
285)	846		Xinjiang-Ug	XJ
112)	846	10	Changzhou 1	JS
2)	846		Haixia 2	FJ
175)	846		Jiaozuo	HEN
216)	850		Zhuhai	GD
1)	855		CPBS 2	
117)	864		Zhejiang	ZJ
128)	864	10	Anhui 1	AH
86)	873	100	Heilongjiang 2	HL, Harbin
271)	873	50	Gansu 1	GS
25)	873v		Xinji	HEB
70)	873		Changchun 1	JL
166)	873		Linyi	SD
172)	873v	1	Kaifeng	HEN
186)	873	10	Wuhan	HB
290)	873		Changji	XJ
142)	882	100	Fujian 1	FJ, Fuzhou
10)	882		Shijiazhuang	HEB
37)	882	10	Hohhot	NM
54)	882	10	Shenyang EC	LN
55)	882	10	Dalian 1	LN
178)	882		Anyang	HEN
278)	882	10	Yushu	QH
287)	882		Karamay-Ch	XJ
280)	891	100	Ningxia 1	NX, Yinchuan
59)	891		Dandong 2	LN
86)	900	10	Heilongjiang 1	HL
4)	900		V.O. Pujiang	SH
51)	900		Dongsheng	NM
104)	900		Nanjing EBS	JS
171)	900		Zhengzhou EBS	HEN
181)	900		Luohe	HEN
261)	900		Dehong	YN, Luxi

1)	909	100	CPBS Taiwan 2	FJ
8)	909	50	Tianjin N	TJ
236)	909	10	Sichuan 1	SC
276)	909	10	Qinghai 1	QH
285)	909	10	Xinjiang-Mo	XJ, Urumqi
74)	909	10	Tonghua	JL
93)	909	5	Yichun	HL
154)	918	10	Shandong 1	SD, Yantai
229)	918	1	Nanning	GX
1)	927		CPBS Taiwan 2	FJ
7)	927	50	Beijing 2	BJ
4)	927		V.O. Pujiang	SH
251)	927	50	Guizhou 1	GZ, Guiyang
71)	927	10	Jilin-shi 1	JL
85)	927		Hunchun	JL
128)	936	200	Anhui 1	AH, Hefei
1)	945		CPBS 2	
8)	954	3.5	Tianjin E	TJ
234)	954	30	Hainan	HAN, Haikou
285)	954		Xinjiang-Kz	XJ
56)	954	1	Anshan 1	LN
119)	954	10	Hangzhou	ZJ
243)	954		Luzhou	SC
272)	954		Lanzhou	GS
52)	963	10	Liaoning	LN, Jinzhou
117)	963	10	Zhejiang	ZJ
285)	963		Xinjiang-Kz	XJ
169)	972	150	Henan EBS	HEN, Zhengzhou
285)	972		Xinjiang-Kz	XJ
87)	972		Harbin 2	HL
1)	981		CPBS 1	
101)	990	50	Shanghai 1	SH
254)	990	20	Yunnan G	YN, Kunming
285)	990		Xinjiang-Kz	XJ
17)	990		Qinhuangdao	HEB
53)	999		Liaoning EBS	LN, Shenyang
212)	999	7.5	Guangdong L	GD, Guangzhou
285)	999		Xinjiang-Ch	XJ
252)	999	10	Guiyang	GZ
237)	1000		Sichuan EBS	SC, Chengdu
141)	1000		Fuyang Xian	AH
1)	1008		CPBS 1	YN, Kunming
266)	1008	10	Shaanxi 1	SN
103)	1008		Nanjing	JS
1)	1017		CPBS 1	
300)	1017	100	R. Beijing	JL, Changchun
212)	1017		Guangdong N Shantou	GD, Shaoguan/
7)	1026	10	Beijing 3	BJ
251)	1026	50	Guizhou 1	GZ
61)	1026		Yingkou	LN
108)	1026		Yancheng	JS
149)	1026	1	Pingxiang	JX
1)	1035		CPBS 1	
300)	1044	300	R. Beijing	JS, Changzhou
285)	1044	10	Xinjiang-Ug	XJ
260)	1044	1	Dali	YN
77)	1053	10	Yanbian-Ch	JL, Yanji
155)	1053	10	Jinan	SD
256)	1053	100	Wenshan	YN, Wenshan
213)	1062	150	Zhujiang EBS	GD, Guangzhou
96)	1062		Qitaihe	HL
300)	1071		R. Beijing	ZJ, Hangzhou
8A)	1071	50	Tianjin EBS	TJ
142)	1071v		Fujian 1	FJ
228)	1071		Guangxi 1	GX
123)	1071		Huzhou	ZJ
269)	1071	10	Baoji	SN
286)	1071		Urumqi-Ug	XJ
300)	1080	200	R. Beijing	YN, Kunming
117)	1080		Zhejiang	ZJ
285)	1080		Xinjiang-Ug	XJ
92)	1080		Daqing	HL
114)	1080	10	Suzhou	JS
217)	1080	5	Shantou	GD
239)	1080	5	Chongqing 2	SC
1)	1089		CPBS Taiwan 2	FJ
52)	1089	200	Liaoning	LN, Shenyang
236)	1089		Sichuan 2	SC, Chengdu
204)	1089	1	Zhuzhou	HN

226)	1098	5	Maoming	GD
289)	1098		Hami-Ug	XJ
34)	1100		Yanbei	SX, Datong
263)	1100		Jiangchuan	YN
69)	1107		Jilin	JL
234)	1107	10	Hainan	HAN
285)	1107	20	Xinjiang-Kz	XJ, Urumqi
90)	1107		Shuangyashan	HL
133)	1107	1	Tongling	AH
144)	1107		Xiamen	FJ
	(r 1110kHz)			
173)	1107		Luoyang	HEN
176)	1107		Hebi	HEN
187)	1112v		Changjiang EBS	HB, Wuhan
	(Nom 1053kHz)			
1)	1116	120	CPBS 2/R. Beijing HL	
1)	1116		CPBS Taiwan 1	FJ
236)	1116	50	Sichuan 1	SC, Chengdu
116)	1116		Changshu 1	JS
9)	1125		Hebei 1	HEB, Handan
117)	1134		Zhejiang	ZJ
225)	1134	10	Zhanjiang	GD
268)	1134		Tongchuan	SN
275)	1134	1	Yumen	GS
292)	1134		Yining-Kz	XJ
117)	1143v		Zhejiang	ZJ
276)	1143		Qinghai 2	QH, Xining
83)	1143		Tumen	JL
112)	1143		Changzhou 2	JS
157)	1143		Zibo	SD
21)	1147v		Xingtai Diqu	HEB
202)	1152		Hunan 1	HN
44)	1152	7.5	Zhelimu-Ch	NM, Tongliao
46)	1152	7.5	Bayannur	NM, Linhe
55)	1152	5	Dalian 1	LN
1)	1161		CPBS 1	
228)	1161		Guangxi 2	GX, Beihai
113)	1161		Wuxi	JS
160)	1161		Weifang	SD
101)	1170		Shanghai 3	SH
42)	1170		Xilingol-Ch	NM, Xilinhot
158)	1170		Zaozhuang	SD
164)	1170		Qingzhou	SD
209)	1170		Changde	HN
185)	1179		Chutian EBS	HB, Wuhan
41)	1179	10	Wulanchabu-Mo	NM, Jining
109)	1179		Yangzhou	JS
287)	1179		Karamay-Ug	XJ
300)	1188	300	R. Beijing	YN
12)	1188		Xingtai	HEB
27)	1188		Botou	HEB
45)	1188	10	Erdox BS-Ch	NM, Dongsheng
48)	1188	10	Xing'an-Ch	NM, Ulanhot
101)	1197		Shanghai 3	SH
88)	1197	10	Qiqihar	HL
49)	1200		Hailar	NM
300)	1206	300	R. Beijing	YN
77)	1206	100	Yanbian-Ko	JL, Yanji
162)	1206v		Weihai	SD
	(Nom 1170kHz)			
1)	1215		CPBS 2	
1)	1224		CPBS Taiwan 2	FJ
228)	1224		Guangxi 2	GX
19)	1224	1	Langfang	HEB
111)	1224		Zhenjiang	JS
202)	1233		Hunan 1	HN
52)	1242		Liaoning	LN
198)	1242		Macheng	HB
126)	1250		Quzhou 1	ZJ
300)	1251		R. Beijing DS	BJ
101)	1251		Shanghai 6	SH
276)	1251	100	Qinghai-Tb	QH, Xining
179)	1251		Puyang	HEB
52)	1260		Liaoning	LN
265)	1260		Shannan	XZ, Nedong
300)	1269		R. Beijing	YN
28)	1269		Shanxi 1	SX
84)	1269		Dunhua	JL
2)	1269	200	V.O. Strait 1	FJ
9)	1278	100	Hebei 1	HEB

100)	1278		Daxing'anling	HL, Jagdaqi
144)	1278	10	Xiamen	FJ
147)	1278	10	Nanchang	JX
247)	1278	1	Neijiang	SC
1)	1287		CPBS 1	
280)	1287		Ningxia 1	NX
63)	1287		Fuxin	LN
180)	1287		Xuchang	HEN
300)	1296	300	R. Beijing	YN
101)	1296	20	Shanghai 6	SH
58)	1296	10	Benxi	LN
246)	1296		Suining	SC
270)	1296		Xianyang	SN
1)	1305		CPBS 2	
300)	1314	100	R. Beijing	NM
102)	1314	10	Jiangsu 1	JS, Huaiyin
161)	1314		Yantai	SD
189)	1314		Xiangfan	HB
239)	1314		Chongqing 1	SC
300)	1323	100	R. Beijing	XZ
266)	1323		Shaanxi 2	SN, Xi'an
67)	1323		Wafangdian	LN
76)	1323		Baicheng	JL
120)	1323	7.5	Ningbo	ZJ
203)	1323	10	Changsha	HN
168)	1332	10	Henan	HEN
70)	1332		Changchun 2	JL
143)	1332	10	Fuzhou	FJ
273)	1332	7.5	Gannan	GS, Xiahe
300)	1341	100	R. Beijing	GD
86)	1341	100	Heilongjiang 1	HL
115)	1341		Taizhou	JS
165)	1341	1	Qufu	SD
199)	1341	1	Puqi	HB
146)	1350v		Jiangxi 1	JX
44)	1350	50	Zhelimu-Mo	NM, Tongliao
106)	1350	5	Lianyungang	JS
255)	1350	50	Kunming	YN
1)	1359		CPBS 1	
183)	1360		Nanyang	HEN
142)	1368v		Fujian 1	FJ
42)	1368		Xilingol-Mo	NM, Xilinhot
91)	1368		Jixi	HL
197A)	1368		Guangshui	HB
2)	1377	200	V.O. Strait 1	FJ
156)	1377	10	Qingdao	SD
284)	1377	1	Qintongxia	NX
8)	1386	20	Tianjin L	TJ
194)	1386	1	Ezhou	HB
200)	1386	1	Shishou	HB
230)	1386	5	Liuzhou	GX
36)	1395		Nei Menggu-Mo	NM
128)	1395	10	Anhui 1	AH
117)	1404		Zhejiang	ZJ
184)	1404		Hubei 1	HB
59)	1404	10	Dandong 1	LN
102)	1413		Jiangsu 1	JS, Wuxi
285)	1413	100	Xijiang-Ug	XJ
65)	1413		Tieling	LN
89)	1413		Hegang	HL
283)	1413		Wuzhong	NX
101)	1422	20	Shanghai 7	SH
29)	1422	10	Taiyuan	SX
241)	1422		Zigong	SC
43)	1431	200	Hulun Bei'er-Mo	NM, Hailar
131)	1431	1	Huaibei	AH
137)	1431	1	Huangshan	AH
201)	1431		Danjiangkou	HB
228)	1440		Guangxi 1	GX
40)	1440	50	Chifeng-Mo	NM
110)	1440		Nantong	JS
145)	1440v		Putian	FJ
146)	1449	20	Jiangxi 1	JX
105)	1449		Xuzhou 1	JS
159)	1449		Dongying	SD
191)	1449		Shashi	HB
36)	1458	150	Nei Menggu-Mo	NM, Hohhot
142)	1467v	50	Fujian 2	FJ, Fuzhou
13)	1467		Baoding	HEB
45)	1467	10	Erdox BS-Mo	NM, Dongsheng

	kHz	kw	Station	Location
48)	1467	10	Xing'an-Mo	NM, Ulanhot
148)	1467	1	Jingdezhen	JX
1)	1476		CPBS 2	
7)	1476	10	Beijing 4	BJ
117)	1476		Zhejiang	ZJ
68)	1476		Haicheng	LN
81)	1476		Qian Gorlos	JL
188)	1476	10	Huangshi	HB
196)	1476		Laohekou	HB
248)	1476		Leshan	SC
277)	1476	1	Xining	QH
288)	1476		Shihezi	XJ
86)	1485		Heilongjiang 2	HL
154)	1485	1	Shandong 1	SD
271)	1485		Gansu 1	GS
11)	1485		Handan	HEB
31)	1485		Yangquan	SX
49A)	1485	1	Manzhouli	NM
62)	1485	1	Panjin	LN
78)	1485		Gongzhuling	JL
122)	1485		Jiaxing	ZJ
134)	1485	1	Bengbu	AH
151)	1485	1	Jiujiang	JX
174)	1485	1	Pingdingshan	HEN
190)	1485	1	Shiyan	HB
231)	1485	1	Guilin	GX
232)	1485		Wuzhou	GX
238)	1485	1	Chengdu 2	SC
257)	1485		Honghe	YN
259)	1485		Chuxiong	YN
262)	1485		Luliang	YN
289)	1485		Hami-Ch	XJ
291)	1485		Kuitun-Ch	XJ
285)	1494	20	Xinjiang-Ch	XJ
50)	1494		Tongliao	NM
71)	1494		Jilin-shi 2	JL
132)	1494		Wuhu	AH
140)	1494		Fuyang	AH
117)	1503		Zhejiang	ZJ
41)	1503	10	Wulanchabu-Ch	NM, Jining
64)	1503	1	Liaoyang	LN
205)	1503		Xiangtan	HN
1)	1512		CPBS Taiwan 1	FJ
40)	1512		Chifeng-Ch	NM
300)	1521	500	R. Beijing	XJ
266)	1521		Shaanxi 1	SN
80)	1521	1	Taonan	JL
192)	1521 (r 1485kHz)		Yichang	HB
257)	1521		Honghe	YN
28)	1530	10	Shanxi 2	SX, Taiyuan
69)	1530		Jilin	JL
118)	1530	10	Zhejiang EBS	ZJ, Hangzhou
206)	1530	10	Hengyang	HN
154)	1548	100	Shandong 1	SD
207)	1548		Shaoyang	HN
18)	1557		Cangzhou	HEB
26)	1557	1	Nangong	HEB
14)	1566		Zhangjiakou	HEB
24)	1566	1	Hengshui	HEB
33)	1566		Jincheng	SX
167)	1566		Liaocheng	SD
197)	1566		Xiaogan	HB
274)	1566		Pingliang	GS
121)	1575	7.5	Wenzhou	ZJ
86)	1584		Helongjiang 2	HL
212)	1584		Guangdong E2	GD, Guangzhou
15)	1584	1	Chengde	HEB
30)	1584		Datong	SX
32)	1584	1	Changzhi	SX
35)	1584		Yuci	SX
79)	1584		Meihekou	JL
123)	1584		Huzhou	ZJ
135)	1584	1	Ma'anshan	AH
136)	1584	1	Anqing	AH
152)	1584		Yingtan	JX
195)	1584	1	Suizhou	HB
86)	1593		Heilongjiang 1	HL
285)	1593	100	Xinjiang-Mo	XJ, Urumqi
97)	1593		Suihua	HL
208)	1593		Yueyang	HN
102)	1602	1	Jiangsu 1	JS, Hongze
193)	1602		Jingmen	HB

Shortwaves: *) inactive.

	kHz	kw	Station	Location	Times
142)	2340		Fujian 1	Fuzhou	2050-2315 1130-1700
121)	2415	1	Wenzhou 1		-1230
146)	2445		Jiangxi 1	Nanchang	as 729kHz
254)	2460		Yunnan G	Kunming	as 990kHz
117)	2475	20	Zhejiang	Hangzhou	as 810kHz
285)	2560	15	Xinjiang-Ch	Urumqi	as 738kHz
2)	3200		Haixia 2		2155-2230
273)	3210*	1	Gannan	Xiahe	alternative
1)	3220*	50	CPBS 1		Winter only
251)	3260	10	Guizhou 1	Guiyang	as 927kHz
4)	3280		V.O. Pujiang	Shanghai	0955-1545
1)	3815*		CPBS	Taiwan 1	
43)	3900		Hulunbei'er-Ch	Hailar	as 603kHz
2)	3900		Haixia 1		2055-2335
254)	3910v*		Yunnan G	Kunming	
36)	3930	50	Nei Menggu-Mo	Hohhot	as 1458kHz
184)	3940v	10	Hubei 1	Wuhan	as 774kHz
276)	3950		Qinghai 1	Xining	as 666kHz
285)	3960*	50	Xinjiang-Ch	Urumqi	Nov-Apr only
36)	3970	50	Nei Menggu-Ch	Hohhot	as 675kHz
285)	3990*	50	Xinjiang-Ug	Urumqi	Nov-Apr only
264)	4035		Xizang-Tb	Lhasa	as 594kHz
1)	4190*		CPBS Minor.		Winter only
285)	4220	15	Xinjiang-Mo	Urumqi	as 909kHz
285)	4330		Xinjiang-Kz	Urumqi	as 1107kHz
1)	4460		CPBS 1		2000-2200 1400-1735
285)	4500		Xinjiang-Ch	Urumqi	as 738kHz
36)	4525*	50	Nei Menggu-Mo	Hohhot	
285)	4735		Xinjiang-Ug	Urumqi	as 558kHz
264)	4750		Xizang-Ch	Lhasa	as 846kHz
43)	4750		Hulunbei'er-Mo	Hailar	as 1431kHz
254)	4760		Yunnan G	Kunming	as 990kHz
117)	4785	20	Zhejiang	Hangzhou	-1130
1)	4800		CPBS 2/Minor.		2200-0200 1000-1600
301)	4830*		China Huayi BC	Fuzhou	
86)	4840		Heilongjiang 1	Harbin	as 621kHz
271)	4865		Gansu 1	Lanzhou	as 873kHz
3)	4875	50	V.O. Jinling	Nanjing	1155-1600
2)	4900		Haixia 2		2155-0030
1)	4905	10	CPBS 2		2100-2330
228)	4915		Guangxi 1	Nanning	as 792kHz
257)	4930v		Honghe	Gejiu	irregular
2)	4940		Haixia 1		2055-0010 1230-1720
4)	4950		V.O. Pujiang	Shanghai	0955-1545
42)	4950*		Xilingol-Mo	Xilinhot	as 1368kHz
285)	4970		Xinjiang-Kz	Urumqi	as 1107kHz
142)	4975		Fujian 1	Fuzhou	as 882kHz
285)	4980		Xinjiang-Mo	Urumqi	as 909kHz
202)	4990	10	Hunan 1	Changsha	as 738kHz
228)	5010		Guangxi 2	Nanning	as 846kHz
146)	5020		Jiangxi 1	Nanchang	as 729kHz
142)	5040		Fujian 1	Fuzhou	2315-1130
5)	5050		Guangxi FBS	Nanning	2300-0100 1000-1600
285)	5060		Xinjiang-Mo	Urumqi	as 909kHz
1)	5090*		CPBS Taiwan 2		Winter only
1)	5125*		CPBS Taiwan 1		Winter only
1)	5163		CPBS 2		2100-2340
264)	5240		Xizang-Ch	Lhasa	as 846kHz
1)	5320		CPBS 1		2000-2310 1330-1735
1)	5420		CPBS Minor.		2130-2230 1230-1300
285)	5440		Xinjiang-Kz	Urumqi	as 1107kHz
285)	5800		Xinjiang-Ug	Urumqi	as 558kHz
1)	5880		CPBS 1		2000-0100 1100-1735
236)	5900		Sichuan 2	Chengdu	as 1089kHz
1)	5915		CPBS 1		2000-0100

228)	5920		Guangxi 3	Nanning	2300-0135
					1000-1235
264)	5935v	50	Xizang-Ch	Lhasa	as 846kHz
86)	5950	50	Heilongjiang 2	Harbin	as 873kHz
264)	5950	50	Xizang-Tb	Lhasa	as 594kHz
254)	5960*	50	Yunnan E	Kunming	0030-0400
					1100-1200
273)	5970	1	Gannan	Xiahe	as 1332kHz
36)	5980	50	Nei Menggu-Ch	Hohhot	as 675kHz
264)	5995	50	Xizang-Tb	Lhasa	as 594kHz
2)	6000	50	Haixia 2		1255-1500
1)	6015*	50	CPBS Taiwan 1		Winter only
47)	6025v	15	Alashan	Bayanhot	irregular
6)	6035	50	Yunnan BS	Kunming	2230-0130
					0900-1100
146)	6040	50	Jiangxi 1		2100-0530
					0950-1505
36)	6045	50	Nei Menggu-Ch	Hohhot	as 675kHz
236)	6060v	50	Sichuan 1	Xichang	as 1116kHz
1)	6095	50	CPBS Taiwan 2		0955-1805
285)	6100	50	Xinjiang-Ch	Urumqi	as 738kHz
1)	6110		CPBS Minor.		2130-2230
					1200-1300
2)	6115	50	Haixia 1		2055-0010
					0855-1720
1)	6125	50	CPBS 1		2000-0100
					1100-1735
271)	6155	15	Gansu 1	Lanzhou	as 873kHz
2)	6170*	15	Haixia 1		
266)	6176	15	Shaanxi 1	Xi'an	0100-0635
301)	6185		China Huayi BC	Fuzhou	0955-1800
36)	6195	50	Nei Menggu-Mo	Hohhot	as 1458kHz
276)	6260		Qinghai 1	Xining	as 666kHz
276)	6500	50	Qinghai-Tb	Xining	as 1251kHz
1)	6750		CPBS 1		2000-0100
					1100-1735
1)	6790*		CPBS Taiwan 2		Winter Only
1)	6840		CPBS 1		2000-1735
1)	6890		CPBS 2		2100-2330
					1100-1600
254)	6940v		Yunnan M	Kunming	2225-0045
					0355-0545
	(nom 6937kHz)				1055-1500
36)	6974	50	Nei Menggu-Mo	Hohhot	as 1458kHz
36)	7105	50	Nei Menggu-Ch	Hohhot	as 675kHz
264)	7110	50	Xizang-Tb	Lhasa	as 594kHz
4)	7115		V.O. Pujiang	Shanghai	0955-1545
264)	7170	50	Xizang-Ch	Lhasa	as 846kHz
285)	7195	15	Xinjiang-Ug	Urumqi	May-Oct only
254)	7210v	15	Yunnan G	Kunming	as 990kHz
3)	7215	50	V.O. Jinling	Nanjing	0055-0300
236)	7225	15	Sichuan 1	Chengdu	as 1116kHz
36)	7270	50	Nei Menggu-Mo	Hohhot	as 1458kHz
251)	7275	7.5	Guizhou 1	Guiyang	as 927kHz
2)	7280	50	Haixia 1		0855-1720
86)	7350		Heilongjiang 1		-1130
285)	7385		Xinjiang-Ch	Urumqi	May-Oct only
1)	7440		CPBS 2		1100-1600
1)	7504		CPBS 1		2000-1735
1)	7516		CPBS 2		2100-2330
					1100-1600
1)	7620*		CPBS Taiwan 1		Winter only
1)	7770		CPBS 2		2100-2340
					1500-1600
1)	7935		CPBS 1		2000-2400
					1410-1735
1)	8566		CPBS Minor.		0430-0530
					1000-1030
					1200-1230
1)	9064		CPBS 2		2100-1600
1)	9080		CPBS 1		2000-2220
1)	9170		CPBS Taiwan 2		2055-0105
1)	9290		CPBS 1		2000-1735
1)	9380*		CPBS Taiwan 1		Winter only
1)	9455		CPBS Taiwan 1		0955-0005
264)	9490*		Xizang-Ch	Lhasa	alternative
2)	9505	50	Haixia 2		0030-0600
					1255-1650
36)	9520	50	Nei Menggu-Ch	Hohhot	0200-0545
285)	9560		Xinjiang-Ch	Urumqi	0430-0730
285)	9595	50	Xinjiang-Ug	Urumqi	0430-0730
36)	9750	50	Nei Menggu-Mo	Hohhot	0130-0550
1)	9755	50	CPBS 2		2340-0200
					1000-1600
1)	9775		CPBS 2/Minor.		2130-0200
					0430-0530
					1000-1600
276)	9780		Qinghai 1	Xining	0355-0630
1)	9800		CPBS 1		1100-1735
1)	9920		CPBS Minor.		0430-0530
					1000-1030
1)	10260		CPBS 2/Minor.		2100-2400
					0030-0130
					0600-0700
					1100-1500
1)	11000		CPBS Taiwan 2		2055-0105
					0355-1805
1)	11040		CPBS 2		2330-1100
1)	11100		CPBS Taiwan 1		0055-0610,
					0955-0005
1)	11330		CPBS 1		2000-1735
1)	11375		CPBS Minor.		0000-0130
					0530-0700
					1030-1200
					1300-1500
2)	11590		Haixia 1		0855-1230
1)	11610		CPBS 2		2100-1600
1)	11630		CPBS 2/Minor.		2100-2340
					1030-1240
					1300-1600etc
36)	11705*	50	Nei Menggu-Ch	Hohhot	0200-0545
1)	11740	120	CPBS 2		2100-2400
					1430-1600
36)	11865*	50	Nei Menggu-Mo	Hohhot	0130-0550
1)	11935		CPBS Taiwan 1		0055-0610
					0955-0005
264)	11950v*	15	Xizang-Tb	Lhasa	
1)	12120		CPBS 1		2200-1400
1)	15390	50	CPBS 1		2220-1735
1)	15500		CPBS 2		2340-1500
1)	15550		CPBS 1		2310-1330
1)	15670		CPBS Minor.		0000-0130
					0530-0700
					1030-1200
					1300-1500
1)	15710		CPBS Taiwan 1		0055-0610
					0955-0005
1)	15880		CPBS Taiwan 2		0355-1805
1)	17605		CPBS 1		0000-1410
1)	17700	50	CPBS 2		0000-1430

Addresses and other information:

1) CENTRAL PEOPLE'S BROADCASTING STATION (CPBS)

ADDR: 2 Fuxingmenwai Dajie, Beijing 100866. **Te:** Directors Office: +86 (1) 8012435 ext 2956. Audience Relations Dept: +86 (1) 863397 — General Dir: Yang Zhengquan.

1st Prgr. 2000-0235 (exc. Tues 0600-0855).
2nd Prgr. 2100-1600 (exc. wed/Fri 0600-0955).
3rd Prgr. 2155-1605 on FM only.
Minorities Prgr on SW — +) relayed by regional sts.

Kazakh		
0100-0130	XJ	15670, 11375, 10260
0630-0700	XJ	15670, 11375, 10260
1130-1200	XJ	15670, 11375, 10260
1400-1430+	XJ	15670, 11375, 10260
Korean		
2130-2200+	JL/HL	9775, 6110, 5420
0430-0500	JL/HL	9920, 9775, 8566
1000-1030	JL/HL	9920, 9775, 8566
1200-1230	JL/HL	9775, 8566, 9775
Mongolian		
2200-2230+	NM	9775, 6110, 5420, 4800
0500-0530+	NM	9920, 9775, 8566, 4800
1230-1300+	NM	9775, 6110, 5420, 4800
1430-1500+	XJ	15670, 11375, 10260
Tibetan		
0000-0030	XZ/QH	15670, 11630, 11375
0530-0600	XZ/QH	15670, 11630, 11375
1030-1100	XZ/QH	15670, 11630, 11375
1300-1330+	XZ/QH	15670, 11630, 11375

Uighur

0030-0100	XJ	15670, 11375, 10260
0600-0630	XJ	15670, 11375, 10260
1100-1130+	XJ	15670, 11375, 10260
1330-1400+	XJ	15670, 11375, 10260

Taiwan Sce:
1st Prgr. in Chinese, Amoy and Hakka: 0055-0610, 0955-0005.
2nd Prgr. in Chinese, Amoy and Hakka: 2055-0105, 0355-1805 (exc. Tues 0605-0955).

BROADCASTS FOR TAIWAN
2) Voice of the Strait (Haixia zhi Sheng), Xindian, Fuzhou, or P.O. Box 187, Fuzhou, Fujian 350012, or P.O. Box 33954, Central, Hong Kong. Operated by the People's Liberation Army of China.
1st Prgr. in Chinese and Amoy: 2055-0010, 0855-1720.
2nd Prgr. in Chinese and Amoy: 2155-0600, 1255-1650.
3) Voice of Jinling (Jinling zhi Sheng), P.O. Box 268, Nanjing, Jiangsu 210002 —0055-0300 on 7215kHz, 1155-1600 on 4875kHz. Also relayed by Jiangsu PBS on 585kHz at 0200-0300 and on 702kHz at 1500-1600.
4) Voice of Pujiang (Pujiang zhi Sheng), 2 Beijing Donglu, Shanghai, or P.O. Box 3064, Shanghai 200002 — 0955-1545. Shanghai dialect: Tues 1130; Thurs 1015, 1130, 1215, 1415; Sat 1130.

BROADCASTS FOR VIETNAM
5) Guangxi Foreign Broadc. St, 12 Minzu Dadao, Nanning, Guangxi 530022 — 2300-0100, 1000-1200, 1400-1600 in Vietnamese, 1200-1400 in Cantonese.
6) Yunnan Broadc. St, Renmin Xilu, Kunming, Yunnan 650031 — 2230-2320, 0030-0120, 0900-0950, 1000-1050 in Vietnamese, 2320-2330, 0120-0130, 0950-1000, 1050-1100 in Ch. Rel. CRI Vietnamese Prgr. 2330-0030.

BEIJING MUNICIPALITY
7) 14 Jianguomenwai Dajie, Beijing 100022.
1st Prgr. on 828kHz 2115-0530(Thurs 0510), 0740-1530.
2nd Prgr. on 927kHz 2225-0500, 0745-1610.
3rd Prgr. (Economic Prgr.) on 1026kHz 2220-0615(Thurs 0530), 0955-1500.
4th Prgr. (Educational Prgr.) on 1476kHz 2128-0600, 0828-1500.
5th Prgr: FM stereo on 94.5MHz 0325(Sun 0225)-1530.

TIANJIN MUNICIPALITY
8) 143 Weijin Lu, Heping Qu, Tianjin 300070.
News Prgr. on 909kHz 2135-0545, 0725-1545.
Literary Prgr. on 1386kHz 2255-0620, 0755-1430.
Educational Prgr. on 954kHz 2225-0230, 1025-1410.
FM stereo Music Prgr. on 99.0MHz 2155-0600, 0900-1500. Rel. CPBS 3: 2200-0400, 0800(Sun/Tue 0900)-1000.
8A) 143 Weijin Lu, Heping Qu, Tianjin 300070 — 2155-0705, 0855-1510.

HEBEI PROVINCE
9) 4 Yuhua Zhonglu, Shijiazhuang, Hebei 050012.
1st Prgr. on MW/FM 2120-0050, 0150-0630, 0830-1530(Sun 1555).
2nd Prgr: FM stereo on 89.1MHz 0355-0600, 0855-1500.
Rel. CPBS 3: 0900-0930.
10) 33 Tiyu Zhongdajie, Shijiazhuang, Hebei 050021 — 2105-0035, 0320-0550, 0920-1440.
11) 246 Renmin Lu, Handan, Hebei 056002 — 2155-?, ?-1305
12) 15 Yejin Lu, Xingtai, Hebei 054000 — 2215-2400, 0255-0535, 1005-?.
13) 30 Wusi Donglu, Baoding, Hebei 071000 — 2220-0125, 0305-0600, 0930-1345.
14) 17 Jianguo Lu, Zhangjiakou, Hebei 075000 — 2220-0025, 0355(Sun 0155)-0525, 0925-1335.
15) 17 Xiaotonggou Nanpo, Chengde, Hebei 067000.
16) Wenhua Lu, Tangshan, Hebei 063000 — 2155-0055 — 0345-0625, 0930-1445.
17) 23 Wenhua Beilu, Qinhuangdao, Hebei 066000 — 2155-0100, 0325-0605, 0925-1400.
18) 229 Jiefang Zhonglu, Cangzhou, Hebei 061001 — 2225-0040, 0320-0530, 0955-1300.
19) 8 Yongfeng Dao, Langfang, Hebei 102800 — 2210-0115, 0340-0610, 0955-1230.
20) 33 Congtai Lu, Handan, Hebei 056002.
21) Nanxiaowang, Xingtai, Hebei 054000 — ?-1245.
22) 25 Wusi Zhonglu, Baoding, Hebei 071051 — 2210-0125, 0355-0540, 0950-1340.
23) 120 Xiaotonggou Nanpo, Chengde, Hebei 067000 — 2200-0020, 0355-0545, 0955-1315.
24) 55 Hongqi Dajie, Hengshui, Hebei 053000 — 2210-0020, 0345-0610, 1015-1315.
25) Xinghua Lu, Xinji, Hebei 052360 — 2225-2355, 0255-0500, 1025-1250.
26) Muchang Jie, Nangong, Hebei 051830 — 2225-0045, 1005-1205.
27) Xiguan Xijie, Botou, Hebei 062150 — 2225-2355, 0345-0450, 1025-1235.

SHANXI PROVINCE
28) 52 Yingze Dajie, Taiyuan, Shanxi 030001.
1st Prgr. on MW/FM 2125-0050, 0255-0650, 0905-1535(SS 1605).
2nd Prgr. on 1530kHz 2240-0240, 0655-1505.
FM stereo: on 98.2MHz 0050-0150.
29) Binhe Xilu, Taiyuan, Shanxi 030024 — on 1422kHz W 2150-0135, Sun 0020-0630, W 0335-0630, D 0900-1440. FM stereo: on 104.4MHz 0700-0800.
30) 2 Xinjian Beilu, Datong, Shanxi 037006 — W 2145-0010, Sun 2215-0205, W 0245-0540, Sun 0350-0630, D0950-1415.
31) 10 Bei Dajie, Yangquan, Shanxi 045000 — 2150-2400, 0300-0535, 0955-1330.
32) 124 Yingxiong Lu, Changzhi, Shanxi 046000 — 2155-0050, 0355-0630, 0945-1335.
33) Xiaozhuang, Jincheng, Shanxi 048000 — 2200-?, 1025-1330.
34) 36 Yantong Donglu, Datong, Shanxi 037044 — 2210-0005, 0345-0555, 0945-1235.
35) 37 Taishan Miao Jie, Yuci, Shanxi 030600 — 2220-0015, 0355-0605, 0955-1400.

NEI MENGGU AUTONOMOUS REGION
36) 19 Xinhua Dajie, Huhhot, Nei Menggu 010058.
Ch Prgr. on MW/SW/FM 2140-0545, 0950-1500(Sat/Sun 1520).
Mo Prgr. on MW/SW/FM 2140-0550, 0950-1500(Sun 1510).
FM stereo: on 99.8MHz 0325-0530.
37) 56 Gongyuan Xilu, Huhhot, Nei Menggu 010035 — 2200-0130, 0300-0600, 0830-1400.
38) 13 Wulan Dao, Baotou, Nei Menggu 014010 — 2145-0500, 0930-1500.
39) Wuhai, Nei Menggu 016000 — W 2225-0020, Sun 0025-0510, W 0325-0515, D 1025-1305(Sat/Sun 1405).
40) Gangtie Jie, Hongshan Qu, Chifeng, Nei Menggu 024001.
Ch Prgr. on MW/FM 2120-0030, 0255-0530, 1000-1500.
Mo Prgr. on MW/FM 2120-?, 0320-0550, 0955-1400.
41) Jining, Nei Menggu 012000.
Ch Prgr. on 1503kHz 2150-0100, 0340-0540, 1005-1500.
Mo Prgr. on 1179kHz 2125-0100, 0330-0600, 1010-1505.
42) Xilin Dajie, Xilinhot, Nei Menggu 026000.
Ch Prgr. on 1170kHz.
Mo Prgr. on 1368/4950kHz 2220-0150, 0255-0500, 1005-1305.
43) Shengli 3 Lu, Hailar, Nei Menggu 021008.
Ch Prgr. on 603/3900kHz 2150-0030, 0355-0550, 0925-1430.
Mo Prgr. on 1431/4750kHz 2150-2400, 0355-0600, 0925-1500.
44) 28 Huolinhe Dajie, Tongliao, Nei Menggu 028001.
Ch Prgr. on 1152kHz 2140-0005, 0320-0530, 0955-1330.
Mo Prgr. on 1350kHz 2120-2335, 0245-0530, 1055-1330.
45) Erdox BS, Dongsheng, Nei Menggu 017000.
Ch Prgr. on 1188/792kHz 2220-0020, 0320-0600, 1000-1320.
Mo Prgr. on 1467kHz.
46) Linhe, Nei Menggu 015000.
47) Bayanhot Zhen, Alashan Zuoqi, Nei Menggu 750306. 2220-?, ?-1335.
48) Aiguo Jie, Ulanhot, Nei Menggu 027400.
Ch Prgr. on 1188kHz 2155-2400, 0955-1400.
Mo Prgr. on 1467kHz.
49) Gongren Xincun, Hailar, Nei Menggu 021000.
49A) Dianshi Jie, Manzhouli, Nei Menggu 021400.
50) Mingren Dajie, Tongliao, Nei Menggu 028001 — 2145-2355.
51) Sujiaqu, Dongsheng, Nei Menggu 017000.

LIAONING PROVINCE
52) 2 Guangrong Jie, Heping Qu, Shenyang, Liaoning 110001 — on MW/FM 2050-0610(Tue 0510), 0825(Tue 0855)-1500. FM stereo: on 104.2MHz 2125-0100, 0155-0500, 0930-1305. Rel. CPBS 3: 2300-0100, 0200-0500, 1100-1300.

53) 2 Guangrong Jie, Heping Qu, Shenyang, Liaoning 110001. 2125-0100, 0225-0540, ?-1500.
54) 9 Nan Dajie, Heping Qu, Shenyang, Liaoning 110003 — on 792kHz 2100-0500, 0825-1430. EC Prgr. on 882kHz 2120-0015, 0315-0820, 0945-1300.
55) 162 Minquan Jie, Shahekou Qu, Dalian, Liaoning 116022.
1st Prgr. on 882/1152kHz W2050-0135, Sun2050-0810, W0315-0730, D0855-1415.
2nd Prgr: FM stereo on 102.4MHz W 2205-0010, Sun 0125-0830, W 0355-0530, W 0855-1500. Rel. R. Beijing DS: W1200-1500.
56) 34 Wudao Jie, Tiedong Qu, Anshan, Liaoning 114001.
1st Prgr. on 954kHz 2100-0030, 0255-0525, 0900-1430.
2nd Prgr. on 96.0MHz 2200-2300, 0400-0500, 1100-1200.
57) 5 Yingbin Jie, Xinfu Qu, Fushun, Liaoning 113008 — 2105-0155, 0345-0640, 0915-1405.
58) Yonghe Jie, Mingshan Qu, Benxi, Liaoning 117000 — 2130-0100, 0325-0535, 0810-1400.
59) 1 Shanshang Jie, Zhenxing Qu, Dandong, Liaoning 118000.
1st Prgr. on 1404kHz 2120-0510, 0850-1300.
2nd Prgr. on 891kHz/106.3MHz 2155-2300, 0455-0600, 0955-1200.
60) 4-3 Beijing Lu, Jinzhou, Liaoning 121001 — 2130-0030, 0315-0540, 0915-1400.
61) 20 Jiankang Li, Bohai Dajie, Yingkou, Liaoning 115000 — 2120-0120, 0325-0520, 0810-1405.
62) Liaohe Lu, Shuangtai Qu, Panjin, Liaoning 124000 — 2115-2350, 0355-0550, 0855-1335.
63) Zhonghua Lu, Haizhou Qu, Fuxin, Liaoning 123000 — on 1287kHz in Ch and Mo. 2110-0005, 0325-0545, 0915-1405.
64) 59 Qingnian Dajie, Liaoyang, Liaoning 111000 — 2130-0005, 0150-0515, 0845-1305.
65) Gongren Jie, Yinzhou Qu, Tieling, Liaoning 112000 — 2145-0055, 0300-0600, 0905-1245.
66) 3-11 Xinhua Lu, Chaoyang, Liaoning 122000 — 2145-2355, 0345-0610, 1015-1445.
67) Delin Jie, Wafangdian, Liaoning 116300 — 2125-0120, 0315-0515, 0925-1305.
68) 15 Huancheng Xilu, Haicheng, Liaoning 114200.

JILIN PROVINCE

69) 2 Xi'an Dalu, Changchun, Jilin 130051 — on MW/FM 2050-0600, 0725(Tue 0820)-1500. FM stereo: on 104.3MHz. Incl. rel. CPBS 3.
70) 63 Jiefang Dalu, Changchun, Jilin 130041.
1st Prgr. on 873kHz 2050-0100, 0300-0730(Tue 0530), 0925-1400.
2nd Prgr. on 1332kHz 2155-0200, 0300-1100.
71) 2 Jiangnan Dajie, Jilin-shi, Jilin 132011.
1st Prgr. on 927kHz 2110-0015, 0340-0615, 0835-1430.
2nd Prgr. on 1494kHz 0125-0430, 0725-1130.
72) 40 Nan Xinhua Dajie, Siping, Jilin 136000 — 2135-0030, 0300-0600, 0855-1300.
73) 15 Hebin Lu, Liaoyuan. Jilin 136200.
74) 24 Cuiquan Lu, Longquan Jie, Tonghua, Jilin 134001 — 2100-0040, 0335-0540, 0825-1405.
75) 36 Hunjiang Dajie, Hunjiang, Jilin 134302.
76) 99 Zhongxing Dong Dalu, Baicheng, Jilin 137000 — 2115-0100, 0945-1430.
77) 166 Juzi Jie, Yanji, Jilin 133000.
Ch Prgr. on 1053kHz 2135-0040, 0305-0525, 0855-1300.
Ko Prgr. on 1206kHz/106.7MHz 2120-0010, 0310-0520, 0900-1320.
78) 1 Dong Toudao Jie, Gongzhuling, Jilin 136100.
79) Henan Jie, Meihekou, Jilin 135000.
80) 29 Gushu Nanjie, Taonan, Jilin 137100.
81) 15 Qingzhen Hutong, Minzhu Jie, Qian Gorlos, Jilin 131100 — on 1476kHz/88.2MHz in Ch and Mo. 2125-2330, 0325-0500, 0955-1230.
82) 1 Ju, Guangming Jie, Yanji, Jilin 133000 — on 576kHz/93.5MHz in Ch and Ko. 2055-2330, 0325-0530, 0925-1230.
83) 12 Xiangshang Jie, Tumen, Jilin 133100 — on 1143kHz in Ch and Ko. 2125-2330, 0325-0505, 0850-1330.
84) 1 Xinhua Xilu, Dunhua, Jilin 133700 — on 1269kHz/93.6MHz in Ch and Ko. 2140-2330, 0320-0525, 0925-1245.
85) Xin'an Jie, Hunchun, Jilin 133300 — on 927kHz in Ch and Ko. 2120-2350, 0300-0520, 0910-1230.

HEILONGJIANG PROVINCE

86) 115 Zhongshan Lu, Nangang Qu, Harbin, Heilongjiang 150001.
1st Prgr. on MW/SW/FM 2055-0550(Fri 0520), 0855-1500. Ko: W 0230-0330.
2nd Prgr. on MW/SW/FM 2125-2400, 0355-1130. Ko: 2200-2400, 0400-0500(Sun 0600), 0900-1100.
FM stereo: on 100.0MHz 0400-0500, 1000-1100.
87) 113 Songhuajiang Jie, Harbin, Heilongjiang 150006.
1st Prgr. on 837kHz 2055-0520, 0855-1445.
2nd Prgr. on 972kHz D 2105-2315, W 0345-0750(Tue/Sat 0500), W 1015-1200(exc. Tues).
88) 99 Yong'an Dajie, Qiqihar, Heilongjiang 161005 — 2100-0015, 0325-0615, 0830-1330.
89) Bama Lu, Xiangyang Qu, Hegang, Heilongjiang 154100 — 2055-2320, 0325-?, 0840-1430.
90) Xinxing Dajie, Shuangyashan, Heilongjiang 155100 — 2050-2400, 0325-0500, 0850-1415.
91) 11 Diantai Lu, Jixi, Heilongjiang 158100 — 2130-2345, 0300-0650, 0850-1350.
92) Jia 1, Dongfeng Lu, Daqing, Heilongjiang 163311 — 2125-0105, 0250-?, 0925-1430.
93) Tonghe Lu, Yichun Qu, Yichun, Heilongjiang 153000 — 2050-0020, 0315-0515, 0925-1400.
94) Da'an Lu, Xi'an Qu, Mudanjiang, Heilongjiang 157000 — 2055-0030, 0300-0620, 0845-1330.
95) 5 Guangfu Lu, Jiamusi, Heilongjiang 154002 — on 666kHz/88.0/101.4MHz 2050-0145, 0255-0515, 0825-1445. FM stereo: on 101.0MHz 0355-?, 0955-?.
96) Shanhu Dajie, Taoshan Qu, Qitaihe, Heilongjiang 154600.
97) 83 Bei Ersi Lu, Suihua, Heilongjiang 152054.
98) Xing'an Jie, Heihe, Heilongjiang 164300.
99) 4 Wenzheng Jie, Dongli Qu, Harbin, Heilongjiang 150040 — 2350-0020, 1130-1200.
100) 47 Xina'an Jie, Jagdaqi Zhen, Heilongjiang 165000 — 2120-0155, 0255-0535, 0825-1430.

SHANGHAI MUNICIPALITY

101) 2 Beijing Donglu, Shanghai 200002.
1st Prgr. on 990kHz 2058-1605.
2nd Prgr. (Economic Prgr.) on 792kHz 2145-1530 (exc. Thurs 0600-0845).
3rd Prgr: Literary Ch and Shanghai dialect on 1197/1170kHz 2128-1500 (exc. Wed 0500-0755).
4th Prgr: FM stereo on 103.7MHz 2200-1600 (exc. Fri 0600-0855).
5th Prgr: FM stereo in Ch and Shanghai dialect on 101.7MHz 0255-1500.
6th Prgr. (Educational Prgr.) on 1296/1251kHz. W: 2128-2400, 0358-0530, 0723-0930, 1028-1400. Sun: 2128-0230, 0428-0900, 0928-1330.
7th Prgr. (Educational Prgr.) on 1422kHz. W: 2128-0330, 0353-0500, 0828-1100, 1153-1400. Sun: 2128-0215, 0353-0845, 0958-1445.
8th Prgr. (**English Prgr. "R. Shanghai"**) on 105.7MHz 2200-0100, 1030-1430. Rel. R. Beijing DS: 2300-2400, 1300-1400.

JIANGSU PROVINCE

102) 8 Xi Citang Xiang, Zhongshan Donglu, Nanjing, Jiangsu 210002.
1st Prgr. on MW/FM 2055-0545, 0745-1455. Rel. V.O. Pujiang on 702kHz: 1500-1600.
2rd Prgr: FM stereo on 95.3MHz 0400-0600, 1000-1300. Rel. CPBS3: D0400-0600, MF1200-1300.
102A) 8 Xi Citang Xiang, Zhongshan Donglu, Nanjing, Jiangsu 210002 — 2105-0030, 0055-0510, 0715-1300. Rel. R. Beijing DS: 1300-1500.
103) 50 Yanling Xiang, Nanjing, Jiangsu 210002 — 2120-0505, 0757-1500.
104) 50 Yanling Xiang, Nanjing, Jiangsu 210002 — ?-1500.
105) 223 Zhongshan Nanlu, Xuzhou, Jiangsu 221003.
1st Prgr. on 1449kHz 2125-0040, 0255-0540, 0900-1455.
2nd Prgr. FM stereo on 93.0/99.2/106.0MHz 2155-2350, 0320-0520, 0955-1405.
106) 6 Jiefanf Xilu, Xinpu, Lianyungang, Jiangsu 222003 — 2125-0005, 0210-0515, 0855-1500.
107) 6 Dazhi Lu, Huayian, Jiangsu 223001 — 2130-?, 0925-1335.
108) 4 Shengyuan Lu, Yancheng, Jiangsu 224001 — 2125-2400, ?-1255.
109) 8 Meiling Donglu, Yangzhou, Jiangsu 225002 — 2140-2400, 0150-0530, 0825-1345.
110) 120 Renmin Zhonglu, Nantong, Jiangsu 226001 — on 1440kHz MF 2135-0030, SS 2135-0520, MF 0155-0520, D 0755-1430. FM stereo: on 99.0MHz. 0300-0515, 1000-1200.

111) 55 Guanyinqiao Xiang, Zhenjiang, Jiangsu 212001 — on 1224kHz 2120-0010, 0215-0520, 0825-1410. FM stereo: on 100.2MHz. 0430-0530, 1030-1130.
112) 30 Damiao Nong, Changzhou, Jiangsu 213003.
1st Prgr. on 846kHz W 2135-0030, Sun 2135-0505, W 0150-0505, D 0750-1400.
2nd Prgr. on 1143kHz/101.4MHz 2200-0200, 0300-0500, 1000-1330. Rel. CPBS 3: 0000-0200.
113) 88 Zhongshan Nanlu, Wuxi, Jiangsu 214002 — on 1161kHz 2120-0015, 0155-0545, 0905-1425. FM stereo: on 99.6MHz.
114) 4 Gongyuan Lu, Suzhou, Jiangsu 215006 — on 1080kHz in Ch and Suzhou dialect. 2130-0010, 0145-0600, 0845-1250. FM stereo: on 98.4MHz W 0258-0555, Sun 0058-0725, D 0858-1140.
115) 20 Qingnian Lu, Taizhou, Jiangsu 225300 — 2155-?, 0900-1300.
116) 43 Shuyuan Nong, Yushan Zhen, Changshu, Jiangsu 215500.
1st Prgr. on 1116kHz 2125-2350, 0155-0555, 0825-1340.
3rd Prgr: FM stereo on 102.7MHz 0340-0450, 0855-1110.

ZHEJIANG PROVINCE

117) 17 Putisi Lu, Hangzhou, Zhejiang 310006 — 2133-0500(Sun 0510), 0758-1510(Sun 1505).
118) 17 Putisi Lu, Hangzhou, Zhejiang 310006 — AM/FM Stereo on 1530kHz/94.8MHz: W 2220-2400, Sun 2320-0100, D 0330-0500, D 0930-1200(Mon/Tues 1300). Rel. CPBS 3: 0000-0330, 0500-0925, 1200-?(on 94.8MHz only).
119) 46 Moganshan Lu, Hangzhou, Zhejiang 310005 — on 954kHz 2130-0445, 0830-1100(Sun 1110). FM stereo: on 102.1MHz 0400-0500, 1000-1100, 1300-1400.
120) 3 Guangji Xiang, Jiefang Nanlu, Ningbo, Zhejiang 315000 — W 2135-0015, Sun 2135-0710, W 0145-0515, D 0915-1405.
121) 19 Xianxue Qianlu, Wenzhou, Zhejiang 325000.
1st Prgr. on 1575/2415kHz in Ch and Wenzhou dialect. 2150-0100, 0220-0450(Sun 0650), 0830-1325.
3rd Prgr. FM stereo on 93.5MHz 0330-0440, 1000-1215.
122) Daoqian Jie, Jiaxing, Zhejiang 314007 — on 1485/657kHz 2150-0010, 0225-0510, 0825-1305. FM stereo: on 103.1MHz.
123) 40 Dongmen Mao'anqian, Huzhou, Zhejiang 313000 — 2210-?, 0915-1345.
124) 2 Banbi Nong, Fushan Zhijie, Shaoxing, Zhejiang 312000 — 2155-0105, 0245-0525, 0910-1335.
125) Xinhua Jie, Jinhua, Zhejiang 321000 — 2145-0025, 0225-0440, 0925-1320.
126) Nanjie, Quzhou, Zhejiang 324000.
1st Prgr. on 711/1250kHz 2150-0025, 0155-0520, 0955-1345.
2nd Prgr: FM stereo on 98.9MHz. H. of tr. as 1st Prgr.
127) 118 Changguo Lu, Dinghai Qu, Zhoushan, Zhejiang 316000. 2140-0015, 0210-0455, 0855-1340.

ANHUI PROVINCE

128) 39 Lujiang Lu, Hefei, Anhui 230065.
1st Prgr. on MW/FM 2120-0530, 0715(Tue 0835)-1500.
2nd Prgr. on 89.5MHz 0325-0600.
3rd Prgr. FM stereo on 98.8MHz. Rel. CPBS 3: 0400-0600.
129) 100 Fuyang Lu, Hefei, Anhui 230001 — on 666kHz 2125-0100, 0230-0600, 0905-1410. FM stereo: on 102.0MHz. 0400-0500.
130) Dongshan, Huainan, Anhui 232001 — 2120-0010, 0255-0510, 0855-1300.
131) Huaihai Donglu, Huaibei, Anhui 235000 — 2155-2350, 0320-0450, 0855-1300.
132) 61 Laodong Lu, Wuhu, Anhui 241000 — 2150-0020, 0155-0510, 0850-1215.
133) Yi'an Beilu, Tongling, Anhui 244000 — 2120-0005, 0155-0510, 0855-1305.
134) 150 Zhongrong Jie, Bengbu, Anhui 233000 — 2155-0005, 0235-0500, 0855-1410.
135) Gongyuan Xincun, Hubei Lu, Ma'anshan, Anhui 243000 — on 1584kHz/90.4MHz 2145-0025, 0250-0530, 0835-1300. FM stereo: on 88.7MHz 0355-0520, 0945-1100.
136) 23 Guanyue Miao, Anqing, Anhui 246004 — 2110-0005, 0220-0535, 0920-1355.
137) Huayuan Nanlu, Tunxi Qu, Huangshan, Anhui 245000 — 2155-0025, 0310-0510, 0925-1300.
138) 123 Huaihai Zhonglu, Suzhou, Anhui 234000.
139) Renmin Donglu, Lu'an, Anhui 237006.
140) Renmin Donglu, Fuyang, Anhui 236014 — 0955-1310.
141) 73 Yingzhou Beilu, Fuyang, Anhui 236001 — 2200-0100, 0255-0540, 0930-1250.

FUJIAN PROVINCE

142) 15 Yongzuoshe, Daoshan Lu, Fuzhou, Fujian 350001.
1st Prgr. on MW/SW/FM in Ch and Amoy. 2050-1700 (exc. Wed/Fri 0600-0850). Prgr. for Taiwan: 0000-0100, 1400-1700.
2nd Prgr. on 1467kHz/FM in Ch and Amoy. 2120-2400, 0250-1600 (exc. Wed 0600-0750). FM stereo: on 102.4MHz 0325(Sun 0155)-0530, 0955-1400.
143) 23 Shengmiao Lu, Fuzhou, Fujian 350001 — on 1332kHz in Ch and Fuzhou dialect. 2130-2335, 0115-0445, 0915-1505.
144) 8 Huyuan Lu, Xiamen, Fujian 361003 — on 1107/1278kHz in Ch and Amoy. 2150-0550, 0810-1505. FM stereo: on 94.9MHz. 0400-0500, 1000-1100.
145) 2 Nanmen Xincun, Putian, Fujian 351100 — on 1440vkHz in Ch and Puxian dialect. 2155-?, 0250-0505, 0855-1330.

JIANGXI PROVINCE

146) 2 Beijing Xilu, Nanchang, Jiangxi 330046.
1st Prgr. on MW/SW/FM 2100-1505 (exc. Tues 0600-0920).
2nd Prgr: FM stereo on 100.5MHz 0155-0600, 0925-1400. Rel. CPBS 3: 0200-0600, 1200-1400.
147) 401 Minde Lu, Nanchang, Jiangxi 330008 — 2120-0005, 0250-0530, 0825-1350.
148) 10 Lianshe Beilu, Jingdezhen, Jiangxi 333000 — 2205-0500, 0940-1300.
149) Yingbin Lu, Pingxiang, Jiangxi 337055 — 2135-0025, 0325-0535, 0935-1425.
150) Ganxin Lukou, Xinyu, Jiangxi 336525 — 2125-2400, 0325-0530, 0925-1310.
151) 14 Jiaotong Lu, Jiujiang, Jiangxi 332000 — 2155-?.
152) 22 Shengli Xilu, Yingtan, Jiangxi 335000 — 2150-0015, 0325-0535, 0930-1300.
153) 74 Hongqi Dadao, Ganzhou, Jiangxi 341000.

SHANDONG PROVINCE

154) 11 Qingnian Donglu, Lixia Qu, Jinan, Shandong 250011.
1st Prgr. on MW/FM W 2050-0130, Sun 2050-0650, W 0245-0650(Tue 0605), D 0845-1530.
3rd Prgr: FM stereo 2200-0530(Tues 0300), 0815-1420. Rel. CPBS3: 1000-1400.
Literary Prgr: FM stereo on 105.0MHz. 2325-0900, 1055-1530.
155) 14 Jing 11 Lu, Lixia Qu, Jinan, Shandong 250014 — 2150-0540, 0850-1505.
156) 30 Shanxian Lu, Qingdao, Shandong 266002 — on 1377kHz 2110-0200, 0255-0540, 0850-1515. FM stereo: on 103.8MHz.
157) 10 Xi'er Xiang, Liuquan Lu, Zhangdian Qu, Zibo, Shandong 255008 — 2150-0125, 0320-0635, 0930-1400.
158) 5 Xinsheng Lu, Zaozhuang, Shandong 277137 — 2050-0100, 0310-0600, 0950-1415.
159) 30 Shenghua Lu, Dongying, Shandong 257031 — on 1449kHz 2150-0030, 0330-0600, 1025-1400. FM stereo: on 105.0MHz.
160) 38 Dongfeng Jie, Weifang, Shandong 261041 — 2145-0130, 0315-0545, 0930-1355.
161) 50 Qingnian Lu, Yantai, Shandong 264000 — 2115-0135, 0255-0600, 0855-1430.
162) 3 Xicheng Lu, Weihai, Shandong 264200 — 2125-0050, 0255-0555, 0855-1350.
163) Changqing Lu, Jining, Shandong 272137.
164) Fangongting Lu, Qingzhou, Shandong 262500 — 2145-?.
165) Gulou Beijie, Qufu, Shandong 273100 — 2155-?, 0900-1300.
166) Jiefang Lu, Linyi, Shandong 270003 — 2120-0020, 0315-0555, 0945-1405.
167) 204 Wenhua Lu, Liaocheng, Shandong 252052 — 2155-?.

HENAN PROVINCE

168) 2 Jing 5 Lu, Zhengzhou, Henan 450003 — 2105-0050, 0330-1510.
169) 2 Jing 5 Lu, Zhengzhou, Henan 450003 — 2155-0600, 0855-1430.
170) Nanduan, Gongren Lu, Zhengzhou, Henan 450007 — 2150-0115, 0300-0555, 0900-1415.
171) Nanduan, Gongren Lu, Zhengzhou, Henan 450007 — 2225-0635, 0855-1500.
172) 27 Gulou Jie, Kaifeng, Henan 475000 — 2155-0105, 0325-0640, 0955-1415.
173) 21 Dingding Nanlu, Luoyang, Henan 471000 — on 1107kHz 2130-0030, 0300-0530, 0925-1440. FM stereo: on 92.25MHz 0400-?, 1130-?.
174) Zhongduan, Kuanggong Lu, Pingdingshan, Henan 467000 — 2150-0005, 0355-0530, 1025-1405.

175) Jiefang Zhonglu, Jiaozuo, Henan 454150 — ?-0030, 0345-?.
176) Zhongduan, Chunlei Lu, Hebi, Henan 456650 — 2155-2400, 0350-0535, 0955-1330.
177) Renmin Lu, Xinxiang, Henan 453000 — 2200-0100, 0355-0550, 0855-1410.
178) 21 Jiefang Lu, Anyang, Henan 455000 — 2155-0025, 0320-0515, 0900-1345.
179) Zhongyuan Lu, Puyang, Henan 457001.
180) 118 Wenfeng Dong Xiang, Xuchang, Henan 461000 — 2150-0005, 0335-0545, 0955-1345.
181) 1 Wenhua Lu, Luohe, Henan 462000 — 2140-0020, 0325-0530, 0920-1325.
181A) Jiadao Jie, Shangqiu, Henan 476000 — 2205-0150, 0355-0555, 0840-1330.
182) Dongduan, Jianshe Lu, Zhoukou, Henan 466001 — ?-1355.
183) Wolonggang Lu, Nanyang, Henan 473081.

HUBEI PROVINCE

184) 563 Jiefang Dadao, Hankou, Wuhan, Hubei 430022.
1st Prgr. on MW/SW/FM 2120-0615, 0850-1530.
2nd Prgr. on 95.0/99.8MHz 2335-0340, 0955-1400.
Literary Prgr. on 103.8MHz 0150-1505 (exc. 0605-0855 on Tue).
185) 342 Jiefang Dadao, Hankou, Wuhan, Hubei 430030 — 2215-0600, 0900-1400.
186) Jianshe Dadao, Hankou, Wuhan, Hubei 430015 — on 873kHz 2159-0710(Wed 0540), 0835(Wed 0855)-1420(Sat 1500). FM stereo: on 90.2MHz 0920-1130.
187) Jianshe Dadao, Hankou, Wuhan, Hubei 430015 — 0350-1420.
188) 38 Wuhan Lu, Huangshi, Hubei 435000 — 2150-0500, 0855-1335.
189) 177 Zhongshan Houjie, Xiangfan, Hubei 441000 — 2120-0005, 0255-0540, 0955-1500.
190) Liuyan Shan, Shiyan, Hubei 442000.
191) 65 Yuanlin Lu, Shashi, Hubei 434000 — 2150-2400, 0255-0545, 0920-1440.
192) Yiling Lu, Yichang, Hubei 443000 — 2155-0015, 0315-0540, 0945-1440.
193) 25 Changning Dadao, Jingmen, Hubei 434500 — 2155-0035, 0325-0600, 0925-1330.
194) 23 Jianshe Jie, Ezhou, Hubei 436000 — 2155-0530, 1000-1420.
195) 193 Lieshan Dadao, Suizhou, Hubei 441300 — 2130-2330, 0330-0500, 1030-1305.
196) Xuefu Lu, Laohekou, Hubei 441800 — 2155-2400.
197) Changzheng Lu, Xiaogan, Hubei 432100 — 2220-0120, 0355-0535, 0957-1355.
197A) 8 Fuqian Jie, Guangshui, Hubei 432700 — ?-1220.
198) 2 Xinjian Jie, Macheng, Hubei 431600.
199) 33 Daqiao Lu, Puqi, Hubei 437300 — 0950-1340.
200) 2 Shannan Xiaoqu, Shishou, Hubei 434400 — 2155-0005, 0955-1235.
201) 4 Renmin Lu, Danjiangkou, Hubei 441900 — 2200-2355, 0955-1155.

HUNAN PROVINCE

202) 27 Yuhua Lu, Changsha, Hunan 410007.
1st Prgr.on MW/SW/FM 2125-1500 (exc. Tues/Fri 0540-0855).
2nd Prgr. on 91.0/94.4MHz. D 2320-0400, Sun 0650-0900, D 1150-1400.
3rd Prgr. on 93.4MHz. 0300-0530, 1030-1400. Rel. CPBS 3: 0300-0400, 1200-1400.
203) 5 Wuyi Xilu, Changsha, Hunan 410005 — 2140-0155, 0325-0540, 0900-1400.
204) Hejiatu, Jianshe Zhonglu, Zhuzhou, Hunan 412000 — ?-0225, 0355-?.
205) 4 Shaoshan Xilu, Xiangtan, Hunan 411100 — 2155-0015, 0325-0555, 0900-1435.
206) 154 Xianfeng Lu, Hengyang, Hunan 421001 — 2155-0100, 0325-0500, 0925-1405.
207) Shazipo, Nanmen Wai, Shaoyang, Hunan 422000. — 2150-0030, 0355-0530, 0930-1415.
208) Zhuyin Jie, Yueyang, Hunan 414000 — 2200-0015, 0330-0530, 0925-1340.
209) Gaoshan Jie, Changde, Hunan 415000 — 2200-2355, 0400-0500, 0925-1100.
210) Jiefang Lu, Yongding Qu, Dayong, Hunan 416600.
211) Renmin Nanlu, Jishou, Hunan 416000.

GUANGDONG PROVINCE

212) 686 Renmin Beilu, Guangzhou, Guangdong 510012.
News Prgr. on MW/FM 2100-1610 (exc. Tues 0015-0355).
Literary Prgr. on 999kHz 2100-1600 (exc. Mon/Thurs 0015-0320)
Music Prgr. on 99.3/98.7MHz W 2200-0630, Sun 2200-1600. W 0930-1600.
English Prgr. "R. Guangdong": on 603kHz 2150-0500, 0845-1100, 1300-1540(Sun 1605).
Educational 1st Prgr. on 103.3MHz 2120-0600, 0925-1540.
Educational 2nd Prgr. on 1584kHz 1055-1505. Rel. R. Beijing DS: 1400-1500.
213) 686 Renmin Beilu, Guangzhou, Guangdong 510012 — on MW/FM stereo in Cantonese. 2100-1610 (exc. 0630-0930 on Thu).
214) Guangzhou, Guangdong 510045 — FM stereo on 96.2MHz in Cantonese.
215) 170 Jiefang Lu, Shenzhen, Guangdong 518001 — FM stereo on 89.8MHz in Cantonese. 2225-?, 0900-1400.
216) 8 Kangning Lu, Zhuhai, Guangdong 519000 — on 850kHz/95.1MHz in Ch and Cantonese. 2205-0600, 0915-1400.
217) 89 Gonghe Lu, Shantou, Guangdong 515031 — on 1080kHz in Ch and Chaozhou dialect. 2125-0055, 0250-0530, 0840-1510. FM stereo: on 95.6MHz. 0330-0505, 1055-1305.
218) Huimin Lu, Shaoguan, Guangdong 512026 — on 765kHz/105.7MHz in Ch and Cantonese. 2225-0135, 0325-0640, 0825-1510.
219) Huizhou, Guangdong 516001.
220) 55 Xizheng Lu, Cheng Qu, Dongguan, Guangdong 511700 — FM stereo on 106.9MHz in Cantonese. ?-0140, 0935-1235.
221) 200 Sunwen Zhonglu, Shiqi Zhen, Zhongshan, Guangdong 528403 — FM stereo on 90.4/101.2MHz in Cantonese. 2225-0605, 0755-1405.
222) 19 Jianshe Lu, Jiangmen, Guangdong 529000 — FM stereo on 100.2MHz in Cantonese. 2200-1600.
223) 98 Yong'an Lu, Foshan, Guangdong 528000 — FM stereo on 94.6MHz in Cantonese. 24h.
224) Wangliaoling, Yangjiang, Guangdong 529500.
225) 93 Yuejin Lu, Chikan Zhen, Zhanjiang, Guangdong 524038 — on 1134kHz in Ch, Cantonese and Leizhou dialect. 2120-2400, 0230-0630, 0850-1505.
226) 95 Renmin Nanlu, Maoming, Guangdong 525000 — on 1098kHz in Ch and Cantonese. 2130-2400, 0330-0600, 0915-1500.
227) 41 Ganyuan Beilu, Zhaoqing, Guangdong 526040 — FM stereo on 92.9MHz in Cantonese. 2200-0440, 0800-1330.

GUANGXI ZHUANG AUTONOMOUS REGION

228) 12 Minzu Dadao, Nanning, Guangxi 530022.
1st Prgr. on MW/SW/FM in Ch, Guangxi and Liuzhou dialect. 2150-0515, 0950-1550.
2nd Prgr. on MW/SW/FM in Guangxi dialect and Zhuang. 2150-0025, 0325-0615(Tue 0605), 0955-1545.
3rd Prgr. on 5920kHz W 2300-0135, W 1000-1235.
FM stereo: on 95.0MHz 0055-0500(exc. Tues), 0930-1035. Rel. CPBS3: 0100-0500(exc. Tues).
229) 7 Chaoyang Lu, Nanning, Guangxi 530012 — on 918kHz/99.2MHz in Ch and Guangxi dialect. 2155-0005(Sun 0030), 0255-0530, 0820-1415.
230) 22 Wuyi Lu, Liuzhou, Guangxi 545001 — on 1386kHz/92.5MHz in Ch and Liuzhou dialect. 2155-2330, 0255-0450, 0920-1405.
231) 5 Minzhu Lu, Guilin, Guangxi 541002. 2225-0525(Sun 0350), W 0910-1400, Sun 1025-1410.
232) Dieshan 2 Lu, Wuzhou, Guangxi 543002 — on 1485kHz in Ch and Guangxi dialect. 2155-2330, 0200-0500, 0830-1330.
233) Guizhou Lu, Beihai, Guangxi 536000 — on 750kHz/101.5MHz in Ch and Guangxi dialect. 2150-2335, 0330-0440, 0950-1330.

HAINAN PROVINCE

234) 1 Haixiu Dadao, Haikou, Hainan 570005 — on 954/1107kHz/91.0MHz in Ch and Hainan dialect — W2200-0100, Sun 2200-0620, W 0250-0620(Mon 0510), D 0850-1530. FM stereo: on 99.0MHz ?-0200, 0355-?.
235) 4 Guangchang Lu, Haikou, Hainan 570005 — on 101.8MHz in Ch and Hainan dialect. 0930-1500.

SICHUAN PROVINCE

236) 119-1, Hongxing Zhonglu, Chengdu, Sichuan 610017.
1st Prgr. on MW/SW/FM 2155-1605.

2nd Prgr. on MW/SW/FM in Ch, Tb and Yi. 2155-0555(Sun 0535), 1025-1345.
237) 119-1, Hongxing Zhonglu, Chengdu, Sichuan 610017 — on 1000kHz/97.8MHz(stereo) 2300-1630.
238) 3 Duan, Dong 1 Huanlu, Chengdu, Sichuan 610051 — 1st Prgr. on 792kHz W 2150-0130, Sun 2225-0320, D 0350-0550(Sun 0610), D0950-1505.
2nd Prgr. on 1485kHz 2155-0030, 0425-0600, 1055-1400.
239) 159 Zhongshan 3 Lu, Chongqing, Sichuan 630015.
1st Prgr. on 1314kHz/105.5MHz 2210-0155, 0320-0655, 0910-1500.
2nd Prgr. on 1080kHz W0055-0230, Sun 0055-0435, W 0355-0435.
240) 159 Zhongshan 3 Lu, Chongqing, Sichuan 630015 — FM stereo: on 101.5MHz 2245-1100(Tue 0600).
241) 87 Tanmulin Jie, Zigong, Sichuan 643000 — 2120-0035, 0320-0530, 0920-1430.
242) Linjiang Lu, Panzhihua, Sichuan 617000 — 2225-0210, 0355-0635, 0810-1355.
243) Jiangyang Lu, Luzhou, Sichuan 646000 — 2205-2400, 0400-0510.
244) 25 Taishan Nanlu, Deyang, Sichuan 618000.
245) 4 Hongxing Nanlu, Mianyang, Sichuan 621000.
246) Dabei Jie, Suining, Sichuan 629000.
247) 8 Xianglong Jie, Xinhua Lu, Neijiang, Sichuan 641000.
248) Haitang Lu, Leshan, Sichuan 614000.
249) 72 Guangchang Lu, Fuling, Sichuan 648000.
250) 109 Renmin Nanlu, Nanchong, Sichuan 637000 — 2155-0015, 0355-0520, 1155-1415.

GUIZHOU PROVINCE

251) 259 Qingyun Lu, Guiyang, Guizhou 550002.
1st Prgr. on MW/SW/FM 2150-0110, 0150-0600, 0850-1605.
2nd Prgr. FM stereo on 91.6/106.2MHz W 0550-0900, Sun 0550-1305, W 0950-1305.
252) Renmin Guangchang Houyuan, Guiyang, Guizhou 550002.
253) 9 Minghu Lu, Liupanshui, Guizhou 553000 — 2225-0030, 0325-0530, 0955-1330.

YUNNAN PROVINCE

254) Renmin Xilu, Kunming, Yunnan 650031.
General Prgr. on MW/SW/FM 2200-0100, 0255-0700(Tue 0600), 0935-1600.
Minority Language Prgr. on 6940kHz in Lahu, Jingpo, Lisu, Dehong Dai and Xishuangbanna Dai. 2225-0045, 0355-0545, 1055-1500.
Educational Prgr. on 5960kHz W 0030-0400, D 1100-1200.
Literary Prgr: FM stereo on 103.4MHz 1200-1600.
255) 610 Beijing Lu, Kunming, Yunnan 650011 — 2255-0030, 0400-0630, 0855-1100.
256) Zhudianpo, Wenshan Xian, Yunnan 663000 — on 1053kHz/96.0MHz in Ch, Zhuang, Miao and Yao. 2225-0100, 0225-0500, 0955-1400.
257) Jianshe Donglu, Gejiu, Yunnan 661400 — on MW/SW/FM in Ch, Hani and Yi. 2225-0030, 0225-0500, 0855-1400.
258) 10 Jinghong Donglu, Jinghong Xian, Yunnan 666100 — on 747kHz in Ch, Xishuangbanna Dai and Hani. 2210-0100, 0250-0600, 1030-1540.
259) 38 Lucheng Beilu, Chuxiong, Yunnan 675000 — W 2225-2400, Sun 2325-0200, D 0325-0600, 0955-1405.
260) Cangshan Lu, Xiaguan Zhen, Dali, Yunnan 671000.
261) Mangshi Zhen, Luxi Xian, Yunnan 678400 — on 900kHz in Ch, Dehong Dai, Jingpo and Zaiwa. 2230-0110, 0330-0700, 1030-1530.
262) Xi Dajie, Luliang Xian, Yunnan 655600.
263) Hubin Lu, Jiangchuan Xian, Yunnan 652600.

XIZANG AUTONOMOUS REGION

264) 146 Beijing Zhonglu, Lhasa, Xizang 850000.
Ch Prgr. on MW/SW W2300-0205, Sun 2300-0750, W 0355-0750, D 1000-1455.
Tb Prgr. on MW/SW W2300-0205, Sun 2300-0750, W 0355-0750, D 1000-1550.
FM stereo: on 98.0MHz 0500-0600, 1100-1200.
265) 25 Nedong Lu, Zetang Zhen, Nedong, Xizang 856000 — on 1260kHz in Ch and Tb. 2335-0135, 0405-0535, 1005-1340.

SHAANXI PROVINCE

266) 119 Bei Dajie, Xi'an, Shaanxi 710003.
1st Prgr. on MW/SW/FM 2140-0635, 0935-1535.
2nd Prgr. on 1323kHz 2155-2400, 0455-0840, 1055-1500.
3rd Prgr. FM stereo on 90.7MHz. W 2258-2400. Rel. CPBS 3: Sun 2330-0600, W 0400-0600, D 0930-1500.
267) 125 Wuyuemiao Menjie, Xi'an, Shaanxi 710003 — 2150-0030, 0325-0600, 0825-1500.
268) Hongqi Jie, Tongchuan, Shaanxi 727000 — 2210-0015, 0330-0515, 0915-1405.
269) 47 Hongqi Lu, Baoji, Shaanxi 721000 — 2145-2400, 0325-0610, 0930-1435.
270) Dong'an Cun, Xianyang, Shaanxi 712000 — 2210-0025, 0345-0640, 0850-1110.

GANSU PROVINCE

271) 146 Donggang Xilu, Lanzhou, Gansu 730000.
1st Prgr. on MW/SW/FM W 2150-0130, Sun 2150-0640, W 0230-0640(Tue/Fri 0610), D 0850-1600.
2nd Prgr. on FM 2200-0255(Sun/Tues 2400), 0430-0855(Sun/Tues 0600), 1050-1400.
3rd Prgr. FM stereo on 104.4MHz. 0355-0600, 0855-1400 (Mon/Wed/Thurs 1300). Rel. CPBS 3: 1100-1300.
272) 34 Qingyang Lu, Lanzhou, Gansu 730030. W 2220-0255, Sun 0000-0530, W 0355-0855, 1000(Sun 0920)-1300.
273) 49 Renmin Xijie, Hezuo Zhen, Xiahe Xian, Gansu 747000 — on 1332/5970kHz/97.2MHz in Ch and Tb. 2220-0030(Sun 0040), 0420-0620, 0950-1355(Sat/Sun 1415).
274) 29 Hongqi Jie, Pingliang, Gansu 744000.
275) Zhongping Qu, Yumen, Gansu 735200.

QINGHAI PROVINCE

276) 96 Kunlun Lu, Xining, Qinghai 810001.
1st Prgr. on MW/SW W 2225-0030, Sun 2225-0630, W 0355-0630, D 0925-1550.
2nd Prgr. on 1143kHz 2225-0030, 0410-0640, 1005-1350(Sun 1405)
3rd Prgr. FM stereo on 97.2MHz. 1155-1500. Rel. CPBS 3: 1200-1400.
Tb Prgr. on 1251/6500kHz 2225-0055, 0355-0545, 1055-1540.
277) 105 Nan Dajie, Xining, Qinghai 810000 — 2225-0530(Tues 0015), 1025-1330(Sun/Mon 1400).
278) 21 Hongwei Lu, Jiegu Zhen, Yushu Xian, Gansu 815000.
279) Renmin Lu, Delingha, Gansu 817000 — on 621kHz 1115-1130 in Ch, 1230-1315 in Mo.

NINGXIA HUI AUTONOMOUS REGION

280) 16 Xixia Donglu, Yinchuan, Ningxia 750021.
1st Prgr. on MW/FM 2210-0150(Sun 0220), 0355-0700(Mon/Tue/wed 0630), 0905-1450.
2nd Prgr. FM stereo on 104.3MHz. 0355-0555, 1025-1400. Rel. CPBS 3: 1200-1400.
281) 122 Zhongshan Nanjie, Yinchuan, Ningxia 750004. W 2225-0145, Sun 000-0500, W 0355-0530, D 1025-1310(Sun 1330).
282) Chaoyang Dongjie, Shizuishan, Ningxia 753000.
283) Yumin Dongjie, Wuzhong, Ningxia 751100.
284) Xiaoba Zhen, Qingtongxia, Ningxia 751600.

XINJIANG UIGHUR AUTONOMOUS REGION

285) 84 Tuanjie Lu, Urumqi, Xinjiang 830044.
Ch Prgr. on MW/SW 2200-0210, 0430-0730, 1030-1650.
Ug Prgr. on MW/SW 2200-0200, 0430-0730, 1030-1650.
Kz Prgr. on MW/SW 0000-0230, 0530-0700, 1200-1645.
Mo Prgr. on MW/SW 2300-0230, 0500-0700, 1130-1645, including Kirghiz Prgr. 2300-2400, 1130-1230.
FM stereo: on 94.9MHz. W 0000-0200, Sun 0230-0530, D 1030-1230.
286) 30 Xinmin Lu Dong, Urumqi, Xinjiang 830002.
Ch Prgr. on 810kHz 2345-0230, 0510-0900, 1050-1615.
Ug Prgr. on 1071kHz 2350-0210, 0500-0730(Sun 1005), 1130-1545.
FM stereo: on 97.9MHz.
287) 42 Tianshan Xilu, Karamay, Xinjiang 834000 — Ch Prgr. on 882kHz 0025-0240, 0600-0800, 1155-1505. Ug Prgr. on 1179kHz.
288) Bei 3 Lu, Shihezi, Xinjiang 832000.
289) 2 Hongxing Xilu, Hami, Xinjiang 839000 — Ch Prgr. on 1485kHz W 2310-0215, Sun 2355-0450, W 0500-0710, Sun 1150-1450, W 1155-1515. Ug Prgr. on 1098kHz.
290) Beijing Beilu, Changji, Xinjiang 831100.
291) Korla Donglu, Kuitun, Xinjiang 833200 — Ch Prgr. on 1485kHz W 2355-0230, Sun 0025-0335, Sun 0528-0720, W 0558-0740, D 1123-1425. Kz Prgr. on 819kHz.
292) 1 Hongqi Lu, Yining, Xinjiang 835000 — Ch Prgr. on 603kHz, Kz Prgr. on 1134kHz.

PRIVATE STATION

301) CHINA HUAYI BROADCAST COMPANY
ADDR: P.O. Box 251, Fuzhou, Fujian 350001 — GM: Cao Jinwang
D.PRGR. on 666/6185kHz + 90.6/99.6/107.1MHz: 0955-1800.
V. by letter.

EXTERNAL SERVICES

300) CHINA RADIO INTERNATIONAL (Zhonggguo Guoji Guangbo Diantai).
ADDR: 2 Fuxingmenwai, Beijing, 100866. **Te:** +86 (1) 868581, +86 (1) 868471. **Fax:** +86 (1) 8013175 — **L.P:** Dir: Mme Cui Yuling.

DOMESTIC SERVICE.

Prgrs in English/French/German/Spanish/Japanese: News, current affairs, features, music & Weather for visitors & resident foreigners (**Te:** +86 (1) 8013135).
Beijing (1251kHz+91.55MHz): English 2300-0500, 0900-1200, 1300-1500; French 0500-0600; German 1500-1530; Spanish 2135-2300; Japanese 1200-1300; Note: no broadcast at 1000-1300 and 0900-1200 on 1251kHz.
Tianjin (801kHz): English 2300-0100, 0400-0500, 1300-1500.
Shanghai (105.7MHz): English 2300-2400, 1300-1400.
Nanjing (585kHz): English 1400-1500.
Guangzhou (603kHz): English 2300-0100, 0400-0500, 1300-1500 (1400-1500 also on 1584kHz).
Shenyang (99.8MHz): English 1300-1320.
Xi'an (603kHz): English 1300-1400 (also on 105.4MHz), 2300-0100, Japanese 1200-1300.
Lanzhou (98.2MHz): English 0900-1200, 1300-1500, 2300-0500.
Jinan (92.8/97.5/99.0/101.7/103.4MHz): English 0000-0100.
Jingdezhen (102MHz): English 0900-1200, 1300-1500, 2300-0500.
Zibo (801kHz): English 0000-0100,1300-1500.
Dalian (102.4MHz): English 1300-1500.
Hangzhou (1071kHz): English 1000-1100, 1300-1500, 2300-0500; German 1500-1530.

OVERSEAS SERVICE

Mediumwave key freqs (kHz): 1017(JL), 1044(ZJ), 1080/1206/1269/1296(YN), 1116(HL), 1188(YN), 1314(NM), 1323(XZ), 1341(GD), 1521(XJ).
Shortwaves: Baoding, HEB (G.C: 115.44E/38.39N): 240/120kW — Beijing, BJ (G.C: 116.27E/39.57N): 240/120/50kW — Jinhua, ZJ (G.C: 119.39E/28.07N) 500kW —Kunming, YN 120/50kW — Shijiazhuang, HEB (G.C:114.28E/38.04N): 500kW — Xi'an, SN (G.C:108.54E/34.12N): 300/150kW. Further trs in border areas are used at times.
Overseas relay st's: CRI exchanges transmitting facilities with R.Canada Int, R.France Int, R. Exterior de España and Swiss Radio Int, and also uses relay facilities in Mali, Brazil and Russia.
Relays: B=Brazil, C=Canada, E=Spain, F=France, G=Fr. Guiana, M=Mali, R=Russia, S=Switzerland.

Albanian	**Area**	**kHz**
1900-1927	Eu	9965, 7470
1930-1957	Eu	15345, 9965, 7470
2100-2127	Eu	9845F
Amoy		
1200-1257	SEAs	15260, 15215, 12015, 11945, 6810, 6010
2330-2357	SEAs	12015, 11685, 11455, 7190, 6165, 6140, 6010, 5220
Arabic		
1600-1657	ME	15540R, 11855, 9440
1830-1927	ME	15170M, 11790M, 12035R
2100-2157	ME	11855, 11515, 9440
Bengali		
1500-1557	SAs	11600, 9900, 9690, 8660, 1206
Bulgarian		
1830-1857	Eu	9945, 9845F, 7385, 6933
2030-2057	Eu	9945, 9845F, 7385
Burmese		
1130-1157	SEAs	11780, 11455, 9880, 1269, 1188
1300-1327	SEAs	11780, 11455, 9880, 1269
1330-1357	SEAs	11780, 9880, 8260, 1269
Cambodian		
1030-1127	SEAs	11780, 11455, 9440, 1080
1200-1257	SEAs	11780, 11455, 9440, 1080
1400-1457	SEAs	9440, 8425, 6010
Cantonese		
0100-0157	NAm/SAm	12055, 11715M, 9770M
0300-0357	NAm	15430
1000-1057	SPac	11695, 11445, 9945, 6825
1100-1157	SEAs	15260, 15215, 12015, 11945, 6810, 6010
1700-1757	SAs/SEAf	11695, 9900, 8345, 7260
1900-1957	Eu	9745, 6974
2300-2327	SEAs	12015, 11685, 11455, 7190, 6165, 6140, 6010, 5220
Chaozhow		
0030-0057	SEAs	12015, 11685, 11455, 7190, 6165, 6140, 6010, 5220
1400-1457	SEAs	15260, 15215, 12015, 11945, 6810
Chinese		
0200-0257	SAm/NAm	15435, 12055, 11715M, 9770M, 9690S
0300-0357	NAm	11695G
0400-0457	NAm	11715M, 9770M
0900-0957	NEAs/SPac	17533, 15260, 15215, 15180, 15100, 12015, 11945, 11695, 9945, 9480, 8425, 6810, 6010
1100-1157	NEAs	11795, 6825,
1300-1357	NEAs/SPac	15260, 15215, 12015, 11945, 6810
1500-1557	SAs/EAf	15455, 11910, 9457
6810, 4020		
1730-1827	WAs/NAf/Eu	15100, 9820, 7800, 7335, 4020
2000-2027	Eu	6165S
2000-2057	WAs/NAf/Eu	11855, 11650, 11445, 9745, 8345, 7660, 5220
2230-2257	NEAs/SEAs	15420, 15170M, 11790M, 12015, 11685, 11455, 7190, 6165, 6140, 6010, 5220
Czech		
1900-1927	Eu	7375, 6933
1930-1957	Eu	9845F, 7375
English		
0000-0057	NAm	11715M, 9770M
0300-0357	NAm	11715M, 9770M, 9690E
0400-0457	WNAm	11680G
0500-0557	NAm	11840C
0900-1057	Pac	17710, 15440, 11755, 8450
1200-1257	ENAm	9655
1200-1357	Pac/SEAs	15450, 15440, 11660, 9715, 8425, 1341
1400-1557	WNAm	7405
	SAs	15165, 11815, 4200
1600-1657	EAf/SAf	15130M, 15170M, 11575, 4130
1700-1757	EAf/SAf	15345, 11575, 9570, 4130
1900-1957	WAf/NAf	11515, 9440
2000-2057	EAf/SAf	15170M, 11715M
	WAf/NAf	9440
	Eu	11500, 9920, 4130
2100-2157	Eu	11500, 9920, 4130
2100-2127	EAf/SAf	15170M, 11715M
2200-2227	Eu	3985S
2200-2257	Eu	7170R
Esperanto		
1100-1125	NEAs	9480, 6955
1300-1327	SEAs	15165, 11650, 4200
2000-2027	Eu	15345, 9965, 7470
2230-2257	SAm	11515, 9480
Filipino/English		
1130-1157	EAs	15135, 12450, 11445, 8425, 1341
1200-1227	EAs	15135, 12450, 12110, 11445
1430-1457	EAs	15135, 11445, 8660
French		
1830-2027	Af/Eu	15100, 9820, 7800, 7335, 4020
2030-2127	Af/Eu	9820, 9640R, 7800, 7700, 7335, 4020
2130-2227	Af/Eu	15170M, 11790M, 15100, 9820, 7800, 7335, 4020
2130-2157	Eu	3985S
German		
1800-1957	Eu	11500, 9920, 4130
2000-2027	Eu	3985S
Hakka		
0000-0027	SEAs	12015, 11685, 11455, 7190, 6165, 6140, 6010, 5220

1000-1057	SEAs/Pac	15260, 15215, 12015, 11945, 8425, 6810, 6010
1600-1657	SEAf	11600, 9900
Hausa		
1730-1757	Af	15170M, 11790M, 11445, 8260
1800-1827	Af	15345, 15170M, 11790M, 11445, 8260
Hindi		
1500-1557	SAs	11675, 9920, 9490, 7590, 7110, 5935, 4035, 1323, 1269, 594
1600-1657	SAs	11675, 9920, 9490, 8345, 5935, 4035, 1269, 594
Hungarian		
1800-1827	Eu	9945, 7385, 6933
2030-2057	Eu	9365, 8450, 6933
2130-2157	Eu	9845F
Indonesian		
0830-0927	SEAs	17680, 15600, 15135, 12450
1030-1127	SEAs	17680, 15135, 12450
1330-1427	SEAs	15135, 11445, 8660
Italian		
1830-1857	Eu	15345, 9965, 7470
2030-2057	Eu	9965, 7470
2100-2127	Eu	9965, 9365, 8450, 7470
2130-2157	Eu	3985S
Japanese		
0930-1527	NEAs	11515, 9855, 4960, 1044
Korean		
1100-1457	NEAs	5965, 4620, 4020, 1017
Lao		
1230-1327	SEAs	8450, 7350, 7190
1430-1527	SEAs	8450, 7350, 7190, 1080
Malay		
0930-1027	SEAs	17680, 15135, 12450
1230-1327	SEAs	15135, 11445, 8660
Mongolian		
1100-1157	CAs	5850, 5145
1200-1257	CAs	5850, 5145, 4883, 4815, 1314
1400-1457	CAs	5850, 5145, 4883, 4815
Nepalese		
1500-1557	SAs	15125, 11445, 9625, 8260, 1269
Pashto		
1500-1557	ME	11975, 11575, 9700, 4130
Persian		
1500-1527	ME	9440, 6955
1800-1827	ME	12055R, 11515, 9440
1830-1857	ME	11855, 11515, 9440
Polish		
2000-2027	Eu	9845F, 7375, 6933
2030-2057	Eu	15345, 7375
Portuguese		
0030-0057	SAm	12055, 11650, 11500, 6825
1900-1957	Af/Eu	15455, 15170M, 11790M, 11445, 9900, 8345
2130-2157	Eu	6165S
2200-2227	SAm	11745, 11650, 11500, 4200
Romanian		
1900-1927	Eu	9945, 9845F
1930-1957	Eu	9945, 6933
Russian		
0000-0057	EAs	4883, 4815
0100-0157	CAs	1521
0300-0357	CAs	15435, 11755, 11515
1000-1057	FE/EAs	15195, 9725, 7820, 7110, 6140, 5145, 4815, 1521, 1116
1100-1157	FE/EAs	9725, 7820, 7110, 6140, 1314, 1521, 1116
1300-1357	Rus/CAs	7820, 6140, 5850, 5145, 4883, 4815, 1521, 1314
1500-1557	Eu/Rus	11650, 9765, 7375, 5850, 4883, 4815, 4620, 1521
1600-1657	Eu/CAs	11745, 11650, 9965, 9945, 9765, 7470, 7375, 4200, 1521
1700-1757	Eu/CAs	11745, 11650, 11500, 9945, 9765, 7470, 7375, 4200, 1521
1800-1857	Eu	11765, 11745, 11650, 9765, 7780, 7660, 7375, 4200, 1521
1900-1957	Eu	11765, 11745, 11650, 9365, 8450, 7660, 4200, 1521
2000-2057	Eu	11765, 11745, 4200
2300-2357	EAs	4883, 4815
Serbo-Croat		
2000-2027	Eu	9945, 9365, 8450, 7385
2030-2057	Eu	6165S
2100-2127	Eu	15345, 15170M, 11790M, 9945, 7385
Sinhalese		
1400-1457	SAs	11780, 9900, 8260, 7260
Spanish		
0000-0057	SAm	15430, 15420, 15405, 12450, 11445, 9945, 5250
0100-0157	SAm	17755B, 15430, 15420, 12450, 11650, 11445, 9945, 8425
0200-0257	SAm	17755B, 15430, 15420, 13685G, 11650, 11445, 9945, 8425
0300-0357	SAm	15445B, 11840C
2100-2127	Eu	6165S
2100-2257	Eu	11765, 9685, 8660, 6933
2300-2357	SAm	15430, 15420, 15170M, 11790M, 12450, 11445, 9945, 5250
Swahili		
1600-1657	Af	9625, 9457, 8660, 7335
1700-1727	Af	15250, 15170M, 11790M, 9625, 8660, 7335
Tamil		
1400-1457	SAs	11975, 11575, 9457, 5220
Thai		
1130-1227	SEAs	8450, 7350, 7190, 6165, 1080
1330-1427	SEAs	8450, 7350, 7190
Turkish		
1400-1427	WAs	9480, 7305R, 6955
1600-1627	WAs	11660, 7480
1900-1927	WAs	15345, 11855
1930-1957	WAs	1170R
Urdu		
1600-1657	SAs	11695, 11445, 9670, 8260, 7590, 1323
Vietnamese		
1100-1457	SEAs	9945, 9605, 8345, 1296
1500-1557	SEAs	9945, 9605, 8345
1600-1657	SEAs	8450, 7360, 6185, 6010

ANN: Arabic: "Idha'atu al-Sini al-Duwaliyah". Chinese: "Zhongguo guoji guangbo diantai". English: "This is China Radio International, CRI, broadcasting from Beijing". Malay: "Inilah Radio Antarabangsa China dalam Bahasa Melayu". Mongolian: "Hyatadyn Olon Ulsyn Radio". Spanish: "Esta es Radio Internacional de China" — **V.** by QSL card.
PUB: Free literature, schedules, calendars and English magazine "The Messenger" in request.

CHINA (Republic of) (Taiwan)

L.T: UTC + 8h — **Pop:** 20.860.000 — **Radio sets:** 8.620.000 — **Pr.L:** Chinese (Mandarin), Taiwanese (Amoy), Hakka — **E.C:** a/c 60, 110/220V.

BROADCASTING CORPORATION OF CHINA (BCC)

(Priv. enterprise under Gov. contract).
ADDR: 53 Jen Ai Road, Sec. 3, Taipei 10628. **Te:** +886 (2) 7710151. **Fax:** +886 (2) 7519277. **Cable:** "Broadcast." **Telex:** 27498 BCCROC.
L.P: Pres: Lee Tsu-yuan, Dir. Eng. Dep: Huang Shuh-shyun.
MEDIUMWAVES: N = Network; N = News Netw, T = Traffic Prgr.

Location	kHz	kW	N				
Ilan	630	10	3	Hsinchu	1017	10	2
Taipei	657	20	1	Chia-i	1035	10	2
Taichung	720	10	1	Taipei	1062	10	3
Tainan	783	10	2	Taichung	1062	10	3
Taitung	819	10	1	Lishan	1071	5	1
Taichung	837	10	2	Jiyuetan	1089	1	FM1
Hualien	855	10	1	Yuli	1116	3.5	1
Kaohsiung	864	10	1	Puli	1152	1	2
Hsinchu	882	10	1	Ilan	1161	10	2
Tainan	891	10	N	Miaoli	1161	10	2
Miaoli	891	3.5	3	Hualien	1188	10	2
Taichung	909	10	T	Jiyuetan	1188	1	2
Kaohsiung	909	10	2	Kaohsiung	1224	10	3
Taipei	954	20	2	Taichung	1242	10	N
Taitung	1008	10	2	Lishan	1251	1	2

Tainan	1296	10	1	Kaohsiung	1449	10	N
Chia-i	1350	10	1	Taipei	1458	10	N
Yuli	1386	3.5	2	Hualien	1467v	10	3
Hsinchu	1386	3.5	3	Chia-i	1467	10	3
Ilan	1404	10	1	Jiyuetan	1467	1	1
Miaoli	1413	10	1	Mobile	1503	1	1
Taitung	1413	3.5	3	Mobile	1530	1	1
Puli	1413	1	1	Mobile	1530	1	2
Unknown*	1422		N	Tainan	1539	10	3

*for Mainland listeners.
+ additional st's of 0.1kW on 1017(1), 1224(1), 1377(1), 1449(2), 1503(1) rel. 1st netw; 1386(1), 1413(1), rel. 2nd netw; 1143(1), 1242(1) rel. 3rd netw; 1143(1), 1242(7), 1449(4) rel. FM 1st netw. Also 1 kW mobile st on 1503 (1st-Taipei), 1530 (2nd-Taichung & So. Tainan).

FM 1st Netw: 102.1MHz Taichung, Hualien, Taitung, Ilan, Taipei (Neihu area); 102.9MHz Miaoli; 103.1MHz Chia-i; 103.3MHz Taipei, Kaohsiung, Yuli.
FM 2nd Netw: 101.5MHz Miaoli; 102.9MHz Ilan; 104.3MHz Chia-i; 105.7MHz Yuli; 105.9MHz Taipei, Kaohsiung; 106.9MHz Taichung, Hualien, Taitung.
FM News Netw: 103.9MHz Taichung, Ilan, Taitung, Huoyanshan; 107.9MHz Taipei, Kaohsiung, Hualien, Chentocisha.
Music Netw: 96.2MHz Ilan, Huoyanshan, Chentochisha; 96.3MHz Taipei, Taichun, Kaohsiung, Hualien, Taitung.
FM: *) relay station, A = Agriculture Netw.

	Network (MHz)				
Location	**FM1**	**FM2/A**	**News**	**Music**	**kW**
Taipei	103.3	105.9	107.8	96.2	35/10
Huoyanshan	102.9	101.5	103.9	96.2	10/01
Taichung	102.1	106.9	103.9	96.3	10
Chentocishan	103.1	104.3	107.8	96.2	10/0.1
Kaohsiung	103.3	105.9	107.8	96.2	10
Taitung	102.1	106.9	103.9	96.3	5/2.5
Ilan	103.1	102.9	103.9	96.2	5/2.5
Hualien	102.1	106.9	107.7	96.3	5/2.5
*Yuli	103.3	105.7			2.5/1
*Puli	107.3				0.1
*Neihu	102.1				0.1

D.PRGR: 1st Netw: 24h in Chinese (2055-1800 on some reg. st's). **2nd Netw:** 2055 (except Miaoli: 2050)-1810 in Taiwanese. **3rd Netw:** 2055-1630 (exc. Taichung-1900). in Chinese (rel. Education Broadc. St: 2200-0000, 0400-0600, 1200-1600). **FM 1st popular Netw:** 24h in Chinese. **FM 2nd Netw:** 2055-1800 in Taiwanese. **Music Netw:** 2155-1900. **AM/FM News Netw:** 2055-1700 (exc. Tainan-1800, Taipei-1600, Taichung 24h). (1st Netw relay 1900-2055) exc. Tainan & Taichung 2055-1705. Also relayed on SW 9610/9765/11725 and via WYFR, USA. **Traffic prgr:** Taichung 1955-1900.
AM stereo (C-QUAM system) on 1st Netw.
Local Prgrs
Addr: 131-1 Kungyuan Rd. Hsinchu 300 — 1 Tientai Str., Fulili, Miaoli 360 — 1 Tientai Str., Pei, Taichung 400 — 121 Dapan South Rd, Chia-i 600 — 38 Nanmen Rd, Tainan 700 — 111 Chungcheng 3rd Rd, Kaohsiung 800 — Ilan 260 — 25 Shuiyüan Rd, Hualien 950 — 36 Chungkuang Str., Tunghai, Taitung, Taitung 930.
Chia-i: 1035kHz: W 0900-1000, 1030-1100, 1130-1200; 1467kHz: W 0730-1000.
Hsinchu: 1017kHz: W 1030-1100, Sun 1100-1200.
Hualien: 1188kHz: W 0400-0415, 0900-1000, 1030-1100, 1467kHz: W 0010-0100, 1100-1200, Sun 0600-0800.
Ilan: 630kHz: W 0010-0100, 0800-0940; 1161kHz: W 1000-1010, 1030-1100.
Kaohsiung: 1224kHz: W 0900-1200, 1600-1700 in Amoy.
Miaoli: 1161kHz: W 0500-0700, 0900-1010, 1030-1100, all in Hakka.
Taichung: 837kHz: W 0900-1000, 1030-1100, 1800-2000, Sun 0415-0500; 1062kHz; W 1100-1200, 1700-1900 in Amoy: W 1000-1210. **Taichung traffic st:** 1955-1900 on 909kHz (10kW). Often rel. 1st netw.
Tainan: 783kHz: W 0700-0900, 1000-1010; 891kHz: 0000-0400, 0530-1000, 1300-1700, 2200-2300, Sun 0000-0400, 0600-1000, 1600-1700, rel. VOFC: W 1000-1300, Sun 1000-1400; 1539kHz: W 1100-1130, 1430-1500, 1530-1600.
Taitung: 1008kHz: W 0900-1000, 1030-1100.
ANN: Chinese: "Chungkuo Kuangpo Kungssu, Ti Hsinwen tiaopin (liuhsing) Kuangpowan" (1st/News/FM (popular) network) or "Chungkuo Kuangpo Kungssu (location) Kuangpo Tientai," Taiwanese: "Tiyon Gok Kon Po Kon Sih, (location) Kon Po Den Tai" — **V.** irr. by QSL-card.

CENTRAL BROADCASTING SYSTEM (CBS)

(Independent broadc. division under BCC. CBS is in charge of the broadcasts to the Mainland.)
ADDR: 55 Pei An Rd, Ta Chih, Taipei 10494. Kuanghua Broadc. St.: P.O. Box 89, Taipei. **Te:** 596-4697. **Telex:** 26484.

Mediumwaves	**kHz**	**kW**	**N**				
Luchiang	603	1000	1	Kuanyin	1053	100	3
Minhsiung	746	100	1	Tanshui	1098	100	1
Kuanyin	801	250	3	Kinmen	1107	—	3/4
Kuanyin	846	250	3	Minhsiung	1206	100	5
Tanshui	900	100	1	Kuanyin	1251	100	3
1)	927	1200	2	Matsu	1341	—	3
Matsu*	961v	—	3	Kinmen	1467	—	3/4
Luchiang	1008	600	4	1)	1521	1200	2

1) reported location: Changchih (Pingtung prov.) — *) inactive.

SHORTWAVES

kHz	**kW**	**N**						
3335	10	5/1	7105	300	2	9690	10	5/1
6040	20	5/1	7150	100	1	11775	300	2
6087	100	4/1	7250	100	1	11905	100	1/6
6180	300	2	9630	300	4/1	15280		J4

J4) Voice of June 4th.
FM: 82.0MHz 3rd Netw.

D.PRGR: 1st Netw. in Chinese (for general public): 24h on 900kHz. Also on 603kHz (0950-1900), 746/1098kHz (2050-0100, 0350-0700, 0950-1900), 7150kHz (2050-0100, 0350-1900), 7250kHz (2050-0700, 0950-1900), 11905kHz (2050-0100, 0350-0950, 1630-1900). At time of editing also 2200-2400 on 3335/6040/6087/9630/9690kHz.
2nd Netw. in Chinese (for members of Communist party): 0955-2155 on 927/1521/6180/11775kHz, 2155-0955 on 7105kHz, 2155-2400 on 6180/11775kHz.
3rd Netw. (Kuanghua Broadc. St.) in Chinese: 0855-1805 on 801/961/1053/1107/1251/1341/1467kHz.
4th Netw. in Chinese (news and musical prgr): 0955-2200 on 1008/6087/9630kHz.
5th Netw: 0950-2200 in Taiwanese on 1206kHz, Hakka on 3335kHz and Cantonese on 6040/9690kHz.
6th Netw. on 11905kHz: 1000-1130 Mongolian, 1140-1310 Tibetan, 1310-1320 Chinese (special messages), 1330-1500 Uighur, 1530-1630 Voice of Free Asia in Chinese (30 min)/Korean (30 min), produced by KBS, Rep. of Korea.
Voice of June 4th in Chinese: On 1st Netw. 0250-0340, 0615-0700, 0915-0955, 1030-1300, 1630-1830 & 2215-0020, also 2100-2200 on 15280kHz — **Addr:** P.O. Box 157939, Chicago, IL 60815.
ANN: "Chungyang Kuangpo Tientai," — exc. 3rd Netw: "Kuang Hua chih Sheng, Kuang Hua Kuangpo Tientai."
V. by QSL-certificate.

"VOICE OF FREE CHINA" (INTERNATIONAL SERVICE, BCC)

ADDR: P.O. Box 24-38, Taipei. **Te:** +886 (2) 771 3849.
L.P: Dir: John C.T. Feng. CE: Huang Shyh Shyan.
STATIONS: (r. sites Anping & Pintung) 10/50/100/250kW—. *) relay via WYFR, USA.

kHz	**kHz**	**kHz**	**kHz**	**kHz**	**kHz**
5950*	9730	9955	11825	15270	17720
7130	9765	11740*	11860	15345	17750*
7445	9845	11745	11915*	15365	17805*
9680*	9850*	11805*	15130*	15370	17845*

Amoy	**Area**	**kHz**
0000-0100	NAm	5950*
0300-0400	SEAs	7130, 11825, 11915
0800-0900	NAm	5950*
1000-1100	EAs	11745
	SEAs	11825, 11915
1500-1600	SEAs	11860, 15370, 17720
Arabic		
2000-2100	Af/ME	9955
Cantonese		
0100-0300	NAm	5950*
0300-0400	SAm	11740*

0500-0600	FE	15345
	NAm	5950*, 9680*
	SEAs	7130, 11745, 11825, 11860 11915, 15270, 15370, 17720
	Pac	9765
0600-0700	NAm	5950*
	SEAs	11745
	FE	15345
1000-1100	SEAs	7445, 11860, 15270, 15370, 17720
	Pac	
1300-1400	SEAs	11860, 15370, 17720

Chaochau

0300-0400	SEAs	11860, 15270, 15370, 17720
1100-1200	SEAs	11860, 15270, 15370, 17720

English

0200-0300	NAm	5950*, 9680*
	FE	15345
	SEAs	11860
	CAm	11740*
	Pac	9765
0300-0400	NAm	5950*, 9680*
	FE	15345
	SEAs	11745
	Pac	9765
0700-0800	NAm	5950*
2200-2300	Eu	9850*, 11915*

French

2000-2100	Eu	9850*, 11915*, 15365
	SEAs	15370

German

2100-2200	Eu	9850*, 11915*, 15365

Hakka

1100-1200	SEAs	11915, 15345
1400-1500	SEAs	11915, 15345

Indonesian

0600-0700	Pac	9765
1300-1400	SEAs	7130,11825, 11915
	SEAs	11915,15345

Japanese

0100-0200	FE	7130, 11745, 15345
1100-1200	FE	7130, 11745
1400-1500	FE	7130, 11745

Korean

1000-1100	FE	7130, 9730
1300-1400	FE	7130, 11745
2200-2300	FE	7130, 9730, 11745

Mandarin

0100-0200	SAm	15215*, 17845*
0400-0500	NAm	5950*, 9680*
	FE	15345
	SEAs	7130, 11745, 11825, 11860, 11915 15270, 15370, 17720
	Pac	9765
0900-1000	FE	7130, 7445, 9730, 11745
	SEAs	11825, 11860, 11915, 15270 15370, 17720
	Pac	11915
1200-1300	FE	7130,11745
	SEAs	11860, 11915, 15270, 15345 15370, 17720
1600-1700	SEAs	15370
1857-1957	Eu	9850*, 17750*
	ME/NAf	9955
2300-2400	FE	7130, 9730, 11745
	Pac	9845

Spanish

0200-0300	SAm	15215*, 17845*
0400-0500	CAm	11740*
2300-2400	CSAm	15130*, 17805*

Thai

0600-0700	SEAs	11860, 15270, 15370, 17720
1400-1500	SEAs	11860, 15370, 17720

Vietnamese

1500-1600	SEAs	11915, 15345

ANN: "This is the Voice of Free China". Chinese: "Tzu Yu Chung Kuo Chih Sheng".
V. By QSL card. Rec. acc—**PUB:** Prgr. schedule free.

VOICE OF ASIA

ADDR: P.O. Box 880, Kaohsiung, Taiwan, ROC.
Te: +886 (2) 771-0151, ext. 283. **Cable:** VOAsia. Kaohsiung.
L.P: Dir: Chen Whai-jyue.
MEDIUMWAVE: Kaohsiung: 585kHz 1200kW — **SW:** 7285/7445/9280kHz 100kW.
Area: E = Philippines, East Malaysia, Eastern & Central Indonesia. W = Hong Kong, Mainland China, Thailand, West Malaysia, Singapore, Western Indonesia.

Mandarin	**Area**	**kHz**
0400-0800	W	7285
0900-1100	W	9280
1050-1100	E	585
English		
1100-1200	E	585, 7445
Indonesian		
1200-1300	E	585, 7445
Mandarin		
1300-1445	E	585, 7445
1450-1500	W	585, 7445
Indonesian		
1500-1600	W	585, 7445
Thai		
1600-1700	W	585, 7445

V. by QSL-card.

OTHER PUBLIC & COMM. ST'S

NATIONAL ASSOCIATION OF BROADCASTERS OF THE REPUBLIC OF CHINA

ADDR: 53 Jen Ai Rd, Sec. 3, Taipei. **Te:** +886 (2) 751-7290.
Cable: Broadcast Taipei. **Telex:** 27498 BCCROC.
L.P: Secr. Gen: Shih Chia Sheng.
Member stations: 1, 3, 6, 7, 12, 13, 14, 15, 17, 18, 19, 20, 21, 22, 23, 24, 25, 28, 30.
Other st's are Public or Military owned. * r. inactive.
Powers: a) 0.5kW, b) 1kW, c) 1.5kW, d) 2kW, e) 3kW, f) 3.5kW, g) 5kW, h) 10kW, i) 20kW, j) 50kW, k) 100kW. Others unknown.
S = AM stereo (C-QUAM system).

	Location	**kHz**			
2)	various	549	12)	Tainan I	837b
1b)	Chingshui	558b	9f)	Penghu (r.840)	846h
2)*	Taipei	558b	1d)	Nanfang'ao	846b
5)	Taipei	576S	2)	various	846
2)	Taipei	594h	1a)	Chia-i	855b
2)	various	594	17)	Taipei II	855b
2)	various	612	18)	Taichung	864h
6c)	Mingchien	630b	19a)	Penghu	882b
6a)	Tahsi	639b	2)	Hsilo	882
1e)	Tawu	648b	2)	Taipei	909g
1b)	Taichung II	657h	2)	various	909
7)	Hsinhua	666	1e)	Chihshang	918b
2)	various	666	2)	Sani	927
2)	various	684	2)	?	936
9)	Taipei I	684	9a)	Taipei II	936
9)	Taoyuan	693h	7)	Hsinying	954h
9g)	Tainan	693e	20)	Wuhaoshan	954
2)	various	693	21a)	Tunglo	950
8b)	Taichung I	702h	6c)	Chunghsing	963h
10)	Hsinfeng	711j	10)	Hsinfeng	981j
11)	Taipei	729b	19)	Kaohsiung II	981b
31)	Kaohsiung ?	738k	27)	Taipei I	990
12)	Penghu II	756b	1b)	Taichung I	990
2)	?	756	8f)	Hualien	990h
2)	various	765	8e)	Ilan	990b
6b)	Taichung	774h	1f)	Kaohsiung	1008b
13)	Taoyuan	774f	22)	Taipei	1008b
9d)	Hualien	792	23)	Sanchung II	1026b
14)	Keelung	792b	21b)	Yuanli	1026
2)	various	792	3)	Kaohsiung	1044g
7)	Hsinhua	801	20)	Hualien	1044b
15)	Changhua	810h	24)	Pingtung	1062g
6a)	Kuanshi	810b	1d)	Ilan	1062b
8d)	Kaohsiung	819h	25)	Tainan	1071b
1)	Taipei I	819g	2)	Taipei	1089g

2)	various	1089	29)	Taipei	1280b
9)	Yunlin	1089a	9b)	Taichung	1287g
16)	Kaohsiung	1089h	24)	Fangliao	1287b
2)	various	1116	17)	Taipei I	1296b
9)	Taipei I	1116c	8c)	Hsinying	1314g
9)	Ilan	1116g	21)	Chunan	1314b
8a)	Hsinchu	1116g	6)	Taipei I	1323b
8d)	Kaohsiung	1116g	6c)	Puli	1323b
1c)	Yünlin	1125g	9e)	Tsaoying	1332
8g)	Taitung	1125g	1d)	Toucheng	1341
26)	Taipei I	1134g	23)	Sanchung I	1350b
31)	Penghu ?	1143	9d)	Hualien	1359b
28)	Taipei I	1152g	1f)	Chishan	1377
19)	Kaohsiung I	1161b	2)	Taipei	1377b
1)	Taipei II	1170g	2)	various	1377
15)	Erhlin	1179b	1c)	Peikang	1386b
6)	Taipei II	1188b	30)	Keelung	1404b
12)	Tainan II	1188	7)	Kuanyin	1422b
6a)	Hsinchu	1206b	10)	Chingshui	1431j
2)	various	1224	10)	Taipei	1440h
28)	Taipei II	1224g	10)	Taichung	1476h
23)	Juifang	1233b	26)	Taipei II	1476
9c)	Kaohsiung	1251g	2)	various	1476
8)	Taipei	1260h	4)	Taipei	1494h
1a)	Putzu	1260b	4a)	Changhua	1494g
9f)	Penghu	1269	8)	Taipei	1512h
1e)	Taitung	1269b	2)	various	1512
1b)	Tungshih	1269b	5a)	Taichung	1566h
2)	various	1269	27)	Taipei II	1575

FM:

	MHz		MHz		MHz		MHz
8d)	93.1	2)	100.7	4a)	103.5	8c)	104.9
2)	93.1	5)	100.7	4d)	103.5	8b)	104.9
26)	93.1	5b)	100.7	4c)	103.7	9)	105.3
8)	94.3	8e)	101.3	2)	103.9	2)	105.9
8b)	94.3	8f)	101.3	9)	104.5	8g)	106.5
16)	94.3	8g)	101.3	9d)	104.5	9)	106.5
5)	100.1	9)	101.3	8)	104.9	9)	107.3
5a)	100.1	4)	101.7	8d)	104.9	2)	107.8
4e)	100.3	4b)	101.7				

NB: Many st's rel. BCC-1 2300-2355.

Addresses and other information (Date and times all acc. to UTC)

1) Cheng Sheng Broadc. Co, 7th Flr, 66-1, Sec. 1, Chungching So. Rd, Taipei. 1st prgr on 819kHz, 2nd prgr on 1170kHz, both 24h (exc. Sun 1800-2100)—1a) 17, Chuiyang Rd, Chia-i 2055 (Sat 2155)—1830 (Sun 1600) 1b) Nanmen Bridge, Taichung (1st prgr on 558, 990, 2nd prgr on 657, 1269, both 24h—1c) 10, Shijuijuan Rd, Huwei Town, Yünlin (sched. as 1a)—1d) 45 Chienchung Rd, Ilan 2055 (Sat 2155)-1800 (Sun 1700) —1e) 21, Lane 380, Shinsheng South Rd, Taitung 2120 (Sat 2125)-1703 (Sun 1503)—1f) 838, Chengching Lake, Chengching Rd, Kaohsiung (24h exc. Sat 2100-2155, Sun 1800-2055).

2) Fuhsing Broadc. Corp, 5, Lane 280, Sec. 5, Chung Shan No. Rd, Taipei (operated by Dept. of the Interior). Most st's: 24h. Fuhsing Broadc. Co. has 26 local st's and is found in most cities and some mountain villages. Local st's have local prgrs mainly during local daytime, using several freq's. Most st's use one freq. on MW for relaying BCC News Netw. **ANN:** "Fuhsing Kuangpo Tientai, (location) Tai".

3) Cheng Kung Broadc. St, 63 Chunghua 3rd Rd, Kaohsiung. 24h (Sun-1600).

4) Educational B'casting System, 41 Nan Hai Rd, Taipei 10728. — 2000-1800 (**F.Pl:** 24h). Local prgr's 1000-1100. **ANN:** "Chiao Yü Chih Sheng, Chiao Yü Kuangpo Tientai." — 4a) Changhua — 4b) 380 Kuangtung 3rd Rd, 80610 Kaohsiung — 4c) 10-5 Linyon Rd, 97052 Hualien — 4d) 30 Lane 76, Shengli Str, 95013 Taitung — 4e) Yuli (relay st.).

5) International Community Radio Taipei, 8 Chung Yung 2nd Rd, Shan Tze Hou, Yangmingshan, Taipei 113. News Dir: Bryan Curtis. 24h in **English** (separate prgrs on MW & FM at times). **N:** on the h. — 5a) Taichung — 5b) Tainan.

6) Taiwan Broadc. Corp, 11th Fl, 21, Chung Ching So. Rd, Sec. 3, Taipei. 24h (exc. Sun 1800-2100). (1st Prgr. on 1323kHz, 2nd Prgr. on 1188kHz)—6a) No. 2, Lane 506, Kao Feng Rd., Hsinchu 300. 2100-1800 (Sun 1600)—6b) 42-1, Sec. 2, Nantun Road, Taichung 40804. 24h—6c) 219 Yuping Road, Tsaotun Town, Nantou 542. 24h (exc. Sun 1800-2050). 1700 s/off on 630kHz.

7) Chien Kuo Broadc. St, 78 Chienkuo Str, Peiliaoli, Hsinying. 2050 (Sat 2150)-1705 (Sun 1605).

8) Public Radio System, 1 Tien Tsin Str, Taipei 10041. 24h on MW/FM (same prgr). Taipei Traffic Prgr 24h on Public Radio System 1512kHz + 94.3MHz — 8a) 126 Minchuan Rd, Hsinchu 30607 — 8b) 99 Poai Str, Nantun, Taichung 40809. Traffic Prgr. on 94.3MHz — 8c) 49 Min Chih Rd, Hsinying, Tainan 73011 — 8d) 246 Chungcheng 4th Rd, Kaohsiung 80116. Traffic Prgr. on 819kHz + 93.1MHz. — 8e) 86 Wusha Rd, Ilan 26021—8f) 21-2 Fuchian Rd, Hualien 97038 — 8g) 3,191, Lane 719, Sec. 1, Chunghua Rd, Taitung 95003 — 8a, 8b, 8c, 8f local prgrs W 0730-0900 on MW freq. — **V.** by QSL-card. **F.Pl:** Sep. prgr on AM & FM. AM will air mainly sports prgr.

9) Han Sheng Broadc. St., 5F, 3, Sec.1, Sinyi: Rd, 10017 Taipei (military broadc. st.), Taipei. 24h [exc. 9a), 9f)]. Mainly separate prgr's on AM & FM. Rel. CBS 1630-1800 on MW—9a) Kuo Kuang Broadc. St, 122, sec. 1, Chung Ching So. Rd., 10016 Taipei. 2045-0230 (Rel. 9) 2045-2255), 0600-0830. Rel. 9) 1000-1800 — 9b) 178 Chensing Rd, 40114 Taichung. Local: W 0100-0200 & Tue/Thu/Sat 1130-1200, Sun 1300-1400 on 1287kHz — 9c) Weiwuying, Fengshan, Kaohsiung 830. Local: W 0310-0400, Sun 0920-1010 — 9d) 643 Chungcheng Rd, 97004 Hualien. Local: W0210-0250, Sun 0130-0200 on 792/1359kHz; W 0210-0250 0800-0850, 1040-1130, Sun 0130-0200 on 104.5MHz — 9e) 40 Mingte-hsing village, 81303 Tsaoying. Local: W 0100-0200, 1130-1200 on 1332kHz — 9f) Chengkungying, Makong, 88000 Penghu. 2000-1900. Local: W 0100-0200, 0900-1000, Sun 0130-0200 on 846/1269kHz — 9g) Fuhsing Rd. Yongkong Village, 71015 Tainan. Local: W 0210-0250, 0900-0930, 1130-1200 on 693kHz.

10) Air Force Broadc. St, 53 Jenai Rd, Sec. 3, Taipei. 2050-?, 0850-1710. **ANN** "Kung Chün Chih Sheng". **N.B:** At time of editing active on 980, 1431kHz.

11) Shi Hsin Broadc. St. (World News Broadc. St), Shichie Hsinwen, 15, Sec. 1, Mushah Rd, Taipei. 0225 (Sun 0255)-0605, 0855-1405.

12) Voice of Victory Broadc. Co, 22 Chienkang Rd, Tainan City. Gen. Mgr: Yuan Hua Lee. St. Mgr: Michael Liu. Dir. Tec: Chen Chi Wu. 1st Prgr. on 837kHz, 2nd Prgr. on 1188kHz, both 2130 (Sat 2220)-1830 (Sun 1600). — **ANN:** "Tainan Shengli chih Sheng Kuangpo Kungssu."

13) Hsien Sheng Broadc. Co, 491 Chungshan Rd, Taoyuan 330. 24h.

14) Cheng Hua Broadc. St, 58-4 Nan Hsin Str, Nan Tsung Rd, Keelung. 2055-1810 (Sun 1710).

15) Kuo Sheng Broadc. Co, 2 Chungshan Vil, Chungshanli, Pakuashan, Changhua 500. 2050 (Sat 2140)-1800 (Sun 1600). Erhlin relay st: 2150-1500.

16) Kaohsiung Broadc. St., 90 Hsinchiang Rd, Kaohsiung 804. (operated by Kaohsiung City Council) — Pres: Wen-ming Chu. 2150-1600.

17) Ming Pen Broadc. Co, 6F, No. 70, Lane 640, Wan-Ta Rd, Taipei 109. Gen. Mgr: Chung-hsin Hu. 1st Prgr on 1296kHz, 2nd Prgr on 855kHz, both 24h exc. Sun 1805-2100.

18) Chung Sheng Broadc. Co, 134 Kuangfu Rd, Taichung. (Rlg. st). 2040-1800.

19) Feng Ming Broadc. Co, 492 Chiu Ju 2nd Rd, Kaohsiung. 1st Prgr. on 1161kHz, 2nd prgr on 981kHz, both 24h exc. Sun 1705-2055 — 19a) Chen Tieh Hsien, Li 38, Makung, Penghu.

20) Yen Sheng Broadc. St, 76 Minchuan Rd, Hualien City. 2100 (Sat 2200)-1605.

21) Tien Sheng Broadc. St, 285 Kongyi Rd, Chunan. 2045 (Sat 2200)-1700 (Sun 1600) — 21a) 8 Kouyla, Chungshiue Rd, Yuanli—21b) 38-1, Tso Dong, Wen Hua Str, Tunglo.

22) Tien Nan Broadc. St, 31, Hangchow So. Rd, Sec. 2, Taipei — Pres: Shang Shaung Ling. 24h.

23) China Broadc. Co, 15, Lane 1, Shuangyuan Str, Sanchung, Taipei — Pres: John Y. Hwang. 1st prgr on 1350kHz, 2nd prgr on 1026kHz, both 2150 (Sat 2250)-1815 (Sun 1600) — **ANN:** "Chung Hua Kuangpo Tientai Ti I/Erh Tai".

24) Min Li Broadc. St, 57-20 Min Sheng Rd, Pingtung. 24h.

25) Tien Sheng Broadc. St, 22 Chienkang Rd, Tainan — Gen. Mgr: Ming W. Lee. 2130 (Sat 2220)-1805 (Sun 1600).

26) Taipei Broadc. St., 62-4, Sec. 3, Chung Shan No. Rd, Taipei. (operated by Taipei City Council). 24h. Prgr 1 on 1134kHz + 93.1MHz, Prgr 2 on 1476kHz.

27) Yu Shih Broadc. St. (China Youth Broadc. St), 16, Sec. 4, Chung Shan No. Rd, Taipei. 2150-0600, 0850-1605. (1st Prgr. on 990kHz, 2nd Prgr on 1575kHz).

28) Hua Sheng Broadc. Co, 18 Hwasheng Str, Shihlin, Taipei. 1st Prgr on 1152kHz, 2nd Prgr on 1224kHz, both 24h.

29) Fu Hsing Kang Broadc. St. (Armed Forces Broadc. St), Fu Hsin Kang, Peitou, Taipei. 2200-0100, 1055-1400.

30) Yi Shih Broadc. St., 75 Bai San Str, Chi-Tu, Keelung (Rlg. st.)— Mgr: Tome Lin. 2130 (Sat 2200)-1700 (1600). **ANN:** "Keelung Yi Shih Kuangpo Tientai".
31) Taiwan Area Fishery Broadc. St., P.O. Box 1110, Kaohsiung. — St. Mgr: Jui-Chang Pai. Dir.Tec: Wen-Yan Hwang. 24h. Weather rpt. at every h. **ANN:** "Taiwan Chü Yüyeh Kuangpo Tientai".

CHRISTMAS ISLAND (Australian Territory)

L.T: UTC + 7h — **Pop:** 2.200 — **Pr.L:** English, Malay, Cantonese, Mandarin — **E.C:** a/c 50, 220V — **ITU:** CHR.

CHRISTMAS ISLAND COMMUNITY RADIO SCE. VLU–2

ADDR: Admin. of the Territory of Christmas Isl, Indian Ocean 6798, Australia. **Te:** 0011 6724 8422 **Fax:** 0011 6724 8425 **L.P:** Co-ord: Mrs. D. Pereira. Superv. VLU–2: Barrie Fullard.
STATION: (G.C: 105.43E/10.25S): VLU–2 1422kHz 0.25kW.
D.PRGR: Local prgrs: 1100-1400. **Continuous music:** 2300 (Sun 0100)-1100. Occ. relay of R. Australia or BBCWS. The carrier remains on the air continuously as the st. is used for shipping, aircraft & weather information.
ANN: "The Power St."—**V.** by letter.

COCOS (Keeling) ISLANDS (Australian)

L.T: UTC + 6.30h — **Pop:** 600 — **Radio sets:** 200 (est.) — **Pr.L:** English, Cocos Malay—**ITU:** ICO.

RADIO VKW (Gov.)

ADDR: P.O. Box 70, Cocos (Keeling) Islands, Indian Ocean 6799. **L.P:** St. Mgr: S. O'Neil. Tec. Dir: K. Beard. Prgr. Mgr: Ms. A. Parker.
STATION (G.C: 96.49E/12.09S): VKW 1404kHz 0.1kW.
D.PRGR: 24h. Rel. BBC World Sce. 1300-1730. Rel. R. Australia 1730-2400 — **V:** by QSL-card.

HONG KONG (British Crown Colony)

L.T: UTC + 8h — **Pop:** 5.890.000 — **Radio sets:** 2.5m — **Pr.L:** Cantonese, English — **E.C:** a/c 50, 200V — **ITU:** HKG.

RADIO TELEVISION HONG KONG (Gov.)

ADDR: Broadcasting House, Broadcast Drive, P.O. Box 70200, Kowloon Central Post Office, Hong Kong. **Te:** +852 339 6300. **Cable:** "Broadcasts". **Telex:** 45568. **Fax:** +852 338 0729. **L.P:** Dir. of Broadc: Miss Man-yee Cheung. JP. Dep. Dir. of Broadc: Chu Pui-hing. Asst. Dir. of Broadc. (Radio): Ng Sek-fai. Asst. Dir. of Broadc. (PATV): Miss Maria Wan. Contr. (Eng. & Prod. Sces): Phillip Manns. Contr. (ETV): Miss Maria Wan. Departmental Secretary: B.S. Pannu.

Mediumwaves:

kHz	Netw.	Location	kW
567	Radio 3	Golden Hill	20
621	Radio 7	Golden Hill	20
675	Radio 6	Peng Chau	10
783	Radio 5	Golden Hill	20
1584	Radio 3	Chung Hom Kok	0.1

FM:

MHz	Netw.	kW	Tr. Location	Target Div.
92.6	Radio 1	1	Mt. Gough	Kowloon
92.9	Radio 1	0.1	Golden Hill	Tsuen Wan
93.2	Radio 1	0.5	Cloudy Hill	Fan Ling
93.4	Radio 1	0.5	Castle Peak	Tuen Mun
93.5	Radio 1	0.1	Beacon Hill	Shatin
93.6	Radio 1	0.3	Lamma Island	HK Island south
94.4	Radio 1	0.5	Kowloon Peak	HK Island north, Sai Kung
94.8	Radio 2	1	Mt. Gough	Kowloon
95.3	Radio 2	0.5	Cloudy Hill	Fan Ling
95.6	Radio 2	0.1	Golden Hill	Tsuen Wan
96.0	Radio 2	0.3	Lamma Island	HK Island south
96.3	Radio 2	0.1	Beacon Hill	Shatin
96.4	Radio 2	0.5	Castle Peak	Tuen Mun
96.9	Radio 2	0.5	Kowloon Peak	HK Island north, Sai Kung
97.6	Radio 4	1	Mt. Gough	Kowloon
97.8	Radio 4	0.5	Cloudy Hill	Fan Ling
97.9	Radio 3		Mt. Nicholson	Jardine's Lookout
98.1	Radio 4	0.1	Beacon Hill	Shatin
98.2	Radio 4	0.3	Lamma Island	HK Island south
98.4	Radio 4	0.1	Golden Hill	Tsuen Wan
98.7	Radio 4	0.5	Castle Peak	Tuen Mun
98.9	Radio 4	0.5	Kowloon Peak	HK Island north, Sai Kung
106.8	Radio 3		Chung Hom Kok	HK Island south

RTHK Radio 1 in Catonese: 24h. **N.** half-hourly 0030-1500. Relay BBC-Cantonese 1300.
RTHK Radio 2 in Cantonese: 24h. **N.** half-hourly 2200-1500, hourly 1500-2200. Relay BBC-Cantonese 1300.
RTHK Radio 3 in English: 24h. AM-Stereo **N.** half-hourly 2200-0200, 0900-1100, hourly 0200-0900, 1100-1600. Relay Radio 4 0200-0400, 1800-2200.
RTHK Radio 4 in English/Cantonese: 24h. **N.** irr. Relay BBC-WS 2300, Relay Radio 3 1000-1200, 1600-1800.
RTHK Radio 5 in Cantonese/Chinese: 24h. AM-Stereo **N.** half-hourly 2300-0200. Hourly 0300-1500. Relay Radio 1 2300-0200. Relay BBC-Cantonese 1300.
RTHK Radio 6 Relay BBC-WS: 24h.
RTHK Radio 7 in Cantonese: 20h 2200-1805 **N.** quarter-hourly.
NB: At 1200-2300, Radios 1, 2, and 5 are in parallel.
V. by QSL-card.

HONG KONG COMMERCIAL BROADC. CO. LTD.

ADDR: 3 Broadcast Drive, GPO Box 3000, Kowloon, Hong Kong. **Te:** +852 336 5111. **Fax:** +852 338 0021. **Cable:** Radioads. (Addr. in Hong Kong: Prince's Bldg. Hong Kong Isl. **Te:** +852 523 6224).
L.P: Chmn. & Man. Dir: G. Ho. Gen. Mgr: Winnie Yu.

Mediumwave

kHz	Location	Prgr.	kW
864	Peng Chau	AM Alive	10

FM:

MHz	Netw.	kW	Tr. Location	Target Div.
88.1	CR1	1	Mt. Gough	Kowloon
88.3	CR1	0.5	Cloudy Hill	Fan Ling
88.6	CR1	0.5	Castle Peak	Tuen Mun
88.9	CR1	0.1	Golden Hill	Tsuen Wan
89.1	CR1	0.3	Lamma Island	HK Island south
89.2	CR1	0.1	Beacon Hill	Shatin
89.5	CR1	0.5	Kowloon Peak	HK Island north, Sai Kung
90.3	CR2	1	Mt. Gough	Kowloon
90.7	CR2	0.5	Cloudy Hill	Fan Ling
90.9	CR2	0.1	Golden Hill	Tsuen Wan
91.1	CR2	0.1	Beacon Hill	Shatin
91.2	CR2	0.5	Castle Peak	Tuen Mun
91.6	CR2	0.3	Lamma Island	HK Island south
92.1	CR2	0.5	Kowloon Peak	HK Island north, Sai Kung

HKCR CR1 in Cantonese: 24h. **N:** half-hourly.
HKCR CR2 in Cantonese: 24h. **N:** half-hourly.
NB: 1700-2200 CR1 and CR2 as common prgr.
HKCR AM Alive in English: 24h. Relay CR1 0412-0900 (weekdays only). **N:** half-hourly 2200-0400, 0900-1600, then hourly.
V. by letter.

METRO BROADCAST CORPORATION LTD.

ADDR: Site 11, Basement 1 Whampoa Gardens Hunghom, Kowloon Hong Kong. **Te:** +852 364 9333. **Fax:** +852 364 2445.
L.P: Gen. Mgr: Craig Quick. Dep. GM: Douglas Gautier. Eng. Dir: Robert N. Palitz. Channel Mgrs: Kelly Dean (Metro News), Joe Chen (Hit Radio), Larry London (Metro Select).

Mediumwave

kHz	Netw.	kW	Location
1044	Metro news	10	Peng Chau

FM:

MHz	Netw.	kW	Tr. Location	Target Div.
99.7	Hit Radio	1	Mt. Gough	Kowloon
100.0	Hit Radio	0.5	Cloudy Hill	Fan Ling
100.4	Hit Radio	0.5	Castle Peak	Tuen Mun
100.5	Hit Radio	0.1	Beacon Hill	Shatin
101.6	Hit Radio	0.1	Golden Hill	Tsuen Wan
101.8	Hit Radio	0.5	Kowloon Peak	HK Island north, Sai Kung
102.1	Hit Radio	0.3	Lamma Island	HK Island south
102.4	FM Select	0.1	Beacon Hill	Shatin
102.5	FM Select	0.5	Castle Peak	Tuen Mun
104.0	FM Select	1	Mt. Gough	Kowloon
104.5	FM Select	0.3	Lamma Island	HK Island south
104.7	FM Select	0.5	Cloudy Hill	Fan Ling
105.5	FM Select	0.1	Golden Hill	Tsuen Wan
106.3	FM Select	0.5	Kowloon Peak	HK Island north, Sai Kung

Metro News in English: 24h. News and Information.
Metro Hit Radio in Cantonese: 24h. Music, conversation and information.
Metro FM Select in English/Cantonese: 24h. Easy listening music.

BRITISH FORCES BROADC. SCE. HONG KONG

ADDR: SSVC-BFBS, Borneo Lines, Sek Kong, Hong Kong (Mil. Addr: SSVC-BFBS, BFPO 1). **Te:** +852 483 7301. **Fax:** +852 488 5455. **Telex:** 34513 BFBS HK.
L.P: GM: Marc Tyley. PD: Allan Miller. Nepali Dir: Kishore Gurung. Nepali PD: Samdupssering Bhutia. Eng. Mgr: Geoff Evans.

Mediumwave

kHz	kW	Location	Prgr.
1251	10	Tam Mei	Nepali/English

FM

MHz	kW	Location	Prgr.
95.4	0.001	Kings Park	Nepali/English
96.6	0.01	Brick Hill	English
102.0	0.01	Sek Kong	English
102.0	0.001	Dills Corner	English
102.0		Hong Lok Yuen	English
104.8	0.001	Tuen Mun	English
104.8	0.01	Kings Park	English
104.8		Happy Valley	English

English Sce: 24h. **N:** Half-hourly 2200-0100 and 0430, otherwise hourly.
Nepali Sce: 2200-0600 & 1100-1500.
V. by letter.

BBC EAST ASIAN RELAY STATION

STATION (G.C: 113.55E/22.55N): 2×250kW tr's.

kHz	kHz	kHz	kHz	kHz	kHz	kHz
7185	11765	11820	11945	15280	17830	21715

D.PRGR: See United Kingdom.

INDIA

L.T: UTC + 5½h — **Pop:** 882.281.000 — **Radio sets:** 55.000.000 (approx) — **ITU:** IND — **Pr.L:** Assamia, Bangla, Dogri, English, Gujarati, Hindi, Kannada, Kashmiri, Marathi, Malayalam, Nepali, Oriya, Punjabi, Sanskrit, Sindhi, Tamil, Telugu & Urdu — **E.C:** a/c 50 + d/c, 230/400V (yes).

MINISTRY OF INFORMATION & BROADCASTING

Main Secretariat, A-Wing, Shari Bhawan, New Delhi-110 001. **Tlx:** 81-31-66349 bib in. **Cable:** Information.
L.P: Minister for Inf. & Broadc: Ajit Panja. Inf. & Broadc. Secr: Ashok Chandra.

AKASHVANI (Gov, semi-comm.)

Administration: Directorate General of All India Radio, Akashwani Bhawan, 1 Sansad Marg, New Delhi-110 001. **Te:** +91 (11) 3710006 (pabx). **Tlx:** 81-31-65585avdg in — **L.P:** Dir. Gen: S. Kapoor. Eng-in-Chief: K.P. Ramaswamy. Asst.Dir. (Freq. Assignments): G. Kuruvilla. (in charge of processing reception reports).
Programming: Broadcasting House, 1 Sansad Marg, New Delhi-110 001. **Te:** +91 (11) 3715411 (pabx). **Cable:** Akashvani — Dir. Gen. (News): B. Bhalla. Dir. Transcription & Prgr. Exchange Sces: G. Raghuram.
Commercial Sce: Vividh Bharati Sce, AIR, 101 M.K. Rd, Bombay-400 020. **Te:** +91 (22) 219415. **Tlx:** 81-11-85414 cbvu in. **Cable:** Vividh Bharati —Dep. Dir: V. Chhabra.
Audience Research: Audience Research Unit, AIR, 2nd Floor, PTI Building, Sansad Marg, New Delhi-110 001. **Te:** +91 (11) 3710033 — Dir: S.K. Khatri. **National Channel:** AIR, Gate 22, Jawaharlal Nehru Stadium, Lodhi Road, New Delhi-110 003. **Tlx:** 81-31-65560 nchjln in — Dir: G.K. Chaturvedi. St. Eng: S.K. Sharma.
Research & Development: Office of the Chief Eng R & D, All India Radio, 14-B Ring Road, Indraprashtha Estate, New Delhi-110 002. **Te:** +91 (11) 3311762. **Tlx:** 81-31-65478 cerd in. **Cable:** Airsearch — Chief Eng: K.M. Paul.
Monitoring: International Monitoring St., All India Radio, Dr. K.S. Krishnan Rd, Todapur, New Delhi-110 097. **Te:** +91 (11) 581461 — Asst. Research Eng (Frequency Planning): M.P. Singh.
Central Monitoring St., AIR, Ayanagar, New Delhi-110 047. **Te:** +91 (11) 666306. **Cable:** Airmonitor — Dir: S. Haider.

Regional Headquarters:
North Zone: Jamnagar House, Shahjahan Road, New Delhi-110 011 — **East Zone:** 4th Floor, Akashvani Bhawan, Eden Gardens, Calcutta-700 001 — **North-East Zone:** AIR Chandmari, Guwahati-781 003, Assam — **West Zone:** AIR, Old CGO Building, 101 M.K. Road, (P.O. Box 11452), Bombay-400 020, Maharashtra — **South Zone:** AIR, Silingi Bldg, 38/39 Greams Road, Thousand Lights, Madras-600 006.

MEDIUMWAVES

kHz	kW	Station	Zone	kHz	kW	Station	Zone
531	100	Jodhpur A	N	963	20	Jalgaon	W
540	20	Aizawl	NE	972	100	Cuttack A	E
549	100	Ranchi	A	981	100	Raipur	N
558	100	Bombay B	W	990	300	Jammu A	N
567	300	Dibrugarh	NE	999	1	Almora	N
576	100	Alappuzha	S	999	10	Coimabatore	S
585	100	Nagpur A	W	1008	100	Calcutta B	E
594*	1000	Mogra	E	1017	1	Mysore	S
603	200	Ajmer	N	1017	10	Delhi D	N
612	200	Bangalore A	S	1026	1	Allahabad	N
621	100	Patna A	E	1035	10	Guwahati B	NE
630	20	Thrisoor	S	1044	100	Bombay A	W
639	50	Kohima	NE	1053	10	Leh	N
648	100	Indore A	N	1062	10	Pasighat	NE
657	100	Calcutta A	E	1071*	1000	Rajkot	W
666	100	Delhi B	N	1089	1	Jammu B	N
675	20	Chhattarpur	N	1089	20	Udipi	S
675	1	Itanagar	NE	1107	10	Gulbarga	S
675c	1	Bangalore B	S	1116	200	Srinagar A	N
684	10	Kozhikode A	S	1125	10	Udaipur	N
684	20	Port Blair A	S	1134*	1000	Mogra	E
702*	300	Jalandhar B	N	1143	20	Ratnagiri	W
711	200	Siliguri	E	1143	20	Rohtak	N
720	200	Madras A	S	1152c	1	Ranchi B	E
729	50	Guwahati A	NE	1161	10	Thiru'puram A	S
738	50	Hyderabad A	S	1170c	1	Hyderabad C	S
747	300	Lucknow A	N	1179	20	Rewa	N
756	20	Jagdalpur	N	1188c	50	Bombay C	W
765	200	Dharwar A	S	1197	1	Shillong A	NE
774	100	Shimla	N	1197	10	Tirunelveli	S
783c	2.5	Madras C	S	1197c	1	Jodhpur B	N
792	100	Pune A	W	1215	1	Pudducheri	S
801	200	Jabalpur	N	1215**	20	Kingsway	N
810	300	Rajkot A	W	1224	10	Calcutta D	E
819	200	Delhi A	N	1224c	1	Srinagar D	N
828	10	Silchar	NE	1233	1	Tura	NE
837	100	Vijayawada A	S	1233*	200	Tuticorin	S
846	200	Ahmedabad A	W	1233c	1	Bhopal B	N
864	100	Shillong B	NE	1242	100	Varanasi	N
873	100	Jalandhar A	N	1251	20	Sangli	W
882	50	Imphal	NE	1260	20	Ambikapur	N
891	10	Rampur	N	1269	10	Madurai	S
891c	1	Tiru'palli B	S	1269	20	Agartala	NE
900	100	Cuddapah	S	1269c	1	Jaipur B	N
909	100	Gorakhpur	N	1278c	1	Lucknow C	N
918	300	Suratgarh	N	1287	100	Panaji A	W
927	100	Visakapatnam	S	1296	10	Darbhanga	E
936*	100	Tiru'palli A	S	1305	10	Parbhani	W
945	20	Sambalpur	E	1314	10	Bhuj	W
954	100	Najibabad	N	1314c	1	Cuttack B	E

1323c	20	Calcutta C	E	1485	1	Adilabad	S
1332	10	Tezu	NE	1485	1	Bhopal A	N
1332	1	Mangalore	S	1485c	1	Allahabad B	N
1341	1	Kohima	NE	1485c	1	Vadodara	W
1350c	1	Dharwar B	S	1494c	1	Thiru'puram B	S
1350c	1	Jalandhar C	N	1503	1	Srinagar C	N
1359	20	Bhadravathi	S	1503c	1	Vijayawada B	S
1368c	10	Delhi C	N	1521	10	Tawang	NE
1377	10	Hyderabad B	S	1521	1	Aurangabad	W
1386	10	Gwalior	N	1530	10	Agra	N
1395	1	Madras B	S	1539	5	Panaji B	W
1395	10	Bikaner	N	1566**	1000	Buttibori	W
1404	10	Gangtok	E	1584	1	Jamshedpur	E
1422c	1	Rajkot B	W	1584	1	Keonjhar	E
1431c	1	Chandigarh	N	1584	1	Kota	N
1431c	1	Kozhikode B	S	1593c	1	Indore B	N
1440	1	Port Blair	S	1602	1	Nagercoil	S
1440c	1	Ahmedabad B	W	1602	1	Nagpur B	W
1449c	1	Kanpur	N	1602	1	Sholapur	W
1458	10	Bhagalpur	E	1602c	1	Patna B	E
1467	20	Jeypore	E	1602c	1	Pune B	W
1476	1	Jaipur A	N				

NB: Ajmer 603kHz relays Jaipur A 1476kHz, Alappuzha 576kHz relays Thiruvananthapuram A 1161kHz.
+) carries Vividh Bharati, *) carries Ext. Sce. **) carries National Channel.
D.Prgr: 0025-1830 (with or without breaks). May be extended for sports coverage and special events. **National Channel:** 1325-0040 in Hindi, Urdu and English.

SHORTWAVES

Regional Sces.

kHz	kW	Station	Zone	Times
3205+	50	Itanagar	NE	1415-1730
3205	50	Lucknow	N	0025-0200, 1215-1740
3223	2.5	Shimla	N	0045-0200 1300-1700/1741
3235	20	Guwahati B	NE	1230-1701(Sat 1700)
3245	50	Itanagar	NE	1230-1701(Sat 1730)
3255	50	Shillong	NE	1100-1630
3268	50	Kohima A	NE	1315-1630/1700
3277	50	Srinagar B	N	0120-0225, 1130-1739 (Sat 1746)
3305	2	Ranchi	E	0055-0200, 1200-1741
3315	10	Bhopal A	N	0025-0215, 1200-1741
3345+	50	Jaipur A	N	0045-0215, 1230-1740
3345	2	Jammu A	N	1130-1741
3355	10	Kurseong	E	1130-1700/1741
3365	10	Delhi	N	1230-1840
3375	50	Guwahati A	NE	1200-1701/1741
4760	10	Leh	N	0225-0430, 1130-1630
4760	10	Port Blair	S	0025-0215
4775	20	Guwahati B	NE	0029-0400, 1030-1215
4790	50	Shillong	NE	0050-0300
4800	10	Hyderabad A	S	0025-0215, 1200-1741
4820	10	Calcutta A	E	0025-0210, 1230-1736
4840	10	Bombay B	W	0025-0215, 1230-1741
4850	10	Kohima A	NE	0025-0215, 0229-0400 1000-1300
4880	50	Lucknow	N	0228-0400
4895	20	Kurseong	E	0100-0400
4910+	50	Jaipur A	N	0015-0400, 1200-1741
4920	10	Madras	S	0015-0215, 1200-1735
4940	50	Guwahati A	NE	0025-0400
4950	2	Jammu A	N	0055-0215
4960	10	Delhi A	N	0025-0435, 1030-1215
4970+	50	Itanagar	NE	0015-0400, 1030-1415
5010+	50	Thiru'puram	S	0015-0215, 1200-1736
5050	50	Aizawl	NE	0030-0400, 1229-1630 (Sat 1741)
5960	2	Jammu A	N	0229-0445(Sun 0400) 0630-0900
5990	10	Bhopal A	N	0228-0405 0700(Sun 0415)-0945
6000	10	Leh	N	0655-0900
6010	10	Calcutta	E	0230-0400, 0429-0443 1015-1215
6020	2.5	Shimla	N	0215)Sun 0415)-0400 0700-0900/0930 1030-1115, 1130-1245
6035	50	Itanagar	NE	0630-0941
6065	50	Kohima A	NE	0229-0400, 0630-0941
6110	50	Srinagar B	N	0245-0453(Sun 0515) 0600-1115
6120	10	Hyderabad	S	0228-0411(Sun 0431)
6140	2	Ranchi A	E	0215-0400(Sun 0430)
6165	50	Jaipur	N	0630-0941
6190	10	Delhi A	N	0734-1031
7105	50	Lucknow B	N	0630(Sun 0415)-0930 0935-0936, 1013-1215
7125	2	Ranchi A	E	0700-0941
7140	10	Hyderabad	S	0530-0610(Sun only) 0610-0840/0910 (Sun 1030, Sat 0930)
7150	10	Guwahati B	NE	0630(Sun 0600)-1000 1010-1011
7160	10	Madras A	S	0230-0415(Sat 0430, Sun 0530) 0610/0630-1030 (Sat 1130, Sun 1145)
7180	10	Port Blair	S	0700-0850(Sun 0950)
7190	50	Shillong	NE	0300-0400
7200	50	Thiru'puram	S	0700-0945
7210	10	Calcutta	E	0630-1031
7230	20	Kurseong	E	0630-0700, 0730-1031
7240	10	Bombay B	W	0229-0430(Sun 0530) 0530-0925/1035
7280	50	Guwahati A	NE	0630-0941(Sun 0430-1000), 1013-1145
7295	50	Aizawl	NE	0700-0930, 1130-1215

Feeder Sces.

kHz	kW	Station	Times
3365	20	Khampur	1230-1840
3905	50	Kingsway	1600-1740/1830
3925	20	Khampur	0030-0040, 1220-1310 1330-1340, 1345-1425 1430-1440, 1445-1740/1830
3945	50	Gorakhpur	0228-0300, 1445-1546
4860	20	Khampur	0125-0135, 0215-0225 0228-0355(Sun 0400)
4860	50	Kingsway	1245-1325, 1330-1340 1345-1400, 1405-1420
4960	50	Kingsway	0025-0229, 0250-0440 1030-1215
6045	250	Khampur	1405-1420
6045	50	Khampur	1220-1255
6045	50	Kingsway	0229-0404
6085	50	Khampur	1400-1425, 1430-1445
6115c	100	Madras	1330-1730
6190	50	Kingsway	0700-1031
7110	50	Gorakhpur	0830-0840
7110	20	Khampur	0125-0135, 0215-0355 (Sun 0400)
7160	50	Khampur	1330-1440, 1445-1740/1830
7210	20	Khampur	0830-0940, 1130-1140
7225	100	Kingsway	0350-0404
7225	100	Khampur	1330-1420
7260c	100	Bombay	0130-0430
7290	100	Khampur	1430-1440, 1445-1740/1830
7412	250	Aligarh	1330-1340, 1445-1730
9535c	100	Madras	0100-0430
9610	50	Kingsway	0800-0940
9615	50	Kingsway	1220-1310
9615c	100	Bombay	1330-1730
9675	100	Kingsway	1130-1140
9950	50	Kingsway	1330-1730
10330	50	Kingsway	1330-1420, 1430-1440 1445-1740/1830
11715	100	Khampur	1115-1140
11735	100	Kingsway	0800-0810
11770	100	Khampur	1220-1245
11830	50	Khampur	0125-0225, 0205-0225 0228-0355(Sun 0400)
11830c	100	Bombay	0700-1230(Sun 0630)
11870	50	Khampur	0125-0200
11920	50	Khampur	0030-0040
11970	100	Khampur	0820-0940
15080	250	Aligarh	1220-1245
15120	50	Kingsway	0030-0040, 0125-0215 0228-0355(Sun 0400)

15125c	100	Madras	0700-1045(Sun 0630)
15145	50	Khampur	1220-1245
15165	100	Kingsway	0730-0740, 0800-0940
15220	50	Kingsway	0125-0200, 0228-0300 0320-0355(Sun 0400)
15250	50	Kingsway	0700-0810, 0820-0940 1130-1140
15420	50	Khampur	1115-1140
17705	50	kingsway	1115-1140
17705	50	Khampur	0800-0940
17795	50	Khampur	0830-0940
17815	50	Khampur	1130-1140
17850	50	Kingsway	0700-0800, 0810-0940
17875	50	Khampur	0228-0300 0320-0355(Sun 0400)

c) carries Vividh Bharati Service.
V. by QSL-card. Re. to Asst. Dir. Eng. (Freq Assignments) at New Delhi addr.

EXTERNAL SERVICES

ADDR: External Services Division, All India Radio, Broadcasting House, 1 Sansad Marg, (P.O. Box 500), New Delhi-110 001. **Te:** +91 (11) 3715411. **Cable:** Airforeign — Dir: P.M. Iyer.
MEDIUMWAVES: See Home Sce.
SHORTWAVES: Aligarh (78.06E/28.00N): 4 x 250kW — Bombay (72.49E/19.11N): 1 x 100kW — Bangalore (G.C: 77.38E/12.58N): 2 x 500kW (plus 4 x 500kW planned) —Delhi-Kingsway (G.C: 77.12E/26.45N): 6 x 100kW, 2 x 50kW — Delhi-Khampur: 2 x 250 kW, 2 x 100kW, 2 x 50kW, 2 x 20kW — Gorakhpur (G.C: 83.24E/26.45N): 1 x 50kW — Kurseong (G.C: 88.20E/26.58N): 1 x 20kW — Madras (G.C: 80.17E/13.04N): 2 x 100kW, 1 x 10kW — Panaji (G.C: 73.52E/15.31N): 2 x 250kW (u.c.) —Thiru'puram (G.C: 66.59E/08.29N): 50kW (u.c.).

kHz	kHz	kHz	kHz	kHz	kHz	kHz
3945	7210	10330	11810	15080	15265	17830
4990	7225	11620	11830	15110	15325	17850
5990	7265	11715	11860	15120	15335	17895
6045	7412	11730	11880	15135	15365	21735
6140	9550	11735	11910	15140	17387	
6165	9610	11745	11935	15145	17705	
7110	9700	11760	11960	15165	17735	
7125	9910	11785	15020	15175	17785	
7140	9950	11805	15050	15185	17805	

Arabic	Area	kHz
0430-0530	ME	17785, 15135
1730-1945	ME	15185, 9910
Baluchi		
1500-1600	PAK	11620, 6165, 1071
Bangla		
0259-0430	BGD	594
0800-1115	BGD	594
1445-1515	BGD	1134
1600-1735	BGD	1134
Cantonese		
1145-1300	CHN	17735, 15050
Dari		
0300-0345	AFG	11910, 9910, 7225
1330-1430	AFG	11960, 7140
English		
1000-1100	NEAs	21735, 17895, 15050
1000-1100	Pac	17387, 15050
1330-1500	SEAs	15120, 11760
1745-1945	EAf	15080, 11935
	NAf	11860, 9950
	NEu	11620, 7412
2045-2230	WEu	11620, 9950, 7412
2045-2230	Pac	15265, 11715, 9910
2245-0045	SEAs	17830, 15145, 15110, 11785 11745, 9910
Farsi		
0400-0430	IRN	17805, 15135, 11730
1615-1730	IRN	11935, 9910, 7265
French		
1945-2030	Af	15185, 11860, 9910
Gujarati		
0415-0430	EAf	17387, 15080
1515-1600	EAf	15120, 11805
Hindi		
0315-0415	EAf	17387, 15080
1615-1700	Af	15080, 11830
1745-1830	ME	11620, 9950, 7412
1945-2045	WEu	11620, 9950, 7412
2300-2400	SEAs	15165, 11880, 9950, 1134
Indonesian		
0845-0945	INS	17387, 15175
Kuoyu		
1300-1315	CHN	17705, 15050
Myanmar		
0100-0130	BRM	15325, 11880, 9950, 1134
1215-1315	BRM	11715, 10330
Nepali		
0130-0229	NPL	9550, 7210, 6045, 3945, 594
0700-0800	NPL	11735, 9610, 7110
1345-1445	NPL	7412, 6140, 3945, 1134
Punjabi		
0800-0830	PAK	702
1230-1430	PAK	702
Pushtu		
0215-0300	AFG	11910, 9910, 7225
1430-1445	AFG	11960, 7140
Russian		
1615-1715	URS	15140, 11810
Sindhi		
0100-0200	PAK	7125, 5990, 1071
1230-1500	PAK	11620, 6165, 1071
Sinhala		
1300-1500	CLN	15020, 9700
Swahili		
0430-0530	EAf	17387, 15080
1515-1615	EAf	15080, 11830
Tamil		
2345-0330	CLN	1233
0000-0045	As	15165, 11880, 11715, 9950, 4990
0700-0930	CLN	1233
1100-1700	CLN	1233
1115-1215	SEAs	17850, 17387, 15335, 15080
Telugu		
2245-2315	SEAs	17830, 15110, 11745
Thai		
1115-1200	THA	17830, 15365, 11620
Tibetan		
0130-0200	CHN	15325, 11880, 9950
1215-1330	CHN	9565, 7412, 1134
Urdu		
0015-0430	PAK	6155, 1071, 702
0130-0430	PAK	9610
0830-1130	PAK	11770, 9675, 1071, 702
1430-1930	PAK	6045, 4860, 702
1600-1930	PAK	1071

News in English: 2300, 0100, 1000, 1055, 1330, 1455, 1800, 1900, 2000, 2100, 2200.
ANN: Home Sce: Hindi: "Ye Akashvani . . . (name of regional st.) hai". English : "This is All India Radio . . . (name of regional st.)". Tamil: "Idi Akashvani . . . (name of regional st.), vanali nilayam".
Ext Sce: English: "This is the General Overseas Service of All India Radio". Hindi: "Yeh Akashvani ke videsh prasaran sewa hai". Tamil: "Idi Akashvani videsh sewa". Sinhala: "Me All India Radio videshiya sevayai". Nepali: "Yo All India Radio ho". Dari: "Injaw Delhi". Pushto: "Da All India Radio de." Persian: "Inja Delhi". Indonesian: "Inilah All India, Radio."
INT-SIG: All India Radio's signature tune is a melody of 8 seconds' duration repeated with intervals of 10 seconds in between. The instruments used are violin, cello and tanpura.
V: Re. to Mr. G. Kuruvilla, Assistant Director (Engineering), Frequency Assignments, Room 207, 1 Sansad Marg, (P.O. Box 500), New Delhi-110 001. QSL cards are issued after checking with station logs of resp stations. Some stations QSL direct via card/letter.
09

INDONESIA

L.T: We. Indonesia (Jawa, Sumatera): UTC + 7h; Ce. Indonesia (Kalimantan, Sulawesi, Bali, Nusa Tenggara): UTC + 8h; Ea. Indonesia (Maluku, Irian Jaya): UTC + 9h — **Pr.L:** Bahasa Indonesia (Indonesian) — **Pop:** 197m — **Radio sets:** 22.000.000 — **E.C:** a/c 50, 127/220V (yes) — **ITU:** INS.

DIRECTORATE GENERAL OF RADIO, TV & FILM

Addr: Jalan Merdeka Barat 9, Jakarta. **Cable:** Radiogen.
L.P: Dir. Gen: Alex Leo Zulkarnain.

DIRECTORATE OF RADIO

Department of Information, Jalan Merdeka Barat 4/5, Jakarta.
Telex: Deppen 44349
L.P: Dep. Dir. for Technical Development: Mr. Sunendra.

FEDERATION OF INDONESIAN COMMERCIAL BROADCASTERS

Persatuan Radio Siaran Swasta Niaga Indonesia (PRSSNI)
Addr: Pengurus Pusat, Persatuan Radio Siaran Swasta Niaga Indonesia, Gedung Dewan Pers Lantai Dasar, Jln Kebon Sirih 32-34, Jakarta Pusat.
L.P: Chmn: Mr. Tagor.

Mediumwaves: v) = varies; r) = reported; u) = unconfirmed but official information, not completely monitored. Commercial st's broadcast between 2200 and 1700 only, with breaks in mid-morning and mid-afternoon.

	kHz	kW	Station
1)	540	10	RRI Bandung
5)	540	0.25	R. Dirgahayu, Barabai
7)	540		R. Deasy, Kendari
8)	540		R. Voice of Yarden, Ambon
9)	558	0.25	RSPDKDT2 Semarang
10)	558	0.25	R. Manggis B.S., Jambi
11)	558	0.25	Suara Beoly Gembira, Telukbetung
12)	558		R. Angkasa B.S., Pandeglang
13)	558		R. Gema Mahasiswa Unsud, Purwokerto
14)	558	0.25	R. Swara Juana Sakti, Pati
15)	558		R. Purnamasidi Broadcast, Wonosobo
17)	558	0.25	R. Rosa, Ujung Pandang
18)	558	0.25	Suara Mesra, Parepare
105)	558		R. Angkasa Medya B.C., Majalengka
540)	558		RSPDT2 Batang
19)	585	50	RRI Surabaya
424)	594		Suara Bhakti, Tanjungkarang
3)	594		R. Pakusarakan Pratita, Bogor
507)	603	0.5	RDKDT2 Sumedang
21)	612		R. Dwikarya 69, Bandung
22)	612	0.25	R. Gema Nusantara, Rangkasbitung
23)	612	0.22	R. Linggarjati, Kuningan
24)	612	0.25	R. Sumasli, Purwokerto
25)	612	0.25	R. Veritas, Semarang
26)	612		R. Asoorkaty B'cast (A.B.C.), Surakarta
27)	612		R. Dhamashinta Rasitania, Pekalongan
28)	612	0.5	R. Aneka Rama Ria, Denpasar
29)	612	0.25	R. Telstar, Ujung Pandang
31)	612		R. Favorite, Ambon
33)	630	100	RRI Ujung Pandang
402)	630		R. Irama Nusantara Ria, Bogor
30)	630		R. Saba Putra, Pringsewu
34)	648	0.35	R. Prima Unisi, Yogyakarta
35)	648	0.25	R. El Florida, Jambi
36)	648	0.25	R. Gema Bundo Kandung, Tanjungkarang
37)	648	0.25	R. Metro Jaya Kartika, Jakarta
38)	648	0.25	R. Swara Duta Citra Irama, Sumedang
39)	648	0.25	R. Gita Kanari Ria, Purwakarta
40)	648		R. Santo Bernadus D.S, Pekalongan
43)	648	0.25	R. Nirwana, Banjarmasin
44)	648		R. Gandaria, Ujung Pandang
46)	648	0.25	R. Veritas, Manado
47)	648		R. Merpati, Ambon
48)	648	0.25	R. Pahlawan Budi Sakti, Serang
501)	648	0.25	R. Gema Merdeka, Denpasar
515)	648		R. Flamboyant Rasistania, Bengkulu
51)	666	0.25	R. Maritim Rasonia, Cirebon
52)	666	0.25	R. Dian Swara, Purwokerto
54)	666	0.25	R. Ramaska B'casting, Surakarta
55)	666	0.25	R. Siebra Patria, Semarang
56)	666		R. Radip, Pati
509)	666		R. Mercidiona, Bandung
50)	666		R. Den Bang, Bd. Jaya
58)	684	0.25	R. Rajawali Terbang, Telukbetung
102)	684		R. Chakti Buddhi Bhakti, Jakarta
510)	684		R. Swara Zenith Angkasa, Salatiga
495)	690	0.5	RKP Kotamadya DT2 Pasuran (R. Gema Suropati)
544)	700		RKPDTI Jawa Timur, Surabaya
62)	702	0.25	R. Mara Ghita, Bandung
63)	702		R. Tjandra Buana, Cianjur
64)	702		R. Leo B.S., Cirebon
65)	702		R. Dhirgantara Permai, Banjarmasin
66)	702	0.25	R. Al Kawakib, Ujung Pandang
512)	702		R. Besakih Rasisonia, Amlapura
267)	702	0.25	R. Siera Alfa Lima, Bantul
574)	702	0.5	RDKDT2 Pandeglang
572)	702	0.5	RSPDT2 Blora
575)	711	0.5	RP Kotamadya DT2 Mojokerto (R. Gelora Mojopahit)
57)	711	0.5	RSPDKDT2 Wonogiri
67)	720	10	RRI Ambon
59)	720		Suara Simpatik Angkasa, Pinrang
243)	720	0.5	RDK Serang
513)	720	0.25	R. Alba Permai, Tabanan
225)	729		R. Swara Mega Mustika, Tanjungkarang
75)	735		R. Prima Sonata, Cirebon
68)	738		R. Walisongo, Pekalongan
69)	738		R. Echo 541, Medan
70)	738	0.25	R. Arbes Rasisonia, Padang
72)	738		R. Nada Kencana Agung, Bandung
74)	738	0.25	R. Galuh, Tasikmalaya
76)	738		R. Konservatori, Surakarta
78)	738	0.45	R. Rinjani Permai, Cakranegara
79)	738	0.25	R. La Fuzsy, Banjarmasin
80)	738	0.5	R. Barata Rasihima, Ujung Pandang
81)	738		R. Rina Bestari, Rantepao
82)	738	0.25	R. Sound of Love, Manado
183)	738	1	RRI Jember
221)	738	0.15	R. Siaran Rezki Kontiki, Banda Aceh
263)	738		R. Aristiara, Bogor
83)	756	10	RRI Purwokerto
84)	756		R. Venus Nusantara, Ujung Pandang
320)	756		R. El Gangga, Bekasi
257)	765	0.2	RRI Banjarmasin
86)	774		Suara Isa, Denpasar
87)	774		R. Antariksa Radang IV, Surabaya
88)	774	0.5	R. Suara Mercurius Kencana, Ujung Pandang
576)	774	0.5	RDK Tangerang
89)	774	0.25	R. Continental Emerald, Manado
136)	774		R. Yusyan Media, Sumedang
226)	774		R. Draba, Jakarta
491)	774	0.2	R. Leonardus Buana Suara, Salatiga
90)	775		R. Pasopati Perkasa, Medan
530)	783	0.3	RDK Lampung Utara, Kotabumi
91)	785	0.5	RKPDT2 Kotamadya Bandung
6)	792		R. Lafayette Studio Pipos, Ujung Pandang
213)	792		R. Gema Warga Karya Satnawa, Bekasi
60)	792		R. Cempaka Angkasa, Ciamis
85)	792		R. Hay Citra Ria, Palangkaraya
92)	801	10	RRI Semarang
101)	801	1	RRI Medan
93)	810	10	RRI Merauke
94)	810	0.25	R. Warastra Bewara Swara, Palembang
95)	810	0.25	R. Romeo and Juliet Rasisonia, Tanjungkarang
97)	810		R. Arvaco Duta Cakrawala, Majalengka
98)	810		R. Gema Kuripan, Amuntai
99)	810		R. Swara Sentosa Pratama, Ujung Pandang
100)	810	0.25	R. Mersí Matuari Waya, Manado
514)	810		R. Nusantara Jaya, Jakarta
577)	810		RDK Bandung
103)	828	0.25	R. Leidya B'casting, Bandung
104)	828	0.25	R. Prabu Kiansantang, Tasikmalaya
106)	828	0.35	R. Gajah Mada Palapa Angkasa, Semarang
107)	828	0.25	R. Swara Citra Suhada Jaya, Pekalongan
108)	828		R. Swara Graha FKUII, Surakarta
109)	828	0.25	R. Persatuan, Bantul
110)	828		R. Merdeka Lokatama, Surabaya
111)	828		R. T.T.-'77, Malang
112)	828	0.25	R. Citra Music Ashbone, Banjarmasin
114)	828		R. Swara Christy Ria, Ujung Pandang
115)	828	0.25	R. Al Chairaat, Manado

116)	828	0.25	R. Yudha, Denpasar
355)	828		R. Almakmuriah Perkasa Angkasa, Jakarta
113)	838	0.5	R. Babagaya, Sampit
97)	840	0.25	R. Arfaco, Majalengka
117)	846	0.25	Suara Gema Atmajaya, Palembang
118)	846	0.25	Suara Cakrawala, Panjang
120)	846	0.25	Swara Fortune Indah, Bandung
121)	846		R. Galunggung 4848, Tasikmalaya
122)	846	0.25	R. Immanuel, Surakarta
123)	846		R. Miniwatt RCA 02, Surabaya
124)	846		R. Andalus Silva Voac, Malang
125)	846	0.1	R. Telerama, Banjarmasin
126)	846	0.25	R. Al-Ikhwan Studio Tamasa, Ujung Pandang
127)	846		R. Swara Caraka Ria, Semarang
132)	846		R. Arief El Kobana, Payakumbuh
262)	846		R. Universitas Darulmukminin, Jakarta.
317)	850	1	RRI Bogor
101)	855	50	RRI Medan (Padang Cermin)
128)	855	10	RRI Mataram
129)	864	10/0.3	RRI Cirebon
16)	864		R. Mataram Buana Suara, Yogyakarta
130)	864	10	RRI Kupang
134)	864	0.25	R. Mercy Jaya Raya, Tanjungkarang
135)	864		R. Assyafiyah, Jakarta
137)	864		R. Menara III, Surabaya
138)	864		R. Sylvania, Banjarmasin
139)	864	0.25	R. Baruna Putera Gita Lestari, Bitung
141)	864	0.25	R. Swara Bayurekhsa Kencana, Ujung Pandang
45)	864		Suara As'Adiyah, Sengkang
578)	879	0.5	RKPKDT2 Sidoarjo
142)	882	0.25	R. Candrika Widya Swara, Bandung
143)	882	0.3	R. Nusantara Bharata Citra (NBC), Garut
144)	882	0.25	R. Chandra Rasisonia, Banjarmasin
145)	882	0.5	R. Bambapuang, Pangkajene
146)	886	0.5	RRI Serui
147)	891	10	RRI Malang
77)	900		R. Gema Cecya Dhaksinarga (G.C.D.), Wonosari
148)	900		R. Elkartika Angkasa Niaga, Padang
149)	900	0.25	R. Jabastigo R. Mercy, Bukittinggi
150)	900	0.25	R. Merpati Dharmawangsa, Jakarta
151)	900		R. Angkatan Muda, Yogyakarta
152)	900		R. M.C.A., Surabaya
314)	900		R. Chandra Taruna, Semarang
389)	900		RPKDT2 Cianjur
463)	900	0.25	R. Gema Megantara Pratama, Tabanan
556)	905	0.25	RPDKDT2 Sumbawa
153)	918	0.075	R. Citra Remaja Maju Jaya, Langsa
154)	918		P.T.D.I., Jakarta
155)	918	0.25	R. Estrelita, Bandung
156)	918		R. El Nury, Bogor
157)	918		R. Pitaloka, Ciamis
232)	918		R. Dhirgantara, Negara
96)	918		R. Gelora Pertiwi, Medan
53)	918	0.25	R. Chandra Kusuma, Pekalongan
579)	918	0.5	RKPKDT2 Gresik
158)	927	25/0.3	RRI Pekanbaru
140)	936		R. Liga Perdana, Balikpapan
159)	936		R. Pasopati, Semarang
160)	936	0.25	R. P.2 S.C., Jakarta
161)	936	0.25	R. Budaya Sari, Bandung
162)	936		R. Bestari B.C., Sukabumi
163)	936	0.25	R. Roiska, Surakarta
164)	936	0.5	R. Yasika AKII, Yogyakarta
165)	936		R. La Victor, Surabaya
166)	936		Suara Mahameru, Kediri
167)	936		R. Niaga Fiska, Bondowoso
168)	936	0.25	R. Gemini Perkasa, Mataram
169)	936		R. Refalado, Banjarmasin
170)	936	0.5	R. Gamasi, Ujung Pandang
197)	950	—	RKPDT2 Blitar
172)	954	10	RRI Kendari
173)	954	0.15	R. Garuda Kenten Jaya, Palembang
174)	954		R. Patrol Radika, Telukbetung
175)	954	0.25	R. Gema Utama Pembangunan, Jakarta
176)	954		R. Contessa Bandung Raya, Bandung
177)	954	0.25	R. Sena Bahana Cakrawala, Sukabumi
178)	954		R. Jaya Pemuda Indonesia, Surakarta

179)	954	0.25	R. B'cast Sahara, Kendal
180)	954	0.25	P. T. D. I. Kota Perak, Yogyakarta
181)	954	0.25	R. El Bayu, Gresik
182)	954	0.25	R. Gema Persada, Banjarmasin
4)	954		R. Nada Berlian, Jambi
131)	954		R. Batur, Bangli
183)	963	10	RRI Jember
185)	966		RPDK Deli Serdang, Medan
186)	972	50	RRI Surakarta
187)	972		R. Attahiriyah, Jakarta
188)	972	0.25	R. Diah Rosanti, Pontianak
189)	972		R. Swara Citra Esa Enang, Langowan
42)	972		R. Guntur IB-29, Singaraja
371)	972		R. Antares, Garut
580)	972	0.25	Suara Atas Roban, Batang
191)	990		R. Cakra Awigra, Surabaya
192)	990		R. K.D.S. VIII, Malang
193)	990		R. Sion, Tomohon
430)	990	0.25	R. Suara Calvary, Klungkung
431)	990		R. Mercu Clan, Martapura
73)	990		R. Trisara Kencana, Cianjur
195)	999	300	RRI Jakarta
196)	1000		R. Angkatan Udara, Medan
198)	1008	10	RRI Madiun
194)	1025	0.25	RSPDT2 Temanggung
200)	1026	0.25	R. Universitas Trisakti, Jakarta
202)	1026		R. Fortuna, Sukabumi
203)	1026		R. Mercy, Semarang
204)	1026	0.25	R. Sumbing Wijaya Kusuma, Klaten
205)	1026	0.25	R. El Victor, Surabaya
206)	1026	0.25	R. Taurus AX-1A, Kediri
207)	1026	0.25	R. Swara Indonusa Perkasa, Gianyar
208)	1026		R. Santanimo, Banjarmasin
383)	1026		R. Rex Bahana Swara, Garut
494)	1029	0.5	RPKDT2 Ciamis
209)	1035	1	RRI Tanjungkarang
533)	1035	0.2	RRI Palangkaraya
210)	1037r		RKPDT2 Malang
171)	1044		R. Sikamoni, Medan
211)	1044	10	RRI Biak
212)	1044	0.25	R. Kayumanis, Jakarta
214)	1044	0.25	R. Purnayudha, Sukabumi
216)	1044	0.25	R. Raka, Tegal
217)	1044	0.25	P.T.D.I. Medari-Sleman
218)	1044		R. Camar, Surabaya
219)	1044		Suara Akbar, Jember
220)	1044	0.25	R. Cakra Swara Perkasa, Singaraja
229)	1044		R. Arya Bomantara, Singkawang
222)	1053	10	RRI Jayapura
186)	1053	1	RRI Surakarta
2)	1062		R. Siaran Cakrawala Utama, Kuala Simpang
223)	1062	0.25	Suara Subuh, Padang
224)	1062	0.25	R. Gema Mutiara, Palembang
227)	1062		R. Maestro, Bandung
230)	1062	0.25	R. Reco Buntung, Yogyakarta
231)	1062		R. Zodiac, Surabaya
233)	1062	0.25	R. Suta Remaja, Mataram
234)	1062		R. Nusantara Antique, Banjarmasin
235)	1062	0.25	R. Memora Anoa Indah, Manado
516)	1062		R. Kalimasadha Sakti, Semarang
506)	1062	0.5	RPKDT2 Tasikmalaya
236)	1071	0.5	R. Volare, Pontianak
237)	1080	0.5/10	RRI Singaraja
239)	1080	0.25	R. Volvo Electronica, Bandung
240)	1080	0.25	R. Young Monarchy, Sukabumi
241)	1080	0.25	R. Carolina Arjuno, Surabaya
242)	1080	0.25	Suara Immanuel, Malang
245)	1097	0.3	RSPKDT2 Kendal
244)	1098		R. Gema Baiturrahman, Banda Aceh
246)	1098	10	RRI Jambi
247)	1098	10	RRI Sumenep
248)	1098		R. Swara No Name, Palembang
249)	1098		R. Universitas Tarumanegara, Jakarta
250)	1098	0.25	R. Paramuda, Bandung
251)	1098		R. Swara Historis Gita, Karawang
261)	1098		R. Diaros Duta Swara, Singkawang
252)	1107	1/10	RRI Yogyakarta
581)	1107	0.25	R. Bintang Niaga, Kisaran
253)	1116	0.5	R. Agustina Yunior, Jakarta

255)	1116	0.25	R. Suzana Bhakti, Surabaya
448)	1116		R. Citra Buana Indah, Medan
517)	1116		R. Mustika Parahiyangan, Bandung
532)	1116	0.25	RPDT2 Kotamadya Binjai
582)	1116	0.25	Suara Batang, Batang
256)	1120v		R. Andalas IX, Banda Aceh
533)	1125	1	RRI Palangkaraya
523)	1125	0.25	RPDT2 Luwu, Palopo
257)	1134	25/1	RRI Banjarmasin
258)	1134	0.1	RSPD Kotamadya Tegal
259)	1134		R. Alpha Romeo, Banda Aceh
260)	1134		R. El Draba Jaya, Medan
265)	1134	0.25	R. Martha, Tasikmalaya
266)	1134		R. Siaran Ichthus, Semarang
264)	1135		R. Ria Cindelaras, Indramayu
268)	1152		R. Alvacos, Medan
269)	1152		R. Febrianta Niaga Angkasa, Padang
270)	1152	0.25	R. Enes Duabelas Ulu, Palembang
271)	1152	0.25	Suara Wajar, Tanjungkarang
272)	1152	0.25	R. Cakrawala Revita, Jakarta
273)	1152	0.25	R. Garuda Tunggal Angkasa, Bandung
274)	1152	0.25	R. Top Media, Serang
275)	1152	0.25	R. Lusiana Namberwan, Semarang
276)	1152	0.35	Suara Istana, Yogyakarta
277)	1152		R. Yasmara CA-27, Surabaya
278)	1152	0.185	R. Wijang Songko, Kediri
280)	1152	0.25	R. Gembira Raya Jaya, Samarinda
92)	1170	50	RRI Semarang
282)	1170		R. Cosanastra, Medan
283)	1170	0.25	R. Ganesha Nada, Bandung
284)	1170	0.25	R. Swara Rama Lokantara, Serang
351)	1170		R. Rajawali, Surabaya
285)	1175v		R. Daerah Perdagangan Bebas Sabang
287)	1180	1/10	RRI Tanjung Pinang
288)	1188	0.25	R. Antar Nusa, Jakarta
289)	1188	0.25	R. Chevy 88, Bandung
290)	1188		R. El Amigo, Sukabumi
291)	1188		R. Widya Bhakti, Magelang
292)	1188	0.25	R. Arma Sebelas, Yogyakarta
293)	1188		R. Swara Perak Jaya, Surabaya
294)	1188		R. Kutilang, Malang
295)	1188	0.25	R. La Barong, Singaraja
299)	1188		R. Dalka Mega Swara, Meulaboh
518)	1188		R. Gema Nirwana, Samarinda
128)	1194	0.5	RRI Mataram
317)	1195	5	RRI Bogor
548)	1197	0.5	RSPDT2 Labuhan Batu, Rantau Prapat
302)	1205		R. Bhaktera Bahana Suara, Pekanbaru
549)	1205		RPKDT2 Lebak, Rangkasbitung
19)	1206	1	RRI Surabaya
20)	1206		R. El Philia, Telukbetung
296)	1206		R. Indah Bahagia Ceria (I.B.C.), Semarang
297)	1206	10/0.5	RRI Denpasar
298)	1206		RSPKDT2 Purbalingga
300)	1206	0.25	R. Istana Merpati Jaya, Medan
301)	1206		R. Dirgan Bravo, Padang
303)	1206		R. Adhika Swara, Bandung
524)	1206	1	RRI Manado
583)	1206	0.25	RSPKDT2 Pekalongan
584)	1206		R. Suara Mitra, Tanjung Gading
304)	1211r		RKPDT2 Bondowoso
519)	1211		R. Kotamadya Cirebon, (R. El Thema), Cirebon
195)	1215		RRI Jakarta
305)	1215	10/0.5	RRI Samarinda
306)	1215		RKPDKDT2 Magetan
307)	1215		RSPDKDT2 Cilacap
308)	1224	0.25	R. Prima Elita, Palembang
309)	1224	0.25	R. Dahlia Flora, Bandung
310)	1224	0.25	R. Buana Jaya, Tasikmalaya
311)	1224	0.25	R. PEEN Rasitania, Surakarta
313)	1224	0.25	R. G.L. 67, Malang
496)	1224	0.25	RKPDT2 Bojonegoro
321)	1224		Suara Tawang Alun, Banyuwangi
585)	1224	0.25	RSPDKDT2 Purworejo
315)	1233	1	RRI Pontianak
316)	1233v	0.3	RPD Kotamadya Sabang
317)	1235	0.5	RRI Bogor
318)	1242	0.25	R. Pelita Kasih, Jakarta
323)	1242		R. Wijaya, Surabaya
184)	1242		R. Airlangga, Sukabumi
520)	1242		R. Swara Rugeri, Garut
527)	1242	1	RRI Bengkulu
317)	1250	1	RRI Bogor
128)	1251		RRI Mataram
325)	1244	0.25	R. Imelda, Semarang
326)	1251	10	RRI Banda Aceh
237)	1251	1	RRI Singaraja
541)	1251	0.25	RSPDT2 Boyolali
327)	1260	0.25	R. Khamasutra, Medan
328)	1260	0.25	R. Lokawisesa, Jakarta
329)	1260	0.25	R. Famor, Bandung
330)	1260		R. Suveri 45, Subang
331)	1260		R. Amatron, Purworejo
332)	1260		R. Kaliwungu Dirgantara, Kaliwungu
333)	1260		R. Gabriel, Madiun
334)	1260		R. R.D.A. 45, Bangil
335)	1260		Suara Mandala, Banyuwangi
336)	1260	0.25	R. Elektronika Gajah Mada (El Gama), Banjarmasin
337)	1260		R. Rajawali Sakti, Balikpapan
433)	1260		Suara Cevillian Antik, Palangkaraya
497)	1270		RKPDT2 Ponorogo
339)	1278		RPDT2 Lombok Timur, Selong
340)	1278v		R. Thosiba, Banda Aceh
341)	1278		R. Didra Nuansa, Medan
342)	1278	0.23	R. Gema Kencana Bahari, Jakarta
343)	1278		R. Ardan Swaratama, Bandung
344)	1278		R. Kusuma Sena, Bogor
345)	1278	0.25	R. Mandalika, Jepara
346)	1278		R. Citra Ikhsaniya, Tegal
347)	1278		R. Swara Kenanga Citra Indah, Purworejo
348)	1278		R. Mercury Madesuk, Surabaya
322)	1278	0.25	RSPD Kotamadya Salatiga
349)	1287	50/0.4	RRI Palembang
586)	1287	0.25	RSPDT2 Pemalang (Suara Widuri)
350)	1296	0.25	R. El Bama, Jakarta
352)	1296		Suara Pioneer, Malang
353)	1296	0.5	R. Duta Gita Bhyomantara Shinta Rama, Cakranegara
498)	1313		RSPD Kotamadya Magelang
354)	1314	0.25	R. Rodhesa, Medan
356)	1314		R. Mutiara, Bandung
357)	1314	0.25	R. Gema Tembang Victory, Bogor
358)	1314		R. Swara Virgin Gemilang, Karawang
359)	1314		R. Sananta Gelora, Tegal
360)	1314		Suara Sion Perdana, Karanganyar
361)	1314		R. Swara Yasfi Indah, Cilacap
362)	1314		R. Bintoro Karya, Demak
363)	1314		R. Kartika Bahari, Sidoarjo
365)	1314	0.25	Suara Casanova, Samarinda
502)	1316		RKPDK Tulungagung
366)	1318		RKPDT2 Lamongan
195)	1332	10	RRI Jakarta
368)	1332		R. Sonya Portibi, Medan
369)	1332		R. Ellusia, Pekanbaru
370)	1332		R. Ozza Mitra Gama, Bandung
504)	1332	0.5	RKPDKDT2 Lumajang
372)	1335		RKPDK Jombang
521)	1339		RDT2 Cirebon
587)	1342		RKPDKDT2 Nganjuk
364)	1350		Suara Kartika, Jember
373)	1350		RKPDT2 Pasuruan
374)	1350	0.25	R. L'Amour Citra Budaya Sriwijaya, Palembang
375)	1350		R. Dei Marganusa, Metro
377)	1350		R. Puspa Jaya, Bojonegoro
338)	1350		R. Mayangkara Ria, Blitar
559)	1350	0.75	RPDT2 Tapanuli Utara, Balige
286)	1350	0.2/10	RRI Padang
545)	1363	0.35	R. Suara Indrakila, Wates
588)	1363	0.3	RKPDT2 Situbondo
378)	1365	0.25	RSPDK Pati
589)	1365	0.25	RKPDKDT2 Pacitan
379)	1365	0.5	RKPKDT2 Trenggalek
380)	1365		RKPD Kotamadya Surabaya
412)	1368		R. Sonya Manis, Bireuen
381)	1368	0.25	R. El Shinta, Jakarta
382)	1368		R. Generasi Muda, Bandung
590)	1368	0.25	RPKDT2 Cirebon

534)	1371	0.25	RDKDT2 Sukabumi
385)	1385		RPKDT2 Majalengka
279)	1386		R. Anka Three's, Kandangan
384)	1386		R. Siaran Expo 70, Banda Aceh
386)	1386		R. Prapanca Buana Suara, Medan
387)	1386	0.25	R. Cendra Wasih Pusat, Jakarta
388)	1386	0.25	R. Kauman, Bogor
390)	1386	0.25	R. Strato, Surabaya
391)	1386	0.25	R. Cakra Bhuwana, Malang
392)	1386		R. Mudhita Buana B.S., Pontianak
591)	1386	0.25	RSPKDT2 Jepara (R. Kartini)
1)	1398	50	RRI Bandung
592)	1400	0.5	RKPDKDT2 Banyuwangi
393)	1404		R. Simponi, Medan
394)	1404	0.25	R. Unasco, Bandung
395)	1404		R. Lesmana, Bogor
396)	1404	0.25	R. Bahtera Yudha, Surabaya
397)	1404		R. Moderato, Madiun
415)	1404		Suara Ramayana Jelita, Palu
398)	1405	0.25	R. Candra Buana, Palembang
522)	1405		R. Swara Nava Ria Gemilang, Palangkaraya
399)	1422		R. El Em Bahama, Padangpanjang
400)	1422	0.25	R. Trijaya Sakti, Jakarta
401)	1422	0.25	R. Shinta Buana, Bandung
432)	1422		R. La Paloma, Pontianak
593)	1422	0.25	R. Citra Kisarannada, Kisaran
133)	1440		R. Gelora Ramona, Palembang
324)	1440		R. Kalimaya Bhaskara, Malang
404)	1440	0.25	RPDT2 Brebes
405)	1440	0.25	R. Safari Bina Budaya, Jakarta
406)	1440	0.25	R. Lita Sari, Bandung
407)	1440	0.25	R. Muria, Kudus
408)	1440		R. Exodus, Temanggung
409)	1440	0.25	R. Double One, Sumenep
410)	1440	0.25	R. Gema Surya, Ponorogo
563)	1440	0.25	R. Indra Kencana, Tembilahan
418)	1440		Suara Totabuan Ria, Kotamobagu
594)	1440		RSPDK Sukoharjo (Suara Raspeda)
550)	1442	0.5	R. Pesona Lematang, Lahat
503)	1445	0.5	RDT2 Indramayu
596)	1450	0.2	RPDT2 Kampar, Bangkinang
254)	1458		R. Swara Cakrawala Sangkuriang, Bandung
411)	1458	0.25	R. Arief Rahman Hakim, Jakarta
414)	1458	0.25	R. Salvatore, Surabaya
475)	1465		R. Kencana Perkasa, Pematang Siantar
453)	1475	1	RKDT2 Karawang
416)	1485		RKPDT2 Kotamadya Kediri
417)	1485		RKPDK Madiun
419)	1485		R. Universitas Sumatera Utara, Medan
420)	1485	0.25	R. Bimantara, Bukittinggi
421)	1485	0.15	R. Swara Arun Cendana, Pekanbaru
422)	1485	0.25	R. Dian Irama, Jambi
423)	1485	0.25	R. Santana Musi Permai, Palembang
425)	1485		R. Angkatan Bersenjata, Jakarta
426)	1485		R. Sindang Kasih, Cirebon
427)	1485	0.25	R. Cendrawasih, Semarang
428)	1485		R. E.M.C., Yogyakarta
429)	1485		R. Angkasa Jaya, Probolinggo
434)	1485	0.25	R. Swara Angela Permai, Manado
553)	1485	0.25	RSPDKDT2 Banjarnegara
555)	1485	0.25	RSPDKDT2 Rembang
435)	1494	0.25	R. Toddo Puli, Palu
436)	1503	0.25	R. Sara 88, Bandung
437)	1503	0.25	R. MC 16, Marcella, Bogor
438)	1503		R. Lebak Ria, Rangkasbitung
439)	1503	0.25	R. Radiks 99, Semarang
440)	1503	0.25	R. Bikima, Yogyakarta
441)	1503	0.25	Suara M.T.B., Surabaya
442)	1503		R. Dardanella, Pontianak
443)	1503	0.25	R. Swara Kasih Sejahtera, Manado
444)	1503		R. Alnora, Medan
446)	1512	0.25	R. Ramona Jelita, Balikpapan
447)	1521		R. Dirgantara, Langsa
449)	1521		R. Cendrawasih Karya Murni, Pematangsiantar
403)	1521	0.35	R. Dharma Cindelaras B.S., Banda Aceh
450)	1521	0.1	RSPDK Klaten
451)	1521	0.25	R. Swara Niaga Hiu Kencana, Semarang
531)	1521	0.3	RSPDKDT2 Wonosobo
595)	1521	0.25	RSPDT2 Kotamadya Pekalongan (Suara Kota Batik)
452)	1539		RSPDKDT2 Magelang
454)	1539	0.2	R. Niaga dan Budaya Simalungan, Pematangsiantar
455)	1539	0.25	R. Esti Elita, Pekanbaru
456)	1539	0.25	R. Hang Tuah Jaya Raya, Palembang
458)	1539		R. Mayapada, Bandung
459)	1539	0.25	R. Vahana Bewara Swara, Cirebon
460)	1539		R. S.A.S., Surakarta
461)	1539		R. Cakrawala Bhakti, Surabaya
462)	1539	0.25	R. Pariwisata Senaputra, Malang
464)	1539	0.25	R. Berita Harapan, Manado
199)	1539		R. Kardopa, Medan
367)	1539		R. Rhodisko Rasisonia, Banda Aceh
465)	1539		R. Swara Manggala Sakti, Kudus
552)	1545	0.75	RKPKDT2 Bekasi
470)	1550	0.06	R. Programa Hiburan dan Informasi DT2 Indragiri Hilir, Tembilahan
524)	1551	0.1	RRI Manado
467)	1557	0.25	R. Miraka Yunior, Bogor
468)	1557	0.25	R. Fantasy 70, Jatiwangi
469)	1566		R. Swara Primadona Mahardika, Cikampek
472)	1566		RKPDT2 Jember
471)	1576		RKPDK Tuban
466)	1584		R. Pangkalan Remaja Don Bosko, Samarinda
473)	1584	0.2	RKPDKDT2 Pamekasan
474)	1584		RSPDKDT2 Demak
476)	1584		R. Dian Erata, Padangpanjang
477)	1584	0.25	R. El Ria Buana, Jambi
478)	1584	0.25	R. La Novia Swara Indah, Palembang
479)	1584		R. Fox Rasikonia B.C., Bandung
480)	1584	0.25	R. Larispol, Magelang
481)	1584		R. Anita, Tegal
482)	1584		R. Istana, Surabaya
483)	1584		Swara Al Karomah Pratama, Martapura
484)	1584		R. Prominda Broadc. Sce, Pontianak
485)	1584	0.25	R. Hex, Balikpapan
554)	1584		R. Sumber Kasih, Manado
486)	1602		R. La Bross, Bukittinggi
487)	1602	0.15	R. Cynthia Rama, Pekanbaru
488)	1602	0.25	Suara Monalisa, Jakarta
489)	1602		R. Paksi, Bandung
490)	1602	0.25	R. Swadaya Cempaka 23, Karawang
492)	1602	0.25	R. Geronimo, Yogyakarta
493)	1602		R. Karya Dharma, Manado
547)	1612		RKIP Surabaya

Shortwaves: r = reported; v = varies; * = inactive.

	kHz	kW	Station
195)	2307	5	RRI Jakarta (Prgr. Nasional)*
252)	2350	1	RRI Yogyakarta: 2100-0100 (**English:** 0030-0045), 0900-1800 (**English:** 1130-1145)
19)	2377	1/0.5	RRI Surabaya: 0200-0730, 0900-2130 (**English:** 0900-0915)
129)	2390	1	RRI Cirebon: 0900-1700, 2145-0300 irr.
326)	2433	1	RRI Banda Aceh: 0700-1000, 1700-2300
349)	2433	1	RRI Palembang*
83)	2473	1	RRI Purwokerto*
33)	2490v	0.5	RRI Ujung Pandang: 1615-2100 (**English:** Sat 1630-1700)
566)	2552	0.175	RRI Wamena: 0345-0615
538)	2582	0.3	RPDT2 Timor Tengah Selatan, Soë*
508)	2694	0.5	RPDK Ende: 2145-2300, 1000-1400
600)	2857		RPDK Bolaang Mongodow, Kotamobagu: 2200-0400, 0800-1400
542)	2904	0.5	RPDT2 Ngada, Bajawa: 1000-1330
543)	2960	0.3	RPDT2 Manggarai, Ruteng: 2145-2300, 1000-1400
537)	3000v	0.25	RPDT2 Bengkalis: 0730-1115, 1155-1600
339)	3109		RPDT2 Lombok Timur, Selong*
536)	3142.9	0.3	RPKDT2 Belitung, Tanjungpandan: 1100-1505
546)	3159	0.03	R. Programa Hiburan dan Informasi DT2 Indragiri Hilir, Tembilahan: 2300-0430, 0630-1100, 1200-1600.
505)	3187		RKPDT2 Bima: 0930-1400

1)	3204	10	RRI Bandung: 2200-0300 (**English:** 0115-0130), 0900-1700
524)	3215	10	RRI Manado: 2100-0200, 0800-1515
128)	3223	5	RRI Mataram: 2100-2400, 0900-1200
287)	3225	10	RRI Tanjung Pinang: 2200-0015, 0700-1215
525)	3232	10	RRI Bukittinggi: 2200-0300, 1030-1605
67)	3241	1	RRI Ambon*
257)	3250	20	RRI Banjarmasin: 2130-0100, 0900-1530
130)	3259.7v	1	RRI Kupang: 0400-0900, 1530-2100
527)	3265	10	RRI Bengkulu: 2100-0300, 0900-1700
528)	3265v	10	RRI Gorontalo: 2100-0015, 0800-1320
195)	3277	1	RRI Jakarta (Prgr. Nasional): 1300-2315 irr.
198)	3286	1	RRI Madiun: 0100-0400, 0700-1000, 1400-2200 (Sat 0700-2200)
529)	3295	0.5	RPDT2 Kotamadya Pematangsiantar*
525)	3305		RRI Bukittinggi: 0430-0715
499)	3306	10	RRI Dili: 2100-0030, 0455-0800, 0855-1550
533)	3325	10	RRI Palangkaraya: 2200-0100, 0800-1500
535)	3345	10	RRI Ternate: 2020-0030 (Sun 2330), 0750-1400
247)	3355	1	RRI Sumenep: 1400-2200
246)	3355	1	RRI Jambi: 0500-0900, 1700-2200
101)	3375	20/7.5	RRI Medan: 1000-1700 (20kW), 1700-0300 (7.5kW)
147)	3381	1	RRI Malanag: 1700-2100
130)	3385	10	RRI Kupang: 2100-0155, 0900-1520
549)	3395	10	RRI Tanjung Karang: 2200-0300, 0820-1700 (Sun 2200)
315)	3445	1	RRI Pontianak: 2200-0115, 1000-1600
551)	3460		RPDT2 Aceh Timur, Langsa*
500)	3488	0.04	RPDT2 Tapanuli Selatan, Padangsidempuan: 2300-0500, 0800-1500
601)	3580	0.35	RPDT2 Poso: 0900-1500
597)	3654	0.2	RPDT2 Buol, Tolitoli: 1000-1400
565)	3644.8v	0.5	RRI Fak-Fak: 1400-2200
93)	3905	0.5	RRI Merauke: 0200-0700
326)	3905		RRI Banda Aceh: 2300-0100, 1000-1610
92)	3934	5/10	RRI Semarang: 2230-0200 (**English:** 0015-0030), 0500-0600, 1000-1700
297)	3945	1	RRI Denpasar: 2100-0030, 1000-1600
209)	3946	2.5	RRI Tanjung Karang: 0900-0200 (rel. Prgr. Nasional 1700-2200)
286)	3960	1	RRI Padang: 1645-2200
560)	3960	10/1	RRI Palu: 0900-1555 (1kW), 1615-2000 (1kW), 2130-0030 (10kW)
19)	3976	10/2	RRI Surabaya: 2130-0130 (**English:** 0100-0115), 0730-1700 (**English:** 0900-0915) irr.
511)	3986	1	RRI Manokwari: 2025-2400, 0225-0625, 1010-1400
315)	3995		RRI Pontianak: 1600-2100
172)	4000	5	RRI Kendari: 2030-0100, 0900-1600
286)	4003	10	RRI Padang: 2200-0300, 0900-1700
146)	4607.4	0.5	RRI Serui: 2000-1520
547)	4697	1.5	RKIP Surabaya: 2200-0200, 0500-0700, 1000-1500
33)	4719	50	RRI Ujung Pandang: 2055-0200
33)	4753	20	RRI Ujung Pandang: 0855-1605
195)	4777v		RRI Jakarta (Prgr. Nasional): 1300-2230 irr.
565)	4789.7v	1	RRI Fak-Fak: 2000-0100, 0700-1400
130)	4805	0.3	RRI Kupang: 2100-0355, 0900-1520
349)	4855.8	10	RRI Palembang: 2200-0300, 0900-1700
67)	4864	10	RRI Ambon: 2000-0015, 0330-0600, 0800-1500
566)	4867		RRI Wamena: 1200-2000
566)	4871		RRI Wamena: alt. to 4867
567)	4875	10	RRI Sorong: 2000-0100 (**English:** 0030-0045), 0800-1355
598)	4881	0.3	R. Suara Kasih Agung, Jayapura: 2030-2230 (Sun 2300), 0800-1200 (Sun 0700-0900)
186)	4900	0.5	RRI Surakarta: 0200-1045, 1700-2200
525)	4901	0.5	RRI Bukittinggi: 1600-2400
246)	4927	7.5	RRI Jambi: 2200-0100 (Sun 0700), 0900-1400
186)	4932	10	RRI Surakarta: 1045-1710
528)	4990v	1	RRI Gorontalo: 0400-0800, 1300-1550
222)	5045	1	RRI Jayapura: 0300-0600, 1300-2100
252)	5046	20	RRI Yogyakarta*
568)	5055	1	RRI Nabire: 2000-0200, 0800-1100
526)	5256	1	RRI Sibolga*
599)	5692	0.5	RPDT2 Berau, Tanjungredeb: 1000-1410
257)	5970	10	RRI Banjarmasin: 0100-0900
158)	5984		RRI Pekanbaru: 2200-1515
524)	5988	10	RRI Manado: 0200-0800
222)	6070	20	RRI Jayapura: 2000-0030 (**English:** Sun 2115-2130), 0600-1400
568)	6127	1	RRI Nabire: 0200-0800, 1100-2000
305)	6135	1	RRI Samarinda: 0200-0515
511)	6189		RRI Manokwari: 0800-1010
286)	6191	10	RRI Padang: 0200-0900
67)	7139	10	RRI Ambon*
146)	7173		RRI Serui
211)	7211	1	RRI Biak: 0300-0800
565)	7230	0.5	RRI Fak-Fak: 0100-0700
252)	7270	50	RRI Yogyakarta: 0300-0900
195)	9525		RRI Jakarta (Prgr. Nasional): 0500-1610
33)	9552	7.5	RRI Ujung-Padang: 0000-0800
222)	9610v		RRI Samarinda: 2300-0210, 0500-0925
305)	9614	0.5	RRI Samarinda: 2300-0210, 0500-0910
195)	9675	250	V. of Indonesia: 0030-0400, 1200-1300, 1730-2100 irr.
195)	9680	50	RRI Jakarta (Prgr. Nasional): 0500-1610
567)	9743		RRI Sorong: 0100-0755
195)	11752v	100	V. of Indonesia: 0030-0400, 0800-1300, 1730-2100
195)	15155v	100	RRI Jakarta (Prgr. Nasional): 2200-0810

FM STATIONS (Commercial Stations only. For RRI FM frequencies, see under Addresses)

	MHz	Station
376)	100.6	R. Mustang Utama, Bogor
557)	100.6	R. KLCBS Ilnafir, Bandung
558)	100.6	R. Fiscaria Jaya Suara, Surabaya
41)	100.9	R. El P. Istara, Surabaya
445)	100.9	R. Sonoramunda, Jakarta
561)	101.1	Suara Irama Indah, Jakarta
201)	101.3	R. Swakarsa Megantara, Bandung
312)	101.3	R. El D.C., Surabaya
562)	101.3	R. Roris Shinta Rama, Medan
564)	101.3	R. Pesona Indah, Palembang
119)	101.6	Suara Kejayaan, Jakarta
319)	101.6	R. Bonsita, Medan
569)	101.6	R. Matra Komersil (Batam), Sekupang
228)	102.0	R. Nasional B.S., Sukabumi
570)	102.0	R. Bahana Pembangunan, Jakarta
49)	102.3	R. Prambors Rasisonia, Jakarta
61)	102.7	R. Camajaya Surya, Jakarta
413)	103.0	R. Jatayu Angkasa, Semarang
571)	103.0	R. Penguin FM Stereo, Denpasar
572)	103.0	Suara Andalas Nada, Bandar Lampung
573)	103.3	Suara Antara Mas, Bogor
238)	103.7	Suara Amigos, Jakarta
215)	105.1	R. Sesanti Suara Sakti, Semarang
281)	105.1	R. Siaran Flamboyant, Banda Aceh
457)	105.8	R. Taman Mini, Jakarta
71)	106.2	R. Ramako Jaya Raya, Jakarta
32)	106.9	R. Queen Stereo, Jakarta
190)	107.2	R. Continental Alpha, Bandung

HOME SERVICES

RADIO REPUBLIK INDONESIA (Gov.)

ADDR: Jln Merdeka Barat 4-5, P.O. Box 356, Jakarta.
Central Station: Jakarta 195)
National Prgr: (Programa Nasional) on 999/3277/4777/9525/9680/15155kHz + FM 100.0/104.0MHz: 24h (SW times: see above). **N:** on the h.
Metropolitan Prgr (Programa Ibukota): 24h on 1332kHz + 93.0/103.0/105.0MHz. Rel. by Programa Nasional at 0400-1000.
FM Stereo Prgr: 0058-1700 on 1215kHz + 89.1/95.7MHz.

FOREIGN SERVICE ("The Voice of Indonesia")
ADDR: P.O. Box 157, Jakarta 10001.
STATIONS: Jakarta (G.C: 106.50E/06.12S): 100kW — *) Padang Cermin (G.C: 98.26E/03.34N): 250kW — **FM:** Jakarta 107.4MHz.

kHz	kHz
9675*	11752v

Arabic	kHz	Indonesian	kHz
0300-0400	9675, 11752	0200-0300	11752
Chinese		1200-1300	9675, 11752
1000-1100	11752	**Japanese**	
English		1130-1200	11752
0100-0200	11752	**Malay**	
0800-0900	11752	0900-1000	11752
2000-2100	9675, 11752	**Spanish**	
French		0030-0100	11752
1900-2000	9675, 11752	1730-1800	9675, 11752
German		**Thai**	
1800-1900	9675, 11752	1100-1130	11752

INT-SIG: Rayuan Pulau Kelapa (the song of the Coconut Islands) played on a Hammond Organ, Flute, Vibe and Piano.
V. by QSL-card or letter. Listeners reports are welcomed.
PUB: Weekly: RRI Radio dan Televisi (Home Prgr.)
Commercial Stations: Permitted power: 0.5kW. Permitted D. Prgrs: 2200-1700.
Radio Angkatan Bersenjata/Radio Angkatan Udara (Military Stations).
Provincial Gov. Station: Radio Khusus Pemerintah Daerah Tingkat Satu.
District Stations: Radio Khusus Pemerintah Daerah Tingkat II (Dua). Sometimes "Khusus" (special) is deleted. "Kabupaten" is occ. used instead of, or as well as, Daerah Tingkat Dua. Both mean "District".
Municipal Stations: Only intended for particular towns; Radio Khusus Pemerintah Daerah (e.g. Kotamadya Surabaya).
Further Radio Pemerintah Daerah (RPD) stations are known to exist, the majority of which operate on MW frequencies. However, outside Java and Bali, there are several such stations which still use SW. Here is a province by province list of those Kabupaten (with capital town indicated in brackets where different from the Kabupaten name) and Kotamadya (KM) whose station frequencies were not available at the time of editing. Station headings (e.g. RPDT2) will vary from station to station.
D.I. Aceh. Aceh Barat (Meulaboh)
Sumatera Utara. Langkat (Binjai), Pulau Nias (Gunungsitoli), KM Tanjung Balai, KM Tebingtinggi, R. Informasi Pertanian (Tg. Morawa), Dairi (Sidikalang), Karo (Kabanjahe), Simalungun (Pematangsiantar).
Riau. Indragiri Hulu (Rengat).
Jambi. Kerinci (Sungaipenuh)
Lampung. Lampung Tengah (Metro)
Jawa Barat. Bogor, Garut, KM Bogor, KM Sukabumi, Kuningan
Jawa Tengah. Grobogan (Purwodadi), Karanganyar, Kebumen, Kudus, Sragen
Jawa Timur. Kediri, Mojokerto, Probolinggo, Sampang, Sumenep, Ngawi, KM Madiun, KM Probolinggo
Bali. Jembrana (Negara), Karangasem (Amlapura)
Nusa Tenggara Barat. Dompu, Lombok Tengah (Praya)
Nusa Tenggara Timur. Belu (Atambua), Timor Tengah Utara (Kefamenanu), Sumba Timur (Waingapu)
Sulawesi Utara. Gorontalo (Limboto), Sangir Talaud (Tahuna), KM Bitung
Sulawesi Tengah. Donggala, Banggai (Luwuk)
Sulawesi Selatan. Soppeng (Watangsoppeng)
Kalimantan Timur. KM Balikpapan, Kutai (Tenggarong), Bulungun (Tg. Selor)
Kalimantan Selatan. Banjar (Martapura), Hulu Sungai Tengah (Barabai), Hulu Sungai Utara (Amuntai), Tabalong (Tanjung), Tanahlaut (Peleihari), KM Banjarbaru, Hulu Sungai Selatan (Kandangan)
Kalimantan Tengah. Kotawaringin Barat (Pangkalanbuun)
Kalimantan Barat. Kapuas Hulu (Putussibau), Pontianak (Mempawah), Sambas (Singkawang) Sintang, Ketapang
Maluku: Maluku Tenggara (Tual)

Indonesian Station Headings:

RDK	Radio Daerah Kabupaten
RKDT2	Radio Kabupaten Daerah Tingkat Dua
RKPD	Radio Khusus Pemerintah Daerah
RKPDT2	Radio Khusus Pemerintah Daerah Tingkat Dua
RKPKDT2	Radio Khusus Pemerintah Kabupaten Daerah Tingkat Dua
RKPDK	Radio Khusus Pemerintah Daerah Kabupaten
RKPDKT2	Radio Khusus Pemerintah Daerah Kabupaten Tingkat Dua
RPD	Radio Pemerintah Daerah
RPDT1	Radio Pemerintah Daerah Tingkat Satu
RPDK	Radio Pemerintah Daerah Kabupaten
RPDKDT2	Radio Pemerintah Daerah Kabupaten Daerah Tingkat Dua
RPDT2	Radio Pemerintah Daerah Tingkat Dua
RPKDT2	Radio Pemerintah Kabupaten Daerah Tingkat Dua
RRI	Radio Republik Indonesia
RS	Radio Siaran

Note: There are many other possible station headings, however, these can be determined by using the appropriate letters/words.
Addresses (Jln = Jalan):
1) Jln Diponegoro 61, Kotak Pos 155, Bandung, Jawa Barat. **FM:** 102.0/104.0MHz.
2) Jln Mayjen Sutoyo 31, Kuala Simpang, D.I. Aceh.
3) Jln Raya Bojongsari 17, Sawangan — Bogor, Jawa Barat
4) Jln Prof. Dr. S. Brojonegoro 24, Jambi, Propinsi Jambi.
5) Jln Murakata 7, Barabai, Kalimantan Selatan.
6) Jln Buru 10A, Ujung Pandang, Sulawesi Selatan.
7) Jln Dahlia 45, Kema Raya - Kendari, Sulawesi Tenggara.
8) Jln Kenanga 15, Ambon, Maluku.
9) Jln Brigjen. Slamet Riyadi, Ungaran — Semarang, Jawa Tengah.
10) Jln Tan Malaka 201C, Jambi, Propinsi Jambi.
11) Jln Laks. Malahayati 1, Telukbetung-Bandar Lampung, Lampung.
12) Jln Pasar Timur 8A, Pandeglang, Jawa Barat.
13) Jln Prof. Dr. Soekarno 6, Purwokerto, Jawa Tengah.
14) Jln Sunan Ngerang 2A, Juwana - Pati, Jawa Tengah.
15) Jln Dieng 1A, Wonosobo, Jawa Tengah.
16) Jln Tegalgendu 12, Tromolpos 6, Kotagede-Yogyakarta, D.I. Yogyakarta.
17) Jln S. Limboto 42, Ujung Pandang, Sulawesi Selatan.
18) Jln Slt. Hasanuddin 14, Parepare, Sulawesi Selatan.
19) Jln Pemuda 82-90, Kotak Pos 239, Surabaya, Jawa Timur. **FM:** 97.0MHz.
20) Jln Dr. Cipto Mangunkusumo 24, Telukbetung-Bandar Lampung, Lampung.
21) Jln Sukarno-Hatta 355, Bandung, Jawa Barat.
22) Jln Sunan Kalijaga 10, Rangkasbitung, Jawa Barat.
23) Jln Siliwangi 35, Cijoho - Kuningan, Jawa Barat.
24) Jln Margantara Tanjung, P.O. Box 45, Purwokerto, Jawa Tengah.
25) Gg. Pinggir 62, Kotakpos 180, Semarang, Jawa Tengah.
26) Jln Kapt. Mulyadi 180, Surakarta, Jawa Tengah.
27) Jln Tondang, Gg. Dhamashinta 3, Pekalongan, Jawa Tengah.
28) Jln Ciung Wenara, Komp. Nitimandala Renon, Denpasar, Bali.
29) Jln Ali Malaka 28-B, Ujung Pandang, Sulawesi Selatan.
30) Jln K.H. Ghalib Raya, Pringsewu-Lampung Selatan, Lampung
31) Jln Kemakmuran 26, Ambon, Maluku.
32) Jln W. Monginsidi 19, Jakarta Selatan.
33) Jln Riburane 2, Kotak Pas 103, Ujung Pandang, Sulawesi Selatan. **FM:** 104.0MHz.
34) Jln Pasar Kembang 41, Yogyakarta, D.I. Yogyakarta.
35) Jln Prof. Dr. Ir. Sutami 62, Jambi, Propinsi Jambi.
36) Jln Kartini 90B, Tanjungkarang-Bandar Lampung, Lampung.
37) Proyek Senen Blok I, Lantai IV/26, Jakarta Pusat.
38) Jln Arief Rachman Hakim 60, Sumedang, Jawa Barat.
39) Jln Pahlawan 313/41, P.O. Box 3, Purwakarta, Jawa Barat.
40) Jln Barito 4, Pekalongan, Jawa Tengah.
41) Jln Panglima Sudirman 72-III, Surabaya, Jawa Timur.
42) Jln Dewi Sartika 52, Singaraja, Bali.
43) Jln Kol. Sugiono 72, Banjarmasin, Kalimantan Selatan.
44) Jln Buru 28, Kotakpos 45, Ujung Pandang, Sulawesi Selatan.
45) Jln Mesjid Raya 5, Sengkang - Wayo, Sulawesi Selatan.
46) Jln Santo Yoseph 17A, Manado, Sulawesi Utara.
47) Jln Patimura 1, Ambon, Maluku.
48) Jln K.H. Abdul Hadi 106, Serang, Jawa Barat.
49) Jln Borobudur 9, Jakarta Pusat.
50) Jln Proklamator Raya 163, Bd. Jaya-Lampung Tengah, Lampung
51) Jln Dr. Sudarsono 20, Cirebon, Jawa Barat.
52) Jln Mas Cilik 45, Purwokerto, Jawa Tengah.

53) Jln Raya Wonopringgo 414, Pekalongan, Jawa Tengah.
54) Jln Purworejo VI/10, Surakarta, Jawa Tengah.
55) Jln Jeruk Raya 27, Semarang, Jawa Tengah.
56) Jln R. Danang 3, Pati, Jawa Tengah.
57) Bagian Humas Setwilda Kab. Dati II Wonogiri, Jln Kabupaten 4, Wonogiri, Jawa Tengah.
58) Jln Terusan Enim/Wai-Ketibung 60, Bandar Lampung, Lampung.
59) Jln Sudirman 73, Pinrang, Sulawesi Selatan
60) Jln Batu Lawang 1, Banjar-Ciamis, Jawa Barat
61) Jln Kelapa Puyuh IV/KB-19, Klp. Gading Permai, Jakarta Utara
62) Jln Guntursari Wetan 27, Bandung, Jawa Barat.
63) Jln Ir. H. Juanda 60, Cianjur, Jawa Barat.
64) Jln By Pass 16, Arjawinangun - Cirebon, Jawa Barat.
65) Jln Dahlia 31, Banjarmasin, Kalimantan Selatan.
66) Jln Ujung Lorong 151/5, Ujung Pandang, Sulawesi Selatan.
67) Jln Jendral Akhmad Yani 1, Ambon, Maluku. **FM:** 105.5MHz.
68) Jln Gajah Mada 9, Pekalongan, Jawa Tengah.
69) Jln Cut Nya Dien 16, Medan, Sumatera Utara.
70) Jln Permindo 57, Padang, Sumatera Barat.
71) Jln Empu Sendok 12, Kebayoran Baru, Jakarta Selatan.
72) Jln Mohammad Toha 146, Bandung, Jawa Barat.
73) Jln Dr. Muwardi 92A, Cianjur, Jawa Barat.
74) Jln R.E. Martadinata 200, Tasikmalaya, Jawa Barat.
75) Jln Pangeran Drajat, Kesambi - Cirebon, Jawa Barat.
76) Jln Kepatihan Tengah, Surakarta, Jawa Tengah.
77) Jln Desa Patuk, Kec. Patuk, Yogyakarta, D.I. Yogyakarta.
78) Jln Sriwijaya 12, Cakranegara, Mataram, Lombok, Nusa Tenggara Barat.
79) Jln Manggis 31, Banjarmasin, Kalimantan Selatan.
80) Jln Rajawali 16, Ujung Pandang, Sulawesi Selatan.
81) Jln Kemakmuran 4, Rantepao, Sulawesi Selatan.
82) Jln Arie Lasut, Manado, Sulawesi Utara.
83) Jln Jendral Soedirman 427, Kotak Pos 5, Purwokerto, Jawa Tengah. **FM:** 94.0/100.0/104.0MHz.
84) Jln Nusantara 261A, Ujung Pandang, Sulawesi Selatan.
85) Jln Haji Ikap 47, Palangkaraya, Kalimantan Tengah.
86) Jln Pantai Sindhu 15, Denpasar, Bali.
87) Jln Anggrek 4, Surabaya, Jawa Timur.
88) Jln Bonto Marannu 11, Ujung Pandang, Sulawesi Selatan.
89) Jln Nusantara II/50, Manado, Sulawesi Utara.
90) Jln Sei Besitang Terusan I, Medan, Sumatera Utara.
91) Jln Jendral A. Yani 296, Bandung, Jawa Barat.
92) Jln Ahmad Yani 144-146, P.O. Box 74, Semarang, Jawa Tengah. **FM:** 94.0/97.0MHz.
93) Jln Jendral A. Yani, P.O. Box 11, Merauke, Irian Jaya. **FM:** 105.0MHz.
94) Jln Pendawa 110/RT.8, 2 Ilir, Palembang, Sumatera Selatan.
95) Jln H.O.S. Cokroaminoto 10, Tanjungkarang-Bandar Lampung, Lampung.
96) Jln Cirebon 3, Belawan-Medan, Sumatera Utara
97) Jln Pramuka 242, Majalengka, Jawa Barat.
98) Jln Candi Agung 88-89, Amuntai, Kalimantan Selatan.
99) Jln Penghibur 92A, Ujung Pandang, Sulawesi Selatan.
100) Jln S. Parman 161, Manado, Sulawesi Utara.
101) Jln Letkol Martinus Lubis 5, Medan, Sumatera Utara. **FM:** 88.0/103.0/105.0MHz.
102) Jln H. Peeng 9, Batusari-Jakarta Barat.
103) Jln Siliwangi 5, Bandung, Jawa Barat.
104) Jln Bojong Tengah 2B, Tasikmalaya, Jawa Barat.
105) Jln Brawijaya 4, Kadipaten-Majalengka, Jawa Barat.
106) Jln M.T. Haryono 161, Lantai II, Semarang, Jawa Tengah.
107) Jln Raya Pekajangan 555, Pekalongan, Jawa Tengah.
108) Jln Bhayangkara 8, Surakarta, Jawa Tengah.
109) Jln Jend. Sudirman A/12, Bantul, D.I. Yogyakarta.
110) Jln Walikota Mustajab 62, Surabaya, Jawa Timur.
111) Jln Dr. Sutomo 26, Malang, Jawa Timur.
112) Jln Belitung Darat 2, Banjarmasin, Kalimantan Selatan.
113) Jln Jendral Achmad Yani (Komplex Nusa Indah 40), Sampit, Kalimantan Tengah.
114) Jln Samiun 17, Ujung Pandang, Sulawesi Selatan.
115) Jln Yos Sudarso VI, Manado, Sulawesi Utara.
116) Jln Raya Sanur 78A, Denpasar, Bali.
117) Jln Bangau 60, Palembang, Sumatera Selatan.
118) Jln Yos Sudarso 17A, Panjang-Bandar Lampung, Lampung.
119) Jln Pakubuwono VI/35, Kebayoran Baru, Jakarta Selatan.
120) Jln L.R.E. Martadinata 229, Bandung, Jawa Barat.
121) Jln Raya Timur 465, Singaparna-Tasikmalaya, Jawa Barat.
122) Jln D.I. Panjaitan 1, Surakarta, Jawa Tengah.
123) Jln Dharmahusada Indah Barat, Blok A/75, Surabaya, Jawa Timur.
124) Jln Probolinggo 4, Malang, Jawa Timur.
125) Jln Bali RT. 08 30C, Banjarmasin, Kalimantan Selatan.
126) Jln Banda Lr. 203/5, Ujung Pandang, Sulawesi Selatan.
127) Jln Kawa V/1, Semarang, Jawa Tengah.
128) Jln Langko 83, P.O. Box 2, Ampenan-Mataram, Lombok, Nusa Tenggara Barat. **FM:** 93.5/102.0/104.0MHz.
129) Jln Brigjen. Dharsono/By Pass, Cirebon, Jawa Barat. **FM:** 93.5/107.2MHz.
130) Jln Tompello 8, Kupang, Timor, Nusa Tenggara Timur. **FM:** 102.0MHz.
131) Jln Merdeka 99, Bangli, Bali.
132) Jln Jendral Sudirman 14, Payakumbuh, Sumatera Barat.
133) Jln Temon 259, 26 Ilir, Palembang, Sumatera Selatan.
134) Jln Mesuji 1, Pahoman - Telukbetung, Bandar Lampung, Lampung.
135) Jln Bali Matraman 17, Jakarta Selatan.
136) Jln Margamulya 128, Cimalaka - Sumedang, Jawa Barat.
137) Jln Simolawang I/96, Surabaya, Jawa Timur.
138) Jln Teluk Tiram Laut 262, Banjarmasin, Kalimantan Selatan.
139) Jln Madidir Weru Lingk. III, Bitung, Sulawesi Utara.
140) Jln Pembangunan 9 R.E. Martadinata, Balikpapan, Kalimantan Timur.
141) Jln S. Limboto 65, Ujung Pandang, Sulawesi Selatan.
142) Jln Paria 8, Bandung, Jawa Barat.
143) Jln Pembangunan, BBK Caringin, Garut, Jawa Barat.
144) Jln Piere Tendean 50, Banjarmasin, Kalimantan Selatan.
145) Jln Andi Cammi 5, Pangkajene, Sulawesi Selatan.
146) Jln Pattimura, Serui, Irian Jaya. **FM:** 93.2MHz.
147) Jln Cerme 16, Kotak Pos 78, Malang, Jawa Timur. **FM:** 102.0/105.0MHz.
148) Jln Sisingamangaraja 1, Padang, Sumatera Barat.
149) Jln Jambu Air, P.O. Box 42, Bukittinggi, Sumatera Barat.
150) Jln Darmawangsa Raya 13, Blok P.2, Kebayoran Baru, Jakarta Selatan.
151) Jln Wates, Kulonprogo, D.I. Yogyakarta.
152) Jln Gadung III/2, Surabaya, Jawa Timur.
153) Jln T. Nya' Arief 2, Langsa, D.I. Aceh.
154) Jln Tebet Timur Dalam I-N/249, Jakarta, Selatan
155) Jln Ir. H. Juanda 52, Bandung, Jawa Barat.
156) Jln Pangkalan Raya 1A, Kd. Halang-Bogor, Jawa Barat.
157) Jln Jend. Sudirman 265, Ciamis, Jawa Barat.
158) Jln Jend. Sudirman 440, Tromolpos 51, Pekanbaru, Riau. **FM:** 88.0MHz.
159) Jln Satria Selatan III/H 262, Semarang, Jawa Tengah.
160) Jln Senen Raya 30-32, Jakarta Pusat.
161) Jln Babakan 35, Majalaya - Bandung, Jawa Barat.
162) Jln Raya Siliwangi 167, Cicurug - Sukabumi, Jawa Barat.
163) Jln Panularan 29/IV, Surakarta, Jawa Tengah.
164) Jln Karangkajen 76, Yogyakarta, D.I. Yogyakarta.
165) Jln Adityawarman 81A, Surabaya, Jawa Timur.
166) Jln Banjaran II/29, Kediri, Jawa Timur.
167) Jln Veteran 6B, Bondowoso, Jawa Timur.
168) Jln Bung Karno 22, Mataram, Lombok, Nusa Tenggara Barat.
169) Jln Letjen. Haryono MT288, Banjarmasin, Kalimantan Selatan.
170) Jln G. Nona 1, Ujung Pandang, Sulawesi Selatan.
171) Jln Monginsidi 89B, Medan, Sumatera Utara.
172) Jln Laute Mandonga 44, Kotak Pos 7, Kendari, Sulawesi Tenggara. **FM:** 107.3MHz.
173) Jln Dr. M. Isa 38, 8 Ilir, Palembang, Sumatera Selatan.
174) Jln Teuku Umar 36, Tanjungkarang-Bandar Lampung, Lampung.
175) Jln Kapuk Muara Raya 53, Jakarta Utara.
176) Jln Burangrang 2, Bandung, Jawa Barat.
177) Jln Perintis Kemerdekaan 86, Cibadak-Sukabumi, Jawa Barat.
178) Jln Mangkubumen Kulon I/2, Surakarta, Jawa Tengah.
179) Jln Makam Pahlawan Kusumajati, P.O. Box 2, Kendal, Jawa Tengah.
180) Jln Mentaok Raya 14, Kotagede-Yogyakarta, D.I. Yogyakarta.
181) Jln Aipda Karel Sasuit Tubun 15, Gresik, Jawa Timur.
182) Jln Brigjen. H. Hasan Basri 64, Banjarmasin, Kalimantan Selatan.
183) Jln. D.I. Panjaitan 61, Jember, Jawa Timur. **FM:** 104.0MHz.
184) Jln Pasundan 81, Sukabumi, Jawa Barat.
185) Jln Brig. Jendral Katamso 43, Lubuk Pakam-Medan, Suma-

tera Utara.
186) Jln Abdul Rahman Saleh 51 (or: Kotakpos No. 40), Surakarta, Jawa Tengah. **FM:** 104.0/105.0MHz.
187) Jln Melayu Besar 68, Jakarta Selatan.
188) Jln Nurali 30, Pontianak, Kalimantan Barat.
189) Jln Desa Koyamas I/36, Langowan, Sulawesi Utara.
190) Jln Salam 44, Bandung, Jawa Barat.
191) Jln Embong Wungu 47, Surabaya, Jawa Timur.
192) Jln Langsep Barat Kav. 30, Malang, Jawa Timur.
193) Jln Bukit Inspirasi, P.O. Box 10, Tomohon-Minahasa, Sulawesi Utara.
194) Jln Jenderal Achmad Yani 32, Temanggung, Jawa Tengah.
195) Tromolpos 157, Jln Medan Merdeka Barat 4-5, Jakarta.
196) Jln Airport Polonia, Medan, Sumatera Utara.
197) Jln Sadanco Supriadi 40, Blitar, Jawa Timur.
198) Jln Mayjen. Panjaitan 10, Madiun, Jawa Timur. **FM:** 99.0/104.0MHz.
199) Jln Iskandar Muda 117, Medan, Sumatera Utara.
200) Jln Kyai Tapa, Grogol, Jakarta Barat.
201) Jln Salendro Timur VII/12A, Bandung, Jawa Barat.
202) Jln Manggis I/14, Sukabumi, Jawa Barat.
203) Jln Dr. Cipto 175, Semarang, Jawa Tengah.
204) Jln Pemuda Selatan 38, Klaten, Jawa Tengah.
205) Jln Darmokali 19A, Surabaya, Jawa Timur.
206) Jln Joyoboyo 77, Kediri, Jawa Timur.
207) Jln Raya Gianyar 100x, Sukawati-Gianyar, Bali.
208) Jln Cempaka Besar 25, Banjarmasin, Kalimantan Selatan.
209) Jln Gatot Subroto 26, Kotakpos 24, Pahoman-Bandar Lampung, Lampung. **FM:** 93.0/98.0MHz.
210) Jln Kawi 37, Kepanjen, Malang, Jawa Timur.
211) Jln Achmad Yani, Kotakpos 505, Biak, Irian Jaya. **FM:** 93.0MHz.
212) Jln Kayumanis V.E./286, Jakarta Timur.
213) Jln Angkasa 1, P.O. Box 44, Bekasi, Jawa Barat.
214) Jln Raya Tipar 16, Sukabumi, Jawa Barat.
215) Jln Kawi 39, Semarang, Jawa Tengah.
216) Jln Lawu 50, Tegal, Jawa Tengah.
217) Jln Raya Medari 81, Medari-Sleman, D.I. Yogyakarta.
218) Jln Pucang Anom Timur II/23, Surabaya, Jawa Timur.
219) Jln Kartini 25, Jember, Jawa Timur.
220) Jln Jendral Sudirman 59, Singaraja, Bali.
221) Jln Rama Setia Lampaseh, Banda Aceh, D.I. Aceh.
222) Jln Tasangkapura 23, Jayapura, Irian Jaya. **FM:** 93.6MHz.
223) Jln Pontianak 22, UKT - Padang, Sumatera Barat.
224) Jln D.I. Panjaitan 3/41, Plaju - Palembang, Sumatera Selatan.
225) Jln Imam Bonjol, Gg. Pelita 45, Tanjungkarang, Bandar Lampung, Lampung.
226) Jln Matraman 39, Jakarta Timur.
227) Jln Kacapiring 12, Bandung, Jawa Barat.
228) Jln Otto Iskandardinata 114, Sukabumi, Jawa Barat.
229) Jln Pelita 34, Singkawang, Kalimantan Barat.
230) Jln Jagalan 36, P.O. Box 104, Yogyakarta, D.I. Yogyakarta.
231) Jln Satelit Indah III/CN.15, Surabaya, Jawa Timur.
232) Jln Gatot Kaca 45, Negara, Bali.
233) Jln Selaparang 60, Mataram, Lombok, Nusa Tenggara Barat.
234) Jln R.K. Ilir 497B, Banjarmasin, Kalimantan Selatan.
235) Jln Mesjid Istiqlal 6, Manado, Sulawesi Utara.
236) Jln Sumatera 28, Pontianak, Kalimantan Barat.
237) Jln Gajah Mada 144, Singaraja, Bali. **FM:** 93.8/103.7MHz.
238) Jln Perdana I/10, Petukangan Selatan, Keb. Lama, Jakarta Selatan.
239) Jln Lengkong Besar 14, Bandung, Jawa Barat.
240) Jln Suryakencana 104, Sukabumi, Jawa Barat.
241) Jln Ngagel Jaya Utara IV/21, Surabaya, Jawa Timur.
242) Jln Dr. Cipto 16, P.O. Box 16, Malang, Jawa Timur.
243) Jln K.H. Abdul Fatah Hasan 9B, Serang, Jawa Barat.
244) Kompleks Mesjid Raya Baiturrahman, Banda Aceh, D.I. Aceh.
245) Sekratariat PKDT2 Kendal, Kendal, Jawa Tengah.
246) Jln Jendral A. Yani 5, Telanaipura-Jambi, Propinsi Jambi. **FM:** 103.7MHz.
247) Jln Urip Sumoharjo 26, Sumenep, Madura, Jawa Timur. **FM:** 93.0/93.5MHz.
248) Jln Puncak Sekuning 462, Palembang, Sumatera Selatan.
249) Jln Letjen. S. Parman 1, Jakarta Barat.
250) Jln Gandapura 67, Bandung, Jawa Barat.
251) Jln K.H.A. Dahlan 1, Karawang, Jawa Barat.
252) Jln Amat Jazuli 4, Tromolpos 18, Yogyakarta, D.I. Yogyakarta. **FM:** 101.0/102.7/107.3MHz.
253) Jln Jepara Lorong 52/6, Tanjung Priok, Jakarta Utara.
254) Jln Van Deventer 5, Bandung, Jawa Barat.
255) Jln Taman Apsari 7, Surabaya, Jawa Timur.
256) Jln Panglima Polim SK 3/2, Banda Aceh, D.I. Aceh.
257) Jln Jenderal A. Yani Km. 3.5, No. 234, Kotak Pos 117, Banjarmasin, Kalimantan Selatan. **FM:** 97.6/105.5MHz.
258) Jln Pemuda 4, "Balaikota", Tegal, Jawa Tengah.
259) Jln Tgk. Cik Ditiro 2, Banda Aceh, D.I. Aceh.
260) Jln Enggang 10, Medan, Sumatera Utara.
261) Jln Gunung Puteng 22, Singkawang, Kalimantan Barat.
262) Jln Bekasi Barat 9, Jakarta Timur.
263) Jln Raya Kav. Pertamina Desa Gandul Cinere, Sawangan-Bogor, Jawa Barat.
264) Jln Olahraga 21, Indramayu, Jawa Barat.
265) Jln R.E. Martadinata 1, Tasikmalaya, Jawa Barat.
266) Jln Mugas Dalam VI/11, Tromolpos 200, Semarang, Jawa Tengah.
267) Jln Kesejahteraan Sosial Wonosewu, Kal. Ngestiharjo, Kec. Kasihan, Bantul, D.I. Yogyakarta.
268) Jln Sei Bohorok 8, Medan, Sumatera Utara.
269) Jln Ratulangi 21B, Padang, Sumatera Barat.
270) Jln K.H.A. Azhari 136, 12 Ulu, Palembang, Sumatera Selatan.
271) Jln Cendana 8, Pahoman - Tanjungkarang, Bandar Lampung, Lampung.
272) Jln Gajahmada 188, Jakarta Barat.
273) Jln Moh. Ramdan 49A, Bandung, Jawa Barat.
274) Jln Jombang Tangsi 64, Cilegon, Serang, Jawa Barat.
275) Jln Raung 7, Candi Baru - Semarang, Jawa Tengah.
276) Jln Puro Pakwalaman, Yogyakarta, D.I. Yogyakarta.
277) Jln Amir Hamzah 18, Surabaya, Jawa Timur.
278) Jln Kilisuci 42, Kediri, Jawa Timur.
279) Jln Panglima Batur 62, Kandangan, Kalimantan Selatan.
280) Jln Iman Bonjol 54, Samarinda, Kalimantan Timur.
281) Jln Diponegoro 4, Banda Aceh, D.I. Aceh.
282) Jln Mataram 3, Medan, Sumatera Utara.
283) Jln Siliwangi 40, Bandung, Jawa Barat.
284) Jln May. Syafe'i 66B, Serang, Jawa Barat.
285) Jln Diponegoro, Sabang, Weh, D.I. Aceh.
286) Jln Jendral Sudirman 12, Kotakpos 77, Padang, Sumatera Barat. **FM:** 89.0/93.0/103.0MHz.
287) Jln St. Abd. Rachman 1, Tanjung Pinang, Propinsi Riau. **FM:** 93.0/97.6MHz.
288) Jln Petojo Utara 7, Jakarta Pusat.
289) Jln Pasirkaliki 59, Bandung, Jawa Barat.
290) Jln Selabintana 146, Sukabumi, Jawa Barat.
291) Jln Pahlawan 131A, Magelang, Jawa Tengah.
292) Jln K.H.A. Dahlan 3, P.O. Box 105, Yogyakarta, D.I. Yogyakarta.
293) Jln Teluk Aru 68, Surabaya, Jawa Timur.
294) Jln Mondoroko 2, Singosari - Malang, Jawa Timur.
295) Jln Jend. A. Yani 123, Singaraja, Bali.
296) Jln Pasir Mas Selatan B.61, Semarang, Jawa Tengah.
297) Jln Hayam Wuruk, P.O. Box 31, Denpasar, Bali. **FM:** 93.5MHz.
298) Jln Dipokusumo, Purbalingga, Jawa Tengah.
299) Jln Teuku Dirundeng 15, Meulaboh, D.I. Aceh.
300) Jln Multatuli 45, Medan, Sumatera Utara.
301) Jln Khairil Anwar 8, Padang, Sumatera Barat.
302) Jln Jend. A. Yani 155, Pekanbaru, Riau.
303) Jln Gembol Wetan 131, Bandung, Jawa Barat.
304) Jln Letnan Karsono 47, Bondowoso, Jawa Timur.
305) Jln M. Yamin 8, P.O. Box 45, Samarinda, Kalimantan Timur. **FM:** 93.5MHz.
306) Jln Basuki Rakhmat Timur, Magetan, Jawa Timur.
307) Jln Jendral Soedirman 16A, Cilacap, Jawa Tengah.
308) Jln Veteran 495/9H, Palembang, Sumatera Selatan.
309) Jln Emur I/2, Bandung, Jawa Barat.
310) Jln Raya Sukamantri 98, Ciawi - Tasikmalaya, Jawa Barat.
311) Jln Kebalen 2, P.O. Box 51, Surakarta, Jawa Tengah.
312) Jln Panglima Sudirman 42-44, Surabaya, Jawa Timur.
313) Jln Let. Jendral S. Parman 115, Malang, Jawa Timur.
314) Jln Kartini Raya 161, Semarang, Jawa Tengah.
315) Jln Jendral Sudirman 7, Kotak Pos 6, Pontianak, Kalimantan Barat. **FM:** 88.0/100.2/104.0MHz.
316) Jln Diponegoro 53, Sabang, Weh, D.I. Aceh.
317) Jln Pangrango 34, P.O. Box 232, Bogor, Jawa Barat. **FM:** 91.3/94.0MHz.
318) Jln Dewi Sartika 136, Jakarta Timur.
319) Jln Babura Baru 2, Medan, Sumatera Utara.

320) Jln Sersan Idris 44, Bekasi, Jawa Barat.
321) Jln Jember 17, Genteng-Banyuwangi, Jawa Timur.
322) Jln Pemuda 3, P.O. Box 43, Salatiga, Jawa Tengah.
323) Jln Perak Timur 20, Surabaya, Jawa Timur.
324) Jln Banten 1, Malang, Jawa Timur.
325) Jln Indragiri 1, Semarang, Jawa Tengah.
326) Jln Sultan Iskandar Muda 13, P.O. Box 112, Banda Aceh, D.I. Aceh. **FM:** 88.5/93.0/97.6/103.7MHz.
327) Jln Hokki 21, Medan, Sumatera Utara.
328) Jln Srengseng Raya 45, Kebon Jeruk, Jakarta Barat.
329) Jln Gamelan 28, Bandung, Jawa Barat.
330) Jln S. Parman 45, Subang, Jawa Barat.
331) Jln Jendral A. Yani 55, Purworejo, Jawa Tengah.
332) Jln Pandean 63, Kaliwungu-Kendal, Jawa Tengah.
333) Jln Pasanggrahan V Taman, Madiun, Jawa Timur.
334) Jln Kauman Dawur 226, Bangil-Pasuruan, Jawa Timur.
335) Jln Jagung A. Suprapto 35, Banyuwangi, Jawa Timur.
336) Jln May. Jendral Sutoyo 5, Komplek Pondok Indah 26, Banjarmasin, Kalimantan Selatan.
337) Jln Karang Bugis SK.XI/82, Balikpapan, Kalimantan Timur.
338) Jln Ciliwung 32A, Blitar, Jawa Timur.
339) Jln Jendral A. Yani, Selong, Lombok, Nusa Tenggara Barat.
340) Jln Taman Siswa 3C, Banda Aceh, D.I. Aceh.
341) Jln Tembakau Deli II/8D, Medan, Sumatera Utara.
342) Jln Warakas VII/11, Jakarta Utara
343) Jln Jurang 74, Bandung, Jawa Barat.
344) Jln Lebak Pasar 17, Bogor, Jawa Barat.
345) Jln Kol. Sugiyono 288, Jepara, Jawa Tengah.
346) Jln Waru 14, Tegal, Jawa Tengah.
347) Jln Brigjen. Katamso 106, Purworejo, Jawa Tengah.
348) Jln Citandui 14, Surabaya, Jawa Timur.
349) Jln Radio 2 Km. 4, Palembang, Sumatera Selatan. **FM:** 88.0/93.0MHz.
350) Jln Taman Bendungan Jatiluhur 13, Jakarta Pusat.
351) Jln Kacapiring 5, Surabaya, Jawa Timur.
352) Jln A. Yani 32, Malang, Jawa Timur.
353) Jln Miru 72, Cakranegara-Mataram, Lombok, Nusa Tenggara Barat.
354) Jln Karantina 8A, Medan, Sumatera Utara.
355) Jln Kayuputih Selatan IIIC/34, Jakarta Timur.
356) Jln Cikamiri 7, Cisadea - Bandung, Jawa Barat.
357) Jln Puter 5, Tanah Sareal - Bogor, Jawa Barat.
358) Jln Ki Hajar Dewantara 69, P.O. Box 15, Karawang, Jawa Barat.
359) Jln Kapten Ismail 120, Tegal, Jawa Tengah.
360) Jln Cangakan, P.O. Box 3, Karanganyar, Jawa Tengah.
361) Koml. Kawasan Industri, Tromolpos 1, Lomanis - Cilacap, Jawa Tengah.
362) Jln Kyai Jebat 1, Demak, Jawa Tengah.
363) Jln Jend. Suparman III/10, Waru-Sidoarjo, Jawa Timur.
364) Jln Kartini 12, Jember, Jawa Timur.
365) Jln Tenggiri 55, Samarinda, Kalimantan Timur.
366) Jln Kombes Pol M. Duryat 20, Lamongan, Jawa Timur.
367) Jln Pemancar, Banda Aceh, D.I. Aceh.
368) Jln Petula 38, Medan, Sumatera Utara.
369) Jln Durian 82, Pekanbaru, Riau.
370) Jln Setrasari II/14, Bandung, Jawa Barat.
371) Jln Merdeka 92A, Garut, Jawa Barat.
372) Jln K.H. Wakhid Hasyim 133, Jombang, Jawa Timur.
373) Jln Hayamwuruk 14, Pasuruan, Jawa Timur.
374) Jln Jend. Sudirman 1025F/RT.33, Palembang, Sumatera-Selatan.
375) Jln Cut Nyak Dien 51, Metro, Lampung.
376) Jln Raya Kav. Pertamina, Desa Gandul, Cinere-Bogor, Jawa Barat.
377) Jln Jaksa Agung Suprapto 85, Bojonegoro, Jawa Timur.
378) Jln Dr. Wahidin 1, Pati, Jawa Tengah.
379) Jln K.H. Wachid Hasyim 1, Trenggalek, Jawa Timur.
380) Humas Gelora 10 Nopember, Jln Tambaksari, Surabaya, Jawa Timur.
381) Jln H.R. Rasuna Said-Setiabudi B17D, Jakarta Selatan.
382) Jln Dr. Hatta 15, Bandung, Jawa Barat.
383) Jln Raya Tarogong 311, Garut, Jawa Barat.
384) Jln Perdagangan Lt. II/12, Banda Aceh, D.I. Aceh.
385) Jln Raya Timur, Majalengka, Jawa Barat.
386) Jln Brigjen. Katamso 41A, Medan, Sumatera Utara.
387) Jln Batu Ceper V/52, Jakarta Pusat.
388) Jln R.E. Abdullah 3, Gn. Batu-Bogor, Jawa Barat.
389) Jln Suroso 46, Cianjur, Jawa Barat.
390) Jln Ngagel Rejo Kidul 61, Surabaya, Jawa Timur.
391) Jln Ciliwung IIIA/26, Malang, Jawa Timur.
392) Jln Jendral Urip, Gg. Kutilang 72, Pontianak, Kalimantan Barat.
393) Jln Medan Area Selatan 820, Gg. Sairin, Medan, Sumatera Utara.
394) Jln Kebon Cau 67, Cimahi - Bandung, Jawa Barat.
395) Jln Sukamulya 15, Bogor, Jawa Barat.
396) Jln Bedadung 6, Surabaya, Jawa Timur.
397) Jln Mayjen. Sungkono 137, Madiun, Jawa Timur.
398) Jln Blabak 30, 3 Ilir, Palembang, Sumatera Selatan.
399) Jln Prof. M. Yamin SH No. 4, Padangpanjang, Sumatera Barat.
400) Jln Ir. H. Juanda II/4A, Jakarta Pusat.
401) Jln Kemuning 1, Bandung, Jawa Barat.
402) Jln Nusantara Raya 3, Depok, P.O. Box 4, Bogor, Jawa Barat.
403) Jln Lampaseh Kota 49, Banda Aceh, D.I. Aceh.
404) Jln Veteran 6, Brebes, Jawa Tengah.
405) Gedung AKA, Jln Bangka II/2, Kebayoran Baru, Jakarta Selatan.
406) Jln Budhi 42, Cilember, Cimahi - Bandung, Jawa Barat.
407) Jln Johar 109, Kudus, Jawa Tengah.
408) Jln Kartini, Temanggung, Jawa Tengah.
409) Jln Diponegoro 115, Sumenep, Madura, Jawa Timur.
410) Jln Merbabu 30A, Ponorogo, Jawa Timur.
411) Taman Ismail Marzuki, Jln Cikini Raya 73, Jakarta Pusat.
412) Jln Pemuda 15, Bireuen, D.I. Aceh.
413) Jln Veteran Kav. 1, Semarang, Jawa Tengah.
414) Jln Embong Tanjung 8, Surabaya, Jawa Timur.
415) Jln Rajawali 25, Palu, Sulawesi Tengah.
416) Jln Jendral Basuki Rakhmad 15, Kediri, Jawa Timur.
417) Jln Alun-Alun Utara 4, Madiun, Jawa Timur.
418) Jln Teuku Umar 155, Kotamobagu, Sulawesi Utara.
419) Jln Kompl. Universitas Sumatera Utara, Medan, Sumatera Utara.
420) Jln Sanjai, Wisma Inkobra, Bukittinggi, Sumatera Barat.
421) Jln Kulim 22A, Pekanbaru, Riau.
422) Jln Prof. Dr. Yamin SH 19, Jambi, Propinsi Jambi.
423) Jln Pabean 232, 28 Ilir, R.T. 1, Palembang, Sumatera Selatan.
424) Jln Letjen. S. Parman 14A, Tanjungkarang-Bandar Lampung, Lampung.
425) Jln Cipinang Cempedak I/51, Polonia Jatinegara, Jakarta.
426) Jln Pilang 109A, Kedawung - Cirebon, Jawa Barat.
427) Jln Wonodri Krajan 677, Semarang, Jawa Tengah.
428) Jln Perintis Kemerdekaan, Yogyakarta, D.I. Yogyakarta.
429) Jln Panglima Sudirman 39C, Probolinggo, Jawa Timur.
430) Jln Flamboyan 55, Klungkung, Bali.
431) Jln Barintik 35, Martapura, Kalimantan Selatan.
432) Jln A. Rahman Saleh, Pontianak, Kalimantan Barat.
433) Jln Jend. A. Yani 23, Palangkaraya, Kalimantan Tengah.
434) Jln Mianggas III, Manado, Sulawesi Utara.
435) Jln Setia Budi 55, Palu, Sulawesi Tengah.
436) Jln Sukarasa 143E, Cicadas - Bandung, Jawa Barat.
437) Jln Cikuray 16, Bogor, Jawa Barat.
438) Jln K.H. Sam'un 2, Rangkasbitung, Jawa Barat.
439) Jln Abdulrachman Saleh 514, Semarang, Jawa Tengah.
440) Jln Let. Jen. S. Parman 8, Yogyakarta, D.I. Yogyakarta.
441) Jln Anjasmoro 37, Surabaya, Jawa Timur.
442) Jln Rajawali 59, Pontianak, Kalimantan Barat.
443) Jln Dr. Sutomo 12, Manado, Sulawesi Utara.
444) Jln Ngalengko 15A, Medan, Sumatera Utara.
445) Jln Gajahmada 109, Lt. IV, P.O. Box 615, Jakarta Barat.
446) Jln D.I. Panjaitan 28A, Balikpapan, Kalimantan Timur.
447) Jln Rantau 19, Langsa, D.I. Aceh.
448) Jln Murai 129, Sei Sikambing-Medan, Sumatera Utara.
449) Jln Simbolon 5, Pematang Siantar, Sumatera Utara.
450) Jln Pemuda Tengah 56, Kotakpos 13, Klaten, Jawa Tengah.
451) Jln Suari 26, Semarang, Jawa Tengah.
452) Jln Pemuda Pucungrejo, Muntilan, Magelang, Jawa Tengah.
453) Jln Brigpol. Nasuha 2, Karawang, Jawa Barat.
454) Jln Bola Kaki 17, Pematangsiantar, Sumatera Utara.
455) Jln Teratai 15A, Pekanbaru, Riau.
456) Jln Sukarjo Harjo Wardoyo 893, 7 Ulu, Palembang, Sumatera Selatan.
457) Jln Kompl. TMII Pd. Gede Kramat Jati, Jakarta Timur.
458) Jln Kaum Kidul 415, Soreang-Bandung, Jawa Barat.
459) Jln Dr. Wahidin S. IV/4, Cirebon, Jawa Barat.
460) Jln Bhayangkara 69, Surakarta, Jawa Tengah.
461) Jln Krembangan Bhakti 86, Surabaya, Jawa Timur.

462) Jln Kahuripan, Malang, Jawa Timur.
463) Jln Parkit 7, P.O. Box 2, Tabanan, Bali.
464) Jln B.W. Lapian 38, Manado, Sulawesi Utara.
465) Jln Sunan Kudus 194, Kudus, Jawa Tengah.
466) Jln Pasundan RT.III, P.O. Box 175, Samarinda, Kalimantan Timur.
467) Jln Raya Puncak 70, Cipayung - Bogor, Jawa Barat.
468) Jln Raya Timur 74, Jatiwangi-Majalengka, Jawa Barat.
469) Jln Jend. A. Yani 33, Cikampek-Purwakarta, Jawa Barat.
470) Jln Veteran 5, Tembilahan, Riau.
471) Komplek Pendopo, Tuban, Jawa Timur.
472) Jln Sudarman 1, Jember, Jawa Timur.
473) Jln Jokotole 57, Pamekasan, Madura, Jawa Timur.
474) Jln Singkil 96, Demak, Jawa Tengah.
475) Jln Kartini 137, Pematang Siantar, Sumatera Utara.
476) Jln Pancasila 18, Padangpanjang, Sumatera Barat.
477) Jln Laksda. Yos Sudarso 34, Jambi, Propinsi Jambi.
478) Jln Segaran 38, 15 Ilir, Palembang, Sumatera Selatan.
479) Jln Jend. A. Yani 34, Purwakarta, Jawa Barat.
480) Jln A. Yani 116, Magelang, Jawa Tengah.
481) Jln Bawal 36, Tegal, Jawa Tengah.
482) Jln Kapasari 1, Surabaya, Jawa Timur.
483) Jln Demang Lehman RT. III/II. 5C, Martapura, Kalimantan Selatan.
484) Jln Akcaya III/7, Pontianak, Kalimantan Barat.
485) Jln Sentosa I/6, Balikpapan, Kalimantan Timur.
486) Jln Veteran 127, Bukittinggi, Sumatera Barat.
487) Jln Melati 82, Pekanbaru, Riau.
488) Jln Dwidaya II/9-Radio Dalam, Jakarta Selatan.
489) Jln Jend. A. Yani 662, Bandung, Jawa Barat.
490) Jln Cempaka 5, Karawang, Jawa Barat.
491) Jln Kemuning 30, P.O. Box 48, Salatiga, Jawa Tengah.
492) Jln Gayam 24, Yogyakarta, D.I. Yogyakarta.
493) Jln Toar 62, Manado, Sulawesi Utara.
494) Jln Ir. H. Juanda 128, Ciamis, Jawa Barat.
495) Jln Pahlawan 20, Pasuruan, Jawa Timur.
496) Jln AKBP M. Soeroko 11, Bojonegoro, Jawa Timur.
497) Jln Kabupaten, Ponorogo, Jawa Timur.
498) Jln Kartini 4, Magelang, Jawa Tengah.
499) Jln Kaikoli, Dili, Timor Timur. **FM:** 93.0/105.0MHz.
500) Jln Willem Iskander, P.O. Box 9, Padangsidempuan, Sumatera Utara.
501) Jln Raya Supratman 90, Denpasar, Bali.
502) Jln Timur Aloon-Aloon, Tulungagung, Jawa Timur.
503) Jln Olah Raga Komplex B.T.N., Indramayu, Jawa Barat.
504) Jln Alun-Alun Utara 7, Lumajang, Jawa Timur.
505) Jln A. Yani 1, Raba, Sumbawa, Nusa Tenggara Barat.
506) Jln Dadaha 17, Tasikmalaya, Jawa Barat.
507) Jln P. Geusan Ulun 125, Sumedang, Jawa Barat.
508) Jln Panglima Sudirman, Ende, Flores, Nusa Tenggara Timur.
509) Jln Cikutra 27, Cibeunying-Bandung, Jawa Barat.
510) Jln Osa Maliki 29, Salatiga, Jawa Tengah.
511) Jln Merdeka 68, Manokwari, Irian Jaya. **FM:** 93.5MHz.
512) Jln Diponegoro 99, Amlapura, Bali.
513) Jln Durian 6, Tabanan, Bali.
514) Jln Kyai Maja 63, Kebayoran Plaza Lt. IV, Jakarta Selatan.
515) Jln K.Z. Abidin 48, Bengkulu, Propinsi Bengkulu.
516) Jln Basudewa 399, Semarang, Jawa Tengah.
517) Jln Gatot Subroto 19/118, Bandung, Jawa Barat.
518) Jln Ruhui Rahayu 45, Samarinda, Kalimantan Timur.
519) Jln Lawang Gada 6, Cirebon, Jawa Barat.
520) Jln Guntur 154, Garut, Jawa Barat.
521) Jln Tujuh Pahlawan Revolusi 69, Cirebon, Jawa Barat.
522) Jln Pangeran Diponegoro 22T, Palangkaraya, Kalimantan Tengah.
523) Jln Manggal, Palopo, Sulawesi Selatan.
524) Jln Radio 1, Tikala Ares — Manado, Sulawesi Utara. **FM:** 102.0MHz.
525) Jln Prof. Muhammad Yamin 199, Kuning, Bukittinggi, Sumatera Barat. **FM:** 93.0/97.7MHz.
526) Jln Ade Irma Suryani Nasution 5, Sibolga, Sumatera Utara.
527) Jln Let. Jend. S. Parman 25, Kotak Pos 13, Bengkulu, Propinsi Bengkulu. **FM:** 90.6/93.0/105.0MHz.
528) Jln Jenderal Sudirman, Gorontalo, Sulawesi Utara. **FM:** 93.6/102.0MHz.
529) Jln Merdeka 6, Pematangsiantar, Sumatera Utara.
530) Komplek Kantor Pemda, Jln. Jend. Sudirman Gapura, Kotabumi, Lampung.
531) Jln Kabupaten 1, Wonosobo, Jawa Tengah.
532) Jln Ismail 5A, Binjai, Sumatera Utara.
533) Jln M. Husni Thamrin 1, Palangkaraya, Kalimantan Tengah. **FM:** 89.4/93.0MHz.
534) Jln R. Syamsudin S.H. 52, Sukabumi, Jawa Barat.
535) Jln Kedaton, Ternate, Maluku. **FM:** 93.5/103.5MHz.
536) Jln Keramika, Tanjungpandan, Belitung, Sumatera Selatan.
537) Jln Jendral Ahmat Yani 74, Bengkalis, Riau.
538) Soë (via Kupang), Timor, Nusa Tenggara Timur.
539) Jln Alun-Alun Utara, Blora, Jawa Tengah.
540) Jln Kartini Raya 1, Batang, Jawa Tengah.
541) Jln Pandanaran 5, Boyolali, Jawa Tengah.
542) Bajawa, Flores, Nusa Tenggara Timur.
543) Ruteng, Flores, Nusa Tenggara Timur.
544) Jln Pahlawan 18, Surabaya, Jawa Timur.
545) Jln Tamtama 3, Wates, D.I. Yogyakarta.
546) Tembilahan, Riau.
547) Jln Jendral Achmad Yani 156, Kotakpos 247, Wonocolo - Surabaya, Jawa Timur.
548) Jln W.R. Supratman 37, Rantau Prapat, Sumatera Utara.
549) Jln Putih Derus 6, Rangkasbitung, Jawa Barat.
550) Jln Raya Bandar Agung 4, Lahat, Sumatera Selatan.
551) Langsa, D.I. Aceh.
552) Gedung Olah Raga, Jln Jendral Akhmad Yani 2, Bekasi, Jawa Barat.
553) Jln Praja 1, Banjarnegara, Jawa Tengah.
554) Jln 14 Februari, Lr. 2 Mei, Manado, Sulawesi Utara.
555) Komplex Kabupaten Rembang, Jln Gatot Subroto 8, Rembang, Jawa Tengah.
556) Jln Kartini, Sumbawa Besar, Sumbawa, Nusa Tenggara Barat.
557) Jln Karang Layung 10, Bandung, Jawa Barat.
558) Jln Wonokitri Besar 40C, Surabaya, Jawa Timur.
559) Komplek Monumen D.I. Panjaitan 1, P.O. Box 4, Balige, Sumatera Utara.
560) Jln R.A. Kartini 39, Palu, Sulawesi Tengah. **FM:** 93.0/105.0MHz.
561) Jln Jend. Sudirman 2 (Sahid Jaya), Jakarta Pusat.
562) Jln Babura 3, Medan, Sumatera Utara.
563) Jln Gerilya 8 Hulu, Tembilahan, Riau.
564) Jln Veteran 757G, Palembang, Sumatera Selatan.
565) Jln Kapt. Tendean, Kotak Pos 54, Fak-Fak, Irian Jaya. **FM:** 93.5MHz.
566) Kotakpos 10, Wamena, Irian Jaya.
567) Jln Jendral Achmad Yani, Klademak II, Sorong, Irian Jaya. **FM:** 97.6MHz.
568) Jln Pepera Tempat, Kotakpos 11, Nabire, Irian Jaya. **FM:** 93.5MHz.
569) Jln Sugiono — Tj. Pinggir, Sekupang — Pulau Batam, Riau.
570) Hotel Jakarta Hilton Lt. III, Jakarta Selatan.
571) Jln Gatot Subroto, Komp. Nitipraj, Denpasar, Bali.
572) Jln Sultan Agung 1, Bandar Lampung, Lampung.
573) Jln Renawijaya 8 (Villa Duta), Bogor, Jawa Barat.
574) Jln Yusuf Martadilaga 55, Pandeglang, Jawa Barat.
575) Jln Gajah Mada 149, Mojokerto, Jawa Timur.
576) Wisma Pemda Tangerang, Jln Sutopo 1, Tangerang, Jawa Barat.
577) Jln Adikusumah, Bale Endah, Dayeuh Kolot — Bandung, Jawa Barat.
578) Jln Gubernur Suryo 1, Sidoarjo, Jawa Timur.
579) Jln Jaksa Agung Suprapto 3, Gresik, Jawa Timur.
580) Jln Raya Tegalsari 7, Batang, Jawa Tengah.
581) Jln Cokroaminoto 171, Kisaran, Sumatera Utara.
582) Jln Pemuda 185, Batang, Jawa Tengah.
583) Jln Raya Wiradesa, Pekalongan, Jawa Tengah.
584) Jln Beringin 1, Tanjung Gading, Kabupaten Asahan, Sumatera Utara.
585) Purworejo, Jawa Tengah.
586) Wisma Klidanggo Lt. 2, Pemalang, Jawa Tengah.
587) Jln Dr. Sutomo 60, Nganjuk, Jawa Timur.
588) Jln P.B. Sudirman 14, Situbondo, Jawa Timur.
589) Jln Jaksa Agung Suprapto 8, Pacitan, Jawa Timur.
590) Jln Tuparev 69, Cirebon, Jawa Barat.
591) Jln Brigjen. Katamso 1, Jepara, Jawa Tengah.
592) Jln Ikan Cakalang 3, Banyuwangi, Jawa Timur.
593) Jln Waja 18, Kisaran, Sumatera Utara.
594) Jln Raya Utama Jend. Sudirman 44B, Sukoharjo, Jawa Tengah.
595) Jln Jatayu 5, Pekalongan, Jawa Tengah.
596) Jln Prof. M. Yamin SH, P.O. Box 21, Bangkinang, Riau.
597) Jln Moh. Ismail Bantilan 4, Tolitoli, Sulawesi Tengah.
598) Jln Trikora 30, Dok V, Jayapura, Irian Jaya.

599) Jln S.A. Maulana, Tanjungredeb, Kalimantan Timur.
600) Jln S. Parman 192, Kotamobagu, Sulawesi Utara.
601) Jln Jenderal Sudirman 7, Poso, Sulawesi Tengah.

JAPAN

L.T: UTC + 9h — **Pop**: 124.513.000 — **Radio Sets:** 97.000.000 — **Pr.L:** Japanese — **EC:** a/c 50 + 60, 110/220V (yes). — **ITU:** J.

RADIO REGULATORY BUREAU MINISTRY OF POSTS & TELECOMMUNICATIONS

ADDR: 3-2, Kasumigaseki 1-chome, Chiyoda-ku, Tokyo 100. **Cable:** Radiogen — **L.P:** Dir. Gen: S. Tanaka.

NIPPON HOSO KYOKAI(NHK)

(The Japan Broadcasting Corporation)

ADDR: 2-2-1 Jinnan, Shibuya-ku, Tokyo 150-01. **Te:** +81 (3) 3465-1111.
Fax: +81 (3) 3481 1576. **Cable:** RADIONHK. **Telex:** J34778 NHKHQ.
L.P: Chmn. (Board of Governors): J. Takemi. Pres: M. Kawaguchi. Exec. VP: Y. Nakamura. Gen. MD & Exec. Dir. Gen. Eng: S. Morikawa. MD's: Y. Horii, S. Oki, M. Suzuki, M. Nakano, Y. Suwa, T. Ando, K. Nakamura, K. Soga, Y. Takenaka. DG. Pub. Rel: I. Harumi. D.G. Int. Rel. & Policy Planning: H. Takashima. DG. Overseas Broadc. Dept. (R. Japan): H. Iwamoto.
STATIONS: Mediumwaves:
1) = 1st Netw.; 2) = 2nd Netw. **Call:** JO —. St's in bold are reg. key st's exc. Tokyo (headquarters).

Name & Prgr.	Call	kHz	kW	Name & Prgr.	Call	kHz	kW
Morioka 1	QG	531	10	**Kumamoto 2**	GB	873	500
Nago 1		531	1	Shizuoka 1	PK	882	10
Yamagata 1	JG	540	5	**Sendai 1**	HK	891	20
Matsumoto 1	SG	540	1	**Nagoya 2**	CB	909	10
Nanao 1		540	1	Wakkanai 1		927	1
Miyazaki 1	MG	540	5	Kofu 1	KG	927	5
Kitakyushu 1	SK	540	1	Fukui 1	FG	927	5
Ishigaki 1		540	1	Tsuyama 1		927	1
Okinawa 1	AP	549	10	Muroran 1	IQ	945	3
Sapporo 1	IK	567	100	Hikone 1	QP	945	1
Kagoshima 1	HG	576	10	Tokushima 1	XK	945	5
Kushiro 1	PG	585	10	Fukue 1		945	1
Tokyo 1	AK	594	300	Aomori 1	TG	963	5
Okayama 1	KK	603	5	Yonago 1	LQ	963	1
Obihiro 1	OG	603	5	Hagi 1		963	1
Fukuoka 1	LK	612	100	**Matsuyama 1**	ZK	963	5
Asahikawa 1	CG	621	3	Saga 1	SP	963	1
Iida 1		621	1	Kochi 1	RK	990	10
Kyoto 1	OK	621	1	Hachinohe 1	TQ	999	1
Nobeoka 1		621	1	Hamamatsu 1	DG	999	1
Shizuoka 2	PB	639	10	Fukuyama 1	DP	999	1
Toyama 1	IG	648	5	Nakamura 1		999	1
Osaka 1	BK	666	100	Fukuoka 2	LB	1017	50
Hakodate 1	VK	675	5	Tsuruoka 2	JD	1035	1
Yamaguchi 1	UG	675	5	Toyama 2	IC	1035	1
Nagasaki 1	AG	684	5	Takamatsu 2	HD	1035	1
Tokyo 2	AB	693	500	**Hiroshima 1**	FK	1071	20
Kitami 2	KD	702	10	**Sendai 2**	HB	1089	10
Hiroshima 2	FB	702	10	Nayoro 2		1125	1
Nagoya 1	CK	729	50	Obihiro 2	OC	1125	1
Sapporo 2	IB	747	500	Muroran 2	IZ	1125	1
Kumamoto 1	GK	756	10	Takayama 2		1125	1
Akita 2	UB	774	500	Tottori 2	LC	1125	1
Enbetsu 1		792	1	Hagi 2		1125	1
Takada 1		792	1	Okinawa 2	AD	1125	10
Takayama 1		792	1	Kushiro 2	PC	1152	10
Naze 1		792	1	Kochi 2	RB	1152	10
Nagano 1	NK	819	5	Kitami 1	KP	1188	10
Oita 1	IP	819	5	Kanazawa 1	JK	1224	10
Sasebo 1	AQ	819	1	Matsue 1	TK	1296	10
Osaka 2	BB	828	300	Fukushima 1	FP	1323	1
Niigata 1	QK	837	10	Yamada 1		1323	1
Nayoro 1		837	1	Iwaki 1	HQ	1341	1
Koriyama 1	CP	846	5	Minamata 1		1341	1
Uwajima 1		846	1	Tsuruoka 1	JP	1368	1
Hitoyoshi 1		846	1	Tottori 1	LG	1368	1

Takamatsu 1	HP	1368	5	Aomori 2	TC	1521	1
Hachinohe 2	TZ	1377	1	Hamamatsu 2	DC	1521	1
Yamaguchi 2	UC	1377	5	Fukui 2	FC	1521	1
Nagasaki 2	AC	1377	1	Yonago 2	LZ	1521	1
Morioka 2	QC	1386	10	Ishigaki 2		1521	1
Kanazawa 2	JB	1386	10	Nakamura 2		1521	1
Okayama 2	KB	1386	5	Niigata 2	QB	1593	10
Kagoshima 2	HC	1386	10	Matsue 2	TB	1593	10
Wakkanai 2		1467	1	Enbetsu 2		1502	1
Hakodate 2	VB	1467	1	Asahikawa 2	CC	1602	1
Nagano 2	NB	1467	1	Kofu 2	KC	1602	1
Oita 2	ID	1467	1	Fukuyama 2	DD	1602	1
Miyazaki 2	MC	1467	1	Uwajima 2		1602	1
Iida 2		1476	1	Kitakyushu 2	SB	1602	1
Akita 1	UK	1503	10	Hitoyoshi 2		1602	1
Koriyama 2	CD	1512	1	Nobeoka 2		1602	1
Matsumoto 2	SC	1512	1	Naze 2		1602	1
Matsuyama 2	ZB	1512	5	Fukushima 2	FD	1602	1
Yamagata 2	JC	1521	1				

SHORTWAVES

Location	kHz	kW	Mode	H. of tr*
Osaka	3377.5	0.3	DSB	0800-1300
Sapporo	3970	0.6	DSB	1100-1200
Matsuyama	3970	0.6	DSB	0600-1100
				1200-1500
Nagoya	3970	0.3	SSB	2000-0600
Sapporo	6005	0.6	SSB	0400-1230
				2000-0030
Nagoya	6008.75	0.3	SSB	0100-0330
Fukuoka	6133.75	0.6	SSB	2000-0400
Tokyo	6178.75	0.9	SSB	2000-0030
Osaka	6190	0.5	DSB	0500-0730
Sapporo	9535	0.6	DSB	0100-0330
Osaka	9535	0.3	DSB	0330-0430
				2000-0300
Fukuoka	9538.75	0.6	SSB	0430-0730
Tokyo	9553.75	0.9	SSB	0100-0730

*) These are standby/emergency tr's, and operate irr. acc. to requirements. H. of tr. will vary, but will normally fall within the **ADDR.** of regional key st's: 3-43, Bamba-cho, Higashi-ku, Osaka 540 — 2-11-10, Otemachi, Naka-ku, Hiroshima 730 — 1, Ohdori Nishi, Chuo-ku, Sapporo 060 — 5, Horinouchi, Matsuyama 790 — 1-11-1, Higashizakura, Higashi-ku, Nagoya 461-91 — 2-7, Chibajo-cho, Kumamoto 860.

FM-Stations: Call: JO — FM. St's below 1kW not mentioned.

Location	Call	MHz	kW	Location	Call	MHz	kW
Utsunomiya	BP	80.3	1	Otsu	QP	84.0	1
Chiba	MP	80.7	5	Fukuoka	LK	84.8	3
Toyama	IG	81.5	1	Urawa	LP	85.1	5
Maebashi	TP	81.6	1	Sapporo	IK	85.2	5
Tsu	NP	81.8	3	Fukushima	FP	85.3	1
Yokohama	GP	81.9	5	Kumamoto	GK	85.4	1
Yamagata	JG	82.1	1	Kofu	KG	85.6	1
Kanazawa	JK	82.2	1	Kagoshima	HG	85.6	1
Niigata	QK	82.3	1	Aomori	TG	86.0	3
Tokyo	AK	82.5	10	Takamatsu	HP	86.0	1
Nagoya	CK	82.5	10	Akita	UK	86.7	3
Sendai	HK	82.5	5	Matsuyama	ZK	87.7	1
Kyoto	OK	82.8	1	Osaka	BK	88.1	10
Morioka	QG	83.1	1	Okinawa	AP	88.1	1
Mito	EP	83.2	1	Hiroshima	FK	88.3	1
Fukui	FG	83.4	1	Okayama	KK	88.7	1
Tokushima	XK	83.4	1	Shizuoka	PK	88.8	1
Gifu	OP	83.6	1	Oita	IP	88.9	1

1st Netw (General prgr): 24h — **N:** W every h (exc. 1300 Sat. only); Sun: Every h exc. 1100, also 1450. **Regional N:** W 2125, 2210, 0155, 0310, 0455, 0555, 0655, 1015, 1255, 1430; Sun 2110, 2255, 0310, 0555, 1010, 1255, 1430 — **Wrp:** W 2030, 2055, 2155, 0351, 0951, 1056, 1156 (exc Sat) & after reg. N; Sun. 2030, 2055, 2155, 2235, 2356, 0056, 0255, 1158, 1358 & after reg. N. **Local Prgr's** (the amount of local prgr's varies between the st's): 2212(W) N, 2356(W) wrp, 0056(W) inf/wrp, 0255(W) inf/wrp, 0313 N, 0455 inf/wrp, 0555 N/wrp, 0655 inf/wrp, 0755 inf/wrp, 0855 inf/wrp, 0950 inf/wrp, 1018 N — **INT-SIG:** Music box, Nat. Anthem at s/on & s/off. **ANN:** "Daiichi Hoso". **Local ID's** with call letters, network & location given by studio st's just before (W): 2000, 0300, 1000; (Sun): 0000, 1200.

2nd Netw (Educational prgr): 2030-1500 — **L.L: Basic English:** W 2105-2125, 0500-0520, 0920-0940 — **Basic English (2nd step):** 2125-2145, 0520-0540, 0940-1000 **Basic English (3rd step):** Mon.–Fri. 0600-0630, 1320-1340 — **English Conversation:** W 2145-2200, 0540-0555, 1000-1015. **Easy English for Businessmen:** Mon.-Fri. 0630-0650, 1340-1400 — **German:** W 2200-2220, 0400-0420. **French:** W 2220-2240, 0420-0440. **Russian:** W 1440-1500, 2350-0010. **Spanish:** W: 2240-2300, 0440-0500. **Chinese:** W 2320-2340, 1420-1440. **Korean:** W 1400-1420, 2300-2320. **English Hour:** Sat 1030-1100, Sun. 1430-1500.
Schools Broadc: Mon-Fri 2045-2105, 0045-0200, 1015-1200; Sat 2045-2105, 0045-0200, 0720-0920, 1015-1200; Sun 2030-2110, 0600-0700, 0720-0900. **Wrp:** 0700, 1300 (both 20 mins.)
INT-SIG: see 1st Netw. — **ANN:** "Daini Hoso". **Local ID's** on certain st's (as 1st Netw.) just before 2030, 2300, 0030, 0720, 1320, 1500.
FM Netw: 2100-1600. **Local ID's** just before (W): 2000, 0200, 1600; (Sun) 0000, 1000.
V: Sapporo & Matsuyama by QSL-card. Other regional key st's have no organised QSL service. However, many local st's verify by QSL-card or letter for DX reports. Rp.

FOREIGN SERVICE (R. JAPAN)

ADDR. & Te: as Home Sce. **Fax:** +81 (3) 3482 1250. **Tlx:** J34179 RADJAPAN. **LP:** DG: Hiroshi Iwamoto. Head of Prgrs: Hirofumi Kato. CE: Toshiro Kawamura. Head of PR: Takao Kiyohara.

Winter Schedule 1992/93

kHz	kHz	kHz	kHz	kHz	kHz	kHz
5960	6185	9590	11840	15230	15430	17845
5970	7140	9640	11850	15270	17765	17860
6005	7180	9645	11865	15280	17775	17890
6025	7210	9670	11875	15300	17785	21575
6030	7230	9675	11880	15325	17810	21610
6050	9535	9685	11925	15340	17820	21635
6085	9560	9770	15170	15350	17825	21640
6120	9580	11815	15195	15355	17835	21700
6125						

Relays: B = Skelton (UK), C = Canada, F = French Guiana, G = Gabon, S = Sri Lanka

Arabic		
0330-0400	ME/NAf	17820S
1930-2000	ME/NAf	17775S, 9560
Burmese		
1115-1145	SEAs	15300, 17820
Bengali		
0200-0230	SAs	11840S
1530-1600	SAs	9535S, 11880
Chinese		
0300-0400	As	17765, 17835
1130-1230	As	7210, 9580
1245-1345	SEAs	15300
1500-1600	As	6185, 7210, 11840
English		
0100-0200	SEAs	17810
	As	15195, 17835
	SAs	17845, 11840S
0300-0330*	Am/HWI	21610, 17825, 15325F
0300-0400	SEAs	17810
	NAm	5960C, 15230
0500-0600	SEAs	17810, 21610
	As	17765
	NAm	15230, 17825
	Oce	17860
	Eu/ME/Af	6085B, 7230B
0700-0800	SEAs	17810
	As	17765
	Eu/ME/Af	5970B, 6025B, 15170, 21575G
0900-1000	SEAs	21610
	As	11840
0900-1000*	Oce	15270, 17860
1100-1200	SEAs	11815
	As	11840
	NAm	6120C
1400-1500	SEAs	11815
	SAs	9535S
	NAm	9505
1500-1600	NAm	9535
	Eu/ME/Af	15355G
1700-1800	SEAs	11815
	As	7140
	NAm	9535
	Eu/ME/NAf	17755S
1900-1930	NAm	9535
	Oceania	9640, 11850
2100-2200	SEAs	11840, 15430, 17810
	As	11815
	Oceania	17890
2300-2400	SEAs	17810, 15430
	As	11815, 15195
	Eu/ME/Af	6050B, 6125B

*) Regional Sce, others Gen. Sce.

French		
0630-0700	Eu	15170, 21690G
1830-1900	ME/NAf	9560
2130-2200	Eu	6050B
German		
0600-0630	Eu	15170, 15355G
0830-0900	Eu	9670B, 9770B
2100-2130	Eu	6005B
Hindi		
0230-0300	SAs	11840S
1600-1630	SAs	9535S, 11880
Indonesian		
1015-1115	SEAs	15300, 17820
2300-2400	SEAs	11840, 17785
Italian		
0545-0600	Eu	15170, 15355G
2015-2030	Eu	15340G
Japanese		
0000-0030	SEAs	17810, 15430
	As	11815, 15195
	SAs	17845
	SAm	21635G
	Eu	6050B, 6125B
0200-0300	SEAs	17810
	As	17835
	SAs	17845
	NAm	5960C, 15230
	CAm	15325F
	SAm	15350F
0200-0300*	Am/HWI	17825, 21610
0400-0500	SEAs	17810
	As	17765
	NAm	15230, 17825
	Eu/ME/Af	5960B, 6030B, 9645G, 17820S
0600-0700	SEAs	17810
	As	17765
	NAm	15230, 17825
	Oceania	17860
	Eu/ME/Af	5970B, 6025B
0800-0900	SEAs	17810, 21610
	As	17765
	SAm	9675F, 11875
	Oceania	17860
	Eu/ME/Af	21575G, 21640G
1000-1100	SEAs	11815
	As	11840
	SAm	15350F
1200-1300	SEAs	11815
	NAm	9535
1300-1400	SEAs	11815
	As	7210
	SAs	9535S
	NAm	9535
1600-1700	SEAs	7210, 11815
	As	6185, 7140
	NAm	9535
	Eu/ME/Af	21700G
1800-1900	SAs	9590
	NAm	9535, 11865
	Oce	9640, 11850
	Eu/ME/Af	17775S
2000-2100	SEAs	11840, 15430
	As	11815
	NAm	17825
	Oce	15280, 17890
	Eu/ME/Af	6005B

2200-2300	SEAs	15430, 17810
	As	11815, 15195
	SAm	9685F
	Eu/ME/Af	6050B, 6125B, 11925G

*) Regional Sce. others Gen. Sce.

Korean
1030-1130 As 7210, 9580
1400-1500 As 6185, 7210
Malay
1145-1215 SEAs 15300, 17820
Persian
1900-1930 ME 9560, 17775S
Portuguese
0900-0930 SAm 9675F, 11875
Russian
0500-0530 Eu 15170, 15355G
0800-0830 Eu 9670B, 9770B
1230-1300 As 7210, 9580
1930-2000 Eu 6005B
Spanish
0330-0400 CAm 15325F
0330-0400 SAm 15350F
CAm/SAm 17825, 21610
0930-1000 SAm 9675F, 11875
2030-2100 Eu 15355G
Swahili
1715-1745 EAf 7180G, 9590
Swedish
0530-0545 Eu 15170, 15355G
2000-2015 Eu 15340G
Thai
1215-1245 SEAs 11875, 15300, 17820
Urdu
0300-0330 SAs 11840S
1630-1700 SAs 9535S, 11880
Vietnamese
1345-1415 SEAs 11875, 15300
2230-2300 SEAs 11840, 17785

ANN: English: "This is Radio Japan, the overseas broadcasting service of NHK in Tokyo" — **French:** "Ici Radio Japon, service des émissions vers l'étranger de la NHK à Tokio" — **German:** "Hier ist Radio Japan. . . ." — **Italian:** "Qui é Radio Giappone, servizio radiofonico della NHK da Tokyo" — **Spanish:** "Aqui Radio Japón, el servicio internacional de la NHK" — **Portuguese:** "Esta é a Rádio Japão, serviço de transmissiões internacionais da NHK" — **Indonesian:** "Inilah Radio Jepang, siaran internasional dari NHK di Tokyo" — **Swedish:** "Här är det skandinaviska programmet fran Radio Japan, NHK i Tokyo" — **Swahili:** "Hii ni Redio Japani, Tokyo" — **Malay:** "Inilah Radio Jepun, Siaran antara bangsa dari NHK di Tokyo".
INT-SIG: Opening music of Reg. Sce's "Sakura". Interval signal music "Kazoe Uta".
V. by QSL-card.

PUB: In Japanese only: Graph: NHK (monthly), Hoso Bunka or Broadcast Culture (monthly), NHK Year Book (annual). In English and others: This is NHK (annual in E), NHK Technical Monograph (irr. in E). NHK Laboratories Note (irr. in E),. The prgr. schedules for Overseas Service (R. Japan), Studies of Broadcasting (annual in E) — **V.** by QSL-card — **FI:** In accordance with the Broadcast Law. NHK's operations are financed by the revenue from receiving fee.

THE NATIONAL ASSOCIATION OF COMMERCIAL BROADCASTERS IN JAPAN (MINPOREN)

ADDR: 5fl Bungei Shunju Bldg, 3-23, Kioi-cho, Chiyodaku, Tokyo 102. **Te:** +81 (3) 3265-7481. **Fax:** +81 (3) 3261-2860. **Cable:** Mimporen Tokyo. **Telex:** 2325163 NABTYO J.
L.P: Pres: Y. Sasaki. Vice-Presidents: S. Hazama, M. Saito, S. Matsumoto. Exec. Dir: T. Matsuzawa. Dir. Int. Rel: Hiaso Aoki.
Mediumwaves: Call: JO-; ID: Company initials usually used as st. identification; °) = key st, other st's which have been assigned call letters have some local prgr's during local daytime, otherwise relay their key st.

Call	kHz	kW	ID	Name, location and h. of tr.
1) CR	°558	20d		R. Kansai, Kobe: 2000-1720 (Fri, Sat 1800, Sun 1700)
2) DF	°684	5	IBC	Iwate Hoso, Morioka: 24h
3) IL	720	1	KBC	Kyushu Asahi Hoso, Kitakyushu: 24h
41)	720	1	RCC	Chugoku Hoso, Mihara: 24h
5) LR	°738	5	KNB	Kita Nihon Hoso, Toyama: 24h
5)	738	1	KNB	Kita Nihon Hoso, Takaoka: 24h
6) RR	°738	5	RBC	Ryukyu Hoso, Naha: 24h
4) JF	°765	5	YBS	Yamanashi Hoso, Kofu: 1930 (Sat, Sun 2000)-1600
15) PF	°765	5	KRY	Yamaguchi Hoso, Tokuyama: 24h
7) XR	°783	5	ROK	R. Okinawa, Naha: 2000-1900 (Sun 1600)
8) SO	864	1	SBC	Shin'etsu Hoso, Matsumoto: 24h
9) HE	864	3	HBC	Hokkaido Hoso, Asahikawa: 24h
9) QF	864	3	HBC	Hokkaido Hoso, Muroran: 24h
10) PR	°864	5d	FBC	Fukui Hoso, Fukui: 24h
48) XN	864	1	CRT	Tochigi Hoso, Nasu: 2100-1500
36)	882	1	STV	Sapporo TV Hoso, Esashi: 24h
9) HO	900	5	HBC	Hokkaido Hoso, Hakodate: 24h
11) HF	°900	5d	BSS	San'in Hoso, Yonago: W 2025-1800, Sun 2055-1600
12) ZR	°900	5d	RKC	Kochi Hoso, Kochi: W 2030, (Sat 2100)-2000, Sun 2100-1600
13) EF	°918	5	YBC	Yamagata Hoso, Yamagata: 24h
13)	918	1	YBC	Yamagata Hoso, Yonezawa: 24h
15) PM	918	1	KRY	Yamaguchi Hoso, Shimonoseki: as 765kHz
15) PN	918	1	KRY	Yamaguchi Hoso, Iwakuni: as 765kHz
36) WS	918	1	STV	Sapporo TV Hoso, Kushiro: 24h
14) TR	°936	5	ABS	Akita Hoso, Akita: 24h
21) NF	°936	5	MRT	Miyazaki Hoso, Miyazaki: 24h
21)	936	1	MRT	Miyazaki Hoso, 4 relay st's: as above
16) KR	°954	100	TBS	Tokyo Hoso, Tokyo: 24h
17) NR	°1008	50	ABC	Asahi Hoso, Osaka: 24h
19) AR	°1053	50	CBC	Chubu Nippon Hoso, Nagoya: 24h
8) SR	°1098	5	SBC	Shin'etsu Hoso, Nagano: 24h
8) SW	1098	1	SBC	Shin'etsu Hoso, Iida: 24h
13) EL	1098	1	YBC	Yamagata Hoso, Tsuruoka: 24h
20) MF	1098	1	NBC	Nagasaki Hoso, Sasebo: 24h
22) GF	°1098	5	OBS	Oita Hoso, Oita: 2030-2000 (Sat 1800, Sun 1500)
23) WO	1098	5	RFC	R. Fukushima, Koriyama: 24h
24) CF	°1107	20	MBC	Minami Nihon Hoso, Kagoshima: Mon-Fri 24h, Sat 2000-1800, Sun 2000-1600
24)	1107	1	MBC	Minami Nihon Hoso, Akune: as above
24)	1107	1	MBC	Minami Nihon Hoso, Okuchi: as above
26) MR	°1107	5	MRO	Hokuriku Hoso, Kanazawa: W 24h, Sat to 1815, Sun 1955-1530
26)	1107	1d	MRO	Hokuriku Hoso, Nanao: as above
27) AF	°1116	5	RNB	Nankai Hoso, Matsuyama: 24h
28) DR	°1116	5	BSN	Niigata Hoso, Niigata: 24h
29) QR	°1134	100	NCB	Bunka Hoso, Tokyo: 24h
25) BR	°1143	20	KBS	KBS Kyoto, Kyoto: 24h
6) RO	1152	0.5	RBC	Ryukyu Hoso, Miyako: 24h
33) OR	°1179	50	MBS	Mainichi Hoso, Osaka: 24h
12)	1197	1	RKC	Kochi Hoso, Nakamura: as 900kHz
30) FO	1197	1	RKB	RKB Mainichi Hoso, Kitakyushu: 24h
31) BF	°1197	5d	RKK	Kumamoto Hoso, Kumamoto: 24h
31)	1197	1	RKK	Kumamoto Hoso, Aso/ Goshinoura/Hitoyoshi: 24h
32) YF	°1197	5	IBS	Ibaraki Hoso, Mito: 2100-1500
36) WL	1197	1	STV	Sapporo TV Hoso, Asahikawa: 24h
36) WN	1197	1	STV	Sapporo TV Hoso, Hakodate: 24h
36)	1197	1	STV	Sapporo TV Hoso, Nayoro/ Wakkanai: 24h
25) BO	1215	1	KBS	KBS Kyoto, Maizuru: 24h
25) BW	1215	1	KBS	KBS Kyoto, Hikone: 24h
20) UR	°1233	5	NBC	Nagasaki Hoso, Nagasaki: 24h
34) GR	°1233	5	RAB	Aomori Hoso, Aomori: 24h
35) LF	°1242	100	NBS	Nippon Hoso, Tokyo: 24h

37) IR	°1260	20	TBC	Tohoku Hoso, Sendai: 24h
9) HW	1269	5	HBC	Hokkaido Hoso, Obihiro: 24h
38) JR	°1269	5	JRT	Shikoku Hoso, Tokushima: 24h
30) FR	°1278	50	RKB	RKB Mainichi Hoso, Fukuoka: 24h
9) HR	°1287	50	HBC	Hokkaido Hoso, Sapporo: 24h
39) UF	°1314	50	OBC	R. Osaka, Osaka: 24h
40) SF	°1332	50		Tokai R. Hoso, Nagoya: 24h
41) ER	°1350	20	RCC	Chugoku Hoso, Hiroshima: 24h
9) JS	1368	1	HBC	Hokkaido Hoso, Wakkanai: 24h
1)	1395	1		R. Kansai, Tajima: as 558kHz
9) QL	1404	5	HBC	Hokkaido Hoso, Kushiro: 24h
42) VR	°1404	10	SBS	Shizuoka Hoso, Shizuoka: 24h
42)	1404	1d	SBS	Shizuoka Hoso, Hamamatsu: 24h
3) IF	°1413	50	KBC	Kyushu Asahi Hoso, Fukuoka: 24h
43) RF	°1422	50d	RF	"RF" R. Nippon, Yokohama: 24h
11) HL	1431	1	BSS	San'in Hoso, Tottori: as 900kHz
11)	1431	1d	BSS	San'in Hoso, Izumo: as 900kHz
20)	1431	1	NBC	Nagasaki Hoso, Fukue: 24h
23) WW	1431	1	RFC	R. Fukushima, Iwaki: 24h
44) VF	°1431	5	WBS	Wakayama Hoso, Wakayama: 24h
45) ZF	°1431	5	GBS	Gifu Hoso, Gifu: 2000-1530
36) WF	°1440	50	STV	Sapporo TV Hoso, Sapporo: 24h
36)	1440	1	STV	Sapporo TV Hoso, Muroran/ Tomakomai: 24h
9) QM	1449	5	HBC	Hokkaido Hoso, Abashiri: 24h
46) KF	°1449	5	RNC	Nishi Nippon Hoso, Takamatsu: 24h
20) UO	1458	1	NBC	Nagasaki Hoso, Saga: 24h
23) WR	°1458	1	RFC	R. Fukushima, Fukushima: 24h
32) YL	1458	1	IBS	Ibaraki Hoso, Tsuchiura: 2100-1500
32)	1458		IBS	Ibaraki Hoso, Sekijo: as above
41)	1458	1	RCC	Chugoku Hoso, Shobara: 24h
15) PL	1485	1	KRY	Yamaguchi Hoso, Hagi: as 765kHz
34) GO	1485	1	RAB	Aomori Hoso, Hachinohe: 24h
9) TL	1494	1	HBC	Hokkaido Hoso, Nayoro: 24h
47) YR	°1494	10	RSK	Sanyo Hoso, Okayama: 24h
47)	1494	1	RSK	Sanyo Hoso, 4 rel. st's: 24h
27) AM	1530	1	RNB	Nankai Hoso, Uwajima: 24h
28) DO	1530	1	BSN	Niigata Hoso, Joetsu: 24h
48) XF	°1530	5	CRT	Tochigi Hoso, Utsunomiya: 2100-1500
36) WM	1530	1	STV	Sapporo TV Hoso, Obihiro: 24h

Low power relays less than 1kW not mentioned.

Addresses and other information:

Schedules: All 24h st's are off the air for 2 to 5 h's until 2000 on Sun. All other days a network prgr is aired 1600 or 1800 to 2000 on most st's. Network prgr's also may be b'cast at other times of the day.

V: All st's verify by QSL-card. Rec. acc. Rp.

1) 1-1-6 Miyuki-cho, Suma-ku, Kobe 654-81 — Pres: R. Matsuda.
2) 6-1 Shike-cho, Morioka 020 — Pres: I. Kohno.
3) 1-1-1 Nagahama, Chuo-ku, Fukuoka 810 — Pres: S. Wakui.
4) 2-6-10 Kitaguchi, Kofu 400 — Pres: Y. Takamuro. Dir. Tec: Nobuo Takeda.
5) 10-18 Ushijima-machi, Toyama 930 — Pres: R. Yokoyama.
6) 2-3-1 Kumoji, Naha 900 — Pres: K. Oroku.
7) 1-4-8 Nishi, Naha 900 — Pres: Y. Aragaki.
8) 1-21-24 Yoshida, Nagano 381 — Pres: S. Hayashi.
9) Kita Ichijo, Nishi 5-2, Chuo-ku, Sapporo 060-01 — Pres: K. Tomihara. Dir. Tec: K. Kashiwakura.
10) 5-105 Itagaki-cho, Fukui 910 — Pres: Y. Ito.
11) 423 Nishi-Fukuhara, Yonago 683 — Pres: T. Fujii.
12) 3-2-15 Honmachi, Kochi 780 — Pres: M. Kanayama.
13) 2-5 Hatago-machi, Yamagata 990 — Pres: K. Yoshida.
14) 7-9-42 Sanno, Akita 010 — Pres: Y. Kudo. Dir. Tec: H. Takahashi.
15) Koen-Ku, Tokuyama, Yamaguchi 745 — Pres: Y. Akao.
16) 5-3-6 Akasaka, Minato-ku, Tokyo 107-06 — Pres: I. Tanaka.
17) 2-2-48 Oyodominami, Kita-ku, Osaka 531-01 — Pres: S. Fujii.
19) 1-2-8 Shinsakae, Naka-ku, Nagoya 460-05 — Pres: K. Takahashi.
20) 1-35 Uwamachi, Nagasaki 850 — Pres: K. Fujiki.
21) 4-6-7 Tachibana Dori Nishi, Miyazaki 880 — Pres: T. Arakawa.
22) 3-1-1 Imazuru, Oita 870 — Pres: T. Mishio.
23) 13-17 Otamachi, Fukushima 960 — Pres: R. Kobari. **V.** Re. to 8 Shimoazako, Fukushima 960.
24) 5-25 Korai-cho, Kagoshima 890 — Pres: H. Taneda.
25) Karasuma, Kamichoja-machi, Kamigyo-ku, Kyoto 602 — Pres: K. Okamura. Dir. Tec: O. Sawada.
26) 3-2-1 Honda-machi, Kanazawa 920 — Pres: I. Saga.
27) 6-24 Dogohimata, Matsuyama 790 — Pres: E. Onishi.
28) 3-18 Kawagishi-cho, Niigata 951 — Pres: M. Takazawa.
29) 1-5 Wakaba, Shinjuku-ku, Tokyo 160-02 — Pres: M. Kobayashi.
30) 4-1-10 Watanabe Dori, Chuo-ku, Fukuoka 810 — Pres: K. Yamamoto.
31) 30 Yamasaki-machi, Kumamoto 860 — Pres: T. Kobori.
32) 2-14 Kitami-cho, Mito-shi, Ibaraki 310 — Pres: Shu Uno. Gen. Mgr: T. Tomioka. Dir. Tec: Junnosuke Nakanishi.
33) 17-1 Chaya-machi, Kita-ku, Osaka 530-04 — Pres: M. Saito. Gen. Mgr: T. Tanno. Dir. Tec: T. Shigezawa.
34) 1-8-1 Matsumori, Aomori 030 — Pres: K. Nara.
35) 1-9-3 Yuraku-cho, Chiyoda-ku, Tokyo 100-87 — Pres: S. Hazama.
36) Kita Ichijo, Nishi 8-1-1, Chuo-ku, Sapporo 060 — Pres: S. Isaka.
37) 26-1 Kazumi-cho, Taihyaku-ku, Yagiyama, Sendai 982 — Pres: K. Ichiriki.
38) 2-5-2 Nakatokushima-cho, Tokushima 770 — Pres: A. Mitani. St. Mgr: A. Fujii.
39) 2-4-9 Umeda, Kita-ku, Osaka 530-13 — Pres: T. Maeda.
40) 1-14-27 Higashi-zakura, Higashi-ku, Nagoya 461-03.
41) 21-3 Moto-machi, Naka-ku, Hiroshima 730 — Pres: T. Hiraoka. Tec. Dir: K. Kosaki.
42) 3-1-1 Toro, Shizuoka 422 — Pres: Jun Matsui. Dir. Tec: Kanae Suzuki.
43) 2-2-1 Azabudai, Minato-ku, Tokyo 106 — Pres: H. Komamura. Dir. Tec: M. Kubo.
44) 3-15 Minato-honmachi, Wakayama 640 — Pres: E. Kitano.
45) 8 Imako-machi, Gifu 500 — Pres: M. Sugiyama.
46) 8-15 Marunouchi, Takamatsu 760 — Pres: T. Hirai.
47) 2-1-3 Marunouchi, Okayama 700 — Pres: S. Kochiyama. (Relay st's: Tsuyama, Bizen, Maniwa, Takahashi).
48) 12-11 Honcho, Utsunomiya-shi, Tochigi 320 — Pres: T. Fukushima; St.Mgr: R. Mori; Dir.Tec: H. Takamatsu.
49) **Nihon Shortwave Broadc. Co. Ltd. (R. Tanpa).**

Call	kHz	kW	Prgr.	Call	kHz	kW	Prgr.
JOZ	3925	50	I	JOZ6	6115	50	II
JOZ4	*3925	10	I	JOZ3	9595	50	I
JOZ5	3945	10	II	JOZ7	9760	50	II
JOZ2	6055	50	I				

*) Sapporo; others Chiba (Tokyo)

ADDR: 1-9-15 Akasaka, Minato-ku, Tokyo 107. **Te:** (03) 3583-8151—Pres: Kinya Sekiguchi.

1st Prgr: 2020-1630 (Sun 1530) on 3925/6055/9595kHz; 2020-2300, 0800-1730 (Sat 1630, Sun 1700) on 3925kHz (Sapporo).

2nd Prgr: 2300-1300 on 3945/6115/9760kHz.

INT-SIG: Slow Tempo chime with Japanese instrument "Koto" at sign on and sign off—**V.** by QSL-card. Rp.

COMMERCIAL FM STATIONS

(Japanese FM-channels: 76-90MHz): Call = JO--

Call	MHz	kW	Name
1) QU	76.1	1	FM Iwate, Morioka
2) LU	76.1	1	FM Fukui
3) JU	77.1	5	FM Sendai
4) SU	77.4	1	FM Naka Kyushu, Kumamoto
5) VU	77.4	0.5	FM San'in, Matsue
35)	77.6	1	Himeji relay
6) XU	77.5	1	FM Niigata
37)	77.9		FM Saga
7) GV	78.0	5	Bay FM, Chiba
8) GU	78.2	1	Hiroshima FM
9) YU	78.6	1	FM Kagawa, Takamatsu
10) NU	78.9	3	FM Mie, Tsu
11) KU	79.2	1	FM Shizuoka, Hamamatsu
12) UU	79.2	1	FM Yamaguchi
13) HU	79.5	1	FM Nagasaki
14) DV	79.5	5	FM Saitama ("Nack Five"), Urawa
17) EU	79.7	1	FM Ehime, Matsuyama
16) ZU	79.7	1	FM Nagano, Matsumoto
38)	79.8		FM Kagoshima
15) WU	80.0	1	FM Aomori, Morioka
18) AU	80.0	10	FM Tokyo

19) FV	80.2	10	FM 802, Osaka
20) FU	80.4	5	FM Hokkaido, Sapporo
21) EV	80.4	1	FM Yamagata
22) HV	80.5	4.1	FM Ishikawa, Kanazawa
23) CU	80.7	10	FM Aichi, Nagoya
24) DU	80.7	3	FM Fukuoka
39)	80.7		FM Tokushima
25) AV	81.3	10	FM Japan ("J-Wave"), Tokyo
40)	81.6		FM Kochi
26) OU	82.7	1	Toyama FM
27) PU	82.8	3	FM Akita
28) CV	83.0	1	FM Fuji, Kofu
29) MU	83.2	1	FM Miyazaki
5)	83.4	3	Chizu relay
6)	83.5	1	Yamato relay
30) TU	84.7	5	FM Yokohama
31) BU	85.1	10	FM Osaka
32) RU	86.3	1	FM Gunma, Maebashi
5)	86.6	1	Hamada relay
33) IU	87.3	1	FM Okinawa
34) JV	88.0	1	FM Oita
35) KV	89.4	1	FM Kyoto
36) IV	89.9	0.5	Hyogo-FM ("Kiss-FM"), Kobe

Repeaters below 1kW not mentioned.

Addresses and other information

1) 8-17 Morioka-Ekimaedori, Morioka 020 — Pres: N. Hatayama; St. Mgr: Y. Sakurai; Dir. Tec: A. Itazawa — 24h.
2) 1-1-Miyuki 1-chome, Fukui 910 — Pres: O. Kuwahara — 1900 (Sat, Sun 2000)-1800 (Fri, Sat 1900).
3) 2-10-28, Hon-cho, Aoba-ku, Sendai 980 — Pres: T. Nakagawa — 2000-1800 (Sun 1700).
4) 5-50 Chibajomachi, Kumamoto 860 — Pres: K. Hasegawa — 24h.
5) Sanin Chuo Bldg, 4th floor, 383 Tonomachi, Matsue 690 — 24h.
6) 2-1-1 Yachiyo, Niigata 950 — Pres: S. Nakano — 24h.
7) 1-11-1, Chuo, Chiba 280 — Pres: S. Nakadai — 24h.
8) 1-8-2, Minamimachi, Minami-ku, Hiroshima 734 — Pres: K. Matsuda — 24h.
9) 1-4-23 Saiho-cho, Takamatsu 760 — Pres: A. Iyama — 24h.
10) 1043-1 Yakio, Kannonji-cho, Tsu-shi, Mie 514 — Pres: B. Koshiba; St. Mgr: H. Kimura; Dir. Tec: N. Ito — 24h.
11) 330-5, Tamachi, Hamamatsu 430 — Pres: S. Kawai — 24h.
12) 3-31 Midori-cho, Yamaguchi 753 — Pres: K. Fukuda — 24h.
13) 5-5, Sakaemachi, Nagasaki 850 — Pres: M. Watanabe — 2030-1800 (Sat 1830).
14) 4-16-2 Tokiwa, Urawa 336 — Pres: I. Nisugi — 24h.
15) 1-7-19 Tsutsumi-cho, Aomori 030 — Pres: K. Sasaki — 24h.
16) 1-13-5 Honjo, Matsumoto 390 — 2100-1800 (Sat 2000, Sun 1700). I. Takizawa.
17) 617-3 Fujiwara-cho, Matsuyama 790 — Pres: N. Kitano — 2057-1803 (Sun 1633).
18) 1-7 Kojimachi, Chiyoda-ku, Tokyo 102-80 — Pres: W. Goto — 24h.
19) 2-2-6, Tenjinbashi, Kita, Kita-ku, Osaka 530-80 — 24h.
20) Kita Ichijo, Nishi 2-1, Chuo-ku, Sapporo 060 — Pres: T. Tajima — 2100-1800 (Sat 2000).
21) 2-3-1, Akoya-cho, Yamagata 990 — S. Tanaka — 24h.
22) 2-1-45, Hikosan-cho, Kanazawa 920 — Pres: K. Hirabayashi — 24h.
23) 2-15-18, Chiyoda, Naka-ku, Nagoya 460 — Pres: S. Ukai — 24h.
24) 2-1-82 Watanabedori, Chuo-ku, Fukuoka 810 — Pres: S. Matsumoto — 24h.
25) 4-17-30 Nishi Azabu, Minato-ku, Tokyo 106-88 — Pres: S. Soyama — 24h.
26) 1-8-10, Marunouchi, Toyama 930 — Pres: Y. Morimoto — 2100-2000.
27) 3-7-10 Yabase-Honcho, Akita 010 — Pres: K. Ito — 2100-1800 (Sun 1000).
28) 2-7-23 Marunouchi, Kofu 400 — Pres: Y. Kanemori — 2000-1900 (Sat 24h).
29) 2-78 Gion, Miyazaki 880 — Pres: Y. Kamino — 2100-1700 (Fri, Sat 2000).
30) 2 Yamashita-cho, Naka-ku, Yokohama 231 — Pres: Y. Ueno — 24h.
31) 3-2-4 Nakanoshima, Kita-ku, Osaka 530-85 — Pres: M. Yamada — 24h.
32) 1-4-8 Wakamiyacho, Maebashi 371 — Pres: M. Sekiguchi — 24h.
33) 40, Kowan, Urasoe, Okinawa 901-21 — Pres: H. Nagahama; Gen.Mgr: J. Yazaki; Dir.Tec: S. Azuma — 2100-1800 (Sun 1600).
34) 17-19 Higashi Kasuga-cho, Oita 870 — Pres: N. Niwa — 24h.
35) 98 Matsumoto-cho, Kamigamo, Kita-ku, Kyoto 603 — Pres: T. Tateishi.
36) 5-4 Hatoba-machi, Chuo-ku, Kobe 650 — Pres: T. Ueshima — 24h.
37) 286-5 Fukuro, Honjo-cho, Saga 840 — 2100-1800 (Sun 1500).
38) 1-38 Higashi Sengoku-cho, Kagoshima 892 — 24h.
39) 1-6 Miyuki-cho, Tokushima 770 — 24h.
40) 3-3-21 Honmachi, Kochi 780.

UNIVERSITY BROADCASTING STATION

ADDR: Hoso Daigaku, 2-11 Wakaba, Chiba 260.
JOUD-FM 77.1MHz 10kW Tokyo, 78.8MHz 1kW Maebashi.
D.PRGR: 2100-1500. Broadcasts texts for distance university studies.

RADIO AUM SHINRIKYO (Rlg.)

ADDR: 3-8-11 Miyamae, Suginami-ku, Tokyo 168. **Fax:** +81 (3) 5370 1604. **Branch Offices:** Aum Supreme Truth, 8 Ea. 48th Str. #2E, New York, NY 10017, USA. **Te:** +1 (212) 421-3687 — Aum Shinrikyo, Auf dem Hügel 48, Endenich 5300, Bonn 1, Germany. **Te:** +45 228 616647 — Aum Sacca Sangha Association, Sora-hena, Watta, Anglugaha, Galle, Sri Lanka.
D. PRGR. via R. Moscow International: English 0530-0600 & 2130-2200. **Japanese:** 1400-1500 on SW, 1500-1600 on 720kHz. **Russian:** via domestic sce. Mayak prgr. (times not known at press time). **Frequencies:** see R. Moscow entry under 'Russia'.

FAR EAST NETWORK (AFRTS)

The network serves the members of the US forces. The st's in Japan broadcast by authority of Commander, US Forces, Japan, in cooperation with the Japanese Radio Regulatory Bureau. St's are linked by land line and microwave.
ADDR: Headquarters, Air Force Pacific Broadcasting Squadron (AFPBS), APO San Francisco 96293. **Te:** Tokyo (0425) 52-2511 ext 5-2374; (Tokyo) AFPBS FEN-Tokyo, APO San Francisco 96328-5000; (Philippines) Det 1, AFPBS, APO San Francisco 96274-5000; (Okinawa) Det 2, AFPBS, APO San Francisco 96239-5000; (Misawa) OLAA, AFPBS, APO San Francisco 96519-5000; (Iwakuni) OLAB, AFPBS, FPO Seattle 98764-5015; (Sasebo) OLAD, AFPBS, FPO Seattle 98766-2903.
L.P: SM: JOCS Robin Barnette. CE: MSgt. John Frost.

Mediumwaves	**kHz**	**kW**		**kHz**	**kW**
Okinawa	684	10	Misawa	1575	0.3
Tokyo	810	50	Iwakuni	1575	1
Sasebo	1575	0.3			

FM: Okinawa 89.1MHz 6kW.
D.PRGR: 24h. **N:** on the h.
Local live prgr's: Mon-Fri 2105-2200, 2230-2400, 0205-0300, 0605-0900.
Live netw. prgr's: D 1505-1800; Sat, Sun 2305-0200, 0805-1105.
ANN: on the h: "FEN Tokyo, the All American Eagle."
V. by QSL-card — **PUB:** Master schedule quarterly.

KAZAKHSTAN

L.T: UTC +6h (ea.), UTC +5h (we.) (Su: UTC +7h/6h) — **Pop:** 16.690.000 — **Pr.L:** Kazakh, Russian — **E.C:** a/c 50, 127/220V — **ITU:** KAZ.

KAZAKH RADIO (Gov.)

ADDR: Zheltoksan Str. 175 A, 480013 Alma-Ata. **Te:** +7 (3272) 634666 (Int. Rel.). **Fax:** +7 (3272) 631207. **Tlx:** 251114 TRVD.
L.P: Chmn: G. M. Shalakhmetov.

Stations*	**kHz**	**kW**	**Prgr.**
Almaty/Alma-Ata	180	250	KR1
Aktöbe/Aktyubinsk	180	150	KR1
Türkistan/Turkestan (1)	180	50	KR1
Almaty/Alma-Ata	243	500	MR1
Karaqandy/Karaganda	243		MR1
Almaty/Alma-Ata	549	1000	RSM
Kostanay/Kustanay	684	30	KR1/Reg
Karaqandy/Karaganda	747	20	KR1/Reg
Oral/Uralsk	747	50	KR1/Reg

Türkistan/Turkestan (1)	783	150	KR2/Reg
Arkalyk/Arkalyk (9)	801	30	KR1/Reg
Öskemen/Ust-Kamenogorsk (7)	801	30	KR1/Reg
Jambyl/Dzhambul	846	50	KR1/Reg
Karaturuk (6)	900	500	RSY/F
Aktau (8)	936	150	RSM
Karaqandy/Karaganda	936		RSM
Kaçiry/Kachiry (2)	945	50	KR1/Reg
Kyzylorda/Kzyl-Orda	954		KR1/Reg
Aktöbe/Aktyubinsk	981		KR2/Reg
Akmola	999	30	KR1/Reg
Jilaydy/Zhilaydy (4)	1008	50	KR1/Reg
Öskemen/Ust-Kamenogorsk (7)	1071	100	KR2
Almaty/Alma-Ata	1098	250	KR2/Reg
Şardara/Chardara (1)	1098		KR2
Jetisay/Dzhetysay (1)	1098		KR2
Karaqayly/Karagayly (3)	1098		KR2
-/Leninskoye (1)	1098		KR2
Şingoja/Shingozha (4)	1098		KR2
Abay/Abay (1)	1188		KR2
Ayaköz/Ayaguz (4)	1188		KR2
-/Yuzhnyy (4)	1188		KR2
Akmola	1197		KR2
Atyrau/Atyrau	1323	20	KR1/Reg
Aktau (8)	1341	25	KR1/Reg
Almaty/Alma-Ata	1341	30	KR2/RL
-/Panfilov (5)	1395		KR2
Bakanas/Bakanas (6)	1440		KR2
Kyzylorda/Kzyl-Orda	1440		KR2
Arkalyk/Arkalyk (9)	1458		KR2
Buran/Buran (7)	1458		KR2
Almaty/Alma-Ata	1494		KR1
Kostanay/Kustanay	1539		KR2
-/Andreyevka (5)	1548		KR2/Reg
Jansügirov/Dzhansugurov (5)	1548		KR2/Reg
Lepsi/Lepsy (5)	1548		KR2/Reg
Tekeli/Tekeli (5)	1548		KR2/Reg
-/Antonovka (5)	1557		KR1/Reg
Kapal/Kapal (5)	1557		KR1/Reg
Uç-Aral/Uch-Aral (5)	1557		KR1/Reg

*) Locations in Kazakh/Russian.

Regions: (1) Şimkent/Chimkent; (2) Pavlodar; (3) Karaqandy/Karaganda; (4) Semey/Semipalatinsk; (5) Taldykorgan/Taldy-Kurgan; (6) Almaty/Alma-Ata; (7) Vistochno-Kazakhstan; (8) Mangistau; (9) Turgay; others = city name.

Shortwaves	**kHz**	**kW**	**H. of tr.**	**Prgr.**
Alma-Ata	4545		0000-2000	KR1
Alma-Ata	4800		0000-2000	KR1
Alma-Ata	5035		0000-2300	KR2
Alma-Ata	5260		0000-2300	KR2
Alma-Ata	5915	100		F
Alma-Ata	5945	20	0000-1530	KR2
Alma-Ata (330°)	5960	100	0000-2300	KR2
Alma-Ata (305°)	5970	50	0000-2300	KR2
Unknown	5985		0000-2300	KR2
Alma-Ata	6060	20	0000-1600	KR2
Alma-Ata	6075	20	0000-1800	KR2
Alma-Ata	6125	20	0000-1500	KR2
Akmola	6130	20	0000-1600	KR2
Alma-Ata	6135	100		F
Alma-Ata (15°)	6180	100	0000-2000	KR1
Unknown	7115	20	0000-1600	KR2
Alma-Ata	7235	20	0000-1600	KR2
Unknown	7240		0000-	KR2
Alma-Ata	7255	100		F
Karaganda	7280	20	0000-1500	KR2
Alma-Ata	9505	20	0000-2300	KR2
Alma-Ata	9550	20	0000-1800	KR2
Alma-Ata	9690	20	0315-1430	KR2
Alma-Ata	9705	20	0000-1430	KR2
Alma-Ata (280°)	9780	50	0000-2000	KR1
Alma-Ata	11825	20	1815-2300	KR2
Alma-Ata	11950	100	0000-2000	KR1
Unknown	15155	20		KR2
Alma-Ata	15215	20	1615-2300	KR2
Unknown	15250	20		KR2
Alma-Ata	15270	20	1515-2200	KR2
Unknown	15285	20		KR2
Alma-Ata	15360	20	1445-2300	KR2
Alma-Ata	15385	20	1545-1800	KR2
Alma-Ata	17605	20	1615-2300	KR2
Alma-Ata	17715	20	1615-2300	KR2
Alma-Ata + 1st	17730	20	2300-0200	KR2
			1445-2300	KR2
Alma-Ata + 1st	17765	20	1815-2200	KR2

Rel. via Balashikha, Russia: 9690kHz daytime, 21490 nighttime, both 20kW. **NB:** Powers are estimates only.

FM: Information not available.

D.PRGR: KR1 in Kazakh/Russian 0000-2000. **German:** Sat 0900-0956 — **KR2** (Şalkar) 0000-2300 in Kazakh incl. prgr's in Uyghur, German— **KR3** (FM only): Details not available — **MR1** = Moscow Ostankino R-1 — **RSY** = Moscow R. St. Yunost — **RSM** = Moscow R. St. Mayak — **RL** = R. Liberty (cf. Germany) 0000-0100, 0200-0300, 0500-0700, 1400-1600 in Russian; 0100-0200, 1300-1400 in Kazakh.

Regional Prgrs: 0115-0200 (ea.), 0215/0230-0300 (we.), 1300-1400 in Kazakh/Russian + minority languages.

Addr. of regional centres: ul. Moskovskaya 49, 473032 **Akmola** — ul. Lenina 43, 463000 **Aktyubinsk** — pr. Gornyakov 36, 459830 **Arkalyk** — pl. Abaya 25, 465017 **Atyrau** — ul. Nekrasova 15, 486024 **Chimkent** — ul. Kumzhata, 484000 **Dzhambul** — ul. Svyazi 14, 470061 **Karaganda** — ul. Pushkina 54, 458000 **Kustanay** — ul. Vokzalnaya 21, 637026 **Pavlodar** — ul. Shugayeva 157, 490018 **Semipalatinsk** — ul. Kirova 432, 488009 **Taldy-Kurgan** — pr. Lenina 204, 417000 **Uralsk** — ul. Stakhanovskaya 70, **Ust-Kamenogorsk.**

FOREIGN SERVICE: Radio Alma-Ata

ADDR: As for Home Sce. (Su: all prgrs 1h earlier).

English
1530-1600 on 900, 5915, 7255
1930-2000 on KR2 freq's
2130-2200 on KR2 freq's
0030-0100 on 900, 5915, 6135 (wi) or 7255 (su)

Kazakh
2300-2330, 0000-0030, 1230-1300, 1300-1330, 1400-1430, 1500-1530 on 900/5915/6135(Wi)/7255(Su)kHz.

ANN: Kazakh: "Almatydan söylep turmiz".

V. by QSL-card. **PUB:** Efirde Kazakstan (weekly in Kazakh for Home Sce.).

RADIO MOSCOW RELAY

Shortwaves: Alma Ata 6 × 100kW — Karaturuk (G.C: 43.39N/77.56E, tentative location) 4 × 500kW. Used for R. Moscow & RSM prgrs.

KOREA (Democratic People's Republic)

L.T: UTC + 9h — **Pop:** 22.230.000 — **Radio sets:** 4.700.000 — **Pr.L:** Korean — **ITU:** KRE.

KOREAN CENTRAL BROADCASTING STATION

(Chosŏn Chung-ang Pangsong)

ADDR: Ministry of Posts & Telecommunications. Chŏnsung-dong, Moron-guyŏk, Pyongyang.

Te: +850 (2) 36344, +850 (2) 812301. **Fax:** +850 (2) 814418.

C = Central Broadcast from Pyongyang, R = Regional Sce, E = rel. Ext. Sce.

Mediumwaves	**kHz**	**kW**	**Prgr.**				
Ch'ŏngjin	702	50	C/R	Shinŭiju	864	250	C/R
Wiwon	720	500	C/R	Hwangju	927	50	C/R
Hyesan	765	50	C/R	Hamhŭng	999	250	C/R
Kaesŏng	810	50	C/R	Haeju	1080	1500	C/R
Pyongyang	819	500	C	Pyongyang	1368	—	E

Shortwaves	**kHz**	**Prgr**			
Sariwon	2350	C/R	Wonsan	3970	C/R
Pyongyang	2850	C	Ch'ŏngjin	3980	C/R
Hamhŭng	3220v	C/R	—	6100	C
P'yŏngsŏng	3350	C/R	—	9665	C
Hyesan	3920v	C/R	—	11680v	C
Kanggye	3960	C/R			

FM: 93.8MHz.
D.PRGR in Korean: 2000-1800 on all freqs exc. 1080 (2000-1500), 6100 (2000-0830 & 1500-1800) and 9665 (2000-0930 & 1730-1800). **N:** 2100, 2200, 0100, 0300, 0400, 0500*, 0700, 0800, 0900, 1100, 1200, 1300, 1500. — *) exc. regionals.
Regional Prgrs: W 0500-0600.
Rel. Pyongyang Broadc. St: 1500-1800 on 702/720/810/864kHz.
ANN: "Chosŏn Chung-ang Pangsong-imnida". Reg. Prgrs: "(location) Pangsong-imnida".
V: does not normally verify.

PYONGYANG BROADCASTING STATION (P'yŏngyang Pangsong)

ADDR: Ministry of Posts & Telecommunications, Pyongyang.
P = Pyongyang Broadc. St, E = Ext. Sce.

Mediumwaves	kHz	kW	Prgr.				
Ch'ŏngjin	621	500	P/E	Sep'o	729	50	P
Kangnam*	657	1500	P	Hwadae	801	500	P
Samgŏ	684	250	P	Sangwon	855	500	P

*) Key st. serving Pyongyang area.
Shortwaves (all freqs variable): Pyongyang 3250/3320/6250kHz, Kanggye 6400kHz.
FM: 89.1/91.1/93.1/96.7/97.3/99.5/101.1/104.5/106.5MHz.
D.PRGR. in Korean: 2100-2030 on 657/729/801/855kHz; 2100-1800 on 684/6400kHz; 2100-1900 on 801/3320kHz; 2100-0930 & 1500-1900 on 6250kHz; 2300-0100 & 0550-0830 on 621/3250kHz; 1500-2030 on 3250kHz; 1500-1800 on 621kHz.
ANN: "P'yŏngyang Pangsong Imnida".
V: does not normally verify.

PYONGYANG FM BROADCASTING STATION (P'yŏngyang FM Pangsong)

STATIONS: Pyongyang 105.2MHz 20kW, +12 relay st's.
D.PRGR: 0700-2000 (National holidays: 2100-2030) (music and information).
ANN: "P'yŏngyang FM Pangsong-imnida".

EXTERNAL SERVICE

ADDR: R. Pyongyang, Pyongyang, D.P.R. of Korea.
MEDIUMWAVE: Ch'ŏngjin 621kHz 500kW — Pyongyang 1368kHz 2kW.
SHORTWAVES: Kanggye (G.C: 126.36E/40.58N), Kujang (G.C: 125.05E/40.05N), Pyongyang (G.C: 125.33E/39.05N): 100/200/400kW.

kHz	kHz	kHz	kHz	kHz	kHz	kHz
3250	7200	9345	9977	11845	13785	15340
6125	7230	9505	11335	11905	15115	17750
6540	7250	9600	11700	13650	15180	17765
6560	7580	9640	11735	13760	15230	17795
6576	9325	9650	11780			

Arabic		
1500-1550*	ME/Af	6540, 9600, 11905
1700-1750	ME/Af	6540, 9600, 11905
1800-1850	ME/Af	6540, 9600, 11905
2000-2050	ME/Af	6540, 9600, 11905
Chinese		
0600-0650	As	15340, 17765
0700-0750*	CHN	13650, 15230
1100-1150*	CHN	6125, 7200, 7230
1300-1350	CHN	9977, 11735
English		
0000-0050	NAm	11335, 13760, 15115
0400-0450	SEAs/IND	15180, 15230, 17765
0600-0650*	SEAs/IND	15180, 15230
0700-0750	SEAs/IND	15340, 17765
0800-0850*	SEAs/IND	15180, 15230
1100-1150	NAm	6576, 9977, 11335
1300-1350*	Eu	9325, 9345
	SEAs/IND	9640, 13650, 15230
1500-1550	Eu/ME/Af	9325, 9640, 9977, 13785
1700-1750	Eu/ME/Af	9325, 9640, 9977, 13785
2000-2050	Eu/ME/Af	6576, 9325, 9640, 9977
2300-2350	NAm	11700, 13650
French		
0000-0050*	SEAs	11845, 15230
0600-0650	SEAs	17750, 17795
1400-1450*	Eu/SEAs	9325, 9345, 11735, 13650
1600-1650	Eu/ME/Af	6576, 9345, 9640, 9977
1900-1950	Eu/ME/Af	9325, 9640, 9977, 13785
2100-2150	Eu/ME/Af	6576, 9345, 9640, 9977
German		
1900-1950	Eu	6576, 9345
2100-2150*	Eu	9325, 11705
Japanese		
0400-0550*	J	9505, 9650, 11780
0600-0750	J	9505, 9650, 11780
0900-1050	J	621, 3250, 6540, 6560, 7580
1100-1250	J	621, 3250, 6540, 6560, 7580
1300-1450	J	621, 3250, 6540, 6560, 7580
2100-2250	J	621, 3250, 9505, 9650, 11780
Korean		
0400-0550	CHN	7200, 9345, 11735
0500-0550	As	15180, 15230, 17765
0800-0850	J	9505
0800-0850	As	7250
1000-1050	Am	6576, 9977
1200-1250	CHN	6125, 7200, 9345
1400-1450	SEAs/IND	9640, 9977, 15230
1600-1650	ME/Af	6540, 9600, 11905
1700-1750	Eu	6576, 9345
1800-1850	Eu/ME/Af	9325, 9640, 9977, 13785
1900-1950	ME/Af	6540, 9600, 11905
2200-2250*	As/Am	11335, 11700, 15115, 15230
2300-2400	CHN	6065, 7200, 9345
Russian		
0900-0950	URS	7200, 9325, 9345, 11335
1000-1050*	URS	7200, 9325, 9345, 11335
1500-1550	Eu	6576, 9345
1600-1650	Eu	9325, 13785
Spanish		
0000-0050	LAm	11700, 13650
0900-0950*	LAm	6576, 9977
1200-1250*	LAm	6576, 9977, 11335
1800-1850	Eu	6576, 9345
2000-2050	Eu	9325, 13785
2300-2350	LAm	13760, 15115

*) also on 1368kHz for Pyongyang area.
INT-SIG: National Anthem.
ANN: "Chosŏn Chung-ang Pangsong imnida".
V. by QSL-card (Ext. Sce. only) — **PUB:** "Radio Pyongyang" upon request.

KOREA (Republic of)

L.T: UTC + 9h (Su: UTC + 10h) — **Pop:** 45.060.000 — **Radio sets:** 42m — **Pr.L:** Korean — **E.C:** a/c 60, 100 / 200V — **ITU:** KOR.

KOREAN BROADCASTING SYSTEM (KBS) (Hanguk Pangsong Kongsa)

(Public Corporation)
ADDR: 18, Yŏ-ŭi-do-dong, Yŏngdŭngp'o-gu, Seoul 150-790.
Te: +82 (2) 781-2410. **Fax:** +82 (2) 781-2499. **Telex:** KBSKB K24599. **Cable:** KBS, Seoul.
L.P: Pres: Suh, Ki-won. Mgr. Technical Planning. Div. & ABU/EBU Tech. Liaison Officer: Lee, Kwang-tae. Dir. Int. Rel. Div. & ABU/EBU Prgr. Liaison Officer: Chang, Myong-Keun.
Mediumwaves: ° = Regional key st, *) Seoul area,[2] = rel N2 exc. for local prgr's (other local st's take N1), + = also used for Ext. sce. **N** = Netw. or local st. area. **Call** = HL. N1 = KBS R. One, N2 = KBS R. Two, L1 = First Liberty (Social Educ.) Prgr, L2 = Second Liberty (Social Educ.) Prgr.

N	Location	Call	kHz	kW	N	Location	Call	kHz	kW
13)	Changsu	SN	540	1	N2)	Kaebong*	CA	639	50
20)	Chŏmch'on	SC	540	1	16)	Posŏng	SL	648	1
11)	Hongsŏng	CZ	540	10	3)	Ch'unch'ŏn°	KM	657	50
16)	Changhŭng	SM	540	1	14)	Kunsan°[2]	AS	675	10
20)	Taegu[2]	QH	558	250	N1)	Sorae*	KA	711	500
13)	Chŏnju°	KF	567	100	20)	Taegu°	KG	738	100
18)	Sunch'ŏn°[2]	KZ	576	1	16)	Kwangju°	KH	747	100
21)	Yŏngju	AG	594	10	N1)	Suwon*	—	756	100
N2)	Namyang*	SA	603	500	6)	Yŏngwol°	CV	783	10
8)	T'aebaek°	SJ	621	10	3)	Yanggu	SY	846	5
26)	Sŏgwip'o	CF	621	10	4)	Kangnŭng°	KR	864	100
9)	Yŏngdong	AY	621	1	11)	Taejŏn°	KI	882	20
3)	Inje	SE	630	5	2)	Pusan°	KB	891	250
19)	Yŏsu°	CY	630	10	20)	Kumi	QY	909	10

N1) Yŏnch'ŏn*	—	918	50	6) P'yŏng-ch'ang	CC	1233	1
12) Puyŏ	QA	927	10	21) Yŏngyang	QG	1233	1
25) Hadong	SU	927	1	15) Namwon	KL	1260	10
7) Hongch'ŏn	QD	927	1	N1) Yangju	—	1269	10
23) Ch'angwon°2	KD	936	10	14) Kurye	SI	1269	1
9) Poŭn	QW	945	10	23) Hapch'ŏn	QV	1278	1
26) Cheju°	KS	963	10	22) Uljin	SV	1305	10
21) Andong°	CR	963	10	16) Yŏng-gwang	QJ	1323	1
L1) Tangjin*	CA	972	1500	22) Ullŭng	CU	1323	1
5) Sokch'o	CS	1008	50	N1) Kimp'o*	—	1341	10
3) Hwach'ŏn	CG	1026	1	13) Muju	KO	1368	1
25) Kŏch'ang	KW	1026	1	1) Ch'ŏlwon*	CO	1395	10
22) P'ohang°	CP	1035	10	24) Ulsan°	QB	1449	10
4) Samch'ŏk	CI	1044	10	25) Hamyang	SH	1458	1
6) Chech'ŏn	CD	1044	10	21) Ponghwa	SD	1458	1
6) Chŏngju°	KQ	1062	50	17) Mokp'o°	KN	1467	50
16) Ch'ungju°	CH	1089	10	18) Kohŭng	QU	1485	1
25) Chinju°	CJ	1098	20	12) Kongju°	QS	1485	1
L1) Hwasŏng*	—	1134	500	20) Kimch'ŏn	SK	1503	1
7) Wonju°	CW	1152	10	8) Kosan	QC	1539	1
L2) Kimje* +	SR	1170	500	26) Tanyang	DK	1584	1
6) Chŏngsŏn	SW	1206	1	25) Sanch'ŏng	—	1584	1
21) Ch'ŏngsong	QR	1206	1	11) Kumsan	QZ	1584	1
6) Chech'ŏn	CD	1215	10	6) Sabuk	QE	1602	1
N2) Kwangju	AA	1224	20				

NOTE: Liberty st's and FM-st's do not use call letters (even if assigned). Other st's without call letters use the calls from their main st's.

Shortwaves: Hwasŏng (G.C: 126.47E/37.13N).

kHz	kW	Prgr.	kHz	kW	Prgr.
3930	10	N1 + L	6135	10	L2 + R. Korea
6015	100	L1			

FM: (MHz) Reg = region in MW section.

Reg. Location	I	II	III	kW
1) Namsan		93.1	89.1a	-/10/10
1) Kwanak-san	97.3			10
1) Yongmun-san	90.3		101.1b	1/-/3
2) Yŏngdo		92.7	97.1c	5/3
3) Hwa-ak-san	99.5	91.1	96.7b	5/5/5
4) Kwaebang-san	98.9	89.1	92.5b	1/5/3
7) Paegun-san		89.5		-/3
7) T'aegi-san	95.5			1
8) Hambaek-san	93.7	97.3		1/3
9) Shikjang-san		102.1		-/3
9) Hŭksŏng-san	89.9			1
9) Uam-san		94.1		-/1
9) Kayŏp-san	90.9			1
10) Kayŏp-san		100.3		-/3
11) Kyeryong-san	94.7	98.5		1/5
13) Mo-ak-san	96.9	100.7		5/5
15) Nogodan	88.3	92.9		1/3
16) Mudŭng-san	90.5	92.3		5/5
17) Yang-ŭl-san		98.3		-/1
19) Mang-un-san	95.7	94.5		1/5
20) P'algong-san	101.3	89.7		5/5
21) Ilwol-san	92.1			1
21) Hakka-san		88.1		-/3
22) Chohang-san	95.9	93.5		1/3
23) Pulmo-san	91.7	93.9		1/5
24) Muryŏng-san	90.7	101.9		1/3
25) Kamak-san		102.5		-/3
25) Mangjin-san		89.3		-/1
26) Kyŏnwolak	99.1	96.3		5/3
26) Sammaebong		99.9		-/3

I-Standard FM; II-KBS FM One; III a = FM-2, b = Nat'l Forces prgr, c = N2.

HOME SERVICES:

KBS R. One (KBS Che-il Radio, HLKA): 24h non-commercial nationwide sce. Key freq's 711/765/3930kHz, 90.3/97.3MHz. Also rel. by Standard FM st's and most reg. st's. All R. One st's except Pusan 891kHz/103.7MHz carry a separate Liberty prgr 1500-1800. Reg. st's may broadcast local prgr's at designated times.

Local N: 2210, 0100, 0305 (Sun), 0315(W), 0900, 1200.

KBS R. Two (KBS Che-i Radio, HLSA): 2000-1800 (558kHz to 1500). Commercial. Key freq's 603/639kHz. Reg. st's may broadcast local prgr's at designated times. All st's except 558/576kHz carry a separate Liberty prgr 1600-1800.

LIBERTY prgr sce's (Sa-hoe Kyo-yuk Pangsong): A sce. for ethnic Koreans living outside of the Rep. of Korea. Two separate sce's + prgr's on KBS R. One and Two.

First Liberty prgr. (Sa-hoe Kyo-yuk Che-il Pangsong, HLCA): 0400-2400 on 972/1134kHz, 0400-2100 on 6015kHz, 1500-1800 on 558kHz.

Second Liberty prgr. (Sa-hoe Kyo-yuk Che-i Pangsong, HLSR): 1000-0400 on 1170kHz, 1400-0000 on 6135kHz. Rebroadcasts First Liberty prgr 1500-1900, 2000-2400 and takes R. Korea 1000-1400, 1900-2000.

KBS FM One (KBS Che-il FM Pangsong, HLQK-FM): 2000-1700. Mainly Korean traditional and western classical music.

KBS FM Two (KBS Che-i FM Pangsong, HLKC-FM) 2000-1700. Mainly Korean and western popular and light classical music. Note: Regional FM One st's air a combination of their own local prgr's, FM One and FM Two prgr's.

KBS-FM Netw. for the National Forces (KBS-FM Kuk-kun Pangsong, HLSE-FM): 24h. Own prgr's 2100-0000, 0300-0400, 0800-1100, other times rel. HLKA 711kHz. Prgr's for soldiers located near the demilitarized zone.

ANN: N1: "Che-il Ladiomnida. HLKA". N2: Che-i Radiomnida HLSA". Liberty prgr: "Yukch'ŏnman-e Kyo-re-e Sudo Sŏ-ur-esŏ Ponaedŭrinŭn Min-joge Sori Kyo-yuk Pangsong-imnida".

Local Identifications: Within local prgrs, also just before the h. at 0300, 0400, 0900, 1000, 1500, 1800, 1900.

V. KBS does not routinely verify reports on domestic prgrs. However, some reg. st's may verify by letter.

Regional Key Station Addresses:
2) 63, Namch'ŏn-dong, Namgu, Pusan 608-790 — 3) 86-1, Nagwon-dong Ch'unch'ŏn 200-100 — 4) 62-5, Yong-gang-dong, Kangnŭng 210-070 — 5) 644-30, Ch'ŏnghak-dong, Sokch'o 217-020 — 6) Yŏnghŭng-7-ri, Yŏngwol-ŭp, Yongwol-gun 230-800 — 7) 79-1, Won-dong 220-060 — 8) 82, Hwangji-1-dong, T'aebaek 235-011 — 9) 604-26, Sajik-dong, Ch'ŏngju 360-070 — 10) 417, Yŏkjun-dong, Ch'ungju 380-790 — 11) 15-30, Mok-dong, Chung-gu, Taejon 301-070 — 12) 571, Shingwan-dong, Kongju 314-040 — 13) 523-3, Kum-am-dong, Chŏnju 560-180 — 14) 796, Naun-dong Kunsan 572-020 — 15) 500, Hyanggyo-dong, Namwon 590-020 — 16) 177-39, Sangu-dong, Sŏ-gu, Kwangju 502-790 — 17) 841-5, Yongdang-dong, Mokp'o 530-360 — 18) 91-3, Sok'yŏn-dong, Sunch'ŏn 540-100 — 19) 8-7, Kwanmun-dong, Yŏsu 550-070 — 20) 100, Shinch'ŏn-dong, Tong-gu, Taegu 701-790 — 21) 666, T'ae-hwa-dong, Andong 760-790 — 22) 483-9, Haedo-1-dong, P'ohang 790-790 — 23) 97-1, Shinwol-dong, Ch'angwon 641-790 — 24) 519-17, Tal-dong, Namgu, Ulsan 680-790 — 25) 13-22, Shinan-dong, Chinju 660-790 — 26) 302-2, Yŏn-dong, Cheju 690-170.

RADIO KOREA (Radio Hanguk)

ADDR: 18, Yŏ-ŭi-do-dong, Yŏngdŭngp'o-gu, Seoul 150-790.

MEDIUMWAVES: See Home Sce.

SHORTWAVES: Kimjae (G.C:126.50E/35.50): 100/250kW — H) Hwasŏng (G.C: 126.47E/37.13N): 10/100kW — S) via Sackville, Canada.

kHz	kHz	kHz	kHz	kHz	kHz	kHz
5975H	6480	9515	9640	9870	11740	15170
6135H	7275	9525	9650S	11715S	11810	15575
6145S	7550	9570	9750	11725	13670	

ND) Omnidirectional.

Arabic	Area	kHz
1645-1730	ND	7275
	Eu	15575
	ME/Af	9515
1900-1945	Eu	6480
		7550
2015-2100	ND	7275
Chinese		
1130-1230	ND	11740
		7275
1300-1400	ND	6135
		1170
1430-1530	ND	7275
		11740
English		
0030-0130	NAm	15575
0600-0700	ND	7275
	NAm	11810
		15170
0800-0900	Eu	7550
		13670
1030-1100	NAm	15575
1100-1200	ME/Af	15575
1215-1315	NAm	9750
1400-1500	SEAs	9750
1600-1700	ND	5975
	ME/Af	9870
1900-2000	ND	1170
		6135
2145-2245	Eu	6480
		7550
		15575
French		
0630-0715	Eu	7550
		13670
	ME/Af	15575
1615-1700	Eu	6480
		7550
1815-1900	ME/Af	9515
		9870
	ND	7275
	Eu	15575
2100-2145	Eu	7550

Time	Area	kHz
German		
0645-0730	Eu	7550
		13670
	ME/Af	15575
1100-1145	Eu	13670
1730-1815	ND	7275
	ME/Af	9515
	Eu	15575
1945-2030	Eu	6480
		7550
		15575
Indonesian		
1115-1200	SEAs	9570
1345-1430	ND	11740
		7275
1500-1545	SEAs	9570
2345-0030	SEAs	9640
	ND	7275
Italian		
0730-0800	Eu	7550
		13670
	ME/Af	15575
1545-1615	Eu	6480
		7550
1900-1930	ME/Af	9515
	ND	7275
2030-2100	Eu	6480
		7550
		15575
Japanese		
0100-0300	ND	7275
0800-1000	ND	7275
1100-1300	ND	6135
		5975
		1170
1600-1700	As	891
2100-2300	ND	5975
Korean		
0300-0400	ND	7275
	NAm	15575
0600-0645	Eu	13670
	NAm	9570
0700-0800	NAm	15170
		11810
	ND	7275
0900-1100	Af/ME/Eu	15575
1000-1100	ND	6135
		5975
		7275
		1170
1100-1130	NAm	6145S
		9650S
0900-0930	Eu/ME/Af	13670
	ND	7275
1100-1130	ND	7275
		11740
	SAm	11725
1400-1430	ND	7275
		11740
1530-1600	ND	7275
1700-1730	Eu/ME/Af	9870
1700-1900	Eu	6480
	ME/Af	7550
	ND	5975
2030-2100	ND	5975
	Eu/ME/Af	9870
2100-2200	ND	7275
		9640
2300-0100	ND	5975
0030-0100	ND	7275
Portuguese		
0930-1015	Eu	13670
	SAm	11725
		9570
2100-2145	Eu	6480
		15575
2330-0015	Eu	15575
Russian		
1230-1315	ND	7275
		11740
1600-1645	Eu	15575
	ND	7275
1900-1945	Eu	15575
	ND	5975
2200-2245	As	9640
	ND	7275
2330-0015	Eu	6480
		7550
Spanish		
0215-0300	NAm	15575
1015-1100	Eu/ME/Af	13670
	SAm	11725
		9570
2245-2330	Eu	6480
	Eu	15575
	ME/Af	7550

INT-SIG: Korean folk music played as "Dawn" with orchestra — **V.** by QSL-card.
PUB: Prgr. schedules, pennant, folder, KBS Handbook (English), KBS Information Sce.

EDUCATION BROADCASTING SYSTEM — EBS (Kyo-yuk Radio)
Programming by Ministry of Education.

ADDR: 926, Umyŏn-dong, Sŏcho-gu, Seoul.
Te: +82 (2) 572-5021, 572-5121, 522-8020.
Call letters HLQL used for all stations.

Service area	Tr. location	MHz	kW
Ch'ŏngju	Kayŏp-san	104.1	5
Wonju	Paegŭm-san	104.9	3
Ch'angwon	Pulmo-san	104.3	5
Seoul	Kwanak-san	104.5	5
Chinju	Kamak-san	104.7	3
Kangnŭng	Kwaebang-san	104.9	3
Taegu	P'algong-san	105.1	5
Kwangju	Mudŭng-san	105.3	5
Taejŏn	Kyeryŏng-san	105.7	5
Ulsan	Muryŏng-san	105.9	3
Yŏsu	Mang-un-san	106.3	1
Ch'unch'ŏn	Hwa-ak-san	106.5	5
P'ohang	Chohang-san	106.7	3
Chŏnju	Mo-ak-san	106.9	5
Taebaek	Hambaek-san	107.1	3
Cheju	Kyŏnwolak	107.3	3
Namwom	Nogodan	107.5	3
Andong	Hakka-san	107.7	3
Pusan	Yŏngdo	107.7	3

D.PRGR: 2000-1600.

MUNHWA YŎ-ŬI-DO-DONG BROADC. CORP.—MBC (Munhwa Pangsong)

Nationwide Commercial Network.
ADDR: 31, Yŏ-ŭi-do-dong, Yŏngdŭngp'o-gu, Seoul 150-728.
Te: +82 (2) 784-2000. **Telex:** MBCNEWS K22203. **Cable:** MBCHLKV SEOUL.
L.P: Pres: Chang-Bong Choi. Vice-Pres: Duk-yul Lee. Man. Dir: Jong-min Lee (corporate planning), Il-Pyung Pyun (corporate adm.), Chang-je Kim (programming), Sang-uk Lee (N. & sports), Sung-hee Lee (production), Jong-han Kim (operation & engineering). Dir. Int. Rel: Young-bok Kim.
Mediumwaves: Call: HL.

	Call	kHz	kW	St. and h. of tr.
1)	CQ	765	10	MBC, Taejŏn: 1955-1700
2)	AJ	774	10	MBC, Cheju: 1955-1700
3)	AN	774	10	MBC, Ch'unch'ŏn: 1955-1800
4)	CT	810	20	MBC, Taegu: 1955-1700
5)	CN	819	20	MBC, Kwangju: 1955-1700
6)	AU	846	10	MBC, Ulsan: 1955-1600
7)	CX	855	10	MBC, Chŏnju: 1955-1700
8)	KV	900	50	MBC, Seoul: 24h
9)	AP	990	10	MBC, Masan: 1955-1600
10)	AW	1017	10	MBC Andong: 1955-1700
11)	AT	1080	10	MBC Yŏsu: 1955-1630
12)	AV	1107	10	MBC P'ohang: 1955-1700
13)	KU	1161	20	MBC Pusan: 1955-1700
14)	AK	1215	10	MBC Chinju: 1955-1700
15)	SB	1242	10	MBC Wŏnju: 1955-1600
16)	AF	1287	10	MBC Kangnŭng: 1955-1608
17)	AX	1287	10	MBC Ch'ŏngju: 1955-1630
18)	AO	1332	10	MBC Ch'ungju: 1955-1700
19)	AQ	1350	10	MBC Samch'ŏk: 1955-1600
20)	AM	1386	10	MBC Mokp'o: 1955-1600

FM Stations

Location	Music FM			General FM		
Studio (Transmitter)	Call	MHz	kW	Call	MHz	kW
8) Seoul (Namsan)	KV-FM	91.9	10			
Seoul (Kwanak-san)				CP-FM	95.9	10
13) Pusan (Yŏngdo)	KU-FM	88.9	5	SX-FM	95.9	3
4) Taegu (P'algong-san)	CT-FM	95.3	5	QQ-FM	96.5	5
5) Kwangju (Mudŭng-san)	CN-FM	91.5	5	SM-FM	93.9	5
Kwangju (Nogo-dan)		95.1	3			
1) Taejŏn (Shikjang-san)	CQ-FM	97.5	5			
7) Chŏnju (Mirŭk-san)	CX-FM	101.7	3			
9) Masan (Pulmo-san)	AP-FM	100.5	1			
3) Ch'unch'ŏn (Pong-ŭi-san)	AN-FM	94.9	1			
17) Ch'ŏngju (Uam-san)	AX-FM	99.7	1			
2) Cheju (Kyŏnwol-ak)	AJ-FM	90.1	3			
Cheju (Sammae-bong)		102.9	3			
6) Ulsan (Muryong-san)	AU-FM	98.7	3			
16) Kangnŭng (Kwaebang-san)	AF-FM	94.3	5			
14) Chinju (Mangjin-san)	AK-FM	97.7	1			
14) Chinju (Kamak-san)		96.1	3			
20) Mokp'o (Yang-ŭl-san)	AM-FM	102.3	1			
11) Yŏsu (Kubong-san)	AT-FM	98.3	2			
Yŏsu (Ch'ŏnhwang-san)				SP-FM	100.7	1
10) Andong (Hakka-san)	AW-FM	91.3	3	SO-FM	100.1	1
15) Wonju (Paegun-san)	SB-FM	98.9	3			
18) Ch'ungju (Kayŏp-san)	AO-FM	88.7	3	QI-FM	96.1	1
Ch'ungju (Check'ŏn)					94.7	0.5
Ch'ungju (Tanyang)					94.1	0.5
19) Samch'ŏk (Hambaek-san)	AQ-FM	98.1	3	AR-FM	101.5	1
Samch'ŏk (Ponghwang-san)		99.9	1			

12) P'ohang (Chohang-san) AV-FM 97.9 3

NOTE: General FM st's simulcast with the MW st in the same city. A seperate sce. is provided to the Music FM st's. All regional st's broadcast a combination of a feed from Seoul and their own local prgr. General FM st's follow the same schedule as their corresponding MW outlet. Music FM sched: 1955-1700.
ANN: "Munhwa Pangsong-imnida (call letters)" or "Munhwa Pangsong" or "MBC". Seoul: "Chungp'a Kubaek (900) kHz, P'yojun FM Kushibo-chŏm'gu (95.9) MHz Munhwa Pangsong."
Addresses:
Add "(location) Munhwa Broadc. Corp." to addr.
1) 381-171, Sŏnhwa-1-dong, Chung-gu, Taejon-shi 301-728 — 2) 321-22, Yŏn-dong, Cheju-shi, Cheju-do 690-170 — 3) 238-3, Samch'ŏn-dong, Ch'unch'ŏn-shi, Kangwon-do 200-200 — 4) 1, Pŏmŏ-dong, Susŏng-gu, Taegu-shi, 706-728 — 5) 300, Wolsan-dong, Sŏ-gu, Kwangju-shi 502-728 — 6) San 7-1, Haksŏng-dong, Chung-gu, Ulsan-shi, Kyŏngsang-namdo 681-180 — 7) 417-114, Chinbuk-dong, Tŏkchin-gu, Chŏnju-shi, Chŏlla-bukto 520-728 — 8) National addr. — 9) 83-9, Sŏsŏng-dong, Happ'o-gu, Masan-shi, Kyŏngsang-namdo 630-728 — 10) 709-1, T'aehwa-dong, Andong-shi, Kyŏngsang-bukdo 760-290 — 11) 101-1, Munsu-dong, Yŏsu-shi, Chŏlla-namdo 550-200 — 12) 907-4, Taejam-dong, P'ohang, Kyŏngsang-bukto 790-728 — 13) 69, Chung-ang-dong-6-ga, Chung-gu, Pusan-shi 600-728 — 14) 47, P'yŏng-an-dong, Chinju-shi, Kyŏngsang-namdo 660-728 — 15) 1023-70, Haksŏng-1-dong, Wonju-shi, Kangwon-do 220-039 — 16) 1091-6, P'onam-dong, Kangnŭng-shi, Kangwon-do, 210-110 — 17) 261-30, Uam-dong, Ch'ŏngju-shi, Ch'ungch'ŏng-bukto 360-728 — 18) 680, Hoam-dong, Ch'ungju-shi, Ch'ungch'ŏng-bukdo 380-130 — 19) 55-6, Namyang-dong, Samch'ŏk-shi, Kangwon-do 245-010 — 20) 1096-1, Yongdang-dong, Mokp'o-shi, Chŏlla-namdo 530-360.
V. Many st's will verify directly. Pusan and Masan have QSL cards. Re. directly to individual st's.

CHRISTIAN BROADCASTING SYSTEM — CBS (Kidokkyo Pangsong)

Mediumwaves: Call: HL.

	Call	kHz	kW	St. and h. of tr.
1)	KY	837	50	°CBS Seoul: 1947-1703
2)	CL	999	10	CBS Kwangju: 1947-1603
3)	KT	1251	10	CBS Taegu: 1947-1603
4)	CM	1314	10	CBS Iri: 1947-1603
5)	KP	1404	10	CBS Pusan: 1947-1603

°key st.
FM: 6) CBS Ch'ongju HLAC-FM 91.5MHz 3kW 1947-1603.
Addresses and other information:
1) CPO Box 2653, Seoul 110 701 — Pres: Lee, Chai-eun; Dir. of Eng.: Cho, Sun-kuk; **ANN:** "Pikkwa Pogŭm-e Sori Chungp'a P'al-baek-sam-ship-ch'il khz Kidokkyo Pangsong-imnida. HLKY."
2) 107-5, Yu-dong, Puk-ku, Kwangju-shi 500-020 — Dir: Lee, Jong-Il.
3) 3-7, Chimsan-dong, Puk-ku, Taegu-shi 702-703 — Dir: Park, Byong-wha.
4) 86-8, Namjung-dong-1-ga, Iri-shi, Chŏlla-bukdo 570-101 — Dir: Song, Taek-ju.
5) 608-1, Chŏnp'a-4-dong, Pusanjin-gu, Pusan 614-044 — Dir: Hong, Yong-pyo.
6) 1408, Pongmyŏng-dong, Ch'ŏngju-shi Ch'ungch'ŏng-bukdo 360-300 — Dir: Cho, Sung-ho.
ANN: St's 2)-6) "Yoginŭn (location) Kidokkyo Pangsong imnida".
V: All st's verify reports for their own prgr's by QSL card.
F.PL: FM st. in Mokp'o.

SEOUL BROADCASTING SYSTEM—SBS (Seoul Pangsong)

Addr: 10-2, Yŏ-ŭi-do-dong, Yŏngdŭng-p'o-gu, Seoul 150-010. **Te:** (02) 786-0792. **Fax:** (02) 786-0785.
Station: HLSQ Kaebong (Seoul) 792kHz 50kW.
D.Prgr: 2000-1700 (commercial).

FAR EAST BROADCASTING CO. (Rlg.)

Mediumwaves	kHz	kW	Mediumwaves	kHz	kW
1) HLKX, Seoul	1188	100	2) HLAZ, Cheju	1566	250/100

FM: 3) HLAD, Taejŏn 93.3MHz 3kW.

Addresses and other information
1) **Far East Broadc. Co,** Map'o P.O. Box 88, Seoul 121-707. **Te:** (02) 320-0114; **Fax:** (02) 333-2627; **Telex:** K22031 — Pres: Dr. Billy Kim. Gen. Mgr: John Song (HLKX), Koh, Sung Cho (HLAZ), Min, San Woong (HLAD). Dir. Tec: Hwang, Young Jim. 2000-1700. **Korean:** 2030-0900, 1100-1230, 1430-1700 **English:** 0900-1100. **Chinese:** 1230-1430. — **ANN:** Korean "Arŭmda-un Ŭmakkwa Kŭrisŭdo Yaesu-e Kippŭn Soshigŭl Chŏnp'ahanŭn Yŏrŏbun-e Kŭkdong Pangsong-imnida". English: "This is HLKX Radio broadcasting with 100,000 watts of power on 1188kHz" — **Fl:** by contributions & free will offerings. **V:** All st's by QSL-card or letter. Rec. acc. (B).
2) **Asia Broadcasting Co.** Addr. & L.P. as 1) — 1000-0800. **Korean:** 1800-2230, 2245-0800. **Chinese:** 2230-2245, 1000-1230, 1345-1500. **Japanese:** 1230-1345. **Russian:** 1500-1800 (incl. languages of the CIS Republics) — **ANN:** Korean: "Kuwŏn-e Kippŭn Soshigŭl Chŏnhanŭn A-se-a Pangsong-imnida".
3) **Kŭkdong Pangsong,** 61-5, Chung-dong, Tong-gu, Taejŏn. **Te:** (042) 253-9330. **D.PRGR:** 2000-1500.

RADIO PACIS (P'yonghwa Pangsong — PBC)

(Endowment by the Catholic Church)
Addr: 2-3, 1-ga, Chŏ-dong, Chung-gu, Seoul 100-031. **Te:** (02) 2702-114; **Fax:** (02) 278-4972.
L.P: Pres: Father Duk-Hyun, Chop; Man. Dir.: Father Won-Haeng, Cho; Prgr. Dir: Hae-Sung, Lee.
Station: HLQP-FM 105.3MHz 5kW: 1957-1702.
Ann: "Chong-hap FM (105.3) Paeg-o-jŏm-sam MHz, Malgŭn Sori, Palgŭn Sesang, PBC, P'yonghwa Pangsong-imnida. HLQP."
F.PL: FM st's in Kwangju, Taegu and Chonju.

BUDDHIST BROADCASTING SYSTEM — BBS (Pulgyo Pangsong)

Owned and operated by the Buddhist Church
Addr: Tabo Building; 140, Map'o-dong, Map'o-gu, Seoul 121-050. **Te:** 704-5114; **Fax:** 704-1878 to 704-1879.
L.P: Pres: Sang-Moon, Chang; Man. Dir: In-Supp, Chung; Dir. of Eng: Chong-gi, Kim; Prgr. Dir: Tae-Haeng, Lee.
Station: HLSG-FM 101.9MHz 5kW: 1956-1700.
Ann: "FM Paeg-il-jŏm-gu (101.9MHz) Kkae-ch'im-e Sori, Nanunŭn Kippum, BBS Pulgyo Pangsong. HLSG."
F.PL: FM Station in Pusan.

TRAFFIC BROADCASTING SYSTEM (Kyot'ong Pangsong — TBS)

Municipal Station
This station is operated by the Seoul Municipal Traffic Broadcast Headquarters to provide traffic information and education of the citizens of Seoul and surroundings.
Addr: 45, Toryŏm-dong, Chongno-gu, Seoul 110-051. **Te:** (02) 724-7114. **Fax:** (02) 724-7239.
L.P: Gen. Mgr: Sok-Ch'un, Paek. Dir. Tec: Pok-Il, Yu, Prgr. Dir: Sang-Ho, So.
Station: HLST-FM 95.1MHz 5kW: 2000-1700.
Ann: "Yŏllin Sori, Yŏllin Sesang. Ku-shib-o-chŏm-il (95.1)MHz, TBS Kyot'ong Pangsong-imnida. HLST."

AMERICAN FORCES KOREA NETWORK (AFKN)

ADDR: As below. **Te:** 7914-6495/6. **L.P:** Commanding Officer: Lt. Col. Harold Wilson.
Stations: F.PL.

	kHz	F.PL. kHz	kW	St. name/location	FM MHz	F.PL. MHz	kW
1)	549	1377	5	Seoul/Yongsan	102.7	88.1	5
2)	576°	1440	5	Munsan/ Camp Pellham	88.5°		0.05
3)	585	1377	5	Taegu/Camp Walker	99.3°	88.5°	1
4)	783	1197	5	Pusan/Camp Hialeah	88.1°		0.25
	1044		1	-/Ch'unch'ŏn (Camp Page)	94.1	88.5	0.1
	1161°		0.25	-/Ŭijŏngbu (Camp Red Cloud)	88.5		0.1
5)	1197°		1	Camp Casey/ Tongduch'ŏn	88.3°		0.25
6)	1359		1	Osan/Osan Air Base	88.5°		0.05
7)	1440°		1	Camp Humphreys/ P'yŏngt'aek	88.3°		0.05
8)	1440		1	Kunsan/Kunsan Air Base	99.9	88.5	0.25
	1440		0.25	-/Wonju (Camp Long)	88.3		0.05
	1440		0.25	-/Waegwan (Camp Carrol)	—		

	1512	0.25	-/P'ohang (Camp Libby)	99.3 88.5	0.05
4)	1512	0.25	-/Chinhae NS	88.5°	0.05

Low power: 1512kHz (Cheju-do Sangdong); 96.1MHz (F.Pl. 88.5MHz) Taejon; 99.1MHz (F.Pl. 88.5MHz) Kwangju Air Base. ° = local prgr's 2005-0000 Sun-Thurs, 0605-0900 Mon-Fri; otherwise rel. 1).

D.PRGR: 24h (MW/FM separate prgr's). **N.** on the h. Formal sign on at 1505.

ANN: AM: "Serving the men and women at (location) on (freq), this is the American Forces Korea Network" (obs: b.cast just before the h over the whole netw, not a local ID). **FM** (Seoul): "Eagle FM" — **V.** by QSL card from addr. 1).

Addresses: 1) Headquarters American Forces Korea Network (HQ-AFKN), Unit #15324, APO AP 96205-0097, USA. **Te:** (822) 7914-6495. Commanding Officer: Lt. Col. Chad C. Starr — 2) Unit #15325, APO AP 96251-0098, USA — 3) Unit #15029, APO AP 96218-5000, USA — 4) Unit #15185, APO AP 96259-0274, USA — 5) Unit #15106, APO AP 96224-0380, USA — 6) Unit #2034, APO AP 96278-5000, USA — 7) Unit #15473, APO AP 96271-0543, USA — 8) Unit #2011, APO AP 96264-5000, USA.

KYRGYZSTAN

L.T: UTC +6h (Su: UTC +7h) — **Pop:** 4.370.000 — **Pr.L:** Kyrgyz, Russian — **E.C:** a/c 50 127/220V — **ITU:** KGZ.

KYRGYZ RADIO (Gov.)

ADDR: ul. Molodoy Gvardii 63, 720300 Bishkek. **Te:** +7 (3312) 253404. **Fax:** +7 (3312) 257930. **Tlx:** 245173 RADIO SU.
L.P: Gen. Dir: Tugelbay Kazakov.

Stations*	kHz	kW	H. of tr.	Prgr.
Bişkek	198	150	0100-1800	MR1
Oş	576	50	0000-2300	RSM
Bişkek	612	300	2300-2200	RSM
Bişkek	882	500	0000-1900	KGR1
Bişkek	1323	30	1000-1300	RL, RSY
Jojomel	1404		0000-1900	KGR1
Haidarkan	1404		0000-1900	KGR1
Orgoçor	1404		0000-1900	KGR1
Talas	1404		0000-1900	KGR1
Sulyukta	1431		0000-1900	KGR1
Falal-abad	1431		0000-1900	KGR1
Naryn	1431		0000-1900	KGR1
Przhevalsk	1431		0000-1900	KGR1
Batken	1467		0100-1800	KGR2
Kara-Kul	1467		0100-1800	KGR2
Bişkek	1467		0100-1800	KGR2
Tas-Kumyr	1530	5		RSM
Bişkek	4010	50	0000-1900	KGR1
Bişkek	4050	50	0100-1800	KGR2

NB: Powers and locations are tentative only. *) names in Kyrgyz.

FM STATIONS**	RL		
Alai-Kuu	69.08	Orgochor	67.82
Bishkek	66.38	Osh	71.69
Dedyumey	71.69	Sulyukta	70.88
Gulcha	69.68	Talas	72.08
Kara-kul	66.86	Terek-Sai	70.64
Naryn	72.39	Vostochnaya	67.04
		Yuzhnaya-2	71.93

**) names in Russian. No further details available.

D.PRGR: KGR1 in Kyrgyz/Russian, incl. rel. of MR1. **N. in English:** 0010-0015. Dungan: W 1615-1700. German: Sat 1530-1615 — **KGR2** in Kyrgyz, Russian and others — **MR1** = Moscow Ostankino R-1 — **RSM** = Moscow R. St. Mayak — **RL** = R. Liberty (cf. Germany) 1000-1200 in Russian, 1200-1300 in Kyrgyz.

ANN: Kyrgyz: "Bişkekden söylöbüz" — **V.** by letter (irr.).

PUB: Efirde Kyrgyzstan (weekly for Home Sce.).

R. ALMAZ (Comm.): r. on 1431 kHz in Bishkek. No further details at press time.

Clandestine broadcasting in Asia:
See *Listen to the World* section.

LAOS (People's Dem. Rep.)

L.T: UTC + 7h — **Pop:** 4.111.000 — **Radio sets:** 425.000 (est.) — **Pr.L:** Lao Soung, Lao Theung — **E.C:** a/c 50, 127/220V — **ITU:** LAO.

NATIONAL RADIO OF LAOS (Gov.)

ADDR: B.P. 310, Vientiane. **Te:** 2457.
L.P: Dir. Gen: Boubane Vorakhoun. Dir. Tec: Bountanh Inthasay. Dir. Foreign Rel: Somsanouk Mixay.

Mediumwaves	kHz	kW	Sce		kHz	kW	Sce
Vientiane	580	150	N	Vientiane	1030	10	E
Vientiane	640	10	N	Vientiane	1190		C
Luang Prabang	705		R	Xieng Khouang	1200	1	R
Oudomxay	800	1	R	Pakse	1350	10	R
Houa Phanh	1000		R	Savannakhet	1430	3	R
Shortwaves	**kHz**	**kW**	**Sce**		**kHz**	**kW**	**Sce**
Oudomxay	4535	1	R	Vientiane	6130	25	N
Houa Phan	4660	1	R	Pakse	6647	1	R
Xieng Khouang	4991*	1	R	Vientiane	7145*	10	E
				Louang Prabang	7164	1	R

NB: SW freq's are variable — *) r. inactive.

FM: Vientiane 97.5MHz 0.05kW (0200-0400, 0700-0900).

D.PRGR: National Sce. (N) on 580/640/6130kHz: 2130-0230, 0400-0700, 0900-1600. **N:** 2300, 0000, 0500, 1200, 1400. **Vientiane Capital R. (C)** on 1190kHz: 0030-1200 approx (exact times irr.).

REGIONAL STATIONS (R) in Laotian/Tribal Languages:
N: Most st's rel. N. from Vientiane in addition to local N.
Houa Phan: 1255-1425 — **Luang Prabang:** 2225-0130, 0355-0600, 1025-1400 —**Oudomxay:** 2300-0200, 1100-1400 — **Pakse:** 2300-0200, 0400-0600, 1000-1400 — **Savannakhet:** 2225-0100, 0425-0600, 1025-1400 — **Xieng Khouang:** 2300-0200, 0430-0560, 1000-1400.

ANN: Nat. Sce: "Thini Withayu heng Sat, krachaisiang chak Wianchan, nakhong-lwang khong sathanalat pasatipatai pasason Lao". Capital City Sce: "Vitthayou Krachaistang Nakhonluang Vientiane". Regional st's : "Thini sathani vithayu krachaisiang heng Luang Prabang", "Thini vithayu krachaisiang heng sat Houa Phan", "Thini vithayu krachaisiang heng sat Ke Thai otaa Pakse", "Thini vithayu krachaisang heng sat kueng Savannakhet".

IS: (National Sce.): Music on Khéne (mouth organ) & Solo (bamboo instrument).

EXTERNAL SERVICE (E) on 1030/7145(announced)kHz:
Cambodian: 0000-0030, 0400-0430, 1230-1300.
English: 0100-0130, 0600-0630, 1330-1400.
French: 0030-0100, 0530-0600, 1300-1330.
Thai: 0500-0530, 1130-1200, 2300-2330.
Vietnamese: 0430-0500, 1200-1230, 2330-2400.

V. by QSL-card. IRC's not acc.

MACAU

L.T: UTC + 8h — **Pop:** 450.725 — **Radio sets:** 75.000 — **Pr.L:** Portuguese, Cantonese — **E.C:** a/c 50, 220V — **ITU:** MAC.

TELEDIFUSAO DE MACAU, SARL (Priv. Comm.)

ADDR: Rua Francisco Xavier Pereira 157-A, Macau. **Te:** +853 550132. **Fax:** +853 595029. **Telex:** 88309 RADIO OM.
L.P: Dir. Gen: K.K. Leung. Contr. Radio: J.N. Nogueira.
D.PRGR. in Portuguese 24h: 98.0MHz (stereo) 2.5kW.
In Cantonese 24h: 900kHz 1kW 100.7MHz (stereo) 2.5kW.

RADIO VILA VERDE (Priv. Comm.)

ADDR: Rua Francisco Xavier Pereira 133, Penthouse, Macau.
L.P: Dir: Simon Leung.
MEDIUMWAVE: 735kHz (10kW) — **FM:** 89.0vMHz.
D.PRGR. in Cantonese: 2230v-1600.

MALAYSIA (Federation of)

L.T: UTC + 8h — **Pop:** 18.413.000 — **Radio sets:** 3.500.000 — **Pr.L:** Bahasa Malaysia, English, Chinese, Tamil — **E.C:** a/c 50, 240V — **ITU:** MLA.

PENINSULAR MALAYSIA

RADIO TELEVISION MALAYSIA — RTM (Gov.)

ADDR: Dept. of Broadc, Angkasapuri, Bukit Putra, 50614 Kuala Lumpur. **Te:** +60 (3) 2745333. **Fax:** +60 (3) 2304735. **Telex:** RTMTV MA31383.
L.P: Dir. Gen: Jafar Kamin. Dep. Dir. Gen. (Programming): Ismail Jah. Dep. Dir. Gen. (Eng): M. Jeganathan. Dir. Eng. (Development): Mohd Nor Abdullah. Dir. Prgr: Saleh Pateh Akhir. Dep. Dir. Eng. (TV): Alimusa Sulaiman. Dir. (Eng.) (Radio): P. Balagonal. Dir. North: Rahman Hamid. Dir. East Reg: Baharin Tahir. Dir. South Reg: Azahar Ahmad. Dir. Sabah: Aziz Abas. Dir. Sarawak: Amran Hamid.

MEDIUMWAVES: JB = Johore Bharu; KL = Kuala Lumpur (Kajang); KB = Kota Bharu; KT = Kuala Terengganu. **Prgrs:** R1 = Bahasa Malaysia, R2 = FM Stereo (musical), R3 = Bahasa Malaysia (State Netw.), R4 = English, R5 = Chinese, R6 = Tamil.

Location	kHz	kW	Sce		kHz	kW	Sce
Tronoh	576	100	R3	KT	765	10	R1
JB	576	50	R1	Kuantan	810	10	R1
Kajang	594	200	R1	Melaka	1008	10	R1
Segamat	621	100	R1	Mersing	1053	20	R1/3
Kuala Lipis	648	20	R1	Grik	1089	20	R5/6
Grik	657	20	R1	Dungun	1260	5	R1/3
Penang	666	10	R1	Mersing	1314	20	R5/6
KB	702	20	R4				

FM: KL 98.3MHz (R1), 95.3MHz (R2), 100.9MHz (R3), 102.3MHz (R3), 100.1MHz (R4), 96.3MHz (R5), 106.7MHz (R6) + relay st's at 13 locations.
SHORTWAVES: Kajang (G.C: 101.46E/03.01N) — OS = Overseas Sce.

kHz	kW	Sce	kHz	kW	Sce
4845	100	R6	7295	100	R4
5965	100	R1	9750	100	OS
6025	100	R5	11885	500	OS
6100	100	OS	15295	500	OS
6175	50	OS			

HOME SERVICES
Radio 1 in Bahasa Malaysia: 24h on MW/FM/5965kHz — **ANN:** "Sepanjang Waktu".
Radio 2 in FM-Stereo 2200-1600.
Radio 3: 12h daily — **ANN:** "Ibu Kota".
Radio 4 in English: 2200-1600. **N:** 2300, 0530, 1030, 1200, 1330, 1500.
Radio 5 in Chinese: 2200-1600.
Radio 6 in Tamil: 2200-1600.
ANN: "Teman Setia Anda".

OVERSEAS SERVICE (Suara Malaysia — Voice of Malaysia)
ADDR: P.O. Box 11272, 50740 Kuala Lumpur. **Te:** +60 (3) 2825333. **Telex:** MA 30283 (Radio).
MEDIUMWAVES: Kota Kinabalu 1476kHz 600kW.
SHORTWAVES: See above.

	kHz		kHz
Arabic		**Indonesian**	
1530-1700	15295	0900-1600	6175, 9750
Bahasa Malaysia		2200-2400	6100, 6175
0830-1030	9750, 15295	**Mandarin**	
1700-1900	6175,9750,15295	1030-1230	11885,15295
Bamar (Burmese)		**Thai**	
1430-1530	6100	1300-1430	6100
English			
0555-0825	6175,9750,15295		
Tagalog			
1030-1300	1476		

N. in English: 0600, 0800.
ANN: "Inilah Suara Malaysia". E: "This is the Voice of Malaysia" — **INT-SIG:** First bar of Nat. Anthem "Negara Ku" (Chimes).
V. by QSL-card. Rec. acc.

JBFM-104 (Comm.)
STATION: Johore Bharu 104.1MHz (MoR music beamed to JB & Singapore area).

HEART 91.3 (Comm.)
STATION: 91.3MHz — 2200-1700.

SABAH

L.T: UTC + 8h — **Pr.L:** Bahasa Malaysia, English, Kadazan, Dusun, Murut, Chinese, Bajau — **E.C:** a/c 50, 230/400V (yes).

RADIO TELEVISION MALAYSIA — KOTA KINABALU (Gov.)

ADDR: 88614, Kota Kinabalu. **Te:** +60 (88) 52711.
Cable: Broadcasts, Sabah. **Telex:** MA 80061 Kota Kinabalu.
STATIONS: Kota Kinabalu (Tuaran) (G.C: 116.14E/06.12N). KK = Kota Kinabalu, LD = Lahad Datu, SM = Suara Malaysia (Overseas Sce).
Languages: BM = Bahasa Malaysia, E = English/local dialects.

Location	kHz	kW	L		kHz	kW	L
Tenom	567	10	BM	LD	828	10	E
KK	603	10	BM	Tawau	927	10	E
LD	675	10	BM	Sandakan	1080	10	E
KK	693	10	E	Kudat	1197	10	E
Tawau	747	10	BM	KK	1476	600	SM
Tenom	774	10	E	KK	4970	10	BM
Sandakan	783	10	BM	KK	5980	10	E
Kudat	801	10	BM				

FM: Layang-Layang 93.5MHz 10kW (E), 97.5MHz 10kW (BM).
Bahasa Malaysia: 2200-1600.
English/local dialects: 2200-0100 & 1000-1600. **Local N. in English:** 2300.
ANN: Bahasa Malaysia: "Inilah R. Malaysia Kota Kinabalu".
INT-SIG: Opens and closes with Nat. Anthem.
V. by QSL-card. Re. to Publicity Officer.

SARAWAK

L.T: UTC + 8h — **Pr.L:** Bahasa Malaysia, English, Chinese and 8 local dialects — **E.C:** a/c 50, 230/400V.

RADIO TELEVISION MALAYSIA-SARAWAK (Gov.)

ADDR: Broadc. House, Jalan Satok, 93614 Kuching. **Te:** +60 (82) 248422. **Cable:** Broadcast Kuching. **Telex:** MA70084.
STATIONS: Kuching (G.C: 110.20E/01.33N) — Sibu (G.C: 111.49E/02.18N) — Limbang (G.C: 115.00E/04.45N) — Miri (G.C: 113.59E/04.23N) — Sri Aman (G.C: 111.28E/01.13N).

Location	kHz	kW	Netw.		kHz	kW	Netw.
Kuching	549	20	Y	Miri	1206	20	G
Miri	576	20	Y	Miri	3385	10	G
Sibu	621	20	R	Kuching	4835	10	Y
Limbang	648	20	Y	Kuching	4895	10	G
Kuching	729	20	Y	Kuching	4950	10	R
Miri	819	20	R	Sibu	5005	10	G
Kuching	846	10	G	Kuching	5030	10	B
Limbang	873	20	R+G	Sibu	6050	10	G
Sibu	909	20	Y	Miri	6060	10	G
Kuching	954	10	B	Kuching	7130	10	B
Sri Aman	1044	20	Y	Kuching	7145	10	Y
Sibu	1062	20	G	Kuching	7160	10	R
Sri Aman	1161	20	Y+G	Kuching	7270	10	G

Red Netw. (R) in English/Chinese: 2200-0600 & 0800-1600 on MW, 2200-0100 & 1000-1600 on 4950kHz, 2300-1500 on 7160kHz. **Local N. in English:** 2330, 0530, 1030,1400.
Yellow Netw. (Y) in Bahasa Malaysia/Melanau: 2100-1200 on MW, 2100-0030 & 1000-1600 on 4835kHz, 2300-1600 on 7145kHz, 1600-2100 on FM 97.7MHz.
Green Netw. (G) in Iban: 2000-2400 & 0400-0500 on MW/4895/5005/7270kHz, 1000-1500 on 6050/6060kHz, 2200-2400 & 1100-1500 on 3385kHz.
Blue Netw. (B) in Bidayuh: 2200-2400 (Sun 0200) & 1030-1530 on 954/5030kHz, 0400-0530 on 954/7130kHz.
NB: All networks have regional prgrs at times.
ANN: Bahasa Malaysia: "Inilah R. Malaysia (location)".
INT-SIG: A musical phrase (played on a native instrument, the Sape), alternating between A and F, followed by the National Anthem. Transmission closes with the National Anthem — **V.** by QSL-card (Miri/Sibu reply direct).

MALDIVES (Rep. of)

L.T: UTC + 5h — **Pop:** 234.600 — **Radio sets:** 25.000 — **Pr.L:** Dhivehi (Maldivian) — **ITU:** MLD.

VOICE OF MALDIVES (Gov.)

ADDR: "Moonlight Higun", Male'. **Te:** +960 322841. **Fax:** +960 328357. **Telex:** 66120 ADU MF — **L.P:** Dir: Ibrahim Manik.
STATION: 1458kHz 5kW — **FM:** 104.2MHz 20W (stereo).
D.PRGR. in Dhivehi (Maldivian): 0030-0400, 0800-1000, 1200-1745. **N:** 0200, 0800, 1200, 1400, 1745.
Radio Eke on 104.2MHz: 0400-0800, 1015-1200, 1500-1900.
ANN: "Mee Dhivehi Raajjeyge Adu" — **V.** by QSL-card (B).
F.PL: SW Sce. to neighboring countries.

MONGOLIA (People's Rep.)

L.T: UTC + 8h (Su: UTC + 9h) — **Pop:** 2.308.000 — **Radio sets:** 186.000 — **Pr.L:** Mongolian — **E.C:** a/c 50, 220V — **ITU:** MNG.

ULAANBAATAR RADIO (Gov.)

ADDR: C.P.O. Box 365, Ulaanbaatar. **Te:** 27900. **Telex:** 223 RTV MH. **Cable:** Mongolian Radio.
L.P: Dir. Foreign Rel: B. Altansukh. Dir. Foreign Sce: Khurel Baatar. Head English Dept: A. Oyuunchimeg.

Longwaves	kHz	kW	Mediumwaves	kHz	kW
Ulaanbaatar	162	1000	Choibalsan	882	75
Ulgii	207	60	Huvsgul	882	150
Dalanzadgad	207	150	Altai*	990	150
Choibalsan	207	150	Ulaanbaatar	990	500
Altai	227	150	Sainshand	1233	75
Murun	882	150	Dalanzadgad	1350	150

*) carries Ext. Sce. (see below).

Shortwaves	kHz	kW		kHz	kW
Dalanzadgad	3960	12	Ulangoom	4865	5
Uliosxai	3960	5	Murun	4897v	12
Ulaanbatar	4081v	50	Choibalsan	4995	12
Ulgii	4750	12	Ulaanbaatar	5960	50
Altai	4828v	12	Ulaanbaatar	6383	50
Ulaanbaatar	4850	50	Ulaanbaatar	7264	50
Sainshand	4865	12	Ulaanbaatar	11855	50

FM: 70.52/72.15MHz (Prgr 1).

HOME SCES. in Mongolian:
Prgr. 1: 2200-1500. **Kazakh:** 1330-1500 on 4750kHz.
Prgr. 2: 0830-1050 on 5960/6383kHz.

EXTERNAL SERVICE

MEDIUMWAVE: Altai 990kHz 150kW.
SHORTWAVES (G.C: 107.00E/47.55N): 500/250/100/50kW.

kHz	kHz	kHz	kHz	kHz
7260	11790	11850	12015	13780

	Days	Area	kHz
Chinese			
0950-1020	1234567	As	12015, 11850, 990
1055-1125	1234567	As	12015, 11850, 990
1325-1355	123456	As	12015, 11850, 990
English			
0910-0940	1234567	Oce	12015, 11850
1200-1230	146	As	12015, 11850
1445-1515	123456	As	13780, 7260
1940-2010	1234567	Eu	11790, 11850
Japanese			
0520-0550	7	J	12015
1530-1600	24	J	12015
2155-2225	36	J	12015
2300-2330	7	J	12015
Mongolian			
0840-0910	1234567	As	12015
1020-1050	1234567	As	12015, 11850, 990
1125-1155	1234567	As	12015, 11850, 990
1255-1325	1234567	As	12015, 11850, 990
1400-1430	12456	As	12015, 11850, 990
Russian			
1200-1230	36	CIS	12015, 11850
1400-1430	36	CIS	12015, 11850
1240-1340	25	CIS	7260
1710-1810	25	CIS	11850, 11790

ANN: "Ulaanbaataraas yaridz baina".
V. by QSL-folder. Rec. acc. (B).

MYANMAR (Union of)

L.T: UTC + 6½h — **Pop:** 48m — **Radio sets:** 3.200.000 — **Pr.L:** Bamar, English — **E.C:** a/c 50, 230V — **ITU:** BRM.

MYANMA RADIO & TV DEPT. (Gov.)

ADDR: GPO Box 1432, Yangon 11181. **Te:** +95 (1) 31355. **Fax:** +95 (1) 30211. **Cable:** Infocast. **Telex:** 21360 BBSXZK BM.
MW: Yangon 576kHz 200kW.
L.P: DG: U. Kyi Lwin. Dir. (Admin): U Khin Maung Htay. Dir. (Broadc): U Ko Ko Htwe. CE: U Tin Wan. Dir (TV): U Phone Myint.
SW (G.C: 96.10E/16.52N): 50kW.

kHz	kHz	kHz	kHz
4725	5990	7185	9730

FM: 98/100/102/104MHz 0.1/0.3/0.5kW.
D.PRGR: 0030-0230 on MW/7185kHz+FM, 0330 (Sat/Sun 0230)-0730 on MW/9730kHz+FM, 0930-1600 on MW/5990kHz +FM. **English:** 0200-0230, 0700-0730, 1430-1600. **N:** 0200, 0700, 1445.
Nationalities Prgr on 4725kHz + FM: 0930-1330. (Languages used: Chin, Kachin, Kayah, Mon, Pow Kayin, Sagaw Kayin, Shan and Rakhine).
ANN: E: "This is R. Myanmar, Yangon". Nationalities Prgr: (Bamar).
INT-SIG: Myanma Orchestral Music — **V.** by letter if IRC is enclosed. Re. to Dir. Gen. Myanma, Radio & TV Dept.

DEFENCE FORCES BROADCASTING UNIT (Mil.)

ADDR: Taunggyi, Shan State.
STATION: 6570kHz (estimated power 1kW).
D.PRGR. in Bamar and minority langs: 0200-0330, 0530-0630, 1030-1330. — **V.** Re. must be in Bamar.

NEPAL

L.T: UTC + 5.45h — **Pop:** 20.100.000 — **Radio sets:** 600.000 — **Pr.L:** Nepali — **E.C:** a/c 50, 220V — **ITU:** NPL.

RADIO NEPAL (Semi-Gov, Comm.)

ADDR: Radio Broadc. Sce., P.O. Box 634, Singha Durbar, Kathmandu. **Te:** +977 (2) 23910. **Cable:** Radionepal. **Telex:** 2590 RNEPAL NP — **L.P:** GM: G.N. Ojha. CE: M.P. Adhikari.

Mediumwaves	kHz	kW	Mediumwaves	kHz	kW
Surkhet	576	100	Kathmandu	792	100
Phankuta	648	100	Dardeldhura	810	10
Pokhara	684	100	Dhalkebar	1143	10

SHORTWAVES: Khumaltar 7165/5005kHz (alt: 3230kHz) 100kW — **Standby tr's:** Jawalakhel 5kW.
D.PRGR: 0015-0515 & 0715-1015 (Sat 0015-1115) & 1145-1715. **N:** 0115, 0245, 0315, 0715, 0845, 1245, 1315, 1545. **N. in English:** 0215, 1415.
EXTERNAL SCE: English: 1115-1145. **N:** 1115.
ANN: Nepali: "Yo Radio Nepal Ho". Ext. Sce: "This is the V. of R. Nepal".
INT-SIG: Instruments used are conch shell, violin, piano and jal tarang — **V.** by QSL-card.

PAKISTAN

L.T: UTC+5h — **Pop:** 120.425.000 — **Radio sets:** 10m (est.) — **Pr.L:** Urdu — **E.C:** a/c 50, 230/400V — **ITU:** PAK.

PAKISTAN BROADCASTING CORPORATION (Gov.)

ADDR: Broadcasting House, Constitution Avenue, Islamabad. **Te:** +92 (51) 829022-39. **Fax:** +92 (51) 811861. **Cable:** DIGBROCAST. **Telex:** 5816 PBCNO PK.
L.P: DG: Abdul Khaliq Awan. Dir. Eng: Malik Ali Chudhry. Contr. (Ext): A.H. Fatmi. Head of Training & Overseas Liaison & ABU Liaison Officer: Nawab Ali Yousafzai. Controller, Frequency Management: Abdul Hafeez Alvi.
R = Region; N = No. We. Frontier Province & Northern tribal areas; F = Federal District of Islamabad; P = Punjab; S = Sind; B = Baluchistan

Mediumwaves	kHz	kW	R	H.of tr
Peshawar	540	300	N	0045-0405, 0600-1808
Khuzdar	567	300	B	1155-1808
Islamabad	585	1000	F	0045-0605, 0800-1900
Lahore-I	630	100	P	0045-0405 (Fri 0820), 0800 (Fri 1000)-1900
Karachi-II	639	10	S	1230-1705
Quetta-I	756	150	B	0045-0805, 1000-1808
Karachi-I	828	100	S	0045-0405, 0600-1900
Quetta-II	855	10	B	0200-0400, 0600-1808
Khairpur	927	100	S	0045-0605, 0800-1808
Hyderabad-I	1008	120	S	0045-0405, 0600-1808
Multan	1035	120	P	0045-0605, 0800-1808
Lahore-II	1080	50	P	1230-1705
Hyderabad-II	1098	10	S	1230-1705
Rawalpindi	1152	10	P	0045-0405, 0600-1808
Bahawalpur	1341	10	P	0850-1808
Peshawar-II	1377	10	N	1230-1705
Dera Ismail Khan	1404	10	N	1055-1808
Faisalabad	1476	10	P	1055-1808
Gilgit (r. 1510)	1512	10	N	1055-1808
Skardu	1557	10	N	1055-1808
Turbat	1584	0.25	B	1300-1808
Sibbi	1584	0.25	B	0755-1108
Abbottabad	1602	0.25	N	0755-1108

D.PRGR: as above. **N.in English** (National): 0300, 0500, 0800, 1100, 1300, 1600.

SHORTWAVES (HOME SERVICES)
Islamabad 10/100kW (G.C: 73.12E/33.27N); Peshawar 10kW (G.C: 71.30E/34.00N); Quetta 10kW (G.C: 67.00E/30.15N).

Islamabad (100kW)
0045-0215 6070
0045-0245 6195
0800-1245 9515
1130-1337 9645
1303-1600 6070
1400-1805 5985

Peshawar (10kW)
1600-1900 7110

Quetta (10kW)
0045-0403 7170
0600-1215 7170
1230-1805 7150

Link Services
Islamabad (10kW)
0230-0245 9460
0700-0800 15190
0900-1000 15190
1230-1300 15190
1600-1630 9460
1700-1800 6070

EXTERNAL SERVICES
MEDIUMWAVES: Islamabad, 1260kHz 500kW — Quetta 756kHz 150kW.
SHORTWAVES: Islamabad 2x250kW & 4x100kW tr's: Karachi 2x50kW tr's.

kHz	kHz	kHz	kHz	kHz
6010	11715	15380	17705	17900
7215	13620	15445	17710	21475
7290	13665	15515	17725	21495
9370	15115	15516	17750	21520
9485	15190	15550	17785	21580
9515	15295	15555	17800	21595
11570	15305	17555	17802	21730

	Area	kHz
Arabic		
1700-1900	ME	13620, 9485
Bamar		
0700-0800	BRM	17785, 21475
Bengali		
0100-0200	SEAs	15190, 17800
1200-1400	SEAs	15295, 17802
Dari		
1600-1800	AFG	1260
English		
0230-0245	SAs	9515, 15115, 17705, 17725, 21730
1100-1120	Eu	17900, 21520
1600-1630	ME	11570, 13665, 15555, 17555
1600-1630	EAf	17725, 21495
Farsi		
0330-0530	IRN	15380, 17710
French		
1930-2030	Eu	11570
1930-2030	Af	15305
Gujrati		
0315-0415	EAf	17725
Hindi		
0030-0130	SAs	7290, 9370
1200-1300	SAs	13665, 15515
Indonesian		
0900-1000	SEAs	17785, 21475
Swahili		
0630-0730	EAf	17750
Tamil		
1030-1130	SAs	13665, 15515
Turki		
1430-1530	AFG	6010, 7215
Turkish		
1700-1800	TUR	13665, 15516
Urdu		
0100-0230	SEAs	17705, 21730
0430-0630	ME	15445, 17555, 21580
0800-1100	Eu	17900, 21520
1330-1530	ME	13665, 15555, 17555, 21595
1700-1900	Eu	11570, 15550
1800-1900	ME	11715

ANN: "This is Radio Pakistan" — **V.** by QSL-card. Re. to Contr. Freq. Management at 1st. addr. (Home Sces: Peshawar & Quetta by letter for direct rpts).
PUB: Ahang; Pakistan Calling (monthly).

AZAD KASHMIR RADIO (Gov.)

Addr: Muzaffarabad, Azad Kashmir, via Pakistan (high power tr's).
L.P: Dep.Controller (Eng): Syed Ahmed.
Stations: Muzaffarabad 792kHz 150kW — 3660kHz 1kW; Trarkhal: 4790/7265kHz 100kW.
D.PRGR: 0045-0405 & 1057-1808 on 792/3660kHz, 0045-0405 & 1430-1808 on 4790kHz, 0930-1415 on 7265kHz.
ANN: 792/3660kHz: "Yeh Azad Kashmir Radio Muzaffarabad Hay". Other freqs: "Yeh Azad Kashmir Radio Trarkhal Hay".
INT. SIG: Azad Kashmir anthem at open and close.
V. by letter. Rp.

PHILIPPINES (Rep. of the)

L.T: UTC + 8h — **Pop:** 67.140.000 — **Radio sets:** 4m (est.) — **Pr.L:** Philippine dialects, Cebuano Ilocano, Bicolano, English — **EC:** a/c 60, 220/110V (no) — **ITU:** PHL.

General Notes: Station identification is made using English alphabet. Generally, station identifications are on the hour and half hour, and consist of callsign, frequency, and station location, and usually, system or network name. Frequency is expressed in Spanish language, English language, or Tagalog (Pilipino) language numerals. When known, individual station addresses should be used in preference to system or network HQ, or KBP repr. Following cities and towns comprise Metropolitan Manila (Metro-Manila), a specially created province; Makati, Malabon, Mandaluyong, Manila, Navotas, Pasay City, Pasig, Quezon City, San Juan, Valenzuela, and Caloocan City.

NB: Cities may be referred to by their simple name, e.g. Baguio City/Baguio, Naga City/Naga, Zamboanga City/Zamboanga. However, Quezon City is **always** referred to by its full name.

Callsign assignments: DU = Shortwave only; DW = Luzon; DX = Mindanao and Sulu; DY = Visayas and Palawan; DZ = Luzon.

NATIONAL TELECOMMUNICATIONS COMMISSION

(Department of Transport and Communications), 865 Vibal Bldg, Esda Corner Times Str, Quezon City.
L.P: Commissioner: Josefina Lichauco. Dep. Commissioners: Aloysius R. Santos, Florentino L. Ampil. Chief Broadcast Sce. Dept: Carlos D. Salivan, Jr.

KAPISANAN NG MGA BRODKASTER SA PILIPINAS (KBP) (Association of Broadcasters in the Philippines).

ADDR: 118 Perea Str, Legazpi Village, Makati, Metro-Manila. **Te:** +63 815-1992.
L.P; Pres: Andre S. Kahn. Chmn: Felipe G. Medina, Jr. Chmn. Metro-Manila Chapter: Mario Garcia.
Member st's: all st's exc. 1), 11), 32), 47), 87), 90), 97x), 99), 100), 102), 103), 105), 109), 111), 113), 114), 115), 117).

PHILIPPINE FEDERATION OF CATHOLIC BROADCASTERS

ADDR: 2307 Pedro Gil Str, Santa Ana, (P.O. Box 3169), Manila 2802. **Te:** +63 58-4828. **Cable:** Reuter Jesuitas Manila.
L.P: Pres: His Excellency Most Rev. Jesus Y. Varela. Chmn: Fr. James Reuter S.J. (14 member st's).

MEDIUMWAVES: For locations, cf. MW Frequency List.
H. of tr: mostly 2100-1600. Variations, where known, are given in the Address section.

	Call	kHz	kW
33)	DZBR	531	1
45)	DZWT	540	10
62)	DYRB	540	10
63)	DZXL	558	10
87)	DWRP	558	10
87)	DZAG	567	2.5
47)	DZYZ	576	1
37)	DYIM	576	10
87)	DYMR	576	50
87)	DYCI	585	10
15)	DZVV	585	5
12)	DXCP	585	2.5
28)	DZBB	594	10
63)	DXPR	603	5
7)	DWLV	603	10
63)	DWHB	612	1
63)	DYHP	612	10
64)	DZTG	612	5
63)	DXDC	621	10
87)	DZFT	621	1
91)	DZMM	630	50
10)	DYAG	639	5
64)	DZRL	639	1
63)	DXKR	639	5
80)	DYMM	648	2.5
114)	DXMB	648	5
103)	DWPS	648	5
84)	DZXC	657	1
12)	DXDD	657	5
57)	DWRN	657	5
111)	DYFL	657	1
39)	DZRH	666	50
87)	DXRP	666	10
32)	DWLW	675	10
64)	DYKC	675	5
75)	DXTP	675	1
32)	DWGW	684	5
41)	DWAU	684	1
25)	DZCV	684	5
64)	DXDX	693	1
18)	DXDN	693	1
63)	DXBC	693	5
24)	DZAS	702	40
47)	DXRD	711	5
21x)	DZYI	711	5
47)	DZYI	711	5
63)	DXIC	711	5
49)	DZSO	720	10
5)	DZJO	720	5
39)	DYEZ	720	5
87)	DZPE	729	10
87)	DWPE	729	10
54)	DZGB	729	2.5
53)	DXOR	729	5
87)	DZSR	738	10
63)	DYHB	747	5
50)	DXND	747	5
39)	DZJC	747	10
87)	DZGC	756	10
32)	DWNW	756	10
61)	DXJM	756	1
6)	DWHL	756	1
47)	DYCB	765	5
47)	DZYT	765	5
62)	DXGS	765	1
87)	DXSO	774	1
87)	DXSO	774	1
40)	DWOO	774	10
39)	DZNL	783	5
52)	DYPR	783	5
42)	DYME	783	5
60)	DXRA	783	10
51)	DYRR	792	5
100)	DZAP	792	1
30)	DWFA	801	1
12)	DYKA	801	5
110)	DYWC	801	5
47)	DXBL	801	1
49)	DZNC	801	5
82)	DXMZ	801	1
71)	DZRJ	810	10
82)	DXUM	819	5
39)	DYVL	819	10
82)	DXMC	819	5
73)	DXSC	819	1
47)	DWRI	819	5
30)	DWZR	828	5
69)	DZTC	828	1
63)	DXCC	828	10
23)	DZXE	837	1
47)	DXRE	837	5
15)	DYFM	837	5
58)	DZNN	846	50
25)	DZGE	855	10
26)	DXLA	855	5
37)	DZPE	855	10
2)	DXCT	855	1
92)	DXWG	855	5
47)	DZSP	864	1
18)	DWSF	864	1
109)	DYRE	864	5
96)	DZPA	873	5
25)	DZRC	873	5
47)	DXRB	873	5
47)	DXRT	873	1
39)	DWIZ	882	20
50)	DXMS	882	10
87)	DYJR	882	10
87)	DXRG	882	1
43)	DYSR	891	10
82)	DXRZ	900	5
82)	DXDN	900	1
72)	DXIP	900	10
86)	DYLA	909	5
87)	DZRB	918	50
75)	DXMM	927	5
99)	DZVS	927	10
61)	DZXT	936	1
64)	DYKW	936	5
63)	DYCC	936	1
87)	DXIM	936	1
1)	DYRO	945	5
47)	DXRO	945	5
14)	DZEM	954	10
107)	DZAL	954	10
47)	DXRI	954	1
39)	DYRC	963	10
47)	DXYZ	963	5
82)	DXMO	972	5
39)	DYSM	972	1
47)	DZRD	981	5
32)	DYBQ	981	10
69)	DXOW	981	10
63)	DXDR	981	5
28)	DWRT	990	15
63)	DXHP	999	1
28)	DYSS	999	5
87)	DZEQ	999	1
64)	DXXX	1008	25/10
32)	DWDW	1017	10
1)	DYRP	1017	5
65)	DXGE	1017	5
9)	DXSN	1017	5
87)	DZLC	1017	1
30)	DWXX	1026	10
82)	DXMI	1026	1
47)	DZAM	1026	10
15)	DZWX	1035	5
49)	DZDR	1044	10
69)	DXCO	1044	5
112)	DXLL	1044	5
39)	DYMS	1044	1
19)	DXML	1044	1
49)	DZVR	1053	10
83)	DYSA	1053	1
64)	DXKD	1053	1
21)	DZEL	1053	5
21)	DZEC	1062	20
24)	DXKI	1062	5
81)	DYXT	1071	5
64)	DXKT	1071	5
101)	DWAM	1080	2.5
21)	DWIN	1080	5
32)	DYSJ	1080	1
64)	DXKS	1080	1
62)	DWRL	1080	1
106)	DWCL	1080	1
30)	DYHR	1089	1
82)	DXCM	1098	5
17)	DWAD	1098	10
47)	DXCL	1098	5
55)	DXMF	1107	10
63)	DYRI	1107	5
24)	DXAS	1116	5
67)	DZLB	1116	5
76)	DYTR	1116	1
24)	DWAS	1125	5
15)	DZWN	1125	10
82)	DXMV	1134	1
57)	DYRM	1134	1
94)	DWDD	1134	10
54)	DZMD	1161	2.5
39)	DWCM	1161	5
8)	DYRD	1161	1
82)	DXDS	1161	1
87)	DZRP	1170	10
87)	DXMR	1170	10
115)	DXMT	1179	10
59)	DYRV	1188	1
61)	DZLT	1188	5
84)	DZXO	1188	5
24)	DXFE	1197	5
1)	DYRH	1197	5
32)	DWAN	1206	10
63)	DXRS	1206	1
97)	DYRF	1215	5
21)	DXED	1224	10
87)	DZMQ	1224	10
87)	DXSM	1224	1
24)	DYVS	1233	5
22)	DWBL	1242	20
77)	DXSY	1242	5
113)	DXZB	1242	5
32)	DYRG	1250	1
47)	DWSI	1251	5
54)	DZMS	1251	2.5
36)	DWMC	1260	5
55)	DYMF	1260	10
54)	DYWB	1269	5
49)	DZVX	1269	5
87)	DZRM	1278	10
32)	DYJJ	1286	1
108)	DZRS	1287	1
117)	DWPR	1296	5
117)	DWLQ	1296	5
21)	DYFX	1305	10
18)	DWXI	1314	10
13)	DXRH	1314	10
3)	DYXX	1323	5
87)	DZRK	1323	1
78)	DYOO	1323	5
64)	DZKI	1332	1
85)	DWAY	1332	5
47)	DWAY	1332	5
57)	DZYS	1341	5
69)	DYRL	1341	5
35)	DZXQ	1350	10
87)	DZER	1350	5
57)	DZYR	1359	5
87)	DYSL	1359	1
87)	DYMR	1359	5
87)	DZRA	1368	1
64)	DZBS	1368	1
64)	DXKO	1368	1
64)	DXKP	1377	10
63)	DYVR	1377	2.5
46)	DXCR	1386	5
84)	DWMG	1395	1
89)	DZOR	1404	2
64)	DYKB	1404	5
97x)	DYDW	1413	5
79)	DWBC	1422	10
20)	DXMU	1422	5
70)	DYRS	1431	1
31)	DWCM	1440	5
104)	DZOM	1440	5
116)	DXRM	1449	5
24)	DWRF	1458	10
69)	DZYA	1476	1
102)	DWRB	1476	1
63)	DYKR	1485	1
62)	DXOC	1494	1
98)	DWLR	1494	10
105)	DYCR	1512	5
11)	DZME	1530	10
57)	DZYM	1539	1
74)	DWGO	1557	2.5
66)	DZHH	1566	10
68)	DZUP	1592	1

Addresses and other information

1) Allied Broadc. Center Inc, 17th Flr, Unit C, Strata 200 Bldg, Emerald Ave, Ortigas Complex, Pasig, Metro Manila — G.M: Atty. E.V. Rosales.

2) Audiovisual Communicators Inc, 17th Floor, Strata 200, Emerald Ave, Ortigas Complex, Pasig, Metro Manila — G.M: A.V. Barreiro Jr.

3) Asian Pacific Broadc. Co. Inc, 5th Flr, Sagittarius Condominium, Officer H. dela Costa Str, Salcedo Village, Makati, Metro Manila — G.M: R.H. Hulog. 2100-1600.

4) Banwag Broadc. System, Rms 402/403, State Condominium I, Salcedo Str, Legaspi Village, Makati, Metro Manila — G.M: Mike Enríquez.

5) Bayanihan Broadc. Corp, 120 J. Arellano Str, San Juan, Metro Manila — Pres. & G.M: Fr. F. Lucas SVD. 2100-1500.

6) Beta Broadc. System Inc, Rm 211, Margarita Bldg, J.P. Rizal, Makati, Metro Manila — G.M: Ms. L. Dela Cruz.

7) Bicol Broadc. Systems Inc, MSI Bldg, 5233 Farenheit Str, Palanan, Makati, Metro Manila — Pres: L. Villafuerte. 2000-1500.

8) Bohol Chronicle Radio, c/o Suite 301, Legaspi Towers 200, Paseo de Roxas, Legaspi Village, Makati, Metro Manila — G.M: Atty. Z. Dejaresco Jr.

9) Silangan Broadc. Corp., 55 Magallanes Str, Surigoo City 8400 — Gen Mgr: J.J. Mendoza Jr. 2100-1400.

10) Cadiz Radio & TV Network Inc, c/o Radio St. DYAG, Cadiz City — G.M: Loel Katalbas.

11) Capitol Broadc. Center, 317 Roosevelt Ave, SFDM, Quezon City — St. Mgr: Joey Luison Jr.
12) Catholic Welfare Organization, 2307 Pedro Gil Str, St. Ana, Metro Manila — G.M: Sr. S. Manapol, SPC.
14) Christian Broadc. Sce, Maligaya Bldg. 2, 887 EDSA, Quezon City — St. Mgr: Nel C. Santiago. Chief Eng: A.H. Ceralde. W 2100-1500, Sun 2200-1400.
15) Consolidated Broadc. System Inc, Ground Flr, Florete Bldg, 2406 Nobel-Edison Str, Makati, Metro Manila — Chief Exec: R.M. Florete.
16) Cotabato Television Corp, ROF Enterprises, Regional Complex, Cotabato City — G.M: Ramon Flores.
17) Crusaders Broadc. System Inc, 209 E. de la Paz Str, Mandaluyong, Metro Manila 3119 — G.M: Atty. F. Borja.
18) Delta Broadc. System, DWXI, Sebastian Rd, Ninoy Aquino Ave, Bo. Sto. Nino, Paranaque, Metro Manila — G.M: G.Z. Velarde. St. Mgr: O.R. Talain.
19) Digos Broadc. System, Associated Labor Unions, Elliptical Rd cor. Maharlika Str, Diliman, Quezon City — S.M: J. Cabatingan.
20) Communication Unit, University Extension, Central Mindanao University, Musuan, Bukidnon — St. Mgr: Prof. Liberty Uego-Josue. Chief Eng: Rainier A. Escarlos. 2100-0100 & 0900-1200.
21) Eagle Broadc. Corp, Maligaya Bldg, No. 887 E. delos Santos Ave, Quezon City — G.M: Eva Santos Vasquez.
21x) East Visayas Bctg, RYGE Specialist, 3rd Flr, Universal-RE Bldg, 106 Paseo de Roxas, Makati, MM — St. Mgr: Florencio Villamor.
22) FBS Radio Network Inc, 18th Flr, Philcomcen Bldg, Ortigas Ave, Pasig, Metro Manila — Pres: Mrs. L. Vera. 24h.
23) Fairwaves Broadc. Network, Rm 601, Allied Bank Bldg, Q. Paredes Str, Binondo, Metro Manila — G.M: Atty. R.S. Encarnacion.
24) **Far East Broadc. Co.** (see end of section).
25) Filipinas Broadc. Network, Rm. 306, Legaspi Towers 200, Paseo de Roxas, Makati, Metro Manila — G.M: Mrs. D. C. Gozum.
26) First United Broadc. Corp, Lorenzo Bldg, 787 Lorenzo Bldg, Vito Cruz Str, Metro Manila — Pres: Maria Clara L. Lobregat. 2100 - 1500.
27) Franciscan Broadc. Corp, 2117 Luna Str, Pasay City — G.M: B. Canoy.
28) GMA Radio-Television Arts, RBS Bldg, E. delos Santos Ave, Diliman, Quezon City — Pres: Menardo Jimenez.
28x) GV Broadc. System, 324-B Natividad Bldg. Cor. T. Pinpin & Escolta Str's, Sta. Cruz, Manila City — G.M: Engr. E. Galang.
29) GVM Radio/Television Corp, Suite 608, Pacific Bank Bldg, Ayala Ave, Makati, Metro Manila — Pres: E.A. Tuason.
29x) Good News Sorsogon Foundation Inc., Rizal Str, Sorsogon — St. Mgr: Rev. Fr. Gerard V. Deyeza. 2100-1500.
30) Hypersonic Broadc. Center, 341 3rd Flr, J&T Bldg, Magsaysay Blvd, Sta. Mesa, Metro Manila — Officer-in-Charge: Hernando E. Baldo.
31) Insular Broadc. System, 6th Flr, 105 Estaban Str, Legaspi Village, Makati, Metro Manila — Pres: E. del Rosario.
32) Intercontinental Broadc. Corp, Broadcast City, Capitol Hills, Quezon City — Chmn: M. Mirasol.
33) Kumintang Broadc. System, Suite 800, LPL Towers, 112 Legaspi Str, Legaspi Village, Makati, Metro Manila — G.M: Antonio Leviste.
35) Mabuhay Broadc. System Inc, 15th Flr, Strata 100, Emerald Cor. Ortigas Complex, Pasig, Metro Manila — Vice Pres: R.M. de Guzman.
36) Magiliw Community Broadc. Co. Inc, 74 F. Blumentritt Str, Mandaluyong, Metro Manila — G.M: R. Estrella Jr.
37) People's Television — Channel 4, Media Center, Bohol Ave, Quezon City — Op's Dir: R. Dalogdog.
38) Makati Broadc. Network Corp, 14th Flr, Insular Life Bldg, Ayala Ave, Makati, Metro Manila — Pres: A.S. Kahn. 2200-1800.
39) Manila Broadc. Co, 4th Flr, FJE Bldg, 105 Esteban Str, Legaspi Village, Makati, Metro Manila — G.M: Atty. E.L. Montilla.
40) Mareco Broadc. Network, 131 Del Monte Ave, Quezon City — G.M: Manuel Villar, Jr. SM: Eng. Terry G. Bondoc. 2100-1700.
41) Masagana Broadc. Corp, 3rd Flr, Admin. Bldg, Araullo Univ, Bitas, Cabanatuan City — Own: Rolan C. Esteban.
42) Masbate Community Broadc. Company Inc, 13 Madelaine Str, Parkway Village, SFDM, Quezon City — G.M: M.E. Manalastas.
43) c/o Intermedia Philippines, P.O. Box 1767, Manila — Gen. Mgr: Ms. A.C. Bangay. 2130-1230.
44) Molave Broadc. Network, Rm. 202, Centrum II Bldg, 150 Valero Str, Salcedo Village, Makati, Metro Manila — G.M: A.M. Carandang Jr.
45) Mountain Province Broadc. Corp, P.O. Box 156, Baguio City — Gen. Mgr: Fr. Herman Godyne. St. Mgr: D. Pineda. 2000-1500.
46) Mountain View College, c/o L. Cruz Compound, 29 Sta. Queteria, Caloocan City, Metro Manila — G.M: R.B. Abordo Jr.
47) Nation Broadc. Corp, Jacinta Bldg, 914 Pasay Rd, Makati, Metro Manila — G.M: A.L. Yabut. 2000-1500.
48) c/o st. 1 — S.M: E. Dejaresco. Chief Eng: Leo Gequinto: 2200-1600.
49) Newsounds Broadc. Network Inc, Ground Flr, Florete Bldg, 2406 Nobel cor. Edison Str, Makati, Metro Manila — G.M: E. Billones.
50) Notre Dame Broadc. Corp, Rm. 302, Gemini Bldg, Sep. G.J. Puyat, Makati, Metro Manila — G.M: Fr. V.P. Quioque, OMI.
51) Ormoc Broadc. Co. Inc, Rms 402/403, State Condominium I, Salcedo Str, Legaspi Village, Makati, Metro Manila — G.M: C.S. Canoy.
52) Palawan Broadc. Corp, No. 6, Rd 1, Highway Hills, Mandaluyong, Metro Manila — Exec. Vice Pres: Ms. L.I. Ilustre.
53) Pedro N. Roa Broadc. System, 4th Flr, National Life Insurance Bldg, Ayala Ave, Makati, Metro Manila — G.M: R.N. Cui.
54) People's Broadc. Network, 64-C Cordillera Str, Quezon City — G.M: J.D. Bayona.
55) People's Broadc. Sce. Inc, Ground Flr, Florete Bldg, 2406 Edison cor. Nobel Str, Makati, Metro Manila — Adm. Officer: G.J. Lucero.
56) Philippine Broadc. Corp, 3rd Flr, FJE Bldg, Esteban Str, Legaspi Village, Makati, Metro Manila — Pres: E.D. Buyco.
57) Philippine Radio Corp, 10th Flr, National Life Insurance Bldg, Ayala Ave, Makati, Metro Manila — G.M: Jose E. Escaner, Jr. Chief Eng: Floriano M. Gloria. 2100-1400.
58) **R. Veritas**—Global Broadc. System. The Penthouse, Century Tower Bldg. Tordesillas Str, cor. dela costa Str, Makati, 1200 Metro Manila — **FM:** DWCS-103.5MHz.
59) RCPI Broadc. Network, RCPI Bldg, E. delos Santos Ave, Quezon City — G.M: O.Q. Pajar.
60) RMC Broadc. Co. Inc, RMC Bldg, Pichon Str, 8000 Davao City — G.M: Dr. Evelyn A. Magno. 1900-1400.
61) Radio Corp. of the Philippines, 10th Flr, National Life Bldg, Ayala Ave, Makati, Metro Manila — Head Tec.Op's: J. Escaner. 2100-1500.
62) Radio Inc, 10th Flr, National Life Bldg, Ayala Ave, Makati, Metro Manila — G.M: Jose E. Escaner, Jr. Chief Eng: Floriano M. Gloria. 24h.
63) R. Mindanao Network Inc, Suite 402/403, State Condominium I, Salcedo Str, Legaspi Village, Makati, Metro Manila — G.M: H.R. Canoy.
64) R. Philippines Network Inc, Broadcast City, Capitol Hills, Old Balara, Quezon City — GM: Felipe G. Medina, Jr.
65) R. St. DXGE, Radio Spectrum Inc, Rm. 601, Campos Rueda Bldg, Tindao Str, Makati, Metro Manila — G.M: Ms. E. Cruz-Rafols.
66) R. St. DZHH (non-comm.), Philippine Air Force, Villamor Air Base, Pasay City —S.M: Maj. V.A. Baes.
67) R. St. DZLB (non-comm.), UP-Los Banos, College Laguna 4031 — S.M: G.L. Paje.
68) R. St. DZUP (non-comm.), Information Office, Quezon Hall, UP, Diliman, Quezon City — G.M: Ms. C.L. Lazaro. 1100-1300.
69) R. Pilipino Corp, 12th Flr, National Life Insurance Bldg, Ayala Ave, Makati, Metro Manila — G.M: J. Escaner Jr.
70) Ragde, Vicente & Sons, Suite 402/403, State Condominium I, Salcedo Str, Legaspi Village, Makati, Metro Manila — G.M: Ret. Gen. H.M. Isleta.
71) Rajah Broadc. Network Inc, Hastings II Bldg, Gen. Luna Cor. Makati Ave, Makati, Metro Manila — Gen. Mgr: A. Austría.
72) Southern Broadc. Network, 3881 E. Vallejo Str. Santol, Sta. Mesa, Metro Manila — G.M: Mrs. C.B. Pacquing.
73) Southern Philippines Mass Communications Center, HQ, Civil Relations Office, Armed Forces of the Philippines — S.M: Col. Buenaventura Ramos.
74) Subic Broadc. Corp, No. 1 National Hwy, Lower Kalaklan, Olongapo City 2200 — Pres. & G.M: J. Gordon Jr. 1930-1400. **F.Pl:** 5kW.
75) Sulu Tawi-Tawi Broadc. Corp. Inc., Suite 4, 3rd Flr, Celta Bldg, Carino Cor. South Superhighway, Makati, Metro Manila — S.M: Sra. Ma. V. Adre, OND.
76) Tagbilaran Broadc. Corp, 301 Cityland III Condominium, Cor. Esteban & Makati Ave, Makati, MM — G.M: N.C. Abacajan.
77) Times Broadc. Corp, 5th Flr, Union Bank Bldg. II, Pasay Rd, Makati, Metro Manila — G.M: Alex V. Sy.

78) Trans-Radio Broadc. Corp, Rm. 608, Pacific Banking Corp. Bldg, Ayala Ave, Makati, Metro Manila — Pres: E. Tuason. 2000-1600 on 594kHz, 2100-1600 on other freqs.
79) United Broadc. Network, 11th Flr, FEMS-Tower I, Zobel Roxas cor. South Superhighway, Metro Manila — G.M: J. Hodreal.
81) DYXT Bldg, A. Luna Str, City of Tagbilaran 6300, Bohol — G.M: R.T. Gatal (St. belongs to R. Mindanao Netw.).
82) Univ. of Mindanao Broadc. Network, Torres Arcade, 1002-J, Pasay Rd. cor. Makati Ave, Makati, Metro Manila — Pres: G. Torres. 2100-1600.
83) 2nd Flr, Univ. of San Agustin Jalandoni Str., Iloilo City 5901 — Gen. Mgr: Miss Marivic Z. Hermano. Chief Eng: Elizer H. Arsenal. 2100-1600.
84) Vanguard R. Network, 208 J & T Bldg, Ramon Magsaysay Blvd, Sta. Mesa, Metro Manila — G.M: M.G. Galvez Jr.
85) Victory Broadc. System Inc, 3rd Flr, Jacinta Bldg, 914 Pasay Rd, Makati, Metro Manila — G.M: J. Yabut.
86) Vimcontu Broadc. System, Broadcast Production & Training Center, ALU-Vimcontu Welfare Center, Port Area, Cebu City — S.M: C.M. Remonde.
87) Bureau of Broadc. Sces (Phill. Broadc. Sce.), PTV. 4 Complex, Sgt. Esquerra Ave, Quezon City — Dir: Francisco M. Batacan Jr; Chief Eng: Jose Q. Borromeo. 24h on 918kHz, 2100-1500 on other freqs.
88) Westwind Broadc. Corp, Cuo-jal Bldg, North Drive, Bacolod City — Pres: L. Ledesma.
89) Zambales Broadc. & Development Corp, DZOR Bldg, 1683 Rizal Avenue, WT — Pres: Dr. H.S. Ruiz.
90) Zamboanga Television Corp, 3616 Bautista Str. cor. Finlandia, Makati, Metro Manila — G.M: P. Rama Jr.
91) ABS-CBN Broadc. Network Inc, Chronicle Bldg, Ortigas Ave, Pasig, Metro Manila — Mgr: L. Balquiedra.
92) Banahaw Broadc. Corp, Broadcast City, Capitol Hills, Quezon City — Repr: M. Mirasol.
93) G.V. Broadc. System, Angeles City, Pampanga — G.M: M. Galang.
94) Ministry of National Defense, Armed Forces of the Philippines, Camp Aguinaldo, Q.C. — S.M: Orly Punzalan.
95) Bo Mayo, Laoag City, Luzon.
96) Abra Community Broad. Corp, Bl. Arnold Janssen Communications Center, Zamora Str, Cor. Rizal Str, Bangued, Abra 2800 — Gen. Mgr: A. Bello.
97) Filipinas Broadc. Assoc. Inc, Univ. of S. Carlos, Pelaez Str, Cebu City — Gen. Mgr: Fr. Mar Alingasa, SVD. 2100-1400.
97x) Filipinas Broadc. Assoc, Baruyan, San Jose, Tacloban City — Gen. Mgr: Fr. Mar Alingasa, SVD. 2200-1300.
98) Ultrasonic Broadc. System, 605-607 Doña Consolacion Bldg, Salcedo Village, Makati, MM.
99) Solid North Broadcasting, Vigan, Ilocos Sur.
100) Angeles Broadc. Network, Angeles City, Pampanga.
101) Radio Bayaniham System, Batangas City, 4200.
102) Ribbon Broadc. Netw, 5th Flr, LCC Bldg, Lipa City.
103) Edelwina F. Pena, Puerto Princesa, Palawan.
104) OMARCO (Ben Viduya), Calapan, Oriental Mindoro.
105) Romblon Broadc. Netw, Odiongan, Romblon.
107) Rinconada Broadc. Corp, UNEP Compound, San Roque, Iriga City.
108) R. Sorsugon Netw, Don Luis Lee Bldg, Plaza Bonifacio, Sorsogon.
109) San Miguel Broadc, Tipdo, Mandawe City.
110) Franciscan Broadc. Corp, Cathedral Compound, Dumaguete — GM: Merlin T. Logronio, P.C. 2100-1630.
111) Eastern Broadc. System, 67 Dr. Alejos Dr, Quezon City.
112) R.T. Broadc. Specialists Philippines, Rm. 9, Poblete Bldg, 17 Sen. Gil J. Puyat Ave, Makati, MM — Pres. & GM: R.L. Tan.
113) DXZB/TV-13 Cooperative for Sce, Campaner Str, Zamboanga City.
114) Ruta Broadc. System, C.M. Recto Str, Malaybalay, Bukidnon.
115) Islamic Dawah Broadc. Netw, Marawi City.
116) Mindanao Broadc. Co. Inc, 7th Flr, Topman Center Bldg, Ayala Ave, Makati, MM.
117) Beacon Communication Systems Inc.
FM: Approx. 100 FM-st's are operating.

EXTERNAL SERVICE

RADYO PILIPINAS

ADDR: Philippine Broadc. Sce, PTV-4 Compound. Sgt. E.A. Esquerra Avenue. Diliman, Quezon City, Metro Manila 1103. **Te:** +63 (2) 969612. **Fax:** +63 (2) 969672.
L.P: MD: Francisco M. Batacan, Jr. PD: Ben Tabisaura. CE: Simeon Salazar, Jr. Head of PR: Lolita Lizano.

STATION: Tinang (see Voice of America Philippines).

English	**kHz**
0230-0330	17760, 17840, 21580
Filipino	**kHz**
1730-1930	15190, 17840, 21455

V. by QSL-card. Rp. (2 IRC's). Rec. acc.

RADIO VERITAS ASIA
Philippine Radio Education & Information Center

ADDR: P.O. Box 939, Manila. **Te:** +63 (2) 900012. **Fax:** +63 (2) 907436. **Cable:** 64420 VERTAS PN. **Tlx:** 632-900014.
LP: Chmn: H.E.J. Cardinal Sin. Mgr: Erlinda G. So. Head of Prgrs: Msgr. Pietro Yan Tai. Tech. Consultant: Fr. H. Delbaere. Prgr. Consultant: Fr. J. Desautels. Tech. Dir: F.L. Kiguchi. Freq. Planning: R.S. Alonzo. Head of PR: Cleote Labindao.
Transmitters: Palauig, Zambales (G.C: 119.90E/15.46N): 2 × 250kW, 2 × 50kW SW-tr's.

kHz	kHz	kHz	kHz	kHz	kHz
7160	9560	11710	11915	15215	15405
7240	9615	11740	11950	15225	15440
9520	9710	11790	15110	15245	17800
9555	11705	11820	15140	15300	21560

Schedule until May 1 1993
Freqs in **bold** are 250kW

Bengali	**kHz**
0100-0130	**15405,** 11820
1400-1430	**9555,** 11740
Hindi	
0130-0155	**15405,** 11820
1330-1400	**9555,** 11740
Indonesian	
2300-2325	**11710,** 15225
1200-1225	**9520,** 15440, 17800
Kachin	
0030-0055	**11790,** 9560
1230-1255	**9555,** 11915
Karen	
0000-0030	**11790,** 9560
1200-1230	**9555,** 11915
Mandarin	
2100-2255	**11790,** 7160
2230-2255	**9560,** 7240
1000-1055	**9520,** 15440
1130-1200	**9520,** 15440
1500-1555	**9615,** 11950
Myan-mar	
2300-2400	**11790,** 9560
1130-1200	**9555,** 11915
Pilipino	
1505-1545	**15140,** 17800
Russian	
2130-2225	**7265,** 9710
1030-1125	**15110,** 17800
Sinhala	
0000-0030	**15300,** 11820
1330-1400	**9520,** 11705
Tamil	
0030-0100	**15300,** 11820
1400-1430	**9520,** 11705
Telugu	
0100-0125	**15300,** 11820
1430-1500	**9520,** 11705
Ukrainian	
2300-2325	**15215,** 11710
1230-1255	**9520,** 15440
Urdu	
0200-0225	**21560,** 15405
1430-1455	**9555,** 11740
Vietnamese	
2300-2355	**11710,** 9710
0130-0225	**15245,** 11790
1000-1025	**9555,** 11790
1100-1125	**9555,** 11790
1300-1325	**9520, 11790**

ANN: "This is R. Veritas Asia, Quezon City, Philippines".
IS: "O Via, Vita Veritas" — **V.** by QSL-card.

FAR EAST BROADCASTING CO. (Rlg.)

ADDR: Box 1, Valenzuela, Metro-Manila 1405. **Te:** +63 (2) 361 1010. **Fax:** +63 (2) 359490. **Cable:** FEBCOM MANILA. **Telex:** 40048 FEBCOM PM.
L.P: MD: Roy J.B. Alvarez (from May 1993 Efren M. Pallorina). Head of Prgrs: Peter McIntyre. CE: Romualdo M. Lintag. Head of PR: Priscilla R. Calica.
PHILIPPINE SCE. in English and Tagalog: On DZAS 702kHz 2130-1600; on 3345kHz: 2100-2320 (Sat 2200-2330), 0800-1100. on 6030kHz; 2200-2315.
FM: Manila 98.7MHz 60kW: 2200-1600; Cebu 98.7MHz 9kW: 2100-0500, 0900-1600.
AM Sts: DXKI 1062kHz (Marbel, So. Catabato), DXAS 1116kHz (Zamboanga), DWAS 1125kHz (Daraga, Albay), DXFE 1197kHz (Davao City), DYVS 1233kHz (Bacolod, C.), DWRF 1458 kHz (Iba, Zambales), (Baco, Mindoro).
D.Prgr: on 702kHz: 2130-1600; on 1062kHz: 2100-0400, 0900-1400; on 1116kHz: 2100-0600, 0900-1400; on 1125/1233kHz: 2200-0500, 0900-1300; on 1197kHz: 2100-0500, 0900-1300; on 1458kHz: 2200-0500, 0900-1100.

OVERSEAS SERVICE (G.C. of the SW-trs: Bocaue, Bulacan 120.55E/14.48N; Iba, Zambales 119.57E/15.22N).

kHz	kHz	kHz	kHz	kHz	kHz
3345	9660	9820	9875	11995	15100
6030	9780	9845	11650	12020	15465
9495	9815	9850	11690	15095	15515

Acehnese	Area	kHz
1045-1115	MLA	15100
Akha		
2300-2315 (Sat/Sun)	LAO/THA	11650
Amoy		
1000-1045	CHN	9820
Ao Naga		
0145-0200 (Sat/Sun)	IND	15465
Burmese*		
1145-1430	BRM	15100
1330-1345	BRM	9660
2330-0130	BRM	15465
*) incl. tribal langs.		
Cambodian		
2230-2315	CBG	9495
1200-1300	CBG	11690
Cantonese		
1130-1330	CHN	9820
English		
0000-0230	IND/SEAs	15450
0900-1100	CHN/PNG	11690
1300-1600	IND/SEAs	11995
Hmong		
2300-2330	LAO/THA	9875
1100-1130	LAO/THA	11690
Indonesian		
2230-2330	INS	9780
0900-1000	INS	11995
1015-1045	INS	15100
Javanese		
0830-0900	INS	11995
Karbi		
0200-0215 (Sun)	IND	15465
Khmu		
2215-2230 (Sat/Sun)	LAO	9875
Kuki		
0130-0145 (Sat/Sun)	IND	15465
Lahu		
2315-2330	BRM	11650
Lao		
2230-2300	LAO	9875
1130-1200	LAO	11690
Mandarin		
0300-0500	CHN	15095
0500-0900	CHN	12020
0900-1600	CHN	9850
1400-1600	CHN	9820
2200-2400	CHN	9815
Meitei		
0130-0145	IND	15465
Minganese		
1000-1030	INS	11995
Russian/Ukrainian		
1600-1900	CIS	9845
Sasak		
1030-1100	INS	11995
Shan		
2345-0030	BRM	15515
Swatow		
1045-1130	CHN	9820
Vietnamese*		
1115-1145	VTN	15100
1115-1330	VTN	9660
*) incl. tribal languages		
Wa		
2330-2400	BRM	11650
Zhuang		
1330-1400	CHN	9820
Zoukam		
0200-0215 (Sat)	IND	15465

ANN: "Broadcasting from the Philippines, this is FEBC Radio International, The Sound Alternative".
INT-SIG: "We have heard the Joyful Sound".
V. by QSL-card (B). Rp. preferred (3 IRCs for Airmail).
PUB: Prgr. schedule free on request. Signal (quarterly).

FAR EAST NETWORK (AFRTS)

ADDR. & L.P: See Far East Network Japan.
STATIONS: Subic Bay 1251kHz 0.25kW, FM 95.1MHz 0.6kW.
D.PRGR: 24h.

VOICE OF AMERICA RADIO STATION PHILIPPINES

MEDIUMWAVES: Poro 1143kHz 1000kW.
SHORTWAVE: Poro (G.C: 120.17E/16.37N): 2×100kW, 3×50kW, 1×35kW SW-tr's — Tinang (G.C: 120.37E/15.21N): 12×250kW, 3×50kW tr's.

kHz	kHz	kHz	kHz	kHz	kHz	kHz
5955	9570	11730	11965	15190	15330	17820
6030	9575	11740	11995	15195	15395	17840
6110	9590	11750	12040	15205	15410	17865
6130	9620	11760	15125	15210	15425	17875
7110	9645	11805	15145	15215	17720	21455
7120	9670	11820	15150	15230	17735	21475
7200	9755	11835	15155	15250	17740	21550
7260	9760	11840	15160	15285	17760	21570
9505	9770	11870	15165	15290	17765	21580
9510	9890	11920	15180	15305	17780	21585
9525	11715	11925	15185	15320	17790	
9545	11720	11930				

D.PRGR: See USA.

SINGAPORE

L.T: UTC+8h — **Pop:** 2.792.000 — **Radio sets:** 800.000 — **Pr.L:** Malay, Mandarin, Tamil, English — **E.C:** a/c 50, 230V — **ITU:** SNG.

SINGAPORE BROADC. CORP. (Statutory Board)

ADDR: Admin: Farrer Rd, P.O. Box 60, Singapore 9128. **English Sce:** Farrer Rd, P.O. Box 42, Singapore 9128. **Malay Sce:** Farrer Rd, P.O. Box 71, Singapore 9128. **Te:** +65 2560401. **Fax:** +65 2538808. **Cable:** Broadcast Singapore. **Telex:** RS 39265 SBCGEN.
L.P: Chmn: Dr. Cheong Choong Kong. Dep. Chmn: Goh Kim Leong. GM: Moses Lee. Asst. GM (Admin): Param Ajeet Singh Bal. Asst. GM (Eng. Sces): Lim Eng Hai. Asst. GM (Ent. Productions): Lee Cheok Yew. Dir. Current Affairs: S. Chandra Mohan. Dir. N: Anthony Chia. Dir. Drama: Kung Lung. Dir. Prod. Sces: Andrew Cheng. Dir. Radio: Mdm. Chua Foo Yong. Dir. Eng: Lam Lychow. Dir. Acquisition & Programming: Sandra Buenaventura. Dir. Public Affairs: Ananda Perera.

Mediumwaves

kHz	kW	Sce.	kHz	kW	Sce.
630	20	Radio One	990	20	Warna
675	20	City Sounds	1368	20	Olikkalanjiam

Shortwaves

kHz	kW	Sce.	kHz	kW	Sce.
5010	10	Radio One	7165	10	Olikkalanjiam
5052	50	Radio One	7250	50	Warna
6000	50	City Sounds	9635	50	City Sounds
6155	50	Warna	11940	50	Radio One

FM (stereo): 89.7MHz (Ria), 90.5MHz (Radio One), 92.4MHz (Symphony), 93.3MHz (Mandarin), 94.2MHz (Warna), 95.0MHz (Class 95 FM), 95.8MHz (City Sounds), 96.8MHz (Olikkalanjiam), 98.7MHz (Perfect 10).
Radio One in English: 2200-1605. **N:** on the h.
Class 95 FM in English: 2200-1800. **N:** 2355, 0055, 0555, 0955, 1055, 1455.
Perfect 10 in English: 24h. **N:** 2200, 0400, 1000, 1600.
Symphony in English: 2200-1600 (Classical music).
Warna in Malay: 21¼h daily.
Ria in Malay: 2200-1600 (Malay pop music).
City Sounds in Mandarin: 2200-1605. **N:** on the h.
FM 93.3 in Mandarin: 2200-1800 (Mandarin pop music).
Olikkalanjiam in Tamil: 2100-1605.
V. by QSL-card. Rp. Rec. acc.

SRI LANKA

L.T: UTC + 5½h — **Pop:** 17.632.000 — **Radio sets:** 2.200.000 — **Pr.L:** Sinhala, Tamil, English — **E.C:** a/c 50, 230V — **ITU:** CLN.

SRI LANKA BROADCASTING CORPORATION (Public Corporation)

ADDR: P.O. Box 574, Torrington Square, Colombo 7.
Te: +94 (1) 697491 (Ext. Sce: 695661). **Cable:** Broadcast, Colombo. **Telex:** 21408 SLABCOR CE.
L.P: Chmn: L.S. Maithrepala. DG: V. Corea. DG (For Rel): M.J.M. Ashroff. Dir. Sinhala Sce: P. Perera. Dir. Tamil Sce: N. Sivarajah. Dir. English Sce: N. Bhareti. Dir. Foreign Sce: N. Jayaweera.
HOME SCES: A=Sinhala Nat. Sce, B=Sinhala Comm. Sce, C=Tamil Sce, D=English Sce, R=Reg. Sce, S=Stereo Sce (English).

Mediumwaves	kHz	kW	Sce.		kHz	kW	Sce.
Ambewela	531	40	B/R	Diyagama	702	25	A
Diyagama	558	20	C	Ratnapura	729	10	A
Kandy	567	10	B/R	Kantale	747	20	B
Kantale	585	20	C	Anuradhapura	774	10	A
Weeraketiya	594	20	R	Maho	801	20	R
Ratnapura	603	10	A	Kandy	819	20	B
Diyagama	621	20	B	Ampara	855	20	B
Maho	639	50	B/R	Diyagama	873	20	D
Ambewela	648	20	A/R	Ampara	972	20	C
Weeraketiya	675	20	B	Galle	1026	10	A
Ampara	693	20	A	Mahiyangana	1485	1	B
				Mahiyangana	1602	1	A

SHORTWAVES: Ekala (G.C: 79.54E/07.06N): 10kW.

kHz	Sce	kHz	Sce.	kHz	Sce.
4870	B	5020	C	6150	C
4902	A	6075	A	6185	B
4940	D	6130	D		

FM: Colombo 88.4/91.7/93.3/98.3/101.35/105.65MHz, Radella 87.5/89.7/94.4/97.0/100.2/103.5MHz, Karagahatenna 92.75/99.6/104.45MHz, Deniyaya 90.8/97.6/107.2/92.8/102.6MHz, Senkadagala 97.6MHz, Palali 102.0MHz.

Nat. Sce. in Sinhala (A): 2330-0200 & 1130-1730 on MW + FM 92.8/98.3/107.2MHz, 0230-1030 on 6075kHz, other times on 4902kHz.
Comm. Sce. in Sinhala (B): 0000-1630 on MW + FM 88.4/97.6/92.75/103.5MHz, 0230-1030 on 6150kHz, other times on 4870kHz.
Comm. Sce. in Tamil (C): 2330-1730 on MW + FM 87.5/104.45/105.65MHz, 0230-1030 on 6185kHz, other times on 5020kHz.
Comm. Sce. in English (D): 0300-0430 (Sat/Sun 1030) & 1130-1630 on MW + FM 90.8/99.6/93.3/100.2MHz, 0300-0430 (Sat/Sun 1030) on 6130kHz, 1130-1630 on 4940kHz.
City FM Sce. on 91.7/94.4/102.6MHz: prgrs in Sinhala, Tamil & English.
Education & Sports Sce: 101.35/89.7MHz.
Regional Sces(R): MW freqs as above + Deniyaya 107.2MHz, Senkadagala 97.6MHz, Palali (Jaffna) 102.0MHz.
ANN: A: "Me Sri Lanka Guwan Viduli Sansthave Welanda Sevaya". B: "Me Sri Lanka Guwan Viduli Sansthave Swadeshiya Sevaya". C: "Illangar Oliparappu Kootuthapanam Tamil Service". D: "This is the Sri Lanka Broadc. Corp." S: "This is City FM Sce. of the SLBC".

EXTERNAL SERVICE

STATION: Ekala (G.C: 79.54E/07.06N): 10/35/100kW.

kHz	kHz	kHz	kHz	kHz
6005	7190	11800	15120	17850
6075	9720	11835	15425	

English	Area	kHz			
0030-0230	As	15425	2330-2400	NAm	15425
		9720	**Hindi* Sce.**		
		6005	0030-0430	As	7190
0230-0430	As	15425			11800
		9720	0630-1030	As	7190
0445-0515	NAm	15425			11800
		9720	1330-1730	As	7190
1030-1130	AUS	11835			11800
	FE	15120	**Sinhalese**		
	SEAs	17850	1745-1900	ME	11800
1230-1730	As	6075	**Tamil**		
		9720	1130-1330	As	7190
		15425			11800
2000-2130	Eu	9720	1900-1950	ME/Af	11800
		15120			

*) incl. Kannada, Telugu, Malayalam, Tamil, Urdu, Nepalese, Sinhalese. **V.** by QSL-card.

TRANS WORLD RADIO (Rlg.)

OP's ADDR: P.O. Box 364, 91 Wijerama Mawatha, Colombo 7. **Te:** +94 (1) 685235/6/7. **Fax:** +94 (1) 685245. **Tlx:** 21654 TWRCEY CE. **Cable:** Votan.
L.P: Dir. of Op's: Mark Blosser. Chief Eng: Robert E. Schultz.
PRGR. ADDR: Trans World Radio-India, Box 4407, L-15, Green Park, New Delhi 110 016. **Te:** +91 (11) 662052. **Fax:** +91 (11) 686 8049. **Tlx:** 031-73095 TWRI IN.
L.P: Exec. Dir: Dr. N. Emil Jebasingh.
MEDIUMWAVE: Puttalam 882 kHz 400kW: 2230-0130, 1330-1700 — **SHORTWAVE:** Puttalam 6035kHz 12.5kW: 1330-1500.

Baduga
1650-1705 (Mon/Tue)
Banjara
0115-0130 (Sat/Sun)
Bengali
2315-2345 (Mon-Fri)
2315-2330 (Sat/Sun)
Chattisgarhi
1650-1705 (Sat/Sun)
Dogri
1635-1650 (Sat/Sun)
Gamit
2330-2345 (Sat/Sun)
Gujarati
2345-0015
1450-1505
Hindi
2230-2315
1405-1420 (Sun)
1420-1450 (Mon-Fri)
1435-1450 (Sat/Sun)
1520-1550 (Sat/Sun)
1535-1550 (Mon-Fri)
Kannada
0100-0130 (Mon-Fri)
0100-1015 (Sat/Sun)
Kashmiri
1620-1635 (Sat/Sun)
Malayalam
0015-0100
Marathi
1550-1620
Oriya
1505-1535 (Mon-Fri)
1505-1520 (Sat/Sun)
Saurashtra
1650-1705 (Wed)
Soura
1420-1435 (Sat/Sun)
Telugu
1335-1420 (Mon-Sat)
1335-1405 (Sun)
Tulu
0650-1705 (Thu/Fri)
Urdu
1620-1650 (Mon-Fri)

ANN: "Vishwa Vani" — **V.** by QSL card (B) — **INT-SIG:** "Send the Light."

R. JAPAN RELAY STATION

STATION: Ekala: 1×300kW SW-tr. operated by SLBC.

kHz	kHz	kHz	kHz
9535	11840	17775	17820

D.PRGR: see Japan.

DEUTSCHE WELLE RELAY SRI LANKA

STATION: Trincomalee (G.C: 81.10E/08.44N): 1548kHz 400kW + 3×250kW SW-tr's.

kHz	kHz	kHz	kHz	kHz	kHz	kHz
6170	9615	11765	11950	15410	17810	21600
7225	9670	11785	15105	15415	17820	21630
9510	9715	11880	15185	17765	17825	21640
9525	11735	11915	15195	17780	21560	21650
9535	11740	11945	15390	17800	21580	

D.PRGR: see Germany.

VOICE OF AMERICA RADIO STATION COLOMBO

STATION: Colombo (G.C: 70.54E/07.06N): 10/35kW.

kHz	kHz	kHz	kHz	kHz	kHz
7115	7125	9645	11705	15250	15395

D.PRGR: see USA.

TADZHIKISTAN

L.T: UTC + 5h — **Pop:** 5.250.000 — **Pr.L:** Tadzhik, Uzbek, Russian — **E.C:** a/c 50, 127/220V - **ITU:** TJK.

GOSUDARSTVENNAYA TELERADIOVESHCHATELNAYA KOMPANIYA TADZHIKISTANA (Gov.)

ADDR: ul. Chapayeva 31, 734005 Dushanbe.
L.P: Head of R. & TV: Mirbobo Mirrahimov.

TADZHIK RADIO (Gov.)

ADDR: ul. Chapayeva 31, 734005 Dushanbe. **Te:** +7 (3772) 276569. **Tlx:** 201392 TELE.

Stations	kHz	kW	H. of tr.	Prgr.
Orzu	270		*	MR1
Yangi-Yul	252	300	0000-1900	TR1
Dushanbe	549	50	2000-1900	RSM
Orzu	648	50	0000-1900	TR1
Orzu	702		0000-2200	RR
Orzu	972	50	0100-1900	TR2,F
Yangi-Yul	1143	150	0100-1900	TR2,F
Orzu	1161		0000-2000	RSY
Dushanbe	1323		0100-1900	TR2
Dushanbe	1502	5	0000-2200	MR1
Yangi-Yul	4635	50	0000-1900	TR1
Yangi-Yul (?)	4740	100	0000-2000	RSY
Yangi-Yul	7245	100	0100-1900	TR2,F

NB: Powers are tentative only. — *) inactive at presstime, under reconstruction.

FM:	TR1	TR3
Dushanbe	70.64	72.20
Tashrabad	67.88	66.32

D.PRGR: **TR1** in Tadzhik, Russian, Uzbek — **TR2** in Tadzhik, Russian: 0500-1500 "Payk-i 'Adzham", at other times FS — **TR3:** music sce. in Tadzhik, Russian. 0700-1600 rel. RSM — **MR1** = Moscow Ostankino R-1 — **RSY** = Moscow R. St. Yunost — **RSM** = Moscow R. St. Mayak — **RR** = Moscow R. Rossii — **F** = Foreign Sce, cf. below.

FOREIGN SERVICE

ADDR: As for Home Sce.

Dari	Area	kHz
0100-0130	As	972, 1143, 7275
0200-0300	As	972, 1143, 7275
0330-0400	As	972, 1143, 7275
1500-1530	As	972, 1143, 7275
1600-1700	As	972, 1143, 7275
1730-1800	As	972, 1143, 7275
Persian		
0400-0430	ME	972, 1143, 7275
1700-1730	ME	972, 1143, 7275

ANN: Tadzhik: "In dzho Dushanbe".
V. Occ. by letter. **PUB:** In dzho Dushanbe.

R. MOSCOW RELAYS

STATIONS: Orzu (G.C: 37.32N/68.42E): 4x250/500/1000kW — Yangi-Yul (?): 4940/4975kHz: 0000-1800. Transmissions included in R. Moscow sched.

THAILAND

L.T: UTC + 7h — **Pop:** 57.610.000 — **Radio sets:** 10.000.000 — **Pr.L:** Thai — **E.C:** a/c 50, 220/380V(no) — **ITU:** THA.

RADIO AND TELEVISION EXECUTIVE COMMITTEE (RTEC)

Constituted under the Broadcasting and TV Rule 1975, this body consists of 17 representatives from 14 Government agencies. It controls Administrative, Legal, Technical and Programming aspects of broadcasting in Thailand.
ADDR: Programme, Administration and Law Section: Div. of RTEC Works, Gov. Public Rel. Dept, Rajchadamnern Klang Rd, Phra Nakhon Region, Bangkok 10200. **Technical Section:** Radio Frequency Management Office, Post & Telegraph Dept, Soi Sai Lom, Phaholyothin Rd, Saam Sen Nai, Phraya Thai Region, Bangkok 10400. **Te:** +66 (2) 710151-60. **Fax:** +66 (2) 713514. **Cable:** Telepost Bangkok. **Telex:** 82503 POSTEL TH.
L.P: Dir. of Int. Sces Div: Somchit Chularat.
STATIONS: All st's are operated by Gov. agencies and/or under the supervision of the Gov.
Mediumwaves: (N = Nakhon, U = Ubon).

	kHz	kW	Location
1)	531	50	Maha Sarakham
13)	540	10	Bangkok
1)	549	10	Krabi
1)	549	100	Lampang (E)
1)	549	10	Mukdahan
1)	558	50	Songkhla (E)
10)	558	10	Chaiyaphum
43)	576	20	Bangkok
8)	585	10	Phrae
31)	585	20	Prachinburi
5)	585	10	Chumporn
8)	585	10	Phitsanulok
5)	594	10	Khon Kaen
40)	603	20	Bangkok
20)	612	20	Lopburi
1)	621	100	Khon Kaen (E)
22)	630	10	Bangkok
1)	639	20	N. Si Thammarat
1)	639	10	Phuket
1)	648	55	Khon Kaen
10)	657	10	Samut Sakhon
11)	657	1	N. Rajchasima
8)	666	10	Tak
17)	675	20	Chieng Mai
4)	684	20	Bangkok
24)	693	20	Saraburi
41)	702	10	Lopburi
19)	711	20	Chieng Mai
19)	711	20	Bangkok
1)	711	20	U. Ratchathani (E)
1)	729	50	N. Rajchasima
5)	738	10	Chieng Mai
18)	738	20	Bangkok
11)	747	1	Chaiyaphum
15)	747	10	Surin
14)	747	1	Prachinburi
12)	747	1	Surat Thani
5)	747	10	Songkhla
11)	747	10	Udon Thani
9)	756	50	Narathivas
6)	765	20	Lampang
25)	774	10	Udon Thani
20)	783	20	Lopburi
13)	792	20	Bangkok
11)	792	10	Kalasin
2)	801	10	Chieng Rai
2)	801	5	U. Ratchathani
2)	801	5	N. Rajchasima
2)	801	5	Lopburi
2)	801	1	Prachuap Khiri Khan
1)	810	20	Nong Kai
1)	810	7	Khon Kaen
1)	810	10	Trang
1)	819	10	Bangkok
1)	819	10	Satoon
8)	828	10	Sukhothai
13)	828	20	Rayong
11)	828	10	Khon Kaen
9)	837	50	Sakon-Nakhon
1)	846	10	Bangkok
1)	864	10	Tak
3)	864	10	Bangkok
1)	873	10	Phattalung
1)	873	10	Phetchabun
1)	891	1000	Saraburi
1)	909	50	Surin
7)	918	10	Chieng Mai
1)	918	1	Chanthaburi (E)
1)	927	10	Bangkok
8)	936	10	Chieng Rai
11)	936	10	N. Rajchasima
12)	936	10	N. Si Thammarat
12)	936	1.5	Pattani
1)	936	50	N. Sawan (E)
2)	954	5	N. Sawan
2)	954	5	Phitsanulok
2)	954	5	Udon Thani
2)	954	1.1	Maha Sarakham
2)	954	5	Bangkok
2)	954	5	Chanthaburi
5)	963	10	N. Rajchasima
1)	963	50	Krabi
9)	972	50	Phetchaboon
1)	981	5	Mae Hong Son
1)	981	10	N. Phanom
3)	981	10	Bangkok
1)	981	50	Yala
7)	990	10	N. Rajchasima
16)	1008	10	Bangkok
1)	1026	50	Phitsanulok
1)	1026	10	Yala
6)	1035	20	Bangkok
15)	1044	10	Khon Kaen
32)	1053	10	N. Sawan
23)	1053	10	Lampang
12)	1053	10	N. Si Thammarat
22)	1062	20	Bangkok
5)	1080	10	N. Sawan
5)	1080	10	Yala
6)	1089	10	Udon Thani
7)	1098	10	Songkhla
1)	1098	10	Tak
4)	1107	20	Chon Buri
17)	1116	20	Samut Sakhon
1)	1125	50	Chanthaburi
1)	1134	10	Lampang
1)	1134	10	Kanchanaburi
1)	1134	10	Phang-Nga
39)	1143	25	Bangkok
18)	1152	20	Chieng Mai
18)	1152	20	Khon Kaen
12)	1152	1	Songkhla
4)	1161	20	N. Phanom
4)	1161	10	Rayong
25)	1170	20	Bangkok
4)	1179	20	Chanthaburi
19)	1179	10	Chieng Rai
15)	1188	10	Sakon Nakhon
8)	1188	10	Phitsanulok
14)	1188	10	Prachin Buri
27)	1197	20	Bangkok
15)	1215	10	Phrae
11)	1215	10	U. Ratchathani
1)	1215	50	Surat Thani
2)	1224	10	Chieng Rai
2)	1224	10	Bangkok
8)	1233	1	Uttaradit
5)	1233	10	Udon Thani
14)	1233	10	Prachuap Khiri Khan
8)	1242	10	Lampang
26)	1242	1	Phayao
8)	1242	10	Phetchabun
10)	1242	10	Roi Et
1)	1242	50	Surat Thani (E)
2)	1251	20	Bangkok
1)	1260	50	Chieng Rai
17)	1269	20	Songkhla
21)	1278	25	Bangkok
5)	1287	10	Chieng Rai
8)	1287	10	Uttaradit
5)	1287	10	U. Ratchathani
5)	1287	10	N. Si Thammarat
1)	1296	10	Pattani
13)	1305	10	Bangkok
17)	1314	20	Khon Kaen
2)	1323	10	Chieng Mai
2)	1323	1	N. Phathom
2)	1323	10	Surat Thani
2)	1323	10	Songkhla
33)	1332	10	Bangkok
1)	1341	10	Loei
1)	1341	50	U. Ratchathani
1)	1341	10	Phang Nga
23)	1350	10	Lampang
5)	1350	10	Trang
16)	1359	10	Bangkok
1)	1368	50	Nan
1)	1368	1	Chumporn
26)	1377	10	Phitsanulok
3)	1386	20	Pathum Thani
9)	1395	25	Chieng Rai
8)	1404	10	Phichit
1)	1404	50	Songkhla
10)	1404	10	Yasothon
6)	1413	10	Tak
7)	1422	10	Phitsanulok
4)	1431	20	Songkhla
8)	1440	1	Nan
11)	1440	10	N. Phanom
5)	1440	10	Samut Sakhon
20)	1449	5	Chumporn
10)	1458	10	Sri Sa Ket
4)	1458	20	Phuket
1)	1467	100	Bangkok (E)
1)	1476	100	Lumphun
1)	1485	1	Surat Thani
12)	1485	1	Trang
12)	1485	1	Krabi
1)	1485	1	Yala
12)	1485	1	Ratchaburi
39)	1494	25	Bangkok
11)	1503	10	Sakon Nakhon
10)	1503	10	Surat Thani
15)	1512	10	Pha Yao
14)	1512	10	Suphan Buri
11)	1512	10	Surin
9)	1521	25	Bangkok
5)	1530	10	Uttaradit
14)	1530	10	Chanthaburi
13)	1539	10	Udon Thani
1)	1539	10	Ratchaburi
16)	1548	10	Bangkok
24)	1557	10	Phetchaboon
37)	1575	1000	Bangkok
3)	1584	1	Chieng Mai
14)	1584	1	Ratchaburi
1)	1584	1	Trad
12)	1584	1	Surat Thani
12)	1584	1	Pattalung
1)	1593	1	Ratchaburi (N)
1)	1593	10	Ranong
1)	1602	1	Burirum
12)	1602	1	N. Si Thammarat

(E) = Educational Sce, (N) = News Sce.
FM: The number of FM-st's operated by each st. is indicated in the address section.

Shortwaves: Pathumthani (G.C: 100.43E/14.03N)

	kHz	kW		kHz	kW		kHz	kW
1)	4830v	10	1)	7115	10	1)	11905	100
1)	6070	10	1)	9655	50			

Other information: Abbreviated short names, e.g. Thor Thor Thor, serve as station ann. if preceded by "Thini. . ." ("Here is . . ."). Full name, given in brackets, is usually preceded by "Thini, Sathani Withayu Krachaisiang . . ." ("Here is R. St . . .").
Numerals, preceded by "Thii" ("number") are: 0 soon, 1 nung, 2 song, 3 saam, 4 sii, 5 ha, 6 hock, 7 jed, 8 pat, 9 kao, 10 sip.

RADIO THAILAND (Gov.)

ADDR: Gov. Public Rel. Dept, Rajchadamnern Klang Rd, Phra Nakhon Region, Bangkok 10200. **Te:** +66 (2) 771814, 771840. (Studio addr: 236 Vibhavadi Rangsit Superhighway, Din Daeng, Huai Khwang, Bangkok 10400).
L.P: Dir. Gen: Manit Varin. Dep. Dir. Gen's: Tavach Meksawan, Arun Ngamdee. Dir. R. Thailand: Chan Phoolsombat. Head Home Sce: Mrs. Chalermsri Hunchareon. Head Ext. Sce: Mrs. Bubpha Laemluang.
HOME SCE: Prgr.I: 2200-1600 on 891/6070/7115kHz + FM 92.5MHz. **N:** On the h. exc. 0500, 0600 & 1100. Also 0530 & 1545.

Prgr.II: 2300-0400 & 0700-1400 on 846kHz. **N:** rel. Prgr. I.
Prgr.III: 2300-1700 on 819kHz(traffic and public affairs information). **N:** Relay Prgr. I.
FM prgrs: "Radio FM 88" on 88.0MHz, Music Prgrs on 95.3/95.5MHz, rel. Foreign sce. on 97.0MHz, "1-0-5 FM" on 105.0MHz.
English: 2300-2400 on 95.5MHz.
Educational Sce: 2200-1640 on all (E) tr's.
Provincial Stations: See MW/FM Frequency List and "Other Stations" (below).
EXTERNAL SCE: On 927/4830/9655/11905kHz + FM97.0MHz: **Burmese:** 1425-1455; **Cambodian:** 1230-1310; **English:** 2300-0430, 0500-0600 (Sat/Sun 0800), 1130-1230; **French:** 0430-0500; **Lao:** 0910-1000; **Mandarin:** 1000-1030; **Thai:** 1030-1100, 1500-1600; **Vietnamese:** 1100-1130; **Malay:** 1310-1400; **Japanese:** 1400-1425.
INT-SIG: Gong.
ANN: E: "This is HSK9, Radio Thailand, External Sce".
V. by QSL-card.

OTHER STATIONS

FM: Total no. of tr's indicated in brackets at end of each entry.

1) **Sathani Withayu Krachaisiang Hang Prathet Thai, Sor. Wor. Tor. (Radio Thailand). HQ addr:** See above; **Provincial addresses:** Most st's can be reached by quoting "Sor. Wor. Tor" and the location given in the freq. list, followed by the phrase "Muang District", and finally the city, which is generally the same as in the freq. list. Exceptions are the following: **Lamphun:** The tr. is in Lamphun, but the studios are in Chieng Mai; **Phang Nga:** The addr. for 1134kHz is Takuapa District, Phang Nga 82110; **Saraburi:** Studios in Bangkok. The 1000kW tr. addr. is: Rim Klong Hog Wa, Moo 4, Nong Rong, Nong Care, Saraburi 18140; **Tak:** The addr. for 1098kHz is Mae-Sod Tak Rd, Mae Sod District, Tak 63110; **Yala:** The addr. for 1026kHz is Baytong District, Yala 95110. (73).

2) **Tor. Or.** (Thahaan Akart), Royal Thai Air Force, Directorate of Communications & Electronics, Bangkok Airport, Wip avadee Rungsit Rd, Don Muang, Bang Khen region, Bangkok 10900. (15).

3) **Por. Chor. Sor.** (Krom Pracha Samphun), Govt. Public Rel. Dept, Rajchadamnern Ave, Phra Nakhon region, Bangkok 10200. (no FM tr's).

4) **Sor. Tor. Ror.** (Siang Chak Thanaan Rua), The Voice of the Navy, Naval Communications Dept, Phra Ratchawang Derm Thonburi, Wat Arun, Bangkok Yai Region, Bangkok 10600 2200-1900. (13).

5) **Wor. Por. Tho.** (Withayu Pracham Thin), Communications Division, Signals Dept, Royal Thai Army. (6).

6) **Nung. Por. Nor.** (Krom Paisanee Thoralek), Post & Telegraph Dept, Sam Sen Nai, Phraya Thai Region, Bangkok 10400. **Te:** 2710151-60 2200-1700 (5).

7) **Sor. Wor. Por.** (Sathani Withayu Pitaksantirat), Police Broadc. Sces, 2nd Communication Div., Directorate of Police Communications, Police Dept, Bang Khen Region, Bangkok 10900. (31).

8) **Tor. Por. Saam (3)** (Kongthap Pak Tii Saam), The 3rd Army Area, Somdej Phranarasuan Maharat Camp, Muang District, Phitsanulok 65000. (1).

9) **Wor. Sor. Kor. Ror. Por. Klang** (Withayu Krachaisiang, Kong Amnuay Kam Klang Raksa Kham Prod Phai (Hang Chat), Central Security Division Radio Station, Radio Broadcasting Division, Headquarters of Supreme Command, Sanam Sua Pa, Rajchadamnern Nok Ave, Dusit Region, Bangkok 10300. (5).

10) **Chor. Sor.** (Krom Chaye Thahaan Süesarn), Broadcasting Control Section, Radio and TV Division, Signals Dept, Royal Thai Army, Phra Ram V Rd, Saphan Daeng, Bangsüe, Dusit Region, Bangkok 10300. (2).

11) **Tor. Por. Song** (Kongthap Pak Tii Song), The 2nd Army Area, Suranaree Camp, Po Klang, Muang District, Nakon Rajchasima 30000. (8).

12) **Tor. Por. Sii** (Kongthap Pak Tii Sii) The 4th Army Area, Vachiravadh Camp, Rajchadamnern Rd, Pak Poon, Muang District, Nakhon Srithammarat 80000 2200-1600. (6).

13) **Yan Kraw,** The 4th Cavalry Batallion (Armoured Unit, Royal Guard), Bangkrabüe Rd, Bangsüe, Dusit Region, Bangkok 10300. (On 792kHz, prgrs produced by Buddhism Diffusion Center, Apitham Mahathat College Foundation). (2).

14) **Tor. Por. Nung.** (Kongthap Pak Thii Nung), The 1st Army Area, Suan Missakawan, Rajchadamnern Nok Ave, Dusit Region, Bangkok 10300. (6).

15) **Kor. Wor. Sor.** (Kitkarn Withayu Krachai Siang), Radio Broadc. & TV Division, Signals Department, Royal Thai Army, Phra Ram V Rd, Saphan Daeng, Bangsüe, Dusit Region, Bangkok 10300. (no FM-tr's).

16) **Phol Nung Ror. Or.** (Kongphol Thii Nung Raksa Phra Ong), The 1st Infantry Division (Royal Guard) HQ, Phitsanulok Rd, Dusit Region, Bangkok 10300. (2).

17) **Mor. Kor.** (Mahavitthayalai Kasetsart), Extension & Training Center, Kasetsart University, Lad Yao, Bang Khen Region, Bangkok 10900. **Te:** 5792294/5790537. (no FM-tr's).

18) **Ror. Dor.** (Kromkarn Raksa Dindan), Territorial Defence Dept, Suan Chaochet, Sanamchai Rd, Wat Po, Phra Nakhon Region, Bangkok 10200. (1).

19) **Wor. Por. Tor.** (Withayu Kromkarn Phalang Ngan Thahaan), Defence Energy Dept., Phetchabari Rd, Pratumwan Region, Bangkok 10500. (7).

20) **Wor. Sor. Por.** (Withayu Soonkarn Thahaan Pin Yai), Artillery Center, Phaholyothin Camp, Koke Kratiem, Muang District, Lopburi 15160. (no FM-tr's).

21) **Kho. Sor. Tor. Bor.** (Kromkarn Khong Song Thahaan), Army Transportation Dept., Bang-Krabüe Rd, Dusit Region, Bangkok 10300. (1).

22) **Mor. Tor. Bor. Nung** (Monthol Thahaan Bok Thii Nung), HQ of the 1st Army Region, at the approach of Kesa Komon, Amnuay Songkram Rd, Dusit Region, Bangkok 10300 2100-1900. (no FM-tr's).

23) **Mor. Tor. Bor. Jed** (Monthol Thahaan Bok Thii Jed), The 7th Army Area HQ, Surasak Montree Camp, Phichai, Muang District, Lampang 52000. (no FM-tr's).

24) **Siang Adison,** The Voice of Adison Camp, Cavalry Center, Adison Camp, Muang District, Saraburi 18000. (no FM-tr's).

25) **Sor. Sor. Sor.** (Siang Samyod), Radio Broadc. Section, The 2nd Communication Division, Directorate of Police Communications, Police Dept., Bangkhen Region, Bangkok 10900. (2).

26) **Wor. Phol. Sii** (Withayu Kong Phol Thii Sii) HQ of the 4th Infantry Division, Somdej Phranarasuan Maharat, Muang District, Phitsanulok 65000. (no FM-tr's).

27) **Wor. Sor. Sor.** (Withayu Suksa), Center for Educational Innovation and Technology, Dept. of Non-Formal Education, Min. of Education, Rajchadamnern Nok Ave, Dusit Region, Bangkok 10300. (1).

28) **Mor. Chor.** (Mahavitthayalai Chieng Mai) Mass Communication Branch, Faculty of Humanities, Chieng Mai University, the approach of Doi Suthep, Huay Krew Rd, Suthep, Muang District, Chieng Mai 50000. (1).

29) **Mor. Kho.** (Mahavitthayalai Khon Kaen), Khon Kaen University, Naimuang, Muang District, Khon Kaen 40000. (1).

30) **Mor. Sor.** (Mahavitthayalai Songkhla Nakarin), Songkhla Nakarin University, Had Yai Campus, Khohong, Had Yai District, Songkhla 90110. (2).

31) **Mor. Tor. Bor. Song** (Monthol Thahaanbock Thii Song), The 2nd Army Region, Chak Krapong Camp, Dong Phra Ram, Muang District, Prachinburi 25000. (1).

32) **Mor. Tor. Bor. Sii** (Monthol Thahaanbock Thii Sii), HQ of the 4th Army Region, Jiraprawat Camp, Muang District, Nakhon Sawan 60000. (no FM-tr's).

33) **Or. Sor.** (Amporn Sathan), Amporn Sathan Throne, Dusit Palace (Chitra Lada Rahothan Palace), Rajchavithee Rd, Chitra Lada, Dusit Region, Bangkok 10300 Tues Sat 0330-0500 & 0900-1200, Sun 0230-0500 (no bc Mon) — **V.** by QSL-Folder. (1).

34) **Thai TV Si Chong Saam** (Thai TV Color Ch. 3 Radio St.), HS-TV3, Lak Song, Nong Khem District, Phetcha Kasem Rd, (Km No.18), Bangkok 10160. (1).

35) **TV Si Chong Jed** (TV7 Color Radio St.), 998/1 Soi Ruam Sirimitra, Phaholyothin Rd, Saam Sen Nai, Phraya Thai Region, Bangkok 10400. **Te:** 2781255. **Cable:** Bebetevee. (1).

36) **Withayu Chula** (Chulalongkorn University Radio St.), Audio-visual Center, Secretariat, Chulalongkorn University, Angreedonang Rd, Nang Mai, Pratumwan Region, Bangkok 10500. (1).

37) **Wor. Or. Sor.** (Withayu Asia Seri), Voice of Free Asia (see below).

38) **Wor. Tor. Or.** (Withayalai Technology Lae Archeva Suska), Vocational and Technology College, Technological Engineering Faculty, Sii Sao Thewes Campus, Sam Sen Rd, Sii Sao Thewes, Dusit Region, Bangkok 10300. (1).

39) **Tor. Tor. Tor.** (Mass Media Organization of Thailand), 222, Mot Building, 6th Floor, Asok-Dindaeng Rd, Huayfang Region, Bangkok 10310. **N.** (R. Thailand): 2300-2330, 0530-0545, 1200-1230, 1300-1330, 1400-1410; Army N. 2345-2400, Navy N. 1245-1300, Police N. 1030-1100. (98).

40) **Por. Tor. Or.** (Kongphol Thahaan Punyai Torsue Akart Yan), Anti-Aircraft Artillery Division, Kiak Kay Junction, Thahaan Rd, Bangsüe, Dusit Region, Bangkok 10300. (1)

41) **Chor. Tor. Ror.** (Changwat Thahaanbock Lopburi), HQ of Lopburi Army Province, Talé Chupson, Muang District, Lopburi 15000. (no FM-tr's).
42) **Tor. Tor. Bor.** (Thorathat Kongthapbock), Army TV-Radio St., 340 Wipavadee Rungsit Rd, Huay Jang Region, Bangkok 10310. (2).
43) **Tor. Chor. Dor.** (Tamruat Trawen Chaidan), HQ of Border Patrol Police Division, Phaholyothin Rd, Saam Sen Nai, Phraya Thai Region, Bangkok 10400. (5).
44) Harbour Dept, Min. of Communications. St. carries maritime and weather information. (1).
45) Meteorological Dept, 612 Sukumvit Rd, Bangkok 10110. **Te:** 2580437-9 (5).
46) Silpakorn University Phitsanulok, Maha Sarakham Campus. (2).
N.B: All st's are required to relay N. from R. Thailand at 0000, 0530, 1200 & 1300 daily.
Reports: Geographical locations given refer to provinces ("Changwat"). St's are not always located in the provincial capitals but in a local district ("Amphur"). However, a report to a provincial station is known to reach its destination even if the "amphur" is unknown.

VOICE OF FREE ASIA (Wor. Or. Sor.)
ADDR: P.O. Box 2-131, Rajdamnoen, Bangkok 10200.
L.P: Dir: Anucha Osathanond.
STATION: Bangkok 1575kHz 1000kW (tr. operated by VOA).
D.PRGR: Mon-Fri: 1030-1130, 1500-1530 & 2230-2400. (1030-1100 Thai, 1100-1130 Malay, 1500-1530 **English,** 2230-2300 Vietnamese, 2300-2330 Cambodian, 2330-2400 Laotian).
INT-SIG: Thai National Anthem.
ANN: "This is the Voice of Free Asia".
V. by QSL-card and letter.

VOICE OF AMERICA RELAY STATION THAILAND
MW: Bangkok 1575kHz 1000kW — **SW:** Udorn Thani: 7×500kW SW-tr's U.C. The first two will come on air in Spring 1993.

kHz	kHz	kHz	kHz	kHz
6045	9560	9680	11705	11855
6070	9615	9760	11785	11905

D.PRGR: United Nations Prgr. beamed to Cambodia on 1575kHz: D 0030-0100 & MF 1200-1230.
OTHER PRGRS: see USA.

TURKMENISTAN

L.T: UTC +5h — **Pop:** 3.630.000 — **Pr.L:** Turkmen, Russian — **E.C:** a/c 50 127/220V — **ITU:** TKM.

TURKMEN RADIO (Gov.)
ADDR: ul. Makhtumkuli 89, 744000 Ashgabat. **Te:** +7 (3632) 251515. **Fax:** +7 (3632) 251421 (manual). **Tlx:** 228125 RADIO.
L.P: Chmn: Anageldi Arazdurdiyev.

Stations*	kHz	kW	H. of tr.	Prgr.
Aşqabat	153	500	0000-2300	RSM
Aşqabat	279	150	0000-2300	TMR1,F
Aşqabat	576	150	0000-2200	MR1
Aşqabat	675	150	0100-2300	TMR2
Ek-Arça (1)	720		0000-2300	TMR1
Esenquli (1)	720		0000-2300	TMR1
Çärýev (2)	765	30	0100-2400	RSM
Çärýev (2)	927	50	0000-2300	TMR1, R
Gyzyletrek (1)	927		0000-2300	TMR1
Krasnovodsk (1)	927		0000-2300	TMR1
Guşgy (3)	945	20		RSM
Krasnovodsk (1)	1080		0000-2200	TMR1
Guşgy (3)	1080	5	0000-2300	TMR1
Aşqabat	1125		0000-2300	TMR1
Syrtaqta (3)	1233	40	0000-2300	TMR1
Aşqabat	4825	50	0000-2300	TMR1,F
Aşqabat	4930	50	0050-2400	RSM
Aşqabat	5015	50	0000-2200	MR1

*) names in Turkmen. **NB:** Powers are estimates only.
(1) Balkan oblast. (2) = Chardzhev. (3) Mary oblast.
FM: No details available.

D.PRGR: TMR1: 0000-2100 in Turkmen, Russian incl. rel MR1. At other times rel. RSM — **TMR2:** 0800-1700 in Turkmen, Russian. At other times rel. RSM — **MR1** = Moscow Ostankino R-1 — **RSM** = Moscow R. St. Mayak.
Regional Prgrs (R): Çärýev: 0500-0530 & 1500-1600 on 927kHz + FM (also r. on 279kHz).

FOREIGN SERVICE: Voice of Turkmen
ADDR: Mening Vatanym Radyo Merkezi, Makhtumkuli 89, 744000 Ashgabat.

Turkmen	Days	Area	kHz
1200-1300	Mon/Wed/Fri	ME	279, 4825
1900-2000	Tues/Thurs/Sat	ME	279, 4825

ANN: Turkmen: "Qepleyär Aşqabat".
V. occ. by letter — **PUB:** Qepleyär Aşqabat (weekly from Home Sce.).

R. MOSCOW RELAY STATION
MW: Ashgabat 1125kHz. R. Moscow prgrs to Afghanistan, India. Cf. R. Moscow schedule.

UZBEKISTAN

L.T: UTC +5h — **Pop:** 20.330.000 — **Pr.L:** Uzbek, Russian — **E.C:** a/c 50 127/220V — **ITU:** UZB.

GOSUDARSTVENNAYA TELERADIOVESHCHATELNAYA KOMPANIYA UZBEKISTANA (Gov.)
ADDR: ul. Khorezmskaya 49, 700047 Tashkent.

UZBEK RADIO (Gov.)
ADDR: ul. Khorezmskaya 49, 700047 Tashkent. **Te:** +7 (3712) 441210. **Fax:** +7 (3712) 440021. **Tlx:** 116062 EFIR.

Stations*	kHz	kW	H. of tr.	Prgr.
Toşkent (11)	162	150	0000-2100	UZR1
Vobkent (2)	576		0000-2300	RSM
Toşkent (11)	576	5	0000-2300	RSM
Toşkent (11)	666	30	0000-2300	UZR2/Reg
Koson (6)	675		0000-2200	MR1
Küngirod (12)	675		0000-2200	MR1
Zarafşon (2)	675		0000-2000	UZR1
Buhoro (2)	711	25	0000-2300	UZR2
Urgonç (4)	711		0000-2300	UZR2
Piskent (11)	756	150	0000-2200	MR1
Andijon (1)	1062	50	0000-2300	UZR2
Nukus (12)	1062		0000-2300	UZR2
Toşkent (11)	1062		0000-2100	UZR1
Nukus (12)	1260	100	0000-2100	UZR1/Reg
Namangan (7)	1269		0000-2200	MR1
Zarafşon (2)	1269		0000-2200	MR1
Koson (6)	1323		0000-2300	UZR2
Kükon (3)	1323			RSM
Muborak (7)	1332		0000-2100	UZR1
Jizzah (5)	1485		0000-2200	MR1
Nurobod (8)	1485			RSM
Samarkand (8)	1539		0000-2300	UZR2
Jar-Kürğon (9)	1593**		0000-2300	UZR2
Farğona (3)	4510		0930-1000	Reg
			1200-1230	Reg
Toşkent (11)	4850	50	0000-2300	UZR2
Toşkent (11)	5995	100	0000-2100	UZR1
Toşkent (11)	9545	100	0000-	UZR2
			1550-2100	UZR2, F
Toşkent (11)	9840	100	1550-2100	UZR2, F
Toşkent (11)	15165	100	-2000	UZR2
Toşkent (11)	15200	100	0400-1400	UZR2, F

*) in Uzbek. **) r. v1583-1585kHz. Further provincial st's may have changed to UZR1. Powers are tentative only.
Regions: (1) Andijon (Andizhan), (2) Buhoro, (3) Farğona (Fergana), (4) Horezm, (5) Jizzah, (6) Koşkodarya, (7) Namangan, (8) Samarkand, (9) Surhandarya, (10) Sirdarya, (11) Toşkent (Tashkent); (12) Karakalpak Rep.

FM (MHz)	UZR1	UZR2	MR1
Toşkent	67.2	68.0	66.3

Freq's are approximate. No further details available.

D.PRGR: UZR1 in Uzbek, Russian. **N. in Russian** (rel. MR1): 0900, 1400 — **UZR2** in Uzbek, Russian, Tadzhik, Crimean Tatar, Kazakh. Includes R. Yoshlik & R. Mash'al. **N. in Russian:** 0215, 0300, 1700 — **MR1** = Moscow Ostankino R-1 (su. 1h earlier) — **RSM** = Moscow R. St. Majak.
Regional st's: ul. Sovetskaya 26, 710000 Andizhan — ul. Kalinina 28, 712000 Fergana: On 4510 in Uzbek, Russian, Tadzhik, Tatar — Karakalpak R, Nukus: On 1260 in Karakalpak, Uzbek, Russian 0215-0300, 0700-0800 (Sun), 1230-1300 (Sun), 1330-1400, 1415-1445 (Sun 1500) — Tashkent: Oblast prgrs on 666kHz, 67.2MHz: 0000-?. On 666kHz: 0730. On 666/4850kHz, 68.0MHz: 1430-2000 (W). City prgrs on 67.2MHz: 0115, 0230, 0730W, 1100W, 1230MF.

FOREIGN SERVICE: Radio Tashkent

ADDR: As above. **Stations:** Tashkent, exc. where indicated.

Arabic	**Area**	**kHz**
1700-1730	ME	7325, 9790, 9840, 9885
1900-1930	ME	7325, 9790, 9840, 9885
English		
0100-0130	As	7235, 7325, 9755, 11975
1200-1230	SAs	5945, 9540, 15470, 17745
1330-1400	SAs	5945, 9540, 15470, 17745
Dari		
0130-0200	ME	7235, 7325, 9740, 11975
1520-1550	ME	7325, 9790, 9840, 9885
Farsi		
1630-1700	ME	7325, 9790, 9840, 9885
1830-1900	ME	7325, 9790, 9840, 9885
Hindi		
1300-1300	SAs	5945, 9540, 15470, 17745
1430-1500	SAs	5945, 9540, 15470, 17745
Pushtu		
0200-0230	As	7235, 7325, 9755, 11975
Turkish		
0400-0430	ME	unconfirmed
1700-1730	ME	unconfirmed
Uighur*		
1230-1330	As	900, 5915, 7255
1430-1500	As	900, 5915, 7255
1530-1600	As	900, 5915, 7255
2330-2400	As	900, 5915, 7255
0030-0100	As	900, 5915, 7255
Urdu		
1230-1300	SAs	5945, 9540, 15470, 17745
1400-1430	SAs	5945, 9540, 15470, 17745
Uzbek		
0230-0330	ME	7235, 7325, 9755, 11975
1550-1630	ME	7325, 9790, 9840, 9885
1730-1830	ME	7325, 9790, 9840, 9885

*) via R. Alma-Ata tr's (Su. 1h earlier).
ANN: Uzbek: "Toşkentdan gapiramiz" — **V.** by QSL-card.
PUB: To kentdan gapiramiz va kürsatamiz (weekly for Home Sce.).

RADIO MOSCOW RELAY STATION

Shortwaves: 10 × 100kW (also used by R. Tashkent).
Transmissions included in R. Moscow schedule.

VOICE OF AMERICA, RADIO LIBERTY relays: Cf. these stations.

VIETNAM (Socialist Rep. of)

L.T: UTC + 7h — **Pop:** 68.980.000 — **Radio sets:** 6m — **Pr.L:** Vietnamese, ethnic — **E.C:** a/c 50, 120/127/230V — **ITU:** VTN.

VIETNAM RADIO AND TELEVISION COMMITTEE (Gov.)

ADDR: 58 Quan Su Street, Hanoi. **Te:** +84 4134.
L.P: Chmn: Tran Lam. Vice Chmn: Ly Van Sau, Le Quy.
Mediumwaves: C = City, t = location tentative.

St.	**Location**	**kHz**	**Times**
2)	Hanoi	570	1020-1430
	Nhatrang (t)	580	1145-
3)	Ho Chi Minh C.	610	0950-1430
	Lang Son (t)	627v	1045-1230
	Nhatrang (t)	670	
1)	Hanoi	680	2145-1600
	Ban Me Thuot (t)	690	
17)	Quang Nam-Da	700	1130-1600
12)	Quy Nhon	740	1045-1230
13)	Tanan	760	1130-1230
	Quy Nhon (t)	770	
14)	Can Tho	780	1000-1100, 1200-1300
1)	Ho Chi Minh C.	870	2145-1600
	Quang Ngai (t)	895	
15)	Rach Gia	975	1145-1230
	Hanoi-FS	1010	1100-1630
	Hanoi	1050	
	?	1068v	-1230
	Hanoi-FS	1240	2230-0100, 0930-1530

SHORTWAVES: All freqs vary widely.

	Station	**kHz**	**Times**
7)	R. Son La	4665	2200-2300, 1130-1230
11)	R. & TV Gia Lai	4787	2200-2300, 1030-1100 1130-1200
4)	Ha Giang BS	4815	2330-2400, 1200-1400 (irr.)
1)	Hanoi Nat. Sce.	4895	W 2200-0100, 0400-0630, 0900-1600. Sun 2200-0630, 0800-1600
1)	Hanoi Hmong Sce.	5020	see below
16)	Radio TV Kontum	5061	1045-1100, 1130-?
5)	R. Lao Cai	5660	1030-1215
1)	Hanoi Nat. Sce.	5925	1100-1600
1)	Hanoi Hmong Sce.	6135	see below
7)	R. Son La	6343	2200-2300, 1130-1230
10)	R. Yen Bai	6370	0000-0030, 1000-1030/1045 1100-1215
6)	R. Lai Chau	6430	0400-0600, 1130-1330
1)	Hanoi Nat. Sce.	6448	as 4895kHz
8)	R. Cao Bang	6477	1200-1400 (irr.)
9)	R. Bac Tai	6525	0300-0500, 1155-1400
1)	Hanoi Nat. Sce.	10059	2300-1600

FM: Hanoi 100MHz.
NB: No official details received. Acc. to recent information, 32 of the 51 provinces have radio st's, incl. more than 20 on SW. However, many of these are low powered and have not been heard outside the country.
Provincial Stations broadcast in Vietnamese and the main tribal languages of their region.
Locations and other information
1) Hanoi — **N:** 2150, 2300, 0500, 1100, 1430. **ANN:** "Da la tieng noi Viet Nam, phat thanh tu Hanoi, thu do nuoc cong hoa xa hoi chu nghia Viet Nam". **Hmong:** 2200-2330, 0430-0600 & 1100-1330 on 5020/6560kHz.
2) Hanoi Broadc. St, 26 pho Hang Dau, Hanoi. **ANN:** "Dáy la dai phát than Hanoi".
3) V. of the People of Ho Chi Minh C, 7 Xo Viet Nge Tinh Str, Ho Chi Minh C. **ANN:** "Dáy la tiend noi Nhan Dan Thanj pho Ho Chi Minh".
4) Ha Giang Province.
5) Lao Cai, Lao Cai Province.
6) Lai Chau Province
7) Son La Province
8) Cao Bang Province
9) Tainguyen, Bac Tai Province.
10) Yen Bai Province.
11) Playcu, Gia Lai Province.
12) Nghia Bin Province.
13) Tanan, Long An Province
14) Can Tho, Hau Giang Province. **ANN:** Khmer: "Thini Sathani Withyu phsay someleng Hau Giang".
15) Rach Gia, Kien Giang Province
16) Kontum Province.
17) Quang Nam-Da Province. **ANN:** "Da lai phat thanh Quang Nam-Da Nan".

FOREIGN SERVICE

ADDR: Voice of Vietnam, 58 Quan Su Str, Hanoi.

kHz	**kHz**	**kHz**	**kHz**	**kHz**	**kHz**	**kHz**
1010	7416	9732	9840	10010	12035	15010
1240						

Cambodian	**Area**	**kHz**
0000-0030	CBG	9732, 7416, 1240
1200-1230	CBG	9732, 7416
1330-1400	CBG	9732, 7416, 1010

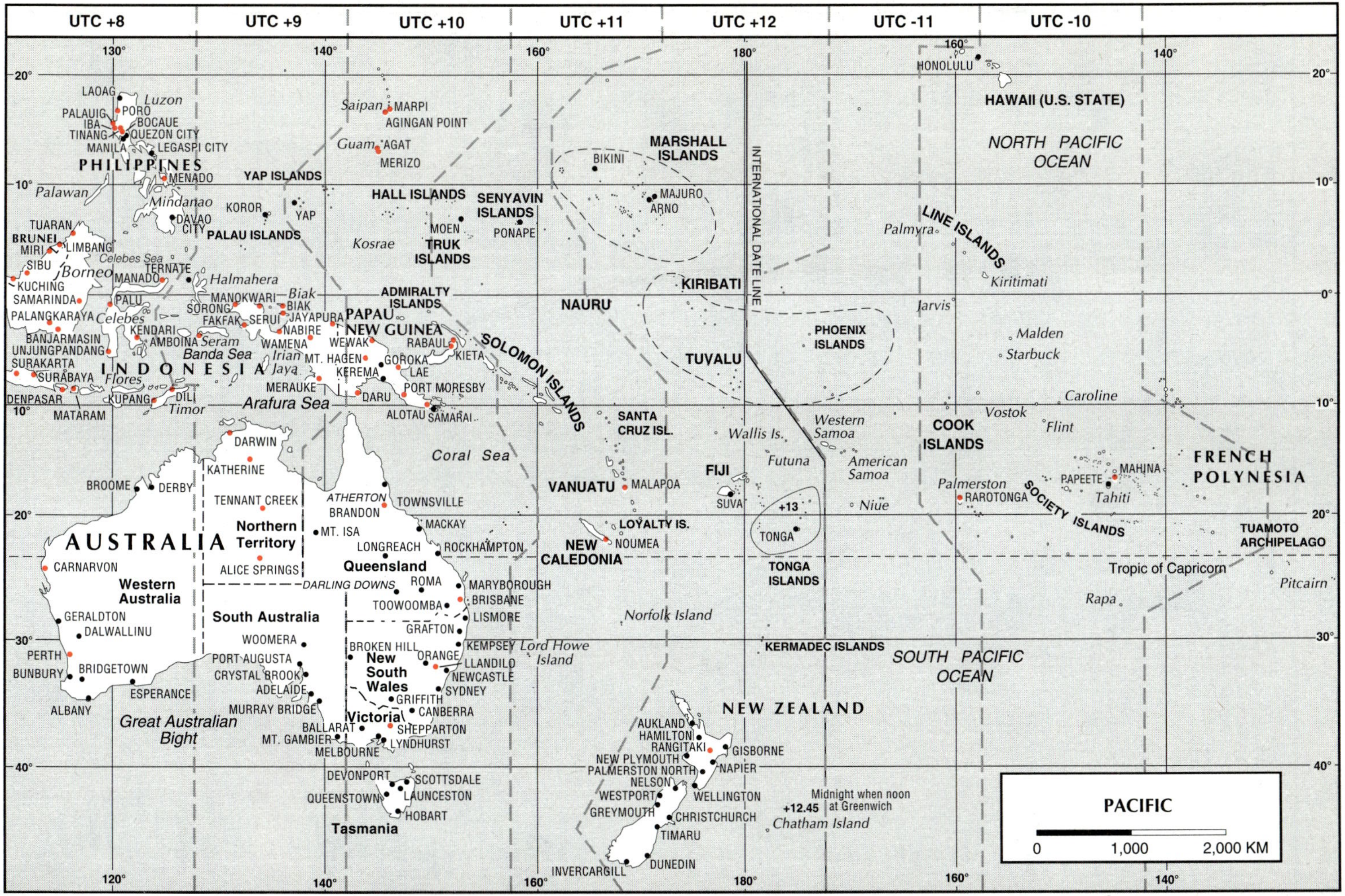
UTC +8
UTC +9
UTC +10
UTC +11
UTC +12
UTC -11
UTC -10
PACIFIC
0 1,000 2,000 KM
HAWAII (U.S. STATE)
HONOLULU
NORTH PACIFIC OCEAN
SOUTH PACIFIC OCEAN
INTERNATIONAL DATE LINE
Tropic of Capricorn
Midnight when noon at Greenwich
+12.45
+13
AUSTRALIA
Western Australia
Northern Territory
South Australia
Queensland
New South Wales
Victoria
Tasmania
NEW ZEALAND
INDONESIA
PHILIPPINES
PAPAU NEW GUINEA
SOLOMON ISLANDS
VANUATU
NEW CALEDONIA
FIJI
TONGA ISLANDS
TUVALU
KIRIBATI
NAURU
MARSHALL ISLANDS
LINE ISLANDS
PHOENIX ISLANDS
COOK ISLANDS
SOCIETY ISLANDS
FRENCH POLYNESIA
TUAMOTO ARCHIPELAGO
KERMADEC ISLANDS
LOYALTY IS.
SANTA CRUZ ISL.
ADMIRALTY ISLANDS
HALL ISLANDS
SENYAVIN ISLANDS
TRUK ISLANDS
YAP ISLANDS
PALAU ISLANDS
BRUNEI
Coral Sea
Arafura Sea
Banda Sea
Celebes Sea
Great Australian Bight
Luzon
Mindanao
Palawan
Borneo
Celebes
Halmahera
Seram
Irian Jaya
Biak
Flores
Timor
Saipan
Guam
Kosrae
Palmyra
Kiritimati
Jarvis
Malden
Starbuck
Caroline
Vostok
Flint
Western Samoa
American Samoa
Wallis Is.
Futuna
Niue
Palmerston
Tahiti
Rapa
Pitcairn
Norfolk Island
Lord Howe Island
Chatham Island
LAOAG
PORO
PALAUIG
IBA
BOCAUE
TINANG
QUEZON CITY
MANILA
LEGASPI CITY
MENADO
DAVAO CITY
KOROR
YAP
TUARAN
LIMBANG
MIRI
SIBU
KUCHING
SAMARINDA
TERNATE
MANADO
PALU
PALANGKARAYA
KENDARI
BANJARMASIN
UNJUNGPANDANG
AMBOINA
SURAKARTA
SURABAYA
DENPASAR
MATARAM
KUPANG
DILI
MANOKWARI
SORONG
BIAK
FAKFAK
SERUI
JAYAPURA
NABIRE
WAMENA
MERAUKE
WEWAK
MT. HAGEN
KEREMA
GOROKA
LAE
RABAUL
KIETA
DARU
PORT MORESBY
ALOTAU
SAMARAI
MARPI
AGINGAN POINT
AGAT
MERIZO
MOEN
PONAPE
BIKINI
MAJURO
ARNO
MALAPOA
NOUMEA
SUVA
TONGA
RAROTONGA
PAPEETE
MAHINA
DARWIN
KATHERINE
BROOME
DERBY
TENNANT CREEK
ALICE SPRINGS
ATHERTON
BRANDON
TOWNSVILLE
MT. ISA
MACKAY
LONGREACH
ROCKHAMPTON
DARLING DOWNS
ROMA
MARYBOROUGH
TOOWOOMBA
BRISBANE
LISMORE
GRAFTON
KEMPSEY
BROKEN HILL
ORANGE
LLANDILO
NEWCASTLE
SYDNEY
GRIFFITH
CANBERRA
SHEPPARTON
LYNDHURST
MELBOURNE
BALLARAT
MT. GAMBIER
MURRAY BRIDGE
ADELAIDE
CRYSTAL BROOK
PORT AUGUSTA
WOOMERA
CARNARVON
GERALDTON
DALWALLINU
PERTH
BUNBURY
BRIDGETOWN
ALBANY
ESPERANCE
DEVONPORT
SCOTTSDALE
LAUNCESTON
QUEENSTOWN
HOBART
AUKLAND
HAMILTON
RANGITAKI
GISBORNE
NEW PLYMOUTH
PALMERSTON NORTH
NAPIER
NELSON
WESTPORT
WELLINGTON
GREYMOUTH
CHRISTCHURCH
TIMARU
DUNEDIN
INVERCARGILL

Cantonese		
1230-1300	CHN	12035, 10010, 1010
1330-1400	CHN	12035, 10010
1430-1500	CHN	9732, 7416
1530-1600	CHN	15010, 12020, 9840
2230-2300	CHN	15010, 12020, 9840, 1240, 1010
Chinese		
0830-0930	CHN	9840, 12020, 15010
1200-1230	CHN	12035, 10010, 1010
1300-1330	CHN	as 1200
1500-1530	CHN	9732, 7416
2200-2230	CHN	15010, 12020, 9840, 1010
English		
1000-1030	As	15010, 12020, 9840, 1240
1100-1130	SEAs	9732, 7416, 1010
1230-1300	As	as 1000
1330-1400	FE	as 1000
1600-1630	Af	15010, 12020, 9840
1800-1830	Eu	as 1600
1900-1930	Eu	as 1600
2030-2100	Eu	as 1600
2330-2400	Eu	as 1000
French		
1300-1330	SEAs	15010, 12020, 9840, 1240
1830-1900	Eu	15010, 12020, 9840
1930-2000	Eu	as 1830
2100-2130	Eu	as 1830
Indonesian		
0930-1000	INS	15010, 12020, 9840, 1240
1030-1100	INS	as 0930
1430-1500	INS	15010, 12020, 9840, 1240, 1010
Japanese		
1200-1230	J	15010, 12020, 9840, 1240
1400-1430	J	as 1200
2130-2200	J	15010, 12020, 9840
Laotian		
0030-0100	LAO	9732, 7416, 1240
1230-1300	LAO	9732, 7416
1400-1430	LAO	9732, 7416, 1010
Russian		
1130-1200	RUS	15010, 12020, 9840, 1240
1300-1330	RUS	9732, 7416
1430-1500	RUS	10010, 12035
1630-1700	RUS	15010, 12020, 9840
Spanish		
1100-1130	LAm	15010, 12020, 9840, 1240
2000-2030	Eu	15010, 12020, 9840
Thai		
1130-1200	THA	9732, 7416, 1010
1400-1430		12035, 10010
1500-1530	THA	15010, 12020, 9840, 1240, 1010
2300-2330	THA	15010, 12020, 9840, 1240
Vietnamese		
0000-0100	As	15010, 12020, 9840
1530-1630	As	1010
1700-1800	Eu	15010, 12020, 9840

V. by QSL-card. Prgr schedule free.

PACIFIC

AUSTRALIA (Commonwealth)

L.T: Vic, NSW, Queensland, Tasmania: UTC + 10h (Vic, NSW, Tas: Oct-March UTC + 11h); So. Australia UTC + 9½h (Oct-March UTC + 10½h); We. Australia: UTC + 8h; No. Territory UTC + 9½h — **Pop:** 17.550.000 — **Radio sets:** 20m — **Pr.L:** English — **E.C:** a/c 240/415V — **ITU:** AUS.

AUSTRALIAN BROADCASTING CORPORATION (ABC)

ADDR: Ultimore Centre, 700 Harris Str, Ultimo, NSW 2007. **Te:** +61 (2) 333 1500. **Fax:** +61 (2) 333 5305.
L.P: MD: David Hill. Asst. MD: M. Long. Dir. Radio: P.L.D. Loxton. Dir. TV: P. Conroy. GM Radio Australia: Richard Broinowski.

MEDIUMWAVES: N = R. National, R = Regional R, M = Metropolitan Sce, P = Parliamentary Netw. **Call letters:** 2 = NSW (exc. Canberra = A.C.T.), 3 = Victoria, 4 = Queensland, 5 = So. Australia, 6 = We. Australia, 7 = Tasmania, 8 = Northern Territory.

Call	kHz	kW	Netw	Location
6DL	531	10	R	Dalwallinu
4QL	540	10	R	Longreach
2CR	549	50	R	Orange
6WA	558	50	R	Wagin
4JK	567	10d	R	Julia Creek
6PU	567	0.1	R	Paraburdoo
6TP	567	0.1	R	Tom Price
6MN	567	0.1	R	Newman
6PN	567	0.1	R	Pannawonica
2RN	567	50	N	Sydney
6PB	585	5	P	Perth
7RN	585	10	N	Hobart
3WV	594	50	R	Horsham
2RN	603	10	N	Nowra
4CH	603	10	R	Charleville
6PH	603	2	R	Port Hedland
4QR	612	50	M	Brisbane
6NM	612	0.2	R	Northam
3RN	621	50	N	Melbourne
2PB	630	10	P	Sydney
4QN	630	50	R	Townsville
6AL	630	0.4	R	Albany
5CK	639	10	R	Port Pirie
5SY	639d	2d	R	Streaky Bay
8KN	639	2	R	Katherine
2NU	648	10	R	Tamworth
6GF	648	2	R	Kalgoorlie
2BY	657	10d	R	Byrock
8RN	657	2	N	Darwin
2CO	675	10	R	Corowa
6BE	675	5	R	Broome
2CN	666	5	M	Canberra ACT
2KP	684	10	R	Kempsey
6BS	684	4	R	Busselton
8RN	684	1	N	Tennant Creek
4MS	693	1	R	Mossman
2BL	702	50	M	Sydney
6KP	702	10	R	Karratha
4QW	711	10d	R	St. George
7NT	711	10d	R	Launceston
2ML	720	0.4	R	Murwillumbah
2AN	720	0.05	R	Armidale
3MT	720	2	R	Omeo
4AT	720	4	R	Atherton
6WF	720	50	M	Perth
5RN	729	50	N	Adelaide
7PB	729	2	P	Hobart
2NR	738	50	R	Grafton
6MJ	738	5	R	Manjimup
4QS	747	10	R	Toowoomba
8JB	747	0.2	R	Jabiru
2TR	756	2d	R	Taree
3RN	756	10	N	Wangaratta
4QA	756	2	R	Mackay
6KW	756	0.1	R	Kununurra
3LO	774	50	M	Melbourne
8AL	783	2	R	Alice Springs
4RN	792	10	N	Brisbane
4QY	801	2	R	Cairns
2BA	810	10	R	Bega
6RN	810	50	N	Perth
2GL	819	10	R	Glen Innes
3GI	828	10	R	Sale
6GN	828	2	R	Geraldton
4RK	837	10	R	Rockhampton
6ED	837	1	R	Esperance
2RN	846	10	N	Canberra
6CA	846	0.2	R	Carnarvon
4QB	855	10d	R	Maryborough
4QO	855	10	R	Eidsvold
6DB	837	2	R	Derby
5AN	891	50	M	Adelaide
4PB	936	10	P	Brisbane
7ZR	936	10	M	Hobart
3RN	990	0.25	N	Albury-Wodonga

8GI	990	0.5	R	Gove
2NB	999	2	R	Broken Hill
6WH	1017	0.1	R	Wyndham
2UH	1044	1d	R	Muswellbrook
4WP	1044	0.5	R	Weipa
6BR	1044	1	R	Bridgetown
4TI	1062	2	R	Thursday Isl.
4MI	1080	0.2	E	Mt. Isa
2RN	1098	0.2	N	Goulburn
5PA	1161	10d	R	Naracoorta
7FG	1161	1d	R	Fingal
6XM	1188	2	R	Exmouth
2NC	1233	10	M	Newcastle
5MV	1305	2	R	Renmark
2LG	1395	0.2	R	Lithgow
2RN	1431	2	N	Wollongong
2PB	1440	2	P	Canberra
2PB	1458	2	P	Newcastle
4HU	1458	0.05	R	Hughenden
5LN	1458	0.2	R	Port Lincoln
5PB	1539	10	P	Adelaide
4QD	1548	50	R	Emerald
4GM	1566	0.2	R	Gympie
2WA	1584	0.1	R	Wilcannia
5MG	1584	0.2	R	Mt. Gambier
5WM	1584	0.05	R	Woomera
7SH	1584	0.1	R	St. Helens
3PB	1593	10	P	Melbourne
2CP	1602	0.05	R	Cooma
3WL	1602	0.2	R	Warnambool
5LC	1602	0.1	R	Leigh Creek So.

Main FM st's (High power tr's and tr's serving main cities)
Networks: N = R. National, R/M = Regional or Metropolitan Netw, FM = ABC FM Netw, J = Triple J (youth) Netw.

Location	State	N	R/M	FM	J
Adelaide	SA		103.9		105.5
Albury-Wondoga	NSW				104.1
Albany	WA		94.5		
Ballarat	VIC		105.5		
Bega	NSW		105.7		
Bendigo	VIC		106.3		
Brisbane	QLD		106.1		107.7
Bunbury	WA		93.3		
Cairns	QLD		105.9		
Canberra	ACT		102.3		101.5
Darwin	NT	105.7		107.3	103.3
Dubbo	NSW	107.1		105.5	
Griffith	NSW		107.5		
Hamilton	VIC		107.3		
Hobart	TAS		93.9	92.9	
Launceston	TAS			93.3	
Loxton	SA		105.1		
Maryborough	QLD	104.1			92.5
Melbourne	VIC		105.9		107.5
Mildura	VIC	104.3		102.7	
Mt. Dundas	QLD	105.7		107.3	
Mt. Gambier	SA			104.1	
Newcastle	NSW		106.1		102.1
Orange	NSW		102.7		
Perth	WA		97.7	99.3	
Pt. Pirie	SA	106.7		104.3	
Rockhampton	OLD	103.1			93.7
Shay Gap	WA	106.1	107.7		104.5
Sydney	NSW		92.9	105.7	
Taree	NSW		107.9		
Toowoomba	QLD	105.7		107.3	
Townsville	QLD	104.7			101.5
Traralgon	VIC		107.1		
Wagga Wagga	NSW		101.9		105.7
Wollongong	NSW		97.3	95.7	

DOMESTIC SHORTWAVE SERVICE

ADDR: VLQ & VLM: ABC, Box 9994, GPO 4001, Brisbane, SA — VLW: ABC, Box 9994, GPO 6001, Perth, WA.
STATIONS: VLQ & VLM Brisbane (G.C: 153.01E/27.19S) — VLW Wannerwoo (G.C: 115.39E/31.51S).

Call	kHz	kW	H. of tr.				
VLM	4910	10	1900-1500			50	0950-2055
VLW	6140	10	0945-0100	VLQ	9660	10	24h
VLW	9610	10	2100-0945	VLW	15425	50	2300-0945

V. by QSL-card Rp.

NORTHERN TERRITORY SHORTWAVE SERVICE

ADDR: Box 9994, Darwin, NT 5750 (CAAMA Radio: P.O. Box 2924, Alice Springs, NT 5740).
STATIONS (50kW):
VL8A Alice Springs: 2310kHz (0830-2130), 4835kHz (2130-0830)
VL8T Tennant Creek: 2325kHz (0830-2130), 4910kHz (2130-0830)
VL8K Katherine: 2485kHz (0830-2130), 5025kHz (2130-0830).
D.PRGR. in English & Aboriginal languages: 24h. Some prgrs produced by ABC, others by the Central Australian Aboriginal Media Association (CAAMA) — **V.** by letter. Rp.

OVERSEAS SERVICE — RADIO AUSTRALIA

ADDR: P.O. Box 755, Glen Waverly, Vic. 3150. **Te:** +61 (3) 881 2222. **Fax:** +61 (3) 881 2346 (Management), +61 (3) 881 2377 (Tech.). **Tlx:** RADAUS 30551. **Cable:** Newscast. **Openline:** + 61 (3) 881 2360.
L.P: GM: Derek White. Asst. GM: Susan Kadar. Contr. Resources & Sces: Howard Deer. CE: Nigel Holmes.
STATIONS : S = Shepparton (GC: 145.25E/36.20S): 6 x 100kW — C = Carnarvon (G.C: 113.43E/24.54S): 1 x 300kW, 1 x 250kW, 1 x 100kW — D = Darwin (G.C: 130.28E/12.25S): 3 x 250kW — B = Brandon (G.C: 147.20E/19.30S): 2 x 10kW.

kHz	kHz	kHz	kHz	kHz	kHz	kHz
5880	7205	9580	11855	15240	17670	21525
5995	7240	9710	11880	15320	17695	21590
6020	7260	9770	11910	15365	17715	21725
6060	9510	9860	13605	15380	17750	21740
6080	9540	11720	13755	15530	17795	25750
7150	9560	11800	15170	15575	17880	

Cantonese	Area	kHz
1300-1430	SPac	6060S
	SEAs	9540C
		9560D
1400-1430	SEAs	13605D
		15170C
Chinese		
0400-0500	CHN	21525D
		17880C
1000-1100	CHN	9510C
		13605D
		15170C
1200-1300	CHN	9540C
1200-1400	CHN	13605D
		15170C
2200-2400	CHN	13605D
2200-2300	CHN	15575D
2300-2400	CHN	15575C
English		
0000-0100	SEAs	13605C
	SEAs	17715D
0000-0400	SEAs	17750C
0030-0100	SEAs	15240C
		17880D
0030-0600	CPac	17795S
0030-0730	CPac	21740S
		15240S
0030-0800	PNG	11880B
	SWPac	11720B
0100-0400	SAs	17670D
0100-0730	CPac	17715S
0100-0800	SEAs	21525D
0100-0900	SAs	21590C
0400-0530	SEAs	17670D
0500-0530	NAs	17880C
0600-0700	SEAs	17880C
	SAs	17670D
0630-0800	SPac	6020S
0700-0900	SAs	17695D
	SEAs	17750C
0730-0830	SPac	15240S
0800-0900*	Pac	11720B
		9710S
		7240B
		6080S
		6020S
0800-0900	NAs	25750D

0900-1000	SEAs	13605D
		9510C
	NAs	15170C
0900-1300	SAs	21725D
0800-2130	CPac	9580S
		5995S
1100-1200	SEAs	13605D
		9510C
	NAs	15170C
1100-1230	PNG	9710S
		6080S
1100-1230	CPac	6020S
1100-2100	PNG	7240B
1300-1400	SEAs	13755C
1300-1630	SEAs	11855S
	CPac	11800S
1430-1800	SAs	13755C
		9510C
	NAs	7260C
1600-2030	CPac	11880S
1630-2030	PNG	6080S
	SPac	11910S
1800-2100	SAs	5880C
		7260C
1900-2100	SEAs	11855C
2100-2230	PNG	9540S
	CPac	11855S
2100-2300	SEAs	9760C
2100-2400	SEAs	11855C
2200-0600	CPac	17795S
2200-0730	SPac	15365S
	PNG	15320S
*) Pacific News Sce.		
French		
2300-2330	SWPac	11720B
	PNG	11880B
	CPac	15240S
		17795S
		21740S
0000-0030	SWPac	11720B
	PNG	11880B
	CPac	15240S
		17795S
		21740S
	SEAs	15240C
		17880D

Indonesian

0900-1230	SEAs	9560C
		9770S
	SEAs	15530D
2100-2300	SEAs	7205S
2100-0030	SEAs	11910S
2200-0030	SEAs	9770D

Khmer

0530-0600	SEAs	17670D
		17880C
1230-1300	SEAs	7150D
		9770S
		13755C

Thai

2300-2400	SEAs	9860C
		15380D

Tok Pisin

0900-1100	PNG	6080S
		7240B
		9710S
	Pac	6020S
		11720B

Vietnamese

1300-1430	SEAs	9770S
		7150D

Reception in No. America: try freqs beamed to CPac.
Reception in Europe/ME: try freqs beamed to SAs.
World N. in English: every odd h. **Current Affairs:** every even h.
ANN: "This is Radio Australia, broadcasting from studios in Melbourne, Victoria".
INT-SIG: "Waltzing Matilda" for 5 mins preceding opening on all freqs. Foreign language b'casts start with the laugh of the Australian bird, the Kookaburra.
PUB: English & Chinese guides twice yearly. Other sces. at irr. intervals. All free on request — **V.** by QSL-card.

PRIVATELY-OWNED BROADCASTING STATIONS

associated in a body called:
Federation of Australian Radio Broadcasters
Addr: P.O. Box 299, St. Leonards, NSW 2065.
Te: +61 (2) 9065944. **Telex:** 25161. **Cable:** FARB, Sydney.
Fax: +61 (2) 9065128.
L.P: Federal Dir: M.J. Hartcher. GM: J.M. Rushton.
Abbreviations: N-1: News on the hour; N-2: News on the half hour; N-3: News on the hour and half hour. The numeral preceding the call letters indicates the state: 2 = New South Wales; 3 = Victoria; 4 = Queensland; 5 = South Australia; 6 = Western Australia; 7 = Tasmania; 8 = Northern Territory.
News: Additional newscasts are often carried during breakfast and drive times — t = translator (rel. main st.).
Stereo: Many st's have, or are planning, AM stereo.

	Call	kHz	kW
1)	2MC	531	5(d)
2)	3GG	531	5(d)
3)	4KZ	531	5(d)
4)	7SD	540	5(d)
5)	4AM	558	5(d)
6)	4GY	558	5(d)
7)	7BU	558	2
8)	2BH	567	0.5
133)	2CS	639	5(d)
132)	6LN	666	1
103)	4LM	666	2
33)	4CC(t)	666	2
9)	4KQ	693	5(d)
103)	4LM(t)	693	0.5
3)	4KZ(t)	693	0.5
144)	3EE	693	5(d)
131)	6SE	747	5
132)	6LN(t)	747	1
10)	2EC	765	3.5(d)
10)	2EC(t)	765	0.5
88)	6KA(t)	765	0.1
134)	5CC	765	5(d)
73)	4GC(t)	765	0.5
11)	4TO	774	5(d)
13)	6VA	783	2
14)	5RM	801	2
15)	4SS	828	5(d)
73)	4GC	828	1
16)	7XS	837	0.5
17)	4CA	846	5(d)
18)	4GR	864	2
19)	6AM	864	2
21)	2GB	873	5
22)	3YB	882	2(d)
23)	6PR	882	2
24)	4BH	882	5(d)
34)	4RR	891	5(d)
25)	2LM	900	5
31)	6BY	900	2
27)	7AD	900	2
28)	8HA	900	2
107)	2LT	900	5(d)
29)	2XL	918	2
31)	6NA	918	2
33)	4CC	927	5(d)
32)	3UZ	927	5
35)	3BO	945	2
69)	4HI(t)	945	1
36)	2UE	954	5
17)	4CA(t)	954	0.5
31)	6TZ(t)	963	2
37)	4WK	963	5(d)
38)	2RG	963	5(d)
93)	5SE	963	5(d)
39)	2MW	972	5(d)
86)	2DU(t)	972	0.3
41)	3HA	981	2
42)	6KG	981	2
112)	2NM	981	5
43)	4RO	990	5(d)
45)	2ST	999	5(d)
46)	4TAB	1008	5(d)
48)	7EX	1008	5(d)
49)	2KY	1017	5
51)	4MK	1026	5(d)
52)	6NW	1026	2
53)	5CS	1044	2
54)	2CA	1053	5(d)
55)	3CV	1071	5
56)	4SB	1071	2
31)	6WB	1071	2
57)	2MO	1080	2
31)	6GL	1080	2
58)	7HT	1080	5
59)	2GZ	1089	5(d)
60)	3WM	1089	5(d)
61)	4LG	1098	2
62)	6MD	1098	2
63)	7LA	1098	5(d)
64)	2UW	1107	5
118)	3AK	1116	5(d)
65)	4BC	1116	5(d)
135)	6MM	1116	2
113)	5MU	1125	5(d)
66)	2AD	1134	2(d)
67)	3CS	1134	5(d)
68)	2HD	1143	2
69)	4HI	1143	5(d)
70)	2WG	1152	2
71)	4MB	1161	2
72)	2CH	1170	5
75)	2NZ	1188	2
78)	2GF	1206	5(d)
79)	6KY	1206	2
80)	2CC	1206	5(d)
45)	2ST(t)	1215	0.35
69)	4HI(t)	1215	0.1
81)	2WS	1224	5(d)
82)	3TR	1242	5(d)
84)	5AU	1242	2(d)
85)	4AK	1242	2
86)	2DU	1251	2
87)	3SR	1260	2
88)	6KA	1260	1
89)	2SM	1269	5
90)	3AW	1278	5
91)	2TM	1287	2
94)	3BA	1314	5(d)
96)	5AD	1323	5
98)	3SH	1332	2
99)	4BU	1332	5(d)
102)	2LF	1350	5(d)
37)	4WK(t)	1359	0.25
104)	2GN	1368	2
105)	3MP	1377	5(d)
106)	5AA	1395	5(d)
108)	2PK	1404	2
111)	2MG	1449	5(d)
114)	3MA	1467	2
115)	4ZR	1476	2
12)	2KA(t)	1476	5
116)	2AY	1494	2
117)	2BS	1503	5(d)
119)	2QN	1521	2
120)	2VM	1530	2
121)	2RE	1557	2
122)	3NE	1566	5(d)
123)	2OO	1575	5(d)
10)	2EC(t)	1584	0.2

FM-Stations (all 24h)

FM	Call	MHz	kW
143)	2WSK-FM	?	25
3)	4KZ(t)	88.5	0.1
122)	3NE(t)	89.3	0.02
138)	4SEA-FM	90.9	24
130)	5SSA-FM(t)	91.1	
51)	4MK(t)	91.4	0.2
78)	4GGG-FM	92.5	5
94)	F.Pl.	92.9	10
35)	F.Pl.	93.5	60
142)	3BAY-FM	93.9	10
101)	3CAT-FM	95.5	35
125)	6NOW-FM	96.1	40
12)	2ONE-FM	96.1	5
29)	2XL(t)	96.3	0.5
95)	2UUL-FM	96.5	5
145)	3SUN-FM	96.9	50
123)	2WIN-FM	98.1	25
120)	2NOW-FM	98.3	10
29)	2XL(t)	98.9	0.2
121)	2RE(t)	100.1	
147)	8HOT-FM	100.1	10
50)	3TTT-FM	101.1	35
140)	2CFM-FM	101.3	10
81)	F.Pl.	101.7	
20)	7HHO-FM	101.7	10
126)	3FOX-FM	101.9	35
40)	5DDN-FM	102.3	5
45)	2ST(t)	102.7	40
109)	2KKO-FM	102.9	10
128)	2DAY-FM	104.1	35
74)	3KKZ-FM	104.3	35
127)	4MMM-FM	104.5	6
138)	2ROC-FM	104.7	10
77)	5KKA-FM	104.7	5
129)	2MMM-FM	104.9	35
124)	3MMM-FM	105.1	35
139)	2NEW-FM	105.3	20
122)	2NE(t)	105.3	0.2
92)	4BBB-FM	105.3	6
137)	2KIX-FM	106.3	5
64)	2WFM	106.5	F.Pl.
45)	2ST(t)	106.7	1
100)	2XXX-FM	106.9	2
8)	2???-FM	106.9	
141)	4QFM-FM	106.9	10
130)	5SSA-FM	107.1	10
16)	7XS(t)	107.1	
97)	2GGO-FM	107.7	10

Addresses and other information:

NB: The term midnight-to-dawn refers to local time. Exact UTC hours vary from st. to st.

1) Midcoast B'casters Pty Ltd, P.O. Box 1611, Port Macquarie NSW 2444 — S.M: G. Hyde. T.D: D. Woodnough. 24h (N. and overnight, rel. st. 36). **F.Pl:** FM license + AM translators.

2) Votraint no. 691 Pty Ltd, P.O. Box 531, Traralgon, Vic. 3844. SM: P. Dethridge. TD: G. Spencer. 24h (N-2) Rel. 35) 1400-0200.

3) Coastal B. Casters Pty. Ltd, P.O. Box 19, Innisfail, Qld. 4860 — S.M: A. Kirton. T.D: J. Roberts. 24h (N-1). Rel. st. 64 0900-1900. Translators: Tully 693kHz 0.5kW, Dunk 1sl 88.5MHz 0.1kW.

4) North East Tasm. R. B. Casters Pty. Ltd, P.O. Box 189, Scottsdale, Tas. 7254 — S.M: S. Johnson. T.D: B. Eyre. W 2000-1300, Sun 2100-1220 (N-1). Rel. 64) overnight.

5) Far Northern Radio Pty. Ltd, P.O. Box 177, Mareeba, Qld 4880 — S.M. & TD: B. Healy. 24h. **F.Pl:** FM st. serving Cairns, AM translator serving Douglasshire.

6) Wesgo Communications p/1, P.O. Box 42, Gympie, Qld 4750 — S.M: G. Walz. T.D: T. Fowler. 24h (N-1). **F.Pl:** AM stereo.

7) Burnie Broadc. Sces. Pty. Ltd, P.O. Box 120, Burnie, Tas. 7320 — S.M: P. Hays . T.D: D. Cables. 2000-1400 (N-1). Rel. 64) midnight to dawn.

8) Far West Radio Pty. Ltd, P.O. Box 699, Broken Hill, NSW 2880 — S.M: R. Daniel. T.D: B. Winn. 24h (N-1.) Rel N. from 21) on the h. **F.Pl:** FM supplementary license.

9) Wesgo Communications Pty Ltd, P.O. Box 693, Stones Corner, Qld 4120 — S.M: H. Roberts. Op's Mgr: M. Spurway. 24h (N-1).

10) Ea. Coast Radio 2EC, P.O. Box 471, Bega, NSW 2550 — S.M: I. Wright. C.E: P. Carson. 24h (Nat. N. from st. 21). **F.Pl:** Move translators to FM. Translators: 765 = Moruya, 1584 = Narooma.
11) 4TO Malbrook p/1, P.O. Box 986, Townsville, Qld 4810 — S.M: P. Thomas. T.D: D. Henry. 24h (N-1). Rel. from st. 9 1400-1900 (Sat/Sun 2000). **F.Pl:** FM.
12) Hayden Nepean Broadc. Pty. Ltd, P.O. Box 145, Penrith, NSW 2750 — S.M: R. Bean. T.D: C. Kruger. 24h (N-1). Translator: Emu Plains.
13) Albany B. Casters Ltd, P.O. Box 293, Albany, WA 6330 — S.M: D. Rutherford. T.D: A. Mead. 24h. (N-1).
14) River Murray B. Casters Ltd, P.O. Box 321, Berri, SA 5343 — S.M: R. Tolano. T.D: B. Wickham. 24h (N-1).
15) Sunshine Coast B. Casters Ltd, P.O. Box 828, Mambour, Qld 4560 — S.M: R. Ferris. T.D: B. Ryan. 24h (N-1).
16) XS West Pty. Ltd, P.O. Box 315, Queenstown, Tas 7467 — S.M. & T.D: S. Adermann. 2000-1400. (N-1). **FM:** 107.1MHz.
17) Greater Cairns Radio, Locked Mailbag 846, Cairns, Qld 4870 — S.M: N. Roberts. T.D: J. Hickey. 24h (N-1). Translator: Gordonvale. Rel. st. 9 midnight-dawn. Rel to 4TO & 4MK 1400-1900. **F.Pl:** Supplementary license.
18) Gold Radio Sces Pty. Ltd, P.O. Box 111, Toowoomba, Qld 4350 — S.M: R. Wilson. T.D: J. Eig. 24h (N-1). Rel. st. 64 midnight to dawn.
19) Radio 6AM Northam Pty. Ltd, P.O. Box 264, Northam, WA 6401 — S.M: J. Ventris. T.D: B. Kent. 24h. Netw. with st. 62) 2200-1000.
20) Comm. B. Casters Pty. Ltd, GPO Box 542F, Hobart, Tas 7001 — S.M: P. Shirley. T.D: Phil McCafferty. 24h (N-1). Rel. st. 64 1400-1930.
21) Harbour Radio. Ltd. GPO Box 4290, Sydney 2001 — S.M: C. Brammall. T.D: K. Lass. 24h (N-3).
22) Regional Communications Pty Ltd, P.O. Box 485, Warrnambool, Vic. 3280 — S.M: M. O'Toole. TD: B. Henderson. 24h (N. from st. 105).
23) Western Broadc. Sces Pty. Ltd, P.O. Box 6072, Hay Str. East, Perth, W.A. 6000 — Joint Chief Execs: B. Martin, J. Watts. S.M: D. Dodds, T.D: A. Mackrill. 24h (N-1). **F.Pl:** new tr. site.
24) Old Radio Ltd, 35 Astor Terrace, Spring Hill, Qld 4000 — S.M: C. Thompson. T.D: T. Stewart. 24h (N-1).
25) Richmond River B. Casters Pty. Ltd, P.O. Box 44, Lismore, NSW 2480 — S.M: C. Lodge. T.D: A. Bone. 2000-1000 (Sat 0800, Sun 1400). Other times rel. st. 36) (N-1). **F.Pl:** translator at Kyogle NSW.
26) Austereo Ltd., P.O. Box 33, Tuart Hill, WA 6060 — 6BY: S.M: P. Barker. T.D: W. Kent. 2200-1000 (N-3). Rel. 6VA 1000 (Sun 2200)-1600. 6IX: S.M: P. Barker. T.D: A. D'mello. 24h (N-1). 6WB: Ellrale Pty Ltd, P.O. Box 148, Bunbury, WA 6230 — SM: C. Mills. TD: P. Browne. **F.Pl:** 6GL move to FM stereo, with new call sign.
27) Northern Tas B. Casters Pty. Ltd., P.O. Box 262, Devonport, Tas 7310 — S.M: Ms. L. Laycock. T.D: B. Eyre. 24h (N-1) rel. 64) overnight.
28) Alice Springs Comm. B. Casters Pty. Ltd, P.O. Box 2106, Alice Springs 5750 — S.M: D. Prior. T.D: J. Farmer. 24h (N-1). Translator at Ayres Rock 101.5 FM. Relays Macsat midnight-to-dawn prgr.
29) Cooma B. Casters Pty. Ltd, P.O. Box 651, Cooma, NSW 2630 — S.M: S. Bow. T.D: I. Stobb. 24h (N-1). **F.Pl:** 5kW (96.3MHz = Jindabyne, 92.1MHz = Thredbo, 98.7MHz = Perisher). Rel. N. from 36).
31) (6NA/6TZ): Radiowest B'casters Pty Ltd, P.O. Box 148, Bunbury, WA 6230 — S.M: M. Spaggs. T.D: R. Bartlett. 24h. Rel. to 6WB/63Y/6NA/6CI.
32) 3UZ Pty Ltd, P.O. Box 927, Carlton, Vic 3053 — S.M: C. Simpson. T.D: N. McCrae. 24h (N-1).
33) Capricorn B. Casters Pty. Ltd, P.O. Box 420, Gladstone, Qld 4680 — S.M: N. Roberts. T.D: R. McCarthy. 24h (N-1). (Translator: Biloela). Rel. 9) midnight-to-dawn.
34) 4RR Reef Radio, P.O. Box 1080, Townsville, Qld 4810 — S.M: B. Martin. 24h (N-1).
35) P.O. Box 108, Goldern Sq. Vic. 3555 — S.M: B. Blacklock. T.D: R. Liersch. Rel. to 2) and 87) overnight. **F.Pl:** move to FM 93.5MHz 60kW.
36) R. 2UE Sydney Pty. Ltd, P.O. Box 950, North Sydney, NSW 2060 — S.M: I. Wallace. T.D: R. Fleming. 24h (N-1).
37) Amalgamated Marketing Pty. Ltd, P.O. Box 403, Toowoomba, Qld 4350 — S.M: T. Rogers. T.D: E. Scoones. 24h (N-1) (Relays st. 91 at 0900-1900). 1359kHz = Toowoomba.
38) 2RG B. Casters Pty. Ltd, P.O. Box 1005, Griffith, NSW 2680 — S.M: C. Corbett. T.D: O. Anderson. 1930-1400 (N-1). **Spanish/Italian:** Wed 1100-1300, Sun 0300-0400.
39) Tweed R. & Broadc. Co. Pty. Ltd, P.O. Box 97, Coolangatta, Qld 4225 — S.M: J. Hurst. T.D: A. Bone. 24h (N-1).
40) First Radio Ltd, P.O. Box 5, North Adelaide, SA 5006 — S.M: M. Wiggins. T.D: E. Tedesco. 24h (N-3). Re. st. 64) 1500-2000. **F.Pl:** 10kW.
41) Western District (3HA) B. Casting, P.O. Box 414, Hamilton, Vic 3300 — S.M: G. Tuckett. T.D: R. Liersch. 24h (N-1). Rel. st. 95) midnight-to-dawn. **F.Pl:** 5kW.
42) Radio 6KG p/1, P.O. Box 440, Kalgoorlie, WA 6430 — S.M: G. Harvey. T.D: R. Holman. 24h (N-1). Rel. 44) 1400-1930.
43) Rockhampton Broadc. Co. Pty. Ltd, P.O. Box 159, Rockhampton, Qld 4700 — G.M: B. Carr. T.D: J. Deakin. 24h (N-1). Rel. Arnsat midnight-dawn.
44) Consolidated Broadc. Syst (WA) Pty. Ltd, P.O. Box 157, Subiaco, WA 6008 — SM: M. Korfe. TD: R. Holman. 24h (N-1). **F.Pl:** 94.5MHz 16kW.
45) South Coast & Tablelands Broadc. Pty. Ltd, P.O. Box 540, Nowra 2540 — S.M: R. Summerall. T.D: D. Munro. 24h (N-1). Translators: Bowrall 1215kHz, Uladulla 106.7MHz 1kW.
46) Broadc. St. 41P Pty Ltd, P.O. Box 275, Albion, Qld 4010 — S.M: J. McCormack. C.E: V.R. Smith. 24h (N-1). Rel. racing from 2KY, 3UZ, 5AA, 6PR, 7HT 2000-1400.
47) Great Northern B. Casters Ltd, P.O. Box 141, Geraldton, WA 6530 — S.M: L. Harris. T.D: T.A. Lennox. W 2200-1500, Sun 2300-1400 (N-1). **F.Pl:** 24h.
48) Tas Radio Pty. Ltd, P.O. Box 812, Launceston, Tas 7250 — S.M: R. Somann-Crawford. T.D: R. Abraham. 24h (N-1). Rel. Arnsat 0900-1900MF & 0800-2000SS.
49) 2KY B. Casters Pty. Ltd, P.O. Box 1303, Parramatta, NSW 2150 — S.M: M. Costa. T.D: M. Carter. 24h (N-1). **F.Pl:** increase power.
50) Doublr T Radio P/L, Private Bag 1011, City Rd. PO, So-Melbourne, Vic. 3205 — S.M: I. Toull. T.D: A. Liddelow. 24h (N-1).
51) Barrier Reef B'casting Pty Ltd, P.O. Box 183, Mackay, Qld 4740 — SM: P.McIntosh. TD: M. Bartlem. 24h (N-1). 91.5MHz = Airlie Beach. Rel. Queensland overnight 1400-1900 (Sat/Sun 2000). **F.Pl:** supplementary license.
52) Northwest Radio Pty. Ltd, P.O. Box 2216, South Hedland, WA 6722 — SM: S. McFarlane. TD: D. Lloyd. 24h. Rel. st. 44 1600-2130, rel. st. 88 0100-0200 & 0400-1600.
53) Coast & Country Broadc. Sces. Pty. Ltd., P.O. Box 481, Pt. Pirie, SA 5540 — S.M: G. Spicer. T.D: J. Carter. 24h (N-1).
54) Austereo Ltd, P.O. Box 163, Canberra City, ACT 2601 — S.M: D. Marshall. T.D: D. Crane. 24h (N. from st. 21).
55) Central Victorian Radio Pty. Ltd, P.O. Box 159, Bendigo, Vic 3550 — S.M: C. Cameron. T.D: B. Lees. 24h (N. from 3EE). **F.Pl:** Conversion to FM.
56) Forsby p/1, P.O. Box 305, Kingaroy, Qld 4610 — S.M: M. Peters. T.D: A. Woodman. 24h (N-3).
57) 2MO Gunnedah Pty. Ltd, P.O. Box 62, Gunnedah 2380 — S.M: B. Kirchner. T.D: B. Holley. 1900-1400 (N-1).
58) Tas Radio Pty. Ltd, GPO Box 572F, Hobart, Tas 7000 — S.M: D. Leitch. T.D: M. Barrenger. 24h (N-1).
59) Aymjit Pty Ltd, P.O. Box 1785, Orange, NSW 2800 — S.M: S. Ward. T.D: O. Anderson. 24h. (n-3).
60) Associated Comm. Enterprises (3WM) Pty. Ltd, P.O. Box 606, Horsham, Vic 3400 — S.M: R. Harding. T.D: R. Barker. 24h. (N-1). Rel. R. Active syndication prgr. 1100-2000MF, 0800-2000SS.
61) Emerald Broadc. Co. Ltd, P.O. Box 20, Longreach, Qld 4730 — S.M: B. Roach. T.D: G. Bonney. 1900-1400. Rel. 4HI Mon-Fri 0300-1930, Sat 2330-1730 & all day Sun. (N-1).
62) Mid-Districts Radio Pty. Ltd, P.O. Box 264, Merredin, WA 6415 — S.M: R. Jerome. 2130-1500 (N-1).
63) 7LA Radio Pty. Ltd, P.O. Box 635G, Launceston, Tas 7250 — S.M: J. Seaman. T.D: M. Nightingale. 24h (N-1). rel. 64) midnight-to-dawn.
64) Commonwealth Broadc. Corp. Pty. Ltd, P.O. Box 1107, Neutral Bay NSW 2089 — S.M: J. Williams. T.D: B. Girdo. 24h (N-1). Key st. in Arnsat satellite netw. to over 40 st's. **F.Pl:** move to FM 106.5MHz.
65) Commonwealth Broadc. Corp (Q'land) Pty. Ltd, P.O. Box 116, Brisbane, Qld 4001 — S.M: J. Musgrove. T.D: B. Griffin. 24h (N-1).
66) New England B. Casters Pty. Ltd, P.O. Box 270, Armidale, NSW 2350 — S.M: D. Thomas. T.D: A. Kollosche. 1900-1400 (N-1). **F.Pl:** 5kW. with new tr. and antenna system.

67) Enterprise B'casters Pty Ltd, P.O. Box 1134, Colac, Vic. 3250 — SM: G. Bratby. TD: M. Gregory. 24h (N-1). Rel. R. Active syndication midnight-to-dawn prgr.
68) 2HD B. Casters Pty. Ltd, P.O. Box 19, Hunter Region Mail Exchange, NSW 2310. S.M: D. Baldwin. T.D: I. Porteous. 24h (N-3).
69) Queensland Regional B'casting Pty. Ltd, P.O. Box 267, Emerald, Qld 4720 — S.M: A. Young. T.D: R. Rumble. 1900-1400 (N-1). Translators: 945 = Dysart, 1215 = Moranban. (**F.Pl:** Moranban to move to FM). 4LG rel. 4HI at times (see 61). Relays to st. 61 0200 (Sun 2000)-1400.
70) Riverina B. Casters (holdings) Pty. Ltd, P.O. Box 480, Wagga Wagga, NSW 2650 — G.M: G. Gross. T.D: P. Combs. 24h (N-1). **F.Pl:** translator at Tumit.
71) Maryborough Broadc. Co. Pty. Ltd, P.O. Box 143, Maryborough, Qld 4650 — S.M: G. Bolderrow. T.D: D. Lewis. 24h. (N-1). Rel. st. 64) Mon-Fri 0900-1900, Sat/Sun 0800-2000.
72) Radio 2CH Pty Ltd. P.O. Box 1170, Nth Sydney, NSW 2059 — SM: C. Joscelyne. TD: J. Oost. 24h (N-1).
73) Radio 4GC Gold City Radio, P.O. Box 381, Charters Towers, Qld 4820 — S.M: B. Barwick. T.D: W. Latham. 24h. Rel. st. 103 0400-0900 then st. 64 overnight. (765kHz = Hughenden).
74) KZFM Pty. Ltd, 24 Victoria Str, Carlton, Melbourne 3053 — S.M: L. Heffer. C.E: B. Perry. 24h (N-1). **F.Pl:** power increase.
75) Northern B'casters Pty. Ltd, P.O. Box 770, Inverell, NSW 2360 — S.M: G. Kachel. T.D: L. Buttsworth. 1930-1405 (N-1). **F.Pl:** apply for supplementary FM license.
76) Gold Coast Radio Broadc. Co. Pty. Ltd, P.M.B. 925, Gold Coast Mail Center Qld. 4217 — GM: N. Scott. TD: P. Cleverley. 24h (N-1).
77) KAFM B. Casters Pty. Ltd, GPO Box 1550, Adelaide, SA 5001 — S.M: S. Robinson. TD: R. Cranwell. 24h (N-1).
78) Westlawn Investment Co. Pty. Ltd, P.O. Box 276, Grafton, NSW 2460 — S.M: R. Bell. T.D: D. Ferry. 24h (National N. from st. 81). **F.Pl:** apply for supplementary FM license.
79) Malberth Pty Ltd, P.O. Box 387, Mt. Hawthorne, WA 6017 — SM: P. Perrin. TD: L. Bordoni. 24h (N-1).
80) Capital City B. Casters Pty. Ltd, P.O. Box 1499, Canberra, ACT 2601 — S.M: M. Baldwin. C.E: B. Klein. 24h (N-1). Rel. Arnsat 1400 (Sat/Sun 0800)-2000.
81) Wesgo Communications Pty. Ltd, P.O. Box 234, Seven Hills, NSW 2147 — S.M: G. Mott. T.D: R. Langhans. 24h (N-1). Midnight to dawn prgrs. to 2GO, 4KQ, 2CS, 2OO, 2AY, 3MP, 5KA-FM, 2MCE-FM, Sunshine-FM. **F.Pl:** move to FM 101.7MHz in early 1993.
82) Southern Cross Radio 85 Hotham Str, Traralgon, Vic. 3844 — SM: D. Broadbent. T.D: S. Mummery. 24h (N-1).
84) 5AU B. Casters Pty. Ltd, P.O. Box 496, Port Augusta, SA 5700 — S.M: G. Kernahan. T.D: B. Wickham. 24h (N-rel. st. 36). Relays 5AD 1430-2000. **Italian:** Sun 1030-1130.
85) Broadc Op's Pty. Ltd, P.O. Box 783, Toowoomba, Qld 4350 — S.M: A. Zastera. T.D: A. Bone. 24h (N-1).
86) Western B. Casters Pty. Ltd, P.O. Box 1221, Dubbo, NSW 2830 — S.M: R. Ferguson. T.D: J. Lane. 24h. Rel. st. 36) N. & midnight to dawn. Translator: Cobar. **F.Pl:** supplementary FM license.
87) Hanor P/L, Ltd, P.O. Box 1260, Shepparton, Vic 3630 — S.M: R. Dunn. T.D: R. Taylor. 24h (rel. st. 35 overnight).
88) Northwest Radio Pty. Ltd, P.O. Box 153, Karratha, WA 6714 — S.M: P. Youd. T.D: M. Lowe. 2130-1600 (N-1). Rel. st. 44) 1400-1930. Translators: Tom Price & Parabourdon on 765kHz. **F.Pl:** 2kW on 1260kHz, translator for 6NW in Broome.
89) Radio 2SM Pty. Ltd, P.O. Box 1270, North Sydney, NSW 2060 — S.M: M. Spurway. T.D: N. Coleman. 24h (N-1). **F.Pl:** FM.
90) Southern Cross 3AW Pty. Ltd, GPO Box 369F, Melbourne 3001 — S.M: T. Bell. T.D: P. Iliff. 24h (N-1).
91) Tamworth Radio Development Co, P.O. Box 497, Tamworth, NSW 2340 — S.M: E. Wilkinson. T.D: R. Holley. 24h (N-1). **F.Pl:** supplementary FM license on 92.9 with 10 or 20kW.
92) B105FM Pty Ltd, P.O. Box 105, Broadway, Brisbane, Qld 4006 — G.M: E. Breslin. T.D: D. Sydney-Smith. 24h (N-1).
93) Southern Cross B'casters Pty. Ltd, P.O. Box 500, Mt. Gambier, SA 5290 — S.M: S. Whelan. T.D: H. Scheggetman. 24h (rel. 5AD 1430-2030) (N-1).
94) Radio Ballarat Pty. Ltd, P.O. Box 360, Ballarat, Vic 3350 — S.M: P. Caligari. T.D: B. Thurnbull. 24h (N. from st. 90). **F.Pl:** supplementary FM license.
95) Associated Communication Enterprises (2WL) P/L, P.O. Box 1234, Wollongong, NSW 2500 — S.M: P. Posetti. T.D: D. Coyle. 24h (National N. from st. 36). **F.Pl:** move to FM.
96) Southern States Broadcasters, GPO Box 518, Adelaide, SA 5001 — S.M: W. Clovten. T.D: G. Wright. 24h (N-1).
97) Wesgo Communications Pty. Ltd, P.O. Box 564, Gosford, NSW 2250 — SM: P. Little. TD: S. Bartlett. 24h (N-1). Rel. st. 81) 1400-1900. **F.Pl:** Move to FM 107.7MHz 10kW from Somersham NSW early 1992.
98) Associated Communications Enterprises (3SH) Pty. Ltd., P.O. Box 504, Swan Hill, Vic 3585 — S.M: G. Rudge. T.D: R. Baker. 24h (N-1).
99) Bundaberg B. Casters Pty. Ltd, P.O. Box 1059, Bundaberg, Qld 4670 — S.M: G. McKennariey. T.D: T. Dekker. W 24h (N-1).
100) Hunter B. Casters Pty. Ltd, P.O. Box 5275D, Newcastle West, NSW 2305 — S.M: D. Heitch. T.D: A. Jobson. 24h (N-3).
101) Geelong B. Casters Pty. Ltd, P.O. Box 9550, Geelong, Vic 3220 — S.M: G. Browne. CE: F. Wilcox. 24h (N-1).
102) Young B. Casters Pty. Ltd, P.O. Box 31, Young, NSW 2594 — S.M: B. Forde. T.D: P. Combs. 24h. Rel. st. 70 0800-1400, rel. st. 64 1400-2000. (N-1).
103) North Queensland Broadc. Co. Pty. Ltd, P.O. Box 780, Mount Isa, Qld 4825 — S.M: T. Lees. T.D: W. Latham. 24h. (N-1). Relays to 4GC 0400-2000. Translator: Cloncurry — **F.Pl:** supplementary FM license.
104) Radio 2GN Pty. Ltd, P.O. Box 115, Goulburn, NSW 2580 — T.D: G. Guyer. 24h (N-1).
105) Radio 3MP, P.O. Box 75, Frankston, Vic. 3001 — S.M: N. Crowe. T.D: K. Brown. 24h (N-3).
106) Festival City B. Casters Pty. Ltd, Pulteny Court, Adelaide SA 5000 — S.M: S. Barrett. T.D: J. Ferguson. 24h (N-1).
107) Midwest Radio Network, Mailbag 90, Lithgow, NSW 2790 — S.M: B. Egan. T.D: P. Barnes. 24h. (N-1).
108) Parkes Broadc. Co. Pty. Ltd, P.O. Box 295, Parkes, NSW 2870 — S.M: K. Birch. T.D: J. Lane 24h (N-3). Rel. 95) overnight.
109) Radio 2KO Newcastle P/L, P.O. Box 606, Charlestown, NSW 2290 — S.M: J. McGahen. T.D: J. Richardson. 24h. rel 36) 1400-1900 (Sat/Sun 2000).
111) Mudgee Broadc. Co. Pty. Ltd, P.O. Box 17, Mudgee, NSW 2850 — S.M: K. Birch. T.D: J. Lane. 24h. (National N. from st. 21).
112) Radio Hunter Valley Pty. Ltd, P.O. Box 600, Muswelbrook, NSW 2333 — S.M: G. Stewart. T.D: M. Reedman. 24h (N-1).
113) Murray Bridge Broadc. Co. Ltd, P.O. Box 470, Murray Bridge, SA 5253 — S.M: R. Kerr. T.D: K. May. 24h (N-1).
114) Sunraysia B. Casters Pty. Ltd, P.O. Box 539, Mildura, Vic 3500 — S.M: R. Harwood. T.D: K. Stephens. 24h (N-1). Greek/Italian/Yugoslavian/Turkish: Sun 1000-1230.
115) Maranoa Broadc. Co. Ltd, P.O. Box 22, Roma, Qld 4455 — Rel. st. 33) overnight.
116) Radio Albury Wodonga Ltd, P.O. Box 670, Albury, NSW 2640 — SM: D. Avery. TD: M. Moses. 24h. Rel. 81) 1400-1900 (N-1). **F.Pl:** FM translator at Falls Creek, Vic.
117) Bathurst B. Casters Pty. Ltd, P.O. Box 310, Bathurst, NSW 2795 — M.D: R. Camplin. T.D: B. Upfold. 24h.
118) 3AK Melbourne B. Casters, P.O. Box 150, Richmond, Vic 3121 — S.M: M. Collier. C.E: M. Dean. 24h (N-1). **F.Pl:** 101.1MHz 10kW.
119) Rich Rivers Broadc. Pty. Ltd, P.O. Box 312, Deniliguin, NSW 2710 — S.M: D. Robertson. T.D: J. Goodall. 24h. (N-1). Rel. Arnsat Mon-Thurs 0900-2000, Fri 1400-2000, Sat/Sun 0800-2000.
120) Moree Broadc. & Development Co. Ltd, P.O. Box 389, Moree, NSW 2400 — SM & TD: S. O'Shea. 1900-1400. (N-1).
121) Manning Valley Broadc. Pty. Ltd, P.O. Box 275, Taree, NSW 2430 — S.M: G. Arklay. T.D: D. Shorter. 1900-1400 (N-1). Translator: Gloucester 100.1MHz.
122) North East B. Casters Pty. Ltd, P.O. Box 449, Wangaratta, Vic 3677 — S.M. & T.D: I. Crockett. 24h (N-1). Rel. Arnsat midnight-dawn. FM translators: Mt. Hotham 89.3MHz 0.02kW, Mt. Buffalo 105.3MHz 0.2kW.
123) Wollongong City R. Ltd, P.O. Box 1800, Wollongong, NSW 2500 — S.M: P. Posetti. T.D: I. Clifford. 24h (N-3).
124) Melbourne F.M. Radio Pty. Ltd, GPO Box 105, Melbourne, Vic 3001 — S.M: G. Buschman. T.D: F. Wilcox (N-1).
125) New Broadc. Pty. Ltd, 111 Wellington Str, Perth, WA 6000 — S.M: G. Roberts. T.D: G. Hately. (N-1).
126) Austereo Ltd, P.O. Box 1019, St. Kilda, Vic. 3182 — G.M: S. Ruhfus. C.E: A. Adler. 24h (N-1).
127) FM 104 Pty. Ltd, GPO Box 1041, Brisbane, Qld 4001 — S.M: P. Butler. T.D: B. Kelly. (N-2).
128) 2DAY-FM Level 2, P.O. Box 920, Crows Nest, NSW 2065 — S.M: R. Scott. T.D: M. Grimes. (N-1).
129) The Triple M Broadc. Co. Pty. Ltd, GPO Box 442, Sydney, NSW 2001 — G.M: G. Smith. T.D: I. Robertson. (N-1).

130) Austereo Ltd, P.O. Box 404, Unley, SA 5061 — S.M: B. Nielsen. T.D: P. Bourke (N-1). Translator = So. Tce, Adelaide.
131) Esperance B'Casters Pty. Ltd, P.O. Box 527, Esperance, WA — S.M: N. McGarath. T.D: R. Davis. 24h. N-1. Rel. 6PPM-FM midnight to dawn.
132) Carnarvon Comm. B. Casters Pty. Ltd, P.O. Box 665, Carnarvon, WA 6701 — G.M: M.J. May. T.D: W. Frost. 2200-1500 (N-1). Translator: Exmouth.
133) Comm. Radio Coffs Harbour Ltd, P.O. Box 1234, Coffs Harbour, NSW 2450 — S.M: M. O'Connor. T.D: C. Molyneaux. 24h (N. rel. st. 81).
134) Coast & Country Broadc. Sces Pty. Ltd, P.O. Box 483, Lincoln, SA 5606 — S.M: D. Prior. T.D: J. Martin. 24h. Rel. st. 36 or 53 0830-1830.
135) West Coast Radio Pty. Ltd, P.O. Box 688, Mandurah, WA 6210 — S.M: D. Wyatt. T.D: K. Bicknell. 24h (N-1). Rel. Arnsat prgr. W 1000-2200, Sun 0400-2000.
136) Austereo Ltd, GPO Box 163, Canberra, ACT 2601 — S.M: D. Marshall. T.D: D. Crane. 24h. Belongs to 54). **F.Pl:** translator for Tuggeranong area.
137) Capital City B'casters Pty. Ltd, P.O. Box 106, Dickson, ACT 2602 — S.M: M. Baldwin. T.D: J. Ockenden. 24h. Belongs to 80).
138) SEA FM P/L, P.O. Box 5910, Gold Coast Mail Centre, Bundall, Qld 4217 — SM: C. Denyer. TD: J. Vasey. (N-1).
139) Newcastle FM P/L. S.M: M. Webb. T.D: R. McClymont. 24h.
140) Gosford Communications Inc, P.O. Box 1997, Gosford, NSW 2250 — SM: J. Eller. TD: G. Bennetts. 24h. (N-3).
141) Ipswich & West Moreston B'casting Corp. Pty Ltd, P.O. Box 7, Ipswich, Qld 4305 — SM: K. Mulcahy. TD: R. Harling. 24h (N-1).
142) Geraldton FM Pty. Ltd, Locked, Bag 98, Geraldton, WA 6530 — S.M: K. Newman. T.D: P. Buzzard. 24 h (N-1).
143) owned by st. 45).
144) AWA Media Pty. Ltd, P.O. Box 7693, Melbourne, Vic 3004. N-3. SM: P. Ramsen. TD: P. Taylor. 24h.
147) P.O. Box 2510, Darwin, NT 0800 — GM: B. Earl.

PUBLIC BROADCASTING ASSN. OF AUSTRALIA

ADDR: Suite One, Level Three, 44-54 Botany Rd. Alexandria, NSW 2015 **Te:** +61 (2) 3102999. **Fax:** +61 (2) 3194545 — Exec. Dir: Ada Hilshoff.

PUBLIC BROADCASTING STATIONS

Medium Waves

	Call	kHz	kW		Call	kHz	kW
1)	5UV	531	0.5	47)	1PPP	1125	0.5
2)	2WEB	585	2(d)	46)	3RPH	1179	5
3)	3CR	855	1	48)	5RPH	1197	2
55)	7RPH	864	0.5	45)	4RPH	1296	5
4)	6NR	927	2	53)	2RPH	1629	*0.5
56)	6RPH	990					
5)	2XX	1008	0.3				
6)	4EB	1053	0.5				

*) **F.Pl:** 1107 or 1224kHz.

FM-Stations

	Call	MHz	kW		Call	MHz	kW
42)	3MFM-FM	88.1	1	58)	3MBR-FM	103.5	3
50)	4CCR-FM	89.1	1	20)	4MBS-FM	103.7	7.5
25)	4CRB-FM	89.3	3	12)	3MBS-FM	103.5	10
22)	3CCC-FM	89.5	2	21)	2NUR-FM	103.7	3
7)	7THE-FM	92.1	10	41)	7LTN-FM	103.7	2
8)	6UVS-FM	92.1	5	26)	2SSS-FM	103.9	10
37)	2ARM-FM	92.1	1	23)	8TOP-FM	104.1	10
9)	2MCE-FM	92.3	1	43)	2CHY-FM	104.1	1
38)	3ZZZ-FM	92.3	10	34)	2BOB-FM	104.5	1.5
10)	2NCR-FM	92.9	3	31)	5TCB-FM(t)	104.5	1
11)	5EBI-FM	92.9	3	39)	3GCR-FM	104.7	5
13)	5MMM-FM	93.7	4	)	4TTT-FM	103.9	1
57)	2OOO-FM	98.5	1	27)	7WAY-FM	105.3	2
33)	6SON-FM	98.5	10	51)	2UUU-FM	105.5	1
56)	3YYR-FM	100.3	1	30)	3MFM-FM	106.1	1.5
40)	4CBL-FM	101.1	2	59)	7DBS-FM	106.1	2.5
29)	2GLA-FM	101.5	5	31)	5TCB-FM	106.1	2
14)	4ZZZ-FM	102.1	7.5	36)	3RPC-FM	106.5	2
15)	2MBS-FM	102.5	5	32)	3PBS-FM	106.7	5
16)	4DDB-FM	102.7	2	44)	3ONE-FM	106.9	5
17)	3RRR-FM	102.7	55	49)	2VOX-FM	106.9	1
18)	2CBA-FM	103.2	5	24)	2SER-FM	107.3	9
19)	7HFC-FM	103.3	3	52)	2REM-FM	107.3	1

+90 additional st's below 1kW not mentioned.

Addresses and other information:
1) R. 5UV, Univ. of Adelaide, 228 North Terrace, Adelaide, SA 5000 — S.M & T.D: J. Langdon. 24h. Rel. BBCWS.
2) Western Region Educational Broadc. Co. Ltd, P.O. Box 426, Bourke, NSW 2840 — S.M: I. Cole. T.D: J. Roberts. 1900-1400. **V.** by QSL-card. Rp. **F.Pl:** 5kW (d).
3) Community R. Federation Ltd, P.O. Box 277, Collingwood, Vic 3066 — S.M: B. Francis. 24h in 17 languages. **F.Pl:** 5kW(d).
4) Curtin Univ. of Technology, GPO Box U1987, Perth, WA 6001 — S.M: I. Hill. T.D: G. Rowley. 24h. Rel. BBCWS Sun-Thurs 1600-1000. Italian Soccer Sun 1400-1530.
5) Community Radio 2XX Inc, P.O. Box 4, Canberra City, ACT 2600 — S.M: E. O'Brien. T.D: K. Vennenon. 1900-1500. 31 non-English prgrs between 1030-1230 (W only). **F.Pl:** Power increase, stereo.
6) P.O. Box 7300, East Brisbane, Qld 4169 — S.M: M. Rogers. T.D: L. Dive. 2000-1400 (Rel. 2EA 2000-2330). 42 different language prgrs.
7) Hobart FM Inc, P.O. Box 1324, GPO Hobart 7001 — Pres: R. Medley. T.D: B. Young. 13 ethnic langs. 2000-1400.
8) Universities Radio Ltd, c/o Univ. of Western Australia, Nedlands, WA 6009 — S.M: A. Tonks. T.D: F. Hayward. 2155-1615/2000v. Rel. BBC-WS N: 2200, 2300, 0000. Prgrs in Italian/French/Russian.
9) Charles Sturt Univ., Mitchell Bathurst, NSW 2795 — S.M: I. Stanistreet. T.D: G. Clulow. 2030-1400. (**N:** rel. comm. st. 81). Dutch Sun 0600-0800, Spanish Sun 0800-0900, German Sun 0900-1000. Rel. N. from comm. st. 81). Translator at Orange on 94.7MHz 0.5kW.
10) Univ. of New England, Northern Rivers, P.O. Box 5123, East Lismore, NSW 2480 — S.M: A. Nicholson. T.D: A. Steward. Rel. N. from comm. st. 72) and BBCWS. Rel. to 2 TEN-FM 1400-1900. Foreign languages W 0300-0400. **F.Pl:** 50kW.
11) Ethnic Broadcasters Inc, 10 Byron Place, Adelaide, S.A. 5000 — S.M: W. De Veer. Prgr. Coordinator: P. Peterejma. T.D: A. Wlodarczak. 24h. 48 foreign languages. **F.Pl:** 103.1MHz 10kW.
12) Music Broadc. Society of Victoria Ltd, 146 Cotham Rd, Kew, Vic 3101 — S.M: S. Worchester. T.D: C. Gliddon. 24h.
13) Progressive Music Broadc. Assoc, 43 Franklin Str, Adelaide, SA 5000 — Administrator: M. White. TD: I. Moore. 24h. **F.Pl:** change freq to 106.3.
14) Creative B.Casters Ltd, P.O. Box 509, Toowong, Qld 4066 — St. Coordinator: G. Fletcher. Tech. Coordinator: S. Brazier. 24h. Limited Aboriginal & Spanish prgrs. **N:** Mon-Fri 2130, 2300, 0800. **F.Pl:** Freq. change.
15) Music Broadc. Society of NSW Co-op Ltd, 76 Candos Str, St. Leonards — S.M: K. Hamilton. T.D: G. Smith. 24h. **F.Pl:** increase power. **V.** by letter. Rp.
16) Darling Downs Broadc. Society, P.O. Box 400, Toowoomba, Qld 4350 — SM: A. Taylor. Prgr. Mgr: D. Boyle. TD: J. Drever. 2000-1400 (Fri 24h). Rel. BBCWS N. **F.Pl:** ethnic prgrs.
17) Triple R.B.Casters Ltd, P.O. Box 304, Fitzroy, Vic 3065 — S.M: L. Rogers. T.D: M. Fisher. 24h. **F.Pl:** power increase.
18) Christian Broadc. Assoc, P.O. Box 54, Five Dock, NSW 2046 — S.M: Rev. V. Turner. T.D: A. Gates. 24h. **V.** by letter. Rp. **F.Pl:** Power increase.
19) Hope Foundation Communications Inc, P.O. Box 281, Moonah, Tas 7009 — S.M: N. Jeanerret. T.D: W. Carpenter. 2000-1300. **F.Pl:** 2000-1400. **V.** by letter.
20) Music Broadc. Society of Q'land, P.O. Box 9, Redhill, Qld 4059 — SM: G. Thorpe. TD: M. Georges. 24h. **V.** by letter.
21) Newcastle Community R, 2NUR-FM, Univ. of Newcastle, NSW 2308 — S.M: S. Ford. T.D: P. Bates. 24h Ethnic prgrs. MF 0830-0930, Sun 0800-0900.
22) Goldfield Community R. Co-op Ltd, Station Str, Harcourt, Vic. 3453 — S.M: G. Pyre. T.D: C. Lomon. 24h. 3 foreign langs. **F.Pl:** power increase.
23) Northern Territory University, P.O. Box 40146, Casuarina, NT 5792 — S.M: C. Millar. T.D: G. Edwards. 24h. 16 foreign language prgrs.
24) Sydney Educational Broadc. Ltd, P.O. Box 123, Broadway, NSW 2007 — S.M: B. Guinn. 24h in 16 languages. Rel. BBCWS N. **V.** by letter. Rp. **F.Pl:** 10kW.
25) Gold Coast Christian & Community Broadc. Assoc, P.O. Box 86, Burleigh Heads, Qld 4220 — S.M: H. Redding. T.D: M. Bullat. 24h. 8 Foreign language prgrs. N. from 4BH-AM.
26) Canberra & District Racing & Sporting B'Casters Ltd, P.O. Box 24, Hackett, ACT 2602 — S.M: J. Paterson. T.D: R. Mallison. 2000-1400. Greek: Thurs 0000-0100.

27) Launceston Christian B'Casters Inc, P.O. Box 1111, Launceston, Tas — S.M: K. Hooper. T.D: R. McLean. 2000-1400.
29) 2LGA-FM, P.O. Box 1015, Foster NSW 2428 — S.M: T. McDermott. T.D: S. Ellis. 24h. Rel. N. from comm. st. 21). **F.Pl:** 2 translators.
31) Tatiana Community FM B'casters Inc, P.O. Box 396, Bordertown, SA 5268 — S.M: G. Williams. T.D: A. Hyslop. 24h. Rel. BBC-WS N. 0300, 0800. (104.5MHz = Keith). **F.Pl:** increase power.
32) Progressive Broadcasting Service Co-op Ltd, P.O. Box 210, St. Kilda Vic. 3182 — S.M: Mrs. M. Knox. T.D: B. Runting. 24h.
33) Good News B'casters Inc, P.O. Box 430, Morley, WA 6062 — S.M: P. Randall. T.D: M. Ewers. 2100 (Sat/Sun 2200)-1600 (N-1).
34) Manning Media Co-op Society Ltd, P.O. Box 4000, Taree, NSW 2430 — S.M: A. Colvin. 24h. German Thurs 0900-1000. **F.Pl:** Power increase.
35) Eurobodalla Access Radio Inc, P.O. Box 86, Moruya, NSW 2537 — 1900-1400. German/French. Fri 0200-0400. **F.Pl:** 2kW & translator at Narooma.
36) P.O. Box 450, Portland, Vic. 3305 — S.M: B. Lander. T.D: G. Denys. 24h.
37) P.O. Box 707, Armidale, NSW 2350 — S.M: S. Lawson. T.D: L. Rau. 24h (N-1). 8 foreign language prgrs. N. from comm. st. 21).
38) Ethnic Public Broadc. Assoc. of Victoria, P.O. Box 422, Carlton South, Vic. 3053 — S.M.: R. Lamshead. T.D: C. Holliday. 47 different languages. 24h.
39) Gippsland Community Radio, P.O. Box 579, Morwell, Vic. 3840 — S.M: P. Strickland, T.D: K. Prestidge. 24h. Foreign language prgrs throughout the week.
40) Radio Logan Inc, P.O. Box 1396, Springwood, Qld 4127 — S.M: A. Anderson. T.D: J. Horrocks. 1800-1400.
41) Launceston Community FM Group Inc, P.O. Box 1501, Launceston, Tas 7250 — SM: D. Sice. MF 2000-2300, SS 2000-1400. Various foreign languages Sat./Tues. **F.Pl:** 24h.
42) South Gippsland FM Radio Inc, P.O. Box 144, Inverloch, Vic. 3996 — SM: D. Ford. TD: A. Hyslop. 24h. (repeaters on 89.1 & 89.5MHz). **F.Pl:** SKW.
43) Community Media CHY Ltd., P.O. Box J233, Coffs Harbour Jetty, NSW 2450 — SM: R. Bunting. TD: J. Williams. 1900-1400.
44) Goulburn Valley Community Radio Inc, P.O. Box 824, Shepparton, Vic. 3630 — S.M: R. Marshman. 24h. 5 foreign languages.
45) Queensland Radio for the Print-Handicapped Ltd, Level 8, Northpoint Bldg., 231 Nth Quay, Brisbane, Qld 4000 — SM: B. Taggart. TD: L. Dive. 24h (N-3). Rel. BBCWS 1300-2009. **V.** by letter. Rp.
46) R. for the Print-Handicapped (Victoria) Co-op Ltd, H.M. Lightfoot Centre, Assoc. for the Blind, 454 Glenferrie Rd, Kooyong 3144 — S.M: J. Hileke. T.D: W. Hartung. 24h. **V.** by QSL-card. Rp.
47) Print-Handicapped Radio of ACT Inc, Barton Highway, Gungahlin, ACT 2912 — 2030-0200 & 0730-1300. **V.** by letter. Rp. **F.Pl:** increase power.
49) P.O. Box 1663, Wollongong, NSW — SM: D. Lucas. TD: L. Stokes. 2000-1400. **F.Pl:** increase power.
50) Cairns FM Broadc. Society, PO Box 300, Manunda, Cairns, Qld 4870
51) Shoalhaven Community R., P.O. Box 884, Nowra, NSW 2541.
52) P.O. Box 1079, Lavington, NSW 2641.
53) R. for the Print-Handicapped (NSW) Co-op Ltd, 186 Blues Pt. Rd, Sydney, NSW — S.M: J. Marchant. **F.Pl:** 1107 or 1224kHz 5kW.
54) Townsville Community Broadc. Co. Ltd, P.O. Box 1033, Townsville, Qld. 4810 — S.M: Lynne Cavanagh. T.D: A. Stephenson. 24h. 17 foreign language prgrs.

SPECIAL BROADCASTING SERVICE (SBS)

SBS RADIO is the radio division of the Special Broadcasting Service. SBS Radio administers two radio stations (2EA in Sydney and 3EA in Melbourne) which broadcast 18 hours each day in upwards of 50 languages each, to cater for the needs of Australians from a non-English speaking background.
ADDR: P.O. Box 0-28, Milsons Point, NSW 2061. **Te:** +61 (02) 964-2828. **Fax:** +61 (02) 956-5393.
L.P: Head of Radio: Tuong Quang Luu.

	Call	kHz	kW		Call	kHz	kW
1)	2EA	1386	5	2)	3EA	1224	5

Addresses and other information:
1) P.O. Box 1386, Bondi Junction, NSW 2022 — S.M: S. Madina. 2000-1400 (Translators: Wollongong 1485kHz. 0.1kW, Newcastle 1584kHz. 0.1kW).
2) P.O. Box 294, South Melbourne, Vic 3205 — St. Mgr: P. Horton. 2000-1400.

COOK ISLANDS

L.T: UTC − 10h (Su: UTC − 9½h) — **Pop:** 18.000 — **Radio sets:** 10.000 — **Pr.L:** English, Maori — **E.C:** a/c 50, 220V — **ITU:** CKH.

COOK ISLANDS BROADCASTING AND NEWSPAPER CORPORATION

ADDR: P.O. Box 126, Avarua, Rarotonga. **Te:** +682 29460. **Fax:** +682 21907. **Cable:** Cooksbroad.
L.P: GM: David G. Nuhn. CE: Ngaire Kavana.
MEDIUMWAVES: ZK1ZC 630kHz 5kW (standby tr: 0.5kW).
SHORTWAVE (G.C: 159.48W/21.11S): 11760kHz 0.5kW.
HOME SCE. in English, Maori: 1600(Sun 1700)-1000 (irr. 24h).
National N. in Maori/English: 1600, 1645, 2245, 0400, 0600.
N. in English (rel. RNZ or ABC): 0100, 0200, 0300, 0500.
ANN: "This is Radio Cook Islands Calling" — **INT-SIG:** Symphony of Drums — **V.** by QSL-card. Rp.

RADIO IKURANGI (Comm.)

ADDR: P.O. Box 521, Avarua, Rarotonga. **Te:** +682 23203.
L.P: Pres. & GM: David Schmidt.
STATIONS: ZK1ZD 95.3/103.3MHz 1kW (stereo).
D.PRGR. in English: 1600 (Sun 1700)-0900 (Sat 1000, Sun 0800). **N.** on the h. (rel. R. Australia or R. NZ).
ANN: KCFM 103 — **V.** by letter.

EASTER ISLAND (Chilean Dependency)

L.T: UTC − 6h (Su: UTC − 5h) — **Pop:** 2.000 — **Pr.L:** Spanish, Rapa Nui.

RADIO MANUKENA (Cult.)

ADDR: Aeropuerto Mataveri, Isla de Pascua, Chile.
L.P: Dir: Wilfredo Reyes Gonzalez.
STATION (G.C: 109W/27S): 580kHz 0.4kW — FM 101.8MHz.
D.PRGR. in Spanish: exact h. of tr. unconfirmed.

FIJI

L.T: UTC + 12h — **Pop:** 750.000 — **Radio sets:** 450.000 — **Pr.L:** English, Fijian, Hindustani — **E.C.:** a/c 50, 240V — **ITU:** FJI.

FIJI BROADC. COMMISSION
(Independent Statutory Body)

ADDR: P.O. Box 334, Suva. **Te:** +679 314333. **Fax:** +679 301643. **Cable:** BROADCOM SUVA. **Telex:** 2142 FBC FJ — P.O. Box 606, Lautoka. **Te:** +679 62121. **Fax:** +679 65855 — P.O. Box 1241, Labasa. **Te:** +679 82888. **Fax:** +679 63450.
L.P: GM: Epeli Kacaimaiwai. Chmn: Ratu Josua Toganivalu. Mgr. Eng./Tech. Resources: Simon Young.

Mediumwaves	kHz	kW	N	Mediumwaves	kHz	kW	N
Suva	558	20	1	Sigatoka	927	2.5	1
Lautoka	639	10	1	Lautoka	990	5	3
Labasa	684	10	1	Suva	1089	1	3
Suva	774	20	2	Sigatoka	1206	2.5	2
Labasa	810	10	2	Rakiraki	1323	2.5	1
Lautoka	891	10	2	Rakiraki	1467	2.5	2

FM: Suva 98.0/104.0MHz 1kW.
R. Fiji 1 in Fijian, **R. Fiji 2** in Hindustani, **R. Fiji 3** in English, **Magic 104** in English on 104.0MHz, **R. Rajdhani** in Hindi on 98.0MHz: 1655-1200 (Fri/Sat 24h). **N. in English** (R. Fiji 3): 1700-2000 hourly, 2200, 0030, 0130, 0400-0700 hourly, 0900, 1000, 1100.
ANN: E: "This is R. Fiji 1/2/3", "The Voice of the Islands".
INT-SIG: Fijian Lali (Log-Drum) — **V.** by QSL-card. Rp.
F.PL: R. Fiji 3 nationwide expansion.

COMMUNICATIONS FIJI (Comm.)

ADDR: 23 Stewart Str, Suva (**Postal addr:** Private Mail Bag, Suva). **Te:** +679 314766. **Fax:** +679 315190. **Telex:** 2496 SUVA FM.
L.P: SM: W. Parkinson. CE: N. Prasad. PD (Hindi): A. Diwarkar.
FM (1kW): Suva 96.0/98.8MHz, Lautoka 95.4/97.4MHz + 3 st's 0.15kW.
FM 96 in English/Fijian: 24h on 96.0/95.4/97.6MHz.
Navtarang in Hindi: 24h on 98.8/97.4/101.6MHz.

GALAPAGOS ISLANDS (Province of Ecuador)

L.T: UTC – 6h — **Pr.L:** Spanish

LA VOZ DE GALÁPAGOS (Cult.)

ADDR: Misión Franciscana, Puerto Baquerizo Moreno, Isla San Cristobal, Galápagos, Ecuador.
L.P: Dir: P. Manuel Valarezo. Gerente: Fr. Manuel Jácome Basantes.
STATIONS: HCVG8 1320kHz 5kW.
D.PRGR in Spanish: 1300-0200 — **V.** by QSL-card.

RADIO SANTA CRUZ (Cult.)

ADDR: Misión Franciscana, Puerto Ayora, Isla de Santa Cruz, Galápagos, Ecuador.
L.P: Dir: P. Nelson Bolaños.
STATION: HCSC8 92.2MHz 1kW — **D.PRGR:** 1300-0100.

GUAM

L.T: UTC + 10h — **Pop:** 120.000 — **Radio sets:** 105.000 (est.) — **Pr.L:** English, Chamorro, Filipino — **E.C:** a/c 60, 110/220V — **ITU:** GUM.

Mediumwaves	**kHz**	**kW**		**kHz**	**kW**
2) KGUM	567	5	6) KTWG	801	10
1) KUAM	612	10			
FM	**MHz**	**kW**		**MHz**	**kW**
5) KSDA-FM	91.9	4	2) KZGZ	97.5	40
1) KUAM-FM	93.9	2	4) KOKU	100.3	5
3) KSTO	95.5	25			

Addresses and other information:
1) Box 358, Agana, Guam 96910 — Pres. & GM: Jon Denight. Asst. GM: Patty Clymer. CE: Dean Reed. PD(AM): Rick Baza. PD(FM): Geoff Elkins. KUAM-AM: "Chamorro Music" st. w/Tagalog music 2h daily M-F. N. in English. KUAM-FM: "Classic 94" – Classic Hits. **V.** by QSL-card.
2) Sorensen Pacific Broadcasting Inc, P.O. Box GM, Agana, Guam 96910 — Chmn: Rex Sorensen. Pres: Jon Anderson. VP: Kethleen Sorensen. Gen. Sales Mgr: Shirley Buetts. Direct Mktg. & Op's: Evan Montuel-Cohen. KGUM: 24h news/talk in English. KZGZ: 24h CHR. **N.** on the h. (rel. CBS) — **ANN:** "K57 Radio", "Power 98". **V.** by letter.
3) Inter-Island Communications Inc, P.O. Box 20249, Guam Main Facility, Barrigada, Guam 96921 — Pres. & GM: E. Poppe. 24h AC in English exc. Chamorro 1900-2000. **N.** on the h. (AP) — **V.** by letter (B).
4) Guam Radio Sces Inc, 530 We. O'Brien, Agana, Guam 96910 — GM: E. Galito. PD: R. Gibson. CE: R. Garman. 24h CHR (0400-0500 relay N. from Guam Cable TV) — **ANN:** "Hit Radio 100".
5) see Adventist World Radio-Asia (below) — **ANN:** "Joy 92".
6) see Trans World Radio Pacific (below).

ADVENTIST WORLD RADIO — ASIA (Rlg.)

ADDR: P.O. Box 319, Hong Kong. **Te:** +1 (671) 565 2000. **Fax:** +1 (671) 565 2983. **Cable:** Adventist. **Tlx:** 721-6548 KSDA GM.
L.P: GM: Max Torkelsen. PD: Greg Scott. CE: Don Myers.
FM: KSDA-FM 91.9MHz 4kW.
SW: KSDA Agat (G.C: 114.39E/13.20N): 2 × 100kW tr's.

kHz	**kHz**	**kHz**	**kHz**	**kHz**	**kHz**	**kHz**	**kHz**
9495	9530	9650	11980	13720	15225	15310	15610

	kHz
Burmese	
0100-0200	15610
1400-1500	13720
Chinese languages	
0200-0900SS	15225
0300-0900SS	13720
0900-1000	13720
1000-1300	11980
1400-1600	11980
2000-2200	9495
2200-0200	15225
English	
0200-0300	13720
1600-1700	11980
1700-1900SS	13720
2300-0100	15610
Indian languages	
1500-1700	13720
Indonesian	
1100-1200	13720
2200-2300	15610
Japanese	
1300-1400	11980
2100-2200	15310
Korean	
1300-1400	9650
2000-2100	15310
Philippine languages	
1200-1300	13720
Russian	
1000-1100	9530
1900-2000SS	11980

IS: "Lift Up the Trumpet" — **ANN:** "This is Adventist World Radio, the Voice of Hope" — **V.** by QSL-card. Rp.
PUB: "AWR Asiawaves", quarterly.

TRANS WORLD RADIO PACIFIC (Rlg.)

ADDR: P.O. Box CC, Agana, Guam 96910. **Te:** +1 (671) 477 9701. **Fax:** +1 (671) 477 2838.
L.P: MD: Edward Stortro. PD (KTWG): Harry Betting. PD (KTWR): Gary Whitmore. CE (KTWG): Robert Chick. CE (KTWR): George Zensen. Head of PR: Wayne Frost.
MW: KTWG 801kHz 10kW: 2000-1300 in English exc. International Hour Mon-Fri 1100-1200 (Mon Tagalog, Tues Mandarin, Wed Korean, Thurs Chamorro, Fri Japanese).
SW: KTWR (G.C: 144/40E/13.17N): 4 × 100kW tr's.

kHz	**kHz**	**kHz**	**kHz**	**kHz**	**kHz**	**kHz**
9590	9870	11650	11665	11805	15200	15610
9785	11580	11660	11700	12025	15485	

	Days	**Area**	**kHz**
Cantonese			
1400-1500	1234567	CHN	9870
2200-2300	1234567	CHN	11660
English			
0800-0915	1234567	FE	15200
0910-1000	1234567	SPac	11805
1500-1535W	123456	SAs	15610
1500-1700	7	SAs	15610
Indian languages			
1400-1530	1234567	IND	11580
Indonesian			
0930-1100	1234567	INS	15200
2100-2230	1234567	INS	12025
Hakka			
0825-0855	1234567	CHN	11665
Japanese			
1200-1330	1234567	J	9870
2100-2200	1234567	J	15485
Kazakh			
1600-1615	567	CIS	9785
Kirghiz			
1600-1615	123	CIS	9785
Korean			
1145-12000	1234567	KOR	11805
1615-1630	123	CIS	9785
Madurese			
0915-0930	56	INS	15200
Mandarin			
0900-1200	1234567 11665	CHN	9785
1100-1330	1234567	CHN	9590
1200-1600	1234567 11700	CHN	9785
2100-2400	1234567	CHN	11580
Russian			
1000-1145	1234567	CIS	11805
1615-1630	4567	9785	
Swatow			
1330-1400	1234567	CHN	9870
2130-2200	1234567	CHN	11660
Tamil			
0915-0930	4567	MLA	15200
Twi			
2030-2100	1234567	Af	11650
Uzbek			
1600-1615	4	CIS	9785

ANN: "This is your station for inspiration, KTWR, Agana" or "This is the voice of Trans World Radio Pacific, KTWR, Agana".
IS: "We've a Story to Tell the Nations" played on an organ.
V. by QSL-card. 3 IRC's for airmail reply, one for surface mail. Rec. acc. **PUB:** Frequency schedule available on request.

See *International Broadcasts in English* for a complete hour-by-hour listing of all English broadcasts.

HAWAII (U.S. State)

L.T: UTC−10h — **Pop:** 1.161.000 — **Radio sets:** 1m (est) — **Pr.L:** English, Japanese, Filipino — **E.C:** 60, mainly 115 V — **ITU:** HWA.

HAWAIIAN ASSOCIATION OF BROADCASTERS

ADDR: 150B Puuhale Rd, Honolulu, HI 96819 — Pres: Dick Shaller.

Mediumwaves: all st's comm. exc. *)non-comm.

	Call	kHz	kW	h. of tr.
1)	KMVI	550	5	24h (exc. Mon 1000-1400)
2)	KQNG	570	1	24h
3)	KSSK	590	7.5	24h
4)	KIPA	620	10	24h
4S)	KIPA-1	620	10	24h
5)	KHNR	650	10	24h
6)	KPUA	670	10	24h (F.Pl. 50kw)
7)	KQMQ	690	10	24h
8)	KUAI	720	5	Mon-Thurs 1500-1000, Fri/Sat 1500-1100, Sun 1500-0900
9)	KGU	760	10	24h
10)	KKON	790	5	24h
11)	KXDS	810	10	F.Pl.
12)	KIKI	830	10	24h
13)	KHLO	850	5	24h
14)	KAIM	870	50	1500-1000
15)	KNUI	900	5	24h (exc. Mon 1000-1430)
16)	KDEO	940	10	24h
17)	KJRI	970	5	F.Pl.
18)	KHVH	990	5	24h
19)	KLHT	1040	5	24h
20)	KAHU	1060	1	1600-0200 (?)
21)	KWAI	1080	5	24h
22)	KAOI	1110	5	silent
23)	KOHO	1170	5	1530-1000
24)	KZOO	1210	5	1530-1005
25)	KNDI	1270	5	24h
26)	KIPO*	1380	5	24h (F.Pl. 10kW)
27)	KCCN	1420	5	24h
28)	KULA	1460	5	24h
29)	KUMU	1500	10	24h
30)	KISA	1540	5	24h
31)	KUAU	1570	1	F.Pl.

Stations on FM only

	MHz	kW	Call		MHz	kW	Call
32)	92.7	3	KHHI	42)	99.5	100	KABG
43)	93.5	25	KPOA	36)	101.1	100	KLHI-FM
33)	93.5	3	KLUA	36a)	101.9	100	KPCY
34)	94.3	3	KDLX	36b)	103.1	3	KAUI
22)	95.1	100	KAOI	35)	103.7	100	KHCR
37)	96.9	100	KFMN	44)	104.3	69.4	KANY
38)	97.1	40	KFSH	45)	105.1	100	KHFX
39)	97.5	80	KPOI-FM	46)	107.9	100	
13)	97.9	35	KKBG				

+ st's below 1kW. Others see below.

Addresses and other information: (9 digit postal zip code is used in this section.) Abbreviations: See under U.S.A. commercial st's.

1) 250 Waiehu Beach Rd, Wailuku, Maui 96793-1462 — G.M: R. Doran; P.D: N. Jackson; C.E: Earl Tulley. ADC, Hawaiian. Japanese 6h, Filipino 11h weekly — **FM**: KMVI-FM 98.3MHz 3kW (Pukalani).

2) Box 1748, Lihue, Kauai 96766-5748. G.M: Rodney T. Sanchez; C.E: E. Nearman. Oldies. (N. at :50). Ilocano: W 1330-1530, Sun 1400-1700. **FM:** KQNG-FM 93.5MHz 100kW.

3) 1505 Colburn St., Suite 208, Honolulu 96817-4905 — G.M: E. McDaniel; P.D: Phil Abbott; C.E: Dale Machado. ADC — **FM**: KSSK-FM 92.3MHz 100kW.

4) 688 Kinoole Str, Hilo, The Big Island 96720-3877 — G.M: H. Gordon; P.D: B. Carnett; C.E: Alan Roycroft. **FM:** KIPA-FM 100.3MHz 74kW.

4S) Kona, Keaau: both synchronized operation with 4).

5) Honolulu. 850 Richard St., Penthouse Ste, Honolulu, HI 96813. Mgr: Bernie Armstrong; All news. "Radio 65."

6) 1145 Kilauea Ave, Hilo, The Big Island 96720 — G.M: John Leonard; P.D: Desiree Berrigan-Douglas; C.E: Glenn Ziegler. MOR; Japanese Mon-Sat 1500-1600. **FM**: KWXX 94.7MHz 100kW.

7) 711 Kapiolani Blvd, Suite 1193, Honolulu 96813 — G.M: Jack Kelly; P.D: Kimo Akane; C.E: R.N. Palitz. CHR, Dance music — **FM**:KQMQ 93.1MHz 54kW.

8) Box 720, Eleele, Kauai 96705-0720 — G.M: W.P. Dahle; P.D: R. De Roos; C.E: David Moore. ADC, Hawaiian.

9) 2153 N. King Str, Suite 376, Honolulu 96819 — G.M: Ron Burley; P.D: Alan Zee; C.E: Martin S. Prater. Newstalk (N-1). **FM:** KTSS 107.9MHz 100kW.

10) Box 845, Kealakekua, The Big Island 96750-0845 — G.M. & P.D: Mark Taylor; C.E: Stan Tomyl — ADC, Hawaiian — **FM**: KAOY 100.5 6kW. N-:56.

11) Princeville (under construction).

12) 345 Queen Str., Suite 601, Honolulu 96813-4715 — G.M: C. Stangl; P.D: J. Christopher; C.E: Jerry Varoujean. Oldies — **FM**: KIKI-FM 93.9MHz 100kW.

13) 913 Kanoelehua Ave, Hilo, The Big Island 96720-4915 — G.M: Jeanine Atebara; C.E: Jeremy Storm. C&W.

14) 3555 Harding Ave, Honolulu 96816-2491 — G.M: D. Gibbs; P.D: Jack Waters; C.E: Ken Wooley. Rlg — **FM**: KAIM-FM 95.5MHz 100kW. (MW tr. in Molokai.)

15) 311 Ano Str, Kahului, Maui 96732-1304 — G.M: Gregg Everett; P.D: Cliff Arquette; C.E: E. Tolley. ADC, oldies — **FM**: KNUI-FM 99.9MHz 100kW.

16) 94-1088 Farrington Hwy, Suite 100, Waipahu, Oahu 96797-3499 — G.M: R.M. Loew; P.D: S. Beach; C.E: D. Moore. CWM. **FM:** KDEO-FM 102.7MHz.

17) Waimea, Kauai (under construction).

18) Suite 660, 1001 Bishop Str, Honolulu 96813-3406 — G.M: Tim Tindall; P.D: Bill Morse; C.E: Sam Hall; News, business, sports. **FM:** KHHH 98.5MHz 100kW.

19) 1190 Nuuanu Ave, Honolulu 96817-5122 — G.M: Derald Skinner; P.D: Naylene Stonebraker; C.E: Ralph Wilson. Christian music. N-1.

20) Box 4454, Hilo, Big Island 96721 — Pres: W. Kaehuaea; G.M. & P.D: C. Awana. All Hawaiian.

21) 1088 Bishop Str, Suite 204, Honolulu 96813 — G.M: Sam Wagenvoord. P.D: Gloria May; C.E: R. Wilson. Talk night (1000-1200 Samoan) N-1.

22) Box 38, Kahului 96732 — Pres: John Detz; P.D: Jack Gist; C.E: Alex Kowalski. Rock. **FM:** KAOI-FM 95.1MHz 100kW.

23) 1142 12th Ave, Honolulu 96816-3712 — G.M: H. Oshita; P.D: I. Tomita; C.E: R. Wilson. Japanese.

24) 250 Ward Ave, Suite 209, Honolulu 96814-4066. G.M: Kozo Suenobu; P.D: Doris Stokstil; C.E: R. Wilson. Japanese/English.

25) 1734 South King Str, Honolulu 96826-2042 — G.M: Leona Jona; P.D: R. Wiorek; C.E: R. Wilson. Variety.

26) 738 Kaheka St, Honolulu 96814-3726 — G.M: Cliff Eblen; P.D: Bob Miller; KIPO-FM soon 75kW. C.E: Mark Wagner. **FM:** KIPO-FM 89.3MHz Honolulu 3.25kW, KKUA 90.7MHz Wailuku 1.3kW, KHPR 88.1MHz Honolulu 25kW, KEAU 88.1MHz Hilo 5kW, KIPU 90.1MHz Lihue 30kW. "Jazz, Rhythm & News." AM will be news; FM all music format.

27) 900 Fort Street Mall, Suite 400, Honolulu 96813-3797 — G.M: Jan Glascock; P.D: Farden Akui; C.E: R. Wilson. All Hawaiian music. Hawaiian Sun 0600-0700. **FM:** KCCN-FM 100.3MHz 5kW.

28)970 N. Kalaheo Ave., Suite C-300, Kailua, Oahu 96734-1873 — G.M. & P.D: Austin Valli; C.E: Scott Jossart. **FM:** KRTR 96.3MHz 75kW.

29) 441 N. Nimitz Hwy, Honolulu 96817-5063 — G.M: Jeff Coelho; S.M: George Rudolph; C.E: E. Nearman. Easy listening. (N. at :55). **FM**: KUMU-FM 94.7MHz 100kW.

30) 904 Kohou Str, Suite 204, Honolulu 96817-4482 — G.M: R.C. Manayan; P.D: Kale Imua; C.E: Jerry Varoujean. Filipino.

32) Hilo (under construction).

33) 74-5605 Luhia St. B7, Kailua-Kona 96740 (under construction).

34) Box 1589, Wailuku 96793.

35) Paauilo, The Big Island (under construction).

36) 840 Waiau Str, Lahaina, Maui 96761-1684.

36a) Pearl City 96782.

36b) Kekaha (under construction).

37) P.O. Box 97, Lihue, Kauai 96766-0097.

38) 47 Punahoa Str, Hilo 96720-6306.

39) 741 Bishop Str, Honolulu 96813-4899.

42) Honolulu (under construction).

43) 505 Front Str, Suite 215, Lahaina, Maui 96761.

44) Kaneohe 96744 (under construction).

45) Penthouse Suite, 850 Richards Str, Honolulu 96813.

KIRIBATI

L.T: UTC – 11h — **Pop:** 72.275 — **Radio sets:** 10.000 — **Pr.L:** I-Kiribati, English — **E.C:** a/c 50, 240V — **ITU:** KIR.

RADIO KIRIBATI (A division of the Broadcasting and Publications Authority).
ADDR: P.O. Box 78, Bairiki, Tarawa. **Te:** +686 21187, 21161, 21162. **Fax:** +686 21096. **Telex:** KI77024.
L.P: Mgr: Tomasi Kei Tauru. Prgr. Organiser: A. Bauro. Eng: B. Reiher.
Mediumwave: Nanikaai 846kHz 10kW.
Shortwave: 17440kHz 0.5kW — **FM:** 98.0MHz 0.1kW.
D.PRGR in I-kiribati (90%)/English (10%): Mon-Fri 1730-1900, 2300-0030 & 0500-0830. Sat 1730-0830 on FM, MW/SW as Mon-Fri. Sun 2300-0030 & 0500-0830. **N. in English:** on the h. Rel. RNZ 1800, rel. R. Australia 0000, rel. BBC 0600.
ANN: "This is R. Kiribati, the national broadc. sce. of Kiribati in the Central Pacific" — "Aio Bwanaan Kiribati te botaki ni kanako bwanaa i bukin Kiribati i nukan te Betebeke".
V. by QSL-card. Enclose US$1 for postage. IRC's not acc.

LORD HOWE ISLAND (Australian Territory)

L.T: UTC + 10½h (UTC + 11h until March 1 and from Oct 25) — **Pr.L:** English — **E.C:** a/c 240V.

RADIO LORD HOWE ISLAND
ADDR: c/o Post Office, Lord Howe Island, NSW 2898, Australia. **Te:** +61 (65) 632123 — **L.P:** SM: Gary Millman — **FM:** 100.1MHz 30W — **D.PRGR:** Thurs 0530-0830. Other times vary seasonally. St. relays 2CS Coffs Harbour when no local announcer available.
V. by letter. Rp.

MARSHALL ISLANDS (Rep. of)

L.T: UTC + 12h (Kwajalein: UTC – 12h) —**Pop:** 50.000 — **Pr.L:** English, Marshallese — **E.C:** a/c 60, 110/220 V— **ITU:** MRL.

WSZO (Gov. Comm.)
ADDR: Dept. of Interior & Outer Isl. Affairs, Majuro, Marshall Isl. 96960. **Te:** 3240. **Fax:** 3413.
L.P: Chief Info Specialist: Billy Sawej.
STATION: 1098kHz 5kW — **D.PRGR:** 1900 (Sun 2000)-1000.
ANN: "This is R. St. WSZO, Majuro, Marshall Isl" — **V.** by letter.

WSZF, Drawer H, Majuro, Marshall Isl. 96960.
Te: 3141 — Rlg.

Marshall Broadc. Co., P.O. Box 19, Majuro, Marshall Isl. 96960. **Te:** 3383/3210 — 97.9MHz.

CENTRAL PACIFIC NETWORK (AFRTS)
ADDR: P.O. Box 1761, APO San Francisco, CA 96555, USA.
L.P: Netw. Mgr: Larry Malinowski; Prgr. Dir: Ms. Kerry Mathieson; Chief Eng: Steve Cole.
MEDIUMWAVE: Kwajalein 1224kHz 1kW — **FM:** 100MHz 1kW.
D.PRGR: 24h (FM automated).
N: hourly via satellite — **ANN:** "The Central Pacific Network."

MICRONESIA (Federated States of)

L.T: Yap, Chuuk UTC +10h; Pohnpei UTC + 11h; Kosrae UTC + 12 h — **Pop:** 125.000 — **Radio sets:** 70.000 — **E.C:** a/c 60, 110/220V — **Pr.L:** Yapese, Trukese, Ponapean, Kosraean, English.

FEDERATED STATES OF MICRONESIA OFFICE OF THE PRESIDENT (Gov.)
ADDR: P.O. Box 34, Palikir Station, Pohnpei 96941. **Te:** +1 (691) 320 2548. **Fax:** +1 (691) 320 2785. **Telex:** 729-6807 FSMGOVT.
L.P: Special Asst. to Pres. for Inf: Terry Gamaburuw. Broadc. Specialist & ABU Liaison Officer: Esikiel Lippwe.

	Call	kHz	kW		Call	kHz	kW
1)	WSZP	1350	—	4)	V6AJ	1584	1
2)	V6AH	1449	10	5)	V6AK	1602	5
3)	VGAI	1494	10				

FM: 6) V6AF 104.0MHz.
Addresses and other information:
1) Moen, Truk State. **Te:** +1 (691) 330 2463. Rlg. (Baptist).
2) P.O. Box 1086, Kolonia, Pohnpei State, 96941, F.S.M. **Te:** +1 (691) 320 2296. 1900-1300 in English (30%)/Ponapean.
3) P.O. Box 119, Colonia, Yap State, 96943, F.S.M. **Te:** +1 (691) 350 2174. 2000-1400 in English/Yapese/Ulithian/Woleaian.
4) P.O. Box 147, Tofol, Kosrae State, 96944, F.S.M. **Te:** +1 (691) 370 3040. 1800-1200 in English/Kosraean.
5) Weno, Chuuk State, 96942, F.S.M. **Te:** +1 (691) 330 2593. 2000-1400 in English Trukese. Rel VOA N: 2200.
6) Nanpohmal, Pohnpei (Addr: c/o Calvary Baptist Church, Drawer H, Kolonia, Pohnpei, FSM 96941. **Te:** +1 (691) 320 2475).
NB: All st's operate 24h during adverse weather conditions.
ANN. ethnic languages: 1) "Met Station V6AH nan Pohnpei" — 2) "Pary e radio station WSZA nu Waab" — 3) "Painge Station V6AJ, fwin an Kosrae" — 4) "Ach nenien appio V6AK lon Chuuk".

MIDWAY ISLANDS (U.S. Terr.)

LT: UTC – 11h — **Pop:** 500 — **Pr.L:** English.

NAVY BROADCASTING SERVICE (AFRTS)
ADDR: Det. 27, U.S. NAF Box 39, San Francisco, CA 96614, USA — **STATION:** KMTH 94MHz 0.05kW.
D.PRGRS: 24h. **N:** on the h. — **ANN:** "The Voice of the Pacific".

NAURU (Republic of)

L.T: UTC + 12h — **Pop:** 9.465 — **Radio sets:** 4.000 — **Pr.L:** English, Nauruan — **E.C:** a/c 50, 220V — **ITU:** NRU.

NAURU BROADCASTING SERVICE (Gov.)
ADDR: Nauru Broadc. Service, Republic of Nauru, Central Pacific. **Te:** 3095. **Cable:** Broadserv, Nauru.
L.P: St. Mgr: Millicent Aroi. Dir. Tec: Anzac Dediya.
STATION: (G.C: 166.56E/00.33S): C2AM 1323kHz 0.2kW.
HOME SCE. in English and Nauruan: 1900-1130.
V. by letter.

NEW CALEDONIA (French)

L.T: UTC + 11h — **Pop:** 175.000 — **Radio sets:** 90.000 (est.) — **Pr.L:** French, Kanak — **E.C:** a/c 50, 220V (yes) — **ITU:** NCL.

RADIODIFFUSION FRANÇAISE D'OUTRE MER (RFO)
ADDR: B.P. G3, Noumea Cédex. **Te:** + 687 274327. **Fax:** + 687 281252. **Telex:** RTOM 3052NM.
L.P: Reg. Dir: Alain Le Garrec. Prgr. Dir: Claude Ruben. Head of Inf: Daniel Franco.
Mediumwave: Sainte-Marie 666kHz 20kW.
Shortwave: (G.C.166.26E/22.16S) 7170.4kHz 20kW.
FM: Noumea 88.9MHz — Mont Do 96.0MHz — Ile de Lifou 93.5MHz — Aoupinie 99.0MHz + 15 repeaters.
D.PRGR: 24h. **N. in French:** 1915, 1945, 2015, 2030, 2230, 2330, 0100, 0530, 0800. **N. in English:** 0700.
ANN: "Bonjour, vous êtes à l'écoute de RFO Nouvelle-Calédonie".

NEW ZEALAND

L.T: UTC+12h (Oct-March UTC+13h) — **Pop:** 3.434.000 — **Radio sets:** 3.100.000 — **Pr.L:** English, Maori — **E.C:** a/c 50, 230V — **ITU:** NZL.

RADIO NEW ZEALAND

(State-owned enterprise)

ADDR: P.O. Box 2092, Wellington. **Te:** +64 (4) 4741555. **Fax:** +64 (4) 4741440. **Cable:** Radionet. **Telex:** NZ 30131.

L.P: Chief Exec: Nigel Milan. GM Eng. & Op's: Norm Collison. GM Metro St's: Ms. Lynne Clifton. GM Provincial Sts: Carl Terry. GM RNZ News: Ray Lilley. GM Public Radio (non-comm.): John Craig

	Call	kHz	kW	Prgr.	Location & H. of tr.
1)	4ZG	558	5	Z	"Hokonui R.", Gore: M: 1200-1700, I: 1800-0600
120)		558	5	A	"Access R.", Gore: Tues-Thurs 0700-1000
2)	2YA	567	50	N	Wellington
3)	4ZF	612	0.4	Z	Te Anau (# 4ZA 864): I: Mon 2105-2110v, Fri 0405-0410v
2)	2YZ	630	20	N	Napier
5)	4YW	639	2	N	Alexandra (# 4ZB 1044): Mon-Fri 1700-2400, Sat 1700-0700, Sun 1800-0800
2)	2YC	657	60	S	Wellington
2)	3YA	675	20	N	Christchurch
2)	4YZ	720	20	N	Invercargill
2)	—	729	—	N	Tokoroa
7)	3ZA	747	2	Z	"R. Scenicland", Greymouth
2)	1YA	756	20	N	Auckland
2)	2YB	783	20	A	"Access R.", Wellington: Ethnic: Fri 0600-1000, Sat & Sun 2100-1000
2)	4YA	810	20	N	Dunedin
110)	IXU	810	5	A	"Access R.", Auckland
2)	1YZ	819	10	N	Rotorua
2)	1YX	837	2	N	Whangarei
2)	1YK	837	2	N	Kaitaia (synchro. with 1YX)
12)	2ZD	846	2	Z	"R. Wairarapa", Masterton
3)	4ZA	864	10	Z	Invercargill
13)	3ZE	873	1	Z	Ashburton # 3ZB 1098 I: 1800-0200 (Sat & Sun 0600)
2)	1YC	882	10	S	Auckland
2)	4YC	900	10	S	Dunedin
113)	4XS	900	10	A	"Access R.", Dunedin: Fri/Sat/Sun 0700-0900
2)	3YZ	918	10	N	Greymouth
14)	2ZA	927	2	Z	Palmerston North
15)	2ZG	945	2	Z	Gisborne
2)	3YC	963	10	S	Christchurch
2)	1YE	981	2	N	Kaikohe
16)	1ZD	1008	10	Z	"R. Bay of Plenty", Tauranga
11)	1ZN	1026	2	Z	"R. Northland", Whangarei
11)	1ZK	1026	2	Z	Kaitaia (synchro with 1ZN)
27)	2ZB	1035	20	Z	Wellington
5)	4ZB	1044	10	Z	Dunedin
17)	2ZP	1053	2	Z	"R. Taranaki", New Plymouth
2)	—	1071	—	N	Masterton
28)	1ZB	1080	10	—	"Newstalk 1ZB", Auckland
6)	3ZB	1098	10	Z	Christchurch
2)	2YX	1116	2	N	Nelson
2)	4YQ	1134	2	N	Queenstown # 4ZB 1044 Mon-Fri 1700-2400, Sat 1700-0700, Sun 1800-0800
2)	1YW	1143	2	N	Hamilton
20)	3ZC	1152	2	Z	"R. Caroline", Timaru
30)	1ZW	1170	0.4	Z	"R. Waitomo", Te Kuiti #ZH-FM 98.6
2)	—	1188	—	N	Rotorua
21)	2ZW	1197	2	Z	"River City R.", Wanganui
130)	—	1206	0.5	A	"Access Community R.", Hamilton
11)	1ZE	1215	2	Z	Kaikohe (# 1ZN 1026)
18)	2ZT	1269	0.4	Z	Takaka # 2ZN 1341
4)	2ZC	1278	2	Z	"Bay City R.", Napier
7)	3ZW	1287	2	Z	Westport # 3ZA 747
19)	1ZH	1296	—	Z	"ZH-AM", Hamilton
2)	2YW	1314	2	N	Gisborne
18)	2ZN	1341	2	Z	"R. Nelson", Nelson
23)	4ZW	1395	2	Z	"R. Waitaki", Oamaru
24)	1ZO	1413	2	Z	"R. Forestland", Tokoroa # 1ZA 1494 I: 1800-
2)	2YM	1449	2	N	Palmerston North
2)	3YW	1458	2	N	Westport
2)	1YT	1494	2.5	N	Taupo
25)	1ZU	1512	1	Z	"King Country R.", Taumaranui (# 1ZA 1494): I: Mon-Fri 1800-0030, Sat 1900-0000
7)	3ZR	1521	0.1	Z	Reefton # 3ZA 747
2)	2YP	1530	2	N	New Plymouth
26)	2ZE	1539	1	Z	"R. Marlborough", Blenheim
31)	2ZH	1557	2	Z	Hawera (# 2ZP 1053): I: Mon-Fri 1800-2200, Sat 2000-2300
26)	2ZF	1584	0.1	Z	Picton #2ZE 1539

FM	Call	MHz	kW	Prgr.	Location
6)	—	89.2	—	M	Sumner (#3ZZM 91.3)
27)	—	89.4	0.3	Z	Kapiti (#2ZB 1035)
2)	2MCP	89.0	40	C	Palmerston North
5)	4ZZB	89.4	3		"ZBFM", Dunedin
2)	3CCP	89.7	40	C	Christchurch
17)	2ZZP	90.0	3	Z	"R. Taranaki," New Plymouth
3)		90.0		C	Invercargill
14)	2QQ	90.6	40	Z	"2 Double Q FM", Palmerston North
32)	2ZZM	90.9	50	Z	"ZM-FM", Wellington
10)	1GEY	90.9	0.2	P	Paeroa Range (#1GEY 97.5)
2)	2HCP	91.1	3	C	Napier
6)	3ZZM	91.3	40	—	"ZM-FM", Christchurch
2)	1WCP	91.4	40	C	Hamilton
2)	2NCP	91.4	3	C	Nelson
2)	2TCP	91.6	3	C	New Plymouth
22)	1TPO	92.4		Z	"R. Lakeland", Taupo
2)	2WCP	92.5	40	C	Wellington
2)	1ACP	92.6	40	C	Auckland
2)	4DCP	92.6	25	C	Dunedin
32)	2ZZM	93.5	0.4	—	Wellington (# 2ZZM 90.9)
28)		94.2	—	Z	"Newstalk 1ZB", Auckland (**F.Pl.**)
)	1BOP	95.0			"BOP FM", Tauranga
2)		95.3	0.8		Banks Peninsula
2)		95.6	0.6	C	Wellington
9)	2ZZK	95.9	3	Z	"Greatest Hits 96FM", Hastings
22)	1TPO	96.7	0.6	Z	"Radio Lakeland", Taupo
2)		96.8	0.1	C	Whangarei
26)		96.9		Z	Marlborough (#2ZE 1539)
61)	3CCT	96.9	3.5	A	"Plains Radio," Christchurch
29)	1ZZM	97.4	40	M	"Classic Hits 97 FM", Auckland
10)	1GEY	97.5	0.4	P	"Geyserland FM", Rotorua
6)		97.7	7.5	Z	"B-FM", Christchurch
14)		97.8		Z	"2ZA-FM", Palmerston North
120)		98.0		A	"Access R.", Invercargill
2)		98.3	0.6	C	Kapiti Coast, Wellington
2)		98.3	0.2	C	Taupo
3)	4ZA	98.5	5	Z	"4ZA-FM," Invercargill
19)	1ZH	98.6	40	Z	"ZH-FM", Hamilton
6)	3ZZM	98.7		M	Sumner
2)		99.0	0.1	C	Wanganui
5)		99.0	0.25	C	Highcliffe (# Dunedin 89.0)
2)		99.2	2.5	C	Masterton
2)		99.5	10	C	Mt. Studholme, Timaru
9)		99.7	0.04	Z	(#2ZZK 95.9), Wairoa
2)		99.9	0.8	C	Sumner/Redcliffs, Christchurch

RNZ Community Net: Commercial. Stations generally have local programming and take the following network schedule: "New Zealand Tonight" 0600-1200, "Shell Allnighter" 1200-1700, "The Early-Bird Show" 1700-1800. Some stations do not join at 0600 e.g. 1ZB, 2ZB, 3ZB, 1ZH. Others will stay local for coverage of sports, horse-racing etc. **N:** on the hour, half-hourly during breakfast and drivetime shows. **NB:** Many RNZ Community Net stations intend to leave AM for FM.

National Radio Net: 24-hour commercial (**NB** mostly in the form of sponsorship). Regional stations may leave the net to join the Concert Programme. Stations often have local sports particularly rugby games. **N:** 1200, 1300, 1400, 1600, 1700, 1800, 1830, 1900, 1930, 2000, 2030, 2100, 2200, 2300, 0000, 0030, 0100, 0200, 0300, 0400, 0500, 0600, 0800, 0900, 1000, 1100. **Regional news:** 1845, 2005, 0034, 0630, Maori News: 1841, 0505. **Main**

weather: 1730, 1838, 1915, 0038, 1015. **"Pacific Islands Magazine":** Mon-Fri 0700-0730 with news in Samoan, Tongan, Cook Islands Maori, Niuean & Tokelauan, with a 12-minute magazine in a different language each week-day.
Concert Programme: 1900-1200. **N:** 1800, 2000 (BBC), 2100, 2300, 0100, 0300 (BBC), 0500, 1100 (BBC).
Parliamentary broadcasts are carried by 1YC, 2YC, 3YC, 4YC.
Note: A feature of New Zealand broadcasting is the high number of short-term stations on both AM & FM permitted by the BCNZ.
Local Addresses
1) P.O. Box 292, Gore — 2) P.O. Box 2092, Wellington — 3) P.O. Box 802, Invercargill — 4) P.O. Box 241, Napier — 5) P.O. Box 888, Dunedin — 6) P.O. Box 1484, Christchurch — 7) P.O. Box 378, Greymouth — 8) Broadcasting House, Auckland — 9) P.O. Box 183, Hastings — 10) P.O. Box 1147, Rotorua — 11) P.O. Box 845, Whangarei — 12) P.O. Box 220, Masterton — 13) P.O. Box 465, Ashburton — 14) P.O. Box 1045, Palmerston North — 15) P.O. Box 1040, Gisborne — 16) P.O. Box 642, Tauranga — 17) P.O. Box 141, New Plymouth — 18) P.O. Box 43, Nelson — 19) P.O. Box 489, Hamilton — 20) P.O. Box 275, Timaru — 21) P.O. Box 632, Wanganui — 22) P.O. Box 740, Taupo — 23) P.O. Box 426, Oamaru — 24) P.O. Box 272, Tokoroa—25) P.O. Box 383, Taumaranui — 26) P.O. Box 225, Blenheim — 27) P.O. Box 2396, Wellington — 28) P.O. Box 2209, Auckland — 29) 1st Floor, David Reid Building, 3-5 Auburn Str., Takapuna, Auckland — 30) P.O. Box 276, Te Kuiti — 31) P.O. Box 341, Hawera — 32) P.O. Box 1991, Wellington.

RADIO NEW ZEALAND INTERNATIONAL

ADDR: P.O. Box 2092, Wellington. **Te:** +64 (44) 741437. **Fax:** +64 (44) 741433. **Telex:** NZ 31031. **Engineering Dept:** P.O. Box 2396, Wellington. **Te:** +64 (4) 4741843. **Fax:** +64 (4) 4741904.
L.P: Mgr: Ian Johnstone. PD: Linden Clark. GM (Eng): Dave Henderson. Freq. Mgr: Adrian Sainsbury. Head of Pub. Rel: Tony King.
STATION: Rangitaki (G.C: 176.25E/38.50S): 100kW.

Times	Days	May-Sept kHz	Oct-April kHz
1650-1849	Sun-Fri	6035	6035
1850-2138	Sun-Fri	11735	15120
2139-0459	Daily	17770	
2139-0658	Daily		17770
0459-0658	Daily	9730	
0659-1206	Daily		9700
1207-1649	Occ. (sports)	6035	9510

Mailbox (incl. SW news): Mon 0430, Thurs 0830, Fri 1930.
ANN: E: "R. New Zealand International". Maori: "Te Reo Irirangi O te Moana-nui-a-kiwa" — **INT-SIG:** Call of the NZ Bellbird.
V. by QSL-card. Rp. (3 IRC's req.). Cassettes not required, and cannot be returned. Send 1 IRC for schedule.

INDEPENDENT BROADCASTERS ASSOCIATION (N.Z.) INC.

ADDR: P.O. Box 3762, Auckland. **Te:** +64 (9) 486 7525. **Fax:** +64 (9) 486 7443. Exec. Dir: B.G. Impey. Secr: Hayley Cornell.

	Call	kHz	kW	Name, location, h. of tr.
40)	4XA	531	2	R. Central, Alexandra (# 4XO 1206): I: W 1800-0600, Sun 2100-2400
42)	1XC	540	5	R. Rhema, Tauranga (# 3XG 1503)
46)	2XR	585	2	Radio Ngati Porou, Ruatoria
42)	3XL	594	2	R. Rhema, St. Andrews (# 3XG 1503)
41)	1XO	603	5	Aotearoa Radio, Auckland: 24h
41)	1XB	603	0.4	Paengaroa (synchro with 1XR)
42)	4XG	621	2	R. Rhema, Dunedin (# 3XG 1503)
42)	2XC	648		R. Rhema, Grisborne
56)	1XP	702	5	R. Pacific, Auckland
79)	2XP	711		"Radio Pacific", Wellington
125)	—	756	0.1	"Goldfields Radio", Lawrence
124)	2XT	765		"Te Toa Takatina", Napier
42)	2XNR	774	4	"Radio Rhema", New Plymouth
42)	2XL	801	1	R. Rhema, Nelson (# 3XG 1503): I: Sun-Thurs 1900-2100 or Mon-Fri 0430-0530
43)	2XS	828	2	"X-5 AM, Palmerston North
42)	1XH	855	2	Hamilton (# 3XG 1503)
44)	2XW	891	5	R. Windy, Wellington
135)	—	909		"Harvest R", Motueka
45)	1XW	954	2	R. Pacific-Waikato, Hamilton
42)	2XG	972	5	R. Rhema, Wellington (#3XG 1503)
126)	4XX	981	1	Ranfurly R, Ranfurly
129)		1017		BBC World Sce, Christchurch
87)	2ZM	1161	5	Te Upoko O Te Ika, Wellington
40)	4XE	1179	0.1	Wanaka (synchro with 4XC)
40)	4XO	1206	2	R. Otago, Dunedin
47)	4XF	1224	2	R. Foveaux, Invercargill
129)		1233		BBC World Sce, Wellington
48)	1XX	1242	2	Whakatane
48)	1XX	1242	0.1	Galatea (synchro with 1XX)
29)	1XG	1251	5	R. Rhema, Auckland (# 3XG 1503)
49)	3XA	1260	2	R. Avon, Christchurch
50)	4XD	1305	2.5	Dunedin R., Dunedin: 24h
41)	2XQ	1323	5	Wellington (# 1XR 603)
41)	3XQ	1323	5	Christchurch (# IXR 603)
51)	1XI	1332	5	R. i, Auckland
40)	4XC	1359	1	Resort R., Queenstown: I: 1800- (# 4XO 1206, #4XA 531)
52)	1XT	1368	1	Village R., Tauranga (Sun 2100-0500)
53)	2XX	1377	2	R. Horowhenua, Levin: I: Mon-Thurs 1800-0700, Fri & Sat 1800-1000, Sun 2000-0400 otherwise relays RNZ Community Netw.
129)	—	1386	5/1	BBC World Sce, Auckland
42)	4XL	1404	2.5	R. Rhema, Invercargill (# 3XG 1503)
108)	3XP	1413	0.1	3XP R. Ferrymead, Christchurch: Sun 2200-0400
112)	1XK	1440		"Te Reo o Tauranga Moana," Tauranga
88)	1XD	1476	5	Infomusic, Auckland Airport
83)		1485	1	"89FM", Gisborne (#89.0)
42)	3XG	1503	2	R. Rhema, Christchurch: 1800-1200
89)	1XN	1548	1	"Classic Country", Rotorua
57)	2XA	1602	1	Print Disabled Radio, Levin: 2200-1000
57)	ZLXA	3935	1	Print Disabled Radio, (# 2XA 1602)
57)	ZLXA	7290	0.25	Print Disabled R, Levin

FM	Call	MHz	kW	Name & Location
73)	2WVU	89.0	0.1	R. Active, Wellington
82)	1WSU	89.0	0.1	Contact FM, Hamilton
83)		89.0		"89 FM", Gisborne
71)		89.0		"Radio Te Arawa", Rotorua
56)		89.1		"Radio Pacific", Hokitika
47)		89.2		"Fouveaux Radio", Invercargill
101)		89.3		"Today FM", Carterton
58)	1MMM	89.4	50	89FM, Auckland
59)	1KCD	89.6	10	Hikurangi # 1KCC 89.3
60)	1KIW	89.8	10	Kiwi FM, Hamilton
27)		89.8	0.1	Towai #2ZB 1035
54)		89.8		"Te Reo Irirangi o Rangitore", Palm. North
56)		89.9		"Radio Pacific", Mt. Studholme (Timaru)
56)		90.1		"Radio Pacific", Westport
128)		90.1		"Baptist Radio", Wairoa
100)		90.2		"Fine Music", Auckland
53)	2XX	90.2		"2XX FM", Forest Heights
59)	1KCC	90.3	0.1	K Double C FM, Whangarei
48)		90.5		Triple X FM, Whakatane
94)		90.5	0.05	"Te Reo Iriraki ki Otautahi", Christchurch
97)		90.6		"Te Reo Irirangi o Ngati Raukawa", Tokoroa
62)	1MJK	91.0	50	91FM, Auckland
75)	4SAO	91.0	1	R. One, Dunedin
59)	1KCF	91.2	5	Maungataniwha # 1KCC 89.3
98)		91.6	0.1	"Te Reo Irirangi o Tauwharetoa", Turangi
136)		91.8	10	"Today FM", Auckland
56)		91.9		"Radio Pacific", Rotorua
85)	4SKI	92.0		Ski FM, Queenstown
56)		92.1		"Radio Pacific", Whakatane
96)		92.1	34.5	"More FM", Christchurch
43)	2XXS	92.2	50	"XS-FM", Palmerston North
80)		92.2	1	"Radio Nga Iwi", Te Aroha
59)	1KCE	92.4	10	Hobson # 1KCC 89.3
63)	2BAY	92.7	2.5	"93-FM", Hastings
93)		92.8	0.2	"Radio Tautoko", Okaihu
42)		92.8		Taumarunui #3XG 1503

102)		92.8		"Star FM", Wanganui
85)	4SKI	92.8	0.25	"Q FM", Queenstown
64)	3FMX	92.9	40	"Classic Rock", Christchurch
74)		92.9		"Sounds FM", Blenheim
107)		93.0		"93 Rock FM", Hamilton
91)	2FFF	93.0	3	"Fifesfire FM", Nelson
48)	1XXX	93.0	.05	Ohope #1XXX 90.5
65)	2NRG	93.2	3	"Energy FM", New Plymouth
90)		93.4	25	"Coastline FM", Tauranga
105)		93.4	1	"Oldies FM", Auckland
40)		93.4		"Rox FM", Dunedin
6)	1KIS	93.5	0.1	"Kis FM", Taupo
59)	1KCR	93.7	0.1	Russell #1KCC 89.6
56)	3XXE	93.7		Christchurch #1XP 702 "R. Pacific"
56)		93.8		"Radio Pacific", Wharite, Palm North
78)		93.9		"Port FM", Twizel #97.9
56)		94.0		"Radio Pacific", Invercargill
44)		94.1	0.5	"Windy FM", Wellington
91)		94.1		"Fifeshire FM", Murchison
103)		94.2		"Ski FM", Ohakune
114)		94.4	1	"Hiku o Te Ika", Kaitaia
74)		94.5		"Sounds FM", Picton
115)		94.6		"Ngati Ranginui", Tauranga
68)		94.6	0.4	Hamilton # "FM Country", 99.8
56)		94.8		"Radio Pacific", Gisborne
69)	1STU	95.0	1	"B-FM", Auckland
42)	1TRR	95.1	0.1	"Radio Rhema", Taupo
53)		95.1		Waikanae #2XX 90.2
56)		95.1		"Radio Pacific", Greymouth
96)		95.3	0.2	"More FM", Southshore
84)		95.4	0.1	"Te Reo Irirangi o Tainui", Ngaruawahia
16)		95.4		"Tumeke FM", Whakatane
117)		95.8	0.1	"Coast FM", Warkworth
67)	1HOT	95.9	0.1	"Lake City FM", Rotorua
53)		95.9		Levin #2XX 90.2
56)		95.9		"Radio Pacific", Alexandra
56)		96.1		"Radio Pacific", Alexandra
118)		96.3		"Kawhia Moana", Kawhia
42)		96.3		Palmerston North #3XG 1503
91)		96.6		Takaka #2FFF 93.0
17)		96.6	0.25	"Coast FM", Orewa
56)		96.6		"Radio Pacific", Dunedin
85)		96.8	0.5	Coronet Peak # "Q FM" 92.8
132)		97.0	50	"Easy Listening FM", Hamilton
119)		97.2		"Te Ika Whenua", Murupara
40)		97.4	1.2	"4XO-FM", Dunedin
56)		97.5		"Radio Pacific", Napier
56)		97.6		"Radio Pacific", Nelson
78)		97.9		"Port-FM", Timaru
44)		98.1	0.1	Towai #"Windy FM" 94.1
51)	1IXI	98.2		"R. i", Auckland
76)	3RDU	98.3	0.05	"Radio U", Christchurch
95)		98.9	0.25	"More FM", Towai (Wellington)
55)		99.0	40	"R. Hauraki", Auckland
71)		99.0		"Urban FM", Rotorua
99)	2XXP	99.1		"Radio Foxton", Foxton
40)		99.2		"Resort Radio", Queenstown
48)		99.3	0.1	Whakatane #1XXX 90.5
96)		99.3	4.5	"More FM", Christchurch
48)	1XXX	99.3	0.25	Ohope #1XXX 90.5 (FPl.)
131)		99.4		"Pare Hauraki", Paeroa
72)	2MMM	99.4	0.1	"Radio Massey", Palmerston North 1900-0100, Weekends 24 hrs
92)		99.4		"Te Reo Irirangi o Maniapoto", Te Kuiti
121)		99.5	0.1	"Ngati Hine", Whangerei
122)		99.6		"More FM", Kapiti
104)		99.7		"Sound Waves", Russell
90)		99.8		"Coastline FM", Tauranga
68)		99.8	2.0	"FM Country", Albany
103)		99.9		Whakapapa #Ski FM 94.2
123)		100.0		"100 FM", Wanganui
95)		100.0	1	"More FM", Wellington
104)		100.0		"New Sound Waves", Hokianga
8)		100.0		"Port FM", Oamaru #97.9

Addresses and other information

40) R. Otago Ltd., Private Bag, Dunedin. (4XA: P.O. Box 143, Alexandra; 4XC: P.O. Box 224, Queenstown).

41) Aotearoa R., P.O. Box 97-254, SAMC, Auckland.

42) R. Rhema, Private Bag, Christchurch. **DX-Prgr:** (Arthur Cushen's DX World) Fri 1103.

43) Manawatu Radio Co., P.O. Box 446, Palmerston North.

44) Capital City R. Ltd., P.O. Box 11-441, Wellington.

45) R. Pacific, Private Bag, Hamilton.

46) R. Ngati Porou, P.O. Box 55, Ruatoria.

47) Foveaux R. Ltd., P.O. Box 1740, Invercargill.

48) R. Bay of Plenty Ltd., P.O. Box 383, Whakatane.

49) R. Avon Ltd., Private Bag, Christchurch.

50) Otago R. Assoc., P.O. Box 404, Dunedin.

51) R. i. Ltd., Private Bag, Auckland.

52) Village R., P.O. Box 597, Tauranga. Station is completely voluntary and operates under the auspices of the Tauranga District Museum as part of the Historic Village complex.

53) R. Horowhenua Ltd., P.O. Box 132, Paraparaumu.

54) Te Reo Irirangi o Rangitore, Palmerston North.

55) Hauraki Enterprises Ltd., P.O. Box 1480, Auckland.

56) R. Pacific, Private Bag, Auckland.

57) P.O. Box 360, Levin 5500. (This station is run on a completely voluntary basis.) — St. Dir: Allen Little. Studio Mgr: Ash Bell. Transmission technician: Steve Jepson. Prgr. Supervisor: Brian Stotoe. On 3935kHz: Local sunset—1000. On 7290kHz: 2200—local sunset.

58) P.O. Box 5089, Auckland.

59) Northland F.M. R. Ltd., P.O. Box 100, Whangarei.

60) Independent Broadc. Co. Ltd., P.O. Box 9540, Hamilton.

61) Canterbury Communications Trust, P.O. Box 22297, Christchurch.

62) Metropolitan F.M., P.O. Box 33-644, Takapuna, Auckland 9.

63) R. Hawkes Bay, P.O. Box 193, Hastings.

64) Canterbury F.M. Ltd., P.O. Box 3930, Christchurch 1.

65) Energy F.M., P.O. Box 869, New Plymouth.

66) Kis F.M., P.O. Box 393, Taupo.

67) Central F.M. Corp., P.O. Box 92, Rotorua.

68) Private Bag, Albany, Auckland.

69) Private Bag, Auckland.

70) Te Upoko O Te Ika, B N Z Building, Lambton Quay, Wellington.

71) Te Arawa Maori Trust Board, P.O. Box 883, Rotorua.

72) Massey Univ. Students Assoc., Massey Univ., Palmerston North.

73) Victoria Univ., Kelburn Parade, Wellington.

74) P.O. Box 930, Blenheim.

75) Otago Univ. Students Assoc., P.O. Box 1437, Dunedin.

76) Canterbury Univ. Students Assoc., Ilam Rd., Ilam, Christchurch.

77) Wellington.

78) P.O. Box 635, Timaru.

79) P.O. Box 11-850, Wellington.

80) P.O. Box 129, Paeroa.

81) P.O. Box 31-497, Lower Hutt.

82) Waikato Student's Union, P.O. Box 13-020, Hamilton.

83) P.O. Box 230, Gisborne.

84) P.O. Box 208, Ngaruawahia.

85) P.O. Box 465, Queenstown. Owner Mgr: Chas Drader.

86) P.O. Box 106, Auckland.

87) P.O. Box 842, Tokoroa.

88) P.O. Box 73-020, Auckland Airport, Auckland.

89) P.O. Box 1007, Rotorua.

90) P.O. Box 2429, Tauranga.

91) P.O. Box 907, Nelson.

92) 4 Corinth Avenue, Te Puke.

93) Mangamauku Bridge, RD2, Okaihu.

94) P.O. Box 22-173, Christchurch.

95) P.O. Box 27-000, Wellington.

96) P.O. Box 25-209, Christchurch.

97) P.O. Box 842, Tokoroa.

98) P.O. Box 209, Turangi.

99) 55 Main St., Foxton.

100) P.O. Box 33-983, Takapuna.

101) P.O. Box 8, Carterton.

102) 53 Ridgway St., Wanganui 5001.

103) P.O. Box 1369, Wellington 6015.

104) P.O. Box 164, Picton.

105) P.O. Box 34-604, Birkenhead.

107) P.O. Box 19-293, Hamilton.

108) 269 Bridal Path Rd., Christchurch.

110) P.O. Box 5609, Wellington.

112) P.O. Box 382, Tauranga.

113) P.O. Box 5323, Dunedin.

114) P.O. Box 458, Kaitaia.
115) P.O. Box 137, Tauranga.
116) P.O. Box 2090, Whakatane.
117) 157 Main Rd., Orewa.
118) P.O. Box 41, Kawhia.
119) P.O. Box 98, Murupara.
120) P.O. Box 1109, Invercargill.
121) P.O. Box 1127, Whangerei.
123) P.O. Box 430, Wanganui.
124) P.O. Box 7010, Taradale.
125) Ross Place, Lawrence.
126) 3 Charlemont St., Ranfurly.
128) Baptist Radio, Wairoa.
129) P.O. Box 2197, Auckland 1015.
130) P.O. Box 15-213, Hamilton.
131) P.O. Box 129, Paeroa.
132) P.O. Box 19-298, Hamilton.
133) Mai FM, Bastion Point, Auckland.
134) Main St., Palmerston North 5301.
135) Motueka Rec Centre, Old Wharf Rd., Motueka.
136) P.O. Box 92, Auckland.

NIUE ISLAND (NZ Dependency)

L.T: UTC – 11h — **Pop:** 2.250 — **Radio sets:** 1.000 (est.) — **Pr.L:** Niuean, English — **E.C:** a/c 50, 220V — **ITU:** NIU.

BROADCASTING CORP. OF NIUE (BCN)

ADDR: P.O. Box 23, Alofi, Niue Isl. **Te:** +683 4026. **Fax:** +683-4217 — **L.P:** GM: Hima Douglas. CE: Trevor Tiakia.
STATION: Alofi ZK2ZN 594kHz 0.25kW — **FM:** 91MHz 0.1kW.
D.PRGR: 1730-2000, 2230-0030, 0500-0830. **N:** on the h.
ANN: "This is R. Sunshine" — **V.** by letter.

NORFOLK ISLAND (Australian)

L.T: UTC + 11½h — **Pop:** 2000 — **Radio sets:** 900 — **Pr.L:** English — **ITU:** AUS.

NORFOLK ISLAND BROADCASTING. SCE. (Gov.)

ADDR: P.O. Box 456, Norfolk Island 2899, Australia. **Te:** +6723 (2) 2137. **Fax:** +6723 (2) 3298.
L.P: Broadc. Officer: Margaret Meadows.
STATION: VL2NI 1566kHz 0.05kW, 93.1/93.9MHz 0.05kW (stereo).
D.PRGR: MF 1930-0230, 0500-1030, Sat 1930-1830, Sun 2030-1830. **FM:** 93.9MHz rel. MW + separate prgrs. at times. Rel. ABC R. national overnight — 93.1MHz rel. ABC-FM 24h.
ANN: "Norfolk Island Community Radio 2NI" — **V:** by QSL-card.

NORTHERN MARIANA ISLANDS (U.S. Commonwealth)

L.T: UTC + 10h — **Pop:** 21.200 — **Radio sets:** 10.500 (est.) — **Pr.L:** English, Chamorro, Carolinian, Filipino — **ITU:** MRA.

INTER ISLAND COMMUNICATIONS INC. (Comm).

ADDR: P.O. Box 914, Saipan, CM 96950. **Te:** +1 (670) 234 7239.
L.P: Gen.Mgr: Hans W. Mickelson. Prgr. Dir: Ken Warnick. Chief Eng: Angel Ocampo. N.Dir: Ken Phillips.
STATION: Chalan Kiya: KCNM 1053kHz 1kW — KZMI 103.9MHz 2kW (stereo).
D.PRGR: 2000-1400. **N:** on the h.
ANN: "KCNM (or) KZMI, Your Music Station" — **V.** by QSL-card & letter (B).

FAR EAST BROADC.CO. (Rlg.)

ADDR: P.O. Box 209, Saipan, CM 96950. **Te:** +1 (670) 234 6520 (KSAI), +1 (670) 322 9088 (KFBS). **Fax:** +1 (670) 322 3060.
L.P: Dir: D. Campbell. Chief Eng (KFBS): Robert Springer.

HOME SERVICE

STATION: Susupe (G.C: 145.50E/15.08N): KSAI 936kHz 10kW.
D.PRGR. in English: 2000-1200 **N.** on the h.

INTERNATIONAL SCE.

STATION: KFBS Marpi (G.C: 145.48E/15.16N): 4x100kW SW-tr's.

kHz	kHz	kHz	kHz	kHz
9465	9495	11650	11705	15380
9475	9670	11665	15320	

	Days	kHz
Bugis		
0930-1000	1234567	15380
Burmese		
1430-1530	1234567	9670
Chinese		
1115-1500	1234567	9495
German		
1500-1530	67	11665
1900-1930	6	9465
Hausa		
1900-1930	1234567	9475
Indonesian		
1100-1230	1234567	15380
2200-2300	1234567	15320
Javanese		
1030-1100	1234567	15380
Kazakh		
1445-1500	123	11665
Kyrgyz		
1445-1500	45	11665
Malay		
1230-1300	1234567	15380
Mongolian		
1130-1145	1234567	11705
Polish		
1915-1930	2	9475
Romanian		
1500-1530	12	11665
Russian		
0900-1400	1234567	11650
1145-1200	1234567	11705
1200-1445	1234567	11665
1400-1830	1234567	9465
1830-1900	34	9465
Sundanese		
1000-1030	1234567	15380
Uighur		
1500-1515	1234567	9495
Ukrainian		
1500-1530	345	11665
1830-1915	2	9465
1830-1930	157	9465
1830-1900	6	9465
1830-1915	34	9465
Uzbek		
1445-1500	1234567	11665
Vietnamese		
1300-1430	1234567	9670
2300-2400	1234567	15320
Yoruba		
1930-2000	1234567	9475

V. by QSL-card (B). Rp. (2 IRC's). Rec. acc.

KHBI — World Sce. of the Christian Science Monitor (Comm.)

ADDR: See WCSN, USA.
L.P: St. Mgr: D. Villar. Others: See WCSN, USA.
STATION: Agingan Point (G.C: 145.41E/15.07N): 2 × 100kW SW-trs.

kHz	Area	Times	kHz	Area	Times
9455	J/KOR	2000-2200	15665	AUS/NZ	0800-1000
9495	Eu/ME	1800-2000		AUS	1200-1400
9530	EAs	1400-1600	17555	EAs	0800-1000
11580	CAs	1600-1800		CAs	1000-1200
13625	SEAs	1000-1200		SEAs	0600-0800
	SAs	1200-1600			0000-0200
	SEAs	2200-2400	17780	EAs/J	0400-0600
13840	AUS	1800-2200		CAs/CHN	0600-0800
15405	CAs	2200-2400	17865	NEAs	0000-0400SS

D.PRGR: see WCSN, USA.

PALAU

L.T: UTC + 9h — **Pop:** 14.500 — **Radio sets:** 9.000 (est.) — **Pr.L:** Palauan, English — **E.C:** a/c 60, 220V.

WSZB BROADCASTING STATION

ADDR: Box 99, Koror State, Republic of Palau 96940. **Te:** +680 488 2417. **Fax:** +688 488 1725. **Telex:** 7288901.
L.P: SM: Salustiano Albert.
MW: WSZB Malakal Island 1584kHz 5kW.
D.PRGR: W 2100-1500, Sun 2200-1400. **N.** on the h.
ANN: "WSZB, the Voice of Palau".

KHBN (Rlg.)

ADDR: P.O. Box 66, Koror, Republic of Palau 96940.
STATION: 1 × 100kW tr — +) using 50kW.

kHz	Area	H. of tr.
9785+	EAs	2000-2400
9830	CAs	0800-1600
11980	CAs	0000-0800

D.PRGR: English, Chinese and Vietnamese.

PAPUA NEW GUINEA

L.T: UTC + 10h — **Pop:** 4.005.000 — **Radio sets:** 235.000 — **Pr.L:** English, Melanesian, Pidgin, Hiri Motu + 30 Vernacular languages — **E.C:** a/c 50, 240V — **ITU:** PNG.

NATIONAL BROADCASTING COMMISSION OF PAPUA NEW GUINEA

ADDR: P.O. Box 1359, Boroko, N.C.D. **Te:** +675 25 5233. **Cable:** Natbroadcast. **Telex:** NE 22112 or NE 22152, Port Moresby. Provincial Prgr. Div (Kundu Sce: **Te:** +675 253022. **Tlx:** NE 23415. Eng. Div: **Te:** +675 253022. **Tlx:** NE 22348.
L.P: Chmn: Sir Alan Tolola. Dir.Eng: Nera Konerus. Prgr.Dir: C. Apelis. Dir. Staff Development & Training: R. Bale. Dir. N. & Current Affairs: E. Tami. Dir. Finance: P. Reptario. Dir. Management: J. Korau. Sen. Tech. Officer: Bob Kabewa.
STATIONS: N=Karai Sce (Nat.): 1900-1400 — P=Kundu Sce (Provincial).

	Call	kHz	kW		St/Times
1)	—	585	2.5	N	Port Moresby
1)	—	675	2	N	Lae*
1)	—	675	2	N	Wewak
1)	—	810	2.5	N	Rabaul*
1)	—	864	2	N	Madang
1)	—	900	2	N	Goroka*
20)	—	1494	10	P	R. Enga: as 2410kHz
8)	—	1593	10	P	R. West Sepik: as 3205kHz
20)		2410	10	P	R. Enga: 1900-2200, 0700-1300
8)	—	3205	10	P	R. West Sepik: 1900-2200, 0700-1300
9)	—	3220	10	P	R. Morobe: 1900-2200, 0700-1300
2)	—	3235	10	P	R. West New Britain: 1900-2200, 0730-1300
10)	—	3245	10	P	R. Gulf: 1900-2200, 0700-1300
11)	—	3260	10	P	R. Madang: 1900-2200, 0700-1300
12)	—	3275	10	P	R. Southern Highlands: 1900-2200, 0700-1300
13)	—	3290	10	P	R. Central: 1900-2200, 0700-1300
14)	—	3305	10	P	R. Western: 1900-2200, 0700-1300
19)	—	3315	10	P	R. Manus: 1900-2200, 0700-1300
15)	—	3325	8	P	R. North Solomons: 1900-2200, 0700-1300
16)	—	3335	10	P	R. East Sepik: 1900-2200, 0700-1300
7)	—	3345	10	P	R. Northern: 1900-2200, 0700-1300
3)	—	3355	10	P	R. Simbu: 1900-2200, 0700-1300
17)	—	3365	10	P	R. Milne Bay: 1900-2200, 0700-1300
6)	—	3375	10	P	R. Western Highlands: 1900-2200, 0700-1300
18)	—	3385	10	P	R. East New Britain: 1900-2200, 0700-1300
4)	—	3395	10	P	R. Eastern Highlands: 1900-2200, 0700-1300
5)	—	3905	10	P	R. New Ireland: 1900-2200, 0700-1300
1)	P2K3	3925	2.5	N	Port Moresby: 1900-2200, 0700-1400
1)	P2T4	4890	5	N	Port Moresby: 1900-1400
6)	—	5965	10	N	NBC Mount Hagen: 2200-0700+
18)	—	5985	10	N	NBC Rabaul: 2200-0700+
17)	—	6040	10	N	NBC Alotau: 2200-0700+
14)	—	6080	10	N	NBC Daru: 2200-0700+
16)	—	6140	10	N	NBC Wewak: 2200-0700+
1)	P2K9	9520	10	N	Port Moresby: 2200-0700 (**F.Pl.**)

*) incl. local prgrs — +) Mon-Fri. **F.PL: FM:** Rabaul 100.8MHz 1kW (Karai Sce).
Kalang Sce (FM): Boroko 100.3MHz 1kW, Kundiawa 101.0MHz 0.05kW +13 st's of 0.1kW — **D.Prgr:** 1900-1400. (**Addr:** Kalang Adv. Pty Ltd, P.O. Box 1534, Port Moresby **Te:** +675 255233. **Fax:** +675 252595. **Tlx:** NE 23404 — MD: J. Malisa).

Addresses and other information:
Karai Service:
Addr. & L.P: see above — 1900-1400 in English, Pidgin, Hiri Motu. **N. in English:** 2000(W), 2045, 2145(W), 2245(Sun), 0230, 0330 (exc. Sat), 0600 (Mon-Fri), 0800 (exc. Sat), 0900, 1100, 1300.
V. by QSL-card. Rp. Specific prgr. details, time, frequency, date required for verifications.

Kundu (Provincial) Stations:
2) P.O. Box 412, Kimbe — Mgr: Valuka Lowa. Prgrs. in Pidgin — **Ann:** "Maus bilong Tavur".
3) P.O. Box 228, Kundiawa — Mgr: R. Teine. Prgrs in Pidgin, Kuman and local dialects — **Ann:** "Karai bilong Mambu".
4) P.O. Box 311, Goroka — Mgr: Tonko Nonao. Prgrs. in Pidgin, Gahuka, Kafe — **Ann:** "Karai bilong Kumul".
5) P.O. Box 140, Kavieng — Mgr: O. Malatana. Prgrs. in Pidgin — **Ann:** "Maus bilong Mai Mai".
6) P.O. Box 311, Mount Hagen — Mgr: P. Ray. Prgrs. in Pidgin, Midwahgi, Melpa — **Ann:** "Nek bilong Tarangau".
7) P.O. Box 137, Popondetta — Mgr: M. Pendaia. Prgrs. in English, Orokaiva, Hiri Motu, Ewage — **Ann:** "Voice of Oro".
8) P.O. Box 37, Vanimo — Mgr: G. Deckwalen. Prgrs. in Pidgin — **Ann:** "Maus bilong Sandaun".
9) P.O. Box 1262, Lae — Mgr: A. Nase. Prgrs. in Pidgin, Kote, Yabim — **Ann:** "Maus bilong Kundu".
10) P.O. Box 36, Kerema — Mgr: M. Daniel. Prgrs. in Hiri Motu, Kamea, Toaripi, Pidgin — **Ann:** "Voice of the Seagull".
11) P.O. Box 2036, Madang — Mgr: S. Tiori. Prgrs. in Pidgin, Graged — **Ann:** "Karai bilong Garamut".
12) P.O. Box 104, Mendi — Mgr: A. Meles. Prgrs. in Pidgin + local vernaculars — **Ann:** "Karai bilong Muruk".
13) P.O. Box 1359, Boroko — Mgr: S. Gamini. Prgrs. in English, Hiri Motu — **Ann:** "Kibi ena Gadona".
14) P.O. Box 23, Daru — Mgr: R. Wainetti. Prgrs. in English, Hiri Motu — **Ann:** "Voice of the Sunrise".
15) P.O. Box 393, Rabaul — Mgr: A. Rumina. Prgrs. in Pidgin — **Ann:** "Maus bilong Sunkamap". **NB:** St. temporarily operating from Rabaul.
16) P.O. Box 65, Wewak — Mgr: L. Umbo. Prgrs. in Pidgin — **Ann:** "Nek bilong Sepik".
17) P.O. Box 111, Alotau — Mgr: S. Muraga. Prgrs. in English, Hiri Motu — **Ann:** "Voice of Kula".
18) P.O. Box 393, Rabaul — Mgr: E. Mael. Prgrs. in Pidgin, Kuana — **Ann:** "Maus bilong Tavuvur".
19) P.O. Box 505, Lorengau — Mgr: E. Sereman. Prgrs. in Pidgin — **Ann:** "Maus bilong Chauka".
20) P.O. Box 196, Wabag — Mgr: J. Kuk. Prgrs. in Pidgin, English, Midwahgi, Medipa, Enga — **Ann:** "Karai bilong Miok".
V. Most st's verify direct. Re. to 3), 18) via National Sce. Addr. in Boroko.

POLYNESIA (French)

L.T: UTC − 10h — **Pop:** 200.000 — **Radio sets:** 84.000 — **Pr.L:** French, Tahitian — **E.C:** a/c 60, 220V (yes) — **ITU:** OCE.

SOCIETE NATIONALE DE RADIO TELEVISION FRANÇAISE D'OUTRE MER (RFO)

ADDR: B.P. 125, Papeete, Tahiti. **Te:** +689 43 05 51. **Fax:** +689 41 31 55. **Cable:** 291FP FROMRPT. **Telex:** 070200.
L.P: Dir: A. Le Garrec. Dir.Inf: B. Minas. Dir.Tec: M. Ollivier.
STATIONS: Mahina (G.C: 149.00W/17.00S): 738kHz 20kW — 6135kHz 4kW — 11825/15170kHz 20kW.
FM: Papeete 89.0MHz + 4 st's.
D.PRGR: 24h — **French:** 1600-2045, 2200-0500, 0800-0930 — other times in Tahitian. **N. in French:** 1600, 1630, 1730, 2000, 2200, 0600.
ANN: F: "Ici Tahiti, Société Nationale de Radio-Télévision Française d'Outre Mer". Tahitian: "O Tahiti teie te RFO" — **INT-SIG:** Tahitian flute (vivo) and drums — **V.** by QSL-card.

PRIVATE STATIONS: Several low power FM-st's are operating.

SAMOA (American)

L.T: UTC − 11h — **Pop:** 44.300 — **Radio sets:** 20.000 (est.) — **Pr.L:** Samoan, English — **E.C:** a/c 60, 120V — **ITU:** SMA.

Mediumwaves	kHz	kW		kHz	kW
1) KBQN	585	50	2) WVUV	648	10
FM	**MHz**	**kW**		**MHz**	**kW**
3) KSBS-FM	92.1	3	2) WVUV-FM	101.1	9.36

Addresses and other information
1) Quality Media Corp, Pago Pago (st. under construction).
2) Box 2567, Pago Pago 96799. **Te:** +684 683 7511 — 24h.
3) Samoa Technologies Inc, Box 793, Pago Pago 96799.

SAMOA (Western)

L.T: UTC−11h — **Pop:** 195.000 — **Radio sets:** 75.000 — **Pr.L:** Samoan, English — **E.C:** a/c 50, 230/410V — **ITU:** SMO.

WESTERN SAMOA BROADC. DEPT. (2AP) (Gov. Comm.)

ADDR: P.O. Box 1868, Apia. **Te:** +685 21420. **Fax:** +685 24000.
L.P: DG: J.K. Brown. PD: L. Polu. CE: T. Tana. Adv. Mgr: I. Solofa. Head of Finance: L. Hunkin.
STATIONS: Apia 540/747kHz 10kW.
Ch. 1 in Samoan/English: 1700-1000. **N:** on the h.
Ch. 2 in English: 0400-1000.
ANN: E: "Samoa Broadc. Sce, 2AP".
INT-SIG: Gong — **AD:** Accepts comm. adv. throughout the day on weekdays — **V.** by QSL-card and letter (B).

SOLOMON ISLANDS

L.T: UTC + 11h — **Pop:** 360.000 — **Radio sets:** 38.000 — **Pr.L:** Pidgin, English — **E.C:** a/c 50, 240V — **ITU:** SLM.

SOLOMON ISLANDS BROADCASTING CORPORATION (Statutory Authority)

ADDR: P.O. Box 654, Honiara. **Te:** +677 20051. **Fax:** +677 23159. **Cable:** Broadcast Honiara. **Telex:** SIBC HQ66406.
Regional Addr: R. Hapi Lagoon, P.O. Box 78, Gizo, Western Province — R. Temotu, Lata, Santa Cruz, Temotu Province.
L.P: GM: J.P. Mae. PD: Johnson Honimae, Ag. CE: Loyley Ngira.
MEDIUMWAVES: Honiara 1035kHz 10kW(d) — Gizo 945kHz 10kW — Lata 1386kHz 5kW.
SHORTWAVES: (G.C: 160.03E/09.25S): 5020/9545kHz 10kW.
D.PRGR: Mon-Fri 1900-2235 & 0455-1130; Sat/Sun 1900-1130. (on 9545kHz: 2055-2235 & 0455-0805, Sat 2055-0805).
Regional Prgrs on 945/1368kHz: 0505-0800. **N. in English:** 2000, 2200 (R. Aust), 0130W (local), 0200 (R. Aust), 0500W (local), 0600 (BBC), 0730 (local), 1000 (R. Aust), 1100 (local).
ANN: "This is the SIBC, Radio Happy Isles". Gizo: "Radio Happy Lagoon". Lata: "This is R. Temotu".
INT-SIG: Drum and Bamboo Pipes — **V.** by QSL-card. Rp. Re. must contain details to establish authenticity.

TONGA (Kingdom of)

L.T: UTC + 13h — **Pop:** 103.200 — **Radio sets:** 66.000 — **Pr.L:** Tongan, English — **E.C:** a/c 50, 240V — **ITU:** TON.

TONGA BROADCASTING COMMISSION (Independent Statutory Board)

ADDR: P.O. Box 36, Nuku'alofa. **Te:** +676 23555/6. **Fax:** +676 24417. **Telex:** 66225 Gentel TS.
L.P: Gen. Mgr: Tavake Fusimalohi. Contr. Eng: Siale Fetu'ufuka. Prgr. Contr: Mrs. L. Vikilani. Contr. N: L. Petelo.
MEDIUMWAVE: 1017kHz 10kW — **SHORTWAVE:** 5030kHz 1kW — **FM:** 94.6/96.0/97.2MHz (all 0.01kW).
R. Tonga 1 on 1017/5030kHz: 1730(Sun 0330)-1000.
N. in English: 1800(BBC), 1900(RA), 2200(local), 0055(VOA), 0900(RA).
R. Tonga 2 on FM: 1730(Sun 0330)-1000.
ANN: E: "This is the call of the Friendly Islands A3Z Nuku'alofa of the Tonga Broadcasting Commission". T: "Ko e Ui 'E ni 'a e 'Otu Felenite, mei he A3Z Nuku'alofa 'a e Komisioni Fakamafolalea Tonga" — **INT-SIG:** Tongan noseflute.

UNITED CHRISTIAN BROADCASTERS (Rlg.)

ADDR: see R.Rhema, New Zealand.
MEDIUMWAVE: 738kHz 5kW — **FM:** 93.0MHz 0.5kW.
D.PRGR: mostly Tongan on MW, English on FM

TUVALU

L.T: UTC+12 — **Pop:** 9.500 — **Radio sets:** 4.000 — **Pr.L:** English, Tuvalu — **E.C:** a/c 50, 240V — **ITU:** TUV.

RADIO TUVALU (Gov.)

ADDR: Broadc. & Inf. Office, Vaiaku, Funafuti. **Te:** +688 732. **Fax:** +688 819. **Telex:** 4800 TUV. **Cable:** Broadcast Tuvalu.
L.P: Broadc. & Inf. Officer: P. Laafai. Technician: P. George.
MEDIUMWAVE: Funafuti T2U2 621kHz 5kW.
D.PRGR: 1852-2000, 2325-0100, W0550-0900. **N. in Tuvalu & English:** 1910. **N. in Tuvalu:** 0730. **N. in English:** 0710.
ANN: E: "This is Radio Tuvalu" — **V.** by letter (B).

VANUATU (Republic of)

L.T: UTC + 11h (Su: UTC + 12h) — **Pop:** 176.000 — **Radio sets:** 20.000 (est.) — **Pr.L:** Bislama, English, French — **ITU:** VUT.

RADIO VANUATU (Gov.)

ADDR: P.O. Box 49, Port Vila. **Te:** +678 2999. **Telex:** 1046.
L.P: Head of Prgrs: A. Thompson.
STATIONS: Malapoa (G.C: 168.33E/17.44S): YJB2 1125kHz 10kW — YBJ4 3945kHz 10kW — YBJ7 7260kHz 10kW.
FM: Port Vila 98.0MHz 280W — Santo 98.5MHz 140W.
D.PRGR: 1900-1115 (Sun 1005).
N. in English: 2000 (rel. R. Australia), 2200 (rel. BBCWS), 0115, 0900. **N. in French:** 2100 & 0000 (rel. R. France Int.), 0120, 1000.
ANN: "Radio Vanuatu" — **V:** by QSL-card (A).

WAKE ISLANDS (U.S. Territory)

L.T: UTC + 12h — **Pop:** 300 — **Radio sets:** 100 (est.) — **Pr.L:** English — **E.C:** a/c 60, 110/220V — **ITU:** WAK.

AFRTS (U.S. Air Force)

ADDR: AFRTS-KEAD, Det.4, 15th Air Base Wing (PACAF), Wake Island, APO San Francisco, CA 96501, USA. **Te:** Autovon 430-0111 Wake ext. 224.
STATION: KEAD 1485kHz 0.05kW, 101.2MHz 0.01kW.
D.PRGR: Tues-Sat 2300-1100 (prerecorded AFRTS prgrs).

WALLIS & FUTUNA (French)

L.T: UTC +12h — **Pop:** 17.100.000 — **Pr.L:** French, Wallisian — **ITU:** WAL.

RADIODIFFUSION FRANÇAISE D'OUTRE MER (RFO)

ADDR: B.P. 102, Mata-Utu, Iles de Wallis et Futuna (par Nouméa, Nouvelle-Caledonie). **Te:** +681 722020. **Fax:** +681 722346. **Fax (Tech):** +681 722526. **Telex:** 250 500.
L.P: SM: A. Quintrie-Lamothe. Head of Info: A. Michon. TD: J. Blasco.
MW: Mata-Utu (Wallis) 1188kHz 2kW — **D.PRGR:** 1800-1000.
ANN: "Bonjour, vous êtes sur RFO Wallis et Futuna," plus French national anthem — **V.** by letter.

NORTH AMERICA

ALASKA (U.S. State)

L.T: UTC – 9h (Western-most Aleutian Is. – 10h) (cf. World Time Table) — **Pop:** 588.000 — **Pr.L:** English — **E.C:** a/c 60, 120/240V — **ITU:** ALS.

ALASKA BROADCASTERS ASSOCIATION

ADDR: c/o KIAK, Box 73410, Fairbanks 99707—Pres: Peter van Nort.

	Call	kHz	kW	h. of tr.
1)	KENI	550	5	24h
2)	KVOK	560	1	1500 (Sun 1600)-0900
3)	KRSA	580	5	24h
4)	KHAR	590	5	24h
5)	KGTL	620	5	W 1455-0906 (Sun 0806)
6)	KJNO	630	5/1	1400-0900 (Fri/Sat 24h)
7)	KIAM	630	5/2.2	1600-0800
8)	KYUK*	640	10	1500 (Sun 1700)-0930 (Sat 1030)
9)	KYAK	650	50	24h
10)	KFAR	660	10	24h
11)	KDLG*	670	10	1500-0900 (June-Aug: 24h)
12)	KBRW*	680	10	1500-0900 (Sat, Sun 1600-1100)
13)	KBYR	700	10	24h
14)	KOTZ*	720	10	1500-0900
15)	KFQD	750	50	24h
16)	KCHU*	770	9.75	1500 (Sat 1700, Sun 1800)-0900
17)	KNOM*	780	10	1455-1110
18)	KCAM	790	5	24h
19)	KINY	800	10/5	24h
20)	KCBF	820	50	24h
22)	KSDP*	830	1	24h
21)	KABN	840	10/5	
23)	KICY	850	10	1455-0930
24)	KSKO	870	10	1457-0910
25)	KBBI*	890	10	1430 (Sat/Sun 1500)-0900
25a)	KZPA*	900	5	F.Pl.
26)	KIYU	910	5	1500-0900, Fri, Sat 24h
27)	KSRM	920	5	24h
28)	KTKN	930	5/1	1400-0900
29)	KNSA*	930	2.5	
30)	KSWD	950	1	24h
31)	KIAK	970	5	24h
32)	KZXX	980	1	24h
33)	KFFR	1020	10	24h
34)	KKSD	1080	10	24h
35)	KSLD	1140	10	24h
36)	KJNP	1170	50/23	1400 (Sat 1600, Sun 1700)-1200 (Sat/Sun 0910)
37)	KVAK	1230	1	24h
38)	KIFW	1230	1	1500-0900
39)	KBCN	1300	1	Silent
40)	KSKM	1420	1	F.Pl.
41)	KLAM	1450	0.25	1500-0700

*) non-comm.

Addresses and other information:
Abbreviations: See under U.S.A. commercial st's.

1) 1777 Forest Park Drive, Anchorage 99517 — G.M: John Ruby; P.D: J.R. Roberts; C.E: Van Craft. News/talk (N-1). F.M: KBFX 100.5 MHz.
2) Box 708, Kodiak 99615 — G.M: Jack Emerson; P.D: Scott Campbell. Variety (N-1).
3) Box 650, St. Petersburg 99833 — Mgr: Donald Johnson; C.E: Daniel Zachary. Rlg. variety (N-1).
4) Box 111566, South Station, Anchorage 99511 — G.M: Craig Harpel; P.D: Dave Stroh; C.E: Walter Sorton. MoR. N:50.
5) Box 103, Homer 99603 — Mgr. & C.E: D.F. Becker; P.D: David Webb. Easy listening (N-1).
6) 3161 Channel Drive, Juneau 99801 — G.M: Steve Rhyner; P.D: Larry Mondelo; C.E: Brian Romeijn. Oldies based ADC.
7) Box 00474, Nenana 99760 — G.M: John Reese; C.E: Tim Zook. Rlg. (N-1).
8) Box 468, Bethel 99559 — G.M: Jerry Brigham; P.D: Chuck Bradley; C.E: B. Humelsine. Variety. (N-1). Bilingual English/Yupik.
9) 2800 East Dowling Rd, Anchorage 99507 — G.M: Gary Donovan; P.D: Steve Chapman; C.E: Derek Edmondson. CWM (N-1).
10) Box 70910, Fairbanks 99707 — G.M: Bill Walley; P.D: Terry Walley; C.E: Lee Santoro. Talk (N-1).
11) Box 670, Dillingham 99576 — Mgr: Les Robinson; P.D: Michelle M. Abrams; C.E: Rob Carpenter. Variety. (N-1).
12) Box 109, Barrow 99723 — Mgr: Don Rinker; P.D: S. Hamlin; C.E: Mel Sather. Variety 20% Eskimo (N-1).
13) Box 10-2200, Anchorage 99510 — Mgr: Ron Moore; P.D: R. Dehn; C.E: Duane Milsap. Oldies (N-1).
14) Box 78, Kotzebue 99752 — G.M: Rob Rawls; P.D: Ed Pentis; C.E: P.A. Lonewolf. Variety (N-1).
15) 9200 Lake Otis Parkway, Anchorage 99507 — G.M: Dennis Bookey; P.D: Michael Rogers; C.E: Jay White. ADC (N-1).
16) Box 467, Valdez 99686 — Mgr: D. Hammock; P.D: Mary Beth Joyce; C.E: M. Pelikan. Variety.
17) Box 988, Nome 99762 — G.M: T.A. Busch; P.D: Katy Clark; C.E: Tim Cochran. Variety (N. at :55).
18) P.O. Box 403, Glennallen 99588 — Stn. Mgr: George Reichman; P.D: Scott Yahr; C.E: Dan J. Zachary. 50% C&W, 50% rlg. (N-1).
19) 1107 8th Str, Suite 2, Juneau 99801 — G.M: D.W. Egan; P.D: Paul Ryder; C.E: C. Gray. ADC.
20) Box 70950, Fairbanks 99707 — G.M: H.H. Hove; P.D: Charlie O'Toole; C.E: Les Secrest. Oldies. (N-1).
21) Box 520368, Big Lake 99652 — G.M: Elaine Welch; P.D: Roger Leff; C.E: C. Plunkett. News, talk, oldies.
22) Box 328, Sand Point 99661 — Variety. (N-1); Also rel. 11), 48).
23) Box 820, Nome 99762 — S.M: John McBride; P.D: Daniel A. Smith. ADC, rlg. (N-1). **Russian** 0800-0930.
24) Box 70, McGrath 99627 — Mgr: Susan Braine; P.D: Maggie Grady. Variety.
25) 215 East Main Court, Homer 99603 — Mgr: Will Peterson; P.D: Kathy Steberl; C.E: S. Morton. Variety, community radio.
25a) Box 126, Fort Yukon 99740 (under construction).
26) Box 165, Galena 99741 — G.M: Robert C. Sommer, Variety (N-1) (mostly relays st. 24).
27) HC 2, Box 852, Soldotna 99669 — G.M: John C. Davis; P.D: Charles Newberg; C.E: Bill Glynn. ADC (N-1). **FM:** KWHQ 100.1 MHz.
28) 526 Stedman, Ketchikan 99901 — G.M: Patti Mackey; P.D: Bob Konet; C.E: Ken Eckland. ADC, MoR. **FM:** KGTW 106.7.
29) Unalakleet 99684 (mostly relays st. 11).
30) Box 405, Seward 99664 — S.M: Dave Kelly; C.E: Van Craft. Oldies/Talk 0600-1500.
31) Box 73410, Fairbanks 99707 — S.M: P. van Nort; C.E: C. Beck. CWM (N-1).
32) 6672 Kenai Spur Hwy, Kenai 99611 — Mgr: Bill Richardson; P.D: Paula Richardson. Oldies (N-1).
33) Box 773527, Eagle River 99577 — S.M: Dean Anderson; P.D: Jim Brewer; C.E: T.K. Reynolds. Cont. Christian.
34) 1300 E. 68th Str, Suite 208, Anchorage 99518 — G.M: Andrew Lohman; P.D: Mike Ford; C.E: Kevin McBride, All News. **FM:** KASH 101.5MHz.
35) 374 Lovers Lane, Soldotna 99669 — G.M: Buzz Barr; C.E: Bill Glynn. Oldies (N-1).
36) Box 0, North Pole 99705 — Pres: Donald L. Nelson; G.M: Yvonne L. Carriker; P.D: B. Olson; C.E: E. Nichols. So. gospel (N-1). **Russian** (Wi only): Mon-Fri 0905-1200; **Eskimo/Indian**: Sun 0600-0845.
37) Box 367, Valdez 99686 — G.M: Joel Meltzer.
38) 611 Lake Str, Sitka 99835 — S.M: Geoffrey A. Brandt; C.E: Brian Romeijn. MoR, Oldies, News.
39) Fairbanks, silent (grant to relocate transmitter site).
40) Bethel (under construction).
41) Box 60, Cordova 99574 — G.M: Elizabeth Walker; C.E: Richard Zook. Cont, C&W (N-1).

ALASKAN FORCES RADIO NETWORK (AFRN)

Netw. Addr: Air Force Arctic Broadcasting Squadron, Attn: Station Manager, Elmendorf AFB, Alaska 99506-5000.
Addr: of AFRN st's with own studios: OLGF, AFABS, Attn: Station Manager, APO Seattle, WA 98733, USA. (Fort Greely).

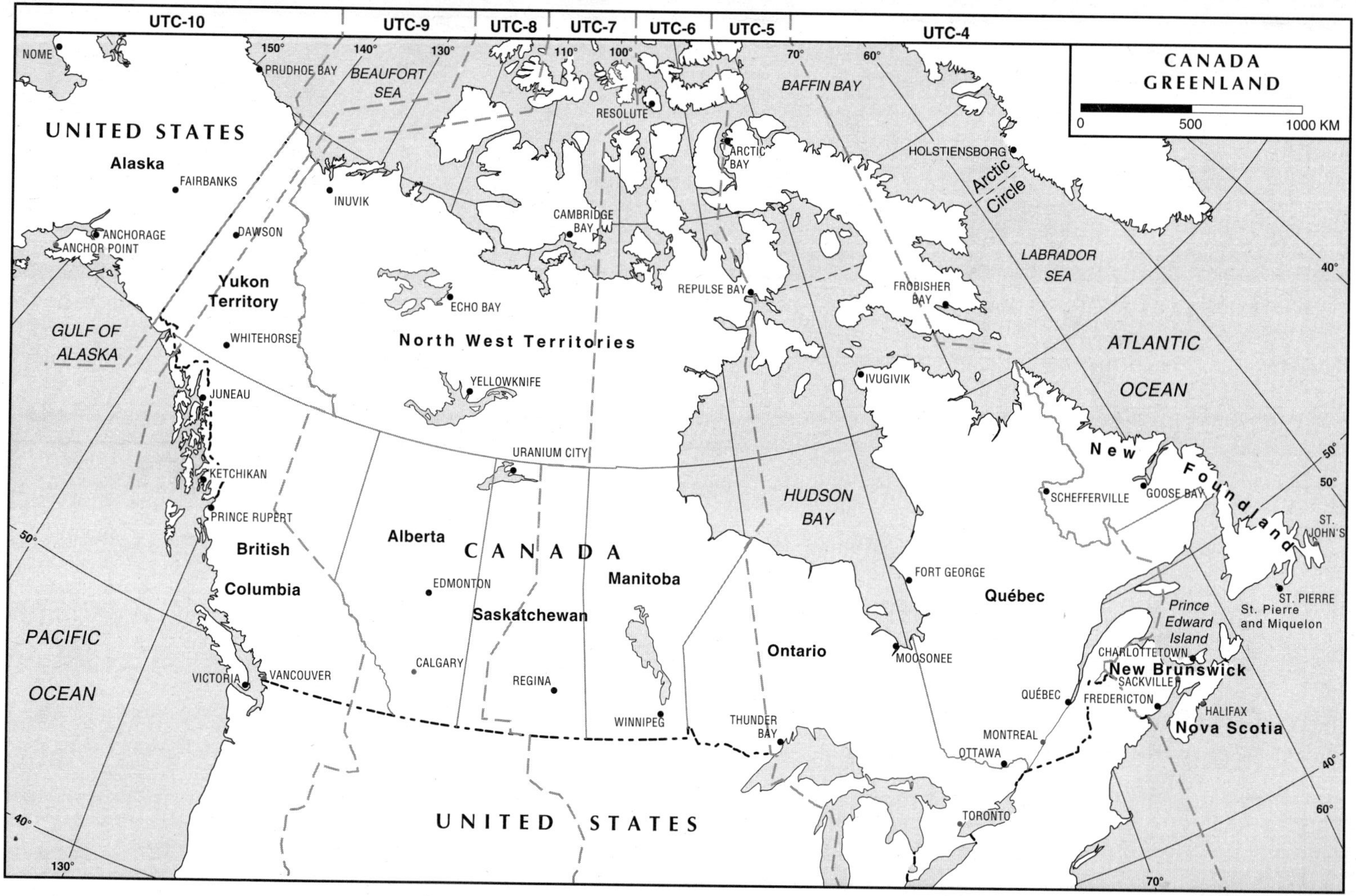
CANADA
GREENLAND
0
500
1000 KM
UTC-10
UTC-9
UTC-8
UTC-7
UTC-6
UTC-5
UTC-4
150°
140°
130°
110°
100°
70°
60°
50°
40°
UNITED STATES
Alaska
NOME
PRUDHOE BAY
FAIRBANKS
ANCHORAGE
ANCHOR POINT
BEAUFORT SEA
GULF OF ALASKA
INUVIK
DAWSON
Yukon Territory
WHITEHORSE
JUNEAU
KETCHIKAN
PRINCE RUPERT
British Columbia
VANCOUVER
VICTORIA
PACIFIC OCEAN
RESOLUTE
CAMBRIDGE BAY
ECHO BAY
North West Territories
YELLOWKNIFE
URANIUM CITY
Alberta
EDMONTON
CALGARY
CANADA
Saskatchewan
REGINA
Manitoba
WINNIPEG
BAFFIN BAY
ARCTIC BAY
REPULSE BAY
HOLSTIENSBORG
Arctic Circle
LABRADOR SEA
FROBISHER BAY
IVUGIVIK
HUDSON BAY
Ontario
THUNDER BAY
MOOSONEE
FORT GEORGE
Québec
QUÉBEC
MONTREAL
OTTAWA
TORONTO
ATLANTIC OCEAN
New Foundland
SCHEFFERVILLE
GOOSE BAY
ST. JOHN'S
ST. PIERRE
St. Pierre and Miquelon
Prince Edward Island
CHARLOTTETOWN
New Brunswick
SACKVILLE
FREDERICTON
HALIFAX
Nova Scotia
UNITED STATES

FM: Fort Greely 93.5MHz 0.3kW.
AFRN Elmendorf feeds relay st's using 90.5, 101.1 and 101.7MHz, all 0.3kW. **MW:** Sweeper Cove, Adak 890 250W.

KNLS — The New Life St. (Rlg)

Owned and operated by World Christian Broadc. Corp.
ADDR: Box 473, Anchor Point, Alaska 99556; **Te:** +1 (904) 235-8262; **Op's Center:** 605 Bradley Court, Franklin, TN 37064. **Te:** +1 (615) 371-8707; **Fax:** +1 (615) 371-8791. **Telex:** 6504393650.
L.P: VP/COO: Edward J. Bailey. PD: Dale Ward. CE: Kevin K. Chambers
STATION: Anchor Point (G.C: 151.44 W/59.45 N); 1 × 100kW tr.

kHz	kHz	kHz
6150	7355	7365

Chinese	Area	kHz	Russian	Area	kHz
1000-1200	CHN	7365	0900-1000	RUS	7365
1500-1700	CHN	7355	1200-1300	RUS	7365
1700-1800	CHN	6150	1400-1500	RUS	7355
English					
0800-0900	EAs	7365			
1300-1400	EAs	7355			

V. by QSL-card. Rec. acc.

BERMUDA

L.T: UTC-4h (Su: UTC-3h) — **Pop:** 59.310 — **Radio sets:** 100.000 (est.) — **Pr.L:** English — **E.C:** a/c 60, 115/230V (yes) — **ITU:** BER.

ZBM-ZFB RADIO & TELEVISION (Comm.)

ADDR: Box HM 452, Hamilton HM BX. **Te:** +1 (809) 295-2828. **Fax:** +1 (809) 295-4282. **Telex:** 3702 ZBM-BA.
L.P: Ag. Gen. Mgr: M.R. Fletcher. Chief Eng: D. Ingham. Prgr. Mgr: B. Lodge.
STATIONS (1kW): ZFB 1230kHz — ABM 1340kHz — **FM:** ZBM-FM 89.1MHz 15kW (stereo) — ZFB-FM 94.9MHz 1kW.
D.PRGR: ZBM: 1000-0300 (Sun 1100-0200). **N, Wrp:** W 1025, 1055, 1125, hourly 1200-0300, also 1608; Sun 1100, 1200, 1300, 1400, 1455, 1615, hourly 1700-0200.
ZFB: 1000 (Sun 1100)-0400. **N:** W 1055, 1130 and every h to 2230, 0230; Sun 1330, 1630, 2230, 0230. **Portuguese:** W 0930-1030, Mon 2330-0100, Fri 2300-0100.
ZBM-FM: 24h music prgr. **N:** 1100, 1200, 1600, 2200.
ZFB-FM: 24h music prgr. **N, Wrp:** 1200, 1700, 2400.
V. by QSL-card or letter. Rp. (B). Rec. acc.

DEFONTES BROADCASTING CO. LTD. (Comm.)

ADDR: P.O. Box 1450, Hamilton 5. **Te:** +1 (809) 295-1450.
L.P: Pres: Kenneth de Fontes. St. Mgr: D.E. Browne. N.Dir: M. De Leon. Dir. Tec: Ed. Tucker.
STATIONS: 1160/1280/1450kHz 1kW — **FM:** 106.2MHz 2.5kW.
D.PRGR: VSB 1 on 1450kHz: 24h (country music). **N, Wrp:** 1000, 1045, 1115, 1145, 1215, 1245, 1415, 1615, 1645, 1715, 1915, 2115, 2215, 0015.
VSB 2 on 1280kHz: 24h (adult contemporary music).
VSB 3 on 1160kHz: 24h Rel. BBC World Sce 1530-1000. Other times tourist info & special prgrs. 1.
VSB V-106 FM on 106.2MHz: 24h (easy listening).
V. by QSL-card or letter.

CANADA

L.T: See Time Table — **Pop:** 27:130.000 — **Radio sets:** 22.600.000 — **Pr.L:** English, French — **E.C:** a/c 60, 115/230V (generally). — **ITU:** CAN.

Abbreviations: NF = Newfoundland, NS = Nova Scotia, PEI = Prince Edward Isl, NB = New Brunswick, PQ = Quebec, ON = Ontario, MB = Manitoba, SK = Saskatchewan, AB = Alberta, BC = British Columbia, NWT = North West Territories, Lab = Labrador, YT = Yukon Territory.

CANADIAN BROADCASTING CORPORATION SOCIÉTÉ RADIO-CANADA (Publicly owned)

ADDR: Head Office: P.O. Box 8478, Ottawa, Ontario K1G 3J5. **Cable:** Broadcasts. **Telex:** 053-4260. **Te:** +1 (613) 724-1200. **Fax:** +1 (613) 738-6749
L.P: President and Chief Executive Officer: Gérard Veilleux — Chairman, Board of Directors: Patrick Watson — Senior Vice-President, Resource Management and Administration: Anthony S. Manera — Vice-President, Planning and Corporate Development: John Shewbridge — Vice-President, Communications: Robert Pattillo — Vice-President, News, Current Affairs & Newsworld, Tim Kotcheff: Vice-President, Regional Broadcasting Operations: Trina McQueen — Executive Director, Business Development and International Relations: Robert O'Reilly — Director of Corporate Communications: Charlotte O'Dea — Vice-President French Television: Guy Gougeon.
English Networks: P.O. Box 500, Station "A", Toronto, ON M5W 1E6. **Cable:** Broadcast. **Telex:** 062-17796. **Te:** +1 (416) 975-3311.
French Networks: P.O. Box 6000, Montreal, PQ H3C 3A8. **Cable:** Radcan. **Telex:** 05-267417. **Te:** +1 (514) 285-3211.
CBC Engineering: 7925 Côte St. Luc Rd., Montreal, PQ H4W 1R5. **Telex:** 055-66437. **Te:** +1 (514) 485-1301.
L.P: Vice-President, Engineering: Brian D. Baldry — Manager, Technical Information: Claudette Lalonde.
English Network (CBC): ° = also on Shortwave.

	Call	kHz	kW		Call	kHz	kW
1)	CBT	540	10	14)	CBZ	970	10
2)	CBK	540	50	15)	CBY	990	10
3)	CFWH	570	5/1	16)	CBW	990	50
4)	CBN	°640	10	17)	CBR	1010	50
5)	CBU	°690	50	18)	CBA	1070	50
6)	CBL	740	50	19)	CBI	1140	10
7)	CBX	740	50	21)	CFFB	1230	1
10)	CFPR	860	10	23)	CFYK	1340	2.5
11)	CHAK	860	1	24)	CBG	1400	1
13)	CBM	940	50	25)	CBE	1550	10

CBC English mono distributed through 416 rebroadcasters.
Shortwaves:
4) CKZN 6160kHz 0.3kW: 0930-0500 (rel. CBN)
5) CKZU 6160kHz 0.5kW: 1400 (Sun 1500)-0905 (rel. CBU).
FM: CFGB-FM (Happy Valley/Goose Bay) 89.5MHz 4.5kW, CBHA-FM (Halifax) 860MHz 90.5kW, CBCT-FM (Charlottetown) 96.1MHz 100kW, CBD-FM (Saint John) 91.3MHz 100kW, CBVE-FM (Quebec) 104.7MHz 100kW, CBO-FM (Ottawa) 920MHz 84kW, CBCS-FM (Sudbury) 99.9Mhz 50kW, CBQT-FM (Thunder Bay) 88.3MHz 10/5kW, CBWK-FM (Thompson) 100.9MHz 0.094kW, CBKA-FM (La Ronge) 105.9MHz 0.08kW, CBQR-FM (Kivalik) 105.1MHz 100.2kW, CBTK-FM (Kelowna) 88.9MHz 4.7kW, CBYG-FM (Prince George) 91.5MHz 100kW.
FM (stereo): CBM-FM (Montreal) 93.5MHz 24.6kW, CBOQ-FM (Ottawa) 103.3MHz 84kW, CBL-FM (Toronto) 94.1MHz 38kW, CBW-FM (Winnipeg) 98.3MHz 160kW, CBU-FM (Vancouver) 105.7MHz 100kW, CBN-FM (St. John's) 106.9MHz 100kW, CBH-FM (Halifax) 102.7MHz 100kW, CBR-FM (Calgary) 102.1MHz 100kW, CBX-FM (Edmonton) 90.9MHz 100kW, CBK-FM (Regina) 96.9MHz 100kW.
French Network (SRC):

	Call	kHz	kW		Call	kHz	kW
25)	CBEF	540	2.5/5	27)	CBV	980	50
7)	CHFA	680	10	16a)	CKSB	1050	10
13)	CBF	690	50	28)	CBGA	1250	10/5
6)	CJBC	860	50	29)	CBJ	1580	50
26)	CJBR	900	10				

CBC French radio is also distributed through 159 rebroadcasters.
FM: CBUF-FM (Vancouver) 97.7MHz 46kW, CBKF-FM (Regina) 97.7MHz 13.7kW, CBON-FM (Sudbury) 98.1MHz 50kW, CBSI-FM (Sept Iles) 98.1MHz 96.7kW, CBAF-FM-5 (Halifax) 92.3MHz 100kW, CBOF-FM 90.7MHz 84kW, CBAF-FM 88.5MHz 22kW.
FM (stereo): CBF-FM (Montreal) 100.7MHz 100kW, CBV-FM (Quebec City) 95.3MHz 100kW, CBJ-FM (Chicoutimi) 100.9MHz 50kW, CBOF-FM (Ottawa) 102.5MHz 84kW, CJBR-FM (Rimouski) 101.5MHz 50kW, CBAF-FM (Moncton) 98.3MHz 77kW.

Addresses:
1) P.O. Box 369, Grand Falls, NF, A2A 2J7.
2) 2440 Broad Str, Regina, SK, S4P 4A1.
3) 3103 — 3rd Avenue, Whitehorse, YT, Y1A 1E5.
4) P.O. Box 12010, St. A, St. John's, NF, A1B 3T8.
5) P.O. Box 4600, Vancouver, BC, V6B 4A2.

6) P.O. Box 500, Station "A", Toronto, ON, M5W 1E6.
7) P.O. Box 555, Edmonton, AB, T5J 2P4.
10) 222 3rd Ave. W, Prince Rupert, BC, V8J 1L1.
11) Mackenzie Rd, Bag Sce. No. 8, Inuvik, NWT, X0E 0T0.
13) P.O. Box 6000, Montreal, PQ, H3C 3A8.
14) P.O. Box 2200 MPO, Fredericton, NB, E3B 5G4.
14a) P.O. Box 2358, Saint John, NB, E2L 3V6.
15) P.O. Box 610, Corner Brook, NF, A2H 6G1.
16) P.O. Box 160, Winnipeg, MB, R3C 2H1.
16a) 607 Langevin Street, St. Boniface, MB, R2H 2W2.
17) P.O. Box 2640, Calgary, AB, T2P 2M7.
18) P.O. Box 950, Moncton, NB, E1C 8N8.
19) P.O. Box 700, Sydney, NS, B1P 6H7.
21) P.O. Box 490, Iqaluit, NWT, X0A 0H0.
23) P.O. Box 160, Yellowknife, NWT, X1A 2N2.
24) P.O. Box 369, Gander, NF, A1V 1W7.
25) P.O. Box 1609, Windsor, ON, N9A 1K7.
26) 273 St. Jean Baptiste Ouest, Rimouski, PQ, G5L 4J8.
27) P.O. Box 10400, Ste-Foy, Quebec, PQ, G1V 2X2.
28) P.O. Box 2000, Matane, PQ, G4W 3P7.
29) P.O. Box 790, Chicoutimi, PQ, G7H 5E7.

V. Reports for st's 1), 15), 24) should be sent to 4). The QSL policy varies from st. to st.

PRIVATELY OWNED STATIONS

Mediumwaves: Stations less than 100W not mentioned. * = also Shortwave, f = mainly French prgrs, e = mainly English prgrs, b = bilingual, m = multilingual. ° = inactive or F.Pl.

	Call	kHz	kW
101a)	CKHL	°530	0.4/0.25e
420)	CIAO	530	1/0.25m
478)	CJSB	540	50/12.5e
17)	CFJC	550	25/5e
1)	CKPG	550	10e
400)	CHNO	550	50/10e
500)	CHLN	550	10/5f
600)	CFNB	550	50e
2)	CHTK	560	1/0.25e
3)	CKNL	560	1e
401)	CJKL	560	5e
402)	CFOS	560	5/1e
501)	CKCN	560	10/5f
901D)	CHVO	560	5e
4A)	CKWL	570	1e
5)	CKEK	570	10/1e
212)	CKSW	570	10e
471)	CHYM	570	10e
601)	CJEM	570	5/1f
900)	CFCB	570	1e
6)	CKXR	580	10/1e
100)	CKUA	580	10e
301)	CKY	580	50e
403)	CFRA	580	50/10e
404)	CKPR	580	5/1e
405)	CKAP	580	10/1e
406)	CKWW	580	0.5e
502)	CHLC	580	10/2.5f
800)	CJFX	580	25e
3A)	CFNL	590	0.25e
7)	CFTK	590	1e
302)	CFAR	590	10/1e
407)	CKYC	590	50e
503)	CKRS	590	25/10f
608A)	CJCW	590	1/0.25e
901)	VOCM	590	25e
8)	CHRX	600	10e
200)	CFQC	600	10e
408)	CFCH	600	10/5e
504)	CIQC	600	10/5e
801)	CKCL	600	10/1e
9)	CHNL	610	25/5e
10)	CJAT	610	10/1e
90)	CKRW	610	1e
101)	CKYL	610	10e
303)	CHTM	610	1e
409)	CKTB	610	10/5e
506)	CFLO	610	1/0.25f
505)	CHNC	610	10/5f
903A)	CKXJ	°610	10e
11)	CJCI	620	10e
201)	CKCK	620	10e
410)	CKOY	620	10/5f
502A)	CFRP	620	1f
901A)	CKCM	620	10e
12)	CKOV	630	5/1e
102)	CHED	630	50e
304)	CKRC	630	10e
411)	CFCO	630	10/1e
412)	CJET	630	10e
507)	CHLT	630	10f
700)	CFCY	630	10e
422)	CFOB	640	1/0.5e
452)	CHOG	640	25e
38)	CISL	650	10/9e
208)	CKOM	650	10e
901c)	CKGA	650	5e
128)	CFFR	660	50e
903DR)	CKXB	670	10/5e
305)	CJOB	680	50e
415)	CFTR	680	50e
903C)	CKXG	680	10e
105)	CKRD	700	50/25e
607)	CHSJ	700	25/10e
416)	CJRN	710	5/2.5e
417)	CHYR	710	10m
501A)	CIPC	710	1f
509)	CKVM	710	10/1f
901E)	CKVO	710	10e
701)	CHTN	720	10/7.5e
13)	CKLG	730	50e
306)	CKDM	730	10/5e
417)	CHYR-7	730	1m
418A)	CJNR	730	1e
510)	CKAC	730	50f
901A)	CHCM	740	10e
211)	CJWW	750	10e
414)	CKGB	750	10/5e
24A)	CFLD	760	1/0.5e
25)	CKQR	760	20/5e
104)	CHQR	770	50e
14)	CKOR	°780	10/2.8e
802)	CFDR	780	50/15e
103)	CFCW	790	50e
419)	CIGM	790	50e
602)	CFAN	790	5/1e
612)	C...	°790	10
900R)	CFNW	790	1e
14)	CKOR	800	10/0.5e
202)	CHAB	800	10e
421)	CKLW	800	50e
423)	CJBQ	800	10e
425)	CKDR	800	1/0.7
511)	CHRC	800	50f
512)	CJAD	800	50/10e
902)	VOWR	800	5/2.5e
130)	C...	°810	5/1e
307)	CKJS	810	10m
603)	CJVA	810	10/5f
450)	CHAM	820	50/10e
127A)	CILW	°830	10/3.5e
466)	CFJR	830	5/1e
4B)	CKBX	840	1/0.5e
124)	CJXX	°840	25/10e
479)	CHUR	840	10e
27A)	CFVR	850	10e
129B)	CKBA	850	1e
513)	CKVL	850	50/35
901D)	CHVO	850	5e
10A)	CKKC	°860	1e
6C)	CKIR	870	1/0.25e
24)	CFBV	870	1/0.5e
900B)	CFSX	870	0.5e
312)	CKLQ	880	25/10e
114)	CHQT	880	50e
30)	CJDC	890	10e
16)	CJVI	900	10e
203)	CKBI	900	10e
400)	CHYC	900	10/1f
424)	CHML	900	50e
514)	CJER	900	1b
516)	CKTS	900	10e
540A)	CKVD	900	10/2.5f
803)	CKDH	900	1e
106)	CKDQ	910	50e
426)	CKLY	910	10/5e
517)	CHRL	910	10f
4)	CKCQ	920	10/1e
308)	CFRY	920	25/15e
428)	CKNX	920	10/1e
604)	CJCJ	920	10/1e
804)	CJCH	920	25e
528)	CFLS	920	10f
107)	CJCA	930	50e
418B)	CKNS	930	10e
605)	CFBC	930	50e
903)	CJYQ	930	50e
18)	CJIB	940	10e
204)	CJGX	940	50/10e
309)	CFAM	950	10e
429)	CKBB	950	10e
606)	CKNB	950	10/1b
805)	CHER	950	10e
108)	CFAC	960	50e
454)	CIAM	960	5/10e
430)	CFFX	960	10/5e
806)	CHNS	*960	10e
109)	CJYR	970	10e
518)	CKCH	970	10/5f
903E)	CFIQ	°970	1/0.5e
19)	CKNW	980	50e
205)	CKRM	980	10/5e
431)	CFPL	980	10/5e
432)	CHEX	980	10/7.5e
519)	CKIS	990	50e
807)	CKBW	1000	10e
555)	CFLP	1000	10f
903D)	CKXD	1010	1e
433)	CFRB	*1010	50e
129C)	CKVH	1020	1/0.4e
15)	CKST	1040	50e
559)	C...	°1040	10/5f
554)	CFZZ	1040	10/1f
37)	CICF	1050	10/1e
110)	CFGP	1050	10e
206)	CJNB	1050	10e
434)	CHUM	1050	50e
111)	CFCN	*1060	50e
520)	CJRP	1060	50/10f
20)	CFAX	1070	10e
436)	CHOK	1070	10e
112)	CKSA	1080	10e
113)	CHEC	1090	10/5e
438)	CKKW	1090	10e
448)	CKTY	1110	10/1e
21)	CKWX	*1130	50e
115)	CFXX	1140	50e
521)	CJTR	1140	50/10f
22)	CKIQ	1150	10e
310)	CKX	1150	50/10e
439)	CJRC	1150	50/10f
440)	CKMO	1150	50e
505A)	CHGM	1150	5f
116)	CKGY	1170	50/15
560)	C...	°1180	1
207)	CFSL	1190	10/5e
23)	CKDA	1200	50e
123)	CHMG	1200	25e
465)	CFGO	1200	50/10
441)	C...	°1200	10/7e
129A)	CKWA	1210	1e
210A)	CFYM	1210	1/0.25e
905)	VOAR	1210	10e
117)	CJOC	1220	10/5e
309A)	CJRB	1220	10e
442)	CHSC	1220	10e
443)	CJRL	1220	5/1e
444)	CJSS	1220	1e
522)	CKSM	1220	10/2.5f
556)	CFVM	1220	10/5f
608)	CKCW	1220	25/10e
7A)	CKTK	1230	1e
9A)	CJNL	1230	1e
109A)	CIYR	1230	1e
118)	CJOK	1230	1e
127A)	CILW	1230	1e
401A)	CJTT	1230	1e
410A)	CHYK	1230	1f
445)	CJLB	1230	4/1e
446)	CKMP	1230	1e
524R)	CJLP	1230	1f
523)	CHVD	1230	1f
514R)	CJSA	1230	1f
900C)	CFGN	1230	0.25e
900D)	CFLN	1230	1e
1A)	CKMK	1240	1e
5A)	CFEK	1240	1e
14A)	CJOR	1240	1e
26)	CJAV	1240	1e
27B)	CKGO	1240	0.25e
33B)	CFNI	1240	1e
95)	CJCD	1240	1e
206A)	CJNS	1240	1e
302A)	CJAR	1240	1/0.5e
423A)	CJNH	1240	1e
427)	CJWA	1240	1e
447)	CJCS	1240	1e
526)	CFLM	1240	1f
527)	CJMD	1240	1f
529)	CKBS	1240	1f
540B)	CKLS	1240	1f
541A)	CJAF	1240	1f
702)	CJRW	1240	1e
901BR)	CKIM	1240	1/0.5e
309B)	CHSM	1250	10e
449)	CHWO	1250	10e
119)	CFRN	1260	50e
610)	CIHI	1260	10e
27)	CHWK	1270	10e
120)	CHAT	1270	10e
423B)	CJTN	1270	1e
530)	CFGT	1270	10/5f
808)	CJCB	1270	10e
122)	CHRB	1280	10e
28)	CHQB	1280	1e
207A)	CJSL	1280	10e
531)	CJMS	1280	50f
532)	CKCV	°1280	50f
311)	CIFX	1290	10e
451)	CJBK	1290	10e
533)	CHRM	1290	10f
209)	CJME	1300	10e

127)	CHLW	1310	10e
453)	CIWW	1310	50e
534)	CHGB	1310	10f
29)	CHQM	1320	50e
449A)	CJMR	1320	10e
809)	CKEC	1320	5e
210)	CJYM	1330	10e
524)	CKLD	1330	10f
6A)	CKCR	1340	1e
9D)	CINL	1340	1e
11A)	CIVH	1340	1e
31)	CKGF	1340	1e
10B)	CFKC	1340	0.25e
106A)	CIBQ	1340	1e
108AR)	CFHC-1	1340	1e
127R)	CHLW-1	1340	0.25e
410R)	CHOH	1340	1f
418)	CKNR	1340	1e
558)	CKMG	1340	1f
509A)	CKVT	1340	1f
524A)	CJAN	1340	1f
527R)	CFED	1340	0.25f
540C)	CHAD	1340	1f
810)	CJLS	1340	5/4e
900E)	CFLW	1340	0.25e
906)	CKXX	1340	10e
907)	CKHV	°1340	1e
39)	CKEG	1350	10e
41)	C...	°1350	1e
456)	CKDO	1350	10/5e
457)	CHVR	1350	1e
534A)	CHAL	1350	1f
537)	CJLM	1350	10/1f
812A)	CKAD	1350	1e
538)	CJVL	1360	10/5f
609)	CKBC	1360	10e
36A)	CHPQ	1370	1e
129)	CFOK	1370	10e
539)	CKOD	1370	10/5f
535)	CFVD	1370	1f
458)	CKLC	1380	10e
459)	CKPC	1380	25e
524B)	CFDA	1380	10f
611)	CHLR	°1380	10
10A)	CKKC	1390	1e
126)	CJCY	1390	10e
460)	CHOO	1390	10e
5R)	CJEK	1400	1e
6B)	CKGR	1400	1e
14D)	CIOR	1400	1e
9C)	CHNL-1	1400	1e
106B)	CKSQ	1400	1e
429A)	CKCB	1400	1e
461)	CHVR-1	1400	1e
514R)	CKSJ	1400	1f
524C)	CKFL	1400	1f
540)	CKRN	1400	10/1f
541)	CJFP	1400	20f
32)	CFUN	1410	50e
462)	CKSL	1410	10e
543)	CFMB	1410	10m
811)	CIGO	1410	10e
213)	CJVR	1420	10e
463)	CKPT	1420	10/5e
524D)	CKTL	1420	1/0.5f
544)	CJMT	1420	10/5f
812B)	CKDY	1420	1e
124)	CJXX	1430	10e
464)	CJCL	1430	50e
33)	CFCP	1440	1e
121)	CJOI	1440	10e
14C)	CHOR	1450	1e
108A)	CFHC	1450	1e
109R)	CKYR	1450	0.5e
467)	CHUC	1450	8/1e
541B)	CHRT	1450	1f
545)	CHEF	1450	20f
812C)	CFAB	1450	1e
120)	CJMH	1460	10e
468)	CJOY	1460	10e
546)	CKRB	1460	10/5f
34)	CJVB	1470	50m
469)	CHOW	1470	10e
547)	CKO	°1470	10/5e
125)	CKER	1480	10m
477)	CKAN	1480	10e
549)	CHRD	1480	50/35f
14AR)	CJOR-1	1490	1e
33A)	CFWB	1490	1e
117A)	CJPR	1490	1e
212A)	CJSN	1490	1e
461A)	CHVR-2	1490	0.25e
402R)	CFPS	1490	1e
506A)	CKLO	1490	1/0.25f
550)	CJMC	1490	1f
551)	CKBM	°1490	1/0.25f
601A)	CKMV	1490	1f
812)	CKEN	1490	1e
35)	CKAY	1500	10/1e
472)	CKOT	1510	10e
552)	CJRS	1510	50/10f
473)	CHIN	1540	50/30m
36)	CHUB	1570	10e
113A)	CFEZ	1570	10/5e
309C)	CKMW	1570	10e
474)	CHLO	1570	10e
475)	CFOR	1570	10/5e
553)	CKLM	1570	50f
476)	CHNR	1600	10e

Shortwaves

	Call	kHz	kW	Relays
504)	CFCX	6005	0.5	CFCF, Montreal: 24h
111)	CFVP	6030	0.1	CFCN, Calgary: 24h
433)	CFRX	6070	1	CFRB, Toronto: 24h
21)	CKFX	6080	0.01	CKWX, Vancouver: 24h
806)	CHNX	6130	0.5	CHNS, Halifax: 24h

Addresses and other information:

Abbreviations: See under U.S.A. comm. st's.

Networks: During recent years many regional networks have been formed. The st's of a network often transmit a common prgr. during part of the broadcast day. The main st has been given a pure number, under which details common to all st's or pertaining to the main st only are found, while the other st's have been given the same number with an individual capital letter added, under which details pertaining only to this st are found. R. means a pure relay st.

CBC Relay: Certain st's are required to carry CBC prgr's for a few hours a day in areas with inadequate coverage by CBC's own tr's. The number of such st's is steadily decreasing.

Télémedia Network (French): St's 410, 500, 501, 505, 506, 507, 509, 510, 511, 517, 518, 524, 526, 527, 532, 533, 535, 540, 541, 544, 546, 549, 551, 555, 556, 558, 603 + satellites. **N:** at :45

Telemedia Network (English): St's 408, 414, 419, 435, 446, 447, 464, 475.

R. Mutuel Network (French): St's 439, 502, 520, 521, 530, 531, 537, 548, 550, 552.

Other Networks: B'cast News, Standard B'cast News, Western Interior Network, Selkirk News Network, Coast to Coast, Star Country, Kiss-97. These networks provide news and/or other prgr's.

Verifications: Many stations verify correct reception reports, especially for exceptional distances. The details of the verification policies often change and thus are omitted here.

News: Extra newscasts are usually aired during prime time (N-3).

British Columbia

1) 1220 6th Ave, Prince George BC V2L 3M8 — G.M: G.M. Leighton; P.D: R. Seabrook. C.E: Ron Kellington. 24h on 550/1240 kHz (N-1).

1A) Box 1210, Mackenzie, BC V0J 2C0 — Mgr: C. Stene. C.E: W. Sanders. Own prgr's: 1500-1900 Mon-Fri.

2) 346 Stiles Place, Prince Rupert, BC V8J 3S5 — Op's Mgr: R. Duck; P.D: Brent Ross; C.E: Harry Nuttma. 24h. (N-1)

3) 10522 Alaska Rd., Fort St. Johns, BC V1T 1B3 — G.M: Mark Tamagi; P.D: R. Beerling, J. Stevens; C.E: F. Roney; 24h on 560/590kHz (N-1).

3A) Box 880, Fort Nelson, BC V0C 1R0 — Mgr: Carl Selzer. Local prgr's: 1300-2100.

4) 160 Front Str, Quesnel, BC V2J 2K1 — G.M: T. Shepherd. P.D: Brian Miles; C.E: F. Lust. 24h on 570/840/920kHz. **German:** Wed 0105-0130. **Italian:** Thurs 0105-0130 (N-1)

4A) 83 S. First Ave, Williams Lake, BC V2G 1H4 — Own prgr's 1400-1800.

4B) Box 939, 100 Mile House, BC V0K 2E0 — P.D: Brian Miles. Own prgr's: Mon-Fri 1500-2100; Sat 1500-1700, 2000-2100.

5) 19 9th Ave. S, Cranbrook, BC V1C 2L9 — G.M: Gary D. Cavers; P.D: R. Ross; C.E: C. Blakely. 24h on 570/1240kHz. (N-1). (CJEK = Sparwood repeater)

5A) Box 1170, Fernie, BC V0B 1M0 — Mgr: H. Ashmore; P.D: R. Robb; C.E: C. Blakey. Own prgr's W 1500-2000, 2300-0200.

6) Box 69, Salmon Arm, BC V1E 4N2 — Mgr: H. Davidson; P.D: A. Laurenson; C.E: G. Hazell. 24h on 580/870/1340/1400kHz (N-1). Ann: "Columbia Shuswap R."

6A) Box 1420, Revelstoke, BC V0E 2S0 — Mgr: T. Biggs; P.D: J. Sousoff. Own prgr's Mon-Fri 1400-2000, Sat 1400-1700.

6B) Golden, BC. Own prgr's: As 6A). Box 1403, Golden, BC V0A 1H0; P.D: Dan Harrington.

6C) Invermere, BC (relay only). P.O. Box 69, Salmon Arm, BC V1E 4N2.

7) 4625 Lazelle Ave, Terrace, BC V8G 1S4 — V.P: B. Edwards; GM: Tim Maclean; P.D: Renita Neys; C.E: K. Barnard. 24h on 590/1230kHz (N-1).

7A) 350 City Centre, Kitimat V8C 1T6 — Mgr: G. Ball. Local prgr's: Mon-Fri 1400-1800, 2300-0100.

8) 1401 West 8th Ave, Vancouver, BC V6H 1C9 — G.M: George Madden; P.D: John Beaudoin; C.E: Alan Kirk. 24h. (N-20).

9) Box 610, Kamloops, BC V2C 5L9 — G.M: R. Dunn; P.D; Jim Reynolds; C.E: Ken Pasolli. 24h on 610/1230/1340/1400kHz (N-1).

9A) Box 1630, Merritt, BC V0K 2B0 — Mgr: R. Chapman. Own prgr's Mon-Fri 1400-2200, Sat 1600-2200.

9C) Clearwater, BC. (no studio)

9D) Ashcroft/Cache Creek, BC. (no studio)

10) 1560 2nd Ave, Trail, BC V1R 1M4 — G.M: D. Gerein; P.D: J. Richardson; C.E: D. Fowler. 24h on 610/1340/1390kHz(N-3). Ann: "KBS".

10A) 1560-2nd Ave., Trail, BC V1R 1M4 — Own prgr's: Mon-Fri 1830-0200.

10B) Box 310, Creston, BC V0B 1G0

11) 1940-3rd Ave, Prince George, BC V2M 1G7 — Mgr: Ron A. East; P.D: W. Russell; C.E: E. Burynuik. 24h on 620/1340kHz (N. at :55). C&W

11A) Box 1370, Vanderhoof, BC V0J 3A0 — G.M: Terry Shepard; P.D: Gary Russell; C.E: Dave Alendal. Own prgr's Mon-Fri 1400-1700, 2300-0200.

12) 3805 Lakeshore Rd., Kelowna, BC V1Y 7V2 — Mgr: Dean Cooper; P.D: Bill Barnes; 24h(N-1).

13) 1006 Richards Str, Vancouver, BC V6B 1S8 — G.M: A. Diehl; P.D: J.J. Johnston; C.E: G. Jardine. 24h(N-1).

14) 33 Carmi Ave, Penticton, BC V2A 3G4 — Mgr: R. Clark; P.D: S. Cook; C.E: H. Nutma. 24h on 800/1240/1450/1490kHz(N-1). Ann: "OR Network."

14A) Box 539, Osoyoos, BC V0H 1V0 — G.M: Ron Clark; P.D: Ron Clark; C.E: L. Gagnon. Own prgr's W 1400-0200 (Sat 2200) (CJOR-1 = Oliver repeater).

14C) 9901 Main Str, Summerland, BC V0H 1Z0 — GM: Bob Duck; P.D: Shannon Cooke; CE: Harry Nutma. Own prgr's 1400-0200 (Sat 2200). 24h(N-3).
14D) Box 1400, Princeton, V0X 1W0 — Mgr: L.M. Currie; Asst. Mgr: W.G. Elliott. Own prgr's Mon-Fri 1400-1800, 2300-0200, Sat. 1400-1800 (N-3).
15) 201-20627 Fraser Hwy, Langley, BC V3A 4G4 — G.M: L. Harvey Gold; P.D: David Marsden; C.E: E. Jurak. 24h(N-1).
16) Box 900, Victoria, BC V8W 2S2 — G.M: Kim Hesketh; P.D: Ken Geiger; C.E: C. Smedley. 24h(N-1).
17) 460 Pemberton Terrace, Kamloops, BC V2C 1T5 — G.M: R. Arnish; P.D: Doug Collins; C.E: V. Gollmar. 24h(N-1).
18) 3313-32nd Ave, Vernon, BC V1T 2E1 — G.M: P. Nicol; P.D: D. Weglo; C.E: D. Schindel. 24h(N-1).
19) 815 McBride Plaza, New Westminster, BC V3L 2C1 — G.M: Rod Gunn; P.D: D. Rutherford. D. Glastetter. 24h(N-1).
20) 825 Broughton Str, Victoria, BC V8W 1E5 — G.M: Mel Cooper; P.D: T. Spence; C.E: Bud Goes. 24h(N-1).
21) 2440 Ash St, Vancouver, BC V5Z 4J6 — GM: Chuck McCoy; P.D: Brian DePoe; C.E: Jack Wiebe. 24h(N-1).
22) 2419 Hwy. 97N, Kelowna, BC V1X 4J2 — G.M: W.E. Gray; C.E: G.J. Young. 24h(N-1).
23) Box 1200, Victoria, V8W 2S5 — G.M: John Sitter; P.D: J. Ferguson; C.E: F. Cole. 24h(N-1).
24) Box 335, Smithers, BC V0J 2N0. P.D: Tom Stokes. 24h on 760/870kHz.
24A) Box 600, Burns Lake, BC V0J 1E0 — own prgr's 1700-2000.
25) 525-11th Ave., Castlegar, BC V1N 1J6 — G.M: Wilfred R. Warner; P.D: Murray Collins; C.E: Gerald Hennig. 24h(N-1).
26) 2970 3rd Ave So, Port Aberni, BC V9Y 7N4 — G.M: Bill Gibson. P.D: Ike Patterson; C.E: Leo Wouters. 24h (N-1).
27) Box 386, Chilliwack, BC V2P 6J7 — Mgr: W.J. Coombes; P.D: B. Laver; C.E: A. Schmidt. 24h on 850/1240/1270kHz (N-1). Ann: "Fraser Valley R."
27A) 2722 Allwood Str, Abbotsford, BC V2T 3R8 — Mgr & P.D: Brian Antonson. Own prgr's 1300-0800.
27B) Box 1600, Hope, BC V0X 1L0 — Mgr: T. Bogge. Own prgr's W1400-2100 (Sat 2000).
28) 6816 Courtenay Str, Powell River, BC V8A 1X1 — G.M: Bob McInnes; C.E: Gary Knodel; P.D: Derek Bouchard. 24h (N-1).
29) 1134 Burrard Str, Vancouver 1, BC V6Z 1Y8 — G.M: Paul Ski; C.E: Dick Dipalma. 24h(N-1).
30) 901-102nd Ave, Dawson Creek, BC V1G 2B6 — G.M: Mike L. Michaud; P.D: Verna Michaud; C.E: Morley Fountain. 24h (N-1).
31) Box 1570, Grand Forks, BC V0H 1H0.
32) 1900 West 4th Ave, Vancouver, BC V6J 1M6 — G.M: P. Ski; P.D: Neil Callagher; C.E: Dave Youell. 24h(N-3).
33) 1595 Cliffe Ave, Courtenay, BC V9N 2K6 — Mgr: Norma Browne; P.D: Ken Armstrong; C.E: Gary Knodell. 24h on 1240/1440/1490kHz(N-1).
33A) 909 Ironwood Str, Campbell River, BC V9W 3E5 — G.M: B. Langston; C.E: Gary Knodel.
33B) Box 1240, Port Hardy, BC V0N 2P0 — G.M: Philip Hicks (also P.D:); CE: Gary Knodel.
34) 814 Richards Str, Vancouver, BC V6B 3A7 — Mgr: Jan van Bruchem; Op's Mgr's: Theo Donnelly, Pat Karl; C.E: Stan Davis. 24h, in E and 21 other languages. Chinese: W0400-1700, English 2000-0200. (N-1) and within foreign language prgrs.
35) 435 Trunk Rd, Duncan, BC V9L 2P5 — Pres: Dick Drew; P.D: Cam Drew; C.E: G. Becker. 24h(N-1).
36) 22 Esplanade, Nanaimo, BC V9R 4Y7 — G.M: Eugene Daniel; P.D. & C.E: Henry Brilz. 24h(N-1).
36A) Box 1370, Parksville, BC V0R 2S0 — G.M: G. Daniel; P.D: Erick Thompson; Own prgr's 1500-0200.
37) 2800 31st, Vernon, BC V1T 5H4 — G.M: Michael Tindall; P.D: J.S. Hewson; C.E: H. Nutma. 24h(N-1).
38) 20-11151 Horseshoe Way, Vancouver, BC V7A 4S5 — G.M: Bill Waddington; P.D: Marty Forbes; C.E: B. Johnston. 24h(N-1).
39) 4550 Wellington Rd., Nanaimo, BC V9T 2H3 — G.M: R. Adshead; P.D: Mike O'Brien; C.E: Bob Calder. 24h(N-3).
41) Williams Lake, BC (under construction)

Yukon Territory
90) 203-4103 4th Ave, Whitehorse, YT Y1A 1H6 — G.M: Glenn Darling; C.E: Bruce Whittington. 24h(N-1).

North West Territories
95) Box 218, Yellowknife, NWT X1A 2N2 — G.M: Chas.Dent. C.E: Ken Pook. 24h(N-1). P.D: Kelly McCrae.

Alberta
100) 10526 Jasper Ave, Edmonton, AB T5J 1Z7 — Mgr: Don E. Thomas; P.D: J. Rollans; C.E: N. Lutes. Mon-Fri 1300-0700, Sat 1400-0800, Sun 1400-0700. (N-1).
101) Bag 300, Peace River, AB T8S 1T5 — G.M: Mary Cambridge; P.D: B. McDonald. 24h(N-1).
101A) High Level, AB
102) 5204-84th Str, Edmonton, AB T6E 5N8 — Mgr: B. Bezpalko; P.D: G. Tougas; C.E: Lome. 24h(N-1).
103) 4752-99th Str, Edmonton, AB T6E 5H5 — G.M: W.H. Holte; P.D: Larry Donohue; C.E: L. Olson. 24h(N-1).
104) 1900, 125-9th Ave S.E., Calgary, AB T2G 0P6 — G.M: Al Anaka; P.D: R. Lewis; C.E: M.B. Hoyme. 24h(N-1).
105) Box 5700, Red Deer, AB T4N 6V5 — Mgr: P. Mason; P.D: Marv Gunderson; C.E: G. Cherepuschak. 24h(N-1).
106) Box 1480, Drumheller, AB T0J 0Y0 — Mgr: J. Blundell; P.D: M. Woods; C.E: G. Umari. 24h on 910/1340/1400kHz(N-1).
106A) Box 180, Brooks, AB T1R 1B3 — G.M: M. Solberg; P.D: John Petrie; C.E: Gary Umari. Own prgr's Mon-Fri 1300-2000, Sat 1400-1900.
106B) 4703-58 Str, Stettler, AB T0C 2L1 — P.D: Russell Thomas; G.M: Jim Blundell; C.E: Gary Umari. Own prgr's: Mon-Fri 1300-0100, Sat 1400-1900).
107) 10250-108 Str, Edmonton, AB T5J 2X3 — G.M: Neil Edwards; P.D: P. Weissbach; C.E: Dave Coulter. 24h(N-1).
108) 3320-17th Ave SW, Calgary, AB T3E 6X6 — G.M: W. Machny; P.D: Bob Spitzer; C.E: Bob MacDonald. 24h on 960/1340/1450 kHz. (N-3).
108A) Box 1450, Canmore, AB T0L 0M0 — Mgr. & P.D: Rob Tyson. Own prgr's: Mon-Fri 1400-0200, Sat 1400-2400, Sun 1830-2330 (1340 Banff relay).
109) Box 6600, Edson, AB T7E 1T9 — G.M: Mel Lazarenko; P.D: D. Schuck; C.E: L. Jesperson. 24h on 970/1450 kHz(N-1) (CKYR=Jasper).
109A) Hinton, AB.
110) 200 Windsor Court, 9835-101 Ave, Grand Prairie, AB T8V 5V4 — G.M. & P.D: Tom Bedore; C.E: Sam Lowe. 24h (N-1).
111) Broadc. House, Box 7060, St. "E",Calgary, AB T3C 3L9 — G.M: V. Di Maggio; P.D: D. Sinclair; C.E: J. Bruins. 24h (N. at :25, :55). Ann: "AM 106".
112) 5026-50th Str, Lloydminster, AB T9V 1P3 — Mgr: Ken Ruptash; P.D: Pat Thomas; C.E: A. Stam. 24h(N-1).
113) Box 1090, Lethbridge, AB T1J 4A3 — G.M: Carmen Toth; P.D: Mike LeBlanc; C.E: Glenn Prins. 24h(N-1).
113A) Box 2199, Taber, AB T0K 2G0 — G.M: Carmen Toth; P.D: Michael LeBlanc; C.E: Glenn Prins. 24h(N-1).
114) 2200-10130-103rd Str, Edmonton, AB T5J 3N9 — G.M: Bruce Davis; P.D: Ed Mason; C.E: Bob Hunter. 24h(N-1).
115) 804-16th Ave. S.W, Calgary, AB T2R 0S9 — G.M: Roy Hennessy; P.D: G. Siemens; C.E: Jerry Pendree. 24h(N-1).
116) Bag 5339, Red Deer, AB T4N 6W1 — G.M: Ron Thompson; P.D: Bob Mills; C.E: Cliff Wheeler. 24h(N-1).
117) Box 820, Lethbridge, AB T1J 3Z9 — Mgr: Brent Seely; P.D: Tom Gillespie; C.E: Kris Rodts. 24h on 1220/1490kHz.
117A) Box 840, Crowsnest Pass, AB T0K 0E0 — Mgr: Daryl Ferguson. Own prgr's: W 1300-1600 (Sat 2000).
118) 9912 Franklin Ave, Ft. McMurray, AB T9H 2K5 — G.M: B. Blackburn; P.D: K. Boyd; C.E: L. Howell. 24h(N-1).
119) P.O. Box 5030, Station "E", Edmonton, AB T5S 1A8 — Mgr: B. Bolli; P.D: W. Gidyk; C.E: S. Knaga. 24h(N-1).
120) Box 1270, Medicine Hat, AB T1A 7H5 — Mgr: B. Ellis; P.D: Jay Hitchen; C.E: J. Simmons. 24h(N-1). (CJMH relays CBR-1010 only).
121) 5220-51st Ave, Wetaskiwin, AB T9A 3E2 — G.M. & P.D: Ray Telford; C.E: Lorne Jesperson. 24h(N-1).
122) #11-5th Ave S.E., High River, AB T1V 1G2 — G.M: Menno H. Friesen; P.D: Peter Gorman; C.E: Jack Hoepener. 24h(N-1).
123) #602, 22 Sir Winston Churchill Ave, St. Albert, AB T8N 1B4 — G.M: Doug Shillington; P.D: Kirk Elliott; C.E: Ray Lawrence. 24h(N-1).
124) #202, 9817-101 Ave, Grande Prairie, AB T8V 0X6 — G.M: Arlene Miller; P.D.: Ken Norman; C.E: Peter Sievert. 24h(N-1).
125) 6005-103rd Str, #200, Edmonton, AB T6H 2H3 — Mgr: Diana Parker; P.D: Roger Charest Jr.; C.E: Ashebir Alemu. 1300-0700 (Fri, Sat 24h). Ethnic prgr's. (N-1).
126) 457-3rd Str. S.E., Medicine Hat, AB T1A 0G8 — G.M: G. Guy; P.D: John Carter; C.E: R. Finlay. 24h(N-1).
127) Box 910, St. Paul, AB T0A 3A0 — Mgr: B. Bezpalko; 24h on 1230/1310/1340kHz. (1340kHz=Grand Centre). Partly in netw. with st. 129).

127A) Box 1476, Wainwright, AB TOB 4P0 — Mgr: B. Hepp; Own prgrs: Mon-Fri 1300-0100, Sat 1300-1900.
128) 2723-37 Ave NE, Station M, Calgary, AB T1Y 5R8 — Op's Mgr: Vince Cowndea; P.D: Gayle Zarbatany; C.E: R. Odegaard. (N-1).
129) Box 1800, Westlock, AB T0G 2L0 — G.M: Ray Telford; P.D: Paul Larson; C.E: Owen Martin. 24h(N-1).
129A) Box 2470, Slave Lake, AB T0G 2A0 — S.M: Mark Edel. Local prgrs: Mon-Fri 1400-2000 2200-0200, Sat 1400-1800.
129B) Box 1800, Athabasca, AB T0G 0B0 — Own prgr's: 1400-0600. Country.
130) Brooks (construction permit, possibly – st 106A).

Saskatchewan

200) 1414-8th St. E., Saskatoon, SK S7H 0T1 — G.M: G.G. Gallagher; P.D: D. Carr. 24h(N-1).
201) Box 6200, Regina, SK S4P 3H7 — G.M: G. Robinson; P.D: Saul Jacobson; C.E: Art Williams. 24h(N-1).
202) 116 Main Str. No, Moose Jaw, SK S6H 3J7 — G.M: Vern Traill; P.D: Gavin Tucker; C.E: M. Soar. 24h(N-1).
203) 1316 Central Ave, Prince Albert, SK S6V 7R4 — Mgr: J. Scarrow; P.D: N. Headrick; C.E: D. Zimmerman. 24h(N-1).
204) 120 Smith Str. E, Yorkton, SK S3N 3V3 — G.M: G.G. Gallagher; P.D: D. Harrison; C.E: Kelly Osmak. 24h(N-1).
205) 2060 Halifax St., Regina, SK S4P 1T7 — G.M: John Huschi; P.D: W. Cole; C.E: D. Ellis. 24h(N-1).
206) Box 1460, North Battleford, SK S9A 2Z5 — Mgr: Harry M. Dekker; P.D: Doug Harrison; C.E: Dave Senft. 24h on 1050/1240kHz(N-1).
206A) Box 1660, Meadow Lake, SK S0M 1V0 — P.D: H. Dekker; C.E: D. Senft. Own prgr's: W1500-1800, 1930-2400.
207) Box 340, Weyburn, SK S4H 2K2 — G.M: J.T. Laing; P.D: W. Todd. 24h on 1190/1280kHz(N-1).
207A) Box 1280, Estevan, SK S4A 2H8 — Mgr: J.R. Empey; P.D: B. Karman; C.E: Jim Hutchings.
208) Box 65000, Saskatoon, SK S7H 0W3 — Mgr: Pam Carley. P.D: Greg Harrison; C.E: Ken Fisher. 24h(N-1).
209) 210-2401 Saskatchewan Dr, Regina, SK S4P 4H8 — G.M: Exec. VP: Douglas E. Rawlinson; P.D: Garth Kalin; C.E: Gordon Stankey. 24h(N1).
210) Box 490, Rosetown, SK S0L 2V0 — Mgr: D. Dyck; P.D: George Leith; C.E: Al Pippin. 1200-0700 (Sun 0600)(N-1).
210A) Kindersley, SK (all correspondence to Rosetown).
211) 345-4th Ave. So, Saskatoon, SK S7K 5S5 — Mgr: Vic Dubois; C.E: A. Pippin. 24h(N-3).
212) 134 Central Ave. N., Swift Current, SK S9H 0L1 — Mgr: W. Neufeld; P.D: R. Cloutier; C.E: D. Funk. 24h on 570/1490kHz(N-1).
212A) Box 1176, Shaunavon, SK S0N 2M0 — Mgr: Joe Gregoire. Own prgr's: Mon-Fri 1330-1800.
213) Box 1420, Melfort, SK S0E 1A0 — Mgr: G.W. Fitz; P.D: B. Wood; C.E: B. Opseth. 24h(N-1).

Manitoba

301) Polo Park, 1440 Rapelje Ave, Winnipeg, MB R3G 0L7 — G.M: Don E. Kay; C.E: George Buzunis. 24h(N-1).
302) Box 430, Flin Flon, MB R8A 1N3 — G.M: Doug O'Brien; P.D: Joe McCormick; C.E: D. Parker. W 1220-0600, Sun 1350-0715 on 590/1240kHz.
302A) Box 2980, The Pas, MB R9A 1R7; Mgr: Jim Hamm.
303) 201 Hayes Rd, Thompson, MB R8N 1M5 — P.D: Ron Krane. 24h.
304) Box 9700, Winnipeg, MB R3C 3E5 — G.M: W. Gorrie; P.D: J. Norris; C.E: R. Patterson. 24h(N-1).
305) 930 Portage Ave, Winnipeg, MB R3G 0P8—G.M: Ralph Warrington; P.D: T. Farr; C.E: D. Sandeman. 24h(N-1).
306) 27-3rd Ave. N.E., Dauphin, MB R7N 0Y5 — G.M: Linus Westberg; P.D: A. Truman; C.E: P. Nykolaishen. 24h(N-3).
307) 520 Corydon Ave, Winnipeg, MB R3L 0P1 — Mgr: T. Carta; C.E: Don Trueman. 24h(N-1).
308) 1500 Sask. Ave West, Portage la Prairie, MB R1N 0N6 — G.M. & C.E: R.D. Hughes; P.D: Bev Edmondson. 24h(N-1).
309) R. Southern Manitoba, Box 950, Altona, MB R0G 0B0 — G.M: D. Wiebe; P.D: Al Friesen; C.E: Ed Siemens. 24h on 950/1220/1250kHz(N-1).
309A) Box 1220, Boissevain, MB R0K 0E0.
309B) Box 1250, Steinbach, MB R0A 2A0. C.E: Randall McLaren.
309C) Box 1570, Winkler, MB R0G 2X0
310) Box 1150, Brandon, MB R7A 6A5 — G.M: Boyd Craig; P.D: Kevin Crexton; C.E: L. Dubois. 24h(N-1).
311) 1445 Pembina Hwy, Winnipeg, MB R3T 5C2 — G.M: Bud Harden; P.D: Al Trickey; C.E: Lorne Anderson. 24h(N-2).
312) Box 880, Brandon, MB R7A 6N6 — G.M: R.W. MacDonald; P.D: S. Antaya; C.E: Steve Ashton. 24h(N-1).

Ontario

400) 295 Victoria Str., Sudbury, ON P3C 1K5 — C.E: D. Elvidge; CHNO: GM: Wayne Berg; P.D: R. Mise. CHYC: G.M: Darcy Kieran — CHYC 1100-0600 on 900kHz, CHNO 24h on 550kHz.
401) Box 430, Kirkland Lake, ON P2N 3J4 — G.M: R. Connelly. P.D: B. Ferguson. C.E: Bob White. 24h on 560/1230kHz.
401A) Box 1058, New Liskeard, ON P0J 1P0 — G.M: Mike Perras. C.E: B. White.
402) P.O. Box 280, Owen Sound, ON N4K 5P5 — Mgr: R. Kentner; P.D: Kevin Kelly; C.E: R. Coyne. 24h on 560/1490kHz (CFPS=Port Elgin repeater). (N-3).
403) 1900 Walkley Rd., Ottawa, ON K1H 8P4 — Mgr: D. Holtby; P.D: D. Barker; C.E: G. Roach. 24h(N-1).
404) 87 N. Hill Str, Thunder Bay, ON P7A 5V6 — G.M: Jerry G. Cox; C.E: M. Volbracht. 24h(N-3). Finnish Sun. 1245.
405) 24 Byng Ave, Kapuskasing, ON P5N 1X5 — Mgr: D. Peplow; P.D: Dave Palmer; C.E: K. Lovquist. 1100-0500(N-3).
406) 300 Cabana Rd. East, Windsor, ON N9G 1A3 — G.M: Al Pervin; P.D: K. Chinnery; C.E: Jim Valvasori. 24h(N-3).
407) 1 Yonge Str, Toronto, ON M5E 1G1 — Mgr: S. Rothschild; P.D:T. Williams; C.E: W.R. Onn. 24h(N-1).
408) Box 3000, North Bay, ON P1B 8K8 — G.M: Gary Duguay. P.D: Jim Hamm; C.E: Wayne Russell. 24h(N-1).
409) P.O. Box 610, St. Catharines, ON L2R 6X7 — G.M: Gene Stevens; C.E: Joe Gurney. 24h(N-3).
410) Box 1340, Timmins, ON P4N 7J8 — G.M: J. Clifford; P.D: L. Roy; C.E: D. Elvidge. 24h on 620/1230/1340kHz. (CHOH-=Hearst).
410A) Kapuskasing, ON. c/o CROY, Box 1340, Timmins, ON P4N 7J8.
411) Box 630, Chatham, ON N7M 5K9 — Mgr: D. Chamberlain; P.D: D. Rollins; C.E: R. Cole. 24h(N-1).
412) Box 630, Smiths Falls, ON K7A 4T4 — G.M: G. Perrin; P.D: B. Perkin; C.E: T. Young. 24h(N-1).
414) Box 1046, Timmins, ON P4N 7H8 — G.M: Chris Ruscica; P.D: Art Pultz; C.E: Scott Sloan. 24h(N-1).
415) 25 Adelaide Str. East, Toronto, ON M5C 1H3 — G.M: S. Sanderson; P.D: Don Stevens; C.E: Kirk Nesbitt. 24h(N-1).
416) Box 710, Niagara Falls, ON L2E 6X7 — Mgr: K. Dancy; C.E: B. McDougall. 24h(N-1).
417) 100 Talbot Str. Ea, Leamington, ON N8H 1L3 — G.M: Lou F. Tomasi; C.E: P. Cribbie.24h. Sunrise to sunset on 710kHz, night on 730kHz.
418) 15 Charles Walk, Elliot Lake, ON P5A 2A2 — G.M: Walter D. Hulme; C.E: Don Elvidge. 24h on 730/930/1340kHz(N-1). Nights in netw. with CHNO-550 "Mid Canada R."
418A) 15 Charles Walk, Elliot Lake, ON P5A 2A2; 24h rel.CKNR.
418B) Box 1910, Espanola, ON P0P 1C0 — G.M: C. Babcock; P.D: J. Reid; C.E: Don Elvidge; 24h(N-1). Own prgr's: M-F 1100-1500.
419) Bag 7900, Stn A, Sudbury, ON P3A 5W7 — Mgr: R. Templeton; C.E: Jim Profit. 24h.
420) 50 Kennedy Rd. South, Brampton, ON L6W 3R7 — Mgr: Angelo Cremisio; P.D: Patrizia di Vincenzo; C.E: Paul Kudla. 24h(N-1).
421) Box 480, Windsor, ON N9A 6M6 — G.M: T.L. Coles; P.D: D. Shafer; C.E: W. Dowhan. 24h.

422) Box 489, Fort Frances, ON P9A 3M8 — Pres: D.A. Fawcett. Mgr: S. Fawcett. 24h.
423) Box 488, Belleville, ON K8N 5B2 — Mgr: W.A. Morton; P.D: Lance Jeffrey; C.E: D.K. Vanderwater. 24h. on 800/1240kHz (N-1).
423A) Box 1240, Bancroft, ON K0L 1C0 — Mgr: Mrs. Nan Whitlaw; P.D: M. Cooper; C.E: Doug Vanderwater. Own prgr's Mon-Fri 1400-1500, 1800-2000, 2300-2400; Sat 1410-1530, 1730-1900; Sun 1330-1500.
423B) Box 9, Trenton, ON K8V 5R1 — Mgr: R.S. Rowbotham; C.E: M. Coffey. 24h(N-1)
424) 875 Main Str. West, Hamilton, ON L8S 4R1 — G.M: Don Luzzi. P.D: Danny Kingsbury; C.E: Ted Townsend. 24h(N-3).
425) Box 580, Dryden, ON P8N 2Z3 — Mgr: B. Walchuk. 24h (N-1).
426) 249 Kent Str W,Lindsay, ON K9V 2Z3—G.M: J.A. McNabb; Op's Mgr: C. Parkhill; C.E: Ray McMurray. 24h(N-1).
427) 53 Broadway Ave, Wava, ON. POS 1K0. Relays CJQM-FM, addr: 642 Great Northern Rd., Sault Ste Marie, ON P6B 4Z9.
428) 215 Carling Terrace, Wingham, ON N0G 2W0 — G.M: Al Skelton; P.D: Matt Miller; C.E: G. Belanger. 1100(Sun 1300)-0408(N-1).
429) Box 950, Barrie, ON L4M 4V1 — G.M: W. Bjorgan; P.D: J. Walther; C.E: J. Evans. 24h on 950/1400 kHz(N-1).
429A) Box 1400, Collingwood, ON L9Y 3Z3.
430) 479 Counter Str, Kingston, ON K7M 7J3 — Mgr: Cam Shillington; P.D: Gord Ambrose; C.E: Larry Cameron. 24h(N-1).
431) Box 2580, London, ON N6A 4H3 — G.M: W.J. Brady; P.D: B.T. Rutledge. C.E: P. Paczynski. 24h(N-1).
432) P.O. Box 4150, Peterborough, ON K9J 6Z9 — G.M: Brian Punchard; P.D: Gordon Gibb; C.E: Ben Wilke. 24h(N-1).
433) 2 St. Clair Ave West, Toronto, ON M4V 1L6 — G.M. & P.D: R. Hennessy; C.E: David Simon. (N-1). CFRX special ID at appr. :22 past each h on 6070kHz.
434) 1331 Yonge Str, Toronto, ON M4T 1Y1 — G.M: Jim Waters; P.D: Ross Davies; C.E: B. Carnegie. 24h(N-1).
436) Box 1070, Sarnia, ON N7T 7K5 — G.M: Wayne Steele; P.D: Paul Godfrey. C.E: R.F. Cooke. 24h(N-1).
438) 864 King St. W., Kitchener, ON N2G 4E9 — G.M: John Spragge; P.D: R. Moss; C.E: J. Brenner. 24h(N-1).
439) 22 rue St. Louis, Gatineau, PQ J8T 2R9. — G.M: Jacques Papin; P.D.: Jean Lemay; C.E: A. Bonneau. 24h (Lic. to Ottawa).
440) Box 1150, Hamilton, ON L8N 3P5 — Mgr: Wolf von Raesfeld; P.D: Nevin Grant; C.E: Gus Sondermeyer; "Oldies 1150". 24h(N-1).
441) Sault Ste., Marie, ON (under construction).
442) Box 3020, St. Catharines, ON L2R 7C7 — Mgr: R. Redmond; P.D: Ted Yates; C.E: L. Garrington. 24h(N-1).
443) 128 Main St. S., Kenora, ON P9N 1S9 — Mgr: H. Syrja. 24h(N-1).
444) Box 969, Cornwall, ON K6H 5V1 — 24h. Mgr: K.H. Clingen; P.D: G. Shaver; C.E: O. Mekitarian.
445) 285-A Memorial Ave, Thunder Bay, ON P7B 6H4 — G.M: Remy C. Gagne; P.D: B. Malcolm; C.E: J. Visser. 24h(N-1).
446) 490 Dominion Ave, Midland, ON L4R 1P6 — G.M: Bob Bennett; P.D: S. Warnock; C.E: Gary Hooper. 1030-0500(N-1).
447) Box 904, Stratford, ON N5A 6W3 — G.M: Doug Ackhurst; P.D: Lynn Riddle; C.E: Bill Tofflemire. 24h(N-1).
448) 1415 London Road, Sarnia, ON N7S 1P6 — Mgr: J. Divinski; P.D: S. Glenn; C.E: Brian Hinz. 24h.(N-1)
449) 284 Church St, Oakville, ON L6J 7N2 — Mgr: M.H. Caine; C.E: Peter Dyck. 24h. 40% ethnic prgr's. (N-1).
449A) Box 1320, Port Credit Postal Sta, Mississauga, ON L5G 4M3. Mgr:/P.D: M. Caine; C.E: Peter Dyck.
450) 151 York Blvd., Hamilton, ON L8R 3M2 — Mgr: K. James; P.D: M. LaPointe; C.E: Rob Meuser. 24h(N-1).
451) P.O. Box 1290, Stn. A, London, ON N6A 5A2 — G.M: Warren Cosford; P.D: Nelson Millman; C.E: Jeffrey Guy. 24h(N-1).
452) 5255 Yonge St., Suite 1400, Toronto, ON M2N 6P4 — Mgr: Don Shafer; P.D: Gary Aube; C.E: Rob Enders. 24h.
453) 112 Kent Str., Suite 1900, Ottawa, ON K1P 6J1 — Mgr: Hal Blackadar; P.D: Jerry Stevens; C.E: S.A. Harrison. 24h(N-1).
454) 46 Main Str, Cambridge, ON N1R 1V4 — G.M: Ray Norris; P.D: Ron Fitzpatrick; C.E: Joe Mignacca. 24h(N-1).
456) 360 King Str. W, Oshawa, ON L1J 2K2 — G.M: Dave Lyman; P.D: Lee Sterry; C.E: Paul Kudla. 24h(N-1).
457) 595 Pembroke St. E., Pembroke, ON K8A 3L7 — GM: Al Kennedy. P.D: Scott Jackson. C.E: A. Gutz. 24h(N-1).
458) Box 1380, Kingston, ON K7L 4Y5 — G.M: J. Wright; P.D: C. St. Clair; C.E: A. Aufleger. 24h(N. at :45).
459) 571 West Str, Brantford, ON N3T 5P8 — G.M: R.D. Buchanan; Op's M, P.D: Gene Stevens; C.E: S. Bayley. 24h(N-1).
460) 339 Westney Rd. South, Suite 201, Ajax, ON L1S 2E6 — Mgr: David Hughes; P.D: S. Kassay; C.E: W. Saikali. 24h(N-1).
461) 690 Stewart St., Renfrew, ON K7V 1Y6 — Mgr: A. Kennedy; P.D: Rick Wyman; C.E: Marc Germain. 24h(N-1).
461A) P.O. Box 1490, Arnprior, ON K7S 3R1 — 24h(N-1).
462) Box 1410, London, ON N6A 5J2 — G.M: Braden Doerr; C.E: W.E. Post. 24h(N-1).
463) Box 177, Peterborough, ON K9J 6Y8 — G.M: Stan Genno; P.D: Dan Gall; C.E: Ed Crompton. 24h(N-1).
464) 40 Holly Str, 7th Floor, Toronto M4S 3C3 — G.M: John Rea; P.D: Allan Davis; C.E: Wally Lennox. 24h(N-1).
465) 1575 Carling Ave, Ottawa ON K1Z 7M3 — G.M: Dianne M. Wilson; P.D: Sharon Taylor; C.E: M. Germain. 24h.
466) Box 666, Brockville, ON K6V 5V9 — P.D: B. Wylie; C.E: Roger Cole. 24h (exc. Mon 0600-1100). (N-1).
467) Box 520, Cobourg, ON K9A 4L3 — G.M: Don T. Conway; P.D: Paul Laing; C.E: Gord Skinner. 24h(N-1).
468) 75 Speedvale Ave, Guelph, ON N1E 6M3 — G.M: Pat St. John; P.D: Guus Hazelaar; C.E: Joe Mignacca. 24h(N-1).
469) Regional Rd. 23, Welland, ON L3B 5R6 — Mgr: G.W. Burnett; P.D: John Marshall; C.E: W. Juchniewicz. 24h(N-1).
471) 305 King Str. W, Kitchener, ON N2G 4E4 — G.M: James T. Webb; P.D: Paul Cugliari; C.E: P. Firminger. 24h(N-3).
472) Box 10, Tillsonburg, ON N4G 4H3 — S.M: J.D. Lamers Jr; P.D: Doug Cooper. C.E: R. Lamers. Daytime only (Jan. 1300-2215, July 1000-0100). Hungarian/Flemish/German: Sun 1800-2100.
473) 622 College Str, Toronto, ON M6G 1B6 — Pres: J. Lombardi; P.D: C. Redhead; C.E: D. Lee. 24h(N-1).
474) Box 160, St. Thomas, ON N5P 3T8 — G.M: V. Furber; P.D: W. Allen; C.E: P.A. Roi. 24h(N-1).
475) Box 550, Orillia, ON L3V 6K2 — G.M: Barry Norman; P.D: Jack Latimer; C.E: Doug Young. 24h(N-1).
476) Box 98, Simcoe, ON N3Y 4K8 — G.M: Clarke Drennen; P.D: Ron Yantho; C.E: Larry Garrington. 24h(N-1).
477) 138 Davis Dr, Newmarket, ON L3Y 2N1 — Mgr: Peter Emmerson; P.D: Terry Thomas; C.E: John Evans. 24h(N-3).
478) 1504 Merivale Rd, Ottawa, ON K2E 6Z5 — G.M: Gary Russell; P.D: Gord Taylor; C.E: Jeff Ruck. 24h (N-20).
479) 245 Oak Str, North Bay, ON P1B 8P8 — G.M: Patty Gysell; P.D: Andy Wilson; C.E: Don Elvidge. 24h(N-1).

Quebec

500) 3550 Boul. Royal, Trois-Rivières, PQ G9A 5G8 — Mgr: M. Cloutier; P.D: P. DeMondehare; C.E: D. Goudreault. 24h(N: at :15, :45).
501) 437 Arnaud, Sept-Iles, PQ G4R 3B3 — Mgr: Y. Bergeron; P.D: R. Savard; 24h (N. at :45).
501A) 8 boul. des Iles, Port-Cartier, PQ G5B 2J4 — G.M: Y. Savoie; P.D: S. Cadieux; C.E: R. Lessard. 24h(N-1).
502) 399 rue De Puyjalon, Baie Comeau, PQ G5C 2Z7 — Mgr: C. St. Pierre; P.D: J.D. Dionne; C.E: G. Paré. 24h on 580/620kHz. (N-1)
502A) Forestville, PQ. See 502.
503) C.P. 59, Jonquière, PQ G7X 7V8 — Pres: Denis Langlois; C.E: D. Dufour. 24h(N-1).
504) 1200 McGill College Ave, Suite 300, Montreal, PQ H3B 4G7 — G.M: P. Arcand; P.D: André Chevalier; C.E: K. Bickerdike. 24h(N-1).
505) C.P. 610, New Carlisle, PQ G0C 1Z0 — G.M: R. Poirier; P.D: Mario Loubier; C.E: A. Houde. 24h N: at :15(local 1200-2300), :45(network).
505A) Location: Gaspé. Studios: New Carlisle.
506) 332 de la Madone, Mont-Laurier, PQ J9L 1R9 — G.M./P.D: Alain Desjardins; C.E: Marcel Bellemare. 24h on 610/1490kHz. (N. at :45. Nights rel. st. 510)
506A) 41 Principale, L'Annonciation, PQ J0T 1T0. Own prgr's 1100-0500.
507) 25 Bryant Str, Sherbrooke, PQ J1J 3Z5 — G.M: M. Fortin; P.D: Alain Campagna; C.E: Michel Laroche. 24h(N: at :15).
509) C.P. 3000, Ville-Marie, PQ J0Z 3W0 — G.M: Yvon Larivière; C.E: Joel Bastien. 24h on 710/1340kHz(N: at :45).
509A) C.P. 388, Temiscaming, PQ. F.Pl: move to FM.
510) 1411 Peel Str, Montreal, PQ H3A 3L5 — Mgr: Liza Frulla-Hebart; P.D: A. Chevalier; C.E: C. Bouchard. 24h(N. at :15, :45).
511) 2136 Chemin Ste-Foy, Ste-Foy, PQ G1V 1R8 — G.M: J. Duhamel; P.D: M. Belanger. C.E: J. Ruel. 24h(N-3).
512) 1411 rue du Fort, Montreal, PQ H3H 2R1 — G.M: Rob Braide; P.D: T. Blackman; C.E: M. Kavanagh. 24h(N-1).
513) 211 Gordon Ave, Verdun, PQ H4G 2R2 — G.M: M.G. Scott; P.D: Bob Beauchamp; C.E: R. Gonneville. 24h(N. at :55, :25).

514) 783, rue Labelle, St.-Jérôme, PQ J7Z 5M4 — Mgr: Pierre Labonté; P.D: Steeve Laplante. C.E: ?. (N-1). CJSA = Ste-Agathe, CKSJ = St-Jovite repeaters.
516) 901 Galt Str. East, Sherbrooke, PQ J1G 1Y6 — G.M: M. Fortin; P.D: T. Silver; C.E: M. Laroche. 24h(N-1).
517) 568 Boul. St-Joseph, Roberval, PQ G8H 2K6 — Mgr: M.-A. Levesque; P.D: G. Gagnon; C.E: M. Bolduc. 24h(N: at :15,:45)
518) 72 rue Laval, Hull, PQ J8X 3H3 — G.M: F. Dunn; P.D: J. Tétrault; C.E: G. Toussignant. 24h(N: at :15,:45).
519) 1310 Greene Ave, Montreal, PQ H3Z 2B5 — G.M: Lee Hambleton; P.D: C. St. Clair. 24h.
520) 1300 Boul. Laurier, Sillery, PQ G1S 1L8 — G.M: A. Gagnon; P.D: Damien Rousseau; C.E: M. Duval. 24h(N-1).
521) 1350 rue Royal, Trois-Rivières, PQ G9A 4J4 — G.M: René Michaus; P.D: Richard Lachance; C.E: Maurice Bouchard. 24h.
522) C.P. 695, Shawinigan, PQ G9N 6V9 — G.M: Marineau Real. P.D: Gille Forcier. C.E: Claude Fitzbay. 1030-0500.
523) 1975 Boul. Wallberg, Dolbeau, PQ G8L 1J5 — Mgr: C. Tremblay; P.D: M. Bouchard; C.E: G. Gosselin. 24h(N-1).
524) Reseau des Appalaches, 327 rue Labbe, Thetford Mines, PQ G6G 5S3 — G.M: François Labbé; P.D: Raymond Cusson; C.E: R. Cloutier. 24h on 1230/1330/1340/1380/1400/1420kHz(N: at :45). Network st's have local prgrs Mon-Fri 1200-1430, 1600-1800, 2200-2230; Sat 1300-1830 (CJLP-Disraeli, studio closed).
524A) 185 Du Roi, Asbestos, PQ J1T 1S4. Pres: Marie-Paule Drouin. C.E: Serge Vachon. 24h(N-1).
524B) C.P. 490, Victoriaville, PQ G6P 6T3 — Mgr: R. Daneau; C.E: S. Vachon; P.D: Raymond Cusson. 24h N:00 + :15.
524C) 5088 rue Frontenac, Lac Mégantic, PQ G6B 1H3 — Mgr: F. Lavallée.
524D) C.P. 160, Plessisville, PQ G6L 2Y7 — S.M: Johanne Talbot.
526) C.P. 850, La Tuque, PQ G9X 3P6 — G.M: L. Leclerc; C.E: Y. Simard. 1130(Sun 1330)-0430.
527) 455-3e rue, Chibougamau, PQ G8P 1N6 — G.M. M. Loubier; P.D: G. Fournier; C.E: Y. Gauthier. Prgr. on 1240/1340kHz. (CFED Chapais).
528) 5 est, Trans-Canada, Levis, PQ G6V 6P5 — G.M: P. Picard; C.E: G. Paré. 24h(N-2).
529) 855 Ave Ste. Marie, St. Hyacinthe, PQ J2S 4R9.
530) 1441 boul Auger Sud, Alma, PQ G8B 5V2 — G.M: G. Pedneault; P.D: M. Larouche; C.E: C. Gagnon. 24h (N-3).
531) 1717 est boul. René-Lévesque, Montreal, PQ H2L 4E8 — G.M: Raynald Brière. P.D: Charles Benoit. C.E: Claude Barnard. 24h(N-1).
532) Quebec, PQ G1R — G.M: Phil Germain; P.D: Roger Bourassa; C.E: Serge Verret. 24h (N. at 1545).
533) 800 Ouest du Phare, Matane, PQ G4W 1V7 — G.M: K. Gagné; P.D: C. Lepage; C.E: René Comeau. 24h(N. at :45).
534) 1000-6e Ave, La Pocatière, PQ G0R 1Z0—G.M: Jean-Marc Belzile; P.D: M. Lévesque; C.E: G. Gélinas. 24h (N-1).
534A) St. Pamphile, PQ (Studio closed). Repeats CHOX-FM.
535) C.P. 670, Ville Dégelis, PQ G0L 1H0 — G.M: G. Caron; C.E: J. Martin. 24h (N-2).
537) C.P. 122, Joliette, PQ J6E 2L5 — G.M., C.E: P. Cardin. 24h(N-3).
538) C.P. 1360, Ste-Marie-de-Beauce, PQ G6E 3C4 — G.M: R. Rhéaume; P.D: J. Lessard; C.E: G. Guay. 24h(N-1). Nights rel. st. 528).
539) 249, rue Victoria, Valleyfield, PQ J6T 1A9 — G.M: Jean-Pierre Major; C.E: Pierre Laberge. 24h (N:30, :45).
540) 380 Ave. Murdoch, Noranda, PQ J9X 1G5 — G.M: Daniel Hamel. P.D: Réal Robert. C.E: Jean-Guy Langevin. 24h on 900/1240/1340/1400kHz. Netw. st's broadcast own prgr's w exc. 2300-1045 (Sat 1100) (N. at :15. :45).
540A) 1729 3e Ave, Val d'Or, PQ J9P 1W3.
540B) 122-5e Ave E, La Sarre, PQ J9Z 2Y1 — Mgr: Sylvie Bourassa.
540C) 751-1 ère Ave Ouest, Amos, PQ J9T 1V7 — P.D: Réal Robert; C.E: Michel Boileau; 24h.
541) 1 rue Frontenac, Rivière-du-Loup, PQ G5R 1R7 — G.M: Guy Simard; P.D: D. St-Pierre; C.E: G. Gélinas. 1045-0500 on 1240/1400/1450kHz.
541A) Cabano, PQ.
541B) Pohénégamook, PQ.
543) R. Montréal, 35 York Str, Westmount, PQ H3Z 2Z5 — G.M./P.D: A. Mielewczyk; C.E: Luigi Valente. 24h in E, F, Italian and 20 other languages.
544) 1200 boul. du Royaume, Chicoutimi, PQ G7H 1T1 — G.M./P.D: P. Tremblay; C.E: S. Villeneuve. W 1030-0800, Sun 1100-0501.
545) 76 rue Dufferin, Granby, PQ J2G 9L4 — G.M: Jacques Bouchard; P.D: Paul Labrecque; C.E: Pierre Racine. 24h (N-3).
546) C.P. 100, St. George-de-Beauce, PQ G5Y 5C4 — G.M: Jacques Petit; P.D: André Leblanc; C.E: G. Paré. 24h (N-1).
547) Pointe Claire, PQ (silent).
549) 2070 St. Georges, Drummondville, PQ J2C 5G6 — G.M: Daniel Lemay; P.D: Michel Duchaine; C.E: M. Courmoyer. 1130-0500 (N. at :15, :45).
550) C.P. 820, St-Anne-des-Monts, PQ G0E 2G0 — Mgr: Jacques Vallée; P.D: Stephane Cyr; 24h(N-1).
551) Montmagny, PQ — (silent).
552) 3395 Ouest rue King, Sherbrooke, PQ J1L 1P8 — G.M: Marlene Simard; P.D: Louis Yves Dubois; C.E: Jean-Pierre Maheux.
553) 3195 Blvd St-Martin ouest, Laval, PQ H7T 1A3 — 24h.
554) 104 rue Richelieu, St-Jean, PQ J3B 6X3 — G.M: Normand Flageole; C.E: Pierre Dauphin. 24h (N-1).
555) 875 boul St. Germain, Rimouski, PQ G5L 3T9 — G.M: G. Rousseau; P.D: E. Bernard; C.E: J. Fournier. 24h(N: at :15, :45).
556) C.P. 1840, Amqui, PQ G0J 1B0 — G.M: Adalbert Lévesque; P.D: Frédérik Bureau; C.E: Jacques Beaudoin. 24h (N. at :15). :30 :00.
558) 163A rue des Oblats, Maniwaki, PQ J9E 1G4 — G.M: Michel Riel; P.D: Jean-Guy Cadieux. C.E: G. Tousignant. 24h (0500-1100 Télémédia prgr). (N. at :45).
559) Dolbeau (Construction permit possibly-st. 523).
560) Perce (under construction).

New Brunswick
600) Box 5555, Fredericton, NB E3B 4Z4 — G.M: John Eddy; P.D: M. Allard; C.E: G. Stone. 24h(N-1).
601) Box 188, Edmundston, NB E3V 3K8 — G.M: Claude Boucher; C.E: M. Soucie. Prgr (French and 5% English): 1000(Sat 1100, Sun 1200)-0330 (Sat 0400, Sun 0300) on 570/1490kHz. (N-1).
601A) Grand Sault, NB (no studio). See 601.
602) Box 338, Newcastle, NB E1V 3M4 — Mgr: Ian Byers; C.E: E. Rogers. 24h(N-1).
603) Box 970, Caraquet, NB E0B 1K0 — Mgr: A. Roussy; C.E: René Lanteigne. 24h(N-1).
604) Box 920, Woodstock, NB E0J 2B0 — G.M: C. Russell; P.D: R. McGuire; C.E: Dick Cleveland. 24h(N-1).
605) Box 930, St. John, NB E2L 4E2 — G.M: Shirley Crawford; P.D: Dave Cochrane; C.E: G. Miller. 24h(N. at :55).
606) Box 340, Campbellton, NB E3N 3G7 — G.M: Dick Alberts; P.D: John Kennedy; C.E: Les Rooke. 24h(N-1).
607) Box 2000, St. John, NB E2L 3T4 — Pres: Larry Nichols; P.D: Bob Henry; C.E: Ken Hauschildt. 24h(N-1). **F.PL:** 50/10 kw.
608) 1000 St. George Blvd, Moncton, NB E1C 2E1 — Mgr: S. Gillis; P.D: D. Pond; C.E: Robert Oke. 24h on 590/1220kHz (N-1).
608A) Box 5900, Sussex, NB E0E 1P0 — Mgr: J.A. MacMullin. Own prgr's Mon-Sat 1000-2200, Sun 1200-1600.
609) Box 1360, Bathurst, NB E2A 4J1 — Mgr: Jim Duncan; P.D: Al Hebert; C.E: K. Chamberlain. (English & 7% French) 24h(N-1).
610) 206 Rockwood Ave, Fredericton, NB E3B 2M2 — Pres: R. Coy; P.D: W. Macdonald; C.E: R. Cleveland. 24h(N-3).
611) Moncton, NB. (Off the air).
612) Bedford-Sackville, NB (under construction)

Prince Edward Island
700) Box 1060, Charlottetown, PEI C1A 7M7 — Mgr: F. Lewis; P.D: M. Brooks; C.E: W. Corney. 24h(N-1).
701) Box 7200, Charlottetown, PEI C1A 8V7 — G.M: Blair Dagget; P.D: Barry Rogers; C.E: G. Mills. 24h(N-1).
702) 763 Water Str. East, Summerside, PEI C1N 4J3 — G.M: P.H. Schurman; P.D: Mike Gallant; C.E: F.L. MacFarlane. 1000-0415(N-1).

Nova Scotia
800) Box 5800, Antigonish, NS B2G 2R9 — Mgr: D. MacLean; P.D: Gus MacKinnon; C.E: Scott MacLeod. 24h(N-1).
801) Box 788, Truro, NS B2N 5E8 — G.M: Roy Publicover; P.D: D. Guy; C.E: H. Thompson. 24h(N-1).
802) Box 1007, Dartmouth, NS B2Y 3Z7 — Mgr: B. Tinkham; P.D: G. Greer; C.E: Kendall Condon. 24h(N-1).
803) Box 670, Amherst, NS B4H 4B8 — G.M: Gary Crowell; P.D: D. March. 24h.
804) Box 1653, Halifax, NS B3J 2Z4 — G.M: Wm. Bodnarchuk; P.D: M. Brookshaw; C.E: W. Labucki. 24h(N-3).

805) #208, 500 King's Rd, Sydney, NS B1S 1B1 — G.M: Eric Morrison; P.D: A. Brocklehurst; C.E: J. McPherson. 24hr(N-1).
806) Box 400, Halifax, NS B3J 2R2 — Mgr: D. Foreman; P.D: J. Keogh; C.E: K. Arsenault. 24h(N-1).
807) 215 Dominion Str, Bridgewater, NS B4V 2G8 — Mgr: R.A. MacLaren; P.D: Gary Richards; C.E: F. Grayney. 24h(N-1, :30).
808) Box 1270, Sydney, NS B1P 6K2 — G.M: Don Brown; P.D: Don Graham. C.E: Roy MacIntosh. 24h(N-1).
809) P.O. Box 519, New Glasgow, NS B2H 5E7 — P.D: Rod Mackey; C.E: P. Lann. 24h(N-1).
810) 328 Main Str. #201, Yarmouth, NS B5A 1E4 — Mgr: G.P. Wyman; P.D: Russ LeBlanc Jr; C.E: G. d'Entremont, 24h(N-1).
811) Box 1410, Port Hawkesbury, NS B0E 2V0 — G.M: Bob MacEachern; C.E: Paul Williams. 24h(N-1).
812) Box 310, Kentville, NS B4N 1H5 — Mgr: Neil MacMullen; P.D: Ken Geddes; C.E: Dave Morrison. 24h. Netw. st's carry local prgrs Mon-Fri 1300-1600(N-1).
812A) Box 550, Middleton, NS B0S 1P0 — Mgr: Christine Morse.
812B) Box 1420, Digby, NS B0V 1A0 — Mgr: Leigh Everett.
812C) Box 278, Windsor, NS B0N 2T0.

Newfoundland
900) Box 570, Corner Brook, NF A2H 6E6 — G.M: James R. O'Rourke; S.M: R. Humber; P.D: Bill Bartlett; C.E: J. Parsons. 24h on 570/790/870/1230/1340kHz (N. at :55). (CFNW Port au Choix repeater).
900B) 30 Oregon Dr., Stephenville, NF A2N 2X9 — Mgr: G. Murphy; P.D: L. Bennett; C.E: J. Goosney. 24h. Own prgr's 0930 (Sat 1030, Sun 1630)-2130. (N:55)
900C) Box 1230, Port aux Basques, NF A0M 1C0 — Mgr: G.N. Critchell; 24h. Own prgr's: Mon-Fri 1230-2130.
900D) Box 4000, Stn "C", Goose Bay, Lab, NF A0P 1C0 — S.M: Paul Saunders. 1000 (Sun 1200)-0500 (N-1). Own prgr's: Mon-Sat 1100-2130 (studio in Corner Brook).
900E) Box 6000, Wabush, Lab, NF A0R 1B0 — S.M: Gary Peckham; C.E: C. Avery. Own prgr's: 1100-2130 (studio in Corner Brook).
901) Box 8-590, St. John's, NF A1B 3P5 — G.M: J. Murphy; C.E: R. McCausland. 24h on 560/590/620/710/730/740/1240kHz (N-2).
901A) Box 560, Marystown, NF A0E 2M0 — S.M: R.J. Murphy; P.D: G. Myles. C.E: H. Steele. Own prgr's: 0930-2130, rel. 901D) 2130-0330.
901B) Box 620, Grand Falls, NF A2A 2K2 — Mgr: J.M. Murdoch; P.D: J. Coady; C.E: H.G. Feltham. Own prgr's: 0930-2100. (CKIM Baie Verte).
901C) Box 650, Gander, NF A1V 1X2 — Mgr: C. Brown; C.E: G. Feltham. Own prgr's: 0930-2130.
901D) Box 560, Carbonear, NF A0A 1T0. G.M: John Harvey; P.D: Tom Ormsby; C.E: H. Steele.
901E) Clarenville, NF — Own prgr's 0930-2130, relays st. 901D) 2130-0330.
902) Box 7430, St. John's, NF A1E 3Y5 (belongs to the Wesley United Church) — Mgr. & P.D: E. Hudson; C.E: Reg McCausland. 1230-0330 (Tues 0500, Thurs 0530, Sat 0430). 30% rlg, non-comm.
903) Box 6180, St. John's, NF A1C 5X8 — Mgr: T. Trainor; C.E: R. Strilec. 24h. CJYQ rel. on 1010kHz. CKIX-FM rel. on 610/670/680/1010kHz.
903A) Grand Bank, NF. Silent.
903C) Box 810, Grand Falls, NF A2A 2M4 — Mgr: Dave Hillier; P.D./C.E: Richard King. Own prgr's 0930-approx. 2000.
903D) 310 Elizabeth Dr., Gander, NF A1V 1J8 — S.M: Alfred Whiffen; P.D: D. Pearcey. Own prgr's 0930-2130. (CKXB = Musgravetown).
903E) Harbour Grace, NF. Silent.
905) Box 2520, Mt. Pearl, NF A1N 4M7 — G.M: Cameron Beierle; C.E: B. Matthews. 1025-0230 (F.Pl. 24h). Rlg (Seventh-Day Adventist Church).
906) Box 1340, Corner Brook, NF A2H 7B2 — G.M: Bob Hicks; P.D: W. Wicks; C.E: Ron Combdon. 24h(N-1).
907) Happy Valley (under construction).

AFRTS
Box 2, Argentia, NF or FPO, New York, NY 09597. 1480kHz 50W: 1000 (Sat/Sun 1030)-0230 (Sun 0130). N. on the h. — Goose Bay, NF. 640kHz. These st's not recently confirmed.

For More Canadian Radio Details, See "ANIK" Satellites

RADIO CANADA INTERNATIONAL
(Canadian Broadcasting Corporation/Société Radio Canada) (Publicly owned)
ADDR: P.O. Box 6000, Montreal, Canada H3C 3A8. **Te:** +1 (514) 597 7555. **Fax:** +1 (514) 284 0891. **Telex:** 0526417.
L.P: Man. Dir: Terry Hargreaves. Prgr. Dir: Allan Familiant. Dir. of Eng: Jacques Bouliane. Head of Pub. Rel: Ousseynou Diop.
STATION: Sackville, New Brunswick (G.C: 64.19W/45.53N): 5 x 250kW, 3 x 100kW tr's. **Relays:** K = Kimje (Korea), M = Moosbrunn (Austria), S = Sines (Portugal), U = Skelton (UK), W = Wertachtal (Germany), X = Xian (China), Y = Yamata (Japan).

kHz	kHz	kHz	kHz	kHz	kHz	kHz
5960	7180	9555	9755	11880	13650	15325
5995	7210	9595	9760	11905	13670	15425
6025	7230	9615	11705	11915	13720	17820
6050	7235	9635	11715	11935	15140	17875
6120	7255	9650	11730	11940	15150	21545
6150	7265	9660	11845	11945	15260	
7150	9505	9670	11850	11955	15270	
7155	9535	9740	11855	11965	15315	

Schedule until March 27th 1993

Arabic	Areas	kHz
0330-0400	ME	6025W, 7210S, 9505M, 9615W
0430-0500	ME	9505M, 11715W, 11965W
Chinese		
1300-1325	CHN	6150K, 11955Y, 15270Y
1530-1555	CHN	6150Y, 7265K, 9660Y
English		
0400-0430	ME	9505M, 9670U
	Eu	6150S
0615-0700MF	Eu	6050U, 6150, 7155U, 9760
	Af	9740U, 11905U
1300-1400MF	Carib	17820
	USA	9635, 11855
1330-1357	EAs	6150K, 9535K
1400-1700Sun	Carib/USA	11955, 17820
1515-1530	CEu/EEu	9555U, 11915S, 11935U, 15315, 15325S, 17820, 21545
1630-1657	IND	7150X, 9555X
1800-1829MF	Af	13670, 15260, 17820
1800-1859SS	Af	13670, 15260, 17820
1900-1930MF	Af	13670, 15260, 17820
2030-2100	CEu/EEu	5995U, 7235U, 11945, 13650, 15140, 15325, 17875
	As	9650Y
2130-2200	Af	11880, 15150, 17820
2200-2230	CEu/EEu	5995D, 7180U, 13650
	SEAs	11705Y
2200-2300	WEu	9760, 11945
2300-2330	Carib	9755, 11730
2300-2330MF	LAm	9535, 11940
2300-2400SS	LAm	9535, 11940
0000-0130MF	USA	5960, 9755
0000-0200SS	USA	5960, 9755
0100-0200SS	LAm	9535, 11845, 11940, 13720
0200-0300MF	LAm	9535, 9755, 11845, 11940, 13720
French		
0300-0330	ME	6025U, 9505M
0600-0615MF	WEu/Af	as English 0615
1300-1400MF	Carib	15425
	USA	9650
1400-1700Sun	Carib/USA	11855
1830-1900MF	Af	13670, 15260, 17820
1900-2000	Eu	5995U, 7235U, 11945, 15325, 17875
1900-2000SS	Af	13670, 15260, 17820
1930-2000MF	Af	13670, 15260, 17820
2100-2130	Af	11880, 15150, 17820
2100-2200	CEu/EEu	5995U, 7230U, 11945, 13650, 15325
2230-2300MF	LAm	11940
	Eu	5995U, 7230U
2230-2300	SEAs	11705Y
	Carib	9755, 11730
0000-0030MF	LAm	9535, 11940
0100-0130MF	LAm	9535, 11845, 11940, 13720
0130-0200MF	USA	5960, 9755
0100-0200SS	LAm	9535, 11845, 11940, 13720
0200-0300SS	LAm	9535, 11845, 11940
	USA	6120, 9755

Spanish

2330-2400MF	LAm	9535, 9755, 11850, 11940
0000-0100SS	LAm	9535, 11940
0030-0100MF	LAm	9535, 11845, 11940, 13720
0130-0200MF	LAm	9535, 11845, 11940, 13720

Russian

1430-1515	RUS	6150Y, 7155K, 7255Y, 9555U, 9595K, 11915S, 11935U, 15325S, 21545
1430-1515W	RUS	15315, 17820
1530-1615	RUS	9555U, 11935U, 21545
1530-1615W	RUS	15325, 17820
1730-1815	RUS	5995U, 7235U, 13650, 15325, 17875

Ukrainian

1615-1700	UKR	5995U, 11935U, 21545
1615-1700W	UKR	13650, 15325, 17820
1815-1900	UKR	5995U, 7235U, 11945, 15325, 17875

ANN: E: "This is Radio Canada International". F: "Ici Radio Canada International".
INT-SIG: first bar of National Anthem.
V: QSL-cards are provided to listeners on the RCI mailing list. They are mailed with the spring/summer programme schedule and may be returned for verification. Other reception reports are not verified at any other time of the year. Records (spoken-word or music), topical discs or tapes can be provided on request to broadcast organizations.
PUB: Radio Canada International programme schedule, published twice a year, is sent free upon request.

CBC NORTHERN QUEBEC SHORTWAVE SERVICE

in English/French/Inuktitut/Cree/Attikamèque: 1155-0610 on 9625kHz.

SACKVILLE RELAY FACILITY

RCI's Sackville facility currently relays broadcasts of the BBC, Deutsche Welle, Radio Japan, Radio Austria International, Radio Beijing and Radio Korea on a reciprocal basis. Details of these broadcasts, and frequencies used, are listed in the schedules of these stations.

GREENLAND

L.T: UTC – 3h (exc. Scoresbysund UTC – 1h). Su: add one hour, exc. Thule area) — **Pop:** 55.000 — **Radio sets:** 25.000 (est.) — **Pr.L:** Greenlandic, Danish — **E.C:** (Nuuk): a/c 50, 220V; US Air Bases: a/c 60, 110/220V — **ITU:** GRL.

KALAALLIT NUNAATA RADIOA (Gov. Comm.)

ADDR: Kalaallit Nunaata Radioa, P.O. Box 1007, DK-3900 Nuuk/Godthåb. **Te:** +299 2 11 72. **Fax:** +299 2 47 03.
L.P: Dir: Jens Lyberth. Dir. Admin: Jørgen Munk. Head N: Naja Paulsen. Head Culture: Juakka Lyberth. Head Eng. Dept: Henrik Jørgensen. Head Tec: Nathan Biilmann.
Regional Addr: Kujataata Radioa (So. Greenland), Box 158, DK-3920 Qaqortoq; Avannaata Radioa (No. Greenland), Box 223, DK-3952 Ilulissat.

Mediumwaves	**kHz**	**kW**		**kHz**	**kW**
Nuuk	570	5	Upernavik	810	5
Qeqertarsuaq	650	5	Uummannaq	900	5
Simiutaq	720	20			

FM: Nuuk stereo, others mono.

	MHz	**W**		**MHz**	**W**
Nuuk	90.5	50	Egedesminde	95.5	50
Thule Air Base	92.0	50	Ilulissat	96.0	50
Holsteinsborg	95.0	50	Sdr. Strømfjord	96.0	10
Uummannaq	95.0	50	Ammassalik	96.0	50

+45 low power st's 50W or less.
D.PRGR: 0930-0300 MF, 1000-0300 Sat, 1100-0210 Sun. **N in Greenlandic:** 1010 MF, 1110 MF, 1515 W, 2130, 0100 MF, 0130 SS. **N in Danish:** 1010 MF, 1110 MF, 1530 W, 2200, 0130 MF, 0145 SS — **Wrp:** 0930 MF, 1000 Sat, 1100 Sun, 1145 Sat, 1150 MF, 1500 Sun, 1545 W, 1615 Sun, 2145 Sun, 2155 MF, 2220 MF, 0200 — **Rlg:** 0940-0950 MF, 1038-1045 Sat, 1130-1135 Sun, 1138-1145 Sat, 1255-1400 Sun — **Faroese:** 2305-2320 Tues — **Reg. Prgr:** 2030-2130 MF — **Prgr's** include commercials.

ANN: "Kallaallit-Nunaata Radioa . . . Grønlands Radio . . ."
INT-SIG: "Sunnia Kalippoq" (The Whaleboat "Sonja" drags whale) played on celeste.
V. by QSL-card. Re to Henrik Jørgensen.

LOCAL STATION (Comm.):
98.5 MHz 10W: Sdr. Strømfjords FM Radio, GC Box C 3, 3910 Kangerlussuaq

AFRTS (US AIR FORCE)

1) Thule — FM: 96.1/102.1MHz.
Addr: Det. 1, AF Arctic Broadc. Squadron (AFABS), APO NY 09023-5000, USA — 24h on 102.1MHz. **N:** on the h. Rel. AFRTS Los Angeles on 96.1MHz.
2) Søndre Strømfjord — **FM:** 96.5/102.1MHz.
Addr: OLA Det. 1, AFABS, APO New York 09121-5000, USA — 24h on 102.1MHz, **N:** on the h. (on 96.5MHz; 24h rel of AFRTS Los Angeles).

ST. PIERRE ET MIQUELON (French)

L.T: UTC – 3h (Su: UTC – 2h) — **Pop:** 6.500 — **Radio sets:** 3.000 (est.) — **Pr.L:** French — **E.C:** a/c 50, 220V — **ITU:** SPM.

RADIO TÉLÉVISION FRANÇAISE D'OUTRE MER (RFO)

ADDR: B.P. 4227, F-97500 St. Pierre et Miquelon. **Te:** + 508 413824. **Telex:** FRTSPM 020428 QN.
L.P: Dir: Alain Quintrie Lamothe. Dir. Tec: Henri Renault. Head of N: Albert Max Briand.
STATIONS: St. Pierre 1375kHz 20kW + FM: 97.9MHz (10W), 99.9MHz (0.5kW) — Miquelon 98.9MHz 50W.
D.PRGR. in French: 0930-0230. **Rel. France-Inter:** 0230-0930.
Local N: 1000, 1530, 2200. **Rel. France-Inter N:** 1100, 1200, 1300, 1400, 1800, 1900 — **Wrp:** 1045, 1128, 1303, 1515, 1715, 2145.
ANN: "Ici RFO, Station de Saint-Pierre et Miquelon".
INT-SIG: La Marseillaise — **V.** by QSL-card. Rec. acc.

UNITED STATES OF AMERICA

L.T: See World Time Table — **Pop:** 254.522.000 — **Radio sets:** 520.000.000 — **Pr.L:** English — **ITU:** USA.

GOVERNMENT AGENCY: THE FEDERAL COMMUNICATIONS COMMISSION

(FCC) — the licensing agency for broadcast stations.
ADDR: 1919 M Street, N.W., Washington, D.C. 20554. **Te:** +1 (202) 632-6460. **Telex:** 7108220160. **Fax:** +1 (202) 653-5402.
L.P: Chmn: Alfred G. Sikes. Man. Dir: Andrew S. Fishel. Chief Mass Media Bureau: Roy J. Stewart.
The FCC is an independent Federal agency composed of five Commissioners appointed by the President with the consent of the Senate. One of its major activities is the general regulation of broadcasting, visual as well as aural. This regulation may be divided into three phases: 1) The allocation of spectrum space to the different types of broadcast services; 2) consideration of applications to build and operate individual stations; 3) regulation of their operations.

Call letters: International agreement provides for the identification of the radio station by the first letter or first two letters of its assigned call signal and for this purpose apportions the alphabet among different nations. USA nations use the initial letters K, N and W exclusively, and part of the A series. For broadcast stations call letters beginning with K are assigned to stations west of the Mississippi River and in the territories and possessions, while W is assigned to broadcast stations east of the Mississippi.
Note: A few exceptions will be noted, namely stations on the East Coast using a "K" callsign. These are old calls retained by special permission.
Stations: More than 11,000 st's. are operating.

UNITED STATES
MEXICO
0
500
1000 KM
CANADA
UNITED STATES
MEXICO
Washington
OLYMPIA
Oregon
SALEM
California
SACRAMENTO
REDWOOD CITY
DIXON
DELANO
RANCHO SIMI
Nevada
CARSON CITY
Idaho
BOISE
Montana
HELENA
Wyoming
CHEYENNE
Utah
SALT LAKE CITY
Colorado
DENVER
Arizona
PHOENIX
New Mexico
SANTA FE
VADO
North Dakota
BISMARCK
South Dakota
PIERRE
Nebraska
TOPEKA
Kansas
Oklahoma
OKLAHOMA CITY
Texas
DALLAS
AUSTIN
Minnesota
ST. PAUL
Iowa
DES MOINES
Missouri
JEFFERSON CITY
Arkansas
LITTLE ROCK
Louisiana
BATON ROUGE
NEW ORLEANS
Wisconsin
MADISON
Michigan
LANSING
IL
SPRINGFIELD
IN
NOBLESVILLE
INDIANAPOLIS
Ohio
COLUMBUS
BETHANY
Kentucky
TN
NASHVILLE
CHATTANOOGA
MS
JACKSON
Alabama
BIRMINGHAM
MONTGOMERY
Georgia
ATLANTA
Florida
TALLAHASSEE
OKEECHOBEE
MIAMI
S. Carolina
COLUMBIA
North Carolina
RALEIGH
GREENVILLE
Virginia
RICHMOND
WV
CHARLESTON
MD
ANNAPOLIS
Del.
DOVER
PA
BETHEL
HARRISBURG
NJ
TRENTON
RED LION
NY
NEW YORK
ALBANY
CT
HARTFORD
RI
PROVIDENCE
MA
BOSTON
NH
CONCORD
VT
MONTPELIER
Maine
AUGUSTA
SCOTTS CORNER
MONTERREY
LINARES
SAN LUIS POTOSI
CIUDAD MANTE
HUAYACOCOTLA
MEXICO CITY
GUADALAJARA
VERACRUZ
MAZATLÁN
LA PAZ
HERMOSILLO
MÉRIDA
BELIZE
BELIZE
TIKAL
GUATEMALA
GUATEMALA CITY
TAPACHULA
HONDURAS
TEGUCIGALPA
LA CEIBA
PUERTO LEMPIRA
PUERTO CABEZAS
CUBA
LA HABANA
BAHAMAS
JAMAICA
HAITI
CAP HAÏTIEN
DOMINICAN REPUBLIC
PACIFIC OCEAN
ATLANTIC OCEAN
GULF OF MEXICO
CARIBBEAN SEA
Tropic of Cancer
140°
130°
120°
110°
100°
80°
70°
60°
40°
30°
20°
UTC-8
UTC-7
UTC-6
UTC-5

Networks providing AM Station Programming

ABC RADIO NETWORKS
ADDR: 125 West End Ave, New York, NY 10023. **Te:** +1 (212) 456-5200. **Fax:** +1 (212) 456-5705 — Exec. VP: Bart Catalane.

CBS RADIO DIVISION
(A division of CBS, Inc.)
ADDR: 51 W. 52nd Str, New York, N.Y. 10019. **Te:** +1 (212) 975-4321.

NBC RADIO
(A division of Westwood One, Inc.)
ADDR: RCA Bldg, 30 Rockefeller Plaza, New York, NY 10020. **Te:** +1 (212) 664-4444.

MUTUAL RADIO NETWORK
(Subsidiary of Westwood One, Inc.)
ADDR: 1775 S. Jefferson Davis Highway, Arlington, VA 22202. **Te:** +1 (703) 685-2090. **Telex:** 4979269. **Fax:** +1 (703) 685-2197.

NATIONAL BLACK NETWORK
(A division of Unity Broadc. Netw, Inc.)
ADDR: 10 Columbus Circle, 10th Floor, New York, N.Y. 10019. 98 affiliate st's coast-to-coast.

SHERIDAN BROADC. NETWORK
ADDR: 411 Seventh Ave, Suite 1500, Pittsburgh, PA 15129. **Te:** +1 (412) 281-6751.

Other National Organizations

NATIONAL PUBLIC RADIO
Non-commercial nationwide system.
ADDR: 2025 M Street NW, Washington, D.C. 20036. **Te:** +1 (202) 822-2000. **Telex:** 440391 npr vi. **Fax:** +1 (202) 822-2329. **NPR** is an interconnected system of member st's serving communities in 48 states, Puerto Rico and the District of Columbia. Educational, cultural and public affairs prgrs are produced, acquired, and distributed by NPR through the two-way system: over one half of prgrs offered by NPR are obtained from member st's, other public radio production sources and foreign sources such as the BBC, CBC and R. Nederland.

C-SPAN AUDIO NETWORKS
ADDR: Suite 650, 400 No. Capital Str. NW, Washington, DC 20001. **Te:** +1 (202) 626-4649.
L.P: Mgr: Beth Talisman.
D.PRGR: Two audio channels available to cable systems nationwide. Ch.1 carries various international broadcasters, Ch.2 is a 24h feed of BBC World Sce. **V:** No QSL's issued.

NATIONAL ASSOCIATION OF BROADCASTERS
ADDR: 1771 N St. NW, Washington, D.C. 20036. **Te:** +1 (202) 429-5300. **Fax:** +1 (202) 429-5343.
Purpose: To foster and promote radio and television broadcasting, ranging from national networks to local service by private commercial stations. NAB conducts the world's largest convention and exhibition of broadcast and production equipment and services in April of each year. NAB holds an HDTV World Convention and Exhibition and a Multimedia World Convention concurrently with its spring convention and in the same city. The 1993-2000 conventions will be in Las Vegas, Nevada. A second annual convention and exhibition each September focuses exclusively on radio. In 1992, NAB Radio Montreux was conducted by the Association in Switzerland for radio broadcasters. Future NAB European Conferences are under study.

COMMERCIAL STATIONS
Especially in multi-station markets, radio stations concentrate their programming to appeal to a given segment of the population or a given listening taste. Many stations devote their entire broadcast day to news and/or talk programs. Others specialize in hit music (adult contemporary, top 40), country music, oldies (e.g. hits of the fifties and sixties), big band/standards, black (urban contemporary, jazz, rhythm & blues), religious services and inspirational music, classical music, ethnic programs (e.g. programs entirely in Spanish).

A new development is the increased use of satellite communications, not only for news/talk programs, but also for musical program networks. Stations making extensive use of network programming pose a special problem to the DXer, since there may be only one local identification an hour, usually on top of the hour. On the technical side several stations have lost their more or less exclusive rights to their respective frequencies ("clear channels"), and such frequencies now are also being allocated to stations requiring a frequency change and/or transmission time extension (i.e. from daytime or limited time to full time) or to new stations in underserved areas. A few stations have been granted temporary licenses for increased powers to combat interference from neighboring countries. Many daytime stations may now operate after local sunset on low or very low powers.

Mediumwaves: Stations less than 10kW not mentioned. For a more comprehensive list of stations please refer to the North America medium wave table of the back of this book.
* = Presunrise Service Authority. A daytime station with a PSA may operate from 0600 local time to local sunrise with low power when local sunrise occurs later than 0600.
Call letters: On stations relaying an FM sister station or carrying certain network programs other call letters than those listed below may frequently be heard. The actual AM call letters are always announced on top of the hour.
Stereo: Roughly 20-25% of all AM stations are, or plan to be, operating in stereo. At present two mutually non-compatible systems are in use.
Obs: Calendar days refer to local time — St. 336) Gov.
Powers: Where a second power is mentioned this refers to early morning/late afternoon operation for daytime st's. For full time stations this appears as the middle one of three power levels.
Transmission hours: For full time stations not operating 24h the number of daily transmission h's is shown in the freq list, and full details are shown in the address section. D = local daytime only.
N.B: Hawaii and Alaska are listed separately.

	Call	kW	H. of tr.
	540kHz		
1)	WGTO	50	24h
2)	KIEZ	50/0.5	24h
3)	KJQI	25/0.36	24h
	640kHz		
4)	KFI	50	24h
5)	KGVW	10/1	17h
6)	WCRV	50/0.48	19h
7)	WLVJ	25/8.2	20h
8)	WDRQ	10/1	F.PL.
9)	WFNC	10/1	24h
10)	WGST	50/1	16h
11)	WNNZ	50/15	24h
12)	WWJZ	50/1	
13	WJTZ	10/0.88	19h
	650kHz		
14)	KRDX	25/1	
15)	KZTU	10/1	F.PL.
16)	WBSO	25/1.6	F.PL.
17)	WKKQ	10/0.5	24h
18)	WSM	50	24h
	660kHz		
19)	KMTI	10/0.9	
20)	WBLX	22/1	
21)	KTNN	50	19h
22)	KGDP	10/1	24h
23)	WFAN	50	24h
24)	WESC	50/5	D*
25)	WTUJ	50/2.5/1	F.PL.
26)	WVAL	10/0.25	24h
27)	KSKY	10/0.5	24h
28)	KAPS	10/1	
	670kHz		
29)	WMAQ	50	24h
30)	WARO	20/5	F.PL.
31)	KBOI	50	24h
32)	WWFE	50/1	
	680kHz		
33)	KKYX	50/10	24h
34)	KNBR	50	24h
35)	WCNN	50/15	24h
36)	WCBM	10/5	24h
37)	WRKO	50	24h
38)	WPTF	50	24h
39)	WEZI	10/5	24h
40)	WCAW	50/0.25	19h
41)	WDBC	10/1	20h
	690 kHz		
42)	WJOX	50/0.5	24h
43)	WPDQ	50/10	24h
44)	KGGF	10/5	19h
45)	KHEY	10	24h
46)	WZAP	10/0.03	24h
47)	WTIX	10/5	24h
	700kHz		
48)	KRMW	50/1	
49)	WLW	50	24h
50)	KFAM	50/1	24h
51)	KGRV	25/0.5	18h
52)	KMJY	10/1	16h
53)	K....	12.5/1	F.PL.
54)	KUET	50/4.2	F.PL.
55)	KMPC	50/10	24h
56)	KFIA	25/5	F.PL.
57)	WAQI	50	24h
58)	KEEL	50/5	24h
59)	WHB	10/5	24h
60)	WOR	50	24h
320)	KBMR	10/4	F.PL.
61)	KGNC	10	24h
62)	WFNR	10	D*
63)	KIRO	50	24h
64)	WDSM	10/5	24h
	720kHz		
65)	WGN	50	24h
66)	KDWN	50	24h
67)	WGCR	10/3	D
68)	WJTO	10/0.5	F.PL.
69)	WRZN	10/0.25	16h
70)	WMXY	10	D
71)	KSAH	10/0.89	24h
72)	WGNZ	10	D F.PL.
	740kHz		
73)	WLWI	50/0.33	
74)	KBRT	10	D
75)	KCBS	50	24h
76)	WWNZ	50	24h
77)	WGSM	25/0.5	24h
78)	WPAQ	10/1	D
79)	KRMG	50/25	24h
80)	KTRH	50	24h
	750kHz		
81)	WSB	50	24h
82)	KMMJ	10	D*
83)	KKNK	10	F.PL.
84)	KXL	50/20	24h
85)	KERR	50/1	24h
86)	KOAL	10/6.8	18h
87)	KAMA	10/1	24h
	760kHz		
88)	WJR	50	24h
89)	KRZN	50/1	24h
90)	KSJL	50/1	24h
91)	KFMB	50	24h
92)	WVNE	10	D
93)	KMTL	10	D
94)	WCHP	25	D
	770kHz		
95)	WVNN	10/0.25	24h
96)	KPLA	50/1	24h
97)	WWCN	10/2.5	24h
98)	KKOB	50	24h
99)	WABC	50	24h
100)	KATL	10/1	24h
101)	KPBC	10/1	24h
102)	KULL	50/5	24h
	780kHz		
103)	WBBM	50	24h
104)	KROW	50	24h
	810kHz		
105)	WJXL	50/0.5	20h
106)	KGO	50	24h
107)	KCMO	50/5	24h
108)	WGY	50	24h
109)	WSJC	50/0.5	16h
110)	KBHB	25	D
111)	WJJQ	10	D*
112)	KTBI	41	D*F.PL.
	820kHz		
113)	WBAP	50	24h
114)	KGNW	50/5	24h
115)	WRFA	50/1	19h
116)	WGGM	10/1	24h
117)	WNYC	50/10	F.PL.
118)	WAPO	10	D
	830kHz		
119)	KFLT	50	D
120)	WCCO	50	24h
121)	WGBV	50/10/5	F.Pl.
122)	KBOA	10	D
123)	KUYO	10	D
124)	WWMO	50/1	
	840kHz		
125)	WHGH	10	D
126)	WHAS	50	24h
127)	KWDF	10	D
128)	KVJY	50/1	
129)	WKTR	10	D
130)	KVEG	50/25	24h
	850kHz		
131)	WYDE	50/1	24h
132)	KOA	50	24h
133)	WHDH	50	24h
134)	WQST	10	D
135)	WUTK	50	D*
136)	WKIX	10/5	24h
137)	WRMR	10/5	24h
138)	WJAC	10	24h
139)	KMTT	10/1	24h
140)	WWJC	10	D
141)	KEYH	10	D*
142)	WNIS	50/25	24h
	860kHz		
143)	KTRB	50/25	24h
144)	KKOW	10/5	24h
145)	KZTW	20/0.5	F.PL.
146)	WOAY	10/0.01	16h
147)	WTEL	10/0.5	D+
148)	KCNR	10/3/1	24h
	870kHz		
149)	WQRX	10/4.7	D
150)	WWL	50	24h
151)	WLAM	10/1	24h
152)	WKAR	10	D
153)	KPRM	25/1	18h
154)	KORD	10/0.25	24h
155)	WZMC	10	D
156)	KOWA	10/1	24h
157)	KIEV	15/1	24h
	880kHz		
158)	WMEQ	10/0.6	
159)	KGHT	50/0.22	
160)	WRFD	50	D
161)	KHAC	10/0.44	16h
162)	KJJR	10/1	24h
163)	WCBS	50	24h
164)	KRVN	50	24h
165)	KJOJ	10/1	19h
166)	KKMC	10/2.5	24h
167)	KIXI	50/10	24h
	890kHz		
168)	KPTO	50/1	
169)	WLS	50	24h
170)	WBMA	10/1	F.PL.
171)	KDXU	10	21h
172)	KVOZ	10/1	24h
173)	WQIS	10	D
174)	WYCV	25	D F.PL
	900kHz		
175)	WSSR	10/1	18h
176)	KTIS	25/0.3	24h
	940kHz		
177)	KFRE	50	24h
178)	WINZ	50	24h
179)	WMAZ	50/10	24h
180)	KBRE	10	D*
181)	KGRL	10	D*+
182)	WCPC	50/0.25	19h
183)	KIOA	10/5	24h
184)	WYLD	10/0.5	24h
185)	WKGM	10/5	
	990kHz		
186)	KTKT	10/1	24h
187)	KWAM	10/0.08	24h
188)	WHOO	50/10/5	24h
189)	WEEB	10/0.5	18h
190)	WVSC	10	D
191)	WZZD	50/10	24h
192)	WALE	50/10	D
193)	WIVK	10	24h
194)	KNIN	10/1	24h
	1000kHz		
195)	WTAK	10/5	D
196)	WLUP	50	24h
197)	WKYB	10/5	D
198)	KIDH	10	D F.PL.
199)	KXRB	10/0.03	D*+2h
200)	KOMO	50	24h
201)	KKIM	10	D*
202)	KIDN	10/5	
	1010kHz		
203)	KBIS	10/5	24h
204)	KIQI	10/2	
205)	KKGZ	10/5	D
206)	WQYK	50/5	24h
207)	WCNU	10	D
208)	WGUN	50/10/	18h
209)	WMOX	10/1	19h
210)	KXEN	50/0.5	D
211)	WINS	50	24h
212)	WFGW	50/10	D
213)	KBBW	50/2.5	18h
214)	KTLE	50/2.5	D
215)	WXTL	25/10	D
	1020kHz		
216)	KTNQ	50	24h
216a)	WART	5	D
217)	WJEP	10	D
218)	WCKW	50/0.35	F.PL.
219)	WNTK	10	D
220)	KCKN	50	20h
221)	KDKA	50	24h
222)	WRIX	10/3	D
	1030kHz		
223)	KFAY	10/1	24h
224)	KKSA	50/1	24h
225)	WTGR	10	D F.PL.
226)	KBUF	25/1	18h
227)	WNTL	50	D
228)	WFTK	35/2.5	D
229)	WBZ	50	24h
230)	KCTA	50	D
231)	KTWO	50	24h
232)	WXSS	50/10/1	24h
233)	WMIN	50/1	18h
234)	WJKZ	10/0.25	19h
235)	KEVT	10/1	F.PL.
236)	KDUN	10/0.63	24h
237)	WNOW	10	D
238)	KMAS	10/1	24h
	1040kHz		
239)	WYFX	10/1	24h
240)	WSKE	10/4	D
241)	WHO	50	24h
242)	KNOB	10	DF.PL.
243)	WSGH	10/0.25	D
244)	KURS	10/5	F.PL.
	1050kHz		
245)	KOFY	50/0.5	24h
246)	WPZA	10/0.5	D*
247)	KMTA	10/0.136	19h
248)	WEVD	50	24h
	1060kHz		
249)	KPAY	10	24h
250)	WNOE	50/5	24h
251)	WBIV	25/2.5	24h
252)	KBGN	10	D*
253)	WAMT	10/5	24h
254)	KYW	50	24h
255)	KLMO	10/0.25	16h
256)	KGFX	10/1	24h
257)	KHYM	10	D*
258)	KKDS	10/1	24h
259)	KFNA	10	D*
	1070kHz		
260)	WIWS	10	D*
261)	WAPI	50/5	24h
262)	WFLI	50/1	24h
263)	KNX	50	24h
264)	WANM	10	D*
265)	WIBC	50/10	24h
266)	KFDI	10/1	24h
267)	WNCT	10	24h
268)	KMOM	10/2.5	24h
269)	WDIA	50/5	24h
270)	WTSO	10/5	24h
271)	WKOK	10/1	20h
272)	WHYZ	50	D*
273)	KRBE	10/5	24h
	1080kHz		
274)	KSCO	10/5	24h
275)	WTIC	50	24h
276)	KWJJ	50/10	24h
277)	WFIV	10	D
278)	KRLD	50	24h
279)	WFTD	10	D*
280)	WDSY	50/25	D*

281)	WVCG	50/20	24h
282)	WDJX	10/1	24h
283)	KVNI	10/1	24h
284)	KRPX	10	D*
	1090kHz		
285)	KAAY	50	24h
286)	KYBG	50/5	24h
287)	KING	50	24h
288)	WGOC	10/1	D
289)	KNCR	10	D
290)	WSLG	10	D
291)	WBAL	50	24h
	1100kHz		
292)	KFAX	50	24h
293)	KCCF	25/1	F.PL.
294)	KNZZ	50/10	19h
295)	KZPM	10/1	F.PL.
296)	WCGA	10	D
297)	WHLI	10	D
298)	WWWE	50	24h
299)	KLWJ	10/1	F.PL.
300)	KDRY	11/1	18h
301)	KWXL	10/0.58	F.PL.
	1110kHz		
302)	KGPL	10	D
303)	KRLA	50/20	24h
304)	WTIS	10	D*
305)	KFAB	50	24h
306)	KBND	25/5	24h
307)	WBT	50	24h
308)	WBCA	10/2.5	D
309)	WJML	10	D*
310)	WZAM	50	D
	1120kHz		
311)	WUST	20	D F.PL
312)	KMOX	50	24h
313)	KPNW	50	24h
314)	KSJI	10/1	F.PL.
315)	WBNM	10/2.5	D*
316)	KANN	10/1	24h
	1130kHz		
317)	KSDO	10	24h
318)	KWKH	50	24h
319)	KTMR	10	D
320)	KBMR	50	D
321)	WCXI	50/10	24h
322)	KFAN	50/30	24h
323)	WBBR	50	24h
324)	WISN	50/10	24h
325)	WLBA	10	D*
	1140kHz		
326)	KRAK	50	24h
327)	KCMJ	10/2.5	24h
328)	WBXR	20	D
329)	WQBA	50/10	24h
330)	KGEM	10	24h
331)	KSOO	10/5	24h
332)	WRVA	50	24h
333)	KLUC	10/2.5	24h
334)	KZMQ	10	D+2
335)	WNHA	10	F.PL.
	1160kHz		
338)	WMLD	10/0.4	
339)	WJJD	50/4.2	24h
340)	WTYN	10/0.5	
341)	KSL	50	24h
342)	WCCS	10/1	24h
343)	WAMB	50/1	24h
344)	KVAR	10/1	24h
345)	WSKW	10/1	24h
346)	WKWA	10/1	F.PL.
	1170kHz		
348)	WACV	10/1	24h
349)	KCBQ	50/1.5	24h
350)	WCXN	10	D
351)	KVOO	50	24h
352)	WWVA	50	24h
353)	KLOK	50/5	24h
354)	WLGO	10	D
355)	KPUG	10/5	24h

	1180kHz		
356)	WHAM	50	24h
357)	VOA	50	17.5h
358)	KOFI	50/10	24h
359)	WMYT	10	D
360)	KYET	10/1	16h
361)	KERI	50/10	24h
362)	WJNT	50/0.5	24h
363)	WHJM	10/2.6	D
	1180		
364)	KTLL	25/1	F.PL.
365)	KGOL	50/10	24h
366)	KKAR	25/1	24h
	1190kHz		
367)	KORG	11/1.4	24h
368)	WGKA	10	D
369)	KEX	50	24h
370)	KUII	50/5	24h
371)	WANN	10	D+2h
372)	KXKS	10/0.03	D*+2h
373)	WLIB	10	D+1h
374)	WOWO	50	24h
375)	WBDY	10	D
	1200kHz		
376)	WFSF	10/2	D
377)	KOQI	25/1	F.PL.
378)	KTCD	10/2.5	F.PL.
379)	KFNW	10/1	17h
380)	WOAI	50	24h
381)	WKOX	10/1	24h
382)	WQDQ	10/0.5	24h
383)	WNSW	10/5	24h
384)	WDCQ	10/1	24h
385)	WSML	10/1	
386)	WGNY	10/2.5	F.PL.
387)	WBZY	10/2.5	24h
388)	WOPA	10/1	24h
389)	WRVX	10/0.25	F.PL.
390)	WCHB	50/1	
391)	WAGE	10/1	F.PL.
	1210kHz		
392)	KQTL	10/1	24h
393)	WDGR	10/2.5	D
394)	WOGL	50	24h
395)	WKNX	10	D*
396)	KGYN	10	24h
397)	KONY	10/0.25	24h
398)	KBSG	50/1	24h
399)	KREW	10/1	16h
400)	KLDI	10/1	
401)	WBIU	10/1	19h
402)	WCMQ	25/2.5	
403)	WGSF	10/0.25	18h
404)	KPRZ	20/5	24h
405)	KUBR	50/1	
	1220kHz		
407)	WREV	10	D
408)	WKNR	50	24h
	1230-1490kHz		
	All st's 5kW or less		
	1500kHz		
409)	KSJX	10/5	24h
410)	WTOP	50	24h
411)	WLQV	50/1	24h
412)	KSTP	50	24h
413)	WSMX	10/1	D
	1510kHz		
414)	KIRV	10	D*
415)	KFNN	10	D*
416)	KDKO	10/5	24h
417)	WNLC	10/5	18h
418)	WLAC	50	24h
419)	KGA	50	24h
420)	WAUK	10	D*
421)	KNSE	10/1	24h
422)	WSSH	50	24h
423)	KIDZ	10	D*
424)	KLLB	10	D
	1520kHz		
425)	KTRO	10/1	24h

426)	WWKB	50	24h
427)	KOMA	50	24h
428)	WGAM	10	D
429)	KFXX	50/10	24h
430)	KACY	10/0.5	24h
431)	KOLM	10	D*
432)	WQWQ	10/5	
433)	KFXE	10	D F.PL.
434)	KJSM	20	D F.PL.
	1530kHz		
435)	KFBK	50	24h
436)	KHPY	50/13	D
437)	WRTP	10	D
438)	WCKY	50	24h
439)	KOPY	10/0.99	D
440)	KGBT	50/10	24h
441)	WTTI	10	D
442)	WRPM	10/1	D*
443)	WCRJ	50	D
	1540kHz		
444)	KSGB	50/5	24h
445)	KLAU	10	24h
446)	KSKQ	50/10	24h
447)	KXEL	50	24h
448)	WPTR	50	24h
449)	WQCC	10	D*
450)	WPGR	50	D*
451)	KASA	10	D*
452)	WRBN	10	D*
453)	WTBI	10/1	D
	1550kHz		
454)	WAAJ	50/0.5	24h
455)	KKHI	10	24h
456)	KRGQ	10/0.5	24h
457)	WRHC	10/0.5	24h

458)	KQXI	10	D*
459)	KVKI	10/0.5	18h
460)	KQWB	10/5	24h
461)	WJRZ	10/3.5	F.PL.
462)	WBSC	10/5	18h
463)	WOKJ	50/10	20h
464)	WAMA	10	D*
465)	WYNX	50/0.5	24h
466)	WARD	10/0.5	20h
467)	KUAT	50	D
478)	WNTN	10	D
469)	WKBA	10	D
470)	KVAN	10	19h
471)	KNTR	10	18h
472)	KSVY	10/2.5	
	1560kHz		
473)	WWGM	10	D*
474)	KNZR	10	24h
475)	WRHC	10/4.4	F.PL.
476)	WQXR	50	24h
477)	WPAD	10/5	19h
478)	KKAA	10/5	20h
479)	WAGL	50	D*
480)	WTRQ	10	D*
481)	KQYX	10	D*
	1580kHz		
482)	KBLA	50	24h
483)	KCWW	50	24h
484)	KWYD	10	D
485)	WTKT	10	D
486)	WPGC	50/0.4	
487)	WSRF	10/5	24h
488)	KZKL	10/0.5	24h
489)	WWOF	10	D

Addresses and other information

G.M = General Manager; P.D = Program Director; C.E = Chief Engineer; S.M = Station Manager; Op's = Operations; N-1 = News on the h; N-2 = News on the half h; N-3 = News on the h and half h; BB = Big Bands; BM = Beautiful Music; MoR = Middle-of-the-Road music; MYL = Music of Your Life (popular music of the 20's, 30's, 40's); C&W = Country & Western music; CWM = Modern country music; ADC = Adult contemporary music; CHR = Contemporary hit radio (Top 40); Oldies = Popular music of the 50's, 60's, 70's; Cont = Contemporary; Pop = popular.

Verifications: The details of the verification policies often vary and thus are omitted here. Most stations will verify correct reports of reception over exceptional distances. Rp is often appreciated if in form of mint stamps or stamped, self-addressed envelope.

News: Additional newscasts often during drive hours e.g. (N-3).

1) 821 Marshall Farms Rd, Ocoee, FL 34761 — G.M: Terry Mason England; C.E: Chuck Jones. Oldies.
2) 1188 Padre Dr, #202, Salinas, CA 93901 — Mgr: Jerry Collins. P.D: Cathy J. Collins; C.E: Gary Beacham; MOR, Oldies.
3) c/o 1500 Cotnor Ave., Los Angeles, CA 90025.
4) Box 76860, Los Angeles, CA 90076 — G.M: Howard Neal; P.D: George Oliva; C.E: Marvin Collins. ADC, talk (L/N-1).
5) 2050 Amsterdam Rd., Belgrade, MT 59714 — G.M: Mark Brashear; C.E: Dale Heidner. Rlg. 1255-0606 (N-1). Easy listening.
6) 4990 Poplar Ave, Memphis, TN 38117 — G.M: Mark Loeffer. Rlg.
7) 1601 Belvedere Rd., Suite 204 E, West Palm Beach, FL 33406 — G.M: Stan Bowman; C.E: Warren Reeves; 1100-0600. Rlg. (N-1).
8) Kingsley, MI 49649.
9) Box 35297, Fayetteville, NC 28305 — G.M: John Dawson; P.D: Jeff Thompson; C.E: Terry Jordan. News, Talk.
10) 550 Pharr Rd, N.E., Atlanta, GA 30363 — G.M: John E. Laver; P.D: Eric Seidel. All News.
11) Box 30064, Springfield, MA 01103 — G.M: C. Hahn; News Dir: Hugh Massey; C.E: W. Weeks. News/talk.
12) P.O. Box 81, Mount Holly, NJ 08060 — MOR.
13) 640 Radio Way, #286-B, Blountville, TN 37617 — G.M: John E. Fry; P.D: Marty Osborne; C.E: Mitch Sandidge. Oldies. (N-3) (1100-0600).
14) Rancho Cordoba, CA. Addr: P.O. Box 417250, Sacramento, CA 95841 —C.E: Gerald Suseoff. Spanish.
15) Junction City, OR 97448.
16) Clinton, MA 01510.

17) Box 1060, Hibbing, MN 55746 — G.M: D. Acaley; P.D: Mike Allen; C.E: Danny Klaysmat. C&W.
18) 2644 McGavock Pike, Nashville, TN 37214 — G.M: Bob Meyer; P.D: Ted Cramer; C.E: Hugh Hickerson. CWM (N-3).
19) Box 1590, Manti, UT 84642 — Mgr: Douglas Barton.
20) Box 1967, Mobile, AL 36608 — G.M: H. Williams; C.E: R. Gallow. Black ADC.
21) Box 2569, Window Rock, AZ 86515 — G.M: Freddie Howard; P.D: Michael D. Moore; C.E: Abram Meria. Navajo 1200-1900, C&W.
22) 3070 Skywave Dr, Santa Maria, CA 93455 — G.M: S.H. Patterson; C.E: N. Patterson. Rlg, beautiful music.
23) 3412 36th Str, Astoria, NY 11106 — G.M: Scott Meier; P.D: Mark Mason; C.E: James Seaman. Sports.
24) Box 660, Greenville, SC 29602 — G.M: Allen Power; P.D: Jeff Blake; C.E: Don Gowens. C&W.
25) Ridgeland, MS 39157.
26) Golden Spike Rd, Sauk Rapids, MN 56379 — G.M: Herb Hoppe; P.D: Bill Zander; C.E: Dave Neeb. Oldies (day), Talk (night).
27) 4144 N. Central Expressway, #266, Dallas, TX 75204 — G.M: Bill Simmons; P.D: Freda Wells; C.E: Mike Vanhooser. Christian. (N-1).
28) Box 70, Mt. Vernon, WA 98273 — Pres: James Keane. C&W. (Currently 1kW only D.)
29) NBC Tower, 455 N. Cityfront Plaza, Chicago, IL 60611 — G.M: Rick Starr; P.D: Jim Frank; C.E: Margaret Bryan. All news.
30) Claremont, VA 23899.
31) Box 1280, Boise, ID 83701 — G.M: A.G. Vuylsteke; P.D: Larry Doss; C.E: W. Frahm. News, ADC. (N: local at :57, CBS N-1),
32) 2381 We. Flagler Str, Miami, FL 33135 — G.M: Emilio M. Milian III. Spanish.
33) 8401 Datapoint Drive, San Antonio, TX 78229 — G.M: Hugh Barr; P.D: Carl Becker; C.E: Ted Ronneburger. C&W (N-2).
34) 55 Hawthorne Str, #1100, San Francisco, CA 94105 — G.M: Tony Salvadore; P.D: Bob Agnew; C.E: Bill Ruck. Sports (N-1).
35) 209 CNN Center, Atlanta, GA 30303 — Pres: Robert D. Jackson; P.D: John Wheeling; C.E: Bob Mayben. All News.
36) 68 Radio Plaza, Owings Mills, MD 21117 — G.M: Michael Plumstead; P.D: Sean Casey; C.E: Dwight Weller. Talk (N-3).
37) 3 Fenway Plaza, Boston, MA 02115 — G.M: Eric Schultz; P.D: Mel Miller; C.E: Larry Bruce. Talk (N-1).
38) 3012 Highwoods Blvd., #200, Raleigh, NC 27604 — G.M: Bryan McIntyre; P.D: Charles Stegall; C.E: Gary Liebisch. News, talk.
39) 5904 Ridgeway Parkway, Memphis, TN 38119 — G.M: Debby Nichols; P.D: Mark Hamlin; C.E: Martinez Gregory. Oldies.
40) Box 4318, Charleston, WV 25364 — G.M: Mike Buxser; C.E: David Higginbotham. ADC. 1100-0600.
41) 604 Ludington Str, Escanaba, MI 49829 — G.M: Alice Sabuco; P.D: Kevin Scannell; C.E: Wayne Peterson. ADC. 1000-0600 (N-1).
42) Penthouse Twin Towers East, 236 Goodwin Crest Dr, Birmingham, AL 35209 — G.M: Mike Hammond; P.D: Jimmy Binyard; C.E: Roger Parker. All sports.
43) 200 Wharfside Way, Jacksonville, FL 32207 — G.M: Bruce C. Maduri; P.D: Jim Shirah; C.E: Jerry Smith. News, talk, all sports.
44) Box 1087, Coffeyville, KS 67337 — P.D: Jim Fischer; C.E: Jack Maxton. News, talk. 1100 (Sun 1200)-0603.
45) 2419 N. Piedras, El Paso, TX 79930 — G.M: Rob Burton; P.D: Brian Kennedy; C.E: H. Rand. C&W (N-1).
46) Box 369, Bristol, VA 24203 — Mgr: R. Al Morris; P.D: Tommy Tester; C.E: J. Faniola. Rlg.
47) 3421 N. Causeway, Suite 800, Metairie, LA 70002 — G.M: Leo F. Holzenthal III; P.D: Jack Wilde; C.E: Barry Chickini. MOR.
48) Silt, CO 81652 — G.M: Jim Seany. Big Band, MoR, Oldies.
49) 1111 St. Gregory Str, Cincinnati, OH 45202 — G.M: J.D. Martin; P.D: D. Reinhart; C.E: Al Kenyon. ADC (N-1). Oldies, country, sports.
50) 1171 S.W. Temple, Salt Lake City, UT 84101 — G.M: D.R. Williams; P.D: Howard Bogart; C.E: Troy German. Beautiful music.
51) Box 1598, Winston, OR 97496 — G.M: Bill Simmons; P.D: S. Welch; C.E: Paul Brown. Rlg. 1330-0800 (N: at :55).
52) Newport, WA. Addr: Box 1740, Oldtown, ID 83822. G.M's: Jim and Helen Stargel. CWM days, CHR nights. (1400-0600).
53) Camas, WA 98607
54) 5637 W. Campo Bello Dr, Glendale, AZ 85308 — Pres: John H. Gates; P.D: Robert O. Stewart; C.E: Evert Fruitman.
55) P.O. Box 710, Los Angeles, CA 90078 — Exec. VP: Bill Ward; P.D: Len Weiner; C.E: Steve Blodgett. All sports.
56) 5705 Marconi Av, Carmichael, CA 95608 — Pres: D. Kahle; G.M: J. Clark; P.D: Steve Gasser. Rlg.
57) 2960 Coral Way, Miami, FL 33145 — G.M: A. Pérez-Roura; C.E: C. Fernández. Spanish "R. Mambí".
58) Box 20007, Shreveport, LA 71120 — G.M: Leeann Lewis; P.D: Howard Clark; C.E: Rick Benson. News, MOR, Sports.
59) 8500 W. 63rd Str, Suite 210, Shawnee Mission, KS 66202 — G.M: Bob Zuroweste; P.D: Don Daniels; C.E: Chris Ostrander. ADC, Oldies. (N-3). (Licensed to Kansas City, MO).
60) 1440 Broadway, New York, NY 10018 — G.M: Bob Bruno; P.D: Ed Walsh; C.E: Paul Stewart. Talk.
61) Box 710, Amarillo, TX 79189 — G.M: Bob Russell; P.D: Bob Reed; C.E: John Harrold. News, talk. (N-1)
62) Box 2100, Christiansburg, VA 24073 — G.M: Robert T.S. Colby; P.D: Bill Lineberry. Bluegrass, Gospel.
63) 2807 3rd Ave, Seattle, WA 98121 — G.M: J.K. Abel; P.D: Andy Ludlum; C.E: Norm Graham. Talk. (N-1).
64) 1105 E. Superior Str, Duluth, MN 55802 — G.M: Ken Buehler; Pres: Patricia McNulty; P.D: Mark Edwards; C.E: P. Maki. C&W, News (N-3).
65) 435 N. Michigan Ave, Chicago, IL 60611 — G.M: Dan Fabian; P.D: Lorna Gladstone; C.E: James Carollo. MoR, Talk (N-3).
66) 1 Main Str, Las Vegas, NV 89101 — Mgr: Clare Reis; P.D: A.J. Williams; C.E: M. Messina. News, Talk (N-1).
67) Box 720, Pisgah Forest, NC 28768 — G.M: Randy C. Barton; P.D: Larry Spears; C.E: Jobie Sprinkle. Rlg. (N. at :55).
68) Box 329, Bath, ME 04530 — G.M: Dan White; P.D: Scott Barrett; C.E: Eugene Terwilliger. Currently on 730kHz.
69) 3988 No. Roscoe Rd, Hernando, FL 32642 — G.M: Frank Watson; P.D: Tony Wallace; C.E: Dave Terry. MOR, Nostalgia (N-1).
70) Box 1114, LaGrange, GA 30241 — G.M: L.A. Wood. Urban cont.
71) 1777 N.E. Loop 410, Suite 803, San Antonio, TX 78217 — G.M: Dennis G. Roberts; P.D: Alberto Martínez; C.E: Bill Haberer. Spanish (R. Festival).
72) Tuckahoe, VA 23229.
73) #300, Colonial Financial Center, Montgomery, AL 36104 — G.M: David Coppock; John Hart; C.E: Larry Wilkins. Nostalgia.
74) 3183 D Airway Ave, Costa Mesa, CA 92626 — Rlg.
75) 1 Embarcadero Center, San Francisco, CA 94111 — G.M: Frank Oxarart; News Dir: Ed Cavagnaro; C.E: Shingo Kamada. All news.
76) Box 740, Orlando, FL 32802.
77) Box 697, Melville, NY 11747 — G.M: Paul Fleischman; P.D: Martin Woolf; Jim Beery. Big bands. (N-1). News 1100-1500.
78) Box 907, Mt. Airy, NC 27030 — G.M. & PD: Kelly Epperson; CE: Ralph D. Epperson. Old Time. (N-1).
79) 7136 So.Yale, Tulsa, OK 74136 — G.M: Gerry Tabio; P.D: Dave Beasing; C.E: Bob Cauthon. ADC, talk.
80) 510 Lovett Blvd, Houston, TX 77006 — S.M: Bill van Rysdam; C.E: Errol R. Coker. News, sports.
81) 1601 W. Peachtree Str. NE, Atlanta, GA 30309 — G.M: Marc W. Morgan; P.D: Jim Ashbery; C.E: Ron Wilson. News (N-3).
82) Box 1847, Grand Island, NE 68802 — G.M: James M. Sander; P.D: Jack Nelson; C.E: Marty Berlinger. CWM (N. at :05, :30).
83) Carson City, NV 89701.
84) Box 14957, Portland, OR 97214 — G.M: Ray Watson; P.D: Jeff Grimes; C.E: Larry Wilson. News, Talk.
85) 581 N. Reservoir Rd, Polson, MT 59860 — G.M: Tom Dwyer; C.E: Jim Bender. Diversified (N-1).
86) Box 875, Price, UT 84501 — G.M: Thomas B. Anderson; C.E: Eric Albrecht. C&W, talk. 1300-0700.
87) 4150 Pinnacle Str, Suite 120, El Paso, TX 79902 — G.M: Greg Heitzman; PD: Carlos Teran; C.E: Cliff Gibson. Spanish.
88) 2100 Fisher Bldg, Detroit, MI 48202 — G.M: Jim Long; P.D: Phil Boyce; C.E: Ed Buterbaugh. MoR, talk, variety (N-1).
89) 5350 So. Roslyn Str, Englewood, CO 80111 — G.M: Gayle Shaw; P.D: Danny Davis; C.E: Bill Harris. Oldies.
90) 217 Alamo Plaza, Suite 200, San Antonio, TX 78205 — G.M: Charles Andrews; P.D: Bill Thorman; C.E: Tom Sittner. Urban AC.
91) 7677 Engineer Rd, San Diego, CA 92111 — G.M: Paul Palmer; P.D: M. Larson; C.E: J. Weigand. News, talk.
92) 70 James St., Suite 140, Worcester, MA 01603 — Rlg.
93) 2808 E. Kiehl Ave, N. Little Rock, AR 72116 — G.M: J.J. Shields; C.E: R. Loewy. Rlg.
94) Box 888, Champlain, NY 12919 — G.M: Teri Billiter; P.D: Teresa Altman; C.E: R.A. Jones. Rlg.
95) Box 389, Athens, AL 35611 — G.M: W. Dunnavant; P.D: Dave Stone; C.E: C. Sampieri. News, talk.

96) P.O. Box 550, Fair Oaks, CA 95628-0550 — G.M: Jack Gardner; P.D: Norman Scott; C.E: Steve Minshall. News, talk, incl. R. Moscow, R. Beijing relays. (N-1).
97) 3440 Marinatown Lane N.W, N. Fort Myers, FL 33903 — G.M: Maureen Knorr; P.D: Lee Gordon; C.E: Richard Gallo. News, talk.
98) 77 Broadcast Plaza, SW, Albuquerque, NM 87103-1351 — C.E: Gary Diamond. MOR (N-3).
99) 2 Penn Plaza, 17th fl., New York, NY 10023 — G.M: Fred Weinhaus; P.D: F.G. Wagstaff; C.E: Bill Krause. Talk.
100) Box 700, Miles City, MT 59301 — G.M., P.D. & C.E: Donald L. Richard. ADC (N-1).
101) 3201 Royalty Row, Irving, TX 75062 — Rlg.
102) Box 91202, Bellevue, WA 98009 — G.M: John Rogers; P.D: Ray Randall; C.E: Robert Trimble. OLD.
103) 630 N. McClurg Court, Chicago, IL 60611 — G.M: G. Peterson; News Dir: Chris Witting; C.E: Mark Williams. All news.
104) Box 11920, Reno, NV 89510 — G.M: John E. O'Brien; C.E: R. Randell; C.E: J. Gradijan. CWM (N-1).
105) Rte 2, Box 1029, Jacksonville, AL 36265 — G.M: Roy Webb. Rlg. 1000-0600.
106) 900 Front Str, San Francisco, CA 94111-1450 — G.M: M. Luckoff; Op's Dir: J. Swanson; C.E: B. Schirmer. Talk, news.
107) 508 Westport Rd, Kansas City, MO 64111 — G.M: Skip Stow; P.D: Brian Wilson; C.E: Lloyd Collins. News, talk.
108) 1430 Balltown Rd, Schenectady, NY 12309 — G.M: Chloe Brothers; P.D: R.E. Brindle; C.E: Roger Bruce. Full Service.
109) 102 Business Park Drive, Suite D, Jackson, MS 39213 — G.M: Bill Clark; Op's Mgr: Joe Bennet. Rlg.
110) Box 99, Hwy 79N, Sturgis, SD 57785 — G.M: Dana Caldwell; PD: Dean Kinny. CE: Steve Neave. C&W (N-1).
111) Box 10, Tomahawk, WI 54487 — Mgr: G. Albert; P.D: M. Everett. C.E: Jim Zastrow. C&W (N-1).
112) Box 8238, Spokane, WA 99203 — G.M: T.W. Read; P.D: N. Bridges; C.E: G. Frese. Big Band, rlg, farm.
113) One B'cast Hill, Fort Worth, TX 76103 — Pres: John Hare; P.D: Ted Stecker; C.E: Clay Steely. C&W (N-1).
114) 2815 2nd Ave, Suite 100, Seattle, WA 98121 — G.M: Richard Ulrich; P.D: Roger Grossenbacher; C.E: R.B. Harris. Rlg (N-1).
115) 800 8th Ave. S.E., Largo, FL 34641 — G.M: Brent Harmon; P.D: Jack Fichter; C.E: Shannon Murdoch; Re. to: David Sharp. Rlg. 1000-0500 (N-1).
116) 10600 Jefferson Davis Hwy, Richmond, VA 23237 — G.M: Paul Scott. C.E. Jeff Loughridge. Rlg. (N-1).
117) 1 Centre Str, New York, NY 10007 — Dir: Thomas B. Morgan; P.D: L.J. Orfaly; C.E: Steve Cellum. News, info.
118) 29 We. Main Str, Jasper, TN 37347 — G.M: Phil Patton. Rlg.
119) Box 36868, Tucson, AZ 85740 — G.M: Ken Hettings; C.E: Jim Bolton; Mgr: Ted Phillips. Rlg.
120) 625 2nd Ave. S, Minneapolis, MN 55402 — G.M: L.S. Goldstein; P.D: Jon Quick; C.E: J. Miller. News, sports, ADC. (N. at :00, :55),
121) Smyrna, GA.
122) Box 509, Kennett, MO 63857 — G.M: Jeff Wheeler; C.E: Larry Anthony. ADC.
123) Box 90395, Casper, WY 82609 — G.M: Steve Stumbo; P.D: Aaron M. Remington; C.E: Kent Abendroth. Rlg. (N-1).
124) Box A, Eden, NC 27288 — G.M: Ron Smith. Variety.
125) Box 2218, Thomasville, GA 31799 — G.M: Curtis Thomas; P.D: James Sherman. C.E: James Devane. Black Gospel (N-1).
126) 520 We. Chestnut, Louisville, KY 40202 — G.M: Robert R. Scherer; P.D: Skip Essick; C.E: Charles Strickland. ADC, talk (N-3).
127) 3735 Rigolette Rd, Pineville, LA 71360 — GM & PD: Richard Dupree. CE: A.T. Moore. So. Gospel.
128) Box 1808, Harlingen, TX 78551 — MOR.
129) Box 339, Quinque, VA 22956 — G.M: Philip Goodwin. Rlg.
130) 3333 Cambridge, Las Vegas, NV 89109 — G.M: Jerry Kutner; P.D: Jack Hayes; C.E: Gordy Aslum. Sports. (N-1).
131) 90 Bagby Dr, #11, Birmingham, AL 35209 — G.M: Jim Romine; P.D: Ken Michaels; C.E: Jim Gray. Easy listening.
132) 1380 Lawrence Str, Suite 1300, Denver, CO 80204 — G.M: Lee Larsen; Kris Ohlinger; C.E: Jan Chadwell. Newstalk.
133) 7 Bullfinch Place, Boston, MA 02114 — G.M: Daniel K. Griffin. P.D: Al Mayers; C.E: Dana Puopolo. Talk.
134) Box 1539, Hwy 80, Forest, MS 39074 — G.M: W.H. Wilbourne. C&W.
135) Univ. of Tennessee, Rm 295, CUEB, P-103 Andy Holt Tower, Knoxville, TN 37996 — G.M: Sam Swan; N. Dir: Pam Doyle; C.E: M. Jones. All news.
136) 5706 New Chapel Hill Rd., Raleigh, NC 27607 — G.M: Mark Warlanmont; P.D: Rick Freeman; C.E: Jon Hardee. Standards.

137) One Radio Lane, Cleveland, OH 44114 — G.M: Rosa R. Turner; P.D: Jim Davis; C.E: D. Szucs. MYL.
138) 109 Plaza Dr, Johnstown, PA 15905 — G.M: Michael F. Brosig, Sr. PD: Jack Michaels. CE: James J. Boxler. CWM (N-1).
139) Box 11335, Tacoma, WA 98411 — G.M: Margaret Dempsey; C.E: Stan MacLafferty. Easy listening. (N-1).
140) 1120 E McCuen Str, Duluth, MN 55808 — G.M. & P.D: Ted Elm; C.E: John Talcott. Gospel (N-2).
141) 10250 Bissonnet #400, Houston, TX 77036 — Pres: D.M. Best; P.D: M. Franco; C.E: D. Armstrong. Spanish, Ranchera music.
142) 999 Waterside Dr, Norfolk, VA 23510
143) Box 3839, Modesto, CA 95352 — Mgr: Bessie Pappas; P.D: K. Roberts; C.E: Steve Menshall. C&W, talk. (N-1).
144) Route 5, Box 45, Pittsburg, KS 66762 — G.M: C.J. Jackson; P.D: S. Passeri; C.E: T. Fast. CWM, Farm (N-2).
145) Troutdale, OR 97060.
146) Box 140, Oak Hill, WV 25901 — G.M: Gene Ellison; P.D: Steve Bush; C.E: Mark Smith. Rlg. (1100-0300, N. at :55.)
147) 1349 Cheltenham Ave, Philadelphia, PA 19126 — G.M: R. Delgado; P.D: Rafael Grullon; C.E: John Faiss. Spanish. (N-1).
148) 180 S. 300 West, Suite 400, Salt Lake City, UT 84101 — G.M: Stuart Stanek; C.E: Chuck Condron. Class. rock.
149) Box 309, Valley Head, AL 35989 — G.M: Tom Crowe; C.E: M. Tolen. ADC/CHR.
150) P.O. Box 53367, New Orleans, LA 70153 — G.M: J.M. Early; P.D: Dave McNamara; C.E: Robert L. Dunn. News, C&W (N-1).
151) 912 Washington St, Auburn, ME 04210 — CE: Armond Girard. News, talk.
152) Michigan State University, East Lansing, MI 48824-1212 — Mgr: S.K. Meuche; P.D: C. Gilleo; C.E: Harold Beer. News, talk.
153) Box 49, Hwy 34 East, Park Rapids, MN 56470 — G.M: E.P. De LaHunt Jr. Ad. Standards.
154) Box 2485, Tri-Cities, WA 99302 — G.M: Jeff Ripley; P.D: Chris Kelly; C.E: Jeff Ripley. Country music (N. at :20).
155) 224 Colonial Heights Rd, Kingsport, TN 37663 — G.M: Grady A. Thomas; P.D: Tom Scott; C.E: Gary Smith. Cont. Christian.
156) 2235 E. Flamingo, #209, Las Vegas, NV 89119.

157) 5900 San Fernando Road, Glendale, CA 91202 — G.M: F.S. Beaton; P.D: R. Sinclair; C.E: H. Williams. Talk.
158) Box 880, Menomonie, WI 54751 — Mgr: Michael E. Phillips. News, talk.
159) Box 580, North Little Rock, AR 72114 — G.M: Wm. E. Hensley; P.D. & C.E: David Jones. C&W, Rlg.
160) Box 802, Columbus, OH 43085. — Mgr: Bill Caridas; C.E: Doug Leonard. Rlg, Farm (N-1).
161) P.O. Drawer F, Window Rock, AZ 86515 — G.M: Lawrence Harper; P.D: Charles Harper; C.E: Wm. Hurne. 1300-0500. (N. at :55, N-1). Rlg, Navajo.
162) Box 880, Whitefish, MT 59937 — G.M: Benny Bee; P.D: Andy Shamely. C&W.
163) 51 W. 52nd Str, New York, NY 10019 — G.M: Chris J. Whitting; News Dir: Harvey Nagler; C.E: Ed Schwartz. All news.
164) Box 880, Lexington, NE 68850-0880 — G.M: Eric Brown; P.D: Charles Brogan, Pam Hird; C.E: Vern Killion. CWM (N-3).
165) Box 73503, Houston, TX 77273 — G.M: Bobby Martinez; C.E: Errol R. Coker. (N-1). Rlg.
166) 8 E. Alisal Str, Suite 501, Salinas, CA 93901 — G.M: John Dick; P.D: Joseph Miller. CE: Ronald Warren. Rlg. 24h.
167) 1100 Olive Way, Suite 1550, Seattle, WA 98101 — G.M: Tim Davidson; P.D: Kevin Cassidy; C.E: Dwight Small. MOR. (N-2).
168) Citrus Heights, CA 95610.
169) 190 N. State St, Chicago, IL 60601 — G.M: Tom Tradup; P.D: Drew Hayes, C.E: Warren Schultz. Talk, info.
170) Dedham, MA 02026. Rlg.
171) Box 1890, St. George, UT 84770 — G.M: B. Philippy; P.D: L. Jewell; C.E: D. Sanford. ADC (N-1); 1300-1000 (Sun 0700).
172) Box 1638, Laredo, TX 78040. G.M: Larry Swikard; P.D: Bobby Pena; C.E: Art Trevino. Bilingual Eng/Sp, CHR.
173) Box 1229, Laurel, MS 39441 — G.M: Leonard Boltun; P.D: Ron Davis; C.E: Dana Walters. Urban cont. (N:35).
174) Box 486, Granite Falls, NC 28630 — Rlg.
175) 701 N. DuPont Hwy, Georgetown, DE 19947 — G.M: Graham Satherley; P.D: Cepth Michaels; C.E: Terry Daulton. Big band/MoR. (N-1). 1100-0500.
176) 3003 N. Snelling Ave, St. Paul, MN 55113 — GM: Wayne Pederson; CE: Don Hayworth. Rlg (N-1).
177) 999 No. Van Ness Ave, Fresno, CA 93728 — G.M: Al Grosby; P.D: Lee Nye; C.E: Hal Torosian. CWM (N-1).
178) 4330 N.W. 207 Dr, Miami, FL 33055 — G.M: Tim Williams; P.D: Marc Kuhn; C.E: Richard Van Hook. News.
179) Box 5008, Macon, GA 31213 — G.M: Dennis DeMichele; P.D: Don King; C.E: W. Sowell. Oldies (N-1).
180) 450W, 400S, Cedar City, UT 84720 — G.M. & P.D: J.C. Hunter. Oldies.
181) Box 5068, Bend, OR 97708 — G.M: Mike Burnette; P.D: Kit Carson; C.E: James Boyd. Old (N-1) (1100-0400).
182) Route 2, Box 10C, Houston, MS 38851 — G.M: Robin H. Mathis; P.D: Melanie Munlin; C.E: J.B. Skelton Jr. Gospel (N-3). 1000-0405.
183) 5161 Maple Dr, Des Moines, IA 50317 — G.M: Kipper McGee; P.D: S. Mathews; C.E: J. Kosobucki. Oldies. (N-1).
184) P.O. Box 19066, New Orleans, LA 70179 — G.M: Penelope Brazile; P.D: Steven Ross; C.E: J.P. Robillard. Gospel. N:55.
185) Box 339, Smithfield, VA 23430 — Mgr: Larry W. Cobb; C.E: H. Wood. Cont. Christian (N-1).
186) 1920 W. Copper, Tucson, AZ 85745 — G.M: Michael Madigan; P.D: Mike Rapp; C.E: Robert Malsbury. All news.
187) Box 11389, Memphis, TN 38111 — G.M: Fred Webb; P.D: Roger Bethel; C.E: Ron Laury. Gospel.
188) No. 1 Radio Rd, Orlando, FL 32808 — G.M: Gary Kimes; P.D: Brian Krysz; C.E: Hal Canaday. Rock.
189) Box 1855, Southern Pines, NC 28387 — G.M: Larry Richardson; P.D: Helen Sansone; C.E: Ed McDabe. ADC. 1100-0500 (N-1).
190) Box 231, Somerset, PA 15501 — G.M: I.R. Adams; P.D: J. Lyons; C.E: H. Showman. C&W.
191) Box 26098, Philadelphia, PA 19128 — G.M: Jennifer Lear Downing; P.D: Carl Dean; C.E: Stu Engelke. Cont. Christian. (N. at :20).
192) 311 Westminster St, Providence, RI 02903 — G.M: Joseph A. Caliri; P.D: Rob Michaels; C.E: Ken Torres. Talk.
193) P.O. Box 10207, Knoxville, TN 37939 — G.M: Bobby Denton; P.D: Les Acree; C.E: Milton Jones. (N-1). News, talk.
194) Box 787, Wichita Falls, TX 76307 — G.M: John Vincent; C.E: V. Beck. (N. at :55). CHR.
195) 6420 Stringfield Rd, Huntsville, AL 35806 — G.M: Virginia Griffith; P.D: Tom Kelley; C.E: Carl Sampieri. Rock.
196) 875 N. Michigan Ave, #3750, Chicago, IL 60611 — G.M: Lawrence J. Wert; S.M: Greg Solk; C.E: Tom Knauss. New, talk.
197) P.O. Box 1006; Hemingway, SC 29554 — G.M. & C.E: Mike Holt; ADC, C&W.
198) Eagle, ID 83616.
199) 3205 So. Meadow, Sioux Falls, SD 57106 — G.M: Don Jacobs; P.D: Charlie Walker; C.E: Tony Randall. CWM (N-1).
200) 100-4th Ave. No, Seattle, WA 98109 — G.M: Rich Robertson; P.D: Norm Gregory; C.E: Jack Barnes. ADC, Full sce. (N-1).
201) 307 Los Ranchos N.W, Albuquerque, NM 87107 — G.M: Michael D. Stitzel; P.D: John Lehman; C.E: Bill Pace. Cont. Christian. (N-1).
202) Box 774743, Steamboat Springs, CO 80477 — G.M: D.H. Gayer; P.D: M. Bachman; C.E: D.H. Gayer. C&W.
203) 2400 Cottondale Lane, Little Rock, AR 72202 — G.M: Philip Jonsson; P.D: John Ramsey; C.E: Norm Laramee. Business. (N-1).
204) 2601 Mission Str, San Francisco, CA 94110 — G.M: R. De la Rosa; P.D: Carlos Rivera; C.E: John Scherer. Spanish popular.
205) Box 707, 1516 Mill St., Brush, CO 80723 — C&W.
206) 9450 Koger Blvd, St. Petersburg, FL 33702 — G.M: Charlie Ochs; P.D: Jay Miller; C.E: Frank Berry. C&W. (N. at :55).
207) Box 1010, Crestview, FL 32536 — C.E: S.L. Atkins. Talk.
208) Box 67, Decatur, GA 30031 — G.M: Dick Schroeder. P.D: Andy Durham. TC.E: George Pass. Rlg.
209) Box 1511, Meridian, MS 39301 — G.M: E. Smith; P.D: Larry Grant; C.E: Jim Johnson. C&W (N-1, at :55). 1030-0600.
210) Box 8085, Mitchell, IL 62040 — G.M: George Spicer; S.M: Daniel A. Allen. C.E: Pete Neichamp. Rlg, Gospel. (N. at :55).
211) 888 7th Ave, New York, NY 10106 — G.M: Warren Maurer; P.D: Steve Swenson; C.E: Ken Nealy. All news.
212) Box 158, Black Mountain, NC 28711 — G.M: Edna Edwards; P.D: Colin O'Brien; C.E: Rob Schermerhorn. Rlg, MoR.
213) 1019 Washington St, Waco, TX 76701 — G.M: Steve Williams; P.D: Robert Huff; C.E: Bernarr Howell. Rlg. (N-1). 1200-0600.
214) Tooele, UT 84074.
215) Box 16656, Jacksonville, FL 32245 — G.M: Karen W. Vincent; C.E: Don Fleming. Gospel.
216) 1645 Vine Str, Los Angeles, CA 90028 — G.M: K.D. Wolt; P.D: J. Liberman; C.E: G. Ogonowski. Spanish.
216a) Port Orange, FL 32019.
217) Box 90, Thomasville, GA 31799 — G.M: L. Birdsong; P.D: Rodney Edmonson; C.E: C. Scott. Cont. Christian. (N-1).
218) Box 5905, Metairie, LA 70002 — G.M. & C.E: S. Levet III. Rock.
219) RR 1, Box 249, New London, NH 03257; Talk.
220) Box 670, Roswell, NM 88202 — G.M: Carl M. Skrobrak. P.D: Wayne Kirpatrick. CE: Donald Niccum. C&W, religion. (N-1). Spanish 0530-0700.
221) One Gateway Center, Pittsburgh, PA 15222 — G.M: Ted Jordan; P.D: Chuck Dickemann; C.E: Jack Layton. News, talk (N-3).
222) Watson Village, Anderson, SC 29624 — G.M: Matt Phillips; P.D: John Woodson; C.E: Bob Bierman. Rlg.
223) Box 878, Fayetteville, AR 72702 — Pres: Pat Demaree. CWM, talk.
224) 1425 River Park Dr, Sacramento, CA 95185-4508.
225) Point Pleasant, WV 25550.
226) Box 798, 1309 Ea. Fulton, Garden City, KS 67846 — G.M: Scott Smith; P.D: Larry Williams. C.E: Don Britnall. CWM, Farm. 1200-0600 (N-1).
227) 10410 Windsor View Dr, Potomac, MD 20854 — G.M. & P.D: Steve Brisker; C.E: P.V. Gurekis. Rlg, ETH, "International Radio".
228) Box 1030, Creedmor, NC 27522 — G.M: Max Mathis; P.D: Robert Reuther; C.E: J.D. Watson. Rlg.
229) 1170 Soldiers Field Rd, Boston, MA 02134 — G.M: John Irwin; P.D: David Bernstein. CE: Paul Donovan. Talk/News (N-1).
230) Box 898, Corpus Christi, TX 78403 — G.M: Bill York; P.D: John Boudresu; CE: Paul Easter. Rlg.
231) 150 N. Nichols, Casper, WY 82601 — G.M: Robert D. Price; PD: Bruce King; C.E: Randy Evans. C&W (N-1).
232) 1188 Minna Place, Suite 214, Memphis, TN 78104 — G.M: Purvis Spann; P.D: Nate Russell; C.E: Jerry Campbell. Rlg.
233) Box 25130, Maplewood, MN 55125 — G.M: Greg Borgen; P.D: Rosann Juenemann; CE: Ken Brenner. Oldies. 1100-0500.
234) 880 Hwy. 70 W., Dickson, TN 37055 — C&W.
235) Cortaro, AZ.
236) Box 147, Reedsport, OR 97467 — G.M: Lyle Irons. C&W. N:3.

237) Box 23509, Charlotte, NC 28212. G.M: Russ Douglass; P.D: Rodney Baucom; C.E Mike Miranda. Rlg music.
238) Box 760, Shelton, WA 98584 — G.M: H.S. Greenberg; P.D: Bill Meyer; C.E: Tom Trotzer. ADC.
239) 400 Gulfstream Blvd, Delray Beach, FL 33444 — G.M: Gary Lewis; P.D: Mike James; C.E: Rick Rieke. Urban Cont.
240) Box 187, Everett, PA 15537 — G.M: Sandra King; P.D: Martin M. Bakner; C.E: R. Resconsin. CWM. (N-1).
241) 1801 Grand Ave, Des Moines, IA 50308 — G.M: Steve Shannon; P.D: Van Harden; C.E: K. Erickson. MoR, farm, talk. (N-3).
242) Frazier Park CA 93225.
243) Box 25368, Winston-Salem, NC 27114 — G.M: Lewis B. Hughes. Rlg.
244) 3511 28th St, San Diego, CA 92104 — G.M: José Mireles. Spanish.
245) 2500 Marine Str, San Francisco, CA 94124 — G.M: Mike Lincoln; P.D: John Warren; C.E: Doug Crall. Spanish.
246) 24 Frank Lloyd Wright, Ann Arbor, MI 48108. News, talk, sports.
247) Box 1426, Miles City, MT 59301 — G.M: C.W. Wilcox; P.D: Rick Tobin; C.E: Bob Parker. CWM, 1200-0700 (N-1).
248) 770 Broadway, New York, NY 10003 — G.M: Thomas Bird; P.D: Bob August; C.E: George Levites. Big Band, Oldies, Standards (N-2).
249) 2654 Cramer Lane, Chico, CA 95928 — G.M: Dave Brower; P.D: Tony Kay; C.E: Dan Butner. News, talk. (N-3).
250) 529 Bienville Str, New Orleans, LA 70130 — G.M: Tom Kennedy. PD: Dave Nicholson. CE: Doug Booth, Jr. C&W (N. at :55 or N-1).
251) 1105 Commonwealth Ave, #204, Boston, MA 02215 — G.M: Robert L. Tarpin; C.E: Geoff McLaughlin. Spanish rlg.
252) 3303 Ea. Chicago Str, Caldwell, ID 83605 — G.M. & C.E: N. Wilson. Rlg.
253) Box 186, Mims, FL 32754 — G.M: Rene Plasensia; C.E: Ron Lykina. Spanish.
254) Independence Mall East, Philadelphia, PA 19106 — G.M: Roy Shapiro; P.D: Mark Helms; C.E: Janet Kowalczyk. News only.
255) Box 799, Longmont, CO 80501 — G.M: Wm. G. Stewart; P.D: Jerry Steffen; C.E: Ron Stephens. C&W (N-1).
256) Box 1197, Pierre, SD 57501 — G.M: M. Swendsen; P.D: Charlie Hale; C.E: C. Hesle. C&W, farm.
257) 104 W. Cass Str, Gilmer, TX 75644 — G.M: Joel Gibbons; P.D: Jim Bowden; C.E: Wayne Shultice. Rlg.
258) Box 57760, Salt Lake City, UT 84157 — G.M: Alan Hague; P.D: Steve Carlson. C.E: Ken Meyer. Childrens prgr's.
259) 5127 Timberwood Dr, El Paso, TX 79903. Spanish.
260) 118 Philpott Lane, Beaver, WV 25813 — G.M: R. Shane Southern; P.D: Bob Spencer; News.
261) Box 10888, Birmingham, AL 35202-0888 — G.M: Bernie Barker; P.D: Frank Giardina; C.E: F. Giardiana. Nostalgia (N-1).
262) 621 O'Grady Dr, Chattanooga, TN 37419 — G.M: Y.H. Benns; P.D: Jim Hill; C.E: Joe Poteet. Rlg.
263) 6121 Sunset Blvd, Los Angeles, CA 91601 — G.M: G. Nicholaw; News Dir: Bob Sims; C.E: Michael P. Smith. All news.
264) Box 10174, Tallahassee, FL 32302 — G.M: Bob Badger; P.D: Gerald Tookes; C.E: H. Haines. News.
265) 9292 No. Meridian, Indianapolis, IN 46260 — G.M: R.G. Cooper; P.D: Jed Duval; C.E: M. Russell. MoR, farm.
266) Box 1402, Wichita, KS 67201 — G.M: Mike C. Oatman; P.D: John Speer; C.E: Craig Maudlin. C&W. (N-3).
267) Box 7167, Greenville, NC 27835 — G.M: Scott Cherney; P.D: W. Baker; C.E: Roy McCoury. News/Talk/Sports (N-1).
268) Box 900, Monticello, MN 55362 — G.M: Greg Davis; News Dir: Jeff McAndrew. MoR.
269) 112 Union Ave, Memphis, TN 38103 — G.M: Ric Caffey; P.D: R. O'Jay; C.E: A. Pendleton. Black (N. at :55).
270) 5721 Tokay Blvd, Madison, WI 53719 — G.M: David Graupner; P.D: J. Tyler; C.E: J.M. Hermanson. C&W (N-3).
271) Box 1070, Sunbury, PA 17801 — G.M: Joseph A. McGranaghan; P.D: Jack Richards; C.E: J.W. Keller Jr. ADC. (N-1). 1100 (Sun 1200)-0703.
272) Box 4309, Greenville, SC 29608 — G.M: Martin McClung; P.D: Janice Henderson; C.E: Ray Slatton. Urban cont. **F.Pl:** 50kW

273) 9801 Westheimer No. 700, Houston, TX 77042 — C.E: John Shadle. Top 40, R&B.
274) 2300 Portola Dr, Santa Cruz, CA 95062 — C.E: Bill Brooks. Talk.
275) 1 Financial Plaza, Hartford, CT 06103 — Mgr: Bob Dunn; P.D: Bill White; CE: Tom Ray. ADC (N-3).
276) 931 S.W. King Ave, Portland, OR 97205 — G.M: Daniel Volz; P.D: Brian Norton; C.E: Chris Cullen. CWM (N-1).
277) Drawer 555519, Orlando, FL 32855 — G.M: E.C. Allmon; C.E: L. Campbell. Farm, C&W (N-1).
278) 7901 Carpenter Fwy, Dallas, TX 75247 — G.M: Michael Ewing; PD: Richard Walker; CE: Charles Staples. News, sport.
279) 774 Roswell Str, Marietta, GA 30060 — GM: Rocky Payne; CE: George Pass. Rlg.
280) 107 6th Str, Pittsburgh, PA 15222 — G.M: Gary Garrett; C.E: C. Bryson. Country.
281) 2100 Salcedo, Suite 203, Coral Gables, FL 33134 — G.M: Matthew Rowe; P.D: Roberto Hernández; C.E: Ron Streeter. Ethnic talk format (E, Sp, Creole).
282) 612 4th Ave, Suite 100, Louisville, KY 40202 — G.M: Wm. V. Wells; P.D: Chris Shebel; C.E: Jerry Snapp. CHR.
283) Box 308, Coeur d'Alene, ID 83814 — G.M: Joe Stanosch; P.D: S. Sibulski; C.E: David Forsman. ADC. (N-1).
284) 163 E. 100 N., Price, UT 84501 — G.M: J.M. Dart, Sr; C.E: J.M. Dart, Jr. ADC. (N. at :30, :55).
285) 7123 I-30, Suite 1, Little Rock, AR 72209 — G.M: Dianne McArthur; C.E: F. McDonald. Rlg.
286) 1776 Grant Str, Denver, CO 80203 — G.M: Ron Jamison; P.D: Rick Crandell; CE: John Vigil. Sports, talk.
287) 333 Dexter Ave No, Suite 400, Seattle, WA 98109 — G.M: Jack Swanson; P.D: Steve Wexler; C.E: Paul Vandegrift. News, talk.
288) Kingsport, TN 37662 — Silent.
289) Box 1090, Fortuna, CA 95540 — G.M: M.K. Allgood; C.E: M. Householter. ADC, oldies.
290) Box 236, Gonzales, LA 70707 — G.M: Anita Burnett; C.E: J. Babin. ADC, oldies, black.
291) 3800 Hooper Ave, Baltimore, MD 21211 — G.M: D.J. Barrett; P.D: J. Beauchamp; C.E: Andy Butler. News, talk.
292) 3106 Diablo Ave, Hayward, CA 94545 — Mgr: Robert Dunker; P.D: Craig Guglielmetti; C.E: Leonard Harris. Rlg. Ethnic. Sat 0030-0700 (I, Gr, Frasi, Sp). (N. at :55).
293) Cave Creek, AZ 85331.
294) Box 969, Grand Junction, CO 81502 — G.M: Jim Terlouw; P.D: Christopher Newby. News, talk.
295) Bakersfield, CA.
296) Rt. 9, Box 280E, St. Simons Island, GA 31522 — G.M: J. Wesley Cox; P.D: Deirdre R. Cox; C.E: Lester Forbes. Visitor info.
297) 1055 Franklin Ave, #306, Garden City, NY 11530 — G.M: K.R. Harris; P.D: D. Anthony; C.E: Bill Schleinitz. Big Bands, MoR.
298) 1468 W. 9th St, 8th fl., Cleveland, OH 44113 — G.M: Roger R. Turner; P.D: Dave Popovich; C.E: Dave Szucs. News, talk, sports.
299) Box 1410, Umatilla, OR 97882 — Pres: D. Marlow; G.M: J. Marlow; C.E: Steve Thompson. Rlg.
300) Box 34478, Alamo Heights, TX 78265 — Pres: Sam Morris. Rlg. 1130 (Sat/Sun 1200)-0600.
301) Hudson, TX.
302) Box 247, Dermott, AR 71638 — GM: P.Q. Gardner. PD: Beth Reilly; CE: Len Rockenbach. CWM (N-1).
303) 3580 Wilshire Blvd, Los Angeles, CA 90010 — G.M: James W. Freeman; P.D: M. Wagner; C.E: D. Ping. Oldies.
304) 311-112th Ave. N.E., St. Petersburg, FL 33716 — G.M: Zog Roseman; P.D: Norm Swenson; C.E: Ed Gorman. Rlg.
305) 5010 Underwood Ave, Omaha, NE 68132 — G.M: Bob Sweet; PD: Kent Pavelka; CE: Jerry McKenna. Talk (N-1).
306) Box 5037, Bend, OR 97708 — G.M: Mike Cheney; C.E. James Boyd. News, Talk.
307) 1 Julian Price Place, Charlotte, NC 28208 — G.M: John Kilgo; P.D: Andy Bickel; C.E: William R. White. MoR, talk (N-1).
308) Box 426, Bay Minette, AL 36507 — G.M: Norman Earls; P.D: Joe Davis; C.E: Lou Stokes. ADC. (N-1).
309) Box 99, Petoskey, MI 49770 — G.M: Cliff Carey; P.D: George Taylor; C.E: Del Reynolds. News.
310) 5520 Greenwich Rd, Suite 100, Virginia Beach, VA 23462 — G.M: Wm. Benns; P.D: Bob Streater. Gospel.
311) 815 "V" Street N.W., Washington, DC 20001; Gospel, UC, talk.
312) 1 Memorial Drive, St. Louis, MO 63102 — S.M: Tim Dorsey; P.D: Bob Osborne; C.E: Paul Grundhauser. News, sports, info, easy listening (nights) (N. at :55, :00).
313) Box 1120, Eugene, OR 97440 — G.M: Dave Woodward; P.D: Jerry Allen; C.E: T. Mulder. News, talk.
314) San Martin, CA.
315) Rt. 6, Box 735, Macon, GA 31201 — G.M: Jerry L. Russell; P.D: K. Burgamy; C.E: Tom McAfee. Business news.
316) Box 3880, Ogden, UT 84409 — G.M: Bob Alzugarat; P.D: Chris Staley; C.E: Tim Hunt. Cont. Christian (N-1).
317) 5050 Murphy Canyon Rd, San Diego, CA 92123 — G.M: Michael Shields; P.D: Kelly Wheeler; C.E: J. Rabell. All news.
318) Box 31130, Shreveport, LA 71130 — G.M: Frank Gunn; P.D: Steve McDonald; C.E: Peter Palagonia. C&W (N-3).
319) Box 188, Edna, TX 77957 — G.M: Johnny Francis; P.D: Joel Hernandez; C.E: Paul Easter. Spanish. Cont. Christian (E,Sp) (N-1).
320) 3500 E. Rosser, Bismarck, ND 58501 — G.M. & C.E: Alvin L. Anderson; P.D: Dean August. C&W (N-1).
321) 2930 E. Jefferson, Detroit, MI 48207 — G.M: P.C. Lamka; P.D: Dave Anthony; C.E: Ralph Hunt. C&W (N-1).
322) 611 Frontenac Place, St. Paul, MN 55104 — G.M: Mick Anselmo; C.E: D. Thompson, Sports.
323) 655 3rd Ave, New York, NY 10017 — G.M: Michael Bloomberg. Financial N.
324) Box 402, Milwaukee, WI 53201 — G.M: Chuck DuCoty; P.D: Gary Jensen; C.E: R.P. Johnson. News, talk, sports.
325) Box 2849, Gainesville, GA 30503 — G.M: A.O. Healan; P.D: Billy Pirkle; C.E: Sandy Griffin. C&W (N. at :20, :50).
326) 1436 Auburn Blvd, Sacramento, CA 95815 — G.M: John A. Winkel; P.D: Don Langford; C.E: Jay Lemmons. C&W (N-3).
327) 490 S. Farrell, #202, Palm Springs, CA 92262 — G.M: S.M. Kiner; P.D: C. West; C.E: S. Tim Cline. C&W (N-1).
328) Box 778, Fayetteville, TN 37334 — G.M: Wm. Malone; C.E: Don Roden. Oldies.
329) 2828 Coral Way, Miami, FL 33145 — G.M: Herb Levin; P.D: Claudia Puig; C.E: M. Triay. Spanish ("La Cubanísima"): News, talk.
330) 5601 Cassia, Boise, ID 83705-0278 — G.M: Ken Koch; P.D: Ty McLean; C.E: Cary Harrison. Adult Standards. (N-1).
331) 2600 So. Spring Ave, Sioux Falls, SD 57105 — Mgr: Thomas J. Ingstad; P.D: Cary Wencil; C.E: Paul Raap. MoR (N-3).
332) Box 1516, Richmond, VA 23212 — G.M: L.G. Jones; P.D: Tim Farley; C.E: J. Francioni. ADC (N-1).
333) 3510 W. Hacienda, Las Vegas, NV 89118 — Mgr: K. Gerst; C.E: Jay Pierce. Top 40, R&B.
334) Box 352, Greybull, WY 82426 — G.M: Jeff Keith. C.E: W.R. Burckhard. CWM (N-1).
335) Concord, NH.
338) East Point, GA 30344.
339) 180 N. Michigan Ave, #1200, Chicago, IL 60601 — G.M: Harvey Pearlman; P.D: Gary Price; C.E: John Hurni. MoR. (N-1).
340) 116 North Trade St., Tryon, NC 28782 — MoR (rel. WHKP-1450). Ops. Mgr: David Bivens.
341) Box 1160, Salt Lake City, UT 84110-1160 — G.M: Russell Wood; P.D: Rod Arguette; C.E: John Dewhurst. News, talk. (N. at :57).
342) Box 1020, Indiana, PA 15701 — Pres: M. E. Harley; P.D: David Gross; C.E: Fred Shetler. ADC.
343) 1617 Lebanon Rd, Nashville, TN 37210 — Mgr: Wm. O. Barry; P.D: Ken Bramming; C.E: G. Brown. Big Bands/Standards (N-1).
344) 501 W. Quincy, San Antonio, TX 78212 — G.M: Gilberto Esquivel; P.D: Saul Mardy Ramírez; C.E: Tom Bray. Spanish "La Fiera".
345) Box 159, Skowhegan, ME 04976 — G.M: A.W. Anderson; C.E: Howie Souze. MoR, Big bands.
346) Mobile, AL 36689 — Urban cont.
348) Box 1669, Montgomery, AL 36102 — G.M: Mark Robbins; P.D. & C.E: Clay Redden. News/talk.
349) 9416 Mission George Rd, Santec, CA 92071 — G.M: Jeff Apregan; PD: Sonny West; C.E: Bill Lipis. Oldies (rel. FM).
350) Box 909, Claremont, NC 28610 — G.M: Donald W. Lee; P.D: Robert Barnette; C.E: Winston Hawkins. Rlg.
351) 4590 Ea. 29th, Tulsa, OK 74114 — G.M: Mike DeMarco; P.D: Mike Adams; C.E: Larry White. C&W (N-3).
352) 1015 Main Str, Wheeling, WV 26003 — G.M: Larry Anderson; P.D: Dave Labrozzi; C.E: J. Lane. C&W (N-3).
353) 2905 S. King Rd, San Jose, CA 95122 — G.M: Carlos Duharte; P.D: Salvador Homero Campos; C.E: Jim Phillips. Spanish (N-1).
354) 145 Branhar View Rd., Lexington, SC 29072 — Oldies.

355) Box 1170, Bellingham, WA 98227 — G.M: M. Pollock; P.D: Mark Edwards; C.E: James S. Kaplan. Oldies (N-3).
356) 207 Midtown Plaza, Box 40400, Rochester, NY 14604 — G.M: J. Murphy; P.D: J. Howlett; C.E: Craig Kincaid. ADC (N-1).
357) For details refer to **R. Martí** (under "Voice of America").
358) Box 608, Kalispell, MT 59903 — G.M: Jerry Adams; P.D: Jack Stevens; C.E: Ron Hirsch. ADC/News. (N-3).
359) 721 Market St, Suite 105, Wilmington, NC — G.M: David H. Moran; P.D: Cheryl Sparks; C.E: Paul Knight. Rlg.
360) Williams, AZ 86046.
361) Box 82225, Bakersfield, CA 93380 — S.M: Mike McCutchan; C.E: T. Geiser. Cont. Christian. (N-1).
362) 2708 Old Brandon Rd, Jackson, MS 39209 — G.M: Jim McAlilly; News Dir: John Hoffman; C.E: Randy Grammar. News, talk.
363) Box 2312, Knoxville, TN 37902 — Pres: Harry J. Morgan; P.D: Tom Burchfield; C.E: Scott Wyrick. Country, Rlg.
364) Placerville, CA 95667.
365) 525 N. Sam Houston Parkway E, Suite 525, Houston, TX 77060 — G.M: Ron Smith; P.D: Bert Salas; C.E: David Anthony. Christian; Sp. Sun 2130-0530.
366) 1001 Farnam Ave, Omaha, NE 68102 — News, talk.
367) 1190 E. Ball Road, Anaheim, CA 92805 — G.M: Milton Sexton; P.D: Chris Cox; C.E: Gary Belzman. Easy listening, NOS, Korean (M-Sat. 1500-0100) (N:23, :54).
368) Box 52128, Atlanta, GA 30355 — G.M: Eathel Holley. Classical. ETH Sunday.
369) 4949 So. We. Macadam Ave, Portland, OR 97201 — G.M: Dave Milner; P.D: Mike Dirkx; C.E: Byron Swanson. ADC, talk (N-3).
370) 3500 Maple Ave., LB-46, #1470, Dallas, TX 75219. C.E: Joshua A. Socolof.
371) Box 631, Annapolis, MD 21404 — G.M: M.H. Blum; P.D: J.S. Blum; C.E: M.W. Pittman. Black, gospel (N: at :55).
372) 6320 Zuni S.E., Albuquerque, NM 87108 — G.M: Kelly Cunningham; P.D: Hector Velez; C.E: Bob Henry. Spanish "La Super X".
373) 801-2nd Ave, New York, NY 10017 — G.M: David Lampel; P.D: Claude Tait; C.E: David Antoine. Black. (Operates to 1h after sunset).
374) 203 W. Wayne, Ft. Wayne, IN 46802 — G.M: Jane E. Bartsch; P.D: Chuck Bailey; C.E: E.B. Culp (N-1). Oldies.
375) Box 509, Bluefield, VA 24605 — Pres: G. Frank Barnes Jr.; P.D: Dean Gatherum; C.E: Jay Belt. CWM (N-1).
376) Box 1259, Ozark, AL 36360 — G.M: Tom O'Brien; P.D: Tracy Hogan; C.E: Neil Riddle. News, talk.
377) Soquel, CA 95073.
378) Eureka, CA 95501. License to Cottonwood, CA.
379) Box 6008, West Fargo, ND 58108 — S.M: Gary Herr; C.E: R. Benjamin. Rlg. 1200-0500.
380) 6222 No. We. Interstate 10, San Antonio, TX 78201 — G.M: William G. Hill; P.D: Pat Rodgers; C.E: Dan Walthers. News, sports.
381) 100 Mt. Wayte Ave, Framingham, MA 01701 — G.M: Jim Cande; P.D: Scott Gibbens; C.E: B. Harden. News, talk, sports.
382) Box 1200, Lebanon, TN 37088 — G.M: William O. Barry; P.D: Kenneth Bramming; C.E: Gary M. Brown. C&W (N-1).
383) Box 2637, Bangor, ME 04401 — G.M: John J. Pineou; P.D: Bobby Russell; C.E: Howie Soles. News.
384) 1227 Del Prado Blvd, Cape Coral, FL 33990 — G.M: John E. Harper; P.D: Glen Yarborough; C.E: Paul Wolf. News, talk.
385) Box 900, Graham, NC 27253 — G.M: Ted J. Gray Jr.; P.D: Olin Campbell; C.E: Jim Davis. Rlg.
386) Box 2307, Newburgh, NY 12550 — G.M: Robert A. DeFelice; P.D: Terry Donovan; C.E: Dave Messing. News. (Still on 1220kHz).
387) 1906 Wilmington Rd, New Castle, PA 16105 — G.M: Robert McCracken; P.D: Carol King; C.E: Bill King. Oldies.
388) 509 We. Roosevelt Rd, Chicago, IL 60607 — G.M: José Antelo. P.D: Guillermo Prince. C.E: John McGuiness. Spanish. (N. at :30, :55).
389) Mt. Carmel, TN 37642.
390) 2994 E. Grand Blvd, Detroit, MI 48202 — Talk.
391) 711 Wage Dr., Leesburg, VA 22075.
392) Box 1511, Tucson, AZ 85701-1511 — G.M: Raul B. Games; P.D: Berta Gallego; C.E: Peter Trowbridge. Spanish (N-1) ("R. Qué Tal").
393) Box 1210, Dahlonega, GA 30533 — Pres: Jay Andrews; S.M: Gigi Andrews; C.E: Dan Davis. ADC (N-3).
394) City Line and Monument Aves, Philadelphia, PA 19131 — G.M: Steve Carver; P.D: Gregory Tantum; C.E: Ken Tankel. Oldies.
395) 306 W. Genesee, Frankenmuth, MI 48734 — G.M: Dana MacVay; P.D: Bob Dyer; C.E: Larry Hilton. Big Bands (N-1).
396) Box 130, Guymon, OK 73942 — G.M: E. Smith; P.D: Bill Weldon; C.E: Jim Stanford. MoR, farm (N-1). 24h.
397) Box 2588, St. George, UT 84770 — G.M: C. Lamar; P.D: Boyd Kelley; C.E: Steve Kimber. CWM (N-1).
398) 1730 Minor Ave, 20th fl, Seattle, WA 98101 — G.M: Bruce Raven-Stark; P.D: David Allen; C.E: Clay Freinwald. ADC.
399) Box 149, Sunnyside, WA 98944 — Mgr: Don Bennett; P.D: Larry Southern; C.E: Carl McDonald. News, ADC, Oldies (Mexican Mon-Fri 0230-0600, Sun 2100-2330). 1400 (Sun 1530)-0606. (N-1).
400) 409 S. 4th St, Laramie, WY 82070 — Oldies.
401) P.O. Box 68, Denham Springs, LA 70727 — G.M: Rip Miller; P.D: Danny Church; C.E: Richard Petty. Rlg (N-1). 1100-0600 News.
402) 1411 Coral Way, Miami, FL 33145 — S.M: Víctor Alemán; P.D: Gilda Miros; C.E: R. Chambers. Spanish.
403) Box 548, Arlington, TN 38002 — G.M: Fred Flinn; P.D: Allen Bray; C.E: Rob Herrin. Cont. Christian Music (N. at :55). 1130-0500.
404) 1635 S. Rancho Sta. Fe Rd, Suite 201, San Marcos, CA 92069 — G.M: Ron Walters; P.D: Steve Sherman; C.E: Doug Schleutker. Rlg. (Sp. Sun 0800-1500, Sat. 0800-1400).
405) Box 252, McAllen, TX 78502. Rlg. in E/Sp.
407) Box 1050, Reidsville, NC 27323 — GM: Tony Mullins. PD: Moon Mullins. CE: Tim Walker. C&W.
408) 9446 Broadview Rd, Broadview Heights, OH 44146 — G.M: John Blassingame; C.E: M. Krieger. News, talk.
409) 1420 Koll Circle, San Jose, CA 95112 — GM: John Sutherland. PD: Dana Jang. CE: David Williams. Asian/Rel.
410) 3400 Idaho Ave N.W., Washington, DC 20016 — G.M: M.L. Douglass; P.D: H. Cooke; C.E: D. Garner. News, talk.
411) 29200 Vassar Dr, #650, Livonia, MI 48152 — G.M: Charlotte Kennedy; P.D: Brian Patrick; C.E: Paul Goodpastor. Rlg.
412) 2792 Maplewood Drive, St. Paul, MN 55109 — G.M: Virginia H. Morris. PD: Allan Searle. CE: Norm Paetznick. News, talk (N-3).
413) Box 16056, Winston Salem, NC 27115 — G.M: Sarah B. Bailey; C.E: Ray Testerman. Rlg, black.
414) 3636 No. First Str, #142, Fresno, CA 93726 — G.M. & P.D: Jim Patterson. Op. Mgr: Gary Tomkins. Christian.
415) 4800 N. Central Ave, Phoenix, AZ 85012 — G.M: Ronald Cohen; C.E: David Dilkson. News, Business.
416) 2559 Welton St, Denver, CO 80205 — G.M: Keith James; P.D: Danny Harris; C.E: Daren McMullin. Urban cont, CHR.
417) Box 1031, New London, CT 06320 — G.M: Richard R. Wolf; C.E: Stephen Keefe. News.
418) 10 Music Circle East, Nashville, TN 37203 — G.M: Elizabeth Yoder; P.D: Alan Eisenson; C.E: Wm. Watt Hairston. News, talk, rlg.
419) Box 8348, Spokane, WA 99203 — G.M: S. Cody; P.D: Art Sanders; C.E: Bill Gott. CWM (N-2).
420) 1021 Whitehall Str, Waukesha, WI 53186 — G.M: Mike S. Saxton; P.D: Joe Till; C.E: Don Hunjadi. CWM (N-1).
421) 8729 E. 9th St, Rancho Cucamonga, CA 91730 — G.M: Humberto Hernández; P.D: H. Hernández; C.E: J. Artal. Spanish "R. Quince".
422) Boston, MA. Addr: 4891 Pacific Highway, San Diego, CA 92110 — G.M: Jennifer McCarn; P.D: Steve Gallagher; C.E: John Kennedy. Spanish.
423) 10841 Ea. 28th Str, Independence, MO 64052 — Chief Operator: Ed Treese. Urban Cont.
424) 6925 Union Park Center, #300 Midvale, UT 84047.
425) Box 1520, Oxnard, CA 93034 — G.M: Harold A. Frank; P.D: Rooster Roads; C.E: Martin Walker. Spanish "R. Tiro".
426) 695 Delaware Ave, Buffalo, NY 14209 — G.M: James H. Meltzer; P.D: Tom Donahue; C.E: Tom Atkins. Business, talk.
427) Box 6000, Oklahoma City, OK 73153 — G.M: Vance Harrison Jr.; P.D: Kent Jones; C.E: Dennis Pharr. Oldies (N-1).
428) Box 910, Greenfield, MA 01302 — G.M: Cheri McSpaden; C.E: Jim Hemingway. Oldies (N-1).
429) 4614 S.W. Kelly Ave, Portland, OR 97201 — G.M: Dan Hern; P.D: Jon Robbins; C.E: L. Holtz. Talk, sports. (N-1).
430) Box J, Lafayette, LA 70502 — G.M: Charles Wood; P.D: Jimmie Cole. Oldies.
431) 1220 4th Ave. S.W, Rochester, MN 55902 — G.M: H.G. Bill; P.D: Jim McCann; C.E: Jim Casey. Oldies.

432) Box 839, Muskegon, MI 49443 — Nostalgia, BB.
433) Camdenton, MO. c/o RR 3, Box 4881, Lake Rd. 54-24, Osage Beach, MO 65065.
434) Sequim, WA 98382.
435) 1440 Ethan Way, Suite 200, Sacramento, CA 95825 — G.M: Rick Eytcheson; P.D: Lucy Thomas; C.E: Dale Harry. News.
436) Box 909, Moreno Valley, CA 92387 — G.M: Doris L. Van Voorhis; C.E: John Patterson. Spanish.
437) 3013 Guess Rd, Durham, NC 27705 — G.M: L.L. Leathers; P.D: Mark Parker; C.E: Jim Davis. Rlg.
438) 219 McFarland Str, Cincinnati, OH 45202 — G.M: Thomas J. Severino; P.D: Bruce Still; C.E: Ted Kendrick. News, talk.
439) Box 6, Georgetown, TX 78626 — G.M: Paul Hrbacek; C.E: Cowboy Otis. Rlg.
440) 1519 W. Harrison, Harlingen, TX 78550 — G.M: T. Solis; P.D: R. Botello; C.E: N. Lindsay. Spanish ("R. Panamericana") (N-1).
441) Box 216, Dalton, GA 30722 — G.M: Devona Poe; C.E: Ed Allen. Rlg.
442) Box 352, Poplarville, MS 39470 — G.M: Wanda Carpenter; C.E: Dick Merrill, C&W.
443) 5900 Picketville Rd, Jacksonville, FL 32205 — G.M: Kim Gatlin; P.D: Fred Rhodes; C.E: Peter V. Gurekis. Rlg.
444) Drawer 711, Harlingen, TX 78551 — G.M: Mary Gaines; P.D: Rogelio Botello Rios; C.E: Nathan Lindsey. Spanish 2000-0030.
445) 8 Harris Court, Suite 5B, Monterrey, CA 93940 — G.M: Stephen Adams; P.D: Ross Davis; C.E: Jim Phillips. Oldies.
446) 5700 Sunset Blvd, Los Angeles, CA 90028 — G.M: José Grimalt; P.D: Cynthia Irene; C.E: Art Pemberton. Spanish. (N-1).
447) Box 1540, Waterloo, IA 50705 — Pres: Cy N. Bahakel; P.D: Ross MacLane; C.E: Leonard Tompkins. C&W, farm, rlg (Spanish). (N. at :55 to :05).
448) Box 12279, Albany, NY 12212 — G.M: J. Kelly; P.D: Mike Morgan; C.E: Paul H. Thurst. News (N-1).
449) 4500 North Tryon Str, Charlotte, NC 28213 — G.M: Wallace Keitt; C.E: David Anthony. Cont. Christian.
450) 1 Bala Plaza, Bala Cynwyd, PA 19004 — G.M: Jeffrey Specter; P.D: Gerry DeFrancesco; C.E: Michael Guidotti. Oldies.
451) 1445 W Baseline Rd, Phoenix, AZ 85041 — G.M: Donna R. Phelps; C.E: Jim Clark. Rlg.
452) Box 25147, Richmond, VA 23260 — G.M: Jon C. King; P.D: Sheri Lynn Stuart; C.E: Jeff Loughridge. Financial. Indian, Korean, Jap. on Sat. 1300-1700.
453) Box 837, Pickens, SC 29671 — G.M: J.C. Clark; S.M: L.K. Wilson; P.D: J. Wint. Rlg (N-1).
454) Box 2466, Huntsville, AL 35804 — G.M: Terry Arnold; P.D: Gary Kaye; C.E: John Hain. Rlg.
455) 335 Powell Str, San Francisco, CA 94102 — G.M: Bruce E. Beebe; P.D: Victor Ledin; C.E: Lou Sinclair. Classical (N. at :55).
456) 5065 W. 2100 S, Salt Lake City, UT 84120 — G.M: Gene Guthrie; C.E: Fritz Ashauer. Spanish.
457) 330 S.W. 27th Ave, Miami, FL 33135 — G.M: George Rodriguez; P.D: Lazaro Asencio. News, talk (Spanish only).
458) 2700 S. Platte River Dr, Englewood, CO 80110 — G.M: G. Spicer; S.M: J. Hull. Gospel, rlg (N. at :55).
459) Box 78277, Shreveport, LA 71137 — G.M: Billy Wilson. P.D: John Brewer. Rlg.
460) Box 1301, Fargo, ND 58107 — G.M: Jim Lakoduk; P.D: Mark Nichols; C.E: Mark Borchart. Oldies (N-1).
461) Box 100, Toms River, NJ 08754 — G.M: Joe Knox; P.D: Lance DeBock; C.E: Mike Morgan. News, talk.
462) Box 629, Bennettsville, SC 29512 — G.M: Ken Harmon. Oldies. W 1100-0500, Sun 1230-2315.
463) Rt 2, 4 Tower Rd, Bolton, MS 39041.
464) 5203 N. Armenia Rd, Tampa, FL 33603 — S.M: R. Cubero; P.D: Silvia Abreu; C.E: Arthur Karmgard. Spanish.
465) 2460 Atlanta Rd. S.E., Smyrna, GA 30080 — G.M: John L. Stirzaker; P.D: Monty Martin; C.E: Skip Marshall. Cont. Christian (N-1).
466) Box 1550, Pittston, PA 18640 — G.M: James Ward; P.D: Rob Neyhard; C.E: Ron Schact. News, talk.
467) Univ. of Arizona, Tucson, AZ 85721 — P.D: E. Kupperstein; C.E: Tom Boone. News, jazz. (Sp. Sun 2000-0230).
468) 143 Rumford Ave, Newton, MA 02166 — G.M: Rob Rudnick; C.E: Leo Sullivan. Talk (Sp. Mon-Fri 1400-2300).
469) 2043 10th Str. N.E., Roanoke, VA 24012 — Pres: D.H. Moran; P.D: Sharon M. Moran; C.E: Ben Girardan. Rlg.
470) Box 4638, Vancouver, WA 98662.
471) Box 308, Ferndale, WA 98248 — G.M: Richard Ellison; P.D: Larry Lee; C.E: Ron Cowell. Rlg. 1400-0800 (N-1).
472) W. 933 3rd Ave, Spokane, WA 99204 — G.M: Dick Wright; P.D: Jerry Anderson; C.E: Ron Mahan. Classical.
473) Box 12040, Nashville, TN 37212 — G.M: F.B. Lindsey; P.D: Jack Hunter; C.E: C. Hooper. Rlg (N-1).
474) 3651 Pegasus Dr, #107, Bakersfield, CA 93308 — G.M: Randy Warwick. News.
476) 122 5th Ave, New York, NY 10011 — G.M: W.G. Bodow; P.D: Tom Bartunek; C.E: Herb Squire. Classical music (N-1).
477) Box 450, Paducah, KY 42001 — C.E: Kirk Harnack. Standards. 1200-0700 (N-1).
478) Box 1770, Aberdeen, SD 57402 — G.M: Deb O'Donnell; P.D. & C.E: Neil Lahammer. CWM, farm (N-3). 1100-0700.
479) Box 28, Lancaster, SC 29720 — G.M: B.L. Phillips, Jr. CWM (N-1).
480) Box 445, Warsaw, NC 28398 — G.M: R.G. Blanchard; P.D: Connie Thigpen; C.E: J. Blanchard.
481) Box 2625, Joplin, MO 64803 — G.M: W.B. Neal; P.D: Steve Scherio; C.E: Jack Lutzinger. Talk.
482) 2049 Century Pk. E., #4390, Los Angeles, CA 90067 — C.E: Ron Russ. Korean.
483) Box 3174, Tempe, AZ 85280 — G.M: M.C. Owens; P.D: Jim Murphy; C.E: Robert A. Van Buhler. C&W.
484) Box 5668, Colorado Springs, CO 80931 — G.M. & P.D: Rick Martin. Country gospel.
485) Box 12890, Lexington, KY 40512 — Oldies.
486) 6301 Ivy Lane, #600, Greenbelt, MD 20770 — G.M: Benjamin Hill; P.D: Steve Chaconas; C.E: T. McGinley. News, talk, business.
487) 3000 SW 60th Ave, Fort Lauderdale, FL 33314.
488) 1651 University Blvd. NE, Albuquerque, NM 87102.
489) 1309 Lejeune Blvd, Jacksonville, NC 28540 — G.M: Randy Field; C.E: Paul Knight. Rlg.

VOICE OF AMERICA (Gov.)

The Voice of America is the worldwide radio broadcasting service of the United States Information Agency, Bureau of Broadcasting.

ADDR: Voice of America, Washington, DC 20547. **Te:** +1 (202) 619 2538. **Fax:** +1 (202) 619 1241. **Tlx:** 7607706 VOA USA. **MCI Mail:** 408-6039. **Frequency Division: Fax:** +1(202) 619 1680. **BBS:** +1 (202) 619 1494.

L.P: Bureau of Broadcasting: Dir: Chase Untermeyer. Dir. of Eng: Walter La Fleur. Dir. of Ext. Affairs: Joseph O'Connell. **Voice of America:** Dir: Chase Untermeyer. Dir. of Prgrs: Sidney Davis. Dir. of B'cast Op's: Edward DeFontaine. Chief, Freq Management & Monitoring Division: William Whitacre.

Transmitters in the USA:

Greenville, No. Carolina (G.C: 77.25W/35.35N): 11 × 500kW, 6 × 250kW and 6 × 50kW tr's.

kHz	kHz	kHz	kHz	kHz	kHz	kHz
5995	9525	11580	11930	15205	15410	17740
6035	9555	11680	11935	15245	15580	17785
6040	9585	11695	13775	15265	17640	17800
6130	9590	11835	15115	15330	17710	21485
6190	9670	11890	15120	15365	17715	21580
7405	9775	11895	15160	15400	17730	21610
9455	9840	11915	15185			

Bethany, Ohio (G.C: 84.21W/39.21N): 6 × 250kW tr's.

kHz	kHz	kHz	kHz	kHz	kHz	kHz
6030	9530	9575	11830	11925	15315	17800
7405	9565	9590	11890	13740	17730	21745
9525						

Delano, California (G.C: 119.17W/35.45N): 4 × 250kW tr's.

kHz	kHz	kHz	kHz	kHz
5985	9465	9590	11945	13740
6130	9515	9690	11965	15120

Dixon, California (G.C: 121.45W/38.23N): 3 × 250kW tr's. This st. currently used as a standby facility only.

Overseas Relay Stations: Refer to listings under Antigua, Ascension Island, Belize, Botswana, Germany, Greece, Kuwait, Morocco, Philippines, Portugal, Sao Tome, Sri Lanka, Thailand, United Kingdom.

SCHEDULE NOV. 1992 – MARCH 1993

Times are UTC, but days of the week refer to local time in the listening area.

Times and frequencies subject to change.

Albanian	Area	kHz
1700-1800	Eu	5965, 9735, 11865, 15195

Amharic

1800-1900	ETH	9650, 11740, 11965, 15165, 15445

Arabic

0330-0430	ME/NAf	1260, 5965, 7130, 9545, 11905, 15160, 17810
0430-0530	ME/NAf	1260, 5965, 7160, 9545, 9670, 11805, 11905, 15160, 17810
0530-0600	ME/NAf	1260, 5965, 9670, 11905, 15160, 17810
0600-0730	ME/NAf	1260, 5965, 7225, 9670, 15160, 15195, 17810
0730-0800	ME/NAf	1260, 5965, 6150, 7225, 9580, 9670, 9705, 11775, 11840, 15160, 15195, 15235, 17810
1330-1400	ME/NAf	1260, 11905, 15400, 21455, 21720
1400-1500	ME/NAf	1260, 11905, 15445, 17885, 21455, 21720
1500-1600	ME/NAf	15305, 15445, 17885, 21570
1600-1800	ME/NAf	1260, 7205, 11960, 15155, 15305, 17885, 21570
1800-1900	ME/NAf	1260, 7205, 9530, 11960, 15155, 15245, 15305, 17885, 21570
1900-2000	ME/NAf	1260, 7205, 9530, 9650, 11960, 15155, 15245, 17740, 17885, 21570
2000-2100	ME/NAf	11895, 15225, 15245, 17740, 9650, 7195
2100-2200	ME/NAf	6160, 7140, 7195, 9530, 9745, 11895, 15225, 15245, 17740

Armenian

1500-1600	CIS	9735, 11805, 15195, 15265, 17805

Azerbaijani

1900-2000	CIS	6160, 7280, 9680, 11835

Bangla

0130-0230	SAs	15210, 17780, 21475
1600-1700	SAs	1575, 9575, 11870, 15185

Bulgarian

0400-0430	Eu	1197, 3980, 6130, 7295, 9530, 9615
1930-2000	Eu	5965, 9735, 11720, 11780
2000-2030	Eu	5965, 9735, 11720, 11780

Burmese

1130-1200	EAs	1575, 9505, 11835, 15145
1200-1230	EAs	9505, 11835, 15145
2330-2400	EAs	7260, 11840, 15215

Cantonese

1500-1600	EAs	1143, 6030

Chinese (Mandarin)

1100-1200	EAs	1143, 5955, 11965, 12040, 15410, 17790, 21585
1200-1300	EAs	5955, 11965, 12040, 15410, 17790, 21585
1300-1330	EAs	5955, 11730, 11965, 12040, 15410, 17790
1330-1400	EAs	1143, 5955, 11730, 11965, 12040, 15410, 17790
1400-1430	EAs	5955, 9590, 11740, 11965, 12040, 15195, 15445, 17790
1430-1500	EAs	5955, 11740, 11965, 12040, 15410, 17790
1500-1600	EAs	11740, 11820, 11965, 12040, 15410, 17790
1600-1700	EAs	1143, 11965, 12040, 15410, 17790
2100-2200	EAs	1575, 7200, 9545, 11925, 15395, 17765
2200-2230	EAs	7200, 9545, 9675, 11820, 11925, 15395, 17765
2230-0100	EAs	7200, 9545, 11925, 15395, 17765

Creole

1230-1300MF	Carib	9525, 11935, 13740
1730-1800	Carib	13740, 17785, 17800
2200-2230	Carib	9670, 11890, 15120

Croatian

2100-2130	Eu	792, 5965, 9735, 11720, 11780
2130-2145	Eu	5965, 9735, 11720, 11780

Czech/Slovak

0430-0500	Eu	1197, 3980
0500-0530	Eu	1197, 3980, 6125, 7295, 9615
1700-1900	Eu	1197, 3980, 6060, 7180, 9585, 11760, 15225

Dari

0215-0300	ME	6015, 9635, 9705, 11825
1515-1600	ME	7295, 11845, 15225, 21520

English

to Ea. Asia/So. Ea. Asia/Pacific

1000-1100	5985, 11720, 15425
1100-1200	5985, 6110, 9760, 11720, 15155, 15425
1200-1230	1143, 6110, 9760, 11715, 15155, 15425
1230-1330	6110, 9760, 11715, 15155, 15425
1330-1400	6110, 9760, 15155, 15425
1400-1500	1143, 6110, 9645, 9760, 15160, 15425
1900-2000	9525, 11870, 15180
2100-2200	11870, 15185, 17735
2200-2400	7120, 9770, 11760, 15185, 15290, 15305, 17735, 17820
0000-0030	1575, 7120, 9770, 11760, 15185, 15290, 17735, 17820
0030-0100	7120, 9770, 11760, 15185, 15290, 17735, 17820

to So. Asia

0100-0300	7115, 7205, 9740, 11705, 15250, 17740, 21550
1400-1500	6110, 7125, 9645, 9760, 15395
1500-1530	6110, 7125, 9645, 9700, 9760, 15395
1530-1600	1575, 6110, 7125, 9645, 9700, 9760, 15395
1600-1700	6110, 7125, 9645, 9700, 9760, 15395
1700-1800	6110, 7125, 9645, 9700, 15395

to Middle East/No. Africa/Europe

0000-0300	792, 1260
0400-0430	792, 5995, 6040, 6140, 7170
0430-0500	5995, 6040, 6140, 7170
0500-0530	792, 5995, 6040, 6140, 7170, 9530, 9700, 11825, 15205
0530-0600	792, 3980, 5995, 6040, 6060, 6140, 7170, 9530, 9700, 11825, 15205
0600-0630	792, 3980, 5995, 6005, 6040, 6060, 7170, 7325, 11805, 11825, 15205
0630-0700	792, 1197, 3980, 5995, 6005, 6040, 6060, 7170, 7325, 11805, 11825, 15205
0800-1000	1260, 11735, 15160, 15195, 17770, 21455
1000-1100	11735, 15160, 15195, 17770, 21455
1400-1500	15205
1500-1600	1260, 9700, 15205
1600-1630	9700, 15205
1630-1700	6180, 9700, 9760, 11855, 15205, 15245
1700-1800	792, 6040, 9700, 9760, 15205
1800-1900	6040, 9700, 9760, 15205
1900-2000	1197, 3980, 6040, 9700, 9760, 15205
2000-2100	3980, 6040, 9700, 9760, 15205
2100-2200	1260, 6040, 9700, 9760, 11960, 15205
2200-2230	6160, 7140, 7195, 9530, 11895, 11960
2230-2300	792, 1260, 6160, 7140, 7195, 9530, 11895, 11960
2300-2330	1260, 6160, 7140, 7195, 9530, 11895, 11960
2330-2400	792, 1260, 6160, 7140, 7195, 9530, 11895, 11960

to Africa

0300-0400	621, 6035, 6090, 7265, 7280, 7405, 9575, 9885, 11835, 11895, 15115
0400-0430	621, 6035, 6090, 6155, 7265, 7280, 7405, 9575, 9885, 11835, 11850, 15115
0430-0500	6035, 7265, 7280, 7405, 9575, 11850, 15115

0500-0600 6035, 7405, 9575, 9885, 11850, 15115, 15600
0600-0700 6035, 7405, 9530, 9575, 9885, 11850, 11925, 15115, 15600
1600-1700 11920, 11995, 13710, 15255, 15320, 15445
1700-1730 11920, 11995, 13710, 15320, 15445
1730-1800 621SS, 11920, 11995, 13710, 15320, 15445
1800-1900 621, 11920, 11995, 13710, 15320, 15410, 15580, 17800
1900-2000 621, 11920, 11995, 13710, 15320, 15410, 15495, 15580, 17800
2000-2030 621, 11820, 13710, 15160, 15320, 15410, 15495, 15580, 17800, 21485
2030-2100 621, 13710, 15320, 15410, 15495, 15580, 17800, 21485
2100-2200 621, 13710, 15290, 15410, 15495, 15580, 17800, 21485

to Caribbean and the American Republics
c = medium wave to northern Central America.
0000-0100 930, 6130, 9455, 11695 (Caribbean program)
0000-0030 5995, 7405, 9775, 11580, 15120, 15205
0030-0100 1530c, 1580c, 5995, 7405, 9775, 11580, 15120, 15205
0100-0200 5995, 6130, 7405, 9455, 9775, 11580, 15120, 15205
0200-0230MF 930, 5995, 7405, 9775, 11580, 15120, 15205
0400-0500MF 1530, 1580 (no. Ce. America)
1000-1100 930, 9590, 11915, 15120 (to the Caribbean)
1100-1200 9590, 11915, 15120 (to the Caribbean)
1200-1230 930 (to the Caribbean)
1700-1730 930 (to the Caribbean)
2100-2200 930 (to the Caribbean)

Estonian
0445-0500 Eu 6015, 6150, 7130, 7210
1930-2000 Eu 6095, 6150, 7180, 11750

Farsi
0300-0430 ME 6060, 9635, 9700, 11805
1700-1900 ME 6160, 7280, 9680, 11835
2300-2330 ME 5970, 7170, 7285, 9680

French
0530-0700 Af 9555MF, 9585MF, 11835MF, 11875, 11890MF, 15375, 17650
1830-2000 Af 12080, 15365, 15625, 17640, 17785, 21485
2000-2030 Af 12080, 15365, 15625, 17640, 17785
2030-2100SS Af 12080, 15365, 15625, 17640, 17785, 17880
2100-2130 Af 12080, 15365, 15625, 17640, 17785
2130-2200 Af 11820, 12080, 15365, 15625, 17640, 17785

Georgian
2000-2100 ME 6140, 7280, 9535

German
2200-2230 Eu 1197, 11845
Also on Radioropa via the Astra and Kopernikus satellites.

Greek
1900-1930 Eu 5965, 5975, 9735, 11760

Hausa
0500-0530 Af 9555, 9775, 11835, 11890, 15375, 17650
2030-2100MF Af 12080, 15365, 15625, 17640, 17785, 17880

Hindi
0000-0100 SEAs 6010, 7105, 7205, 11825
1600-1700 SEAs 6160, 7280, 9675, 11945

Hungarian
0530-0600 Eu 6125, 7295, 9615

Indonesian
1130-1230 EAs 9620, 11930, 15125
2200-2330 EAs 9620, 11805, 15205

Khmer
1330-1500 EAs 1575, 9575, 11930, 15285
2200-2230 EAs 1575, 7260, 9510, 11720

Korean
1300-1330 EAs 11930, 15195, 17720
2130-2200 EAs 6130, 7110, 9755, 11750

Kurdish
0230-0330 ME 1260, 5965, 5995, 7270, 9760

Lao
1230-1300 EAs 1575, 9645, 11930, 15230
1300-1330 EAs 1575

Latvian
2000-2030 Eu 6095, 6150, 7180

Lithuanian
0500-0515 Eu 6015, 6150, 7130, 7210
2030-2100 Eu 6095, 6150, 7180

Pashto
0130-0215 ME 6015, 9635, 9705, 11825
1430-1515 ME 7295, 11845, 15225, 21520

Polish
0530-0630 Eu 1197, 5955, 6160, 7130
2100-2200 Eu 1197, 3980, 6095, 6150, 9650
2200-2230 Eu 3980, 6095, 6150, 7210, 9650
2230-2400 Eu 1197, 3980, 6095, 6150, 7210, 9650

Portuguese
0430-0500 Af 9555, 9775, 11835, 11890, 15375, 17650
1730-1800 Af 621MF, 15150, 15625, 17705, 17740, 21485
1800-1830 Af 15150, 15625, 17705, 17740, 21485
2300-2400 B 9455, 11680, 15330, 17740

Romanian
0430-0445 Eu 792, 6130, 7295, 9530, 9615
1800-1900 Eu 792, 5965, 9735, 11780, 15125
1900-1930 Eu 792, 11780, 15125
1930-2000 Eu 792

Russian (E = to eastern Russia)
0400-0500 RUS 5985, 6160, 7270, 9760, 9770
0900-1100E RUS 11930, 11965, 15150, 15410, 17865
1600-1800 RUS 7245, 9530, 9660, 11710, 11740, 11805, 15235, 17865
1800-1900 RUS 6140, 7170, 7270, 9690, 11710, 11805, 11825, 15235
1900-2000 RUS 5985, 6060, 6140, 7170, 7270, 9575, 9690, 11710, 11825
2000-2100 RUS 6060, 7170, 7270, 9575, 9585, 9690, 11710, 11825
2100-2200 RUS 6060, 6085, 6140, 7170, 7270, 7280, 9575, 11710, 11825
2200-2300 RUS 6060, 6085, 6140, 7170, 7270, 7280, 11710, 11825

Serbian
0445-0500 Eu 792, 6130, 7295, 9530, 9615
0630-0700 Eu 6125, 6160, 7285
2030-2100 Eu 792, 5965, 9735, 11720, 11780
2300-2330 Eu 792

Slovene
0600-0630 Eu 6125, 7285, 9715

Spanish (C = primarily to Central America)
0100-0300 LAm 1530C, 1580C, 6030, 6190, 9465, 9670, 9840, 11895, 13740, 15400
0300-0400 LAm 1530C, 1580C, 6030C, 9465C, 9670C, 9840C, 13740C
0930-1100MF LAm 6040, 7405, 9465, 9525, 11580, 11935, 15265, 17810, 21490
1100-1130MF LAm 6040, 7405, 9465, 9525, 11580, 11935, 15265, 17810
1200-1300MF LAm 11890C, 11945C, 13775C, 21580C
1200-1400SS LAm 11890, 11945, 13775, 15265, 17730, 17830, 21490, 21580, 21610
1400-1500SS LAm 11890, 11945, 13775, 15265, 17730, 21580, 21610
1700-1730MF LAm 15185, 17710, 17785, 21745
2130-2200MF LAm 11890, 15120, 17715, 17730

Swahili
1630-1700MF Af 15150, 15165, 15330, 17705, 17740, 21485
1700-1730 Af 15150, 15165, 15330, 17705, 17740, 21485

Tibetan
0100-0130 EAs 6090, 7180, 15320, 17875, 21570
1500-1530 EAs 9575, 11705, 15255, 15285

Turkish
2000-2100 Eu/ME 1260, 6160, 7205, 9525, 11740

Ukrainian
0400-0600 CIS 7200, 9570, 9715, 9740
2100-2200 CIS 7210, 9715, 11805

Urdu
0100-0130 SEAs 6010, 7105, 9635, 9705, 11825
0130-0200 SEAs 7105, 9530
1330-1430 SEAs 11825, 15145, 17875, 21520

Uzbek

0100-0130	CIS	5985, 6195, 7130, 7190
1500-1600	CIS	11835, 15185, 15235, 17780, 17855, 21625

Vietnamese

1230-1330	EAs	1143, 9890, 11835, 15250
2230-2330	EAs	7260, 9510, 11720

ANN: E: At start and end of transmission period on each frequency: "This is the Voice of America, Washington DC, signing on/off". Before all foreign language prgrs: "This is the Voice of America. The following program is in . . . (language)".
INT-SIG: Musical signature, "YANKEE DOODLE", is played at the beginning & conclusion of transmissions & within prgs.
PUB: Prgr. schedule available upon request by E-mail from Dan Ferguson at Internet address df @ voa3.voa.gov. — **V.** by QSL-card (B). Address technical reception reports to Frequency Division.

VOA EUROPE

ADDR: 330 Independence Avenue, S.W, Washington, D.C. 20547, USA. **Te:** +1 202 619 2167. **Munich Office:** P.O. Box 221 220, D-8000 Munich 22, Germany. **Te:** +49 89 286 091. **Fax:** +49 89 280 9210 or +1 (202) 619 0918. **Telex:** (Germany) 523737.
L.P: Dir. VOA Europe: Terry Hourigan. Affiliate Relations Officer (Munich): Dick Bertel
VOA Europe, a 24-hour English-language service of contemporary music, news, and features, is available on cable systems and FM stations in many cities of Europe. For more information, listeners in Europe may write to VOA Europe, P.O. Box 221 220, 8000 Munich 22, Germany. VOA Europe can also be heard in some parts of Europe via VOA medium wave and shortwave transmitters, as follows:

Times	kHz
0700-0800	1197(W)
0800-1000	1197(W), 1260, 11735, 15160, 15195, 21455
2000-2100	1197

RADIO MARTI (Gov.)

Radio Marti is the Cuban radio broadcasting service of the United States Information Agency, Bureau of Broadcasting (see Voice of America).

ADDR: Washington, D.C. 20547, USA. **Te:** +1 202 401 7013. **Fax:** +1 202 401 3340.
L.P: Dir, Office of Cuba Broadcasting: Antonio Navarro. Dir, Radio Marti: Dr. Rolando Bonachea.

Times	kHz
0600-1200	1180, 6030
1200-1400	1180, 9590
1400-2300	1180, 11930
2300-0600	1180, 9525

Radio Marti does not broadcast UTC Monday 0300-0900.

UNITED STATES ARMED FORCES RADIO AND TELEVISION SERVICE

ADDR: AFRTS-BC, 10888 La Tuna Canyon Rd, Sun Valley, CA 91352-2098, USA.
L.P: PD: Gerald M. Fry. Radio Prgr. Mgr: Emanuel J. Levy. Public Affairs Mgr: Keith L. Anderson.
In service since 1942, the mission of the Armed Forces Radio and Television Service (AFRTS) is to inform and entertain US military personnel and their families stationed in more than 120 countries. Television and radio programming consisting of news, sports, public affairs and other timely information is relayed from Los Angeles, CA to overseas locations by satellite. Programming is derived from the US commercial radio and television networks, and other programming sources. Radio entertainment programs, which include a wide range of popular music formats, are transcribed or otherwise duplicated for shipment to stations. Television entertainment and news programs are videotaped for distribution to locations

KCBI INTERNATIONAL (Rlg.)

ADDR: 22720 SE 410th Str, Enamclaw, WA 98002. **Te:** +1 (206) 825 1099. **Fax:** +1 (206) 825 4517.
STATION: KCBI, Denton (G.C: 96.52W/33.13N): 50kW.
Consulting Eng: George Jacobs & Assoc. Inc.

Times	Area	kHz
0230-1400	CAN	9815
1400-0230	CAN	15375

INTERNATIONAL BROADCAST STATION KGEI (La Voz de la Amistad, The Voice of Friendship)

ADDR: Friendship Station, Redwood City, CA 94065. **Te:** +1 (415) 591 7374. **Cable:** FEBCOM RDCY.
STATION: Redwood City (G.C: 120.13W/27.38N): 50/250kW.

kHz	kW	kHz	kW	kHz	kW	kHz	kW
7365	250	9615	50	15280	50	15355	50

Spanish	Area	kHz	Portuguese	Area	kHz
0315-1400	LAm	9615	2300-2330	LAm	15280
1400-1600	LAm	15355	**Russian**		
2200-2300	LAm	15280	0200-0600	As	7365
2330-0315	LAm	15280			

ANN: (sign-on): "This is the Voice of friendship, International Broadcast St. KGEI, San Francisco, California, the United States of America, broadcasting to . . . on . . . kHz". Sp: "Esta es KGEI en San Francisco, California, Estados Unidos, La Voz de la Amistad, transmitiendo para Mexico, Centro America, El Caribe y America del Sur, en los . . . metros, banda de . . . kHz".
V. by QSL-card. Rp. (2 IRC's for surface mail, 4 IRC's for airmail".

RADIO STATION KJES (Rlg.)

ADDR: P.O. Box 1422, El Paso, TX 79948-1422. **Te:** +1 (915) 533 9122.
STATION: Vado, NM (G.C: 106.35W/32.08N): KJES 50kW

kHz	Area	H. of tr.	kHz	Area	H. of tr.
9510	CAN	1800-1900	11715	eCAN	1300-1400
		2000-2100		wCAN	1400-1600

V. by letter. Rp.

KTBN (Rlg.)

ADDR: P.O. Box A, Santa Ana, CA 92711. **Te:** +1 (714) 665 2145. **Fax:** +1 (714) 730 0661
L.P: Pres: Paul F. Crouch. V.P. (Eng): Ben Millar. Consulting Eng: George Jacobs & Assoc. Inc.
STATION: KTBN (G.C: 122.03W/40.39N): 100kW.

Times	Area	kHz
0200-1600	NAm	7510
1600-0200	NAm	15590

D.PRGR: audio from TBN tlevision netw.
V. by QSL-card. Rp.

KVOH (Rlg.)

ADDR: High Adventure Ministries Inc, Box 7466, Van Nuys, CA 91409. **Te:** +1 (805) 520 9460. **Fax:** +1 (805) 520 7823.
L.P: Founder: George Otis. **V.P:** John Tayloe. Consulting Eng: George Jacobs & Assoc. Inc.
STATION: KVOH Rancho Simi, CA (G.C: 118.38W/34.15N): 50kW.

kHz	kHz	kHz
9785	13695	17775

Times	Area	kHz	Times	Area	kHz
0100-0400*	NAm	13695	1200-0400	NAm	17775
0400-0800	NAm	9785			

*) occ. use.
Spanish: 1400-1900, 2100-0200. Other times in English.

RADIO EARTH INTERNATIONAL, INC. (Priv, Comm.)

ADDR: 1724 Sherman Avenue, Evanston, IL 60201. **Te:** +1 (708) 492 9300. **Fax:** +1 (708) 492 9812.
L.P: Chmn. & MD: Michael D. Poulos. PD's: Michael D. Poulos, Suzanne M. Poulos. Head of PR: Suzanne M. Poulos. CE's: J.D. Freberg, J.L. Gorski.
Schedule: call for current details.
ANN: "Presenting 'The World' to the world, this is R. Earth, worldwide commercial radio".
V. by QSL-card. Rp — **F.PL:** resume European Sce.

RADIO MIAMI INTERNATIONAL (Comm.)

ADDR: P.O. Box 526852, Miami, FL 33152. **Te:** +1 (305) 267 1728. **Fax:** +1 (305) 267 7375.
L.P: GM: Jeff White. CE: Indalecio Espinosa.
STATION: WRMI (G.C: 80.21W/25.54N): 1 × 50kW SW-tr. on 9955kHz.
D.PRGR: At press time, st. was expected to sign on in late 1992. The freq of 9955kHz was registered by WRMI for use 2h a day. Prgrs were expected to be in Spanish, English & Creole. At press time, st. was broadcasting via tr's of WRNO and WHRI acc. to the following schedule:

Times	Days	kHz	St.
1100-1300	Sun	9850	WHRI
2100-2300	Mon-Fri	17830	WHRI
2300-0100	Daily	7355	WRNO
0100-0300	Mon-Sat	7355	WRNO
0300-0500	Tues-Sun	7395	WRNO

V. by QSL-card. Prgr. schedules free on request.

WEWN CATHOLIC RADIO SERVICE (Rlg.)

ADDR: P.O. Box 100279, Birmingham, AL. **Te:** +1 (205) 956 9537. **Fax:** +1 (205) 951 0340. **Tlx:** 3762848 AT: 70413,40. **E-mail:** CompuServe 70413,40.
L.P: Pres: Bill Steltemeier. Senior VP: Matt Scalici. Head of Prgrs: Glen Tapley. Project Dir: Bob German. Head of PR: Marynell Ford. Consulting Eng: George Jacobs & Assoc. Inc.
STATION: WEWN (G.C: 86.28W/33.30N): 4 × 500kW SW-tr's.

kHz	kHz	kHz	kHz	kHz	kHz
5825	7540	9870	11885	13710	18930
7465	9350	9985	11970	17760	21670
7520	9410	11735	13615	17890	21735

Times	Area	kHz		Area	kHz
0000-0100	EAs	17890	1600-1800	NAf	21670
0000-0400	SAm	9410	1600-2400	SAm/NAm	13615
0100-0200	SAs	9985	1700-2200	SAm	21735
0200-0300	Eu/ME	9985	1800-2000	EAf/SAf	21670
0300-0500	Eu	7520	1800-2000	Eu	13710
0400-0800	SAm	9350	2000-2200	Eu	7540
0500-1000	Eu	7465	2000-2200	Eu/NAf	11970
0800-1000	SAm	9985	2000-2400	CAm	13710
0800-1600	NAm	9870	2200-0200	SAm	11885
1300-1600	NAf/EAf	21670	2200-0800	NAm	7540
1300-1700	MEX	11735	2200-1000	Eu	5825
1300-1800	Eu	18930	2200-2400	SAm	17760
1400-1800	SAm	13710			

V. by QSL-card. Rp (3 IRC's).

WHRI, WORLD HARVEST RADIO (Rlg./Comm.)

ADDR: P.O. Box 12, South Bend, IN 46624. **Te:** +1 (219) 291 8200. **Tlx:** 810-299-2595.
L.P: Chmn: Dr. Lester Sumrall. GM: Peter Sumrall. Sales Mgr: Tom Lewis. CE: Douglas W. Garlinger. Consulting Eng: George Jacobs & Assoc. Inc.
STATION: WHRI, Noblesville, IN (G.C: 85.57W/40.01N): 2 × 100kW SW-tr's.

kHz	kHz	kHz	kHz	kHz	kHz
7315	9465	9850	13760	15105	17830
7355	9495	11790			

Times	Area	kHz		Area	kHz
0000-1300	eNAm	7315	1400-1800	CAm/SAm	15105
0800-1100	CAm/SAm	7355	1600-2400	eNAm/Eu	13760
1000-1300	CAm/SAm	9850	1800-2300	CAM/SAm	17830
1100-1400	CAm/SAm	11790	2300-0800	CAm/SAm	9495
1300-1700	eNAm	9465			

Radio Miami International: Spanish: Mon-Fri 2100-2300 on 17830kHz, Sun 1100-1300 on 9850kHz.
Rel. Hrvatski Radio (Croatia): Croatian: 0000-0100 & 0500-0600 on 7315/9495kHz.
IS: "Onward Christian Soldiers".
ANN: "This is World Harvest Radio, the Int. Sce. of LeSea Broadc. Corp — **V.** by QSL-card. Reports & Monitors wanted.
PUB: World Harvest magazine.

WMLK (Rlg.)

Operated by the Assemblies of Yahweh.
ADDR: P.O. Box C, Bethel, PA 19507. **Te:** +1 (717) 933 4518.
L.P: Dir: Elder Jacob O. Meyer. CE: Gary McAvin.
STATION (G.C: 76.17W/40.29N): 1 × 50kW SW-tr.

Times	Area	KHz
0400-0900a	Eu/ME	9465
1700-2200a	Eu/ME	9465

a) exc. Sat — **V.** by QSL-card. Rp — **F.PL:** 100kW.

WORLD INTERNATIONAL BROADCASTERS INC. (Comm.)

ADDR: P.O. Box 88, Red Lion, PA 17356. **Te:** +1 (717) 244 5630.
L.P: SM: John H. Norris. CE: Fred Wise.
STATION: WINB (G.C: 76.34W/39.54N): 1 × 50kW tr.

kHz	kHz	kHz
15145	15185	15295

Times	Area	kHz	Times	Area	kHz
1600-2100	Eu/NAf	15295	2247-2345	Eu/NAf	15145
2100-2245	Eu/NAf	15185	2347-0330	SAm	15145

D.PRGR: Mostly English rlg. prgrs. and music.
ANN: "You are listening to World International Broadcasters, radio station WINB, broadcasting from Red Lion, Pennsylvania, the United States of America, on a frequency of . . . Megacycles, with transmitting power of 50,000 watts" — **V.** by QSL-card and prgrs. schedule.

WORLDWIDE CHRISTIAN RADIO (Comm.)

ADDR: 1300 WWCR Avenue, Nashville, TN 37218. **Te:** +1 (615) 255 1300 (in USA: 1 800 238 5576) — **L.P:** MD: George McClintock. Asst. MD: Joe Brashier. PD: Adam W. Lock, Sr. CE: Watt Hairston.
STATION: WWCR 2 × 100kW tr's (third tr. expected early 1993).

kHz	kHz	kHz	kHz
5920	7435	13815	15690

Times	kHz	Tr.	Times	kHz	Tr.
0000-1000	7435	1	1000-2400	15690	1
0100-1200	5920	2	1200-0100	13815	2

D.PRGR: 24h (different prgrs. on each tr.).
V. by QSL-card. Rp. Rec. acc.

WRNO WORLDWIDE (Comm.)

ADDR: 4539 I-10 Service Road, Metairie, LA 70002. **Te:** +1 (504) 889 2424. **Fax:** +1 (504) 889 0602 — **L.P:** Owner: Joseph Costello III. CE: Jack Bruce. Consulting Eng: George Jacobs & Assoc.
STATION: WRNO (G.C: 90.07W/29.50N): 1 × 100kW SW-tr.

kHz	kHz	kHz	kHz
7355	7365	7395	15420

Times	Area	kHz	Times	Area	kHz
1600-2300	NAm/Eu	15420	0300-0400	eNAm	7395*
2300-0400	eNAm	7355	0400-0500	eNAm	7365

*) occ. use.
Spanish (R. Miami Internacional): 0200-0500.
V. by QSL-card. Rp.(2 IRC's). Rec. acc.

WORLD SERVICE OF THE CHRISTIAN SCIENCE MONITOR (Comm.)

ADDR: P.O. Box 860, Boston, MA 02123. **Te:** +1 (617) 450 2929.
Recorded schedule (available 24h): +1 (617) 450 2060.
L.P: Pres: Al Carnesciali. Dir. of Int. Radio Broadc: Kate Dearborn. Sen. Producer: Kim Shippey. Sen. SM: Ed Evans.
Consulting Eng: George Jacobs & Assoc. Inc.
STATION: WCSN, Scotts Corners, ME (G.C: 68.34W/45.09N): 1 × 500kW.
L.P: Robert Stessel. CE: Bruce Clark. Sys. Eng: Ken Fox.

kHz	kHz	kHz	kHz	kHz	kHz	kHz
5850	7510	9350	9840	13770	17510	21640

H.of tr.	Area	kHz	H.of tr.	Area	kHz
0000-0200	Af	9850	**1200-1400**	Eu/ME	17510
0200-0400	ME/EAf	9350	1400-1600	Eu/ME	17510
0400-0600	SAf/WAf	9840	1600-2000	Af/ME	21640
0600-0800	Eu	5850	2000-2200	Eu/ME	7510
0800-1000	Eu/NAf	9840	2200-2400	Eu/Af	7510
1000-1200	Eu/ME	13770			

N.B.: Times in bold Sat & Sun only.
D.PRGR: Mon-Fri **World Sce.** - a 2h block of news, analysis & features in English. Sat/Sun rlg. prgrs. in French, Spanish, German, Portuguese, Czech, Russian and English.
ANN: "This is the World. Sce. of the Christian Science Monitor, coming to you from Boston, Massachusetts, USA".
STATION: WSHB, Cypress Creek, SC (G.C: 87.01W/32.41N): 2 × 500kW.
L.P: SM: Ed Evans. CE: Tony Kobatake. Sys. Eng: Michael Batchelor.

kHz	kHz	kHz	kHz	kHz	kHz
5850	9455	9495	9870	13760	15665
7395	9465	9840	13615	13770	17555

Times	Area	kHz	Times	Area	kHz
0000-0200	eNAm/Carib	5850	1400-1600	wNAm/MEX	13760
0000-0400	SAm	13760	**1400-1800**	eNAm/Carib	15665
0200-0400	wNAm/MEX	5850	**1600-2000**	wNAm/MEX	17555
0400-0600	MEX	9870	1800-2200	eNAm/Carib	15665
0400-0800	wNAm	5850	1800-2000	NAm/Eu	13770
0600-0800	Eu	9840	2000-2200	NAm/Eu	13770
0800-1000	AUS	13615	2000-2200	eSAm/B	17555
0800-1200	cSAM/B	9455	2200-2400	cSAm/B	17555
1000-1400	eNAm	9495	2200-2400	eNAm/Carib	9465
1200-1400	CAm	13760			

N.B.: Times in bold Sat & Sun only.

D.PRGR: Mon-Fri **World Sce.** in English & Spanish. Sat/Sun rlg. prgrs. in French, Spanish, Portuguese, German & English.
ANN: same as WCSN.
KHBI, Saipan : see 'Northern Marianas' in Pacific section.

WORLD WIDE GOSPEL RADIO INC. (Rlg.)

ADDR: P.O. Box 91, Upton, KY 42784.
L.P: Pres: Pastor John Powell.
STATION: WJCR, Millerstown, KY : 2 × 100kW SW-tr's.

kHz	kHz
7465	7490

D.PRGR: 24h.

WYFR-FAMILY RADIO (Rlg.)

ADDR: 290 Hegenberger Rd, Oakland, CA 94621. **Te:** +1 (510) 568 6200. **Fax:** +1 (510) 633 7983. **Tlx:** 703259 FAMILY RAD. UD.
L.P: Pres: Harold Camping. Dir. Tec: Wesley D. Becker. Prgr. Mgr: Thomas A. Schaff. Dir. Int. Rel: Richard H. Homeres.

kHz	kHz	kHz	kHz	kHz	kHz	kHz
5275+	6175	9680	9955+	11830	15170	17805
5950	6300+	9705	11550+	11855	15215	17845
5985	7355	9715	11580	11875	15355	21500
6065	9280+	9770	11715	11915	15400	21525
6085	9465+	9775	11725	13695	15440	21615
6105	9505	9850	11740	15130	15566	21720
6130	9605	9852.5	11775	15145	17750	

Arabic	Areas	kHz
0500-0600	Eu/ME	17725
1600-1700	Eu/ME	21500
1700-1800	Eu/ME	21525
English		
0000-0100	NAm	6085*
0100-0300	NAm	6065*, 9505*, 15440
0300-0500	NAm	6065*, 9505*
0500-0600	Eu/Af	9880, 11915, 13695
0500-0700	NAm	5985
0600-0800	Eu/Af	7355, 9680, 13695
1000-1100	NAm	5950*
1100-1200	NAm	5950*, 7355*
1200-1300	NAm	5950*, 7355, 11830*, 17760*
1300-1400	NAm	5950*, 7355, 11830*, 13695, 17760*
1302-1502	IND	11550+
1400-1500	NAm	9705, 11830*, 17760*
1500-1700	NAm	11830*, 15215, 17760*
1600-1700	Eu/Af	15355, 21525, 21615
1700-1900	Eu/Af	21500
1900-2000	Eu/Af	15355, 21615
2000-2200	Eu/Af	7355, 15566, 17750, 21525
2200-2300	Eu/Af	7355, 17750, 21525

*) Netw. prgrs Mon-Fri.

French		
0600-0700	Eu/Af	9770, 11580, 15170
1000-1100	Caribb	15130, 15170
1100-1300	CAN	13695
1800-1900	Eu/Af	17750, 21525, 21720
2000-2100	Eu/Af	11580
2300-2400	CAN	6085
2300-0100	Caribb	15400
German		
0400-0500	Eu/Af	11915
0500-0600	Eu/Af	7355, 9770
1700-1800	Eu/Af	15355, 21615
1900-2000	Eu/Af	21500
Hindi		
1502-1602	As	11550+
Italian		
0600-0700	Eu/Af	11915
0700-0800	Eu/Af	9770, 11580
1800-1900	Eu/Af	15355, 21615
2100-2200	Eu/Af	11580
Mandarin		
1102-1605	CHN	5275+, 9280+
2100-2200	CHN	6300+, 9280+, 9955+
2200-2300	CHN	6300+, 9280+, 9465+
2300-2400	CHN	6300+, 9465+, 11550+
Portuguese		
0100-0200	B	11775, 15130
0400-0500	Eu/Af	15170
0800-0900	B	6175, 9605
0900-1000	B	6175, 9605, 11830, 15130
1900-2000	Eu/Af	21525
2200-2300	B	15130
2300-0100	B	13695
Russian		
0305-0350	Eu/Af	6130, 9775
0400-0500	Eu/Af	7355, 9850
1505-1705	Eu/Af	9955+
1700-1800	Eu/Af	17750, 21720
Spanish		
0000-0100	CAm	9715
0100-0200	SAm	15170
0100-0300	Caribb	5985, 11715
0200-0500	CAm	9715, 11855
0300-0500	Caribb	5985
0500-0600	CAm	9705, 11855
0800-1000	SAm	6105, 9555
1000-1100	SAm	6105, 9555, 11830
1000-1300	Caribb	6085, 6170
1100-1200	CAm	9605, 11725
1100-1200	SAm	15130, 15145
1200-1400	CAm	11725, 15130, 15145
1300-1400	Caribb	5985
1400-1500	CAm	15130
2200-0100	Caribb	5985
2200-2300	Eu/Af	7355, 11580
2300-0100	SAm	15170, 15215, 17845

ANN: "This is your Family Radio, International Broadcast Station WYFR, Okeechobee, Florida, the United States of America". G: "Dies ist Ihr Familien Radio, die Internationale Radio-Station WYFR, in Okeechobee, Florida, der Vereinigten Staatan von Amerika". Sp: "Esta es Family Radio WYFR con el Sonido de la Nueva Vida transmitiendo desde Okeechobee, Florida, y con estudios de produccion ubicados en Oakland, California, Estados Unidos de Norte America".
IS: "To God be the Glory" — **V.** by QSL-card. No rec.
PUB: International Program and Frequency Guide.

ORGANIZATION OF AMERICAN STATES (OAS)

ADDR: 17th Str. & Constitution Ave. NW, Washington, DC 20006.
L.P: OAS Radio & TV Unit Chief: Ivan Silva. Principal Specialist: Rene Olivares. CE: Sergio Bonilla. Asst. CE: Mario Martinez. Asst. PR: Delia Levandoski.
D.PRGR. (via Voice of America tr's): **Spanish** 2345-0030 on 9565/11830kHz (Bethany) and 15160kHz (Greenville).
Purpose: "La Voz de la OEA" (the Voice of the OAS) promotes a better understanding of the OAS member countries through reports, newscasts, commentaries, music, interviews, pen pal section (international mailbox), interviews etc — **V.** by QSL-card & letter.

UNITED NATIONS RADIO

ADDR: Room S850, New York, NY 10017. **Fax:** +1 (212) 963 0765.
United Nations radio prgrs. can be heard on radio st's throughout the world. Produced at UN HQ in New York, they comprise transcriptions as well as news and current affairs magazines which by special arrangement are fed by radio/telephone circuits to a number of international broadcasting organizations, mostly on shortwave. UN Radio currently broadcasts on SW as follows:

Arabic	Days	Areas	Station	kHz
1645-1700	Sun	ME/NAf	R.Cairo	9700, 11665
Bangla				
1650-1700	Sat	SAs	AIR	4820, 1134, 594
Chinese				
0900-1000	Wed	SEAs	China R.	15165, 9590
1100-1200	Wed	NEAs	China R.	17715, 15110, 9590
1300-1400	Wed	SEAs	China R.	15165, 9945, 9590
English				
2150-2200	M-F	NAm/LAm	RFPI	7375, 13630, 15030, 21465
2350-2400	M-F	NAm/LAm	RFPI	7375, 13630, 15030, 21465
2100-2115	M/W/F	NAm/LAm	RFPI	7375, 13630, 15030, 21465
1845-1900	Sun/Mon	NAm/LAm	RFPI	7375, 13630, 15030, 21465

CENTRAL AMERICA
& THE CARIBBEAN
0
150
300 KM
GULF OF MEXICO
UNITED STATES
MIAMI
BAHAMAS
ATLANTIC
OCEAN
Tropic of Cancer
LA HABANA
PINAL DEL RIO
CUBA
SANTIAGO DE CUBA
CAP HAÏTIEN
HAITI
PORT-AU-PRINCE
DOMINICAN REP.
SANTO DOMINGO
PUERTO RICO
SAN JUAN
VIRGIN ISLANDS (U.S. & Br.)
ANTIGUA
GUADELOUPE
DOMINICA
MARTINIQUE
ST. LUCIA
ST. VINCENT
BARBADOS
GRENADA
TOBAGO
TRINIDAD
JAMAICA
KINGSTON
MÉRIDA
MÉXICO
BELIZE
TIKAL
BARILLAS
COBÁN
GUATEMALA
JOCOTÁN
SANTIAGO ATITLÁN
ESCUINTLA
GUATEMALA CITY
SAN LUIS
LA CEIBA
SANTA BÁRBARA
HONDURAS
TEGUCIGALPA
SAN SALVADOR
EL SALVADOR
PUERTO LEMPIRA
PUERTO CABEZAS
NICARAGUA
MANAGUA
BLUEFIELDS
CARIBBEAN SEA
COSTA RICA
ALAJUELA
PUNTARENAS
SAN JOSÉ
LIMÓN
COLÓN
PANAMÁ
PACIFIC
OCEAN
ARUBA
BONAIRE
CURAÇAO
CARACAS
MARACAIBO
TOVAR
VALENCIA
OCUMARE DEL TUY
BARCELONA
MATURÍN
EL TIGRE
CIUDAD BOLÍVAR
VALERA
BARINAS
MÉRIDA
SAN CRISTÓBAL
VENEZUELA
COLOMBIA
90°
85°
80°
75°
70°
60°
25°
20°
15°
10°
UTC -6
UTC -5
UTC -4

2330-2345	Sat/Sun	NAm/LAm	RFPI	7375, 13630, 15030, 21465
2330-2345U	Mon/Thu	NAm/LAm	RFPI	7375, 13630, 15030, 21465
2100-2115U	Tues	NAm/LAm	RFPI	7375, 13630, 15030, 21465
2000-2030U	Thurs	NAm/LAm	RFPI	7375, 13630, 15030, 21465
0200-0230	M-F	NAm/Eu	IRRS	7125
0730-0800	M-F	NAm/Eu	IRRS	7125
1700-1730	M-F	NAm/Eu	IRRS	7125
1330-1400	Sat	NAm/Eu	IRRS	7125
0730-0800	Sun	NAm/Eu	IRRS	7125
1700-1730	Sun	NAm/Eu	IRRS	7125
1630-1645	Mon	As	Bhutan	5025
1430-1445	Thurs	As	Bhutan	5025
2115-2130	Fri/Sun	Af	SLBS	3316
1900-1915	Sun	Af	SLBS	3316
1346-1401	Sat	As	AIR	9565, 11760, 11810, 15110
1930-1945	Sat	As	AIR	9565, 11760, 11810, 15110
2030-2200	Sun	WAf	Cairo	15335
French				
2200-2215	Mon	WAf/CAf	Cairo	15335
0830-0800	Sun	Eu	IRRS	7125
KiSwahili				
1650-1705	Mon	EAf/CAf	Cairo	11975
1730-1745	Sun	EAf/CAf	TZA	11734
Russian				
1845-1900	Sat	Eu	Poland	6095, 9540, 11840
0645-0715	Sun	EEu	IRRS	7125
0230-0300	Sun	EEu	IRRS	7125
Spanish				
1400-1415	M-F	Am	RFPI	7375, 13630, 15030, 21465
1530-1600	Mon	Am	RFPI	7375, 13630, 15030, 21465
0015-0020	M-F	LAm	OAS	11830, 15160
1430-1445	M-F	Am	HCJB	6050, 9765, 11910, 690
1930-1945	M-F	Am	HCJB	6050, 15230, 17875, 690
2130-2145	M-F	Am	HCJB	6050, 15230, 690
0300-0315	M-F	Am	HCJB	3220, 6080, 11960, 15230, 690
2100-2115	M-F	Am	Mexico	5985, 9705, 11770, 15430
0900-0930	Sun	Eu	IRRS	7125

Abbreviations: U = UNESCO Radio (re. to 7 Place de Fontenoy, F-75007 Paris, France), AIR = All India Radio, IRRS = Italian Radio Relay Sce, OAS = Voice of the OAS, RFPI = R. for Peace Int. (Costa Rica), SLBS = Sierra Leone Broadc. Sce, TZA = R. Tanzania Zanzibar

BBC RELAYS VIA VOA TRANSMITTERS

B = Bethany, D = Delano, G = Greenville

kHz	kHz	kHz	kHz
9515D	9690D	15315D	17715G
9590D	11965D	17850B	

D.PRGR: see United Kingdom.

CENTRAL AMERICA & THE CARIBBEAN

ANGUILLA (British West Indies)

L.T: UTC-4h — **Pop:** 7.000 — **Radio sets:** 2.500 (est.) — **Pr.L:** English — **E.C:** a/c 50, 230V — **ITU:** IOB.

RADIO ANGUILLA (Gov.)

ADDR: The Valley, Anguilla, BWI. **Te:** +1 (809) 497 2218. **Fax:** +1 (809) 497 2751.
L.P: Dir: V. Menes Hodge. Prgr. Contr. & N. Editor: N. A. Hodge. CE: O. Gumbs —**MW:** 1505kHz 1kW.
D.PRGR: W 0930-1800 & 2100-0210, Sun 1600-0210. **N: BBC:** 1100W, 1200, 1300W, 1600, 2300, 0200. **VOA:** 0100. **Local:** 1115W, 1615Sun, 2315 —**ANN:** "1505 Radio" — **V.** by letter.

ZJF-FM (Comm.)

ADDR: Sachasses, Anguilla, BWI. **Te:** +1 (809) 497 3157.
FM: 105.3MHz.

THE CARIBBEAN BEACON (Rlg.)

ADDR: Box 690, Anguilla, BWI. **Te:** +1 (809) 497 4340. **Fax:** +1 (809) 497 4311 — Owner: Dr. Gene Scott.
MW: 690kHz 15kW, 1610kHz 50kW — **FM:** 100.1MHz 0.1kW.
D.PRGR: 0900-0600. **N:** 0100MF — **V.** by QSL-card.

ANTIGUA & BARBUDA

L.T: UTC-4h — **Pop:** 64.200 — **Radio sets:** 75.000 (est.) — **Pr.L:** English — **E.C:** a/c 60, 220V — **ITU:** ATG.

ANTIGUA & BARBUDA BROADC. SCE. (Gov. Comm.)

ADDR: P.O. Box 590, St.John's, Antigua. **Te:** +1 (809) 462 3602.
Cable: Broacast Antigua. **Telex:** 2090 TELCO AK.
L.P: GM: Hollis Henry. PD: D.L. Payne. CE: Dennis Leandro.
MEDIUMWAVE: 620kHz 10kW — **D.PRGR:** 0925-0500.
N: 1100(BBC), 1600, 2300(BBC), 2310, 0400(BBC).
ANN: "This is the Antigua & Barbuda Broadc. Sce." or "You are listening to 620 - ABS" — **V.** by letter.

RADIO ZDK (Priv. Comm.)

ADDR: Grenville Radio Ltd, P.O. Box 1100, St.John's, Antigua.
Te: +1 (809) 462 1100. **Cable:** Grenville Radio Limited - Antigua.
L.P: Chief Exec: Mrs. Lydia Bird. PD: Ivor Bird. GM (Admin): Clayton Hinds. CE: E.Phillip.
MEDIUMWAVE: 1100kHz 10kW — **FM:** 99.0MHz.
D.PRGR: Mon-Fri 0900-0504, Sat 0956-0504, Sun 1426-0204.
N: 1000W, 1400W, 1600, 1800.
ANN: "Magic Radio" — **V.** by QSL-card.

CARIBBEAN RADIO LIGHTHOUSE (Rlg.)

ADDR: Box 1057, St. Johns', Antigua. **Te:** +1 (809) 462 1454.
L.P: Dir: Curtis L. Waite. Asst. Dir. & Dir. Tec: Jerry Baker.
MEDIUMWAVE: 1165kHz 10kW — **FM:** 90.0MHz 0.01kW.

D.PRGR: 0925-0230. **N.** (BBC): 1100, 1200, 1300, 1600, 2000, 2300.
ANN: "This is the Caribbean Radio Lighthouse, broadcasting from Antigua" —**V.** by QSL-card

CARIBBEAN RELAY CO. LTD. (BBC/DW)
ADDR: P.O. Box 1203, St. John's.
FM: 98.1MHz (rel. BBWs 24h incl. Caribbean Report)
SHORTWAVE (G.C: 61.48W/17.06N): 2 × 250kW SW-tr's jointly operated by BBC & Deutsche Welle.
Frequencies (BBC)

kHz	kHz	kHz	kHz	kHz	kHz
5975	6110	6195	9640	15220	17840

D.PRGR: see United Kingdom
Frequencies (Deutsche Welle)

kHz	kHz	kHz	kHz	kHz	kHz
6040	9545	11795	11965	15275	17810
6075	9670	11810	15105	15410	17860
6160	9690	11960	15205	17715	

D.PRGR: see Germany.

ARUBA

L.T: UTC – 4h — **Pop:** 63.000 — **Radio sets:** 40.000 (est.) — **Pr.L:** Papiamento, English, Dutch — **E.C:** a/c 50+60, 127/220V.

	kHz	KW	MHz	KW	Station
1)	960	10	93.1	1	Radio Victoria, Oranjestad (Rlg.)
2)	1270	1.5	—	—	R. 1270, San Nicolaas
3)	1320	1	89.9	10	Voice of Aruba, Oranjestad
4)	1440	1	106.7	6	R. Kelkboom, Oranjestad
5)	—	—	97.9	—	R. Caruso Booy, Oranjestad
6)	—	—	103.5	—	R. Carina FM, Oranjestad

Addresses and other information:
1) Pos Chiquito, P.O. Box 410, Oranjestad. **Te:** +2978 47090. **Fax:** +2978 49052 — SM: E. Parsons. CE: W. Gandre. **MW:** Mon-Fri 1000-2300, Sat 1100-2300, Sun 1130-0100. **N. in English:** 1130W, 1630W, 2230W. **FM:** 1200-0300W (music). **V.** by letter. Rp.
2) Bernardstraat 138, P.O. Box 28, San Nicolaas. **Te:** +2978 45602. **Fax:** +2978 27753 — SM: J.A.C. Alders. Dir: F.A. Leauer.
3) Van Leeuwenhoekstraat 26, P.O. Box 219, Oranjestad. **Te:** +2978 24134 (admin.) 21601 (studios) — Dir: A.M. Arends Jr. **MW:** 1000-0500 (Sat 0700). **FM:** ("**Channel 90**"): 1000 (Sun 1200)-0500. **V.** by QSL-card or letter.
4) Bloemond 14, P.O. Box 146, Oranjestad. **Te:** +2978 21899. **Fax:** +2978 34825 — GM: C.A. Kelkboom. CE & PD: E.A.M. Kelkboom. 1000 (Sun 1200)-0400. **Dutch:** 1030-1100, 1230-1300, 2315-0000.
5) De Bruynewijk 49, Savaneta. **Te:** +2978 47752. **Fax:** +2978 43351 — Owner: Ms. S. Booy.
6) Emanstraat 49A, Oranjestad. **Te:** +2978 26433/21450.

BAHAMAS

L.T: UTC – 5h — **Pop:** 255.000 — **Radio sets:** 200.000 — **Pr.L:** English — **E.C:** a/c 60, 120/220V — **ITU:** BAH.

BROADC. CORP. OF THE BAHAMAS (Comm.)
ADDR: P. O. Box N-1347, Nassau. **Te:** +1 (809) 322 4623. **Fax:** +1 (809) 322 3924. **Tlx:** ZNS NASSAU 253.
L.P: GM: C. Johnson. Dep. GM: L. Hanchell. PD: A. Foster. Dir. Eng: M. Tompson. SM (Freeport): A. Adderley.
MW: Nassau ZNS1 1540kHz 20kW (d), ZNS2 1240kHz 1kW — Freeport ZNS3 810kHz 1kW.
D.PRGR: 24h — **N** (ZNS1): 0800, 1300, 1830, 0000, 0300.
ANN: ZNS1: "This is R. Bahamas". ZNS2: "This is R. New Providence". ZNS3: "This is Northern Sce, R. Bahamas" — **V.** by letter.

BARBADOS

L.T: UTC – 4h — **Pop:** 261.352 — **Radio sets:** 200.000 — **Pr.L:** English — **E.C:** a/c 50, 110/220V — **ITU:** BRB.

CARIBBEAN BROADC. CORP. (Gov. Comm.)
ADDR: P. O. Box 900, Pine Hill, Bridgetown. **Te:** +1 (809) 429 2041. **Fax:** +1 (809) 429 4795. **Tlx:** 2560 CABROCO WB.
L.P: GM: Sam Taitt. Dep. GM (Broadc.): Tony Cumming. CE (Radio): G. Husbands. Prgr. Mgr. (Radio): C. Graham.
MW: 900kHz 10kW — **FM:** 98.1MHz 5kW (stereo).
D.PRGR. on 900kHz: 0855-0500. **Local N:** 1000, 1045, 1200W, 1500W, 1700, 2100W 2200. **BBC N:** 1100. **RCI N:** 1300 — **on FM:** 0900-2300 (Fri/Sat 0400). **N:** 1055, 1155W, 1455W, 1700, 2055W, 2155Sat — **ANN:** "CBC Radio 900", "Liberty FM 98.1". **V.** by letter.

BARBADOS REDIFFUSION SCE. (Comm.)
ADDR: River Rd, Bridgetown. **Te:** +1 (809) 426 0820. **Fax:** +1 (809) 429 8093. **Cable:** Rediffusion.
L.P: Dir. & GM: V. Fernandez. Dep. GM (Finance & Admin): St. Elmo Walker. Dir. N. & Public Affairs: D. Ellis. CE: J. S. W. Marshall. PD: L. Johnson.
MW: 790kHz 20kW — **FM:** 104.1MHz 3.5kW (stereo) + Cable Sce.
V. of Barbados on MW: 0900-0500. **N:** 0930, 1030, 1130W, 1230Mon-Thurs, 1330Fri-Sun, 1630, 1830MF, 2030, 2130, 0130, 0230MF.
Yess FM: 24h. **N:** 0930W, 1030, 1130, 1630, 2030MF, 2130, 0130.
ANN: MW: "This is the V. of Barbados", "This is 790 VOB".

BARBADOS BROADC. SCE. LTD.
ADDR: Astoria, St. George. **Te:** +1 (809) 437 9550. **Fax:** +1 (809) 437 9554 — **FM:** 90.7/102.1MHz (stereo).
BBS-FM (Comm.): 24h on 90.7MHz. **N:** 1100 (BBC), 2200MF.
Faith FM (Rlg.) 0900-1500, 1900-0400 on 102.1MHz.

BELIZE

L.T: UTC-6h — **Pop:** 236.000 — **Radio sets:** 100.000 — **Pr.L:** English, Spanish — **E.C:** a/c 60, 110/220V — **ITU:** BLZ.

BELIZE BROADC. NETW. (Gov. Semi-Comm.)
ADDR: Box 89, Belize City. **Te:** +501 (2) 77246. **Fax:** + 501 (2) 75040. **Cable:** Radio Belize.
L.P: GM: R. Villanueva. CE: M. J. Gundy. PD (AM): A. Nicholson. PD (FM): T. Jeffries. CE: E. Rosado.

MW	kHz	kW		kHz	kW
Ladyville	730	10	Corozal	930	1
Punta Gorda	910	1	San Ignacio	940	1

SW: Belmopan 3285kHz 1kW (inactive).
FM: Ladyville 88.9/91.1MHz + 4 low power st's.
Belize R. One on MW + 91.1MHz: 1100-0600. **N: English:** 1340, 1500, 1700, 1830, 2100, 2300, 0100, 0300. **Spanish:** 1845, 0115.
Friends FM on 88.9MHz + repeaters: 1600-1100.
ANN: "R. Belize, the V. of Belize" or at 1800/0100 : "This is the Belize National Radio Netw, R. Belize & Friends FM".
V. by QSL-card. Rec. acc.

BRITISH FORCES BROADC. SCE.
ADDR: BFBS Belize, BFPO 12 — SM: Alan Jennings.
FM: 93.1/99.1MHz 24h.

VOICE OF AMERICA RELAY STATION BELIZE
MW: 1530/1580kHz 50kW — **D.PRGR: English:** D 0030-0100, Tues-Sat 0400-0500. **Spanish:** 0100-0400.

CAYMAN ISLANDS (British)

L.T: UTC+5h — **Pop:** 28.640 — **Radio sets:** 18.000 — **Pr.L:** English — **E.C:** a/c 60, 110V — **ITU:** IOB.

RADIO CAYMAN (Gov. Comm.)
ADDR: Box 1110, George Town, Grand Cayman, BWI. **Te:** +1 (809) 949 7799. **Fax:** +1 (809) 949 7544. **Cable:** Broadcast Grandcayman. **Telex:** 4260 CIGOVT.
L.P: Dir. of Broadc: Loxley Banks. Dep. Dir: Doren Miller.
MW: Gun Bluff 1555kHz 10kW — George Town 1205kHz 1kW.
FM: 105.3MHz 5kW, 89.9MHz 0.25kW — Brac 91.9MHz 0.25kW.

D.PRGR: 1100-0500. **N:** 1100W*, 1200W, 1300W, 1300Sun*, 1500W, 1700, 1900MF, 2000SS, 2100MF, 2300MF*, 2300SS, 0000W, 0300*.
*) BBC — **V.** by letter. Rec. acc.

COSTA RICA

L.T: UTC – 6h (SU: UTC — 5h) — **Pop:** 3.190.000 — **Radio sets:** 255.000 — **Pr.L:** Spanish — **E.C:** a/c 60, 120/220V — **ITU:** CTR.

CONTROL NACIONAL DE RADIO (CNR)

ADDR: Ap. 8000, 1000 San José. **Te:** + 506 257364. **Cable:** Radcontrol. **Tlx:** 3420 MIGOP CR — Dir: Warren Murillo Martínez.

CAMARA NACIONAL DE RADIO (CANARA)

ADDR: Ap. 1583, 1002 San José. **Te:** + 506 331845. **Fax:** +506 554483 — Exec. Dir: Luzmilda Vargas Gonzáles.
Most st's carry the prgr. "Panorama" (produced by CANARA) at 1300-1315.

Mediumwaves: Call: TI. °) also on Shortwave, (r:) = repeater. *) inactive.

	Call	kHz	kW	Name and h. of tr.
1)	CAL	°v530	18	R. Rumbo, Cartago: 1100-0600v (r. 530.4)
2)	SCL	550	20/2	R. Santa Clara, Cd, Quesada: 1100-0300
3)	CDL	570	20	R. Fundación, Desamparados: 1015-0600
4)	RN	590	40	R. Nacional, San José: 1030-(Sun 1300)-0600
6)	RSU	610	4	R. Sistema Universal, Guadalupe: 1100-0700
5)	JJ	640	1	R. Musical, Moravia: 24h
7)	TNT	670	10	R. Monumental, San José: 1030-0615
12)	JC	700	20	R. Sonora, Guadalupe: 1000-0600
8)	HB	°730	25	R. Reloj, San José: 24h
9)	W	760	20	R. Columbia, San José: 24h
10)	RA	v780	2	R. América Latina, San José: 1000-0500 (Sun 0400)
9)	W	800	5	R. Uno, San José: 24h.
11)	GC	820	2.5	R. Centro, Tibás: 1130-0600
52)	XEW	850	20	XEW La Exitosa, San José: 1100-0500
13)	UCR	°870	12/6	R. UCR, San José: 1258-0400
7)	HOT	890	10	R. Fabulosa, San José: 1200-0500
14)	QM	910	1	R. Metrópolis Norte, Cd. Quesada: 1200-0600
14)	QM	910	1	(r. TIQM 910), Tibás
53)	RCR	930	5	R. Saprissa "La V. del Monstruo," Zapote: 1200-0400
16)	RI	°980	20	R. Cordillera, Esparza: 1030-0600
17)	MIL	1000	5	R. Mil, San José: 24h
29)	TIC	1020	2	R. Mil Veinte, San José: 1200-0400
18)	AC	1040	12	R. Fides, San José: 1000-0400
9)	LX	1060	1	(r: TILX 760), San Isidro de El General
9)	MS	1060	1	(r: TILX 760), Liberia
19)	FC	°1080	19	Faro del Caribe, San José: 24h
19)	FC	1080	1	(r: TIFC 1080), Chomes
19)	FC	1080	1	(r: TIFC 1080), Liberia
19)	FC	1080	1	(r: TIFC 1080), La Marina
20)	SCR	1100	20	R. Chorotega, Santa Cruz: 1100-0300
15)	VAL		2	R. Guápiles: 1100-0200
8)	SHB	1120	10	Sonido 11-20, San José: 1200-0500
7)	BAS	1140	1	(r: TITNT 670), San Isidro de El General
7)	BAS	1140	1	(r: TITNT 670), San Carlos
9)	LXX	1160	3	(r: TILX 760), Puntarenas
22)	PJ	1180	10	R. Victoria, Heredia: 1200-0500
23)	TQ	1200	2	Cadena KW, San José: 1130-0600
23)	TG	1200	0.25	(r: TITQ 1200), San Isidro del General
21)	Q	°1220	1	R. Casino, Limón: 1100-0600
9)	WC	1240	1	(r: TILX 760), Nicoya
24)	HN	1260	6	R. Emaus, San Vito de Coto Brus
25)	HT	1280	2	R. Alajuela, Alajuela: 1000-0600
26)	LC	1300	7.5	R. La Fuente Musical, Cartago: 1030-0500
9)	LX	1320	1	(r: TILX 760), San Carlos
27)	HR	1340	6	R. Sideral, San Ramón: 1000-0400
7)	LX	1360	1	(r: TITNT 670), Limón
7)	TNT	1360	1	(r. TNT 670) Puntarenas
28)	MS	1380	1	R. Guanacaste, Liberia
28)	MS	1380	1	(r: TIMS 1380), San José: 1000-0500
24)	CJ	1400	12	R. Sinaí, San I. de El General: 1000-0400
7)	BAS	1420	1	(r: TITNT 670), Nicoya
30)	RSC	1440	5	R. San Carlos, Cd. Quesada: 1000-0400
9)	LX	1460	2	(r: TILX 760), Limón
9)	AW	1480	1	R. Puntarenas, Puntarenas: 1100-0600
30)	ASF	1500	12	R. Cima, Cd. Quesada: 1000-0400
9)	LX	1520	0.5	(r: TILX 760), Turrialba
33)	—	1540	—	R. Mía, San José (F.Pl.)
32)	JM	1560	6	R. Nicoya, Nicoya: 1100-0400
21)	Q	1580	0.5	(r: TIQ 1220), Siquirres + 9 other locations
34)	CLS	1580	0.25	R. Cultural Los Santos, San Marcos
35)	RSCLC	1580	0.25	R. Sist. Cultural, La Cruz
37)	RCC	1580	0.25	R. Cultural de Corredores, Cd. Neily
38)	RSCM	1580	0.25	R. Sist. Cult. Maleku, Tonjibe
39)	RCT	1600	0.25	Em. Cultural, Turrialba
40)	RCBA	1600	0.25	R. Cultural de Buenos Aires, Bs. As.
41)	RCP	1600	0.25	R. Cultural, Pital
42)	RSCN	1600	0.25	R. Sist. Cultural Nicoyano, Nicoya
43)	RCU	1600	0.25	R. Cultural, Upala
44)	JV	1600	1.5	R. Quepos, Quepos
45)	LGJ	1600	1.5	R. 16, Grecia: 1100-0400 (r. 1590)
46)	MMCC	1600	2/1	R. Golfito, Pto. Golfito: 24h
47)	MQ	1600	1.5	R. Pococí, Guápiles: 1100-0400

Shortwaves:

	Call	kHz	kW	Name and h. of tr.
8)	HB	4831.5	3	R. Reloj, San José: 24h
19)	FC	5055	5	Faro del Caribe, San José: 1000-2000, 2300-0600 (**English:** 0300-0400)
31)	AWR	5030	20	AWR Latin America
21)	Q	5953.5	1	R. Casino, Limón: 1030-0600 (**English:** 0400-0600, 1100-1200)
31)	AWR	5970	50	AWR Latin America: 2300-0500
8)	HB	6006	3	R. Reloj, San José: 24h
1)	CAL	*6075	1	R. Rumbo, Cartago: 1100-0530
13)	UCR	6105	2	R. UCR, San José: 1258-0400
31)	AWR	6150	50	AWR Latin America: 1200-1500, 2300-0500
19)	FC	6175	2.5	Faro del Caribe: 1000-2000
48)		v7375	30	R. for Peace International, Cd. Colón: 1400-2400 (AM), 0000-1300 (USB)
19)	FC	9645	0.5	Faro del Caribe: 1000-1800
31)	AWR	9725	50	AWR: 2300-0500
31)	AWR	11870	50	AWR: 1100-1500
48)		13630v	2	R. for Peace International, Cd. Colón: 24h (USB)
31)	AWR	13750	20/50	AWR Latin America
48)	—	15030v	5	R. for Peace International, Cd. Colón: 24h
31)	ASD	15460	50	AWR: 1100-1500
48)		21465v	2	R. for Peace International, Cd. Colón: 1400-2400 (USB)

Addresses and other information:

1) Ap. 140, 7050 Cartago — Dir: Carlos Lafuente Fernández.
2) Ap. 221, Cd. Quesada, San Carlos — Gte: Pbro. Marcos Solís.
3) Ap. 301, 2400 Desamparados — Gte: V. Julio Bermúdez.
4) Ap. 7-1980, 1000 San José — Dir: Luis B. Chávez R.
5) Ap. 13, 2150 Moravia — Gte: Rigoberto Urbina P.
6) Ap. 4666, 1000 San José — Dir: Edgar Alfaro.
7) Ap. 800, 1000 San José — D.G: José Luis Ortiz Araya. All news (on 670kHz). Fax: +506 331021.
8) Ap. 341, 1000 San José — Gte: Hernán Barquero.
9) Ap. 708, 1002 San José — Dir: C. Alfaro McAdams.
10) Ap. 2525, 1000 San José — DG: J. Sepúlveda T.
11) Ap. 6133, 1000 San José — Owner: Roberto Hernández R.
12) Ap. 1999, 1002 San José — DG: Fernando Naranjo.
13) Cd. Universitaria Rodrigo Facio, 1000 Jan José.
14) Ap. 7849, 1000 San José — Dir. Limberth Quesada.
15) Ap. 266, Guápiles — Dir: Eduardo Solano S.
16) Ap. 497, 2050 San Pedro de Montes de Oca, or: Ap. 80, Esparza — Own: Parmenio Medina.
17) Ap. 10.001, 1000 San José — Dir: R. Giralt C.
18) Ap. 5079, 1000 San José — Dir: Jorge Campos Sandí.

19) Ap. 2710, 1000 San José — Mgr: Juan J. Ochoa F.
20) Ap. 92, 5175 Santa Cruz — Gte: Pbro. Carlos M. Castillo.
21) Ap. 287, 7301 Puerto Limón — Fax: 583029 — Dir: L. Grau V.
22) Ap. 298, 3000 Heredia — Gte: Emilio Piedra G.
23) Ap. 1128, 1000 San José — Gte: José R. Traube Z.
24) Ap. 262, San Isidro del General — Dir: Pbro. Alvaro Coto O.
25) Ap. 122, 4050 Alajuela — Dir: Flora Herrera.
26) Ap. 555, Cartago — Dir: F. Lafuente C.
27) Ap. 73, 4250 San Ramón — Mgr: Róger Retana C.
28) Ap. 27, Liberia, or: Ap. 6462, 1000 San José — Dir: Mario Salgado S.
29) Ap. 722, 1011 San José — Dir: Alfredo Aspitia.
30) Ap. 259, 4400 Cd. Quesada — Mgr: Carlos Hernández P.
31) Ap. 1177, 4050 Alajuela — Mgr: David L. Gregory (owned & operated by AWR-Latin America). **English:** 1100-1300 (Sat 1230), 2300-0100. **Spanish:** 1300 (Sat 1330)-1500, 0100-0500. **Dutch:** Sat 1300-1330. **Papiamento:** Sat 1230-1300. **Ann:** "R. Mundial Adventista".
32) Ap. 50, Nicoya, or: Ap. 184, Cd. Quesada — Dir: Tony Díaz.
33) Ap. 1583, 1002 San José — Repr: Fernando López Fallas.
34-43) St's are affiliated to **Instituto Costarricense de Enseñanza Radiofónica**, Ap. 132, 2050 San Pedro de Montes de Oca. A joint project of the Principality of Liechtenstein and Costa Rica — Coordin: Marco Antonio Gómez V.
44) Ap. 99, Quepos — Gte: Juan Vega Q.
45) Ap. 16, Grecia — Own: Luis G. Jiménez R.
46) Ap. 11, Golfito, or: Ap. 5362, San José — Dir: Marco A. Muñoz C.
47) Ap. 160, Guápiles — Gte: Jorge Quesada B.
48) Ap. 88, Santa Ana, or: Ap. 199, Escazú. **Te:** +506 491821. **Fax:** +506 491929. (**Addr. in USA:** P.O. Box 10869-B, Eugene, OR 97440. **Te:** +1 (503) 741 1794. **Fax:** +1 (503) 741 1279.) — Dir. Gen: Debra Latham. A joint project of World Peace University, Eugene, OR, USA and the UN affiliated University for Peace, in Costa Rica. **D. Prgr:** Eng., exc 1400-1700 Sp., Tues., Fri. 2300-2330 German.
50) Ap. 400, 2150 Moravia.
51) Ap. 10270, 1000 San José.
52) Ap. 606, 1250 Escazú — Own: Federico Sosto Peralta.
53) Ap. 365, 1009 San José — Gte: Fernando A. Almodóvar.
FM in San José and vicinities: 48) 88.3 — 31) 88.7 R. Lira — 14) 89.1 R. Mundo — 89.5 R. Sendas de Vida — 50) 89.9 R. Continental — 6) 90.7 — 9) 91.1 R. Uno — 3) 91.5 — 9) 92.7 — 18) 93.1 — 7) 93.5 — 50) 93.9 Sonido Latino — 8) 94.3 — 14) 94.7 — 7) 95.1 R. Fabulosa — 17) 95.9 — 13) 96.7 — 19) 97.1 — 5) 97.5 — 51) 97.9 R. Titania — 9) 98.7 — 9) 99.1 R. Sabrosa — 99.9 Azul — 23) 100.3 — 4) 101.5 — 50) 102.3 Súper R. — 103.1 103 R. — 103.5 R. Popular — 24) 103.9 — 51) 104.3 R. Sensación — 104.7 R. Emperador — 105.0 R. Omega — 12) 105.9 — 106.3 — 106 — 8) 106.7.

CUBA

L.T: UTC – 5h (Mar-Oct: UTC – 4h) — **Pop:** 10.840.000 — **Radio sets:** 2.140.000 — **Pr.L:** Spanish — **E.C:** a/c 60, 115/120 V — **ITU:** CUB.

MINISTERIO DE COMUNICACIONES (MC)

Dirección General de Telecomunicaciones
ADDR: Plaza de la Revolución; Ciudad de la Habana.
Cable: GENPOSTEL — **L.P:** Dir. Tec: Carlos Martínez Albuerne.

INSTITUTO CUBANO DE RADIO Y TELEVISION (ICRT)

ADDR: Edif. Radiocentro, 23 y L, Vedado, Habana 4. **Te:** +53 (32) 3369, +53 (32) 6940 — **L.P:** Dir: Rodobaldo Díaz O. Dir. Int. Rel: Benjamín Rodríguez.

MEDIUMWAVES: Call: CM--.
NOTE: Callsigns used over the air differ from those registered with the IFRB in Geneva; the latter are herein suffixed with a "-".

	Call	kHz	kW	Name & Location
N1)	HV	540	1	R. Rebelde, Sancti Spíritus, SS
N1)	—	550	30	R. Rebelde, Pinar del Río, PR
N1)	—	550	1	R. Rebelde, Manzanillo, GR
N1)	DC	570	1	R. Rebelde, Pilón, GR
N1)	KF	580	5	R. Rebelde, Baracoa, GU
N1)	AN	580	10	R. Rebelde, Mantua, PR
N1)	W	590	150	R. Rebelde, La Julia, HA
N1)	DN	590	10	R. Rebelde, Guantánamo, GU
N3)	HI	590	60	R. Reloj, Santa Clara, VC
N1)	KV	600	150	R. Rebelde, Urbana Noris, HO
N1)	AN	610	1	R. Rebelde, Bahía Honda, PR
N3)	HI	610	1	R. Reloj, Trinidad, SS
N1)	KF	620	1	R. Rebelde, Moa, HO
N1)	GN	620	30	R. Rebelde, Colón, MA
N2)	BC	640	50	R. Progreso, Guanabacoa, HA
N2)	DQ	640	10	R. Progreso, Las Tunas, LT
N1)	BC	640	5	R. Rebelde, Las Mercedes, GR
N1)	KU	650	1	R. Rebelde, Santiago de Cuba, SC
N1)	DC	650	1	R. Rebelde, Media Luna, GR
L31)	—	v650	—	R. Sandino, Pinar del Río: 2300-0300
N2)	HG	660	30	R. Progreso, Santa Clara, VC
N2)	—	670	—	R. Progreso, ?
N3)	Q	670	50	R. Reloj, Arroyo Arenas, CH
N2)	HN	680	1	R. Progreso, Cienfuegos, CI
N1)	JV	680	10	R. Rebelde, Ciego de Avila, CA
N2)	DB	680	1	R. Progreso, Santiago de Cuba, SC
N2)	—	690	20	R. Progreso, Jovellanos, MA
N5)	DU	700	1	R. Enciclopedia, Guantánamo, GU
N1)	GA-	700	1	R. Rebelde, Sancti Spíritus, SS
N1)	KJ	710	10	R. Rebelde, Holguín, HO
N1)	HQ	710	50	R. Rebelde, Santa Clara, VC
N1)	JN	710	30	R. Rebelde, Camagüey, CM
N1)	FC-	720	1	R. Rebelde, Cienfuegos, CI
N2)	OB-	730	10	R. Progreso, La Fé, IJ
N2)	JL	740	10	R. Progreso, Camagüey, CM
N2)	KP	740	20	R. Progreso, Sagua de Tánamo, HO
N1)	HV	750	1	R. Rebelde, Trinidad, SS
N3)	CD	760	5	R. Reloj, Guanabacoa, HA
N3)	DE	760	20	R. Reloj, Las Mercedes, GR
N3)	—	780	—	R. Reloj, ?
N3)	AQ	790	30	R. Reloj, Pinar del Río, PR
N2)	DT	v800	1	R. Progreso, Manzanillo, GR (r. 792)
N2)	DW	810	10	R. Progreso, Guantánamo, GU
L30)	CA	820	10	R. Ciudad de la Habana, "la emisora joven," Santa Catalina, CH: 1000-0500
N2)	JT	820	10	R. Progreso, Morón, CA
N3)	DE	820	10	R. Reloj, Contramaestre, GR
L06)	—	830	300	R. Taíno, La Julia, HA
P05)	HW	840	10	"W", Santa Clara, VC: 1100-0600
P10)	KC	840	1	R. Revolución, Santiago de Cuba, SC
N2)	DQ	840	1	R. Progreso, Las Tunas, LT
N5)	BQ	840	1	R. Enciclopedia, La Fé, IJ
N2)	HL	850	1	R. Progreso, Trinidad, SS
N2)	KB	850	5	R. Progreso, Baracoa, GU
L01)	BL	860	1	R. Metropolitana, Arroyo Arenas, CH
N2)	HL	870	10	R. Progreso, Sancti Spíritus, SS
N2)	DT	870	1	R. Progreso, Pilón, GR
N2)	AF	880	30	R. Progreso, Pinar del Río, PR
N4)	HD	890	1	R. Musical, Santa Clara, VC
N4)	DZ	890	1	R. Musical, Santiago de Cuba, SC
N2)	KP	900	50	R. Progreso, Cacocum, HO
N3)	GL	910	10	R. Reloj, Bolondrón, MA
P08)	FA	910	10	R. Cadena Agramonte, Camagüey, CM
N3)	GL	920	1	R. Reloj, Unión de Reyes, MA
N3)	GB	930	1	R. Reloj, Cárdenas, MA
N3)	JS	930	10	R. Reloj, Ciego de Avila, CA
N3)	KN	930	1	R. Reloj, Santiago de Cuba, SC
N3)	DP	930	10	R. Reloj, Las Tunas, LT
N3)	GU	940	1	R. Reloj, Colón, MA
N3)	KD	940	10	R. Reloj, Holguín, HO
N4)	BF	950	5	R. Musical, Arroyo Arenas, CH
N3)	HI	950	1	R. Reloj, Sancti Spíritus, SS
N5)	GF	960	1	R. Enciclopedia, Matanzas, MA
N3)	HJ	960	1	R. Reloj, Cienfuegos, CI
N4)	JD	960	0.25	R. Musical, Ciego de Avila, CA
N3)	DJ	960	10	R. Reloj, Guantánamo, GU
L02)	CK	980	1	El Periodico del Aire, Sapo, CH
L03)	KR	980	1	La Voz del Níquel, Moa, HO
N3)	DE	980	1	R. Reloj, Bayamo, GR
P01)	AS	990	1	R. Guamá, Pinar del Río, PR
P01)	AC	1000	10	R. Guamá, Los Palacios, PR
N4)	HB	1000	1	R. Musical, Sancti Spíritus, SS
N4)	JB	1000	1	R. Musical, Camagüey, CM
P01)	AP	1010	10	R. Guamá, Guane, PR
N4)	KM	1010	1	R. Musical, Holguín, HO

P01)	AP	1020	1	R. Guamá, Bahía Honda
P01)	AX	1030	1	R. Guamá, La Palma, PR
P12)	KT	1040	1	R. Victoria, Puerto Padre, LT
L06)	—	1040	—	R. Taíno, ? (irr.)
P12)	KT	1050	10	R. Victoria, Las Tunas, LT
P12)	KT	1060	1	R. Victoria, Amancio Rodríguez, LT
P13)	DX	1060	5	CMKS, Baracoa, GU
P03)	—	1070		
P13)	KS	1070	10	CMKS, Guantánamo, GU
P03)	CH	1080	10	R. Cadena Habana, Güines, HA
P01)	AP	1090	1	R. Guamá, Minas de Matahambre, PR
P03)	CH	1090	—	R. Cadena Habana, San Antonio de los Baños
P11)	KO	1090	1	R. Angulo, Moa, HO
P03)	CH	1100	1.5	R. Cadena Habana, La Salud, HA
P11)	KO	1100	1	R. Angúlo, Banes, HO
L06)	—	1100	—	R. Taíno, ? (irr.)
P11)	KO	1110	10	R. Angulo, Holguín, HO
P03)	CH	1120	5	R. Cadena Habana, Artemisa, HA
P11)	KO	1120	1	R. Angúlo, Mayarí, HO
N5)	HA	1130	1	R. Enciclopedia, Santa Clara, VC
P11)	KO	1130	5	R. Angulo, Sagua de Tánamo, HO
P03)	CG	1140	5	R. Cadena Habana, Loma de la Cruz, HA
P14)	KX	1140	1	R. Bayamo, Media Luna, GR
P14)	KX	1150	10	R. Bayamo, Entronque Bueycito, GR
P14)	KX	1160	—	R. Bayamo, Pilón, GR
P13)	KS	1170	10	CMKS, Maisí, GU
L06)	—	1180	300	R. Taíno, Villa María, CH
N2)	DB	1180	1	R. Progreso, Mayarí Arriba, SC
N3)	BM	v1180	1	R. Reloj, Nueva Gerona, IJ
P07)	HT	1190	1	R. Sancti Spíritus, Yaguajay, SS
P04)	—	1190	1	R. 26, La Caridad, MA
L07)	BS	1200	0.25	R. Ariguanabo, San Antonio de los Baños, HA
P07)	HT	1200	1	R. Sancti Spíritus, Trinidad, SS
P10)	KC	1200	10	R. Revolución, Palma Soriano, SC
P07)	HT	1210	10	R. Sancti Spíritus, Sancti Spíritus, SS
P10)	KC	1210	1	R. Revolución, Chivirico, SC
P10)	KC	1210	1	CMKC, Mayarí Arriba, SC
P04)	GY	1220	10	R. 26, Central España, MA
P04)	GJ	1230	3	R. 26, Unión de Reyes, MA
P04)	GW	1240	10	R. 26, Bolondrón, MA
P05)	HW	1250	1	"W", Caibarién, VC
P14)	KS	1250	0.25	CMKS, Imías, GU
P04)	GH	1260	1	R. Victoria de Girón, Torriente
N3)	JG	1270	10	R. Reloj, Camagüey, CM
N5)	GF	1280	1	R. Enciclopedia, Varadero, MA
L04)	KW	1280	1	R. Mambí, Santiago de Cuba, SC
L08)	JC	1280	1	R. Rectángulo, Guáimaro, CM
N5)	DS	1290	5	R. Enciclopedia, La Pastora, HA
P05)	HW	1290	1	"W", Rancho Veloz, VC
N5)	DA	1300	1	R. Enciclopedia, Las Tunas, LT
P05)	HR	1310	1	"W", Sagua La Grande, VC
P15)	BN		1	R. Caribe, Nueva Gerona, IJ
N5)	DA	1320	1	R. Enciclopedia, Santiago de Cuba, SC
N5)	HA	1320	1	R. Enciclopedia, Sancti Spíritus, SS
L09)	AD	1330	1	R. Artemisa, Artemisa, HA
L10)	DO	1330	1	R. Banes, Banes, HO
P06)	HU	1340	10	R. Ciudad del Mar, Palmira, VC: 1100-0600
P06)	HU	1350	1	R. Ciudad del Mar, Aguada de Pasajeros, CI
L11)	KY	1350	1	R. Libertad, Puerto Padre, LT: 1200-2200
P08)	FA	1360	10	R. Cadena Agramonte, Rodolfo Ramírez Esquivel, CM
P08)	FA	1370	1	R. Cadena Agramonte, Nuevitas, CM
L12)	DV	1370	1	R. Siboney, Santiago de Cuba, SC
P08)	FA	1380	10	R. Cadena Agramonte, Central Brasil, CM
L13)	BT	1390	1	R. Jaruco, Jaruco, HA
N4)	GX	1400	1	R. Musical, Matanzas, MA
P08)	FA	1400	1	R. Cadena Agramonte, Guaimaro, CM
N5)	AL	1410	1	R. Enciclopedia, Pinar del Río, PR
L14)	JW	1410	1	R. Santa Cruz, Santa Cruz del Sur, CM
L16)	HS	1420	1	R. Caibarién, Caibarién, VC
L15)	—	1420	0.25	R. Llanuras, Colón, MA
P09)	JY	1430	10	R. Surco, Morón, CA
L17)	BU	1440	0.25	R. Güines, Güines, HA
P09)	JP	1440	10	R. Surco, Ciego de Avila, CA
L18)	JF	1450	1	R. Maboa, Amancio Rodríguez, LT
L28)	HZ	1460	1	R. Cubitas, Sola, CM (r. 1550)
L20)	GE	1470	1	R. Ciudad Bandera, Cárdenas, MA
L21)	JI	1480	0.25	R. Florida, Florida, CM
L22)	BW	1490	0.25	R. Camoa, San José de las Lajas, HA
L23)	DH	1490	1	R. Mayarí, Mayarí, HO
L24)	KQ	1500	0.25	R. Majaguabo, San Luis, SC
N5)	DA	1500	1	R. Enciclopedia, Holguín, HO
N5)	DA	1510	1	R. Enciclopedia, Moa, HO
L25)	KZ	1520	1	R. Baragua, Palma Soriano, SC
L26)	JX	1530	1	R. Morón, Morón, CA
L27)	—	1540	1	R. Sagua, Sagua La Grande, VC
L19)	JQ	1550	1	R. Nuevitas, Nuevitas, CM
N5)	—	1560	1	R. Enciclopedia, Ciego de Avila, CA
P08)	FA	1580	1	R. Cadena Agramonte, Santa Cruz del Sur, CM
L29)	DF	1590	0.5	R. Granma, Manzanillo, GR: 1000-0500

SHORTWAVES: N1) Bauta 5025kHz 10kW, 3365v 0.5kW.

FM: in Habana 90.3 MHz N2) — 92.1 MHz N1) — 93.7 MHz N5) — 95.1 MHz N4) — 95.9 MHz N3) — 97.7 MHz L02) — 101.0 L30) — 102.1 L01) — 104.0 P3).

PROVINCES: CA = Ciego de Avila — CH = Ciudad Habana — CI = Cienfuegos — CM = Camagüey — GR = Granma — GU = Guantánamo — HA = Habana — HO = Holguin — IJ = Isla de la Juventud — LT = Las Tunas — MA = Matanzas — PR = Pinar del Río — SC = Santiago de Cuba — SS = Sancti Spiritus — VC = Villa Clara.

NATIONAL NETWORKS

N1) **R. Rebelde**, Edif. Radiocentro, Av. 23 N° 258, Vedado, Habana 4 (Ap. 6277, Habana 6) — Head of Public Rel. Dept: Jorge Luís Mas Zabala. **D.Prgr**: Actualities; "Haciendo Radio" 1100-1400; **N:** 1900, 0100.

N2) **R. Progreso, "La Onda de la Alegría",** Infanta 105, Habana 3 (Ap 30-42) — **D.Prgr:** Actualities; "Ayer, hoy y mañana" 1100-1400; **N:** "Su reportero RP" 27 minutes past each hour.

N3) **R. Reloj (CMCD),** P y 23, Vedado, Habana 4 — **D.Prgr:** Continuous news & commentary with background of clock-like pulses marking the seconds; "Radio Reloj", chime, and timecheck on the minute.

N4) **CMBF, "R. Musical Nacional",** Av. 23 N° 258, Vedado, Habana 4 — **D.Prgr:** 1200-0400. Classical music and programming of a cultural nature.

N5) **R. Enciclopedia,** Calle N N° 266, Vedado, Habana 4 — **D.Prgr:** Light instrumental music with timechecks and encyclopedic "facts" by female announcer.

PROVINCIAL NETWORKS

P01) R. Guamá, Ap. 14, Pinar del Río.

P03) CMCH, Radio Cadena Habana, "La Frecuencia Popular": Av. 51 el 128 y 130, La Habana.

P04) R. 26, Matanzas.

P05) CMHW ("W"), Ap. 376, or: Parque Vidal el Martha Abreu y Pao Chao, Santa Clara.

P06) R. Ciudad del Mar, Ap. 290, or: Av. 58 No. 3311, Cienfuegos 55100.

P07) CMHT, R. Sancti Spíritus, Céspedes 411, San José de Socorro, Sancti Spíritus.

P08) R. Cadena Agramonte, Ap. 140, Camagüey. **Te:** +53 5210.

P09) R. Surco, Ciego de Avila.

P10) CMKC, R. Revolución, Ap. 232, Santiago de Cuba.

P11) CMKO, R. Angulo, Holguín.

P12) CMKT, R. Victoria, Las Tunas.

P13) CMKS, Trinchera antiimperialista, Donato Mármol 409, 95100 Guantánamo.

P14) R. Bayamo, Bayamo GR.

P15) R. Caribe, La Emisora del Archipiélago de los Canarreos, Nueva Gerona, Isla de la Juventud.

LOCAL STATIONS (additional information only)

L01) R. Metropolitana, Av. 26 No. 877, Nuevo Vedado, La Habana.

L02) CMCK El Periódico del Aire, Av. 15 No. 210, Vedado, La Habana.

D.Prgr: news, music & sports in "magazine" format.

L06) R. Taíno, "Tour" radio in Havana; Calle "0" No. 216, Vedado, (Postal addr: P.O. Box 3040), La Habana 3 — **D.Prgr:** 1100-0600. **English:** 2000-2200, 0100-0200, 0300-0430. Bilingual ID's. Music, features, & "commercials" for tourists & visitors to Cuba.

L07) Av. 41 No. 5614, San Antonio de las Baños 32500.
L11) Av. de la Libertad 95, Puerto Padre, LT.
L12) R. Siboney, Santiago de Cuba — **D.Prgr:** instrumental music with time-checks by female announcer.
L13) Av. 25 No. 1810, Jaruco 32800.
L15) Ap. 97, Colón, MA.
L17) Calle 76 No. 7707, Güines 33900.
L19) R. Nuevitas, Ap. 46, Nuevitas, CM.
L20) Calle 13 y 5a. Av., Cárdenas, MA.
L22) Av. 47 No. 10202, San José de las Lajos 32700.
L28) R. Cubitas, Sola, MA.
L29) Ap. Postal 220, Manzanillo, GR.
L30) Addr. as N5).
L31) Cd. Sandino, Pinar del Río, PR.

EXTERNAL SERVICE

RADIO HABANA, Cuba

ADDR: Apartado 6240, La Habana. **Te:** +28 74954. **Telex:** 1344. **L.P:** Man. Dir: Milagro Hernandez Cuba. Head of Prgrs: Ignacio Canel. Chief Eng: Luis Pruna.
STATIONS (G.C: 82.30W/23.00N): 50/75/100/250kW.

kHz	kHz	kHz	kHz	kHz	kHz	kHz
5965	9505	9655	11875	15105	15340	17705
6000	9550	9815+	11950	15230	15415	17750
6010	9620	11760	11970	15300	15425	17770
6180						

+) SSB.

Arabic	**Area**	**kHz**	**Guarani**	**Area**	**kHz**
1900-2000	Eu/Af/ME	17770	2300-2400	SAm	15340
		15425	**Portuguese**		
2200-2300	SAm	11970	0730-0830	Eu	15105
0100-0200	SAm	15340	1000-1100	SAm	15340
Creole			2000-2100	Eu	9590
2100-2200	CAm	9620	2200-2300	SAm	15340
0200-0300	CAm	11760	**Quechua**		
English			1000-1050	SAm	11760
0000-0200	ENAm	9815	0000-0100	SAm	15340
0200-0500	ENAm	6010	**Spanish**		
0200-0430	CNAm	9655	0630-0730	Eu	15105
0400-0500	ENAm	6180	0900-1100	SAm	9505
0500-0700	WNAm	6000	0900-1145	SAm	15300
2100-2200	Eu/Af/ME	17705	1100-1400	NAm/CAm	5965
2200-2300	CAm/USA	6180	1100-1650	Carib	9505
Esperanto*			1100-1700	Am	11760
1700-1730	Am	11760		SAm	11875
	SAm	11875			15230
		15230		CAm/SAm	15340
		15340	1200-1700	SAm	17750
	CAm	15415	1400-1700	CAm	15415
1830-1930	Eu/Af/ME	17770	1800-2000	Eu	9590
1830-1900	Eu	17705	2000-2200	SEu/NAf	17770
		15425			15425
2200-2230	Eu/Af/ME	17770	2300-0200	NAm/CAm	5965
		15425		Am	11760
2230-2300	NAm/CAm	5965	2300-0600	Carib	9620
		17750		SAm	11875
	Carib	9620			11970
	SAm	11875			15230
		15230	0000-0400	CAm	9550

*) Sun only.

French		
0300-0400	Carib	6180
	Am	11760
2100-2240	Eu	17705
2100-2200	Af	9600

V. by QSL-card. Rec. acc. Rp — **PUB:** Prgr. schedule free on request.

AFRTS STATIONS (US Navy)

ADDR: FPO New York, NY 09593, USA.
STATION: Guantánamo Bay: 1340kHz 0.25kW.
FM: 102MHz 0.4kW (stereo) — 103MHz 0.05kW.
D.PRGR: 24h on 1340kHz/102MHz. Rel. AFRTS satellite sce. on 103MHz.

DOMINICA

L.T: UTC − 4h — **Pop:** 88.000 — **Radio sets:** 45.000 — **Pr.L:** English, Patois — **E.C:** a/c 50, 240V — **ITU:** IOB.

DOMINICA BROADC. CORP. (Gov. Comm.)

ADDR: P. O. Box 1, Victoria Str, Roseau. **Te:** +1 (809) 448 3282. **Fax:** +1 (809) 448 2918. **Cable:** RADOM Dominica — **L.P:** GM: G. R. John. Dep. GM: Miss J. Pascal. PD: Mrs. S. Brown. CE: G. James.
MW: Hillsborough 595kHz 10kW — **FM:** 88.1MHz.
D.PRGR: 0930(Sun 0955)-0230. **N:** 1000, 1030, 1100(BBC), 1400, 1715, 2000, 2100, 2200. **Patois:** 1800-1930MF.
ANN: "DBS Radio" — **V.** by QSL-card — **F.PL:** 20kW.

GOSPEL BROADC. CORP. (Rlg.)

ADDR: P. O. Box 205, Roseau. **Te:** +1 (809) 448 4391.
L.P: GM: W. K. DeBoer. CE: F. R. Guiste.
MW: Stow Estate ZGBC 740kHz 10kW — **FM:** 102.1MHz.
D.PRGR: 0900-0300. **BBC N:** 1100, 1200W, 1300W, 2000W. **Local N:** 1755, 2300, 0230. **Patois:** Sun 1800-1830, Wed 1215-1230.
ANN: "Caribbean V. of Life from Dominica" — **V.** by letter.

VOICE OF THE ISLANDS (Rlg.)

ADDR: P. O. Box 402, Roseau. **Te:** +1 (809) 448 4042.
L.P: Exec. Dir: Rev. Raymond Conrad.
MW: 860kHz 10kW — **FM:** 96.1MHz 1kW (stereo).
D.PRGR: 1015-0300. **BBC N:** 1100, 1200.
ANN: "V. of the Islands – VOI" — **F.PL:** additional st's.

R. Caribbean Int: FM 98.1MHz 24h (see under St. Lucia).

DOMINICAN REPUBLIC

L.T: UTC − 4h — **Pop:** 7.532.000 — **Radio sets:** 1.150.000 — **Pr.L:** Spanish — **E.C:** a/c 60, 110V — **ITU:** DOM.

DIRECCION GENERAL DE TELECOMUNICACIONES

ADDR: Calle Isabel la Católica 73, Santo Domingo. **Tel.** +1 (809) 689-4161 — **Cable:** Telecom — **L.P:** Ramón Soto Then, Rear Admiral.

ASOCIACION DOMINICANA DE RADIODIFUSORAS (ADORA)

ADDR: Calle Paul Harris 3, Centro de los Héroes, Santo Domingo. **Te:** +1 (809) 535-4057. **L.P:** Pres. Teo Veras.
STATIONS: Mediumwaves: Call: HI—

	Call	**kHz**	**kW**	**Name and h. of tr.**
1)	CM	540	5	R. ABC, Sto. Domingo: 0900-0400
2)	AA	560	5	R. Ritmos, Santiago: 1000-0400
3)	MS	570	5	R. Cristal, Sto. Domingo: 1000-0400
4)	DV	590	10/5	R. Santa María, La Vega: 0900-0200
6)	JR	610	5/1	R. Acción, Santiago: 24h.
7)	SD	620	10	R. Televisión Dominicana, Santo Domingo: 0900-0500
8)	AF	630	1	R. Monte Cristi: 1100-0300
7)	SD	640	1	(r: HISD 620), Santiago: 0900-0500
9)	AT	650	20	R. Universal, Sto. Domingo: 24h
10)	AM	660	5	R. Quisqueyana, Santiago: 1000-0400
59)	BS	670	5	R. Dial (R. 670), San Pedro de Macorís: 0900-0400
11)	JX	680	3/2	R. Zamba, Santiago Rodríguez: 0930-0300
12)	AW	690	10	R. Guarachita, Sto. Domingo: 0900-0400
13)	DC	700	1,5	R. Mao, Valverde: 1000-0400
14)	AQ	°720	5	R. Norte, Santiago: 0900-0500
15)	Z	730	10	R. HIZ, Broadcasting Nacional, Sto. Domingo: 0930-0500
16)	DB	750	5	R. Alegre, Santiago: 0900-0500
17)	CO	760	5	R. Cordillera, Sto. Domingo: 1000-0600
18)	MD	770	1	R. Aguila, Santiago: 1000-0400
19)	BO	780	0.5	R. Constanza: 1100-0200
20)	L	790	5	La Voz del Trópico, Sto. Domingo: 0930-0400
21x)	AV	810	5	R. Baní: 1100-0300.
21)	AZ	820	5	R. Santiago: 1000-0400
22)	JB	830	10	R. HIJB, Sto. Domingo: 1000-0500

23) AB	840	5	R. Isabel de Torres, Puerto Plata: 1000-0400
72) GA	850	5	R. Guarocuya, Barahona: 1000-0400
107) LR	850	5	R. Clarín, Santiago: 0930-0400
107) LR	860	50	R. Clarín, Sto. Domingo: 0900-0500
25) VG	870	5	R. La Vega: 1000-0400
26) OR	880	1	R. Santa Cruz, Mao Valverde: 0900-0300
27) PJ	890	4/5	R. Continental, Sto. Domingo: 0930-0600
28) BN	900	5	R. Puerto Plata: 0900-0400
29) LB	910	5	R. Noventiuno (R.91), Bonao: 0930-0300
30) BA	*920	5	R. Diamante, Sto. Domingo
31) CK	930	5	Ondas del Yaque, Santiago: 0945-0400
32) G	950	10	R. Popular, Sto. Domingo: 0900-0600
33) FF	960	5	La Voz del Atlántico, Puerto Plata: 1000-0500
34) VP	970	1	R. Olímpica, La Vega: 1000-0400
35) FA	980	5	La Voz de las Fuerzas Armadas, Sto. Domingo: 0900-0400
36) SA	990	5	R. Cibao, Santiago: 1000-0400
37) HG	1000	5/1	R. Beller, Dajabón: 1000-0300
38) JP	1010	10	R. Comercial, Sto. Domingo: 0900-0400
38) JP	1010	1	(r: HIJP 1010), Salcedo: 0900-0400
38) JP	1010	1	(r: HIJP 1010), San Juan: 0900-0400
102) TS	1020	10	R. Enriquillo, Tamayo: 0900-0300
39) DL	*1030	10	R. Novedades, La Vega: 1000-0400
40) ON	1040	10	R. Central, Sto. Domingo: 24 h.
41) CB	1050	10	R. Hispaniola, Santiago: 0900-0400
42) KF	1060	1	R. Azua: 1000-0200
43) RV	1060	1	R. Mar, San Pedro de Macorís: 0900-0300
44) BI	1070	5	HIBI R. AM, San Francisco de Macorís: 0900-0400
45) MC	1080	0.5	R. Ambar, Sto. Domingo: 1000-0300
46) JM	1090	3	R. Amistad, Santiago: 0930-0300
50) RB	1090	1/0.25	R. Jimani: 0900-0300
47) HD	1100	1	R. Oriente, San Pedro de Macorís: 0900-0400
48) MP	1100	1	R. Ocoa, San José de Ocoa: 1200-0200
49) PS	1100	1	R. Nagua: 0900-0200
51) JJ	*1110	1	R. Jarabacoa: 1000-0300
95) OS	1110	1/0.5	R. Marién, Dajabón: 0930-0200
53) RL	1130	10	R. Exitos, Santiago: 24h.
54) RA	1140	5	R. Anacaona, San Juan de la Maguana: 1100-0400
55) AS	1150	5	Onda Musical, Sto. Domingo: 0930-0400
56) BG	1160	5/1	Radiolandia, Santiago: 0900-0400
57) BE	1180	10	R. Mil, Sto. Domingo: 0930-0500
58) AG	1190	5	R. Azul, Santiago: 1000-0400
50) MR	1200	1/0.25	R. Caracol, Azua: 1000-0400
61) CJ	1210	5	R. Merengue, San Francisco de Macorís: 1000-2200
62) N	1220	25	R. HIN, Sto. Domingo: 1000-0400
63) MB	1230	1	R. Ideal, Moca: 1000-0300
64) AU	1240	1	R. Revelación, Puerto Plata: 0900-0300
50) CV	1240	5/0.5	R. Barahona: 0900-0400
66) BC	1250	5	La Voz del Progreso, San Francisco de Macorís: 1000-0400
67) RJ	1250	1	R. Juventud, La Romana: 18h.
38) T	1260	1	R. Visión, Sto. Domingo: 0900-0400
69) TA	1270	1	R. Ambiente, Baní: 1000-0400
71) HZ	1280	1	R. Clave, Monte Plata: 1000-0300
70) VM	1290	1	R. Bonao: 1000-0400
74) KQ	1300	1	R. Radio, Sto. Domingo: 0900-0600
75) MH	1310	1	R. Real, La Vega: 1100-0400
7) SD	1310	1	(r: HISD 620), El Seibo
35) FA	*1330	5	La Voz de las Fuerzas Armadas, Moca: 0900-0400
76) BZ	1340	1	R. Centro, San Juan de la Maguana: 1000-0300
7) SD	1350	1	(r: HISD 620), Monte Cristi: 0900-0500
77) MG	1360	0.5	R. Rutas Musical, La Romana: 1000-0400
7) SD	1360	1	(r: HISD 620), La Vega: 0900-0500
7) SD	1370	1	(r: HISD 620), Barahona: 0900-0500
35) FA	1370	5	La Voz de las Fuerzas Armadas: Elías Piña: 0900-0400
79) RP	1380	0.5	R. Seybo: 0900-0200
80) SC	1380	5	R. Nacional, Santiago: 1000-0400
81) AR	1390	1	R. San Cristóbal: 1100-0300
82) AL	1400	1	Ondas del Valle, La Vega: 1100-0200
7) SD	1400	1	(r. HISD 620), San Juan: 0900-0500
7) SD	1400	1	(r. HISD 620), Pedernales: 0900-0500
83) AE	1410	1	R. Revelación en América, Sto. Domingo: 24h
84) RM	1410	3	R. Sol, Higüey: 1000-0300
85) GG	1410	1	R. Gri-Gri, Río San Juan: 1000-0300
50) CH	1410	1/0.25	R. 14-10, Barahona: 0900-0400
86) ED	1420	1	R. Oro, Cotuí: 1000-0200
87) LF	1430	1	R. Cayacoa, Higüey: 0900-0400
103) JC	1430	5	R. Maibá, Santiago: 1000-0400
88) FS	1440	1	R. Bahía, Nagua: 1000-0400
89) AD	1440	5	R. San Juan: 1000-0300
90) AK	1440	5	R. Renuevo, La Voz de La Verdad, Sto. Domingo: 1000-0400 (rel.)
91) AC	1450	10	R. Util, Salcedo: 0900-0400
92) AN	1460	0.5	R. Renacimiento, Hato Mayor
93) DE	1470	5	R. Super 56, San Francisco de Macorís: 1000-0400
98) AH	1480	5	R. VEN, Sto. Domingo: 1000-0400
7) SD	1490	1	(r: HISD 620), Puerto Plata: 0900-0500
96) AP	1490	2	La 54 AM, Moca: 1000-0300
97) PA	1500	0.5	R. Color, Higüey: 0900-0400
98) BL	1510	10	R. Pueblo, Sto. Domingo: 1000-0400
99) WJ	1520	1	R. Samaná, "R. 15-20,": 0930-0400
100) BU	1540	1	La Voz de La Romana: 0930-0400 (r. 1545)
38) FB	1540	1	R. Recuerdos, Sto. Domingo: 0900-0400
50) PZ	1560	1/0.25	R. Pedernales: 0900-0300
104) AJ	°1570	5	R. Amanecer Internacional, Sto. Domingo: 1000-0400
50) PK	1580	1/0.25	R. Neiba: 0900-0400
105) DA	*1590	3	R. Hit Musical, Santiago: 1100-0300
106) EG	1600	5	R. Revelación, Sto. Domingo: 24 h.

SHORTWAVES: Call HI—

Call	kHz	kW	Name and h. of tr.
14) AQ	4799.8	1	La N-103, Santiago: 1000-0600*
50) 5V	4930	5	R. Barahona: 0000-0400 (irr.)
108) VR	4962	1	R. Cima 100, Santo Domingo: 2100-1000
104) 1J	v6025	5	R. Amanecer Internacional, Sto. Domingo: 1000-0400
21) AZ	9877.5v	1	R. Santiago: irr.

Addresses and other information:

1) Avenida Rómulo Betancourt 2078, Sto. Domingo — Dir: Padre Teófilo de la Cruz.

2) Ap. 581, Santiago — Owner: Johannes W. Streese.

3) Ap. 819, Sto. Domingo — Owner: Elizardo Dickson. Dir: Fabio C. Herrera.

4) Ap. 55, La Vega — Dir: Antonio Llluberes.

6) Ap. 556, Santiago — Dir: Jaime Nelson Rodríguez.

7) Ap. 965, Sto. Domingo — Dir: Adriano Rodríguez. Asst. Dir: George Rodriguez. Chief Eng: Julio Moscoso.

8) Ap. 52, Monte Cristi — Dir: R. O. Moscoso C.

9) Ap. 2000, Sto. Domingo — Dir: Hector J. Torres.

10) Ap. 26, Santiago — Dir: Marcos Tulio Cepeda.

11) Ap. 2, Santiago Rodríguez — Dir: Manuel de Js. Thomas B.

12) Calle Palo Hincado 302, Sto. Domingo — DG: R.A. Aracena.

13) Ap. 20, Valverde-Mao (or: Ap. 789, Santiago) — Pres: José Raposo P.

14) Pedro Francisco Bonó 58, Santiago — Acting Dir: Lic. Luis José Rodríguez.

15) Ap. 68, Sto. Domingo — Dir: Oscar Ureña

16) Ap. 112, Santiago — Dir: A. Vanderhorst.

17) Calle Emilio A. Morel esq. Luis Pérez, Ensanche La Fe — Own: J. Jiménez Maxwell.

18) Calle El Sol 51, 3a. Planta, Edif. Lamarche Alvarez, or: Ap. 1636, Santiago.
19) Calle Duarte 17, Constanza — Dir: R.M. Tactuk.
20) Ap. 335, Sto. Domingo — Owner: J. Custals.
21) Ap. 282, Santiago — Owner: Luis F. Moscoso F.
21x) Calle Pres. Billini esq, Duarte, Bani — DG: Manuel E. Bello.
22) Edif. Teleantillas, Km. 7½, Carr. Duarte — Dir: Héctor Olivo.
23) Ap. 146, Puerto Plata — Own: Antonio Damián Brisso B.
25) Autopista Duarte, La Vega — Dir: Gabriel Grullón A.
26) Ap. 80, Valverde — Dir: Guarionex Aquino R.
27) Ap. 156, Sto. Domingo — Dir: José Lluberes.
28) Calle Beller 35, Puerto Plata — Dir: Waldo R. Musa.
29) Calle Mella 50, Bonao — Dir: Salvador Chestaro.
30) Calle Arzobispo Meriño 208, Apto. 205, Sto. Domingo — Dir: Pedro J. Santana, hijo.
31) Ap. 225, Santiago — Dir. Roque Candelario. Ll.
32) Ap. 928, Sto. Domingo — Dir: R. Corporán de los Santos.
33) Duarte 65, altos, Puerto Plata — Dir: Antonio Badui.
34) Calle Mella esq. Colón, La Vega — DG: M. Fernández M. — **F.Pl:** SW.
35) Ap. 1350, Sto. Domingo — Dir: Osvaldo Cepeda C.
36) Ap. 141, Santiago — Owner: July Bermúdez.
37) Hno. Martin Juffermans, Dajabón — Dir: Nelson Ricards O.
38) Ap. 1322, Sto. Domingo — Dir. J. Francisco N.
39) Las Arboledas 5, La Vega — Dir: M.C. Fernández.
40) Avenida 27 de Febrero 350, Sto. Domingo — Pres. Adm: Ing. Irvin Pérez P.
41) Ap. 454, Santiago — Dir: Patria Villamil de Candelario.
42) Calle Emilio Prud'homme 17A, Azua — Dir: J.A. Félix A.
43) Ap. 476, San Pedro de Macorís — Dir: D. Aguiló Hidalgo.
44) Ap. 201, San Francisco de Macorís — Owner: J.A. González B.
45) Edif. Jaar, Calle El Conde esq. Espaillat, Sto. Domingo — Owner: V.M. Furment U.
46) Ap. 561, Santiago — Dir: José E. McDougal Jr.
47) Ap. 64, San Pedro de Macorís — Dir: Agustín Santana.
48) Calle Canada, San José de Ocoa — Dir: Felipe A. Isa P.
49) Calle Colón 66, Nagua — Pres: P. Riggio.
50) Empresas Radiofonicas, S.A., Ap. 20339, Santo Domingo. Mgr: Rodolfo Z. Lama Jaar. C.E: Roberto Lama Sajour.
51) Ap. 10, Domingo Savio 1, Jarabacoa — Dir: J.J. Batista.
53) Calle El Sol 42, Santiago — Dir: Moraldo Ovalle.
54) Ap. 37, San Juan de la Maguana — Dir: M.A. Rojas F.
55) Ap. 860, Sto. Domingo — Mgr: M. Báez A.
56) Ap. 187, Santiago — Dir: Lic. Américo Rodríguez A.
57) Ap. 1372, Sto. Domingo — **Te:** 565-5366 — Own: Manuel Ma. Pimentel — **F.Pl:** react. 4940 kHz.
58) Ap. 79, Santiago — Dir: Leonardo A. Atiles R.
59) Ap. 142, San Pedro de Macorís — Dir: O. Cepeda.
61) Ap. 57, San Francisco de Macorís — Dir: Ernesto A. Grullón.
62) Ap. 1220, Sto. Domingo — Dir: Fco. Suárez.
63) Corazón de Jesús 61, Moca — Dir: P. Mirelis.
64) Av. Circunvalación Norte, Puerto Plata — Dir: L.A. Pelegrín.
66) Calle San Francisco 50, or: Ap. 264, San Francisco de Macorís — Dir: Fabio Raposo Tabar.
67) Ap. 151, La Romana — Own: Fco. A. Micheli, hijo.
68) E. A. Morel 27, Sto. Domingo — Dir: Jose Fco. Núñez.
69) Sánchez esq. Mella, Baní — Dir: Pedro C. Guerrero.
70) Calle Libertad 97, Bonao — Dir: José A. Aquino.
71) Miguel A. Monclús, Monte Plata — Dir: Ricardo Acosta.
72) Padre Billini esq, Jaime Mota — Dir: Alci Pimentel.
74) Ap. 209, Sto. Domingo — Dir: R.A. Martínez G.
75) Juan Rodríguez 76-A, La Vega — Dir: Cuto F. Holguín.
76) Ap. 65, San Juan de la Maguana — Dir: Iván A. Ramírez.
77) Ap. 207, La Romana — Dir: Jorge Garib M.
79) Ap. 266, El Seibo — Dir: Pablo Puerto P. — **F.Pl:** 5kW.
80) General López 80, Santiago — Dir: Nombrito Wehbe.
81) Calle Padre Borbón 16, San Cristóbal — Dir: Marcos A. Bello.
82) Restauración 64, La Vega — Dir: Dionisio Moya.
83) Calle Arturo Logroño 158, Ensanche La Fe, Sto. Domingo — Dir: Inés Abreu.
84) Carr. Mella, Km. 1.5, Higüey — Dir: Juan Julio Campos.
85) Calle Gri-Gri, Río San Juan — Dir: Estela Ovalle Vda. Abud.
86) María Sánchez 68, Cotuí — Dir: P.S. Inoa.
87) Calle General Santana 65, Higüey — Dir: Modesto Amado Mercedes.
88) Calle Duarte 76, Nagua — Dir: Juan Fco. Santos H.
89) Ap. 88, San Juan de la Maguana — Dir: L.P. González.
90) Prol. Av. Bolívar 49, Santo Domingo — Dir: Lic. Braulio Portes.
91) Ap. 2, Salcedo — Dir: Eusebio H. Bencosme.
92) Calle Felipe de Castro, Hato Mayor — Dir: Domingo A. Ubiera.
93) Ap. 123, San Francisco de Macorís — Own: Pedro R. Fernández.
95) Henríquez 53, Dajabón — Dir: Roberto Alonso, S.J.
96) Ap. 213, Moca — Dir: Gabino Nūñez Rosa.
97) Calle 16 de Agosto, Higüey — Dir: Ramón Ant. Lescaille.
98) Ap. 2217, Santo Domingo — DG: Ezequiel Molina R.
99) Av. Malecón, Samaná — Dir: Miguel Castillo.
100) Ap. 213, La Romana — Pres: N. Maríñez.
101) Calle Emilio A. Morel 27, Ensanche La Fe, Sto. Domingo — Dir: Augustín Ferrer.
102) Ap. 99, Tamayo — Dir: Gerardo Rodman.
103) Ap. 42, Santiago — Own: Julio Caballero.
104) Ap. 1500, Sto. Domingo — Dir: Pastor Alexis Muñoz D., C.E: Ing. Sócrates Domínguez.
105) Ap. 513, Santiago — Dir: Pedro Manuel Jorge.
106) Av. 25 de Febrero 144, Ens. Las Américas, Sto. Domingo — DG: Ing. Julio Soto Medina.
107) Apartado 205-2, Santo Domingo. **Te:** +1 (809) 565-7676. Pres: José Penzo. Administrator: Maria Elena J. de Read.
108) Ap. 804, Santo Domingo — Own: Agr. Roberto Vargas. Fax: 809 541 1088.

FM in Santo Domingo:
88.5 La Brava — 15) 89.1 R. WAO — 89.5 Super 89 — 12) 90.1 — 90.5 R. Eco — 91.1 La 91 — 38) 91.7 Tuya — 92.1 Hits 92 — 40) 92.5 Mágica — 106) 92.9 — 93.5 R. Alfa Omega — 62) 94.1 Viva — 94.5 KQ-94 — 38) 95.1 — 22) 95.7 La Nota Diferente — 3) 96.5 — 9) Galaxia — 97.5 Clásica — 9) 98.1 — 20) 98.5 R. Prisma — 99.1 Sonido Suave — 99.5 R. Listín — 7) 99.9 — 108) 100.5 R. Cima — 100.9 Super Q — 101.3 Z-101 — 9) 102.1 La X — 57) 103.1 — 30) 103.7 — 17) 104.1 — 104.5 La Super Potente — 1) 105.1 — 107) 105.5 — 106.1 R. Disco — 38) 106.5 Super Rock — 62) 107.9 — Numerous outlets in the rest of the country: Santiago alone has 19.

EL SALVADOR

L.T: UTC – 6h — **Pop:** 5.527.000 — **Radio sets:** 1.935.000 — **Pr.L:** Spanish — **E.C:** a/c 60, 110/220V — **ITU:** SLV.

ADMINISTRACION NACIONAL DE TELECOMUNICACIONES, ANTEL

ADDR: Centro de Gobierno, San Salvador. **Cable:** Presantel. **Fax:** 226849.
L.P: Pres: Lic. Saúl Suster. Head, Div. Radioeléctrica: Ing. Carlos Pineda R.

ASOCIACION SALVADOREÑA DE EMPRESARIOS DE RADIODIFUSIÓN (ASDER)

ADDR: Ap. Postal 210, San Salvador — Pres: Manuel A. Flores B.
STATIONS: Call: YS—.
H. of tr: 1100-0400 generally. (r) = repeater.

	Call	kHz	kW	Name, location and h. of tr.
1)	HV	540	10	R. Restauración, San Salvador: 1200-0500
2)	FG	550	2	R. Variedades, Sonsonate
3)	DR	570	1	(r: YSKT 570), Santa Ana
3)	KR	570	1	(r: YSKT 570), San Miguel
3)	KT	570	10	R. Cadena Central, San Salvador: 1000-0600 (weekends 0800).
4)	NK	600	3	R. Exitos, San Salvador
5)	SS	610	10	(r: YSSS 655), Morazán
6)	LN	630	10	La Monumental/San Salvador: 1200-0500
5)	SS	655	10	R. El Salvador, San Salvador (n. 660)
5)	SS	680	10	(r: YSSS 655), Cabañas
7)	UU	700	5	R. Cadena YSU, San Salvador
7)	UU	700	1	(r: YSU), Ahuachapán
7)	UU	700	1	(r: YSU), Usulután
7)	UU	700	1	(r: YSU), San Miguel
7)	UU	700	1	(r: YSU), Santa Ana
7)	US	700	1	(r: YSU), Sonsonate
9)	RA	720	1	(r: YSR 720), Santa Ana
9)	RA	720	5	(r: YSR 720), San Miguel
9)	RA	720	5	Cir. YSR, San Salvador: 1100-0500
5)	SS	750	1	(r: YSSS 655), Santa Ana

10)	KL	780	10	R. Cadena YSKL, San Salvador: 1030-0530
10)	KL	780	5	(r: YSKL 780), San Miguel
10)	KL	780	1	(r: YSKL 780), Sonsonate
10)	KL	780	1	(r: YSKL 780), Usulután
10)	KL	780	1	(r: YSKL 780), Santa Ana
11)	AX	800	20	La Voz Panamericana, San Salvador
12)	FA	810	2	R. Lorenzana, San Vicente: (n 840)
13)	PX	830	4	R. Pax, San Miguel
14)	FB	840	5	R. Vanguardia, San Salvador: 1000-0430 (n 810)
15)	RC	860	1	R. Tecana, Santa Ana
16)	AR	870	1	R. 10, Cadena de Oro, San Salvador
16x)	CD	880	1	R. Ritmo, Santiago de María
17)	LA	890	3	R. Musical, Santa Ana: 1000-0500
18)	QJ	900	10	Radiópolis, San Salvador: 1000-0300
19)	TG	930	5	R. Cadena Sonora, San Salvador: 1100-0500
19)	TG	930	1	(r: YSTG 930), Sonsonate
19)	TG	930	1	(r: YSTG 930), San Miguel
19)	TG	930	1	(r: YSTG 930), Santa Ana
19)	TG	930	4	(r: YSTG 930), Ahuachapán
19)	TG	930	1	(r: YSTG 930), Usulután
20)	HG	950	1	R. Minuto, San Miguel
21)	TW	960	0.5	R. Centro, Sonsonate
18)	MS	970	1	R. UTEC, San Salvador
23)	TA	990	5	R. Clásica, San Salvador
24)	HH	1000	1	Estación "H", Santa Ana
25)	CA	*1020	10	R. Internacional, San Salvador
27)	RM	1030	1	R. Frontera, Ahuachapán
7)	QR	1050	3	Auto R., San Salvador
26)	AN	r1070	1	La Voz de los Ausoles, Ahuachapán
3)	EM	1080	1	(r: YSME 1080), San Miguel
3)	EM	1080	10	R. Cadena 1080, San Salvador
28)	MG	v1090	3	R. 1090, Atiquizaya
29)	RF	1100	3	R. Ranchera, San Salvador
30)	CL	1110	1	(r: YSCL 1160), Santa Ana
30)	CL	1110	2.5	(r: YSCL 1160), San Miguel
10)	LR	1120	3	La Romántica, San Salvador: 1100-0500
20)	LG	1130	1	R. Chaparrastique, San Miguel
23)	TS	1140	10	R. El Mundo, San Salvador
31)	AJ	1130	2	R. Moderna, Santa Ana
32)	CF	1150	3	Ondas Orientales, "La Romántica de Oriente," San Miguel: 1100-0400
30)	CL	1160	3	R. Horizonte, San Salvador
2)	CB	r1175	0.25	La Voz del Pacífico, Sonsonate (n. 1170)
33)	VE	1180	5	R. VEA, San Salvador
29)	MM	1200	10	R. Fiesta, San Salvador
34)	KJ	1200	1	R. Sirama, San Miguel
22)	CG	1210	0.5	R. La Paz, Zacatecoluca
3)	NB	1230	1	R. Uno Dos Tres, San Salvador
	MT	1240	0.5	R. Metapan, Santa Ana
	QR	1240	1	R. Norteña, San Miguel
35)	AA	1260	3	R. América, San Salvador
34)	QZ	1270	1	R. Mía, San Miguel
36)	QV	1280	1	R. Galaxia, Santa Ana
37)	MA	1290	1	R. Chalatenango, Chalatenango
38)	LV	1300	5	"La Versatil", San Salvador: 1145-0400
	RV	1310	5	R. Veritas, Usulután
	AH	1320	1	R. Emanuel, La Unión
39)	HQ	1330	5	R. Progreso, San Salvador: 1100-0500
40)	XW	1340	1	R. Novedades, Usulután
41)	JI	1390	1.5	La Voz del Litoral, Usulután
42)	UCA	1420	1	R. Universitaria, San Salvador
46)	FM	1450	5	Super R., San Salvador
	KR	1450	1	R. Renovación, San Miguel
43)	CS	1500	1	R. Fides, Usulután: 1100-0400
44)	DA	1500	1	R. Imperial, Santo Tomás: 1100-0400
45)	CZ	1530	1	(r: YSCZ 1550, San Miguel
45)	CZ	1550	10	Cadena Cuscatlán, San Salvador
45)	CZ	1570	0.5	(r: YSCZ 1550), Santa Ana

Addresses and other information:

1) Iglesia Elim, Col. Sta. Lucía, Ilopango, San Salvador.
2) Av. Sta. Mónica 9, Sonsonate — DG: Larry A. Zedán B.
3) 25 Calle Poniente No. 113, San Salvador — DG: José R. Durán Pacheco — Ranchera.
4) Edif. TV2, Alameda Dr. Manuel E. Araujo, San Salvador.
5) Parque Venustiano Carranza, San Salvador — DG: L. Moreira.
6) 4a Calle Oriente 224, Edif. Comercial, San Salvador.
7) Av. Olímpica y Alameda Dr. M.E, Araujo, Colonia Escalón, San Salvador — DG: Julio Suvillaga Zaldívar.
9) Ap. Postal 720, San Salvador. **Te:** +503 23-6744 — **L.P:** DG: Andrés Rovira.
10) 4a Calle Oriente 528, San Salvador. **Te:** +503 81-4111. **Fax:** +503 22-8809 — Pres: Manuel Antonio Flores.
11) 1a Calle Poniente 3412 or: Ap. 1835, San Salvador — Dir: Pbro. Roberto A. Torruella.
12) Costado Norte Pargue Central, San Vicente.
13) 10 Calle Oriente 102 Bis, San Miguel.
14) Pasaje Vilanova 114, entre 10 y 8 Av. Nte., San Salvador.
15) Altos del Cine Tecana, Santa Ana.
16) Edif. Palomo, C. Rubén Darío y 23 Av. Sur, San Salvador.
16x) 2a Av. Nte. No. 24, Santiago de María, Usulután.
17) 4a Av. Sur entre 7a. y 9a., Calle Poniente, Edif. Plaza de Vidrio, Santa Ana — DG: Ma. Emma de Peñate.
18) Av. Las Acacias 230, Col. San Benito, San Salvador.
19) Urb. La Esperanza 1322, San Salvador — DG: Roberto Castañedo.
20) Atrás de Jardines de Bolonia, San Miguel — DG: Joaquín Aparicio Borja.
21) 5a Calle Oriente, Sonsonate — Dir: J.L. Gudiel G.
23) Calle y Colonia Universitaria Norte, Mejicanos, San Salvador.
24) 9a Calle Poniente 25, Santa Ana — Dir: M. Montes M.
25) 29a Calle Oriente 218, San Salvador — DG: Enrique Ochoa.
27) Av. 2 de Abril y 8a. C. Pte., Ahuachapán.
26) Av. Morazán km 101, Ahuachapán.
28) Esq. 2a. Av. Sur y 1a. C. Ote. 3-93, Atiquizaya.
29) Calle y Col. Roma, Edif. 3-B, San Salvador.
30) 4a. C. Pte. entre 43 y 45 Av. Sur, San Salvador — Dir: Julio René Vargas.
31) 4a Av. Norte y Calle Don Bosco 14, Santa Ana.
32) Ap. 19, San Miguel — DG: Oscar René Polío.
33) Km 18½ Carretera Panamericana, Cantón la Palma, San Martín — DG: Enríque Ramírez Castillo.
34) Km. 134, Carr. Panam., Cantón Jalacatal, San Miguel.
35) Col. San Benito, Av. Las Palmas, Pje. 6 No. 109, San Salvador.
36) 2a Calle Poniente 43-B, Colonia Santa Lucía, Santa Ana.
37) 4a. C. Pte. No. 11, Chalatenango — DG: Mario A. Espinoza.
38) 5te. Av. Norte 1803, San Salvador — DG: M. de Jesús S.
39) Ap. 855, San Salvador — Dir: José M. Cárcamo.
40) 4a Av. Sur 302, Usulután — DG: Oscar Aparicio Borjas.
41) 4a Av. Norte 5, Usulután — DG: José Infantozzi.
42) UCA, Autopista Sur, San Salvador — Dir: Mario Torres C.
43) 4a. C. Ote. No. 2, Usulután — Dir: René R. Rivera A.
44) Ap. 56, Sonsonate — Gte: Daniel E. Cardona Lara.
45) Estado Mayor de la F.A., Carr. a Santa Tecla, San Salvador — Dir: Eugenio Acosta R.
46) 12 Av. Norte 1712, San Salvador — Dir: Fco. A. Melara L.

FM in San Salvador.
St. names: Estéreo . . . exc. where indicated. All 1-5 kW.
9) 88.9 — 89.7 R. Vida — 90.5 R. 2000 — 91.3 Club — 92.1 "92" — 92.9 R. Láser — 23) 93.7 — 94.5 Mi Preferida — 95.3 Doble L — 96.1 Scan — 96.9 Yess — 97.7 R. Agape Metropolitana — 45) 98.5 — 29) 99.3 — 100.1 Juvenil — 100.5 R. Venceremos — 100.9 Amor — 3) 101.7 — 102.1 R. Farabundo Martí — 29) 102.5 R. Femenina — 23) 103.3 — 10) 104.1 — 29) 104.9 R. Fiesta — 105.7 Súper — 16) 106.5 — 7) 107.3 — 107.9 R. Cadena La Exitosa.

GRENADA

L.T: UTC − 4h — **Pop:** 83.000 — **Radio sets:** 80.000 — **Pr.L:** English — **E.C:** a/c 50, 220V — **ITU:** GRD.

GRENADA BROADC. CORP. — G. B. C. Radio (Statutory Body)

ADDR: P. O. Box 535, Morne Jaloux, St. George's. **Te:** +1 (809) 440 3033. **Fax:** +1 (809) 440 4180. **Cable:** Broadcast Grenada W. I.
L.P: GM: Lew Smith. PD: Eugenie Mason. CE: Allan Palmer.
MW: 535kHz 20kW — **D.PRGR:** 0957-0400(Sun 0200) — **N: BBC:** 1200, 1600, 0200. **VOA:** 1000, 2400.
ANN: "This is GBC AM 535" or "This is 535 Radio".
V. by letter. Rp.

HARBOUR LIGHT OF THE WINDWARDS (Rlg.)

ADDR: Carriacou. **Te:** +1 (809) 443 7628 — SM: Randy Cornelius — **MW:** 1400kHz 5kW.
D.PRGR: 0930-0200 — **N:** rel. BBC & VOA.
ANN: "This is the Harbour Light of the Windwards broadcasting from beautiful and friendly Carriacou" — **V.** by QSL-card.
F.PL: broadcasts in French & Spanish.

GUADELOUPE (French)

L.T: UTC-4h — **Pop:** 350.000 — **Radio sets :** 100.000 (est) — **Pr.L:** French, Creole patois — **E.C:** a/c 50, 220V — **ITU :** GDL.

RADIODIFFUSION FRANCAISE D'OUTRE-MER (RFO)

ADDR: B.P. 402, 97163 Point à Pitre Cédex. **Te:** +590 939696. **Fax:** +590 939682. **Telex:** 919064.
L.P: Dir: R. Surjus. St. Mgr: E. Roumiguieres. Editor-in-Chief: R. Sargenton. Dir. Prgr: G. César.

Mediumwaves	kHz	kW		kHz	kW
Point à Pitre	640	40	Morne à Louis	1420	5

FM: Mourne à Louis 97.0MHz 3kW + 8 relay st's.
D.PRGR: 24h. **N:** 1100, 1700, 2230 + rel. France Inter.
ANN: "Ici Point à Pitre, RFO Guadeloupe" or "RFO".
INT-SIG: Biguin, played on guitar — **V.** by QSL-card. Rp.
Private stations: Over 30 FM st's are operating.

GUATEMALA

L.T: UTC−6h (Su: UTC−5h) — **Pop:** 9.500.000 — **Radio sets:** 400.000 — **Pr.L:** Spanish — **EC:** a/c 60, 110/120/127/220V — **ITU:** GTM.

MINISTERIO DE TRANSPORTES Y OBRAS PUBLICAS
Dirección General de Radiodifusión y Televisión Nacional

ADDR: 18 de Septiembre 6-72, Z-1, Edif. Tipografía Nacional, Nivel 3, Guatemala. **Te:** +502 (2) 532539.
L.P: Dir. Gen: Ricardo Gómez Flor. Dep. Dir: Enrique Hernán.

Mediumwaves: Call; TG —, ° = Shortwave, (r) = repeater, * = inactive.

Call	kHz	kW	Name, Location and h. of tr.
1) RV	560	10	R. 560, Guatemala: 1200-0500
1x)	r560	1	R. Quetzal, Malacatán
2) PA	570	1	R. Palmeras, Escuintla
3) Y	580	3	R. Progreso "La Onda Alegre", Guatemala: 1100-0700
4) RQ	590	5	R. Quiché, Sta. Cruz: 1200-0400
5) RC	600	1	R. Campesina, Escuintla Tiquisate: 1100 (Sun 1200)-0500
6) GA	610	0.5	R. Señorial 610, Guatemala: 1200-0700
7) PQ	620	5	R. 6-20, San Cristóbal: 1200-0400
7x) EL	630		R. El Porvenir, Sta. Elena: 1000-0400 (Sun. 1100-0200)
8) W	*640	11.5	R. Nal "La Voz de Guatemala", Guatemala: *
9) Q	660	10	R. Nacional, La Voz de Quetzaltenango: 1100-0400
10) RT	670	10	R. Fabulosa, Guatemala: 1000-0400
11) VP	680	10	R. Norte, Cobán: 1100-0400
12) VB	690	1	R. Tamazulapa, Jutiapa
13) HR	700	15	R. Mundial, Guatemala: 1000-0600
14) XL	710	1	R. Tecún Umán, Quetzaltenango: 1100-0430
15) RO	720	1	R. Corona, Morales: 24h
16) N	°730	10	R. Cultural, Guatemala: 0955-0630
17) HF	740	1	R. Tacaná, San Marcos: 1100-0500
18) AJ	750	1	R. Tropicana, Escuintla
19) HB	760	5	Nueva R. Super, Guatemala: 1000-0500
20) BX	770	1	R. Fraternidad, Quetzaltenango: 1000-0600
21) O	790	6	R. Festival, Guatemala: 1100-0500
79) YZ	800	1	R. Rosa, Chiquimulilla: 24h
22) TO	820	10	R. Internacional, Guatemala: 1000-0600
23) AV	830	5	R. Satélite, Mazatenango: 1100-0400
24) SM	840	0.35	La Voz de San Marcos, San Marcos: 1300-0400
49) X	850	5	R. Ranchera, Guatemala: 24h
26) FP	*860	1	R. Nal. Tikal, Flores Petén: *
59) L	870	0.5	R. Victoria, Mazatenango
27) J	880	10	R. Nuevo Mundo, Guatemala: 1030-0500
28) HU	890	1	R. Escuintla, Escuintla
29) MA	900	1	R. Amatique, Pto. Barrios: -0500
13) KL	910	10	R. Emperador, Guatemala: 1100-0500
30) RS	920	0.2	R. Sur, Escuintla
31) JL	v925	5	R. Imperial, San Pedro Carchá (n. 930): 1000-0400
32) TL	940	5	La Voz del Hogar, Guatemala: 1200-0500
33) AF	950	1	R. Indiana, Mazatenango
34) RU	960	1	R. Utatlán, Sta. Cruz del Quiché: 1100-0300
35) AX	970	5	R. Continental, Guatemala: 1200-0430
36) MQ	980	1	R. Retama, San Marcos: 1200-0500
37) AL	990	1	R. Perla de Oriente, Chiquimula: 24h
8) WC	1000	10	(r: TGW 640), Guatemala: irr.
39) RX	1010	1	R. Miramundo, Zacapa
61) —	1010	1	R. Ixil, Nebaj
40) CM	1020	5	R. Frontera, Pajapita: 1100-0400
41) UX	1030	10	R. Panamericana, Guatemala: 1200-0500
42) JP	1040	1	R. Oriental, Jalapa
43) SL	1050	5/1	La Voz de los Cuchumatanes, Huehuetenango: 1100-0600
44) XA	1060	10	Disco R., Guatemala: 24h
45) D	1070	3/2.5	La Voz de Occidente, Quetzaltenango: 1130-0500
45a) LU	v1080	1	R. Novedad, Zacapa
10) Z	1090	10	R. Mil Noventa, Guatemala
46x) SR	1100	1	R. Superior, Coatepeque
46) MK	1110	1	R. Marianalá, Santa Lucía: 1100-0400
47) C	1120	0.5	R. Poderosa, Guatemala
48) VR	v1135	1	La Voz de la Costa Sur, Retalhuleu
49) RR	1150	1	R. Fiesta, Guatemala: 24h
50) RI	1160	1	R. Izabal, Morales: 1100-0130
51) RL	1170	5	R. Cadena Landívar, Quetzaltenango: 0900-0300
52) T	1180	25/10	R. Sonora, Guatemala: 1130-0600
53) LU	1190	5	R. Novedad, Zacapa (r1079)
54) RJ	1200	1	R. Jutiapa, Jutiapa
55) MX	1210	5	R. Rumbos, Guatemala: 24h
56) MT	1220	1	R. Amiga, Antigua: 1100-0300
57) AT	1230	1	R. Atlántida, Puerto Barrios: 1100-0400
58) K	1240	5	R. Sensación, Guatemala
87) PY	1250	1	R. Payakí, Esquipulas: 1100-0300
60) PA	1260	1	R. Monumental, Antigua: 1100-0300
41) CQ	1270	2.5	Saturno AM, Guatemala: 1100-0400
62) TU	1290	0.5	La Voz de Totonicapán, Totonicapán: 0000-0400
63) TL	*1300	1	R. Tropical, Guatemala
64) AN	1310	1	La Voz de los Altos, Quetzaltenango: 1100-0300
65) ME	1320	1	R. Quezada, Jutiapa
66) MU	°1330	10	Unión Radio, Guatemala: 1100-0300
67) CO	1340	10	La Voz del Trópico, Coatepeque: 24h
68) MC	1350	1	R. Municipal Cultural, Cobán
41) LK	1360	10	R. 9-80, Guatemala: 1100-0500
69) AC	1370	1	La Voz de Colomba, Colomba: 1100-0300
70) EB	1380	1	R. Momostenango Educativa, Momostenango: 1100-1900
71) YC	1390	5	R. Estrella, Guatemala: 1200-2300
72) RB	1400	1	R. Porteña, Pto. Barrios: 24h
73) GH	1410	3/5	R. Xelajú, Quetzaltenango: 1155-0600
74) RP	1420	1	R. Capital, Guatemala: 1100-0400
75) AG	1430	1.2	La Voz de Huehuetenango: 1100-0400
76) MS	1440	0.5	La Voz de Mazatenango, Mazatenango: 0000-0400
13) LG	1450	1	R. Epoca, Guatemala: 24h

78)	RN	1460	5	R. Petén, Flores
80)	BH	1480	1	R. Horizontes, Guatemala: 1030-0500
81)	RE	1490	1	R. Modelo, Retalhuleu: 0900-0300
82)	DX	1510	5	R. Centroamericana, Guatemala: 1130-0500
83)	RS	1520	1	R. Superior, Coatepeque
85)	CK	1550	1	R. Sultana de Oriente, Zacapa: 1100-0500
86)	VE	1570	10	R. Voz Evangélica de America, (R. VEA), Guatemala: 1030-0430
88)	XC	1590	1	R. Triunfadora, Chimaltenango
8)	WC	*1600	10	(r: TGE 640), Guatemala

Shortwaves: Call: TG —. Location: Guatemala City exc. where indicated. All st's are non-commercial.

89)	BA	2360	1	R. Maya, Barillas: 1030-1400, 2230-0330
90)	DS	2389.9	1	R. La Voz de Atitlán, Santiago Atitlán: W 2200-0115, Sun 1800-2400 (n. 2390)
16)	NC	3300	10	R. Cultural: 0955-0630
89)	BA	3324.8	1	R. Maya, Barillas: 1030-1400, 2230-0330
91)	VN	3360.4	1	La Voz de Nahualá, Nahualá: 1100-1400, 2100-0300
92)	TZ	3369.9	1	R. Tezulutlán, Cobán: 1030-1600, 2045-0300, 2130-0230 (n. 3370)
93)	CH	3380	1	R. Chortis, Jocotán: 1100-1300, 2100-0300
95)	MI	4799.8	1	R. Buenas Nuevas, San Sebastián H.: 1100-1400, 2200-0230 (n. 4800)
94)	MN	4825.3	1	R. Mam, Cabricán: 1300-1700, 2000-2330
92)	TZ	4834.7	5	R. Tezulutlán, Cobán: as 3370 (n. 4835)
96)	VC	4844.3v	2.5	R. K'ekchi', Fray Bmé.de las Casas: 1100-1700, 2100-0300 (n. 4845)
91)	VN	5040	1	La Voz de Nahualá: 1100-1400, 2100-0300
16)	NA	5955.5	0.25	R. Cultural: 1000-0630 (n. 5955)
66)	UA	5981.7v	5	La Voz de la Iglesia Adventista en Latinoamérica, Guatemala: 1200-1500, 0000-0200 (n. 5980)

Addresses and other information:

1) 8 Calle 1-11, Z-11, Guatemala — Dir: Edna Castillo Obregón.
1x) Malacatán, San Marcos.
2) 15 Calle 2-43, Z-3, Escuintla — Dir: F. Tres G.
3) 9a Av. 0-32, Z-2, Guatemala — Dir: G. Humberto Andrino.
4) 7 Calle 3-67, Z-5, Sta. Cruz del Quiché — Dir: Edgar A. López — D. Prgr: Vernaculars only.
5) Col. 15 de Junio, Z-3, Tiquisate, Escuintla — Dir: José M. Alvarez C.
6) 15 Calle "A" 0-81, Z-1, Guatemala — DG: Haroldo Castillo O.
7) Barrio La Cienaga, San Cristóbal, Totonicapán — Dir: José F. Herrera Nelson.
8) 18 Calle 6-70, Z-1, Guatemala — DG: Arturo Soto Echeverría.
9) 13 Av. 8-19, Z-1, or: Ap. 113, Quetzaltenango — DG: Francisco Rojas López.
10) 7a Ave. 6-45, Z-9, Guatemala — Dir: Rolando Archila Marroquín. (various st's rel. **N.** from this st. under slogan "Emisoras Unidas").
11) 2 Calle 557, Z-3 Cobán, Alta Verapaz — DG: Guillermo Casa F.
12) Calle 15 Septiembre, Santa Cruz, Jutiapa — DG: Silvia L. de Azurdia.
13) 6a Av. 2-80, Z-1, Guatemala — Dir: Federico Azurdia.
14) 6a Av. 6-41, Z-1, Quetzaltenango — Dir: M. Sosa de Cifuentes.
15) Calle Principal, Morales, Izabal (or: cf. 37) — Dir: Ana de Liu.
16) Ap. 601, 01901 Guatemala — Dir: Stephen R. Sywulka. **English:** 0300 (Sun 2345)-0430. Vernaculars: 1100-1130.
17) 8 Calle 8-01, Z-2, San Marcos — DG: Sandra E. Rubio.
18) 6 Av. 2-45, Z-1, Escuintla — DG: Carlos Echeverría.
19) 30 Av. 3-86, Z-11, Utatlán II, 01001 Guatemala — Dir: Dr. Ernesto Ponce Bedoya.
20) Ap. 90, Quetzaltenango — DG: P. Edgar A. Mazariegos.
21) 11 Calle 2-43, Z-1, Guatemala — DG: Pablo F. Nuiza S.
22) 37 Av. 1-15, Z-7, 1007 Guatemala — Dir: José Antonio Gonzáles A.
23) Mazatenango, Such — Dir: Herberto Díaz G.
24) Palacio Maya, San Marcos.
26) Flores Petén.
27) 6a Av. 10-45, Z-1, Guatemala — Dir: Alfredo González G. Various st's relay **N.** from this st. under slogan "Sistema de Emisoras Nuevo Mundo".
28) Masagua, Escuintla — Dir: Freddy Azurdia.
29) Km 291 Ruta Atlántico, Puerto Barrios, Izabal — Gte: J. Aquino R.
30) 4a Av. 7-30, Z-1, Escuintla — Dir: Violeta Gonzales G.
31) 5a Calle 7-53, Z-1, San Pedro, Carchá, A.V — DG: Rolando Archila Marroquín.
32) Ap. 31, Guatemala — Dir: Carlos Martínez.
33) 6a Av. 10-54, Z-1, Mazatenango — Dir: Guillermo Alcázar.
34) 5 Av. 4-22, Z-5, Sta. Cruz del Quiché — Dir: C.I. de Archila.
35) 15 Calle 3-45, Z-1, Guatemala — Dir: Roberto Vizcaíno R.
36) 5a Calle 8-21, Z-1, San Pedro, San Marcos — Dir: Heberto Díaz M.
37) 7a Av. 4-00, Z-1, Chiquimula (or: 6a. Av. 0-60, Z-4, Torre Prof. II, of. 904, 01004 Guatemala) — Gte: Eduardo Alfonso Liu.
39) Palacio Nal. Ministerio de la Defensa Nacional, Guatemala.
40) Pajapita, San Marcos — Dir: Carlos Alcázar F.
41) 1a Av. 35-48, Z-7, Guatemala — Dir: María V. de Paniagua.
42) Av. Chipilapa "A" 1-03, Z-2, Jalapa — Dir: A. Morales H.
43) 2a Calle 4-42, Z-1, Huehuetenango — Dir: Héctor L. Alvarez G.
44) 10 Calle 5-20, 21, Guatemala.
45) 7a Av. 0-26, Z-2, Quetzaltenango — Dir: María Cristina R. de Ríos.
46) 5a Av. 5-122, Z-1, Santa Lucía — Dir: V.M. Marroquín.
46x) Coatepeque — Dir: Marie P. Quintana.
47) Edif. El Centro, Nivel 16, Z-1, Guatemala.
48) Ap. 84, Retalhuleu — Dir: José E. Archila M.
49) Grupo Radial El Tajín, 2a Av. "A" 13-45, Z-1, or: Ap. 1526, Guatemala. Fax: 81419 — Pres: Eleonora Girón de Rubens.
50) Barrio El Carrizal, Morales, Izabal — Dir: Carlos Enrique Molina.
51) 14 Av. "A" 0-78, Z-1, Quetzaltenango — Dir: Herberto Díaz M.
52) 2a Calle 18-07, Z-15, Vista Hermosa I 3er. Nivel, Guatemala — DG: Lic. Eduardo Mendoza.
53) 4 Calle 10-34, Z-1, Zacapa — DG: Eduardo Liu.
54) Km 117, Carr. Interamericana, Jutiapa — Dir: C. Archila.
55) 4a Av. 1-13, Z-1 Guatemala — Dir: Roberto Bocaletti L.
56) Av. del Desengaño 20, Antigua — Dir: C. Echeverría.
57) Ap. 425, Puerto Barrios — Dir: Jaime Archila.
58) Ap. 281, Guatemala — Dir: Humberto Gonzales G.
59) La Libertad 9-91, Z-1, Mazatenango — Dir: Alfonso Medina.
60) 7a. Av. Norte No. 8, Antigua — Dir: E. Cifuentes.
61) Nebaj, Quiché — C.E: Benjamín García. Prgr. in Sp., Quiché and Ixil.
62) Palacio Municipal, Totonicapán.
63) Guatemala — DG: Rafael Matta R.
64) Ap. 107, Quetzaltenango — Dir: J. César García P. — Rlg.
65) Quezada, Jutiapa.
66) Ap. 51-C, Guatemala — Dir: Alberto Ixcat. **English:** 0200-0400. Owned & operated by AWR-Latin America.
67) 3 Calle 2-06, Z-1, Coatepeque — Dir: Gilberto Calderón V.
68) Edif. Municipalidad, 5a. Calle 1-06, Cobán, Alta Verapaz — DG: Mauro R. Acchun.
69) Colomba, Quetzaltenango — Coord: José Molina.
70) Momostenango, Totonicapán — DG: José F. Guinea Y.
71) Ap. 1405, Guatemala — Dir. Gen: Jośe E. Pérez M.
72) 8 Av. 15 y 16 Calle, Pto. Barrios (or: cf. 37) — Dir: Mario A. Miranda M.
73) 4 Calle 15A-62, Z-1, Quetzaltenango — Dir: Rolando Oliva V.
74) 11 Calle 32-15, Z-7, Guatemala — Dir: Eduardo Soria S.
75) Ap. 13, Huehuetenango — Dir: Mauro Guzmán M.
76) 2a Av. 25 Col. 30 de Junio, Mazatenango.
78) Isla Santa Bárbara, 17001 Flores (Petén) — Dir: Mariano Colmenares.
79) Edif. Municipal, Chiquimulilla, Santa Rosa (or: cf. 37) — Dir: José L. Castro.
80) 17 Ave. No. 20-25, Com. Las Pérgolas, Res. Granaí III, Guatemala Z-11 — DG: Joel Melgar.
81) Retalhuleu (Ap. 183-A, Guatemala) — Dir: G. Andreu C.
82) 3 Av. 6-32, Z-1, Guatemala — Dir: Carlos Peynado.
83) 3 Calle 3-38, Z-1, Coatepeque, Retalhuleu — Dir: M. Plinio Q.
84) 10a Calle 7-65, Z-9, Guatemala — Dir: Mario Estupinian R.
85) Calz. M. García, Centro de Oriente, Zacapa.
86) Ap. 1213, 01901 Guatemala — Dir: Efraín Gutierrez L. F.Pl: SW.
87) 5a Av. 9-02, Z-1, 20007 Esquipulas — Dir: Luis Felipe Paz R.
88) 2 Calle 3-53, Z-3, Chimaltenango.

89) 13026 Barillas, Huehuetenango — Mgr: José Castañeda — Prgr. in Sp. and 5 vernaculars.
90) Santiago de Atitlán, Depto. Sololá — Mgr: Juan Ajtzip A.
91) Nahualá, Depto. Sololá — Dir: Sister Janet Druffel.
92) Ap. 19, 16901 Cobán, A.V. — Dir: Mons. Gerardo Flores R. Exec. Dir: Hno. Antonio Jacobs, O.P.
93) Centro Social, 20004 Jocotán, Chiq. — Mgr: P. Juan María Boxus.
94) Acu'Mam, Cabricán, Dept. Quetzaltenango — Dir: José B. Escalante Ramos — F.Pl: 5kW, FM.
95) 13020 San Sebastián H., Huehuetenango — Mgr: Israel Rodas M. — Prgr in Mam, Aguacatec, Quiché, Jacaltec and Ch'ol.
96) Fray Bartolomé de las Casas, 16015 Alta Verapaz (or: Medios Educativos, A.C., Ap. 310, Cd. Satélite, Edo. de México 53100, México) — Mgr: G. Sun Xicol.

FM in Guatemala City: St. names: Estéreo . . . exc. where indicated. If available, MW fq's relay FM.
10) 88.1 Fabuestéreo — 88.5 Galaxia — 89.5 R. Stereo — 10) 90.9 Cadena Sideral — 49) 91.9 R. Exitos — 92.3 R. Universidad — 92.7 Cristal — 1) 93.5 Amor — 93.9 Jazz 94 — 55) 94.5 — 37) 94.9 "FM 95" — 49) 95.9 R. Ranchera — 97.3 Alfa Super — 98.1 Doble S — 37) 98.9 Globo — 1) 99.7 Conga — 16) 100.5 — 1) 101.0 Fresca — 102.0 "102" — 102.3 Eco "la Cariñosa" — 102.9 Metroestéreo — 49) 103.7 R. Fiesta — 105.0 Tropicálida — 66) 105.7 Advenesléreo — 49) 106.1 Máxima — 106.5 Ejecutiva — 8) 107.1 — 37) 107.5 Fama.

HAITI

L.T: UTC – 5h (Apr–Oct UTC – 4h) — **Pop:** 6.431.000 — **Radio sets:** 3.000.000 — **Pr.L:** Créole, French — **E.C:** a/c 50/60, 110/220V — **ITU:** HTI

CONSEIL NATIONAL DES TELECOMMUNICATIONS (CONATEL)

ADDR: Cité de l'Exposition 16 (B.P. 2002), Port-au-Prince. **Fax:** +1 (809) 23-0579, **Te:** +1 (809) 22-0300
Cable: CONATEL — **L.P:** Dir. Gen: Samuel Dubois, M.S.
STATIONS: all comm. except 1), 2), 3), 4), 14) Cult./gov., 16), 19) Cult/rlg.
Mediumwaves: Call: 4V—.

	Call	kHz	kW	Name and h. of tr.
1)	JS	640	0.4	La Voix du Seigneur, Port-au-Prince: 1600-1800, 2300-0300
2)	I	660	5/1	R. Lumière, Port-au-Prince: 1000-0100
2)	IA	720	1	R. Lumière, Carrefour Paye: 1000-0100
2)	IE	740	1	R. Lumière, Pignon-le-Jeune: 1000-0100
2x)	—	r740	1	R, Reveille, Cap Haïtien
2)	U	760	5	R. Lumière, Les Cayes: 1000-0100
2)	O	780	0.5	R. Lumière, Jérémie: 1000-0100
3)	EF	840	10	R. 4VEH, Cap-Haïtien: 1000-0300
4)	MK	860	3	R. Min Kontre, Les Cayes: 1000-0300
5)	JV	870	1	R. Express Continentale, Jacmel: 1030-0300
6)	PM	880	0.8	R. Indépendence Nouvelle, Gonaïves: 1000-0300
7)	VB	895	0.5	R. Trans-Artibonite, Gonaïves: 1000-0300
8)	KB	920	1	(r: KB 930), Petite Anse
8)	KB	930	10/3	R. Cap-Haïtien, Cap-Haïtien: 24h.
9)	LF	940	0.4	R. St. Marc Haïti, Saint-Marc: 1200-0200
10)	FF	940	0.2	R. Diff. Jacmélienne, Jacmel: 1100-0400
11)	CD	960	0.3	R. Carillon, Port-au-Prince: 1100 (Sun 1200)-0300
12)	CPS	990	0.2	R. Cacique, Port-au-Prince: 1100-0300
13)	JH	1020	1	R. Pétion-Ville, Pétion-Ville: 1100-0300
14)	DN	1080	20	R. Nationale, Port-au-Prince: 1000-0500
15)	AB	1150	5	R. Caraïbes, Port-au-Prince: 1000-0400 (r.1146)
16)	RS	1170	10	R. Soleil, Port-au-Prince: 1000-0200
17)	AF	1190	0.2	La Voix de la Grande-Anse, Jérémie
30)	LS	1210	1	R. Plus, Port-au-Prince: 24h
18)	AV	1230	1	R. La Voix de l'Avé-Maria, Cap-Haïtien: 1000-0400
19)	SJ	1240	10	R. Antilles Internationales, Port-au-Prince: 24h
31)	S	1250	1	La Voix du Plateau Central, Hinche
20)	AM	1280	10	R. Métropole, Port-au-Prince: 1100-0500
21)	JLD	1330	10	R. Haïti-Inter, Port-au-Prince: 1030-0400
32)	RL	1360	5	R. Liberté, Port-au-Prince: 24h
22)	EE	1370	1	R. Citadelle, Cap-Haïtien: 1000-0400
23)	MM	1370	1	R. Diffusion Cayenne, Les Cayes: 1100-0300
24)	SS	1380	0.7	R. Port-au-Prince, Port-au-Prince: 1030-0400
25)	TS	1410	1	La Voix du Nord Ouest, Port-de-Paix: 1100-0300
26)	GM	1430	10	M B C, Port-au-Prince: 1030 (Sun 1330)-0400
27)	EA	1460	0.2	La Voix du Nord, Cap-Haïtien: 1000-0400
28)	AA	1470	1	R. Arc-en-Ciel, (R. Lakansyel), Port-au-Prince: 0900-0500
33)	OC	1500	0.5	R. Select, Port-au-Prince
29)	VE	1560	10	La Voix de L'Espérance (R. Adventiste), Port-au-Prince: 0900-1700, 2100-0200

Addresses and other information: P-au-P = Port-au-Prince.
1) B.P. 1164, P-au-P — Dir: J. Simon Jean-Baptiste.
2) B.P. 1050, P-au-P — Dir: Robinson Joseph.
3) B.P. 1, Cap-Haïtien (B.P. 1739, P-au-P) or: c/o MFI, Box 15665, West Palm Beach, FL 33416 — Dir: Gaudin L. Charles. Eng: Claude Beachy — **D.Prgr: 840kHz** (DA 210°): Créole & French 1000-0300.
4) 137 Rue Antoine Simon, Les Cayes — Dir: Mgr Alix Verrier.
5) Rue de l'Eglise et rue Magloire Ambroise, Jacmel — Dir: Jean Verdier.
6) 116 Rue Egalité, Gonaïves — Dir: Fred Brutus.
7) 32 rue Anténor Firmin, Gonaïves — Dir: V. Bastien.
8) B.P. 64, Cap-Haïtien — Dir: Kalil Bitar.
9) Rue Armand Thoby, 20, St. Marc — Dir: L. S. Fleury.
10) 32 Rue Dorleáns, Jacmel — Dir: Fresnel François.
11) Bourdon, Ave. John Brown 356, P-au-P — Dir: Mme. C. Desmangles.
12) B.P. 1480, P-au-P — Dir: Mme. Jean-Claude Carrié.
13) Rue Rigaud 71, Pétion-Ville — Mgr: Mme. Carlo Hubert.
14) B.P. 1143, P-au-P — Dir: Pierre Raymond Dumas.
15) Ruelle Chavannes 23, P-au-P. — Dir: Jacques G. Simeon.
16) B.P. 1362, P-au-P — Dir: Père Arnoux Chéry.
17) 52 Rue Dr. Hyppolite, Jérémie — Dir: Héritos Felix.
18) B.P. 22, Cap-Haïtien — Dir: Rev. Marcel Bussels.
19) B.P. 81, P-au-P — Prés: J. Sampeur.
20) B.P. 62, P-au-P — Dir: Herbert Widmaier. MW: N. (:55) and sports. FM 99.1 and 100.1.
21) B.P. 737, P-au-P — Dir: Jean L. Dominique.
22) Rue 16-A, Cap-Haïtien — Dir: E. C. Eugène.
23) 55 Rue Duvivier Hall, Les Cayes — Dir: M. Mattieu.
24) B.P. 81, P-au-P — Dir: Georges L. Hérard.
25) Rue B. Sylvain, Port-de-Paix — Dir: St.-Aubain Saintil.
26) B.P. 367, P-au-P — Dir. Gen: F. C. Magloire.
27) Rue 20 B, Cap-Haïtien — Dir: Daniel Dupuy.
28) 21 Bizoton, P-au-P — Dir: Alex St.-Surin.
29) B.P. 1339, P-au-P — Dir: Wébert Lahens.
30) B.P. 1174, Port-au-Prince — Dir: Lionel Benjamin.
31) Rue Bon Cœur 10, Hinche — Dir: Jude Simon.
32) B.P. 1485, Port-au-Prince — Dir: Serge Beaulieu
33) Port-au-Prince — Adm: William Bonhomme.
34) Fort-Liberté — Own: Eric Lamothe.

HONDURAS (Rep.)

L.T: UTC – 6h — **Pop:** 5.100.000 — **Radio sets:** 1.800.000 — **Pr.L:** Spanish — **E.C:** a/c 60, 110V — **ITU:** HND.

EMPRESA HONDUREÑA DE TELECOMUNICACIONES (HONDUTEL)

ADDR: Ap. Postal 1794, Tegucigalpa. **Te:** +504 22-2101.
Telex: 1220 HT — **L.P:** Dir: Ing. Camilo A. Pon Z. Jefe Depto. de Frec: Ing. Emilio A. Montesi P.
Mediumwaves: Call HR—; °)also on SW; (r) = repeater. *)inactive — **H. of tr.** 1100-0500 exc. where indicated.

	Call	kHz	kW	Name, location and h. of tr.
2)	XD	550	3	R. Manantial, Santa Rosa de Copán: 1115-0300
2x)	JO	560	1	R. Campeona, Comayagua
3)	RZ2	560	1.5	R. Juticalpa, Juticalpa: 1100-0400
4)	PX	560	1	R. Tropical, San Pedro Sula
5)	LP2	570	1	(r: HRLP 610), Tela
5)	LP10	570	1	R. La Ceiba, La Ceiba
6)	ZQ	580	3	R. Tegucigalpa, Tegucigalpa
7)	VW2	590	3	(r: HRVW 650), Siguatepeque
5)	LP3	590	1	(r: HRLP 610), San Pedro Sula
7)	VW	600	1	(r: HRVW 650), La Ceiba
5)	LP	610	10	R. América, Tegucigalpa
5)	LP4	610	1	(r: HRLP 610), Sta. Rosa de Copán
5)	LP6	620	1	(r: HRLP 610), Comayagua
5)	LP5	620	1	(r: HRLP 610), Siguatepeque
5)	LP9	620	1	(r: HRLP 610), Juticalpa
28)	LP17	620	1	R. Continental, San Pedro Sula
5)	LP3	630	1	(r: HRLP 610), Choluteca
5)	LP7	630	1	(r: HRLP 610), La Ceiba
5)	LP13	630	1	(r: HRLP 610), San Lorenzo
8)	NN4	640	3	R. Exitos, Tegucigalpa
7)	VW	640	10	(r: HRVW 650), Santa Rosa de Copán
7)	VW	650	50	La Voz de Centroamérica, San Pedro Sula: 1100-0500
8)	NN18	660	3	(r: HRN 670), La Ceiba
8)	N	670	1	La Voz de Honduras, "la N", Tegucigalpa
8)	NN20	670	1	(r: HRN 670), Sta. Rosa de Copán
8)	NN3	680	1	(r: HRN 670), San Pedro Sula
8)	NN7	680	1	(r: HRN 670), Danlí
8)	NN10	680	1	(r: HRN 670), Juticalpa
8)	NN2	680	1	(r: HRN 670), Siguatepeque
8)	NN3	690	1	(r: HRN 670), Choluteca
8)	NN9	690	1	(r: HRN 670), Tela
8)	KL	700	1	R. Reloj, Tegucigalpa
9)	RH	710	5	La Voz de Occidente, Sta. Rosa de Copán: 1000-0500 (n. 700)
14)	NN	710	1	Estéreo Rey, San Pedro Sula
10)	LK	v718	1	R. Comayagua, Comayagua (n710)
11)	KN	710	2	La Voz de Olancho, Catacamas: 1100-0300
8)	NN13	710	1	(r: HRN 670), Olanchito
8x)	NN3	720	1	R. Caribe, La Ceiba
8)	TG	730	1	R. Televisión, Tegucigalpa
12)	QQ	730	1	R. Intibucá, La Esperanza (n. 740)
13)	IH	740	1.2	R. 7-40, Juticalpa: 1200-0400
14)	NN23	740	1	R. Eco, San Pedro Sula (r. 770.6)
15)	JH	750	1	R. Trujillo, Trujillo
16)	XK	°750	1	La Voz de la Mosquitia, Pto. Lempira
18)	XW	760	5	R. Comayagüela, Comayagüela: 1200-0600
14)	NN21	770	10	R. Norte, San Pedro Sula (r. 770.6)
19)	MV	770	1	R. Aguán, Olanchito
135)	RO	770	2.5/1	R. Majestad, Juticalpa: 1200-0400
20)	IR	790	1	R. Relámpago, Sta. Bárbara: 1000-0400
8)	TG2	790	3	R. Satélite, Tegucigalpa
17)	MA	800	1	R. Mundial, San Pedro Sula
21)	DL	r805	2	R. Corporación, Comayagua (n800): 1100-0400
22)	LP26	800	1	R. Sonora, Danlí
23)	MD	*800	1	R. Yoro, Yoro
90)	VC	810	1	(r: HRVC 1390), La Ceiba
25)	LP24	810	3	R. Valle, Choluteca: 1000-0400
5)	LP16	820	5	R. Moderna, Tegucigalpa
25x)	KW	820	3	R. Sultana, Sta. Rosa de Copán
24)	XS	830	2	R. Estéreo Sula, San Pedro Sula
26)	JB	830	3	Cadena Radial Impacto, Comayagua
27)	VQ	830	1	R. Excelsior, Juticalpa: 1100-0400
8)	UP	850	10	R. Centro, Tegucigalpa: 1100-0400
28)	BS	860	10	R. San Pedro, San Pedro Sula
8)	UP6	860	1	(r: HRUP 850), Danlí
8)	UP13	860	1	(r: HRUP 850), Juticalpa
1)	H9	860	1	(r: HRH 880), La Ceiba
1)	H4	870	1	(r: HRH 880), Choluteca
1)	H	880	10	R. Honduras "V. Oficial del Estado"
1)	H7	880	10	(r: HRH 880), Juticalpa
1)	H5	880	10	(r: HRH 880), Santa Rosa de Copán
1)	H3	890	10	(r: HRH 880), San Pedro Sula
1)	H2	890	10	(r: HRH 880), Siguatepeque
1)	H2	890	1	(r: HRH 880), Comayagua
1)	H6	890	1	(r: HRH 880), El Paraíso
1)	H8	890	1	(r: HRH 880), Olanchito
8)	UP7	900	1	(r: HRUP 850), Choluteca
8)	UP6	900	1	(r: HRUP 850), La Ceiba
29)	VS	910	10	R. Católica,"La Voz de Suyapa", Tegucigalpa: 1000-0500
30)	RM	920	1	R. Sistema, Comayagua: 1300-0300
31)	ZV	920	5	R. Fabulosa, San Pedro Sula: 1200-0500
32)	SK	920	5	R. Catacamas: 1200-0400
18)	CR	940	5	R. Dif. Cristiana, Comayagüela:
15)	BO	940	1	R. Cadena Occidental, Sta. Rosa de Copán
34)	QL	950	3	R. Centro Radial, Siguatepeque: 1300-0300
35)	ZE	950	1.2	R. Cortés, Puerto Cortés: 1200-0400
36)	YF	960	1	R. Fergusón, Choluteca
37)	RD3	960	1	R. Sangrelaya, Sangrelaya, Colón
38)	TL	970	3	Estéreo Tic Tac, Tegucigalpa: 24h
39)	ZC	r975	2	R. Monumental, San Pedro Sula (n.980): 1200-0500
40)	CD	r980	1	R. Novedades, Comayagua: 1200-0400
41)	YG	980	1	R. Tocoa, Tocoa
42)	RD	980	1	R. Emperador, Campamento Olancho
44)	RP	1000	2.5	Unión R., Tegucigalpa: 1200-0600
43)	MH	°1000	3	La Voz del Junco, Santa Bárbara: 1100-0400
46)	LP23	1010	1	R. Moderna, El Progreso
47)	LC	1010	5	R. Proyección, Minas de Oro
47x)	LS	1010	0.5	R. Dinorama, La Paz
48)	UW	1020	1	R. Michele, La Ceiba
9)	RH3	1030	1	R. Ticante, Ocotepeque
8)	UP3	1030	1	Estéreo Mil, Tegucigalpa
11)	FX	1040	1	R. Musical, Catacamas: 1200-0200
14)	NN22	1040	2	R. Exitos, San Pedro Sula
33)	MJ	1040		R. Minerva, Comayagua
49)	ZX	1040	10/5	R. La Primerísima, Olanchito: 1200-0500
7)	VW	1060	25	(r: HRVW 650), Tegucigalpa
50)	FA	1060	1	R. Minería, El Mochito
52)	GR	1070	3	R. Guaymuras, El Paraíso: 1100-0400
53)	LE	1070	1	R. Mi Favorita, San Pedro Sula
54)	LP26	1070	1	R. Siguatepeque, Siguatepeque
55)	XM	1080	1	R. Meridiano, Choluteca
56)	ID	1080	1	R. Miramar, Tela: 1200-0400
57)	WC	1090	1	R. Aeropuerto, Tegucigalpa: 24h
58)	ND	1100	1	R. Esperanza, La Esperanza: 1100-0300
59)	VA	1100	1	R. Tiempo, San Pedro Sula: 1100-0400
60)	VL	1100	1	R. Lux, Olanchito: 1100-0400
5)	LP25	1110	1	R. Sur, Choluteca
38)	TL2	1120	3	R. Fiesta, Tegucigalpa: 24h
62)	D4	1120	5	R. Oriental "RCO", Danlí: 1100-0600
66)	AV	1120	1.5	(r: HRAV1150), ?
61)	PL	1130	5	R. Progreso, El Paraíso: 1000-0500
63)	BT	1130	1	R. San Francisco, San Francisco de la Paz: 1100-0200
63x)	HP	1130	1	R. Pinares, Siguatepeque
64)	AP	1140	1	R. Mercurio, Nacaome
65)	UN	1140	5	R. Palmeras, La Ceiba
66)	AV	1150	10/5	Ondas del Ulúa, "La V. de Sta. Bárbara," Sta. Bárbara: 1000-0400
5)	LP12	1150	1	R. Universal, Tegucigalpa
67)	QN	1160	3	La Voz del Atlántico, Puerto Cortés: 1130-0400
68)	GF	r1163	1	R. El Paraíso, El Paraíso (n1160)
34)	QL2	1160	1	R. Sensación, Siguatepeque: 1100-0500
4)	AZ2	1170	1	R. Hits, La Ceiba: 24h
45)	AF	1170	1.5	R. Atenea, Choluteca
57)	AZ3	1180	1	R. Taxi, Tegucigalpa
70)	PO	1190	1	R. Santa María de la Luz, Gualaco
113)	WM	1190	1	R. Alfa Stereo, San Pedro Sula: 1200-0700
134)	GK	1190	0.5	R. Brassavola, Minas de Oro: 1000-0300

71)	DS	1200	1	R. Nacaome, Nacaome
72)	SI	1200	3	R. Impacto, Roatán
73)	RO	1210	1	R. Capital, Comayagüela
114)	HO	1210	1	R. Maya, La Entrada
74)	HH2	1220	1	La Voz de la Costa, San Pedro Sula
75)	YS	1220	1	R. Suari, Marcala
56)	QW	1230	10	R. Tela, Tela: 1200-0200
133)	ZC	1240	1	R. Monumental, Tegucigalpa
4)	AT	1250	10	Super R., San Pedro Sula
51)	CC	1250	1	R. Cadena Continental, Comayagua
115)	YN	v1250	1	R. Latina, Danlí (r. 1255)
77)	YF2	1260	1	R. San Marcos, San Marcos de Colón
116)	2R	1260	1	R. 1260, La Ceiba
5)	NQ	1270	1	R. Sonora, Tegucigalpa: 1100-0500
117)	OF	1270	1	Ecos del Celaque, Gracias
78)	AM	1280	1	R. Olanchito, Olanchito
136)	CG	1280	0.5	R. Columbia, Pto. Cortés: 1200-0400
80)	YT	1280	1	R. Estrella de Oro, San Pedro Sula: 1030-0430 (n 1400)
107)	VN	1280	1	R. San Miguel, Marcala: 1100-0300
81)	NN26	1290	2	R. Choluteca, Choluteca
118)	GS	1290	0.5	Bay Island Christian Network, Roatán
82)	LR	1300	3	R. Santa Rosa, Sta. Rosa de Copán: 1100-0300
83)	LH	1300	1	Stereo Amistad, Tegucigalpa
90)	VC	1310	1	R. Rumbo, (r: HRVC 1390), San Pedro Sula
85)	AD	v1310	2	R. Majestad, Juticalpa (n770): 1000-0400
103)	RL	1310	1	R. Libertad, Marcala, La Paz
119)	MG	1320	1	R. Bahía, La Ceiba
126)	GM	1320	1	R. Juvenil, Choluteca
86)	SW	1330	1	La Nueva R. Cadena, Comayagüela
74)	HH	v1338	10	R. El Mundo, San Pedro Sula: 1100-0500
57)	RF	1340	2.5	R. Cadena Nal. de Noticias, Tegucigalpa: 24h
120)	JC	1340	1	R. Colonial, Comayagua
87)	ME	r1354	1	R. La Voz del Patio, La Ceiba (n1350)
28)	LP18	1360	1	(r: HRLP21 860), Tegucigalpa
121)	CP	1360	1	R. Continente, Trinidad
66)	BH	1360	3/1.5	R. Santa Bárbara "la B grande": 1200-0400
69)	TR	r1372	1.5	R. Danlí, Cd. Danlí (n1370)
88)	ST	v1370	1	R. Fraternidad, San Pedro Sula
89)	CD	v1370	2	R. Constelación, Juticalpa (n1010): 1200-0400
127)	AH	1380	0.5	R. Jutiapa, Jutiapa
90)	VC	°1390	10/5	La Voz Evangélica, Tegucigalpa: 1100-0500
91)	JU	1400	1	R. Estéreo Punto, Comayagua
92)	DD	1410	1	La Voz de Atlántida, La Ceiba: 1100-0500
93)	SY	1410	1	R. La Voz del Pacífico, San Lorenzo
94)	SL	1420	1	R. Actualidad, Trinidad
95)	IC	1430	1	La Radio del 80, Puerto Cortés
96)	SJ	1420	2	R. Mundial, Tocoa: 1100-0200
97)	VM	1430	1	R. Maranatha, La Paz
98)	RD	1440	1	R. San Isidro, La Ceiba: 1100-0500
44)	XZ2	1450	0.5	R. Titania, Tegucigalpa: 1200-0600
128)	GC	1460	1	R. Conga, San Pedro Sula
100)	QX	1460	5	Radiolandia, Comayagua
122)	CX	1460	5	La Voz de Patuca, Catacamas: 1000-0400
123)	SA	1470	1	R. Luz y Verdad, La Ceiba
101)	BR	1480	1	R. Cultural de Copán, La Entrada
102)	EZ	1480	1	R. Cadena Caracol, Tegucigalpa
35)	GO	1490	1	R. Porteña, Puerto Cortés: 1200-0100
104)	MO	1490	2	R. Omega, La Esperanza: 1100-0400
129)	RA	1490	1	R. Juventud, Sonaguera
105)	TX	1500	1	R. Victoria, Choluteca
124)	AN	1510	1	R. Novedades, Tegucigalpa
106)	EM	1510	1	R. Emanuel, Ocotepeque
130)	RG	1520	1	R. Providencia, Danlí
108)	FU	1520	1	R. Futura, San Pedro Sula: 1200-0400
131)	HJ	1520	1	R. Santiago, Yoro
109)	YK	1540	1	R. Nuevo Mundo, Tegucigalpa
132)	AO	1550	1	R. Original, Cofradía
111)	PC	°1600	1	R. Luz y Vida, San Luís: 2200-0400
125)	XN	1600	1	R. Solaris, Tegucigalpa

Shortwaves: Call HR—

111)	CP	3249.5v	1	R. Luz y Vida, San Luís: 1230-1630, 2200-0400 (n. 3250)
112)	RI	*4755	10	Sani R., Pto Lempira
90)	VC	4820.2	5	La Voz Evangélica, Tegucigalpa: 1030-0500 (n. 4820)
16)	XK	*4910.4	0.5	La Voz Evang. de la Mosquitia, Pto. Lempira (n. 4910)
43)	MH5	6075	1	La Voz del Junco, Santa Bárbara: irr.
112)	RI	6299.4	—	Sani R., Pto. Lempira 1200-?, ?-2400

Addresses and other information:

1) Ap. 403, Tegucigalpa — Dir: Miguel Rafael Zavala. Fax: +504 379721.

2) 1a. Av. 439, Bo. San Martín, Santa Rosa de Copán — Dir: J.R. Bueso P.

2x) Av. El Calvario 229, Comayagua.

3) Ap. 3, Juticalpa — Dir: Victor Rubí Z.

4) Ap. 24, San Pedro Sula.

5) Ap. 259, Tegucigalpa — DG: Rodrigo Wong Arévalo.

6) Calle la Fuente, Tegucigalpa — DG: Antonio Masariego.

7) Ap. 120, San Pedro Sula — Pres: Blanca Sikaffy.

8) Emisoras Unidas, Blvd. Suyapa, or: Ap. 642, Tegucigalpa — DG: Nahum Valladares.

8x) Bo. Solares Nuevos, Av. República, La Ceiba.

9) Cas. 206, Santa Rosa de Copán — Dir: A. Rendón P.

10) Frente Ant. Local C. Hispano, Comayagua.

11) 3a. C. SO No. 46, Catacamas, Olancho — DG: Raúl Zaldívar G.

12) La Esperanza, Intibucá — Gte: Ismael Martínez A.

13) Ap. 9, Juticalpa — DG: Alexis Salinas.

14) Ap. 163, San Pedro Sula.

15) Bo. El Triángulo, Sta. Rosa de Copán—Dir: B. Rivera D.

16) Puerto Lempira (or: Global Outreach, Box 1, Tupelo MS 38802, USA). Dir: L. Wilkinson.

17) Av. New Orleans 20-C, San Pedro Sula.

18) 9a. C. No. 322, Comayagüela, or: Ap. 3448, Tegucigalpa — DG: Humberto Andino N.

19) Coyoles Contral, Olanchito Yoro — Gte: Rodolfo Irías.

20) Ap. 26, Sta. Bárbara — C.E: Angel Leo D. Rodriguez — FM: 90.9.

21) Barrio San Francisco, Comayagua — Gte: René Martínez.

22) Danlí, El Paraíso — Dir: H. Mendoza.

23) Yoro, Yoro — Dir: I. Martínez.

24) Ap. 271, San Pedro Sula — DG: Roosevelt Assaf A.

25) Ap. 29, Choluteca — Dir: Sra. Lety Stercke de Madragón.

25x) Calle Centenario, Santa Rosa de Copán.

26) Ap. 33, Comayagua — Dir: Juan Bosco Campos.

27) Ap. 28, Juticalpa, Olanchito — DG: Orlando Aguirre.

28) Ap. 364, San Pedro Sula.

29) Atrás del Palacio Arzobíspal, Tegucigalpa — Dir: Pbtro. Aureliano Santaolaya.

30) Bo. San Francisco, Comayagua — Dir: R. Martínez V.

31) 5 Calle S.O. 10- y 11 Ave. 91, San Pedro Sula — DG: Mario R. Zavala M.

32) Ap. 50, Catacamas — Own: Roger Mario Brown.

33) Casa No. 220, Av. Ppal., Comayagua—Dir: Jesus Martínez V.

34) 2 Ave. 2-3 Calle 269, Siguatepeque — DG: José A. Baires.

35) 3a. Av. No. 772, Puerto Cortés—Dir: Mario E. Prieto — FM: 102.7 and 89.1.

36) C. Vic. Williams, Edif. Fergusón, Choluteca.

37) Muncipio de Iriona, Sangrelaya, Colón.

38) Ap. 771, Tegucigalpa, DC — Dir: Elias J. Chahin Davila.

39) Ap. 996, San Pedro Sula — Dir: Fausto Mejía.

40) Comayagua — DG: Héctor Machado.

41) Tocoa, Colón — Dir: José F. Mejía H.

42) Campamento-Olancho — Dir: H. Robilio R.

43) Ap. 6, Santa Bárbara — Gte: Dra. Carlota Hasbun C. C.E: Miguel E. Hasbun C. — Te: +504 642693.

44) Ap. 614, Tegucigalpa, D.C. — DG: Fernando Lardizabal.

45) Bo. Guadalupe, Av. Herrera, Choluteca—Dir: Samuel A. Aguilera.

46) Col. Brisas del Ulúa, El Progreso, Yoro.

47) Minas de Oro, Comayagua.

48) Av. 14 de Julio, 9 Calle, La Ceiba — Dir: Gregorio Irías N.

49) Av. Francisco J. Mejía, Bo. Sofoco, Olanchito, Yoro.

50) Barrio La Hoya, Santa Bárbara, Juticalpa — Dir: Isidro R. Pineda.

51) Bo. Costado Nte. Cine Valladolid, Comayagua—Dir: Oscar I. Murillo B.

52) Bo. Santa Clara, El Paraíso — DG: M.A. Ordóñez G.

53) Ap. 1576, San Pedro Sula — DG: Salvador A. Handal.

55) Barrio La Esperanza, Choluteca — DG: A.D. Lardizábal.
56) Edif. Canales, Av. Panamá, Tela — Dir: Elida Canales de Molina — FM: 96.3 10kW.
57) Ap. 1082, Tegucigalpa, D.C. — Dir: Ubaldo García A. & Hubert Bodden.
58) Ap. 25, La Esperanza — Dir: José M. del Cid.
59) Ap. 906, San Pedro Sula — Dir: A. Roldán F.
60) Ap. 109, Olanchito, Yoro — DG: Erlin E. Jiminez.
61) Ap. 20, El Progreso, Yoro — Gte: A. Ocaña.
62) Ap. 21, Danlí, El Paraíso — Dir: M. Mendoza T.
63) San Francisco de la Paz, Olancho — DG: N. De Montalvan.
63x) Barrio Abajo, Siguatepeque. — Dir: Helda Olinda Pinel.
64) Contiguo Cine Rey, Nacaome, Valle — Dir: L.A. Pavón G.
65) Av. San Isidro 3 Calle, La Ceiba. — Dir: Oscar L. Irias.
66) Ap. 004, Santa Bárbara — DG: Benjamín E. Handal — Fax (504) 642940.
67) 12 Calle 2a Av. No. 206, Puerto Cortés — DG: A. Griffin C. — F.Pl: Move to 1430kHz.
68) Bo. San Isidro, El Paraíso — Gte: Roy S. Villafranca.
69) Barrio El Centro, Danlí.
70) Iglesia Católica, Gualaco, Olancho.
71) Bo. El Centro, Nacaome, Valle.
72) Km 1, Crtra. a La Ceiba, Tela — Dir: R. Irías.
73) Colonia El Prado 1C-107A, Comayagüela.
74) Ap. 210, San Pedro Sula Own: F. Battistello S., Joel López P.
75) Calle Principal, Marcala, La Paz.
77) San Marcos de Colón — Dir: A. Fergusón.
78) Olanchito, Yoro — Dir: C.A. Muñoz.
79) Minas de Oro, Comayagua — Dir: D. Donaire F.
80) Ap. 303, San Pedro Sula — DG: Donald Moore.
81) Bo. Campo Luna, Choluteca — Dir: M. Villeda T.
82) Ap. 203, Santa Rosa de Copán — Gte: P. Iván de Jesus A.
83) Ap. 955, Tegucigalpa, DC — DG: J.A. Padilla H.
84) Blvd. Concepción 4 Av. entre 1 y 2 Calle, San Pedro Sula.
85) Ap. 15, Juticalpa, Olancho — DG: Héctor Robylio Romero P.
86) Ap. 3405, Tegucigalpa — Dir: Roberto Durón C.
87) Av. San Isidro, La Ceiba — Dir: R. Irías.
88) Colonia Fesitran, San Pedro Sula.
89) Juticalpa, Olancho — Dir: Roberto J. Torres.
90) Ap. 3252 Tegucigalpa, DC. Te: +504 333933 — Dir: Venancio Mejía Cartagene (Owned by: Conservative Baptist Home Mission Society, Box 828, Wheaton, IL 60187, USA — **English**: Mon 0300-0500).
91) 1a Av. No. 189, Comayagua — Dir: Denis Roberto Velásquez.
92) Ap. 17, La Ceiba — Dir: Miguel R. Moncada R.
93) San Lorenzo, Valle — Dir: E. Pineda Hernández.
94) 2a Av. No. 38, Trinidad, Sta. Bárbara.
95) 12 Calle 2a Ave 206, Barrio La Curva, Puerto Cortés.
96) Bo. La Esperanza, Tocoa, Colón — Dir: J.A. González.
97) Santiago de la Paz, La Paz — Dir: Eva A. Garcia V.
98) Ap. 51, La Ceiba — DG: José Luis Asenjo.
100) Calle Bulevar, Comayagua — Dir: Rolando Barahona.
101) Tomas López Moreno, La Entrada, Copán.
102) R.A. Banegas, Barrio los Dolores, Tegucigalpa, D.C — DG: E. Zúñiga.
103) Bo. San Miguel, Calle Ppal., Marcala, La Paz — Dir: Luis Enrique Guzmán.
104) La Esperanza — Dir: Miguel A. Oreppana M.
105) Bo. La Cruz, Calle Chorotega, Choluteca — DG: J.R. Aguilar B.
106) Ocotepeque — Dir: José R. Rodríguez O.
107) Marcala, La Paz — Dir: P. Lucio Núñez C.
108) 5a Av. N.E. No. 1303, San Pedro Sula — Dir: José A. Peña.
109) Col. Miramontes, 2 C. 1a Av., Tegucigalpa.
111) San Luis, Santa Bárbara (Ap. 303, San Pedro Sula).
112) Ap. 113, La Ceiba — Gte: Edward A. Pfister.
113) Col. Los Laureles, 2a C., San Pedro Sula.
114) Col. Progreso 2 y 3 C. 2 y 3 Av. S.O., La Entrada de Copán.
115) Barrio El Centro, Danlí, El Paraíso — Dir: Hernán Mendoza M.
116) 4a Calle No. 1185, La Ceiba.
117) 2a Av. 9C. No. 9, Gracias, Lempira.
118) Islas de La Bahía, Roatán — Dir: Glenn R. Solomon.
119) Bo. La Isla, 4C. Av. Juan Ramón M., La Ceiba.
120) Calle del Comercio 12A, Comayagua.
122) Av. Independencia 210, Catacomas.
123) Bo. Loma Jackson No. 3, La Ceiba.
125) Bo. El Bosque, 9 Av. No. 1420, Tegucigalpa.
126) 1 Calle No. 197, Choluteca — Dir: Guillermo Moreno H.
127) Jutiapa, Atl. — Dir: Angel A. Martínez.
128) Edif. Bendeck 1 Calle 4 y 5 Av., San Pedro Sula — Dir: Victor A. Handal H.
129) 1 Calle 2a Av. 197, Sonaguera, Colón — Dir: Raúl Anael Funes.
130) Danlí, El Paraíso — Dir: Roberto Gamero.
131) Yoro — Dir: Jamil N. Hawil Castro.
132) Cofradía, Cortés — Dir: Carlos R. Pineda R.
133) Ap. 914, Tegucigalpa — Dir: Fernando H. Gómez H.
134) Camino Malcotal, Minas de Oro, Com. — Dir: Gary Kesler. F.Pl: SW 60m.
135) Bo. El Centro 56, Juticalpa, Olancho.

FM in Tegucigalpa.
St. names — Estéreo..., exc. where indicated. MW fq's usually relay FM.
57) 88.3 — 83) 89.3 — 89.9 Saturno — 8) 90.5 R. Exitos — 18) 91.1 R. Comayagüela — 38) 91.7 R. Fiesta — 8) 92.9 La "N" — 6) 93.5 — 8) 94.1 Eco — 5) 94.7 — 95.3 Centaury — 18) 95.9 R. Panamericana — 96.5 Mágica — 38) 97.1 Tic Tac — 57) 97.7 Azul — 98.8 R. Taxi — 18) 99.5 Suprema — 100.1 Morazán — 44) 100.7 — 1) 101.3 — 8) 101.9 R. Satélite — 8) 102.5 R. Exitos — 90) 103.7 — 104.3 R. Nuevo Mundo — 104.9 Amor — 8) 105.5 R. Centro — 8) 106.1 Mil — 7) 106.7 — 107.3 Energía — 6) 107.9.

JAMAICA

L.T: UTC − 5h — **Pop:** 2.540.000 — **Radio sets:** 1.5m — **Pr. L:** English — **E.C:** a/c 50, 110/220V (yes) — **ITU:** JMC.

JAMAICA BROADCASTING CORPORATION (Gov/Comm.)

ADDR: P.O. Box 100, Kingston 10. **Te:** +1 (809) 926-5620-9. **Fax:** +1 (809) 929-1029. **Cable:** JARAD Jamaica. **Telex:** 2218 BROADCORP JA.
L.P: DG: Claude Robinson. Dep. GM: B. C. Barovier. Dir. Radio: Gladstone Wilson. Dir. Eng: Martin Darien.

Radio One	**kHz**	**kW**		**kHz**	**kW**
Kingston	560	5	Port Maria	750	10
Mandeville	620	5	Savanna-la-Mar	850	10
Montego Bay	700	10	Morant Bay	1090	1

FM: Coopers Hill 97.1MHz 3.5kW — Half Way Tree 99.7MHz 1 kW — Spur Tree 93.3MHz 35kW — Flower Hill 97.3MHz 35 kW — Oracabessa 103.9MHz 35kW — Port Antonio 98.7MHz 2kW.
D.PRGR: 24h. **N:** 1030, 1100, 1130, 1200, 1230, 1300, 1530, 1731, 1830, 2030, 2130, 2230, 0100, 0300, 0310, 0430(W) — **Wrp. and Fishermen's Forecast:** W 2248.
JBC-FM (Stereo): 24h. Coopers Hill 91.1MHz 5kW; Montego Bay 92.1MHz 1kW, Oracabessa 100.3MHz 1kW, Half Way Tree 105.7MHz (0.25kW).
ANN: "This is Jamaica Broadc. Corpo" or "This is JBC Radio Two (or JBC-FM)" — **IS:** Guitar Chi Chi Bud (Jamaica folksong) — **V.** by letter. Rp.

RADIO JAMAICA LTD. (Comm.)

ADDR: P.O. Box 23, Kingston 5. **Te:** +1 (809) 926-1100. **Cable:** Broadco. **Fax:** +1 (809) 929-7467.
L.P: Man. Dir: J.A. Lester Spaulding. Dir. Prgr: D. Topping. Prgr. Mgr "FAME FM": Norma Brown-Bell. Chief Eng: Carroll Lawrence. Mktg. Mgr: Michael Johnston. News & Current Affairs Editor: Jennifer Grant.

RJR	**kHz**	**kW**	**RJR**	**kHz**	**kW**
Montego Bay	550	5	Kingston	720	10
Galina	580	10	Spur Tree	770	5

FM: Kingston 94.5MHz 5kW, 104.5MHz 0.6kW; Spur Tree 90.5MHz 20kW, Montego Bay 104.5MHz 11.41kW, 92.9MHz 2kW; Oracabessa 101.3MHz 5kW.
D.PRGR: 24h. **N, Wrp:** W 1100, 1200, 1230, 1300(BBC), 1500, 1700, 1800, 2000, 2100MF, 2200, 2300, 0000, 0300; Sun 1200, 1300(BBC), 1400, 1500, 1700, 2200, 0100, 0300.
"FAME FM": Kingston 92.7MHz 5kW, 95.7MHz 0.5kW; Oracabessa 91.5MHz 5kW; Montego Bay 95.7MHz 11.41kW; Spur Tree 98.1MHz 10kW — **D.PRGR:** 24h.
Imminent Hurricane Warnings: On the h (both st's).
ANN: "This is RJR". "This is FAME FM". — **V.** by QSL-card.

ISLAND BROADC. SERVICES LTD. (Comm.)

ADDR: 19 Caledonia Rd, Mandeville. **Te:** +1 (809) 962-2215-7. **Fax:** +1 (809) 962-2004. (Addr. in Kingston: 41b Half Way Tree Rd, Kingston 5) — Chief Exec: Neville James.
STATION: KLAS-FM 89.3MHz — **D.PRGR:** 24h.

MARTINIQUE (French)

L.T: UTC – 4h — **Pop:** 366.500 — **Radio sets:** 60.000 — **Pr.L:** French, Creole patois — **E.C:** a/c 50, 220V (no) — **ITU:** MRT.

SOCIÉTÉ NATIONALE DE RADIO-TÉLÉVISION FRANÇAISE D'OUTRE MER (RFO)

ADDR: B.P. 662, F–97263 Fort de France. **Te:** +596 595200. **Telex:** 029659 — Dir: Fred Jouhoud. CE: Jean Claude Arrivé.
MEDIUMWAVE: (G.C: 61.00W/14.30N): 1310kHz 2x20kW.
FM: Fort de France 92MHz 0.3kW/95MHz 0.05kW + 4 relays.
D.PRGR: 0800-0400. **N:** 1000, 1030, 1200+, 1300+, 1400+, 1700+, 2000, 2300 — +) rel. France-Inter.
ANN: "Radio Télévision Française pour l'Outre Mer, station de Martinique" or "RFO" — **INT-SIG:** First eight notes of "Adieu Foulard, Adieu Madras" on piano — **V.** by QSL-card.

RADIO CARAIBES INTERNATIONAL (Comm.)

ADDR: B.P. 111, F-97209 Fort-de-France Cédéx. **Te:** +596 731212. **Telex:** 912579 MR.
STATIONS: 1090kHz 20kW (tr. located on St. Lucia) + FM 89.9/92.5/98.7/106.1MHz.
D.PRGR: 24h. **N:** on the h. (rel. Europe 1).
ANN: F: "R. Caraibes International, Fort de France".
V. by letter. Rp.

PRIVATE STATIONS

Approx. 40 st's are licenced to broadcast on FM.

MEXICO

L.T: UTC – 6h in all states except: UTC – 7h (Sonora, Sinaloa, Nayarit, Baja California Sur; and during summer Baja California Norte; UTC-8h (Baja California Norte, winter) — **Pop:** 91.990.000 — **Radio sets:** 16.325.000 — **Pr.L.:** Spanish — **EC:** a/c 60, 110/220V — **ITU:** MEX.

SECRETARIA DE COMUNICACIONES Y TRANSPORTES

Dirección General de Telecomunicaciones

ADDR: L. Cárdenas 567-11° Ala Nte., Narvarte, 03020 México — **Te:** +52 (5) 5199734.
L.P: Dir. Gen: Ing. Enrique Luengas H.
Departamento de Frecuencias, Av. Niño Perdido y Cumbres de Acultzingo, México 12, D.F. **Te:** +52 (5) 5303060/9.
L.P: Dir: Ing. Luis Valencia P.

DIRECCIÓN GENERAL DE CONCESIONES Y PERMISOS DE TELECOMUNICACIONES

Depto de Asignación de Frecuencias
ADDR: Unidad Contel Sga-2, Av de las Telecomunicaciones s/n, Ixtapalapa 09310, Mexico, D.F. **Te:** +52 (5) 6920077.
L.P: Dir: Ing. Sergio Cervantes.

CAMARA NACIONAL DE LA INDUSTRIA DE LA RADIO Y TELEVISION DE MÉXICO

ADDR: Horacio 1013, Col. Polanco, 11550 México, D.F. **Te:** +52 (5) 2502122.
L.P: Pres: Lic. Casio Carlos Narváez. Gte: César Hernández Espejo.

Note: Most stations carry "La Hora Nacional" Sundays 0400-0500 (Mon UTC).

STATIONS: Mediumwaves: call XE-.

The letters preceding the st. number indicate the State. Addresses are listed by State in alphabetical order.

	Call	kHz	kW	Name and h. of tr.
NL01)	WA	540	150	Super Estelar, Monterrey: 24h
BC19)	TIN	540	—	R. 540, Tijuana: 1300-0400
CH01)	PL	550	55	La Super Estación, Cd. Cuauhtémoc: 1200-0800
GR01)	ACD	550	0.5	La Máquina Musical, Acapulco: 1200-0600
JL01)	MZA	550	0.5d	R. Gallito, Cd. Guzmán: 1300-0100
NA13)	TNC	550	10	R. Atlán, Tepic: 1300-0500
OX01)	UC	550	1	R. Felicidad, Tehuantepec: 1200-0500
VE01)	KL	550	1/0.25	Dimensión 550, Jalapa: 1300-0400
YU01)	QW	550	2	La Poderosa, Mérida: 1130-0830
DF01)	OC	560	5	R. Chapultepec, México: 24h
DG05)	SRD	560	2.5	R. Sierra, Santiago Papasquiaro
SO31)	YO	560	1	R. 560, Huatabampo
ZC07)	XZ	560	2	Zacatecas: 1200-0700
DG04)	TJ	570	1/0.25	R. Alegría, Gómez Palacio: 1200-0600
MI01)	LQ	570	1	La Tigresa, Morelia: 1200-0600
NL02)	BJB	570	5	La Metrópoli Norteña, Monterrey: 24h
OX02)	OA	570	5	R. Oaxaca: 1200-0500
PU13)	VJP	570	1	Xicotepec de Juárez
SO26)	UK	570	1	La UK, Caborca: 1400-0609
TB01)	VX	570	1	R. Voz, Comalcalco: 1200-0400 (A)
YU09)	ME	570	5	R. Valladolid: 1200-0400
CO02)	MU	580	5/0.25	Rancherita del Aire, Piedras Negras: 24h
CS01)	UE	580	5	La Voz de Chiapas, Tuxtla Gutiérrez: 1100-0600
CH02)	FI	580	5	R. Mexicana, Chihuahua: 24h
JL02)	AV	580	10	R. Guadalajara: 1200-0700
QR01)	YI	580	1/0.25	R. Festival, Cancún: 1200-0600 (A)
S006)	HO	580	1	R. Centro, Cd. Obregón: 24h
TM01)	HP	580	1	Estudio 5-80, Cd. Victoria: 1200-0400
VE02)	DZ	580	1	R. Ondas, Córdoba: 1200-0400
DF02)	PH	590	5	R. Alicia (R. 590), México: 1200-0800
DG01)	E	590	1	R. Festival, Durango: 1300-0400 (A)
SO01)	HQ	590	1	R. Mexicana, Hermosillo
TM02)	FD	590		La Consentida, Reynosa: 1300-0400
VE17)	OM	590	1	(r: XEL 1260), Coatzal coalcos: 24h
CO01)	DN	600	1	La Ranchera de la Laguna, Torreón: 1200-0500
GJ01)	GTO	600	5	R. Capital, Guanajuato: (A)
GR02)	BB	600	3	La Grande, Acapulco: 1200-0400 (A)
MI02)	TA	600	0.5	La Divertida, Zitácuaro: 1200-0400
NL03)	MN	600	1	R. Variedades, Monterrey
YU02)	Z	600	5	R. 600, Mérida: 1200-0400
CO03)	BX	610	5/0.5	La Primera Estación, Sabinas: 1100-0600
MI03)	UF	610	5	Radio Variedades, Uruapan: 1200-0600
OX03)	KZ	610	1	La Voz del Istmo, Tehuantepec: 1200-0300
SL02)	CV	610	1/0.25	La Gran Compañía, Cd. Valles: 1200-0500
SN02)	GS	610	1	R. Guasave: 1230-0600
VE01)	JA	610	1	R. Fiesta, Jalapa: 1200-0500
ZC01)	EL	610	5	R. Progreso, Fresnillo: 1200-0500
CH03)	BU	620	5/0.25	La Norteña, Chihuahua: 1200-0700
DF03)	NK	620	10/5	R. 620, México: 1200-0600
DG02)	CK	620	1/0.5	R. Variedades, Durango: 1300-0500
NA01)	OO	620	1	Onda 6-20, Tepic: 24h
TB05)	ACM	620	1	R. Pueblo, Villahermosa: 1200-0600
CS02)	TS	630	3	R. Tapachula: 1100-0400 (A)
GR02)	ACA	630	5	R. Voz, Acapulco: 1200-0400 (A)
JL02)	JB	630	1/5	R. Gobernación, Guadalajara: 1300-0700
NL03)	FB	630	10	La Estación de las Noticias, Monterrey
QR08)	CCQ	630	1	R. Sensación, Cancún: 1200-0300
SN03)	TK	630	4	R. Hit, Mazatlán: 24h
SO02)	FX	630	1/0.25	Los Profesionales de la R., Guaymas: 1100-0700
VE03)	FU	630	5/0.75	La Voz Amiga, Cosamaloapan: 1100-0500
HG04)	NQ	640	10	La Voz de la Provincia, Tulancingo: (A)
ZC09)	YQ	640	10	R. Fresnillo: 24h (n. 890)
CS03)	WM	640	1d	R. 640, San Cristóbal de las Casas

JL03)	EJ	650	1/0.5d	R. Paraíso, Puerto Vallarta: 1200-0200 (A)
OX12)	PX	650	1	
MI04)	ZM	650	5d	La Expresión de Zamora: 1230-0100
SN04)	TNT	650	1	R. 65, Los Mochis: 24h (A)
YU01)	VG	650	1	Mérida
CH04)	ACB	660	1d	R. 6-60, Cd. Delicias: 1300-0100
DF04)	RIN	660	50	R. 660, México: 1200-0800
OX04)	YG	660	0.25d	R. 6-60, Matías Romero: 1200-0100
NL02)	FZ	660	10	Metrópoli, Monterrey (n. 740)
TM20)	AR	660	5	La Mexicana, Tampico: 24h
ZC02)	EY	660	10d	La Consentida, Xalpa: 1100-0100
QR09)	CPR	660	1	R. Chan Santa Cruz, Felipe Carrillo Puerto
DG06)	WX	660	1	R. Mexicana, Duranga
CO04)	TOR	670	1d	R. Ranchito, Torreón: 1200-0230
JL01)	IS	670	5d	La Rancherita Contenta, Cd. Guzmán: 1300-0100
CS04)	KQ	680	5	Tapachula: 1200-0500
CH02)	FO	680	1d	R. Capital, Chihuahua: 1200-0100
GJ02)	LG	680	10	La Grande, León: 24h
PU01)	FJ	680	1	R. Teziutlán: 1200-0400
SN05)	ORO	680	1	Canal 6-80, R. Oro, Guasave: 1200-0600
SO22)	SON	680	1	R. Capital, Hermosillo: (A)
BC01)	TRA	690	50	Newsradio 6-90, Tijuana: 24h **(English)**
CL03)	CS	690	5	La Grande, Manzanillo: 1100-0600
DF05)	N	690	20	R. Mundo, México: 1200-0600
NL04)	RG	690	55	La Deportiva 690, Monterrey: 1100-0600
SN22)	ST	690	2.5	R. Sensación, Mazatlán: 1300-0500
ZC03)	MA	690	5	La Onda Musical del Altiplano, Fresnillo: 1200-0600
VE07)	UY	690	2.5	Coatzacoalcos: 24h
JL04)	DKR	700	1	R. Red, Guadalajara: 1200-0100
CH25)	GD	700	1	R. Novedades, Hidalgo del Parral: 1200-0500
VE02)	VC	700	10	R. Cordobesa, Córdoba: 1300-0100
YU01)	YK	710	5	R. Mágica, Mérida
CO05)	PQ	710	1/0.1	La Estación que Piensa Joven, Cd. Muzquiz: 24h (A)
CL01)	RL	710	1	Colima: 1100-0600 (A)
CS05)	ON	710	3	La Nueva XEON, Tuxtla Gutiérrez: 24h
CH05)	DP	710	5d	La Ranchera de Cuauhtémoc: 24h
DF04)	MP	710	1	R. 710, México: 1200-0800
GR02)	MAR	710	1	R. Felicidad, Acapulco (A)
NA02)	RK	710	1	R. Korita, Tepic: 1200-0800
OX05)	RPO	710	1	R. Variedades, Oaxaca: 24h
SN01)	BL	710	5	R. Bonita, Culiacán: 1300-0900
TM10)	OLA	710	1	R. Ola, Altamira
SO03)	PS	710	1	R. Amistad, Empalme: 1300-0700 (A)
QR05)	CPQ	720	1	La Estrella Maya, Felipe Carrillo Puerto
JL05)	QZ	720	1d	San Juan de los Lagos: 1300-0100
SN19)	VU	720	1	La Grande de Mazatlán, Mazatlán (A)
VE37)	AVR	720	5	Alvarado
DF06)	X	730	100	La Super Grande, México: 24h
CO04)	QN	740	5	Torreón
TB02)	KV	740	1	R. Sensación, Villahermosa: 24h
AG02)	LTZ	740	1	R. Regional, Aguascalientes
JL25)	VAY	740	1	La Mexicana, Pto. Vallarta
GR03)	KOK	750	5	R. Sensación, Acapulco: 1200-0100
SN23)	CSI	750	1	R. Sensación, Culiacán: 24h
CH21)	OH	750	1d	R. Camargo: 1300-0100
OX06)	QF	750	5	R. Loma, Loma Bonita: 1200-0400
VE31)	TI	750	1	R. Tempoal: 1200-0600
DF07)	ABC	760	20/5	ABC R., "la estación de la palabra," México: 24h
DG03)	DGO	760	5	R. Capital, Durango: 1300-0500
JL06)	ZZ	760	5d	R. Gallito, Guadalajara
SL03)	EQ	760	0.25d	La Pantera, San Luis Potosí: 1200-0100
MI25)	LAC	760	5	La Voz de las Balsas, Cd. Lázaro Cárdenas
CS15)	RA	760	1	R. Chiapas, S. Cristóbal de las Casas

SO07)	EB	760	1	La Rancherita, Cd. Obregón: 1300-0800 (A)
SO17)	NY	760	1	R. Geny, Nogales: 1200-0700
YU01)	YW	760	5	Tropi Ranchera, Mérida
NL14)	ACH	770	5	La Regiomontana, Monterrey: 1100-0600
CH06)	JZ	770	1d	Z-7-7, Cd. Jiménez: 1300-0200
CH07)	HB	770	0.5d	R. Jilguerita, San Francisco del Oro: 1300-0200
MI05)	ML	770	1d	R. Tropical, Apatzingán: 1200-0100
SN04)	ZA	770	1	Z-77, Los Mochis
VE41)	QRV	770	1	R. Costera, Cd. Cardel: 1135-0300
ZC01)	IH	770	10	Nueva Dimensión en Radio, Fresnillo: 1200-0600
GJ03)	ZN	780	0.5d	R. Mundo, Celaya
GR04)	XY	780	1d	La Voz del Balsas, Cd. Altamirano: 1300-0100
JL07)	LD	780	5d	R. Costa, Autlán
TM20)	MTS	780	1	La Mexicana, Tampico: 24h
TM25)	SFT	780	5/1	La Bronca Norteña, San Fernando
AG01)	BI	790	1	R. B-I, Aguascalientes: 1200-0400
BC02)	SU	790	1	R. 7-90, Mexicali
BS01)	NT	790	5/0.75	La Paz: 1300-0700
CH02)	RPC	790	5	Chihuahua: 24h
DF08)	RC	790	10	R. El Fonógrafo, México: 24h
DG04)	GZ	790	1d	La Pantera, Gómez Palacio: 1300-0100
TB03)	VA	790	5/0.25	La Emisora del Hogar, Villahermosa: 1000-0600
TM03)	FE	790	1/0.5	R. Fiesta, Nuevo Laredo: 1230-0200
YU06)	UP	790	1	La Bonita del Oriente, Tizimin
BC03)	MMM	800	0.5	R. 80, Onda Ranchera, Tijuana: 1400-0800
CH08)	ROK	800	150	R. Cañón, Cd. Juárez: 24h
JL08)	AN	800	1d	Súper Sonido, Ocotlán: 1300-0100
CL05)	MAX	810	3	R. Felicidad, Armería
CO06)	IM	810	0.5	R. Capital, Saltillo (A)
CS06)	IN	810	1	La Voz del Valle, Cintalapa: 1100-0500 (A)
CS04)	OE	810	1	R. Amistad, Tapachula: 1100-0400
NA03)	UX	810	10/0.25	La Gran Máquina Musical, Tuxpan: 1300-0100
SO04)	RSV	810	5d	R. Alegría, Cd. Obregón: (A)
TM04)	RI	810	0.5	R. Rey, Reynosa: 24h
TM05)	FW	810	50	Tampico: 1200-0400
TX01)	HT	810	1d	R. Huamantla: 1200-0100
QR04)	RB	810	5	La Onda Musical del Caribe, Cozumel
YU03)	MQ	°810	2	Tus Panteras, Mérida: 1230-0600
ZC05)	ZC	810	1/0.125	R. Cañón, Río Grande: 1200-0500
BC07)	YX	820	3,5	La Gigante, Mexicali: 1100-0300
CA01)	ESC	820	0.75d	Escársega: 1300-0100
CH09)	SB	820	1d	R. Sensación, Santa Bárbara: 1200-0200
DG06)	DRD	820	1	La Poderosa, Durango
JL06)	BA	820	10d	La Consentida, Guadalajara: 24h
TB09)	ZQ	820	1/0.1	R. Diversión, Huimanguillo
VE02)	KG	820	10	R. Variedades, Córdoba: 1300-0100
DF09)	LA	830	10	Buena Música en México: 1200-0600
NL13)	LN	830	1	R. Linares: 1300-0100
ZC07)	LK	830	5	Onda Juvenil, Zacatecas: 1300-0800
SN06)	VQ	830	5d	La Grande de Sinaloa, Culiacán (A)
CS01)	IO	840	10d	Emisora de Chiapas, Tuxtla Gutiérrez: 1130-
GJ04)	FG	840	1d	R. Juventud, Celaya: (A)
NA13)	TEY	840	1	R. Sensación, Tepic
OX07)	ACC	840	0.25d	La Voz del Puerto, Puerto Escondido: 1300-0100 (n. 860)
JL16)	XXX	840	5	Triple X, Tamazula
TM06)	MY	840	0.5d	R. Tropical, Cd. Mante: 1200-0100
BC04)	ZF	850	0.5d	La Rancherita Contenta, Mexicali: 1300-0300
CO07)	ZR	850	1/0.25	ZR-85, Zaragoza: 1200-0400
CH03)	M	850	5d	R. Exitos, Chihuahua: 1200-0200
JL08)	MIA	850	1d	R. Felicidad, Guadalajara: (A)

SO05)	US	°850	1d	R. Universidad, Hermosillo: 24h
VE04)	TQ	850	10	R. Sensación, Orizaba: 1300-0500
AG02)	PLA	860	1	R. Mexicana, Aguascalientes
BC05)	MO	860	5	R. 86, Tijuana: 1330-0930
CL02)	AL	860	5	Onda Internacional Juventud, Manzanillo: 24h
CS08)	DB	860	1	Canal 86, Tonalá: 1100-0500 (A)
CH10)	ZOL	860	1	R. Zol, Cd. Juárez: 1200-0600
DF10)	UN	860	50/25	R. UNAM, México
DG03)	DU	860	5	DU la que le gusta, Durango: 1200-0500
QR07)	CNN	860	5	R. Caribe, Cancún: 1200-0600
NL04)	NL	860	5	R. Recuerdo, Monterrey: 1200-0600
SN07)	NW	860v	1	La Estación del Silbidito, Culiacán: 1200-0700 (r. 859.7)
TB04)	ZX	860	1/0.15	Voz del Usumacinta, Tenosique: (A)
TM05)	TW	860	1d	R. Fiesta, Tampico: 1200-0100
CH28)	TAR	870	10	La Voz de la Sierra Tarahumara, Guachochi
GR14)	RO	870	5	R. Guerrero, Chilpancingo
GJ05)	AMO	870	1d	Estación del Amor, Iraputo: 1300-0100
SN22)	FIL	870	1	R. Festival, Mazatlán: (A)
MI06)	LY	870	1d	R. Sensación, Morelia: 1200-0130
CO16)	TC	880	10	Estéreo Mayrán, Torreón: 1300-0500
CS10)	OB	880	5	Pichucalco
CH02)	V	880	1	R. Sensación, Chihuahua: 24h
JL09)	AAA	880	10	Románticos AM, Guadalajara: 24h
SL04)	EM	880	1d	R. Alegría, Río Verde: 1300-2400
QR06)	CTL	880	5	R. Quintana Roo, Chetumal
SN04)	PNK	880	5d	Canal 88, Los Mochis: 1200-0300
TB10)	QQQ	880	3	Súper Q, Macuspana (A)
GR10)	IG	880	1	RCN de Iguala
CS04)	TAP	890	5d	Tapachula: 1200-0100
SN01)	NZ	890	50	Energía 8-90, Culiacán: 24h
YU07)	UL	890	2.5	Tu nueva Onda, Progreso
ZC06)	PC	890	5	Sonido Estrella, Zacatecas: 1200-0800
DF06)	W	900	250	La Voz de la América Latina desde México: 24h
VE05)	WB	900	50/5	(r: XEW 900), Veracruz: 24h
BC06)	AO	910	1	La Voz de Mexicali, Mexicali: 24h
GJ06)	ACN	910	5	La Nueva Onda, León: 1300-0500
BC09)	SDA	920	2.5	Rock 92, Ensenada: 1400-0800
CO07)	OP	920	1	La Campeona, Monclova: 1200-0600
CO08)	MJ	920	1	R. Exitos, Piedras Negras: 1200-0500
CO04)	TAA	920	1/0.2	R. Exitos, Torreón: 24h
CS05)	VV	920	5/0.25	Doble V Juvenil, Tuxtla Gutiérrez: 1200-0600
CH03)	QD	920	1/0.1	La Divertida, Chihuahua: 1200-0700
GJ07)	RE	920	5d	R. Cañón, Salvatierra: (A)
JL06)	LT	920	10	La Sabrosita, Guadalajara: 1200-0500
NL05)	OK	920	1	El Radiazo Mexicano, Monterrey: 1200-0600 (A)
PU14)	ZAR	920	1	Digital 9-20, Puebla
SL03)	BM	920	5	La Rancherita, San Luis Potosí: 1200-0600
SN08)	CQ	920	5	La Ranchera de Culiacán, 1300-0800
SO01)	BH	920	5/1	La Correcaminos, Hermosillo: 1300-0600
CL01)		930	1	Festival 93, Colima (A)
CS12)	MK	930	5	R. Progreso, Huixtla
VE06)	U	°930	5	La "U" de Vera Cruz, "El Eco de Sotavento": 1130-0600
BC07)	WV	940	1d	Onda 94, La Gigante del Cuadrante, Mexicali: 1300-0755
DF06)	Q	°940	50	La Q Mexicana, México: 1200-0600
TM07)	RKS	940	1d	R. Cañón, Reynosa: 1230-0100
AG02)	CAA	950	1	La Doble A, Pedernal: 1200-0400
BC08)	KAM	950	10	R. 95, Tijuana: 24h
CA02)	MAB	950	1	Canal Internacional, Cd. del Carmen: 1200-0400
CS01)	TUG	950	1	Sonovida, Tuxtla Gutiérrez
GJ22)	CEL	950	1	R. Fiesta, Celaya
CO09)	YJ	950	5	La Nueva Imagen del R., Nueva Rosita: 24h
CH02)	FA	950	1	La Nueva FA, Chihuahua: 24h
JL22)	MEX	950	5	La Mexicana, Cd. Guzmán
NL06)	RN	950	1d	R. Naranjera, Montemorelos: (A)
SN09)	ORF	950	5	R. Voz, Los Mochis: 1100-0600
SO22)	PB	950	10	R. Amor, Hermosillo: 1300-0800 (A)
CO10)	KS	960	1	La Voz del Tiempo, Saltillo: 1300-0600
CH11)	CC	960	0.25	Cd. Camargo: 1300-0400
GR05)	UQ	960	1	R. Variedades, Zihuatanejo: 1200-0300
JL10)	HK	960	1/0.25	Voz de Guadalajara: 24h
MI07)	MM	960	1	R. Capital, Morelia: 1200-0600 (A)
QR02)	ROO	960	1	La Pirata del Caribe, Chetumal: 1200-0500
SL03)	CZ	960	0.5	R. Capital, San Luis Potosí: 1200-0600
SO07)	IQ	960	1/0.5	R. Felicidad, Cd. Obregón: 24h
TM08)	K	960	5/1	La Estación Grande de Nuevo Laredo: 1130-0400
VE07)	GB	960	1	R. Fiesta, Coatzacoalcos: 1200-0600
VE08)	OZ	960	1	R. Lobo, Jalapa: 1200-0600 (A)
CO11)	QX	970	0.5	Musicalísima, Monclova: 1200-0600
CH12)	J	970	10/5	R. Mexicana, Cd. Juárez: 24h
DF11)	DF	970	10	R. Noticias, México: 24h
GJ08)	UG	970	0.5d	R. Universidad de Guanajuato
MI08)	CJ	970	1	R. Apatzingán: 1200-0400
SN10)	VOX	970	5	R. Voz, "Fiesta Mexicana," Mazatlán: 24h
SO08)	EZ	970	5	La Super Z, Caborca: 1300-0600 (A)
TB05)	VT	970	5	R. Festival, Villahermosa: 1100-0600
TM01)	BJ	970	1	R. 9-70, Cd. Victoria: 1200-0600
TM09)	O	970	1	R. Gallito, Matamoros: 24h (A)
YU03)	MH	970	5	La Grande, Mérida: 1200-0600
CS04)	TAP	970	5	Tapachula: 1100-0500
CO12)	NR	980	5/0.5	Nueva Rosita: 1200-0600
MI09)	LC	980	5	"98", La Piedad: 1230-0500
NA04)	XT	980	1	XT Internacional, Tepic: 1300-0600
QE01)	QG	980	1/0.2	Canal 98, Querétaro: 1200-0400
SO09)	FQ	980	1	R. 98, Cananea: 1400-0500
SO10)	KE	980v	1/0.25	R. 980 Musical, Navojoa: (A) (r. 980.4)
TM10)	TU	980	10	R. Tampico: 1200-0600
VE03)	QO	980	5	R. Romance, Cosamaloapan: 1200-0400
ZC03)	QS	980	1	Romance en Radio, Fresnillo: 1200-0400
BC04)	CL	990	5	La "L" de Mexicali: 1355-0800
BS02)	HZ	990	1	R. 990, La Paz: 1300-0700
CS09)	TG	990	10	La Grande del Sureste, Tuxtla Gutiérrez: 24h (A)
CH13)	ER	990	0.5	R. Formula, Cd. Cuauhtémoc: 1130-0600
JL01)	BC	990	1	La Divertida, Cd. Guzmán: 1200-0600
NL04)	T	990	50	La "T" Grande de Monterrey: 24h
OX08)	IU	990	1	R. Felicidad, Oaxaca: 1200-0400 (A)
PU02)	OL	990	1	R. Impacto, Teziutlán: 1200-0400 (A)
YU04)	UM	990	1	R. 99, Valladolid: 1200-0400 (A)
ZC10)	FP		10	R. Alegría, Xalpa: 24h
CS02)	TAC	1000	1d	R. Mil, Tapachula: 1200-0100
CH10)	FV	1000	1d	La Rancherita, Cd. Juárez: 1200-0600
DF02)	OY	1000	10	R. Mil, México: 24h
SN11)	MIL	1000	1	R. Mil, Los Mochis: 24h
SN23)	MMS	1000	1	R. Mil, Mazatlán
QR08)	QOO	1000	1	R. Mil, Cozumel
YU02)	MYL	1000	1d	R. Mil, Mérida: 1300-0100
TM24)	NLT	1000	0.5	R. Mil del Norte, Nuevo Laredo
BC09)	DX	1010	1	R. Variedades, Ensenada: 24h
CO13)	KD	1010	0.5/0.25	La Norteña, Cd. Acuña: 1200-0600
CH14)	LO	1010	1	R. Lobo, Chihuahua: 1300-0400
CH15)	TX	1010	1/0.25	Voz de la Sierra Madre Occidental, Nuevo Casas Grandes: 1200-0500
CO01)	VK	1010	5	R. Juventud, Torreón: 1200-0600
JL06)	HL	1010	50	La Poderosa, Guadalajara: 24h
SN12)	WS	1010	5d	R. Capital, Culiacán: (A)

SO11)	XN	1010	0.5	R. Ures: 1200-0400
VE09)	FM	1010	1	R. Moderna, Veracruz: 1200-0600
MI10)	XL	1020	1d	R. Láser, Pátzcuaro: 1300-2400
QR03)	WO	1020	1d	R. Chetumal: 1200-0000
QE01)	KH	1020	0.5	R. Centro, Querétaro
VE29)		1020	0.75	Gutiérrez Zamora: 1200-0300
DF08)	QR	1030	10	R. Centro, México: 24h
JL11)	LJ	1030	1d	R. 10-30, Lagos de Moreno: 1300-0100
GR09)	VP	1030	1	Máximo Sentido, Acapulco
SL07)	IE	1030	5	R. Alegría, Matehuala
SN13)	MPM	1030	1	R. Fama, Los Mochis: 1300-0800
JL09)	BBB	1040	10	R. Fama "Energía 10-40," Guadalajara
CS01)	PLA	1040	1	R. Palenque, Palenque:1000-0400
QE02)	JAQ	1040	0.25d	R. Felicidad, Jalpan (A)
VE01)	GR	1040	1d	La Pantera, Jalapa: 1300-0100
CH02)	HES	1040	1	La Poderosa, Chihuahua; 24h
BC02)	D	1050	10/1	R. Mexicali, "La Gran D," Mexicali: 24h
NA05)	LUP	1050	5	R. Exitos, Compostela
NL08)	G	1050	100	La Ranchera de Monterrey: 24h
BS05)	BCS	1050	10/1	R. Cultural Surcaliforniana, La Paz: 1300-0500
DF12)	EP	1060	50/20	R. Educación, México: 24h
CA02)	IT	1070	1	R. Hit, Cd. del Carmen: 1200-0400
JL12)	SP	1070	5	R. Juventud, San Pedro Tlaquepaque: 1300-0700
PU03)	GY	1070	1/0.25	R. Fiesta, Tehuacán: 1200-0600
SL10)	EI	1070	1	R. Mexicana, San Luís Potosí
SO18)	OBS	1070	1	R. Mexicana, Cd. Obregón
VE10)	MI	1070	1	Minatitlán: 1100-0600
BC10)	DY	1080	0.5d	R. Gallo "Canal 108," Cd. Morelos: 1300-0900
CL04)	UU	1080	1	R. Variedades, Colima: 1100-0500
GJ05)	CN	1080	1d	La Juvenil de Irapuato: 1300-0100
VE11)	XK	1080	0.5	R. Mundo, Poza Rica: 1200-0200
CH27)	DT	1080	1	La Divertida, Cd. Cuahtémoc
BC11)	PRS	1090	50	La Gigante Musical, Rosarito: 24h
JL13)	LB	1090	0.25	R. La Barca: 1300-0300
NL04)	AU	1090	2.5	R. AU, Monterrey: 1200-1600
PU04)	HR	1090	0.5	R. Variedades, Puebla: (A)
TM11)	WL	1090	1	R. Cañón, La Internacional: Nuevo Laredo
VE12)	MCA	1090	5d	La Emisora de las Huastecas, Pánuco: 1300-0100
VE13)	IL	1090	3/1	R. Festival, Veracruz: (A)
GJ09)	BV	1100	5	R. Alegría, Moroleón: (A)
VE14)	BD	1100	5d	R. Ambiente, Perote: 1200-0100
CH10)	WR	1110	1.5d	R. Hits, Cd. Juárez: 1200-0200
CH03)	ES	1110	1d	R. Alegría, Chihuahua: 1200-0200
DF13)	RED	1110	50	R. Red, México: 1200-0600
GJ02)	LEO	1110	5d	La Rancherita, León: 1300-0100
JL25)	PVJ	1110	1	R. Vallarta: 24h
SO12)	VS	1110	1d	R. Festival, Hermosillo: 24h
TB06)	JAC	1110	1d	Cárdenas: 1200-0100
TM02)	OQ	1110	1	La Pantera Cumbiambera, Reynosa
JL14)	UNO	1120	0.5d	R. Exitos, Guadalajara: 1200-0100
OX09)	ZB	1120	0.5d	R. Fono, Oaxaca: 1200-0400.
QE01)	GV	1120	0.5d	R. Felicidad, Querétaro: 1200-0200
SL05)	TR	1120	1d	R. Panorámica, Cd. Valles: 1200-0100
BC19)	MX	1120	1	Sonido de Mexicali: 1300-0700 (A)
GR15)	CHG	1130	1	R. Fiesta, Chilpancingo
MI11)	FN	1130	1d	R. Moderna, Uruapan: 1300-0100
SO13)	HN	1130	1d	La Pantera, Nogales: 1200-0200
VE08)	ZL	1130	10	Su Gallo, Jalapa: 1200-0600 (A)
GJ10)	XF	1140	5d	R. Felicidad, León: 1200-0100 (A)
NL02)	MR	1140	50	La Nueva MR, Monterrey: 24h
PU04)	TE	1140	1d	R. Manantial, Tehuacán: (A)
BC06)	RM	1150	1	R. Sensación, Mexicali: 1400-0900
CO14)	BF	1150	1	R. Mexicana, San Pedro: 1200-0400
CH16)	JS	1150	1	R. Exitos, Hidalgo del Parral: 1245-0500
DF08)	JP	1150	10	R. Variedades, México: 24h
JL06)	AD	1150	20	R. Metrópoli, Guadalajara: 24h
OX10)	XP	1150	0.5d	R. XP, Tuxtepec: 1200-0100
TB07)	RTM	1150	1	R. Variedades, Macuspana: 1200-0200
YU01)	RRF	1150	5	R. Lobo, Mérida: 1200-0400
GJ11)	VW	1160	1d	R. Sensación, Acámbaro: 1300-0130
MI12)	IW	1160	1d	R. 160, Uruapan
VE16)	BE	1160	1d	R. Perote: 1200-0100
AG01)	UVA	1170	0.5d	R. Uva, Aguascalientes: 1200-0100
ME01)	RLU	1170	1	R. Miled, Atlacomulco
MI09)	LP	1170	2/0.1	Onda 1170, La Piedad: 1300-0300
PU05)	CD	1170	1	11-70, La Divertida, Puebla: 1200-0600
SO32)	YH	1170	1	R. Mil, Hermosillo
SO33)	IB	1170	0.25	R. Sensación, Caborca
TM07)	RT	1170	5d	La Rancherita Contenta, Reynosa: 1230-0100
VE17)	ZS	1170	0.5	R. Felicidad, Coatzacoalcos: 1200-0600
VE18)	PV	1170	0.25d	R. Variedades, Papantla: 1200-0100
DF14)	FR	1180	5	R. Felicidad, México: 24h (A)
VE19)	YV	1180	1d	R. Acción, Huatusco: 1200-0100
BC07)	MBC	1190	1	R. Variedades, Mexicali: 1400-0600
CH12)	PZ	1190	1d	R. Norteña, Cd. Juárez: 1200-0200
HG01)	PK	1190	1	R. Alegría, Pachuca: 1200-0100
JL15)	WK	1190	50	La "W" de Guadalajara: 1200-0600
MI13)	SOL	1190	0.5d	R. Sol, Cd. Hidalgo: 1300-0100
TM20)	TOT	1190	5	La Poderosa, Tampico: 1200-0100
MO01)	DO	1190	1d	R. 11-90, Apotla: 1200-0200
NL08)	CT	1190	1d	R. Melodía, Monterrey: 1230-0100
ME02)	QY	1200	1	R. Fiesta, Toluca
MI14)	ZI	1200	1d	Canal Festivo 120, Zacapu: 1200-0100
SN23)	WT	1200	1	Culiacán: 1200-0300
VE20)	PW	1200	0.25d	R. XEPW, Poza Rica: 1200-0200
GJ12)	EMM	1210	0.25d	R. Salmantina, Salamanca: 1300-0100
CL07)	BCO	1210	50	R. Occidente, Colima
JL26)	ZAJ	1220	2.5	Sonoritmo 1220, Guadalajara
DF04)	B	1220	100	La B Grande de México: 24h.
TB11)	—	1230	—	R. Tabasco, Villahermosa: 1100-0500
VE15)	ID	1230	1	R. Alamo: 1200-0400
CO15)	VM	1240	0.25	Canal 124, La Juvenil, Piedras Negras: 1200-0600
CH17)	BN	1240	1/0.25	Cd. Delicias: 1300-0400
CH08)	WG	1240	0.25	Mágico 12-40, Cd. Juárez: 1200-0600
CS01)	LM	1240	1	R. Sensación, Tuxtla Gutiérrez
GJ10)	RZ	1240	1	R. Amistad, León: 1200-0600 (A)
MI15)	RPA	1240	0.5	R. Ranchito, Morelia: (A)
NA06)	SI	1240	1	R. Felicidad, Santiago Ixcuintla: 1300-0500
NL04)	IZ	1240	1	R. Exitos, Monterrey: 1200-0600
OX08)	CE	1240	0.5	R. Hit, Oaxaca: 1100-0600
PU06)	NG	1240	1	R. Nueva Generación, Huauchinango: 1100-0300
SO14)	BQ	1240	1/0.25	Estelar 12-40, R. Mexicana, Guaymas: 1300-0500
SO13)	CG	1240	1	R. Mexicana, Nogales: 1300-0100
TM10)	S	1240	1	R. Jet, Tampico 1200-0600
VE04)	OV	1240	1	R. Moderna, Orizaba: 1200-0400
CO03)	SC	1250	1	R. 1250, Sabinas
CO06)	SJ	1250	1	R. Saltillo: 24h
CS11)	MG	1250	1/0.25	Arriaga: 1200-0400
CH19)	AT	1250	1	R. Onda, Hidalgo del Parral: 1300-0300
GR06)	PI	1250	1	La Consentida, Chilpancingo: 1200-0200
JL04)	DK	1250	5	La Grande, Guadalajara: 24h
PU05)	ZT	1250	1	La Máquina Musical, Puebla: 1200-0600
QE01)	JX	1250	1	R. Dólar, Querétaro: 1200-0600
SO12)	DL	1250	1	R. Ambiente, Hermosillo: 24h
SO31)	SN	*1250	—	R. Sonora, Caborca
VE21)	TF	°1250	1	La Jarocha, Veracruz: 1130-0600
CO11)	MF	1260	1	R. Fiesta, Monclova: 1200-0600
CH20)	OG	1260	1	R. Ranchito, Ojinaga: 1200-0400
DF14)	L	1260	10	R. ACIR, México: 24h (A)
GJ13)	ZH	1260	0.5d	Vibraciones en DO, RE, MI, Salamanca
NL13)	R	1260	1	R. Recuerdo de Linares: 1200-0300

SL02)	XR	1260	1d	R. Mensajera, Cd. Valles: 1300-0100
SN08)	SA	1260	5	R. Hits, Culiacán: 1300-0500
SO15)	MW	1260	1	R. San Luis: 24h
VE22)	MTV	1260	0.25	R. Lobo, Minatitlán: 1200-0100
BC12)	AZ	1270	1	Zeta 13, Tijuana: 24h
DG04)	WN	1270	0.5	R. Variedades, Gómez Palacio: 1200-0600
GJ06)	RPL	1270	5	R. Poderosa, León: 24h
MI16)	ZU	1270	0.5	R. Moderna, Zacapu: 1300-0400
OX09)	AX	1270	5/0.5	R. Festival, Oaxaca: 1200-0500
SO16)	GL	1270	1	La Pionera de la R. en Navojoa: 1200-0700
SO17)	NY	1270	1	R. Geny, Nogales
TM10)	RRT	1270	2	R. Ranchito, Cd. Madero: 24h
VE20)	RRR	1270	1	R. Alegría, Poza Rica: 1200-0300
HG05)	QH	1270	1	Ixmiquilpán
YU05)	YW	1270	5	Tu Chica, Mérida: 1200-0400
CA03)	CAM	1280	1	La Voz de las Murallas, Campeche: 1200-0500
CS12)	KY	1280	1	La Voz de la Costa de Chiapas, Huixtla: 1200-0200
CH03)	BW	1280	1/0.6	Canal 12-80, Chihuahua: 24h
GJ14)	SQ	1280	0.5	R. San Miguel, S.M. de Allende: 1200-0300
JL14)	QP	1280	0.5/0.25	R. Variedades, Guadalajara: 1300-0700
NL04)	AW	1280	5	Música Inolvidable, Monterrey: 24h
PU05)	EG	1280	1	R. Alegría, Puebla: 1200-0700
SN14)	HS	1280	1	R. Ambiente, Los Mochis: 1300-0600
VE02)	AG	1280	1	R. Juventud, Córdoba: 1200-0400
CA04)	TH	1290	0.25	R. 12-90, Palizada: 1200-0400
DF09)	DA	1290	10	R. 13, México: 24h.
MI17)	IX	1290	1	R. Ritmo, Jiquilpan: 1300-0200
SL04)	IY	1290	1d	R. Juventud, Río Verde: 1200-2200
SN10)	NX	1290	10	La Poderosa NX, Mazatlán: 24h
SO18)	AP	v1290	1	R. Sensación, Cd. Obregón: 1300-0500 (r. 1292)
CH12)	P	1300	1/0.5	R. 13, Cd. Juárez: 24h
CH22)	SW	1300	1d	R. Madera, Cd. Madera: 1300-0100
GJ15)	XV	1300	1d	Serenata 13, León: 1300-0100
MI07)	KW	1300	0.5	R. Festival, Morelia: 1200-0400 (A)
SN15)	JL	1300	1	R. Guamuchil: 1200-0400 (A)
SO13)	XW	1300	1	R. Sensación, Nogales: 1200-0600
TM13)	LE	1300	1	Radio 13, Tampico: 1200-0600
VE23)	HU	1300	1	R. Tropical, M. de la Torre: 1200-0600
QE01)	HY	1310	1	La Poderosa, Querétaro
BC13)	C	1310	1	R. Enciso, Tijuana: 1400-0900
CH23)	RU	1310	1/0.25	R. Universidad, Chihuahua
DG04)	BP	1310	10	R. Sensación, Gómez Palacio: 24h
GR07)	HJ	1310	0.25d	R. Petatlán: 1200-0100
JL06)	TIA	1310	10	La Cotorra de tu Radio, Guadalajara: 1200-0700
NL02)	VB	1310	4	R. 13, Monterrey: 1200-0600
PU14)	HIT	1310	1	Tropifiesta, Puebla: 1200-0600
SL03)	PO	1310	0.5	R. Voz, San Luis Potosí: (A)
SO19)	FH	1310	1	La FH, Agua Prieta: 1400-0300
TM14)	AM	1310	5	La Norteña, Matamoros: 1200-0600
VE06)	HV	1310	1	R. Trópico, Veracruz: 1200-0600
DG04)	BP	1310	1	R. Sensación, Gómez Palacio
BS03)	SR	1320	0.5	R. Cachanía, Santa Rosalía: 1300-0700
CS13)	UI	1320	1	R. Comitán: 1200-0400
DF08)	CMQ	1320	10	R. Sensación, México: 24h
HG02)	CY	1320	1/0.25	La Voz de la Huasteca, Huejutla: 1200-0100
MI18)	NI	1320	1	La Rancherita Bonita, Nueva Italia: 24h
OX10)	UH	1320	1	R. UH, Tuxtepec: 1200-0400
SN03)	RJ	1320	5	R. Mazatlán: 1200-0800
CO18)	WQ	1330	1	La Superestación, Monclova: 24h
CO19)	AJ	1330	0.5	R. Bonita, Saltillo: 1200-0600
GJ16)	BO	1330	5	R. Variedades, Irapuato: 1100-0600
OX11)	AH	1330	0.5	R. Hit, Juchitán: 1130-0630
PU07)	EV	1330	0.5d	Alma Musical de Izúcar de Matamoros: 1200-0100
SN16)	UAS	*1330	5/1	R. Universidad, Culiacán
TM10)	RP	1330	1	R. Sensación, Cd. Madero: 24h
VE23)	UZ	1330	1	R. Veracruzana, M. de la Torre: 1200-0600
YU01)	FC	1330	5/1	La Romántica, Mérida
BC06)	AA	1340	1	R. Idolos, Mexicali: 24h
CA05)	BCC	1340	0.25	Es Su Estación, Cd. del Carmen: 1200-0600 (A)
CO13)	DH	1340	1	R. Tropical, Cd. Acuña: (A)
CH24)	BZ	1340	1	La Rancherita, Meoqui
CH20)	RCH	1340	0.25	La Pantera, Ojinaga: 1300-0300
GR03)	CI	1340	1	La Nueva Radiorama, Acapulco
HG03)	QB	1340	1	La Divertida, Tulancingo
JL04)	DKT	1340	1	R. Ranchito, Guadalajara: 1200-0600
MI19)	CR	1340	1	R. Variedades, Morelia: 1200-0530
MO01)	JC	1340	1	R. Sensación, Cuernavaca: 1200-0600
NA07)	ZE	1340	1	La Cotorra de Tu Radio, Stgo. Ixcuintla: 1400-0200
NL02)	NV	1340	5	R. Alegría, Monterrey: 24h
PU08)	LU	1340	0.5d	R. Esmeralda, Cd. Serdán: 1300-0100
SL05)	SL	1340	0.5/0.25	R. Variedades, San Luis Potosí: 1200-0600
SN17)	QE	*1340	1d	La Perla Camaronera, Escuinapa: 1300-0200
SN18)	CW	1340	1/0.2	R. Variedades, Los Mochis: 24h
SO06)	OS	1340	1	R. Variedades, Cd. Obregón: 1300-0500
TB08)	YR	1340	1	R. 13, Teapa: 1100-0300
TM01)	RPV	1340	1	La "V" de la Victoria: 1300-0200
TM14)	MT	1340	1	La Tremenda, Matamoros: 1200-0300
TM15)	BK	1340	5	El Norteñazo, Nuevo Laredo: 1200-0600
VE24)	BY	1340	0.25	R. Nivel del Mar, Tuxpam: 1200-0400
CO04)	TB	1350	5/0.5	R. Laguna, Torreón: 24h
CS16)	CAH	1350	5	R. Chiapas "La Voz de Soconusco," Cacaohuatán
DF04)	QK	1350	1	La Hora Exacta, "XE-Cuca", México: 24h
JL24)	JY	1350	1d	La Voz del Valle, El Grullo: 1300-0100
SO20)	TM	1350	1d	El Heraldo de la Frontera, Naco: 1300-0200
SO15)	LBL	1350	0.5d	R. Centro, San Luis: 1300-0200
TM16)	ZD	1350	0.25	La Doña de la Frontera, Camargo
VE25)	JF	1350	1	R. Variedades, Tierra Blanca: 1300-0400
CO15)	IK	1360	0.5	La Primera, Piedras Negras: 1200-0600
CS09)	UD	1360	5	La "U" de Tuxtla Gutiérrez: 1100-0700 (A)
CH02)	DI	1360	1	La Pantera, Chihuahua: 24h
GJ04)	Y	1360	1/0.25	R. Celaya: 1300-0600
GR08)	KF	1360	1	R. Independiente, Iguala: 1200-0500
VE26)	DQ	1360	1	R. Alegría, San Andrés Tuxtla: 1200-0600 (A)
ZC08)	XM	1360	1	R. Jerez, Jerez de García Salinas: 1200-0400
BC02)	HG	1370	0.5	R. Ranchito, Mexicali: 1400-0800
CA06)	A	1370	1	Radiorama del Sureste, Campeche: 1200-0600
DG02)	RPU	1370	1	R. Sensación, Durango: 1300-0500
GJ17)	JE	1370	0.5d	La Consentida, Dolores Hidalgo: 1300-0100
SL10)	SMR	1370	1	La Poderosa, San Luis Potosí
JL08)	PJ	1370	5	R. Capital, Guadalajara: (A)
MI20)	SV	1370	0.5d	R. Unicolás, Morelia
NL14)	MON	1370	10	Poder 13-70, Monterrey: 1100-0600
PU05)	PA	1370	5	La Poblanita, Puebla: 1100-0600
SO13)	HF	1370	5	R. Norteña, Nogales: 1400-0700
TM17)	GNK	1370	0.25d	R. Ritmo, Nuevo Laredo: 1215-0100
BC14)	KT	1380	0.5	Tu Chica Musical, Tecate: 1300-0500
CO20)	VD	1380	1/0.1	R. Allende: 1200-0400

DF02)	CO	1380	10	R. Dimensión 13-80, México: 24h.
DG04)	RS	1380	1	La Divertida, Gómez Palacio: 1300-0400
GJ11)	AK	1380	1.5	R. Consentida, Acámbaro: 1200-0400
TM01)	GW	1380	5	R. 13-80, Cd. Victoria: 1200-0400
VE27)	TP	1380	5	R. Sensación, Jalapa: 1300-0100
CL06)	TY	1390	5	R. Tecomán: 1300-0400
GJ06)	RW	1390	10	Sonido 13-90, León: 24h
SO21)	QC	1390	1d	La Reina del Mar, Puerto Peñasco
TM18)	XO	1390	5	La Diferente, Cd. Mante: 1200-0600
TM19)	OR	1390	1	R. Gallito, Reynosa: 1100-0600 (A)
VE24)	TL	1390	5/1	La Voz de la Huaxteca, Tuxpam: 1200-0400
YU08)	RUY	1400	10	R. Universidad, Mérida: 1300-0600
AG03)	AC	1400	0.5	R. Cordialidad, Aguascalientes: 1300-0600
BC15)	PF	1400	1	La Rancherita, Ensenada: 1400-0730
CO19)	DE	1400	1/0.25	La Divertida, Saltillo: 1200-0600
GR09)	KJ	1400	0.5/0.25	R. Alegría, Acapulco: 1200-0400
ME03)	XI	1400	1	La "I" de Ixtapan, Ixtapan de la Sal: 1200-0400
MI21)	OJ	1400	1	R. Horizonte, Cd. Lázaro Cárdenas: 1200-0400
MI22)	I	1400	1/0.25	R. Morelia: 24h
NA08)	LH	1400	1d	La Gardenia Musical, Acaponeta: 1300-0100
NA09)	NAY	1400	0.25d	La Romántica, Ahuacatlán: 1400-0100 (n. 1450)
NL09)	SH	1400	5	R. Alegría, Cd. Sabinas Hidalgo: 1200-0600 (n. 1450)
PU09)	FS	1400	0.25	R. Matamoros, Izúcar de Matamoros: 1200-0200 (A)
QE03)	VI	1400	0.5	R. Ambiente, San Juan del Río: 1200-0400
SL06)	WU	1400	0.25	Matehuala: 1200-0400
SN12)	EX	1400	0.25	R. Felicidad, Culiacán: 1300-0500 (A)
SO23)	AB	1400	1	Santa Ana: 1400-0500
TM20)	TO	1400	1	Radiorama, Tampico: 24h
DF02)	BS	1410	5/1	R. Sinfonola "La Est. del Barrilito," México: 1200-0800
JL17)	KB	1410	8	Canal 14-10, Guad: 24h
SL11)	IR	1410	5	Estelar 14-10, Cd. Valles: 24h
SN13)	CF	1410	1/0.5	Canal 14-10, R. Mexicana, Los Mochis: 1300-0600
TB03)	TAB	1410	5/0.5	R. Juvenil, Villahermosa: 1200-0400
TM11)	AS	1410	1	R. Jalapeño, Nuevo Laredo: 1200-0400
CO25)	YD	1410	1	La Norteña, Torreón
BC16)	XX	1420	2	R. Ranchito, Tijuana: 1300-0800
CH10)	F	1420	5	R. Centro, Cd. Juárez: 1000-0700
GJ05)	WE	1420	10	La Estación Familiar, Irapuato: 1100-0500
HG01)	RD	1420	0.25	R. Lobo, Pachuca: 1300-0500
JL18)	KMX	1420	1d	Canal 1420, Sayula: 1200-2400
MO01)	WF	1420	0.5	R. Fiesta, Cuernavaca: 24h
NL03)	H	1420	1	La Tremenda Tropical, Monterrey: 1200-0500
PU10)	WJ	1420	1	R. Popular, Tehuacán: 1200-0600
TM21)	EW	1420	1	La Voz del Bajo Bravo, Matamoros: 1200-0600
VE30)	JV	1420	1/0.25	Jaltipan: 1200-0100
CA06)	RAC	1430	1	R. RAC, Campeche: 1200-0600
ME04)	XOO	1430	1	La "O" de Oro, El Oro: 1200-0400
OX13)	CA	1430	1	R. Exitos, Ixtepec: 1200-0200
SO24)	OX	1430	5/1	Canal 14-30, Cd. Obregón: 24h
TM22)	WD	1430	2	La Grande, Cd. Miguel Alemán: 1200-0400
TX02)	TT	1430	0.7d	R. Tlaxcala: (A)
VE06)	LL	1430	1	R. Onda, Veracruz: 1200-0600
BS04)	VSD	1440	1	La Señal del Progreso, Cd. Constitución: 1300-0700
DF08)	EST	1440	10	R. Alegría, México: 24h
JL09)	CCC	1440	5	R. Fama, Guadalajara: 1200-0600
SN20)	HW	1440	0.25d	R. Provincia, Rosario: 1300-0200
AG04)	YZ	1450	1	R. Variedades, Aguascalientes: 1200-0400
BC17)	SS	1450	1	R. Fiesta, Ensenada: 1400-0800
CH26)	ARE	1450	1	R. Pegüis, Ojinaga: 1200-0400
GR11)	RY	1450	1/0.1	La Poderosa Voz del Sur, Arcelia
JL19)	JTF	1450	1	La Voz de la Provincia Jalisciense, Zacoalco de Torres: 1200-0400 (A)
MI17)	GC	1450	1	R. Continental, Sahuayo: 1200-0500
NA10)	TD	1450	0.25d	R. Sensación, Tecuala: 1300-0600
NL04)	JM	1450	1	La Tropicalísima, Monterrey: 1200-0600
PU11)	PUE	1450	1	R. Cañón, Atlixco: 1200-0600
QE01)	NA	1450	1/0.5	R. Alegría, Querétaro: 1200-0400
SO25)	DJ	1450	1	Magdalena: 1300-0300
SO15)	CB	1450	1	R. Ranchito, San Luis: 24h
TM06)	CM	1450	1	Música Romántica, Cd. Mante: 1200-0400
TM14)	VH	1450	1	R. Galáctica, Matamoros: 1200-0400
VE22)	KM	1450	1/0.5	R. Mina, Onda 14-50, Minatitlán: 1100-0600
VE32)	PP	1450	5	La Voz de Orizaba: 1200-0800
VE20)	JD	1450	1/0.25	R. Tropicana, Poza Rica: 1200-0600
YU03)	PY	1450	0.5	La Tropical Ardiente, Mérida: 1200-0500
CH12)	YC	1460	1	R. Sensación, Cd. Juárez: 24h
JL20)	HE	1460	1d	La Voz de Jalisco, Atotonilco el Alto: 1300-0100
MI02)	LX	1460	10	La Grande de Michoacán, Zitácuaro: 1200-0300
OX05)	KC	1460	1	R. Exitos, Oaxaca: 1200-0600
SL08)	XQ	1460	0.25	R. Universidad Potosina, San Luis Potosí
SO05)	HX	1460	1/0.25	R. Fama, Cd. Obregón: 1200-0400
VE08)	JH	1460	0.5d	R. Capital, Jalapa: (A)
BC18)	RCN	1470	5	RCN de Tijuana: 1300-0800
CA08)	BAL	1470	1	R. Voz Maya de México, Bécal
CO01)	YD	1470	1/0.1	R. Norteña, Fco. I. Madero: 1200-0200
DF11)	SM	1470	10	R. Cañon, "Rockañón," México: 24h
DG03)	CAV	1470	1	R. Internacional, Durango: 1200-0500
GJ05)	YA	1470	1d	La Super Estación, Irapuato: 1300-0100
SN18)	CU	1470	5	La Rancherita, Los Mochis: 24h
SN21)	ACE	1470	1	R. Exitos, Mazatlán: 24h
TM23)	HI	1470	3	El Heraldo Internacional, Cd. Miguel Alemán: 1200-0430
CO21)	XU	1480	1	R. Variedades, Monclova: 24h
CH17)	HM	1480	1d	HM R. Cd. Delicias: 1300-0100
GJ19)	GX	1480	0.5d	R. La Paz, San Luis de la Paz: 1300-0100
GR12)	XC	1480	1	La Voz de Plata, Taxco: 1200-0500
JL10)	ZJ	1480	1	R. Selecciones, Guadalajara: 1100-0700
NL07)	TKR	1480	1	La TKR, Rancherita y Regional, Monterrey: 1200-0600
OX14)	OU	1480	0.5d	R. Joya, Huajuapan de León: 1200-2400
SO10)	NS	1480	5/1	La NS, Navojoa: 1200-0700
VE20)	PR	1480	5	R. Poza Rica: 1200-0600
AG01)	RO	1490	1	La Nueva R-O, Aguascalientes: 1130-0330
CH12)	CJC	1490	1	La Pantera, Cd. Juárez: 24h
JL21)	ED	1490	1/0.25	XEED Ameca: 1200-0400
ME05)	CH	1490	0.5	R. Estrella, Toluca: 1100-0500
MI04)	KN	1490	1d	R. Variedades, Huetamo: 1300-0100
MI23)	GT	1490	1	R. Alegría, Zamora: 1230-0430
NA11)	SK	1490	0.25	R. Esparcimiento, Cd. Ruiz: 1300-0300
PU05)	POP	1490	1	R. Capital, Puebla: 1230-0630 (A)
QE01)	XE	1490	1/0.2	R. Capital, Querétaro: 1200-0600
SL06)	FF	1490	1	Matehuala: 1230-0100
SO27)	AQ	1490	1	R. Juventud, Agua Prieta: 1700-0400 (A)
SO28)	DR	1490	1/0.25	R. Marinera, Guaymas: 1300-0500 (A)
TM14)	MS	1490	1	R. Mexicana "La 'M' Grande," Matamoros: 24h

VE33)	VZ	1490	1	R. Sensación, Acayucan: 1200-0600
VE28)	YT	1490	1	Centro de Promoción Social y Cultural, Teocelo
CO22)	JQ	1500	0.25d	R. Parras
DF11)	AI	1500	20	R. AI "Canal Tropical," México: 24h
GJ20)	FL	1500	0.25d	R. Santa Fé de Guanajuato: 1300-0100
VE34)	GN	1500	1	R. Triunfadora, Piedras Negras: 1200-0100
GJ04)	OF	1510	5d	La Mexicana, Cortázar: 1200-0600 (A)
CH12)	JCC	1520	1	R. Paso del Norte, Cd. Juárez
CO23)	VUC	1520	1d	La Est. de los Super Grupos, Allende: 1200-0200
MI06)	CGP	1520	0.5d	R. Fiesta, Morelia: 1200-0100
SO15)	EH	1520	1d	R. Exitos, San Luis: 1300-0200
TM18)	YP	1520	1d	La Ranchera del Mante, El Limón: 1200-0100
VE35)	VO	1520	1d	R. San Rafael: 1300-0100
DF02)	UR	1530	5	R. Onda, México: 24h.
GJ21)	SD	1530	1	R. Silao: 1300-0400
MI24)	GQ	1530	1d	R. Esmeralda, Los Reyes
CS14)	VF	1540	5	R. Variedades, Villaflores
GJ04)	NC	1540	1/0.25	La Voz Musical, Celaya: 1300-0500
NL15)	STN	1540	—	R. Red, Monterrey
SO29)	HOS	1540	5	La Poderosa, Hermosillo: 24h
PU12)	RTP	1540	1	R. Impacto, San Martín Texmelucán
BC03)	BG	1550	1	15-50, Tijuana: 24h
TM11)	NU	1550	5d	La Rancherita Contenta, Nuevo Laredo: 1200-0100
VE36)	RUV	1550	0.35	R. Universidad Veracruzana, Jalapa: 1100-0700
CA07)	SE	1560	5d	Champotón: 1200-2400
CH10)	JPV	1560	1d	La Supersónica, Cd. Juárez: 1200-0200
DF08)	FAJ	1560	20	R. Consentida, México: 24h
GJ12)	MAS	1560	1	Música de Pegue, Salamanca: 1300-0500
NA12)	RIO	1560	5d	R. Río, Ixtlán del Río: 1300-0200
NL10)	DD	1560	0.5	R. Amistad, Montemorelos: 24h
SL09)	ZW	1560	1d	R. Diversión, Cerritos: 1200-0100
CO24)	RF	1570	250	Stereo Frontera, La Voz de la Amistad, Cd. Acuña: 24h
GJ04)	AF	1580	1d	Super AF, Celaya: (A)
GR13)	LI	1580	1	La Estación de la Familia Guerrerense, Chilpancingo: 1200-0400
HG06)	MQ	1580	5	La V. de Hidalgo, Tulancingo
ME06)	VAB	1580	0.25	R. Miled, Valle del Bravo: 1200-0400
MI26)	QL	1580	1	Sonido Brillante, Zamora: 1200-0400
SO30)	DM	1580	50	La Grande de Sonora, Hermosillo: 1200-0700 (A)
BC15)	HC	1590	1	R. Bahía, Ensenada: 1400-0700
CH24)	JK	1590	1	R. Amistad, Cd. Delicias: (A)
DF14)	VOZ	1590	10	Capital R., "La R. de la Capital," México: 24h (A)
MO02)	ART	1590	1	La Señal de la Estrella, Zacatepec: 1200-0400
VE39)	PT	1590	1	R. Misantla: 1200-0200
BC14)	TKT	1600	1	Rock 16, Tecate: 1400-0200
CO13)	AE	1600	1	R. Variedades, Cd. Acuña: 1300-0400
JL23)	ZK	1600	0.5	R. Alteña, Tepatitlán: 1200-0300
ME07)	GEM	1600	5	R. Mexiquense, Mectepec: 1200-0600
SL12)	GI	1600	1	Tamazunchale: 1200-0400
VE09)	QT	1600	0.5	La Pantera, Veracruz: 1200-0600

Shortwaves: (location = México City unless otherwise stated)

VE40)	JN	2390	0.8	R. Huayacocotla, Huayacocotla: W: 1200-1500, 2100-0100
NL12)	UJ	5980	0.5	R. XEUJ, Linares: 1100-0300
DF15)	RMX	5985	10	R. México Int: 1300-1700
DF02)	OI	6009.9	0.25	R. Mil: 1200-0600 (0800 Sat, Sun)
VE06)	UW	v6020	0.25	La "U" de Vera Cruz, "El Eco de Sotavento,": 1130-0600 (r. 6016.5v)
SL08)	XQ	6045	0.25	Radio Universidad de San Luis Potosí: 1300-0500
TM06)	CMT	*6090	1	Música Romántica, Cd. Mante: 1200-0400
YU03)	QM	6105v	0.25	Tus Panteras, Mérida: 1200-0600 (r. 6105.2)
SO05)	UDS	6114.9	1	R. Universidad de Sonora, Hermosillo: 1500-0715
DF06)	WW	6165	0.5	(r: XEW 900): 24h
DF12)	PPM	6185	5	R. Educación: 24h
DF06)	WW	9515	0.5	(r: XEW 900): 24h
VE21)	FT	9545v	0.5	La Jarocha, Vera Cruz (r. 9546.3v)
DF04)	QK	9555	0.5	La Hora Exacta: 24h
DF10)	YU	9600	1	R. UNAM: 1400-2400
DF06)	QQ	*9680	0.5	La Q Mexicana, México
DF15)	RMX	9705	10	R. México Int: 2000-0445
DF15)	RMX	11769.7	10	R. México Int: 1300-1700 (n. 11770)
DF06)	WW	15160	0.5	(r: XEW 900): 24h (alt. 15177)
DF15)	RMX	15430	50	R. México Int: 2000-0500
DF15)	RMX	17765	10	R. México Int: 1300-1700, 2000-0500

GRUPO ACIR, Pirineos 700, 11000 México, D.F. **Te:** (52) 52590001. Originates daily newsreel prgrs at 1330, 2030 and 0330. Relaying st's indicated by (A) in fq list.

AG00) AGUASCALIENTES (Agsc.)

AG01) Ap. 173, 20000 Aguascalientes — DG: Ing. A. Rivas G.
AG02) Madero 333-501, 20000 Aguascalientes — GG: A. Morales P.
AG03) Ap. 467, 20230 Aguascalientes — Gte: J. Ramírez G.
AG04) Ap. 232, 20000 Aguascalientes — GG: R. Zúñiga C.

BC00) BAJA CALIFORNIA NORTE (B.C.N.)

BC01) Ap. 100, 22000 Tijuana (or: c/o Noble Broadc. of San Diego, 4891 Pacific Highway, San Diego, CA 92110-4082) — Gte: Ing. E. Liaño G.
BC02) Ap. 666, 21000 Mexicali — Gte: A. Ruiz de Anda.
BC03) Av. de los Olivos 305, 22410 Tijuana — DG: Lic. M. Enrique M.
BC04) Alfareros 1301, 21010 Mexicali (or: Box 1014, Calexico, CA 92231) — Gte: E. Bonillas.
BC05) Ap. 860, 22540 Tijuana (or: 713 Broadway Suite "F", Chula Vista, CA 91910) — DG: G. E. Astiazarán.
BC06) Ap. 1500, Mexicali — Gte: A. Ruiz de Anda.
BC07) Prolong. Alfareros 253, Centro Cívico, 21000 Mexicali (or: Box 1123, Calexico, CA 92331) — Gte: M. Hurtado S. C.E: Luis F. García H.
BC08) Km. 19 Carr. S. Ant. de los Buenos, Tijuana — GG: C.E.J. Alvarado.
BC09) Ap. 526, 22800 Ensenada — Gte: F. Ruanova S.
BC10) Ap. 148, Morelos — Gte: Elisa Eguía Tello de Brassea.
BC11) Ap. 1090, 22420 Tijuana — Gte: Ing. J. Wilkins G.
BC12) Ap. 375, Tijuana — DG: J. Sánchez M.
BC13) Ap. 23, Tijuana — GG: G. Márquez M.
BC14) Av. Hidalgo 1193, 21450 Tecate — Gte: Fco. J. Fimbres.
BC15) Ap. 123, 22800 Ensenada — Gte: A. Cañas M.
BC16) Carlos Robirosa 142, 22000 Tijuana — Gte: G. López M.
BC17) Ap. 287, 22800 Ensenada — Gte: G. Reyes J.
BC18) Ap. 1470, 22000 Tijuana — Gte: R. Navarro A.

BS00) BAJA CALIFORNIA SUR (B.C.S.)

BS01) Ap. 105, 23010 La Paz — Gte: Francisco King.
BS02) Hidalgo 314, 23000 La Paz — DG: Raúl A. Aréchiga E.
BS03) Av. Las Flores 1, 23920 Santa Rosalía — Gte: I. Tonche P.
BS04) Ap. 279, 23600 Cd. Constitución — Gte: F. Riestra C.
BS05) Univ. Autónoma de Baja California Sur, 23000 La Paz.

CA00) CAMPECHE (Camp.)

CA01) Calle 44 y 21, 24350 Escársega — Gte: R. Palma P.
CA02) Calle 22 N° 147, Cd. del Carmen — Adm: H. Muñoz L.
CA03) Ap. 33, 24000 Campeche — Gte: G. Ortiz A.
CA04) Ap. 22, 24200 Palizada — R. Palma P.
CA05) Calle 30 N° 86-A, 24100 Cd. del Carmen — Adm: M. A. Canepa C.
CA06) Tamaulipas 15, 24050 Campeche — Gte: R. Palma P.
CA07) 24400 Champotón — Gte: R. Palma P.
CA08) 24930 Bécal.

CH00) CHIHUAHUA (Chih.)

CH01) Ap. 412, 31500 Cd. Cuauhtémoc — GG: A. Moreno S.
CH02) Ap. 9, 31000 Chihuahua — Gte: J. L. Medina R.
CH03) Ap. 220, 31000 Chihuahua — GG: R. Salmón de Boone.
CH04) Calle 4 Pte. 606, 33000 Cd. Delicias — Gte: R. Díaz G.
CH05) Ap. 264, 31500 Cd. Cuauhtémoc — Gte: G. Loyo E.
CH06) Allende 613, 33980 Cd. Jiménez — Gte: R. Montes A.
CH07) Ap. 8, 33500 S. Fco. del Oro — Gte: A. Gutiérrez M.

CH08) Ap. 266, Cd. Juárez — GG: J. Sánchez C.
CH09) Coronado 71, Santa Bárbara — Gte: R. Serrano Q.
CH10) 16 de Septiembre Ote. 337-305, 32000 Cd. Juárez.
CH11) Comonfort 301, 33700 Cd. Camargo — Gte: H. Piñera L.
CH12) Profra. E. Dosamantes 400, 32380 Cd. Juárez — Gte: José Pérez Michel
CH13) Ap. 177, 31500 Cd. Cuauhtémoc — Gte: G. López de la Rocha.
CH14) Cuauhtémoc 2000, 31020 Chihuahua — GG: G. López B.
CH15) Ap. 59, 31700 Nvo. Casas Grandes — Gte: L. F. Molinar.
CH16) Ap. 125, 33800 Hgo. del Parral — Gte: M. Orpinel M.
CH17) Ap. 222, 33000 Cd. Delicias — Gte: H. L. Pérez B.
CH19) Ap. 122, 33800 Hgo. del Parral — Gte: A. Gutiérrez M.
CH20) Ap. 29, 32880 Ojinaga — Dir: S. Bustamante V.
CH21) Ap. 4, 33700 Cd. Camargo — GG: J. Padilla C.
CH22) Calle 3 N° 1204, 31940 Cd. Madera — Gte: A. Muñoz E.
CH23) Univ. de Chihuahua — Dir: A. Varona T.
CH24) Ap. 250, 33000 Cd. Delicias — Gte: A. Juarico L.
CH25) Ap. 190, 33800 Hgo. del Parral — Gte: D. Salayandía N.
CH26) Ap. 94, 32880 Ojinaga — Gte: Lic. A. Orpinel R.

CL00) COLIMA (Col.)

CL01) 27 de Septiembre 328, 28000 Colima — DG: Roberto F. Levy.
CL02) Balbino Dávalos 17, 28200 Manzanillo — Dra: S.E. Godoy C.
CL03) Ingenieros 14, 28200 Manzanillo — Dir: E. Romo R.
CL04) Zaragoza 67 Altos, 28000 Colima — Gte: E. Padilla V.
CL05) Allende 408-101, 28100 Tecomán — Gte: Ing. H. Cobián C
CL06) Ap. 2, Tecomán — Gte: Dr. M. Alberto M.
CL07) Ap. 2-1690, Suc. A, 28000 Colima.

CO00) COAHUILA (Coah.)

CO01) (See DG04)
CO02) Ap. 3, 26000 Piedras Negras — Gte: C. M. Bres G.
CO03) Ap. 60, 26700 Sabinas — Adra: M. Mercedes G.
CO04) Ap. 225, 27140 Torreón — Gte: A. Brassea E.
CO05) Ap. 74, 26340 Cd. Múzquiz — Dir: L. Isaac E.
CO06) Am. Latina y Alaska, 25220 Saltillo — Gte: Lic. J. Luis S.
CO07) Ap. 285, 25700 Monclova — Gte: Roberto Apud A.
CO08) Ap. 57, 26000 Piedras Negras — Gte: Arq. F. Elizondo C.
CO09) Ap. 16, 26850 Nva. Rosita — Gte: H. H. Martínez T.
CO10) Pérez Treviño Pte. 839, 25000 Saltillo — Dir: J. López C.
CO11) Ap. 181, 25700 Monclova — DG: H. Ainslie de Medina.
CO12) Ap. 32, 26850 Nva. Rosita — Gte: G. Briseño G.
CO13) Ap. 10, 26200 Cd. Acuña — GG: I. Mendoza S.
CO14) Ap. 78, 27800 San Pedro — Gte: C. Enríquez F.
CO15) Emilio Carranza 1104, 26000 Piedras Negras — GG: R. M. González.
CO16) Desp. 601, Morelos y Cepeda, 27000 Torreón — G. Adm: R. Ortega A.
CO18) Ap. 89, 25700 Monclova — Gte: Norberto R. Carrizales L.
CO19) Universidad 205, 25280 Saltillo — Gte: A. Jaubert A.
CO20) Juárez 307 Sur, 26530 Allende — Gte: H. Moreno V.
CO21) Ap. 91, 25600 Monclova — GG: J. F. Elizondo V.
CO22) Madero 501 Pte., 27980 Parras — Pte: Fco. J. Sánchez C.
CO23) Arteaga 106 Pte., 26530 Allende — GG: R. M. González.
CO24) Ap. 98, 26200 Cd. Acuña — Gte: Lic. A. C. González.
CO25) Matamoros Pte. 655, 27000 Torreón.

CS00) CHIAPAS (Chis.)

CS01) Ap. 59, 29000 Tuxtla Gutiérrez — Gte: R. Olvera P.
CS02) Ap. 57, 30700 Tapachula — GG: G. Flores S.
CS03) Ap. 74, 29200 S.C. Las Casas — Gte: Fco. Narváez R.
CS04) Ap. 76, 30700 Tapachula — Gte: K. Ordaz E.
CS05) Av. Central Pte. 554, 29000 Tuxtla Gutiérrez.
CS06) Ap. 60, 30400 Cintalapa — Gte: G. Orozco M.
CS08) Ap. 19, 30500 Tonalá — Gte: Lic. C. León C.
CS09) Ap. 600, 29000 Tuxtla Gutiérrez — GG: J. C. Chanona A.
CS10) 29520 Pichucalco — Gte: Ricardo Eslava.
CS11) Ap. 28, 30450 Arriaga — Adm: C. Sanz L.
CS12) Av. Central Norte 8, 30640 Huixtla — Gte: C. Baxín A.
CS13) Ap. 99, 30000 Comitán — Gte: R. Avendaño V.
CS14) Ap. 100, 30470 V. Flores — GG: O. Fonseca A.

DF00) DISTRITO FEDERAL (D.F.)

DF01) Chapultepec 473, 06600 México — GG: O. Obregón M.
DF02) Núcleo Radio Mil, Insurgentes Sur 1870, 01030 México — Pte: Lic. E. G. Salas P.
DF03) Bartolache 1918, 03100 México — DG: V. Blanco R.
DF04) Instituto Mexicano de la Radio (IMER), Margaritas 18 Col. Florida, 01030 México — DG: T. Rentería A.
DF05) Insurgentes Sur 1871-8°P., 01020 México — DG: E. Nassar H.
DF06) Ayuntamiento 52, 06070 México. **Te:** +52 (5) 518 1220 — DG: Lic. J. Almeida P. C.E: Ing. Miguel A. Barrientos V.
DF07) Eugenia 1462, 03020 México — DG: E. Gandarilla A.
DF08) Organización Radio Centro, Artículo 123 N° 90, 06050 México — Pta: Ma. E. Gómez de Aguirre.
DF09) Trabajadoras Sociales 309 Eje 6, 09410 México — Pte: Lic. J. L. Fernández H.
DF10) Adolfo Prieto 133 Col. del Valle, México — DG: A. Zayas.
DF11) Privada de Horacio 10 Polanco, 11560 México — DG: Julio E. Velarde A.
DF12) Angel Urraza 622, 03100 México — C.E: Gustavo Carreño L.
DF13) Radio Programas de México, La Presa 212 San Jerónimo Lídice, 10200 México — Gte: Roberto Ordorica C.
DF14) Pirineos 770, 11000 México — DG: F. Ibarra L.
DF15) Ap. 19-737, 03900 México — Gte: Lic. Beatriz Villa Medina G. — DX Prgr in Sp. Sun 0300, Wed & Fri 1500, 2200, Thurs. 0300, 0400.
DF16) Blvd. de los Virreyes 1030, Lomas de Chapultepec, 11000 México.
DF17) Mariano Escobedo 532, Col. Anzures, 11300 México.

FM in Mexico City:
DF13) 88.1 R. VIP; DF16) 88.9 Azul 89; DF02) 89.7 — Estereomil — DF09) 90.5 Imagen — DF13) 91.3 Estación Alfa — 92.1 Crystal — DF06) 92.9 La Tropi Q — DF08) 93.7 R. Joya — DF04) 94.5 Opus 94; — DF14) 95.3 R. Amistad — DF10) 96.1, DF06) 96.9 Magia Digital — DF08) 97.7 Stereo 97.7 FM — DF09) 98.5 Classic — DF16) 99.3 Digital 99 — 100.1 Stereo Cien — DF02) 100.9 Rock 101 — DF06) 101.7 Estéreo 102; — DF17) 102.5 Stereorey — DF11) 103.3 Kosmo Estéreo — DF11) 104.1 Jazz — DF17) 104.9 Globo — DF04) 105.7 Joven — DF16) 106.5 Amor 106 — DF08) 107.3 R. Universal. — Prgr. different from MW. All st's 30-180 kW ERP.

DG00) DURANGO (Dgo.)

DG01) Ap. 174, 34000 Durango — Gte: H. R. Piña S.
DG02) Ap. 405, 34000 Durango — Gte: Alejandro Stevenson B.
DG03) Negrete 405-8 Nte., 34000 Durango — DG: C. Armas H.
DG04) González de la Vega y Piedras Negras Z.I., 35070 Gómez Palacio — Gte: C.P. C. Enríquez F.
DG05) Madero y Col. Militar, Santiago Papasquiaro.
DG06) Av. 20 de Noviembre 210 Ote., Edif. Morosán, 34000 Durango.

GJ00) GUANAJUATO (Gto.)

GJ01) Alhóndiga 25-C, 36080 Guanajuato — Gte: D. Castrojón T.
GJ02) Ap. 301, 37160 León — Gte: J. Torres M. de Campo.
GJ03) Ap. 528, 38000 Celaya — Gte: L. E. Rodríguez M.
GJ04) López Mateos 402-B, 38000 Celaya — GG: J. Miranda.
GJ05) Ap. 61, 36500 Irapuato — GG: M. Ant. H. Contreras S.
GJ06) Ap. 392, 37480 León — Gte: D. Pérez V.
GJ07) Fray Andrés de San Miguel 97, 38900 Salvatierra — GG: R. Sandovar A.
GJ08) Universidad de Guanajuato, 36080 Guanajuato.
GJ09) Padre Quintana 10, 38800 Moroleón — GG: G. F. Ortiz M.
GJ10) López Mateos 434 Ote., 37000 León — GG: D. Castrejón T.
GJ11) Allende 16, 38600 Acámbaro — Gte: G. Argueta S.
GJ12) Ap. 300, 36700 Salamanca — GG: M. Ant. H. Contreras S.
GJ13) Ap. 24, 36700 Salamanca — Gte: Lic. E. Silva E.
GJ14) Sollano 4, 37700 San Miguel de Allende — Gte: J. M. Zavala Z.
GJ15) Juárez 783, León — GG: M. Vázquez O.
GJ16) Ap. 72, 36500 Irapuato — Gte: A. Martínez D.
GJ17) Ap. 43, 37800 Dolores Hidalgo — Gte: J. R. Reyna L.
GJ19) Ap. 67, 37900 S.L. de la Paz — Gte: M. González S.
GJ20) Municipio Libre 8, 36080 Guanajuato — GG: G. Hernández C.
GJ21) Ap. 60, 36100 Silao — Gte: J. Sánchez R.
GJ22) Priv. V. Carranza 119, 38000 Celaya.

GR00) GUERRERO (Gro.)

GR01) C. de la Paz 190-6, 39300 Acapulco — Cdor: H. Núñez F.
GR02) Ap. 60, 39300 Acapulco — Gte: M. Hernández T.
GR03) Av. del Tanque Col. Lázaro Cárdenas, Acapulco.
GR05) Ignacio Altamirano y 5 de Mayo, 40880 Zihuatanejo — Gte: M. Morales V.
GR06) Ap. 2, 39170 Tixtla — DG: A. Peyrón G.
GR07) Ap. 31, 40830 Petatlán — Gte: J. Orta Z.
GR08) Ap. 2, 40000 Iguala — DG: J. S. Garrido H.
GR09) Ap. 21, 39300 Acapulco — CG: H. Núñez F.
GR10) Ap. 52, 40000 Iguala — Gte: J. N. González R.
GR11) 40500 Arcelia.
GR12) Calle del Arco 1, 40200 Taxco — Gte: B. García A.
GR13) Ap. 40, 39000 Chilpancingo — Gte: H. Castorena N.

GR14) Arturo Martínez Adame s/n, Col. Cuauhtémoc Nte., 39000 Chilpancingo.
GR15) Col. Eduardo Neri, Chilpancingo — Gte: Lic. Abel Sánchez S.
HG00) HIDALGO (Hgo.)
HG01) Ap. 123, 42000 Pachuca — Gte: F. Sierra.
HG02) Ap. 35, 43000 Huejutla — Gte: J. A. Reyes F.
HG03) Hidalgo Ote. 209, 43600 Tulancingo — Gte: H. Segura O.
HG04) Ap. 96, 43600 Tulancingo — GA: L. Castelán L.
HG05) Carr. a Cardenal km. 2, Ixmiquilpan.
HG06) Pza. Constitucíon y Miguel F. Soto, Tulancingo.
JL00) JALISCO (Jal.)
JL01) Ap. 60, 49000 Cd. Guzmán — Pte: Fco. J. Díaz R.
JL03) Austria 238 Col. Versailles, 48310 Pto. Vallarta — GG: R. López F.
JL04) Lázaro Cárdenas 3126, 45040 Guad. — GG: Lic. J. del Pozo M.
JL05) 44100 S.J. de los Lagos.
JL06) Ap. 1333, 44100 Guadalajara — DG: Enrique Pérez Maestre.
JL07) Ap. 7, 48900 Autlán — DG: J. A. Perales G.
JL08) Juárez 127, 47800 Ocotlán — Dir: J. Raúl N.
JL09) Mariano Otero 3405 Fracc. Verde Valle, 45560 Guad. — DG: J. Pérez R.
JL10) Vidrio 2056, 44100 Guad. — GG: G. Fregoso M.
JL11) Ap. 32, 47400 Lagos de Moreno — GG: D. S. de Sánchez.
JL12) Pablo Casals 567 Prados Providencia, Guad. — Dir: L. F. Aguirre.
JL13) Morelos 27, 47910 La Barca — Gte: R. Alvarez L.
JL14) Organización OIR, T.V. Gómez 47 Col. Arcos Sur, 44500 Guad. — Dir: Lic. J. Jiménez M.
JL15) Lerdo de Tejada 2186, 44160 Guadalajara — Gte: Lic. P. Moreno V.
JL16) Finca El Cerrito, 49650 Tamazula — GG: Fco. Uribe C.
JL17) 16 de Septiembre 426, 44100 Guad. — Gte: D. Martínez.
JL18) Ap. 36, 49300 Sayula — GG: J. A. Aguilar H.
JL19) Madero 77, 45750 Zacoalco de Torres — GG: J. Toscano F.
JL20) Zaragoza 59 Altos, 47750 Atotonilco El Alto — GG: S. Rubio P.
JL21) Ap. 16, 46600 Ameca — Gte: A. Navarro A.
JL22) Pascual Galindo Ceballos 188, 49000 Cd. Guzmán.
JL23) Hidalgo 78-4, 47600 Tepatitlán.
JL24) Ap. 5, 48740 El Grullo — Adm: R. Pérez G.
JL25) Manuel M. Diéguez 314, 48380 Pto. Vallarta.
JL26) Niños Héroes No. 1555, Edif. Plaza Tolsá, 44100 Guadalajara.
ME00) ESTADO DE MEXICO (Edo. Méx.)
ME01) Isidro Fabela 42-A, 50450 Atlacomulco — GG: M. Libien K.
ME02) Independencia 406 Ote., 50000 Toluca — GG: L. F. Nieto V.
ME03) 51900 Ixtapan de la Sal — DG: R. Ordorica C.
ME04) Independencia 19, El Oro — DG: R. Ordorica C.
ME05) Ap. 37, 50000 Toluca — GG: O. Azcue R.
ME06) 51200 Valle del Bravo.
ME07) Av. Edo. de México Km. 1, Mectepec — DG: Ing. J. Muñoz E.
MI00) MICHOACAN (Mich.)
MI01) Acueducto 992, 58000 Morelia — Dir: J. Laris I.
MI02) Ap. 50, 61500 Zitácuaro — GG: M. Jiménez R.
MI03) Ap. 61, 60000 Uruapan — GG: M. Treviño N.
MI04) Ap. 37, 59670 Zamora — Gte: Fco. Elizalde G.
MI05) Constitución Norte 10 Altos, 60600 Apatzingán.
MI06) Acueducto 992, 58000 Morelia — Gte: D. López I.
MI07) Laguna de Parras 630, 58020 Morelia — Gte: R. Rodríguez M.
MI08) Constitución Norte 2 Altos, 60600 Apatzingán — Gte: C. Hernández R.
MI09) Ap. 10, 59300 La Piedad — Dir: H. Guízar C.
MI10) Ap. 10, 61600 Pátzcuaro — Gte: H. Guido S.
MI11) Ap. 132, 60000 Uruapan — Gte: J. Lira R.
MI12) Mazatlán 30, 60050 Uruapan — Gte: F. de Jesús Flores L.
MI13) Ap. 27, 61100 Cd. Hidalgo — Adm: Ing. J. Robledo.
MI14) Ap. 65, 58600 Zacapu — Gte: J. Ortiz A.
MI15) Av. Madero Pte. 398-10, 58000 Morelia — Gte: J. Gpe. Muñoz M.
MI16) Ap. 50, 58600 Zacapu — Gte: E. González M.
MI17) Ap. 60, 59000 Sahuayo — Gte: O. Castro.
MI18) Ap. 49, 61760 Nueva Italia.
MI19) Ap. 275, 58020 Morelia — GG: J. M. Treviño N.
MI20) Universidad de San Nicolás, 58000 Morelia.
MI21) Río Balsas 7, 60950 Cd. Láz. Cárdenas — DG: Fco. Bautista V.
MI22) Ignacio López Rayón 334, 58000 Morelia — GG: E. Salgado B.
MI23) Ap. 68, 59670 Zamora — Gte: M. A. Sánchez G.
MI24) Ap. 45, 60300 Los Reyes — Gte: E. Pérez Z.
MI25) Ap. 430, 60950 Cd. Láz. Cárdenas — Gte: D. Mendoza Q.
MI26) Ap. 40, 59600 Zamora — Gte: J. Martínez G.
MO00) MORELOS (Mor.)
MO01) Ap. 720, 62000 Cuernavaca — GG: D. López I.
MO02) Pasaje Robles, 62900 Jojutla — Gte: A. E. Martínez S.
NA00) NAYARIT (Nay.)
NA01) Ap. 296, 63000 Tepic — GG: J. de Jesús Macías C.
NA02) Ap. 95, 63000 Tepic — GG: A. Núñez I.
NA03) Independencia 1330 Ote., 63200 Tuxpan — Gte: L. Herena M.
NA04) Ap. 68, 63000 Tepic — GG: L. O. Bertrand P.
NA05) Ap. 41, 63700 Compostela — Gte: J. Mondragón G.
NA06) Ap. 22, 63300 Stgo. Ixcuintla — Gte: J. Mondragón G.
NA07) Ap. 4, 63310 Stgo. Ixcuintla — GG: J. de Jesús Macías C.
NA08) Ap. 49, 63400 Acaponeta — GG: H. J. Aguiar A.
NA09) Juárez 67, 63900 Ahuacatlán — GG: J. de Jesús Macías C.
NA10) Ap. 7, 63440 Tecuala — DG: P. Aguiar V.
NA11) Juárez 702, 63600 Cd. Ruiz — GG: R. Abud Q.
NA12) Ap. 33, 63940 Ixtlán del Río.
NL00) NUEVO LEON (N.L.)
NL01) Nadadores 153, 64700 Monterrey — (Transmitters at Río Verde, SLP).
NL02) Grupo R. Alegría S.A., Madero Ote. 1110, 64700 Monterrey — Pres: Gonzalo Estrada Cruz.
NL03) 5 de Mayo 110 Ote., 64000 Monterrey — DG: F. Aguirre G.
NL04) Ap. 628, Monterrey — Gte: J.M. Díaz.
NL05) Ap. 1111, Monterrey — Adm. Francisca Arredondo.
NL06) 67500 Montemorelos — DG: J. A. Gómez G.
NL07) Ap. 1430, Monterrey — Gte: Fco. A. González.
NL08) Ap. 118, 64000 Monterrey — GG: Lic. Teófilo Zimeri Bichara.
NL09) Ap. 38, 65290 Cd. Sabinas Hgo. — Gte: Lic. A. Ríos L.
NL10) Ap. 45, 67500 Montemorelos — GG: A. J. Pezzino G.
NL12) Ap. 62, 67700 Linares — DG: M. Becerra G.
NL13) Ap. 81, 67700 Linares — Pte: C. Serna M.
NL14) Juan Ignacio Ramón 506 Ote., 20° piso Edif. Latino, 64000 Monterrey.
OX00) OAXACA (Oax.)
OX01) Ap. 46, 70760 Tehuantepec — Gte: H. López L., Jr.
OX02) Ap. 175, 68000 Oaxaca — Gte: W. Hernández M.
OX03) Ap. 21, 70760 Tehuantepec — Gte: E. Velázquez T.
OX04) Ap. 66, 70300 Matías Romero — Gte: H. López L.
OX05) Bustamante 112, 68000 Oaxaca — GG: Fco. Pliego G.
OX06) 16 de Septiembre 23, Loma Bonita — GG: E. Haaz D.
OX07) 71980 Pto. Escondido — Gte: I. Orihuela F.
OX08) Ap. 229, 68000 Oaxaca — Gte: S. J. Alcázar.
OX09) Gómez Farías 113, 68000 Oaxaca — DG: A. Márquez R.
OX10) Ap. 48, 68300 Tuxtepec — Gte: A. Gallegos M.
OX11) Ap. 60, 70030 Juchitán — Gte: H. López L., Jr.
OX12) Ap. 35, Pto. Angel — Gte: I. Orihuela F.
OX13) 70110 Ixtepec — Gte: H. López L.
OX14) Ap. 48, 69000 Huajuapan de León — GG: R. Martínez O.
PU00) PUEBLA (Pue.)
PU01) Allende 507, 73800 Teziutlán — Gte: J. Sánchez T.
PU02) Ap. 176, 73800 Teziutlán — Gte: R. Romero R.
PU03) Ap. 84, 75700 Tehuacán — Gte: Fco. Pérez N.
PU04) 2 Pte. 101, 72000 Puebla — DG: Lic. R. Cañedo B.
PU05) Juárez 2108, 72400 Puebla — Pte: A. Grajales S.
PU06) Ap. 54, 73160 Huauchinango — Gte: M. R. Rojano S.
PU07) Ap. 49, 74400 Izúcar de Matamoros — Gte: Sra. A. Rodríguez.
PU08) Ap. 21, 75520 Cd. Serdán — Gte: R. Bautista H.
PU09) Corregidora 4, 74400 Izúcar de Matamoros — Gte: A. Tavares H.
PU10) Ap. 44, 75700 Tehuacán — Gte: Fco. Sánchez T.
PU11) Independencia 104 Altos, 74200 Atlixco — DG: E. E. Cabrera S.
PU12) Ap. 4, 74000 S.M. Texmelucan — Gte: R. Franco P.
PU13) Reforma No. 120, 73080 Xicotepec de Juárez.
PU14) Blvd. Héroes del 5 de Mayo No. 9, Col. Las Palmas, 72550 Puebla.
QE00) QUERETARO (Qro.)
QE01) Zaragoza Pte. 15, 76000 Querétaro — DG: E. Nassar H.
QE02) Independencia 2, 76340 Jalpan — Gte: M. Ugalde G.
QE03) Juárez 38 Pte., 76800 S.J. Del Río — Gte: E. Morales G.
QR00) QUINTANA (Q. Roo)
QR01) Ap. 506, 77500 Cancún — DG: R. López M.

QR02) Ap. 96, 77010 Chetumal — DG: A. García L.
QR03) Héroes 232, 77010 Chetumal — Gte: L. Pavia M.
QR04) Rafael Melgar 15 Nte. Altos, 77600 Cozumel — Gte: L. A. Pavia M.
QR05) Ap. 13, 77200 Felipe Carrillo Puerto — GG: S. Uc Yam.
QR06) Av. Hidalgo 201, 77000 Chetumal — DG: Roberto Coral G.
QR07) Av. Uxmal 30, 77500 Cancún.
QR08) L. 21-23, Mz. 62, Rg. 92, Z-7, 77500 Cancún.
SL00) SAN LUIS POTOSI (S.L.P.)
SL01) (See DF06 and NL01)
SL02) Ap. 36, 79000 Cd. Valles — Gte: M. Castro E.
SL03) Madero 307 Altos, 78000 S.L.P. — DG: J. Briones R.
SL04) Hidalgo 7-A, 79600 Río Verde — Gte: M. I. Martínez S.
SL05) Ap. 160, 79050 Cd. Valles — Gte: R. Aguilar G.
SL06) Ap. 80, 78700 Matehuala — Gte: M. Monlui.
SL07) 78700 Matehuala — Gte: N. Rueda L.
SL08) Ap. 456, 78000 S.L.P. — Dir: M. Carrillo G.
SL09) Martín de Turubiartes 500, 79400 Cerritos — Gte: S. Ramírez H.
SL10) González Bocanegra 500, 78000 San Luis Potosí.
SL11) Av. Hidalgo 1, 79000 Cd. Valles.
SL12) Km 5.5 Crtra. a Cd. Valles, 79960 Tamazunchale.
SN00) SINALOA (Sin.)
SN01) Ap. 231, 80000 Culiacán — DG: R. Campos P.
SN02) Ap. 61, 81000 Guasave — Gte: C. Chávez L.
SN03) Ap. 60, 82010 Mazatlán — DG: O. Pérez E.
SN04) Ap. 965, 81200 Los Mochis — Gte: R. Chávez L.
SN05) Ap. 68, 81000 Guasave — Gte: L. Chávez R.
SN06) Benito Juárez 855, 80000 Culicán — GG: M. H. López A.
SN07) Juan Carrazco 107, 80000 Culiacán — Conc: H. Ramos R.
SN08) Ap. 233, 80000 Culiacán — Pte: O. Pérez.
SN09) Ap. 1, 81200 Los Mochis — Gte: H. Uribe L.
SN10) Angel Flores 412 Pte., 82000 Mazatlán — GG: Lic. J. Manuel R.
SN11) Ap. 965, 31200 Los Mochis — Dir: R. Chávez C.
SN12) Ap. 120, 80000 Culiacán — Gte: L. Carrillo B.
SN13) Ap. 971, 81240 Los Mochis — Gte: Ing. J. Pérez R.
SN14) Ap. 1380, 81220 Los Mochis — DG: Ing. M. F. Pérez M.
SN15) 81400 Guamuchil — Gte: J. L. Arce A.
SN16) Rosales 284 Pte., 80000 Culiacán.
SN17) Hidalgo 408, 82400 Escuinapa — CCnc: Fco. M. Ramos.
SN18) Hidalgo 61 Pte., 81200 Los Mochis — Gte: J. L. Muñoz E.
SN19) H. Frías 2-22, 82000 Mazatlán — Dir: J. Horacio S.
SN20) Ap. 35, 82800 Rosario — GG: M. Morales S.
SN21) H. Frías 2-27 Sur, 82000 Mazatlán — Gte: P. Xibillé P.
SO00) SONORA (Son.)
SO01) Plutarco Elías Calles y Pedro Moreno, 83000 Hermosillo — Dir: A. Hoeffer O.
SO02) Ap. 630, 85480 Guaymas — Gte: L. Gallardo V.
SO03) Ap. 58, 85340 Empalme — Gte: Lic. H. Lizárraga Z.
SO04) Durango No. 901 Sur, 85160 Cd. Obregón.
SO05) Ap. 1817, 83000 Hermosillo — Dir: A. Merino M.
SO06) Rodolfo Elías Calles 252 Ote., 85000 Cd. Obregón — GG: C. E. Gálvez R.
SO07) Sinaloa Sur No. 408, 85000 Cd. Obregón — Gte: A. Castañeda S.
SO08) Av. Morelos y Obregón, Col. Centro, 83600 Caborca — GG: J. de Jesús P.
SO09) Ap. 213, 84620 Cananea — Gte: J. L. Trueba L.
SO10) Juárez 206, 85800 Navojoa — Gte: J. S. Terminel U.
SO11) Ap. 6, 84900 Ures — Gte: Fco. Vidal E.
SO12) Angel Flores 12 Pte., 83000 Hermosillo — GG: C. Aparicio R.
SO13) Ap. 744, 84000 Nogales — Gte: H. M. Mireles A.
SO14) Ap. 371, 85400 Guaymas — GG: A. A. Padilla.
SO15) Obregón Ote. 1051, 83400 S.L.R.C. — Gte: O. Navarro Z.
SO16) Ap. 9, 85800 Navojoa — Gte: R. Gómez B.
SO17) Ap. 256, 84000 Nogales — Gte: Ma. del Carmen Guzmán de Ojeda.
SO18) California y Guerrero, Loc. E-17, 85140 Cd. Obregón.
SO19) 84200 Agua Prieta — Gte: H. Rivera E.
SO20) Ap. 7, 84180 Naco — Gte: J. M. Franco M.
SO21) Ap. 66, 83550 Puerto Peñasco.
SO22) Blvd. Las Quintas, Hermosillo — Gte: M. Torres R.
SO23) Ap. 44, 84600 Santa Ana — Gte: A. Marín C.
SO24) Ap. 158, 85000 Cd. Obregón — DG: L. F. García de León M.
SO25) Ap. 63, 84160 Magdalena — Gte: J. de Jesús Rochín D.
SO26) 83600 Caborca — GG: J. de Jesús P.
SO27) Ap. 28, 84200 Agua Prieta — Gte: C. Morales F.
SO28) Ap. 62, 85400 Guaymas — Gte: I. C. Verdugo R.
SO29) Ap. 1519, 83000 Hermosillo — Adm: I. Miranda R.
SO30) Ap. 285, 83000 Hermosillo — DG: J.C. Romero M.
TB00) TABASCO (Tab.)
TB01) Morelos 311, 86300 Comalcalco — GG: M. Dagdug L.
TB02) J. Alvarez 301, 86000 Villahermosa — Gte: E. Lodoza.
TB03) Ap. 270, 86000 Villahermosa — Gte: V. Gómez B.
TB04) Calle 28 N° 117 Altos, 86900 Tenosique — Adm: Ant. Nazar B.
TB05) Ap. 48, 86000 Villahermosa — Dir: Lic. G. Gaudiano P.
TB06) 86500 Cárdenas — GG: A. Calderón L.
TB07) Melchor Ocampo 412-6, 86700 Macuspana — GG: M. Huerta M.
TB08) José Julián Dueñas 201 Parque Hidalgo, 86800 Teapa.
TB09) Allende 61, 86400 Huimanguillo — GG: O. Zerecero D.
TB10) Ap. 397, 86000 Villahermosa.
TM00) TAMAULIPAS (Tamps.)
TM01) Gaspar de la Garza 170 Sur, 87000 Cd. Victoria — Gte: B. Zurita M.
TM02) Ap. 134, 88500 Reynosa — DG: N. Salem G.
TM03) Ap. 4, 88000 Nvo. Laredo — DG: R. Esparza G.
TM04) Ap. 246, 80500 Reynosa — GG: Ant. Gallegos G.
TM05) Ap. 797, 89160 Tampico — Gte: V. J. Flores.
TM06) Ap. 79, 89800 Cd. Mante — Gte: S. Braña A.
TM07) Ap. 52, 80500 Reynosa — Dir: E. Góngora O.
TM08) Ap. 99, 88000 Nvo. Laredo — Gte: Lic. E. Villareal M.
TM09) Ap. 735, 87300 Matamoros — GG: N. Salem G.
TM10) Ap. 414, 89150 Tampico — GG: G. Téllez O.
TM11) Ap. 200, 88000 Nuevo Laredo — GG: O. Maltos M.
TM13) Héroes del Cañonero Tampico 306 Ote., Tampico — Gte: Ma. E. Barrón G.
TM14) 7a Abasolo y Gonzáles 150, 87300 Matamoros — DG: Ant. S. Gallegos E.
TM15) Ap. 232, 88000 Nuevo Laredo — Gte: M. Cortés.
TM16) Ap. 20, 88440 Cd. Camargo — Gte: A. Hinojosa V.
TM17) Ap. 110, 88000 Nuevo Laredo — Gte: C. Rangel O.
TM18) Juárez 703 Ote., 89800 Cd. Mante — Gte: S. Braña A.
TM19) Ap. 134, 88500 Reynosa — GG: N. Salem G.
TM20) B. Juárez 506-A, 89160 Tampico — Gte: G. Moses N.
TM21) Ap. 465, 87350 Matamoros — DG: J. Cárdenas G.
TM22) Ap. 13, 83000 Cd. M. Alemán — Gte: A. Hinojosa V.
TM23) Ap. 1, 83000 Cd. M. Alemán — Gte: Ma. A. Gómez S.
TM24) V. Carranza 3151, 88000 Nuevo Laredo.
TM25) II Centenario 52, Col. Centro, San Fernando.
TX00) TLAXCALA (Tlax.)
TX01) Juárez Norte 203, 90500 Huamantla — GG: R. Romero R.
TX02) Mariano Sánchez 11-B, 90000 Tlaxcala — GG: A. Hernández C.
VE00) VERACRUZ (Ver.)
VE01) Plaza Cristal Local 26, 91000 Jalapa — Gte: E. García B.
VE02) Av. 2 N° 720, 94500 Córdoba — Dir: C. Ferráez M.
VE03) Ap. 18, 95400 Cosamaloapan — Gte: A. Aguirre.
VE04) Oriente 6 N° 261, 94300 Orizaba — Gte: D. Marín M.
VE05) (See DF06)
VE06) Ocampo 119, 91700 Veracruz — DG: Ing. B. Pazos de la Torre.
VE07) Av. Hidalgo 1117-A Altos, 96400 Coatzacoalcos.
VE08) Pje. Revolución 5, 91000 Jalapa — Dir: R. Cañedo B.
VE09) B. Franklin 4, 91700 Veracruz — GG: O. E. Pérez R.
VE10) Hidalgo 48 Altos, 96700 Minatitlán — Gte: A. Girón M.
VE11) Altamira 214, Col. Palma Sola, 93320 Poza Rica.
VE12) Ignacio de la Llave 36, 92000 Pánuco — Gte: Ant. P. de Barrenechea.
VE13) Ap. 482, 91700 Veracruz — Dir: J. de Dios Rodríguez D.
VE14) Gral. Rincón 41, 91270 Perote — GG: Prof. G. Morales C.
VE15) Alamo 1, 92730 Alamo — GG: D. Ruiz M.
VE16) Humboldt 36, 91270 Perote — Gte: Prof. R. Molina R.
VE17) R. Malpica 1414, 96400 Coatzacoalcos — Gte: Lic. O. Bejar G.
VE18) Ap. 15, 93400 Papantla — Gte: J. Vázquez G.
VE19) Av. 4 Pte. 421-17, 94100 Huatusco — Gte: H. M. Gómez C.
VE20) Ap. 4, 93300 Poza Rica — DG: Artemio López S.
VE21) Ap. 21, 91700 Veracruz — Gte: J. Rodríguez S.
VE22) Eulalio Vela 17, 96700 Minatitlán — Gte: María E. González.
VE23) Ap. 80, 93600 M. de la Torre — GG: E. Sainz Vda. de Manterola.
VE24) Juárez 13, 92800 Tuxpam — Asesor: Lic. C. Almazán F.
VE25) Libertad y Morelos, 95100 Tierra Blanca — Pte: E. Haaz P.
VE26) Constitución 7, 95700 S.A. Tuxtla — Gte: I. Fariña G.
VE27) Jalapeños Ilustres 3, 91000 Jalapa — DG: J. L. Oliva M.
VE29) Camacho 11, Gtz. Zamora — DG: L. Pintado C.
VE30) Ap. 50, 96200 Jaltipan — GG: A. Girón.
VE31) Ap. 1, 92060 Tempoal — Gte: I. Aradillas A.

VE32) Ap. 18, 94300 Orizaba — GG: E. Hernández M.
VE33) Ap. 26, 96000 Acayucan — Gte: C. de los Santos M.
VE34) Libertad 201, 95220 Piedras Negras — Gte: G. Alarcón R.
VE36) Ap. 426, 91000 Jalapa.
VE37) 95250 Alvarado — Gte: A. Mendoza M.
VE39) Zaragoza 105, 93820 Misantla — GG: E. Sainz Vda. de Manterola.
VE40) Ap. 13, 92600 Huayacocotla — Dir: J. Maldonado G.
VE41) Ap. 75, 91680 Cd. Cardel — GG: E. Landa S.
YU00) YUCATAN (Yuc.)
YU01) Calle 62 No. 465 x 55, 97000 Mérida — DG: L. A. Rivas A.
YU02) Ap. 152, 97000 Mérida — DG: A. García L.
YU03) Ap. 217, 97000 Mérida — DG: A. Iglesias V.
YU04) 97780 Valladolid — DG: R. López M.
YU05) Calle 56 N°447, 97000 Mérida — Gte: R. Rodríguez R.
YU06) Ap. 5, 97700 Tizimín — Gte: J. Jiménez P.
YU07) Ap. 78, 97320 Progreso — GG: A. Igelsias V.
YU09) Calle 42 N° 194-A, 97780 Valldolid — DG: L. A. Rivas A.
ZC00) ZACATECAS (Zac.)
ZC01) 99030 Fresnillo — GG: J. Torres G.
ZC02) Porfirio Díaz 6, 99600 Xalpa — GG: A. Rodríguez Z.
ZC03) Ap. 41, 99000 Fresnillo — Gte: A. García V.
ZC04) (See AG02)
ZC05) Delgadillo 18, 98400 Río Grande — GG: A. Rodríguez Z.
ZC07) Ap. 78, 98000 Zacatecas — Gte: E. Llamas F.
ZC08) Ap. 198, 99300 J. de García S.
ZC09) Ap. 324, 99000 Fresnillo — GG: A. Rodríguez Z.
ZC10) Ocampo 22, 99600 Xalpa — GG: G. Díaz A.

MONTSERRAT (British West Indies)

L.T: UTC – 4h — **Pop:** 12.630.000 — **Radio sets:** 8.500 — **Pr.L:** English — **E.C:** a/c 60, 220V — **ITU:** MSR.

RADIO MONTSERRAT (Gov.)

ADDR: P.O. Box 51, Plymouth, Montserrat. **Te:** +1 (809) 491 2885/2451. **Fax:** +1 (809) 491 6780. **Telex:** 5720. **Cable:** RADMON.
L.P: GM: Ms. Rosalind Willock. Asst. Mgr: Ms. Juliette Brade. CE: Lowell Mason.
STATION: 885kHz 1.5kW.
D.PRGR: 0900 (Sun 1100)-0200.
ANN: "ZJB Radio, the Voice of Montserrat".

RADIO ANTILLES (Gov. Comm.)

ADDR: P.O. Box 930, Montserrat, West Indies.
Te: +1 (809) 491 2755. **Fax:** +1 (809) 491 2724.
STATION: 930kHz 100kW (beam 215°).
D.PRGR: 0900 (Sun 0930)-0400. **Rel. V. of America:** 0000-0100, 0200-0230 (Tues-Sat), 1000-1100, 1200-1230, 1700-1730, 2100-2200. **Rel. BBC:** 0300-0400, 1100-1130, 1660-1615, 1800-1830, 2230-2245 (MF–Caribbean Report), 2245-2400.

GEM RADIO NETWORK (Comm.)

ADDR: P.O. Box 488, Plymouth, Montserrat.
Te: +1 (809) 491 3601. **Fax:** +1 (809) 491 2505.
L.P: Gen. Mgr: Wilsie White.
STATIONS: Garibaldi Hill 94.4MHz 20 kW, Ferrels 93.8MHz 1kW.

NETHERLANDS ANTILLES

L.T: UTC – 4h — **Pop:** 184.000 — **Radio sets:** 125.000 (est.) — **Pr.L:** Papiamento (Leeward Antilles), English (Windward Antilles), Spanish — **E.C:** a/c 50 + 60, 127/220V — **ITU:** ATN.

LANDSRADIO, TELECOMMUNICATION ADMINISTRATION

ADDR: Schouwburgweg 22, P.O. Box 103, Curaçao.
Telex: 1075 LRDIR NA. **Te:** +599 (9) 631111.

STATIONS: Mediumwaves

	Call	kHz	kW	MHz	kW	Name and location
1)	PJB	800	500			TWR, Bonaire
2)	PJZ-86	860	10	95.5	4	R. Curom, Willemstad
3)	PJC-7	1010	3	101.9	5	R. Hoyer I, Willemstad
4)	PJL-3	1100	0.25			R. Caribe, Willemstad
5)	PJE-3	1120	1			R. Statia, St. Eustasius
6)	PJD-2	1300	1	102.7	3.5	PJD-2 Radio, Philipsburg
7)	PJF-1	1410	5	93.9	??	Voice of Saba, Saba
3)	PJC-9	1500	3	105.1	5	R. Hoyer II, Willemstad
8)				93.9	20	R. Korsou FM, Willemstad
9)				94.7	1	Voz di Bonaire, Bonaire
10)				97.1		Flamingo Network, Bonaire
8)				101.1	5	R. Korsou FM/Laser 101, Willemstad
9)				101.1	0.4	Voz di Bonaire, Bonaire
11)				103.1		R. Paradise, Willemstad

Addresses and other information
All stations commercial, except 1) which is religious (see separate listing under Trans World Radio)
2) R. Curom, Roodeweg 64, P.O. Box 2169, Willemstad, Curaçao. **Te:** +599 (9) 626586 — Dir. & Gen. Mgr: O. Cuales. Chief Eng: C. Siegenthaler. 1000-0400 in **Papiamento**. Separate music prgrs. ("**Z-FM**") on 95.5 MHz — **V.** by letter. Rp.
3) Plasa Hoyer 21, Willemstad, Curaçao. **Te:** +599 (9) 611678, 612713. **Fax:** +599 (9) 616528 — Man. Dir: Ms. H. Hoyer. W 0930-0430, Sun 1230-0330. R. Hoyer I in **Papiamento** on 1010 kHz + 101.9 MHz, R. Hoyer II in **Dutch + English** (5%) on 1500 kHz + 105.1 MHz. **V.** by letter. (B). **N.B.** FM-transmitters are fully fed by solar energy
4) Ledaweg, Brievengat, Willemstad, Curaçao. **Te:** +599 (9) 611780. **Cable:** R. Caribe — Dir. Gen. C.R. Heilegger. Dir. Tec: G.A. Heilegger — 1000 (Sun 1200)-0400. W: **Papiamento:** 1000-1700, 1900-2200, 0100-0400. **N:** 1615, 1700 — **Spanish:** 1700-1900, 2200-0100. **N:** 1700, 1800 (VOA), 2200, 0015. Sun: **English** (rlg. prgr's): 1200-1400, 2200-2300. **Papiamento:** 1400-1700, 1830-2200, 2300-0400. **Spanish:** 1700-1830. **N:** 1815. **V.** by letter. Rec. acc.
5) St. Eustasius Broadc. N.V., Korthals Weg, St. Eustasius. **Te:** +599 (3) 82262.
6) Plaza 21 Shopping Centre, P.O. Box 366, Philipsburg, St. Maarten. **Te:** +599 (5) 22580, 22764. **Fax:** +599 (5) 24905 — Gen. Mgr/Dir: Donald R. Hughes. AM: 0930-0600 (Mon 0400). **N:** 1030, 1700, 0000. FM: 0930-0400. **N:** 1200, 1800, 0000.
7) P.O. Box 1, The Bottom, Saba. (Studios in St. Maarten). **Te:** +599 (5) 63213 — Owner: Max Nicholson. On 1410 kHz: 1000-2330 (Sun 1700). **N:** on the h. also 1030. On 93.4 MHz???: rel. ZGM-FM 94 Montserrat with local commercials inserted. **V.** by letter.
9) Kaya Gob. N. Debrot 2, Kralendijk, Bonaire. **Te:** +599 (7) 7220 (offices), 8273 (studios). **Fax:** +599 (7) 8220 — Dir: Mr. F. Piloto. 1000-0400 in **Papiamento and Dutch. N:** 2200. **N: in Dutch:** 1900, 2000
8) Bataljonweg 7, P.O. Box 3250, Willemstad, Curaçao. **Te:** +599 (9) 373934, 371501 (studios), 373012 (offices). **Fax:** +599 (9) 372888 — Owner: H. Oosterhof. Prgr. Dir: Alan H. Evertsz. 24h. Separate prgrs ("**Laser 101**") on 101.1 MHz. 24h. **N in Dutch:** 1000, 1600, 2100. **N in Papiamento:** 2200. **English:** Tues 0000. **Portuguese:** Wed 2330. **Sranan Tongo:** Fri 0000. **Indonesian:**
10) Bonaire. Fully automated music st. **Te:** +599 (7) 5971.
11) ITC Building, P.O. Box 6103, Willemstad, Curaçao. **Te:** +599 (9) 636103, 636403. **Fax:** +599 (9) 636404 — MD: J.A. Visser. 1000-0400. **Dutch:** 1000-1300. **N. and tourist information: Dutch:** 1430, 1630, 1830, 2030, 0030. **English:** hourly 1300-2300. **Spanish:** 1530. **Portuguese:** 1330.

TRANS WORLD RADIO (Cultural, Educational, Religious)

ADDR: Trans World Radio, Bonaire, Netherlands Antilles. **Te:** +599 (7) 8800. **Fax:** +599 (7) 8808. **Cable:** VOTAN.
L.P: MD: Thomas F. Corcoran. PD: McDaniel E. Phillips. CE: David J. Magnuson. Freq. Coordinator: Chuck Roswell. Head of PR: A. Cort Tangeman.
STATIONS (G.C: 68.28W/12.11N): 800kHz 500kW.
Shortwaves: 50/100kW.

kHz	kHz	kHz	kHz	kHz
9515	11815	11885	15345	15375
9535	11865	11930	15355	15445

D.PRGRS on 800kHz: 2159-2400 Spanish, 0000-0100 Portuguese, 0100-0200 Spanish, 0200-0400 English, 0400-0500 (0435SS) Spanish, 0659-0900 Portuguese, 0900-1100 Spanish, 1100-1300 (1330SS) English. **N. in English:** 1230W, 1300SS, 0300D.

Shortwave Schedule

English	Area	kHz	Portuguese	Area	kHz
0255-0430	NAm	11930	0655-0940	SAm	9515
0300-0400	NAm	9535	0655-0940	SAm	11885
1055-1300	NAm	11815	**Spanish**		
		15345	0000-0050	SAm	15355
German			0030-0230	SAm	15375
2325-2400	SAm	15355	0055-0230	CAm	11930
0955-1030	SAm	11885	0405-0500	CAm	9535
Portuguese			0955-1045	SAm	11865
2155-2400	SAm	15375			
2155-2315	SAm	15445			

ANN: "This is the international sound of the Caribbean, Trans World Radio, Bonaire" — **INT.SIG:** "Stand up, Stand up for Jesus" played with various instruments.

V. by QSL-card. Rp. preferred. IRC's requested(B). All correspondence replied by airmail. Re. in E and Sp, P & G, no tapes. Do not address reports to individuals — **PUB:** Prgr. schedule free — **F.Pl:** Spanish satellite sce. to LAm.

RADIO NEDERLAND RELAY STATION

Bonaire RNW (G.C: 68.19W/12.13N): 2 × 250kW tr's.

kHz	kHz	kHz	kHz	kHz	kHz
6020	9630	9775	11835	15315	17605
6165	9715	11660	11895	15330	21515
9590	9720	11720	15120	15560	21590

ST. MARTIN (French)

RFO: 99.2MHz (rel. RFO-Guadeloupe)
Radio Caraibes International (Comm.), B.P. 173, Marigot, F-97150 St. Martin. **Te:** +590 875406. **Fax:** +590 878887 — Dir: Patrice Toure. Head of Prgrs: Christophe Hubert — **FM:** 104.7MHz (relay st's: Guadeloupe 95.0MHz, Martinique 105.6MHz) — **D.Prgr:** 24h (0400-1000 rel. Europe I).
Radio St. Martin (Comm.), Port de Marigot, 97150 St. Martin — Mgr: H. Cocks — **FM:** 95.3MHz — **D.Prgr. In French/English:** 1000-0500(Sun 0400). **Spanish:** 2000-2100W.
Radio Voix Chretiennes de St. Martin (Rig.), B.P. 103, Marigot, 97150 St. Martin. **Te:** +590 873159 — Mgr: Father Cornelius Charles — **FM:** 106MHz 0.25kW — **D.Prgr.In English/French:** 0845-0530.

NICARAGUA

L.T: UTC −5h — **Pop:** 3.857.000 — **Radio sets:** 880.000 — **Pr.L:** Spanish — **E.C:** a/c 60, 120V — **ITU:** NCG.

DIVISION DE FRECUENCIAS ELECTRICAS

ADDR: Instituto Nicaragüense de Telecomunicaciones y Correos (TELCOR), Ap. 2664, Managua. **Te:** +505 (2) 673333.
L.P: Dir: Reinerio E. Montiel D.
STATIONS: 4) Gov, — 11), 35) Rlg.
Mediumwaves: Call YN, ° = Shortwave, (r:) = repeater.

	Call	kHz	kW	Name, location and h. of tr.
1)	ARC	v540	1	R. Corporación, Managua: 1000-0600
2)	CDJ	555	10	R. 19 de Julio, Chinandega: 1100-0400
6)	ALP	570	2.5	(r: YNALP 680), Managua
26)	NCN	590	1	La Voz de Chontales, Juigalpa: 1000-0300
41)	ARY	v600	10	R. Ya, Managua: 1000-0600 (r. 601)
4)	ALN	620°	50	R. Nicaragua, Managua: 24h
4)	ALN	640°	10	(r: YNALN 620), Managua
3)	LRI	640	12	R. Insurreción, Matagalpa: 1000-0600
5)	ELS	650	10	R. Sport, Granada: 1000-0600
4)	ALN	650°	10	(r: YNALN 620), Bluefields
4)	ALN	660°	75	(r: YNALN 620), Estelí
22)	PLC	670	10	R. Caribe, Puerto Cabezas
28)	ELD	670	10	R. Diriangen, Granada
6)	ALP	680*	10	R. La Primerísima, Managua: 24h
7)	ALI	700	10	R. Istmo, Managua: 1100-0600
8)	ACE	720	10	R. Católica, Managua: 1200-0400
23)	JSR	730	10	R. Segovia, Ocotal: 1000-0550
9)	ARS	760	50	R. Sandino "La S Grande", Managua: 1000-0500
10)	GFS	810	10	R. Rivas, Rivas: 1000-0600
11)	AOL	820	20	R. Ondas de Luz, Managua: 1000-0400 (r. 820.5)
40)	PRZ	830	10	R. Zinica, Bluefields: 1000-0600
12)	ARN	840	5	R. Noticias, Managua: 1030-0400
13)	ARP	880	5	R. El Pensamiento, Managua: 1000-0400
15)	ART	900	1	R. Tiempo, Managua: 1100-0400
16)	ARM	920	1	R. Mundial, Managua: 1100-0400
17)	OTO	960	5	R. 13 de Octubre, San Carlos: 1000-0500
18)	ARF	1000	10	R. Fiesta, Managua: 1200-0400
20)	DMY	1030	1	R. Masaya, Masaya
21)	HPR	1040	2.5	R. Pancasán, Jinotega: 1000-0600
24)	ALD	1130	1	R. Reloj, Managua: 1130-0030
25)	BRD	1170	1.5	R. Darío, León
13)	ADM	1230	5	R. Democracia, Managua: 1030-0300
29)	ARG	1260	2	R. Güegüence, Managua: 1100-0330
30)	BRA	1270	3	R. Atenas, León
31)	ACT	1300	1	Canal 130 AM, Managua: 1200-2400
32)	CSC	1310	3	R. San Cristóbal, Chinandega: 1100-0300
33)	IST	1330	1	R. Satélite, Estelí
34)	FRL	1360	1	R. Libertad, Jinotepe: 1000-0600
35)	MRP	1370	1	R. Patria, Boaco: 1230-0130
36)	LLM	1370	1	R. Matagalpa, Matagalpa
37)	BRV	1410	3	R. Venceremos, León: 1000-0400
38)	IRL	1430	5	R. Liberación, "La Tayacana", Estelí: 1000-0100
39)	AMT	1500	1	R. Minuto, Managua: 1100-2400

	Shortwaves:	kHz	kW	Name, location & h.of tr.
14)	—	v4920	1.2	R. RICA, Managua (v. to 4900)
4)	ALN	5950*	50	R. Nicaragua, Managua
19)	PMK	5970	1	R. Miskut, Puerto Cabezas: 1100-2330v (r. 4560, 5560, 4690)

Addresses and other information:
1) Ap. 2442, Managua — DG: José Castillo O. **F.Pl:** 10kW.
2) Ap.12, Chinandega — Dir: Alberto Jarquín.
3) Costado Norte de la Iglesia Catedral, Matagalpa.
4) Contiguo a Telcor, Villa Panamá (Ap. 3170), Managua — Dir: Frank Arana. **F.Pl:** Reactivate SW (last r. on 6001.9v and 6098.7v).
5) Ap. 64, Granada — Dir: Manuel Arana.
6) Del Cine Dorado, 1c al sur y 3c arriba, Managua — Dir: William Grigsby.
7) Ap. 700, Managua — Dir: Jesús Miguel Blandon.
8) Ap. 2183, Managua — Dir: Mons. Bismarck Carballo.
9) Ap. 4776, Managua — Fax: (505) 2 624002 — Dir: Conrado Pineda Aguilar.
10) Carretera San Jorge, Rivas.
11) Ap. 607, Managua — DG: Juan A.Ríos.
12) Ap. A-150, Managua. Te: (505) 2 673052 — Dir: A. Fuentes S.
13) Altamira No. 73, Managua — Dir: Allan Tefel-Alba.
14) Ap. 38, Suc. 14 de Septiembre, Managua — Dir: Digna Bendaña.
15) Ap. 2776, Managua — Dir: O. Hunguia.
16) Ap. 3170, Managua — Owner: Manuel Arana Valle. F.Pl: 10kW.
17) Costa norte de la Iglesia Católica, San Carlos.
18) De la Vicky, 6½c al lago, Altamira Este, Managua — Owner: Alejandro Mendoza.
19) Puerto Cabezas — Dir: Evaristo Mercado Pérez — Prgr. in Miskut & Sp.
20) Calle Carlos Ulloa, del Parque 1½c al oeste, Masaya — Dir: Jorge Correa Montiel.
21) Del Cine Betti 2½c al norte, Jinotega.
22) Puerto Cabezas.
23) Costado oeste del Catedral 2c al sur, Ocotal.
24) Ap. 2839, Managua — Dir: Francisco Rodriguez.
25) Residencial Posada del Sol No. 93, León — Dir: Juan Toruno Jiron.
26) Contiguo al BANIC, Juigalpa.
28) Piedra Bocona, ½c abajo, Granada,
29) Ap. 1929, Managua — Dir: Lorenzo Cardenal V. (classical music).
30) Ap. 51, León — Dir: Anabell Reyes Lanzas.
31) Carretera a Masaya, Km 12 3/4, 450 mts. al Este, Managua — Dir: Reinerio E. Montiel B.
32) Ap. 59, Chinandega — Dir: José Gutierrez.

33) Avenida Sandino, contiguo al restaurante Churrasco, Estelí — Dir: Santiago Hudiel Espinoza.
34) Fte. al Rest. Casa Blanca, Jinotepe.
35) Ap. 18, Boaco — Dir: Maria Auxiliadora Gomez L.
36) Sector 7, Del Parque Central 1½c al sur, Matagalpa — Dir: Eusebio Alfaro.
37) 1a. Av. NO No. 413, León — Dir: Pablo Medina Cardoza.
38) De la Shell, 1c al norte, Estelí — DG: Wilfredo Rodríguez.
39) Ciudad Jardín, Q-20, Managua — Dir: Carlos Gadea.
40) Ap. 25, Bluefields — Dir: Arturo J. Váldez R.
41) Fte. a la UCA, Managua — Dir: Carlos Guadamuz.

FM in Managua:
24) 89.9 — 91.0 R. Futuro — 16) 91.8 — 11) 94.3 — 1) 97.5 — 9) 98.3 — 98.9 Azul — 13) 99.1 La Grande — 12) 100.5 — 29) 101.3 — 102.5 R. Deportes — 15) 105.7 — 14) 106.1 — 106.7 R. Mundo — 9) 107.5.

PANAMA

L.T: UTC – 5h — **Pop:** 2.528.000 — **Radio sets:** 450.000 — **Pr.L:** Spanish — **E.C:** a/c 60, 110/115/120/126V — **ITU:** PNR (Canal Zone = PNZ).

DIRECCION NACIONAL DE MEDIOS DE COMUNICACION SOCIAL

ADDR: Ap. 1628, Panamá 1. **Telex:** 2746 MINGO PG. **Te:** (507) 227300.
L.P: Dir: Edwin Cabrera. Asesor: Alfredo de Souza.

ASOCIACION PANAMEÑA DE RADIODIFUSION

ADDR: Ap. 7387, Panamá — Pres: Fernando Eleta Casanovas.

STATIONS: Mediumwaves: Call HO-. * = inactive, (r) = repeater.

	Call	kHz	kW	Name and h. of tr.
1)	U23	540	10	R. Mía de Ciriquí (r:HOS 22 650), David: 24 h
2)	PU	550	—	La Emisora Líder, Panamá
3)	H2	560	1	(r:HOHM 610), Colón
4)	S	570	1	R. Soberana Civilista, Panamá: 1000-0300
3)	H4	580	10	(r:HOHM 610), David
3)	H3	590	10	(r:HOHM 610), Chitré
3)	HM	610	10	RPC Radio, Panamá: 24h
5)	J35	630	1	R. Provincias, Chitré
6)	K22	640	1	CPR, Colón: 1000-0300
1)	S22	650	5	R. Mía, "Cadena Nal.", Panamá: 24h
3)	F33	660	1	(r:HOHM 610), Bocas del Toro: 24h
7)	LY	670	5	R. Hogar, Panamá: 1100-0300
8)	F32	680	5	Super Z Estéreo, David
9)	R43	690	10	R. Veraguas, Santiago
10)	Q51	700	5	KW Continente, Panamá
11)	B52	710	3.5	Ondas del Caribe, Bocas del Toro: 1000-0400
12)	B50	720	5	R. República, Chitré: 1100-0300
13)	N26	740	10	R. Cristal, David
14)	R44	740	10	CMQ, La Exitosa Libre, Panamá: 24h
15)		750	5	R. Inolvidable, Chitré: 1130-2100
16)	XO	760	5	La Voz del Istmo, Panamá: 1145-0300
17B)	L83	770	1.5	La Voz de Herrera, Chitré: 1100-0100
18)	B35	780	10	R. Chiriquí, David: 1100-0300
19)	ACA20	790	10	Southern Command Network, Ft. Clayton, Canal Zone
20)	L60	800	15	La Exitosa Provincias Centrales, Chitré: 24h
21)	G	810	10	R. Mundial, Panamá
22)	F28	820	10	R. Ritmo, David: 1100-0400
23)	R56	830	1	R. Península, Macaracas
17A)	L80	*840	10	R. Nacional de Panamá, Panamá
14)	T61	850	10	La Nueva Exitosa Libre de Chiriquí, David
14)		850	3	(r:HOR44 740), Colón
24)	L55	860	10	R. Reforma, Chitré: 1000-0300v
25)	HO	870	5	R. Libre, Panamá: 24h
26)	B51	880	1	R. Hit, Colón
12)	Q62	890	10	R. Ritmo, Chitré
27)	HA	900	1	La Voz del Pueblo, Panamá: 1100-0500
17D)	L85	910	1	R. Nal. Cristóbal Colón, Colón
17C)	L81	910	10	R. Nal. Guaymí, David
1)	S56	920	1	(r: HOS22 650), Los Santos
14)	R43	930	3	(r: HOR44 740), La Chorrera
29)	K85	930	2.5	Mi Preferida Estéreo, Pto. Armuelles: 1100 (Sat/Sun 1200)-0300 (Sun 2400)
30)	X	*940	5	R. TV2, Panamá
17E)	L84	950	1	R. Nal. Victoriano Lorenzo, Penonomé
60)	M33	950	0.5	R. Cadena Azul "RCN", La Concepción 1100-0300 (Sun 1200-1800)
31)	K71	960	10	Onda Popular, Panamá: 1000-0500
32)	S97	970	1	Ondas Centrales, Santiago
1)	S57	980	10	(r:HOS22 650), David
33)	U44	*990	3	R. Monumental, San Miguelito: 24h
34)	K36	1000	10	R. Poderosa, Aguadulce: 1000-0400
17F)	L86	1015	3	La V. del Teribe, Bocas del Toro
35)		1020	10	R. Ancón, Panamá: 1000-0400
36)	J2	1040	1	La Voz del Canajagua, Las Tablas
37)	J60	1060	3	La Voz de Panamá, Panamá: 1100-0300
38)	J24	*1080	1	R. Tic Tac, Panamá
17G)	L82	1090	10	R. Urracá, Civilista, Santiago: 1000-0500
39)		1100	10	R. Suave, Panamá: 24h
10)	M21	*1120	1	R. Sonora, Panamá
40)	U80	1130	1	R. Sensación, Aguadulce
41)	B49	1140	1	R. Juvenil, Panamá: 1100-0500
1)	S25	1150	1	(r:HOS22 650), Colón
42)	C20	1160	10	Ondas Chiricanas, David
43)		*1160	3	R. Rock, Panamá: 1000-0400
44)	U84	1180	10	R. Belén, Santiago
21)	E91	1210	4	R. Diez, Panamá
47)	M56	1240	1	R. Boquete, Boquete
48)		1240	5	R.BB, Panamá: 24h
14)	J22	1270	0.5	R. Alegre "CMQ", Panamá
43)	S23	1290	10	R. Guadalupe, Panamá
49)	I417	1300	1	R. Bahá'i, Boca del Monte: 1045-2400
10)	M92	*1330	1	R. Fiesta, Panamá
50)	Z38	1350	10	R. Cadena Millonaria, San Miguelito: 24h
61)	B64	1370	1	R. Sitrachilco, Pto. Armuelles: 1000-0100 (Sun. 1200-2200)
13)	PJ60	1370	1	R. Colosal, David
35)		1380	10	R. América, Panamá: 1130-0400
51)	L	1390	1	R. Super Sol, Colón
52)	T40	1400	5	R. Luz, La Chorrera
53)	H779	1410	5	R. Mensabé, Las Tablas: 1100-0500
19)		1420	10	SCN, Southern Command Netw., Ft. Davis: 24h
56)	D42	1460	0.5	La Voz de Almirante, Bocas del Toro 1400-0400
14)	R43	1490	3	(r:HOR44 740), La Palma
57)	A95	v1510	1	Hosana R., Panamá
59)	E35	1570	1/2.5	La Voz del Trópico, "La Grande", Colón: 24h

Addresses and other information
1) Ap. 5117, Panamá 5 — DG: M. González Natera. (540 own pgr: 1230-1730, 1900-0200).
2) Ap. 7751, Panamá.
3) Ap. 1795, Panamá 1 — DG: Fernando Eleta C.
4) Calle 58, 13 Obarrio, or: Ap. 6-2323, El Dorado, Panamá. **Te:** 645136. **Fax:** 645042.
5) Ap. 423, Chitré — Gte: Gaspar Reyes.
6) Ap. 33, Colón — DG: Jacobo L. Salas.
7) Ap. 102, Panamá 9A — Dir: Fernando Guardia.
8) Ap. 991, David — Own: S. Gozaine.
9) Ap. 48, Santiago — Dir: Régulo José Franco.
10) Vía España, Edif. Regency, Panamá.
11) Finca 13, Empalme, Changuinola (Bocas del Toro), or: Ap. 375 — Dir: F. Artola.
12) Paseo Enrique Geenzier, or: Ap. 191, Chitré — Dir: Elías S. Solís C.
13) Ap. 540, David — Pres: Daniel Morhaim R.
14) Ap. 7462, Panamá 5. **Te:** +507 255858 — Dir: A. de Icaza.
15) Ap. 375, Chitré — DG: Rogelio Herrera.
16) Ap. 6-1192, El Dorado, Panamá.
17A) Closed since Dec. 21, 1989; 17B) Av. Pérez, Chitré; 17C) Av. 5E, David; 17D) Av. Bolívar Calle 9, Colón; 17E) Villa Interamericana, Penonemé; 17F) Av. Central, Bocas del Toro; 17G) Barriada La Hilda, Santiago de Veraguas — Gte: Carlos Hernández.

18) Ap. 43, David — Dir: Tomás Guerra.
19) (AFRTS) Drawer 919, APO Miami, FL 34004, USA).
20) Ap. 38, Chitré — Dir: Oriel E. Jaramillo.
21) Vía Argentina 33, Panamá — Dir: Carlos I. Zúñiga.
22) Calle D Norte, David — Dir: Denis Arce.
23) Macaracas, Los Santos.
24) Ap. 194, Chitré — Dir: Roberto Solís Villalaz.
25) Ap. 10577, Panamá 4 — DG: N. de Icaza.
26) Av. Meléndez 15119, Colón — Dir: D. Caballero.
27) Ap. 6-3045, Panamá — Dir: Modesto Lombardo Vega.
29) Ap. 44, Puerto Armuelles — Dir: José A. Mora.
30) Ap. 6-3092, El Dorado, Panamá — Own: Carlos Duque. Gte: A. Ayala.
31) Ap. 6-4747, Panamá — DG: Héctor González A.
32) Av. Centrales, Santiago — Dir: Héctor A. Santacoloma.
33) Ap. 7471, Panamá 5 — Dir: Víctor R. Vásquez.
34) Ap. 090, Aguadulce (Coclé) — Dir: Italo E. Rojas.
35) Ap. 6-5519, El Dorado, Panamá — Dir: René A. Hernández.
36) Ap. 10, Las Tablas — Own: S. E. Cohen.
37) Calle Belén 42, Panamá 7 — Dir: J. Díaz.
38) Ap. 3115, Panamá 3.
39) Calle 74, San Francisco, Edif. El Golf, Panamá — Dir: Delsa L. Fábrega.
40) Av. Arosemena, Aguadulce — Dir: N. González.
41) Ap. 6956, Panamá 5 — Dir: R. de Icaza.
42) Ap. 172, David — Dir: E. Calvo.
43) Vía España y Calle 45, Edif. El Conquistador, Panamá — Dir: F. Nuñez Fábrega.
44) Av. Central y Calle 8, Santiago — Dir: Vicente Caballero.
47) Crtra. Principal, Volcancito, Boquete — Dir: S. Cordero.
48) Ap. 5316, Panamá 5 — Dir: Andrés Vega C.
49) Ap. 1187, David — DG: Nasrollah Hashemí.
50) Ap. 6-8868, El Dorado, Panamá — Dir: Nino Macías.
51) Ap. 5011, Colón — Dir: Estéban Lam.
52) Ap. 473, La Chorrera — DG: F. & E. Brands.
53) Ap. 20, Las Tablas — Dir: Rigoberto Amaya M.
56) Calle 6 y Av. Almirante, Bocas del Toro. — Dir: Nicolás O. Dosman
57) Comunidad Misionera Hosana, via Ricardo J. Alfaro, or: Ap. 7671, Panamá 9 — Own: Olmedo Russel.
59) Ap. 506, Colón — DG: David A. Simons. Hill.
60) Entrega Gral., La Concepción, Chiriquí.
61) Ap. 03, Pto. Armuelles — Dir: Benigno Araúz.

FM in Panamá City. All st's less 5kW. St. names: Estéreo. . . exc. where indicated.
19) 87.9 — 16) 88.5 — 89.3 "RM" — 89.9 "89" — 14) 90.5 R. Super Q — 3) 90.9 — 19) 91.5 — 48) 92.5 — 43) 92.9 "F" — 25) 93.5 — 93.9 Universal — 7) 94.3 — 94.9 Tricolor — 14) 95.3 R. Alegre — 95.9 Festival — 96.3 Selecta — 1) 96.7 — 3) 97.3 — 98.5 Oeste — 3) 99.3 FM 99 — 100.1 R. El Mundo — 30) 100.5 — 101.1 Azul — 38) 101.7 — 102.6 Rey — 38) 103.1 FM Latina — 38) 103.5 Continental — 38) 103.9 Mil — 39) 104.7 — 16) 105.1 R. Vida — 105.7 Bahía — 27) 106.7 Panamá — 16) 107.3 — 107.7 Universidad.

PUERTO RICO

L.T: UTC – 4h — **Pop:** 3.300.000 — **Radio sets:** 2m (est) —**Pr.L:** Spanish, English — **EC:** a/c 60, 120V — **ITU:** PTR.

BROADCASTERS ASSOCIATION OF PUERTO RICO

ADDR: Suite 212, Cobians Plaza, Ave. Ponce de Léon 1607, Santurce, 00912.
Stations: All st's comm. exc. 17, 51). Most st's broadcast in Spanish only. *English or mainly English.
Locations: Hato Rey and Santurce are districts in San Juan.
Networks
NotiUno: News 0930-1300, 1900-2100 on st's 10), 16), 25), 32), 39), 41), 43), 44), 52), 55), 63), 64).
R.Reloj: 0930-1300, 1900-2100 on st's 2), 6), 11), 28), 38).

	Call	kHz	kW	Name, location and h. of tr.
1)	WPAB	550	5	R.5-50, Ponce: W 0900-0400 Sun 1100-0300
2)	WKAQ	580	10d	R. Reloj, San Juan: 24h
3)	WAEL	600	1d	R. WAEL, Mayagüez: 0830-0400
3a)	WEXS	610	1/0.5d	Patillas.
4)	WSKN	630	5d	La Gran Cadena, San Juan: 24h
5)	WAPA	680	10	Guapa R, San Juan: 24h
6)	WKJB	710	10/0.75	KJB, Mayagüez: 0830-0600
7)	WIAC	740	10d	WIAC, San Juan: 0859-0400(Fri/Sat 0600)
8)	WORA	760	5d	La Gran Cadena, Mayagüez: 24h
9)	WKVM	810	50d	AM-81, San Juan: 0900-0400
10)	WXEW	840	5/1d	R. Victoria, Yabucoa: 0900-0200
11)	WABA	850	5/1	Super 85, Aguadilla: 24h
12)	WQBS	870	5d	R. Voz, San Juan: -0300
13)	WYKO	880	1/0.5	R. 8-80, Sabana Grande
14)	WFAB	890	1	Las Piedras
15)	WPRP	910	5	La Gran Cadena, Ponce: 24h
16)	WEKO	930	2.5	NotiUno, Cabo Rojo: 0900-0500
17)	WIPR	940	10d	La Emisora del Pueblo, San Juan: W 0930-0230, Sun 1130-0400
18)	WJYT	960	1	R. Salvación, Quebradillas
19)	WPRA	990	1	R. Mil, Mayagüez: 0900-0400
20)	WPJC	1020	1	R. Gigante, Adjuntas: 1000-0200
21)	WOSO*	1030	10d	El Oso, San Juan: 24h
22)	WCXQ	1040	0.25	Super Q, Moca: Daytime
23)	WCGB	1060	5/0.5	R. Iniciativa, Juana Díaz: 0900-0300
24)	WMIA	1070	0.5/2.5	R. Arecibo, Arecibo: 0900-0400
25)	WLEY	1080	0.25	R. Ley, Cayey: 24h
26)	WSOL	1090	0.25	R. Sol, San Germán: 0930(Sun 1000)-0400
27)	WVJP	1110	2.5/0.5	R. Caguas, Caguas: 0930-0300
28)	WMSW	1120	5	La Gran W, Hatillo: 1000-0200
29)	WOIZ	1130	2.5	R. Antillas, Guayanilla (daytime)
30)	WQII	1140	10d	Once Q, San Juan: 24h
31)	WBQN	1160	5/2.5d	Super Borinquén, Barceloneta: 1030-0100
32)	WLEO	1170	0.25	R. Leo, Ponce: 24h
33)	WBMJ	1190	10/5d	WBMJ, San Juan: 0930 (Sat. Sun 1100)-0300
34)	WGDL	1200	0.25	R. Grito, Lares: Daytime
35)	WHOY	1210	5d	R. Hoy, Salinas: 0900-0300
36)	WNIK	1230	1	Arecibo
37)	WALO	1240	1/5	Walo (R. Oriental), Humacao: 0900-0300
	W...	1250	1/0.25d	Sabana
38)	WISO	1260	1	R. Wiso, Ponce: 1000-0100
39)	WCMN	1280	5/1	R. Centro, Arecibo: 0900-0400
40)	WTIL	1300	1	R. Util, Mayagüez: 1000-0400
41)	WUNO	1320	5	Noti-Uno, San Juan: 24h
42)	WENA	1330	1	La Buena del Sur, Peñuelas: Daytime
43)	WNOZ	1340	1	R. Nosotros, Aguadilla: 0900-0300
44)	WEGA	1350	2.5d	Vega Baja: 1000-2200
45)	WCHQ	1360	5/1d	C-H-Q, Camuy
46)	WIVV*	1370	5/1	WIVV Missionary R. St., Vieques Isl: 0925 (Sun 1000)-0230
47)	WOLA	1380	1	R. Prócer, Barranquitas 0900-0200
48)	WISA	1390	1	R. Noroeste, Isabela: 0930-0300
49)	WIDA	1400	1	R. Vida, Carolina: 24h
50)	WRSS	1410	2.5	R. Cristal, San Sebastian: 0900-0200
51)	WEUC	1420	1	R. Univ. Católica, Ponce: 0900-0200
52)	WNEL	1430	5	R. Tiempo, Caguas: 24h
53)	WCPR	1450	1	R. Coamo, Coamo: 1000-0200
54)	WRRE	1460	5d	La Fabulosa, Juncos
55)	WLRP	1460	0.5	R. Raíces, San Sebastian: 0900-0400
56)	WKCK	1470	1/2.5d	R. Cumbre, Orocovis: 0900-0200
57)	WMDD	1480	5	Sonido 14-80. (R. El Conquistador), Fajardo: 1000-0400(Sun 0200)

58)	WZBS	1490	5/1	R. Señorial, Ponce: 0900-0400
59)	WMNT	1500	1/0.25	R. Atenas, Manatí: 1000-0200
60)	WAVB	1510	1	Super B, Lajas
61)	WRAI	1520	10d	R. Aeropuerto Int, San Juan: 24h
62)	WUPR	1530	1/0.25	Exítos 1530, Utuado: 1000-0230
63)	WIBS	1540	1	R. Caribe, Guayama: Daytime
64)	WKFE	1550	0.25	R. Café, Yauco: 24h
65)	WRSJ	1560	5/0.75	R. San Juan, San Juan: 24h
66)	WPPC	1570	1	R. Felicidad, Peñuelas: Daytime
67)	WMTI	1580	5/2.5d	R. Jefe, Morovis: 1000-0200
		1580	0.25/0.5	Manatí (synchr. WMTI)
		1580	0.1/0.5	Arecibo (synchr. WMTI)
68)	WXRF	1590	1	R. 15-90, Guayama: 24h
69)	WLUZ	1600	5d	R. Luz, Santurce: 1000-0400

Addresses and other information:
Abbreviations: See USA comm. st's. Obs: New zip codes are being introduced outside the greater San Juan area.
1) Box 7243, Ponce 00732 — G.M: Alfonso Giménez-Porrata. P.D: José Rivera. C.E: William Batista — Talk. **FM:** 93.3MHz WOQ1.
2) Box 4668, San Juan 00936 — G.M: H. Biaggi. N.Dir: W. García. C.E: C. Valladares — News. **FM:** 104.7MHz.
3) Box 1370, Mayagüez 00709 — Mgr: María del Pilar Pirallo. C.E: José Marty — ADC. **FM:** 96.1MHz.
3a) Patillas 00723.
4) Box 13427, Santurce 00908 — V.P: S. Robiou. C.E: R. Hernández — News, talk. **FM:** 99.9MHz.
5) Box 4563, San Juan 00905 — C.E: W. Blanco P — Cont (N-2).
6) Box 1293, Mayagüez 00680 — G.M: Mabel Santos. P.D: Alan Corales. C.E: Pedro Vélez, Jr. — Top 40. **FM:** 99.1MHz.
7) Box 4504, San Juan 00905 — G.M: L.A. Mejía. C.E: Juan Rivera. — Nostalgia. **FM:** 102.5MHz.
8) Box 43, Mayagüez 00708 — G.M: G. Bonet. C.E: N. Nin — News, talk. **FM:** 97.5MHz.
9) Calle Ing. J.A. Carbonell #415, Hato Rey 00918 — G.M: L.A. Rivera. P.D: E. Liboy Shaleck. C.E: J. Gómez R. — Secular Catholic prgr's. **F.M:** 105.7MHz.
10) Box 42, Yabucoa 00767 — G.M: V. Vargas. P.D: V.M. Calderón. C.E: M. Laboy — MoR.
11) Box 188, Aguadilla 00605 — G.M: Rosa M. Pellot. C.E: Eduardo Ramírez. ADC.
12) Penthouse, Darlington Bldg, Río Piedras 00925 — Pres: Pichin Román Collazo. G.M: M. Nazario. — Salsa. **FM:** 107.7MHz.
13) Box 1373, Sabana Grande 00919 — Rlg.
15) Box 7302, Ponce 00732 — S.M: M.A. Borrero. P.D: P. García, C.E: R. Guzmán — News, talk. **FM:** 105.1MHz.
16) Box 681, Cabo Rojo 00623 — G.M: David Ortiz. P.D: Mrs. Jean Avilés. C.E: Max Pérez — All news. Mostly parallel with 41).
17) Ave. Hostos esq. Tous, Urb. Baldrich, Hato Rey 00919 — G.M: A. Mercado R. P.D: F. Percy. C.E: G. Albizu. **FM:** 91.3MHz.
18) Box 960, Quebradillas 00742 — G.M: Idalia Arrieta.
19) Box 1869, Mayagüez 00708 — G.M: E.C. de Tamayo. P.D: C.A.R. Camara. C.E: S. Pérez Jr — Ballads.
20) Box 982, Adjuntas 00601 — G.M: Jorge I. Quiles. P.D: José F. Pérez. C.E: Victor M. Pérez — Variety.
21) Box 4349, San Juan 00905 — G.M: A.L. Cavallaro. C.E: Rodolfo Rivas. — News/talk. All English.
22) 182-A Baldorioty Str, Moca 00716 — G.M: L.A. Rosario S. C.E: Maximino Pérez.
23) Box 248, Juana Díaz 00665 — G.M: J. Rodríguez. C.E: René Guzmán. — MoR.
24) Box 1055, Arecibo 00613 — G.M. & C.E: E. Rodríguez-Vélez. P.D: Joe Mercado. ADC, news.
25) Box 1186, Cayey 00634 — G.M: Martín Colón, Jr. C.E: Rafael Acosta. — News, MoR.
26) Box 154, San Germán 00753 — Mgr: J.M. Miranda. S.M: Ada Acosta. C.E: S. Pérez — Top 40.
27) Box 207, Caguas 00625 — G.M: H. Pereira. P.D: N. Rodríguez T. C.E: J.R. Gómez. **FM:** 103.3MHz.
28) Box 1652, Arecibo 00613 — G.M: H. Lloyd Santiago S. — P.D: Manolo Santiago — News, talk.
29) Calle Jobos 7, Ponce 00731.
30) Box 3779, Hato Rey 00919 — G.M: G.A. Tryon. P.D: R. Moreira — Adult cont.
31) Box 993, Manatí 00674 — Pres: A.R. Rivera. G.M. & P.D: Luis R. Rivera. C.E: J. Gómez — Top 40.
32) Box 7213, Ponce 00732 — G.M: J.H. Conesa Jr. P.D: Guillermo Droing. C.E: Rafael Acosta — ADC.
33) Box 367000, San Juan 00936-7000 — G.M: Janet Luttrell. P.D: Chuck Newby. C.E: William Heffley. — Rlg. (English/Spanish. Often in parallel with 46).
34) Box 872, Lares 00669 — G.M: Pedro Hernández.
35) Box 1148, Salinas 00751 — Pres. & C.E: Johnny Ortíz. P.D: Zulma Rodríguez — MoR.
36) Box 556, Arecibo 00613. **FM:** 106.5MHz.
37) Apartado 1240, Humacao 00661 — G.M: Efraín Archilla D. P.D: Luis A. Gándara-G. C.E: Oswald Pinott. Talk, music.
38) Box 7251, Ponce 00732 — G.M: Mrs. N.P. Freyre. P.D: Luis F. Vélez. C.E: Modesto Delgado — MoR, talk.
39) Box 436, Arecibo 00612 — G.M: B. Mitchell. C.E: R. González — MoR. **FM:** 107.3MHz.
40) Box 1360, Mayagüez 00709 — Pres. & G.M: Gilbert Mamery. P.D: Lidia Basora S. C.E: G. Bonet — Nostalgia.
41) Box Q, Hato Rey 00919 — G.M: R. Royo Jr. C.E: T. Gómez — All news.
42) Box 1330, Peñuelas 00724 — G.M: N. Rodríguez. P.D: I.M. Vásquez — Adult Cont, Oldies. **F.Pl:** Move to Yauco with 2kW full time.
43) Apartado #1, Aguadilla 00605 — G.M: Luís Omar Ortiz. P.D: David Ortiz A. C.E: Héctor Sáncheza — News, Nostalgia.
44) Box 1488, Vega Baja 00693 — G.M: Carmelo A. Santiago. P.D: Alberto Adorno. C.E: Max Pérez.
45) Box 629, Camuy 00627 — Pres: Dr. J. Rodríguez García. C.E: E. Rodríguez — Top 40. **FM:** 102.9MHz.
46) Box 338, Vieques Island 00765 — G.M. & P.D: Janet Luttrell. C.E: William Heffley. **Spanish** 0100-0230, otherwise **English.** All rlg. Also owns 33).
47) Box 669-A, Barranquitas 00618 — G.M: Edwin Aponte M. P.D: Orlando Morales. C.E: Jesús R. Gómez. Top 40.
48) Box W, Isabela 00662 — C.E: J. Santiago — MoR. **FM:** 101.5MHz.
49) Box 188, Carolina 00689 — G.M: William Lebrón; P.D: Josue Salgado; C.E: Alberto Pereira. Rlg. **FM:** 90.5MHz.
50) Ap. 1410, San Sebastian 00755 — G.M: Eulogio Cardona. P.D: Axel Toro. C.E: G. Bonet. — Variety.
51) Box 529, Sta. 6, Catholic Univ. of Puerto Rico, Ponce 00732 — G.M: Juan L.R. Ricart. C.E: R. Acosta. **FM:** 88.9MHz.
52) Box 487, Caguas 00626 — G.M: Carmen Pagan. P.D: Jacinto Ruiz. C.E: L.A. Soto — ADC. **FM:** 98.5MHz.
53) Box 316, Coamo 00769 — G.M. & C.E: José D. Soler — Top 40, oldies.
54) Box WW, Juncos 00666 — Pres: A.E. Román. G.M: A.O. Román. C.E: A. Flores.
55) Box 1670, San Sebastian 00755. — G.M: Ramón Edwin Colón P. P.D: Alexis Vásquez.; C.E: Guillermo Bonet. MoR, news.
56) Apartado 1210, Orocovis 00720 — G.M: Arnaldo García R. C.E: Carlos Arroyo — Tropical.
57) Box 948, Fajardo 00648 — G.M: R.J. Friedman. P.D: A.L. Vásquez. C.E: C. Soto P.
58) Box 7612, Ponce 00732 — G.M: Wilfredo Marrero. C.E: José L. Torres — Top 40.
59) Box 6, Manatí 00701 — Pres: José A. Ribas; P.D: Wilton Collazo; C.E: Arcadio Rodríguez. Top 40 — MoR. F.M.: 96.9MHz.
60) Puerto Rico Rd., Lajas 00667.
61) 74 Mayagüez Str, San Juan 00917 — G.M: C.M. Pirallo. P.D: N. Acevedo. C.E: R. Rivas — Rlg.
62) Box 868, Utuado 00641 — G.M: José A. Martínez. C.E: Epifanio Rodríguez — Top 40.
63) Box 1540, Guayama 00655 —
64) Box 324, Yauco 00768 — G.M: Jaime L. Bermúdez. P.D: Guillermo Valls Colón. C.E: Julio H. Conesa — News, ADC, Oldies.
65) Box 1560, Bayamón 00619 — ADC.
66) Box 9064, Ponce 00732 — G.M: Rev. Carlos Morales — Rlg.
67) Box 995, Manatí 00701 — Owner: Ing. W.G. Blanco P. G.M: C. Galdo. P.D: J. Blanco.
68) Box 1590, Guayama 00655 — G.M: José Raúl Fuster; C.E: Alberto Pereira. — Popular.
69) Box 9394, Santurce 00908 — G.M: Julia Acosta. C.E: R. Rivas — MoR.

AFRTS (US Air Force)
STATIONS: 1) AHD 780kHz 0.05kW (Ramey Air Force Base, Aguadilla) — 2) ACE 1040kHz 0.05kW (Buchanan, San Juan) — 3) AFCN 1200kHz 0.05kW (Roosevelt Roads) — 4) AFCN 1460kHz.
ADDR: 1), 2) Box 43, APO New York 09503; 3) FPO New York 09503.

ST. KITTS & NEVIS

L.T: UTC − 4h — **Pop:** 40.000 — **Radio sets:** 25.000 — **Pr.L:** English — **E.C:** a/c 60, 220V.

RADIO ZIZ (Gov. Comm.)
ADDR: P. O. Box 331, Springfield, Basseterre. St. Kitts. **Te:** +1 (809) 465 2555. **Fax:** +1 (809) 465 5202. **Cable:** ZIZ St. Kitts.
L.P: Ag. GM: Mrs. C. Manchester. CE: V. Grey.
MW: ZIZ 555kHz 20kW — **FM:** 90MHz — **D.PRGR:** 1000-0400. **BBC N:** 1100, 2300, 0200 — **Local N:** 1115, 1615W, 1755, 1855, 1955, 2055, 2155, 2315 — **ANN:** "The V. of the Federation of St. Kitts & Nevis, R. ZIZ" —**V.** by letter.

RADIO PARADISE (Rlg.)
ADDR: P. O. Box 423, Basseterre, St. Kitts. **Te:** +1 (809) 465 8500. **Fax:** +1 (809) 465 1700 — **Studio:** Nevis. **Te:** +1 (809) 469 5425 (St. owned by Trinity Broadc. Netw, Sta. Ana, CA, USA).
MW: 825kHz 50kW — **D.PRGR:** 0900-0400(Sat 0500).
ANN: "This is R. Paradise, 825 on your AM dial, broadcasting from St. Kitts —**V.** by QSL-card.

VOICE OF NEVIS (Comm.)
ADDR: P. O. Box 195, Charlestown, Nevis. **Te:** +1 (809) 465 1616. **Fax:** +1 (809) 465 1700 — **MW:** 895kHz 10kW.
ANN: "This is VON Radio on 895 AM".

GEM RADIO NETWORK (Comm.)
FM: 92.9MHz 1kW (rel. prgrs from Montserrat).

ST. LUCIA

L.T: UTC − 4h — **Pop:** 158.000 — **Radio sets:** 90.000 — **Pr.L:** English, Creole — **E.C:** a/c 50, 440V.

ST. LUCIA BROADC. CORP. (Gov. Comm.)
ADDR: P. O. Box 660, Castries. **Te:** +1 (809) 45 22337. **Fax:** +1 (809) 45 31568. **Cable:** Sunradio — **L.P:** Chmn: H. Bristol. Mgr. & PD: W. Springer. Asst. Mgr: H. George.
MW: 660kHz 10kW, 625kHz 1kW — **FM:** 99.5MHz + s relay st's.
D.PRGR: 0800-0300. **Local N:** 1030, 1200, 1500, 1700, 2000, 2100, 2200, 2345. **BBC N:** 1100, 1600, 2300, 0200.
ANN: "You are listening to R. St. Lucia".
V. by QSL-card — **F.PL:** 10kW on 625kHz.

RADIO CARIBBEAN INTERNATIONAL (Comm.)
ADDR: P. O. Box 121, Castries. **Te:** +1 (809) 45 22636. **Fax:** +1 (809) 45 22637. **Telex:** 6240 RADOCAR.
L.P: Pres: H. Coquerelle. PD: C. Bonnel.
MW: 840kHz 20kW — **FM:** 101.1MHz + 3 relays (1 in Dominica).
D.PRGR: 24h on FM, 0500-1802 on 840kHz. **N:** 0600, 0700, 1230.
ANN: "This is R. Caribbean International, the Caribbean Sound".
V. by letter.

RADIO KOULIBWI (Comm.)
ADDR: Box 20, Marchand PO, Castries. **Te:** +1 (809) 45 30874.
FM: 105.0MHz.

CATHOLIC TV BROADC. SCE. (Rlg.)
ADDR: Micoud Str, Castries. **Te:** +1 (809) 45 27050.
FM: 87.8MHz (rel. TV sound of EWTN, USA).

ST. VINCENT & THE GRENADINES

L.T: UTC − 4h — **Pop:** 115.820 — **Radio sets:** 55.000 — **Pr.L:** English — **E.C:** a/c 50, 230V.

ST. VINCENT & THE GRENADINES NAT. BROADC. CORP. (Gov. Comm.)
ADDR: P. O. Box 705, Kingstown, St. Vincent. **Te:** +1 (809) 45 71111. **Fax:** +1 (809) 45 62749. **Cable:** Broadcast St. Vincent. **Telex:** 7473 RADIO VQ.
L.P: GM: E. B. John. PD: Ms. P. Barbour. CE: L. McKie.
MW: 705kHz 10kW — **D.PRGR:** W 0930-0300 (Fri/Sat 0400), Sun 1000-0200. **N:** 1630, 2230, 2345W.
ANN: "NBC 705 Radio" — **V.** by QSL-card. Rec. acc.

TRINIDAD & TOBAGO

L.T: UTC − 4h — **Pop:** 1.300.000 — **Radio sets:** 700.000 — **Pr.L:** English — **E.C:** a/c 60, 115V (yes) — **ITU:** TRD.

NATIONAL BROADCASTING SCE. (Gov.)
ADDR: P.O. Box 610, Port of Spain. **Te:** +1 (809) 623 2529. **Fax:** +1 (809) 624 2045. **Cable:** Voice Trinidad. **Telex:** Sixten WG 22222.
L.P: Gen. Mgr: F. P. Thompson. Prgr. Dir: H. Clement. CE: B. Rampersand.
STATION: 610kHz 50kW — **FM:** 98.9/100.0MHz — **Tobago relay:** 91.1MHz 20kW.
D.PRGR. in English: 24h on MW + 91.1/98.9MHz. **N.** On the h (BBC 2000, 0200) — **Wrp:** On the half h.
R. 100: 24h on 100.0MHz.
ANN: "This is NBS Radio, 610 AM, 98.9 FM stereo, and 91.1MHz on the FM band serving Tobago" — **INT-SIG:** At sign-on: The tune "Yellow Bird, Choucoun Destiny" played on Guitar and Xylophone.
V. by letter. Rp.

TRINIDAD BROADCASTING CO. (Comm.)
ADDR: P.O. Box 716, Port of Spain. **Te:** +1 (809) 622 1151-7. (R. Tempo: +1 809 628 6118, Rhythm R. 95: +1 809 622 2744). **Fax:** +1 (809) 622 2380. **Cable:** Broadcast Trinidad. **Telex:** 22239 BROCA WG.
L.P: Gen. Mgr: E.R. McDowell. Dir. of Prgrs: A. Henry. Sales Mgr: M. Mahabir. Chief Eng: R. Rambaran.
MEDIUMWAVE: 730kHz 20kW.
FM: 95.1MHz 20kW, 105.1MHz 20kW (stereo).
R. Trinidad: 24h on 730kHz.
N: 0200+, 1000, 1030, 1100+, 1110, 1200, 1300+, 1500, 1600, 1630, 1700, 1800, 2000, 2200 — +) BBC.
ANN: "This is R. Trinidad, first with the news".
Rhythm Radio FM95: 24h on 95.1MHz.
R. Tempo: 0925-0400 on 105.1MHz.
V. by letter. Rec. acc.

MUSIC RADIO 97 FM STEREO (Comm.)
ADDR: Level 4, Long Circular Mall, Long Circular Rd, St. James. **Te:** +1 (809) 622 4124. **Fax:** +1 (809) 622 6693.
FM: 97MHz — **D.PRGR:** 0900-0500.

PRIME RADIO 106 FM (Comm.)
ADDR: 35-37 Independence Square, Port of Spain. **Te:** +1 (809) 627 7463. **Fax:** +1 (809) 627 2721.
FM: 106MHz — **D.PRGR:** 24h. **BBC N:** 1100, 2300.

TURKS & CAICOS ISLANDS (British)

L.T: UTC − 5h (Su: UTC − 4h) — **Pop:** 10.200 — **Radio sets:** 5000 (est.) —**Pr.L:** English — **E.C:** ac 60, 110/220V — **ITU:** IOB.

RADIO TURKS & CAICOS (Gov. Comm.)
ADDR: P. O. Box 69, Grand Turk. **Te:** +1 (809) 946 2007. **Fax:** +1 (809) 946 2777. **Tlx:** 946 8212 GOVTCI TQ.
L.P: Ag. Mgr: Ms. Lynette Smith — **MW:** 1460kHz 2.5kW.
D.PRGR: 1100-0400(Sun 0315) — **ANN:** "This is R. Turks & Caicos on Grand Turk, Turks & Caicos Islands" — **V.** by letter.

ATLANTIC BEACON (Rlg.)
ADDR: So. Caicos. **Te:** +1 (809) 946 3311. **USA Addr:** Christian Radio Productions, P. O. Box 5312, Ft. Lauderdale, FL 33310. **Te:** +1 (305) 748 9500.
L.P: Overseas Op's Mgr: Buddy Tucker. CE: Jerry Kilfer.
MW: So. Caicos 1570kHz 50kW — **D.PRGR:** 24h — **V.** by QSL-card.

SOUTH AMERICA
0
750
1500 KM
CARIBBEAN SEA
NORTH
ATLANTIC
OCEAN
SOUTH
PACIFIC
OCEAN
SOUTH
ATLANTIC
OCEAN
Tropic of Capricorn
PANAMA
COLOMBIA
VENEZUELA
GUYANA
SURINAM
FRENCH GUIANA
ECUADOR
PERU
BRAZIL
BOLIVIA
PARAGUAY
CHILE
ARGENTINA
URUGUAY
FALKLAND
ISLANDS
Stanley
MARACAIBO
TOVAR
CARACAS
VALENCIA
VALERA
BARINAS
MÉRIDA
SAN CRISTÓBAL
OCUMARE
DEL TUY
BARCELONA
MATURÍN
EL TIGRE
CIUDAD BOLÍVAR
SPARENDAAM
PARAMARIBO
MONTSINERY
CAYENNE
MEDELLIN
TUNJA
PEREIRA
BOGOTA
IBAGUE
VILLAVICENCIO
ESPINAL
CALI
NEIVA
POPAYAN
FLORENCIA
TUMACO
PIFO
IQUITOS
TUMBES
TALARA
TARAPOTO
PUCALLPA
CERRO DE PASCO
LA OROYA
HUANCAYO
LIMA
ICA
JULIACA
PUNO
AREQUIPA
TACNA
LLALLAGUA
ARICA
LA PAZ
COCHABAMBA
ORURO
HUANUNI
SANTA CRUZ
POTOSI
SUCRE
ANIMAS
TARIJA
CALAMA
CHUQUICAMATA
COPIAPO
SANTIAGO
CONCEPCIÓN
TEMUCO
PUERTO MONTT
COYHAIQUE
PTO. AYSEN
MANAUS
SAO LUIZ
FORTALEZA
RECIFE
SALVADOR
BRASILIA
GOIANIA
CACHOEIRA PAULISTA
BELO HORIZONTE
RIBEIRAO PRETO
APARECIDA
SAO GOANCALO
SAO PAULO
RIO DE JANEIRO
CURITIBA
FLORIANOPOLIS
PORTO ALEGRE
PELOTAS
ARTIGAS
PEDRO JUAN
CABALLERO
CONCEPCIÓN
FOZ DO IGUACU
ASUNCIÓN
VILLARRICA
ENCARNACIÓN
GRAL PACHECO
MONTEVIDEO
UTC -5
UTC -4
UTC -3

CORAL RADIO (Comm.)

ADDR: Butterfield Sq, Providenciales. **Te:** +1 (809) 946 4496.
L.P: SM: Bob Cooper — **FM** (0.25kW): 89.3/89.9/90.5/92.5MHz.
D.PRGR: 24h (4 separate pgrs).

WPRT RADIO (Comm.)

ADDR: Leeward Hwy, Providenciales. **Te:** +1 (809) 946 4267.
L.P: SM: Peter Stubbs — **FM:** WPRT-FM 88.7MHz 0.25kW.

VIRGIN ISLANDS (American)

L.T: UTC−4h — **Pop:** 100.100 — **Radio sets:** 90.000 — **Pr.L:** English, Spanish, Creole — **E.C:** a/c 60, 110V — **ITU:** VIR.

	Call	kHz	kW		Call	kHz	kW
1)	WSTX	970	5/1	3)	WRRA	1290	0.5
2)	WVWI	1000	5/1	4)	WSTA	1340	1
6)	WGOD	1090	0.25				
	FM	**MHz**	**kW**		**FM**	**MHz**	**kW**
5)	WIUJ	88.9	0.15	7)	WSTX-FM	100.3	50
7)	WDCM	92.3	50	12)	WTBN	102.1	50
11)	WAVI	93.5	6	10)	WIYC	104.3	44.8
8)	WJKC	95.1	50	9)	WVGN	105.3	50
6)	WGOD	97.9	50	13)	WVIS	106.1	50
7)	WVIQ	99.5	10.5				

Addresses and other information:

1) Family B'casting Inc., Box 3279, Christiansted, St. Croix 00822.
2) Box 5678, St. Thomas 00801. — Mon-Fri 24h, Sat/Sun 1000-0400. **N:** on the h — **Ann:** "Radio One".
3) Box 277, Frederiksted, St. Croix 00840 (Daytime only).
4) Box 1340, St. Thomas 00801 — 24h. **Ann:** "Lucky 13".
5) Virgin Isl. Youth Development Radio, GPO, Govt. of Virgin Isl., St. Thomas 00801.
6) Box 5012, St. Thomas 00801 (Rlg).
7) St. John FM Inc., 350 Cruz Bay 00830.
8) Suite V-2, Caravelle Arcade, Christiansted, St. Croix 00820.
9) Buccaneer Mall No. 14, St. Thomas 00802.
10) Box 5234, St. Thomas 00801.
11) Box 25016, Christiansted, St. Croix 00824.
12) Trans Caribbean Broadc. Co., Haversight Exec. Tower, Ste. 1033 No. 19 Estate Thomas, St. Thomas 00801.
13) Box 487, Frederiksted, St. Croix 00840.

VIRGIN ISLANDS (British)

L.T: UTC−4h — **Pop:** 12.550 — **Radio sets:** 7.000 (est.) — **Pr.L:** English — **ITU:** IOB.

VIRGIN ISLANDS BROADC. Ltd. (Comm.)

ADDR: Box 78, Roadtown, Tortola. **Te:** +1 (809) 4942430. **Fax:** +1 (809) 4942010.
L.P: GM: Meritt Herbert. Sen. N. Editor: Michael Jarvis.
STATION: ZBVI 780kHz 10kW.
D.PRGR. in English: 1000-0200 (Sun 1100-0100).
N: W on the h. (exc. 1000, 1700, 1900, 0100). Sun 1600, 2000.
Local N: 1110, 2230.
ANN: "This is ZBVI Radio from Tortola" — **V.** by letter.

CARIBBEAN BROADC. SYSTEM (Comm.)

ADDR: Box 3059, Roadtown, Tortola. **Te:** +1 (809) 44990.
L.P: Gen. Mgr: Alvin Korngold.
STATIONS (10kW): Z-Wave 97.3MHz, Z-Hit 94.3MHz, Z-Gold 91.7MHz — **D.PRGR:** 1000-0400 (three different formats).

**Discovered a new station?
Please inform the
WRTH Editorial Office.**

SOUTH AMERICA

ARGENTINA

L.T: UTC − 3h, − 4h in Ma., SJ, SL, LR, Ca. and SE (21 Oct.-03 Mar.) **Pop:** 33.023.000 — **Radio sets:** 21.582.456 — **Pr.L:** Spanish — **E.C:** a/c 50, 220V — **ITU:** ARG.

MINISTERIO DE OBRAS Y SERVICIOS PUBLICOS SECRETARIA DE COMUNICACIONES

Dirección Nacional de Telecomunicaciones
Departamento de Operaciones
ADDR: Secretaría de Comunicaciones, Sarmiento 151, 1000 Buenos Aires. **Te:** Sector Radiodifusión: +54 (1) 30-5031.
L.P: Ing: Roberto Hugo Zubieta.
Departamento de Operaciones: Ayacucho 1556, 1112 Buenos Aires. **Te:** +54 (1) 84-0504.
Dirección Nal. de Telecomunicaciones:
L.P: Subsecr: Ing. Di Benedetto.

COMITE FEDERAL DE RADIODIFUSION (COMFER)

ADDR: Suipacha 765, 1008 Buenos Aires. Dir: León Guinzburg.
L.P: Pres: Pedro Sánchez. Programación y Desarrollo: Dr. Mario Heffes. Secr.: Ing. Juan C. Magliano.
COMFER controls certain technical aspects of broadcasting along with Secretaria de Comunicaciones, and also controls the prgrs. transmitted over all kinds of broadcasting stations.
STATIONS:
Mediumwaves

	Call	kHz	kW	Name, location and h. of tr.
1)	LRA14	540	25/8	R. Nac. Santa Fé: 1000-0300
2)	LRA25	540	5	R. Nac. Tartagal: 1100-0500
25)	LU17	540	10/5	R. Golfo Nuevo, Pto. Madryn: 0930-0330
3)	LRA9	560	15/7.5	R. Nac. Esquel: 1000-0400
4)	LRA13	560	25/5	R. Nac. Bahia Blanca: 1000-0500
5)	LT15	560	10/3	R. del Litoral, Concordia: 0900-0300
6)	LV1	560	25/15	R. Colón, San Juan: 24h
7)	LRA16	560	12.5	R. Nac. La Quiaca: 1000-0400
8)	LU20	580	12	R. Chubut "La 20," Trelew: 0930-0330
9)	LW1	580	25/5	R. Universidad Nacional de Córdoba, Córdoba: 0900-0400
10)	LS4	590	25/10	R. Continental, Buenos Aires: 24h
11)	LRA30	590	25	R. Nac. San Carlos de Bariloche: 1000-0400
12)	LV12	590	25/5	R. dif. Independencia, San Miguel de Tucumán: 1000-0500
94)	LV4	620	25/5	R. San Rafael: 1000-0400
13)	LRA18	620	25/5	R. Nac. Río Turbio: 0945-0400
14)	LT17	620	25/5	R. Provincia de Misiones, Posadas: 0900-0500
85)	LRA26	620	10/5	R. Nac. Chaco, 0900-0400
15)	LRA28	620	25/5	R. Nac. la Rioja: 0900-0400
16)	LS5	630	25/5	R. Rivadavia, Buenos Aires: 24h
65)	LU4	630	25/5	R. Dif. Patagonia Argentia, Comodoro Rivadavia: 0900-0500
17)	LRA24	640	25/5	R. Nac. Río Grande: 24h
18)	LU18	640	10/1	R. El Valle, General Roca: 0900-0300
19)	LV15	640	10/1	R. Villa Mercedes: 0900-0300
20)	LRA11	670	25/5	R. Nac. Comodoro Rivadavia: 24h
21)	LRA52	670	1	R. Nac. Chos Malal: 0945-0200
22)	LT4	670	25	R. Dif. Misiones, Posadas: 0800-0200
23)	LRI209	670	20/16	R. Mar del Plata, Mar del Plata: 24h
110)	LT3	680	10	R. Cerealista, Rosario: 24h
24)	LU12	680	30/1	R. Río Gallegos: 1000-0300
26)	LV6	680	25/5	R. Nihuil, Mendoza: 1000-0700
27)	LRA4	690	20	R. Nac. Salta: 0900-0500
28)	LU19	690	10/5	La Voz de Comahue, Cipoletti: 0900-0500
29)	LRA17	710	25/1	R. Nac. Zapala: 24h

29x)LRA19	710	25/5	R. Nac. Pto. Iguazú: 0900-0300
30) LS1	710	37/5	R. Municipal de la Ciudad de Buenos Aires: 24h
31) LRA59	720	1	R. Nac. Governador Gregores: 1100-2300
32) LV10	720	25/5	R. de Cuyo, Mendoza: 24h
33) LRA3	730	5	R. Nac. Santa Rosa: 0900-0430
34) LU23	730	10/1	R. Lago Argentino, El Calafate: 1000-0400
35) LRA27	730	25/5	R. Nac. Catamarca: 0900-0400
36) LRA55	740	1	R. Nac. Alto Río Senguerr: 1100-0300
36x)LRI200	740	1	R. Puerto Deseado: 24h
37) LRA7	750	100/10	R. Nac. Córdoba: 24h
38) LRA10	780	5/1	R. Nac. Ushuaía e Islas Malvinas, Ushuaía: 24h
39) LRA12	780	5	R. Nac. Santo Tomé: 0900-0300
40) LV8	780	8	R. Libertador, Mendoza: 0930-0600
83) LV19	°790	5	R. Malargüe: 1000-0400
41) LR6	790	25/5	R. Mitre "AM 80," Buenos Aires: 24h
42) LRA22	790	25/5	R. Nac. San Salvador de Jujuy: 0945-0500
44) LU15	800	25/2	R. Viedma: 0900-0300
45) LV23	800	1/0.25	R. Rio Atuel, General Alvear: 1000-0400
46) LRA8	820	25/5	R. Nac. Formosa: 0900-0400
48) LU24	820	5/3	R. Tres Arroyos: 0900-0400
49) LT8	830	10/1.5	R. Rosario: 24h
50) LT43	830	1	R. Mocovi, Charata: 0900-0300
114) LV18	830	0.25	R. Municipal San Rafael: 1030-0230
118) LT21	830	0.5	R. Municipal, Alvear: 1000-0200
51) LU14	830	5	R. Provincia de Santa Cruz, Las Heras: 1000-0600
52) LW8	830	25/5	R. San Salvador de Jujuy: 0900-0400
165) —	830	—	R. Federal, Lanús
53) LRK200	840	10/1	R. Dif. El Clavillo, Concepción: 1000-2300
54) LT12	840	10/5	R. General Madariaga, Paso de los Libres: 0900-0300
56) LU2	840	25/10	R. Bahía Blanca: 0900-0400
57) LRA56	860	1	R. Nac. Perito Moreno: 0955-0300
57x) —	860	5	R. Municipal, Chilecito: 1000-0400
58) LRA1	870	100	R. Nac. Buenos Aires: 24h
51) LU14	°880	5	R. Provincia de Santa Cruz, Santa Cruz: 1000-0600
59) LU33	890	20/1	Em. Pampeana, Santa Rosa: 24h
60) LV11	890	25/5	Em. Santiago del Estero: 0930-0330
61) LT7	900	25/2	R. Provincia de Corrientes: 0900-0300
62) LRA23	910	50	R. Nac. San Juan: 24h
63) LR5	910	25/5	La Red, Buenos Aires: 24h
144) LRH200	940	3/0.5	R. Chajarí: 0900-0300
64) LR3	950	25/5	R. Libertad, Buenos Aires: 24h
66) LRA6	960	25/5	R. Nac. Mendoza: 24h
68) LU13	960	10/1	R. Necochea: 0900-0400
69) LU5	970	20/10	R. Neuquén: 0900-0500
70) LV2	970	25/10	R. General Paz, Córdoba: 24h
109) LU37	980	10/6	R. Gral. Pico: 0930-0330
146) LT25	980	0.25	R. Guaraní, Curuzú Cuatiá: 0900-0300
136) LT39	980	1/0.25	Em. Victoria: 0900-0300
63) LR4	990	25/5	R. Splendid, Buenos Aires: 24h
70x) LRJ201	990	1	R. Calingasta, Barreal: 1000-0300
71) LU16	1000	25/1	R. Río Negro, Villa Regina: 0900-0300
72) LV16	1010	25/5	R. Rio Cuarto: 0830-0430
130) LW2	1010	1	Radioem. Tartagal: 1000-0300
73) LT10	1020	10/5	R. Universidad Nacional del Litoral, Santa Fé: 0800-0500
15) LRA28	1020	1	R. Nac. La Rioja, Chilecito
74) LRA58	1020	1	R. Nac. Río Mayo: 1100-2300
76) LS10	1030	25/5	R. del Plata, Buenos Aires: 24h

75) LRH252	*1050	1	R. Pucará, Reconquista
77) LR1	1070	25/5	R. El Mundo, Buenos Aires: 24h
78) LU3	1080	25/10	R. del Sur, Bahía Blanca: 0900-0500
79) —	1080	0, 25	R. Departamento Minas, Andacollo: 1000 (Sat/Sun 1200)-2300
79x) LR2	1110	15/5	R. Argentina, Buenos Aires: 24h
80) LV7	1110	10/5	R. Tucumán, San Miguel de Tucumán: 0900-0400
81) LU6	1120	25/5	Em. Atlántica, Mar del Plata: 24h
82) LV5	1120	25/5	R. Sarmiento, San Juan: 1000-0500
84) LRA21	1130	25/5	R. Nac. Santiago del Estero: 0900-0400
84x) LRA2	°1150	10	R. Nac. Viedma: 0900-0400
86) LRA51	1150	1	R. Nac. Jáchal: 1030-0300
87) LT9	1150	10/5	R. Brigadier Gral. Estanislao López, Santa Fé: 0930-0400
88) LRA57	1160	1	R. Nac. El Bolsón: 1000-0400
89) LRH253	1160	1.8	R. Cataratas del Iguazú, Pto. Iguazú: 1000-1600
90) LU32	1160	10/5	R. Coronel Olavarría, Olavarría: 0900-0300
91) LV9	1160	25/5	R. Salta: 1000-0500
92) LRA29	1170	25/3	R. Nac. San Luis: 0900-0400
93) LT16	1170	25/3	R. Esmeralda, Presidencia Roque Sáenz Peña: 0900-0300
95) LRA15	1190	50/5	R. Nac. San Miguel de Tucumán: 1000-0400
96) LR9	1190	25/5	R. América, Buenos Aires: 24h
66) —	1200	1	(r. LRA6 960), Valle de Uspallata: 24h
97) LT6	1200	10/1.5	R. Genaro Berón de Astrada de Goya, Goya: 0900-0200
98) LT18	1210	10/5	R. Eldorado: 0930-2130
99) LV3	1210	25/10	R. Córdoba: 0900-0500
100) LT2	1230	20/5	R. Gral. San Martín, "R.2", Rosario: 24h
101) LW5	1230	5/1	R. Libertador General San Martín, Ledesma: 1000-0300
102) LT14	1260	10/5	R. General Urquiza, Paraná: 0900-0500
103) LRA20	1270	5	R. Nac. Las Lomitas: 0900-0300
104) LS11	1270	25/5	R. Provincia de Buenos Aires, La Plata: 24h
163) —	1290	5/1	Em. del Lago, Junín: 0900-0300
105) LU22	1300	5/1.5	R. Tandil: 0930-0330
106) LRA5	1300	10/1	R. Nac. Rosario: 0900-0300
106x) LRA42	1310	1	R. Nac. Gualeguaychú (r. 1309.5)
106z) —	1310	0.5	R. Dr. Gregorio Alvarez, Piedra de Aguila: 24h
107) LRH251	1320	25/1.5	R. Chaco, Resistencia: 0900-0300
108) LU10	1320	5	R. Azul: 0900-0300
111) LS6	1350	25/10	R. Buenos Aires, "RBA": 24h
112) LT46	1360	1/0.25	R. Provincia Bernardo de Irigoyen: 1000-1700, 1900-0300
113) LRA54	1370	1	R. Nac. Ingeniero Jacobacci: 1000-0400
113x) LRJ241	1370	20/3	R. Dimensión, San Luis: 0900-0400
66) —	1390	1	R. Nac. Mendoza, Valle de Uspallata: 24h
115) LR11	1390	10/2.5	R. Universidad-Nacional de La Plata: 0900-0300
116) LU11	1400	10	R. Em. del Oeste, Trenque Lauquen: 0900-0300
116x)	r1400	0.45	R. Municipal Catriel, Catriel
166) —	1410	0.1	R. Fortín Federación, Junín: 1000-0400
167) —	1415	—	R. Cultura, Lanús
117) LRK221	1420	1/0.25	R. Ciudad Perico, Perico: 1000-0200 (r. 1400)
176) —	1420	—	R. Club, Lomes de Zamora, BA
119) LT24	1430	1	R. San Nicolás: 0900-0300
120) LV26	1430	1/0.25	R. Río Tercero: 0930-0330
168) —	1430	—	R. Los Alamos, Glew
121) LRI235	1430	1	R. Balcarce, Balcarce: 0900-0300
43) LRA53	1440	1	R. Nac. San Martín de los Andes: 1000-0400

122)	LT41	1440	1	R. La Voz del Sur Entrerriano, Gualeguaychú: 0900-0300
123)	LU36	1440	0.25	R. Coronel Suárez: 1000-0400
124)	LV20	1440	1/0.25	R. Laboulaye: 1000-0400
125)	LV27	1440	1/0.25	R. San Francisco: 0900-0300
177)	—	1440	—	R. Chascomús "RCH," Chascomús, BA
126)	LV28	1450	1/0.25	R. Villa María: 0900-0300
67)	LT44	1450	1/0.25	R. Fortín Yunka, Formosa: 0900-0400
127)	LT29	1460	1/0.25	R. Venado Tuerto: 0930-0330
128)	LU30	*1460	0.25	R. Maipú: 0900-2400
129)	LU34	1460	0.25	R. Pigüé: 1000-0300
130)	LW2	1460	0.25	R. Emis. Tartagal: 1000-0300
131)	LT20	v1470	1/0.25	R. Junín: 0900-0300
132)	LT26	1470	1/0.25	R. Nuevo Mundo, Colón: 0900-0300
133)	LT28	1470	1	R. Rafaela: 0900-0300
134)	LU26	1470	1.5/0.6	Emis. Coronel Dorrego: 1030-0300 (r. 1475 & 1480)
135)	—	1470	1	R. Municipal, Luis Beltrán: 1200-0100
55)	LT13	1470	1	R. Oberá: 0800-0200
141x)	LU27	1480	0.25	R. Dolores: 1100-0300
178)	—	1480	—	AM Santa María, San Justo, BA
137)	LT42	1490	1/0.25	R. del Iberá, Mercedes: 0900-2400
138)	LV22	1490	1	R. Huinca Renancó: 1000-0300
138x)	—	1500	1	R. Municipal, Gral. Conesa: 1000-2400
139)	LT22	1500	0.25	R. Nueva Era, Pehuajó: 1000-0300
169)	—	1500	—	R. El Sol, Lanús: 1100-0300
140)	LT34	v1500	0.25	R. Nuclear, Zárate: 0900-0000
141)	LT45	1500	1/0.25	R. San Javier: Mon-Fri: 1100-0500; Sat/Sun 1100-2100
142)	LV25	1500	5/1	R. Unión, Bell Ville: 0900-0400
170)	—	1500—		"AM 1500" R. Chascomús
143)	LV21	1510	1/0.25	R. San Joaquín, Villa Dolores: 1000-0100
145)	LW4	1510	0.25	R. Orán: 0900-0300
179)	—	1520	—	R. Güemes, Quilmes Oeste, BA
147)	LT37	1520	0.25	Em. Rufino: 1000-0400
148)	LT38	1520	0.25	R. Gualeguay: 0900-0300
149)	LV24	1520	0.25	R. Manantiales, Tunuyán: 0900-0400
171)	—	1530	—	La R. Gaucha, Glew
180)	—	1530	—	R. Cd. de San Martín, Villa Ballester, BA
150)	LRJ200	1530	1	R. Centro Morteros, Morteros: 0900-0300
151)	LU25	1530	0.1	R. Carhué: 0900-0330
152)	LT35	1540	0.25	R. Mon, Pergamino: 1000-0400
153)	LU28	1540	0.25	R. Tuyú, General Madariaga: 1000-0300
154)	LT23	1550	10/1	R. San Jenaro Norte: 0900-2300 (Sat 0300)
155)	LT32	1550	0.25	R. Chivilcoy: 1000-0400
156)	LT40	1550	0.25	R. La Voz de la Paz, La Paz: 0900-0100
172)	—	1550	—	R. La Cruz del Sur, Guernica
181)	—	1550	—	R. Contacto, San Antonio de Padua, BA
157)	LT11	1560	5/1	R. Gral. Francisco Ramírez, Concepción del Uruguay: 0900-0500
173)	—	1560	—	R. Cd. de Libertad, Libertad
158)	LT33	1560	0.25	R. Nueve de Julio: 0900-0300
159)	LRI229	1570	1	R. Las Flores, Las Flores: 0900-0200
174)	—	1580	—	R. Gral. Martín Miguel de Güemes, Guernica
160)	LT27	1580	0.25	R. La Voz del Montiel, Villaguay: 0900-0300
161)	LT36	1580	0.25	R. Chacabuco: 1000-0400
162)	LR14	1600	1	R. Ciudad Aluminé, Aluminé
175)	—	1600	—	R. Esperanza, Gral. Viamonte

Shortwaves

	Call	kHz	kW	Name, h. of tr
58)	LRA31	6060	50	R. Nac. Buenos Aires: 2100-1500
51)	LU14	*6100	0.5	R. Provincia de Santa Cruz
83)	LV19	6160.7	0.5	R. Malargüe: 1100-2200
66)	LRA34	*6180v	1	R. Nac. Mendoza: 24h
58)	LRA31	9690	25	R. Argentina al Exterior: Tue/Sat 0100-1100
58)	LRA35	11710	100	R. Argentina al Exterior: 1000-1400, Sat/Sun 1900-2200
64)	LRY2	11781.2	1	R. Libertad: 1500-2100
58)	LRA31	15345.2	100	R. Argentina al Exterior: Ext. Sce: 1105-1300, 1700-0100 r. of LRAI Weekends 1100-2300

ASOCIACION DE RADIODIFUSORAS PRIVADAS ARGENTINA (ARPA)

ADDR: Pte. Perón 1561, 1037 Buenos Aires. **Te:** (01) 35-4412.
L.P: Pres: Carlos Alberto Mirson. Gte: Jorge J. Zaffore.
ARPA is an association of all privately owned commercial st's exc. 5), 61), 134).

SERVICIO OFICIAL DE RADIODIFUSION (SOR)

SOR is operated by Dirección Nacional de Radiodifusión which is responsible to the Subsecretaria de Comunicaciones.
ADDR: Sarmiento 151, 1041 Buenos Aires. **Te:** (01) 30-2121.
L.P: Dir. Nac: Julio Maharbiz.
All LRA st's belong to SOR (also includes st. LRA36 in Antarctica). Common prgrs. (originated from LRA1).

SECRETARIA DE INFORMACIÓN PUBLICA SUBSECRETARIA OPERATIVA

Controls all national state owned commercial st's.

STATE ABBREVIATIONS (Provincias): BA = Provincia de Buenos Aires; C = Catamarca; Ca = Córdoba; Cs = Corrientes; Ch = Chaco; Ct = Chubut; ER = Entre Ríos; F = Formosa; J = Jujuy; LP = La Pampa; LR = La Rioja; Ms = Misiones; Ma = Mendoza; N = Neuquén; RN = Río Negro; S = Salta; SC = Santa Cruz; SE = Santiago del Estero; SF = Sante Fé; SJ = San Juan; SL = San Luis; T = Tucumán; (Territorio): TFAIA = Territorio Nacional de la Tierra del Fuego, Antártida e Islas del Atlántico Sur.
NB: These abbreviations should not be used for postal purposes; instead use full name.

Addresses and other information:

1) Mendoza 2430, 3000 Santa Fé, SF — Dir: Héctor J. Perino.
2) Ruta Nac. 34,km. 1433, 4560 Tartagal, S — Dir: E. Esper.
3) Av. Alvear 1180, 9200 Esquel, Ct — DG: Raúl H. Pasarín.
4) Moreno 36, 8000 Bahía Blanca, BA — Dir: Juan J. Calloni.
5) Cas. 21, 3200 Concordia, ER — Dir: Hugo H. Soria.
6) Mendoza 169 Sur, 5400 San Juan, SJ — Dir: J. Barassi.
7) Av. España No. 700 Sur, 4650 La Quiaca, J — DG: Jesús Alberto Guerrero.
8) Hipólito Yrigoyen 1735, 9100 Trelew, Ct. — Mgr: Carlos Pérez Luces.
9) Rivera Indarte 170, Pje. Muñoz, 5000 Córdoba — Mgr: Lidia López Paz de Devoto.
10) Rivadavia 835, 1002 Buenos Aires. Te: (54) 1 331-7311 — Dir: Alfredo Ojea.
11) Cas. 108, 8400 San Carlos de Bariloche, RN. Dir: Sebastián A. Tetelboin.
12) Rivadavia 118-120, 4000 San Miguel de Tucumán, T — DG: Ing. Eduardo J. Bader.
13) Casa 16, Las Heras s/n, Barrio Las Lengas, 9407 Río Turbio, SC — DG: Ramón Lozano.
14) Cas. 372, 3300 Posadas, Ms — Dir: Pedro D. Cabrera.
15) Yrigoyen 324, 5300 La Rioja, LR — DG: N.A. Flores.
16) Arenales 2467, 1124 Buenos Aires — **Te:** 824-1087. **Tlx:** 22012 RIVAD AR — DG: Luis María Cetra.
17) Monseñor Fagnano y Leonardo Rosales, 9420 Río Grande, TF — Dir: Miguel Bersier.
18) Cas. 77, 8332 General Roca, RN — DG: H.X. de Cozzi.
19) Salta 24, 5730 Villa Mercedes, SL — Dir: Hector E. Echandía.
20) 25 de Mayo 453, 9000 Comodoro Rivadavia, Ct — DG: J. Horat.
21) Gral. Paz 536, 8353 Chos Malal, N — Dir: Luis José Castro.
22) Bolívar 318, 3300 Posadas, Ms — Pte: Pedro Warenycia.
23) La Rioja 2371, 7600 Mar del Plata, BA — Dir: Marcelo Freites.
24) Zapiola 25, 9400 Río Gallegos, SC — Dir: Alberto R. Segovia.
25) Estivariz 226, 9120 Pto. Madryn, Ct — DG: H. Castro.

26) Echeverría 144, 5500 Mendoza, Ma — Dir: Daniel E. Vila.
27) Mitre 294, 4400 Salta, S — DG: Carlos Skaf.
28) Cas. 14, 8324 Cipoletti, RN — Dir: H. Rodríguez S.
29) San Martín y Chanetón, 8340 Zapala, N — DG: Antonio O. Vecchiarelli.
29x) Av. Victoria Aguirre 1556, 3370 Puerto Iguazú, Ms. — Dir: Alfredo Gil.
30) Sarmiento 1551, 1042 Buenos Aires — Dir: Marcelo A. Simón.
31) Av. San Martín 114, 9311 Governador Gregores, SC — DG: O. Muñoz.
32) San Martín y Gral. Paz, 5500 Mendoza, Ma — DG: M. Fernandez R.
33) Rivadavia 402, 6300 Santa Rosa, LP — DG: Eduardo N. Nicoletti.
34) Cas. 2, 9405 El Calafate, SC — Dir: Raúl R. Miño.
35) Rivadavia 741, 4700 San Fernando del Valle de Catamarca, C — Dir: Raúl J. Acuña.
36) Av. Comandante Fontana y Dr. Mariano Moreno, 9033 Alto Río Senguerr, Ct — Dir: Roxana B. de Jensen.
36x) 12 de Octubre 633, 9050 Puerto Deseado, SC — Mgr: Mario dos Santos Lopes.
37) Santa Rosa 241, 5000 Córdoba, Ca — Dir: José O. Gaidon.
38) Av. San Martín 351, 9410 Ushuaía, TFAIA — Dir: E.J. Mayor.
39) Chacra 46, La Tablada, 3340 Santo Tomé, Cs — DG: H. Centeno.
40) Rioja 1484, 5500 Mendoza, Ma — DG: Norma S. Fontemachi.
41) Mansilla 2668, 1425 Buenos Aires — Dir: Dr. Jorge H. Santos.
42) Olavarría y Río Bermejo, 4600 San Salvador de Jujuy, J. — Mgr: José Ramón Casas. C.E: Roberto Pemberton.
43) Villegas 251, 8370 San Martín de los Andes, N — DG: Norberto Hilzerman.
44) Av. Alvaro Barros 1148, 8500 Viedma, RN — Dir: Alberto Andria.
45) Patricias Mendocines 72-78, 5620 General Alvear, Ma — Dir: José M. Ferraro.
46) Junín 665, 3600 Formosa, F — Dir: Gustavo dos Santos.
48) Belgrano 457, 7500 Tres Arroyos, BA — DG: Cr. Domingo F. Leonel Elías.
49) Córdoba 1843, 2000 Rosario, SF — DG: Manuel A. Fernández.
50) Cas. 83, 3730 Charata, Ch — Dir: Tratis R. Politis, C.E: Adolfo Zinzer.
51) Av. Roca 823, 1p, 9400 Río Gallegos, SC — DG: F. Santa Cruz.
52) Paraguay y Honduras, 4600 San Salvador de Jujuy, J — DG: Ricardo Marcalaín.
53) San Martín 1471, 4146 Concepción, T — DG: H. Rearte.
54) Av. Lib. Gral. San Martín 2031, 3230 Paso de los Libres, CS — Dir: E. Nicholas.
55) Larrea 886, 3360 Oberá, Ms — Dir: Hugo W.R. Amable — F.Pl: Move to 840kHz 10kW.
56) Sarmiento 64, 8000 Bahía Blanca, BA — Dir: Diana Julio de Massot.
57) Saavedra 1318, 9040 Perito Moreno, SC — DG: Ing. Hugo Santos Rojas.
57x) 25 de Mayo 25, 5360 Chilecito, LR — Dir: Miguel O. Barberis.
58) Maipú 555, 1000 Buenos Aires — Dir: Vicente di Leo.
International Sce: see end of section.
59) Av. San Martín 227, 6300 Santa Rosa, LP — DG: M. Loggia.
60) 9 de Julio 390, 4200 Santiago del Estero, SE — Dir: Eduardo Maidana.
61) Rioja 743, 3400 Corrientes, Cs — Dir: Jorge Félix Gómez.
62) Av. Ignacio de la Roza 293 este, 5400 San Juan, SJ — Dir: Ali R. Mattar.
63) Arenales 1925, 1124 Buenos Aires — Dir: Alberto Mirson.
64) Uruguay 1237, 1016 Buenos Aires — Int: J.C. Vilanova. "La Voz de Sudamerica": Mon-Fri 2300-0300.
65) Cas. 229, 9000 Comodoro Rivadavia, Ct — DG: N. Dames.
66) Emilio Civit 460, 5500 Mendoza, Ma — Dir: Lic. J. Parvanoff.
67) España 376, 3600 Formosa, F — Dir: M.F. Bedoya.
68) Calle 64 No. 2946, 7630 Necochea, BA — DG: Dra. Julia J. Rico de Coupau.
69) Cas. 204, 8300 Neuquén, N — Dir: Jorge Ocampo.
70) 25 de Mayo 424, 5000 Córdoba, Ca — DG: José O. Gaido.
70x) Garibaldi 7, 5405 Barreal, SJ — DG: Victor H. Echegaray.
71) Reconquista 135, 8336 Villa Regina, RN — DG: R.A. Musso.
72) Constitución 399, 5800 Río Cuarto, Ca — DG: J. Castellina.
73) 9 de Julio 3560, 3000 Santa Fé, SF — Dir: Jorge Ricci.
74) Ruta 40 s/n, 9030 Río Mayo, Ch — DG: Hugo R. Font.
75) Gral. Oblegado 1023, 3560 Reconquista, SF — Dir: José C. Odasso.
76) Av. Santa Fé 2043, 1123 Buenos Aires — DG: Armando V. Pérez. (AM Stereo)
77) Pte. Perón 646, 1038 Buenos Aires — Dir: Guillermo L. Higuy.
78) Cas. 67, 8000 Bahía Blanca, BA — DG: P.B. Serrat.
79) Cas. 2, 8353 Andacollo, N — Dir: Delia Ma. Callieri.
79x) Tacuarí 2035/37, 1009 Buenos Aires — DG: R. Gangeme.
80) Mendoza 273, 4000 San Miguel de Tucumán, T — DG: Victor Manuel Cuevas.
81) Córdoba 1865, 7600 Mar del Plata, BA — Carlos Lucchetti.
82) Mendoza 452 Sur, 5400 San Juan, SJ — Dir: Emilio Ventura.
83) Esquivel Aldao 350, 5613 Malargüe, Ma — DG: D. Porras.
84) Urquiza 332, 4200 Santiago del Estero, SE — DG: Antonio Medel.
84x) Cas. 424, Viedma, RN — Dir: Carlos E. Reyes.
85) Av. 9 de Julio 1855, 3500 Resistencia, Ch — DG: Alfredo H. Norniella — Occ. prgr. in Toba and Guaraní.
86) Ruta Nal. 150, Km. 3,5460 San José del Valle de Jáchal, SJ — DG: C. Paez.
87) 4 de Enero 2153, 3000 Santa Fé, SF — DG: N. Vittori.
88) Cas. 176, 8430 El Bolsón, RN — DG: Carlos Pogliano.
89) Cas. 55, 3370 Puerto Iguazú, Ms — Dir: E. Gurovici.
90) Alsina 3377, 7400 Olavarría, BA — Dir: Juan A. del Bono.
91) Cas. 113, 4400 Salta, S — Dir: CPN Alberto V. Verón.
92) Rivadavia 563, 5700 San Luis, SL — DG: Hugo Sudai.
93) Justo José de Urquiza 585, 3700 Presidencia Roque Sáenz Peña, Ch — Dir: Miguel L. Braccone.
94) Hipólito Yrigoyen y Godoy Cruz, 5600 San Rafael, Ma.
95) San Martín 251, Piso 4, 4000 San Miguel de Tucumán, T — DG: Adrián Lomello.
96) Honduras 5665, or: Bonpland 1745, 1414 Buenos Aires, BA — Fax: (54) 1 775 3694 — Mgr: César Carabelli. (All news)
97) Dr. Mariano L. Lloza 231, 3450 Goya, Cs — DG: José Escobar.
98) Cas. 218, 3384 Eldorado, Ms — Dir: J.E. Gamas.
99) Av. Colón 119, P. 4, 5000 Córdoba, Ca — Fax: (54) 51 241080. DG: (Cr.) Carlos M. Molina. DP: Mario Pereyra.
100) Dorrego 1042, 2000 Rosario, SF — Dir: Alberto C. Gollán.
101) Victoria 472, 4512 Lib. Gral. San Martín, J — Dir: Héctor Anauti.
102) Rivadavia 126, 3100 Paraná, ER — DG: R. Galanti.
103) Intersección Rutas Nacional 81 y Provincial 32, 3630 Las Lomitas, F — DG: María Dolores Ramírez.
104) Av. 53 No. 810, 1900 La Plata, BA — Dir: R. Barreiros.
105) Cas. 231, 7000 Tandil, BA — Dir: Alcides Fortunato.
106) Córdoba 1331, 2000 Rosario, SF — Dir: Jorge Vidal.
106x) Urquiza al Oeste, 2820 Gualeguaychú, ER.
106z) Casa 16, Barrio Jardín, 8315 Piedra del Aguila, N — Dir: Oscar I. Lillo.
107) Córdoba 710, 3500 Resistencia, Ch — Dir: Raúl Puigbó.
108) Av. Mitre 819, 7300 Azul, BA — DG: D. Santamauro.
109) Calle 29 y 40, 6360 General Pico, LP — DG: J. Diván.
110) Balcarce 840, 2000 Rosario, SF — DG: Carlos McGuire.
111) Av. Belgrano 270, 1092 Buenos Aires, BA — DG: Jorge A. Civit — DP: Eduardo Haida.
112) Av. Independencia s/n, 3366 Bernardo de Irigoyen, Ms.
113) Martín Coronado y José Hernández, 8418 Ingeniero Jacobacci, RN — Dir: Abel H. Chaina.
113x) Av. Pte. Illia 128, 5700 San Luís — Dir: Alberto E. Cometto.
114) Cdte. Salas 287, 5600 San Rafael, Ma — DG: Juan C. Rivas.
115) Plaza Rocha 133, 1900 La Plata, BA. — Dir: Silvina Fernández Cortés.
116) Cas. 36, 6400 Trenque Lauquen, BA — Dir: Andrés E. Fernández V. F.Pl: Move to 1280kHz.
116x) Av. San Martín 558, 8307 Catriel, RN.
117) Villafañe y Calilegua, 4608 Perico, J — DG: G. Jenefes.
118) Gral. Paz esq. 1. Abitbol, 3344 Alvear, Cs. — Mgr: Magda E. Trindade F.Pl: 3 kw.
119) Av. Moreno 124, 2900 San Nicolás, BA — DG: A.A. Ramini.
120) Libertad 455, 5850 Río Tercero, Ca — DG: E. Yantorno.
121) Cas. 77, 7620 Balcarce, BA — DG: Juan H. Orofino.
122) Urquiza y Gualeguay 1399, 2820 Gualeguaychú, ER — DG: L. Mettler de Fariña.
123) Avellaneda 1144, 7540 Coronel Suárez, BA — Dir: Eduardo D. Alvarez.
124) Tucumán 159, 6120 Leboulaye, Ca — Dir: José C. Alisio.
125) Colón 26, Piso 1, 2400 San Francisco, Ca — DG: Dionisio E. Peretti.
126) Santa Fé 1490, 5900 Villa María, Ca — Dir: Roberto Kfuri.

127) Cas. 340, 2600 Venado Tuerto, SF — Dir: Hilmar H. Long.
128) Lavalle Sud 312, 7160 Maipú, BA — Dir: Luis A. Tavela.
129) Cas. 101, 8170 Pigüe, BA — Dir: Isidro A. Bros.
130) Gorriti 524, 4560 Tartagal, S — Dir: Olimpia P. del Busto.
131) Hipólito Yrigoyen 86, 6000 Junín, BA — Dir: Raúl A. Diez.
132) Pte. Perón 117, 3280 Colón, ER — Dir: Edwin Benítez.
133) Av. Lehman 245, 2300 Rafaela, SF — DG: A. Robert.
134) San Martín y Uslenghí, 8150 Coronel Dorrego, BA — DG: Salvador E. Lavios.
135) 8361 Fray Luis Beltrán, RN — Dir: D. Natali.
136) Sarmiento 474, 3153 Victoria, ER — Dir: Guido A. Anghilante.
137) Cas 33, 3470 Mercedes, Cs — Dir: José M. Enciso.
138) Santa Fé y La Pampa, 6270 Huinca Renancó, Ca — DG: O. Berlaffa.
138x) Roca 570, 8503 General Conesa, RN — Dir: Jorge O. Monteoliva.
139) Perito Moreno 700, 6450 Pehuajó, BA — Dir: Raúl Negreira.
140) Independencia 501, 2800 Zárate, BA — Dir: Jorge E. Lynch.
141) Ordóñez 737, 3357 San Javier, Ms — Dir: H. Andersson.
141x) Rico 424, 7100 Dolores, BA — Dir: N.G. Peñoñori.
142) Villalba s/n, 2550 Bell Ville, Cs — Dir: Osvaldo Etrat.
143) Belgrano 33, 5870 Villa Dolores, Ca — Dir: Héctor O. Escudero.
144) Pablo Stampa 2430, 3228 Chajarí, ER — DG: J. Ponzoni.
145) 9 de Julio 163, 4530 Orán, S — DG: A. Bernard.
146) San Martín 1380, 3460 Curuzú Cuatiá, Cs. — DG: M. Bermudez.
147) Chacabuco 128, 6100 Rufino, SF — Dir: Clara R. Vda. de Valeri.
148) Chacabuco 38, 2840 Gualeguay, ER — DG: A. Carbone.
149) Alem y Juan B. Justo, 5560 Tunuyán, Ma — Dir: F. Rossi.
150) 25 de Mayo 133, 2421 Morteros, Ca — Dir: José L. Brunone.
151) Av. Colón y Rivadavia, 6430 Carhué, BA — DG: Mario V. Fernández.
152) Dr. Alem 340, 2700 Pergamino, BA — DG: C. Trincavelli.
153) Cas. 123, 7163 General Madariaga, BA — DG: Hìpólito Maceira.
154) Calle Juan Chavarri 452, 2147 San Jenaro Norte, SF — Dir: Rubén Cavallera.
155) Cas. 179, 6620 Chivilcoy, BA — Dir: Rafael E. Mónaco.
156) Sáenz Peña 1082, 3190 La Paz, ER — Dir: Juan Malvasio.
157) Onésimo Leguizamón 269, 3260 Concepción del Uruguay, ER — Dir: Carlos Godein.
158) Continuación Av. Bartolomé Mitre, Sección A, 6500 9 de Julio, BA — Dir: Héctor H. Tinetti.
159) Avellaneda 773, 7200 Las Flores, BA — DG: César A. Dantas.
160) Cas. 40, 3240 Villaguay, ER — Dir: Padre José Lavini.
161) Almirante Brown 135, 6740 Chacabuco, BA — DG: Juan P. Martini.
162) 8345 Aluminé, N.
163) Av. Rep. Argentina 737, 6000 Junín, BA — Dir: Jorge Andreollo.
165) 1824 Lanús, BA.
166) Além 469, 6000 Junín, BA.
167) 1824 Lanús, BA.
168) 1856 Glew, BA.
169) Sarmiento 1436, 1824 Lanús, BA.
170) Libres del Sud 132, 7130 Chascomús, BA.
171) 1856 Glew, BA.
172) Av. Eva Perón esq. Calle 113, 1862 Guernica, BA.
173) 1716 Libertad, BA.
174) Calle 32 e/ll y ll, 1862 Guernica, BA.
175) 6015 Gral. Viamonte, BA.

FM in Buenos Aires: 30) 92.7 Frec. Municipal; 77) 94.3 Horizonte; 76) 95.1 "FM95"; 63) 95.9 Tango; 58) 96.7 FM Folklore y Tango Nacional; 79x) 97.5 Clásica; 96) 98.3 Sport; 58) 99.1 R. Nacional; 41) 99.9 FM Cien; 64) 100.7 FM Feeling; 63) 101.3 "FM 101.3"; 96) 102.3 Aspen 102; 16) 103.1 Viva; 10) 105.5 FM Hit; 111) 106.3 Rock and Pop.
In the City area there are over 150 unlicensed LP FM st's, about 900 in the rest of the country.

RADIODIFUSION ARGENTINA AL EXTERIOR (RAE)

Operated by Servicio Oficial de Radiodifusión (SOR).
ADDR: Maipú 555, 1000 Buenos Aires (postal addr: Casilla de Correo 555O **L.P:** Dir: Marcela G.R. Campos 1000 Buenos Aires).
Te: +54 (1) 3256368. **Tlx:** 21250 LRA AR.
STATION: (G.C: 58.22W/34.36S): 50/100kW.

kHz	kHz	kHz
9690	11710	15345.2v

	Areas	kHz
Arabic		
1800-1900	ME/Af	15345
English		
1900-2000	Eu/Af	15345
0200-0300	NAm	11710
French		
2100-2200	Eu/Af	15345
0300-0400	NAm	11710
German		
2200-2300	Eu/Af	15345
Italian		
2000-2100	Eu/Af	15345
Japanese		
1000-1055	FE	11710
Portuguese		
0100-0200	SAm	9690
1100-1200	SAm	15345
Spanish		
1200-1400	Am	15345
2305-2400	Eu/Af	11710
0005-0200	Am	11710

Broadcasts from Mo to Fri only. On Sat & Sun relays from LRA1 R. Nacional.

DX-Prgr: Arabic: Tues 1820; **English:** Wed 1930; **French:** Thurs 2115, Fri 0315; **German:** Fri 2215; **Italian:** Fri 1930; **Japanese:** Thurs 1015, Fri 1015; **Portuguese:** Thurs 1120, Fri 0120; **Spanish:** Tues 1230, 2330, Fri 0030, 1230. **DX-Editor:** Gabriel Iván Barrera.

ANN: English: "This is R.A.E., the International Sce. of the Argentine Radio" — **V.** by QSL-card.

BOLIVIA

L.T: UTC – 4h — **Pop:** 7,330.000 — **Radio sets:** 4m — **Pr.L:** Spanish, Quechua, Aymará — **E.C:** a/c 50, La Paz 110/220V, Santa Cruz 220/380V — **ITU:** BOL.

DIRECCION GENERAL DE TELECOMUNICACIONES

ADDR: Av. Mariscal Santa Cruz, Edif. Palacio de Comunicaciones, P. 14, (Cas. 4475), La Paz.
Cable: Gentel. **Te:** +591 (2) 368789. **Telex:** 2595 Gentel BV.
L.P: Dir. Gen: Dr. Héctor Guzmán Hinojosa. Dir. de R. y TV: Ing. Fabián Tito Luque.

CAMARA NACIONAL DE MEDIOS DE COMUNICACION

ADDR: Casilla 2431, La Paz.
STATIONS: Mediumwaves: ° = shortwave. *) = inactive.

	Call	kHz	kW	Name, location and h. of tr.
1)	CP91	°580	25	R. Panamericana, La Paz: 0930-0500 (Sun 1030-0100)
2)	CP190	600	8	R. ACLO, Sucre: 0930-1230 (Sat 1430), 1800-0200, Sun 1000-1500
3)	CP63	°620	20/5	R. San Gabriel, La Paz: 0830 (Sat 0930, Sun 1000)-0230, (Sun 0100)
5)	CP204	640	10	R. Tarija, San Lorenzo: 0920-1300, 1600-0100
17)	CP263	660	—	R. Dif. Integración, El Alto
7)	CP34	680	10	R. Méndez Agricultura, La Paz: 0900-0130
139)	CP50	710	10	R. Pío XII, Siglo Veinte: 0900-0300 (Sun 2300)
8)	CP148	720	2.5	R. Yungas, Chulumani: 0900-1400, 1945-0200
9)	CP165	730	3	R. Mensaje, Montero: 0930-0030
10)	CP27	°730	10	R. La Cruz del Sur, La Paz: 0900-0030
11)	CP29	°760	16	R. Fides, La Paz: 1000-0300 (Sun 0600)
12)	CP116	°770	10	R. Cosmos, Cochabamba: 0900-0200
13)	CP46	°790	10	R. Libertad, La Paz: 1100-0400 (Sun 0500; 2400)
14)	CP157	800	0.25	R. Santa Clara, Sorata: 0900-2400
15)	CP188	815	0.25	R. Warnes, Warnes: 1000-0400
16)	CP35	°820	10	R. Altiplano, La Paz: 0950-0400 (Sun 1050-0300)
18)	CP126	°840	1.5/3	R. Juan XXIII, San Ignacio de Velasco: 1000-1430, 1930-2400
19)	CP160	°850	1	R. 21 de Diciembre, Mina Catavi: 1000-0100
20)	CP210	°850	1	R. María Auxiliadora, Montero: 1000-0200

21)	CP8	°860	10	R. Nueva América, La Paz: 1000-0300 (Sun 0130)
22)	CP185	°860	5	R. Paitití, Guayaramerín: 1030-0300
23)	CP20	880	0.18	R. El Cóndor, La Paz: 0800-0230 (Sun 0900-0100) (n. 900)
24)	CP79	900	1	R. Em. La Voz Nacional, Tarija: 1000-0200
25)	CP145	930	10	R. Metropolitana, "la radio popular," La Paz: 0900-0400
26)	CP-	940	0.8	R. San Lorenzo, Colcapirhua: 1000-0200 (n. 960)
27)	CP93	960	1	R. Kollasuyo, Potosí: 0930-2400 (Sun 2200)
28)	CP30	°970	10	R. Santa Cruz, Santa Cruz: 0900-0100 (Sun 1000-2200)
29)	CP118	980	3	R. Mar AM 980, La Paz: 0930-0300
30)	CP192	990	5	R. Esperanza, Aiquile: 0930-1400, 2145-0145
31)	CP119	°1000	3	R. Dif. Trópico, Trinidad: 1300-0030/0100
32)	CP288	1000	3	R. Cochabamba, "CBA": 0945-0400 (n. 900)
33)	CP-	1000	0.15	R. Em. Poopó, Poopó: 1000-1800, 2100-0200
34)	CP220	1010	1	R. Baha'i de Bolivia, Caracollo: 0700-2400
35)	CP4	°1020	10	R. Illimani, La Paz: 1000-0400
36)	CP173	1030	1	R. Agricultura, Portachuelo: 1100-0300
37)	CP113	1040	1	R. Villazón, Villazón: 1100-0200 (Sun 2200)
116)	CP208	1040	5	R. Sipe, Sipe: 0900-2300
38)	CP233	°1050	5	R. El Mundo, Santa Cruz: 1000-0100
39)	CP-	1060	0.25	R. Splendid, Vinto: 1000-2330 (Sat/Sun 1830)
40)	CP181	1060	0.5	R. La Voz de la Frontera, Pto. Suárez: 0900-0300 (Sat 0230, Sun 2300)
42)	CP-	1070	0.5	R. Quillacollo, Quillacollo: 0900-0100
43)	CP-	1080	1	R. Dif. Colosal, Sucre: 0900-0300 (n. 1070)
45)	CP45	1090	0.3	R. Cultura, Cochabamba: 1100-1800, 2000-0300
48)	CP55	1100	1	R. Universidad de Oruro: 1200-0200
46)	CP184	1100	5	R. El Dorado, Trinidad: 1000-0400
47)	CP194	°1100	1	R. Em. Chaco, Yacuíba: 0900-0400
6)	CP137	1100	4	R. Mundial, La Paz: 0900-0330 (Sun 0530)
49)	CP83	°1115	1.5	R. Norte, Montero: 1100-0400 (r. 893)
44)	CP-	1125	0.5	R. Jenecheru "La V. de Cotoca": 1100-0200
50)	CP-	*1130	—	R. Em. 27 de Julio, Huanuni
51)	CP-	1135	1	Kanata Radiodifusión, Cochabamba: 0900-0300 (r. 1125)
52)	CP19	1145	1	R. Agustín Aspiazu, La Paz: 0900-0300 (Sun 0100)
53)	CP71	°1150	0.5	R. El Cóndor, Oruro: 1100-2030
54)	CP221	1150	1	R. Horizontes, Cobija: 0930-0400
55)	CP25	°1160	5	R. Centenario la Nueva, Santa Cruz: 0900-2330
56)	CP78	1160	1.5	RTC, La Deportiva, Cochabamba: 1000-0100 (Sun 2200)
57)	CP98	°1170	1	R. Nuevo Mundo, Sucre: 1030-0300 (Sun 1100-0200)
58)	CP235	1180	1	R. Em. Ingavi, Viacha: 1000-1800, 2000-0100
59)	CP108	1200	1	R. Independencia, Quillacollo: 1000-0100
60)	CP31	1200	3	R. Oriental, Santa Cruz: 0900-0400
61)	CP37	°1200	1	R. Santa Ana, Santa Ana de Yacuma: 1000-0230v
62)	CP171	1200	0.25	R. 24 de Noviembre, Arani: 1030-0400
63)	CP67	1220	1	R. Splendid, La Paz: 1000-0200
64)	CP162	1220	3	R. Batallón Topáter, Oruro: 1000-0400 (Sun 1300-2400)
65)	CP-	r1220	—	R. El Cóndor, Arque
66)	CP35	°1235	1	Indoamérica R., Potosí: 1030-2400 (n. 1250)
67)	CP180	1240	1	R. San Miguel, Arani: 1000-0300
68)	CP54	°1250	2.5	R. La Plata, Sucre: 1000-0230 (Sun 2200)
69)	CP17	°1250	0.5	R. Sararenda, Camiri: 1000-0400
70)	CP16	°1250	2	R. Los Andes, Tarija: 1000-0300 (Sun 2400)
71)	CP26	1250	1	R. Amboró, Santa Cruz: 0950-0400 (Sun 1100-2400)
72)	CP69	°1250	5	R. Nacional, Cochabamba: 1000-0300 (Sun 1200-2200)
73)	CP14	1250	2	Emisoras Unidas, La Paz: 1000-0300 (Sun 2400)
74)	CP65	1250	0.4	R. Oruro, Oruro: 1300-0400
75)	CP100	1250	1	R. Ibaré, Trinidad: 1100-0200
76)	CP47	°1250	0.3	R. Frontera, Cobija: 1000-2400
77)	CP-	1260	0.4	R. Uncía, Uncía: 1100-0300
78)	CP134	1270	1	R. Vanguardia, Colquiri: 1000-0300
79)	CP187	1275	0.5	R. Chané, Mineros: 1000-0400 (Sat/Sun 1100-2300)
80)	CP112	°1280	1.5	R. Frontera, Yacuíba: 1000-0300
81)	CP212	°1290	1	R. Dif. Minería, Oruro: 1000-0230
82)	CP51	°1300	1	R. Loyola, Sucre: 0930-0400 (Sun 1100-2230)
83)	CP127	*1300	0.15	R. Juan XXIII, Uyuni:*
84)	CP82	1300	1	Coop. Radial Electra, Potosí: 1000-0100 (Sun 1100-0500)
85)	CP168	1300	0.3	R. Chichas, Siete Suyos: 1100-0400
86)	CP-	*1300	0.5	R. Coronel Eduardo Avaroa, Santa Cruz: 1000-0330
97)	CP-	1300	—	R. Sol, Viacha
87)	CP68	1310	10	R. San Rafael, Cochabamba: 0900 (Sun 1000)-0300
88)	CP176	1330	1.5	R. América, Oruro: 1145-0100
89)	CP106	1330	0.25	R. Trinidad, Trinidad
90)	CP24	°1340v	1	R. Grigotá, Santa Cruz: 1000 (Sun 1100)-0100v
91)	CP146	1340	0.35	R. San Francisco, Apolo
92)	CP214	1350	1	R. Dif. Ichilo, Yapacaní, San Carlos: 0945-0100
93)	CP-	1350	1	América Radiodifusión, Sucre: 1000-2300, 0230-0600 (Sun 1100-2200)
94)	CP154	1355v	0.25	R. Armonía, Cliza: 0900-0400v (n. 1350)
95)	CP143	1360	1	R. Libertad, Villazón: 1100-0300v (Sun 2200)
41)	CP-	1360	—	Difusoras Ximénez, La Paz
96)	CP158	1370	0.5	R. La Voz de Minero, Siglo XX: 0900-1700, 2200-2300
74)	CP248	1370	3/1.5	R. Dif. Coral, Oruro: 1000-0400
98)	CP33	1370	1	R. Agricultura, Achacachi
99)	CP186	1370	0.15	R. Libertad, Cliza: 0930-0300
100)	CP221	1380	0.5	R. 16 de Noviembre, Sacaba: 1030-0300
101)	CP227	°1380	0.5	R. Em. Luis de Fuentes, Tarija: 0930-0400
102)	CP3	°1390	3	R. Nacional de Bolivia, La Paz: 0900-0300 (Sun 0200)
103)	CP169	1390	0.25	R. La Voz Minera del Sud, Mina Telamayu: 1100-0300
104)	CP174	1400	1	R. Libertador, Santa Cruz: 1000-0400 (Sun 1100-2300)
105)	CP124	1410	0.25	R. Atlántida, Oruro: 1200-2400
106)	CP-	1410	0.25	R. Roboré, Roboré: 1000 (Sun 1100)-0400
121)	CP-	1420	—	R. Chaka, Pucarani
108)	CP49	1420	1	R. Centro, Cochabamba: 1000 (Sun 1100)-0400
109)	CP141	1430	0.25	R. Nuestro Señor de Burgos, Mizque: 1000-0200
110)	CP193	1430	0.25	R. 23 de Marzo, Tupiza: 1300-0200
111)	CP61	°1440	0.6	R. Batallón Colorados, La Paz: 1030-0200 (Sun 1200-2400)
112)	CP107	1440	1.5	R. Yaguarí, Vallegrande: 1500-0200
88)	CP61	1450	1	R. Em. Bolivia, Oruro: 0900-0400 (Sun 1000-0300)
51)	—	1455	2	R. Estévan Arze, Tarata: 0900-0200
117)	CP152	r1465	2.5	El Mundo Radiodifusión, Sacaba: 0900-0200 (n. 1495)

118)	CP215	°1470	0.25	R. CORDECH, Alcalá: 1100-1600, 1900-0200
119)	CP-	r1475	—	R. Tiraque, Tiraque
120)	CP-	1475	1	Emisoras Verde y Blanco, Santa Cruz: 1030-0300
115)	CP-	1480	0.1	Patrimonio R., Potosí: 1000-0100
114)	CP-	r1480	—	R. Cristal, La Paz
122)	CP135	1485	1	R. La Voz del Valle, Punata: 1000-0300 (n. 1500)
123)	CP172	°1490	1	R. San José, San José, Oruro: 1000-2300
124)	CP196	1490	0.25	R. Moxos, San Ignacio de Moxos
125)	CP198	1490	0.35	R. Pedro Domingo Murillo, Quime: 1000-1400
126)	CP-	1490	0.25	R. Mairana, Mairana: 1100-0300
127)	CP238	1500	0.3	R. Sagrado Corazón, Mineros
128)	CP80	1510	0.25	R. Litoral, Cochabamba: 0930-0200 (Sun 2400)
129)	CP102	°1510	0.25	R. 27 de Diciembre, Villamontes: 1000-0200 (Sun 1100-0100)
130)	CP1	1510	5/1	R. Chuquisaca, La Paz: 1000-0500
131)	CP179	*1520	1	R. Espectador, Santa Cruz
132)	CP207	1520	0.25	La Vnz del Cobre, Corocoro: 1100-0200
133)	CP111	°1530	1	R. Em. Ballivián, San Borja: 1100-1800, 2300-0400 (r. 1540)
134)	CP237	1530	0.3	R. San Juan Bosco, Kami
135)	CP200	°1530	0.25	R. Litoral, Llica: 1400-0300
136)	CP191	1545	0.35	R. Mejillones, Tarata: 1900-0400
137)	CP115	1550	5/1	R. Em. Caranavi, Caranavi: 0900-1200, 1530-1800, 2200-0200
138)	CP205	1550	1	R. Tamengo, Pto. Quijarro (n. 1495)
140)	CP-	1560	0.5	1° de Octubre "la radio," Capinota: 1030-0100
141)	CP132	1570	10	R. Continental, La Paz: 0900-0300
142)	CP-	1570	—	R. 1° de Mayo, 1° de Mayo: 1130-2400 (Sun 2200)
143)	CP-	1570	0.3	R. Urkipiña, Quillacollo: 1000-0200 (Sun 0900-0300)
144)	CP166	1580	0.8	R. Llallagua, Llallagua: 1100-0400
107)	CP155	1590	0.1	R. Bermejo, Bermejo: 0830-1900
145)	CP153	1600	0.5	R. Continental, Punata: 1000-0200

Shortwaves

	Call	kHz	kW	Name, location and h. of tr.
146)	CP-	*3200.3	0.5	R. 9 de Abril, Pulacayo: 1000-1200, 2330-0230v
47)	CP-	*3280	—	R. Em. Chaco, Yacuíba
147)	CP114	3310.3	1	R. San Miguel, Riberalta: 0900-1800, 2000-0300
148)	CP206	*3330	0.3	R. Charcas, Monteagudo
149)	CP99	3340.2	1	R. Viloco, Viloco: 1000-?, 2300-0200v (n. 3350)
129)	CP103	3350.1	1	R. 27 de Diciembre, Villamontes: 2230-0200v (n. 3350)
150)	CP209	3370.5	1.2	R. Florida, Samaipata: 1100-0300
57)	CP231	*3376	1	R. Nuevo Mundo, Sucre
151)	CP167	*3380.3	1	R. Cumbre, Tazna: 1030-1300, 1500-1700, 2100-0300v (Sun 0130)
152)	CP175	3390.3	1	R. Emisoras Camargo, Camargo: 1500-0145v
153)	CP-	v3471.5	0.5	R. Dif. Alfonso Padilla, Padilla: 2300-0200 (v. up to 3479)
154)	CP-	v3754.3	—	R. 20 de Septiembre, Bermejo: -0200 (n. 3700)
121)	—	4183.2v	—	R. 11 de Octubre, Cobija?
180)	—	4211	—	R. Perla del Oriente, Roboré
155)	CP-	4409.2	1	R. Eco, Reyes: 1500-1700, 2230-0300
156)	CP-	4420.7	1	R. Santa Rosa, Santa Rosa de Yacuma: 1400-1900, 2200-0130
163)	CP-	4435.1	—	R. CVU, "La Voz del Trópico," Villa Tunari: 1000-1400, 2100-0200
76)	CP59	4449.9	-	R. Frontera, Cobija: 1000-2400
157)	CP142	4472.5	1	R. Movima, Santa Ana de Yacuma: 1030-1900v, 2200-0230v (n. 4835)
160)	CP-	4530.1	—	R. Hitachi, Guayaramerín
54)	CP-	4542.5	—	R. Horizontes, Cobija (n. 4540)
31)	CP120	4549.3	0.3	R. Dif. Trópico: 1300-?, -0030/0100
158)	CP-	4599.7	—	R. Em. Villamontes, Villamontes: 1000-
159)	CP-	4600.1	—	R. Perla del Acre, Cobija: 1100-0300v
174)	CP116	4625.7	0.2	R. Dif. Machupo, San Ramón
61)	CP89	4649.2	1	R. Santa Ana, Santa Ana de Yacuma: 1100-1800, 2100-0230v (n. 4805)
22)	CP185	4681.4	5	R. Paitití, Guayaramerín: 1030-1900, 2130-0300 (n. 4865)
161)	CP136	4712.3	0.5	R. Abaroa, Riberalta: 1030-0200v (n. 4760)
162)	CP147	4739.6	0.5	R. Mamoré, Guayaramerín: 1000-, 2200-2400v (n. 4815) (v. up to 4739.9)
164)	CP219	4765.3	0.3	R. Guanay, Guanay: 1030-1600, 1800-2200v
70)	CP184	4775.2	3	R. Los Andes, Tarija: 1000-0300 (Sun 2400)
133)	CP152	4785	1	R. Em. Ballivián, San Borja: irr.
21)	CP73	*4795.9	10	R. Nueva América, La Paz
80)	CP131	4808	0.25	R. Frontera, Yacuiba (n. 4805)
165)	CP-	4808.9	—	R. Dif. Libertad, Centro Mínero Santa Fé: 1000-1700 (Sun 2100), 2100-0100.
102)	CP144	4815	1	R. Nacional de Bolivia, La Paz: irr.
90)	CP70	4830	5	R. Grigotá, Santa Cruz: 1000-1830, 2200-0030v (Sun 1100-2400)
11)	CP72	4845.0	5	R. Fides, La Paz: 1000-1730, 2200-0300 (Sun 0600; 2400)
55)	CP66	4855	5	R. Centenario "la Nueva," Santa Cruz: 0900-2330
166)	CP-	*4856.7	0.75	R. El Cóndor, Uyuni: (n. 4860)
167)	CP-	4864.5	1	R. Em. 16 de Marzo, Centro Polimetalúrgico Bolívar: 1000-0100v (Sat/Sun 2330)
10)	CP75	4875.1	10	R. La Cruz del Sur, La Paz: 0900-1300, 1500-1700, 2200-0030
69)	CP77	4886	1	R. Sararenda, Camiri: 1000-0400
168)	CP-	v4901.5	0.5	R. San Ignacio, San Ignacio de Moxos: 1030-?, 2230-2400v
147)	CP114	v4922	5	R. San Miguel, Riberalta: 0900-1800, 2000-0300
49)	CP110	4938	1.5	R. Norte, Montero (n. 4935)
35)	CP7	*4945	10	R. Illimani, La Paz
18)	CP90	4965	3/1	R. Juan XXIII, San Ignacio de Velasco: 1000-1430, 1930-2400
20)	CP236	4975	1	R. María Auxiliadora, Montero: 1000-0200
81)	CP213	4980.9	1	R. Dif. Minería, Oruro: 1000-1330, 0000-0230 (n. 4985)
170)	CP163	4991.2	1	R. Animas, Chocaya: 1030-1800, 2100-0130 (Sun 1200-2200)
13)	CP48	5005.8	5	R. Libertad, La Paz: 1100-0400 (Sun 0500; 2400)
135)	CP-	*5010	0.1	R. Litoral, Llica
16)	CP-	5044.8	5	R. Altiplano, La Paz: 0950-, -2400v
172)	CP218	5156.2	0.2	R. Galaxia, Guayaramerín: 1000-0300v
173)	CP-	*5217	—	R. Em. Tarabuco, Tarabuco
113)	CP-	5500.4	—	R. Luz del Oriente, Roboré?
175)	CP109	v5505.3	0.5	R. 2 de Febrero, Rurrenabaque: 1200-1500, 2130-0300 (Sun 2330-0200) (n. 5020)
176)	CP-	v5580.2	1	R. San José, San José de Chiquitos: 1100-1700, 2100-0200v
178)	CP-	v5925	—	R. Centinela, Tupiza: 1100-2200
163)	CP-	593	—	R. CVU "La V. del Trópico," Villa Tunari
139)	CP60	5948.7	5	R. Pío XII, Siglo Veinte: 0900-0200 (Sun 2300)
179)	CP177	5964.8	1	R. Nacional de Huanuni, Huanuni: 0930 (Sun 1100)-0300
72)	CP200	v5974	1	R. Nacional, Cochabamba: 1000-1500, 2000-2300
123)	CP222	*5985	2.5	R. San José, San José, Oruro
82)	CP41	*5995	0.8	R. Loyola, Sucre
38)	CP234	6015.8	10	R. El Mundo, Santa Cruz: 1000-0100 (n. 6015)
35)	CP5	6025	10	R. Illimani, La Paz: 0930-0400
53)	CP18	*6069	2	R. El Cóndor, Oruro (n. 6070)
3)	CP229	6085	5	R. San Gabriel, La Paz: 0830 (Sun 1000)-1800, 2000-0230

19)	CP159*6082	0.85		R. 21 de Diciembre, Mina Catavi
12)	CP216	6094.2	1	R. Cosmos, Cochabamba: 0930-0200 (n. 6095)
1)	CP92	6105.5	7.5	R. Panamericana, La Paz: 0930-0030v
28)	CP32	6135	10	R. Santa Cruz, Santa Cruz: 0900-0100 (Sun 1000-2200)
101)	CP81	6140.7	1	R. Em. Luis de Fuentes, Tarija: 0930-1300, 1600-1800, 2200-0200
11)	CP12	6155	10	R. Fides, La Paz: 1000-0300 (Sun 0600; 2400)
66)	CP74	*6175	1	R. Indoamérica, Potosí
111)	CP-	6185	3	R. Batallón Colorados, La Paz
74)	CP249	6330	1	R. Dif. Coral, Oruro
68)	CP21	9717.2	1	R. La Plata "La Voz de la Capital," Sucre: 1000-1900 (Sun 2200) (n. 9715)

Addresses and other information:

ERBOL (Escuelas Radiofónicas de Bolivia), Cas. 5946, La Paz. **Te:** +591 354142 — Pte: Jorge Trias S.J.
Affiliated st's: 2), 3), 5), 8), 14), 18), 28), 30), 82), 87), 139), 147).
UNESBO (Unión de Emisoras Sindicales de Bolivia), Yanacocha 689, La Paz. **Te:** 341881 Pte: Jorge Bustillo Burgos.
Affiliated st's: 4), 19), 23), 53), 60), 65), 74), 78), 85), 88), 103), 123), 132), 141), 151), 165), 166), 167), 170), 179).

1) Potosí 1320, or: Cas. 503, La Paz. **Te:** +591 (2) 354215 — Own: Miguel Dueri. Dir: Daniel Sánchez Rocha. **N:** "El Panamericano" relayed by many st's.
2) Acción Cultural Loyola, Cas. 538, Sucre — DG: P.J. Javier Velazco. Dir: Prof. René Santillán. Prgr in Quechua.
3) Gral. Lanza 2001, Cas. 4792, La Paz — Dir: José Canut Saurat. Prgr in Aymara, exc. N. in Sp. at 0000.
5) Obispado de Tarija, Cas. 1003, Tarija — Dir: M. van den Valck.
6) Av. Sánchez Lima 2554, P. 2, La Paz — Dir: Sonia D. de Arévalo.
7) Cas. 8122, La Paz — Dir: Abelardo Murillo.
8) Cas. 4535, La Paz — Dir: R.P. José Geldens.
9) Igl. Metodista de Bolivia, Cas. 434, Santa Cruz — Dir: Marcelo Zanabria.
10) Nicaragua 1759, Cas. 1408, La Paz — DG: Pastor Rodolfo Moya Jiménez. Prgr. in Aymara 0900-1110, 2200-0030, otherwise in Sp.
11) Jenaro Sanjinés 799, Cas. 9143, La Paz. **Fax:** 379030 — Dir: R.P. Eduardo Pérez. **N:** "La hora del país", relayed by many st's, at 1000, 1630, 2200, 0130. F.Pl: SW 9625 kHz.
12) Sucre 3509, Cas. 5303, Cochabamba — Gte: Laureano Rojas.
13) Cas. 5324, La Paz.
14) Cas. 2329, La Paz — Dir: Fr. Krystyn Kurka.
15) Cas. 830, Santa Cruz — Gte: Lino Azurduy.
16) Abdón Saavedra 2170, Cas. 1081, La Paz — Dir: Eduardo Ibáñez.
17) El Alto, La Paz. Te: 810048.
18) Vicariato Apostólico, San Ignacio de Velasco (Santa Cruz).
20) Congreg. Salesiana Don Bosco, Cas. 507, Montero (Santa Cruz) — DG: Mario Pani. Dir: Hna. Rosa Gallardo.
21) Indaburo 184, Cas. 2431, La Paz — Coord. Gral: Jorge Rocha Siles.
22) Cas. 321, Guayaramerín (Beni) — Dir: A. Mollinedo B.
23) Yanacocha 689, La Paz — Dir: Jorge Bustillo Burgos.
24) Cas. 404, Tarija — Dir: J. Jaime Fernández Mogro.
25) Cas. 8704, La Paz. **Te:** 356785 — Dir: Carlos Palenque.
26) Cajón 21, Cochabamba — Dir: J. Valverdo Leiza.
27) Cobija 19, Potosí — Gte: Angel Moscoso M.
28) Inst. Radiofónico Fé y Alegría, Cas. 672, Santa Cruz — DG: Víctor Blajot, S.I. DP: Dr. Alvaro Puente. Prgr. in Sp., Quechua and Guaraní.
29) Cas. 11228, La Paz — Dir: Dr. Fernando Salazar P.
30) Cas. 5, Aiquile, or: Cas. 4430, Cochabamba — DG: P. Floriano Weiss. Prgr. in Quechua; Sp. 3½ h.
31) Cas. 60, Trinidad (Beni) — Dir: Eduardo Avila Alberdi.
32) Cas. 2787, Cochabamba — Own: Edgar Barriga Ch.
33) Cooperativa Minera Poopó, Poopó (Oruro) — Dir: Marcial Socompi Juárez.
34) Cas. 1019, Oruro — Dir: Augusto E. Costas M. F.Pl: 5 kW.
35) Ayacucho 476, Cas. 1042, La Paz. **Te:** 376364. Dir: Lic. Fernando Lozada.
36) Calle Warnes, Portachuelo (Sta. Cruz) — Dir: Franklin Gutiérrez.
37) Cas. 58, Villazón (Potosí) — Dir: G. Cocca A.
38) Cas. 1984, Santa Cruz. **Te:** (33) 464646 — DG: Hugo Paz Méndez.
39) Av. Patino Km. 17, Vinto (Cbba.) — Gte: María P. Avila A.
40) Cas. 18, Pto. Suárez (Sta. Cruz) — Dir: Leonardo Arteaga.
42) Quillacollo (Cochabamba) — Dir: Jaime Valdivia Z.
43) Cas. 335, Sucre — Dir: Prof. Moisés Torres R.
44) V. Gutiérrez s/n, Cotoca (Depto. de Santa Cruz).
45) Cas. 422, Cochabamba — Dir: José Maldonado.
46) Cas. 720, Trinidad (Beni) — Dir: Julio Ferrer S.M.
47) Inst. Politécnico Campesino, Cas. 42, Yacuíba (Tarija) — Dir: Dr. Armando Vidal Soruco.
48) Universidad Técnica de Oruro, Oruro.
49) Cas. 19, Montero (Santa Cruz) — DG: Leonardo Arteaga R.
50) Lizárraga 120, Huanuni (Oruro).
51) Cas. 3475, Cochabamba — Dir: Silvio Vallejos.
52) Edif. Batallón Colorados, Cas. 8084, La Paz — Dir: Hugo A. Aspiazu
53) Cas. 620, Oruro.
54) Cas. 196, Cobija (Pando) — Dir: Alberto Lazarta C.
55) Cas. 818, Santa Cruz — Dir: Pedro Salces Ruiz.
56) Cas. 846, Cochabamba — Dir: Carlos Dalence Loayza.
57) Cas. 25, Sucre — Dir: Víctor Hugo Hevia R.
58) Viacha (Depto. de La Paz) — Dir: Mario Mamani.
59) Cas. 108, Quillacollo (Cbba.) — Dir: René Rojas B.
60) Cas. 186, Santa Cruz — Dir: Luis A. Mendívil.
61) Cobija 285 esq. Bolívar, Santa Ana de Yacuma — Dir: Delicia Guardia Lima.
62) Arani (Cochabamba) — Dir: Oscar J. López A.
64) Cas. 1539, La Paz — Dir: Guido Velasco A. Prgr in Aymara.
64) Calle Junín y Velasco Galvarro, Oruro — Dir: J.A. de Tizanos P.
65) Arque (Cochabamba)
66) Cas. 572, Potosí — Dir: Agustín Vera Duarte.
67) Plaza Progreso 20, Arani (Cbba.) — Dir: Wilfredo Rojas.
68) Avaroa 422, Cas. 276, Sucre. **Te:** +591 (64) 23231 — Dir: Gregorio Donoso Daza.
69) Plaza 12 de Julio, Camiri (Sta. Cruz) — Dir: Freddy Lara A.
70) Cas. 34, Tarija — Dir: Jaime Rollano Monje.
71) Cas. 697, Santa Cruz — Dir: Hugo R. Tarabini.
72) Ayacucho 5900, Cas. 2429, Cochabamba — Gte: Glady A. de Peredo.
73) Cas. 3250, La Paz — Dir: Alberto Cajías C.
74) Cas. 845, Oruro — Dir: Alfredo Luján M.
75) Univ. Técnica del Beni, Cas. 38, Trinidad (Beni).
76) Cas. 179, Cobija (Pando) — Dir: L. Miahuchi von Ancken.
77) Cas. 15, Uncía (Potosí) — Dir: Roberto Velásquez.
78) Cas. 154, Colquiri (La Paz) — Dir: Alfredo Murillo M.
79) Mineros (Santa Cruz) — Dir: Manuel Emilio Pena.
80) Cas. 24, Yacuíba (Tarija) — Dir: Juan Castillo R.
81) Cas. 247, Oruro — Dir: Dr. José Carlos Gómez Espinoza.
82) Cas. 40, Sucre — Gte: Renato Hurtado Chumacero.
83) Cas. 28, Uyuni (Potosí) — Dir: Juan López Claros.
84) Cas 328, Potosí — Dir: J. Antonio González B.
85) Campamento Minero, Siete Suyos (Potosí).
86) Cas. 1745, Santa Cruz — DG: Tcnel. Inf. Jesús Cuéllar N.
87) Cas. 546, Cochabamba — DG: Salim Sauma. **F.Pl:** react. sw.
88) Velasco Galvarro 5156, Oruro — Dir: Hilarión Berrios.
89) Cas. 90, Trinidad (Beni) — Gte: Juan L. Farrel.
90) Abaroa 223, Cas. 203, Santa Cruz. **Te:** +591 (33) 22142 — Dir: Daniel Arteaga Farrell.
91) Apolo, Pcia. Franz Tamayo (La Paz).
92) Congreg. Salesiana Buena Vista, Cas. 463, Santa Cruz — Dir: Gheraldina Anjarí.
93) Eduardo Berdecio 568, Sucre — Dir: Jorge Poppé Avilés.
94) Cliza (Cochabamba) — Dir: Francisco Quiroz J.
95) Cas. 40, Villazón (Potosí) — Dir: Jaime Néstor Lima.
96) Campamento Minero Siglo XX (Potosí) — Dir: Teodoro Orosco.
98) Achacachi, Pcia. Omasuyos (La Paz).
99) Cliza (Cochabamba) — Dir: Crescencio Escobar A.
100) Cas. 2522, Cochabamba — Dir: Luis Camacho C.
101) Cas. 125, Tarija — Own: Jimmy Borda.
102) Isac Tamayo 640, La Paz — Dir: Jorge Antureano. C.E: Ing. Carlos Belmonte. Prgr. mainly in Quechua.
103) Mina Telamayu (Potosí) — Dir: Raúl Chorocoro.
104) Arenales 630, Cas. 1333, Santa Cruz — Dir: Juan Flores Ortiz.
105) Linares 1160, Oruro — Dir: Federico García S.
106) Roboré (Santa Cruz) — Gte: Ciro Rivera C.

107) Campamento Central IAB, Bermejo (Tarija) — Dir: Manuel Barja P.
108) Cas. 839, Cochabamba — Dir: Antonio Torrico Navia.
109) Cas. 893, Cochabamba — Dir: P. Mario Comi.
110) Cochabamba s/n, Tupiza (Potosí) — Dir: Mario Loza F.
111) Edo. Mayor Gral., Final Av. Saavedra, La Paz — Dir: Tcnl. Com: Carlos Polo M. DP: Félix Machicado.
112) Correo Central, Vallegrande (Sta. Cruz) — Dir: Prof. Edgar Bonilla M.
114) Cas. 5303, La Paz — Dir: Mario Castro.
115) Victor Flores 410, Potosi — Dir: Carlos Velástegui.
116) Cochabamba s/n, Quillacollo (Cbba.) — Dir: Eduardo Murillo Orellana.
117) Plaza 6 de Agosto, Sacaba; Cas. 3230, Cochabamba — Dir: Víctor Gutiérrez Iriarte.
118) Cas. 156, Sucre — Dir: Jorge Crespo Avilés.
119) Tiraque (Cochabamba)
120) Av. Landívar 242, Santa Cruz.
121) Pucarani (La Paz).
122) Cas. 1361, Cochabamba — Dir: Roberto Mendoza.
123) Cas. 433, Oruro — Dir: Roberto Ramírez Guzmán.
124) San Ignacio de Moxos (Beni) — Dir: Mina Bauer de Suárez.
125) Quime, Inquisivi (La Paz) — Dir: Oscar Terrazas B.
126) Mairana, Florida (Depto. de Santa Cruz).
127) Congregación Salesiana, Mineros (Santa Cruz).
128) Cas. 1009, Cochabamba — Dir: Marcos Orgaz Escobar.
129) Plaza 15 de Abril, Villamontes (Tarija) — Dir: J. Maldonado T.
130) Cas. 3123, La Paz — Dir: Mercedes Camacho de Kuncar.
131) Cas 3556, Santa Cruz — Dir: G. Antonio Paredes.
132) Corocoro, Pacajes (La Paz).
133) Oruro s/n, San Borja (Beni) — Dir: Ignacio Viruez.
134) Congregación Salesiana, Kami (Cochabamba)
135) Llica, Pcia. Daniel Campos (Potosí).
136) Tarata (Cochabamba) — Dir: Filiberto Rodríguez.
137) Liga de Oración, Misión Mundial, Cas. 266, La Paz — Dir: Felipe Estéban Kittelson H.
138) Quijarro, Puerto Suárez (Santa Cruz).
139) Cas. 434, Llallagua (Potosí) — Dir: Pbro. Roberto Durette OMI. Te: 591 52 53163.
140) Capinota (Cochabamba) — Dir: Benjamín Rojas F.
141) Av. República 872, La Paz — Dir: José Gróver F.
142) Calle 5, Cd. Satélite 1° de Mayo (Santa Cruz).
143) Av. Blanco Galindo 800, Quillacollo (Cbba.) — Dir: J. Salazar.
144) Llallagua (Potosí) — Dir: Juvenal Veizaga.
145) Punata (Cbba.) — Own: Fermín Grágeda O.
146) Planta Industrial de Pulacayo, Pcia. Quijarro (Potosí).
147) Vicariato Apostólico de Pando, Cas. 102, Riberalta (Beni) — Dir: Héctor Salas Takaná.
148) Correo Central, Monteagudo (Chuquisaca).
149) Centro Minero Viloco, Cas. 1414, Viloco (La Paz) — Dir: Félix Barrios.
150) Cas. 5901, Santa Cruz de la Sierra — Dir: Felipe Aponte. Fax: (591) 33 60649.
151) Campamento Minero Tazna, Pcia. Nor Chichas (Potosí).
152) Cas. 99, Camargo, Pcia. Nor Cinti (Chuquisaca) — Dir: Pablo García Benito.
153) Alcaldía Municipal, Padilla (Chuquisaca) — Dir: A. Paredes.
155) Reyes, Pcia. Ballivián (Beni) — Dir: Carlos Espinoza Cortez.
156) C.C., Sta. Rosa del Yacuma (Beni) — Gte: Carlos Baeny.
157) Santa Ana de Yacuma (Beni) — Dir: Rubén Serrano López.
158) Villamontes, Pcia. Gran Chaco (Tarija).
159) Cobija (Pando) — Dir: Rafael Vidal.
160) Guayaramerín — Gte: Jéber Hitachi Banegas.
161) Cas. 136, Riberalta (Beni) — Dir: René Arias P.
162) 25 de Mayo s/n Guayaramerín (Beni) — Dir: Carlos Pinedo S.
163) Cas. 2494, Cochabamba — Dir: Oscar Ustáriz Aranda.
164) Calle Boston, Guanay, Pcia. Larecaja (La Paz).
165) Cas. 568, Zona Postal 5, Oruro — Dir: Aurelio Núñez.
166) Teatro Municipal 2 do. piso, Uyuni (Potosi) — Dir: David Huaylla.
167) Cas. 821, Zona Postal No. 5, Oruro — Dir: Víctor E. Flores Salas.
168) San Ignacio de Moxos (Beni).
170) Chocaya, Pcia. Animas (Potosí) — Dir: Rómulo Lampa F.
171) Cas. 61, Tupiza (Potosí) — Dir: A. Córdova S.
172) Guayaramerín (Beni) — Dir: Carlos Arteaga Tacaná.
173) Otero 98, Sucre — DP: Irma Ortega D. — (Closed in Aug. 1990).
174) San Ramón, Mamoré (Beni) — Dir: Farid Raslán A.
175) Vacadiez 400, Rurrenabaque (Beni) — Dir: Roger Arce Salmón.
176) Cas. 15, San José de Chiquitos (Santa Cruz).
178) Cas. 180, Tupiza (Potosí) — Dir: Israel Jurado G.
179) Cas. 681, Oruro (or: Huanuni, Dto. Postal No. 5, Oruro) — Dir: Rafael Linneo Morales.
180) Roboré, Chiquitos (Santa Cruz).

NB: Whenever listed, Casilla addresses should preferably be used for mailing purposes.

FM in La Paz.
16) 90.1 FM 90 — 21) 91.0 — 141) 93.1 — 35) 94.1 — 10) 95.0 — 52) 96.0 — 1) 97.1 — 11) 98.1 FM Láser — 25) 100.0 Melodía — 3) 101.0 Color y servicio — 102.9 Cristal — 130) 103.7 Contemporánea — 111) 105.4 — 102) 106.3.

BRAZIL

L.T: Eastern (Coastal); UTC – 3h. Western: UTC – 4h. Acre: UTC – 5h (until Jan 28th, 1991, UTC – 2h, 3h and 5h respectively) — **Pop:** 161.654.000 — **Radio sets:** 60.000.000 — **Pr.L:** Portuguese — **E.C:** a/c 50-60, 110-127/220V — **ITU:** B.

DEPARTAMENTO NACIONAL DE TELECOMUNICAÇÕES (DENTEL)

Ministério das Comunicações, Esplanada dos Ministérios, 70340-000 Brasília, DF — **Cable:** Dentel.
L.P: Dir. Gen. Dr. Rubens Bussacos. Dir. of Radio: Roberto Blois Montes de Souza.

ASSOCIAÇÃO BRASILEIRA DE EMISSORAS DE RÁDIO E TELEVISÃO (ABERT)

ADDR: Hotel Nacional, Salas 5a8, 70340-000 Brasília, DF (C.P. 08780, 70340-970). **Te:** +55 (61) 224-4600. **FAX:** +55 (61) 321-7583. **Telex:** (061) 2001.
L.P: Joaquim Mendonça. Exec. Dir: Antônio Abelin.
N.B: all st's carry "A Voz do Brasil" Mon-Fri: 2200-2300 (official programme); 2300-2330 "Projeto Minerva" (educational) (Sat/Sun: 2230-2300).

STATIONS: Mediumwave: Call ZY----, ° = also shortwave. * = inactive.
The letters preceding the stn. number indicate the state or territory. Addresses are listed by state in alphabetical order. H. of tr. usually 0800-0300. Larger st's may operate 24h.

	Call	kHz	kW	Name and location
BA01)	H481	540	1/0.25	R. Regional, Irecê
CE01)	H610	540	1/0.25	R. Jornal, Canindé
GO01)	H755	540	10/1	R. Riviera, Goiânia
MA01)	H894	540	1/0.25	R. Guajajara, Barra do Corda
MG151)	L331	540	1/0.25	Sist. Ipanema de Comunicação, Ipanema
MG146)	L...	540	1/0.25	R. Meridional, Uberaba (CP)
MS37)	I...	540	10	R. Nacional, Corumbá (CP)
PI31)	I914	540	1/0.25	R. Primeiro de Julho, Agua Branca
PR110)	J322	540	1/0.25	R. Flor do Café Borrazópolis
RJ01)	J450	540	10/2.5	R.AM Fluminense, Niterói
RS01)	K226	540	1/0.5	R. Real, Canoas
RS02)	K322	540	10/1	R. Sepe Tiaraju, Santo Angelo
SC01)	J778	540	10/1	R. Mirador, Rio do Sul
SE01)	J924	540	20/2.5	R. Jornal de Sergipe, Aracaju
SP01)	K697	540	1/0.25	R. Uirapuru, Biriguí
SP02)	K734	540	1/0.25	R. Nova Sumaré, Sumaré
CE59)	H644	550	1/0.25	R. Em. Acopiara, Acopiara
MG01)	L225	550	5/1	R. Cataguases, Cataguases
MG02)	L263	550	5	R. Soc. Norte de Minas, Montes Claros
MT29)	I429	550	1	R. Cultura, Sinop
PE01)	I796	550	5/1	R. Meridional, Garanhuns
PI01)	I902	550	1/0.25	R. Serra da Capivara, São Raimundo Nonato
PI22)	I907	550	1/0.25	R. Igaracu, Parnaíba
PR139)	J331	550	5/0.5	R. Estação da Luz, Curitiba
RS03)	K287	550	2.5/0.25	R. Santa Cruz do Sul, Santa Cruz do Sul
SP03)	K578	550	5/0.5	R. Mantiqueira Globo, Cruzeiro
SP04)	K696	550	5/0.25	R. Educação e Cultura, Sertãozinho

BA02)	H456	560	5/1	R. Jornal, Itabuna
CE41)	H604	560	1/0.25	R. Educadora Jaguaribana, Limoeiro do Norte
MA02)	H887	560	25/5	R. Educadora do Maranhão, São Luís
MG05)	L277	560	5/0.25	R. Difusora, Patrocinio
MT01)	I395	°560	10/2.5	R. Araunã, Barra do Garças (F.PI.) (still on 1460)
MT24)	I419	560	1/0.25	R. Pioneira, Tangara da Serra
PB16)	I695	560	1/0.25	R. Potiguara, Mamanguape
PR01)	J214	560	1/0.5	R. Londrina, Londrina
PR02)	J281	560	5/1	R. Cultura, Guarapuava
RJ02)	J496	560	5/1	R. AM-560, Araruama
RS04)	K231	560	5/1	R. São Francisco, Caxias do Sul
SP213)	K761	560	25/10	R. Paulista AM, São Paulo
AL01)	H244	570	5/1	R. Novo Nordeste, Arapiraca
AM21)	H...	570	1	Empresa de Comunicação da Amazônia ("R. Tupinabarana"), Parintins (CP)
CE02)	H613	570	5/0.25	R. Vale do Cariri, Juazeiro do Norte
CE03)	H614	570	1/0.25	R. Uirapuru, Itapipoca
GO02)	H750	570	1/0.25	R. Cultura, Catalão
MA06)	H890	570	10/1	R. Imperatriz, Imperatriz
MA19)	H...	570	10/0.25	Sist. Timon de Radiodifusão, Timon
MG03)	L261	570	25/5	R. Capital de Minas, Belo Horizonte
MT30)	I...	570	1/0.25	R. Jornal, São José dos Quatro Marcos (CP)
PR146)	J349	570	1/0.5	R. Continental, Palotina
RS05)	K267	570	1/0.25	R. Passo Fundo, Passo Fundo
SC02)	J735	570	5/0.25	R. Eldorado AM, Criciúma
SC99)	J794	570	1/0.25	R. Fronteira Oeste, Dionísio Cerqueira
SP05)	K595	570	1/0.25	R. Clube, Itapeva
SP06)	K672	°570	5/1	R. Difusora, Taubaté
SP195)	K698	570	1/0.25	R. Jornal, Nhandeara
SP150)	K717	570	1/0.25	Bariri R. Clube, Bariri (F.PI.) (still on 1370)
AM01)	H290	580	10	R. Nacional, Tefé
BA03)	H477	580	1/0.25	R. Sobral, Teixeira de Freitas
MG04)	L328	580	5/1	R. Visão, Uberlândia
MS01)	I387	°580	10/1	R. Educação Rural, Campo Grande
PE02)	I776	580	20/10	R. Globo, Recife
PR03)	J330	580	1/0.25	R. Grande Lago, Santa Helena
RJ03)	J465	°580	20/5	R. Relógio Federal, Rio de Janeiro
RS06)	K299	580	2/0.25	R. São Gabriel, São Gabriel
RS07)	K318	580	10/5	R. Fátima, Vacaria
SP07)	K540	580	1/0.25	R. Jornal, Americana
SP08)	K724	580	1/0.25	R. Regional, Palmital
BA04)	H445	590	50/5	R. Cruzeiro da Bahia, Salvador
CE04)	H627	590	5/0.25	R. Vale do Rio Poty, Crateús
ES01)	I213	590	5/0.5	R. Tribuna, Vitória
GO03)	H798	590	10/1	R. Manchester, Anápolis
MT03)	I420	590	10/5	R. Globo Vila, Cuiabá
PB25)	I692	590	2/0.25	R. Serrana, Araruna
PR04)	J234	590	10/5	R. Difusora, Curitiba
RR01)	O700	590	10	Super R. Roraima, Boa Vista
RS08)	K210	590	5/0.5	R. Alegrete, Alegrete
SC03)	J809	590	2/1	R. Progresso, Descanso
SP09)	K534	590	10/1	R. Atlântica, Santos
SP11)	K643	°590	5/1	R. Ribeirão Preto, Ribeirão Preto
AM02)	H287	°600	10	R. Nacional, São Gabriel da Cachoeira
BA05)	H486	600	10/1	R. Vale do Rio Grande, Barreiras
BA64)	H...	600	1/0.25	R. Difusora, Rio Real (CP)
MA16)	H...	600	10/1	R. Litoral Maranhense, São Luís (CP)
PE03)	I789	600	1/0.25	R. Cardeal Arcoverde, Arcoverde
RS09)	K278	°600	100	R. Gaúcha, Porto Alegre
AL10)	H249	610	10/2	R. Cidade Imperial, Marechal Deodoro
AM18)	H...	610	10/2.5	R. Grande Rio, Iranduba
GO10)	H786	610	25/2	R. Jornal, Luziânia
MG06)	L268	°610	50/10	R. Itatiaia, Belo Horizonte
MT04)	I425	610	10/5	R. Celeste, Sinop
PB01)	I678	610	1/0.25	R. Progresso, Sousa
PI02)	I899	610	10/1	R. Poty, Teresina
SC04)	J746	610	5/0.5	Radiodifusao Indio Condá, Chapecó
SP12)	K532	610	1/0.25	R. Chamonix, Mogi Mirim
SP13)	K577	610	1/0.25	R. Em. A Voz de Catanduva, Catanduva
SP14)	K589	610	1/0.25	R. Piratininga, Guaratinguetá
SP15)	K726	610	1/0.25	R. Paranapanema, Piraju
TO01)	H785	610	1/0.25	R. Anhanguera, Porto Nacional
CE05)	H590	620	10	R. Assunção Cearense, Fortaleza
GO04)	H777	620	1/0.25	R. Cristã Educativa, Pires do Rio
MG125)	L320	620	10/0.25	R. Educadora, Porteirinha
MG147)	L348	620	1/0.25	R. Ibiá, Ibiá
MG145)	L357	620	5/0.25	R. Catuí, Manhuaçu (F.PI.) (still on 1310)
PR05)	J332	620	1/0.25	R. Cidade Jandaia, Jandaia do Sul
RS10)	K270	620	10/1	R. Pelotense, Pelotas
RS11)	K315	620	1/0.25	R. Municipal, Tenente Portela
SC05)	I779	620	5/0.25	R. Difusora Alto Vale, Rio do Sul
SP16)	K521	620	50/10	R. Jovem Pan, São Paulo
AP01)	H422	°630	10	R. Difusora, Macapá
MG07)	L299	630	1/0.5	R. Difusora, Uberaba
MS32)	N603	°630	10/1	R. IPB AM, Campo Grande
MT05)	I384	630	10/5	R. Difusora Bom Jesús, Cuiabá
PI03)	I904	630	1/0.25	R. Difusora da Barra, Barras
PR06)	J284	630	10/0.5	R. Estadual do Paraná, Curitiba
PR07)	J300	630	5/0.25	R. Educadora Marechal, Marechal Cândido Rondón
RJ04)	J466	630	25/10	R. Roquette Pinto, Rio de Janeiro
RS12)	K259	630	1/0.5	R. Cacique, Lagoa Vermelha
RS13)	K289	630	5/1	R. Santamariense, Santa Maria
SC06)	J800	630	1/0.25	R. Doze de Maio, São Lourenço d'Oeste
SE02)	J920	630	50/5	R. Aperipe, Aracaju
SP17)	K613	630	1/0.25	R. Difusora, Mirassol
SP18)	K635	630	5/0.25	R. Cidade Globo, Presidente Prudente
BA12)	H458	640	10/0.5	R. Difusora Sul da Bahia, Itabuna
ES02)	I204	°640	10/0.5	R. Vitória, Vitória
GO05)	H757	640	25/5	R. Difusora, Goiânia
MG08)	L308	640	.5/0.25	R. Santa Cruz, Pará de Minas
MT06)	I406	640	1	R. Progresso, Alta Floresta
MT18)	I424	640	10/1	R. Tangará, Tangará da Serra
PI28)	I924	640	1	R. Cruzeiro, Pedro II (F.PI.) (still on 1460)
PR08)	J262	640	5/1	R. Auri Verde, Londrina
RJ05)	J489	640	5/0.25	R. Agulhas Negras, Resende
RN01)	J590	640	10/5	R. Cabugi, Natal
RS14)	K277	640	50	R. Bandeirantes, Porto Alegre
SP19)	K547	640	5/1	R. A Morada do Sol, Araraquara
BA06)	H462	650	5/0.5	R. Clube, Valença
GO06)	H790	650	1/0.25	R. Cultural do Araguaia, Jussara
MA16)	H924	650	1/0.25	R. Macaru, Viana
MG09)	L200	650	5/0.5	R. Princesa, Lagoa Formosa
MG85)	L309	650	5/0.5	R. Veredas, Unai
MG142)	L...	650	10/1	R. Itatiaia, Timóteo
MT19)	I414	650	5	R. Educadora, Colider
PA20)	I540	650	10/1	R. Tropical, Santarém
PB02)	I672	650	1/0.25	R. Alto Piranhas, Cajazeiras
PI26)	I925	650	1/0.25	R. Tapuio, Miguel Alves
PR91)	J250	650	5/1	R. Colméia, Cascavel
RS15)	K238	650	5/0.5	Radiodifusão Sul Riograndense, Erechim
SP20)	K508	650	5/1	R. Andradina, Andradina
SP21)	K524	650	5/1	R. Difusora, Piracicaba
SP232)	K...	650	25/0.5	R. Difusora, São Paulo
BA07)	H465	660	5/0.25	R. Jornal, Itapetinga
BA36)	H480	660	10/ 0.25	R. Bom Jesus da Lapa, Bom Jesus da Lapa (F.PI.) (still on 1390)
BA65)	H...	660	1/0.25	R. Planalto, Euclides da Cunha (CP)
BA66)	H...	660	1/0.25	R. Tribuna do Vale do São Francisco, Xique-Xique (CP)
CE06)	H619	660	1/0.25	R. Rio das Garças, Itarema (Acaraú)
G007)	H778	660	5/0.25	R. Primavera, Itapuranga
G051)	H794	660	5/0.5	R. Alvorada, Quirinópolis
MG11)	L206	660	10/0.25	R. Clube, Curvelo
MT07)	I401	660	10/0.5	R. Amorim, Rondonópolis
PA21)	I...	660	1/0.25	R. Xinguara, Xinguara
PE04)	I787	660	5/1	R. Difusora, Limoeiro
PE05)	I795	660	1/0.25	R. da Grande Serra, Araripina
PI34)	I925	660	1/0.25	R. Tacarijus, São Miguel do Tapuio
PR09)	J202	660	1/0.5	R. Cultura, Cambará
RJ06)	J472	660	1/0.5	R. Sociedade de Friburgo, Nova Friburgo
R001)	J673	660	10/5	R. Eldorado, Porto Velho
RS16)	K286	660	1/0.25	R. Marajá, Rosário do Sul
RS17)	K319	660	10/0.25	R. Esmeralda, Vacaria
SP22)	K518	660	5/0.5	R. A Tribuna, Santos
SP23)	K639	°660	10/0.5	R. Clube, Ribeirão Preto

AM04)	H288	°670	10	R. Cabocla, Benjamin Constant
AM03)	H297	°670	1/0.25	R. Vale do Rio Madeira, Humaitá
CE07)	H606	670	1/0.25	R. Cultura, Várzea Alegre
G008)	H747	670	5/2.5	R. São Francisco, Anápolis
MG12)	L310	670	10/2.5	R. Educadora, Montes Claros
MG126)	L347	670	1/0.25	R. Visão, Ponte Nova
MG123)	L361	670	5/0.25	R. Uberaba, Uberaba
MG135)	L3..	670	1/0.25	R. Difusora, Bambuí
MS28)	N600	670	10	R. Fronteira OM, Ponta Porã
PA02)	I537	670	1/0.25	R. Rural, Altamira
PA22)	I539	670	1/0.25	R. Atalaia, Óbidos
PA27)	I546	670	1/0.25	R. Tropical AM, Paragominas
PI32)	I927	670	1/0.25	R. Livramento, José de Freitas
PR10)	J231	670	2.5/ 0.25	R. Soc. Nova Esperança, Nova Esperança
PR11)	J248	670	10/2	R. Cidade, Curitiba
RS18)	K296	670	2.5/ 0.25	R. Cultura Jaguarão, Santa Vitória do Palmar
RS19)	K370	670	10/0.5	R. Gazeta, Carazinho
SE03)	J921	670	10/5	R. Cultura de Sergipe, Aracaju
SP24)	K574	670	1/0.5	R. Oceânica, Caraguatatuba
SP25)	K585	670	1/0.25	R. Centro Oeste AM, Garça
SP26)	K598	670	1/0.5	R. Emissora Convenção, Itu
TO02)	H774	670	1/0.25	R. Independência do Tocantins, Paraíso do Tocantins
BA08)	H471	680	10/0.5	R. Clube, Sto. Antônio de Jesús
GO09)	H765	°680	1/0.25	R. Difusora, Jataí
MA03)	H885	°680	10/5	R. Difusora do Maranhão, São Luís
MG13)	L326	680	1/0.25	R. União, João Pinheiro
MS02)	I389	680	5/1	R. Cultura, Campo Grande
PB26)	I683	680	2.5/ 0.25	R. Integração do Brejo, Bananeiras
PE06)	I793	680	10/1	R. do Grande Rio, Petrolina
RJ07)	J452	°680	20/5	R. Copacabana, Rio de Janeiro
RS69)	K275	680	50	R. Farroupilha, Porto Alegre
SP27)	K576	680	1/0.25	R. Difusora, Catanduva
SP28)	K628	680	2/0.25	R. Piratininga, Piraju
BA09)	H453	690	5/0.25	R. Cultura, Ilhéus
CE08)	H587	690	25/10	R. Dragão do Mar, Fortaleza
MG14)	L228	690	10/3	R. Mineira, Belo Horizonte
MS03)	I402	690	5/0.25	R. Cultura, Naviraí
MT31)	I451	690	5/1	R. Parecis, Diamantino
PA03)	I532	°690	20/5	R. Clube do Pará, Belém
PR13)	J229	°690	5/0.25	R. Difusora, Londrina
PR14)	J252	690	1/0.25	R. Difusora, Ponta Grossa
RS21)	K252	690	5/0.25	R. Progresso, Ijuí
SC07)	J772	690	5/0.5	R. Clube, Lages
SP29)	K561	690	1/0.25	R. Bebedouro, Bebedouro
SP30)	K588	690	1/0.25	R. Clube, Guaratinguetá
SP31)	K625	690	1/0.25	R. Cidade Pereira Barreto, Pereira Barreto
SP220)	K646	690	1/0.25	R. Brasil, Santa Bárbara d'Oeste
BA10)	H500	700	50/1	R. Cultura AM, Feira de Santana
MG136)	L322	700	1/0.25	R. Mucuri, Teófilo Otoni
PI04)	I890	700	10/5	R. Clube, Teresina
RS22)	K356	700	1/0.25	R. Em. Batovi, São Gabriel
SP32)	K686	700	50	R. Nova Eldorado AM, São Paulo
AL02)	H240	710	5/1	R. AM 710, Maceió
BA46)	H490	710	10/0.25	R. Jacarandá, Sta. Cruz Cabrália
CE09)	H628	710	1/0.25	R. Difusora Asa Branca, Boa Viagem
DF07)	H...	710	10/2.5	R. Nova Aliança, Brasília (DF)
MA12)	H891	710	1	R. Verdes Campos, Pinheiro (F.Pl.) (still on 1450)
MA27)	H910	710	1/0.25	R. Verdes Vales, Grajaú
MG79)	L219	710	5/0.5	R. Cancella, Ituiutaba
MG15)	L258	710	5/0.25	R. Manhuaçu, Manhuaçu
MG16)	L333	710	1/0.25	R. Difusora, Pouso Alegre
MT08)	I386	°710	5/0.5	R. Cultura, Cuiabá
MT23)	I436	710	5/1	R. Nova Xavantina, Nova Xavantina (F.Pl.) (still on 1110)
PA04)	I534	°710	10/5	R. Rural, Santarém
PB03)	H685	710	5/0.25	R. Educadora, Conceição
PI19)	I901	710	1/0.25	R. Alvorada do Sertão, São João do Piauí
PI23)		710	1/0.25	R. Clube, Barras
RJ09)	J451	710	10/0.25	R. Difusora Carioca, Rio de Janeiro
SC08)	I793	710	1/0.25	R. Fraiburgo, Fraiburgo
SP33)	K559	710	10.0.25	R. Globo, Bauru
AC01)	H202	°720	10	R. Integração, Cruzeiro do Sul
AM05)	H281	720	1	R. Difusora, Itacoatiara
MG28)	L330	720	25/0.5	R. Divinópolis AM, Divinópolis
MS04)	I390	°720	5/1	R. Clube, Dourados
MT20)	I411	720	5/1	R. Difusora, Barra do Garças
PE07)	I770	720	100	R. Clube de Pernambuco, Recife
RS23)	K276	°720	50/100	R. Guaíba, Porto Alegre
SP34)	K575	720	1/0.25	R. Difusora, Casa Branca
SP35)	K701	720	1/0.25	R. Sentinela, Ourinhos
SP36)	K718	720	5/0.25	R. Cruzeiro, Cruzeiro
SP37)	K722	720	1/0.25	R. Menina, Olímpia
CE45)	H640	730	5/0.5	R. Sinal, Aracati
ES16)	I...	730	10/0.5	Sistema Canaã de Comunicação, Cariacica (CP)
G031)	H759	°730	50/5	R. Clube, Goiánia
MA04)	H896	730	1/0.25	R. Eldorado, Codó
MG17)	L287	730	5/0.25	R. Soc. Triângulo Mineiro, Uberaba
MG18)	L297	730	5/1	R. Nova Amizade AM, Juiz de Fora
MS38)	I452	730	1/0.25	R. Princesa do Vale, Camapuã
MT09)	I410	730	10/2.5	R. Jornal, Cáceres
PE08)	I780	°730	10/5	Em. Rural, A Voz do São Francisco, Petrolina
PR15)	J208	°730	5/0.25	R. Marumby, Curitiba
PR16)	J323	730	1/0.25	R. Humaitá, Campo Mourão
RS24)	K268	730	5/1	R. Planalto, Passo Fundo
SC09)	J787	730	1/0.25	R. Sociedade Tubá, Tubarão
SP38)	K523	730	5/0.25	R. Cidade, Jundiaí
SP39)	K610	730	10/1	R. Dirceu, Marília
AC02)	H203	740	20/10	R. Progresso do Acre, Rio Branco
BA11)	H446	740	100	R. Soc. da Bahia, Salvador
MT25)	N403	740	1/0.5	R. Cidade, Alto Araguaia
PR17)	J259	740	1/0.25	R. Goio-Erê, Goio-Erê
PR135)	J354	740	1/0.25	R. Placar, Ortigueira
RS25)	K265	°740	2.5/ 0.25	R. Palmeira, Palmeira das Missões
RS26)	K283	740	5/0.25	R. Cultura Riograndina, Rio Grande
SC10)	J753	740	5/1	R. Diário da Manhã, Florianópolis
SP93)	K519	740	5/0.25	R. Assunção, Jales
SP40)	K553	740	1/0.25	R. Cultura, Bariri
SP41)	K650	740	20/0.5	R. 740 AM, Santo André
DF01)	H709	750	50/25	R. Globo, Brasília
MG19)	L213	750	50/5	R. América, Belo Horizonte
PA28)	I...	750	1/0.25	R. Ximango, Alenquer (CP)
PB04)	I682	750	1/0.25	R. Panati, Patos
PI05)	I897	750	1/0.25	R. Heróis do Jenipapo, Campo Maior
PI35)	I...	750	1/0.25	R. Cantonense, Canto do Buruti (CP)
RS27)	K264	750	5/0.25	R. Osório, Osório
SC11)	J815	750	5/0.25	R. Aliança, Concórdia
SE04)	J927	750	10/0.25	R. Progresso, Lagarto
SP42)	K516	750	1/0.25	R. Clube Osvaldo Cruz
SP43)	K642	750	10/1	R. Renascença, Ribeirão Preto (nf 1380)
SP44)	K661	750	2.5/ 0.25	R. Piratininga, São José dos Campos
TO03)	H792	750	1/0.25	R. Tocantins, Tocantinópolis
AL12)	H...	760	1/0.25	R. Pioneira, Delmiro Gouveia (CP)
AP03)	H...	760	20/10	Rede Amanpaense de Radiodifusão, Macapá (CP)
CE10)	H588	760	10	R. Record, Fortaleza
GO11)	H783	760	10/1	R. Pousada do Rio Quente, Caldas Novas
MG83)	L257	760	1/0.25	R. Difusora, Machado (F.Pl.) (still on 1370)
MG137)	L360	760	10/0.5	R. Terra, Monte Claros
MT32)	I...	760	10/5	R. Atividade, Chapada dos Guimarães (CP)
PR12)	J...	760	1/0.25	R. Cacique, Guarapuava (CP)
RJ11)	J478	760	25/1	R. Manchete AM, Niterói
RS28)	K222	760	1/0.25	R. Jornal do Comércio, Candelária
RS29)	K351	760	2.5/ 0.25	R. Ametista, Planalto
SC12)	J742	760	1/0.25	R. Nereu Ramos, Blumenau
SP149)	K541	760	1/0.25	R. Urubupungá, Andradina (F.Pl.) (still on 1370)
SP45)	K560	760	10/0.5	R. Auri Verde, Bauru
BA51)	H491	770	5/0.5	R. Rio Corrente, Santa Maria da Vitória
CE11)	H609	770	10/0.25	R. Vale do Salgado, Lavras da Mangabeira
ES03)	I211	770	5/1	R. Difusora Princesa do Sul, Cachoeiro de Itapemirim
GO12)	H745	°770	5/1	R. Carajá, Anápolis

MA28)	H922	770	1/0.25	R. Jornal, Coelho Neto
MG20)	L209	770	1/0.25	R. Cultura d'Oeste, Lavras
MG21)	L302	770	10/0.5	R. Clube de Patos, Patos de Minas
MG108)	L315	770	5/0.5	R. Iturama, Iturama
MG22)	L337	770	5/0.25	R. Itabira, Itabira
MS11)	I412	770	10/5	R. Caiuás, Dourados (F.Pl.) (still on 1250)
MT21)	I428	770	1	R. Sorriso, Sorriso
MT28)	N404	770	1	E. Cidade de Matupá, Colider (Matupá) (F.Pl.) (still on 1380)
PA29)	I...	770	10/ 0.25	SNC—Sistema Norte de Comunicações ("R. Eldorado AM") Marabá (CP)
PR131)	J344	770	1/0.5	R. Metropolitana, Cambé
SE05)	J922	770	10/5	R. Atalaia de Sergipe, Aracaju
SP46)	K506	770	5/0.5	R. Jornal do Povo, Limeira
AM22)	H291	780	10	R. Nacional, Eirunepê
GO13)	H789	780	10/1	R. Soc. Vera Cruz, Goianésia
MA24)	H919	780	10/5	R. Alvorada, Zé Doca (F.Pl.) (still on 1480)
MG23)	L246	780	1	R. Educadora, Uberlândia
PE09)	I771	780	20/5	R. Jornal do Comércio, Recife
PR18)	J247	780	1/0.25	R. Porta Voz, Cianorte
PR19)	J305	780	5/0.25	R. Chopinzinho, Chopinzinho
RS30)	K229	780	5/2	R. Carazinho, Carazinho
RS31)	K279	780	2.5	R. Princesa, Porto Alegre
SC54)	J751	780	1/0.25	R. Brasil Novo, Jaraguá do Sul
SC13)	J788	780	1	R. Marconi, Urussanga
SP161)	K619	780	1/0.25	R. Difusora, Monte Aprazível
SP47)	K695	°780	50/10	R. CBN (Globo), São Paulo
BA13)	H484	790	10/1	R. Barreiras, Barreiras
BA14)	H505	790	1/0.25	R. Regional, Serrinha
CE12)	H629	790	1/0.25	R. Jornal Centro Sul, Iguatu
GO28)	H761	790	5/0.5	R. Xavantes, Ipameri
GO14)	H771	790	1/0.5	R. Eldorado, Mineiros
MA29)	H904	790	1/0.25	R. Pérola do Turi, ("R. Rio Turiaçu"), Santa Helena
MA20)	H915	790	1/0.25	R. Cultura, Açailândia
MA30)	H...	790	1/0.25	Sistema Clube de Comunicação, Tuntum (CP)
MG24)	L279	790	5/0.25	R. Soc. Ponte Nova, Ponte Nova
MG25)	L311	790	1/0.25	R. Treze de Junho, Mantena (nf 1190)
MG26)	L314	790	5/1	R. Tropical, Lagoa da Prata
MT22)	I456	790	1	R. Difusora, Nortelândia
PB05)	I679	790	2.5/1	R. Cultura, Guarabira
PI36)	I931	790	1/0.25	R. Mafrense, Simplício Mendes
PR20)	J337	790	1	R. Panorama, Mandirituba
RS32)	K285	790	1/0.25	R. Rio Pardo, Rio Pardo
SC14)	J789	790	1/0.25	R. Videira, Videira
SP48)	K538	790	1/0.25	R. Brasil, Adamantina (nf 1270)
SP49)	K546	°790	5/1	R. Cultura, Araraquara
DF02)	H705	800	10/1	R. MEC, Brasília
PI08)	I921	800	10	R. Antares, Teresina
RJ12)	J457	°800	100	R. MEC, Rio de Janeiro
RS33)	K292	800	10	R. Universidade, Santa Maria
CE13)	H589	810	50/5	R. Verdes Mares, Fortaleza
GO15)	H767	810	5/0.5	R. Alvorada, Rialma
MG27)	L202	810	1	R. Aimorés, Aimorés
MG76)	L266	810	1/0.25	R. Clube, Nepomuceno
MG138)	L354	810	1/0.25	R. Capinópolis, Capinópolis
MT33)	N402	810	5/0.25	Gaspar Radiodifusão, São José do Rio Claro
MT34)	I...	810	1/0.25	Magalhães Barros Radiodifusão, Alta Floresta (CP)
PR111)	J336	810	5/0.5	R. Esperança, Prudentópolis
SP50)	K604	810	5/0.25	R. Difusora Jundiaiense, Jundiaí
SP89)	K655	810	1/0.5	R. Universal, Santos
SP51)	K732	810	5/0.5	R. Centro-América, São José do Rio Preto
AC03)	H...	°820	1/0.25	R. Educadora 6 de Agosto, Xapuri
AM06)	H294	820	1/0.25	R. Princesa do Solimões, Manacapuru
CE14)	H624	820	1/0.25	R. União, Camocim
CE60)	H655	820	1/0.25	R. Sul Cearense, Brejo Santo
ES04)	I212	820	10/2.5	R. Gazeta AM, Vitória
GO16)	H752	820	10/5	R. Jornal de Goiás, Goiânia
MG29)	L255	820	5/0.25	R. Barbacena, Barbacena
MG30)	L291	820	1/0.25	R. Difusora Paraisense, São Sebastião do Paraíso
MT10)	I400°	°820	10/1	R. Difusora, Cáceres
PA05)	I543	820	1	R. Regional do Araguaia, Conceição do Araguaia
PE10)	I775	820	5/1	R. Universitária, Recife
PI06)	I912	820	5/0.25	R. Cacique Bruenque, Regeneração
PR21)	J238	°820	10/5	R. Cultura, Foz do Iguaçu
RJ13)	J477	820	5/0.25	R. Macaé, Macaé
RS34)	K241	820	5/1	R. Taquari, Estrela
SC15)	J738	820	10/5	R. União, Blumenau
SP52)	K542	°820	10/0.5	R. Aparecida, Aparecida
SP53)	K602	820	1/0.25	R. Jauense, Jaú
SP54)	K622	820	1/0.25	R. Clube, Ourinhos
SP55)	K624	820	1/0.25	R. Difusora, Penápolis
BA67)	H506	830	1/0.25	R. Extremo Sul da Bahia, Itamaraju
CE65)	H659	830	1/0.25	R. Pioneira, Forquilha
GO50)	H...	830	1/0.25	R. Princesa do Sul, Goiatuba
MA26)	H905	830	10/1	R. Mirante do Maranhão, Imperatriz
MA21)	H...	830	1/0.25	R. Boas Novas, Esperrantinópolis (CP)
MG86)	L207	830	5/0.25	R. Imbiara, Araxá
MG31)	L244	830	10/5	Cultura R. Sucesso, Belo Horizonte
MG99)	L265	830	1/0.25	R. Difusora, Nanuque
MT11)	N430	830	5/0.25	R. Xavantes, Jaciara
MT26)	N401	830	1/0.25	R. Educadora, Juina
PA24)	I556	830	1/0.25	R. Guaraní de Marajó, Soure
PI07)	I906	830	1/0.25	R. Primeira Capital, Oeiras
PI37)	I...	830	1/0.25	R. União, União (CP)
PR22)	J224	830	5/0.5	R. Iguaçu, Araucária
PR24)	J266	830	10/0.25	R. Tabajara, Londrina
PR23)	J311	830	1/0.25	R. Progresso, Clevelândia
RN02)	J595	830	1/0.5	Emiss. de Educação Rural, Caicó
RS35)	K332	830	5/0.25	R. Independente, Cruz Alta
SC16)	J773	830	2.5/ 0.25	R. Soc. Cruz da Malta
SE06)	J926	830	20/1	R. Princesa da Serra, Itabaiana
SP56)	K681	830	10/1	R. Clube, Votuporanga
SP227)	K746	830	1/0.25	R. dos Bons Tempos, Nova Odessa
AM07)	H298	840	1/0.25	R. Rio Madeira, Manicoré
BA16)	H447	840	25/5	R. Excelsior da Bahia, Salvador
CE51)	H648	840	1/0.25	R. Campo Maior, Quixeramobim
PI38)	I930	840	1/0.25	R. Ribeirão, Demerval Lobão
PI39)	I...	840	1/0.25	R. Vitória, Batalha (CP)
RO13)	J679	*840	50	R. Nacional, Porto Velho
RS36)	K248	840	10	R. Capital, Porto Alegre
SC17)	J750	840	10/1	R. Rural, Concórdia
SP57)	K687	°840	50	R. Bandeirantes, São Paulo
BA17)	H474	850	5/0.25	R. Caraibas, Senhor do Bonfim
CE15)	H599	850	1	R. Iracema de Juazerio, Juazeiro do Norte
GO17)	H776	850	10/5	R. Tropical, Porangatu
MA31)	H...	850	1/0.25	R. Vitória, VItória do Mearim (CP)
MG32)	L233	850	1/0.25	R. Difusora Formiguense, Formiga
MG33)	L254	°850	1/0.25	R. Por Um Mundo Melhor, Governador Valadares
MG34)	L295	850	1/0.25	R. Tupaciguara, Tupaciguara
MS30)	I438	850	1/0.25	R. Difusora Nor'estado, São Gabriel d'Oeste
MT02)	I416	850	10/2	R. Cultura, Poxoréo
PA17)	I538	850	10/1	R. Itacaíunas, Marabá
PA06)	I555	850	1/0.25	R. Tocantins, Cametá
PA18)	I557	850	5/1	R. Itaituba, Itaituba
PB22)	I693	850	5/1	R. Rural, Guarabira
PR50)	J254	850	5/0.25	R. Dif. Colméia, Campo Mourão (F.Pl.) (still on 1090)
RO03)	J675	850	5/1	R. Ariquemes, Ariquemes
SC20)	J808	850	1/0.25	R. Cidade, Brusque
SP59)	K563	850	1/0.25	R. Clube, Biriguí
SP58)	K644	850	1/0.25	R. Jornal, Rio Claro
CE16)	H592	860	25/10	R. Cidade, Maracanaú
RJ14)	J459	860	100	R. Mundial, Rio de Janeiro
RS37)	K288	860	10/0.25	R. Guarathan, Santa Maria
AL03)	H245	870	5/1	R. Educadora Sampaio, Palmeira dos Indios
AM19)	H...	870	1/0.25	R. Jutanópolis, Manacapuru
BA18)	H457	870	10/0.25	R. Clube, Itabuna
BA56)	H499	870	5/0.25	R. Rio Vale AM, Juazeiro
CE17)	H591	870	1/0.25	R. Iracema, Iguatu
CE66)	H658	870	1/0.25	R. Tabajara, São Benedito
GO18)	H749	870	5/0.5	R. Educadora, Uruaçu
GO32)	H754	870	20/0.5	R. Universitária, Goiânia (F.Pl.)

MA05)	H903	870	10/0.5	R. Mirante, Codó
MG78)	L304	870	10/0.5	R. Juriti, Paracatu
MG38)	L318	870	1/0.25	R. Cultura, Diamantina
MG128)	L349	870	5/0.25	R. Princesa da Mata, Muriaé
MG127)	L...	870	1/0.25	R. A Voz do São Francisco, Januária
NT35)	I...	870	1/0.25	R. Garça Branca, Guiratinga (CP)
PI40)	I...	870	1/0.25	R. Vale do Pajeú, Canto do Buriti (CP)
PR25)	J243	870	5/0.25	R. Metropolitana, Maringá
SC96)	J784	870	5/0.25	R. Difusora, São Francisco do Sul
SP60)	K620	870	1/0.25	R. Novo Horizonte, Novo Horizonte
SP61)	K705	870	5/1	R. Central, Campinas
T004)	H762	°870	1/0.25	R. Anhanguera, Araguaína
MG35)	L275	°880	100	R. Inconfidência, Belo Horizonte
BP18)	I680	880	1/0.25	R. Maringá, Pombal (F.Pl.) (still on 1490)
RS38)	K249	880	10/2.5	R. Itaí, Porto Alegre
RS87)	K317	880	2.5/ 0.25	R. São Miguel, Uruguaiana
RS20)	K...	880	1/0.25	R. Seberi, Seberi
CE46)	H642	890	1/0.25	R. Itatiaia, Santa Quitéria
DF03)	H706	890	50/10	R. Planalto, Brasília
MG36)	L250	890	10/0.25	R. Santa Cruz, Jequitinhonha
PA13)	I536	890	5/1	R. Ponta Negra, Santarém
PE11)	I772	890	20/10	R. Tamandaré, Recife
PR117)	J287	890	1/0.25	R. Ubá, Ivaiporã
PR26)	J338	890	5/0.25	R. Itapuã, Pato Branco
RJ10)	J499	890	10/0.5	R. Musical, Cantagalo (F.Pl.) (still on 1520)
RO16)	J685	890	10/2.5	R. Planalto, Vilhena (F.Pl.) (still on 1530)
RS39)	K215	890	5/0.25	R. Viva, Bento Gonçalves
RS40)	K295	890	1/0.5	R. Noroeste, Santa Rosa
SC18)	J755	890	1/0.25	R. Santa Catarina, Florianópolis
SP62)	K690	°890	25/10	R. Gazeta, São Paulo
BA19)	H488	900	5/1	R. Sisal, Conceição do Coité
GO41)	H768	900	10/1	R. Sudoeste AM, Rio Verde
MG124)	L338	900	1/0.25	R. Vinícola, Andradas
MG148)	L...	900	2.5/ 0.25	R. Carangola, Carangola (CP)
MT36)	I455	900	10/2.5	R. Difusora Arco Iris, Araputanga
PR27)	J272	900	1/0.25	R. Emiss. Sant'Ana, Ponta Grossa
PR28)	J295	900	5/0.25	R. União, Toledo
RJ15)	J454	900	50/10	R. Tamoio, Rio de Janeiro
RN03)	J591	900	20/10	R. Nordeste, Natal
RO04)	J672	900	5/1	R. Alvorada de Rondonia, Ji-Paraná
RS41)	K211	900	2.5/0.5	R. Aratiba, Aratiba
RS42)	K263	900	5/0.5	R. Progresso, Novo Hamburgo
RS164)	K303	900	1/0.25	R. Mun. Sampedrense, São Pedro do Sul
SP63)	K511	°900	0.5/ 0.25	R. Difusora, Presidente Prudente
SP64)	K664	900	10/0.25	R. Brasil Nôvo, São José do Rio Preto
SP65)	K742	900	1/0.25	R. Clube, Itapetininga
TO07)	H797	900	10/5	R. Cristal, Cristalândia
CE61)	H645	910	1/0.25	R. Assunção Cearense, Sobral
GO20)	H763	910	5/1	R. Paranaíba, Itumbiara
GO23)	H	910	1/0.25	R. Cidade, Jaraguá
MG37)	L292	910	1/0.25	R. Teófilo Otoni, Teófilo Otoni (nf 990)
MG132)	L346	910	1/0.25	R. Difusora Industrial, Nova Serrana
MG149)	L...	910	5/0.5	R. Jornal Cidade, Juiz de Fora (CP)
PE12)	I785	910	5/1	R. Liberdade, Caruaru
PI41)	I...	910	10/1	R. Chapada do Corisco, Teresina (CP)
PR29)	J207	910	1/0.25	R. Difusora, Apucarana
RS43)	K320	910	5/0.5	R. Venâncio Aires, Venâncio Aires
SC19)	J811	910	1/0.25	R. Difusora, Içara
SC90)	J824	910	1/0.25	R. Rainha das Quedas, Abelardo Luz
SP66)	K536	910	5/0.25	R. Alvorada, Piracicaba (ff 950)
SP228)	K763	910	1/0.25	R. Princesa, Monte Azul Paulista
BA57)	H...	920	10/2	R. Globo AM 920, Salvador
BA57)		910	10/2	R. Globo, Salvador
ES05)	I207	920	5/1	R. Cultura, Linhares
GO33)	H788	920	1/0.25	R. Vale da Serra, São Luís de Montes Belos (F.Pl.) (still on 1450)
MG39)	L271	920	5/0.5	R. Cultura, Visconde do Rio Branco
MG40)	L276	920	5/0.25	R. Mineira do Sul, Passa Quatro
PI09)	I895	920	1/0.25	R. Difusora, Picos
PI12)	I905	920	10/0.5	R. Itamaraty, Piripiri (F.Pl.) (still on 1330)
RN04)	J600	920	1/0.25	R. Currais Novos, Currais Novos
RS44)	K348	920	20/2	R. Tramandaí, Tramandaí
SP67)	K584	920	5/0.25	R. Franca do Imperador, Franca
SP69)	K652	920	2.5/ 0.25	R. Cultura, São Vicente
SP221)	K...	920	2.5/1	R. Nacional, Cotia
AM16)	H296	930	10/5	R. Ajuricaba, Manaus
CE18)	H605	930	1/0.25	R. Salamanca, arbalha
CE52)	H646	930	10/0.5	R. Metropolitana, Caucaia
GO45)	H802	930	10/1	R. Caraiba AM, Aparecida de Goiânia
MG41)	L220	930	1/0.5	R. Clube, Campo Belo
MG42)	L229	930	5	R. Araguari, Araguari
MG87)	L237	930	10/1	R. Ibituruna, Governador Valadares
MS05)	I454	930	10/ 0.25	Radiosul Emissoras Integradas ("AM Capital"), Campo Grande
MT12)	I423	°930	5/0.25	R. Clube, Rondonópolis
PA07)	I600	930	5/1	R. Rauland, Castanhal
PR30)	J227	930	1/0.25	R. Cultura, Rolândia
PR31)	J232	930	10/1	R. Cultura, Curitiba
PR69)	J235	930	10/1	R. Princesa, Francisco Beltrão
RO05)	J676	930	10/1	R. Soc. Rondónia, Cacoal (F.Pl.) (still on 1470)
RS45)	K230	930	20/2.5	R. Caxias, Caxias do Sul
RS46)	K298	930	10/0.25	Super R. Sto. Angelo, Santo Angelo
SE07)	J923	930	20/5	R. Liberdade de Sergipe
SP71)	K500	930	1/0.25	R. Dinâmica de Santa Fé, Santa Fé do Sul
SP68)	K503	930	1/0.25	R. Clube, Itapira
SP69)	K652	930	10/1	Super R. Cultura, Santos
SP70)	K713	930	5/1	R. Universal, Agudos
SP214)	K747	930	1/0.25	R. Jóia, Adamantina
AC04)	H204	940	5/1	R. Verdes Florestas, Cruzeiro do Sul
PI25)	I911	940	10/0.25	R. Sete Cidades, Piracuruca
RJ16)	J453	940	100	R. Jornal do Brasil, Rio de Janeiro
BA50)	H489	950	1/0.25	R. Bahia Noroeste, Paulo Afonso
CE19)	H593	950	5/1	R. Educadora do Nordeste, Sobral
MG43)	L212	950	10/5	R. Atalaia, Belo Horizonte
MG44)	L281	950	5/0.5	R. Jornal Sul de Minas, Bueno Brandão
MT17)	I439	950	1	R. Tucunare ("A Voz do Sertão"), Juara
PB17)	I681	950	1/0.25	R. Jornal, Sousa
PE13)	I782	950	1/0.25	R. Planalto, Carpina
PI20)	I915	950	10/0.25	R. São José dos Altos, Altos
PI42)	I923	950	1/0.25	R. Boa Esperança, Padre Marcos
PR114)	J239	950	5/0.25	R. Difusora Cultural, Irati (F.Pl.) (still on 1510)
RS47)	K260	950	5/0.25	R. Independente, Lajeado
SC21)	J736	950	1/0.25	R. Vale Tijucas, Tijucas
SP72)	K510	950	5/0.25	R. 950, Marília
SP66)	K536	950	5/0.5	R. Alvorada, Piracicaba (F.Pl.) (still on 910)
AL04)	H241	960	10	R. Difusora de Alagôas, Maceió
CE37)	H618	960	1/0.25	R. Cultura dos Inhamuns, Tauá
ES15)	I216	960	10/0.25	R. Aquidaban, Cachoeiro de Itapemirim (F.Pl.) (still on 1500)
GO21)	H764	960	5/0.25	R. Difusora, Itumbiara
PA26)	I551	960	5/1	R. Clube, Itaituba
PR32)	J257	960	1/0.25	R. Difusora, Maringá
RS48)	K291	960	10/1	R. Imembuí, Santa Maria
SC22)	J733	960	5/0.25	R. Guarujá, Orleãns
SC23)	J813	960	5/0.5	R. Difusora, Xanxerê
SP73)	K689	960	50/10	R. São Paulo, São Paulo
BA20)	H451	°970	10/5	R. Sociedade, Feira de Santana
CE20)	H612	970	5/0.25	R. Monólitos, Quixadá
MG45)	L243	970	2.5/ 0.25	R. Sociedade Caratinga, Caratinga
MG46)	L285	970	1/0.25	R. São João Del Rei, São João Del Rei
MG47)	L321	970	1/0.25	R. Central, Monte Alegre de Minas
MS18)	I399	970	5/1	R. Vale do Taquari, Coxim
PB06)	I684	970	1/0.25	R. Princesa Isabel, Princesa Isabel
PI46)	I910	970	10/0.5	R. Vale do Parnaíba, Luzilândia
PR33)	J260	°970	5/1	R. Alvorada, Londrina
PR34)	J277	970	5/0.25	R. Difusora Paraná, Marechal Cândido Rondón
RS49)	K201	°970	50/10	R. Pampa, Porto Alegre
RS50)	K349	970	1/0.5	R. do Povo, Humaitá
SC24)	J730	970	1/0.25	R. Araguaia, Brusque
SP74)	K505	970	2/0.25	R. Difusora, Itapetininga
SP75)	K529	970	1/0.25	R. Piratininga, São João da Boa Vista

SP76)	K684	970	1/0.25	R. Hertz, Franca
SP215)	K744	970	5/0.25	R. Alvorada, Estrela d'Oeste
DF04)	H707	°980	50/600	R. Nacional, Brasília
AM23)	H299	990	1/0.25	Rede Amazonense de Comunicação, Maués
BA21)	H483	990	1/0.25	R. Alvorada de Teixeira de Freitas, Caravelas
PI44)	I922	990	1/0.25	R. Vale do Canindé, Oeiras
RJ17)	J461	990	100/10	R. Ipanema, Rio de Janeiro
RN05)	J596	990	10/1	R. Rural, Mossoró
RS51)	K314	990	1/0.25	R. Tupanciretã, Tupanciretã
RS52)	K335	990	1/0.25	R. Sananduva, Sananduva
RS154)	K360	990	1/0.25	R. Clube, Pedro Osório
SC25)	J763	990	1/0.25	R. Peperi, Itapiranga
SP239)	K579	990	0.5/0.25	R. Cultura, Dois Corregos
PE14)	I791	1000	5/0.25	R. Princesa Serrana, Timbaúba
SP77)	K522	°1000	200	R. Record, São Paulo
BA22)	H448	1010	25/5	R. Bahia, Salvador
CE21)	H625	1010	50	R. O Povo, Fortaleza
MG50)	L230	1010	10/0.5	R. Educadora, Coronel Fabriciano
MG48)	L264	1010	5/0.5	R. Solar AM, Juiz de Fora
MG49)	L325	1010	1/0.25	R. Estância, Jacutinga
MT13)	I421	1010	5/0.25	R. Difusora, Mirassol d'Oeste
PR35)	J263	1010	25/5	R. Celinauta, Pato Branco
RS53)	K232	1010	1/0.5	R. 1010 Caxias do Sul, Caxias do Sul
RS54)	K344	1010	2.5/0.25	R. Missioneira Sete Povos, São Luís Gonzaga
SC71)	J764	1010	5/0.5	R. Jaraguá, Jaraguá do Sul (F.Pl.) (still on 1510)
SC26)	J807	1010	1/0.25	R. Verde Vale, Braço do Norte
SP78)	K556	1010	1/0.25	R. Independente, Barretos
SP79)	K611	1010	1/0.25	R. Tuiuti, Presidente Prudente
AL05)	H247	1020	25/1	R. Clube, Rio Largo
CE22)	H600	°1020	5/1	R. Educadora do Cariri, Crato
ES06)	I205	1020	20/10	R. Difusora, Colatina
GO52)	H781	1020	1/0.25	R. Maranata, Firminópolis
MG55)	L224	°1020	5/0.25	R. Congonhas, Congonhas
MG51)	L260	1020	10/0.25	R. Cultura, Uberlândia
MS06)	I381	1020	10/0.25	R. Independente, Aquidauana
PB19)	I686	1020	1/0.25	R. Cenecista, Picuí
PR36)	J244	1020	10/0.25	R. Colombo do Paraná, Curitiba
PR37)	J307	1020	1/0.25	R. Independência, Medianeira
RJ42)	J484	1020	5/0.25	R. A Noticia AM, Campos (F.Pl.) (still on 1510)
RO02)	J680	1020	5/1	R. Educadora, Rolim de Moura
RS49)	K202	1020	10/5	R. Caiçara, Porto Alegre
SC27)	J805	1020	1/0.25	R. Continental, Coronel Freitas
SP80)	K513	°1020	10/0.25	R. Canção Nova, Cachoeira Paulista
SP81)	K515	1020	5/0.25	R. Cultura, Assis
SP82)	K531	°1020	2.5/0.5	R. Educadora, Limeira
SP83)	K600	1020	5/0.25	R. Cultura, Jales
BA40)	H475	1030	10/1	R. Bahiana, Itaberaba (F.Pl.) (still on 1480)
GO22)	H746	1030	10/1	R. Imprensa, Anápolis
MA07)	H892	1030	10/1	R. Jaínara, Bacabal
MS07)	I396	1030	1/0.25	R. Cidade, Maracaju
PE15)	I777	1030	20/5	R. Olinda, Olinda
PR38)	J240	1030	1/0.25	R. Difusora, Cruzeiro do Oeste
PR39)	J271	1030	1/0.25	R. Atalaia, Londrina
PR40)	J329	1030	1/0.25	R. Difusora do Xisto, São Mateus do Sul
RJ18)	J467	°1030	50/5	R. Capital, Rio de Janeiro
RO10)	J...	1030	5/1	R. Colina de Machadinho, Ariquemes (CP)
RS55)	K253	1030	10/1	R. Repórter, Ijuí
SC28)	J771	1030	5/0.5	R. Princesa, Lages
SP84)	K525	1030	2.5/0.5	R. Difusora, Franca
SP85)	K554	1030	1/0.25	R. Emissora da Barra, Barra Bonita
SP86)	K606	1030	1/0.25	Lins Rádio Clube, Lins
SP87)	K537	1040	200	R. Capital, São Paulo
TO05)	H791	1030	5/0.25	R. Siqueira Campos, Colinas de Tocantins
BA47)	H494	1050	10/0.5	R. Metropolitana da Bahia, Camaçari
CE54)	H644	1050	5/0.5	R. Primeira Capital, Aquiraz
ES07)	I203	°1050	50/5	R. Capixaba, Vitória
MG52)	L236	1050	1/0.25	R. Rural, Tupaciguara
MS20)	I391	1050	10/0.5	R. Difusora Paranaibense, Paranaíba
PB07)	I676	1050	5/1	R. Caturité, Campina Grande
PI45)	I...	1050	1/0.25	Jet Radiodifusão, Teresina (CP)
PR66)	J226	1050	1/0.25	R. Dif. Platinense, Santo Antônio da Platina
RJ19)	J497	1050	10/0.5	R. Angra, Angra dos Reis
SP160)	K601	1050	5/0.25	R. Super Tupi, Jardinópolis
BA23)	H460	1060	5/1	R. Clube de Conquista, Vitória da Conquista
BA68)	H...	1060	2.5/0.25	R. Clube, Itapicuru (CP)
MG53)	L278	1060	25/5	R. Grande Belo Horizonte ("R. Grande Be Aga"), Pedro Leopoldo
MG54)	L306	1060	1/0.25	R. Itajubá, Itajubá
MS39)	N604	1060	5/0.5	R. Tupinambás, Dourados
PR42)	J246	1060	10/0.5	R. Paraná, Curitiba
PR43)	J298	1060	1/0.25	R. Colorado, Colorado
PR44)	J306	1060	10/0.25	R. Educadora, Francisco Beltrão
RJ20)	J495	1060	1/0.25	R. Tropical, Miguel Pereira
RN06)	J597	1060	5	R. Tapuyo, Mossoró
RS56)	K220	1060	5/0.25	R. Camaqüense, Camaquã
RS57)	K302	1060	2/0.25	R. São Luís, São Luís Gonzaga
RS81)	K307	1060	1/0.5	R. Cristal, Soledade
SP88)	K533	1060	5/0.5	R. Educadora, Piracicaba
SP229)	K765	1060	1/0.25	R. Universitária, Garça
BA48)	H492	1070	5/0.5	R. Rural, Ipiaú
CE47)	H643	1070	1/0.25	R. Cultura, Paracuru
MG56)	L316	1070	1/0.25	R. Rural, Muzambinho
MG150)	L355	1070	10/0.25	Super R. Patos, Patos de Minas
MT14)	I427	1070	10/2.5	R. Industrial, Várzea Grande
PB08)	I673	1070	20/2.5	Difusora R. Cajazeiras, Cajazeiras
PR45)	J203	1070	2/0.25	R. Difusora União, União da Vitória
PR46)	J319	1070	1/0.25	R. Guaraniaçu, Guaraniaçu
RJ21)	J483	1070	10/0.25	R. Cidade, Campos
RS58)	K218	1070	2/0.25	R. Caçapava, Caçapava do Sul
RS59)	K343	1070	1/0.25	R. Metrópole, Crissiumal
RS60)	K357	1070	1/0.25	R. Revista, Bento Gonçalves
SC91)	J747	1070	1/0.25	R. Gralha Azul, Urubici
SP91)	K603	1070	1/0.25	R. Piratininga, Jaú
SP145)	K615	1070	10/0.25	R. Metropolitana, Mogi das Cruzes
SP92)	K633	1070	5/1	R. Presidente Prudente, Pres. Prudente
SP212)	K758	1070	5/0.25	R. Jornal, Barretos
AP04)	H...	1080	1/0.25	R. Cultura de Quixadá, Mazagão (CP)
BA24)	H470	1080	20/0.5	R. Subaé AM, Feira de Santana
BA25)	H485	1080	1/0.25	R. Fascinação, Itapetinga
DF05)	H708	1080	10/5	R. Capital, Brasília
MG57)	L251	1080	5/0.5	R. Capital, Juiz de Fora
MT27)	I437	1080	1/0.25	R. Gaspar, Itiquira
PA09)	I530	1080	10/2.5	R. Guajará, Belém
PE16)	I784	1080	5/0.5	R. Difusora, Caruaru
PE33)	I824	1080	1/0.25	R. Voluntários da Pátria, Ouricuri
PR47)	J201	1080	1/0.5	R. Clube Pontagrossense, Ponta Grossa
PR48)	J245	1080	1/0.25	R. Cultura do Norte, Paranavaí
PR49)	J261	1080	1/0.25	R. Educadora, Cornélio Procópio
RS61)	K254	1080	1/0.25	R. Marabá, Iraí
RS62)	K280	1080	10	Radiodifusão da Univ. Federal, Porto Alegre
SC29)	J759	1080	1/0.25	R. Clube, Indaial
SP94)	K557	1080	5/1	R. Difusora, Batatais
SP95)	K607	1080	1/0.25	R. Alvorada, Lins
SP96)	K669	1080	5/0.5	R. Clube, Sorocaba
SP97)	K704	1080	1/0.25	R. Monumental, Aparecida
SP190)	K710	1080	1/0.25	R. Alvorada, Cardoso
BA26)	H455	1090	10/1	R. Santa Cruz, Ilhéus
GO24)	H758	1090	10/1	R. Universal, Goiânia
MA08)	H893	1090	10/1	R. Rio Balsas AM, Rio Balsas
PR50)	J254	1090	5/0.25	R. Difusora Colméia, Campo Mourão (ff 890)
PR51)	J283	1090	1/0.25	R. Vicente Pallotti, Coronel Vivida
RJ22)	J468	1090	25/5	R. Metropolitana, Rio de Janeiro
RN07)	J592	1090	10/5	R. Rural de Natal, Natal
RS63)	K216	1090	5/0.25	R. Cachoeira do Sul, Cachoeira do Sul
RS64)	K262	1090	1/0.25	R. Salette, Marcelino Ramos
RS65)	K341	1090	1/0.25	R. Giruá, Giruá

SC30)	J732	1090	1/0.25	R. Colón, Joinville
SC31)	J786	1090	1/0.25	R. Tabajara, Tubarão
SP98)	K609	°1090	1/0.5	R. Clube, Marília
SP99)	K618	1090	1/0.25	R. Cultura, Monte Alto
SP233)	K768	1090	1/0.25	R. Em. do Planalto, Paulinía
CE67)	H638	1100	1/0.25	R. Difusora dos Inhamuns, Tauá
RN18)	J607	1100	1/0.25	R. A Voz do Seridó, Caicó
SP100)	K694	°1100	150	R. Globo, São Paulo
CE32)	H620	1110	5/0.25	R. Litoral, Cascavel
GO46)	H782	1110	25/2	R. Redentor, Santo Antônio do Descoberto
MG58)	L205	1110	1/0.25	R. Planalto, Araguari
MG59)	L267	1110	1	R. Aurilândia, Nova Lima
MS08)	I392	1110	1	R. Ponta Porã, Ponta Porã
MT23)	I436	1110	5/1	R. Nova Xavantina, Nova Xavantina (ff 710)
PB09)	I786	1110	10	R. Tabajara, João Pessoa
PR52)	J241	1110	10/1	R. Paiquerê, Londrina
PR53)	J253	1110	5/0.5	R. Eldorado, São José dos Pinhais
PR151)	J356	1110	1/0.25	R. Clube, Ubiratã
RJ23)	J471	°1110	50/5	R. Cultura, Campos
RS66)	K257	1110	2.5/ 0.25	R. Cultura, Jaguarão
RS67)	K306	1110	1/0.25	R. Sobradinho, Sobradinho
RS68)	K325	1110	2.5/ 0.25	R. Cruzeiro do Sul, Itaqui
RS152)	K364	1110	1/0.25	R. Solaris, Antônio Prado
SC74)	J743	1110	1/0.25	R. Caçanjure, Caçador (F.Pl.) (still on 1520)
SC32)	J752	1110	1/0.5	R. Cultura, Florianópolis
SC33)	J812	1110	1/0.25	R. São Carlos, São Carlos
SP101)	K544	1110	2.5/0.5	R. Luz, Araçatuba
SP102)	K592	1110	1/0.25	R. Ibitinga, Ibitinga
SP103)	K617	1110	1/0.25	R. Cultura, Mogi Mirim
CE23)	H598	1120	1	R. Tupinambá, Sobral
ES14)	I215	1120	1	R. Cricaré-AM, São Mateus
MG10)	L272	1120	2.5/0.5	R. Ouro Preto, Ouro Preto
MG60)	L301	1120	2.5/ 0.25	R. Sete Colinas, Uberaba
MG61)	L332	1120	1/0.25	R. Serra da Boa Esperança, Boa Esperança
MS40)	I...	1120	25/1	Empresa de Radiodifusão Campograndense, Campo Grande (CP)
PB10)	I687	1120	1/0.25	R. Independência, Catolé do Rocha
PE17)	I778	1120	5/1	R. Paulista, Paulista
RS69)	K274	1120	50	R. 1120 AM, Porto Alegre
SP104)	K631	1120	1/0.25	R. Em. Portofelicense, Porto Feliz
SP105)	K660	1120	10/1	R. Clube AM, São José dos Campos
SP106)	K671	1120	1/0.25	R. Clube Imperial, Taquaritinga
GO25)	H769	1130	5/0.25	R. Emissora Sul Goiana, Quirinópolis
PA08)	I531	°1130	10	R. Marajoara, Belém
PE18)	I783	1130	5/1	R. Cultura do Nordeste, Caruaru
PR55)	J220	1130	1/0.25	R. Castro, Castro
PR54)	J333	1130	5/0.25	R. Ingamar, Marialva
RJ17)	J460	1130	100/50	R. Nacional, Rio de Janeiro
R006)	J677	1130	5/0.25	R. Ji-Paraná, Ji-Paraná
RS70)	K290	1130	5/1	R. Medianeira, Santa Maria
SC34)	J790	1130	5/1	R. Princesa d'Oeste, Xanxerê
SP107)	K676	1130	1/0.25	R. Tupã, Tupã
BA81)	H449	1140	10	R. Cultura da Bahia, Salvador
CE24)	H607	1140	1/0.25	R. Progresso, Russas
GO26)	H751	1140	1/0.25	R. Tapirapés, Formosa
GO47)	H801	1140	5/0.25	R. Pouso Alto, Piracanjuba (F.Pl.) (still on 1480)
MG62)	L204	1140	1/0.5	R. Minas, Divinópolis
MG63)	L248	1140	1/0.25	R. Difusora, Campanha
MG64)	L253	1140	1/0.5	R. Sociedade Muriaé, Muriaé
MG129)	L362	1140	5/0.5	R. Clube, Bocaiuva
MS36)	I418	1140	1/0.25	R. Cidade, Aparecida do Taboado
MS33)	I453	1140	10/0.5	R. Guaicurus, Fátima do Sul
RS71)	K228	1140	2.5/ 0.25	R. Cruz Alta, Cruz Alta
RS72)	K316	1140	5/0.5	R. Charrua, Uruguaiana
RS73)	K330	1140	1/0.25	R. Sobral, Butiá
SC35)	J748	1140	10/0.5	R. Coroado, Curitibanos
SP108)	K550	1140	10/0.5	R. Difusora, Assis
SP109)	K555	1140	2/0.5	R. Barretos, Barretos
SP110)	K645	1140	1/0.5	R. Educação e Cultura, Rio Claro
SP169)	K708	1140	1/0.25	R. Regional AM, Registro

SP111)	K709	1140	5/0.25	R. Costa Azul, Ubatuba
AL11)	H250	1150	20/1	R. Cultura, Arapiraca
MG65)	L283	1150	20/5	R. Globo de Minas, Belo Horizonte
PI10)	I891	°1150	10/5	R. Pioneira, Teresina
RJ24)	J456	1150	10/0.5	R. Três Rios, Três Rios
RN25)	J617	1150	1/0.25	R. Cabugi do Seridó, Jardim do Seridó
SP112)	K656	1150	100/50	R. Tupi, São Paulo
AM24)	H...	1160	1/0.25	R. Sociedade TV Manauara, Boca do Acre (CP)
CE25)	H634	1160	2/0.5	R. Guarany, Pacajus
CE62)	H652	1160	5/0.25	R. Vale do Coreaú, Granja
ES08)	I202	1160	20/10	R. Espírito Santo, Vitória
MG66)	L324	1160	1/0.25	R. Sacramento, Sacramento
MT15)	I3850	°1160	10/1	R.A. Voz d'Oeste, Cuiaba
PA30)	I...	1160	1/0.25	R. Guamá, São Miguel do Guamá (CP)
PB11)	I674	1160	1	R. Soc. Campina Grande, Campina Grande
PI46)	I...	1160	1/0.25	R. Cidade Modelo, Bocaina (CP)
PR56)	J258	1160	1/0.25	R. Norte, Londrina
RS74)	K242	1160	5/0.5	R. Miriam, Farroupilha
RS75)	K245	1160	5/0.5	R. Luz e Alegria, Frederico Westphalen
RS76)	K256	1160	1/0.5	R. Jaguari, Jaguari
RS77)	K273	1160	1	R. Universidade Católica, Pelotas
SC36)	J741	1160	1	R. Globo, Blumenau
SC37)	J767	1160	1/0.25	R. Eldorado, Laguna
SC38)	J776	1160	2.5/ 0.25	R. Difusora Colmeia, Porto União
SP113)	K502	1160	1/0.5	R. Presidente Venceslau, Presidente Venceslau
SP114)	K517	°1160	5/0.25	R. Cacique, Sorocaba
SP115)	K582	1160	1/0.25	R. Difusora, Fernandópolis
SP237)	K673	1160	1/0.25	R. Cacique, Taubaté
SP116)	K685	1160	1/0.25	R. Clube, Mococa
AM08)	H284	1170	5/2.5	R. Guaranópolis, Maués
AM25)	H...	1170	1/0.25	R. Difusora, Anori (CP)
AM26)	H...	1170	1/0.25	Juruá Comunicações, Carauari (CP)
BA27)	H473	1170	5/0.25	R. Jornal, Eunápolis
MG152)	L234	1170	1/0.25	R. Clube Fronteira, Fronteira (F.Pl.) (still on 1590)
MG75)	L327	1170	10/ 0.25	R. Vanguarda do Vale do Aço, Ipatinga (F.Pl.) (still on 1270)
MG67)	L336	1170	5/0.25	R. Cidade, Araxá
PR57)	J273	1170	20/10	R. Record, Curitiba
RN08)	J598	1170	10/1	R. Difusora, Mossoró
RS78)	K207	1170	1/0.25	R. Itapuí, Santo Antônio da Patrulha
RS79)	K213	1170	5/1	R. Difusora A Voz de Bagé, Bagé
RS80)	K359	1170	5/0.25	R. Uirapuru, Passo Fundo
RS155)	K380	1170	2/1	R. Pitangueira, Itaqui
SP117)	K569	1170	10/5	R. Educadora, Campinas
AL06)	H248	1180	1/0.25	R. Correio do Sertão, Santana do Ipanema
AM20)	H280	°1180	10/2.5	R. Difusora do Amazonas, Manaus
MA09)	H889	°1180	10/5	R. Ribamar, São Luís
MS34)	I602	1180	25/10	R. Guanandi AM, Campo Grande
MT38)	I...	1180	5/0.25	Sociedade R. Educadora Nova Xavantina, Guarantã do Norte (CP)
PB20)	I690	1180	1/0.25	R. Bonsucesso, Pombal
PE19)	I797	1180	1/0.25	Radio Cultural de Vitória, Vitória de Santo Antão
PR58)	J237	1180	5/0.25	R. Guaçu, Toledo
RJ25)	J463	1180	50/10	R. CBN (Globo), Rio de Janeiro
RS128)	K340	1180	10/0.5	R. Gazeta, Santa Cruz do Sul
SC39)	J737	1180	1/0.25	R. Integração d'Oeste, São José do Cedro
SC72)	J770	1180	1/0.5	R. Difusora, Lages
SP179)	K647	1180	1/0.5	R. Difusora, Santa Cruz do Rio Pardo
SP217)	K749	1180	5/0.25	R. Nova Bebedouro, Bebedouro
BA28)	H459	1190	10/1	R. Juazeiro, Juazeiro
CE68)	H663	1190	1/0.25	R. Guaraciaba, Guaraciaba do Norte
GO35)	H800	1190	1/0.25	R. Rio Vermelho, Silvânia
MG68)	L221	1190	10/1	R. Guarani, Belo Horizonte
PR59)	J309	1190	1/0.25	R. Pontal, Nova Londrina
RN09)	J594	1190	10/1	R. Tropical, Natal
RS82)	K234	1190	5/0.5	R. Soc. Cerro Azul, Cerro Largo
RS102)	K301	1190	2.5/ 0.25	R. São Lourenço, São Lourenço do Sul
RS83)	K354	1190	1	R. Rosário, Serafina Corrêa

SC40)	J783	1190	1/0.25	R. Clube, São João Batista
SC87)	J817	1190	1/0.25	R. Planalto, Major Vieira
SC41)	J820	1190	1/0.25	R. Clube, São Domingos
SP199)	K512	1190	2.5/0.25	R. Clube Marconi, Paraguaçu Paulista
SP118)	K700	1190	5/0.25	R. Cidade AM, Votuporanga
SP119)	K729	1190	10/1	R. 31 de Março, Santa Crus das Palmeiras
SP120)	K741	1190	1/0.25	R. Regional, Taquarituba
BA29)	H482	1200	1/0.25	R. Clube Rio do Ouro, Jacobina
CE26)	H585	°1200	10	Ceará R. Clube, Fortaleza
R007)	J	1200	5/1	R. Tucumã, Ji-Paraná
RS84)	K239	1200	5/1	R. Erechim, Erechim
RS85)	K342	1200	1/0.5	R. Cotrisel, São Sepé
SP121)	K520	°1200	50/25	R. Cultura, São Paulo
BA30)	H452	1210	5/1	R. Carioca, Feira de Santana
BA58)	H498	1210	10/0.25	R. Bandeirantes, Vitória da Conquista
CE48)	H641	1210	5/0.25	R. Boa Esperança, Barro
CE55)	H657	1210	1/0.25	R. Difusora, Nova Russas
DF08)	H...	1210	50/2.5	R. Brasília, Brasília (CP)
ES09)	I200	1210	25/1	R. Cachoeiro, Cachoeiro de Itapemirim
MG69)	L238	°1210	1/0.5	R. Clube, Varginha
MG70)	L305	1210	1/0.5	R. Difusora, Uberlândia
PE20)	I786	1210	5/1	R. Difusora, Garanhuns
PI11)	I893	1210	10/1	R. Educadora, Parnaíba
PR60)	J219	°1210	10/5	R. Universo, Curitiba
PR140)	J325	1210	2.5/1	R. Brotense, Porecatu
RS86)	K240	1210	10/5	R. Record AM, Porto Alegre
RS88)	K353	1210	1/0.5	R. Blau Nunes, Santa Bárbara do Sul
SC42)	J785	1210	1/0.25	R. Santa Catarina, Tubarão
SP122)	K509	1210	1/0.25	R. Nova Vida, Jaboticabal
SP123)	K545	1210	5/0.25	R. Difusora, Araçatuba
SP124)	K558	1210	1/0.25	R. Bandeirantes, Bauru
SP125)	K668	1210	5/0.25	R. Vanguarda, Sorocaba
RJ25)	J458	°1220	150	R. Globo, Rio de Janeiro
BA59)	H...	1230	1/0.25	R. Jornal, Ubatã
GO27)	H756	°1230	10/2.5	R. Anhanguera, Goiânia
MA23)	H...	1230	1/0.25	R. Alecrim, Caxias
MG105)	L208	1230	1/0.25	R. Correio da Serra, Barbacena
MG102)	L216	1230	2.5/0.25	R. Sociedade, Passos
MG71)	L296	1230	5/0.25	R. Educadora Rio Doce, Governador Valadares
PB12)	I670	1230	10/1	Super R. Correio, João Pessoa
PR92)	J225	1230	1/0.25	R. Soc. Monte Alegre, Telêmaco Borba
RS146)	K297	1230	1/0.25	R. Santiago, Santiago
RS89)	K326	1230	1/0.25	R. Nonoai, Nonoai
RS90)	K333	1230	1/0.25	R. Prata, Nova Prata
RS91)	K352	1230	1/0.25	R. Encruzilhadense, Encruzilhada do Sul
SC88)	J816	1230	10/1	R. Guararema, São José
SP126)	K573	1230	1/0.25	R. Cidade, Capão Bonito
SP127)	K703	1230	2.5/0.25	R. Noticias, Matão
SP128)	K716	1230	5/0.5	R. Jequitibá AM, Campinas
BA31)	H463	1240	10/0.5	R. Emiss. de Alagoinhas, Alagoinhas
CE49)	H654	1240	1/0.25	R. São Francisco, Canindé
MG97)	L294	1240	5/0.25	R. Três Pontas, Três Pontas
MG84)	L298	1240	1/0.25	Soc. R. Ubaense, Ubá
MG72)	L303	1240	1/0.25	R. Platina, Ituiutaba
MG116)	L317	1240	2.5/0.25	R. Pirapora, Pirapora
MG153)	L367	1240	5/0.25	R. Metropolitana, Vespasiano (F.Pl.) (still on 1560)
MG154)	L...	1240	1/0.25	R. Clube, Inhapim (CP)
MS09)	I388	1240	25/10	R. Difusora, Campo Grande
PE21)	I774	1240	5	R. Capibaribe, Recife
PR112)	J215	1240	5/0.25	R. Arapongas, Arapongas
PR61)	J280	1240	2/0.25	R. Matelândia, Matelândia
PR94)	J301	1240	1/0.25	R. Difusora Ubiratanense, Ubiratã (F.Pl.) (still on 1450)
R008)	J...	1240	1	R. Verde, Jaru (CP)
RS92)	K200	1240	1/0.25	R. Aparados da Serra, Bom Jesus
RS93)	K251	1240	5/0.25	R. Ibirubá, Ibirubá
RS94)	K355	1240	1/0.25	R. São Jerônimo AM, São Jerônimo
SC43)	J774	1240	5/0.25	R. Em. São José, Mafra
SC44)	J810	1240	1/0.25	R. Iracema, Cunha Porã
SP129)	K565	1240	2.5/0.25	R. Municipalista, Botucatu
SP130)	K621	1240	5/0.25	Orlândia R. Clube, Orlândia
SP131)	K653	1240	10/2.5	R. Clube, Santos
SP132)	K711	1240	1/0.25	R. Vale do Rio Tietê, José Bonifácio
AM09)	H289	°1250	1/0.25	R. Educação Rural, Coari
CE27)	H594	1250	1	R. Educadora, Crateús
CE69)	H...	1250	1/0.25	R. Liberdade, Itarema (CP)
GO29)	H748	1250	1/0.25	R. Difusora São Patricio, Ceres
MG73)	L282	°1250	5	R. Difusora, Poços de Caldas
MS10)	I394	1250	1/0.25	R. Difusora, Três Lagoas
MS11)	I412	1250	5/0.25	R. Caiuás, Dourados (ff 770)
PB27)	I...	1250	1/0.25	R. Sociedade de Soledade, Soledade (CP)
PI47)	I932	1250	1/0.25	R. João de Paivo, Altos
PR62)	J211	1250	5/0.5	R. Difusora, Guarapuava
PR63)	J233	1250	1/0.25	R. Paranavaí, Paranavaí
PR64)	J313	1250	1/0.25	R. Danúbio Azul, Santa Isabel do Oeste
RJ50)	J500	1250	5/0.5	R. Litoral, Casimiro de Abreu
RS95)	K233	1250	10/0.5	R. Difusora Caxiense, Caxias do Sul
RS96)	K272	1250	1	R. Tupanci, Pelotas
RS142)	K361	1250	1/0.25	R. Aguas Claras, Catuípe
SC45)	J766	1250	5/0.25	R. Cultura, Joinville
SE08)	J925	1250	10/1	R. Esperança, Estância
SP133)	K702	1250	5/0.5	R. Universal, Caçapava
AL07)	H242	1260	50/5	R. Gazeta de Alagoas, Maceió
CE28)	H596	1260	1/0.25	R. Vale do Jaguaribe, Limoeiro do Norte
RO09)	J670	°1260	5	R. Educadora, Guajará Mirim
RS97)	K204	1260	5/0.25	R. Cultura, São Borja
RS98)	K327	1260	5/0.25	R. Em. Fandango, Cachoeira do Sul
RS99)	K345	1260	1/0.25	R. Gaurama, Gaurama
SC46)	J740	1260	5/0.5	R. Jornal de Santa Catarina, Blumenau
SP134)	K688	1260	50/25	R. Morada do Sol, São Paulo
AM10)	H282	°1270	2.5	R. Educação Rural, Tefé
GO30)	H753	°1270	50/10	R. Brasil Central, Goiânia
MG74)	L227	1270	5/0.25	R. Carijós, Conselheiro Lafaiete
MG107)	L300	1270	5/0.25	R. Estância, São Lourenĉo
MG75)	L327	1270	5/1	R. Vanguarda do Vale do Aço, Ipatinga (ff 1170)
MG155)	L...	1270	5/1	R. Libertas do Vale do Aço, Ipatinga (CP)
PB28)	I...	1270	1/0.25	R. Cidade, Sumé (CP)
PR65)	J222	1270	5/0.5	R. Atual Guairaçá, Mandaguari
PR67)	J236	1270	5/0.5	R. Capital, Curitiba
PR68)	J289	1270	5/0.5	R. Cidade, Cascavel
RJ26)	J474	1270	10/0.5	R. Continental, Campos
RN10)	J593	°1270	10/1	R. Poti, Natal
RS131)	K206	1270	1/0.5	R. América do Rio Grande do Sul, Montenegro
RS100)	K244	1270	1/0.5	R. São Roque, Faxinal do Soturno (ff 1480)
RS101)	K250	1270	5/0.5	R. Vera Cruz, Horizontina
SC47)	J765	1270	5/0.25	R. Soc. Catarinense, Joaçaba
SC48)	J768	1270	1/0.25	R. Garibaldi, Laguna
SP135)	K640	1270	2/0.5	R. Tropical, Ribeirão Preto
SP136)	K678	°1270	5/1	R. Brasil 5000, Campinas
PB21)	I688	1280	10/1	R. Sinhauá, Bayeux
RJ27)	J455	1280	100	Super R. Tupi, Rio de Janeiro
AM11)	H286	°1290	10/2.5	R. Rio Mar, Manaus
BA32)	H450	1290	10/1	R. Clube, Salvador
MA10)	H888	°1290	50/5	R. Timbira do Maranhão, São Luís
MG77)	L273	1290	10/5	R. Record, Uberlândia
PR73)	J310	1290	10/0.5	R. Brasil Sul, Londrina
RN26)	J...	1290	1/0.25	Sistema Potiguar de Informação, Caicó (CP)
RS103)	K331	1290	2.5/2	R. Planetário, Espumoso
SC49)	J804	1290	5/0.5	R. Camboriú, Balneário Camboriú
SP240)	K662	1290	1/0.25	R. Difusora, São José do Rio Pardo
SP137)	K663	1290	5/1	R. Independência, São José do Rio Preto
SP216)	K745	1290	5/0.5	R. Metropolitana, São José dos Campos
CE29)	H586	1300	10	R. Iracema, Fortaleza
ES11)	I210	1300	1	R. Novo Tempo — Cidade, Afonso Cláudio
MG143)	L339	1300	5/1	R. Eldorado, Sete Lagoas
PE22)	I799	1300	1/0.25	R. Guarany AM Stereo, Camarajibe
PR71)	J278	1300	1/0.25	R. Pitangui AM, Ponta Grossa
PR127)	J288	1300	5/0.25	R. Educadora, Dois Vizinhos

RS104)	K203	1300	10	Super R. Tupi, Porto Alegre
RS105)	K337	1300	1/0.25	R. Regional, Santo Cristo
RS106)	K347	1300	5/0.5	R. Marathan, Santana do Livramento
SC89)	J819	1300	1/0.25	R. Alvorada, Santa Cecília
SP138)	K535	1300	10/5	R. Diário do Grande ABC, São Bernardo do Campo
BA33)	H454	1310	1/0.25	R. Bahiana, Ilhéus
BA42)	H476	1310	5/1	R. Educadora Santana de Caetité, Caetité
BA63)	H501	1310	5/0.25	R. Canto da Sereia, Jacobina (F.Pl.) (still on 1590)
CE63)	H656	1310	1/0.25	R. Liberdade, Boa Viagem
CE30)	H602	1310	1	R. Progresso de Juazeiro, Juazeiro do Norte
MG145)	L357	1310	1/0.25	R. Catuí, Manhuaçu (ff 620)
MG144)	L359	1310	1/0.25	R. Montanhesa, Vazante
MS31)	I426	1310	5/0.25	R. Panorama, Sidrolândia (F.Pl.) (still on 1370)
PB23)	I691	1310	10/0.5	R. Cidade Esperança, Esperança
PR70)	J274	1310	10/0.5	R. Atalaia, Maringá
RJ28)	J504	1310	1/0.25	R. Difusora Coroados, São Fidélis
RO17)	J...	1310	10/1	Rede Sanmori de R. e TV ("R. Tropical AM"), Porto Velho (CP)
RS107)	K305	1310	5/1	R. Sarandi, Sarandi
RS124)	K329	1310	2.5/0.25	R. Integracâo, Restinga Seca
RS160)	K371	1310	2.5/0.5	R. Horizonte, Capão da Canoa
SC85)	J801	1310	10/0.25	R. Ituporanga, Ituporanga
SP141)	K566	1310	2.5/0.25	R. Bragança, Bragança Paulista
SP139)	K596	1310	2/1	R. Difusora, Itápolis
AL08)	H243	1320	1/0.25	R. Progresso de Alagoas, Maceió
BA69)	H503	1320	5/0.5	R. AM Cicero Dantas, Cicero Dantas
CE31)	H597	1320	1	R. Cidade, Sobral
CE70)	H...	1320	1/0.25	R. Jaguaribana, Aracati (CP)
MG80)	L319	1320	5/0.25	R. Planeta, Carmo do Paranaíba
PE31)		1320	1/0.25	R. Cultura, São José do Egito
PR72)	J255	1320	5/1	R. Porto Alegre AM, Curitiba
RJ29)	J475	1320	25/5	R. Difusora Boas Novas, Rio de Janeiro
RS108)	K223	1320	5/0.25	R. Clube de Canela, Canela
RS109)	K266	1320	10/0.5	R. Panambi, Panambi
RS110)	K271	1320	5/1	R. Cultura, Pelotas
SP140)	K630	1320	1/0.25	R. Difusora, Pirassununga
SP241)	K675	1320	1/0.25	R. Clube, Tupã
BA34)	H468	1330	5/1	R. Difusora, Serrinha
PA10)	I533	°1330	25/5	R. Liberal, Belém (vf)
PI12)	I905	1330	1/0.25	R. Itamaraty, Piripiri (ff 920)
PR74)	J264	1330	10/0.5	R. Jaguariaíva, Jaguariaíva
PR75)	J320	1330	5/0.25	R. Educadora Inconfidência, Umuarama
RN27)	J...	1330	10/0.5	R. Eldorado, Natal (CP)
RS111)	K236	1330	1/0.25	R. Upacaraí, Dom Pedrito
RS112)	K323	1330	1/0.5	R. Diplomata, São Marcos
SC50)	J739	1330	5/0.5	R. Clube, Blumenau
SC51)	J749	1330	5/1	R. Chapecó, Chapecó
SP142)	K638	1330	1/0.25	R. Regente, Regente Feijó
SP143)	K641	1330	1	R. Cultura, Ribeirão Preto
SP187)	K736	°1330	10/1	R. Iguatemi, Osasco
CE71)	H661	1340	1/0.25	R. Planalto, Maracanaú
MA11)	H886	1340	50/1	R. São Luís, São Luís
MG81)	L241	1340	5/0.25	R. Mundo, Itabirito
MG139)	L352	1340	1/0.25	R. Minas Liberdade, Passos
MG156)	L...	1340	1/0.25	R. Capital do Triângulo, Patrocínio (CP)
MS12)	I380	°1340	1/0.25	R. Nova Difusora, Aquidauana
PB13)	I671	1340	5/1	R. Arapuan, João Pessoa
PR76)	J205	1340	2.5/0.25	R. Rio Negro, Rio Negro
PR41)	J...	1340	1/0.25	R. Cascavel, Cascavel (CP)
PR77)	J249	1340	1/0.25	R. Cultura, Arapongas
RJ40)	J490	1340	10/0.25	R. Jornal Contémporânea, Rio Bonito
RS113)	K227	1340	10/1	R. Educadora, Porto Alegre
SP144)	K543	1340	5/0.5	R. Cultura, Araçatuba
SP146)	K738	1340	1/0.25	R. Canoa Grande, Igaraçu do Tietê
AC05)	H201	1350	5	R. Capital, Rio Branco
AP02)	H420	°1350	5	R. Equatorial, Macapá
BA70)	H...	1350	50/2.5	R. Cristal, Salvador (CP)
CE56)	H662	1350	1/0.25	R. Liberal, Morada Nova
MG82)	L214	1350	10/5	R. Cultura, Poços de Caldas
PB14)	I675	°1350	0.5	Super R. Borborema, Campina Grande
RS114)	K205	1350	1/0.25	R. Aurora, Guaporé
RS115)	K313	1350	1	R. Difusora, Três Passos
RS116)	K336	1350	1/0.25	R. Agudo, Agudo
SC81)	J734	1350	1/0.25	R. Araranguá, Araranguá
SC52)	J745	1350	1/0.25	R. Clube, Canoinhas
SC53)	J760	1350	1/0.25	R. Clube, Itajaí
BA35)	H469	1360	10/1	R. Cultura, Paulo Afonso
CE57)	H650	1360	5/0.25	R. Iracema, Ipu
MS14)	I383	1360	2	R. Difusora Matogrossense, Corumbá
PR12)	J265	1360	10/0.25	R. Cidade, Pato Branco
PR78)	J268	1360	1/0.25	R. Pepita de Ouro, Assaí
PR79)	J302	1360	5/0.25	R. Cultura, Iporã
RJ30)	J464	1360	50/10	R. Guanabara, Rio de Janeiro
RN11)	J605	1360	1/0.25	R. Ouro Branco, Currais Novos
RS117)	K261	1360	2/0.5	R. Alvorada, Marau
SP147)	K581	1360	5/1	R. Cultura, Fernandópolis
SP148)	K739	1360	1/0.25	R. Regional, Dracena
SP235)	K759	1360	1/0.25	R. Luzes da Riberalta, Santa Bárbara d'Oeste
CE81)	H...	1370	1/0.25	R. Vanguarda, Caridade (CP)
MG83)	L257	1370	1/0.25	R. Difusora, Machado (ff 760)
MS31)	I426	1370	1/0.25	R. Panorama, Sidrolândia (ff 1310)
PE34)	I800	1370	1/0.25	R. Vale do Capibaribe, Santa Cruz do Capibaribe
PI13)	I892	1370	10/1	R. Difusora, Teresina
PR80)	J267	1370	50/2	R. Independência, Curitiba
RN28)	J...	1370	1/0.25	Sistema Potiguar de Informação, São Miguel (CP)
RS118)	K243	1370	10/0.25	R. Regional do Vêneto, Flores da Cunha
RS119)	K334	1370	1/0.25	R. Gazeta, Alegrete
SC55)	J782	1370	5/0.25	R. Peperí, São Miguel do Oeste
SE09)	J929	1370	1/0.25	R. Capital do Agreste, Itabaiana
SP149)	K541	1370	1/0.25	R. Urubupungá, Andradina (ff 760)
SP150)	K717	1370	1/0.25	Bariri R. Clube, Bariri (ff 570)
SP223)	K766	1370	100/20	R. Atual AM, São Paulo
AM12)	H283	°1380	5/1	R. Alvorada, Parintins
MG130)	L323	1380	1/0.25	R. Gorutubana AM, Janaúba
MT28)	N404	1380	1/0.25	R. Cidade de Matupá, Colider (Matupá) (ff 770)
PA11)	I547	1380	1/0.25	R. Marajó, Breves
PA31)	I...	1380	1/0.25	R. Tocantins, Tucuruí (CP)
PE23)	I773	1380	10/5	R. Continental, Recife
PR122)	J276	1380	2/0.25	R. Bom Jesus, Siquiera Campos
PR81)	J314	1380	1/0.25	R. Educadora, São José do Ivaí
PR153)	J...	1380	1/0.25	Rede Integração de Comunicação, Toledo (CP)
RJ31)	J470	1380	10/1	Campos R. Difusora, Campos
RS120)	K293	1380	1/0.25	R. Cultura, Santana do Livramento
RS134)	K311	1380	1/0.25	R. Maristela, Torres (F.Pl.) (still on 1460)
RS121)	K350	1380	1/0.25	R. Gazeta, Tapera
RS165)	K372	1380	1/0.25	Chiru Comunicações, Palmitinho
SC56)	J821	1380	1/0.25	R. Cidade, Itaiópolis
SC93)	J...	1380	1/0.25	R. Barriga Verde, Capinzal (CP)
SP151)	K507	1380	1/0.25	R. Difusora, Lençóis Paulista
SP152)	K616	1380	1/0.5	R. Difusora, Mogi Guaçu
SP224)	K751	1380	5/0.25	R. Diário, Presidente Prudente
SP234)	K...	1380	1/0.25	R. República, Morro Agudo
BA36)	H480	1390	10/0.25	R. Bom Jesus da Lapa, Bom Jesus da Lapa (ff 660)
CE72)	H...	1390	1/0.25	R. Paty, Senador Pompeu (CP)
ES10)	I201	1390	10/0.5	R. Cariacica AM, Cariacica
MA22)	H907	1390	10/0.5	R. Cultura do Rio Jordão, Coroatá
MG157)	L358	1390	2.5/0.25	R. Ouro Verde, São Sebastião do Paraíso
PA12)	I535	°1390	10/1	R. Educadora, Bragança
PE24)	I788	1390	1	R. Difusora, Pesqueira
PR82)	J242	1390	10/1	R. Cultura, Maringá
PR83)	J335	1390	1/0.25	R. Independência, Salto do Lontra
RJ32)	J473	1390	5/0.5	R. Sul Fluminense, Barra Mansa
RN21)	J614	1390	5/1	R. Santa Cruz AM, Santa Cruz (F.Pl.) (still on 1590)
RO18)	J...	1390	5/1	Sistema Itapirema de Comunicações, Ji-Piraná (CP)
RR02)	O701	1390	10/5	R. Roraima, Caracaraí

RS122)	K209	1390	10	R. Esperança, Porto Alegre
RS166)	K...	1390	1/0.25	R. Atlântica, Constantina (CP)
SC57)	J769	1390	1/0.25	R. Araucária, Lages
SP153)	K570	1390	5/0.25	R. Cultura, Campinas
SP90)	K594	1390	2.5/ 0.25	R. Anchieta, Itanhaém
SP154)	K636	1390	1/0.25	R. Cultura, Promissão
AC06)	H200	°1400	10	R. Difusora Acreana, Rio Branco
BA71)	H...	1400	1/0.25	R. Vale do Vasa Barris, Jeremoabo (CP)
GO32)	H754	1400	5/0.25	R. Universitaria, Goiânia (ff 870)
PB15)	I677	1400	5/1	R. Espinharas, Patos
PI27)	I926	1400	1/0.25	R. Cantagalo, Jaicós
PR84)	J256	1400	1/0.25	R. Cruzeiro do Sul, Londrina
PR85)	J285	1400	2.5/ 0.25	R. Educadora, Laranjeiras do Sul
PR86)	J291	1400	1/0.25	R. Alvorada do Sul, Reboucas
PR87)	J299	1400	1/0.25	R. Fronteira d'Oeste, Terra Roxa
RJ33)	J462	°1400	50/5	R. Rio de Janeiro ("R. Rio AM"), Rio de Janeiro
RS123)	K247	1400	1/0.25	R. Sideral, Getúlio Vargas
SC58)	J775	1400	1/0.25	R. Entre Rios, Palmitos
SP155)	K527	1400	1/0.25	R. Difusora, Lucélia
SP156)	K658	1400	5/0.25	R. Progresso, São Carlos
SP157)	K682	1400	5/0.25	R. Metrópole AM, São José do Rio Preto
TO08)	H...	1400	5/1	Soc. de Radiodifusão Guaraí, Guaraí (CP)
BA37)	H467	1410	10/0.5	R. São Gonçalo, São Gonçalo dos Campos
GO19)	H...	1410	10/1	R. Fraternidade Universal, Santo Antônio do Descoberto (CP)
MS15)	I382	1410	5/1	R. Clube, Corumbá
PR141)	J328	1410	1/0.25	R. Cultura, Cândido de Abreu
RJ34)	J486	1410	10/0.5	R. Itaperuna, Itaperuna
RS137)	K246	1410	5/0.5	R. Garibaldi, Garibaldi
RS125)	K284	1410	1/0.25	R. Minuano, Rio Grande
RS126)	K294	1410	5	R. Santa Rosa, Santa Rosa
SP158)	K691	1410	50/10	R. América, São Paulo
BA60)	H504	1420	1/0.25	R. Difusora, Irecê
CE73)	H...	1420	1/0.25	R. Difusora do Vale Acaraú, Acaraú
MG88)	L286	1420	1/0.25	R. Difusora, São João Nepomuceno
MG89)	L288	1420	5/0.25	R. Cultura, Sete Lagoas
MG90)	L313	1420	1/0.25	R. Montanhês, Botelhos
MS16)	I397	1420	1/0.25	R. Difusora Cacique, Nova Andradina
PA32)	I...	1420	10/5	R. Maguary, Belém (CP)
PR88)	J269	1420	5/0.25	R. Cultura, Umuarama
PR89)	J282	1420	1/0.25	R. Educadora, Jacarezinho
PR90)	J334	1420	1/0.25	R. Entre Rios, Santo Antônio do Sudoeste
RN12)	J609	1420	1/0.25	R. Tapuyo, Alexandria
RS149)	K258	1420	10/0.25	R. 14 de Júlho, Julio de Castilhos
SC60)	J744	1420	5/0.25	R. Cultura, Campos Novos
SC59)	J754	°1420	5/2.5	R. Guarujá, Florianópolis
SP159)	K597	1420	1/0.25	R. das Nações, Itatiba
SP10)	K612	1420	1/0.25	R. Clube, Mirandópolis
SP162)	K674	1420	1/0.25	R. Cultura, Taubaté
SP163)	K733	1420	1/0.25	R. Nova São Manoel, São Manoel
CE50)	H637	1430	5/1	R. Principe Imperial, Crateús
MG91)	L239	1430	1/0.25	R. Clube, Guaxupé
MG92)	L252	1430	1/0.25	R. Educadora, Ubá
MG158)	L...	1430	1/0.25	R. Planalto, Perdizes (CP)
PE32)	I826	1430	5/0.25	R. Independência, Goiana
PR134)	J200	°1430	25/10	R. Clube Paranaense, Curitiba
RN13)	J604	1430	10/0.5	R. Libertadora Mossoroense, Mossoró
RO11)	J671	1430	1	R. Caiari, Porto Velho
RS129)	K224	1430	1/0.25	R. Cultura, Canguçu
RS167)	K366	1430	1/0.25	R. Querência, Coronel Bicaco
SP164)	K666	1430	1/0.25	R. Serra Negra, Serra Negra
AM13)	H285	1440	10	R. Baré, Manaus
BA38)	H466	1440	10/0.5	R. Independência, Santo Amaro
CE33)	H603	1440	10/1	R. Cidade, Crato
MG159)	L...	1440	1/0.25	Empreendimentos Guimarães e Franzão ("R. Som 200"), Santa Vitória (CP)
MS41)	I407	1440	1/0.25	R. Bela Vista, Bela Vista
PR147)	J353	1440	1/0.25	R. Integração Oeste, Corbélia
RJ35)	J469	1440	20/5	R. 1440 AM ("R. AM O Día"), Rio de Janeiro
RS168)	K221	1440	1/0.25	R. Ceres de Campo Real, Naõ Me Toque
RS130)	K328	1440	1/0.25	R. Excelsior, Gramado
RS153)	K362	1440	1/0.25	R. Caibaté, Caibaté
SC61)	J792	1440	1/0.25	R. Difusora, Maravilha
SE11)	J930	1440	5/0.25	R. Educadora, Frei Paulo
SP165)	K634	1440	5/0.25	R. Comercial, Presidente Prudente
SP218)	K752	1440	2.5/0.5	R. Azul Celeste, Americana
BA72)	H...	1450	1/0.25	R. Belo Campo, Belo Campo (CP)
BA73)	H...	1450	1/0.25	R. Ipirá, Ipirá (CP)
CE34)	H601	1450	1/0.25	R. Difusora Cristal, Quixeramobim
CE35)	H623	1450	1/0.25	R. Pinto Martins, Camocim
ES12)	I208	1450	1/0.5	R. Gaeta, Guarapari
GO33)	H788	1450	1/0.25	R. Vale da Serra, São Luís de Montes Belos (ff 920)
MA12)	H891	1450	1/0.25	R. Verdes Campos, Pinheiro (ff 710)
MA32)	H900	1450	1/0.25	R. Boa Esperança, São João dos Patos
MA13)	H901	1450	1/0.25	R. Cultura, Pedreiras
MG93)	L249	1450	1/0.25	R. Globo de Minas, João Monlevade
MG94)	L312	1450	1/0.25	R. Diamante, Coromandel
MG160)	L...	1450	1/0.25	R. Princesa do Vale, Itaobim (CP)
MS13)	I417	1450	1/0.25	R. Difusora, Rio Brilhante
PA33	I...	1450	1/0.25	R. Juruá, São Felix do Xingu (CP)
PB29)	I...	1450	1/0.25	R. Itatiunga, Patos
PE25)	I794	1450	1/0.25	R. Cultura, Palmares
PI21)	I908	1450	1.5/1	R. Cultura do Gurguéia, Bom Jesus
P148)	I917	1450	1/0.25	R. Confederação Valenciana ("R. Valenciana"), Valença do Piauí
PR93)	J279	1450	1/0.25	R. Cabiúna, Bandeirantes
PR128)	J293	1450	1/0.25	R. Najuá, Irati (F.Pl.) (still on 1580)
PR94)	J301	1450	1/0.25	R. Difusora Ubiratanense, Ubiratã (ff 1240)
PR95)	J317	1450	1/0.25	R. Rainha d'Oeste, Altônia
PR148)	J346	1450	1/0.25	R. Jornal São Miguel, São Miguel do Iguaçu
RJ36)	J480	1450	5/0.25	R. do Comércio, Barra Mansa
RJ37)	J503	1450	1/0.25	R. Feliz, Santo Antônio de Pádua
RO12)	J674	1450	1/0.25	R. Vilhena, Vilhena
RS132)	K346	1450	1/0.25	R. Cassino, Rio Grande
SC62)	J802	1450	1/0.25	R. São Bento, São Bento do Sul
SC63)	J822	1450	5/0.25	R. Soc. Hulha Negra, Criciúma
SC97)	J...	1450	1/0.25	R. Belos Montes, Seara (CP)
SE12)	J...	1450	1/0.25	R. Abais, Estância (CP)
SP238)	K526	1450	1/0.25	R. Cultura, Ituverava
SP166)	K587	1450	1/0.25	R. Difusora, Guararapes
SP167)	K591	1450	20/0.5	R. Difusora Boa Nova, Guarulhos
SP168)	K657	°1450	1/0.25	R. São Carlos, São Carlos
AL13)	H...	1460	1/0.25	R. Canavieiro, União dos Palmares (CP)
AM17)	H300	1460	5	R. Clube, Parintins
BA39)	H472	1460	1/0.25	R. Bahiana, Jequié
BA54)	H...	1460	1/0.25	Arcofaba Radiodifusão, Riacho de Santana (CP)
BA74)	H...	1460	1/0.25	R. Alvorada, Cruz das Almas (CP)
CE36)	H616	1460	1/0.25	R. Uirapuru, Morada Nova
CE53)	H	1460	1/0.25	R. Dragão do Norte, Massapé
GO34)	H766	1460	1/0.25	R. Morrinhos, Morrinhos
MA33)	H...	1460	1/0.25	R. Vanguarda, Santa Luzia (CP)
MG95)	L201	1460	1/0.25	R. Cultura de Porto Novo, Além Paraíba
MG161)	L356	1460	1/0.25	R. Buritis, Buritis
MG131)	L363	1460	1/0.25	R. Entre Rios, Raul Soares
MT01)	I395	1460	1/0.25	R. Araunã, Barra do Garças (ff 560)
PB30)	I...	1460	1/0.25	R. Oeste da Paraíba, Cajazeiras (CP)
PI14)	I903	1460	1/0.25	R. Cultura, Amarante
PI28)	I924	1460	1/0.25	R. Cruzeiro, Pedro II, (ff 640)
PR96)	J204	1460	1/0.25	R. Difusora, Paranaguá
PR97)	J228	1460	1/0.25	R. Central do Paraná, Ponta Grossa
PR98)	J251	1460	1/0.25	R. Cultura, Apucarana
PR99)	J286	1460	2.5/ 0.25	R. Clube de Palmas, Palmas
PR100)	J297	1460	2/0.25	R. Guaíra, Guaíra
PR101)	J308	1460	1/0.25	R. Ampere, Ampere
PR102)	J318	1460	1/0.25	R. Educadora, Loanda
RN20)	J615	1460	1/0.25	R. Agreste, Santo Antônio
RS133)	K214	1460	1/0.25	R. Cultura, Bagé
RS134)	K311	1460	1/0.25	R. Maristela, Torres (ff 1380)
RS135)	K312	1460	1/0.25	R. Colonial, Três de Maio
SC64)	J756	1460	1/0.25	R. Sentinela do Vale, Gaspar

SP242)	K548	1460	1/0.25	R. Clube Ararense, Araras
SP170)	K608	1460	1/0.25	R. Cultura, Lorena
SP171)	K707	1460	1/0.25	R. Universal, São Roque
CE64)	H665	1470	1/0.25	R. Guanacés de Itapagé, Itapagé
ES17)	I214	1470	1/0.25	R. São Francisco, Barra de São Francisco
ES18)	I...	1470	1/0.25	Sistema Norte de R., Serra (CP)
GO36)	H773	1470	1/0.25	R. Dif. Serra dos Cristais, Cristalina
GO37)	H779	1470	1/0.25	R. Cidade, Goiás
MA34)	H...	1470	1/0.25	Sistema Maranhense de R. e TV, Presidente Dutra (CP)
MG96)	L247	1470	1/0.25	R. Difusora, Ituiutaba
MS29)	I413	1470	1/0.25	R. Alvorada, Dourados
MT16)	I422	1470	1/0.25	Emissoras Reunidas, Poconé
PA25)	I548	1470	1/0.25	R. Moreno Braga, Vigia
PE35)	I822	1470	1/0.25	R. Educadora de Belém Belém de São Francisco
PE37)	I...	1470	1/0.25	R. Bom Conselho, Bom Conselho (CP)
PI15)	I900	1470	1/0.25	R. Difusora Vale do Uruçuí, Uruçuí
PI24)	I913	1470	1/0.25	R. Ingazeira, Paulistana
PI33)	I928	1470	1/0.25	R. AM Cidade, Castelo do Piauí
PR103)	J294	1470	1/0.25	R. Novo Horizonte, Ibaiti
PR104)	J304	1470	2/0.25	R. Jornal, Assis Chateaubriand
PR105)	J327	1470	1/0.25	R. Auri Verde, Pitanga
RJ08)	J476	1470	1/0.25	R. Jornal Fluminense, Campos
RJ38)	J481	1470	1/0.25	R. Vale do Paraíba, Barra do Piraí
RN23)	J616	1470	1/0.25	R. AM de Parelhas, Parelhas
RO05)	J676	1470	1/0.25	R. Soc. Rondônia, Cacoal (ff 930)
RS169)	K219	1470	1/0.25	R. Cultura Cacequiense, Cacequi
RS136)	K324	1470	1/0.25	R. Cinderela, Campo Bom
SC65)	J781	1470	2.5/1	R. Gazeta, São José
SC66)	J798	1470	1/0.25	R. Líder do Vale, Herval d'Oeste
SP172)	K586	1470	1/0.25	R. Cultura, Guaíra
SP173)	K599	1470	1/0.25	R. Clube, Jacareí
SP174)	K632	1470	1/0.25	R. Primavera, Porto Ferreira
SP175)	K712	1470	1/0.25	R. Jornal, Indaiatuba
SP243)	K771	1470	1/0.25	R. Bastos AM, Bastos
BA40)	H475	1480	1/0.25	R. Bahiana, Itaberaba (ff 1030)
BA61)	H...	1480	5/1	R. Alvorado, Guanambi
BA75)	H...	1480	1/0.25	R. Santana, Santana (CP)
CE58)	H636	1480	1/0.25	R. Cidade, Campos Sales
CE74)	H...	1480	1/0.25	R. Universal, Morrinhos (CP)
GO47)	H801	1480	1/0.25	R. Pouso Alto, Piracanjuba (ff 1140)
MA14)	H897	1480	1/0.25	R. Itapecuru, Colinas
MA24)	H919	1480	1/0.25	R. Alvorada, Zé Doca (ff 780)
MG98)	L235	1480	1/0.25	Soc. R. Frutal, Frutal
MG100)	L307	1480	1/0.25	R. Emboabas, Tiradentes
MS17)	I393	1480	1/0.25	R. Caçula, Três Lagoas
PE26)	I790	1480	1/0.25	R. A Voz do Sertão, Serra Talhada
PE36)		1480	1/0.25	R. Vale do Ipojuca, Gravatá
PI29)	I929	1480	1/0.25	R. Vale do Coroata, Elesbão Veloso
PR153)	J221	1480	1/0.25	R. Piraí do Sul, Piraí do Sul
PR106)	J230	1480	1/0.25	R. Astorga, Astorga
PR107)	J270	1480	1/0.25	R. Educadora, União da Vitória
PR142)	J359	1480	1/0.25	Radiodifusão Campo Aberto, Laranjeiras do Sul
RJ39)	J488	1480	1/0.25	R. Mauá Solimões, Nova Iguaçu
RJ53)	J...	1480	1/0.25	R. Soc. Fraiburgo, Cordeiro (CP)
RN14)	J601	1480	1/0.25	R. Princesa do Vale, Açu
RO15)	J681	1480	1/0.25	R. Soc Rondônia, Pimenta Bueno
RS100)	K244	1480	1/0.25	R. São Roque, Faxinal do Soturno (F.Pl.) (still on 1270)
RS127)	K321	1480	5/0.25	R. Veranense, Veranópolis
SC67)	J731	1480	1/0.25	R. Difusora, Joinville
SC68)	J762	1480	1/0.25	R. Difusora 26 de Abril, Imaruí
SC94)	J826	1480	1/0.25	R. Caibi, Caibi
SE10)	J928	1480	1/0.25	R. Atalaia, Simão Dias
SP176)	K539	1480	1/0.25	R. Clube, Altinópolis
SP177)	K551	1480	2.5/0.5	R. Técnica, Atibaia
SP178)	K562	1480	1/0.25	R. Clube, Bilac
TO06)	H795	1480	1/0.25	R. Anhanguera, Miracema do Tocantins
AL09)	H246	1490	5/1	Em. Rio São Francisco, Penedo
BA41)	H478	1490	1/0.25	R. Educadora, Ipiaú
BA76)	H...	1490	1/0.25	Radiodifusão Sertaneja, Brumado (CP)
BA77)	H...	1490	1/0.25	Rio São Francisco Radiodifusão, Bom Jesus da Lapa (CP)
CE38)	H617	1490	1/0.25	R. Cultura, Aracati
MG162)	L231	1490	1/0.25	R. Cacique, Araguari
MG163)	L274	1490	1/0.25	R. Paraisópolis, Paraisópolis
MG164)	L345	1490	1/0.25	R. Cidade, Arcos
MG165)	L353	1490	1/0.25	R. Pirapetinga, Pirapetinga
MG166)	L...	1490	1/0.25	R. Matinada, Paraguaçu (CP)
MS19)	I404	1490	1/0.25	R. Paiaguás, Glória de Dourados
PB18)	I680	1490	1/0.25	R. Maringá, Pombal (ff 880)
PB31)	I...	1490	1/0.25	R. Aliança, João Pessoa (CP)
PI49)	I...	1490	1/0.25	R. Lagoa, Buriti dos Lopes (CP)
PR108)	J210	1490	1/0.25	R. Cornélio Procópio, Cornélio Procópio
PR109)	J217	1490	1/0.25	R. Legendaria, Lapa
PR149)	J347	1490	1/0.25	R. Difusora, São Jorge d'Oeste
RS138)	K208	1490	1/0.25	Rádiodifusão Assisense, São Francisco de Assis
RS139)	K225	1490	1/0.25	R. Liberdade, Canguçu
RS140)	K309	1490	1/0.25	R. Taquara, Taquara
SC69)	J757	1490	1/0.25	R. Sentinela de Alto Vale, Ibirama
SC70)	J791	1490	1/0.25	R. Cultura, Xaxim
SP180)	K530	1490	1/0.25	R. Difusora, Olímpia
SP181)	K580	1490	1/0.25	R. Nova Dracena, Dracena
SP182)	K583	1490	1/0.25	R. Educadora, Fernandópolis
SP244)	K680	1490	1/0.25	R. Cultura, Vargem Grande do Sul
SP183)	K764	1490	1/0.25	R. Mauá, Mauá
SP222)	K769	1490	1/0.25	R. Icatu, Penápolis
AL14)	H...	1500	1/0.25	R. Manguaba, Pilar (CP)
BA49)	H487	1500	5/1	R. Jacuípe, Riachão do Jacuípe
CE39)	H615	1500	1/0.25	R. Macico, Baturité
ES15)	I216	1500	1/0.25	R. Aquidaban, Cachoeiro de Itapemirim (ff 960)
MG101)	L215	1500	5/0.25	R. Montanhesa, Viçosa
MG140)	L340	1500	1/0.25	R. Aparecida do Sul, Ilicínea
PA19)	I542	1500	1/0.25	R. Floresta, Tucuruí
PE27)	I779	1500	1/0.25	R. Pajeu, Afogados da Ingazeira
PI50)	I919	1500	1/0.25	R. A Voz do Longá Esperantina
PR143)	J360	1500	1/0.25	R. A Voz do Sudoeste, Coronel Vivida
RJ41)	J494	1500	1/0.25	R. Stereo Sul, Volta Redonda
RN29)	J...	1500	1/0.25	R. Vale do Potengi, São Paulo do Potengi (CP)
RS161)	K365	1500	1/0.25	R. Simpatia Chapada
SP184)	K549	1500	1/0.25	R. Centenário, Araras
SP185)	K626	1500	1/0.25	R. Difusora, Pindamonhangaba
SP186)	K706	1500	1/0.25	R. Vale do Rio Grande, Miguelópolis
SP236)	K...	1500	1/0.25	R. Cidade, Apiaí
SP237)	K...	1500	1/0.25	R. Cumbica AM, Guarulhos
BA52)	H493	1510	1/0.25	R. Difusora do Descobrimento, Porto Seguro
CE40)	H608	1510	1/0.25	R. Planalto da Ibiapaba, São Benedito
GO38)	H770	1510	1/0.25	R. Goiatuba, Goiatuba
GO48)	H780	1510	1/0.25	R. Sociedade, Ceres
MG103)	L259	1510	2.5/0.25	R. Manhumirim, Manhumirim
MG104)	L269	1510	1/0.25	R. Oliveira, Oliveira
PA01)	I544	1510	1/0.25	R. Oriente de Redenção, Redenção
PE38)	I...	1510	1/0.25	R. Difusora Vale do Rio Una, São Bento do Una (CP)
PI16)	I894	1510	1/0.25	R. Floriano, Floriano
PI17)	I896	1510	1/0.25	R. Araucária, Corrente
PI30)	I909	1510	1/0.25	R. Grande de Picos, Picos
PR113)	J216	1510	1/0.25	R. Educadora, Venceslau Brás
PR114)	J239	1510	1/0.25	R. Difusora Cultural, Irati (ff 950)
PR115)	J326	1510	1/0.25	R. União, Céu Azul
PR155)	J362	1510	1/0.25	R. Poema, Pitanga
RJ43)	J492	1510	1/0.25	R. Teresópolis, Teresópolis
RN15)	J602	1510	1/0.25	R. Centenário, Caraúbas
RS170)	K...	1510	1/0.25	R. Educadora, São João da Urtiga (CP)
SC71)	J764	1510	1/0.25	R. Jaraguá, Jaraguá do Sul (ff 1010)
SC73)	J795	1510	1/0.25	R. Centro Oeste, Pinhalzinho
SP188)	K654	1510	1/0.25	R. Cacique, Santos
SP189)	K665	1510	1/0.25	R. Clube, São Manoel
SP230)	K...	1510	1/0.25	R. Rural, Rinópolis
BA78)	H...	1520	5/1	R. Bela Vista, Poções (CP)
CE75)	H...	1520	1/0.25	R. Araripe, Ipu (CP)
GO39)	H772	1520	1/0.25	R. Santalenense, Santa Helena de Goiás
GO53)	H...	1520	1/0.25	R. Campos Belos, Campos Belos (CP)
MA15)	H899	1520	1/0.25	R. Ribamar, Pindaré Mirim

MA17)	H916	1520	1/0.25	R. Karajás, João Lisboa
MA35)	H...	1520	1/0.25	R. Educadora Vale do Munin, Chapadinha (CP)
MG106)	L245	1520	1/0.25	R. Clube, Itaúna
MS21)	I405	1520	1/0.25	R. Jornal, Amambaí
MS42)	I...	1520	1/0.25	R. Campo Alegre, Rio Verde de Mato Grosso (CP)
PE28)	I801	1520	1/0.25	R. Educadora, Surubim
PR116)	J218	1520	1/0.25	R. Antoniense, Antonina
PR118)	J292	1520	1/0.25	R. Cultura Palotinense, Palotina
PR132)	J340	1520	1/0.25	R. Internacional, Quedas do Iguaçu
PR156)	J358	1520	1/0.25	Soc. de Radiodifusão Padre Eduardo ("R. Guairacá"), Terra Rica
RJ44)	J491	1520	10/0.5	R. Continental, Rio de Janeiro
RJ10)	J499	1520	1/0.25	R. Musical, Cantagalo
RN22)	J610	1520	1/0.25	R. Salinas, Macau
RS141)	K217	1520	1/0.25	R. Vale do Jacuí, Cachoeira do Sul
RS171)	K308	1520	1/0.25	R. Tapense, Tapes
SC74)	J743	1520	1/0.25	R. Caçanjurê, Caçador (ff 1110)
SC75)	J758	1520	1/0.25	R. Difusora, Imbituba
SC76)	J806	1520	1/0.25	R. Cultura, Timbó
SE13)	J931	1520	1/0.25	R. Imperatriz dos Campos, Tobias Barreto
SP191)	K614	1520	2.5/0.25	R. Diário, Mogi das Cruzes
SP192)	K627	1520	1/0.25	Pinhal R. Clube, Espirito Santo do Pinhal
SP225)	K760	1520	1/0.25	R. Manchester, Votorantim
BA43)	H479	1530	10/0.5	R. Cultura, Guanambi
CE76)	H...	1530	1/0.25	R. das Tres Fronteiras, Campos Sales (CP)
MG109)	L232	1530	1/0.25	R. Cultura, Dores do Indaiá
MG110)	L280	1530	1/0.25	R. Clube, Pouso Alegre
MG167)	L...	1530	1/0.25	R. Bom Sucesso, Minas Novas (CP)
MS22)	I398	1530	1/0.25	R. Uberlim, Fátima do Sul
PE29)	I781	1530	1/0.5	R. Bituri, Belo Jardim
PR120)	J312	1530	1/0.25	R. Clube, Realeza
PR121)	J321	1530	1/0.25	R. Regional, Cianorte
PR119)	J339	1530	5/0.5	R. Difusora, Balsa Nova
PR157)	J...	1530	1/0.25	R. Vale do Iguaçu, Verê (CP)
RJ45)	J502	1530	1/0.25	R. Princesinha do Norte, Miracema
RN16)	J603	1530	1/0.25	R. Curimataú, Nova Cruz
RO16)	J685	1530	1/0.25	R. Planalto, Vilhena (ff 890)
RS143)	K235	1530	1/0.25	R. Sulina, Dom Pedrito
RS144)	K300	1530	1/0.25	R. São Leopoldo, São Leopoldo
RS145)	K304	1530	1/0.25	R. Tapejara, Tapejara
RS172)	K...	1530	1/0.25	R. Itaqui, Itaqui
SC77)	J761	1530	1/0.25	R. Difusora, Itajaí
SC78)	J780	1530	1/0.25	R. Difusora, São Joaquim
SC79)	J796	1530	1/0.25	R. Porto Feliz, Mondaí
SP193)	K677	1530	1/0.25	R. Brasil, Tupi Paulista
SP194)	K699	1530	1/0.25	R. Noticias, Tatuí
SP231)	K755	1530	1/0.25	R. Universal, Teodoro Sampaio
BA79)	H...	1540	1/0.25	R. Jornal, Souto Soares (CP)
CE42)	H611	1540	1/0.25	R. Sant'Ana, Tianguá
CE77)	H631	1540	1/0.25	R. Sertões, Mombaça
CE78)	H...	1540	1/0.25	R. Aratanha, Pacatuba (CP)
ES19)	I206	1540	1/0.25	R. Agricultura, Santa Teresa
GO40)	H760	1540	1/0.25	R. Jornal, Inhumas
MA36)	H...	1540	1/0.25	R. Santa Maura, Lago da Pedra (CP)
MG111)	L217	1540	1/0.25	R. Bomdespachense, Bom Despacho
MG112)	L226	1540	1/0.25	R. Clube de Minas Gerais, Conselheiro Lafaiete
MG113)	L293	1540	1/0.25	R. Tropical, Três Corações
MG168)	L351	1540	1/0.25	R. Difusora, Salinas
MS35)	N601	1540	1/0.25	R. Regional do Piraveve, Ivinhema
PA14)	I545	1540	1/0.25	R. Boa Vista, São Sebastião da Boa Vista
PB24)	I694	1540	1/0.25	R. Santa Maria, Monteiro
PR144)	J352	1540	1/0.25	R. San Thiago Dantas, Chopinzinho
PR150)	J357	1540	1/0.25	R. Central, Roncador
RJ54)	J...	1540	1/0.25	R. Matozinho AM, Paraíba do Sul (CP)
RN30)	J...	1540	1/0.25	R. Baixa Verde, João Câmara (CP)
RS157)	K282	1540	1/0.25	R. Quaraí, Quaraí
RS156)	K367	1540	1/0.25	R. Querência, Santo Augusto
SC80)	J803	1540	1/0.25	R. Capinzal, Capinzal
SP245)	K514	1540	1/0.25	R. Cultura, Leme
SP196)	K564	1540	2/0.25	R. Em. de Botucatu, Botucatu
SP197)	K723	1540	5/0.5	R. Difusora Oeste, Osasco
SP246)	K727	1540	1/0.25	R. Central, Pompéia
AL15)	H...	1550	1/0.25	Radiodifusão Eldorado, Pão de Açucar (CP)
BA44)	H461	1550	1/0.25	R. Regional de Conquista, Vitória da Conquista
BA80)	H518	1550	2.5/0.25	R. Independência do São Francisco, Juazeiro
BA15)	H...	1550	1/0.25	R. Cultura, Utinga (CP)
CE79)	H664	1550	1/0.25	R. Macambira, Ipueiras
GO44)	H799	1550	1/0.25	R. Serra Azul, Caiapônia
MA37)	H...	1550	1/0.25	Sistema Janaina de Radiodifusão, Vargem Grande (CP)
MG114)	L211	1550	1/0.25	R. Cultura, Monte Carmelo
MG169)	L222	1550	1/0.25	R. Difusora, Carmo do Rio Claro
MG115)	L289	1550	1/0.25	R. Difusora Santarritense, Santa Rita do Sapucaí
PA34)	I550	1550	1/0.25	R. Cabano, Maracanã
PB32)	I...	1550	1/0.25	R. Bruxaxá, Areia (CP)
PR133)	J213	1550	1/0.25	R. Ipiranga, Palmeira
PR123)	J303	1550	1/0.25	R. Pioneira de Formosa, Formosa do Oeste
PR124)	J315	1550	1/0.25	R. Cristal, Marmeleiro
PR136)	J355	1550	1/0.25	R. Cidade, Palmital
PR154)	J363	1550	1/0.25	R. Colméia, Mandaguaçu
RJ46)	J479	1550	1/0.5	R. Imperial, Petrópolis
RN19)	J606	1550	1/0.25	R. Gazeta d'Oeste, Areia Branca
RS162)	K...	1550	1/0.25	R. Soc. Zona Sul, Capão do Leão
SC92)	J814	1550	1/0.25	R. Imigrantes, Turvo
SP198)	K501	1550	1/0.25	R. Clube, Itararé
SP200)	K572	1550	1/0.25	R. Cacique, Capivari
SP201)	K590	1550	5/1	R. Guarujá Paulista, Guarujá
SP247)	K623	1550	1/0.25	R. Cultura, Pederneiras
SP202)	K659	1550	1/0.25	R. São Joaquim, São Joaquim da Barra
SP219)	K740	1550	1/0.25	R. Auriflama, Auriflama
CE43)	H622	1560	1/0.25	R. Difusora Vale de Curu, Pentecoste
MA18)	H902	1560	1/0.25	R. Agua Branca, Vitorino Freire
MG117)	L256	1560	1/0.25	R. Jornal, Leopoldina
MG153)	L367	1560	5/0.25	R. Metropolitana, Vespasiano (ff 1240)
PR125)	J275	1560	1/0.25	R. Capanema, Capanema
RJ47)	J501	1560	5/0.25	R. Grande Rio, Itaguaí
RN17)	J608	1560	1/0.25	R. Cultura do Oeste, Pau dos Ferros
RS158)	K310	1560	2.5/0.25	R. Açoriana, Taquari
RS173)	K...	1560	1/0.25	R. Globo, Ijuí (CP)
SC82)	J797	1560	1/0.25	R. Educadora, Taió
SC95)	J825	1560	1/0.25	R. Cidade, São Miguel d'Oeste
SP203)	K571	1560	1/0.25	R. Em. Campos do Jordão, Campos do Jordão
SP248)	K679	1560	1/0.25	R. Valparaíso, Valparaíso
SP249)	K725	1560	1/0.25	R. Cidade, Pedreira
SP250)	K...	1560	1/0.25	R. Vale do Rio Paraná, Presidente Epitácio (CP)
BA56)	H496	1570	1/0.25	R. Vale Aprazível, Jaguaquara
CE44)	H621	1570	1/0.25	R. Sertão Central, Senador Pompeu
GO42)	H784	1570	1/0.25	R. Itaberaí, Itaberaí
GO49)	H787	1570	1/0.25	R. Mantiqueira, Niquelândia
MG118)	L203	1570	1/0.25	Nova R. Cultura, Alfenas
MG119)	L210	1570	1/0.25	R. Difusora, Itapecerica
MG141)	L344	1570	1/0.25	R. Cidade, Corinto
MG170)	L364	1570	1/0.25	R. Difusora, Piranga
MS23)	I408	1570	1/0.25	R. Patriarca, Cassilândia
MS24)	I409	1570	1/0.25	R. Caarapó, Caarapó
PE30)	I798	1570	1/0.25	R. Asa Branca, Salgueiro
PR126)	J223	1570	1/0.25	R. Atalaia, Guarapuava
PR158)	J324	1570	1/0.25	R. Brasileira, Bela Vista do Paraíso
PR137)	J341	1570	1/0.25	R. Clube, Nova Aurora
PR145)	J351	1570	5/0.5	R. Foz, Foz do Iguaça
PR159)	J...	1570	1/0.25	R. Arapoti, Arapoti (CP)
RJ48)	J493	1570	1/0.25	R. Cultura, Valença
RJ49)	J498	1570	1/0.25	R. Bom Jesus, Bom Jesus de Itabapoana
RS147)	K358	1570	1/0.25	R. Vale do Gravataí, Gravataí
SC83)	J777	1570	1/0.25	R. Rio Negrinho, Rio Negrinho
SC98)	J...	1570	1/0.25	R. Modelo, Modelo (CP)
SP204)	K552	1570	1/0.25	R. Avaré, Avaré
SP205)	K605	1570	1/0.25	R. Junqueirópolis, Junqueirópolis
SP206)	K651	1570	1/0.25	R. Emissora ABC, Santo André
SP207)	K667	1570	1/0.25	R. Socorro, Socorro

SP208)	K670	1570	1/0.25	R. Clube, Tanabi
BA45)	H464	1580	1/0.25	R. Emissora Radiovox, Muritiba
BA53)	H497	1580	1/0.25	R. Barra do Mendes, Barra do Mendes
BA62)	H...	1580	1/0.25	R. Atalaia, Canavieiras
ES13)	I209	1580	1/0.25	R. Educadora, Afonso Cláudio
GO43)	H775	1580	1/0.25	R. Rio Claro, Iporã
MG120)	L284	1580	1/0.25	R. Paranaíba, Rio Paranaíba
MG121)	L290	1580	1/0.25	R. Cultura, Santos Dumont
MG122)	L329	1580	1/0.25	R. Educadora, Espinosa
MG133)	L335	1580	1/0.25	R. Rural Nova Guaranésia, Guaranésia
MS25)	I415	1580	1/0.25	R. Laguna, Jardim
MS43)	I...	1580	1/0.25	R. Difusora, Ivinhema (CP)
PI18)	I898	1580	1/0.25	R. Irapuá, Floriano
PR128)	J293	1580	1/0.25	R. Najuá, Irati (ff 1450)
RJ51)	J487	1580	1/0.25	R. Popular Fluminense, Conceição de Macabu
RN24)	J613	1580	1/0.25	R.Novos Tempos, Ceará Mirim
RS148)	K237	1580	1/0.25	R. Encantado, Encantado
RS150)	K339	1580	1/0.25	R. Difusora Fronteira, Arroio Grande
RS159)	K...	1580	1/0.25	R. Difusora das Missões, Palmeira das Missões (CP)
SC84)	J818	1580	1/0.25	R. Pomerode, Pomerode
SP251)	K504	1580	1/0.25	R. Difusora, Amparo
SP209)	K743	1580	1/0.25	R. Trans Universal, Itaporanga
BA63)	H501	1590	5/0.25	R. Canto da Sereia, Jacobina (ff 1310)
CE80)	H600	1590	1/0.25	R. Araripe, Cedro
MG152)	L234	1590	1/0.25	R. Clube Frontiera, Fronteira (ff 1170)
MG171	L368	1590	1/0.25	Rede Dinâmica de Comunicação, Ubá
MG134)	L...	1590	10/1	Rima Comunicações, Várzea da Palma (CP)
MS26)	I403	1590	1/0.25	R. Independência, Eldorado
PR129)	J290	1590	1/0.25	R. Cultura, Andirá
PR160)	J296	1590	1/0.25	R. Havaí, Capitão Leônidas Marques
PR130)	J316	1590	1/0.25	R. Clube, Faxinal
RJ55)	J485	1590	1/0.25	R. Duque de Caxias, Duque de Caxias
RJ52)	J506	1590	1/0.25	R. Resende AM, Resende
RN21)	J614	1590	1/0.25	R. Santa Cruz AM, Santa Cruz (ff 1390)
RS174)	K212	1590	1/0.25	R. Clube, Bagé
RS151)	K281	1590	1/0.25	R. Navegantes, Porto Lucena
SP226)	K762	1590	1/0.25	R. Realidade AM, São Carlos

Shortwaves: * = inactive.

	Call	kHz	kW	Name and Location, h. of tr.
SP82)	ZYG852	2380	0.25	R. Educadora, Limeira: 24h
AC07)	ZYF202	2410	5	R. Transamazônica, Senador Guiomard: 1000-0100 (ff 3255)
SP168)	ZYG862	2420	0.5	R. São Carlos, São Carlos: 0800-0300
AC02)	ZYF204	2460	1	R. Progresso do Acre, Rio Branco (CP)
SP114)	ZYG851	*2470	0.25	R. Cacique, Sorocaba
SP210)	ZYG866	2490	0.25	R. Oito de Setembro, Descalvado: 0900-0200
AM03)	ZYF279	*3205	5	R. Vale do Rio Madeira, Humaitá
SP11)	ZYG861	3205	1	R. Ribeirão Preto: 0700-0400
SP98)	ZYG860	3235	0.5	R. Clube, Marília: 0800-0400
MG69)	ZYG203	3245	1	R. Clube, Varginha: 0900-0100
AC07)	ZYF202	3255	5	R. Transamazônica, Senador Guiomard: 1000-0100 (F.Pl.) (still on 2410)
CE22)	ZYF535	3255	1	R. Educadora Cariri, Crato: 0900-0200
PA23)	ZYG366	3285	1	R. Sentinela da Amazônia, Óbidos: 24h
PA10)	ZYG430	3325	5	R. Liberal, Belém: 24h
SP112)	ZYG867	3325	2.5	R. Tupí, São Paulo: 24h
PR33)	ZYG641	3335	5	R. Alvorada, Londrina: 0800-0300 (ff 4865)
SP49)	ZYG855	3365	1	R. Cultura, Araraquara: 0700-0300
AM02)	ZYF276	v3375	5	R. Nacional, São Gabriel da Cachoeira: 0800-0300
MS04)	ZYF907	v3375	5	R. Dourados, Dourados: 0800-0400
AP02)	ZYF361	*v3375	5	R. Equatorial, Macapá
RO09)	ZYG792	3375	5	R. Educadora, Guajará Mirim: 0900-0200
AM10)	ZYF271	3385	5	R. Educação Rural, Tefé: 0900-0200
AC03)	ZYF...	v3401	2	R. Educadora 6 de Agosto, Xapuri: 1000-0200
AC08)	ZYF	v3568.7	1.5	R. Educadora 3 de Julho, Brasiléia: 0900-0300
AC09)	ZYF	v4118	0.25	R. Difusora, Sena Madureira: 1000-0230
MA03)	ZYF810	4755	5	R. Difusora do Maranhão, São Luíz: 24h
MS01)	ZYF904	4755	10	R. Educação Rural, Campo Grande: 0800-0500
AC01)	ZYF200	v4765	10	R. Integração, Cruzeiro do Sul: 0900-0300
PA04)	ZYG363	4765	10	R. Rural, Santarém: 0800-0300
MG55)	ZYG207	4775	1	R. Congonhas, Congonhas: 0800-0100
MT15)	ZYF902	4775	1	R. Portal da Amazônia, Cuiabá: 0900-0300
RO14)	ZYG794	4775	5	R. Amarela, Rolim do Moura (CP)
MA09)	ZYF812	*4785	5	R. Ribamar, São Luís
RO11)	ZYG790	4785	10	R. Caiari, Porto Velho: 0800-0400
SP136)	ZYG857	4785	1	R. Brasil, Campinas: 0730-0030
MS12)	ZYF900	4795	1	R. Nova Difusora, Aquidauana: 0830-0300
AM08)	ZYF273	4805	10	R. Difusora do Amazonas, Manaus: 0900-0300
MG06)	ZYG209	*4805	0.5	R. Itatiaia, Belo Horizonte
AM04)	ZYF277	v4815	10	R. Cabocla, Tabatinga: 0800-0300
PR13)	ZYG640	v4815	10	R. Difusora, Londrina: 0800-0400
PA12)	ZYG364	4825	10	R. Educadora, Bragança: 0830-0200
SP80)	ZYG868	4825	10	R. Canção Nova, Cachoeira Paulista: 24h
MS27)	ZYF908	*4835	5	R. Atalaia, Corumbá
AM14)	ZYF278	4845	250	R. Cabocla, Manaus: 0800-0300
SP102)	ZYG869	4845.1	1	R. Meteorologia Paulista, Ibitinga: 0830-2300
MG33)	ZYG202	*4855	1	R. Por Um Mundo Melhor, Governador Valadares
MT01)	ZYF905	4855	2.5	R. Aruanã, Barra do Garças: 0800-0300
AC04)	ZYF203	v4865	5	R. Verdes Florestas, Cruzeiro do Sul: 1000-0300
BA20)	ZYF390	*4865	1	R. Sociedade, Feira de Santana
PR33)	ZYG641	4865	5	R.Alvorada, Londrina: 0800-0300 (F.Pl.) (still on 3335)
RR01)	ZYG810	4875	10	Super R. Roraima, Boa Vista: 0900-0300
AC06)	ZYF201	4885	5	R. Difusora Acreana, Rio Branco: 0900-0530
GO12)	ZYF692	v4885	1	R. Carajá, Anápolis: 0730-0330
PA03)	ZYG362	4885	5	R. Clube do Pará, Belém: 24h
AM13)	ZYF270	4895	5	R. Baré, Manaus: 0730-0300
MS32)	ZYR200	4895	5	R. IPB AM, Campo Grande: 0800-0500
RJ03)	ZYG683	4905	5	R. Relógio Federal, Rio de Janeiro: 0730-0330
TO04)	ZYF693	4905	1	R. Anhanguera, Araguaína: 0800-0300
AP01)	ZYF360	4915	10	R. Difusora, Macapá: 0800-0300
GO27)	ZYF691	4915	10	R. Anhanguera, Goiânia: 24h
SP06)	ZYG684	4925	1	R. Difusora, Taubaté: 0700-0300
ES07)	ZYF641	4935	1	R. Capixaba, Vitória: 24h
GO09)	ZYF694	4935	2.5	R. Difusora, Jataí: 0900-0300
MG73)	ZYG201	4945	1	R. Difusora, Poços de Caldas: 0800-0400
PE08)	ZYG525	4945	1	Emiss. Rural A Voz do São Francisco, Petrolina: 0755-0300
RO13)	ZYG791	4945	50	R. Progresso, Porto Velho: 0900-0200
MT12)	ZYF906	*4955	2.5	R. Clube, Rondonópolis
PA08)	ZYG361	4955	10	R. Marajoara, Belém: 0830-0300
RJ23)	ZYG682	4955	1	R. Cultura, Campos: 0700-0300
AM12)	ZYF275	4965	5	R. Alvorada, Parintins: 0900-0200

RN10)	ZYG761	*4965	1	R. Poti, Natal
MA10)	ZYF813	4975	5	R. Timbira do Maranhão, São Luís: 0800-0300
SP112)	ZYG865	4975	1	Super R. Tupí, São Paulo: 24h
SP187)	ZYG685	*4976	1	R. Iguatemi, Osasco
GO30)	ZYF690	4985	10	R. Brasil Central, Goiânia: 24h
MT08)	ZYF903	5015	1	R. Brasil Tropical, Cuiabá: 24h
PI10)	ZYG595	5015	1	R. Pioneira, Teresina: 0800-0300
RJ07)	ZYG685	5015	1	R. Copacabana, Rio de Janeiro: 24h
PA15)	ZYG365	5025	5	R. Jornal da Transamazônica, Altamira: 0900-0200
PB14)	ZYG481	*5025	1	Super R. Borborema, Campina Grande
RO06)	ZYG793	*5025	5	R. Morimoto, Ji-Paraná
AM09)	ZYF272	5035	5	R. Educação Rural, Coari: 0900-0230
SP52)	ZYG853	5035	10	R. Aparecida, Aparecida: 0800-0300
PA16)	ZYG360	5045	10	R. Cultura do Pará, Belém: 0800-0300
SP63)	ZYG850	5045	1	R. Difusora, Presidente Prudente: 0800-0400
AM15)	ZYF280	*5055	5	R. Jornal A Crítica, Manaus
MT10)	ZYF901	5055	1	R. Difusora, Cáceres: 0800-0400
SP62)	ZYE965	5955	10	R. Gazeta, São Paulo: 24h
RS163)	ZYE857	5965	7.5	R. Transamérica, Santa Maria: 1000-0200
MG06)	ZYE523*	5970	10	R. Itatiaia, Belo Horizonte
SC59)	ZYE891	5980	10	R. Guarujá, Florianópolis: 0900-0300
RJ12)	ZYE766	*5990	7.5	R. MEC, Rio de Janeiro
MA25)	ZYE481	6000	5	R. Boas Nova, São Luís (CP)
RS23)	ZYE852	6000	10	R. Guaíba, Porto Alegre: 0830-0400
MG35)	ZYE521	6010	25	R. Inconfidência, Belo Horizonte: 24h
BA55)	ZYE299	*6020	10	R. Educadora da Bahia, Salvador
RS09)	ZYE851	6020	10	R. Gaúcha, Porto Alegre: 24h
RJ25)	ZYE770	6030	10	R. Globo, Rio de Janeiro: 0900-0300
PR134)	ZYE725	6040	7.5	R. Clube Paranaense, Curitiba: 0800-0300
MG68)	ZYE520	6050	10	R. Guarani, Belo Horizonte: 0800-0300
PR60)	ZYE726	6060	15	R. Universo, Curitiba: 24h
RJ18)	ZYE765	6070	7.5	R. Capital, Rio de Janeiro: 0800-0300
GO27)	ZYE442	6080	5	R. Anhanguera, Goiânia
PR138)	ZYE729	6080	10	R. Novas de Paz, Curitiba: 0900-0300
SP57)	ZYE956	6090	10	R. Bandeirantes, São Paulo: 0700-0500
PR21)	ZYE728	6105	5	R. Cultura, Foz do Iguaçu
SP80)	ZYE971	6105	5	R. Canção Nova, Cachoeira Paulista: 0800-0300
SP100)	ZYE968	6120	10	R. Globo, São Paulo: 0900-2200
SP52)	ZYE954	6135	25	R. Aparecida, Aparecida: 0800-0300
SP77)	ZYE950	6150	7.5	R. Record, São Paulo: 1000-2200
AM11)	ZYE246	6160	10	R. Rio Mar, Manaus: 0800-2200
RS49)	ZYE854	*6160	10	R. Nova Esperança, Porto Alegre
SP121)	ZYE959	6170	7.5	R. Cultura, São Paulo: 0800-0400
DF06)	ZYE365	6180	250	R. Nacional da Amazônia, Brasília: 0900-0200
SP77)	ZYE951	9505	7.5	R. Record, São Paulo: 1000-2200
PR138)	ZYE729	9515	10	R. Novas de Paz, Curitiba: 0900-0300
BA55)	ZYE300	9540	10	R. Educadora da Bahia, Salvador: 0800-0300
RS49)	ZYE855	9550	5	R. Nova Esperança, Porto Alegre
PR60)	ZYE727	9565	20	R. Universo, Curitiba: 24h
SP48)	ZYE969	9585	10	R. CBN (Globo), São Paulo: 0900-2200
RJ12)	ZYE772	*9600	7.5	R. MEC, Rio de Janeiro
SP121)	ZYE960	9615	7.5	R. Cultura, São Paulo: 0800-0400
SP52)	ZYE954	9630	10	R. Aparecida, Aparecida: 0800-0300
SP57)	ZYE957	9645	7.5	R. Bandeirantes, São Paulo: 0700-0500
SC86)	ZYE890	9665	10	R. Marumby, Florianópolis: 0900-0100
SP80)	ZYE971	9675	10	R. Canção Nova, Cachoeira Paulista: 24h
SP62)	ZYE963	9685	7.5	R. Gazeta, São Paulo: 0700-2200
AM11)	ZYE245	9695	7.5	R. Rio Mar, Manaus: 0800-2200
RJ17)	ZYE768	9705	7.5	R. Nacional, Rio de Janeiro: 0800-0300
PR134)	ZYE725	9725	7.5	R. Clube Paranaense, Curitiba: 0800-0300
RS163)	ZYE857	11705	10	R. Transamérica, Santa Maria: 1000-2200
DF06)	ZYE365	11780	250	R. Nacional da Amazônia, Brasília: 0900-0200
RS23)	ZYE853	11785	7.5	R. Guaíba, Porto Alegre: 0830-0400
RJ25)	ZYE775	11805	10	R. Globo, Rio de Janeiro: 0900-0300
GO30)	ZYE440	11815	7.5	R. Brasil Central, Goiânia: 0800-0300
GO27)	ZYE441	11830	1	R. Anhanguera, Goiânia: 0800-0300
SP52)	ZYE954	11855	5	R. Aparecida, Aparecida: 0800-0300
RS49)	ZYE856*	11895	1	R. Nova Esperança, Porto Alegre
PR60)	ZYE726	11905	10	R. Universo, Curitiba: 24h
RS09)	ZYE850	11915	7.5	R. Gaucha, Porto Alegre: 24h
SP57)	ZYE958	11925	10	R. Bandeirantes, São Paulo: 0700-0500
PR134)	ZYE725*	11935	7.5	R. Clube Paranaense, Curitiba
RJ12)	ZYE773	11950	7.5	R. MEC, Rio de Janeiro: irr.
SP77)	ZYE952	11965	7.5	R. Record, São Paulo: 1000-2200
SP77)	ZYE953	15135	7.5	R. Record, São Paulo: 1000-2200
MG35)	ZYE522	15190	5	R. Inconfidência, Belo Horizonte: 0800-0300
DF06)	ZYE365	15200	250	R. Nacional da Amazônia, Brasília
MA10)	ZYE480*	15215	2.5	R. Timbira, São Luís
SP100)	ZYE970*	15265	10	R. Globo, São Paulo
SP62)	ZYE964*	15325	1	R. Gazeta, São Paulo
SP23)	ZYE955*	15415	1	R. Clube, Ribeirão Preto
DF06)	ZYE365*	15445	1	R. Nacional da Amazônia, Brasília
SP121)	ZYE961	17815	7.5	R. Cultura, São Paulo: 0800-0400
RJ12)	ZYE774*	17875	7.5	R. MEC, Rio de Janeiro

AC00) ACRE

AC01) Rua Alagoas, 270, 69980-000 Cruzeiro do Sul — Gte: Claudio Onofre Ferreira.
AC02) Av. Getúlio Vargas 752, 69900-000 Rio Branco.
AC03) Rua Pio Nazario 31, 69930-000 Xapuri.
AC04) C.P. 53, 69981-970 Cruzeiro do Sul — DG: Rui Matos Said.
AC05) Rua Epaminondas Jacome 236, 69908-420 Rio Branco.
AC06) Rua Benjamim Constant 161, 69900-160 Rio Branco.
AC07) Av. Castelo Branco 287, 69925-000 Senador Guiomard.
AC08) Rua Genni Assis s/n, 69932-000 Brasiléia.
AC09) Rua Avelino Chaves 707, 69940-000 Sena Madureira.

AL00) ALAGOAS

AL01) Av. Coronel Wilson Santa Cruz 6, 57314-000 Arapiraca.
AL02) Via Expressa 4360, 57080-000 Maceió — DG: Waldenir S. Rodrigues.
AL03) Rua José Maria Passos 25, 57600-030 Palmeiras dos Indios.
AL04) Av. Fernandes Lima 1047, 57055-000 Maceió.
AL05) Av. Getúlio Vargas 111, 57100-000 Rio Largo.
AL06) Praça Senador Eneas Araújo 61, 57500-000 Feira do Ipanema.
AL07) Rua do Comércio 515, 57020-000 Maceió.
AL08) Rua Barão de Penedo 258, 57020-000 Maceió.
AL09) C.P. 6, 57201-970 Penedo — Dir: Raul Rodrigues de Lima.
AL10) 57160-000 Marechal Deodoro.
AL11) Rua Porcos s/n, 57300-000 Arapiraca
AL12) Parque Industrial Delmiro Gouveia ao lado da Rodovia BR-423, 57480-000 Delmiro Gouveia.
AL13) Praça Parque Antenor Uchoa 72, 57800-000 União dos Palmares.
AL14) BR-316, dm. 213, Margem Direita, 57150-000 Pilar.
AL15) 56798-000 Pão de Açucar.

AM00) AMAZONAS

AM01) Rua Brasilia s/n, 69470-000 Tefé — DG: F. H. de Santos.
AM02) Av. Pres. Costa e Silva s/n, 69750-000 São Gabriel da Cachoeira — Gte: V.S. Marques.
AM03) Av. Rio Madeira 1940, 69800-000 Humaitá.
AM04) Gleba Tocantins, lote 15, 69640-000 Tabatinga — Gte: Jaime A. Silva.
AM05) Rua Eduardo Ribeiro esq. Solimões, 69100-000 Itacoatiara.
AM06) Rua Joana D'Angelo 2255, 69400-000 Manacapuru.
AM07) Av. Major Santana 2502, 69280-000 Manicoré — DG: Hamilton Maia C.
AM08) Estrada dos Morães s/n, 69190 Maués.
AM09) Praça São Sebastião 228, 69460-000 Coari — Gte: José C. Martins Cabral.
AM10) Praça Santa Tereza 283, 69470-000 Tefé — Gte: Thomas Schwamborn.
AM11) Rua José Clemente 500, 69010-070 Manaus — DG: Dom Clovis Frainer.
AM12) Travessa Leopoldo Neves 503, 69151-460 Parintins — Dir: Raimunda Ribeiro da Motta.
AM13) Av. Santa Cruz Machado 170A, 69010-070 Manaus. —Dir: Clodoaldo J. Guerra
AM14) C.P. 2929, 69001-970 Manaus — DG: Francisco Puga.
AM15) C.P. 2250, 69001-970 Manaus — DG: Rui Alencar.
AM16) Alameda Cosme Ferreira s/n, 69083-000 Manaus.
AM17) Av. Amazonas 1958, 69151-000 Parintins.
AM18) Rodovia Manoel Urbano, km. 02, 69405-000 Iranduba.
AM19) Av. Ribeiro Junior 381, 69400-000 Manacapuru.
AM20) C.P. 311, 69011-970 Manaus — Dir: José J. Marinho.
AM21) Rua Eduardo Ribeiro 2050, 69151-000 Parintins.
AM22) 69880-000 Eirunepê.
AM23) Estrada dos Morães, km. 05, 69190-000 Maués.
AM24) Av. Leopoldo Neves 360, 69850-000 Bôca do Acre.
AM25) Rua Anori/Anama 327, 69440-000 Anori.
AM26) Estrada do Gavião, km. 05, 69500-000 Carauari.

AP00) AMAPA

AP01) C.P. 2929, 68901-970 Macapá — Gte: Antônio Filho.
AP02) Rua Major Eliezer Levy 684, 68900-140 Macapá.
AP03) Av. Nações Unidas 256, 68900-140 Macapá.
AP04) 68930-000 Mazagão.

BA00) BAHIA

BA01) Rua M. Dourado Sobrinho 78, 44900-000 Irecê.
BA02) Praça João Bastos 2, 45600-000 Itabuna — DG: Waldery Andrade.
BA03) Av. Marechal Castelo Branco 193, 45995-000 Teixeira de Freitas.
BA04) Rua Senador Costa Pinto 925, 40230-040 Salvador.
BA05) Av. Amazonas 306, 47800-000 Barreiras.
BA06) Av. Nilo Peçanha 572, 45400-000 Valença.
BA07) Praça Duque Caxias 3, 45700-000 Itapetinga.
BA08) Rua dos Sururu s/n, 44570-000 Santo Antônio de Jesus.
BA09) Rua Joana Angélica 125, 45660-000 Ilhéus — Gte: Robério Meneses.
BA10) Rua Germiniano Costa 47, 44025-070 Feira de Santana.
BA11) Rua Ferreira Santos 5, 40230-040 Salvador.
BA12) C.P. 19, 45601-970 Itabuna — Gte: P. Nunes E.
BA13) Rua 24 de Outubro 167, 47800-000 Barreiras.
BA14) Praça Luiz Nogueira 385, 48700-000 Serrinha.
BA15) Rodovia BA-052, km. 48, 46810-000 Utinga.
BA16) Praça da Sé 26/28, 40020-210 Salvador.
BA17) Av. Visconde do Rio Branco 633, 48970-000 Senhor do Bonfim.
BA18) Av. Juarez Tavora s/n, 45600-000 Itabuna.
BA19) C.P. 15, 48730-000 Conceição do Coité — Dir: Edvaldo de Santiago.
BA20) C.P. 32, 44001-970 Feira de Santana.
BA21) Gleba Fazenda Ouro Verde, 45900-000 Caravelas.
BA22) Rua Gabriel Soares 23, 40060-040 Salvador — DG: Gilmar Teixeira.
BA23) Praça Barão do Rio Branco 42, 45100-000 Vitória da Conquista — Gte: A. Ribeiro N.
BA24) Av. Maria Quitéria 223, 44070-100 Feira de Santana.
BA25) Rua Santos Dumont 22, 45700-000 Itapetinga.
BA26) Rua Marquês de Paranaguá 259, 45660-000 Ilhéus.
BA27) C.P. 29, 45825-000 Eunápolis — DG: José Silvano F.
BA28) Rua Aprigo Duarte 4, 48900-000 Juazeiro.
BA29) Morro do Peru Pelado, 44700-000 Jacobina.
BA30) Rua Comandante Almiro 465, 44010-010 Feira de Santana — DG: Julio E. dos Santos.
BA31) Rua Dom Pedro II 98, 48100-000 Alagoinhas.
BA32) Av. Joana Angélica 251, 40050-001 Salvador.
BA33) Av. 2 de Julho 14, 45660-000 Ilhéus — DG: Tony Neto
BA34) Praça Luiz Nogueira 385, 48700-000 Serrinha.
BA35) Rua São Francisco 159, 48600-000 Paulo Afonso.
BA36) C.P. 11, 47601-970 Bom Jesus da Lapa — DG: José N. Grossi.
BA37) Praça Ruy Barbosa s/n, 44330-000 São Gonçalo dos Campos — DG: Margarida Alves B.
BA38) Rua Marechal Deodoro 16, 44200-000 Santo Amaro.
BA39) Rua 2 de Julho 2, 45200-000 Jequié — DG: Cid Teixeira.
BA40) Rua Sítio Escurinha s/n, BR-242, km. 90, 46880-000 Itaberaba — DG: João Leão Carneiro.
BA41) Rua Castro Alves 191, 45570-000 Ipiaú.
BA42) Av. Contorno s/n, 46400-000 Caetité — Dir: F. H. Negretros.
BA43) Rua Rui Barbosa 128, 46430-000 Guanambi.
BA44) Praça Marcelino Mendes 430, 45100-000 Vitória da Conquista.
BA45) Rua Dr. Teixeira 16, 44340-000 Muritiba.
BA46) Av. Dom Pedro II 672, 45810-000 Santa Cruz Cabrália — DG: Paulo H. Santos Soares.
BA47) Rua da Bandeira 27, 42800-000 Camaçari.
BA48) Travessa do Contorno 26, 45570-000 Ipiaú.
BA49) Pe. A. Guimarães 32, 44640-000 Riachão do Jacuípe.
BA50) Av. Getúlio Vargas s/n, 48600-000 Paulo Afonso.
BA51) Praça da Bandeira 102-1E, 47640-000 Santa Maria da Vitória.
BA52) Travessa Luis Viana Filho s/n 45820-000 Porto Seguro.
BA53) Rua Alvaro Campos s/n, 44990-000 Barra do Mendes.
BA54) Rua Sitio São Felix s/n, 46470-000 Riacho de Santana.
BA55) Rua Pedro Gama 413/E, Alto Sobradinho Federação, 40230-290 Salvador — Dir: Walter Sequieros R. Tanure.
BA56) Praça Guilherma Silva 85, 45343-000 Jaguaquara — Dir: Italo Rabelo de Amaral.
BA57) Rua Ferreira Santos 5, 40230-040 Salvador.
BA58) Centro Industrial Imbares, BR-116 Km 837, 45100-000 Vitória da Conquista.
BA59) Rua Gonçalo Martins 19, 45550-000 Ubatã.
BA60) Rua Regio Emilia 32, 44900-000 Irecê.
BA61) C.P. 22, 464300-000 Guanambi — Dir: Clóvis F. Filho.
BA62) Rua General Pederneiras s/n, 45860-000 Canavieiras.
BA63) Rua Margem do Rio do Ouro 115, 44700-000 Jacobina.
BA64) Rua Farias Goes 164, 48330-000 Rio Real.
BA65) Rua Otavio Mangabeira 13, 48500-000 Euclides da Cunha.
BA66) Rua Custodio Morais 342, 47400-000 Xique-Xique.
BA67) Praça Castelo Branco 294 — Centro, 45830-000 Itamaraju.
BA68) Praça Barão de Gerenoabo s/n, 48475-000 Itapicuru.
BA69) Rua Barão do Rio Branco 183, 48410-000 Cicero Dantas.
BA70) Av. Suburbana 7000, 40230-040 Salvador.
BA71) Praça Coronel Antonio Louren O. 48, 48540-000 Jeremoabo.
BA72) Rua Almiro Ferraz 30, 45160-000 Belo Campo.
BA73) 44600-000 Ipirá.
BA74) 44380-000 Cruz das Almas.
BA75) Rua Teixeira de Freitas s/n, 47700-000 Santana.
BA76) Rua Virgilio Ataide s/n — Bairro Pedra do Conselho, 46100-000 Brumado.
BA77) 47600-000 Bom Jesus da Lapa.
BA78) Periferia da Cidade, Margem da BR-116, 45260-000 Poções.
BA79) 46990-000 Santo Soares.
BA80) Rodovia BR-407 Km 1, Estrada da Quidaela, 48700-000 Juazeiro.
BA81) Rua Machado de Assis 16 — Brotas, 40230-040 Salvador.

CE00) CEARA

CE01) C.P. 45, 62701-970 Canindé.
CE02) Rua São Pedro 918, 63010-010 Juazeiro do Norte.
CE03) Av. Monsenhor Tabosa 2514, 62500-000 Itapipoca — DG: José I. Magalhães.
CE04) Rua Carlos Rolim/Praça da Matriz, 63700-000 Crateús.
CE05) C.P. 635, 60001-970 Fortaleza.
CE06) Rodovia CE-161, alt. do KM 2.5, 62580-000 Acaraú.
CE07) Rua Major Joaquim Alves 221, 63540-000 Várzea Alegre.
CE08) C.P. 2651, 60121-970 Fortaleza.
CE09) Rua Agranomando Rangel, 63870-000 Boa Viagem.

CE10) C.P. 1515, 60001-970 Fortaleza — DG: Carlúcio G. de Assis.
CE11) Rua Monsenhor Macena 51, 63300-000 Lavras da Mangabeira.
CE12) Av. Agenor Araujo 1194, 63500-000 Iguatu.
CE13) C.P. 851, 60001-970 Fortaleza — Dir: E. Gomes de Souza.
CE14) Praça Vicente Aguilar s/n, 62400-000 Camocim.
CE15) Rua São Luís 68, 63010-350 Juazeiro do Norte.
CE16) C.P. 2683, 60121-970 Fortaleza — DG: Miguel D. de Souza.
CE17) Rua Coronel Virgilio Correia s/n, 63500-000 Iguatu.
CE18) Praça Figueira Sampaio 32, 63180-000 Barbalha.
CE19) Av. Dom José 947, 62030-630 Sobral — DG: E. Rodrigues A.
CE20) C.P. 87, 63901-970 Quixadá — DG: Heilton Dantas.
CE21) Av. Aguanambi 282, 60055-402 Fortaleza.
CE22) C.P. 57, 63101-970 Crato — Gte: G. Farias F.
CE23) Rua Conselheiro José Júlio 126, 62010-820 Sobral.
CE24) Rua Raul Vieira 562, 62900-000 Russas.
CE25) Rua Conego Eduardo Araripe 1692, 62870-000 Pacajus.
CE26) C.P. 222, 60001-970 Fortaleza — Dir: M. Pinheiro C.
CE27) Rua Coronel Zezé, 63700-000 Crateús.
CE28) Rua Coronel Malveira 122, 62930-000 Limoeiro do Norte.
CE29) Av. Barão de Studart 1864, 60120-001 Fortaleza.
CE30) C.P. 34, 63011-970 Juazeiro do Norte — DG: Geraldo Meneses.
CE31) Praça Duque de Caxias 552, 62011-300 Sobral.
CE32) Av. Dr. Pedro de Queiroz Ferreira 2129, 62850-000 Cascavel.
CE33) Rua São Francisco 139, 63100-000 Crato.
CE34) Rua Monsenhor Salviano Pinto 71, 63800-000 Quixeramobim.
CE35) Praça Pinto Martins, 62400-000 Camocim.
CE36) Av. Manoel Castro 815, 62940-000 Morada Nova.
CE37) Rua Moacir Pereira Gondim s/n, 63660-000 Tauá.
CE38) Rua Coronel Alexanzito 835, 62800-000 Aracati — DG: Fátima Costa.
CE39) Rua Hildo Furtado s/n, 62760-000 Baturité.
CE40) Rua Italiano Júlio Filizola 512, 62370-000 São Benedito — DG: Francisco de Araujo Leite.
CE41) Rua Coronel Antônio Joaquim 2148, 62930-000 Limoeiro do Norte.
CE42) Rua Maestro Quincas Bizerril s/n, 62320-000 Tianguá.
CE43) Rua Poeta A. Martins, 62640-000 Pentecoste.
CE44) Av. Santos Dumont 414, 63600-000 Senador Pompeu.
CE45) Praça Adolfo Caminha 247, 62800-000 Aracati.
CE46) M. Lourdes 545, 62280-000 Santa Quitéria.
CE47) Rua Coronel Meireles 35, 62680-000 Paracuru.
CE48) Rua Firmino Tavares 246, 63380-000 Barro.
CE49) Travessa Mercedes Santos 762, 62700-000 Canindé.
CE50) Rua Coronel Lucio 489, 63700-000 Crateús — DG: José M. B. do Bonfim.
CE51) Rua Dr. Monteiro Filho 10, 63800-000 Quixeramobim.
CE52) Visconde de Cauípe 695, 61600-000 Caucaia.
CE53) Rodovia Massapê/Sobral (Mumbaba), 62140-000 Massape.
CE54) Rua Tibúrcio Targino 155, 61700-000 Aquiraz — DG: Iva Lioba.
CE55) Rua Dr. Almir Farias 446, 62200-000 Nova Russas.
CE56) Rua Raimundo Nonato 81 — Centro, 62940-000 Morada Nova.
CE57) Rua Coronel Felix 1237 altos, 62250-000 Ipu.
CE58) Travessa 10 de Novembro 198, 63150-000 Campos Sales.
CE59) Rua Cazuzinha Marques 87, 63560-000 Acopiara.
CE60) Rua Manoel Inacio de Lucena 249, 63260-000 Brejo Santo.
CE61) BR-222 Km 220, 62100-000 Sobral.
CE62) Rua Coronel Porto 410, 62430-000 Granja.
CE63) Rua Antônio Queiroz 343, 63870-000 Boa Viagem.
CE64) Rua Major Barreto 3000, 62600-000 Itapagé — DG: Eldo R. Louzada.
CE65) Loteamento Pioneiro, Entrada Sobral Santa Quitéria, 62100-000 Forquilha.
CE66) Rua Paulo Marques 374 — Centro, 62370-000 São Benedito.
CE67) Rua Monsenhor Joviniano Barreto s/n, 63660-000 Tauá.
CE68) Rua Monsenhor Eurico s/n, 62380-000 Guaraciaba do Norte.
CE69) Av. Rios 92, 62580-000 Itarema.
CE70) 62800-000 Aracati.
CE71) Av. José de Borba Vasconcelos 3589, 61940-000 Maracanaú.
CE72) Fazenda Veneza — Distrito de Veneza, 63600-000 Senador Pomeu.
CE73) Rua Dom Pedro II 141, 62580-000 Acaraú.
CE74) 62550-000 Morrinhos.
CE75) Rua Major Liberalino s/n, 62250-000 Ipu.
CE76) 63150-000 Campos Sales.
CE77) Praça Padre Pedro Leão 110, 63610-000 Mombaça.
CE78) Rua Caio Prado 406 — Centro, 61800-000 Pacatuba.
CE79) Rua Coronel Eufrasino de Pinho s/n, 62230-000 Ipueiras — Dir: Francisco Airton de Pinho.
CE80) Margem Direita da Ferrovia Cedro — Fortaleza, 63400-000 Cedro.
CE81) 62730-000 Caridade.

DF00) DISTRITO FEDERAL

DF01) SRTV, Q-702, bloco P, salas 3109, 70340-000 Brasília — DG: Neder Adib A.
DF02) Setor de Rádio e TV Sul, Q-2 Edif. Palácio do Rádio I, 70340-000 Brasília — Dir: Antônio Moraes e Souza.
DF03) Av. W-3, Q-16, lotes 26/28-Gr. 314, 70340-000 Brasília.
DF04) Av. W2 Sul Q-701, bloco A SRTV, 70340-000 Brasília — DG: Eduardo da Veiga R. Fajardo.
DF05) SRTV/Sul, Q-701, bloco E, Térreo, 7034-000 Brasília.
DF06) **Radiobras.** C.P. 070.747, 70359-970 Brasília. **Te:** +55 (61) 321 3949. **Telex:** 061-1682. **L.P:** — Pres: Marcelo Netto; Dir. Radio: Iolando Lourenço; Dir. News: Miriam Moura; Dir. Adm: Januário Procópio; Dir. Tec: Toshihiro Kanagae; Dir. SW: Luiza Alves.
DF07) SRTS SL 316, Edif. Palacio do Radio II, 70340-000 Brasília.
DF08) SCS-Quadra 05, Bl. B, Lotes 47 a 57, Ns. 39/40, 70340-000 Brasília.

ES00) ESPIRITO SANTO

ES01) Av. Alberto Torres 345, 29040-700 Vitória.
ES02) C.P. 700, 29001-970 Vitória — Dir: Leo Amorim.
ES03) Rua Coronel Guardia 11, 29300-070 Cachoeiro de Itapemirim — DG: Lucia H.C. Martins.
ES04) C.P. 275, 29001-970 Vitória — DG: Carlos Lindenberg. Neto.
ES05) C.P. 125, 29900-971 Linhares — DG: Paulo Sérgio Gava.
ES06) C.P. 178, 29700-971 Colatina — DG: G. Pereira.
ES07) C.P. 509, 29001-970 Vitória — Dir: Jairo Gouvea Maia.
ES08) C.P. 809, 29001-970 Vitória — DG: J.A. Figueiredo.
ES09) C.P. 134, 29300-971 Cachoeiro de Itapemirim.
ES10) Rua Alberto de Oliveira Santos 42, 29140-000 Cariacica — DG: Romulo Conde.
ES11) C.P. 37, 29601-970 Afonso Cláudio — DG: Zaqueu K. Hein.
ES12) Rua da Matriz 86, 29200-000 Guarapari.
ES13) Av. Presidente Vargas 439, 29600-000 Afonso Cláudio.
ES14) C.P. 132, 29930-000 São Mateus — DG: Paulo Sérgio Gava.
ES15) Av. Dr. Jeremias Sandoval s/n, 29308-240 Cachoeiro de Itapemirim.
ES16) Rua 09 esquina com Rua Lauro Cunha Freire, 29140-000 Cariacica.
ES17) Rua Astrogildo Romão dos Anjos 189, 29800-000 Barra de São Francisco.
ES18) Av. Presidente Castelo Branco 905, 29160-000 Serra.
ES19) 29650-000 Santa Teresa.

GO00) GOIÁS

GO01) Av. Goiás 636, 74010-010 Goiânia — DG: M. Barbosa L.
GO02) C.P. 41, 75701-970 Catalão — DG: Verônica F. Margon.
GO03) Rua Coronel Barbosa 420, 75070-110 Anápolis.
GO04) C.P. 117, 75201-970 Pires do Rio — DG: U. de Oliveira.
GO05) C.P. 15200, 74501-970 Goiânia — DG: Pe. Cesar Moreira.
GO06) Rua Prof. Ferreira 169, 76270-000 Jussara — DG: M. Soares.
GO07) Rua 48 No. 288, 76680-000 Itapuranga.
GO08) C.P. 670, 75001-970 Anápolis — DG: Mária Lúcio Gomes Santos.
GO09) C.P. 33, 76800-000 Jataí — DG: Zacharias Falleiros.
GO10) Rua Evangelino Meireles 26, 72800-000 Luziânia.
GO11) C.P. 150, 75691-970 Caldas Novas — DG: Aldo Moreira.
GO12) C.P. 520, 75001-970 Anápolis — DG: Nelson Silva Rosa.
GO13) Av. Brasil 272, 76380-000 Goianésia.
GO14) Praça Coronel Carrijo, 75830-000 Mineiros — DG: Elias de Oliveira.

GO15) Av. Bernardo Sayao 371, 76310-000 Rialma — DG: José R. Rego.
GO16) Av. Anhanguera 3511, 74610-010 Goiânia.
GO17) Av. Floriano Peixoto 17, 76550-000 Porangatu — DG: Agenor H. Manzi.
GO18) Av. Tocantins 2011, 76400-000 Uruaçu.
GO19) Praça da Imortalidade 05, 77227-000 Santo Antonio do Descoberto.
GO20) C.P. 149, 75503-970 Itumbiara — DG: Anair G. Machado.
GO21) Praça da Bandeira 94, 75500-000 Itumbiara.
GO22) C.P. 501, 75001-970 Anápolis — DG: José Cunha Gonçalves.
GO23) Rua de Contorno 702, 76330-000 Jaraguá.
GO24) Rua 8 No. 497, 74010-010 Goiânia — Gte: A. Cozar.
GO25) Rua Jatai 18, 75860-000 Quirinópolis.
GO26) Rua Herculano Lopo 80, 73800-000 Formosa.
GO27) C.P. 13, 74001-970 Goiânia.
GO28) C.P. 34, 75781-970 Ipameri — Gte: W. Naves.
GO29) Rua 15 No. 83, 76300-000 Ceres.
GO30) C.P. 330, 74001-970 Goiânia — DG: Weber Borges.
GO31) C.P. 62, 74001-970 Goiânia — Dir: Luis G. de Araújo.
GO32) Lagos das Rosas, setor Oeste, 74320-060 Goiânia — DG: Wilmar Antônio Alves.
GO33) Av. Amazonas 367, 76100-000 São Luíz de Montes Belos — DG: João P. de Souza.
GO34) Rua Barão do Rio Branco 999, 75650-000 Morrinhos.
GO35) Rua Couto Magalhães s/n, 75180-000 Silvânia.
GO36) Rua Kisleu Maciel 113, 73850-000 Cristalina — DG: Waldenir Figueiredo.
GO37) C.P. 60, 76601-970 Goiás — DG: Wilians de Souza P.
GO38) C.P. 70, 75601-970 Goiatuba.
GO39) Praça Emilio Garrastazu Médici s/n, 75920-000 Santa Helena de Goiás.
GO40) C.P. 117, 75401-970 Inhumas — DG: Lúsio de Freitas B.
GO41) C.P. 131, 75901-970 Rio Verde — DG: Iris C. de Freitas.
GO42) Rua Major Garcia 555, 76630-000 Itaberaí.
GO43) Rua Catalão 671, 76200-000 Iporã — Dir: S. Alves C.
GO44) Av. Lindolfo Alves Dias 571 — Centro, 75850-000 Caiapônia.
GO45) Rua Goiás 400, 75300-000 Aparecida de Goiânia.
GO46) Quadra 33, Lote 23 No. 24, 72900-000 Santo Antônio do Descoberto.
GO47) Rua Percival Rebelo 375, 75640-000 Piracanjuba.
GO48) C.P. 185, 76301-970 Ceres — Dir: Marcela J. Silva.
GO49) Praça Silva Junior 184, 76420-000 Niquelândia — DG: Sebastião da Silva Rocha Vidal.
GO50) Praça Lucio Prado s/n, 75600-000 Goiatuba.
GO51) Av. Brasil 100, 76260-000 Quirinópolis.
GO52) Av. Joaquim Davi Ferreira 1390, 76515-000 Firminópolis.
GO53) Rua das Industrias s/n, 77335-000 Campos Belos.

MA00) MARANHÃO

MA01) Rua Frederico Figueira s/n, 65950-000 Barra do Corda.
MA02) Praça Dom Pedro II s/n, 65030-000 São Luíz.
MA03) C.P. 152, 65001-970 São Luíz — Dir: C.M. Duque B.
MA04) Rua Henrique de Figueiredo 485, 65400-000 Codó.
MA05) Av. São Benedito 1075, Bairro São Benedito, 65400-000 Codó — DG: Raimundo Leonel M. Araujo.
MA06) Rua Dorgival Pinheiro de Sousa s/n, 65900-000 Imperatriz.
MA07) Rua Manoel Alves de Abreu s/n, 65700-000 Bacabal.
MA08) Av. Coronel Fonseca 200, 65800-000 Balsas — DG: Waldemar C. de Paula.
MA09) Parque do Bom Menino, 65025-180 São Luís.
MA10) Rua do Correio s/n, Bairro Fátima, 65030-340 São Luís.
MA11) Av. Presidente Médici 77, 65031-410 São Luís.
MA12) Rua Albino Paiva 655, 65200-000 Pinheiro.
MA13) Rua Manoel Trindade 410, 65725-000 Pedreira.
MA14) Av. Dr. Osano Brandão s/n, 65690-000 Colinas.
MA15) BR-316 km 115, 65370-000 Pindaré Mirim.
MA16) Fazenda São João, BR-14 Km 37, 65215-000 Viana.
MA17) Av. Babaculândia Km 08, 65922-000 João Lisboa.
MA18) Rua A. Bandeira 831, 65320-000 Vitorino Freire.
MA19) Av. Belo Horizonte s/n, 65630-000 Timon.
MA20) Rua T. Caldas s/n, 65926-000 Açailândia.
MA21) Rua Costa e Silva s/n, Fazenda Nova, 65750-000 Esperantinópolis.
MA22) Travessa Tiradentes 338 — Centro, 65415-000 Coroatá.
MA23) Rua Aarão Reis s/n, 65600-000 Caxias.
MA24) Av. Coronel Stanley Fortes Batista 454, 65365-000 Zê Doca.
MA25) Av. Marechal Castelo Branco 605, sala 204, 65075-000 São Luís.
MA26) 65700 Imperatriz.
MA27) Av. Amaral Raposo s/n, 65940-000 Grajaú.
MA28) Rua Rui Barbosa s/n, 65620-000 Coelho Neto.
MA29) Rua Dr. Paulo Ramos 495, 65208-000 Santa Helena.
MA30) Rua Frederico Coelho esquina com Av. Frei Aniceto, 65763-000 Tuntum.
MA31) Rua do Puraqueo s/n, 65350 Vitória do Mearim.
MA32) Rua 15 No. 72, 65665-000 São João dos Patos.
MA33) Rua Mangueira s/n, 65309-000 Santa Luzia.
MA34) Rua Terra esquina com Rua Jupiter s/n, 6570-0000 Presidente Durtra.
MA35) Rua do Comércio s/n, 65500-000 Chapadinha.
MA36) Travessa Aristides Lourenço s/n, 65715-000 Lago da Pedra.
MA37) Rua Hemeterio Leitão 103, 65430-00 Vargem Grande.

MG00) MINAS GERAIS

MG01) C.P. 123, 36771 Cataguases — DG: Mauro Rios Valentim.
MG02) C.P. 109, 39310-970 Monte Claros — DG: Elias Siufi.
MG03) Rua Cafelândia 161, 31015-020 Belo Horizonte — DG: Terezinha Jesus M. Vasconcelos.
MG04) Rua Cruzeiro dos Peixotos 499, 38406 38406-004 Uberlândia — DG: Maurilio Catito Oliveira.
MG05) Praça Honoratio Borges 787, 38740-000 Patrocínio.
MG06) Rua Itatiaia 117, 31210-170 Belo Horizonte — DG: Emanuel.
MG07) C.P. 24, 38001-970 Uberaba.
MG08) Rua dos Expedicionários 67, 35660-040 Pará de Minas.
MG09) Rua Euripides Ribeiro 739, 38720-000 Lagoa Formosa — DG: Silvio Ribeiro.
MG10) Rua Padre Antônio Gabriel Carvalho 17, 35400-000 Ouro Preto — DG: Januário Carneiro.
MG11) C.P. 30, 35791-970 Curvelo — DG: Nelson Rocha F.
MG12) Rua Prof. Monteiro Fonseca 119, 39400-149 Montes Claros.
MG13) Praça Coronel Hermógenes 292, 38770-000 João Pinheiro.
MG14) Rua José de Alencar 700, 30480-500 Belo Horizonte.
MG15) Praça 5 de Novembro 339, 36900-000 Manhuaçu.
MG16) C.P. 37, 37551-970 Pouso Alegre — DG: Milton L. de Paula.
MG17) C.P. 276, 38001-970 Uberaba — DG: Antonio S. Lazenda.
MG18) Rua Dr. João Penido Filho 269, 36021-600 Juiz de Fora.
MG19) Av.Itaú 515, 30730-280 Belo Horizonte.
MG20) Rua Dr. Getúlio Vargas 63, 37200-000 Lavras.
MG21) C.P. 69, 38701-970 Patos de Minas.
MG22) C.P. 10, 35802-970 Itabira.
MG23) C.P. 401, 38409-970 Uberlândia — DG: A. de Castro.
MG24) Praça Getúlio Vargas 10, 35430-002 Ponte Nova.
MG25) C.P. 153, 35291-970 Mantena — DG: U.A. Moreira.
MG26) C.P. 34, 35591-970 Lagoa da Prata — DG: Samuel M. de Almeida.
MG27) Av. Dr. Américo Martins da Costa 92, 35200 Aimorés.
MG28) Rua Minas Gerais 655, 35500-007 Divinópolis.
MG29) Praça Dom Pedro Teixeira 49, 36200-000 Barbacena.
MG30) C.P. 3, 37951-970 São Sebastião do Paraíso — DG: Sebastião Lopes Sousa.
MG31) Rua Itatiaia 117, 31210-170 Belo Horizonte — DG: Emanuel Soares Carneiro.
MG32) Rua Barão de Piunhi 31, 37290-000 Formiga.
MG33) Av. Brasil 2770, 35020-070 Governador Valadares — DG: Odilon Lagares.
MG34) C.P. 65, 38431-970 Tupaciguara — DG: A.M. Netto.
MG35) C.P. 1027, 30161-970 Belo Horizonte — DG: Manuel E. de Lima Torres.
MG36) Rua Dr. Olinto Martins 207, 39960-000 Jequintinhonha.
MG37) Rua Benedito Valadares 139, 39800-000, Teófilo Otoni.
MG38) Praça Couto Magalhães 300, 39100-000 Diamantina.
MG39) Rua Aldo Dato de Almeida 191, 36520-000 Visconde do Rio Branco — DG: Adilson Cunha Honório.
MG40) Rua Tenente Viotti 131, 37460-000 Passo Quatro.
MG41) C.P. 62, 37271-970 Campo Belo — DG: Eugênio Cardoso Furtado.
MG42) Rua Brasil Acioli 28, 38440-000 Araguari.
MG43) C.P. 890, 30161-970 Belo Horizonte — DG: Carlos A. Rodrigues Pinto.
MG44) Rua Dr. Vicente Chirico 46, 37578-000 Bueno Brandão — DG: Jair Asbahr.
MG45) Rua dos Viajantes 81, 35300-000 Caratinga.

MG46) C.P. 34, 36301-970 São João Del Rei — Gte: Weber Neder Issa.
MG47) Rua Antenor Airosa Machado 64, 38420-000 Monte Alegre de Minas — DG: Carlos Nogueira M.
MG48) C.P. 353, 36001-970 Juiz de Fora.
MG49) C.P. 88, 37591-970 Jacutinga — DG: João Alves.
MG50) C.P. 143, 35135-970 Coronel Fabriciano.
MG51) C.P. 557, 38409-970 Uberlândia.
MG52) Rua Bueno Brandão 27, 38430-000 Tupaciguara.
MG53) C.P. 34, 33601-970 Pedro Leopoldo.
MG54) Rua Olegário Maciel 200, 37500-000 Itajubá — DG: José L. Marcondes S.
MG55) C.P. 5, 36404-970 Congonhas — Gte: Lúcio de Sousa C.
MG56) Av. Dr. Américo Luz 153, 37890-000 Muzambinho — DG: Anirce Aparecida Guida.
MG57) C.P. 262, 36001-970 Juiz de Fora — DG: G. Mendes.
MG58) C.P. 85, 38441-970 Araguari — Gte: W. Gebrim.
MG59) Rua Areião Domata Douro s/n, 34000-000 Nova Lima.
MG60) C.P. 253, 38101-970 Uberaba — DG: J. Curi Peres.
MG61) C.P. 34, 37171-970 Esperança — DG: José do Carmo.
MG62) C.P. 10, 35501-970 Divinópolis — DG: Mayrinck Pinto de Aguiar.
MG63) Rua Saturno 360, 37400-000 Campanha.
MG64) C.P. 41, 36881-970 Muriaé — DG: Paulo R. Barros.
MG65) Rua Pernambuco 1077, 30130-151 Belo Horizonte.
MG66) C.P. 61, 38191-970 Sacramento — Dir: José Inácio Medeiros.
MG67) Rua Cassiano Lemos 87, 38180-000 Araxá.
MG68) Av. Assis Chateaubriand 499, Floresta, 30150-101 Belo Horizonte — Dir: Camilo Teixeira da Costa F.
MG69) C.P. 102, 37002-970 Varginha — Dir: S. Sampaio M.
MG70) C.P. 173, 38409-970 Uberlândia — Dir: E. Costa.
MG71) Av. Brasil 3049, 35020-070 Governador Valadares.
MG72) C.P. 110, 38301-970 Ituiutaba.
MG73) C.P. 937, 37701-970 Poços de Caldas — DG: Marcelo de Oliveira.
MG74) Praça Getúlio Vargas 81, 36400-000 Conselheiro Lafayete — DG: Alfredo Ganaime.
MG75) Rua Itajubá 80, 35160-035 Ipatinga.
MG76) Rua Francisco Ribeiro 59, 37250-000 Nepomuceno.
MG77) Rua Cruzeiro dos Peixotos 343, 38400-050 Uberlândia — DG: Paulo dos Santos Rodriques.
MG78) Travessa Juriti 59, 38600-000 Paracatu.
MG79) C.P. 40, 38301-970 Ituiutaba — Dir: Ricardo Alves.
MG80) C.P. 43, 38841-970 Carmo do Paranaíba.
MG81) Rua Dr. Guilherme 286, 35450-000 Itabirito.
MG82) C.P. 143, 37701-970 Poços de Caldas — Dir: F. de Araújo.
MG83) Rua Prof. Maria Justiana 26, 37750-000 Machado.
MG84) Praça Guido Marliere 30, 36500-000 Ubá.
MG85) Rua Canabravo 403, 38610-000 Unaí.
MG86) Rua Calimério Guimarães 308, 38180-000 Araxá.
MG87) C.P. 44, 35001-970 Governador Valadares — DG: José L. Barbosa.
MG88) Rua Dr. Péricles de Mendoça 10, 36680-000 São João Nepumuceno.
MG89) Rua Niquel 457, 35701-107 Sete Lagoas — DG: Gerardo Padrão.
MG90) Av. Major Antônio Alberto Fernandes 445, 37720-000 Botelhos.
MG91) Av. Conde Ribeirão do Vale 661, 37800-000 Guaxupé.
MG92) Rua XV de Novembro 62, 36500-000 Ubá.
MG93) Praça Aires Quaresma s/n, 35930-000 João Monlevade.
MG94) Rua Gerson Coutinho da Silva 1001, 38550-000 Coromandel — Dir: Carlos Amado de Oliveira.
MG95) Praça Pres. Vargas 9, 36660-000 Além Paraíba.
MG96) Av. 17 No. 1045, 38300-000 Ituiutaba.
MG97) Praça Conego Victor 73, 37190-000 Três Pontas.
MG98) Praça da Matriz 5, 38200-000 Frutal.
MG99) C.P. 63, 39861-970 Nanuque — DG: José W. Teixeira L.
MG100) Av. 7 de Setembro s/n, 36325-000 Tiradentes.
MG101) Rua Peixoto 31, 36570-000 Viçosa — DG: J. Bosco T.
MG102) C.P. 28, 37901-970 Passos — Dir: M.A. Soares M.
MG103) C.P. 61, 36971-970 Manhumirim — DG: Antônio Pio L.
MG104) C.P. 044, 35541-970 Oliveira — DG: Wander Assis Ribeiro.
MG105) Rua 13 de Maio 42, 36200-000 Barbacena.
MG106) C.P. 54, 35402-970 Itaúna — DG: Afonso H.S. Lima.
MG107) C.P. 189, 37471-970 São Lourenço — DG: Fernando Ribeiro Jardim.
MG108) Rua Ribeirao São Domingos 689, 38280-000 Iturama.
MG109) Praça São Vicente 64, 35610-000 Dores do Indaiá.
MG110) Praça Senador Eduardo Amaral 39, 37550-000 Pouso Alegre — DG: Murilio Maia.
MG111) Rua Dr. José Gonçalves 17, 35600-000 Bom Despacho.
MG112) Praça Getúlio Vargas 81, 36400-000 Conselheiro Lafayette.
MG113) Av. Getúlio Vargas 422, 37410-000 Três Corações — DG: Victor Cunha.
MG114) Praça Nossa Senhora do Carmo 224, 38500-000 Monte Carmelo.
MG115) C.P. 20, 37541-970 Santa Rita do Sapucaí.
MG116) Praça Coronel Ramos 3, 39270-000 Pirapora.
MG117) C.P. 35, 36701-970 Leopoldina — DG: José A. Barcellos.
MG118) Rua Coronel L. Ribeiro 6, 37130-000 Alfenas — Dir: Albany Gandya.
MG119) Av. Ribeiro Pena 35, 35560-000 Itapecerica — DG: Wantuil Rodrigues Nascimento.
MG120) Rua Anastácio José Gonçalves 139, 38810-000 Rio Paranaíba — DG: João G. da Costa.
MG121) Alto da Boa Vista s/n, 36240-000 Santos Dumont.
MG122) Av. Minas Gerais 584, 39510-000 Espinosa — DG: Antonio P. Tércio.
MG123) Praça Nossa Senhora D'Abadia, 38025-430 Uberaba — DG: Luíz Carlos G. Andrade.
MG124) Av. Hermenegildo Donatti 199, 37795-000 Andradas — DG: José V. Sales.
MG125) Praça Coronel Odilon Coelho s/n, 39520-000 Porteirinha.
MG126) Rua Ordalino Rodrigues 351, 35430-148 Ponte Nova.
MG127) Av. Itapiraçaba 26, 39480-000 Januária.
MG128) Av. Constantino Pinto 90, 36880-000 Muriaé.
MG129) Rua Padre Pedro 53, 39391-000 Bocaiuva.
MG130) Rua Rui Barbosa 74, 39440-000 Janaúba — Gte: Odilon Coelho.
MG131) Av. Getúlio Vargas 25, 35350-000 Raul Soares.
MG132) Rua Divinópolis 619, Ap. 01, 35519-000 Nova Serrana.
MG133) Rua Quintino Bocaiuva 735, 37810-000 Guaranésia.
MG134) BR-496 Km 103, 39260-000 Várzea da Palma.
MG135) Rua dos Quarteis s/n, 38900-000 Bambuí.
MG136) C.P. 318, 38801-970 Teófilo Otoni.
MG137) Av. Dulce Sarmento 767, 39400-000 Montes Claros.
MG138) Rua 102 No. 498, 38360-000 Capinópolis.
MG139) Praça Monsenhor Messias Bragança 80, 37900-000 Passos.
MG140) Rua Padre João Lourenço Leite 100, 37175-000 Ilicínea.
MG141) Rua José Brigido 70, 39200-000 Corinto.
MG142) 35180-000 Timóteo.
MG143) Av. Presidente Vargas 111, 35700-000 Sete Lagoas.
MG144) Rua Guarda Mor 470, 38780-000 Vazante.
MG145) Praça C.P. Coelho, cj. 1003, 36900-000 Manhuaçu.
MG146) 38100-000 Uberaba.
MG147) Rua 20 esquina com Rua 03, 38950-000 Ibiá.
MG148) Praça Getúlio Vargas 108, 36800-000 Carangola.
MG149) Av. dos Andradas 63, 36100-000 Juiz de Fora.
MG150) Rua Ceará esquina com Rua Alagoas, 38700-000 Patos de Minas.

MS00) MATO GROSSO DO SUL

MS01) C.P. 261, 79002-970 Campo Grande — GTE: Ailton Guerra.
MS02) C.P. 315, 79002-970 Campo Grande.
MS03) C.P. 104, 79951-970 Naviraí.
MS04) C.P. 68, 79804-970 Dourados.
MS05) Av. Dr. Mario de Freitas 120, 79115-800 Campo Grande.
MS06) Rua José Bonifácio 616, 79200-000 Aquidauana.
MS07) Av. Mario Correia s/n, 79150-000 Maracaju.
MS08) Rua Joaquim Pereira Teixeira 135, 79900-000 Ponta Porã.
MS09) Rua Maracaju 858, 79002-212 Campo Grande.
MS10) C.P. 37, 79601-970 Três Lagoas — DG: O. Tácio de Lima.
MS11) Av. Marcelino Pires s/n, 79830-000 Dourados.
MS12) C.P. 18, 79201-970 Aquidauana — DG: Also Bertoni.
MS13) Rua Antônio Lino Barbosa 270, 79130-000 Rio Brilhante.
MS14) C.P. 138, 79301-970 Corumbá — DG: Uriel Raghiant.
MS15) C.P. 217, 79301-970 Corumbá — DG: Fause Anache.
MS16) Rua Walther Hubachi 643, 79750-000 Nova Andradina.
MS17) Av. Aldair Rosa de Oliveira 1045, 79640-100 Três Lagoas.
MS18) Rua Ferreira 69 B, Piracema, 79400-000 Coxim — DG: Moacir Moura.
MS19) Av. Pres. Vargas 1893, 79730-000 Glória de Dourados.
MS20) Rua 13 de Maio 249, 79500-000 Paranaíba.
MS21) C.P. 126, 79991-970 Amambaí — Gte: Eldir Vieira N.
MS22) Rua Tenente Antônio João 1592, 79700-000 Fátima do Sul.

MS23) C.P. 200, 79541-970 Cassilândia — DG: João Lourenço Girotto.
MS24) Av. da Saudade, Q-146, 79940-000 Caarapó.
MS25) Rua Dr. A. Coelho de Oliveira 549, 79240-000 Jardim.
MS26) Rua Ponta Porã 433, 79970-000 Eldorado — DG: Valério de Medeiros.
MS27) C.P. 129, 79301-970 Corumbá.
MS28) C.P. 199, 79901-970 Ponta Porã — DG: Elite José Sandri.
MS29) C.P. 68, 79804-970 Dourados.
MS30) Rua São Paulo 1359, 79460-000 São Gabriel d'Oeste.
MS31) Rua Marques de Tamandaré 349, 79170-000 Sidrolândia.
MS32) Rua Itajai 473, Bairro Antonio Vendas, 79041-270 Campo Grande.
MS33) Rua Severino de Araujo Ferreira 1375, 79700-000 Fátima do Sul.
MS34) Av. Catógeros 1932, 79005-000 Campo Grande.
MS35) Av. Costa Rica 654, 79740-000 Ivinhema.
MS36) Av. João Pedro Pedrossian 1545, 79570-000 Aparecida do Tabuado.
MS37) 79300-000 Corumbá.
MS38) Rua Candido Severino 462, 79420-000 Camapuã.
MS39) Rua Major Capile 1564, 79800-000 Dourados.
MS40) Av. Bandeiras s/n, Prolongamento, 79100-000 Campo Grande.
MS41) Rua Coronel Comisão 621, 79260-000 Bela Vista.
MS42) Rua Barão do Rio Branco 245, 79480-000 Rio Verde de Mato Grosso.
MS43) Rua Atilio Reginato 355, 79740-000 Ivinhema.

MT00) MATO GROSSO

MT01) C.P. 214, 78601-970 Barra do Garças.
MT02) Av. Brasil 27, 78600-000 Poxoréo — Gte: Edisio Rodrigues Rocha.
MT03) Praça Alice Garias 117, 78020-630 Cuiabá — DG: Onofre Ribeiro.
MT04) C.P. 133, 78551-970 Sinop — DG: Ascancio B. Carvalho.
MT05) C.P. 22, 78005-970 Cuiabá — Gte: A. Chaves Vasconcellos.
MT06) Av. Leste s/n, 78580-000 Alta Floresta.
MT07) C.P. 401, 78701-970 Rondonópolis — DG: Edson A. Pires.
MT08) C.P. 405, 78005-970 Cuiabá — Gte: Antônio Fauser.
MT09) Rua São Pedro 806, 78200-000 Cáceres.
MT10) C.P. 297, 78201-970 Cáceres — DG: Ivo Vignardi.
MT11) C.P. 227, 78821-970 Jaciara — DG: Francisco A. de Moura.
MT12) Rua Fernando Correia da Costa 651, 78740-000 Rondonópolis.
MT13) Rua 28 de Outubro 761, 78280-000 Mirassol D'Oeste.
MT14) Rua Benedito Monteiro 68, 78110-390 Várzea Grande.
MT15) C.P. 277, 78005-970 Cuaibá.
MT16) Rua 2 No. 32, 78175-000 Poconé.
MT17) Chacara 70/71, 78575-000 Juara.
MT18) Av. Brasil 1830, 78300-000 Tangará da Serra.
MT19) Av. Principal 1632, 78500-000 Colider.
MT20) Rua Valdir Rabelo s/n, 78600-000 Barra do Garças.
MT21) Av. Natalino J. Brescancin 355, 78890-000 Sorriso.
MT22) Praça Brigadeiro Eduardo Gomes 28, 78430-000 Nortelândia.
MT23) Av. Mato Grosso 320, 78690-000 Nova Xavantina.
MT24) Rua 6 No. 498, 78300-000 Tangará da Serra.
MT25) Rua Benjamim Constant s/n, 78780-000 Alto Araguaika.
MT26) Av. Barra do Garças s/n, 78320-000 Juina.
MT27) Av. Mario Correa 350, 78530-000 Itiquira.
MT28) Av. Agricola Pães de Barros 924, 78500-000 Colider (Matpuá).
MT29) Rua das Primaveras 3574, 78260-000 Sinop.
MT30) Rua Niterói 584, 78735-000 São José dos Quatro Marcos.
MT31) Rua 2 esquina com Rua 6, 78860-000 Diamantino.
MT32) Rua 7, Quadra 37, 78260-000 Chapada dos Guimarães.
MT33) Rua Leonildo Bernardes Quadra 06, lote 9, 78870-000 São José do Rio Claro.
MT34) Rua 02 — Canteiro Central, 78890-000 Alta Floresta.
MT35) Av. Bahia 430, 78400-000 Guiratinga.
MT36) Rua Joaquim Nabuco s/n, 78700-000 Araputanga.
MT37) Rua São Paulo 1440, 78770-000 Pontes e Lacerda.
MT38) Margem Esquerda BR-163 Km 725, 79640-000 Guarantã do Norte.

PA00) PARÁ

PA01) Av. Araguaia s/n, 68551-000 Redenção.
PA02) C.P. 119, 68371-970 Altamira — Gte: Ivanese Braga.
PA03) C.P. 533, 66017-970 Belém — DG: Edyr Paiva Proença.
PA04) Rua Floriano Peixoto 632, 68105-060 Santarém — DG: Evaldo P. Azevedo.
PA05) Rua V. Virgulina Ferreira s/n, 68540-000 Conceição do Araguaia.
PA06) Praça dos Notáveis 1006, 68400 Cametá — DG: Gerson dos S. Peres.
PA07) Rodovia BR-316 Km 58, 68741-740 Castanhal.
PA08) Tavessa Campos Sales 370, 66019-050 Belém — DG: Jones Lara Tavares.
PA09) Av. Governador José Malcher 1332, 66060-230 Belém.
PA10) C.P. 498, 66017-970 Belém.
PA11) C.P. 038, 68801-970 Breves — DG: Adilson Almeida.
PA12) Rua Barão do Rio Branco 1151, 68600-000 Bragança.
PA13) Rua Galdino Veloso 786, 68005-070 Santarém.
PA14) Av. Coronel Monfredo 42, 68820-000 São Sebastião da Boa Vista.
PA15) C.P. 226, 68371-970 Altamira.
PA16) Av. Almirante Barroso 735, 66090-000 Belém — Dir: Ronald Pastor.
PA17) Rodovia PA-150 Km 8e9, 68500-000 Marabá.
PA18) 68180-000 Itaituba.
PA19) Av. Lauro Sodré 722, 68456-000 Tucuruí — DG: José A. Costa
PA20) Av. Rui Barbosa 136, 68100-000 Santarém.
PA21) Av. Xingu s/n, 68555-000 Xinguara.
PA22) Travessa E. Simões 230, 68250-000 Óbidos.
PA23) C.P. 8, 68251-970 Óbidos — Dir: Max Hamoy.
PA24) Travessa 18 No. 1863, entre 4 e 5 ruas, 68870-000 Soure.
PA25) Rua Visconde de Souza Franco 116, 68780-000 Vigia.
PA26) Rodovia BR-230 (Transamazônica) Km 01, 68180-000 Itaituba.
PA27) Rodovia PA-125 Km 15, 68630-000 Paragominas.
PA28) Rua 2 de Outobro s/n, 68200-000 Alenquer.
PA29) Rodovia Transamazônica Km 04, 68500-000 Marabá.
PA30) Rua Francisco Chaves s/n, 68660-000 São Miguel do Guamá.
PA31) 68460-000 Tucuruí.
PA32) 66240-000 Belém.
PA33) Av. Beira Rio s/n, 68380-000 Sáo Felix do Xingu.
PA34) Av. Augusto Montenegro s/n, 68720-000 Maracanã.

PB00) PARAÍBA

PB01) Rua Pres. João Pessoa 25, 58800-010 Sousa.
PB02) C.P. 26, 58900-970 Cajazeiras — Francisco A. de Albuquerque.
PB03) Rua Epitácio Pessoa 10, 58970-000 Conceição.
PB04) Rua Presidente Epitácio Pessoa 242, 58700-020 Patos.
PB05) C.P. 40, 58200-970 Guarabira — DG: H. Camilo.
PB06) Praça Pres. Epitácio Pessoa 167, 58755-000 Princesa Isabel.
PB07) C.P. 134, 58101-970 Campina Grande.
PB08) Rua Coronel Juvêncio Carneiro 160, 58900-000 Cajazeiras — DG: Francisca Zélia Ribeiro.
PB09) C.P. 1089, 580001-970 João Pessoa — DG: Deodato Borges.
PB10) Rua Manoel Pedro s/n, 58884-000 Catolé do Rocha.
PB11) Rua Venâncio Neiva 287, 58100-060 Campina Grande — DG: Luiz Aguacar.
PB12) Av. Pedro II 657, 58013-040 João Pessoa.
PB13) Rua das Trincheiras 198, Centro, 58011-000 João Pessoa — DG: Alberto P. de Castro.
PB14) C.P. 160, 58100-970 Campina Grande — DG: L. Barbosa A.
PB15) C.P. 57, 58701-970 Patos — DG: Gerardo Andrade Ponte.
PB16) Rua Antônio Martins s/n, 58290-000 Mamanguape.
PB17) Rua Dr. Carlos Pires 17, 58804-200 Sousa.
PB18) Rua Monsenhor Valeriano s/n, 58840-000 Pombal.
PB19) Rua Antônio Firmino s/n, 58187-000 Picuí — DG: A. Ferreira D.
PB20) Rua Cândido de Assis 421, 58840-000 Pombal.
PB21) Rua Oswaldo Cruz 161, 58305-000 Bayeux.
PB22) Rua Epitácio Pessoa 8, 58200-000 Guarabira.
PB23) Rua Monsenhor Palmeira s/n, 58135-000 Esperança.
PB24) Av. Getúlio Vargas 129, 58500-000 Monteiro.
PB25) Rua Coronel Pedro Targino s/n, 58233-000 Araruna.
PB26) Rodovia PB-105 Km 33 Anel do Brejo, 58220-000 Bananeiras.
PB27) 58155-000 Soledade.
PB28) Rua J. Sabia s/n, 58540-000 Sumé.

PB29) Rua Luíz Felix esquina com Francisco Rodrigues, 58700-000 Patos.
PB30) Rua Epifanio Sobreira 14, 58900-000 Cajazeiras.
PB31) Av. Almirante Barros 918, 58000-000 João Pessoa.
PB32) Rua Epitácio Pessoa 92, 58397-000 Areia.

PE00) PERNAMBUCO

PE01) Rua Mauricio de Nassau 317, 55290-000 Garanhuns.
PE02) Rua Floriano Peixoto 780, 50020-060 Recife — DG: Fernando Espinoza.
PE03) Av. Joaquim Nabuco 322, 56500-000 Arcoverde — DG: Áureo Bradley Netto.
PE04) Praça da Bandeira, 55700-000 Limoeiro.
PE05) Rodovia Araraipina — Picos Km 3, 56280-000 Araripina.
PE06) Travessa 7 de Setembro s/n, 56300-000 Petrolina.
PE07) Rua do Veiga 590, 50040-110 Recife — Gte: Marcos Leite.
PE08) C.P. 8, 56300-970 Petrolina — DG: Frei Paulo Cardoso.
PE09) Rua Capitão Lima 250, 50040-080 Recife — DG: Paulo F. Neto.
PE10) C.P. 1076, 50001-970 Recife — DG: Carlos A. Pacheco.
PE11) C.P. 292, 50001-970 Recife — DG: J.C. Poncel Neto.
PE12) Rua Conceição 16, 55000-000 Caruaru — Dir: J. del Lacerda.
PE13) Av. Padre Rocha s/n, 55810-000 Carpina — Dir: V. Alves C.
PE14) Rua João Belarmino Guerra s/n, 55870-000 Timbaúba.
PE15) Estrada do Passarinho 1415, 53170-110 Olinda — Dir: R. Neto.
PE16) C.P. 88, 55001-970 Caruaru — Gte: Liezed Alvara.
PE17) Rua Floriano Peixoto 780, 1° andar, 50020-060 Recife.
PE18) Rua do Expedicionário 30, 55000-000 Caruaru.
PE19) Rua dos Ferreiros s/n, Granja Fazenda Nova, 55600-000 Vitória de Santo Antão — DG: Eduard Glaser.
PE20) C.P. 13, 55301-970 Garanhuns — DG: Aluizio Alves de Melo.
PE21) Rua Coronel Urbano Ribeiro de Sena 956, 52221-000 Recife — DG: Humberto de Mota Pinto.
PE22) 54750-000 Camarajibe.
PE23) Rua Pajussara 225, Bairro Tigipió, 55010-460 Recife — Dir: M.A. da Cunha.
PE24) Av. F. Pessoa de Queiróz s/n, 55200-000 Pesqueira — Gte: Geraldo Santana.
PE25) Bairro São Manoel a margem da BR-101, 55540-000 Palmares.
PE26) Rua Inocencio Gomes de Andrade 619, 56900-000 Serra Talhada.
PE27) Rua 3 de Maio s/n, 56800-000 Afogadas da Ingazeira.
PE28) Rua Benjamin Constant 16, 55750-000 Surubim.
PE29) Rua Siqueira Campos 108, 55150-000 Belo Jardim.
PE30) Rua Antônio Figueira Soares s/n, 56000-000 Salgueiro.
PE31) Av. C. Colombo 16, 56700-000 São José do Egito.
PE32) Praça Duque de Caxias 818, 55900-000 — Gte: José Daniel da Silva.
PE33) Av. Fernando Bezerra s/n, 56200-000 Ouricuri.
PE34) Rua Manoel Balbino 184, 55190-000 Santa Cruz do Capibaribe.
PE35) Av. Coronel Trapia s/n, 56440-000 Belém de São Francisco — Dir: Rosimary G. Torres.
PE36) Rua Santo Amaro 26, 56440-000 Gravatá.
PE37) Rua do Cigna s/n, 55330-000 Bom Conselho.
PE38) Osvaldo Celso Maciel s/n, 55370-000 São Bento do Una.

PI00) PIAUÍ

PI01) Praça Júlio Paixão s/n, 64770-000 São Raimundo Nonato.
PI02) Av. Centenário 1200, 64003-700 Teresina.
PI03) Rua Taumaturgo de Azevedo 995, 64100-000 Barras — DG: E. Cronemberger.
PI04) C.P. 209, 64001-970 Teresina — Gte: Valter Alencar.
PI05) Av. Heróis do Jenipapo 37, 64280-000 Campo Maior.
PI06) Praça Pres. Kennedy 233, 64490-000 Regeneração.
PI07) Praça Coronel O. Carvalho 400, 64500-000 Oeiras.
PI08) Av. Prof. Valter Alencar 2021, 64017-500 Teresina — DG: José D. da Silva Lima.
PI09) Rua Joaquim Baldoíno 40, 64600-000 Picos – DG: José Nunes de Barros.
PI10) Rua 24 de Janeiro 150, 64001-230 Teresina — Dir: Tony Batista.
PI11) Av. Pres. Vargas 266, 64200-000 Parnaíba.
PI12) Rua Prof. Bem 838, 64260-000 Piripiri.
PI13) C.P. 125, 64001-970 Teresina — DG: V. Machado.
PI14) Av. Prefeito J. de Carvalho, 64400-000 Amarante — Gte: José A. Nunes N.
PI15) Av. Rio Branco s/n, 64860-000 Uruçuí.
PI16) C.P. 41, 64800-970 Floriano — DG: J. do Vale Reis.
PI17) Praça Emilio Cavalcante 29, 64980-000 Corrente.
PI18) Rua Antônio Neto 1065, 64800-000 Floriano.
PI19) Rua São Paulo 696, 64760-000 São João do Piauí.
PI20) Av. João de Paiva 94, 64290-000 Altos — DG: Elvira Raulino.
PI21) Rua Arsénio Santos 555, 64900-000 Bom Jesus — DG: Dantas Lima.
PI22) Rua Riachelo 770, 64200-280 Parnaíba.
PI23) T. Azavedo 800, 64100-000 Barras.
PI24) Praça Presidente Castelo Branco s/n, 64750-000 Paulistana.
PI25) Rua Fernando Bacelar 480, 64240-000 Piracuruca.
PI26) Av. José de Deus Lacerda 584, 64130-000 Miguel Alves.
PI27) Praça Getúlio Vargas, 64680-000 Jaicós.
PI28) Rua Corrinto de Andrade s/n, 64255-000 Pedro II.
PI29) Praça da Independência 69, 64325-000 Elesbão Veloso.
PI30) Rua Joaqium Baldoíno 40, 64600-000 Picos.
PI31) Av. João Fereira 199, 64460-000 Agua Branca.
PI32) Rua Hugo Napoleão 940, 64110-000 José de Freitas.
PI33) Rua Pedro II 695, 64340-000 Castelo do Piauí — Gte: Alcides L. Martins.
PI34) Rua Pedro II s/n, 64330-000 São Miguel do Tapuio.
PI35) Rua Merechal Dutra 366, 64810-000 Canto do Buriti.
PI36) Rua Matias Gomes 510, 64700-000 Simpício Mendes.
PI37) Rua Coronel Narciso 728, 64120-000 União.
PI38) Rua Padre Joaquim Nonato s/n, 64390-000 Demerval Lobão.
PI39) Rua Coronel Messias Melo 766A, 64190-000 Batalha.
PI39) Praça Marechal Deodoro da Fonseca 349, 64810-000 Canto do Buriti.
PI41) Rua Antônio Freire 1356, 64000-000 Teresina.
PI42) Praça Waldemar Leal 42, 64660-000 Padre Marcos.
PI43) Rua Sete de Setembro 471, 64160-000 Luzilândia.
PI44) Praça da Bandeira 91, 65400-000 Oeiras.
PI45) 64000-000 Teresina.
PI46) Estrada Picos/Bocaina Km 22, 64615-000 Bocaina.
PI47) Rodovia BR-343 s/n, 64290-000 Altos.
PI48) Rua Coronel Anibal Martins 581, 64300-000 Valença do Piauí.
PI49) Estrada Barra do Longa, Periferia da Cidade, 64220-000 Buriti dos Lopes.
PI50) Rua Coronel José Fortes 549, 64189-000 Esperantina.

PR00) PARANÁ

PR01) C.P. 44, 86101-970 Londrina — Dir: R. dal Coineto.
PR02) C.P. 157, 85100-970 Guarapuava — DG: Luis V. Sanquetta.
PR03) C.P. 10, 85892-970 Santa Helena — DG: Naudé Pedro Prates.
PR04) C.P. 202, 80001-970 Curitiba — Dir: João L. Seiter B.
PR05) Av. Anunciato Sonni 1673, 86900-000 Jandaia do Sul.
PR06) Rua Cruz Machado 66, 80410-170 Curitiba — DG: Raul Varassin.
PR07) Rua 7 de Setembro 520, 85960-000 Marechal Cândido Rondón — DG: Werner Wanderer.
PR08) C.P. 337, 86001-970 Londrina.
PR09) C.P. 78, 86390-970 Cambará.
PR10) C.P. 218, 87600-970 Nova Esperança.
PR11) Av. Marechal Humberto de Alencar Castelo Branco 69, 82530-020 Curitiba — Dir: A. R. de Assis.
PR12) Prolongamento da Rua XV de Novembro s/n, 85100-000 Guarapuava.
PR13) C.P. 1870, 86101-970 Londrina — DG: C.A. Gomes.
PR14) Rua XV de Novembro 439, 84010-020 Ponta Grossa.
PR15) Av. Paraná 1896, Bairro Boa Vista, 82510-000 Curitiba.
PR16) C.P. 528, 87300-970 Campo Mourão — DG: Nelson Amaral.
PR17) Av. Daniel Portela 614, 87360-000 Goio-Erê.
PR18) C.P. 209, 87201-970 Cianorte — DG: S. Garcia P.
PR19) Rua Diego A. Feijo s/n, 85560-000 Chopinzinho.
PR20) Av. Brasil 178, 83800-000 Mandirituba.
PR21) C.P. 312, 85890-970 Foz do Iguaçu — DG: Ennes Mendes da Rocha.
PR22) C.P. 60, 83700-970 Araucária — DG: J. Ukachenski.
PR23) Rua Coronel Manoel Ferreira Bello 64, 85530-000 Clevelândia.

PR24) C.P. 255, 86001-970 Londrina — DG: Amarildo Lopes.
PR25) Av. Getúlio Vargas 268, 87013-130 Maringá.
PR26) C.P. 200, 85500-970 Pato Branco.
PR27) Praça Marechal Floriano Peixoto 581, 84010-680 Ponta Grossa — DG: Nilson de Oliveira.
PR28) Rua Guarani 1479, 85900-901 Toledo.
PR29) Rua Prof. João Cândido Ferreira 315, 86800-100 Apucarana.
PR30) C.P. 178, 86600-970 Rolândia — DG: A. Pinceli.
PR31) Rua João Negrão 554, 80010-200 Curitiba.
PR32) Rua Joubert de Carvalho 623, 87013-200 Maringá.
PR33) C.P. 414, 86001-970 Londrina — DG: Pe. José Guidoreni.
PR34) Rua Santa Catarina 1020, 85960-000 Marechal Cândido Rondón — DG: Elio E. Winter.
PR35) C.P. 591, 85501-970 Pato Branco — Gte: Lindolfo Schmitz.
PR36) Praça Generoso Marques 90, 80020-030 Curitiba.
PR37) C.P. 219, 85884-970 Medianeira — DG: Irineu Pelissari.
PR38) Rua João de Rezende 259, 87400-000 Cruzeiro do Oeste.
PR39) Av. São Paulo 165, 86010-040 Londrina.
PR40) Rua Barão do Rio Branco 404, 83900-000 São Mateus do Sul.
PR41) Av. Carlos Gomes 1600, 85800-000 Cascavel.
PR42) Rua Dr. Muricy 926, 80020-040 Curitiba.
PR43) Praça Alvorada, 86690-000 Colorado.
PR44) Av. Julio Assis Cavalheiro 162, 85601-000 Francisco Beltrão.
PR45) Av. Manoel Ribas 110, 84600-000 União da Vitória.
PR46) C.P. 09, 85400-970 Guaraniaçu — DG: Caetano B. Filho.
PR47) C.P. 114, 84001-970 Ponta Grossa.
PR48) Rua Getúlio Vargas 1050, 87772-000 Paranavaí.
PR49) C.P. 101, 86300-000 Cornélio Procópio — DG: Santiago Reghin.
PR50) Av. Capitão Indio Bandeira 1115, 87300-000 Campo Mourão — DG: Irineu Luiz F. Lima.
PR51) C.P. 149, 85500-970 Coronel Vivida — DG: M. Cassol.
PR52) C.P. 596, 86001-Londrina — DG: J.B. Faria.
PR53) C.P. 107, 83005-970 São José dos Pinhais.
PR54) Av. Cristóvão Colombo 1055, 86990-000 Marialva.
PR55) Rua Dr. Jorge Xavier da Silva 540, 84165-680 Castro — DG: Ronie Cardoso Filho.
PR56) C.P. 56, 86001-970, Londrina — DG: Toshiro Okada.
PR57) Rua Presidente Beaurepaire Rohan 44, 80050-340 Curitiba — DG: João B. Ramos da Silva.
PR58) C.P. 258, 85900-970 Toledo — DG: Avelino Campagnolo.
PR59) Av. Londrina 500, 87970-000 Nova Londrina — DG: Joaquim de Paula.
PR60) C.P. 7133, 80011-970 Curitiba — Dir: Luiz Andreu Rúbio.
PR61) Av. Paraná 596, 85887-000 Matelândia.
PR62) Rua Dr. Vicente Machado 1384, 85010-260 Guarapuava.
PR63) Av. Getúlio Vargas 500, 87702-000 Paranavaí.
PR64) Av. dos Pinheiros 1970, 85650-000 Santa Isabel do Oeste.
PR65) C.P. 239, 86975-970 Mandaguari — Dir: Cylleneo Junior.
PR66) C.P. 91, 86430-970 Santo Antônio da Platina.
PR67) Rua Pedro Eloy de Souza 51, 82820-130 Curitiba.
PR68) Rua Rio Grande do Sul 1110, 85806-010 Cascavel.
PR69) C.P. 71, 85600-970 Francisco Beltrão — DG: A. Seleski.
PR70) C.P. 1300, 87100-970 Maringá.
PR71) C.P. 700, 84001-970 Ponta Grossa — Dir: Helio Porto.
PR72) Rua Desembargador Clotário Portugal 222, 84010-220 Curitiba.
PR73) Av. Paraná 343, sala 1-006, 86100-000 Londrina.
PR74) C.P. 26, 84201-970 Jaguariaíva — DG: A. da Silva C.
PR75) C.P. 13, 87500-970 Umuarama.
PR76) C.P. 45, 83880-970 Rio Negro — Dir: A. de Aquinno.
PR77) Rua Aguias 27, 86707-190 Arapongas — DG: J. Dutra.
PR78) Av. Paula Harris 50, 86220-000 Assaí.
PR79) Rua Pedro Alves Cabral 1609, 87560-000 Iporã.
PR80) Rua Amauri Lange Silvério 608, 82120-000 Curitiba.
PR81) Rua Meron Heuko 501, 86930-000 São João do Ivaí.
PR82) C.P. 76, 87001-970 Maringá — Dir: Joaquim Dutra.
PR83) Rua Vereador Idanir Canello 18, 85670-000 Salto do Lontra.
PR84) Av. Paraná 590, 86010-370 Londrina.
PR85) Av. Dep. Ivan Ferreira do Amaral Filho 1496, 85303-000 Laranjeiras do Sul — DG: Ângelo Manoel da Cunha.
PR86) Rua Simão Domingues 26, 84550-000 Rebouças.
PR87) Rua Vereador Vicente Balan 12, 85990-000 Terra Roxa.
PR88) C.P. 263, 87500-970 Umuarama — DG: Marcia Mori.
PR89) C.P. 381, 86400-970 Jacarezinho — Gte: Padre João Chiaroy.
PR90) Av. Brasil 702, 85710-000 Santo Antônio do Sudoeste.
PR91) C.P. 66, 85801-970 Cascavel — DG: B.J. Motto.
PR92) C.P. 307, 84260-970 Telêmaco Borba — DG: Horacio Klabin.
PR93) Av. Bandeirantes 958, 86360-000 Bandeirantes.
PR94) C.P. 04, 87350-970 Ubiratã — DG: A. Zamprôno.
PR95) C.P. 171, 87550-970 Altônia — DG: Antônio Duma.
PR96) Rua Prof. Viana Cleto 281, 83221-320 Paranaguá.
PR97) Rua XV de Novembro 452, 84010-020 Ponta Grossa.
PR98) C.P. 777, 86800-970 Apucarana — DG: Cynyra Woiski Ignácio.
PR99) C.P. 66, 84670-970 Palmas — DG: Miraldo de Carli.
PR100) C.P. 217, 85980-970 Guaíra — DG: Edson Luiz da Silva.
PR101) Rua Londrina 410, 85640-000 Ampere.
PR102) C.P. 171, 87900-970 Loanda — DG: Theodoro H. Neto.
PR103) C.P. 16, 86590-970 Ibati.
PR104) C.P. 54, 85935-970 Assis Chateaubriand — DG: Afonso C.R. do Prado.
PR105) Rua Ebano Pereira 157, 85200-000 Pitanga.
PR106) C.P. 45, 86730-970 Astorga — DG: O. Bassan.
PR107) Rua Ipiranga 91, 84600-000 União da Vitória — Dir: Adelino Ritzmann.
PR108) C.P. 230, 86300-000 Cornélio Procópio.
PR109) Rua 7 de Setembro 42, 83750-000 Lapa.
PR110) C.P. 195, 86925-970 Borrazópolis.
PR111) C.P. 121, 84400-970 Prudentópolis.
PR112) C.P. 304, 86700-970 Arapongas.
PR113) C.P. 72, 86500-970 Venceslau Brás.
PR114) Rua XV de Novembro 253, 84500-000 Irati.
PR115) Rua Curitiba 1534, 85840-000 Céu Azul.
PR116) Travessa Vale Porto 240, 83370-000 Antonina.
PR117) C.P. 267, 86870-970 Ivaiporã — DG: Maria das Dores Moreira.
PR118) Rua 5 de Julho s/n, 85950-000 Palotina.
PR119) C.P. 10, 83650-970 Balsa Nova.
PR120) Rua Mauá 2518, 85770-000 Realeza.
PR121) C.P. 358, 87200-970 Cianorte — DG: A. Prado Ruiz.
PR122) Rua Marechal Deodoro 414, 86530-000 Siqueira Campos.
PR123) C.P. 11, 85830-970 Formosa do Oeste.
PR124) C.P. 10, 85615-970 Marmeleiro — DG: Augustinho Seleski.
PR125) C.P. 100, 85761-970 Capanema — Gte: M. Cruz.
PR126) Rua XV de Novembro 3544, 85010-000 Guarapuava.
PR127) C.P. 134, 85590-970 Dois Vizinhos —DG: Valdir Luiz P.
PR128) C.P. 199, 84500-970 Irati — DG: Nagib Harmuch.
PR129) Rua São Paulo 180, 86380-000 Andirá — Dir: V. de Resende.
PR130) Rua São Paulo 489, 86840-000 Faxinal.
PR131) Rua Noruega 98, 86182-000 Cambé — DG: Antônio E. Casaroto.
PR132) Praça São Pedro 1210, 85460-000 Quedas do Iguaçu.
PR133) C.P. 46, 84330-970 Palmeira — DG: Euclides A. de Paula F.
PR134) C.P. 448, 80001-970 Curitiba — DG: Flavio Augusto Dallegrave.
PR135) Av. Brasil 746, 84350-000 Ortigueira.
PR136) Av. Maximiliano Vicentim 240, 85270-000 Palmital — DG: Edmilson Cecura.
PR137) C.P. 101, 85410-970 Nova Aurora.
PR138) C.P. 934, 82001-970 Curitiba — Gte: João Falavinha Ienze.
PR139) Rua Oyapoque 649, 80050-000 Curitiba.
PR140) Rua Urbano Lunardelli 875, 86160-000 Porecatu.
PR141) Av. Paraná 126, 84470-000 Cândido de Abreu.
PR142) Av. Santos Funont 971, Laranjeiras do Sul.
PR143) Rua Marechal Deodoro 175, 85550-000 Coronel Vivida.
PR144) Av. 15 de Novembro s/n, 85560-000 Chopinzinho.
PR145) Rua Almirante Barroso 1215, 85851-010 Foz do Iguaçu — DG: Nilton Vitorassi.
PR146) Av. Presidente Kennedy 170, 85950-000 Palotina.
PR147) Rua Amor Perfeito 1827, 85420-000 Corbélia.
PR148) Rua Souza Naves 102, 85877-000 São Miguel do Iguaçu.
PR149) Av. Iguaça 257 — Centro, 85575-000 São Jorge d'Oeste.
PR150) Av. Santo Antônio 826, 87320-000 Roncador.
PR151) Av. Yolanda Loureiro 1021, 87350-000 Ubiratã.

RJ00) RIO DE JANEIRO

RJ01) Rua Visconde de Itaborai 184, 24030-091 Niterói — Gte: Alberto F. Torres.

RJ02) C.P. 496, 28970-970 Araruama — Dir: Ruy Carvalho da Silva.
RJ03) Av. Presidente Vargas 409, Crupo 2201, 20071-003 Rio de Janeiro — DG: Leonardo Abbud.
RJ04) Av. Erasmo Braga 118, 20020-000 Rio de Janeiro.
RJ05) Rodivia Presidente Dutra Km 303, 27540 Resende.
RJ06) Praça Demerval Barbosa, Moreira 28, 28610-000 Novo Friburgo.
RJ07) Rua Visconde de Inhaúma 37, 20091-000 Rio de Janeiro — Dir: Marilea Sales.
RJ08) Av. 24 de Outubro 201, 28100-000 Campos.
RJ09) Rua México III, Sobraloja, 20031-145 Rio de Janeiro.
RJ10) Av. Djalma Beda Coube s/n, 28500-000 Cantagalo — DG: Nelson Leão Nascimento.
RJ11) Rua do Russel 766, 22210-010 Rio de Janeiro.
RJ12) Praça da República 141-A, 3° andar, sala 306, 20211-350 Rio de Janeiro — Dir: Paulo Henrique Cardoso.
RJ13) Rua Teixeira de Gouveia 431, 27913-120 Macaé.
RJ14) Rua do Russel 434, Glória, 22210-010 Rio de Janeiro.
RJ15) Av. Portugal 96, Urca, 22291-050 Rio de Janeiro.
RJ16) Av. Brasil 500, sala 729, São Cristovão, 20940-070 Rio de Janeiro — DG: Geraldo Leite.
RJ17) Praça Mauá 7, 20081-240 Rio de Janeiro — Gte: Luis E. Porto.
RJ18) Rua do Bispo, 20261-060 Rio de Janeiro — DG: W. Luiz.
RJ19) Rua Coronel Carvalho 173, 23900-000 Angra dos Reis.
RJ20) Rua Machado Bittencourt 300, 26900-000 Miguel Perreira.
RJ21) Av. Sete de Setembro 380, 28013-000 Campos.
RJ22) Rua México 111, 20031-145 Rio de Janeiro — Dir: S. Pires C.
RJ23) C.P. 79, 28100-970 Campos — DG: Salvador Macedo.
RJ24) Rua Duque de Caxias 627, 25802-120 Três Rios — DG: Edson Elias B. Jorge.
RJ25) Rua do Russel 434, 22210-010 Rio de Janeiro.
RJ26) Rua Carlos de Lacerda 75, 28100-000 Campos — DG: Diva Arreu B.
RJ27) Rua do Livramento 189, 20221-191 Rio de Janeiro — DG: Alfredo R. Filho.
RJ28) Rua Alberto Torres 410, 28400-000 São Fidelis — DG: Amary de Almeida.
RJ29) Av. 28 de Setembro 258, 20551-031 Rio de Janeiro.
RJ30) Av. Treze de Maio 23, 20031-000 Rio de Janeiro — DG: Laércio Alves.
RJ31) Rua Carlos Lacerda 52, 28100-000 Campos.
RJ32) C.P. 85919, 27401-970 Barra Mansa — Dir: O. Marcui N.
RJ33) Av. Teodoro da Silva 371, 20560-000 Rio de Janeiro.
RJ34) Av. Cardoso Moreira 422, 28300-000 Itaperuna — DG: José G. Leonardo.
RJ35) Rua Riachuelo 359, 20230-012 Rio de Janeiro.
RJ36) Av. Joaquim Leite 279, 2° andar, Centro, 27330-042 Barra Mansa — DG: Ildeu Nodge A. da Silva.
RJ37) Rua Dr. Temistocles de Almeida 97, 28470-000 Santo Antônio de Pádua.
RJ38) Praça Nilo Peçanha 42, 27123-020 Barra do Piraí.
RJ39) Praça Procópio Ferreira 22, 26000-000 Novo Iguaçu.
RJ40) Rua Eugênio Cordeiro 28, 28800-000 Rio Bonito.
RJ41) Rua 100 No. 01, Bairro Laranjal, 27255-000 Volta Redonda — Gte: Joel de Mattos Alvarenga.
RJ42) Rua Barão de Cotegipe 90, 28100-000 Campos.
RJ43) Rua José Augusto da Costa 14, 25953-160 Teresópolis.
RJ44) Rua do Livramento 189, 20221-191 Rio de Janeiro.
RJ45) Rua Paulino Padilha 80, 28460-000 Miracema — DG: Luciano B. Mercanti.
RJ46) Rua Alencar Lima 26, 25600-000 Petrópolis — DG: Roberto Santos Ribeiro.
RJ47) Rua Vereador Darey Teixeira Fontes 556, 23800-000 Itaguaí — DG: Adilson Santana B.
RJ48) Av. Nilo Peçanha 572, 27600-000 Valença.
RJ49) Praça Governador Portela 18, 28360-000 Bom Jesus do Itabapoana — Dir: F. R. Nevei Xavier.
RJ50) Av. Amaral Peixoto 366, 28860-000 Casimiro de Abreu.
RJ51) Rua Frei Valerio 58, 28740-00 Conceição de Macabu.
RJ52) Rua Nilo Piçanha 320, 27542-210 Resende — Dir: Luiz Oliveira Fontes.
RJ53) 28540-000 Cordeiro.
RJ54) 28850-000 Paraíba do Sul.
RJ55) Av. Presidente Kennedy 1763, 25000-000 Duque de Caxias.

RN00) RIO GRANDE DO NORTE

RN01) Av. Duque de Caxias 106, Ribeira, 59010-200 Natal — DG: Ricardo Luiz V. Alves.
RN02) Av. Otávio Lamartine 603, 59300-000 Caicó.
RN03) C.P. 625, 59022-970 Natal — Dir: F. Rodrigues.
RN04) Rua Presidente João Pessoa 26, 59380-000 Currais Novos.
RN05) C.P. 112, 59600-970 Mossoró — DG: Emery J. Costa.
RN06) C.P. 60, 59600-970 Mossoró — Dir. Francisco B. Neto.
RN07) Praça Pio X 335, 59000-000 Natal — Dir: Angelo D. Barreto.
RN08) C.P. 101, 59600-970 Mossoró — DG: Angelo A. Fernandes.
RN09) Rua Romualdo Galvão 973, Tirol, 59056-100 Natal — DG: Isidro José Moro.
RN10) C.P. 145, 59001-970 Natal — Gte: Cid Lobo.
RN11) Praça Tomaz Salustriano 42, 59380-000 Currais Novos.
RN12) Rua Francisca Delfina 30, 59965-000 Alexandria — Gte: Duilio Gutemberg A.
RN13) Praça Bento Praxades 104, 59600-000 Mossoró — Gte: José G. Leite M.
RN14) Rua Otávio Amorim 643, 59650-000 Açu.
RN15) Estrada Caraúbas Janduis Km O — Liberdade, 59780-000 Caraúbas.
RN16) Rua Frei Alberto Cabral s/n, 59215-000 Nova Cruz.
RN17) Rua Getúlio Vargas 1296, 59900-000 Pau dos Ferros — DG: Raimundo Nonato O.
RN18) Av. Coronel Martiniano 847, 59300-000 Caicó.
RN19) Av. Rio Branco 173, 59665-000 Areia Branca — Gte: Jaltion Rodrigues.
RN20) Rua Ana de Pontes s/n, 59225-000 Santo Antônio.
RN21) Rua Camilo José Rocha 93, 59200-000 Santa Cruz.
RN22) Rua Martins Ferreira 183, 59500-000 Macau.
RN23) Rua Cícero Tomaz de Acevedo s/n, 59360-000 Parelhas — Gte: Sergio de Miranda Silva.
RN24) Usina São Francisco s/n, 59570-000 Ceará Mirim.
RN25) Rua Sebastião Guilherme Caldas s/n, 59343-000 Jardim do Seridó.
RN26) 59300-000 Caicó.
RN27) Av. Nascimento de Castro 1597, 59000-000 Natal.
RN28) 59920-000 São Miguel.
RN29) 59460-000 São Paulo do Potengi.
RN30) Praça Monsenhor Freitas 336, 59550-000 João Câmara.

RO00) RONDÔNIA

RO01) Rua Joaquim Nabuco 1573, 79815-300 Porto Velho.
RO03) Av. Jamari 4218, 78932-000 Ariquemes.
RO04) Rua Dr. Fiel s/n, 78958-000 Ji-Paraná.
RO05) Rua Rui Barbosa 3375, 78975-000 Cacoal — Dir: Herivelton J. Bernard.
RO06) Rua Costa e Silva 1113, Vila Jotão, 78958-000 Ji-Paraná — Dir: Joel Nogueira.
RO07) Rua Dom Augusto 681, Centro, 78958-000 Ji-Paraná.
RO08) Rua Ricardo Catanhede esquina com a Rua Goiás s/n, 78941-000 Jaru.
RO09) C.P. 51, 78957-970 Guajará Mirim — DG: E. Oliveira R.
RO10) Rua Dourados 4, Setor de Areas Especiais, 78932-000 Ariquemes.
RO11) C.P. 104, 78900-970 Porto Velho — DG: Carlos A. Diniz Martins.
RO12) Rua Princesa Isabel 128, 78995-000 Vilhena — DG: Ruth Morimoto.
RO13) C.P. 005, 78900-970 Porto Velho — DG: Angela Xavier.
RO14) 78987-000 Rolim de Moura.
RO15) Rua Carlos Donege s/n, 78984-000 Pimenta Bueno.
RO16) Rua Gonçalves Dias 227, 78995-000 Vilhena.
RO17) Rodovia BR-364 Km 3 (Estradas de acesso do Cenetron), 78900-000 Porto Velho.
RO18) Rua 6 de Maio esquina com Av. Rio Jaru, 78958-000 Ji-Paraná.

RR00) RORAIMA

RR01) C.P. 225, 69301-970 Boa Vista — DG: Geraldo Franca.
RR02) Rua Sebastião Diniz 363, 69350-000 Caracaraí.

RS00) RIO GRANDE DO SUL

RS01) Av. Victor Barreto 3056/207, 92010-000 Canoas — DG: Amabilio J. Castro.
RS02) Av. Antunes Ribas 1535, 98801-630 Santo Angelo.
RS03) Rua Marechal Deodoro 1157, 96810-110 Santa Cruz do Sul.
RS04) C.P. 719, 95001-970 Caxias do Sul — Dir: Pedro D. Fattori.
RS05) C.P. 89, 99001-970 Passo Fundo — DG: Waldir R. Sudbrack.

RS06) Praça Fernando s/n, 97300-000 São Gabriel.
RS07) C.P. 67, 95201-970 Vacaria — DG: Antoninho Pasqualon.
RS08) Praça Oswaldo Aranha 39, 97540-000 Alegrete.
RS09) Av. Ipiranga 1075, 90060-000 Porto Alegre.
RS10) C.P. 284, 96001-970 Pelotas — Gte: Gilberto Gomes.
RS11) Rua Suécia 255, 98500-000 Tenente Portela — DG: Claudio N. Gádens.
RS12) Rua 14 de Julho 588, 95300-000 Lagoa Vermelha.
RS13) Praça Saldanha Marinho s/n, 97010-540 Santa Maria.
RS14) C.P. 1474, 90001-970 Porto Alegre — DG: Gudy Emunds.
RS15) C.P. 195, 99701-970 Erechim — DG: Idyllio S. Badalotti.
RS16) C.P. 80, 97590-970 Rosário do Sul — Gte: Moacir Guazina.
RS17) Rua Farroupilha 110, 95200-000 Vacaria.
RS18) Rua 7 de Setembro 1713, 96230-000 Santa Vitória do Palmar.
RS19) C.P. 55, 99500-970 Carazinho — DG: I.B. Albuquerque.
RS20) Travessa 4 de Junho 84, 98380-000 Seberi.
RS21) C.P. 31, 98700-970 Ijuí — DG: José L. Bonamigo.
RS22) Rua Marechal Mascarenhas de Morães 298, 97300-000 São Gabriel — DG: Nissio Eskenazi Gianfranco Trevisan.
RS23) Rua Caldas Jr. 219, 90010-260 Porto Alegre — DG: Carlos A. Bastos Ribeiro.
RS24) Av. Scarpelini Ghezzi 353, 99072-000 Passo Fundo — DG: Pe. Anacleto J. Zaffare.
RS25) C.P. 75, 98300-970 Palmeira das Missões.
RS26) C.P. 147, 96500-970 Rio Grande — DG: P. Hahuys C.
RS27) Av. Marechal Floriano 920, 95520-000 Osório — DG: Pedro Edir D. Farias.
RS28) C.P. 69, 96930-970 Candelária.
RS29) Rua Duque de Caxias 736, 98470 Planalto — DG: Moacir Zilio.
RS30) Av. Flores da Cunha 1596, 99500-000 Carazinho.
RS31) Rua Marechal Floriano 38, 90020-060 Porto Alegre.
RS32) Rua Andrade Neves 379, 96640-000 Rio Pardo — DG: Norberto J. Frantz.
RS33) Cidade Universitária, 97100-000 Santa Maria — DG: Sergio de Assis B.
RS34) Rua Fernando Abott 42, 95880-000 Estrela — DG: Frederico A. Ballve.
RS35) Rua Voluntários da Pátria 805, 98025-770 Cruz Alta.
RS36) Rua Casimiro de Abreu 1233, 90420-001 Porto Alegre.
RS37) Rua Bolzano 1336, Edif. Segala, 97100-000 Santa Maria — Dir: Cláudio Zappe.
RS38) Rua Andradas 1155, 90020-007 Porto Alegre.
RS39) C.P. 72, 95700-970 Bento Gonçalves — DG: Carlos Domingos P.
RS40) C.P. 121, 98900-970 Santa Rosa — DG: Sergio Ambros M.
RS41) Rua 15 de Novembro 336, 99770-000 Aratiba.
RS42) Rua Joaquim Nabuco 957, 93310-002 Nôvo Hamburgo.
RS43) Rua Tirandentes 710, 95800-000 Venâncio Aires.
RS44) Rua Saihydi Abrahão 35, 95590-000 Tramandaí.
RS45) Rua Garibaldi 789, 95084-060 Caxias do Sul — Dir: Carlos A. Riegel.
RS46) Av. Brasil 523, 98801-590 Santo Angelo.
RS47) C.P. 49, 95900-970 Lajeado.
RS48) Rua Venâncio Aires 1980, 97010-004 Santa Maria — DG: Rudy Seibel.
RS49) Rua dos Andradas 756, 90020-004 Porto Alegre.
RS50) C.P. 48, 98670-970 Humaitá.
RS51) Rua Otacilio Azevedo 2, 98170-000 Tupanciretã.
RS52) Av. Florentina Bacchi 791, 99840-000 Sananduva.
RS53) Rua Garibaldi 789, 95084-060 Caxias do Sul.
RS54) Rua 13 de Maio 1871, 97800-000 São Luís Gonzaga.
RS55) C.P. 155, 98700-970 Ijuí — DG: Mathilde Salete Mânica.
RS56) Rua General Zaca Neto 1396, 96180-000 Camaguã.
RS57) C.P. 172, 97800-970 São Luís Gonzaga.
RS58) Rua 15 de Novembro 236, 96570-000 Caçapava do Sul.
RS59) Rua Tucunduva 758, 98640-000 Crissiumal.
RS60) C.P. 72, 95701-970 Bento Gonçalves — DG: Carlos Domingos P.
RS61) Rua João Carlos Machado 645, 98460-000 Iraí.
RS62) Rua Sarmento Leite 426, 90050-170 Porto Alegre — DG: Carlos Urbim.
RS63) Rua Ramiro Barcelos 258, 96508-070 Cachoeira do Sul.
RS64) C.P. 71, 99800-970 Marcelino Ramos — DG: Noemi M. Basso.
RS65) Rua Alcides Filau 71, 98870-000 Giruá.
RS66) Rua Galvão Osório 1160, 96300-000 Jaguarão.
RS67) C.P. 11, 96901-970 Sobradinho — DG: João V.Z. de Castro.
RS68) C.P. 80, 97650-970 Itaqui — Gte: José I. Dornelles.
RS69) Rua João Pessoa 919, 90040-000 Porto Alegre — DG: Ricardo Ferro Gentelini.
RS70) C.P. 1277, 97001-970 Santa Maria — DG: E. Botton.
RS71) Rua Pinheiro Machado 628, 98050-000 Cruz Alta — DG: Sidney Amaral.
RS72) Rua Domingos de Almeida 2194, 97500-004 Uruguaiana — DG: João R. Cobelli.
RS73) C.P. 50, 96750-970 Butiá.
RS74) C.P. 200, 95180-970 Farroupilha — DG: B. Zorzi.
RS75) Rua Tenente Lyra 950, 98400-000 Frederico Westphalen.
RS76) Rua General Osório 1160, 97760-000 Jaguari — DG: Silvia B. Bertoncheli.
RS77) C.P. 100, 96001-970 Pelotas.
RS78) Av. Coronel Victor Villa Verde 491, 95500-000 Santo Antônio da Patrulha.
RS79) C.P. 157, 96400-970 Bagé — DG: J.H. Biancullo G.
RS80) Rua Sete de Setembro 366, 99010-121 Passo Fundo.
RS81) Rua Mauricio Cardoso 697, 99300-000 Soledade.
RS82) Rua Senador Pinheiro Machado 383, 97900-000 Cerro Largo — DG: H. Dias B.
RS83) C.P. 65, 99250-970 Serafina Corrêa.
RS84) Rua Torres Gonçalves 33, 99700-000 Erechim.
RS85) Próximo ao Km 233 da BR-392, 97340-000 São Sepé — DG: Enilton J. Bolzan.
RS86) 90020-060 Porto Alegre.
RS87) Rua Domingos de Almeida 1722, 97500-001 Uruguaiana — Dir: Antônio Everardo Pinto Bermúdez.
RS88) Rua Victor Dumonal 1756, 98240-000 Santa Bárbara do Sul.
RS89) C.P. 89, 99600-970 Nonoai — DG: José Reck.
RS90) Av. Fernando Luzatto 42, 95320-000 Nova Prata.
RS91) Av. Rio Branco s/n, 96610-000 Encruzilhada do Sul.
RS92) Rua Júlio de Castilhos 430, 95290-000 Bom Jesus.
RS93) C.P. 11, 98200-970 Ibirubá — DG: Saulo A. Stefanello.
RS94) C.P. 82, 96700-970 São Jerônimo — DG: Gilton Paschoal.
RS95) C.P. 707, 95001-970 Caxias do Sul.
RS96) Rua 15 de Novembro 717, 96015-000 Pelotas.
RS97) Rua Riachuelo 928, 97670-000 São Borja.
RS98) Rua Moron 1520, 96508-030 Cachoeira do Sul.
RS99) Rua José Sponchiado 418, 99830-000 Guarama.
RS100) Rua Júlio de Castilho 1038, 97220-000 Faxinal do Soturno.
RS101) C.P. 112, 98920-970 Horizontina — DG: Luiz C. Pilau.
RS102) Rua Coronel Alfredo Born 447, 96170-000 São Lourenço do Sul.
RS103) Av. Angelo Macalos 490, 99400-000 Espumoso — DG: Verdi Ugiratan.
RS104) Av. Farrapos 146, 90220-000 Porto Alegre — Gte: G. Gianuca.
RS105) Rua 15 de Julho 39, 98960-000 Santo Cristo.
RS106) Av. João Belchior Goulart, 97574-330 Santana do Livramento.
RS107) Rua Duque de Caxias 1503, 99560-000 Sarandi — DG: Francisco A. Rosin.
RS108) Av. Júlio de Castilhos 232, 95680-000 Canela.
RS109) Rua Sete de Setembro 457, 98280-000 Panambi.
RS110) Rua Sete de Setembro 353, 96015-300 Pelotas.
RS111) Av. Rio Branco 401, 96450-000 Dom Pedrito — Dir: O. Silva.
RS112) C.P. 22, 95190-970 São Marcos — DG: Nelson Tomiello.
RS113) Av. Bastian 286, 90130-020 Porto Alegre — DG: Erika A.W. Coestel K.
RS114) Praça Vespaziano Correa 1180, 99200-000 Guaporé — DG: Valmir Baldo.
RS115) Rua Daltro Filho 417, 98600-000 Três Passos.
RS116) Av. Concordia 1013, 96540-000 Agudo — DG: Leo Pachaly.
RS117) C.P. 10, 99150-970 Marau — DG: Alécio Turcatel.
RS118) Rua São José 2500, 95270-000 Flores da Cunha.
RS119) Rua Gaspar Martins 55, 94820-380 Alegrete — DG: Samuel M. da Silva.
RS120) Rua Conde de Porto Alegre 135, 97573-581 Santana do Livramento.
RS121) Rua Tiradentes 35, 99490-000 Tapera.
RS122) C.P. 844, 90001-970 Porto Alegre — DG: Antonio F. Marin.

RS123) Rua Irmao Gabriel Leão 358, 99900-000 Getúlio Vargas.
RS124) C.P. 19, 97200-970 Restinga Seca.
RS125) Rua Marechal Floriano 176, 96205-190 Rio Grande.
RS126) C.P. 79, 98900-970 Santa Rosa — Dir: R. Donadel.
RS127) Rua 24 de Maio 181, 95330-000 Veranópolis.
RS128) C.P. 118, 96800-970 Santa Cruz do Sul — DG: André L. Jungblut.
RS129) Rua General Osório 943, 96600-000 Canguçu.
RS130) Av. Coronel Diniz 78, 95670-000 Gramado — DG: Sergio Lima.
RS131) C.P. 120, 95780-970 Montenegro — DG: Amabilio J. Cabral.
RS132) Rua General Osório 531, 96200-400 Rio Grande.
RS133) Av. Sete de Setembro 672, 96400-000 Bagé.
RS134) Av. das Hortencias 78, 95560-000 Torres — DG: Amabilio J. Castro.
RS135) Rua Dr. Bruno Dockhorn 18, 98910-000 Três de Maio.
RS136) C.P. 22, 93700-970 Campo Bom — DG: M. Vergara G.
RS137) C.P. 64, 95720-970 Garibaldi — DG: Nestor Ferronato.
RS138) Rua João Moreira 1733, 97610-000 São Francisco de Assis.
RS139) Rua Silveira Martins 153, 96600-000 Canguçu.
RS140) Rua Tristão Monteiro 2140, 95600-000 Taquara.
RS141) Rua Sete de Setembro 960, 96508-010 Cachoeira do Sul.
RS142) C.P. 706, 98770-970 Catuípe.
RS143) C.P. 144, 96450-970 Dom Pedrito — DG: Pascoal A. Brandi.
RS144) Rua Independência 192, 93010-001 São Leopoldo.
RS145) Rua Coronel Amancio Cardoso 596, 99950-000 Tapejara — DG: Aleen Basso.
RS146) Rua Benjamin Constant 126, 97700-000 Santiago.
RS147) Av. José Loureiro da Silva 1727, 94010-001 Gravataí.
RS148) C.P. 20, 95960-970 Encantado.
RS149) Rua Assis Brasil 83, 98130-000 Júlio de Castilhos.
RS150) Rua José Bonifácio 29, 96330-000 Arroio Grande.
RS151) Rua Marechal Floriano Peixoto 331, 98980-000 Porto Lucena.
RS152) Rua Valdomiro Bocchese 872, 95250-000 Antônio Prado.
RS153) Av. Padre Reus 324, 97930-000 Caibaté.
RS154) Rua Presidente Vargas 176, 96360-000 Pedro Osório.
RS155) Rua Borges de Medeiros 1462, 97650-000 Itaqui.
RS156) C.P. 241, 98590-970 Santo Augusto.
RS157) Rua Brum 943, 97560-000 Quaraí.
RS158) Rua Leandro Ribeiro 28, 95860-000 Taquari.
RS159) Rua Rui Barbosa s/n, 98300-000 Palmeira das Missões.
RS160) Rodovia RS-407 No. 1019, 95555-000 Capão da Canoa — Pres: Egon Birlim.
RS161) Rua 7 de Setembro 601, 99530-000 Chapada.
RS162) Rua Edilio Vitória 833, 96160-000 Capão do Leão.
RS163) C.P. 551, 97001-970 Santa Maria (C.P. 6084, 90031-970 Porto Alegre).
RS164) Rua Coronel Scherer 195, 97400-000 São Pedro do Sul.
RS165) Rua Duque de Caxias 255, 98430-000 Palmitinho.
RS166) Av. Cantidio Rodrigues de Almeida 107-02, 99680-000 Constantina.
RS167) Av. Presidente Vargas 1443, 98580-000 Coronel Bicaco.
RS168) Av. Alto Jacuí 435, Terreo, 99470-000 Não Me Toque.
RS169) Rua Tiradentes 35, Próximo a Praça Getúlio Vargas, 97450-000 Cacequi.
RS170) Av. João XXIII s/n, 99855-000 São João da Urtiga.
RS171) Av. Borges de Medeiros 407, 96760-000 Tapes.
RS172) 97650-000 Itaqui.
RS173) 98700-000 Ijuí.
RS174) Rua General Sampaio 18, Edificio Consorcio, Bl. Central, Conj. 09, 96400-000 Bagé.

SC00) SANTA CATARINA

SC01) C.P. 61, 89161-970 Rio do Sul — DG: Osny J. Gonçalves.
SC02) Morro Cechinel s/n, 88800-000 Criciúma — DG: Pedro V. Barcellos.
SC03) C.P. 8, 89911-970 Descanso — Dir: Wolmir Hübner.
SC04) Rua J. Goulart 1020, 89800-000 Chapecó.
SC05) C.P. 313, 89161-970 Rio do Sul — Dir: Edison Andrade.
SC06) C.P. 25, 89960-970 São Lourenço d'Oeste.
SC07) Rua Carlos Jofre de Amaral 67, 88501-010 Lages — DG: Eni Maria Ranzan.
SC08) Av. Rio Grande do Sul 149, 89580-000 Fraiburgo.
SC09) C.P. 71, 88701-970 Tubarão — Dir: Raimundo Ghizoni.
SC10) Rua Allan Kardec 12, 88025-100, Florianópolis — DG: Gilberto Benati.
SC11) Rua Dr. Maruri 1204, 89700-000 Concórdia.
SC12) Rua Angelo Dias 251, 89010-020 Blumenau — DG: E.P. Vieira.
SC13) Rua da Criança s/n, 88840-000 Urussanga — Gte: Osmar Nunes.
SC14) Rua Pedro Andreazza 49, 89560-000 Videira.
SC15) C.P. 209, 89010-971 Blumenau — DG: Walter U. Mick.
SC16) Rua Orleañs 54, 88880-000 Lauro Müller.
SC17) C.P. 71, 89700-970 Concórdia — DG: Octaviano Zandorai.
SC18) Rua Jaú Guedes da Fonseca 17, 88080-080 Florianópolis — DG: Luiz Carlos da Silva.
SC19) Margens da Rodovia SC-444, Linha Tres Ribeirões, 88820-000 Içara.
SC20) Rua Rodrigues Alves 111, 88350-000 Brusque — DG: César Moritz.
SC21) Loteamento Jardim Portobello, Rua A, 88200-000 Tijucas — Dir: Walmor Silva Telles.
SC22) Rua Barão de Rio Branco 229, 88870-000 Orleañs.
SC23) Rua José de Miranda Ramos 100, sala 105, 89820-000 Xanxerê.
SC24) C.P. 96, 88350-970 Brusque.
SC25) C.P. 63, 89896-970 Itapiranga — DG: A. Baldissera.
SC26) C.P. 67, 88750-970 Braço do Norte.
SC27) Av. Santa Catarina Quadra 623, 89840-000 Coronel Freitas.
SC28) Rua Otacilio Vieira da Costa 40, 88501-050 Lages —DG: Gilberto Berati.
SC29) C.P. 34, 89130-970 Indaial — DG: Aroldo Laemmel.
SC30) C.P. 25, 89200-970 Joinville.
SC31) Rua Princesa Isabel 300, 88701-010 Tubarão.
SC32) C.P. 1477, 88001-970 Florianópolis.
SC33) C.P. 89, 89835-970 São Carlos — DG: Mario Luiz Rauber.
SC34) C.P. 15, 89820-970 Xanxerê.
SC35) C.P. 30, 89520-970 Curitibanos — DG: Frei E. Tambosi.
SC36) Rua 15 de Novembro 600, sala 401, 89010 Blumenau.
SC37) Rua Oswaldo Cabral 145, Ap. 7, 88790-000 Laguna.
SC38) Rua Siqueira Campos 33, 89400-000 Porto União.
SC39) C.P. D4, 89930-970 São José do Cedro.
SC40) Rua Joaquim Geraldo 169, 88240-000 São João Batista.
SC41) C.P. 59, 89835-970 São Domingos — DG: Valentim Bigolin.
SC42) C.P. 312, 88700-970 Tubarão — DG: Carlos A. Kürten.
SC43) Rua Tenente Ari Rauen 1361, 89300-000 Mafra.
SC44) Rua Rui Barbosa 136, 89860-000 Cunha Porã — DG: Edsom Zanetti.
SC45) C.P. 335, 89200-970 Joinville — DG: R. da Silva.
SC46) Rua São Paulo 1120, 89012-000 Blumenau.
SC47) C.P. 30, 89600-970 Joaçaba — DG: Albino B. Scanzeria.
SC48) C.P. 72, 88790-970 Laguna — Gte: Paulo G. da Silva.
SC49) Rua 400 No. 585, 88330-000 Balneário Camboriú — Gte: Marinho Lopes Sringari.
SC50) Rua Búenos Aires 131, 89051-050 Blumenau — DG: Mario E. Binder.
SC51) C.P. 44, 89800-970 Chapecó — Dir: J.F. Müller B.
SC52) Rua Princesa Isabel 311, 89460-000 Canoinhas.
SC53) Av. Gov. Adolfo Konder 1500, Bairro São Vicente, 88308-000 Itajaí.
SC54) Rua Reinoldo Rau 400, 89250-000 Jaraguá do Sul.
SC55) C.P. 001, 89900-000 São Miguel do Oeste — DG: Ademar P. Baldissera.
SC56) Rua José Gonçalves 333, 89340-000 Itaiópolis — DG: Darcy Zanghelini.
SC57) Rua Carlos Jofre do Amaral 67, 88501-000 Lages.
SC58) Rua Visconde do Rio Branco 1028, 89887-000 Palmitos.
SC59) C.P. 45, 88010-970 Florianópolis — Dir: Alberto Gonçalves de Souza.
SC60) Praça Lauro Müller 255, 89620-000 Campos Novos.
SC61) Av. Araucaria 228, 89874-000 Maravilha.
SC62) C.P. 307, 89290-970 São Bento do Sul.
SC63) Morro Cechinel, 88800-000 Criciúma.
SC64) C.P. 63, 89110-970 Gaspar.
SC65) Rua Ademar da Silva 1209, Bairro Kobrassol, 88117-400 São José — DG: Daniel Espírito Santo.
SC66) C.P. 91, 89610-970 Herval d'Oeste — Gte: Elmir Bender.
SC67) Rua Dr. João Colin 572, 89204-000 Joinville.
SC68) Rua Izau Luiz de Bittencourt s/n, 88770-000 Imarui.
SC69) Rua Getúlio Vargas 183, 89140-000 Ibirama.
SC70) Rua do Comércio 476, 89825-000 Xaxim.
SC71) C.P. 86, 89250-970 Jaraguá do Sul — DG: W. Luiz F.
SC72) C.P. 104, 88502-970 Lages — DG; J.P. Bággio.

SC73) C.P. 74, 89870-970 Pinhalzinho.
SC74) Rua A. Guimarães 230, 89500-000 Caçador.
SC75) Av. Dr. João Rimse 797, 88780-000 Imbituba.
SC76) C.P. 234, 89120-970 Timbó — DG: J. Reinart.
SC77) C.P. 122, 88300-970 Itajaí.
SC78) Rua Lauro Müller 109, 88600-000 São Joaquim.
SC79) Rua Colombo 151, 89893-000 Mondaí — Dir: Romiro A. Utzig.
SC80) C.P. 160, 89665-970 Capinzal.
SC81) Av. Getúlio Vargas 429, 88900-000 Araranguá.
SC82) C.P. 3, 89190-970 Taió — DG: R. Amaral.
SC83) C.P. 103, 89295-970 Rio Negrinho — DG: Goldwin Meier.
SC84) Rua 15 de Novembro 39, 89107-000 Pomerode.
SC85) Rua José Steffens 260, 88400-000 Ituporanga — nDG: Solange G. Maciel.
SC86) C.P. 62, 88010-970 Florianópolis — DG: Matheus Iensen.
SC87) Rua Argemiro Borges 11, 89480-000 Major Vieira.
SC88) C.P. 1477, 88103-970 São José — DG: Ismael Fabião.
SC89) Montese 45, 89540-000 Santa Cecília.
SC90) Av. Getúlio Vargas 974, 89830-000 Abelardo Luz.
SC91) Av. Antônio Francisco Guizoni s/n, 88650-000 Urubici.
SC92) Rua Rui Barbosa 1321, 88930-000 Turvo.
SC93) Rua Professor João Sobotka 222, Bairro São Cristovão, 89665-000 Capinzal.
SC94) Av. Progresso 374, 89888-000 Caibi.
SC95) C.P. 001, 89900-970 São Miguel d'Oeste.
SC96) Rua Major Lucio Caldeira 54, 89230-000 São Francisco do Sul.
SC97) Rua Vicente Rech 399, 89770-000 Seara.
SC98) Rua do Comércio 214, 89872-000 Modelo.
SC99) Rua 7 de Setembro 495, 89950-000 Dionísio Cerqueira.

SE00) SERGIPE

SE01) Av. Barão de Maroim 66, 49010-340 Aracaju.
SE02) C.P. 172, 49001-970 Aracaju — Dir: Pedro P. Valverde.
SE03) Rua Simão Dias 643, 49010-430 Aracaju — DG: Marcus Pinto Aguiar.
SE04) Praça Dr. Filomeno Mora 4, 49400-000 Lagarto.
SE05) C.P. 409, 49001-970 Aracaju — DG: E. Ferreira S.
SE06) Rua João Pessoa 85, 49500-000 Itabaiana.
SE07) Rua Itabaianinha 162, 49010-190 Aracaju.
SE08) C.P. 6, 49200-970 Estância — DG: Jorge Prado L.
SE09) Av. Dr. Luíz Magalhães 346, 49500-000 Itabaiana.
SE10) Rua Presidente Vargas 280, 49480-000 Simão Dias.
SE11) Av. José da Cunha 6, 49510-000 Frei Paulo.
SE12) Rua Barão do Rio Branco 262, 49200-000 Estância.
SE13) Travessa Riachão s/n, 49300-000 Tobias Barreto.

SP00) SÃO PAULO

SP01) Rua Padre Geraldo Goseling 798, 16200-000 Biriguí — DG: Leonardo Sabioni.
SP02) Av. 7 de Setembro 100, 13170-440 Sumaré.
SP03) C.P. 62, 12700-970 Cruzeiro — Gte: Carlos R. da Silva Reimer.
SP04) Rua José Bonini 1415, 14160-000 Sertãozinho.
SP05) Rua Cafeiá 161, 18400-000 Itapeva — Dir: A. Lerma.
SP06) Rua Dr. Sousa Alves 960, 12020-030 Taubaté.
SP07) Rua 7 de Setembro 425, 13465-000 Americana.
SP08) C.P. 400, 19970-970 Palmital.
SP09) Rua General Câmara 5, 11010-121 Santos.
SP10) Rua Nações 127, 16800-000 Mirandópolis.
SP11) C.P. 814, 14001-970 Ribeirão Preto — DG: O. Sousa S.
SP12) Av. Luíz Gonzaga de Amoedo Campos 28, 13800-000 Mogi Mirim.
SP13) C.P. 169, 15800-970 Catanduva — DG: Gerson J. de C. Gabas.
SP14) Rua Comendador João Galvão 58, 12500-000 Guaratinguetá.
SP15) C.P. 150, 18800-970 Piraju — DG: Orley Amaral de Souza.
SP16) Av. Paulista 807, 01311-100 São Paulo.
SP17) Rua Capitão Neves 18128, 15130-000 Mirassol.
SP18) Av. Marcondes Filho 1130, 19013-160 Presidente Prudente.
SP19) Praça José Palomone Lepre 99, 14804-010 Araraquara — DG: Roberto Montoro P.
SP20) Rua Homero Rodrigues 1090, 16900-000 Andradina — DG: Nivaldo Bueno F. Rocha.
SP21) Praça José Bonifácio 815, 13400-430 Piracicaba — Dir: José R. Soave.
SP22) Rua João Pessoa 129, 11013-001 Santos — DG: R.M. Santini.
SP23) C.P. 1252, 14001-970 Ribeirão Preto — DG: P. de Tarso P.
SP24) Rua Teotino Tibiriçá Pimenta 380, 11660-230 Caraguatatuba.
SP25) Rua Heitor Penteado 173, 17400-000 Garça.
SP26) C.P. 107, 13300-970 Itu — DG: Horlimar P. de Almeida.
SP27) Rua 13 de Maio 720, 15800-000 Catanduva.
SP28) Av. Dr. Alvaro Schmidt Gallo 317, 18800-000 Piraju — Dir: Nelson Paula Meira.
SP29) Rua Francisco Inácio 257, 14700-000 Bebedouro.
SP30) Praça Cons. Rodrigues Alves de Andrade 182, 12500-000 Guaratinguetá — DG: J.M. Oliveira.
SP31) Rua Campos Salles, 15370-000 Pereira Barreto — DG: Nivaldo B.F. Rocha.
SP32) Rua Major Quedinho 76, 01050-902 São Paulo — Dir: J. Mendonça.
SP33) Rua 1 de Agosto 927, 17010-011 Bauru — DG: Tobias Ferreira.
SP34) Rua dos Pelegrine s/n, Bairro Desterro, 13700-000 Casa Branca.
SP35) C.P. 355, 19900-970 Ourinhos — DG: Ruy S. Karpstein.
SP36) C.P. 25, 12700-970 Cruzeiro — DG: Carlos C. Fernandes.
SP37) C.P. 361, 15400-970 Olímpia — DG: Silvio R. Mathas N.
SP38) Rua Barão de Jundiaí 954, 13200-002 Jundiaí.
SP39) C.P. 324, 17500-970 Marília — DG: Edson J. de Souza.
SP40) C.P. 88, 17250-970 Bariri — DG: Osvaldo Orides Gimenez.
SP41) Travessa São João 47, 09090-540 Santo André.
SP42) C.P. 171, 17700-970 Osvaldo Cruz — DG: B. Boroni.
SP43) Rua Visconde do Rio Branco 401, 14015-000 Ribeirão Preto — Dir: Carlos Orlandi.
SP44) Rua Euclides Miragaia 394, 12245-550 São José dos Campos.
SP45) C.P. 621, 17001-970 Bauru — DG: Tobias Ferreira G.
SP46) Praça Toledo Barros 280, 13480-008 Limeira — DG: Orlando J. Zovico.
SP47) Rua dos Palmeiras 315, 01288-900 São Paulo.
SP48) Av. Armando de Sales Oliveira 575, 17800-000 Adamantina.
SP49) Av. Espanha 284, 14801-130 Araraquara — DG: Ricardo Lupo.
SP50) C.P. 30, 13200-970 Jundiaí — DG: Hélio L. Lorencini.
SP51) Rua Benjamin Constant 3327, 15015-600 São José do Rio Preto.
SP52) Av. Getúlio Vargas 185, 12570-000 Aparecida — DG: Antônio C. Moreira.
SP53) Rua Tenente Lopes 191, 17201-460 Jaú.
SP54) Rua José Galvão 359, 19900-000 Ourinhos.
SP55) Rua Anchieta 154, 16300-000 Penápolis.
SP56) C.P. 145, 15500-970 Votuporanga — DG: Nelson Camargo.
SP57) C.P. 372, 01059-970 São Paulo — DG: Salomão Esper.
SP58) Av. Visconde do Rio Claro 2128, 13500-580 Rio Claro.
SP59) C.P. 154, 16200-970 Biriguí — Dir: Libério Telies.
SP60) Rua Antônio Sabino 840, 14960-000 Novo Horizonte.
SP61) Rua Luzitana 717, 13015-121 Campinas.
SP62) Av. Paulista 900, 01310-100 São Paulo — DG: Alberto Helena Jr.
SP63) C.P. 5, 19001-970 Presidente Prudente.
SP64) C.P. 746, 15001-970 São José de Rio Preto.
SP65) C.P. 281, 18200-970 Itapetininga — DG: Francisco Carlos Silva Janez.
SP66) Rua Alferes José Caetano 1039, 13400-120 Piracicaba — DG: Orlando J. Zovico.
SP67) Rua Major Claudiano 1393, 14400-690 Franca — Gte: Maria Abadiade Almeida.
SP68) C.P. 52, 13970-970 Itapira — DG: Luís Antônio R. Nucci.
SP69) C.P. 2011, 11051-970 Santos.
SP70) Rua Rio Branco 719, 17015-310 Bauru — Gte: Carlos R. Morgado.
SP71) C.P. 150, 15775-970 Santa Fé do Sul — DG: Helena Coelho Rubinho.
SP72) Av. Sampaio Vidal 185, 17501-040 Marília — Dir: Wilson N. Matos.
SP73) C.P. 57, 01059-970 São Paulo — DG: Paulo Coev.
SP74) C.P. 56, 18200-970 Itapetininga — DG: José Abrão.
SP75) Praça Roque Fiore 18-B, 13870-000 São João da Boa Vista.
SP76) C.P. 34, 14400-970 Franca — DG: Sidnei Franco Rocha.

SP77) Indianópolis, 04084-900 São Paulo — DG: Paulo Machado de Carvalho.
SP78) Praça Joel Waldo Dal Moro 1, 14781-574 Barretos.
SP79) C.P. 15, 19500-970 Martinópolis — DG: Sebastião Rodrigues N.
SP80) C.P. 57, 12630-970 Cachoeira Paulista — DG: Wellington Silva Jardim.
SP81) Rua Capitão Francisco Rodrigues Garcia 187, 19800-000 Assis.
SP82) C.P. 105, 13480-970 Limeira — DG: Vitorio Bortolan Filho.
SP83) C.P. 16, 15700-970 Jales — DG: Alexandrina C. Garcia.
SP84) C.P. 284, 14400-970 Franca — DG: Ricardo Pucci Pieri.
SP85) Rua Francisco Pompeu s/n, 17340-000 Barra Bonita.
SP86) C.P. 310, 16400-970 Lins — Dir: A. Zago.
SP87) Av. 9 de Julho 3939, 01488-900 São Paulo — DG: Antonio Aggio Jr.
SP88) C.P. 85, 13400-970 Piracicaba — DG: Jairo Mattos.
SP89) Av. Ana Costa 90, 11060-000 Santos.
SP90) C.P. 31, 11740-970 Itanhaém — Dir: Ademir Bispo.
SP91) C.P. 237, 17200-970 Jaú — DG: Waldemar Baub.
SP92) C.P. 804, 19901-970 Presidente Prudente — DG: Carlos A. Arruda Campos.
SP93) C.P. 109, 15700-970 Jales — DG: Luis Demétrio V.
SP94) C.P. 85, 14300-970 Batatais — DG: João C. Cangussú.
SP95) Rua Olavo Bilac 693, 16400-000 Lins — Gte: Humberto Abdo.
SP96) C.P. 565, 18001-970 Sorocaba — Dir: Gastão Lima N.
SP97) Av. Monumental Papa João Paulo II 221, 12570-000 Aparecida — DG: Liliane Freitas.
SP98) C.P. 326, 17500-970 Marília — Gte: José Marques Beato.
SP99) Rua Jeremias de Paulo Eduardo 916, 15910-000 Monte Alto.
SP100) Rua das Palmeiras 315, 01288-900 São Paulo.
SP101) Av. Brasília 1324, 16015-460 Araçatuba — Dir: Oscar L. Piconez.
SP102) C.P. 91, 14940-970 Ibitinga — Dir: Roque de Rosa.
SP103) Rua Dr. José Alves 33, 13800-000 Mogi Mirim.
SP104) C.P. 95, 18540-970 Porto Feliz — DG: Danilo C. Casteluici.
SP105) C.P. 151, 12201-970 São José dos Campos — DG: Ronaldo Wagner Mimessi.
SP106) Rua Duque de Caxias 260, 15900-000 Taquaritinga.
SP107) C.P. 258, 17600-970 Tupã — Gte: J.A. Neto.
SP108) Rua Gonçalves Dias 208, 19800-000 Assis.
SP109) C.P. 115, 14780-970 Barretos.
SP110) C.P. 139, 13500-970 Rio Claro — DG: Orlando José Zovico.
SP111) C.P. 153, 11680-970 Ubatuba — Gte: Eduardo de Souza Neto.
SP112) Rua Nadir Dias de Figueiredo 1329, 02110-000 São Paulo.
SP113) Rua Almirante Barroso 456, 19400-000 Presidente Venceslau.
SP114) C.P. 486, 18001-970 Sorocaba.
SP115) Rua São Paulo 731, 15600-000 Fernandópolis.
SP116) Rua 15 de Novembro 539, 13730-000 Mococa.
SP117) C.P. 11, 13001-970 Campinas — DG: João R. Pinheiro.
SP118) C.P. 380, 15500-970 Votuporanga — DG: Oliveira S. Prates.
SP119) C.P. 75, 13650-970 Santa Cruz das Palmeiras — DG: Joaquim M. Ramos Neto.
SP120) Rodovia SP-253 Km 384, 18740-000 Taquarituba — DG: Francisco Rodrigues.
SP121) Rua Cenno Sbrighi 378, 05036-900 São Paulo — DG: Roberto Muylaert.
SP122) Rua São Sebastião 145, 14870-000 Jaboticabal.
SP123) C.P. 684, 16001-970 Araçatuba — DG: José C.S. Hernandes.
SP124) C.P. 209, 17001-970 Bauru — DG; Maximo Duarte.
SP125) C.P. 250, 18001-970 Sorocaba — DG: Maria App. F. Pavlovsky.
SP126) Rua Floriano Peixoto 310, 18300-000 Capão Bonito.
SP127) Rua João Pessoa 1990, 15990-000 Matão — Dir: Benedito J. Fernandes.
SP128) Rua Santo Antônio Claret 193, 13073-450 Campinas.
SP129) Praça Comendador Emilio Pedutti 28, 18600-410 Botucatu.
SP130) Rua 8 No. 472, 14620-000 Orlândia — DG: Chéster A. Martins.
SP131) Rua José Cabalero 60, 11055-000 Santos — DG: José Fornos Rodrigues.
SP132) Av. Nove de Julho s/n, 15200-000 José Bonifácio.
SP133) Av. da Saudade 200, 12280-000 Caçapava.
SP134) C.P. 18384, 04699-970 São Paulo.
SP135) Rua Cerqueira César 35, 14010-130 Ribeirão Preto.
SP136) Av. Benjamin Constant 1214, 5° andar, Centro, 13013-000 Campinas — DG: Synésio Pedroso Jr.
SP137) Rua Bernardino de Campos 3180, 15015-300 São José do Rio Preto.
SP138) C.P. 966, 09701-970 São Bernardo do Campo — DG: Gisele P. Catalde.
SP139) C.P. 66, 14900-970 Itápolis — DG: Acácio Batista.
SP140) Rua Duque de Caxias 154, 13630-000 Pirassununga.
SP141) C.P. 75, 12900-970 Bragança Paulista — Dir: J.F. Ferreira.
SP142) Rua Brigadeiro Tobias 911, 19570-000 Regente Feijó.
SP143) Rua Visconde do Rio Branco 401, 14015-000 Ribeirão Preto — Dir: Carlos Orlandi.
SP144) C.P. 246, 16001-970 Araçatuba — DG: M. Pavanello C.
SP145) Rua Barão de Jaceguai 468, 08710-160 Mogi das Cruzes.
SP146) C.P. 20, 17350-970 Igaraçu do Tietê.
SP147) C.P. 173, 15600-970 Fernandópolis — Dir: A. Vidale.
SP148) Av. José Bonifácio 1229, 17900-000 Dracena — DG: Belmiro Borini.
SP149) C.P. 376, 16900-970 Andradina — DG: Nivazdo Bueno Franco da Rocha.
SP150) Av. Antônio J. de Carvalho 1671, 17250-000 Bariri — DG: Carlos E. Marcos Galizia.
SP151) Rua 15 de Novembro 843, 18680-030 Lençóis Paulista.
SP152) Rua Guanabara 144, 13840-000 Mogi Guaçu — Gte: Cleuzio Fonseca.
SP153) Rua Benjamin Constant 1214, 13010-141 Campinas.
SP154) Rua Dr. Erico de Abreu Sodré 542, 16370-000 Promissão.
SP155) Av. Brasil 1170, 17780-000 Lucélia.
SP156) C.P. 96, 13560-970 São Carlos — DG: Dr. Afranio R. Zambel.
SP157) C.P. 755, 15001-970 São José do Rio Preto.
SP158) C.P. 7200, 01064-970 São Paulo — DG: António C. d'Elboux.
SP159) C.P. 29, 13250-970 Itatiba — DG: Paulo Abreu.
SP160) Rua Cerqueira Cesar 481, 14010-130 Ribeirão Preto.
SP161) Rua Oswaldo Aranha 1040, 15150-000 Monte Aprazível.
SP162) Rua Dr. Souza Alves 960, 12020-030 Taubaté.
SP163) Rua Coronel Joaquim Floriano 287, 18650-000 São Manoel — Gte: José A. di Santis.
SP164) Praça Lourenço Franco de Oliveira 81, 13930-000 Serra Negra.
SP165) C.P. 1087, 19001-970 Presidente Prudente.
SP166) Praça Nossa Senhora da Conceição 434, 16700-000 Guararapes.
SP167) Av. André Luís 723, 07082-050 Guarulhos.
SP168) C.P. 115, 13560-970 São Carlos — Dir: G. Rossi.
SP169) Av. Dr. Fernando Costa 192, 11900-000 Registro.
SP170) C.P. 71, 12600-970 Lorena — DG: Jose A. dos Santos.
SP171) Rua Professor Germano Negrini 62, 18130 São Roque.
SP172) Av. 15 No. 225, 14790-000 Guaíra.
SP173) Rua Alfredo Schuring, 12300-000 Jacareí.
SP174) C.P. 40, 13660-970 Porto Ferreira — Dir: José Wanderlei Klein.
SP175) C.P. 297, 13330-970 Indaiatuba — DG: Antônio R. Geiss.
SP176) Rua Renato Jardim 511, 14350-000 Altinópolis — DG: Mauro Matsuo.
SP177) C.P. 161, 12940-970 Atibaia — DG: J. Adibabi C.
SP178) Praça Oswaldo Martins 218, 16210-000 Bilac.
SP179) Rua Conselheiro Antônio Prado 518, 18900-000 Santa Cruz do Rio Pardo — Gte: Mauricio Azevedo F.
SP180) Rua Marechal Floriano Peixoto 1149, 15400-000 Olímpia.
SP181) Rua Monte Castelo 941, 17900-000 Dracena.
SP182) C.P. 63, 15690-970 Fernandópolis — DG: L.F. de Oliveira.
SP183) Rua Princesa Isabel 227, 09310-010 Mauá.
SP184) Praça Barão de Araras 201, 13600-000 Araras.
SP185) Rua Rubião Júnior 411, 12400-000 Pindamonhangaba.
SP186) Av. Leopoldo Carlos de Oliveira 1038, 14530-000 Miguelópolis.
SP187) C.P. 66, 06001-970 Osasco — DG: J. Gutier Navarro.
SP188) Av. Washington Luis 314, 11055-000 Santos.
SP189) C.P. 13, 18650-970 São Manoel — DG: Daniel O. Neves.
SP190) Av. Romeu Viana Romaneli 1510, 15570-000 Cardoso.
SP191) Rua Dr. Ricardo Vilela 568, 08710-150 Mogi das Cruzes.
SP192) C.P. 66, 13990-970 Pinhal — DG: José V. Sales.

SP193) C.P. 125, 17930-970 Tupi Paulista — Gte: J. Matias.
SP194) C.P. 64, 18270-970 Tatuí — DG: Luiz G.V. Camargo.
SP195) Rua Edmilson Pessoa Cavalcante 833, 15190-000 Nhandeara.
SP196) C.P. 4, 18603-970 Botucatu — DG: E. Placio Paganni.
SP197) C.P. 403, 06001-970 Osasco — DG: Francisco Rossi.
SP198) Rua Dom José Carlos Aguirre 567, 18460-000 Itararé.
SP199) Rua Pedro de Toledo 205, 19700-000 Paraguaçu Paulista.
SP200) Rua Regente Feijó 121, 13360-000 Capivari — Dir: Antônio Mattar Jr.
SP201) Rua José Vaz Porto 175, Vila Santa Rosa, 11431-190 Guarujá — DG: Orivaldo R. Rampazo.
SP202) C.P. 135, 14600-970 São Joaquim da Barra.
SP203) Av. Dr. Januario Miraglia 1337, 12460-000 Campos do Jordão.
SP204) C.P. 199, 18700-970 Avaré — DG: Marilia P. de Almeida.
SP205) C.P. 221, 17890-970 Junqueirópolis — DG: J. Sinicato.
SP206) Rua Tatui 321, 09015-620 Santo André — DG: A.C. Netto.
SP207) Rua Campos Sales 94, 13960-000 Socorro — Dir: Assir Bujato.
SP208) C.P. 68, 15170-970 Tanabi — DG: Adirley Lustri.
SP209) Rua Dr. Felipe Vita 1616, 18480-000 Itaporanga — DG: Arnoldo Krubniki.
SP210) C.P. 8, 13690-970 Descalvado — Dir: Pedro Gaspar.
SP211) Rua Dr. Timóteo Penteado 2207, 07094-000 Guarulhos.
SP212) Av. 17 No. 560, 14780-000 Barretos.
SP213) Rua Cardoso de Almeida 1932, 01251-000 São Paulo.
SP214) Rodovia João Ribeiro de Barros, Vila Jamil de Lima, 17800-000 Adamantina.
SP215) Av. São Paulo 788, 15650-000 Estrela d'Oeste.
SP216) Av. João Guilhermino 429, 12245-000 São José dos Campos — Dir: Mauro Brandão.
SP217) Rua Brandão Veras 1274, 14700-000 Bebedouro — Gte: Auro Ferreira Leite.
SP218) Rua Antônio Lobo 237, 13465-000 Americana.
SP219) Estrada Serrinha Km 200, 15350-000 Auriflama.
SP220) Rua Duque de Caxias 520, 13450-000 Santa Bárbara d'Oeste.
SP221) Rua Lions, junto ao No. 646, 06700-000 Cotia.
SP222) Rua Dr. Mário Sabino 131, 16300-000 Penápolis.
SP223) Rua Jacofer 615, 02712-070 São Paulo — DG: José M. de Abreu.
SP224) Rua José Bongiovani 691, 19050-680 Presidente Prudente — DG: Antônio C. Sishieri F.
SP225) Rua Albertino Nasciamento 207, 18110-000 Votorantim.
SP226) Rua Bento Carlos 61, 13560-600 São Carlos — Dir: Vanderlice de Melo.
SP227) Av. Tres, Quadra L, Lote 7, 13460-000 Nova Odessa — Gte: Yara Regina Ayres.
SP228) Rua Américo Vespúcio 20, 14730-000 Monte Azul Paulista.
SP229) Av. Dr. Labiano da Costa Machado 1735, 17400-000 Garça.
SP230) Rua Professor Sud Menucci 464, 17740-000 Rinópolis.
SP231) Alameda Alvares Machado 1211, 19280-000 Teodoro Sampaio.
SP232) Av. Consolação 2570, 15° andar, 01416-000 São Paulo.
SP233) Av. 9 de Julho 304, 13140-000 Paulinía.
SP234) Rua Capitão Francisco Marcelino Junqueiram 126, 14640-000 Morro Agudo.
SP235) Rua General Câmara 733, 13450-000 Santa Bárbara d'Oeste.
SP236) Rua Carlos Achinitz 313, 18320-000 Apiai.
SP237) Rua 5 No. 170, Bairro Cristo Redentor, 12100-000 Taubaté.
SP238) Rua Ademar de Barros 275, 14500-000 Itaverava.
SP239) Rua Frederico Ozanan 554, 17300-000 Dios Corregos.
SP240) Rua 13 de Maio 195, 13720-000 São José do Rio Pardo.
SP241) Rua Bororos 344, 17600-000 Tupã.
SP242) Rua Coronel Justiniano 200, 13600-000 Araras.
SP243) Av. 18 de Junho 367, 17690-000 Bastos.
SP244) Rua Santana 440, Centro, 13880-000 Vargem Grande do Sul.
SP245) Rua Rafael de Barros 126, 13610-000 Lemé.
SP246) Rua Getúlio Vargas 203, 17580-000 Pompéia.
SP247) Rua 7 de Setembro S-73, 17280-000 Pederneiras.
SP248) Rua Tenente Adolfo Padilha 157, 16880-000 Valparaíso.
SP249) Rua 15 de Novembro 52, 13920-000 Pedreira.
SP250) Rua Vitória 01-62, 19470-000 Presidente Epitácio.
SP251) Av. Dr. Carlos Burgos 1680, 13900-000 Amparo.

TO00) TOCANTINS

TO01) Av. Joaquim Aires 2393, 77500-000 Porto Nacional.
TO02) Praça José Torres Q-44, 77495-000 Paraiso do Norte.
TO03) Av. Nossa Senhora de Fátima 894, 77900-000 Tocantinópolis.
TO04) Rua Bela Vista 1079, Bairro Saõ João, 77807-040 Araguaína.
TO05) C.P. 67, 77761-970 Colinas de Tocantins — DG: Dirce Noda R.
TO06) Av. Tocantins 422, 77560-000 Miracema do Norte.
TO07) Rua Francisco Lucena Barbosa s/n, 77480-000 Cristalândia.
TO08) Chacara Manduca, Estrada para Saneago, 77640-000 Guaraí.

EXTERNAL SERVICE

RADIOBRAS

ADDR. & L.P: see under DF06.
SHORTWAVES: 250kW.

kHz	kHz	kHz	kHz
9745	11745	15265	17755

English	Area	kHz	Portuguese	Area	kHz
1200-1250	NAm	11745	1630-1720	Eu	15265
1800-1850	Eu	15265	1800-1850	Af	17755
German			**Spanish**		
1930-2020	Eu	15265	1000-1050	SAm	9745

R. Bras SW-tr's also relay certain prgrs of the BBC, Swiss Radio International & Deutsche Welle. See respective entries for details.

FM in Brasília:
89.1 Sigma Radiodifusão — 89.9 Brailia Super FM — 91.7 Brasília Comunicação — 93.7 Atlântida FM — DF01) 95.3 — DF04) 96.9 — 97.7 Manchete FM — 99.3 Antena 9 — 100.1 Transamérica — DF01) 100.9 Cultura FM — 101.7 R Jornal de Brasília — DF01) 105.5 Atividade — 107.9 Universitária.

FM in Rio de Janeiro:
RJ11) 89.3 — 90.3 Panorama FM — RJ25) 92.5 — 93.3 Nacional FM — RJ04) 94.1 — RJ01) 94.9 — 95.7 Alvorada FM — RJ27) 96.5 — 97.3 Melodia FM — RJ25) 98.1 FM 98 — RJ12) 98.9 — RJ16) 99.7 — RJ35) 100.5 — 101.3 Transamérica FM — 102.1 Imprensa FM — 102.9 Cidade — 103.7 Antena Um — 104.5 Tropical — RJ16) 105.1 105 FM — 106.3 Universidade — 106.7 Catedral — 107.1 107 — 107.9 Universidade.

FM in São Paulo:
SP62) 88.1 — 89.1 FM 89 — SP77) 89.7 — SP100) 90.5 X(Xis) — 91.3 Manchete FM — SP32) 92.9 — 93.7 Universidade FM — 94.7 Antena Um FM — SP57) 96.1 — 96.9 Cidade FM — 97.7 FM 97 — 98.3 Metropolitana FM — 99.3 Scala FM — 100.1 Transamérica — SP16) 100.9 — 101.7 Alpha FM — 102.5 Imprensa — SP121) 103.3 — 104.1 Líder FM — 105.7 Musical FM — 106.3 Nova Omega FM — 107.3 Brasil 2000 — 107.9 Tropical FM.

CHILE

L.T: UTC – 4h (UTC – 3h until 7 Mar.) — **Pop:** 13.486.000 — **Radio sets:** 4.250.000 — **Pr.L:** Spanish — **E.C:** a/c 50, 220/380V (yes) — **ITU:** CHL.

SUBSECRETARIA DE TELECOMUNICACIONES

ADDR: Amunátegui 139, Santiago. **Postal Addr:** Clasificador 120, Correo 21, Santiago. **Te:** + 56 (2) 672 6503. **Fax:** +56 (2) 699 5138. **Telex:** 341 156.
L.P: Subsect of Telecomm's: Roberto Pliscoff Vásquez.

ASOCIACIÓN DE RADIODIFUSORES DE CHILE

ADDR: Cas. 10476, Santiago de Chile. **Te:** +56 (2) 56398755.
Cable: ARCHI.
L.P: Pres: Oscar Pizarro R. Secr. Adm: Horacio Guajardo Contreras de Villaseca.

STATIONS: SW: Call letters CE are used for all zones.
MW: Call letters CA, CB, CC and CD indicate: A=No. Zone, B=Central Zone, C=So. Zone and D=Antarctic Zone. The figures indicate the freq. in kHz minus one cipher, f. inst. CB82 = Central Zone 820kHz.

Mediumwaves: ° = Shortwave, v = varying freq.

	Call	kHz	kW	Name, location and h. of tr.
1)	CB54	540	10	R. Ignacio Serrano, Melipilla
2)	CD54	540	1	R. Calle Calle, Valdivia: 1000-0400
2x)	CC55	550	1	R. V. de la Tierra, Angol
3)	CB57	°570	50	R. Agricultura, Santiago: 1000-0400
4)	CA59	590	1	R. Minería, Antofagasta: 1000-0430
5)	CC59	590	1	R. Caracol, Concepción: 24h
6)	CD59	590	10	R. Pres. Ibáñez, Punta Arenas: 1000-0430
6x)	CB60	600	10	R. Monumental, Santiago: 24h
6y)	CD60	600	10	R. Tricolor, Osorno: 1000-0500
7)	CA62	620	1	R. Norte Verde, Ovalle: 1100-0400
8)	CC62	620	10	R. Bío-Bío, Concepción: 24h
9)	CB63	630	10	R. Fundación, Valparaíso
10)	CD64	640	10	R. Cooperativa, Temuco: 1000-0430
11)	CB66	660	50	R. Chilena, Santiago: 1000-0700
12)	CA68	680	10	R. El Loa, Chuquicamata: 24h
13)	CC68	°680	10	R. Universidad de Concepción: 1030-0400
14)	CB69	690	10	R. Santiago, Santiago: 1000-0600
15)	CD69	690	10	R. Estrella del Mar, Ancud: 1100-0400
15x)	CA70	700	1	R. Amancay, Copiapó: 1030-0300
28)	CD70	700	1	R. Valdivia, Valdivia: 1030-0500
16)	CB73	730	50	R. Cooperativa, Valparaíso: 1000-0415
17)	CD73	730	1	R. Angelina, Los Angeles
18)	CD73B	730	1	R. Aysén, Pto. Aysén: 1100-0400
16)	CB76	760	75	R. Cooperativa, Santiago: 1000-0430
19)	CD78	780	10	R. Sago AM, Osorno: 1030-0430
20)	CB80	800	5	R. Minería, Viña del Mar: 1000-0500
21)	CA82	820	10	R. Gabriela Mistral, La Serena: 24h
22)	CC82	820	1	R. Almirante Latorre, Talcahuano: 1000-0430
23)	CD82	820	1	R. Concordia, La Unión: 1100-0400
25)	CA82	820	0.25	R. Pampa, Pedro de Valdivia
26)	CB84	840	10	R. Portales, Valparaíso: 1000-0430
27)	CD84	°840	10	R. Santa María, Coyhaique: 1000-0400
95)	CC86	860	5	R. Inés de Suárez, Concepción
29)	CB88	880	10	R. Corporación, Santiago: 1030-0400
30)	CA89	890	10	R. León Trece, Pozo Almonte: 1100-0200
31)	CC89	890	5	R. Interamericana, Concepción: 1100-0430
32)	CD89	890	20	R. Nal., Punta Arenas: 1000-0400
33)	CA90	900	5	R. Univ. de Atacama, Copiapó: 1100-0400
9)	CB90	900	1	R. Amapola, Valparaíso: 1100-0400
34)	CC90	900	1	R. Ñuble, Chillán: 1130-0400
35)	CD90	900	1	La Voz de la Costa, Osorno: 1100-0400
35x)	CD92	920	1	R. CD92, Temuco.
36)	CA93	930	10	R. Nal., Antofagasta: 1130-0400
37)	CB93	930	10	R. Nuevo Mundo, Santiago: 1000-0530
38)	CD93	930	10	R. Reloncaví, Puerto Montt: 1000-0500
48)	CB94	940	1	R. Valentín Letelier, Valparaíso
38x)	CA94	940	1	R. 9-40, Copiapó
39)	CB96	960	10	R. Carrera, Santiago: 1100-0400
40)	CD96	960	10/5	R. Polar, Punta Arenas: 24h
41)	CA97	970	1	R. Calama, Calama: 1000-0400
42)	CC97	970	1	R. Lautaro, Talca: 1000-0500
43)	CD97B	°970	1	R. Patagonia Chilena, Coyhaique: 1130-0400
44)	CD97A	970	1	R. Austral, Valdivia: 1030-0400
45)	CA98	980	1	R. Univ. de Tarapacá, Arica: 1100-0400
3)	CB98	980	10	R. Agricultura, Valparaíso
145)	CC99	990	1	R. El Roble, Parral
46)	CB100	1000	10	R. Cien AM, Santiago: 1030-0500
47)	CD101	1010	10	Emisoras Ñielol, Temuco: 1015-0400
49)	CC103	1030	10	R. Chilena, Concepción: 1100-0400
49x)	CB103	1030	1	R. Progreso, Talagante
50)	CD103A	1030	1	R. Chiloé, Castro: 1100-0330
51)	CD103	1030	1	R. Paine, Puerto Natales
52)	CD104	1040	1	R. Raíces, Curacautín: 1100-0100
52x)	CD105	1050		R. Bío Bío, Osorno: 0930-0500
53)	CB106	°1060	100	R. Minería, Santiago: 1000-0500
54)	CA108	1080	1	R. Río Elqui, Vicuña
55)	CD108	1080	1	R. Los Confines, Angol: 1100-0400
56)	CC109	1090	5	R. Chilena, Talca: 1000-0400
58)	CA110	1100	1	R. La Portada, Antofagasta
59)	CB110	1100	10	R. Recreo, Viña del Mar: 1100-0500
60)	CD111	1110	10	R. La Frontera, Temuco: 1030-0400
61)	CB114	°1140	40/75	R. Nal., Santiago: 1000-0600
62)	CC116	1160	1	R. Ancoa, Linares: 1000-0600
62x)	CD116	1160	1	R. Baha'i, Labranza: 1100-1800 2130-0100
29)	CB118	°1180	50	R. Portales, Santiago: 1030-0430
64)	CA120	*1200	1	R. Almirante Blanco Encalada, Tocopilla
3)	CD120	1200	10	R. Agricultura, Los Angeles: 1000-0400
65)	CA121	1210	1	R. Universidad de Antofagasta "R. 121": 1100-0400
66)	CB121	1210	5	R. Valparaíso, Viña del Mar: 0930-0600
67)	CC121	1210	1	R. Universidad de Talca, Talca: 1100-0400
67x)	CD121	1210	10	R. Nal, Puerto Montt
67z)	CD122	1220	10	R. Minería, Temuco
68)	CB124	1240	25	R. Universidad de Santiago, Santiago: 1100-0400
68x)	CA124	1240	0.25	R. Principal Chuquicamata, Calama
69)	CA124	1240	1	R. del Salitre, Iquique
70)	CA125	1250	5	R. América, La Serena
71)	CD125	1250	10	R. Universidad Austral de Chile, Valdivia: 1100-0400
72)	CA126	1260	10	R. Nal., Arica: 10-04
73)	CC126	1260	1	R. Condell, Curicó: 1100-0500
57)	CA126	1260	10	R. Minería, Punta Arenas: 1100-0400
74)	CB127	1270	10	R. Festival, Viña del Mar: 1000-0700
75)	CD127	1270	10	R. Agricultura, Temuco: 1000-0400
151)	CC128	1280	1	R. Arturo Prat Chacón, San Carlos
150)	CA129	1290	0.25	R. Coya, María Elena
164)	CD128	1280	10	R. del Sur, Osorno
165)	CD129	1290	1	R. Mulchen, Mulchen
76)	CB130	1300	5	R. Tierra, Santiago: 1100-0400
77)	CD130	1300	10	R. Pilmaiquén, Valdivia: 1000-0430
78)	CD130	1300	1	R. Cabo de Hornos, Pto. Williams: 1100-0400 (r. 1350)
79)	CA132	1320	1	R. Estrella del Norte, Vallenar
80)	CD132	1320	1	R. Lincoyan, Mulchén: 1100-0300
46)	CB133	1330	3	R. Metropolitana, Santiago
83)	CD133	1330	1	R. Vicente Pérez Rosales, Puerto Montt.
84)	CA134	*1340	0.25	R. El Toco, Tocopilla
85)	CB134	1340	10	R. Colo Colo del Pacífico, Valparaíso: 0930-0300
86)	CC134	v1340	1	R. Doña Isabel Riquelme, Chillán: 1000-0430
87)	CD134	1340	1	R. Panguipulli, Panguipulli: 1200-0100
89)	CD135	1350	1	R. San Carlos, Ancud: 1130-0400
90)	CA135	1350	1	R. Guardiamarina Riquelme, Coquimbo: 1030-0430
78)	CD135	1350	1	R. Cabo de Hornos, Pto. Williams: 1100-0400
92)	CC136	1360	1/5	R. Universidad del Bío Bío, "136 AM", Concepción: 1100-0300
153)	CD137	1370	1	Conunhueno, R. La Popular, Temuco: 1100-0400 (Sat 0800)
93)	CB138	1380	50	R. Colo Colo, Santiago: 24h
154)	CD140	1400	5	R. La Amistad, Los Angeles: 24h
94)	CA140	v1400	1	R. Nal., Iquique: 1045-0600

166)	CD140	1400	5	R. Llanquihue, Puerto Montt
96)	CB141	1410	1	R. Litoral, Valparaíso: 24h
97)	CD141	1410	1	R. Loncoche, Loncoche: 1100-0400
82)	CB142	1420	1	R. Panamericana, Santiago: 1100-0430
100)	CC142	1420	1	R. Nueva Maule, Cauquenes: 1050-0430
101)	CC142	v1420	1	R. Formidable, Rancagua: 1100-0400
102)	CA144	1440	1	R. Arica, Arica: 1006-0430
103)	CA144	1440	1	R. Agricultura, La Serena: 1000-0400
104)	CC144	1440	1	R. El Sembrador, Chillán: 1000-0430
105)	CB145	1450	5	R. Universidad Técnica "Federico Santa María", Valparaíso: 1100-0300
106)	CC145	1450	5	R. Libertad, Curicó: 1000-0600
108)	CD145	v1450	1	R. Turismo, Pto. Varas (r 1445)
109)	CA146	1460	10	R. Antofagasta, Antofagasta: 1130-0300
110)	CB146	1460	1	R. Yungay, Santiago: 1100-0500
111)	CC146	1460	1	Nueva R. Talcahuano, Talcahuano: 18h
112)	CB147	1470	1	R. Sargento Aldea, San Antonio: 1100-0500
113)	CA148	1480	1/ 0.25	R. Amanecer, Ovalle: 1000-0305
114)	CC148	1480	1	R. Tomé, Tomé: 1100-0230
115)	CD148	1480	1	R. General Baquedano, Valdivia
116)	CA149	1490	1	R. Alicanto, El Salvador
117)	CB149	*1490	1	R. Chena, San Bernardo
88)	CD149	1490	5	R. Malleco, Victoria: 0300
118)	CA150	1500	10	R. Almirante Patricio Lynch, Iquique: 1100-0500
119)	CC150	1500	1/ 0.25	R. Centenario, San Javier: 1100-0100
120)	CD150	1500	1	R. Tierra del Fuego, Puerto Porvenir: 1200-0400
136)	CB150	1500	1	R. Trasandina, Los Andes: 1100-0400
149)	CA151	1510	1	R. Illapel, Illapel
121)	CC151	1510	1	R. Rancagua, Rancagua: 1100-0400
122)	CD151	1510	1	R. Em. Teniente Merino, Lebu: 1130-0300
123)	CC152	1520	1	R. Nueva Soberanía, Linares: 1030-0430
124)	CD152	1520	0.1	R. Aníbal Pinto, Lautaro
155)	CB152	1520	1	R. Integración, Llolleo: 0900-0500
125)	CA153	1530	1	R. Juan Godoy, Copiapó
156)	CB153	1530	1	R. Nexo, Quillota: 1100-0600
126)	CC153	1530	1	R. El Carbón, Lota
167)	CD153	1530	1	R. Calbuco, Calbuco
128)	CB154	1540	1	R. Sudamérica, Santiago: 1200-0030
157)	CC154	1540	1	R. Central, Chillán
129)	CD154	1540	1	R. San José de Alcudia, Río Bueno: 0955-0300
129x)	CD154	1540	1	R. Central, Chillán
130)	CB155	1550	1	R. Provincial, Putaendo
131)	CC155	v1550	1	R. Manuel Rodríguez, San Fernando (r 1555): 1100-0400
132)	CD155	1550	0.25	R. Regional, Traiguén
133)	CA156	v1560	5/3	R. Parinacota, Putre: 0955-0430 (r. 1565)
134)	CB156	1560	1	R. Manantial, Talagante: 1100-0400
135)	CD156	1560	1	R. Turismo Internacional, Villarica: 1100-0300
137)	CC157	v1570	1	R. Nueva O'Higgins, Rancagua: 24h
159)	CC157	1570	1	R. Sonorama, Talca: 24h.
138)	CD157	1570	0.25	R. Acuarela, Nueva Imperial: 1155-1810, 2200-0200
140)	CC158	v1580	1	R. Colchagua, Santa Cruz
141)	CD158A	1580	0.25	R. Millaray, Cañete
142)	CD158B	1580	0.5	R. Continental, Collipulli
143)	CB159	1590	1	R. Aconcagua, San Felipe: 1100-0400
144)	CC159	v1590	0.25	R. Rengo, Rengo: 1000-0400
145)	CC159	1590	0.1	R. El Roble, Parral
168)	CD159	1590	5	R. Llanquihue, Llanquihue
146)	CB160A	1600	0.25	R. 2000, Puente Alto: -0000
147)	CB160B	1600	0.25	R. Reloj, Viña del Mar
148)	CC160A	v1600	0.25	R. Regional, Concepción: 1000-0430

Shortwaves: Location: Santiago exc. where indicated.

27)	CE603	6029.7	1	R. Santa María, Coyhaique: 1030-0400 (n. 6030)
43)	CE608	6080.2	1	R. Patagonia Chilena, Coyhaique: W1100-0400
158)	CE609	6088.7	1	R. Esperanza, Temuco: W. 1300-2300, Fri-Sat. 1200-0400, Sun. 24h.
3)	CE963	*9630	10	R. Agricultura, Santiago:
53)	CE975	*9750	1	R. Minería, Santiago
61)	CE1514	*15139.4	60	R. Nal. de Chile: (n. 15140)

Addresses and information:

1) Cas. 110, Melipilla — Mgr: José Massoud S.
2) Cas. 20-D, Valdivia — Gte: Gastón G. Zaffaroni C.
2x) O'Higgins 297, or: Cas. 268, Angol — Dir: Nelson Quesada.
3) CB57: Cas. 40-D, Santiago — Gte: Ricardo Hoffmann L.; CD120: Cas. 327, Los Angeles — Gte: Sergio Jiménez R; CB98: Cas. 90, Valparaíso — Gte: Carlos Prieto.
4) Cas. 1060, Antofagasta — Gte: Rodolfo Pizarro N.
5) Cas. 486, Concepción — DG: Hugo Díaz Uribe.
6) Cas. 97-D, Punta Arenas — Dir: Alfredo Fonseca M.
6x) Condell 910, (or Cas. 183), Santiago — Gte: Claudio Jaque.
6y) Cas. 923, Osorno — Gte: G. Kittsteiner R.
7) Cas. 355, Ovalle — Mgr: Francisco Morales C.
8) O'Higgins 680, Concepción — Gte: Nibaldo Mosciatti M.
9) Cas. 4218, Valparaíso — Gte: Guillermo Morera C.
10) Cas. 41-D, Temuco — Gte: Hernán Espina Rojas.
11) Phillips 40, 2° piso, Santiago — DG: Ernesto Corona B.
12) Domeyo con Avda. Los Pimientos, Chuquicamata — Gte: Federico Gana Johnson.
13) Cas. 2337, Concepción — Dir: Hernán Miller Mansilla.
14) Cas. 10195, Santiago — Mgr: Waldo Mora Longa.
15) Ramírez 207, or Cas. 260, Ancud — DG: Carlos González S.
15x) Vallejos 650, Copiapó — Gte: Freddy Arancibia F.
16) CB76: Cas. 16367, Correo 9, Santiago — Mgr: Luis Ajenjo Isasi; CB73: Cas. 37-V, Valparaíso — Mgr: Carlos Spahie Nasser.
17) Cas. 51, Los Angeles — Dir: Julio Cerda C.
18) Cas. 295, Pto. Aysén — Gte: Carlos Alarcón C.
19) Cas. 35-0, Osorno — DG: Miguel Loayza V.
20) 5 Norte 168, Viña del Mar — Gte: Manuel Miranda Jerez.
21) Cordovez 490 piso 3, La Serena — Dir: Claudio Gómez Arceu.
22) Cas. 358, Talcahuano — Dir: Tamara Deij S.
23) Cas. 312, La Unión — Mgr: César Figueroa F.
25) Of. SOQUIMICH, Pedro de Valdivia.
26) Cas. 89-V, Valparaíso — Mgr: Leandro Pons P.
27) Cas. 1, Coyhaique — DG: Rocco Martiniello Avila.
28) Caupolicán 597, of. 31, Valdivia — Dir: Rubén Rivas A.
29) Cas. 13962, Santiago — Gte: Ignacio Astete.
30) Ap. 6, Pozo Almonte, Iquique — Mgr: Jean Marie Tremblay.
31) Cas. 3115, Concepción — Dir: Manuel J. Fonseca A.
32) I. Carrera Pinto 718, Punta Arenas — DG: Roberto Bravo.
33) Cas. 240, Copiapó — Gte: Juan Soto Meneses.
34) Cas. 267, Chillán — Mgr: Manuel Acevedo A.
35) Cas. 5-0, Osorno — DG: Humberto R. Cárcamo V.
35x) Gral. Cruz 56-A, Temuco — Dir: Sergio Marín R.
36) Cas. 1298, Antofagasta — Dir: Patricio Herrera.
37) Estado 115, Santiago — Gte: Jorge Carrasco T.
38) Cas. 67, Puerto Montt — Gte: René Salinas P.
38x) Maipú 370, Copiapó — Gte: Carlos Porchille.
39) Arzobispo Vicuña 15 Dto. 111, Santiago — Gte: Oscar Pizarro R.
40) Cas. 599, or: Bories 871, Punta Arenas — Gte: René Venegas O.
41) Granaderos 3659, Calama — DG: Raúl Gallardo V.
42) Cas. 214, Talca — Gte: Juan Enrique Concha V.
43) Cas. 30-D, Coyhaique — Mgr: Paulo Ojeda T.
44) Cas. 65-D, Valdivia — Dir: Augusto E. Olave Pávez
45) Cas. 6-D, Arica — DG: Ricardo Castillo C.
46) Cas. 1102, Santiago — Gte: Uros Domic Besic.
47) Cas. 640, Temuco — Dir: Miguel Díaz R.
48) Cas. 3325, Correo 3, Valparaíso — Gte: Alvaro Quintanilla P.
49) Aníbal, Pinto 215, of 801, Concepción — Dir: Héctor Alarcón M.
49x) José Leyán 1233, Talagante.

50) Cas. 106, Castro — Dir: A. Arcanio Velásquez C.
51) Cas. 97, Puerto Natales — Mgr: Juan Navarro V.
52) Cas. 136, Curacautín, Malleco — Gte: Marina E. Sánchez T.
52x) Bilbao 1141, Osorno.
53) Miguel Claro 161, Providencia, Santiago — Dir: Jaime Herrera Ramírez.
54) Cas. 34, Vicuña — Mgr: Fernando Zambra A.
55) Cas. 359, Angol — Gte: Patricia Merino Z.
56) Cas. 516, Talca — Dir: Manuel Yáñez G.
57) Cas. 3-D, Punta Arenas — Gte: Humberto Ramón Utz M.
58) Cas. 410, Antofagasta — Gte: Carlos Brown P.
59) Uno Oriente 938, or: Cas. 596, Viña del Mar — Gte: Uros Domic B.
60) Cas. 910, Temuco — Gte: Eduardo Díaz H.
61) Cas. 244-V, Santiago — Dir: Oscar Sesnic M.
62) Cas. 7-D, Linares — Gte: Herbert Krebs Rosenberg.
62x) Cas. 56-D, Temuco — DG: Roberto Jara B.
64) Cas. 2001, Tocopilla — Gte: Mario V. Reyes R.
65) Cas. 170, Antofagasta — Mgr: Humberto Cortés I.
66) Cas. 1-D, Viña del Mar — Gte: María Cecilia Venegas.
67) Cas. 450, Talca — Gte: César Hernández B.
67x) Cas. 808, Puerto Montt. — Dir: Lorenzo Miranda Y.
67z) Antonio Varas 920, Temuco
68) Cas. 442, Correo 2, Santiago — Dir: Erico Carrasco.
68x) Cas. 127, Calama — Dir: Iris Contreras M.
69) Cas. 290, Iquique — Gte: Elsa Díaz Benavides.
70) Pje. A. Risopatrón 725, La Serena. — Gte: Fernando Zambra A.
71) Cas. 694, Valdivia — Dir: Arnaldo A. Cerda C.
71x) J. Nogueira 1277, Punta Arenas
72) Baquedano 575, or: Cas. 49-D, Arica — DG: Alejandro Ortega.
73) Cas. 492, Curicó — Gte: Juana Cervera T.
74) Cas. 337, Viña del Mar — Gte: Santiago Chiesa H.
75) Cas. 400, Temuco — Dir: Sergio Jiménez R.
76) Dardignac 97, Santiago — DG: Carmen Gloria Quiroz.
77) Cas. 597, Valdivia — Mgr: Luis Papic D.
78) Calle sur s/n, Base Naval, Puerto Williams — Mgr: Marco Arellano.
79) Cas. 13, Vallenar — Gte: Gretta Pizarro B.
80) Gana 360, Mulchén — Mgr: Pedro Amigo A.
82) Gran Avenida 5848, P. 4, La Cisterna (Santiago) — Gte: Julio Vizacarra S.
83) Cas. 166, Puerto Montt — Gte: Rodrigo Muñoz.
84) Cas. 2098, Tocopilla.
85) Plaza de la Justicia 45, Of. 702, Valparaíso Dir: Alfredo Barra B.
86) 18 de Septiembre 721, Chillán — Gte: Daniel Sepúlveda.
87) Bernard O'Higgins 792, Panguipulli. — Gte: J. Figueroa.
88) Cas. 257, Victoria — Gte: Roberto Muñoz Barra
89) Cas. 283, Ancud, Isla de Chiloé — Ricardo Wagner.
90) Cas. 92, Coquimbo — Dir: Juan Ramírez P.
92) Cas. 5-C, Concepción — Dir: Alejandro Salazar.
93) Phillips 56, p. 6, or: Cas. 56042, Santiago — Mgr: Fernando Malafesta.
94) Cas. 614, Iquique — Gte: Néstor Jofré N.
95) Cas. 862, Concepción — Gte: Rafael Arjona M.
96) Cas. 1096, Valparaíso — Gte: María Eliana Durán L.
97) Cas. 61, Loncoche — Mgr: Osvaldo Núñez H.
100) Victoria 495, Cauquenes — Gte: Víctor Chamorro.
101) Brasil 825, Rancagua — Dir: Roberto Rojas.
102) Cas. 225, Arica — Gte: Lucy Rodríguez.
103) Cas. 536, La Serena — Gte: Rodrigo Jara.
104) Cas. 336, Chillán — Gte: José Fco. Arrau U.
105) Cas. 110-V, Valparaíso — Gte: David Dahma B.
106) Yungay 737, Curicó — Gte: Nelson Tarud A.
108) Cas. 328, Pto. Varas — Mgr: Einar Lindh G.
109) Cas. 248, Antofagasta — Mgr: Noemí Solar de García.
110) José Domingo Cañas 1061, Santiago — Gte: Abel Velázquez A.
111) Aníbal Pinto 399, Talcahuano — Mgr: Rubén Sáez S. (Evangelical).
112) Cas. 68, Correo 2, San Antonio — Gte: G. González Vera.
113) Cas. 34, Ovalle — Gte: Sergio Peralta P.
114) Cas. 148, Tomé — Gte: Patricia Torres.
115) Cas. 758, Valdivia — Mgr: Miriam Rosas Wagenknecht.
116) Montandón 916, El Salvador — Gte: Pedro Seracci.
117) Cas. 386, or: Av. Cristóbal Colón 625, San Bernardo — Dir: E. Peñaloza G.
118) Cas. 290, Iquique — Dir: Luis Roldán M.
119) Cas. 18-D, San Javier — Gte: Mario Gaete L.
120) Cas. 89, Puerto Porvenir — Gte: Juan Nesbet G.
121) Cas. 8-D, Rancagua — Gte: Jorge Romero.
122) Cas. 76, Lebu — Gte: Abel Peña Peña.
123) Cas. 11-D, Linares — Gte: Alfonso Astete B.
124) Cas. 242, Lautaro — Gte: Enrique Devaud R.
125) Colipí 371, Copiapó — Gte: Héctor Galarce.
126) Cas. 66, Lota — Mgr: Nibaldo Mosciatti M.
128) Cas. 1346, Santiago — Gte: Manuel Casabianca L.
129) Pedro Lagos 295, Río Bueno — Gte: Nelson Gormaz C.
130) Cas. 75, Putaendo — Gte: Hernán Arancibia C.
131) Cas. 147, San Fernando — Mgr: Flor Quiroz S.
132) Cas. 293, Traiguén — Gte: Rosa Valenzuela V.
133) Calle José Miguel Carrera s/n, Putre (or: Cas 82, Arica) — DG: Tomislav E. Simunovic G.
134) Cas. 223, Talagante — Gte: Francisco Leyton F.
135) Cas. 110, Villarica — Gte: Anita María Ketterer V.
136) Cas. 307, Los Andes — Gte: Carlos Grez T.
137) Cas. 309, Rancagua — Gte: Sergio Díaz C.
138) Cas. 18, Nueva Imperial — Gte: Luis E. Rodríguez C.
139) Aviador Aracena 850, Huasco — Gte: Gretta Pizarro B.
140) Cas. 170, Santa Cruz — Gte: Jorge Villalón G.
141) Arturo Prat 366, Cañete — Gte: Carmen Vallejos D.
142) Cas. 3, Collipulli — Gte: José Urrutia B.
143) Cas. 100, San Felipe — Gte: Raúl Grez G.
144) Cas. 147, Rengo—Gte: Arturo Vergara.
145) Cas. 37, Parral — Gte: Juan Rogazy C.
146) Arturo Prat 53, Puente Alto (Santiago) — Gte: Hernán Claro.
147) Cas. 872, Viña del Mar — Gte: Salvador Cortés P.
148) Cas. 2311, Concepción — DG: Juan Antonio Ríos C.
149) Constitución 420, Illapel — Gte: Sergio Guzmán Leiva.
150) Of. SOQUIMICH, María Elena.
151) Cas. 265, San Carlos — Dir: Hernán Vacaro C.
153) Cas. 12, Temuco — Dir: Jorge Figueroa M.
154) Cas. 541, Los Angeles — Dir: Jaime Sala S.
155) Cas. 33, Llolleo, San Antonio — Gte: Emilio Alcaíno.
156) Cas. 529, Quillota — Gte: Luis Pardo S.
157) Constitución 681, Of. 6, Chillán — Gte: Orlando Villamán.
158) Cas. 830, Temuco — Gte: Eleazar J. Salazar.
159) 4 Oriente 738, Talca — Dir: Francisco Burgos.
165) Aníbal Pinto 720, Mulchen (Bío Bío).
166) Benavente 385, Puerto Montt.
167) Camino Pte. Ibáñez, Calbuco (Llanquihue).
168) E. Ramirez y Pte. Ibáñez, Llanquihue (Puerto Montt).

FM in Santiago. Power 1-10 kW. Slogans: Name + "FM". 11) 88.1 Aurora — 61) 88.9 Futuro — 11) 89.7 Galaxia — 90.5 Pudahuel — 91.3 El Conquistador — 3) 92.1 San Cristóbal — 92.9 La Clave — 93.7 Universo — 16) 94.1 — 68) 94.5 — 95.3 Umbral — 76) 95.5 Tiempo — 96.5 Beethoven — 97.1 Clásica — 97.7 Viva — 98.5 Andrés Bello — 29) 99.3 Carolina — 76) 100.1 Infinita — 100.9 La Ciudad — 101.7 Concierto — 102.5 Univ. de Chile — 103.3 Horizonte — 104.1 Romantica — 104.9 Niña — 105.7 Manquehue — 46) 106.9 Sintonia.

COLOMBIA

L.T: UTC – 5h — **Pop:** 33.236.000 — **Radio sets:** 34.487.000 — **Pr.L:** Spanish — **E.C:** a/c 60, 110/120V — **ITU:** CLM.

MINISTERIO DE COMUNICACIONES: Dirección General de Radiocomunicaciones, Edificio Murillo Toro, Santa Fé de Bogotá, D.C.
L.P: Jefe División Radio: Ing. José Genaldo Cespedes S.

STATIONS: Mediumwaves: Call: HJ--
° = shortwaves, * = inactive, v = varying fq, (r: ...) = repeater.

	Call	kHz	kW	Name, location and h. of tr.
5A)	KA	540	15	R. Auténtica, Bogotá: 24h.
2)	HF	550	50	(r: HJND 570), Marinilla
3)	GS	560	10	(r: HJND 570), Tunja
4)	PF	560	25	"R. 5-60," La Voz de la Pampa, Maicao: 0900-0300
2)	ND	°570	100	R. Dif. Nacional de Colombia, Bogotá: 0930-0500
2)	HP	*580	50	(r: HJND 570), Cali
22A)	CR	590	100	Caracol Antioquia, Medellín
6A)	HJ	600	50	R. Libertad, Barranquilla: 24h
1A)	KL	610	30	RCN 6-10 Internacional, Bogotá: 24h (partly in **English**)

8A)	EL	620	50	Colmundo R., Cali: 24h
9)	VP	620	15	Colmundo R., Cartagena
10A)	FD	630	20	R. Manizales, Manizales: 24h
10X)	WB	°630	15	La Voz del Guainía, Puerto Inírida
11A)	BJ	640	10	RCN, Santa Marta: 24h
1B)	KH	650	30	Antena 2 RCN, Bogotá: 24h
12A)	QS	660	25	Colmundo R., Cúcuta: 24h
5C)	JM	660	10	R. Auténtica, Cali: 24h
14A)	PL	670	10	Antena 2 RCN, Medellín: 24h
2)	BU	v680	100	(r: HJND 570), Zambrano (r. 695)
15B)	CZ	690	30	R. Deportes, Bogotá: 24h
28A)	CX	700	120	Caracol Occidente, Cali: 24h
16)	NX	°710	50	R. Super, Medellín: 24h
6B)	AN	720	30	Emisoras Unidas, Barranquilla
18)	VO	720	25	Transm. Quindío, Armenia
19)	CU	°730	50	Cadena Melodía, Bogotá: 24h
20)	HB	740	15	Ecos de Pasto, Pasto: 1000-0300
21)	NS	740	50	R. Guatapurí, Valledupar
22B)	DK	750	50	Caracol, Medellín: 24h
23A)	AJ	760	30	RCN, Barranquilla: 24h
1C)	JX	°770	50	RCN, Bogotá: 24h
24A)	ZG	780	30	La Voz del Valle, Cali: 24h
25)	ZW	780	15	R. Almirante, Riohacha: 1000-0500
22C)	DC	790	10	La Voz de Colombia, Medellín: 24h
26)	BW	800	10	La Em. de Atalaya, Bucaramanga: 0900-0300
15A)	CY	°810	250	Caracol Colombia, Bogotá: 24h
27A)	AD	820	15	R. Vigía, Cartagena: 1000-0500
28A)	ED	820	50	Caracol, Cali: 24h
22D)	DM	830	15	R. Deportes, Medellín: 24h
29)	BI	840	30	Ondas del Caribe, Santa Marta: 1100-0500
30)	KK	840	20	H J Doble K, Neiva: 24h
15C)	KC	850	50	R. Recuerdos, Bogotá: 24h
32)	EP	860	10	Voces de Occidente, Buga: 24h
33)	NJ	860	50	La Voz del Cañaguate, Valledupar
35)	LA	870	15	La Voz del Tolima, Ibagué: 24h
36A)	SB	870	30	R. Mar Caribe Internacional, Barranquilla
112A)	GE	°880	10	Caracol Bucaramanga: 1100-0400
47)	CS	890	25	Todelar Bogotá: 24h
39)	PM	890	20/10	R. Galeón, Sta. Marta: 24h
40)	DD	900	30	R. 900, Cúcuta
24B)	EY	900	20	La Voz de Cali: 24h
41)	SS	900	10	R. Malecón, Montería
42A)	DO	910	30	La Voz del Río Grande, Medellín: 24h
43)	MY	910	1	RCN, San Andrés, Isla de San Andrés
44)	AA	920	10	Emisoras Fuentes, Cartagena: 1000-0420 (r. 930)
45)	JN	920	30	Ondas del Mayo, Pasto: 1000-0400
46)	SJ	920	15	Colmundo R., Ibagué: 24h
38A)	CE	930	15	La Voz de Bogotá: 24h
24C)	GB	940	15	R. Calima, Cali: 24h
48A)	TL	940	25	RCN, Cúcuta: 24h
49A)	FN	950	15	Caracol, Pereira: 24h
70A)	OP	960	120	Caracol Caribe, Magangué: 24h
50)	CH	970	10	R. Super, Bogotá: 24h
51)	VK	v°970	30	Armonías del Caquetá, Florencia: 1000-0300 (r972)
53A)	ME	970	15	RCN Guajira, Maicao: 1100-0300
13B)	ES	980	25	R. El Sol, Antena 2 RCN, Cali: 24h
52A)	JV	980	30	R. Punto, Cúcuta
14B)	DB	990	150	RCN, Medellín: 24h
54A)	AQ	1000	15	RCN, Cartagena: 24h
2)	JG	v1000	10	(r: HJND 570), Manizales
303)	—	1000	—	R. Cultural, Cajibío
55)	BN	v1010	10/5	La Voz del Galeras, Pasto: 24h (r.1000)
56)	CC	1010	15	Colorín Color R., Bogotá
57A)	JR	v°1010	20	Caracol, Neiva: 24h (r1005)
70C)	FF	1010	15	R. Reloj, Barranquilla
58)	IX	1010	10	R. Yarima, Barráncabermeja
59)	ZD	v1010	15	R. Panzenú, Montería (r.1011)
42B)	DQ	1020	15	Emisora Claridad, Medellín: 24h
61)	DZ	v1020	15	R. Primavera, Bucaramanga: 0900-0400 (r1015)
62)	KS	°1020	15	La Voz del Llano, Villavicencio: 24h
63A)	FQ	1020	15	RCN, Pereira: 24h
64)	FT	°1020	10	Ecos del Combeima, Ibagué: 24h

65)	DJ	1030	10	La Voz de los Libertadores, Duitama: 24h
66)	—	1030	5	Ondas del Vaupés, Mitú
13C)	ER	1030	30	RCN, Cali: 24h
68)	GX	1030	1	R. Progreso de Córdoba, Lorica: 1030-0300
69)	RF	1030	15	Ondas del Cesar, Aguachica
71)	BF	1040	15	La Voz del Norte, Cúcuta
72)	CJ	1040	15	Colmundo R., Bogotá: 1100-0600
73)	CQ	1040	15	La Voz de Armenia, Armenia: 1100-0500
74)	SY	1040	15	R. Mil 40, Popayán: 0900-0400
75)	UB	1040	15	Colmundo R., Pasto
76)	BB	1050	15	R. Valledupar: 0900-0500
42C)	DR	1050	15	R. Sonorama, Medellín: 24h
77)	IO	1050	5	La Voz de la Conquista, Granada: 24h
78)	LZ	v°1050	10	La Voz del Cinaruco, Arauca: 24h
79)	NG	1050	5	R. Palmira, Palmira: 24h
80)	FZ	1050	10	La Voz del Centro, Espinal: 24h
81)	TJ	1050	15	Ondas de Urrá, Montería: 1000-0600
82A)	FJ	1060	15	RCN Caldas, Manizales: 24h
83A)	GU	1050	10	R. Bucarica, Bucaramanga
84)	OV	1060	15	R. Surcolombiana, Neiva: 0955-0530 (r.1050)
85)	LY	1060	15	R. Delfín, Riohacha: 0900-0300
86)	MG	1060	1	R. Prosperidad, Turbo: 1000-0400
87)	MV	1060	10	R. Furatena, Chiquinquirá: 0900-0600
88)	YX	1060	1	R. Caracoli, Sincelejo: 24h
89A)	AH	v1070	20	Emis. Atlántico, Barranquilla: 24h
90)	CG	°1070	30/15	R. Santa Fé, Bogotá: 24h
91)	VR	1070	15	Ondas del Puracé, Popayán: 1100-0500
92)	AW	v1080	10	La Voz de Montería, Montería: 1000-0300 (r1095)
22E)	AX	v1080	10	R. Reloj, Medellín: 24h
24D)	JF	1080	15	R. Eco, Cali: 24h
93)	KT	°1080	10	R. Macarena, Villavicencio: 0900-0500
94)	JS	1080	2	R. Pontoná, La Dorada: 1000-0500
95)	MH	1080	10	R. Melodía, Bucaramanga
96)	IG	°1090	10	La Voz de la Selva, Florencia 1000-0200
97A)	BC	1090	15	Caracol, Cúcuta: 24h
98)	IH	1090	10	Caracol, Sogamoso: 24h
10B)	IA	1090	10	Ondas del Nevado, Manizales: 24h
99)	OM	1090	5	R. Bucanero, Cartagena
15D)	CN	1100	10	R. Reloj, Bogotá: 24h
70B)	AT	1100	15	Caracol, Barranquilla: 24h
100)	EF	1100	2	Brisas del Orinoco, Pto. Carreño
103)	MK	1100	5	Emisora Ideal, Planeta Rica
101)	GQ	1100	5	Transmisora Surandes, Andes: 24h
102)	GI	1100	5	La Voz de Colombia, Socorro: 0930-0330
104)	YZ	1100	15	R. Futuro, Neiva
105)	DI	1110	10	R. Bolivariana, Medellín: 24h
28B)	EW	1110	15	R. Reloj, Cali: 24h
106)	GP	°1110	5	La Voz del Río Arauca, Arauca
108)	JP	1110	10	RCN, Villavicencio: 24h
109)	PA	1110	1	La Voz de las Islas, Isla San Andrés: 1100-0500
110)	ZE	1110	20	R. Piragua, Sincelejo
111)	NC	1110	10	La Voz del Nevado, Ibagué: 1000-0400
12B)	TI	1120	10	R 24, Cúcuta
112B)	GH	1120	15	Corazón AM, Bucaramanga: 24h
113A)	KQ	1120	10	R. Tunja, Tunja: 1100-0400
115A)	JC	1120	5	R. Matecaña, Pereira: 24h
116A)	AC	1130	15	Em. Riomar, Barranquilla: 0930-0400
107)	VA	1130	20	R. "K", Bogotá: 1100-0500
120A)	QQ	1130	15	R. Reloj, Pasto
186)	NN	1130	1	Ondas del Río, Magangué
14C)	DL	1140	10	R. Paisa, Medellín: 24h
123)	RW	°1140	20	La Voz de los Centauros, Villavicencio: 0900-0500
124)	KO	1140	10	La Voz de la Victoria, Cartagena
125)	RN	1140	10	La Voz de la Prov. Colombiana, Barbosa: 1000-0400
126)	CL	1140	10	R. Girardot, Girardot

127)	FH	1140	10	La Voz de Anserma, Anserma: 1030-0500
131)	SL	1150	5	RCN, Sincelejo
119)	FP	1150	15	R. Neiva: 24h
128A)	GJ	1150	1	R. Vencedores, Duitama: 1100-0300
129A)	FI	1150	15	Caracol, Armenia: 24h
130)	BT	1150	10	R. Catatumbo, Ocaña
132)	TE	°1150	1	La Voz del Chocó, Quibdó
133)	AZ	1160	5	La Voz de Sinú, Montería: 24h
6C)	BL	1160	10	R. Aeropuerto, Barranquilla
135)	EC	1160	10	R. San José de Cúcuta, Cúcuta: 0900-0500
136)	AU	°1160	15	Ondas del Orteguaza, Florencia: 1030-0300
118)	CO	1160	30	R. Total, Bogotá: 24h
137)	ZV	1160	10	R. Las Lajas, Ipiales: 1100-0200
24E)	EV	1160	15	La Caleñísima, Cali: 24h
355)	—	1160	1	La Voz de Samaniego, Samaniego: W. 1300-1700, 1900-0100 (n. 1540)
113B)	GA	1170	10	R. Boyacá, Tunja
138)	JE	1170	5	RCN, Tuluá: 24h
139)	JY	1170	1	Ondas del Folcklor, Guamo: 1000-0300
140)	BX	°1170	15	Ondas del Meta, Villavicencio
141)	PB	1170	10	Ondas de Macondo, Valledupar
142)	IZ	°1170	15	Meridiano 70, Arauca: 0900-0500
143A)	NW	1170	10	Caracol, Cartagena: 24h
144)	WA	°1170	5	La Voz del Guaviare, San José del Guaviàre
42D)	KW	1170	15	R. Nutibara, Medellín: 24h
145)	—	1180	15	Em. Coorpurabá, Apartadó
146A)	FX	1180	15	Caracol, Manizales: 24h
147A)	GK	1180	20	RCN Santander, Bucaramanga: 24h
148A)	JT	1180	10	RCN, Ibagué: 24h
149)	CT	1190	10	La Voz de la Costa, Barranquilla: 1030-0400
38B)	CV	1190	15	R. Cordillera, Bogotá: 24h
150)	EO	1190	15	Ondas del Valle, Cartago: 24h
151)	KI	1190	5	R. Barají, Sahuguín: 1000-0500
152)	KG	°1190	10	R. Mira, Tumaco: 1100-0500
154)	CD	1200	15	Emis. Nueva Epoca, Fusagasugá: 0900-0400
155A)	IJ	1200	15	R. 1200, "La Voz de la Raza," Medellín: 24h
156A)	LR	1200	10	Antena 2 RCN, Sogamoso: 1100-0500
157)	NF	1200	10	R. Super, Cali: 24h
158)	BZ	1200	10	Ondas de Riohacha, Riohacha
159)	BV	1200	1	R. Príncipe, Cartagena
57B)	FR	1210	10	R. Reloj, Neiva: 24h
63B)	BQ	1210	15	La Voz Amiga, Pereira: 24h
48B)	BE	1210	30	Antena 2 RCN, Cúcuta: 24h
162)	AV	1220	10	RCN, Montería
163)	KR	1220	15	R. Coral, Bogotá: 24h
164)	NM	v1220	10	Emisora Cultural Bolívar, Ipiales: 1100-0200
165)	MT	1220	10	RCN, San Gil: 24h
167)	IL	1230	10	La Voz de las Américas, Medellín: 24h
169)	BR	1230	10	R. Unica, Tunja: 1100-0500
170)	EH	1230	15	Colmundo R., Bucaramanga
13D)	LK	1230	10	R. Calidad, Cali: 24h
53B)	MJ	1230	5	R. Tribuna, Maicao
172)	TP	1230	5	R. Colina, Girardot: 18h
173)	FG	1240	10	RCN, Armenia: 24h
174)	GN	1240	5	Emisora Barrancabermeja: 1000-0400
175)	JA	°1240	5	R. Buenaventura: 24h
176)	GO	1240	1	R. Caribabare, Sarabena
116B)	OK	1250	1	Emisoras ABC, Barranquilla: 24h
38C)	CA	1250	10	R. Capital, Bogotá: 24h
177)	EM	1250	5	La Voz de Corozal, Corozal: 1100-0500
97B)	HS	1250	15	R. Reloj, Cúcuta: 24h
120C)	FV	1250	1.5	R. Viva, Pasto: 1100-0500
5D)	DA	1260	5	R. Mundial, Medellín: 24h
179)	DV	1260	5	Caracol, Ibagué: 24h
28C)	ET	1260	5	R. Deportes, Cali: 24h
180)	HU	1260	1	R. Morgan, Isla San Andrés
181)	LX	°1260	5	R. Cinco, Villavicencio: 24h
182)	OH	1260	10/5	RCN Cesar, Valledupar: 24h
128B)	NO	1260	2.5	R. Duitama, Duitama: 1000-0300
183)	OU	1260	2	Ondas del Amazonas, Leticia:

184)	QN	1260	5	R. Preferida, Montería
185)	TM	1260	5	R. Sonar, Ocaña: 1000-0500
186X)	XQ	1270	1	R. Melodía, Ubaté
187)	IM	1270	1	Colmundo R., Pereira: 0930-0515
112C)	TX	1270	5	R. Deportes, Bucaramanga: 24h
54B)	AR	1270	1	Antena 2 RCN, Cartagena: 24h
187X)	SV	1270	1	La Voz de Orito, Orito: 1100-2300
188)	HO	°1280	5	Impacto Popular, San Juan del Cesar
120B)	JI	1280	5	Caracol, Pasto: 24h
38D)	KN	1280	5	R. Musical, Bogotá: 24h
189)	RP	1280	5	Ecos de Tibú, Tibú: 1100-2300
190)	TK	1280	5	R. Super, Caicedonia
191)	MB	1280	1	R. Suroeste, Concordia
192)	NQ	1280	2.5	La Voz del Río Suárez, Barbosa: 0930-0200
6D)	SO	1280	5	R. Playa Mendoza, Barranquilla
194)	EB	1290	5	La Voz del Turismo, Sta. Marta
195)	CM	1290	10	R. Sur, Pitalito: 0845-0500 (n.1280)
196)	OI	1290	5	R. Chacurí, Sampués: 1100-0300
197)	MC	1290	5	R. Impulso, Cali: 1100-0500
198)	NE	1290	15/5	La Voz del Ariari, Granada: 1000-0400
199)	HI	1290	5/1	La Voz de Garagoa, Garagoa: 0930-0330
200)	KY	1290	1	Ondas del Puerto, Girardot: 1000-0100
201)	TH	1290	5	La Voz de las Estrellas, Medellín: 24h
201x)	SZ	1290	5	Emisora del Ejército, Saravena
202)	UA	1300	5	R. Sindamanoy, Mocoa: 0900-0300
202x)	RQ	1300	2.5	R. Austral, Túquerres: 0900-0400 (n. 1540)
203)	EA	1300	5	R. Lumbí, Mariquita
204)	IN	1300	5	R. Eucha, Belalcázar: 1200-2330
49B)	LD	1300	5	R. Reloj, Pereira: 24h
112C)	NB	1300	5	R. Reloj, Bucaramanga: 24h
27B)	OG	1300	15	R. La Voz de las Antillas, Cartagena: 24h
205)	RB	1300	10	R. Super, Tunja (r.1320)
206)	DG	1310	5	Caracol, Montería (Transmisora Río): 1000-0500
207)	AK	1310	5	La Voz de la Patria, Barranquilla: 0900-0500
208)	LM	1310	10	R. Santa Bárbara, Santa Bárbara: 1000-0500
361)	—	1310	0.5	R. Cd. Paisaje, Samaniego: 1200-0100
209)	IR	1310	5	RCN Urabá, Apartadó: 24h
15E)	JZ	1310	5	Corazón AM, Bogotá: 24h
210)	WD	1310	15	Micrófono Civico, Palermo
211)	TQ	1310	5	R. Tasajero, Cúcuta: 1000-0500
212)	HT	1320	5	R. Guateque, Guateque: 1000-0300
213)	MS	1320	10	R. El Sol, Barrancabermeja: 0900-0400
214)	NK	1320	1	R. Luna, Palmira
215)	ON	1320	1	R. El Sol, Girardot: 24h
216)	LV	1320	5	Esquina del Progreso, Fundación
217)	QI	1320	10	R. Leda Internacional, San Andrés Isla: 1100-0500
218)	TA	1320	1	La Voz de la Independencia, Medellín: 24h
63C)	FE	1330	10	Antena 2 RCN, Pereira: 24h
221)	NR	1330	10	R. Reloj, San Gil: 0900-0200
222)	RD	1330	1	R. Coopeñol, El Peñol
321)	LS	1330	8	Caracol Popayán
223)	IS	1340	5	R. El Sol, Buenaventura: 24h
89B)	FA	1340	1	R. Olímpica, Barranquilla: 24h
1D)	FB	1340	5	R. Uno, Bogotá: 24h
224)	HY	1340	5	RCN Sucre, Sincelejo: 24h
225)	KD	1340	5	RCN, Neiva: 24h
226)	NP	1340	1	R. Comunal, Nariño: 1100-0100
83B)	NY	1340	5	R. Sonorama, Bucaramanga
52B)	PY	1340	5	R. Mil, Cúcuta
229A)	HA	1340	10	RCN Nariño, Pasto: 24h
227)	VL	1340	2	La V. de los Héroes, Belén de los Andaquíes
230)	DS	1350	10	Ondas de la Montaña, Medellín
231)	EN	1350	10	R. Melodía, Cali: 24h
232)	HL	1350	1	R. Festival, Ibagué
233)	HW	1350	5	Ecos del Río, Puerto Boyacá
234)	MN	1350	1	R. Perijá, Codazzi
235)	LO	1350	5	R. El Sol, Caucasia: 24h
11B)	OC	1350	5	Antena 2 RCN, Sta. Marta: 24h
237)	OZ	°1350	3	La Voz del Yopal: 0900-0200

238)	KV	1360	1	R. Laser, Zapatoca
239)	RA	1360	1	Ecos del Risaralda, Pereira
143B)	UT	1360	5	R. Reloj, Cartagena
240)	SD	1360	1	R. Morichal, San Martín
280)	PK	1360	5	La Voz de Abejorral: 1000-0100
241)	JQ	1370	1	RCN Antena 2, Zarzal: 1100-0500
242)	NI	1370	5	Caracol, Sincelejo: 24h
243)	NU	1370	1	RCN, Rionegro
244)	BD	1370	5	Nueva R. Guaimaral, Cúcuta: 1000-0500
245)	BO	1370	5	R. Piloto, Barranquilla: 24h
246)	EQ	1370	10	RCN Cauca, Popayán: 24h
5B)	KX	1370	5	R. Mundial, Bogotá: 24h
247)	EJ	1380	1	Armonías del Palmar, Palmira: 1130-0300
248)	ID	1380	3	Em. Villa de Páez: 1100-0200
362)	—	1380	1	Ondas del Sur, Túquerres
249)	LG	1380	5	La Voz de la Dorada, La Dorada: 1000-0500
250)	MM	1380	5	R. Reloj, Valledupar: 24h
251A)	JD	*1380	2.3	Colmundo R., Medellín
252)	EE	1380	5	RCN, Tunja: 24h
253)	FO	1390	5	La Voz de los Andes, Manizales: 24h
254)	FY	1390	5	R. Avenida, Espinal: 1000-0400
255A)	ZY	v1390	5	R. Lemas, Bucaramanga: 1000-0500
256)	YW	1390	10	R. Auténtica, Pacho: 24h
23C)	AS	1400	5	Antena 2 de RCN, Barranquilla
258)	BK	1400	1	La Voz de la Gran Colombia, Cúcuta: 24h
360)	WY	1400	1	La Voz del los Samanes, Quilichao: 1100-0000
259)	DF	1400	5	La Voz de Níquel, Montelíbano
261)	HM	1400	5	Antena 2 RCN, Calarcá: -0300
262)	IT	1400	2	Ecos del Atrato, Quibdó
263)	JJ	1400	5	R. Ipiales, Ipiales
264)	KM	1400	5	Emisora Mariana, Bogotá: 1100-0130
265)	LL	1400	5	Antena 2 RCN, Santa Bárbara: 24h
266)	DU	1410	5	Emisora Cultural Universidad de Antioquia, Medellín: 1130-0330
267)	EI	1410	10	R. Guadalajara, Buga: 1100-0300
268)	FS	1410	10	Antena 2 RCN, Honda: 24h
269)	TY	v1410	1	Ondas del Carare, Vélez
270)	AP	1420	5	R. Bahía, Cartagena: 24h
271)	BH	1420	5	Caracol, Santa Marta: 24h
146B)	HK	1420	15	R. Reloj, Manizales: 24h
148B)	LE	1420	10	Antena 2 RCN, Ibagué
272)	SN	1420	1	R. Lenguerque, Zapatoca
274)	QX	1430	5	R. Majagual, Sincelejo
275)	BP	1430	3	R. Cariongo, Pamplona: 0900-0330
276)	EG	1430	2	La Voz de Belalcázar, Popayán: 1000-0500
277)	IU	1430	1	Armonías del Ingrumá, Riosucio: 1100-0200
278)	KU	1430	7.5	La Voz de María "R. Kennedy", Bogotá: 1100-0300
279)	MF	1430	1	R. Venus, Puerto Berrío
281)	PW	1430	5	Colmundo R., Barranquilla
156B)	GM	1440	5	RCN, Sogamoso
283)	EK	1440	1	R. Reloj, Tuluá: 24h
284)	BM	1440	1	La Voz de Honda, Honda
285)	IB	°1440	5	RCN Caquetá, Florencia: 24h
251B)	NZ	1440	5	R. 24, Medellín: 1100-0600
286)	TT	1450	5	Ondas del Porvenir, Samacá
255B)	HH	1450	1	R. Metropolitana, Bucaramanga
82B)	NL	1450	5	Antena 2 de RCN, Manizales: 1000-0500
288)	MX	1450	1.5	R. Mancomoján, Carmen de Bolívar: 1030-0400
288X)	—	1450	1	R. Libanense, Líbano
289)	VY	1450	5	R. Ciudad de Flandes, Flandes
290)	IW	1460	1	R. Monumental, Villa del Rosario
291)	AL	1460	1	R. Sincelejo, Sincelejo: 1000-0400
292)	FL	1460	5	Atalaya Agustiniana, San Agustín: 1000 (Sun 1100)-2300
293)	JH	1460	10	R. Ciudad Milagro, Armenia: 24h
294)	JW	1460	5	Nuevo Continente, Bogotá: 24h
295)	TF	°1460	1	Ondas del Darién, Turbo: 1000-0400
229B)	ZU	1460	5	Antena 2 RCN, Pasto: 24h
296)	MU	1460	1	La Voz de Amalfi, Amalfi: 1100-0300
23D)	VH	1460	5	R. Multicolor, Barranquilla

298)	HQ	v1470	5	R. Futurama, Pacho: 0900-0300
299)	IF	°1470	1	R. Tres Fronteras, Puerto Assís
300)	TB	1470	5	Ondas de Ibagué, Ibagué: 1000-0400
42E)	II	1470	5	R. Musical, Medellín: 24h
8B)	NT	1470	5	R. Total, Cali: 1100-0500
301)	OD	1480	5	R. Rodadero, Santa Marta: 24h
361)	KB	1480	1	R. Guayabal, Guayabal
304)	PX	1480	10	La Voz de las Estrellas, Cartagena: 24h (n1470)
115B)	FC	1480	5	R. Sonorama, Pereira: 1000-0500
305)	TC	1480	1	R. Sonsón, Sonsón (n.1490)
306)	YD	1490	10/5	R. La Paz, Paipa: 0930-0300
147B)	TZ	1480	5	Antena 2 RCN, Bucaramanga: 24h
307)	BS	1490	4	R. Punto Cinco, Bogotá
308)	AG	1490	5	R. Garzón, Garzón: 24h
309)	AY	1490	10	Onda Nueva, Barranquilla: 1100-0500
310)	JO	1490	1	La Voz de San Marcos, San Marcos
311)	ZB	1490	15	La Voz de los Robles, Tuluá: 0900-0500
353)	—	°1500	2	R. Cultural "La Voz del Telembí," Barbacoas: 1300-2300
302)	LJ	1500	10	R. Mil 500, Cali: 24h
312)	MP	1500	1	La Voz de Aguachica, Aguachica
313)	SH	1500	10	R. Reloj, Moniquirá: 0900-0100
314)	TW	1500	10/5	R. Sumapaz, Fusagasugá: 24h
315)	UW	1500	5	La Voz de la Feria, Manizales: 24h
155B)	ZH	1500	5	R. Disco, Medellín: 24h
316)	HX	1510	1	La Voz Panamericana, Bucaramanga: 1000-0500
316X)	ZA	1510	1	La Voz del Centenario, Armenia
317)	UH	1510	5	La Voz de Ciénaga Espectacular, Ciénaga
318)	A22	1510	1	La Voz de San Luis, San Luis Gaceno
319)	AM	1520	1	R. Altamizal, Dolores: 1000-0300
320)	LI	1520	10	Ecos del Palmar, Bogotá: 1100-0500
36B)	LQ	v1520	1	R. Minuto, Barranquilla: 24h
322)	MZ	1520	5	Ecos de la Sierra Flor, Sincelejo: 1030-0430
323)	MA	1520	5	La Voz de Suroeste, Jericó: 1000-0500
282)	RL	1520	5	Antena de los Andes, Santa Rosa de Cabal: 24h
324)	GD	1530	5	Emisora Reina de Colombia, Chiquinquirá: 0800-0200
325)	OZ	1530	5	La Voz de la Prov. de Padilla, San Juan del Cesar: 1000-0400
326)	MI	1530	5	R. Auténtica, Melgar: 0900-0500
363)	—	1530	0.5	La Voz Comunal de la Sabana, Túquerres
251C)	DN	1530	5	R. Total, Medellín: 24h
328)	JB	1530	1	R. Sevilla, Sevilla
329)	ZF	1540	1	R. Cóndor, "Em. Universitaria", Manizales: 1100-0500
330)	HD	1540	2	La Voz del Petróleo, Barrancabermeja: 1000-0300
129B)	QD	1550	5	R. Reloj, Armenia: 24h
333)	UN	1550	1	La Voz de Aguadas, Aguadas: 1000-0300
23B)	CB	1550	5	R. El Sol, Barranquilla: 24h
334)	GY	*1550	5	R. Foro Popular, Lorica: 1000-0300
8C)	LT	1550	5	R. 24, Cali: 1040-0500
336)	ZI	1550	10	R. Fantasía, Bogotá: 24h
362)	UJ	1560	1	Armonías Boyacenses, Motavita
337)	CP	1560	5	Antena 2 RCN, Arbeláez: 24h
338)	HE	1560	5	Voces Rovirenses, Málaga: 1000-0100
339)	LP	1560	5	Antena 2 RCN, Tuluá: -0300v
340)	PZ	1560	1	R. Codazzi, Codazzi
251D)	XZ	1560	5	Colmundo R., El Poblado
342)	ZT	1570	1	La Voz de los Fundadores, Manizales
342X)	TG	1570	1	R. Camino, Machetá
343)	LC	1580	2.5	La Voz de El Banco, El Banco: 1000-0400
344)	KB	1580	5	R. Zulima, Villa del Rosario
345)	OE	1580	1	R. Miraflores, Rovira
346)	QT	1580	1	R. Cachaca, Bogotá
89C)	QZ	1580	1	La Voz de las Estrellas, Barranquilla: 1000-0400
347)	RM	1580	10	R. Costanera, Sincelejo: 1000-0430
348)	SQ	1580	5	R. Robledo, Cartago: 0900-0500

349)	IP	1590	5	R. Metropolitana, Envigado
350)	QM	1590	1	Ecos de la Miel, Samaná
359)	WB	1590	5	R. Ntra. Sra. del Socorro: 0930-0400
351)	HB	1600	1	Armonías Zipaquireñas, Zipaquirá
352)	SP	1600	5	R. Saratama, Pereira

Shortwaves: Call: HJ--, * = inactive, v = varying frequency, h. of tr. 1100-0500 exc. where indicated.

	Call	kHz	kW	Name and h. of tr.
10X)	WA	3500	5	La Voz del Guainía, Puerto Inírida (n. 3315): 1000-2400
353)	—	r3805v		R. Cultural "La Voz del Telembí," Barbacoas: 1300-2300
64)	LW	v4785	5	Ecos del Combeima, Ibagué: -0100v (r. 4785.6v)
175)	—	4833	1	R. Buenaventura
112)	GF	4845	1	Caracol, Bucaramanga
78)	LZ	4865.1	5	La Voz del Cinaruco, Arauca: 0900-0030
140)	IE	4884.9	5	Ondas del Meta, Villavicencio: 0945-2400
106)	—	4895	10	La Voz del Río Arauca, Arauca: 1000-0300
51)	—	4915	3	Armonias del Caquetá, Florencia: 1000-2400
142)	IZ	4925.1	2	Meridiano 70, Arauca: irr.
90)	AE	*4965	2	R. Santa Fé, Bogotá
136)	QA	v4976.2	1	Ondas del Orteguaza, Florencia: 1000-2330v (n. 4975)
262)	PC	5019.8	1	Ecos del Atrato, Quibdó: 1000-0400
237)	PV	5050.1	1	La Voz de Yopal, Yopal: 0900-0200
15A)	GC	5075	50	Caracol Colombia, Bogotá: 2300-1100
358)	—	v5535.5	—	Ecos Celestiales: 0945-1200, 0045-0400
332)	—	r.5555	1.6	La Voz del Valle, Sibundoy (n. 5550)
356)	—	v5569.7	0.2	R. Nueva Vida, Tibú: 1030-1330, 2300-0200
123)	KW	5955	5	La Voz de los Centauros, Villavicencio: 0900-0600
93)	HZ	5975	5	R. Macarena, Villavicencio: 0900-2300
2)	GV	r5985	5	R. Dif. Nacíonal, Tunja: -2300
152)	OW	6015.3	3	R. Mira, Tumaco
144)	OY	6035.2	5	La Voz del Guaviare, San José del Guaviare
19)	QE	6045.5	1	R. Melodía, Bogotá: irr.
72)	CH	6065	1	R. Super, Bogotá: 24h
15A)	IA	*6075	50	Caracol Colombia, Bogotá
295)	TF	v6085	1	Ondas del Darién, Turbo: 1000-0500
62)	IQ	6116.5	10	La Voz del Llano, Villavicencio: 0900-0400v
15A)	OR	6150	10	Caracol Colombia, Bogotá: 1100-2300
96)	KF	6169.2	2.5	La Voz de la Selva, Florencia: 0900-2200 (n. 6170)
2)	ZO	v11821.5	20	R. Dif. Nacional, Bogotá: 0930-0500
2)	ZQ	v17865U	20	R. Dif. Nacional, Bogotá

CARACOL (Primera Cadena Radial Colombiana)

ADDR: Ap. A. 9291 or 8700 (for reports), (Cra. 39A No. 15-81), Bogotá, D.C. **Te:** +57 (1) 2685200. Fax: 2681582.
L.P: Pres: Ricardo Alarcón Gaviria. Vice-Pres: Jorge Eduardo Correa. Dir. Tec: Luis Hernán Camacho.
STATIONS: Basic network: 4), 15), 22), 25), 28), 35), 49), 56), 57), 61), 68), 69), 70), 76), 78), 79), 86), 96), 97), 98), 102), 103), 109), 110), 112), 120), 123), 129), 139), 143), 146), 150), 152), 154), 158), 160), 168), 174), 175), 176), 177), 178), 179), 183), 185), 188, 189), 191), 198), 199), 200), 202), 206), 212), 221), 222), 233), 237), 238), 242), 248), 249), 250), 259), 262), 263), 267), 269), 271), 282), 283), 284), 292), 296), 299), 306), 308), 310), 313), 319), 321), 324), 332), 333), 338), 343), 351). The net has numerous additional affiliates relaying N. only.

RCN (Radio Cadena Nacional)

ADDR: Calle 37 No. 13A-19, (or: Ap. A. 4984) Bogotá, D.C. — **Telex:** 43279. **Te:** +57 (1) 2882288. **Fax:** +57 (1) 2886130.
L.P: Pte: Ricardo Londoño L. Gte Nal: Antonio Pardo García. Dir. Tec: Hermán Medina.
STATIONS: 1), 11), 13), 14), 23), 43), 48), 53A), 54), 63), 65), 80), 82), 108), 137), 138), 147), 148), 156), 162), 165), 173), 182), 209), 213), 215), 223), 224), 225), 229), 235), 241), 243), 246), 252), 257), 261), 265), 268), 285), 337), 339).

TODELAR (Circuito Todelar de Colombia)

ADDR: Av. 13 No. 84-42. Bogotá. **Te:** +57 (1) 2364742.
L.P: Pte: Bernardo Tobón de la Roche. Mgr: Jimmy García Camargo. Dir. Tec: Guillermo Escobar B.
STATIONS: 10), 24), 27), 30), 32), 33), 38), 42), 47), 55), 73), 77), 83), 84), 93), 94), 101), 115), 116), 119), 130), 132), 133), 136), 141), 142), 144), 164), 172), 180), 192), 195), 203), 211), 214), 234), 237), 244), 248), 254), 272), 275), 276), 277), 286), 291), 300), 301), 311), 312), 318), 325), 328), 330), 340), 348), 349).

SUPER RADIO

ADDR: Ap. A. 23316 (Calle 39A No. 18-12), Bogotá, D.C. **Te:** +57 (1) 2877777
L.P: Pte: Jaime Pava N. Gte: Alvaro Pava C.
STATIONS: 16), 40), 50), 53B), 59), 61), 62), 64), 99), 111), 126), 140), 157), 186), 190), 194), 205), 207), 239), 280), 293), 322), 342).

COLMUNDO RADIO (La Cadena de la Paz)

ADDR: Ap. A. 36750 (Diag. 58 No. 26A-29), Bogotá. **Te:** +57 (1) 2178911. **L.P:** Pte: Carlos Humberto Sierra T. Gte: Néstor Chamorro.
STATIONS: 8), 9), 12), 46), 75), 92), 118), 170), 187), 251), 281).

State abbreviations: (Departmentos, Intendencias y Comisarías). Ant = Antioquia; Atl = Atlántico; Bol = Bollívar; Boy = Boyacá; Caq = Caquetá; Ces = Cesar; Cor = Córdoba; Cun = Cundinamarca; D.C. = Distrito Capital (Bogotá renamed to Santa Fé de Bogotá); Gua = Guajira; Hui = Huila; Mag = Magdelena; Nar = Nariño; NS = Norte de Santander; Put = Putumayo; Qui = Quindío; Ris = Risaralda; SS = Santander del Sur; Tol = Tolima; Valle = Valle del Cauca.
N.B: D.C. and Valle can be used for postal purposes, also "Islas" for San Andrés y Providencia. In all other cases full name is preferred.
Ap. A. = Apartado Aéreo. (A.A. also currently used).

Addresses and other information:
1A-D) Calle 37 No. 13A-19, Bogotá, D.C. — Gte: A. Pardo G.
2) CAN, Av. El Dorado, Bogotá, D.C. — Dir: Luz Amanda Rodríguez.
3) Calle 19 No. 9-50, Tunja (Boy) — Gte: Alberto Upegui A.
4) Cra. 9 No. 12-29, Maicao — Gte: Carlos Serrano
5AB) Ap. A. 18350, or: Calle 32 No. 16-12, Bogotá — Gte: J. Enrique Gómez Jr.
5C) Calle 5C No. 41-30, Cali — Gte: Oswaldo Arias.
5D) Cra. 74 No. 49-97, Medellín — Gte: Rafael Linares.
6A-D) Ap. A. 3143, or: Cra. 53 No. 55-166, Barranquilla (Atl) — Gte: Roberto Esper R.
8ABC) Cra. 26 No. 5C-25, Cali — Gte: Gloria Roa de Martínez.
9) Av. Venezuela, Edif. Banco Intern., La Matuna 8B-05, Cartagena (Bol) — Dir: Valerio Mejía.
10AB) Ap. A. 67, Manizales (Caldas) — Gte: Carlos A. Mejía.
10X) Calle 6 con Cra.3, Pto. Inírida (Guainía) — Gte: Luis Fernando Román.
11AB) Ap. A. 540, Santa Marta (Mag) — Gte: Javier Orozco R.
12AB) Calle 5 No. 3-26, Cúcuta (NS) — Gte: Luis Blanco.
13A-D) Ap. A. 1972, Cali (Valle) — Gte: Jairo Domínguez A.
14A-D) Calle 44 No. 57A-21, Medellín (Ant) — Gte: Mario Ceballos Z.
15A-E) Ap. A. 9291, Bogotá, D.C. — Gte: A) Darío Arizmendi; B) Hernán Peláez; D) Pastor Vesga.
16) Ap. A. 3399, Medellín (Ant) — Dep. Mgr: Henry Pava C.
18) Cra. 16 No. 19-23, Armenia (Qui) — DG: Aníbal Mejía.
19) Ap. A. 19823, Bogotá, D.C. — DG: Fernando Páez Majía — AM Stereo.
20) Ap. A. 375, Pasto (Nar) — Gte: E. del Castillo C.
21) Ap. A. 51, Valledupar (Ces) — Gte: Gustavo Cuello D.
22A-E) Cra. 79A No. 39-45, Medellín — Gte: Dr. Luis Salazar Gómez.
23AB) Ap. A. 1883, Barranquilla (Atl) — Gte: Rubén D. Garcia.
24A/E) Ap. A. 4666, Cali (Valle) — Gte: Dr. Eduardo Rueda S.
25) Cra. 8 No. 3-27, Riohacha (Gua) — Gte: Roberto Gutiérrez.
26) Ap. A. 915, Bucaramanga (SS) — Own: Gustavo Serrano F.

27AB) Calle 35 No. 8B-10, Edif. Banco Internacional, Cartagena.
28ABC) Av. 4N No. 4-65, Cali (Valle) — Dir: María del Socorro Valencia. Gte: B,C) Sammy Jalil.
29) Ap. A. 757, or: Cra. 5 No. 18-32, Santa Marta (Mag) — Gte: Roberto Esper R.
30) Ap. A. 727, Neiva (Hui) — DG: Jorge H. Charry A.
32) Ap. A. 96, Buga (Valle) — Gte: Sonia Govea.
33) Cra. 5 No. 13-52, Valledupar (Ces).
35) Ap. A. 780, Ibagué, (Tol) — Gte: Carmen Laserna.
36AB) Cl. 72 No. 41C-64 (or: Ap. A. 5402), Barranquilla (Atl).
37) Ap. A. 572, Bucaramanga (SS) — Gte: Jorge Serrano.
38AB) Ap. A. 056654, Bogotá, D.C. — Gte: Germán Tobón M.
38C) Cra. 28 No. 95-27, Bogotá.
38D) Calle 68 No. 14-76, Bogotá.
39) Ap. A. 103, Sta. Marta (Mag) — Own: Rodrigo Ahumada B.
40) Av. O No. 8-45, Cúcuta (NS) — Gte: J. Duque Castrillón.
41) Calle 66 No. 3-171, Montería (Cor) — Gte: Jaime Cabrales P.
42A-E) Ap. A. 1431, Medellín (Ant) — DG: Héctor J. Velásquez.
43) Ap. A. 354, San Andres (Isla) — DG: Humberto Paffen G.
44) Ap. A. 1771, or: Calle Mayor No. 6-34, Cartagena (Bol.).
45) Ap. A. 635, Pasto (Nar) — Gte: Alirio Ortega P.
46) Calle 14 No. 2A-14, P. 2, Ibagué (Tol) — Gte: Alberto Viancha.
47) Calle 48 No. 18-77, Bogotá, D.C. — Gte: Germán Tobón M.
48AB) Ap. A. 400, Cúcuta (NS) — Gte: Fdo. Fonseca R.
49AB) Ap. A. 354, Pereira (Ris) — Gte: A) Erney Ocampo; B) Antonio José Uribe.
50) Ap. A. 23316, or: Calle 39A No. 18-12, Bogotá, D.C.
51) Florencia (Caq) — Dir: Padre Alvaro Serna A.
52AB) Calle 5 No. 0-45, Cúcuta (NS) — Dir: Luis E. Mantilla.
53AB) Ap. A. 125 & 246, Maicao (Gua) — Gte: William Gómez.
54AB) Ap. A. 246, Cartagena (Bol) — Gte: Luis Lopera.
55) Ap. A. 454, Pasto (Nar) — Gte: Felipe Chaves.
56) Ap. A. 10556, Bogotá, D.C. — Gte: Adriana Giraldo.
57AB) Ap. A. 150, Neiva (Hui) — Gte: Luis F. Escobar.
58) Calle 10 No. 17-73, Barrancabermeja (SS).
59) Cra. 3 No. 30-12, Montería (Cor) — Gte: José V. Muskus.
61) Cra. 27 No. 45-80, Bucaramanga (SS).
62) Cra. 31 No. 38-07, piso 2, Villavicencio (Meta) — Adm: Gloria Inés Rodriguez.
63ABC) Ap. A. 045, Pereira (Ris) — Gte: Augusto Torres G.
64) Pq. Murillo Toro 3-31, Ibagué (Tol) — Gte: Henry Pava C.
65) Ap. A. 1030, Duitama (Boy) — Gte: Gloria Guatibonza.
66) Mitú (Vaupés).
68) Calle 1 No. 16A-03, Lorica (Cor) or Ap. A. 337 Montería — Own: Freddy Sánchez A. F.Pl: 5kW.
69) Calle 7 No. 16-39, Aguachica (Ces). — Gte: Alberto Cabrales.
70A/D) Ap. A. 1688, Barranquilla (Atl) — Gte: Alberto Díaz M.
71) Ap. A. 624, Cúcuta (NS) — Gte: Gustavo Sánchez Chacón.
72) Ap. A. 23316, Bogotá, D.C. — Gte: Alvaro Pava C.
73) Calle 9 No. 13-50, or: Ap. A. 2361, Armenia, Armenia (Qui).
74) Cra. 8 No. 3-17, or: A.A. 1321, Popayán (Cauca) — Gte: Marco A. Segura.
75) Calle 20 No. 24-37, P. 6, Pasto (Nar) — Gte: Ovidio Palacio C.
76) Ap. A. 22, Valledupar (Ces) — Gte: Eduardo Dangond.
77) Calle 16 No. 15-90, Granada (Met) — Gte: Javier Ruiz A.
78) Calle 19 No. 19-62, Arauca — Gte: Efrahim Varela N.
79) Ap. A. 280, Palmira (Valle) — Gte: Fernando Burbano.
80) Cra. 7 x Calle 10, El Espinal (Tol) — Dir: Carlos Barón A.
81) Calle 23 No. 1-53, Montería (Cor).
82AB) Ap. A. 244, Manizales (Caldas) — Gte: Alvaro Fonseca C.
83AB) Calle 39 No. 22-54, Bucaramanga (SS).
84) Ap. A. 496, Neiva (Hui) — DG: Samuel Sánchez C.
85) Calle 7 No. 14-16, Riohacha (Gua).
86) Cra. 19 No. 20-66, Turbo (Ant) — DG: Fabio Patiño G.
87) Ap. Nal. 13, Chiquinquirá (Boy) — Gte: Alejandro Rocha Z.
88) Ap. A. 750, Sincelejo (Suc) — Gte: J. L. Feris.
89ABC) Ap. A. 51266, Barranquilla (Atl) — Gte: José A. Restrepo.
90) Ap. A. 9339, Bogotá, D.C. — DG: Blanca Bernal M.
91) Ap. A. 1700, Popayán (Cauca).
92) Ap. A. 497, Montería (Cor) — Own: Magola Gómez G. **F.Pl:** 15kW.
93) Ap. A. 2484, Villavicencio (Meta) — Gte: Enrique Zambrano
94) Edif. Banco Ganadero, La Dorada (Caldas) — Gte: F. Ordóñez M.
95) Calle 35 No. 18-20, Girón (SS).
96) Ap. A. 465, Florencia (Caq) — Dir: Eduardo Flores C.
97AB) Ap. A. 519, Cúcuta (NS) — Dir: Carlos Pérez.
98) Ap. A. 282, Sogamoso (Boy) — Dir: Miguel Cristancho.
99) Edif. Banco Popular (or: Ap. A. 2215), Cartagena (Bol).
100) Calle 8 No. 1-24, Puerto Carreño — Mgr: Sergio Cardona.
101) Av. Santander, 47A-84, Andes (Ant) — DG: J.O. Martinez H.
102) Calle 16 No. 15-01, El Socorro (SS) — Gte: Helio Zambrano.
103) Cra. 8 No. 17-56, Planeta Rica (Cor) — Gte: A. Bula C.
104) Cra. 6 No. 6-51 P. 2, Neiva (Hui).
105) Ap. A. 1178, Medellín (Ant) — Gte: Humberto Mesa R.
106) Cra. 22 No. 18-21, Arauca (or: Ap. A. 16555, Bogotá) — Dep. Mgr: Gloria Jeanne Silvia R.
107) Cra. 16 No. 43-09, Bogotá.
108) Cra. 30 No. 36-14, P. 4, Villavicencio (Meta) — Dir: J.A. Martinez.
109) Ap. A. 1034, San Andrés Isla — Gte: Mauro Gallardo C.
110) Ap. A. 448, Sincelejo (Sucre) — Gte: Juan Guerra T.
111) Ap. A. 926, Ibagué (Tol) — Gte: Antonio Rocha P.
112A-D) Cl. 35 No. 16-24, P. 8 (or: Ap. A. 223), Bucaramanga (SS) — Gte: Germán Gómez Vahos.
113AB) Calle 20 No. 8-54, Tunja (Boy) — Gte: Victor Umaña.
115ABC) Cra. 7 No. 16-25, Pereira — DG: Hernán Ospina G.
116AB) Ap. A. 2010, Barranquilla (Atl) — DG: Wenceslao Herrera.
118AB) Ap. A. 36750, Bogotá, D.E. — Gte: Fabio Restrepo G.
119) Ap. A. 137, Neiva (Huila) — Gte: Jorge Enrique Sánchez.
120AB) Cra. 27 No. 19-30, Pasto (Nar) — Dir: Gonzalo Ardila.
120C) Cra. 24 No. 17-75, Pasto (Nar) — DG: Manuel Eraso.
123) Cra. 31 No. 37-71, Of 1001, Villavicencio (Meta) — Gte: Guillermo Torres B.
124) Ap. A. 2456, Cartagena (Bol) — Gte: J. Franco E.
125) Transv. 6 No. 9-56, Barbosa (San) — DG: Henry Cortés T.
126) Cra. 10 No. 13-36, Girardot (Cun) — Gte: José C. Cifuentes G.
127) Cra. 4 No. 8-58, Anserma (Caldas) — Gte: J.D. Giraldo.
128AB) Ap. A. 1064, Duitama (Boy) — Gte: G. Benavides G.
129AB) Ap. A. 2481, Armenia (Qui) — Gte: César Marulanda.
130) Calle 9 No. 13-10, Ocaña (NS) — Gte: Carlos Navarro.
131) Cra. 16 No. 23-40, Sincelejo (Suc) — DG: Juan A. Navarro.
132) Cl. 23 No. 3-36, Quibdó (Chocó) — Gte: Alfonso Agudelo.
133) Ap. A. 148, Montería (Cor) — Gte: A. Gómez D.
135) Ap. A. 2284, Cúcuta (NS) — DG: Wm. Javier Gómez.
136) Ap. A. 209, Florencia (Caq) — Gte: J. D. Santos C.
137) Ap. A. 1005, Ipiales (Nar) — Gte: Fidencio A. Erazo R.
138) Ap. A. 126, Tuluá (Valle) — Gte: Luis F. Ossa P.
139) Cra. 9a No. 9-56, Guamo (Tol) — Gte: Camilo Sarmiento G.
140) Ap. A. 2196, Villavicencio (Meta) — Adm: Yolanda P. de Lozada.
141) Ap. A. 492, Valledupar (Ces) — Gte: Alfonso Campos S.
142) Ap. A. 18227, Arauca — Gte: Alvaro Pérez G.
143AB) Calle 36 No. 9-89 (or: Ap. A. 97), Cartagena — Dir: Humberto González Kerguelen.
144) Cra. 3 Calle 2, San José del Guaviare.
145) Apartadó (Ant).
146AB) Ap. A. 2000, Manizales (Caldas) — Gte: Dr. J. Giraldo Neira.
147AB) Ap. A. 1100, Bucaramanga (SS) — Gte: Fdo. Ardila.
148) Ap. A. 2419, Ibagué (Tol) — Gte: Alonso Botero Palacio.
149) Ap. A. 50906, Barranquilla (Atl) — **Te:** +57 340420 — Dir: G. Forero Sanmiguel. Prgr. all news.
150) Ap. A. 145, Cartago (Valle) — Gte: Cristóbal de los Rios.
151) Calle de Comercio, Sahagún (Cor) — DG: B.E. Nader.
152) Ap. A. 165, Tumaco (Nar) — Dir: Julio Cortés B.
154) Calle 8 No. 5-08, Fusagasugá (Cun). Own: F. Carrera Otero.
155AB) Ap. A. 4552, Medellín (Ant) — DG: Aida Luz Herrera.
156AB) Ap. A. 019, Sogamoso (Boy) — Dir: E. Cubillos C.
157) Calle 15 No. 8-75, 7° piso, Cali (Valle) — Dir: H. Pava C.
158) Ap. A. 3, Riohacha (Gua) — Gte: José I. Vives.
159) Av. 3 No. 21-62, La Manga, Cartagena (Bol).
162) Calle 27 No. 8-25, Montería (Cor) — Dir: H. Rangel R.
163) Ap. A. 10951, Bogotá, D.C — Gte: Guillermo Romero.
164) Cra. 8 No. 4-48, Ipiales (Nar) — Gte: Alvaro A. Enríquez.
165) Calle 10 No. 9-71, San Gil (SS) — Dir: T. Rueda R.
167) Ap. A. 3998, Medellín (Ant) — Gte: Carlos Nicholls Posada.
168) Calle 15 No. 13-56, Acacias (Meta) — DG: Eduardo Riveros.
169) Calle 19 No. 9-35, Tunja (Boy) — DG: Antonio Piedrahita.
170) Cra. 19 No. 33-52, Bucaramanga — Gte: Omar Martínez.
172) Calle 14 No. 9-10, Of. 302, Girardot — Gte: Germán Rodríguez P.
173) Ap. A. 556, Armenia (Qui) — Gte: Jorge Orozco.
174) Ap. A. 23, Barrancabermeja (SS) — Gte: Xiomara Domínguez G.
175) Ap. A. 383, Buenaventura — Gte: Mauricio Castaño.
176) Ap. A. 6558, Sarabena, Arauca — Own: Mario Fric.
177) Ap. A. 100, Corozal (Sucre) — Gte: Nelson Martelo O.
178) Calle 38 No. 28-87, Pasto (Nar) — Gte: R.P. Jaime Alvarez.
179) Ap. A. 1094, Ibagué (Tol) — Gte: Fabio Zúñiga.
180) Edif. Bermuda, piso 2, Av. de las Américas, San Andrés Isla.

181) Ap. A. 2284, Villavicencio (Meta) — Gte: Omar López R.
182) Ap. A. 250, Valledupar (Cesar) — Gte: Santiago Marmolejo M.
183) Ap. A. 236, Leticia (Amazonas) — DG: Luis M. Usoga.
184) Calle 73 No. 7-17, Montería (Cor) — DG: Jaime Gonzáles V.
185) Ap. Nal. 19, Ocaña (NS) — Gte: Myriam M. Cabrales C.
186) Edif. Cure, or: Ap. A. 180, Magangué (Bol).
186X) Cra. 6 No. 6-38, Ubaté (Cun).
187) Cra. 7a. No. 18-21, of. 1103, Pereira — Gte: Luis Alberto Delgado.
187X) Calle Ppal., Orito (Put) — Dir: Jeppy Navarro.
188) Cra. 6 No. 5S-37, San Juan del Cesar (Gua).
189) Curia Episcopal, Tibú (NS) — Gte: J.H. Puentes Torres.
190) Cra. 16 No. 6-22, Calcedonia (Valle) — Gte: Carlos Alberto Orozco.
191) Cra. 3 Calles 2 y 3, Concordia (Ant).
192) Calle 7, 7-44, Barbosa (SS) — Gte: Rafael Fonseca L.
194) Ap. A. 1529, Sta. Marta (Mag) — Gte: P. de Andreís A.
195) Calle 6 No. 5-36, Pitalito (Huila) — Pte: Manuel Castro T.
196) Ap. A 191, Sincelejo — Gte: Marino Diconza.
197) Calle 22N No. 5N-46, Cali.
198) Cra. 13 No. 28-05, Granada (Met) — DG: Néstor G. Padilla.
199) Cra. 9 No. 8-65, Garagoa (Boy) (or: Ap. A. 13729, Bogotá) — DG: H. Gámez R.
200) Ap. A. 416, Girardot (Cun) — Gte: Jairo E. Cifuentes.
201) Ap. A. 4300, Medellín (Ant) — Gte: Marco F. Eusse.
201x) Saravena (Arauca).
202) Ap. A. 011, Mocoa — Own: Jaime Herrera A.
202x) Calle 20 No. 15-02, Túquerres (Nar) — Gte: J. Aguirre B.
203) Calle 3 No. 3-62, Mariquita (Tol) — Gte: Lisimaco Parra.
204) Casa Cural, Parque Ppal., Belalcázar-Páez (Cauca) — Gte: P. Luis Eduardo Quíroga T. Prgr. in Páez vernacular.
205) Calle 20 No. 10-64, Tunja (Boy) — DG: Victor Umaña.
206) Ap. A. 364, Montería (Cor) — Gte: Francisco T. Zubiría R.
207) Ap. A. 231, Barranquilla (Atl) — Own: Clemente Vasallo G.
208) Ap. A. 3854, Medellín (Ant) — Gte: Ignazio Ramírez B.
209) Calle 94 No. 99-51, Apartadó (Ant) — Dir: Carlos Vélez.
210) Edif. Antiguo de la Alcaldía, Palermo (Hui) (or: Ap. A 446, Neiva) — Gte: Alvaro Falla R.
211) Ap. A. 473, Cúcuta (NS) — Gte: Carlos Ernesto Díaz M.
212) Cra. 7 No. 9-57, Guateque (Boy) — Gte: Guillermo Quintero.
213) Ap. A. 578, Barrancabermeja (SS) — Gte: Enrique Fuentes S.
214) Cra. 30 No. 29-09, Palmira (Valle) — Gte: Ma. Consuelo Valdez.
215) Ap. A. 457, Girardot (Cun) — Gte: Carlos A. Barón.
216) Cra. 5 No. 6-12, Fundación (Mag) — Gte: R. Arango Galvis.
217) Ap. A. 665, San Andrés Isla — Own: León Cybul.
218) Ap. A. 4910, Medellín (Ant) — Gte: Enrique Hincapié B.
221) Calle 12 No. 10-30, San Gil (SS) — Gte: Mario G. Rojas.
222) Centro Cooperativo, El Peñol (Ant) — Dir: Edgar A. Garcia.
223) Cra. 6 No. 2-16, Buenaventura — Gte: C. Astorquiza.
224) Calle 20 No. 24-93, Av. las Penitas, Sincelejo (Sucre).
225) Ap. A. 192, Neiva (Huila) — Gte: Ernesto Cruz D.
226) Casa Cural, Nariño (Ant) — Gte: Rodolfo Salazar Z.
227) Belén de los Andaquíes (Caq).
229AB) Ap. A. 516, Pasto (Nar) — Dir: Carlos E. Pérez L.
230) Ap. A. 7870, Medellín (Ant) — Gte: Nora Roldán.
231) Av. 3Bis No. 23CN-71, Cali (Valle) — DG: Efraín Páez E.
232) Edif. Banco del Comercio, Ibagué (Tol).
233) Cra. 3 No. 13-74, Puerto Boyaca (Boy) — Gte: M. Caraballo.
234) Cra. 16 No. 11-102, Codazzi (Ces).
235) Cra. 2 No. 21-54, Caucasia (Ant) — Dir: Hdo. García.
237) Cra. 9 No. 7-17, Yopal (Casanare) — Dir: P.A. Socha.
238) Calle 16 No. 4-47, Zapatoca (SS) — DG: Angel J. Jaimes.
239) Cra. 7 No. 15-10, Pereira (Ris) — Gte: Rosas Salazar R.
240) Km. 3 Vía Villav/o, San Martín (Meta) — Gte: Hernando Sánchez.
241) Cra. 11 No. 9-43, Zarzal (Valle) — Dir: A. Jurado.
242) Cra. 18 No. 18-31, Sincelejo (Sucre) — Gte: N. Miserque.
243) Cra. 51 No. 49-09, Rionegro (Ant) — Gte: A. Arenas B.
244) Ap. A. 2582, Cúcuta (NS) — Dir: Carlos Ernesto Díaz.
245) Ap. A. 320, Barranquilla — Gte: Dorian Fayán N.
246) Ap. A. 535, Popayán (Cauca) — Dir: Javier Perdomo F.
247) Ap. A. 201, Palmira (Valle) — Gte: Jaime Bejarano D.
248) Cra. 4 No. 4-12, La Plata (Huila) — Pbro. Fernando Floriano C.
249) Ap. A. 34, La Dorada (Caldas) — Gte: Javier Comas P.
250) Cra. 9 No. 5-02, Valledupar (Ces).
251A/D) Calle 48B No. 79-38, Medellín (Ant) — Gte: Rogelio Tangarife.
252) Ap. A. 1186, Tunja (Boy) — Dir: Alfonso Escobar F.
253) Ap. A. 1585, Manizales (Caldas) — Gte: Artura Gallego E.
254) Ap. A. 64, Espinal (Tol) —Gte: Antonio Forero B.
255AB) Ap. A. 1620, Bucaramanga (SS) — Gte: A. Sanmiguel.
256) Calle 7 No. 14-66, Pacho (Cun) — Gte: Gloria E. Ortiz.
258) Ap. A. 1303, Cúcuta (NS) — Gte: José Neira Rey.
259) Cra. 5 No. 14-35, Montelibano (Cor) — DG: Demetrio Nader.
261) Cra. 25 No. 41-38, Calarcá (Qui) — Dir: Jorge Orozco.
262) Ap. A. 78, Quibdó (Chocó) — Gte: Nancy Hinestroza P.
263) Cra. 6 No. 10-60, Ipiales (Nar) — Gte: Roberto Ramírez.
264) Calle 6 No. 7-22, Bogotá, D.C. — Gte: Luis A. Monroy B.
265) Cra. Bolívar, Calle López, Santa Bárbara (Ant) — Dir: H. Gómez C.
266) Ap. A. 1226, Medellín (Ant) — Dir: Luis Alfonso Yepes.
267) Cra. 14 No. 5-81, Buga (Valle) — Gte: José M. González.
268) Ap. A. 536, Honda (Tol) — Dir: Rafael Pabón B.
269) Cra. 3 No. 3-42, Vélez (SS) — Gte: Rodrigo Correa F.
270) Calle Román 5-45, Cartagena — Gte: Juan Gallo.
271) Ap. A. 1240, Santa Marta (Mag) — Gte: Gustavo Morales.
272) Calle 20 No. 7-18, Zapatoca (SS) — Gte: Alvara Grass.
274) Ap. A. 542, Sincelejo (Sucre) — Gte: Antonio Turbay.
275) Ap. A. 1074, Pamplona (NS) — Dir: María Socorro Luna S.
276) Ap. A. 759, Popayán (Cauca) — Gte: G.A. Valencia M.
277) Cra. 5 No. 11-102, Riosucio (Caldas) — Gte: O. Cuesta S.
278) Ap. A. C.K. 72825, Bogotá, D.C. — DG: Padre José V. Echeverri.
279) Calle 6 No. 1-23, Puerto Berrío (Ant).
280) Calle 50 No. 50-53, Abejorral (Ant) — Gte: B. García R.
281) Cra. 44 No. 70-61, Barranquilla (Alt) —Dir: Margarita Bermúdez.
282) Cra. 15 No. 11-80, Santa Rosa de Cabal (Ris) (or: Ap. A. 2718, Pereira) —DG: J.G. Cardona R.
283) Ap. A. 130, Tuluá (Valle) — Gte: Eduardo Ruega S.
284) Edif. Banco Central Hipotecario, P. 3, Honda (Tol) — Dir: José Eddy Galindo.
285) Ap. A. 150, Florencia (Caq) — Gte: Carlos Moya.
286) Cra. 6, Pg. Principal, Samacá (Boy).
288) Calle 25 No. 49-33, Carmen de Bolívar (Bol) — Dir: E. Sanabria T.
289) Cra. 2 No. 11-27, Flandes (Tol) — DG: Antonio Forero.
290) Cra. 1 Cl. 7 y 8, Villa del Rosario (NS).
291) Ap. A. 307, Sincelejo (Sucre) — Gte: A. Gómez Jiménez.
292) Calle 3 Cra. 4, San Agustín (Huila) — Dir: H. Paredes C.
293) Cra. 14 No. 21-26, Armenia (Qui) — Gte: Sabogal Vásquez.
294) Ap. A. 10900, Bogotá, D.C. — Gte: Antonio Neira V.
295) Ap. A. 1289, Medellín (Ant) — Dir: Manuel Cardona A.
296) Calle 19 No. 18-33, Amalfi (Ant) — Gte: Ma. Helena González.
298) Calle 3 No. 16-39, Pacho (Cun) — Gte: Luis A. Rocha.
299) Ap. A. 9, Puerto Assís (Put) — Gte: Carlos A. Herrera R.
300) Ap. A. 589, Ibagué (Tol) — Gte: Melyda T. de Molina.
301) Ap. A. 052, Santa Marta (Mag) — Gte: Edgard Polo F.
302) Calle 3A No. 42-78, Cali, or: Calle 7 No. 6-73, El Cerrito (Valle) — Gte: Ernesto Quijano.
303) Cajibío (Cauca).
304) Ap. A. 2254, Cartagena — DG: Ubaldo Morales M.
305) Calle 7 No. 7-54, Sonsón (Ant) — Gte: J. Cardona H.
306) Calle 5 No. 5-65, Paipa (Boy) — Dir: Abraham Casteblanco.
307) Av. 15 No. 123-71, Of. 408, Bogotá, D.C. — Dir: Hugo Prieto R.
308) Cra. 7 No. 7-05, Garzón (Huila) — Gte: A. Losada L.
309) Ap. A. 2647, Barranquilla (Atl) — Gte: Jaime Jiménez V.
310) Cra. 28 Calle 18, San Marcos (Sucre).
311) Calle 27 No. 33-35, Tuluá (Valle) — DG: E. Castro L.
312) Cra. 10a No. 4-24, Aguachica (Ces) — Gte: Carlos Navarro.
313) Calle 7 No. 3-69, Moniquirá (Boy) — Admor: G. Pardo A.
314) Ap. A. 006, Fusagasugá (Cun) — Gte: J. Munevar B.
315) Ap. A. 990, Manizales (Caldas) — Gte: Myriam Uribe B.
316) Ap. A. 432, Bucaramanga (SS) — Gte: Gonzalo Ayala O.
316X) Ap. A. 617, Armenia (Qui) — Gte: Leonardo Marín I.
317) Calle 18 No. 5-58, Ciénaga (Mag). —Gte: Anibal S. Ceballos
318) Av. Durán Quintero y Calle 3, San Luis Gaceno (Boy).
319) Cra. 7a. No. 6-58, Dolores (Tol) — Dir: Héctor Hernández N.
320) Cra. 41C No. 1A-14, Bogotá — Dir: José Paipilla P.
321) Calle 5A No. 11-25, Popayán (Cauca) — Gte: Carlos Lamprea R.
322) Calle 25A No. 18-18, Sincelejo (Sucre) — Gte: Roberto Angulo M.
323) Calle 7, Cras. 3 y 4, Jericó (Ant) — DG: P. Sánchez P.
324) Calle 18 No. 12-81, Chiquinquirá (Boy) —Gte: Jorge Enrique Flores N.
325) Calle 1 No. 5-63, San Juan del Cesar (Cesar).
326) Calle 7 No. 20-70, Melgar (Tol) — Gte: Gernando Sierra R.
328) Cra. 51 No. 49-21, Sevilla (Valle) — Gte: Oscar Salazar J.
329) Ap. A. 441, Manizales (Caldas) — Dir: Gabriel Jaramillo.

330) Ap. A. 250, Barrancabermeja (SS) — Gte: J. García R.
332) Sibundoy (Putumayo).
333) Cra. 3 No. 7-25, Aguadas (Boy) — Dir: J.F. Gómez E.
334) Lorica (Cor) — Dir: Gustavo Gómez R.
336) Calle 48 No. 18-28, Bogotá, D.C. — Dir: César Castellanos D.
337) Cra. 3 No. 2-36, Arbeláez (Cun) or Av. 37 No. 75-84, Bogotá.
338) Pje. Carrillo Casas, piso 3, Málaga (SS) — Gte: P. Constantino Páez.
339) Ap. A. 126, Tuluá (Valle) — Gte: Wilson Olave L.
340) Calle 14 No. 13A-04, Codazzi (Ces).
342) Cra. 23 No. 25-61, Edif. Don Pedro, Manizales (Caldas).
342x) Cra. 7 No. 8-06, Machetá (Cun)) Gte: José Sánchez.
343) Ap. A. 45, El Banco (Mag) — DG: J. Vanegas T.
344) Edif. Juana Paula, B1.5, Apto. 151, Villa del Rosario (NS).
345) Cra. 2 No. 3-74, Rovirá (SS) — Gte: Luis F. Tovar.
346) Av. 39 No. 15-42, Bogotá, D.E.
347) Ap. A. 167, Sincelejo — Gte: C. Cohén.
348) Ap. A. 127, Cartago (Valle) — Dir: Edgar Arroyave L.
349) Ap. A. 81095, Envigado (Ant) — Gte: Aurelio Toro.
350) C.A.M., Samaná (cds.).
351) Cra. 8 No. 3-35, Zipaquirá (Cun) — Gte: Pedro S. Rozo S.
352) Calle 7 No. 16-26, Pereira.
353) Fundación Tomás Cipriano de Mosquera, Calle Polo, Barbacoas (Nar) — Gte: Amado Waldemar Ortiz.
355) Cra. 5 No. 3-15, Samaniego (Nar) — DG: William Montenegro Chamorro.
356) Ap. A. 402, Cúcuta (NS) — Dir: Marco Antonio Caicedo.
357) Casa Cural, Paipa (Boy) — Dir: Pbro. Abraham Casteblanco V.
358) Ap. A. 6447, Medellín — Dir: Arnulfo Villada A.
359) Socorro (SS).
360) Calle 3 No. 11-60, Quilichao (Cauca).

FM in Bogotá.
50) 88.9 — 89.9 HJCK El Mundo en Bogotá — 1) 90.9 Bogotá — 91.9 Emisora Javierana — 38) 92.9 Sonorama — 1) 93.9 Cerros Stereo — 1) 94.9 Rumba — 19) 96.9 — 97.4 R. Univ. Nal. — 15) 97.9 Musicar — 2) 98.9 — 15) 99.9 — 15) 100.9 Bienvenida — 118) 101.9 Candela Stereo — 15) 102.9 Radioactiva — 38) 103.9 Todelar Stereo — 15) 104.9 Acuario — 105.9 Olimpica — 106.9 Universidad Jorge Tadeo Lozano — 107.9 Minuto de Dios.

ECUADOR

L.T: UTC – 5h — **Pop:** 11m — **Radio sets:** 3m — **Pr.L:** Spanish, Quechua — **E.C:** a/c 60, 110/220V — **ITU:** EQA.

INSTITUTO ECUATORIANO DE TELECOMUNICACIONES (IETEL)

Addr: Casilla 3066, Quito.
Cable: IETEL. **Telex:** 2150, 2202.
L.P: Dir. Nal. de Frecuencias: Ing. Mario Cuesta Bastidas. Jefe Admón. Tec: Osvaldo Benavides.

ASOCIACION ECUATORIANA DE RADIODIFUSION

Addr: Cas. 6014, CCI, Quito.
L.P: Pres. Edgar Yánez V.
STATIONS: Call HC-
Mediumwaves: °) Shortwaves, v) varying fq., *) inactive

222)	FA2	540	25	R. Tropicana, "Canal 540," Guayaquil: 1100-0600 +
144)	GM1	550	50	R. Reloj, Quito: 1100-0600
222)	AK2	560	25	C.R.E. Guayaquil: 0900-0430
165)	CE1	570	10	R. El Sol, Quito: 24h
223)	PC2	580	10	R. Uno, Guayaquil: 24h
146)	SP1	590	10	R. Carrousel, Quito: 24h
224)	XY2	600	50	R. Nal. del Ecuador, Guayaquil
147)	MJ1	°610	25	Em. Gran. Colombia, Quito: 1100-0545
360)	XY3	620	50	R. Nal. del Ecuador, Loja: 1100-0400
280)	HA2	°630	10	Ondas Quevedeñas, Quevedo: 2300-0300
148)	XY1	°640	50	R. Nal. del Ecuador, Quito: 1000-0430
461)	FD4	650	5	R. Visión de Manta, Manta: 0900-0500
225)	LG2	660	25	R. Carrousel, Guayaquil
149)	FF1	°670	12.5	R. Jesús del Gran Poder, Quito: 0945-0100
226)	VP2	°680	25/12	Sistema de Emis. Atalaya, Guayaquil: 0900-0500 (Sun. 1000-0300)
462)	FA4	690	5	R. Sucre, Portoviejo
150)	JB1	°690	50d	La Voz de los Andes, Quito: 0830-0500
227)	RS2	700	50	R. Sucre, Guayaquil: 24h
565)	EP5	°710	10	Escuelas Radiofónicas Populares, Riobamba: 0900-0200
463)	GB4	v720	5	La Voz de Portoviejo, Portoviejo: 1000-0400 (r. 715v)
151)	IC1	720	5	R. Municipal, Quito: 1045-0400
371)	MO3	720	15	R. Matovelle, Loja: 0930-0200
228)	MG2	730	10	R. Guayaquil: 24h
144)	GC1	740	10	R. Melodía, Quito: 1100-0600
464)	SE4	v740	10	R. Libertad, Chone (r767v): 1100-0600
229)	RC2	750	30	R. Caravana, Guayaquil: 1100-0600
152)	RQ1	°760	12.5	R. Quito, Quito: 1000-0500
230)	MF2	770	20	R. El Telégrafo, Guayaquil: 0930-0500
465)	RG4	780	2.5	R. Mía, Manta: 1000-0500
153)	AA1	780	0.5	R. Noticia, Quito: 1000-0300
225)	ML2	800	25	R. Super K, Guayaquil: 1200-0500
146)	FV1	800	1	R. Uno más uno, Quito
232)	VT2	815	5	R. Atalaya, El Milagro: 2300-0300
122)	CR1	820	3	Super R. F2, San Pablo del Lago
154)	FB1	820	1	R. Monumental, Sto. Domingo Cds.
460)	RF4	820	1	Canal Manabita, Portoviejo
233)	RM2	830	25	R. Huancavilca, Guayaquil: 24h
566)	BP5	830	2.5	R. Promoción, Riobamba: 1000-1400, 2200-0100 (Sat/Sun 1000-1800)
155)	PN1	840	50	R. Vigía, La Voz del Tránsito Nal., Quito: 1100-0500
466)	EM4	840	1	R. Costa Azul, Portoviejo: 1100-0500
234)	VS2	850	50	R. San Francisco, Guayaquil: 0945-0300
777)	GB7	855	0.5	R. Nal. Espejo, El Puyo
156)	PC1	860	10	R. Visión, Quito: 1015-0400
666)	GS6	870	0.4	R. Píllaro, Píllaro: 1130-1930, 2100-0330
235)	LY2	870	25	R. Cristal, Guayaquil: 24h
157)	RP1	°880	100	R. Católica Nal. del Ecuador, Quito: 1015-02(Sun 1155-0100)
567)	TL5	890	1	Ondas del Chimborazo, Riobamba: 1100-0500
333)	RS3	890	30	R. Superior, Machala: 1000-0700
158)	VA1	900	10	R. Sucre, Quito
467)	OF4	v900	5	R. Chone, Chone: 1100-0400
535)	RR5	900	1	R. Carrousel, Cuenca: 1100-0300
236)	BO2	*905	2	R. Espectáculo, Guayaquil (n. 910)
568)	GE5	910	1	R. Mundial, Riobamba
334)	RU3	920	12.5	C.R.O., Machala: 1030-0500
159)	CM1	920	10	R. Colón, Quito: 1100-0600
233)	VI2	930	5	Ondas Antillanas, Guayaquil: 24h
667)	BA6	930	5	R. Ambato, Ambato: 0900-0500
160)	BZ1	940	10	R. Dif. de la Casa de la Cultura Ecuatoriana Quito: Mo-Fri 1100-0200, Sat/Sun 1300-2300
569)	UE5	950	3	R. Colta, La Voz de la Asociación, Colta: 0900-0200
267)	DE2	950	10	R. Latina, Guayaquil: 1100-0700
668)	JX6	955	3	La Voz del Santuario, Baños: 1100-0100 (n. 970)
536)	SA5	960	1	Sono Onda Internacional, Cuenca: 1100-0500 (Sun 1200-0400)
161)	NC1	960	1	R. Cosmopolita, Quito: 1050-0330
162)	OT1	°965	25	R. Católica, Sto. Domingo de los Colorados: 1100-0200
135)	MB1	970	1	R. Imperio, Ibarra: 1130-2000, 2200-0400
231)	AW2	970	20	R. Católica Nal. del Ecuador, Guayaquil: 1000-0500
570)	JI5	980	1	R. El Prado, Riobamba: 1200-0400
163)	GH1	v990	25	R. Tarqui, Quito: 1115-0400
537)	AO5	990	5	R. América, Cuenca: 1030-0600
239)	EW2	990	15	Frecuencia Mil, Guayaquil: 24 h
164)	CR1	1000	1	R. Alegría, Sto. Domingo Cds.: 1000-0500
538)	RV5	1000	5	R. Visión, Cuenca: 24h
468)	RC4	1010	1	R. Cenit, Manta: 1200-0500
669)	NR6	1010	10	R. Variedades "Onda Mil", Ambato: 1000-0500
240)	RZ2	*1010	3	R. Victoria, Guayaquil: 1100-0500

335)	GO3	1020	3	R. Estelar, Santa Rosa
606)	CR6	1020	5	R. Surcos, Guaranda: 1030-0100
163)	HR1	v1020	5	R. Hit, Quito: 24h
469)	KL4	v1030	5	R. Unión, Chone: 1100-0500 (r.1059)
241)	RF2	1030	5	R. Horizonte, Guayaquil: 1100-0400
773)	F07	1030	3	R. Francisco de Orellana, Coca
166)	CW1	1040	3	La Voz del Valle, Machachi: 1100-1700, 2130-0300
545)	EV5	1040	10	R. Splendit, "RS", Cuenca: 24h
670)	GB6	1040	1	R. Colosal, Ambato: 0930-0500
124)	IM1	1050	5	R. Municipal La Voz de Imbabura, Ibarra: 0900-1900
222)	RQ2	1050	5	Súper R., Guayaquil: 1030-0400
626)	MG6	v1055	5	R. Ecos del Pueblo, Saquisilí: 1045-0330 (Sun. 0200) (n. 1060)
167)	RS1	1070	1	R. Lubacán, Sto. Domingo Cds.: 0950-0200 (Sun. 2300)
539)	RT5	1070	5	R. La Voz de Tomebamba, Cuenca: 1045-0400
168)	VP1	1070	1	R. Libertad, Quito: 1100-0300v
242)	KD2	1070	10	R. Tigre, Guayaquil: 1100-0500
627)	VH6	1080	10	R. Latacunga, Latacunga: 1000-0300
470)	AB4	1080	1	R. Eco, Manta: 1000-0400 (r. 1085)
180)	VI1	1090	5	R. Deporte: "Estación 1090," Quito
628)	GR6	1100	5	R. Novedades, Latacunga: 1100-0400
243)	FW2	v1100	10	R. Alegría, Guayaquil: 24h
671)	RP6	v1105	3	R. Pelileo, Pelileo: 1000-0300
540)	JC5	1110	3	Ondas Azuayas, Cuenca: 1030-0500
169)	EG1	1110	1	R. Exito, Quito: 1100-0300
111)	EB1	1120	2	Ecos de Montúfar, San Gabriel: 1200-0200
174)	LE1	1120	10	R. Dif. Marañón, Santo Domingo Cds: 1300-0200
362)	NT3	1120	5	R. "Once" Catamayo, La Toma: 1100-0300
244)	LA2	1120	5	R. América, Guayaquil: 1100-0500
780)	AS7	1120	3	R. Variedades del Puyo, El Puyo
672)	PV6	°1130	3	R. Centro, Estación 112, Ambato: 0800-0600
123)	RD1	1130	2	R. Punto, Ibarra: 1100-0500
170)	IR1	1140	5	R. Amazonas, Pomasqui: 1100-0400
541)	AZ5	1140	1	R. Alfa Musical, Cuenca: 1100-0600
245)	FB2	1140	1.5	R. Cóndor, Guayaquil: 1130-0500
483)	MF4	1140	—	R. Momento, Calceta
363)	AV3	°1150	10	R. Luz y Vida, Loja: 1000 (Sun 1140)-0330 (Sat 0400, Sun 0700)
571)	GB5	v1150	5	La Voz de Riobamba: 0900-0300 (r. 1145)
774)	VC7	1150	3	R. El Cisne, Lago Agrio
471)	WD4	1155	1	R. Cenit, Portoviejo: 1200-0500 (n. 1160)
171)	CP1	v1160	2	R. Presidente, Quito: 24h
336)	VR3	1160	2	R. Vía, Machala
629)	UR6	1160	1	R. Runatacuyaj, La Voz de la Asociación, Latacunga
243X)	RV2	1170	10	R. Filadelfia, Guayaquil: 24h
337)	AH3	1170	3	R. Trébol, Zaruma
444)	JM4	°1170	10	R. Antena Libre, Esmeraldas: 1300-0100
572)	JV5	1170	5	Em. Central, Riobamba: 0900-0500
542)	DP5	1180	4	R. Cuenca, La Voz de los 4 Ríos, Cuenca: 1000-0500
172)	LR1	1180	12.5	Nueva Em. Central, Quito: 1045-0400
112)	RT1	1180	1	R. Familiar, Julio Andrade: 1030-1800, 2200-0200
246)	DE2	1190	2	R. Excelsior, Guayaquil: 24h
630)	RF6	1190	5	R. El Sol, Pujilí: 1100-0200
173)	CS1	v1200	7.5	R. Súper K, Sangolquí: 18h
473)	MP4	1200	1	R. La Grande, Bahía de Caráquez
543)	RM5	°1200	5	R. El Mercurio, Cuenca: 1300-0500
281)	RE2	1200	5	La Voz del Trópico, Quevedo: 1030-0400
364)	VC3	°1210	10	CDS, R. Centinela del Sur, Loja: 1100-0400
247)	BJ2	1210	20	R. El Mundo, Guayaquil: 1200-0300 (Sat 0100, Sun 0400)
174)	AP1	1220	10	Radiodif. Marañón, Quito: 1000-0200
673)	JM6	v1225	3	R. Sira, Ambato: 1000-0700
544)	MV5	°1230	3	R. Popular Independiente, Cuenca: 1045-0500
445)	FG4	1230	5	R. Sucre, Esmeraldas: 1000-0500 (Fri/Sat 24h)
125)	RI1	1230	3	CRI, Centro Radiofónico de Imbabura, Ibarra: 1100-0300
264)	GT2	1230	5	R. Galáctica, Guayaquil
631)	RL6	°1235	1	La Voz de Saquisilí y Libertador, Saquisilí: 1045-0300 (n. 1230)
580)	--5	1240		R. Sonorama, Riobamba
338)	RF3	1240	0.5	R. Fénix, Zaruma: 1100-0400
175)	PA1	1240	1	R. Metropolitana, Quito: 1200-0300
248)	HB2	1250	2	R. Linda, Guayaquil: 1100-0300
113)	EM1	1250	5	Ondas Carchenses, Tulcán: 1000-0200
176)	MY1	1250	3	La Voz del Triunfo, Sto. Domingo Cds.: 1000-0500
554)	PB5	1260	2	R. Stereo AM, Cuenca
347)	RB3	1260	1	R. Benemérita, Machala
674)	RO6	1260	3	R. Calidad, Ambato: 0930-0600
177)	MO1	1260	10	R. HCM-1, La Voz del Santuario del Quinche, Quito: 1030-0330
474)	LD4	1270	5	R. Junín, Junín: 1100-0500
242)	UM2	1270	15	R. Universal, Guayaquil: 0900-0500
178)	AB1	1280	10	R. Democracia, Quito: 1000-0400
339)	RP3	1280	2	R. Continental, Arenillas
577)	WW5	1280	3	R. Canal Tropical, Riobamba: 0900-0500
475)	IN4	1280	1	La Voz del Sur de Manabí, Jipijapa: 1000-0500
250)	OF2	1290	1	Canal Milagreño, El Milagro
546)	JA5	°1290	2	La Voz del Rio Tarqui, Cuenca: 1100-0300
126)	NS1	1290	1	R. Popular, Atuntaqui: 1100-0300
633)	VM6	1295	0.5	R. Once de Noviembre, Latacunga: 1200-0400
179)	RU1	1300	2.5	R. Festival, Santo Domingo Cds: 1030-0500
251)	DC2	1300	5	R. Cenit, Guayaquil: 1200-0400
144)	GB1	°1310	20	R. Nal. Espejo, Quito: 24h (r. 1309.7)
340)	CP3	1310	1	La Voz de El Oro, Pasaje: (r. 1315)
675)	JD6	1320	10	R. Continental, Ambato: 0930-0400
560)	CI5	1320	3	T.V.O., Biblián
747)	OB7	°1320	0.5	R. Nal. Limón, Limón Indanza
282)	FR2	1320	5	R. Guayaquil, Babahoyo: 1100-0400
114)	OV1	1330	3	Ondas del Volante, El Angel: 1200-2400
547)	LW5	1330	2	R. La Voz de la Cordillera, Cuenca: 1130-0300
341)	RV3	1330	5	R. Nal. El Oro, Machala: 1000-0600
192)	JA1	v1330	5	R. Sideral, Quito: 24h
287)	SF2	1340	1	R. Bodegas, Babahoyo
476)	DR4	*1340	1	R. Cenit, Bahía de Caráquez: 1000-0500
581)	VP5	1340	2.5	R. San Francisco, Penipe: 0900-0300
676)	PB6	°1340	10/5	R. Paz y Bien, Ambato: 0900-0200
115)	PZ1	1350	5	R. Rumichaca, Tulcán: 1045-0415
193)	PU1	1350	1	La Voz de Santo Domingo, Sto. Domingo de los Colorados
237)	VP2	1350	3	R. Reloj, Guayaquil: 24h
548)	SF5	1350	0.3	R. San Fernando, San Fernando: 1100-0300
345)	HG3	1360	5	R. El Sol, Machala: 1045-0500
181)	MT1	v1360	1	R. dif. Oyambaro, Tumbaco (r. 1360.4)
477)	EG4	1360	3	La Voz del Carmen, El Carmen
767)	LE7	1360	1	R. Oriental, Tena: 1000-0400
575)	RJ5	1360	3	R. América, Riobamba: 1100-0300
365)	ER3	°1370	5	Sistema de Emisoras Progreso, Loja: 1000-0315v
127)	JS1	1370	2	Ecos Andinos, Pimampiro (r.1365)
778)	RP7	°1370	2	R. Pastaza, El Puyo: 1100-1800, 2300-0400
252)	VO2	1370	5	La Voz del Milagro, El Milagro
677)	JR6	v1380	5	R. Mera, Ambato: 24h
182)	CV1	v1380	5	R. Cristal, "RCQ", Quito: 24h
342)	OA3	1380	5	R. Impacto, Piñas: 1100-0500
752)	WV7	1380	3	R. Morona, Macas
446)	VE4	1390	10/5	La Voz de Esmeraldas: 1000-0500
128)	IE1	1390	1.5	R. Uno, Urcuquí: 0800-0200
549)	EA5	1390	5	R. Tropicana, Cuenca: 1130-0500
576)	DN5	1390	3	R. Atenas, Riobamba: 0900-0500
223)	FL2	1400	10	R. Z-1, Guayaquil: 24h

787)	VZ7	1400	1	La Voz de Zamora, Zamora: 1100-0100
117)	VL7	1400	7	R. Multicolor, Tulcán
147)	EC1	1410	1	R. El Tiempo "la em. del amor," Quito: W. 1130-1900; at other times r. HCAP1 1220
253)	CQ2	1410	1	R. Presidente "La Voz del Pueblo," El Milagro
447)	FR4	1410	1	La Voz de Quinindé, Quinindé
550)	KD5	1410	1	R. Centro Gualaceo, Gualaceo
768)	MS7	1410	1	La Voz del Río Coca, El Coca
635)	MA6	v1420	5	R. Nuevos Exitos, Salcedo: 1000-0400
129)	RN1	°1420	3	R. Bahá'í, Otavalo: 0900-2400
348)	NR3	1420	1	La Voz de Huaquillas, Huaquillas
233)	MB2	1430	10	R. Caribe, Guayaquil: 24h
607)	JC6	1430	5	R. Guaranda, Guaranda: 1100-0300
366)	CV3	1430	5	R. Musical, "Punto 14-30", Loja: 1130-0330
194)	GF1	1430	1	R. Continental, Quito
130)	DF1	1440	5	R. Panorama, Ibarra: 1030-0400
555)	OV5	1440	3	Ondas del Volante, Azogues: 1230-
448)	DY4	1440	5	R. Iris, Esmeraldas: 1000-0400
779)	MD7	1440	5	R. Puyo, El Puyo: 17h
577)	SC5	1445	3	R. Calidad, Riobamba: 0800-0400 (n. 1450)
195)	SC1	1450	1	R. Sensación, Cayambe
283)	HW2	1450	3	R. Rey, Quevedo: 1000-0500
254)	SE2	1450	1	R. Sta. Elena, Sta. Elena: 2200-0200
255)	DR2	1450	1	R. Minutera, Pascuales (r.1460)
772)	RI7	°1450	1	R. Interoceánica, Sta. Rosa de Quijos
367)	CL3	1460	5	R. Cariamanga, Cariamanga: 1100-0300
636)	IC6	1460	5	R. Nuevos Horizontes, Latacunga: 1000-0200
183)	ED1	1470	1.5	Ecos de Cayambe, Cayambe: 0845-0400
256)	LD2	1470	1	R. Ecos de Naranjito, Naranjito: 1200-0300
484)	RD4	1470	1	R. Richi, Tosagua (Calceta)
343)	BS3	v1480	3	Ondas del Jubones, Pasaje: 1000-0500
478)	JB4	1480	3	R. La Voz de Jipijapa: 1100-0400
131)	MC1	1480	1	R. Municipal, Cotacachi: 1000-0100
578)	WP5	1480	5	R. Atlántida, Alausí: 1100-0300
637)	CY6	1480	5	R. Popular de La Maná, La Maná
257)	VY2	1490	1	La "R" Dinámica, Guayaquil
449)	AE4	1490	5	R. Unión, Esmeraldas
556)	SM5	1490	5	R. Santa María, Azogues: 1100-0100
678)	AI6	1490	0.5	R. Moderna, Píllaro: 1230-0200
574)	WN5	1500	2	R. Puruhá, Riobamba: 0945-0500
284)	HG2	1500	5	La Voz del Río Vinces, Vinces: 1100-0500
132)	RO1	1500	3	R. Otavalo, Otavalo: 1200-0300
479)	AD4	1500	5	R. Satélite, El Carmen: 1000-0500
344)	UE3	1500	10	R. Sucre, Machala: 24h
788)	OY7	1505	1.5	Ondas del Río Yacuambí, Yacuambí: 1000-1400, 2000-2400
258)	—	°1510	2	Inst. Oceanográfico de la Armada, Guayaquil: time signals 24h
184)	DC1	°1510	1	R. Iris, Quito: 1000-0500
372)	UC3	1510	3	R. Unión Calvense, Cariamanga
557)	RC5	1510	0.5	La Voz de la Juventud, Cañar
608)	RY6	1510	1	R. Runacunapac Yachana, Simiátug: 7h
769)	VL7	°1510	3	R. Ecos del Oriente, Lago Agrio: 1100-0400
265)	RN2	1520	1	La Voz de Naranjal, Naranjal
285)	JJ2	1520	3	R. Punto Verde, Quevedo
579)	RI5	1520	2.5	La Voz de Guamote, Guamote
133)	TI1	1520	1	R. Ibarra, Ibarra: 1000-0400
480)	EB4	v1525	0.35	R. Manta, Manta: 1100-0500 (n.1520)
259)	MP2	v1530	3	La Voz de la Península, La Libertad: 1100-0400 (r. 1529.8)
679)	MZ6	1530	1	La Voz del Dorado, Pelileo: 1130-0230
558)	CC5	1530	5	Ondas Cañaris, Azogues: 1030-0400
582)	VP5	1530	1	La Voz de Pallatanga, Pallatanga
185)	JY1	1530	5	R. Uno, La Concordia: 1000-2400
638)	MH6	1535	0.5	O.R.C., Latacunga: 1000-0400 (n.1540)
116)	PV1	1540	1	R. Mira, Mira
286)	FM2	1540	3	R. Cristal, Ventanas
748)	VB7	°1540	2.5	R. dif. Católica Cultural "Voz del Upano", Macas: 1045-0300 (Prgr A)
186)	DP1	v1545	1	R. Caracol, Quito: 1000-0400 (r. 1545.7)
551)	AB5	1550	1	La Voz de Chaguarurco, Santa Isabel
260)	AD2	1550	3	La Voz de El Triunfo, El Triunfo: 1100-0400
680)	EI6	1550	2	R. Montalvo, Ambato: 1130-0300
770)	RA7	1550	1	R. Amazonas, Archidona
346)	TR3	1560	2	La Voz del Guabo, El Guabo: 1100-1300, 2300-0400
261)	CS2	1560	2	R. Sideral, Daule: 1300-0500
559)	VI5	1560	5	R. Ingapirca, Cañar: 0900-0330
187)	HG1	1560	1	La Voz de El Dorado, Puembo
134)	ZD1	1560	1.5	Ecos Culturales de Urcuquí, Urcuquí
583)	AI5	1570	0.5	Eco de los Andes, Chunchi: 1000-0200
188)	PG1	*1570	10	R. Ñucanchic, P.V. Maldonado
189)	LF1	1580	1	R. Orellana, Machachi: 1030-0230
266)	CP2	1580	0.5	Canal del Pueblo, Sanborondón
368)	AB3	1580	0.25	Ondas de Paltas, Catacocha
552)	AE5	1580	3	Ecos del Portete, Girón: 1200-0330
450)	VA4	1580	5	R. Bolívar, La Estación de la Alegría, Esmeraldas: 24h
609)	--6	1580	1.5	La Voz de Libertador, Chimbo (n. 1600)
190)	RZ1	1590	1	R. Mensaje, Tabacundo: 1000-1400, 2130-0230
262)	AS2	1590	0.25	R. Record, La Libertad
639)	CS6	1590	1	La Voz de La Maná, La Maná
681)	QT6	1590	1	R. Panamericana, Quero: 1000-0200 (Sun. 2400)
191)	UP1	*1600	5	R. Unión, Quito: 1100-2400
263)	JP2	v1600	1	R. Consular, Playas: 1200-0100 (r. 1605)
553)	PB5	1600	3	R. Intiñán, Girón: 1100-0300

Shortwaves: Call: HC —

	Call	kHz	kW	Name and h. of tr.
150)	JB	3220	10	La Voz de los Andes, Quito: 0830-1430, 2100-0200
444)	JM4	3240.2	1	R. Antena Libre, Esmeraldas: 1300-1500, 2100-0100
769)	JV7	3270	1	R. Ecos del Oriente, Lago Agrio: 2000-0400
771)	VN7	3279.8	2.5	La Voz del Napo, Tena: 1000-1215, 2200-0230 (n. 3280)
546)	JA5	3285.3	0.35	La Voz del Río Tarqui, Cuenca
672)	PV6	3289.9	0.5	R. Centro, "Estación 112", Ambato: 0800-0600v
778)	RP7	*3315.1	2	R. Pastaza, El Puyo
749)	RS7	*3322	1	R. Dif. Sangay, Macas
280)	HA2	3325	1.5	Ondas Quevedeñas, Quevedo: irr.
775)	JQ7	3332	3	R. Cumandá, Coca: 1100-0300v (n. 3350)
750)	SK7	*3360	5	R. Federación Shuar, Sucúa
747)	OB7	3370	0.25	R. Nacional Limón, Limón Indanza: 1100-0400
162)	OT1	3394.8	10	R. Católica, Sto. Domingo Cds: 1100-1300, 2200-0200 (n. 3395)
370)	AA3	4270.6	1	R. Gonzanamá: 1200-1400, 1700-1900, 0000-0300
226)	VP2	*4760	5	Sistema de Emis. Atalaya, Guayaquil
482)	AS4	4795.2	5	La Voz de los Caras, Bahía de Caráquez: 2300-0300
544)	MV5	4799.8	5	R. Popular Independiente, Cuenca: 1045-0500 (n. 4800)
676)	RT6	4819.7	3	R. Paz y Bien, Ambato: 0900-1430, 1700-2200 (n. 4820)
772)	RI7	4840	1	R. Interoceánica, Santa Rosa de Quijos: 1100-1500, 2200-0200
363)	AV3	4850.6	2	R.Luz y Vida, Loja: 1000-0330 (Sat 0400, Sun 0700) (n. 4850)
144)	WE1	4879.6	1	R. Nal. Espejo, Quito: irr.
364)	VC3	v4890	5	CDS, R. Centinela del Sur, Loja: 1100-0400
631)	VS6	4900	1	La Voz de Saquisilí y Libertador, Saquisilí: local mornings
147)	MJ1	4911	5	Em. Gran Colombia, Quito (n. 4910)

152)	QR1	4920	5	R. Quito, Quito: 1045-1430, 1645-1900, 2245-0500
129)	RN1	4950	1	R. Bahá'í, Otavalo: 0900-2400
750)	SE7	4960.9	5	R. Federación Shuar, Sucúa: 1100-1300, 2100-0100
565)	ER5	5010.3	1	Escuelas Radiofónicas Populares, Riobamba: 0900-1400, 2100-0200 (n. 5010)
748)	VB7	5020	10	R. dif. Católica Cultural "Voz del Upano," Macas: 1045-0300 (Prgr B)
157)	RP1	5030.1	10	R. Católica Nal. del Ecuador, Quito: 1020-1400, 2245-0200
748)	VB7	5040	10	R. dif. Católica Cultural "Voz del Upano," Macas: 1045-0300v (Prgr A)
149)	FF1	5049.8	5	R. Jesús del Gran Poder, Quito: 0945-1500, 2300-0100 (n. 5050)
365)	EH3	v5062	2	Sistema de Emisoras Progreso, Loja: 1000-0315v (v. ± 3kHz)
748)	VB7	5964.8	10	R. dif. Católica Cultural "Voz del Upano," Macas: 1045-0300 (Prgr B)
750)	SK7	5980	5	R. Federación Shuar, Sucúa: 1100-1300, 2100-0100
748)	VB7	5999.3	10	R. dif. Católica Cultural "Voz del Upano," Macas: 1045-0300 (Prgr B)
150)	JB	6050	50	La Voz de los Andes, Quito: 0830-0500
150)	JB	6080	50	La Voz de los Andes, Quito: 0830-1430, 2100-0500
148)	—	15350	100	R. Nal. del Ecuador, Quito (via HCJB): Mon-Fri: 1730-1800

Addresses and other information

Name of province should be added to names of minor towns. In ZIP code for P.O. Boxes under implementation the first two digits refer to province, the second two digits to local P.O.

REGION 1

Carchi (04)

111) Colón 619, San Gabriel — Dir: Yolando Pozo P.
112) Convento Parroquial, Julio Andrade — Gte: Mons. José A. Fuentes.
113) Ap. 30, Tulcán — Dir: Fausto Almeida C.
114) Calle Salinas, El Angel — Dir: Ignacio Mena.
115) Casilla 42, Tulcán — Dir: Efraín Cabezas M.
116) G. Suárez 824, Mira — Dir: Leopoldo V. Padilla.
117) Cas. 60, Tulcán — Dir: Ruth Muñoz L.

Imbabura (10)

122) San Pablo del Lago — Dir: F. Beltrán Proaño
123) Cas. 10-01-0099, Ibarra — Dir: Augusto Báez R.
124) Casilla 179, Ibarra — Dir: Ricardo R. Ruiz.
125) Casilla 181, Ibarra — Dir: Silvio Morán Madera.
126) Casilla 3, Atuntaqui — Dir: Carlos A. Calderón.
127) Olmedo 181, Pimampiro — Dir: José Valle N.
128) Antonio Ante 100, Urcuquí — Dir: G. Ruiz S.
129) Ap. 14, Otavalo — Dir: Nooshin Bushwell. **F.Pl:** SW 8 kW.
130) Casilla 694, Ibarra — Dir: Cristóbal Bonifaz.
131) Calle García Moreno, Cotacachi — Dir: Ulpiano Saltos T.
132) Casilla 1474, Otavalo — Dir: Marco A. Chicaiza T.
133) Av. Pérez Guerrero 812, Ibarra — Gte: Carlos Arciniega.
134) Antonio Ante s/n, Urcuquí — Dir: Fausto Dueñas C.
135) Casilla 413, Ibarra — Dir: Luis Viteri.

Pichincha (17)

144) Cas. 352, Quito — Dir: Marco Caicedo.
145) Cas. A-57, Quito — Dir: Marilou L. Parra.
146) Amazonas 1538, Quito — Dir: Byron Meza.
147) Cas. 2246, Quito. **Te:** 211670. — Dir: Grace Cevallos de Reyes.
148) Cas. 60, or: Mariano Echeverría y Brasil, Quito — Dir: Gustavo Cevallos.
149) Cas. 133, Quito — Dir: R.P. Jorge Enríquez Silva.
150) cf. separate listing.
151) Antigua Alcaldía, Quito — Dir: Sonny de la Torre.
152) Ap. 57, Quito — Dir: Gonzalo Ruiz. **D.Prgr:** Ecuadoradio network **N.** 1145 and 0310 (Sun 1145 and 2005).
153) Cas. 2938, Quito — Dir: Arcesio Arcentales R.
154) Cas. 038, Santo Domingo de los Colorados.
155) Ramírez Dávalos 612 y 10 de Agosto, Quito — Dir: My. Edgar Noboa.
156) Casilla 251-A, or: Checoeslovaquia 323, Quito.
157) Cas. 540-A, Quito — DG: R.P. Antonio Arregui Y. Dir. Prgr: René Torres.
158) Palacios 303 y La Gasca, Quito — Dir: L. Pumagualle.
159) Cas. 8167, Quito — Own: A. Granda C. Dir: Atahualpa Ruiz R.
160) Ap. 67, Quito — Dir: Lourdes Mendoza.
161) 24 de Mayo 1224, Quito — Dir: Numa P. Castro.
163) Cas. 2558, Quito — Dir: Hernan Herdoiza L.
164) Ap. 045, Santo Domingo de los Colorados — Dir: C. Reinoso M. **F.Pl:** SW 3440kHz.
165) Cas. 8270, or: 9 de Octubre 215 y 18 de Septiembre, Quito — Dir: Jaime Bowen.
166) García Moreno 486, Machachi — Gte: S.B. Cueva S.
167) Guayaquil 124, Santo Domingo Cds. — Dir: M. Zavala V.
168) Tarqui 785 y Estrada, Quito — Dir: Raúl Armendáriz.
169) Cas. 17-03-0736, or: Eustorgio Salgado 442 y Armero, Quito.
170) Cas. 17-01-638, Quito — Dir: Dr. Raúl Izurieta.
171) Cas. 2493, or: Marquesa Dosantos y Montúfar, Quito.
172) Ap. 6110, Quito — Dir: Carlos E. Machado.
173) Av. Gral. Enríquez 714, 3er. P. de Almacenes Salvador, (or: Cas. 47), Sangolquí — Dir: Abdón Calderón G.
174) Cas. 17-01-2263, Quito — Dir: Ligia Wilches de Garcés.
175) 12 de Octubre 227, Quito — Dir: Eduardo Loza.
176) Cas. 17-24-0043, Santo Domingo Cds. — Dir: M. Yánez O.
177) Cas. 17-01-3386, Quito — Dir: P. Jesús Palomino I.
178) Edif. Doral Mariscal, Of. 86, Páez y Mercadillo, Quito — DG: Gonzalo Rosero
179) San Miguel y 29 de Mayo, Santo Domingo Cds. — Dir: Rómmel Velástegui I.
180) Ap. 6014, CCI, Quito — Dir: Jorge Aguilar V.
181) Pizarro y Bolívar, Tumbaco — Dir: Saulo Barros.
182) Av. de la Prensa Mz. 1 C-2 y Av. del Maestro, Quito — Dir: Ricardo Romero R.
183) Terán 527, Cayambe — Dir: Juan M. Córdoba Solá.
184) Cas. 1018, Quito — Dir: Daniel Alvarez T.
185) La Concordia (or: via st. 175) — Dir: Jesús Yánez B.
186) Venezuela 701 y Espejo, Quito — Dir: J.A. Buenaño.
187) Av. 24 de Mayo s/n, Puembo — Dir: G. Hernández N.
188) Pedro Vicente Maldonado (or: Concejo Provincial de Pichincha, Manuel Larrea y Antonio Ante, Cas. 298, Quito).
189) Luis Cordero 226, Machachi — Dir: José J. Quinga.
190) Pza. Gutiérrez 255, Tabacundo (or: Ap. 7179, Suc. 10, Quito) — Dir: Isaías Barriga N.
191) Unión Nacional de Periodistas, Iñaquito y NN. UU., or: Cas. 5184, Quito.
192) Madrid 1234, or: Cas. 17-12-861, Quito — Dir: Enrique Grímbal.
193) Edif. Vidal Gómez, P.3, Sto. Domingo Cds — Dir: Joseph Sánchez.
194) Olmedo 122 y Ríos, Quito.
195) Bolívar y Rocafuerte esq., Cayambe — Own: Ramiro Almeida.

FM in Quito.
88.1 Líder — 150) 89.3 — 90.5 Concierto — 156) 91.7 — 92.1 Ecuatoriana — 92.9 Cordillera — 157) 94.1 — 95.3 Teleonda Musical — 180) 96.5 Bolívar — 97.7 Centro — 159) 98.9 — 99.7 Añoranza — 162) 100.5 — 101.3 Onda Azul — 149) 102.5 Francisco — 103.7 Sonorama — 104.9 Ecuashyri — 106.1 Pichincha — 107.3 JC R. "La Bruja".

REGION 2

Guayas (09)

222) Ap. 4144, Guayaquil — Dir: Rafael Guerrero G. (Tropical).
223) Cas. 2119, Guayaquil (or: Amazonas 743 y Veintimilla, P.8, Quito) — Dir: Raúl Marcos Salcedo C.; Quito: Dr. D. Troya B.
224) V. M. Rendón y Garaycoa, Edif. Huancavelica, Guayaquil.
225) Av. de Las Américas junto a Canal 10, or: Cas. 9974, Guayaquil — Dir: Kléber Chica.
226) Cas. 204, Guayaquil — Gte: A. Mendoza Paladines.
227) Ap. 11714, Guayaquil — Dir: Vicente Arroba D.
228) Cas. 2440 u, or: Escobedo 1504 y Aguirre, piso 9, Guayaquil.
229) P. Icaza 437, or: Cas. 716, Guayaquil — Dir: Ignacio Dueñas.
230) Colón 548 y Boyacá, P. 7, or: Cas. 7377, Guayaquil — DG: Roberto Hanze Salem.
231) García Avilés 606 y Aguirre, Guayaquil — Dir: Mons. Elías Dávila.
232) Juan Montalvo 1042, El Milagro — Gte: Ecuador Martínez C.
233) Edif. Gran Pasaje, Of. 906/908 (or: Cas. 856), Guayaquil — Dir: A. Feraud M.

234) Cas. 5762, Guayaquil — Dir: Padre Jorge González P.
235) Cas. 5662 or: Laque 1407 y Antepara, Guayaquil — Dir: Armando Romero Rodas.
236) Baquerizo Moreno 803, Guayaquil — Dir: C. Córdova V.
237) Cas. 2088, Urdesa Central, Guayaquil.
238) Cas. 1166, Guayaquil — Gte: David Huerta Castelló.
239) 1° de Mayo 1006, or: Cas. 10863, Guayaquil — Dir: Guillermo Rendón.
240) Esmeraldas 902, Guayaquil — Dir: J. Llerena M.
241) Casilla 4719, or: García Moreno 913, Guayaquil. Mgr: Ilona Chávez.
242) Cas. 5177, Guayaquil — Dir: Gabriel Arroba D.
243) Cas. 856, Guayaquil — Dir: Fausto Feraud Manssur.
243X) Cas. 8729, Guayaquil — Dir: Mario José Zurita F.
244) Ap. 4839, Guayaquil — Dir: Luis Albán.
245) Cas. 1054, Guayaquil — Dir: V. Villacís G.
246) Casilla 5177, or: Higueras 101 y Costanera, Guayaquil.
247) Cas. 3311, Guayaquil — Dir: Gabriel Vergara J.
248) L. de Garaycoa 2615, Guayaquil — Dir: H. Alvarado P.
250) Laurel y Guayacanes, El Milagro — Dir: Francisco Ortiz G.
251) Ap. 4065, Guayaquil — Dir: W. Delgado C.
252) Av. 17 de Septiembre, El Milagro — Dir: Carlos H. Vega V.
253) Av. Cuba y Ecuador, El Milagro — Dir: Stelvio H. Centenario Q.
254) Guayaquil y 9 de Octubre, Santa Elena — Dir: Walter Gellibert L.
255) Km. 1 Vía Guayaquil, Pascuales — Dir: F. Dávila de la R.
256) Av. 5 de Octubre 150, Naranjito — Dir: Leoncio Delgado H.
257) Alianza 1204 y Laureles, Guayaquil — Dir: Inocente Yance P.
258) Cf. listing under Time Signal St's. — Cas. 5940, Guayaquil.
259) 4a. Av. 619 y Robles, La Libertad — Own: M. Pinoargote P.
260) Av. 8 de Abril y Roldós Aguilera, El Triunfo — Gte: A. Torres.
261) Sitio El Belén, Daule — Dir: Francisco Dávila de la R.
262) 12 de Octubre 1032, La Libertad — Dir: A.N. Suárez B.
263) Gral. Villamil, Playas (or: Cas. 10983, Guayaquil) — Dir: Jorge Pazos S.
264) Edif. El Forum, Of. 508/9, or: Cas. 2119, Guayaquil — Gte: Galo Verduga V.
265) Pastaza y Guayaquil, Naranjal — Dir: José D. Toapanta Ch.
266) El Oro y Los Ríos, Sanborondón — Dir: W. Gellibert L.
267) Cas. 11548, or: 9 de Octubre 2009 y Los Rios, Guayaquil — Dir: Wilson Chávez.

Los Ríos (12)

280) 12va. Calle No. 207, Quevedo — Dir: Humberto Alvarado P.
281) Cas. 104, Quevedo — Dir: Rosendo V. Escobar C.
282) Cas. 77, Babahoyo — Dir: Fernando E. Ronquillo F.
283) Cas. 84, Quevedo — Dir: Benito Wong Y.
284) Olmedo 109, Vinces — Dir: Selvi Elizalde C.
285) 7 de Octubre 917, Quevedo — Dir: Julio César Jalón F.
286) 28 de Mayo y 6 de Octubre, Ventanas — Dir: Arnaldo Rivas.
287) Cas. 31, Babahoyo — Gte: Gustavo von Lippke.

REGION 3

El Oro (07)

333) Cas. 221, Machala — Dir: Rodrigo Pineda I.
334) Cas. 473, Machala — Dir: José Ugarte Aguilar
335) Cdla. Marlene Nieto, Calle Eloy Alfaro, Santa Rosa — Dir: A. Acosta.
336) 9 de Octubre y Paéz or: Cas. 07-01-0086, Machala — Dir: Héctor Bravo.
337) Sucre 153, Zaruma — Dir: José Valdivieso L.
338) Cas. 23, Zaruma — Dir: Lautaro Barrezueta J.
339) Arenillas — Dir: Franco Ordóñez.
340) Ochoa León 505, Pasaje — Dir: César Prado A.
341) Cas. 895, Machala — Dir: Vicente Serrano Zambrano.
342) Sucre 1105, Piñas — Dir: G. Olmedo Torres. F.Pl: Move to 1385 or 1390kHz.
343) Rocafuerte 619, Pasaje — Dir: Alfredo H. Sarmiento.
344) 9 de Mayo y Pasaje, P. 2, Machala — Dir: Olmedo Minuchi.
345) Cas. 474, Machala — Dir: Silvio Pinos G.
346) Av. del Ejército, El Guabo — Dir: Rosa R. Vda. de Ramón
347) Chávez Franco y Guayas, Edif. Sind. de Choferes, Machala — Dir: Joffre Unda Ch.
348) Cdla. Jaime Roldós, Calle Martha Bucaram, Huaquillas — Gte: Carlos Naranjo M.

Loja (11)

360) Cas. 521, Loja — Dir: A. Maldonado.
362) 24 de Mayo y Espejo, La Toma — Dir: Norberto A. Torres C.
363) Cas. 222, Loja — Dir: Eloy Torres P.
364) Cas. 196, Loja — Dir: José Coronel Illescas.
365) Cas. Letra V. Loja — Dir: Efraín Herrera G.
366) Ap. 431, Loja — Dir: Víctor M. Sánchez.
367) 18 de Noviembre s/n, Cariamanga — Dir: Jorge Montero.
368) Cdla. Progreso, Catacocha — Dir: Artemio Ríos G.
370) Cas. 379, Loja — DG: Miguel A. Mendieta.
371) El Cisne, Casa Parroquial, Loja.
372) R. Carrasco y Paz, Cariamanga.

REGION 4

Esmeraldas (08)

444) Cas. 65, Esmeraldas — Dir: Luis E. Velasco León.
445) Mejía 325 y Olmedo, Esmeraldas — Dir: Rónald Murillo C.
446) Cas. 108, Esmeraldas — Dir: Héctor Endara E.
447) Plata Torres, Quinindé — Dir: Fulton Villagómez.
448) Cas. 116, Esmeraldas — Dir: Homero López.
449) Cas. 155, Esmeraldas — Dir: Raúl Delgado G.
450) Cas. 379, Esmeraldas — Own: Jaime Recalde B.

Manabí (13)

460) Cas. 69, Portoviejo — Dir: Lauro Fernández.
461) Cas. 4817, Manta — Dir: César Dávalos H.
462) Olmedo 426 y 10 de Agosto, Portoviejo — Dir: Jorge Gutiérrez.
463) Cas. 89, Portoviejo — Dir: Guido Bonilla.
464) Cas. 6, Chone — Dir: Stalin Espinel A.
465) Ap. 4810, Manta — Dir: Enrique Quevedo A.
466) Colón 180, Portoviejo — Dir: Sergio Enrique Morlas A.
467) Cas. 8, Chone — Dir: Argemiro Andrade.
468) P. Solano 701, Manta — Dir: W. López Mero.
469) Cas. 05, Chone — Dir: Alberto E. Espinel A.
470) Av. 24 No. 1266, Manta — Dir: Angel Delgado.
471) Cas. 145, Portoviejo — Dir: Dúmar Iglesias M.
473) Bolívar 1209, Bahía de Caráquez — Dir: Trajano Velástegui.
474) 10 de Agosto 180, Junín — Dir: Julio Alvarez M.
475) Cas. 705, Jipijapa — Own: José R. Falconí Y. **F.Pl:** 5kW.
476) Bolívar 920, Bahía de Caráquez — Dir: W. Delgado C.
477) El Carmen — Dir: Enrique García E.
478) Noboa y Colón, Jipijapa — Dir: Ulbio Delgado H.
479) 4 de Dicbre. y Alfaro, El Carmen — Dir: Enrique García E.
480) Cas. 4869, Manta — Dir: Efraín Loor.
481) 10 de Agosto 156, Calceta — Dir: Roque Navarrete V.
482) Cas. 628, Bahía de Caráquez — Own: A. Nevárez Pinto. Dir: Marcelo Nevárez F.
483) Salinas 119, Calceta.
484) Av. Principal, Tosagua — Dir: Luis Grijalva D.

REGION 5

Azuay (01)

535) Cordero 944, Cuenca — Dir: Jorge Antonio Chamoun.
536) Larga 1043, Cuenca — Dir: Vicente Serrano A.
537) Luis Cordero 944, Cuenca — Gte: Fabián Molineros N.
538) Cas 198, Cuenca — Dir: Paúl Rodas Moscoso.
539) Cas. 493, Cuenca — Dir: Jorge Piedra C.
540) Cas. 01-01-4980, Cuenca — DG: Gustavo Cardoso.
541) Cas. 1583, Cuenca — Dir: Oswaldo Flores de la T.
542) Cas. 1598, Cuenca — Dir: Daniel Pinos Guaricela.
543) Cas. 60, Cuenca — Dir: Germán Morales Campoverde.
544) Av. Loja 2408, Cuenca — Dir: Manena E. de Villavicencio.
545) Cas. 1352, Cuenca — Dir: Edmundo Valencia V.
546) La Mar y Montalvo, or: Cas. 877, Cuenca — Dir: Manuel Pulla C.
547) Gran Colombia 817 y Cordero, or: Cas. 877, Cuenca — Dir: Manuel Pulla Cornejo.
548) Av. José María Quito, San Fernando — Dir: Humberto Romero V.
549) Pumapungo E-25, or: Cas. 830, Cuenca — Dir: Lautaro Pinos A.
550) 9 de Octubre 3102, Gualaceo — Gte: José Toral L.
551) 3 de Novbre. y Sucre, Santa Isabel — Dir: Rodrigo Palacios.
552) Antonio Flor 657, Girón — Dir: Telmo Pinos Guaricela.
554) J. Dávila y C. Marchán, Cuenca.

Cañar (03)

555) Cas. 720, Azogues — Dir: Lic. Wilson G. León.
556) Ap. 730, Azogues — Dir: Fray Raúl Pinto E.
557) Av. San Antonio y Panam, Cañar — Dir: Manuel Córdova E.
558) Cas. 731, Azogues — Dir: Humberto Vicuña N.
559) Av. Ingapirca s/n, Cañar (or: Cas. 447, Cuenca) — Dir: Martín Samaniego C.
C.E: Juan K. Tommerbakk. **F.Pl:** SW 5025 kHz 3.5kW.
560) Tomás Sacoto 161 y Sucre, Biblián (or: Cas. 729, Azogues) — Gte: Telmo V. Ortega.

Chimborazo (06)

565) Cas. 06-01-4755, Riobamba — Dir: Juan Pérez Sarmiento. Prgr. in Quichua 0900-1100, 2330-0200.

566) Cas. 242, Riobamba — Dir: Guillermo Burbano.
567) Pichincha 1363, Riobamba — Dir: Luis A. Tasambay.
568) Cas. 1244, Riobamba — Dir: Galo Encalada R.
569) Cas. 87-A, Riobamba — Prgrs in Quichua only.
570) Cas. 20, Riobamba — Dir: Marcelo Vizcaíno M.
571) Cas. 348, Riobamba — Dir: Patricio Pumagualle.
572) 10 de Agosto 1742, Riobamba — Dir: J. Segundo León V.
573) Cas. 4747, Riobamba — Dir: Luis Aliaga.
574) Cas. 1226, Riobamba — Dir: N. Fausto M. **F.Pl:** 10kW.
575) Cas. 82, Riobamba — Dir: Eloy Lomas C. **F.Pl:** 10kW.
576) Cas. 287, Riobamba — Dir: Galo Silva.
577) Pichincha y Argentinos esq., piso 3, Riobamba — Dir: Luz Luzuriaga.
578) Cas. 805, Alausí — Dir: Lic. Julio Martínez.
579) Soc. Unión Obrera, Guamote — Dir: Miguel Alcócer.
580) M.E. Flor 4099, Riobamba — Dir: Luis F. Aliaga Dávalos.
581) Av. Amazonas y Chávez, Penipe — Dir: Pedro V. Padilla.
582) Pallatanga — Gte: Enrique Naranjo H.
583) Ricaurte 529, Chunchi.

REGION 6
Bolívar (02)
606) Cas. 21, Guaranda — Gte: P. Fausto Gaybor G.
607) Cas. 86, Guaranda — Dir: Jorge W. Caravajal G.
608) Cas. 23, Simiátug — Dir: Mariano Poaquiza.
609) Guayas 194, San José de Chimbo — Dir: Oswaldo Jiménez.
Cotopaxi (05)
626) Cas. 251, Saquisilí — Dir: Galo Morocho T.
627) Ap. 392, Latacunga — Dir: Mons. Mario Ruiz.
628) Cas. 504, Latacunga — Gte: Jorge E. Erazo H.
629) Ap. 208, Latacunga — Prgrs in Quichua only.
630) B. Quevedo 551, Pujilí — Dir: Jorge Barreno A.
631) Cas. 669, Saquisilí — Dir: Arturo Mena Herrera.
633) Cas. 286, Latacunga — Dir: Hilda Nieto de Muñoz.
635) Cas. 294, Latacunga — Dir: Wilson R. Abril G.
636) Cas. 252, Latacunga — Dir: Isaías Carrera N.
637) Calle Calavi, La Maná — Dir: José María Lema Yugcha.
638) Cas. 474, Latacunga — Gte: Carlos Morán M. **F.Pl:** 5 kW.
639) B. León y La Prensa, La Maná — Dir: José Gonzalo S.
Tungurahua (18)
666) Bolívar 537, Píllaro — Dir: Angel Salazar.
667) Cas. 181, Ambato — Dir: Mario Barona Andrade.
668) Cas. 25, Baños — Dir: Luis A. Sevilla.
669) Cas. 541, Ambato — Dir: Alberto Salazar.
670) Chayán y Daquilema, Cdla. Presidencial, Ambato — Dir: Ramón Salazar C.
671) Juan Montalvo y Padre Chacón, Pelileo — Dir: José Rodríguez — F.Pl: Leave MW for FM.
672) Cas. 18-01-0574, Ambato — Dir: Luis Gamboa T.
673) Cas. 818, Ambato — Dir: Angel Núñez.
674) Cas. 18-01-0198, Ambato — Dir: GG: Luis Ramiro Vela G.
675) Cas. 368, Ambato — Dir: Gerardo Velástegui G.
676) Cas. 94, Ambato — Dir: Padre Angel Falconí. Quichua at 0100.
677) Cas. 618, Ambato — Dir: Isabel Yolanda Andaluz O.
678) Bolívar y Tamayo, Píllaro — Dir: Andrés E. Ibarra.
679) Cas. 146, Pelileo — Dir: Luis Jordán M.
680) Cas. 173, Ambato — Dir: Pedro Calvache.
681) Montalvo 106, Quero — Dir: Fauri Llerena.

REGION 7
Morona Santiago (14)
747) Quito s/n, Limón Indanza — Dir: Antonio Ochoa.
748) Misión Salesiana, 10 de Agosto s/n, Macas (or: Cas. 692, Quito) — Dir: Sor Dolores Palacios C. — Prgr A: commercial, Prgr B: educational.
749) Macas — Dir: Simón Rivadeneira R.
750) Federación de Centros Shuar, Domingo Comín 1738, Sucúa (or: Ap. 4122, Quito) — Dir: Albino Ututia J. C.E: Johannes Regniet. Prgr in Shuar only.
Macas (or: 10 de Agosto 4106 y Rumipamba, Quito) — Dir: Eduardo Romero.
Napo (15)
767) Cas. 260, Tena — Dir: Luis Enrique Espín E. F.Pl: SW.
768) El Coca — Dir: Gonzalo Mansilla Santos.
769) Mcal. Sucre 148 y 12 de Febrero, Lago Agrio — Dir: Fausto Velástegui F.
770) Archidona — Dir: Aurelio Rosero Castillo.
771) Misión Josefina, Tena — Dir: P. Rostagno.
772) Santa Rosa de Quijos (or: Ap. 11294, Quito) — Dir: Walther Manzo B. DP: Byron Medina.
773) El Coca — Dir: Leonor Cargua de Jiménez.
774) Av. Principal 425, Lago Agrio — Gte: Víctor J. Campoverde.
775) Principal y Espejo, Coca (or: via st. 189).
Pastaza (16)
777) Cas. 744, El Puyo — Dir: Julio Zamora.
778) Cas. 728, El Puyo — Dir: Orlando Tamayo Santos.
779) Cas. 777, El Puyo — Dir: Padre Frumen Escudero A.
780) Km 1.5 Vía Macas, El Puyo — Dir: Alberto Salvador S.
Zamora Chinchipe (—)
787) Sevilla de Oro 312, Zamora — Dir: Mons. Serafín Cartagena.
788) Parroquia 28 de Mayo, Cantón Yacuambí.

REGION 8
Galápagos (—)
(Cfr. separate listing in Pacific section)

LA VOZ DE LOS ANDES (HCJB)

ADDR: Casilla 17-17-691, Quito. **Te:** +593 (2) 433512. **Fax:** +593 (2) 447263. **Cable:** Vozandes (USA Office: PO Box 39800, Colorado Springs, CO 80949-9800).
L.P: MD: Glenn Volkhardt. PD: John Beck. CE: Jim Heck. Head of PR: Harold Goertzen.

kHz	kHz	kHz	kHz	kHz	kHz	kHz
690	6110	9745	11835	11960	15250	17790
3220	6205	9765	11845	15115	15270	17875
6050	9600	11730	11910	15140	15295	21455U
6080	9670	11735	11925	15155	15350	21480

Arabic	Area	kHz
1630-1700Sat	ME	17790, 21455U, 21480
1745-1800SS	ME	17790, 21455U, 21480
Czech		
0500-0530	Eu	6205, 11835
1800-1830	Eu	17790, 21455U, 21480
English		
0030-0430	NAm	9745, 15155, 21455U
0500-0700	NAm	11925
0500-1600	*	21455U
0700-0830	Eu	6205, 11735
0730-1130	SPac	9745, 11925
1130-1430	NAm/SAm	15115
1130-1600	NAm	11925, 17890
1630-1700MF	ME	17790, 21455U, 21480
1630-1745Sun	ME	17790, 21455U, 21480
1700-1745Sat	ME	17790, 21455U, 21480
1900-2000	Eu	17790, 21455U, 21480
2130-2200	Eu	15270, 17790, 21455U
French		
0000-0030	NAm	15155, 21455U
0600-0630	SPac	6050
0630-0700	Eu	6205, 11835
1600-1630	ME	17790, 21455U, 21480
2030-2100	ME	15270, 17790, 21455U
Georgian		
0400-0430a	As	9600, 11735
German		
0000-0030	SAm	15250
0430-0500	Eu	9600, 11735
0600-0630	Eu	6205, 11835
0930-1000	SAm	6110, 9670
1830-1900	Eu	15270, 17790, 21455U
	Af	15295
2100-2130	Eu	15270, 17790, 21455U
2300-2330	SAm	15250, 21455U
Japanese		
0430-0500	NAm	9745, 15155, 21455U
1130-1200	J	9745, 11845
2200-2230	SAm	15295, 17875
Kikongo		
0500-0515MF	Af	9765
Nordic		
0530-0600	Eu	6205, 11835
2000-2030	Eu	15270, 17790, 21455U
Portuguese		
0515-0600MF	AF	9765
0730-1030	SAm	11845
1000-1030	SAm	6110, 9670

1500-1800	SAm	15295
1600-1800	SAm	11925
2300-2400	SAm	17875
2300-0200	SAm	15295
Quechua		
0830-1030	EQA	690, 3220, 6050, 6080
0900-0930	SAm	6110, 9670
1030-1130	SAm	6110, 9670
1030-1430	EQA	3220, 6080
2100-0200	EQA	3220, 6080
2200-0030	SAm	9745
Russian		
0150-0200a	As	9600, 11735
0200-0400	As	9600, 11735
0400-0430b	As	9600, 11735
0500-0600SS	As	11735
0500-0700	As	9600
0600-0700	As	11735
1030-1130	As	11730
2230-2300	SAm	15250, 21455U
2330-2400	NAm	15250, 21455U
Spanish		
0130-0150b	NAm	11910
1030-0500	EQA	690, 3220, 6050, 6080
1030-1300	SAm	11960
1030-1500	NAm	9765
1200-1430	SAm	11910
1300-2030	SAm	17875
1500-0500	NAm	15140
1730-1800	NAm	15350
2030-0500	SAm	11960
2200-2230	Eu	15270, 17790, 21455U
b) Fri/Sat		
Tadjik		
0115-0130a	As	9600, 11735
Uzbek		
0130-0145a	As	9600, 11735

a) Sun-Thurs, B) Fri/Sat

SSB test transmissions: 24h on 21455kHz.

R) Regional, a) Mon-Fri 0645-0700, b) Mon-Fri only, c) Sat/Sun to 0500, d) Sat & Sun to 0500, e) Sat & Sun only, f) Finnish Saturdays, 0630 & 2000, g) Sat/Sun to 1500, h) Tues/Thurs/Sat, j) Sun/Mon/Wed/Fri, k) MF to 0645.

DX-Prgrs: English: Sat during "Studio 9" Prgr. at 0730, 1000, 2130; Sun 0030, 0230, 0500. **German:** Sat 0600, 1830, 2100; **Swedish:** Fri 0530, 2100; **Japanese:** Tues/Fri 1130; **French:** Mon/Fri 0630, 2030; Tues/Sat 0015.

N. in English: Latin American N. at 0730, 1000, 2130, 0030, 0230, 0500. World N. at end of most Studio 9 releases.

ANN: E: "This is the Voice of the Andes, HCJB, Quito, Ecuador in South America, The time at the tone will be . . . , Eastern Standard Time, the same as . . . Coordinated Universal Time. A programme follows in . . ."

V. Listener reports are welcomed. QSL-cards (a new one every 2 months) sent upon receipt of complete reception reports. Rp. Write for International Programme Schedule or request a specific Language Division Schedule. Other explanatory literature available by writing to HCJB, Casilla 691, Quito, Ecuador. Reception tapes are appreciated but cannot be returned to the sender.

FALKLAND ISLANDS

L.T: UTC – 3h — **Pop:** 2.000 (excl. British Forces) — **Pr.L:** English — **ITU:** FLK.

FALKLAND ISLANDS BROADC. STATION (Gov.)

ADDR: John Str, Stanley. **Te:** +500 27277/8. **Fax:** +500 27279.
L.P: Broadc. Officer: Patrick Watts, MBE. Secr: Steff Hanlon.

BRITISH FORCES BROADCASTING SERVICE

ADDR: BFBS Falkland Islands, Mount Pleasant, BFPO 655. **Te:** 32179. **Fax:** 32193.
L.P: St. Mgr: Roger Wood. Sen. Eng: Adrian Almond.

MW: Bush Rincon 550kHz 10kW — **FM:** March Ridge 90 MHz 2kW, Mount Alice 92.5MHz 0.3kW, Port Stanley 96.5 MHz 2kW, Mount Maria 103.4 MHz 2kW, Mount Byron 104.7MHz 0.3kW.

D.PRGR: 24h (FIBS: approx 1000-1215 and 1715-2200; other times BFBS).
N: 0900, 1000, 1100, 1200, 1500, 1600, 1700, 1900, 2000 (relays of BBCWS or BBC News fed from BFBS London by satellite).
V. by QSL-card. Rp.

GUIANA (French)

L.T: UTC-3h — **Pop:** 105.000 — **Pr.L:** French — **EC:** a/c 50, 127/220V (no) — **ITU:** GUF.

R.F.O. GUYANE (RADIODIFFUSION FRANÇAISE D'OUTRE-MER)

ADDR: B.P. 336, F-97305 Cayenne. **Te:** +594 30 15 00. **Fax:** +594 302649. **Telex:** 910526 FG.
L.P: Dir: Henri Neron. Head of N: Anastasie Bourquin. Dir Tec: Daniel Beugin. PD: Jean-Pierre Karam.
MEDIUMWAVES: Matoury 1070kHz 10kW — St. Laurent du Maroní 1060kHz 0.05kW.
SHORTWAVE: Matoury (G.C: 52.20W/04.54N): 3385kHz 4kW — 5055kHz 10kW.
FM: Cayenne 91.5/95.2MHz 0.05kW, 94.0MHz 0.3kW, Korou 96.3/Sinnamary 99.2/Iracoubou 97.0/Mana 99.5/Saint-Laurent 92.5MHz (all 0.05kW).
HOME SCE: in French: 24h on all freqs (exc. 1070kHz + 91.5MHz).
N: 1030, 1630, 2230.
Rel. France-Inter: 24h on 1070kHz + FM91.5MHz.
ANN: "Ici Cayenne, R.F.O. Guyane" — **INT-SIG:** "Nos richesses" on guitar — **V.** by QSL-folder. Rec. acc.

Local Radio: 6 private FM-st's are operating.

RADIO FRANCE INTERNATIONALE RELAY ST.

STATION: Montsinery (G.C: 53.00W/05.00N): 3 × 500kW SW-tr's.

kHz	kHz	kHz	kHz	kHz	kHz
9675	9800	11995	15365	17620	21645
9685	11670	13685	15435	17720	21685
9715	11680	15200	17560	17860	21765
9790	11685	15300	17575		

D.PRGR: See RFI listing under "France", also R. Japan & China R. Int.

GUYANA (Republic)

L.T: UTC – 3h — **Pop:** 749.000 — **Radio sets:** 310.000 — **Pr.L:** English, Amerindian dialects — **E.C:** a/c 50/60, 120/240V — **ITU:** GUY.

GUYANA BROADCASTING CORPORATION (Gov/Comm.)

ADDR: P.O. Box 10760, Georgetown. **Te:** +592 (2) 58734.
Cable: GUYCAST GEORGETOWN.
L.P: GM: Ms. A. Brewster. PD: Ms. M. Lawrence. CE: S. Goodman.

MEDIUMWAVES	kHz	kW	Ch		kHz	kW	Ch
Georgetown	560	10	2	Georgetown	760	10	1
Linden	700	1	1				

SHORTWAVES (G.C: 58.05W/06.49N): 5950kHz 10kW (Ch. 2).
FM: Georgetown 100MHz 0.2kW (Ch. 1), 102MHz 0.05kW (Ch. 2); Linden 104MHz 0.05kW (Ch. 2).
D.PRGR: R. Roraima (Ch. 1): 0800-0200. **N:** 0900, 1000, 1100, 1330, 1500, 1900, 2100, 2230 (Sun), 2300 (W), 0100.
V. of Guyana (Ch. 2): 0730-0300 (irr. 24h). **N:** as Ch. 1.
V: by letter.

PARAGUAY

L.T: UTC – 4h Mar 1-Sep 30 (UTC – 3h Oct 1-Mar 31) — **Pop:** 4.938.000 — **Radio sets:** 775.000 — **Pr.L:** Spanish, Guaraní — **E.C:** a/c 50, 220V — **ITU:** PRG.

ADMINISTRACION NACIONAL DE TELECOMUNICACIONES
ADDR: Administración General, Cas. Correo 84, Asunción.
Cable: Antelco. **Te:** +595 (21) 44001.
L.P: Pres: Ing. M.C. Guanes. Dir. Adm: F.F. Duarte. Dir. of Tech. Dept: Ing. Osmar G. López R. Dir. Int. Affairs: Ing. M.H. Gini. Dir. Radio: Ing. Angel Barboza.
STATIONS: All st's comm. exc. R. Nacional.
Mediumwaves: °) Shortwave. *) inactive.

	Call	kHz	kW	Name, location and h. of tr.
1)	ZP16	550	25	R. Parque, Cd. del Este: 0900-0500
2)	ZP15	v°572	30	La Voz del Amambay, Pedro Juan Caballero: 0950-0200 (n.570)
3)	ZP32	590	5	R. Ycuámandyyú, Villa de San Pedro: 0900-0200
4)	ZP30	610	10/1	R. La Voz del Chaco Paraguayo, Filadelfia: 0930-0215
5)	ZP40	v620	1	R. Ñassaindý, San Estanislao (r. 619-616.6v)
6)	ZP19	640	10	R. Caaguazú, Coronel Oviedo: 0700-0300
8)	ZP26	660	10	R. Itapirú, Cd. del Este: 0855-0100
29)	ZP11	680	10	R. Cáritas, Asunción: 0830-0400
9)	ZP41	680	2	R. Ypacaraí, Ypacaraí: 0920-0200
10)	ZP12	v700	10	R. Carlos Antonio López, "La Voz del Neembucú," Pilar: 0900-0200 (r. 699.6)
11)	ZP-	730	50	Cardinal AM Stéreo, Asunción
12)	ZP5	760	25	R. Encarnación: 0800-0400
13)	ZP70	780	25	R. Primero de Marzo, Asunción: 0900-0430
14)	ZP27	800	5	R. Mbaracayú, Salto de Guairá
15)	ZP6	v°840	5	R. Guairá, Villarica: 0900-0300
16)	ZP28	v860	25	La Voz de la Cordillera, Caacupé: 0900 (Sat/Sun 1000)-0400 (r. 860.4)
17)	ZP33	890	1	R. Tres de Febrero, Itá: 1000-0200
18)	ZP1	°920	100	R. Nacional del Paraguay, Asunción: 0800-0400
19)	ZP9	970	10	R. Comuneros, Asunción
20)	ZP31	980	1	R. Mburucuyá, Pedro Juan Caballero: 0855-0600
21)	ZP36	v996	10	R. Ypoá, Paraguari: 0855-0200 (n. 1000)
22)	ZP14	1020	10	R. Ñandutí, Asunción: 24h.
23)	ZP43	1040	5	R. Arapizandú, San Ignacio (Misiones): 0900-0200
24)	ZP13	v1060	3	R. Boquerón, Alberdi: 1000-0300 (r. 1060.5)
25)	ZP25	1080	10	Radiodif. Nanawa, Luque: 0800-0330
26)	ZP71	1100	5	R. Ñu Verá, Capitán Badó: 0900-0200
27)	ZP24	1120	10	R. Nuevo Mundo, San Lorenzo: 1000-0300
28)	ZP22	1140	5	R. Panambí Verá, Villarica: 0900-0200
28x)	ZP52	1180	3	R. Coronel Oviedo, Coronel Oviedo
31x)	ZP34	1210		R. Cultura, Villarica
30)	ZP21	1230	3	R. Centenario, Caaguazú: 0800-0300
31)	ZP3	1250	10	R. Asunción: 0830-0400
32)	ZP10	°1300	5	Em. Paraguay, Asunción: 0800-0400
41)	ZP53	1310		La Voz del Este, Cd. del Este
33)	ZP4	1330	10	R. Chaco Boreal, Asunción
34)	ZP44	1360	1	R. Arasapotý, Itaguá: 0900 (Sun 1000)-0200
36)	ZP37	1360	1	R. Yby Yau, Carapeguá: 0900-0300
35)	ZP8	1380.7	2	R. Concepción: 0930-0200
37)	ZP35	v1411	1	R. Dif. Mangore, San Juan Bautista Misiones: 1000-0130 (n. 1430)
38)	ZP42	1417	5	R. Güyrá Campana, Horqueta (n. 1420)
7)	ZP29	1450	5	R. Vallemí
39)	ZP23	1480	1	R. Mariscal Francisco Solano López, Bella Vista: 1000-0100
40)	ZP20	v1480	1	R. America, Ñembý: 0900-0300

Shortwaves:

	Call	kHz	kW	Name and location, h. of tr.
15)	A6	5935v	5	R. Guairá, Villarica: 0900-1800 (irr.) (n5975)
2)	A15	*5995	2	La Voz del Amambay, Pedro Juan Caballero
32)	A10	6014.4v	0.3	Em. Paraguay, Asunción: 1100-1500, 1900-2300
18)	A1	6025	0.7	R. Nacional del Paraguay, Asunción
18)	A1	9735	100	R. Nacional del Paraguay: 0800-0400
12)	A5	11939.5	0.5	R. Encarnación: 0800-0400

Addresses and other information:
1) km 10, Ruta Internacional, Cd. del Este.
2) 14 de Mayo 475, Pedro Juan Caballero — DG: L.C. Rolón P.
3) Villa de San Pedro — Dir: Ing. Juan B. Ibáñez.
4) Casilla 984, Filadelfia, Chaco — Dir: Andreas Friesen. Prgr. in Guaraní & vern., German, E. and Sp. operated by Mennonite Mission.
5) San Estanislao — Dir: Demetrio Sánchez Candia.
6) Ruta Mcal. Estigarribia Km 6½, Coronel Oviedo.
7) Gral. Santos y Manuel Domínguez, 5° piso, Asunción.
8) Nanawa y Adrian Jara, Edif. Saba Cd. del Este — Dir: Pablo Porfirio Cilía.
9) Calle Paraguay, Ruta 2, km 36, Ypacaraí.
10) Alberdi 998, Pilar — Gte: Miguel Angel Rodríguez Hermosa.
11) 8va. e/2a. y3a., Lambaré (or: Cas. 247), Asunción.
12) Gral. Artigas 724, Encarnación — Dir: Eugenio Pérez.
13) Av. General Máximo Santos y Felicidad, Asunción.
14) Salto del Guairá — Dir: Víctor Pérez Ávalos.
15) Alejo García y Presidente Franco, Villarica — Own: Lídice R. de Traversi. DG: Enrique Traversi R.
16) Dr. Venancio Pino y 3ra. Proyectada, Caacupé.
17) Av. Enrique Doldán Ibieta y Presidente Franco, Itá.
18) Montevideo y Estrella, Asunción — DG: Antolín Leguizamón, C.E: Carlos Montaner V.
19) Estados Unidos 2019-2022, Asunción.
20) Barrio María Victoria, Pedro Juan Caballero.
21) Radio Parque San Francisco, Paraguarí — Dir: Carmelo Ruggilo.
22) Choferes del Chaco 1194, Asunción — Dir: Humberto Rubín.
23) Mariscal López y Cerro Corá, San Ignacio (Misiones).
24) Curupaytý 288, Alberdi — DG: J. Isidro Gamarra
25) Av. General Aquino — km. 12, esq, José Bonifacio, Luque — Dir: Silvio Noguera.
26) Capitán Badó — Dir: Víctor Castro Monte.
27) Coronel Romero y Florida, San Lorenzo (Asunción: General Díaz 488, Ofic. 34) — Dir: Pablo Pereira.
28) General Caballero 748, Villarica — Dir: Dr. Nery González.
29) Azara y Kubitschek (or: Cas. 1313), Asunción. **Te:** +595 204161 — Dir: Pbro. Cristóbal López.
30) Caaguazú — Dir: Robustiano Cabrera J.
31) Av. Artigas y Capitán Lombardo, Asunción.
32) Gral. Santos y 18 de Julio, Asunción — Dir: Roque A. Fleytas.
33) Av. Mcal. López 2948, Asunción — Own: Humberto Domínguez D.
34) Ruta Mariscal Estigarribia — Km 29, Itauguá.
35) Casilla de Correo 78, Concepción — DG: S.E. Dacak.
36) Carapeguá — Dir: Alfonzo Romero.
37) San Juan esq. Presidente Franco, San Juan Bautista Misiones — Dir: Johnny Sánchez Martínez.
38) Horqueta — Dir: Dr. Salvador R. Paredes.
39) Calle Iturbe 146, Bella Vista — Dir: Blanca Cardoso.
40) Cas. 2220, Ñembý.

FM in Asunción.
89.0 R. Reloj — 90.7 Ysapý — 91.6 R. 15 de Agosto — 11) 92.3 — 18) 95.0 — 13) 97.1 — 98.0 R. Curupaytý — 100.0 Canal 100 — 100.5 R. Itá Pirú — 100.9 R. Cristal — 102.1 R. Tayý — 33) 103.1 — 32) 105.0 — 22) 108.0.

PERU

L.T: UTC – 5h (UTC-4h. Jan. 15-Apr. 15) — **Pop:** 22.810.000 — **Radio sets:** 4.400.000 — **Pr.L:** Spanish, Quechua — **E.C:** a/c 60, 220V — **ITU:** PRU.

MINISTERIO DE TRANSPORTES Y COMUNICACIONES, Dirección General de Telecomunicaciones.
ADDR: Av. 28 de Julio 800, Lima 1. **Cable:** Digecom.
L.P: Dir. de Telecomunicaciones: Ing. Carlos A. Romero Sanjinés. Dir. Freq. Div: Ing. Raúl Gómez.

ASOCIACION DE RADIOIFUSORAS DEL PERU (ARP)
ADDR: Manuel Corpancho 208, Lima 1.
L.P: Pres: Fernando Gonzales del Campo.

INSTITUTO NACIONAL DE COMUNICACION SOCIAL
ADDR: Jr. de la Unión 264, Lima — **L.P:** Hernán Valdizán C. Dir. Bc: Sra. Clarisa P. de Olivera.

UNION DE RADIOEMISORAS DE PROVINCIAS DEL PERU (UNRAP)
ADDR: Mariano Carranza 754, Santa Beatriz, Lima 1.
L.P: Pres: Humberto Maldonado Balbín.
Mediumwaves: ° = shortwave, * = inactive; v = varying fq.

	Call	kHz	kW	Name and h. of tr.
400)	OBX4E	540	10	R. Inca del Perú, Lima: 24h
260)	OCX2D	540	1	R. San Antonio, El Porvenir
401)	OBZ4I	v560	2	R. Oriente, Lima: 1000-0600 (Sun 2300) (r. 564)
333)	OAX3X	570	1	R. Progreso, Chimbote: 1100-0500
200)	OAX2E	°580	10	R. Marañón, Jaén: 1100 (Sun 1200)-0100
434)	OAX4M	580	15	R. Star, Lima: 0900-0300
261)	OBX2N	590	0.5	R. Esperanza 2000, Esperanza
650)	OAX6X	*600	0.3	R. Cultural, Toquepala
403)	OBZ4W	°600	10	R. Cora del Perú, Lima: 0930-0500
262)	OAX2M	620	1	R. Chepén, Chepén
404)	OCY4A	620	10	R. Estrella del Sur, Cañete: 1000-0430
750)	OBX7B	°640	5	R. Onda Azul, Puno: 0900-0100
405)	OAZ4K	°640	10	R. Del Pacífico, Lima: 1100-0500 (Sun 0300)
540)	OAX5R	650	1	R. Regionalista, Ica: 1030-0530
263)	OAX2N	*650	1	R. Ondas del Norte, Trujillo
406)	OCX4R	660	10	R. Omega, Lima: 1100-0700
618)	OBX60	660	1	R. Internacional, Arequipa
460)	OCX4L	°660	1	R. Chinchaycocha, Junín
751)	OAX7H	670	10	R. Nacional Pachacútec, Puno
407)	OBX4A	680	5	RBC, la estación: Lima: 24h
264)	OBX2L	680	0.5	R. Amauta, Chocope
541)	OAX5E	680	10	R. Independencia, Ica: 1100-0600
600)	OBX6Q	690	0.5	Radiodif. La Joya, La Joya
334)	OBX3C	690	1	R. El Puerto, Chimbote
408)	OBZ4H	700	1	R. 700, Lima
131)	OBX1U	700	10	R. Cutivalú "La Voz del Desierto", Castilla: 0930-0300
543)	OBX5Q	710	5	(r. RPP 730), Ica
265)	OAX2B	720	2.5	R. Nacional Pachacútec, Trujillo
800)	OAX8A	720	10	R. Nacional Pachacútec, Iquitos
409)	OAX4G	730	10	R. Programas del Perú, "RPP", Lima: 24h
132)	OAX1D	740	10	R. Progreso, Piura: 1100-0500
266)	OAX2J	*740	10	R. Sudamericana, Trujillo
601)	OAX6C	°740	10	R. Continental, Arequipa: 0900-0600
335)	OAX3T	750	1	R. El Mundo, Chimbote: 1100-0500
490)	OCX4B	°750	5	R. Altura, Cerro de Pasco: 1100-0500
267)	OBX2K	760	0.5	R. Andino, Otuzco
410)	OBZ4X	760	10	Radiomar, Lima: 24h
122)	OCX1T	°770	—	R. Horizonte, Chiclayo
801)	OAX8M	°770	5	La Voz de la Selva, Iquitos: 1000-0300 (Sun 1100-1700)
151)	OAX1K	780	10	R. Nacional del Perú, Tumbes
411)	OAX4X	°780	10	R. Victoria, Lima: 1000-0600
268)	OAX2I	790	10	R. CRU, (r: RPP 730), Trujillo: 24h
700)	OCX7G	790	1	R. Canchis, Sicuani
412)	OAZ4U	800	0.5	R. Sol, Huaral
544)	OBX5B	800	0.5	R. Sur Medio, Ica
640)	OBX6A	800	1	R. La Porteña, Ilo
602)	OBX6A	800	0.75	R. Texao, Arequipa: 0900-0430
752)	OAX7V	810	10	(r. RPP 730), Juliaca
413)	OAX4O	820	10	R. Libertad, Lima: 24h
360)	OAX3Y	830	1	R. La Selva, Tingo María: (n. 800)
701)	OCX7H	830	1	R. Inti, Santiago
297)	—	r835	—	R. Fraternidad, El Porvenir: 0950-
651)	OAX6D	°830	10	R. Nacional del Perú, Tacna
603)	OBX6Y	840	1	R. Onda Azul, Paucarpata
510)	OAX5H	840	10	R. Ayacucho, Ayacucho
336)	OAX3S	840	4	R. Casma, Casma
414)	OAX4A	850	40	R. Nacional Pachacútec, Lima: 24h.
461)	OBX4V	860	10	R. Junín, Huancayo
111)	OBX1F	870	10	R. CRU, (r: RPP 730), Chiclayo
298)	—	r870	—	R. Vernacular, Trujillo: (n. 900)
702)	OCX7R	°870	1	R. Mundo, Cusco: 0900-0100
415)	OBZ4N	°880	10	R. Unión, Lima: 24h
753)	OBX7S	900	1	R. Bahá'í, Chucuíto
416)	OBX4X	900	10	R. El Sol, Lima: 0930-0530
361)	OAX3E	900	0.5	R. Ribereña, Aucayacu: 1230-0300
604)	OBX6K	900	3/1	Nevada R., Arequipa: 24h
299)	—	r905	—	La Súper R., San Antonio
269)	OAX2H	910	1	R. Huamachuco: 1000-0300
270)	OBX2S	920	1	R. Ollantay, Virú
133)	OBX1J	920	10	(r: RPP 730), Piura
727)	OCX7M	*920	1	R. Andina, Cusco
417)	OAX4E	930	10	Moderna "R.papá", Lima: 24h
703)	OBX7L	940	5/3	R. Willkamayu, Cuzco: 24h
605)	OBX6T	940	3	R. Onda Sideral, Arequipa: 24h
134)	OBX1C	940	1	R. Presidente, Paita
337)	OBX3S	950	1	(r: RPP 730), Chimbote
201)	OBX2G	°950	1	R. Cutervo, Cutervo
300)	—	r960	—	R. Est. Caracol. Trujillo: 1100-
418)	OAX4D	960	10	R. Panamericana AM, Lima
462)	OCY4V	°960	1.5	Constelación Radial, Huancayo (r 1000): 24h
237)	—	960	—	R. Lider del Norte, Cajamarca: 1100-0300
606)	OBX6S	°960	3	R. Hispana, Arequipa: 0900-0500
546)	OBX5A	970	1	Latinoamericana R., Ica: 1100-0500
135)	OBX1V	970	1.5	R. La Capullana, Sullana
607)	OAX6I	980	1	R. Universidad, Arequipa: 1200-0200
271)	OAX2F	980	2	R. La Hora, Trujillo: 1100-0455
511)	OAX5T	°980	2.5	R. La Voz de Huamanga, Ayacucho: 0930-0300
419)	OBX4J	985	10	R. Latina, Lima: 1000-0500
463)	OBZ4R	990	1	R. La Voz del Valle de Mantaro, Jauja
652)	OAX6K	990	10	R. Bicolor, Tacna: 0900-0500
202)	OBX2M	v990	0.5	R. Contumazá: 1100-0400
920)	—	°1000	1.5	R. Central, Bellavista: 18 h
921)	—	1000	1	R. El Sol, Saposoa: 1000-0400
608)	OBX6R	1000	1	R. Endesa, Arequipa
704)	OAX7V	1000	5	R. Machu Picchu, Cusco
203)	OBX2P	*1010	2	R. San Francisco, Cajamarca
420)	OAX4U	°1010	50	R. América, Lima: 24h
112)	OBX1G	1020	1	R. Heroica, Chiclayo: 1000-0500
338)	OBX3L	1020	1	R. Onda Nueva, Chimbote
754)	OAX7N	°1030	5	R. La Voz del Altiplano, Puno: 1000-0400
296)	OAX2U	°1030	3	R. Los Andes, Huamachuco: 1000-0200
705)	OCX7O	1030	1	Super R., Cusco
242)	OBX2O	1040	1	R. Nor Oriente, Jaén
421)	OBX4O	1040	10	R. Red, Lima: 24h
609)	OBX6B	°1050	1	R. Arequipa: 1000-0400
274)	OCX2B	1050	1	R. San Sebastián, Chepén: 0930-0200
339)	OAX3A	1050	1	R. Alpamayo, Huaraz
464)	OBZ4J	1050	5	R. Bolognesi, Huancayo
421)	OCY4D	1060	1	Exito Amplitud Modulada, Lima
500)	OBX5D	°1060	2	R. Andahuaylas: 1000-0400
706)	OCX70	1060	1	R. Amauta, Cusco
275)	OAX2B	1070	2	R. Trujillo "La Voz del Norte del Perú": 0950-0500
547)	OAX5A	1070	0.25	R. Centinela, Pto. de San Juan
136)	OBX1D	1080	5	R. San Miguel, Piura: 1000-0500
707)	OAX7S	1080	1	R. Salcantay, Cusco: 0900-0400 (Sun 1000-0200)
802)	OAX8N	1080	1	R. Amazonas, Iquitos: 1130-0500
491)	OCX4P	°1080	1	R. Huayllay, Huayllay
820)	OAX8C	°1080	1	R. Sideral, Pucallpa (n 1100)
204)	OBX2A	1090	1.2/3	R. Cajabamba, Cajabamba: 0900-0500

362)	OAX3I	°1090	1	R. Tingo María, Tingo María (n. 1100)
610)	OBX6X	1090	1	R. Amistad, Arequipa
365)	OBX3M°	1100	1.5	Emisoras JSV, Huánuco: 1100-0200
424)	OAZ4W	1100	1	R. Musical (r: RPP 730), Barranca
425)	OCX4S	1100	0.2	R. Imperial, "La Voz de la Provincia", Cañete: 0900-0300
548)	OAX5Z	1100	1	R. Oasis, Ica: 1100-0200
465)	OCY4G	1100	1	Sonorama R., Huancayo: 1000-0300
113)	OBX1L	1100	1	R. Star, Chiclayo: 1000-0600
755)	OBX7Z	1100	1.5	R. Collao, Juliaca: 1000-0400
922)	OAX9J	1100	0.5	R. Lamas, Lamas
708)	OCX7T	1100	5	R. Lider, Cusco: 1000-0430
340)	OBX3B	1110	1	R. Heróica, Huaraz
276)	OBX2I	1120	2	R. Dínámica, Trujillo: 0900-0500
427)	OAX4N	1130	3	R. Selecta, Lima: 1000-0500
466)	OAZ4S	°1130	0.4	R. Chanchamayo, La Merced: 1400-0600
924)	OAX9Q	°1130	1	R. San Martín, Tarapoto: 0930-0400
205)	OAX2V	1130	1	R. Los Andes (r: RPP 730), Cajamarca
611)	OAX6L	°1140	12	R. Concordia, Arequipa: 0900-0500
467)	OCY4C	1140	10	R. Centro, Huancayo
558)	OAX5W	1140	0.5	R. Chinchaysuyo, Chincha
236)	—	1150	1	R. Naylamp "Estación Pucará, Pucará
709)	OCX7Q	°1150	2.5	R. Universal, Cusco: 1000-0400
803)	OAX8F	1150	10	R. Loreto, Iquitos: 1000-0005
238)	—	1150	—	R. Colonial, Cajamarca: 0900-0300
428)	OAX4C	1160	10	Once Sesenta AM, Lima: 1100-0700
641)	OBX6G	1160	1	R. Hogar, Moquegua
271)	OAX2C	1160	5	R. Libertad, Trujillo: 1100-0400
516)	OBX5O	°1160	1	R. Huanta 2000, Huanta: 0900-0400
342)	OAX3B	1160	1	R. Huaraz, Huaraz: 1100-0300
468)	OCX4Y	°1170	1	R. COSAT, Satipo: 1100-0200
612)	OBX6L	1170	10	R. Andina (r: RPP 730), Arequipa
469)	OCY4Z	°1180	1	R. Libertad, Junín
925)	OAX9V	°1180	1	R. Marginal, Tocache (n 920)
277)	OCX2A	1180	1.5	R. América, Quiruvilca
114)	OAX1E	1190	10	R. Chiclayo, Chiclayo: 1100-0600
549)	OBX5M	1190	1	R.Lurén, Ica (n 1060)
343)	OBX3D	°1190	3	R. Ancash, Huaraz: 0930-0500 (r. 1150)
248)	—	°1190	—	R. Membrillar, Cascas
946)	—	1200	—	R. Las Palmas, Nuevo Bambamarca: -0200
923)	OAX9C	1200	2	R. Rioja, Rioja: 0900-0500
756)	OCX7S	1200	1	R. San Ramón, Juliaca
926)	OBX9C	°1200	1	R. Estación "C", Moyobamba
642)	OAX6N	1200	1	R. El Faro, Ilo: 1000-0500
367)	—	1200	1	R. Satélite (r: RPP730), Tingo María
429)	OAX4B	1200	4	R. Cadena Informativa, Lima: 1000-0500
725)	OAX7B	°1200v	5	R. Tawantinsuyo, Cusco
247)	—	1200	—	R. San Mateo, Contumazá
206)	OCX2T	1200	1	R. Noticias, Cajamarca: 1100-0300
470)	OCY4T	1210	1	R. Galaxia, Satipo
430)	OBZ4A	°1210	1	R. Paramonga, Barranca: 0900-0300
278)	OAX2Q	1210	1	R. Universo, Trujillo: 1000-
710)	OAX7M	°1210	1	R. Quillabamba: 1000-0200
550)	OAX5G	1220	1	R. San Pedro, Grocio Prado
613)	OAX6X	°1220	3	R. Melodía, Arequipa: 24h
900)	OAX9T	°1220	3	R. Bagua, Bagua: 1000-0300
207)	OCX2H	°1220	1	R. Wuaira, Chota
279)	OAX2T	1230	1	R. Albújar, Guadalupe
351)	—	1230	1	R. Santa, Chimbote
471)	OBZ4Y	1230	1	R. Selecciones, Tarma: 1100-0500
740)	OBX7I	°1230	0.5	R. Madre de Dios, Puerto Maldonado: 1000-0200
280)	OCX2K	1240	1.5	R. Indoamérica, Trujillo: 1100-0500 (r.1250)
137)	OAX1L	°1240	1	R. Huancabamba, Huancabamba
431)	OBX4Q	1240	5	R. Oro, Huacho
711)	OBX7M	1240	1	R. Túpac Amaru, Sicuani: 24h
123)	—	r1250	—	R. Sensacional, Chiclayo
821)	OAX8P	°1250	2	R. Pucallpa, Pucallpa: 1030-0500
432)	OAX4L	1250	10/5	R. Miraflores, Lima: 1100-0500
927)	OAX9K	°1250	0.7	R. Estación Tarapoto, Tarapoto: 1000-0400
281)	OBX2C	1260	1	R. Otuzco, Otuzco
614)	OBX6D	1260	1	R. Red, Arequipa
301)	—	r1270	—	R. Imperial, Laredo: 1200-
433)	OAZ4H	1270	1	R. Huacho, Huacho: 1100-0300
551)	OAX5B	1270	1	R. Ica, Ica
804)	OAX8T	°1270	1	R. Eco, Iquitos: 0900-0400
472)	OBZ4T	1270	0.5	R. La Merced, La Merced
615)	OBX6H	1270	0.35	R. Mollendo, Mollendo
712)	OBX7A	1270	0.7	R. Continente, Cusco: 1000-0500
757)	OBX7G	°1270	1	Ondas del Titicaca, Puno: 0930-0300
901)	OAX9F	°1270	1	R. Nor Peruana, Chachapoyas
138)	OAX1H	°1270	1.5/3	R. Grau, Piura: 24h
208)	OBX2F	1280	1	R. Moderna, Cajamarca: 1000-0400
616)	OBX6P	1280	1	R. Fénix, Camaná
154)	—	1290	1	(r: RPP 730), Tumbes
282)	OBX2Z	1290	1	R. Casa Grande, Casa Grande
908)	OAX9N	1290	10	R. Dif. Nor Oriental, Bagua
302)	—	r1295	—	R. San Juan, Trujillo: 1030-
115)	OAX1I	1300	1	R. Difusora Comercial, Chiclayo
344)	OAX3O	1300	0.5	R. Huascarán, Huaraz: 1100-0400
341)	OAX3R	1300	2	R. Bahía, Chimbote
822)	OAX8B	°1300	3	R. Nuevo Mundo, Pucallpa: 24h
402)	OAX4S	1300	5	R. Comas, Comas: 1000-0500
473)	OAZ4B	°1300	2	R. Andina, Huancayo: 0900-0500v
653)	OAX6P	1300	2	R. Latina (r: RPP 730), Tacna
758)	OAX7X	°1300	1.5	R. Juliaca, Juliaca: 1000-0300
713)	OAX7P	°1300	10	R. Onda Imperial, Cusco: 1100-0100
552)	OBX5K	1310	1	R. San Martín, Pisco
231)	OBX2D	°1310	1	R. Chota, Chota: 1200-0400
414)	OAX4I	°1320	10	R. La Crónica, Lima: 1100-0500
209)	OBZ2Z	°1320	1	R. La Voz de Cutervo, Cutervo: 1100 (Sun 1000)-0100 (r. 1200)
210)	—	*1320	—	R. Libertad, Cajamarca
617)	OBX6W	1320	0.8	R. Premier, Camaná
492)	OBX4Z	°1320	5/1	R. La Voz de Oxapampa: 1000-0430
139)	OAX1F	1320	0.35	R. Talara, Talara
947)	OCZ9K	1320	0.25	R. Pajatén, Juanjuí
352)	—	r1324	—	R. Continental, Caraz
244)	—	*1330	—	R. Universal, Cajamarca
283)	OAX2P	1330	0.5	R. Paiján, Paiján: 1100-0400
714)	OCX7K	°1330	1	R. San Miguel, Cusco: 1000-0100 (Sun 1700)
353)	—	1330	1	R. Nal. del Perú, Chimbote
929)	OAX9C	°1340	3	R. Tropical, Tarapoto: 0945-0400
474)	OAZ4Q	1340	1	R. Jauja, Jauja
553)	OAX5D	1340	0.5	R. Chincha, Chincha Alta
140)	OBX1K	1340	1.5	R. San Francisco, Piura: 24h
436)	OAZ4M	1340	0.2	R. Huaral, Huaral
823)	OAX8D	1350	1.5	R. Super AM, Pucallpa
363)	OAX3N	°1350v	1	R. Ondas del Huallaga, Huánuco: 1000-0500
211)	—	*1350	—	R. San José, Cajamarca
303)	—	r1350	—	R. Sintonía, Trujillo: 1150-
345)	OAX3U	1350	1	R. Miramar, Chimbote
643)	OBX6F	1350	1	R.Ilo, Ilo
124)	—	r1350	—	R. Continental, Chiclayo
284)	OBX2W	1360	1	R. Super Uno, Cartavio Ascope
407)	OAX4Y	1360	5	R. Excelsior, Lima: 18h
715)	OAX7R	°1365	1.5	R. Sicuani, Sicuani: 1000-0300
346)	OAX3H	1370	1	R. Interamericana, Chimbote
644)	OAX6T	1370	0.5	R. Moquegua, Moquegua: 0900-0500
304)	—	r1370	—	R. Marítima, Salaverry: 1300-
475)	OAZ4F	1370	0.35	R. La Oroya, La Oroya (n 1490)
152)	OAX1G	1370	1	R. Tumbes, Tumbes
728)	OAZ7J	1370	1	R. Santa Mónica, Cusco
930)	—	1370	—	R. Tocache, Tocache
438)	OAZ4O	1370	1	R. Cosmos, Huacho
212)	OAX2W	°1380	0.3	R. Atahualpa, Cajamarca: 24h
554)	OBX5C	1380	1	R. El Pueblo, Ica: 18 h
122)	OBX2V	1380	1	Sistema Radial, Ferreñafe
439)	OCY4U	1380	1	R. ABC, Chosica: 1000-0300
619)	OAX6O	1380	5	R. San Martín, Arequipa: 0900-0330
759)	OCX7U	1390	1	R. Cultura, Yunguyo
250)	—	°1400	—	R. La Merced, Tongod
760)	—	1400	—	R. Constelación, Puno: 1000-0200
245)	—	1400	0.8	R. Layzón, Cajamarca: 1100-0300
354)	—	1400	2	R. Chimbote, Chimbote
285)	OAX2O	1400	1	R. Perú, Chepén

476)	OAZ4P	°1400	0.5	R. San Juan, Tarma: 1100-0300
931)	—	°1400	0.2	R. Juanjuí, Juanjuí
932)	OBX9A	1400	1.5	R. Palmera, Bellavista (r. 1398.2)
948)	OBX9E	1400	1	R. El Dorado, Villa Picota
440)	OBX4W	1400	2.5	Callao Super R., Callao: 1100-0500
620)	OAX6J	°1400	3	R. Landa, Arequipa
716)	OAX7I	°1400	3	R. La Hora, Cusco: 1000-0300 (r. 1390v)
729)	—	°1400	—	Ondas del Suroriente, Quillabamba.
347)	OBZ3F	1400	1	R. Willaq, Huaraz
116)	OAX1N	1410	1	R. Ideal, Lambayeque: 24h
286)	OAX2Y	1410	2	R. Heróica, Trujillo: 24h
441)	OBZ4V	1410	1	R. Universal, Huacho
761)	OCX7E	1420	1	R. Qollasuyo, Juliaca: 1000-0400
805)	OAX8Z	°1420	1	R. Oriente, Yurimaguas: 0900-0300 (Sat/Sun 0400)
501)	OBX5H	1420	1	R. Apurímac, Abancay: 1100-0300
429)	OBZ4G	1420	1	R. San Isidro, Lima: 1000-0600 (Sun 0400)
291)	—	1430	—	R. El Puerto, Pacasmayo
287)	OBX2T	1430	1	R. Santa Bárbara, Ascope
355)	—	1430	1	R. Chavín, Chimbote
717)	—	°1430	1	R. Espinar, Espinar
477)	OAZ4V	1430	1.5	R. Universal, Huancayo
221)	—	°1430	—	R. Frecuencia Líder, Bambamarca
621)	OAX6R	1440	1	R. El Tiempo, Arequipa: 1000-0400
117)	OBX1T	v1440	2	R. Cooperativa Tumán, Chiclayo: 1100-0300 (r. 1440.3)
762)	—	1440	—	R. San Francisco, Yunguyo
443)	OAX4K	1444	5	R. Imperial 2, Lima: 0900-0530 (n 1440)
444)	OBX4K	1450	1	R. Fortaleza, Barranca
718)	OCX7W	1450	1	R. Santa Rosa, Quillabamba
622)	OAX6C	1460	1	R. Bahía, Mollendo
141)	OAX1V	1460	1	R. Sullana "La Voz de Chira", Sullana (r. 1465)
214)	—	1460	0.5	Ondas del Perú, Cajamarca: 1000-0300
935)	OBX9B	°1460	1	R. Frec. Popular, Rioja (r: 1600)
364)	OAX3C	°1460	0.5	R. Estrella, Huánuco
763)	OAX7W	°1460	1	R. El Sol de los Andes, Juliaca
555)	OAX5K	1460	2.5	R. Internacional, Pisco
288)	OBX2Y	1460	1	R. Estelar, Guadalupe
478)	OCY4I	1460	0.5	R. Imperial, Junín
348)	OBX3A	1460	1	R. Chimú (r: RPP 730), Chimbote
452)	—	1470	25	Antena 1, Lima
213)	OBX2U	°1470	1	R. Ilucán, Cutervo: 24h
479)	OCX4D	°1470	1	R. Huancayo, Huancayo: 24h
933)	OAX9A	°1470	5	R. Imagen, Tarapoto: 1000-0500 (n 1500)
654)	OAX6M	°1470	1	R. Tacna, Tacna: 0845-0455 (Sun 0300)
719)	OAX7G	°1470	1	R. Cusco: 1000-0300
142)	—	1480	1	R. Frecuencia Popular, Piura: 1100-0500
289)	OCX2C	1480	0.6	R. San Pedro, Virú
118)	OAX1L	1490	1	R. Imperio "R. 1460", Chiclayo
475)	OAZ4F	°1490	1	R. La Oroya, La Oroya: 1000-0500 (n. 1460)
556)	OAX5N	1490	0.5	R. Nasca, Nasca
623)	OAX6Q	1490	1.2	R. Surperuana-Fidelidad," Arequipa: 1100-0500
806)	OAX8J	°1490	1	R. Atlántida, Iquitos
502)	OAX5L	°1490	0.25	R. Abancay, Abancay
764)	OCX7P	1500	2	R. Em. Frontera, Puno: 0900-0500
290)	—	1500	—	R. El Porvenir, El Porvenir
305)	OBX2Q	1500	1	R. JCL, Ascope: 0845-0300
766)	OAX7F	°1500	1	R. Ayaviri: 1000-0300 (n. 1550)
512)	OBX5J	°1500	1	R. San Cristóbal de Huamanga, Ayacucho: 1100-0400
445)	OBX4I	°1500	10	R. Santa Rosa, Lima: 1000-0400
624)	OBX6V	1500	1	Nueva R. 2000, Orcopampa
480)	OCX4F	1500	1	R. Mundo, Huancayo (n 1560)
218)	OCX2Q	°1510	2	R. Inca, Baños del Inca: 0900-0100
446)	OCX4B	*1510	0.3	R. Fessa, Cañete
481)	OCX4J	°1510	3	R. Tarma, Tarma: 1000-0500
557)	OAX5F	1510	1	R. La Voz de Nasca, Nasca: 24h
720)	OBX7P	1510	1	R. Red, Huánchac
934)	OAX9X	°1520	0.25	R. Visión, Juanjuí (r. 1545)
482)	OAZ4S	1520	0.5	R. Minería, La Oroya
119)	OAX1C	1520	1	R. Délcar, Chiclayo
292)	OCX2G	1520	1.2	R. Santiago, Santiago de Chuco: 0900-0500
447)	OBX4D	*1520	5	R. Reloj, Lima
483)	OBZ4S	1530	1	R. 15-50, "La Nueva Voz del Centro", Huancayo: 1000-0600
356)	—	1530	—	Antena 1, Chimbote
215)	—	°1530	—	R. San Ignacio, San Ignacio
216)	—	1530	—	R. Oriental, Jaén
721)	OAZ7F	1530	0.5	R. Confraternidad, Yauri
494)	OBX4N	1540	1	R. Corporación, Cerro de Pasco: 18h
293)	OBX2B	1540	1	R. Star, Trujillo: 1100-
448)	OBZ4U	1540	1	R. Barranca, Barranca: 1100-0500
153)	OBX1B	1540	1	La Voz de la Frontera, Tumbes
722)	OCX7V	1540	1	Onda Tropical, Cusco
449)	OBX4P	1550	1	R. Independencia, Lima: 1000-0530
765)	OCX7A	1550	1	R. Cultura, Puno
120)	OAX1O	1550	1	R. Em. Superior, Monsefú: 1000-0500
294)	OAX2X	v1560	1	R. Primavera, Trujillo (r. 1562)
723)	—	1560	—	R. Municipal de Cusco, Cusco
217)	OBX2J	°1560	1.5	R. Nuevo Continente, Cajamarca: 0900-0400
525)	OAX5M	°1580	2	R. Villa Rica, Huancavelica
121)	OBX1M	°1580	1	R. Naylamp, Lambayeque: 0900-0500
295)	OBX2H	°1580	1	R. Tayabamba, Tayabamba: 1000-0500 (Sun 1100-1500)
724)	OBX7Q	1580	1	R. Triunfo (r: RPP 730), Cusco
513)	—	°1580	1	R. Cultural Amauta, Huanta: 1000-0200
450)	OAZ4Z	1590	5	R. Agricultura, Lima: 1100-0500
514)	OBX5C	1590	1	R. Esmeralda, Huanta
246)	—	1598	1	R. La Luz de la Vida, Cajamarca: 1000-0300 (n. 1600)
625)	—	1600	—	R. Camaná, Camaná
307)	OBX2Q	1600	1	R. Popular, Chepén

Shortwaves

430)	OBZ4E	3204.6	1	R. Paramonga, Barranca
763)	OBX7D	3230.2	0.4	R. El Sol de los Andes, Juliaca: 0930-0300
761)	OCX7C	3250	1	R. Qollasuyo, Juliaca: 0900-0300
492)	OAZ4I	3260	1	R. La Voz de Oxapampa, Oxapampa
515)	OAZ5D	3280.2	1	Estación Wari, Ayacucho: 1000-0400
900)	OAX9U	*3310	1	R. Bagua, Bagua: 1000-0300
363)	OAX3Q	3330.3	5	R. Ondas del Huallaga, Huallaga: 1100-0500 (v. ±0.8)
490)	OBZ4B	3339.1v	1	R. Altura, Cerro de Pasco: 1100-0500 (v. up to 3339.9)
903)	—	*3464	—	R. Reina de la Selva, Chachapoyas: 0930-0400
306)	—	v3502.1	—	R. Chilia, Chilia: 2300-0200
451)	—	3800.3	1	R. Oyón, Oyón: 0930-1300, 2100-0100
495)	—	*3810	—	R. Centenario, Cerro de Pasco
349)	—	r3981	—	R. Caraz, Caraz
910)	—	4002.2	—	R. San Nicolás, Rodriguez de Mendoza: 2330-0200
935)	—	4012.1	0.6	R. Frecuencia Popular, Rioja: 1030-0400 (Sun 1200-0100)
925)	—	4039.1	1	R. Marginal, Tocache: 1000-0230 (Sun 0200)
121)	—	v4108.5	0.5	R. Naylamp, Lambayeque
220)	—	*4141	—	R. Gran Pajatén, Celendín
248)	—	v4180	0.1	R. Membrillar, Cascas (n. 4190, v. ±5kHz)
218)	—	4299.6	1.3	R. Inca, Baños del Inca: 0900-0300
221)	—	v4419	0.85	R. Frecuencia Líder, Bambamarca: 1000-1400, 2300-0430
222)	—	4461.8	0.18	R. Nor Andina, Celendín: 2300-0400
239)	—	4485	—	R. Frecuencia VH, Celendín
247)	—	4495	—	R. San Mateo, Contumazá: 1100-1300, 2300-0400
122)	—	4509.3	—	R. Horizonte, Chiclayo
766)	OBX7F	4606.5	1	R. Ayaviri "La Voz de Melgar", Ayaviri: 1100-0100 (n. 5035)

295)	OAZ2D	4629.6	1	R. Tayabamba: 1000-1600, 2330-0400
207)	OCX2I	4700.1	1	R. Waira, Chota: 1200-0500 (n 2475)
938)	—	4705	0.25	Estación Láser, Rioja: 1000-2400v
350)	—	v4731.5	1	R. San Juan, Caraz: 1300-2300v (n 4780)
516)	OAZ5B	v4752	0.5	R. Huanta 2000, Huanta: 1000-1400, 2200-0400 (n 5030)
362)	OAX3P	v4759	0.5	R. Tingo María: 0900-1200, 2300-0400
481)	OCX4E	4774.9	1	R. Tarma, Tarma: 1000-1400, 2000-0500
468)	OCX4Z	4784.9	1	R. COSAT, Satipo: 1000-0200
503)	—	4785.4	—	R. Chincheros, Chincheros: 1000-1400, 2300-0300
806)	OAX8F	v4790	3	R. Atlántida, Iquitos: 0900-0500 (Sun 1100-0400)
750)	OBX7C	4800.7	1.5	R. Onda Azul, Puno: 0900-0300 (n 4800)
525)	OAX5X	4804.5	0.8	R. Villa Rica, Huancavelica: 0940-0315v (n. 4885)
924)	OAX9R	4810.2	3	R. San Martín, Tarapoto: 0930-0400
212)	OAZ2F	4820.8	1	R. Atahualpa, Cajamarca: 0900-0500 (n 4820)
801)	OAX8S	4824.5	10	La Voz de la Selva, Iquitos: 1000-0300
715)	OAX7T	4826.3	1.5	R. Sicuani, "La Voz de Canchis", Sicuani: 1000-0300 (n 4835)
519)	—	r4830	—	Estación Sivia, Huanta
200)	OCX2E	4834.8	1	R. Marañón, Jaén: 1100 (Sun 1200)-0100
500)	OBX5E	4840	2	R. Andahuaylas: 1000-0300
706)	OAZ7A	*4850	2.5	R. Amauta, Cusco (F.Pl.)
904)	OAX9W	4850	1	R. Comercial Educativa, La Peca: 1300-0300
716)	OAX7J	4860	1	R. La Hora, Cusco: 1000-0100
460)	OBZ4Z	*4860	1	R. Chinchaycocha, Junín: 0900-0400
531)	—	4870.7	—	R. Cosmos, Lircay
822)	OAZ8H	4881.5	1	R. Nuevo Mundo, Pucallpa: 24h (n 4880)
517)	OAZ5C	v4890	0.5	Dif. R. Huanta, Huanta: 1000-0200
231)	OAZ2B	4890.3	1	R. Chota: 1200-0300v
466)	OAZ4T	4895	0.4	R. Chanchamayo, La Merced: 1000-0500 (Sun 1100-0300)
475)	OAZ4G	4905	—	R. La Oroya, La Oroya
725)	OBX7V	4908	5	R. Tawantinsuyo, Cusco
271)	OAX2K	4910.6	1	R. Libertad de Trujillo, Trujillo: 6h./week
403)	OAZ4N	4914.5	10	R. Cora del Perú, Lima: 0930-0500
920)	OAZ9A	v4920	1	R. Central "La Voz del Huallaga Central", Bellavista: 1000-0400
757)	OBX7H	4924.1	1	R. Ondas del Titicaca, Puno: 0930-0300
527)	OAZ5A	4925.8	0.5	R. Cobriza 2000, San Pedro de Coris: 1000-0300
929)	OAX9E	v4935	3	R. Tropical, Tarapoto: 0945-0400
476)	OAZ4R	4940	0.5	R. San Juan, Tarma
824)	OAX8Y	4950	0.5	R. Ucayali, Contamana
740)	OBX7J	4950.1	5	R. Madre de Dios, Puerto Maldonado: 1000-0200
513)	OAX5S	4955	5	R. Cultural Amauta, Huanta: 1000-0200
472)	OCX4T	4960.3	1	R. La Merced, La Merced: irr
714)	OAZ7B	v4966	1	R. San Miguel, Cusco: 0930-0300
933)	OAX9M	4970.4	1	R. Imagen, Tarapoto
405)	OCX4H	4974.9	4	R. del Pacífico, Lima: 2300-0430
343)	OAZ3B	4990.9	5	R. Ancash, Huaraz: 0900-0500 (n 4990)
473)	OAZ4G	4995.8	2	R. Andina, Huancayo: 0900-0500v
902)	—	5008.7	5	R. Horizonte, Chachapoyas: 1100-0115
758)	OAX7Z	n5015	0.5	R. Juliaca, Juliaca
927)	OAZ9B	5015.5	0.7	R. Estación Tarapoto, Tarapoto: 1000-0400
512)	OAX5Y	5020	1	R. San Cristóbal de Huamanga, Ayacucho: 1100-1300, 1700-2100
710)	OAX7Q	5025	5	R. Quillabamba, Quillabamba: 1000-0200
296)	OAZ2A	5030.6	5	R. Los Andes, Huamachuco

469)	OCY4Y	5039.3	1	R. Libertad "RLJ," Junín: 1000-0400
713)	OBX7O	5055	—	R. Onda Imperial, Cusco. irr.
923)	OAX9L	5046	3	R. Rioja, Rioja: (n. 5045)
518)	—	5050	0.5	R. Municipal, Cangallo: 1100-1400, 2330-0300
528)	OBX5G	5059	0.5	R. Lircay, Lircay: -0300
730)	—	5068.7	—	Ondas del Suroriente, Quillabamba: -0100
905)	OAZ9C	5081	0.5	R. Huayabamba, Huayabamba
702)	OAZ7D	5082.6	0.5	R. Mundo, Cusco: 0900-0100 (n 5150)
243)	—	5085	0.1	R. Celendín: 2300-0500
804)	OAX8V	v5097.3	1	R. Eco, Iquitos (n. 5010): 0900-0400
362)	—	v5147.3	—	"RQ", Tingo María
217)	OAZ2E	*5190	2	R. Nuevo Continente, Cajamarca
224)	—	5186.8	0.5	R. Onda Popular, Bambamarca: 2330-0400v
626)	—	5235	0.3	R. San Antonio de Padua, Arequipa: Rosary at 1000, 1700 and 2300
246)	—	r5265	—	R. Estación "A," Sócota, Cutervo
909)	—	5271.8	—	R. Nororiental, Rodríguez de Mendoza: -0300
143)	—	5277.6	1.5	R. Grau, Huancabamba
529)	—	5323.5	0.5	R. Origen, Huancavelica
225)	—	5340	0.1	R. La Voz de los Andes, San Miguel: 1145-0300v
351)	—	5344	0.5	R. Pomabamba (n. 5370)
934)	—	5359.8	0.1	R. Visión, Juanjuí
949)	—	v5419.1	—	R. Sonorama, Saposoa
937)	—	5445.2	—	R. Alto Huallaga, Uchiza: 1100-2300
226)	—	v5525	—	R. Comercial, Chota: 2300-0300v (n 5600)
227)	—	5538.7	—	Frecuencia Modular "la radio", Celendín: -0400 (n. 5540)
213)	OAX2G	5619.3	0.1	R. Ilucán, Cutervo: Sun. 1100-1400
228)	—	v5645	0.1	R. Bambamarca: 1400-1700v
209)	—	5660.5	0.6	R. La Voz de Cutervo, Cutervo: 1100 (Sun 1000)-0100v
215)	—	5700	0.1	R. San Ignacio, San Ignacio: 2300-0315
939)	—	5800	—	R. Nuevo Cajamarca, Nueva Cajamarca: 1145-1500, 1800-0200 (Sun 0100)
754)	OAX7L	5816	1	R. La Voz del Altiplano, Puno: 1000-0400
230)	—	5825	—	R. Comercial Cosmos, Bambamarca: 2300-0300
366)	—	v5833	—	R. Uno, Huánuco
767)	—	r5835.3	—	R. Tinajani, Ayaviri: 2100-0300
606)	OAX6Z	5895	0.1	R. Hispana, Arequipa: Sat/Sun 0900-0300
609)	OAX6A	5951.3	1	R. Arequipa, Arequipa: 1000-0400
479)	OAX4P	5955.3	1	R. Huancayo, Huancayo: 24h
416)	OBX4Q	5969.7	10	R. El Sol, Lima: irr. 1200-1300
409)	OBX4M	*5980	5	R. Programas del Perú, Lima
613)	OBX6L	5995.2	1	R. Melodía, Arequipa: 24h
420)	OAX4W	6010.9	2.5	R. América, Lima: irr. (n 6010) (r. 6042)
411)	OAX4Q	*6020	5	R. Victoria, Lima: 1100-1400, 2300-0200
620)	—	6033.1	—	R. Landa, Arequipa
445)	OCY4H	6045.5	3	R. Santa Rosa, Lima: 1000-0400
601)	OAX6D	6055.3	2	R. Continental, Arequipa: 0900-0600
365)	OAX3L	°6061	3	R. JSV, Huánuco: 1100-0100 (Sat & Sun 1200-2300) (n 6060)
709)	OAZ7C	6093.6	2	R. Universal, Cusco
415)	OBZ4O	6115	10	R. Unión, Lima: 0900-0600 (occ. 24h)
491)	OBZ4P	6140	2.1	R. Huayllay, Huayllay: 1000-0230v
611)	OBX6B	v6140.2	1	R. Concordia, Arequipa: 1000-2200 (Sun 1000-1700) (n. 6150)
821)	OAX8Q	v6155	0.5	R. Pucallpa, Pucallpa: 1030 (Sun 1130)-0300 (Sun 0500)
725)	OAX7C	6173.8	5	R. Tawantinsuyo, Cusco: 0930-0200
805)	OAX8I	6187.8	3	R. Oriente, Yurimaguas: 0900-0300 (Sat/Sun 0400) (n 6190)
719)	OAX7A	6191.7	1	R. Cusco "La Voz de la Capital Arqueológica de América," Cusco: 1000-0300 (n 6195)

511)	OAX5U	6200.6	2	R. La Voz de Huamanga, Ayacucho: 0930-0300 (n 6070)
726)	—	v6243.3	0.12	R. Calca, Calca: 1000-1500
931)	—	6261.1	—	R. Juanjuí, Juanjuí: 2300-0100v
137)	—	v.6281.7	1.5	R. Huancabamba, Huancabamba: 1030-2300v (n 6290)
627)	—	6304.7	—	R. Acarí, Acarí: 2300-0300
502)	OAX5Q	6325.8	0.25	R. Abancay
926)	—	6329.6	0.8	R. Estación "C", Moyobamba: 1030-0200
229)	—	6339.5	—	R. San Miguel Arcángel, San Miguel
940)	—	6363	0.2	R. Huallaga, Saposoa: 1130-1700, 2330-0400
768)	—	6390.9	—	R. San Miguel, San Miguel de llave
717)	—	6429	1	R. Espinar, Espinar: 1030-1330, 2300-0300v (n 6425)
654)	—	v6571.5	0.17	R. Tacna, Tacna: Mon-Fri 1800-1900v
520)	—	v6585	—	R. Tradición, Huanta
729)	—	v6587	—	R. Santa Fé, Santa Fé de Urubamba
249)	—	6628.5	0.5	R. La Voz de San Antonio, Bambamarca
232)	—	*6655	—	R. Santa Cruz, Santa Cruz
941)	—	r6666	0.01	R. Jepelacio, Jepelacio: 2300-0400
308)	—	6670.2	—	R. Santa Mónica, Santiago de Chuco
201)	OAX2R	v6690.5	1	R. Cutervo, Cutervo: 1400-0100 (Sun 1000-1500)
233)	OAX2L	v6724.5	0.15	R. Satélite, Santa Cruz: 2300-0330 (n 4780)
250)	—	6754.7	—	R. La Merced, Tongod
234)	—	6781.7	—	R. Tigre, Sorochuco: 0000-0430
942)	—	6803.2	—	Ondas del Mayo, Nueva Cajamarca
235)	OAX1K	v6815	0.1	R. Universo, Pandalle: Sun 2200-0300v
627)	—	6851.2	—	R. Cultural, Cailloma
144)	—	6895.2	—	R. Sensación, Huancabamba: 1100-0030 (n. 6890)
767)	—	v7032	—	R. Azángaro, Azángaro (v. +20kHz) Progreso
241)	—	8065	—	R. El Paraíso de los Andes: 0100-0430
943)	—	v8515	0.2	R. Amistad, Soritor: 2100-0230
944)	—	v8930	0.15	R. Continente, Juanjuí: 1100-0300 (n 8900)
138)	OAX1T	9462	3	R. Grau, Piura
654)	OAX6H	9486	0.2	R. Tacna, Tacna: Sun 1000-2400
654)	OAX6H	9505	0.25	R. Tacna: 0900-0100
420)	OAX4W	v9510	5	R. América, Lima: irr. 1000-1800
414)	OAX4J	9520	5	R. La Crónica, Lima: 1100-0500
611)	OBZ6A	9580	1	R. Concordia, Arequipa: 1000-2200 (Sun 1700)
901)	OAX9G	9655	1	R. Nor Peruana, Chachapoyas: 1000-0200v
405)	OAZ4L	9675	5	R. del Pacífico, Lima: 1100-2300
601)	OAX6E	9685	2	R. Continental, Arequipa
820)	OAX8W	9755	1.2	R. Sideral, Pucallpa

Addresses and other information:
(Names of **departamentos** listed under each region (cf. call-sign numbers) should be added to address info, main capitals excluded).

REGION 1

Lambayeque
111) Urb. Patazoa Mz. 41, Chiclayo
112) Ap. 490, Chiclayo — Dir: H. Hubert Limo Mendoza.
113) Cas. 448, Chiclayo — Dir: Jorge E. Alfaro R.
114) Cas. 210, Chiclayo — Pte: Carlos Montjoy S.
115) Av. Balta 413, Chiclayo
116) San Martín 440, Lambayeque
117) Ap. 77, Chiclayo — Dir: Víctor Sandoval G.
118) Jr. Arica 972, Chiclayo — Gte: Mesías Sánchez F.
119) Cas. 9, Chiclayo — Dir: Ricardo Montjoy O.
120) Av. Mcal. Castilla 839, Monsefú, Chiclayo — Gte: Víctor Yaipen U. **F.Pl:** 3 kW
121) Av. Huamachuco 1080, Lambayeque — Dir: Juan J. Grández V.
122) Jr. Próceres 745, Urb. Latina, Chiclayo.

Piura
131) Cas. 165, Piura — Dir: Rolando Ojeda S. **Fax:** +51 (74) 323115.
132) Cas. 254, Piura — Dir: Raquel la Madrid J.
133) Cayetano Heredia 498, Piura
134) Cas. 55, Paita — Dir: César Minaya V.
135) Esq. Grau y Piura 1084, Sullana
136) Loreto 114, Piura — Dir: Geraldo Garrido P.
137) Unión 610, Huancabamba — Dir: César Colunche B.
138) Callao 557, Piura — DG: Walter Ruiz Zelada
139) Cas. 74, Talara — Gte: Max H. Cornejo P.
140) Moquegua 458, Piura — Dir: Cecilia Cruzado Solari
141) Jr. Sucre 556, Sullana — Dir: Jorge Geller G.
142) Jr. Blas Atienza, Galería Anita, Piura — Gte: Segundo N. Rojas E.
143) Grau 403, Huancabamba — Dir: Ignacio León Alva
144) Av. Ramón Castilla 331, Huancabamba — Dir: Justo E. Carrasco M.

Tumbes
151) Pza. Alipio Rosales, Tumbes — Admor: C. Fernández P.
152) Bolívar 111, Tumbes
153) Piura 1010, Tumbes

REGION 2

Cajamarca
200) Ap. 50, Jaén (via Chiclayo) — DG: Ubaldo Ramos Cismeros, S.J. Fax: (51) 74 731245.
201) Jr. Comercio 725, Cutervo
202) Benjamín Gálvez 118, Contumazá
203) Ap. 245, Cajamarca — Dir: Rosa Estela de Hoyos
204) Ap. 61, Cajabamba — Own: Salustiano Asmad Vergara
206) Jr. Amalia Puga 673, Cajamarca — Dir: Eduardo Lezama
207) Jr. Gregorio Malca 141, Chota — Adm: Eladio Cusma G.
208) Jr. 2 de Mayo 484, Cajamarca — Dir: Segundo Lozano F.
209) Jr. Comercio 624, Cutervo — Dir: Julio C. Sánchez C.
210) Jr. José Gálvez 273, Cajamarca
211) Barrio San José, Cajamarca
212) Pza. Bolognesi s/n, Cajamarca — Gte: Luis Cruzado Ossio
213) Jr. Lima 228, Cutervo — Dir: José Gálvez S.
214) Jr. Angamos 889, Cajamarca — Dir: Tito Chávarry Delgado.
215) Jr. El Carmen 618, San Ignacio — Dir: Franklin Hoyos C.
216) Av. Mesones Murón 157, Jaén
217) Ap. 219, Cajamarca — Dir: Eduardo Cabrera U.
218) Cas. 418, Baños del Inca — Dir: Enrique Ocas S.
220) Jr. Amazonas 710, Celendín.
221) Jr. Jorge Chávez 416, Bambamarca — Dir: Valentín Peralta D.
222) Jr. Pardo 579, Celendín — Dir: Roberto Alcántara G.
224) Alfonso Ugarte 463, Bambamarca — DG: Carlos Acuña S.
225) Jr. José Gálvez 689, San Miguel de Pallaques — Gte: José Belaunde Villoslada R.
227) Ayacucho 101, Celendín — Admor: Jorge Chaves V.
228) Bambamarca — Dir: Augusto Arana M.
229) San Miguel de Pallaques
230) Jr. Iván Cnel. Nogueira 362, Bambamarca
231) Ap. 3, Chota — Dir: Mauro Vásquez G.
233) Jr. Cutervo 543, Santa Cruz — Dir: Sabino Llamo Ch.
234) Distrito Sorochuco, Provincia Celendín
235) Jr. 22 de Octubre 1809, Cutervo — Own: Hnos. Lozano Tantaleán
236) Pucará, Jaén, or: via st. 121)
237) Jr. Huánuco 2361, Cajamarca — Gte: E. Machuco
238) Av. Rafael Hoyos 342, Cajamarca
239) Jr. Unión 311 y Pza. de Armas, Celendín — Dir: Fernando Vásquez Castro
241) Niepos, San Miguel
242) Jr. Bolívar 1020, Jaén
243) Jr. 2 de Mayo 501, Celendín — Gte: José Camacho V.
245) Pje. Atahualpa 643, Cajamarca — Dir: Marcos Manay M.
246) Jr. Contamana 315, Cajamarca — Dir: W. Esquivel
247) Av. Benjamín Gálvez 118, Contumazá
248) Jr. San Martín 610, Cascas, Contumazá
249) Jr. Mariscal Sucre 731, Bambamarca — Dir: Valentin Mejía Vásquez
250) Distrito de Tongod, Santa Cruz

La Libertad
260) Av. González Prada 695, piso 12, Trujillo
261) Condorcanqui 1757, La Esperanza
262) Jr. Lima 578, Chepén — Dir: José A. Tejada T.
263) Ap. 186, Trujillo — Dir: Víctor Meléndez S.
264) Casagrande 218, Chocope
265) Pizarro 532, Trujillo — Dir: Carlos Martínez P.

266) Pizarro 574, Of. 205, Trujillo
267) Jr. Trujillo 579, Otuzco
268) Km 548 Panamericana Norte, Urb. Larrea, Trujillo
269) Bolívar 937, Huamachucho — Dir: Manuel D. Gil G. **F.Pl:** 3980kHz, at Trujillo
270) Alfonso Ugarte s/n, Virú
271) Cas. 712, Trujillo — Dir: Carlos Burmester B.
272) Jr. Balta 252, Huamachuco
274) Jr. Atahualpa 795, Chepén — Dir: Jorge Acevedo Rojas
275) Ap. 1215, Trujillo — Pte: Carmela Castro de Manrique.
276) Grau 439, Of. 213, Trujillo — DG: Jorge L. Mantilla O.
277) Jr. Progreso 121, Quiruvilca — Dir: Walter Díaz Ramos
278) Cas. 1029, Trujillo — Pte: Luis F. Cook L.
279) Victoria 229, Guadalupe
280) Bolívar 780, Trujillo — Dir: Pedro Avalos G.
281) San Antonio 880, Otuzco
282) Urb. Miguel Grau, Casa Grande, Chocope
283) Jr. Chocope 684, Paiján
284) Camal 13, Cartavio Ascope
285) Lima 392, Chepén
286) Gamarra 713, Trujillo — Dir: Carlos Casanova B.
288) Jr. Unión 669, Guadalupe, Pcia. Pacasmayo
289) Grau s/n, Virú
290) José E. Sánchez Carrión 500, Of. 534, El Porvenir
292) Jr. Bolognesi 1247, Santiago de Chuco — Dir: Saturnino Asmad Vergara
293) Pasaje San Luis 149, Trujillo
294) Bolívar 679, Trujillo
295) Bolívar 565, Tayabamba — DP: Israel Alva L.
296) Pje. Damián Nicolau s/n, Huamachuco — Dir: Mons. Sebastián Ramis Torrens
301) San Ignacio 601, Laredo
302) Jr. Bolívar 627, of 105, Trujillo
303) Av. del Ejército 717, Trujillo
306) Jr. Miraflores y San Cristóbal, Chilia, prov. Pataz
307) Pacasmayo 136, Chepén
308) Mcal. Cáceres 453, Santiago de Chuco — Gte: Faustino L. Rodríguez Rebaza.

REGION 3
Ancash
333) Cas. 117, Chimbote — Dir: Darío C. Moreno
334) Panam. Nte. Km 440 (or: Leoncio Prado 660), Chimbote
335) Ap. 294, Chimbote — Gte: Humberto Luna L.
336) Barrionuevo 623, Casma — Gte: Fausto León R.
337) Aguirre y Pizarro, Chimbote
338) Urb. El Trapecio 2a. etapa, Mz G, Lote 18, Chimbote
339) Ap. 240, Huaraz — Dir: Antonio Camones Blas
340) Luzuriaga 1239, Huaraz
341) Av. Alfonso Ugarte 311, Chimbote — Dir: F. Cruz M.
342) Jr. José de Sucre 1124, Huaraz — Dir: Máximo Janampa
343) Ap. 210 or 221, Huaraz — Dir: Armando Moreno Romero
344) Av. Luzuriaga 646, Huaraz — Dir: Francisco Sotelo López
345) San Pedro 246, Chimbote
346) Unión 1100, Miramar Bajo, Chimbote
347) Ap. 348, Huaraz — Dir: Lorgio Sánchez G. **F.Pl:** sw (in Quechua)
348) Barrio 1 de Mayo, Chimbote
350) Jr. Pumacahua 528, Caraz — Dir: Víctor Morales G.
351) Jr. José Olaya 917 of 200, Chimbote
Huánuco
360) Av. Raimondi 432, Tingo María
361) Malecón Huallaga 1038, Aucayacu, José Crespo y Castillo
362) Ap. 25, Tingo María — DG: Ricardo Abad Vásquez
363) Ap. 343, Huánuco — Gte: Flaviano Llanos M. Prgr. in Quechua 1800-1900, 2300-2400
364) Ap. 180, Huánuco — Gte: Jorge Díaz S.
365) Ap. 250, Huánuco — Gte: Jesús A. Santa Cruz V.
366) Sector 5, San Luis, Huánuco

REGION 4
Lima
400) Bernardo Alcedo 375, Lima 14 — Dir: Augusto S. Irei S.
401) Jr. Camaná 615, Of. 605, Lima 1 — Dir: Enrique Delgado M.
402) Av. Estados Unidos 327, Urb. Huaquillay, Comas, Lima 7 — Dir: Julio Saldaña G.
403) Paseo de la República 144, Centro Cívico, Of. 5, Lima 1 — Dir: Juan Ramírez Lazo
404) Cas. 55, San Vicente de Cañete
405) Ap. 4236, Lima 1 — Gte: Pedro Ferreyra G.
406) Av. Javier Prado Este 309, Lima 27 — Gte: Alberto Carreño B.
407) Av. Manco Cápac 333, La Victoria, Lima 33 — Gte: Gonzalo Vallarino E.
408) Av. de la Marina 2099, Lima 32 — Gte: Martha de Láinez
409) Jr. Alejandro Tirado 217, Santa Beatriz, Lima 1. **Te:** +51 (14) 246042 — Gte: Rolando Estremadoyro R. **F.Pl:** 50kW
410) Parque Fátima 403, Lima 9 — DG: Abraham Zavala Falcón. Fax: 14 675557
411) Cas. 3639, Lima 1 — Gte: Santiago Salcedo G.
412) Ap. 8, Huaral — Gte: Santiago Kong M.
413) Av. Salaverry 1082, Lima 11 — DG: Humberto Panfichi
414) Av. Petit Thouars 447, Lima 1 — Gte: (850kHz) David Odría; (1320kHz) Juan Silva Vidaurre
415) Ap. 6205, Lima — DG: Miguel Palacios R.
416) Av. Uruguay 355, 7 piso, Lima — Gte: Emilio García Lara
417) Av. Arenales 431, Of. 404, Lima 1
418) Av. Mariano Carranza 126, Santa Beatriz, Lima 1 — DG: Mauricio Alcántara & Fernando Torres D.
419) Ignacio Merino 230, Lima 18 — DG: Delia Ll. A. de Caballero
420) Leon Velarde 1140, Lima 14, or: Cas. 1192, Lima 1 — Pres: Luis Umbert D.
421) La Torre de Lima 3201, Centro Cívico, Lima 1 — DG: Pedro Tello B.
425) 2 de Mayo 573, San Vicente de Cañete — Own: Roberto Pachas
427) Jr. Manuel A. Fuentes 860, San Isidro, Lima 27 — Gte: Oscar Artacho S.
428) Ap. 2355, Lima 21 — DG: Augusto Lanatta B.
429) Av. Arequipa 4130, Of. 1001, Miraflores, Lima — Gte: Julio Schiappa-Petra C.
430) Av. Francisco Vidal 921, Paramonga, Prov. Barranca — Gte: Francisco Niquín Quezada
431) Ap. 156, Huacho — DG: Jorge H. Bisso J.
432) Av. Manco Cápac 495, Lima 18 — DG: Ing. Ricardo Palma Michelsen
433) Av. Echenique 140, Huacho — Gte: Angel Flores I.
434) Mariano Carranza 754, Santa Beatriz, Lima — DG: Humberto Maldonado B.
436) Cas. 66, Huaral.
438) Av. Grau 538, Huacho — Gte: M. Arakaki O.
439) Cas. 40, Chosica, Lima 15 — Gte: Luis Bordo Rosell
440) Av. Alte. Miguel Grau 487, Callao 1 — DG: Bruno Espósito M.
441) Bolognesi 163, Huacho — DG: César H. Minaya
443) Pq. Industrial, Lote U, Mz. 5, Parcela 1, Villa El Salvador, Lima — Gte: Aníbal Torres A.
444) Alfonso Ugarte 153, Barranca — Dir: Santiago Jau G.
445) Ap. 4451, Lima 1 — Gte: R.P. Juan Sokolich A.
446) Av. Mariscal Benavides 614, San Vicente de Cañete
447) Cádiz 285, Lima 27 — Gte: Hernán Balderrama. **F.Pl:** 10 kW
448) Ap. 23, Barranca — Gte: Wesley Portnoy R.
449) Tiahuanaco 107, Independencia, Lima 28
450) Ap. 1200, Lima 100 — DG: Bartolomé Dextre F.
451) Av. Huánuco 144, Oyón
452) Paseo de la República 3662, Lima

FM in Lima.
88.3 Telestereo 88 — 416) 88.9 Sol y Armonía — 409) 89.7 — 452) 90.3 — 91.1 R. San Borja — 407) 91.9 — 92.5 Studio 92 — 410) 93.1 R. Ritmo — 420) 94.2 — 95.1 R. "A" — 432) 96.1 — 97.1 R. Omega — 428) 98.1 — 98.1 R. Doble 9 — 100.1 Stereolima — 418) 101.1 — 405) 102.1 — 415) 103.1 — 407) 104.7 — 429) 105.5 — 410) 106.3 Radiomar Plus — 107.1 Turbo — 107.9 R. 1.

Junín
460) Ap. 16, Junín — Dir: Julio O. Baldeón M.
461) Esq. Ancash e Ica s/n, Huancayo
462) Cas. 668, Huancayo — DG: Ramón Chang P.
463) Jr. Junín 1008, Jauja — Gte: Jonás Barreto P.
464) Cas. 560, Huancayo — DG: Eduardo Sanjinés A.
465) Real 270, El Tambo, Huancayo — DG: Guido Barreto
466) Ap. 96, La Merced — Gte: Augusto Espejo P.
468) Jr. Manuel Prado 459, Satipo — Dir: Fermín Zúñiga T.
469) Ap. 2, Junín — Dir: Mauro Chaccha G.
470) Av. Colones Fundadores 396, Satipo
471) Jr. Moquegua 648, Tarma — Own: César Anaya Chagua
472) Junín 171, La Merced — Dir: Teodoro Fidehl F.
473) Jr. Real 175, Chilca, Huancayo — Dir: Fernando Cruz M.
474) Jr. Junín 843, Jauja
475) Calle Lima 190, 3er piso, La Oroya — Dir: Jacinto Figueroa Y.
476) Ap. 214, Tarma — Admor: César E. Zevallos L.

477) Jr. Mariátegui 699, El Tambo, Huancayo
478) Jr. Bolívar 481, Junín — DG: Aníbal Torres A.
479) Jr. Real 517, Huancayo — Gte: Nicolás Bernal R.
480) Jr. Real 455, Huancayo — Gte: Max Guannay G.
481) Cas. 167, Tarma — DG: Mario Monteverde P.
482) Darío 425, La Oroya — Gte: Emilio Navarro C.
483) Ap. 230, Huancayo — Dir: Horacio Gago E. **F.Pl:** react. SW 4800 kHz

Pasco

490) Ap. 140, Cerro de Pasco — Gte: O. de la Cruz Vásquez
491) Huaral 129, Huayllay
492) Jr. Mullenbruck 469, Oxapampa — Own: Pascual Villafranca G. — Eng. & German 1800-1830.
494) Ap. 19, Cerro de Pasco — Gte: Nicolás Herrera A.
495) Jr. Puno 122, Cerro de Pasco

REGION 5

Apurímac

500) Jr. Ayacucho 248 (or: Cas. 92), Andahuaylas — Gte: Daniel Andreu C.
501) Jr. Cuzco 206, Abancay — Gte: Raúl Montesinos E.
502) Jr. Cuzco 300, Abancay — Gte: Luis R. Bedoya C.

Ayacucho

510) 28 de Julio 211, Ayacucho
511) Pasaje El Nazareno 138, Ayacucho — DG: F. Cruz M.
512) Ap. 30, Ayacucho — Dir. Tec: Julio A. Julca Tineo
513) Cas. 24, Huanta — Gte: Abdilio E. González, C.E: Vicente Saico. Quechua at 1000-1215, 2100-2345.
514) Jr. Ayacucho 248, Huanta — Dir: A. Liberatto A.
515) Calle Nazareno 108, Ayacucho — Dir: Walter Muñoz Ynga I. **F.Pl:** MW 1 kW, SW 2kW.
516) Jr. Gervasio Santillana 455, Huanta (or: Jr. Galena 112, Urb. Millotingo, Tambo, Huancayo) — Dir: Ronald Sapaico M.
517) Cas. 43, Huanta — Dir: Clemente Ramos Q.
518) Alcaldía Muncipal, Cangallo — Dir: T. Felipe Fajardo

Huancavelica

525) Virrey Toledo 468 (or: Ap. 92), Huancavelica — Dir: Fernando Cruz M. v/s: Fidel Hilario H.
527) Campamento Cobriza, San Pedro de Coris, Tayacaja (or: alt. listed under 516)) — Dir: Humberto Sapaico S.
528) Jr. Libertad 188, Lircay, Angaraes — v/s: Gilmar Zorrilla Ll.
529) Av. Augusto B. Leguía 126, Huancavelica — Dir: Jesús Acuña Quispe
531) Av. Centenario 225, Bo. Pueblo Nuevo, Lircay.

Ica

540) Alvarez de Arenales 1160, Ica — Dir: Juana F. de Nieri
541) Fray Ramón Rojas 120, Urb. Luren, Ica — DG: Luis A. Umbert
543) Km 308, Panamericana Sur, Los Aquijes, Ica
544) Av. Arenales 1050, Ica — DG: Oscar F. Drago V.
546) Jr. Cajamarca 195, Ica — DG: Julio Huayanca P.
547) Av. Los Incas 130, Puerto de San Juan
548) Jr. Salaverry 410, Ica — DG: Lucio H. Fox M.
549) Grau 520, Ica — DG: Fernando Cruz Mendoza
550) Av. Progreso 400, Grocio Prado, Chincha
551) Jr. San Martín 351, Ica
552) Barrionuevo 142, Pisco
553) Av. Progreso s/n, Chincha Alta
554) Ap. 91, Ica — DG: Rosa Ch. de Flores
555) Cas. 24, Pisco — DG: Wilfredo Villa M.
556) Av. Los Incas 117, Nasca
557) Ap. 57, Nasca — DG: Fernando Cruz R.
558) Ap. 54, Chincha Alta — DG: Augusto Córdova B.

REGION 6

Arequipa

600) Lote 64-5-C, La Joya
601) Cas. 856, Arequipa — DG: J. Antonio Umbert D.
602) Alto de la Luna 334, Cercado, Arequipa — Dir: Victor Apaza B.
603) Francia 120, Urb. Satélite Chico, Paucarpata
604) Victor A. Belaúnde C-8, Umacollo, Arequipa — Dir: Leopoldo Rondón F. **F. Pl:** SW 3kW
605) Pizarro 320, P. 3 (or: Ap. 2580), Arequipa — Dir: Celestino Mamani P.
606) Bolivar 118, Arequipa — Dir: Ing. David Rivera R.
607) Cas. 23, Arequipa — Dir: Alejandra Cornejo
608) Av. Jorge Chávez 201, Arequipa
609) Cas. 657, Arequipa — Dir: Pedro Rufino Zúñiga G.
610) Av. Volívar 317, Arequipa
611) Av. La Paz 512-A, Arequipa — Dir: Victor R. Fernández B.
613) San Camilo 501, Arequipa — Dir: Hermógenes Delgado T. — F.Pl: SW 10kW.
614) Jr. Victor Lira 105, Arequipa
615) Puno 801, Mollendo
616) Av. Lima 481, Camaná
617) Av. Lima 301, Camaná
618) Av. Parra 239, Arequipa
619) Cas. 56, Arequipa — Dir: P. Arnulfo Salinas A., O.P.
620) Sucre 409, Arequipa — Dir: Máximo Landa D.
621) Cas. 451, Arequipa — Dir: Tirso Borja Vizcarra
622) Cas. 128, Mollendo — Gte: J.A. Bernedo Ll.
623) Ap. 2330, Arequipa — DG: Gustavo Quintanilla P.
624) Urb. Buena Ventura AN-7, Orcopampa, Castilla
625) Av. Progreso 932, Miraflores, Arequipa — Dir: Héctor Rozand M.
626) Acari, Caravelí
627) Comunidad Minera de Cailloma, Cailloma

Moquegua

640) Zepita 743, Ilo — Dir: Guillermo Fernández P.
641) Municipalidad de Moquegua
642) Ap. 31, Ilo — Dir: Javier Valencia Alponte. (Ann. OAX70).
643) Los Geranios 105, Agrup. Huáscar, Ilo
644) Ap. 22, Moquegua — Dir: Guillermo Kuong F.

Tacna

650) Cas. 303, Tacna — Dir: Lester E. Kitch
651) Cas. 113, Tacna — Adm: Royce Zeballos R.
652) Cas. 281, Tacna
653) Ap. 115, Tacna — Dir: Rosa L. R. de Liendo
654) Ap. 370, Tacna — DG: Yolanda Vda. de Cáceres. Dir. Tec: Alfonso Cáceres Contreras

REGION 7

Cusco

700) Calle Sucre s/n, Sicuani
701) Matto Turner 340, Santiago
702) Calle Tecte 245, Cusco — DG: Valentín Olivera P.
703) Av. Infancia 527, Cusco — DG: Julio C. Tello A.
704) Av. Tacna 320, Cusco — Own: J. Luis Castro G.
705) Belén 886, Cusco
706) Calle Triunfo 363, Cusco
707) Cuesta del Almirante 209, Cusco
708) Lote E-11, Urb. Bancopata, Cusco
709) Ap. 808, Cusco — Dir: Luis Villasante C.
710) Ap. 76, Quillabamba — Dir: P. Francisco Panera. Quechua at 1100-1400, 2030-2300
711) Ap. 50, Sicuani — Dir: Manual Infantes Q.
712) Centro Comercial, Ollanta 346 — Gte: Gabino Pérez C. **F.Pl:** 5 kW
713) Cas. 40, Cusco — Dir: Alberto Barrionuevo S.
714) Av. Garcilaso 604, Huánchac, Cusco — Gte: Germán Alencastre D.
715) Cas. 45, Sicuani — Gte: Raúl Siu Almonte. Prgr. in Sp., Quechua at 1000-1200, 0100-0300 Fax: 84 223308
716) Ap. 540, Cusco — Dir: Edmundo Montesinos G.
717) Av. Sol 150, Espinar — Dir: Marcelino Taco Quispe
718) Independencia 143, Quillabamba
719) Cas. 251, Cusco — DG: Raúl Siu Almonte. Prgr. in Sp., Quechua only 1000-1200.
720) Ap. 85, Cusco — Dir: Carola M. de Ramón Flores
721) Pza. de Armas, Yauri, Espinar
722) Jr. Tandapata 145, Barrio San Blas, Cusco
725) Cas. 39, Cusco — Dir: Raúl Montesinos Espejo
726) Edif. Municipal, Calca — José Victor del Castillo Farfán & J. Serapio Surco Ramírez
727) Calle "C" 13, Urb. Jardin, Cusco
728) Urb. Marcavalle, P-20, Huánchac, Cusco
729) Santa Fé de Urubamba
730) Edgar de la Torre, Quillabamba

Madre de Dios

740) Ap. 37, Puerto Maldonado — Admor: Fco. Javier Aniz B., O.P.

Puno

750) Cas. 210, Puno — Dir: Raúl Rodríguez Chalco
751) Jr. Arequipa 387, Puno
753) Asociación Cultural Nueva Era, Chucuíto
754) Ap. 130, Puno — Dir: G. Humpiri C.
755) Cas. 163, Juliaca — Dir: Francisco Aramayo P.
756) Jr. 2 de Mayo 151, piso 2, Juliaca
757) Ap. 3, Puno — Dir: Alberto Lescano R.
758) Ap. 67, Juliaca — Own: Alberto Quintanilla Ch. **F.Pl:** MW 5 kW
759) San Martín 134, Yunguyo

760) Cas. 543, Puno — Dir: Víctor H. Avila G.
761) Cas. 13, Juliaca — Dir: David Gamero Alatrista
762) Jr. San Martín 125, Yunguyo
763) 2 de Mayo 257, Juliaca — Dir: Pedro Cáceres V.
764) Av. Titicaca 160-C, Puno — Dir: Atalo Gutiérrez V.
765) Cas. 340, Puno — Gte: Jorge Aramayo G.
766) Ap. 8, Ayaviri — Gte: J. Solórzano M. (Prgr. in Sp., Quechua only 0930-1050)
767) Ap. 24, Ayaviri
768) San Miguel de llave, Provincia de Collao

REGION 8
Loreto
800) Av. Raimondi 143, Iquitos
801) Ap 207, Iquitos — Gte: Julia Jáuregui R.
802) Jr. Tacna 146, Iquitos — Dir: Teddy R. Bendayan D.
803) Arica 228, Iquitos — Dir: Manuel Santos V.
804) Cas. 174, or: Jr. Próspero 645, piso 2, Iquitos — Dir: Marcos Panduro R.
805) Av. Progreso 112, Yurimaguas — Dir: Prof. Ricardo Arévalo F.
806) Ap. 786, (Reports to: Pablo Rojas Bordales, Jr. Bermúdez 445), Iquitos — Dir: Julio Reátegui B.
Ucayali
820) Av. Cnel. Portillo 542, Pucallpa
821) Cas. 263, Pucallpa — Dir: Julio R. Zevallos M.
822) Av. 9 de Diciembre 642, Pucallpa — Dir: Luis Salas B.
823) Av. Cnel. Portillo 448, Pucallpa
824) Alfredo Vargas 40, Contamana — Dir: L. Vargas S.

REGION 9
Amazonas
900) Ap. 4, Bagua — Dir: Julio Cardozo G. **F.Pl:** 5 kW
901) Junín 819, Chachapoyas
902) Ap. 69, Chachapoyas — Dir: Santiago Izuzquiza, S.J. F.Pl: Change of freq.
903) Jr. Ayacucho 944, Chachapoyas — Gte: José David Reina N.
904) Jr. Trujillo 422, La Peca, Bagua
905) Jr. Huayabamba 513, Santa Rosa de Huayabamba, Rodríguez de Mendoza — Dir: José M. Calderón S.
908) Simón Bolívar 433, Bagua Grande
909) Jr. Matiaza Rimache 201, Rodriguez de Mendoza — Luis Elber Julca Ll.
San Martín
920) Jr. Progreso 389, Bellavista — Dir: Francisco González Meléndez
921) Av. Lima 715, Saposoa
922) San Martín 317, Lamas
923) Jr. Maldonado 930-934, Rioja — Dir: Teófilo Juan Mori M.
924) Ap. 169, Tarapoto — Gte: José Manuel Chong Velasco
925) Jr. San Martín 257, Tocache — Dir: Nélson Chaves S.
926) Ap. 210, Moyobamba — Dir: Porfirio Centurión Tapia
927) Jr. Federico Sánchez 720, Tarapoto — Dir: Luis H. Hidalgo
929) Cas. 31, Tarapoto — Gte: Luis F. Mori R. **F.Pl:** 10kW
932) Jr. Huallaga, cuadra 5, Bellavista — Dir: Severo Ruiz R.
933) Ap. 42, Tarapoto — Dir: Víctor D. Coral P.
934) Jr. Triunfo 600, Juanjuí — Admor: Heber Arévalo Ríos
935) Jr. Alte. Grau 703, P.3, Rioja — Dir: Tito Pilco M.
937) Av. Atahualpa 1140, Uchiza — Dir: Antenor Cantuta V.
938) Jr. Amargura 962, Rioja — Dir: Juan Bolaños C.
939) Av. Cajamarca 126, Nueva Cajamarca
940) Av. Lima 814, Saposoa — Gte: M. Pinedo S.
941) Jr. Lima 755, Jepelacio — Gte: Fernando Rojas G.
942) Jr. Huallaga 350, Nueva Cajamarca — Gte: Edilberto Peralta Lozada
943) Jr. Ramón Castilla 892, Soritor — Dir: Enedina Vda. de Rojas
944) Jr. E. Peña Meza 419, Juanjuí — Gte: A. N. Amara P.
947) Jr. Progreso s/n, Bo. La Victoria, Juanjuí
948) Jr. Túpac Amaru s/n, Villa Picota, Picota

SURINAME

L.T: UTC – 3h — **Pop:** 414.650 — **Radio sets:** 250.000 — **Pr.L:** Dutch, Sranan Tongo, Sarnami Hindi — **E.C:** a/c 60, 110/115/127/220V — **ITU:** SUR.

TELESUR N.V. (Gov. Administration)
ADDR: P.O. Box 1839, Paramaribo. **Te:** +597 473944. **Cable:** Latel. **Telex:** 131.

STICHTING RADIO-OMROEP SURINAME (SRS-Gov, Comm.)
ADDR: P.O. Box 271, Paramaribo. **Te:** +597 498115.
STATIONS: 725kHz 5kW (inactive) — **FM:** Paramaribo 96.3MHz 1kW (stereo), 93.1MHz 0.1kW — Nieuw Nickerie 95.7MHz 0.01kW.
D.PRGR: 0730-0400 (Sat 0430, Sun 0300) — **Sarnami Hindi:** 0730-0830 on all freqs, W 1900-2030 & 2315-2400 on 600kHz + 93.1MHz. **Javanese:** 0830-0930. Other times in Dutch. **N. in Dutch:** W 1000, 1100, 1200, 1730, 2000, 2345, 0100; Sun 1600, 1900, 0000. **N. in Sranan Tongo:** 0900, 2100 (W) — **Ann:** "Dit is de Stichting Radio Omroep Suriname, de SRS in Paramaribo" — **V.** by QSL-folder or letter. Rec. acc.

PRIVATE COMMERCIAL STATIONS

	kHz	kW	Name and Location
4)	600	10	R. Paramaribo, Paramaribo
2)	820	1	Radio Apintie, Paramaribo
3)	914	3	Radio Nickerie (RANI), Nieuw Nickerie

FM: 2) 97.1MHz 1kW, Nieuw Nickerie 92.1MHz 0.01kW, 4) 89.7MHz 1.5kW, Nieuw Nickerie 103.5MHz 0.01kW, 5) 98.3MHz 1kW, 6) 99.3MHz 1.5kW, 7) 103.1MHz 1.5kW, Nieuw Nickerie 101.1MHz 0.01kW.

SHORTWAVE: 2) 5005.7 kHz 0.35kW.

Addresses and other information:
2) P.O. Box 595, Paramaribo. **Te:** +597 497950. **Cable:** Apintie — Mgr: Ch. Vervuurt. 0800-0400 (Sun 0300). **Sarnami Hindi:** W 1900-2030, Sun 1800-1900. **Javanese:** W 2030-2100, Sun 1900-1930. Other times in Dutch. **N. in Dutch:** 1730, 2200 — **Ann:** "Ulvistert naar R. Apintie op AM-FM stereo en special voor het binnenland op de kortegolf in de zestig meter band". **Int-Sig:** The beat of the Apintie drum — **V.** by letter.
3) Waterloo straat 3, Nieuw Nickerie. **Te:** +597 231462 — Dir: L. Bissesar. 0900-1700, 2000-2300. **Dutch:** Sat 1630-1900. **Javanese:** 2130-2300. Other times in Sarnami Hindi — **Int-Sig:** "Surinam hamara pyara desh" sung by G. Kallasing.
4) P.O. Box 975, Paramaribo. **Te:** +597 499995 — Dir: R. Pierkhan. Mgr: F. Pierkhan. 24h. **Sarnami Hindi:** W 0845-0930, 1500-1600; Mon-Thurs 2100-0100; Fri 2200-0200; Sat 2200-0300; Sun 0845-1330, 1930-2100. **Javanese:** W 0800-0845, 1900-2100. Other times in Dutch. **N. in Dutch:** 1830 — **Ann:** "Dit is R.P. Internationaal, the hot one", or "dit is R. Paramaribo, RAPAR N.V. op de 500 meter AM Band en 89.7 FM".
5) RADIKA, Indira Ghandiweg 165, Paramaribo. **Te:** +597 481175, +597 482910 — Mgr: Mrs. Radhakishun. 0800-0300 in Sarnami Hindi plus some Dutch.
6) R. Sangeet Mala, Indira Gandhiweg 40, Paramaribo **Te:** +597 42390/2 — Dir: Radjen Soekhradj. 0800-0400 (Sat 0500) in Sarnami Hindi. **N. in Dutch:** 0830, 1830. **N. in Hindi:** 0915, 2300.
7) Kara's Broadc. Corp, Verlengde Keizerstraat 5-7, Paramaribo. **Te:** +597 475032 — Mgr: Orlando Karamat Ali. 0800-0400 (Fri/Sat 0600). **Ann:** "This is the spirit of the nation".

URUGUAY

L.T: UTC – 3h (UTC –2h until 28 Feb) — **Pop:** 3.140.000 — **Radio sets:** 1.800.000 — **Pr.L:** Spanish — **E.C:** a/c 50, 220V — **ITU:** URG.

MINISTERIO DE DEFENSA NACIONAL
ADDR: Av. 8 de Octubre 2628, Montevideo.

DIRECCION NACIONAL DE COMUNICACIONES
ADDR: Sarandi 472, 11000 Montevideo. **Te:** +598 (2) 960215. **Tlx:** 23213 Dinaco Uy — **L.P:** Cnel. Estéban J. Hackembruch.

ASOCIACION NACIONAL DE BROADCASTERS URUGUAYOS (ANDEBU)
ADDR: Calle Yi, 1264, 11100 Montevideo. **Te:** + 598 (2) 900053. **Cable:** Andebu.
L.P: Pres: — All st's affiliated exc. 4), 22), 32), 80); All affiliated st's carry "Cadena Andebu" prgr. D 1455-1500.
STATIONS: All st's comm. exc. 4) CX6; ° = shortwave, * = inactive.

	Call	kHz	kW	Name and h. of tr.
1)	CW1	551	10	R. Colonia, Colonia: 24h (n. 550)

2) CX58	580	2.5/1	R. Clarín, Montevideo: 24h
3) CX4	610	50	R. Capital, Montevideo: 0800-0600 (Sat/Sun 0300)
4) CX6	°650	100	S.O.D.R.E. Montevideo: 0950-0400 (classical)
5) CW27	680	5/1	R. Tabaré, Salto: 0900 (Sun 1100)-0300
6) CX8	690	10/5	R. Sarandí, Montevideo: 0800-0500
7) CX10	°730	5/2.5	R. Continente, Montevideo: 24h
8) CX12	°770	100/35	R. Oriental, Montevideo: 0925-0300 (Sat/Sun 0500)
9) CX14	°810	50	R. El Espectador, Montevideo: 24h
10) CW23	820	1/0.5	R. Cultural, Salto: 0800-0300
11) CX16	850	82	R. Carve, Montevideo: 0830-0400
49) CX18	°890	25/10	R. Libertad Sport "la pantera," Montevideo: 0900-0530
12) CW17	900	2.5	R. Frontera, Artigas: 0900-0300
8) CX20	°930	25/10	R. Monte Carlo, Montevideo: 24h
13) CW96	960	1	Radio Yi, Durazno: 1000-0200
14) CX22	970	10/5	R. Universal, Montevideo: 0900-0530
11) CX24	1010	20	R. El Tiempo, Montevideo: 0900-0330
4) CX26	°1050	6	S.O.D.R.E. Montevideo: 0950-0400
15) CX28	1090	25/15	R. Imparcial, Montevideo: 24h
16) CW31	1120	5/10	R. Salto, Salto: 1000 (Sun 1100)-0300
17A) CX30	°1130	20	R. Nacional, Montevideo: 24h, Sun 0800-0500
18) CW116	1160	0.5	R. Agraria del Uruguay, Cerro Chato: 0830-0130
19) CX32	°1170	10/5	Radiomundo, Montevideo: 1000 (Sun 1200)-0500
43) CX118	°1180	10/5	La Voz de Artigas, Artigas: 0930-0300
20) CW33	1200	2	La Nueva Radio, Florida: 24h
21) CX120	1200	1	R. Reconquista, Rivera: 0900-0300
22) CX121	1209.9	3/1	Difusora Soriano, Mercedes: 24h
23) CV121	1210	0.25	Em. RBC, Piriápolis: 0900-0400 Summer, 1000-0200 Winter
24) CW121	1210	0.25	Em. El Libertador, Villa Vergara: 0830-0300
25) CW35	1240	5/1.5	R. Paysandú, Paysandú: 0900-0400
26) CX36	1250	5/10	R. 36, Montevideo: 0900-0500
27) CW125	1250	5	R. Bella Unión, Bella Unión: 0930-0300
82) CW37	1260	4.5/1	Dif. Rochense, Rocha: 0900-0300
28) CV127	1270	3/1	R. Cuareim, Artigas: 1000-0300
4) CX38	1290	10	S.O.D.R.E. Montevideo: (relays 1050 kHz)
4) CX159	1290	1/0.5	S.O.D.R.E. Carmelo: 0950-0400
30) CW39	1320	1/0.5	R. La Voz de Paysandú, Paysandú: 0858-0400
31) CW132	1320	0.3	R. Fortaleza, Rocha: 0900-0300
32) CX40	1330	5	R. Fenix, Montevideo: 1000-0600
33) CW53	1340	2.5/1.4	La Voz de Melo, Melo: 0800-0300
34) CW136	1360	1/0.5	R. Rio Branco, Rio Branco: 0930-0230
35) CW41	1360	2.5/0.5	Broadcasting San José, San José: 0900-0300
36) CX42	1370	7.5	Emis. Ciudad de Montevideo: 1100-0600
37) CV137A	1370	0.5/0.25	Emis. Real, Minas de Corrales: 0930-0130
39) CW45	v1390	2.5/1	Dif. Treinta y Tres, Treinta y Tres: 0800 (Sun 0900)-0300
40) CX140	v1400	25	R. Zorrilla de San Martin, Tacuarembó: 0900-0300
41) CW141	1410	1/0.5	R. Belén, Belén: 1030-2400
42) CX44	1410	5	R. Panamericana, Montevideo: 0900-0310
43) CW142	°1420	1/0.25	La Voz de Artigas, Artigas: 0900-0400
44) CW43	1420	5	R. Lavalleja, Minas: 0830-0300
45) CX142	1420.1	0.1	R. Felicidad, Paysandú: 1000-0300
46) CW25	1430	5/1	R. Durazno, Durazno: 0900-0300
47) CV144	1440	1	R. Chuy, Chuy: 0900-0300
48) CX144	1440	3/0.5	R. Rivera, Rivera: 0830 (Sun 1000)-0300 (r.1445.3v)
17B) CX46	1450	1.5	R. América, Montevideo: 0900-0500
50) CX146	1460	1	R. Carmelo, Carmelo: 0900-0300
29) CV146	1460	0.25	R. José Battle y Ordoñez, José Battle y Ordoñez: 1100-0300
51) CX147	1470	3/0.5	R. Cristal del Uruguay, Las Piedras: 0955-0500 (Sun 0130)
52) CW147	1470	1/0.5	R. Tacuarí, Melo: 0830-0330
53) CW148	1470	3/1	R. Atlantica, Castillos: 1000-0300
54) CW43B	1480	3/1.5	R. Internacional, Rivera: 0900-0300
55) CX149	1490	5/1	R. del Oeste, Nueva Helvecia: 0930-0300
56) CX151	1510	0.25	R. Rincón, Fray Bentos: 1000-0300
57) CW57	1510	1/0.75	R. San Carlos, San Carlos: 0900-0300
58) CW151	v1510	0.25	R. Ibirapitá, San Gregorio de Polanco: 1000-0200
59) CX152	v1520	2	R. Cerro Largo, Melo: 0925-0300
60) CV152	1520	0.1	R. Guichón, Guichón: 1000-2300
61) CX50	v1530	1.2/0.25	R. Independencia, Montevideo: 1000-0300
62) CW153	1530	0.25	Emisora Cono Sur, Nueva Palmira: 0900-0300
63) CW154	1540	0.1	R. Charruá, Paysandú: 1000-0300
64) CX154	1540	0.25	R. Patria, Treinta y Tres: 0800 (Sun 0900)-0300
65) CV154	1540	0.1	R. Centro, Cardona: 0900-0200
66) CV155	1550	1/0.25	R. Agraciada, Mercedes: 0800-0300
67) CW155	1550	0.25	R. Sarandí del Yí, Sarandí del Yí: 1000-0200
68) CW51	1560	2/0.5	R. Maldonado, Maldonado: 24h
69) CX156	1560	0.25	Dif. Americana, Trinidad: 0930-0130
70) CV156	1560	1/0.25	R. Vichadero: 1000-0200
71) CW157	1570	0.25	R. Young, Young: 0900-0300
72) CX157	1570	1	R. Canelones: 1055-0200
73) CW54	v1580	0.5	Emisora del Este, Minas: 0830-0300
74) CW158	1580	0.25	R. San Salvador, Dolores: 0900-0300
75) CW159	1590	0.25	R. Regional, Lascano: 0900-0300
76) CV159	1590	0.25	R. Regional, Constitución: 1000-0200
77) CX159	v1590.5	2.5	R. Real, Colonia: 0900-0300
78) CX159A	1590	0.1	R. Paso de los Toros, Paso de los Toros: 1100-0200
79) CV160	1600	5	Emisora Continental, Pando: 0900 (Sun 1000)-0300
80) CX160	1600	0.5	R. Litoral, Fray Bentos: 1000-0300
81) CW160	v1600	0.25	R. Tacuarembó, Tacuarembó: 0900-0300

Shortwaves: Call CX—

19) A2	*6000	1.5	Radiomundo
17) A30	*6035	1	R. Nacional, Montevideo: 1300-0500
49) A61	6045.1	1	R. Integración Americana, 1300-1400, 0100-0200
43) A3	v6073.3	2.5	La Voz de Artigas (n. 6075): 1100-1500
8) A20	6139.9	1.5	R. Monte Carlo: 1000-1630v
8) A72	9595	1.5	R. Monte Carlo: 2330-0300
4) A6	9620.3	1	S.O.D.R.E: 0950-0500
8) A7	11735	1	R. Oriental, Montevideo: 1630-2330 (irr.)
9) A19	v11835.7	0.5	R. El Espectador: 1000-0300v

Addresses and other information:

1) Rivadavia 383, 70000 Colonia — Dir: Tabaré Grazioso.

2) Av. 18 de Julio 1516, 11200 Montevideo — DG: F.D. Tourreilles.

3) Colonia 2212, 11200 Montevideo — Dir: Cr. P. Burunsuzian.
4) Sarandí 430, 11000 Montevideo — Dir: Luis Alberto Zavala. CX6: classical music; CX26 music & spoken word. DX Prgr. Sun 0300, produced by DX Club del Uruguay, on 1050kHz.
5) Larrañaga 37, 50000 Salto — Dir: Carlos Gelpi.
6) Enriqueta Compte y Rique 1250, 11800 Montevideo. **Te:** 598 (2) 282615 — Dir: Jorge N. Mullins S — News and talk.
7) German Barbato 1472, 11200 Montevideo — Own: Heber Pinto.
8) Av. 18 de Julio 1224, 11100 Montevideo — DG: Daniel Romay S.
9) Río Branco 1483, 11100 Montevideo — Dir: Luis De María.
10) Lavalleja 48, 50000 Salto — Dir: Leonardo Vinci — FM 106.5 25kW "Em. del Exodo."
11) Mercedes 973, 11100 Montevideo — Own: Pablo Fontaina. (N.:30).
12) Av. Lecueder 544, 55000 Artigas — Dir: Basilio Borgato.
13) Zorrilla de San Martín 875, 97000 Durazno — Dir: T.B. Rodriguez C.
14) Av. 18 de Julio, 1220, 3er piso, 11100 Montevideo — DG: R. Baleato.
15) Av. Libertador Brg. Gral. Lavalleja 1708, 11800 Montevideo — DG: José Félix Iglesias.
16) Brasil 715, 50000 Salto — Gte: Enrique Pera Erro.
17) Pza. Independencia 846, 11100 Montevideo — Dir: 17A): A. Grille; 17B): Antonio Iglesias.
18) Dr. Luis A. de Herrera s/n, 30204 Cerro Chato.
19) Convención 1343, 9°, Montevideo — Dir: Pablo Lecueder.
20) Antonio Ma. Fernández 800, 94000 Florida — DG: W. Pastorini C.
21) Figueroa 887, 40000 Rivera — DG: Ever Viera.
22) Castro y Careaga 568, 75000 Mercedes — Dir: Miguel Racioppi.
23) Chacabuco y Moreno, 20200 Piriápolis — Dir: C. Repetto F.
24) Basilio Araujo 273, Villa Vergara, 33000 Treinta y Tres — DG: Geraldo Alzugaray.
25) 18 de Julio 1044, 60000 Paysandú — DG: Aldo Rubens Bayce.
26) 18 de Julio 1357, 11200 Montevideo — Dir: Leonardo Nicolini.
27) Enrique Ferreira 1550, 55100 Bella Unión — Dir: L. Murillo M.
28) Av. Lecueder 167, 55000 Artigas — Dir: José H. Meirelles Zabalegui.
29) Cno. Nacional s/n, 30200 José Battle y Ordoñez — Dir: Elbio Erazu P.
30) 18 de Julio 614, 60000 Paysandú — Dir: E. Estefanell.
31) Zorrilla de S. Martin 200, 27000 Rocha — Dir: Homero Casals O.
32) Canelones 1969, 11200 Montevideo — Dir: Miguel M. Racioppi.
33) Remigio Castellanos 721, 37000 Melo — Gte: Luis A. Bengochea.
34) Virrey Arredondo 986, 37100 Rio Branco — Dir: Raquel Balero González.
35) Treinta y Tres 890, 80000 San José — Dir: Marcos Marin.
36) Canelones 2061, 11200, Montevideo — Dir: Aramazd Yizmeyián.
37) Av. Davinson 320, 40002 Minas de Corrales — Dir: R.C.A. Andina.
39) Pablo Zufriátegui 1076, 33000 Treinta y Tres — Dir: A. Pinho.
40) 18 de Julio 302, 45000 Tacuarembo — Dir: Luis O. Dini.
41) Av. A. Sánchez s/n, 50001 Belén — Dir: Juan Arredondo.
42) Av. 18 de Julio 965, 11100 Montevideo — Dir: C. Casares.
43) Av. Lecueder 483, 55000 Artigas — Dir: Solange Murillo R.
44) José E. Rodó 530, 30000 Minas — Dir: Juan José Volante.
45) Calle 18 de Julio 817, 60000 Paysandú — Gte: Nery R. Cabillon.
46) Dr. E. Penza 805, 97000 Durazno — Dir: Sandy S. de Lerena.
47) Laguna Negra 174, 27100 Chuy — Gte: A. Francois.
48) Florencio Sanchez 1093, 40000 Rivera — Dir: E. Pedezert.
49) Eduardo Acevedo 1469, P. 9, 11100 Montevideo — Dir: Luis E. Parodi S. Sep. prgr. on SW for MERCOSUR, from Rivera, at 1300-1400, 0100-0200.
50) 19 de Abril 444, 70100 Carmelo — Dir: A. Lobecio P.
51) Av. Artigas 781, 90200 Las Piedras, Canelones — Dir: Julio Olivar.
52) Treinta y Tres 949, 37000 Melo, Own: Myrian Suárez G.
53) Ferrer 1404, 27000 Castillos — Dir: Julio Gómez T.
54) Av. Sarandí 792, 40000 Rivera — Dir: Patricia I. de Yanuzzo.
55) Calle Berna 1375, 70201 Nueva Helvecia — Dir: Rafael M. Ravazzani.
56) Av. Trienta y Tres 3223, 65000 Fray Bentos — Dir: Omar Donato.
57) Sarandí 775, 20400 San Carlos, — DG: Enrique Lamaison.
58) Artigas 145, 42500 San Gregorio de Polanco — Own: Luis A. Gancio.
59) Aparicio Saravia 571, 37000 Melo — Dir: Mario Barrios Viana.
60) Trabajo 360, 60008 Guichón — Dir: Rebeca O. de Luces.
61) Paysandú 1186, 11100 Montevideo — Gte: Antonio Rupenian.
62) Chile esq. Gral. Artigas, 70101 Nueva Palmira — Dir: Lorenzo Pagés.
63) Leandro Gómez 1089, Local 18, 60000 Paysandú — Dir: Lucio F. Francolino.
64) Lavalleja 1530, 33000 Treinta y Tres — Dir: José Eduardo Rodríguez.
65) Lavalleja s/n, 75200 Cardona — Dir: A. Martínez D.
66) E. José Rodó 781, 75000 Mercedes — Dir: Nestor Grana.
67) Sosa Bernadet 619, 97100 Sarandí del Yí — Dir: Arnol Franco.
68) Zelmar Michelini 819, 20000 Maldonado — Dir: Enrique Lamaison.
69) 25 de Agosto 724, 85000 Trinidad — Dir: R. Estrada B.
70) Bld. Artigas s/n, 40003 Vichadero — Dir: José E. Mariño.
71) Rincón 1689, 65100 Young — Dir: Gerardo Arévalo.
72) T. González 434, 90000 Canelones — Dir: Walther Espiga.
73) Treinta y Tres 632, 30000 Minas — Dir: C.&E. Falco.
74) Av. Asencio, 1695, 75100 Dolores — DG: Osmar Andriolo.
75) Ituzaingó 1170, 27300 Lascano — DG: A. A. Punales L.
76) Av. D. Pérez s/n, 50002 Constitución — Gte: Carlos A. Gelpi.
77) Av. Gral. Flores 468, 70000 Colonia — Gte: Elbio Geymonat.
78) 18 de Julio 743, 45100 Paso de los Toros — Dir: Nelson Gómez.
79) Av. Artigas 932, 91000 Pando — DG: Elías A. Alves.
80) 18 de Julio y 25 de Agosto, 65000 Fray Bentos — DG: M. Racioppi.
81) Ituzaingó 246, 45000 Tacuarembó — Dir: Lauro Siqueira.
82) 25 de Agosto 108, 27000 Rocha — Dir: Edgardo R. Ramírez G.

FM in Montevideo. All st's 10-100 kW. St. name Emisora (del) . . . 19) 93.9 Océano — 94.7 Concierto — 95.5 Plata — 96.3 Alfa — 4) 97.1 — 98.7 Diamante — 99.5 Sol — 100.3 El Dorado — 101.9 Azul — 105.9 Galaxia — 106.7 Cabildo.
In the rest of the country there are well over 30 FM outlets.

VENEZUELA

L.T: UTC – 4h — **Pop:** 20.675.000 — **Radio sets:** 8.100.000 — **Pr.L:** Spanish — **E.C:** a/c 50 + 60, 120/240V — **ITU:** VEN.

MINISTERIO DE TRANSPORTE Y COMUNICACIONES

ADDR: Dir. Gral. Sectorial de Comunicaciones, Torre Este, piso 35, Parque Central, Caracas. **Cable:** Diteleco. **Fax:** 58 (2) 5740753.
L.P: Dir. Gral. Sectorial de Comunicaciones: Luis E. García. Head, International Affairs: Nidya Crocker.

CAMARA VENEZOLANA DE LA INDUSTRIA DE RADIODIFUSIÓN

ADDR: Ap. 3955, Caracas. **Cable:** Camradio. **Te:** +58 (2) 2611651. Fax: 261 4783.
L.P: Pres: Marietta Hernández de Gómez.

STATIONS: All st's comm. exc. 1) and 11) (Gov); 4), 65), 75), 85), 131) (Cult), **H. of tr:** 1000-0400 exc. where indicated.

Mediumwaves: Call YV — °) shortwave; v = varying fq. *) inactive.

	Call	kHz	kW	Name, location and h. of tr.
1)	UR	*540	25	La Voz de Manapiare, S. Juan de Manapiare
2)	OY	540	120/50	Mundial Perijá, Villa del Rosario: 0900-0400
3)	KE	550	100/50	YVKE Mundial, Caracas: 24h
11)	RH	560	100/50	(r: YVKA 630), Cd. Guayana
114)	PJ	560	20	R. Exitos "Latina 5-60", Rubio: 1000-0300
5B)	LX	570	100/50	R. Rumbos, Villa de Cura: 24h
6)	MJ	580	50/10	La Voz de la Fé, Maracaibo: 24h
7)	KL	°590	20	R. (Reloj) Continente, Caracas: 0900-0500
8)	QB	600	10	R. Sucre, Cumaná: 0900-0400

149)	SW	600	15	R. Alto Llano, Sta. Bárbara de Barinas: 0900-0500
9)	SE	610	10	R. Cristal, Barquisimeto
4)	XY	610	10	R. Centro 610, Cantaura
10)	NO	620	10	R. Libertad, Cabimas: 0900-0400
11)	KA	°630	50	R. Nacional "Canal Ligero," Caracas: 24h
12)	QO	640	30	Porteñas 6-40, Puerto La Cruz: 24h
13)	LH	650	50/20	R. Aragüeña 650, Maracay: 0900-0500
14)	NA	660	10	Ondas de los Médanos, Coro: 0900-0400
5A)	LL	°670	100/50	R. Rumbos, Caracas: 24h
15)	QR	680	10	Cumaná 680 R., Cumaná
9)	MR	690	50/20	R. Barquisimeto, Barquisimeto
16)	MH	700	10	R. Popular, Maracaibo: 1000-0400
160)	PQ	700	5/2	R. Sur, Puerto Ordaz: 24h
17)	KY	°710	50/20	R. Capital, Caracas: 1000-0600
18)	QE	720	100/50	R. Felicidad "Super Criolla", Porlamar: 0900-0500
30)	XE	720	10	R. Elorza, Elorza
28)	OO	°730	10	R. Frontera, San Antonio del Táchira
9)	MT	730	10	R. Universo, Barquisimeto
19)	NC	°740	10	R. Maracaibo, Maracaibo: 0900-0400
55)	NQ	740	50	R. Caroní, Pto. Ordaz
20)	KS	750	100/50	R. Caracas R. "Red Nal. de Noticias", Caracas: 0900-0400
21)	QQ	760	10	Doble Q, Pto. La Cruz: 0955-0300
118)	SO	760	10	Mundial Primera, Trujillo
11)	KK	770	50/20	(r: YVKA 630), Valencia
22A)	OD	°780	50/20	Ecos del Torbes, San Cristóbal: 0900-0400 (Weekends 0600)
87)	MN	780	10	R. Coro, Coro: 24h
11)	—	790	—	(r. YVKA 630), Cd. Bolívar
23)	KC	790	10	R. Dif. Venezuela, "AM 790", Caracas
107)	XM	790	50	R. Minuto "La Barquisimetana", Barquisimeto: 24h
71)	—	800	50	R. Fé y Alegría, Guasdualito: 1100-0200
25)	LP	810	50	R. 810, Valencia: 0900-0500
169)	XG	820	50	R. Guadalupana, Coro
26)	SH	820	50	R. Guayana, Upata: 0900-0500
142)	KU	820	10	R. Altura, La Grita: 1000-0400
27)	LT	830	50	R. Sensación, Caracas: 0900-0500
9)	MY	840	10	Unión R. 840, Barquisimeto
175)	RV	850	10	RV-850, Valencia
173)	YE	860	20/10	R. Enlace 860, Valle de la Pascua
29)	OL	860	10	R. Rumba 860 AM, San Cristóbal: 0900-0500
163)	RU	870	10	R. Especial 870, Pto. La Cruz
9)	MP	870	10	R. Lara, Barquisimeto: 0958-0400
135)	YM	880	20/10	Mundial 880, Puerto Ordaz
23)	KV	°880	10	R. Libertador, Caracas
31)	LW	890	10	R. América, Valencia: 0900-0400
45)	VO	890	1	R. Oriente, El Tigre
32)	MD	°900	10	R. Mara "Ritmo 900", Maracaibo
33)	RQ	910	50/20	RQ 910, Caracas: 24h
34)	QX	920	20	R. Nueva Esparta, Porlamar: 1000-0355
116)	QU	920	10/5	R. San Carlos, San Carlos: 0955-0300
35)	LJ	930	10	R. Maracay, Maracay: 1000-0600
36)	NN	940	10	R. Punto Fijo, Punto Fijo: 0900-0500
110)	ZR	°940	15	R. Continental, Barinas
153)	LU	940	20	R. Fé y Alegría, Campo Mata: 19h
37)	KG	950	100/50	Informativa 950, Caracas: 24h
38)	RB	960	50/20	R. Monagas, Maturín
176)	—	960	10	Mundial 960, Acarigua
39)	SS	960	10	R. San Sebastián, San Cristóbal: 1000-0500
40)	LR	970	10	R. Central Continente, Maracay: 0900-0400
41)	SD	°970	15	R. Turismo, Valera: 0855-0355
177)	—	970	10	Mundial Oriental, Barcelona: 24h
42)	QM	°980	10	"R. 980" La Voz de El Tigre, El Tigre: 1000-0300
11)	—	*980	10	(r: YVKA 630), Maracaibo
43)	RT	990	20	R. Tropical "99-0", Caracas
44)	TA	990	10	Mundial Tricolor, Barquisimeto: 24h
22B)	OA	°1000	10	R. Táchira, San Cristóbal: 0900-0400
46)	NM	1000	10	Mundial Mil, Morón: 24h
47)	PC	1010	10	R. Aragua, Cagua: 1000-0400
48)	QF	°1010	10	Mundial Bolívar, Cd. Bolívar: 0900-0400
49)	MX	1020	50/10	R. Calendario, Maracaibo: 24h
50)	RS	1020	10	Mundial Margarita, La Asunción: 0955-0500
124)	TW	1020	25	R. Alegría Chivacoa
51)	TD	°1030	20/10	R. Valles del Tuy, Ocumare del Tuy: 1000-0300
52)	QY	1030	20/5	R. Onda, Guanare: 0900-0500
53)	ON	°1040	20/10	R. Los Andes "1040", Mérida
54)	LB	°1040	20	La Voz de Carabobo, Valencia: 0900-0400
11)	KZ	1050	20/10	R. Nacional "Canal Clásico," Caracas: 1000-0400
56)	LN	1060	10	R. Guárico, San Juan de los Morros: 1030-0330
57)	OE	1060	10	R. San Cristóbal "RSC", San Cristóbal: 1000-0600
81)	PX	1070	5	R. El Sol, La Fría
59)	MA	1070	10	Mundial Zulia, Maracaibo: 24h
60)	NR	1080	10	Mundial 1080, Maracay: 24h
61)	QJ	1080	10	R. Barcelona, Barcelona
62)	PB	1090	10	R. Yaracuy 1090 AM, San Felipe
63)	SZ	1090	20	Exitos 1090 Stereo, Caracas: 24h
64)	TG	1090	3	Melódica 1090, Machiques: 0900-0500
65)	OP	°1100	15	R. Occidente, Tovar: 1000-0400
66)	SV	1100	10	R. Angostura, Cd. Bolívar: 0900-0430
67)	QT	1110	10	R. Carúpano, Carúpano
155)	MF	1120	10	R. Super Ondas Cristianas, Maracaibo: 24h
171)	SK	1120	20/10	R. Dif. del Sur, San Fernando de Apure
70)	XZ	1120	5	R. República "La Estación Feliz", Maturín
44)	KQ	1130	10	11-30 Musical, Barquisimeto: 0930-0400
72)	RL	1130	20/10	R. Ideal, Macuto
73)	PY	1130	1	R. Amazonas, Puerto Ayacucho: 1100-0200
74)	MU	1140	10	R. Carora, Carora: 0955-0400
76)	MV	1150	10	Mundial Caribe, Punto Fijo: 24h
77)	QD	1150	10	Ecos del Orinoco, Cd. Bolívar: 24h
78)	OK	1160	10/1	R. Universidad, Mérida
79)	RR	1160	20/10	R. Industrial "Doble Uno 6-0", Guarenas
80)	QV	1170	20/10	R. Acarigua, Acarigua
58)	KW	1170	10	R. 1070, "Señal 11-70", La Guaira
82)	OR	°1180	10	R. Maturín, Maturín: 0900-0400
94)	LQ	1180	10	Super Suave 11-80, La V. de la Victoria, La Victoria
83)	NJ	1180	10	R. Petrolera, Cd. Ojeda: 0900-0700
84)	RE	1190	25	R. Barinas 1190 AM Estéreo, Barinas
85)	ZD	1190	25	R. Dif. Cultural del Táchira, San Cristóbal: 24h
105)	PF	1190	20/10	R. Canaima, San Félix: 0900-0300
95)	SF	1200	10	R. Dimensión, Caripito: 1000-0300
33)	OZ	1200	10	R. Tiempo, Caracas: 24h
86)	NH	1200	1	Ondas del Escalante, Sta. Bárbara del Zulia: 1000-0300
88)	ZT	°1210	10	R. Anzoátegui, Barcelona: 24h
89)	RD	1220	10	La Voz de Apure, San Fernando de Apure
178)	VM	1220	10/5	Mundial, Valencia
90)	ZO	1220	20/10	R. Aeropuerto 1220, Maracaibo: 24h
91)	NT	1230	10	R. Barlovento, Caucagua
92)	OH	°1230	10	R. Valera, Valera: 0900-0400
96)	PZ	1250	20/10	R. Puerto Ordaz: 0900-0400
97)	ML	1250	5	R. Cabimas, Cabimas: 1000-0300
98)	RM	1260	10	R. Super Suave, Caracas
99)	RY	1260	10	R. Horizonte, Nirgua: 1000-0200
100)	OU	1270	10	Ondas Panamericanas, El Vigía
101)	TR	1270	1	R. Tucupita, Tucupita: 0900-0500
102)	OF	°1280	10	R. Trujillo, Trujillo
103)	QS	1280	10/5	R. Zaraza, El Tamarindo: 1000-0300
104)	LF	1290	10	R. Puerto Cabello, Puerto Cabello
106)	KH	1300	10	R. 1300, Caracas: 1000-0400
16)	NS	1300	10	R. Reloj, Maracaibo: 1000-0400
109)	TS	1310	5	R. Sonido 13-10, Isnotú: 0900-0500

11)	SM	1310	20/10	(r: YVKA 630), Barcelona: 24h
11)	SL	1310	10	(r: YVKA 630), Guri
111)	WP	1320	10/5	R. Apolo, Turmero: 24h
112)	SG	1320	10	R. Colonial "Canal 13", El Tocuyo
113)	NX	1330	5	R. Los Llanos, Calabozo: 0900-0300
115)	TU	1330	10	R. Regional, Ciudad Ojeda: 0900-0500
117)	NE	1340	50/20	R. Uno "AM 1340," Caracas: 24h
119)	ZZ	1350	5	R. Guanipa, "R. 1350", El Tigrito: 24h
121)	TJ	1350	5	R. Falcón, Puerto Cumarebo
122)	TI	1360	10	R. Internacional, Maracaibo: 24h
123)	TZ	1360	5	R. Armonía, Charallave: 1000-0300
125)	OQ	1370	5	R. La Pascua, Valle de la Pascua: 1000-0300
126)	LG	1370	5	R. Cumbre, Ejido
127)	ST	1370	10	R. Portuguesa, Araure: 1000-0400
128)	ME	1380	5	R. Fantasía, Soledad
129)	NG	1380	10	Ondas del Mar, Puerto Cabello
120)	TL	1380	1	R. Triunto 1380, Caja Seca: 0900-0500
108)	ZC	1390	10	R. Selecta, Maracaibo: 1100-0100
130)	TT	1390	10	R. Super Criolla, Cabudare: 24h
131)	ZA	1390	12	R. Fé y Alegría, Caracas: 1000-0300
132)	NF	1400	1	R. Sabana, El Sombrero: 1000-0300
134)	QZ	1400	5	R. Anaco: 1000-0300
133)	ZJ	1400	10	R. 1400, Barinas: 0900-0500
136)	SP	1410	10	R. Simpatía, Valera
137)	ST	1410	5	R. Turén, Turén
138)	TP	1420	10	R. Bahía, Puerto la Cruz
139)	NZ	1420	5	Super R. Marabina, Maracaibo: 0900-0500
140)	RW	1420	10/5	R. Cardenal, Carora: 1000-0400
180)	—	1420	10	R. Sintonía, San Antonio de los Altos
141)	NB	1430	10	R. Satélite, Guacara: 1000-0600
143)	TM	1430	10/5	R. Caicara, Caicara del Orinoco
144)	RF	1440	5	Super Suave Orituco, Altagracia del Orituco
145)	TY	1440	20	R. Sucesos, Táriba: 0950-0400
146)	ZI	1440	5	R. Estelar, Guanare: 1000-0300
146x)	XC	1450	10/5	R. Cachamay, "RC 14-5-0", Puerto Ordaz
147)	ZQ	1450	10	R. Triunto, Puerto de Altagracia: 0900-0500
148)	KJ	1450	10	Sonera 14-50, Catia la Mar
150)	RJ	1460	5	R. Jardín, Boconó
151)	JW	1470	10	R. Latina, Valencia: 24h
152)	SY	1470	10	R. Vibración, Carúpano
174)	—	1480	—	R. Agrícola, Acarigua (CP)
154)	SQ	1490	1	R. Mérida 1490, Mérida: 24h
155)	RP	1490	10/5	R. Favoritas El Sol, Maracaibo
158)	XD	1490	10	Dinámica 14-90, Caracas
156)	RZ	1500	10/5	R. Dos Mil, Cumaná
11)	PO	1500	20	(r: YVKA 630), Cabudare
179)	—	1510	20	La V. del Centro, Güigüe
159)	ND	1520	10	R. Bonita "La Guapa", Guatire: 0930-0400
161)	NP	1530	10	R. San Felipe, San Felipe:
170)	XO	1550	10/5	R. Ojeda, Cd. Ojeda.
162)	MW	1550	20/5	R. Metropolitana, Los Teques: 0900-0400, 24h weekends
164)	LZ	1560	10/5	R. Dif. Andina, "R. 1560", Mérida
166)	UZ	1570	10/5	R. Guarapiche, "RG 15-70", Maturín: 24h
167)	TK	1580	10	Manzanares 15-80, Cumaná
168)	YV	1580	10	R. Dif. Venezolana, Calabozo: 1000-0200
172)	YO	1580	10	R. Occidental, Santa Rita
98)	—	1590	10	R. Deporte, Caracas

Shortwaves:

	Call	kHz	kW	Name and h. of tr.
23)	KT	*3244.9	1	R. Libertador, Caracas
32)	MC	*3274.9	1	R. Mara, Maracaibo
28)	PP	4761	1	R. Frontera, S. Antonio del Táchira (n. 4760): 1000-0300
48)	NW	4770	1	R. Mundial Bolívar, Cd. Bolívar: 0900-0400 (irr.)
54)	LA	*4780	1	La Voz de Carabobo, Valencia
22B)	OB	4830	10	R. Táchira, San Cristóbal: 0900-0400
92)	OI	4840.4	1	R. Valera: 0900-0400 (n. 4840)
19)	RN	*4859.9	1	R. Maracaibo: 0900-0400
110)	PA	4939.6	1	R. Continental, Barinas: 0900-1100, 2200-0400
5A)	LK	v4970	10	R. Rumbos, Villa de Cura: 24h
22A)	OC	4980	10	Ecos del Torbes, San Cristóbal: 0900 (Sun 1000)-0400 (Weekends 0600)
7)	—	5029.9	7.5	R. (Reloj) Continente, Caracas: 24h
82)	QH	5040.1	1	R. Maturín: irr.
53)	SB	*6010.6	1	R. Los Andes "1040", Mérida
88)	WK	*6020	10	R. Anzoátegui, Barcelona
11)	NV	9540	25	R. Nal. de Venezuela: 1100-1200, 1400-1500, 1800-1900, 2100-2200, 0000-0100, 0300-0400
22A)	SC	9640	10	Ecos del Torbes: as 4980
5A)	LM	9659.4	10	R. Rumbos, Caracas: 24h

Networks

State abbreviations (Estados): Ama = Amazonas; Amro = Amacuro; Anz = Anzoátegui; Apu = Apure; Ara = Aragua; Bar = Barinas; Bol = Bolívar; Car = Carabobo; Coj = Cojedes; Fal = Falcón; Guá = Guárico; Lar = Lara; Mér = Mérida; Mir = Miranda; Mon = Monagas; Nes = Nueva Esparta; Por = Portuguesa; Suc = Sucre; Tách = Táchira; Tru = Trujillo; Yar = Yaracuy; Zul = Zulia.
N.B: These abbreviations are not officially recognised by the Venezuelan P.O. Letters should therefore carry full name.

Addresses and other information:

1) San Juan de Manapiare, Ama.
2) Calle Central, Villa Rosario, Zul (or: Edif. La Palmera, Piso 1, Las Delicias, Maracaibo) — Dir: Ali Rachid. L.
3) Ap. 1684, Caracas 1010A — Dir: Antonio Mejías Jaimes.
4) Cantaura, Anz — Dir: Pedro Fermín V.
5AB) Ap. 2618, Caracas 1010A (or: Multicentro Empresarial del Este, Edif. Libertador, P.7, Núcleo A, Chacao, Caracas 1060). 5B) Villa de Cura, Ara — Pte: Andrés Serrano T. — SW relays 570/670 at 0400-1200, 570 at 1200-0400.
6) Ap. 459, Maracaibo 4002, Zul — Dir: Padre Angel Rodríguez del Palacio.
7) Ap. 866, Caracas 1010A — DG: Rafael Enrique Mena.
8) Ap. 26, Cumaná 6101A, Suc — Dir: Jaime Muñoz.
9) Ap. 567 y 576, Barquisimeto, Lara — DG: Ligia Villanueva A.
10) Ap. 99, Cabimas 4013, Zul — Dir: Dario E. Layba.
11) Ap. 3979, Caracas 1010A — **Te:** +58 (2) 746666. Dir: Jaime Alsina. Dir. Foreign. Sce: Gregorio Montiel. Gen. Mgr. of Eng: Ing. A. Miranda B. (630 kHz + relays; Light music 1050kHz Classical music). **International Sce** on SW: as above. 1 h. blocks in Sp., Eng., Fr. and, on Su., Creole.
12) Ap. 4123, 6023 Puerto La Cruz, Ang — DG: Jaime Ross. Fri. 0200-0300 "Sintonía DX", host José Elías Díaz.
13) Hotel Maracay, Maracay 2101, Ara — DG: Gilberto Mewa M.
14) Edif. Don Beto, Ca. Bolívar, Coro, Fal — Dir: José Jatar Sr.
15) Ap. 114, Cumaná, Suc — DG: Yolanda Marín Rojas.
16) Ap. 347, Maracaibo, Zul — Dir: José Higuera M.
17) Centro Comercial Los Ruices, Av. Francisco de Miranda, Caracas.
18) Hotel Bella Vista, Calle Velázquez, Porlamar, Nes — Dir: Calasanz Guzmán.
19) Calle 67 No. 24-88, Maracaibo, Zul — Dir: Rafael E. Mena.
20) Av. Páez, El Paraíso, Caracas 1021 — Dir: Arnaldo Salazar. Format: All news.
21) Ap. 4082, Puerto La Cruz 6023, Anz — Dir: Sergio Ramos.
22AB) Ap. 152, San Cristóbal 5001, Tách — Own: Gregorio González L. C.E: Iván Escobar S. 22A) **Te:** +58 (76) 438244. Dir: Jaime E. Villamizar. 22B) **Te:** +58 (76) 430015. Dir: Eleázar Silva M.
23) Edif. Los Jabillos, Of. 4AB, Av. Los Jabillos, Sabana Grande, Caracas — Gte: Mercedes de Vallejo. Prgr. in various lang. for foreign communities in Caracas.
24) Calle Garcés, Edif. Sabino, Coro 4101, Fal.
25) C.C. Big Low Center, Nave H, Valencia — Dir: Alfonso López.
26) Av. Raúl Leoni, Edif. Antonelly, Upata, Bol. — Dir: Edgar H. Rodríguez.
27) Av. Stgo. de Chile 07-29, Los Caobos, Caracas 1050 — DG: Luis Armando Arreaza.
28) Cra. 5 entre 4a. y 5a., Edif. Radio, San Antonio del Táchira, Tách — Dir: M. Marchena.
29) Ap. 220, San Cristóbal, Tách — Dir: Nelson V. Chacín.
30) Av. Aeropuerto esq. Calle 9, Municipio Rómulo Gallegos, Elorza 7007, Apu. — Dir: Luis H. Borjas.
31) Ap. 117, Valencia, Car — Mgr: Manolo Fachín V.

32) Ap. 1969, Maracaibo 4001A, Zul — Dir: Gustavo E. Vecino.
33) C. Comercial Concresa P.B., Prados del Este, Caracas 1080.
34) Ap. 58, Porlamar, Nes — Dir: Luis J. Arreaza A.
35) Av. Boyacá, Edif. Centro, Of. 01, Maracay 2101, Ara — Dir: G. Mewa M.
36) Ap. 25, Punto Fijo, Fal — Dir: Román O. Aguilar.
37) Final Av. LaSalle, Colinas de los Caobos, Caracas — Dir: Sixto Martínez.
38) Ap. 140, Maturín, Mon — Dir: Víctor Ramón Cordova.
39) Ap. 338, San Cristóbal 5001, Tách — Mgr: Jesús Contreras.
40) Ap. 4545, Maracay 2101, Ara — Dir: Freddy Contreras.
41) Ap. 12, Valera, Tru. — Own: Pedro José Fajardo
42) Ap. 430, El Tigre 6034, Anz — Dir: Fco. González P.
43) Torre Oeste, Puente Nuevo a Puerto Escondido, Caracas 1010 — Dir: Maria Vittoria de Grasso. **F.Pl:** 50kW
44) Edif. Núñez, Av. Vargas Carrera 16, Barquisimeto, Lar. — Dir: Argenis Lugo.
45) Av. Táchira 3, El Tigrito 6035, Anz — Dir: Calazans Guzmán.
46) Ed. Cumboto, Calle Comercio, Pto. Cabello, Car.
47) Calle Boyacá Nte. 9 (or: Calle Vargas Norte 21, Edif. Tufano 4to piso, Of. 16) Cagua, Ara — Dir: J.M. Labrador U.
48) Ap. 123, Cd. Bolívar, Bol — Dir: Luvén Rossi V.
49) Ap. 789, Maracaibo, Zul — Mgr: León Magno Montiel.
50) Ap. 145, La Asunción, Nes — Mgr: Bernardo Pirela.
51) Calle Urdaneta, Edif. Radio, Ocumare del Tuy 1209, Mir — Dir: Eleázar Ramos B.
52) Calle 15 esq. Cra. 9, Guanare 3310, Por —Pte: Rafael Vicente Camacho. **F.Pl:** 50kW.
53) Ap. 40, Mérida, Mér — Dir: Nergio Peña.
54) Torre Trébol, Urb. Lomas del Este, Valencia — Dir: Guillermo Degwitz.
55) Altavista, Calle Caura, Edif. Los Bancos, Puerto Ordaz, 8015.
56) Ap. 25, S. Juan de los Morros 2301, Guá — DG: J. Ghersy.
57) Ap. 374, San Cristóbal 5001, Tách — Own: G. González Lovera.
58) Av. Soublette, Edif. Américas-B, piso 16, Maiquetía 1160.
59) Av. 26, No. 22-23, Maracaibo 4001, Zul — DG: Pedro Carreño.
60) Calle 98, Nte. 6, Calicanto, Maracay 2101, Ara — Dir: Abraham Granadillo.
61) Ap. 27, 6001 Barcelona 6001, Anz — Dir: Ruffo José Vivas.
62) 5a. Av. con Calle 15, San Felipe 3201, Yar — Dir: Orión Oviedo Pinto.
63) Ap. 65598, Caracas 1061 — Pte: Enrique Cusco — **AM Stereo**.
64) Av. Gral. Trias, Machiques, Zul — Gte: Antonio Perozo.
65) Cra. 4 No. 6-46, Tovar, Mer — DG: Pedro R. Silguero.
66) Ap. 31, Bolívar 8001, Bol — Dir: Tomás León.
67) Ap. 590, Carúpano 6124, Suc — Dir: Pedro Luis Regnault Villalba.
68) Edif Ramos, Calle Páez No. 97-53, Valencia, Car.
70) Edif. Isnotú, Calle Monagas esq. Mariño, Maturín.
71) Barrio La Arenosa, Guasdualito 5063, Apu.
72) Av. Soublette, Calle El Trébol, Qta. Anna, Maiquetía 1160 — Dir: Luciano Landaeta L.
73) Av. Bolívar 4, Puerto Ayacucho, Ama — Dir. Gen: A.A. Pérez.
74) La Casita, Sucre e/Cras. 7 y 8, Carora 3040, Lar — DG: Marconi González C. **F.Pl:** 25kW, fq. move to 640 kHz.
75) Ap. 23, San Cristóbal, Tách — DG: Modesto Marchena.
76) Calle Ecuador, Punto Fijo, Fal — Dir: Bartolo Cárdenas.
77) Ap. 47, Cd. Bolívar, Bol — Dir: Luis Galeano Martínez.
78) Av. 2 No. 21-22, Mérida, Mér — DG: Lourdes E. Dubuc.
79) Calle Régulo Franquiz, Edif. La Electricidad, Guarenas, Mir.
80) Av. 35 con Calle 29, Acarigua, Por — Gte: Oswaldo A. Ramírez.
81) Calle 4, Edif. Illinois, La Fría, Tách — Dir: Luis Orlando Soto.
82) Calle Sucre, Maturín 6201 — Dir: Alonso Hurtado.
83) Ap. 201, Cd. Ojeda 4019, Zul — Dir: H. Zabala P.
84) Final Av. Sucre, Barinas, Bar — Dir: Rubén Darío Rodríguez. **AM Stereo.**
85) Ap. 150, San Cristóbal, Tách — Dir: Elio R. Ríos R.
86) Av. 5 No. 2-21, Sta. Bárbara del Zulia, Zul — Dir: J. Alvarado S.
87) Ap. 7421, Coro, Fal — Dir: José G. Polanco.
88) C.C. Tricentenaria, P.A., Barcelona 6001, Anz. **Te:** (81) 771732 — Gte: Dally V. Arreaza G.
89) Ap. 14, San Fernando de Apure, Apu — Dir: J.R. Barbella.
90) Ed. Plaza, Local 3, Calle 78 con 3H, Maracaibo 4002 — Dir: José Bonacia.
91) Calle Pantoja, Caucagua, Mir — Dir: Atilano Raymundi.
92) Av. 10 No. 9-31, Valera, Tru — Dir: José Silvino Pérez.
94) Edif. Belis, Blvd. Aldao, La Victoria, Ara — Dir: R. Vásquez.
95) Av. Jerusalem al lado de CANTV, Caripito 6211, Mon.
96) Centro Comercial S. Vicente, Ctra. El Tocuyo, Pto. Ordaz, Bol — Dir: Angel de Jesús Coa.
97) Av. Andrés Bello, Cabimas, Zul — DG: Harold J. Zabala P.
98) Qta. Radio Suave, Av. Los Mangos, La Florida, Caracas 1050. Dir: Samuel Robinson.
99) Urb. Las Tunitas, Nirgua, Yar — Dir: Alejandro Arzola.
100) Av. Bolívar 10-87, El Vigía, Mér — Dir: Dr. Serafín González Alvarez.
101) Calle Petión esq. La Paz, Tucupita, DAm.
102) Av. Independencia 10-74, Trujillo, Tru — DG: L. Torres.
103) Calle Concordia 35, Qta. Pto. Arturo, Zaraza, Gua — DG: Luìs Solórzano F.
104) Ap. 4, Pto. Cabello, Car — Gte: Juan Manuitt.
105) Av. Antonio de Berrios, San Félix, Bol — Dir: L.J. Pastrano.
106) Av. Libertador, Edif. Perú, El Bosque, Caracas — Dir: Héctor Mata.
107) C.C. Ven Rol, Av. Pedro León Torres, Barquisimeto — Own: Freddy Andrade.
108) Av. 3E No. 63-50 (or: Ap. 10247), Maracaibo — Coord: Lic. Gerardo Lombardi.
109) Ap. 63, Valera, Tru (or: Calle Iglesia J.G. Hernández, Isnotú) — Dir: Leonardo Salas.
110) Av. Marqués del Pumar, Barinas, Bar — Dir: Angel M. Pérez — "América en antena" Mo. 0000-0400, host: Santiago San Gil.
111) Calle Bermúdez, Edif. Torre Apolo, Turmero, Ara — DG: Mercedes de Salazar.
112) Av. Lisandro Alvarado, El Tocuyo, Lara — DG: O. Ramírez.
113) Ed. Paez, Carr. 12, Calabozo, Guá — Dir: Eleázar Ramos B.
114) Av. 9 No. 13-61, Rubio, Tách — Dir: Josefina da Rocha H.
115) Calle Vargas 141, Cd. Ojeda 4019, Zul — Dir: Héctor Toyo.
116) Edif. Gral. Manrique, Av. Sucre, San Carlos, Coj.
117) Torre Central, Av. Sucre, Cen. Pq. Boyacá, Los Dos Caminos (or: Ap. 76335, Caracas 1070A) — DG: Alberto Paruto.
118) Ap. 4, Trujillo 3102 (Tru) — Dir: Brixio Oliveros.
119) Av. Fernández Padilla 26-80, San José de Guanipa 6035 (El Tigrito), Anz — Mgr: José Luis Tineo L.
120) C.C. Nuevo Mundo, Loc. 2, Caja Seca — Dir: A. Granadillo.
121) Urb. Alta Vista, Pto. Cumarebo, Fal — DG: J. Montes I.
122) Calle 89, Av. 15, Edif. Occimaquinaria, P.B., Maracaibo 4001, Zul — DG: Gerardo Pozo.
123) Calle Zamora 8, Charallave 1210, Mir — Dir: A. Sarmiento C.
124) Av. 7 No. 61-1, Chivacoa, Yar — Dir: Pedro Luis Sánchez — AM stereo.
125) Calle 5 de Julio 20, Valle de La Pascua 2307, Guá.
126) Calle Industria, Ejido, Mér — Dir: Ender Ramírez.
127) Av. 26, Centro Radial, Araure 3303, Por — DG: Armando Roque López.
128) Ap. 240, Cd. Bolívar, Bol — Dir: Marcos Pitter.
129) Av. Bolívar, Edif. Sabatinos, Pto. Cabello, Car. Own: Carlos Rodríguez.
130) Av. Libertador, Altos Farmacia San Rafael, Cabudare, Lar — Dir: Luis Hidalgo.
131) Ap. 20484, San Martín, Caracas 1020A — DG: J. Martínez Terrero, S.J.
132) Calle Alegría, Qta. Galia, El Sombrero 2319, Guá.
133) Qta. Clari, Av. Sucre, Barinas 5201, Bar — Dir: Ing. Rafael González V. F. Pl: Move to 680kHz.
134) Calle Cajigal 39, Anaco, Anz — DG: Otman H. Quiñones
135) C.C. Venezuela, Local 15, piso 1, Urb. Villa Colombia, Pto. Ordaz 8015 — Dir: Luiz Alexander Pallares.
136) Av. 11 No. 12-59, Valera, Tru — Dir: Eladio Pacheco.
137) Ed. Los Andes, Av. Peñalver, Turén, Por — DG: N. Pérez L.
138) Av. Mpal., Torre Pelicano P.8, Puerto la Cruz — Dir: Fernando Zaurín
139) Calle 85 × Av. 18, Edif. Marilú, Loc. 2 No. 8-81, Maracaibo — Dir: Agustín Arteaga.
140) Calle 3 No. 6-100, Carora 3040, Lara — Dir: Jorge D. González L.
141) Av. Jacinto Lara, Edif. R.S., Guacara, Car. — Dir: S. Hidalgo F.
142) Av. Fco. de Cáceres, 9-88, Qta. Delia Mercedes, La Grita 5022, Tach — DG: Prof. Luis A. Aguilar Ch.
143) Caicara del Orinoco, Bol — DG: A. Castillo.
144) Calle Andrés Eloy Blanco, Altagracia de Orituco, Gua. — DG: Enrique Morón A.
145) Carrera 4 No. 1-35, Tariba, Tach. — DG: Eliseo A. Suárez U. **F.Pl:** 20kW
146) Av. Circunvacación, Edif. Gómez Galerío, Guanare, Por.
146x) Av. Ppal. Castillito, Edif. Arosa, Puerto Ordaz — Dir: Luis A. Pallares.

147) Sector La Salina, Puertos de Altagracia, Zul — Dir: Luis Enrique Torres.
148) Av. Tacagua, Qta. Sola, Urb. La Atlántida, Catia La Mar 1162.
149) Ap. 325, San Cristóbal (or: Cra. 3 No. 7-39, Santa Bárbara de Barinas) — Dir: R. Carvajal.
150) Av. Bolívar, Plaza la Alameda, Boconó, Tru. — DG: Fermín Mejías.
151) Ap. 1624, Valencia, Car — Dir: Nelson Villalba.
152) Av. Independencia 141, Edif. Plaza, Carupano 6124 — Dir: A. Hernández.
153) Campo Mata, Dtto. Freites, Cantaura (Anz) — DG: Gerardo Monreal.
154) Av. 2 No. 36-18, Mérida, Mér — Gte: Gustavo Arévalo P.
155) Calle 74, Av. 3D y 3E, Edif. Superondas, Sector Lalago, Maracaibo — Dir: Danilo Bautista F.
156) Av. Santa Rosa 18, Cumaná 6101 — Dir: J. Alfonzo.
158) Av. Lecuna esq. Cipreses, Edf. Corp. Felman, PH, Caracas — Dir: Enrique Bolívar Navas.
159) Ap. 009, Guatire, Mir — DG: G. Obelmejias.
160) Calle Guasipati, C.C. Artigas, Of. 2, Puerto Ordaz 8015, Bol — DG: Jesús R. Sifontes.
161) Av. Cartagena, entre 19 y 20, San Felipe, Yar.
162) Calle Ribas, Centro Empresarial P.7, Los Teques, Mir — DG: Omar A. Garcia — **AM Stereo.**
163) Torre Porteña, Of. 2-4, Av. Municipal, Puerto la Cruz 6023 — Dir: José Monga.
164) Ap. 265 Mérida — DG: Gildardo Moreno G.
166) Carrera 5 No. 33, Maturín, Mon — Dir: Rubén León M.
167) Av. Miranda, Qta. Tere, Cumaná 6101 — DG: Roselia G. de López.
168) Cra. 12 No. 3-57, Calabozo, Guá — DG: G. Helimenas Rojo P. **F.Pl:** Move to 1270kHz.
169) Ap. 7342, Coro 4101A, Falcón.
170) Crtra. N, Av. 51, Pq. Industrial, Cd. Ojeda 4019, Zul.
171) Gobernación del Edo. Apure, San Fernando 7001, Apu. Dir: Francisco Javier Loreto.
172) Av. Pedro Lucas Uribarri, Santa Rita 4020, Zul.
173) Av. Rómulo Gallegos, Res. Flor de Pascua, Loc. 2., Valle de la Pascua 2307, Guá.
175) Av. Mérida 107, Edf. Felpo, Valencia — Dir: Armando Moure G.
176) Barrio Reja, Acarigua 3301, Por.
177) Av. Intercomunal, C.C. Géminis, P. 3, Barcelona.
178) Urb. Lomas del Este, Av. Rotaria, Ed. El Parque, PB2, Valencia — Dir: Alfonso Nava M.
179) Av. Bolívar, Edif. Padre Cecilio Avila, PH, Güigüe, Car. — Dir: José Jiménez.
180) C.C. El Picacho, P. 8, San Antonio de los Altos, Mir. — Gte: Luis A. Alarcón.

FM in Caracas.
88.1 Imagen — 88.9 Stereo 88.9 — 89.7 Super Suave — 11) 91.1 — 91.9 Avila — 20) 92.9 — 94.1 Hot 94 — 95.5 Ritmo 95 — 97.7 Cultural de Caracas — 99.1 Mágica — 63) 99.9 Hits 100 — 101.5 KYS — 117) 102.3 Super Uno — 3) 103.3 — 17) 104.5 La Capital del Sonido — 37) 106.5 Fiesta 106 — 148) 106.9 Playa 107 — 63) 107.3 Exitos 107.
Prgr. different from MW, exc. st. 11).

ANTARCTICA

AMERICAN FORCES ANTARCTIC NETWORK (AFAN McMurdo)

ADDR: AFAN McMurdo, US Naval Support Force Antarctica, 651 Lyons Str, Port Hueneme, CA 93043-4345, USA.
L.P: St. Mgr: Michael Ray.
STATION: McMurdo 6012kHz 1kW — **FM:** 100.1MHz 0.05kW.
D.PRGR: 2000-1200 (on 6012kHz AM or SSB may be used).
ANN: "AFAN Radio International, Sixty-Twelve on your dial".
V. by QSL-card.

R. NACIONAL ARCANGEL SAN GABRIEL

ADDR: Base Antártida Esperanza, C.P. 9411, Antártida Argentina — Dir: Jefe de la Base Esperanza, Ejército Argentino.
STATION: (G.C: 63.24S/56.59W): LRA 36 15474kHz 1kW.
D.PRGR. In Spanish: 2130-0030.
V. by QSL-card & letter. Reports to Comando de Comunicaciones, Comando en Jefe del Ejército.

WORLD TELEVISION

This section is designed to give an accurate impression of the state of development of TV services in each country. It is not our intention to list all transmitters operated by each organisation, but main stations are included and wherever possible the total number of relay or repeater stations is given. Terrestrial programmes that are also relayed via satellite are indicated with the symbol .

TELEVISION STANDARDS

Recently a lot has been accomplished to facilitate easy reception of foreign broadcasts by simplifying television standards. The discontinuation of the high-definition French 819 lines standard and the rather poor British 405 lines standard left the world with just 2 standards; 525 lines (generally accompanied by the American NTSC colour system) and 625 lines (usually in PAL or SECAM colour).

The technical characteristics of the various transmission systems are given in the tables below. The colour standard (PAL, SECAM, NTSC) and television system used by each country are indicated at the top of the country entry. Antenna polarization varies from one transmitter to another and is either (H)orizontal or (V)ertical.

NEW STANDARDS

New television standards are being used for some satellite TV transmissions. The MAC (Multiplexed Analogue Components) system comes in 5 varieties; A-, B-, C-, D-, D2-MAC. B-MAC is currently used in North America, Australia and South Africa, C-MAC/D-MAC in the United Kingdom and some parts of Scandinavia and D2-MAC in the rest of Western Europe. The major benefit of the MAC system is that it enables programme providers to use a "hacker-proof" encryption system.

TELETEXT

Many TV broadcasters nowadays use a few lines of their transmission to send out a text service, which can be decoded by anyone having a suitable TV set. This service is generally known as TELETEXT, although the advertised name varies from country to country. TELETEXT services normally are available throughout the regular programme and often when a test card is shown.

N.B: The information used in this section comes from a variety of sources. Despite all our efforts to obtain complete details, there may be some omissions or contradictions in this section. We would welcome comments and observations that will help us to make this section as comprehensive as possible.

CHARACTERISTICS OF TELEVISION SYSTEMS

(as indicated in CCIR Report 624-3, XVIth Plenary Assembly, Dubrovnik, 1986)

System	Number of lines	Channel width MHz.	Vision band-width MHz.	Vision/Sound separation MHz.	Vestigial side-band MHz.	Vision mod.	Sound mod.
B	625	7	5	+5.5	0.75	Neg.	FM
D	625	8	6	+6.5	0.75	Neg.	FM
G	625	8	5	+5.5	0.75	Neg.	FM
H	625	8	5	+5.5	1.25	Neg.	FM
I	625	8	5.5	+5.996	1.25	Neg.	FM
K	625	8	6	+6.5	0.75	Neg.	FM
L	625	8	6	+6.5	1.25	Pos.	AM
M	525	6	4.2	+4.5	0.75	Neg.	FM
N	625	6	4.2	+4.5	0.75	Neg.	FM

N.B: Channels L2, L3, L4, Vision/Sound separation is −6.5 MHz. (France)

CHANNEL INFORMATION

(frequencies in MHz.)

VHF Channels:

West European "E" Channels

2 = 48.25	5 = 175.25	9 = 203.25
2A = 49.75	6 = 182.25	10 = 210.25
3 = 55.25	7 = 189.25	11 = 217.25
4 = 62.25	8 = 196.25	12 = 224.25

Italy

A = 53.75	D = 175.25	G = 201.25
B = 59.75	E = 183.75	H = 210.25
C = 82.75	F = 192.25	H1 = 217.25

Ireland

A = 45.75	E = 183.75	H = 207.25
B = 53.75	F = 191.25	I = 215.25
C = 61.75	G = 199.25	J = 223.25
D = 175.25		

France

2 = 55.75	5 = 176	8 = 200
3 = 60.50	6 = 184	9 = 208
4 = 63.75	7 = 192	10 = 216

East European "R" Channels

1 = 49.75	5 = 93.25	9 = 199.25
2 = 59.25	6 = 175.25	10 = 207.25
3 = 77.25	7 = 183.25	11 = 215.25
4 = 85.25	8 = 191.25	12 = 223.25

North/South America

2 = 55.25	6 = 83.25	10 = 193.25
3 = 61.75	7 = 175.25	11 = 199.25
4 = 67.25	8 = 181.25	12 = 204.25
5 = 77.25	9 = 187.25	13 = 211.25

Japan

1 = 91.25	5 = 177.25	9 = 199.25
2 = 97.25	6 = 183.25	10 = 205.25
3 = 103.25	7 = 189.25	11 = 211.25
4 = 171.25	8 = 193.25	12 = 217.25

Australia

0 = 46.25	5 = 102.25	8 = 189.25
1 = 57.25	5A = 138.25	9 = 196.25
2 = 64.25	6 = 175.25	10 = 209.25
3 = 86.25	7 = 182.25	11 = 216.25
4 = 95.25		

New Zealand

1 = 45.25	4 = 175.25	7 = 196.25
2 = 55.25	5 = 182.25	8 = 203.25
3 = 62.25	6 = 189.25	9 = 210.25

China (P.R.)

1 = 49.75	5 = 85.25	9 = 192.25
2 = 57.75	6 = 168.25	10 = 200.25
3 = 65.75	7 = 176.25	11 = 208.25
4 = 77.25	8 = 184.25	12 = 216.25

South Africa

4 = 175.25	7 = 199.25	10 = 223.25
5 = 183.25	8 = 207.25	11 = 231.25
6 = 191.25	9 = 215.25	13 = 247.43

Morocco

4 = 163.25	7 = 187.25	9 = 203.25
5 = 171.25	8 = 195.25	10 = 211.25
6 = 179.25		

French Overseas Territories

4 = 175.25	6 = 191.25	8 = 207.25
5 = 183.25	7 = 199.25	9 = 215.25

UHF Channels:

North/South America

14 = 471.25	33 = 585.25	52 = 699.25
15 = 477.25	34 = 591.25	53 = 705.25
16 = 483.25	35 = 597.25	54 = 711.25
17 = 489.25	36 = 603.25	55 = 717.25
18 = 495.25	37 = 609.25	56 = 723.25
19 = 501.25	38 = 615.25	57 = 729.25
20 = 507.25	39 = 621.25	58 = 735.25
21 = 513.25	40 = 627.25	59 = 741.25
22 = 519.25	41 = 633.25	60 = 747.25
23 = 525.25	42 = 639.25	61 = 753.25
24 = 531.25	43 = 645.25	62 = 759.25
25 = 537.25	44 = 651.25	63 = 765.25
26 = 543.25	45 = 657.25	64 = 771.25
27 = 549.25	46 = 663.25	65 = 777.25
28 = 555.25	47 = 669.25	66 = 783.25
29 = 561.25	48 = 678.25	67 = 789.25
30 = 567.25	49 = 681.25	68 = 795.25
31 = 573.25	50 = 687.25	69 = 801.25
32 = 579.25	51 = 693.25	

UHF TV channels 70-83 were discontinued. (to be used by radio)

Europe/Africa

21 = 471.25	38 = 607.25	54 = 735.25
22 = 479.25	39 = 615.25	55 = 743.25
23 = 487.25	40 = 623.25	56 = 751.25
24 = 495.25	41 = 631.25	57 = 759.25
25 = 503.25	42 = 639.25	58 = 767.25
26 = 511.25	43 = 647.25	59 = 775.25
27 = 519.25	44 = 655.25	60 = 783.25
28 = 527.25	45 = 663.25	61 = 791.25
29 = 535.25	46 = 671.25	62 = 799.25
30 = 543.25	47 = 679.25	63 = 807.25
31 = 551.25	48 = 687.25	64 = 815.25
32 = 559.25	49 = 695.25	65 = 823.25
33 = 567.25	50 = 703.25	66 = 831.25
34 = 575.25	51 = 711.25	67 = 839.25
35 = 583.25	52 = 719.25	68 = 847.25
36 = 591.25	53 = 727.25	69 = 855.25
37 = 599.25		

Australia

28 = 527.25	42 = 625.25	56 = 723.25
29 = 534.25	43 = 632.25	57 = 730.25
30 = 541.25	44 = 639.25	58 = 737.25
31 = 548.25	45 = 646.25	59 = 744.25
32 = 555.25	46 = 653.25	60 = 751.25
33 = 562.25	47 = 660.25	61 = 758.25
34 = 569.25	48 = 667.25	62 = 765.25
35 = 576.25	49 = 674.25	63 = 772.25
36 = 583.25	50 = 681.25	64 = 779.25
37 = 590.25	51 = 688.25	65 = 786.25
38 = 597.25	52 = 695.25	66 = 793.25
39 = 604.25	53 = 702.25	67 = 800.25
40 = 611.25	54 = 709.25	68 = 807.25
41 = 618.25	55 = 716.25	69 = 814.25

China (P.R.)

13 = 471.25	21 = 534.25	29 = 637.25
14 = 479.25	22 = 543.25	30 = 645.25
15 = 487.25	23 = 551.25	31 = 653.25
16 = 495.25	24 = 559.25	32 = 661.25
17 = 503.25	25 = 605.25	33 = 669.25
18 = 511.25	26 = 613.25	34 = 677.25
19 = 519.25	27 = 621.25	36 = 693.25
20 = 527.25	28 = 629.25	

N.B: Japan Channel 13-62 = No./So. America Channel 14-63

WRTH ON LINE

If you have a computer & modem, you can contact the WRTH editors through the international Fidonet network. Ask your SysOp for the "Shortwave Echo."

Philips anticipated the new 16:9 screen format by launching the new Matchline 100Hz wide screen color TV.

EUROPE

ALBANIA

TV-sets: 3m — **Systems:** B & G — **Colour:** PAL.

RADIOTELEVISIONE SHQIPTAR (Gov.)

Addr. & L.P: see radio.

Stations:	ch	kW/Pol	Stations:	ch	kW/Pol
Tirana	5	60H	Butrint	12	5H
Elbasan	6	10	Kükes	12	100H
Gjirokaster	7	10H	Vlora	12	10H
Peshkopi	8	2H	Peskopi	32	10H
Berat	9	10H	Letaj	39	25H
Tirana	10	0.2V	Tirana	57	800H
Pogradec	11	100H			

D.Prgrs: 1700-2100 (Sun also 0900-1430).

AUSTRIA

TV-sets: 2.706.000 — **Colour:** PAL — **Systems:** B & G.

ÖSTERREICHISCHER RUNDFUNK

Addr: ORF-Zentrum Wien, A-1136 Wien, Würzburggasse 30. **Te:** 43 (1) 87 8780. **Fax:** +43 (1) 87 878 2250. **Tlx:** 133601.
L.P: Dir. Gen: Gerd Bacher. Dir. of TV: Joh. Kunz, E. Wolfram Marboe. Tech.Dir: H. Doucha.
Stations: System B ch1-12, System G ch21-68.

Location	Prgr 1	kW	Prgr 2	kW	Pol
St. Polten	2A	60	21	500	H
Innsbruck	4	80	23	800	H
Neumarkt	4	1	47	10	H
Bad Ischl	5	2	38	5	H
Bregenz	5	100	24	350	H
Kufstein	5	3	24	50	H
Wien	5	100	24	1000	H
			34*	50	H
Linz	6	100	43	500	H
Rechnitz	6	3			V
			43	55	H
Graz	7	100	23	800	H
Weitra	7	5	58	90	H/V
Salzburg	8	100	32	800	H
			36	300	H
Wofsberg	8	2	28	20	H
Bruck/Mur	9	20	41	200	H
Bludenz	9	2	33	30	H
Klagenfurt	10	150	24	1250	H
Semmering	10	10	36	65	H/V
Schladming	11	10	40	80	H
Spittal/Drau	12	2	54	20	H
Mattersburg	38	30	52	30	V
St. Pölten	38	150	21	600	H
Poysdorf	51	10	43	10	H
Lienz			41	15	H
Gmunden			59	10	H

+ 867 low power tr's — *) also carries local prgr.
N.B: ORF 1&2 partly stereo.
D.Prgr: Prgr. 1: W 0800-2345, Sun 0800-2330 (all times approx). **Prgr. 2:** Mon-Fri 1530-2330, Sat 1245-2400, Sun 0800-2345 (all times approx).

AZORES (Portuguese)

TV-sets: 3380—**Colour:** PAL—**System:** B.

RADIOTELEVISÃO PORTUGUESA (RTP)

Addr: Ponta Delgada, S. Miguel.
Stations: Pico da Barrosa chE7 150kW H, Santa Barbara chE9 100kW H, Lages (Isle of Terceira) chE4 1 kW + 8 repeaters (only relay of RTP's I st. prgr.).
D. Prgrs: 2000-2400.

AFRTS (US Air Force)

Addr: Detachment 3, Air Force European Broadc. Squadron, APO New York, NY. 09406-5000, USA
Station: (System M): Lajes Field chA8 0.1kW H.
D. Prgr: 0900-0200 (Fri/Sat 0400).

BELARUS

TV-sets: N/A — **Colour:** SECAM — **Systems:** D&K.

BELARUSKAJE TELEBAČANNE (Gov.)

Addr: Makajonka 9, 220807 Miensk. **Te:** +7 (0172) 649286. **Fax:** +7 (0172) 648182. **Telex:** 252267 TV SU.
L.P: See radio.

Stations	ch	Pol		ch	Pol
Smjataničy	5	H	Pinsk	21	H
Miensk	6	H	Žlobin	22	H
Heraneny	7	V	Krascjukovičy	27	H
Mahilеŭ	7	H	Brest	30	H
Homiel	10	H	Bragin	39	H
Słonim	10	H	Babrujsk	40	H
Ušačy	11	H	Mjadzel	41	H
Viciebsk	12	V			

+ 5 low power repeators.
D.Prgr: 0630 (Mon 1205)-2200.

OSTANKINO KANAL 1 (OK-1)

Stations	ch	Pol		ch	Pol
Miensk	1	H	Smjataničy	8	H
Viciebsk	2	H	Ušačy	9	H
Homiel	3	H	Krascjukovičy	9	V
Hrodna	3	H	Salihorsk	11	V
Mahilеŭ	4	H	Babrujsk	12	V
Pinsk	4	H	Słonim	12	H
Brest	7	H	Bragin	27	H
Mjadzel	8	V	Heraneny	29	H

+ 6 low power repeators.
D.Prgr: Relays of OK-1 from Moscow. Further details see under Russia.

ROSSIJSKOJE TELEVIDENIJE (RTV)

Stations	ch	Pol		ch	Pol
Miensk	3	H	Žlobin	34	H
Viciebsk	31	H	Hrodna	34	H

+ 13 low power repeators.
D.Prgr: Relays of RTV from Moscow. Further details see under Russia.

TV PETERBURG (TV-P)

Stations	ch	Pol		ch	Pol
Brest	40	H	Hrodna	51	H
Miensk	46	H			

+ 1 low power repeator.
D.Prgr: Relays of TV-P from Sankt-Peterburg. Further details see under Russia.

BELGIUM

TV sets: 4.200.000 — **Colour:** PAL — **Systems:** B & H.

BELGISCHE RADIO EN TELEVISIE (BRTN)

Addr: see radio. **Cable:** BRT-TV. **Te:** +32 (2) 741 3111.
L.P: Dir. Gen: J. Ceuleers. Dir. Prgr. Planning: J. Bauwens.
Stations: ch2-11 System B, ch21-68 system H. NICAM stereo audio.

Location	Prgr 1	kW	Prgr 2	kW	Pol
Antwerpen	2			0.1	V
Waver/Overijse	10	100	25	10	H
Brussel	11	5W			V
Egem	43	1000	46	1000	H
Genk	44	200	47	200	H
Oostvleteren	49	20	55	20	V
Schoten			62	200	H

D.Prgr: BRT1: 1400-2200. **BRT2:** 1800 (Sun 1300)-2230.

RADIO TELEVISION BELGE DE LA COMMUNAUTE CULTURELLE FRANCAISE

Addr: see radio. **Cable:** RTBF-TV.
L.P: Prgr. Dir: G. Konen.
Stations: ch2-11 System B, ch21-68 system H. NICAM stereo audio.

Location	Prgr 1	kW	Télé 21	kW	Pol
Liège	3	100	42	1000	H
Wavre	8	100	28	500	H
Léglise	11	10	60	200	H
Bruxelles			45	0.5	H
Profondeville	52	200	49	200	H
Tournai	57	20	63	20	V
Anderlues			61	200	H

\+ 8 low power relay st's.
D.Prgr: RTBF1: 1545-2230. **Télé 21:** 1800 (Sat 1600, Sun 1300)-2200.

VTM (Vlaamse Televisie Maatschappij)

Addr: c/o Perexma, Slankrijkel 86b, B-2000 Antwerpen. **Te:** +32 (3) 231 47 90 — Man. Dir: J. Merckx. Commercial TV Service in Dutch on cable only.

CANAL PLUS (Comm.)

Addr: Chaussee de Lauvain 656, 1050 Brussels. **Te:** +32 (2) 730 0222. **Fax:** +32 (2) 732 1848. **Station:** ch 39 (Liège), 58 (Anderlues), 50 (Brussels).

Location	ch	kW	Pol
Liège	39	200	H
Wavre	50	500	H
Anderlues	58	200	H

RTL–TVi (Comm.)

Addr: Ave. Franklin Roosevelt 67, B-1050 Bruxelles. **Te:** +32 (2) 640 51 50. **Telex:** 64430.
Station: ch27.
D.Prgr in French: 12h daily.
N.B: ch27 tr located in Dudelange, Luxembourg.

TV-5 — Europe

Station: Bruxelles ch56 1kW H.
Rebroadcasts the TV-5 satellite sce.

AMERICAN FORCES NETWORK - SHAPE

Colour: NTSC — **System:** M.
Station: chE33 1kW V, ch 34 4.5kW (rel. AFN Germany).

BOSNIA/HERCEGOVINA

TV-sets: N/A — **Colour:** PAL — **Systems:** B&H.

TELEVIZIJA SARAJEVO

Addr: VI Proleterske brigade 4, 71000 Sarajevo. **Te:** +38 (71) 522333.
NB: Due to the continuing unrest, the actual situation may vary.
Stations: System B=ch2-12, System H=ch21-68. **ChE.** audio powers 1/10 of vision powers indicated.

TV Sarajevo 1st Prgr.	ch	kW (ERP)		ch	kW (ERP)
Majevica	5	10	Hum	8	6
Bjelašnica	5	6	Trovrh	9	10
Kozara	6	18	Leotar	10	6
Tusnica	6	10	Plješevica-B	10	2
Velež	7	10	Vlašić	11	100

\+ 152 low power sts.
D.Prgr: Mon-Fri 0500-1130, 1345-0030. Sat 1100-0015, Sun 0700-0015.

TV Sarajevo 2nd Prgr.					
Trovrh	21	100	Tusnica	43	7
Velez	26	1000	Majevica	46	750
Kozara	27	1000	Bjelašnica	47	60
Vlašić	29	1000	Plješvica	47	125
Leotar	37	100	Hum	52	22

\+ 140 repeaters.
D.Prgr: Mon-Fri 1355-1600, 1730-2230. Sat 0900-2230, Sun 0635-2200.

TV Sarajevo 3rd Prgr.		
Hum	37	100

D.Prgr: 1700-2400.

BULGARIA

TV-sets: 3.127.000 — **Colour:** SECAM — **Systems:** D & K.

BÂLGARSKA TELEVIZIJA (Gov.)

Addr: Ul. San Stefano 29, 1504 Sofia 4. **Te:** +359 (2) 652870. **Fax:** +359 (2) 871871. **Cable:** TV-Bulgar. **Telex:** 22581 bt sof bg.
L.P: DG: Assen Agov; TD: Ivan Dantchev.
Stations: (System D ch1-12, System K ch21-60):

Prgr. 1	ch	kW		ch	kW
Shumen	R5	100	Kjustendil	R10	50
Smoljan	R6	5	Botev Vrâh	R11	250
Burgas	R7	100	Sliven	R12	20
Sofija	R7	10	Belogradčik	R12	50
Silistra	R8	1	Dobrich (ex-Tolbuhin tr)	R12	5
Zelena Glava	R9	50	Ruše	R9	20
Kârdjali	R9	10			
Varna	R9	5			

D.Prgr: Rel. TSS-1 (Russian tv).

Prgr. 2	ch	kW		ch	kW
Botev Vrâh	24	1000	Sliven	31	200
Gotze Delchev	25	?	Varna	33	100
Burgas	26	1000	Kârdjali	34	100
Mihajlovgrad	26	1000	Kjustendil	34	100
Ruše	27	250	Silistra	35	100
Dobrich (ex-Tolbuhin tr)	28	100	Stara Zaora	37	?
			Smoljan	38	100
Plovdiv	28	?	Shumen	39	1000
Sofija	29	?	Belogradčik	46*	1000
Belogradčik	30	?			

\+ 250 low power tr's (Prgr. 1 + 2)
D.Prgr: Mon-Fri 1750-2130, Sat 1715-2200, Sun 1740-2200.
F.Pl: 3rd Netw using ch R 48 250kW. Broadcasting via satellite.
*Future Prgr. 3 trs.

RUSSIAN TELEVISION RELAY

Station: Sofia ch36 2kW, Ruše ch32 100kW.
D.Prgr: Mon-Fri 1230-2300; Sat/Sun 0630-2330. Relays of OK-1 via satellite from Moscow.

TV 5 EUROPE

Station: Plovdiv chR6, Sofija ch41 0.2kW, Burgas ch32.

CROATIA

TV-sets: N/A — **Colour:** PAL — **Systems:** B&H.

HRVATSKA TELEVISIJA (HTV)

Addr: Setaliste K. Marxa bb, 41000 Zagreb. **Te:** +38 (44) 61885
Fax: +38 (41) 510093 **Telex:** 21477
L.P: Dir: Branco Lentić
NB: Due to the continuing unrest, the actual situation may vary.
Stations: System B=ch2-12, System H=ch21-68, POL H (unless indicated otherwise).

Transmitter	HTV1	HTV2	HTV3	kW
Labištica	4	23	34	15/2×500
Psunj	4	50	58	80/2×1000
Plještivica–H	5	53	—	30/300
Kalnik	5	43	—	25/10
Srdj	6	28	22	37/2×250
Rota	5	38	—	1/5
Mirkovica	7	43	46	25/2×500
Ćelevac	8HV	31	(21)	80/1000(×2)
Belje	8	38	51	50/2×500
Sljeme	9	28	25	100/2×1000
Biokovo	9	41	(45)	110/1000(×2)
Slavonski Brod	9	42	48	1/2×5
Razromir	9	55	—	1.6/10
Učka	11	29	39	100/2×1000
Vidova Gora	11	29	53	8/2×10
Stipanov Grič	12	24	29	1/2×10
Ulijenje	12	35	—	0.6/5

Podsused	12	39	54	0.5/2×5
Moslavačka Gora	34	(21)	—	50(×2)
Uglijan	34	37	28	5
Veleb. Plješivica	34	44	—	2.5
Ivančica	36V	40V	—	35
Papuk	53	56	—	500
Županja	49	57	—	5
Liske	(7)	(44)	—	(40/200)

+ 106 lp's for HTV 1, 98 lp's for HTV2 and 20 lp's for HTV3. Channels and powers in brackets are planned.

Dprgr:
HTV1: Mon/Thu 0810-0000, Tue/Wed 0810-0045, Fri 0810-0030 (or 0215), Sat 0710-0030, Sun 0825-0015.
HTV2: Mon-Thu 1520-2330 (Thu 0000), Fri 1430-2330, Sat 1300-0515 (or 0015), Sun 1000-0020.
HTV3: 0800-2340 (Fri 0000, Sat 0100).

CZECH & SLOVAK REPUBLICS

TV-sets: 5.720.000 — **Colour:** SECAM — **Systems:** D & K.

ČESKOSLOVENSKÁ TELEVIZE (Gov.)

Addr: Jindrisska 16, 11150 Praha 1. **Te:** +42 (2) 221247. **Fax:** +41 (2) 425484. **Cable:** Televize. **Telex:** 121800.
L.P: GM: Jiri Kanturek. TD: Milan Bauman. Film Buyer: Jaroslav Bazant.
F1: Pol: H.

Stations	ch	kW	Pol	Stations	ch	kW	Pol
Praha	1	150	H	Trencín	10	1.6	V
Ostrava	1	100	H	Jihlava	11	2	H
Č. Budejovice	2	100	H	Zilina	11	100	V
Bratislava	2	150	H	Trutnov	11	1	V
Námestovo	4	0.8	H	Usti n. Labem	12	100	circ.
Bardejov	4	1.2	H	Domalize	24	100	H
Jesenik	4	10	H	M. Kamen	12	15/6	V
Poprad	5	80	V	N. Mesto n. Váhom	12	2	V
Košice	6	100	V	Uherský Brod	21	8	H
Hradec Kralove	6	100	H	Cheb	26	40	H
Klatovy	6	1.6	H	Bratislava	31	2	H
B. Bystrica	7	100	H	Č. Budejovice	36	20/6	H
Jachymov	7	0.8	H	Banska Stiavnica	40	9	H
Praha (city)	7	3.5	H	Zilin	41	10	H
Liberec	8	35	V	Ostrava	42	0.8	H
Brno	9	150	H	Borsky Mikules	42	1	H
Ruzomberok	9	1	H	Brno (city)	49	24	H
Susice	9	1.6	H				
Stúrovo	9	1.5	V				
Plzeň	10	100	H				

+ 1087 low power sts not mentioned.
D.Prgr: Mon-Fri 0900-1200, 1600-2330; Sat 0830-0045; Sun 0825-2400.
CTV & ST: Pol: H (Audio power 1/5 of vision power). * means H/V polarized.

Stations	ch	kW	Stations	ch	kW
Domazlice	12	1.6	Rychnov	28	100
Modrý Kamen	21	100	Brno	29	600
Kosice (city)	21	2	Námestovo	29	100
Klatovy	22	100	Poprad	30	600
Hradec Kralove	22	600	Ostrava	31	600
Zilin	22	100	Plzeň	31	600
Roznava	22	100	Liberec*	31	100
Trutnov	23	1000	Sturovo	31	100
Trencín	23	300	B. Brystica	32	600
Snina	23	15	Vimperk	32	100
Praha (mesto)	24	60	Zdar N. Sazavou	32	100
Svitavy	24	100	Lucenec	33	100
Jihlava	25	1	Olomouc	33	300
V. Klobouky	25	10	Usti n. Labem	33	600
Kosice	25	600	Nový Jicin	34	100
Praha	26	1000	Plzeň (city)	34	100
Mikulov	26	300	Zilina	35	1000
Banska Stiavnica	26	300	Brno (city)	35	20
Tabor	27	7.8	Susice	35	100
Ruzomberok	27	18	Cheb	36	100
Bratislava	27	1000	Pacov	36	1.6
Stara Lubovna	27	100	Jesenik	36	600
Trebíc	28	100	Frýdek-Mistek	37	300

Borsky-Mikulás	37	100	Č. Budejovice	39	600
Bardejov	37	100	N. Mesto n. Váhom	39	600
Jáchymov	38	300			

+ 484 repeater stations not mentioned.
D.Prgr: Mon-Fri 1530-2330; Sat/Sun 1000-1230, 1530-2330.

TA-3. Colour: PAL K, Pol: H.

Stations	ch	kW	Stations	ch	kW
Ruzomberok	22	0.15	Cheb	51	0.1
Plzen	27	2	Bratislava	50	0.5
Kosice	27	0.2	Liberec	51	0.1
Praha (mesto)	41	2	Zilin	51	0.2
Ceske Budejovice	49	2	Ostrava	51	2
Banska Bystrica	49	2	Brno—mesto	52	2
Trutnov	40	0.1	Banska Stiavnica	52	0.1
Usti Nad. Labem	50	0.1	Hradec Kralove	56	0.15

+ 15 low power stations not mentioned.
D.Prgr: 24h. relays satellite prgrs (Screensport, CNN, La7, TV5, Worldnet, TSS1, RTL Aktuell).
NB: Possible discontinuation after 01/01/93.

DENMARK

TV-sets: 2.500.000 — **Colour:** PAL — **System:** B.

DANMARKS RADIO (Gov.)

Addr: Danmarks Radio, TV-Byen, DK-2860 Søborg. **Te:** +45 (35) 203040. **Fax:** +45 (35) 202644. **Telex:** 22695.
L.P: Dir. Prgr: Finn Rowold. Head TV-Fakta (culture): Kjeld Veirup. Head of News & Actualities: Ole Sippel. Head of TV-Fiktion (Music, entertainment & drama): Henrik Iversen. Head of TV-International: Bo Lynnerup. Head of TV-sport: Per Krogh Simonsen. Head of the Province Dept (Århus): Peter Roland Larsen.
Stations: Pol H exc. Ølgod: V — Stereo: NICAM digital system, only on Copenhagen-Hove ch 31.

Location	ch	kW		ch	kW
Sdr. Højrup (Fyn)	3	50/5	Rangstrup (S. Jyll.)	7	60/6
Copenhagen-Glads	4	50/5	Århus	8	60/6
Ålborg	5	50/5	Holstebro	10	60/6
Ølgod	5	5/0.5	Copenhagen-Hove	31	600/60
Bornholm	5	10/1	Mosbjerg (Vendsyssel)	57	22/2.2
Næstved	6	60/6			

+ 27 low-powered repeaters not mentioned.
D.PRGR: 1500-2230 MF, 1300-2300 SS (all colour). **N. & Wrp:** 1700-1715 & 1830-1900 (Sun 1830-1930). **Teletext:** 0500-end of prgr's.
ALF: DKK 974 (b/w), DKK 1514 (colour) — both including radio & TV2.

TV2 (Comm.)

Addr: Rugaardsvej 25, DK-5100 Odense C. **Te:** +45 (65) 91 12 44. **Fax:** +45 (65) 91 33 22.
L.P: Dir. Gen: Jørgen Schleimann. Head of News: Ulla Terkelsen. Head of Prgr's: Jørgen Steen Nielsen. Head of Facts Dept: Lally Hoffmann. Head of Fiction Dept: Lone Bastholm. Head of Tech. Dept: O. B. Larsen.
Stations: Pol H, exc. Thisted: V — Stereo: NICAM digital system.

Location	ch	kW		ch	kW
Odense (Tommerup)	22	500	Ringkøbing (Videbæk)	40	600
Århus (Hadsten)	26	600	Jyderup	48	600
Åbenrå	27	600	Nakskov	52	100
Thisted	28	250	Copenhagen-Hove	53	600
Vejle (Hedensted)	30	500			
Svendborg	32	150	Bornholm (Rø)	56	600
Varde	33	500	Viborg	56	500
Nibe	35	600	Vordingborg	58	600
Vendsyssel (Tolne)	37	50			

+18 low-power translators.
D.PRGR: 1600-2300 MF, 1300-2300 SS (all colour) — **N:** 1800-1825 & 2050-2105 MF. **Comm:** between prgr's.
Regional Prgr's: 1830-1900 (exc. Sat).

Regional Addr's:
TV2/Bornholm, Ravnsgade 5, 3720 Aakirkeby. **Te:** +45 (53) 97 57 10. **Fax:** +45 (53) 97 45 85. **Dir:** Bent Nørby Bonde. **D.Prgr.** on ch. 56.
TV/Fyn, Kongensgade 68 — I, 5000 Odense C. **Te:** +45 (66) 14 18 18. **Fax:** +45 (66) 12 44 24. **Dir:** Ebbe Larsen. **D.Prgr.** on ch. 22 & 32.
TV2/Lorry, Allegade 7-9, 2000 Frederiksberg. **Te:** +45 (38) 88 00 11. **Fax:** +45 (38) 88 31 11. **Dir:** Dan Tschernia. **D.Prgr.** on ch. 53.
TV/Midt-Vest, Søvej 2, P.O. Box 1460, 7500 Holstebro. **Te:** +45 (97) 40 33 00. **Fax:** +45 (97) 40 14 44. **Dir:** Ivar Brændgaard. **D.Prgr.** on 28, 40 & 56.
TV/Nord, Søparken 4, 9440 Åbybro. **Te:** +45 (98) 24 46 00. **Fax:** +45 (98) 24 46 98. **Dir:** Bent Bjørn. **D.Prgr.** on ch. 35 & 37.
TV/Syd, Laurids Skausgade 12, 6100 Haderslev. **Te:** +45 (74) 53 05 11. **Fax:** +45 (74) 53 23 16. **Dir:** Helge Lorenzen. **D.Prgr.** on ch. 27, 30, 33 & 42.
TV/Øst, Kildemarksvej 7, 4760 Vordingborg. **Te:** +45 (55) 34 02 00. **Fax:** +45 (55) 34 01 34. **Dir:** Ole Dalgaard. **D.Prgr.** on ch. 48, 52 & 58.
TV2/Østjylland, Niels Brocks Gade 16, 8900 Randers. **Te:** +45 (86) 40 40 00. **Fax:** +45 (86) 40 56 86. **Dir:** Erling Bundgaard. **D.Prgr.** on ch. 26.
ALF: DKK 195 of the Radio Denmark license income, plus commercial income.

TV3 (Comm.)

Addr: Overgaden oven Vandet 10, DK-1415 København K. — **Te:** +45 (31) 57 30 80. **Fax:** +45 (31) 57 80 81.
L.P: Dir: Klaus Fog. Head of prgr's: Susanne Teilmann.
Station: Via Intelsat VA-F11 — 27.5 degr. west, 11.135 GHz, using D2MAC and digital sound.
D.PRGR: 1600-2400. Commercials interrupt prgr's.
Local TV Stations: Approx. 35 organizations are on the air on UHF with 0.2-3 kW ERP. Details are included in the list of local stations issued by Hartvig Media Aps., as mentioned in the radio section.

ESTONIA

TV-sets: 600.000 — **Colour:** SECAM — **Systems:** D&K.

TALLINN EESTI TELEVISIOON

Addr: 12 Faehlmanni Street, Tallinn EE 0100. **Te:** +7 (014) 2-434113. **Fax:** +7 (014) 2-434155. **Telex:** 173 869 ETV.
L.P: St. Mgr: Mart Siimann; T.D: Leiger Põldma; Film Buyer: Tõnis Värnik.
TV-stations

Estonian TV	**ch**	**ERP kW**	**Pol**
Tallinn	R2	150 (a)	H
Pärnu	R4	15	H
Valgjärve	R6	40*	H
Kohtla-Nómme	R11	40	H
Orissaare	R11	40	H

+5 low power st's not mentioned.
*radiated power restricted to north & north-west.

Ostankino TV (Moscow)

Tallinn	R12	30	H
Kohtla-Nómme	29	400	H
Valgjärve	30	400	H

+16 low power st's not mentioned.

Russian TV (Moscow)

Tallinn	28	400	H

+3 low power st's not mentioned.

St Peterburg's TV

Tallinn	R3	1	H

+3 low power st's not mentioned.
D.PRGR: Relays of Centralnoje televidenije **(CT)** and Rossijskoje televidenije **(RTV)** from Moscow, as well as Sankt-Peterburgskoje televidenije **(SPTV)** from Sankt-Peterburg. Further details see under USSR.

FAROE ISLANDS (Danish)

TV-sets: 14.000 — **System:** B&G — **Colour:** PAL.

SJÓNVARP FØROYA

Addr: M.A. Winthersgøta, 100 Tórshavn. **Te:** +298 17780.
L.P: Gen. Mgr: J.A. Skaale. Dir. of Adm: F. Lómstein. Head of Technical Dept: R.C. Joensen.
Stations: Tórshavn ch6 145kW H, Suduroy ch9 6.4kW H, Eysturoy ch10 1kW V + 29 low power repeaters. **Sat. relays:** ch22/28 MTV Europe, ch30/37 BBC Europe, ch34/44 Scansat TV3, Eurosport.
D.Prgrs: Tues 1930-2300; Thurs 2100-2200; Fri 1930-2330; Sat 1930-2400; Sun 1500-1700 & 1930-2300.

FINLAND

TV-sets: 1.900.000 — **Colour:** PAL — **Systems:** B & G.

OY YLEISRADIO AB (Non-comm, Public Broadc. Sce.)

Addr: TV-1, Box 10, SF-00241 Helsinki. **Te:** +358 (0) 14801. **Cable:** YLEtv. **Telex:** 121270. **Fax:** +358 (0) 14805148 TV-2, Box 196, SF-33101 Tampere. **Te:** +358 (31) 456111. **Telex:** 22749. **Fax:** +358 (31) 456892.
L.P: Dir. TV-1: A. Kaila. TV-news Co-ord: L. Toivola. Acq.: J. Sorjonen, Co-prod & Sales: J. Vakkuri. Dir. TV-2: A. Wessberg. Acq.: T. Tuovinen. For other personnel: See radio.
Stations: (System B ch 2-12, System G ch 21-60):

Prgr. I	**ch**	**kW**	**Pol**	**Prgr. I**	**ch**	**kW**	**Pol**
Tervola	3	20	H	Tampere	8	80	H
Ruka	4	3	V	Vuotso	8	0.25	H
Vuokatti	4	40	H	Haapavesi	9	60	H
Kuttanen	5	3	H	Kerimäki	9	10	H
Lapua	5	80	H	Lahti	9	80	H
Pyhätunturi	5	60	H	Joutseno	10	10	H
Anjalankoski	5	20	H	Pihtipudas	10	60	H
Ahvenanmaa	5	10	H	Posio	10	10	H
Espoo	6	180	H	Tammela	10	10	H
Iisalmi	6	1	H	Inari	10	30	H
Mikkeli	6	10	H	Eurajoki	11	20	H
Pyhävuori	6	20	H	Jyväskylä	11	10	H
Taivalkoski	6	60	H	Vaasa	11	1	H
Utsjoki	6	1	H	Ylläs	11	60	H
Karigasniemi	7	1	H	Ähtäri	26	100	H
Kruunupyy	7	10	H	Kiihtelysvaara	35	50	H
Kuopio	7	80	H	Fiskars	31	100	H
Oulu	7	60	H	Pernaja	39	100	H
Turku	7	60	H	Rovaniemi	53	100	H
Koli	8	20	H				

+ 127 low power repeaters.
YLE-prgrs: approx 65h weekly.

Prgr. II	**ch**	**kW**	**Pol**	**Prgr. II**	**ch**	**kW**	**Pol**
Espoo	8	60	H	Utsjoki	34	20	H
Turku	9	30	H	Vuotso	35	1	H
Vaasa	9	1	H	Kuopio	36	600	H
Pyhävuori	10	20	H	Pernaja	36	100	H
Oulu	11	20	H	Tammela	37	600	H
Iisalmi	22	20	H	Mikkeli	38	600	H
Tervola	22	1000	H	Ähtäri	39	100	H
Taivalkoski	23	600	H	Lahti	40	600	H
Karigasniemi	23	20	H	Inari	40	500	H
Lapua	24	1000	H	Fiskars	41	100	H
Jyväskylä	25	600	H	Kiihtelysvaara	41	50	H
Haapavesi	28	1000	H	Anjalankoski	49	600	H
Ruka	28	300	H	Koli	51	600	H
Ahvenanmaa	28	300	H	Tampere	53	1000	H
Kruunupyy	30	200	H	Ylläs	55	900	H
Posio	30	500	H	Kerimäki	55	600	H
Joutseno	32	600	H	Rovaniemi	56	100	H
Pihtipudas	32	1000	H	Vuokatti	56	1000	H
Pyhätunturi	32	650	H	Kuttanen	60	20	H
Eurajoki	33	600	H				

+ 126 low power repeaters.
D.Prgr: approx 54h weekly.
Relay st. FINLANDS TV in Stockholm 55h weekly of Finnish prgr from all networks. See Sweden.

Prgr. III	ch	kW	Pol	Prgr. III	ch	kW	Pol
Espoo	24	600	H	Vaasa	51	20	H
Oulu	33	40	H	Turku	54	1000	H
Jyväskylä	35	600	H	Pernaja	55	100	H
Lapua	40	1000	H	Fiskars	58	100	H
Kuopio	49	600	H	Tampere	59	1000	H
Lahti	51	600	H	Eurajoki	36	600	H
Mikkeli	28	600	H	Joutseno	60	600	H
Anjalankoski	56	600	H	Kerimäki	58	600	H
Pyhävuori	12	20	H	Koli	57	600	H
Iisalmi	29	20	H	Vuokatti	59	1000	H
Tammela	30	600	H	Rovaniemi	49	100	H
Kruunupyy	33	200	H				

+ 85 low power repeaters.
MTV-3 rgrs: approx. 70h weekly.

Prgr. IV	ch	kW	Pol	Prgr. IV	ch	kW	Pol
Fiskars	23	100	H	Espoo	35	600	H
Pernaja	26	100	H	Turku	57	1000	H

+6 low power repeaters.
Prgrs: (produced by Swedish television, SVT) approx 55h weekly.
Fi: By license fee and adv. Yle owns the nation's television networks and rents air time to MTV Oy and Channel Three Finland.

MTV Oy (Comm.)

Addr: Ilmalantori 2, SF-00240 Helsinki 24. **Te:** +358 015001. **Fax:** +358 0 1500707. **Cable:** Camtele. **Telex:** 121544 & 124319.
L.P: Pres: E. Pilkama. Exec. Vice Pres: J. Paavela. Prgr. Dir: T. Äijälä. News Dir. (Editor-in-chief): Pekka Hyvärinen. Tec. Dir: H. Marsalo. Marketing Dir: Eero Aalto. Sales Dir: Heikki Rotko. Pub. Rel. Mgr: M. Paaso. Ass. Dir./Acquisitions: Jorma Sairanen.
Prgrs: Own production & acquisition programmes on III network (MTV3).

CHANNEL THREE FINLAND/OY KOLMOSTELEVISIO AB (Comm., subs. of MTV Oy)

Addr: Ilmalankatu 2 C, SF-00240 Helsinki. **Te:** +358 (0) 150 01/150 0833. **Telex:** 126068 CHANT SF. **Fax:** +358 (0) 150 0677/150 0715/150/0757.
L.P: Man. Dir: H. Lehmusto, Prgr. Dir: H. Seppälä. Marketing Dir: J-P. Louhelainen. Prgr. Mgr: S. Kievari. Producer: R. Haavisto.
Prgrs: Sport and other subcontracted programming on III network (MTV3).

FRANCE

TV-sets: 29.300.000 — **Colour:** SECAM — **System:** L.

TELEVISION FRANÇAISE 1 (TF1) (Priv, Comm).

Addr: Société TF1, 1 quai du Point du Jour 92100 Boulogne-Billancourt. **Te:** +33 (1) 41 41 12 34.
L.P: Pres. & Dir. Gen: P. Le Lay. Dir. Tec: F. Hericourt.

FRANCE 2 (Public Television)

Addr: Société Nationale Antenne 2, 22 Ave Montaigne, F-75008 Paris. **Te:** +33 (1) 44214242. **Fax:** +33 (1) 44215145. **Telex:** 204 068. **L.P:** Pres. & Dir. Gen: Hervé Bourges.

FRANCE 3 (FR3) (Public Television)

Addr: 116 Avenue du President Kennedy, 75016 Paris. **Te:** +33 (1) 42 30 22 22. **Fax:** +33 (1) 42890327.
L.P: Pres. & Dir. Gen: Hervé Bourges. Dir. Gen: Dominique Alduy.

CANAL PLUS (Private)

Addr: 85/89 Quai Andre Citroën, 75015 Paris. **Te:** +33 (1) 44251000. **Fax:** +33 (1) 44251234. **Telex:** CPLUS 201 141 F.
L.P: Man. Dir: Pierre Lescure, Tec. Dir: Lucien Banton, Film Buyer: Rene-Bonnell.
Canal Plus is a subscription network. The signals are mostly coded and subscribers need a decoder. Times for uncoded signals: W 0600-0630, 1130-1230, 1715-1930.

Stations (Pol = H, exc. where indicated)
*) Lower power applies to Canal Plus only.

Location	TF1	F2	F3	Canal* Plus	kW(ERP)
Abbeville	63	57	60		250
Ajaccio	31	21	24	4	500/8
Albertville	45	39	42		5
Alençon	48	51	54		100
Ales	27	21	24	65	100/3
Amiens	41	47	44	10	500/0.01
Angers	47	44	41	10	20/0.04
Annemasse	—	—	66	—	1.75
Argenton	46	40	43		80
Aurillac	59	65	62	9V	500/4
Aurillac			54		350
Autun	48	51	54	10	500/0.1
Auxerre	37	31	34	6	300/0.03
Avignon	42	45	39a		125
Avignon			33		40
Bar-le-Duc	54	51	48	6	200/0.05
Bastia	41	47		2V	500/17
Bayonne	64	58	61	7V	500/2.5
Bergerac	37	34	31		250
Besançon	47	41	44	3V	500/60
Besançon	29	23	26	10	250/0.02
Bordeaux	63	57	60	8	1000/50
Boulogne	29	34	37	10	100/0.5
Bourges	23	26	29	8	1000/211
			43		34
Brest	27	21	24	10	1000/230
Brive	23	29	26	6	150/0.5
Caen	22	25	28	9	1000/198
Carcassonne	64	58	66/61	3V	1000/100
Chambery	29	23	26	8	100/1
Chambery			49		4
Chamonix	25b	28	22	7	50/0.02
Champagnole	58	61	64		80
Chartres	55	50	53	9	250/0.18
Chartres			34		30
Chaumont	52&62	49	55		80
Cherbourg	65	59	62	6	100/8
Clermont-Fd.	22	28	25	5	1000/200
Cluses	56	50	53	6	8/0.1
Corte	51	61	54	7	100/6
Dijon			56		30
Dijon	59	62	65	9	1000/30
Dunkerque	42	39	45		200
Epinal	65	60	63		100
Forbach	47	22	25	28	20/2
Gap	27	21	24	9	8/0.3
Gex	27	21	24	5V	1000/30
Grenoble	56	50	53	6	50/0.6
Gueret	64	58	61	9	80/0.3
Hirson	54	48c	51	32	200/0.25
Hyères	65	59	62	6	35/2
Laval	63	57d	60	8	80/0.02
Le Creusot	35	33	30	67	20/10
Le Havre	46	43	40	5	100/2
Le Havre			35		2
Le Mans	24	27	21	5V	1000/200
LePuy	63	57	60	6	100/0.13
Lesparre	39	45	42		10
Le Vilhain			36		3.6
Lille	27	21	24	5	1000/200
Limoges			32		450
Limoges	56	50	53	10	1000/260
Longwy	52	47	44	8	100/0.1
Lyon (Mt. Pilat)	46	40	43	10	1000/400
Lyon	61	58	64	66	10/10
Macon	57	55	49		5
Mantes	64	58	61	9	80/0.02
Marseille	29	23	26	5	1000/200
Marseille	40	46	43	57	22/4.5
Maubeuge	39	42	45	29	5/2.5
Mende	37	31	34	68	80/1
Menton	62	50	56	68	50/17
Metz	37	34	31	5V	1000/45
Mezières	29	23	26	36	500/?
Millau	47	44	41	10	20/0.15
Montmelian	64	58	61	9	3/?
Montpellier	56	50	53	9	1000/2.5
Mortain	50	52	55		10
Morteau	48	54	51	10	20/0.01
Mulhouse	27	21	24	5	1000/300
Nancy	23	29	26	8	500/30

Nantes	23	29	26	9V	1000/300
Neufchatel	51	48	54	65	80/5
Nice	64	58	61	66	10/6
Niort	28	22	25&58e	6V	1000/405
Orleans	42	39	45		100
Paris	25	22	28	6	700/104
Paris Est	43	46	40	53	5/5
Paris Nord	45	39	56	59	10/3
Paris Sud	49	52	62	65	3/3
Parthenay	52	49	55		80
Perpignan	22	25	28	7	10/3
Pignans	46	43	40	56	2/3
Porto-Vecchio	40	37	34	6	25/?
Privas	64	58	61	6	50/0.02
Reims	43	46	40	9	1000/36
Rennes	39	45	42	7	1000/200
Rouen	23	33	26	7	500/65
Sarrebourg	50	40	53		250
Sens	57	63	50/60	5	100/?
Serres	50	53	56	4	2/1
St. Etienne	35	30	33	38	10/12
St. Flour	52	49	55	7	6/0.4
St. Martin	48	51f	54	8	5/0.03
St. Raphael	25	28	22	10V	1000/70
Strasbourg	62	56	43	10V	1000/20
Tarascon	52	55	49	8	5/0.5
Toulon	51	48	54	9	100/1.6
Toulouse	27g	21	24	5	500/100
Toulouse			47		600
Toulouse	45	39	42	7	10/1
Tours	65	59	62	10	200/0.1
Troyes	27	24	38/21	7	1000/100
Ussel	42	45	39	4	50/?
Utelle	47	44	41		10
Vannes	50	56	53	5	500/16
Verdun	65	59	62	8	500/0.08
Villers-Cotterets	65	59	62		4
Vittel	30	35	32		80
Wissembourg	54	48	51		50

+ low power st's not mentioned.
a = 300kW; b = 50kW; c = 500kW; d = 100kW; e = 310kW; f = 3kW; g = 1000kW.
D.Prgrs: TF1: 24h — **F2:** 24h — **FR3:** 0630-2300 — **CANAL Plus:** 0600-0200, Sat/Sun 24h. **F.Pl:** F2 & FR3: 24 h.

ARTE (Cultural, Public Television)

Addr: 2A rue de la Fonderie, 67000 Strasbourg. **Te:** +33 88 52 22 22.
L.P: Jérôme Clément.
D.Prgrs: 1800-2300.
F.Pl: A private network is planned to use the transmitters before 1800.

M6 Metropole TV (Priv. Comm.)

Addr: 16 Cours Albert 1er, 75008 PARIS. **Te:** +33 (1) 44 21 66 66. **Fax:** +33 (1) 45 63 78 52.
L.P: Dir. Gen: Jean Drucker.
D.Prgrs: 24h.

	ARTE	kW (ERP)	M6	kW (ERP)
Abbeville	45	80	63	80
Ajaccio	41	110	44	110
Alencon	42	12	45	12
Amiens	49	5	52	(5)
Angers	50	20	53	(15)
Angouleme	31	5	34	(5)
Annemasse	54	3.8	51	3.8
Argenton	38	18	51	18
Aurillac	57	80	67	80
Autun	45	300	42	(300)
Auxerre	55	50	49	(100)
Avignon	47	40	36	40
Bastia	49	5	35	5
Bayeux	52	1.6	49	1.6
Bayonne	56	40	45	(40)
Beauvais	49	100	52	(100)
Bergerac	66	90	58	90
Bordeaux	65	130	43	10
Boulogne	59	3.2	62	(3.2)
Bourges	21	130	56	(130)
Brest	34	100	60	(100)
Brive	21	2	48	(2)
Caen	64	100	61	100
Carcassonne	46	56	43	56
Chalon-Sur-Saone	44V	2	41V	(2)
Chambery	55	4	52	(4)
Charleville	44	5	41	(5)
Chartres	47	50	44	(50)
Cherbourg	35	10		
Cluses	32	1.6	63	1.6
Clermont-Ferrand	30	100	33	(100)
Dijon	46V	1		
Dunkerque	59	1	62	1.8
Gueret	30	10	33	8
Hyeres	67	15	38	10
La Rochelle	48	1	51	1
Laval	33	20	30	(20)
Le Creusot	38	2.5	60	(2.5)
Le Havre	53	20	56	(20)
Le Mans	32	150	35	(150)
Lens	51	9.5	54	9.5
Les Sables D'Olonne	61	1	63	1
Le Touquet	55	2.1	32	2.1
Lille	65	6	53	(6)
Limoges	38	1.4	35	1.4
Lorient	62	1	65	1
Lyon	28	10	22	(10)
Lyon	59	250	62	(250)
Mantes	55	10	53	(10)
Marseille	32	150	38	(100)
Marseille	54	22	49	(22)
Maubeuge	32	3	37	(3.5)
Mende	50	1.5	53	(1.5)
Metz	39	200		
Montpellier	48	250	40	100
Nancy	55	7	43	(5)
Nantes	21	50	65	(50)
Neufchatel	—	—	31	10
Nice	51	2	45	2
Niort	38	200	64	(200)
Orleans	52	20	36	(20)
Paris	30	100	33	(100)
Paris-Est	48	5	58	(5)
Paris-Nord	65	3	62	(3)
Paris-Sud	59	3	42	(3)
Parthenay	60	30	67	(30)
Pau	60	2	63	(2)
Perpignan	38	2	35	(2)
Poitiers	41	1	44	1
Reims	53	25	56	(25)
Rennes	34	75	31	(75)
Rouen	59	100	62	(100)
Royan	—	—	33	2.1
Saint-Etienne	65	10	55	10
Saint-Nazaire	55	4	52	(4)
Saint-Raphael	36	80	39	(4)
Sens	47	40	44	(40)
Toulon	57	50	60	(50)
Toulouse	32	5	34	(5)
Toulouse	29	80	38	(80)
Tours	57	80	54	(80)
Troyes	29	250	50	(250)
Valenciennes	49	6	34	(6)
Vannes	58	75	48	(75)
Villers Cotterets	39V	25	57V	(25)

+ 474 stations under 1kW.

TELE MONTE CARLO (relays in France)

Addr: 16 boulevard Princesse Charlotte, Monte Carlo 98090, France — **Te:** +33 93 15 14 15.
Stations: Toulon ch33 1kW, Marseille ch35 10kW, Marseille (town) ch51 0.25 all using L/SECAM.
D.Prgr: 1100-0000.

PRIVATE LOCAL STATIONS

TELE TOULOUSE (Private, Comm.)

Addr: 3 Place Alphonse Jourdain, 31069 Toulouse Cedex — **Te:** +33 61 23 65 65.
L.P: Etienne Mallet.
Station: Toulouse ch37 2kW ERP.
D.Prgr: 0630-0000.

TELE BLEUE (Private, Comm.)

Addr: Rue Brousan, 30128 Garons. **Te:** +33 66 70 01 23.
Station: NIMES ch61 250 W ERP.

TELE LYON METROPOLE (Priv., Comm.)

Addr: 21 bd Yves Farge, 69007 Lyon — **Te:** +33 78727373.
L.P: Jerome Bellay.
Station: Lyon ch25 2kW ERP. Lyon (south) ch56 1kW. St Etienne ch52 10 kW
D.Prgr: 0630-0000. rel.

8 Mont Blanc (Priv., Comm.)

Addr: Rue du Centre, 74330 Epagny. **Te:** +33 50 22 66 00.
L.P: André Campana.
Stations: Cluses ch47 1.8kW, Annemasse ch57 3.8kW ERP, Araches ch 60 1kW ERP. +28 low power stations under 1 kW. + Cluses ch47 1.8kW.

AQUI TV (Priv. Comm.)

Addr: BP 104, 24202 Sarlat Cedex. Tel. (33) 53 31 21 05. + (33) 53 03 50 01.
Stations: Bergerac Ch25 3.5kW. Perigueux Ch59 0.075kW. Perigueux Ch21 0.04kW. Sarlat Ch47 0.06kW.

GERMANY (Federal Republic)

TV-sets: 30.500.000 — **Colour:** PAL — **System:** B.

ARD (PROGRAMMDIREKTION DEUTSCHES FERNSEHEN)

Addr: Arnulfstraβe 42, D-8000 München 2. **Te:** +49 (89) 59 00 01. **Fax:** +49 (89) 5900 3249.
L.P: Prgr. Dir: Dr. Günter Struve; TD. Chairm.: Ingo Dahrendorf; Film Buyer: Klaus Lackschewitz.
NB: The ARD is an umbrella organisation representing regionalized German public radio- and tv broadcasters. The ARD is responsible for the first public tv network (ARD Eins) and third public tv programs. A cooperative ARD/SRG satellite channel is called **"Eins Plus"** (see satellite section: Kopernikus-, ASTRA- and TV-SAT satellites), but the regular **"ARD Eins"** prgrs are also relayed via satellite. (ECS II f1 13° East).
ARD members:
— **Bayerischer Rundfunk Fernsehen,** Rundfunkplatz 1, 8000 München 2. **Te:** +49 (89) 59 00 01 **Fax:** +49 (89) 5900 2375.
— **Hessischer Rundfunk Fernsehen,** Bertramstraβe 8, 6000 Frankfurt 1. **Te:** +49 (69) 1551 **Fax:** +49 (69) 1552 900.
— **MDR Fernsehen,** Kantstraβe 71-73, O-7030 Leipzig. **Te:** +49 (341) 22760 **Fax:** +49 (341) 5663 544.
— **NDR Fernsehen,** Rothenbaumchaussee 132, D-2000 Hamburg 13. **Te:** +49 (40) 4131 **Fax:** +49 (40) 4476 02.
— **ORB Fernsehen,** August-Bebel-Straβe 25-53, O-1591 Potsdam-Babelsberg. **Te:** +49 (331) 72 36 00 **Fax:** +49 (331) 77395.
— **Radio Bremen Fernsehen,** Hans-Bredow-Straβe 10, 2800 Bremen 44. **Te:** +49 (421) 2460 **Fax:** +49 (421) 246 2010/1010.
— **Saarländischer Rundfunk Fernsehen,** Funkhaus Halberg, P.O. Box 1050, 6600 Saarbrücken. **Te:** +49 (681) 6020 **Fax:** +49 (681) 6023 874.
— **SDR Fernsehen,** Neckarstraβe 230, 7000 Stuttgart 1. **Te:** +49 (711) 929 1 **Fax:** +49 (711) 929 2600.
— **SFB Fernsehen/B 1,** Masurenallee 8-14, W-1000 Berlin 19. **Te:** +49 (30) 3031 0 **Fax:** +49 (30) 301 50 62.
— **SWF Fernsehen,** Hans-Bredow-Straβe, P.O. Box 820, 7570 Baden-Baden. **Te:** +49 (7221) 92 0 **Fax:** +49 (7221) 92 20 13.
— **WDR Fernsehen,** Appelnhoffplatz 1, D-5000 Köln 1. **Te:** +49 (221) 22 01 **Fax:** +49 (221) 2204 800 **D.Prgr:** ARD Eins, Germany's first public tv network, broadcasts nationwide on average from 0800-0030 UTC (except for 2½ hrs. in the evening when regional tv from 13 tv centers is relayed). In 1992 ARD together with Germany's second public tv network, ZDF, started broadcasting a common breakfast tv program called "Morgenmagazin" (Mon-Fri 0500-0800). Germany's third program comprises a combination of seven separate channels: **NDR/RB** in **"N 3"**; **MDR** in **"MDR 3"**; **ORB** in **"ORB 3"**; **WDR** in **"West 3"**; **HR** in **"HR 3"**; **SWF/SDR/Saarländischer Rundfunk** in **"Südwest 3"**; **Bayerischer Rundfunk** in **"BR 3"**.

ZWEITES DEUTSCHES FERNSEHEN (ZDF)

Addr: P.O. Box 4040, 6500 Mainz. **Te:** +49 (6131) 70 1. **Fax:** +49 (6131) 7021 57. **L.P:** Gen. Dir: Prof. Dr. Dieter Stolte; TD: Dr. Albert Ziemer; Film Buyer: Dr. Hans-Jürgen Steimer.
D.Prgr: The ZDF is Germany's second public tv network, broadcasting nationwide on average from 0800-0030 UTC.
F.Pl: One main st. and ±130 rel. sts. opening 1993. Relay via satellite.

COMMERCIAL TV STATIONS

SAT EINS

Addr. Hegelstraβe 61, Mainz, 6500, Germany. **Te:** +49 (6) 131 380. **Fax:** +49 (6) 131 380 100. **Telex:** 176131959.
D.Prgr: This commercial satellite tv station, set up by a number of German publishing houses, broadcasts on average from 0430-0030. Reg. prgr. from Hannover at 1645-1715 UTC.

RTL PLUS

Addr: Aachenerstraβe 1036, Cologne 40, 5000, Germany. **Te:** +49 (221) 48 950. **Fax:** +49 (221) 48 95 490.
see satellite section.
Dr. Prgr: Originally Luxembourg's German language tv channel, but due to German legal rules currently operating from Cologne. RTL Plus broadcasts 24 hrs. with loc. prgrs. (Hamburg, Essen, Frankfurt, Mannheim, München, Berlin) from 1700-1745.
NB: Late 1992 RTL started up a new station called **"RTL2"**. (details see satellite section: ASTRA 19.2° East/ECS II f1 13° East).

DEUTSCHES SPORTFERNSEHEN (DSF)

(formerly Tele Fünf)
NB: Since January 1st 1993 former Tele Fünf changed its format to a full 24h. sports & leisure channel under the new name "DSF".
Addr: Broadcast Center Munich, Bahnhofstraβe 28, D-8043 Unterföhring.
L.P: Man. Dir: Donald P.T. McLoughlin; Press & PR: Margit Lehmann.
D.Prgr: 24h. Sports & Leisure.

VOX (Westschienenkanal)

NB: This new "Infotainment" channel, broadcasting on satellite and terrestrial trs. (mostly lps) started on January 1st 1993. The freqs. of SAT 1 in North Rhine-Westphalia will eventually be taken over by VOX.
Addr: Richard-Byrd-Straβe 6, W-5000 Köln 30. **Te:** +49 (221) 160 210.
D.Prgr: N. every h. on the h. (10 min). Every half h. "Headline News".
Stations: Dortmund ch26 (100W); Gevelsberg ch5 (2.5W); Iserlohn ch7 (2W); Herdecke ch22 (20W); Siegen ch36 30W); Unna ch51 (200W); Lippstadt ch24 (30W); Hamm ch? (?W); Wuppertal ch10 (1000W); Bonn ch27 (60W); Remscheid ch? (?W); Köln ch27 (200W); Stolberg ch29 (1500W); Düren ch22 (400W); Siegburg/Troisdorf ch51 (200W); Bonn ch51 (200W); Köln/Bonn ch40 (3000W); Bielefeld ch36 (500W); Münster ch34 (200W); Paderborn ch22 (100W); Rheine ch57 (1000W); Burscheid ch41 (4000W); Bochum ch50 (100W); Mülheim ch27 (250W).

STATIONS:

1) BAYERISCHER RUNDFUNK FERNSEHEN

BR. third program = "Bayerisches Fernsehen", at some moments split up in an "Altbayern" and "Frankenland" version.

Area	ARD	ZDF	3	SAT1	RTL	DSF	POWER kW
Amberg	—	37	43	50	52	—	350/320
Augsburg	—	23	44	38	30	58	310/430
Bamberg	52	24	56	45	48	—	50/85/90
Bayreuth	—	30	54	39	46	—	98/96
Brotjacklr./D.	7	33	40	22	35	52	100/380/270
Büttelberg	55	—	—	—	—	—	400
Coburg	—	22	41	—	—	—	190/41
Dillb./Nürnb.	6	34	59	40	36	53	100/400/492
Grünten-Allgäu	2	28	46	—	—	—	100/470/470
H.Linie/Rgnsb.	53	21	42	38	34	46	75/370/400
H.Peiβenberg	25	22	—	42	33	—	10/350
Högl	—	42	50	—	—	—	40
Hoher Bogen	55	28	59	—	—	—	200
Hühnrb./Hsslb.	60	32	47	—	—	—	400/140/200
Kreuzb./R.	3	29	49	44	—	21	100/350/270

Area	ARD	ZDF	3	SAT1	RTL	DSF	POWER kW
Landshut	—	39	58	—	—	—	27/55
Ochsenk./Hof	4V	23	57	—	—	—	100/500/500
Passau	—	30	60	—	—	—	41/34
Pfaffenb./S.	59	35	51	—	—	—	100/230/250
Pfaffenhofen	—	31	41	—	—	—	480/420
Pfarrkirchen	—	27	57	—	—	—	250
Schnaitsee	—	26	54	—	—	—	360/280
Wendelstein/	10	35	56	59	24	37	100/200/200
München	—	—	—	—	—	38	—
Würzburg	10	25	45	21/	57/	—	5/410/360
	—	—	—	38	60	—	

2) HESSISCHER RUNDFUNK FERNSEHEN

HR. third program = "Hessen 3".

Area	ARD	ZDF	3	SAT1	RTL	DSF	POWER kW
Biedenk./A.	2	24	52	—	—	—	100/500/500
Fulda/Rhön	47	29	37	60	41	—	10
G. Feldberg	8	34	54	—	—	—	100/500/500
Habichtswald	56	28	41	—	—	—	50/100/100
Hardb./Krehb.	5	33	43	—	—	—	10/100/100
Hoher Meißner	7	32	55	—	—	—	100/390/470
Hohes Lohr	—	32	45	—	—	—	63/63
Kassel	—	—	—	42	35	58/60	
Rimberg	57	25	39	—	—	—	400/400/332
Würzberg	56	—	—	—	—	—	100/350/350

3) MITTELDEUTSCHER RUNDFUNK FERNSEHEN

MDR. third program = "MDR 3"

Area	ARD	ZDF	3	SAT1	RTL	DSF	POWER kW
Brocken	6H	—	34H	—	—	—	100/1000
Chemnitz	8	—	32	—	—	—	—
Dequede	12V	—	21H	—	—	—	5/1000
Dresden	10V	46	29H	—	—	—	100/100/1000
Halle	—	—	—	57	—	—	?
Inselberg	5H	—	31H	—	—	—	100/1000
Leipzig	9V	42	22H	—	—	—	100/500/1000
Löbau	27H	56	39H	—	—	—	500/500
Magdeburg	—	—	—	47	—	—	?
Sonneberg	12H	—	33H	—	—	—	30/500

D.Prgr: MDR3: Mon-Fri 1300-2330 approx. (includes relays of SAT1, and also local "Fenster" (Window) programmes. Sat/Sun: 1200-0100 approx.

4) NDR FERNSEHEN/B1/RB FERNSEHEN

NDR/RB. third program = "N 3". NDR1 Hamburg on ch E56 (500kW), NDR1 Cuxhaven on ch E51 (250kW).

NB: On October 1st 1992 Sender Freies Berlin Fernsehen started up a new regional prgr. under the name **"B 1"**. This prgr. replaces "N 3" in the Berlin area.

Area	ARD	ZDF	3	SAT1	RTL	DSF	POWER kW
Aurich	53	33	43	—	—	—	230/410/400
Bannenb./Höhb.	43	21	45	—	—	—	250/500/440
Berlin	7	33	39	25	22	9V	100/430/200
Bremen	22	32	42	29	46/49	—	100/500/500
	—	—	—	—	49	—	
Bremerh./Cuxh.	45	24	48	5	8/11	—	30/330/330
Bungsb./Eutin	50	21	47	31	44	—	260/235/280
Flensburg	4	39	57	28	24	—	50/160/150
Hamburg	9/26	30	40	48	46	34	20/100/460/500
Hannover	8	24	44	40	36	—	5/500/500
Harz/Göttingen	10	21	59	39	29	—	100/160/250
Heide/Eiderst.	10V	31	44	—	—	—	25/500/380
Helpterberg	37	52	22	—	—	—	500/200/1000
Kiel	5	35	55	53	24	—	2/250/200
Lingen	41	24	59	30V	—	—	400/500/400
Lübeck	—	23	33	42	36	—	240
Marlow	8	43V	24V	—	—	—	100/100
Minden	—	26	54	—	—	—	500/20
Neumünster/S.	28	26	45	—	—	—	500/100/100
Niebüll	29	34	60	—	—	—	100/200/150
Osnabrück	50	39	56	44V	36	—	160/250/138
Schwerin	11	—	29	—	—	—	100
Stadthagen	47	—	—	—	—	—	100
Steinkimmen/C.	55	37	40	27V	35V	—	400/47/47
Torfhaus	—	23	53	—	—	—	500
Ülzen	—	27	58	—	—	—	500/400
Visslh./Verden	7V	25	60	—	—	—	20/66/50

5) OSTDEUTSCHER RUNDFUNK BRANDENBURG FERSEHEN

ORB. third program = "ORB 3" (broadcast to Brandenburg)

Area	ARD	ZDF	3	SAT1	RTL	DSF	POWER kW
Berlin	5	—	27	—	—	—	100
Cottbus	53	57	23	—	—	—	100
Höhbeck	—	—	51	—	—	—	100

6) SWF FERNSEHEN/SDR FERNSEHEN/ SR FERNSEHEN

SWF/SDR/SR. third program = "Südwest 3" (further regional split-ups as indicated under a-d).

a) SWF-Mainz:

Area	ARD	ZDF	3	SAT1	RTL	DSF	POWER kW
Ahrweiler	—	33	56	—	—	—	170/210
Bd. Marienberg	47	21	44	—	—	—	50/190/165
Boppard	—	28	41	—	—	—	160/120
Donnersberg	10	37	60	—	—	—	100/250/250
Eifel	23	30	40	—	—	—	20/195/180
Haardtkopf	25	35	55	—	—	—	400
Hohe Derst	23	44	46	—	—	—	1
Kaiserslautern	3	22	44	50	33	—	25/25
Koblenz	6	31	51	57	36	—	50/85/83
Mainz	11	—	—	36	—	44	1
Saarbrücken/	29	45	53	26/	—	—	20/50/50
Teufelskopf	—	—	—	56	—	—	
Trier	5	22	48	56	—	—	0.6/41/22
Weinbiet	6	—	—	—	—	—	30

b) SR-Saarbrücken

Area	ARD	ZDF	3	SAT1	RTL	DSF	POWER kW
Göttelb.H./	2	32	42	—	—	—	100/470/500
Saarbrücken	—	45	—	—	—	—	500
Nelb./Piesb.	28	36	44	—	—	—	?

c) SWF-Baden-Baden

Area	ARD	ZDF	3	SAT1	RTL	DSF	POWER kW
Brandenkopf	48	28	45	—	—	—	2.5/25/50
Eggb./Hochrhn.	22	39	52	—	—	—	0.8/225/295
Feldberg/	8	22	57	50/	36/	—	100/100/100
Donausch.	—	—	—	60	38	—	
Freib. im Brsg.	7	33	58	—	38	—	0.9/500/500
Grünten	43	—	—	—	—	—	500
Hornisg./B-B	9	31	41	49	—	55	80/220/290
Raichen-bach/R.	4	35	55	—	—	—	100/260/330
Ravensburg	—	37	40	27	—	30V	270/290
Waldshut	21	—	—	—	—	—	1
Wannenberg	30	—	—	—	—	—	5

d) SDR-Stuttgart

Area	ARD	ZDF	3	SAT1	RTL	DSF	POWER kW
Aalen/Heubach	8	29	52	—	—	—	50/209/135
Bd. Mergenth	48	—	—	—	—	—	10
Heidelberg	7/50	27	53	—	—	—	100/500/440/492
Heilb./Eberb.	49	30	58	—	—	—	10/215/195
Langen-brand/P.	21	34	59	23	—	—	100/170/250
Stuttgart	11	26	39	—	—	—	100/404/234
Ulm	—	33	54	36	—	48/	370/330
	—	—	—	—	—	51	
Waldenburg/L.	9	28	42	—	—	—	100/165/185

7) WESTDEUTSCHER RUNDFUNK FERNSEHEN

WDR. third program = "West 3".

Area	ARD	ZDF	3	SAT1	RTL	DSF	POWER kW
Aachen	24	37	58	27	26	26	200/320/400
Bonn	43	26	49	—	5	—	100/81/100
Dortmund	—	25	53/	47	58	58	500/470
	—	—	43*	—	—	—	
Ederk./Hochs.	50	27	40	—	—	—	50/250/250
Eggebirge	—	31	48	—	—	—	250
Kleve/	46	35	48/	—	52	52	100/500/475/80
Wesel	—	—	59	—	—	—	
Köln	11	—	—	—	—	—	10
Langenberg/	9	29	39/	44	36	36	100/500/100/450
Düsseldorf	—	—	55	—	—	—	
Minden	—	26	57	—	—	—	500/200
Münster	32	21	45	51	38	38	500/170/250
Nordh./Lüdens.	30	37	60	—	—	—	240/200/250
T. Wald/Bielef.	11	33	46	38	59	59	100/310/450
Wuppertal	6	22	42	—	—	—	0.6/100/85

*) possible new frequency to be used by VOX.

OTHER TV STATIONS

CFN/RFC-TV (Canada)

Addr: 9 Jammstraβe, D-7630 Lahr. **Te:** +49 (7821) 20 98 92.
L.P: Dir: K. Randall; Tec. Dir: G. Campbell; Studio Mgr: P. Bisson.
Stations: Lahr ch60V (6W); ch38 (6W)
D.Prgr: English (ch60) 1315-2300, French (ch38) 1300-2300.
System: SECAM (G).

AMERICAN FORCES TELEVISION GERMANY

Addr: Headquarters AFN Europe, Bertramstraβe 6, D-6000 Frankfurt a/M. (or APO New York, NY 09757, USA) **Te:** +49 (611) 1516 101.
Stations:

Site	ch	kW	Pol		ch	kW	Pol
Wiesbaden AB	22	2	H	Karlsruhe	39	0.1	H
Berlin	29	0.5	H	Ulm	42	0.1	H
Kaiserslautern	30	3	H	Bitburg	51	5	V
Garlstedt	35	0.2	H	Mannheim	56	0.15	V
H. Oldendorf	38	0.1	V	Kitzingen LB	58	0.2	V

+ many lps less than 0.1kW
Satellite: Intelsat 604 (60° East) **D.Prgr:** 24h.

SSVC TELEVISION (U.K.)

Addr: BFPO 40
System: PAL (I)
Stations:

Site	ch	kW	Pol		ch	kW	Pol
Münster	23	0.8	H	Hameln	44	0.1	V
Lemgo	37	0.15	V	Osnabrück	48	0.9	H
Geilenkirchen	40	0.4	V	Paderborn	49	0.2	H
Hannover	40	0.4	V	Wassenberg	49	0.4	H
Bielefeld	41	0.1	V	Celle	51	0.2	V
Hildesheim	41	0.2	V	Herford	52	0.1	V
Mülheim	41	0.1	H	Niederkrüchten	53	1	H
Berlin Westend	41	0.2	V	Hamm	55	0.1	V
Minden	41	0.4	H	Wolfenbüttel	56	0.14	V
Gütersloh	44	0.4	V	Düsseldorf	57	0.12	H

+ many lps less than 100W
D.Prgr in English: 1100-2200. Special service intended solely for H.M. Forces.

BELGIUM FORCES TELEVISION

Stations: Bensberg ch39 (600W); N.-Hinsbeck ch47H (200W); Bensberg & N.-Hinsbeck ch51H (800/200W); Titz ch39H (600W); Arolsen ch9 (500W)

FRENCH FORCES TELEVISION

Addr: see radio.
Stations: Berlin ch31H (85kW).
System: SECAM (G).
D.Prgr: relays French TV5.

RUSSIAN TELEVISION RELAYS

Stations:

Loc	ch	Loc	ch
Frankfurt a/O	6	Neuruppin	12
Wittenberge	7V	Magdeburg	12
Cottbus	8	Wünsdorf	21
Ludwigslust	8	Karl–Marx–St.	27
Nauen	10	Rochlitz	28
Premnitz	11	Dresden	32
Rosslau	11	Erfurt	39
Bad Düben	12		

D.Prgr: Mon-Fri 0500-0830, 1130-2300; Sat-Sun 0500-2300. Relays of TSS-1 via satellite from Moscow.

GIBRALTAR

TV-sets: 7.500 — **Colour:** PAL — **System:** B & G.

GBC TELEVISION (partly comm.)

Addr: Broadcasting House, 18 So. Barrack Rd, Gibraltar.
Te: +350 79760. **Cable:** Broadcasts. **Fax:** +350 78673. **Telex:** 2229 GBCEE GK.
L.P: GM: George Valarino. Dir: Jue Adambery.
Station: chE6 0.2/0.4kW H (+ low power repeaters ch12, 53, 56).
D.Prgr: 1600-2300.

GREECE

TV-sets: 2.300.000 — **Colour:** SECAM — **Systems:** B & G.

ELLINIKI TILEORASSI-1 (ET-1)

Addr: Leophoros Messogeion 432, GR-153 42 Aghia Paraskevi Attikis. **Te:** +30 (1) 63 95 970. **Fax:** +30 (1) 63 92 263. **Telex:** 216066.
L.P: Dir. Gen: M. Demetriou. (Other personnel, see radio).

Stations	ch	kW	Pol		ch	kW	Pol
Akarnanika	3	1.5/		Rhodes Isl.	9	10/2	V
		0.3	H	Mytilini	9	10/2	H
Thessaloniki	5	30/6	H	Tripolis	10	30/6	H
Pilion (Volos)	6	30/6	H	Heraklion	10	10/2	H
Alexandroupolis	6	30/6	V	Ioannina	10	10/2	H
Kalamata	6	10/2	H	Parnis (Athens)	11	30/6	H
Chania	7	10/2	H	Hymettos			
Kavala	7	30/6	H	(Athens)	21	15/3	H
Kastoria	7	30/6	V	Tholopotamos	22	30/6	H
Kefalinia Isl.	8	30/6	V	Alexandroupolis	33	15/3	H
Thira	8	30/6	V	Thassos Isl.	39	550/55	H
Gerania	9	10/2	H	Euros	50	18/7	H
Corfu Isl.	9	10/2	V				

+ 700 low power repeaters.
D.Prgr: W 0500-1315, 1630-2220; Sat 1200-2215; Sun 1100-2200.

ELLINIKI TILEORASSI-2 (ET-2)

Addr: Leophoros Messogeion 136, GR-115 25 Athens. **Te:** +30 (1) 7701911. **Fax:** +30 (1) 77 97 776. **Telex:** 210886.
L.P: Gen. Dir: D. Yiannarakos.

Stations	ch	kW	Pol		ch	kW	Pol
Hymettos				Pilion (Athens)	41	750	H
(Athens)	5	30/7.5	V	Rhodes Isl.	42	500	H
Alexandroupolis	21	170	H	Tripolis	42	500	H
Thassos Isl.	23	1000	H	Akarnanika	43	1000	H
Ioannina	25	250	H	Mytilini	48	200	H
Thira Isl.	29	500	H	Vitsi (Kastoria)	49	250	H
Thessaloniki	30	1000/		Corfu Isl.	50	530	H
		200	H	Kefalinia Isl.	57	430	H
Parnis (Athens)	34	450	H	Kavala	59	650	H

+300 low power repeaters

ET-3

Addr: Aggelaki 2. GR-546 21 Thessaloniki. **Te:** +30 (31) 27 87 84. **Fax:** +30 (3) 23 64 66.
L.P: Dir. Gen: G. Nikolopoulos.
Station: Thessaloniki ch23/27 H (local prgrs), Pilion ch 44H, Paggaion ch 35H, Parnis ch 52H, Hymettos ch 31H, Thassos Isl. ch 26H, Florina ch 39H, Polygiros ch 21H.

AFN TV

Station: chA2 Iraklion, A6.

HUNGARY

TV-sets: 4.261.600 — **Colour:** PAL — **Systems:** D & K.

MAGYAR TELEVÍZIÓ

Addr: Szabadság ter 17, 1810 Budapest, 5. **Te:** +36 1111 4059. **Fax:** +36 1115 74979. **Telex:** 325558.
L.P: Pres: Gyula Berecky.
Stations: (Prgr. I System D):
D.Prgr: MTV1— ch33 (Vasvár), MTV2 — ch31 (Nagykanizsa). MTV2 — ch46 (Vasvár), MTV2 — ch32 (UNID).

Prgr. I	ch	kW (ERP)	Pol				
Budapest	R1	150/50	H	Kékes	R8	30/3	H
Nagyka-				Sopron	R9	5/1	V
nizsa	R1	50/5	V	Miskolc	R9	5/1	H
Pécs	R2	60/6	V	Szentes	R10	200/20	V
Tokaj	R4	80/20	H	Kabhegy	R12	150/15	H
Komádi	R7	30/3	V	Csávoly	28	80/8	H
Györ	R8	10/1	V				

Prgr. II (System K) **kW**

	ch	kW (ERP)	Pol		ch	kW (ERP)	Pol
Csávoly	R7	9/0.9	V	Tokaj	26	420/42	H
Kabhegy	22	760/76	H	Komádi	32	200/20	H
Szentes	23	480/48	H	Pécs	32	400/40	H
Budapest	24	1000/100	H	Györ	35	105/10	H
				Kékes	36	880/88	H

low power st's not mentioned.

D.Prgr: Prgr. I: Tue/Thu 0655-1045, 1440-2245; Mon/Wed 0800-1045, 1440-2245; Fri 0800-1045, 1440-2345; Sat 0530-1140, 1415-2300; Sun 0700-1140, 1250-2210. Prgr. II: 1600-2215; Sat 1300-2320; Sun 1300-2300.

ICELAND

TV-sets: 76.250 — **Colour:** PAL — **Systems:** B&G.

RÍKISÚTVARPID — SJÓNVARP ICELANDIC NATIONAL BROADCASTING SERVICE — TELEVISION

Addr: Laugavegur 176, 105 Reykjavik. **Te:** +354 (1) 693900. **Fax:** +354 (1) 693789. **Telex:** 051 94012601 RUVI G.
L.P: MD: Petur Guafinnsson. CE: E. Valdimarsson.
Stations: Pol H; Audio power 10% of visual power indicated.

Location	chE	kW (ERP)		chE	kW (ERP)
Stykkishólmur	3	90	Girdisholt	7	12.5
Skálafell	4	300	Hegranes	8	12
Gagnheidi	4	80	Hnjúkar	9	20
Vestmannaeyjar	5	39	Djúpivogar	10	25
Heidarfjall	5	10	Vatnsendi	10	20
Vadlaheidi	6	490	Vidarfjall	11	8.4
Háfell	7	1100			

+ 132 low power relay st's.

D.Prgr: Mon-Thurs 1800-2300; Fri 1800-2400; Sat 1600-0100; Sun 1600-2330.

STÖD 2 (Comm.)

Addr: Krókhálsi 6, 110 Reykjavík. **Te:** +354 (1) 672255/3777. **Fax:** +354 (1) 673011. **Telex:** 3111.
L.P: Dir. Gen: Jón Ó. Ragnarsson. Dir. Prgr: Jónas R. Jónsson.
Station: Vatnsendi ch12 10kW.
D.Prgr: 1730-0100, also Sat/Sun 0800-1100. Scrambled prgr. from 2100 and on Sat/Sun mornings.

IRELAND

TV-sets: 1m — **Colour:** PAL — **Systems:** I.

RADIO TELEFIS EIREANN (Statutory Corporation)

Addr: Donnybrook, Dublin 4. **Te:** +353 (1) 643111. **Telex:** 93700. **Fax:** +353 (1) 643080.
L.P: Contr. of TV. Prod: Liam Miller.

Stations	I	II	kW	Pol		I	II	kW	Pol
Maghera	B		100	H	Mt.Leinster	F	I	100	V
		H	100	V	Kippure	H	J	100	H
Mullaghanish	D	g	100	V	Truskmore	I	G	100	H
N.E.Donegal	23	26	20	H	Longford	40	43	800	H
Dublin	29	33	25	H	Co.Louth	52	56	250	V

+36 low power transposer st's.

D.Prgr: RTE One, Mon-Fri 1100-2345 (Fri 0100 appr.), Sat 0900-0100, Sun 1005-2345. RTE Two, Mon-Fri 1430-2330, Sat 1230-0030, Sun 1015-2345.

ITALY

TV-sets: 17m — **Colour:** PAL — **Systems:** B & G.

RADIOTELEVISIONE ITALIANA

Addr: Direzione Centrale TV, Viale Mazzini 14, 00195 Roma. **Te:** +39 (6) 36861. **Fax:** +39 (6) 3612012.
L.P: MD: Carlo Fuscagni. Prgr. Dir: Mario Maffucci.
RAI Uno (System B).
Audio powers 1/10 of vision powers; Pol = H exc. where indicated.

Location	ch	kW (ERP)	Location	ch	kW (ERP)
Monte Caccia(a)	A	34	Udine	F	55
Monte Cammarata(b)	A	35	Pescara	F	14
Monte Nerone(c)	A	34	Roseto Capo Spulico(p)	F	12V
Monte Penice(d)	B	100	San Cerbone(q)	G	48
Monte Falto(e)	B	40	Roma-Monte Mario	G	36
Crotone(f)	B	5V	Monte Serpeddi(r)	G	35
Torino	C	16	Milano	G	24
Monte Serra(g)	D	270	Pantelleria	G	24V
Martina Franca(h)	D	220	Trieste	G	15
Punta Badde Urbara(i)	D	145	Monte Scuro(s)	G	8
Monte Venda(j)	D	100	Portofino	H	127
Gambarie(k)	D	19	Monte Sambuco(t)	H	35
Monte Conero(l)	E	100	Monte Peglia(u)	H	32
Monte Soro(m)	E	5	Monte Pellegrino(v)	H	8
Malonno	E	4V	Col Visentin(x)	H	3.5
Golfo di Salerno(f)	E	4V	Monte Limbara(y)	H	3
Monte Argentario(n)	E	2.5	Monte Favone (ø)	H	2
Monte Lauro(o)	F	200			

+ many low power st's.

D.Prgr: Mon-Fri 1130-2300; Sat/Sun 0900-2300.

Rai Due (System G).

Location	ch	kW (ERP)	Location	ch	kW (ERP)
Paganella(aa)	21	10	Bologna	28	10
Udine	22	150	Portofino	29	1000
Monte Penice(d)	23	1800	Monte Favone	29	20
Monte Faito(e)	23	1400V	Torino	30	200
Capo Spartivento(z)	23	14.6V	Pescara	30	30V
Monte Luco(bb)	23	20	Monte Serpeddi(r)	30	25
Monte Argentario(n)	24	250V	Bertinoro	30	20
Monte Lauro(o)	24	200	Monte Peglia(u)	31	200
Monte Caccia(a)	25	400	Trieste	31	25
Monte Venda(j)	25	1000	St. Vincent	31	16.5
Gambarie(l)	26	200V	Monte Vergine	31	15
Velletri	26	200	Trapani-Erice	31	10
Monte Conero(l)	26	20	Martina-Franca(h)	32	700
Milano	26	15	Mont-Beigua(cc)	32	60
Monte Sambuco(t)	27	200	Monte Limbara(y)	32	20
Monte Serra(g)	27	200	Monte Soro(dd)	32	15
Punta Badde Urbara(i)	27	200	Monte Nerone(c)	33	200
			Monte Cammarata(b)	34	200
Monte Pellegrino(v)	27	40V	Salento(ee)	34	16
Crotone(f)	27	11	Col Visentin(x)	34	10
Monte Scuro(s)	28	30	Monte Cavo	35	20
Roma-Monte Mario	28	150	Vibo Valentia(ø)	35	16

+ st's below 10kW not mentioned.

D.Prgr: Mon-Fri 1130-2300; Sat/Sun 0900-2300.

TV-3

Location	ch	kW (ERP)	Location	ch	kW (ERP)
Monte Venda(j)	32	1000	Portofino	45	1000
Milano	33	26	Monte Argentario	46	500V
Monte Penice	35	2000	Monte Scuro(s)	46	60
Monte Faito(e)	39	1250V	Pescara	46	60V
Monte Luco	39	40	Monte Sambuco	47	400
Torino	40	400	Udine	47	300
Monte Pellegrino(v)	40	75V	Punta Badde		
Bertinoro	40	40	Urbara(i)	47	400
Crotone	40	20	Monte Paganella	47	30
Monte Lauro(o)	41	400	Bologna	48	200
St. Vincent	41	20	Monte Cammarata	48	400
Monte Conero	42	400	Martina Franca	49	1000
Monte Beigua	42	120	Monte Serpeddi	49	500
Monte Serra(g)	43	400	Monte Peglia	49	400
Roma	43	300	Monte Sambuco(t)	50	400
Monte Caccia(a)	44	610	Gambarie	51	400V
Gamborie	44	400V	Monte Nerone(c)	54	400
Trieste	44	50	Campo de Fiori	59	20

+ st's below 10kW(ERP) not mentioned.

D.Prgrs: W 1600-2300; Sun 1130-1230, 1400-2300.

a) Bari — b) Ribera, Sicilia — c) Arezzo — d) Piacenza — e) Salerno — f) Calabria — g) Lucca — h) Taranto — i) Suni, Sardegna — j) Rovigo — k) Messina — l) Ancona — m) Taormina — n) Grosseto — o) Siracuso — p) Taranto — q) Firenze — r) Cagliari — s) Cosenza — t) Campobasso — u) Perugia — v)

Palermo — x) Udine — y) Sargono, Sardegna — z) Reggio Calabria — ø) Catanzaro — aa) Bolzano — bb) Firenze — cc) Genova — dd) Taormina — ee) Otranto.

RUNDFUNKANSTALT SÜDTIROL (RAS)

(Public Statutory Body of the Autonomous Province of Southern Tyrol)
ADDR: Europaalee 164A, I-39100 Bozen.
L.P: See radio.
Stations: RAS1 (relay ORF-FS1); RAS2 (relay ZDF); RAS3 (relay ORF-FS2+SRG).

Location	ch	kW	Netw.	Location	ch	kW	Netw.
Perdonig	23	2.5	III	Penegal	55	10	II
Penegal	48	10	III	Perdonig	59	7.5	I
Vinschgau	50	3	III	Vinschgau	63	3	II
Penegal	52	10	I	Perdonig	65	7.5	II
Vinschgau	53	3	I				

+ 122 low power st's not mentioned.

PRIVATE TV STATION NETWORKS

There are about 840 privately operated tv sts in Italy, mostly on local service. Due to space limitations, only those sts with nationwide networks are mentioned. Readers requiring more information should write (with Rp pls) to: D. Monferini, Via Davanzati 8, I-20158 Milano, Italy.

1)	Canale 5	24h	(key st. Milano)
2)	Circuito 5 Stelle	0600-0100	(key st. Milano)
3)	Cons. Italia 9 Netw.	24h	(key st. Firenze)
4)	Junior TV	24h	(key st. Firenze)
5)	Italia 7	24h	(key st. Milano)
6)	Italia 1	24h	(key st. Milano)
7)	Odeon	0600-0100	(key st. Rho-Milano)
8)	Rete A	24h	(key st. Milano)
9)	Rete Quattro	24h	(key st. Milano)
10)	Tele Montecarlo	0710-0130	(key st. Roma)
11)	Tele Piú 2 Sport	24h	(key st. Milano)
12)	Video Music	24h	(key st. Castelv. Pascoli)

Addresses and other information

1) Società Canale 5, Via Marconi 27, I-20090 Segrate (MI). **Te:** +39 (2) 216001. **Fax:** +39 (2) 2138019; 20 tv-sts affiliated.
2) Società Circuito Cinque Stelle, Via del Caravaggio 4, I-20144. **Te:** +39 (2) 433889. **Fax:** +39 (2) 4985514; 13 tv-sts affiliated.
3) Società Consorzio Italia 9 Network Television, Via Settala 57, I-20124 Milano. **Te:** +39 (2) 221964. **Fax:** +39 (2) 29404892; 20 tv-sts affiliated.
4) Junior TV, Corso Garibaldi 35, I-20121, Milano. **Te:** +39 (2) 801370. **Fax:** +39 (2) 71001165; 20 tv-sts affiliated.
5) Società Rete 10 s.r.l., Via Marconi 27, I-20090 Segrate. **Te:** +39 (2) 216001. **Fax:** +39 (2) 2138019; 20 tv-sts affiliated.
6) Italia 7 c/o Tele Radio City, Via Lavagello 31, I-15060 Castelletto d'Orba (AL). **Te:** +39 (2) 45701747. 19 tv-sts affiliated.
7) Odeon, Via Tavecchia 43/45, I-20017 Rho (MI). **Te:** +39 (2) 935151. **Fax:** +39 (2) 93504423. 18 tv-sts affiliated.
8) Reta A, Viale E. Marcelli 165, I-20099 Sesto S. Giovanni (MI). **Te:** +39 (2) 2428175. **Fax:** +39 (2) 2401630. 18 tv-sts affiliated.
9) Società Rete Quattro, Via Marconi 27, I-20090 Segrate (MI). **Te:** +39 (2) 216001. **Fax:** +39 (2) 2138019; 19 tv-sts affiliated.
10) Tele Montecarlo, Piazza della Balduina 59, I-00136 Roma. **Te:** +39 (6) 35584271. **Fax:** +39 (6) 3454043. 16 tv-sts affiliated.
11) Tele Piú, Via Piranesi 16, I-20137 Milano. **Te:** +39 (2) 700271. **Fax:** +39 (2) 70027201. 20 tv-sts affiliated. **FPI:** To become PYE-TV.
12) Società Elefante CTV c/o Video Music, P.O. Box 1, I-55020 Castelvecchio Pascoli (LU). **Te:** +39 (583) 7191. **Fax:** +39 (583) 723197; 20 tv-sts affiliated.

LATVIA

TV-sets: 1.2m — **Colour:** SECAM — **System:** D&K

LATVIJAS TELEVIZIJA (LTV) (Gov)

Addr: 3, Zakusala, Riga LV 1018 **Te:** +97 (132) 200 314/227 906
Tlx: 161188 video su **Fax:** +358 (49) 348 655
L.P.: Dir Gen: Imants Rakins; Ch. Eng: Ilmars Muuls; Hd. PR: Inese Vitkus

Location	Netw I ch/kW	Netw II ch/kW	Netw III ch/kW	ch/kW
Cesvaine	8/5	5/5	41/20	36/0.1
Dagda	36/0.01	24/0.01	—	—
Daugavpils	7/4.5	10/0.1	40/20	21/0.1
Dundaga	25/5	30/5	8/0.03	—
Jekabpils	—	—	29/0.1	—
Kandava	12/0.01	—	—	—
Kraslava	—	4/0.01	—	—
Kuldiga	6/5	1/5	40/0.1	—
Liepaja	33/5	21/5	35/2	5/0.1
Preili	—	4/0.01	2/0.01	—
Rezekne	39/20	27/20	6/0.3	10/0.1
Riga*	3/25	10/20	28/20	31/5
	—	—	—	7/0.3
Sabile	8/0.01	5/0.01	11/0.01	—
Valmiera	33/20	21/18	11/0.3	9/?
Ventpils	5/0.1	9/0.1	12/0.1	7/0.03
Vilani	—	—	11/0.01	—

*N.B: Riga ch7 relays foreign satellite prgrs. (Mon-Sun 0600-2200 UTC). Riga ch31 is leased to private companies

D.Prgr: Netw I (LTV national ch) Mon-Fri 0700-0900/1600-2200, Sat&Sun 0700-2200; Netw II (LTV 2nd ch) Mon-Fri 1600-2200 (LTV prgrs in Russian and retransmitted prgrs of Russian Ostankino TV); Netw III (relays of Russian tv) Mon-Sun 0600-2200
N: LTV I Mon-Fri 1600-1610/1830-1900, Sat&Sun 1830-1900; LTV II (in Russian) Mon-Fri 2100-2110

NON-GOVERNMENT STATIONS

There are over 100 regional, commercial and municipal tv stations and studios in the main centers of Latvia.
Stations: Latgales TV (Rezekne), Ogres TV (Ogre), Liepajas TV (Liepaja), Rojas TV (Roja), Daugavpils TV (Daugavpils), Malpils TV (Malpils), Rudbarzu TV, Vidzemes TV (Valmiera)

LITHUANIA

TV-sets: 1.400.000 — **Colour:** SECAM — **Systems:** D&K.

LIETUVOS TELEVIZIJA (Gov.)

ADDR: see radio. **Te:** +7012 (2) 534471. **Telex:** 261151 litvr. **Fax:** +7012 (2) 221571. **L.P:** Dir: Algirdas Kaušpedas.

STATE OWNED TRANSMITTERS

Location	ch R-	Pol	kW(ERP)
Taurage	1	H	0.2
Pažagieniai	1	H	0.2
Druskininkai	1	H	0.316
Vilnius	2	V	177
Raseiniai	2	H	0.2
Jurbarkas	3	H	0.2
Vilnius	4	V	177
Mažeikiai	4	H	0.2
Ignalina	5	H	0.2
N. Akmene	5	H	0.2
Neringa	5	H	0.2
Pažagieniai	5	H	0.2
Plunksniai	6	H	0.2
Utena	6	H	0.63
Marijampole	6	H	0.2
Salčininkeliai	6	H	0.035
Alytus	6	H	0.19
Mažeikiai	7	H	0.4
Kaunas	7	H	0.2
Klaipeda	8	H	50
Druskininkai	8	H	0.32
Juragiai	9	H	2300
Klaipeda	10	H	0.2
Pažagieniai	10	H	0.2
Ignalina	10	H	0.19
Bubiai	11	H	70
Vilnius	11	H	1
Alytus	11	H	0.2
Viešintos	12	H	63
Plunksniai	12	H	0.2
Juragiai	21	H	800

Bubiai	22	H	500
Neringa	22	H	0.355
Truikiai	24	H	0.2
Raseiniai	25	H	0.2
Vilnius	26	H	11
Viešintos	26	H	0.2
N. Akmene	27	H	0.2
Taurage	27	H	600
Juragiai	28	H	800
Klaipeda	29	H	700
Utena	29	H	0.36
Sniečkus	30	H	500
Vilnius	31	H	800
Viešintos	32	H	700
Sniečkus	33	H	125
Bubiai	34	H	500
Sniečkus	35	H	125
Taurage	39	H	600
Neringa	40	H	0.32
Mažeikiai	41	H	0.92

*) stn's planned to carry other prgrs in future.
D.PRGR: 0545-0800 (Tues-Fri), 1600-2230 (Mon-Fri), 0700-2230 (Sat/Sun). Closing times are approximate. **Russian:** 0915-0935 (Sat), 1650-1700 (Mon-Sat). **Polish:** 1400-1500 (Sun), 1700-1710 (Mon-Sat). **Rel. CNN International:** Mon-Fri 1500-1600.

LRT KAUNO DIREKCIJA (Gov.)

ADDR: see radio.
Prgrs: are broadcast at various times on Lietuvos Televizija.

RYTU LIETUVOS TELEVIZIJA (Gov.)

ADDR: Tumo-Vaižganto 2, LT-2000 Vilnius. **Te:** +7012 (2) 629472, 625723. **L.P:** Dir: Igoris Šileikis.

Stations	ch R-	Pol	kW(ERP)
Šalčininkėliai	6	H	0.035
Vilnius	38	H	500

D.PRGR: in Lithuanian/Russian/Polish for national minorities in Eastern Lithuania: 0500-0530 (Sun-Fri), 0615-1830 (Mon-Fri), 2030-2150 (Mon-Fri). Other times relays Lietuvos Televizija (ch R-6) or TP-1 from Poland (ch 38). **F.Pl:** Time expansion on weekends. Tr in Šalčininkėliai to be dismantled and replaced by a more powerful one.

PANEVĖŽIO TELESTUDIJA (Priv., Comm.)

ADDR: Angariečio 44-24, LT-5300 Panavėžys. **Te:** +7012 (54) 30830. **L.P:** Dir: S. Bukelis.

Station	ch R-	Pol	kW(ERP)
Pažagieniai	8	H	0.2

D.PRGR: test transmissions at various times. Regular broadcasts to begin on 1.1.1992 (times unknown).

PRIVATELY OWNED TRANSMITTERS

PANEVEZIO TELESTUDIJA:	Pažagieniai	ch R-8H	0.2 kW ERP
"BLOKAS":	Taurage	ch R-3H	0.01 kW ERP

PIRATE TRANSMITTERS known to be operating:

Elektrenai	ch R-6 unkn. power (10 W or so) — relays satellite prgrs.
Palange	— no further details known.

KAUNO TELESTUDIJA PLIUS (Priv.)

ADDR: Laisvės al. 86, LT-3000 Kaunas. **Te:** +7012 (7) 207220. **Telex:** 269860 apa. **Fax:** +7012 (7) 207220. **L.P:** Editor-in-chief: Arvydas Linartas.

Stations	ch R-	Pol	kW(ERP)
Kaunas	7	H	0.2

D.PRGR: 1300-0000 (closing time approximate). **Rel. SAT. 1** (Mon/Wed/Fri/Sun) with own voice-over translation of some of the movies into Lithuanian, and **Super Channel** (Tues/Thurs/Sat). **F.Pl:** morning broadcasts (0600-0900). Own prgrs.

RUSSIAN TELEVISION RELAYS

Stations	CH1	CH2	Pol.	kW
Pažagieniai	1	5	H	0.2
Vilnius	2	31	V/H	177/800
Raseiniai	2	25	H	0.2
Neringa	5	40	H	0.2/0.32
Mažeikiai	7	41	H	0.4/0.92
Ignalina	10	—	H	0.19
Alytus	11	6	H	0.2
Plunksniai	12	—	H	0.2
Jarugiai	12	21	H	800
Kleipeda	29	10	H	700/0.2
Viešintos	32	26	H	700/0.2
Bubiai	34	22	H	500
Sbiečkus	35	33	H	125/0.32
Mariampolė	—	6	H	0.2
Druskininkai	—	8	H	0.32
Truikiai	—	24	H	0.2
Naujoji Akmenė	—	27	H	0.2
Tauragė	—	28	H	600
Utena	—	29	H	0.36

Ostankino on Ch1; **Russia Channel** on Ch2.

SATELLITE TELEVISION RELAYS
Station only relaying satellite TV:

Station	ch R-	Pol	kW(ERP)
Vilnius	11	H	1

D.PRGR: relays **Sky News, TeleClub, MTV, Screensport** (times vary).

LUXEMBOURG

TV-sets: 100.500 — **Colour:** PAL & SECAM — **Systems:** B, L & G.

TELE LUXEMBOURG (Comm.)

ADDR: Compagnie Luxembourgeoise de Télédiffusion, Villa Louvigny, Luxembourg.
Te: +352 47661. **Fax:** +352 4766 2730.
L.P: Dir. Gen: G. Thorn. Dir. Prgr: J.C. de Keyser & J.C. Thieltgen. Dir. Adm: P. Ulveling. Chief Eng: P. Faber.
Stations: Dudelange ch7 (System B) 100/25kW H, Dudelange ch21 (System L) 1000/100kW, Dudelange ch27 (System G) 1000/100kW (2 studios), ch24 100kW.
D.Prgr. in French: W 1045-1210, 1600-2230 (Sat 1400-2240); Sun 1100-2200 on ch21 (SECAM) for French viewers and ch27 (PAL) for Belgium & Luxembourg. **Italian/Luxembourgian:** Sun 1100-1300.
"RTL Plus" in German: 1727-2130 on ch7 (PAL). (See W. Germany)
RTL-4 in Dutch & Multilingual. Luxembourg ch. 41 (30kW).

MACEDONIA

TV-sets: N/A — **Colour:** PAL — **Systems:** B&H

TELEVIZIJA MAKEDONIJE

Addr: Dolno Nerezi bb, 91000 Skopje. **Te:** +38 (91) 258 230.
Stations:
Macedonia
TV Makedonije
1st Prgr.

Pelister	4	30	Mali Vlaj	9	10
Golak	5	10	Vodno	9	10
Cnri Vrv	6	100	Popova Sapka	10	10
Stogovo	6	10	Turtel	11	50
Belasica	7	6.6	Vodno	12	10
Boskija	8	10			

+58 low power sts.
D.Prgr: Mon-Thu 1110-2200, Fri 1445-2200, Sat 0820-2145, Sun 0830-2100.
TV Skopje
2nd Prgr.

Ohrid	21	4	Oteseva	37	10
Turtel	22	1000	Golak	38	200
Pelister	29	600	Popova Sapka	38	10
Crn Vrv	30	600	Mali Vlaj	44	200
Stogovo	31	10	Boskija	57	20
Belasica	35	13			

+56 low power sts.
D.Prgr: Mon-Fri 1355-2230, Sat 0900-2230, Sun 0650-2200.

MALTA

TV-sets: 133.000 — **Colour:** PAL — **System:** B.

XANDIR TELEVISION (TV division of Telemalta Corp.)
Addr: See radio.
L.P: Pres: Manni Spiteri; MD: Roland Flamini; Dir. of Prgr: George Peresso.
Station: chE10 10/2.5kW H.
D. Prgr: 1700-2230 (Sun also 1200-1530).

MONACO

TV-sets: 20.000 — **Colour:** SECAM & PAL — **Systems:** L & G.

TELE MONTE CARLO (Comm.)
Addr: 16 Blvd. Princess Charlotte, MC 98090 Monaco-Cedex.
Cable: Tele-Carlo. **Te:** +3393 505940. **Fax:** +3393 250109.
Telex: 469823 Tele Carlo.
L.P: Pres: Jean-Louis Medecin. Dir. Prgrs: George Giaveret.
Stations: chF8 (L/SECAM) 50kW(ERP), ch30(L/SECAM) 500kW, ch33(G/PAL) 0.05kW, ch35(G/PAL) 40kW, ch39(L/SECAM) 0.02kW.
D.Prgr: French 1800-2400 on ch 8, 30, 39; **Italian** 1000-2400 on ch 33, 35.
(See also listing of relay st's in France).

NETHERLANDS

TV-sets: 6m — **Colour:** PAL — **System:** B & G.

NEDERLANDSE OMROEPPROGRAMMA STICHTING (NOS)
Addr: see radio
L.P: Dir: E.J.W. van Westerloo; Pub.Rel: A. de Vries; Prod: J.D.P. Meijer; Styl./Prgr. Coord: H.G. Kuperus; Curr. Aff: T. Kamlag; Hd. News Dept: G. van der Wulp.
Stations:

Loc.	Ned. 1	Ned. 2	Ned. 3	kW
Arnhem	50	53	43	30/30/30
Eys	51	54	48	1/1/1
Goes	29	32	35	250/250/250
Hulsberg	57	60	43	0.1/0.1/0.1
Lopik	4	27	30	100/1000/1000
Losser	—	26	34	3/3
Maastricht	53	56	59	1/1/1
Markelo	7	54	51	30/300/300
Noorbeek	46	49	52	0.01/0.01/0.01
Roermond	5	31	34	50/250/250
Slenaken	29	35	32	0.02/0.02/0.02
Smilde	6	47	44	40/1000/1000
St.Pietersberg	26	33	23	0.25/0.25/0.25
Wieringermeer	39	45	42	300/300/300
Wijk aan Zee	33	49	21	0.004/0.004/0.004

+ 5 low power repeaters.
D.Prgr: New broadcasting format as of October 1st 1992:

	Mon	Tue	Wed	Thu	Fri	Sat	Sun
Ned. 1	NCRV	AVRO	AVRO/ KRO/ NCRV	KRO	NCRV	AVRO	KRO
Ned. 2	EO/ TROS	VOO	EO/ NOS	EO	TROS	VOO	TROS/ VOO
Ned. 3	VPRO	NOS	VARA	VARA	NOS	VARA	VPRO

AMERICAN FORCES NETWORK
Addr: see Belgium
System: M. **Colour:** NTSC
Station: Soesterberg chA18 (European UHF ch24) 20kW
D.Prgr: Mon-Fri 1400-2200; Sat/Sun 1000-2200.

NORWAY

TV-sets: 1.500.000 — **Colour:** PAL — **Systems:** B & G.

NORSK RIKSKRINGKASTING
Addr: Bjoernstjerne, Bjoernson Plass 1, 0340 Oslo 3. **Te:** +47 (2) 459050. **Fax:** +47 (2) 459 618. **Telex:** International Relations 76820. Eurovision: 71794. TV News: 18530. Prgr. Purchases: 19676.
L.P: Dir. Gen (Radio & TV): Bjartmar Gjerde. Dir. Adm & Finance: Oddvar Svanevik. Techn. Dir: Kjell Lövaas. Dir. Prgr. TV: T. Strand Head of Depts: News: H. Sörebö. Cultural & Current Affairs: O. Foss. Sports and O.B: Rolf Johnsen. Films: B. Rinnan. International Rel: A. Amlie (Ag), Eurovision: S. Johanssen. Sales: S. Torjusson (see also Radio).

Location	ch	kW (ERP)	Location	ch	kW (ERP)
Greipstad(a)	2	60	Alta	7	5
Gulen	2	30	Mosvik	7	20
Melhus	2	100	Bokn(p)	8	100
Steigen	2	60	Gausta	8	3
Varanger(b)	2	30	Jetta	8	30
Bagn(c)	3	10	Kappfjell	8	2
Gamlemsveten(d)	3	60	Kistefjell	8	30
Hemnes(e)	3	60	Tana(q)	8	30
Kautokeino	3	8	Bergen	9	140
Bremanger	4	30	Hammerfest	9	30
Hadsel(f)	4	30	Kongsvinger(r)	9	60
Kongsberg(g)	4	100	Lyngdal(s)	9	30
Førde	5	1.2	Tron	9	30
Grong(h)	5	30	Vega	9	30
Karasjok	5	10	Narvik	10	30
Lyngen(i)	5	3	Norfjordeid	10	3
Nordhue(j)	5	60	Skien	10	60
Stord(k)	5	30	Halden	11	60
Bjerkreim(l)	6	15	Kopparen	11	20
Hestmannen	6	4.6	Lønahorgi	11	25
Nordkapp	6	30	Trolltind(t)	11	60
Oslo	6	100	Store Jekkir	11	4
Andenes(m)	6	4	Fister	11	1.2
Reinsfjel(n)	6	60	Ullandhaug	35	1.1
Hovdefjell	7	30	Ringevike	41	2
Salten	7	30	Mistberget	44	1.7
Sogndal(o)	7	20	Lifjell	52	1

+ 2100 low power repeaters less than 1kW not mentioned.
a) Kristiansand — b) Finnmark — c) Lillehammer — d) Ålesund — e) Trondheim — f) Vesterålen — g) Telemark — h) Namsos — i) Tromsö — j) Hamar — k) Haugesund — l) Stavanger — m) Narvik — n) Kristiansund — o) Leikanger — p) Stavanger — q) Finnmark — r) Hamar — s) Mandal — t) Tromsö.
D.Prgr: W 1655-2200; Sat 1655-2300; Sun 0955-1115, 1330-1445, 1700-2200 (times are approx.) School prgrs not included.

PHILIPS PETROLEUM 66
TV1: Ekofisk ch55. V. relays NRK.
TV2: Ekofisk ch52. V. relays Film Net.

POLAND

TV-sets: 10m — **Colour:** PAL — **Systems:** B & G.

TELEWIZJA POLSKA (Gov.)
Addr: ul.J.P. Woronicza 17, P-35,00950 Warszawa. **Te:** +48 (22) 478191. **Fax:** +48 (22) 440206. **Cable:** Telewar. **Telex:** 813421.
L.P: Pres: Andrej Drawicz.
Power = ERP Pol H exc. where indicated.

Prgr I	chR	kW/Pol	Prgr 1	chR	kW/Pol
Bydgoszcz	1	120	Kraków	10	200
Pila	2	50	Zamošć	10	100V
Warszawa	2	90	Rzeszów	12	100V
Kielce	3	100	Szczecin	12	100
Zielona Góra	3	200	Wroclaw	12	150
Suwalki	5	100	Przemyšl	24	100
Lódź	7	20	Warszawa	27	100
Bialystok	8	100	Plock	29	1000
Katowice	8	250	Jelenia Góra	30	200
Koszalin	8	100V	Rabka	31	2
Lublin	9	100	Opole	40	250
Poznań	9	150	Kudowa	34	?
Olsztyn	9	100V	Siedlce	52	300
Gdańsk	10	100	Klodzko	52	200

+124 lp's

Prgr II	chR	kW/Pol	Prgr II	chR	kW/Pol
Wroclaw	2	1	Kielce	28	1000
Lublin	2	0.4V	Zielona Góra	29	1000
Gdynia	3	2	Rzeszów	29	50
Kielce	5	2	Krakow	33	250
Lódź	10	1V	Szczecin	30	700
Warszawa	11	30	Świnoujscie	33	?
Bialystok	11	1V	Jelenia Góra	35	200
Koszalin	11	0.5V	Lobez	35	25
Bydgoscz	12	0.6V	Bydgoszcz	36	2
Rzeszów	7	0.5V	Suwalki	36	160
Katowice	21	400	Rabka	36	10
Elblag	21	20	Zamošć	36	250
Tarnów	22	15	Siedlce	37	50
Opole	23	250	Klodzko	38	200
Pila	24	140	Lomza	38	7
Lebork	25	50	Czluchów	39	13
Wroclaw	25	1000	Plock	39	1000
Olsztyn	26	270	Slupsk	40	12
Poznań	27	250	Przemyšl	41	100

+ 87 lp's.

Prgr. III (local on Progr. II transmitters): Katowice ch 21. **D. Prgr:** W 0800-1000, 1500-1530.

Satellite TV Relay: Katowice ch R6 20kW — **Echo (priv):** Wroclaw ch 28 1kW — **RAI Uno Relay:** Krakow ch 50 50kW — **Soviet TV Relay (TSS CT1):** Szczecin ch R3, Warszava ch 51 2kW.

TELE-9

Kielce ch R5, Tarnow ch R22, Kielce ch R28, Rzeszow ch R29, Kraków ch R33, Przemyśl ch R41.

PORTUGAL

TV-sets: 1.686.513 — **Colour:** PAL — **System:** B&G.

RADIOTELEVISÃO PORTUGUESA (RTP) (Comm.)

Addr: Head Office: Av. 5 de Outubro 197, 1094 Lisboa. **Te:** +351 (1) 7931 774. **Cable:** Televisão. **Telex:** 14527-RTP RE P. **Fax:** +351 (1) 7931 758. **Mailing addr:** Apartado 2934, Lisboa.

L.P: Stat. Mgr. José Eduardo Moniz; TD: Ismael Augusto; Film Buyer: José Eduardo Moniz.

Pol H unless stated otherwise.

RTP Channel I

Stations	Ch	kW	Stations	Ch	kW
Muro	2	10/67V	Valença	7	1/7
Lousa	3	12/60	Foia	8	2/20
Marofa	5	2/16	Leiranco	8	2/4
Mendro	5	5/30	Gardunha	8	1/3
Montejunto	6	5/22	Porto	9	10/100
S. Miguel	6	2/20	Palmela	22	2/115
Marao	6	5/40	Mosteiro	24	1/9
Bornes	7	5/12	S. Macario	47	2/67V
Lisboa	7	10/100			

+42 repeaters.

D.Prgr: Mon-Fri 0900-0030, Sat 0900-0230, Sun 0900-0030.

RTP Channel II

Stations	Ch	kW	Stations	Ch	kW
Mosteiro	21	1/9	Gardunha	34	2/18
Lisboa	25	10/405	Marao	35	10/300
Bornes	25	10/200	Porto	41	10/100
Lousa	26	20/540	Montejunto	46	10/200
Mendro	27	20/560	Valença	46	2/70
Muro	27	10/500	Foia	47	10/550
S. Miguel	31	10/250	Marofa	48	10/300
Palmela	32	2/128	S. Macario	50	2/67V
Leiranco	34	5/40			

+20 repeaters.

D.Prgr: Mon-Fri 1200-0050/0130, Sat 0900-0130, Sun 0900-0100.

ROMANIA

TV-sets: 4m — **Colour:** PAL — **System:** D&K.

RADIOTELEVIZIUNEA ROMÂNĂ (Gov.)

Addr: 191 Calea Dorobanti, Bucharest. **Te:** +40 (0) 331092. **Fax:** +40 (0) 337544. **Telex:** 10182 T/R. **Cable:** Radio-televiziunea —

Stations: Pol = H (exc. where indicated).

	ch	kW		ch	kW
Bucuresti II	R2	150	Vascau	R8	2.5V
Turnu	R2	15	Iasi	R9	25
Oradea	R3	120	Timisoara	R9	50
Semenic	R3	15V	Magura	R9	10
Dobrogea	R3	15	Pitesti II	R9	500
Bistrita	R3	2.5	Baia Mare	R10	15
Bucuresti	R4	50	Constanta	R10	50
Suceava	R4	100	Brasov	R10	10
Birlad	R5	50V	Petrosani	R10	10
Gheorghieni	R5	50	Bacau	R10	2
Bucegi	R6	150	Cluj	R11	25
Delta	R6	50V	Mangalia	R11	1
Piatra	R6	5	Arad	R12	50
Zalau	R6	2	Comanesti	R12	20
Galati	R7	2V	Deva	R12	20
Sibiu	R7	50	T. Mures	R12	2
Varatec	R7	2	Cozia	R12	2
Constanta	R8	10V	Tulcea	R12	0.5
Craiova	R8	150	Timisoara (II)	21	50
Cimpulung	R8	15	Brasov (II)	34	50

D.Prgr: Prgr. I: Mon-Fri 0800-2200, Sat/Sun 0700-2300.
Prgr. II: 1500-2200.

ANTENA INDEPENDENTA

First independent TV station — **D.Prgr:** 24 hrs. UHF ch 34.

RUSSIA

TV-sets: N/A — **Colour:** SECAM — **Systems:** D&K.

ROSSIJSKAJA GOSUDARSTVENNAJA TELERADIOKOMPANIJA "OSTANKINO" (Gov.)

Addr: ul. Akad. Korol'ova 19, 127427 Moskva.

L.P: See radio.

NB: Teleradiokompanija "Ostankino" is producing two national TV-channels: OSTANKINO KANAL 1 (OK-1) and KANAL 4 (OK-4).

OSTANKINO KANAL 1 (OK-1)

Stations: OK-1 is broadcast over a vast number of main and relay transmitters throughout Russia and the memberstates of the CIS. Since complete transmitter-lists are not available, a list of OK-1 transmitters has been omitted in this edition of the WRTH.

D.Prgr: 0255-2400 (European edition). Timeshifted editions of OK-1 are aired via satellite to the Centralasian and Far East parts of Russia and the CIS. There are 4 editions: "Orbita 4" to "Orbita 1" respectively.

OSTANKINO KANAL 4 (OK-4)

Stations: Since complete details are currently not available, a list of OK-4 transmitters has been omitted in this edition of WRTH.

D.Prgr: 0600-2200. Educational programmes ("Telekanal Rossijskije universitety"): Mon-Fri 0600-1900.

OK-4 is only transmitted in the European part of Russia.

VSEROSSIJSKAJA GOSUDARSTVENNAJA TELERADIOKOMPANIJA (VGTRK) (Gov.)

Addr: Jamskogo polja 5-ja ul. 19/12, 125124 Moskva.

Te: +7 (095) 2179981. **Fax:** +7 (095) 2500506.

L.P: See radio.

NB: VGTRK produces one national TV-channel: ROSSIJSKOJE TELEVIDENIJE (RTV).

ROSSIJSKOJE TELEVIDENIJE (RTV)

Stations: RTV is broadcast over a large number of main and relay stations throughout Russia and most memberstates of the CIS. Since current complete details are not available, a list of RTV transmitters in Russia has been omitted in this edition of WRTH.

D.Prgr: 0500-1115, 1300-2130 (Sat/Sun 0500-2200) (European edition). Timeshifted editions of RTV are aired by satellite to the Centralasian and Far East parts of Russia and the CIS. These editions are known as "Dubl 4" to "Dubl 1" respectively.

TELERADIOKOMPANIJA "PETERBURG" (Muncip.)

Addr: ul. Ital'anskaja 27, 1910011 Sankt-Peterburg.

TV-PETERBURG (TV-P)

Stations:	ch	kW (not ERP)
Sankt-Peterburg	3	50

NB: TV-P is aired via various main and relay transmitters in the European part of Russia, as well as the Republic of Belarus.
D.Prgr: 0430 (Sat/Sun 0500) — 2200.

SLOVENIA

TV-sets: N/A — **Colour:** PAL — **Systems:** B&H

TELEVIZIJA SLOVENIJA

Addr: Tavcarjeva 17, 61000 Ljubljana. **Te:** +38 (61) 311922.
NB: Due to the continuing unrest, the actual situation may vary.
Stations: System B=ch2-12, System H=ch21-68. **ChE.** audio powers 1/10 of vision powers indicated.

TV Slovenija
1st Prgr.

Kum	3	50	Plesivec	6	31.6
Krvavec	5	50	Pohorje	11	10
Nanos	6	31.6	Kum	38	20

+165 low power sts.
D.Prgr: Mon-Fri 0750-2215 (Thu 2300, Fri 2330), Sat 0700-1115, 1330-2345, Sun 0725-2230.

TV Slovenija
2nd Prgr.

Krvavec	21	25	Plesivec	49	32
Kum	32	20	Pohorje	56	400
Nanos	41	398			

+150 low power sts.
D.Prgr: Mon-Fri 1430-2300, Sat 1530-0030, Sun 0900-2300.

TV Koper-Capodistria

Nanos	27	398

+33 low power sts.
D.Prgr: Mon-Fri 1245-2400, Sat 0900-2400, Sun 0930-2400.

SPAIN

TV-sets: 17.000.000 — **Colour:** PAL — **Systems:** B&G.

RADIOTELEVISION ESPAÑOLA (RTVE) (Gov.)

Addr: Prado del Rey, 280023 Madrid. Torrespaña C/ O'Donnell 77, 28007 Madrid. **Te:** +34 (1) 346 8754. **Fax:** +34 (1) 581 7125. **Telex:** 22053. **L.P:** Dir. Gen: Jordi Garcia Candau. Ex.Dir.TVE, S.A: Ramón Colom Esmatges. Dir. of Int. Coop: Alfonso Callego. TD: Francisco Baquedano. Film Buyer: Fernando Moreno. **Stations:** (System B&G): Powers = ERP. Audio powers 10% of vision powers indicated.
N.B: TVE's 1st. and 2nd. prgr. are also relayed via satellite. The 1st. prgr is relayed under the name TVE Internacional. All Pol H. unless otherwise indicated.

TVE 1st Prgr.

Station	ch	(kW)
Pechina	41	100
Jerez de la Frontera	8	6
	26	32
Jerez de la Frontera		
Cordoba	11	10
Parapanda	9	24
Sierra de Lujar	7	150
Huelva	39	250
Sierra Almaden	5	40
Mijas	65	200
Guadalcanal	4	120
Valencina	63	79
Arguis	64	36
Camarena de la Sierra	9	53V
Inoges	5	24
La Muela	3	35
Gamoniteiro	3	50
Alfabia	6	50
Alfabia	54	119
Lierganes	40	100
Villadiego	8	13
Villadiego	47	40
Castropodame	7	22V
Matadeon	8	12
El Cabaco	9	63
Soria	39	16
Zamora	31	160
Chinchilla	11	40
La Mancha	9	60
Isleta	6	4
Montaña Mina	32	6
Pozo de las Nieves	10	10
Fuencaliente	7	10
Izaña	3	200
Tibidabo	4	150
Gerona	49	126
Alpicat	5	15
Soriguera	39	40
La Musara	8	22
Monte Caro	10	60
Aitana	3	60
Benicasim	7	10V
Torrente	5	20
Montanchez	11	45
Ares	5	50
Santiago de Compostela	4	112
Paramo	9	9
Parada del Sil	5	40V
Domayo	10	10
Madrid	4	15
Navacerrada	2	250
Navacerrada	34	604
Torrespaña	49	117
Murcia	59	160
Monreal	29	158
Sollube	4	60
Jaizquibel	54	158
Zaldiaran	45	15

TVE 2nd Prgr.

Station	ch	(kW)
Pechina	47	100
Jerez de la Frontera	23	32
Cordoba	21	200
Parapanda	23	160
Sierra de Lujar	57	250
Huelva	45	250
Sierra Almaden	39	260
Mijas	59	200
Guadalcanal	40	32
Valencina	52	79
Alpicat	55	112
Arguis	58	36
Camarena de la Sierra	41	200
Monte Caro	43	60
Inoges	57	200
La Muela	46	158
La Muela	33	158
Gamoniteiro	39	158
Alfabia	48	119
Lierganes	46	100
Villadiego	41	200
Castropodame	21	200
Matadeon	33	28
El Cabaco	39	315
Soria	45	16
Zamora	37	160
Chinchilla	43	158
La Mancha	58	200
Isleta	28	20
Montana Mina	35	6
Pozo de las Nieves	59	48
Temejereque	52	25
Fuencaliente	27	148
Izaña	45	87
Tibidabo	41	28
Tibidabo	31	144
Gerona	55	126
Alpicat	49	160
Soriguera	45	40
La Musara	57	300
Monte Caro	26	225
Aitana	32	200
Benicasim	59	200
Monte Caro	23	232
Torrente	22	160
Montanchez	23	158
Ares	22	200
Santiago de Compostela	45	316
Santiago de Compostela	2	40
Paramo	57	158
Parada del Sil	47	400
Domayo	39	141
Madrid	21	100
Navacerrada	24	870
Navacerrada	27	604
Torrespaña	55	117
Murcia	65	160
Monreal	23	158
Archanda	22	50
Jaizquibel	48	158
Zaldiaran	39	15

D.Prgr: TVE 1 0800-0230 (24 H. at weekends), TVE 2 0800-0200.

TV Networks in autonomous areas
TV3–TELEVISIO DE CATALUNYA (Aut)

Televisió de Catalunya, S.A. (TV3 and Canal 33) Appartat de Correus 20300, 08080 Barcelona. **Te:** +34 4730333. **Telex:** 53280. **Fax:** +34 4730671. **L.P:** Dir: Jaume Ferrús i Estopa. T.D: Pere Vila. Film Buyer: Jaume Santacana.

Canal 33

Stations:	ch	kW
Tibidabo	23	10
Alpicat	38	8
Rocacorba	45	6
La Mussara	35	4
Montcaro	36	1
Montserrat	40	0.1
Collsuspina	60	0.1
Pic de L'orri	56	2
Sant Pere de Ribes	32	0.1

TV3

Stations:	ch	kW
Tibidabo	44	20.000
Alipicat	52	10.000
Rocacorba	52	10.000
La Mussara	63	10.000
Montcaro	29	5.000
Ager	35	20
Aiguafreda	42	5
Alcanar	39	8
Alins	36	8
Almenar	35	20
Arbucies	42	20
Arenys de Munt	26	10
Areu	52	1
Argentona	25	5
Artesa de Segre	46	20
Bagergue	40	5
Baix Llobregat	50	20
Bausen	53	1
Begues	21	0
Bellmunt d'Osona	48	90
Boixadors	36	20
Boldis	31	0
Bossost	43	20
Cabrils-Mataro	53	100
Cadaques	42	5
Calaf	50	20
Calella	36	100
Calonge	62	100
Camprodon	38	20
Canet	65	2
Cap de Vaqueira	30	20
Cardona	48	5
Castellfollit Riubregos	57	1
Collsuspina	26	100
Comiols	61	100

El Vendrell	27	100	Os de Civis	30	1
Escalo	23	2	Osor	28	1
Espinabell	21	1	Palafrugell	25	20
Esterri de	53	1	Pic de L'Orri	42	5.000
Cardos			Pobla de Segur	24	8
Flix I	61	20	Pont de Suert	31	100
Gallifa	39	1	Portbou	37	20
Garraf	61	5	Puig-Reig	36	20
Gerri de la Sal	48	6	Rialp	48	1
Girona	40	20	Ribes de Freser	54	20
Gombren	33	1	Ripoll I	39	100
Guardiola	0	0	Ripoll II	21	1
Igualdada	61	20	Sa Nela	62	8
La Cerdanya	62	200	St Carles Rapita	25	20
La Coma	40	5	St Hilari Secalm	27	5
La Molina	24	1	St Llorenç Morunys	38	5
La Pobla de	46	4	St Llorenç Savall	36	1
Claramunt			St Pere de Ribes	42	150
L'Estartit	60	20	Senterada	53	5
Llavaneras	28	1	Solsona	36	8
Llavorsi	31	1	Taga	25	6
Llimiana	21	6	Taull	43	1
Maçanet de	62	200	Tornafort	25	5
Cabrenys			Torre de Capdella	47	8
Mangraners	25	40	Tossa de Mar	24	5
Molins de Rei	64	20	Ulldecona	0	0
Monistrol	50	5	Ulldemolins	25	5
Monistrol de	50	1	Unarre	61	8
Calders			Vall de Bohi	53	20
Montserrat	28	200	Vallirana	51	5
Montblanc	43	100	Valls de Valira	22	20
Montcada	65	100	Vandellos	33	6
Oliana	34	5	Vielha	52	2
Olot	21	40	Viladrau	30	1
Orcau	27	5	Vilaller	21	5
Organya	47	200	Xerallo	21	1

D.Prgr: TV3 1115-0030, Canal 33 1900-2400.

TV DE GALICIA (TVG)

Addr: Apt. 707, San Marcos (Santiago de Compostela). **Te:** +34 (81) 565141. **Telex:** 97012. **Fax:** +34 (81) 562886 — Dir: Xerardo R. Rodríguez. **Station:** ch 42. In Galician: 1200 (Sun 1030)-2300 (approx).

CANAL SUR (Gov.)

Addr: (See Satellite Section)

TELEVISIO VALENCIANA (Gov.) CANAL 9

Addr: 46100 Burjassot, Valencia. **Te:** +34 (6) 364 1100. **Fax:** +34 (6) 363 9516. **L.P:** GM: Amaden Fabregat-Manes. MD: Rafael Cano-Baron.

TV VASCA-EUSKAL TELEBISTA (Gov.)

Addr: Barria Lurreta, 48200 Durango (Vizcaya). **Te:** +34 (94) 6816600. **Telex:** 34441. **Fax:** +34 (94) 6816526 — Dir: Koldo Anasagasti — **Stations:** ch 42 (ETB-1) & ch 49 (ETB-2) +160 lp's. **D.Prgr:** ETB-1 1400-2000 — ETB-2 1430-2100.

COMMERCIAL STATIONS:

ANTENA TELEVISION (Antena 3)

Addr: (See Satellite Section)
L.P: Pres: Javier de Godó; G.M: Manuel Martín Ferrand; Film Buyer: Condorcet da Silva Costa.

CANAL PLUS

Addr: (See Satellite Section)

TELE 5

Addr: (See Satellite Section)
D.Prgr: 1100-0300 (CET).
Projected Autonomous Networks:
Radiotelevision Madrid, Radiotelevision Navarra, TV Andalucia, TV Cantabria & TV Valenciana.
Stations:

Loc.	Tele-5	Antena 3	C. Plus	kW
Albacete-Chinchilla	50	53	56	50
Alicante-Aitana	50	53	60	100
Almería-Pechina	58	61	64	100
Avila-San Mateo	59	62	65	1
Barcelona-Tibidado	27	34	47	100
Bilbao-Archanda	59	62	65	50
Burgos	30	33	36	1
Cádiz-San Cristobal	49	53	55	30
Castellón-Desierto	49	52	55	50
Ciudad Real-Atalaya	23	26	29	1
Córdoba-Lagar	58	61	64	100
Cuenca-San Cristóbal	37	60	63	1
Gerona-Rocacorba	32	35	38	50
Granada-Parapanda	50	53	56	100
Guadalajara	31	37	51	1
Huelva-Punta Umbría	32	35	56	100
Huesca-Arguis	21	24	27	30
Jaen-Sierra Almadén	32	35	49	100
La Coruña-Ares	35	62	65	100
Las Palmas-La Isleta	32	35	38	20
León-El Portillo	49	52	55	1
Lérida-Alpicat	59	62	65	50
Logroño-Moncalvillo	40	46	48	20
Lugo-Páramo	23	41	44	100
Madrid-Torrespaña	59	62	65	100
Málaga-Mijas	39	42	45	100
Mallorca-Alfabia	58	61	64	100
Montánchez	59	62	65	100
Murcia-Carrascoy	38	42	44	50
Orense-Barbadanes	26	40	43	1
Oviedo-Gamoniteiro	28	32	35	100
Palencia-Villamuriel	48	51	54	1
Pamplona-S. Cristóbal	49	52	55	1
Pontevedra-Tomba	31	35	41	1
Salamanca-Teso	29	60	63	1
San Roque-Carbonaras	21	24	27	20
San Sebastián-Ulia	31	41	44	1
Santander-Peña Cabarga	29	60	63	50
Santiago-Pedroso	38	56	59	100
Segovia	48	51	54	1
Sevilla-Valencia	38	41	44	50
Soria-Santa Ana	21	24	27	1
Tarragana-La Musara	37	50	53	20
Tenerife-Izaña	23	26	29	100
Teruel-Santa Bárbara	26	30	33	1
Toledo-Los Palos	30	53	56	1
Valencia-Torrente	40	43	46	100
Valladolid-Contienda	50	53	56	1
Vigo-Domayo	54	61	64	100
Vitoria-Zaldiarán	29	32	35	15
Zamora-El Viso	58	61	64	100
Zaragoza-La Muela	22	30	54	100

TELEVISION MURCIANA

Addr: Plateraa 44, 230001 Murcia
Te: +34 (68) 212 224. **Fax:** +34 (68) 214 673.

AFRTS (US Air Force Europe)

Addr. Det. 1 AFEBS, Zaragoza AB, c/o APO New York 09286-5000, USA.
Station: Zaragoza chA4 (Power: cable distribution), Torrejon chA4 2kW.
D.Prgrs: W 12h; Sat/Sun 15h.
NBS TV: Rota chA2 2kW.

SWEDEN

TV-sets: 3.750.000 — **Colour:** PAL — **Systems:** B & G.

SVENSK RUNDRADIO AB (Teracom Swedish Broadcasting)

This company has the responsibility for the distribution of the prgrs produced by Sveriges Television AB and TV4 Nordisk TV.
Addr. and L.P: See Radio.

SVERIGES TELEVISION AB (Non-comm.)

Addr: Oxenstiernsgatan 26-34, Stockholm.
Postal Addr: S-105 10 Stockholm. **Te:** +46 (8) 7840000 & 7847400. **Fax:** +46 (8) 7841500. **Telex:** 10000. **Cable:** Broadcast.

L.P: Chmn. Board of Governors: L. Sandgren. Man. Dir: S. Nilsson. Dir. Prgr. (SVT1): I. Bengtsson. Dir. Prgr: (SVT2): O. Hansson. Dir. Eng: S.O. Ekholm. Dir. Finance & Adm: B. Böhlin. Dir. Staff: S. Wahlström. Contr. News and Curr. Aff. (SVT1). S. Strömberg. Head News. and Curr. Aff. (SVT2): E. Winblad. Contr. SVT International: B. Nordell. Legal Adviser. U. Peyron. Head International Rel: Å. Iwarsson. Press and Information: S. Carlsson.

Stations (Systems B ch2-11, System G ch21-68): Sound power 1/20 of vision power stated.
Prgr. I (Systems B and G, NICAM stereo):

Name	ch	Pol	kW (ERP)	Name	ch	Pol	kW (ERP)
Örebro	2	H	60	Kalix	8	H	60
Vännäs	2	H	60	Lycksele	8	H	15
Skövde	3	H	60	Ånge	8	H	15
Sveg	3	H	60	Mora	8	H	20
Stockholm	4	H	60	Västerås	8	H	10
Östersund	4	H	100	Filipstad	9	H	5
Älvsbyn	4	H	60	Gällivare	9	H	60
Arvidsjaur	5	H	60	Gävle	9	H	60
Norrköping	5	H	60	Göteborg	9	H	60
Sundsvall	5	H	60	Tåsjö	9	H	60
Karlstad	5	H	1	Visby	9	H	60
Bollnäs	6	H	60	Karlskrona	9	V	3
Kiruna	6	H	60	Helsingborg	9	H	1
Västervik	6	H	60	Borlänge	10	H	60
Skellefteå	6	H	35	Nässjö	10	H	60
Borås	6	H	3	Storuman	10	H	60
Uppsala	6	H	10	Malmö	10	H	3
Örnsköldsvik	6	H	7	Varberg	10	V	1
Halmstad	7	H	60	Överkalix	10	H	5
Pajala	7	H	60	Hudiksvall	11	H	4
Sollefteå	7	H	60	Kisa	11	H	30
Sunne	7	H	60	Östhammar	11	H	30
Motala	7	H	10	Karishamn	11	H	30
Trollhättan	7	H	1	Uddevalla	33	H	1000
Väddö	8	V	200	Vislanda	39	H	1000
Bäckefors	8	H	60	Finnveden	41	H	1000
Emmaboda	8	H	60	Hörby	43	H	1000
Jönköping	8	H	1				

+ 471 low power trs.
D.Prgr (School prgrs not included): Mon-Fri 1650-2300; Sat 0810-2300; Sun 1330-2300.

Prgr. 2 (System G, NICAM stereo) (all sts with Pol H).

Name	ch	kW (ERP)	Name	ch	kW (ERP)
Arvidsjaur	21	1000	Karlskrona	34	20
Sveg	21	1000	Pajala	34	1000
Mora	22	1000	Kalix	35	1000
Nässjö	22	1000	Älvsbyn	36	1000
Stockholm	23	1000	Skövde	37	500
Uddevalla	23	1000	Tåsjö	37	1000
Halmstad	24	1000	Örnsköldsvik	39	400
Bäckefors	26	1000	Visby	41	1000
Karlshamn	26	1000	Borås	42	1000
Västervik	26	1000	Ånge	42	250
Östhammar	26	1000	Karlstad	43	200
Gävle	27	1000	Västerås	44	1000
Malmö	27	150	Motala	45	1000
Östersund	27	1000	Lycksele	45	1000
Jönköping	28	15	Överkalix	45	1000
Bäckeforg	29	1000	Skellefteå	46	1000
Bollnäs	29	1000	Sollefteå	46	1000
Kiruna	29	1000	Borlänge	47	1000
Göteborg	30	1000	Sundsvall	47	1000
Helsingborg	30	10	Vännäs	47	1000
Emmaboda	31	1000	Finnveden	48	1000
Hudiksvall	31	1000	Örebro	48	1000
Västerås	31	1000	Kisa	49	1000
Norrköping	32	1000	Uppsala	49	200
Vislanda	32	1000	Varberg	49	200
Filipstad	33	1000	Sunne	50	1000
Hörby	33	1000	Trollhättan	51	20
Gällivare	33	1000	Matala	52	1000
Storuman	33	1000			

+ 610 low power trs.
D.Prgr: (School prgrs not included). Mon-Fri 1645-2300, Sat/Sun 0800-2300.

TV4 NORDISK TELEVISION Co. (Comm.)

Addr: S-115 79 Stockholm. **Te:** +46 (8) 644 4400 **Fax:** +46 (8) 644 4440 **Tlx:** 14124.
L.P: Pres: C.E.O. Björn Nordstrand; Hd. of Prgrs: Lars Weiss; Hd. of Inf. & Pub. Rel: Helga Baagøe; Marketing Dir: Tony Michaelsen: Dir. of News & Curr. Aff: Kerstin Persdotter; Hd. of Eng: Olle Mossberg.
Stations (System G, NICAM stereo) (All sts Pol H)

Name	ch	kW	Name	ch	kW
Mora	25	1000	Emmaboda	47	1000
Nässjö	25	1000	Malmö	47	150
Kalix	29	1000	Skövde	47	1000
Gävle	30	1000	Bäckefors	49	1000
Jönköping	31	15	Skellefteå	49	1000
Motala	39	1000	Sollefteå	49	1000
Helsingborg	41	10	Bollnås	49	1000
Stockholm	42	1000	Horby	50	1000
Örnsköldsvik	42	400	Sondsvall	50	1000
Gällivore	43	1000	Vännäs	50	1000
Vadevalla	43	1000	Västerås	51	1000
Varberg	43	200	Oppsala	52	200
Hudiksvall	44	1000	Alvsbyn	52	1000
Karlshamn	44	1000	Sonne	53	1000
Halmstad	45	1000	Norrköping	54	1000
Östersund	45	1000	Borås	55	1000
Göteborg	46	1000	Örebro	58	1000
Karlstad	46	200	Borlänge	60	1000

+ 17 lps

FINNISH TELEVISION RELAY

Stockholm (Nacka) ch39 1000kW(ERP). Pol H.
Relay of Finnish Prgr II: Mon-Fri 1530-2115, Sat 1200-2200, Sun 1200-2115.
+ 2 lp st.

SWITZERLAND

TV-sets: 2.2m — **Colour:** PAL — **Systems:** B&G.

SBC—SWISS BROADCASTING CORPORATION

The SBC's Television programme services are an integral part of the Swiss Broadcasting Corporation.

Addr: SBC, Giacomettistrasse 3, CH-3000 Berne 15. **Cable:** Radif. **Te:** +41 (31) 439111. **Telex:** 911590 SSr ch. **Fax:** +41 (31) 43 92 56.
L.P: see radio section.
Prgr. Sce. in German: TV Prgr. Dir: Peter Schellenberg, Schweizer Fernsehen DRS, Fernsehstrasse 1-4, 8052 Zürich. **Te:** +41 (1) 3056611. **Telex:** 823823 TVZ. **Fax:** +41 (1) 305 57 11.
Prgr. Sce. in French: Radio-TV Dir: Jean-Jacques Demartines, Radio-télévision suisse romande, RTSR, 20 Quai Ernest Ansermet. B.P. 234, 1211 Geneva 8. **Te:** +41 (22) 293333. **Fax:** +41 (22) 204049. TV Prgr. Dir: G. Chenevière, Télévision suisse romande, 20 quai Ernest Ansermet, CH-1211 Geneva 8.
Te: +41 (22) 293333. **Telex:** 427 701 TVR. **Fax:** +41 (22) 28 34 30.
Prgr. Sce. in Italian: RTSI, Radiotelevisione della Svizzera italiana, Via Canevascini, P.O. Box 6903, Lugano-Besso. Radio-TV Dir: M. Blaser, Radio-TV Prgr. Dir: D. Balestra, 6903 Lugano-Besso. **Te:** +41 (91) 585111. **Fax:** +41 (91)589150.
Studio Televisione, Casella postale, CH-6949 Comano. **Te:** +41 (91) 585111. **Fax:** +41 (91) 585355.

Stations (Systems B&G)

German Prgr:	ch	kW (ERP)			
Bantiger	2	47.9	P. Matro	29	13.8
Uetliberg	3	56.2	La Dôle	31	234
Rigi	6	38.5	Les Ordons	31	14
Säntis	7	30	La Chaux-		
Celerina	9	2.8	de Fonds	32	18
Valzeina	10	5	Piz Lagalb	37	1.1
Gebidem	11	2.5	Castel		
St. Chrischona	11	38	San Pietro	39	5.4
Niederhorn	12	5	Olten	42	14
Chasseral	25	12	Haute Nendaz	43	30

Mt. Pelerin	44	18
Gruyères	46	1
Monte Ceneri	46	35
Vallée de Joux	46	1.0
Schaffhausen	47	14
Savièse	50	1.1
Ravoire	51	8
Monte San Salvatore	54	100
Chamossaire	57	2
Chasseral	59	15.0
Monte Morello 1	62	5.0
Stans 2	64	1.2
St. Niklaus/ VS 2	64	1.0
Fanas 2	67	1.2

+ approx 450 low power st's.

French Prgr:	**ch**	**kW**
La Dôle	4	264
Haute Nendaz	7	3
Les Ordons	7	4.5
Monthey 2	8	5.0
La Chaux-de Fonds	9	4
Ravoire	9	1.8
Bantiger	10	111
Mont Pelerin	11	5.8
Chasseral	22	12
Feldis	24	1.7
Uetliberg	23	93.3
Niederhorn	27	15
Säntis	31	158
Pizzo Matro	32	13.8
Rigi	32	70.8
Ausserberg	37	1
Piz Lagalb	40	1.1
Castel San Pietro	42	6.1
Olten	45	14
St. Chrischona	46	105
Celerina	49	4.3
Monte Ceneri	49	35
Pfänder (A)	49	2.0
Schaffhausen	50	14
Bantiger	50	234.4
Schüpfheim	51	1.0
Ziegelbrücke	51	1.5
Gebidem	52	10.2
Wattwil	53	1.0
Klewenalb	55	2.0
Sedrun	55	1.0
Valzeina	56	8.5
Monte San Salvatore	57	100
Wattenwil	57	6.0
Monte Morello 1	66	5.0
Stans 2	67	1.2

+ over 400 low power st's.

Italian Prgr:	**ch**	**kW(ERP)**
Pizzo Matro	6	0.3
Celerina	7	2.8
Monte Ceneri	5	4.5
Monte San Salvatore	10	10
Feldis	21	1.94
Uetliberg	26	93.3
Rigi	29	70
Niederhorn	30	15
Celerina	33	4.3
La Dôle	34	234
Les Ordons	34	14
Säntis	34	158
La Chaux-de-Fonds	35	18
Ausserberg	40	1
Bantiger	40	234.4
Haute Nendaz	46	30
Mt. Pelerin	47	18
St. Chrischona	49	105
Gruyères	49	1
Vallée de Joux	49	1.0
Sedrun	52	1.0
Savièse	53	1.1
Ofenpass	53	0.8
Valzeina	53	8.5
Ravoire	54	8
Schüpfheim	54	1.0
Ziegelbrücke	54	1.5
Gebidem	55	10.2
Piz Lagalb	55	1.1
Chasseral	56	15
Castel San Pietro	56	7.7
Wattwil	56	1.0
Chamossaire	60	2
Schaffhausen	60	14
Wattenwil	60	6.0
Pfänder	62	2.0
Olten	63	14
Monte Morello	68	5

+ over 400 low power st's (TV studios 3).

D.Prgr: German: Sun 0630-2230; Mon 1200-2130; Tues/Fri 0700-0930, 1200-2230; Wed/Thurs 1200-2200; Sat 0800-2300. **French:** Sun 0630-2230; Mon 0845-2100; Tues/Wed 0845-2200; Thurs 0900-2200; Fri 0845-2300; Sat 0915-2300. **Italian:** Sun 0800-2200; Mon 1400-2130; Tues-Fri 0700-0830 & 1400-2130; Sat 1100-2230.

TELECINE

Station: ch 69/250kW La Dôle.

TELECLUB

Pay TV Sce. see Satellite Section.

UKRAINE

TV-Sets: N/A — **Colour:** SECAM — **Systems:** D&K.

UKRAJINSKA TELEBAČENNJA (Gov.)

Addr: vul. Chreščatik 26, 252001 Kyjiv. **Te:** +7 (044) 2290638. **Fax:** +7 (044) 2296945.
L.P: See radio.

UT-1 (*Not ERP)

Stations	**ch**	**kW***
Sovetskij	2	5
Izmajil	3	0.1
Komiš Zorja	3	5
Nikopol'	3	2.5
Kyjiv	4	50
Kovel'	5	5
Sevastopol'	5	5
Ternopil'	5	25
Kirovograd	6	5
Simferopol'	6	5
Ivano-Frankivs'k	7	5
Kam'jans'ke	7	5
L'viv	8	25
Kryvyj Rig	9	5
Krasnogorivka	10	25
Mykolajiv	10	5
Krasnoperekops'k	12	5
Odesa	12	5
Kam'janac'-Podil'c'kyj	21	20
Olevs'k	22	20
Pryluky	22	5
Kerč	25	1
Rivne	27	25
Starobil's'k	28	25
Izjum	29	25
Melitopol'	29	20
Pervomejc'k	29	5
Mariupol'	31	5
Kramators'k	34	20
Kotovs'k	39	5

+various low power translators.
NB: The above list of main transmitters is not complete.
D.Prgr: 0600-2200.

UT-2 (*Not ERP)

Stations	**ch**	**kW***
Kramators'k	1	0.1
Ivano-Frankivs'k	2	0.1
Simferpol'	3	5
Kirovograd	5	0.1
L'viv	6	5
Sevastopol'	8	0.1
Kyjiv	9	50
Odesa	9	5
Krasnoperekops'k	10	0.1
Pervomejs'k	12	0.1
Pryluky	12	0.1
Kirovograd	21	5
Mariupol'	25	5
Krasnogorivka	28	—
Komiš Zorja	28	0.1
Kam'jans'ke	29	—
Kerc	30	1
Kovel'	31	0.1
Nikopol'	31	1
Sevastopol'	31	20
Mykolajiv	33	5
Kramators'k	37	20
Ternopil'	38	20
Krasnoperekops'k	39	20
Izmajil	39	—
Rivne	39	20
Ivano-Frankivs'k	40	5
Izjum	40	20
Kryvyj Rig	40	5
Melitopol'	41	20
Pryluky	41	—

*various low power translators.
NB: The above list of main transmitters is not complete.
D.Prgr: 0500-2200. Includes relays of Rossijskoje televidenije (RTV) from Moscow.

OSTANKINO KANAL 1 (OK-1) (*Not ERP)

Stations	**ch**	**kW***
L'viv	1	40
Simferopol'	1	25
Kryvyj Rig	1	5
Kyjiv	2	50
Mykolajiv	2	5
Rivne	3	5
Kam'jans'ke	4	5
Mariupol'	5	5
Odesa	5	5
Kovel'	6	25
Kotovs'k	6	5
Pervomejs'k	7	5
Sovetskij	7	5
Komiš Zorja	8	5
Krasnogorivka	8	25
Starobil's'k	8	5
Ivano-Frankivs'k	9	5
Izjum	11	25
Kirovograd	11	5
Melitopol'	11	5
Sevastopol'	11	5
Ternopil'	11	25
Kerč	12	5
Kramators'k	22	20
Izmajil	27	1
Krasnoperekops'k	32	25
Kam'janac'-Podil'c'kyj	33	20
Olevs'k	34	20
Pryluky	34	25
Nikopol'	41	0.1

+various low power translators.
NB: The above list of main transmitters is not complete.
D.Prgr: Relays of OK-1 from Moscow. Further details see under Russia.

UNITED KINGDOM

TV-sets: appr. 20 m — **Colour:** PAL — **System:** I.

BRITISH BROADCASTING CORPORATION

Addr: Television Centre, London W12. **Te:** +44 (81) 743 8000. **Telex:** 265781. **Fax:** +44 (81) 749 7520.
L.P: MD: Will Wyatt. Asst. MD: Jane Drabble. Contr. BBC-1: J. Powell. Contr. BBC-2: Alan Yentob. CE: Peter Marchant. Publ: Keith Samuel.

Stations: See below.
D.Prgr: BBC1: Mon-Fri 0600-2400; Sat/Sun 0900-2400 (closing times vary). **BBC2:** 1500-2400 (Sat/Sun 0100) approx.
N.B: BBC started broadcasting in NICAM-stereo as of August 1991 (cnly main trs).

INDEPENDENT TELEVISION COMMISSION

Addr: 70 Brompton Rd, London SW3 1EY. **Te:** +44 (71) 581 7011. **Fax:** +44 (71) 589 5533.
L.P: Chief Exec: David Glencross; Dir of Prgrs; Clare Mulholland; Dir of Advert: Frank Willis; Head of Press: James Conway.
Program contractors:
(*new license holders as of Jan. 1st 1993)
Anglia Television (1), Anglia House, Norwich NR1 3JG. **Te:** +44 (603) 615151. **Fax:** +44 (603) 631032.
Border Television (3), The Television Centre, Carlisle, CA1 3NT. **Te:** +44 (228) 25101. **Fax:** +44 (228) 41384.
***Carlton Television Ltd.** (8), 101 St. Martin's Lane, London WC2N 4AZ. **Te:** +44 (71) 240 4000. **Fax:** +44 (71) 240 4171.
Central Independent Television (2), Central House, Broad Street, Birmingham, B1 2JP. **Te:** +44 (21) 643 9898. **Fax:** +44 (21) 616 4766.
Channel 4 Television Company Ltd, 60 Charlotte Street, London W1P 2AX. **Te:** +44 (71) 631 4444. **Fax:** +44 (71) 637 1495.
Channel Television (4), The Television Centre, St. Helier, Jersey, Channel Islands. **Te:** +44 (534) 68999. **Fax:** +44 (534) 59446/24770.
***Data Broadcasting International Ltd,** Allen House, Station Road, Egham, Surrey, TW20 9NT. **Te:** +44 (784) 471515.
***GMTV Ltd,** The London Television Centre, London SE1 9LT. **Te:** +44 (71) 827 7000. **Fax:** +44 (71) 827 7001.
Grampian Television (5), Queen's Cross, Aberdeen, AB9 2XJ. **Te:** +44 (224) 646464. **Fax:** +44 (224) 635127.
Granada Television (6), Granada Television Centre, Manchester, M60 9EA. **Te:** +44 (61) 832 7211. **Fax:** +44 (61) 839 0454.
HTV Wales (7), The Television Centre, Culverhouse Cross, Cardiff, CF5 6XJ. **Te:** +44 (222) 590590. **Fax:** +44 (222) 597183.
HTV West (7a), The Television Centre, Bath Road, Bristol, BS4 3HG. **Te:** +44 (272) 778366. **Fax:** +44 (272) 722400.
London Weekend Television (LWT) (8), South Bank Television Centre, London, SE1 9LT. **Te:** +44 (71) 620 1620. **Fax:** +44 (71) 928 6948.
***Meridian Broadcasting Ltd** (10), Television Centre, Southampton, SO9 5HZ. **Te:** +44 (703) 222 555. **Fax:** +44 (703) 335050.
S4C, Welsh Fourth Channel Authority, Clos Sophia, Cardiff, CF1 9XY. **Te:** +44 (222) 747 444. **Fax:** +44 (222) 75 4444.
Scottish Television (9), Cowcaddens, Glasgow, G2 3PR. **Te:** +44 (41) 332 9999. **Fax:** +44 (41) 332 6982.
***Teletext UK Ltd,** 101 Farm Lane, London SW6 1QJ. **Te:** +44 (71) 386 5000. **Fax:** +44 (71) 386 5002.
Tyne Tees Television (11), The Television Centre, City Road, Newcastle upon Tyne, NE1 2AL. **Te:** +44 (91) 261 0181. **Fax:** +44 (91) 261 2302.
Ulster Television (12), Havelock House, Ormeau Road, Belfast, BT7 1EB. **Te:** +44 (232) 328122. **Fax:** +44 (232) 246695.
***Westcountry Television Ltd** (13), Western Wood Way, Language Science Park, Plymouth, PL7 5BG. **Te:** +44 (752) 333333. **Fax:** +44 (752) 333 444.
Yorkshire Television (14), The Television Centre, Leeds, LS3 1JS. **Te:** +44 (532) 438283. **Fax:** +44 (532) 445107.

BRITISH SKY BROADCASTING

(see satellite section, ASTRA satellite 19.2° East)

INDEPENDENT TELEVISION NEWS (ITN)

Addr: 200 Gray's Inn Road, London WC1X 8XZ. **Te:** +44 (71) 833 3000.

INDEPENDENT TELEVISION ASSOCIATION (ITVA)

Addr: Knighton House, 56 Mortimer Street, London W1N 8AN. **Te:** +44 (71) 612 8000. **Fax:** +44 (71) 580 7892.

Stations: (Pol = H)	**BBC1**	**BBC2**	**ITV**	**Ch4**	**kW (ERP)**
So. Ea. England (+ 66 relays)					
Bluebell Hill (10)	40	46	43	65	30
Crystal Palace (8)	26	33	23	30	1000
Dover (10)	50	56	66	53	100
Heathfield (10)	49	52	64	67	100
Oxford (2)	57	63	60	53	500
So. We. England & Channel Isl. (+ 81 relays)					
Beacon Hill (13)	57	63	60	53	100
Caradon Hill (13)	22	28	25	32	500
Huntshaw Cross (13)	55	62	59	65	100
Redruth (13)	51	44	41	47	100
Stockland Hill (13)	33	26	23	29	250
Fremont Point (4)	51	44	41	47	20
So. England (+ 38 relays)					
Hannington (10)	39	45	42	66	250
Midhurst (10)	61	55	58	68	100
Rowridge (10)	31	24	27	21	500
We. England (+ 51 relays)					
Mendip (7a)	58	64	61	54	500
Ea. England (+ 18 relays)					
Sandy Heath (1)	31	27	24	21	1000
Sudbury (1)	51	44	41	47	250
Talcolneston (1)	62	55	59	65	250
Ce. England (+ 60 relays)					
Ridge Hill (2)	22	28	25	32	100
Sutton Coldfield (2)	46	40	43	50	1000
The Wrekin (2)	26	33	23	29	100
Waltham (2)	58	64	61	54	250
No. England (+ 52 relays)					
Belmont (14)	22	28	25	32	500
Emley Moor (14)	44	51	47	41	870
No. We. England (+ 66 relays)					
Winter Hill (6)	55	62	59	65	500
No. Ea. England (+ 64 relays)					
Bilsdale (11)	33	26	29	23	500
Caldbeck (3)	30	34	28	32	500
Chatton (11)	39	45	49	42	100
Pontop Pike (11)	58	64	61	54	500
Scotland (+ 188 relays)					
Angus (5)	57	63	60	53	100
Black Hill (9)	40	46	43	50	500
Sandale	22	—	—	—	500
Craigkelly (9)	31	27	24	21	100
Darvel (9)	33	26	23	29	100
Durris (5)	22	28	25	32	500
Eitshal (5)	33	26	23	29	100
Keelylang Hill (5)	40	46	43	50	100
Knock More (5)	33	26	23	29	100
Rosemarkie (5)	39	45	49	42	100
Rumster Forest (5)	31	27	24	21	100
Selkirk (5)	55	62	59	65	50
No. Ireland (+ 41 relays)					
Brougher Mt. (12)	22	28	25	32	100
Divis (12)	31	27	24	21	500
Limavady (12)	55	62	59	65	100
Wales (+ 162 relays)	**BBC1**	**BBC2**	**ITV**	**S4C**	**kW (ERP)**
Blaen-plwyf (7)	31	27	24	21	100
Carmel (7)	57	63	60	53	100
Llanddona (7)	57	63	60	53	100
Moel-y-Parc (7)	52	45	49	42	100
Presely (7)	46	40	43	50	100
Wenvoe (7)	44	51	41	47	500

NB: The number in brackets indicates the ITC company responsible for prgrs on ITV (3rd column).

YUGOSLAVIA (Federal Republic of)

TV-sets: 1.642.522 — **Colour:** PAL—**System:** B&G.

UDRUŽENJE JUGOSLOVENSKIH RADIOTELEVIZIJA d.O.O. (JRT)

Addr: JRT-Permanent Services — Television Department: Hartvigova 70/I, 11000 Beograd. **Te:** +38 (11) 434-910. **Telex:** 11469 yu jurate — **Televizija Srbije-Televizija Beograd:** Takovska 10, 11000 Beograd. **Te:** +38 (11) 342-001. **Telex:** 11884 — **Televizija Srbije-Televizija Novi Sad:** Kamenički put 45, 21000 Novi Sad. **Te:** +38 (21) 56-855. **Televizija Srbije-Televizija Priština:** Beogradska 66, 38000 Priština. **Te:** +38 (38) 31 211 — **Televizija Crne Gore:** Cetinski put bb, 81000 Podgorica. **Te:** +38 (81) 41 529. **Stations** (System B) (Ch 2-12), System G (Ch 21-69). **Ch E.** Audio powers 1/10 of vision powers indicated.
St's below 3kW not mentioned.

MONTENEGRO

TV Crne Gore	I	II	III	IV	kW
Bjelasica	6	12	37	62/43	100/100/1/500/1
Sudjina Glava	12	6	24	39	5/5/15/15
Sjenica	6	33	23	25/29	1.5/15/15/15
Luštica	4	26	33	39/42	1/10/10/10
Lovćen	8	31	10	35/67	100/1000/20/1000/500
Volujica	6	12	24	38	1/1/10/10
Mužura	12	23	33	43/53	1/10/10/10/10

+321 low power st's.

SERBIA

TV Srbije-TV Beograd	I	II	III	kW
Kopaonik	3	41		50/1000
Jastrebac	5	27	33	100/1000/1000
Avala	6	22	28	100/1000/30
Deli Jovan	6	23	43	10/50/100
Tornik	7	53	59	1/50/50
Besna Kobila	8	49	59	10/300/300
Ovčar	8	42	56	10/400/10
Maljen	9	26	32	5/2.5/250
Crveni Čot	10	24	30	100/1000/1000
Tupižnica	10	25	31	35/500/500
Crni Vrh (Svetozarevo)	11	35	38	35/500/500
Cer	7	37	34	3/300/300

+288 low power st's.

TV Srbije-TV Novi Sad	I	II	III	kW
Subotica	5	43		35/1000
Venac	41		48	1000/1000
Vršac	39	56		1000/1000

+13 low power st's.

TV Srbije-TV Priština	I	II	III	kW
Goleš	7	44		35/600
Cviljen	9	21		10/400

+63 low power st's.

TV POLITIKA: Makedonska 29, 11000 Beograd — Local st's Ch: 43, 47, 60 1kW.

NTV Studio "B": Masarikova 5, 11000 Beograd-Local and Ch: 32, 49, 51, Regional st. 53, 1 kW.

AFRICA

ALGERIA

TV-sets: 2m — **Colour:** PAL — **System:** B.

ENTREPRISE NATIONALE DE TÉLÉVISION (E.N.T.V.) (Gov.)

Addr: 21 Blvd. des Martyrs, Algiers 16000. **Te:** +213 (2) 62300. **Telex:** 66101 or 65282. **Fax:** +213 (2) 601042.
L.P: Dir. Gen: A. Brahimi. Dir. Gen. Adj: M. Malaika. Dir. Ext. Rel: R. Hamini. Dir. of Admin: Y. Nader. Dir. of N: M. Benabi. Dir. Prgr: B. Azibi. Dir. Prod: K. Bendissari. Dir. Tec: A. Bouhired.
Stations: Pol: H.

	ch	kW (ERP)		ch	kW (ERP)
M. Cid	5	150/30	Adrar	7	17/1.7
Ain-n-Sour	5	100/20	Mecheria	8	30/6
Reggane	5	25/2.5	Chrea	9	120/24
Ghardaia	5	11/1.1	Touggourt	9	12/1.2
Akfadou	6	100/20	Metlili	10	150/30
Nador	6	150/30	Tessala	10	150/30
Ouargla	6	5/0.5	Alger	11	20/4
Constantine	7	100/20	Aflou	11	150/30
Aouilef	7	19/1.9	In Amenas	11	10/1

Low power st's not mentioned.
D.Prgr: 1500-2300 (Thurs/Sat/Sun 1300-2400).

ANGOLA

TV-sets: 50.500 — **Colour:** PAL — **System:** I.

TELEVISÃO POPULAR DE ANGOLA (Gov.)

Addr: Avenida Ho-Chi-Min, P.O. Box 2604, Luanda. **Te:** +244 (2) 320025. **Fax:** +244 (2) 391091. **Telex:** 3238, 4153, 4157 TPA-AN. **Cable:** TPA.
L.P: MD: Carlos Chunha. Head of Prgrs: António Pedreira.
Station: ch9 13kW (ERP) + st's at Benguela, Huambo, Lubango, Namibe, Cabinda, Bié — **D.Prgrs:** 1730-2300 (Mon-Fri), 1400-2300 (Sat), 0900-2300 (Sun).

BENIN

TV-sets: 20.000 — **Colour:** SECAM. — **System:** K.

OFFICE DE RADIODIF. ET TV DE BENIN (ORTB) (Gov.)

Addr: P.O. Box 366, Cotonou. **Te:** +229 3010628.
L.P: Dir. Gen: Nicolas Benon. Dir. TV: Michèle Badarou. Chief of Sce. (TV): Marcellin Illougbade. Head of Prgrs: Didier Falde.
Station: ch4 20kW.
D.Prgr: Mon-Fri 1800-2100, Sat/Sun 1700-2200.

BOTSWANA

TV-sets: 13.800 — **Colour:** SECAM — **System:** K.

GABORONE TELEVISION CORP.

Addr: Private Bag 0060, Gaborone. **Te:** +267 352541. **Fax:** +267 357 138.
L.P: Dep. Dir. of Prgrs: Mrs. B. Tafa; Dep. Dir. of Prod: Mr. S. Moribame.
Station: ch. not known.
D.Prgr. in English: 1700-2000.

SATELLITE MARKETING BOTSWANA PTY. LTD.

L.P: Raymond Smith.
F.Pl: Seven or eight overseas channels to be made available to southern Africa by 1992.

BURKINA FASO

TV-sets: 45.500 — **Colour:** SECAM — **System:** K.

TÉLÉVISION NATIONALE BURKINA (Gov.)

Addr: B.P. 2530, Ouagadougou. **Te:** + 226 354773/306621. **Telex:** 5327.
L.P: DG: Seydou Azad Sawadogo.
Station: Ouagadougou & Bobodioulasso ch6 50/10W V.
D.Prgr: Tues-Fri 1833-2220; Sat 1508-2000; Sun 1205-1300, 1833-2200 — **Projected:** Ouagadougou 10kW H.

BURUNDI (Rep.)

TV sets: 4500 — **Colour:** SECAM — **System:** K.

TÉLÉVISION NATIONALE DU BURUNDI (Gov.)

Addr: B.P. 1900, Bujumbura. **Te:** +257 2247 60. **Telex:** 5119.
L.P: DG: Didace Barandaraetse; Hd. of Prgr: Leonidas.
Station: ch25 0.5kW — **D.Prgr:** 1600 (Sat/Sun 1400)-2200.

CAMEROON

TV-sets: 15.000 — **Colour:** PAL — **System:** B

RADIODIFFUSION CAMEROUNAISE (Gov.)

Addr: P.O. Box 1634, Yaoundé. **Te:** +237 234088. **Fax:** +237 204340. **Telex:** 8888 KN.
L.P: Dir. Gen: Pr. Gervais Mendoze.

CANARY ISLANDS

TV-sets: 240.000 — **Colour:** PAL — **Systems:** B & G.

TELEVISION ESPAÑOLA EN CANARIAS

Addr: 69 Calle Buenos Aires, Santa Cruz de Tenerife. **Te:** +34 (22) 216200.
Stations: Netw. I: Santa Cruz de Tenerife (Izana) chE3 350/35kW H, Fuencaliente chE7 4.7/0.47 H, Poso de las Nieves ch10 10/1H, Arrecife ch 32 6/0.6 kW H + 38 relays.
Netw. II: Fuencaliente ch27 170/17kW H, La Isleta ch28 18.6/1.86kW H, Arrecife ch38 6/0.6kW H, Izana ch45 87/8.7kW H, Puerto Rosario ch52 23.8/2.38kW H, Los Christianos ch 57 6/0.6kW H, Pozo de las Nieves ch59 48/4.8kW H + 13 relays.
D.Prgr. I: W 0615-0020 (Sun 0745-0020); II: W 1745-2300; Sun 0845-2325 (closing time may vary).

CENTRAL AFRICAN REPUBLIC

TV-sets: 7.500 — **Colour:** SECAM — **System:** K.

RADIODIFFUSION — TÉLÉVISION CENTRAFRIQUE

Addr: P.O. Box 940, Bangui. **Te:** +236 613242.
L.P: MD: Paul Service; Hd. of Prgrs: Henri-Gustav Hytayu.
Stations: N/A.

CHAD

TV-sets: N/A — **Colour:** SECAM — **System:** D.

TÉLÉTCHAD (Gov.)

ADDR: B.P. 74, N'Djamena. **Te:** +235 51 29 23. **Telex:** 5307.
L.P: Dir: Hourmadji Houssa Doumgor. Adj. Dir: Houssa Dago.
Station: N'Djamena ch7 (offset) 100W.
D.Prgr. in French/Arabic: 1800-2100 4 days per week.

CONGO (People's Rep.)

TV-sets: 8.500 — **Colour:** SECAM — **System:** D.

RADIODIFFUSION TELEVISION CONGOLAISE

Addr: B.P. 2241, Brazzaville. **Te:** +242 814574/814273/814030.
L.P: DG: Guy Noel Fam Ovey.
Station: ch 7 10/20kW H.
D.Prgr: 1730-2300

COTE D'IVOIRE

TV-sets: 810.000 — **Colour:** SECAM — **System:** K.

TELEVISION IVOIRIENNE (Gov.)

Addr: 08 B.P. 883, Abidjan 08. **Te:** +225 439039. **Fax:** +225 222297. **Telex:** 26110 ditele.
L.P: Dir. Gen: Mamadou Berté.

Stations	**chK'**	**kW**	**Pol**				
Koun	4	10/2.5	V	Touba	6	1/1	H
Tiémé	4	10/2.5	V	Man	7	10/2.5	H
Bouake	4	0.1	V	Dabakala	7	2	H
Abobo*	4	0.1	H	**Abidjan**	8	10/2.5	H
Séguéla	5	10/2.5	H	Niangbo	8	10/2.5	H
Digo	5	2/1	H	Niangué	8	10/2.5	H
Dimbroko	6	2/1	H	Bouaflé	9	10/2.5	V

*) 2nd prgr for Abidjan only.
D.Prgrs: 1st Prgr: Mon-Wed 1200-1330 & 1900-2300, Thurs 1200-1300 & 1600-2300, Sat 1200-0130, Sun 1030-2330. **2nd Prgr:** Mon-Fri 2030-2300, Sat 1600-2030 (Sun no transmissions).
Bouaké Regional Prgrs: Thurs 1200-1300, Fri 1700-1830.

DJIBOUTI (Rep.)

TV-sets: 17.000 — **Colour:** SECAM — **System:** K.

RADIO TELEVISION DE DJIBOUTI (Gov.)

Addr: P.O. Box 97, Djibouti. **Te:** +253 352294. **Telex:** 5863 DJ.
L.P: DG: Mohamed Tara Moussa.
Station: Djibouti chK'6 0.05kW H.
D.Prgr: 1500-2000. **F.Pl:** 5 relay st's.

EGYPT

TV-sets: 5m — **Colour:** SECAM — **System:** B.

EGYPTIAN RADIO AND TV UNION (Gov.)

Addr: TV Bldg, Cornish El-Nil, Maspero Cairo. **Te:** +20 (2) 757155. **Cable:** Cibrotev. **Telex:** 92152 Karadun.
L.P: Chmn. of TV Prgr. Sector: Abd. El-Salam El-Nady. Chmn. of Economic Sector: Hassan Abd El Fatah. Head of Int. Rel: Kamal Osman.

Stations	**I**	**II**	**kW**	**Pol**					
Dumyat	4	2	0.9	H	Idfu	8	11	165	H
Asswan	5	9	67	H	Mahalla	8	10	1600	H
Cairo	5	9	200	H	Abu Simbil	9	11	0.05	H
Hurghada	5	7	89	V	Ras Gharib	9	11	66	H
Nag Hamadi	5	8	17	H	Salum	9	11	6.9	H
Port Said	5	7	10	V	Qena	9	6	80	H
Sidi Barani	5	7	11.2	H	Assiut	10	6	67.2	H
El Farafra	5	7	10	V	El Kharga	10	8	40	V
Alexandria	6	11	110	H	Kom Ombo	10	7	40	H
Beni Ali	6	9	5	V	Matruh	10	8	39.2	H
El Arish	6	10	182	V	Bawiti	10	8	21.9	H
Isna	6	9	17	H	Beni Suef	11	7	110	V
Siwa	6	8	10	V	Ismailia	11	9	260	V
Dahab	6	8	9.33	V	Luxor	11	7	19	H
El Daba	7	5	2.95	V	Safaga	11	9	50	V
Quseir	7	5	50	H	Negala	22	25	5	H
Sohag	7	11	52	H	Nuweiba	26	35	25.7	V
Suez	7	5	20	H	Sharm				
Managem	7	5	2.5	H	El-Sheikh	27	33	8.91	H
Baharia					Taba	32	37	25.7	H
Baris	7	5	10	V	Barnis	24	29	830	H
El Dakhla	8	6	22.4	H	El Hamam	48	51	69.2	H
El Minya	8	5	165	V					

D.Prgr: Prgr. I: 0800-2330.
Prgr. II: 0800-2300.
Prgr. III: Cairo ch7 200kW H, ch21 100kW H.
D.Prgr: 1200-2100.
Prgr. IV: Suez ch54 20kW H, Ismailia ch31 260kW V, Port Said ch42 10kW V.
D.Prgr: 1100-2100 (Thurs: 1100-2200).
Prgr. V: Alexandria ch36 670kW H.
D.Prgr: 0800-1030, 1430-2200.
Esat Prgr: Shalatin ch96 10kW H, Cathrine ch12 1kW H.
D.Prgr: 0800-2300.

EQUATORIAL GUINEA

TV-sets: 2.500 — **Colour:** SECAM — **System:** B.

TELEVISION NACIONAL (Gov.)

Addr: Malabo Bioko Norte.
L.P: Dir: Antonio Nkulu Oye.
Station: chE2.
No further information available.

ETHIOPIA

TV Sets: 100.000 — **Colour:** PAL — **System:** B (Pol=H).

ETHIOPIAN TELEVISION (Gov.)

Addr: P.O. Box 5544, Addis Ababa. **Te:** +251 116701. **Telex:** 21429 ETV ET.

L.P: St. Mgr: Wole Gurmu. TD: Gessese Abay. Film Buyer: Mrs. Almaz Degene.

City	ch	kW		ch	kW
Asmara	5	5	Mekele	7	4
Bahir Dar	5	1.5	Gondar	7	2.5
Dire Dawa	5	1	Dessie	9	5
Addis Ababa	7	5	Deber	9	1
Harar	7	5	Jima	11	2.5

D.Prgr: Mon-Fri 1900-2230, Sat/Sun 1800-2400 (Sun 2300). **N:** 2000, 2100, 2200.

GABON

TV-sets: 40.000 — **Colour:** SECAM — **System:** K.

RADIO TELEVISION GABONAISE (Gov.)

Addr: B.P. 150, Libreville. **Te:** +241 732152. **Fax:** +241732153.
Cable: Radiotelegabon.
L.P: Pres: William Oyonne.
Stations: Libreville chK'4 & K'8 2kW H, Port Gentil chK10 0.1kW H, Moanda 1kW (relay) + 5 low power relay sts.
D.Prgr: 1800-2200; ch10: 1800-2100 (relay Libreville).

CANAL FRANCE INTERNATIONAL

Prgrs would be sent from France via Intelsat-V as part of the "Canal France" project targeted on Africa.

GAMBIA

TV-Sets: N/A — **Colour:** PAL — **System:** B.

INFORMATION & BROADCASTING Sce.

Addr: 19 Hagan Street, Banjul. **Te:** +220 27230.
Stations: N/A.

GHANA

TV-sets: 250.000 — **Colour:** PAL — **System:** B.

GHANA BROADCASTING CORPORATION

Addr: P.O. Box 1633, Accra. **Te:** +233 (21) 221161. **Fax:** +233 (21) 773227. **Telex:** 2114.
L.P: Dir of Broadc: C.K.O. Nyamboki. Dir. of Prgrs: Mrs. E. Namai.
Stations: *) colour; others monochrome.
Accra ch4 (5kW) & 9* (10kW), Kisi ch2 (5kW) & 11* (10kW), Jamasi ch3 (5kW) & 5* (10kW), Bolgatanga ch6 (5kW) + 8 translators — **D.Prgr:** W 1845-2230; Sat/Sun 0645-0945, 1530-2230.

GUINEA (Republic)

TV sets: 65.000 — **Colour:** PAL — **System:** K.

RADIODIFFUSION TELEVISION GUINÉENNE (Gov.)

Addr: B.P. 391, Conakry. **Te:** +224 442205. **Telex:** 22341 rtg ge conakry - Guinée.
L.P: Dir. Gen: B. Camara. Dir. of Prgrs: B. Kaba.
Stations: Conakry ch 5 1kW + Kindia ch4 0.2 kW, Faranah ch5 0.5kW, Labé ch7 8kW, Mamou/Mali ch9 0.2kW, Kankan ch9 1kW.
D.Prgr: 1700-2000 (approx).

KENYA

TV-sets: 260.000 — **Colour:** PAL — **System:** B.

VOICE OF KENYA TELEVISION (Gov, Comm.)

Addr: Box 30456, Nairobi. **Te:** +254 (2) 334567. **Telex:** 25361 VOK KE.
L.P: Dir. of Broadc: C.K.O. Nyamboki. Dir. of Prgrs: Mrs. E. Namai.
Stations: ch4 (Nairobi) 10/1kW, Timboroa ch2 10/1 kW, Mombasa ch6 10/1kW, Mazeras ch6 5/1kW, Nyeri ch10 10/1kW, Nyambene ch11 10/1kW—**D.Prgrs:** 1400 (Sat/Sun 1100)—2100.

COMMERCIAL STATIONS:

KENYA TELEVISION NETWORK (KTN-TV)

Station: UHF ch 62 (Nairobi).
D.Prgr: W0000-0505, 1000-1800, 2000-2400. SS 0515-0715, also included relays of US Cable News Network (CNN).

LESOTHO

TV-sets: 50.000 — **Colour:** PAL — **System:** I.

LESOTHO NATIONAL BROADCASTING SERVICES

Addr: P.O. Box 36, Maseru. **Te:** +266 323561. **Fax:** +266 310003. **Telex:** 4340.
L.P: Dir: T. Ntasane. CE: Mr. Letele.
D.Prgr: No information available.

LIBERIA

TV-sets: 45.000 — **Colour:** PAL — **System:** B.

LIBERIAN BROADCASTING CORPORATION (Gov. Comm.)

Addr: P.O. Box 594, Monrovia. **Te:** +231 271250. **Telex:** 44249 BROADCAST.
L.P: GD: Alhaji GV. Kromah.
Station: ELTV chE6 1/0.1kW + 4 low power repeaters.
D.Prgr: W 1815 (Sat 1615)-2300; Sun 1415-2230.

LIBYA

TV-sets: 500.000 — **Colour:** PAL — **System:** B.

PEOPLES REVOLUTION BROADCASTING — TELEVISION (Gov.)

Addr: P.O. Box 333, Tripoli.
L.P: Dir: Youssif Debri.

Stations:	chE	kW (ERP)	Pol		chE	kW (ERP)	Pol
Benghazi	5	10/2	V	Khoms	8	5/1	V
Tripoli	6	20/4	H	Tobruk	8	5/1	H
Derna	6	5/1	V	Yefren	9	20/4	H
Elmarj	7	5/1	V	El Beida	9	5/1	H
Houn	7	1/0.2	H	Misurata	10	5/1	H
Sirte	7	1/0.2	H	Egdabia	11	15/3	H

+ 1 low power repeater.
D.Prgr's: 1700-2230.

MADAGASCAR

TV-sets: 130.000 — **Colour:** SECAM — **System:** K.

RADIO TELEVISION MALAGASY (Gov.)

Addr: P.O. Box 442, Antananarivo. **Te:** +261 (2) 21784. **Telex:** 22506.
L.P: Dir: M. Rabesahala.
Stations: Antananarivo ch5 1kW H + 36 low power st's.
D.Prgr: 1600-1930.

MADEIRA

TV-sets: 80.000 — **Colour:** PAL — **System:** B.

RADIO TELEVISÃO PORTUGUESA, E.P. (Comm.)

Addr: Rua das Maravilhas 42, 9000 Funchal. **Te:** +351 45197/8.
Fax: +351 48859. **Telex:** 72478.
L.P: Gen. Mgr: A. Abreu.

Stations: Funchal (Pico do Silva) chE5 20/4kW H, with repeaters on ch5/100kW H, 6/20kW H, 8/100kW H (+4 low power st's not mentioned.)
N.B: only RTP's 1st. prgm. is broadcasted.
D.Prgr: 1800 (Sat/Sun 1400)-2300.

MALI

TV-sets: 10.000 — **Colour:** SECAM — **System:** K.

RADIODIFFUSION TÉLÉVISION DU MALI

Addr: B.P. 171, Bamako. **Te:** +223 (22) 2019 2243 08.
L.P: see radio.
Stations: Bamako ch5 10kW + 2 repeaters.
D.Prgr: 1854 (Sat/Sun 1454)-2300.

MAURITANIA (Islamic Rep.)

TV-sets: 1100 — **Colour:** SECAM — **System:** B.

TELEVISION NATIONALE (Gov.)

TVM (Television du Mauritanie)
Addr: B.P. 5522
Te: 53303, **Telex:** 5817.
Station: Nouakchott chE5 (2x1kW).
D.Prgr: 2000-2245.

MAURITIUS

TV-sets: 138.111 — **Colour:** SECAM — **System:** B.

MAURITIUS BROADCASTING CORP. (Comm.)

Addr: 1 Louis Pasteur Str, Forest Side. **Te:** +230 86 5001. **Fax:** +230 86 7332. **Telex:** 4320.
L.P: Dir: Dilliparsad Suraj Bali. Head of Prgrs: Mrs. M. Boolell.
Stations: Malherbes chE4 15/3kW H, + 3 repeaters st's (all 0.4kW ERP)
D.Prgr: 1200 (Sat 0530/Sun 0600)-1900.

MAYOTTE

TV-sets: 3.500 — **Colour:** SECAM — **System:** K.

R.F.O.—MAYOTTE

ADDR. & L.P: see radio.
Stations: Lavigie ch 9H 100W, Mamaidzou ch 7H 50W, Lima Combanich 4H 200W.
D.Prgr: Mon-Fri 1100-2100, Sat/Sun 0800-2100.

MOROCCO

TV-sets: 1.210.000 — **Colour:** SECAM — **System:** B.

RADIODIFFUSION TÉLÉVISION MAROCAINE (Gov.)

Addr: 1, rue Al Brihi, Rabat. **Te:** +212 (7) 64951.
L.P: Dir. Gen: Mohamed Issari.
Stations: Pol: H.

	ch	kW(ERP)		ch	kW(ERP)
Zerhoun	4	120/12	Essaouira	11	20/2
Zaio	4	9/1.8	Tan Tan	8	11/1.1
Laayoune	E4	316/31.6	Safi	8	20/2
Oujda	5	267/26.7	Tazerkount	8	90/18
Boukhouali	5	150/15	Tazekka	8	4/0.8
Tanger	5	20/4	Touzarine	9	9/1.8
Sidi Bounouara	5	11/1.1	Tiguelmamine	9	9/1.8
Oukaimeden	6	18/3.6	S. Bounouara	9	11/1
Azougar	6	9/1.8	Biougra	9	4/0.8
Dakhla	6	11/1.1	Casablanca	10	180/18
Figuig	6	9/1.8	Hafa Safa	10	9/1.8
Rabat	7	180/18	Ourzazate	10	267/26.7
Izeft	7	14/2.8	Bouarfa	10	267/26.7

+ 35 low power relay sts.
N.B: All ch's M except where otherwise indicated.
D.Prgr: 1800-2400 in Arabic.

2M INTERNATIONAL (Comm.)

Addr: 52 Ave. Hassan II, Casablanca. **Te:** +212 (35) 4444.
Fax: +212 223955.
L.P: Dir. Gen: Antonio Filali.
D.Prgr: 15h daily, partly subscription.
F.Pl: Pan-African satellite sce.

MOZAMBIQUE

TV sets: 35.000 — **Colour:** PAL — **System:** B.

TELEVISAO EXPERIMENTAL DE MOCAMBIQUE (Gov.)

Addr: C.P. 2675, Maputo. **Te:** +258 744788 or 741395. **Telex:** 6-346 TEVEMP MO.
L.P: Dir: Botelho Moniz. Dt.Tec: Jaime Ferreira. Film Buyer: Arlando Tembe.
Station: Maputo ch33 1kW.
D.Prgr: 24h weekly.

NAMIBIA

TV-sets: 27.000 — **Colour:** PAL — **System:** I.

NAMIBIA BROADCASTING CORPORATION

Addr: P.O. Box 321, Windhoek 9000. **Te:** +264 (61) 215811. **Fax:** +264 (61) 2912291. **Telex:** 50908 – 622/708. **Cable:** Broadcast.
L.P: see radio.

Stations	ch		ch
Rundu	4	Signalberg	9
Keetmanshoop	4	Rösing	10
Paresis	5	KleinWindhoek	10
Windhoek	6	Mariental	10
Erongo	7	Waterberg	10

+ 3 low power st's and approx. 20 private low power relay st's.
D.Prgrs: Mon-Sat: 1400-2100, Sun: 0600-1100, 1400-2100 in English.

NIGER

TV-sets: 25.000 — **Colour:** SECAM — **System:** K.

TELE-SAHEL (Gov.)

Addr: B.P. 309, Niamey. **Te:** +227 723155. **Telex:** 5229 NI.
L.P: Dir. Gen: DG: Abdou Souley. TD: Zoudi Issouf.

Stations	ch	kW	Pol		ch	kW	Pol
Agadez	4	10	H	Gaya	8	1	V
Dosso	4	10	H	Niamey	9	10	H
Zinder	5	10	H	Konni	9	10	H
Arlit	6	1	H	Diffa	9	10	H
Maradi	7	10	V	Tessaoua	9	1	V
Dogondoutchi	7	1	H				

+ 7 low power relay st's.
D.Prgr: 0630-1130, 1430-1730.

NIGERIA

TV-sets: 6.100.000 — **Colour:** PAL — **Systems:** B.

NIGERIAN TELEVISION AUTHORITY (Gov.)

Addr: Television House, PMB 12036, Victoria Island, Lagos. **Te:** +234 (1) 614966/615154/612529. **Telex:** 22536 NTA HQ.
L.P: Dir. Gen: Mallam Mohammed Ibrahim. Dir. Tec: I. Wakombo. Dir. of Prgrs & Film Buyer: Dr. Victoria Ezeokoli.
National TV Production Centre.
TV House, Victoria Island, PMB 12005, Lagos.
L.P: Man. Dir: A. Micika.

Stations: (Figures in brackets indicate ERP in kW); P = Polarization.

	Location	ch/kW/P			
2)	Jaredi	3(250) H	2)	Gusau	9(100) H
1)	Ibadan	4(60) H	7)	Ilorin	9(80) H
3)	Jaji	4(40) H	7)	Chokochoko	9(40) H
6)	Lagos 2	5(120) H	6)	Tejuoso	10(100) H
1)	Idominasi	5(60) H	17)	P. Harcourt	10(120) H
4)	Auchi	5(5) V	14)	Makurdi	10(66.7) V
8)	Mongum	5(56.5) V	15)	Maiduguri	10(60) H
13)	Azare	5(100) H	5)	Kano	10(100) H
5)	Kano	5 (100) H	3)	Kachia	10(100) H
2)	Koko	5(10) H	16)	Minna	10(100) H
19b)	Suleja	5(100) H	18)	Iju	11(10) H
11)	Aba	6(100) H	4)	Iseleuku	11(100) H
14)	Ayangba	6(100) H	8)	Keffi	11(100) H
12)	Jalingo	6 (100) H	12)	?	11(25) H
7)	Egbe	6(40) V	19)	Abeokuta	12(100) H
7)	Kiama	6(40) V	4)	Ughelli	12(100) H
19a)	Ikom	6(10) H	14)	Kasina Ala	12(100) H
1)	Ashileke	7(60) H	8)	Langtang	12(100) H
8)	Jos	7(82.2) H	16)	Bida	12(100) H
13)	Gombe	7(100) H	1)	Oyo	22(500)
15)	Gashua	7(100) H	20)	Akure	23
5)	Arbus	7 (37.5) H	21)	Abeokuta	25(400) H
3)	Funtua	7(100) H	27)	Ibadan	28
2)	T. Amodu	7(100) H	25)	Jos	29(300) H
16)	Kontangora	7(100) H	21a)	Lagos	35
9)	Ikeja	8(10)	9)	Ikeja	35(40)
10)	Enugu	8(100) H	22)	Enugu	50(15)
12)	Yola	8(25) H	25)	Shendam	53(300) H
12)	Wukari	8(10) H	23)	Benin City	55
5)	Birninkudu	8(50) H	28)	Port Harcourt	55
4)	Eku	9(50) V	24)	Owerri	59(100)
19a)	Calabar	9(100) H	26)	Kano	67(100) V
13)	Bauchi	9(120) H			

Addresses & other information:

1) NTA-Ibadan, P.O. Box 1460, Ibadan, Oyo State. 1500-2300 (Sat/Sun 0730-2300).
2) NTA-Sokoto, P.M.B. 2351, Sokoto, Sokoto State. 1530-2315 (Sat/Sun 1130-2400).
3) NTA-Kaduna, P.O. Box 250, Kaduna. 1500-2300.
4) NTA-Benin, P.M.B. 1593, Benin, Edo State. 1700-2200 (Sun 1000-2100).
5) NTA-Kano, P.M.B. 3343, Kano, Kano State. — St. Mgr: M.G. Abubakar. Dir. Tec: A. Aduma.
6) NTA-Lagos, Victoria Island, P.M.B. 12005, Lagos — ch10: 1400-2245. Sat 1000-2315; Sun 1000-2245; ch5: 1600 (Sat/Sun 0800)-2300.
7) NTA-Ilorin, P.M.B. 1470, Ilorin, Kwara State.
8) NTA-Jos, P.M.B. 2134, Jos, Plateau State.
9) NTA-Ikeja, Broadcasting House, Obafemi Awolowoway, PBM 210 35, Ikeja.
10) NTA-Enugu, P.M.B. 1440, Enugu, Enugu State — 1345-2300.
11) NTA-Owerri, P.M.B. 7126, Abia, Abia State. — St. Mgr: Ph.O. Okereke. Dir. Tec: J.O.N. Ezekoka — 1355(Sat/Sun 1255)-2315; 1400-2320 (Sat/Sun 1300-2300).
12) NTA-Yola, P.M.B. 2197, Yola, Adamawa State. 1400 (Sat/Sun 0800)-2300.
13) NTA-Bauchi, P.M.B. 0146, Bauchi, Bauchi State.
14) NTA-Makurdi, P.M.B. 2044, Makurdi, Benue State.
15) NTA-Maiduguri, P.M.B. 1487, Maiduguri, Borno State. Mon-Fri 1500-2310; Sat/Sun 0800-1100, 1500-2310.
16) NTA-Minna, TV House, P.M.B. 79, Minna, Niger State — 1500 (Sat/Sun 1400)-2300.
17) NTA-Port Harcourt, P.M.B. 5797, Port Harcourt, Rivers State — 1400-2325.
18) NTA-Akure, P.M.B. 794, Akure, Ondo State. **L.P:** Gen. Man: Victor Folivi.
19) NTA-Abeokuta, P.M.B. 2190, Abeokuta, Ogun State. 1500 (Sat/Sun 0700)-2300.
19a) NTA-Calabar, P.M.B. 1299, Calabar, Cross River State. 1400-2230 (Sat/Sun 2400).
19b) NTA-Abuja, Abuja, FCT. 8h daily.

Educ. Prgrs on all NTA-st's: 1415-1500.
F.Pl: St's under construction at Katsina & Uyo.

State TV-Services

20) Ondo State TV (ODTV), PMB 709. Akure. St. Mgr: A. Ayemi.
21) Ogun State Television, P.M.B. 2239, Abeokuta. 1500 (Sat/Sun 0600)-2400.
21a) Lagos TV, P.M.B. 12036, Lagos.
22) ESBS, P.M.B. 1600, Enugu, Enugu State. **Te:** +234 (42) 337983. W 1500-2310, Sat/Sun 0800-2400.
23) Edo State TV, P.M.B. 1012, Benin City. 1500 (Sat 1400)-2000.
24) Imo Broadc. Corp., P.O. Box 329, Owerri—1450 (Sat/Sun 1100)—2300.
25) Plateau State TV, P.O. Box 2043, Jos. Plateau State. St. Mgr: Susan R. Janta.
26) CTV — Kano Broadc. Corp, P.O. Box 10094, Kano. 10h Mon-Fri, 12h Sat/Sun. **Te:** 234 (64) 622322.
27) Broadc. Corp. of Oyo State (BCOS), Ibadan.
28) Rivers State TV, Port Harcourt.

REUNION

TV-sets: 90.500 — **Colour:** SECAM — **System:** K.

SOCIÉTÉ NATIONALE DE RADIO-TÉLÉVISION D'OUTRE-MER (RFO)

Addr. & L.P: See radio.
Stations: P. Textor chK'9 0.5kW + 18 low power repeaters.
D.Prgr: Prgr.I: 1300 (Sat/Sun 1100)-1930; Prgr.II: no times available.

SENEGAL

TV-sets: 61.000 — **Colour:** SECAM — **System:** K.

SENEGAL TELEVISION (Gov.)

Addr: Office de Radiodiffusion Television de Senegal. 58 Blvd. de la Republique. B.P. 2735. Dakar. **Te:** +221 (21) 5689/215366.
L.P: DG: Babacar Diagne.
Station: Dakar chK'7 10kW H.
D.Prgrs: 1900-2300 (Sat/Sun 1200-2330).

SEYCHELLES

TV-sets: 8.200 — **Colour:** PAL — **System:** B.

SEYCHELLES BROADCASTING CORP. (Gov.)

Addr & LP: see radio. **Te:** + 248 24161.
Stations: La Misère chE2 1kW, St. Louis chE7 6kW + 9 low power repeaters.
D.Prgr: Mon-Thurs 1355-1800, Fri 1355-1700, Sat 1200-1730, Sun 1200-1800.

SIERRA LEONE

TV-sets: 25.000 — **Colour:** PAL — **System:** B.

SIERRA LEONE TELEVISION (Gov. Comm.)

Addr: Private Mailbag, Freetown. **Te:** + 232 (22) 40403/40906.
Telex: RADTEX 3334 SL.
L.P: Ag. Head of Prgrs: G. Felix-George. Sen Eng. (TV): W.A.I. George.
Station: chE2 1kW H, chE7 126kW H.
D.Prgr: 1745-2330.

SOMALIA (Rep)

TV sets: N/A — **Colour:** PAL — **System:** B.

MINISTRY OF INFORMATION

Addr: P.O. Box 1748. **Telex:** 999621 Mogadishu.
L.P: Dir. of TV project: A. Ali Askar. Dir. Tec: A. Hassan.
Station: ch6 1kW.
D.Prgrs: 1700-2000 (Fri 2100).

SOUTH AFRICA

TV-sets: 3.445.000 — **Colour:** PAL — **System:** I.

SOUTH AFRICAN BROADCASTING CORPORATION (SABC-TV)

Addr: see Radio.
L.P: Dep. Dir. Gen: T.v.Heerden. Gen. Mgr. TV1: H.P.N. Jordaan. CCIR System 1 (PAL colour) used on bands III/IV/V using ch4-13 (174-254MHz) and 21-68 (470-845MHz), chs 12/38 not used. Sound/Vision spacing is + 6MHz.
N.B: As of June 1st 1992 TV 2/3/4 no longer exist. These sts are combined in "The Contemporary Community Value TV".
TV-1:
Stations: 100kW (ERP) exc. *) 1-10kW

Location	ch	Pol	Location	ch	Pol
Durban	4	H	Ladismith (C.P.)*	22	H
Kimberley	4	H	Bedford*	23	H
Potgietersrus	4	H	Boesmanskop*	23	H
De Aar	5	H	Nelspruit	24	H
George*	5	V	Petrus Steyn*	24	H
Pretoria	5	V	Kareedouw*	25	H
Theunissen	5	H	Glencoe	27	H
Donnybrook*	6	H	Schweizer Reneke	33	H
Graaff-Reinet*	6	V	Eshowe	36	H
Prieska*	6	V	Mooi River*	37	H
Springbok*	6	V	Springfontein*	37	H
Volkrust*	6	V	Hoedspruit*	39	H
Walvis Bay*	6	V	Matjiesfontein*	39	H
Port Elizabeth	7	H	Vryheid*	39	H
Queenstown	7	H	Carnarvon*	40	H
Welverdiend	7	H	Matatiele*	40	H
Villiersdorp	7	H	Cradock*	40	H
Cape Town	8	V	Middelburg (Tvl)	41	H
Garies*	8	H	Senekal*	42	H
Grahamstown	8	H	Williston*	42	V
Kuruman	8	H	Klerksdorp	45	H
Louis Trichardt*	8	V	Blouberg*	45	V
Port Shepstone	8	V	Ubombo*	45	H
Bloemfontein	9	H	Carolina*	50	H
East London	9	H	Piet Plessis*	50	H
Napier*	9	V	Zeerust	52	H
Piketberg	9	H	Dullstroom*	53	H
Oudtshoorn	9	H	Greytown*	53	H
Victoria West	9	H	Noupoort*	54	H
Vanrhynsdorp	10	H	Bethlehem	55	H
Beaufort West*	10	H	King William's Town*	56	H
Pofadder	10	V			
Upington	10	H	Ladybrand*	56	H
Piet Retief*	11	H	Rustenburg*	56	H
Pomfret*	13	H	Douglas*	57	H
Johannesburg	13	H	Kroonstad	57	H
Thabazimbi	13	V	Willowmore*	57	H
Riversdale	13	H	Suurberg*	59	H
Davel	22	H	Aliwal North	61	H
Calvinia*	22	H	Christiana*	62	H
Enzelsberg	22	H	Tzaneen	64	H

+ 63 gap fillers and estimated 400 privately owned low power tr's.
D.Prgr: 0400-0615 (Mon-Fri), 1300 (Sat 0400, Sun 1130)-2200 (Sat 2230).

CONTEMPORARY COMMUNITY VALUE TV

Addr: P.O. Box 2491, Johannesburg 2000.
Stations:

Loc.	ch	Pol			
Bethlehem	63	H	Kimberley	7	H
Bloemfontein	13	H	King William's Town	60	H
Cape Town	5	V	Klerksdorp	41	H
Christiana	58	H	Kroonstad	61	H
Davel	30	H	Kuruman	11	H
Donnybrook	13	H	Louis Trichardt	11	V
Durban	7	H	Middelburg	45	H
East London	13	H	Nelspruit	32	H
Eshowe	28	H	Nongoma	58	H
George	11	V	Oudtshoorn	6	H
Glencoe	31	H	Piketberg	6	H
Grahamstown	5	H	Port Elizabeth	4	H
Johannesburg	6/9	H	Port Shepstone	5	V
Potgietersrus	7	H	Theunissen	8	H
Pretoria	11/8	V	Villiersdorp	10	H
Queenstown	4	H	Volksrust	9	V
Rustenburg	64	H	Welverdiend	10/4	H
Schweizer-Reneke	25	H	Zeerust	44	H

D.Prgrs: broadcasts in various languages.

M-NET TELEVISION (Pay channel, Comm, Private).

Addr: P.O.Box 4950, Randburg 2125. **Te:** +27 (11) 8891251.
L.P: Gen. Mgr: T. Vosloo. Dir. Prgrs: T. Ellis.
Stations: Bloemfontein ch6 (H), Alverstone & Pt. Elizabeth ch10 (H), Constantiaberg ch11 (V), Pretoria ch21 (H), Johannesburg ch39 (H) + 16 repeaters, Durban ch10, Cape Town ch11, George ch7, Newcastle ch62.
D.Prgr: Mon-Fri 0830-1030 & 1300-2300; Sat/Sun 0500-2300. **Indian Prgr:** Sun 0530-0830. **Portuguese Prgr:** Sun 0830-1130. **Oper time:** 1600-1700.

BOPHUTHATSWANA

BOPHUTHATSWANA TELEVISION (Bop-TV) (Para-statal, Comm.)

Addr: P/Bag X 2150, Mmbatho 8681. **Te:** +27 (140) 289111. **Fax:** +27 (140) 26009. **Telex:** (0937) 3017 BP.
L.P: Dir: R. Minton. Head of Prgrs: R. Magau.

Stations	ch	Pol	kW	Stations	ch	Pol	kW
Mmabatho	24	V	1	Kameeldrift	36+	V	0.1
Itsoseng	59	H	0.5	Madibogo	55	H	1
Makadima	54	H	0.25	Ganyesa	42+	H	1
Motswedi	45	V	0.5	Taung	39−	H	1
Madikwe	26−	V	0.5	Kuruman	56−	H	1
Pilanesberg	57+	V	0.25	Thaba Nchu	67+	H	1
Klipvoordam				Mogwase	66+	V	0.5
Mabopane	44−	V	0.025	Pankop	64+	V	0.2
Garankuwa				Tihabane	40−	V	0.01
Brixton	37	V	0.1	Bethanie	44−	V	0.01

+ rel. st's in Johannesburg (ch36) & Pretoria (ch37) both 0.25kW.
D.Prgr: 1300 (Sat/Sun 1200)-2200 (Fri/Sat 2300).

CISKEI

TRINITY BROADC. NETW. CISKEI (TBNC)

(owned by the Rhena Church)
Station: Bisho ch24 2kW
D.Prgr. in English: 1400-2030.

TRANSKEI (Republic)

TV Sets: 40.000 — **Colour:** PAL — **System:** I.

TRANSKEI BROADCASTING CORPORATION

TRANSKEI TV (Govt. Comm.)

Addr: P.O. Box 794, Umtata, 5100 Transkei. **Te:** +27 (471) 312611.
L.P: Same as for Radio.
D.Prgr: 16H00-16H30.
Ann: This is Transkei TV.

M.N.E.T. TRANSKEI (Pay TV Service)

Addr: MNET Transkei: Metropolitan Building, Leeds Street, Umtata,100. **Te:** +27 (471) 312222.
D.Prgr: Scrambled Service except between 15H00-17H00 When programmes are unscrambled. Transkei TV uses ½ hour of this unscrambled time for its programmes.
Relay S.A.B.C. TV1 & TV2/4 Programmes by agreement.

Location	TV1	TV2/4	TKEITV/MNET	
Butterworth	ch 29	ch 33	ch 21	1kW
Mt. Ayliff	ch 35±	ch 31±	ch 23	1kW
Umtata	ch 59	ch 63	ch 55	1kW
Port St. Johns	ch 57+	ch 61+	ch 53	1kW

Power: ± 220 Watt, + 100 Watt
F.Pl: Relay stations at Mt. Fletcher.

TRINITY BROADCASTING NETWORK TRANSKEI (Religious)

Addr: Private Bag X 5130, Umtata, 5100, Transkei. **Te:** +27 (471) 312582. **Fax:** +27 (471) 312581.
L.P: Station Sup: Stephan Wessels.

Location: Umtata, Ch 67, Power 1kW.
D.Prgr: 1330-2030 daily and special programming during school holidays.
Ann: "This is the Trinity Broadcasting Network, Channel 67, Transkei." or "This is the Trinity Broadcasting Network beaming the Gospel of Jesus Christ from Umtata, Capital of Transkei."
F.Pl: A 1kW Transposer at Butterworth and Mt. Ayliff.

SUDAN

TV-sets: 250.000 — **Colour:** PAL — **System:** B.

SUDAN TELEVISION (Gov, Comm.)

Addr: P.O. Box 1094, Omdurman. **Te:** +249 (11) 55022. **Telex:** 28002, 28053. **Cable:** Sudan TV.
L.P: Head of Directorate: Hadid al-Sira.
Stations: Omdurman chE5 5kW H, Gezira chE7 10kW, Atbara chE9 0.5kW.
D.Prgr: 1500-2200.

SWAZILAND

TV sets: 12.500 — **Colour:** PAL — **Systems:** B & G.

BROADCASTING AUTHORITY OF SWAZILAND

Addr: Swazi TV, P.O. Box A146, Mbabane. **Te:** +268 43036/7. **Telex:** 2138 WD. **Fax:** +268 42093.
L.P: Man. Dir: Dan S. Lamini.
Stations: Bulembu ch5 1.5kW H, Ntondozi ch15 15kW H + 7 relay st's.
D.Prgr. in English: 1600-2100.

TANZANIA

TV-sets: 80.000 — **Colour:** PAL — **System:** B.

TELEVISION ZANZIBAR TVZ (Gov.)

Addr: P.O. Box 314, Zanzibar. **Te:** +255 (54) 32816/7. **Telex:** 57200 TVZ TZ.
L.P: MD: Jama A. Simba. TD: George H. Majaliwa. Film Buyer: Jaffar S. Kassingo; Prod. Mgr: Abdulhamid H. Dau.
Stations: Unguja chE21 40kW, Pemba chE9 40kW (ERP).
D.Prgr: 1645-1900 (Sat/Sun/National Holidays 1645-2000).

TOGO

TV sets: 23.000 — **Colour:** SECAM — **System:** K.

TELEVISION TOGOLAISE (Gov.)

Addr: B.P. 3286, Lomé. **Te:** +228 21-53-57. **Telex:** 5320 ATOP.
L.P: Dir. of TV: Koffi Panou. Head of Prgrs: Eho K. Vioto.
Stations: Mt. Agou ch6 10KW V, Lomé ch8 1kW H, Aldjo-Kadara ch8 10kW H + 2 relay st's.
D.Prgr: Mon-Fri 1830-2230, Sat/Sun 1230-2400.

TUNISIA

TV-sets: 650.000 — **Colour:** SECAM — **System:** B&G.

RADIODIFFUSION TELEVISION TUNISIENNE (Gov.)

Addr: 71, Ave de la Liberté, 1070 Tunis. **Te:** +216 (1) 780088. **Fax:** +216 (1) 781 058. **Telex:** 14960.
L.P: Dir: Abdeh Afidh Hardudm.
Stations: Network I (Arabic)

	ch	kW (ERP)	Pol		ch	kW (ERP)	Pol
Remada	4	32/6.4	H	Gafsa	10	400/80	V
Kasserine	6	340/68	H	Zaghouan	11	280/56	H
Boukornine	7	12.5/1.25	V	Ain Draham	12	32/6.4	H
Sfax	8	365/73	H	Zarzis	12	180/36	H
Goraa	9	32/6.4	H				

Network II: (French) Pol = H.

	ch	kW (ERP)		ch	kW (ERP)
Zarzis	22	1500	Ain Drahan	35	87
Biadha	23	780	Chambi	40	1410
Boukornine	26	44	Ghraba	51	1800
Zaghouan	33	1425	Bizerte	53	40
Remada	33	89	Goroa	55	89

low power st's not mentioned.
D.Prgr: Netw. I & II: 1600-2300.

UGANDA

TV-sets: 115.000 — **Colour:** PAL — **System:** B (Pol = H).

UGANDA TELEVISION (Gov.)

Addr: P.O. Box 7142, Kampala. **Te:** +256 (41) 254461. **Telex:** KNOLLEDGE 61084.
L.P: Ag. Dir: J.M.A. Obo.
Stations: Kampala chE5, Lira chE7, Masaka chE8, Mbale chE8, Mbarara chE10, Soroti chE10. Gulu chE9, Jinja chE11 + low power repeater at Kabale.
D.Prgr: Mon-Fri 1500-2100; Sat/Sun 1200-2100.

ZAIRE (Rep. of)

TV-sets: 22.000 — **Colour:** SECAM — **System:** K.

OZRT (Gov.)

Addr: B.P. 3171, Kinshasa, Gombe 3164. **Te:** +243 23171. **Telex:** 21583.Vozakin ZR.
L.P: Dir. Gen: B. Dongo. Dir. Tec: S. Lepamabla
Stations: a) regular; b) regular but not fulltime; c) irregular.

Location	ch	kW	Location	ch	kW
Kinshasa	5a	27	Uvira	6c	24
Kananga	4b	2	Kambove	7a	2
Kamina	4b	2	Goma	7a	2
Gbadolite*	4a	2	Gemena	8c	24
Kolwezi	5a	1	Kisangani*	8a	2
Mbuji Mayi	6b	2	Bandundu	8b	2
Mbandaka	6b	2	Lubumbashi*	9a	24
Kalemie	6a	2	Bukavu	9a	2
Kindu	6b	2	Isiro	9c	2

*) own studio facilities.
D.Prgr: Mon/Tues/Thurs/Fri 1130-1330 & 1630-2300, Wed 1130-2300, Sat/Sun 0900-2300. Relayed on Intelsat 66°E, C-band. Tr's from inland towns are dependent on power supplies (i.e. fuel availability).

ZAMBIA

TV-sets: 200.000 — **Colour:** PAL — **System:** B.

TELEVISION-ZAMBIA (Gov.)

Addr: Broadc. House, P.O. 50015, Lusaka 229648. **Te:** +260 219673; P.O. Box 20748, Kitwe. **Te:** +260 254989. **Telex:** 41221 ZA. **Cable:** Broadcasts TV.
L.P: Ag. Contr: Ch. Mando.

Stations:	chE	kW (ERP)	Pol		chE	kW (ERP)	Pol
Solwezi	3	0.075	H	Kitwe	9	200	H
Kapiri Mposhi	6	200	H	Lusaka	10	200	H
Pemba	8	200	V	Senkobo	10	200	V
Kasama	8	16	H	Chipata	11	16	H
Mubwa	8	0.1	H				

D.Prgrs: Tues-Fri 0800-1400 (Educational), 1500 (Sat/Sun 1400)-2100 (Fri/Sat 2200).

ZIMBABWE

TV-sets: 137.090 — **Colour:** PAL — **System:** B.

ZIMBABWE TELEVISION (Gov, Comm.)

Addr: See radio.
L.P: DG: Pavia Tsimba; Hd. of Prgrs: Tomi Mandndigora.

Stations:	chE	kW (ERP)	Pol		chE	kW (ERP)	Pol
Gweru	2	17.6	H	Bulawayo	6	6	V
Bulawayo	3	3	H	Mutare	7	12	V
Harare	4	20	H	Masvingo	7	6	H
Harare	5	30	H	V. Falls	7	1	H
Gwendingwe	5	7	H	Gwanda	9	6	H
Rukotso	5	6	V	Gweru	11	25	H
Kariba	5	1	H	Mutoro-shanga	12	8	H
Kamativi	5	16	V				
Kadoma	6	4	H				

D.Prgr: English/Shona/Ndebele 0700-2200.

NEAR & MIDDLE EAST

AFGHANISTAN

TV-sets: 100.000 — **Colour:** PAL & SECAM — **System:** B.

PEOPLES RADIO & TV AFGHANISTAN (Gov.)

Addr: P.O. Box 544, Kabul. **Te:** 25241. **Telex:** 232 KHAREJA AF.
Station: Kabul ch5 10/1kW H.
D.Prgr: 1330-1830.
Reg. st's are operating in 9 different provinces at 1330-1530 or 1430-1630 (approx.).

ARMENIA

TV-sets: N/A — **Colour:** SECAM — **Systems:** D&K.

ARMENIAN TELEVISION (Gov.)

Addr: Alek Manukyan 5, 375025 Yerevan. **Te:** +7 (8852) 552502. **Fax:** +7 (8852) 551513.
L.P: See radio.

Stations (Prgr. 1)	**ch**	**Stations** (Prgr. 2)	**ch**
Yerevan	1	Yerevan	11
D.Prgr: 1730-2100.		**D.Prgr:** 1730-2030.	

NB: Prgr. 1 is carried by various main and relay transmitters throughout the country. Further details not known.

OSTANKINO KANAL 1 (OK-1)

Stations	**ch**
Yerevan	8

D.Prgr: Relays of OK-1 from Moscow. Further details see under Russia.

ROSSIJSKOJE TELEVIDENIJE (RTV)

Stations	**ch**
Yerevan	5

D.Prgr: Relays of RTV from Moscow. Further details see under Russia.
NB: OK-1 and RTV are carried by various main and relay transmitters throughout the country. Further details not known.

BAHRAIN

TV-sets: 270.000 — **Colour:** PAL — **Systems:** B&G.

BAHRAIN TELEVISION (Gov, Comm.)

Addr: P.O. Box 1075, Bahrain. **Te:** +973 781888/686000. **Fax:** +973 681544.
L.P: Dir: Dr. H. Al-Umran. Head of Prgrs: Fowzia Zainal.
Station: chE4, chE55, ch57.
D.Prgr: ch4: Main Arabic ch. loc. & int. news, cultural and variety prgrs, Arabic & English feature films.
ch44: Satellite broadcasts, loc. & int. sports. Also rel. Egypt satellite ch.
ch55: Main English ch. Also rel. CNN.
ch57: Transmits BBC World Sce. TV, 24h.

CYPRUS

TV-sets: 234.000 — **Colour:** PAL — **Systems:** B & G.

CYPRUS BROADCASTING CORPORATION

Addr: P.O. Box 4824, Nicosia. **Cable:** Broadcasts. **Te:** +357 (2) 422231. **Fax:** +357 (2) 314050. **Telex:** 2333CYBC.
L.P: Head of TV prgr: P. Ioannides. Other personnel: See radio.
Stations: Mt. Olympus chE6 40/4kW V + 35 low power st's.
D.Prgr. in Greek, Turkish, English: 1700-2300 in Winter, 1800-2330 from June-September.

BFBS (SSVC)

Addr: BFPO 58, Dhekelia Mil 381. **Te:** +357 (47) 23326/7. **Fax:** +357 (47) 23565.
Stations: ch 69 and 66 in ESBA, ch 60 and 68 in the WSBA.
D.Prgrs: relays of English prgrs. + live prgrs. from satellite.

NORTHERN CYPRUS

TV-sets: 75.000 — **Colour:** PAL — Systems: B & G.

BAYRAK RADIO & TELEVISION CORP.

Addr. see radio.
L.P: Mgr. Taylan Kav. Eng. Mgr: Kadri Balli.
Stations: Sinan daği ch8 100kW (Prgr. 2) — Selvilitepe ch40 450kW (Prgr. 3), ch44 450kW (Prgr. 1) + 4 relay st's.
D.Prgrs: no details received.

IRAN

TV-sets: 2.250.000 — **Colour:** SECAM — **System:** B.

ISLAMIC REPUBLIC OF IRAN TELEVISION (Gov.)

Addr: P.O. Box 19395, 1774 Tehran. **Te:** +98 (21) 294024. **Cable:** "IRIB". **Telex:** 213910.
L.P: See radio.
Stations: Two production centers in Tehran producing for two networks.
Network I: 1330 (Holidays 1030)-2030.
Network II: 1330 (Holidays 1300)-2030. (During summer also 0630-0830, Fri 0430-0630).
Local st's (28) Abadan, Ahwaz, Alamdeh, Ardebil, Baharlo, Bakhtaran, Bandar-Abbas, Booshehr, Esfahan, Hamedan, Kerman, Khoramabad, Kohe Genon, Kohe Noor, Kosangar, Mashhad, Oromieh, Rasht, Sanandaj, Sari, Shahrkord, Shiraz, Tabriz, Tehran, Yazd, Zahedan, Zanjan, Zibakenar + 450 low power repeaters.

IRAQ

TV-sets: 1 m — **Colour:** SECAM — **System:** B.

BAGHDAD TELEVISION (Gov, Limited Comm.)

Addr: See radio.
Te: +964 (1) 5371151. **Telex:** 212246 IDAH.
L.P: GM: Saad Al Bazzaz.
Stations: 1 = 1st Prgr; 2 = 2nd Prgr; R = Regional Prgr.

Location	chE	kW/Pol	Location	chE	kW/Pol
Misan (1)	5	144H	Baghdad (1)	9	360H
Muthanna (1)	6	144V	Basrah (1)	9	360V
Al-Taamin (R)	6	144H	Nenavah (1)	9	180V
Baghdad (2)	7	288H	Al-Taamin (2)	10	144H
Nenavah (2)	7	144V	Misan (2)	11	180H
Al-Taamin (1)	8	180H	Basrah (2)	12	144V
Muthanna (2)	8	180V			

D.Prgrs: Prgr I: 1400 (Fri 0700)-1930 (Fri 2100); Prgr II: 1400-2100.

ISRAEL

TV-sets: 1.500.000 — **Colour:** PAL — **Systems:** B&G.

ISRAEL BROADCASTING AUTHORITY (IBA)

Addr: P.O. Box 7139, Jerusalem 91071. **Te:** +972 (2) 291888. **Fax:** +972 (2) 292944. **Telex:** 25301 — **L.P:** DG: Ayre Mekel.

ISRAEL EDUCATIONAL TELEVISION

Addr: 14 Klausner Str, Tel Aviv. **Te:** +972 (3) 5434343. **Telex:** 342325 ITCIL.
L.P: Gen. Mgr: Y. Lorberbaum. Dir. of Eng: A. Kaplan. Dir. of Prgrs: Mrs. Y. Prener.

Stations	chE	kW (ERP)	Stations	chE	kW (ERP)
M. Ramon	5	15/1.5	Jerusalem	27	1/0.2
Acco	7	20/2	Tel Aviv	28	60/6
Eilat	7	0.15/0.015	Zefat	34	22/2.2
Eitanim	8	200/20	Haifa	36	40
Manara	8	1.5/0.15	Karmiel	39	1.5/0.15
Arad	9	10/1	Jerusalem	40	3/0.3
Grofit	10	15/1.5	Manara	44	20/2
Haifa	10	100/10	Haifa	46	200/20
Bar Jehuda	10	1.5/0.15	Jerusalem	56	1.5/0.15
Beersheba	11	100/10	K. Hayarden	60	20/2
Acco	24	50/5			

+31 low power repeaters.
D.Prgr. Educational: 0600-1530 (Fri 1430). (Sat: no trs).
IBA: Arabic: 1630 (Fri 1430, Sat 1400)-1800.
Hebrew: 1430-1530 (exc. Fri/Sat), 1800-2300v.

CHANNEL 2

Addr: 97 Jaffa St, Jerusalem 94340. **Te:** +972 (2) 242776/242750. **Telex:** 25678 ITRA. **Fax:** +972 (2) 242720.
L.P: Man. Dir: Oren Tokatly.
Stations: Eitanim chE22 50kW (ERP), Acco ch E27 2kW, Beersheba chE35 6kW — **D.Prgr:** 1830-2230.

JORDAN

TV-sets: 250.000 — **Colour:** PAL — **Systems:** B.

JORDAN RADIO & TELEVISION CORP.

Addr: P.O. Box 1041, Amman. **Te:** +962 (6) 638760. **Fax:** +962 (6) 788 115. **Cable:** Television. **Telex:** 24213 RTVENG JO.
L.P: St.Man: M. Amin, TD: O. Asfoura, Film Buyer: K. Ferray.
1st Prgr (in stereo): Suweilih chE3 104kW H, Aqaba ch9 5kW, Ras Munif-Ajlun ch9 500kW H, Deir Alla ch26 6kW H + low power repeaters.
2nd Prgr: Aqaba ch5 5kW H, Suweilih ch6 108kW H, Ras Munif-Ajlun ch11 500kW H, Deir Alla ch29 6kW H + low power repeaters.
1st Prgr in Arabic (in stereo): 1330 (Fri 0800)-2200.
2nd Prgr: 1600-2200. **French:** 1600-1730. **N:** 1700. **Hebrew:** 1730-1800. **Arabic:** 1800-1830 (rel. 1st Prgr). **English:** 1830-2200. **N:** 2000.
+ 43 low power sts.

KUWAIT

TV sets: 800.000 — **Colour:** PAL — **Systems:** B & G.

KUWAIT TELEVISION (Gov.)

Addr: P.O. Box 621 SAFAT, 13007 Safat. **Te:** +965 242 3774. **Telex:** (496) KTV 22169 KT (Dir. of TV), (496) MI 46285 KT (Dir. of Eng.). **Fax:** +965 245 6660.
L.P: Dir. of TV: Salem Al-Fahed. Chief Eng: Majid Al-Baghdadi. Others: see radio.
Stations: Mutlaa ch5 & ch12 40kW, ch45 & ch47 110kW; Rawdatein ch6 10kW; Failaka ch8 & ch10 10kW; Min. of Inf. complex ch24 & ch39 0.5kW.
D.Prgr: 1st Prgr: Sat-Wed 1100 (June-Sept 0600)-2100, Thurs/Fri 0600-2300. 2nd Prgr: 1300 (Thurs/Fri 1500)-2100.

LEBANON

TV-sets: 1.100.000 — **Colour:** SECAM — **Systems:** B & G.

TELE-LIBAN (Gov.)

Addr: B.P. 115054, Hazmieh, Beirut. **Te:** +961 450100.

Stations	ch	kW (ERP)	Prgr	Stations	ch	kW (ERP)	Prgr
Fih	2	1/0.2	I	Fih	8	10/1	II
Jounieh	2	1/0.2	II	Zahle	8	10/1	II
Beit Mery	2	1/0.2	III	Beyrouth	9	5/0.5	III
M. el Chouf	4	60/6	II	Fih	10	10/1	I
Beyrouth	5	50/5	I	Zahle	10	10/1	I
Fih	6	10/1	III	Beit Mery	11	10/1	II
Beyrouth	7	5/0.5	II				

+ 4 low power repeaters
D.Prgrs: 1st and 2nd Prgr (Arabic/French/English): 1500-2230; **N. in English:** 1600 (Prgr. 1). **N. in French:** 1615 (Prgr. 2). 3rd Prgr. (French/English) 1630-2200.

MIDDLE EAST TELEVISION

Addr: P.O. Box 5689, Nicosia, Cyprus (Studios in Marjayoun).
Addr. in USA: 1000 Centreville Turnpike, Virginia Beach, VA 23464, USA. **Te:** +1 (804) 523-7085. **Fax:** +1 (804) 523-7990.
L.P: Man. Dir. Int: Dr. Daniel Olsen. Head of Prgrs: Mary St. John.
Station: chE12 100kW (Pol:V) — chE5 10W.
D.Prgrs: 1130-2200 (approx) in English & Arabic.
N. in English: 1700.

LEBANESE BROADC. CORP. INT.

Addr: P.O. Box 16-5853 Beirut. **Te:** +961 938938. **Fax:** +961 937919.
L.P: St. Mgr: Piere El-Daher; TD: Nasim Boustany; Film Buyer: Selim El-Azar.
Stations: ch12H 60kW; ch33H 325kW; ch10H 35kW, ch9H 35kW; ch5H 35kW.

MOLDOVA (Moldavia)

TV-sets: N/A — **Colour:** SECAM — **Systems:** D&K.

TELEVIZIUNEA NAŢIONALĂ (Gov.)

Addr: şoseaua Hînceştilor 64, 277028 Chişinău. **Te:** +7042 (2) 721881. **Fax:** +7042 (2) 723537.
L.P: See radio.

Stations	ch	Stations	ch
Bălţi	2	Ungheni	32
Straşeni	3	Cimişlia	33
Cahul	8		

+ 6 low power translators.
D.Prgr: 0500-0800, 1500-2230 (Sun 0700-2230).

TELEVIZIUNEA ROMÂNĂ (TVR-1)

Stations	ch	Stations	ch
Cahul	1	Rezina	25
Cimişlia	2	Bălţi	41
Straşeni	11		

+ 6 low power translators.
D.Prgr: Relays of TVR-1 from Bucureşti/Romania. Further details see under Romania.

OSTANKINO KANAL 1 (OK-1)

Stations	ch	Stations	ch
Bălţi	8	Straşeni	30
Rezina	27	Edineţ	31
Ungheni	29	Cahul	31

+ 8 low power translators.
D.Prgr: Relays of OK-1 from Moscow. Further details see under Russia.

OMAN (Sultanate of)

TV-sets: 1.500.000 — **Colour:** PAL — **Systems:** B & G.

OMAN TELEVISION (Gov.)

Addr: P.O. Box 600, Muscat. **Te:** +968 603222. **Fax:** +968 602381. **Telex:** 5454/5151.
L.P: See radio.

Location	ch	kW	Pol	Location	ch	kW	Pol
Qurum	6	120/24	H	Salala	10	300/60	V
Sur	7	6/1.2		Saham	10	25/5	
Buraimi	8	160/32		Nizwa	10	200/20	
Thumrait	8	140/28	V				

+ 25 low power relay st's.
D.Prgr: 1100-2000.

QATAR

TV-sets: 250.000 — **Colour:** PAL — **System:** B.

QATAR TELEVISION SERVICE

Addr: Min. of Information, QTV, P.O. Box 1944, Doha.
Te: +974 894444. **Fax:** +974 864511.
Cable: Television, Doha. **Telex:** 4040 TeeVee DH.
L.P: Dir. TV: Mana-Al-Hajri. Head of Eng: H Abdullah Jaffar.
Stations: Doha chE9 200kW (ERP), Jamiliya chE11 (repeater) 600kW (ERP), Doha chE37 695kW (ERP), ch 49 (high power repeater for ch 37), (+ 5 repeaters of 1kW each).
D.Prgr: 1200-2130 (Fri also 0600-0900).

SAUDI ARABIA

TV-sets: 4,500,000 — **Colour:** SECAM & PAL — **System:** B & G.

SAUDI ARABIAN TELEVISION (Gov.)

Addr: P.O. Box 570, Riyadh 11421. **Te:** +966 (1) 4014440. **Fax:** +966 (1) 4044192/4054176. **Telex:** 401030 SJ.

1st Prgr:	ch	kW (ERP)	Pol		ch	kW (ERP)	Pol
Abu Qumais	5	200	H	Ranyah	10	7	H
Riyadh	5	45	H	Al Ardiyah	10	7	H
Madinah	5	170	H	Bani Malik	10	2	H
Abha	5	170	V	Ar Rass	10	5	H
Hail	5	42	H	Wadi Al Freah	10	9	H
Zilfi	5	9	V	Jebel Garah	10	3	H
Arar	5	14	H	Al Summan	10	29	H
Tayma	5	5	H	Tathlith	10	9.5	H
Bijadiah	5	9	V	Al Howaidh	10	4.7	H
Harradh	5	5	V	Al Kharj	11	5	V
Halat Ammar	5	8	H	Al Majmaah	11	160	H
Al Hariq	5	2	H	Al Sulaiyal	11	5	H
Linah	5	5	H	Bani Saad	12	6	H
Dammam	6	800	H	Owegliyah	12	26	H
Taif	6	47	H	Al Harjah	21	2	H
Yanbu	6	46	H	Al Ahmer	22	16	H
Baqa	6	5	V	Shaqra	22	100	H
Sharourah	6	42	V	Durma	22	8	H
Al Quwaiyah	6	5	V	Khybar	22	7	H
Hanakiyah	6	5	H	Al Qunfudah	22	7	H
Athnen	6	6	H	Al Nuariyah	22	8	H
Mahd Ad Dahab	6	5	H	Al Sudah	22	19	H
Khamasin	6	6	H	Artawiyah	23	9	H
Makkah	7	52	H	Turaif	24	3	H
Al Baha	7	94	H	Hufuf	25	25	H
Jizan	7	160	H	Al Musayjid	25	11	H
King Khaled City	7	2	H	Al Khurma	26	7	H
Al Qurayyat	7	1	H	Dammam	27	44	H
Al Ula	7	5	H	Layla	27	7	H
Buraydah	8	98	H	Haql	27	7	H
Tabarjal	8	12	V	Rafha	28	7	H
Hafr Al Batin	9	16	V	Al Nimas	30	6	H
Al Hassi	9	8	H	Al Alaya	31	10	H
Uqlat Assuqr	9	9	H	Bishah	39	6	H
Al Muwayh	9	9	V	Turabah	40	7	H
Nabq Abu Qasr	9	2	H	Umm Lajj	41	7	H
Skaka	9	5	H	Rabigh	42	8	H
Badr Hunayn	9	5	H	Al Hawtah	45	7	H
Sajir	9	13	V	Qaryat Al Ulya	46	17	H
Jebel Khasheb	9	10	H	Tuwal	50	100	H
Al Bad	9	1	H	Al Khafji	51	40	H
Jeddah	10	27	H	Al Wajh	53	7	H
Tabuk	10	15	V				

+ 25 Stations not mentioned.

2nd Prgr:	ch	kW (ERP)	Pol		ch	kW (ERP)	Pol
Makkah	5	52	H	Hail	7	42	H
Skaka	5	5	H	Zilfi	7	9	V
King Khaled City	5	2	H	Arar	7	14	H
Buraydah	6	98	H	Taif	8	47	H
Riyadh	7	45	H	Tabuk	8	15	V
Madinah	7	170	H	Yanbu	8	51	H
Sharourah	8	42	H	Al Howaidh	12	4.7	H
Khamasin	8	6	H	Shaqra	25	100	H
Abha	9	170	V	Hufuf	28	25	H
Al Baha	11	106	H	Dammam	29	890	H
Hafr Al Batin	11	17	V	Tuwal	53	100	H
Jeddah	12	27	H	Al Khafji	53	40	H

+ 6 Stations not mentioned.
D.Prgr: 1st Prgr: (Arabic): Sat-Wed 0700-0930, 1400-2130; Thurs/Fri: 0700-2130. **2nd Prgr:** (English): 1400-2130 (Thurs/Fri also 0600-0900).

CHANNEL 3 TV (Non-Comm, Private Co.)

Addr: Bldg. 3030 LIP, Dhahran.
Te: 875-4634. **Cable:** Aramco, Dhamedia.
Station: chE3 Dhahran 5kW.
D.Prgr: 0600-2030 (or 2100). **Colour:** PAL.

SYRIAN ARAB REP.

TV-sets: 700.000 — **Colour:** SECAM on ch's 2-12 (System B), PAL on ch26 (System G).

SYRIAN ARAB TELEVISION (Gov.)

Addr: Ommayyad Square, Damascus. **Te:** +963 (11) 720700. **Fax:** +963 (11) 720700. **Telex:** 411223.
L.P: Dir: Abd-Al-Salam Hejab. For other personnel see radio.
Stations: Pol = H. +) Prgr 2, others Prgr 1.

Location	chE	kW (ERP)		chE	kW (ERP)
Holms	2	?+	Aein-Al Arab	7	100/10
Abou-Kmal	3	200/20	Tabqua	8	100/10
Nabi-Saleh	3	100/10	Kaldoun	8	10/1
Hassakeh	4	200/20	Slenfeh	9	200/20
Aleppo	5	200/20	Salhieh	11	30/3
Damascus	5	100/10	Afrien	11	10/1
Deir-Al-Zoor	6	100/10	Palmyra	11	10/1
Soueida	7	350	Al-Malkieh+	12	200/20
Homs	7	200/20	Lattakia	26	60

+ 36 low power & repeater st's.
Prgr 1: 1300 (Fri 0800)-2200.
Prgr 2 in Arabic/English/French: 1700-2130.

TURKEY

TV-sets: 10.530.000 — **Colour:** PAL — **System:** B.

TURKISH RADIO TELEVISION CORPORATION

Addr: TRT-TV Department, Nevzat Tandogan Caddesi 2, Kavaklidere, Ankara. **Te:** +90 (41) 280546. **Fax:** +90 (41) 42767.
L.P: Head of TV Dept: M. Turan Akköprülü. Dep. Head Prgr: Teoman Ertan. Dir. Ankara TV: S. Alpdogan. Dir. of Tech. Affairs: E. Gönüllü.

TV-1

Location	chE	kW (ERP)	Location	chE	kW (ERP)
Ankara	5	100	Hatay	8	10
Istanbul	5	100	Canakkale	8	30
Antalya	5	30	Uşak	8	30
Agri	5	30	Ersincan	8	30
Giresun	5	30	Nigde	8	30
Cizre	5	30	Afyon	9	30
Adana	6	30	Amasya	9	100
Erzurum	6	100	Edirne	9	30
Bursa	6	100	Diyarbakir	9	30
Kastamonu	6	30	Trabzon	9	30
Aksehir	6	30	Kars	9	100
Malatya	7	30	Zonguldak	9	30
Eskisehir	7	100	Izmir	10	100
Kirsehir	7	30	Gaziantep	10	30
Samsun	7	30	Konya	10	100
Silifke	7	30	Adapazari	10	30
Van	7	30	Bingöl	10	30
Aydin	7	30	Tokat	10	30

+ 476 low power repeaters not mentioned.

TV-2*	chE				
Elbistan	5	Izmir	23	Ersincan	42
Adapazari	21	Kayseri	26	Kastamonu	48
Cizre	21	Gaziantep	27	Mus	48
Kars	21	Denizli	28	Istanbul	51
Adana	22	Edirne	28	Samsun	52
Canakkale	22	Trabzon	28	Bursa	55
Eskisemir	22	Agri	32	Diyarbakir	56
Van	22	Elbistan	34	Giresun	57
Zonguldak	23	Ankara	37	Mugla	55
Erzurum	23	Isparta	40		

*) all tr's 450kW (ERP).
D.Prgr: TV-1: Mon-Fri 0330-0700, 1200-2200v; Sat/Sun 0600-2200v. **TV-2:** Mon-Fri 1700-2200, Sat/Sun 1200-2200.
TV-3, TV-4, TV-5
Stations: (Istanbul area) TV-3 ch48, TV-4 ch54, TV-5 ch24.
D.Prgr: TV-3 1800-2130, TV-4 1700-2130.

UNIVERSITE TECHNIQUE

Stations: Istanbul chE4, 0.5kW.
NB: Turkish Private Comm. TV Stations see Satellite section: Intelsat 604 at 60° East; ECS II at 16°, 13°, 10°, 7° East; Intelsat 601 at 27.5° W.

TURKMENISTAN

TV-sets: N/A — **Colour:** SECAM — **Systems:** D&K.

TURKMEN TELEVISION (Gov.)

Addr: Machtumkuli 89, 744000 Aşgabat. **Te:** +7 (3632) 251515. **Fax:** +7 (3632) 251421 (manual).
L.P: See radio.
Stations and **D.Prgr:** Current details not available.

TV RELAYS

Various main and relay transmitters throughout the country relay programmes of OK-1 and RTV from Moscow, as well as TRT from Turkey. Further details not known.

UNITED ARAB EMIRATES

TV-sets: 170.000 — **Colour:** PAL — **Systems:** B & G.

UNITED ARAB EMIRATES TELEVISION SERVICE (Gov.)

Addr: P.O. Box 637, Abu Dhabi. **Te:** +971 (2) 452000. **Fax:** +971 (2) 461823. **Telex:** 22557 teevee em.
L.P: Asst. Under Secr: A.M. Al-Hamili. Dir. Gen: Ali Obaid.
Stations:

Loc	prgr.I	prgr.II	kW
Abu Dhabi	7/11	48	40/5
Al-Ain	5	35	40/5
Umm Al Quwain	6	43	40
Ras Al Khaimah	11	—	10
Fujairah	9	36	10
Habshan	3	—	40
Dibba	21	23	0.5
Jabel Al Dhanna	31	—	40
Khorfakkan	28	—	0.02

D.Prgr: Prgr. I (Arabic main prgr), Prgr. II (Arabic/English/French): 0600-1005, 1200-2030 (Fri & holidays 0600-2030).
F.Pl: Mohawi ch's 7, 48, 51 (instead of Abu Dhabi existing sts.)

U.A.E. RADIO AND TELEVISION-DUBAI (Gov.)

Addr: P.O. Box 1695, Dubai. **Te:** +971 (4) 470255. **Telex:** 45605 DRCTV EM.
L.P: Dir. Gen: Abdul G. Al Sayeed Ibrahim. Contr. of Prgrs: Nasib Bitar. Contr. of Eng: Ahmed Najeeb.
Stations (Systems B & G): Audio powers are 1/10 of the vision powers indicated.

Location	ch	kW (ERP)	Pol	Prgr
Trade Centre	2	150	H	1
Zabeel	10	455	H	1
Zabeel	33	1700	H	2
Jebel Hatta	41	1600	H	1

D.Prgrs: Arabic (prgr. 1) on chs 2, 10 & 41: 1200 (Fri 1000)-2130 (or 2030). English (prgr. 2) on ch33: 1300 (Fri 1200)-2000 (approx).

SHARJAH TV (Gov.)

Addr: P.O. Box 111, Sharjah. **Te:** +971 (6) 547755.
Stations: ch22, 28, 54, 57.
D.Prgr: 1300-2000 (in Arabic).

YEMEN (Rep. of)

TV-sets: 100.000 — **Colour:** PAL/NTSC — **System:** B.

YEMEN RADIO & TV CORP. (Gov.)

Addr. & L.P: See radio. **Te:** +967 (2) 231181.
Stations: chE5 (0.2kW), chE6 (2 st's, 4 & 1kW), chE7 (2kW), chE8 (4kW), chE10 (4kW), chE11 (1kW), chE12 (2kW) + low power repeaters.
D.Prgr: 1300-2100.

ASIA

AZERBAIDZHAN

TV-sets: N/A — **Colour:** SECAM — **Systems:** D&K.

AZERBAIDZHANI TELEVISION (Gov.)

Addr: Mehdi Huseyn küçäsi 1, 370000 Baki. **Te:** +7 (8922) 398585. **Fax:** +7 (8922) 395452.
L.P: See radio.
Stations + **D.Prgr:** Details not available.

TV RELAYS

Various main and relay transmitters throughout the country relay programmes of OK-1 and RTV from Moscow, as well as TRT from Turkey. Further details not known.

BANGLADESH

TV-sets: 350.000 — **Colour:** PAL — **System:** B.

BANGLADESH TELEVISION (Gov, Comm.)

Addr: Rampura, Dhaka-17. **Te:** +880 (2) 400131/9. **Cable:** BTV. **Telex:** 65624 BTV BJ DHAKA.
L.P: Chmn: S. Bari. Dep. Dir. Gen. (Prgrs): Ms. K. Fahmi. CE: S.M. Nausher Ali.

Stations	chE	kW (ERP)			
Chittagong	5	15	Dhaka I	9	60
Rangpur	6	360	Cox's Bazar	10	7
Sylhet	7	136	Khulna	11	91
Satkhira	7	250	Mymensingh	12	390
Rangamati	8	55	Noakhali	12	390
Natore	8	90			

D.Prgr: 1100 (Fri 0900)-1740.

BRITISH INDIAN OCEAN TERRITORY

TV-sets: 650 — **Colour:** NTSC — **System:** M.

AF DIEGO GARCIA TELEVISION (AFRTS)

Addr: US Navy Broadcasting Service, Detachment, Box 14, US Navy Support Facility, F.P.O. San Francisco, Calif. 96685, USA.
L.P: St. Mgr: J.A. Riccio. Dir. Tec: R.L. Newton.
Station: chA8 0.25kW H, ch 10 0.1 kW H.
D.Prgr: 0600 (Sat/Sun 0300)-2200; ch 10 24h.

BRUNEI DARUSSALAM

TV-sets: 67.000 — **Colour:** PAL — **System:** B.

RADIO TELEVISYEN BRUNEI (Gov.)

Addr: Bandar Seri Begawan, 2042, Negara Brunei Darussalam. **Te:** +673 (2) 222190/243008. **Fax:** +673 (2) 27204. **Telex:** 2720. **Cable:** Broadcasts.
L.P: Dir: Hj Alidin Hj Othman. CE: Hj Md Yusof Hj Abd Rahman.
Stations (System B): Brunei ch5 10kW H, ch8 20kW H.
D.Prgrs: 0800 (Fri/Sun 0400)-1600.

CAMBODIA

TV-sets: 70.000 — **Colour:** PAL — **System:** B/G.

CAMBODIAN TELEVISION (Gov.)

Addr: 19, Street 242, Chaktomuk, Daun Penh. Phnom Penh.
Te: +2 (2983) 22349/(4449) 24149
L.P.: St. Man: Tan Yan. T.D: Uy Thuon.
Stations: ch7 1kW.
D.Prgr: 0715 PM-0745 PM.

CHINA (People's Republic of)

TV sets: 26.000.000 — **Colour:** PAL — **System:** D.

CHINA CENTRAL TELEVISION (CCTV)

Addr: 11 Fuxinglu Road, Beijing 100859, China. **Te:** +86 (1) 801 1149. **Fax:** +86 (1) 801 1149. **Telex:** 222299 CCTVE CN.
L.P: Dir: Huang Huiquan. CE: Sun Tonggeng.
Stations: Beijing ch2, ch8, ch6, ch15.
1st Prgr (National): Approx 0200-1600 on ch2 (Beijing), ch5 (Shanghai).
2nd Prgr (National): Approx 0030-1500 on ch8 (Beijing), ch4 (Shanghai).
English: 1430-s/off. **N:** 1410.
3rd Prgr: Approx 0945-1430 on ch15 to the Beijing area.
Educational Sce: Approx 1230-1530 on ch15.
Shanghai TV: ch8 & ch20. **N. in English:** approx 1500 on ch20.
Other TV Centres: 29 Provincial, municipal and autonomous regional centres with 361 local st's.

CHINA (Republic of) TAIWAN

TV-sets: 7m — **Colour:** NTSC — **System:** M.

CHINA TELEVISION COMPANY (Comm.)

Addr: No. 120 Chung-Yang Road, Nankang District, Taipei. **Te:** +886 (2) 783 8308. **Fax:** +886 (2) 782 6007. **Telex:** 25080 CHINA TV.
L.P: Pres: Hu Ping Chung. CE: Mr. Chen.
Stations: BEF21 chA9 180/40kW (No. Taiwan), BEF22 chA10 150/17kW (Ce. Taiwan), BEF23 chA9 180/80/20kW (So. Taiwan).
D.Prgr: W 0400-0500, 0930-1515; Sat 0450-1615; Sun 0250-1515.

CHINESE TELEVISION SERVICE (Comm.)

Addr: 100 Kuang Fu South Road, Taipei. **Te:** +886 (2) 751 0321. **Fax:** +886 (2) 751 6019.
L.P: Chmn: Chien-Chiu Yee. Pres: Shih-shung Wu. CE: Shiao Ho Whu.
Stations: BET31 chA11 200kW (No.We.), BET32 chA8 190kW (Ce.), BET33 chA11 131kW (So.We.), BET34 chA8 24.5kW (No.Ea.), BET35 chA11 18kW (Ce.Ea. Taiwan), BET36 chA8 9.5kW (So.Ea. Taiwan), BET37 ch35 1702kW (No.We.), BET39 ch36 1383kW (Ce.), BET40 ch34 1383kW (So.We.), BET41 ch33 316kW (No.Ea.), BET42 ch35 316kW (Ce.Ea.), BET43 ch33 316 kW (So.Ea.), BET38 ch33 283kW (We.) + 2 relay st's.
D.Prgr: W 2320-0120, 0340-1600; Sun 2320-1600.

EDUCATIONAL TV

BET37 chA35 1717kW (No.We. Taiwan), BET38 chA33 285kW (Ce.No. Taiwan), BET39 chA36 1500kW (Ce. Taiwan), BET40 chA34 1500kW (So.We. Taiwan), BET41 chA33 354kW (No.Ea. Taiwan), BET42 chA35 343kW (Ce.Ea. Taiwan), BET43 chA33 334kW (So.Ea. Taiwan).
D.Prgr: W 1030-1435, Sun 1130-1505.

TAIWAN TELEVISION ENTERPRISE Ltd. (Comm.)

Addr: No. 10, Pa Te Rd, Section 3, Taipei 10560. **Te:** +886 (2) 7711515. **Fax:** +886 (2) 7413626. **Cable:** Television. **Telex:** 25714 TV TAIWAN.
L.P: Chmn: Ching-Teh Hsu. Pres: Walter C.H. Wang. Executive vice president: Wei-Yung Lee. Vice president: Hsiang-chuan Hsiung. Vice president: Shen-Wen Lee. Manager, Engineering Department: To-Hui Yang. Manager, Business Department: Wen-Lung Liu. Film buyer: Nancy Hu. Manager, News Department: Victor S.T. Chang. Manager, Sports Department: Jason Liao. Manager, Program Department: Sheng Chu-yu. Dir. Inf. Center: Ko-Jan Hwang.
Stations: BET21 chA7 282/95kW (No. Taiwan), BET22 chA12 162/136/6.5kW (Ce. Taiwan), BET23 chA7 136/162/5.5kW (So. Taiwan), BET24 chA10 4.0/0.4kW (I-Lan), BET25 chA7 12/36kW (Hua Lien), BET 26 chA10 1.9kW (Taitung) + 7 Transposer stations.
D.Prgr: Mon-Fri 2159-0030, 0311-0810, 0839-1610; Sat 0244-1740; Sun 2339-1610.

GEORGIA

TV-sets: N/A — **Colour:** SECAM — **Systems:** D&K.

GEORGIAN TELEVISION (Gov.)

Addr: M. Kostavas 68, 380071 Tbilisi. **Te:** +7 (8832) 362294. **Fax:** +7 (8832) 362319 (manual).
L.P: See radio.

Stations (Prgr. 1)	**ch**		**ch**
Tbilisj	4	Kutaisi	10
Suchumi	5	Gori	12
Batumi	8		

+ various translators.
D.Prgr: Mon-Fri 1000-1100, 1200-2100. Sat/Sun 0900-2100.

Stations (Prgr. 2)	**ch**
Tbilisi	28

D.Prgr: Current schedule not available.

OSTANKINO KANAL 1 (OK-1)

Stations	**ch**		**ch**
Batumi	3	Tbilisi	9
Kutaisi	7	Gori	31
Suchumi	9		

+ various translators.
D.Prgr: Relays of OK-1 from Moscow. Further details see under Russia.

ROSSIJSKOJE TELEVIDENIJE (RTV)

Stations	**ch**
Tbilisi	6

+ various translators.
D.Prgr: Relays of RTV from Moscow. Further details see under Russia.

HONG KONG

TV-sets: 1.749.000 — **Colour:** PAL — **System:** I.

TELEVISION BROADCASTS LTD.

Addr: TVB Complex Building, Lot no. 214, Clearwater Bay Road, Kowloon. **Te:** +852 (3) 7194828. **Fax:** +852 (3) 581300. **Telex:** 43596 HX.
L.P: TD: Kelly KL Cheng, Film Buyer: Sophia Chan (Jade Channel), Musetta Wu (Pearl Channel).
Stations: chE21 & chE25, 10kW.
D.Prgr: Jade network on ch21: 2245-2030 (approx). Pearl network on ch25: 0000 (Sat 0100, Sun 0030)-1730 (approx).

ASIA TELEVISION LIMITED

Addr: 81 Broadcast Drive, Kowloon. **Te:** +852 (3) 399111. **Fax:** +852 (3) 380438. **Telex:** HX44680. **Cable:** ASIATV.
L.P: Dir: Carl Tong. CE: K.C. Fong.
Stations: chE23 & chE27, 10kW.
D.Prgr: Home Channel in Chinese on ch23: 2230-1930. World Channel in English on ch27: 24h.

INDIA

TV-sets: 20M (est.) — **Colour:** PAL — **System:** B.

DOORDARSHAN INDIA (government, semi-commercial)

Addr. I: Directorate General of Doordarshan, Mandi House, Copernicus Marg, New Delhi-110 001. **Te:** +91 (11) 382094/99. **Fax:** +91 (11) 386507. **Telex:** 81-31-65290 dgdd in, 81-31-66413 dgdd in. **Cable:** TVgeneral.
L.P: DG: S. Krishnan, **Te:** +91 (11) 387786. TD: N.J. Nair, **Te:** +91 (11) 388113.
Addr. II: Doordarshan Kendra, Akashvani Bhavan, 1 Sansad Marg, New Delhi-110 001. **Te:** +91 (11) 3710006/3715411(pabx).
D.Prgrs: National: 1510-1800v (extended telecasts during live coverages) — Network: 0130-0315, 0830-0940, Suns 0330-0800/0830-1400/1415-1510 — General/Regional: Mons-Sats 1130-1510, Suns 0315-0330/1400-1415 — Countrywide Classroom: Weekdays 0730-0830, 1030-1130 — Distance Education: 0100-0130 (Mons, Weds, Fris) — School Telecasts: Weekdays 0305-0715, 0905-1115 (except summer holidays) — Channel II: 1230-1630.
F.Pl: HPTs (1 kW) to be setup at Itanagar, Mokokchung, Lunglei and Churachandpur, a 10kW hpt at Ramgarh and an undisclosed number of repeater LPTs in various places in the country.
Stations:
Only HPTs (1 kW and above) are listed below.

Station	Ch	kW		Ch	kW
Bombay	4	10	Nagpur	7	10
Calcutta	4	10	Panaji	7	10
Delhi	4	10	Sambalpur	7	1
Hyderabad	4	10	Vijayawada	7	10
Lucknow	4	10	Visakapatnam	7	10
Madras	4	10	Agartala	8	10
Srinagar	4	10	Cuttack	8	1
Ahmedabad	5	10	Kohima	8	1
Bangalore	5	10	Kurseong	8	10
Bhopal	5	10	Varanasi	8	10
Jaipur	5	10	Aizawl	9	1
Kanpur	5	10	Agra	9	10
Pune	5	6	Gorakhpur	9	10
Raipur	5	1	Guwahati	9	10
Silchar	5	10	Indore	9	10
Bombay II	6	1	Jalandhar	9	10
Calcutta II	6	10	Thiru'puram	9	10
Imphal	6	1	Murshidabad	10	10
Kasauli	6	10	Mussoorie	10	10
Madras II	6	10	Ranchi	10	10
Muzaffarpur	6	1	Dibrugarh	11	10
Rajkot	6	10	Patna	11	10
Allahabad	7	10	Poonch	11	1
Amritsar	7	10	Bathinda	12	10
Asansol	7	10	Kochi	12	10
Delhi II	7	10	Dwarka	12	10
Gulbarga	7	1	Jammu	12	10
Kodaikanal	7	10	Shillong	12	1

Total transmitters = 540+ including low power relays.

INDONESIA

TV sets: 11.000.000 — **Colour:** PAL — **System:** B.

TELEVISI REPUBLIK INDONESIA TVRI (Gov.)

Addr: J1 Gerbang Pemuda-Senayan, Jakarta. **Te:** +62 (21) 5704 720. **Fax:** +62 (21) 5821 40/5821 09. **Cable:** TELVIRI Jakarta. **Telex:** 46154 Jakarta.
L.P: Man. Dir: Azis Husein — TD: Djoko Widayat — Film Buyer: Adi Kasno.

Key Stations

Location	chE	kW(ERP)			
Ujung Padang	4	1	Pontianak	7	10
Medan	5	10	Jakarta	8	10
Banda Aceh	5	10	Yogyakarta	8	10
Padang	5	5	Denpasar	8	5
Banjarmasin	5	10	Kupang	8	5
Jakarta	6	5	Palembang	9	5
Bandung	6	10	Balikpapan	9	1
Semarang	6	5	Surabaya	9	10
Ambon	7	5	Manado	9	1

+ 10 relay st's.
D.Prgr: W 0930-1630 (Sat 1700), Sun & Holidays 0100-0630, 0930-1630. **English:** 1130-1200. **N:** 1130.

Pay Television

RCTI

(PT Rajawali Citra Televisi Indonesia)
Addr: Jalan 3 Raya, Pjuangan Aban Jerua, Jakarta 11530.
Te: +62 (21) 549 3847/3853/3855/3860. **Fax:** +62 (21) 549 3452.
L.P: Pres: Joni Sobandono. GM: Peter Langlois. Hd. of Prgrs: Ms. Ratner Mahadi.
D.Prgr: 90 hrs per week.

JAPAN

TV-sets: 100 m — **Colour:** NTSC — **System:** M.

NIPPON HOSO KYOKAI (Corporation, non Gov, non Comm.)

Addr: 2-2-1 Jinnan, Shibuya-ku, Tokyo 150. **Te:** +81 (3) 465 1111. **Fax:** +81 (3) 481 1576.
1) = General Sce; 2) = Educational Sce.

Main st's	Call	chkW		P					
Tokyo	JOAK	1	50	1	Hiroshima	JOFB	7	10	2
Osaka	JOBK	2	10	1	Niigata	JOQK	8	5	1
Matsuyama	JOZB	2	5	2	Akita	JOUK	9	5	1
Okinawa	JOAP	2	5	1	Nagoya	JOCB	9	10	2
Sapporo	JOIK	3	10	1	Matsuyama	JO	10	23	1
Sendai	JOHK	3	10	1	Sapporo	JOIB	12	10	2
Aomori	JOTG	3	5	1	Niigata	JOQB	12	5	2
Tokyo	JOAB	3	50	2	Osaka	JOBB	12	10	2
Nagoya	JOCK	3	10	1	Okinawa	JOAD	12	5	2
Hiroshima	JOFK	3	10	1	Kobe	JOPP	28	10	1
Okayama	JOKB	3	10	2	Tsu	JONP	31	5	1
Fukuoka	JOLK	3	10	1	Kyoto	JOOK	32	10	1
Kagoshima	JOHG	3	5	1	Takamatsu	JOHP	37	10	1
Sendai	JOHB	5	10	2	Tokushima	JOXB	38	10	2
Aomori	JOTC	5	5	2	Saga	JOSP	38	5	1
Okayama	JOKK	5	10	1	Takamatsu	JOHD	39	10	2
Kagoshima	JOHC	5	5	2	Gifu	JOOP	39	5	1
Matsuyama	JOZK	6	5	1	Saga	JOSD	40	5	2
Fukuoka	JOLB	6	10	2					

+ approx. 3500 st's.
D.Prgr: 1) Gen. TV: 2100-1500. **2) Educ. TV:** 2100-1500.

COMMERCIAL TV-STATIONS

Japan Satellite Broadcasting (JSB)

Addr: see satellite section.
Call: JO33-BS-TV ch3 120W — **D. Prgr:** General (approx.) 2100 (or 2130) -1700 (or 1800) in Tokyo, Osaka and other major cities; in smaller cities generally 2200, 2300 or 0000-1400 (or 1500).

The National Association of Commercial Broadcasters in Japan

Addr: Bungei-Shunju Building, 3-23, Kioi-Cho, Chiyoda-ku, Tokyo. **Te:** +81 (3) 3265 7481. **Fax:** +81 (3) 3261 2860.

Key Networks:

The Nippon Television Network Corp. (NTV)

Addr: 14, Niban-cho, Chiyodaku, Tokyo 102. **Te:** +81 (3) 265-2111. **Fax:** +81 (3) 263-6008.
Telex: J24566 JOAXNTV — **L.P:** Pres: Y. Sasaki.
Station: JOAX-TV ch4 50kW. **D.Prgr:** 24h (+83 relay sts).

Tokyo Broadcasting System Inc.

Addr: 5-3-6 Akasaka, Minato-ku, Tokyo 107-06. **Te:** +81 (3) 224 250411. **Fax:** +81 (3) 505 1584 — **L.P:** Exec. Vice Pres: K. Takagaki.
Station: JOKR-TV ch6 370kW. **D.Prgrs:** 2100-1600 (70 relay sts).

Fuji Television Netw. Ltd.

Addr: 3-1, Kawada-cho, Shinjuku-ku, Tokyo 162. **Te:** +81 (3) 353-1111. **Telex:** J22560. **Fax:** +81 (3) 358 8038.
L.P: Pres: H. Hieda.
Station: JOCX-TV ch8 370kW. **D.Prgrs:** 24h (85 relay sts).

Asahi National Broadc. Co. Ltd.

Addr: 1-1-1, Roppongi, Minato-ku, Tokyo 106 **Te:** +81 (3) 587 5412. **Fax:** +81 (3) 505 3539. **Telex:** 22520 — **L.P:** Pres: K. Tashiro. Film Buyer: K. Kikuya. Dir. Tec: S. Tomita.
Station: JOEX-TV ch10 50kW. **D.Prgr:** 2100-1800 (+7 relay sts).

Hokkaido Broadc. Co.

Addr: Kita 1 Nishi 5-2, Chuo-ku, Sapporo 060. **Te:** +81 (11) 231 5800. **Fax:** +81 (11) 232 3120.
L.P: St. Mgr: K. Tomihara; TD: K. Kashiwagura.
Station: ch1 (+6 rel. sts.)

ARMED FORCES RADIO & TV SERVICE (U.S. Mil.)

Stations: Misawa ch66 1kW, Iwakuni ch66 0.4kW — Okinawa ch8 5kW — Sasebo ch3 & Tokyo ch11 (cable) — **Philippines:** Olongapo ch14 0.25kW, Angeles City ch17 1kW, San Miguel ch40 0.03kW, Baguio ch14 0.3kW, San Fernando ch17 0.25kW, Capas ch17 25W.
Addr: (Misawa) OLAA, AFPBS, APO San Francisco 96519 — (Iwakuni) OLAB, AFPBS, FLO Seattle 98764 — (Philippines) Det 1, AFPBS, APO San Francisco 96274 — (Okinawa) Det 2, AFPBS, APO San Francisco 96239 — (Guam) OLAC, AFPBS, APO San Francisco 96334 — (Tokyo) AFPBS, APO San Francisco 96238 — (Sasebo) OLAD, AFPBS, FPO Seattle 98776.

KAZAKHSTAN

TV-sets: N/A — **Colour:** SECAM — **Systems:** D&K.

KAZAKH TELEVISION (Gov.)

Addr: Jeltoksan köşesi 175, 480013 Almati. **Te:** +7 (3272) 695188. **Fax:** +7 (3272) 631207.
L.P: See radio.

Stations	ch		ch
Karağandi	1	Akmola	10
Semey	4	Öskemen	10
Aktöbe	8	Oral	11
Almati	10	Petropavlovsk	12

+ further main transmitters and translators. Complete details not known.
D.Prgr: Mon-Fri 0255-0800, 1200-1830. Sat/Sun 0300-1830.

OSTANKINO KANAL 1 (OK-1)

Stations	ch		ch
Aktöbe	1	Akmola	7
Öskemen	1	Semey	8
Oral	2	Karağandi	12
Almati	3		

+ further main transmitters and translators. Complete details not known.
D.Prgr: Relays of OK-1 from Moscow. Further details see under Russia.

ROSSIJSKOJE TELEVIDENIJE (RTV)

Stations	ch
Almati	?
Karağandi	9

+ further main transmitters and translators. Complete details not known.
D.Prgr: Relays of RTV from Moscow. Further details see under Russia.

TV RELAYS

Stations	ch
Almati	7
Almati	40

D.Prgr: The above transmitters carry relays of private programme producers, Kyrgyz and Turkish TV.

KOREA (Rep.)

TV-sets: 8.700.000 — **Colour:** NTSC — **System:** M.

KOREAN BROADCASTING SYSTEM — TV (Publicly owned).

ADDR: see radio.
L.P: Pres: Suh, Ki-Won. V.P: Sohn, Young-Ho.
Stations: I = Prgr. 1; II = Prgr. 2; III = Prgr. 3 (Educational).

Location	ch	kW	Prgr.		ch	kW	Prgr.
Yŏsu	4	10	I	Hwangji	25	10	III
Seoul	5	10	I	Kwangju	25	30	II
Taejŏn	6	10	I	P'ohang	26	10	III
Masan	6	10	I	Chinju	27	5	II
Kangnŭng	6	10	II	Ulsan	27	10	II
Seoul	7	50	II	Ch'unch'ŏn	28	30	III
Chŏnju	7	10	I	Kangnŭng	28	5	II
Pusan	7	10	II	Sŏsan	29	5	III
Taegu	8	10	I	Andong	29	10	III
Ch'unch'ŏn	8	10	I	Yŏsu	30	30	III
Seoul	9	50	I	Ch'ungju	30	10	II
Pusan	9	10	I	Wŏnju	31	10	II
Kangnŭng	9	10	I	Kangnŭng	34	5	III
Cheju	10	5	II	Chinju	36	5	III
Kwangju	11	10	I	Hwangji	37	5	II
Cheju	12	5	I	Seoul	37	10	II
Seoul	13	10	III	Taegu	38	30	II
Kwangju	19	30	III	Taejŏn	39	30	III
Kangnŭng	19	10	III	Masan	39	30	III
Cheju	20	10	III	Wŏju	40	10	III
Pohang	20	10	II	Taejŏn	42	10	II
Ulsan	21	10	III	Chinju	42	5	II
Ch'ungju	21	30	III	Seoul	43	30	III
Ch'unch'ŏn	22	30	II	Taegu	44	30	III
Kurye	22	10	III	Chŏnju	45	30	III
Pusan	23	30	III	Masan	45	30	II
Andong	23	10	II	Ponghwa	47	5	III
Yŏsu	24	30	II	Taejŏn	54	5	III
Ch'ungju	24	5	II				

+ approx 600 st's less than 5kW.
D.Prgr: 1st prgr: W 2100-0100, 0830-1500; Sat/Sun 2100-1500; 2nd Prgr; W 2100-0900, 0830-1500; Sat/Sun 2100-1500; 3rd Prgr: W 0730-1500; Sat 0700-1500; Sun 2200-1500.

MUNHWA TV-RADIO BROADC. CORP. (Comm. & Public Corp.)

Addr: 31, Yŏ-ŭido-dong, Yŏngdŭngp'o-gu, Seoul, 150-728. **Te:** +82 (2) 784 2000. **Fax:** +82 (2) 784 0888. **Telex:** MBCNEWS K22203. **Cable:** MBCHLKV SEOUL.
L.P: Pres: Chang-Bong Choi. CE: Jung-Soo Oh.

Stations:	ch	kW			
Andong	5	1	Samch'ŏk	22	5
Pohang	6	1	Ch'unch'ŏn	24	30
Cheju	7	5	Ch'ŭngju	26	3
Mokp'o	7	0.5	Yŏsu	28	2
Yŏsu	7	0.5	Kangnŭng	31	10
Taejon	8	5	Ulsan	33	10
Chinju	8	1	Ch'ŏngju	33	5
Kwangju	9	10	Wŏnju	34	10
Taegu	10	5	Ch'ŭngju	36	10
Chŏnju	10	5	Samch'ŏk	43	5
Ch'unch'ŏn	10	1	Taejon	45	10
Seoul*	11	50	Wŏnju	49	5
Cheju	11	1	Masan	48	5
Pusan	11	10	Kwangju	51	10
Masan	13	5	Taegu	59	5

*) Key st.
D.Prgrs: Mon-Fri 2100-0100, 0830-1500; Sat 2100-0100, 0430-1500; Sun 2100-1500.

KOREA (Dem. Peoples Rep.)

TV-sets: 2.000.000 — **Colour:** PAL — **System:** D & K.

KOREAN CENTRAL TELEVISION STATION (Gov.)

Addr: Ministry of Posts and Telecommunications, Pyongyang.

Stations	chR	kW	Prgr.				
Hamhŭng	3	10	1	Pyongyang	9	20	2
Pyongyang	5	50	1	Haeju	11	10	1
Sep'o 6	10	1		Pyongyang	12	50	3
Sinŭiji	6	10	1	Chóngjin	12	10	1
Hanggye	8	10	1	Kaesong	12	10	1
Hysesan	9	10	1				

Prgrs. 1 & 2: 0000-0300, 0600-1400.
Prgr. 3: 0100-0400, 0700-1300.

MANSUDAE TELEVISION STATION
Addr: Mansudae, Pyongyang. **Telex:** Mansudae, Pyongyang.

KYRGYZSTAN

TV-sets: N/A — **Colour:** SECAM — **Systems:** D&K.

KYRGYZ TELEVISION (Gov.)
Addr: pr. Molodoj Gvardii 63, 720885 Bişkek. **Te:** +7 (3312) 253404. **Fax:** +7 (3312) 257930.
L.P: See radio.
Stations: current details not known.
D.Prgr: 1300-1600.

TV RELAYS
Various main and relay transmitters throughout the country relay programmes of OK-1 and RTV from Moscow, as well as Uzbek and Kazakh TV, and programmes of TRT from Turkey. Further details not known.

LAOS

TV-sets: 32.000 — **Colour:** PAL — **System:** M.

LAO NATIONAL TV (Gov.)
Addr: B.P. 310, Vientiane. **Te:** +856 5609.
L.P: See radio.
Stations: ch8 0.1kW, ch10 10W.
D.Prgrs: ch8 (colour) 1330-1500; ch10 1130-1300.

MACAU

TV-sets: 70.000 — **Colour:** PAL — **System:** I.

TELEDIFUSAO DE MACAU (TDM SARL)
Addr. & L.P: see radio.
Te: +853 519188. **Fax:** +853 520208. **Telex:** 88309.
L.P: Man. Dir: Dr. Leung Kwok-Kwong.
Stations: ch30 0.2kW + 3 repeaters of 10W.
D.Prgr. (bilingual): 2330-0100, 0930-1600.

MALAYSIA (Federation of)

TV-sets: 2m* — **Colour:** PAL — **System:** B. — *) incl. Sabah & Sarawak.

Peninsular Malaysia

RADIO TELEVISION MALAYSIA (Gov.) (TV1 & 2)
Addr: Dept. of Broadc, Angkasapuri, Kuala Lumpur 50614.
Te: +60 (3) 282 5333. **Fax:** +60 (3) 282 4735. **Cable:** Tivimalsia, Kuala Lumpur. **Telex:** MA 31383 Kuala Lumpur.
L.P: See radio.

	Network I		Network II	
Location	ch	kW(ERP)	ch	kW(ERP)
Genting Sempah	2	112	10	112
Johor Baru/Gunong Pulai	3	38	10	142.5
Tangkak	4	154	7	77.4
Kuala Lumpur	5	100	8	100
Alor Setar	5	100	8	100
Kuantan	5	127	8	127
Kuala Pilah	5	7.95	8	7.95
Kuala Lipis	5	20	8	20
Mersing	6	100	9	100
Ipoh	6	190	9	190
Melaka	6	185	9	185
Kota Bharu	6	250	9	250

+ low power relay st's.
D.Prgr: Netw. I: Mon-Wed 0900-1600, Thurs/Sat 0700-1600; Fri/Sun 0200-1600. Netw. II: 0900 (Fri/Sun 0700)-1600.

SYSTEM TV MALAYSIA BERHAD (TV3)
Addr: 7–9th Floor, KUB Bldg. 1 Lorong Kapar, P.O. Box 11124, Kuala Lampur. **Te:** +60 (3) 274 3111. **Fax:** +60 (3) 274 8896. **Telex:** 33014 STMB MA.
L.P: Chief Ex. Off: Mohammed Noor Salleh.

Stations: Kuala Lumpur ch12 124/25kW H.
Repeater st's of high power in Kuantan (ch 11), Ipoh (ch 11), Gunung Ledang (ch 12), Johore Bahru (ch 26), Kedah (ch 29), Kota Bharu (ch 27), Kuala Terengganu (ch 11), Dungun (ch 27), Tampin (ch 23), Taiping (ch 41), Kuching (ch 12), Sibu (ch 11), Kota Kinabalu (ch 29), Ulu Kali (ch 29) & Machang (ch 11).
D. Prgrs: Mon-Wed 0830-1615, Thurs-Fri 0630-1615, Sat/Sun 0100-1615.

East Malaysia

TV MALAYSIA SABAH AND SARAWAK (Gov.)
Addr. P.O. Box 1016, 88614 Kota Kinabalu. **Te:** +60 (88)52711. **Cable:** Broadcasts, Sabah. **Telex:** MA 80061 Kota Kinabalu.
L.P: Prgr. Contr: M.A. Mahmood.
Stations: Pol = H.

Location	Prgr.I chE	Prgr.II chE	kW (ERP)
Sipitang	11	3	13.6
Lahad Datu	7	10	100
Sandakan	8	5	100
Tawau	9	6	33
Tambunan/Keningau	9	12	36
Kudat	9	6	100
K.K. (Lawa Mandau)	10	7	100
Limbang	2	4	13
Sibu	8	5	100
Bintulu	9	6	25
Simangang (Sri Aman)	9	6	120
Miri	10	12	55
Kuching	10	7	220
Kapit	4	10	18

D.Prgrs: as for Peninsular Malaysia.

MALDIVES (Rep. of)

TV sets: 4.750 — **Colour:** PAL — **System:** B.

TELEVISION MALDIVES (Gov.)
Addr: 20-04 Male'. **Te:** +960 (32) 2106. **Fax:** +960 (32) 5083. **Telex:** 66085 HOADHUN MF.
L.P: St. Mgr: Ali Shareef. Dir.Tec: Esa Ibrahim. Film Buyer: Shafeega Ali.
Station: ch7 1kW H.
D.Prgr: 0300-0500 (Fri), 1200-1620.

MONGOLIA

TV-sets: 120.000 — **Colour:** SECAM — **System:** D.

MONGOL TELEVIZ
Addr: Mongolian TV Centre Ulaanbaatar.
L.P: Dir. TV: Mr. Altai.

Stations: chR2 5kW, chR4 0.5kW.
+ relay tr's (0.5kW each) in all 20 provincial centres and 60 small towns along the relay route.
D.Prgrs: ch.2: 7th daily. Ch.4: rel. Moscow 1st Prgr. 50h weekly.
Orbit st. receivers in provinces: 310.

MYANMAR (Union of)

TV-sets: 1m (est.) — **Colour:** NTSC — **System:** M.

TV MYANMAR

Addr.: GPO Box 1432, Yangon. **Te:** +95 (1) 31355.
Stations: Yangon ch6 10/2.5kW H + 7 relay st's.
D. Prgrs: 1300-1530 (Sat & Sun also 0630-0830).

NEPAL

TV-sets: 250.000 — **Colour:** PAL — **System:** B.

NEPAL TELEVISION CORPORATION (Gov.)

Addr: P.O. Box 3826, Singha Durbar, Kathmandu. **Te:** +977(1) 227290. **Fax:** +977(1) 228312. **Telex:** NP 2548 NEPTV.
L.P: Chmn & Gen. Mgr: N.B. Shah. Dir. Tec: R.S. Rana.
Station: Phulchowki chE5 0.1kW H.
D.Prgr: 0730-0830, 1830-2200.

PAKISTAN

TV-sets: 2.080.000 — **Colour:** PAL — **System:** B.

PAKISTAN TELEVISION CORPORATION LTD.

Addr: P.O. Box 1221, Islamabad. **Te:** +92 (51) 828651. **Fax:** +92 (51) 823406. **Cable:** Pakteevee. **Telex:** 5833 PPV RP.
L.P: Man. Dir: A.R. Siddiqi. TD: Akhtar Mehmood Dad. Film Buyer: Fazal Kamal.
Stations: (Prgr. Centres at Lahore, Karachi, Peshawar, Quetta and Islamabad).

Location	ch	kW		ch	kW
Karachi	4	60	Mangora	7	10
Sakessar	4	114	Muree	8	180
Lak Pass	4	8	Shikarpur	8	213
Kala Shah Kaku	5	400	Quetta	8	1.25
Nurpur	5	170	Shujabad	8	178
Ghazaband	5	10	Thana Bola Khan	9	205
Quetta	6	30	Cherat	10	170
Thandiani	6	5	Sahiwal	10	277
Faisalabad	6	20	J. din Wali	10	200
Sibi	6	6	Pasrur	10	6.8
Sakessar	7	168	Tando Allahyar	11	400

+7 low power st's.
D.Prgr: 1130-1930. **N:** 1400, 1600.

PHILIPPINES (Rep. of the)

TV-sets: 7m — **Colour:** NTSC — **System:** M.

NATIONAL TELECOMMUNICATIONS COMMISSION

For full particulars, see Philippines entry in Radio section.
Stations C = City.

	Call	chA	kW	Location
1)	DWWX-TV	2	35	Quezon C.
2)	DXRV-TV	2	1	Ihgan C.
3)	D-3-Z0-TV	3	100	Baguio, C.
12)	—	3	1	Metro Manila
1)	DYCB-TV	3	10	Cebu C.
11)	DYLL-TV	3	500	Zamboanga C.
3)	DWGT-TV	4	25	Metro Manila
1)	DYXL-TV	4	2	Bacolod C.
1)	DXAS-TV	4	10	Davao C.
2)	DYXX-TV	6	1	Iloilo C.
2)	DZBB-TV	7	25	Quezon C.
2)	DYSS-TV	7	5	Cebu C.
5)	DXSS-TV	7	1	Davao C.
2)	DWAI-TV	7	1	Naga C.
4)	DYKB	8	5	Bacolod C.
4)	DZKB-TV	9	25	Quezon C.
4)	DYKC-TV	9	5	Cebu C.
6)	DXLA-TV	9	2.5	Zamboanga C.
2)	DYAF-TV	10	5	Bacolod C.
2)	DWMJ-TV	10	1	Baguio C.
2)	DWMT-TV	10	25	Benguet
2)	DWLA-TV	12	10	Legaspi C.
2)	D-12-ZB	12	1	Batangas
4)	DZBS-TV	12	5	Baguio C.
8)	PR-TV	12	?	Tacloban C.
9)	DXNS-TV	12	1	Cotabato C.
3)	—	13	20	Metro Manila
10)	DZTV-TV	13	6.25	Quezon C.
7)	DXGL-TV	13	2	Butuan C.
10)	DXTV-TV	13	5	Davao C.
10)	DYTV-TV	13	12.5	Cebu C.

low power st's not mentioned.

Addresses & other information:
1) ABS-CBN Broadc. Corp, Mother Ignacia Ave, 1100 Quezon C. — Mon-Fri 2300-1530 (Sat/Sun 0000-1730).
2) Republic Broadc. System Inc. EDSA, Diliman, Quezon City, Metro Manila. Mon-Fri 0800-0130, Sat 0730-0130, Sun 0730-0200.
3) PTV Channel 4, Media Center, Bohol Ave, Quezon City — 0425 (Sat/Sun 0300)-1730.
4) RPN Channel 9, Broadcast City, Capitol Hills, Quezon City — 0230-1600.
5) Southern Broadc. Netw, 3881 E. Vallejo Str, Santol Sta. Mesa, Metro-Manila.
6) First United Broadc. Corpo, Lozenzo Bldg, 787 Vito Cruz Str, Metro-Manila. **Te:** +63 583082. **Fax:** +63 583082.
7) Philippine Electronic & Communications Institute, Montilla Blvd, Butuan C — Mon-Fri 1000-1500, 0500-2200.
8) East Visayas Broadc, 2647 Donada Str, Malate, Metro-Manila.
9) Cotabato TV Corpo, Regional Complex, Catabato City.
10) Intercontinental Broadc. Corp., Quezon C.
11) Zamboanga TV Corp., Zomboanga C.
12) RT Broadcasting Spec. Rm.9, RC Poblete Bld., 17 Sen. Gil Y. Puyat Av. Makati, Metro Manila.

SINGAPORE

TV-sets: 550.000 — **Colour:** PAL— **System:** B.

SINGAPORE BROADC. CORP (SBC)

Addr: Caldecott Hill, Andrew Rd. Singapore, 1129. **Te:** +65 (256) 0401. **Fax:** +65 (253) 8808. **Telex:** 39265.
L.P: see radio.
Stations: Singapore chE5/chE8/chE12 120kW.
D.Prgr: chE5: Mon-Fri 0700-1625; Sat 0500-1700; Sun 0100-1630; chE8: Mon-Fri 0940-1540; Sat & Sun 0700-1630; chE12: Daily 1125-1710.

SRI LANKA

TV-sets: 700.000 — **Colour:** PAL — **System:** B.

INDEPENDENT TELEVISION NETWORK (ITN)

Addr: Wickramasinghepura, Battaramulla. **Te:** +94 (1) 564591. **Fax:** +94 (1) 564591. **Telex:** 22445.
Station: Wickramasinghepura ch12 10kW (ERP) ch9.
D.Prgr: 1800-2300 (approx).

SRI LANKA BROADC. CORP.

Addr: Independence Square, P.O. Box 574, Colombo 7. **Te:** +94 (1) 597491.

SRI LANKA RUPAVAHINI CORPO.
Addr: P.O. Box 2204, Colombo 7. **Te:** +94 (1) 580136. **Fax:** +94 (1) 580929. **Cable:** Rupavahini. **Telex:** 22148 SLTV CE.
L.P: St.Man: Asoka Malimage. TD: Hartley Fernando. Film Buyer: Asoka Malimage.
Stations: Kokavil ch8 20 kW, ch11 1 kW Sooriyakanda +2 low power repeaters.
D.Prgrs: 1130-1700. Sun 1000-1700.

TADZHIKISTAN

TV-sets: N/A — **Colour:** SECAM — **Systems:** D&K.

TADZHIK TELEVISION (Gov.)
Addr: Behzod küça 7, 734013 Dushanbe. **Te:** +7 (3772) 224357.
L.P: See radio.
Stations and **D.Prgr:** current details not available.

TV RELAYS
Various main and relay transmitters throughout the country relay programmes of OK-1 from Moscow, as well as TRT from Turkey and IRIB from Iran. Further details not known.

THAILAND

TV-sets: 3.300.140 — **Colour:** PAL — **Systems:** B&M.

TELEVISION OF THAILAND (Gov.)
Addr: Public Rel. Dept., Ratchadamoen Rd, Bangkok 10200.
Te: +66 (2) 2228821. **Telex:** 72243 PRDTHAI TH.
Stations: Bangkok ch11 200kW (ERP) + 27 relay st's.
D.Prgrs: 0930-1400.

BANGKOK ENTERTAINMENT CO. Ltd.
(Licensed through Mass Communications Organisation of Thailand).
Addr: 1126/1 New Petchbury Road, Bangkok 10400.
Te: +66 (2) 2539970-3. **Fax:** +66 (2) 2539978. **Telex:** 82616 BECOM TH. **Cable:** BANGERTAIN.
L.P: Prgr. Dir: Pravit Maleenont. TD: Manoontham Thachai.
Stations: Bangkok ch3 650 kW + 32 relay st's not mentioned.
D.Prgr: Mon-Fri 1400-2400, SS 0800-2400.

MASS COMMUNICATIONS ORG. OF THAILAND
Addr: 222 Thanon Ysok Asok-Dindaeng, Bangkok 10210.
Te: +66 (2) 2450700. **Telex:** 84577 MOT BKK TH.
Stations: Bangkok ch9 20/4kW + 32 relay st's not mentioned.
D.Prgrs: 0850 (Sat/Sun 0025)-1700.

THE ARMY TELEVISION HSA-TV (Gov, Comm.)
Addr: Phaholyothin Rd, Sanampao. Bangkok 10400. **Te:** +66 (2) 2710069. **Fax:** +66 (2) 2712510. **Telex:** 81080 ATV TH.
L.P: DG: Maj. Gen. Vijit Junapart.
Stations: Bangkok ch5 20/4kW + 18 relay st's not mentioned.
D.Prgr: 0900 (Sat/Sun 0100)-1700.

BANGKOK BROADCASTING & TV CO. LTD. (Comm.)
Addr: P.O. Box 4-56, Bangkok 10900. **Te:** +66 (2) 2781255. **Fax:** +66 (2) 2701976. **Telex:** 82730 BBTV TH.
L.P: St. Man: Chatchur Karnasuta. TD: Supoch Sangsayan.
Stations: Bangkok ch7 20kW + 22 relay st's not mentioned.
D.Prgr: 0900 (Sat/Sun 0100)-1700.

UZBEKISTAN

TV-sets: N/A — **Colour:** SECAM — **Systems:** D&K.

UZBEK TELEVISION (Gov.)
Addr: Navoii küçasi 69, 700011 Toşkent. **Te:** +7 (3712) 495214.
L.P: See radio.

Stations: (Prgr. 1)	**ch**
Toşkent	5

D.Prgr: Current schedule not available.

Stations: (Prgr. 2)	**ch**
Toşkent	11

D.Prgr: Current schedule not available.

OSTANKINO KANAL 1 (OK-1)

Stations	**ch**
Toşkent	3

D.Prgr: Relays of OK-1 from Moscow. Further details see under Russia.

ROSSIJSKOJE TELEVIDENIJE (RTV)

Stations	**ch**
Toşkent	9

D.Prgr: Relays of RTV from Moscow. Further details see under Russia.

TV RELAYS

Stations	**ch**
Toşkent	40

D.Prgr: Relays of Kazakh, Kyrgyz and Tadzhik TV, as well as TRT from Turkey.

VIETNAM

TV-sets: 2.500.000 — **System:** M — **Colour:** NTSC/SECAM

TELEVISION VIETNAM (Gov.)
Addr: 59 Giang Vo, Hanoi. **Te:** 43748/43188. **Telex:** 412279 THVN-VT.
Stations: Hanoi ch2, ch6 (SECAM); Ho Chi Minh City ch9 240kW; Hue, Cantho & Quinhon ch7 and DaNang ch13 (NTSC). Details of st's in Vinh & Nha Trang not available. 13 additional relay tr's.
D.Prgr: Central Prgr. from Hanoi: Mon-Fri 90 min's a day, Sat/Sun 3h (prgrs in black & white only).
Ch 6: Hanoi City Sce. **Ch. 2:** Provincial Sce. for Red River delta.
Other st's: 25% in colour (no further details available).

PACIFIC

AUSTRALIA

TV-sets: 8m — **Colour:** PAL. **System:** B.

AUSTRALIAN BROADCASTING CORPORATION (ABC)
The 'National Television Service' is controlled by the Australian Broadcasting Corporation which is responsible to Parliament through the Minister for Transport and Communications.

In addition to the terrestrial transmitters listed below, ABC also broadcasts to remote areas using five transponders of the Australian domestic satellite system to cover the five main geographic regions and time zones. This service is known as the 'Homestead and Community Broadcasting Satellite Service' (HACBSS) and is carried in the frequency band 12 250 to 12 750 MHz on satellites which are owned and operated by AUSSAT Proprietary Limited. Except in the States of Victoria and Tasmania, most transmitters outside the State capital cities are fed by satellite. The satellite signal is 'B-MAC' encoded and carries three sound-only programs, one of which is stereophonic, in addition to the television signal and its associated sound channels.

Addr: Box 9994, GPO, Sydney 2001 — **L.P:** See radio.
Stations: Audio powers are 1/10 of vision powers when a single sound channel is used. Stations below 10 kW ERP are not listed. Pattern: 'O': omnidirectional, 'D': directional. Channels 3, 4, 5 and 5A are progressively being phased out to make way for FM radio and other services — the expected new channel numbers and other information covering changes arising from the first stages of the clearance of stations from 'Band II' are given in parentheses in the table.

Call	ch	ERP(kW)	POL	Pattern	Location
Australian Capital Territory					
ABC	3	100	V	O	Canberra
ABN	2	100	H	O	Sydney
New South Wales					
ABMN	0	100	H	O	Wagga Wagga
ABCN	1	100	V	O	Orange
ABTN	1	100	V	D	Taree
ABDN	2	100	H	D	Grafton/Kempsey
ABQN	5(11)	100(200)	V	O(D)	Dubbo
ABHN	5A	100	H	O	Newcastle
ABWN	5A(56)	100(600)	H	D	Illawarra
ABRN	6	100	H	O	Lismore
ABGN	7	100	H	O	Griffith
ABUN	7	100	H	D	Tamworth
ABSN	8	100	V	D	Bega/Cooma
Victoria					
ABAV	1	100	H	D	Albury
ABEV	1	100	V	D	Bendigo
ABV	2	100	H	O	Melbourne
ABSV	2	100	V	O	Swan Hill
ABGV	3(11)	100	V	D	Shepparton
ABRV	11	100	H	D	Ballarat
ABLV	40	100(1000)	H	D	Traralgon
ABMV	4(6)	100	H	D	Mildura
ABWV	5A	66	H	D	Western Victoria
ABWV	45	100	H	D	Horsham/Dimboola
Queensland					
ABSQ	1	100	H	D	Warwick
ABQ	2	100	H	O	Brisbane
ABDQ	3(32)	100(1000)	H	O(D)	Toowoomba
ABRQ	3(9)	100	H	D	Rockhampton
ABTQ	3	100	H	D	Townsville
ABMQ	4(8)	100(180)	H	D	MacKay
ABWQ	6	150	V	D	Maryborough
ABNQ	9	100	H	D	Cairns
ABQ	49	50	H	D	Gold Coast
South Australia					
ABGS	1	100	H	D	Mount Gambier
ABNS	1	100	V	D	Port Pirie
ABS	2	100	H	D	Adelaide
ABRS	3	150	V	D	Loxton
ABS	30	10	H	D	Wudinna
Western Australia					
ABW	2	100	H	O	Perth
ABAW	2	100	V	O	Albany
ABSW	5	100	H	O	Bunbury
ABCW	5A	100	H	O	Central Agricultural
ABGW	6	10	H	O	Geraldton
ABW	8	60	H	D	Wagin
ABCMW	8	10	H	D	Morawa
ABMW	46	10	H	D	Dalwallinu
ABMW	60	60	H	D	Moora
Tasmania					
ABT	2	100	H	O	Hobart
ABNT	3(11)	100(300)	H	D	Launceston
Northern Territory					
ABD	6	30	H	D	Darwin

+ 343 low power services with ERPs below 10kW, made up of 110 translators and 233 satellite-fed transmitters.
+ 46 (approx.) low power (up to 150 watts ERP), privately owned transmitters/translators which are licensed under a 'self-help' scheme to carry ABC television programs.
Daily Prgrs: Mon-Fri 2100-1400; Sat/Sun 24h.

SPECIAL BROADCASTING SERVICE (SBS)

Addr: 4 Cliff Str, Milsons Point, Sydney, NSW 2061. **Te:** +61 (2) 964 2828. **Fax:** +61 (2) 957 3571. **Telex:** AA120944.
Stations: Sydney ch28 300kW, Canberra ch28 200kW, Melbourne ch28 300kW, Brisbane ch28 300kW, Adelaide ch28 300kW, Perth ch28 300kW, Hobart ch28 225kW.

IMPARJA TELEVISION PTY LTD.

Addr: 14 Leichhardt Terrace, Alice Springs, NT 0871. **Te:** +61 (89) 530300 — **Fax:** +61 (89) 530322. **Telex:** AA81166.
This st. is owned by the Central Australian Aboriginal Media Association (CAAMA).
Station: ch9 250kW.

COMMERCIAL TV STATIONS

Federation of Australian Commercial Television Stations
Addr: 44A Avenue Rd, Mossman, NSW 2088. **Te:** +61 (2) 960 2622. **Fax:** + 61 (2) 969 3520. **Telex:** 121542.

Main Networks

6TH METROPOLITAN TV CHANNEL (planned)
NB: The Australian government is considering issuing a license for a 6th TV channel to operate in each capital city. The station is expected to be a public (non-commercial) station on UHF channel 31. Transmission is expected to commence as early as late 1992. At present time, the plans were to allow temporary low-power transmissions by various organizations until 1997, when the future of the channel will be reviewed. Such transmissions have already taken place on this channel.

The Seven Network

Addr: Television Centre, Mobbs Lane, Epping, NSW 2121. **Te:** +61 (2) 858 7777. **Fax:** +61 (2) 858 7888. **Telex:** AA20250. (5 owned and 9 affiliated st's).

The Nine Network

Addr: P.O. Box 27, Willoughby, NSW 2068. **Te:** +61 (2) 430 0444. **Fax:** +61 (2) 436 2193.

Network Ten Australia

Addr: P.O. Box 10, Lane Cove, NSW 2066. **Te:** +61 (2) 887 0222.
Stations: The two first letters are an abbreviation of the name of the licence, the third indicates the state and the numeral signifies the channel — Aural power 10% of the visual power.

	Call	kW(ERP)		Call	kW(ERP)
1)	ATV10	100(H)	26)	SEQ8	100(V)
2)	TVQ0	100(H)	27)	SES8	100(H)
3)	RVN2	100(H)	28)	STV8	100(H)
4)	NBN3	100(H)	29)	NRL8	100(H)
5)	BTW3	50(H)	5)	VEW8	4(H)
7)	AMV4	100(H)	31)	ITQ8	0.5(H)
8)	GTS4	100(V)	32)	NTD8	20(H)
9)	SDQ4	100(H)	35)	ECN8	100(V)
9a)	RTS5A	80(V)	33)	TCN9	100(H)
10)	BTV6	100(H)	34)	MTN9	100(H)
11)	GMV6	100(V)	35)	NEN9	100(H)
12)	TVT6	100(H)	36)	GTV9	100(H)
13)	CWN6	200(V)	37)	QTQ9	100(H)
14)	MVQ6	100(H)	38)	NSW9	100(H)
15)	ATN7	100(H)	39)	STW9	100(H)
16)	CTC7	100(V)	25)	TNT9	300(H)
17)	HSV7	100(H)	5)	GSW9	50(V)
18)	BTQ7	100(H)	41)	TEN10	100(H)
19)	RTQ7	100(H)	9)	DDQ10	100(H)
20)	TNQ7	100(H)	16)	CTC10	50(V)
21)	SAS7	100(H)	20)	FNQ10	10(H)
22)	TVW7	100(H)	25)	BCV11	20(V)
23)	BKN7	100(H)	29)	NRN11	100(H)
13)	CBN8	100(V)	43)	ADS10	100(V)
24)	RTN8	100(H)	44)	GTW11	100(H)
25)	BCV8	125(V)	45)	NEW10	100(H)
25)	GLV8	100(V)			

1) Austarama TV Pty. Ltd, P.O. Box 42, Nunawading, Vic 3131. **Cable:** Austarama — St. Mgr: C. O'Connell. Dir. Tec: J. Crane. Film Buyer: R. Burns. 24h.
2) Universal Telecasters Qld. Ltd, G.P.O. Box 751, Brisbane 4001. **Cable:** Channel 0 — Gen. Mgr: B. Campbell. Dir. Tec: J.C. Stevenson. 2000 (Sat/Sun 2100)-1430.
3) Riverina & North East Victoria TV Ltd, P.O. Box 2, Kooringal NSW 2650. **Cable:** Rixtel — GM: A.J. Cook. TD: N. Baggio. 2100-1400.
4) NBN Ltd, P.O. Box 750, Newcastle 2300. **Cable:** Chanthree — Man. Dir: J. Sweeney. Dir. Tec: M. Lewis. 2100 (Sat 2130, Sun 2030)-1400 (Sat 1500).
5) Golden West Netw, P.O. Box 112, Bunbury, W.A. 6230. **Cable:** Swetel — Man. Dir: W. G. Rayner. Dir. Tec: H. James. 0300-1600 (relay GSW9, Bunbury & VEW8, Tuart Hill). DSB Sce to We. Australia, with teletext.
7) Television Centre, Wagga Wagga, NSW 2650.

8) Spencer Gulf Telecasters Ltd, P.O. Box 305, Port Pirie, SA 5540. **Cable:** "GTS4" — Man. Dir: R. Davis. Generally 0530 (Sun 2330)-1330.
9) Darling Downs TV Ltd, P.O. Box 670, Toowoomba, Qld. 4350. **Cable:** Deedecue — Gen. Mgr: B.M. Finlay. Dir. Tec: G. Walburn. Relay SDQ4 (Toowoomba) 2100 (Sun 2030)-1400.
9a) Riverland TV Pty. Ltd, P.O. Box 471, Loxton 5333 — G.M: Wendy Mudge. Dir. Tec: K. Garrow. 0130-1430.
10) Ballarat & Western Victoria TV Ltd, P.O. Box 464, Ballarat, Vic. 3350. **Cable:** Beeteesix — Gen. Mgr: J. Malone. Chief Eng: H.G. Riley. 0100 (Sat 2100, Sun 2145)-1400.
11) Goulburn Murray TV Ltd, P.O. Box 666, Shepparton, Vic. 3630. **Cable:** Geemveesix — Man. Dir: J. Malone. Dir. Tec: B. Knight. 2030 (Sat 2130)-1430 (Fri 1500, Sat 1530).
12) Tasmanian TV Ltd, G.P.O. Box 1209M, Hobart, Tasmania 7001. **Cable:** Tasteevee — G.M: D. Rouse. Dir. Eng: R. Hernan. 0100-1400.
13) Country TV Services Ltd, P.O. Box 465, Orange 2800. **Cable:** Cebtes — Man. Dir: Ian Ridley. Dir. Tec: J. Vanderstock. 2100-1400 (Sat 1430). Relay st. CWN6 Dubbo.
14) MacKay TV Ltd, P.O. Box 496, MacKay, Qld 4740. **Cable:** Emveecue — Gen. Mgr: R.J. Cox. Dir. Tec: D. Brooks. 2100 (Sat/Sun 0030)-1330.
15) Amalgamated TV Sces Pty. Ltd, Television Centre, Epping 2121. **Cable:** Telecentre, Sydney — Gen. Mgr: E. Thomas. Dir. Tec: G. Healy. Film Buyer. G. Kinging. Continuous exc. Sat 1600-2000 & Sun 1430-2000.
16) Australian Capital TV Ltd, Private Mailbag 10, Dickson, A.C.T. 2602. **Cable:** Ceeteecee — St. Mgr: & Film Buyer: P.F. Hunniford. Dir. Tec: A.R. Martin. 2030-1530 (Sat 1630).
17) HSV Channel 7, Pty. Ltd, Box 215D, G.P.O. Melbourne 3000. **Cable:** Hertel — Gen. Mgr: G. Carrington. Dir. Tec: L. De Stefano. Mon-Fri 24h; Sat/Sun 2100-1500.
18) Brisbane TV Ltd, GPO Box 604, Brisbane, Qld. 4001. **Cable:** Beeteecue — Man. Dir: Les Riley. Dir. Tec: Andrew Robertson. 24h.
19) Rockhampton TV Ltd, P.O. Box 568, Rockhampton, Qld. 4700. **Cable:** Arteecue — G.M: A.C. Shields. Dir. Tec: P. Deavin. 2100-1400.
20) Telecasters North Queensland, P.O. Box 1016, Townsville, Qld 4810. **Cable:** Teenq — St. Mgr: R. Power. Dir. Tec: P. Ament. 2000 Sat/Sun 2100)-1400.
21) TV Broadcasters Ltd, P.O. Box 7, North Adelaide, SA 5006. **Cable:** Adseven — Dir: S. Pippos, Gen. Mgr. Chief Eng: P. Hartmann. 2130-1600.
22) TVW Enterprises Ltd, P.O. Box 77, Tuart Hill, W.A. 6060. **Cable:** Teleview — Gen. Mgr: K. Campbell. Dir. Tec: T. O'Dea. 2200-1830 (Sun 2100-1630).
23) Broken Hill TV Ltd, P.O. Box 472, Broken Hill 2880 — Man. Dir: J.M. Sturrock. 0830 (Wed 0230)-1230 (Wed/Fri 1300); Sun 0830-1200 (relay of st. 8).
24) Richmond Tweed TV Ltd, P.O. Box 15, Lismore Heights 2480. **Cable:** Richteeve. 2100 (Sat 0000, Sun 2200)-1330 (relays st. 29).
25) Southern Cross Communications Ltd, P.O. Box 888, Bendigo, Vic. 3550. **Cable:** Beneight — Chief Exec: G.L. Eddy. Dir. Tec: P.R. Long.
26) Wide Bay Burnett TV Ltd, P.O. Box 30, Maryborough, Qld. 4650. **Cable:** Wideteve — Man. Dir: G.J. McKean. Dir. Tec: B. Hunter. 2100 (Sun 2000)-1330.
27) South East Telecasters Ltd, P.O. Box 821, Mt. Gambier, SA 5290. **Cable:** Setel — Man. Dir: G.J. Gilbertson. Dir. Tec: G.D. Smythe. Mon-Fri 2030-1430, Sat 2300-1500, Sun 2130-1430.
28) Sunraysia Television Ltd, P.O. Box 1157, Mildura, Vic 3500 — St. Mgr: A. Sharpe. Dir. Tec: D. Carson. 2100 (Sun 2000)-1500.
29) Northern Rivers TV Ltd, P.O. Box 920, Coffs Harbour 2450. **Cable:** Enareleven — Gen. Mgr: J. Davis. Chief Eng: R. Higgens. 2000-1500.
31) Mt. Isa Television Pty. Ltd, P.O. Box 1557, Mt. Isa, Qld. 4825 — Gen. Mgr: B. Hiller. Dir. Tec: G. Dent. 0200 (Sat/Sun 0330)-1300.
32) Territory Television Pty. Ltd, P.O. Box 1764, Darwin, NT 5794 — Gen. Mgr: P. Dooley. Dir. Tec: J. Wilkinson. 2130-1430.
33) TCN Channel 9 Ltd, P.O. Box 27, Willoughby 2068. **Cable:** Diffusion, Sydney — Gen. Mgr: D. Leckie. Dir. Tec: Bruce Robertson. 24h.
34) Broadcast Operations Pty. Ltd, P.O. Box 493, Griffith 2680 — Man. Dir: R. Gamble. Dir. Tec: Ove Andersen. 2100 (Sat 2200)-1400; Sun 2030-1330.
35) TV New England Ltd, P.O. Box 317, Tamworth 2340. **Cable:** Nenine — Gen. Mgr. M. Moroney. Dir. Tec: R. Barrett. 2100 (Sat 0000, Sun 2200)-1400. Relay ECN8 (Middle Brother).
36) General TV Corporation Pty. Ltd, P.O. Box 100, Richmond, Vic. 3121. **Cable:** Geeteevee — Man. Dir: G.W. Rice. Dir. Tec: I. Hodgson. 24h.
37) Queensland TV Ltd, Box 72, GPO, Brisbane 4001. **Cable:** Queteenine — Gen. Mgr: I.R. Müller. Dir. Tec: M.J. Bartlett. 24h.
38) Southern TV Corp. Pty. Ltd, P.O. Box 9, North Adelaide, SA 5006. **Cable:** Newsnine — Gen. Man: Tyrrell Talbot. Ch. Eng: Robert Miers. Film Buyer: R. Boland. 24h.
39) Swan Television and Radio Broadcasters Ltd, P.O. Box 99, Tuart Hill, WA 6060. **Cable:** Swantel — Gen. Mgr: K. Gannaway. Dir. Tec: T. Provan. 24h.
41) United Telecasters Sydney Ltd, P.O. Box 10, Lane Cove, NSW 2066. **Cable:** Unitelcast, Sydney — Man. Dir: I. Gow. Dir. Eng: B. Robertson. 2000-1530 (Fri 2000-Sun 1530 continuous).
43) South Australian Telecasters Ltd, 45-49 Park Terrace, Gilberton, SA 5081. **Cable:** Saseven — Gen. Mgr: D.R. Earl. Dir. Tec: D. Caddy. 2030-1530 (Sat/Sun 24 h).
45) West Coast Telecasters Ltd (NEW 10), P.O. Box 1010, Mirrabooka, WA 6061. **Fax:** (09) 3448076 — Man. Dir: W. McKenzie. Chief Eng: P. Scotland.

COOK ISLANDS

TV-sets: 3500 — **Colour:** PAL — **System:** B.

COOK ISLAND BROADCASTING CORPORATION (Gov.)

Addr: P.O. Box 126, Avarua, Rarotonga. **Te:** (682) 29460. **Fax:** (682) 21907 Cook Island Television (CITV).
Stations: VHF ch1 & 6.

EASTER ISLAND

TV-sets: N/A — **Colour:** PAL — **System:** B.

TV RAPANUI, Isla de Pascua.

L.P: Dir. Gen: J. Edmund Paoa. Head Tec. Sce's: J. Pont Chavez.
Station: ch (unknown). **D.Prgr:** 0000-0600.

FIJI

TV-sets: N/A — **Colour:** NTSC — **System:** M.

Two TV st's, at Suva and Monsavu, are under construction with Chinese aid.

FIJI BROADCASTING COMM.

Addr: P.O. Box 334, Broadcasting House, Suva — **Te:** 314333.
Telex: 314333.

GALAPAGOS ISLANDS

TV-sets: 4.000 — **Colour:** NTSC — **System:** M.

TELEGALAPAGOS (Cult)

ADDR: see radio. **LP:** Dir. Gen: P. Victor Maldonado. Dir. Tec: Fernando Mendez. Film Buyer: Remigio Andrade.
Station: chA13 — **D.Prgr:** 2000-0400.

GUAM (US Terr.)

TV-sets: 75.000 — **Colour:** NTSC — **System:** M.

KUAM TELEVISION (Comm.)

Addr: P.O. Box 368, Agana 96910. **Te:** +1 (671) 477-9861.
Cable: Kuam.

L.P: Pres: L.S. Berger. Gen. Mgr: Greg Perez. Dir. Tec: K. Tydingco.
Station: chA8 21.9/3.8kW
D.Prgr: Mon-Fri 2000-1400 (also rel. CBS/NBC prgrs).

KGTF TELEVISION (Educ.)
Addr: Guam Educational Telecommunications Corporation, P.O. Box 21449, Guam, Marianas Is. 96921. **Te:** +1 (671) 734-3030. **Fax:** –1 (671) 734-5483. **Telex:** 6467 KGTF TV GM.
L.P: Gen. Mgr: Joseph E. Tighe. TD: Edmond Cheung. Film Buyer: Doris Gallo.
Station: chA12 27.6/5.5kW. **F.Pl.** ch 14 (Saipan & Tinian); ch 16 (Rota).
D.Prgr: 2000-1300.

HAWAII (US State)

TV-sets: 552.500 — **Colour:** NTSC — **System:** M.
All st's comm. exc. 7) and 9) Educational, 18) religious.

	Call	ch	kW (ERP)		Call	ch	kW (ERP)
1)	KHON-TV	2	100/20	7)	KMEB-TV	10	5/0.5
1a)	KHBC-TV	2	2.29/0.458	9)	KHET	11	150/30
2)	KGMV	3	14.1/2.82	9a)	KHAW-TV	11	2.09/0.27
3)	KITV	4	100/20	10)	KMAU-TV	12	30/4.36
16)	KFVE	5	100/20	11)	KHNL-TV	13	316/63.2
14)	KVHF	6	52.6/6.7	12)	KHVO	13	4.68/0.59
4)	KAII-TV	7	29.8/5.9	13)	KHAI-TV	20	537/63
5)	KGMB-TV	9	209/41.8	14)	KMGT	26	100/10
6)	KGMD	9	9.68/1.71	15)	KBFD	32	146/14.6

1) Burnham Broadc. Co., 1170 Auahi Str, Honolulu 96814-4975.
1a) Hilo Broadc. Corp, Box 4250 Hilo, Big Island 96720-0520.
2) Wailuku (Relay of st. 5).
3) TAK Communications, 1290 Ala Moana Blvd, Honolulu 96814-4299.
4) Wailuku (Relay of st. 1).
5) Lee Enterprises, 1534 Kapiolani Blvd,honolulu 96814-3799. Mon-Fri 1600-1130; Sat/Sun 24h.
6) Hilo (Relay of st. 5).
7) Wailuku (Relay of st. 9).
9) Hawaii Public Broadc. Authority, 2350 Dole Str, Honolulu 96822-2495.
9a) Hilo (Relay of st. 1).
10) Wailuku (Relay of st. 3).
11) King Broadc. Co, 150-B Puuhale Rd, Honolulu, HI 96819.
12) Hilo (Relay of st. 3).
13) Honolulu Family TV Ltd, 735 Sheridan Str, Honolulu 96814-3095.
14) Mauna Kea Broadc. Co, 970 N. Kalaheo Ave, Honolulu 96734-1892.
15) The Allen Broadc. Corp, 1188 Bishop Str, Honolulu 96813-3314.
16) 315 Sand Island, Access Rd, Honolulu 96819-2245.

MARSHALL ISLANDS

TV-sets: N/A — **Colour:** NTSC — **System:** M.

Adele TV Station: Te: 3550.

AFRTS TELEVISION (Department of Defense)
Addr. & L.P: as radio.
Stations: ch9, ch13 0.25kW (24 h to Kwajalein Island & Roi-Namur Island).

MICRONESIA (Fed. States Of)

TV-sets: 7000 — **Colour:** NTSC — **System:** M.

TV STATION POHNPEI (Comm.)
Addr: KPON-TV, Central Micronesia Communications, P.O. Box 460, Kolonia, Pohnpei 96941.
L.P: Pres: Bernard Hegenberger. Dir. Tec: David Cliffe.
Station: Pohnpei chA7 1kW + cable TV on ch's 4, 5, 9.
D.Prgr: 0400-1200.

NEW CALEDONIA

TV-sets: 35.500 — **Colour:** SECAM — **System:** K.

RFO-TV (Gov.)
Addr: & **L.P:** See radio.
Stations: Mont Do chK4 2x1kW H, Noumea chK8 0.4kW H, Lifou chK7 0.4kW H (+ 25 low power repeaters).
D.Prgr: W 0630-1130, Sun 0515-1115.

NEW ZEALAND

TV-sets: 1.100.000 — **Colour:** PAL — **System:** B.

TELEVISION NEW ZEALAND
Addr: P.O. Box 3819, Auckland. **Te:** +64 (9) 770630. **Fax:** +64 (9) 750979. **Telex:** NZ 60056.
L.P: Dir. Gen: J.D. Mounter. Contr. Eng: P.W. Sandlant. Dir. Prgrs and Production: D.J. Monaghan.
Stations: Channel One: Wellington ch1 100/20 kW H, Auckland ch2 100/20 kW H, Christchurch ch3 100/20 kW H, Dunedin ch2 50/10 kW H + relay st's at Hamilton ch1 100/20 kW V, Palmerston North ch2 100/20 kW V, Invercargill ch1 1/0.1 kW H + 18 medium and 412 low powered relay st's.
D.Prgr: 2100 (Sat 1900, Sun 2000)-1200 (Fri 1300, Sat 1400).
Channel Two: Auckland ch4 300/30 kW H, Wellington ch5 300/30 kW H, Christchurch ch3 300/30 kW H, Dunedin ch4 300/25 kW H + repeaters at Hamilton ch3 100/20 kW H, Palmerston North ch4 300/30 kW V, Invercargill ch3 1/0.1 kW H + 17 medium and 240 low powered repeaters.
D.Prgr: 1830-1200.

TV3 (Comm.)
Addr: P.O. Box 5185, Auckland. **Te:** +64 (9) 779 730. **Fax:** +64 (9) 366 7029.
L.P: Head of Netw. Programming: Kel Geddes.
D.Prgr: Auckland ch7 H, Christchurch ch6 H, Hamilton ch9 V, Invercargill ch7 H, Wellington ch11 H, Dunedin ch10 H, Palmerston North ch7 V.

CANTERBURY TELEVISION (CTV)
Addr: 196 Gloucester Street, Christchurch.
D.Prgr: 0500-1200 UTC — **Station:** ch48 65kW.

NORFOLK ISLAND

TV-sets: 900 — **Colour:** PAL — **System:** B.

NORFOLK ISLAND TELEVISION SCE. (Gov.)
Addr: New Cascade Rd, Norfolk Island 2899, Australia. **Te:** 2137.
Station: Mt. Pitt ch7 0.02kW V.
D.Prgr: rel. ABC Australia from Aussat Satellite.

NORTHERN MARIANA ISLANDS (US Commonwealth)

TV-sets: 4.100 — **Colour:** NTSC — **System:** M.

MICRONESIA BROADC. CORPO. (Comm.)
Addr: c/o KUAM, Box 368, Agana, Guam 96920.
L.P: Pres: H. Scott Killgore. Gen. Mgr: T. Dickey. Asst. Gen. Mgr: A. Ocambo. Technician: M. Madaing.
Stations: WSZE-TV ch10 0.5kW (Saipan)
D.Prgr: 0600-1400.

PALAU (Rep.)

TV-sets: 1.600 — **Colour:** NTSC — **System:** M.

STV-TV KOROR (Comm.)

Addr: Koror, Palau 96940.
L.P: Mgr: David Nolan. Technician: Ray Omelen.
Station: Ngermit, Koror ch7 0.1kW.
D.Prgr: 0400-1400.

PAPUA NEW GUINEA

TV-sets: 10.000 — **Colour:** PAL — **Systems:** B&G.

EMTV (Comm.)

Addr: Media-Niugini Pty. Ltd, P.O. Box 443, Boroko. **Te:** +675 257322. **Fax:** +675 254450.
L.P: CEx. John Taylor. CE: Rod Tindale.
Stations: Burns Peak ch9 1.1kW, Air Niugini Hill ch31 0.17kW, Garden City ch68 0.02kW (all Port Moresby area) — **F.Pl:** st's at Mt. Hagen, Goroko, Lae, Rabaul.

POLYNESIA (French)

TV-sets: 26.500 — **Colour:** SECAM — **System:** K.

SOCIÉTÉ NATIONALE DE RADIO TÉLÉVISION D'OUTRE MER (RFO)

Addr: & L.P: See radio.
Stations: Papeete chK4 0.1kW H, Mont Marau chK8 0.5kW H, Vaitape chK7 0.2kW V, Taravao chK4 0.1kW H (+ 8 low power repeaters).
D.Prgr: 0400 (Sat/Sun 0200)-0830.

SAMOA (American)

TV-sets: 8.000 — **Colour:** NTSC — **System:** M.

KVZK-TV (Gov.)

Addr: American Samoa Government, Pago Pago 96799.
Te: + 684 6334191.
L.P: Gen. Mgr: Fulifuli Taveuveu. Dir. Tec: Robert Blauvelt.
Stations: chA2/A4/A5 (72kW).
D.Prgr: ch2 & 5: 1830-1000; ch4: 1830-1100.

WALLIS & FUTUNA

TV-sets: N/A — **Colour:** SECAM — **System:** K.

RADIODIFFUSION FRANCAISE D'OUTRE-MER (RFO)

Addr & L.P: see radio.
Stations: ch6 & ch9.
D.Prgr: 5-6 hours each evening.

NORTH AMERICA

ALASKA (US State)

TV-sets: N/A — **Colour:** NTSC — **System:** M.

TV-STATIONS: All comm. exc. 2, 4, 5, 7 (Educational).

	Call	ch	kW(ERP)		Call	ch	kW(ERP)
1)	KTUU-TV	2	100/10	6)	KJUD-TV	8	0.24/0.05
1a)	KATN-TV	2	28.2/5.5	7)	KUAC-TV	9	46.8/8.33
2)	KTOO-TV	3	2.45/0.49	8)	KTVA-TV	11	50.7/5.07
3)	KJNP-TV	4	19.1/3.31	9)	KTVF-TV	11	27/5.5
4)	KYUK-TV	4	4.67/1.16	10)	KIMO-TV	13	39/7.8
4a)	KTBY-TV	4	42.5/8.5	11)	KTNL-TV	13	0.2/0.03
5)	KAKM-TV	7	163/16.3				

1) Box 102880, Anchorage, AK 99510 — Gen. Mgr: Al Bramstedt Jr. 2230 (Sat/Sun 0000)-0900.
1a) P.O. Box 74730, Fairbanks, AK 99701.
2) Capital Community Broadc. Inc, 224 Fourth Str, Juneau 99801 — Pres. & Gen. Mgr: D. Rinker. Dir. Tec: J.W. Foster. 1930 (Sat/Sun 1530)-0800.
3) Evangelistic Alaska Missionary Fellowship, Box "O", North Pole 99705 — Pres. & Dir: D.L. Nelson. Dir. Tec: E. Nichols.
4) Bethel Broadc. Inc, Box 468, Bethel 99559 — Dir. Gen: J. Brigham.
4a) KTBY Inc, 1840 S. Bragnaw Str, Anchorage 99508 — Gen. Mgr: R.V. Bradley. Dir. Tec: E. Gjernes.
5) Alaska Public Television Inc, 2677 Providence Dr, Anchorage 99508 — Gen. Mgr: E. Sackett. Dir. Tec: F. Mengel.
6) 1107 West Eighth St., Suite 2, Juneau, AK 99801.
7) University of Alaska, Fairbanks 99701 — St. Mgr: Kathryn Jensen. Dir. Tec: David L. Walstad. 1800-0900.
8) Northern TV Inc, Box 102200, Anchorage 99510 — Dir. Tec: D. Milsap. 1700-1100.
9) Northern TV Inc, Box 950, Fairbanks 99707 — 1700-1100.
10) Alaska 13 Corp, 2700 Tudor Rd., Anchorage 99507 — Pres: D.L. Triplett.
11) Sitka Broadc. Co. Inc, Box 2668, Sitka 99835 — Gen. Mgr: D. Etulain. Mon-Fri 1900-0900, Sat 1700-0900, Sun 1400-0800.

ARMED FORCES RADIO & TV SCE. (US Mil.)

Adak ch 8 0.12kW, Officer in Charge, NAVBCSTSVC Det. Adak, Navsta Box 14, FPO Seattle, WA 98791 — St. Mgr: E.C. Miller, MSgt, USAF.

BERMUDA

TV-sets: 30.000 — **Colour:** NTSC — **System:** M.

ZBM-ZFB RADIO & TELEVISION (Comm.)

Addr: P.O. Box HM 452, Hamilton. **Te:** +1 (809) 295-2828. **Fax:** +1 (809) 295-4282. **Telex:** 3702 ZBMBA.
L.P: Prgr. Mgr: Bryan Lodge.
Station: ZBM-TV chA10 17.5/8.75kW H.
D.Prgr: 1100-0430.

CANADA

TV-sets: 17.400.000 — **Colour:** NTSC — **System:** M.

CANADIAN BROADCASTING CORPORATION SOCIÉTÉ RADIO-CANADA (Publicly owned)

ADDR: Head Office: P.O. Box 8478, Ottawa, Ontario K1G 3J5. **Cable:** Broadcasts. **Telex:** 053-4260. **Te:** +1 (613) 724-1200. **Fax:** +1 (613) 738-6887.
L.P: President and Chief Executive Officer: Gérard Veilleux — Chairman, Board of Directors: Patrick Watson — Senior Vice-President, Resource Management and Administration: Anthony S. Manera — Senior Vice-President, Television Services: (vacant) — Senior Vice-President, Radio Services: Michael McEwen — Vice-President, Planning and Corporate Development: John Shewbridge — Vice-President, Communications: Robert Pattillo — Vice-President, News, Current Affairs & Newsworld: Tim Kotcheff — Vice-President, Regional Broadcasting Operations: Trina McQueen — Executive Director, Business Development and International Relations: Robert O'Reilly — Director of Corporate Communications: Charlotte O'Dea — Vice-President French Television: Guy Gougeon.
English Networks: P.O. Box 500, Station "A", Toronto, ON M5W 1E6. **Cable:** Broadcast. **Telex:** 062-17796. **Te:** +1 (416) 975-3311.
French Networks: P.O. Box 6000, Montreal, PQ H3C 3A8. **Cable:** Radcan. **Telex:** 05-267417. **Te:** +1 (514) 285-3211.
CBC Engineering: 7925 Côte St-Luc Road, Montreal, PQ H4W 1R5. **Telex:** 055-66437. **Te:** +1 (514) 485-1301.
L.P: Vice-President, Engineering: Brian D. Baldry — Manager, Technical Information: Claudette Lalonde.

Affiliated Stations: CKVR Barrie, Ont; CKX Brandon, MB; CJDC Dawson Creek, BC; CFJC Kamloops, BC; CHBC Kelowna, BC; CKWS Kingston, Ont; CKSA Lloydminister, AB; CHAT Medicine

Hat, AB; CHNB North Bay, Ont; CHEX Peterborough, Ont; CKBI Prince Albert SK; CKPG Prince George BC; CDMI Quebec, PQ; CKRD Red Deer, AB; CHSJ Saint John, NB; CJIC Sault St. Marie, Ont; CKNC Sudbury, Ont; CJFB Swift Current, SK; CFTK Terrace, BC:CKPR Thunder Bay, Ont; CFCL Timmins, Ont.
Stations: Pol: H.

English Network	Call	chA	kW
St. John's, Nfld.	CBNT	8	196
Charlottetown, P.E.Q.	CBCT	13	178
Halifax, N.S.	CBHT	3	56
Montreal, Que	CBMT	6	100
Ottawa, Ont.	CBOT	4	100
Toronto, Ont.	CBLT	5	84
Winnipeg, Man.	CBWT	6	100
Regina, Sask.	CBKT	9	140
Edmonton, Alta.	CBXT	5	318
Vancouver, B.C.	CBUT	2	50
Yellowknife, N.W.T.	CFYK	8	2.4

CBC English TV is also distributed through 200 rebroadcasters
Addresses:
P.O. Box 12010 Stn. "A", St. Johns, Nfld. A1B 3T8.
P.O. Box 2230, Charlottetown, P.E.I. C1A 8B9.
P.O. Box 3000, Halifax, N.S. B3J 3E9.
P.O. Box 6000, Montreal, Que. H3C 3A8.
P.O. Box 3220, Stn. "C", Ottawa, Ont. K1Y 1E4.
P.O. Box 500, Stn. "A", Toronto, Ont. M5W 1E6.
P.O. Box 160, Winnipeg, Man. R3C 2H1.
2440 Broad Str, Regina, Sask. S4P 4AI.
P.O. Box 555, Edmonton, Alta. T5J 2P4.
P.O. Box 4600, Vancouver, B.C. V6B 4A2.
P.O. Box 160, Yellowknife, N.W.T. X1A 2N2.

French Netw.	Call	chA	kW
Moncton, N.B.	CBAFT	11	163
Montreal, Que.	CBFT	2	100
Quebec, Que.	CBVT	11	252
Ottawa, Ont.	CBOFT	9	128
Winnipeg, Man.	CBWFT	3	59.0
Regina, Sask	CBKFT	13	140
Edmonton, Alta.	CBXFT	11	90
Vancouver, B.C.	CBUFT	26	105

CBC French TV is also distributed through 200 rebroadcasters

Addresses:
P.O. Box 950, Moncton, N.B. E1C 8N8.
P.O. Box 10400, St-Foy, Que G1V 2X2.
Others: See English Netw.

Private TV Networks

CITY-TV

Addr: 299 Queen Str. W. Toronto, Ontario, M5V 2Z5. **Te:** +1 (416) 591-5757. **Fax:** +1 (416) 591-7791. **Telex:** 06218283.
L.P: GM: Dennis Fitz-Gerald.

CANWEST GLOBAL SYSTEM

Addr: 81 Barber Greene Rd, Don Mills, ON, M3C 2A2. **Te:** +1 (416) 446-5311. **Fax:** +1 (416) 446-5371/5490.
Stations: CKVU Vancouver, BC; CKKX Calgary, AB; CISA Lethbridge, AB; CFRE Regina, SK; CFSK Saskatoon, SK; CKND Winnipeg, MB; CIII Toronto, Ont; MITV Saint John, NB; MITV Halifax/Dartmouth, NS.

CTV TELEVISION NETWORK LIMITED

Head Office: 42 Charles Str. E., Toronto, Ontario M4Y 1T5.
Te: +1 (416) 928-6000. **Telex:** 06-22080. **Fax:** +1 (416) 928-0907.
L.P: Pres. & C.E.O: Murray H. Chercover; Chief Financial Officer and Treasurer: Duncan Morrison; VP Network Relations: Marge Anthony; VP Gov't Relations & Corporate Planning: John T. Coleman; VP Operations: Joseph A. Colson; VP Sports: Johnny Esaw; VP News, Features & Information Programming: Tim Kotcheff; VP Finance: Peter O'Neill; VP Sales: Peter Sisam; VP Programming: Philip Wedge; VP Entertainment Programming: Arthur Weinthal.
Affiliated Stations: NTV St. John's, Nfld; CJCB Sydney, N.S; CJCH Halifax, N.S.; CKCW Moncton, N.B.; CKNY North Bay, Ont.; CITO Timmins, Ont.; CJOH Ottawa, Ont.; CFTO Toronto, Ont.; MCTV Sudbury, Ont.; CKCO Kitchener, Ont.; CFCF Montreal, Que.; CKY Winnipeg, Man.; CKTV Regina, Sask.; CFQC Saskatoon, Sask.; CFRN Edmonton; Alta.; CFCN Calgary, Alta.; BCTV Vancouver, B.C.; CHEK Victoria, B.C.; CIPA Prince Albert, Sask.; CHBX Sault Ste. Marie, Ont.; CHFD Thunder Bay, Ont.; CICC Yorkton, Sask.; CITL Lloydminster, Alta.; CJBN Kenora, Ont.; CHRO Pembroke, Ont.

LE RESEAU DE TELEVISION (TVA)

Addr: 2600 boul. de Maisonneuve, Montreal, PQ, H2L 4P2. **Te:** +1 (514) 526-9251.
Stations (ch no in brackets): CFCM Quebec (4), CJPM Chicoutimi (6), CHLT Sherbrooke (7), CHEM Trois-Rivières (8), CIMT Rivière d. Loup (9), CFTM Montreal (10), CFER Rimouski (11), CFEM Rouyn-Noranda (13), CHOT Hull/Ottawa (40), CHAU-TV Carleton (5).

TÉLÉVISION QUATRE SAISONS

Addr: 405 Oglivy Ave., Montreal, PQH3N 1M4. **Te:** +1 (514) 495-6884. **Fax:** +1 (514) 495-6231.
Stations: CFJP Montreal (35); CFGS Hull (49); CFRS Jonguiêre (4); CFAP Quebec (2); CFPC Rimouski (18); CFTF Rivière-du-Loup (29); CFKS Sherbrooke (30); CFKM Trois Rivières (16); CFVS Val d'Or (20/25).

Public TV Networks

SOCIETE DE RADIO-TELEVISION DU QUEBEC

Addr: 800 rue Fullum, Montreal, Quebec H2K 3L7. **Te:** +1 (514) 521-2424. **Telex:** 05-25808.
L.P: Dir. Gen: Ms. Françoise Bertrand. Head of Prgrs: Pierre Roy.
Affiliated Stations: CKRS Jonguière; CKRN Rouyn-Noranda; CKRT Rivière-du-Loup; CKSM Sherbrooke; CKTM Troit-Rivières.

TV ONTARIO

Ontario Educational Communications Authority.
Addr: Box 200, Station Q, Toronto, Ont. M4T 2T1.
Te: +1 (416) 484-2600. **Fax:** +1 (416) 484-7771. **Telex:** 06-23547.
L.P: Chmn. & Chief Exec. Officer: B. Ostry. Chief Operating Officer: P. Bowers. Man. Dir. French Prgr. Sces: Jacques Bensimon. Man. Dir. English Prgr. Sces: Don Duprey. MD. Ext. Rel: Bill Roberts/Judith Tobin.
Station: CICA-TV chA19 1080/108kW + low power relays.

GREENLAND

TV-sets: 21.000 — **Colour:** PAL — **System:** B.

KALAALLIT NUNAATA RADIOA (Gov.)

ADDR: Kalaallit Nunaata Radioa TV, P.O. Box 1007, DK-3900 Nuuk/Godthåb. **Te:** +299 25333. **Fax:** +299 25042.
L.P: Head of Prgr's: H.P. Møller Andersen. Head of Production: Niels Pavia Lynge. Head of News: Maliinannguaq M. Mølgaard.
Stations: The KNR TV prgr's are aired locally using band 3 channels with 1 to 20 watts tr. power.
D.Prgr: 2000-2100, 2225-0230. **N: Greenlandic:** 2230-2255 MF. **Danish:** 2300-2330 (Wed 2400): rebroadcasts from the Danish TV aired 3 days after the actual broadc. date in Denmark. Prgrs mainly originate from Radio Denmark TV, plus 10 hours a week from the Danish TV2 + local commercials.
Text-TV: 24 hrs.
F.Pl: Prgr's distributed via satellite from 1992.
Local TV organizations mainly distribute via closed networks, and include some satellite relays.

AFRTS (US Air Force)

Addr: OLA Det 1, AFABS, APO New York, NY 09121-5000—Thule Air Base, Det 1, AFABS, APO New York, NY 09023-5000.
Stations: Søndre Strømfjord chA8 0.1kW H; Thule chA8, chA13 0.1kW (using NTSC colour).
D.Prgr: 1330-0630 (Sat 1300-0700).

ST. PIERRE ET MIQUELON

TV-sets: 2.000 — **Colour:** SECAM — **System:** K.

SOCIETE NATIONALE DE RADIO TÉLÉVISION FRANÇAISE D'OUTRE MER (RFO)

Addr: & L.P: See radio.

Stations (Pol H)
1st Prgr: St. Pierre chK4 0.1kW, chK8 0.5kW, chK39 5W — Miquelon chK6 0.1kW.
2nd Prgr: St. Pierre chK31 0.5kW, chK55 0.05kW — Miquelon chK56 0.2kW.
D.Prgr: 1st Prgr: W 1815-0230, Sun 1730-0300. **2nd Prgr:** Mon-Fri 2115-0330, Sat 1730-0400, Sun 1830-0400.

USA

TV-sets: 215m — **Colour:** NTSC — **System:** M.

ABC TELEVISION DIVISION (Comm.)
Addr: 1330 Ave. of the Americas, New York, NY 10019. **Te:** +1 (212) 887-7777. **Fax:** +1 (212) 887-6850. **Telex:** 422003.
Owned stations: WABC-TV/New York — WLS-TV/Chicago — KGO-TV/San Francisco — KABC-TV/Los Angeles — KTRK-TV/Houston — KFSN-TV/Fresno, CA — WPVI-TV/Philadelphia — WTVD-TV/Durham, North Carolina. **Affiliates:** approx. 220.

CBS, Inc. (Comm.)
Addr: 51 West 52nd Str, New York, NY 10019. **Te:** +1 (212) 975-4321.
Owned Stations: WCBS-TV, New York; WBBM-TV, Chicago; KCBS-TV, Los Angeles; WCAU-TV, Philadelphia; KMOX-TV, St. Louis. **Affiliated Stations:** 200.

NATIONAL BROADCASTING COMPANY (Comm.)
Addr: 30 Rockefeller Plaza, New York, NY 10012. **Te:** +1 (212) 664-6606. **Fax:** +1 (212) 333-7546. **Telex:** 662131.
NBC Television Stations: WNBC-TV, New York; WRC-TV, Washington; WMAQ-TV, Chicago; WKYC-TV, Cleveland; KNBC-TV, Los Angeles — KCNC-TV, Denver; WTVJ-TV, Miami. **Affiliated stations:** 208.

THE FOX TELEVISION NETWORK
Addr: 205 E. 67th Str., New York, NY 10021. **Te:** +1 (212) 452-5555.

PUBLIC BROADCASTING SERVICE (Non-comm, Educ)
Addr: 1320 Braddock Place, Alexandria, VA 22314-1698. **Te:** +1 (703) 739-5000. **Fax:** +1 (703) 739-0775.
Member Stations: 314.

TV Marti (Gov.)
N.B: TV Marti is the Cuban television broadcasting service of the United States Information Agency, Bureau of Broadcasting (see Voice of America).
ADDR: Washington, D.C. 20547, USA. **Te:** +1 (202) 501-7210. **Fax:** +1 (202) 208-7808.
L.P.: Dir., Office of Cuba Broadcasting: Antoniao Navarro, Dir., TV Marti: Antonio Dieguez.
Stations: VHF ch13 (0830-1100).

All major U.S. cities have three network affiliates at least one PBS st. & several additional comm. st's.
Satellite — delivered tv services: see Satellite Section.

SEE SATELLITE SECTION FOR DETAILED AMERICAN SATELLITE INFORMATION

CENTRAL AMERICA & THE CARIBBEAN

ANTIGUA & BARBUDA

TV-sets: 28.000 — **Colour:** NTSC — **System:** M.

ANTIGUA & BARBUDA BROADC. SERVICE (Gov.)
Addr: see radio. **Te:** + 1 (809) 462-0010. **Fax:** +1 (809) 462-1622.
L.P: Prgr. Mgr. (TV): James Tanny Rose. CE: Denis Leandro. Film Buyer: J.T. Rose.
Stations: ABS-TV chA10 50/20kW H.
Relay: Montserrat ch13 10kW.
D.Prgr: 1900-0700.

CTV ENTERTAINMENT SYSTEMS (Comm.)
Addr: 25 Long Str, St. Johns. **Te:** + 1 (809) 462-0346. **Fax:** + 1 (809) 462-4211.
L.P: Prgr. Dir: J. Cox.

ARUBA

TV-sets: 19.000 — **Colour:** NTSC — **System:** M.

TELE ARUBA (Comm.)
Addr: P.O. Box 392, Oranjestad. **Te:** + 297 (8) 47302. **Fax:** + 297 (8) 41683. **Telex:** 5195 TELAR.
L.P. Gen. Mgr: Mrs. Jane Lampkin. CE: Miguel Roga.
Station: chA13 3/0.6kW H.
D.Prgr: 2030-0400.

BAHAMAS

TV-sets: 60.000 — **Colour:** NTSC — **System:** M.

BAHAMAS TELEVISION (owned and operated by the Broadc. Corp. of the Bahamas).
Addr: see radio. **Te:** +1 (809) 322 4623. **Fax:** +1 (809) 322 3924.
L.P: Ag. Prgr. Dir (TV): Ms. R. Simmons.
Station: ZNS ch13 50kW (ERP).
D.Prgr: 2200 (Sat 2100, Sun 2000)-0400.

BARBADOS

TV-sets: 69.350 — **Colour:** NTSC — **System:** M.

CARIBBEAN BROADCASTING CORP. (Gov, Comm.)
Addr: see radio. **Te:** +1 (809) 429 2041. **Fax:** +1 (809) 429 4795. **Telex:** 2560.
L.P: Prgr. Mgr (TV): Mrs. O. Cumberbatch.
Station: CBC-TV ch8 60/30kW, ch 9, ch 14, ch 18, ch 22, 26.
D.Prgr: Mon-Fri 0940-1400 and 2000-0330, Sat 1200-0430, Sun 1200-0330.
Cable TV: A Gov. Sce, STV, provides two additional subscription ch's.
N.B: Wednesday and Friday closing may be after 0400 as movies are shown on these days. During close times classified commercials and coming attractions are broadcast continuously.

BELIZE

TV-sets: 27.048 — **Colour:** NTSC — **System:** M.

TROPICAL VISION (Comm.)

Addr: 48 Albert Str, P.O. Box 89, Belize City. **Te:** +501 77246-8 — **Fax:** +501 (2) 75040 — **Station:** ch 7.

CTV (Comm.)

Addr: 27 Dayman Av., Belize City. **Te:** + 501 (2) 44400 — **Station:** ch 9.

COSTA RICA

TV-sets: 340.000 — **Colour:** NTSC — **System:** M.

Stations	chA	kW	Stations	chA	kW
2)	4	10	2)	9	10
1)	6	300	4)	13	2.5
3)	7	360	5)	15	0.1

Addresses & other information:
1) Corporación Costaricense de Televisión (Comm.), P.O. Box 2860, 1000 San José. **Te:** +506 312222 — Dir. Gen: M. Sotela B. — 1530-0700.
2) Multivision, Apt 4666, 1000 San José. **Te:** +506 334444. **Telex:** 3043 Televi. **Fax:** +506 211734 — Dir: Arnaldo Vargas V. — 1600-0600.
3) Televisora de Costa Rica, Apt 3876, San José. **Te:** +506 322222. **Telex:** 2220 Teletica — 1730-0600 (+ 9 repeaters).
4) Rede Nacional de Televísion, Apt 7-1980, San José. **Te:** +506 200071 — Dir. Gen: Dr. Ch. Zelaya Goodman — 2200-0400 (+ 2 repeaters).
5) Universidad de Costa Rica, San Pedro, Montes de Doa, San José — Dir. Gen: Dr. Sergio Guevara Fallas — 1600-2200. **Te:** +506 340463. **Fax:** +506 256950.

CUBA

TV-sets: 2.500.000 — **Colour:** NTSC — **System:** M.

INSTITUTO CUBANO DE RADIODIFUSION (Gov.)

Addr: Television Nacional, Calle M No. 313, Vedado, La Habana.

CANAL TELE REBELE/CUBAVISION

Addr: Mazón No. 52, Vedado, La Habana. **Te:** +537 (32) 3369. **Telex:** 511661.
L.P: VP: Gary Gonzalez.
Studios in Santiago de Cuba, Holguín & La Habana.
Stations: Pol H (no calls used).

ch	Location	kW	ch	Location	kW
2	Babiney, P.Río	8	11	Cienfuegos	8
2	La Habana	132	11	Salón, P.Río	32
4	Camagüey	53	12	Prov. Granmá	54
5	La Palma	8	12	Chivirico	56
5	Stgo. de Cuba	77.5	12	Ciego de Avila	56
5	Santa Clara	60	13	Pinar del Río	63
8	Holguín	27.5	13	Guantánamo	79
9	Baracoa	12	13	Prov. Matanzas	120
10	I. Juventud	100			

\+ 12 st's below 4kW.
D. Prgr: 1200-0500.

TELEVISION CUBANA (Canal 6) (Gov.)

Addr: Calle M No. 313, Vedado, La Habana.

ch	Location	kW	ch	Location	kW
2	Stgo. de Cuba	32	7	Salón, P.Río	32
3	La Palma	8	8	Ciego de Avila	51
3	Holguín	27.5	8	I. Juventud	100
3	Santa Clara	60	9	Pinar del Río	63
6	Babiney, P.Río	6	9	Prov. Matanzas	120
6	Camagüey	56	10	Prov. Granma	54
6	La Habana	129	10	Chivirico	56
7	Baracoa	12	11	Guantánamo	79
7	Cienfuegos	8			

\+ 12 st's below 4kW.
D.Prgr: 1200-0500.

AFRTS (US Navy)

Addr: US Naval Base, P.O. Box 22, FPO New York, NY 09406.
Station: Guantanamo Bay chA8 0.485kW.

DOMINICA (Commonwealth of)

TV-sets: 5.200 — **Colour:** NTSC — **System:** M.

DOMINICA BROADCASTING CORPORATION

Addr: P.O. Box 1, Victoria Street, Roseau. **Te:** +1 (809) 448-3282/3. **Fax:** +1 (809) 448-2918.

MARPIN-TV (Comm.)

Addr: P.O. Box 382, Roseau. **Te:** +1 (809) 448- 4107. **Fax:** +1 (809) 448- 2965.
L.P: Prgr. Mgr: Ron Abraham.

DOMINICAN REPUBLIC

TV-sets: 728.000 — **Colour:** NTSC — **System:** M.

	VHF Stations	chA			chA
1)	Sto. Domingo	2	7)	Sto. Domingo	11
2)	Cibao/Costa Norte	2	5)	Cibao/Costa Norte	11
3)	Sto. Domingo	4	3)	Barahona	12
3)	Cibao/Costa Norte	4	3)	Dajabon	12
3)	Alto Bandera	5	3)	Descubierta	12
4)	Sto. Domingo	6	3)	El Cercado	12
4)	Cibao/Costa Norte	6	3)	Enriquillo	12
5)	Sto. Domingo	7	3)	La Romana	12
6)	Santiago	7	3)	Pedernales	12
2)	Sto. Domingo	9	3)	Puerto Plata	12
2)	La Romana	9	8)	Sto. Domingo	13
7)	Cibao/Costa Norte	9	1)	Cibao/Costa Norte	13
	UHF Stations	**chA**			
5)	La Naviza	70			

N.B: Because of contradicting information, no transmitter power figures are given.

Addresses and other information
1) Teleantillas, Autopista Duarte Km.7, Sto. Domingo. **Te:** +1 (809) 567-7751.
2) Color Vision, Corporación Dominicana de Radio & TV, Emilio Morel esq. Lulu Perez, Ens. La Fé, Sto. Domingo. **Te:** +1 (809) 556-5876.
3) RadioTelevisión Dominicana, Dr. Tejeda Florentino 8, Sto. Domingo. **Te:** +1 (809) 689-2120.
4) Canal 6, Circuito Independencia C-A, Mariano Cestero esq. Enrique Henriquez, Sto. Domingo. **Te:** +1 (809) 689-8151.
5) Rahintel, Av. Indepencia, Centro de los Heroes, Sto. Domingo. **Te:** +1 (809) 532-2531.
6) Canal 7 Cibao, Edificio Banco Universal, Calle del Sol, Santiago. **Te:** +1 (809) 583-0421.
7) Telesistema, 27 de Febrero, Sto. Domingo. **Te:** +1 (809) 567-1251.
8) TV 13, Av. Pasteur esq., Santiago, Sto. Domingo. **Te:** +1 (809) 687-9161.

EL SALVADOR

TV-sets: 500.700 — **Colour:** NTSC — **System:** M.

	chA	kW (ERP)		chA	kW (ERP)		chA	kW (ERP)
1)	2	100/25	3)	6	150/30	4)	10	109/5
2)	4	75/37.5	4)	8	109/5			

Addresses & other information:
1) Canal Dos SA(Comm.), Ap. Postal 720, San Salvador. **Te:** +503 236744 — 1730-0530.
2) Canal Cuatro(Comm.), Ap. Postal 720, San Salvador. **Te:** +503 244633 — 2100(Sat/Sun 1500)-0600.

3) Canal Seis(Comm.), Km. 6, Carretera Panamericana a Santa Tecla, San Salvador. **Te:** +503 235122 — 2300 (Sat/Sun 1700)-0600.
4) Television Cultural Educativa, Ap. Postal 4, Santa Tecla. **Te:** +503 280499 — Dir. Gen: Maura Echaverria — 1300-0500.

GRENADA

TV-sets: 30.000 — **Colour:** NTSC — **System:** M.

GRENADA TELEVISION SERVICE, LTD.

Addr: P.O. Box 535, St. George's, Grenada. **Te:** +1 (809) 444 5521/22 (PBX). **Fax:** +1 (809) 444 5054.
Station: ch7 (4kW), ch11 (5kW).

GUADELOUPE

TV-sets: 150.000 — **Colour:** SECAM — **System:** K.

RFO-GUADELOUPE

Addr. & L.P: see radio.
Stations: Basse Terre chK5 2kW + low power repeaters.
D.Prgr: 1900-0230.

SAINT MARTIN

An RFO st. is operating on chK7 0.1kW relaying RFO Guadeloupe.

GUATEMALA

TV-sets: 475.000 — **Colour:** NTSC — **System:** M.

Stations: 2)Gov, others comm.

	Call	chA	kW/Pol		Call	chA	kW/Pol
1)	TGV-TV	3	240H	1)	TGV-TV	10	25H
2)	TGCE-TV	5	?	4)	TGMO-TV	11	316/63
3)	TGVG-TV	7	108/22H	5)	TGSS-TV	13	25/5

Addresses & other information:
1) Radio-Television Guatemala, Apt. 1367, Guatemala. **Te:** +502 (2) 922491. **Cable:** Teletenango — Pres: Lic. M. Kestler F. Dir. Tec: E. Sandoval — 1200-0600 (+ 2 repeaters).
2) Television Cultural Educativa, 4a Calle 18-38, Zona 1, Guatemala. **Te:** +502 (2) 531913 — 2200-0500.
3) Televisiete, Apt. Postal 1242, Guatemala. **Te:** +502 (2) 62216. **Cable:** TV siete — 1800-0600(+ 3 repeaters).
4) Teleonce, Ca. 20, 5-02, Zona 10, Guatemala. **Te:** +502 (2) 682165 — 1800-0600 (+ 2 repeaters).
5) Trecevision, 3a Calle 10-70, Zona 10, Guatemala. **Te:** +502 (2) 63266. **Telex:** 6070 Trece GU — 1800-0600 (+ 12 repeaters).

HAITI

TV-sets: 25.000 — **Colour:** NTSC — **System:** M.

TELEVISION NATIONALE D'HAITI (Gov., Cult.)

Addr: P.O. Box 13400 Delmas 33, Port-au-Prince. **Te:** +509 (1) 63324/64049/62202.
L.P: Dir: Mme. Jacqueline André.
Stations: Port-au-Prince chA8 0.3kW, chA10 5kW.
D.Prgr: Mon-Fri 2100-0400 in French & Creole.
F.Pl: chA12 Cap. Haïtien.

TÉLÉ HAITI S.A. (Comm.)

Addr: B.P. 1126, Port-au-Prince. **Te:** +509 (1) 23000.
L.P: Dir: Walter Bussenius.
Operates cable TV 24h in Port-au-Prince area.

HONDURAS (Rep.)

TV-sets: 160.000 — **Colour:** NTSC — **System:** M.

Stations (all comm.)

	Call	chA	kW		Call	chA	kW
1)	HRJS-TV	2	25	2)	HRTS-TV	7	10/2
2)	HRCV-TV	3	2/0.4	4)	HRLP-TV	7	2/0.4
4)	HRLP-TV	4	2/0.4	1)	HRJS-TV	9	25
3)	HRTG-TV	5	10/2	3)	HRTG-TV	9	5/1
3)	HRTG-TV	5	5/1	5)	HRNQ-TV	13	2/0.4
6)	HRGJ-TV	6	10/2				

Addresses & other information:
1) Corp. Centroamericana de Comunicaciones, S.A. de C.V, Apt. Postal 120, San Pedro Sula — Pres: J.J. Sikaffy. Exec. Vice Pres: F.J. Sikaffy. Dir. Tec: R. Beurket & A. Pinto — 1530-0400 (+ 2 relays).
2) Telesistema Hondureño, Apt. Postal 642, Tegucigalpa — 1530-0400 (+ 6 relays).
3) Compañia Televisora Hondureña, Apt. Postal 734, Tegucigalpa — 1230-0400 (+ 9 relays). **Te:** +504 (32) 7835. **Fax:** +504 (32) 0097.
4) Compañia Centroamericana de TV, Apt. Postal 68, Tegucigalpa — 1430-0400.
5) Cruceña de Televísíon, Casilla 3424, Tegucigalpa — Pres: Lic. Ivo Kuljis F. Dir. Gen: Lic. Walter Gasser Diaz C. — 1700-0800.
6) Compañia Broadcasting, Apt. Postal 882, Barrio Rio Piedras — 1130-0500.

JAMAICA

TV-sets: 484.000 — **Colour:** NTSC — **System:** M.

JAMAICA BROADCASTING CORPORATION (Comm.)

Addr: P.O. Box 100, Kingston 10. **Te:** +1 (809) 926-8683. **Fax:** +1 (809) 926-8669. **Telex:** 2218.
L.P: Prgr. Mgr: D. Bucknor.
Station: HWT ch 7, 8, 9, 10, 11, 12, 13.
D.Prgr: 2200(Sat 1930, Sun 1900)-0500.

MARTINIQUE

TV-sets: 65.000 — **Colour:** SECAM — **System:** K.

SOCIÉTÉ NATIONAL DE RADIO-TÉLÉVISION D'OUTRE MER (RFO)

Addr. and L.P: See radio.
Stations: Fort de France chK4 1kW (+ 9 relay st's). Pol: V.
D. Prgr: 2200 (Sun 2000)-0300.

MEXICO

TV-sets: 56.000.000 — **Colour:** NTSC. — **System:** M.

INSTITUTO MEXICANO DE TELEVISIÓN (IMEVISION) (State Television)

Addr: Periférico Sur 4121, Col. Fuentes del Pedregal, 14140 México, D.F. **Te:** +52 (5) 652-0973 — Dir.Gral: Lic. Jose Antono Alvarez Lima.
STATIONS: México, D.F; IMT(7), DF(13), IMT(22) — Chihuahua: CH(2) — Monterrey: FN(8).
Red Nacional 7: Aguascalientes: LGA(10), Cd. Obregón: BK(10), Culiacán: BL(13), Guadalajara: SFJ(11), Hermosillo: TH(10), Irapuato: CCG(7), Jalapa: IC(13), Matamoros: OR(14), Mérida: DH(11), México DF: IMT(7), Nogales: FA(2), Oaxaca: PSO(10), Tampico: WT(12) + 51 others.
Red Nacional 13: Aguascalientes: JCM(4), Cd. Obregón: CSO (6), Cuernavaca: CUR(13), Guadalajara: JAL(13), Hermosillo: HSS(4), Irapuato: MAS(12), Mérida: MEY(7), Mexicali: AQ(5), México DF: DF(13), Monterrey: WX(4), Oaxaca: IG(12), Puebla: TEM(12), Saltillo: SAO(11), S.L. Potosi: CLP(6), Tuxtla Gutiérrez: AO(4) + 39 others. ch7 & 13 are national, 11 & 22 regional.

TELEVISA, S.A. (Priv, Comm.)

Addr: Av. Chapultepec 28, 06724 México, D.F — **Te:** +52 (5) 709-3333. **Telex:** XEWTM 77-3154 — **Fax:** +52 (5) 709-3021 —

Chmn. of the Adm. Council: R. O'Farrill Jr. Pres. TV: Emilio Azcárraga.
STATIONS: XEW-TV(2), XHTV(4), XHGC(5), XEQ(9).
Cadena Canal 2: Acapulco: AP(2), Cd. Juárez: EPM(2), Cd. Obregón: BS(4), Córdoba/Orizaba: AH(8), Culiacán: BT(7), Chihuahua: FI(5), Guadalajara: EWO(2), Hermosillo: XEWH(6), Jalapa: AH(8), León: L(10), Los Mochis: BS(4), Mazatlán: OW(12), Matamoros: AB(7), Mérida: TP(9), México DF: XEW(2), Mexicali: BM(14), Monterrey: XHX(10), Morelia: KW(10), Nuevo Laredo: BR(11), Oaxaca: BN(7), Saltillo: AE(5), Tampico: XHD(4), Torreón: O(11), Tuxtla Gutiérrez: TX(8), Villahermosa: LL(13) + 28 others.
Cadena Canal 4: Córdoba/Orizaba: AI(10), Jalapa: AI(10), México DF: TV(4), Puebla: XEX-TV(7) + 2 others.
Cadena Canal 5: Aguascalientes: AG(13), Guadalajara: GA(9), Hermosillo: TH(10), Jalapa: AJ(5), México DF: GC(5), Monterrey: XET-TV(6), Saltillo: AD(7) + 3 others.
Cadena Canal 9: Mexico City.
Ch2 & 5 are national, 4 & 9 regional.
Part-time affiliates (after 2200) noted below.

STATIONS originating local programming.
Call: XH— (unless otherwise stated).
*) see Televisa above — **)see Imevision above.

No.	Call	ch	kW	No.	Call	ch	kW
1)	IA	2	30	**)	IMT	7	?
2)	CH	2	6	24)	GEM	7	1
*)	XEW-TV	2	100	25)	MZ	7	72
3)	AP	2	1.2	26)	GO	7	50
4)	KG	2	0.5	27)	FN	8	17.5
5)	FB	2	100	28)	US	8	1.6
6)	I	2	86.5	29)	TVL	9	250
7)	RIO	2	100	30)	FW	9	50
8)	FE	2	60	*)	Q	9	100
9)	FM	2	14.2	31)	K	10	8
10)	BC	3	100	32)	A	10	0.85
11)	PN	3	5.9	33)	L	10	150
12)	JMA	3	13	34)	IPN	11	300
13)	P	3	30	35)	WT	12	66
14)	Q	3	100	36)	MH	12	1
15)	Y	3	100	37)	ND	12	1
16)	LN	4	4	38)	AW	12	100
*)	TV	4	100	39)	CG	12	7
17)	AL	4	2	40)	AK	12	5
18)	G	4	88	41)	DE	13	50
19)	CC	5	3.5	42)	UT	13	2
20)	XEJ-TV	5	67.4	43)	ST	13	54
*)	GC	5	100	**)	IMT	22	?
21)	ETV	6	100	44)	S	23	18
22)	EDK	6	84	45)	IJ	44	330
23)	EWH	6	1.6				

MONTSERRAT

TV-sets: 1.625 — **Colour:** NTSC — **System:** M.

ANTILLES TV LIMITED (Comm.)

Addr: P.O. Box 342, Plymouth, Montserrat. **Te:** +1 (809) 491-2226. **Fax:** +1 (809) 491-4511.
L.P: Gen. Mgr: K. Osborne. Dir. Tec: Z.A. Joseph.
Station: Chance Pic chA7 (48kW towards Dominica, 3kW towards Antigua & St. Kitts).
D.Prgrs: 1000-1110, 2000-0400 (Fri/Sat 0700).

NETHERLANDS ANTILLES

TV-sets: 35.000 — **Colour:** NTSC — **System:** M.

TELE CURAÇAO (Gov, Comm.)

Addr: P.O. Box 415, Curaçao. **Te:** +599 (9) 61288. **Fax:** +599 (9) 614138.
L.P: Gen. Mgr: Norbert Hendrikse. Prgr. Mgr: H. van der Beist. Dir. Tec: J. Rufina.
Station: chA8 20/5kW H/A6 — **D.Prgr:** 2000-0345.
Cable TV: Telecuraçao also provides a cable service relaying various U.S. satellite networks and two Venezuelan channels.

LEEWARD BROADC. CORP. — TELEVISION

Addr: P.O. Box 375, Philipsburg, St. Maarten.
Station: chA7 5kW.
D.Prgr: 2030 (Sun 1800)-0300 (approx); Sat also 1200-1300.

A Gov. st. on chA6 is reported operating from Saba.

NICARAGUA

TV-sets: 210.000 — **Colour:** NTSC — **System:** M.

SISTEMA SANDINISTA DE TELEVISIÓN (Gov.)

Addr: Km 3½ Carretera Sur, Contig o Shell, Las Palmas, Managua. **Te:** +505 (2) 660028/660879. **Fax:** +505 (2) 662411. **Telex:** 1226 Sandino.
L.P: Hd. of Sales: Miguel Chivel.
Stations: chA2 25kW, chA6 25kW H, Chg 1 kW.
+ 7 low power repeaters.
D.Prgr: 2030-0600 (approx.)

PANAMA

TV-sets: 204.539 — **Colour:** NTSC — **System:** M.

	Location	chA	kW		Location	chA	kW
1)	Cerro Azul	2	650	6)	Fort Davis	10	5
2)	Cerro C. Chilibre	4	32	4)	Cerro Oscuro	11	110
3)	Cerro Azul	5	20	2)	El Valle	12	5
6)	Cerro Ancón	8	5	5)	Cerro Oscuro	13	10
1)	El Valle	9	130				

Addresses and other information
1) Televisora Nacional, Apt. 6-3092, El Dorado, Panamá — Dir. Gen: Lic. Alejandro Ayala V. — 1700-0500 (+6 relays).
2) RPC Television, Apt. 1795, Panamá 1 — 1330-0600 (+8 relays).
3) Panavision, Apt. 6-2605, El Dorado, Panamá (+ 1 relay).
4) Sistema de TV Educativa, Estafeta Universitaria, Universidad de Panamá — Dir. Gen: I. Velasquez de Cortes — Mon-Fri 2200-0400, Sat 1500-2200. **F.Pl:** 1 relay st at Ninguna in 1992.
5) Telemetro, P.O. Box 8-116, Panamá 8 — Dir. Gen: B. Marques — 1730-0530 (+ 1 relay).
6) Armed Forces Television (US Mil), Drawer 919, Bldg. 209, Fort Clayton — 24h.

PUERTO RICO

TV-sets: 830.000 — **Colour:** NTSC — **System:** M.

	Call	ch	kW		Call	ch	kW
1)	WKAQ-TV	2	53.7/10.5	9a)	WPRV-TV	13	170/17
2)	WIPM-TV	3	72.4/7.24	9b)	WSJU-TV	18	759
3)	WAPA-TV	4	53.7/8.1	9c)	WKPV	20	100/10
4)	WORA-TV	5	95.5/19.1	8b)	WJNX	2	?
5)	WIPR-TV	6	53.7/5.37	9d)	WSJN-TV	24	4384/490
6)	WLUZ-TV	7	166/22.4	12)	WMTY	40	—
7)	WSUR-TV	9	57.5/7.24	10)	WVEO-TV	44	993/198
8)	WLII	11	200/39.8	11)	WATX-TV	54	11.78/6.35
9)	WOLE-TV	12	316/31.6	11a)	WUJA	58	55/5.5

1) Telemundo of Puerto Rico Inc., 383 Roosevelt Av., Hato Rey, PR 00918 — GM: Jose Ramos. CE: Jose Medina. 1210-0500 (+ 8 relay sts).
2) Dept. of Education, Box 909, Hato Rey 00919 — 1800-0500 (Sat/Sun 1330-0400).
3) SFN Communications, Inc, GPO Box 2060, San Juan 00938.
4) Telecinco Inc, Box 43, Mayagüez 00708 — St. Mgr. & Film Buyer: E. Bado. Dir. Tec: G.A. Bonet. 1200-0400.
5) Avda. Hostos Hato Rey. Esquina Tous Urbanization Baldrich, 00919 Puerto Rico.
6) Ponce TV Corpo, Isabel Esq Montaner, Ponce 00731 — Pres: L.T. Muniz.
7) La Ramble, Ponce (belongs to 8).
8) American Colonial Broadc. Corp, Box S-4189, San Juan 00905. 1230-0430 — 8b) rel. 8.
9) Western Broadc. Corpo. of Puerto Rico, Box 1200, Mayagüez 00709 — 1245-0500 (partly rel. st. 3).

9a) No. 10 Simon Madeira Rio Piedras, 00929 — GM: Nacha Rivera.
9b) Three Star Telecast Inc, Box 18, Carolina 00628 — Dir. Gen.: Barakat Saleh. 1030-0400.
9c) Multi Media TV, Box 2556, San Juan 00936.
9d) belongs to 9c) 24h (mostly rel. st's 1 and 4). All News.
10) Siglares Iglesia Catolica Inc, Buzon C-339, Quebradillas 00742.
11) Arecibo Video Corp, Arecibo 00612. Pres: F. Velasquez.
11a) Community TV of Cagua, Box 6556, Caguas 00626.
12) Ana G. Mendez Foundation, Box 21345, Rio Piedras 00928 — GM: Gloria Hernandez. CE: Ariel Diaz.

AFRTS (US Mil.)
Addr: US Naval Station, Box 3029, FPO Miami 34051.
Stations: Roosevelt Roads chA40 1kW, Ft. Allen ch40 0.05kW, Viques Isl. ch56 0.05kW.

ST. KITTS & NEVIS

TV-sets: 9500 — **Colour:** NTSC — **System:** M.

ZIZ TELEVISION (Gov, Comm.)
Addr: P.O. Box 331, Basseterre, St. Kitts. **Te:** +1 (809) 465 2621. **Fax:** +1 (809) 465-5202.
L.P: Gen. Mgr: Mrs. Claudette Manchester. Acting Producer TV: Barry Thomas.
Stations: Basseterre chA5 20/5kW H. + 3 repeaters.
D.Prgr: 2000-0430.

ST. LUCIA

TV-sets: 25.000 — **Colour:** NTSC — **System:** M.

HELEN TV (Comm.)
Addr: P.O. Box 621, Le Morne Castries. **Te:** +1 (809) 4522 693. **Fax:** +1 (809) 454 1737. **Telex:** 6254 HTSTV.
L.P: MD: Linford Fevrier. CE: Stephenson Anius.
Station: ch's 4 & 11.

CABLEVISION
Addr: George Gordon Bld., Bridge Str. Castries, St. Lucia. P.O. Box 111, Castries. **Te:** (809) 452 3301. **Fax:** +1 (809) 453 2544. **Telex:** 6362.
Station: ch's 16 on CATV.

ST. VINCENT

TV-sets: 17.700 — **Colour:** NTSC — **System:** M.

ST. VINCENT & THE GRENADINES BROADC. CORPORATION Ltd.
Addr: P.O. Box 617, Kingstown. **Te:** +1 (809)-4561078. **Fax:** +1 (809) 4561015.
L.P: Chief Eng: R.P. MacLeish.
Station: ch9. Relay sts. ch's 7, 11, 13 varying between 5-30 W. 1500-0900.
Coverage area: St. Vincent, Grenadines, St. Lucia & Grenada.

TRINIDAD & TOBAGO

TV-sets: 250.000 — **Colour:** NTSC — **System:** M.

TRINIDAD & TOBAGO TELEVISION CO. LTD.
Addr: Television House. 11A Maraval Rd., P.O. Box 665, Port of Spain.
Te: +1 (809) 622-4141-4. **Fax:** +1 (809) 622-0344. **Cable:** Television Trinidad.
L.P: GM: Grenfell Kissoon. Chief Eng: D. Dhani.
Stations: chA2 12/6kW, chA13 2/0.2kW (ch's 9 & 14 since 1983).
D.Prgr: 0940-0400.

VIRGIN ISLANDS (American)

TV-sets: 31.500 — **Colour:** NTSC — **System:** M.

WBNB-TV
Addr: Box 1947, Charlotte Amalie, St. Thomas 00801.
Te: +1 (809) 774-0300. **Fax:** +1 (809) 776-3511.
L.P: Sen. Vice Pres: J. Potter. St. Mgr: P. Stull.
Station: WBNB-TV chA10 113/76kW.

CARIBBEAN COMM. CORP.
Addr: 1 Beltjen Place, St. Thomas, V1 00802. **Te:** +809 77621 50. **Fax:** +809 774 5029.
L.P: Hd. of Sales: Randolph H. Knight; MD: Andrea L. Martin.

VIRGIN ISLANDS PUBLIC TV-SYSTEM
Addr: Box 7879, Charlotte Amalie, St. Thomas 00801. **Te:** +1 (809) 774-6255. **Fax:** +1 (809) 774 7092.
L.P: TD: Leslie Hayes; Film Buyer: Lori Elskoe.
Station: WTJX-TV chA12 31.6/6.32kW.

VIRGIN ISLANDS (British)

TV-sets: 3.000 — **Colour:** NTSC — **System:** M.

TELEVISION WEST INDIES LTD. (Comm.)
Addr: P.O. Box 34, Broadcast Peak, Chawell, Tortola, BVI.
Te: +1 (809) 409 43332
Station: ZBTV (Tortola) chA5 30/3kW.

BVI CABLE TV
Addr: P.O. Box 694, Road Town, Tortola. **Te:** (809) 495 3205.
L.P: MD: Todd Klindworth.

SOUTH AMERICA

ARGENTINA

TV-sets: 7.165.000 — **Colour:** PAL — **System:** N.

ASOCIACIÓN DE TELERADIODIFUSORAS ARGENTINAS (ATA)
ADDR: Av. Córdoba 323, 6to., 1054 Buenos Aires. **Te:** +54 (1) 312-4208/4219/4533. **Fax:** +54 (1) 312-4208. **Telex:** 17253 ATA AR. **Cable:** Teleradio Baires — **L.P:** Pres: Alejandro Enrique Massot. CE: Enrique Parodi.

	Call	chA	kW (ERP)	City of location
1)	LS86	2	100/10	La Plata
2)	LT83	3	90/20	Rosario
3)	LU89	3	30/3	Santa Rosa
4)	LT84	5	36/6	Rosario
6)	LV84	6	1.5/0.15	San Rafael
5)	LU93	6	1.5/0.15	San Carlos de Bariloche
7)	LS82	7	212/20	Buenos Aires
8)	LU81	7	28/3	Bahía Blanca
9)	LU84	7	24/3	Neuquén
10)	LU90	7	30/3	Rawson
11)	LV89	7	45/5	Mendoza
12)	LW80	7	10/1	San Salvador de Jujuy
13)	LW81	7	28/3	Santiago del Estero
14)	LRI486	8	75/7.5	Mar del Plata
15)	LV82	8	10/1	San Juan
16)	LV85	8	18/1.8	Córdoba
17)	LRK458	8	5/0.5	San Miguel de Tucumán
18)	LS83	9	62/6	Buenos Aires
19)	LU80	9	43/4	Bahía Blanca
20)	LU83	9	6/0.6	Comodoro Rivadavia

21) LV83	9	52/5	Mendoza
22) LU85	9	1.5/0.15	Río Gallegos
23) LT81	9	1.5/0.15	Resistencia
24) LV91	9	5/0.5	La Rioja
25) LU82	10	295/30	Mar del Plata
26) LV80	10	80/8	Córdoba
27) LW83	10	6/0.6	San Miguel de Tucumán
28) LU92	10	50/5	General Roca
29) LRH450	10	22/2.2	Junín
30) LS84	11	180/20	Buenos Aires
31) LT88	11	230/20	Formosa
32) LU87	11	15/1.5	Ushuaia
33) LW82	11	11/1	Salta
34) LV81	12	170/20	Córdoba
35) LT85	12	20/2	Posadas
36) LU91	12	4/0.4	Trenque Lauquén
37) LS85	13	116/20	Buenos Aires
38) LT80	13	65/6	Corrientes
39) LT82	13	80/8	Santa Fé
40) LU88	13	15/1.5	Río Grande
41) LV86	13	5/0.5	Río Cuarto
42) LV90	13	12/1.2	San Luis

Associated to A.T.A.: 2), 4), 5), 8), 9), 12), 13), 14), 15), 16), 17), 19), 20), 21), 23), 25), 33), 36), 38), 39), 41).
Operated by State Administration: 3), 7), 10), 22), 24), 28), 29), 30), 31), 32), 34), 35), 37), 40), 42).
Addresses and other information:
1) Radiodifusora El Cármen S.A., Calle 27 e/530 y 531, 1900 La Plata. Pcia. Buenos Aires. (+5 relays).
2) Televisión Litoral S.A., Av. Godoy 8101., 2000 Rosario. Pcia. Santa Fé. (+ 5 relays).
3) Ruta 35 Km.322, Casilla Correo 139, 6300 Santa Rosa. Pcia. de La Pampa. (+7 relays).
4) Rader S.A., Av. Belgrano 1055, 2000 Rosario. Pcia. de Santa Fé. (+5 relays).
5) TV Río Diamante, Luzuriaga 360, 5600 San Rafael. Pcia. de Mendoza. (+3 relays).
6) Bariloche Televisión SRL., Elflein 251, 8400 San Carlos de Bariloche. Pcia. de Río Negro. (+5 relays).
7) Argentina Televisora Color S.A., Figueroa Alcorta 2977, 1425 Buenos Aires. Capital Federal. (+163 relays).
8) Telba S.A., Blandengues 225, 8000 Bahía Blanca. Pcia. de Buenos Aires. (+4 relays).
9) Neuquén TV S.A., Av. Argentina 1700, 8300 Neuquén. Pcia. de Neuquén. (+30 relays).
10) Av. Fontana 50, 9103 Rawson. Pcia. de Chubut. (+9 relays)
11) Garibaldi 7 Piso 5, 5500 Mendoza. Pcia. de Mendoza. (+8 relays).
12) R. Visión Jujuy S.A., Av. 19 de Abril 749, 4600 San Salvador de Jujuy. Pcia. de Jujuy. (+8 relays).
13) Pellegrini 345, 4200 Santiago del Estero. Pcia. de Santiago del Estero. (+7 relays).
14) Emisora Arenales de Radiodifusión S.A., Av. Luro 2907, 7600 Mar del Plata. Pcia. de Buenos Aires. (+ 8 relays).
15) Mitre 59 Oeste, 5400 San Juan. Pcia. de San Juan. (+5 relays)
16) DICOR S.A., Vélez Sarsfield 4300, 5000 Córdoba. Pcia. de Córdoba. (+16 relays)
17) Televisora Tucumana Color S.A., Av. Salta y Delfín Gallo. 4000, San Miguel de Tucumán. Pcia. de Tucumán.
18) Telearte S.A., México 990, 1097 Buenos Aires. Capital Federal.
19) Telenueva S.A., Sarmiento 64, 8000 Bahía Blanca. Pcia. de Buenos Aires. (+13 relays)
20) Comodoro Rivadavia TV, Rawson 1459, 9000 Comodoro Rivadavia. Pcia. de Chubut. (+7 relays).
21) Cuyo Televisión S.A., San Martín 1027, Galería Piazza, Local 10. 5500 Mendoza. Pcia. de Mendoza. (+5 relays)
22) H. Irigoyen 250, 9400 Río Gallegos. Pcia. de Santa Cruz. (+15 relays).
23) TV Resistencia S.A., Av. Alvear 50, 3500 Resistencia. Pcia. de Chaco. (+9 relays).
24) Av. Ortiz de Ocampo 1700, 5300 La Rioja. Pcia. de La Rioja. (+6 relays).
25) TV Mar del Plata S.A., Independencia 1163, 7600 Mar del Plata. Pcia. de Buenos Aires. (+4 relays).
26) SRTV S.A., Rivera Indarte 170, 5000 Córdoba. Pcia. de Córdoba. (+24 relays)
27) Televisora de Tucumán S.A., Av. Buenos Aires 296, 4000 San Miguel de Tucumán. Pcia. de Tucumán. (+2 relays).
28) Mitre y Sarmiento Piso 1, 8332 General Roca. Pcia. de Río Negro. (+10 relays).
29) Junín TV S.A., Belgrano 84, 6000 Junín. Pcia. de Buenos Aires. (+1 relay).
30) Telefé S.A., Pavón 2444, 1248 Buenos Aires. Capital Federal.
31) Tucumán 56, 3600 Formosa. Pcia. de Formosa. (+2 relays).
32) Magallanes 1310, 9410 Ushuaia. Tierra del Fuego. (+1 relay).
33) CorTe S.A., España 475, 4400 Salta. Pcia. de Salta. (+4 relays).
34) Telecor S.A., Av. Fader 3469, Cerro de las Rosas. 5000 Córdoba. Pcia. de Córdoba. (+21 relays).
35) Rioja 161, 3300 Posadas. Pcia. de Misiones. (+11 relays).
36) Av. Leandro N. Alem 351, 6400 Trenque Lauquén. Pcia. de Buenos Aires.
37) ArTeAr S.A., Cochabamba 1153, 1147 Buenos Aires. Capital Federal.
38) Río Paraná TV SRL., Calle 13 s/n., 3400 Corrientes. Pcia. de Corrientes. (+3 relays).
39) Televisora SantaFesiona S.A., Blvd. Gálvez 840, 3000 Santa Fé. Pcia. de Santa Fé. (+6 relays).
40) Alberti 739, 9420 Río Grande. Tierra del Fuego.
41) Imperio Televisión S.A., Alberdi 823, 5800 Río Cuarto. Pcia. de Córdoba. (+6 relays).
42) Colón 925, 5700 San Luis. Pcia. de San Luis. (+6 relays).

BOLIVIA

TV-sets: 50.000 — **Colour:** NTSC — **System:** M&N.

	Location	chA	kW(ERP)		Location	chA	kW(ERP)
1)	La Paz	2		13)	Cobija	9	
2)	Cochabamba	2		16)	Cochabamba	9	
3)	Santa Cruz	2		17)	Santa Cruz	9	
4)	Trinidad	2		18)	Sucre	9	
4a)	Oruro	3		20)	Potosi	9	0.1
5)	La Paz	4	5	1)	Oruro	10	
5a)	Cochabamba	4		19)	La Paz	11	
6)	Santa Cruz	4	8	13)	Tarija	11	
13)	Trinidad	4		20)	Cochabamba	11	1
7)	Santa Cruz	5	2	20)	Santa Cruz	11	10
8)	Oruro	5		13)	Potosi	11	
7)	Trinidad	5	2	20)	Trinidad	11	0.1
10)	La Paz	6		20)	Sucre	12	0.1
11)	Cochabamba	6		1)	Trinidad	12	
10)	Oruro	6		20)	La Paz	13	10
1)	Cochabamba	7		7)	Cochabamba	13	5
1)	Sucre	7		22)	Santa Cruz	13	2
12)	La Paz	7		20)	Oruro	13	2
1)	Potosi	7		20a)	Trinidad	13	
1)	Tarija	7		23)	La Paz	15	
13)	Oruro	8		24)	Cochabamba	15	
14)	Trinidad	8		24a)	La Paz	15	
20)	Tarija	8	1	25)	Santa Cruz	15	1
15)	La Paz	9		26)	La Paz	20	

Addresses & other information
1) Television Boliviano, Cas. 4837, La Paz.
2) CCA-TV, Av. Heroinas 467, Cochabamba. 1500-0430.
3) Cristal de TV, Cas. 4399, Santa Cruz.
4) Benivision, Cas. 54, Trinidad.
4a) TV O Ltda, Cas 631, Oruro.
5) TV-Popular, Calle Juan de la Riva 1527, Casilla #8704, La Paz — 1200-0600 (+76 relays).
5a) Casilla 4573, Cochabamba.
6) Galavision, Cas. 495, Santa Cruz — 1000-0600.
7) TV-Universo, Av. Circunvalación, Santa Cruz — 24h.
8) Orureña TV, Cas. 14, Oruro.
10) America TV, Cas. 10076, La Paz. 24h.
11) TV Cochabamba, Cas. 1009, Cochabamba.
12) Empresa Nacional de TV, Cas. 900, La Paz. 24h.
13) Illimani de Comunicaciones, Oruro.
14) Trinivision, Cas. 333, Trinidad.
15) Rede ATB, Av. 6 de Agosto 2972, La Paz.
16) Cochabamba TV, Cas. 4545, Cochabamba.
17) Teleoriente, Cordillera 550, Santa Cruz.

18) Chuquisaqueña TV, Cas. 187, Sucre.
19) Teleandina, Cas. 1665, La Paz.
20) TV Universitaria, Cas. 21982, La Paz. 1900-0300.
20a) F.Pl. (Trinivision Ltda).
22) Cruceña TV, Cas. 3424, Santa Cruz.
23) Sonomac/Tricolor, Av. 16 de Julio 1810, La Paz.
24) TV Integral, Cochabamba.
24a) SONOMAC, Cas. 21375, La Paz.
25) Grigota TV, Santa Cruz.
26) Tecnitron, Cas. 4410, La Paz.

BRAZIL

TV-sets: 30.000.000 — **Colour:** PAL — **System:** M.
All st's comm. exc. where indicated.

ASSOCIACÃO BRASILEIRA DE EMISSORAS DE RADIO E TELEVISÃO (ABERT).

Addr: Hotel Nacional, s/5 e 7, C.P. 040-280, 70322 Brasilia, DF. **Te:** +55 (61) 2238113.
SEARA — Serviços Associados de Rádio Ltda, Rua do Livramento 189, 20021 Rio de Janeiro, RJ. **Te:** +55 (61) 2432225.
STATIONS (relay st's omitted):

	ch	kW	Name and location
1)	2	13.17	TV Educativa, Manaus
2)	2	70/7	TV Guaiba, Porto Alegre
3)	2	100/10	TV Anhanguera, Goiânia
5)	2	78.25/8	TV Jornal do Comercio, Recife
6a)	2	6	TV Globo, Bauru
7)	2	57	TV Educativa, Rio de Janeiro.
8)	2	282/60	TV Cultura, São Paulo
8)	2	282	TV Geradora
9)	2	10	RBS TV, Erexim
9a)	2	120	TV Manchete, Fortaleza
10)	2	42	TV Educativa, São Luiz
11)	2	37.8	TV Educativa da Bahia, Salvador
12)	2	10.6	TV Educativa Espirito Santo, Vitória
63)	2	1	TV Liberal Marabé
12a)	2	8.5	TV Curitiba, Curitiba
13)	3	30.4	TV Nacional de Brasilia
6j)	3	10	TV Coroados, Londrina
16)	3	50	RBS TV, Blumenau
18)	3	1.6	TV Studios Rio de Janeiro, Nova Friburgo
108)	3	10	RBS TV, Cruz Alta
108)	4	5	Santa Angelo
19)	4	7.2	TV À Critica, Manaus
20)	4	100/10*	TV Aratu, Salvador
21)	4	10.31	TV Gazeta, Vitória
22)	4	10/1*	TV Record, Franca
6)	4	100	TV Globo, Rio de Janeiro
23)	4	14.77	TV Goyá, Goiânia
106)	4	3.16	TV Paranaiba-Prata
24)	4	1.8	TV Montes Claros, Montes Claros
25)	4	12.3	TV Difusora do Maranhão, São Luis
27)	4	120	TV Manchete, Belo Horizonte
29)	4	15.6	TV Guajará, Belém
30)	4	10	TV Iguaçu, Curitiba
31)	4	3.1	TV Tapajós, Santarém
32)	4	6.7	TV Radio Clube Teresina, Teresina
33)	4	15	RBS TV, Pelotas
34)	4	30/6	TV Rio Grandense, Porto Alegre
38a)	4	116*	TVS, São Paulo
39)	4	0.5	TV Roraima
39)	4	4	TV Rondônia
39)	4	0.5	TV Acre
92a)	4	100	TV Santa Cruz
85)	4	1	TV O Estado Florianopolis
88)	4	—	TV Bandeirantes, Brasilia
4)	5	100	TV Alterosa, Belo Horizonte
39)	5	13	TV Amazonas, Manaus
40)	5	2/0.5	TV Universitaria, Natal
40a)	5	5	TV Educativa, Teresina
41)	5	100/10*	TV Itapoan, Salvador
42)	5	10	RBS TV, Joinville
43)	5	55	TV Educativa, Fortaleza
6f)	5	2	TV Globo, Juiz de Fora
46)	5	38/1	TV Alagoas, Maceio
6b)	5	75	TV Globo, São Paulo
38b)	5	100*	TVS, Porto Alegre
38c)	5	100*	Sistema B de TV, Belem
92a)	5	100	TV Sudoeste
39)	6	1	TV Amapá
47)	6	24	TV Brasilia, Brasilia
48)	6	8	TV Morena, Campo Grande
49)	6	12/2	TV Parana, Curitiba
17)	6	30/3*	TV Cruz Alta, Cruz Alta
50)	6	70*	TV Tarobá, Cascavel
96)	6	20	RBS TV, Santa Cruz (F.Pl.)
51)	6	8.5	TV Cultura, Florianópolis
52)	6	400	TV Manchete, Rio de Janeiro
52a)	6	180	TV Manchete, Recife
53)	6	10	RBS TV, Bagé
54)	6	14	TV Ribamar, São Luiz
54a)	6	27.6	TV Vitoria, Vitoria
55)	7	189/18.9*	TV Bandeirantes, Salvador
56)	7	14/1.2	TV Tropical, Londrina
57)	7	20	TV Esplanada, Ponta Grossa
58)	7	31.6*	TV Sudoeste do Parana, Patc Branco
59)	7	316	TV Record, São Paulo
88)	7	316/31.6	TV Bandeirantes, Belo Horizonte
6h)	7	5	TV Ribeirão Preto, Ribeirão Preto
62)	7	12	TV Gazeta de Alagoas, Maceió
63)	7	60	TV Liberal, Belém
64)	7	57.48	TV Bandeirantes, Rio de Janeiro
65)	7	100	TV Educativa, Porto Alegre
66)	7	40	RBS TV, Passo Fundo
66a)	7	7.7	TV Uberaba, Uberaba
99)	7	?	TV Independência, Curitiba
67)	8	15.8	TV Capital, Brasilia
68)	8	3.16/0.36*	TV Sul Fluminense, Barra Mansa
70)	8	9.6	TV Brasil Oeste, Cuiabá
71)	8	5	TV Campo Grande
72)	8	31.6	RBS TV, Caxias do Sul
73)	8	20/4	TV Uberlândia, Uberlândia
74)	8	80 /8*	TV Record, São José do Rio Preto
75)	8	9.2	TV Ajuricaba, Manaus
76)	8	87.9	TV Atalaia, Aracaju
77)	8	8.91	TV de Fortaleza, Fortaleza
6k)	8	5	TV Cultura, Maringá
108)	9	5	Santa Rosa
79)	9	2/0.5	TV Borborema, Campina Grande
80)	9	31.6	RBS TV, Rio Grande
81)	9	21.6	TV Educativa, Belo Horizonte
82)	9	76	TV Morado do Sol, Araraquara
83)	9	1.32	TV Eldorado, Criciuma
106)	9	316	TV Paranaiba—Frutal
83a)	9	320	TV Manchete, São Paulo, SP
83b)	9	16.2	TV Record, Rio de Janeiro
84)	10	138	TV Verde Mares, Fortaleza
85)	10	2	TV O Estado Chapeco
85)	10	36	TV Planalto, Lages
106)	10	3.16	TV Paranaiba—Araxa
106)	10	31.6	TV Paranaiba—Uberlândia
6c)	10	47	TV Globo Capital, Brasilia
87)	10	30/6	TV Paranaiba, Uberlândia
87a)	10	?	TV Carima, Cascavel
88)	10	126	TV Bandeirantes, Porto Alegre
88)	10	8.8	TV Bandeirantes, Pres. Prudente
88b)	11	70	TVS, Rio de Janeiro
63)	11	2	TV Liberal—Castanhal
105)	11	10	TVU
90)	11	5.6	TV Tibagi, Apucarana
91)	11	5/1.5	TV Universitaria, Recife
92)	11	10	TV Gazeta, São Paulo
92a)	11	316	TV Bahia, Salvador
108)	11	1	São Luiz Gonzaga
6d)	12	35	TV Globo, Belo Horizonte
6i)	12	25	TV Paranaense, Curitiba
95)	12	5	TV Norte Fluminense, Campos
96)	12	520	RBS TV, Porto Alegre
97)	12	31.6	RBS TV, Santa Maria
6g)	12	1	TV Campinas, Campinas
99)	12	13	TV Independência, Cornelio Procopio
99)	12	?	TV Independência, Maringá
100)	12	31.6	TV Cultura, Chapeco
101)	12	316	RBS TV, Florianópolis

38e)	12	10	TVS, Belem
105)	12	7.5	Televisão Imembui
102)	13	200	TV Bandeirantes, São Paulo
6e)	13	330	TV Globo, Recife
103)	13	80	TV Brasil Central, Goiânia
104)	13	31.6	RBS TV, Uruguaiana

*) ERP.

Addresses and other information:

1) Rua Major Gabriel s/n, 69000 Manaus, AM — 2200-0400.
2) Rua Caldas Jr. 219, 90000 Porto Alegre, RS — 1200-0400.
3) Rua Thomaz Edson, Qd.7, 74000 Goiânia, GO—0900-0600.
4) Av. A. Chateaubriand 499, 30150 Belo Horizonte, MG — 1100-0600.
5) Rua do Lima 250, Recife, PE — 1200-0430.
6) Rua Lopez Quintas 303 (Studio: Rua Von Martins), Jardim Botanico, 22463 Rio de Janeiro, RJ. Dir. Tec: A. Pontes Malta. 0900-0600.
6a) Rua Padre Anchieta 941, 17100 Bauru, SP.
6b) Alameda Santos 1893, 01419 São Paulo, SP — 0930-0300.
6c) SCS Q2, Bl. B. no 81, Edif. Bradesco, 70300 Brasilia, DF.
6d) Rua Rio de Janeiro 1279, 30000 Belo Horizonte. 1200-0300.
6e) Av. Dantes Barreto 1186, 50000 Recife, PE.
6f) Rua Ewbank da Camara 46, 36100 Juiz de Fora, MG — 0930-0530.
6g,h) Rua Javari 3099, 14110 Ribeirão Preto, SP.— Dir. Gen: A.C. Coutino Nogueira. Dir. Tec: A. João Filho. 0900-0600.
6i) Av. Batel 1393, 80000 Curitiba, PR — 1430-0500.
6j) Av. Tirandentes 1370, 86100 Londrina, PR — 2000-0300.
6k) Rua Sta. Joaquina de Vedruna 625, 87100 Maringá, PR.
7) Av. Gomes Freire 474-B, Rio de Janeiro, RJ.— 0930-0345.
8) Rua Cenno Sbrighi 378, Agua Branca, 05099 São Paulo, SP — St. Mgr: R. Muylaert. Dir. Tec: J. Munhoz — 1100-0300.
9) Rua Soledade 277, Erexim, RGS.
9a) Av. Antonio Sales Esq, 60000 Fortaleza — 1100-0500.
10) Av. Kennedy s/n, São Luis, MA — 1000-0300.
11) Rua Pedro Gama 413/E, 40000 Salvador, BA — 1330-0300.
12) Rua Pedro Palacio 99, 29000 Vitoria, ES.
12a) Rua Francisco Caron 29, Curitiba, PR.
13) Praça 31 de Marco s/n, 70000 Brasilia — 1400-0300.
16) Rua Getulio Vargas 32, 89100 Blumenau, SC.
17) Rua Jango Vidal 427, 98100 Cruz Alta, RS — 1000-0500.
18) Rua E. Brasilio 30, 28600 Nova Friburgo.
19) Estrada do Aleixo Km 3, 69060 Manaus, AM — 0900-0400.
20) Rua Pedro Gama 31, 40230 Salvador, BA — 0900-0500.
21) CP. 1070, 29050 Vitória, ES — 0900-0600.
22) Rua José Maria Medeiros 5120, 14400 Franca, SP — 1100-0500. **Te:** +55 (16) 727 0400.
23) Av. Goiás 187, 74000 Goiânia, GO.
24) Praça dos Morrinhos S/N, Morrinhos. 0930-0530 (Sat/Sun 24h).
25) Av. Camboa do Maio 120, 65000 São Luis, Maranhão — 1000-0600.
29) Av. Governador José Malcher 1332, Belém, PR — 1400-0400.
30) Rua João Tscharnell 800, Jardim Merces, 80000 Curitiba, PR.
31) Av. Ismael Aranju 266, 68100 Santarem, PA.
32) C.P. 209, 64000 Teresina, PI—Pres: S.F. de Alénear. Dir Tec: H.P. de Carvalho. 0900-0600.
33) Rua 15 de Novembro 612, Pelotas, RGS. 1200-0600.
34) Rua Orfanotrôfio 711, Alto Teresópolis, 90000 Porto Alegre, RS — 1000-0500.
35) Rua Alto do Morro, 49000 Aracaju, SE.
38a-e) Rua Dona Santa Velozo 535, 02050 São Paulo — 1000-0500 (on all st's).
39) Ave. Carvalho Leal 1270, Cachoeirinha, Manaus, AM — 0930-0500.
40) Rua Princesa Isabel 758, 59020 Natal, RN — Dir: C.A. Martins. Dir. Tec: R. de Andrade Martins — 1000-0300.
40a) Ave. Prof. V. Alencar s/n, 64065 Teresina, PI — Dir. Tec: F.J. de Paiva Ribeiro — 1000-0400.
41) Rua Ferreira Santos 5, 40000 Salvador, BA — Dir: A. Moraes. 0930-0430.
42) Rua Saguaçu s/n, Joinville, SC.
43) Rua Oswaldo Cruz, 1985, 60000 Fortaleza, CE — 1000-0200.
46) Rua Cel. Paranhos 305, Jacintinho, 57000 Maceió, AL.
47) Av. W-3, Setor de Radio e TV, Brasilia — 1300-0500.
48) Av. Eduardo E. Zahran s/n, Campo Grande, MS — 2030-0300.
49) C.P. 7061, 80000 Curitiba, PR — 1000-0500.
50) C.P. 1169, 85800 Cascavel, PR — 1030-0600.
51) Alto do Morro Antão, Florianópolis, SC.
52) Rua do Russel 766-804, 20000 Rio de Janeiro — Dir. Gen: R. Furtado. Dir. Tec: F. Cavalcanti — 1100-0500.
52a) 1100-0500.
53) Rua do Acampamento 2550, 96400 Bage, RGS.
54) Praça Bom Menino s/n, São Luiz, Maranhão.
54a) Av. Presidente Florentino Avidos, 350 7° anolar Vitoria ES.
55) Largo do Candomblé 19A, Salvador, B.A — Dir. Gen: F.H. Chagas. Dir. Tec: R. Blum. 0900-0500.
56) Rodovia Celso Garcia, Londrina, PR — 1300-0500.
57) Rua João França Silva 2885, Ponta Grossa, PR. 1030-0300.
58) C.P. 591, 85500 Pato Branco, PR — Dir: V. Hillesheim. Dir. Tec: L. Schmitz — 1000-0300.
59) Av. Miruna 713, Aeroporto, 01000 São Paulo, SP — 1100-0500. **Te:** +55 (11) 542 9000.
59a) Campo de São Cristovão 105, Rio de Janeiro, RJ.
60) Rua Tomé de Souza 1251, 30000 Belo Horizonte, MG — Dir. Gen: M. Pereira Leite. Dir. Tec: O. Dominco Dalip. Film Buyer: R. Hachich Maluf. 1200-0600.
62) Av. Aristeu de Andrade s/n, Maceió, AL — 1830-0330.
63) Av. Nazaré, 350 Belem Pará, PA — 1200-0300.
64) Rua Alvaro Ramos 492, Bota fogo, Rio de Janeiro, RJ — 1230-0300.
65) Rua Correa Lima 2118, Morro Santa Tereza, 90640 Porto Alegre, RS — St. Mgr: A.C. Fedrizzi — 1145-0330.
66) Rua Princesa Isabel, 99100 Passo Fundo, RGS.
66a) Rua General Osorio 755, 38100 Uberaba, MG — 2000-0400.
67) Torre de Televisão de Brasília, Box 1, Brasilia, DF — 1440-0200.
68) C.P. 85919, 27400 Barra Mansa, RJ—Dir: O.M. Nora. Dir. Tec: C. Pina. 1000-0500.
70) Rua Giboia, s/n°, Bairro Concil, 78000 Cuiaba, MT—1030-0600.
71) Av. Calogeras 315, 79300 Campo Grande, MT — 0600-0200.
72) Av. Rio Grande do Sul, Caxias do Sul, RS — 1530-1730, 1900-0500.
73) Rua R.G. do Norte 1069, 38400 Uberlândia, MG.
74) Via Washington Luiz, km 436, 15100 São José do Rio Preto, SP — Dir: P.M. de Carvalho F. Dir. Tec: C. Victor Donato. 1100-0500.
75) Rua O.G. 18 Santo Antonio, 69000 Manaus.
76) Rua Claudio Batista 122 (ex Cláudio Batista S/N), 49045 Aracaju, SE.
77) Av. Desembargador Moreira 2565, Fortaleza, CE. 1300-0400.
79) Rua Venâncio Neiva 287, 2° andar, 58100 Campina Grande, PB — 1355-0400.
80) Rua Duque de Caxias 63, 1° e 2° andares, Rio Grande, RGS.
81) Av. Assis Chateaubriand 167, 30150 Belo Horizonte, MG — 0900-0500.
82) Praça José Palmores Lepre 99, 14800 Araraquara, SP — 24h.
83) Rua Silva Jardim 216, Cricuma, SC.
83a) Rua Bruxelas 193 Sumare, 01000 São Paulo, S.P.
83b) Rua G. Padilha 144, 20000 Rio de Janeiro.
84) Av. Desembargador Moreira 2430, Fortaleza, CE. 1405-0220.
85) Rua Carlos Jofre do Amaral 67, 88500 Lages, SC.
87) C.P. 210, 38400 Uberlândia, MG — St. Mgr: A. de Castro jun. Dir. Tec: M.J. Rodrigues dos Reis — 0900-0500.
87a) Av. Barao do Rio Branco, 1960, Sao Paulo.
88) Rua Delfino Riet 183, C.P. 1474, Porto Alegre, RS — 1300-0400.
88) Rua Radiantes 13, Sao Paulo, Brazil 05699.
90) Av. Santos Dumont 11, 86800 Apucarana, PR — 1200-0400.
91) Av. Norte 68, Santo Amaro, Recife, PE. 1900-0300.
92) Av. Paulista 900, São Paulo, SP. 1500 (Sat/Sun 1200)-0300.
92a) Estrada de São Lázaro 540, Federação, Salvador-BA, Brazil.
95) Av. 24 de Outubro 201, 28100 Campos, RJ — 0945-0530.
96) Rua Radio y TV Gaucha 189, 90650 Porto Alegre, RS — 0900-0500.
97) Av. 2 de Novembro s/n, 97100 Santa Maria, RS.
99) SSC — Sistema Sul de Comunicacão, R. Amaury Lange Silvério, 450 Pilarzinho, 82000 Curitiba, PR.
100) Estrada de Seara Km 3, 89800 Chapeco, SC.
101) Morro da Cruz, 88000 Florianópolis, SC.
102) Rua Radiantes 13, 05699 São Paulo, SP — 1300-0400.
103) Rua 201 no. 430, Vila Nova, Goiânia, GO.

104) Rua Domingos de Almeida 1722, 97500 Uruguaiana, RGS. 24h.
105) Avenida Norte, 68 Sante Amaro, Recife, Pernambuco.
106) Av. Prof. José Ignácio de Souza, 2710, 38400, Uberlandia.
107) Av. Mauricio Sirotsky Sobrinho, 25.
108) Caixa Postal 324, 98100, Cruz Alta.

CHILE

TV-sets: 2m — **Colour:** NTSC — **System:** M.

TELEVISION NACIONAL DE CHILE (Gov.)

Addr: Bellavista 0990 (P.O. Box 16104, Correo 9), Santiago. **Te:** +56 774552. **Telex:** 241375 TVNCH CL. **Cable:** TV Chile. **Fax:** +56 353000. — Dir. Gen: Jorge Navarrete Martinez, GM: Bartolome Dezerega Salgado, CE: Jaime Sancho Martinez.

Stations Location	ch	kW (ERP)	Stations Location	ch	kW (ERP)
El Roble	2	175	Pto. Williams	8	1
Valdivia	3	20	Ancud	8	2
Puerto Montt	4	20	Santiago II	9	40
Coquimbo	4	20	Iquique	10	30
San Fernando	5	25	Castro	10	50
Antofagasta	6	20	Salvador	10	4
Osorno	6	20	Talca	10	20
Punta Arenas	6	20	Ovalle	10	3
Cayumanqui	6	175	Vallenar	12	5
Santiago	7	340	Valparaíso	12	100
Temuco	7	115	San Antonio	12	10
Copiapo	7	4			

+ 96 low power repeaters less than 1 kW.
D.Prgr: 1200 (Sat/Sun 1400)-0500.

CORPORACION DE TELEVISION UNIVERSIDAD DE CHILE (Comm.)

Addr: Ines Matte Urrejola, 0825, Santiago. **Te:** +56 (2) 377851. **Telex:** 340 492 TVCH. **Fax:** 56 (2) 377923.
Stations: chA11 60/30kW (+ relay st. at Valparaíso ch10).
D.Prgr: 2145-0430.

CORPORACION DE TELEVISION UNIVERSIDAD CATOLICA DE CHILE

Addr: Inés Matte Urrejola 0848, Santiago. **Te:** +56 (2) 514000. **Telex:** 440182 TRECE CZ. **Fax:** +56 (2) 377044.

Location	ch	kW	Location	ch	kW
Arica	8	21.7	Talca	8	27
Iquique	8	28.8	Chillán	13	245
María Elena	11	20	Concepción	5	56
Chuquicamata	12	20	Angol	10	1.2
Antofagasta	13	41.6	Victoria	10	1.2
Copiapó	11	9.5	Temuco	4	45
La Serena	13	21.6	Villarrica	9	2
Ovalle	5	2	Valdivia	12	45
San Felipe	7	21	Osorno	9	47
Valparaíso	8	137	Puerto Montt	13	66.4
Santiago	13	220	Ancud	5	7
San Antonio	10	21.7	Castro	12	73.8
San Fernando	5	37			

+ 25 low power tr's less than 1 kW.

RED DE RADIOTELEVISION DE LA UNIVERSIDAD DEL NORTE (TELENORTE)

Addr: Carrera 1625, Antofagasta. **Te:** +56 222496 — **L.P:** GM: Juan Carlos Salas Floras.
Stations: ch3 Antofagasta, ch11 Arica, ch12 Iquique +7 low power sts.
D. Prgr: 1735-0435.

UNIVERSIDAD CATOLICA DE VALPARAÍSO

Addr: Agua Santo Alto 2455 (Casilla 4059), Viña del Mar. **Te:** +56 (32) 611675. **Fax:** +56 (32) 610505.
Stations: chA4 (Valparaíso), chA5 (Santiago), ch7 (Puerto Montt), chA8 (La Serena).

COLOMBIA

TV-sets: 5.500.000 — **Colour:** NTSC — **System:** M.

INSTITUTO NACIONAL DE RADIO Y TELEVISION (INRAVISION)

Addr: Centro Administrativo Nacional, Via Eldorado, Bogota. **Te:** +57 (1) 2220700. **Fax:** +57 (1) 222 1426. **Telex:** 43311 INRACO.
Inravision rents airtime to 26 comm. companies. The three largest are:

Caracol

Addr: Calle 19, 3-16 Bogota. **Te:** +57 (1) 282 2088. **Fax:** +57 (1) 286 3160. **Telex:** 41480 — Head of Prgrs: Rut Echeverdi.
Key st: Manjui ch7.

Punch

Addr: Carrera 28, 49-98 Bogota. **Te:** +57 (1) 2174750.
Key st: Manjui ch9.

RTI

Addr: Calle 19 N 4-56 Piso 2, Bogota. **Te:** +57 (1) 282 7700. **Fax:** +57 (1) 284 9012. **Telex:** 43294 — Head of Prgrs: Patricio Wills.
Key st: Manjui ch11.

Station/Area	Netw I ch	Netw I kW (ERP)	Netw II ch	Netw II kW (ERP)
Manjui (1)	7	668	9	668
La Rusia (2)	10	54	8	54
C. Oriente (3)	8	336	3	336
Tasajero (4)	5	150	11	100
Jurisdicciones (5)	6	468	4	372
Saboya (6)	13	27	11	20
Chigorodo (7)	7	15	11	15
El Ruiz (8)	12	54	10	54
Padre Amaya (9)	3	152	13	252
Galeras (10)	12	393	9	393
Planadas (11)	4	16	2	125
Munchique (12)	10	54	3	54
Paramo de Dominguez (13)	8	426	13	426
La Popa (14)	7	10	9	10
La Pita (15)	10	?	2	?
Monteria (16)	7	27	9	25
Cerro Kennedy (17)	13	700	11	700
El Alguacil (18)	7	270	12	270
Arauca (19)	11	1.2		
Gabinete (20)	6	10	4	10
Buenavista (21)	9	5	7	5
Leticia (22)	8	54	13	54
Nieva (23)	12	54	10	54
Alto del Tigre (24)	12	16	2	10
San Andres (25)	11	27	9	15

+ 7 low power stations not mentioned.
Areas (Departments) served: 1) Bogota (Sabana); Tolima, Huila; 2) Boyaca, Santander (South); 3) Santander (North), Arauca; 4) Santander (North & Central); 5) Santander, Magdalena (South); 6) Boyaco; 7) Uraba; 8) Manizales, Caldas; 9) Medellin, Antioquia (Central); 10) Nariño (Central); 11) Quindio; 12) Popayan; 13) Cauca & Cauca Valley; 14) Cartagena, Bolivar (North); 15) Bolivar, Sucre; 16) Cordoba; 17) Atlantico, Magdalena (North); 18) Guajira; 19) Arauca; 20) Caqueta, Huila; 21) Huila (South); 22) Leticia, Amazonas; 23) Huila (Central & North); 24) Meta; 25) San Andres.
Network III: Manjui (1) ch A 11 668kW.
D. Prgrs: 1630-1830 (Comm), 1830-2130 (Educ.), 2100-0500 (Comm).

ECUADOR

TV-sets: 900.000 — **Colour:** NTSC — **System:** M.

Stations	chA	kW	Stations	chA	kW
1) Guayaquil	2	6/0.6	9) Portoviejo	7	5
2) Quito	2	1/0.1	7) Quito	8	10/1
3) Cuenca	2	8	11) Quito	8	10
4) Quito	4	5/0.5	8) Portoviejo	9	3
5) Guayaquil	4	3/0.3	9) Guayaquil	10	10/1
) Quito	5	?	9) Quito	10	10
6) Esmeraldas	6	0.5/0.05	10) Quito	13	1/0.1

Addresses and other information:
1) Corporación Ecuatoriana de Televisión, Casilla 1239, Guayaquil. **Te:** +593 (4) 300150. **Fax:** +593 (4) 303677. **Telex:** 3409 TVDOSG. **Cable:** Teledos — 1200-0600.
2) Canal 2 Quito, Murgeon 732, Quito. **Te:** +593 (2) 540877 — 1930-0730 (Sun 1330-0700).
3) Telecuenca, Canal Universitaria Catolica, Casilla 400, Cuenca. **Te:** +593 827862. **Telex:** 4775 TELCUE ED — 1600-0400.
4) Teleamazonas, Av. Diguja 529 y Brazil, Quito. **Te:** +593 (2) 430313.
5) Canal 4 Guayaquil S.A, 9 de Octubre 1200, Guayaquil. **Te:** +593 (4) 308194.
6) Canal 6 Esmeraldas, Cas. 108, Esmeraldas. **Te:** +593 (2) 710090. **Cable:** TECEM — 2300-0400.
7) Canal 8 Quito, Cas. 3888, Quito. **Te:** +593 (2) 244888 — 1730-0530.
8) Manavision S.A, Apt. 50, Portoviejo — 1930-0530.
9) Canal 10, Guayaquil, Casilla 673, Guayaquil. **Te:** +593 (4) 391555 — 1130-1430, 1630-0500 (+14 relays).
10) Canal 13 Quito, Rumipampa 1039, Quito. **Te:** +593 (2) 242758.
11) Televisora Nacional, Bosmediano 447 y José Carbo, P.O. Box 6615, Quito.

FALKLAND ISLANDS

TV-sets: N/A — **Colour:** PAL — **System:** I.

BRITISH FORCES BROADCASTING SERVICE (BFBS)

ADDR: BFBS Falkland Islands, Mount Pleasant, BFPO 655. **Te:** 32179. **Fax:** 32193.
L.P: St.Mgr: Richard Jones. Gen. Eng: Julian Brower. **Station:** Mount Pleasant ch 24 UHF/100 W. Port Stanley ch 30 UHF/15 W.
D.Prgr: 4 h. of taped broadcasts from BBC and ITV London.

GUIANA (French)

TV-sets: 6.500 — **Colour:** SECAM — **System:** K.

RFO-GUYANE

Addr: See radio.
L.P: Dir: Yves Rambeau. T.D: Gérard Malhache.
Stations: Cayenne chK'4 0.1kW, + 8 low power repeaters.
D.Prgr: 2100-0130.

GUYANA

TV-sets: 15.000 — **Colour:** NTSC — **System:** M.

GUYANA TELEVISION (Gov.)

Addr: 68 Hadfield St., Lodge Georgetown. **Te:** +592 (2) 69231-4/62691-4/58584.
L.P: GM: A. Brewster; CE: S. Goodman.
Station: Georgetown ch10 0.04kW.
Prgrs: Sun 1500-1600, repeated 2100-2200.
Two private TV stations relay U.S. satellite sces.

PARAGUAY

TV-sets: 350.000 — **Colour:** PAL — **System:** N.

	Stations	chA	kW		Stations	chA	kW
1)	Encarnación	7	60/12	4)	Asunción	9	5/1
2)	Pto. Stroessner	8	25/5	4)	Asunción	13	30/6
3)	Asunción	9	60/12				

1) Television Itapua, Avda Irrazabal y 25 de Mayo, Encarnación — Gen. Mgr: J. Mateo G. — 1600-0400.
2) Televisora del Este, Area 5, Cd. Puerto Stroessner — Gen. Mgr: A. Villalba V. — 2000-0300 (Sat/Sun 1100-0330).
3) Canal 9 TV Cerro Cora SA. **Addr:** Avenida Carlos Antonio Lopez 572, Asuncion. **Te:** +595 (84) 22226. **Fax:** +595 (84) 498911. **L.P:** Hd. of Sales: Hugo Montgomery.
4) Teledifusora Paraguaya S.A., 8 Proy Lambare, Asunción — 1600-0400. **Te:** +595 (21) 443093.

PERU

TV-sets: 2m — **Colour:** NTSC — **System:** M.

Stations: *) in stereo.

	chA	kW (ERP)		chA	kW (ERP)
1)	2	22.54/2.25	5)	8	5
5)	3	20	5)	9	315
2)	4	25/12.5	6)	11	30/6
5)	4	20	1)	13	?
3)	5	290/29	7)	27*	7
5)	6	2	8)	33*	10
4)	7	10/5			

Addresses & other information:
1) Compañia Latinoamericana de Radiodifusion S.A. Av. San Felipe 968, Jesús Mariá, Lima 11. **Te:** +51 (14) 707272. **Fax:** +51 (14) 712688.
2) Compañia Peruana de Radiodifusion, Cas. 1192, Lima. Dir. Gen: M. M. Arbulu B. Man. Dir: N. Gonzalez U. Dir. Tec: D. Capella. Operates st's in Piura(ch2 2 kW), Chiclayo(ch 4 5 kW), Trujillo(ch 6 2 kW), Tacna(ch 9 2 kW), Huancayo(ch 4 2 kW) + 59 repeaters — 1000-0600.
3) Panamericana de Television, Av. Arequipa 1110, Lima. 1500-0600 (+ 14 relay st's).
4) Empresa de Cine, Radio y Television Peruana, José Galvez 1040, Lima. Pres: Carlos Guillen B. 1900-0400 (+ 39 repeaters).
5) Andina de Television, Arequipa 3570, San Isidro, Apartado 270077, Lima. 1900-0400.
6) RBC Television, Juan de la Fuente 453, Miraflores, Lima.
7) Difusora Universal de Television, Paseo de la República 6099, San Antonio, Miraflores, Lima — Exec. Pres: J.L. Banchero H. 2200-0400.
8) Empresa Radiodifusora 1160 TV, Apt. Postal 2355, Lima. 800-0500.
N.B: ch 2, 4, 5, 13.

SURINAME

TV sets: 43.000 — **Colour:** NTSC — **System:** M.

SURINAAMSE TELEVISIE STICHTING (STVS Gov./Comm.)

Addr: P.O. Box 535, Paramaribo. **Te:** +597 473100. **Fax:** +597 477216. **Telex:** 271 STVS. **Cable:** Surteve. **L.P:** Dir: Frits J. Pengel.
Stations: Paramaribo chA8 6kW + 5 relay st's.
D.Prgr: 2130-0300 (+ Sat 1500-2130, Sun 1400-1700).

ALTERNATIEVE TELEVISIE VERZORGING (ATV Telesur)

Addr: Adrianusstraat 1, Paramaribo. **Te:** +597 410027, +597 470425. **Fax:** +597 479260. **Telex:** 488 ATV TLS SN. **L.P:** Mgr: Roy Doorson.
Stations: Paramaribo chA2 + chA12 + 2 relay st's.
D.Prgr: 1115-0320.
(They started official broadcasts on August 15, 1987).

URUGUAY

TV-sets: 600.000 — **Colour:** PAL — **System:** N.

St's 1), 3), 8), 9), 13), 15), 16), 17), 18), 20), 23), 24) & 26) are affiliated to ANDEBU (Associación Nacional de Broadcasters Uruguayos) — All st's are comm.

Location	ch	kW(ERP)		ch	kW(ERP)
1) Artigas	3	0.5	15) Maldonado	9	12
2) Colonia	3	1/0.5	16) Bella Unión	10	0.1
3) Paysandú	3	1/0.5	17) Montevideo	10	600
4) Rio Branco	3	0.1	18) Rivera	10	1/0.1
5) Montevideo	4	300	19) Punta del Este	11	5/0.5
6) Montevideo	5	96/9	20) Treinta y Tres	11	1/0.5
7) Rivera	5	7/0.7	21) Durazno	11	0.1
8) Rocha	7	16	22) Fray Bentos	12	30/3
9) Tacuarembó	7	75	23) Melo	12	1/0.5
10) M. de Corrales	7	0.1	24) Montevideo	12	600/180
11) Melo	8	0.5/0.05	25) Chuy	12	0.1
12) Rosario	8	75/7.5	26) Minas	13	1/0.5
13) Salto	8	2/1			
14) P. de los Toros	9	0.1			

+10 st's 0.1kW (all rel. st 6).

Addresses and other information:

1) Tele-Artigas, Lecueder 291, 55000 Artigas — Dir. Gen: Carlos F. Falco. 1800-0300.
2) Televisora Colonia, W. Barbot 172, 70000 Colonia — Dir. Gen. G. de Gonzáles.
3) Río de los Pájaros TV, Av. España 1629, 60000 Paysandú — Dir Gen: A. Davison. Dir. Tec: H. Caporale.
4) 37100 Río Branco, Cerro Largo.
5) Monte Carlo TV Color, Paraguay 2253, 11800 Montevideo — Dir. Gen: H. Romay. Film Buyer: M. Fonticiella. Dir. Tec: J. Spinella — 1830/0400 (Sat/Sun 1300-0400)
6) S.O.D.R.E., Bul Artigas 2552, 11600 Montevideo.
7) Canal 5 S.O.D.R.E, Bulevar Artigas 2552, Montevideo. Gen. Man: Julio Frade. Dir. Tec: Pedro Narancio.
8) Tele-Rocha, Av. O. de los Santos 105, 27000 Rocha — Dir. Gen: M. Scherchener. Dir. Tec: J. Regalo. Film Buyer: L. Castillo.
9) Radiotelevisión "Zorilla de San Martín", 18 de Julio 302, 45000 Tucuarembó — Dir. Gen: D. Dini. S. St. Mgr: G. Valdés G. 2030 - 0330. Dir. Tec: G. Acosta. Film Buyer: Jose Abbondanza.
10) 40002 Minas de Corrales, Rivera.
11) Canal 8 TV Melo, 18 de Julio 572, 37000 Melo, Cerro Largo — Dir: Raul Figueredo. Dir. Tec: Eduardo Baptista.
12) Canal 8 Rosario TV Color, Ruta 2 Km. 136.500, 70200 Rosario, Colonia — Dir: H. Fripp.
13) Televisora Salto Grande, Av. Viera 1280, 50000 Salto — Dir. Gen: Dr. Carlos A. Gelpi. Dir. Tec: K. Muguerza.
14) 45100 Paso de los Toros, Tacuarembó.
15) Canal 9 del Este TV Color, Av. Artigas 879, 20000 Maldonado — Dir. Gen: M. Scherschener. Dir. Tec: Fernando Bareño. Film Buyer: J. López — 2000-0400.
16) Telediez, General Rivera s/n, 55100 Bella Unión, Artigas — Dir. Gen: C. Gelpi.
17) SAETA TV Canal 10, Dr. Lorenzo Carnelli 1234, 11200 Montevideo — Dir. Gen: J. de Feo. Dir. Tec: Oscar Inchausti. Film Buyer: H. Villar — Mon-Fri 1930.
18) Tevediez, Sarandí 705, 40000 Rivera — Dir. Gen: A. Pereira. 2000-0300.
19) Canal 11 Punta del Este, Cantegril Country Club, 20100 Punta del Este, Maldonado — Dir. Gen: D. Romay — 1830-0400.
20) Televisora Treinta y Tres, Pablo Zufriategui 226, Treinta y Tres — Dir. Gen: A. Pinho. St. Mgr: A. Lagos. Dir. Tec: D. Ponce.
21) 97000 Durazno.
22) Canal 12 Río Uruguay, Cno. San Salvador s/n, 65000 Fray Bentos, Río Negro — Dir. Gen: D. Romay.
23) Melo TV, Castellanos 723, 37000 Melo, Cerro Largo — Dir. Gen: R. Lucas. Dir. Tec: C. Britos. Film Buyer: J. Lucas.
24) Teledoce Televisora Color, Enriqueta Compte y Riqué 1276, 11800 Montevideo — Dir. Gen: H. Scheck. Dir. Tec: M. Donnangelo. Film Buyer: C. Restano — 1810 (Sat 1600, Sun 1330)-0400.
25) 27100 Chuy, Rocha.
26) TV Cerro del Verdún, Treinta y Tres 632, 30000 Minas, Lavalleja — Dir. Gen: C. Falco. Dir. Tec: J. Rodriguez.

VENEZUELA

TV-sets: 3.700.000 — **Colour:** NTSC — **System:** M.

CAMARA VENEZOLANA DE LA TELEVISION
(Organization for private TV stations).

Addr: Ap. 60423, Chacao, Caracas 1050. **Te:** +58 (2) 7814608. **Cable:** CAVETEL. **Telex:** 21144 — **L.P:** Pres. H. Ponsdomenech.

VENEZOLANA DE TELEVISION "5" (Gov.)

Addr: Ap. 3979, Caracas. **Te:** +58 (2) 2399811. **Fax:** +58 (2) 35734. **Telex:** 25401.
Stations: Caracas (Central St): chA5 210/105kW.

Repeaters	chA	kW(ERP)		chA	kW(ERP)
Margarita	5	99.6/49.8	El Tocuyo	9	2/1
La Grita	6	75.9/37.9	Terepalma	13	72/36
Los Olivos	5	40.9/20.9	Pto. La Cruz	13	72/36
Maracaibo	6	108/54	Táchira	13	72/36
Aleton	6	67.2/38.6	Litoral Central	13	60.25/30.12
Güigüe	6	67.2/38.6	Escuque	13	26.25/13.12
Cd. Bolivar	6	12.5/6.25	Caricuao	13	0.631

D.Prgr: 1000-0400 (approx.)

VENEZOLANA DE TELEVISION "8" (Gov.)

Addr: Ap. 2739, Caracas. **Te:** 349571. **Cable:** VTV. **Telex:** 25401-25412.
Stations: Caracas (Central st): chA8 190/95kW.

Repeaters	chA	kW(ERP)		chA	kW(ERP)
Mérida	5	5/2.5	Valencia	11	151/75.5
Barquisimeto	7	107.5/53.5	Litoral	11	79/49.5
Maracaibo	8	157/78.5	Pto. La Cruz	11	23/11.5
Boconó	8	2/1	Cd. Bolivar	11	23/11.5
Cd. Piar	8	0.15	Vidoño	11	23/11.5
Anaco	9	12.5/6.25	Pto. Cabello	12	20/10

D.Prgr: 1600-0600 (approx.)

RADIO CARACAS TELEVISION (Comm.)

Addr: Ap. 2057, Caracas. **Te:** +58 (2) 239 8722. **Fax:** +58 (2) 340123. **Telex:** 21527.
Stations: Caracas (Central St): chA2 132/61kW.

Repeaters	chA	kW(ERP)		chA	kW(ERP)
Maracaibo	2	105.6/52.8	Valencia	7	550/275
Barcelona	3	108/54	Táchira	7	75/37.5
Barquisimeto	3	40.2/20.1	Pto. Cabello	10	10/5
Litoral Central	7	200/100	Falcón	10	330/165

+ 5 low power repeaters less than 1kW not mentioned
D.Prgr: 1030-0500.

VENEVISION (Comm.)

Addr: Av. La Salle, EDF, Venevision, Los Caobos, Caracas. **Te:** +58 (2) 782 0111. **Cable:** Ventel.
Stations: Caracas (Central st): chA4 132/66kW.
Repeaters: Trujillo Ch3(6kW), Valencia ch4(20), Puerto Cabello ch4(20), Maracaibo ch4(336), Puerto Ordaz ch4(20), El Tigre ch4(12), Barquisimeto ch6(198), Barcelona ch7(25), Ci. Bolivar ch7(10), Maturin ch7(25), La Guaira ch9(21), Maracay ch9(476), Mérida ch9(16), San Cristobal ch9(40), Valera ch9(12), Anyarito ch9(10), Caricuao ch9 (11), Curimagua ch12(40), Guaramacal ch12(13kW).
D.Prgr: 1330-0430.

TELEVISORA ANDINA MERIDA (Cult & Rlg)

Addr: Calle 23 Entre Avenidas 4 y 5, Merida CP5101 Edo Merida. **Te:** +58 (74) 525758. **Fax:** +58 (74) 520098. **Telex:** 74287.
Station: Merida chA6 1kW.
D.Prgr: 2000-0200.

NCTV (Comm.)

Addr: Urv. La Paz, Avenida 57 y Maracaibo, Maracaibo. **Te:** +58 (61) 512662. **Fax:** +58 (61) 512729.
L.P: Dir: Gustavo Ocando Yamarte.
Station: Maracaibo chA11 108/54kW (est).
D.Prgr: 1600-0400.

CANAL 10

Addr: Av. Francisco de Miranda, con Principal de los Ruices, Centro Empresarial Miranda PHD, Caracas. **Te:** +58 (2) 239 8679. **Fax:** +58 (2) 239 7757.
Station: Caracas ch10.

CANAL 12-OMNIVISION

Addr: Edificio Omnivision, La California Sur, Caracas. **Te:** +58 (2) 256 5011. **Fax:** +58 (2) 256 3586. **Telex:** 21024.
Station: Caracas ch12. **F.Pl:** stations in other major cities.

WORLD SATELLITE BROADCASTS

This section comprises Radio and TV signals distributed via satellite. In some countries free reception of satellite transmissions may be restricted by the law. The information supplied in this section was compiled without taking restrictive legislation into account. Although some attempts are made to indicate target areas, space limitations prevent us from being able to give detailed information about footprints and signal strength.

R = circular polarized signals turning Right
L = circular polarized signals turning Left
H = Horizontally polarized signals
V = Vertically polarized signals
SCPC = Single Channel Per Carrier (audio)
t.. = transponder number
sc = subcarrier frequency (audio)
DSR = Digital Satellite Radio
(≡) = encryption system
[16:9] = wide screen

S-Band	**2535-2655 MHz**
C-Band	**3700-4200 MHz (USA & CIS)**
Ku1-Band	**10.9-11.75 GHz (FSS)**
Ku2-Band	**11.75-12.5 GHz (DBS)**
Ku3-Band	**12.5-12.75 GHz (Télécom)**
Ka-Band	**18.00-20.00 GHz**

NB: Users of this section are cordially invited to send in comments and observations to Bart Kuperus

The following countries operate Satellite News Gathering (SNG) mobile uplink units. These units can be recognized by their colorbar testcard + ID.

Id	Ku-Band	User	Country
AUT-1	high/low	PTT	Switzerland
SUI-1N	high/low	PTT	Switzerland
SUI-2N	high/low	PTT	Switzerland
SUI-4	high/low	PTT	Switzerland
D1 to D10	high/low	DBP	Germany
D12 to D14	high/low	DBP	Germany
E1 to E5	low	Telefonica	Spain
F7 & F8	high/low	France Télécom	France
F12 & F13	high/low	France Télécom	France
F15	high/low	France Télécom	France
UKI-1	low	BBC	UK
UKI-16 & 17	low	BBC	UK
UKI-20 & 21	low	BBC	UK
UKI-23	low	BBC	UK
UKI-35	low	BBC	UK
UKI-7	low	IBA/NTL/ITN	UK
UKI-11	low	IBA/NTL	UK
UKI-31	low	IBA/NTL	UK
UKI-3 to 5	low	BT	UK
UKI-8 to 10	low	BT	UK
UKI-12 to 15	low	BT	UK
UKI-18 & 19	low	BT	UK
UKI-27 to 30	low	BT	UK
UKI-22	low	WTN/BEA Comm.	UK
UKI-24 & 25	low	WTN/BEA Comm.	UK
UKI-26	high/low	Croydon Uplink	UK
UKI-32	high/low	Croydon Uplink	UK
UKI-34	low	BSkyB	UK
ITA-1 & ITA-9	low	Telespazio	Italy
ITA-17 to 21	low	Telespazio	Italy
ITA-30	low	Telespazio	Italy
ISR-1	low	Bezeq	Israel
HOL-1	high	PTT	The Netherlands
HOL-2	low	PTT	The Netherlands
SWE-1 to 4	low	Televerket	Sweden

*Ku-band low = 10.05 GHz - 11.7 GHz, Ku-band high = 12.5 - 12.75 GHz

N.B: Commercial SNG units not listed.

INTERNATIONAL TELECOMMUNICATION UNION REGION 1:

EUROPE, AFRICA, NEAR & MIDDLE EAST

63° E. Intelsat 602 (Int)

This spacecraft coming from 60° East replaces the Intelsat 511, which was redeployed at 177° East. The payload is similar to all other Intelsat 600 satellites.

C-Band (MHz)
Global Beam:

4166/ 4188 R	occasional feeds PAL/SECAM/NTSC audio: 6.60 or 6.65 MHz

S. Zone Beam:

3910 R	**M-Net** (≡Irdeto)
3780 R	**SABC-TV**, Johannesburg. (B-MAC) sc: 6.58 MHz **Radio MF-702** (Johannesburg)

E. hemisph. Beam:

3885 R	**Bangkok Entertainment Co., Channel 3**, Thailand sc: 6.20/6.80 MHz stereo TV
3920 R	**Mass Communications Org. of Thailand, Channel 9**, Bangkok, sc: 6.20/6.80 MHz stereo TV

Ku-Band (GHz)
Spot Beam East:

10.990 V	**IRIB TV 2**. (SECAM, neg.) audio: 6.80 MHz sc: 6.20 MHz **IRIB Radio 1**
11.155 V	**IRIB TV 1**. (SECAM, neg.) audio: 6.80 MHz sc: 6.20 MHz **IRIB Radio 2**
11.470 H	**AFRTS-E2**

Spot Beam West:

10.971 H	occasional feeds
11.011 H	**Rete 4** (Italy) audio: 6.60 MHz
11.137 H	**Italia 1** (Italy) audio: 6.60 MHz
11.173 H	**Canale 5** (Italy) audio: 6.60 MHz

60° E. Intelsat 604 (Int)

This satellite replaces the Intelsat 515 which was redeployed at 18.5° West. The Intelsat 604 satellite has 38 C-Band and 10 Ku-Band transponders.

C-Band (MHz)
Global Beam:

4166/ 4188 R	occasional news & sports feeds PAL/ SECAM/NTSC, audio: 6.60 or 6.65 MHz

Ku-Band (GHz)
Spot Beam West:

10.974 V	occasional video
11.010 V	**Wir in Bayern** **Bayern Journal**, audio: 6.65 MHz
11.495 V	**AFN-Television**, Frankfurt (≡525-line B-MAC)
11.590 V	**interSTAR** feeds (neg. video)

Spot Beam East:

10.974 H	**TRT TV 4**, audio: 6.80 MHz sc: 8.28 MHz Turkish radio
11.138 H	**TRT TV 3**, audio: 6.80 MHz sc: 8.28 MHz Turkish radio
11.472 H	EBU
11.647 H	**TRT TV 1**, audio: 6.80 MHz sc: 8.28 MHz Turkish radio
11.683 H	**TRT TV 2** **TV Gap**, audio: 6.80 MHz sc: 8.28 MHz Turkish radio

57° E. Intelsat 507 (Int)

Launched in 1983, this satellite has 38 C-Band channels of 36 MHz, four Ku-Band channels of 72 MHz and two of 241 MHz bandwidth. In addition two transponders around 1540 MHz for maritime communications. This satellite has gone into an inclined orbit at the end of 1991.

C-Band (MHz)

3900 R — **EET** Ethiopia (PAL)
sc: 6.60 MHz TV audio

3967 R — **Sudanese TV** (PAL) Khartoum
sc: 6.20 MHz TV audio
SCPC: 3940.2 MHz **Sudanese radio**
SCPC: 3940.7 MHz TV audio.

Ku-Band (GHz)

Sporadic TV feeds.

53° E. Gorizont 11 (CIS)

Also known as Statsionar 5. This spacecraft was launched in 1985 and carries the usual payload of six 34 MHz wide C-Band and experimental Ku-Band transponders.

C-Band (MHz)

Spot Beam:

3675 R — **Ostankino TV 1**
sc: 7.0 MHz TV audio
sc: 7.5 MHz **Radio Mayak**

3825 R — **MIR-Station**

Ku-Band (GHz)

Spot Beam:

11.512 L — feeds

11.525 L — feeds

43° E. Arabsat 1B (Arab League Countries)

This spacecraft was launched in 1985 and has a payload of 25 C-band transponders as well as one very strong S-Band transponder. The Arabsat 1B satellite was drifting eastwards from 26° East at time of editing. Last known position was 43° East. The spacecraft is also in inclined orbit due to reduced station keeping, making reception quite difficult.

NB: All TV PAL, unless stated otherwise.

C-Band (MHz)

Spot Beam: (to the Mediterranean)

3830 R t7 — occasional video relay, audio: 6.60 MHz

4010 R t17 — **RTM-1**, Morocco. SCPC audio: 4029.5 MHz
RTM radio 4029.2 and 4030 MHz
Radio RTM-International 4030.3 MHz

4070 L t20 — **Oman TV** SCPC audio: 4078.5 MHz,
radio program 4078.9 MHz

4089 R t21 — **TV Mauretania** (French & Arab services)
sc: 6.6 MHz TV audio

4126 R t23 — Inter-Arab news exchanges via Tunis.
(PAL/SECAM), audio: 6.60 MHz

4163 R t25 — **Saudi-Arabian 2nd program** (SECAM),
sc: 6.60 MHz TV audio (English)

40° E. Gorizont 12 (CIS)

Also known as Statsionar 12. This Indian Ocean reserve satellite has a rather large orbital inclination of +/-2° of arc. Telecommunication payload comprises 6 transponders within 6-4 GHz range, 1 transponder in the 14-11 GHz and 1 in the 1.6-1.5 GHz range.

C-Band (MHz)

N. hemisph. Beam:

3675 R — **Rossia TV 2**
sc: 7.50 MHz **Radio Junost**

3730 R — **TV 5 Europe** (French)

Global Beam:

3825 R — former CNN transponder

3930 R — **RTP International** (Portuguese)

33.5° E. Kopernikus (DFS-3) FRG

This satellite was launched in September 1992 and will serve as a backup satellite to the other two Kopernikus spacecraft. Its payload comprises seven 12 GHz transponders with 44 MHz and three 11 GHz transponders with 90 MHz bandwith.

28.5° E. Kopernikus (DFS-2) FRG

This satellite was launched in July 1990 and is identical to the DFS-1 at 23.5° E. DFS-2 is supposed to serve telecommunications purposes only, but occasionally some video activity can be witnessed.

Ku-Band (GHz)

11.475 tA1H — occasional feeds

11.525 tA2H — occasional feeds

11.555 tB1V — occasional feeds

11.600 tB2V — occasional feeds

12.580 tK3H — occasional feeds

12.625 tK4H — **Wir in Niedersachsen** (1600-1700 UTC)
sc: 6.65 MHz TV audio

12.658 tK5H — ARD occasional feeds (sound-in-sync)

12.692 tK6H — **RTL Nord** (1600-1700 UTC)
sc: 6.65 MHz TV audio

12.725 tK7H — ARD occasional feeds (sound-in-sync)

26° E. Arabsat 1C (Arab League Countries)

Launched in March 1992 this spacecraft replaces the Arabsat 1A (used to be at 19° East). Its payload consist of 25 C-Band transponders and 1 strong S-Band transponder. This satellite is used for both communications and TV purposes. S-Band signals are also receivable throughout Europe. Expected lifetime is approx. 7 years. After the elderly Arabsat 1B started drifting eastwards in Autumn 1992, this satellite was moved from 31° East and repositioned at this orbital position.

NB: All TV PAL.

S-Band (MHz)

2635 V t1b — **Middle East Broadcasting Centre** (MBC)

C-Band (MHz)

3790?R t9 — **The Kuwait Satellite Channel**
sc: 6.60 MHz TV audio

3790 R t9 — **Canal France Internationale**

3829 R — **CNN**

3974 R t15 — **Saudi Arabian TV 1** (in Arabic)
sc: 6.60 MHz TV audio

4049 R t19 — **Saudi Arabian TV 2** (in English)
sc: 6.60 MHz TV audio

4062 L — **Oman TV**, audio: 4078/4076 MHz

4078 L SCPC — **Oman Radio**

4162 R — **The Egyptian Space Channel**

23.5° E. Kopernikus (DFS-1) (FRG)

Launched in 1989, this satellite carries seven 44 MHz and three 90 MHz Ku-Band channels as well as one 90 MHz wide Ka-Band transponder (19.75 GHz). DFS-1 Kopernikus was designed to mainly serve communications purposes, but was transformed into a de-facto broadcasting satellite lateron. Frequencies are in the 11- and 12 GHz range meaning that for proper reception 2 LNBs or a dualband LNB is needed.

NB: All TV PAL, audio: 6.65 MHz.

Ku-Band (GHz)

Spot Beam:

11.475 tA1H — **SAT-1** (German commercial TV)
sc: 7.38/7.56 MHz **Star*Sat Radio**

11.500 tA1H — occasional video

11.525 tA2H — **3-SAT** (ZDF-ORF-SRG)

11.550 tB1V — **ARTE** (French/German cultural program)
sc: 7.02/7.20 MHz stereo TV

11.600 tB2V — **VOX** (German "Westschienen" program)

11.625 tC1H — **Eins Plus** (ARD/SRG coop.)

11.675 tC2H — **RTL Plus** (Cologne)
sc: 7.02/7.20 MHz **RTL** (German sce)

12.524 tK1V — **n-tv** (German news channel)
sc: 7.02/7.20 MHz stereo TV

12.559 tK2H — **PRO-7** (German commercial TV)
sc: 7.38/7.56 MHz stereo TV

12.591 tK3V — **Premiere** (≡Syster)
sc: 7.02/7.20 MHz stereo TV
sc: 7.38/7.56 MHz **Klassik Radio Hamburg**

12.625 tK4H — DSR, 16 channels:
***Bayern 4 Klassik** (public, classical)
***Deutschlandfunk** (public, news/talk)
***DS Kultur** (public, classical)
***HR2 Kultur** (public, classical)
***Klassik Radio** (comm., classical)
***NDR 3** (public, classical)
***Radio ffn** (comm., pop)
***Radioropa** (comm., news/talk)
***RB2 Kulturell** (public, classical)
***RIAS Eins** (public, pop)
***RTL Radio** (comm., pop)
***SDR/SWF2 Kultur** (public, classical)

	***SR1 Saarland** (public, classical)
	***Star*Sat Radio** (comm., pop)
	***WDR3 Köln** (public, classical)
12.658 tK5V	**West-3** (regional German TV)
	sc: 6.65 MHz **WDR 3** (when no TV is broadcast)
	sc: 7.02/7.20 MHz **Radioropa**.
	sc: 7.38 MHz relay of radio commercials
	sc: 7.56 MHz non-stop music
	sc: 7.74/7.92 MHz **Deutschlandfunk**
12.692 tK6H	**DSF** (Deutsches Sportfernsehen)
12.725 tK7V	**Bayerisches Fernsehen** (regional German TV)
	sc: 7.68 MHz non-stop music
	sc: 7.90/8.05 MHz non-stop music
19.700 H	DBS experiments

21.5° E. Eutelsat I f5

This spacecraft was redeployed at this position after being replaced by a new Eutelsat satellite (ECS II f2) at 10° E.
NB: All TV PAL using sound-in-sync technique, unless stated otherwise.
Ku-Band (GHz)

10.974 V	occasional feeds
10.992 H	telecommunications
11.010 V	occasional feeds (EBU channel A)
11.136 H	occasional feeds (EBU channel B)
11.177 H	occasional feeds (EBU channel D)
11.471 H	occasional feeds
11.501 H	occasional feeds (audio: 6.5/6.2/6.8 MHz)
11.512 H	occasional feeds
11.555 H	occasional TV (TV channel 5A)
11.595 H	occasional TV (TV channel 5B)
11.658 H	telecommunications

19.2° E. Astra 1A & 1B (Luxembourg)

The ASTRA 1A satellite was launched in 1988, the ASTRA 1B in 1991. Both spacecraft have a payload of sixteen 27 MHz wide Ku-Band transponders, each within a densely packed frequency scheme. Société Européenne des Satellites plans to launch a third satellite **"ASTRA 1C"** in April or May 1993.
NB: All TV PAL, audio 6.50 MHz, unless stated otherwise.
Ku-Band (GHz)
ASTRA 1C

10.964 t33H	11.023 t37H	11.082 t41H	11.141 t45H
10.979 t34V	11.038 t38V	11.097 t42V	11.156 t46V
10.993 t35H	11.052 t39H	11.111 t43H	11.170 t47H
11.008 t36V	11.067 t40V	11.126 t44V	11.185 t48V

NB: tentative list of ASTRA 1C program providers: *Two further Spanish channels; *n-tv (German news channel), *VOX (German Infotainment), *RTL-2 (German entertainment), *FilmNet Plus, *The Discovery Channel Europe, *Bravo, *The Disney Channel, *Nickelodeon, *TV Asia, *The Children's Channel, *Thames Channel 5, *The Games Channel, *Budapest-1 (BP-1).

ASTRA 1A

11.214 t1H	**Screen Sports**
	sc: 7.02/7.20/7.38/7.56 MHz (multilingual)
11.229 t2V	**RTL Plus**
	sc: 7.38 MHz **Deutsche Welle**
	sc: 7.56 MHz **Deutsche Welle** (English)
	sc: 7.74 MHz **RTL** (German sce)
11.244 t3H	**Scansat TV3** Sweden (≡D2MAC EuroCrypt)
11.259 t4V	**Eurosport**
	sc: 6.65/7.02/7.20/7.38 MHz multilingual
11.273 t5H	**Lifestyle TV**
	(possible new user after Nov. 28th 1992: RTL 2)
	sc: 7.02/7.20 MHz stereo TV/multilingual
11.288 t6V	**Sat-1**
	sc: 7.38/7.56 MHz **Deutschlandfunk**
	sc: 7.74/7.92 MHz **Deutschlandfunk** (foreign)
11.303 t7H	**TV 1000 Succekanalen**
	(Scandinavian)(≡D2MAC EuroCrypt)
11.318 t8V	**SKY One** (commercial U.K.)
	sc: 7.02/7.20 MHz stereo TV.
	sc: 7.38/7.56 MHz **SKY Radio** (light music)
	sc: 7.74/7.92 MHz **Hit Radio**
11.332 t9H	**Teleclub** (≡Payview)
	sc: 7.20 MHz **Swiss Radio International (SRI)**
	sc: 7.74/7.92 MHz **Radio E Viva!**
11.347 t10V	**3-SAT** (ZDF-ORF-SRG)
	sc: 7.02/7.20 MHz stereo TV
11.362 t11H	**FilmNet** (Scandinavian)(≡D2MAC EuroCrypt)
11.377 t12V	**SKY News** (British)
	sc: 7.92 MHz **Supergold** (oldies)
11.391 t13H	**RTL-4** (Dutch, commercial TV)
	sc: 7.02/7.20 MHz stereo TV
	sc: 7.38/7.56 MHz former R. Luxembourg
	sc: 7.74/7.92 MHz **RTL Radio** (Dutch)
11.406 t14V	**Pro-7** (German)
	sc: 7.02/7.20 MHz stereo TV
	sc: 7.38/7.56 MHz **Star*Sat Radio**
	sc: 7.74/7.92 MHz **Radioropa**.
11.421 t15H	**MTV** (video clips, commercial)
	sc: 7.02/7.20 MHz stereo TV
	sc: 7.38/7.56 MHz **Power FM**
11.436 t16V	**SKY Movies Plus** (≡VideoCrypt)
	sc: 7.02/7.20 MHz stereo TV
	sc: 7.38/7.56 MHz **QEFM/ERF** (1h)
	sc: 7.74 MHz **ASDA**

ASTRA 1B

11.464 t17H	**Premiere** (≡Syster)
	sc: 7.02/7.20 MHz stereo TV
11.479 t18V	**The Movie Channel** (≡VideoCrypt)
	sc: 7.02/7.20 MHz stereo TV
	sc: 7.56 MHz **Holland FM**
	sc: 7.38 MHz **Sunrise Radio** (ethnic)
11.493 t19H	**Eins Plus** (ARD/SRG coop.)
	sc: 7.02/7.20 MHz stereo TV
	sc: 7.38/7.56 MHz **SWF-3** (German)
11.508 t20V	**Sky Sports** (≡VideoCrypt)
	sc: 7.02/7.20 MHz stereo TV
	sc: 7.56 MHz **EuroNet**
11.523 t21H	**DSF** (Deutsches Sportfernsehen)
	sc: 7.02/7.20 MHz stereo TV
	sc: 7.38/7.56 MHz **RTL Radio** (German sce)
11.538 t22V	**MTV** (video clips, commercial)
	sc: 7.02/7.20 MHz stereo TV
11.553 t23H	**UK Gold** (Thames TV & BBC)(≡VideoCrypt)
11.567 t24V	**JSTV** (≡VideoCrypt)
	NHK (Japanese)(≡VideoCrypt)
	TCC (The Children's Channel)
	sc: 7.20 MHz occasional English language
11.582 t25H	**N 3** (NDR, SFB, RB Germany)
	sc: 7.02/7.20 MHz **N 4**
11.597 t26V	**Sky Movies Gold** 1600-2400 UTC (≡VideoCrypt)
	Home Video Channel 0000-0400 UTC (≡VideoCrypt)
	TV Asia 0600-1600 UTC (≡VideoCrypt)
	sc: 7.74 MHz **Radio Sweden**
	sc: 8.01 MHz **Radio Asia**
11.611 t27H	**Scansat TV3** Denmark (≡D2MAC EuroCrypt)
11.626 t28V	**CNN International**
	sc: 7.02/7.20 MHz TV audio
	sc: 7.74 MHz **CNN Radio**/Spanish audio
11.641 t29H	**BSkyB** (≡[16:9] D2MAC Videoguard)
11.656 t30V	Spanish movie channel (≡Nagravision)
11.670 t31H	**Scansat TV3** Norway (≡D2MAC EuroCrypt)
11.685 t32V	Spanish cultural channel (≡Nagravision)

16° E. Eutelsat II f3 (European)

This satellite, launched December 7th 1991, is the third of a new generation Eutelsat satellites. It replaces the old ECS I f1 wich was directed to the "Junk Orbit". The payload is similar to the other ECS II spacecraft, currently stationed at 13° and 10° East. Expected lifetime is approx. 12 years.
NB: All TV PAL, unless stated otherwise.
Ku-Band (GHz)

10.986 t20H	**Eurostep** (Dutch educational)(D2MAC)
10.971 t25V	**Antena Tres** (Spanish)(≡Nagravision)
	sc: 7.02/7.20 MHz **Onda Cero**
	sc: 7.38/7.56 MHz **Los 40 Principales**
	sc: 7.74 MHz **Cadena SER Convencional**
10.987 t20H	**RTV Zagreb** (Hvratska Televisija) Croatia
	sc: 6.65 MHz TV audio
	sc: 7.02 MHz **Radio Zagreb** (Hvratski Radio)
10.991 t25V	
11.075 t21H	
11.095 t26V	**Tele 5** (Spanish)(≡Nagravision)
	sc: 7.02 MHz **Antena Tres**
	sc: 7.20 MHz **COPE Convencional**
	sc: 7.38/7.56 MHz **Agencia EFE**

11.162 t27V **The Egyptian Space Channel** (planned)
11.174 t27V **Canal Plus Español** (Spanish)(≡Nagravision)
sc: 7.02/7.20 MHz **Popular FM**
sc: 7.38 MHz **Radio Olé**
sc: 7.56 MHz **Radio 80**
11.554 t32H **Polish Television's Third Program**
sc: 6.65 MHz TV audio
sc: 7.20 MHz Polish Radio
11.595 t33H **Hungaria TV** (occasional)
11.575 t37V **RTP International**
sc: 7.02/7.20 MHz **Radio Renascenca**
sc: 7.74/7.92 MHz **RFM** (R. Renascenca 2)
sc: 8.10/8.28 MHz **Antena I** (Portuguese)
sc: 8.46 MHz **Antena I Int.**
11.616 t38H **HBB-TV/HAS-TV** (Turkish)
sc: 7.02/7.20 MHz **RFM** (Turkish)
sc: 7.74/7.92 MHz **Radio Number One**
11.658 t39V **TV7 Tunesia** (8 hrs a day)
sc: 7.38/7.56 MHz Tunesian radio
11.638 t34H PTT (Dutch)
11.675 t34H **TV PLUS** (Dutch)(≡[16:9] D2MAC EuroCrypt)
12.522 t40H occasional use (British Telecom)
12.563 t41H
12.604 t42H
12.646 t43H
12.708 t44H (72MHz)
12.542 t45V occasional VTM
12.583 t46V
12.625 t47V
12.667 t48V
12.708 t49V

13° E. Eutelsat II f1 (European)

This satellite, launched in August 1990, is the first of a new generation Eutelsat satellites. It replaces one of the first generation Eutelsat satellites, the ECS I f4, which was moved to 7° E. replacing the even older ECS I f2. The Eutelsat II f1 satellite is more powerful than its predecessor and has a payload of 16 Ku-Band channels of 31, 36 and 72 MHz bandwidth. Frequencies are in the 10.95-12.75 GHz range, meaning that for proper reception 2 LNBs or a dual-band LNB is needed. Expected lifetime approx. 10 years.

NB: All TV PAL, unless stated otherwise.

Ku-Band (GHz)

10.971 t20H **Eurosport**, audio: 6.60 MHz
sc: 7.02/7.20/7.38 MHz various languages
10.990 t20H PTT/Intrax (Dutch)
10.987 t25V **Super Channel/Channel e**
sc: 6.60 MHz TV audio
sc: 7.02/7.20 MHz multi lingual
sc: 7.38 MHz **BBC World Service** (English)
sc: 7.56 MHz **BBC World Service** (multilingual)
11.055 t21H **Der Kabelkanal** (D2MAC)
11.080 t26V **TV 5 Europe**
sc: 6.6 MHz TV audio/**RFI**
11.095 t21H **RTL-2** (German), audio: 6.60 MHz
sc: 7.74 MHz **RFE** (Czech/Slovak)
sc: 7.92 MHz **RFE** (Baltic)
sc: 8.10 MHz **RFE** (Polish)
11.148 t22H PTT (Dutch)
11.162 t27V **Deutsche Welle Fernsehen**
WorldNet (Mon-Fri 0800-1600 UTC)
sc: 7.02 MHz **Deutsche Welle** (German)
sc: 7.20 MHz **Deutsche Welle** (foreign)
sc: 7.38/7.56 MHz **Voice of America Europe**
sc: 8.10 MHz **Deutsche Welle** (Asian)
11.181 t22H **TRT** (Turkish)
Red Hot Dutch (pornography)(≡SAVE)
sc: 6.60 MHz TV audio
11.554 t32H **Middle East Broadcasting Centre** (Arabic)
sc: 6.60 MHz TV audio
sc: 7.02/7.20 MHz stereo TV
11.575 t37V **Euronews** (planned)
sc: 7.38 MHz **Show Radio**
11.595 t33H **ARD Eins** (Germany's 1st channel)
sc: 7.02/7.20 MHz stereo TV
11.616 t38V occasional use
11.638 t34H **FilmNet Plus** (Belgium)(≡SATPAC)
sc: 7.01 MHz NICAM TV audio
sc: 7.02/7.92 MHz **Radio Contact** (2 sces)
11.658 t39V former TV 5
11.678 t38H **FilmNet** (Dutch)(SATPAC)
sc: 7.01 MHz NICAM TV audio
sc: 7.92/8.10 MHz **Radio 10 Gold**
12.522 t40H **VisEurope** (clear/≡VideoCrypt)
BrightStar (feeds PAL/NTSC)
sc: 6.60/7.20 MHz main audio
12.542 t45V digital radio services (French)
12.555 t45V occasional feeds
12.563 t41H **WTN/ITN** (Maxat feeds)
EuroPace (D2MAC)
12.584 t46V **TV Sport** (France Télécom)
sc: 5.80 MHz TV audio
12.625 t47V occasional use
12.667 t48V occasional use
12.708 t49V occasional use

10° E. Eutelsat II f2. (European)

This spacecraft is the second of a new generation Eutelsat satellites and was launched in January 1991. Payload is similar to the other ECS II satellites.

NB: All TV PAL, unless stated otherwise.

Ku-Band (GHz)

10.971 t25V **RAI Uno** (Italy)(≡Discrete-2)
sc: 6.60 MHz TV audio
10.990 t20H occasional video
11.012 t25V occasional RAI
11.055 t26V occasional RAI
11.077 t21H CSS GUADALAJARA (Spain)
11.095 t26V **RAI Due** (Italy)(≡Discrete-2)
sc: 6.60 MHz TV audio
11.136 t27V occasional video
11.149 t22H **TVE-Internaçional** (Spain)
sc: 6.60 MHz TV audio
sc: 7.38 MHz **RNE Radio Una**
sc: 7.56 MHz **RNE Radio Cinco**
sc: 7.74 MHz **REE** (Radio Exterior España)
11.174 t27V occasional video
11.183 t27V occasional video
11.575 t37V **Show TV** (Turkey), audio: 6.65 MHz
sc: 7.38/8.20 MHz **Show Radyo**
11.596 t33H **TELEON** (Turkey), audio: 6.65 MHz
11.617 t38V **interSTAR** (AKK Magic Box, Turkey)
sc: 6.65 MHz TV audio
sc: 7.02 MHz **Radio Liberty** (Russian)
sc: 7.20 MHz **Radio Liberty** (Ukrainian)
sc: 8.10 MHz **RFE** (Bulgarian)
12.542 t45V digital services. (a.o. Spanish radio)
12.583 t46V **SIP Canal Courses** (B-MAC)
TV Campus (occasional)

7° E. Eutelsat II f4. (European)

This spacecraft was launched in the summer of 1992 and put into use in September. Its Payload is similar to the other ECS II satellites.

NB: All TV PAL, unless stated otherwise.

Ku-Band (GHz)

10.971 t20H occasional video
10.987 t25V occasional video
10.990 t20H occasional video
11.075 t26V occasional video
11.146 t22H **ET 1** (Greece)(PAL/SECAM)
sc: 6.60 MHz TV audio
11.163 t27V **Kanal 6** (Turkish)
sc: 7.02/7.20 MHz **Radyotek**
11.178 t22H **CYBC** (Cyprus)
sc: 6.60 MHz TV audio
11.596 t33H **RTV Beograd** Serbia (1800-2400 CET)
sc: 6.60 MHz TV audio
sc: 7.02 MHz **Radio Beograd**
sc: 7.56 MHz occasional use

5° E. Tele-X DBS slot 3

Tele-X (Scandinavia): Launched in 1989, this satellite operates two DBS-frequencies and one wide-band Business-Band transponder.

Assigned channels 26, 32 and 40 + 12.5-12.75 GHz for business purposes.

Channel:

26	**TV4** (Swedish commercial)(PAL) sc: 6.50 MHz TV audio sc: 7.02/7.20 MHz stereo TV sc: 7.38 MHz **Radio Sweden** sc: 7.74/7.92 MHz **The Voice**
32	**NRK** (Norwegian TV)(D-MAC) extra audio: **NRK-1 & NRK-2/Radio ROKS** (rock-music beamed to FM transmitters in Russia)

Ku-Band (GHz)

12.475 L	**TV5 Nordic** (PAL) **Kompetenskanalen** (1155-1300 UTC)
12.607 L	occasional use
12.640 L	occasional use
12.673 L	**SSC Tele-X/1** (occasional)
12.720 L	occasional use
12.755 L	occasional use

European DBS-frequency allocations:

Slot 1 (44° East) and 2 (25° East) are assigned to the CIS.
Slot 5 (7° West) is only assigned to former Yugoslavia ch 21, 25, 29, 33 and 37 all R and Albania ch 22, 26, 30, 34 and 38 all L

Frequency		Slot 3	Slot 4	Slot 6	Slot 7	Slot 8
CH.	(GHz.)	5° East	1° West	19° West	31° West	37° West
1	11.72748	Turkey	Poland	France		San Marino
2	11.74666	Finland	Romania	F.R.G.	Ireland	
3	11.76584	Greece	Cz-Slov	Luxemburg	Portugal	Liechtenst
4	11.78502	Sweden	Bulg.	Austria	U.K.	Andorra
5	11.80420	Turkey	Poland	France		San Marino
6	11.82338	Finland	Roman.	F.R.G.	Ireland	
7	11.84256	Greece	Cz-Slov	Luxemburg	Portugal	Liechtenst
8	11.86174	Sweden	Bulg.	Austria	U.K.	Andorra
9	11.88092	Turkey	Poland	France		San Marino
10	11.90010	Finland	Romania	F.R.G.	Ireland	
11	11.91928	Greece	Cz-Slov	Luxemburg	Portugal	Liechtenst
12	11.93848	Denmark	Bulg.	Austria	U.K.	Andorra
13	11.95764	Turkey	Poland	France		San Marino
14	11.97682	Norway	Romania	F.R.G.	Ireland	
15	11.99600	Greece	Cz-Slov	Luxemburg	Portugal	Liechtenst
16	12.01518	Denmark	Bulg.	Austria	U.K.	Andorra
17	12.03436	Turkey	Poland	France		San Marino
18	12.05354	Norway	Romania	F.R.G.	Ireland	
19	12.07272	Greece	Cz-Slov	Luxemburg	Portugal	Liechtenst
20	12.09190	Denmark	Bulg.	Austria	U.K.	Andorra
21	12.11108	Cyprus	F.R.G.	Belgium	Iceland	Monaco
22	12.13026	Finland	Hungary	Switzerl		
23	12.14944	Iceland		Netherl.	Spain	Vatican
24	12.16862	Denmark		Italy		
25	12.18780	Cyprus	?	Belgium	Iceland	Monaco
26	12.20698	Finland	Hungary	Switzerl		
27	12.22616	Iceland		Netherl.	Spain	Vatican
28	12.24534	Norway		Italy		
29	12.26452	Cyprus	?	Belgium	Iceland	Monaco
30	12.28370	Sweden	Hungary	Switzerl		
31	12.20288	Iceland		Netherl.	Spain	Vatican
32	12.32206	Norway		Italy		
33	12.34124	Cyprus	?	Belgium	Iceland	Monaco
34	12.36042	Sweden	Hungary	Switzerl		
35	12.37960	Iceland		Netherl.	Spain	Vatican
36	12.39878	Denmark		Italy		
37	12.41796	Cyprus	?	Belgium	Iceland	Monaco
38	12.43714	Norway	Hungary	Switzerl		
39	12.45632	Iceland		Netherl.	Spain	Vatican
40	12.47550	Sweden		Italy		

NB: Polarizations are decided upon as follows:
For Slot 3, 5, 6 and 8: even channels are L, odd channels R.
For slot 4 and 7: even channels are R, odd channels L.

3° E. Télécom 1C (France)

This satellite was redeployed at this position after being replaced by the new Télécom 2B at 5° West.

Ku-Band (GHz)

Spot Beam V:

12.522	French DSR
12.564	occasional use
12.606	occasional use
12.648	occasional use
12.690	occasional use
12.732	occasional use

1° W. Thor (formerly Marcopolo II)

In July 1992 the Marcopolo II satellite was purchased by The Norwegian Telecoms Administration, which redeployed it at 1° West. Program providers not known at time of editing.

Ku-Band (GHz)

1° W. Intelsat 512 (Int)

Launched in 1989, this satellite has 42 C-Band channels of 36 MHz, four Ku-Band channels of 72 MHz and two of 241 MHz bandwidth.

C-Band (MHz)

Global Beam:

4175 R	**AFRTS** (USA)(≡B-MAC)

NE. Zone Beam:

3900 L	**AFRTS-SEB** (USA)(≡B-MAC)

E. hemisph. Beam:

3725 R	**RTG-TV 1 Gabon** extra SCPC audio: 3728.5 MHz TV audio 3728.9 MHz extra audio carrier 3729.2 MHz **Radio Gabon** (French) 3729.6 MHz popmusic channel.
3770 R	**RTG-TV 2 Gabon**
3912 R	**ORTN Tele-Sahel**, Niger. extra SCPC audio: 3930 MHz TV audio 3939.7 & 3940.5 MHz radio channels.
3968 R	**RTA-TV Algerian Television** (PAL) sc: 6.2 MHz TV audio SCPC: 3895 MHz **Algerian radio** SCPC: 3895.4 MHz 2nd TV audio.

Ku-Band (GHz)

Spot Beam West:

10.969 t61aH	test pattern
11.016 t61bH	**TVN** comm. Norwegian TV (PAL) sc: 6.60 MHz TV audio sc: 7.38 MHz **Radio Nettverk** sc: 7.56 MHz **Radio Atlantic** sc: 7.74 MHz **Radio Nyheter/R. Noordsee Int.**
11.132 t63aH	**Nordisk TV 4**, Norway (D-MAC). extra audio: **NRK 1&2/Radio ROKS**
11.177 t63bH	**SVT-2**, Sweden (≡D-MAC) also 2nd Swedish radio program.
11.467 t67bH	sports feeds
11.683 t69bH	**SVT-1**, Sweden (≡D-MAC) also 1st and 3rd Swedish radio program.

Spot Beam East:

11.019 t71V	**IBA TV 3** (PAL testpattern)
11.174 t73V	**IBA TV 2** (PAL), audio: 6.60 MHz sc: 7.20 MHz **Kol Israel Radio 1**.(Hebrew)
11.597 t77V	**IBA TV 1** (PAL), audio: 6.60 MHz

5° W. Télécom 2B (France)

This satellite took over from the Télécom 1C in July 1992. Its payload is similar to the Télécom satellite at 8° West. The old 1C was redeployed at 3° East.

NB: All TV SECAM.

C-Band (MHz)

Semi-Global Beam:

3710 L	**RFO/France-2**
3731 L	**RFO**/various broadcasters, audio: 6.60 MHz target areas: Caribbean, Africa and Indian Ocean region.

Ku-Band (GHz)

Spot Beam V:

12.522 tR1	**Métropole-6**, audio: 5.8 MHz sc: 6.40/7.20 MHz **Frequence Mousquetaire**

	sc: 6.85/8.20 MHz **Europe 1**
	sc: 7.75 MHz **UNICO**
12.564 tR2	**France-2**, audio: 5.8 MHz
	sc: 6.40 MHz **Rires & Chansons**
	sc: 7.75/8.65 MHz **AFP Radio**
	sc: 8.20 MHz **Ondes Latines**
12.606 tR3	**ARTE**
	sc: 6.85/8.20 MHz **RTL Luxembourg**.
12.648 tR4	**Canal Plus** (≡Syster)
	sc: 5.8 MHz (encrypted TV audio)
	sc: 6.40 MHz **Couleur FM**
	sc: 6.85/8.20 MHz **Radio Monte Carlo**
	sc: 7.25 MHz **Mood Music 4**
12.690 tR5	**TF 1**, audio: 5.8 MHz
12.732 tR6	**Canal J**(eunesse)
	Canal Jimmy
	sc: 5.8 MHz TV audio
	sc: 6.40 MHz **Mont Martre**
	sc: 6.85 MHz **Chérie-FM**
	sc: 7.25/8.20 MHz **M-40**
	sc: 7.75 MHz **SER Rockland**
	sc: 8.65 MHz **Fourviere FM**

Spot Beam H:

12.543 tR7	test pattern
12.585 tR8	test pattern
12.627 tR9	test pattern (HD-MAC)
12.669 tR10	**Cine Cinema** (≡D2MAC)
12.711 tR11	DSR

8° W. Télécom 2A (France)

This satellite took over from the aging Télécom 1A in April 1992 and has a payload of 10 C-Band semi-global and hemispherical beams as well as 11 Ku-Band spot beam transponders. All Ku-Band transponders are centered on France with an EIRP of 52.4 dBW.

C-Band (MHz)

3725 L tC1	3965 R tC5b
3783 L tC2	4049 L tC6a (7)
3849 L tC3	4049 R tC6b (8)
3907 L tC4	4154 L tC7a (9)
3965 L tC5a	4154 R tC7b (10)

Ku-Band (GHz)

Spot Beam V:

12.522 tR1	**Canal Plus** (≡[16:9]D2MAC)
12.564 tR2	**France 2** (≡[16:9]D2MAC)
12.606 tR3	**Cine Cinefil** (≡[16:9]D2MAC)
12.648 tR4	**Canal Plus** (≡Syster)
	sc: 5.8 MHz TV audio
12.689 tR5	**Cine Cinema** (≡[16:9]D2MAC)
12.732 tR6	**Canal J(eunesse)/Canal Jimmy** (≡Syster)
	sc: 5.8 MHz TV audio

Spot Beam H:

12.543 tR7	**MCM Euromusique** (≡Syster)
	sc: 5.8 MHz TV audio
12.585 tR8	**Planete** (≡Syster)
	sc: 5.8 MHz TV audio
12.627 tR9	**Cine Cinefil** (≡Syster)
	sc: 5.8 MHz TV audio
12.669 tR10	**Cine Cinema** (≡Syster)
	sc: 5.8 MHz TV audio
12.711 tR11	**TV Sport** (≡Syster)

NB: Transmissions on this satellite are a mixture of SECAM and D2MAC. All SECAM broadcasts will be Syster encrypted, all D2MAC broadcasts will be Eurocrypt encrypted.

11° W. Gorizont 7 (CIS)

Launched in 1983 with a standard payload of 6 C-Band 34 MHz wide transponders and one experimental Ku-Band transponder, this satellite occupies the Statsionar 11 position. On July 15th 1992 another Gorizont satellite was successfully launched from Baykonur atop a Proton rocket and put at this position.

C-Band (MHz)

Spot Beam:

3675 R	no video activity

Global Beam:

3825 R	**EBU Moscow** (PAL)

Ku-Band (GHz)

11.525 R	occasional use

14° W. Gorizont 15 (CIS)

Launched in 1988 with the usual six 34 MHz C-Band transponders and two strong experimental Ku-Band carriers. This spacecraft occupies the Statsionar 4 position.

C-Band (MHz)

Spot Beam:

3675 R	**Ostankino TV 1/BrightSide TV**
	sc: 7.0 MHz TV audio
	sc: 7.5 MHz **Radio Mayak**, Moscow.

Varied beams:

3725 R	**RTM Malagasy** (Madagascar)(SECAM)
	sc: 7.50 and/or 6.65 MHz TV audio
3825 R	Intersputnik, news & sports exchanges.
	Cubavision (NTSC)(2230-0030 UTC)
	sc: 6.85 MHz TV audio
3875 R	**RFO** (French), audio: 6.60 MHz

Ku-Band (GHz)

Spot Beam:

11.525 R	**Moscow Visnews** (feeds)

18.5° W. Intelsat 515 (Int)

Launched in 1989, this satellite has 42 C-Band channels of 36 MHz, four Ku-Band channels of 72 MHz and two of 241 MHz bandwidth. This spacecraft was initially stationed at 60° E. but was redeployed at this position at the end of February 1992, replacing the Intelsat 506. The latter was redeployed over the Atlantic at 50° W.

C-Band (MHz)

Global Beam:

4166.5/ 4188.5 R	occasional news and sports feeds.

Ku-Band (GHz)

Spot Beam East V:

10.975 t71a	former Rete 4
11.005 t71b	former Telespacio II
11.135 t73a	former Rete Uno
11.170 t73b	former Canale 5

Spot Beam West H:

10.969 t61a	**STV 1** (D-MAC)
11.016 t61b	**TV-Norge**
	sc: 7.38 MHz **Radio Nettverk**
	sc: 7.56 MHz **Radio Atlantic**
	sc: 7.74 MHz **R. Nyheter/R. Noordsee Int.**
11.132 t63a	**TV-4 Norway** (D-MAC)
11.177 t63b	**STV 2** (D-MAC)
11.545	**TV2** (Norway)(≡D2MAC)

18.8° W. Olympus 1 (DBS slot 6) (ESA)

This satellite was launched in January 1989 and has 2 DBS-channels of 27 MHz bandwidth R, 4 low power Ku-Band channels of 18 MHz bandwidth and 3 low power Ka-Band of 240 and 1700 MHz bandwidth. Expected lifetime was approx. 10 years, but after some faulty manoeuvres were carried out in summer 1991 expected lifetime has decreased dramatically.

Channel:

24	**RAISat** (PAL/D2MAC)

Ku-Band (GHz)

12.525 H/V	occasional use
12.528 H/V	**CBC** (Canadian, occasional NTSC)
12.550 H/V	occasional use
12.570 H/V	occasional use

19° W. TDF 1A/1B (DBS slot 6) (France)

These satellites have 5 transponders of 27 MHz bandwidth and R polarization each. Broadcasting frequencies are between 11.7 and 12.1 GHz Expected lifetime is approx. 9 years. TV system D2MAC. Malfunctioning transponders (two permanent transponder failures so far) have made TDF 1A unreliable. TDF 1B has taken over most of the broadcasting activities.

Channel:

1	**Monte Carlo Music**
	extra audio (1): **Victor**
	extra audio (2): **RFI**
	extra audio (3): **RFI** (English sce)
5	**Canal Plus** (French)(≡EuroCrypt)
9	**ARTE** (French/German)
	extra audio: **Hector**
13	defunct
17	**France-2** [16:9]
	extra audio: **RFI**

19.2° W. TV-Sat 1/2 (DBS slot 6) (FRG)

Launched autumn 1987 (TV-Sat 1) and 1989 (TV-Sat 2). Expected lifetime approx. 8 years. These satellites have 4 DBS channels of 27 MHz bandwidth and 1 channel for 16 stereo radio programs. TV system D2MAC. In order to save energy only 4 channels can be used simultaneously.

Channel:

2	**RTL Plus**
	extra audio: **RTL Radio**
6	**Sat-1**
	extra audio: **Radioropa**
10	**3Sat**
	extra audio: **Deutsche Welle**
14	DSR, 16 channels.
	***Bayern 4 Klassik** (public, classical)
	***Deutschlandfunk** (public, news/talk)
	***DS Kultur** (public, classical)
	***HR2 Kultur** (public, classical)
	***Klassik Radio** (comm., classical)
	***NDR 3** (public, classical)
	***Radio ffn** (comm., pop)
	***Radioropa** (comm., news/talk)
	***RB2 Kulturell** (public, classical)
	***RIAS Eins** (public, pop)
	***RTL Radio** (comm., pop)
	***SDR/SWF2 Kultur** (public, classical)
	***SR1 Saarland** (public, classical)
	***Star*Sat Radio** (comm., pop)
	***WDR3 Köln** (public, classical)
18	**Eins Plus** (ARD/SRG)
	extra audio: **Deutschlandfunk**

21.5° W. Intelsat 502 (Int)

Launched in 1980, this aged satellite has 38 C-Band channels of 36 MHz, four Ku-Band channels of 72 MHz and two of 241 MHz bandwidth. Due to reduced stationkeeping, constant ground tracking is necessary for proper reception.

Ku-Band (GHz)

11.133 H
11.170 H

21.5° W. Intelsat K (Int)

This medium-powered spacecraft was launched on June 10th 1992 and is the first Ku-Band only satellite operated by Intelsat. The payload consists of 16 Ku-Band transponders of 54 MHz bandwidth, operating in the 11- and 12 GHz range. Expected lifetime approx. 10-12 years. Spot Beams: WNAm (47 dBW), CAm (45 dBW), SAm (45 dBW), CEu (50 dBW). As of October 1992 the Italian TV **RAI** broadcasts to the Americas and the Caribbean in both PAL and NTSC.

Ku-Band (GHz)

Spot Beam to USA and South America:

11.735 t1H/V
11.795 t2H/V
11.855 t3H/V
11.915 t4H/V

Spot Beams to all target areas:

11.465 t5V	**BrightStar** (feeds)
11.536 t6V	**BrightStar** (feeds PAL/NTSC)
11.500 t7V	**BrightStar** (feeds PAL/NTSC)
11.565 t8V	**BrightStar** (≡B-MAC)

Spot Beam to Europe:

12.536 t9 V
12.595 t10V
12.655 t11V
12.715 t12V
t13-16 double use Europe/USA

24.5° W. Intelsat 605 (Int)

This spacecraft was launched on 14th of August 1991 and replaces the Intelsat 602. Its payload is similar to the other Intelsat 600 satellites. The Intelsat 602 has been redeployed at 63° East.

C-Band (MHz)

4166 R	**Channel Africa** (SABC)
4188 R	occasional video

27.5° W. Intelsat 601 (Int)

The Intelsat 600 spacecraft are spin-stabilized satellites with one-half times the capacity of the 500-series satellites. This increased capacity was achieved by adding two more zone beam transponder sets and 2 additional 72 MHz wide hemispherical beam transponders. Late 1991 Intelsat announced that the Intelsat 604, previously at this position, was to be replaced by the Intelsat 601. The Intelsat 604 is redeployed at 60° East, replacing the Intelsat 602 (now 63° E).

C-Band (MHz)

Global Beam:	
4135 R	**FBIS** Caversham (USA/UK)(B-MAC)
E. hemisph. Beam:	
3650 R	**M-Net** (South African)(≡Irdeto)
	BBC World Service TV (≡SAT-TEL)
	(0000-0900 UTC, clear PAL)
	sc: 6.60 MHz TV audio
3743 R	**USIA/WorldNet/C-Span/BBC/CNN/DW Fernsehen**
	sc: 6.60 MHz TV audio
	sc: 7.03 MHz **VOA Europe**
	sc: 7.21 MHz **VOA**
3790 L	**WorldNet**, audio:6.60 MHz
3883 R	**Canale Horizons** (Senegal)(≡SECAM-Nagravision)
	sc: 6.60 MHz TV audio
3915 R	**Canal France Internationale** (SECAM)
	sc: 5.80 MHz TV audio
3931 L	**NMABATHO** (Africa)
4005 L	**CNN International** (NTSC), audio: 6.60 MHz
4023 R	**Libyan TV** (Tripoli)(PAL), audio: 6.20 MHz
4065 R	**NTA Channel 10** (Nigeria)(≡PAL-NIL),
	audio: SCPC 4037.2 MHz
E. Zone:	
3770 L	**BrightStar**, audio: 6.60 MHz
N.E. Zone:	
4048 L	French TV from USA (irregular, SECAM)
	sc: 5.80 and/or 6.60 MHz TV audio
4050 R	**CNN International** (PAL)
W. Zone:	
3860 L	**Televen-Venez**, audio: 6.65 MHz
3895 L	**VTV Venez**, audio: 6.65 MHz
W. hemisph. Beam:	
3740 R	**TVE** (Spain), audio: 6.65 MHz
3780 R	**Inravision** (Colombia), audio: SCPC
3995 R	**C-Span/WorldNet**, audio: 6.8 MHz
	sc: 5.94/6.12/7.34/7.43/7.52/7.61 MHz **VOA**

Ku-Band (GHz)

Spot Beam West H:	
10.975 t61	occasional use
11.015 t61	occasional use
11.055 t62	**Bravo** (≡SAVE)(*)
	sc: 6.65 MHz TV audio
	sc: 7.02/7.20 MHz stereo TV
	sc: 7.74/7.92 MHz **Electric FM**
11.090 t62	**The Parliamentary Channel**
11.135 t62	**The Children's Channel** (*)
	sc: 6.65 MHz TV audio
	sc: 7.38 MHz Dutch TV audio
11.175 t63	**Discovery Channel** (*)
	KinderNet (*)
	The Learning Channel (*)
	sc: 6.65 MHz TV audio
	sc: 7.02/7.20 MHz stereo TV
11.475 t69	occasional use
11.515 t69	occasional use
11.591 t69	**Ladbroke Horse Racing** (≡B-MAC)
Spot Beam East V:	
10.995 t71	**BBC World Service TV** (≡SAT-TEL)
	sc: 6.65 MHz TV audio
	sc: 7.02 MHz **BBC 648**
	sc: 7.20 MHz **BBC Foreign Service** (multil.)
11.080 t72	occasional feeds
11.155 t73	**CNN** (*)
11.470 t79	**EBU** (New York)
11.515 t79	**Country Music TV Europe**
	sc: 7.02/7.20 MHz stereo TV audio
11.565 t79	**SSVC** (≡Cryptovision)
	sc: 6.30 MHz **BBC Radio 5**
	sc: 7.20 MHz **BFBS/BBC World Sce.**
	sc: 7.38 MHz TV audio (digital Musicam)
	sc: 7.56 MHz **BBC World Sce** (multilingual)
	sc: 7.74 MHz **BBC for Europe**
	sc: 7.92 MHz **BBC World Sce** (multilingual)

	sc: 8.10 MHz **BBC World Sce** (multilingual)
11.646	occasional feeds
11.664 t79	**Kanal Market** (Turkish)

NB: Services indicated with (*) will move or have moved to the ASTRA satellites at 19.2° E.

31° W. Marcopolo I (DBS slot 7)

In July 1992 the Marcopolo II satellite was purchased by The Norwegian Telecoms Administration which redeployed it at 1° West. The future of the other spacecraft was unknown at time of editing.

31° W. Hispasat (Spain) (DBS slot 7)

This spacecraft was built by MATRA and British Aerospace and has a payload of two spot beams (Spain and Canary Islands), one hemispherical beam (Americas), one global- and one omni-directional beam. It was launched atop an ARIANE rocket in September 1992. The expected lifetime is approx. 10 years.
DBS channels (Spain):
23, 27, 31, 35, 39 R
Ku-Band

12.540 t1 V	12.015 t6 V	12.540 t10H	12.078 t15H
12.591 t2 V	11.517 t7 V	12.591 t11H	12.517 t16H
12.631 t3 V	11.597 t8 V	12.631 t12H	12.597 t17H
12.671 t4 V	11.668 t9 V	12.671 t13H	12.668 t18H
12.711 t5 V		12.711 t14H	

34.5° W. Intelsat 603 (Int)

After an unsuccessful launch in 1990 this satellite was considered to be lost, but a special Space Shuttle re-boost mission in may 1992 salvaged the spacecraft. It now replaces the older Intelsat 504, which was redeployed at 40.5° West.

C-Band (MHz)
Global Beam:

4166.5/	
4188.5 R	occasional News and Sports feeds.
4142 R	radio feeds **Kalaallit Nunaata Radioa** (Greenland)
E. hemisph. Beam:	
3980 R	Spanish video relay channel.
N.E. Zone:	
3908 L	**TVE-Internaçional** (Relay for Canary Islands)
3932 L	**TVE-2** (Relay for Canary Islands)

All three Spanish channels in (PAL) audio: 6.60 MHz
Ku-Band (GHz)
Spot Beam:

10.975 V	occasional use BBC (Britain)

41° W. TDRSS-41 (NASA, USA)

Both Columbia Communications and NASA use this spacecraft for telecommunications and video purposes. This satellite covers the US, Western Europe and areas over the Atlantic. The payload consists of 12 powerful C-qBand, as well as S- and Ku-Band transponders (used by NASA). On August 12th 1992, Columbia Communications Corp. fired up tr. 6 V.

C-Band (MHz)

3800 t6V	**Columbia Comm.** (NTSC)
	sc: 6.2/6.8 MHz TV audio

45° W. Panamsat f1 (USA)

Launched in 1987, this satellite has six horizontally polarized Ku-Band transponders of 72 MHz bandwidth centered on Europe and 18 C-Band transponders (12 of 36 MHz and 6 of 72 MHz bandwidth). Another Ku-Band Conus Beam is centered on North America. **RTV Beograd** (see 7° East) is relaying its European satellite program on one of the North American Conus Beam transponders. Because of the relatively wide bandwidth some stations share one transponder.
(broadcasts on "half-transponder" basis are indicated with a/b).

C-Band (MHz)
Spot Beams to Latin, South, North and Central America. Polarization: Latin and South America V, North and Central America H

Spot Beam: (to Latin America)	
3723 t1	**CNN** (B-MAC)
3798 t3	**TNT International** (B-MAC)
3883 t5	**ECO-Televisia** (Mexico)
4119.5 t15a	**ESPN International** (B-MAC)
4159.5 t15b	data
Spot Beam: (to South America)	
3750 t2a	**TV Nacionale de Chile** (ch7)(≡Leitch)
3769 t2b	**TV Nacionale de Chile** (≡Leitch)
3840 t4	**HBO Olé** (B-MAC)
3910 t6a	TELEFE (Argentina)
3929 t6b	Space (Argentina)
4060 t14	CTC-Voice
Spot Beam: (to North America)	
3720 t7a	future video
3750 t7b	future video
3770 t9a	data
3790 t9b	Hondutel-Voice
3900 t11	**HBO-Olé** (B-MAC)
4000 t16	data
4140 t18a	**Canal Sur** (Peru)
4181 t18b	future video
Spot Beam: (to Centr. America)	
3771 t8a	**Canal 5** Lima
3789 t8b	**Canal 4** Lima
3871 t10a	**Canal 13** Lima
3889 t10b	**Canal 2** Lima
3940 t12	Peru (educational)
3955.5 t13a	SCPC **BBC World Sce**
3998 t13b	**RAI/WTN/VisNews**

Ku-Band (GHz)

Spot Beam H: (to Europe)	
11.515 t19B	**Galavisión** (PAL), audio: 6.20/6.80 MHz
11.546 t20	SCPC **EuroJazz** (digital radio service for European cable headends)
11.595 t20B	**NHK** Tokyo (NTSC), audio: 6.60 MHz
11.639 t21A	**CBS News/ABC World News** (PAL/NTSC) sc: 6.20/6.80 MHz TV audio
11.676 t21B	occasional use
Conus Beam H: (to North America)	
11.724 t22A	
11.760 t22B	**RTV Beograd** (NTSC), audio: 6.20 MHz
	sc: 5.80 MHz English TV audio
	sc: 6.80 MHz **Radio Beograd I**
11.8058 t23A	SCPC **BBC World Sce** (Eng., Span, Por.)
11.840 t23B	
11.884 t24A	
11.920 t24B	**TV Japan/NHK** (feeds)

50° W. Intelsat 506 (Int)

Launched in 1983, this satellite has 38 C-Band channels of 36 MHz, four Ku-Band channels of 72 MHz and two of 241 MHz bandwidth. In addition two transponders around 1540 MHz for maritime communications. Expected lifetime is approx. 7-10 years. This satellite was redeployed at this position, after being replaced by the Intelsat 515 (18.5° W).

53° W. Intelsat 513 (IBS) (Int)

Launched in 1989, this Intelsat also has 46 C-Band channels of 36 MHz, four Ku-Band channels of 72 MHz and two of 241 MHz bandwidth. Special C-Band configuration allows for distribution of ten high powered channels into South America.

C-Band (MHz)

W. hemisph. Beam:	
3915 R	occasional video
Zone Beam:	
3975 R	occasional video
4015 L	occasional video
E. hemisph. Beam:	
3743 R	two half-transponders **RTP** feeds Lisbon-Madeira.
3798 R	**RTM-1** (Morocco)
3990 R	**M-2** (Morocco)(≡Discrete-12)
S.W. Spot Beam: (to South America)	
4055 R	**ENTV** (Ch.7) Bolivia, audio: 6.6 MHz
4086 R	**ATC** (Ch.7) Buenos Aires, audio: 6.6 MHz
4086 L	**RTP** (Ch.7) Lima, audio: 6.6 MHz
4140 R	**Omnivision** Venezuela, audio: 6.6 MHz
4140 L	**VCC** Argentina, 6.6 MHz
4166 L	**Liberta** (Ch.9) Argentina, audio: 6.6 MHz
4184 L	**CableVision** Argentina, audio: 6.6 MHz

Ku-Band (GHz)
Spot Beam H/V: Business traffic on Ku3-Band, further occasional video on regular Ku1-Band frequencies.

For Satellite Receivers Test
See "Listen To The World" Section

INTERNATIONAL TELECOMMUNICATION UNION REGION 2:

THE AMERICAS & THE CARIBBEAN

North American C-Band Channel Frequency Chart. (standardized)

Ku-Band frequency chart, based on 16-channel subdivision. For 32-channel format use half-spacing. No universal standard is set.

Channel 1 = 3720 / 11730 MHz.	Channel 13 = 3960 / 12096 MHz.
Channel 2 = 3740 / 11743 MHz.	Channel 14 = 3980 / 12109 MHz.
Channel 3 = 3760 / 11791 MHz.	Channel 15 = 4000 / 12157 MHz.
Channel 4 = 3780 / 11804 MHz.	Channel 16 = 4020 / 12170 MHz.
Channel 5 = 3800 / 11852 MHz.	Channel 17 = 4040 MHz.
Channel 6 = 3820 / 11865 MHz.	Channel 18 = 4060 MHz.
Channel 7 = 3840 / 11913 MHz.	Channel 19 = 4080 MHz.
Channel 8 = 3860 / 11926 MHz.	Channel 20 = 4100 MHz.
Channel 9 = 3880 / 11974 MHz.	Channel 21 = 4120 MHz.
Channel 10 = 3900 / 11987 MHz.	Channel 22 = 4140 MHz.
Channel 11 = 3920 / 12035 MHz.	Channel 23 = 4160 MHz.
Channel 12 = 3940 / 12048 MHz.	Channel 24 = 4180 MHz.

NB: All TV in 525-lines NTSC, audio 6.20 / 6.80 MHz. unless stated otherwise. (VC2) = VideoCipher II encrypted video.

69° W. Spacenet 2 (S2) (USA)

Launched in 1984, having 12 C-Band transponders of 36 MHz, 6 C- and 6 Ku-band transponders of 72 MHz bandwidth.

C-Band
Channel:

1 H	occasional video (news)
2 V	**Dr. Gene Scott**
3 H	**USIA/WorldNet/Deutsche Welle Fernsehen**
	sc: 7.605/7.515/7.425/7.355 MHz **VOA**
	sc: 5.95/6.10 MHz **VOA Europe** (in stereo)
4 V	occasional video
5 H	**Channel America**
7 H	**Main Street TV**
9 H	**Telemundo**
10 V	**Eco/Galavision**
11 H	**ESPN International**
12 V	**Eco/Galavision**
14 V	occasional video
17 H	**Telemundo** (Miami)
18 V	occasional video
19 H	**KBL** (Pittsburg)
20 V	**AFRTS** occasional video (≡B-MAC)
21 H	occasional video
22 V	**Canal Sur** (Spanish news from Lat. Am.)
23 H	**SCOLA** (international news)
	sc: 5.8 MHz **Radio France Internationale**

Ku-Band
Channel:

2 H	digital (SCPC)
4 H	occasional video
6 H	**TI-INN** (≡B-MAC)
8 H	occasional video
12 V	**CNN** (The Airport Channel)

70° W. Brasilsat (SBTS-A1/A2) (Brazil)

Both spacecraft were launched in 1986. The A2 was at first a spare satellite but was put into regular use in 1991.
NB: All TV in M-PAL (525 lines)
audio 5.8/6.6 MHz (unless stated otherwise).

C-Band MHz
Spot Beam:

3720 H	**Rede Globo TV**
3740 V	**SBT TV**.
3760 H	**TV Educativa** (Brazil)
3780 V	**Rede Manchete**.
3800 H	**Bandeirantes**.
3810 V	**Rede Amazonica TV**.
3820 H	**TV Amazona**
3830 V	**TV Record**, Brazil
3850 H	occasional video
3910 H	**Kay TV/Jockey Club** (≡B-MAC)
4140 V	**Joven Pan TV**
4170 V	regional TV
4180 H	occasional video
4190 V	regional TV

72° W. Satcom F-2R (F2) (USA)

Launched in 1983, having 24 C-Band transponders of 36 MHz bandwidth.

C-Band
Channel:

1 V	**The Arab Network**, audio: 6.8 MHz
2 H	**WRAL (CBS)** (≡VC2)
3 V	news & sports feeds
4 H	**WABC-New York (ABC)** (≡VC2)
5 V	occasional video (NASA)
7 V	**All News Channel** (≡VC2)
8 H	SCPC (19 channels)
9 V	SCPC (26 channels)
10 H	SCPC (19 channels)
11 V	**Home Shopping SPREE**
12 H	**WXIA-Atlanta (NBC)** (≡VC2)
13 V	**NASA Select TV**
14 H	SCPC (26 channels incl. **VOA**)
15 V	SCPC (15 channels)
16 H	occasional video
17 V	occasional video

74° W. Galaxy 2 (G2) (USA)

Lauched in 1983, having 24 C-Band transponders of 36 MHz bandwidth. Used primarily as a general communications satellite.

C-Band
Channel:

1 H	occasional video
2 V	occasional video
3 H	SCPC (30 sts a.o. **Peoples Network**)
4 V	news & sports
5 H	ecucational feeds
6 V	**FDMA**
7 H	occasional video
8 V	occasional video
9 H	occasional video
10 V	occasional video
11 H	occasional video
12 V	SCPC (9 channels)
13 H	**Virginia State Educational Network**
16 V	horse racing
17 H	news & sports
18 V	**University of Virginia**/horse racing
19 H	**Virginia Technological University**
20 V	horse racing (B-MAC)
21 H	horse racing/news & sports
22 V	horse racing from New York (B-MAC)
23 H	horse racing from the Poconos (B-MAC)
24 V	horse/dog racing

81° W. Satcom K-2 (USA)

Launched in 1986, mainly used as TV broadcaster's satellite. 16 Ku-Band transponders of 54 MHz bandwidth offer up to 32 TV channels in half2Dtransponder operation.

Ku-Band
Channel:

1 H	**NBC** (East)
2 V	SCPC
4 V	**CycleSat** (occasional)
5 H	**NBC** (Central)
6 V	news and sports feeds
7 H	**NBC Skypath NY/NBC A-News**
8 V	**USTV**
9 H	**NBC Skypath NY** (feeds)
11 H	occasional video
12 V	SCPC
13 H	**NBC News Channel** (feeds)
14 V	**The Asia Network**
	NAm Chinese TV
15 H	**NBC** (feeds)

82° W. Anik D2 (A2) (Canada)

Launched in 1984, having 12 horizontally and 12 vertically polarized C2DBand transponders of 36 MHz bandwidth. The ANIK D2 became surplus to requirements following the deployment of ANIK E1 at 111.1° West. It took over from the Satcom F-4 after it was withdrawn from service. Eventually the ANIK D2 will be replaced by new Satcom C-4 (currently at 135° W).

C-Band

Channel:	
1 V	**Sports Channel Philadelphia**
2 H	**Action Pay-Per-View**
3 V	**Sports Channel America**
4 H	**Cable Video Store** (≡VC2)
5 V	**New England Cable News Network**
6 H	**Madison Square Garden Network** (≡VC2)
7 V	**Sports Channel New York** (≡VC2)
8 H	**PreVue Guide/Superstar Guide** sc: 5.58/5.76 MHz **Tidewater Radio Network** sc: 7.38/7.56 MHz **KILA** (Las Vegas)
9 V	**Sports Vision Chicago** (≡VC2)
10 H	**Valu Vision** (shopping channel)
11 V	**Sports Channel New York**
12 H	**The International Channel**
13 V	**NESN**/occasional advertising, audio: 6.8 MHz
14 H	**Sports Channel Florida** (≡VC2)
15 V	**Shop At Home/Silent Network** sc: 5.80 MHz **Lighthouse Inspirational** sc: 6.20 MHz **Yesterday USA Superstation** (oldies)
16 H	**Sports Channel Ohio** (≡VC2)
17 V	**Shop At Home/Sports Channel Los Angeles**
18 H	**SPICE PPV** (≡VC2)
19 V	**Digital Music Express**
20 H	occasional video
21 V	**The Nostalgia Channel**, audio: 6.8 MHz sc: 6.20 MHz **Yesterday USA Superstation** (oldies)
22 H	**Playboy at Night** (≡VC2)
23 V	**Sports Channel New England** (≡VC2)
24 H	**Home Shopping Network**

85° W. Satcom K-1 (K1) (USA)

This Ku-Band only DBS-satellite is mainly used by PrimeStar & Partners, for delivering TV services to cable head end stations. The Broadcasting system used is B-MAC.

Ku-Band

channel:	
2 V	**Viewer's Choice I** (PPV)(≡B-MAC)
3 H	**Super Station WWOR** (≡B-MAC)
4 V	**Super Station KTVU** (≡B-MAC)
5 H	**TV Japan** (≡B-MAC)
6 V	**Viewer's Choice II** (PPV)(≡B-MAC)
7 H	SCPC
8 V	**Super Station WSBK** (≡B-MAC)
9 H	occasional video
10 V	**Request TV I** (PPV)(≡B-MAC)
11 H	**Super Station WPIX** (≡B-MAC)
12 V	**Super Station KTLA** (≡B-MAC)
13 H	**Super Station WTBS** (≡B-MAC)
15 H	occasional video
16 V	**Super Station WGN** (≡B-MAC)

85° W. Telstar 302 (T2) (USA)

Launched in 1984, having 24 C-Band transponders.

C-Band

Channel:	
5 V	**HBO-2** (tests)(≡VC2)
7 V	**CBS**
8 H	occasional video
9 V	**HBO** (East)(≡VC2)
10 H	**ABC-Network West** (stereo 6.2/6.8 wide discrete)
11 V	**ABC-Network** feed
14 H	**HBO** (West)(≡VC2)
15 V	**CBS News East** (stereo 6.2/6.8 wide discrete)
16 H	**CBS Network West**
17 V	feeds
18 H	feeds
19 V	feeds
20 H	**CBS Network East**
21 V	**Cinemax** (≡VC2)
22 H	occasional video
23 V	occasional video

87° W. Spacenet 3 (S3) (USA)

Launched in 1988, having 12 C-Band transponders of 36 MHz and 6 C- and 6 Ku-Band transponders of 72 MHz bandwidth. Ku-Band transponders are used in split beam configuration.

C-Band

Channel:	
1 H	SCPC
2 V	**The Nebraska Educational TV Network**
3 H	**WSBK** Boston (≡VC2) sc: 6.12 MHz **WROL** Boston.
4 V	**The Nebraska Educational TV Network (ETN-2)**
5 H	**KTVT Dallas-Ft.Worth** (≡VC2) sc: 6.16 MHz **Pan American Network**. (Spanish) sc: 5.76/5.94 MHz **USA Radio Network** (religious) sc: 6.8 MHz **Sun Radio Network** (talk)
9 H	**WPIX-New York** (≡VC2) sc: 5.76/5.93 MHz **Modern Country** sc: 6.30/6.48 MHz **National Black Network** sc: 6.80 MHz **IBS/Talkradio**.
10 V	**FDMA** (digital)
11 H	occasional video
12 V	**FDMA** (digital)
13 H	SpaceCom Systems (audio carriers) a.o. **Z-Rock** (SCPC)
14 V	**CNN** (≡Leitch)
15 H	**KTLA** (Los Angeles)(≡VC2) sc: 5.58/5.76 MHz **KKJZ** (Jazz)/**KKGO** Los Angeles sc: 5.94/6.48 MHz **Ambassador Inspirational Network** sc: 6.12 MHz **FBN/Financial News** sc: 6.16 MHz **Radio Sedeye** (Iran) sc: 6.30 MHz **Talk Radio 1** sc: 6.48 MHz **Talk Radio 2**
16 V	**Video Catalog**
17 H	SpaceCom Systems (audio subcarriers)
18 V	occasional sports & news.
19 H	**Sports South** (≡VC2), audio: 6.8 MHz
20 V	occasional video
21 H	**PASS** (≡VC2) sc: 6.2 MHz **Let's Talk Radio** (a.o. Gary Bourgois)
22 V	**Las Vegas Network**
23 H	**Home Team Sports** (≡VC2)
24 V	**Satellite Market USA**

Ku-Band

Channel:	
2 H	telecommunications
5 H	telecommunications
7 H	occasional video
9 H	occasional video
10 H	occasional video
13 H	occasional video
15 H	occasional video

91° W. SBS 4 (USA)

This is the oldest US Ku-Band only satellite and it was launched in 1984. It has a payload of 10 Ku-Band transponders of 43 MHz bandwidth. Expected lifetime is approx. 9 years.

Ku-Band

channel:	
1 V	TOYOTA
2 H	V-SAT Network
3 V	occasional video
4 H	occasional video
5 V	Automobile Sales Training Network
6 H	V-SAT Network
7 V	General Motors Satellite Network
8 H	V-SAT Network
9 V	V-SAT Network
10 H	V-SAT Network

93.5° W Galaxy 3 (G3) (USA)

Launched in 1984, having 24 C-Band transponders of 36 MHz bandwidth.

C-Band

Channel:	
1 H	**Encore** (≡VC2)
2 V	**FoxNet** feeds
5 H	DCR (Digital Cable Radio)
7 H	**ACTS** (christian) sc: 5.58 MHz **Gospel Radio Net.** sc: 5.76 MHz **Yesterday USA** sc: 5.98 MHz Spanish radio sc: 7.56 MHz religious radio sc: 7.74 MHz religious radio

	sc: 7.95 MHz **West & Wild Rock**
8 V	**Home Shopping Club 1**
	sc: 7.84 MHz **WCCO** Minneapolis
	sc: 8.05 MHz **Business Radio Net.**
	subc. 8.23 MHz **Minneapolis Public Radio**
9 H	**QVC Shopping Network**
	sc: 5.94/6.12/7.50 MHz **WCCO-AM** Minneapolis
10 V	**EWTV** (Eternal Word TV Network)
11 H	**Mind Extension University**
	sc: 5.04/7.74 MHz **Galactic Country**
	sc: 5.22/5.40 MHz **Galactic POP**
	sc: 5.58/5.76 MHz **Soft Sounds**
	sc: 5.94/6.12 MHz **Galactic Light Rock.**
	sc: 6.30/6.48 MHz **Galactic Classical**
	sc: 7.38/7.56 MHz **Galactic Jazz.**
	sc: 7.875 MHz **In Touch** (reading for the blind).
12 V	**Showtime** promos (only nights)
13 H	**The Weather Channel** (≡VC2)
14 V	**C-Span II** (US-Senate)
15 H	**Video Hits I** (≡VC2)
16 V	**Viewer's Choice II** (PPV)(≡VC2)
17 H	**MTV** (East)(≡VC2)
18 V	**MTV** (West)(≡VC2)
19 H	**Nickelodeon** (East)(≡VC2)
20 V	**Lifetime** (West)(≡VC2)
21 H	**Viewer's Choice I** (≡VC2)
22 V	**Nickelodeon** (West)(≡VC2)
23 H	**The Comedy Channel West** (≡VC2)
	sc: 5.58/5.76 MHz stereo TV
	sc: 7.38 MHz **WRVL-FM**, VA. (religious)
24 V	**C-Span I** (US-House of Representatives)
	sc: 5.2 MHz C-Span audio I (various SW)
	sc: 5.4 MHz **BBC World Service** (C-Span audio II)
	sc: 5.6 MHz audio program guide

95° W. SBS 3 (USA)

Launched in 1982, having 10 horizontally polarized Ku-Band transponders. Used primarily for video and telecommunication, both in entertainment and professional services, such as The Private Satellite Network, The American Law Network and Conus communications. All broadcasts in B-MAC.

Ku-Band
channel:

1 H	**Satellite Cinema 4** (≡B-MAC)
3 H	**Satellite Cinema 1** (≡B-MAC)
4 H	**Showtime 1** (≡B-MAC)
5 H	**WTBS/Satellite Cinema 2** (≡B-MAC)
6 H	**CNN/ESPN** (≡B-MAC)
8 H	**ESPN** (≡B-MAC)
9 H	**Satellite Cinema 3/Showtime 2** (≡B-MAC)
10 H	occasional video
14/16 H	**CONUS/HUBCOM**

96° W. Telstar 301 (T1) (USA)

Launched in 1983, having 24 C-Band channels.
Mainly in use for linking of Networks such as **CBS, Fox, Hughes and ABC**, as well as independent TV producers organized in **"Wold"**.

C-Band
Channel:

1 V	sports/occasional feeds
2 H	**CBS** (West)(≡VC2)
3 V	**Warner Brothers** (amusement)
4 H	**ABC** Los Angeles (occasional)
5 V	**CBS/Pro Sports** (occasional
6 H	**ABC** (occasional)
7 V	**CBS** (East)(≡VC2)
9 V	**Keystone Communications**
10 H	**ABC** (East)
12 H	**ABC**
13 V	occasional video
14 H	college and professional sports feeds
16 H	**ESPN** college sports feeds
17 V	college and professional sports feeds
19 V	recurring feeds
20 H	**Tokyo Broadcasting Service (TBS)**
21 V	variety
22 H	recurring feeds
23 V	**ESPN Computer Show**
24 H	various users

99° W. Galaxy 6 (G6) (USA)

G6 took over this orbital position at the end of 1991 after Westar 4 ran out of station-keeping fuel. G6 has a payload of 24 C-Band transponders of 36 MHz bandwidth.

C-Band
Channel:

1 H	SCPC (22 channels a.o. **Family Radio**)
2 V	SCPC (14 channels)
3 H	SCPC (50 channels, a.o. **BBC World Sce**)
4 V	SCPC (21 channels)
5 H	occasional video
6 V	**NCN Shepherd's Chapel**
	audio: 6.53 MHz **WCRP-FM**, Puerto Rico
7 H	occasional video
8 V	**CTNA** (religious)
10 V	occasional video
11 H	feeds
12 V	**ITN World News** (BBC News relay)
13 H	occasional video
14 V	**Family Net**
15 H	**World Harvest TV**
16 V	occasional video
17 H	**Opportunity Channel**
18 V	**ABC** backhaul to Tokyo/**NHK**
19 H	college sports, Oklahoma S.U.
20 V	various users
21 H	feeds
23 H	**Bore Sight** (only Thursday nights)
24 V	**ESPN** feeds.

99° W. SBS 6 (USA)

Launched on October 12th 1990, this Ku-Band only satellite has 19 transponders of 27 MHz bandwidth. This satellite will be replaced by the Galaxy VI satellite some time in 1993. All transponders are in use for news-, sports-, teleconference- and educational feeds.

Ku-Band
Channel:

4 H	occasional video	14 H	occasional video
6 H	occasional video	15 V	occasional video
8 H	occasional video	16 H	occasional video
9 V	occasional video	17 V	**All News Channel**
10 H	occasional video	20 H	occasional video
12 H	occasional video		

101° W. Spacenet IV (USA)

This spacecraft was launched on April 12th. 1991 and has a payload of six 27 MHz wide Ku-Band, six 27 MHz wide C-Band channels and twelve 36 MHz C-Band channels. Unfortunately at least three Ku-Band transponders have shown permanent failure so far.

C-Band
Channel:

9 H	**Three Angels** (religious)
10 V	occasional video
11 H	occasional video
18 V	occasional video

Ku-Band
Channel:

2 H	defunct	13 H	defunct
5 H	occasional video	15 H	test pattern
7 H	defunct	24 H	test pattern

103° W. GSTAR A1 (USA)

Launched in 1985, having 16 Ku-Band transponders of 54 MHz bandwidth, capable of delivering up to 32 video transmissions simultaneously. Two transponders are configured to cover the USA including Hawaï and Alaska with extra strong signals.
NB: all services use digital transmission techniques.

105° W. GSTAR 2 (USA)

Launched in 1986, having 16 Ku-Band transponders of 54 MHz bandwidth, capable of delivering up to 32 video transmissions simultaneously. Two transponders are configured to cover the USA including Hawaï and Alaska with extra strong signals.

Ku-Band
Channel:

11 H	occasional video
12 V	occasional video
13 H	occasional video

107.3° W. Anik E2 (Canada)

This is the first of a fifth generation ANIK satellites and was launched on April 5th 1991. The satellite's payload consists of 24 C-Band and 16 Ku-Band transponders. After suffering a lot of problems caused by jammed antennas the spacecraft finally became operational in July 1991. Late 1991 this satellite took over all services of the ANIK D1 (104.5° W).

C-Band

Channel:

1 H	**Meteomedia** (Canadian weather channel)
2 V	**TSN Sports Network** (≡Oak Orion)
	sc: 5.40/5.58 MHz talk/music
	sc: 5.76/5.94 MHz music/TV audio
4 V	**First Choice** (movies)(≡VC2)
5 H	CBC feeds (occasional)
6 V	**MuchMusic**
	sc: 5.41/6.2 stereo TV
	sc: 6.67 MHz TV audio
	sc: 7.50 MHz **Background Music**.
	sc: 6.84 MHz **CHIN** (Toronto)
7 H	**CBC Labrador**, Newfoundland
8 V	**CHCH Hamilton** (≡Oak Orion)
	sc: 5.40/5.58 MHz **FM 98** (Toronto)
9 H	**WDIV** (NBC), Detroit (≡Oak Orion)
	sc: 6.2 MHz Canadian radio feeds
10 V	**WXYZ** (ABC), Detroit (≡Oak Orion)
11 H	**CBC** (North Pacific)(occasional)
	sc: 6.12 MHz **CBC Radio**
12 V	**Global Satellite Services**
13 H	**Super Channel** (≡Oak Orion)
14 V	**CFTM**, Montreal
	sc: 5.76 MHz **CITE**, Montreal
	sc: 6.255 MHz **CKAC**, Montreal. (French)
	sc: 6.435 MHz **CKAC Radio**, Montreal. (French)
	sc: 7.605 MHz **CJMS-AM**, Montreal. (French)
15 H	**CBC-French Network**
	sc: 5.40/5.76 MHz **Radio Canada East** (variety)
	sc: 5.58 MHz **CBC Radio** (French, ethnic)
16 V	**CBC House of Commons** (French)(≡Oak Orion)
	sc: 5.40/5.58 MHz **CBC Stereo F** (classical)
	sc: 5.76/5.94 MHz **CBC Stereo E** (classical)
	sc: 6.12/6.3 MHz **Vancouver Radio**
	sc: 7.44 MHz **Voice Print Reading**
17 H	CBC feeds
	sc: 6.17 MHz **CBC-FM** (English/Ind.)
18 V	**CITV Edmonton** (≡Oak Orion)
	sc: 5.41 MHz **Canadian Country**
	sc: 6.44 MHz **CFNR-FM**
	sc: 6.8 MHz **CKRW** Whitehorse
19 H	**CBC-North** (Atlantic)
20 V	**CFTM Montreal**
	sc: 6.12 MHz **CBM** Vancouver
	sc: 5.4/5.58 MHz **CBC** (French)
	sc: 5.76 MHz **CBU** Vancouver
21 H	**WTVS-(PBS)** Detroit (≡Oak Orion)
	sc: 5.41 MHz **CKNM-FM** Yellowknife (C&W)
22 V	**CHAN** Vancouver (≡Oak Orion)
	sc: 5.76/5.94 MHz **CBU-FM** Vancouver
	sc: 6.8 MHz **CFMI-FM** New Westminster
	sc: 7.38 MHz **CBU-AM** Vancouver
23 H	**WJBK** (CBS), Detroit (≡Oak Orion)
	sc: 5.41 MHz **CKRW-AM/CHON-FM** (C&W)
	sc: 6.17 MHz **VOCM-AM** (rock)
24 V	**CBC House of Commons** (English)
	sc: 5.40/5.58 MHz **CBC Stereo F** (classical)
	sc: 5.76/5.94 MHz **CBC Stereo E** (variety)
	sc: 6.12/6.30 MHz **CBC Stereo E** (classical)

Ku-Band

channel:

3 H	**Canal Famille** (≡VC2)
4 H	**Musique Plus**
	sc: 5.76/5.94 MHz stereo TV
5 H	**Radio Quebec** audio: 6.41 MHz
5 V	occasional video
6 H	**Quatre Saisons**
	sc: 5.76/5.94 MHz stereo TV
6 V	occasional video
7 V	occasional video
8 H	**La Réseau des Sports** (≡VC2)
9 H	**La Chaine** audio: 5.41 MHz
9 V	**Alberta Access Network** (educational)
10 H	**TV Ontario** (educational)
11 H	**Super Ekran** (≡VC2)
12 H	**Ontario Legislative Assembly**
	sc: 5.41 MHz TV audio
13 H	**Youth TV**
	sc: 5.76/5.94 MHz stereo TV
13 V	occasional video
14 H	**The Family Channel** (≡VC2)
15 H	**Atlantic Satellite Network**
	sc: 5.41 MHz TV audio

110° W. Anik C1 & C2 (Canada)

These two satellites were co-located at this position in 1992.

Ku-Band

video activity on channels 6, 8, 18 H, 5 and 7 V

111.1° W. Anik E1 (A1) (Canada)

This satellite was launched late 1991 and took over from the ANIK D2, which moved to 82° West.

C-Band

Channel:

7 H	occasional video
8 V	occasional video
10 V	SCPC
12 V	**CTV** (feeds)
13 H	occasional video
15 H	**Vision TV Network/Musique Plus**
	sc: 5.58/5.94 MHz stereo TV
16 V	**CTV** (≡Leitch)
17 H	**TV5 Quebec**
	sc: 5.41/6.12 MHz **Radio France Internationale**
18 V	occasional video
19 H	**TVNC (TV Northern Canada)**
20 V	**Canadian Home Shopping Club**
	sc: **Q-107** (rock)
22 V	occasional video
23 H	occasional video
24 V	**CTV** (≡Leitch)

Ku-Band

channel:

14 H	testpattern
15 H	**CBC World News** (feeds)

113.5° W. Morelos 1 (F1) (Mexico)

Launched in 1985, having 18 C-Band and 4 Ku-Band transponders. Six of the C-Band and all Ku-Band transponders are 72 MHz wide, the others have a 36 MHz bandwidth. A second, identical satellite (F2) is positioned at 116.5° West. Both spacecraft will eventually be replaced by the new "Solidarity 1 and 2" satellites.

C-Band

Channel:

2 V	**XHIMT**, Mexico City.
	sc: 7.40 MHz **Radio Felicidad**, Mexico City.
4 V	SCPC
8 V	**XHDF**, Mexico City, audio: 6.2 MHz
9 H	occasional video
10 V	**XHGC**, Mexico City, audio: 6.2 MHz
11 V	occasional video
13 V	occasional video
14 V	**XHTV** (Channel 4, Mexico City) audio: 6.2 MHz
17 V	occasional video
18 H	occasional video
20 V	occasional feeds
21 V	occasional video
23 V	occasional video

Ku-Band

Channel:

4 H	**XHIMT**, Mexico City.

114.9° W. Anik C3 (Canada)

This satellite is used for telecommunications purposes only

116.7° W. Morelos 2 (F2) (Mexico)

This spacecraft is identical to the one at 113.5° W.

C-Band

Channel:

8 V	**XHGC TV**, audio: 6.20 MHz
9 H	**Super Canal**, audio: 6.20 MHz

10 V	occasional video
11 H	occasional video
12 V	**TNT International** (B-MAC)
13 H	**Canal 11**, audio: 6.20 MHz
14 V	**XEW-TV**, audio: 6.20 MHz
	sc: 7.38 MHz Mexican Radio
15 H	**TVC** (B-MAC)
16 V	occasional video
17 H	occasional video
18 V	occasional video
19 H	CNI text service
20 V	occasional video
21 H	occasional video
23 H	occasional video

120° W. Spacenet 1 (S1) (USA)

Launched in 1984, this satellite has six C-Band and six Ku-Band transponders of 72 MHz bandwidth. In addition there are also 12 C2DBand transponders of 36 MHz bandwidth available. Split Ku-Band beam configuration is standard.

C-Band

Channel:

1 H	**Digital Planet** (audio scs.)
2 V	**Univision** (feeds)
3 H	**PBS-C**
4 V	**Shop At Home**
5 H	**PBS-B**
	sc: 5.40 MHz **SUN Radio Net** (News, talk)
	sc: 5.58/5.76 MHz **WCIE-FM** (AP News)
7 H	**Three Angels Broadcasting** (religious)
9 H	**PBS-D** (≡VC2)
11 H	occasional video
13 H	**Keystone Inspirational**
12 V	data
14 V	data
15 H	**ESPN Basketball**
16 V	data
17 H	**SpaceCom Systems** (audio)
	sc: 5.04 MHz **Muzak Service**
	sc: 5.2 MHz data
	sc: 5.38/5.74 MHz **Safeway In Store Music**
	sc: 5.58 MHz data
	sc: 5.94 MHz **Grocery Muzak Channel**
	sc: 6.12 MHz **Contemporary Country**
	sc: 6.3 MHz **Muzak Service**
	sc: 6.48 MHz **Safeway In Store Music**
	sc: 6.62 MHz **Music Service**
	sc: 7.18/7.72 MHz **Safeway In Store Music**
	sc: 7.36 MHz unidentified
	sc: 7.56 MHz **KISN** Salt Lake City
	sc: 7.92 MHz **Muzak Service**
19 H	sports feeds
21 H	occasional video
22 V	occasional video
23 H	**PBS** (children's programming)
24 V	**SCOLA** (educational)
	sc: 5.8 MHz **Radio France Internationale**
	sc: 6.2 MHz secundary TV audio

Ku-Band

10 V	**CNN** (The Checkout Channel)

123° W. Telstar 303 (T3) (USA)

Launched in 1985, having only 24 C-Band channels. Telstar 303 (coming from 125° W) took over this orbital slot in May 1992.

NB: Transponders 1,2,3,4,5,6,7,8,9,20: Pay-Per-View by **TVN**.

C-Band

channel:

18 H	**Fox** (East), audio: 5.8/6.2/6.8 MHz
19 H	**Showtime Multiplex 2** (≡VC2)
23 V	**Fox** (West)
24 H	preview

125° W. GSTAR IV (USA)

This spacecraft has a payload of 16 Ku-Band transponders of 54 MHz bandwidth. The GSTAR VI was launched in 1990 and has a maximum EIRP of 45.9 dBW.

Ku-Band

channel:

3 H	occasional video
4 H	occasional video
8 V	occasional video
13 H	test pattern
15 H	occasional video

125° W. Galaxy 5 (USA)

Galaxy V is a new Hughes Communications satellite and was launched in March 1992. It replaces the older Telstar 303 satellite, which was redeployed at 123° West.

C-Band

channel:

1 H	**The Disney Channel** (≡VC2)
3 H	**Trinity Broadcasting**
4 V	**Sci-Fi Channel** (≡VC2)
5 H	**CNN** (≡VC2)
	sc: 6.3 MHz **CNN Radio**
6 V	**WTBS-Atlanta** (≡VC2)
7 H	**WGN-Chicago** (≡VC2)
	sc: 6.075 MHz **Seeburg Lifestyle 1**
	sc: 6.165 MHz **Seeburg Lifestyle 2**
	sc: 6.3/6.48 MHz **WFMT** (Chicago)
8 V	**HBO** (West)(≡VC2)
9 H	**ESPN** Primary feed (≡VC2)
10 V	**Shop At Home**
11 H	**The Family Channel** (≡VC2)
	sc: 6.3/6.48 MHz **CBN Radio Net.**
12 V	**Discovery Channel** (≡VC2)
13 H	**CNBC** (≡VC2)
14 V	**The Movie Channel** (≡VC2)
15 H	**HBO** (East)(≡VC2)
16 V	**Cinemax** (West)(≡VC2)
17 H	**TNT** (≡VC2)
18 V	**The Nashville Network** (≡VC2)
	sc: 7.38/7.56 MHz **WSM-FM** (Nashville)
19 H	**USA Network** (East)(≡VC2)
20 V	**Black Entertainment TV**
21 H	**Mind Extension University**
	sc: 5.04/7.74 MHz **Galactic Country**
	sc: 5.22/5.40 MHz **Galactic POP**
	sc: 5.58/5.76 MHz **Soft Sounds**
	sc: 5.94/6.12 MHz **Galactic Light Rock.**
	sc: 6.30/6.48 MHz **Galactic Classical**
	sc: 7.38/7.56 MHz **Galactic New Age of Jazz.**
	sc: 7.875 MHz **In Touch** (reading for the blind).
	sc: 8.11/8.38 MHz **Golden Oldies**
22 V	**Headline News** (≡VC2)
	sc: 6.3 MHz **CNN Radio**
23 H	**Arts & Entertainment** (≡VC2)
24 V	**Showtime** (East)(≡VC2)

128° W. ASC 1 (USA)

Launched in 1985, this satellite has twelve 36 MHz and six 72 MHz wide C-Band, as well as six 72 MHz wide Ku-Band transponders. Sole user is Global Access Telecommunications Services (GATS).

131° W. Satcom C-3 (USA)

This spacecraft, the second of two advanced cable TV satellites, was launched atop a Ariana 4 rocket on September 10th 1992. The Satcom C-3 has a payload of 24 linear C-Band channels of 36 MHz bandwidth.

C-Band

Channel:

1 V	**The Family Channel** (East)
2 H	**The Learning Channel**
3 V	**Viewers Choice I**
4 H	**Lifetime** (West)
5 V	**Vision Interfaith**
6 H	**Courtroom TV**
7 V	**C-Span I**
8 H	**QVC Fashion**
9 V	DSR
10 H	**Home Shopping Network II**
11 V	**Prime Network**
12 H	**Arts & Entertainment**
13 V	**The Weather Channel**
14 H	**NESN**
15 V	**Showtime** (East)
16 H	**MTV** (West)
17 V	**The Movie Channel** (East)
18 H	**Nickelodeon** (West)

19 V	**Viacom Growth**
20 H	**Jones Intercable**
21 V	**Comedy Central** (East)
22 H	**TBA**
23 V	**E! Entertainment**
24 H	**TBA**

133° W. Galaxy 1 (G1) (USA)

Launched in 1983, this satellite has a standard 24 C-Band channels of 36 MHz bandwidth payload. No Ku-Band transponders.
NB: during 1992 most services moved to Galaxy V (125° W)
C-Band
Channel:

1 H	**The Comedy Central** (East)
6 V	**Univision**
10 V	**The Movie Channel** (East)(≡VC2)
11 H	**Eternal Word**
13 H	**Country Music TV** (≡VC2)
15 H	**WWOR-NY**, Secaucus, New Jersey (≡VC2) sc: 6.30/6.48 **WQXR-FM** (classical)
16 V	**Showtime** (West)(≡VC2)
17 H	**The New Inspirational Network** sc: 5.58/5.76 MHz stereo TV sc: 6.20 MHz **Satellite Radio Network** (religious) sc: 7.92 MHz **USA Radio Network** (religious)
19 H	**Cinemax** (East)(≡VC2)
20 V	**Galavision** (Spanish)
21 H	**USA** (East)(≡VC2)
22 V	**Discovery Video Mall**
23 H	**HBO** (East)(≡VC2)
24 V	**The Disney Channel** (West)(≡VC2)

135° W. Satcom C-4 (USA)

This new satcom satellite was launched on August 31st 1992, from Cape Canaveral Air Force Station. Late 1992/early 1993, it will take over all video services from the ANIK D2 at 82° West.

137° W. Satcom C-1 (F1) (USA)

Launched in 1983, this satellite has a standard 24 C-Band-only channels of 36 MHz bandwidth payload.
C-Band
Channel:

1 H	occasional video
2 V	**KUSA TV** (ABC) Denver (≡VC2)
3 H	**KRMA** Denver (≡VC2)
4 V	**Pacific Sports Channel** (≡VC2)
6 V	**KMGH** (CBS) Denver (≡VC2)
7 H	**Prime Ticket Sports** (≡VC2) sc: 5.58 MHz **KKJZ/KSHO-AM** sc: 5.76 MHz **KSUR-FM** Monterey CA.
8 V	**NBC Network** (≡Leitch)
9 H	**NBC TV Network**
10 V	**American Agriculture Network/HSE** (≡VC2) sc: 6.80 MHz **KKBQ-FM** (contemporary)
11 H	occasional video
13 H	occasional video
14 V	**KCNC** Denver (≡VC2)
16 V	occasional video
17 H	**Prime Sports Network** Feed
18 V	**Prime Sports Network** Feed
20 V	**Sport Channel L.A.** (≡VC2)
21 H	NHK backhaul (occasional)
22 V	**Prime Sports Northwest** (≡VC2)
23 H	**KWGN** Denver (≡VC2)
24 V	**Sunshine Network** (≡VC2)

139° W. Satcom C-5 (C5) "Aurora II" (USA)

This C-Band only satellite was launched in 1982, carrying a standard 24 channels of 36 MHz bandwidth payload. It has reached its projected end-of-life period, and is primarily used for non-video purposes. Transponders 1-23 are used for DATS (Digital Audio Transmission System).
C-Band
Channel:

24 H	**Alaska Satellite TV** sc: 6.80 MHz **KABN-AM**, Long Island, Alaska. sc: 7.38/7.56 MHz **KSKA-FM**

INTERNATIONAL TELECOMMUNICATION UNION REGION 3:

ASIA, AUSTRALIA, JAPAN & THE PACIFIC

180° E. Intelsat 508 (Int)

Launched in 1984, this satellite has 42 C-Band channels of 36 MHz bandwidth, four Ku-Band channels of 72 MHz and two of 241 MHz bandwidth. In addition two transponders around ~540 MHz for maritime communications.
C-Band (MHz)
Global Beam:

4014 t13R	**NHK/TV Tokyo/ABC**
4045 t18R	**RFO** (Tahiti)
4052 t18R	**Nine Network Australia**
4166 R	occasional use
4187 t24R	**WTN** (Pacific feed)

W. hemisph. Beam:

3719 t1 R	**ESPN** (USA)(≡B-MAC)
3720 t2 L	occasional use
3766 t3 R	**Network 10** (Australia) sc: 6.60 MHz **Today's Japan** (NHK)
3840 t4 L	occasional use
3845 t7 R	**CNN International** sc: 6.65 TV audio
3878 t9 R	**NBC/CNBC**
3898 t10R	occasional use
3931 t12R	**ABC/CBS**
3974 t14R	**WorldNet/C-Span/Deutsche Welle Fernsehen** sc: 6.6 MHz TV audio sc: 7.02/7.20/7.335/7.425/7.515/7.605 MHz **VOA**

Ku-Band
E. Spot Beam:

11.480 t79V	**TBS**/CBS/EBN, audio 5.8/6.8 MHz
11.510 t79V	**Fuji TV**/ABC, audio 5.8/6.8 MHz

177° E. Intelsat 511 (Int)

Launched in 1985, this satellite has 46 C-Band channels of 36 MHz, 4 Ku-Band transponders of 72 MHz and 2 of 241 MHz bandwidth. It was first positioned at 63° East, but redeployed at this position early 1992. Previously at this position was the Intelsat 503, which was moved to 183° East in September 1992.
C-Band (MHz)
Global Beam:

4135 t22R	**AFRTS** Pacific relay (≡B-MAC) sc: 6.80 MHz TV audio
4167 t23R	sports & news feeds PAL/SECAM/NTSC sc: 6.60/6.65 MHz TV audio
4187 t24R	**TV New Zealand** (news and sports feeds)

NW Zone Beam:

3799 t5 R	**CNN International** sc: 6.80 MHz TV audio

174° E. Intelsat 510 (Int)

Launched in 1985, the Intelsat 510 satellite has 46 C-Band channels of 36 MHz, four Ku-Band channels of 72 MHz and two of 241 MHz bandwidth. This satellite was moved in April 1990 from 24.5° W. to serve the Pacific Ocean Region (POR).
C-Band (MHz)
Global Beam:

4166 R	occasional feeds
4177.5 R	occasional news and sports feeds.

NB: This orbital position will be taken over by the first Intelsat 700 satellite, to be launched late 1993.

162° E. Superbird B1 (Japan)

This spacecraft was launched in 1992 and has 19 linear polarized Ku-Band and 10 circular Ka-Band channels. It will mainly serve telecommunications purposes, but a number of Ku-Band transponders are already active carrying video and audio transmissions. Expected lifetime is approx. 10 years.
Ku-Band
Spot Beam:

12.290 t1 V	**TV Tepco**
12.320 t11H	**TBS**

12.340 t12H **NIHON TV Netw.**
12.360 t13H **NIHON TV Netw.**
12.370 t3 V **The Power Channel**
12.380 t14H **Fuji TV**
12.400 t15H **Fuji TV**
12.410 t4 V **One World Television** (≡SkyPort)
Universal One Television
12.430 t16H **Space Wave**
12.450 t5 V **Eisei Channel**
12.470 t17H occasional video
12.490 t6 V **Nihon Cable TV Net** (≡SkyPort)
12.530 t7 V **Space Vision Net**
12.550 t19H **Let's Try/Life Design**
12.570 t8 V **Cable Soft Network**
12.610 t9 V **Super Channel/Weather Channel** (≡SkyPort)
12.650 t10V **Japan Cable Television** (≡SkyPort)
12.670 t22H test pattern
12.690 t10V **The Music Channel**
12.730 t12V **Star Channel**

164° E. Aussat A3 (AUS)
160° E. Aussat A1 (AUS)
156° E. Aussat A2 (AUS)

Launched in 1985 (1 & 2) and 1987 these three satellites provide an extensive television network for the Australian "outback" and Pacific territories in a score of different arrays. Aussat deploys B-MAC standard video signals and the radio signals are "piggy-backed" onto the TV signal. Each satellite has 15 Ku-Band channels of 45 MHz bandwidth. Frequencies are in the 12.25-12.75 GHz range and receivers should be based on the specification DOC 512.

NB: On August 14th 1992 the first satellite of the AUSSAT B program was successfully launched. The **AUSSAT B1** satellite will take over all activities form the A1 satellite (160° East) some time in 1993. A second AUSSAT B satellite will replace the A2 (156° East). The AUSSAT A3 will remain in service until 1997.

footprints:
Central serving South Australia and the Northern Territory.
North Eastern serving Queensland
South Eastern serving New South Wales, Australian Capital Territory, Victoria and Tasmania
Western serving Western Australia

HACBSS (Homestead and Community Broadcasting Satellite Service) refers to the provision of **ABC** (Australian Broadcasting Corporation) TV and radio broadcasts via satellite.

RCTS (Remote Commercial Television Service) refers to the broadcast of commercial TV programs to remote areas via satellite.

RCRS (Remote Commercial Radio Service) refers to the broadcast of commercial radio programs to remote areas.

AUSSAT A3
Ku-Band
to North Australia
12.405 t3 V **Network 9 Astralia** (≡VAMP E-PAL)
to Central Australia
12.629 t14H **Sky Channel** (≡B-MAC)

AUSSAT A1
Ku-Band
to North Eastern Australia:
12.277 t1 V **TCN** Channel 9 (≡VAMP E-PAL) feeds
12.405 t3 V **ATN 7** (≡VAMP E-PAL)
12.469 t4 V **TV 10** feed, audio: 6.6 MHz (companded)
12.597 t6 V **TV 10** feed (≡VAMP E-PAL)
sc: 7.38/7.56 MHz TV audio
12.309 t9 H **TCN** Channel 9 (PAL) interchange.
Companded audio: 6.8 MHz Stereo 7.38/7.56 MHz
12.373 t13H SCPC (ABC)
12.629 t14H **HACBSS** (≡B-MAC)
sc: **ABC** National Radio
sc: **ABC** Regional Radio
12.693 t15H **RCTS** (≡B-MAC)
to South Eastern Australia:
12.553 t5 V **SBS** (Special Broadcasting Service)(≡B-MAC)
12.661 t7 V **HACBSS** (≡B-MAC)
sc: **ABC** National Radio
sc: **ABC** Regional Radio
12.725 t8 V AUSSAT interim use B-MAC/PAL teleconferencing.
RCTS B-MAC (future).
12.501 t12H **ABC** interchange (PAL), audio: 6.8 MHz

AUSSAT A2
Ku-Band
to Central Australia:
12.501 t12H **HACBSS** (ABC SA)
sc: **ABC** National Radio
sc: **ABC** Regional Radio
12.629 t14H **IMPARJA** RCTS/RCRS
12.693 t15H **HACBSS** (ABC NT)
sc: **ABC** National Radio
sc: **ABC** Regional Radio
to Western Australia:
12.611 t7 V **Golden West Network/RCRS** (≡B-MAC)
12.725 t8 V **HACBSS**
sc: **ABC** National Radio
sc: **ABC** Regional Radio
to South East Australia:
12.533 t5 V **Techn. Advanced Further Educ.** (TAFE)
The Enterprise Channel

154° E. JCSat 2 (Japan, JCSC)

This satellite was launched in 1989, together with the JCSat 1 and is used to deliver TV, telephone, facsimile and high speed data services to Japanese businesses and consumers. The communications payload consists of 32 Ku-Band transponders of 27 MHz bandwidth. All audio 6.20 MHz, unless stated otherwise.

Ku-Band (GHz)
12.283 t2 H **Japan Religious Broadcasting Channel Co. Ltd**
12.313 t4 H **Nikkei Satellite News**
12.343 t6 H occasional video
12.373 t8 H **Satellite ABC**
12.403 t10H occasional video
12.433 t12H occasional video
12.448 t13V **Fantasy Channel** (≡B-MAC)
High Tech Shower (≡M-System)
12.463 t14H **Rainbow Channel** (≡M-System)
12.478 t15V **The Book Channel**
12.493 t16H **Satellite Culture Japan/TCC** (≡B-MAC)
12.524 t18H **Space Shower TV**
12.538 t19V test pattern
12.568 t21V **Channel NECO**
12.583 t22H **Japan Sports Channel**
12.613 t24H **Satellite Theatre**

150° E. JCSat 1 (Japan, JCSC)

This spacecraft was launched in 1989 together with the JCSat 2 and carries the same payload. All audio 6.20 MHz, unless stated otherwise.

Ku-Band (GHz)
12.283 t2 H **TV ASAHI**
12.313 t4 H occasional video
12.328 t5 V test pattern
12.373 t8 H occasional video
12.403 t10H occasional video
12.433 t12H **AUCNET**
12.463 t14V **P-Sat (MPT-Japan)**
Nihon Mercantile TV (≡)
12.583 t22H occasional video
12.643 t26H occasional video
12.688 t29V occasional video
12.733 t32H occasional video

132° E. CS-3a Sakura (Japan)

Launched in 1988, having 2 C-Band and 10 Ka-Band (17.7-19.45 GHz) transponders. Mainly used for the relay of domestic TV programs. Details of users unknown.

113° E. Palapa B2P (Indonesia)

Launched in 1987, this satellite replaces Palapa B2 which appeared unusable after ending up in a faulty orbit. It has 24 C-Band transponders of 36 MHz bandwidth. Late 1992 it was announced that the Australian Broadcasting Organisation **ABC** began to lease a transponder on this satellite to support its international transmission network to 15 Asian countries.

NB: All TV PAL, unless stated otherwise.

C-Band (MHz)
Spot Beam:
3800 t5 H **RCT1** Indonesia
3820 t6 V **TV 3** (commercial TV, Malaysia)
sc: 6.60 MHz TV audio

3840 t7 H **CNN International** (B-MAC)
3850 t8AV **Army TV Channel 5**, Bangkok, audio SCPC
3870 t8BV **BBTV channel 7**, Bangkok, audio SCPC
3880 t9 H SCPC ch 5 & 7
3900 t10H **TV 1 Malaysia**
3980 t14V occasional video (B-MAC)
4000 t15H **ABS/CBN** (Philippines), audio: 5.0 MHz
4020 t16V **ASEAN/VIZ** news feeds, audio: 7.55 MHz
4042 t17V **ESPN International** (B-MAC)
4060 t18H SCPC (7 carriers Radio Thailand)
4180 t24H occasional video

110.5° E. DFH2-A2 (China)

This spacecraft was launched in 1988 and payload is similar to the DFH2-A1 at 87.5° East.
C-Band (MHz)
3848 t7 H **CETV** (China)
sc: 6.7 MHz TV audio
4178 t24H **CETV** (China)
sc: 6.54 MHz TV audio

110° E. BS-3A & BS-3B (Japan)

Two broadcasting satellites also known as Yuri-3A/B. BS-3A was successfully launched in August 1990 and started broadcasting in November of that year. BS-3B was launched in August 1991. Both satellites are equipped with three 12 GHz transponders for broadcasting and one wide-band data channel with 37 dB EIRP over the main part of Japan.
Program providers: **NHK** (government) and the commercial **JSB** (Japan Satellite Broadcasting)
Ku-Band (GHz)
11.765 t11R **DBS-2 (NHK)**
11.804 t5 R **JSB** (≡COATEC)
11.842 t7 R **DBS-1** (Joho)
11.880 t9 R HDTV tests

108° E. Palapa B2R (Indonesia)

This satellite was launched on April 13th 1990 and has an expected lifetime of approx. 8 years. The payload comprises 24 C-Band channels of 36 MHz bandwidth.
C-Band (MHz)
Spot Beam:
4000 H **TVRI**, Djakarta, Indonesia (PAL)
sc: 6.80 MHz TV audio
4140 V **Indonesian TV** feeds (PAL)
sc: 6.80 MHz TV audio

105.5° E. AsiaSat 1 (Asia)

This spacecraft was launched in April 1990 by a Chinese Long March rocket. Originally, this satellite was called Westar 6 and was launched by an American space shuttle in 1984. It failed to lift into a geostationary orbit and was retrieved by a special shuttle mission in 1985. Afterwards it was sold to the AsiaSat consortium. Payload consists of 24 C-Band transponders of which 15 have been leased by Hong Kong, Myanmar (Burma), Pakistan and Mongolia. **BBC World Service Television** provides a 24 hours-a-day Asian news service from Hong Kong. Other channels are operated by **Star TV**, Hong Kong (see address section).
North Beam audio subcarrier frequencies: 5.58 MHz, 5.76 MHz, 5.94 MHz, 6.12 MHz, 6.30 MHz. South Beam audio subcarrier frequencies: 6.3 MHz, 6.48 MHz, 6.66 MHz, 6.84 MHz, 7.02 MHz, 7.20 MHz.
C-Band (MHz)
North Beam: (NTSC, unless stated otherwise)
3720 H tN1
3760 H tN2 **Mongolian Television** (SECAM)
3800 H tN3 **Sports Channel**
sc: 5.58/5.76 MHz Eng. stereo TV
sc: 5.94 MHz (Mandarin)
3840 H tN4 **Music Channel** (24 hrs-a-day)
sc: 5.58/5.76 MHz Eng. stereo TV
sc: 6.48 MHz **BBC World Sce** (Engl.)
3880 H tN5 **BBC World Service TV**
sc: 5.58/5.76 MHz Eng. stereo TV
sc: 5.94 MHz (Mandarin)
3920 H tN6 **Chinese Channel**
sc: 5.58/5.76 MHz Mandarin stereo TV
sc: 5.94 MHz **BBC World Sce** (Asian)
3960 H tN7 **Entertainment Channel**
sc: 5.58/5.76 MHz Eng. stereo TV
sc: 6.48 MHz **BBC World Sce** (Asian)
3980 V tN8 **Chinese Channel**
sc: 6.30/6.48 MHz Mandarin stereo TV
4040 H tN9 **Chinese TV**
sc: 7.50 MHz Chinese local radio
4080 H tN10
4120 H tN11 **CCTV 1** (China)
4160 H tN12
South Beam: (PAL, unless stated otherwise)
3740 V tS1
3760 H tS2
3820 V tS3
3860 V tS4 **Sports Channel**
sc: 6.30/6.48 MHz Eng. stereo TV
sc: 7.20 MHz Mandarin
3900 V tS5 **Music Channel** (24 hrs)
sc: 6.30/6.48 MHz Eng. stereo TV
sc: 7.20 MHz **BBC World Sce** (Eng.)
3940 V tS6 **BBC World Service TV**
sc: 6.30/6.48 MHz Eng. stereo TV
sc: 7.20 MHz (Mandarin)
3980 H tS7 **Zee TV** (Indian entertainment in Hindi)
sc: 6.8 MHz TV audio
sc: 7.20 MHz **BBC World Sce** (Asian)
4020 V tS8 **Entertainment Channel**
sc: 6.30/6.48 MHz Eng. stereo TV
sc: 7.20 MHz **BBC World Sce** (Asian)
4060 V tS9
4100 V tS10 **Pakistan TV**
audio: 6.60 MHz
4140 V tS11 **Myanmar TV** (Burma)(NTSC)
audio: 6.60 MHz
4180 V tS12

103° E. Statsionar 21 (CIS)

C-Band (MHz)
3876 R **Azerbaidzhan TV**

99° E. Ekran 19 (CIS)

Aside from this one, launched in 1987, Several satellites with the same purpose stationed here to secure uninterrupted service. Domestic UHF relay of Orbita III, Siberian version of 1st Russian TV program on 714 MHz Powerful signal, available over all of Asia.
C-Band (MHz)
3714 R **Moscow 1**

96.5° E. Gorizont 19 (CIS)

Also known as Statsionar 14, this spacecraft was launched in September 1989. This satellite has a payload of six C-Band transponders and one Ku-Band transponder of 34 MHz bandwidth.
C-Band (MHz)
Spot Beam East:
3800 R **ATN** (Asian TV Network in Hindi)
3825 R **CCTV 1** (China)(PAL)

91.5° E. Intelsat 501 (Int)

This orbital position is a newly established fourth service region, to provide expanded coverage and improved connectivity to the Asian and Pacific countries. The Intelsat 501 spacecraft was moved from 183° East to provide new telecommunication- and TV services at this position. Frequencies not known at time of editing.

90° E. Gorizont 13 (CIS)

Launched in 1987, with the usual six C-Band transponders and experimental 11 GHz transponder.
C-Band (MHz)
Spot Beam:
3675 R **Vostok Orbita III**, CT-1 version for C.-Siberia
sc: 7.0 MHz TV audio
sc: 7.5 MHz **Radio Orbita 3**
N. hemisph. Beam:
3875 R **The Asian Network** (India), audio 5.8 MHz

87.5° E. DFH2-A1 (China)

This satellite was launched in March 1988 and carries a payload of 4 C-Band channels. Expected lifetime is approx. 10 years.

C-Band (MHz)
transp:
1 **CCTV 1**
2 **CCTV 2**
3 currently in use by **CETV 1** (educational)
4 spare

83° E. Insat 1D (India)

This satellite was launched from Cape Canaveral on June 12th 1990 after almost a year's delay. The Insat 1D satellite has 12 transponders for telecommunication and 2 high-power national coverage TV broadcast transponders. Expected lifetime is approx. 7 years.

S-Band (MHz)
2575 V **AIR** (India)
2615 V **Community TV**

80° E. Gorizont 10 (CIS)

This satellite was launched in July 1991 on top of a Proton rocket carrier. It replaces the aging Statsionar 13 and will provide communications and TV services troughout the Urals and Western Siberia.

C-Band (MHz)
3875 R feeds in SECAM/PAL/NTSC
THVN (Vietnam), audio: 7.50 MHz
Ku-Band (GHz)
11.525 L (a.o) **Kasachstan TV** (Alma Ata)
sc: 7.00 MHz TV audio
sc: 7.50 MHz **Radio Alma Ata**

74° E. Insat 2A (India)

This is the first of the second generation Insat satellites and was launched in the summer of 1992. It will provide telecommunication services as well as TV and radio. All India (Doordarshan) TV claimes for itself the distinction of being the first user agency to use the C-Band transponders of this satellite. In August 1992 the Doordarshan regional services of Karnataka (Bangalore TV), Tamil Nadu (Madras TV) and Maharashtra (Bombay TV) were shifted from the Arabsat 1C satellite at 31° East to this new spacecraft. Further moves will follow.

C-Band (MHz)
4115 H/V **Doordarshan TV** (Madras TV)
4135 H/V **Doordarshan TV** (Bangalore TV)
4175 H/V **Doordarshan TV** (Bombay TV)

66° E. Intelsat 505 (Int)

Launched in 1982, this satellite has 42 C-Band channels of 36 MHz, four Ku-Band channels of 72 MHz and two of 241 MHz bandwidth. In addition two transponders around 1540 MHz for maritime communications.

C-Band (MHz)
Global Beam:
4175 R **USIA/WorldNet/Deutsche Welle Fernsehen**
sc: 6.80 MHz TV audio
sc: 7.02/6.20/7.335/7.425/7.515/7.605 MHz **VOA**
E. hemisph. Beam:
3760 R **CCTV 3**, China. (PAL) audio: 6.65 MHz
3800 R **CCTV 7**, China. (PAL) audio: 6.60 MHz
3925 R **CCTV 11**, China. (PAL) audio: 6.60 MHz
Zone Beam:
3835 L **RTM2 TV**, Malaysia (PAL)
sc: 6.60 MHz TV audio
W. hemisph. Beam:
3725 L **SAIS** South African Information Sce. (PAL Video-Crypt)
sc: 7.02/7.20 MHz **Radio Metro**
sc: 7.56/7.92 MHz **Radio 5**
sc: 7.38 MHz **Radio 2**
sc: 7.74 MHz **Radio SABC**
3793 R **Zaïre TV (ORTZ)**
3823.7 SCPC TV audio
3824 SCPC radio
3997 R **Canal France Internationale**
4055 R **BOP TV1**
Ku-Band (GHz)
no video activity.

Non-geostationary: Molniya (CIS)

Three Molniya satellites are used to supply those areas of the CIS that are too close to the North Pole for reception of standard Gorizont TV signals. Payload is similar to Gorizont/Raduga satellites: C-Band frequencies including capacity to broadcast on frequencies as low as 3425 MHz. Broadcasts cover a large part of the world, from the North pole down to a line going across Paraguay, Nigeria, Egypt, Iran, China and Japan. A second loop is made from India across to the most eastern parts of Mongolia into the CIS, but it is unknown if any broadcasting takes place from there. Aside from TV broadcasting many non2Dvideo services are operated, a.o. the famous Washington-Moscow "Hot Line". A fourth Molniya satellite, Molniya-1, was launched on August 6th 1992 from the Plesetsk cosmodrome. Maybe broadcasts into Canada will resume with this new satellite. The latest Molniya was launched on October 14th 1992.

C-Band (MHz)
Global Beam:
3875 MHz **Double II** (East-Siberian version of the second TV program)

EUROPE

Addresses

3Sat, Postfach 4040, Mainz, D-6500, Germany. **Te:** +49 (6131) 70-1 **Fax:** +49 (6131) 705412 **Telex:** 4187930 zdf d

Antena 3 TV, Ctra Madrid-Irun Km 19.3, 28700 San Sebastian de los Reyes (Madrid), Spain. **Te:** +34 (1) 623 0500 **Fax:** +34 (1) 652 7144 .

ARTE, 50 avenue Theopile Gautier, 75016 Paris, France. **Te:** +33 (1) 44 14 77 77 **Fax:** +33 (1) 44 14 77 00

BBC-TV Europe, Woodlands, 80 Wood Lane, London, W12 0TT, United Kingdom. **Te:** +44 (81) 576 2248 **Fax:** +44 (81) 743 9256

BFS3 (Bayerischer Rundfunk), Rundfunkplatz 1, Munich 2, D-8000, Germany. **Te:** +49 (89) 5900 01 **Fax:** +49 (89) 5900 2375 **Telex:** 522480

BrightStar Communications, Room 723 PD PS 211, 120 Holborn, London, EC1N 2TE, United Kingdom. **Te:** +44 (71) 936 2000

British Sky Broadcasting, 6 Centaurs Business Park, Grant Way, Off Syon Lane, Isleworth,Middx, TW7 5QD, United Kingdom. **Te:** +44 (71) 782 3000 **Fax:** +44 (71) 782 3030

Canal Courses (PMU), 83 rue de la Boetie, 75008 Paris, France **Te:** +33 (1) 4953 3900 **Fax:** +33 (1) 4289 0895

Canal J, 91 rue de Cherche Midi, Paris, 75006, France. **Te:** +33 (1) 49 54 54 54 **Fax:** +33 (1) 49 42 22 87

Canal Jimmy, 45 Place Abel Gance, 92654 Boulogne-Cedex **Te:** +33 (1) 4761 9955 **Fax:** +33 (1) 4761 9400

Canal Plus España, Torre Picasso, Plaza Pablo Picasso s/n, 28020 Madrid, Spain **Te:** +34 (1) 396 5500 **Fax:** +34 (1) 556 4259

Canal Plus Satellite, 85-89 Quai André Citroën, 75015 Paris, France **Te:** +33 (1) 44 25 1000 **Fax:** +33 (1) 44 25 12 34

Canal Sur, Ctra San Juan de Aznalfarache Tomares, km. 1300, 41920 San Juan de Aznalfarache, Sevilla, Spain. **Te:** +34 (54) 763 111 **Fax:** +34 (54) 761 630

Children's Channel, 9/13 Grape Street, London, WC2H 8DR, United Kingdom. **Te:** +44 (71) 240 3422 **Fax:** +44 (71) 497 9113 **Telex:** 28622 child g

Country Music TV Europe, look under North America.

CNN News-Europe, CNN House, 19-22 Rathbone Place, London, W1P 1DF, United Kingdom **Te:** +44 (71) 637 6700 **Fax:** +44 (71) 637 6768

Deutsche Welle Fernsehen (formerly RIAS TV), Voltastraße 5, 1000 Berlin 65, Germany. **Te:** +49 (30) 46 990 **Fax:** +49 (30) 46 31 998

Deutsches Sportfernsehen (DSF) (fomerly Tele Fünf), see Germany, TV section

Discovery Channel Europe, Twyman House, 16 Bonny Street, London, NW1 9PG, United Kingdom. **Te:** +44 (71) 284 1570 **Fax:** +44 (71) 284 2042

EBC, Wagistrasse 4, Schlieren/Zürich, CH-8952, Switzerland. **Te:** +41 (1) 730 4144 **Fax:** +44 (1) 730 3877

Eins Plus (ARD), Arnulfstrasse 42, Munich 2, D-8000, Germany. **Te:** +49 (89) 5900 3280/1 **Fax:** +49 (89) 5900 3249 **Telex:** 522480

EuroPACE, Boîte Privée 529, 2 Place de la Défence, Paris, 92053, France. **Te:** +33 (1) 4776 3335 **Fax:** +33 (1) 4676 4272

European Television Networks (European Sports Network, Lifestyle, Kindernet) The Quadrangle, 180 Wardour Street, London, W1V 4AE, United Kingdom. **Te:** +44 (71) 439 1177 **Fax:** +44 (71) 439 1415 **Telex:** 255871 whcwo2

Eurosport, 17 rue de l'Arrivée, 75015 Paris, France **Te:** +33 (1) 4275 1234 **Fax:** +33 (1) 4275 2890
EuroStep, P.O. Box 11112, 2301 EC Leiden, The Netherlands
FilmNet, Archimedesbaan 21, Nieuwegein, 3439 ME, The Netherlands. **Te:** +31 (3402) 50 120
France-2, 22 avenue Montaigne, Paris, 75008, France. **Te:** +33 (1) 4421 4242 **Fax:** +33 (1) 4421 5145 **Telex:** 642313
Galavision, Televicine SA, Arrastaria S/n, Pol Las Mercedes, Ctra Nacional II km 11.300, 28022 Madrid, Spain **Te:** +34 (1) 329 3917/329 3873 **Fax:** +34 (1) 747 6463
Independent Television News (ITN), ITN House, 200 Grays Inn Road, London WC1, UK. **Te:** +44 (71) 83 33 000 **Fax:** +44 (71) 43 04 344
Intersputnik, 2nd Smolensky Blvd 1-4, Moscow, 121099, CIS. **Te:** +7 (95) 244 0333 **Telex:** 411 288
interSTAR (AKK Magic Box), Divan Yolu Turbedar Sok 22, Cagaloglu, Istambul, Turkey. **Te:** +90 (1) 511 4328 **Fax:** +90 (1) 511 8999
Japan Satellite TV Ltd (JSTV), Unit 17 Kings exchange, Tileyard Rd., London N7 9AH, United Kingdom **Te:** +44 (71) 607 7677 **Fax:** +44 (71) 607 7442
Kabelkanal Der, Martin Kollar-Straße 13, 8000 München 82, Germany
Kindernet, P.O. Box 7146, Amsterdam, 1007 JC, The Netherlands. **Te:** +31 (20) 665 9576 **Fax:** +31 (20) 692 87 72
M6 (Métropole Télévision), 16 cours Albert Premier, Paris, 75008, France. **Te:** +33 (1) 44 21 6666 **Fax:** +33 (1) 45 63 7822
MBC Ltd.(Middle East Broadcasting Centre), 10 Heathmans Road, London SW6 40J, United Kingdom. **Te:** +44 (71) 37 19 59 7 **Fax:** +44 (71) 37 19 60 1
MCM Euromusique, Immeuble Le Levant, 305 avenue le Jour Se Lève, 92100 Boulogne Billancourt, France **Te:** +33 (1) 4694 6161 **Fax:** +33 (1) 4694 0015/34
MTV Europe, Centre House, 20-23 Mandela Street, London, NW1 0DU, United Kingdom. **Te:** +44 (71) 383 4250 **Fax:** +44 (71) 388 2064 **Telex:** 929580
N 3 (see Germany, TV Section).
n-tv, Taubenstraße 1, O-1086 Berlin. **Te:** +49 (30) 6093 820
Nordic Channel, Stockholmsvägen 30, 18281 Stocksund, Norway. **Te:** +47 (8) 624 4400 **Fax:** +47 (8) 854606
Nor-Net, Nordre gate 20, Oslo 5, N-0551, Norway. **Te:** +47 (2) 38 38 40 **Fax:** +47 (2) 38 43 97
NRK, Norsk Rikskringkasting, Bj Bjoenssons Plass 1, N-0340 Oslo 3, Norway **Te:** +47 (2) 459050 **Fax:** +47 (2) 459645
Pro 7, 8 Leopoldstrasse 145, Munich 40, D-800, Germany. **Te:** +49 (89) 360 30 **Te:** +49 (89) 361 7792 **Telex:** 52113943
Rai Uno, Rai Due, Radiotelevisione Italiana, Via le Mazzini 14, Rome, 00195, Italy. **Te:** +39 (6) 3878/350304 **Fax:** +39 (6) 361 2012 **Telex:** 614432
Red Hot Dutch (Continental TV Plc.), I. Mex House, 40 Princess Street, M1 6DE Manchester, UK. **Te:** +44 (61) 237 1017 **Fax:** +44 (61) 237 1018
RTL Plus/RTL-2, see Germany TV section
RTL-4, Villa Louvigny, Allée Marconi, Luxembourg, L-2850, Luxembourg. **Te:** +352 47661 **Fax:** +352 4766 2730
Sat 1, see Germany TV section
Scansat TV3, P.O. Box 2094, Stockholm, S-10313, Sweden. **Te:** +46 (8) 403 333 **Fax:** +46 (8) 216 564 **Telex:** 810 6269
Scansat TV 1000 Succé Kanalen, P.O. Box 2133, Stockholm, 10314, Sweden. **Te:** +46 (8) 100 019 **Fax:** +46 (8) 243 840
Super Channel, Melrose House, 14 Lanark Square, Lime Harbour, London. E14 9QD, United Kingdom. **Te:** +44 (71) 41 89 418. **Fax:** +44 (71) 41 89 420/21
Te: +44 (71) 637 6700 **Fax:** +44 (71) 637 6768 **Telex:** 269169 tbsi g
Tele-5 (Telecinco), Plaza Pablo Picasso, S/N Planta 14, 28020 Madrid, Spain. **Te:** +34 (1) 396 6100 **Fax:** +34 (1) 555 0044
Teleclub, Fluelastrasse 7, Zürich, 8048, Switzerland. **Te:** +41 (1) 492 4433 **Fax:** +41 (1) 492 0122 **Telex:** 823516 fcag ch
TF1, 17 rue de l'Arrivee, 750015, Paris, France. **Te:** +33 (1) 42 75 1234 **Fax:** +33 (1) 42 75 2688
TRT International, Nevsat Tandogan Caddesi 2 Kavaklidere, Ankara, Turkey **Te:** +90 (4) 190 4491/19044
TV Asia, 9 Belvue Business Centre, Belvue Road, Northolt, Middlesex UB5 5QQ, **Te:** +44 (81) 841 5112 **Fax:** +44 (81) 845 2266
TV 5, 174 rue de la Université, Paris, 75007, France. **Te:** +33 (1) 45 56 00 80 **Fax:** +33 (1) 45 56 00 04 **Telex:** 203303
TV 5 Nordic, Stockholmsvågen 30, 18271 Stocksund, Stockholm, Sweden. **Te:** +46 (8) 850085 **Fax:** +46 (8) 854606
TV Norge, Sagvm 17, Oslo 4, 0458, Norway. **Te:** +47 (2) 3533 79 **Fax:** +47 (2) 3543 74
TV Sport, Immeuble Le Levant, 305 avenue Le Jour Se Lève, 92100 Boulogne Billancourt, France. **Te:** +33 (1) 4694 6100 **Fax:** +33 (1) 4694 0030
TVE Internaçional, Television Española, SA, O'Donnell 77, 28007 Madrid, Spain. **Te:** +34 (1) 409 5773 **Fax:** +34 (1) 574 2015 **Telex:** 43281
UK Gold, Thames Television House, 306-316 Euston Road, London NW1 3BB. **Te:** +44 (71) 388 8877 **Fax:** +44 (71) 383 3986
US Airforces Europe, Broadcast, DBS & Cable Office, Bldg 548, Flpltz Ramstein, D-672, Germany. **Te:** +49 (6371) 6565 **Fax:** +49 (6371) 43469 **Telex:** 45958
VOX (Westschienenkanal), P.O. Box 30 14 33, Richard-Byrd-Straße 6, 5000 Cologne 30, Germany. **Tel:** +49 (211) 16 02 10
West 3, see German section.

CIS

OSTANKINO TV (ROSSIJASKAJA GOSUDARSTVENNAJA TELERADIOKOMPANIJA), Akademika Korolova 19, 127 427 Moskva (headquarters), CIS. **Te:** +7 (095) 2337860 (Platnickaja office)

NORTH AMERICA

ABC, 77 West 66th Street, New York NY 10023.
ACTION PPV, 100 Wilshire Blvd., ste. 400, Santa Monica CA 90401 **Te:** +1 (213) 394-0477
ACTS, 6350 West Freeway, Fort Worth TX 76150.
AFRTS, 1735 N. Lunn, Suite 210, Arlington VA 22209
AgriSat, P.O.Box 314, Belmont WI 53510.
Alberta Access Network, 295 Midpark Way S.E., Calgary, Alberta T2X 2A8 Canada
All News Channel, 3415 University Minneapolis MN 55414 **Te:** +1 (612) 642-466
American Extasy, 419 Park Avenue South, Suite 1400, New York NY 10016.
American Movie Classics, 150 Cross Way Park West, Woodbury NY 11797 **Te:** +1 (516) 364 2222
American Satellite TV Alliance, 16 Broadway, Ste. 400, Valhalla NY 10595
AMCEE, 500 Tech. Parkway NW Atlanta, Georgia 30313 **Te:** +1 (404) 894 3362
Arts & Entertainment, 555 Fifth Avenue, 10th floor, New York NY 10017.
Atlantic Satellite Network, 285 Robie Street, Halifax, Nova Scotia, Canada B3J 2Z4
BET, 1899 Ninth Street NE, Washington DC 20018.
Biznet, 1615 H Street, Washington D.C. 20062 **Te:** +1 (202) 463 5817
Bravo, 150 Crossways Park West, Woodbury NY 11797.
Brightstar, International Building Rockefeller Center 630, 5th Avenue 22th floor, New York NY 10111 **Te:** +1 (212) 582 8578
BTN, 127 9th Avenue North, Nashville TN 27234 **Te:** +1 (615) 251 2000
C-Span, 400 N. Capitol Street NW, Suite 650, Washington DC 20001.
Cable Radio Network, 10487 Sunland Blvd, Sunland CA 91040 **Te:** +1 (818) 352 7152
Cable Value Network, 14605 28th Avenue N., Minneapolis MN 55441.
Canadian Broadcasting Corporation (CBC), P.O. Box 8478, Ottawa, Ontario K1G 3J5, Canada
Caribbean Super Station (CSS), 333 Roosevelt, Hollywood Hills, FL 33021 **Te:** +1 (305) 985 2641
CBN, CBN Centre, Virginia Beach VA 23463.
CBS, 51 West 52nd Street, New York NY 10019.
CNN, One CNN Centre, Box 105366, Atlanta GA 30348.
Continental Broadcasting Network, Virginia Beach, VA **Te:** +1 (804) 424 7777
Conus Communications, 3415 University Avenue St. Paul, MN 55114 **Te:** +1 (612) 642 4645
Country Music TV, 704 18th Avenue South, Nashville TN 37203.
CTV Television Network Ltd., 42 Charles St. East, Toronto, Ontario M4Y 1T4, Canada
Discovery Channel, 8201 Corporate Dr., Suite 1260, Landover MD 20.
Disney Channel, 3800 W. Alameda Avenue, Burbank CA 91505.
E! Entertainment, 1800 North Vine Street, Hollywood, CA 90028
ESPN, ESPN Plaza, 935 Middle St., Bristol CT 06010.

Eternal Word TV Network, 5817 Old Leeds Road, Birmingham AL 35210.
Family Net, P.O. Box 196, Forest VA 24551-0196.
First Choice, 98 Queen St. East, Suite 300, Toronto, Ontario M5C-1S6, Canada.
First Run, 7270 NW 25th Street, Miami FL 33122.
Florida News Network (FNN), 1851 Southhampton Jacksonville, 32207
Galavision, 2121 Avenue of the Stars, Suite 2300, Los Angeles CA 90067.
HBO/Cinemax, 1100 Avenue of the Americas, New York NY 10036.
Hit Video USA, 1000 Louisiana, Suite 3500, Houston TX 77002.
Home Shopping Network, 1529 US Highway 19 South, Clearwater FL 33546.
Home Team Sports, 1111 18th Street NW., Suite 200, Washington DC 20036.
HSE, 5251 Gulfton Street, Houston TX 77081.
KCNC, 1044 Lincoln Street, Denver CO 80203.
KMGH, 123 Speer Blvd., Denver CO 80203.
KRMA, 1261 Glenarm Place, Denver CO 80204.
KTVT, P.O. Box 2495, Fort Worth TX 76103.
KUSA, 1089 Bannock Street, Denver CO 80217.
KWGN, P.O. Box 5222, Denver CO 80155.
Learning Channel, 1525 Wilson Blvd, Suite 550, Rosslyn VA 22209.
Lifetime, 36-12 35th Avenue, Astoria NY 11106.
Madison Square Garden, 4 Pennsylvania Plaza, New York NY 10001.
MTV, 1775 Broadway, New York NY 10019.
Much Music, 299 Queen West, Toronto,Ontario M5V-2Z5, Canada.
Nashville Network, 2806 Opryland Dr., Nashville TN 37214.
National Christian Netw., P.O. Box 416, Gravette AR 72736.
National Jewish TV, 2621 Palisades Avenue, Riverdale NY 10463.
NBC, 30 Rockefeller Plaza, New York NY 10020.
Netlink, 5806 Lake Washington Blvd, Ste 400, Kirkland WA 98033.
New England Sports Netw., 70 Brookline Avenue, Fenway Park, Boston MA 02215.
Nickelodeon, 1775 Broadway, New York NY 10019.
North America One (KSBN), Highway 14 East, Suite 300, Box 642, Richland Center WI 53581.
Nostalgia Channel, Royal Commerce Centre, 2025 Royal Ln, Suite 300, Dallas TX 75229-3220.
PBS, 1320 Braddock Place, Alexandria VA 22314.
Playboy Channel, 8560 Sunset Blvd., Los Angeles CA 90069.
Premier Choix, 666 Sherbrooke W, 15th floor, Montreal PQ H3A 1E7, Canada.
Prime Time 24, 342 Madison Avenue, Suite 1520, New York NY 10173.
Pro Am Sports Systems, 500 Stevenson St., Suite 204, Troy MI 48084.
PTL, Charlotte NC 28279.
QVC Network, Goshen Corp. Park, West Chester PA 19380.
SBCA, 300 N. Washington St., Ste 208, Alexandria VA 22314.
SCOLA, 2500 California St, Omaha NE 68178-0778.
Showtime/The Movie Channel, Paramount Plaza Bldg, 1633 Broadway, New York NY 10019.
Silent Network, P.O. Box 1902, Beverly Hills CA 90213.
Sports Channel Florida, 1951 NW 19th Street 102, Boca Raton FL 33431.
Sports Channel New England, 150 Crossways Park West, Woodbury NY 11797.
Sportsvision, 875 N. Michigan Avenue, Suite 1540, Chicago 60611.
Superstar Connection, (United Video), P.O. Box 35728, Tulsa OK 74153-07278.
TBS, One CNN Center, P.O. Box 105366, Atlanta GA 30348.
Telesat (Canadian satellite operator), 1601 Telesat Court, Gloucester, Ontario K1B 5P4, Canada
Three Angles Broadcasting, P.O. Box 582, West Frankfort IL 62896.
Travel Channel, 1370 Avenue of the Americas, 27th floor, New York NY 10019.
Trinity Broadcasting Netw., 2442 Michelle, Tustin CA 92680.
TSN, 1155 Leslie Street, Don Mills, Ontario M3C-2J6, Canada.
University Network, 1615 South Glendale Avenue, Glendale CA 91205-3395.
Univision, 460 W 42nd Street, New York NY 10036.
USA Network, 1230 Avenue of the Americas, New York NY 10020.
VH-1, 1775 Broadway, New York NY 10019.
Viewer's Choice, 1633 Broadway, New York NY 10019.
Weather Channel, 2840 Mount Wilkinson Pkwy, Atlanta GA 30339.
WorldNet, Patrick Henry Building, 601 D Street N West, Washington DC 20547

ASIA

Japan Satellite Broadcasting Inc. (JSB), No. 6 Central Building, 1-19-10 Toranomon, Minato-ku, Tokyo, Japan.
National Space Development Agency (NASDA), World Trade Center Building, 2-4-1 Hamamatsu-cho, Minato-ku, Tokyo, Japan. **Te:** +81 (3) 5470 4111 **Fax:** +81 (3) 3436 2928 (International Relations Dept.)
NHK Tokyo, see Japan, Television Section.
STAR TV (Satellite Television Asian Region Ltd), 12th Floor Hutchison House, 10 Harcourt Road, Hong Kong. **Te:** +852 532 1888 **Fax:** +852 524 4093

Broadcasts In English *(Cont'd from page 574)*

2230-2400	VOA	ME/NAf/Eu	11960, 11895, 9530, 7195, 7140, 6160, 1260, 792
2240-2250	Greece	CAm/SAm	9395, 7450 (exc. Sat)
2241-2251	Armenia	Eu	12060, 9480, 7440
2245-0015	Sofia	NAm/Eu	11720, 9700, 7225
2245-0045	India	As	15145, 15110, 11745, 11715, 9910
2245-2305	Vatican	As/AUS/NZ	11830, 9600
2247-2305	WINB	Eu/NAf	15145
2300-0030	BBC WS	As	11945, 11955, 15280
2300-0100	AWR-CR	CAm/SAm	11870, 9725
2300-0100	KSDA	SEAs	15610
2300-0400	WRNO	NAm/CAm	7355 (Spanish Tues-Fri)
2300-0430	Thailand	As	11905, 9655, 4830, 927
2300-0800	WHRI	SAm/CAm	9495
2300-2315	BBC WS	Eu	1296
2300-2330	Canada	CAm	11730, 9755
2300-2330MF	Canada	LAm	11940, 9535
2300-2330SS	Norway	NAm/CAm	11795
2300-2350	N. Korea	Am	13650, 11700
2300-2350	Turkey	Eu/NAm/ME	11895, 9445, 7185
2300-2400	R. Austr.	As/Pac/PNG	17795, 15365, 15320, 11855
2300-2400	BBC WS	As	9570
2300-2400	BBC WS	NAm/CAm	15070, 12095, 9590, 6175, 5975
2300-2400	Japan	As/Eu/ME/Af	17810, 15430, 15195, 11815, 6160, 6025
2300-2400	Moscow	WS	21790, 21770, 21690*, 21530*, 21480, 17890*, 17665, 17655, 17570, 15425, 15350, 15130, 12050, 11775, 11655, 9885, 9870, 9755, 9735, 9725, 9715*, 9695, 9625, 9530, 9480, 9450, 7390, 7355*, 7315*, 7295, 7285, 7255, 7220, 7170*, 7145, 7135, 7120, 7115, 7110, 6045, 5985, 5895, 648, 585
2300-2400	WEWN/4	WEu	5825
2300-2400SS	Canada	LAm	11940, 9535
2315-0215	Iraq	SAm	17740, 15180 (Arabic/Eng)
2330-0025	Holland	ENAm	6165 6020
2330-0100	KHBN	As/Pac	9830
2330-2400	Austria	SAm	13730, 9870
2330-2400	Colombia	NAm/CAm	17862v, 11822v
2330-2400	Sweden	Eu	6065, 1179
2330-2400	Vietnam	Eu	15010, 12020, 9840
2330-2400Mon	SriLanka	WNAm	15425
2335-2345	Greece	SAm/CAm	11645, 9425, 7450 (exc. Sat)
2341-2351	Armenia	Eu	11980, 9480
2345-0450	BBC WS	Eu	198
2347-0330	WINB	SAm	15145

LONG & MEDIUM WAVE STATIONS

Compiled by
Olle Alm, Bengt A. Ericson
and Torgeir Woxen,
members of the Arctic Radio Club, Sweden
November 1st, 1992

CONVERSION TABLE

Frequency kHz = Wavelength meters

kHz m	kHz m	kHz m	kHz m	kHz m	kHz m	kHz m
153-1961	560-535.7	735-408.2	910-329.7	1085-276.5	1260-238.1	1435-209.1
162-1852	565-531.0	740-405.4	915-327.9	1090-275.2	1265-237.2	1440-208.3
171-1754	570-526.3	745-402.7	920-326.1	1095-274.0	1270-236.2	1445-207.6
173-1734	575-521.7	750-400.0	925-324.3	1100-272.7	1275-235.3	1450-206.9
180-1666	580-517.2	755-397.4	930-322.6	1105-271.5	1280-234.4	1455-206.2
189-1587	585-512.8	760-394.7	935-320.9	1110-270.3	1285-233.5	1460-205.5
198-1515	590-508.5	765-392.2	940-319.0	1115-269.1	1290-232.6	1465-204.8
207-1449	595-504.2	770-389.6	945-317.5	1120-267.9	1295-231.7	1470-204.1
216-1389	600-500.0	775-387.1	950-315.8	1125-266.7	1300-230.8	1475-203.4
225-1333	605-495.9	780-384.6	955-314.0	1130-265.5	1305-229.9	1480-202.7
234-1282	610-491.8	785-382.2	960-312.5	1135-264.3	1310-229.0	1485-202.0
245-1224	615-487.8	790-379.7	965-310.9	1140-263.2	1315-228.1	1490-201.3
254-1181	620-483.9	795-377.4	970-309.3	1145-262.0	1320-227.3	1495-200.7
263-1140	625-480.0	800-375.0	975-307.7	1150-260.9	1325-226.4	1500-200.0
272-1103	630-476.2	805-372.7	980-306.0	1155-259.7	1330-225.6	1505-199.3
281-1067	635-472.4	810-370.4	985-304.6	1160-258.6	1335-224.7	1510-198.7
340-882	640-468.8	815-368.1	990-303.0	1165-257.5	1340-223.9	1515-198.0
362-829	645-465.1	820-365.9	995-301.5	1170-256.4	1345-223.0	1520-197.4
385-779	650-461.5	825-363.6	1000-300.0	1175-255.3	1350-222.2	1525-196.7
388-773	655-458.0	830-361.4	1005-298.5	1180-254.2	1355-221.4	1530-196.1
400-750	660-454.5	835-359.3	1010-297.0	1185-253.2	1360-220.6	1535-195.4
433-692	665-451.1	840-357.1	1015-295.6	1190-252.1	1365-219.8	1540-194.8
	670-447.8	845-355.0	1020-294.0	1195-251.0	1370-219.0	1545-194.2
	675-444.4	850-352.9	1025-292.7	1200-250.0	1375-218.2	1550-193.5
	680-441.2	855-350.9	1030-291.3	1205-249.0	1380-217.4	1555-192.9
Mediumwave	683-438.0	860-348.8	1035-289.9	1210-247.9	1385-216.6	1560-192.3
kHz m	690-434.8	865-346.8	1040-288.4	1215-246.9	1390-215.8	1565-191.7
520-576.9	695-431.7	870-344.8	1045-287.1	1220-245.9	1395-215.1	1570-191.1
525-571.4	700-428.6	875-342.9	1050-285.7	1225-244.9	1400-214.3	1575-190.5
530-566.0	705-425.5	880-340.9	1055-284.4	1230-243.9	1405-213.5	1580-189.9
535-560.7	710-422.5	885-339.0	1060-283.0	1235-242.9	1410-212.8	1585-189.3
540-555.6	715-419.6	890-337.0	1065-281.7	1240-241.9	1415-212.0	1590-188.7
545-550.5	720-416.7	895-335.2	1070-280.4	1245-241.0	1420-211.3	1595-188.1
550-545.5	725-413.8	900-333.3	1075-279.0	1250-240.0	1425-210.5	1600-187.5
555-540.5	730-411.0	905-331.5	1080-277.8	1255-239.0	1430-209.8	1605-186.9

EUROPE, AFRICA, NEAR & MIDDLE EAST

Abbreviations: AFN = American Forces Network — AFRTS = US Armed Forces Radio & TV Service — AWR = Adventist World Radio — CI = Cadena Ibérica — COPE = Cadena de Ondas Populares Españolas — DLF = Deutschlandfunk — DS = Deutschlandsender Kultur — EI = R. Euskadi/Euskadi Irratia — ERF = Europe No. 1 — HR = Hessischer Rundfunk — IND = Emissores Independentes — NDR = Norddeutscher Rundfunk — NRK = Norsk Rikskringkasting — OCR = Onda Cero Radio — RA = Radio Altitude — RFI = Radio France International — RFE = Radio Free Europe — RMC = Radio Monte Carlo — RMR = Radio Moscow-relay station — RNE = Radio Nacional de España — RRE = Rádio Renascença — SDR = Süddeutscher Rundfunk — SER = Sociedad Española de Radiodifusion — SFB = Sender Freies Berlin — SWF = Südwestfunk — TWR = Trans World Radio — VOA = Voice of America — WDR = Westdeutscher Rundfunk.

Country	kW	Location
153 kHz		
Germany	500	Donebach (DLF)
Romania	1200	Brasov
Russia	500	Ufa
Algeria	1000	Bechar
162 kHz		
France	2000	Allouis
Turkey	1000	Agri
171 kHz		
Belarus	1000	Osipovičy
Russia	1200	Bolshakovo
	1000	Kaliningrad
	1000	Krasnodar
	500	Noginsk
	300	Syktyvkar
Morocco	2000	Nador
Ukraine	500	Lviv
180 kHz		
Germany	100	Oranienburg (DS) (177)
	100	Kónigs-Wusterhausen (DS) (177)
	2000	Saarlouis (ERF) (183)
Turkey	1200	Polatli
189 kHz		
Georgia	500	Dusheti
Italy	10	Caltanissetta
198 kHz		
Russia	300	Ufa
	150	Sankt-Peterburg
	100	Noginsk
United Kingdom	500	Droitwich
	50	Burghead
	50	Westerglen
Algeria	1000	Ouargla
Turkey	120	Etimesgut

207 kHz		
Germany	500	Aholming (DLF)
Iceland	100	Vatnsendi
	10	Eidar
Ukraine	500	Kiev
Morocco	800	Azilal
Jordan	600	Al Karanah
216 kHz		
Azerbaidzhan	500	Gyanca
France	1400	Roumoules (RMC)
Norway	200	Oslo
225 kHz		
Poland	?	Konstantynow 1
Turkey	600	Van
234 kHz		
Armenia	100	Kamo
Luxembourg	2000	Junglinster
Moldavia	1000	Grigoriopol
Russia	1200	Samara
	1000	Sankt-Peterburg
	150	Arkhangelsk
243 kHz		
Denmark	300	Kalundborg
Turkey	200	Erzurum
252 kHz		
Armenia	150	Yerevan
Finland	200	Lahti
Ireland	500/100	Clarkestown
Russia	150	Kazan
Algeria	1500/750	Tipaza
261 kHz		
Bulgaria	60	Sofia
Germany	200	Burg (RMR/Volga)
Russia	2000	Taldom
270 kHz		
Czech & Slovak Rep.	1500/750	Topolná
279 kHz		
Belarus	500	Osipovičy
520 kHz		
Germany	0.2	Hof (BR)
	0.2	Würzburg (BR)
Spain	20	Bilbao
526 kHz		
Vatican	5	Vatican City
531 kHz		
Faroe Isl.	100 (200)	Tórshavn
Germany	100	Leipzig (MDR)
Romania	15	Petrosani
Russia	30	Cheboksary
Spain	10	Pontevedra (RNE5)
	20	Oviedo (RNE5)
	10	Pamplona (RNE5)
	5	Córdoba (RNE5)
Switzerland	500	Beromünster
Yugoslavia	10	Titovo Užice
	1	Vranje
Algeria	600/300	Ain-El-Beida
Madeira	10	Porto Santo
Nigeria	50	Akure
Tanzania	10	Dar-es-Salaam
Iran	20	Iranshahr
Israel	100	Jerusalem
540 kHz		
Belgium	150	Wavre-Overijse
Hungary	2000	Solt
Ireland	2	Conamara
Russia	—	Unknown E.
Spain	10	Vitoria (EI)
Bophuthatswana	100/50	Ga-Rankuwa
Kenya	100	Voi
Malawi	10	Mangochi
Mali	100	Bamako
Morocco	600	Sidi Bennour
Nigeria	50	Sokoto
Sudan	50	Nyala
Iran	200	Mashhad
Kuwait	1500	Sulaibiyah
549 kHz		
Azerbaidzhan	65	Gyanca

Belarus	1000	Osipovičy
Albania	—	Unknown loc.
Germany	200	Thurnau (DLF)
	100	Nordkirchen (DLF)
Russia	1000	Sankt-Peterburg
	150	Chkalovskaya
	25	Kaliningrad
	—	Rostov-na-Donu
Slovenia	20	Beli Kriz
Ukraine	150(?)	9 st's (Mayak)
Yugoslavia	100	Pristina
Algeria	600/300	Les Trembles
Gabon	20	Oyem
Nigeria	50	Kano
Zambia	100	Kitwe
Saudi Arabia	2000	Qurayyat
	20	Rafha
	1	Gizan
558 kHz		
Czech & Slovak Rep.	2	Broumov
	1	2 st's
Finland	50	Espoo
Germany	20	Rostock (NDR)
Portugal	10	Guarda
	10	Faro
Romania	200	Tirgu Jiu
Slovenia	20	Maribor
Spain	20	Valencia (RNE5)
	10	La Coruña (RNE5)
	10	San Sebastián (RNE5)
Switzerland	300	Cima di Dentro
United Kingdom	0.8	London
Egypt	40	Cairo
Kenya	20	Kisumu (557)
Liberia	10	Gbarnga (inactive ?)
Malawi	10	Karonga
Mozambique	5	Inhambane (557)
Nigeria	50	Calabar
Transkei	100	Umtata
Cyprus	10	Paphos
Iran	1000	Gheslagh
Iraq	300	Rutba
Saudi Arabia	1	Jeddah
Turkey	600	Denizli
567 kHz		
Czech & Slovak Rep.	14	R. Sobota
	14	Zilina
Germany	100	Berlin (SFB1)
Ireland	500	Tullamore
Italy	20	Bologna + 4 st's
Portugal	10	Valença
Romania	50	2 st's
Russia	250	Volgograd
Spain	50	Murcia (RNE5)
	10	Socuellamos (RNE5)
	5	Marbella (RNE5)
Yugoslavia	10	Strumica
	10	Pljevlja
Morocco	50	Laayoune
Nigeria	50	Owerri
Zaire	2	Lubumbashi
Zambia	10	Kasama
Saudi Arabia	15	Afif
	5	Abha
Syria	1000	Damas-Adra
576 kHz		
Bosnia/Hercegovina	2	Prijedor
Bulgaria	500	Vidin
Germany	500	Mühlacker (SDR)
	250	Schwerin (NDR)
Latvia	500	Riga
	7	Daugavpils
Portugal	10	Braga (RRE)
Russia	50	Astrakhan
	25	Nalchik
Spain	50	Barcelona (RNE5)
Algeria	300	Bechare
Canary Islands	10	Las Palmas (RNE1)
Nigeria	50	Ibadan
South Africa	100	Bloemendal

Sudan	100	Soba + 1 st.
Uganda	100	Mawagga
Iran	600	Abadan
Israel	200	Tel Aviv
Saudi Arabia	1	Gizan
580 kHz		
Iran	600	Abadan
582 kHz		
Iran	100	Tehran
585 kHz		
Austria	600/240	Wien-Bisamberg
France	8	Paris
Russia	40	Perm
Spain	200	Madrid (RNE 1)
	5	Cádiz (RNE 1)
United Kingdom	2	Dumfries
Liberia	10	Voinjama (inactive ?)
Nigeria	50	Ugwuaji
Tanzania	10	Chumbum
Tunisia	350	Gafsa
Saudi Arabia	1200	Riyadh
594 kHz		
Bulgaria	250	Pleven + 2 st's
Croatia	20	Osijék
Germany	1000/400	Frankfurt (HR)
	200	Hoher Meissner (HR)
Moldovia	500	Grigoriopol
Portugal	100	Muge (RRE)
Russia	150	Izhevsk
	22	Vladikavkaz
Slovenia	1	Cerkno
Yugoslavia	10	Nikšić
	1	3 st's
Malawi	50	Lilongwe
Morocco	100	Oujda
Namibia	100	Windhoek
Nigeria	100	Kaduna (593)
Iran	100	Zahedan + 1 st.
Saudi Arabia	2000	Duba
Turkey	600	Malatya
603 kHz		
Czech & Slovak Rep.	25	Brno-Komárov
France	300	Lyon-Tramoyes
Romania	50	Botosani
	30/2	3 st's
Spain	50	Sevilla (RNE5)
	10	Palencia (RNE5)
United Kingdom	2	Newcastle
	0.1	Littlebourne
	0.1	Cheltenham
Guinea	25	Kankan
Madeira	10	Arieiro
Nigeria	30	Ibese
Reunion	4	Le Port
Tanzania	100	Dodoma
Transkei	100	Umzimkulu
Tunisia	10	Monastir
Zambia	10	Mongu
Cyprus	100	Nicosia
Iraq	300	Nineva
Saudi Arabia	10	Afif
612 kHz		
Azerbaidzhan	25	Baky
Bosnia/Hercegovina	600	Sarajevo
Estonia	100	Tallinn
Germany	10	Kiel (NDR)
Ireland	100	Tullamore
Lithuania	50	Vilnius
Russia	100	Petrozavodsk
	50	Moskva
	30	Murmansk
Spain	10	Lérida (RNE 1)
	10	Vitoria (RNE1)
Kenya	100	Nairobi
Morocco	300	Sebaa-Aioun
Nigeria	50	Ilorin
Zimbabwe	2	Gweru
Bahrain	100	Manama
Jordan	200	Amman
Saudi Arabia	15	Khamasin
	5	Hail
Syria	300	Homs (617v)
621 kHz		
Albania	—	?
Belgium	300	Wavre-Overijse
Czech & Slovak Rep.	14	Orava
	10	Jaén (RNE 1)
	10	Avila (RNE 1)
Spain	50	Barcelona (OCR)
Botswana	50	Selebi-Phikwe
Canary Isl	200	Santa Cruz, Tenerife (RNE 1)
Egypt	1000	Batrah + 1 st
Tanzania	50/10	Mbeya
Iran	20	Birjand
630kHz		
Bosnia/Hercegovina	2	Livno + 1 st
Germany	80	Dannenberg (NDR)
Norway	100	Vigra
Portugal	50	Sta. Isabel
	1	Chaves
	1	Miranda do Douro
Romania	400	Timisoara
Russia	50	Saratov
United Kingdom	2	Redruth
	0.2	Luton
Madagascar	20	Tamatave
Nigeria	50	Maiduguri
	50	Pt. Harcourt
Tunisia	600	Tunis-Djedeida
Zambia	500	Lusaka
Afghanistan	10	Herat
Iraq	20	Kirkuk
Saudi Arabia	20	Gizan
	10	Najran
Turkey	300	Tarsus
639 kHz		
Bosnia/Hercegovina	1	Gračanica
Czech & Slovak Rep.	1500	Praha-Liblice
Spain	100	LaCoruña (RNE 1)
	20	Almería (RNE 1)
	20	Bilbao (RNE 1)
	50?	Zaragoza (RNE 1)
	10	Albacete (RNE 1)
Yugoslavia	1	Drniš
	1	Štip
	1	Soko Banja
Nigeria	25	Kaduna
	5	Warri
Sudan	5	Obeid
Uganda	50	Kampala
Tanzania	100	Dar-es-Salaam
Cyprus	500	Zakaki
Iran	400	Bonab
648 kHz		
Albania	15	Rrogozhina (v. freq.)
Azores	1	Lajes
Bulgaria	30	Plovdiv
Slovenia	10	M. Sobota
Spain	10	Palma de Mallorca (RNE 1)
	10	Badajoz (RNE 1)
Ukraine	150	Oktyabrskoye
United Kingdom	500	Orfordness
Botswana	50	Orapa
Gambia	10	Bonto
Libya	300	Tobruk
Tanzania	100/10	Nachingwea
Afghanistan	10	Taluqan
Saudi Arabia	2000	Jeddah
657 kHz		
Germany	40	Neubrandenburg (NDR)
Italy	120	Napoli
	100	Firenze
	50/20	3 st's
Russia	150	Murmansk
	50	Groznyy
Spain	20	Madrid (RNE5)
Ukraine	25	Chernivtsi
	2	Wrexham
United Kingdom	0.5	Bodmin
Nigeria	100	Ibadan

Tanzania	50	Dar-es-Salaam
Afghanistan	25	Kabul
Israel	200	Tel Aviv
Saudi Arabia	20	Rafha
UAE	2×50	Sadiyat
666 kHz		
Germany	300	Rohrdorf (SWF)
Greece	15	Athens
Iceland	0.5	Höfn
Lithuania	500	Sitkūnai
Portugal	135	Lisboa
	70	Vila Real
Russia	30	Maykop
	—	Unk. So. E.
Spain	10	Barcelona (COPE)
United Kingdom	0.34	Exeter
	0.5	York
Yugoslavia	10	Sombor
Algeria	5	Tindouf
Egypt	2	El Minya
Nigeria	50	Benin
Reunion	20	St. Pierre
Sudan	5	Kassala
Zambia	10	Chipata
Syria	600	Damas-Saboura
675 kHz		
France	600	Marseille
Netherlands	120	Lopik
Norway	10	Bodø
Russia	50	Ukhta
Ukraine	50	Uzhgorod
Yugoslavia	10/5	Bosilegrad + 1 st
Kenya	20	Nairobi
Libya	100	Benghazi
Malawi	50	Lusangazi
Morocco	20	Tan-Tan
Nigeria	10	Aha
Iran	10	Hamedan
Israel	50	Jerusalem
Qatar	100	Doha
Saudi Arabia	20	Afif
	5	Abha
684 kHz		
Germany	40	Hof (RIAS)
Lithuania	200	Palanga (F.Pl.)
Russia	10	Sankt-Peterburg
Spain	500	Sevilla (RNE 1)
Yugoslavia	2000	Beograd
Mauritius	10	Curepipe
South Africa	5	East London (683)
Tunisia	10	Kairouan
Iran	100	Mashhad
Saudi Arabia	50	Jeddah
	10	Riyadh
693 kHz		
Albania	10	Pogradec
Azores	10	Santa Barbara
Georgia	150(?)	Batumi
Germany	40	Berlin (ORB)
Italy	20	Potenza
Portugal	10	Viseu
Russia	30	Ufa
Spain	2	Tortosa (RNE 1)
United Kingdom	150	Droitwich
	50/1	10 st's
Algeria	5	Ain-el-Hamam + 1 st
Madagascar	2x75	Antananarivo
Senegal	20	Ziguinchor
Sudan	100	Juba
Zaire	600	Kinshasa
Cyprus	20	Nicosia (692)
Iraq	600	Basrah
702 kHz		
Armenia	—	Yerevan
Bulgaria	40	Pirin
Czech & Slovak Rep.	400	Bánská Bystrica
	400	Prešov
Germany	3	Bonn/Kleve (WDR)
	5/1.2	Flensburg (NDR)
	2	Herford/Siegen (WDR)

Monaco	300	Monte Carlo
Norway	20	Vadsø
Russia	—	Unk. E
Spain	10	Zamora (RNE 1)
Yugoslavia	10/0.1	Šabac + 1 st
Bophuthatswana	100	Ga-Rankuwa
Angola	10	Luanda
Benin	1	Parakou
Egypt	10	El Kharga
Kenya	100	Meru
Morocco	740	Sebaa-Aioun
Nigeria	10	Bida
	25	Wukari
Iran	800	Rasht
Oman	1500	Masirah Island (BBC)
Saudi Arabia	40	Duba
Turkey	150	Catalca
711 kHz		
Estonia	50	Tallin + 3 LP st's
France	300	Rennes
Germany	5	Heidelberg (SDR) + 4 st's
Romania	30	Sighet
Russia	7	Naryan-Mar
Spain	5	Murcia (COPE)
Ukraine	150	Donetsk
Yugoslavia	100	Niš
Egypt	100	Cairo
Libya	50	Ghadames
	50	Jefren
	50	Sebha
Morocco	600	Laayoune
Tanzania	100/10	Kigoma
Zimbabwe	2	Gweru
Israel	1	Jerusalem
Yemen	200	Sana'a
720 kHz		
Germany	200	Langenberg (WDR)
	150	Holzkirchen
Poland	—	Warsaw
Portugal	100	Norte
	10	4 st's
Romania	2/1	3 st's
United Kingdom	10	Lisnagarvey
	0.5	London + 1 st.
Yugoslavia	1	Djakovo + 3 st's
Canary Isl	20	Santa Cruz Tenerife (RNE5)
Chad	0.1	Sarh
Kenya	10	Garissa
Nigeria	20	Bali (722)
Tanzania	5/10	Mwanza
Tunisia	100	Sfax-Sidi Mansour
Cyprus	500	Zakaki
Iran	400	Tayebad
729 kHz		
Bosnia/Hercegovina	1	Maglaj
Germany	12	Putbus (NDR)
Greece	150	Athens
Ireland	10	Cork
Spain	50	Oviedo (RNE 1)
	20	Logroño (RNE 1)
	20	Malaga (RNE 1)
	10	Alicante (RNE 1)
	10	Valladolid (RNE 1)
	5	Cuenca (RNE 1)
United Kingdom	0.2	Manningtree
Nigeria	25	Dutshe
Reunion	20	St. André
Uganda	100	Butebo
Zambia	10	Livingstone
UAE	750	Sadiyat
738 kHz		
France	4	Paris-Romainville (RFI)
Iceland	5	Skjaldarvik
Poland	300	Poznan
Russia	7.5	Moskva
Spain	500	Barcelona (RNE 1)
United Kingdom	0.37	Worcester
Yugoslavia	25	Deanovac
	10	Krusevac
	2	Sibenik

	1	Tetovo
Algeria	5	Ain-Amenas
Mozambique	50	Maputo (737)
Iran	10	Mashhad
Israel	1200	Tel-Aviv
747 kHz		
Bulgaria	500	Petrich
Estonia	—	Tallinn
Holland	400	Flevoland
Russia	10	Sankt-Peterburg
Spain	10	Cádiz (RNE5)
Burkina Faso	100	Ouagadougou
Canary Isl.	20	Las Palmas (RNE5)
Gambia	1	Basse
Kenya	100	Nairobi (746)
Namibia	100	Gobabis
Nigeria	20	Bali (744)
Iran	800	Bandar Turkanmen
Saudi Arabia	1	Najran
Syria	100	Sarakeb
756 kHz		
Bosnia/Hercegovina	10	Mostar
Germany	800/200	Braunschweig (DLF)
	100	Ravensburg (DLF)
Portugal	5	Lisbon (R.C.)
	1	Lamego (R.AD)
Romania	400	Lugoj
Spain	5	Bilbao (EI)
United Kingdom	2	Redruth
	1	Montgomeryshire
	1	Carlisle
Egypt	10/1	Qena + 1 st.
Malawi	50	Blantyre
Nigeria	10	Ibadan
	50	Minna
	25	Damagun
Lebanon	—	Marjayoun
765 kHz		
Greece	20	Ioannina
Russia	150	Petrozavodsk
Switzerland	600	Sottens
United Kingdom	0.5	Chelmsford
Ukraine	50	Odesa
Yugoslavia	2/1	Zadar + 2 st's
Senegal	200/300	Dakar
Sudan	50	Arda
Iran	1200	Chahbahar
Saudi Arabia	1000	Dammam
Turkey	600	Gaziantep
774 kHz		
Bosnia/Hercegovina	2	Bihac
Bulgaria	2×50	Sofia
	50	Varna
Croatia	50	Hvar
Czech & Slovak Rep.	40	Hradec Králové
Germany	5	Bonn (WDR)
Russia	100	Voronezh
Spain	50	Caceres (RNE 1)
	50	Valencia (RNE 1)
	50	San Sebastian (RNE 1)
	20	Orense (RNE 1)
	10/4	4 st's (RNE 1)
United Kingdom	1	Enniskillen + 1 st
	0.7	Littlebourne
	0.5	Leeds
	0.14	Gloucester
Egypt	500	Abis
Morocco	50	Agadir
Nigeria	10	Bali
Iran	10	Arak
783 kHz		
Germany	100	Burg (MDR)
Portugal	50	Miramar
Yugoslavia	10	Zagreb-Buje
Algeria	5	Djanet
Mozambique	5	Lichinga (v.f.)
Nigeria	25	Anyagba
South Africa	5	Bloemfontein
Sudan	5	Atbara
Iran	20	Iranshahr

Saudi Arabia	100	Dammam
Syria	600	Tartus
792 kHz		
Bosnia/Hercegovina	1	Banoviči
	1	Čapljina
Czech & Slovak Rep.	40	Praha
	10	Bratislava
France	300	Limoges
Germany	5	Lingen (NDR)
Greece	500	Kavalla
Russia	50	Astrakhan
	5	Moskva
Spain	20	Sevilla (SER)
United Kingdom	1	Londonderry
	0.2	Bedford
Yugoslavia	2	Arandjelovac
Kenya	100	Mombasa
Libya	20	Sirte
Nigeria	10	Idominasi
Zimbabwe	1	Kadoma
Saudi Arabia	50	Jeddah
Yemen	750	Al-Hiswah
801 kHz		
Azerbaidzhan	150	Baky + 1 st
Germany	300/550	München-Ismaning (BR)
	50	Dillberg (BR)
Russia	1000	Sankt-Peterburg
Spain	20	Lugo (RNE 1)
	10	Castellón (RNE 1)
	10	Ciudad Real (RNE 1)
	10	Burgos (RNE 1)
United Kingdom	2	Barnstaple
Egypt	10	Assiut
Niger	0.1	N'Guigmi
Nigeria	20	Damaturu
South Africa	2	Durban
Bahrain	100	Manama
Jordan	2000	Ajlun + 1 st.
810 kHz		
Estonia	5	Vöru/Haapsalu
Georgia	150	Tbilisi
Germany	5	Berlin (DLF)
Hungary	12	Siofok
Russia	150	Volgograd + 2 st's
Spain	20	Madrid (SER)
Ukraine	6	Kharkiv + 2 st's
United Kingdom	100	Burghead
	100	Westerglen
	5	Redmoss
Yugoslavia	1000	Skopje
Senegal	1	Podor
South Africa	5	Grahamstown
Uganda	100	Bobi
UAE	50	Maqtaa
819 kHz		
France	20	Toulouse
Italy	20	Trieste
Poland	300	Warszawa
Spain	5	San Sebastian (EI)
United Kingdom	1	Ludlow
Egypt	1000	Batra/Cairo
Mauritius	10	Forest Side
Morocco	25	Rabat
Sudan	5	Dongola
Zambia	200	Lusaka
Iran	20	Sari
828 kHz		
Azores	1	Flores
Bulgaria	500	Shumen
	50	Sofia
Germany	100/5	Hannover (NDR)
	40	Freiburg (SWF)
Ireland	1	Corca Dhuibhne
Portugal	10	S. Isabel + 5 st's
Russia	50	Nizhniy Novgorod
	—	Unk. E
Spain	50	Barcelona (SER)
United Kingdom	0.3	Bournemouth
	0.12	Leeds
	0.2	Luton

	0.2	Sedgley
Yugoslavia	1	Vrbovec
Libya	300	Sebha
Morocco	100	Oujda
Nigeria	25	Enugu
South Africa	5	Welgedacht
Zambia	10	Mongu
Saudi Arabia	20	Medinah
Syria	1500	Deir-Ez-zoor
UAE	1	Al Ain
837 kHz		
Azores	10	Barossa
France	200	Nancy
Spain	10	Sevilla (COPE)
	10	Burgos (COPE)
	2	El Ferrol (COPE)
	1	Ibiza (COPE)
Ukraine	150	Kharkiv
United Kingdom	1.5	Barrow
	0.45	Leicester
Yugoslavia	50	Novi Sad + 1 st.
Algeria	5	Ben-Abbas
Canary Isl	10	Las Palmas
Mozambique	5	Chimoio
Tanzania	1	Dar-es-Salaam
Iran	100	Shiraz
Lebanon	10	Beirut
Yemen	30	Sana'a
840 kHz		
Chad	20	N'Djamena
846 kHz		
Czech & Slovak Rep.	30	Ostrava
Italy	540	Roma
Russia	50	Noginsk
	30	Elista
Kenya	100	Kisumu
Nigeria	10	Azare
South Africa	100	Komga
Iran	2	Damghan
Iraq	300	Nasiriya
Israel	5	Safed
UAE	20	Umm-al-Qiwain
850 kHz		
Chad	1	Sarh
855 kHz		
Germany	100	Berlin (RIAS)
Romania	1500	Bucuresti
Russia	—	Kamenka
Spain	125	Murcia (RNE 1)
	20	Santander (RNE 1)
	20	Pontevedra (RNE 1)
	10/5	7 st's (RNE 1)
United Kingdom	1.5	Postwick
	1	Plymouth
	1.5	Preston
	0.2	Ludlow
Ethiopia	100	Harar
Sudan	7	Port Sudan (860)
Zimbabwe	1	Masvingo
Zambia	10	Chipata
Cyprus	10	Cape Greco
Jordan	10	Amman
864 kHz		
Albania	5	Saranda
Armenia	150	Yerevan
Bosnia/Hercegovina	1	Sokolac
Bulgaria	50	Blagoevgrad
	10	Suvorovo
	5	Samuil
Czech & Slovak Rep.	10	Olomouc
	6	Ústi n. Labem
	1	Snina
	1	Čadca
France	300	Paris
Spain	2	Socuellamos (RNE 1)
Yugoslavia	10/1	Ucka
Afghanistan	10	Kandahahr
Congo	10	Ft Rousset (863)
Egypt	500	Santah
Morocco	15	Errachidia

Sierra Leone	50	Kenema
Saudi Arabia	500	Dammam
873 kHz		
Bulgaria	50	Stara Zagora
Germany	150	Frankfurt (AFN)
Hungary	20	Lakihegy
	15	Pecs
Moldovia	150	Chişinău
Russia	150	Sankt-Peterburg + 3 st's
Spain	20	Zaragoza (SER)
Ukraine	50	Kyyiv + 4 st's
United Kingdom	1	Enniskillen
	0.3	West Lynn
Algeria	5	Ghardaia
Ethiopia	100	Addis Ababa
Mozambique	50	Beira (866 r)
Sudan	5	Medani
Lebanon	80	Beirut
Saudi Arabia	10	Qassim
882 kHz		
Germany	250	Wachenbrunn (MDR)
Russia	50	Stavropol
Spain	5	Málaga (COPE)
	2	Alicante (COPE)
	2	Valladolid (COPE)
	2	Gijón (COPE)
	2	Sabadell (COPE)
United Kingdom	100/1	Washford + 3 st's
Yugoslavia	600/300	Titograd
	10	Pljevlja + 4 st's
Canary Isl	20	La Laguna
Egypt	10	Matruh
Tunisia	1	Remada
Zambia	10	Kasama
Iran	10	Mahabad
Israel	1	Beit Hillel
Saudi Arabia	100	Dammam
891 kHz		
Azerbaidzhan	30	Baky
Cyprus	—	Unk.
Netherlands	20	Hulsberg
Portugal	10	Vilamoura (RRE)
Russia	—	Unk. E
Ukraine	150	Uzhgorod
Algeria	600/300	Alger
Lesotho	100	Maseru
Nigeria	10	Ogbomosho
Zimbabwe	2	Harare
Iran	20	Yasuj
Turkey	600	Antalya
900 kHz		
Czech & Slovak Rep.	40	Plzeň
	20	Karlovy Vary
	14	Poprad (Tatry)
Italy	600	Milano
Russia	50	Yoshkar-Ola
Spain	10	Bilbao (COPE)
	5	Vigo (COPE)
	5	Granada (COPE)
Yugoslavia	2	Beograd
Cameroon	20	Yaoundé (899)
Kenya	100	Meru
Nigeria	25	Abeokuta
Iran	50	Tehran
Saudi Arabia	1000	Qurayyat
909 kHz		
Albania	15	Korçë
Azores	5	Angra do Heroísmo
Romania	50	Cluj + 1 st
Russia	—	Unk, So. E.
Spain	10	Palma de Mallorca (RNE5)
United Kingdom	200/1	Moorside Edge + 10 st's
Algeria	5	Tamanrasset
Gambia	2.5	Banjul
Libya	20	Giaghboub
	10	Kufra
Morocco	5	Safi C
Nigeria	50	Abuja
Uganda	20	Kampala
Zambia	10	Solwezi

Afghanistan	10	Mazar-I-Sharif
Iraq	300	Hurriyah
918 kHz		
Russia	100	Arkhangelsk
	75	Moskva
	50	Makhachkala
Spain	20	Madrid (IND)
Slovenia	600	Ljubljana-Domžale
Egypt	10	Bawiti
	10	Hurghada
Kenya	20	Nakuru
Nigeria	50	Makurdi
	50	Ikeja
Zambia	10	Mansa
Iran	20	Jiroft
Syria	200	Al-Hassake
927 kHz		
Belgium	150	Wolvertem
Czech & Slovak Rep.	40	Trnava
Greece	50	Zakynthos
Portugal	1	Evora
Spain	10	Lérida-Lleida (SER)
Algeria	5	Timimoun
Zambia	10	Livingstone
Israel	50	Haifa
	1	Eilat
Turkey	200	Izmir
936 kHz		
Germany	100	Bremen (RB)
Italy	20	Venezia
	10	Trapani/Genova
Russia-Matveyevka	20	Valladolid (RNE 5)
	10	Alicante (RNE 5)
	10	Bilbao (RNE 5)
	10	Malaga (RNE 5)
	10	Zaragoza (RNE 5)
	5	Vigo (RNE 5)
Spain	2	Lérida (SER)
Ukraine	300	Lviv
United Kingdom	0.2	W. Wiltshire
Yugoslavia	10/1	Djakovica + 3 st's
Benin	16/10	Parakou
Egypt	10	Salum
Mozambique	1	Beira (F.Pl.)
Morocco	600	Agadir
Zambia	1	Kabwe
Afghanistan	10	Baghlan
Iran	10	Shahar-e-Kord + 1st
Lebanon	—	Rumaylah
Saudi Arabia	5	Riyadh
945 kHz		
Bosnia/Hercegovina	100	Sarajevo
Czech & Slovak Rep.	6	Liberec
France	300	Toulouse
Greece	5	Larissa
Latvia	150	Riga (VOA Europe)
Romania	14	Miercurea Ciuc
Russia	300	Rostov-na-Donu
United Kingdom	0.2	Derby
Yugoslavia	2/1	4 st's
Angola	25	Luanda (944)
Ethiopia	50	Asmara
São Tomé e P.	20	São Tomé
Sudan	100	Kadguli
Afghanistan	10	Farah
Iran	20	Marivan
Lebanon	12	Marjayoun
Saudi Arabia	5	Hail
954 kHz		
Bulgaria	50	Shumen
Czech & Slovak Rep.	200	Prostějov
	7	Strakonice
Greece	20	Heraklion
Spain	20	Madrid (CI)
United Kingdom	0.32	Torquay
	0.16	Hereford
Yugoslavia	2	Gnjilane
Kenya	100	Kisumu
Nigeria	20	Awka
Swaziland	50/30	Sidvokodvo

Israel	100	Jerusalem
Lebanon	100	
Qatar	1500	Al Arish
Turkey	300	Trabzon
963 kHz		
Bulgaria	50	Shumen
	2 × 50	Sofia
	2 × 20	Pirin
Finland	600	Pori
France	8	Paris
Ireland	10	Tir Chonaill
Portugal	10	Seixal (RRE)
Russia	15	Moskva
Slovenia	2	Celje
Egypt	10	Salum
Nigeria	10	Katsina
Senegal	1	Matam
Somalia	150	Mogadishu (962)
Sudan	100	Soba
Tunisia	200	Tunis-Djedeida
Cyprus	100	Nicosia
Iran	20	Birjand
Lebanon	100	Jounieh
972 kHz		
Albania	20	Puke
Germany	300	Hamburg (NDR)
Spain	2	Monforte de Lemos (RNE 1)
	5	Córdoba (RNE 1)
Ukraine	500	Mykolayiv
Botswana	50	Gaborone
Cameroon	20	Bertoua (971)
Libya	100	—
Morocco	1	Marrakech
Nigeria	25	Lokoja
	—	Katsina
Iran	10	Ilam
UAE	2 × 50	Fujairah
981 kHz		
Czech & Slovak Rep.	30	Česke Budějovice
	2	Ostrava
Greece	200	Megara
Italy	10	Trieste
Portugal	10	Coimbra + 3 st's (RRE)
Yugoslavia	10/1	Čačak + 4 st's
Algeria	600/300	Alger
Egypt	10	Assiut + 2 st's.
Kenya	100	Voi
Nigeria	10	Port Harcourt (r. inactive)
Zimbabwe	1	Mutare
Saudi Arabia	20	Medinah
990 kHz		
Albania	8	Kukes (996)
Bosnia/Hercegovina	1	Foča
Germany	300	Berlin (RIAS)
Italy	10	Potenza
Spain	10	Bilbao (SER)
United Kingdom	1	Exeter
	1	Tywyn
	1	Redmoss
	0.1	Wolverhampton
	0.3	Doncaster
Yugoslavia	1	1 st
Ceuta	5	Ceuta
Ethiopia	1	Addis Ababa
Gabon	20	Epila
Liberia	5	Enoanyi
Nigeria	50	Bauchi
	20	Ikeja
	2	Pietersburg
Iran	400	Shiraz (vf)
Lebanon	10	Amchit (989)
999 kHz		
Germany	20	Schwerin (RIAS)
Italy	50/0.5	Torino + 3 st's
Malta	5	Delimara
Moldovia	300	Grigoriopol
Spain	50	Madrid (COPE)
United Kingdom	1	Fareham
	0.25	Nottingham
	0.8	Preston

USSR	300	Grigoriopol
Yugoslavia	1	Kladovo
Angola	1	Lobito (1000)
Cameroon	10	Yaoundé
Morocco	10	Ad-Dakhla
Uganda	100	Kabale
Zimbabwe	100	Gweru
Afghanistan	20	Jalalabad
Iran	20	Baneh
Qatar	50	Al Khaisah
Saudi Arabia	20	Tabuk
1008 kHz		
Belarus	50	Slonim + 4 st's
Greece	50	Kerkyra
Netherlands	400	Flevoland
Russia	—	Unk. So. E.
Spain	5	Badajoz (SER)
	5	Gerona (SER)
	3	Vigo (SER)
Yugoslavia	400/120	Aleksinac
	200/120	Beograd
Angola	1	Huambo (1010)
Burkina Faso	10	B. Dioulasso
Canary Isl	10	Las Palmas
Egypt	100	El Arish + 2 st's
Mozambique	50	Maputo
Nigeria	50	Kontagora
	10	Ibadan
Iran	20	Semnan
Yemen	600	Sana'a
	30	Taiz (1012)
1017 kHz		
Bosnia/Hercegovina	1	Gazin
Bulgaria	2×20	Vidim
Czech & Slovak Rep.	50/2	Nitra + 4 st's
Germany	800	Rheinsender (SWF)
		Mnichovo Hrodiště
Spain	10	Burgós (RNE5)
	5	Granada (RNE5)
United Kingdom	2	Shrewsbury
Yugoslavia	10/1	Crna Trava
Algeria	2	Batna
Madeira	1	Santana
Morocco	1	Tanger
Nigeria	10	Yola
Iran	20	Zahedan
Lebanon	—	Ba'labekk
Turkey	1200	Mundanya
1026 kHz		
Austria	100/50	Graz-Dobl + 2 st's
Belarus	15	Mahileŭ + 5 st's
Russia	5	Unk. No. E. (3 st's)
Spain	10	Réus (SER)
	5	Oviedo (SER)
	3	Alicante (SER)
	2	Salamanca (SER)
	2	Jaén (SER)
	2	Jerez (SER)
United Kindgom	1	Jersey
	1.7	Belfast
	0.5	Cambridge
Yugoslavia	10	Kragujevac
	5	Bar
	5	Skopje
	1	Nova Gradiška
Algeria	5	Hassi Messaoud
Kenya	5	Malinda
Morocco	1	Rabat
Iran	100	Tabriz
	20	Chalus
Israel	50	Tel Aviv
1035 kHz		
Bosnia/Hercegovina	1	Bos Šamac
Czech & Slovat Rep.	7	Banska Bystrica
Estonia	500	Tallinn
Italy	50	Milano + 7 st's
Portugal	135	Lisbon
United Kingdom	1	Sheffield
	0.5	Hoo
	0.3	Ayr
	0.8	Aberdeen
Yugoslavia	10	T. Mitrovica + 1 st
Egypt	10	Luxor
Guinea Bissau	5	Bissau (1034)
Tanzania	10	Dar-es-Salaam
Venda	100	Tshisahulu
Iran	20	Yazd
Oman	100	Salalah
1044 kHz		
Georgia	100	Tbilisi
Germany	250	Dresden (DT 64)
Greece	150	Thessaloniki
Spain	10	San Sebastian (SER)
Yugoslavia	1	P. Slatina + 1 st.
Egypt	7	Shalatin
Morocco	300	Sebaa-Aioun
Niger	0.1	Sahel (1041)
Cyprus	10	Limassol
1050 kHz		
Lebanon	—	Beirut
1053 kHz		
Romania	1000	Iasi
Russia	10	Sankt-Peterburg
United Kingdom	150/1	Droitwich + 13 st's
Libya	50	Tripoli
Morocco	600	Tanger
Spain	10	Zaragoza (COPE)
Iran	20	Khorramabad
1062 kHz		
Denmark	250	Kalundborg
Italy	25	Cagliari
	25	Squinzano
	2/1	5 st's
Portugal	100	Norte
Russia	150	Saransk
Yugoslavia	10/1	Sljeme + 1 st
	10	Svetozarevo
Nigeria	10	Onitsha
Iran	20	Kerman
Turkey	300	Diyarbakir
1071 kHz		
Bosnia/Hercegovina	40	Mostar
	20	Banja Luka
Czech & Slovak Rep.	50/7	M. Budějovice + 3 st's
France	40	Lille + 4 st
Latvia	75	Riga + 3 st's
Spain	5	Bilbao (EI)
Ukraine	20	Dnipropetrovsk
Egypt	60	Cairo
Guinea-Bissau	100	Bissau
Zambia	100	Kitwe
1080 kHz		
Greece	20	Orestiada
Lebanon	—	Beirut
Poland	1500	Katowice
Spain	5	Granada (SER)
	5	Palma de M. (SER)
	2	La Coruña (SER)
	3	Huesca (SER)
	5	Toledo (OCR)
Egypt	10	El Minya + 1 st
Kenya	100	Mombasa
Libya	40	Ajedabia + 1 st
Morocco	1	Casablanca
Mozambique	5	Maputo (1079)
Nigeria	50	Minna
Iran	20	Abadan
Israel	5	Jerusalem
Lebanon	50	Alayh
1089 kHz		
Albania	150	Durrës
Germany	20	Weimar (T1)
Russia	300	Krasnodar
United Kingdom	150/1	Brookmans Park + 9 st's
Yugoslavia	10	Subotica
Algeria	5	Adrar
Angola	25/100	Luanda (1088)
Comoros	40	Haboho
Nigeria	50	Yola

	20	Lagos
Saudi Arabia	10	Qurayyat
Syria	200	Al-Hassake
1098 kHz		
Czech & Slovak Rep.	1500	Nitra
Russia	5	Vologda + 4 st's
	—	Unk. E
Spain	10	Lugo (RNE5)
	10	Almería (RNE5)
	5	Huelva (RNE5)
	5	Avila (RNE5)
Algeria	5	Ouargla
Bophuthatswana	100	Ga-Rankuwa
Canary Isl	2	S. Cruz de La Palma (RNE 5)
Cyprus	100	Yeni Iskele
Iran	200	Zabol
Saudi Arabia	5	Dammam
1107 kHz		
Germany	10	Berlin (AFN)
	10	Grafenwöhr (AFN)
	10	Nürnberg (AFN)
Italy	6	Roma
Lithuania	150	Sitkúnai + 6 st's
Russia	25	Nalchik + 2 st's
Spain	5	Cáceres (RNE5)
	25	Logroño (RNE 5)
	10	Santander (RNE5)
	10	Ponferrada (RNE5)
	5	Teruel (RNE5)
United Kingdom	1.5	Inverness
	0.5	Wallasey
Yugoslavia	150	Novi Sad
Cameroon	5	Douala (1106)
Egypt	600	Batrah
Malawi	1	Nkhota Kota
Nigeria	25	Kaduna
Afghanistan	1000	Kabul
UAE	10	Dubai
1116 kHz		
Germany	5	Chemnitz (S1)
Hungary	12	Miskolc
	2	Mosonmagyaróvár
Italy	150	Bari
	60	Bologna
	25	Pisa
	10	Palermo
	2	Aosta
Netherlands	0.05	Bloemendaal
Russia	30	Kaliningrad
	30	Sochi
	5	Moskva
Spain	5	Pontevedra (SER)
	5	Albacete (SER)
United Kingdom	1.2	Derby
	0.5	Guernsey
Angola	10	Sumbe
	5	M'B Kongo
Morocco	15	Ouarzazate
South Africa	2	Pietersburg
Iran	125	Bandar Lengeh
Iraq	300	Rutba
1125 kHz		
Belgium	20	La Louvière
Belarus	—	Miensk + 1 st
Bulgaria	2 × 500	Kalyakra Cape (F.Pl.)
Croatia	100	Deanovec
Russia	150	Sankt-Peterburg
Spain	10	Castellon (RNE5)
	10	Soria (RNE5)
	10	Vitoria (RNE5)
United Kingdom	1	Llandrindod Wells
Libya	500	El Beida (1124)
Niger	20	Niamey
Zambia	10	Solwezi
Syria	200	Al-Hassake
1134 kHz		
Lithuania	1	Pazagieniai
Russia	—	Volgodonsk + 2 st's
Spain	2/1	7 st's (COPE)
Ukraine		Kyyiv + 1 st
Yugoslavia	600/1200	Zadar
Nigeria	10	Ogoja
Iran	20	Bojnurd
Kuwait	750	Sulaibiyah
Lebanon	—	Beirut
1143 kHz		
Bulgaria	2×20	Varna
Croatia	300	Tovarnik (inactive)
Germany	10	Stuttgart (AFN)
	10	Kaiserslautern (AFN)
	5	Bremerhaven (AFN)
	1	5 low powers (AFN)
	0.25	10 low powers
Ireland	5	Cork
	2	Dublin
Italy	6	Messina
Russia	150	Bolshakovo
	100	Samara
Spain	2	Reus (COPE)
Egypt	10	Sohag
Nigeria	50	Bida
Zambia	1	Mbala
Iran	20	Yasuj
Int. Waters	10	Arutz Sheva
1152 kHz		
Romania	950	Cluj
Russia	50	Moskva
Spain	10	Lérida-Lleida (RNE5)
	10	Zamora (RNE5)
	10	Cartagena (RNE5)
	10	Albacete (RNE5)
United Kingdom	23.5	London
	1.8	Newcastle
	3.6	Glasgow
	3.0	Birmingham
	1.5	Manchester
	0.8	Norwich
	0.3	Plymouth
Angola	10	Menongue (1148 r)
Cameroon	300/1	Bamenda
Liberia	10	Voinjama
Mozambique	1	Maputo (1148)
Iran	100	Tabriz
UAE	50	Ras Al Khaima
1161 kHz		
Bosnia/Hercegovina	2	Zenica
Bulgaria	500/10	Stara Zagora + 3 st's
France	200	Strasbourg
	100	Toulouse
	10	Ajaccio
Russia	—	Volgograd
Spain	50	San Sebastian (EI)
United Kingdom	1	Bexhill
	1.4	Dundee
	0.2	Swindon
	0.4	Hull
	0.1	Bedford
Algeria	5	Ain-Salah
Angola	10/1	Huambo (1160)
Egypt	60	Tanta
Libya	—	?
Mozambique	5	Tete (1160)
Zaire	50	Matadi (1160)
Iran	10	Abadan (1160)
1170 kHz		
Albania	1	Berati
Belarus	150	Osipovičy
Portugal	10	Vila Real
Russia	500	Krasnodar
Slovenia	300	Beli Križ
United Kingdom	0.3	Stockton
	0.6	Swansea
	0.3	Ipswich
	0.1	Portsmouth
	0.2	Stoke-on-Trent
Yugoslavia	2	V. Banja
Djibouti	20	Djibouti
Nigeria	40	Abeokuta
Swaziland	50	Mpangela Range

Iran	2	Damghan
Israel	10	Jerusalem
UAE	10	Ras Al Khaima (1175)
1179 kHz		
Greece	50	Thessaloniki
Romania	200	Bacau
	5	Vascau
Spain	10	Santiago (SER)
	5	Murcia (SER)
	2	Logroño (SER)
Sweden	600	Sölvesborg
Yugoslavia	2	Pula
	1	Samobor
Canary Isl.	20	Sta. Cruz, TF
Egypt	10	Qena
1188 kHz		
Belgium	5	Kuurne
Germany	5	Reichenbach (MDR)
Hungary	135	Szombathely
Italy	6	San Remo
Russia	10	Sankt-Peterburg
Angola	10	Malanje
Egypt	10	Ras Gharib
Morocco	1	Casablanca
Iran	100	Tehran
Lebanon	—	Rumaylah
Yemen	400	Al-Hiswah
1197 kHz		
Belarus	50	Miensk + 2 st's
Bosnia/Hercegovina	2	Bjeljina + 1 st
Germany	300	München (VOA)
Romania	14	Bod
Spain	5	Vitoría (El)
United Kingdom	1/0.2	Enniskillen + 3 st's
Angola	0.25	Benguela
Egypt	25	Alexandria
Lesotho	100	Lancer's Gap (BBC)
Morocco	20	Agadir
Iran	20	Moghan
Iraq	300	Nineva
1206 kHz		
Albania	10	Korçë
Bosnia/Hercegovina	1	Bos. Brod
France	100	Bordeaux
Italy	0.5	Castel S. Pietro
Poland	200	Wroclaw
	60	Lublin
	10/5	4 st's
Russia	10	Moscow (Ekho Moskvy)
Yugoslavia	1	Majdanpek
Gabon	1	Oyem
Kenya	20	Nakuru
Nigeria	10	Hong
Sierra Leone	2x50	Goderich
Rodrigues Isl.	10/2	Rodrigues
Afghanistan	500	Mazar I Sharif
Israel	50	Haifa
1215 kHz		
Albania	500	Lushnje
Estonia	50	Tartu + 1st
Russia	1200	Bolshakovo
Spain	2/1	5 st's (COPE)
United Kingdom	100/0.3	Moorside Edge + 16 st's
Yugoslavia	1	Djurdjevac
	1	Mladenovac
Niger	0.1	Tahoua
Tanzania	50/10	Arusha
Bahrain	0.1	Awali
1224 kHz		
Bulgaria	2 × 500	Vidin
Spain	5/1	9 st's (COPE)
Yugoslavia	1	Krapina
Kenya	10	Garissa
Nigeria	50	Jos
Mozambique	5	Nampula (1223)
Senegal	20/1	Ziguinchor
Israel	20	Beersheba
1233 kHz		
Belgium	5	Liège
Bosnia/Hercegovina	1	Iljaš
Czech & Slovak Rep.	50/2	Brno + 7 st's
Russia	—	Moskva
	—	Unk, No. E
Yugoslavia	1	Našice
Angola	10	Lubango (1228)
	1	Huambo (1232)
Gabon	5	Pt Gentil
Madagascar	0.25	Maintirano
Morocco	200	Tanger
Venda	100	Sibasa
Cyprus	600	Cape Greco
Qatar	100	Al Khaisah
1242 kHz		
France	150	Marseille
Ukraine	150	Kyyiv + 9 st's (UR2)
United Kingdom	0.3	Maidstone
	0.5	Isle of Wight
Yugoslavia	10	Ohrid
	10	Kraljevo
Angola	5	Saurimo
Ethiopia	25	Harar (Taque)
Côte d'Ivoire	1	Bondoukou (1240)
Zimbabwe	1	Bulawayo (1241)
Iran	10	Kerman
Oman	200	Muscat
1251 kHz		
Bosnia/Hercegovina	1	Goražde
Hungary	500	Nyiregyhaza
Netherlands	10	Hulsberg
Portugal	10	Porto
	10	Viseu
	1	2 low power st's
Russia	5	Urup
	—	2 st's unk., E.
United Kingdom	0.8	Bury St. Edmunds
Libya	500	Tripoli
Iran	20	Sari
United Arab Emirates	600	Dubai
1260 kHz		
Albania	10(?)	Fier
Azores	1	Faial (1259)
Greece	500	Rhodos
Poland	10	Szczecin + 1 st
Russia	10	Moskva
	10	Sankt-Peterburg
Spain	20	Valencia (SER)
	5	Algeciras (SER)
Ukraine	50	Kharkiv
United Kingdom	1.6	Bristol
	0.3	Leicester
	0.64	Wrexham
	0.5	Scarborough
Vatican	5	Citta del Vaticano
Angola	10/1	N'Dalatando
Cameroon	20	Garoua
Egypt	10	Asswan
Liberia	10	Zwedru (inactive ?)
Nigeria	10	Jalingo
1269 kHz		
Germany	600	Neumünster (DLF)
Spain	5	León (COPE)
	3	Badajoz (COPE)
	2	Figueras (COPE)
	10	Ciudad Real (COPE)
Yugoslavia	750	Novi Sad
Kenya	20	Nyeri (1268)
Mali	4	Mopti
Zaire	2	Mbandaka
Zambia	10	Mansa
1278 kHz		
Belarus	150	Osipovičy
France	300	Strasbourg
Greece	20	Florina
Ireland	10	Dublin
	10	Cork
Russia	100	Nizhniy Novgorod + 8 st's
Ukraine	150	Odesa + 1 st
United Kingdom	0.4	Bradford
Comoros	10	Moheli
Egypt	10	Asswan

Malawi	1	Bangula
	1	Kasungo
Afghanistan	100	Kabul
Iran	100	Kermanshah
1287 kHz		
Bulgaria	2	Smolyan
Czech & Slovak Rep.	400	Melnik (RFE)
	400	Trnava (RFE)
	300	Litomyšl (RFE)
	14	2 st's (RFE)
Portugal	1	Portoalegre
Russia		4 st's E
Spain	10	Lérida-Lleida (SER)
	2	Lugo
Yugoslavia	1	Kočani
Algeria	5	El Golea
Cameroon	1	Yaoundé (1286)
Ceuta	5	Ceuta
Madagascar	1	Diego Suarez
Senegal	4	Kaolack
South Africa	2	Welgedacht
Israel	100	Tel Aviv
Saudi Arabia	5	Makkah
1296 kHz		
Azerbaidzhan	150	Pirsagat
Bulgaria	150	Kârdžali
	30	Pleven
Italy	5	Toranto + 2 sts.
Spain	10	Valencia (COPE)
United Kingdom	500	Orfordness
Yugoslavia	10	Loznica
	10	Vranje
Angola	1	Uige (1295)
Guinea	1	Conakry (1295)
Morocco	1	Rabat C
Mozambique	10	Quelimane (1295)
Nigeria	20	Maiduguri
Sudan	1500 (?)	Rebia
1305 kHz		
Belgium	10/5	Marche
Bosnia/Hercegovina	1	B. Krupa
Italy	3	Pisa
Poland	100	Rzeszów
	60	Bialystok
	60	Gdansk
	60	Lódz
Russia	10	Moskva
Spain	10	Orense (RNE5)
	10	León (RNE5)
	10	Ciudad Real (RNE 5)
United Kingdom	0.2	Newport
	0.2	Barnsley
Yugoslavia	1	Cakovec + 2 st's
Algeria	20	Constantine
Egypt	10	Assiut
Senegal	20	Dakar
Iran	20	Bushehr
Israel	20	Haifa
	5	Eilat
Saudi Arabia	1	Taif
1314 kHz		
Armenía	250	Yerevan
Austria	0.05	Scharnitz
Greece	20	Tripolis
Italy	6/1	Ancona + 3 st's
Norway	1200	Kvitsøy
Romania	30	Timisoara
	14	Constantza
	7	Craiova
Spain	10	Valladolid (RNE5)
	5	Cuenca (RNE5)
	4	Cabra (RNE5)
	2	Tarrega (RNE5)
Yugoslavia	100	Skopje
	10	Ohrid
Angola	1	Lubango (1313)
Egypt	10	Hurghada + 2 st's
Kenya	10	Garissa
South Africa	5	Port Elizabeth
Iran	20	Ardebil

Lebanon	—	Beirut
Syria	10	Aleppo
UAE	1000	Dabiya
1318 kHz		
Italy	0.5	Campogalliano
1323 kHz		
Albania	15	Shkoder
Georgia	30	Tskhinvali
Germany	1000/150	Wachenbrunn (RMR)
Romania	2	Sulina
Russia	—	Unk, Ural area
	—	Unk, So. E.
United Kingdom	0.63	Taunton
	0.5	Brighton
Yugoslavia	10/1	Bitola + 3 st's
Madagascar	1	Tuléar
Morocco	5	Safi C (1325)
Senegal	1	Linguére
Cyprus	200	Zyyi (BBC)
Iran	20	Jolfa
1332 kHz		
Czech & Slovak Rep.	50	M. Budějovice
	40	Prostejov
	14	Jihlava
Estonia	30	Yöru + 2 st's
Italy	300	Rome
	50	Bari
	25	Pescara
	12.5	Palermo
Portugal	10	Elvas
Romania	15	Galatzi
Russia	—	Moskva
United Kingdom	0.6	Peterborough
	0.3	Lacock
Yugoslavia	2	Gnjilane
Angola	1	Namibe (1331)
Ivory Coast	0.1	Maine
	0.1	Soroa
Madeira	1	Funchal
Iran	100	Tehran
1341 kHz		
Bosnia/Hercegovina	1	Prnjavor
Hungary	300	Lakihegy
Spain	5	Tarrasa (SER)
	2	Almería (OCR)
United Kingdom	100	Lisnagarvey
Yugoslavia	10	Zaječar
Burkina Faso	1	Ouagadougou
Egypt	100	Cairo + 3 st's
Zimbabwe	1	Bulawayo
Iran	1	Bam
Kuwait	200	Magwa
1350 kHz		
France	300	Nice
	100	Nancy
Georgia	50	Sukhumi
Greece	4	Pyrgos
Hungary	10	Pecs
	5	Györ/Szolnok
Latvia	50	Cesvaine/Kuldiga
Yugoslavia	10	Beograd
	1	Zabok
Egypt	10	Queseir
Madagascar	0.1	Fianarantsoa
Mauritania	20	Nouakchott (1349)
Iran	10	Bandar-Abbas
1359 kHz		
Albania	50	Tirana
Germany	250/100	Berlin (DLF)
Russia	50	Perm
	15	Chkalovskaya
Spain	600	Arganda (RNE-FS)
Ukraine	—	Simferopol
United Kingdom	0.85	Bournemouth
	0.2	Cardiff
	0.3	Chelmsford
	0.3	Coventry
Yugoslavia	2	Titov Vrbas-Kula
Kenya	5	Maralal
Melilla	5	Melilla

Nigeria	10	Iwo
1368 kHz		
Bosnia/Hercegovina	1	Sanski Most
Italy	20/1	Venezia + 10 st's
Poland	60	Kraków
United Kingdom	20	Foxdale
	2	Lincoln
	0.5	Duxhurst
	0.1	Swindon
Yugoslavia	10	Valjevo
Angola	100	Luanda
Egypt	10	El Kharga + 1 st
Kenya	20	Nyeri
Nigeria	10	Calabar
Senegal	20	St Louis
Seychelles	10	Victoria
Zimbabwe	1	Kwekwe
1377 kHz		
Armenia	—	Sisian
France	300	Lille
Portugal	10	Porto
Ukraine	50/25	5 st's (UR2)
Yugoslavia	10	Zadar
	10	Prizren
Swaziland	50	Sandlane
Iraq	1000	Tanaf
Lebanon	—	Murabitun
1386 kHz		
Greece	50	Athens
Iceland	0.002	Hellissandur
Russia	2500	Bolshakovo
Yugoslavia	1	Kratovo
Angola	1	Bié (alt. 1404)
Egypt	10	Luxor + 1 st
Guinea	50	Labe
Iran	400	Ahwaz
1395 kHz		
Albania	1000	Lushnje
Armenia	—	Yerevan
Poland	1	Slupsk
Russia	—	Petrozavodsk + 6 st's
Spain	2	Calahorra (RNE5)
	2	Tortosa (RNE5)
Ukraine	—	Oktyabrskoye
Yugoslavia	1	Bugojno
	1	Doboj
Madagascar	1	Antananarivo
Nigeria	20	Uyo
	10	Abak
Togo	20/1	Lomé
1404 kHz		
France	20/1	4 st's
Greece	50	Komotini
Romania	15	Baia Mare
Ukraine	50	Kotovsk + 7 st's (UR2)
Angola	1	Bié (alt. 1386)
Guinea	200	Kipe
Libya	20	Tripoli (1402)
Nigeria	10	Gombe
South Africa	2	Welgedacht
Iran	800	Rasht
Israel	10	Jerusalem
1413 kHz		
Germany	0.2	Wertheim (SDR) + 1 st
Russia	—	Volgograd
Spain	10	Salamanca (RNE5)
	10	Jaén (RNE5)
	5	Gerona (RNE5)
United Kingdom	0.13	Southall
Yugoslavia	1000	Priština
Egypt	1	Queseir
Oman	1500	Masirah Isl (BBC)
1422 kHz		
Georgia	30	Batumi
Germany	1200/600	Heusweiler
Latvia	50	Valmiera + 2 st's
Romania	7	Vilcea
Algeria	50	Alger
Egypt	10	Salum + 1 st
Iran	100	Kermanshah

Saudi Arabia	20	Riyadh
1431 kHz		
Italy	2	Foggia
	2	Pesaro
	1	Taranto
Portugal	10/5	Caramulo
United Kingdom	0.35	Southend
	0.14	Reading
Ukraine	400	Mikolayiv
Yugoslavia	1	Križevci
	1	Probištip
Mozambique	7	Beira
Iran	200	Isfahan
Iraq	1000	Missan
1440 kHz		
Luxembourg	1200	Marnach
Russia	7.5	Moskva
Yugoslavia	20 × 10	Svetozarevo
Central African Rep.	20/50	Bangui
Saudi Arabia	1600	Dammam + 1 st
1449 kHz		
Germany	5	Berlin (SFB)
Italy	50	Squinzano
	10	Sassari
	6	Catania
	2/0.5	20 low power st's
Moldovia	30	Chişinău
Russia	25	Kushalino + 3 st's
Ukraine	—	Chernivtsi + 1 st
United Kingdom	2	Redmoss
	0.15	Gunthorpe
Yugoslavia	2	Karlovac
Cameroon	20	Bafoussam
Libya	20	Misurata
Zaire	2	Kinshasa (1448)
Iran	400	Bandar Turkaman
1458 kHz		
Albania	500	Lushnje
Germany	5	Weida (MDR)
Gibraltar	0.5/2	Wellington Front
Lithuania	12	Kaunas
Romania	50	Constantza
Russia	7	Kudymkar + 2 st's
United Kingdom	50/0.5	London + 5 st's
USSR	30	Kudymkar
	—	3 st's Europe
Yugoslavia	1	Kladovo + 1 st
Mayotte	100	Pamandzi
Nigeria	5	Lagos
Zambia	1	Kabwe
·Israel	10	Jerusalem/Eilat
1467 kHz		
Bosnia/Hercegovina	1	Zvornik
France	1000	Romoules (RMC/TWR)
Moldovia	500	Grigoriopol
Russia	5	Moskva + 1 st
Ukraine	—	Yalta
Yugoslavia	2/1	Zrenjanin + 1 st
Iran	200	Isfahan
1476 kHz		
Austria	600	Wien-Bisamberg
Azerbaidzhan	25	Baky + 2 st's
Russia	—	Aleksin
Spain	2	Toledo (RNE5)
Ukraine	120	Lviv
United Kingdom	1	Guildford
Benin	50/20	Cotonou
Congo	20	Brazzaville
Egypt	10	El Minya + 1 st
	10	Sohag/Aswan
Ethiopia	1	Asmara
Iran	200	Sanandaj
1481 kHz		
U.A.E.	1500	Dubai (1476 r)
1485 kHz		
Bosnia/Hercegovina	1	26 st's
Bulgaria	3	Haskovo
Czech & Slovak Rep.	2	Plzen
	1	Brno (R.F.E.)
	1	Volary

Germany	0.2/0.1	Buchen-Walldürn (SDR)+1 st
	1	Baden-Baden (SWF)
	1/0.25	7 low powers (AFN)
	1	LP st's (RMV + SR)
Greece	1	Volos
	1	Orestiada
	1	Patras
	0.25	Kato Souli (AFRTS)
Italy	2/0.1	7 low powers
Iceland	0.002	2 low powers
Lithuania	3.5	Vilnius
Norway	1	Longyearbyen (NRK)
Poland	1	Gizycko + 5 low powers
Russia	1	Oktyabrski
	—	Unk. Ural
Slovenia	1	4 st's
Spain	5/2	7 st's (SER, OCR)
Switzerland	1	Savièse
Ukraine	2	Kyyiv + 2 st's
United Kingdom	2/0.5	Wallasey + 4 st's
Yugoslavia	1	Sarajevo (202)
	1/0.05	41 low powers
Angola	1	Luachimo (1484)
Ascension Island	0.5	English Bay
Azores	1	S. Cruz
Egypt	1	El Tur
Ethiopia	1	3 st's
Libya	1	Brak
Madeira	1	Funchal
Madagascar	0.25	Antalaha
Malta	1	Bugibba (1480)
Morocco	1	Casablanca
Niger	0.1	Gaya
Sudan	1	Kassala
Uganda	10	Arua
Bahrain	0.25	Manama
Iran	2	Abadan + 2 st's
Jordan	1	Aqaba
Lebanon	5	Marjayoun
Qatar	1	Al-Khaisah
Syria	10	Homs
Saudi Arabia	1	Jeddah
1494 kHz		
Albania	1	Prenjas
France	20	Clermont-Ferrand
	4	Bayonne
	1	Besançon
Greece	50	Rhodos
Moldovia	30	Cahul
	30	Edinet
Russia	1200	Sankt-Peterburg
Côte d'Ivoire	20	Abidjan (1493)
Egypt	1	El Arish
Kenya	5	Kitale
Mozambique	5	Pemba (1493)
Cyprus	10	Yeni Iskele
Jordan	100/10	Al Karanah
1503 kHz		
Azores	0.1	Lajes
Bosnia/Hercegovina	1	Zavidovići
Greece	0.25	Iraklion (AFRTS)
Italy	2	Roma
Poland	300	Stargard Szczecinski
Russia	—	Belgorod
Spain	5/2	4 st's (RNE5)
United Kingdom	1	Stoke-on-Trent
Yugoslavia	10	Beograd "202"
	10	Ulcinj
	1	Kutina
Angola	1	Benguela (1502)
Egypt	25	El Arish
Libya	10	Zelten
Madagascar	4	Antananarivo
Senegal	20	Tambacounda
Togo	10	Kara (1502)
Iran	100	Bushehr
1512 kHz		
Belgium	25/300	Wolvertem
Estonia	30	Tallinn
Greece	50	Chania

Iceland	0.002	3 low power st's
Italy	2	Palermo 3
	1	Pavullo
Russia	30	Staryy Oskol
	7	Ibresi
	5	Chaykovskiy
	—	Sochi
	5	Chaykovskiy
Yugoslavia	10	Priština
Niger	1	Zinder (1511)
Saudi Arabia	1000	Jeddah
1521 kHz		
Armenia	5	Kamo
Bosnia/Hercegovina	1	Derventa
Czech & Slovak Rep.	600	Košice
	50	M. Hradiště
	40	Karlovy Vary
	40/10	6 st's
Italy	0.25	Sicily (AWR)
Russia	5	Unk., E
—		Unk., Ce. E
Spain	2	Manresa (SER)
United Kingdom	0.64	Reigate
Yugoslavia	1	1 low power
Congo	0.25	Pointe Noire (1520)
Egypt	10	Queseir
Gabon	1	Port Gentil (1520)
Iran	600	Kiashahr
Saudi Arabia	2000	Duba
1530 kHz		
Azerbaidzhan	5	Baky
Bosnia/Hercegovina	1	Donji Vakuf
Iceland	0.25	Keflavik (AFRTS)
Romania	15	Mahmudia
	15	Mihaileni
Ukraine	5	Trostyanets + 1 st
United Kingdom	0.74	Huddersfield
	0.5	Worcester
	0.15	Southend
	—	Kettering/Corby
Vatican	150/450	Citta del Vaticano
Yugoslavia	1	3 low powers
Madeira	1	Funchal
Iran	20	Yazd
1539 kHz		
Germany	700	Mainflingen (DLF)
Latvia	7	Cesvaine + 2 st's
Russia	—	Moskva
Spain	5	Valladolid (SER)
Ukraine	—	Izyum
Yugoslavia	10/1	Pec + 2 st's
Int. Waters	75	Voice of Peace (1540)
Djibouti	20	Djibouti
Madagascar	1	Majuuga
Melilla	4	Melilla
Senegal	10	Dakar
Iraq	1000	Arbat
U.A.E.	2×50	Sadiyat
1544 kHz		
Morocco	—	L.V.D. de Sahara
1548 kHz		
Moldavia	500	Grigoriopol
United Kingdom	97.5	London
	2	Edinburgh
	5	Liverpool
	0.74	Sheffield
	5	Bristol
	1	Stockton
Yugoslavia	1	Dragaš
Algeria	20	Laghouat
St Helena	0.45	Jamestown
Iran	20	Marivan
	10	Ardebil (1551)
1554 kHz		
Gabon	20	Melene
1557 kHz		
France	300	Nice
Lithuania	25	Vilnius + 2 st's
Malta	600	Cyclops
United Kingdom	0.8	Northampton

	0.5	Southampton
	0.25	Oxcliffe
	—	Clacton
Yugoslavia	300 (?)	Tovarnik
Madagascar	1	Diego-Suarez (1554)
Iran	10	Ardabil
1566 kHz		
Armenia	5	Yerevan
Belarus	—	Pinsk + 5 st's
Azores	10	Vila de Porto
Estonia	5	Tartu
Portugal	1	Covilhã (1562)
Slovenia	2	Smarje
Switzerland	300	Monte Ceneri
Ukraine	5	Odesa + 1 st
Yugoslavia	1	Vrgin Most
Madagascar	1	Tamatave
Tunisia	1200	Sfax
Iran	100	Bandar Abbas
1570 kHz		
Angola	10	Cabinda
Turkey	0.01	Diyarbakir
1575 kHz		
Germany	250	Burg (DLF)
Italy	50	Genova
	20	Perugia
	10	Sassari
	2/1	14 st's
Portugal	10	Porto-Canidelo
Spain	5	Córdoba (SER)
	5	Vilanova (OCR) + 1 st (SER)
Ukraine	—	Botanichnoye
Yugoslavia	0.1	Bać
Côte d'Ivoire	0.1	Bouaké (1578)
Egypt	10	Queseir
Gabon	15	Tichibanga
Mauritius	1	Forest Side
Niger	1	Niamey
Israel	5	Haifa
U.A.E.	50	Sharjah
1580 kHz		
Turkey	0.01	Adana
1584 kHz		
Azores	0.3	Lajes Field
Bosnia/Hercegovina	2/1	2 st's
Czech & Slovak Rep.	2	Tachov
	1	Lipno
Greece	1	Athens
	1	Serres
	1	Amaliás (1590)
Italy	1	Verona
	0.5	Bologna
Poland	1	Ostróda
Portugal	1	Guarda (R.A.)
Russia	—	3 st's (RR)
Slovenia	1	2 st's
Spain	5/2	5 st's (SER) (IND)
Ukraine	—	Simferopol + 1 st
United Kingdom	1	Clipstone
	0.2	Perth
Yugoslavia	1	11 low power st's
Angola	10	Luena (1586)
Egypt	10	Idfu + 2 st's
Niger	0.1	2 st's
Sudan	1	2 st's
Bahrain	1	Manama
Iran	1	Hamadan
	1	Sirjan
1593 kHz		
Czech & Slovak Rep.	6	Liberec
Germany	800/400	Langenberg (WDR)
Moldovia	5	Chişinău
Romania	14	M. Ciuc
	7	Oradea
	7	Sibiu
Russia	—	Unk, No. E
Ukraine	5	Dnipropetrovsk + 1 st
Egypt	10	Matruh
Niger	0.1	Téra
Iran	100	Isfahan
Yemen	1	Beihan
1602 kHz		
Czech & Slovak Rep.	1	Volary
Estonia	1	Viljandi
Germany	1	LP st's
Greece	1	Kavala/Kozani/Samos
Italy	2	Bolzano + 7 st's
Poland	1	Lidzbark + 3 st's
Russia	5	Gayny + 2 st's
Slovenia	1	2 st's
Spain	10	Vitoria (EI)
	5/2	5 st's (SER)
Ukraine	—	Sevastopol + 1 st
United Kingdom	0.25	Rusthall
Yugoslavia	1	Negotin + 8 st's
Ascension Island	1	ZD8VR (AFRTS)
Egypt	10	Al Dakhla + 2 st's
Gabon	1	Franceville
Madagascar	1	Diego-Suarez
Sudan	1	Al-Ubayyid
Iran	2	Zanjan
Qatar	1	Al Khaisah
1611 kHz		
Italy	15	Vatican City
1620 kHz		
Greece	1	Ierapetra
Italy	5	Verona

E. ASIA & PACIFIC

Abbreviations peculiar to the E.Asia/Pacific section of MW freq. lists: AF = allocated freq. C. = City. PO = Present operation on---. Proj. = Projected station. Rptr. = repeater. Trtr = translator. Unk. = Unknown location.

Australia: The numeral preceding the call letters indicates the state: 2 = New South Wales. 3 = Victoria. 4 = Queensland. 5 = South Australia. 6 = Western Australia. 7 = Tasmania. 8 = Northern Territory. ACT = Australian Capital Territory.

China, P.R: Only main stations, generally observed overseas, are listed by name. If there is no dominant station on a frequency, only the number of stations reported on the frequency is listed. For full details see text section. A 'v' in brackets under location: variable frequency.

Indonesia: Only RRI stns included. For details of other stns see text section.

Philippines: Province Abbreviations: Ak = Aklan; Ant = Antique; Boh = Bohol; Bat = Batangas; Buk = Bukidnon; Bul = Bulacan; Cag = Cagayan; Cam Nte = Camarines Norte; Cam Sur = Camarines Sur; Dvo Nte = Davao Norte; Dvo Sur = Davao Sur; Isa = Isabela; IN = Ilocos Norte; IS = Ilocos Sur; LU = La Union; Mas = Masbate; MM = Metropolitan Manila; M Octal = Mindoro Occidental; M Ortal = Mindoro Oriental; Pang = Pangasinan; Que = Quezon; Riz = Rizal; S Sur = Surigao del Sur; Sor = Sorsogon; Tar = Tarlac; Z Nte = Zamboanga del Norte; Z Sur = Zamboanga del Sur.

Russia: Regions in the Asian parts of Russia: Sib. = Siberia. FE = Far East.

Country	kW	Call	Location
153 kHz			
Russia	1000		Komsomolsk, FE
Turkmenistan	500		Ashgabat
162 kHz			
Mongolia	1000		Ulaan baatar
Uzbekistan	150		Tashkent, Uz.
Russia			Norilsk, Sib.
171 kHz			
Russia			Tomsk, Sib.
	500		Yakutsk, Sib.
180 kHz			
Kazakhstan	250		Alma Ata
	150		Aktyubinsk
Russia	50		Turkestan
	150		Chita, Sib.
	150		Petropavlovsk-Kamch, FE
189 kHz			
Russia	1000		Blagoveshchensk, Sib.
198 kHz			
Kirghizia	150		Bishkek
Russia	50		Aleksandrovsk-Sakhalinskiy, FE
207 kHz			
Mongolia	60		Ulgii, Bayan-Ulgii
	150		Dalanzadgad, So. Gobi
	150		Choibalsan, Ea. Aimak
Russia	30		Blagoveshchensk, Sib.
	30		Skovorodino, Sib.
216 kHz			
Russia	300		Yeniseysk, Sib.
	30		Birobidzhan, FE
225 kHz			
Russia	150		Altai, Gobi
USSR	150		Surgut, Sib.
234 kHz			
Russia	500		Angarsk, Sib.
	1000		Magadan, FE
243 kHz			
Kazakhstan	500		Alma Ata
			Karaganda
Russia	1000		Vladivostok, FE
252 kHz			
Russia			Unk, W. Sib.
Tadzhikistan	300		Yangi-Yul, Tad.
261 kHz			
Russia	1000		Chita, Sib.
270 kHz			
Russia	150		Novosibirsk, Sib.
	100		Khabarovsk, FE
Uzbekistan	500		Tashkent
279 kHz			
Russia	150		Ulan Ude, Sib.
Turkmenistan	150		Ashgabat
	50		Yekaterinburg, Sib.
	150		Yuzhno-Sakhalinsk, FE
	50		Barnaul, Sib.
531 kHz			
Australia	5	2MC	W. Kempsey
	5	3GG	Warragul
	5	4KZ	Innisfail
	0.5	5UV	Adelaide (Univ.)
	10	6DL	Dalwallinu (ABC3)
China, PR	10		ZJ
India	100		Jodhpur A
Japan	10	JOQG	Morioka (NHK1)
	1		Nago (NHK1)
New Zealand	2	4XA	Alexandra (rel 4XO)
Philippines	5	DZBR	Lipa C., Bat.
Sri Lanka	40		Ambewela
Thailand	50		Maha Sarkham
540 kHz			
Australia	10	4QL	Longreach (ABC3)
	5	7SD	Scottsdale
China, PR			(C1) Synchro
India	20		Aizawl
Indonesia	10		Bandung (RRI)
Japan	5	JOJG	Yamagata (NHK1)
	5	JOMG	Miyazaki (NHK1)
	1		4 sts (NHK1)
Korea, Rep.	10	HLCZ	Hongsŏng (KBS)
	1	HLSC	Chŏmch'on (KBS)
	1	HLSM	Changhŭng (KBS)
	1	HLSN	Changsu (KBS)
Pakistan	300		Peshawar I
Philippines	10	DYCU	Cebu C.
	10	DZWT	Baguio C., Benguet
New Zealand	5	1XC	Tauranga (R. Rhema)
Samoa, W.	10	2AP	Apia
Thailand	10		Bangkok
549 kHz			
Australia	50	2CR	Orange (ABC3) V;
China, PR			SX
China (Taiwan)			Tainan (Fuhsing)
India	100		Ranchi A
Japan	10	JOAP	Okinawa (NHK1)
Kazakhstan	1000		Alma Ata
Korea, Rep.	5		Seoul (AFKN) (F. pl. 1377)
Malaysia	20		Kuching, Sarawak
Russia	50-150		5 sts (sync.)
Tadzhikistan	50		Yangi-Yul
Thailand	10		Krabi
	100		Lampang
	10		Mukdahan
550 kHz			
Hawaii	5	KMVI	Wailuku, Maui
558 kHz			
Australia	5	4AM	Atherton (Mareeba)
	5	4GY	Gympie
	50	6WA	Wagin (ABC3)
	2	7BU	Burnie
Bangladesh	100		Khulna
China, PR			FJ;XJ;NM
China (Taiwan)	1	BEV58	Chingshui (Cheng Sheng)
	1		Taipei (Fuhsing)
Fiji	20	VRH	Suva (Naulu)
India	50		Bombay B
Japan	20	JOCR	Kobe (R. Kansai)
Korea, Rep.	250	HLQH	Taegu (KBS)
New Zealand	5	4ZG	Gore (Dacre)
Philippines	10	DZXL	Makati, MM
	10	DWRP	Naga C. Cam. Sur
Russia			Unk., Ce. Sib.
Sri Lanka	20		Diyagama (Colombo)
Thailand	10		Chaiyaphum
	50		Songkhla
567 kHz			
Australia	0.5	2BH	Broken Hill
	10	4JK	Julia Creek (ABC3)
	0.1		4 sts W.A. (ABC3)
China, PR			(C1)
Guam	5	KGUM	Agana
Hong Kong	20		Golden Hill (RTHK3)
India	300		Dibrugarh
Japan	100	JOIK	Sapporo (NHK1)
Korea, Rep.	100	HLKF	Chŏnju (KBS)
Malaysia	10		Tenom, Sabah
New Zealand	50	2YA	Wellington
Pakistan	300		Khuzdar
Philippines	2.5	DZAG	Agoo, LU
Russia	50		Kyzyl, Sib.
Sri Lanka	10		Senkadagala (Kandy)
570 kHz			
Hawaii	1	KQNG	Lihue, Kauai (FP: 3 kW)
Vietnam			Hanoi (Hanoi B.S.)
576 kHz			
Australia	50	2RN	Sydney (ABC2)
China (Taiwan)	10		Taipei (ICRT)
China, PR			YN;ZJ(v);JL
India	100		Alappuzha
Japan	10	JOHG	Kagoshima (NHK1)
Kirghizia	50		Osh
Korea, Rep.	1	HLKZ	Sunch'ŏn (KBS)
	5		Munsan (AFKN) (F. pl. 1440)
Malaysia	100		Tronoh
	50		Johore Bharu
	20		Miri, Sarawak
Myanmar	200		Yangon
Nepal	100		Surkhet
Philippines	50	DYMR	Cebu City

	1	DZYZ	Laoag C, 1N
Russia	100-500		4 sts. (sync)
Thailand	20		Bangkok
Turkmenistan	150		Ashgabat
Uzbekistan	5		Tashkent
580 kHz			
Easter Isl.	0.4		Mataveri Airport
Laos	150		Vientiane
Pakistan	1000		Lawrencepur (AF585)
Vietnam	50		Nhatrang?
585 kHz			
Australia	5	6PB	Perth
	10	7RN	Hobart (ABC 2)
	2	2WEB	Bourke (Educ)
China, PR			JS;HEB(v);JX
China (Taiwan)	1200		Kaohsiung (VO Asia)
India	100		Nagpur A
Indonesia	50		Surabaya (RRI)
Japan	10	JOPG	Kushiro (NHK1)
Korea, Rep.	5		Taegu (AFKN) (F. pl. 1377)
New Zealand	2	2XR	Ruatoria
Pakistan	1000		Lawrencepur (r. 580)
Philippines	2.5	DXCP	Makati, MM
	10	DYCI	Iloilo C
	5	DZVV	Vigan, I.S.
Papua N. Guinea	5		Port Moresby
Russia	500		Svobodnyy, Sib. (FS)
Samoa, Am.	50	KBQN	Pago Pago under constr.
Sri Lanka	20		Kantale
Thailand	10		Chumporn
	20		Prachinburi
	10		Phitsanulok
	100		Phrae
590 kHz			
Hawaii	7.5	KSSK	Honolulu
594 kHz			
Australia	50	3WV	Horsham (ABC3)
Brunei	200		Tutong
China, PR			SD;FJ(T2);XZ
China (Taiwan)	1	BEH2	Taipei + 11 sts (Fuhsing)
India	1000		Calcutta (Magra)
Japan	300	JOAK	Tokyo (NHK1)
Korea, Rep.	10	HLAG	Yŏngju (KBS)
New Zealand	2	3XL	St. Andrews (R. Rhema)
Niue	0.4	ZK2ZN	Alofi
Philippines	10	DZBB	Quezon C, MM
Russia	500		Surgut, Sib.
	150		Yeniseysk, Sib.
			Unk., Sib.
Sri Lanka	20		Weeraketiya
Thailand	10		Khonkaen
600 kHz			
China, PR			SC
603 kHz			
Australia	10	2RN	Nowra (ABC)
	5	4CH	Charleville (ABC3) (rptr)
	2	6PH	Port Hedland (ABC3)
China, PR			4 sts
China (Taiwan)	1000	BED76	Luchiang (CBS)
India	200		Ajmer
Japan	5	JOKK	Okayama (NHK1)
	1	JOCY	Nayoro (NHK1)
Korea, Rep.	500	HLSA	Namyang (Seoul) (KBS)
Malaysia	10		Batu Pahat
	10		Kota Kinabalu, Sabah
New Zealand	5	1XO	Auckland
New Zealand	0.4	1XB	Paengaroa
Philippines	10	DWLV	Naga C.
	10	DXPR	Pagadian C, Zamb. Sur
Russia	50		Salekhard, Sib.
Sri Lanka	10		Ratnapura
Thailand	20		Bangkok
610 kHz			
Vietnam	20		Ho Chi Minh C.
612 kHz			
Australia	50	4QR	Brisbane (ABC1)
	0.2	6NM	Northam (ABC3)
China, PR			FJ + 2 sts
China (Taiwan)	600	BEG50	Kaohsiung (V. of Asia)
			3 sts (Fuhsing)
Guam	10	KUAM	Agana
India	200		Bangalore A
Japan	100	JOLK	Fukuoka (NHK1)
Kirghizia	300		Bishkek
Malaysia	10		Kangar
New Zealand	0.4	4ZF	Te Anau (rel. 4ZA)
Philippines	5	DWHB	Baguio C., Benguet
	10	DYHP	Cebu C. (r 615)
	5	DZTG	Teguegarao, Cag.
Russia			Kzyl, Sib.
Thailand	20		Lopburi
620 kHz			
Hawaii	10	KIPA	Hilo
	10	KIPA-1	Kona/Keaan (rptr KIPA)
621 kHz			
Australia	50	3RN	Melbourne (ABC2)
China, PR			HL + 1 st
China (Taiwan)			Haituan (Cheng Sheng)
Hong Kong	20		Golden Hill (RTHK7)
India	100		Patna A
Japan	3	JOCG	Asahikawa (NHK1)
	3-1		3 sts (NHK1)
Korea, DPR	500		Ch'ŏngjin
Korea, Rep.	10	HLCF	Sŏgwip'o (KBS)
	10	HLSJ	T'aebaek (KBS)
	1	HLAY	Yŏngdong (KBS)
Malaysia	20		Sibu, Sarawak
	100		Segamat
New Zealand	2	4XG	Dunedin (R.Rhema)
Philippines	10	DXDC	Davao C.
	1	DZFT	Legaspi City, Albay
Russia			Khabarovsk, FE
			Yakutsk, Sib.
Sri Lanka	20		Diyagama (Colombo)
Thailand	100		Khon Kaen
Tuvalu	10	T2U2	Funafuti
625 kHz			
Vietnam	50?		Lang Son? (V. of Vietnam) (v)
630 kHz			
Australia	10	2PB	Sydney
	50	4QN	Townsville (ABC3)
	0.4	6AL	Albany (ABC3)
	0.4	7QN	Queenstown (ABC3)
China, PR			(C2) synchro
China (Taiwan)	10	BEG51	Ilan (BCC3/Chiao Yu)
	1		Chungshing (Min Chien)
Cook Is.	5	ZK1ZC	Rarotonga (Black Rock)
India	20		Thrisoor
Indonesia	100		Ujung Pandang (RRI)
Korea, Rep.	10	HLCY	Yŏsu (KBS)
	5	HLSE	Inje (KBS)
New Zealand	20	2YZ	Napier (Opapa)
Pakistan	100		Lahore I
Philippines	50	DZMM	Quezon C., MM
Russia	100		Komsomolsk, FE (FS)
Singapore	50		Singapore
Thailand	10		Bangkok
639 kHz			
Australia	5	2CS	Coff's Harbour
	1	4MS	Mossman (ABC3)
	10	5CK	Port Perie (ABC3)
	2	8KN	Katherine (ABC)
China, PR			(C1)
China (Taiwan)			Taipei + 3sts (Fuhsing)
			Tahsi (TBC)
Fiji	10		Lautoka (Drasa)
India	50		Kohima
Japan	10	JOPB	Shizuoka (NHK2)
Korea, DPR	500		Chongjin (FS)
Korea, Rep.	50		Kaebong (Seoul) (EBS)
New Zealand	2	4YW	Alexandra
Pakistan	10		Karachi II
Philippines	5	DYAG	Cadiz, Neg. Occ.
	5	DXKR	Marbel, Cot Sur
	1	DZRL	Batac, IN
Russia	50		Omsk, Sib.
Sri Lanka	50		Maho
Thailand	20		N. Si Thammarat
	10		Phuket

640 kHz			
Laos	10		Vientiane
648 kHz			
Australia	10	2NU	Tamworth (ABC3)
	2	6GF	Kalgoorlie (ABC3)
China, PR			GD + 2 sts.
China (Taiwan)	1		Tawu (Cheng Sheng)
India	100		Indore A
Japan	10		Okinawa C. (FEN)
	5	JOIG	Toyama (NHK1)
Korea, Rep.	1	HLSL	Posŏng (KBS)
Malaysia	20		Kuala Lipis
	20		Limbang, Sarawak
Nepal	100		Dhankuta
New Zealand		2XC	Gisborne (R. Rhema)
Philippines	5	DXMB	Malaybalay, Buk.
	2.5	DYMM	Tacloban C, Leyte
	5	DWPS	Puerto Princesa, Palawan
Russia	1000		Vladivostok, FE (FS)
Samoa, Am.	10	WVUV	Pago Pago (Vailoa)
Sri Lanka	20		Ambewela
Thailand	55		Khon Kaen
650 kHz			
Hawaii	10	KHNR	Honolulu
657 kHz			
Australia	10	2BY	Byrock (ABC3)
	2	8RN	Darwin (ABC2)
China, PR			HEN + 2 sts.
China (Taiwan)	20	BED34	Taipei (BCC1)
	10	BEV59	Taichung (Cheng Sheng)
India	100		Calcutta A
Japan	5	JOOG	Obihiro (NHK1)
Korea, Rep.	50	HLKM	Ch'unch'ŏn (KBS)
Korea, DPR	1500		Kangnam
Malaysia	20		Grik
New Zealand	60	2YC	Wellington
Philippines	5	DXDD	Ozamis C., Mis. Occ.
	5	DWRN	Naga C., Cam. Sur
	1	DYFL	Borongan, E. Samar
	1	DZXC	San Fernando, LU
Russia	50		Verkhoyansk, Sib.
Thailand	10		Samut Sakhon
	1		N. Rajchasima
666 kHz			
Australia	5	2CN	Canberra, ACT
	2	4LM	Mt. Isa
	2	4CC	unk. (trtr)
	1	6LN	Carnarvon
China, PR			QH;FJ + 6 sts
China (Taiwan)		BEH71	Taitung + 5 sts (Fuhsing)
			Hsinhua (Chien Kuo)
India	100		Delhi B
Japan	100	JOBK	Osaka (NHK1)
	0.1		Miki (NHK1)
Malaysia	10		Penang (r 665)
New Caledonia	20		Noumea (RFO)
Philippines	10	DXRP	Davao C.
	25	DZRH	Navotas, MM
Russia	100		Unk., Yakut SSR
			Komsomolsk-na-Amure, FE
Thailand	10		Tak
Uzbekistan	30		Tashkent
			Urgenc
670 kHz			
Hawaii	10	KPUA	Hilo
Vietnam	50?		Nhatrang (?)(r)
675 kHz			
Australia	10	2CO	Corowa (ABC3) (FP 684)
	10	2KP	Smithtown (PO 684)
	5	6BE	Broome (ABC3)
Brunei	200		Serasa
China, PR	150		1 st
Hong Kong	10		Peng Chan (RTHK6)
India	20		Chhattarpur
	1		Bangalore B
	1		Itanagar
Japan	5	JOUG	Yamaguchi (NHK1)

	5	JOVK	Hakodate (NHK1)
Korea, Rep.	10	HLAS	Kunsan (KBS)
Malaysia	10		Lahad Datu, Sabah
New Zealand	20	3YA	Christchurch
Papua N. Guinea	2		Lae/Wewak
Philippines	5	DYKC	Mandawe (Cebu)
	5	DWLW	Laoag C., IN
	1	DXTP	Bongao, Sulu
Russia	150		Novosibirsk, Sib.
Singapore	50		Singapore
Sri Lanka	20		Weeraketiya
Thailand	20		Chieng Mai
Turkmenistan			Ashgabat
680 kHz			
Vietnam			Hanoi
684 kHz			
Australia	10	2CO	Corowa (PO 675)
	10	2KP	Kempsey (ABC3) (FP 675)
	4	6BS	Busselton (ABC3)
	1	8RN	Tennant Creek (ABC)
China, PR			FJ (T2) + 6 sts
China (Taiwan)		BEC23	Taipei (Han Sheng)
			several (Fuhsing)
Fiji	10		Labasa
India	10		Kozhikode
	20		Port Blair
Japan	5	JOAG	Nagasaki (NHK1)
	5	JODF	Morioka + rptr (IBC)
Kazakhstan	30		Kustanay
Korea, DPR	250		Samgŏ
Nepal	100		Pokhara
Philippines	5	DWGW	Legaspi C., Albay
	1	DWAU	Cabanatuan, Nva. Ecija
	5	DZCV	Tuguegarao, Cag.
Thailand	20		Bangkok
690 kHz			
Hawaii	10	KQMQ	Honolulu
Vietnam	55		Ban Me Thuot
693 kHz			
Australia	5	3EE	Melcourne
	5	4KQ	Brisbane (Newstead)
	0.5	4KZ	"3) Private" (trtr)
	0.5	4LM	Cloncurry (trtr)
	2	5SY	Streaky Bay (ABC3)
	0.5		Yulara (trtr 8HA) (FP)
Bangladesh	1000		Dhaka A
China, PR			SN; FJ (T1)
China (Taiwan)	10	BEC25	Taoyuan (Han Sheng)
	3	BEC32	Tainan (Han Sheng)
			several (Fuhsing)
Japan	500	JOAB	Tokyo (NHK2)
Malaysia	10		Kota Kinabalu, Sabah (r690)
Philippines	5	DXBC	Butnan, Agusau Nte
	1	DXDX	Gen. Santos C., Cot Sur
Russia	25		Anadyr, FE
Sri Lanka	20		Amparai
Thailand	20		Saraburi
700 kHz			
China, PR			LN
Vietnam			Quang Nam-Da
702 kHz			
Australia	50	2BL	Sydney (ABC1)
	10	6KP	Karratha (ABC)
China, PR			JS + 2 sts
China (Taiwan)	10	BEP24	Taichung (Chingcha)
India	300		Jalandhar
Japan	10	JOFB	Hiroshima (NHK2)
	10	JOKD	Kitami (NHK2)
Korea, DPR	50		Ch'ŏngjin
Malaysia	20		Kota Bharu
New Zealand	5	1XP	Auckland
Philippines	40	DZAS	Valenzuela, MM
Russia			Bratsk, Sib.
Sri Lanka	25		Diyagama (Colombo)
Tadzhikistan	50		Orzy
Thailand	10		Lopburi

705 kHz			
Laos			Luang Prabang
711 kHz			
Australia	10	2CU	Cootamundra (ABC3)
	10	4QW	St. George (ABC3)
	10	7NT	Launceston (ABC3)
Brunei	20		S. Hanching
China, PR			QH + 5 sts
China (Taiwan)	50	BEC72	Hsinfeng (Kung Chun)
India	200		Siliguri
Korea, Rep.	500	HLKA	Seoul (Sorae) (KBS)
New Zealand		2XP	Wellington
Philippines	5	DXIC	Iligan C., Lanao Nte
	5	DXRD	Davao C.
	5	DZYI	Tacloban C, Leyte
Russia	50		Khabarovsk, FE
Thailand	20		Bangkok
	20		Chieng Mai
	20		U. Ratchathani
Uzbekistan			Bukhara, Uz.
720 kHz			
Australia	0.05	2AN	Armidale (ABC3)
	0.4	2ML	Murwillumbah (ABC3)
	2	3MT	Omeo (ABC3)
	4	4AT	Atherton (ABC3)
	50	6WF	Perth (ABC1)
China, PR			(C2) 1st
China (Taiwan)	10	BED58	Taichung (BCC1)
Hawaii	5	KUAI	Eleele, Kauai
India	200		Madras A
Indonesia	10		Ambon (RRI)
Japan	1	JOIL	Kitakyushu (KBC)
	1		Mihara (RCC)
	0.1		2 sts (Comm.)
Korea, DPR	500		Wiwon
New Zealand	20	4YZ	Invercargill
Philippines	5	DYEZ	Bacolod, Neg. Occ.
	5	DZJO	Infanta, Que.
	5	DZLU	San Fernando, LU
Russia	500		Yuzhno-Sakhalinsk, FE
Turkmenistan			Gasan-Kuli/Ek-Archa
729 kHz			
Australia	50	5RN	Adelaide (ABC2)
	2	7PB	Hobart
China, PR			JX + 1 st
China (Taiwan)	1		Taipei (Shi Hsin)
India	50		Guwahati A
Japan	50	JOCK	Nagoya (NHK1)
Korea, DPR	50		Sep'o
Malaysia	20		Kuching, Sarawak
New Zealand			Tokoroa
Philippines	10	DWPE	Tuguegarao, Cag
	5	DXOR	Cagayan de Oro, Mis. Or.
	2.5	DZGB	Legaspi C., Albay
Sri Lanka	10		Ratnapura
Thailand	50	HSKT	N. Rajchasima
735 kHz			
Macau	10		Em. Vila Verde
738 kHz			
Australia	50	2NR	Grafton (ABC3)
	5	6MJ	Manjimup (ABC3)
China, PR	100		HN; JL; XJ + 1 st
China (Taiwan)	100	BEL2	Kaohsiung? (Fishery BS)
India	50		Hyderabad A
Japan	5	JOLR	Toyama (KNB)
	1		Takaoka (KNB)
	5	JORR	Naha, Okinawa (RBC)
Korea, DPR	100		Sinsang
Korea, Rep.	100	HLKG	Taegu (KBS)
Philippines	10	DZFM	Quezon C. MM
Polynesia (Fr.)	20		Papeete, Tahiti
Thailand	20		Bangkok
	10		Chieng Mai
Tonga	5	A3R	Tangatapy (UCB)
Russia	50		Chelyabinsk, Sib.
	50		Palana, FE
740 kHz			
Cambodia	120		Phnom Penh
Vietnam	50		Quy Nhon
747 kHz			
Australia	10	4QS	Toowoomba (ABC3)
	1		Exmouth (trtr 6LN)
	5	6SE	Esperance
	0.25	8JB	Jabiru (ABC/R. Austr)
China, PR			4 sts
China (Taiwan)			Pingtung (Fuhsing)
	100	BED2	Minhsiung (CBS)
India	300		Lucknow A
Japan	500	JOIB	Sapporo (NHK2)
Kazakhstan	20		Karaganda
Korea, Rep.	100	HLKH	Kwangju (KBS)
Malaysia	10		Tawau, Sabah
New Zealand	2	3ZA	Greymouth (Kumara)
Philippines	1	DXND	Kidapawan, Cotab. Nte
	5	DYHB	Bacolod, Neg. Occ.
	5	DZJC	Laoag C, IN
Samoa, W.	10	2AP	Apia (Afiamalu)
Sri Lanka	20		Kantalai
Thailand	10		Songkhla
	10		Surin
	10		Udon Thani
	1		3 sts
	50		Uralsk
750 kHz			
China, PR			LN;GX
756 kHz			
Australia	2	2TR	Taree (ABC3)
	10	3RN	Wangaratta (ABC2) (rptr)
	2	4QA	Mackay (ABC3)
	0.1	6KW	Kununurra (ABC3)
China, PR			(C1)
China (Taiwan)	1		Penghu (Shengli Chih Sheng)
India	20		Jagdalpur
Indonesia	10		Purwokerto (RRI)
Japan	10	JOGK	Kumamoto (NHK1)
Korea, Rep.	100		Suwon (KBS)
New Zealand	20	1YA	Auckland
	0.1		Lawrence (F.Pl.)
Pakistan	150		Quetta 1
Philippines	1	DWHL	Olongapo C, Zamb.
	10	DZGC	Tayug, Pang. (inactive?)
	10	DWNW	Naga C., Cam. Sur.
	1	DXJM	Butuan C., Agusan Nte
Thailand	50		Narathiwat
Uzbekistan			Tashkent
760 kHz			
Hawaii	10	KGU	Honolulu
Philippines	5	DYCB	Cebu C. (AF 765)
Vietnam			Tanan. (Long An BS)
765 kHz			
Australia	3.5	2EC	Bega
	5	5CC	Port Lincoln
	0.1-0.5		4 trtrs
China (Taiwan)	5		Taitung (Fuhsing)
China, PR			GZ; FJ (T1); NM + 1 st
India	200		Dharwar A
Indonesia	1/0.15		Banjarmasin (RRI)
Japan	5	JOJF	Kofu (YBS)
	5	JOPF	Tokuyama (KRY)
Korea, Rep.	10	HLCQ	Taejŏn (MBC)
Korea, DPR	50		Hyesan
Malaysia	10		Kuala Trengganu
New Zealand		2XT	Taradale
Philippines	1	DXGS	Gen. Santos C., S.Cotab.
	5	DYCB	Cebu C. (r 760)
	5	DZYT	Tuguegarao, Cag.
Russia			Unk., FE
Thailand	20		Lampang
Turkmenistan	30		Chardzhou
770 kHz			
Vietnam			Quy Nhon
774 kHz			
Australia	50	3LO	Melbourne (ABC1)
	5	4TO	Townsville
China, PR	100		HB
China (Taiwan)	3.5	BEV88	Taoyuan (Hsien Sheng)
	10	BEV94	Taichung (TBC)

Fiji	20		Suva (Naulu)
India	100		Shimla
Japan	500	JOUB	Akita (NHK2)
Korea, Rep.	10	HLAJ	Cheju (MBC)
	10	HLAN	Ch'unch'ŏn (MBC)
Malaysia	10		Sungai Petani
	10		Tenom, Sabah
New Zealand	5	2XN	New Plymouth (R. Rhema)
Philippines	1	DXSO	Marawi C., Lanao del Norte
	10	DWOO	Quezon C., Manila
Sri Lanka	10		Anuradhapura
Thailand	10		Udon Thani
780 kHz			
Vietnam			Can Tho (Hau Giang B.S.)
783 kHz			
Australia	2	6VA	Albany
	2	8AL	Alice Springs (ABC/RA)
China PR	100		HEB
China (Taiwan)	10	BED92	Tainan (BCC2)
Hong Kong	20		Golden Hill (RTHK5)
India	2.5		Madras C
Japan	10	JOXR	Naha, Ok (ROK) (F. pl.)
Kazakhstan	150		Turkestan
Korea, Rep.	10	HLCV	Yŏngwol (KBS)
	5		Pusan (AFKN) (F. pl. 1197)
Malaysia	10		Sandakan, Sabah
Philippines	10	DXRA	Davao C.
	5	DYME	Masbate, Mas
	5	DYPR	Pto. Princesa, Pal
	1	DZNL	San Fernando, LU
Russia	75		Spassk Dalniy, FE
Thailand	20		Lopburi
Turkmenistan	300		Ashgabat
790 kHz			
Hawaii	5	KKON	Kealakekua
792 kHz			
Australia	10	4RN	Brisbane (ABC2)
China, PR			FJ (T1); GX + 4 sts
China (Taiwan)			Hualien (Han Sheng)
	1	BEV79	Keelung (Cheng Hua) + 3 sts (Fuhsing)
India	100		Pune A
Japan	0.1-1		8 sts (NHK1)
Korea, Rep.	50	HLSQ	Kaebong (Seoul) (SBS)
Nepal	100		Kathmandu
Pakistan	150		Muzaffarabad, Kashmir
Philippines	5	DYRR	Ormoc C.
	1	DZAP	Angeles C, Pampanga
Russia	30		Abakan, Sib.
			Aleksandrovsk-Sakhalinskiy, FE
Thailand	20		Bangkok
	20		Kalasin
800 kHz			
Laos	1		Oudomxay
801 kHz			
Australia	2	4QY	Cairns (ABC3)
	2	5RM	Berri
China, PR	50		GD + 4 sts
China (Taiwan)			Hsinhua (Chien Kuo)
	250	BEC50	Kuanyin (CBS/Kuang Hua)
Guam	10	KTWG	Agana
India	200		Jabalpur
Indonesia	10		Semarang (RRI)
	1		Meclan (RRI)
Japan	1	JOFL	(RFC)
	0.1		9 sts (Comm.)
Kazakhstan	30		Arkalyk
			Ust Kamenogorsk
Korea, DPR	500		Hwadae
Malaysia	10		Kudat, Sabah
New Zealand	1	2XL	Nelson (rel. 3XG)
Philippines	1	DWFA	Sorsogon, Sor.
	1	DXBL	Bislag, Surigao S.
	1	DXMZ	Gen. Santos C., S.Cotab.
	5	DYKA	San José, Ant.
	5	DYWC	Dumaguete, Neg. Or.
	5	DZNC	Cauayan, Isa.

Russia	1000		Chita, Sib. (FS)
Sri Lanka	20		Maho
Thailand	10		Chieng Mai
	5-1		4 sts.
810 kHz			
Australia	10	2BA	Bega (ABC3)
	2	2MY	Moruya (ABC3) (FP)
	10	6RN	Perth (ABC2)
China, PR	200		ZJ + 5 sts
China (Taiwan)	1		Kuanhsi (TBC)
	10	BEV54	Changhua (Kuo Sheng)
Fiji	10		Labasa
Hawaii	10	KXDS	Princeville (F. pl.)
India	300		Rajkot
Indonesia	10		Merauke (RRI)
Japan	50		Tokyo (FEN)
Korea, DPR	50		Kaesŏng
Korea, Rep.	20	HLCT	Taegu (MBC)
Malaysia	10		Kuantan
Nepal	10		Dandeldhura
New Zealand	20	4YA	Dunedin (Highcliff)
Papua N. Guinea	2		Rabaul
Philippines	10	DZRJ	Quezon C, MM
	1	DXMV	Cotabato C
Russia	30		Sverdlovsk, Sib.
	150		Vladivostok, FE (FS)
Thailand	7		Khon Kaen
	20		Nong Kai
	10		Trang
819 kHz			
Australia	10	2GL	Glen Innes (ABC3)
Bangladesh	100		Dhaka B (r 820)
China, PR	200		SX + 1 st
China (Taiwan)	10	BED28	Taitung (BCC1)
	10	BEP28	Kaohsiung (Chingcha)
	5	BEV34	Taipei (Cheng Sheng)
India	200		Delhi A (F. pl. 810)
Japan	5	JONK	Nagano (NHK1)
	5	JOIP	Oita (NHK1)
	1	JOAQ	Sasebo (NHK1)
Korea, Rep.	20	HLCN	Kwangju (MBC)
Korea, DPR	500		Pyongyang
Malaysia	20		Miri, Sarawak
New Zealand	10	1YZ	Rotorua (Paengaroa)
Philippines	5	DXMC	Davao C
	1	DXSC	Zamboanga C, Z Sur
	5	DYVL	Tacloban, Leyte
Sri Lanka	20		Senkadagala (Kandy)
Thailand	10		Bangkok
	10		Satoon
820 kHz			
Easter Isl.	0.25	CB82	R. Manukena
828 kHz			
Australia	10	3GI	Sale (ABC3)
	5	4SS	Nambour
	5	4GC	Charters Towers
	2	6GN	Geraldton (ABC3)
China, PR			BJ; GD; HEN (2 sts)
India	10		Silchar
Japan	300	JOBB	Osaka (NHK2)
	0.1		Miki (NHK2)
Malaysia	10		Lahad Datu, Sabah
New Zealand	2	2XS	Palmerston No.
Pakistan	100		Karachi I
Philippines	5	DWZR	Legaspi C., Albay
	1	DXCC	Cagayan de Oro Mis. Or.
	1	DZTC	Tarlac, Tar.
Russia	50		Kyzyl, Sib.
Singapore	50		Singapore
Thailand	20		Royang
	10		Khon Kaen
	10		Sukhothai
830 kHz			
Hawaii	10	KIKI	Honolulu
831 kHz			
China (Taiwan)			Unk. (Fuhsing)
837 kHz			
Australia	10	4RK	Rockhampton (ABC3)
	1	6ED	Esperance (ABC3)
	0.5	7XS	Queenstown

	0.1		Yulara (trtr 8HA) (F.PL.)
China, PR			(C1) + 1 st
China (Taiwan)	1	BEV56	Tainan (Shengli Chih Sheng)
	10		Taichung (BCC2)
India	100		Vijayawada A
Japan	10	JOQK	Niigata (NHK1)
Korea, Rep.	50	HLKY	Seoul (CBS)
New Zealand	2	1YK	Kaitaia (rel. 1 YX)
	2	1YX	Whangarei
Pakistan	10		Gilgit (Kashmir)
Philippines	5	DXRE	Gen. Santos C., S.Cotab.
	5	DYFM	Iloilo C.
	1	DZXE	Vigan, IS
Russia			Unk., FE
Thailand	50		Sakon Nakhon
846 kHz			
Australia	10	2RN	Canberra, ACT (ABC2)
	5	4CA	Cairns
	0.2	6CA	Carnarvon (ABC3)
Bangladesh	100		Rajshahi
China, PR			FJ (H2); SX + 5 sts
China (Taiwan)	250		Kuanyin (CBS/Kuang Hua)
	10	BEC38	Penghu (Han Sheng) (r 840)
			Nanfangao/Ilan? (Cheng Sheng)
			3 sts (Fuhsing)
India	200		Ahmedabad A
Japan	5	JOCP	Koriyama (NHK1)
	1		2 sts (NHK1)
	0.5-0.05		5 sts (NHK1)
Kazakhstan	50		Dzhambul
Kiribati	10	T3K1	Naranakei, Betio I. (v)
Korea, Rep.	10	HLAU	Ulsan (MBC)
	5	HLSY	Yanggu (KBS)
Malaysia	10		Jerantut
	20		Kuching, Sarawak
New Zealand	2	2ZD	Masterton
Philippines	50	DZNN	Quezon C., MM
Thailand	10		Bangkok
850 kHz			
China (PR)			GD
Hawaii	5	KHLO	Hilo
Indonesia	1	RR1	Bogor
Laos			Luang Prabang
855 kHz			
Australia	1	3CR	Melbourne
	10	4QB	Maryborough (ABC3)
	10	4QO	Eidsvold (ABC3)
China, PR			(C2) Synchro
China (Taiwan)	1	BEV24	Taipei (Min Pen)
	1	BEV72	Chia-i (Cheng Sheng)
	10	BED27	Hualien (BCC1)
Indonesia	50		Medan (RRI)
	10		Mataram (RRI)
Korea, Rep.	10	HLCX	Chŏnju (MBC)
Korea, DPR	500		Sangwon
New Zealand	2	1XH	Hamilton (rel.3KG)
Pakistan	10		Quetta II
Philippines	5	DXWG	Ilagan C. Lanao Nte.
	5	DXLA	Zamboanga c, Z sur
	10	DZGE	Naga C., Cam. Sur
Sri Lanka	20		Amparai
860 kHz			
Pakistan	10		Quetta (r)
864 kHz			
Australia	2	4GR	Toowoomba
	2	6AM	Northam
	0.5	7RPH	Hobart
China, PR			AH; ZJ (C1)
China (Taiwan)	10	BED25	Kaohsiung (BCC1)
	10	BEV52	Taichung (Chung Sheng)
Hong Kong	10		Peng Chau (Comm. R.)
India	100		Shillong
Indonesia	10		Cirebon/Kupang (RRI)
Japan	5	JOPR	Fukui (FBC)
	3	JOHE	Asahikawa (HBC)
	3	JOQF	Muroran (HBC)
	1	JOSO	Matsumoto (SBC)
	0.1	JOSM	Toyohashi (Tokai)
	1	JOXN	Nasu (CRT)
	10	JOXR	Naha, Okinawa (ROK)
Korea, DPR	250		Shinŭiju
Korea, Rep.	100	HLKR	Kangnŭng (KBS)
Malaysia	10		Penang
New Zealand	10	4ZA	Invercargill (Dacre)
Papua N. Guinea	2		Madang
Philippines	1	DWSF	San Fernando, Pampanga
	5	DYRE	Mandaue C, Cebu
	1	DZSP	San Pablo C., Laguna
Russia			Unk., Amur oblast, Sib.
Sri Lanka	20		Mannar
Thailand	10		Bangkok
	10		Tak
870 kHz			
Hawaii	50	KAIM	Honolulu (temp: 5 kW)
Vietnam	100		Ho Chi Minh C.
873 kHz			
Australia	5	2GB	Sydney
	2	6DB	Derby (ABC3)
Bangladesh	10		Chittagong
China, PR			HL; GS + 6 sts
India	100		Jalandhar A
Japan	500	JOGB	Kumamoto (NHK2)
Malaysia	20		Limbang, Sarawak
New Zealand	1	3ZE	Ashburton
Philippines	5	DXRB	Butuan C., Agusan Nte
	1	DXRT	Tawitawi, Sulu
	5	DZPA	Bangued, Abra
	5	DZRC	Legaspi C., Albay
Sri Lanka	20		Diyagama (Colombo)
Thailand	10		Phattalung
	10		Phetchabun
882 kHz			
Australia	2	3YB	Warrnambool
	5	4BH	Brisbane
	2	6PR	Perth
China, PR	100		FJ + 7 sts
China (Taiwan)	1		Hsilo (Fuhsing)
	10	BEG77	Hsinchu (BCC1)
	1		Penghu (Feng Ming)
India	50		Imphal
Japan	10	JOPK	Shizuoka (NHK1)
	1		Esashi (STV)
Kirghizia	500		Bishkek
Korea, DPR	250		Wonsan (v)
Korea, Rep.	20	HLKI	Taejŏn (KBS)
Mongolia	150		Murun, Hövsgöl
	75		Choibalsan, Ea. Aimak
New Zealand	10	1YC	Auckland
Philippines	10	DWIZ	Navotas, MM
	10	DXMS	Cotabato C., Cot. Nte
	1	DXRG	Gingoog, Misamis Or.
	10	DYJR	Calbayog, W. Samar
Sri Lanka	400		Puttalam
891 kHz			
Australia	5	4RR	Townsville (Ayr)
	50	5AN	Adelaide (ABC1)
China, PR			NX + 1 st
China (Taiwan)	10	BED24	Tainan (BCC)
	3.5	BEG78	Miaoli (BCC3/Chiao Yu)
Fiji	10		Lautoka (Drasa)
India	10		Rampur
	1		Tiruchirapalli B
Indonesia	10		Malang (RRI)
Japan	20	JOHK	Sendai (NHK1)
Korea, Rep.	250	HLKB	Pusan (KBS)
New Zealand	5	2XW	Wellington (R. Windy)
Philippines	10	DYSR	Dumaguete C., Neg. Or.
Russia	100		Pravaya Khetta, Sib.
Thailand	1000		Saraburi
895 kHz			
Vietnam			Quang Ngai (r)
900 kHz			
Australia	2	2LM	Lismore
	5	2LT	Lithgow
	2	6BY	Bridgetown
	2	7AD	Devonport

	2	8HA	Alice Springs
China, PR	10		HL + 6 sts
China (Taiwan)	100	BED96	Tanshui (CBS)
Hawaii	5	KNUI	Kahului, Maui
India	100		Cuddapah
Japan	5	JOHF	Yonago (BSS)
	5	JOZR	Kochi (RKC)
	5	JOHO	Hakodate (HBC)
Kazakhstan	500		Alma Ata (FS)
Korea, Rep.	50	HLKV	Seoul (MBC)
Macau	1		Teledif. de M.
New Zealand	10	4YC	Dunedin (Highcliff)
	10	4SX	Dunedin (Access R)
Papua N. Guinea	2		Goroka
Philippines	1	DXDN	Davao C, Dvo Nte
	10	DXIP	Davao C.
	5	DXRZ	Zamboanga C.

909 kHz

China, PR			FJ (T2);TJ + 5 sts
China (Taiwan)	10	BEG52	Kaohsiung (BCC2)
	10	BEG38	Taichung (BCC)
			Taipei + 5 sts (Fuhsing)
India	100		Gorakhpur
Japan	10	JOCB	Nagoya (NHK2)
Korea, Rep.	10	HLQY	Kumi (KBS)
Malaysia	20		Sibu, Sarawak
New Zealand			Motueka
Philippines	5	DYLA	Cebu C.
Russia			Unk, FE
Thailand	50	HSKS	Surin

918 kHz

Australia	2	2XL	Cooma (FP 5kW)
	2	6NA	Narrogin
China, PR	10		SD + 1 st
China (Taiwan)	1		Chihshang (Cheng Sheng)
India	300		Suratgarh
Japan	5	JOEF	Yamagata (YBC)
	1		Yonezawa (YBC)
	1	JOPM	Shimonoseki (KRY)
	1	JOPN	Iwakuni (KRY)
	1	JOWS	Kushiro (STV)
Korea, Rep.	50		Yŏnch'ŏn (KBS)
New Zealand	10	3YZ	Greymouth (Kumara)
			Timaru
Philippines	50	DZRP	Quezon C., MM
Russia	5		Shadrinsk/Shumikha, S.
			Unk., E. Sib.
Thailand	10		Chieng Mai
	1		Chanthaburi

920 kHz

China, PR			Fuzhou, FJ

927 kHz

Australia	5	3UZ	Melbourne
	5	4CC	Gladstone
	2	6NR	Perth
China, PR			BJ; GZ; FJ (T2) + 3 sts
China (Taiwan)	1		Sani (Fuhsing)
	1200		Matsu? (CBS)
Fiji	2.5		Sigatoka
India	100		Visakapatnam
Indonesia	25		Pekanbaru (RRI)
Japan	5	JOFG	Fukui (NHK1)
	5	JOKG	Kofu (NHK1)
	1		2 sts (NHK1)
Korea, Rep.	10	HLQA	Puyŏ (KBS)
	1	HLQD	Hongch'ŏn (KBS)
	1	HLSU	Hadong (KBS)
Korea, DPR	50		Hwangju
Malaysia	10		Tawau, Sabah
New Zealand	2	2ZA	Palmerston No.
Pakistan	100		Khairpur
Philippines	5	DXMM	Sulu
	10	DZVS	Vigan, IS
Russia			Komsomolsk, FE
Thailand	10		Bangkok
Turkmenistan	50		Chardzhev
			Kizyl-Atrek/Krasnovodsk

936 kHz

Australia	5	4BK	Brisbane (PO 1296)
	10	4PB	Brisbane
	10	7ZR	Hobart (ABC)
China, PR	200		AH
China (Taiwan)			Taipei (Han Sheng)
India	100		Tiruchirapalli A
Japan	5	JOTR	Akita (ABS)
	5	JONF	Miyazaki (MRT)
	1-0.1		4 sts (MRT)
Kazakhstan	150		Shevchenko
			Karaganda
Korea, Rep.	10	HLKD	Ch'angwon (KBS)
N. Mariana Is.	10	KSAI	Saipan
Philippines	5	DXUM	Davao C.
	1	DXIM	Cagayan de Oro C, Mis. Or.
	1	DYKW	Binalgaban, Neg. Occ.
	1	DYOG	Calbayog C, W. Samar
	1	DZXT	Tarlac, Tar.
Thailand	10		Chieng Mai
	10		N. Rajchasima
	50		N. Sawan
	10		N. Si Thammarat
	1.5		Pattani

940 kHz

Hawaii	10	KDEO	Waipahu, Oahu

945 kHz

Australia	2	3BO	Bendigo
	1		Dysart (trtr 4HI)
China, PR			(C1) synchro
India	20		Sambalpur
Japan	1	JOQP	Hikone (NHK1)
	3	JOIQ	Muroran (NHK1)
	5	JOXK	Tokushima (NHK1)
	1		Fukue (NHK1)
Kazakhstan	50		Kachiry
Korea, Rep.	10	HLQW	Poŭn (KBS)
Malaysia	100		Tronoh
	50		Johore Bharu
New Zealand	2	2ZG	Gisborne
Philippines	5	DXRO	Cotabato C., Cot Nte
	5	DYRO	Roxas C., Capiz
Solomon Is.	10	H4B	Gizo
Turkmenistan	20		Kushka

954 kHz

Australia	5	2UE	Sydney
	0.25	4VL	Cunnamulla (trtr)
	0.5		Gordonvale (trtr 4CA)
China, PR			HAN + 6 sts
China (Taiwan)	20	BED55	Taipei (BCC2)
			Wuhaoshan (Yen Sheng)
	10	BEV85	Hsinying (Chien Kuo)
			Yuanli/Tunglo (Tien Sheng)
India	100		Najibabad
Indonesia	10		Kendari (RRI)
Japan	100	JOKR	Tokyo (TBS)
Kazakhstan			Kzyl-Orda
Malaysia	20		Kuching, Sarawak (r 960)
New Zealand	2	1XW	Hamilton
Philippines	10	DZEM	Quezon City, MM
	1	DXRI	Iligan C, Misamis Or
Thailand	5		7 sts.

960 kHz

Malaysia	20		Kuching, Sarawak (AF954)

961 kHz

China (Taiwan)		BEC53	Matsu (CBS/Kuang Hua)

963 kHz

Australia	5	2RG	Griffith
	5	4WK	Warwick
	5	5SE	Southern Cross Br. c.
	2	6TZ	Bunbury (trtr)
Bangladesh	20		Sylhet
China, PR			LN + 2 sts
China (Taiwan)	10	BEV84	Chunghsing (TBC)
India	20		Jalgaon
Indonesia	10		Jembar (RRI)
Japan	5	JOTG	Aomori (NHK1)
	5	JOZK	Matsuyama (NHK1)
	1	JOSP	Saga (NHK1)
	1	JOLQ	Yonago (NHK1)

	1		Hagi (NHK1)
Korea, Rep.	10	HLKS	Cheju (KBS)
	10	HLCR	Andong (KBS)
Malaysia	10		Kuala Trengganu
New Zealand	10	3YC	Christchurch
Philippines	10	DYRC	Cebu C.
	5	DXYZ	Zamboanga C.
Russia			Blagoveschensk, Sib.
			Khabarovsk, FE
Thailand	50		Krabi
	10		N. Ratchasima
970 kHz			
Hawaii	5	KJRI	Waimea (CP)
972 kHz			
Australia	0.3		Cobar (trtr 2DU)
	5	2MW	Murwillumbah
Brunei	1		Kuala Belait
China, PR	150		HEN + 2 sts
India	100		Cuttack A
Indonesia	50		Yogyakarta (RRI)
Korea, Rep.	1500	HLCA	Tangjin (KBS)
New Zealand	5	2XG	Wellington (R. Rhema)
Philippines	5	DWTI	Lucena C, Que
	5	DXMO	Cagayan de Oro, Mis.Or.
	1	DYSM	Catarman, N. Samar
Russia	50		Orzu, Tad. (FS)
	50		Yuzhno-Sakhalinsk, FE
Sri Lanka	20		Amparai
Thailand	50		Phetchaboon
975 kHz			
Vietnam			Rach Gia (Kien Giang B.S.)
981 kHz			
Australia	5	2NM	Muswellbrook
	2	3HA	Hamilton
	2	6KG	Kalgoorlie
China, PR			(C1) synchro
China (Taiwan)	1	BEV68	Kaohsiung (Feng Ming)
	50		Hsinfeng (Air Force)
India	100		Raipur
Kazakhstan	30		Aktyubinsk
New Zealand	2	1YE	Kaikohe
	1	4XX	Ranfurly
Philippines	10	DXOW	Davao C.
	5	DXDR	Dipolog, Zamb. Nte
	20	DYBQ	Iloilo C.
	5	DZRD	Dagupan C., Pang.
Thailand	50	HSKZ	Yala
	10		Bangkok
	10		Nakhon Phanom
	5	HSKH	Mae Hong Son
990 kHz			
Australia	0.25	3RN	Albury/Wodonga (ABC)
	5	4RO	Rockhampton
		6RPH	
	0.5	8GO	Gove (ABC/R. Austr)
China, PR			SH; YN + 2 sts
China (Taiwan)	1	BEG22	Taipei (Yu Shih)
	10	BEP36	Hualien (Chingcha)
	1	BEP38	Ilan (Chingcha)
	10	BEV58	Taichung (Cheng Sheng)
Fiji	5		Lautoka (Vomo)
Hawaii	5	KHVH	Honolulu
India	300		Jammu A (Kashmir)
Japan	10	JORK	Kochi (NHK1)
Korea, Rep.	10	HLAP	Masan (MBC)
Mongolia	150		Altai
	500		Ulaanbaatar
Philippines	15	DWRT	Makati, MM
Singapore	20		Singapore
Thailand	10		N. Ratchasima
999 kHz			
Australia	2	2NB	Broken Hill (ABC3)
	5	2ST	Nowra
Bangladesh	10		Thakurgaon
China, PR.			XJ + 3 sts
India	10		Coimbatore
	1		Almora
Indonesia	300		Jakarta (RRI)
Japan	1	JODG	Hamamatsu (NHK1)
	1		Nakamura (NHK1)
	1	JOTQ	Hachinohe (NHK1)
	0.1		4 sts (NHK1)
Kazakhstan	30		Akmola
Korea, Rep.	10	HLCL	Kwangju (CBS)
Korea, DPR	250		Hamhŭng
Philippines	1	DXHP	Bislig, Surigao S.
	5	DYSS	Cebu C.
	1	DZEQ	Baguio C., Benguet
Russia			Birobidzhan, FE
1000 kHz			
Laos			Houa Phan
1008 kHz			
Australia	0.3	2XX	Canberra, ACT
	5	4TAB	Brisbane (Ipswich)
	5	7EX	Launceston
China, PR			YN (C1) + 2 sts
China (Taiwan)	600	BED62	Luchiang (CBS)
	10	BED88	Taitung (BCC2)
	1	BEV60	Kaohsiung (Cheng Sheng)
	1	BEV92	Taipei (Tiennan)
India	100		Calcutta B
Indonesia	10		Madiun (RRI)
Japan	50	JONR	Osaka (ABC)
Kazakhstan			Semipalatinsk
Korea, Rep.	50	HLCS	Sokch'o (KBS)
New Zealand	10	1ZD	Tauranga (Paengaroa)
Pakistan	120		Hyderabad I
Philippines	25/10	DXXX	Zamboanga C.
Russia	50		Unk., Kamchatka, FE
			Khabarovsk, FE
Thailand	10		Bangkok
1010 kHz			
Vietnam	100		Hanoi (FS)
1017 kHz			
Australia	5	2KY	Sydney
	0.1	6WH	Wyndham (ABC3)
China, PR			(C1); JL (FS) + 1 st
China (Taiwan)	10	BED53	Hsinchu (BCC2)
	0.1	BED71	Kuanshan (BCC1)
India	10		Delhi D
	1		Mysore
Japan	50	JOLB	Fukuoka (NHK2)
Korea, Rep.	10	HLAW	Andong (MBC)
Malaysia	100		Segamat
New Zealand			Christchurch (BBCWS)
Philippines	10	DWDW	Dagupan C., Pang.
	10	DWLC	Lucena C., Que.
	5	DXGE	Davao C.
	5	DXSN	Surigao, S. Nte
	5	DYRP	Iloilo C. (v)
Tonga	10	A3Z	Nuku'alofa
1026 kHz			
Australia	5	4MK	Mackay
	2	6NW	Port Hedland
China, PR			GZ;BJ + 3 sts
China (Taiwan)			Yuanli (Tien Sheng)
	1	BEV51	Sanchung (Chung Hua)
India	1		Allahabad
Japan	0.1		24 sts (NHK1)
Korea, Rep.	1	HLKW	Kŏch'ang (KBS)
	1	HLCG	Hwach'ŏn (KBS)
New Zealand	2	1ZK	Kaitaia (rel. 1ZN)
	2	1ZN	Whanggarei
Philippines	10	DZAM	Quezon C, MM
	1	DXMI	Iligan C., Lanao Nte
Russia	200		Novosibirsk, Sib.
Sri Lanka	10		Galle
Thailand	50		Phitsanulok
	10		Yala
1030 kHz			
Laos	10		Vientiane (FS)
1035 kHz			
China, PR			(C1)
China (Taiwan)	10	BED26	Chia-i (BCC2)
India	10		Guwahati B
Indonesia	0.2		Palangkaraya (RRI)
	1		Tanjungkarang (RRI)
Japan	1	JOJD	Tsuruoka (NHK2)
	1	JOIC	Toyama (NHK2)
	1	JOHD	Takamatsu (NHK2)

	0.1		4 sts (NHK2)
Korea, Rep.	10	HLCP	P'ohang (KBS)
Malaysia	10		Penang
New Zealand	20	2ZB	Wellington
Pakistan	120		Multan
Philippines	2.5	DXCP	Gen. Santos C., S. Cotabato
	5	DZWX	Baguio C., Benguet
Solomon Is.	10	H4B1	Honiara
Thailand	20		Bangkok
1040 kHz			
Hawaii	5	KLHT	Honolulu
1044 kHz			
Australia	1	2UH	Muswellbrook (ABC3)
	0.5	4WP	Weipa (ABC3)
	2	5CS	Port Pirie
	1	6BR	Bridgetown (ABC3)
China, PR	300		JS (FS) +2 sts
China (Taiwan)	5		Kaohsiung (Cheng Kang)
		BEV78	Hualien (Yen Sheng)
Hongkong	10		Peng Chau (Metro Broadc.)
India	100		Bombay A
Indonesia	10		Biak (RRI)
Korea, Rep.	10	HLCD	Chech'ŏn (CKBS)
	10	HLCI	Samch'ŏk (CKBS)
	1		Ch'unch'ŏn (AFKN)
Malaysia	20		Sri Aman, Sarawak
New Zealand	10	4ZB	Dunedin (Highcliff)
Philippines	1	DXML	Digos, Dvo. Sur
	2.5	DXLL	Zamboanga C.
	1	DYMS	Catbalogan C., W. Samar
	5	DZDR	Naga C., Cam. Sur
Thailand	10		Khon Kaen
1050 kHz			
Vietnam			Hanoi
1053 kHz			
Australia	5	2CA	Canberra, ACT
	0.5	4EB	Brisbane (West End)
Bangladesh	10		Rangpur
China P.R.	100		YN + 2 sts
China (Taiwan)	100	BEC49	Kuanyin (CBS/Kuang Hua)
India	10		Leh (Kashmir)
Indonesia	10		Jayapura (RRI)
	1		Surakarta (RRI)
Japan	50	JOAR	Nagoya (CBC)
Korea, DPR	1500		Haeju (Kuguk-e Sori)
Malaysia	20		Mersing (r 1055)
N. Mariana Is.	1	KCNM	Navy Hill, Saipan
New Zealand	2	2ZP	New Plymouth
Philippines	5	DXCO	Cayagan de Oro, Mis. Or.
	1	DXKD	Dipolog, Zamb. Nte
	5	DYSA	Iloilo C.
	5	DZEL	Lucena, Que. (v)
	5	DZVR	Dagupan C, Pang.
Thailand	10		N. Si Thammarat
	10		Nakhon Sawan
	10		Lampang
1060 kHz			
Hawaii	1	KAHU	Hilo (CP)
1062 kHz			
Australia	2	4TI	Thursday Isl (ABC3)
	10	4--	Charleville (FP)
	2	5AD	Adelaide (PO 1323)
China, PR	150		GD + 1 sts
China (Taiwan)	10	BED85	Taipei (BCC3/Chiao Yu)
	1	BEV82	Ilan (Cheng Sheng) (r. 1064)
	5	BEV74	Pingtung (Min Li)
India	10		Pasighat
Japan	0.1		25 sts (Comm.)
Korea, Rep.	50	HLKQ	Ch'ŏngju (KBS)
Malaysia	20		Sibu, Sarawak
Philippines	5	DXKI	Marbel, S. Cotabato
	10	DZEC	Quezon C., MM
Thailand	20		Bangkok
Uzbekistan			Andizhan
			Karshi
1068 kHz			
Vietnam			unk. (r)
1071 kHz			
Australia	5	3CV	Maryborough
	2	4SB	Kingaroy
	2	6WB	Katanning
China, PR			TJ + 6 sts
China (Taiwan)	10	BED57	Taipei (BCC)
	5	BED8	Lishan (BCC1)
	1	BEV96	Tainan (Den Sheng)
			Yuli (Fuhsing)
India	1000		Rajkot
Japan	20	JOFK	Hiroshima (NHK1)
Kazakhstan	100		Ust-Kamenogorsk
Philippines	5	DXKT	Davao C.
	5	DYXT	Tagbilaran C., Bohol
	5	DZAL	Iriga C., Cam Sur
Russia			Unk., Ea. Sib.
1080 kHz			
Australia	2	2MO	Gunnedah
	0.2	4MI	Mt. Isa (ABC3)
	2	6GL	Perth (FP 5kW)
	5	7HT	Hobart
Bangladesh	10		Rajshahi
China, PR	200		YN (FS) + 6 sts
Hawaii	5	KWAI	Honolulu
Indonesia	0.5/10		Singaraja (RRI)
Japan	0.1-1		9 sts (NHK2)
Korea, DPR	1500		Haeju
Korea, Rep.	10	HLAT	Yŏsu (MBC)
Malaysia	10		Sandakan, Sabah
New Zealand	10	1ZB	Auckland
Pakistan	50		Lahore II
Philippines	5	DWAM	Batangas C
	5	DWIN	Dagupan C., Pang.
	1	DWRL	Albay
	1	DXKS	Surigao, S. Nte
	1	DYSJ	Kalibo, Ak
Russia	500		Angarsk. S. (FS)
Thailand	10		N. Sawan
	10		Yala
Turkmenistan	50		Krasnovodsk
	5		Gushgy
1089 kHz			
Australia	5	2GZ	Orange
	5	3WM	Horsham
China, PR			LN; FJ (T2) + 2 sts
China (Taiwan)	1	BED56	Jiyuetan (BCC)
	5	BEH5	Taipei + 14 sts (Fuhsing)
	0.5	BEC31	Yunlin (Han Sheng)
	10	BEG28	Kaohsiung (City Council)
Fiji	1		Suva/mobile
India	20		Udipi
	1		Jammu B (Kashmir)
Japan	10	JOHB	Sendai (NHK2)
Korea, Rep.	10	HLCH	Ch'ungju (KBS)
Malaysia	20		Grik
Philippines	5	DXCM	Cotabato C., Cot. Nte
	1	DYHR	Calbayog C., W. Samar
Thailand	10		Udon Thani
1090 kHz			
Wallis & Futuna	1		Alu, Futuna (RFO)
1098 kHz			
Australia	0.2	2RN	Goulburn (ABC)
	2	4LG	Longreach
	2	6MD	Merredin
	5	7LA	Launceston
China, PR			GD; XJ 2 sts (C2)
Fiji	1		Suva (mob./spec.prgr.)
Indonesia	10		Jambi/Sumenep (RRI)
Japan	1	JOEL	Tsuruoka (YBC)
	1		Sakata (YBC) (FP)
	5	JOGF	Oita (OBS)
	1	JOMF	Sasebo (NBC)
	5	JOSR	Nagano (SBC)
	1	JOSW	Iida (SBC)
	5	JOWO	Koriyama (RFC)
	0.1		3 sts (Comm.)
Kazakhstan			Alma Ata/Shingoza
			Ekardara/Dzhetysay
			Karagayly/ Leninskaya
Korea, Rep.	20	HLCJ	Chinju (KBS)
Marshall Is.	5	WSZO	Majuro
New Zealand	10	3ZB	Christchurch

Pakistan	10		Hyderabad II
Philippines	10	DWAD	Mandaluyong, MM
	5	DXCM	Cotabato C. (1089?)
	5	DXCL	Cagayan de Oro, Mis. Or.
Russia	150		Tilichiki, FE
Thailand	10		Songkhla
	10		Tak
1100 kHz			
China, PR			SX; YN
China (Taiwan)	100	BED97	Tanshui (CBS)
1107 kHz			
Australia	5	2UW	Sydney
China P.R.			XJ + 6 sts
China (Taiwan)	1	BEC47	Kinmen (Kuang Kua)
India	10		Gulbarga
Indonesia	1/10		Yogyakarta (RRI)
Japan	5	JOMR	Kanazawa (MRO)
	1	JOMO	Nanao (MRO)
Japan	1		Akuma (MBC)
	20	JOCF	Kagoshima (MBC)
	1		Okuchi (MBC)
Korea, Rep.	10	HLAV	Po'hang (MBC)
Papua N.G.	5		Kavieng (Proj.)
Philippines	10	DXMF	Davao C.
	5	DYRI	Iloilo C.
Thailand	20		Chon Buri
1110 kHz			
China, PR			FJ (nf 1107)
Hawaii	5	KAOI	Kihei (Pukalani), Maui
1116 kHz			
Australia	5	3AK	Melbourne
	5	4BC	Mandurah
China P.R.			HL (C2/FS); FJ (T1) + 2 sts
China (Taiwan)	1.5	BEC22	Taipei (Han Sheng)
	5	BEC30	Ilan (Han Sheng)
	5	BEP26	Hsinchu (Chingcha)
	5	BEP25	Kaohsiung (Chingcha)
	3.5	BED72	Yuli (BCC1)
India	200		Srinagar A (Kashmir)
Indonesia	10		Biak (RRI)
Japan	5	JOAF	Matsuyama (RNB)
	5	JODR	Niigata (BSN)
	0.1		3 sts (Comm.)
New Zealand	2	2YX	Nelson
Philippines	5	DXAS	Zamboanga C.
	1	DYTR	Tagbilaran C, Bohol
	5	DZLB	Los Banos, Laguna
Russia			3 st's, FE
			Chita, Sib.
Thailand	20		Samut Sakhon
1125 kHz			
Australia	0.5	1PPP	Canberra
	5	5MU	Murray Bridge
China, PR			HEB 3 sts
China (Taiwan)	5	BEV36	Yünlin (Cheng Sheng)
			Taitung (Chingcha)
India	10		Udaipur
Indonesia	1		Palangkaraya
Japan	10	JOAD	Naha, Okinawa (NHK2)
	1-0.1		7 sts (NHK2)
Philippines	5	DWAS	Daraga, Albay
	5	DXAS	Zamboanga C.
	10	DZWN	Dagupan C., Pang.
Russia			Unk., C. Asia
Thailand	50	HSKI	Chanthaburi
Vanuatu	10	YJB2	Port Vila (Malapoa)
1130 kHz			
Hawaii	10	KLEI	Kailua (inact.)
1134 kHz			
Australia	2	2AD	Armidale (FP 5kW)
	5	3CS	Colac
China, PR			5 sts
China (Taiwan)	5	BEG26	Taipei (Taipei Bc. Stn.)
India	1000		Calcutta (Magra)
Indonesia	25/1		Banjarmasin (RRI)
Japan	100	JOQR	Tokyo (Bunka Hoso)
Korea, Rep.	500		Hwasŏng (KBS)
New Zealand	2	4YQ	Queenstown
Philippines	1	DYRM	Dumaguete, Neg. Or.
	10	DWDD	Camp Aguinaldo, MM
	1	DXMV	Valencia, Buk.
Thailand	10		Kanchanaburi
	10		Lampang
	10		Pang-Nga
1140 kHz			
Malaysia		10	Kuantan (AF 1143)
1143 kHz			
Australia	2	2HD	Newcastle
	5	4HI	Emerald
China, PR			QH ZJ (V) + 3 sts
China (Taiwan)	0.1		2 sts (BCC)
	100	BEL3	Penghu (Fishery BS)
Fiji	2.5		Labasa ? (r)
India	20		Ratnagiri/Rohtak
Japan	20	JOBR	Kyoto (KBS)
Malaysia	10		Kuantan (r 1140)
Nepal	10		Dhalkebar
New Zealand	2	1YW	Hamilton
Philippines	1000	DWVA	Poro Pt, LU (VOA)
Russia			Unk., C. Sib.
Tadzhikistan	150		Yangi-Yul
Thailand	25		Bangkok
1147 kHz			
China, PR			HEB (v)
1152 kHz			
Australia	2	2WG	Wagga Wagga
China, PR			HN + 3 sts
China (Taiwan)	1	BED68	Puli (BCC2)
	5		Taipei 1 (Hua Sheng)
India	1		Ranchi B
Japan	10	JOPC	Kushiro (NHK2)
	10	JORB	Kochi (NHK2)
	0.5	JORO	Miyako, Okinawa (RBC)
	0.1		Tsuyama (NHK2)
Korea, Rep.	10	HLCW	Wonju (KBS)
New Zealand	2	3ZC	Timaru
Pakistan	10		Rawalpindi
Russia	25		Khanty-Mansiysk, Sib.
	30		Komsomolsk, FE
			Unk., C. Asia
Thailand	20		Chieng Mai
	20		Khon Kaen
	1		Songkhla
1161 kHz			
Australia	2	4MB	Maryborough
	10	5PA	Naracoort (ABC3)
	1	7FG	Fingal (ABC3)
China, PR			(C1) + 3 sts
China (Taiwan)	10	BED86	Ilan (BCC2)
	10	BED89	Miaoli (BCC2)
	1	BEV67	Kaohsiung (Feng Ming)
India	10		Thiruananthapuram
Japan	0.1		24 sts (NHK1)
Korea, Rep.	20	HLKU	Pusan (MBC)
	0.25		Uijŏngbu (AFKN)
Malaysia	20		Sri Aman, Sarawak
	10		Melaka
New Zealand	5	2XM	Wellington
Philippines	1	DYRD	Tagbilaran C, Bohol
	1	DXDS	Digos, Dvo. Nte
	2.5	DZMD	Daet, Cam Nte
Thailand	20		Nakhon Phanom
	10		Rayong
Tadzhikistan			Orzu
1170 kHz			
Australia	5	2CH	Sydney
China, PR			5 sts
China (Taiwan)	5	BEV35	Taipei (Cheng Sheng)
Hawaii	5	KOHO	Honolulu
India	1		Hyderabad C
Indonesia	50		Semarang (RRI)
Korea, Rep.	500	HLSR	Kimje (KBS)
New Zealand	0.4	1ZW	Te Kuiti
Philippines	10	DXMR	Zamboanga C.
	10	DZCA	Quezon C, MM
Russia	250		Angarsk, S.
Thailand	20		Bangkok
1179 kHz			
Australia	5	3RPH	Melbourne
China, PR			HB + 3 sts

China (Taiwan)	1	BEV54	Erhlin (Kuo Sheng)
India	20		Rewa
Indonesia			Padang (RRI)
Japan	50	JOOR	Osaka (MBS)
New Zealand	0.1	4XE	Wanaka (rel. 4XC)
Philippines	10	DXMT	Marawi C., Lanao Sur
Thailand	20		Chanthaburi
	10		Chiangrai
1180 kHz			
Indonesia	1/10		Tanjung Pinang (RRI)
1188 kHz			
Australia	2	2NZ	Inverell
	2	6XM	Exmouth (ABC3)
China, PR	300		YN (FS) + 4 sts
China (Taiwan)	10	BED32	Hualien (BCC2)
	1	BED46	Jiyuetan (BCC2)
	1	BEV46	Taipei (TBC2)
		BEV57	Tainan (Shengli Chih Sheng)
India	50		Bombay C
Japan	10	JOKP	Kitami (NHK1)
Kazakhstan			Abay/Ayaguz/Yuzhnyy
Korea, Rep.	100	HLKX	Inchŏn (FEBC)
New Zealand			Rotorua
Philippines	1	DYRV	Catbalogan, W. Samar
	5	DZLT	Lucena C, Que.
	5	DZXO	Cabanatuan, Nva. Ecija
Thailand	10		Sakon Nakhon
	10		Phitsanulok
	10		Prachin Buri
Wallis & Futuna	2		Mata Utu, Wallis (RFO)
1190 kHz			
Laos			Vientiane
1194 kHz			
Indonesia	0.5		Mataram (RRI)
1195 kHz			
Indonesia	5		Bogor (RRI)
1197 kHz			
Australia	2	5RPH	Adelaide
China, PR			HL; SH
India	10		Tirunelveli
	1		Jodhpur B
	1		Shillong A
Japan	5	JOBF	Kumamoto (RKK)
	1-0.1		3 sts. (RKK)
	5	JOYF	Mito (YBS)
	1	JOFO	Kitakyushu (RKB)
	1	JOWL	Asahikawa (STV)
	1	JOWN	Hakodate (STV) + 2 st's
	1		Nayori (STV)
	1		Nakamura (RKC)
	1-0.1		4 sts (Comm.)
Kazakhstan			Akmola
Korea, Rep.	1		Tongduch'on (AFKN)
Malaysia	10		Kudat, Sabah
New Zealand	2	2ZW	Wanganui
Philippines	5	DXFE	Davao C.
	5	DYRH	Bacolod, Neg. Occ.
Thailand	20		Bangkok
1200 kHz			
China, PR			HL
Laos	1		Xieng Khouang
1206 kHz			
Australia	5	2CC	Canberra, ACT
	5	2GF	Grafton
	2	6KY	Perth
China, PR			YN (FS); JL; SD (V) (nf 1170)
China (Taiwan)	100	BED50	Minhsiung (CBS)
	1	BEV62	Hsinchu (TBC)
Fiji	2.5		Sigatoka
Indonesia	10/0.5		Denpasar (RRI)
	1		Surabaya (RRI) (r)
	1		Manado (RRI)
Korea, Rep.	1	HLSW	Chŏngsŏn (KBS)
	1	HLQR	Ch'ŏngsong (KBS)
Malaysia	20		Miri, Sarawak
New Zealand	2	4XO	Dunedin
	0.5		Hamilton (Access CR.)
Philippines	10	DWAN	Quezon C, MM
	1	DXRS	Surigao C, S. Nte
1210 kHz			
Hawaii	1	KZOO	Honolulu
1215 kHz			
Australia	0.35		Bowral (trtr 2ST)
	0.1		Moranbah (trtr 4HI)
China, PR			(C2) synchro
India	20		Delhi (Kingsway)
	1		Pudducheri
Indonesia	10/0.5		Samarinda (RRI)
			Jakarta (RRI)
Japan	1	JOBO	Maizuri (KBS)
	1	JOBW	Hikone (KBS)
Korea, Rep.	10	HLCD	Chech'ŏn (KBS)
	10	HLAK	Chinju (MBC)
New Zealand	2	1ZE	Kaikohe (rel. 1ZN)
Philippines	10	DYRF	Cebu C.
Thailand	50	HSKV	Surat Thani
	10		Phrae
	10		U. Ratchathani
1224 kHz			
Australia	5	2WS	Sydney
	5	3EA	Melbourne (SBS)
China, PR			FJ (T2) + 3 sts
China (Taiwan)	10	BED79	Kaohsiung (BCC)
	0.1	BED98	Fengping (BCC1)
	5	BEV71	Taipei (Hua Sheng)
			Taitung/Matan (Fuhsing)
India	1		Srinagar D (Kashmir)
	10		Calcutta D
Japan	10	JOJK	Kanazawa (NHK1)
Kazakhstan			Dzhambul
Korea, Rep.	20		Kwangju (KBS)
Malaysia	50		Johore Bharu
	100		Tronoh
Marshall Is.	1		Kwajalein (AFRS)
New Zealand	2	4XF	Invercargill (Foveaux R.)
Philippines	10	DXED	Davao C.
	1	DXSM	Jolo, Sulu
	10	DZMQ	Dagupan C, Pang.
Russia			Unk, C. Sib.
Thailand	10		Bangkok
	10		Chieng Rai
1233 kHz			
Australia	10	2NC	Newcastle (ABC[illegible])
China, PR	100		HN
China (Taiwan)			Juifang (Chung Hua)
			2 sts (Fuhsing)
India	1		Bhopal B
	1		Tura
	200		Tuticorin
Indonesia	1		Pontianak (RRI)
Japan	5	JOGR	Aomori (RAB)
	5	JOUR	Nagasaki (NBC)
	0.1		2 st's (RNC)
Korea, Rep.	1	HLCC	P'yŏngch'ang (KBS)
	1	HLQG	Yŏngyang (KBS)
Mongolia	75		Sainshand
New Zealand			Wellington (rel. B3CWS)
Philippines	5	DYVS	Bacolod, Neg. Occ.
Thailand	10		Udon Thani
	10		Prachuap Khiri Khan
	1		Uttaradit
Turkmenistan	40		Syrt-Takhta
1235 kHz			
Indonesia	0.5	RRI	Bogor
1242 kHz			
Australia	5	3TR	Sale
	2	4AK	Toowoomba (FP: 5kW)
	2	5AU	Port Augusta
	2/1	SMU	Murray Bridge
China, PR			LN; HB
China (Taiwan)	10	BED 77	Taichung + 7 sts (BCC)
India	100		Varanasi
Indonesia	1		Bogor (RRI)
	1		Bengkulu (RRI)
Japan	100	JOLF	Tokyo (Nippon Hoso)
Korea, Rep.	10	HLSB	Wŏnju (MBC)
New Zealand	2	1XX	Whakatane
	0.1	1XX	Galatea (rptr)

Philippines	10	DWBL	Pasig, MM
	5	DXZB	Zamboanga C
	5	DXSY	Ozamis C, Mis. Occ.
Thailand	10		Roi Et
	10		Lampang
	10		Phetchabun
	1		Phayao
	50		Surat Thani
1250 kHz			
China, PR		ZJ	
Indonesia	1	RRI	Bogor
Philippines	1	DYRG	Kalibo, Ak
1251 kHz			
Australia	2	2DU	Dubbo
China, PR			BJ; QH; SH + 1 sts
China (Taiwan)	100	BEC22	Kuanyin (CBS/Kuang Hua)
	5	BEC29	Kaohsiung (Han Sheng)
	1	BED61	Lishan (BCC2)
Hong Kong	10		Tam Mei (BFBS)
India	20		Sangli
Indonesia	10		Banda Aceh (RRI)
			Mataram (RRI)
	1		Singaraja (RRI)
Korea, Rep.	10	HLKT	Taegu (CBS)
New Zealand	5	1XG	Auckland (R. Rhema)
Philippines	5	DWSI	Santiago, Isa.
	2.5	DZMS	Sorsogon, Sor
	1		Subic Bay (Olongapo C.) (FENP)
Russia	1000		Vladivostok, FE (FS)
Thailand	20		Bangkok
1260 kHz			
Australia	2	3SR	Shepparton
	0.5		Biloela (trtr 4CD) (PO 927)
	1	6KA	Karratha
China, PR			LN; XZ + 3 sts
China (Taiwan)	10	BEP22	Taipei (Chingcha)
	1		Putzu (Cheng Sheng)
India	20		Ambikapur
Japan	20	JOIR	Sendai (TBC)
Korea, Rep.	10	HLKL	Namwon (KBS)
Malaysia	5		Dungun
New Zealand	2	3XA	Christchurch
Philippines	5	DWMC	Rosales, Pang.
	10	DYMF	Cebu C.
Thailand	50		Chieng Rai
Uzbekistan	100		Nukus
1269 kHz			
Australia	5	2SM	Sydney
China, PR			FJ (H1); YN (FS) + 2 sts
China (Taiwan)	1	BEV37	Taitung (Cheng Sheng)
	1		Tungshih (Cheng Sheng)
			6 sts (Fuhsing)
			Penghu (Han Sheng)
India	20		Agartala
	1		Jaipur B
	10		Madurai
Japan	5	JOHW	Obihiro (HBC)
	5	JOJR	Tokushima (JRT)
Korea, Rep.	1	HLSI	Kurye (KBS)
	10		Yangju (KBS)
New Zealand	0.4	2ZT	Takata (rptr 2ZN)
Philippines	5	DZVX	Daet, Cam. Nte
	5	DYWB	Bacolod, Neg. Occ
Thailand	20		Songkhla
1270 kHz			
Hawaii	5	KNDI	Honolulu
1278 kHz			
Australia	5	3AW	Melbourne
China, PR			HEB + 4 sts
China (Taiwan)	1	BEC40	Taipei (Fu Hsing Kang) :(r1280)
India	1		Lucknow C
Japan	50	JOFR	Fukuoka (RKB)
Korea, Rep.	1	HLQV	Hapch'ŏn (KBS)
New Zealand	2	2ZC	Napier
Philippines	10	DZRM	Quezon C, MM
Russia			Unk, C. Sib.
Thailand	25		Bangkok

1287 kHz			
Australia	2	2TM	Tamworth
China, PR			(C1) + 3 sts
China (Taiwan)	5	BEC27	Taichung (Han Sheng)
			Fangliao (Min Li)
India	100		Panaji A
Indonesia	50/0.4		Palembang (RRI)
Japan	50	JOHR	Sapporo (HBC)
Kazakhstan			Unk.
Korea, Rep.	10	HLAF	Kangnŭng (MBC)
	10	HLAX	Ch'ŏngju (MBC)
New Zealand	2	3ZW	Westport (rel. 3ZA)
Philippines	1	DXRS	Sorsogon
Russia			Novosibirsk oblast, Sib.
			Unk., C. Sib.
			Unk., FE
Thailand	10		Chieng Rai
	10		U. Ratchathani
	10		Uttradit
	10		N. Si Thammarat
1296 kHz			
Australia	5	4RPH	Brisbane
China, PR	300		YN (FS) + 4 sts
China (Taiwan)	10	BED47	Tainan (BCC1)
	1	BEV23	Taipei (Min Pen)
India	10		Darbhanga
Japan	10	JOTK	Matsue (NHK1)
New Zealand		1ZH	Hamilton
Philippines	5	DWPR	Dagupan C, Pang.
	1	DYJJ	Roxas C.
	5	DWLQ	Lucena C., Que. (v;r 1293)
Thailand	10		Pattani
1305 kHz			
Australia	2	5MW	Renmark
China, PR			(C2)
India	10		Parbhani
Korea, Rep.	10	HLSV	Uljin (KBS)
New Zealand	2	4XD	Dunedin
Philippines	10	DYFX	Cebu C. (v)
Philippines	10	DYFX	Cebu C. (v)
Russia			Unk, W. Sib.
Sri Lanka	1		Welikada
Thailand	10		Bangkok
1310 kHz			
Hawaii	0.5	KKAI	Honolulu (CP)
Malaysia	20		Mersing (AF 1314)
1314 kHz			
Australia	5	3BA	Ballarat
China, PR	100		NM (FS) + 4 sts
China (Taiwan)	1	BEV76	Chunan (Tien Sheng)
	5	BEP32	Hsinging (Chingcha)
India	10		Bhuj
	1		Cuttack B
Japan	50	JOUF	Osaka (OBC)
Korea, Rep.	10	HLCM	Iri (CBS)
Malaysia	20		Mersing (r1310)
New Zealand	2	2YW	Gisborne
Philippines	10	DXRH	Davao C.
	10	DWXI	Paranaque, MM
	0.25		Baguio (FENP)
	0.25		Wallance (S. Fernando) (FENP)
Thailand	20		Khon Kaen
1320 kHz			
Galapagos Is.	5	HCVG8	Puerto Baqueriz
1323 kHz			
Australia	0.1		Hughenden (trtr 4GC)
	5	5AD	Adelaide
China, PR	100		XZ (FS); SN + 4 sts
China (Taiwan)		BEV45	Taipei/Puli (TBC)
Fiji	2.5		Rakiraki
India	20		Calcutta C
Japan	1	JOFP	Fukushima (NHK1)
	1		Yamada (NHK1)
	0.1		4 sts (NHK1)
Kazakhstan	20		Atyran + unk
Kirghizia			Bishkek
Korea, Rep.	1	HLQJ	Yŏng-gwang (KBS)
	1	HLCU	Ullŭng (KBS)
Nauru I.	0.2	C2AM	Nauru

New Zealand	5	2XQ	Wellington
	5	3XQ	Christchurch
Philippines	1	DYXX	Iloilo C.
	1	DZRK	Kalinga Apayao, Beng.
Tadzhikistan			Yangi-Yul
Thailand	10		Chieng Mai
	10		Songkhla
	1		N. Pathom
	10		N. Surat Thani
1332 kHz			
Australia	2	3SH	Swan Hill
	5	4BU	Bundaberg
China, PR			4 sts
China (Taiwan)	1	BEC36	Tsaoying (Han Sheng)
India	10		Tezu
	1		Mangalore
Indonesia	10		Jakarta (RRI)
Japan	50	JOSF	Nagoya (Tokai R.)
Kazakhstan	30		Alma Ata
Korea, Rep.	10	HLAO	Ch'ungju (MBC)
New Zealand	5	1XI	Auckland (R.I)
Philippines	5	DWAY	Cabanatuan, Nva. Ecija
	5	DYBB	Baybay, Leyk
	1	DZKI	Iriga C., Cam. Sur
Thailand	10		Bangkok
1335 kHz			
Indonesia			Pekanbaru (RRI) (r)
1341 kHz			
China, PR	100		GD (FS); HL + 3 sts
China (Taiwan)			Toucheng (Cheng Sheng)
			Ilan (Cheng Sheng)
China (Taiwan)		BEC48	Matsu (Kuang Hua)
India	1		Kohima
Indonesia	5		Tanjung Pinang
Japan	1	JOHQ	Iwaki (NHK1)
	1		Minamata (NHK1)
	0.1		20 sts (NHK1)
Kazakhstan	25		Aktan
			Alma Ata
Korea, Rep.	10		Kimp'o (KBS)
New Zealand	2	2ZN	Nelson
Pakistan	10		Bahawalpur
Philippines	5	DYRL	Bacolod, Neg. Occ.
	5	DZYS	Santiago, Isa.
Thailand	50		Ubon Ratchathani
	10		Loei
	10		Pang Nga
1350 kHz			
Australia	5	2LF	Young
China, P.R.	50		YN; + 2 sts
China (Taiwan)	10	BEP63	Chia-i (BCC1)
	1	BEV50	Sanchung 1 (Chung Hua)
India	1		Dharwar B/Jalandhar C
Japan	20	JOER	Hiroshima (RCC)
Korea, Rep.	10	HLAQ	Samch'ŏk (MBC)
Laos	10		Pakse
Mongolia	150		Dalanzadgad
Philippines	10	DZXQ	Makati, MM
	5	DZER	Boac, Marinduque
Russia			Onguday/Ust-Kan/ Ust-Alagan, Sib.
Thailand	10		Lampang
	10		Trang
1359 kHz			
Australia	0.25	4WK	Toowoomba (trtr)
China, PR			(C1) Synchro
China (Taiwan)	10	BED23	Taichung (BCC3/Chiao Yu)
	1	BEC33	Hualien (Han Sheng)
India	20		Bhadravathi
Japan	0.1		26 sts (NHK2)
Korea, Rep.	1		Osan (AFKN)
New Zealand	1	4XC	Queenstown (Resort R.)
Philippines	1	DYSL	Sogod, So. Leyte
	5	DYMR	Mandaue C, Cebu
	5	DZYR	S. Fernando, LU
Russia			Unk, C. Sib.
Thailand	10		Bangkok
1360 kHz			
Cambodia	20		Phnom Penh (FS)
China, PR			HEN
1368 kHz			
Australia	2	2GN	Goulburn
China, PR			FJ (V) + 3 sts
India	10		Delhi C
Japan	5	JOHP	Takamatsu (NHK1)
	1	JOJP	Tsuruoka (NHK1)
	1	JOLG	Tottori (NHK1)
	1	JOJS	Wakkanai (HBC)
	0.1		5 sts (NHK1)
Korea, Rep.	1	HLKO	Muju (KBS)
Korea, PDR			Pyongyang
New Zealand	1	1XT	Tauranga (Village R.)
Philippines	1	DZRA	Virac, Catanduanes
	5	DXKO	Cagayan de Oro, Mis. Or.
	1	DZBS	Baguio C., Benguet
Singapore	20		Singapore
Thailand	50		Nan
	1		Chumporn
1377 kHz			
Australia	5	3MP	Mornington
China, PR	200		FJ (H1) + 2 sts
China (Taiwan)			Chihsan (Cheng Sheng)
	0.1	BED41	Chengkung (BCC1)
			Taipei + 4 sts (Fuhsing)
			Unk. (TBC)
India	10		Hyderabad B
Japan	5	JOUC	Yamaguchi (NHK2)
	1	JOTZ	Hachinohe (NHK2)
	1	JOAC	Nagasaki (NHK2)
New Zealand	2	2XX	Levin
Pakistan	10		Peshawar II
Philippines	10	DXKP	Pagadian C., Zamb. Sur
	2.5	DYVR	Roxas C., Capiz
Russia	30		Vladivostok, FE
			Okha, FE
			Yekaterinburg, Sib.
Thailand	10		Phitsanulok
1380 kHz			
Hawaii	5	KIPO	Pearl City
1386 kHz			
Australia	5	2EA	Sydney (SBS)
China, PR	20		TJ + 3 sts
China (Taiwan)	3.5	BED87	Hsinchu (BCC3/Chiao Yu)
	3.5	BED74	Yuli (BCC2)
	0.1	BEG36	Hehuanshan (BCC2)
	0.5		Peikang (Cheng Sheng)
India	10		Gwalior
Japan	10	JOQC	Morioka (NHK2)
	10	JOJB	Kanazawa (NHK2)
	10	JOHC	Kagoshima (NHK2)
	5	JOKB	Okayama (NHK2)
Korea, Rep.	10	HLAM	Mokp'o (MBC)
New Zealand	0.2	1ZT	Turangi (rel. 1ZA)
	5/1		Auckland (rel. BBCWS)
Philippines	5	DXCR	Malaybalay, Buk.
Solomon Isl.	5		Lata, Sta. Cruz I.
Thailand	20		Pathum Thani
1395 kHz			
Australia	0.2	2LG	Lithgow
	5	5AA	Adelaide
China, PR			2 sts
India	10		Bikaner
	1		Madras B
Kazakhstan			Panfilov
Korea, Rep.	10	HLCO	Ch'ŏlwon (KBS)
New Zealand	2	4ZW	Oamaru
Philippines	1	DWMG	Solano, Nva. Visc.
Thailand	25		Mae Chau Chiangrai
1398 kHz			
Indonesia	50	RRI	Bandung
1404 kHz			
Australia	2	2PK	Parkes
China, PR	10		LN + 2 sts
China (Taiwan)	10	BED65	Ilan (BCC1)
China (Taiwan)	5	BEV78	Keelung (Yi Shih)
Cocos Is.	0.1	VKW	Keeling
India	10		Gangtok, Sikkim
Japan	10	JOVR	Shizuoka (SBS)
	1		Hamamatsu (SBS)
	0.1		Gotomba (SBS)

	5	JOQL	Kushiro (HBC)
Kirghizia			4 sts
Korea, Rep.	10	HLKP	Pusan (CBS)
New Zealand	2.5	4XL	Invercargill (R. Rhema)
Pakistan	10		Dera Ismail Khan
Philippines	5	DYKB	Bacolod, Neg. Occ.
	2	DZOR	Olongapo, Zamb.
Samoa. W	2.5	2AP	Unk.
Thailand	10		Yaso Thon
	10		Phichit
	50		Songkhla
1410 kHz			
Galapagos	1	HCSC8	Puerto Ayora
1413 kHz			
Bangladesh	10		Comilla
China, PR	100		XJ + 4 sts
China (Taiwan)	10	BED54	Miaoli (BCC1)
	1	BED67	Puli (BCC1)
	3.5	BED80	Taitung (BCC3/Chiao Yu)
Japan	50	JOIF	Fukuoka (KBC)
New Zealand	2	1ZO	Tokoroa
	0.1	3XP	Christchurch
Russia			Sangar, Sib.
Thailand	10		Bangkok
1420 kHz			
Hawaii	5	KCCN	Honolulu
1422 kHz			
China, PR	20		SH + 2 sts
China (Taiwan)	1		Kuanyin (Chien Kuo)
			Unknown (BCC)
Christmas I.	0.25	VLU2	Phosphate Hill
India	1		Rajkot B
Japan	50	JORF	Yokohama (R. Nippon)
Philippines	15	DWBC	Quezon C, MM
	5	DXMU	Musuan, Buk.
Russia			Unk., C. Sib.
Thailand	10		Phitsanulok
1430 kHz			
Laos			Sowanuakhet
1431 kHz			
Australia	2	2RN	Wollongong (ABC3)
China P.R.	200		NM + 3 sts
China (Taiwan)	50	BEC73	Chingshui (Kung Chun)
India	1		Kozhikode B/Chandigarh
Japan	5	JOVF	Wakayama (WBS)
	5	JOZF	Gifu (GBS)
	1	JOWW	Iwaki (RFC)
	1		Fukue (NBC)
	1	JOHL	Tottori (BSS)
	1		Izumo (BSS)
	0.1		4 sts (Comm.)
Philippines	5	DYRS	San Carlos, Neg. Occ.
Thailand	20		Songkhla
1440 kHz			
Australia			(rel. 2ST)
	2	2PB	Canberra
China, PR	50		NM + 3 sts
China (Taiwan)	10	BEC60	Taipei (Kung Chun)
India	1		Ahmedabad B
	1		Pt. Blair (standby tr.)
Japan	50	JOWF	Sapporo (STV)
	1		2 sts (STV)
Kazakhstan			Kzyl-Orda/Bakanas
Korea, Rep.	1-0.25		4 sts (AFKN)
New Zealand		1XK	Taurango
Philippines	5	DWCM	Dagupan C., Pang.
	1	DZOM	Calapan, M Ortal
Russia			Ust-Koksha/
			Kosh-Agach, Sib.
Thailand	10		Samut Sakhon
	10		N. Phanom
			Nan
1449 kHz			
Australia	5	2MG	Mudgee
China, PR	20		JX + 3 sts
China (Taiwan)	10	BED48	Kaohsiung
			(BCC3/Chiao Yu)
	0.1		8 sts (BCC3)
India	1		Kanpur
Japan	5	JOKF	Takamatsu (RNC)
	5	JOQM	Abashiri (HBC)
	0.1		Isahaya (NBC)
Korea, Rep.	10	HLQB	Ulsan (KBS)
Micronesia	10	V6AH	Kolonia, Pohnpei
New Zealand	2	2YM	Palmerston No.
Philippines	5	DXRM	Marawi C, Lanao Sur
Thailand	5		Chumporu
1453 kHz			
Cambodia			Unk. in Batambang Province
1458 kHz			
Australia	2	2PB	Newcastle
China, PR	150		NM ZJ + 1 st
China (Taiwan)	10		Taipei (BCC)
India	10		Bhagalpur
Japan	1	JOUO	Saga (NBC)
	1	JOYL	Tsuchiura (IBS)
	1	JOWR	Fukushima (RFC)
	1		Shobara (RCC)
			Sekijo (IBS)
	0.1		3 sts (Comm.)
Kazakhstan	15		Arkalyk Buran
Korea, Rep.	1	HLSH/ST	Hamyang/Ponghwa (KBS)
Maldives	5		Malé
New Zealand	0.4	3YW	Westport (rptr)
Philippines	10	DWRF	Iba, Zamb.
Russia	15		Arkalyk, Kaz
			Tara, Sib.
			Unk, E. Sib.
Thailand	20		Phuket
	10		Sri Sa Ket
1460 kHz			
Hawaii	5	KULA	Kailua
1467 kHz			
Australia	2	3MA	Mildura
China, PR	50		FJ (V) + 4 sts
China (Taiwan)	10	BED81	Hualien (BCC3/Chiao Yu)
	10	BED82	Chia-i (BCC3/Chiao Yu)
	1	BED29	Jiyuetan (BCC1)
China (Taiwan)		BEC47	Kinmen (Kuang Hua)
Fiji	2.5		Rakiraki
India	20		Jeypore
Japan	1		5 sts (NHK2)
	0.5-0.1		5 sts (NHK2)
Kirghizia	30		Bishkek, Kir.
	15		Osh, Kir.
			Batken/Kara-Kul, Kir.
Russia			Unk., Ce. Asia
Korea, Rep.	50	HLKN	Mokp'o (KBS)
Thailand	100		Bangkok
1476 kHz			
Australia	5		Penrith (trtr 2KA)
	2	4ZR	Roma
China, PR			(C2) + 9 sts
China (Taiwan)	10	BEC74	Taichung (Kung Chun)
	25	BEG27	Taipei (Taipei Bc. Stn.)
			4 sts (Fuhsing)
India	1		Jaipur A
Japan	1	JOSD	Iida (NHK2)
	0.1		Ozu (NHK2)
Malaysia	600		Kota Kinabalu, Sabah
			(r 1475)
New Zealand	5	1XD	Auckland, Airport
Philippines	1	DZYA	Angeles C, Pamp.
	1	DWRB	Lipa C, Bat
		DYWR	Tacloban C, Leyte
Russia	150		Vladivostok, FE (FS)
			Onguday, Sib.
Thailand	100		Chiengmai
1485 kHz (Low-power ch.)			
Australia	0.1-0.2		4 sts
Br.Ind.Oc.Terr.	0.25		Diego Garcia (AFRTS)
China, PR			Many sts
India	1		4 sts
Japan	1	JOGO	Hachinohe (RAB)
	1	JOPL	Hagi (KRY)
	0.1		24 sts (Comm.)
Korea, Rep.	1	HLQS	Kongju (KBS)
	1	HLQU	Kohŭng (KBS)
New Zealand	1		Gisborne
Pakistan	10		Faisalabad

Philippines	1	DYKR	Kalibo, Aklan
Russia			Unk, Yakut ASSR, Sib.
			Kamenskoya, FE
			Chelyabinsk oblast, Sib.
			Tazovskiy/Krasnoselkup, Sib.
Sri Lanka	1		Mahiyangana
Thailand	1		5 sts
Wake I.	0.05	KEAD	Wake (AFRTS)
1494 kHz			
Australia	2	2AY	Albury
China, PR	120		XJ + 4 sts
China (Taiwan)	10	BEE32	Taipei (Chiao Yu)
	5	BEE34	Changhua (Chiao Yu)
India	1		Thiruananthapuram B
Japan	1	JOTL	Nayoro (HBC)
	5	JOYR	Okayama (RSK)
	1-0.1		6 sts (RSK)
Kazakhstan			Alma Ata
Micronesia	10	V6A	Colonia, Yap
New Zealand	2.5	1YT	Taupo
Papua N.Guinea	10		Wabag
Philippines	1	DXOC	Ozamis C., Mis.Occ.
	10	DWLR	Quezon C, MM
	1	DYKR	Kalibo, Ak
Thailand	25		Bangkok
1500 kHz			
Hawaii	10	KUMU	Honolulu
1503 kHz			
Australia	5	2BS	Bathurst
China, PR	10		NM + 3 sts
China (Taiwan)	0.1	BED64	Hehuanshan (BCC1)
	1	BED83	Mobile (BCC1)
India	1		Srinagar D/Vijayawada B
Japan	10	JOUK	Akita (NHK1)
Korea, Rep.	1	HLSK	Kimch'on (KBS)
New Zealand	2	3XG	Christchurch (R. Rhema)
Russia			Salekhard, Sib.
Thailand	10		Surat Thani
	10		Sakon Nakhon
1512 kHz			
Australia	10	2NA	Newcastle (ABC2)
China, PR			FJ (T1); NM
China (Taiwan)	10	BEP27	Taipei (Ching Cha)
			11 sts (Fuhsing)
Japan	5	JOZB	Matsuyama (NHK2)
	1	JOCD	Koriyama (NHK2)
	1	JOSC	Matsumoto (NHK2)
	0.5-0.1		5 sts (NHK2)
Korea, Rep.	0.25-0.1		4 sts (AFKN)
New Zealand	1	1ZU	Taumarunui
Pakistan	10		Muzaffarabad,
	10		Gilgit, Kashmir (r 1510)
Philippines	1		Clark (Angeles C., Pamp) (FENP)
	0.25		S. Miguel (FENP)
Thailand	10		Suphan Buri
	10		Pha Yao
	10		Surin
1521 kHz			
Australia	5	2QN	Deniliquin
China, PR	500		XJ (FS) + 3 sts
China (Taiwan)	1200		Changchih (CBS) (reported)
India	1		Aurangabad
	10		Tawang
Japan	1		7 sts (NHK2)
	0.1		4 sts (NHK2)
New Zealand	0.2	3ZR	Reefton (rptr)
Russia			Boguchany, Sib.
Thailand	25		Bangkok
1530 kHz			
Australia	2	2VM	Moree
China, PR	10		SX; ZJ; + 1 st
China (Taiwan)	1	BEG37	Mobile (BCC2)
	1	BEG41	Mobile (BCC1)
India	10		Agra
Japan	1	JODO	Joetsu (BSN)
	1	JOAM	Uwajima (RNB)
	1	JOWM	Obihiro (STV)
	5	JOXF	Utsunomiya (CRT)
	0.1		2 sts (Comm.)

Kirghizia			Tash-Kumyr
New Zealand	2	2YP	New Plymouth
Philippines	10	DZME	Quezon C., MM
Thailand	10		Uttaradit
	10		Chanthaburi
1539 kHz			
Australia	10	5PB	Adelaide
China (Taiwan)	10	BED78	Tainan (BCC3/Chiao Yu)
India	5		Panaji (Goa) B
Japan	0.1		20 sts (NHK2)
Kazakhstan			Kustanay
Korea, Rep.	1	HLQC	Kosan (KBS)
New Zealand	1	2ZE	Blenheim
			Mangere
Philippines	1	DZYM	San José, Mind. Occ.
Thailand	10		Udon Thani
	10		Ratchaburi
Uzbekistan			Samarkand
1540 kHz			
Hawaii	5	KISA	Honolulu
1548 kHz			
Australia	50	4QD	Emerald (ABC3)
China, PR	100		SD + 1 st
Kazakhstan			Tekeli/Dzhansugurov
			Lepsy/Andreyevka
New Zealand		1XD	Pukekohe (Counties R.)
	1	1XN	Rotorua
Sri Lanka	600		Trincomalee (DW relay)
Thailand	10		Bangkok
1551 kHz			
Indonesia	0.1		Manado (RRI) (r)
1557 kHz			
Australia	2	2RE	Taree
China, PR			2 sts
Japan	0.1		19 sts (Comm.)
Kazakhstan			Ucharal/Kapal/Antonovka
New Zealand	2	2ZH	Hawera (rel. 2ZP)
Pakistan	10		Skardu (Kashmir)
Philippines	2.5	DWGO	Olongapo C., Zamb.
Thailand	10		Lom Sak, Phetchaboon
1566 kHz			
Australia	5	3NE	Wangaratta
	0.2	4GM	Gympie (ABC3)
China, PR			HEB + 6 sts
China (Taiwan)	10		Taichung (ICRT)
India	1000		Nagpur (Buttibori)
Korea, Rep.	250/100	HLAZ	Cheju (FEBC)
Norfolk I.	0.05	VL2NI	Kingston
Philippines	10	DZHH	Pasay C, MM (r 1570) (Time-share)
Russia			Unk., FE
1570 kHz			
Hawaii	1	KUAU	Kaihua-Kona (F.p.)
1575 kHz			
Australia	5	2OO	Wollongong
China, PR	7.5		Wenzhou, ZJ
China (Taiwan)			Taipei (Yu Shih)
Japan	1		Iwakuni (FEN)
	0.25		Misawa/Sasebo (FEN)
Kazakhstan	10		Semipalatinsk, Kaz.
Russia	5		Yuzhno-Sakhalinsk, FE
Thailand	1000		Bangkok (Ayutthaya)
1584 kHz (Low-power ch.)			
Australia	0.1-0.2		6 sts
China, PR			Many sts
Hong Kong	0.1		Chung Hom Kok (RTHK3)
India	1		Jamshedpur
	1		Keonjhar
	1		Kota
	1		Mathura
Japan	0.1		24 sts (NHK1)
Korea, Rep.	1	HLDK	Tanyang (KBS)
	1	HLQZ	Kumsan (KBS)
	1		Sanch'ŏng (KBS)
Micronesia	1	V6AJ	Tofol, Kosrae
New Zealand	0.4	2ZF	Picton
Pakistan	0.25		Turbat
Palau	3	WSZB	Koror (Malakai Is.)
Russia			Klyuchi/Tigil, FE
Thailand	1		5 sts

1593 kHz			
Australia	10	3PB	Melbourne
	2	5MV	Renmark (ABC3)
China, PR	100		XJ + 3 sts
India	1		Indore B
Japan	10	JOQB	Niigata (NHK2)
	10	JOTB	Matsue (NHK2)
Papua N.Guinea	10		Vanimo
Philippines	1	DZUP	Quezon C, MM
Russia			Angarsk, S.
			Unk., FE
Thailand	10		Ranong
	1		Ratchaburi
Uzbekistan			Termez
1600 kHz			
Hawaii	5	KHNL	Honolulu (CP)
1602 kHz (Low-power ch.)			
Australia	0.05-0.5		3 sts (ABC3)
China, PR			2 sts
India	1		5 sts (B)
Japan	1		10 sts (NHK2)
	0.1		16 sts (NHK2)
Korea, Rep.	1	HLQE	Sabuk (KBS)
Micronesia	5	V6AK	Weno, Chunk
New Zealand	1	2XA	Levin
Russia			Kargasok, Sib.
Sri Lanka	1		Mahiyangana
Thailand	1		Buriram
	1		N. Si Thammarat
1629 kHz			
Australia	0.5	2RPH	Sydney (F. Pl. 1107 or 1224)
1656 kHz			
Australia			Gold Coast (TAFE R.)
1700 kHz			
China (Taiwan)			Keelung (Fishery & Weather BS)
1720 kHz			
Australia	0.2		Kensington (trtr VL2UV)

NORTH AMERICA (excl Mexico)

Explanation of listing:—
FIRST COLUMN—Call letters. These are required identification for US sts, usually on top of the hour.

SECOND COLUMN—Location. The official Post Office abbreviations are used for US states, and Canadian provinces. For US st's the state abbreviation is followed by the Post Office zip code (cf. "Addresses" below).

THIRD COLUMN—Transmitter powers. A daytime station listed with two powers uses the second one during critical hours (cf. D and U6 below). For stations using a directional antenna the ERP in most cases varies greatly with the direction from the station.

FOURTH COLUMN—Contains authorized operating hours and directional antenna information. The symbols mean as follows: **D** is daytime operation (between local sunrise and local sunset), **D1** without directional antenna, **D3** with directional antenna, **D4** with directional antenna during critical hours only, **D5** with directional antenna except during critical hours, **D6** with directional antenna, different patterns during critical and non-critical hours. **L** is limited time, and means a st. West of the dominant st. can operate from as early as sunrise at the dominant sts location; A st. East of the dominant st. can operate as late as the dominant sts sunset. Number indicates directional pattern as under "U" below. **SH** equals "Specified Hours" and usually means a st. that would normally operate unlimited time files a schedule of reduced hours with the FCC because of economic factors. **ST** means "Shares Time". This usually finds a college, university, or church st. on the air two or three hours a day, during which time the commercial st. signs off. **U** means the st. is authorized to operate any time up to 24 hours a day. **U1** means without directional antenna; **U2** means with directional antenna at night; **U3** means directional antenna, same pattern, day and night; **U4** means directional antenna, different pattern day and night; **U5** means directional antenna daytime, non-directional at night; **U6** means directional antenna at night and during critical hours (until two hours after sunrise and from two hours before sunset). **U7** means different directional patterns for day, critical hours and night. **U8** as U7 but non-directional day. **U9** means directional day and night (different patterns), but non-directional during critical hours (usually on reduced power).

FIFTH COLUMN—Gives an indication of actual operating hours as follows: **NH** normal hours, usually from 18 to 20h a day. **AN** all night operation, i.e. 24h a day most days. **AN7** continuous operation. **AN6** 24h six days a week, off a few hours during the early morning local time on the seventh day for maintenance, usually Sun or Mon morning. **AN5** off both Sun and Mon morning, etc. **D** daytime, i.e. local sunrise to sunset only. **D*** 0600 local time to local sunset, but with low power (from a few watts to a maximum of 500 watts) before local sunrise if sunrise is later than 0600. A similiar extension of evening hours also exists. **SH** specified hours as fourth column; or special hours, i.e. less than NH, often with breaks during the day.

SCOPE—Because of the vast number of stations in operation in North America it has been found necessary to limit the scope of this list. All US and Canadian st's on the "local" channels (1230, 1240, 1340, 1400, 1450, 1490kHz), except those having more than 1kW power, have been omitted. For the remaining frequencies a lower power limit of 5kW day and/or 1kW night has been set for US st's and 1kW day for Canadian st's. For each frequency the approximate number of omitted low power **(lp)** stations is mentioned. Full details of these stations may be obtained from specialized publications.

ADDRESSES—For USA stations not listed elsewhere in this book, city, state and zip code as listed below may do. The US Post Office is, however, becoming more reluctant to deliver incompletely addressed letters. Complete addresses for all US MW stations may be obtained from the National Radio Club AM Radio Log (see club list for address) and from trade publications. In some cases the address location differs from the licence location. Where necessary for the sake of clarity a differing licence location is shown separately, with the address within brackets

530kHz				
CKHL	High Level, AB	0.4/0.25	U1	AN
CIAO	Brampton, ON	1/0.25	U1	AN
540kHz				
KJQ1	Costa Mesa, CA	25/0.36	U7	
KIEZ	Salinas, CA	50/0.5	U3	AN7
WGTO	Pine Hills, FL	50	U4	AN6
WSTH	Columbus, GA (Addr: Alexander City, AL 35010)	5/0.5	U4	AN6
KWMT	Fort Dodge, IA 50501	5/0.2	U4	AN
KNOE	Monroe, LA 71201	5/1	U4	AN7
KNMX	Las Vegas, NM 87701	5	D3	D
WETC	Wendell, NC 27591	8/0.5	U4	
WWCS	Canonsburg, PA (Addr: Pittsburgh, PA 15241)	7.5/0.25		AN
CBT	Grand Falls, NF	10	U1	NH
CBEF	Windsor, ON	2.5/5	U3	AN
CBGA1	New Carlisle, PQ	10	U4	NH
CJSB	Ottawa, ON	50/12.5	U1	AN7
CBK	Watrous, SK	50	U1	NH
+lp: USA 12				
550kHz				
KENI	Anchorage, AK 99517	5	U1	AN7
KOY	Phoenix, AZ 85004	5/1	U1	AN7
KCWR	Bakersfield, CA 93368	5	U4	AN7
KRAI	Craig, CO 81625	5/0.5	U2	AN7
WAYR	Orange Park, FL 32073	5	D3	D
WDUN	Gainesville, GA 30503	5/2.5	U2	AN7
KFRM	Salina, KS (Addr: Belleville, KS 66935)	5	D3	D

KUSA	St. Louis, MO 63132	5	U2	AN7
KBOW	Butte, MT 59701	5/1	U2	AN6
WGR	Buffalo, NY 14202	5	U2	AN7
KFYR	Bismarck, ND 58502	5	U2	AN6
WKRC	Cincinnati, OH 45219	5/1	U4	AN7
KOAC	Corvallis, OR 97331	5	U4	NH
WJMW	Bloomsburg, PA 17815	1	U4	AN
KCRS	Midland, TX 79701	5/1	U4	AN
KTSA	San Antonio, TX 78218	5	U2	AN7
WDEV	Waterbury, VT 05676	5/1	U4	NH
WSVA	Harrisonburg, VA 22801	5/1	U2	AN
KARI	Blaine, WA 98230	5/2.5	U4	AN7
WSAU	Wausau, WI 54402	5	U4	AN7
CFJC	Kamloops, BC	25/5	U4	AN
CKPG	Prince George, BC	10	U2	AN7
CFNB	Fredericton, NB	50	U4	AN7
CHNO	Sudbury, ON	50/10	U4	AN
CHLN	Trois-Rivières, PQ	10/5	U4	AN7

+lp: USA 5

560kHz

WOOF	Dothan, AL 36302	5	D1	D*
KVOK	Kodiak, AK 99615	1	U1	NH
KBLU	Yuma, AZ 85364	1	U2	AN
KSFO	San Francisco, CA 94133	5	U2	AN7
KLZ	Denver, CO 80222	5	U3	AN7
WQAM	Miami, FL 33140 (Addr: Hollywood, FL 33024)	5/1	U1	AN7
WIND	Chicago, IL 60611	5	U4	AN6
WGAN	Portland, ME 04104	4.8	U3	AN7
WFRB	Frostburg, MD 21532	5	D1	D
WHYN	Springfield, MA 01104	5/1	U3	AN7
WHND	Monroe, MI 48220	5	D3	D*
WEBC	Duluth, MN 55805	5	U4	AN
KWTO	Springfield, MO 65807	5/4	U2	AN6
KMON	Great Falls, MT 59403	5	U2	AN6
WEAZ	Philadelphia, PA (Addr: Bala Cynwyd, PA 19004)	5	U4	AN7
WVOC	Columbia, SC 29221	5	U2	AN
WHBQ	Memphis, TN 38111	5/1	U4	AN7
KLVI	Beaumont, TX 77702	5	U2	AN
KPQ	Wenatchee, WA 98801	5	U2	AN7
WJLS	Beckley, WV 25802	5/0.5	U2	AN7
CKNL	Ft. St. John, BC	1	U2	AN7
CHTK	Prince Rupert, BC	1/0.25	U1	AN7
CHVO	Carbonear, NF	5	U2	AN7
CJKL	Kirkland Lake, ON	5	U2	NH
CFOS	Owen Sound, ON	5/1	U4	NH
CKCN	Sept-Iles, PQ	10/5	U4	AN

+lp USA 4/Can 2

570kHz

WAAX	Gadsden, AL 35902	5/0.5	U2	AN
KCNO	Alturas, CA 96101	5	U4	
KLAC	Los Angeles, CA (Addr. Burbank, CA 91510-7806)	5	U2	AN7
WHNZ	Pinellas Park, FL (Addr: St. Petersburg, FL 33743)	5	U4	AN7
WACL	Waycross, GA 31501	5/1	U2	NH
WTEM	Bethesda, MD 20014	5/1	U4	AN6
WVMI	Biloxi, MS 39531	5/1	U4	AN7
KGRT	Las Cruces, NM 88001	5	D1	D*
WMCA	New York, NY 10019	5	U3	AN7
WSYR	Syracuse, NY 13202	5	U4	AN7
WWNC	Asheville, NC 28816	5	U2	AN6
WKBN	Youngstown, OH 44501	5	U2	AN7
WNAX	Yankton, SD 57078	5	U2	AN6
KLIF	Dallas, TX 75219	5	U4	AN7
KISN	Salt Lake City, UT 84116	5	U3	AN7
KVI	Seattle, WA 98101	5	U1	AN7
CKEK	Cranbrook, BC	10/1	U3	AN6
CKWL	Williams Lake, BC	1	U4	AN6
CJEM	Edmundston, NB	5/1	U2	NH
CFCB	Corner Brook, NF	1	U1	AN7
CKGL	Kitchener, ON	10	U3	AN7
CKSW	Swift Current, SK	10	U4	AN
CFWH	Whitehorse, YT	5/1	U1	AN7
	Nuuk (Godthåb), Greenland	25	U1	SH

+lp: USA 4, Can 3

580kHz

KRSA	Petersburg, AK 99833	5	U1	NH
KSAZ	Tucson, AZ 85718	5/0.5	U2	AN7
KMJ	Fresno, CA 93726	5	U1	AN6
KUBC	Montrose, CO 81401	5/1	U2	NH
WDBO	Orlando, FL 32804	5	U2	AN7
WGAC	Augusta, GA 30907	5/1	U2	AN7
KFXD	Nampa, ID (Addr: Meridian, ID 83642)	5	U2	AN6
WILL	Urbana, IL 61801	5/0.1	U4	AN
KKSU	Manhattan, KS 66506	5/0.5	ST1	SH
WIBW	Topeka, KS 66601	5	ST2	AN
KALB	Alexandria, LA 71301	5/1	U2	NH
WTAG	Worcester, MA 01602	5	U4	AN6
WTCM	Traverse City, MI 49685	5/0.5	U4	AN
KCMX	Ashland, OR 97520	1	U4	AN6
WHP	Harrisburg, PA 17105	5	U2	AN6
KRLB	Lubbock, TX 79408	5	D3	D*
WCHS	Charleston, WV 25324	5	U2	AN
WKTY	LaCrosse, WI 54601	5/0.74	U4	AN
CKUA	Edmonton, AB	10	U4	NH
CKXR	Salmon Arm, BC	10/1	U2	AN7
CKY	Winnipeg, MB	50	U4	AN7
CJFX	Antigonish, NS	25	U4	AN7
CKAP	Kapuskasing, ON	10/1	U2	NH
CFRA	Ottawa, ON	50/10	U4	AN7
CKPR	Thunder Bay, ON	5/1	U1	AN
CHLC	Hauterive, PQ	10/2.5	U4	AN2

+lp: USA 8/Can 2

590kHz

KHAR	Anchorage, AK 99511	5	U1	AN
KBHS	Hot Springs, AR 71901	5	D	D*
KRSO	San Bernardino, CA 92412	1	U4	AN7
KTHO	South Lake Tahoe, CA 95705	2.5/0.5	U3	AN7
KCSJ	Pueblo, CO 81002	1	U4	AN6
WGNE	Panama City, FL 32401	1.7/2.5	U2	AN
WKHX	Atlanta, GA 30339	5/4.5	U2	AN7
KID	Idaho Falls, ID 83401	5/1	U2	NH
KEZK	Wood River, IL (Addr: St. Louis, MO 63105)	1	U4	AN7
WVLK	Lexington, KY 40592	5/1.6	U4	AN6
WEEI	Boston, MA 02199	5	U3	AN7
WJMS	Ironwood, MI 49938	5/1	U2	NH
WKZO	Kalamazoo, MI 49008	5	U2	NH
WOW	Omaha, NE 68114	5	U1	AN7
WROW	Albany, NY 12204	5/1	U4	AN
WGTM	Wilson, NC 27893	5	U4	NH
KUGN	Eugene, OR 97402	5/1	U2	AN6
WARM	Scranton, PA 18501	5	U4	AN6
WMBS	Uniontown, PA 15401	1	U2	AN
KLBJ	Austin, TX 78753	5/1	U2	AN7
KSUB	Cedar City, UT 84720	5/1	U2	NH
WLVA	Lynchburg, VA 24501	5/1	U4	AN6
KAQQ	Spokane, WA 99203	5	U1	AN7
CFTK	Terrace, BC	1	U3	AN7
CFAR	Flin Flon, MB	10/1	U5	NH
CJCW	Sussex, NB	1/0.25	U4	AN6
VOCM	St. John's, NF	20/10	U2	AN7
CKYC	Toronto, ON	50	U4	AN7
CKRS	Jonquière, PQ	25/10	U4	AN

+lp: USA 4/Canada 1

600kHz

KCLS	Flagstaff, AZ 86001	5/0.5	U2	NH
KHTE	Redding, CA 96001	5	U3	AN
KKLQ	San Diego, CA 92123	5	U3	AN7
KIIX	Fort Collins, CO 80524	5/0.5	U4	
WOKV	Jacksonville, FL 32236	5	U2	AN7
WMT	Cedar Rapids, IA 52406	5	U2	AN6
WKLW	Paintsville, KY 41240	5	D1	D
WFST	Caribou, ME 04736	5	D1	D*
WCAO	Baltimore, MD 21208	5	U3	AN7
KGEZ	Kalispell, MT 59903	5/1	U4	NH
WSJS	Winston-Salem, NC 27101	5	U4	AN7
KSJB	Jamestown, ND 58402	5	U3	AN7
WREC	Memphis, TN 38103	5	U4	AN7
KROD	El Paso, TX 79902	5	U2	AN7
KTBB	Tyler, TX 75711	5/2.5	U2	AN
CHRX	Vancouver, BC	10	U3	AN7
CBNA	St. Anthony, NF	10	U4	NH
CKCL	Truro, NS	10/1	U3	AN
CFCH	North Bay, ON	10/5	U4	AN
CIQC	Montreal, PQ	10/5	U3	AN7
CFQC	Saskatoon, SK	10	U4	AN7

+lp: USA 14/Canada 3

610kHz				
WZZK	Birmingham, AL 35209	5/1	U2	AN7
KAVL	Lancaster, CA 93534	5/4	U4	AN
KFRC	San Francisco, CA 94111	5	U1	AN7
KSKE	Vail, CO 81657	5	D1	D
WIOD	Miami, FL (Addr: N. Bay Village, FL 33141)	10	U4	AN7
WRUS	Russellville, KY 42276	2.5/0.5	U2	
KDAL	Duluth, MN 55802	5	U2	AN5
WDAF	Kansas City, MO 64108	5	U1	AN7
KOJM	Havre, MT 59501	1	U4	NH
WGIR	Manchester, NH 03105	5/1	U4	NH
KZSS	Albuquerque, NM 87103	5	U2	AN7
WAQS	Charlotte, NC 28216	5/1	U4	AN7
WTVN	Columbus, OH 43215	5	U2	AN7
KYJC	Medford, OR 97501	2.5/5	U4	AN7
WIP	Philadelphia, PA 19103	5	U3	AN7
KILT	Houston, TX 77006	5	U4	AN7
KVNU	Logan, UT 84321	5/1	U2	AN6
WSLC	Roanoke, VA 24017	5/1	U4	AN6
KONA	Tri-Cities, WA 99302	5	U4	AN7
CKYL	Peace River, AB	10	U2	AN7
CHNL	Kamloops, BC	25/5	U4	AN7
CJAT	Trail, BC	10/1	U5	AN
CHTM	Thompson, MB	1	U1	AN7
CKTB	St. Catherines, ON	10/5	U3	AN6
CFLO	Mont-Laurier, PQ	1	U2	NH
CHNC	New Carlisle, PQ	10/5	U3	AN7
CKRW	Whitehorse, YT	1	U1	AN7
+lp: USA 7/Canada 2				
620kHz				
WKNI	Lexington, AL (Addr: Killen, AL 35645)	5	D1	D
KGTL	Homer, AK 99603	5	U1	NH
KTAR	Phoenix, AZ 85013	5	U2	AN6
KIGS	Hanford, CA 93232	1	U2	AN
KSTR	Grand Junction, CO 81501	5/0.8	U2	D*
WSUN	St. Petersburg, FL 33702	10/5	U2	AN7
KWAL	Wallace, ID (Addr: Osburn, ID 83849)	1	U2	NH
KMNS	Sioux City, IA 51101	1	U4	AN
WZON	Bangor, ME 04401	5	U2	NH
WJDS	Jackson, MS 39206	5/1	U2	AN7
WSKQ	Newark, NJ (Addr: Livingston, NJ 07039)	5	U4	AN7
WHEN	Syracuse, NY 13217	5/1	U2	AN7
WDNC	Durham, NC 27702	5/1	U4	AN7
KGW	Portland, OR 97201	5	U2	AN7
WRJZ	Knoxville, TN 37917	5	U2	AN
KWFT	Wichita Falls, TX 76307	5	U2	NH
WVMT	Burlington, VT (Addr: Colchester, VT 05446)	5	U4	AN6
WTMJ	Milwaukee, WI 53201	5	U2	AN7
CJCI	Prince George, BC	10	U2	AN7
CKCM	Grand Falls, NF	10	U3	AN7
CKOY	Timmins, ON	10/5	U4	NH
CFRP	Rivière Portneuf, PQ	1	U3	AN2
CKCK	Regina, SK	10	U4	AN7
+lp: USA 6				
630kHz				
KJNO	Juneau, AK 99801	5/1	U1	NH
KIAM	Nenana, AK 99760	5	U1	NH
KIDD	Monterey, CA 93940	1	U4	AN6
KHOW	Denver, CO 80237	5	U4	AN7
WMAL	Washington, DC 20015	5	U4	AN7
WBMQ	Savannah, GA 31402	5	U2	AN
KIDO	Boise, ID 83707	5	U4	AN7
WLAP	Lexington, KY 40577	5/1	U4	AN6
KDWB	St. Paul, MN 55119	5/0.5	U4	AN7
KXOK	St. Louis, MO 63105	5	U4	AN6
KOH	Reno, NV 89505	5/1	U2	AN6
WMFD	Wilmington, NC 28403	1	U4	AN
KWRO	Coquille, OR 97423	5	D1	D
WPRO	East Providence, RI 02915	5	U2	AN7
KSLR	San Antonio, TX 78229	5	U4	AN7
KCIS	Seattle, WA 98133	3.9/2.5	U2	
CHED	Edmonton, AB	50	U4	AN7
CKOV	Kelowna, BC	5/1	U1	AN7
CKRC	Winnipeg, MB	10	U4	AN7
CFCO	Chatham, ON	10/1	U4	AN7

CJET	Smiths Falls, ON	10	U4	AN7
CFCY	Charlottetown, PEI	10	U4	AN7
CHLT	Sherbrooke, PQ	10	U4	NH
+lp: USA 9				
640kHz				
KYUK	Bethel, AK 99559	10	U1	NH
KFI	Los Angeles, CA	50	U1	AN7
WLVJ	Royal Palm Beach, FL 33411	25/8.2	U4	NH
WGST	Atlanta, GA	50/1	U4	AN7
WOI	Ames, IA 50011	5/1	U2	AN7
KTIB	Thibodeaux, LA 70301	5/1	U4	
WNNZ	Westfield, MA 01085	50/15	U4	AN
WROR	Zeeland, MI 49464	1	U1	AN7
KGVW	Belgrade, MT 59714	10/1	U2	NH
WWJZ	Mt. Holly, NJ 08060	50/1	U4	
WFNC	Fayetteville, NC 28303	10/1	U1	AN
WHLO	Akron, OH 44702	5/0.5	U4	AN7
WWLS	Moore, OK (Addr: Norman, OK 73071)	1	U2	NH
WJTZ	Blountsville, TN 37617	5/0.81	U4	AN7
WCRV	Collierville, TN	50/0.48	U2	NH
CBN	St. John's, NF	10	U1	NH
CFOB	Ft. Frances, ON	1	U2	AN
CHOG	Toronto, ON	50/25	U4	AN7
+lp: Can 1/USA 2				
650kHz				
KYAK	Anchorage, AK 99507	50	U4	AN7
KRDX	Rancho Cordova, CA	25/1	U5	
WORL	Christmas, FL	5	D3	D
WBSO	Clinton, MA 01510	25/1.6	U4	F.Pl.
WKKQ	Hibbing, MN 55746	10/0.5	U2	AN
KZTU	Junction City, OR 97448	10/1	U4	F.Pl.
WLEM	Emporium, PA 15834	5	D1	D
WSM	Nashville, TN	50	U1	AN6
KMTI	Manti, UT 84642	10/0.9	U4	AN
KUUY	Cheyenne, WY 82003	8.6/0.5	U2	
CISL	Richmond, BC	10/9	U4	AN7
CKGA	Gander, NF	5	U4	AN7
CKOM	Saskatoon, SK	10	U2	AN7
	Qeqertarsuaq (Godhavn), Greenland	5	U1	SH
+lp: USA 2				
660kHz				
WBLX	Fairhope, AL	22/0.85	U1	
KFAR	Fairbanks, AK 99707	10	U1	AN
KTNN	Window Rock, AZ 86515	50	U2	NH
KGDP	Orcutt, CA	10/1	U4	AN7
WVAL	Saulk Rapids, MN 56379	10/0.25	U4	
WFAN	New York, NY	50	U1	AN7
KEYZ	Williston, ND 58802	5	U4	F.Pl.
WESC	Greenville, SC 29602	50/5	SH3	D*
KSKY	Dallas, TX 75235	10/0.5	U2	AN
KAPS	Mt. Vernon, WA 98273	10/1	U1	F.Pl.
CFFR	Calgary, AB	50	U4	AN7
C...	Red Lake, ON	1/0.7	U2	F.Pl.
+lp: USA 2				
670kHz				
WYLS	York, AL 36925	5	D1	D
KDLG	Dillingham, AK 99576	10	U1	NH
KWXI	Glenwood, AR 71943	5	D1	D*
KWNK	Simi Valley, CA 93063	5/1	U3	
KMVP	Commerce City, CO	5/1	U2	
WWFE	Miami, FL	50/1	U4	
KBOI	Boise, ID 83701	50	U2	AN6
WMAQ	Chicago, IL 60654	50	U1	AN7
WIEZ	Lewiston, PA17044	6	D1	D
WARO	Claremont, VA 23899	20	D3	F.Pl.
CKXB	Musgravetown, NF	10/5	U4	AN7
+lp: USA 2				
680kHz				
KBRW	Barrow, AK 99723	10	U1	NH
KNBR	San Francisco, CA 94111	50	U1	AN7
WCNN	Atlanta, GA 30303	50/10	U4	AN
WCBM	Baltimore, MD	10/5	U4	AN7
WRKO	Boston, MA 02115	50	U4	AN7
WDBC	Escanaba, MI 49829	10/1	U4	NH
KFEQ	St. Joseph, MO 64502	5	U4	AN7
KVCM	Helena, MT 59604	5	D1	F.Pl.
WPTF	Raleigh, NC	50	U2	AN7
WRGC	Sylva, NC 28779	1	U2	NH

WEZI	Memphis, TN	10/5	U2	AN7
KKYX	San Antonio, TX 78229	50/10	U2	AN7
KOMW	Omak, WA 98841	5	D1	D
WCAW	Charlestown, WV 25364	50/0.25	U4	AN
CHFA	Edmonton, AB	10	U3	NH
CJOB	Winnipeg, MB	50	U4	AN7
CKXG	Grand Falls, NF	10	U4	AN7
CFTR	Toronto, ON	50	U4	AN7

+lp: USA 10/Can 1

690kHz

WJOX	Birmingham, AL 35209	50/0.5	U2	AN
WPDQ	Jacksonville, FL	50/10	U2	AN7
KGGF	Coffeyville, KS 67337	10/5	U4	NH
WTIX	New Orleans, LA	10/5	U4	AN6
KHEY	El Paso, TX 79930	10	U4	AN
WZAP	Bristol, VA 24203	10/0.03	U1	AN
CBU	Vancouver, BC	50	U3	NH
CBF	Montreal, PQ	50	U1	NH
CBKF-1	Gravelbourg, SK	5	U2	NH

+lp: USA 17/Can 6

700kHz

KBYR	Anchorage, AK	10	U1	AN7
WWTL	Walkersville, MD	5	D3	F.Pl.
WLW	Cincinnati, OH	50	U1	AN7
KGRV	Winston, OR 97496	25/0.5	U1	NH
KSEV	Tomball, TX 77375	2.5/1	U4	AN
KFAM	Salt Lake City, UT 84101	50/1	U4	AN7
K...	Camas, WA 98607	12.5/1	U4	F.Pl.
KJMY	Newport, WA (Addr: Old Town, ID 83822)	10/1	U2	NH
CKRD	Red Deer, AB	50/25	U2	AN7
CHSJ	St John, NB	50/10	U2	AN7

+lp: USA 3

710kHz

WKRG	Mobile, AL 36606	1	U2	AN6
KUET	Black Canyon City, AZ 85324	50/4.2	U3	F.Pl.
KFIA	Carmichael, CA 95608	25/5	U4	AN
KMPC	Los Angeles, CA 90078	50/10	U2	AN6
KNUS	Denver, CO 80202	9/5	U3	AN6
WAQI	Miami, FL 33145	50	U4	AN6
KEEL	Shreveport, LA 71120	50/5	U4	AN7
WREM	Monticello, ME 04760	5	D1	D
WHB	Kansas City, MO	10/5	U4	AN7
WOR	New York, NY 10018	50	U3	AN7
KBMR	Bismarck, ND	50/4	U7	F.Pl.
KGNC	Amarillo, TX 79189	10	U4	AN6
WFNR	Christiansburg, VA 24065	5	D3	D*
KIRO	Seattle, WA 98121	50	U2	AN7
WDSM	Superior, WI	10/5	U2	AN6
CKVO	Clarenville, NF	10	U3	AN7
CHYR	Leamington, ON	10	D3	D
CJRN	Niagara Falls, ON	10/5	U4	AN6
CIPC	Port-Cartier, PQ	1	U3	NH
CKVM	Ville-Marie, PQ	10/1	U2	NH

+lp: USA 14/Can 3

720kHz

KOTZ	Kotzebue, AK 99752	10	U1	NH
WRZN	Hernando, FL 32642	10/0.25	U2	F.Pl.
WMXY	Hogansville, GA	10	D3	D
WGN	Chicago, IL	50	U1	AN7
WWDF	Richland, MS	5	D1	D
KDWN	Las Vegas, NV 89101	50	U2	AN6
KSAH	Universal City, TX	10/0.89	U4	AN7
WGNZ	Tuckahoe, VA	D3	10	F.Pl.
CHTN	Charlottetown, PEI	10/7.5	U2	AN7
	Simiutaq, Greenland	20	U1	SH

+lp: USA 1

730kHz

WSNI	Thomasville, GA 31792	5	D1	D*
WMTC	Vancleve, KY 41385	5	D3	D
WACE	Springfield, MA 01101	5	D1	D*
KURL	Billings, MT 59107	5/0.23	U1	NH
WPIT	Pittsburgh, PA 15222	5/0.02	U1	NH
WCPT	Alexandria, VA 22304	5/0.02	U1	AN
CKLG	Vancouver, BC	50	U4	AN7
CKDM	Dauphin, MB	10/5	U2	AN
CJNR	Blind River, ON	1	U2	AN
CHYR-7	Leamington, ON	1	N2	SH
CKAC	Montreal, ON	50	U3	AN7

+lp: USA 25/Can 4

740kHz

WLWI	Montgomery, AL 36198	50/0.17	U3	
KBRT	Avalon, CA	10	D3	D
KCBS	San Francisco, CA 94111	50	U4	AN6
KSSS	Colorado Springs, CO 80906	3.3/1.5	U4	AN7
KVFC	Cortez, CO 81321	1	U2	AN
WWNZ	Orlando, FL 32802	5/1	U2	AN7
KATK	Carlsbad, NM 88220	1	SH1	SH
WGSM	Huntington, NY 11747	25/1	D5	D
WPAQ	Mt. Airy, NC 27030	10/1	D1	D
KRMG	Tulsa, OK 74136	50/25	U4	AN5
KTRH	Houston, TX	50	U4	AN6
KCMC	Texarkana, TX 75501	1	U3	AN7
KWNT	Buckley, WA 98321	5/0.5	U4	F.Pl.
CBX	Edmonton, AB	50	U4	NH
CHCM	Marystown, NF	10	U2	AN7
CBL	Toronto, ON	50	U1	NH

+lp: USA 16/Can 15

750kHz

KFQD	Anchorage, AK 99507	50	U1	AN6
WSB	Atlanta, GA 30309	50	U1	AN7
KJEL	Lebanon, MO 65536	5	D1	D
KERR	Polson, MT 59860	50/1	U2	NH
KMMJ	Grand Island, NE 68801	10	L3	D*
KKNK	Carson City, NV 89701	10	U2	F.Pl.
WHEB	Portsmouth, NH 03801	1	L1	D
WYHV	Canton, NY (Addr: Canandaigua, NY 14424)	5/1	U4	
KXL	Portland, OR 97214	50/20	U4	AN7
KAMA	El Paso, TX 79902	10/1	U4	AN
KOAL	Price, UT 84501	10/6.8	U2	NH
CBGY	Bonavista, NF	10	U4	NH
CKGB	Timmins, ON	10/5	U2	AN7
CJWW	Saskatoon, SK	10	U4	AN7

+lp: USA 7

760kHz

KMTL	Sherwood, AR	10	D1	D
KFMB	San Diego, CA 92111	5	U3	AN7
KRZN	Thornton, CO	50/1	U1	NH
WEND	Brandon, FL	5/1	U4	F.Pl.
KBCB	Overland Park, KS (Addr: Independence, MO 64052)	6	D3	D
WVNE	Leicester, MA (Addr: Holden, MA 01520)	5	D1	D
WJR	Detroit, MI 48202	50	U1	AN7
WCHP	Champlain, NY 12919	25	D3	D
KSJL	San Antonio, TX 78216	50/1	U4	AN
CFLD	Burns Lake, BC	1/0.5	U1	AN
CKQR	Castlegar, BC	20/5	U2	NH

+lp: USA 3

770kHz

WVNN	Athens, AL 35611	10/0.5	U2	AN
KCHU	Valdez, AK 99686	9.75	U1	NH
KPLA	Riverbank, CA	50/1	U4	AN6
WWCN	North Fort Myers, FL	10/1	U4	AN7
WFBN	Lynn Haven, FL	5/0.5	U4	
KJCB	Lafayette, LA 70501	5/1	U4	AN7
KUOM	Minneapolis, MN 55455	5/2.2	ST2	SH
WCAL	Northfield, MN 55057	5	ST1	D
KATL	Miles City, MT 59301	10/1	U2	NH
KKOB	Albuquerque, NM 87103	50	U2	AN7
WABC	New York, NY 10019	50	U1	AN7
WTOR	Youngstown, NY	5	D3	F.Pl.
WLWL	Rockingham, NC 28379	5	D1	D
KPBC	Garland, TX	10/1	U4	AN7
WYRV	Cedar Bluff, VA 24609	5	D1	D
KULL	Seattle, WA	50/5	U4	AN7
CHQR	Calgary, AB	50	U2	AN7

+lp: USA 5

780kHz

WZZX	Lineville, AL 36266	5	D1	D
KNOM	Nome, AK 99762	10	U1	NH
KAZM	Sedona, AZ 86336	5/0.5	U2	NH
WJKB	Siesta Key, FL	4/1	U4	
WBBM	Chicago, IL 60611	50	U1	AN7
WLRM	Ridgeland, MS (Addr: Jackson, MS 39213)	5	D1	D
WJAG	Norfolk, NE 68701	1	L1	D*
KROW	Reno, NV 98510	50	U2	AN7
WSEN	Baldvinsville, NY	7.5/5	D3	F.Pl.

KRDA	Springville, UT	5/0.45	U4	F.Pl.
WABS	Arlington, VA 22207	5	D1	D
CKOK	Penticton, BC	20/2.8	U1	F.Pl.
CFDR	Dartmouth, NS	50/15	U4	AN7
+lp: USA 4				
790kHz				
WTSK	Tuscaloosa, AL 35405	5	D1	D
KCAM	Glenallen, AK 99588	5	U1	AN
KWFM	Tucson, AZ 85745	5/0.5	U3	AN7
KURM	Rogers, AR 72756	5/0.5	U2	
KOQO	Clovis, CA (Addr: Fresno, CA 93794)	5/2.5	U4	
KFLI	Eureka, CA 95501	5	D1	D
KABC	Los Angeles, CA 90016	5	U2	AN7
WLBE	Leesburg, FL 32748	5/1	U2	AN
WMRZ	South Miami, FL 33169	25	U4	AN7
WQXI	Atlanta, GA 30326	5/1	U2	AN7
KBRV	Soda Springs, ID 83276	5	D1	D*
KXXX	Colby, KS 67701	5	D1	D*
WWKY	Louisville, KY 40202	5/1	U4	AN7
WSGW	Saginaw, MI 48605	5/1	U4	AN6
KGHL	Billings, MT 59102	5	U2	AN6
WTNY	Watertown, NY 13601	1	U2	NH
KFGO	Fargo, ND 58108	5	U2	AN7
KWIL	Albany, OR 97321	1	U4	AN6
WAEB	Allentown, PA 18103	3.8/5	U4	AN6
WLKW	Providence, RI 02903	5	U2	AN7
WETB	Johnson City, TN 37601	5	D1	D*
WMC	Memphis, TN 38104	5	U2	AN7
KKBQ	Houston, TX 77046	5	U4	AN6
KFYO	Lubbock, TX 79464	5/1	U4	NH
WTAR	Norfolk, VA 23510	5	U3	AN7
KGMI	Bellingham, WA 98225	5/1	U2	AN6
KJRB	Spokane, WA 99203	5	U2	AN7
WEAQ	Eau Claire, WI 54702	5	U2	AN6
CFCW	Camrose, AB	50	U4	AN7
CFAN	Newcastle, NB	5/1	U3	AN
C...	Bedford-Sackville, NB	10	U4	F.Pl.
C...	Yarmouth, NS	25/10	U4	F.Pl.
CFNW	Port au Choix, NF	1	U3	AN7
CIGM	Sudbury, ON	50	U4	AN6
CKBH	Baie Comeau, PQ	1/0.25	U3	
+lp: USA 17				
800kHz				
KINY	Juneau, AK 99801	10/5	U1	AN
WTMR	Camden, NJ 08104	5/0.5	U4	
WSVS	Crewe, VA 23930	5/0.3	U1	AN
WKEE	Huntington, WV 25724	5	D1	D*
WDUX	Waupaca, WI 54981	5/0.5	U4	
CKOR	Penticton, BC	10/0.5	U1	AN6
VOWR	St. John's NF	10/2.5	SH1	SH
CJBQ	Belleville, ON	10	U4	AN6
CKLW	Windsor, ON	50	U4	AN7
CJAD	Montreal, PQ	50/10	U4	AN7
CHRC	Quebec, PQ	50	U3	AN7
CHAB	Moose Jaw, SK	10	U2	AN7
+lp: USA 21/Can 1				
810kHz				
WJXL	Jacksonville, AL 36265	50/0.5	U4	NH
KGO	San Francisco, CA 94111	50	U3	AN7
WEKG	Jackson, KY 41339	5	D1	D*
KCMO	Kansas City, MO	50/5	U2	AN7
WSJC	Magee, MS	50/0.5	U4	NH
KSWV	Santa Fe, NM 87501	5	D1	D*
WGY	Schenectady, NY 12309	50	U1	AN6
WQIZ	St. George, SC (Addr: Charleston, SC 29418)	5	D1	D*
KBHB	Sturgis, SD 57785	5	D1	D*
WMTS	Murfreesboro, TN 37130	5	D1	D*
KTBI	Ephrata, WA	41	D3	F.Pl.
WJJQ	Tomahawk, WI 54487	10/0.5	U3	NH
C...	Brooks, AB	5/1	U2	F.Pl.
CKJS	Winnipeg, MB	10	U3	AN
CJVA	Caraquet, NB	10/5	U2	AN
	Upernavik, Greenland	5	U1	SH
+lp: USA 15				
820kHz				
KCBF	Fairbanks, AK 99707	50	U1	AN6
WRFA	Largo, FL	50/1	U2	
WSCR	Chicago, IL 60611	5	D1	D

WQSI	Frederick, MD 21701	5/1	U4	
WIQT	Horseheads, NY 14845	5/1	U4	AN
WNYC	New York, NY 10007	50/10	U4	F.Pl.
WOSU	Columbus, OH 43210	5/1	U2	NH
WBAP	Forth Worth, TX	50	U1	AN7
WGGM	Chester, VA	10/1	U4	AN
KGNW	Seattle, WA 98121	50/5	U4	AN7
CHAM	Hamilton, ON	50/10	U4	AN
+lp: USA 4				
830kHz				
KSDP	Sand Point, AK 99661	1	U1	F.Pl.
KFLT	Tucson, AZ	50/1	U2	AN
KNCO	Grass Valley, CA 95945	5/1	U2	AN
KPLS	Orange, CA 92666	2.5/1	U2	
WRFM	Hialeah, FL (Addr: Miami, FL 33148)	1	U1	
WGBV	Lithia Springs, GA 30057	50/5	U4	F.Pl.
WADU	Norco, LA (Addr: La Place, LA 70068)	5/0.75	U4	AN
WCRN	Cherry Valley, MA	3/1	U1	F.Pl.
WCCO	Minneapolis, MN 55402	50	U1	AN7
KBOA	Kennett, MO 63857	10	D1	D
WNYC	New York, NY 10007	1	L3	NH
WWMO	Eden, NC 27288	50/1	U2	F.Pl.
KUYO	Evansville, WY	10	D1	D
CILW	Wainwright, AB	10/3.5	U4	AN7
CFJR	Brockville, ON	5/1	U2	AN7
+lp: USA 3				
840kHz				
WBHY	Mobile, AL 36633	10	D1	D
KABN	Big Lake, AK 99652	10	U1	NH
KSDP	Sand Point, AK	1	U1	AN
WHGH	Thomasville, GA 31799	10	D1	D
WHAS	Louisville, KY 40201	50	U1	AN7
KWPN	West Point, NE 68788	5	D1	D
KUIP	Gardnerville, NV	10/1	U2	F.Pl.
KVEG	North Las Vegas, NV	50/25	U3	AN
WCTG	Columbia, SC	50	D3	F.Pl.
KIKN	Pharr, TX (Addr: Harlingen, TX 78551)	5/1	U3	
WKTR	Earlysville, VA 22936	8.2	D3	
KHDL	Opportunity, WA (Addr: Spokane, WA 99214)	50/0.25	U2	F.Pl.
WTLI	Plover, WI 54467	5	D3	F.Pl.
CJXX	Grande Prairie, AB	25/10	U4	AN7
CKBX	100 Mile House, BC	1/0.5	U1	AN
CHUR	North Bay, ON	10	U2	AN
+lp: USA 9				
850kHz				
WYDE	Birmingham, AL	50/1	U4	AN
KICY	Nome, AK 99762	10	U1	NH
KOA	Denver, CO 80236	50	U1	AN6
WREF	Ridgefield Center, CT (Addr: Ridgefield, CT 06877)	5	D1	D
WRUF	Gainesville, FL 32604	5	U2	AN7
WEAT	West Palm Beach, FL 33406	5/1	U4	AN7
WPTB	Statesboro, GA 30458	1	U2	NH
WHDH	Boston, MA 02116	50	U4	AN6
WKBZ	Muskegon, MI 49443	1	U3	NH
WWJC	Duluth, MN 55808	10	D1	D
WQST	Forest, MS 39074	10	D3	D
KFUO	Clayton, MO 63105	5	L1	D
WKIX	Raleigh, NC 27605	50/5	U2	AN6
WRMR	Cleveland, OH 44114	10/5	U4	AN7
WJAC	Johnstown, PA 15907	10	U3	AN
WEEU	Reading, PA 19601	1	U2	AN6
WUTK	Knoxville, TN	50	D3	D
KEYH	Houston, TX 77098	10	D3	D*
WNIS	Norfolk, VA	50/25	U4	AN7
KMTT	Tacoma, WA 98411	10/1	U4	AN7
CKBA	Athabasca, AB	1	U1	AN7
CFVR	Abbotsford, BC	10	U4	AN6
CKVL	Montreal, PQ	50/35	U4	AN7
+lp: USA 10				
860kHz				
KVVA	Phoenix, AZ 85016	1	U2	
KTRB	Modesto, CA 95352	50/10	U2	AN7
WAEC	Atlanta, GA 30309	5/0.5	U2	
WDMG	Douglas, GA 31533	5	U2	NH
KKOW	Pittsburg, KS 66762	10/5	U2	AN7

KZTW	Troutdale, OR 97060	20/0.5	U4	F.Pl.
WTEL	Philadelphia, PA 19126	10/0.5	U3	NH
KONO	San Antonio, TX 78215	5/1	U2	AN7
KLZX	Salt Lake City, UT	10/0.2	U1	AN
WOAY	Oak Hill, WV 25901	10/0.01	U1	NH
CKKC	Nelson, BC	1	U3	F.Pl.
CFPR	Prince Rupert, BC	10/2.5	U3	NH
CHAK	Inuvik, NWT	1	U1	AN
CJBC	Toronto, ON	50	U1	AN7
CBKF-2	Saskatoon, SK	10	U4	AN

+Ip: USA 26/Can 26

870kHz

WQRX	Valley Head, AL 35989	10/4.7	D1	D
KSKO	McGrath, AK 99627	10	U1	NH
KIEV	Glendale, CA 91206	15/1	U4	AN7
WWL	New Orleans, LA 70116	50	U3	AN7
WLAM	Gorham, ME (Addr: Westbrook, ME 04092)	5	U4	AN
WKAR	East Lansing, MI 48824	10	D3	D
KPRM	Park Rapids, MN 56470	25/1	U2	NH
KOWA	East Las Vegas, NV	5/0.3	U2	An
KROL	Laughlin, NV	10/1	U4	AN
WHCU	Ithaca, NY 14851	5/1	L2	
WZMC	Colonial Heights, TN 37663	10	D3	D
KORD	Pasco, WA	10/0.25	U1	AN
WRPQ	Baraboo, WI 53913	5	U2	F.Pl.
CKIR	Invermere, BC	1/0.25	U3	AN7
CFBV	Smithers, BC	1/0.5	U3	AN7

+Ip: USA 5/Can 1

880kHz

KGHT	Sheridan, AR	50/0.2	U2	
KKMC	Gonzales, CA 93926	10/1	U4	AN
WBKZ	Jefferson, GA 30549	5	D3	D
WINU	Highland, IL 62249	2.5	U4	F.Pl.
KJJR	Whitefish, MT 59937	10/1	U2	AN6
KRVN	Lexington, NE 68850	50	U4	AN
KHAC	Tse Bonito, NM	10/0.44	U2	NH
WCBS	New York, NY 10019	50	U1	AN7
WRFD	Columbus, OH 43085	9/5	D3	D
KWIP	Dallas, OR 97338	5/1	U1	NH
KTMT	Phoenix, OR (Addr: Medford, OR 97501)	1	U1	NH
KJOJ	Conroe, TX 77305	10/1	U4	AN
KWVG	Hamby, TX	2.5/1	U4	F.Pl.
KIXI	Seattle, WA 98109	50/10	U4	AN7
WMEQ	Menomonie, WI	10/0.25	U2	
CHQT	Edmonton, AB	50	U2	AN7
CKLQ	Brandon, MB	25/10	U2	AN

+Ip: USA 2/Can 1

890 kHz

KBBI	Homer, AK 99603	10	U1	NH
KPTO	Citrus Heights, CA	50/1	U4	F.Pl.
WLS	Chicago, IL 60601	50	U1	AN7
WBMA	Dedham, MA 02026	10/1	U4	F.Pl.
WQIS	Laurel, MS 39441	10	D1	D
WYCV	Granite Falls, NC	25/6	D3	F.Pl.
WFQA	Nicholson, PA 18446	10	D3	F.Pl.
KVOZ	Laredo, TX	10/1	U1	AN
KDXU	St. George, UT 84770	10	U2	NH
CJDC	Dawson Creek, BC	10	U2	AN

+Ip: USA 5

900kHz

KZPA	Fort Yukon, AK	5	U1	F.Pl.
WSSR	Georgetown, DE 19947	10/1	U4	F.Pl.
WEAS	Savannah, GA 31414	5/0.15	U1	NH
WLSI	Pikeville, KY 41501	5/0.15	D1	D*
KTIS	Minneapolis, MN	25/0.3	U4	D
CJVI	Victoria, BC	10	U3	AN7
CKDH	Amherst, NS	1	U4	AN7
CHML	Hamilton, ON	50	U4	AN7
CHYC	Sudbury, ON	10/1	U4	AN7
CJBR	Rimouski, PQ	10	U2	NH
CJER	Saint-Jerome, PQ	1	U3	AN
CKTS	Sherbrooke, PQ	10	U4	AN7
CKVD	Val-d'Or, PQ	10/2.5	U3	AN
CKBI	Prince Albert, SK	10	U2	AN7
	Uummannaq, Greenland	5	U1	SH

+Ip: USA 51/Can 6

910kHz

KIYU	Galena, AK	5	U1	NH
KFYI	Phoenix, AZ 85003	5	U2	AN7
KLCN	Blytheville, AR 72315	5	D1	D*
KAMD	Camden, AR 71701	5/0.5	U4	NH
KECR	El Cajon, CA (Addr: San Diego, CA 92112)	5	U4	AN7
KNEW	Oakland, CA 94607	5	U2	AN7
KOXR	Oxnard, CA 93030	5/1	U4	NH
KPOF	Denver, CO (Addr: Westminster, CO 80030)	5/1	U1	AN7
WNEZ	New Britain, CT (Addr: Farmington, CT 06032)	5	U2	AN7
WFNS	Plant City, FL 33566	1	U3	NH
WFVR	Valdosta, GA 31601	5	U2	AN7
WSUI	Iowa City, IA 52242	5	U4	24h
WNDC	Baton Rouge, LA 70892	1	U3	AN6
WABI	Bangor, ME 04401	5	U2	NH
WFDF	Flint, MI 48502	5/1	U4	AN
WALT	Meridian, MS 39302	5/1	U1	AN
KBIM	Roswell, NM 88201	5/0.5	U2	NH
WRKL	New City, NY 10956	1	U	
WLAS	Jacksonville, NC 28541	5	U2	AN7
KCJB	Minot, ND 58701	5/1	U4	AN6
WBRJ	Marietta, OH 45750	5	D3	D*
KVIS	Miami, OK 74355	1	U3	AN7
WAVL	Apollo, PA 15613	5	D3	D*
WSBA	York, PA 17405	5/1	U4	AN6
WORD	Spartanburg, SC 29302	5/1	U4	AN7
KJJQ	Volga, SD (Addr: Brookings, SD 57006)	0.5/1	U3	AN
WJCW	Johnson City, TN 37602	5/1	U4	AN6
WEPG	South Pittsburg, TN 37380	5	D1	D
KRIO	McAllen, TX (Addr: Edinburg, TX 78539)	5	U4	AN6
KBLN	Sherman, TX (Addr: Ft. Worth, TX 76106)	1	U3	AN7
KALL	Salt Lake City, UT 84111	5/1	U2	AN6
WRNL	Richmond, VA 23228	5	U2	AN
KKSN	Vancouver, WA (Addr: Portland, OR 97204)	5	U4	AN7
WHSM	Hayward, WI 54843	5	D1	D*
CKDQ	Drumheller, AB	50	U4	AN7
CKLY	Lindsay, ON	10/5	U4	AN6
CHRL	Roberval, PQ	10/2.5	U2	AN7

+Ip: USA 12/Can 1

920kHz

WKYD	Andalusia, AL 36420	5/0.5	U2	NH
KSRM	Soldotna, AK 99669	5	U1	NH
KARN	Little Rock, AR 72204	5	U2	AN6
KLOC	Modesto, CA 95353	0.5/2.5	U4	
KDES	Palm Springs, CA 92263	5/1	U4	AN6
KLMR	Lamar, CO 81052	5/0.5	U1	NH
WMEL	Melbourne, FL 32936	1	U2	AN7
WAFS	Atlanta, GA 30309	5/1	U1	AN7
WBAA	West Lafayette, IN 47907	5/1	U2	NH
KYFR	Shenandoah, IA 51601	5/2.5	U2	NH
WTCW	Whitesburg, KY 41858	5	D1	D*
WBOX	Bogalusa, LA 70427	1	U2	AN
WPTX	Lexington Park, MD 20653	5/1	U4	NH
KDHL	Faribault, MN 55021	5	U4	NH
KWAD	Wadena, MN 56482	1	U2	NH
KORK	Las Vegas, NV 89109	5/0.5	U2	AN7
KQLO	Reno, NV 89503	5/1	U2	AN6
WTTM	Trenton, NJ 08618	1	U3	AN7
WGHQ	Kingston, NY 12401	5/0.5	U3	
WIRD	Lake Placid, NY 12946	5	D1	D*
WBBB	Burlington, NC 27215	5	D1	D*
KSHO	Lebanon, OR (Addr: Albany, OR 97321)	1	U3	AN6
WHJJ	Providence, RI 02914	5	U2	AN7
WYMB	Manning, SC 29102	5/1	U2	F.Pl.
WAZQ	Orangeburg, SC 29115	5	D1	D
KKLS	Rapid City, SD 57709	5	D3	D*
KYST	Texas City, TX 77054	5/1	U4	AN
KVEL	Vernal, UT 84078	5/1	U2	
KQEU	Olympia, WA 98507	5/0.5	U4	AN7
KXLY	Spokane, WA 99201	5	U1	AN6
WMMN	Fairmont, WV 26554	5	U2	NH
WOKY	Milwaukee, WI 53220	5/1	U4	AN7
CKCQ	Quesnel, BC	10/1	U2	AN6
CFRY	Portage la Prairie, MB	25/10	U4	AN6

CJCJ	Woodstock, NB	10/1	U4	AN7
CJCH	Halifax, NS	25	U4	AN7
CKCY	Sault Ste. Marie, ON	10/5	U4	
CKNX	Wingham, ON	10/1	U4	AN
CFLS	Lévis, PQ	10	U2	AN7
+Ip: USA 19/Can 4				
930kHz				
WYNI	Monroeville, AL 36461	5	D1	D
WJBY	Rainbow City, AL	5/0.5	U2	
	(Addr: Gadsden, AL 35902)			
KTKN	Ketchikan, AK 99901	5/1	U1	NH
KNSA	Unalakleet, AK 99684	2.5	U1	F.Pl.
KAFF	Flagstaff, AZ 86002	5	D1	D*
KKHJ	Los Angeles, CA 90034	5	U2	AN6
KIUP	Durango, CO 81302	5/1	U2	AN6
WNZS	Jacksonville, FL 32201	5	U2	AN7
WKXY	Sarasota, FL 33578	5/2.5	U4	AN7
WMGR	Bainbridge, GA 31717	5/0.5	U2	NH
KSEI	Pocatello, ID 83204	5	U2	AN6
WTAD	Quincy, IL 62301	5/1	U2	NH
WAUR	Sandwich, IL	4.2/2.2	U4	AN
	(Addr: Aurora, IL 60506)			
WKCT	Bowling Green, KY 42102	5/0.5	U2	NH
WFMD	Frederick, MD 21701	5/2.5	U4	AN
WBCK	Battle Creek, MI 49015	5/1	U4	AN
WSLI	Jackson, MS 39205	5	U2	NH
KWOC	Poplar Bluff, MO 63901	5/0.5	U2	NH
KLCY	East Missoula, MT 59807	5/1	U2	
KOGA	Ogallala, NE 69153	2.5/0.5	U4	NH
WZNN	Rochester, NH 03867	5	U2	NH
WPAT	Paterson, NJ	5	U4	AN7
	(Addr: Clifton, NJ 07013)			
WBEN	Buffalo, NY 14207	5	U2	AN7
WYFQ	Charlotte, NC 28234	5/1	U2	AN6
WRRF	Washington, NC 27889	5/1	U2	AN7
WEOL	Elyria, OH 44036	1	U4	AN6
WKY	Oklahoma City, OK 73113	5	U2	AN6
KAGI	Grants Pass, OR 97526	5/1	U2	AN7
WCNR	Bloomsburg, PA 17815	1/0.5	U2	
KSDN	Aberdeen, SD 57401	5/1	U4	AN
WSEV	Sevierville, TN 37862	5	D1	D
KLUP	San Antonio, TX 78218	5/1	U2	AN7
WLLL	Lynchburg, VA 24506	5	D1	D*
KZTA	Yakima, WA 98907	5/0.5	U2	
WTKZ	Huntington, WV 25713	5/1	U2	AN7
WLBL	Auburndale, WI	5	D1	D
	(Addr: Madison, WI 53713)			
KROE	Sheridan, WY 82801	5	D1	D*
CJCA	Edmonton, AB	50	U4	AN7
CFBC	St. John, NB	50	U4	AN7
CJYQ	St. John's, NF	50	U4	AN7
CKNS	Espanola, ON	10	U2	AN6
+Ip: USA 14				
940kHz				
KNST	Tucson, AZ 85702	5/1	U4	AN7
KFRE	Fresno, CA 93728	50	U4	AN7
WINZ	Miami, FL 33055	50/10	U4	AN7
WMAZ	Macon, GA	50/10	U2	AN
WMIX	Mount Vernon, IL 62864	5/1.5	U4	AN7
KIOA	Des Moines, IA	10/5	U4	AN7
WYLD	New Orleans, LA 70119	10/0.5	U4	AN6
WIDG	St. Ignace, MI 49781	5	D1	D*
WCPC	Houston, MS 38851	50/0.25	U4	NH
KVSH	Valentine, NE 69201	5	D1	D*
WKYK	Burnsville, NC 28714	5/0.25	U2	NH
WLNC	Laurinburg, NC 28352	5/1	U2	F.Pl.
KGRL	Bend, OR 97708	10/0.5	U3	NH
WECO	Wartburg, TN 37887	5	D1	D*
KIXZ	Amarillo, TX 79106	5/1	U4	AN7
KBRE	Cedar City, UT 84720	10	D1	D*
WNRG	Grundy, VA 24614	5	D1	D*
WKGM	Smithfield, VA 23430	10/3.1	U2	
CJIB	Vernon, BC	10	U2	AN7
CBM	Montreal, PQ	50	U3	NH
CJGX	Yorkton, SK	10	U2	AN6
+Ip: USA 14/Can 3				
950kHz				
WSYA	Montgomery, AL 36101	1	U2	AN7
KSWD	Seward, AK 99664	1	U1	AN
KXJK	Forrest City, AR 72335	5	D1	D*
KFSA	Fort Smith, AR 72902	1/0.5	U4	NH
KAHI	Auburn, CA 95603	5	U4	AN
KYGO	Denver, CO 80224	5	U3	AN7
WOMX	Orlando, FL	5	U2	AN7
	(Addr: Winter Park, FL 32790)			
WGTA	Summerville, GA 30747	5	D1	D*
WGOV	Valdosta, GA 31601	5/1	U2	
KOZE	Lewiston, ID 83501	5/1	U4	AN
WJPC	Chicago, IL 60605	1/5	U2	AN7
WXLW	Indianapolis, IN 46222	5	D3	D*
KOEL	Oelwein, IA 50662	5/0.5	U4	NH
WROL	Boston, MA 02116	5	D1	D*
WWJ	Detroit, MI	5	U2	AN7
	(Addr: Southfield, MI 48086)			
KJJO	St. Louis Park, MN	1	U4	AN7
	(Addr: Eden Prairie, MN 55343)			
WBKH	Hattiesburg, MS 39401	5	D1	D*
KLIK	Jefferson City, MO 65102	5/0.5	U2	NH
KMTX	Helena, MT 59624	5	U2	NH
KNFT	Bayard, NM	5	D1	D*
	(Addr: Silver City, NM 88061)			
WBBF	Rochester, NY 14604	1	U4	AN7
WIBX	Utica, NY 13503	5	U3	AN
WPEN	Philadelphia, PA	5	U2	AN7
	(Addr: Bala Cynwyd, PA 19004)			
WSPA	Spartanburg, SC 29304	5	U2	AN7
KWAT	Watertown, SD 57201	1	U2	NH
KPRC	Houston, TX 77074	5	U2	AN7
KXTQ	Lubbock, TX 79408	5/0.5	U4	AN6
WJDK	Richmond, VA 23225	5	D1	D*
KJR	Seattle, WA 98124	5	U2	AN7
WQBE	Charleston, WV 25323	5/1	U2	AN7
KMER	Kemmerer, WY 83101	5	D1	D*
CFAM	Altona, MB	10	U4	AN7
CKNB	Campbellton, NB	10/1	U4	AN6
CHER	Sydney, NS	10	U3	AN6
CKBB	Barrie, ON	10	U4	AN6
+Ip: USA 14				
960kHz				
WERC	Birmingham, AL 35202	5	U2	AN6
WLPR	Pritchard, AL 36612	2.5/1	U2	
KOOL	Phoenix, AZ 85016	5	U2	AN7
KCGS	Marshall, AR 72650	5	D1	D
KZXY	Apple Valley, CA	5	D1	D*
	(Addr: Victorville 92392)			
KABL	San Francisco, CA 94111	5	U3	AN7
WELI	New Haven, CT 06501	5	U2	AN6
WGRO	Lake City, FL 32055	0.5/1	U2	NH
WJCM	Sebring, FL 33870	5/1	U3	NH
WJYZ	Albany, GA 31702	5	D3	D*
WRFC	Athens, GA	5/2.5	U2	AN6
	(Addr: Atlanta, GA 30309)			
WSBT	South Bend, IN 46601	5	U4	AN
KMA	Shenandoah, IA 51601	5	U2	NH
WPRT	Prestonsburg, KY 41653	5	D1	D*
WLVW	Salisbury, MD 21801	5	U4	AN7
WFGL	Fitchburg, MA 01420	1	U4	NH
WHAK	Rogers City, MI 49779	5	D1	D*
KLTF	Little Falls, MN 56345	5	D1	D*
KZIM	Cape Girardeau, MO 63701	5/0.5	U2	NH
KFLN	Baker, MT 59313	5	D1	D*
KNDN	Farmington, NM 87401	5	D1	D*
WEAV	Plattsburgh, NY	5	U4	NH
	(Addr: Peru, NY 12972)			
WRNS	Kinston, NC 28501	5/1	U2	NH
KGWA	Enid, OK 73702	1	U3	NH
KLAD	Klamath Falls, OR 97601	5	U2	NH
WHYL	Carlisle, PA 17013	5	D3	D*
WATS	Sayre, PA 18840	5	D	D*
KGKL	San Angelo, TX 76902	5/1	U2	AN
KFMY	Provo, UT 84601	5/1	U2	AN7
WFIR	Roanoke, VA 24002	5	U2	AN
KALE	Richland, WA	5/1	U2	AN6
	(Addr: Tri-Cities, WA 99302)			
WTCH	Shawano, WI 54166	1	U2	NH
CFAC	Calgary, AB	50	U2	AN7
CHNS	Halifax, NS	10	U2	AN7
CIAM	Cambridge, ON	1	U4	AN
CFFX	Kingston, ON	10/5	U4	AN6
+Ip: USA 15				

970kHz

WERH	Hamilton, AL 35570	5	D1	D*
WTBF	Troy, AL 36081	5/0.5	U2	NH
KIAK	Fairbanks, AK 99707	5	U1	AN7
KVWM	Show Low, AZ 85901	5	D1	D*
KAFY	Bakersfield, CA 93302	1/5	U4	AN7
KVIM	Coachella, CA (Addr: Indio, CA 92201)	5/1	U4	AN7
KBEE	Modesto, CA 95353	1	U4	AN
WFLA	Tampa, FL 33602	5	U4	AN7
WNIV	Atlanta, GA 30309	5/0.04	U3	NH
WVOP	Vidalia, GA 30474	5	D1	D*
WFSR	Harlan, KY 40831	5	D1	F.Pl.
WAVG	Louisville, KY 40233	5	U4	AN6
KSYL	Alexandria, LA 71306	1	U4	AN
WYNZ	Portland, ME 04104	5	U2	AN
WMVN	Ishpeming, MI 49849	5	D1	D*
WKHM	Jackson, MI 49201	1	U4	AN7
KQAQ	Austin, MN 55912	5/0.5	U4	AN
KCTR	Billings, MT 59101	5	U2	AN7
KJLT	North Platte, NE 69103	5	D1	D*
KNUU	Paradise, NV (Addr: Las Vegas, NV 89119)	5/0.5	U4	
WWDJ	Hackensack, NJ 07602	5	U4	NH
WEBR	Buffalo, NY 14202	5	U3	AN
WWIT	Canton, NC 28716	5	D1	D*
WDAY	Fargo, ND 58108	5	U2	AN7
WFUN	Ashtabula, OH 44004	5/1	U4	NH
KCFO	Tulsa, OK 74107	2.5/1	U4	AN6
KBBT	Portland, OR 97201	5	U2	AN7
WWSW	Pittsburgh, PA 15212	5	U4	AN6
WJMX	Florence, SC 29501	5/3	U2	NH
KIXL	Del Valle, TX (Addr: Austin, TX 78703)	1	U4	
WANV	Waynesboro, VA 22980	5/1	U4	NH
KTRW	Spokane, WA 99203	5/1	U2	AN7
WHA	Madison, WI 53713	5	D1	D*
CJYR	Edson, AB	10	U3	AN
CKCH	Hull, PQ	10/5	U4	AN6

+lp: USA 21/Can 1

980kHz

KZXX	Kenai, AK 99611	1	U1	AN7
KCAB	Dardanelle, AR 72834	5	D1	D
KINS	Eureka, CA 95501	5/0.5	U2	NH
KFWB	Los Angeles, CA 90028	5	U1	AN6
WWRC	Washington, DC (Addr: Silver Spring, MD 20910)	5	U4	AN7
WLUS	Gainesville, FL 32602	5	D1	D*
WFXP	Pensacola, FL 32575	2.5/1	U4	
WWNN	Pompano Beach, FL 33060	5/1	U4	AN7
WPGA	Perry, GA 31069	5	D1	D*
KUPI	Ammon, ID (Addr: Idaho Falls 83402)	5/1	U4	AN
WITY	Danville, IL 61832	1	U4	NH
KOKA	Shreveport, LA 71161	5	D1	D*
WCAP	Lowell, MA 01852	5/0.5	U4	AN
KMZZ	Richfield, MN (Addr: Eagan, MN 55122)	5	U3	AN7
WAPF	McComb, MS 39648	5	D1	D*
KMBZ	Kansas City, MO (Addr: Shawnee Mission, KS 66205)	5	U2	AN7
KVLV	Fallon, NV 89406	5	D1	D*
KICA	Clovis, NM 88101	1	U2	NH
KMIN	Grants, NM 87020	1	U1	AN
WTRY	Troy, NY (Addr: Latham, NY 12110)	5	U2	AN7
WAAV	Wilmington, NC 28401	5	U2	AN
WONE	Dayton, OH 45402	5	U2	AN7
WILK	Wilkes-Barre, PA 18701	5/1	U2	AN
KDSJ	Deadwood, SD 57732	5/1	U2	NH
WYFN	Nashville, TN 37203	5	U2	AN6
KSVC	Richfield, UT 84701	5/1	U2	
WXBQ	Bristol, VA 24201	5/1	U2	AN7
KUTI	Selah, WA (Addr: Yakima, WA 98907)	5/0.5	U2	AN
WCUB	Manitowoc, WI 54221	5	U4	AN
CKNW	New Westminster, BC	50	U3	AN7
CFPL	London, ON	10/5	U4	AN7
CHEX	Peterborough, ON	10/5	U4	AN7
CBV	Quebec, PQ	50	U3	NH
CKRM	Regina, SA	10/5	U4	AN7

+lp: USA 27

990kHz

KTKT	Tucson, AZ	10/1	U4	AN7
KKIS	Pittsburg, CA (Addr: Concord, CA 94520)	5	U4	AN7
KKJZ	Santa Barbara, CA 93101	5/0.5	U4	NH
KRKS	Denver, CO 80226	5/0.4	U2	
WFBA	Miami, FL 33131	5	U4	F.Pl.
WHOO	Orlando, FL	50/5	U4	AN7
KSVP	Artesia, NM 88210	1	U1	NH
WRMM	Rochester, NY 14604	5/2.5	U4	AN
WEEB	Southern Pines, NC 28387	10	D1	D
WZZD	Philadelphia, PA 19128	50/10	U4	AN7
WVSC	Somerset, PA 15501	10	D3	D
WALE	Providence, RI	50	D3	D
WIVK	Knoxville, TN	10	U2	AN7
KWAM	Memphis, TN	10/0.45	U4	AN
KZZB	Beaumont, TX 77702	1	U3	AN
KNIN	Wichita Falls, TX 76307	10/1	U4	AN6
WNRV	Narrows, VA 24124	5	D1	D
CBW	Winnipeg, MB	50/46	U1	NH
CBY	Corner Brook, NF	10	U1	NH
CKIS	Montreal, PQ	50	U4	AN7

+lp: USA 25/Can 21

1000kHz

WTAK	Huntsville, AL 35806	5	D1	D
WZTN	Montgomery, AL 36101	5	D1	D
KFLG	Bullhead City, AZ 86442	5	D1	D
KIDN	Hayden, CO 81689	10/5	D1	D
KIDH	Eagle, ID 83616	10	D3	F.Pl.
WLUP	Chicago, IL	50	U4	AN6
WXTN	Lexington, MS 39095	5	D1	D
KKIM	Albuquerque, NM 87112	10	D1	D*
WLNL	Horseheads, NY 14845	5	D1	D
KTOK	Oklahoma City, OK 73101	5	U4	AN7
WKYB	Hemingway, SC 29554	10/5	D1	D
KXRB	Sioux Falls, SD 57106	10	D3	D*
WMUF	Paris, TN 38242	5/2.5	D3	D
KOMO	Seattle, WA 98109	50	U2	AN7
CKBW	Bridgewater, NS	10	U2	AN
CFLP	Rimouski, PQ	10	U4	AN7

+lp: USA 21

1010kHz

WPYK	Dora, AL 35062	5	D1	D
KXEG	Tucson, AZ 85710	7.5/1	U4	AN7
KBIS	Little Rock, AR 72202	10/5	U2	AN6
KCHJ	Delano, CA 93216	5/1	U4	NH
KPSL	Thousand Palms, CA (Addr: Palm Springs, CA 92262)	3.6/0.4	U4	F.Pl.
KIQI	San Francisco, CA 94111	10/0.5	U3	
KKGZ	Brush, CO 80723	10/5	D1	D
WCNU	Crestview, FL 32536	10	D1	D
WXTL	Jacksonville, FL	10	D3	D
WXTL	Baldwin, FL	25/10	U4	F.Pl.
WQYK	Seffner, FL	50/5	U4	AN7
WGUN	Atlanta, GA	50/0.07	U1	AN7
WMOX	Meridian, MS 39301	10/1	U4	NH
KXEN	St. Louis, MO 63166	50/0.5	U4	D
WINS	New York, NY	50	U3	AN7
WFGW	Black Mountain, NC 28711	50/19/0.5	U3	D
KZRC	Milwaukie, OR (Addr: Lake Oswego, OR 97034)	8/0.5	U4	
WHIN	Gallatin, TN 37066	5	D1	D
KDJW	Amarillo, TX 79117	5/0.5	U3	AN6
KLAT	Houston, TX 77008	5	U4	AN7
KBBW	Waco, TX 76703	10/2.5	U4	NH
KTLE	Tooele, UT 84074	50/2.5	D1	D
WPMH	Portsmouth, VA 23705	5	D3	D
CBR	Calgary, AB	50	U4	NH
CKXD	Gander, NF	1	U2	AN7
CFRB	Toronto, ON	50	U4	AN7

+lp: USA 20/Can 7

1020kHz

KFFR	Eagle River, AK	10	U2	AN
KTNQ	Hollywood, CA	50	U4	AN6
WART	Port Orange, FL 32019	10	D3	F.Pl.
WJEP	Ochlocknee, GA	10	D1	D
WCKW	Garyville, LA 70051	50/0.35	U4	F.Pl.

WNTK	Newport, NH	10	D1	D
KCKN	Roswell, NM 88202	50	U4	NH
KKWC	Watford City, ND	5/1	U2	F.Pl.
KDKA	Pittsburgh, PA 15222	50	U1	AN6
WRIX	Anderson, SC 29624	10/3	D1	D
KYXE	Selah, WA 98942	5/0.5	U5	AN7
KWIQ	Moses Lake, WA 98837	5/0.5	U5	AN
CKVH	High Prairie, AB	1/0.4	U1	AN7

+lp: USA 2

1030kHz

KFAY	Fayetteville, AR 72701	10/1	U2	AN
KEVT	Cortaro, AZ	10/1	U4	F.Pl.
KKSA	Folsom, CA 95630	50/1	U4	AN7
KJDJ	San Luis Obispo, CA 93401	2.5/0.8	U4	AN
WCAG	Oviedo, FL 32765	25/0.5	U4	F.Pl.
KBUF	Holcomb, KS	25/1	U4	AN
WBZ	Boston, MA 02134	50	U1	AN7
WNTL	Indian Head, MD	50	D3	D
WUFL	Sterling Heights, MI	5	D3	
WMIN	Maplewood, MN 55109	50/1	U5	
WNOW	Mint Hill, NC	10	D3	D
WFTK	Wake Forest, NC	35/2.5	D3	
KDUN	Reedsport, OR	10/0.63	U1	AN7
WXSS	Memphis, TN 38104	50/1	U2	AN
WJKZ	White Bluff, TN 37187	10/1	U4	NH
KCTA	Corpus Christi, TX 78403	50	L1	D
KMAS	Shelton, WA 98584	10/1	U1	NH
WTGR	Point Pleasant, WV	10	D3	F.Pl.
KTWO	Casper, WY 82601	50	U2	AN6

+lp: USA 8

1040kHz

KURS	San Diego, CA	9.5/4.5	U2	
KCBR	Monument, CO 80132	5/2.5	D1	D
WYFX	Boynton Beach, FL 33435	10/1	U4	
WHO	Des Moines, IA 50308	50	U1	AN7
WJHR	Flemington, NJ 08822	4.7/1	U4	F.Pl.
WSGH	Lewisville, NC 27114	9.1/0.25	D3	D
WJTB	North Ridgeville, OH 44039	5/2.5	D1	D
WSKE	Everett, PA 15337	10/4	D1	D
WQBB	Powell, TN	10/3	D3	D
CKST	Langley, BC	50/25	U4	AN
C...	Dolbeau, PQ	10/5	U2	F.Pl.
CFZZ	St-Jean, PQ	10/2.5	U4	AN

+lp: USA 11

1050kHz

KTBA	Tuba City, AZ 86045	5	D1	D
KNOB	Frazier Park, CA	10	D3	D
KOFY	San Mateo, CA	50/10	U4	AN
WROS	Jacksonville, FL 32205	5	D1	D
WFAM	Augusta, GA 30901	5	D1	D*
WPZA	Ann Arbor, MI 48108	10/0.5	D3	D*
KLOH	Pipestone, MN 56164	9/0.4	U4	D
KMTA	Miles City, MT	10/0.13	U1	
WEVD	New York, NY	50	U3	AN7
KORE	Springfield, OR 97477	5	D1	D
WJXY	Conway, SC 29526	5	D1	D*
WCMS	Norfolk, VA (Addr: Virginia Beach, VA 23464)	5/0.35	U3	NH
KEYF	Dishman, WA (Addr: Spokane, WA 99203)	5/0.36	U1	AN
KBLE	Seattle, WA 98122	5/0.45	U1	NH
WADC	Parkersburg, WV 26101	5	D1	D*
CFGP	Grande Prairie, AB	10	U3	AN7
CICF	Vernon, BC	10/1	U2	NH
CKSB	Winnipeg, MB	10	U2	NH
CFYN	Sault Ste. Marie, ON	10/5	U2	
CHUM	Toronto, ON	50	U4	AN7
CJNB	North Battleford, SK	10	U2	AN

+lp: USA 44

1060kHz

KUKQ	Tempe, AZ (Addr: Phoenix, AZ 85016)	5/0.5	U2	AN7
KPAY	Chico, CA 95926	10	U2	AN6
KLMO	Longmont, CO 80501	10/0.25	U1	NH
WAMT	Titusville, FL 32781	10/5	U4	AN6
WKNG	Tallapoosa, GA 30176	5	D1	D*
KBGN	Caldwell, ID 83605	10	D1	D*
WNOE	New Orleans, LA 70130	50/5	U4	AN7
WBIV	Natick, MA	25/5	U4	NH
WHFB	Benton Harbor, MI 49022	5/2.5	D1	D*
KKVV	Las Vegas, NV 89109	5	D1	F.Pl.
WRCW	Canton, OH 44711	5	D3	D*
KYW	Philadelphia, PA 19106	50	U3	AN7
KGFX	Pierre, SD 57501	10/1	U4	AN
KFNA	El Paso, TX	10	D1	D*
KHYM	Gilmer, TX 75644	10	D1	D*
KIJN	Farwell, TX 79325	5	D1	D
KKDS	Salt Lake City, UT 84107	10/1	U4	AN
CFCN	Calgary, AB	50	U2	AN7
CJRP	Quebec, PQ	50/10	U4	AN7

+lp: USA 20

1070kHz

WAPI	Birmingham, AL	50/5	U2	AN
KNX	Los Angeles, CA 90028	50	U1	AN7
WANM	Tallahassee, FL 32302	10	D1	D
WIBC	Indianapolis, IN 46260	50/10	U4	AN7
KFDI	Wichita, KS 67201	10/1	U2	AN6
KMOM	Monticello, MN 55362	10/2.5	U4	NH
KHMO	Hannibal, MO 63401	5/1	U4	NH
KATQ	Plentywood, MT 59254	5	D1	D
WKDR	Plattsburgh, NY 12901	5	D1	D
WNCT	Greenville, NC 27834	10	U4	AN6
WKOK	Sunbury, PA 17801	10/1	U2	NH
WHYZ	Greenville, SC 29608	50/1.5	U3	AN
WFLI	Lookout Mountain, TN	50/2.5	U4	AN7
WDIA	Memphis, TN 38103	50/5	U4	AN6
KDSI	Alice, TX 78332	1	U2	AN
KRBE	Houston, TX 77042	10/5	U4	AN7
WINA	Charlottesville, VA 22902	5	U2	NH
WIWS	Beckley, WV	10	D1	D*
WTSO	Madison, WI	10/5	U4	AN6
CFAX	Victoria, BC	5	U3	AN7
CBA	Moncton, NB	50	U1	NH
CHOK	Sarnia, ON	10	U4	AN7

+lp: USA 6/Can 1

1080kHz

WKAC	Athens, AL 35611	5/1	D1	D
KKSD	Anchorage, AK 99507	10	U1	AN7
KSCO	Santa Cruz, CA 95061	10/5	U2	AN7
WTIC	Hartford, CT 06103	50	U2	AN7
WVCG	Coral Gables, FL 33134	50/20	U4	AN7
WFIV	Kissimmee, FL	10	D3	D
WFTD	Marietta, GA	10	D3	D
KVNI	Coeur d'Alene, ID 83814	10/1	U2	AN7
WDJX	Louisville, KY 40202	10/1	U4	AN6
WSSI	Carthage, MS 39051	5	D3	D
WKGX	Lenoir, NC 28645	5/2.5	D1	D*
WNCR	St. Pauls, NC 28384	50/25	D3	D
KWJJ	Portland, OR 97205	50/10	U3	AN6
WDSY	Pittsburgh, PA 15222	50/25	D3	D*
KRLD	Dallas, TX 75247	50	U2	AN6
KRPX	Price, UT 84501	10/5	U4	
WVKV	Hurricane, WV 25526	5	D3	D*
CKSA	Lloydminster, AB	50/10	U2	AN6

+lp: USA 17/Can 4

1090kHz

WRBK	Flomaton, AL 36441	5	D1	D
KAAY	Little Rock, AR 72209	50	U2	AN7
KNCR	Fortuna, CA 95540	10	D1	D
KJRC	S. Lake Tahoe, CA	5	D1	F.Pl.
KYBG	Aurora, CO	50/1	U4	AN7
WKII	Port Charlotte, FL	4.8	D1	D
WSLG	Gonzales, LA 70737	10	D3	D
WBAL	Baltimore, MD 21211	50	U2	AN7
WILD	Boston, MA 02219	5/1	D1	D
KBOZ	Bozeman, MT 59715	5	U2	AN7
KRZY	Albuquerque, NM 87110	50/2.5	U4	F.Pl.
WKSP	Kingstree, SC 29556	5	D1	D*
WGOC	Kingsport, TN 37662	10/1	D1	D
KKYN	Plainview, TX 79072	5/0.5	U4	AN
KING	Seattle, WA 98109	50	U4	AN6
WAQE	Rice Lake, WI 54868	5	D3	D
CHEC	Lethbridge, AB	10/5	U4	AN7
CKKW	Kitchener, ON	10	U4	AN7

+lp: USA 23/Can 8

1100kHz

KCCF	Cave Creek, AZ 85331	25/1	U4	F.Pl.
KZPM	Bakersfield, CA 93302	10/1	U4	F.Pl.
KFAX	San Francisco, CA	50	U4	AN6
KNZZ	Grand Junction, CO 81502	50/10	U6	NH

WCGA	Woodbine, GA	10	D1	F.Pl.
KKLL	Webb City, MO 64870	5	D1	D
WHLI	Hempstead, NY 11550	10	D3	D
WWWE	Cleveland, OH 44114	50	U3	AN6
KLWJ	Umatilla, OR 97682	10/1	U2	F.Pl.
KDRY	Alamo Heights, TX 78265	11/1	U2	NH
WYKR	Wells River, VT 05081	5/2	D1	D
+lp: USA 1				
1110kHz				
WBCA	Bay Minette, AL 36507	10/2.5	D5	D
KGFL	Clinton, AR 72031	5/1	D1	D
KGPL	Dermott, AR 71638	10	D3	D
KRLA	Pasadena, CA	50/20	U4	AN6
KRCX	Roseville, CA 95678	5/1	U4	NH
WTIS	Tampa, FL	10	D3	D*
WMBI	Chicago, IL 60610	5	D1	D
WUHN	Pittsfield, MA 01202	5	D3	D
WJML	Petoskey, MI 49770	10	D3	D
KFAB	Omaha, NE 68132	50	U2	AN6
WNNW	Salem, NH 03079	5	D3	D
KYKK	Humble City, NM (Addr: Hobbs, NM 88240)	5	D1	D*
WBT	Charlotte, NC 28208	50	U2	AN7
KEOR	Atoka, OK 74525	5	D3	D
KBND	Bend, OR 97701	25/5	U2	AN7
WWRX	East Providence, RI 02914	5	D3	D
WZAM	Norfolk, VA	50	D3	D
CKTY	Sarnia, ON	10/1	U4	AN7
+lp: USA 22/Can 4				
1120kHz				
KSJI	San Martin, CA 95046	10/1	U4	F.Pl.
WUST	Washington, DC 20001	20/2.5	D1	D
WBNM	Gordon, GA 31031	10/2.5	D1	D*
WADN	Concord, MA 01742	5/1	U4	
WTWZ	Clinton, MS 39056	5/2.5	D1	D
KMOX	St. Louis, MO 63102	50	U1	AN7
KPNW	Eugene, OR 97440	50	U3	AN7
KANN	Roy, UT	10/1	U2	
+lp: USA 13				
1130kHz				
KRDU	Dinuba, CA 93618	5/6	U4	AN7
KSDO	San Diego, CA	50/10	U4	AN6
KHTH	Dillon, CO 80498	5	D1	D
WWBF	Bartow, FL 33830	2.5/0.5	U2	NH
WLBA	Gainesville, GA	10	D1	D*
WMGA	Moultrie, GA 31768	10/0.25	U6	NH
WSJP	Murray, KY 42071	2.5/0.5	U4	AN
KWKH	Shreveport, LA 71130	50	U2	AN7
WCXI	Detroit, MI	50/10	U4	AN7
KFAN	Minneapolis, MN	50/30	U4	AN6
KOFK	Milan, NM	5/2.5	D1	D
WNEW	New York, NY 10017	50	U2	AN7
WEDE	Eden, NC 27288	5/0.25	U1	
KBMR	Lincoln, ND	50/5	U4	
WASP	Brownsville, PA 15417	5/1	D6	D*
KTMR	Edna, TX 77957	10	D3	D
WISN	Milwaukee, WI 53201	50/10	U4	AN7
CKWX	Vancouver, BC	50	U4	AN7
+lp: USA 21/Can 1				
1140kHz				
WBXR	Hazel Green, AL	15.6	D3	D
KSLD	Soldotna, AK 99669	10	U1	AN7
KCMJ	Palm Springs, CA 92263	10/2.5	U4	AN
KRAK	Sacramento, CA 95860	50	U4	AN6
WQBA	Miami, FL 33145	50/10	U4	AN6
KGEM	Boise, ID 83705	10	U2	AN6
WVEL	Pekin, IL 61554	5/1	D1	D*
WKWM	Kentwood, MI 49508	5	D3	D
WSAO	Senatobia, MS 38668	5	D1	D
KLUC	Las Vegas, NV	10/2.5	U2	AN6
WNHA	Concord, NH	10	D3	F.Pl.
KSOO	Sioux Falls, SD 57105	10/5	U2	AN
KSSQ	Conroe, TX 77301	5	D1	D
WRVA	Richmond, VA 23212	50	U3	AN7
KZMQ	Greybull, WY 82426	10	D1	D
CFXX	Calgary, AB	50	U2	AN6
CBI	Sydney, NS	10	U2	NH
CJTR	Trois-Rivières, PQ	10	U4	AN7
+lp: USA 14/Can 4				
1150kHz				
WSPZ	Tuscaloosa, AL 35401	5/1	U2	AN7
KCKY	Coolidge, AZ 85228	5/1	U4	AN7
KLRG	Little Rock, AR 72203	5/1	U2	AN7
KIIS	Los Angeles, CA 90028	5	U2	AN7
KBAI	Morro Bay, CA 93442	5	U2	NH
KMXN	Santa Rosa, CA 95406	5/1	U4	AN6
KCUV	Englewood, CO (Addr: Denver, CO 80203)	5/1	U2	AN
WDEL	Wilmington, DE 19803	5	U4	AN7
WNDB	Daytona Beach, FL 32015	1	U2	AN7
WTMP	Tampa, FL 33601	5/2.5	U4	AN7
WJEM	Valdosta, GA 31601	5	D3	D*
WGGH	Marion, IL 62959	5	D3	D*
KWKY	Des Moines, IA 50303	1	U4	NH
KSAL	Salina, KS 67402	5	U2	AN7
WJBO	Baton Rouge, LA 70821	5	U3	AN
WMEX	Boston, MA 02116	5	U4	AN7
KSEN	Shelby, MT 59474	5	U4	NH
KDEF	Albuquerque, NM 87109	5/0.5	U2	AN7
WRUN	Utica, NY (Addr: Oriskany, NY 13424)	5/1	U4	AN
WGBR	Goldsboro, NC 27530	5/1	U4	AN
WIMA	Lima, OH 45802	1	U2	AN7
KNED	McAlester, OK 74502	1	U2	AN
KAGO	Klamath Falls, OR 97601	5/1	U2	AN
KKEY	Portland, OR 97228	5	D3	D*
WHUN	Huntingdon, PA 16652	5	D1	D*
WJZS	Orangeburg, SC 29115	5/0.5	U4	NH
KIMM	Rapid City, SD 57701	5/0.5	U2	
WGOW	Chattanooga, TN 37401	5/1	U2	AN7
WCRK	Morristown, TN 37814	5/0.5	U2	NH
KEZX	Seattle, WA 98103	5	U2	AN6
WELC	Welch, WV 24801	5	D1	D*
WAYY	Chippewa Falls (Addr: Eau Claire, WI 54702)	5	D1	D*
WHBY	Kimberly, WI (Addr: Neenah, WI 54956)	5	U3	AN7
CKIQ	Kelowna, BC	10	U2	AN7
CKX	Brandon, MB	50/10	U4	AN
CKMO	Hamilton, ON	50	U4	AN7
CJRC	Ottawa, ON	50/5	U4	AN
CHGM	Gaspé, PQ	5	U2	NH
+lp: USA 31/Can 7				
1160kHz				
WKWA	Mobile, AL 36689	10/0.8	U2	F.Pl.
WELX	Callahan, FL 32011	50/0.5	U4	F.Pl.
WJJD	Chicago, IL 60601	50/5	U4	AN7
WMLX	Florence, KY (Addr: Cincinnati, OH 45202)	5/1	U4	F.Pl.
WKCM	Hawesville, KY 42348	2.5/1	U2	AN
WSKW	Skowhegan, ME 04976	10/1	U1	AN
WPVG	Funkstown, MD 21713	1	U1	
WACY	Fenton, MI 48430	1	U1	NH
KNKQ	Virginia City, NV	5	D1	F.Pl.
WOBM	Lakewood, NJ 08701 (Addr: Toms River, NJ 08753)	5/8.9	U4	
WVNJ	Oakland, NJ	10/2.5	U4	
WMVI	Mechanicville, NY 12118	5/0.47	U1	NH
WPIE	Trumansburg, NY	5/0.31	U4	F.Pl.
WYRU	Red Springs, NC 28377	5/0.25	U2	NH
WTYN	Tryon, NC 28782	10/0.5	U2	
WCCS	Homer City, PA (Addr: Indiana, PA 15701)	10/1	U4	AN7
WYNS	Lehighton, PA 18235	4.4/1	U4	AN6
WAMB	Donelson, TN	50/1	U2	AN
KVAR	San Antonio, TX 78212	10/1	U3	AN7
KSL	Salt Lake City, UT	50	U1	AN7
WCBX	Fieldale, VA	5/0.25	U4	
VSB3	Hamilton, Bermuda	1	U1	SH
+lp: USA 3/Can 1				
1170kHz				
WACV	Montgomery, AL 36102	10/1	U4	AN6
KJNP	North Pole, AK 99705	50/23	U2	NH
KCBQ	San Diego, CA	50/1.5	U4	AN
KLOK	San Jose, CA	50/5	U4	AN7
WAVS	Davie, FL 33314	5/0.5	U2	NH
WHNE	Cumming, GA 30130	5/2.5	D1	D
WLBH	Mattoon, IL 61938	5	D3	D*

KSTT	Davenport, IA 52808	1	U4	AN7
WOBM	Lakewood, NJ (Addr: Toms River, NJ 08753)	5	D3	D
WCXN	Claremont, NC 28810	7.7/1	D1	D
WCLN	Clinton, NC 28328	5/1	D1	D
KVOO	Tulsa, OK 74101	50	U2	AN7
WLGO	Lexington, SC 29072	10	D1	D
KPUG	Bellingham, WA	10/5	U2	AN6
WWVA	Wheeling, WV 26003	50	U2	AN7
CKGY	Red Deer, AB	50/14	U4	AN7

+ lp: USA 15

1180kHz

KYET	Williams, AZ 86046	10/1	U1	NH
KTLL	Placerville, CA	25/1	U4	F.Pl.
KERI	Wasco, CA	50/10	U4	AN
"VOA"	R Marti, Marathon Key, FL	50	U3	NH
WSAF	Trion, GA 30753 (Addr: Summerville, GA 30747)	5	D1	D
WJNT	Pearl, MS	10/0.5	U2	AN
KOFI	Kalispell, MT 59901	50/10	U2	AN6
KKAR	Bellevue, NE (Addr: Omaha, NE 68102)	5/1	U4	AN
WHAM	Rochester, NY 14604	50	U1	AN6
WMYT	Carolina Beach, NC	10	D3	F.Pl.
WHJM	Knoxville, TN 37901	10/2.6	D1	D
KGOL	Humble, TX 77338	10/5	U4	
KLAY	Lakewood, WA	2.4/1	U2	F.Pl.

+ lp: USA 7

1190kHz

KRDS	Tolleson, AZ (Addr: Phoenix, AZ 85021)	5/0.25	U4	AN7
KMOA	Kensett, AR 72006	10/1.9	D1	D
KORG	Anaheim, CA 92805	11/1.4	U4	AN7
KBCO	Boulder, CO 80301	5	U2	
WAJL	Pine Castle, FL (Addr: Orlando, FL 32859)	5	D1	D
WOEQ	Royal Palm Beach, FL	0.89/1	U2	
WGKA	Atlanta, GA 30355	10/2.5	D1	D
WOWO	Fort Wayne, IN 46802	50	U2	AN7
WANN	Annapolis, MD 21404	10	D3	D
KKOJ	Jackson, MN 56143	5	D3	D
WBSL	Bay St. Louis, MS 39520	5	D1	D
KHAD	De Soto, MO 63020	5	D3	D*
KJLA	Kansas City, MO 64111	5/0.25	U2	AN7
KXKS	Albuquerque, NM	10	D1	D
WLIB	New York, NY 10017	10	L3	D*
KEX	Portland, OR 97201	50	U2	AN7
WSDQ	Dunlap, TN 37327	5/1	D1	D
KGBS	Dallas, TX	50/5	U4	AN7
WBDY	Bluefield, VA	10	D3	D
CFSL	Weyburn, SK	10/5	U2	AN

+ lp: USA 18

1200kHz

WFSF	Ozark, AL 36361	10/2	D1	D
KCPC	Cathedral City, CA	5/1	U4	F.Pl.
KTCD	Cottonwood, CA	10/2.5	U4	F.Pl.
KRDE	Pismo Beach, CA 93449	5/1	U4	F.Pl.
KOQI	Soquel, CA	25/1	U2	F.Pl
WDCQ	Pine Island Center, FL	10/1	U4	AN7
WFLZ	Thonotosassa, FL	10/1	U4	F.Pl.
WOPA	Chicago, IL	10/2.5	U4	AN
WANG	Radcliff, KY 40160	5/0.5	U2	
WNSW	Brewer, ME 04412,	10	U3	AN
WKOX	Framingham, MA 01701	10/1	U2	AN
WCHB	Taylor, MI	50/1	U4	
WGNY	Newburgh, NY 12550	10/1	U4	F.Pl.
WNSS	North Syracuse, NY (Addr: Clay, NY 13041)	1	U2	
WSML	Graham, NC 27253	10/1	U2	F.Pl.
KFNW	West Fargo, ND	10/1	U2	NH
WBZY	New Castle, PA 16101	5/1	U2	
WQDQ	Lebanon, TN 37088	10/0.5	U2	AN
WRVX	Mt. Carmel, TN	10/0.25	U4	F.Pl.
WOAI	San Antonio, TX 78201	50	U1	AN6
WAGE	Leesburg, VA 22075	10/1	U2	NH
CHMG	St. Albert, AB	25	U4	AN7
CKDA	Victoria, BC	50	U3	AN7
CFGO	Ottawa, ON	50	U4	AN7
C . . .	Sault Ste. Marie, ON	10/7	U4	F.Pl.
CKOT	Tillsonburgh, ON	10	U4	F.Pl.

+ lp: Can 1

1210kHz

KQTL	Sahuarita, AZ	10/1	U1	AN
KRGO	Fowler, CA 93625	10/0.37	U5	AN
KEBR	Rocklin, CA (Addr: Sacramento, CA 95821)	5/0.5	U5	
KPRZ	San Marcos, CA 92069	20/5	U4	AN
WCMQ	Miami Springs, FL	25/2.5	U4	
WDGR	Dahlonega, GA 30533	10/2.5	D1	D
WBIU	Denham Springs, LA 70726	10/1	U2	AN
WKNX	Frankenmuth, MI	10	D3	D*
KGYN	Guymon, OK 73942	10	U2	AN
WOGL	Philadelphia, PA 19131	50	U1	AN7
KOKK	Huron, SD	10/1	U4	
WGSF	Arlington, TN 38002	10/0.25	U3	NH
KUBR	San Juan, TX	50/1	U1	
KONY	Washington, UT	10/0.25	U1	AN
KBSG	Auburn, WA	10/1	U5	AN7
KREW	Sunnyside, WA 98944	10/1	U1	NH
KRSV	Afton, WY 83110	5/0.25	U1	
KLDI	Laramie, WY 82070	10/1	U2	
CKWA	Slave Lake, AB	1	U1	AN7
VOAR	St. John's, NF	10	U4	NH
CFYM	Kindersley, SK	1	U1	AN

+ lp: USA 3/Can 1

1220kHz

KDFC	Palo Alto, CA (Addr: San Francisco, CA 94109)	5/0.14	U1	NH
WJAX	Jacksonville, FL 32211	5	D3	D*
WSLM	Salem, IN 47167	5	D3	D*
WTCN	Stillwater, MN 55082	5	D1	D*
WGNY	Newburgh, NY 12550	5	D3	D*
WREV	Reidsville, NC 27323	10	D1	D
WENC	Whiteville, NC 28472	5	D1	D*
WKNR	Cleveland, OH 44147	50	U3	AN7
WFAX	Falls Church, VA 22046	5	D1	D*
CJOC	Lethbridge, AB	10/5	U2	AN7
CJRB	Boissevain, MB	10	U4	AN7
CKCW	Moncton, NB	25/10	U2	AN6
CJSS	Cornwall, ON	1	U4	AN
CJRL	Kenora, ON	1	U1	AN
CHSC	St. Catharines, ON	10	U4	AN7
CFVM	Amqui, PQ	10/5	U2	AN
CKSM	Shawinigan, PQ	10/2.5	U4	AN

+ lp: USA 42

1230kHz

USA 172 st's mostly 1kW, Canada's 33 st's 40W to 1kW

CJLB	Thunder Bay, ON	4/1	U1	AN
CHVD	Dolbeau, PQ	10/1	U1	AN7
ZFB	Hamilton, Bermuda	1	U1	NH

1240kHz

USA 156 st's mostly 1kW, Canada 37 st's 40W to 1kW

CJCD	Yellowknife, NWT	4/1	U1	NH

1250kHz

WZOB	Fort Payne, AL 35967	5	D1	D
WAPZ	Wetumpka, AL 36092	5	D1	D*
KHIL	Willcox, AZ 85643	5	D1	D
KURB	Little Rock, AR 72207	2/1.2	U2	AN7
KTMS	Santa Barbara, CA 93102	2.5/1	U3	AN7
KLLK	Willits, CA 95490	5/2.5	U4	
KGRJ	Fraser, CO 80442	5/0.5	U4	F.Pl.
WDAE	Tampa, FL 33609	5	U3	AN7
WGL	Fort Wayne, IN 46819	1	U4	NH
KFKU	Lawrence, KS 66045	5	ST2	SH
WREN	Topeka, KS 66601	5	ST2	AN7
WARE	Ware, MA 01082	5/2.5	U4	NH
KBRF	Fergus Falls, MN 56537	2.2/1	U2	AN
WHNY	McComb, MS 39648	5/1	U2	NH
KIKC	Forsyth, MT 59327	5	D1	D*
WKBR	Manchester, NH 03102	5	U4	AN
WMTR	Morristown, NJ 07960	5	U4	AN
WGHB	Farmville, NC 27828	5/2.5	U4	
WBRM	Marion, NC 28752	5	D1	D
WTAE	Pittsburgh, PA 15221	5	U2	AN7
WTMA	Charleston, SC 29417	5/1	U2	AN7
KALO	Port Arthur, TX 77640	5/1	U2	AN
KZEP	San Antonio, TX 78210	1	U2	AN
KNEU	Roosevelt, UT 84066	5	D1	D*
WDVA	Danville, VA 24541	5	U2	AN6
WPRZ	Warrenton, VA 22186	5	D3	D
KWSU	Pullman, WA 99164	5	U1	NH

KKFX	Seattle, WA 98121	5	U2	AN7
WYKM	Rupert, WV 25984	5	D1	D*
WEMP	Milwaukee, WI (Addr: Hales Corners, WI 53130)	5	U4	AN7
CHSM	Steinbach, MB	10	U4	AN7
CHWO	Oakville, ON	10	U4	AN7
CBGA	Matane, PQ	10/5	U2	NH
+lp: USA 33				

1260kHz

WCEO	Birmingham, AL 35209	5/1	U2	
KGIL	San Fernando, CA (Addr: Mission Hills, CA 91345)	5	U4	AN6
KOIT	San Francisco, CA 94108	5/1	U1	AN7
KRKE	Aspen, CO 81611	5	D1	D*
WWDC	Washington, DC 20036	5	U4	AN7
WSUA	Miami, FL 33145	5	U4	AN6
WUFE	Baxley, GA 31513	5	D1	D*
WTJH	East Point, GA 30364	5	D1	D*
KICN	Idaho Falls, ID 83401	5	D1	D*
WIBV	Belleville, IL 62221	5	U4	AN6
WNDE	Indianapolis, IN 46220	5	U2	AN6
KFGQ	Boone, IA 50036	5	D3	D
WEZE	Boston, MA (Addr: Milton, MA 02186)	5	U2	AN6
WWJQ	Zeeland, MI (Addr: Holland, MI 49423)	5/1	U4	
KTTS	Springfield, MO 65805	5	U2	AN7
WBUD	Trenton, NJ 08638	5/2.5	U4	NH
KVSF	Santa Fe, NM 87504	5/1	U1	F.Pl.
WNDR	Syracuse, NY 13277	5	U2	AN7
WKXR	Asheboro, NC 27203	5/0.5	U4	NH
WRDZ	Cleveland, OH 44141	5	U4	AN7
WNXT	Portsmouth, OH 45662	5/1	U4	NH
KWSH	Wewoka, OK 74884	1	U2	NH
KLCY	McMinnville, OR 97128	1	U2	NH
WRIE	Erie, PA 16509	5	U4	AN7
WPHB	Philipsburg, PA 16866	5	D1	D*
WMUU	Greenville, SC 29609	5	D1	D*
WRIP	Lake City, SC 29560	5	D3	D
KWYR	Winner, SD 57580	5	D1	D*
WNOO	Chattanooga, TN 37406	5	D1	D
WQZQ	Dickson, TN 37055	5	D1	D*
WCHV	Charlottesville, VA 22905	5/2.5	U4	AN6
WXCE	Amery, WI 54001	5	U4	NH
KPOW	Powell, WY 82435	5/1	U2	NH
CFRN	Edmonton, AB	50	U2	AN7
CIHI	Fredericton, NB	10	U2	AN7
+lp: USA 34				

1270kHz

WKSJ	Prichard, AL (Addr: Mobile, AL 36608)	5	D1	D*
KDJI	Holbrook, AZ 86025	5	D1	D*
KPBA	Pine Bluff, AR 71601	5	D1	D*
KNWZ	Thousand Palms, CA	5/0.75	U4	
KJUG	Tulare, CA 93274	5/1	U2	AN
WNOG	Naples, FL 33940	5/2.5	U4	NH
WHBS	Eatonville, FL (Addr: Orlando, FL 32808)	5	U3	AN
WNOG	Naples, Fl 33940	1/2.5	U4	NH
WNLS	Tallahassee, FL 32303	5	U2	NH
WHYD	Columbus, GA 31906	5	D1	D*
WJJC	Commerce, GA 30529	5	D1	D
KTFI	Twin Falls, ID 83303	5/1	U1	AN
WKBF	Rock Island, IL 61204	5	U2	AN7
WCMR	Elkhart, IN 46515	5/0.5	U4	AN
WWCA	Gary, IN 46402	2.5	U3	NH
WCBC	Cumberland, MD 21501	5/1	U4	AN
WLAW	Fairhaven, MA 02719	5	U4	F.Pl.
WSPR	Springfield, MA 01102	5/1	U3	AN7
WMKT	Charlevoix, MI 49720	5	U2	AN
WXYT	Detroit, MI (Addr: Southfield, MI 48037)	5	U2	AN7
WJJY	Baxter, MN 56401	5	U4	F.Pl.
KWEB	Rochester, MN 55901	5/1	U4	AN6
KPLY	Sparks, NV 89431	5	U2	
WTSN	Dover, NH 03820	5	U4	AN5
WHLD	Niagara Falls, NY (Addr, Grand Island, NY 14072)	5	D3	D*
WDLA	Walton, NY 13856	5	D1	D*
WCGC	Belmont, NC 28012	5/0.5	U4	NH
WMPM	Smithfield, NC 27577	5	D1	D*
KTRT	Claremore, OK 74017	1	U4	
KAJO	Grants Pass, OR 97526	5	D1	D*
WLBR	Lebanon, PA 17042	5/1	U4	NH
WYAK	Surfside Beach, SC 29577	5	D3	D
KNWC	Sioux Falls, SD 57106	2.5	U2	NH
WLIK	Newport, TN 37821	5/0.5	U2	NH
KIOX	Bay City, TX 77414	1	U2	NH
KESS	Fort Worth, TX (Addr: Dallas, TX 75247)	5	U3	AN7
WTJZ	Newport News, VA (Addr: Hampton, VA 23669)	1	U2	AN
WHEO	Stuart, VA 24171	5	D1	D*
KBAM	Longview, WA 98632	5	D1	D*
KIML	Gillette, WY 82716	5/1	U2	AN7
CHAT	Medicine Hat, AB	10	U2	AN7
CHWK	Chilliwack, BC	10	U2	AN6
CJCB	Sydney, NS	10	U2	AN7
CJTN	Trenton, ON	1	U1	AN
CFGT	Alma, PQ	10/5	U2	AN7
+lp: USA 30/Can 1				

1280kHz

WNPT	Tuscaloosa, AL 35401	5/0.5	U2	AN
KKAL	Arroyo Grande, CA 93420	5/2.5	U4	AN
KFRN	Long Beach, CA 90802	1	U1	AN7
KJAX	Stockton, CA 95201	1	U2	AN7
KXKL	Denver, CO 80204	5	U4	AN7
WGTX	DeFuniak Springs, FL 32433	5	D1	D*
WSVE	Jacksonville, FL 32211	5	D1	D*
WIBB	Macon, GA 31208	5	D1	D*
WBIG	Aurora, IL 60504	1	U4	AN6
WWOK	Evansville, IN 47702	5/1	U2	AN7
WQUE	New Orleans, LA 70112	5	U3	AN7
WABK	Gardiner, ME 04345	5	U2	NH
WEIM	Fitchburg, MA 01420	5/1	U4	NH
WWTC	Minneapolis, MN 55403	5	U2	AN6
KVOX	Moorehead, MN 56560	5/1	U4	AN7
KDOL	Henderson, NV 89015	5	D1	D
KRZE	Farmington, NM 87401	5	D1	D*
WADO	New York, NY 10017	5	U3	AN6
WPXY	Rochester, NY 14604	5	U2	AN6
WSAT	Salisbury, NC 28144	1	U2	NH
WYAL	Scotland Neck, NC 27874	5	D1	D*
KDUK	Eugene, OR 97401	5/1	U2	AN7
WHVR	Hanover, PA 17331	5/0.5	U4	NH
WKST	New Castle, PA 16101	5/1	U2	NH
WANS	Anderson, SC 29622	5/1	U2	AN7
WJAY	Mullins, SC 29574	5	D1	D*
WMCP	Columbia, TN 38401	5/0.5	U4	
KDYL	Salt Lake City, UT 84101	5/0.75	U2	AN7
KUDY	Spokane, WA 99203	5	D3	D*
KIT	Yakima, WA 98907	5/1	U1	AN7
WNAM	Neenah, WI 54957	5/1	U4	AN7
CHRB	High River, AB	10	U4	AN7
CHQB	Powell River, BC	1	U3	NH
CJMS	Montreal, PQ	50	U4	AN7
CKCV	Quebec, PQ	50	U4	
CJSL	Estevan, SK	10	U4	AN6
VSB2	Hamilton, Bermuda	1	U1	AN
+lp: USA 32/Can 1				

1290kHz

WOPP	Opp, AL 36467	2.5/0.5	U4	NH
KCUB	Tucson, AZ 85703	5/1	U5	AN7
KDMS	El Dorado, AR 71730	5	D1	D*
KUOA	Siloam Springs, AR 72761	5	D1	D*
KHSL	Chico, CA 95927	5	U2	AN
KAZA	San Jose, CA 95118	5	D3	D*
KMEN	San Bernardino, CA 92402	5	U4	AN7
WTMC	Ocala, FL 32670	5/1	U2	AN
WBZT	West Palm Beach, FL 33416	5	U4	AN
WCHK	Canton, GA 30114	5/0.5	U3	NH
WCHY	Savannah, GA 31402	5	U2	AN7
WIRL	Peoria, IL 61614	5	U4	AN7
WDGS	New Albany, IN (Addr: Clarksville, IN 47130)	0.5/1	U4	
KWLS	Pratt, KS 67124	5/0.5	U4	NH
WCBL	Benton, KY 42025	5	D1	D*
WKLB	Manchester, KY 40962	5	D1	D
WHGR	Houghton Lake, MI 48629	5	D3	D
KGVO	Missoula, MT 59806	5	U2	AN7

KOIL	Omaha, NE 68102	5	U2	AN6
WKNE	Keene, NH 03431	5	U3	AN
WNBF	Binghamton, NY 13902	5	U2	AN7
WHKY	Hickory, NC 28601	5/1	U2	AN
WHIO	Dayton, OH 45401	5	U2	AN7
KUMA	Pendleton, OR 97801	5	U2	NH
KPHP	Lake Oswego, OR 97034	5	U3	
WFBG	Altoona, PA 16603	5/1	U2	AN
WCRP	Providence, RI 02904	5	U4	NH
WQMC	Sumter, SC 29151	1	U2	AN
WTNX	Lynchburg, TN 37352	5	D1	D
WATO	Oak Ridge, TN 37830	5/0.5	U4	NH
KRGE	Weslaco, TX 78596	5	U2	NH
KLLF	Wichita Falls, TX 76308	5/1	U2	AN6
KKZN	Kanab, UT 84741	5	D1	F.Pl.
WSTK	Colonial Heights, VA 23834	5	D1	D
WPVA	Petersburg, VA 23803	5	D1	D*
WVOW	Logan, WV 25601	5/1	U2	AN
WMVP	Milwaukee, WI 53216	5	U4	AN
WKLJ	Sparta, WI 54656	5	D1	D*
KOWB	Laramie, WY 82070	5/1	U4	NH
CIFX	Winnipeg, MB	10	U4	AN7
CJBK	London, ON	10/5	U3	AN7
CHRM	Matane, PQ	10	U4	AN7

+ Ip: USA 29

1300kHz

WKXM	Winfield, AL 35594	5	D1	D*
KBCN	Fairbanks, AK	1	U1	F.Pl.
KWCK	Searcy, AR 72143	5	D1	D*
KYNO	Fresno, CA 93703	5/1	U2	AN7
KPMO	Mendocino, CA 95460	5	D3	D*
KAZN	Pasadena, CA 91107	5/1	U2	AN7
KVOR	Colorado Springs, CO 80901	5/1	U1	AN7
WAVZ	New Haven, CT (Addr: North Haven, CT 06473)	1	U2	AN6
WXXU	Cocoa, FL 32922	5/1	U4	NH
WFFG	Marathon, FL 33050	2.5	U3	NH
WQBN	Tampa, FL 33606	5/1	U4	NH
WMTM	Moultrie, GA 31768	5	D1	D*
KLER	Orofino, ID 83544	5/1	U2	NH
WTAQ	La Grange, IL 60525	5/4	U4	AN2
KGLO	Mason City, IA 50401	5	U2	NH
WLXG	Lexington, KY 40578	2.5/1	U2	AN7
WIBR	Baton Rouge, LA 70821	5/1	U4	AN7
KFLO	Shreveport, LA 71101	5	D1	D
WJFK	Baltimore, MD 21204	5	U4	AN7
WOOD	Grand Rapids, MI 49503	5	U2	AN
WKXI	Jackson, MS 39206	5/1	U1	AN
KBRL	McCook, NE 69001	5	D3	D*
KPTL	Carson City, NV 89701	5/0.5	U2	AN
WPNH	Plymouth, NH 03264	5	D1	D*
WIMG	Trenton, NJ 08608	3.2/1.3	U4	
WXRL	Lancaster, NY 14086	2.5	U4	AN
WQBK	Rensselaer, NY (Addr: Albany, NY 12201)	5	U4	AN
WSYD	Mount Airy, NC 27030	5/1	U2	NH
WERE	Cleveland, OH 44114	5	U3	AN7
KAKC	Tulsa, OK 74135	5/1	U4	AN7
KDOV	Phoenix, OR (Addr: Ashland, OR 97520)	5	U2	
WXPX	West Hazelton, PA 18201	5/0.5	U4	
KOLY	Mobridge, SD 57601	5	D1	D*
WMTN	Morristown, TN 37814	5	D1	D*
WNQM	Nashville, TN 37203	5	U2	AN7
KVET	Austin, TX 78767	5/1	U4	AN7
WKCY	Harrisonburg, VA 22801	5	D1	D*
KMPS	Seattle, WA 98124	5	U2	AN7
CJME	Regina, SK	10	U4	AN7

+ Ip: USA 30

1310kHz

WAJO	Marion, AL 36756	5	D1	D*
KXAM	Mesa, AZ (Addr: Scottsdale, AZ 85252)	5/0.5	U2	AN6
KIQQ	Barstow, CA 92311	5/0.5	U4	
KDIA	Oakland, CA 94662	5	U3	AN7
KFKA	Greeley, CO 80632	5/1	U2	NH
WICH	Norwich, CT 06360	5	U4	AN
WYND	DeLand, FL 32720	5	D1	D*
WAUC	Wauchula, FL 33873	5	D3	D*
KLIX	Twin Falls, ID 83301	5/2.5	U2	AN7
WTUX	Indianapolis, IN 46208	5/1	U2	AN7
WTTL	Madisonville, KY 42431	1/1.5	U2	AN
WDOC	Prestonsburg, KY 41653	5	D1	D
KMBS	West Monroe, LA 71291	5	D1	D*
WLOB	Portland, ME 04104	5	U4	NH
WORC	Worcester, MA 01608	5/1	U4	NH
WMTG	Dearborn, MI 48126	5	U4	AN6
WCCW	Traverse City, MI 49684	5	D1	D*
WHLV	Hattiesburg, MS 39401	5	D1	D*
KFSB	Joplin, MO 64802	5/1	U4	NH
KEIN	Great Falls, MT 59403	5/1	U1	AN7
WJLK	Asbury Park, NJ 07712	2.5/1	U4	AN7
KIVA	Corrales, NM (Addr: Albuquerque, NM 87102)	5/0.5	U2	AN
WCGR	Canandaigua, NY 14424	2.5	U4	F.Pl.
WVIP	Mount Kisco, NY 10549	5	D3	D*
WTLB	Utica, NY 13503	5/0.5	U4	AN6
WISE	Asheville, NC 28804	5/1	U2	AN6
WGSP	Charlotte, NC 28203	1/1.6	U2	
WTIK	Durham, NC 27702	5/1	U4	AN6
KNOX	Grand Forks, ND 58206	5	U2	AN6
KNPT	Newport, OR 97365	5/1	U2	NH
WBFD	Bedford, PA 15522	5	D1	D*
WGSA	Ephrata, PA 17522	5	D3	D*
WNAE	Warren, PA 16365	5	D1	D*
WDKD	Kingstree, SC 29556	5	D1	D*
WDOD	Chattanooga, TN 37405	5	U2	NH
WDXI	Jackson, TN 38301	5/1	U2	NH
KAAM	Dallas, TX 75248	5/1	U4	AN7
KZVE	San Antonio, TX 78217	5/0.25	U3	
WDCT	Fairfax, VA 22030	5/0.5	U4	NH
WRAP	Newport News, VA (Addr: Hampton, VA 23666)	5	U2	AN7
KARY	Prosser, WA 99350	5	D1	D*
WSLW	White Sulphur Springs, WV 24986	5	D1	D*
WIBA	Madison, WI 53701	5	U2	AN6
CHLW	St. Paul, AB	10	U2	AN
CIWW	Ottawa, ON	50	U4	AN7
CHGB	La Pocatière, PQ	10	U2	NH

+ Ip: USA 26

1320kHz

WAGG	Birmingham, AL 35203	5	D1	D*
WAGF	Dothan, AL 36301	1	U2	AN
KWHN	Fort Smith, AR 72901	5	U2	NH
KQIQ	Farmerville, CA	5/3.1	U3	F.Pl.
KCTC	Sacramento, CA 95818	5	U4	AN
WATR	Waterbury, CT 06706	5/1	U4	NH
WLQY	Hollywood, FL 33020	5	U4	AN7
WQIK	Jacksonville, FL 32205	5	U2	AN7
WAMR	Venice, FL 33595	5/1	U2	AN7
WHIE	Griffin, GA 30224	5/0.08	U1	NH
KNCB	Vivian, LA 71082	5	D1	D*
WARA	Attleboro, MA 02703	5	U4	NH
WILS	Lansing, MI 48910	5/1	U4	AN7
WDMJ	Marquette, MI 49855	5/1	U2	NH
KOZY	Grand Rapids, MN 55744	5	U2	NH
WRJW	Picayune, MS 39466	5	D1	D*
KSIV	St. Louis, MO 63144	5	D1	D*
KOLT	Scottsbluff, NE 69361	5/1	U2	AN
WDER	Derry, NH 03038	5/1	U4	
WHHO	Hornell, NY 14843	5	D1	D*
WGLD	Greensboro, NC (Addr: High Point, NC 27260)	5	U4	AN7
WKRK	Murphy, NC 28906	5	D1	D*
WOBL	Oberlin, OH 44074	1	U4	AN
WKAP	Allentown, PA (Addr: Whitehall, PA 18052)	5/1	U4	AN7
WJAS	Pittsburgh, PA 15220	5	U2	AN7
WOMG	Columbia, SC 29202	5/2.5	U2	AN
KELO	Sioux Falls, SD 57102	5	U2	AN
WKIN	Kingsport, TN 37662	5/0.5	U2	
WMSR	Manchester, TN 37355	5	D1	D*
KXYZ	Houston, TX 77017	5	U2	AN7
KCNR	Salt Lake City, UT 84115	5	U3	AN7
WLEE	Richmond, VA 23225	5	D3	D*
KXRO	Aberdeen, WA 98520	5/1	U2	AN6
WFHR	Wisconsin Rapids, WI 54494	5/0.5	U2	NH
CHQM	Vancouver, BC	50	U4	AN7
CKEC	New Glasgow, NS	25	U2	AN7

CJMR	Mississauga, ON	20/2	U4	AN

+lp: USA 31

1330kHz

WPRN	Butler, AL 36904	5	D	D
WZCT	Scottsboro, AL 35768	5/0.5	U2	
KHYT	Tucson, AZ 85713	2/5	U2	AN
KWKW	Los Angeles, CA 90028	5	U2	AN7
KLBS	Los Banos, CA 93635	0.5/5	U2	
KRDG	Redding, CA 96002	5	U4	
WJNX	Fort Pierce, FL 33454	5/1	U2	AN
WEBY	Milton, FL 32570	5	D1	
WCVC	Tallahassee, FL 32315	5	D1	D*
WMLT	Dublin, GA 31040	3.4/0.4	U2	NH
KSKI	Hailey, ID 83333	5/1	U2	
WSSY	Evanston, IL (Addr: Northbrook, IL 60062)	5/0.5	U4	
WVHI	Evansville, IN 47735	5/1	U2	AN
KWLO	Waterloo, IA 50704	5	U4	AN7
KFH	Wichita, KS 67202	5	U2	AN7
WKDP	Corbin, KY 40702	5	D3	D*
KVOL	Lafayette, LA 70502	5/1	U2	AN7
WASA	Havre de Grace, MD 21078	5/0.5	U2	
WRCA	Waltham, MA 02254	5	U4	AN7
WTRX	Flint, MI 48529	5/1	U4	AN7
KNOW	St. Paul, MN 55101	5	U4	AN6
WFTO	Fulton, MS 38843	5	D1	D*
KGAK	Gallup, NM 87301	5/1	U2	AN
WWRV	New York, NY 10312	5	U3	AN
WEBO	Owego, NY 13827	5	D1	D*
WFWC	Springville, NY 14141	1	U4	NH
KDRQ	Wishek, ND 58495	5	U4	
WZKC	Youngstown, OH 44505	0.5/1	U4	AN7
KUPL	Portland, OR 97221	5	U3	AN7
WEYZ	Erie, PA 16512	5	U4	AN7
WADJ	Somerset, PA 15501	5	D3	D*
WPJS	Conway, SC 29526	5/0.5	U2	NH
WFBC	Greenville, SC 29602	5	U2	AN7
KLBO	Monahans, TX 79756	5/1	U2	NH
WBTM	Danville, VA 24543	5/1	U2	NH
WOLD	Marion, VA 24354	5	D1	D
WESR	Onley, VA 23418	5	D1	D*
KMBI	Spokane, WA 99203	5	D1	D*
WHBL	Sheboygan, WI 53081	5/1	U4	NH
KOVE	Lander, WY 82520	5/1	U2	NH
CKLD	Thetford Mines, PQ	10	U3	AN7
CJYM	Rosetown, SK	10	U3	NH

+lp: USA 30

1340kHz

USA 174 st's, mostly with 1kW, Canada 41 st's 40W to 1kW.

CKXX	Cornerbrook, NF	10/2.5	U	AN
CKHV	Happy Valley, NF	1	U1	F.Pl.
CFYK	Yellowknife, NWT	2.5	U1	AN
CJLS	Yarmouth, NS	5/4	U4	NH
ZBM	Hamilton, Bermuda	1	U1	NH

1350kHz

WGAD	Gadsden, AL 35902	5/1	U2	NH
KCKC	San Bernardino, CA 92406	5/0.5	U4	AN6
KSRO	Santa Rosa, CA 95403	5	U2	NH
KGHF	Pueblo, CO 81002	5/1	U2	AN7
WNLK	Norwalk, CT 06852	2.5/0.5	U2	NH
WINY	Putnam, CT 06260	5	D1	D8
WWHL	Cocoa, FL 32922	1	U2	AN7
WCRM	Fort Myers, FL 33916	5/0.15	U3	
WGIA	Blackshear, GA 31516	5	D1	D*
WCOP	Warner Robins, GA 31093	5/0.5	U2	
KRLC	Lewiston, ID 83501	5/1	U2	AN7
WXCL	Peoria, IL 61604	1	U4	AN7
WIOU	Kokomo, IN 46902	5/1	U4	AN6
KRNT	Des Moines, IA 50309	5	U2	AN7
WLOU	Louisville, KY 40208	5/0.5	U2	AN
WSMB	New Orleans, LA 70113	5	U2	AN7
WLNH	Laconia, NH 03246	5	D1	D*
WHWH	Princeton, NJ 08542	5	U4	AN6
KABQ	Albuquerque, NM 87106	5/0.5	U2	NH
WSLR	Akron, OH 44313	5	U3	AN7
WOYK	York, PA 17404	5/1	U2	AN7
KTXJ	Jasper, TX 75951	5	D1	D*
KCOR	San Antonio, TX 78207	5	U2	AN7
WNVA	Norton, VA 24273	5	D1	D*
WBSK	Norfolk, VA 23501	5	U4	AN7
CKEG	Nanaimo, BC	10	U3	AN7
C...	Williams Lake, BC	1	U2	F.Pl.
CKAD	Middleton, NS	1	U3	NH
CKDO	Oshawa, ON	10/5	U4	AN7
CHRO	Pembroke, ON	1	U3	NH
CJLM	Joliette, PQ	10/1	U4	AN7
CHAL	St-Pamphile, PQ	1	U3	NH

+lp: USA 37/Can 5

1360kHz

WMOB	Mobile, AL 36601	5	D3	D*
KNNS	Glendale, AZ (Addr: Phoenix, AZ 85003)	5/0.8	U2	AN7
KFFA	Helena, AR 72342	1	U2	NH
KFIV	Modesto, CA 95353	5/1	U4	AN7
KPOP	San Diego, CA 92111	5/1	U1	AN7
WDRC	Hartford, CT (Addr: Bloomfield, CT 06002)	5	U2	NH
WYXY	Cypress Gardens, FL (Addr: Winter Haven, FL 33880)	5/2.5	U4	NH
WCGL	Jacksonville, FL 32207	5	D1	D
WKAT	Miami, FL 33181	5/1	U1	AN7
WYSE	Bainbridge, GA 31717	5/1	U2	AN7
KSCJ	Sioux City, IA 51104	5	U2	AN
WEBB	Baltimore, MD 21215	5/1.5	U2	
WKYO	Caro, MI 48723	1	U2	NH
WKMI	Kalamazoo, MI 49005	5/1	U2	AN6
KKBJ	Bemidji, MN 56601	5/2.5	U2	AN6
WVSJ	Vineland, NJ 08360	5/1	U4	AN7
KBUY	Ruidoso, NM 88345	5	D1	D*
WKOP	Binghamton, NY 13902	5/0.5	U4	AN7
WCHL	Chapel Hill, NC 27514	5/2/5	U2	AN7
WSAI	Cincinnati, OH 45204	5	U2	AN7
KOHU	Hermiston, OR 97838	4.3/1	U2	NH
KUIK	Hillsboro, OR 97123	5	U2	AN
WIXZ	McKeesport, PA 15035	5/1	U2	AN6
WPPA	Pottsville, PA 17901	5/0.5	U4	NH
WWHY	Milan, TN 38358	5/0.5	U4	
KWWJ	Baytown, TX 77520	5/1	U4	AN7
KRYS	Corpus Christi, TX 78408	1	U1	AN7
KNRB	Fort Worth, TX 76107	5/1	U2	AN7
WBOB	Galax, VA 24333	5	D1	D*
WHBG	Harrisonburg, VA 22812	5	D1	D*
KKMO	Tacoma, WA 98401	5	U1	AN6
WGEE	Green Bay, WI 54305	5	U2	AN7
KRKK	Rock Springs, WY 82902	5	U2	AN7
CKBC	Bathurst, NB	10	U2	AN
CJVL	Ste-Marie-Beauce, PQ	10/5	U4	AN

+lp: USA 41/Can 1

1370kHz

KWRM	Corona, CA 91720	5/2.5	U4	AN6
KPCO	Quincy, CA 95971	5/0.5	U4	
KEEN	San Jose, CA 95150	5	U4	AN7
KTMG	Deer Trail, CO 80105	5	D3	
WOCA	Ocala, FL 32678	5	D1	D*
WCOA	Pensacola, FL 32573	5	U2	AN6
WLOP	Jesup, GA 31545	5	D1	D*
WGCL	Bloomington, IN 47401	5/0.5	U4	NH
WLTH	Gary, IN 46409	1/0.5	U2	NH
KDTH	Dubuque, IA 52001	5	U2	AN
KGNO	Dodge City, KS 67801	5/1	U2	NH
WGOH	Grayson, KY 41143	5	D1	D
WDEA	Ellsworth, ME 04605	5	U4	NH
WKIK	Leonardtown, MD 20650	1	U2	
WKJF	Cadillac, MI 49601	5/1	U4	NH
KSUM	Fairmont, MN 56031	1	U4	AN6
KXTL	Butte, MT 59702	5	U1	AN7
WFEA	Manchester, NH 03101	5	U3	AN7
WELV	Ellenville, NY 12428	5	D1	D*
WXXI	Rochester, NY 14601	5	U2	AN7
WLTC	Gastonia, NC 28052	5	D1	D*
WLLN	Lillington, NC 27546	5	D3	D*
WTAB	Tabor City, NC 28463	5	D1	D*
WSPD	Toledo, OH 43602	5	U2	AN7
KAST	Astoria, OR 97103	1	U2	NH
WWCB	Corry, PA 16407	1/0.5	U2	NH
WKMC	Roaring Spring, PA (Addr: Altoona, PA 16603)	5	D3	D*
WMNY	Santee, SC 29142	5	D3	D
WDEF	Chattanooga, TN 37402	5	U2	NH
KFRO	Longview, TX 75601	1	U2	AN7

KSOP	Salt Lake City, UT 84125	5/0.5	U4	AN7
WHEE	Martinsville, VA 24115	5	D1	D*
WJWS	South Hill, VA 23970	5	D1	D*
WVMR	Frost, WV	5	D1	D*
	(Addr: Dunmore, WV 24934)			
WZAO	Moundsville, WV	5	D1	D*
	(Addr: Glen Dale, WV 26038)			
WCCN	Neillsville, WI 54456	5	D1	D*
CFOK	Westlock, AB	10	U2	AN
CHPQ	Parksville, BC	1	U3	AN7
CKOD	Valleyfield, PQ	10/5	U3	NH
CFVD	Dégelis, PQ	1	U3	AN

+lp: USA 37/Can 2

1375kHz

RFO	St. Pierre & Miquelon	20	U1	NH

1380kHz

WVSA	Vernon, AL 35592	5	D1	D*
KPAL	Little Rock, AR 72203	5/2.5	U4	
KSMJ	Sacramento, CA 95825	5	U4	AN7
KTOM	Salinas, CA 93912	5	U4	AN
WFNW	Waterbury, CT 06721	5/0.5	U4	NH
WAMS	Wilmington, DE 19807	5/1	U4	AN6
WELE	Ormond Beach, FL 32074	5/2.5	U4	AN7
WRBQ	St. Petersburg, FL	5	U2	AN7
	(Addr: Tampa, FL 33609)			
WAOK	Atlanta, GA 30365	5/4.2	U2	AN7
WSIZ	Ocilla, GA 31774	5	D1	D*
WBEL	South Beloit, IL	5	U2	NH
	(Addr: Beloit, WI 53511)			
WQHK	Fort Wayne, IN 46816	5	U4	AN7
KCIM	Carroll, IA	1	U4	NH
	(Addr: Village Park, IA 51401)			
KCNW	Fairway, KS	5/1	U4	AN
	(Addr: Kansas City, KS 66106)			
WYNK	Baton Rouge, LA 70821	5	D3	D*
WPHM	Port Huron, MI 48060	5	U4	NH
KLIZ	Brainerd, MN 56401	5	U2	AN6
KASP	St. Louis, MO 63139	5/1	U4	AN
WCQL	Portsmouth, NH 03801	5/1	U2	AN
WAHB	Bath, NY 14810	5/0.35	U1	D
WKDM	New York, NY	5	U3	AN
	(Addr: Carlstadt, NJ 07072)			
WTOO	Asheville, NC 28806	5/1	U2	NH
WSFL	New Bern, NC 28560	5	D1	D
WTOB	Winston-Salem, NC 27106	5/2.5	U2	AN6
KSWO	Lawton, OK 73502	1	U4	NH
KSRV	Ontario, OR 97914	5/1	U2	NH
WGUS	No. Augusta, SC	5	D1	D*
	(Addr: Augusta, GA 30903)			
KOTA	Rapid City, SD 57709	5	U2	NH
WIZO	Franklin, TN 37064	5/0.5	U2	AN
WMPS	Millington, TN	2.5/1	U4	NH
	(Addr: Memphis, TN 38053)			
KTSM	El Paso, TX 79925	5/0.5	U1	AN
WSYB	Rutland, VT 05701	5/1	U2	AN7
WTVR	Richmond, VA 23230	5	U4	AN6
KRKO	Everett, WA 98206	5	U1	AN6
WFCL	Clintonville, WI 54929	3.9/1.8	U4	
CKPC	Brantford, ON	10	U4	AN7
CKLC	Kingston, ON	10	U4	AN6
CFDA	Victoriaville, PQ	10	U4	AN7

+lp: USA 36/Can 4

1390kHz

WHMA	Anniston, AL 36202	5/1	U2	NH
KKDV	Fields Landing, CA 95537	5	D1	F.Pl.
KGER	Long Beach, CA 90807	5	U4	AN6
KMIX	Turlock, CA 95381	5	U4	AN6
KJME	Denver, CO 80209	5/0.5	U4	NH
WAJD	Gainesville, FL 32605	5	D1	D*
WISK	Americus, GA 31709	5	D1	D*
WGCI	Chicago, IL 60602	5	U4	AN7
KKSO	Des Moines, IA 50312	1	U3	AN6
WKIC	Hazard, KY 41701	5	D1	D*
WTMS	Presque Isle, ME 04769	5	U2	AN
WPLM	Plymouth, MA 02360	5	U4	NH
WNNY	Charlotte, MI 48813	5	D3	D*
KXSS	Waite Park, MN	2.5/1	U4	AN
	(Addr: St. Cloud, MN 56302)			
WROA	Gulfport, MS 39503	5	U4	AN7
WMER	Meridian, MS 39301	5	D1	D*
KJPW	Waynesville, MO 65583	5	D1	D
KENN	Farmington, NM 87401	5/1.3	U2	AN
KHOB	Hobbs, NM 88240	5/0.5	U2	AN7
WEOK	Poughkeepsie, NY 12602	5	D3	D
WFBL	Syracuse, NY 13204	5	U4	AN6
WEED	Rocky Mount, NC 27801	5/2.5	U2	NH
KRRZ	Minot, ND 58702	5/1	U1	AN6
WMPO	Middleport, OH 45760	5	D1	D*
WHOT	Youngstown, OH 44512	5	U2	NH
WLAN	Lancaster, PA 17603	5/1	U4	AN
WRSC	State College, PA 16801	2/1	U2	AN
WXTC	Charleston, SC 29403	5	U2	AN7
WTJS	Jackson, TN 38301	5/1	U2	NH
KLGN	Logan, UT 84321	5/0.5	U2	AN
WDOT	Burlington, VT 05401	5	U2	AN
WMZQ	Arlington, VA	5	U4	AN
	(Addr: Washington, DC 20015)			
WWOD	Lynchburg, VA 24501	5/1	U2	NH
KBBO	Yakima, WA 98909	5/0.5	U2	NH
WRIG	Schofield, WI	5	U4	NH
	(Addr: Wausau, WI 54402)			
CJCY	Medicine Hat, AB	10	U2	AN7
CKKC	Nelson, BC	1	U3	AN
CHOO	Ajax, ON	10	U4	AN7

+lp: USA 26

1400kHz

USA 176 st's, mostly 1kW; Canada 27 st's, 40W-1kW.

CBG	Gander, NF	4	U1	NH
CJFP	Rivière-du-Loup, PQ	20	U2	AN7
CKRN	Rouyn, PQ	10/1	U2	AN7

1410kHz

WLVV	Mobile, AL 36652	5	U2	AN6
WRNB	Prattville, AL 36067	5/1	U4	NH
KERN	Bakersfield, CA 93303	1	U1	AN7
KRML	Carmel, CA 93922	5	U4	
KMYC	Marysville, CA 95901	5/1	U4	NH
KCAL	Redlands, CA 92373	5/4	U4	NH
KCOL	Fort-Collins, CO 80522	5/1	U4	AN7
WPOP	Hartford, CT	5	U4	AN6
	(Addr: Newington, CT 06111)			
WDOV	Dover, DE 19903	5	U4	AN7
WMYR	Fort Myers, FL 33902	5	U2	AN
WQBQ	Leesburg, FL 34788	5	D1	D*
WHBT	Tallahassee, FL 32303	5	D1	D*
WLAQ	Rome, GA 30161	1	U2	AN
KKLO	Leavenworth, KS 66048	5/0.5	U4	NH
KQAM	Wichita, KS 67204	5/1	U4	AN7
WLBJ	Bowling Green, KY 42101	5/1	U2	NH
WHLN	Harlan, KY 40831	5	D1	D*
KRWB	Roseau, MN 56751	1	U2	NH
WDTL	Boyle, MS	5/1	U4	
	(Addr: Cleveland, MS 38732)			
KGNN	Cuba, MO 65453	5	D3	D
KOOQ	North Platte, NE 69101	5/0.5	U2	NH
KFMS	Las Vegas, NV 89114	5	U4	AN7
WELM	Elmira, NY 14901	5/1	U4	AN6
WNCQ	Watertown, NY 13601	5/1	U2	AN6
WSRC	Durham, NC 27702	5	D3	D*
WING	Dayton, OH 45429	5	U2	AN7
KBNP	Portland, OR 97201	5/1	U2	
WLSH	Lansford, PA 18232	5	D3	D*
KQV	Pittsburgh, PA 15219	5	U4	AN7
KLEV	Cleveland, TX 77327	1	U4	NH
KRIL	Odessa, TX 79760	1	U2	AN7
WRIS	Roanoke, VA 24017	5	D1	D*
WSCW	South Charleston, WV 25303	5	D1	D*
WIZM	LaCrosse, WI 54602	5	U2	AN6
KWYO	Sheridan, WY 82801	5/0.5	U1	NH
CFUN	Vancouver, BC	50	U4	AN7
CIGO	Port Hawkesbury, NS	10	U4	NH
CKSL	London, ON	10	U4	AN7
CFMB	Montreal, PQ	10	U3	AN7

+lp: USA 38

1420kHz

WACT	Tuscaloosa, AL 35402	5	D1	D*
KSKM	Bethel, AK 99559	1	U1	F.Pl.
KXOW	Hot Springs, AR 71901	5	D1	D*
KSTN	Stockton, CA 95206	5/1	U4	AN7
WLIS	Old Saybrook, CT 06475	5/0.5	U2	NH
WDBF	Delray Beach, FL 33444	5/0.5	U4	NH

WBRD	Palmetto, FL	2.5/1	U4	
	(Addr: Bradenton, FL 33506)			
WRCG	Columbus, GA 31994	5	U2	AN
WLET	Toccoa, GA 30577	5	D1	D*
WIMS	Michigan City, IN 46360	5	U4	NH
WOC	Davenport, IA 52807	5	U4	AN6
WVJS	Owensboro, KY 42302	5/1	U4	AN7
WBSM	New Bedford, MA 02740	5/1	U3	AN7
WBEC	Pittsfield, MA 01201	1	U2	AN
WFLT	Flint, MI 48506	5	U4	
KTOE	Mankato, MN 56001	5	U2	AN7
WIGG	Wiggins, MS 39577	5	D1	D*
WASR	Wolfeboro, NH 03894	5	D1	D*
WACK	Newark, NY 14513	5/0.5	U4	NH
WLNA	Peekskill, NY 10566	5/1	U4	AN
WHK	Cleveland, OH 44115	5	U2	AN7
WCOJ	Coatesville, PA 19320	5	U2	NH
WCED	DuBois, PA 15801	5/0.5	U2	NH
WEMB	Erwin, TN 37650	5	D1	D*
WKSR	Pulaski, TN 38478	1	U2	NH
KSRK	Lufkin, TX 75901	5/1	U2	AN
WKCW	Warrenton, VA 22186	5	D1	D*
KITI	Centralia, WA 98531	5	U4	AN6
KUJ	Walla Walla, WA 99362	5	U2	AN7
WTCR	Kenova, WV	5/0.5	U2	AN
	(Addr: Catlettsburg, KY 41129)			
CKDY	Digby, NS	1	U3	AN
CKPT	Peterborough, ON	10/5	U4	AN7
CJMT	Chicoutimi, PQ	10/5	U2	AN
CKTL	Plessisville, PQ	1/0.5	U2	AN7
CJVR	Melfort, SA	10	U2	AN

+lp: USA 44/Can 1

1430kHz

WFHK	Pell City, AL 35125	5	D1	D
KWDO	Fresno, CA 93744	5	U4	AN7
KALI	San Gabriel, CA	5	U4	AN7
	(Addr: Hollywood, CA 90038)			
KNTA	Santa Clara, CA	1	U4	AN6
	(Addr: San Jose, CA 95150)			
KEZW	Aurora, CO	5	U2	AN7
	(Addr: Denver, CO 80231)			
KMIY	Grand Junction, CO	5	D1	F.Pl.
WOIR	Homestead, FL 33030	5/0.5	U2	AN
WLKF	Lakeland, FL 33801	5/1	U1	AN
WLTG	Panama City, FL 32401	5	U4	AN7
WGFS	Covington, GA 30209	5	D1	D*
WWGS	Tifton, GA 31794	5/1	U2	NH
WEEF	Highland Park, IL 60035	1	U4	
WCKN	Indianapolis, IN 46208	5	U2	AN7
WYMC	Mayfield, KY 42066	1	U1	AN
WNAV	Annapolis, MD 21404	5/1	U2	NH
WTTT	Amherst, MA 01002	5	D3	D*
WXKS	Everett, MA	5/1	U2	NH
	(Addr: Medford, MA 02155)			
WION	Ionia, MI 48846	5	D3	D*
WLAU	Laurel, MS 39440	5	D1	D*
WRTH	St. Louis, MO 63101	5	U4	AN7
KRGI	Grand Island, NE 68801	5/1	U2	NH
WNJR	Newark, NJ	5	U2	AN6
	(Addr: Hillside, NJ 07205)			
KCRX	Roswell, NM 88201	5/1	U2	NH
WMRV	Endicott, NY	5	U2	AN7
	(Addr: Endwell, NY 13760)			
WDEX	Monroe, NC 28110	2.5	U4	AN7
WMNC	Morganton, NC 28655	5/1	U2	NH
KTYN	Minot, ND 58701	5	D1	D*
WFOB	Fostoria, OH 44830	1	U4	NH
KSKS	Tulsa, OK 74133	5	U2	AN7
KYKN	Keizer, OR	5	U2	AN
	(Addr: Salem, OR 97308)			
WVAM	Altoona, PA 16602	5/1	U2	AN6
WBLR	Batesburg, SC 29006	5	D1	D*
WNWZ	Germantown, TN 38117	2.8/2.5	U4	
WEMG	Knoxville, TN 37917	5	D1	
WYZO	Madison, TN	5/1	U2	AN
	(Addr: Nashville, TN 37203)			
KEES	Gladewater, TX	5/1	U2	NH
	(Addr: Longview, TX 75604)			
KCOH	Houston, TX 77004	5	D3	D*
KLO	Ogden, UT 84401	5	U4	AN7
WPES	Ashland, VA 23005	5	D3	D*
WKEX	Blacksburg, VA 24060	5	U4	
WDIC	Clintwood, VA 24228	5	D1	D*
KBRC	Mount Vernon, WA 98273	5/1	U2	NH
KCLK	Asotin, WA			
	(Addr: Clarkston, WA 99403)	5/1	U4	
WEIR	Weirton, WV 26062	1	U4	AN
WBEV	Beaver Dam, WI 53916	1	U2	NH
CJCL	Toronto, ON	50	U4	AN7

+lp: USA 32

1440kHz

WHHY	Montgomery, AL 36105	5/1	U2	AN7
KOPA	Scottsdale, AZ 85252	5	D1	D*
KITA	Little Rock, AR 72202	5/0.24	U2	
KVON	Napa, CA 94558	5/1	U4	AN
KDIF	Riverside, CA 92502	1	U1	AN
KUHL	Santa Maria, CA 93454	5/1	U2	AN6
KRDZ	Wray, CO 80758	5	D1	D
WWCL	Lehigh Acres, FL	5/1	U4	AN
	(Addr: Ft. Myers, FL 33907)			
WPRD	Winter Park, FL	5/1	U2	AN6
	(Addr: Orlando, FL 32854)			
WGIG	Brunswick, GA 31523	5/1	U2	AN
WGEM	Quincy, IL 62306	5/1	U4	NH
WROK	Rockford, IL 61125	5/0.5	U5	AN7
KMAJ	Topeka, KS 66604	5/1	U3	AN7
WCDS	Glasgow, KY 42141	5	D3	D*
KMLB	Monroe, LA 71211	5/1	U2	AN
WWGT	Portland, ME 04101	5	U2	AN
WVEI	Worcester, MA 01608	5	U4	AN7
WUNI	Bay City, MI 48707	5/2.5	U4	AN6
WMKM	Inkster, MI 48141	1	U4	AN7
KQRS	Minneapolis, MN 55422	5/0.5	U2	AN6
WRBE	Lucedale, MS 39452	5	D1	D
WBLA	Elizabethtown, NC 28337	5	D1	D*
WLXN	Lexington, NC 27292	5/1	U2	NH
WRRO	Warren, OH 44482	5	U4	AN6
KMED	Medford, OR	5/1	U1	AN6
	(Addr: Gold Hill, OR 97525)			
KODL	The Dalles, OR 97058	5/1	U2	NH
WCDL	Carbondale, PA 18407	5	D1	D*
WSSL	Greenville, SC 29602	5	U2	AN7
WZYX	Cowan, TN 37318	5	D1	D*
WHDM	McKenzie, TN 38201	5	D3	D*
KPUR	Amarillo, TX 79102	5/1	U2	AN7
KEYS	Corpus Christi, TX 78469	1	U2	AN
KDNT	Denton, TX 76201	5/0.5	U2	NH
KETX	Livingston, TX 77351	5	D1	D*
WBBC	Blackstone, VA 23824	5	D1	D*
KJUN	Puyallup, WA 98371	5/2	U4	F.Pl.
WHIS	Bluefield, WV 24701	5/0.5	U1	AN6
WAJR	Morgantown, WV 26505	5/0.5	U4	AN6
WNFL	Green Bay, WI 54307	5/0.5	U4	AN
CJOI	Wetaskiwin, AB	10	U2	NH
CFCP	Courtenay, BC	1	U2	AN7

+lp: USA 26/Can 1

1450kHz

USA 178st's mostly with 1kW, Canada 34 st's 40W-1kW

CHEF	Granby, PQ	20	U4	NH
VSB1	Hamilton, Bermuda	1	U1	AN

1460kHz

WFMH	Cullman, AL 35055	5/0.5	U2	AN
WPNX	Phenix City, AL	5/1	U2	AN6
	(Addr: Columbus, GA 31902)			
KTYM	Inglewood, CA 90302	5/0.5	U4	AN7
KRQC	Salinas, CA 93912	5	U3	AN7
KKCS	Colorado Springs, CO 80904	5/0.5	U2	AN7
WZEP	DeFuniak Springs, FL 32433	5	D1	D
WZNZ	Jacksonville, FL 32204	5	U2	AN7
WXEM	Buford, GA 30518	5	D1	D*
KGGO	Des Moines, IA 50317	5	U2	AN7
WBPA	Elkhorn City, KY 41522	5	D1	D
WXOK	Baton Rouge, LA 70896	5/1	U2	AN6
WBET	Brockton, MA 02403	5/1	U2	NH
WBRN	Big Rapids, MI 49307	5/2.5	U2	NH
KDMA	Montevideo, MN 56265	1	U2	NH
KIRL	St. Charles, MO 63301	5/0.5	U4	AN
KKPR	Kearney, NE 68847	5	D1	D*
KENO	Las Vegas, NV 89103	5/1	U2	AN7

WIFI	Florence, NJ (Addr: Burlington, NJ 08016)	5	D3	D*
WGNA	Albany, NY 12201	5	U2	NH
WWWG	Rochester, NY 14618	5	U2	AN7
WNBR	Fuquay-Varina, NC 27526	5	D3	D*
WEWO	Laurinburg, NC 28352	5	U4	NH
KLTC	Dickinson, ND 58601	5	U2	AN
WBNS	Columbus, OH 43215	5/1	U2	AN6
WIMX	Harrisburg, PA 17105	5	U2	AN7
WEMR	Tunkhannock, PA 18657	5/1.2	U4	NH
WBCU	Union, SC 29379	1	U2	NH
WACO	Waco, TX 76714	1	U2	AN7
WPRW	Manassas, VA 22110	5	U4	NH
WRAD	Radford, VA 24141	5/0.5	U2	NH
KARR	Kirkland, WA 98033	5/2.5	U4	AN
KMWX	Yakima, WA 98907	5	U2	AN7
WBUC	Buckhannon, WV 26201	5	D1	D*
CJMH	Medicine Hat, AB	10	U2	NH
CJOY	Guelph, ON	10	U4	AN7
CKRB	St-George-de-Beauce, PQ	10/5	U2	AN7

+lp: USA 36/Can 1

1470kHz

KUTY	Palmdale, CA 93550	5	U4	AN7
KXOA	Sacramento, CA 95808	5/1	U4	AN7
WMMW	Meriden, CT 06450	2.5	U4	AN
WLVU	Dunedin, FL 33528	5	D1	D*
WRBD	Fort Lauderdale, FL 33319	5/2.5	U4	AN
WRGA	Rome, GA 30161	5	U2	NH
WCFJ	Chicago Heights, IL 60611	1	U4	AN
WMBD	Peoria, IL 61604	5	U4	AN7
KWSL	Sioux City, IA 51102	5	U4	AN6
KERE	Atchison, KS 66002	1	U3	NH
WFSR	Harlan, KY 40831	5	D1	D*
KLCL	Lake Charles, LA 70602	5	U2	AN
WKZN	Lewiston, ME 04240	5	U3	AN7
WJDY	Salisbury, MD 21801	5	D3	D*
WTTR	Westminster, MD 21157	1	U2	NH
WSRO	Marlboro, MA 01752	5/0.5	U2	AN
WKMF	Flint, MI 48501	5/1	U4	AN6
WQSN	Kalamazoo, MI 49007	5/1	U2	
KBCW	Brooklyn Park, MN (Addr: Anoka, MN 55303)	0.5/5	U4	NH
WTKO	Ithaca, NY 14850	5/1	U2	AN6
WWBG	Greensboro, NC 27420	5/3.4	U2	
WPNC	Plymouth, NC 27962	5	D1	D*
WTOE	Spruce Pine, NC 28777	5	D1	D*
WDIX	Yadkinville, NC 27055	5/3.5	U4	F.Pl.
WWWM	Toledo, OH 43616	1	U4	AN7
KDUN	Reedsport, OR 97467	5	D1	D*
WXKW	Allentown, PA (Addr: Whitehall, PA 18052)	5	U2	AN7
WQXL	Columbia, SC 29230	5	D1	D*
WVOL	Berry Hill, TN 37207	5/1	U4	NH
KNTS	Abilene, TX 79604	5/1	U2	AN7
KWRD	Henderson, TX 75653	5	D1	D
KUFA	Tremonton, UT 84337	5	D1	D
WBTX	Broadway, VA 22815	5	D1	D*
WTZE	Tazewell, VA 24651	5	D1	D*
KELA	Chehalis, WA 98532	5/1	U1	NH
KBSN	Moses Lake, WA 98837	5/1	U4	NH
WHRD	Huntington, WV 25701	5	D1	D*
WBKV	West Bend, WI 53095	2.5	U4	NH
CJVB	Vancouver, BC	50	U4	AN
CHOW	Welland, ON	10	U4	AN7

+lp: USA 39/Can 1

1480kHz

WLPH	Irondale, AL (Addr: Birmingham, AL 35210)	5	D1	D*
WABB	Mobile, AL 36652	5	U2	AN7
KPHX	Phoenix, AZ 85034	5/0.5	U4	AN
KTHS	Berryville, AR 72616	5	D1	D*
KWUN	Concord, CA 94521	0.5/5	U4	AN
KRED	Eureka, CA 95501	5/0.68	U1	AN
KYOS	Merced, CA 95340	5	U2	AN6
KWIZ	Santa Ana, CA 92703	5	U4	AN6
WMIB	Marco Island, FL (Addr: Naples, FL 33942)	1	U4	NH
WYZE	Atlanta, GA 30312	5	D1	D*
WAKB	Augusta, GA 30903	5	U2	AN7
WTHI	Terre Haute, IN 47808	5/1	U4	AN7
KZSN	Wichita, KS 67220	5/1	U4	AN6
WNKY	Neon, KY 41840	5	D1	D*
WSAR	Fall River, MA 02722	5	U3	NH
WMAX	Grand Rapids, MI 49508	5	U2	NH
WSDS	Ypsilanti, MI 48197	0.5/3.8	U4	
KAUS	Austin, MN 55912	1	U4	AN6
KKCQ	Fosston, MN 56542	5/2.5	U2	
KGCX	Sidney, MT 59270	5	U3	NH
KFMQ	Lincoln, NE 68506	5/1	U4	AN7
KKEL	Hobbs, NM 88240	5/1	U2	AN
WZRC	New York, NY 10021	5	U4	AN7
WADR	Remsen, NY 13438	5	D1	D*
WCNT	Charlotte, NC 28232	5	U4	AN7
WAJA	Franklin, NC 28734	5	D1	D
WHBC	Canton, OH 44711	5	U2	AN6
WCIN	Cincinnati, OH 45217	5/0.5	U4	AN7
WDAS	Philadelphia, PA 19131	5/1	U4	AN7
WISL	Shamokin, PA 17872	1	U2	NH
WBBP	Memphis, TN 38104	5	D1	D*
KCMZ	Dallas, TX 75238	5/1.9	U4	AN7
WCFR	Springfield, VT 05156	5	D1	D*
WBBL	Richmond, VA 23220	5	U4	
WTOY	Salem, VA 24153	5	D1	D*
KBMS	Vancouver, WA 98668	1/2.5	U2	AN7
WTDY	Madison, WI 53701	5	U4	AN7
CKAN	Newmarket, ON	10	U4	AN6
CKER	Edmonton, AB	10	U2	AN2
CHRD	Drummondville, PQ	10	U4	NH

+lp: USA 39/Can 3

1490kHz

USA 177st's, mostly 1kW, Canada 22 st's 40W-1kW

1500kHz

KRCK	Los Angeles, CA	50/14	U4	AN
KSJX	San Jose, CA 95112	50/5	U4	AN7
WFIF	Milford, CT 06460	5	D3	D
WTOP	Washington, DC 20016	50	U4	AN7
WCHM	Clarkesville, GA 30523	5/0.5	D1	D
WBRI	Indianapolis, IN 46220	5	D3	D
WLQV	Detroit, MI	50/5	U4	AN7
KSTP	St. Paul, MN	50	U2	AN7
WSMX	Winston-Salem, NC 27115	10/1	D6	D
KXVQ	Pawhuska, OK 74056	5/0.5	D5	D*
CKAY	Duncan, BC	10/1	U3	AN

+lp: USA 30

1510kHz

KFNN	Mesa, AZ 85202	10	D1	D*
KIRV	Fresno, CA 93726	10	D3	D*
KNSE	Ontario, CA	10/1	U4	AN6
KDKO	Littleton, CO	10/1.3	U4	AN7
WNLC	New London, CT	10/5	U4	NH
WSSH	Boston, MA	50	U4	AN7
WJCO	Jackson, MI 49203	5	D3	D*
KIDZ	Independence, MO 64052	10	D3	D*
WMHQ	Dover, NJ	10/1	U4	AN6
WAHT	Annville, PA (Addr: Lebanon, PA 17042)	5	D3	D
KMSD	Milbank, SD 57252	5/1	D1	D*
WLAC	Nashville, TN 37203	50	U2	AN7
KQHN	Nederland, TX 77627	10	D3	D
KLLB	West Jordan, UT 84084	10	D1	D
KGA	Spokane, WA 99203	50	U2	AN7
WAUK	Waukesha, WI 53186	10	D3	D*
CKOT	Tillsonburg, ON	10	D3	D
CJRS	Sherbrooke, PQ	10	U4	AN7

+lp: USA 36/Can 1

1520kHz

WZMG	Opelika, AL 36801	5	D3	D
KMPG	Hollister, CA 95024	5/3.5	D3	D
KTRO	Oxnard, CA 93034	10/1	U4	AN6
WTLN	Orlando, FL 32860	5	D3	D
WTIN	Seminole, FL (Addr: St. Petersburg, FL 33708)	1	U4	
WEXY	Wilton Manors, FL (Addr: Ft. Lauderdale, FL 33309)	3.5/1	U2	AN7
WHOW	Clinton, IL 61727	5/1	D1	D
WLGC	Greenup, KY 41144	5/0.5	D1	D
KACY	Lafayette, LA 70502	10/0.5	U6	AN
WTRI	Brunswick, MD 21716	9.3	D3	D
WGAM	Greenfield, MA 01302	10	D3	D
WQWQ	Muskegon, MI 49443	10/5	U4	

WMLM	St. Louis, MI 48880	1	U4	
KOLM	Rochester, MN 55901	10	D4	D*
KFXE	Camdenton, MO	10/4.2	D3	F.Pl.
KMPL	Sikeston, MO 63801	5/0.5	U7	NH
WWKB	Buffalo, NY 14209	50	U3	AN7
WDSL	Mocksville, NC 27028	5/1	D1	D
WARR	Warrenton, NC 27589	5	D1	D
WVOI	Toledo, OH 43613	1	U4	24h
KOMA	Oklahoma City, OK	50	U2	AN7
KFXX	Portland, OR	50/10	U4	AN7
WKZQ	Myrtle Beach, SC 29577	5	D3	D
WREA	Dayton, TN 37321	5/0.5	D1	D
KJSM	Sequim, WA	20	D3	F.Pl.

+Ip: USA 32

1530kHz

KFBK	Sacramento, CA 95825	50	U3	AN7
KHPY	Moreno Valley, CA 92399	50/13	D3	F.Pl.
WDJZ	Bridgeport, CT 06605	5	D3	D
WCRJ	Jacksonville, FL 32205	50	D3	D
WTTI	Dalton, GA 30722	10	D6	D
WWGZ	Lapeer, MI 48446	5	D3	D
KKCM	Shakopee, MN 55379	8.6	D3	D
WRPM	Poplarville, MS 39470	10/1	D1	D*
KHAT	Lincoln, NE 68506	5	D3	D*
WRTP	Durham, NC 27707	10	D3	D
WCKY	Cincinnati, OH 45202	50	U2	AN7
KXTD	Wagoner, OK (Addr: Broken Arrow, OK 74012)	5	D3	D
KOPY	Georgetown, TX	10/0.99	D4	D
KGBT	Harlingen, TX 78551	50/10	U2	AN
KCLR	Ralls, TX 79357	5/1	D1	D*
KSHY	Fox Farm, WY	10/1	U4	F.Pl.

+Ip: USA 31/Can 2

1540kHz

KASA	Phoenix, AZ 85009	10	D3	D
KLAU	Aptos, CA	10	U3	AN
KXED	Los Angeles, CA	50/10	U4	AN7
KXEL	Waterloo, IA	50	U2	AN7
WMDO	Wheaton, MD 20902	5	D1	D
WMYF	Exeter, NH 03833	5/2.5	D1	D
WPTR	Albany, NY	50	U3	AN7
WPGR	Philadelphia, PA	50	D3	D*
WECZ	Punxsutawney, PA 15767	5/1	D1	D
WTBI	Pickens, SC 29671	10/1	D1	D*
KSGB	Fort Worth, TX	50/5	U4	AN6
KEDA	San Antonio, TX 78204	5/1	U4	AN7
WRBN	Richmond, VA 23227	10	D3	D*
KLSY	Bellevue, WA 98005	5	U2	AN7
CHIN	Toronto, ON	50/30	U4	AN7

+Ip: USA 37

1550kHz

WAAJ	Huntsville, AL 35804	50/0.5	U4	AN7
KUAT	Tucson, AZ 85721	50	D1	D
KAPL	Apple Valley, CA 92307	5/0.5	U2	
KXEX	Fresno, CA 93777	5/2.5	U4	AN
KKHI	San Francisco, CA 94102	10	U4	AN7
KQXI	Arvada, CO	10	U2	D*
WLVX	Bloomfield, CT (Addr: West Hartford, CT 06110)	5/2.5	U4	NH
WRHC	Miami, FL 33135	10/0.5	U4	AN
WAMA	Tampa, FL 33601	10	D1	D*
WTHB	Augusta, GA 30903	5	D1	D
WYNX	Smyrna, GA 30080	50/0.5	U4	AN7
WLUX	Baton Rouge, LA 70821	5/0.5	U2	
KVKI	Shreveport, LA 71137	10/0.5	U2	AN6
WNTN	Newton, MA 02166	10	D1	D
WOKJ	Jackson, MS	50/10	U4	NH
KLFJ	Springfield, MO 65802	5	D	D*
KAPE	Cape Girardeau, MO 63701	5	D3	D*
KSFT	St. Joseph, MO 64508	5	U2	AN6
KSWV	Santa Fe, NM 87504	50/0.25	U2	F.Pl.
WJRZ	Toms River, NJ	6/3	U4	F.Pl.
KQWB	Fargo, ND	10/5	U2	AN
WARD	Pittston, PA 18640	10/0.5	U4	
WBSC	Bennettsville, SC 29512	10/5	U2	NH
WBCV	Bristol, TN 37621	5	D1	D*
WCTZ	Clarksville, TN 37041	2.5/0.25	U4	NH
KRGQ	West Valley City, UT	10/0.5	U2	AN7
WKBA	Vinton, VA	10	D3	D
WVAB	Virginia Beach, VA 23451	5	D1	D
KNTR	Ferndale, WA 98248	50/10	U2	NH
KSVY	Spokane, WA 99204	10/2.5	U2	AN
KVAN	Vancouver, WA 98662	10	U2	NH
WXVA	Charles Town, WV 25414	5	D1	D*
WHIT	Madison, WI 53711	5	D3	D*
CBE	Windsor, ON	10	U3	NH

+Ip: USA 43/Can 4

1560kHz

WTKN	Daleville, AL 36322	5	D1	D
KNZR	Bakersfield, CA	10	U3	AN
WRHC	Miami, FL 33135	10/4.4	U4	F.Pl.
WINV	Inverness, FL 32652	5/2.3	U2	
WTAI	Melbourne, FL 32901	5	D1	D*
KCJJ	Iowa City, IA 52244	1	SH4	AN7
WPAD	Paducah, KY 42001	10/5	U4	NH
WSLA	Slidell, LA (Addr: New Orleans, LA 70130)	1	U2	
WHEZ	Portage, MI (Addr: Kalamazoo, MI 49002)	5	D3	D*
KQYX	Joplin, MO	10	D3	D*
WQXR	New York, NY	50	U4	AN6
WMJO	Webster, NY 14580	10	D3	F.Pl.
WTRQ	Warsaw, NC 28398	10	D4	D
WCNW	Fairfield, OH 45014	5/1	D6	D*
WTOD	Toledo, OH 43614	5	D3	D*
KWCO	Chikasha, OK 73023	1	U2	NH
WAGL	Lancaster, SC 29720	50	D6	D*
KKAA	Aberdeen, SD	10/5	U4	AN2
WWGM	Nashville, TN 37212	10	D3	D*
KTXZ	West Lake Hills, TX (Addr: Austin, TX 78746)	2.5	U4	AN7

+Ip: USA 36

1570kHz

WTQX	Selma, AL 36701	5	D1	D*
KCVR	Lodi, CA 95241	5/0.5	U4	AN
KPRO	Riverside, CA 92504	5	D3	D*
KTGE	Salinas, CA 92905	5/0.5	U4	
WTWB	Auburndale, FL 33823	5	D1	D*
WQAI	Fernandina Beach, FL 32034	5	D1	D
WSSA	Morrow, GA 30260	5/0.05	U1	
WFRL	Freeport, IL 61032	5/0.5	U3	AN
WZCC	New Albany, IN 47150	2/0.4	U1	NH
WFEL	Baltimore, MD 21204	5	D1	D*
KYCR	Golden Valley, MN (Addr: Minneapolis, MN 55422)	2.5/0.24	U1	NH
WIZK	Bay Springs, MS 39422	5	D1	D*
WFLR	Penn Yan, NY 14527	5	D1	D*
WBUX	Doylestown, PA 18901	5/0.05	U3	
WNKX	Centerville, TN 37033	5	D1	D*
WCLE	Cleveland, TN 37311	5	D1	D*
WTRB	Ripley, TN 38063	28/0.5	U4	D
WMQA	Minocqua, WI 54548	5	D1	D
CFEZ	Taber, AB	10/5	U2	AN7
CHUB	Nanaimo, BC	10	U4	AN7
CKMW	Winkler, MB	10	U4	AN7
CFOR	Orillia, ON	10/5	U2	AN6
CHLO	St. Thomas, ON	10	U4	AN7
CKLM	Montreal, PQ	50	U4	AN7

+Ip: USA 63

1580kHz

KCWW	Tempe, AZ 85281	50	U2	AN7
KBLA	Santa Monica, CA	50	U4	AN6
KWYD	Colorado Springs, CO 80931	10	D1	D
WTCL	Chattahoochee, FL 32324	5	D1	D
WSRF	Fort Lauderdale, FL	10/5	U4	AN7
WBGB	Mount Dora, FL 32757	5	D1	D
WEAM	Columbus, GA 31902	2.3/1	U2	AN6
WTKT	Georgetown, KY	10	D3	D
KXZZ	Lake Charles, LA 70602	1	U2	AN7
WPGC	Morningside, MD	50/0.4	U4	
WAMY	Amory, MS 38821	5	D1	D
WESY	West Greenville, MS 38701	1	U2	
WZZJ	Pascagoula, MS 39567	5	D3	D
KZKL	Albuquerque, NM	10/0.05	U1	AN
WWOF	Camp LeJeune, NC	10	D1	D
WUIV	Icard Township, NC 28666	5	D3	D
WBBR	Travelers Rest, SC 29690	5/1	D1	D
WMRE	Knoxville, TN 37901	5/1	D1	D
WPUV	Pulaski, VA 24301	5/1	D1	D
CBJ	Chicoutimi, PQ	50	U3	NH

+Ip: USA 50

1590kHz

WGYJ	Atmore, AL 36504	5/1	U2	NH
WVNA	Tuscumbia, AL 35674	5/1	U2	AN7
KSSU	St. Johns, AZ 85936	5	U2	F.Pl.
KYDE	Pine Bluff, AR 71601	5/0.5	U4	
KLIV	San Jose, CA 95108	5	U2	AN7
KOGO	Ventura, CA 93003	5	U4	AN7
WQQW	Waterbury, CT 06704	5	U3	AN6
WPSL	Port St. Lucie, FL 33452	5	D1	D
WRXB	St. Petersburg Beach, FL 33711	5/1	U4	AN
WALG	Albany, GA 31702	5/1	U4	AN7
WQCH	La Fayette, GA 30728	5	D1	D*
WONX	Evanston, IL 60202	3.5/2.5	U2	NH
WAIK	Galesburg, IL 61401	5	D3	D*
WNTS	Beach Grove, IN (Addr: Indianapolis, IN 46203)	5/0.5	U4	NH
KVGB	Great Bend, KS 67530	5	U2	NH
WASY	Gorham, ME 04038	5/2.5	U4	
WJRO	Glen Burnie, MD 21061	1	U4	AN7
WTVB	Coldwater, MI 49036	5/1	U2	AN
KCNN	East Grand Forks, MN 56721	5/1	U2	NH
WZRX	Jackson, MS 39204	5/1	U2	AN7
KHIT	Sun Valley, NV (Addr: Reno, NV 89502)	5	D1	D*
WSMN	Nashua, NH 03061	5	U3	AN
WAUB	Auburn, NY 13021	0.5/1	U4	AN
WASB	Brockport, NY 14420	1	U4	NH
WGGO	Salamanca, NY 14779	5	D1	D*
WHPY	Clayton, NC 27520	5	D3	D*
WAKR	Akron, OH 44320	5	U2	AN7
KTIL	Tillamook, OR 97141	5/1	U2	NH
WCBG	Chambersburg, PA 17201	5/1	U2	AN
WCZN	Chester, PA 19014	2.5/1	U2	NH
WARV	Warwick, RI 02886	5	U4	
WKTP	Jonesborough, TN (Addr: Kingsport, TN 37662)	1.6/5	U3	AN
KELP	El Paso, TX 79905	5	D	D
KYOK	Houston, TX 77004	5	U2	AN
KLLL	Lubbock, TX 79412	1	U4	AN7
WFTH	Richmond, VA 23221	5	D1	D*
KZOK	Seattle, WA 98119	5	U2	AN7
WIXK	New Richmond, WI 54017	5	D1	D*

+lp: USA 50

1600kHz

WEUP	Huntsville, AL 35806	5/0.5	U2	AN7
WXVI	Montgomery, AL 36195	5/1	U4	AN
KXEW	So. Tucson, AZ 85713	2.5/1	U4	AN7
KNWA	Bellefonte, AR 72601	5	D1	D
KGST	Fresno, CA 93775	5	U2	AN
KMNY	Pomona, CA (Addr: Anaheim, CA 92802)	5	U2	AN7
KUBA	Yuba City, CA 95991	5/2.5	U2	AN
KRXY	Lakewood, CO (Addr: Denver, CO 80227)	5	U2	AN7
WKEN	Dover, DE 19901	5/1	U4	NH
WQBR	Atlantic Beach, FL 32233	5	D1	D*
WPOM	West Palm Beach, FL 33407	5/4.6	U3	AN7
WXTO	Orlando, FL 32804	5	U3	AN6
WAOS	Austell, GA 30001	5	D1	D
KLGA	Algona, IA 50511	5/0.5	U4	
KCRG	Cedar Rapids, IA 52401	5	U2	AN6
WINX	Rockville, MD 20850	1/0.5	U2	NH
WUNR	Boston, MA 02114	5	U3	AN7
WIXY	East Longmeadow, MA 01028	5/2.5	U4	AN7
WAAM	Ann Arbor, MI 48106	5	U4	AN
WSFN	Muskegon, MI 49444	5	U2	AN
KWOM	Watertown, MN 55388	5	U3	F.Pl.
KATZ	St. Louis, MO 63101	5	U2	AN7
WWRL	New York, NY (Addr: Woodside, NY 11377)	5	U4	AN7
WGIV	Charlotte, NC 28236	1	U2	AN7
WTZQ	Hendersonville, NC 28739	5	D3	D*
WRNC	Reidsville, NC 27320	1	U2	
KEED	Eugene, OR 97440	5/1	U2	AN7
WAYC	Bedford, PA 15522	5	D1	D*
WWBR	Harriman, TN 37748	5	D1	D*
KBBB	Borger, TX 79007	5	D1	D*
KBOR	Brownsville, TX 78520	1	U4	NH
KSSA	Plano, TX (Addr: Dallas, TX 75201)	5/1	U4	AN
KOGT	Orange, TX 77630	1	U4	AN6
KBBX	Centerville, UT (Addr: Salt Lake City, UT 84110)	5/1	U2	NH
WJQI	Chesapeake, VA 23324	5	D1	D*
WXMY	Saltville, VA 24370	5	D1	D
WNST	Milton, WV 25541	5	D1	D*
WBBD	Wheeling, WV 26003	5	D1	D*
WCWC	Ripon, WI 54971	5	U4	NH
CHNR	Simcoe, ON	10	U4	NH

+lp: USA 37/Can 1

CENTRAL AMERICA AND THE CARIBBEAN

Abbreviations particular to this section of the medium wave tables: CP = Construction Permit; Dom. Rep. = Dominican Republic; LV = La Voz; Neth. Ant. = Netherlands Antilles; nf = nominal frequency; vf = varying frequency. Others, see Main Abbreviations list.

The sign o indicates that location is not given where company name includes the name of the town concerned.

Where several tr's are indicated for the same st. or network, a full list is given in the country section.

México: Stations with a power of 1 kW or less not mentioned. Please see México pages.

kHz	kW	Call	Country	Station/Location
530	18	TICAL	Costa Rica	R. Rumbo, Cartago (vf)
535	20		Grenada	GBC AM 535, St. George's
540	1	CMHV		Cuba R. Rebelde, Sancti Spíritus
	5	HICM	Dom. Rep.	R. ABC, Sto Domingo
	10	YSHV	El Salvador	R. Restauración, San Salvador
	150	XEWA	México	Super Estelar/La Voz de la América Latina, Monterrey
	5	XETIN		R. 5-40, Tijuana
	1	YNARC	Nicaragua	R. Corporación, Managua
	10	HOU23	Panamá	R. Mía, David (rel 650)
550	20/2	TISCL	Costa Rica	R. Sta Clara, Cd. Quesada
	10	CMAA	Cuba	R. Rebelde, Pinar del Río
	1	CMNA		R. Rebelde, Manzanillo
	2	YSFG	El Salvador	R. Variedades, Sonsonate
	3	HRXD	Honduras	R. Manantial, Sta Rosa de Copán
	5		Jamaica	RJR/Montego Bay
	5	XEPL	México	La Super Estación, Cd. Cuauhtemoc
	2	XEQW		La Poderosa, Mérida
	10	XETNC		R. Atlán, Tepic
	50	HOPU	Panamá	La Emisora Líder, Panamá
	5	WPAB	Puerto Rico	R. 5-50, Ponce
555	10	YNCDJ	Nicaragua	R. 19 de Julio, Chinandega
	20	ZIZ	St. Kitts &	R. ZIZ, St. Kitts
560	5	HIAA	Dom. Rep.	R. Ritmos, Stgo
	10	TGRV	Guatemala	R. 560, Guatemala
	1	TG..		R. Quetzal, Malacatán
	1	HRJO	Honduras	R. Campeona, Comayagua
	1	HRPX		R. Tropical, San Pedro Sula
	1.5	HRRZ2		R. Juticalpa, o
	5		Jamaica	JBC. R. One, Kingston
	5/0.75	XEOC	México	R. Chapultepec, México
	2.5	XESRD		R. Sierra, Stgo Papasquiaro

	2	XEXZ		Zacatecas
	1	HOH2	Panamá	RPC Radio, Colón (rel 610)
570	20	TICDL	Costa Rica	R. Fundación, Desamparados
	1	CMDC		R. Rebelde, Pilón
	5	HIMS	Dom. Rep.	R. Cristal, Sto Domingo
	10	YSKT	El Salvador	R. Cadena Central, San Salvador +2 repeaters
	1	TGPA	Guatemala	R. Palmeras, Escuintla
	1	HRLP2	Honduras	R. América, Tela (rel 610)
	1	HRLP10		R. Ceiba, La Ceiba
	5	XEBJB	México	La Metrópoli Norteña, Monterrey
	5	XEOA		R. Oaxaca, o
	5	XEME		R. Valladolid, o
	2.5	YNALP	Nicaragua	R. La Primerísima, Managua (rel 680)
	5	HOS	Panamá	R. Soberana Civilista, Panamá
580	5	CMKF	Cuba	R. Rebelde, Baracoa
	10	CMAN		R. Rebelde, Mantua
	3	TGY	Guatemala	R. Progreso/La Onda Alegre, Guatemala
	3	HRZQ	Honduras	R. Tegucigalpa, o
	10		Jamaica	RJR, Galina
	10/1	XEAV	México	R. Guadalajara/Canal 58, o
	5	XEFI		R. Mexicana, Chihuahua
	5/0.25	XEMU		R. Rancherita del Aire, Piedras Negras
	5	XEUE		LV de Chiapas, Tuxtla Gutiérrez
	10	HOH4	Panamá	RPC Radio, David (rel 610)
	10	WKAQ	Puerto Rico	R. Reloj, San Juan
590	40	TIRN	Costa Rica	R. Nacional, San José
	150	CMW		R. Rebelde, La Julia
	5	CMDN		R. Rebelde, Guantánamo
	60	CMHI		R. Reloj, Sta Clara
	10/5	HIDV	Dom. Rep.	R. Sta María, La Vega
	5	TGRQ	Guatemala	R. Quiché, Sta Cruz
	1	HRLP3	Honduras	R. América, San Pedro Sula (rel 610)
	3	HRVW		La Voz del Centroamérica, Siguatepeque (rel 650)
	5/0.25	XEFD	México	La Consentida, Reynosa
	5	XEPH		R. 590/R. Alicia, México
	1	YNNCN	Nicaragua	LV de Chontales, Juigalpa
	10	HOH3	Panamá	RPC Radio, Chitré (rel 610)
595	10		Dominica	DBS Radio, Hillsborough
600	150	CMKV	Cuba	R. Rebelde, Urbano Noris
	3	YSNK	El Salvador	R. Exitos, San Salvador
	1	TGRC	Guatemala	R. Campesina, Tiquisate
	1	4VIA	Haiti	R. Lumière, Carrefour Paye (cfr 720)
	1	HRVW	Honduras	LV del Centroamérica, La Ceiba (rel 650)
	3	XEBB	México	La Grande, Acapulco
	5	XEGTO		R. Capital, Guanajuato
	5	XEZ		R. 600, Mérida
	10	YNART	Nicaragua	R. Ya, Managua (vf)
	1	WAEL	Puerto Rico	R. 600 AM, Mayagüez
610	4	TIRSU	Costa Rica	R. Sistema Universal, San José
	1	CMAN	Cuba	R. Rebelde, Bahia Honda
	1	CMHI		R. Reloj, Trinidad
	5/1	HIJR	Dom. Rep.	R. Acción, Stgo
	1	YSSS	El Salvador	R. El Salvador, Morazán (rel 655)
	0.5	TGGA	Guatemala	R. Señorial 610, Guatemala
	10	HRLP	Honduras	R. América, Tegucigalpa
	1	HRLP4		R. América, Sta Rosa de Copán (rel)
	5/0.5	XEBK	México	La Primera/Estación, Sabinas
	5	XEEL		R. Progreso, Fresnillo
	5/1	XEUF		R. Variedades, Uruapan
	10	HOHM	Panamá	RPC Radio, Panamá
	1/0.5	WEXS	Puerto Rico	WEXS, Patillas (F.Pl.)
	50	VPL6	Trinidad & Tobago	NBS Radio 610, Trinidad
620	10		Antigua & Barbuda	Antigua & Barbuda Bc. Sce, St. Johns, Antigua
	30	CMGN	Cuba	R. Rebelde, Colón
	1	CMKF		R. Rebelde, Moa
	10	HISD	Dom. Rep.	R. TV Dominicana, Sto Domingo
	5	TGPQ	Guatemala	R. Occidental/R. 6-20, San Cristóbal
	1		Honduras	R. América, 3 st's (rel. 610)
	1	HRLP17		R. Continental, San Pedro Sula
	5		Jamaica	JBC. R. One, Mandeville
	5/0.25	XEBU	México	La Norteñita/La B-U, Chihuahua
	10/5	XENK		R. 6-20, México
	50	YNALN	Nicaragua	R. Nicaragua, Managua
625	1		St. Lucia	R. St. Lucia (rel 660)
630	1	HIAF	Dom. Rep.	R. Monte Cristi, o
	10	YSLN	El Salvador	La Monumental, San Salvador
		TGEL	Guatemala	R. El Porvenir, Sta. Elena
	1		Honduras	R. América, 3 st's (rel. 610)
	5/0.25	XEACA	México	R. Voz, Acapulco
	10/1	XEFB		La Estación de las Noticias, Monterrey
	5/0.75	XEFU		LV Amiga, Cosamaloapan
	4/0.25	XETK		R. Hit/La T-K, Mazatlán
	3/0.5	XETS		R. Tapachula, o
	1	HOJ35	Panamá	R. Provincias, Chitré
	5	WSKN	Puerto Rico	La Gran Cadena, San Juan
640	1	TIJJ	Costa Rica	R. Musical, Moravia (rel TIJC 700 kHz)
	50	CMBC	Cuba	R. Progreso, Guanabacoa
	10	CMDQ		R. Progreso, Las Tunas
	5	CMBC		R. Rebelde, Las Mercedes
	1	HISD	Dom. Rep.	R. TV Dominicana, Santiago
	2x20		Guadeloupe	RFO, Pointe á Pitre
	11.5	TGW	Guatemala	R. Nacional/La Voz de Guatemala (inactive)
	0.4	4VJS	Haiti	La Voix du Seigneur, Port-au-Prince
	3	HRNN4	Honduras	R. Exitos, Tegucigalpa
	10	HRVW		La Voz del Centroamérica, Sta Rosa de Copán (rel 650)
	10	XEYQ	México	R. Fresnillo, o (n. 890)
	10	XENQ		LV de la Provincia, Tulacingo
	10	YNALN	Nicaragua	R. Nicaragua, Managua (rel 620)
	12	YNLRI		R. Insurreción, Matagalpa
	1	HOK22	Panamá	CPR–Cadena Panameña de Radiodifusión, Colón
650	1	CMKU	Cuba	R. Rebelde, Stgo de Cuba
	1	CMDC		R. Rebelde, Media Luna
				R. Sandino, Pinar del Río
	20	HIAT	Dom. Rep.	R. Universal, Sto Domingo
	50	HRVW	Honduras	La Voz de Centroamérica, San Pedro Sula
	5	XEZM	México	R. La Expresión de Zamora, o
	10	YNALN	Nicaragua	R. Nicaragua, Bluefields (rel 620)
	10	YNELS		R. Sport, Granada
	5	HOS22	Panamá	R. Mía/Cadena Nacional, Panamá
655	10	YSS	El Salvador	R. El Salvador, San Salvador (nf 660)
660	30	CMHG	Cuba	R. Progreso, Sta Clara
	5	HIAM	Dom. Rep.	R. Quisqueyana, Stgo
	10	TGQ	Guatemala	R. Nacional/LV de Quetzaltenango, o
	5/1	4VI	Haiti	R. Lumière, Port-au-Prince
	3	HRNN18	Honduras	LV de Honduras, La Ceiba (rel 670)
	10	XEEY	México	La Consentida, Jalpa
	10	XEFZ		R. Metrópoli, Monterrey (n. 740)
	10/50	XERIN		R. 660, México
	5	XEAR		La Mexicana, Tampico

	75	YNALN	Nicaragua	R. Nicaragua, Estelí (rel 620)
	1	HOF33	Panamá	RPC Radio, Bocas del Toro (rel 610)
	10		St. Lucia	R. St. Lucia, Babonneau
670	10	TITNT	Costa Rica	R. Monumental, San José
	50	CMQ	Cuba	R. Reloj, Arroyo Arenas
	5	HIBS	Dom. Rep.	R. Dial, San Pedro de Macorís
	10	TGRT	Guatemala	R. Fabulosa, Guatemala
	10	HRN	Honduras	LV de Honduras, Tegucigalpa +1 repeater
	5	XEIS	México	Onda 6-70/La Rancherita Contenta, Cd. Guzmán
	10	YNELD	Nicaragua	R. Diriangen, Granada
	10	YNPLC		R. Caribe, Pto Cabezas
	5	HOLY	Panamá	R. Hogar, Panamá
680	1	CMHN	Cuba	R. Progreso, Cienfuegos
	1	CMDB		R. Progreso, Stgo de Cuba
	10	CMJN		R. Rebelde, Ciego de Avila
	3/2	HIJX	Dom. Rep.	R. Zamba, Stgo Rodriguez
	10	YSSS	El Salvador	R. El Salvador, Cabañas (rel 655)
	10	TGVP	Guatemala	R. Norte, Cobán
	1		Honduras	LV de Honduras, 4 st's (rel 670)
	5/0.5	XEKQ	México	El Eco de Tacana, Tapachula
	10/5	XELG		L-G La Grande, León
	10	YNALP	Nicaragua	R. La Primerísima, Managua
	5	HOF32	Panamá	Super Z Estéreo, David
	10	WAPA	Puerto Rico	R. Guapa, San Juan
690	15		Anguilla	The Caribbean Beacon
	20	CM..	Cuba	R. Progreso, Jovellanos
	10	HIAV	Dom. Rep.	R. Guarachita, Sto Domingo
	1	TGVX	Guatemala	Tamazulapa Radio, Jutiapa
	1	HRNN19	Honduras	LV de Honduras, 2 st's (rel 670)
	20/5	XEN	México	R. Mundo, México
	5	XERG		La Deportiva 690, Monterrey
	2.5	XEUY		Coatzacoalos
	2.5	XEST		R. Sensación, Mazatlán
	50	XETRA		Extra Newsradio, Tijuana
	5	XECS		La Grande, Manzanillo
	10	HOR43	Panamá	R. Veraguas, Stgo
700	20	TIJC	Costa Rica	R. Sonora, San José
	1	CMDU	Cuba	R. Enciclopedia, Guantánamo
	1	CMGA		R. Rebelde, Sanctí Spíritus
	1.5	HIDC	Dom. Rep.	R. Mao, o
	5	YSUU	El Salvador	R. Cadena YSU, San Salvador +5 repeaters
	15	TGHR	Guatemala	R. Mundial, Guatemala (vf)
	1	HRKL	Honduras	R. Reloj, Tegucigalpa
	10		Jamaica	JBC. R. One, Montego Bay
	10	XEVC	México	R. Cordobesa, Córdoba
	5	XEYK		R. Mágica, Mérida
	10	YNALI	Nicaragua	R. Istmo, Managua
	5	HOQ51	Panamá	KW Continente, Panamá
705	10		St. Vincent & Grenadines	NBC 705 Radio, Kingstown St. Vincent
710	30	CMJN	Cuba	R. Rebelde, Camagüey
	50	CMHQ		R. Rebelde, Sta Clara
	10	CMHQ		R. Rebelde, Holguín
	1	TGXL	Guatemala	R. Tecún Umán, Quetzaltenango
	2	HRKN		LV de Olancho, Catacamas
	1	HRLK		R. Comayagua, o (vf)
	1	HRNN		Estéreo Rey, San Pedro Sula
	1	HRNN13		LV de Honduras, Olanchito
	5	HRRH		LV de Occidente, Sta Rosa de Copán
	5	XEBL	México	R. Bonita, Culiacán
	5	XEDP		R. D-P/La Ranchera de Cuauhtémoc, o
	3	XEON		R. Mexicana/La Nueva XEON, Tuxtla Gutiérrez
	3.5	HOB52	Panamá	R. Ondas del Caribe, Changuinola
	10/0.75	WKJB	Puerto Rico	KJB, Mayagüez
720	1	CMFC	Cuba	R. Rebelde, Cienfuegos
	5	HIAQ	Dom. Rep.	R. Norte, Stgo
	10	YSR	El Salvador	Circuito YSR, San Salvador
	5	YSRA		Circuito YSR, San Miguel
	1	YSRA		Circuito YSR, Sta Ana
	1	TGRO	Guatemala	R. Corona, Morales
	1	4VIA	Haiti	R. Lumière, Carrefour Paye (nf 600)
	1	HRNN3	Honduras	R. Caribe, La Ceiba
	10		Jamaica	RJR/Kingston
	5	XEAVR	México	Alvarado
	10	YNACE	Nicaragua	R. Católica, Managua
	5	HOB50	Panamá	R. República, Chitré
730	10		Belize	R. Belize, Ladyville
	25	TIHB	Costa Rica	R. Reloj, San José
	1	CMOB	Cuba	R. Progreso, La Fé
	10	HIZ	Dom. Rep.	Emisoras Nacionales, Sto Domingo
	10	TGN	Guatemala	R. Cultural, Guatemala
	1	HRQQ	Honduras	R. Intibucá, La Esperanza (nf 740)
	1	HRTG		R. Televisión, Tegucigalpa
	100	XEX	México	La Super Grande, México
	5	XEQN		Torreón
	10	YNJSR	Nicaragua	R. Segovia, Ocotal
	20		Trinidad & Tobago	R. Trinidad, Port of Spain, Trinidad
740	20	CMKP	Cuba	R. Progreso, Sagua de Tánamo
	10	CMJL		R. Progreso, Camagüey
	10	ZGBC	Dominica	Caribbean Voice of Life
	1	TGHF	Guatemala	R. Tacaná, San Marcos
	1	4VIE	Haiti	R. Lumière, Pignón-le-Jeune
	1	4V..		R. Reveille, Cap Haïtien
	1	HRNN23	Honduras	R. Eco, San Pedro Sula
	1.2	HRIH		R. 7-40, Juticalpa
	10	HON26	Panamá	R. Cristal, David
	10	HOR44		CMQ-La Exitosa, Panamá
	10	WIAC	Puerto Rico	R. Internacional, San Juan
750	1	CMHV	Cuba	R. Rebelde, Trinidad
	5	HIDB	Dom. Rep.	R. Alegre, Stgo
	1	YSSS		R. El Salvador, Sta Ana (rel 655)
	1	TGAJ	Guatemala	R. Tropicana, Escuintla
	1	HRJH	Honduras	R. Trujillo, o
	1	HRXK		LV de la Mosquitia, Pto Lempira
	10		Jamaica	JBC. R. One, Port Maria
	5	XEKOK	México	R. Sensación, Acapulco
	5	XEQF		R. Loma, Loma Bonita
	50	YNARS	Nicaragua	R. Sandino, Managua
	5	HO..	Panamá	R. Inolvidable, Chitré
760	20	TILX	Costa Rica	R. Columbia, San José
	5	CMCD	Cuba	R. Reloj, Guanabacoa
	20	CMDE		R.Reloj, Las Mercedes
	5	HICO	Dom. Rep.	R. Cordillera, Sto Domingo
	10	TGHB	Guatemala	Nueva Super R, Guatemala
	5	4VU	Haiti	R. Lumière, Les Cayes
	5	HRXW	Honduras	R. Comayagüela, Tegucigalpa
	20/5	XEABC	México	ABC R, "la estación de la palabra", México
	5	XELAC		LV de las Balsas, Cd. Lázaro Cárdemas
	5	XEYW		Tropi Ranchera, Mérida
	5	XEZZ		R. Gallito, Guadalajara
	10	YNARS	Nicaragua	R. Sandino, Managua (irr)
	5	HOXO	Panamá	La Voz del Istmo, Panamá
	5	WORA	Puerto Rico	La Gran Cadena, Mayagüez
770	1	HIMD	Dom. Rep.	R. Aguila, Stgo
	1	TGBX	Guatemala	R. Fraternidad, Quetzaltenango
	2.5/1	HRRO	Honduras	R. Majestad, Juticalpa (cfr 1310)

	1	HRMV		R. Aguán, Olanchito
	10	HRNN21		R. Norte, San Pedro Sula (r. 770.6)
	5		Jamaica	RJR/Spur Tree
	5	XEACH	México	La Regiomontana, Monterrey
	10/0.2	HEIH		Nueva Dimension en Radio, Fresnillo
	10	HOL83		LV de Herrera, Chitré
780	10	ZBVI	British Virgin Isl.	ZBVI Radio, Tortola
	2	TIRA	Costa Rica	R. América Latina, San José
	0.5	HIBO	Dom. Rep.	R. Constanza, o
	10	YSKL	El Salvador	R. Cadena YSKL, San Salvador +4 repeaters
	0.5	4VO	Haiti	R. Lumière, Jérémie
	5	XELD	México	R. Costa, Autlán
	5/1	XESFT		La Bronca Norteña, San Fernando
	10	HOB35	Panamá	R. Chiriquí, David
	0.05	AHD	Puerto Rico	AFRTS, Ramey Air Force Base, Aguadilla
790	20		Barbados	V. of Barbados, Bridgetown
	30	CMAQ	Cuba	R. Reloj, Pinar del Río
	5	HIL	Dom. Rep.	La Voz del Trópico, Sto Domingo
	6	TGO	Guatemala	R. Festival, Guatemala
	1	HRIR	Honduras	R. Relámpago, Sta Bárbara
	3	HRTG3		R. Satélite, Tegucigalpa
	5/0.75	XENT	México	R. La Paz, o
	10/1	XERC		Expresión 7-90, México
	5/1	XERPC		R. Ranchito/R-P-C, Chihuahua
	5/0.25	XEVA		La Emisora del Hogar, Villahermosa
	10	ACA20	Panamá (Canal Zone)	SCN–Southern Command Network, Ft. Clayton
800	3	TIW	Costa Rica	R. Uno, San José
	1	CMDT	Cuba	R. Progreso, Manzanillo (vf, r. 792)
	20	YSAX	El Salvador	La Voz Panamericana, San Salvador
	1	TGYZ	Guatemala	R. Rosa, Chiquimulilla
	1	HRLP26	Honduras	R. Sonora, Danlí
	1	HRMA		R. Mundial, San Pedro Sula
	1	HRMD		R. Yoro, o
	150	XEROK	México	R. Cañón, Cd. Juárez
	500	PJB	Neth. Ant.	Trans World Radio, Bonaire
	15	HOL60	Panamá	La Exitosa, Chitré
810	1	ZNS3	Bahamas	R. Bahamas, Freeport
	10	CMDW	Cuba	R. Progreso, Guantánamo
	5	HIAV	Dom. Rep.	R. Baní, o
	1	YSFA	El Salvador	R. Lorenzana, San Vincente (nf 840)
	2	HRDL	Honduras	R. Corporación, Comayagua
	3	HRLP24		R. Valle, Choluteca
	1	HRVC		LV Evangélica, La Ceiba (rel 1390)
	50	XEFW	México	Voz e Imagen de Tampico, o
	3	XEMAX		R. Felicidad, Armería
	2	XEMQ		Tus Panteras, Mérida
	5	XERB		R-B/La Onda Musical del Caribe, Cozumel
	5/0.25	XERSV		R. Alegría, Cd. Obregón
	10	XEUX		La Gran Máquina Musical, Tuxpán
	10	YNGFS	Nicaragua	R. Rivas, o
	10	HOG	Panamá	R. Mundial, Panamá
	50	WKVM	Puerto Rico	AM-81, San Juan
820	2.5	TIGC	Costa Rica	R. Centro, San José
	10	CMCA	Cuba	R. Ciudad de Habana, o
	10	CMJT		R. Progreso, Morón
	10	CMDE		R. Reloj, Contramaestre
	5	HIAZ	Dom. Rep.	R. Stgo, o
	10	TGTO	Guatemala	R. Internacional, Guatemala
	5	HRLP16	Honduras	R. Moderna, Tegucigalpa
	1	HRKW		R. Sultana, Sta Rosa de Copán
	10	XEBA	México	La Consentida, Guadalajara
	3.5	XEYX		La Gigante del Cuadrante, Mexicali
	10	XEKG		R. Variedades, Córdoba
	10	YNAOL	Nicaragua	R. Ondas de Luz, Managua
	10	HOF28	Panamá	R. Ritmo, David
825	50		St. Kitts & Nevis	R. Paradise, Basseterre, St. Kitts
830	10		Belize	Belize R. One, Ladyville
	300	CM..	Cuba	R. Taino, La Julia
	10	HIJB	Dom. Rep.	R. HIJB, Sto Domingo
	4	YSPX	El Salvador	R. Pax, San Miguel
	5	TGAV	Guatemala	R. Satélite, Mazatenango
	1	HRJB	Honduras	R. Cadena Radial Impacto, Comayagua
	1	HRVQ		R. Excelsior, Juticalpa
	2	HRXS		R. Estereo Sula, San Pedro Sula
	10	XELA	México	Buena Música en México, México
	5	XELK		Onda Juvenil, Zacatecas
	5	XEVQ		La Grande de Sinaloa, Navolato
	10	YNPRZ	Nicaragua	R. Zinica, Bluefields (cfr 1090)
	1	HOR56	Panamá	R. Peninsula, Macaracas
840	10	CMHW	Cuba	Doblevé, Sta Clara
	1	CMKC		R. Revolución, Stgo de Cuba
	1	CMBQ		R. Enciclopedia, La Fé
	1	CMDQ		R. Progreso, Las Tunas
	5	HIAB	Dom. Rep.	R. Isabel de Torres, Pto Plata
	1	YSFA	El Salvador	R. Lorenzana, San Vicente (cfr 810)
	5	YSFB		R. Vanguardia, S. Salvador
	0.35	TGSM	Guatemala	LV de San Marcos, o
	10	4VEF	Haiti	4VEH-La Voix Evangelique d'Haïti, Cap Haïtien
	10	XEIO	México	R. Capital/La Emisora de Chiapas, Tuxtla Gutiérrez
	5	XEXXX		Triple X, Tamazula
	5	YNARN	Nicaragua	R. Noticias, Managua
	10	HOL80	Panamá	R. Nacional de Panamá, o
	5/1	WXEW	Puerto Rico	R. Victoria, Yabacua
	10		St. Lucia	R. Caribbean International.
850	20	TIXEW	Costa Rica	XEW La Exibosa, S. José
	1	CMHL	Cuba	R. Progreso, Trinidad
	5	CMKB		R. Progreso, Baracoa
	5	HIGA	Dom. Rep.	R. Guarocuya, Barahona
	5	HILR		R. Clarín, Stgo (rel 860)
	5	TGX	Guatemala	R. Ranchera, Guatemala
	0.8	4VGM	Haiti	MBC-Magloire Bc. Circuit, Port-au-Prince (nf 1430)
	10	HRUP	Honduras	R. Centro, Tegucigalpa
	10		Jamaica	JBC. R. One, Savanna-la-Mar
	5	XEM	México	R. Exitos/El Ritmo 8-50, Chihuahua
	10	XETQ		R. Sensación, Orizaba
	10	HOT61	Panamá	La Nueuva Exitosa Libre de Chiriquí, David
	3	HO..		CMQ-La Exitosa, Colón (rel 740)
	5/1	WABA	Puerto Rico	Super 85, Aguadilla
860	1	CMBL	Cuba	R. Metropolitana, Arroyo Arenas
	10		Dominica	Voice of the Islands/VOI
	50	HILR	Dom. Rep.	R. Clarín, Sto Domingo
	1	YSRC	El Salvador	R. Tecana, Sta Ana
	1	TGFP	Guatemala	R. Nal. Tikal (inactive)
	3	4VMK	Haiti	R. Min Kontré, Les Cayes
	10	HRBS	Honduras	R. San Pedro, San Pedro Sula
	1			Circuito R. Centro, 2 st's (rel. 850)

	1	HRH9		R. Nacional de Honduras, La Ceiba (rel 880)
	5	XEAL	México	La Onda Internacional de la Juventud, Manzanillo
	5	XECCN		R. Caribe, Cancún
	5	XECTL		R. Quintana Roo, Chetumal
	5	XEDU		DU la que le gusta, Durango
	5	XEMO		La Poderosa/R. 86, Tijuana
	5/2	XENL		R. Recuerdo, Monterrey
	50/25	XEUN		R. Unam, México
	10	PJZ86	Neth. Ant	R. Curom/Z 86, Willemstad, Curaçao
	10	HOL55	Panamá	R. Reforma, Chitré
870	12/6	TIUCR	Costa Rica	R. UCR, San José
	10	CMHL	Cuba	R. Progreso, Sancti Spíritus
	1	CMDT		R. Progreso, Pilón
	5	HIVG	Dom. Rep.	R. La Vega, o
	1	YSAR	El Salvador	R. 10/Cadena de Oro, San Salvador
	0.5	TGL	Guatemala	R. Victoria, Mazatenango
	1	4VJV	Haiti	R. Express Continentale, Jacmel
	1	HRH4	Honduras	R. Nacional de Honduras, Choluteca (rel 880)
	10	XETAR	México	LV de la Sierra Tarahumara, Guachachi
	5	XERO		R. Guerrero, Chilpancingo
	10	XETC		Estéreo Mayrán, Torreón
	5	HOHO	Panamá	R. Libre, Panamá
	5	WQBS	Puerto Rico	R. Voz, San Juan
880	30	CMAF	Cuba	R. Progreso, Pinar del Río
	1	HIOR	Dom. Rep.	R. Sta Cruz, Valverde
	1	YSCD	El Salvador	R. Ritmo, Stgo de María
	10	TGJ	Guatemala	R. Nuevo Mundo, Guatemala
	0.8	4VPM	Haiti	R. Indépendence Nouvelle, Gonaïves
	10	HRHQ	Honduras	R. Nacional de Honduras, Tegucigalpa +2 repeaters
	10	XEAAA	México	Románticos AM, Guadalajara
	5	XECTL		R. Quintana R. o, Chetumal
	5	XEOB		Pichucalco
	5	XEPNK		Canal 88/La Jovén, Los Michis
	3	XEQQQ		Super Q, Macuspana
	10	XETC		XETC-La Imagen del Sonido, Torreón
	5	YNARP	Nicaragua	R. El Pensamiento, Managua
	1	HOB51	Panamá	R. Hit, Colón
	1/0.5	WYKO	Puerto Rico	Radio 8-80, Sabana Grande
885	1	ZJB	Montserrat	R. Montserrat, Plymouth
890	10	TIHOT	Costa Rica	R. Fabulosa, San José
	1	CMHD	Cuba	R. Musical, Sta Clara
	1	CMDZ		R. Musical, Stgo de Cuba
	4/5	HIJP	Dom. Rep.	R. Continental, Sto Domingo
	3	YSLA	El Salvador	R. Musical, Sta Ana
	1	TGHU	Guatemala	R. Escuintla, o
	10/1	HRH2	Honduras	R. Nacional de Honduras, (rel 880)
	50	XENZ	México	Energía 8-90, Culiacán
	5	XEPC		Sonido Estrella, Zacatecas
	5	XETAP		R. Tapachula, o (cfr 970)
	2.5	XEUL		Tu Nueva Onda, Progresso
	10	HOQ62	Panamá	R. Ritmo, Chitré
	1	WFAB	Puerto Rico	WFAB, Las Piedras
895	0.5	4VVB	Haiti	R. Trans-Artibonite, Gonaïves (nf 890)
	10		St. Kitts & Nevis	VON Radio 895, Charlestown, Nevis
900	10		Barbados	CBC R. 900, Bridgetown
	50	CMKP	Cuba	R. Progreso, Cacocum
	5	HIBN	Dom. Rep.	R. Pto Plata, o
	10	YSQJ	El Salvador	Radiópolis, San Salvador
	1	TGMA	Guatemala	R. Amatique, Pto Barrios
	1		Honduras	Circuito R. Centro, 2 st's (rel 850)
	250	XEW	México	LV de la América Latina, México
	50/5	XEWB		LV de la América Latina, Veracruz (rel)
	1	YNART	Nicaragua	R. Tiempo, Managua
	1	HOHA	Panamá	LV del Pueblo, Panamá
910	1		Belize	R. Belize, Punta Gorda
	1	TIQM	Costa Rica	R. Metrópolis Norte, Cd. Quesada
	1	TIQM		R. Metrópolis, Tibás
	10	CMFA	Cuba	R. Cadena Agramonte, Camagüey
	10	CMGL		R. Reloj, Bolondrón
	5	HILB	Dom. Rep.	R. 91, Bonao
	10	TGKL	Guatemala	R. Emperador, Guatemala
	10	HRVS	Honduras	R. Católica/LV de Suyapa, Tegucigalpa
	5	XEACN	México	La Nueva Onda, León
	10	HOL81	Panamá	R. Nacional Guaymí, David
	1	HOL85		R. Nacional Cristóbal, Colón
	5	WPRP	Puerto Rico	La Gran Cadena, Ponce
920	1	CMGL	Cuba	R. Reloj, Unión de Reyes
	5	HIBA	Dom. Rep.	R. Diamante, Sto Domingo (inactive)
	0.2	TGRS	Guatemala	R. Sur, Escuintla
	1	4VKB	Haiti	R. Cap Haïtien, Petite Anse (rel 930)
	1	HRRM	Honduras	R. Sistema, Comayagua
	5	HRSK		R. Catacamas, o
	5	HRZV		R. Fabulosa, San Pedro Sula
	5/1	XEBH	México	XEBH-La Correcaminos, Hermosillo
	5/1	XEBM		La Rancherita, San Luís Potosí
	5	XECQ		C-Q La Ranchera de Culiacán, o
	10	XELT		La Sabrosita, Guadelajara
	5	XERE		R. Cañon, Salvatierra
	5/0.25	XEVV		Doblevé Juvenil, Chiapa de Corzo
	2.5	XESDA		Rock 92, Ensenada
	1	YNARM	Nicaragua	R. Mundial, Managua
	1	HOS56	Panamá	R. Mía, Los Santos (rel 650)
930	1		Belize	R. Belize, Corozal
	5	TIRCR	Costa Rica	R. Saprissa, "LV del Monstruo", Zapote
	1	CM..	Cuba	R. Reloj, 4 st's
	5	HICK	Dom. Rep.	Ondas del Yaque, Stgo
	5	YSTG	El Salvador	R. Sonora, San Salvador +5 repeaters
	5	TGJL	Guatemala	R. Imperial, San Pedro Carchá (vf)
	10/3	4VKB	Haiti	R. Cap Haïtien, o
	5	XEMK	México	R. Progreso, Huixtla
	5/1	XEU		La U de Veracruz/El Eco de Sotavento, Veracruz
	100		Montserrat	R. Antilles, Plymouth
	2.5	HOK85	Panamá	Mi Preferida Estéreo, Pto Armuelles
	3	HOR43		CMQ-La Exitosa, La Chorrera (rel 740)
	2.5	WEKO	Puerto Rico	WEKO/NotiUno, Cabo Rojo
940	1		Belize	R. Belize, San Ignacio
	10	CMGU	Cuba	R. Reloj, Colón
	10	CMKD		R. Reloj, Holguín
	5	TGTL	Guatemala	LV del Hogar, Guatemala
	0.2	4VFF	Haiti	R. Diff. Jacméllienne, Jacmel
	0.4	4VLF		R. Saint Marc Haïti, Saint Marc
	1	HRBO	Honduras	R. Cadena Occidental, Sta Rosa de Copán
	5	HRCR		R. Dif. Cristiana, Tegucigalpa

	50	XEQ	México	La Q Mexicana, México
	5	HOX	Panamá	R. TV2, Panamá (inactive)
	10	WIPR	Puerto Rico	La Emisora del Pueblo, San Juan
950	5	CMBF	Cuba	R. Musical, Arroyo Arenas
	1	CMHI		R. Reloj, Sancti Spiritus
	10	HIG	Dom. Rep.	R. Popular, Sto Domingo
	1	YSGH	El Salvador	R. Minuto, San Miguel
	1	TGAF	Guatemala	R. Indiana, Mazatenango
	3	HRQL	Honduras	R. Centro Radial, Siguatepeque
	1.2	HRZE		R. Cortés, Pto Cortés
	5	XECEL	México	R. Fiesta, Celaya
	10/5	XEKAM		R. 95 de California, Tijuana
	5	XEMEX		La Mexicana, Cd. Guzmán
	5/0.5	XEORF		R. Voz, El Fuerte
	10	XEPB		R. Amor, Hermosillo
	5	XEYJ		La Nueva Imagen del R, Nueva Rosita
	1	HOL84	Panamá	R. Nacional Victoriano Lorenzo, Penonomé
	0.5	HOM33		R. Cadena Azul "RCN", La Concepción
960	10	P4A6	Aruba	R. Victoria, Oranjestad
	1	CMHJ	Cuba	R. Reloj, Cienfuegos
	1	CMGF		R. Enciclopédia, Matanzas
	0.25	CMID		R. Musical, Ciego de Avila
	10	CMDJ		R. Reloj, Guantánamo
	5	HIFF	Dom. Rep.	LV del Atlántico Pto Plata
	0.5	YSTW	El Salvador	R. Centro, Sonsonate
	1	TGRU	Guatemala	R. Itatlán, Sta Cruz del Quiché
	0.3	4VCD	Haiti	R. Carillon, Port-au-Prince
	1	HRRD3	Honduras	R. Sangrelaya, o
	1	HRYF		R. Ferguson, Choluteca
	5/1	XEK	México	La Estación Grande de Nuevo Laredo, o
	5	YNOTO	Nicaragua	R. 13 de Octubre, San Carlos
	10	HOK71	Panamá	Onda Popular, Panamá
	1	WJYT	Puerto Rico	R. Salvación, Quebradillas
970	5/1	WSTX	American Virgin Isl.	Virgin Isl Bc Corp., Christiansted, St. Croix
		HIVP	Dom. Rep.	R. Olímpia, La Vega
	1	YSMS	El Salvador	R. UTEC, San Salvador
	5	TGAX	Guatemala	R. Continental, Guatemala
	3	HRTL	Honduras	Estéreo Tic Tac, Tegucigalpa
	10/5	XEDF	México	D-F/R. Noticias, México
	5	XEEZ		La Super Z, Caborca
	10/5	XEJ		La Jota/La Mexicana, Cd. Juarez
	5	XEMH		La Grande, Mérida
	5	XETAP		R. Tapachula, o (cfr 890)
	5/0.25	XEVOX		R. Voz, Mazatlán
	5/0.4	XEVT		R. Festival, Villahermosa
	1	HOS97	Panamá	Ondas Centrales, Stgo
975	2	HRZC	Honduras	R. Monumental, San Pedro Sula (nf 980)
980	20	TIRI	Costa Rica	R. Cordillera, Esparza
	1	CMCK	Cuba	El Periódico del Aire, Sapo
	1	CMKR		La Voz del Niquel, Moa
	0.25	CMDE		R. Reloj, Bayamo
	5	HIFA	Dom. Rep.	La Voz de las Fuerzas Armadas, Sto Domingo
	1	TGMQ	Guatemala	R. Retama, San Marcos
	1	HRCD	Honduras	R. Novedades, Comayagua
	1	HRRD		R. Emperador, Campamento
	1	HRYG		R. Tocoa, o
	5/0.1	XELC	México	Canal 98, La Piedad
	5/0.5	XENR		La N-R/La Emisora Grande, Nueva Rosita
	5	XEQQ		R. Romance, Cosamaloapán
	10/1	XETU		R. Tampico, o
	10	HOS57	Panamá	R. Mía, David (rel 650)
990	1	CMAS	Cuba	R. Guamá, Pinar del Río
	5	HISA	Dom. Rep.	R. Cibao, Stgo
	5	YSTA	El Salvador	R. Clásica, San Salvador
	1	TGAL	Guatemala	R. Perla de Oriente, Chiquimula
	0.2	4VCPS	Haiti	R. Cacique, Port-au-Prince
	5	XECL	México	R. Mexicali, o
	5	XEER		R. Formula, Cd. Cuauhtemoc
	10	XEFP		R. Alegría, Xalpa
	50	XET		La T Grande de Monterrey, o
	10	XETG		La Grande del Sureste, Tuxtla Gutiérrez
	3	HOU44	Panamá	R. Monumental, San Miguelito (inactive)
	1	WPRA	Puerto Rico	R. Mil, Mayagüez
1000	5/1	WVWI	American Virgin Isl.	R. One, St. Thomas Charlotte Amalie
	5	TIMIL	Costa Rica	R. Mil, San José
	10	CMAC	Cuba	R. Guamá, Los Palacios
	1	CMJB		R. Musical, Camagüey
	1	CMHB		R. Musical, Sancti Spíritus
	5/1	HIHG	Dom. Rep.	R. Beller, Dajabón
	1	YSHH	El Salvador	Estación H, Sta Ana
	10	TGWC	Guatemala	R. Nacional/LV de Guatemala, o (rel 640)
	2.5	HRXZ2	Honduras	Unión Radio, Tegucigalpa
	3	HRMH		LV del Junco, Sta. Bárbara
	10	XEOY	México	R. Mil, México
	10	YNARF	Nicaragua	R. Fiesta, Managua
	10	HOK36	Panamá	R. Poderosa, Aguadulce
1010	10	CMAP	Cuba	R. Guamá, Guane
	1	CMKM		R. Musical, Holguín
	10	HIJP	Dom. Rep.	R. Comercial, Sto Domingo +2 repeaters
	1	TGRX	Guatemala	R. Miramundo, Zacapa
	1			R. Ixil, Nebaj
	2	HRCD	Honduras	R. Constelación, Juticalpa (cfr 1370)
	5	HRLC		R. Proyección, Minas de Oro
	1	HRLP23		R. Moderna, El Progreso
	0.5	HRLS		R. Dinorama, La Paz
	50	XEHL	México	H-L Onda 10-10/ La Poderosa, Guadalajara
	5/1	XEWS		R. Capital, Culiacán
	5/0.1	XEVK		La Primera V-K, Torreón
	3	PJC7	Neth. Ant.	R. Hoyer 1, Willemstad, Curaçao
1015	3	HOL86	Panamá	LV del Teribe, Bocas del Toro
1020	2	TITIC	Costa Rica	R. 1020, San José
	1	CMAP	Cuba	R. Guamá, Bahia Honda
	10	HITS	Dom. Rep.	R. Enriquillo, Tamayo
	10	YSCA	El Salvador	R. Internacional, San Salvador
	5	TGCM	Guatemala	R. Frontera, Pajapita
	1	4VJH	Haiti	R. Pétion-Ville, o
	1	HRUW	Honduras	R. Michelle, La Ceiba
	10	HO..	Panamá	R. Ancón, Panamá
	1	WPJC	Puerto Rico	R. Gigante, Adjuntas
1030	1	CMAX	Cuba	R. Guamá, La Palma
	10	HIDL	Dom. Rep.	R. Novedades, La Vega (inactive)
	1	YSRM	El Salvador	R. Frontera, Ahuachapán
	10	TGUX	Guatemala	R. Panamericana, Guatemala
	1	HRRH3	Honduras	R. Ticante, Ocotepeque
	1	HRUP3		R. Estero Mil, Tegucigalpa
	5	XE1E		R. Alegría, Matehuala
	10/5	XEQR	México	R. Centro, México
	1	YNDMY	Nicaragua	R. Masaya, o
	10	WOSO	Puerto Rico	El Oso, San Juan
1040	12	TIAC	Costa Rica	R. Fides, San José
		CM..	Cuba	R. Taino, (irr.)
	1	CMKT		R. Victoria, Pto Padre
	10	HION	Dom. Rep.	R. Central, Sto Domingo
	10/5	HRZX	Honduras	R. LaPrimerísima, Olanchito
	1	TGJP	Guatemala	R. Oriental, Jalapa
	1	HRFX	Honduras	R. Musical, Catacamas
	1	HRMJ		R. Minerva, Comayagua
	2	HRNN22		R. Exitos, San Pedro Sula
	10	XEBBB	México	R. Fama/Triple B, Guadalajara
	2.5	YNHPR	Nicaragua	R. Pancasán, Jinotega

	1	HOJ2	Panamá	LV del Canajagua, Las Tablas
	0.25	WCXQ	Puerto Rico	Super Q, Moca
1050	10	CMKT	Cuba	R. Victoria, Las Tunas
	10	HICB	Dom. Rep.	R. Hispañola, Stgo
	3	Y5QR	El Salvador	Auto R, San Salvador
	5/1	TGSL	Guatemala	LV de los Chuchumatanes, Huehuetenango
	10/1	XEBCS	México	R. Cultural Surcalifornia, La Paz
	10/1	XED		R. Internacional/La Gran D, Mexicali
	100	XEG		La Ranchera de Monterrey, o
1060	1	TILX	Costa Rica	R. Columbia, San Isidro del General (rel 760) +1 relay
	1	CMKT	Cuba	R. Victoria, Amancio Rodriguez
	5	CMDX		CMKS, Baracoa
	1	HIRV	Dom. Rep.	R. Mar 1060, San Pedro de Macorís
	1	HIKF		R. Azua, o
	10	TGLR	Guatemala	Disco Radio, Guatemala
	1	HRFA	Honduras	R. Minería, El Mochito
	10	HRVW		LV del Centroamérica, Tegucigalpa (rel 650)
	50/20	XEEP	México	R. Educación, México
	3	HOJ60	Panamá	LV de Panamá, o
	5/0.5	WCGB	Puerto Rico	R. Iniciativa, Juana Díaz
1070	10	CMKS	Cuba	CMKS, Guantánamo
	10	CM..		R. Guamá, Pinar del Río
	10	HIBI	Dom. Rep.	HIBI Radio AM, San Francisco de Macorís
	1	YSAN	El Salvador	LV de los Ausoles, Ahuachapán
	3/2.5	TGD	Guatemala	LV del Occidente, Quetzaltenango
	3	HRGR	Honduras	R. Guaymuras, El Paraíso
	1	HRLE		R. Mi Favorita, San Pedro Sula
	1	HRLP26		R. Siguatepeque, o
	5/1	XESP	México	R. Juventud Unica, Guadalajara
	0.5/2.5	WMIA	Puerto Rico	R. Arecibo/R. Mía, Arecibo
1080	19	TIFC	Costa Rica	Faro del Caribe, San José + 3 repeaters
	10	CMCH	Cuba	R. Cadena Habana, Güines
	0.5	HIMC	Dom. Rep.	R. Ambar, Sto Domingo
	10	YSME	El Salvador	R. 1080, San Salvador
	1	YSME		R. 1080, San Miguel (rel)
	5	TGLU	Guatemala	R. Novedad, Zacapa (vf), (cfr 1190)
	20	4VRD	Haiti	R. Nationale, Port-au-Prince
	1	HRID	Honduras	R. Miramar, Tela
	1	HRXM		R. Meridiano, Choluteca
	1	HOJ24	Panamá	R. Tic Tac, Panamá
	0.25	WLEY	Puerto Rico	NotiUno, Cayey
1090	0.25	WGOD	American Virgin Isl.	WGOD, Charlotte Amalie St. Thomas
	1	CMAP	Cuba	R. Guamá, Minas de Matahambre
		CMCH		R. Cadena Habana, S. Antonio de los Baños
		CMKO		R. Angulo, Moa
	3	HIJM	Dom. Rep.	R. Amistad, Stgo
	1/0.25	HIRB		R. Jimaní, o
	3	YSMG	El Salvador	R. 1090, Atiquizaya
	10	TGZ	Guatemala	R. 1090, Guatemala
	1	HRWC	Honduras	R. Aeropuerto, Tegucigalpa
	1		Jamaica	JBC One, Morant Bay
	20		St. Lucia	R. Caraïbes International, St. Lucia (rel. Martinique)
	2.5/ 0.25	XEAU	México	R. AU, Monterrey
	3/1	XEIL		R. Festival, Veracruz
	5	XEMCA		La Emisora de las Huastecas, Pánuco
	50	XEPRS		1090 Express/La Gigante Musical, Rosarito
	10	YNPRZ	Nicaragua	R. Zinica, Bluefields (cfr 830)
	10	HOL82	Panamá	R. Urracá, Civilista Veraguas
	0.25	WSOL	Puerto Rico	R. Sol, San Germán
1100	10	ZDK	Antigua & Barbuda	"Magic Radio", St. Johns, Antigua
	2	TINAL	Costa Rica	R. Guápiles, o
	20	TISCR		R. Chorotega, Sta Cruz
	1.5	CMCH	Cuba	R. Cadena Habana, La Salud
	1	CMKPO		R. Angúlo, Banes
		CM..		R. Taíno (irr.)
	1	HIMP	Dom. Rep.	R. Ocoa, San José de Ocoa
	1	HIPS		R. Nagua, o
	1	HIHD		R. Oriente, San Pedro de Macorís
	3	YSRF	El Salvador	R. Ranchera, San Salvador
	1	TGSR	Guatemala	R. Superior, Coatepeque
	1	HRDN	Honduras	R. Esperanza, La Esperanza
	1	HRVA		R. Tiempo, San Pedro Sula
	1	HRVL		R. Lux, Olanchito
	5	XEBD	México	R. Ambiente, Perote
	5	XEBV		R. Algría, Moroleón
	0.25	PJL3	Neth. Ant.	R. Caribe, Willemstad, Curaçao
	10	HO..	Panamá	R. Suave, Panamá
1110	10	CMKO	Cuba	R. Angulo, Holguín
	1/0.5	HIOS	Dom. Rep.	R. Marién, Dajabon
	1	HIJJ		R. Jarabacoa (inactive)
	2.5/1	YSCL	El Salvador	R. Cadena Libertad, 2 st's (rel 1160)
	1	TGMK	Guatemala	R. Marinalá, Sta Lucia
	1	HRLP25	Honduras	R. Sur, Choluteca
	5	XELEO	México	La Rancherita Consentida, León
	50	XERED		R. Cadena Nacional/ R. Red, México
	1.5	XEWR		La Poderosa/R. Hits, Cd. Juárez
	2.5/0.5	WVJP	Puerto Rico	R. Caguas, o
1120	10	TISHB	Costa Rica	Sonido 11-20, San José (vf)
	5	CMCH	Cuba	R. Cadena Habana, Artemisa
	1	CMKO		R. Angúlo, Mayari
	3	YSLR	El Salvador	R. La Romántica, San Salvador
	0.5	TGC	Guatemala	R. Poderosa, Guatemala
	5	HRD4	Honduras	R. Oriental/RCO, Danlí
	1	HRTL1		R. Fiesta, Tegucigalpa
	1.5	HRAV	Honduras	Ondas del Ulúa (rel. 1120)
	1	PJE3	Neth. Ant.	R. Statia, St. Eustatius
	1	HOM21	Panamá	R. Sonora, Panamá
	5	WMSW	Puerto Rico	La Gran W, Hatillo
1130	1	CMHA	Cuba	R. Enciclopedia, Sta Clara
	5	CMKO		R. Angúlo, Sagua de Tánamo
	10	HIRL	Dom. Rep.	R. Exitos, Stgo
	1	YSLG	El Salvador	R. Chaparrastique, San Miguel
	2	YSAJ		R. Moderna, Sta Ana (cfr 1150)
	1	TGVR	Guatemala	R. La Voz de la Costa Sur, Retalhuleu (vf. r1135)
	1	HRBT	Honduras	R. San Francisco, San Francisco de la Paz
	1	HRHP		R. Pinares, Siguatepeque
	5	HRPL		R. Progreso, El Progreso
	5	XELUP	México	R. Exitos, Compostela
	10	XEZL		Su Mero Gallo, Jalapa
	1	YNALD	Nicaragua	R. Reloj, Managua
	1	HOU80	Panamá	R. Sensación, Aguadulce
	2.5	WOIZ	Puerto Rico	R. Antillas, Guayanilla
1140	1	TITNT	Costa Rica	R. Monumental, 2 st's (rel 570)
	5	CMCG	Cuba	R. Cadena Habana, Loma de la Cruz
	1	CMKX		R. Bayamo, Media Luna
	5	HIRA	Dom. Rep.	R. Anacaona, San Juan de la Maguana
	10	YSTS	El Salvador	R. El Mundo, San Salvador

	1	HRAP	Honduras	R. Mercurio, Nacaome
	5	HRUN		R. Palmeras, La Ceiba
	50	XEMR	México	La Nueva M-R, Monterrey
	5	XEXF		R. Felicidad, León
	1	HOB49	Panamá	R. Juvenil, Panamá
	10	WQII	Puerto Rico	WQII/Once-Q, San Juan
1146	5	4VAB	Haiti	R. Caraïbes, Port-au-Prince (nf 1150)
1150	10	CMKX	Cuba	R. Bayamo, Entronque Bueycito
	5	HIAS	Dom. Rep.	Onda Musical, Sto Domingo
	2	YSAJ	El Salvador	R. Moderna, Sta Ana (nf 1130)
	3	YSCF		Ondas Orientales/ La Romántica de Oriente, San Miguel
	1	TGRR	Guatemala	R. Fiesta, Guatemala
	10/5	HRAV	Honduras	Ondas del Ulúa, ("LV de Sta. Barbara"), o
	1	HRLP12		R. Universal, Tegucigalpa
	20/0.5	XEAD	México	R. Metrópol, Guadalajara
	10	XEJP		R. Variedades, México
	1	HOS25	Panamá	R. Mía, Colón (rel 650)
		WCRD	Puerto Rico	WCRD, Cabo Rojo
1160	3	TILX	Costa Rica	R. Columbia, Puntarenas (rel 760)
	5	XERRF		R. Lobo, Mérida
	10	CMKX		R. Bayamo, Pilón
	5/1	HIBG	Dom. Rep.	Radiolandia, Stgo
	3	YSCL	El Salvador	R. Horizonte, San Salvador
	1	TGRI	Guatemala	R. Izabal, Morales
	1	HRQL2	Honduras	R. Sensación, Siguatepeque
	3	HRQN		LV del Atlántico, Pto Cortés (F.Pl. 1430 kHz)
	10	HOC20	Panamá	Ondas Chiricanas, David
	3	HO..		R. Rock, Panamá (inactive)
	5/2.5	WBQN	Puerto Rico	Super Borinquén, Barceloneta
1163	1	HRGF	Honduras	R. El Paraíso, o (nf 1160)
1165	10		Antigua & Barbuda	Caribbean R. Lighthouse, St. Johns, Antigua
1170	10	CMKS	Cuba	CMKS, Maisí
	5	TGRL	Guatemala	R. Cadena Landívar, Quetzaltenango
	10	4VRS	Haiti	R. Soleil, Port-au-Prince
	1	HRAZ2	Honduras	R. Hits, La Ceiba
	1.5	HRAF		R. Atena, Choluteca
	2/0.1	XELP	México	Onda 11-70, La Piedad
	5	XERT		La Rancherita del Valle, Reynosa
	1.5	YNBRD	Nicaragua	R. Darío, León
	0.25	WLEO	Puerto Rico	R. Leo, Ponce
1175	0.5	YSCB	El Salvador	LV del Pacífico, Sonsonate (nf 1170)
1180	10	TIPJ	Costa Rica	R. Victoria, Heredia
	300	CM..	Cuba	R. Taíno, Villa María
	1	CMDB		R. Progreso, Mayarí Arriba
	1	CMBM		R. Reloj, Nueva Gerona (vf)
	10	HIBE	Dom. Rep.	R. Mil, Sto Domingo
	5	YSVE	El Salvador	LV Evangélica de América/R. Vea, San Salvador
	25/10	TGT	Guatemala	R. Sonora, Guatemala
	1	HRAZ3	Honduras	R. Taxi, Tegucigalpa
	5	XEFR	México	R. Felicidad, México
	10	HOU84	Panamá	R. Belén, Stgo
1190	1	CMHT	Cuba	R. Sancti-Spíritus, Yaguajay
	5	HIAG	Dom. Rep.	R. Azul, Stgo
	1	R26		La Caridad
	5	TGLU	Guatemala	R. Novedad, Zacapa (cfr 1080)
	1	HRPO	Honduras	R. Sta María de la Luz, Gualaco
	5	HRWM		R. Alfa Stereo San Pedro Sula
	0.5	HRGK	Honduras	R. Brassavola, Minas de Oro
	5	XETOT	México	La Poderosa, Tampico
	50	XEWK		La Doblevé, Guadalajara
	10	HO..	Panamá	R. Bolívar, Panamá (F.Pl.)
	10/5	WBMJ	Puerto Rico	WBMJ, San Juan
1195	0.2	4VAF	Haiti	La Voix de la Grande Anse, Jérémie (nf 1190)
1200	2	TITQ	Costa Rica	Cadena KW, San José +1 repeater
	0.25	CMBS	Cuba	R. Ariguanabo, San Antonio de los Baños
	1	CMHT		R. Sancti-Spíritus, Trinidad
	10	CMKC		R. Revolución, Palma Soriano
	1/0.25	HIMR	Dom. Rep.	R. Caracol, Azua
	1	YSKJ	El Salvador	R. Sirama, San Miguel
	10	YSMM		R. Fiesta, San Salvador
	1	TGRJ	Guatemala	R. Jutiapa, o
	1	HRDS	Honduras	R. Nacaome, o
	3	HRSI		R. Impacto, Roatán
	0.25	WGDL	Puerto Rico	R. Grito, Lares
1205	1		Cayman Isl.	R. Cayman, George Town
	1	4VLS	Haiti	R. Plus, Port-au-Prince (nf 1210)
1210	10	CMHT	Cuba	R. Sancti-Spíritus, o
	1	CMKC		R. Revolución, 2 st's
	5	HICJ	Dom. Rep.	R. Merengue, San Francisco de Macorís
	0.5	YSCG	El Salvador	R. La Paz, Zacatecoluca
	5	TGMX	Guatemala	R. Rumbos, Guatemala
	1	HRHO	Honduras	R. Maya, La Entrada
	1	HRRO		R. Capital, Tegucigalpa
	50	XEBCO	México	R. Occidente/La Voz de Colima, o
	4	HOE91	Panamá	R. 10, Panamá
	5	WHOY	Puerto Rico	R. Hoy, Salinas
1220	1	TIQ	Costa Rica	R. Casino, Limón
	10	CMGY	Cuba	R. 26, Central España
	25	HIN	Dom. Rep.	R. HIN, Sto Domingo
	1	TGMT	Guatemala	R. Amiga, Antigua
	1	HRHH2	Honduras	LV de la Costa, San Pedro Sula
	1	HRYS		R. Suari, Marcala
	100	XEB	México	La B Grande de México, o
	2.5	XEZAJ		Sororitmo 1220, Guadalajara
1230	10	CMGT	Cuba	R. 26, Unión de Reyes
	1	HIMB	Dom. Rep.	R. Ideal, Moca
	5	YSNB	El Salvador	R. 1-2-3, San Salvador
	1	TGAT	Guatemala	R. Atlántida, Pto Barrios
	1	4VAV	Haiti	R. La Voix de l'Avé-Maria, Cap Haïtien
	10	HRQW	Honduras	R. Tela, o
	5	YNADM	Nicaragua	R. Democracia, Managua
	1	WNIK	Puerto Rico	R. Unica, Arecibo
1240	1	ZNS2	Bahamas	R. Bahamas
	1	TILX	Costa Rica	R. Columbia, Nicoya (rel 760)
	10	CMGW	Cuba	R. 26, Bolondrón
	1	HIAU	Dom. Rep.	R. Revelación, Pto Plata
	5/0.5	HICV		R. Barahona, o
	1	YSQR	El Salvador	R. Metapan, Sta. Ana
	0.5	YSMT		R. Norteña, S. Miguel
	5	TGK	Guatemala	R. Sensación, Guatemala
	10	4VSJ	Haiti	R. Antilles Internationales, Port-au-Prince (vf)
	1	HRZC	Honduras	R. Monumental/La Voz del Recuerdo, Tegucigalpa
	1	HOM56	Panamá	R. Bouquete, o
	5	HO..		R. BB, Panamá
	1/5	WALO	Puerto Rico	R. Walo/R. Oriental, Humacao
1250	1	CMHW	Cuba	Doblevé, Caibarién
	5	CMKS		CMKS, Imías
	5	HIBC	Dom. Rep.	La Voz del Progreso, San Francisco de Macorís
	1	HIRJ		R. Juventud, La Romana
	1	TGPY	Guatemala	Payakí
	1	4VS	Haiti	La Voix du Plateau Central, Hinche

	10	HRAT	Honduras	Super R, San Pedro Sula
	1	HRCC		R. Cadena Continental, Comayagua
	1	HRYN		R. Latína, Danlí
	5/1	XEDK	México	La Grande, Guadalajara
	1/0.25	W...		Sabana
1260	6	TIHN	Costa Rica	R. Emaus, San Vito de Coto Brus
	1	CMGH	Cuba	R. Victoria de Girón, Torriente
	1	CM..		R. Reloj
	1	HIT	Dom. Rep.	R. Visión, Sto Domingo
	3	YSAA	El Salvador	R. América, San Salvador
	1	TGPA	Guatemala	R. Monumental, Antigua
	1	HRYF2	Honduras	R. San Marcos, San Marcos de Colón
	1	HR2R		R. 1260, La Ceiba
	10/5	XEL	México	R. ACIR, México
	5/0.5	XESA		S-A/R. Hits, Culiacán
	2	YNARG	Nicaragua	R. Güegüence, Managua
	1	WISO	Puerto Rico	R. Wiso/La Emisora del Pueblo, Ponce
1270	1.5	P4A8	Aruba	R. 12-70, St. Nicolaas
	10	CMJG	Cuba	R. Reloj, Camagüey
	1	HITA	Dom. Rep.	R. Ambiente, Baní
	1	YSQZ	El Salvador	R. Mía, San Miguel
	2.5	TGCQ	Guatemala	R. Saturno AM, Guatemala
	1	HRNQ	Honduras	R. Sonora, Tegucigalpa
	1	HROF		Ecos del Celaque, Gracias
	5/0.5	XEAX	México	R. Festival, Oaxaca
	5	XERPL		R. Poderosa, León
	2	XERRT		R. Ranchito, Cd. Madero
	2.5/0.5	XEYW		Tu Chica, Mérida
	3	YNBRA	Nicaragua	R. Atenas, León
	0.5	HOJ22	Panamá	R. Alegre/CMQ Panamá
1280	2	TIHT	Costa Rica	R. Alajuela, o
	1	CMGF	Cuba	R. Enciclopedia, Varadero
	1	CMKW		R. Mambí, Stgo de Cuba
	1	CMJC		R. Rectángulo, Guáimaro
	1	HIHZ	Dom. Rep.	R. Clave, Monte Plata
	1	HIVM		R. Bonao, o
	1	YSQV	El Salvador	R. Galaxia, Sta Ana
	10	4VAM	Haiti	R. Métropole, Port-au-Prince
	1	HRAM	Honduras	R. Olanchito, o
	1	HRYT		R. Estrella de Oro, San Pedro Sula (nf 1400)
	1	HRVN		R. San Miguel, Marcala
	0.5	HRCG		R. Columbia, Pto. Cortés
	5/0.5	XEAW	México	Música Inolvidable, Monterrey
	5/1	WCMN	Puerto Rico	R. Centro, Arecibo
1290	1	WRRA	American Virgin Isl.	WRRA, Frederiksted, St. Croix
	5	CMDS	Cuba	R. Enciclopedia, La Pastora
	1	CMHW		Doblevé, Rancho Veloz
	0.5	HIVD	Dom. Rep.	R. Janico, o (inactive)
	1	YSMA	El Salvador	R. Chalatenango, o
	0.5	TGTU	Guatemala	LV de Totonicapán, o
	0.5	HRGS	Honduras	Bay Island Christian Network, Roatán
	2	HRNN26		R. Choluteca, o
	10/1	XEDA	México	R. 13/Sonido Internacional, México
	10/0.25	XENK		La Poderosa N-X, Mazatlán
	10	HOS23	Panamá	R. Guadalupe/La Pantera, Panamá
1300	7.5	TILC	Costa Rica	R. La Fuente Musical, Cartago
	1	CMDA		R. Enciclopedia, Las Tunas
	1	HIKQ	Dom. Rep.	R. Radio, Sto Domingo
	5	YSLV	El Salvador	R. Versatil, San Salvador
	1	TGTL	Guatemala	R. Tropical, Guatemala
	1	HRLH	Honduras	Stéreo Amistad, Tegucigalpa
	3	HRLR		R. Sta Rosa, Sta Rosa de Copán
	1	PJD2	Neth. Ant.	PJD2 Radio, Philipsburg, St. Maarten
	2	YNACT	Nicaragua	Canal 130 AM, Managua
	1	HOI417	Panamá	R. Baha'í, Boca del Monte
	1	WTIL	Puerto Rico	R. Util, Mayagüez
1310	1	CMHR	Cuba	Doblevé, Sagua la Grande
	1	CMBN		R. Caribe, Nueva Gerona
	1	HIMH	Dom. Rep.	R. Real, La Vega
	1	HISD		R. TV Dominicana, El Seibo (rel 620)
	5	YSRV	El Salvador	R. Veritas, Usulután
	1	TGAN	Guatemala	LV de los Altos, Quetzaltenango
	2	HRAD	Honduras	R. Majestad, Juticalpa (nf 770)
	1	HRRL		R. Libertad, Marcala
	1	HRVC		R. Rumbo, San Pedro Sula (rel 1390)
	2×20		Martinique	RFO, Fort de France
	5	XEAM	México	La Norteña, Matamoros
	10	XEBP		R. Sensación, Torreón
	10	XETIA		La Cotorra de tu Radio, Guadalajara
	4	XEVB		R. 13, Monterrey
	3	YNCSC	Nicaragua	R. San Cristóbal, Chinandega
1315	0.5	HIAV	Dom. Rep.	R. Baní, o (nf 1310)
1320	1	P4A10	Aruba	V. of Aruba, Oranjestad
	1	TILX	Costa Rica	R. Columbia, San Carlos
	1	CMDA	Cuba	R. Enciclopedia, Santiago de Cuba
	1	CMHA		R. Enciclopedia, Sancti Spíritus
	1	YSAH	El Salvador	R. Emanuel, La Unión
	1	TGME	Guatemala	R. Quesada, Jutiapa
	1	HRGM	Honduras	R. Juvenil, Choluteca
	1	HRMG		R. Bahía, LaCeiba
	10/1	XECMQ	México	R. Sensación, México
	5/0.5	XERJ		R. Mazatlán, o
	5	WUNO	Puerto Rico	NotiUno, San Juan
1330	1	CMAD	Cuba	R. Artemisa, o
	1	CMDO		R. Banes, o
	5	HIFA	Dom. Rep.	LV de las Fuerzas Armadas, Moca (inactive)
	5	YSHQ	El Salvador	R. Progreso, San Salvador
	10	TGMU	Guatemala	Unión Radio, Guatemala
	10	4VJLD	Haiti	R. Haïti Inter, Port-au-Prince
	1	HRSW	Honduras	R. Cadena, Tegucigalpa
	5	XEBO	México	R. Variedades, Irapuato
	5/1	XEFC		La Romántica, Mérida
	5/1	XEUAS		R. Universidad, Culiacán
	1	YNIST	Nicaragua	R. Satélite, Estelí
	1	HOM92	Panamá	R. Fiesta, Panamá
	1	WENA	Puerto Rico	La Buena del Sur, Peñuelas
1340	1	WSTA	American Virgin Isl.	WSTA/Lucky 13, Charlotte Amalie, St. Thomas
	6	TIHR	Costa Rica	R. Sideral, San Ramón
	10	CMHU	Cuba	R. Ciudad del Mar, Palmira
	0.25	WGBY		AFRTS, Guantánamo Bay
	1	HIBZ	Dom. Rep.	R. Centro, San Juan de la Maguana
	1	YSXW	El Salvador	R. Novedades, Usulután
	10	TGCO	Guatemala	R. LV del Trópico, Coatepeque
	10	HRHH	Honduras	R. El Mundo, San Pedro Sula (vf)
	2.5	HRRF		R. Cadena Nal. de Noticias, Tegucigalpa
	1	HRJC		R. Colonial, Comayagua
	1	XENV	México	R. Alegría, Monterrey
	5	XEBK		El Norteñazo, Nuevo Laredo
	1	WNOZ	Puerto Rico	R. Nosotros/NotiUno, Aguadilla
1350	1	CMHU	Cuba	R. Cd. del Mar, Aguada de Pasajeros
	1	CMHU		R. Libertad, Pto Padre
	1	HISD	Dom. Rep.	R. TV Dominicana, Monte Cristi (rel 620)
	1	TGMC	Guatemala	R. Municipal Cultural, Cobán
	1	HRME	Honduras	R. LV del Pueblo, La Ceiba

	5/0.5	XETB	México	R. Laguna, Torreón
	10	HOZ38	Panamá	R. Cadena Millonaria, San Miguelito
	2.5	WEGA	Puerto Rico	R. Las Vegas, Vega Baja
1360	1	TITNT	Costa Rica	R. Monumental, Limón (rel 670)
	1	TITNT		R. Monumental, Puntarenas, (rel 670)
	10	CMFA	Cuba	R. Cadena Agramonte, Rodolfo Ramírez Esquivel
	1	HISD	Dom. Rep.	R. TV Dominicana, La Vega (rel 620)
	0.5	HIMG		R. Rutas Musical, La Romana
	10	TGLK	Guatemala	R. 9-80, Guatemala
	5	4VRL	Haiti	R. Liberté, Port-au-Prince
	5/1	HRBH	Honduras	R. Sta. Barbara, "La B Grande", o
	1	HRCP		R. Continente, Trinidad
	1	HRLP18		R. San Pedro, Tegucigalpa (rel 860)
	5	XEUD	México	La U de Tuxtla, Tuxtla Gutiérrez
	1	YNFRL	Nicaragua	R. Libertad, Jinotepe
	5/1	WCHQ	Puerto Rico	R. C-H-Q, Camuy
1370	1	CMFA	Cuba	R. Cadena Agramonte, Nuevitas
	1	CMDV		R. Siboney, Stgo de Cuba
	5	HIFA	Dom. Rep.	LV de la Fuerzas Armadas, Elías Piña
	1	HISD		R. TV Dominicana, Barahona (rel 620)
	1	TGAC	Guatemala	LV de Colombá, o
	1	4VMN	Haiti	Radiodiff. Cayenne, Les Cayes
	1	4VEE		R. Citadelle, Cap Haïtien
	2	HRCD	Honduras	R. Constelación, Juticalpa (nf 1010)
	1	HRST		R. Fraternidad, San Pedro Sula
	1.5	HRTR		R. Danlí, o
	5	XEHF	México	La Norteña, Nogales
	10	XEMON		Poder 13-70, Monterrey
	5/1	XEPA		R. 13-70/La Poblanita, Puebla
	5	XEPJ		R. Capital, Guadalajara
	1	YNLLM	Nicaragua	R. Matagalpa, o
	1	YNMRP		R. Patria, Boaco
	1	HOB64	Panamá	R. Sitrachilco, Pto. Armuelles
	1	HOPJ60		R. Colosal, David
	5/1	WIVV	Puerto Rico	WIVV Missionary, Vieques Isl.
1380	1	TIMS	Costa Rica	R. Guanacaste, Liberia
	1	TIMS		R. Guanacaste, San José (rel)
	10	CMFA	Cuba	R. Cadena Agramonte, Central Brasil
	5	HIRP	Dom. Rep.	R. El Seibo, o (nf 1490)
	5	HISC		R. Nacional, Stgo
	1	TGEB	Guatemala	R. Momostenango Educativa, Momostenango
1380	0.7		Haita	R. Port-au-Prince, o
	0.5	HRAH	Honduras	R. Jutiapa, o
	1.5	XEAK	México	R. Consentida, Acámbaro
	10	XECO		R. Dimensión 13-80, México
	5	XEGW		R. 13-80, Cd. Victoria
	5	XETP		R. Sensación, Jalapa
	10	HO..	Panamá	R. América, Panamá
	1	WOLA	Puerto Rico	R. Procér, Barranquitas
1390	1	CMBT	Cuba	R. Jaruco, o
	1	HIAR	Dom. Rep.	R. San Cristóbal, o
	1.5	YSJI	El Salvador	La Voz del Litoral, Usulután
	5	TGYC	Guatemala	R. Estrella, Guatemala
	10	HRVC	Honduras	LV Evangélica, Tegucigalpa
	10/0.25	XERW		Sonido 13-90, León
	5/1	XETL		LV de la Huasteca, Tuxpan
	5/0.25	XETY		R. Tecomán, o
	5/0.1	XEXO		X-E Exito/La Diferente, Cd. Mante
	1	HOL	Panamá	R. Super Sol, Colón
	1	WISA	Puerto Rico	R. Noroeste, Isabela
1400	12	TICJ	Costa Rica	R. Sinai, San Isidro del General
	1	CMGX	Cuba	R. Musical Matanzas
	1	CMFA		R. Cadena Agramonte, Guáimaro
	1	HIAL	Dom. Rep.	Ondas del Valle, La Vega
	1	HISD		R. TV Dominicana, San Juan (rel 620)
	1	HISD		R. TV Dominicana, Pedernales (rel 620)
	5		Grenada	Harbour Light, Carriacou
	1	TGRB	Guatemala	R. Porteña, Pto Barrios
	1	HRJU	Honduras	R. Punto/ Onda Popular, Comayagua
	1	HRYT		R. Estrella de Oro, San Pedro Sula (cfr 1280)
	10	XERUY	México	R. Universidad, Mérida
	5	XESH		R. Alegría, Ct. Sabinas Hidalgo (n. 1410)
	1	HOT40	Panamá	R. Luz, La Chorrera
	1	WIDA	Puerto Rico	R. Vida, Carolina
1410	1	CMAL	Cuba	R. Enciclopedia, Pinar del Río
	1	CMJW		R. Sta Cruz, Sta Cruz del Sur
	1	HIAE	Dom. Rep.	R. Revelación en América, Sto Domingo
	1/0.25	HICH		R. 14-10 AM, Barahona
	1	HIGG		R. Gri Gri, Río San Juan
	1/0.25	HIPK		R. Neiba, o (nf 1580)
	1	HIRM		R. Sol, Higüey (vf)
	3/5	TGGH	Guatemala	R. Xelajú, Quetzaltenango
	1	4VTS	Haiti	La Voix de Nord'Ouest/ R. Familial, Port-de-Paix
	1	HRDD	Honduras	LV de Atlántida, La Ceiba
	1	HRSY		R. LV del Pacífico, San Lorenzo
	5/1	XEBS	México	R. Sinfonola/R. Mil, México
	5	XEIR		Estelar 14-10, Cd. Valles
	10/0.25	XEKB		Canal 14-10, Guadalajara
	5/0.5	XETAB		R. Tabasco/La Juvenil, Villahermosa
	5	PJF1	Neth. Ant.	14-10 The Voice of Saba, Bottom, Saba
	1	YNBRV	Nicaragua	R. Venceremos, León
	5	HOH779	Panamá	R. Mensabé, Las Tablas
	1	WRSS	Puerto Rico	R. Cristal, San Sebastián
1420	1	TITNT	Costa Rica	R. Monumental, Nicoya (rel 670)
	0.25		Cuba	R. Llanuras de Colón, o
	1	CMHS		R. Caibarién, o
	1	HIED	Dom. Rep.	R. Oro, Cotuí
	1	YSUCA	El Salvador	R. Universitaria, San Salvador
	5		Guadeloupe	RFO, Morne à Louis
	1	TGRP	Guatemala	R. Capital, Guatemala
	1	HRSL	Honduras	R. Actualidad, Trinidad
	2	HRSJ		R. Mundial, Tocoa
	5/0.5	XEF	México	R. Centro, Cd. Juárez
	10/0.2	XEWE		La Estación Familiar, Irapuato
	2	XEXX		La Ranchito/La Doble X, Tijuana
	10	ACB20	Panamá (Canal Zone)	SCN-AM Radio Southern Command Netw, Ft. Davis
	1	WEUC	Puerto Rico	R. Universidad Católica, Ponce
1430	10	CMOY	Cuba	R. Surco, Morón
	1	HILF	Dom. Rep.	R. Cayacoa, Higüey
	1	HIJC		R. Maíba, Stgo
	1	TGAG	Guatemala	LV de Huehuetenango, o
	1	TGRN		R. Petén, Flores Petén
	10	4VGM	Haiti	MBC-Magloire Bc. Circuit, Port-au-Prince (cfr 850)
	1	HRIC	Honduras	La Radio del 90, Pto Cortés
	1	HRVM		R. Maranatha, La Paz
	5/1	XEOK	México	Canal 14-30, Cd. Obregón
	2/0.15	XEWD		La Grande, Cd. Miguel Alemán

	5	YNIRL	Nicaragua	R. Liberación/La Tayacana, Estelí
	5	WNEL	Puerto Rico	R. Tiempo/NotiUno, Caguas
1440	1	P4A5	Aruba	R. Kelkboom, Oranjestad
	5	TIR5C	Costa Rica	R. San Carlos, Cd. Quesada
	0.25	CMBU	Cuba	R. Güines, o
	10	CMJP		R. Surco, Ciego de Avila
	5	HIAD	Dom. Rep.	R. San Juan, San Juan de la Maguana
	5	HIAK		R. Renuevo/LV de la Verdad, Sto Domingo
	1	HIFS		R. Bahía, Nagua
	0.5	TGMS	Guatemala	LV de Mazatenango, o
	1	HRRD	Honduras	R. San Isidro, La Ceiba
	5	XECCC	México	R. Cañon, Guadalajara
	10	XEEST		R. Alegría, México
1450	1	CMJF	Cuba	R. Maboa, Amancio Rodríguez
	10	HIAC	Dom. Rep.	R. Util, Salcedo
	5	YSFM	El Salvador	Super Radio, San Salvador
	1	YSKR		R. Renovación, 5 Miguel
	1	TGLG	Guatemala	R. Epoca, Guatemala
	0.5	HRXZ2	Honduras	R. Titania, Tegucigalpa
	5	XEPP	México	LV de Orizaba, o
	1	WCPR	Puerto Rico	R. Coamo, o
1460	1	TILX	Costa Rica	R. Colombia, Limón
	1	CMHZ	Cuba	R. Cubitas, Sola (r. 1550)
	0.5	HIAN	Dom. Rep.	R. Renacimiento, Hato Mayor
	1	HIAR		R. San Cristóbal, o
	5	TGRN	Guatemala	R. Petén, Flores
	0.2	4VEA	Haiti	R. La Voix du Nord, Cap Haïtien
	1	HRGC	Honduras	R. Conga, San Pedro Sula
	5	HRQX		Radiolandia, Comayagua
	5	HRCX		LV de Patuca, Catacamas
	10	XELX	México	La Grande de Michoacán, Zitácuaro
	0.5	HOD42	Panamá	LV de Almirante, Bocas del Toro
	5	WRRE	Puerto Rico	La Fabulosa, Juncos
	0.5	WCRP		R. Raices, San Sebastián
	0.25			AFCN, Camp García, Viequez Island
	2.5		Turks & Caicos Isl.	R. Turks & Caicos, Grand Turk
1470	1	CMGE	Cuba	R. Cd Bandera, Cárdenas
	5	HIDE	Dom. Rep.	Super 56, San Francisco de Macorís
	1	HIEH		R. Sur, Barahona (vf)
	1	4VAA	Haiti	R. Arc-en-Ciel, Port-au-Prince
	1	HRSA	Honduras	R. Luz y Verdad, La Ceiba
	3	XEHI	México	El Heraldo Internacional, Cd. Miguel Alemán
	50	XERCN		RCN de Tijuana, o
	5	XECU		La Rancherita de los Mochis, o
	10/5	XESM		R. Cañon, México
	1/2.5	WKCK	Puerto Rico	R. Cumbre, Orocovis
1480	1	TIAW	Costa Rica	R. Puntarenas, o
	0.25	CMJI	Cuba	R. Florida, o
	5	HIAH	Dom. Rep.	R. VEN-Voz Evangélica Nacional, Sto Domingo
	1	TGBH	Guatemala	R. Horizonte, Guatemala
	1	HRBR	Honduras	Cultura de Copán, La Entrada
	1	HREZ		R. Cadena Caracol, Tegucigalpa
	5/1	XENS	México	La NS, Navojoa
	5	XEPR		R. Poza Rica, o
	5	WMDD	Puerto Rico	Sonido 14-80/R. El Conquistador, Fajardo
1490	0.25	CMBW	Cuba	R. Camoa, San José de las Lajas
	1	CMDH		R. Mayarí, o
	2	HIAP	Dom. Rep.	La 54 AM, Moca
	1	HISD		R. TV Dominicana, Pto Plata (rel 620)
	1	HIRP		R. El Seibo, o (cfr 1380)
	1	TGRE	Guatemala	R. Modelo, Retalhuleu
	1	HRGO	Honduras	R. Porteña, Pto Cortés
	2	HRMO		R. Omega, La Esperanza
	1	HRRA		R. Juventud, Sonaguera
	3	HOR43	Panamá	CMQ-La Exitosa, La Palma (rel 740)
	5/1	WZBS	Puerto Rico	R. San Miguel, Ponce
1500	12	TIASF	Costa Rica	R. Cima, Cd. Quesada
	0.25	CMKQ	Cuba	R. Majaguabo, San Luis
	1	CMDA		R. Enciclopedia, Holguín
	0.5	HIPA	Dom. Rep.	R. Color, Higüey
	1	YSCS	El Salvador	R. Fides, Usulután
	1	YSDA		R. Imperial, Santo Tomás
	0.5	4VOC	Haiti	R. Select, Port-au-Prince
	1	HRTX	Honduras	R. Victoria, Choluteca
	20/5	XEAI	México	R. A-I/Canal Tropical, México
	3	PJC9	Neth. Ant.	R. Hoyer 2, Willemstad, Curaçao
	1	YNAMT	Nicaragua	R. Minuto, Managua
	1/0.25	WMNT	Puerto Rico	R. Atenas, Manatí
1505	1		Anguilla	R. Anguilla/The Valley
1510	1	CMDA	Cuba	R. Encíclopedía, Moa
	10	HIBL	Dom. Rep.	R. VEN-Voz Evangélica Nacional, Sto Domingo
	5	TGDX	Guatemala	R. Centroamericana, Guatemala
	1	HREM	Honduras	R. Emanuel, Ocotepeque
	1	HRAN		R. Novedades, Tegucigalpa
	5	XEOF	México	O-F La Mexicana, Cortáza
	1	HOA95	Panamá	Hosana R, Panamá (VF)
	1	WAVB	Puerto Rico	Super B, Lajas
1520	0.5	TILX	Costa Rica	R. Columbia, Turrialba (rel 760)
	1	CMKZ	Cuba	R. Baraguá, Palma Soriano
	1	HIWJ	Dom. Rep.	R. Samaná/R. 15-20, Samaná
	1	TGRS	Guatemala	R. Superior, Coatepeque
	1	HRFU	Honduras	R. Futura, San Pedro Sula
	1	HRHJ		R. Santiago, Yoro
	1	HRRG		R. Providencia, Danlí
	10	WRAI	Puerto Rico	R. Aeropuerto/R. Deportes, San Juan
1530	50		Belize	V. of America, Punta Gorda
	1	CMJX	Cuba	R. Morón, o
	1	YSCZ	El Salvador	R. Cadena Cuscatlán, San Miguel
	5	XEUR	México	R. Onda, México
	1/0.25	WUPR	Puerto Rico	Exitos 15-30, Utuado
1540	20	ZNS1	Bahamas	R. Bahamas, Nassau
		TI..	Costa Rica	R. Mía, San José (F.Pl.)
	1		Cuba	R. Sagua, Sagua la Grande
	1	HIBU	Dom. Rep.	LV de La Romana, La Romana (vf)
	1	HIFB		R. Recuerdos, Sto Domingo
	1	HRYK	Honduras	R. Nuevo Mundo, Tegucigalpa
	5	XEHOS	Mexico	La Poderosa, Hermosillo
	5	XESTN		XESTN, Sta Catarina
	5	XEVF		R. Variedades, Villaflores
	2.5	XEYK		R. Magica, Motul
	1	WIBS	Puerto Rico	R. Caribe, Guayama
1550	1	CMJQ	Cuba	R. Nuevitas, o
	10	YSCZ	El Salvador	R. Cadena Cuscatlán, San Salvador
	1	TGCK	Guatemala	R. Sultana de Oriente, Zacapa
	1	HRAO	Honduras	R. Original, Cofradía
	5	XENU	México	La Rancherita Contenta, Nuevo Laredo
	10	XERUV		R. Universidad Veracruzana, Jalapa
	0.25	WKFE	Puerto Rico	R. Café, Yauco
1555	10		Cayman Isl.	R. Cayman, Gun Bluff
1560	6	TIJM	Costa Rica	R. Nicoya, o
	1		Cuba	R. Enciclopedia, Ciego de Avila
	1/0.25	HIPZ	Dom. Rep.	R. Pedernales, o

kHz	kW	Call	Country	Station/Location
	10	4VEE	Haiti	La Voix de l'Espérance, Port-au-Prince
	20/10	XEFAJ	México	R. Consentida, México
	5	XERIO		R. Río, Ixtlán del Río
	5	XESE		LV de Campeche, Champotón
	5/0.75	WRSJ	Puerto Rico	R. San Juan, o
1570	5	HIAJ	Dom. Rep.	R. Amanacer Internacional Sto. Domingo
	0.5	YSCZ	El Salvador	Cadena Cuscatlán, Sta. Ana
	10	TGVE	Guatemala	R. Voz Evangélica de América/R. VEA, Guatemala
	250	XERF	México	R. Frontera/La Voz de la Amistad, Cd. Acuna
	1/2.5	HOE35	Panamá	LV del Trópico/ La Grande, Colón
	1	WPPC	Puerto Rico	R. Felicidad, Peñuelas
	50		Turks & Caicos Isl.	Atlantic Beacon, South Caicos
1580	50		Belize	V. of America, Punta Gorda
	0.5	TIQ	Costa Rica	R. Casino, 10 st's (rel 1220)
	0.25	TICLS		R. Cultural Los Santos, San Marcos
	0.25	TIRCC		R. Cultural de Corredores, Cd. Neily
	0.25	TIRSCLC		R. Sistema Cultural La Cruz, o
	0.25	TIRSCM		R. Sistema Cultural Maleku, Tonjibe
	1	CMFA	Cuba	R. Cadena Agramonte, Sta Cruz del Sur
	0.25	CM..		R. Reloj
	1/0.25	HIPK	Dom. Rep.	R. Neiba, o (cfr 1410)
	50	XEDM	México	La Grande de Sonora, Hermosillo
	5	XEMQ		LV de Hidalgo, Tulancingo
	5/2.5	WMTI	Puerto Rico	R. Jefe, Morovis
	0.25/0.5	WMTI		R. Jefe, Manatí (rel)
	0.1/0.5	WMTI		R. Jefe, Arecibo (rel)
1590	0.5	CMDF	Cuba	R. Granma, Manzanillo
	3	HIDA	Dom. Rep.	R. Hit Musical, Stgo (inactive)
	1	TGXC	Guatemala	R. Triunfadora, Chimaltenango
	10/5	XEVOZ	Mexico	Capital R. Voz, México
	1	WXRF	Puerto Rico	R. 15-90, Guayama
1600	1.5	TIJV	Costa Rica	R. Quepos, o
	1.5	TILGJ		R. 16, Grecia (r. 1590)
	2/1	TIMMCH		R. Golfito, Pto Golfito
	1.5	TIMQ		R. Pococí, Guápiles
	0.25	TIRCBA		R. Cultural de Buenos Aires, o
	0.25	TIRCP		R. Cultural de Pital, Pital de San Carlos
	0.25	TIRCT		Emisora Cultural, Turrialba
	0.25	TIRCU		R. Cultural de Upala, o
	0.25	TIRSCN		R. Sistema Cultural, Nicoyano, Nicoya
	5	HIEG	Dom. Rep.	R. Revelación en América, Sto Domingo
	10	TGWC	Guatemala	R. Nacional/LV de Guatemala, Guatemala (rel 640)
	1	HRPC	Honduras	R. Luz y Vida, San Luís
	1	HRXN		R. Solaris, Tegucigalpa
	5/0.2	XEAK	México	R. Consentida, Acámbaro
	2	XEGEM		R. Mexiguense, Mectepec
	5	WLUZ	Puerto Rico	R. Luz, Santurce
1610	50		Anguilla	The Caribbean Beacon

SOUTH AMERICA

Abbreviations particular to this section of the medium wave tables: CP = Construction Permit, ff = future frequency, Mun. = Municipal, nf = nominal frequency, Prov. = Provincia, Radiodif. = Radiodifusora, R. Dif. = Radio Difusora, RJ = Rio de Janeiro, RPP = Radio Programas del Perú, ti = temporarily inactive (may reopen shortly), ts = temporarily suspended (may reopen shortly), vf = varying frequency.

The sign o indicates that the location is not given where company name includes the name of the town concerned.

Brazil: Stations with a power of less than 1 kW are not mentioned.

kW	Call	Country	Station/Location
540 kHz			
25/8	LRA14	Argentina	R. Nacional, Santa Fé, o
5	LRA25		R. Nacional Tartagal, o
10/5	LU17		R. Golfo Nuevo, Puerto Madryn
50	ZYH295	Brazil	R. Nacional, Manaus, AM
1/0.25	ZYH894		R. Guajajara, Barra do Corda, MA
1/0.25	ZYI914		R. Primeiro de Julho, Agua Branca, PI
1/0.25	ZYH610		R. Jornal, Canindé, CE
20/2.5	ZYJ924		R. Jornal de Sergipe, Aracaju, SE
1/0.25	ZYH481		R. Regional, Irecê, BA
1/0.25	ZYL331		Sistema Ipanema de Comunicacão, Ipanema, MG (F.Pl.) (still on 1080)
1/0.25	ZYL...		R. Meridional, Uberaba, MG (CP)
10/2.5	ZYJ450		R. Fluminense AM, Niterói, RJ
1/0.25	ZYK697		R. Uirapuru, Biriguí, SP
1/0.25	ZYK734		R. Nova Sumaré, Sumaré, SP
1/0.25	ZYJ322		R. Flor do Café, Borrazópolis, PR
10/1	ZYJ778		R. Mirador, Rio do Sul, SC
1/0.5	ZYK226		R. Real, Canoas, RS
10/1	ZYK322		R. Sepe Tiaraju, Santo Angelo, RS
10	ZYI...		R. Nal. de Corumbá, o, MS (CP)
10/1	ZYH755		R. Riviera, Goiânia, GO
10	CB54	Chile	R. Ignacio Serrano, Melipilla
1	CD54		R. Calle-Calle, Valdivia
15	HJKA	Colombia	R. Auténtica, Bogotá
25	HCFA2	Ecuador	R. Tropicana, ("Canal 540"), Guayaquil
10	OBX4E	Peru	R. Inca del Perú, Lima
120/50	YVOY	Venezuela	Mundial Perijá, Villa del Rosario
25	YVUR		La Voz de Manapiare, San Juan de Manapiare
548 kHz			
1	OCX2D	Peru	Radiodif. San Antonio, El Porvenir (nf 540) (vf)
550 kHz			
1/0.25	ZYI902	Brazil	R. Serra da Capivara, São Raimundo Nonato, PI
1/0.25	ZYI907		R. Igaracu, Parnaíba, PI
1/0.25	ZYH644		R. Em. Acopiara, o, CE
5/1	ZYI796		R. Meridional, Garanhuns, PE
5/1	ZYL225		R. Cataguases, o, MG
5	ZYL263		R. Soc. Norte de Minas, Montes Claros, MG
5/0.5	ZYK578		R. Mantiqueira Globo, Cruzeiro, SP
5/0.25	ZYK696		R. Educação e Cultura de Sertãozinho, o, SP
5/0.5	ZYJ331		R. Estação da Luz, Curitiba, PR
2.5/ 0.25	ZYK287		R. Santa Cruz do Sul, o, RS
1	ZYI429		R. Cult. de Sinop, o, MT
1	CC55	Chile	R. V. de la Tierra, Angol
50	HJHF	Colombia	R. Dif. Nacional, Medellín (Marinilla) (relay 570)
10		Falkland Islands	Falkland Isl. Bc. Station/ BFBS, Bush Rincon
50	HCGM1	Ecuador	R. Reloj, Quito

25	ZP16	Paraguay	R. Parque, Ciudad del Este
100/50	YVKE	Venezuela	R. Mundial, ("YVKE Mundial"), Caracas

551 kHz

10	CW1	Uruguay	R. Colonia, o (nf 550)

560 kHz

15/7.5	LRA9	Argentina	R. Nacional Esquel, o
25/5	LRA13		R. Nacional Bahía Blanca, o
125	LRA16		R. Nacional La Quiaca, o
10/3	LT15		R. del Litoral, Concordia
25/15	LV1		R. Colón, San Juan
25/5	ZYH887	Brazil	R. Educ. do Maranhão, São Luís, MA
1/0.25	ZYH604		R. Educ. Jaguaribana, Limoeiro do Norte, CE
1/0.25	ZYI695		R. Potiguara, Mamanguape, PB
5/1	ZYH456		R. Jornal, Itabuna, BA
5/1	ZYL277		R. Dif. de Patrocinio, o, MG
25/10	ZYK761		R. Paulista AM, São Paulo, SP
1/0.5	ZYJ214		R. Londrina, o, PR
5/1	ZYJ281		R. Cult. de Guarapuava, o, PR
5/1	ZYK231		R. São Francisco, Caxias do Sul, RS
10/2.5	ZYI395		R. Aruanã, Barra do Garças, MT (F.Pl.) (still on 1460)
1/0.25	ZYI419		R. Pioneira, Tangará da Serra, MT
10	HJGS	Colombia	R. Dif. Nal., Tunja (rel 570)
25	HJPF		La Voz de la Pampa ("R. 5-60"), Maicao
25	HCAK2	Ecuador	C.R.E. Radiodif. del Ecuador, Guayaquil
10		Guyana	GBC, Georgetown
20	YVPJ	Venezuela	R. Exitos, "Latina 5-60" Rubio
100/50	YVRH		R. Nal, Cd Guayana (rel 630)

564 kHz

2	OBZ4I	Peru	R. Oriente, Lima (nf 560)

570 kHz

1	ZYH...	Brazil	ECOS Empresa de Comunicação da Amazônia ("R. Tupinabarana"), Parintins, AM (CP)
10/1	ZYH890		R. Imperatriz, o, MA
10/0.25	ZYH...		Sistema Timon de Radiodifusão, Timon, MA (CP)
5/0.25	ZYH613		R. Vale do Cariri, Juazeiro do Norte, CE
1/0.25	ZYH614		R. Uirapuru, Itapipoca, CE
5/1	ZYH244		R. Novo Nordeste, Arapiraca, AL
25/5	ZYL261		R. Capital de Minas, Belo Horizonte, MG
1/0.25	ZYK595		R. Cl. de Itapeva, o, SP
5/1	ZYK672		R. Dif. de Taubaté, o, SP
1/0.25	ZYK698		R. Jornal, Nhandeara, SP
1/0.25	ZYK717		Bairiri R. Clube, o, SP (F.Pl.) (still on 1370)
1/0.5	ZYJ349		R. Continental, Palotina, PR
5/1	ZYJ735		R. Eldorado AM, Criciúma, SC
1/0.25	ZYJ794		R. Fronteira Oeste, Dionísio Cerqueira, SC
1/0.25	ZYK267		R. Passo Fundo, o, RS
1/0.25	ZYI416		R. Cult. de Poxoréo, o, MT (ff 850)
1/0.25	ZYI...		R. Jornal, São José dos Quatro Marcos, MT (CP)
1/0.25	ZYH750		R. Cult. de Catalão, o, GO
50	CB57	Chile	R. Agricultura, Santiago
100	HJND	Colombia	R. Dif. Nacional, Bogotá
10	HCCE1	Ecuador	R. El Sol, Quito
1	OAX3X	Peru	R. Progreso, Chimbote (ti)
100/50	YVLX	Venezuela	R. Rumbos, Villa de Cura (rel 670)

572 kHz

30	ZP15	Paraguay	R. La Voz del Amambay, Pedro Juan Caballero (nf 570)

580 kHz

12	LU20	Argentina	R. Chubut, "La 20" Trelew
25/5	LW1		R. Univ. Nal. de Córdoba, o
25	CP91	Bolivia	R. Panamericana, La Paz
10	ZYH290	Brazil	R. Nal. de Tefé, o, AM
20/10	ZYI776		R. Globo, Recife, PE
1/0.25	ZYH477		R. Sobral, Teixeira de Freitas, BA
5/1	ZYL328		R. Visão Uberlândia, MG
20/5	ZYJ465		R. Relógio Federal, RJ, RJ
1/0.25	ZYK540		R. Jornal, Americana, SP
1/0.25	ZYK724		R. Regional, Palmital, SP
5/0.25	ZYJ330		R. Grande Lago, Santa Helena, PR
2/0.25	ZYK299		R. São Gabriel, o, RS
10/5	ZYK318		R. Fátima, Vacaria, RS
10/1	ZYI387		R. Educ. Rural, Campo Grande, MS
50	HJHP	Colombia	R. Dif. Nacional, Cali (rel 570) (ti)
10	HCPC2	Ecuador	R. Uno ("La Primerisima"), Guayaquil
10	OAX2E	Peru	R. Marañón, Jaén
15	OAX4M		R. Star, Líma
2.5/1	CX58	Uruguay	R. Clarín, Montevideo
50/10	YVMJ	Venezuela	La Voz de la Fé, Maracaibo

590 kHz

25/1	LRA30	Argentina	R. Nacional Bariloche, San Carlos de Bariloche
25/10	LS4		R. Continental, Buenos Aires
25/5	LV12		R. Dif. Independencia, San Miguel de Tucumán
10	ZYJ700	Brazil	Super R. Roraima, Boa Vista, RR
5/0.25	ZYH627		R. Vale do Rio Poty, Crateús, CE
2/0.25	ZYI692		R. Serrana, Araruna, PB
50/5	ZYH445		R. Cruzeiro da Bahia, Salvador, BA
1/0.25	ZYL277		R. Dif. de Patrocínio, o, MG (ff 560)
5/0.5	ZYI213		R. Tribuna, Vitória, ES
10/1	ZYK534		R. Atlântica de Santos, o, SP
5/1	ZYK643		R. Ribeirão Preto ("R. 79"), o, SP
10/5	ZYJ234		R. Dif. de Curitiba, o, PR
2/1	ZYJ809		R. Progresso, Descanco, SC
5/0.5	ZYK210		R. Alegrete, o, RS
10/5	ZYI420		R. Globo Vila, Cuiabá, MT
10/1	ZYH798		R. Manchester, Anápolis, GO
1	CA59	Chile	R. Minería, Antofagasta
1	CC59		R. Caracol, Concepción
10	CD59		R. Pres. Ibañez, Punta Arenas
100	HJCR	Colombia	CARACOL Antioquia, Medellín
10	HCSP1	Ecuador	R. Carrousel, Quito
5	ZP32	Paraguay	R. Ycuámandyyú, San Pedro de Ycuámandyyú
0.5	OBX2N	Peru	R. Esperanza 2000, La Esperanza
20	YVKL	Venezuela	R. (Reloj) Continente, Caracas

600 kHz

8	CP150	Bolivia	R. ACLO, Sucre
10	ZYH287	Brazil	R. Nal. de São Gabriel da Cachoeira, o, AM
10/1	ZYH...		R. Litoral Maranhense, São Luís, MA (CP)
1/0.25	ZYI...		R. Boa Esperança, Padre Marcos, PI (CP)
1/0.25	ZYI789		R. Dif. Cardeal Arcoverde, Arcoverde, PE
10/1	ZYH486		R. Vale do Rio Grande, Barreiras, BA
1/0.25	ZYH...		R. Dif. de Rio Real, o, BA (CP)
100	ZYK278		R. Gaúcha, Porto Alegre, RS
10	CB60	Chile	R. Monumental, Santiago
10	CD60		R. Tricolor, Osorno
50	HJHJ	Colombia	R. Libertad, Barranquilla
50	HCXY2	Ecuador	R. Nal. del Ecuador, Guayaquil
10	OBZ4W	Peru	R. Cora del Perú, Lima
0.25	OAX6S		R. Cult. Toquepala, Tacna (ti)
10	PZX26	Suriname	SRS, Paramaribo
10	YVQB	Venezuela	R. Sucre, Cumaná
15	YVSW		R. Alto Llano, Santa Bárbara de Barinas

610 kHz

10/2.5	ZYH...	Brazil	R. Grande Rio ("R. Sucesso"), Iranduba, AM
10/1	ZYI899		R. Poty, Teresina, PI
1/0.25	ZYI678		R. Progresso, Sousa, PB
10/2	ZYH249		R. Cidade Imperial, Marechal Deodoro, AL
50/10	ZYL268		R. Itatiaia, Belo Horizonte (Nova Lima), MG
1/0.25	ZYK532		R. Chamonix ("R. Alvorada"), Mogi-Mirim, SP
1/0.25	ZYK577		R. Em. A Voz de Catanduva, o, SP
1/0.25	ZYK589		R. Piratininga, Guaratinguetá, SP
2/0.25	ZYK726		R. Paranapanema, Piraju, SP

5/0.5	ZYJ746		Radiodifusão Indio Condá, Chapecó, SC
10/5	ZYI425		R. Celeste, Sinop, MT
25/2	ZYH786		R. Jornal, Luziânia, GO
1/0.25	ZYH785		R. Anhanguera, Porto Nacional, TO
30	HJKL	Colombia	R. Internacional, Bogotá
25	HCMJ1	Ecuador	Em. Gran Colombia, Quito
10/1	ZP30	Paraguay	R. La Voz del Chaco Paraguayo, Filadelfia
50	CX4	Uruguay	R. Rural, Montevideo
10	YVSE	Venezuela	R. Cristal, Barquisimeto
10	YVXY		R. Centro 610, Cantaura

619 kHz

1	ZP40	Paraguay	R. Ñasaindý, Santaní (San Estanislao) (vf)

620 kHz

25	LRA18	Argentina	R. Nacional, Río Turbio, o
10/5	LRA26		R. Nacional Chaco, o
25/5	LRA28		R. Nacional, La Rioja, o
25/5	LT17		R. Prov. de Misiones, Posadas
25	LV4		R. San Rafael, o
20/5	CP63	Bolivia	R. San Gabriel, La Paz
10	ZYH590	Brazil	R. Assunção Cearense, Fortaleza, CE
10/ 0.25	ZYL320		R. Educ. de Porteirinha, o, MG
1/0.25	ZYL348		R. Ibja, O, MG
5/0.25	ZYL357		R. Catuí, Manhuaçu, MG, (F.Pl.) (still on 1310)
50/10	ZYK521		R. Panamericana, ("Jovem Pan"), São Paulo, SP
1/0.25	ZYJ332		R. Cidade Jandaia, Jandaia do Sul, PR
5/0.25	ZYJ779		R. Dif. Alto Vale, Rio do Sul, SC
10/1	ZYK270		R. Pelotense, Pelotas, RS
1/0.25	ZYK315		R. Municipal, Tenente Portela, RS
1/0.25	ZYH777		R. Cristã Educativa, Pires do Rio, GO
1	CA62	Chile	R. Norte Verde, Ovalle
10	CC62		R. Bío-Bío, Concepción
15	HJVP	Colombia	Colmundo R, Cartagena
50	HJEL		Colmundo R, Cali
50	HCXY3	Ecuador	R. Nal. del Ecuador, Loja
1	OAX2M	Peru	R. Chepén, o
10	OCY4A		R. Estrella del Sur, Cañete
10	YVNO	Venezuela	R. Libertad, Cabimas

630 kHz

25/5	LS5	Argentina	R. Rivadavia, Buenos Aires
25/5	LU4		Radiodif. Patagonia Argentina, Comodoro Rivadavia
10	ZYH422	Brazil	R. Dif. de Macapá, o, AP
1/0.25	ZYI904		R. Dif. da Barra, Barras, PI
50/5	ZYJ920		R. Aperipe, Aracaju, SE
1/0.5	ZYL299		R. Dif. de Uberaba, o, MG
25/10	ZYJ466		R. Roquette Pinto, RJ, RJ
1/0.25	ZYK613		R. Dif. de Mirassol, o, SP
5/0.25	ZYK635		R. Cidade Globo, Pres. Prudente, SP
10/0.5	ZYJ284		R. Estadual do Paraná, Curitiba, PR
5/0.25	ZYJ300		R. Educ. Marechal, Marechal Cândido Rondón, PR
1/0.25	ZYJ800		R. Doze de Maio, São Lourenço d'Oeste, SC
1/0.5	ZYK259		R. Cacique, Lagoa Vermelha, RS
5/1	ZYK289		R. Santamariense, Santa Maria, RS
10/5	ZYI384		R. Dif. Bom Jesús, Cuiabá, MT
10/1	ZYN603		R. IPB AM, Campo Grande, MS
10	CB63	Chile	R. Fundación, Valparaíso
20	HJFD	Colombia	R. Manizales, o
15	HJWB		LV de Guainia, Pto. Inírida
10	HCHA2	Ecuador	Ondas Quevedeñas, Quevedo
50	YVKA	Venezuela	R. Nacional, Caracas

640 kHz

25/5	LRA24	Argentina	R. Nacional Rio Grande, o
10/1	LU18		R. El Valle, General Roca
10/5	LV15		R. Villa Mercedes, o
10	CP204		R. Tarija, San Lorenzo
10/5	ZYJ590	Brazil	R. Cabugi, Natal, RN
10/0.5	ZYH458		R. Dif. Sul da Bahia, Itabuna, BA
5/0.25	ZYL308		R. Santa Cruz, Pará de Minas, MG
10/0.5	ZYI204		R. Vitória, o, ES
5/0.25	ZYJ489		R. Agulhas Negras, Resende, RJ
5/1	ZYK547		R. A Morada do Sol, Araraquara, SP
5/1	ZYJ262		R. Auri Verde, Londrina, PR
50/10	ZYK277		R. Bandeirantes, Porto Alegre, RS
10/5	ZYI406		R. Progresso, Alta Floresta, MT
10/1	ZYI424		R. Tangará, Tangará da Serra, MT
1	ZY1924		R. Cruzeiro, Pedro II (F.Pl.) (still on 1460)
25/10	ZYH757		R. Dif. de Goiânia, o, GO
10	CD64	Chile	R. Cooperativa, Temuco
10	HJBJ	Colombia	RCN Santa Marta, o
50	HCXY1	Ecuador	R. Nal. del Ecuador, Quito
10	ZP19	Paraguay	R. Caaguazú, Coronel Oviedo
10	OAZ4K	Peru	R. del Pacífico, Lima
5	OBX7C		R. Onda Azul, Puno
0.5		Surinam	Kara's Broadcasting Corp., Paramaribo
50/20	YVQO	Venezuela	Porteñas 640, Puerto La Cruz

650 kHz

10/1	ZYI540	Brazil	R. Tropical, Santarém, PA
1/0.25	ZYH924		R. Macaru, Viana, MA
1/0.25	ZYI925		R. Tapuio, Miguel Alves, PI
1/0.25	ZYI672		R. Alto Piranhas, Cajazeiras, PB
5/0.5	ZYH462		R. Cl. de Valença, o, BA
5/0.5	ZYL200		R. Princesa, Lagoa Formosa, MG
10/1	ZYL272		R. Ouro Preto, o, MG (ff 1120)
5/0.5	ZYL309		R. Veredas, Unaí, MG
10/1	ZYL...		R. Itatiaia, Timóteo, MG
1/0.25	ZYK508		R. Andradina, o, SP
5/1	ZYK524		R. Dif. de Piracicaba, o, SP
25/0.5	ZYK...		R. Difusora, São Paulo, SP (CP)
5/1	ZYJ250		R. Colméia, Cascavel, PR
5/0.5	ZYK238		R. Dif. Sul Riograndense, Erechim, RS
5	ZYI414		R. Educ. de Colider, o, MT
1/0.25	ZYH790		R. Cultural do Araguaia, Jussara, GO
30	HJKH	Colombia	Antena 2 de RCN, Bogotá
5	HCFD4	Ecuador	R. Visión, Manta
1	OAX2N	Peru	Ondas del Norte, Trujillo (ti)
1	OAX5R		R. Regionalista, Ica
100	CX6	Uruguay	S.O.D.R.E., Montevideo
50/20	YVLH	Venezuela	R. Aragüeña 650, Maracay

660 kHz

	CP263	Bolivia	R. Dif Integración, El Alto
10/5	ZYJ673	Brazil	R. Eldorado, Porto Velho, RO
1/0.25	ZYI...		R. Xinguara, o, PA
1/0.25	ZYI925		R. Tacarijus, São Miguel do Tapuio, PI
1/0.25	ZYH619		R. Rio das Garças, Itarema, (Acaraú), CE
5/1	ZYI787		R. Dif. de Limoeiro, o, PE
1/0.25	ZYI795		R. da Grande Serra, Araripina, PE
5/0.25	ZYH465		R. Jornal, Itapetinga, BA
10/0.25	ZYH480		R. Born Jesus de Lapa, o, BA (F.Pl.) (still on 1390)
1/0.25	ZYH...		R. Planalto, Euclides da Cunha, BA (CP)
1/0.25	ZYH...		R. Tribuna do Vale do São Francisco, Xique-Xique, BA (CP)
5/0.25	ZYL206		R. Cl. de Curvelo, o, MG
1/0.25	ZYL320		R. Educ. de Porteirinha, o, MG (ff 620)
1/0.5	ZYJ472		R. Soc. de Friburgo, Nova Friburgo, RJ
5/0.5	ZYK518		R. A Tribuna, Santos, SP
10/0.5	ZYK639		R. Cl. de Ribeirão Preto, o, SP
1/0.5	ZYJ202		R. Cult. de Cambará, o, PR
1/0.25	ZYK286		R. Marajá, Rosário do Sul, RS
10/ 0.25	ZYK319		R. Esmeralda ("R. Cidade"), Vacaria, RS
10/0.5	ZYI401		R. Amorim, Rondonópolis, MT
5/0.25	ZYH778		R. Primavera, Itapuranga, GO
5/0.5	ZYH794		R. Alvorada, Quirinópolis, GO
50	CB66	Chile	R. Chilena, Santiago
10	HJJM	Colombia	R. Auténtica, Cali

25	HJQS		R. Ciudad de Cúcuta, o
25	HCLG2	Ecuador	R. Carrousel, Guayaquil
10	ZP26	Paraguay	R. Itapirú, Ciudad del Este
1	OCX4L	Peru	R. Chinchaycocha, Junín
10	OCX4R		R. Omega, Lima
1	OBX60		R. Internacional, Arequipa
10	YVNA	Venezuela	Ondas de los Médanos, Coro

670 kHz

25/5	LRA11	Argentina	R. Nacional Comodoro Rivadavia, o
1	LRA52		R. Nacional Chos Malal, o
20/10	LRI209		R. Mar del Plata, o
25	LT4		Radiodif. Misiones, Posadas
10	ZYH288	Brazil	R. Cabocla, Benjamin Constant, AM
1/0.25	ZYH297		R. Vale do Rio Madeira, Humaitá, AM
1/0.25	ZYI537		R. Rural, Altamira, PA
1/0.25	ZYI539		R. Atalaia, Óbidos, PA
1/0.25	ZYI546		R. Tropical AM Paragominas, o, PA
1/0.25	ZYI927		R. Livramento, José de Freitas, PI
1/0.25	ZYH606		R. Cult. de Várzea Alegre, o, CE
10/5	ZYJ921		R. Cult. de Sergipe, Aracaju, SE
5/2.5	ZYL310		R. Educ. de Montes Claros, o, MG
1/0.25	ZYL347		R. Visão, Ponte Nova, MG
5/0.25	ZYL361		R. Uberaba, o, MG
1/0.25	ZYL...		R. Dif. de Bambuí, o, MG
1/0.5	ZYK574		R. Oceânica, Caraguatatuba, SP
1/0.25	ZYK585		R. Centro Oeste Paulista, AM, Garça, SP
1/0.5	ZYK598		R. Em. Convenção de Itu, Itu, SP
2.5/ 0.25	ZYJ231		R. Soc. Nova Esperança, o, PR
10/2	ZYJ248		R. Cidade de Curitiba, Curitiba, PR
1/0.25	ZYK296		R. Cult. Jaguarão, Santa Vitória do Palmar, RS
10/ 0.25	ZYK370		R. Gazeta ("R. Tangara"), Carazinho, RS
10	ZYN600		R. Fronteira OM, Ponta Porã, MS
5/1	ZYH747		R. São Francisco, Anápolis, GO
1/0.25	ZYH774		R. Independência do Tocantins, Paraíso do Tocantins, TO
10	HJPL	Colombia	Antena 2 de RCN, Medellín
12.5	HCFF1	Ecuador	R. Dif. Jesús del Gran Poder, Quito
10	OAX7H	Peru	R. Nal. Pachacútec, Puno
100 /50	YVLL	Venezuela	R. Rumbos, Caracas

680 kHz

5	LT3	Argentina	R. Cerealista, Rosario
30/1	LU12		R. Río Gallegos, o
25/5	LV6		R. Nihuil, Mendoza
10	CP34	Bolivia	R. Méndez Agricultura, La Paz
10/5	ZYH885	Brazil	R. Dif. do Maranhão, São Luís, MA
2.5/ 0.25	ZYI683		R. Integração do Brejo, Bananeiras, PB
10/1	ZYI793		R. do Grande Rio, Petrolina, PE
10/0.5	ZYH471		R. Cl. de Santo Antônio de Jesús, o, BA
1/0.25	ZYL326		R. União, João Pinheiro, MG
20/5	ZYJ452		R. Copacabana, São Gonçalo, RJ
1/0.25	ZYK576		R. Dif. de Catanduva, o, SP
2/0.25	ZYK628		R. Piratininga, Piraju, SP
50	ZYK275		R. Farroupilha, Porto Alegre, RS
5/1	ZYI389		R. Cult. de Campo Grande, o, MS
1/0.25	ZYH765		R. Dif. de Jataí, o, GO
10	CA68	Chile	R. El Loa, Chuquicamata
10	CC68		R. Universidad, Concepción
100	HJBU	Colombia	R. Dif. Nacional, Zambrano (relay 570) (r. 695)
25/12	HCVP2	Ecuador	Sistema de Em. Atalaya, Guayaquil
10	ZP11	Paraguay	R. Cáritas, Asunción
2	ZP41		R. Ypacaraí, o
0.5	OBX2L	Peru	R. Amanta, Chocopé
5	OBX4A		RBC, Lima
10	OAX5E		R. Independencia, Ica
5/1	CW27	Uruguay	R. Tabaré, Salto
10	YVQR	Venezuela	R. Cumaná 680 R, o

690 kHz

20	LRA4	Argentina	R. Nacional Salta, o
10/5	LU19		LV del Comahue, Cipolletti
20/5	ZYI532	Brazil	R. Cl. do Pará, Belém, PA
25/10	ZYH587		R. Dragão do Mar, Fortaleza, CE
5/0.25	ZYH453		R. Cult. de Ilhéus, o, BA
10/3	ZYL228		R. Mineira, Belo Horizonte, MG
1/0.25	ZYK561		R. Bebedouro, o, SP
1/0.25	ZYK588		R. Cl. de Guaratinguetá, o, SP
1/0.25	ZYK625		R. Cidade Pereira Barreto, o, SP
1/0.25	ZYK646		R. Brasil, Santa Bárbara d'Oeste, SP
5/0.25	ZYJ229		R. Dif. Londrina, o, PR
1/0.25	ZYJ252		R. Dif. de Ponta Grossa, o, PR
5/0.5	ZYJ772		R. Cl. de Lages, o, SC
5/0.25	ZYK252		R. Progresso, Ijuí, RS
5/1	ZYI451		R. Parecis, Diamantino, MT
5/0.25	ZYI402		R. Cult. de Naviraí, o, MS
10	CB69	Chile	R. Santiago, o
10	CD69		R. Estrella del Mar, Ancud
30	HJCZ	Colombia	R. Deportes, Bogotá
50	HCJB1	Ecuador	La Voz de los Andes, Quito
10	HCFA4		R. Sucre, Portoviejo (ts)
1	OBX3C	Peru	R. El Puerto, Chimbote (ti)
0.5	OBX6Q		R. La Joya, o
10/5	CX8	Uruguay	R. Sarandí, Montevideo
50/20	YVMR	Venezuela	R. Barquisimeto, o

700 kHz

10/5	ZYI890	Brazil	R. Cl. de Teresina, o, PI
50/1	ZYH500		R. Cult. AM, Feira de Santana, BA
1/0.25	ZYL322		R. Mucuri, Teófilo Otoni, MG
50	ZYK686		Nova Eldorado AM, São Paulo, SP
1/0.25	ZYK356		R. Em. Batovi, São Gabriel, RS
1/0.75	CA70	Chile	R. Amancay, Copiapó
1	CD70		R. Valdivia, o
120	HJCX	Colombia	CARACOL Pacífico, Cali
50	HCRS2	Ecuador	R. Sucre, Guayaquil (ts)
10	ZP12	Paraguay	R. Carlos Antonio López, ("La Voz del Neembucú"), Pilar (vf)
10	OBX1U	Peru	R. Cutivalú ("La Voz del desierto"), Castilla
1	OBZ4H		R. 700, Lima
10	YVMH	Venezuela	R. Popular, Maracaibo
5/2	YVPQ		R. Sur, Puerto Ordaz

710 kHz

25/1	LRA17	Argentina	R. Nacional Zapala, o
25/5	LRA19		R. Nacional Iguazú, Puerto Iguazú
37/5	LS1		R. Municipal de la ciudad de Buenos Aires, o
10	CP50		Bolivia R. Pío XII, Siglo Veinte
10/5	ZYI534	Brazil	R. Rural, Santarém, PA
1	ZYH891		R. Verdes Campos, Pinheiro, MA (F.Pl.) (still on 1450)
1/0.25	ZYH910		R. Verdes Vales, Grajaú, MA
1/0.25	ZYI901		R. Alvorada do Sertão, São João do Piaui, PI
1/0.25	ZYI933		R. Cl. de Barras, o, PI
1/0.25	ZYH628		R. Dif. Asa Branca, Boa Viagem, CE
5/0.25	ZY1685		R. Educ. de Conceição, o, PB
5/1	ZYH240		R. AM 710, Maceió, AL
10/ 0.25	ZYH490		R. Jacarandá, Santa Cruz Cabrália, BA
5/.05	ZYL219		R. Cancella, Ituiutaba, MG
5/0.25	ZYL258		R. Manhuaçu, o, MG
1/0.25	ZYL333		R. Dif. de Pouso Alegre, o, MG
10/0.25	ZYJ451		R. Dif. Carioca, RJ, RJ
10/0.25	ZYK559		R. Globo, Bauru, SP
1/0.25	ZYJ793		R. Fraiburgo, o, SC
5/0.5	ZYI386		R. Cult. de Cuiabá, o, MT
10/2.5	ZYH...		R. Nova Aliança, Brasília, DF
50	HJNX	Colombia	R. Super, Medellín
10	HCEP5	Ecuador	Escuelas Radiofonicas Populares, Riobamba
5	OBX5Q	Peru	R. Programas del Perú, Ica (rel 730)
50/20	YVKY	Venezuela	R. Capital, Caracas

715 kHz

10	HCGB4	Ecuador	La Voz de Portoviejo, o (nf 720) (vf)

720 kHz

1	LRA59	Argentina	R. Nacional Gobernador Gregores, o
25/10	LV10		R. de Cuyo, Mendoza
2.5	CP148	Bolivia	R. Yungas, Chulumani
10	ZYH202	Brazil	R. Integração, Cruzeiro do Sul, AC
1	ZYH281		R. Dif. de Itacoatiara, o, AM

100	ZYI770		R. Cl. de Pernambuco, Recife, PE
25/0.5	ZYL330		R. Divinópolis AM, o, MG
1/0.25	ZYK575		R. Dif. de Casa Branca, o, SP
1/0.25	ZYK701		R. Sentinela, Ourinhos, SP
5/0.25	ZYK718		R. Cruzeiro, o, SP
1/0.25	ZYK722		R. Menina, Olímpia, SP
50/10	ZYK276		R. Guaíba, Porto Alegre, RS
5/1	ZYI411		R. Dif. de Barra do Garças, o, MT
5/1	ZYI390		R. Cl. de Dourados, o, MS
30	HJAN	Colombia	Em. Unidas, Barranquilla
25	HJVO		Tr. del Quindío, Armenia
5	HCIC1	Ecuador	R. Municipal de Cultura, Quito
15	HCMO3		R. Matovelle, Loja
2.5	OAX2B	Peru	R. Nal. Pachacútec, Trujillo (vf)
10	OAX8A		R. Nal. Pachacútec, Iquitos
100/50	YVQE	Venezuela	R. Felicidad "Super Criolla," Porlamar
10	YVXE		R. Elorza, o

730 kHz

5	LRA3	Argentina	R. Nacional Santa Rosa, o
25/5	LRA27		R. Nacional Catamarca, San Fernando del Valle de Catamarca
10/1	LU23		Em. Lago Argentino, El Calafate
10	CP27	Bolivia	R. La Cruz del Sur, La Paz
3	CP165		R. Mensaje, Montero
1/0.25	ZYH896	Brazil	R. Eldorado, Codó, MA
5/0.5	ZYH640		R. Sinal, Aracati, CE
10/5	ZYI780		Em. Rural ("A Voz do São Francisco"), Petrolina, PE
5/0.25	ZYL287		R. Soc. Triângulo Mineiro, Uberaba, MG
5/1	ZYL297		R. Nova Amizade AM, Juiz de Fora, MG
10/0.5	ZYI...		Sistema Canaã de Comunicação, Cariacica, ES (CP)
5/0.25	ZYK523		R. Cidade, Jundiaí, SP
10/1	ZYK610		R. Dirceu de Marília, o, SP
5/0.25	ZYJ208		R. Marumby, Campo Largo (Curitiba), PR
1/0.25	ZYJ323		R. Humaitá, Campo Mourão, PR
1/0.25	ZYJ787		R. Soc. Tubá, Tubarão, SC
5/1	ZYK268		R. Planalto, Passo Fundo, RS
10/2.5	ZYI410		R. Jornal, Cáceres, MT
1/0.25	ZYI452		R. Princesa do Vale, Camapuã, MS
50/5	ZYH759		R. Cl. de Goiânia, o, GO
10/1	ZYH786		R. Jornal, Luziânia, GO (ff 610)
50	CB73	Chile	R. Cooperativa, Valparaíso
1	CD73		R. Angelina, Los Angeles
1	CD73B		R. Aysen, Puerto Aysen
50	HJCU	Colombia	R. Melodía, Bogotá
10	HCMG2	Ecuador	R. Guayaquil, o
50	ZP..	Paraguay	R. Cardinal AM Estéreo, Asunción
1	OBX2J	Peru	R. Sudamericana, Trujillo (nf 740) (ti)
10	OAX4G		R. Programas del Perú, ("RPP"), Lima
5/2.5	CX10	Uruguay	R. Continente, Montevideo
10	YVMT	Venezuela	R. Universo, Barquisimeto
10	YVOO		R. Frontera, San Antonio del Táchira

740 kHz

1	LRA55	Argentina	R. Nal. Alto Río Senguerr, o
10	LRI200		R. Puerto Deseado, o
20/10	ZYH203	Brazil	R. Progresso do Acre, Rio Branco, AC
100	ZYH446		R. Soc. da Bahia, Salvador, BA
5/0.25	ZYK519		R. Assunção de Jales, o, SP (F.Pl.) (still on 1080)
1/0.25	ZYK553		R. Cult. de Bariri, o, SP
5/0.5	ZYK650		R. 740 AM, Santo André, SP
1/0.25	ZYJ259		R. Goio-Erê, o, PR
1/0.25	ZYJ354		R. Placar, Ortigueira, PR
5/1	ZYJ753		R. Diário da Manhã, Florianópolis, SC
2.5/0.25	ZYK265		R. Palmeira, Palmeira das Missões, RS
5/0.25	ZYK283		R. Cult. Riograndina, Rio Grande, RS
1/0.5	ZYN403		R. Cidade, Alto Araguaia, MT
15	HJHB	Colombia	Ecos de Pasto, o
50	HJNS		R. Guatapurí, Valledupar
25	HCGC1	Ecuador	R. Melodía, Quito
10	OAX1D	Peru	R. Progresso, Piura
10	OAX6C		R. Continental, Arequipa
10	YVNC	Venezuela	R. Maracaibo, o
50	YVNQ		R. Caroní, Puerto Ordaz

748 kHz

10	HCSE4	Ecuador	R. Libertad, Chone (nf 740) (vf 747 - 767)

750 kHz

100/10	LRA7	Argentina	R. Nacional, Córdoba, o
1/0.25	ZYI...	Brazil	R. Ximango, Alenquer, PA (CP)
1/0.25	ZYI897		R. Heróis do Jenipapo, Campo Maior, PI
1/0.25	ZYI...		R. Cantonense, Canto do Buriti, PI (CP)
1/0.25	ZYI682		R. Panati, Patos, PB
10/0.25	ZYJ927		R. Progresso, Lagarto, SE
10/1	ZYH458		R. Dif. Sul da Bahia, Itabuna, BA (ff 640)
50/5	ZYL213		R. América, Belo Horizonte, MG
1/0.25	ZYK516		R. Cl. de Osvaldo Cruz, o, SP
10/1	ZYK642		R. Renascença, Ribeirão Preto, SP (nf 1380)
2.5/0.25	ZYK661		R. Piratininga, São José dos Campos, SP
5/0.25	ZYJ815		R. Aliança, Concórdia, SC
5/0.25	ZYK264		R. Osório, o, RS
1/0.25	ZYH792		R. Tocantins, Tocantinópolis, TO
50/25	ZYH709		R. Globo, Brasília, DF
50	HJDK	Colombia	CARACOL Medellín, ("La Voz de Antioquia"), o
30	HCRC2	Ecuador	R. Caravana, Guayaquil
1	OAX3T	Peru	R. El Mundo, Chimbote
5	OCX4B		R. Altura, Cerro de Pasco
100/50	YVKS	Venezuela	R. Caracas R. ("Red Nal. de Noticias"), o

760 kHz

16	CP29	Bolivia	R. Fides, La Paz
10	ZYH588	Brazil	R. Record, Fortaleza, CE
1/0.25	ZYH...		R. Pioneira, Delmiro Gouveia, AL (CP)
10/0.5	ZYL360		R. Terra, Montes Claros, MG
25/1	ZYJ478		R. Manchete AM, Niterói, RJ
10/0.5	ZYK560		R. Auri Verde, Bauru, SP
1/0.25	ZYJ...		R. Cacique, Guarapuava, PR (CP)
1/0.25	ZYJ742		R. Nereu Ramos, Blumenau, SC
1/0.25	ZYK222		R. Jornal do Comércio, Candelária, RS
2.5/0.25	ZYK351		R. Ametista, Planalto, RS
5/0.25	ZYI411		R. Dif. de Barra do Garças, o, MT (ff 720)
10/5	ZYI...		R. Atividade, Chapada dos Guimarães, MT (CP)
10/1	ZYH783		R. Pousada do Rio Quente, Caldas Novas, GO
20/10	ZYH...		Rede Amapaense de Radiodifusão, Macapá, AP (CP)
1/0.25	ZYK541		R. Urubupungá, Andradina (F.Pl.) (still on 1370)
1/0.25	ZYL257		R. Difusora, Machado, MG (F.Pl.) (still on 1370)
75	CB76	Chile	R. Cooperativa, Santiago
30	HJAJ	Colombia	RCN Barranquilla, o
12.5	HCRQ1	Ecuador	R. Quito, o
10	8RG	Guyana	GBC, Georgetown
25	ZP5	Paraguay	R. Encarnación, o (vf)
0.5	OBX2K	Peru	R. Andino, Otuzco
10	OBZ4X		Radiomar, Lima
10	YVQQ	Venezuela	R. Doble Q, Puerto La Cruz
10	YVSO		Mundial Primera, Trujillo

770 kHz

10	CP116	Bolivia	R. Cosmos de Bolivia, Cochabamba
10/0.25	ZYI...	Brazil	SNC—Sistema Norte de Comunicações ("R. Eldorado AM"), Marabá, PA (CP)
1/0.25	ZYH922		R. Jornal, Coelho Neto, MA
10/0.25	ZYH609		R. Vale do Salgado, Lavras da Mangabeira, CE

10/5	ZYJ922		R. Atalaia de Sergipe, Aracaju, SE
5/0.5	ZYH491		R. Rio Corrente, Santa Maria da Vitória, BA
1/0.25	ZYL209		R. Cult. d'Oeste, Lavras, MG
10/0.5	ZYL302		R. Cl. de Patos, Patos de Minas, MG
5/0.5	ZYL315		R. Iturama, o, MG
5/0.25	ZYL337		R. Itabira, o, MG
5/1	ZYI211		R. Dif. Princesa do Sul, Cachoeiro de Itapemirim, ES
5/0.5	ZYK506		R. Jornal do Povo, Limeira, SP
1/0.5	ZYJ344		R. Metropolitana, Cambé, PR
1	ZYI428		R. Sorriso, o, MT
10/5	ZYI412		R. Caivás, Dourados, (F.Pl.) (still on 1250)
1	ZYN404		R. Cidade de Matupá, Colider (Matupá), MT (F.Pl.) (still on 1380)
5/1	ZYH745		R. Carajá, Anápolis, GO
50	HJJX	Colombia	RCN Bogotá, o
20	HCNF2	Ecuador	R. El Telégrafo, Guayaquil
5	OAX8M	Peru	La Voz de la Selva, Iquitos
	OCX1T		R. Horizonte, Chiclayo
100/25	CX12	Uruguay	R. Oriental, Montevideo
50/20	YVKK	Venezuela	R. Nal., Valencia (rel. 630)

780 kHz

5/1	LRA10	Argentina	R. Nacional Ushuaia e Islas Malvinas, Ushuaia
5	LRA12		R. Nacional, Santo Tomé, o
8	LV8		R. Libertador, Mendoza
10	ZYH291	Brazil	R. Nal. de Eirunepê, o, AM
20/5	ZYI771		R. Jornal do Comércio, Recife, PE
1/0.25	ZYL246		R. Educ. de Uberlândia, o, MG
1/0.25	ZYK619		R. Dif. de Monte Aprazível, o, SP
50/10	ZYK695		R. CBN (Globo), São Paulo, SP
1/0.25	ZYJ247		R. Porta Voz de Cianorte, Cianorte, PR
5/0.25	ZYJ305		R. Chopinzinho, o, PR
1/0.25	ZYJ751		R. Brasil Novo, Jaraguá do Sul, SC
1	ZYJ788		R. Marconi, Urussanga, SC
5/2	ZYK229		R. Carazinho, o, RS
2.5	ZYK279		R. Princesa, Pto. Alegre, RS
10/1	ZYH789		R. Soc. Vera Cruz, Goianésia, GO
10/5	ZYH919		R. Alvorada, Zé Doca, MA (F.Pl.) (still on 1480)
10	CD78	Chile	R. Em. S.A.G.O. AM, Osorno
30	HJZG	Colombia	La Voz del Valle, Cali
30	HJZW		R. Almirante, Riohacha
0.5	HCAA1	Ecuador	R. Noticia, Quito
2.5	HCRG4		R. Mía, Manta
25	ZP70	Paraguay	R. Primero de Marzo, Asunción
10	OAX1K	Peru	R. Nal. del Perú, Tumbes (vf)
10	OAX4X		R. Victoria, Lima
10	YVMN	Venezuela	R. Coro, o
50/20	YVOD		Ecos del Torbes, San Cristóbal

790 kHz

25/5	LRA22	Argentina	R. Nacional San Salvador de Jujuy, o
25/5	LR6		R. Mitre, ("AM 80"), Buenos Aires
5	LV19		R. Malargüe, o
10	CP46	Bolivia	R. Libertad, La Paz
1/0.25	ZYH904	Brazil	R. Pérola do Turi ("R. Rio Turiaçu"), Sta. Helena, MA
1/0.25	ZYH915	Brazil	R. Cult. de Açailândia, o, MA
1/0.25	ZYH...		Sistema Cl. de Comunicação, Tuntum, MA (CP)
1/0.25	ZYI931		R. Mafrense, Simplício Mendes, PI
1/0.25	ZYH629		R. Jornal Centro Sul, Iguatu, CE
2.5/1	ZYI679		R. Cult. de Guarabira, o, PB
10/1	ZYH484		R. Barreiras, o, BA
1/0.25	ZYH505		R. Regional, Serrinha, BA
5/0.25	ZYL279		R. Soc. Ponte Nova, o, MG
1/0.25	ZYL311		R. Treze de Junho, Mantena, MG (nf 1190)
5/0.25	ZYL314		R. Tropical, Lagoa da Prata, MG
1/0.25	ZYK538		R. Brasil, Adamantina, SP (nf 1270)
5/1	ZYK546		R. Cult. de Araraquara, o, SP
1	ZYJ337		R. Panorama, Mandirituba, PR
1/0.25	ZYJ789		R. Videira, o, SC
1/0.25	ZYK285		R. Rio Pardo, o, RS
1	ZYI456		R. Dif. de Nortelândia, o, MT
5/0.5	ZYH761		R. Xavantes, Ipameri, GO
1/0.5	ZYH771		R. Eldorado, Mineiros, GO
20	HJDC	Colombia	La Voz de Colombia, Medellín
10	OAX2I	Peru	R. C.R.U., Trujillo (rel RPP 730)
1	OCX7C		R. Canchis, Sicuani
10	YVKC	Venezuela	Radiodif. Venezuela, ("AM 790"), Caracas
50	YVXM		R. Nal, Cd. Bolívar (rel 630) R. Minuto "La Barquisimetana", Barquisimeto

800 kHz

25/2	LU15	Argentina	R. Viedma, o
1/0.25	LV23		R. Río Atuel, General Alvear
0.25	CP157	Bolivia	R. Santa Clara, Sorata
10	ZYI...	Brazil	R. Cult. do Pará, Belém, PA
10	ZYI921		R. Antares, Teresina, PI
100	ZYJ457		R. MEC, RJ, RJ
10	ZYK292		R. Universidade de Santa Maria, o, RS
10/1	ZYH705		R. MEC, Brasília, DF
5	CB80	Chile	R. Minería, Viña del Mar
10	HJBW	Colombia	Em. Atalaya, Bucaramanga
1	HCFV1	Ecuador	R. Uno Más Uno, Quito
25	HCML2		R. Super K, Guayaquil
0.5	OAZ4U	Peru	R. Sol, Huaral
0.5	OBX5B		R. Sur Medio, Ica
1	OAX6A		R. La Porteña, Ilo
0.75	OBX6A		R. Texao, Arequipa
50	YV..	Venezuela	R. Fé y Alegría, Guasdualito

810 kHz

50/5	ZYH589	Brazil	R. Verdes Mares, Fortaleza, CE
1/0.25	ZYH491		R. Rio Corrente, Santa Maria da Vitória, BA (ff 770)
1	ZYL202		R. Aimorés, o, MG
1/0.25	ZYL266		R. Cl. Nepomuceno, o, MG (F.Pl.) (still on 1290)
1/0.25	ZYL330		R. Divinópolis AM, o, MG (ff 720)
1/0.25	ZYL354		R. Capinópolis, o, MG
5/0.25	ZYK604		R. Dif. Jundiaiense, Jundiaí, SP
1/0.5	ZYK655		R. Universal, Santos, SP
5/0.5	ZYK732		R. Centro-América, São José do Rio Preto, SP
5/0.5	ZYJ336		R. Esperança, Prudentópolis, PR
5/0.25	ZYN402		Gaspar Radiodifusão, São José do Rio Claro, MT
1/0.25	ZYI...		Magalhães Barros Radiodifusão, Alta Floresta, MT (CP)
5/0.5	ZYH767		R. Alvorada, Rialma, GO
250	HJCY	Colombia	CARACOL Colombia, Bogotá
10	OAX7V	Peru	R. Radial Andina, Juliaca (relay RPP 730)
50	CX14	Uruguay	R. El Espectador, Montevideo
50	YVLP	Venezuela	R. 810, Valencia

815 kHz

0.25	CP188	Bolivia	R. Warnes, o
5	HCVT2	Ecuador	R. Atalaya, El Milagro (nf 810)

820 kHz

25/5	LRA8	Argentina	R. Nacional Formosa, o
5/3	LU24		R. Tres Arroyos, o
10	CP35	Bolivia	R. Altiplano, La Paz
1/0.25	ZYH294	Brazil	R. Princesa do Solimões, Manacapuru, AM
1	ZYI543		R. Regional do Araguaia, Conceição do Araguaia, PA
5/0.25	ZYI912		R. Cacique Bruenque, Regeneração, PI
1/0.25	ZYH624		R. União, Camocim, CE
1/0.25	ZYH655		R. Sul Cearense, Brejo Santo, CE
1/0.25	ZYH...		R. Educadora G de Agosto, Xapuri, AC
5/1	ZYI775		R. Universitária, Recife, PE
5/0.25	ZYL255		R. Barbacena, o, MG
1/0.25	ZYL291		R. Dif. Paraisense, São Sebastião do Paraíso, MG
10/2.5	ZYI212		R. Gazeta AM, Vitória (Serra), ES
5/0.25	ZYJ477		R. Macaé, o, RJ
10/0.5	ZYK542		R. Aparecida, o, SP
1/0.25	ZYK602		R. Jauense, Jaú, SP
1/0.25	ZYK622		R. Cl. de Ourinhos, o, SP
1/0.25	ZYK624		R. Dif. de Penápolis, o, SP
10/5	ZYJ238		R. Cult. de Foz do Iguaçu, o, PR

10/5	ZYJ738		R. União, Blumenau, SC
5/1	ZYK241		R. Alto Taquari, Estrela, RS
10/1	ZYI400		R. Dif. de Cáceres, o, MT
10/5	ZYH752		R. Jornal de Goiás, Goiânia, GO
10	CA82	Chile	R. Gabriela Mistral, La Serena
0.25	CA82		R. Pampa, Pedro de Valdivia
1	CC82		R. Almirante Latorre, Talcahuano
1	CD82		R. Concordia, La Unión
15	HJAD	Colombia	R. Vigía, Cartagena
50	HJED		CARACOL Cali, o
3	HCCR1	Ecuador	Super R. F2, San Pablo del Lago
2	HCFB1		R. Monumental, Santo Domingo de los Colorados
1	HCRF4		Canal Manabita, Portoviejo
5	ZP27	Paraguay	R. Mbaracayú, Salto del Guairá (nf 800) (vf)
10	OAX40	Peru	R. Libertad, Lima
1	PZX7	Suriname	R. Apintie, Paramaribo
1/0.5	CW23	Uruguay	R. Cultural, Salto
10	YVKU	Venezuela	R. Altura, La Grita
50	YVSH		R. Guayana, Upata
50	YXG		R. Guadalupana, Coro
830 kHz			
10/1.5	LT8	Argentina	R. Rosario, o
0.5	LT21		R. Mun. Alvear, o
1	LT43		R. Mocovi, Charata
25	LU14		R. Prov. de Santa Cruz, Las Heras (relay 880)
0.25	LV18		R. Mun. de San Rafael, o
25/5	LW8		R. San Salvador de Jujuy, o R. Federal, Lanús
1/0.25	ZYI556	Brazil	R. Guarani de Marajó, Soure, PA
10/1	ZYH905		R. Mirante do Maranhão, Imperatriz, MA
1/0.25	ZYH...		R. Boas Novas, Esperantinópolis, MA (CP)
1/0.25	ZYI906		R. Primeira Capital, Oeiras, PI
1/0.25	ZYI...		R. União, o, PI (CP)
1/0.25	ZYH659		R. Pioneira, Forquilha, CE
1/0.5	ZYJ595		Em. de Educação Rural, Caicó, RN
20/1	ZYJ926		R. Princesa da Serra, Itabaiana, SE
1/0.25	ZYH506		R. Extremo Sul da Bahia, Itamaraju, BA
5/0.25	ZYL207		R. Imbiara, Araxá, MG
10/5	ZYL244		Cultura R. Sucesso, Belo Horizonte, MG
1/0.25	ZYL265		R. Dif. de Nanuque, o, MG
10/1	ZYK681		R. Cl. de Votuporanga, o, SP
1/0.25	ZYK746		R. dos Bons Tempos, Nova Odessa, SP
5/0.5	ZYJ224		R. Iguaçu, Araucária, PR
10/0.25	ZYJ266		R. Tabajara, Londrina, PR
1/0.25	ZYJ311		R. Progresso, Clevelândia, PR
2.5/ 0.25	ZYJ773		R. Soc. Cruz de Malta, Lauro Müller, SC
5/0.25	ZYK332		R. Independente, Cruz Alta, RS
5/0.25	ZYI430		R. Xavantes, Jaciara, MT
1/0.25	ZYN401		Soc. R. Educ. de Juina, o, MT
1/0.25	ZYH...		R. Princesa do Sul, Goiatuba, GO
15	HJDM	Colombia	R. Deportes, Medellín
25	HCRM2	Ecuador	R. Huancavilca, Guayaquil
2.5	HCBP5		R. Promoción, Riobamba
1	OAX3Y	Peru	R. La Selva, Tingo María
10	OAX6D	Peru	R. Nal. del Perú, Tacna
1	OCX7H		R. Inti, Santiago
50	YVLT	Venezuela	R. Sensación, Caracas
835 kHz			
5	ZP6	Paraguay	R. Guairá, Villarrica (vf)
	O..2.	Peru	R. Fraternidad, El Porvenir
840 kHz			
10/1	LRK200	Argentina	Radiodif. el Clavillo, Concepción
10/5	LT12		R. General Madariaga, Paso de los Libres
25/10	LU2		R. Bahía Blanca, o
1.5/3	CP126	Bolivia	R. Juan XXIII, San Ignacio de Velasco
1	CP210		R. Maria Auxiliadora, Montero (nf 850) (vf)
50	ZYJ679	Brazil	R. Nal. de Porto Velho, o, RO (ti)
1/0.25	ZYH298		R. Rio Madeira, Manicoré, AM
1/0.25	ZYI930		R. Ribeirão, Demerval Lobão, PI
1/0.25	ZYI...		R. Vitória, Batalha, PI (CP)
1/0.25	ZYH648		R. Campo Maior, Quixeramobim, CE
25/5	ZYH447		R. Excelsior da Bahia, Salvador, BA
50	ZYK687		R. Bandeirantes, São Paulo, SP
1/0.25	ZYJ266		R. Tabajara, Londrina, PR (ff 830)
10/1	ZYJ750		R. Rural, Concórdia, SC
10	ZYK248		R. Capital, Gravataí (Porto Alegre), RS
10	CB84	Chile	R. Portales, Valparaíso
10	CD84		R. Santa María, Coyhaique
30	HJBI	Colombia	Ondas del Caribe, Sta. Marta
20	HJKK		HJ Doble K, Neiva
50	HCPN1	Ecuador	R. Vigía, La Voz de la Policía Nal., Quito
1	HCEM4		R. Costa Azul, Portoviejo
4	OAX3S	Peru	R. Casma, o
10	OAX5H		R. Ayacucho, o
1	OBX6Y		R. Onda Azul, Paucarpata
10	YVMY	Venezuela	Unión R. 840, Barquisimeto
850 kHz			
1	CP160	Bolivia	R. 21 de Diciembre, Mina Catavi
5/1	ZYJ675	Brazil	R. Ariquemes, o, RO
10/1	ZYI538		R. Itacaíunas, Marabá, PA
1/0.25	ZYI555		R. Tocantins, Cametá, PA
5/1	ZYI557		R. Itaituba, o, PA
1/0.25	ZYH...		R. Vitória, Vitória do Mearim, MA (CP)
5/1	ZYH599		R. Iracema de Juazeiro, Juazeiro do Norte, CE
5/1	ZYI693		R. Rural, Guarabira, PB
5/0.25	ZYH474		R. Caraibas, Senhor do Bonfim, BA
1/0.25	ZYL233		R. Dif. Formiguense, Formiga, MG
5/0.25	ZYJ254		R. Dif. Colméi, Campo Mourão (F.Pl.) (still on 1090)
1/0.25	ZYL254		R. Por Um Mundo Melhor, Governador Valadares, MG
1/0.25	ZYL295		R. Tupaciguara, o, MG
1/0.25	ZYK563		R. Cl. de Biriguí, o, SP
1/0.25	ZYK644		R. Cl. de Rio Claro, o, SP
1/0.25	ZYJ808		R. Cidade, Brusque, SC
10/2	ZYI416		R. Cult. de Pcxoréo, o, MT
1/0.25	ZYI438		R. Dif. Nor estado, São Gabriel do Oeste, MS
10/5	ZYH776		R. Tropical, Porangatu, GO
50	HJKC	Colombia	R. Recuerdos, Bogotá
50	HCVS2	Ecuador	R. San Francisco, Guayaquil
40	OAX4A	Peru	R. Nal. Pachacútec, Lima
	O..5.		R. Democracia, Ayacucho
82	CX16	Uruguay	R. Carve, Montevideo
10	YVLD	Venezuela	RV-850, Valencia
855 kHz			
0.5	HCGB7	Ecuador	R. Nal. Espejo, El Puyo (nf 850)
860 kHz			
1	LRA56	Argentina	R. Nal. Perito Moreno, o
5			R. Mun. de Chilecito, o
			R. Del Lago, Villa Carlos Paz
10	CP8	Bolivia	R. Nueva América, La Paz
5	CP185		R. Paitití, Guayaramerín
25/10	ZYH592	Brazil	R. Cidade de Fortaleza, Maracanaú (Fortaleza), CE
100	ZYJ459		R. Mundial, RJ, RJ
10/ 0.25	ZYK288		R. Guarathan, Santa Maria, RS
5	CC86	Chile	R. Inés de Suárez, Concepción
10	HJEP	Colombia	Voces de Occidente, Buga
50	HJNJ		La Voz del Cañaguate, Valledupar
10	HCPC1	Ecuador	R. Visión, Quito
25	ZP28	Paraguay	R. La Voz de la Cordillera, Caacupé (vf 860-861)
10	OBX4V	Peru	R. Junín , Huancayo
10	YVOL	Venezuela	R. Rumba 860 AM, San Cristóbal
20/10	YVYE		R. Enlace 8-60, Valle de La Pascua
870 kHz			
100	LRA1	Argentina	R. Nacional, Buenos Aires
1/0.25	ZYH...	Brazil	R. Jutanópolis, Manacapúru, AM
10/1	ZYH903		R. Mirante, Codó, MA
1/0.25	ZYI...		R. Vale do Pajeú, Canto do Buriti, PI (CP)
1/0.25	ZYI...		R. Garça Branca, Guiratinga, MT (CP)

1/0.25	ZYH591		R. Iracema de Fortaleza, Iguatu, CE
1/0.25	ZYH658		R. Tabajara, São Benedito, CE
5/1	ZYH245		R. Educ. Sampaio, Palmeira dos Indios, AL
5/0.25	ZYH457		R. Cl. de Itabuna, o, BA
5/0.25	ZYH499		R. Rio Vale AM, Juazeiro, BA
5/0.25	ZYJ784		R. Difusora, São Francisco do sol, SC
10/0.5	ZYL304		R. Juriti de Paractu, Paracatu, MG
1/0.25	ZYL318		R. Cult. de Diamantina, o, MG
5/0.25	ZYL349		Multisom R. Princesa da Mata, Muriaé, MG
1/0.25	ZYL...		R. A Voz do São Francisco, Januária, MG
1/0.25	ZYK620		R. Novo Horizonte, o, SP
5/1	ZYK705		R. Central, Campinas, SP
5/0.25	ZYJ243		R. Metropolitana, Maringá, PR
1/0.25	ZYI...		Empresa Guiratinguense de Radiodifusão, Guiratinga, MT (CP)
5/0.5	ZYH749		R. Educ. de Uruaçu, o, GO
20/0.5	ZYH754		R. Universitária, Goiânia, GO (F.Pl.) (still on 1400)
1/0.25	ZYH762		R. Anhanguera, Araguaína, TO
15	HJLA	Colombia	R. Reloj ("La Voz del Tolima"), Ibagué
30	HJSB		R. Mar Caribe Internacional, Barranquilla
25	HCLY2	Ecuador	R. Cristal, Guayaquil
0.4	HCGS6		R. Píllaro, o
10	OBX1F	Peru	R. C.R.U., Chiclayo (rel RPP 730)
	O..2.		R. Vernacular, Trujillo
1	OCX7R		R. Mundo, Cuzco
10	YVMP	Venezuela	R. Lara, Barquisimeto
10	YVRU		R. Especial, Puerto La Cruz

880 kHz

5	LU14	Argentina	R. Prov. de Santa Cruz, Río Gallegos
0.18	CP20	Bolivia	R. El Cóndor, La Paz
100	ZYL275	Brazil	R. Inconfidência, Belo Horizonte, MG
1/0.25	ZYI680		R. Maringá, Pomball, PB (F.Pl.) (still on 1490)
2.5/0.25	ZYK317		R. São Miguel, Uruguaira, RS
10/2.5	ZYK249		R. Itaí, Pto. Alegre, RS
1/0.25	ZYK...		R. Seberi, o, RS (CP)
10	CB88	Chile	R. Corporación, Santiago
15	HJGE	Colombia	CARACOL Bucaramanga, o
100	HCRP1	Ecuador	R. Católica Nal. del Ecuador, Quito
50	OBZ4N	Peru	R. Unión, Lima
10	YVKV	Venezuela	R. Libertador, Caracas
20/10	YVYM		Mundial 880, Puerto Ordaz

890 kHz

20/1	LU33	Argentina	Em. Pampeana, Santa Rosa
25/5	LV11		Em. Santiago del Estero, o
10/2.5	ZYJ685	Brazil	R. Planalto, Vilhena, RO (F.Pl.) (still on 1530)
5/1	ZYI536		R. Ponta Negra ("R. Cl. de Santarém"), Santarém, PA
1/0.25	ZYH642		R. Itatiaia, Sta. Quitéria, CE
20/10	ZYI772		R. Tamandaré, Recife, PE
10/0.25	ZYL250		R. Santa Cruz, Jequitinhonha, MG
25/10	ZYK690		R. Gazeta, São Paulo, SP
10/0.5	ZYJ499		R. Musical, Cantagalo, RJ (F.Pl.) (still on 1520)
1/0.25	ZYJ287		R. Ubá, Ivaiporã, PR
5/0.25	ZYJ338		R. Itapuã, Pato Branco, PR
1/0.25	ZYJ755		R. Santa Catarina, Florianópolis, SC
5/0.25	ZYK215		R. Viva, Bento Gonçalves, RS
1/0.5	ZYK295		R. Noroeste, Santa Rosa, RS
50/10	ZYH706		R. Planalto, Brasília, DF
10	CA89	Chile	R. León Trece, Pozo Almonte
5	CC89		R. Interamericana, Concepción
20	CD89		R. Nal., Punta Arenas, o
15	HJCE	Colombia	R. Continental, Bogotá
20/10	HJPM		R. Galeón, Santa Marta
50	HCFP3	Ecuador	R. Superior, Machala
1	HCLT5		Ondas del Chimborazo ("Canal 890"), Riobamba
1	ZP33	Paraguay	R. Tres de Febrero, Itá
25/10	CX18	Uruguay	R. Libertad Sport "lapantera," Montevideo
10	YVLW	Venezuela	R. América, Valencia
1	YVVO		R. Oriente 890 AM, El Tigre

900 kHz

25/5	LT7	Argentina	R. Prov. de Corrientes, o
1	CP79	Bolivia	Radioem. LV Nal, Tarija
5/1	ZYJ672	Brazil	R. Alvorada de Rondônia, Ji-Paraná, RO
20/10	ZYJ591		R. Nordeste, Natal, RN
5/1	ZYH488		R. Sisal, Conceição do Coité, BA
1/0.25	ZYL338		R. Vinícola, Andradas, MG
50/10	ZYJ454		R. Tamoio, RJ, RJ
5/0.25	ZYK511		R. A Voz do Sertão, Pres. Prudente, SP
10/0.25	ZYK664		R. Brasil Novo, São José do Rio Preto, SP
1/0.25	ZYK742		R. Cidade, Itapetininga, SP
1/0.25	ZYJ272		R. Em. Sant'Ana, Ponta Grossa, PR
5/0.25	ZYJ295		R. União, Toledo, PR
2.5/0.5	ZYK211		R. Aratiba, o, RS
5/0.5	ZYK263		R. Progresso, Nôvo Hamburgo, RS
1/0.25	ZYK303		R. Mun. Sampedrense, São Pedro do Sul, RS
10/1	ZYH768		R. Sudoeste AM, Rio Verde, GO
10/5	ZYH797		R. Cristal, Cristalândia, TO
2.5/0.25	ZYL...		R. Carangola, o, MG (CP)
10/2.5	ZYI455		R. Dif. Arco Iris, Araputanga, MT
5	CA90	Chile	R. Univ. de Atacama, Copiapó
1	CB90		R. Amapola, Valparaíso
1	CC90		R. Ñuble, Chillán
1	CD90		La Voz de la Costa, Osorno
30	HJDD	Colombia	R. 900, Cúcuta
20	HJEY		La Voz de Cali, o
10	HJSS		R. Malecón, Montería
10	HCVA1	Ecuador	R. Sucre, Quito
1	HCRR5		R. Carrousel, Cuenca
5	HCOF4		R. Chone, o (vf)
0.5	OAX3E	Peru	R. Ribereña, Aucayacu
10	OBX4X		R. El Sol, Lima
3/1	OBX6K		Nevada R., Arequipa
1	OBX7S		R. Baha'í, Chucuito
2.5	CW17	Uruguay	R. Frontera, Artigas
10	YVMD	Venezuela	R. Mara ("Ritmo 900"), Maracaibo

905 kHz

1	HCBO2	Ecuador	R. Espectáculo, Guayaquil (nf 910) (ti)
	O..2.	Peru	R. San Antonio ("La Super R."), o

910 kHz

50	LRA23	Argentina	R. Nacional San Juan o
25/5	LR5		La Red, Buenos Aires
10/1	ZYI...	Brazil	R. Chapada do Corisco, Teresina, PI (CP)
1/0.25	ZYH645		R. Assunção Cearense, Sobral, CE
5/1	ZYI785		R. Liberdade, Caruaru, PE
1/0.25	ZYL292		R. Teófilo Otoni, o, MG (nf 990)
1/0.25	ZYL346		R. Difusora Industrial, Nova Serrana, MG (CP)
5/0.5	ZYL...		R. Jornal Cidade, Juiz de Fora, MG (CP)
5/0.25	ZYK536		R. Alvorada, Piracicaba, SP (ff 950)
1/0.25	ZYK763		R. Princesa, Monte Azul Paulista, SP
1/0.25	ZYJ207		R. Dif. Apucarana, o, PR
1/0.25	ZYJ811		R. Dif. de Içara, o, SC
1/0.25	ZYJ824		R. Rainha das Quedas, Abelardo Luz, SC
5/0.5	ZYK320		R. Venâncio Aires, o, RS
5/1	ZYH763		R. Paranaíba, Itumbiara, GO
1/0.25	ZYH...		R. Cidade, Jaraguá, GO
30	HJDO	Colombia	LV del Río Grande, Medellín
30	HJMY		RCN San Andrés, Isla San Andrés
1	HCGE5	Ecuador	R. Mundial, Riobamba
1	AX2H	Peru	R. Huamachuco, o
50/20	YVRQ	Venezuela	RQ-910, Caracas

914 kHz

3		Suriname	RANI, Nieuw Nickerie

920 kHz

1/0.25	ZYI895	Brazil	R. Dif. de Picos, o, PI

10/0.5	ZYI905		R. Itamaraty, Piripiri, PI (F.Pl.) (still on 1330)
1/0.25	ZYJ600		R. Currais Novos, o, RN
10/2	ZYH...		R. Globo, AM 920, Salvador, BA
1/0.25	ZYH788		R. Valle da Serra, São Luís de Montes Bellos (F.Pl.) (still on 1450)
5/0.5	ZYL271		R. Cult. Rio Branco, Visconde do Rio Branco, MG
5/0.25	ZYL276		R. Mineira do Sul, Passa Quatro, MG
5/1	ZYI207		R. Cult. de Linhares, o, ES
5/0.25	ZYK584		R. Franca do Imperador, Franca, SP
20/2	ZYK348		R. Tramandaí, o, RS
2.5/1	ZYK...		R. Nacional, Cotia
1	CD92	Chile	R. CD 92, Temuco
10	HJAE	Colombia	Em. Fuentes, Cartagena (vf)
30	HJJN		Ondas del Mayo, Pasto
15	HJSJ		Colmundo R, Ibagué
10	HCCM1	Ecuador	R. Colón, Quito
12.5	HCRU3		C.R.O., Machala
100	ZP1	Paraguay	R. Nal. del Paraguay, Asunción
10	OBXIJ	Peru	R. Programas del Perú, Piura (relay RPP 730)
1	OBX2S		R. Ollantay de Virú, Virú
1	OCX7M		R. Andina, Cusco (ti)
1	OAX9V		R. Marginal, Tocache (also reported on 1180)
10/5	YVQU	Venezuela	R. San Carlos, o
20	YVQX		R. Nueva Esparta, Porlamar

930 kHz

10	CP145	Bolivia	R. Metropolitana ("R. Popular"), La Paz
10/1	ZYJ676	Brazil	R. Soc. Rondônia, Cacoal, RO (F.Pl.) (still on 1470)
10/5	ZYH296		R. Ajuricaba, Manaus, AM
5/1	ZYI600		R. Rauland, Castanhal, PA
1/0.25	ZYH605		R. Salamanca, Barbalha, CE
10/0.5	ZYH646		R. Metropolitana, Caucaia, CE
20/5	ZYJ923		R. Liberdade de Sergipe, Aracaju, SE
1/0.5	ZYL220		R. Cl. de Campo Belo, o, MG
5	ZYL229		R. Araguari, o, MG
10/1	ZYL237		R. Ibituruna, Governador Valadares, MG
1/0.25	ZYK500		R. Dinâmica de Santa Fé, Santa Fé do Sul, SP
1/0.25	ZYK503		R. Cl. de Itapira, o, SP
2.5/ 0.25	ZYK652		Super R. Cult. de Santos, o, SP
5/1	ZYK713		R. Universal, Agudos (Bauru), SP
1/0.25	ZYK747		R. Jóia, Adamantina, SP
10/2	ZYK...		R. Sistema São Paulo ("R. Atual"), Cotia, SP
1/0.25	ZYJ227		R. Cult. de Rolândia, o, PR
10/1	ZYJ232		R. Cult. de Curutiba, o, PR
10/1	ZYJ235		R. Princesa, Francisco Beltrão, PR
20/2.5	ZYK230		R. Caxias, Caxias do Sul, RS
10/0.5	ZYK298		R. Santo Ângelo, o, RS
5/0.25	ZYI423		R. Cl. de Rondonópolis, o, MT
1/0.25	ZYN400		R. Jornal, Pontes e Lacerda, MT
10/ 0.25	ZYI454		Radiosul Emissoras Integradas ("AM Capital"), Campo Grande, MS
10/1	ZYH802		R. Caraíba, AM, Aparecida de Goiânia, GO
10	CA93	Chile	R. Nal., Antofagasta, o
10	CB93		R. Nuevo Mundo, Santiago
10	CD93		R. Reloncaví, Puerto Montt
15	HJCS	Colombia	La Voz de Bogotá, o
5	HCVI2	Ecuador	Ondas Antíllanas, Guayaquil
5	HCBA6		R. Ambato, o
10	OAX4E	Peru	R. Moderna ("R. Papá"), Lima
25/10	CX20	Uruguay	R. Monte Carlo, Montevideo
10	YVLJ	Venezuela	R. Maracay, o

935 kHz

	O..3.	Peru	R. Pilcomoto, Cauri

940 kHz

3/0.5	LRH200	Argentina	R. Chajarí, o
0.8	CP...	Bolivia	R. San Lorenzo, Colcapirhua
5/1	ZYH204	Brazil	R. Verdes Florestas, Cruzeiro do Sul, AC
10/ 0.25	ZYI911		R. Sete Cidades, Piracuruca, PI
100	ZYJ453		R. Jornal do Brasil ("R. JB"), RJ, RJ
1	CA94	Chile	R. 9-40, Copiapó
1	CB94		R. Valentín Letelier, Valparaíso
15	HJGB	Colombia	R. Calima, Cali
25	HJTL		RCN Cúcuta, o
20	HCBZ1	Ecuador	Radiodif. de la Casa de la Cultura Ecuatoriana, "Benjamín Carrión", Quito
1	OBX1C	Peru	R. Presidente, Paita
3	OBX6T		R. Onda Sideral, Arequipa
5/3	OBX7L		R. Willkamayu, Cuzco
10	YVLU	Venezuela	R. Fé y Alegría Oriente, Cantaura
10	YVNN		R. Punto Fijo, o
15	YVZR		R. Continental, Barinas

950 kHz

50	LR3	Argentina	R. Libertad, Buenos Aires
10/0 0.25	ZYI915	Brazil	R. São José dos Altos, Altos, PI
1/0.25	ZYI923		R. Boa Esperança, Padre Marcos, PI
5/1	ZYH593		R. Educ. do Nordeste, Sobral, CE
1/0.25	ZYI681		R. Jornal, Sousa, PB
1/0.25	ZYI782		R. Planalto, Carpina, PE
1/0.25	ZYH489		R. Bahia Nordeste, Paulo Afonso, BA
10/5	ZYL212		R. Atalaia, Belo Horizonte, MG
5/0.5	ZYL281		R. Jornal Sul de Minas, Bueno Brandão, MG
5/0.25	ZYK510		R. 950, Marília, SP
5/0.5	ZYK536		R. Alvorada, Piracicaba, SP (F.Pl.) (still on 910)
5/0.25	ZYJ239		R. Difusora Cultural, Irati, PR (F.Pl.) (still on 1510)
1/0.25	ZYJ736		R. Vale Tijucas, o, SC
5/0.25	ZYK260		R. Independente, Lajeado, RS
1	ZYI439		R. Tucunare ("A Voz do Sertão"), Juara, MT
15	HJFN	Colombia	CARACOL Pereira, o
10	HCDE2	Ecuador	R. Latina, Guayaquil
3	HCUE5		R. Colta, LV de la Asociación, Colta
1	OBX2G	Peru	R. Cutervo, o
1	OBX3S		R. Programas del Perú, Chimbote
100/50	YVKG	Venezuela	Informativa 950, Caracas

953 kHz

3	HCJX6	Ecuador	R. La Voz del Santuario, Baños (nj 970) (vf)

960 kHz

25/5	LRA6	Argentina	R. Nacional Mendoza, o
10/1	LU13		R. Necochea, o
10	CP30	Bolivia	R. Santa Cruz, o (nf 970)
1	CP93		R. Kollasuyo, Potosí
5/1	ZYI551	Brazil	R. Cl. de Itaituba, o, PA
1/0.25	ZYH618		R. Cult. dos Inhamuns, Tauá, CE
10	ZYH241		R. Dif. de Alagôas, Maceió, AL
10/ 0.25	ZY1216		R. Aquídaban, Cachoeira de Itapemirim, ES (F.Pl.) (still on 1500)
50/10	ZYK689		R. São Paulo, o, SP
1/0.25	ZYJ257		R. Dif. de Maringá, o, PR
5/0.25	ZYJ733		R. Guarujá, Orleãns, SC
5/0.5	ZYJ813		R. Dif. de Xanxerê, o, SC
10/1	ZYK291		R. Imembuí, Santa Maria, RS
5/0.25	ZYH764		R. Dif. de Itumbiara, o, GO
10	CB96	Chile	R. Carrera, Santiago
10/5	CD96		R. Polar, Punta Arenas
120	HJOP	Colombia	CARACOL Caribe, Magangué
1	HCNC1	Ecuador	R. Cosmopolita, Quito
1	HCSA5		R. Unión ("Sono Onda Internacional"), Cuenca
1	OBX1Y	Peru	R. Amor, José Leonardo Ortiz (CP)
	0..2.		R. Estación Caracol, Trujillo
10	OAX4D		R. Panamericana, Lima
1.5	OCY4V		R. Constelación Radial, Huancayo (r. 1000)
3	OBX6S		R. Hispana, Arequipa (vf)
			R. Lider del Norte, Cajamarca
1	CW96	Uruguay	R. Yi, Durazno
50/20	YVRB	Venezuela	R. Monagas, Maturín
10	YVSS		R. San Sebastian, San Cristóbal (ff 590)
	YV..		Mundial 960, Acarigua

965 kHz

25	HCOT1	Ecuador	R. Católica, Santo Domingo de los Colorados

970 kHz

25/5	LU5	Argentina	R. Neuquén, o
25/10	LV2		R. General Paz, Córdoba
1/0.25	ZYI910	Brazil	R. Vale do Parnaíba, Luzilândia, PI
5/0.25	ZYH612		R. Monólitos de Quixadá, Quixadá, CE
1/0.25	ZYI684		R. Princesa Isabel, o, PB
10/5	ZYH451		R. Soc. de Feira de Santana, o, BA
2.5/0.25	ZYL243		R. Soc. Caratinga ("R. Sistec"), o, MG
2.5/0.25	ZYL285		R. São João del Rei, o, MG
1/0.25	ZYL321		R. Central, Monte Alegre de Minas, MG
2/0.25	ZYK505		R. Dif. de Itapetininga, o, SP
1/0.25	ZYK529		R. Piratininga, São João de Boa Vista, o, SP
1/0.25	ZYK684		R. Hertz de Franca, Franca, SP
5/0.25	ZYK744		R. Alvorada, Estrela d'Oeste, SP
5/1	ZYJ260		R. Alvorada, Londrina, PR
5/0.25	ZYJ277		R. Dif. Paraná, Marechal Cândido Rondon, PR
1/0.25	ZYJ730		R. Araguaia, Brusque, SC
50/10	ZYK201		R. Pampa, Porto Alegre, RS
1/0.5	ZYK349		R. Do Povo, Humaitá, RS
5/1	ZYI399		R. Vale do Taquari, Coxim, MS
1	CA97	Chile	R. Calama, o
1	CC97		R. Lautaro, Talca
1	CD97A		R. Austral, Valdivia
1	CD97B		R. Patagonia Chilena, Coyhaique
10	HJCH	Colombia	R. Super, Bogotá
15	HJME		RCN Guajira, Maicao
30	HJVK		Em. Armonías del Caquetá, Florencia
1	HCMB1	Ecuador	R. Imperio, Ibarra
20	HCAW2		R. Católica Nacional, Guayaquil
10	ZP9	Paraguay	R. Comuneros, Asunción
1.5	OBX1V	Peru	R. La Capullana, Sullana
10	OAX2D		R. Cajamarca, o
1	OBX5A		Latinoamericana R., Ica
10/5	CX22	Uruguay	R. Universal, Montevideo
10	YVLR	Venezuela	R. Central Continente, Maracay
15	YVSD		R. Turismo, Valera
10	YV..		Mundial Oriental, Barcelona

980 kHz

0.25	LT25	Argentina	R. Guaraní, Curuzú Cuatiá
1/0.25	LT39		Em. Victoria, o
10/6	LU37		R. General Pico, o
3	CP118	Bolivia	R. Mar AM 980, El Alto de La Paz
50/600	ZYH707	Brazil	R. Nal. de Brasília, o, DF
1	CA98	Chile	R. Univ. de Tarapacá, Arica
10	CB98		R. Agricultura, Valparaíso
50	HJES	Colombia	Antena 2 de RCN ("R. El Sol"), Cali
30	HJJV		R. Punto, Cúcuta
0.5	HCJI5		R. El Prado ("La Voz de la ciudad"), Riobamba
1	ZP31	Paraguay	R. Mburucuyá, Pedro Juan Caballero
2	OAX2F	Peru	R. La Hora, Trujillo
2.5	OAX5T		R. La Voz de Huamanga, Ayacucho
1	OAX6I		R. Universidad, Arequipa
10	YVQM	Venezuela	LV de El Tigre ("R. Novecientos Ochenta"), El Tigre
10	YV..		R. Nacional Zulia, Maracaibo (rel 630)

985 kHz

10	OBX4J	Peru	La Nueva Estación X ("R. Latina"), Lima

990 kHz

25/5	LR4	Argentina	R. Splendid, Buenos Aires
1	LRJ201		R. Calingasta, Barreal
5	CP192	Bolivia	R. Esperanza, Aiquile
1/0.25	ZYH299	Brazil	Rede Amazonense de Comunicação, Maués, AM
1/0.25	ZYI922		R. Vale do Canindé, Oeiras, PI
10/1	ZYJ596		R. Rural, Mossoró, RN
1/0.25	ZYH483		R. Alvorada de Teixeira de Freitas, Caravelas, BA
100/10	ZYJ461		R. Ipanema, RJ, RJ
1/0.25	ZYJ763		R. Peperi, Itapiranga, SC
1/0.25	ZYK314		R. Tupanciretã, o, RS
1/0.25	ZYK335		R. Sananduva, o, RS
1/0.25	ZYK360		R. Cl. de Pedro Osório, o, RS
0.5/0.25	ZYK579		R. Cultura, Dois Corregos
30	HJDB	Colombia	RCN Medellín, o
25	HCGHI	Ecuador	R. Tarqui, Quito (vf)
15	HCEW2		Frecuencia Mil, Guayaquil
5	HCAO5		R. América, Cuenca
0.5	OBX2M	Peru	R. Contumaza, o (vf)
1	OBZ4R		Radiodif. del Centro del País, ("La Voz del Valle de Mantaro"), Jauja
8	OAX6K		R. Bicolor, Tacna
20	YVRT	Venezuela	R. Tropical ("Tropical 9-90"), Caracas
10	YVTA		R. Mundial Tricolor, Barquisimeto

996 kHz

10	ZP36	Paraguay	R. Ypoá, Paraguarí (nf 1000) (vf)

1000 kHz

25	LU16	Argentina	R. Río Negro, Villa Regina
3	CP28	Bolivia	R. Cochabamba ("CBA"), o (nf 900)
0.15	CP...		R. Em. Poopó, o
3	CP119		R. Dif. Trópico, Dif. Trinidad
5/0.25	ZYI791	Brazil	R. Princesa Serrana, Timbaúba, PE
200	ZYK522		R. Record, São Paulo, SP
10	CB100	Chile	R. Cien AM, Santiago
15	HJAQ	Colombia	RCN Cartagena, o
10	HJJG		R. Dif. Nacional, Manizales (rel 570)
5	HJ..		R. Cultural, Cajibío
1	HCCR1	Ecuador	R. Alegría, Santo Domingo de los Colorados
5	HCRV5		R. Visión, Cuenca
1.5	OCY4V	Peru	R. Constelación Radial, Huancayo (nf 960)
1	OBX6R		R. Endesa, Arequipa
5	OAX7V		R. Macchu Picchu, Cuzco
1.5	O..9.		R. Central, Bellavista
1	O..9.		R. El Sol, Saposoa
10	YVNM	Venezuela	R. Mundial Mil, Morón
10	YVOA		R. Táchira, San Cristóbal

1010 kHz

25/5	LV16	Argentina	R. Río Cuarto, o
1	LW2		Radioemisora Tartagal, o
1	CP220	Bolivia	R. Baha'í de Bolivia, Caracollo
50	ZYH625	Brazil	R. O Povo, Fortaleza, CE
25/5	ZYH448		R. Bahia, Salvador, BA
10/0.5	ZYL230		R. Educ. de Coronel Fabriciano, o, MG
5/0.5	ZYL264		R. Solar AM, Juiz de Fora, o, MG
1/0.25	ZYL325		R. Estância, Jacutinga, MG
1/0.25	ZYK556		R. Independente, Barretos, SP
1/0.25	ZYK611		R. Tuiuti, Pres. Prudente, SP
25/5	ZYJ263		R. Celinauta, Pato Branco, PR
1/0.25	ZYJ807		R. Verde Vale, Braço do Norte, SC
1/0.5	ZYK232		R. 1010 Caxias do Sul, o, RS
2.5/0.25	ZYK344		R. Missioneira 7 Povos, São Luís Gonzaga, RS
5/0.5	ZYJ764		R. Jaraguá, Jaraguá do Sul, SC (F.Pl.) (still on 1510)
5/0.25	ZYI421		R. Dif. de Mirassol d'Oeste, o, MT
10	CD101	Chile	Em. Ñielol, Temuco
10/5	HJBN	Colombia	La Voz del Galeras, Pasto (vf)
15	HJCC		Colorín Color R, Bogotá
10	HJFF		R. Reloj, Barranquilla
10	HJIX		R. Yarima, Barrancabermeja
20	HJJR		CARACOL Neiva, o (vf)
10	HCRZ2	Ecuador	R. Victoria, Guayaquil (ti)
1	HCRC4		R. Cenit, Manta
12.5	HCNR6		R. Variedades, Ambato
2	OBX2P	Peru	R. San Francisco, Cajamarca (ti)
50	OAX4U		R. América, Lima
20	CX24	Uruguay	R. El Tiempo, Montevideo
10	YVPC	Venezuela	R. Aragua, Cagua
10	YVQF		R. Mundial Bolívar, Ciudad Bolívar

1011 kHz			
20	HJZD	Colombia	R. Panzenú, Montería
1020 kHz			
1	LRA28	Argentina	R. Nacional La Rioja, Chilecito (rel 620)
1	LRA58		R. Nacional, Rio Mayo, o
10/5	LT10		R. Univ. Nal. del Litoral, Santa Fé
10	CP4	Bolivia	R. Illimani, La Paz
5/1	ZYJ680	Brazil	R. Educ. de Rolim de Moura, o, RO
5/1	ZYH600		R. Educ. do Cariri, Crato, CE
25/1	ZYH247		R. Cl. de Rio Largo, o, AL
5/ 0.25	ZYL224		R. Congonhas, o, MG
10/ 0.25	ZYL260		R. Cult. de Uberlândia, o, MG
20/10	ZYI205		R. Dif. de Colatina, o, ES
5/0.25	ZYJ484		R. Sucesso ("R. A Noticia AM"), Campos, RJ (F.Pl.) (still on 1510)
10/ 0.25	ZYK513		R. Canção Nova, Cachoeira Paulista, SP
5/0.25	ZYK515		R. Cult. de Assis, o, SP
2.5/0.5	ZYK531		R. Educ. de Limeira, o, SP
5/0.25	ZYK600		R. Cult. de Jales, o, SP
10/ 0.25	ZYJ244		R. Colombo do Paraná, Colombo (Curitiba), PR
1/0.25	ZYJ307		R. Independência, Medianeira, PR
1/0.25	ZYJ805		R. Continental, Coronel Freitas, SC
10/5	ZYK202		R. Caiçara, Porto Alegre, RS
10/ 0.25	ZYI381		R. Independente, Aquidauana, MS
1/0.25	ZYI686		R. Cenecista, Picuí, PB
1/0.25	ZYH781		R. Maranata, Firminópolis, GO
15	HJDQ	Colombia	Em. Claridad, Medellín
15	HJDY		R. Primavera, Bucaramanga
15	HJFQ		RCN Pereira, o
5	HJFT		Ecos del Combeima, Ibagué
10	HJKS		LV del Llano, Villavicencio (vf)
5	HCHR1	Ecuador	R. Hit (vf), Quito
3	HCGO3		R. Estelar, Santa Rosa
5	HCCR6		R. Surcos, Guaranda
10	ZP14	Paraguay	R. Ñandutí, Asunción (vf)
1	OBX3L	Peru	R. Onda Nueva, Chimbote
1	OBX1G		R. Heroíca, Chíclayo
50/10	YVMX	Venezuela	R. Calendario, Maracaibo
10	YVRS		Mundial Margarita, La Asunción
25	YVTW		R. Alegría, Chivacoa
1030 kHz			
25/5	LS10	Argentina	R. del Plata, Buenos Aires
5/1	ZYJ...	Brazil	R. Colina de Machadinho, Ariquemes (Machadinho), RO (CP)
10/1	ZYH892		R. Jaínara, Bacabal, MA
20/5	ZYI777		R. Olinda de Pernambuco, Olinda, PE
10/1	ZYH475		R. Bahiana, Itaberaba, BA (F.Pl.) (still on 1480)
50/5	ZYJ467		R. Capital, RJ, RJ
5/0.5	ZYK525		R. Dif. de Franca, o, SP
1/0.25	ZYK554		R. Em. da Barra, Barra Bonita, SP
1/0.25	ZYK606		Lins R. Cl., Lins, SP
1/0.25	ZYJ240		R. Dif. Cruzeiro do Oeste, o, PR
1/0.25	ZYJ271		R. Atalaia, Londrina, PR
1/0.25	ZYJ329		R. Dif do Xisto, São Mateus do Sul, PR
5/0.5	ZYJ771		R. Princesa, Lages, SC
10/1	ZYK253		R. Repórter, Ijuí, RS
1/0.25	ZYI396		R. Cidade, Maracaju, MS
10/ 0.25	ZYH746		R. Imprensa, Anápolis, GO
5/0.25	ZYH791		R. Siqueira Campos, Colinas do Tocantins, TO
1	CB103	Chile	R. Progreso, Talagante
10	CC103		R. Chilena, Concepción
1	CD103A		R. Chiloé, Castro
1	CD103		R. Payne, Puerto Natales
10	HJDJ	Colombia	La Voz de los Libertadores, Duitama
30	HJER		RCN Cali, o
1	HJGX		R. Progreso de Córdoba, Lorica
15	HJOL		Ondas del César, Aguachica
5	HJ..		Ondas del Vaupés, Mitú
5	HCRF2	Ecuador	R. Horizonte, Guayaquil
5	HCKL4		R. Unión, Chone (vf)
3	HCFO7		RR Francisco de Orellana, Coca
3	OAX2U	Peru	R. Los Andes, Huamachuco
5	OAX7N		La Voz del Altiplano, Puno
1	OCX7O		Super R., Cusco
5		Surinam	R. Paramaribo, o
20/5	YVQY	Venezuela	R. Onda, Guanare
20/10	YVTD		R. Valles del Tuy, Ocumare del Tuy
1040 kHz			
1	CP113	Bolivia	R. Villazón, o
500	CP208		R. Sipe Sipe, o
200	ZYK537	Brazil	R. Capital, São Paulo, SP
1	CD104	Chile	R. Raíces, Curacautín
15	HJBF	Colombia	La Voz del Norte, Cúcuta
15	HJCJ		Colmundo R, Bogotá
15	HJCQ		La Voz de Armenia, o
15	HJSY		R. Mil 40, Popayán
15	HJUB		Colmundo R, Pasto
3	HCCW1	Ecuador	LV del Valle, Machachi
5	HCEV5		R. Splendit ("RS"), Cuenca
1	HCGB6		R. Colosal, Ambato
5	ZP43	Paraguay	R. Arapizandú, San Ignacio de Misiones
1	OBX2O	Peru	R. Nor Oriente, Jaén
10	OBX4O		R. Red, Lima
20	YVLB	Venezuela	LV de Carabobo, Valencia
25	YVON		R. Los Andes ("R. Mil Cuarenta"), Mérida
1050 kHz			
1	LRH252	Argentina	R. Pucará, Reconquista
5	CP233	Bolivia	R. El Mundo, Santa Cruz
1/0.25	ZYI...	Brazil	Jet Radiodifusão, Teresina, PI (CP)
5/0.5	ZYH644		R. Primeira Capital, Aquiraz, CE
5/1	ZYI676		R. Caturité, Campina Grande, PB
10/0.5	ZYH494		R. Metropolitana da Bahia, Camaçari, BA
50/5	ZYI203		R. Capixaba, Vitória, ES
1/0.25	ZYL236		R. Rural, Tupaciguara, MG
10/0.5	ZYJ497		R. Angra, Angra dos Reis, RJ
5/0.25	ZYK601		R. Super Tupi, Jardinópolis, SP
1/0.25	ZYJ226		R. Dif. Platinense, Santo Antônio da Platina, PR
10/0.5	ZYI391		R. Dif. Paranaibense, Paranaíba, MS
	CD105	Chile	R. Bío Bío, Osorno
15	HJBB	Colombia	R. Valledupar, o
15	HJDR		R. Sonorama, Medellín
10	HJFZ		La Voz del Centro, El Espinal
10	HJGU		R. Bucarica, Bucaramanga
5	HJIO		LV de la Conquista, Granada
10	HJLZ		La Voz del Cinaruco, Arauca
5	HJNG		R. Palmira, o
15	HJOV		R. Surcolombiana, Neiva
15	HJTJ		Ondas de Urrá, Montería
5	HCIM1	Ecuador	R. Municipal ("La Voz de Imbabura"), Ibarra
5	HCRQ2		R. Super, Guayaquil
1	OCX2B	Peru	R. San Sebastian, Chepén
1	OAX3A		R. Alpamayo, Huaraz
5	OBZ4J		R. Bolognesi ("La nueva Bolognesi"), Huancayo
1	OBX6B		R. Arequipa, o
6	CX26	Uruguay	S.O.D.R.E., Montevideo
20/10	YVKA	Venezuela	R. Nacional, Caracas
1055 kHz			
5	HCMG6	Ecuador	R. Ecos del Pueblo, Saquisilí (nf 1060) (vf)
1060 kHz			
0.5	CP181		R. La Voz de la Frontera, Puerto Suarez
0.25	CP...		R. Splendid, Vinto,
5	ZYJ597	Brazil	R. Tapuyo, Mossoró, RN
5/1	ZYH460		R. Cl. de Conquista, Vitória da Conquista, BA
25/5	ZYL278		R. Grande Belo Horizonte ("R. Grande Be Aga"), Pedro Leopoldo, MG
1/0.25	ZYL306		R. Itajubá, o, MG
1/0.25	ZYJ495		R. Tropical, Miguel Pereira, RJ

5/0.25	ZYK533		R. Educ. de Piracicaba, o, SP
1/0.25	ZYK765		R. Universitária, Garça, SP
10/0.5	ZYJ246		R. Paraná, Curitiba, PR
1/0.25	ZYJ298		R. Colorado, o, PR
10/ 0.25	ZYJ306		R. Educ. de Francisco Beltrão, o, PR
5/0.25	ZYK220		R. Camaqüense, Camaquã, RS
2/0.25	ZYK302		R. São Luís, São Luís Gonzaga, RS
1/0.5	ZYK307		R. Cristal, Soledade, RS
1/0.25	ZYN604		R. Tupinambás, Dourados, MS
2.5/ 0.25	ZYH...		R. Clube, Itapicuru, BA (CP)
100	CB106	Chile	R. Minería, Santiago
15	HJFJ	Colombia	RCN Caldas, Manizales
15	HJLY		R. Delfín, Riohacha
1	HJMG		R. Prosperidad, Turbo
10	HJMV		R. Furatena, Chiquinquirá
1	HJYX		R. Caracolí, Sincelejo
0.05		Fr. Guiana	R.F.O., Guyane, St. Laurent du Maroni
3	ZP13	Paraguay	R. Boquerón, Alberdi (vf)
1	OCY4D	Peru	R. Exito 1060, ("Exito Amplitud Modulada"), Lima
2	OBX5D		R. Andahuaylas, o
1	OAX7O		R. Amauta, Cusco (ti)
10	YVLN		R. Guárico, San Juan de los Morros
10	YVOE		R. San Cristóbal, "RSC", o

1070 kHz

25/5	LR1	Argentina	R. El Mundo, Buenos Aires
	CP...	Bolivia	R. Em. Quillacollo, o
1/0.25	ZYH643	Brazil	R. Cult. de Paracuru, o, CE
20/2.5	ZYI673		Dif. R. Cajazeiras, o, PB
5/0.5	ZYH492		R. Rural, Ipiaú, BA (F. Pl.) (still on 1450)
5/0.25	ZYL224		R. Congonhas, o, MG (ff 1020)
1/0.25	ZYL316		R. Rural, Muzambinho, MG
5/0.25	ZYL355		Super R. Patos, Patos de Minas, MG
10/ 0.25	ZYJ483		R. Cidade, Campos, RJ
1/0.25	ZYK603		R. Piratininga, Jaú, SP
10/ 0.25	ZYK615		R. Metropolitana, Moji das Cruzes, SP
5/1	ZYK633		R. Pres. Prudente, o, SP
5/0.25	ZYK758		R. Jornal, Barretos, SP
2/0.25	ZYJ203		R. Dif. União, União da Vitória, PR
1/0.25	ZYJ319		R. Guaraniaçu, o, PR
1/0.25	ZYJ747		R. Gralha Azul, Urubici, SC
2/0.25	ZYK218		R. Caçapava, Caçapava do Sul, RS
1/0.25	ZYK343		R. Metrópole, Crissiumal, RS
1/0.25	ZYK357		R. Revista, Bento Gonçalves, RS
10/2.5	ZYI427		R. Industrial, Várzea Grande, MT
1/0.25	ZYH...		R. Cultura de Quixadá, Mazagão, AP (CP)
20	HJAH	Colombia	Em. Atlántico, Barranquilla
30/15	HJCG		R. Santa Fé, Bogotá
15	HJVR		Ondas del Puracé, Popayán
1	HCRS1	Ecuador	R. Lubacan, Santo Domingo de los Colorados
0.5	HCVP1		R. Libertad, Quito (relay Radiodif. Tarqui 980)
10	HCKD2		R. Tigre, Guayaquil (ts)
5	HCRT5		R. La Voz del Tomebamba, Cuenca
10		Fr. Guiana	R.F.O. Guyane, Matoury
2.5	OAX2B	Peru	R. Trujillo ("La Voz del Norte del Perú"), o
0.25	OAX5A		R. Centinela, Puerto San Juan
10	YVMA	Venezuela	Mundial Zulia, Maracaibo
5	YVPX		R. El Sol, La Fría

1075 kHz

1	CP173	Bolivia	R. Agricultura, Portachuelo (nf 1030)

1080 kHz

25/10	LU3	Argentina	R. del Sur, Bahía Blanca
0.25			R. Departamento Minas, Andacollo
1	CP...	Bolivia	R. Dif. Colosal, Sucre (nf 1070)
10/2.5	ZYI530	Brazil	R. Guajará, Belém, PA
1/0.25	ZYH...		R. Cult. de Quixadá, Mazagão, AP (CP)
5/0.5	ZYI784		R. Dif. de Caruaru, o, PE
1/0.25	ZYI824		R. Voluntários da Pátria, Ouricuri, PE
20/0.5	ZYH470		R. Subaé AM, Feira de Santana, BA
1/0.25	ZYH485		R. Fascinação, Itapetinga, BA
5/0.5	ZYL251		R. Capital, Juiz de Fora, MG
1/0.25	ZYL331		Sistema Ipanema de Comunicação, Ipanema, MG (ff 540)
5/0.25	ZYK519		R. Assunção de Jales, o, SP (ff 740)
5/1	ZYK557		R. Dif. Batatais, o, SP
1/0.25	ZYK607		R. Alvorada, Lins, SP
5/0.5	ZYK669		R. Cl. de Sorocaba, o, SP
1/0.25	ZYK704		R. Monumental Aparecida, SP
1/0.25	ZYK710		R. Alvorada, Cardoso, SP
1/0.5	ZYJ201		R. Cl. Pontagrossense, Ponta Grossa, PR
1/0.25	ZYJ245		R. Cult. do Norte, Paranavaí, PR
1/0.25	ZYJ261		R. Educ. de Cornélio Procópio, o, PR
1/0.25	ZYJ759		R. Cl. de Indaial, o, SC
1/0.25	ZYK254		R. Marabá, Iraí, RS
10	ZYK280		Radiodifusão da Univ. Federal, Porto Alegre, RS
1/0.25	ZYI437		R. Gaspar, Itiquira, MT
10/5	ZYH708		R. Capital, Brasília, DF
1	CA108	Chile	R. Río Elqui, Vicuña
1	CD108		R. Los Confines, Angol
10	HJAW	Colombia	La Voz de Montería, o (vf)
10	HJAX		R. Reloj, Medellín
2	HJIS		R. Pontoná, La Dorada
15	HJJF		R. Eco, Cali
10	HJKT		R. Macarena, Villavicencio
10	HJMH		R. Melodía, Bucaramanga
10	HCVH6	Ecuador	R. Latacunga, o
10	ZP25	Paraguay	Radiodif. Nanawa, Luque
5	OBX1D	Peru	R. San Miguel, Piura
1	OCX4P		R. Huayllay, o
2.5	OAX7S		R. Salcantay, Cusco
1	OAX8N		R. Amazonas, Iquitos
1	OAZ8C		R. Sideral, Pucallpa
10	YVNR	Venezuela	Mundial 1080, Maracay
10	YVQJ		R. Barcelona, o
10	YV..		R. Nacional, Mérida

1085 kHz

1	HCAB4	Ecuador	R. Eco, Manta (nf 1080)

1090 kHz

0.3	CP45	Bolivia	R. Cultura, Cochabamba
10/1	ZYH893	Brazil	R. Rio Balsas AM, Balsas, MA
10/5	ZYJ592		R. Rural, Natal, RN
10/1	ZYH455		R. Santa Cruz, Ilhéus, BA
25/5	ZYJ468		R. Metropolitana, RJ, RJ
1/0.5	ZYK609		R. Cl. de Marília, o, SP
1/0.25	ZYK618		R. Cult. de Monte Alto, o SP
1/0.25	ZYK768		R. Em. do Planalto, Paulinía, SP
5/0.25	ZYJ254		R. Dif. Colméia, Campo Mourão, PR (ff 890)
1/0.25	ZYJ283		R. Vicente Pallotti, Coronel Vivida, PR
1/0.25	ZYJ732		R. Colón, Joinville, SC
1/0.25	ZYJ786		R. Tabajara, Tubarão, SC
5/0.25	ZYK216		R. Cachoeira do Sul, o, RS
1/0.25	ZYK262		R. Salete, Marcelino Ramos, RS
1/0.25	ZYK341		R. Giruá, o, RS
10/1	ZYH758		R. Universal, Goiânia, GO
5	CC109	Chile	R. Chilena, Talca
15	HJBC	Colombia	CARACOL Cúcuta, o
10	HJIA		Ondas del Nevado, Manizales
10	HJIG		LV de la Selva, Florencia
10	HJIH		CARACOL Sogamoso, o
5	HJOM		R. Bucanero, Cartagena
5	HCVI1	Ecuador	R. Deporte, "Estación 1090", Quito
1.2/3	OBX2A	Peru	R. Cajabamba, o
1	OAX3I		R. Tingo María, o (nf 1100)
1	OBX6X		R. Amistad, Arequipa (vf)
25/15	CX28	Uruguay	R. Imparcial, Montevideo
10	YVPB	Venezuela	R. Yaracuy 1090 AM, San Felipe
20	YVSZ		Exitos Mil Noventa, Caracas
10	YYTG		R. Melódica 1090, Machiques

1100 kHz

1	CP55	Bolivia	R. Universidad de Oruro, o
5	CP184		R. El Dorado, Trinidad
1	CP194		R. Em. Chaco, Yacuiba

4	CP246		R. Mundial, La Paz
1/0.25	ZYH638	Brazil	R. Dif. dos Inhamuns, Tauá, CE
1/0.25	ZYJ607		R. A. Voz do Seridó, Caicô, RN
150	ZYK694		R. Globo de São Paulo, o, SP
10	CA110	Chile	R. La Portada, Antofagasta
10	CB110		R. Recreo, Viña del Mar
15	HJAT	Colombia	CARACOL Barranquilla, o
10	HJCN		R. Reloj, Bogotá
2	HJEF		Brisas del Orinoco, Puerto Carreño
5	HJGI		LV de Colombia, El Socorro
5	HJGQ		Tr. Surandes, Andes
5	HJMK		Em. Ideal, Planeta Rica
15	HJYZ		R. Futuro, Neiva
10	HCFW2	Ecuador	R. Alegría, Guayaquil (vf)
5	HCGR6		R. Novedades, Latacunga (vf)
5	ZP71	Paraguay	R. Ñu Verá, Capitán Badó
1	OBX1L	Peru	R. Star, Chiclayo
1.5	OBX3M		R. JSV, Huánuco
1	OAZ4W		R. Musical, Barranca (relay RPP 730)
0.2	OCX4S		R. Imperial, Cañete
1	OCY4G		Sonorama R., Huancayo
1	OAX5Z		R. Oasis, Ica
1.5	OBX7Z		R. Collao ("R. Comercial Collao"), Juliaca
15	YVOP	Venezuela	R. Occidente, Tovar
10	YVSV		R. Angostura, Ciudad Bolívar
1105 kHz			
3	HCRP6		R. Pelileo, o (vf)
1110 kHz			
15/5	LR2	Argentina	R. Argentina, Buenos Aires
10/5	LV7		R. Tucumán, San Miguel de Tucumán
5/0.25	ZYH620	Brazil	R. Litoral, Cascavel, CE
10	ZYI786		R. Tabajara, João Pessoa, PB
1/0.25	ZYL205		R. Planalto, Araguari, MG
1	ZYL267		R. Aurilândia, Nova Lima, MG
50/5	ZYJ471		R. Cult. de Campos, o, RJ
2.5/0.5	ZYK544		R. Luz, Araçatuba, SP
1/0.25	ZYK592		R. Ibitinga, o, SP
1/0.25	ZYK617		R. Cult. de Moji-Mirim, o, SP
10/1	ZYJ241		R. Paiquerê, Londrina, PR
5/0.5	ZYJ253		R. Eldorado, São José dos Pinhais, PR
1/0.5	ZYJ356		R. Cl. de Ubiratã, o, PR
1/0.5	ZYJ752		R. Cult. de Florianópolis, o, SC
1/0.25	ZYJ812		R. São Carlos, o, SC
2.5/ 0.25	ZYK257		R. Cult. Jaguarão, o, RS
1/0.25	ZYK306		R. Sobradinho, o, RS
2.5/ 0.25	ZYK325		R. Cruzeiro do Sul, Itaqui, RS
1/0.25	ZYK364		R. Solaris, Antonio Prado, RS
5/1	ZYI436		R. Nova Xavantina, o, MT (ff 710)
1	ZYI392		R. Ponta Porã, o, MS
25/2	ZYH782		R. Redentor, Santo Antônio do Descoberto, GO
1/0.25	ZYJ743		R. Caçanjure, Caçador, SC (F.Pl.) (still on 1520)
10	CD111	Chile	R. La Frontera, Temuco
10	HJDI	Colombia	R. Bolivariana, Medellín
15	HJEW		R. Reloj, Cali
5	HJGP		LV del Río Arauca, Arauca
10	HJJP		RCN Villavicencio, o
15	HJNC		La Voz del Nevado, Ibagué
1	HJPA		CARACOL San Andrés ("La Voz de las Islas"), Isla San Andrés
20	HJZE		R. Piragua, Sincelejo
1	HCEG1	Ecuador	R. Exito, Quito
3	HCJC5		Ondas Azuayas, Cuenca
1	OBX3B	Peru	R. Heróica, Huaraz
5	OCX7T		R. Lider, Cusco
10	YVQT	Venezuela	R. Carúpano, o
1115 kHz			
1.5	CP83	Bolivia	R. Norte, Montero (vf)
1120 kHz			
25/5	LU6	Argentina	Em. Atlántica, Mar del Plata
25/5	LV5		R. Sarmiento, San Juan
1	ZYH598	Brazil	R. Tupinambá, Sobral, CE
1/0.25	ZYI687		R. Independência, Catolé do Rocha, PB
5/1	ZYI778		R. Paulista, Paulista (Recife), PE
2.5/ 0.5	ZYL272		R. Ouro Preto, o, MG
2.5/ 0.25	ZYL301		R. Sete Colinas, Uberaba, MG
1/0.25	ZYL332		R. Serra da Boa Esperança, Boa Esperança, MG
1	ZYI215		R. Dif. de São Mateus, ("R. Cricaré"), o, ES
1/0.25	ZYK631		R. Em. Portofelicense, Porto Feliz, SP
10/1	ZYK660		R. Clube AM São José dos Campos, SP
1/0.25	ZYK671		R. Cl. Imperial, Taquaritinga, SP
50	ZYK274		R. 1120 AM, Porto Alegre, RS
25/1	ZYI...		Empresa de Radiodifusão Campograndense, Campo Grande, MS (CP)
15	HJGH	Colombia	Corazón AM, Bucaramanga
5	HJJC		R. Matecaña, Pereira
10	HJKQ		R. Tunja, o
10	HJTI		R. 24, Cúcuta (nf 1130)
2	HCEB1	Ecuador	R. Ecos del Montúfar, San Gabriel
1	HCLE1		Radiodif. Marañón, Santo Domingo de los Colorados (rel 1220)
5	HCLA2		R. América, Guayaquil
5	HCNT3		R. Once Catamayo, La Toma
3	HCAS7		R. Variedades del Puyo, El Puyo
10	ZP24	Paraguay	R. Nuevo Mundo, San Lorenzo
2	OBX2I	Peru	R. Dinámica, Trujillo
0.5	OAX9J		R. Lamas, o (nf 1100)
5/10	CW31	Uruguay	R. Salto, o
10	YVMF	Venezuela	R. Super Ondas Cristianas, Maracaibo
20/10	YVSK		Radiodif. del Sur, San Fernando de Apure
5	YVXZ		R. República, Maturín
1125 kHz			
0.5	CP...	Bolivia	R. Jenecheru, "LV de Cotoca", o
1	CP...		Kanata Radiodifusión, Cochabamba (nf 1135)
1130 kHz			
25/5	LRA21	Argentina	R. Nacional Santiago del Estero, o
0.1	CP...	Bolivia	R. Em. 27 de Julio, Huanuni (ti)
5/0.25	ZYJ677	Brazil	R. Ji-Paraná, o, RO
10	ZYI531		R. Marajoara, Belém, PA
5/1	ZYI783		R. Cult. do Nordeste, Caruarú, PE
100/50	ZYJ460		R. Nacional, RJ, RJ
1/0.25	ZYK676		R. Tupã, o, SP
1/0.25	ZYJ220		R. Castro, o, PR
5/0.25	ZYJ333		R. Ingamar, Marialva, PR
5/1	ZYJ790		R. Princesa d'Oeste, Xanxerê, SC
5/1	ZYK290		R. Medianeira, Santa Maria, RS
5/0.25	ZYH769		R. Em. Sul Goiana, Quirinópolis, GO
15	HJAC	Colombia	Emisoras Riomar, Barranquilla
1	HJNN		Ondas del Río, Magangué
15	HJQQ		R. Reloj, Pasto
20	HJVA		R. K. 11-30, Bogotá
2	HCRD1	Ecuador	R. Punto, Ibarra
3	HCPV6		R. Centro ("Estación 112"), Ambato
1	OAX2V	Peru	R. Los Andes, Cajamarca (relay RPP 730)
5	OAX4N		R. Selecta 2000, Lima
0.4	OAZ4S		R. Chanchamayo, La Merced
0.25	OAX7H		R. Puno, o
1	OAX9Q		R. San Martín, Tarapoto
20	CX30	Uruguay	R. Nacional, Montevideo
10	YVKQ	Venezuela	R. 11-30 Musical, Barquisimeto
1	YVPY		R. Amazonas, Puerto Ayacucho
20/10	YVRL		R. Ideal ("Ideal 11-30"), Macuto
1140 kHz			
1/0.25	ZYH607	Brazil	R. Progresso, Russas, CE
10	ZYH449		R. Cult. da Bahia, Salvador, BA
5/0.25	ZYH801		R. Pouso Alto, Piracanjuba, GO (F.Pl.) (still on 1480)
1/0.5	ZYL204		R. Minas, Divinópolis, MG
1/0.25	ZYL248		R. Dif. da Campanha, o, MG
1/0.5	ZYL253		R. Soc. Muriaé, o, MG

5/0.5	ZYL362		R. Cl. de Bocaiuva, o, MG
5/0.5	ZYK550		R. Dif. de Assis, o, SP
2/0.5	ZYK555		R. Barretos, o, SP
1/0.5	ZYK645		R. Educ. e Cult. de Rio Claro, o, SP
1/0.25	ZYK708		R. Regional AM, Registro, SP
5/0.25	ZYK709		R. Costa Azul, Ubatuba, SP
10/0.5	ZYJ748		R. Coroado, Curitibanos, SC
2.5/ 0.25	ZYK228		R. Cruz Alta, o, RS
5/0.5	ZYK316		R. Charrua, Uruguaiana, RS
1/0.25	ZYK330		R. Sobral, Butiá, RS
5/0.25	ZYI418		R. Cidade, Aparecida do Taboado, MS
10/0.5	ZYI453		R. Guaicurus, Fátima do Sul, MS
1/0.25	ZYH751		R. Tapirapés, Formosa, GO
40/75	CB114	Chile	R. Nal., Santiago, o
10	HJDL	Colombia	R. Paisa, Medellín
10	HJFH		La Voz de Anserma, o
10	HJKO		LV de la Victoria, Cartagena
10	HJRN		La Voz de la Provincia Colombiana, Barbosa
20	HJRW		CARACOL Villavicencio ("La Voz de los Centauros"), o
10	HJCL		R. Girardot, o
5	HCIR1	Ecuador	R. Amazonas, Quito (Pomasqui)
5	HCFB2		R. Cóndor, Guayaquil
	HCMF4		R. Momento, Calceta
1	HCAZ5		R. Alfa Musical, Cuenca
5	ZP22	Paraguay	R. Panambí Verá, Villarrica
10	OAX4C	Peru	R. Centro, Huancayo (relay RPP 730)
0.5	OAX5W		R. Chinchaysuyo, Chincha
1	OAX6L		R. Concordia, Arequipa
10	YVMU	Venezuela	R. Carora, o
1145 kHz			
1	CP19	Bolivia	R. Agustín Aspiazu, La Paz (ti)
1150 kHz			
10	LRA2	Argentina	R. Nacional Viedma, o
1	LRA51		R. Nacional Jáchal, o
10/5	LT9		R. Brigadier General Estanislao López, Santa Fé
0.5	CP71	Bolivia	R. El Cóndor, Oruro
1	CP221		R. Horizontes de Campi, Cobija
10/5	ZYI891	Brazil	R. Pioneira de Teresina, o, PI
1/0.25	ZYJ617		R. Cabugí do Seridó, Jardim do Seridó, RN
20/1	ZYH250		R. Cult. de Arapiraca, o, AL
20/5	ZYL283		R. Globo de Minas, Belo Horizonte, MG
10/0.5	ZYJ456		R. Três Rios, o, RJ
100/50	ZYK656		R. Tupi, São Paulo, SP
10	HJBT	Colombia	R. Catatumbo, Ocaña
15	HJFI		CARACOL Armenia, o
15	HJFP		R. Neiva, o
1	HJGJ		R. Vencedores, Duitama
5	HJSL		RCN Sincelejo, o
1	HJTE		La Voz del Chocó, Quibdó
10	HCAV3	Ecuador	R. Luz y Vida, Loja
5	HCGB5		La Voz de Riobamba, o (vf)
3	HCVC7		R. El Cisne, Lago Agrio
2.5	OCX7Q	Peru	R. Universal, Cusco
10	OAX8F		R. Loreto, Iquitos
1			R. Naylamp, "Estación Pucara," Pucara R. Colonial, Cajamarca
10	YVMV	Venezuela	Mundial Caribe, Punto Fijo
10	YVQD		Ecos del Orinoco, Ciudad Bolívar
1155 kHz			
1	HCWD4	Ecuador	R. Cenit, Portoviejo (nf 1160)
1160 kHz			
1	LRA57	Argentina	R. Nacional El Bolsón, o
1.8	LRH253		R. Cataratas del Iguazú, Puerto Iguazú
10/5	LU32		R. Coronel Olavarría, Olavarría
25/5	LV9		R. Salta, o
5	CP25	Bolivia	R. Centenario "la Nueva", Santa Cruz
1.5	CP78		R. RTC La Deportiva, Cochabamba
0.25	CP200		R. Litoral, Llica (nf 1530)
1/0.25	ZYH...	Brazil	Soc. de TV Manauara, Boca do Acre, AM (CP)

1/0.25	ZYI...		R. Guamá, São Miguel do Guamá, PA (CP)
1/0.25	ZYI...		R. Cidade Modelo, Bocaina, PI (CP)
2/0.5	ZYH634		R. Guarany, Pacajus, CE
1/0.25	ZYH652		R. Vale do Coreaú, Granja, CE
1	ZYI674		R. Soc. Campina Grande ("R. Cariri"), o, PB
1/0.25	ZYL324		R. Sacramento, o, MG
20/10	ZYI202		R. Espírito Santo, Vitória, ES
1/0.5	ZYK502		R. Pres. Venceslau, o, SP
5/0.25	ZYK517		R. Cacique, Sorocaba, SP
1/0.25	ZYK582		R. Dif. de Fernandópolis, o, SP
1/0.25	ZYK673		R. Cacique de Taubaté, o, SP
1/0.25	ZYK685		R. Cl. de Mococa, o, SP
1/0.25	ZYJ258		R. Norte, Londrina, PR
1	ZYJ741		R. Globo, Blumenau, SC
1/0.25	ZYJ767		R. Eldorado ("R. Dif. Laguna"), Laguna, SC
2.5/ 0.25	ZYJ776		R. Dif. Colméia, Porto União, SC
5/0.5	ZYK242		R. Miriam, Farroupilha, RS
5/0.5	ZYK245		R. Luz e Alegria, Frederico Westphalen, RS
1/0.5	ZYK256		R. Jaguari, o, RS
1	ZYK273		R. Univ. Católica de Pelotas, o, RS
10/5	ZYI385		R. A Voz d' Oeste, Cuiabá, MT
1	CC116	Chile	R. Ancoa, Linares
1	CD116		R. Baha'í, Labranza
10	HJAU	Colombia	Ondas del Orteguaza, Florencia
5	HJAZ		La Voz de Sinú, Montería
10	HJBL		R. Aeropuerto, Barranquilla (Soledad)
30	HJCO		R. Total, Bogotá
10	HJEC		R. San José de Cúcuta, Cúcuta
15	HJEV		La Caleñísima, Cali
10	HJZV		R. Las Lajas, Ipiales
2	HCCP1	Ecuador	R. Presidente, Quito (vf)
2	HCVR3		R. Vía, Machala
1	CHUR6		R. Runatecuyaj ("R. Cultural, La Voz de la Asociación"), Latacunga
5	OAX2C	Peru	R. Libertad, Trujillo
1	OAX3B		R. Huaraz, o
10	OAX4C		R. Once Sesenta, ("Once Sesenta AM"), Lima
1	OBX50		R. Huanta 2000, Huanta
1	OBX6G		R. Hogar, Moquegua
0.5	CW116	Uruguay	R. Agraria, Cerro Chato
10/1	YVOK	Venezuela	R. Universidad, Mérida
20/10	YVRR		R. Industrial, Guarenas
1170 kHz			
25/3	LRA29	Argentina	R. Nacional San Luís, o
25/3	LT16		R. Esmeralda ("La Nueva Voz"), Pres. Roque Sáenz Peña
1	CP98	Bolivia	R. Nuevo Mundo, Sucre
5/2.5	ZYH284	Brazil	R. Guaranópolis, Maués, AM
1/0.25	ZYH...		R. Dif. de Anori, o, AM (CP)
1/0.25	ZYH...		Juruá Comunicações, Carauari, AM (CP)
10/1	ZYJ598		R. Dif. de Mossoró, o, RN
5/0.25	ZYH473		R. Jornal, Eunápolis, BA
5/0.25	ZYL336		R. Cidade, Araxá, MG
10/5	ZYK569		R. Educ. de Campinas, o, SP
20/10	ZYJ273		R. Record, Curitiba, PR
1/0.25	ZYK207		R. Itapuí, Santo Antonio da Patrulha, RS
5/1	ZYK213		R. Dif. A Voz de Bagé, o, RS
5/0.25	ZYK359		R. Uirapuru, Passo Fundo, RS
2/1	ZYK380		R. Pitangueira, Itaqui, RS
1/0.25	ZYL234		R. Cl. Fronteira, o, MG (F.Pl.) (still on 1590)
10/0.25	ZYL327		R. Vanguarda do Vale do Aço, Ipatinga, MG (F.Pl.) (still on 1270)
15	HJBX	Colombia	Ondas del Meta, Villavicencio (nf 1200)
10	HJGA		R. Boyacá, Tunja
5	HJJE		RCN Tuluá, o
1	HJJY		Ondas del Folcklor, Guamo
15	HJKW		R. Nutibara, Medellín
15	HJ1Z		Meridiano 70, Arauca

10	HJNW		CARACOL Cartagena, o
10	HJPB		Ondas de Macondo, Valledupar
5	HJWA		La Voz del Guaviare, San José del Guaviare (ti)
10	HCRB2	Ecuador	R. Filadelfia, Guayaquil
3	HCAH3		R. El Trebol, Zaruma
10	HCVC3		R. Centinela del Sur ("CDS Internacional"), Laja (nf 1210)
10	HCJM4		R. Antena Libre, Esmeraldas
5	HCJV5		R. Central, Riobamba
1	OCX4Y	Peru	R. COSAT, Satipo
10	OBX6L		R. Radial Andina, Arequipa (relay RPP 730)
10/5	CX32	Uruguay	Radiomundo, Montevideo
10	YVKW	Venezuela	R. Mil Setenta ("Señal 11-70"), La Guaira
20/10	YVQV		R. Acarigua, o

1180 kHz

1	CP235	Bolivia	R. Ingavi, Viacha
10/2.5	ZYH280	Brazil	R. Dif. do Amazonas, Manaus, AM
10/5	ZYH889		R. Ribamar, São Luís, MA
1/0.25	ZYI690		R. Bonsucesso, Pombal, PB
1/0.25	ZYI797		R. Cultural, de Vitória, Vitória de Santo Antão, PE
1/0.25	ZYH248		R. Correio do Sertão, Santana do Ipanema, AL
50/10	ZYJ463		R. CBN (Globo), RJ, RJ
1/0.5	ZYK647		R. Dif. de Santa Cruz do Rio Pardo, o, SP
5/0.25	ZYK749		R. Nova Bebedouro, Bebedouro, SP
5/0.25	ZYJ237		R. Guaçu, Toledo, PR
1/0.25	ZYJ737		R. Integração do Oeste, São José do Cedro, SC
1/0.5 0.25	ZYJ770		R. Dif. de Lages, o, SC
10/0.5	ZYK340		R. Gazeta, Santa Cruz do Sul, RS
5/0.25	ZYI...		Soc. R. Educ. Nova Xavantina, Guarantã do Norte, MT (CP)
25/10	ZYI602		R. Guanandi AM, Campo Grande, MS
50	CB118	Chile	R. Portales, Santiago
15	HJFX	Colombia	CARACOL Manizales, o
20	HJGK		RCN Santander, Bucaramanga
10	HJJT		RCN Ibagué, o
15	HJ...		Em. Coorpurabá, Apartadó
20	HCLR1	Ecuador	Nueva Em. Central, Quito
1	HCRT1		R. Familiar, Julio Andrade
4	HCDP5		R. Cuenca ("La Voz de los Cuatro Rios"), Cuenca (vf)
3	ZP52	Paraguay	R. Coronel Oviedo, o
1.5/1	OCX2A	Peru	R. América, Quiruvilca
1	OCY4Z		R. Libertad, Junín
1	OAX9V		R. Marginal, Tocache (nf 920)
10/5	CX118	Uruguay	La Voz de Artigas, o
10	YVLQ	Venezuela	Super Suave 11-80, LV de La Victoria, o
10	YVNJ		R. Petrolera, Ciudad Ojeda
10	YVOR		R. Maturín, o

1190 kHz

50/5	LRA15	Argentina	R. Nacional San Miguel, San Miguel de Tucumán
25/5	LR9		R. América, Buenos Aires
1/0.25	ZYH663	Brazil	R. Guaraciaba, Guaraciaba do Norte, CE
10/1	ZYJ594		R. Tropical, Natal, RN
10/1	ZYH459		R. Juazeiro, o, BA
10/1	ZYL221		R. Guarani, Belo Horizonte, MG
2.5/ 0.25	ZYK512		R. Cl. Marconi, Paraguaçu Paulista, SP
5/1	ZYK700		R. Cidade AM, Votuporanga, SP
5/0.25	ZYK729		R. 31 de Março, Santa Cruz das Palmeiras, SP
1/0.25	ZYK741		R. Regional, Taquarituba, SP
1/0.25	ZYJ309		R. Pontal, Nova Londrina, PR
1/0.25	ZYJ783		R. Cl. de São João Batista, o, SC
1/0.25	ZYJ817		R. Planalto, Major Vieira, SC
1/0.25	ZYJ820		R. Cl. São Domingos, o, SC
5/0.5	ZYK234		R. Soc. Cerro Azul, Cerro Largo, RS
2.5/ 0.25	ZYK301		R. São Lourenço, São Lourenço do Sul, RS
1	ZYK354		R. Rosário, Serafina Corrêa, RS
1/0.25	ZYH800		R. Rio Vermelho, Silvânia, GO
10	HJCT	Colombia	LV de la Costa, Barranquilla
10	HJCV		R. Cordillera, Bogotá
15	HJEO		Ondas del Valle, Cartago
5	HJKI		R. Barají, Sahagún
10	HJKG		R. Mira, Tumaco
2	HCDE2	Ecuador	R. Excelsior, Guayaquil
5	HCRF6		R. El Sol, Pujilí
10	OAX1E	Peru	R. Chiclayo, o
	O..2.		R. Membrillar, Cascas
3	OBX3D		R. Ancash, Huaraz
1	OBX5M		R. Luren, Ica (nf 1060)
20/10	YVPF	Venezuela	R. Canaíma, San Félix
25	YVRE		R. Barinas, 1190 AM Estéreo, o
25	YVZD		Radiodif. Cultural del Táchira, San Cristóbal

1200 kHz

1	LRA6	Argentina	R. Nacional Mendoza, Valle de Uspallata (relay 960)
10/1.5	LT6		R. Genaro Berón de Astrada, Goya
3	CP31	Bolivia	R. Oriental, Santa Cruz
1	CP37		R. Santa Ana, Santa Ana del Yacuma
1	CP108		R. Independencia, Quillacollo
0.25	CP171		R. 24 de Noviembre, Arani
5/1	ZYJ...	Brazil	R. Tucumã, Ji-Paraná, RO (CP)
10	ZYH585		Ceará R. Cl. Fortaleza, CE
1/0.25	ZYH482		R. Cl. Rio de Ouro, Jacobina, BA
50/20	ZYK520		R. Cult. de São Paulo, o, SP
5/1	ZYK239		R. Erexim, o, RS
1/0.5	ZYK342		R. Fundação Cotrisel, São Sepé, RS
1	CA120	Chile	R. Almirante Blanco Encalada, Tocopilla (ti)
10	CD120		R. Agricultura, Los Angeles
10	HJBZ	Colombia	Ondas del Riohacha, o
15	HJCD		Em. Nueva Epoca, Fusagasugá
15	HJIJ		R. 1200 ("La Voz de la Raza"), Medellín
10	HJLR		Antena 2 de RCN, Sogamoso
10	HJNF		R. Super, Cali
10	HJBV		R. Principe, Cartagena
5	HCCS1	Ecuador	R. Súper K, Sangolquí (vf)
5	HCRE2		La Voz del Trópico, Quevedo
1	HCMP4		R. La Grande, ("La Grande"), Bahía de Caráquez (vf)
5	HCRM5		R. El Mercurio, Cuenca
	O..2.	Peru	R. San Mateo, Contumaza
5	OAX4B		R. Cadena, Lima
1	OAX6N		R. El Faro, Ilo
5	OAX7B		R. Tawantinsuyo, Cuzco (vf)
1	OCX7S		R. Cult. de Juliaca ("R. San Ramón"), o
2	OAX9C		R. Rioja, o
1			R. Satelite, Tingo María (rel. RPP 730)
1	OCX2T		R. Noticias, Cajamarca
1	OBX9C		Estación C, Moyobamba
	O..9.		R. Las Palmas, Nuevo Bambamarca
2	CW33	Uruguay	La Nueva R., Florida
1	CX120		R. Reconquista, Rivera
1	YVNH	Venezuela	Ondas del Escalante, Santa Bárbara del Zulia
10	YVOZ		Radiotiempo, Caracas
10	YVSF		R. Dimensión, Caripito

1210 kHz

10/5	LT18	Argentina	R. Eldorado, o
25/10	LV3		R. Córdoba, o
10/1	ZYI893	Brazil	R. Educ. de Parnaíba, o, PI
5/0.25	ZYH641		R. Boa Esperança, Barro, CE
1/0.25	ZYH657		R. Dif. de Nova Russas, o, CE
5/1	ZYI786		R. Dif. de Garanhuns, o, PE
5/1	ZYH452		R. Carioca, Feira de Santana, BA
1/0.25	ZYH498		R. Bandeirantes, Vitória da Conquista, BA
1/0.5	ZYL238		R. Cl. de Varginha, o, MG
1/0.5	ZYL305		R. Dif. de Uberlândia, o, MG
25/1	ZYI200		R. Cachoeiro do Itapemirim, o, ES
1/0.25	ZYK509		R. Nova Vida, Jaboticabal, SP
5/0.25	ZYK545		R. Dif. de Araçatuba, o, SP

1/0.25	ZYK558		R. Bandeirantes, Bauru, SP
5/0.25	ZYK668		R. Vanguarda, Sorocaba, SP
10/5	ZYJ219		R. Universo, Curitiba, PR
2.5/1	ZYJ325		R. Brotense, Porecatu, PR
1/0.25	ZYJ785		R. Santa Catarina, Tubarão, SC
10/5	ZYK240		R. Record AM, Porto Alegre, RS
1/0.5	ZYK353		R. Blau Nunes, Santa Bárbara do Sul, RS
50/2.5	ZYH...		R. Brasília, o, DF (CP)
1	CA121	Chile	R. Univ. de Antofagasta ("R. 121"), o
5	CB121		R. Valparaíso, Viña del Mar
1/0.8	CC121		R. Univ. de Talca, o
10	CD121		R. Nal., Puerto Montt, o
30	HJBE	Colombia	Antena 2 de RCN, Cúcuta
10	HJBQ		La Voz Amiga, Pereira
10	HJFR		R. Reloj, Neiva
20	HCBJ2	Ecuador	R. El Mundo, Guayaquil
10	HCVC3		R. Centinela del Sur ("CDS Internacional"), Loja (also on 1170)
1	ZP34	Paraguay	R. Cult. de Villarrica, o (vf)
1	OAX2Q	Peru	R. Universo, Trujillo
1	OBZ4A		R. Paramonga, o
1	OCY4T		R. Galaxia, Satipo
1	OAX7M		R. Quillabamba, o
0.25	CV121	Uruguay	R. RBC del Este, Piriápolis
0.25	CW121		R. El Libertador, Villa Vergara
3/1	CX121		Dif. Soriano, Mercedes
10	YVZT	Venezuela	R. Anzoátegui, Barcelona

1219 kHz

10	HJNM	Colombia	Em. Cultural Bolívar, Ipiales

1220 kHz

1	CP67	Bolivia	R. Splendid, La Paz
3	CP162		R. Batallón Topater, Oruro
	CP...		R. El Cóndor, Arque
150	ZYJ458	Brazil	R. Globo, RJ, RJ
10	CD122	Chile	R. Minería, Temuco
10	HJAV	Colombia	RCN Montería, o
15	HJKR		R. CORAL, Bogotá
10	HJMT		RCN San Gil, o
10	HCAP1	Ecuador	Radiodif. Marañón, Quito
1	OCX2H	Peru	R. Wuaira, Chota
1	OAX5G		R. San Pedro, Grocio Prado
3	OAX6X		R. Melodía, Arequipa
3	OAX9T		R. Bagua, o
10	YVRD	Venezuela	La Voz de Apure, San Fernando de Apure
10/5	YVVM		Mundial Valencia, o
20/10	YVZO		R. Aeropuerto 1220, Maracaibo

1226 kHz

3	HCJM6	Ecuador	R. Sira, Ambato (nf 1225) (vf)

1230 kHz

20/5	LT2	Argentina	R. General San Martín ("R. Dos"), Rosario
5/1	LW5		R. Libertador General San Martín, Ledesma
1/0.25	ZYH. . .	Brazil	R. Alecrim, Caxias, MA
10/1	ZYI670		Super R. Correio, João Pessoa, PB
1/0.25	ZYH...		R. Jornal, Ubatã, BA
1/0.25	ZYL208		R. Correio da Serra, Barbacena, MG
2.5/0.25	ZYL216		R. Soc. Passos, o, MG
5/0.25	ZYL296		R. Educ. Rio Doce, Governador Valadares, MG
1/0.25	ZYK573		R. Cidade, Capão Bonito, SP
2.5/0.25	ZYK703		R. Noticias, Matão, SP
5/0.5	ZYK716		R. Jequitibá AM, Campinas, SP
1/0.25	ZYJ225		R. Soc. Monte Alegre, Telêmaco Borba, PR
10/1	ZYJ816		R. Guararema, São José, SC
1/0.25	ZYK297		R. Santiago, o, RS
1/0.25	ZYK326		R. Nonoaí, o, RS
1/0.25	ZYK333		R. Prata, Nova Prata, RS
1/0.25	ZYK352		R. Encruzilhadense, Encruzilhada do Sul, RS
10/2.5	ZYH756		R. Anhanguera, Goiânia, GO
10	HJBR	Colombia	R. Unica, Tunja
10	HJGV		Colmundo R. Bucaramanga
10	HJIL		LV de las Américas, Medellín
10	HJLK		R. Calidad, Cali
5	HJMJ		R. Tribuna, Maicao
5	HJTP		R. Colina, Girardot
3	HCRI1	Ecuador	Centro Radiofonico de Imbabura, ("C.R.I."), Ibarra
20	HCGT2		R. Galactica, Guayaquil
5	HCFG4		R. Sucre, Esmeraldas (ts)
3	HCMV5		R. Popular de Cuenca, o
3	ZP21	Paraguay	R. Centenario, Caaguazú (vf)
1	OBX2T	Peru	R. Albujar de Guadalupe, Guadalupe (Chepén)
1	O..3.		R. Stana, Chimbote
1	OBZ4Y		R. Selecciones, Tarma
0.5	OBX7J		R. Madre de Diós, Puerto Maldonado
10	YVNT	Venezuela	R. Barlovento, Caucagua
10	YVOH		R. Valera, o

1235 kHz

1	CP35	Bolivia	Indoamérica R., Potosí (vf)
1	HCRL6	Ecuador	La Voz de Saquisilí y Libertador, Saquisilí

1237 kHz

1.5	OCX2T	Peru	R. Noticias, Cajamarca

1240 kHz

1	CP180	Bolivia	R. San Miguel, Arani
1	ZYJ...	Brazil	R. Verde, Jaru, RO (CP)
1/0.25	ZYH654		R. São Francisco, Canindé, CE
5	ZYI774		R. Capibaribe do Recife ("Jovem Cap"), o, PE
10/0.5	ZYH463		R. Em. de Alagoinhas, o, BA
5/0.25	ZYL294		R. Três Pontas, o, MG
1/0.25	ZYL298		R. Soc. Ubaense, Ubá, MG
1/0.25	ZYL303		R. Platina, Ituiutaba, MG
2.5/0.25	ZYL317		R. Pirapora, o, MG
1/0.25	ZYL...		R. Cl. de Inhapim, o, MG (CP)
5/0.25	ZYL367		R. Metropolitana, Vespasiano (F.Pl.) (still on 1560)
2.5/0.25	ZYK565		R. Municipalista, Botucatu, SP
5/0.25	ZYK621		Orlândia R. Cl., o, SP
10/2.5	ZYK653		R. Cl. de Santos, o, SP
1/0.25	ZYK711		R. Vale do Rio Tietê, José Bonifácio, SP
5/0.25	ZYJ215		R. Arapongas, o, PR
2/0.25	ZYJ280		R. Matelândia, o, PR
1/0.25	ZYJ301		R. Dif. Ubiratanense, Ubiratã, PR (F.Pl.) (still on 1450)
5/0.25	ZYJ774		R. Em. São José ("R. Nova Era"), Mafra, SC
1/0.25	ZYJ810		R. Iracema, Cunha Porã, SC
1/0.25	ZYK200		R. Aparados da Serra, Bom Jesus, RS
5/0.25	ZYK251		R. Ibirubá, o, RS
1/0.25	ZYK355		R. São Jerônimo, o, RS
25/10	ZYI388		R. Dif. de Campo Grande, o, MS
5/0.25	ZYH761		R. Xavantes, Ipameri, GO (ff 790)
1	CA124	Chile	R. del Salitre, Iquique
0.25	CA124		R. Principal Chuquicamata, Calama
25	CB124		R. Univ. de Santiago de Chile, Santiago
10	HJFG	Colombia	RCN Armenia, o
5	HJGN		Em. Barrancabermeja, o
1	HJGO		R. Caribabaré, Sarabena
5	HJJA		R. Buenaventura, o
1	HCPA1	Ecuador	R. Metropolitana, Quito
0.5	HCRF3		R. Fénix, Zaruma
5	HCLA5		R. Sonorama, Riobamba
1	OAX1L	Peru	R. Huancabamba, o
5	OBZ4Q		R. Oro de Huacho, Huacho
1	OBX7M		R. Túpac Amaru, Sicuaní
5/1.5	CW35	Uruguay	R. Paysandú, o

1250 kHz

2	CP14	Bolivia	Em. Unidas, La Paz
2	CP16		R. Los Andes, Tarija
0.5	CP17		R. Sararenda, Camiri
1	CP26		R. Amboró, Santa Cruz
0.3	CP47		R. Frontera, Cobija
2.5	CP54		R. La Plata, Sucre
0.4	CP65		R. Oruro, o
5	CP69		R. Nacional, Cochabamba

1	CP100		R. Ibaré, Trinidad
1/0.25	ZYH289	Brazil	R. Educ. Rural, Coari, AM
1/0.25	ZYI932		R. João de Paiva, Altos, PI
1	ZYH594		R. Educ. de Crateús, o, CE
1/0.25	ZYH...		R. Liberdade, Itarema, CE (CP)
1/0.25	ZYI...		R. Soc. de Soledade, o, PB (CP)
10/1	ZYJ925		R. Esperança, Estância, SE
5	ZYL282		R. Dif. de Poços de Caldas, o, MG
5/0.5	ZYJ500		R. Litoral, Casimiro de Abreu, RJ
5/0.5	ZYK702		R. Universal, Caçapava, SP
5/0.5	ZYJ211		R. Dif. Guarapuava, o, PR
1/0.25	ZYJ233		R. Paranavaí, o, PR
1/0.25	ZYJ313		R. Danúbio Azul, Santa Isabel do Oeste, PR
5/0.25	ZYJ766		R. Cult. de Joinville, o, SC
10/0.5	ZYK233		R. Dif. Caxiense, Caxias do Sul, RS
1	ZYK272		R. Tupanci, Pelotas, RS
1/0.25	ZYK361		R. Aguas Claras, Catuípe, RS
1/0.25	ZYI394		R. Dif. de Três Lagoas, o, MS
5/0.25	ZYI412		R. Caiuás, Dourados, MS (ff 770)
1/0.25	ZYH748		R. Dif. São Patrício, Ceres, GO
1/0.25	ZYH...		R. Princesa do Sul, Goiatuba, GO (ff 830)
5	CA125	Chile	R. América, La Serena
10	CD125		R. Univ. Austral de Chile, Valdivia
10	HJCA	Colombia	R. Capital, Bogotá
5	HJEM		La Voz de Corozal, o
1.5/1	HJFV		R. Viva, Pasto
15	HJHS		R. Reloj, Cúcuta
10	HJOK		Em. A.B.C., Barranquilla
5	HCEM1	Ecuador	Ondas Carchenses, Tulcán
3	HCMY1		LV del Triunfo, Santo Domingo de los Colorados
5	HCHB2		R. Linda, Guayaquil
10	ZP3	Paraguay	R. Asunción, o
	O..1.	Peru	R. Sensacional, Chiclayo
1.5	OCX2K		R. Indoamérica, Trujillo
10/5	OAX4L		R. Miraflores, Lima
1	OAX8P		R. Pucallpa, o
1	OAX9K		R. Estación Tarapoto, o
5	CW125	Uruguay	R. Bella Unión, o
5/10	CX36		R. Centenario ("R. 36"), Montevideo
5	YVML	Venezuela	R. Cabimas, o
20/10	YVPZ		R. Puerto Ordaz, o

1260 kHz

10/5	LT14	Argentina	R. General Urquiza, Paraná
0.4	CP...	Bolivia	R. Uncía, o
5	ZYJ670	Brazil	R. Educ. de Guajará-Mirim, o, RO
1/0.25	ZYH596		R. Vale do Jaguaribe, Limoeiro do Norte, CE
10/5	ZYH242		R. Gazeta de Alagôas, Maceió, AL
50/25	ZYK688		R. Mulher ("R. Morada do Sol"), São Paulo, SP
5/0.5	ZYJ740		R. Jornal de Santa Catarina, Blumenau, SC
5/0.25	ZYK204		R. Cult. de São Borja, o, RS
5/0.25	ZYK327		R. Em. Fandango, Cachoeira do Sul, RS
1/0.25	ZYK345		R. Gaurama, o, RS
10	CA126	Chile	R. Nal., Arica, o
1	CC126		R. Condell, Curicó
10	CA126		R. Minería, Punta Arenas
5	HJDA	Colombia	R. Auténtica ("R. Mundial"), Medellín
5	HJDV		CARACOL Ibagué, o
5	HJET		R. Deportes, Cali
1	HJHU		R. Morgan, Isla San Andrés (vf)
5	HJLX		R. Cinco, Villavicencio
2.5	HJNO		R. Duitama, o
10/5	HJOH		RCN César, Valledupar
2	HJOU		Ondas del Amazonas, Leticia
5	HJQN		R. Preferida, Montería
15	HJTM		R. Sonar, Ocaña
10	HCMO1		R-HCM1, LV del Santuario del Quinche, Quito
1	HCRB3		R. Benémerita, Machala
2	HCPB5		R. Stereo AM, Cuenca
3	HCR06		R. Calidad, Ambato
1	OBX2C	Peru	R. Otuzco, o
1	OBX6D		R. Red, Arequipa
4.5/1	CW37	Uruguay	Dif. Rochense, Rocha
10	YVRM	Venezuela	R. Super Suave, Caracas
10	YVRY		R. Horizonte, Nirgua

1270 kHz

5	LRA20	Argentina	R. Nacional Las Lomitas, o
25/5	LS11		R. Prov. de Buenos Aires, La Plata
1	CP134	Bolivia	R. Vanguardia, Colquiri
2.5	ZYH282	Brazil	R. Educ. Rural, Tefé, AM
10/1	ZYJ593		R. Poti, Natal, RN
1/0.25	ZYI...		R. Cidade, Sumé, PB (CP)
5/0.25	ZYL227		R. Carijós, Conselheiro Lafaiete, MG
5/1	ZYL327		R. Vanguarda do Vale do Aço, Ipatinga, MG (ff 1170)
5/1	ZYL...		R. Libertas do Vale do Aço, Ipatinga, MG (CP)
10/0.5	ZYJ474		R. Continental, Campos, RJ
2/0.5	ZYK640		R. Tropical, Ribeirão Preto, SP
5/1	ZYK678		R. Brasil 5000, Campinas, SP
1/0.25	ZYJ222		R. Atual Guairacá, Mandaguari, PR
5/0.5	ZYJ236		R. Capital, Piraquara (Curitiba), PR
5/0.5	ZYJ289		R. Cidade, Cascavel, PR
5/0.25	ZYJ765		R. Soc. Catarinense, Joaçaba, SC
1/0.25	ZYJ768		R. Garibaldi, Laguna, SC
1/0.5	ZYK206		R. América do Rio Grande do Sul, Montenegro, RS
1/0.5	ZYK244		R. São Roque, Faxinal do Soturno, RS (ff 1480)
5/0.5	ZYK250		R. Vera Cruz, Horizontina, RS
50/10	ZYH753		R. Brasil Central, Goiânia, GO
1	CA127	Chile	R. Universo, Antofagasta
10	CB127		R. Festival, Viña del Mar
1	CD127		R. Agricultura, Temuco
1	HJAR	Colombia	Antena 2 de RCN, Cartagena
1	HJIM		Colmundo R. Pereira
1	HJSV		La Voz de Orito, o
5	HJTX		R. Deportes, Bucaramanga
1	HJXQ		R. Melodía, Ubaté
15	HCUM2	Ecuador	R. Universal, Guayaquil
5/4.8	HCLD4		R. Junín, o
1.5/3	OAX1H	Peru	R. Grau, Piura
	O..2.		R. Imperial, Laredo (nf 1260)
1	OAZ4H		R. Huacho, o
1	OBZ4T		R. La Merced, o
1	OAX5B		R. Ica, o
0.5	OBX6H		R. Mollendo, o
1	OBX7A		Radiodif. Continente, Cuzco
1	OBX7G		R. Ondas del Titicaca, Puno
1	OAX8T		R. Eco, Iquitos
1	OAX9F		R. Nor Peruana, Chachapoyas
3/1	CV127	Uruguay	R. Cuareim, Artigas
10	YVOU	Venezuela	Ondas Panamericanas, El Vigía
1	YVTR		R. Tucupita, o

1275 kHz

0.5	CP187	Bolivia	R. Chané, Mineros

1280 kHz

1.5	CP112		R. Frontera, Yacuiba
10/1	ZYI688	Brazil	R. Sinhauá, Bayeux, PB
100	ZYJ455		Super R. Tupí, RJ, RJ
1	CC128	Chile	R. Arturo Prat Chacón, San Carlos
10	CD128		R. del Sur, Osorno
5	HJCM	Colombia	R. Sur, Pitalito
5	HJHO		Impacto Popular, San Juan del César
5	HJKN		R. Musical, Bogotá
5	HJLR		CARACOL Pasto, o
1	HJMB		R. Suroeste, Concordia
2.5	HJNQ		LV del Río Suarez, Barbosa
5	HJRP		Ecos de Tibú, o
15	HJSO		R. Playa Mendoza, Barranquilla
5	HJTK		R. Super, Caicedonia
10	HCAB1	Ecuador	R. Democracia, Quito
2	HCRP3		R. Continental, Arenillas
1	HCIN4		R. La Voz del Sur de Manabí, Jipijapa
1	HCWW5		R. Canal Tropical, Riobamba
1	OBX2F	Peru	R. Moderna, Cajamarca (vf)
1	OBX6P		R. Fenix, Camaná
10	YVOF	Venezuela	R. Trujillo, o

10/5	YVQS		R. Zaraza, o

1290 kHz

5/1		Argentina	Em. del Lago, Junín
2	CP212	Bolivia	R. Dif. Minería, Oruro
10/2.5	ZYH286	Brazil	R. Rio Mar, Manaus, AM
50/5	ZYH888		R. Timbira do Maranhão, São Luís, MA
1/0.25	ZYJ...		Sistema Potiguar de Informação, Caicó, RN (CP)
10/1	ZYH450		R. Cl. de Salvador, o, BA
1/0.25	ZYL266		R. Cl. Nepomuceno, o, MG (ff 810)
10/0.5	ZYL273		R. Record, Uberlândia, MG
10/1	ZYL304		R. Juriti de Paracatu, Paracatu, MG (ff 870)
1/0.25	ZYK662		R. Difusora, São José do Rio Pardo
5/1	ZYK663		R. Independência, São José do Rio Preto, SP
5/0.5	ZYK745		R. Metropolitana, São José dos Campos, SP
10/0.5	ZYJ310		R. Brasil Sul, Ibiporã, PR
5/0.5	ZYJ804		R. Camboriú, Balneário Camboriú, SC
2.5/2	ZYK331		R. Planetário, Espumoso, RS
1	CD129	Chile	R. Mulchén, o
0.25	CA129		R. Coya, María Elena
5	HJEB	Colombia	La Voz del Turismo, Santa Marta (El Rodadero)
5/1	HJHI		La Voz de Garagoa, del Valle de Tenza, Garagoa
5	HJKY		Ondas del Puerto, Girardot
5	HJMC		R. Impulso, Cali
15/5	HJNE		La Voz del Ariari, Granada
8	HJOI		R. Chacurí, Sampués
5	HJSZ		Em. del Ejército, Sarabena
5	HJTH		LV de las Estrellas, Medellín
1	HCNS1	Ecuador	R. Popular, Atuntaqui
1	HCOF2		Canal Milagreño, El Milagro
2	HCJA5		LV del Río Tarqui, Cuenca
1	OBX2Z	Peru	R. Casa Grande, o
10	OAX9N		R. Dif. Nor Oriental, Bagua
1			R. Programas del Perú, Tumbes
10	CX38	Uruguay	S.O.D.R.E., Montevideo (relay 1050)
1/0.5	CX129		S.O.D.R.E., Carmelo
10	YVLF	Venezuela	R. Puerto Cabello, o

1295 kHz

0.5	HCVM6	Ecuador	R. Once de Noviembre, Latacunga (nf 1290)
	0..2.	Peru	R. San Juan, Trujillo

1300 kHz

10/1	LRA5	Argentina	R. Nacional Rosario, o
5/1.5	LU22		R. Tandil, o
1	CP51	Bolivia	R. Loyola, Sucre
1	CP82		Coop. Radial Electra, Potosí
0.15	CP127		R. Juan XXIII, Uyuni
0.3	CP168		R. Chichas, Siete Suyos
0.5	CP...		Radioemisora Militar Coronel Eduardo Abaroa ("Radioemisora Eduardo Abaroa"), Santa Cruz
	CP...		R. Sol, Viacha
10	ZYH586	Brazil	R. Iracema, Fortaleza, CE
1/0.25	ZYI799		R. Guarany AM Stereo, Camarajibe (São Lourenço da Mata), PE
1/0.25	ZYL219		R. Cancella, Ituiutaba, MG (ff 710)
5/1	ZYL339		R. Eldorado, Sete Lagoas
5/0.25	ZYI210		R. Novo Tempo—Cidade, Afonso Cláudio, ES
10/5	ZYK535		R. Diario do Grande ABC ("Grande ABC AM"), São Bernardo do Campo, SP
1/0.25	ZYJ278		R. Pitangui AM, Ponta Grossa, PR
5/0.25	ZYJ288		R. Educ. de Dois Vizinhos, o, PR
1/0.25	ZYJ819		R. Alvorada, Santa Cecília
10	ZYK203		Super R. Tupi, Porto Alegre, RS
1/0.25	ZYK337		R. Regional, Santo Cristo, RS
5/0.5	ZYK347		R. Marathan, Santana do Livramento, RS
5	CB130	Chile	R. Tierra ("R. Antena Uno"), Santiago
10	CD130		R. Pilmaiquén, Valdivia
1	CD130		R. Cabo de Hornas, Pko. Williams
5	HJUA	Colombia	R. Sindamanoy, Mocoa
5	HJEA		R. Lumbí, Mariquita
5	HJLD		R. Reloj, Pereira
5	HJNB		R. Reloj, Bucaramanga
15	HJOG		LV de las Antillas, Cartagena
10	HJRB		R. Super, Tunja
2.5	HJRQ		R. Austral, Túquerres
5	HJIN		R. Eucha, Belalcázar
10	HCRU1	Ecuador	R. Festival, Santo Domingo de los Colorados
5	HCDC2		R. Cenit, Guayaquil
5	ZP10	Paraguay	Em. Paraguay, Asunción
1	OAX1I	Peru	R. Dif. Comercial, Chiclayo
0.5	OAX3O		R. Huascarán, Huaraz
2	OAX3R		R. Bahía Chimbote
2	OAX4B		R. Andina, Huancayo
5	OAX4S		R. Comas, o
2	OAX6P		R. Latina, Tacna (rel RPP 730)
10	OAX7P		R. Onda Imperial, Cusco
1.5	OAX7X		R. Juliaca, o
3	OAX8B		R. Nuevo Mundo, Pucallpa
10	YVKH	Venezuela	R. 1300, Caracas
10	YVNS		R. Reloj, Maracaibo

1310 kHz

1	LRA42	Argentina	R. Nal. Gualeguaychú, o (r. 1309.5)
0.5			R. Dr. Gregorio Alvarez, Piedra del Aguila
0.17			R. Horizonte, Villa Carlos Paz
10	CP68	Bolivia	R. San Rafael, Cochabamba
10/1	ZYJ...	Brazil	Rede Sanmori de R. e TV ("R. Tropical AM"), Porto Velho, RO (CP)
1	ZYH602		R. Progresso de Juazeiro, Juazeiro do Norte, CE
1/0.25	ZYH656		R. Liberdade, Boa Viagem, CE
10/0.5	ZYI691		R. Cidade Esperança, Esperança, PB
1/0.25	ZYH454		R. Bahiana, Ilhéus, BA
5/0.25	ZYH501		R. Canto da Sereia, Jacobina (F.Pl.) (still on 1590)
5/1	ZYH476		R. Educ. Santana de Caetité, Caetité, BA
5/0.25	ZYI426		R. Panorama, Sidrolândia, M5 (F.Pl.) (still on 1370)
1/0.25	ZYL357		R. Catuaí, Manhuaçu, MG (ff 620)
1/0.25	ZYL359		R. Montanhesa, Vazante, MG
1/0.25	ZYJ504		R. Dif. Coroados, São Fidélis, RJ
2.5/ 0.25	ZYK566		R. Bragança AM, Bragança Paulista, SP
2/1	ZYK596		R. Dif. de Itápolis, o, SP
10/0.5	ZYJ274		R. Atalaia, Maringá, PR
10/ 0.25	ZYJ801		R. Ituporanga, o, SC
5/1	ZYK305		R. Sarandi, o, RS
2.5/ 0.25	ZYK329		R. Integração, Restinga Seca, R5
2.5/	ZYK371		R. Horizonte, Capão da Canoa, RS
5	HJAK	Colombia	LV de la Patria, Barranquilla
5	HJDG		CARACOL Montería ("Tr. Río"), o
5	HJIR		RCN Urabá, Apartadó
5	HJJZ		R. Corazón AM, Bogotá
10	HJLM		R. Santa Bárbara, o
5	HJTQ		R. Tasajero, Cúcuta
1	HJWD		Micrófono Cívico, Palermo
	HJ..		R. Ciudad Paisaje, Samaniego
20	HCGB1	Ecuador	R. Nal. Espejo, Quito (r. 1309.7)
3	HCCI5		R. TVO, Biblián
1	ZP53	Paraguay	LV del Este, Ciudad del Este
1	OBX2D	Peru	R. Chota, o
1	OBX5K		R. San Martín, Pisco
10	YVSL	Venezuela	R. Nacional, Guri (rel 630)
20/10	YVSM		R. Nal. de Venezuela Oriente, Barcelona (rel 630)
5	YVTS		R. Sonido 13-10, Isnotú

1315 kHz

1	HCCP3	Ecuador	La Voz de El Oro, Pasaje

1320 kHz

25/1.5	LRH251	Argentina	R. Chaco, Resistencia
5	LU10		R. Azul, o
1	ZYH597	Brazil	R. Cidade, Sobral, CE
1/0.25	ZYH...		R. Jaguaribana, Aracati, CE (CP)

1/0.25	ZYI...		R. Cult. de São José de Egito, o, PE
1/0.25	ZYH243		R. Progresso de Alagôas, Maceió, AL
5/0.5	ZYH503		R. AM Cicero Dantas, o, BA
5/0.25	ZYL319		R. Planeta, Carmo do Paranaíba, MG
1/0.25	ZYL322		R. Mucuri, Teófilo Otoni, MG (ff 700)
25/5	ZYJ475		R. Dif. Boas Novas, Petrópolis (RJ), RJ
1/0.25	ZYK630		R. Dif. de Pirassununga, o, SP
1/0.25	ZYK675		R. Cl. de Tupã, o, SP
5/1	ZYJ255		R. Porto Alegre AM, Curitiba, PR
5/0.25	ZYK223		R. Cl. de Canela, o, RS
10/0.5	ZYK266		R. Panambi, o, RS
5/1	ZYK271		R. Cult. de Pelotas, o, RS
1	CA132	Chile	R. Estrella del Norte, Vallenar
1	CD132		R. Lincoyán, Mulchén
5	HJHT	Colombia	R. Guateque, o
5	HJLV		Em. La Esquina del Progreso, Fundación
10	HJMS		R. El Sol, Barrancabermeja
1	HJNK		R. Luna, Palmira
5	HJON		R. El Sol, Girardot
5	HJQI		R. Leda Internacional, Isla San Andrés
5	HJTA		La Voz de la Independencia, Medellín
5	HCFR2	Ecuador	R. Guayaquil, Babahoyo
10	HCJD6		R. Continental, Ambato
0.25	HCOB7		R. Nal. Limón, Limón Indanza
0.35	OAX1F	Peru	R. Talara, o
1	OBX2Z		R. La Voz de Cutervo, o
	O..2.		R. Libertad, Cajamarca (ti)
10	OAX4I		R. La Cronica, Lima
5/1	OBX4Z		R. La Voz de Oxapampa, o
0.8	OBX6W		R. Premier, Camaná
0.25	OCZ9K		R. Pajatén, Juanjuí
1/0.5	CW39	Uruguay	La Voz de Paysandú, o
0.3	CW132		R. Fortaleza, Rocha
10	YVSG	Venezuela	R. Colonial ("Canal 13"), El Tocuyo
10/5	YVWP		R. Apolo, Turmero
1324 kHz			
	O...3.	Peru	R. Continental, Caraz
1330 kHz			
0.25	CP106	Bolivia	R. Trinidad, o
1.5	CP176		R. América, Oruro
25/5	ZYI533	Brazil	R. Liberal, Belém, PA (vf)
1/0.25	ZYI905		R. Itamaraty, Piripiri, PI (ff 920)
10/0.5	ZYJ...		R. Eldorado, Natal, RN (CP)
5/1	ZYH468		R. Dif. de Serrinha, o, BA
1/0.25	ZYK638		R. Regente, Regente Feijó, SP
1	ZYK641		R. Cult. de Ribeirão Preto, o, SP
10/1	ZYK736		R. Iguatemi, Osasco, SP
10/0.5	ZYJ264		R. Jaguariaíva, o, PR
5/0.25	ZYJ320		R. Educ. Inconfidência, Umuarama, PR
5/0.5	ZYJ739		R. Cl. de Blumenau, ("R. Alegría O Povo"), o, SC
5/1	ZYJ749		R. Chapecó, o, SC
1/0.25	ZYK236		R. Upacaraí, Dom Pedrito, RS
1/0.5	ZYK323		R. Diplomata, São Marcos, RS
3	CB133	Chile	R. Metropolitana, Santiago
1	CD133		R. Vicente Pérez Rosales, Puerto Montt
10	HJFE	Colombia	Antena 2 de RCN, Pereira
8	HJLS		CARACOL Popayán, o
10	HJNR		R. Guanentá, San Gil
1	HJRD		R. Copeñol, El Peñol
5	HCJA1	Ecuador	R. Sideral, Quito (vf)
3	HCOV1		Ondas del Volante, El Angel
5	HCRV3		R. Nal. El Oro, Machala (vf)
2	HCLW5		R. La Voz de la Cordillera, Cuenca
10	ZP4	Paraguay	R. Chaco Boreal, Asunción
0.5	OAX2P	Peru	R. Paiján, o
	O..2.		R. Universal, Cajamarca (ti)
1	O..3.		R. Nal. del Perú, Chimbote
1	OCX7K		R. San Miguel, Cusco
5	CX40	Uruguay	R. Fenix, Montevideo
5	YVNX	Venezuela	R. Los Llanos, Calabozo
10	YVTU		R. Regional, Ciudad Ojeda
1340 kHz			
1	CP24	Bolivia	R. Grigotá, Santa Cruz (vf)
0.35	CP146		R. San Francisco, Apolo
50/1	ZYH886	Brazil	R. São Luís, o, MA
1/0.25	ZYH661		R. Planalto, Maracanaú, CE
5/1	ZYI671		R. Arapuan, João Pessoa, PB
5/0.25	ZYL241		R. Mundo, Itabirito, MG
1/0.25	ZYL352		R. Minas Liberdade, Passos, MG
1/0.25	ZYL...		R. Capital do Triângulo, Patrocínio, MG (CP)
10/ 0.25	ZYJ490		R. Jornal Contemporânea, Rio Bonito, RJ
5/0.5	ZYK543		R. Cult. de Araçatuba, o, SP
1/0.25	ZYK738		R. Canoa Grande, Igaraçu do Tietê, SP
2.5/ 0.25	ZYJ205		R. Dif. de Rio Negro, o, PR
1/0.25	ZYJ249		R. Cult. de Arapongas, o, PR
1/0.25	ZYJ...		R. Cascavel, o, PR (CP)
10/1	ZYK227		R. Educadora, Porto Alegre, RS
1/0.25	ZYI380		R. Nova Difusora Aquidauana, o, MS
0.25	CA134	Chile	R. El Toco, Tocopilla (ts)
10	CB134		R. Colo Colo del Pacífico, Valparaíso
1	CC134		R. Doña Isabel Riquelme, Chillán
1	CD134		R. Panguipulli, o
1	HJFA	Colombia	R. Olímpica, Barranquilla
5	HJFB		R. Uno, Bogotá
5	HJHY		RCN Sucre, Sincelejo
5	HJJS		R. El Sol, Buenaventura
5	HJKD		RCN Neiva, o
10	HJHA		RCN Nariño, Pasto
1	HJNP		R. Comunal, Nariño
5	HJNY		R. Sonorama, Bucaramanga
5	HJPY		R. Mil, Cúcuta
2	HJVL		La Voz de los Héroes, Belén de los Andaquíes
2	HCSF2	Ecuador	R. Bodegas, Babahoyo
1	HCDR4		R. Cenit, Bahia de Caráquez (ti)
2.5	HCVP5		R. San Francisco, Penipe
5/10	HCPB6		R. Paz y Bién, Ambato (vf)
1.5	OBX1K	Peru	R. San Francisco, Piura
0.2	OAZ4M		R. Huaral, o
1	OAZ4Q		R. Jauja, o
0.5	OAX5D		R. Chincha, Chincha Alta
3	OAX9C		R. Tropical, Tarapoto
2.5/1.4	CW53	Uruguay	La Voz de Melo, o
50/20	YVNE	Venezuela	R. Uno, Caracas
1350 kHz			
25/10	LS6	Argentina	R. Buenos Aires ("R.B.A."), o
1.5	CP124	Bolivia	R. Dif. Ichilo, San Carlos (also reported on 1510)
1	CP...		América Radiodifusión, Sucre
5	ZYH201	Brazil	R. Capital, Rio Branco, AC
5	ZYH420		R. Equatorial, Macapá, AP
1/0.25	ZYH662		R. Liberal, Morada Nova
5/1	ZYI675		Super R. Borborema, Campina Grande, PB
50/2.5	ZYH...		R. Cristal, Salvador, BA (CP)
10/5	ZYL214		R. Cult. de Poços de Caldas, o, MG
1/0.25	ZYJ734		R. Araranguá, o, SC
1/0.25	ZYJ745		R. Cl. de Canoinhas, o, SC
1/0.25	ZYJ760		R. Cl. de Itajaí, o, SC
1/0.25	ZYK205		R. Aurora, Guaporé, RS
1	ZYK313		R. Dif. Três Passos, o, RS
1/0.25	ZYK336		R. Agudo, o, RS
1	CA135	Chile	R. Guardia Marina Riquelme, Coquimbo (vf)
1	CD135		R. Cabo de Hornos, Puerto Williams
1	CD135A		R. San Carlos, Ancud
15	HJDS	Colombia	Ondas de la Montaña, Medellín
10	HJEN		R. Melodía, Cali
1	HJHL		R. Festival, Ibagué
5	HJHW		Ecos del Río, Puerto Boyacá
5	HJLO		R. El Sol, Caucasia
1	HJMN		R. Perijá, Codazzi
5	HJOC		Antena 2 de RCN, Santa Marta
5	HJOZ		La Voz del Yopal, o
1	HCPU1	Ecuador	LV de Sto. Domingo, Sto. Domingo de los Colorados
5	HCPZ1	Ecuador	R. Rumichaca, Tulcán

3	HCVP2		R. Reloj, Guayaquil
0.3	HCSF5		R. San Fernando, o
	O..1.	Peru	R. Continental, Chiclayo
	O..2.		R. San José, Cajamarca
	O..2.		R. Sintonía, Trujillo
1	OAX3N		Ondas del Huallaga, Huánuco
1	OAX3U		R. Miramar, Chimbote
1	OBX6F		R. Ilo, o
1.5	OAX8D		Radiodif. del Ucayali ("R. Super AM"), Pucallpa
5	YVTJ	Venezuela	R. Falcón, Puerto Cumarebo
5	YVZZ		R. Guanipa ("R. 1350"), San José de Guanipa

1355 kHz

0.25	CP154	Bolivia	R. Armonía, Cliza (nf 1350) (vf)

1360 kHz

1/0.25	LT46	Argentina	R. Provincial Bernardo de Irigoyen, o
1	CP143	Bolivia	R. Libertad, Villazón
	CP...		Difusoras Ximénez, La Paz
1/0.25	ZYH620	Brazil	R. Litoral, Cascavel, CE (ff 1110)
5/0.25	ZYH650		R. Iracema, Ipu, CE
1/0.25	ZYJ605		R. Ouro Branco, Currais Novos, RN
10/1	ZYH469		R. Cult. de Paulo Afonso, o, BA
50/10	ZYJ464		R. Guanabara, RJ, RJ
5/1	ZYK581		R. Cult. de Fernandópolis, o, SP
1/0.25	ZYK739		R. Regional, Dracena, SP
1/0.25	ZYK759		R. Luzes da Ribalta, Santa Bárbara d'Oeste, SP
10/ 0.25	ZYJ265		R. Cidade, Pato Branco, PR
1/0.25	ZYJ268		R. Pepita de Ouro, Assaí, PR
5/0.25	ZYJ302		R. Cult. de Iporã, o, PR
2/0.5	ZYK261		R. Alvorada, Marau, RS
2	ZYI383		R. Dif. Matogrossense, Corumbá, MS
1/5	CC136	Chile	R. Univ. del Bío-Bío ("136 AM"), Concepción
1	HJKV	Colombia	R. Laser, Zapatoca
5	HJPK		La Voz de Abejorral, o
1	HJRA		Ecos del Risaralda, Pereira
1	HJRK		HJRK, Miraflores (CP)
1	HJSD		R. Morichal, San Martín
5	HJUT		R. Reloj, Cartagena
1	HCMT1	Ecuador	R. Dif. Oyambaro, Tumbaco (vf)
5	HCHG3		R. El Sol, Machala
3	HCEG4		La Voz del Carmen, El Carmen
1	HCRJ5		R. América, Riobamba
1	HCLE7		R. Oriental, Tena
1	ZP37	Paraguay	R. Yby Yau ("R. Arapoty"), Carapeguá (also r. 1400)
1	ZP44		R. Arazá Potý, Itauguá
1	OBX2W	Peru	R. Super Uno, Cartavio Ascope
5	OAX4Y		R. Excelsior, Lima
2.5/ 0.5	CW41	Uruguay	Bc. San José, o
1/0.25	CW136		R. Río Branco, o
10	YVTI		Venezuela R. Internacional, Maracaibo
5	YVTZ		R. Armonia, Charallave

1365 kHz

2	HCJS1	Ecuador	Ecos Andinos, Pinampiro
1.5	OAX7R	Peru	R. Sicuaní, o

1370 kHz

1	LRA54	Argentina	R. Nacional, Ingeniero Jacobacci, o
20/3	LRJ241		R. Dimensión, San Luís
1	CP133	Bolivia	R. Agricultura, Achacachi
0.5	CP158		R. La Voz del Minero, Siglo Veinte
0.25	CP186		R. Libertad, Cliza
3	CP248		Dif. R. Coral, Oruro
10/1	ZYI892	Brazil	R. Dif. de Teresina, o, PI
1/0.25	ZYH		R. Vanguarda, Caridade, CE (CP)
1/0.25	ZYJ...		Sistema Potiguar de Informação, São Miguel, RN (CP)
1/0.25	ZYI800		R. Vale do Capibaribe, Santa Cruz do Capibaribe, PE
1/0.25	ZYJ929		R. Capital do Agreste, Itabaiana, SE
1/0.25	ZYL257		R. Dif. do Machado, o, MG (ff 760)
1/0.25	ZYL298		R. Soc. Ubaense, Ubá, MG (ff 1240)
1/0.25	ZYK541		R. Urubupungá, Andradina, SP (ff 760)
1/0.25	ZYK717		Bariri R. Cl., o, SP (ff 570)
100/20	ZYK766		R. Atual AM, Itapevi, SP
50/2	ZYJ267		R. Independência do Paraná, Curitiba, PR
5/0.25	ZYJ782		R. Peperí, São Miguel do Oeste, SC
10/ 0.25	ZYK243		R. Regional do Vêneto, Flores da Cunha, RS
1/0.25	ZYK334		R. Gazeta, Alegrete, RS
1/0.25	ZYI426		R. Panorama, Sidrolândia, MS
1	CD137	Chile	Conunhueno, R. La Popular, Temuco
5	HJBD	Colombia	R. Guaimaral, Cúcuta
5	HJBO		R. Piloto, Barranquilla
10	HJEQ		RCN Cauca, Popayán
1	HJJQ		Antena 2 de RCN, Zarzal
5	HJKX		R. Auténtica ("R. Mundial"), Bogotá
5	HJNI		CARACOL Sincelejo, o
2.3	HJNU		RCN Rionegro, o
5	HCVM2	Ecuador	LV del Milagro, El Milagro
5	HCER3		Sistema de Emisoras Progreso, Loja
2	HCRP7		R. Pastaza y Voz de Oriente, El Puyo
1	OAX1G	Peru	R. Tumbes, o
	O..2.		R. Marítima, Salaverry
1	OAX3H		R. Interamericana, Chimbote
1	OAZ4F		R. Cosmos, Huacho
1	OAX6T		R. Moquegua, o
1	OAZ7J		R. Santa Monica, Cusco
1	OBX9A		R. Palmera, Bellavista (also r. on 1450)
	O..9.		R. Tocache, o
0.5/ 0.25	CV137A	Uruguay	Em. Real, Minas de Corrales
7.5	CX42		Em. Ciudad de Montevideo, o
5	YVLG	Venezuela	R. Cumbre, Ejido (Mérida)
5	YVLO		R. La Pascua, Valle de La Pascua
10	YVST		R. Portuguesa, Araure

1380 kHz

0.5	CP221	Bolivia	R. 16 de Noviembre, Sacaba
0.5	CP227		R. Luís de Fuentes, Tarija
5/1	ZYH283	Brazil	R. Alvorada, Parintins, AM
1/0.25	ZYI...		R. Tocantins, Tucuruí, PA (CP)
10/5	ZYI773		R. Continental, Recife, PE
1/0.25	ZYL309		R. Veredas, Unaí, MG (ff 650)
1/0.25	ZYL323		R. Gorutubana AM, Janaúba, MG
10/1	ZYJ470		Campos Dif., o, RJ
1/0.25	ZYK507		R. Dif. de Lençóis Paulista, o, SP
1/0.5	ZYK616		R. Dif. de Mogi-Guaçu, o, SP
5/0.25	ZYK751		R. Diário, Pres. Prudente, SP
1/0.25	ZYK...		R. República, Morro Agudo, SP
2/0.25	ZYJ276		R. Bom Jesus ("RBJ-O Som Lider Norte Pioneiro"), Siqueira Campos, PR
1/0.25	ZYJ314		R. Educ. de São João do Ivaí, o, PR
1/0.25	ZYJ...		Rede Integração de Comunicação, Toledo, PR (CP)
1/0.25	ZYK311		R. Maristela, Torres, RS (F.Pl.) (still on 1460)
1/0.25	ZYJ821		R. Cidade, Itaiópolis, SC
1/0.25	ZYJ...		R. Barriga Verde, Capinzal, SC (CP)
1/0.25	ZYK293		R. Cult. de Santana do Livramento, o, RS
1/0.25	ZYK350		R. Gazeta, Tapera, RS
10/ 0.25	ZYK371		R. Horizonte, Capão da Canoa, RS (ff 1310)
1/0.25	ZYK372		Chiru Comunicações, Palmitinho, RS
1/0.25	ZYN404		R. Cidade de Matupá, Colider (Matup), MT (ff 770)
50	CB138	Chile	R. Colo Colo, Santiago
5	HJEE	Colombia	RCN Tunja, o
1	HJEJ		Armonías del Palmar, Palmira
1	HJID		Em. Villa del Páez, La Plata
10	HJLG		La Voz de La Dorada, o
5	HJLM		Colmundo R, Medellín (ti)
5	HJMM		R. Reloj, Valledupar
1	HJ..		Ondas del Sur, Túquerres

5	HCCV1	Ecuador	R. Cristal ("RCQ"), Quito
5	HCOA3		R. Impacto, Piñas
5	HCJR6		R. Mera, Ambato (vf)
3	HCWV7		R. Morona, Macas
1	OAX2W	Peru	R. Atahualpa, Cajamarca
1	OBX2V		Sistema Radial, Ferreñafe
1	OCY4U		R. ABC, Chosica
1	OBX5C		R. El Pueblo, Ica
5	OAX6O		R. San Martín, Arequipa
5	YVME	Venezuela	R. Fantasía, Soledad
10	YVNG		Ondas del Mar, Puerto Cabello
1	YVTL		R. Triunfo 1380, Caja Seca

1381 kHz

1	ZP8	Paraguay	R. Concepción, o (nf 1380)

1390 kHz

10/2.5	LR11	Argentina	R. Univ. Nal. de La Plata, La Plata
3	CP3	Bolivia	R. Nal. de Bolivia, La Paz
0.25	CP169		R. La Voz Minera del Sud, Mina Telemayu
5/1	ZYJ...	Brazil	Sistema Itapirema de Comunicações, Ji-Paraná, RO (CP)
10/5	ZYO701		R. Roraima, Caracaraí, RR
10/1	ZYI535		R. Educ. de Bragança, o, PA
10/0.5	ZYH907		R. Cult. do Rio Jordão, Coroatá, MA
1/0.25	ZYH...		R. Paty, Senador Pompeu, CE (CP)
1	ZYI788		R. Dif. de Pesqueira, o, PE
10/0.25	ZYH480		R. Bom Jesus da Lapa, o, BA (ff 660)
1/0.25	ZYL358		R. Ouro Verde, São Sebastíão do Paraíso, MG
10/0.5	ZYL...		R. Educ. de Uberlândia, o, MG (CP)
10/0.5	ZYI201		R. Cariacica AM, o, ES
5/0.5	ZYJ473		R. Sul Fluminense, Barra Mansa, RJ
5/0.25	ZYK570		R. Cult. de Campinas ("R. JB AM"), o, SP
2.5/0.25	ZYK594		R. Anchieta, Itanhaém, SP
1/0.25	ZYK636		R. Cult. de Promissão, o, SP
10/1	ZYJ242		R. Cult. de Maringá, o, PR
1/0.25	ZYJ335		R. Independência, Salto do Lontra, PR
5/1	ZYJ614		R. Santa Cruz AM, o, RN (F.Pl.) (still on 1390)
1/0.25	ZYK...		R. Atlântica, Constantina, RS (CP)
1/0.25	ZYJ769		R. Araucária, Lages, SC
10	ZYK209		R. Esperança, Porto Alegre, RS
5	HJFO	Colombia	LV de los Andes, Manizales
5	HJFY		R. Avenida, El Espinal
5	HJYW		R. Auténtica, Pacho
5	HJZY		R. Lemas, Bucaramanga (vf)
1.5	HCIE1	Ecuador	R. Uno, Urcuquí
5/10	HCVE4		La Voz de Esmeraldas, o
3	HCDN5		R. Atenas, Riobamba
5	HCEA5		R. Tropicana, Cuenca
1	OCX7U	Peru	R. Cult. de Yunguyo, o
2.5/1	CW45	Uruguay	Dif. Treinta y Tres, o (vf)
10	YVTT	Venezuela	R. Super Criolla, Cabudare (Barquisimeto)
20	YVZA		R. Fé y Alegría, Caracas
10	YVZC		R. Selecta, Maracaibo

1400 kHz

1/0.25	LRK221	Argentina	R. Ciudad Perico ("La Voz de Perico"), o (nf 1420)
10	LU11		Radioemisora del Oeste, Trenque Lauquen
0.45			R. Mun. Catriel, Catriel
1	CP174	Bolivia	Radiodifusoras Libertador, Santa Cruz
10	ZYH200	Brazil	R. Dif. Acreana, Rio Branco, AC
1/0.25	ZYI926		R. Cantagalo, Jaicós, PI
5/1	ZYI677		R. Espinharas, Patos, PB
1/0.25	ZYH...		R. Vale do Vasa Barris, Jeremoabo, BA (CP)
50/5	ZYJ462		R. Rio de Janeiro ("R. Rio AM"), RJ, RJ
1/0.25	ZYK527		R. Dif. de Lucélia, o, SP
5/0.25	ZYK658		R. Progresso, São Carlos, SP
5/0.25	ZYK682		R. Metrópole AM, São José do Rio Preto, SP
1/0.25	ZYJ256		R. Cruzeiro do Sul, Londrina, PR
2.5/0.25	ZYJ285		R. Educ. de Laranjeiras do Sul, o, PR
1/0.25	ZYJ291		R. Alvorada do Sul, Rebouças, PR
1/0.25	ZYJ299		R. Fronteira d'Oeste, Terra Roxa, PR
1/0.25	ZYJ775		R. Entre Rios, Palmitos, SC
1/0.25	ZYK247		R. Sideral, Getúlio Vargas, RS
5/0.25	ZYH754		R. Universitária, Goiânia, GO (ff 870)
5/1	ZYH...		Soc. de Radiodifusão, Guaraí, o, TO (CP)
1	CA140	Chile	R. Nal., Iquique, o (vf)
5	CD140		R. Llanquihue, Puerto Montt
5	CD140		R. La Amistad, Los Angeles
10	HJAS	Colombia	Antena 2 de RCN, Barranquilla
1	HJBK		La Voz de la Gran Colombia, Villa del Rosario (Cúcuta)
5	HJDF		LV del Niquel, Montelíbano
5	HJHM		Antena 2 de RCN, Calarcá
2	HJIT		Ecos del Atrato, Quibdó
5	HJJJ		R. Ipiales, o
5	HJKM		Em. Mariana, Bogotá
5	HJLL		Antena 2 de RCN, Santa Bárbara
1	HJWY		LV de los Samanes, Quilichao
1	HCVL1	Ecuador	R. Multicolor, Tulcán
10	HCFL2		R. Z-1, Guayaquil
1	HCVZ7		La Voz de Zamora, o
1	ZP37	Paraguay	R. Yby Yau ("R. Arapoty"), Carapeguá (nf 1360)
1	OAX2O	Peru	R. Perú, Chepén
0.8	O..2.		R. Layzón, Cajamarca
	O..2.		R. La Merced, Tongod
	O..2.		R. Quilavó, Limosulpi
1	OBZ3F		R. Willaq, Huaraz
2	O..3.		R. Chimbote, o
0.5	OAZ4P		R. San Juan, Tarma
2.5	OBX4W		Callao Super R., o
3	OAX6J		R. Landa, Arequipa
	O..7.		R. Constelación, Puno
1.5	OBX9A		R. Palmera, Bellavista (r. 1398.2)
1	OBX9E		R. El Dorado, Villa Picota
0.2	O..9.		R. Juanjuí, o
			Ondas del Suroriente, Quillabamba
25	CX140	Uruguay	R. Zorrilla de San Martín, Tacuarembó (vf)
1	YVNF	Venezuela	R. Sabana, El Sombrero
5	YVQZ		R. Anaco, o (ff 660)
10	YVZJ		R. 1400, Barinas

1410 kHz

0.1		Argentina	F. Fortín Federación, Junín
0.3	CP124	Bolivia	R. Atlántida, Oruro
0.1	CP...		R. Melodía, Bermejo
0.25	CP...		R. Roboré, o
10/0.5	ZYH467	Brazil	R. São Gonçalo, São Gonçalo dos Campos, BA
10/0.5	ZYJ486		R. Itaperuna, o, RJ
50/10	ZYK691		R. América, São Paulo, SP
1/0.25	ZYJ328		R. Cult. de Candido de Abreu, o, PR
5/0.5	ZYK246		R. Garibaldi, o, RS
1/0.25	ZYK284		R. Minuano ("R. Pampa"), Rio Grande, RS
5	ZYK294		R. Santa Rosa, o, RS
5/1	ZYI382		R. Cl. de Corumbá, o, MS
10/1	ZYN603		R. IPB AM, Campo Grande, MS (ff 630)
10/1	ZYH...		R. Fraternidade Universal, Santo Antônio do Descoberto, GO (CP)
1	CB141	Chile	R. Litoral, Valparaíso
1	CD141		R. Loncoche, o
5	HJDU	Colombia	Em. Cultural, Medellín
10	HJEI		R. Guadalajara, Buga
10	HJFS		Antena 2 de RCN, Honda
1	HJTY		Ondas del Carare, Vélez (vf)
1	HCEC1	Ecuador	R. El Tiempo, Quito
1	HCCQ2		R. Presidente ("La Voz del Pueblo"), El Milagro
1	HCFR4		La Voz de Quininde, o
1	HCKD5		R. Centro Gualaceo, Gualaceo
1	HCMS7		La Voz del Río Coca, El Coca
1	OAX1N	Peru	R. Ideal ("Frecuencia 14-10"), Lambayeque
2	OAX2Y		Heróica R., Trujillo

1	OBZ4V		R. Universal, Huacho
1	OAX7I		R. La Hora, Cusco (nf 1400)
1/0.5	CW141	Uruguay	R. Belén, o
5	CX44		R. Panamericana, Montevideo
10	YVSP	Venezuela	R. Simpatía, Valera
5	YVST		R. Clarín, Turén (ff 1180)
1411 kHz			
1	ZP35	Paraguay	R. Dif. Mangore, San Juan Bautista, Misiones
1415 kHz			
		Argentina	R. Cultura, Lanús
1417 kHz			
5	ZP42	Paraguay	R. Güyrá Campana, Horqueta
1419 kHz			
2.5	ZP35		R. La Voz de Misiones, San Juan Bautista de las Misiones (nf 1430)
1420 kHz			
		Argentina	R. Club, Lomas de Zamorra
1	CP49	Bolivia	R. Centro, Cochabamba
	CP..		R. Chaka, Pucarani
10/5	ZYI...	Brazil	R. Maguary, Belém, PA (CP)
1/0.25	ZYH...		R. Dif. do Vale Acaraú, Acaraú, CE (CP)
1/0.25	ZYJ609		R. Tapuyo, Alexandria, RN
1/0.25	ZYH504		R. Dif. de Irecê, o, BA
5/0.25	ZYL207		R. Imbiara, Araxá, MG (ff 830)
10/ 0.25	ZYL237		R. Ibituruna, Governador Valadares, MG (ff 930)
1/0.25	ZYL286		R. Dif. São João Nepomuceno, o, MG
5/0.25	ZYL288		R. Cult. de Sete Lagoas, o, MG
1/0.25	ZYL313		R. Montanhês, Botelhos, MG
1/0.25	ZYK597		R. das Nações, Itatiba, SP
1/0.25	ZYK612		R. Cl. de Mirandópolis, o, SP
1/0.25	ZYK619		R. Dif. de Monte Aprazível, o, SP (ff 780)
1/0.25	ZYK674		R. Cult. de Taubaté, o, SP
1/0.25	ZYK733		R. Nova São Manoel, São Manoel, SP
5/0.25	ZYJ269		R. Cult. de Umuarama, o, PR
1/0.25	ZYJ282		R. Educ. Rural, Jacarezinho, PR
1/0.25	ZYJ334		R. Entre Rios, Santo Antônio do Sudoeste, PR
5/0.25	ZYJ744		R. Cult. de Campos Novos, o, SC
5/2.5	ZYJ754		R. Guarujá, Florianópolis, SC
10/ 0.25	ZYK258		R. 14 de Julho, Júlio de Castillos, RS
1/0.25	ZYI397		R. Dif. Cacique, Nova Andradina, MS
1	CB142	Chile	R. Panamericana, Santiago
1	CC142		R. Nueva Maule, Cauquenes
1	CC142		R. Formidable, Rancagua
5	HJAP	Colombia	R. Anténtica ("R. Bahia"), Cartagena
5	HJBH		CARACOL Santa Marta, o
15	HJHK		R. Reloj, Manizales
10	HJLE		Antena 2 de RCN, Ibagué
1	HJSN		R. Lenguerque, Zapatoca
3	HCRN1	Ecuador	R. Bahá'í, Otavalo
1	HCNR3		La Voz de Huaquillas, o
5	HCMA6		R. Nuevos Exitos, Salcedo (vf)
1	OBZ4G	Peru	R. San Isidro, Lima
1	OBX5H		R. Apurímac, Abancay
1	OCX7E		R. Qollasuyo, Juliaca
1	OAX8Z		R. Oriente, Yurimaguas
5	CW43	Uruguay	R. Lavalleja, Minas
0.1	CX142		R. Felicidad, Paysandú (vf)
5	YVNZ	Venezuela	Super R. Marabina, Maracaibo
10/5	YVRW		R. Cardenal, Carora
10	YVTP		R. Bahía, Puerto La Cruz
10	YV..		R. Sintonía, San Antonio de los Altos
1430 kHz			
1	LRI235	Argentina	R. Balcarce, o
1/0.25	LT24		R. San Nicolás, o
1/0.25	LV26		R. Río Tercero, o R. Los Alamos, Glew
0.25	CP141	Bolivia	R. Nuestro Señor de Burgos, Mizque
0.25	CP193		R. 23 de Marzo, Tupiza
1	ZYJ671	Brazil	R. Caiari, Porto Velho, RO
5/1	ZYH637		R Principe Imperial, Crateús, CE
10/0.5	ZYJ604		R. Libertadora Mossoroense, Mossoró, RN
5/0.25	ZYI...		R. Independência AM, Goiana, PE
1/0.25	ZYL239		R. Cl. de Guaxupé, o, MG
1/0.25	ZYL252		R. Educ. de Ubá, o, MG
1/0.25	ZYL...		R. Planalto, Perdizes, MG (CP)
1/0.25	ZYK666		R. Serra Negra, o, SP
25/10	ZYJ200		R. Cl. Paranaense, Curitiba, PR
1/0.25	ZYK224		R. Cultura, Canguçu, RS
1/0.25	ZYK366		R. Querência, Coronel Bicaco, RS
1	CC143	Chile	R. Nal. de Chile, Rancagua (vf)
3	HJBP	Colombia	R. Cariongo, Pamplona
5	HJEG		LV del Belalcázar, Popayán
1	HJIU		Armonías del Ingruma, Riosucio
7.5	HJKU		Em. Kennedy, Bogotá
1	HJMF		R. Venus, Puerto Berrío
5	HJQX		R. Majagual, Sincelejo (nf 1420)
15	HJPW		Colmundo R, Barranquilla
	HJ..		R. Ciudad del Tambo ("Ecos del Guáitara"), Sandoná
1	HCGF1	Ecuador	R. Continental, Quito
10	HCMB2		R. Caribe, Guayaquil
5	HCCV3		R. Musical, ("Punto 14-30"), Loja
5	HCJC6		R. Guaranda, o
1	OBX2T	Peru	R. Santa Bárbara, Ascope
	O..2.		R. El Puerto, Pacasmayo
1	O..3.		R. Chavín, Chimbote
1.5	OAZ4V		R. Universal, Huancayo
1	O..7.		R. Espinar, o
5/1	CW25	Uruguay	R. Durazno, o
10	YVNB	Venezuela	R. Satélite, Guacara
10/5	YVTM		R. Caicara, Caicara del Orinoco
1440 kHz			
1	LRA53	Argentina	R. Nacional San Martín de los Andes, o
1/0.25	LT41		R. La Voz del Sur Entrerriano, Gualeguaychú
0.25	LU36		R. Coronel Suárez, o
1/0.25	LV20		R. Laboulaye, o
1/0.25	LV27		R. San Francisco, o
			R. Chascomús, "RCH", o
0.6	CP72	Bolivia	R. Batallón Colorados ("Voz del Ejército Boliviano"), La Paz
1.5	CP107		R. Yaguarí, Vallegrande
10	ZYH285	Brazil	R. Baré, Manaus, AM
10/1	ZYH603		R. Cidade, Crato, CE
5/0.25	ZYJ930		R. Educ. de Frei Paulo, o, SE
10/0.5	ZYH466		R. Independência, Santo Amaro, BA
1/0.25	ZYL...		Empreendimentos Guimarães e Franzão ("R. Som 200"), Santa Vitória, MG (CP)
20/5	ZYJ469		R. 1440 AM, RJ, RJ
5/0.25	ZYK634		R. Comercial, Pres. Prudente, SP
2.5/0.5	ZYK752		R. Azul Celeste, Americana, SP
1/0.25	ZYJ353		R. Integração Oeste, Corbélia, PR (F.PI.) (still on 1490)
1/0.25	ZYJ792		R. Dif. Maravilha, o, SC
1/0.25	ZYK221		R. Ceres de Campo Real, Não Me Toque, RS
1/0.25	ZYK328		R. Excelsior, Gramado, RS
1/0.25	ZYK362		R. Caibaté, o, RS
1/0.25	ZYI407		R. Bela Vista ("A Voz do Apa"), o, MS
1	CA144	Chile	R. Arica, o (vf)
1	CA144	Chile	R. Agricultura, La Serena
1	CC144		R. El Sembrador, Chillán
5/1	HJBM	Colombia	La Voz de Honda, o
1	HJEK		R. Reloj, Tuluá
5	HJGM		RCN Sogamoso, o
5	HJIB		RCN Caquetá, Florencia
5	HJNZ		R. 24, Medellín
5	HCDF1	Ecuador	R. Panorama, Ibarra
5	HCDY4		R. Iris, Esmeraldas
5	HCOV5		Ondas del Volante, Azogues
5	HCMD7		R. Puyo, o
2	OBX1T	Peru	R. Corporación Tumán, Chiclayo (vf)
1	OAX6R		R. El Tiempo, Arequipa
	O..7.		R. San Francisco, Yunguyo
1	CV144	Uruguay	R. Chuy, o

5	YVRF	Venezuela	Super Suave Orituco, Altagracia del Orituco
20	YVTY		R. Sucesos, Táriba
5	YVZI		R. Estelar, Guanare

1443 kHz

10	HCSC5	Ecuador	R. Calidad, Riobamba (nf 1450) (vf)

1444 kHz

5	OAX4K	Peru	R. Imperial Dos, Lima (nf 1440) (vf)

1445 kHz

1	CD145	Chile	R. Turismo, Puerto Varas (nf 1450)
3/0.5	CX144	Uruguay	R. Rivera, o (nf 1440)

1450 kHz

1/0.25	LT44	Argentina	R. Fortín Yunká, Formosa
1/0.25	LV28		R. Villa María, o
1	CP61	Bolivia	R. Em. Bolivia, Oruro
1/0.25	ZYJ674	Brazil	R. Vilhena, o, RO
1/0.25	ZYI...		R. Juruá, São Felix do Xingu, PA (CP)
1/0.25	ZYH891		R. Verdes Campos, Pinheiro, MA (ff 710)
1/0.25	ZYH900		R. Boa Esperança, São João dos Patos, MA
1/0.25	ZYI901		R. Cult. de Pedreiras, o, MA
1/0.25	ZYI908		R. Cult. de Gurguéia, Bom Jesus, PI
1/0.25	ZYI917		R. Confederação Valenciana ("R. Valenciana"), Valença do Piauí, PI
1/0.25	ZYH601		R. Dif. Cristal, Quixeramobim, CE
1/0.25	ZYH623		R. Pinto Martins, Camocim, CE
1/0.25	ZYI...		R. Itatiunga, Patos, PB (CP)
1/0.25	ZYI794		R. Cultura Palmares, o, PE
1/0.25	ZYJ...		R. Abais, Estância, SE (CP)
1/0.25	ZYH492		R. Rural, Ipiaú, BA (ff 1070)
1/0.25	ZYH...		R. Belo Campo, o, BA (CP)
1/0.25	ZYH...		R. Ipirá, o, BA (CP)
1/0.25	ZYL249		R. Globo de Minas, João Monlevade, MG
1/0.25	ZYL312		R. Diamante, Coromandel, MG
1/0.25	ZYL...		R. Princesa do Vale, Itaobim, MG (CP)
1/0.25	ZYI208		R. Gaeta, Guarapari, ES
1/0.25	ZYJ480		R. do Comércio, Barra Mansa, RJ
1/0.25	ZYJ503		R. Feliz de Santo Antônio, Santo Antônio de Pádua, RJ
1/0.25	ZYK526		R. Cult. de Ituverava, o, SP
1/0.25	ZYK587		R. Dif. de Guararapes, o, SP
5/0.5	ZYK591		R. Difusora Boa Nova, Guarulhos, SP
5/0.25	ZYK621		Orlândia R. Cl., o, SP (ff 1240)
1/0.25	ZYK657		R. São Carlos, o, SP
1/0.25	ZYK708		R. Registro, o, SP (ff 1140)
1/0.25	ZYJ225		R. Soc. Monte Alegre, Telêmaco Borba, PR (ff 1230)
1/0.25	ZYJ279		R. Cabiúna, Bandeirantes, PR
1/0.25	ZYJ293		R. Najuá, Irati (F.PI) (still on 1580)
1/0.25	ZYJ301		R. Dif. Ubiratanense, Ubiratã, PR (ff 1240)
1/0.25	ZYJ317		R. Rainha d'Oeste, Altônia, PR
1/0.25	ZYJ346		R. Jornal São Miguel, São Miguel do Iguaçu, PR
1/0.25	ZYJ802		R. São Bento, São Bento do Sul, SC
1/0.25	ZYJ...		R. Belos Montes, Seara, SC (CP)
1/0.25	ZYJ...		R. Abais, Estância, SE (CP)
5/0.25	ZYJ822		R. Soc. Hulha Negra, Criciúma, SC
1/0.25	ZYK346		R. Cassino, Rio Grande, RS
1/0.25	ZYI455		R. Dif. Arco Iris, Araputanga, MT
1/0.25	ZYI417		R. Dif. Rio Brilhante, o, MS
1/0.25	ZYH788		R. Vale da Serra, São Luís de Montes Belos, GO (ff 920)
5	CB145	Chile	R. Univ. Técnica "Federico Santa María", Valparaíso
5	CC145		R. Libertad, Curicó
1	HJHH	Colombia	R. Metropolitana, Bucaramanga
1	HJMX		R. 1450 Mancomoján, El Carmen de Bolívar
5	HJNL		Antena 2 de RCN, Manizales
5	HJTT		Ondas del Porvenir de Boyacá, Samacá
5	HJVY		R. Ciudad Flandes, Flandes
1	HJ..		R. Libanense, Líbano
1	HCSC1	Ecuador	R. Sensación, Cayambe
5	HCHW2		R. Rey, Quevedo (vf)
1	HCSE2		R. Santa Elena, o
1	HCRI7		R. Interoceánica, Santa Rosa de Quijos
5	ZP29	Paraguay	R. Vallemí, o
1	OBX4K	Peru	R. Fortaleza, Barranca
1	OCX7W		R. Santa Rosa, Quillabamba
1.5	CX46	Uruguay	R. América, Montevideo
10	YVKJ	Venezuela	R. Sonera 14-50, Catia La Mar
10/5	YVXC		R. Cachamay ("RC 14-5-0"), Puerto Ordaz
10	YVZQ		R. Triunfo, Puerto de Altagracia

1455 kHz

2	CP..		Bolivia R. Estévan Arce, Tarata

1460 kHz

1/0.25	LT29	Argentina	R. Venado Tuerto, o
0.25	LU30		R. Maipú, o (ti)
0.25	LU34		R. Pigüé, o
1	CP212		R. Minería, Capinota (ti)
1/0.25	ZYH300	Brazil	R. Cl. de Parintins, o, AM
1/0.25	ZYH...		Rede Amapaense de Radiodifusão, Macapá, AP (CP)
1/0.25	ZYH...		R. Vanguarda, Santa Luzia, MA (CP)
1/0.25	ZYI901		R. Alvorada do Sertão, São João do Piaui, PI (ff 710)
1/0.25	ZYI903		R. Cult. de Amarante, o, PI
1/0.25	ZYI924		R. Cruzeiro, Pedro II, PI (ff 640)
1/0.25	ZYH616		R. Uirapuru, Morada Nova, CE
1/0.25	ZYH...		R. Dragão do Norte, Massapé, CE
1/0.25	ZYJ615		R. Agreste, Santo Antônio, RN
1/0.25	ZYH...		R. Canavieiro, União dos Palmares, AL (CP)
1/0.25	ZYI...		R. Oeste da Paraíba, Cajazeiras, PB (CP)
1/0.25	ZYH472		R. Bahiana, Jequié, BA
1/0.25	ZYH...		R. Alvorada, Cruz das Almas, BA
1/0.25	ZYH...		Arcofaba Radiodifusão, Riacho de Santana, BA (CP)
1/0.25	ZYL201		R. Cult. de Porto Novo, Além Paraíba, MG
1/0.25	ZYL356		R. Buritis, o, MG
1/0.25	ZYL363		R. Entre Rios, Raul Soares, MG
1/0.25	ZYK548		R. Cl. Ararense, Araras, SP
1/0.25	ZYK608		R. Cult. de Lorena, o, SP
1/0.25	ZYK707		R. Universal, São Roque, SP
1/0.25	ZYJ204		R. Dif. de Paranaguá, o, PR
1/0.25	ZYJ228		R. Central do Paraná, Ponta Grossa, PR
1/0.25	ZYJ251		R. Cult. de Apucarana, o, PR
2.5/ 0.25	ZYJ286		R. Cl. de Palmas, o, PR
2/0.25	ZYJ297		R. Guaíra ("A Voz das 7 Quedas"), Guaíra, PR
1/0.25	ZYJ308		R. Ampere, o, PR
1/0.25	ZYJ318		R. Educ. de Loanda, o, PR
1/0.25	ZYJ756		R. Sentinela do Vale ("R. Cl. de Gaspar"), Gaspar, SC
1/0.25	ZYK214		R. Cult. de Bagé, o, RS
1/0.25	ZYK311		R. Maristela, Torres, RS (ff 1380)
1/0.25	ZYK312		R. Colonial, Três de Maio, RS
1/0.25	ZYI395		R. Aruanã, Barra do Garças, MT (ff 560)
1/0.25	ZYH749		R. Educ. de Uruaçu, o, GO (ff 870)
1/0.25	ZYH766		R. Morrinhos, o, GO
1/0.25	ZYH774		R. Independência do Tocantins, Paraíso do Tocantins, TO (ff 670)
10	CA146	Chile	R. Antofagasta, o
1	CB146		R. Yungay, Santiago
1	CC146		Nueva R. Talcahuano, o
1	HJAL	Colombia	R. Sincelejo, o
5	HJFL		Atalaya Agustiniana, San Agustín
1	HJIW		R. Monumental, Villa del Rosario
10	HJJH		R. Ciudad Milagro, Armenia
5	HJJW		Em. Nuevo Continente, Bogotá
1	HJMU		La Voz de Amalfi, o
1	HJTF		Ondas del Darién, Turbo
10	HJVH		R. Multicolor, ("La R. Jalisco"), Barranquilla
5	HJZU		Antena 2 de RCN, Pasto
1	HCDR2	Ecuador	R. Minutera, Pascuales

5	HCCL3		R. Cariamanga, o
5	HCIC6		R. Nuevos Horizontes, Latacunga
1	OBX2Y	Peru	R. Estelar, Guadalupe
0.5	OAX3C		R. Estrella, Huánuco
1	OBX3A		R. Chimú, Chimbote (relay RPP 730)
1	OAZ4F		R. La Oroya, o
0.5	OCY4I		R. 1460, Junín
2.5	OAX5K		R. Internacional, Pisco
1	OAX6C		R. Bahía, Mollendo
1	OAX7W		R. El Sol de los Andes, Juliaca
1	OBX9B		R. Frecuencia Popular, Rioja (also r. 1602)
0.5			Ondas del Perú, Cajamarca
0.25	CV146	Uruguay	R. José Batlle y Ordóñez, o
1	CX146		R. Carmelo, o
5	YVRJ	Venezuela	R. Jardín, Boconó
1465 kHz			
2.5	CP152	Bolivia	El Mundo Radiodifusión, Sacaba (nf 1495)
1	OAX1V	Peru	R. Sullana ("La Voz de Chira"), Sullana (nf 1460)
1468 kHz			
	HJ..	Colombia	R. El Sol, San Jeronimo
1470 kHz			
1	LT13	Argentina	R. Oberá, o (ff 840)
1/0.25	LT20		R. Junín, o (vf)
1/0.5	LT26		R. Nuevo Mundo, Colón
1/0.25	LT28		R. Rafaela, o
1.5/0.6	LU26		Em. Coronel Dorrego, o (r. 1475/1480)
1			R. Luís Beltrán, Fray Luís Beltrán
0.25	CP215	Bolivia	R. Cordech, Alcalá
1/0.25	ZYJ676	Brazil	R. Soc. Rondônia, Cacoal, RO (ff 930)
1/0.25	ZYI548		R. Moreno Braga, Vigia, PA
1/0.25	ZYH...		Sistema Maranhense de R. e TV, Pres. Dutra, MA (CP)
1/0.25	ZYI900		R. Dif. Vale do Uruçuí, Uruçuí, PI
1/0.25	ZYI913		R. Ingazeira, Paulistana, PI
1/0.25	ZYI928		R. AM Cidade, Castelo do Piauí, PI
1/0.25	ZYH665		R. A Voz de Itapagé, o, CE
1/0.25	ZYJ616		R. AM de Parelhas, o, RN
1/0.25	ZYI822		R. Educ. de Belém, Belém de São Francisco, PE
1/0.25	ZYI...		R. Bom Conselho, o, PE (CP)
1/0.25	ZYL247		R. Dif. de Ituiutaba, o, MG
1/0.25	ZYL294		R. Três Pontas, o, MG (ff 1240)
1/0.25	ZYL...		R. Libertas do Vale do Aço, Ipatinga, MG (CP)
1/0.25	ZYI214		R. São Francisco, Barra do São Francisco, ES
1/0.25	ZYI...		Sistema Norte de R., Serra, ES (CP)
1/0.25	ZYJ476		R. Jornal Fluminense, Campos, RJ
1/0.25	ZYJ481		R. Dif. Vale do Paraíba ("R. de Barra do Piraí"), Barra do Piraí, RJ
1/0.25	ZYK586		R. Cult. de Guaíra, o, SP
1/0.25	ZYK599		R. Cl. de Jacareí, o, SP
1/0.25	ZYK632		R. Primavera, Porto Ferreira, SP
1/0.25	ZYK771		R. Bastos AM, o, SP
1/0.25	ZYK712		R. Jornal, Indaiatuba, SP
1/0.25	ZYK...		Org. Kimura Nakaya de Radiodif., Bastos, SP (CP)
1/0.25	ZYJ294		R. Novo Horizonte, Ibaiti, PR
2/0.25	ZYJ304		R. Jornal, Assis Chateaubriand, PR
1/0.25	ZYJ327		R. Auriverde, Pitanga, PR
1/0.25	ZYJ781		R. Gazeta, São José, SC
1/0.25	ZYJ798		R. Líder do Vale, Herval d'Oeste, SC
1/0.25	ZYK219		R. Cult. Cacequiense, Cacequi, RS
1/0.25	ZYK324		R. Cinderela, Campo Bom, RS
1/0.25	ZYI422		Em. Reunidas, Poconé, MT
1/0.25	ZYI413		R. Alvorada, Dourados, MS
1/0.25	ZYH773		R. Dif. Serra dos Cristais, Cristalina, GO
1/0.25	ZYH779		R. Cidade, Goiás, GO
1	CB147	Chile	R. Sargento Aldea, San Antonio
10	HJHQ	Colombia	R. Futurama, Pacho (vf)
1	HJIF		R. Tres Fronteras, Puerto Assís
5	HJII		R. Musical, Medellín
5	HJNT		R. Total, Cali
10	HJPX		LV de las Estrellas, Cartagena (also r. 1480)
5	HJTB		Ondas de Ibagué, o
1.5	HCJC1	Ecuador	Ecos de Cayambe, o
1	HCLD2		R. Ecos de Naranjito, o
1	HCRD4		R. Richi, Tosagua (Calceta)
1	OBX2U	Peru	R. Ilucán, Cutervo
25	O..4.		Antena Uno R., Lima
1	OCX4D		R. Huancayo, o (vf)
1	OAX6M		R. Tacna, o
1	OAX7G		R. Cusco, o
5	OAX9A		R. Imagen, Tarapoto (nf 1500)
3/1	CV147	Uruguay	Em. Atlántica, Castillos
1/0.5	CW147		R. Tacuarí, Melo
3/0.5	CX147		R. Cristal del Uruguay, Las Piedras
10	YVJW	Venezuela	R. Latina, Valencia
10	YVSY		R. Vibración, Carúpano
1475 kHz			
	CP...	Bolivia	R. Tiraque, o
	CP..		Emisoras Verde y Blanco, Santa Cruz
1480 kHz			
0.25	LU27 AM	Argentina	R. Dolores, o Santa María, San Justo
0.1	CP..	Bolivia	Patrimonio R, Potosí
	CP..		R. Cristal, La Paz
1/0.25	ZYJ681	Brazil	R. Soc. Rondônia, Pimenta Bueno, RO
1/0.25	ZYH897		R. Itapecuru, Colinas, MA
1/0.25	ZYH919		R. Alvorada, Zé Doca, MA (ff 780)
10/ 0.25	ZYI911		R. Sete Cidades, Piracuruca, PI (ff 940)
1/0.25	ZYI929		R. Vale do Coroatá, Elesbão Veloso, PI
1/0.25	ZYH636		R. Cidade, Campos Sales, CE
1/0.25	ZYH...		R. Universal, Morrinhos, CE (CP)
1/0.25	ZYJ601		R. Princesa do Vale, Açu, RN
1/0.25	ZYI790		R. A Voz do Sertão, Serra Talhada, PE
1/0.25	ZYI...		JMB Empreendimentos ("R. Vale deo Ipojuca"), Gravatá, PE
1/0.25	ZYJ928		R. Atalaia, Simão Dias, SE
1/0.25	ZYH475		R. Bahiana de Itaberaba, BA (ff 1030)
5/1	ZYH...		R. Alvorada, Guanambi, BA
1/0.25	ZYH...		R. Santana, o, BA (CP)
1/0.25	ZYL235		Soc. R. Frutal, o, MG
1/0.25	ZYL307		R. Emboabas de Minas Gerais, Tiradentes, MG
1/0.25	ZYL...		R. Carangola, o, MG (CP)
1/0.25	ZYI210		R. Cidade, Afonso Cláudio, ES (ff 1300)
1/0.25	ZYJ488		R. Mauá Solimões, Nova Iguaçú, RJ
1/0.25	ZYJ...		R. Soc. Fraiburgo, Cordeiro, RJ (CP)
1/0.25	ZYK539		R. Cl. Altinópolis, o, SP
2.5/0.5	ZYK551		R. Técnica, Atibaia, SP
1/0.25	ZYK562		R. Cl. de Bilac, o, SP
1/0.25	ZYK647		R. Dif. de Santa Cruz do Rio Pardo, o, SP (ff 1180)
1/0.25	ZYJ221		R. Piraí do Sul, o, PR
1/0.25	ZYJ230		R. Astorga, o, PR
1/0.25	ZYJ270		R. Educ. União da Vitória, o, PR
1/0.25	ZYJ359		Radiodifusão Campo Aberto, Laranjeiras do Sul, PR
1/0.25	ZYJ731		R. Dif. de Joinville, o, SC
1/0.25	ZYJ762		R. Dif. 26 de Abril, Imaruí, SC
1/0.25	ZYJ826		R. Caibi, o, SC
5/0.25	ZYK321		R. Veranense, Veranópolis, RS
1/0.25	ZYI393		R. A Voz da Caçula, Três Lagoas, MS
1/0.25	ZYI399		R. Vale do Taquari, Coxim, MS (ff 970)
1/0.25	ZYH801		R. Pouso Alto, Pirancanjuba, GO (ff 1140)
1/0.25	ZYH795		R. Anhanguera, Miracema do Tocantins, TO
1/0.25	ZYK244		R. São Roque, Faxinal do Soturno (F.Pl.) (still on 1270)

1/0.25	CA148	Chile	R. Amanecer, Ovalle
1	CC148		R. Tomé, o
1	CD148		R. General Baquedaño, Valdivia
5	HJFC	Colombia	R. Sonorama, Pereira
1	HJKB		R. Guayabal, o
5	HJOD		R. Rodadero, Santa Marta
10	HJPX		La Voz de las Estrellas, Cartagena (nf 1470)
1	HJTC		R. Sonsón, o (nf 1490)
5	HJTZ		Antena 2 de RCN, Bucaramanga
1	HCMC1	Ecuador	R. Municipal, Cotacachi
5	HCBS3		Ondas del Jubones, Pasaje
3	HCJB4		La Voz de Jipijapa, o
5	HCWP5		R. Atlántida, Alausí
5	HCCY6		R. Popular de la Maná,
1	ZP20	Paraguay	R. América, Villeta (vf)
1	ZP23		R. Mariscal Francisco Solano López, Bella Vista
	ZP..		R. América, Ñembý
	O..1.	Peru	R. Frequencia Popular, Piura
0.6	OCX2C		R. San Pedro, Virú
1	O..4.		R. K'ler, Paramonga-Barranca
3/1.5	CW43B	Uruguay	R. Internacional, Rivera
	YV..		Venezuela R. Agrícola, Acarigua (CP)

1485 kHz

1	CP135	Bolivia	Radiodifusoras la Voz del Valle, Punata (nf 1500)

1490 kHz

1/0.25	LT42	Argentina	R. del Iberá, Mercedes
1	LV22		R. Huinca Renancó, o
1	CP172	Bolivia	R. San José, o
0.25	CP196		R. Moxos, San Ignacio de Moxos
0.35	CP198		R. Pedro Domingo Murillo, Quime
0.25	CP...		R. Mairana, o
1/0.25	ZYI...	Brazil	R. Lagoa, Buriti dos Lopes, PI (CP)
1/0.25	ZYI...	Brazil	R. Aliança, João Pessoa, PB (CP)
1/0.25	ZYH617		R. Cult. de Aracati, o, CE
1/0.25	ZYI680		R. Maringá, Pombal, PB (ff 880)
1/0.25	ZYI...		R. Aliança, João Pessoa, PB (CP)
5/1	ZYH246		Em. Rio São Francisco, Penedo, AL
1/0.25	ZYH478		R. Educ. de Ipiaú, o, BA
1/0.25	ZYH...		Radiodifusão Sertaneja, Brumado, BA (CP)
1/0.25	ZYH...		Rio São Francisco Radiodifusão, Bom Jesus da Lapa, BA (CP)
1/0.25	ZYL231		R. Cacique, Araguari, MG
1/0.25	ZYL274		R. Paraisópolis, o, MG
1/0.25	ZYL345		R. Cidade, Arcos, MG
1/0.25	ZYL353		R. Pirapetinga, o, MG
1/0.25	ZYL...		R. Matinada, Paraguaçu, MG (CP)
1/0.25	ZYK530		R. Dif. de Olímpia, o, SP
1/0.25	ZYK580		R. Nova Dracena, Dracena, SP
1/0.25	ZYK583		R. Educ. de Fernandópolis, o, SP
1/0.25	ZYK680		R. Cult. de Vargem Grande do Sul, o, SP
1/0.25	ZYK764		Sistema Mauá de Comunicação ("R. Mauá"), Mauá, SP
1/0.25	ZYK769		R. Icatu, Penápolis, SP
1/0.25	ZYJ210		R. Cornélio Procópio, o, PR
1/0.25	ZYK217		R. Legendária, Lapa, PR
1/0.25	ZYJ347		R. Dif. de São Jorge d'Oeste, o, PR
1/0.25	ZYJ353		R. Integração Oeste, Corbélia, PR (ff 1440)
1/0.25	ZYJ757		R. Sentinela do Alto Vale, Ibirama, SC
1/0.25	ZYJ791		R. Cult. de Xaxim, o, SC
1/0.25	ZYK208		Radiodifusão Assisense, São Francisco de Assis, RS
1/0.25	ZYK225		R. Liberdade, Canguçu, RS
1/0.25	ZYK309		R. Taquara, o, RS
1/0.25	ZYI404		R. Paiaguás, Glória de Dourados, MS
1	CA149	Chile	R. Alicanto, El Salvador
1	CB149		R. Chena, San Bernardo (ts)
5	CD149		R. Malleco, Victoria
5	HJAG	Colombia	R. Garzón, o
10	HJAY		Onda Nueva, Barranquilla (vf)
4	HJBS		Punto Cinco, Bogotá
1	HJJO		La Voz de San Marcos, o
15	HJYD		R. La Paz, Paipa
5	HJZB		La Voz de Los Robles, Tuluá
1	HCVY2	Ecuador	R. Dinámica, ("La R. Dinámica"), Guayaquil
5	HCAE4		R. Unión, Esmeraldas
5	HCSM5		R. Santa Maria, Azogues
0.5	HCAI6		R. Moderna, Pillaro
1	OAX1L	Peru	R. Imperio, Chiclayo
0.25	OAX5L		R. Abancay, o
0.5	OAX5N		R. Nazca, o
1.2	OAX6Q		R. Surperuana-Fidelidad, Arequipa
1	OAX8J		R. Atlántida, Iquitos
5/1	CX149	Uruguay	R. del Oeste, Nueva Helvecia
5/10	YVRP	Venezuela	R. Favoritas El Sol, Maracaibo
1	YVSQ		R. Mérida 1490, o
10	YVXD		Dinámica 14-90, Caracas

1495 kHz

1		Argentina	R. Mun. de General Conesa, o (nf 1500)
5	HCHG2	Ecuador	La Voz del Río Vinces, Vinces (nf 1500)

1500 kHz

0.25	LT22	Argentina	R. Nueva Era, Pehuajó
0.25	LT34		R. Nuclear, Zárate
1/0.25	LT45		R. San Javier, o
5/1	LV25		R. Unión, Bell Ville R. El Sol, Lanús AM 1500 R Chascomús, Chascomús
0.3	CP238	Bolivia	R. Sagrado Corazón, Mineros
1/0.25	ZYI542	Brazil	R. Floresta, Tucuruí, PA
1/0.25	ZYI919		R. A Voz do Longá, Esperantina, PI
1/0.25	ZYH615		R. Maciço, Baturité, CE
1/0.25	ZYJ...		R. Vale do Potengi, São Paulo do Potengi, RN (CP)
1/0.25	ZYI779		R. Pajeu de Educ. Popular, Afogados da Ingazeira, PE
1/0.25	ZYH...		R. Manguaba, Pilar, AL (CP)
5/1	ZYH487		R. Jacuípe, Riachão do Jacuípe, BA
5/0.25	ZYL215		R. Montanhesa, Viçosa, MG
1/0.25	ZYL340		R. Aparecida do Sul, Ilicínea, MG
1/0.25	ZYI216		R. Aquidaban, Cachoeiro do Itapemirim, ES (ff 960)
1/0.25	ZYJ494		R. Stereo Sul, Volta Redonda, RJ
1/0.25	ZYK549		R. Centenário, Araras, SP
1/0.25	ZYK626		R. Dif. de Pindamonhangaba, o, SP
1/0.25	ZYK706		R. Vale do Rio Grande, Miguelópolis, SP
1/0.25	ZYK...		R. Cumbica AM, Guarulhos, SP
1/0.25	ZYK...		R. Cidade, Apiaí, SP
1/0.25	ZYJ322		R. Flor do Café, Borrazópolis, PR (ff 540)
1/0.25	ZYJ360		R. A Voz do Sudoeste, Coronel Vivida, PR
1/0.25	ZYJ...		R. Colméia, Mandaguaçu, PR (CP)
1/0.25	ZYK365		R. Simpatia, Chapada, RS
10	CA150	Chile	R. Almirante Lynch, Iquique
1	CB150		R. Trasandina, Los Andes
1/0.25	CC150		R. Centenario, San Javier
1	CD150		R. Tierra de Fuego, Puerto Porvenir
10	HJLJ	Colombia	R. Mil 500, Cali
1	HJMP		La Voz de Aguachica, o
10	HJSH		R. Reloj, Moniquirá
10/5	HJTW		R. Sumapaz, Fusagasugá
10	HJUW		La Voz de la Feria, Manizales
5	HJZH		R. Disco, Medellín
2	HJ..		R. Cultural ("La Voz del Telembi"), Barbacoas
3	HCRO1	Ecuador	R. Otavalo, o
5	HCHG2		La Voz del Río Vinces, Vinces
10	HCUE3		R. Sucre, Machala
5	HCAD4		R. Satélite, El Carmen
2	HCWN5		R. Puruhá ("Canal 150"), Riobamba
1	OBX2X	Peru	R. JCL, Ascope
	O..2.		R. El Porvenir, o
10	OBX4I		R. Santa Rosa, Lima
1	OCX4F		R. Mundo, Huancayo (nf 1560)
0.5	OBX5J		R. San Cristóbal de Huamanga, Ayacucho
1	OBX6V		Nueva R. 2000, Orcopampa
1	OCX7F		R. Ayaviri, o (nf 1550)

2	OCX7P		R. Em. Frontera, Puno
20	YVPO	Venezuela	R. Nal. de Venezuela, Cabudare (rel 630)
10/5	YVRZ		R. Dos Mil, Cumaná

1505 kHz

1.5	HCOY7	Ecuador	Ondas del Río Yacuambí, Cantón Yacuambí

1506 kHz

	YV..	Venezuela	R. La Voz del Centro, Valencia (vf)

1510 kHz

1/0.25	LV21	Argentina	R. San Joaguín, Villa Dolores
0.25	LW4		R. Orán, San Ramón de la Nueva Orán R. Mau, Charata
5/1	CP1	Bolivia	R. Chuquisaca, La Paz
10	CP80		R. Litoral, Cochabamba
0.25	CP102		R. 27 de Diciembre, Villa Montes
1.5	CP124		R. Dif. Ichilo, San Carlos (nf 1350)
1/0.25	ZYI544	Brazil	R. Oriente de Redenção, Redenção, PA
1/0.25	ZYI894		R. Floriano, o, PI
1/0.25	ZYI896		R. Araucária, Corrente, PI
1/0.25	ZYI909		R. Grande de Picos, Picos, PI
1/0.25	ZYI925		R. Tapuio, Miguel Alves, PI (ff 650)
1/0.25	ZYH608		R. Planalto da Ibiapaba, São Benedito, CE
1/0.25	ZYJ602		R. Centenário, Caraúbas, RN
1/0.25	ZYI...		R. Dif. Vale do Rio Una, São Bento do Una, PE (CP)
1/0.25	ZYH476		R. Educ. Santana de Caetité, Caetité, BA (ff 1310)
1/0.25	ZYH493		R. Dif. do Descobrimento, Porto Seguro, BA
1/0.25	ZYL216		R. Soc. Passos, o, MG (ff 1230)
2.5/ 0.25	ZYL259		R. Manhumirim, o, MG
1/0.25	ZYL269		R. Oliveira, o, MG
1/0.25	ZYJ492		R. Teresópolis, o, RJ
1/0.25	ZYK654		R. Cacique, Santos, SP
1/0.25	ZYK665		R. Cl. de São Manoel, o, SP
1/0.25	ZYK...		R. Rural, Rinópolis, SP
1/0.25	ZYJ215		R. Arapongas, o, PR (ff 1240)
1/0.25	ZYJ216		R. Educ. de Venceslau Brás, o, PR
1/0.25	ZYJ239		R. Dif. Cultural, Irati, PR (ff 950)
1/0.25	ZYJ326		R. União, Céu Azul, PR
1/0.25	ZYJ362		R. Poema, Pitanga, PR
1/0.25	ZYJ764		R. Jaraguá, Jaraguá do Sul, SC (ff 1010)
2.5/ 0.25	ZYJ770		R. Dif. de Lages, o, SC (ff 1180)
1/0.25	ZYJ795		R. Centro Oeste, Pinhalzinho, SC
1/0.25	ZYK...		R. Educ. de São João de Urtiga, o, RS (CP)
1/0.25	ZYI391		R. Dif. Paranaibense, Paranaíba, MS (vf) (ff 1050)
1/0.25	ZYI396		R. Cidade, Maracaju, MS (ff 1030)
1/0.25	ZYH770		R. Goiatuba, o, GO
1/0.25	ZYH780		R. Soc. de Ceres, o, GO
1	CA151	Chile	R. Illapel, o
1	CC151		R. Rancagua, o
1	CD151		R. Teniente Merino, Lebú
1	HJHX	Colombia	La Voz Panamericana, Floridablanca (Bucaramanga)
5	HJUH		La Voz de Ciénaga, o
1	HJZA		LV del Centenario, Armenia
1	HJA22		La Voz de San Luis, San Luis de Gaceno
1	HCDC1	Ecuador	R. Iris, Quito
2	HD21OA		Instituto Oceanográfico de la Armada, Guayaquil
3	HCUC3		R. Unión Calvense, Cariamanga
0.5	HCRC5		La Voz de la Juventud, Cañar
1	HCRY6		R. Runacunapac Yachana Huasi, Simiátug
3	HCVL7		R. Ecos del Oriente, Lago Agrio
1	OCX2Q	Peru	R. Inca, Los Baños del Inca
0.3	OCX4B		R. Fessa, Cañete
3	OCX4J		R. Tarma, o
1	OAX5F		R. La Voz de Nasca, o
1	OBX7P		R. Red, Huánchac
	O..9.		R. Colonial, Nuevo Cajamarca
0.25	CX151	Uruguay	R. Rincón, Fray Bentos
1/0.75	CW57		R. San Carlos, o
0.25	CW151		R. Ibirapitá, San Gregorio de Polanco (vf)
20	YV..		Venezuela LV del Centro, Güigüe

1520 kHz

0.25	LT37	Argentina	Em. Rufino, o
0.25	LT38		R. Gualeguay, o
0.25	LV24		R. Manantiales, Tunuyán
			R. Güemes, Quilmes Oeste
1	CP179	Bolivia	R. El Espectador, Santa Cruz (vf)
0.25	CP207		La Voz de Cobre, Corocoro
1/0.25	ZYH899	Brazil	R. Ribamar, Pindaré- Mirim, MA
1/0.25	ZYH916		R. Karajás, Joáo Lisboa, MA
1/0.25	ZYH...		R. Educ. Vale do Munin, Chapadinha, MA (CP)
1/0.25	ZYH...		R. Araripe, Ipu, CE (CP)
1/0.25	ZYJ610		R. Alagamar ("R. Salinas"), Macau, RN
1/0.25	ZYI801		R. Educ. de Surubim, o, PE
1/0.25	ZYJ931		R. Imperatriz dos Campos, Tobias Barreto, SE
5/1	ZYH...		R. Bela Vista, Poções, BA (CP)
1/0.25	ZYL208		R. Correio da Serra, Barbacena, MG (ff 1230)
1/0.25	ZYL245		R. Cl. de Itaúna, o, MG
1/0.25	ZYL315		R. Pontal do Triângulo Mineiro, ("R. Iturama"), Iturama, MG (ff 770)
1/0.25	ZYL340		R. Aparecida do Sul, Ilicínea, MG (ff 1500)
10/0.5	ZYJ491		R. Continental, Rio de Janeiro (São João de Meriti), RJ
2.5/ 0.25	ZYK614		R. Diário de Moji, Moji das Cruzes, SP
1/0.25	ZYK627		Pinhal R. Cl., Espiri to Santo do Pinhal, SP
1/0.25	ZYK760		R. Manchester, Votorantim, SP
1/0.25	ZYJ218		R. Antoniense, Antonina, PR
1/0.25	ZYJ292		R. Cult. Palotinense, Palotina, PR
1/0.25	ZYJ340		R. Internacional, Quedas do Iguaçu, PR
1/0.25	ZYJ358		Soc. de Radiodifusão Padre Eduardo ("R. Guairacá"), Terra Rica, PR
1/0.25	ZYJ743		R. Caçanjurê, Caçador, SC (ff 1110)
1/0.25	ZYJ758		R. Dif. de Imbituba, o, SC
1/0.25	ZYJ806		R. Cult. de Timbó, o, SC
1/0.25	ZYK217		R. Vale do Jacuí, Cachoeira do Sul, RS
1/0.25	ZYK308		R. Tapense, Tapes, RS
1/0.25	ZYI405		R. Jornal, Amambaí, MS
1/0.25	ZYI...		R. Campo Alegre, Rio Verde de Mato Grosso, MS (CP)
1/0.25	ZYH772		R. Santalenense, Santa Helena de Goiás, GO
1/0.25	ZYH...		R. Cidade, Jaraguá, GO (ff 910)
1/0.25	ZYH...		R. Campos Belos, o, GO (CP)
1/0.25	ZYH797		R. Cristal, Cristalândia, TO (ff 900)
1/0.25	ZYJ499		R. Musical, Catagallo (ff 890)
1	CB152	Chile	R. Integración, Llolleo
1	CC152		R. Soberanía, Linares
0.1	CD152		R. Aníbal Pinto, Lautaro
1	HJAM	Colombia	R. Altamizal, Dolores
10	HJLI		R. Ecos del Palmar, Bogotá
1	HJLQ		R. Minuto, Barranquilla (vf)
5	HJMA		La Voz del Suroeste, Jericó
5	HJMZ		Ecos de la Sierra Flor, Sincelejo
15	HJRL		Antena de los Andes, Santa Rosa de Cabal
1	HCTI1	Ecuador	R. Ibarra, o
3	HCJJ2		R. Punto Verde, Quevedo
1	HCRN2		La Voz de Naranjal, o
2.5	HCRI5		La Voz de Guamote, o
1	OAX1C	Peru	R. Délcar, Chiclayo
1.2	OCX2G		R. Santiago, Santiago de Chuco
0.5	OAZ4D		R. Minería, La Oroya
5	OBX4D		R. Reloj, Lima (ti)
0.1	CV152	Uruguay	R. Guichón, o
2	CX152		R. Cerro Largo, Melo (vf)

10	YVND	Venezuela	R. Bonita ("R. Bonita La Guapa"), Guatire

1525 kHz

0.35	HCEB4	Ecuador	R. Manta, o (nf 1520)

1530 kHz

1		Argentina	R. Centro Morteros, o
0.1	LU25		R. Carhué, o
			La R. Gaucha, Glew R. Ciudad de San Martín, Villa Ballester
1	CP111	Bolivia	Radioemisora Ballivián, San Borja (r. 1540)
0.3	CP237		R. San Juan Bosco, Kami
1/0.25	ZYJ685	Brazil	R. Planalto, Vilhena, RO (ff 890)
1/0.25	ZYH604		R. Educ. Jaguaribana, Limoeiro do Norte, CE (ff 560)
1/0.25	ZYH...		R. das Tres Fronteiras, Campos Sales, CE (CP)
1/0.25	ZYJ603		R. Curimataú, Nova Cruz, RN
1/0.25	ZYI781		R. Bituri, Belo Jardim, PE
10/0.5	ZYH479		R. Cult. de Guanambi, o, BA
1/0.25	ZYL232		R. Cult. de Dores do Indaiá, o, MG
1/0.25	ZYL280		R. Cl. de Pouso Alegre, o, MG
1/0.25	ZYL...		R. Bom Sucesso, Minas Novas, MG (CP)
1/0.25	ZYJ502		R. Princesinha do Norte, Miracema, RJ
1/0.25	ZYK677		R. Brasil, Tupi Paulista, SP
1/0.25	ZYK698		R. Jornal, Nhandeara, SP (ff 570)
1/0.25	ZYK699		R. Notícias, Tatuí, SP
1/0.25	ZYK703		R. Notícias, Matão, SP (ff 1230)
1/0.25	ZYK755		R. Universal, Teodoro Sampaio, SP
1/0.25	ZYJ312		R. Cl. de Realeza, o, PR
1/0.25	ZYJ321		R. Regional, Cianorte, PR
5/0.5	ZYJ339		R. Dif. de Balsa Nova, o, PR
1/0.25	ZYJ...		R. Vale do Iguaçu, Verê, PR (CP)
1/0.25	ZYJ761		R. Dif. Itajaí, o, SC
1/0.25	ZYJ780		R. Dif. São Joaquim, o, SC
1/0.25	ZYJ796		R. Porto Feliz, Mondaí, SC
1/0.25	ZYK235		R. Sulina, Dom Pedrito, RS
1/0.25	ZYK300		R. São Leopoldo, o, RS
1/0.25	ZYK304		R. Tapejara, o, RS
1/0.25	ZYK...		R. Itaqui, o, RS
1/0.25	ZYI398		R. Uberlim, Fátima do Sul, MS
1	CA153	Chile	R. Juan Godoy, Copiapó
1	CB153		R. Nexo, Quillota
1	CC153		R. El Carbón, Lota
1	CD153		R. Calbuco, o
5	HJDN	Colombia	R. Total, Medellín
5	HJGD		Em. Reina de Colombia, Chiquinquirá
1	HJJB		R. Sevilla, o
5	HJMI		R. Auténtica, Melgar
5	HJOZ		La Voz de la Prov. de Padilla, San José del Cesar
0.5	HJ..		La Voz Comunal de la Sabana, Túquerres
5	HCJY1	Ecuador	R. Uno, La Concordia
5	HCMC2		La Voz de la Península, La Libertad (vf) (r. 1529.8)
5	HCCC5		Ondas Cañaris, Azogues
1	HCVP5		La Voz de Pallatanga, o
1	HCMZ6		La Voz del Dorado, Pelileo
	O..2.	Peru	R. Frecuencia San Ignacio, San Ignacio
3	O..2.		R. Oriental, Jaén
	O..3.		Antena Uno, Chimbote
1	OBZ4S		R. La Voz Quince Cincuenta ("La Nueva Voz del Centro"), Huancayo
0.5	OAZ7F		R. Confraternidad, Yauri
0.25	CW153	Uruguay	Em. Cono Sur, Nueva Palmira
1.2/ 0.25	CX50		R. Independencia, Montevideo (vf)
10	YVNP	Venezuela	R. San Felipe, o

1535 kHz

0.5	HCMH6	Ecuador	Organización Radiofónica del Cotopaxi ("R.O.R.C."), Latacunga (nf 1540)

1540 kHz

0.25	LT35	Argentina	R. Mon, Pergamino
0.25	LU28		R. Tuyú, General Madariaga
1/0.25	ZYI545	Brazil	R. Boa Vista, São Sebastião da Boa Vista, PA
1/0.25	ZYH...		R. Santa Marua, Lago da Pedra, MA (CP)
1/0.25	ZYH611		R. Sant'Ana, Tianguá, CE
1/0.25	ZYH631		R. Sertões, Mombaça, CE
1/0.25	ZYH...		R. Aratanha, Pacatuba, CE (CP)
1/0.25	ZYJ...		R. Baixa Verde, João Câmara, RN (CP)
1/0.25	ZYI694		R. Santa Maria, Monteiro, PB
1/0.25	ZYH...		R. Jornal, Souto Soares, BA (CP)
1/0.25	ZYL217		R. Dif. Bomdespachense, Bom Despacho, MG
1/0.25	ZYL226		R. Cl. de Minas Gerais, Conselheiro Lafaiete, MG
1/0.25	ZYL293		R. Tropical, Três Corações, MG
1/0.25	ZYL351		R. Dif. de Salinas, o, MG
1/0.25	ZYI206		R. Agricultura, Santa Teresa, ES
1/0.25	ZYJ...		R. Matozinho AM, Paraíba do Sul, RJ (CP)
1/0.25	ZYK514		R. Cult. de Leme, o, SP
2/0.25	ZYK564		R. Em. de Botucatu, o, SP
5/0.5	ZYK723		R. Dif. Oeste, Osasco, SP
1/0.25	ZYK727		R. Central, Pompéia, SP
1/0.25	ZYJ276		R. Bom Jesus, ("RBJ-O Som Lider Norte Pioneiro"), Siqueira Campos, PR (ff 1380)
1/0.25	ZYJ352		R. San Thiago Dantas, Chopinzinho, PR
1/0.25	ZYJ357		R. Central, Roncador PR
1/0.25	ZYJ803		R. Capinzal, o, SC
1/0.25	ZYK282		R. Quaraí, o, RS
1/0.25	ZYK297		R. Santiago, o, RS (ff 1230)
1/0.25	ZYK367		R. Querência, Santo Augusto, RS
1/0.25	ZYN601		R. Regional do Piraveve, Ivinhema, MS
1/0.25	ZYH760		R. Jornal, Inhumas, GO
1	CB154	Chile	R. Sudaméricana, Santiago
1	CC154		R. Central, Chillán
1	CD154		R. San José de Alcuida, Río Bueno
2	HJHD	Colombia	La Voz del Petróleo, Barrancabermeja
1	HJZF		R. Cóndor ("Em. Universitaria"), Manizales
1	HCPV1	Ecuador	R. Mira, o
3	HCFM2		R. Cristal, Ventanas
10	HCVB7		La Voz del Upano, Macas
1	OBX1B	Peru	LV de la Frontera, Tumbes
1	OBX2B		R. Star, Trujillo
1	OBX4B		R. Corporación, Cerro de Pasco
1	OBZ4U		R. Barranca, o
1	OCX7V		R. Los Andes ("Onda Tropical"), Cusco
0.1	CV154	Uruguay	R. Centro, Cardona
0.1	CW154		R. Charrúa, Paysandú
0.25	CX154		R. Patria, Treinta y Tres

1545 kHz

0.35	CP191	Bolivia	R. Mejillones, Tarata
1	HCDP1	Ecuador	R. Caracol, Quito (vf) (nf 1547.7)
0.5	OAX9X	Peru	R. Visión, Juanjuí (nf 1520)

1550 kHz

10/1	LT23	Argentina	R. San Jenaro Norte, o
0.25	LT32		R. Chivilcoy, o
0.25	LT40		R. La Voz de La Paz, o R. La Cruz del Sur, Guernica R. Contacto, San Antonio de Padua
10	CP50	Bolivia	R. Pio Doce, Siglo Veinte
5/1	CP115		Radioemisoras Caranavi, o
1	CP205		R. Tamengo, Puerto Quijarro (nf 1495)
1/0.25	ZYI550	Brazil	R. Cabano, Maracanã, PA
1/0.25	ZYH...		Sistema Janaina de Radiodifusão, Vargem Grande, MA (CP)
1/0.25	ZYH664		R. Macambira, Ipueiras, CE
1/0.25	ZYJ606		R. Gazeta d'Oeste, Areia Branca, RN

1/0.25	ZYI...		R. Bruxaxá, Areia, PB (CP)
1/0.25	ZYH...		Radiodifusão Eldorado, Pão de Açucar, AL (CP)
1/0.25	ZYH461		R. Regional de Conquista, Vitória da Conquista, BA
1/0.25	ZYH518		R. Independência do São Francisco, Juazeiro, BA
1/0.25	ZYH...		R. Cult. de Utinga, o, BA (CP)
1/0.25	ZYL211		R. Cult. de Monte Carmelo, o, MG
1/0.25	ZYL222		R. Dif. de Carmo do Rio Claro, o, MG
1/0.25	ZYL289		R. Dif. Santarritense, Santa Rita do Sapucaí, MG
1/0.25	ZYL317		R. Pirapora, o, MG (ff 1240)
1/0.25	ZYJ479		R. Imperial, Petrópolis, RJ
1/0.25	ZYK501		R. Cl. de Itararé, o, SP
1/0.25	ZYK572		R. Cacique, Capivari, SP
5/0.25	ZYK590		R. Guarujá Paulista, Guarujá, SP
1/0.25	ZYK623		R. Cult. de Pederneiras, o, SP
1/0.25	ZYK659		R. São Joaquim, São Joaquim da Barra, SP
1/0.25	ZYK740		R. Auriflama, o, SP
1/0.25	ZYJ213		R. Ipiranga, Palmeira, PR
1/0.25	ZYJ303		R. Pioneira de Formosa, Formosa do Oeste, PR
1/0.25	ZYJ315		R. Cristal, Marmeleiro, PR
1/0.25	ZYJ354		R. Placar, Ortigueira, PR (ff 740)
1/0.25	ZYJ355		R. Cidade, Palmital, PR
1/0.25	ZYJ363		R. Colméia, Mandaguaçu
1/0.25	ZYJ814		R. Imigrantes, Turvo, SC
1/0.25	ZYK259		R. Cacique, Lagoa Vermelha, RS (ff 630)
1/0.25	ZYK...		R. Soc. Zona Sul, Capão do Leão, RS
1/0.25	ZYH799		R. Serra Azul, Caiapônia, GO
1	CB155	Chile	R. Provincial, Putaendo
1	CC155		R. Manuel Rodríguez, San Fernando
0.25	CD155		R. Regional, Traiguén
5	HJCB	Colombia	R. El Sol, Barranquilla
5	HJGY		R. Foro Popular, Lorica
5	HJLT		R. 24, Cali
5	HJQD		R. Reloj, Armenia
1	HJUN		La Voz de Aguadas, o
10	HJZI		R. Fantasía, Bogotá
3	HCAD2	Ecuador	La Voz de El Triunfo, o
1	HCAB5		La Voz de Chaguarurco, Santa Isabel
2	HCEI6		R. Montalvo, Ambato
1	HCRA7		R. Amazonas, Archidona
1	OAX10	Peru	Radioemisora Superior, Monsefú
1	OBX4P		R. Independencia, Lima
1	OCX7A		R. Cult. de Puno, o
1/0.25	CV155	Uruguay	R. Agraciada, Mercedes
0.25	CW155		R. Sarandí del Yi, o
20/5	YVMW	Venezuela	R. Metropolitana, Los Teques
10/5	YVXO		R. Ojeda, Ciudad Ojeda

1559 kHz

		Argentina	R. Libre, Berazategui

1560 kHz

5/1	LT11		R. General Francisco Ramirez, Concepción del Uruguay
0.25	LT33		R. Nueve de Julio, o R. Ciudad de Libertad, o
0.5	CP156	Bolivia	1° de Octubre "la Radio", Capinota
	CP...		R. Occidental, Oruro
1/0.25	ZYH902	Brazil	R. Agua Branca, Vitorino Freire, MA
1/0.25	ZYH622		R. Dif. Vale do Curu, Pentecoste, CE
1/0.25	ZYJ608		R. Cult. do Oeste, Pau dos Ferros, RN
1/0.25	ZYI686		R. Cenecista de Picuí, Picuí, PB
1/0.25	ZYL256		R. Jornal, Leopoldina, MG
5/0.25	ZYL367		R. Metropolitana, Vespasiano, MG (ff 1240)
1/0.25	ZYI208		R. Gaeta, Guarapari, ES (ff 1450)
5/0.25	ZYJ501		R. Comunicadora Grande Rio, ("R. Grande Rio"), Itaguaí, RJ
1/0.25	ZYK571		R. Em. de Campos do Jordão, o, SP
1/0.25	ZYK679		R. Valparaíso, o, SP
1/0.25	ZYK725		R. Cidade, Pedreira, SP
1/0.25	ZYK...		R. Vale do Rio Paraná, Pres. Epitácio, SP (CP)
1/0.25	ZYJ275		R. Capanema, o, PR
1/0.25	ZYJ325		R. Brotense, Porecatu, PR (ff 1210)
1/0.25	ZYJ797		R. Educ. de Taió, o, SC
1/0.25	ZYJ825		R. Cidade, São Miguel d'Oeste, SC
2.5/0.25	ZYK310		R. Açoriana, Taquari, RS
1/0.25	ZYK...		R. Globo, Ijuí, RS (CP)
1/0.25	ZYH768		R. Sudoeste AM, Rio Verde, GO (ff 900)
5/3	CA156	Chile	R. Parinacota, Putre (vf)
1	CB156		R. Manantial, Talagante
1	CD156		R. Turismo Internacional, Villarica
5	HJCP	Colombia	Antena 2 de RCN, Arbeláez
5/4	HJHE		Voces Rovirences, Málaga
5	HJLP		Antena 2 de RCN, Tuluá
1	HJPZ		R. Codazzi, o
1	HJUJ		Armonías Boyacenses, Motavita
5	HJXZ		Colmundo R, El Poblado
1	HCHG1	Ecuador	R. La Voz de El Dorado, Puembo
1.5	HCZD1		Ecos Culturales, Urcuquí
2	HCCS2		R. Sideral, Daule
2	HCTR3		La Voz del Guabo, El Guabo
5	HCVI5		R. La Voz de Ingapirca, Cañar
1.5	OBX2J	Peru	R. Nuevo Continente, Cajamarca
	O..7.		R. Municipal de Cusco, o
1/0.25	CV156	Uruguay	R. Vichadero, o
2	CW51		R. Maldonado, o
0.25	CX156		Dif. Americana, Trinidad
10/5	YVLZ	Venezuela	Radiodif. Andina ("Exitos 15-60"), Mérida

1562 kHz

1	OAX2X	Peru	R. Primavera, Trujillo

1570 kHz

1	LRI229	Argentina	R. Las Flores, o
10/0.25	CP132	Bolivia	R. Continental, La Paz
0.3	CP...		R. Urkupiña, Quillacollo
	CP...		R. Primero de Mayo, Santa Cruz de la Sierra
1/0.25	ZYH621	Brazil	R. Sertão Central, Senador Pompeu, CE
1/0.25	ZYI798		R. Asa Branca, Salgueiro, PE
1/0.25	ZYH496		R. Vale Aprazivel, Jaguaquara, BA
1/0.25	ZYL203		Nova R. Cult. de Alfenas, o, MG
1/0.25	ZYL210		R. Difusora, Itapecerica, MG
1/0.25	ZYL364		R. Difusora, Piranga, MG
1/0.25	ZYL344		R. Cidade, Corinto, MG
1/0.25	ZYL346		R. Dif. Industrial, Nova Serrana, MG
1/0.25	ZYJ493		R. Cult. de Valença, o, RJ
1/0.25	ZYJ498		R. Bom Jesus, Bom Jesus de Itabapoana, RJ
1/0.25	ZYK552		R. Avaré, o, SP
1/0.25	ZYK605		R. Junqueirópolis, o, SP
1/0.25	ZYK651		R. Em. A.B.C., Santo André, SP
1/0.25	ZYK667		R. Nossa Senhora, Socorro, SP
1/0.25	ZYK670		R. Cl. de Tanabi, o, SP
1/0.25	ZYJ223		R. Atalaia, Guarapuava, PR
1/0.25	ZYJ288		R. Educ. de Dois Vizinhos, o, PR (ff 1300)
1/0.25	ZYJ324		R. Brasileira, Bela Vista do Paraíso, PR
1/0.25	ZYJ341		R. Cl. de Nova Aurora, o, PR
5/0.5	ZYJ351		R. Foz, Foz do Iguaçu, PR
1/0.25	ZYJ...		R. Arapoti, o, PR (CP)
1/0.25	ZYJ...		R. Modelo, o, SC (CP)
1/0.25	ZYJ777		R. Rio Negrinho, o, SC
1/0.25	ZYJ...		R. Modelo, o, SC (CP)
1/0.25	ZYK358		R. Vale do Gravataí, Gravataí, RS
1/0.25	ZYI408		R. Patriarca, Cassilândia, MS
1/0.25	ZYI409		R. Caarapó, o, MS
1/0.25	ZYI418		R. Cidade, Aparecida do Taboado, MS (ff 1140)
1/0.25	ZYH784		R. Itaberaí, o, GO
1/0.25	ZYH787		R. Mantiqueira, Niquelândia, GO
1	CC157	Chile	R. Sonorama, Talca
1	CC157		Nueva R. Libertador Bernardo O'Higgins, Rancagua (vf)
0.25	CD157		R. Acuarela, Nueva Imperial
1	HJTG	Colombia	R. Camino, Machetá

1	HJZT		La Voz de los Fundadores, Manizales
10	HCGP1	Ecuador	R. Ñucanchic, P.V. Maldonado (ti)
0.5	HCAI5		R. Ecos de los Andes, Chunchi
0.25	CW157	Uruguay	R. Young, o
1	CX157		R. Canelones, o
10/5	YVUZ	Venezuela	R. Guarapiche ("RG 15-70"), Maturín
1580 kHz			
0.25	LT27	Argentina	R. LV del Montiel, Villaguay
0.25	LT36		R. Chacabuco, o R. General Martín Miguel de Güemes, Guernica
0.8	CP166	Bolivia	R. Llallagua, o
1/0.25	ZYI898	Brazil	R. Irapuá, Floriano, PI
1/0.25	ZYJ613		R. Novos Tempos, Ceará Mirim, RN
1/0.25	ZYH464		R. Em. Radiovox, Muritiba, BA
1/0.25	ZYH497		R. Barra do Mendes, o, BA
1/0.25	ZYH...		R. Atalaia, Canavieiras, BA (CP)
1/0.25	ZYL284		R. Paranaíba, Rio Paranaíba, MG
1/0.25	ZYL290		R. Cult. de Santos Dumont, o, MG
1/0.25	ZYL329		R. Educ. de Espinosa, o, MG
1/0.25	ZYL335		R. Rural Nova Guaranésia, Guaranésia, MG
1/0.25	ZYI209		R. Educ. de Afonso Cláudio, o, ES
1/0.25	ZYJ487		R. Popular Fluminense, Conceição de Marabu, RJ
1/0.25	ZYK504		R. Dif. de Amparo, o, SP
1/0.25	ZYK743		R. Trans Universal, Itaporanga, SP
1/0.25	ZYJ293		R. Najuá, Irati, PR (ff 1450)
1/0.25	ZYJ818		R. Pomerode, o, SC
1/0.25	ZYK207		R. Itapuí, Santo Antonio da Patrulha, RS (ff 1170)
1/0.25	ZYK237		R. Encantado, o, RS
1/0.25	ZYK339		R. Dif. Fronteira, Arroio Grande, RS
1/0.25	ZYK...		R. Dif. das Missões, Palmeira das Missões, RS (CP)
1/0.25	ZYI415		R. Laguna, Jardim, MS
1/0.25	ZYI...		R. Dif. de Ivinhema, o, MS (CP)
1/0.25	ZYH775		R. Rio Claro, Iporá, GO
1	CC158	Chile	R. Colchagua, Santa Cruz
0.25	CD158A		R. Millaray, Cañete
0.5	CD158B		R. Continental, Collipulli
5	HJKD	Colombia	R. Zulima, Villa del Rosario
2.5	HJLC		La Voz del Banco, El Banco
1	HJOE		R. Miraflores, Rovira
1	HJQT		R. Cachaca, Bogotá
1	HJQZ		La Voz de las Estrellas, Barranquilla
10	HJRM		R. Costanera, Sincelejo
5	HJSQ		R. Robledo, Cartago
10	HCLF1	Ecuador	R. Orellana, Machachi
0.5	HCCP2		Canal del Pueblo, Sanborondón
0.25	HCAB3		Ondas de Paltas, Catacocha
5	HCVA4		La Estación de la Alegría, Esmeraldas
3	HCAE5		Ecos del Portete, Girń
1.5	HC..6		LV del Libertador, Chimbo (n. 1600)
1	OBX1M	Peru	R. Naylamp, Lambayeque (also r. 1590)
1	OBX2H		R. Tayabamba, o
2	OAX5M		R. Villa Rica, Huancavelica
	O..5.		R. Cultural Amauta, Huanta
2	OBX7Q		R. El Triunfo, Cusco (relay RPP 730)
0.5	CW54	Uruguay	Em. del Este, Minas (vf)
0.25	CW158		R. San Salvador, Dolores
10	YVTK	Venezuela	R. Manzanares ("Manzanares 15-80"), Cumaná
10	YVYO		R. Occidental, Santa Rita
10	YVYV	Venezuela	R. Dif. Venezolana ("La Em. de la Diócesis de Calabozo"), Calabozo (ff 1270)
1582 kHz			
	O..4.	Peru	R. R & R, Villa Maria del Triunfo
1590 kHz			
0.25	CP155	Bolivia	R. Bermejo, o
1/0.25	ZYH600	Brazil	R. Araripe, Cedro, CE
1/0.25	ZYJ614		R. Santa Cruz AM, o, RN (ff 1390)
1/0.25	ZYH489		R. Bahia Nordeste, Paulo Afonso, BA (ff 950)
5/0.25	ZYH501		R. Canto da Sereia, Jacobina, BA (ff 1310)
1/0.25	ZYL234		R. Cl. de Fronteira, o, MG (ff 1170)
1/0.25	ZYL368		Rede Dinâmica de Comunicação, Ubá, MG (CP)
10/1	ZYL...		Rima Comunicações, Várzea da Palma, MG (CP)
1/0.25	ZYJ296		R. Havaí, Capitão Leônidas Marques, PR
110.25	ZYJ485		R. Dif. Duque de Caxias, o, RJ
1/0.25	ZYJ506		R. Resende AM, o, RJ
1/0.25	ZYK762		R. Realidade AM, São Carlos, SP
1/0.25	ZYJ290		R. Cult. de Andirá, o, PR
1/0.25	ZYJ316		R. Cl. de Faxinal, o, PR
1/0.25	ZYK212		R. Cl. de Bagé, o, RS
1/0.25	ZYK281		R. Navegantes, Pôrto Lucena, RS
1/0.25	ZYI403		R. Independência, Eldorado, MS
1	CB159	Chile	R. Aconcagua, San Felipe
0.25	CC159		R. Rengo, o (vf)
0.1	CC159		R. El Roble, Parral
5	CD159		R. Llanquihue, o
5	HJIP	Colombia	R. Metropolitana, Envigado (Medellín)
1	HJQM		Ecos de la Miel, Samaná
5	HJWB		R. Ntra. Señora del Socorro, El Socorro
1	HCRZ1	Ecuador	R. Mensaje, Tabacundo
0.25	HCAS2		R. Record, La Libertad
1	HCCS6		La Voz de La Maná, o
1	HCQT6		R. Panamericana, Quero
1	OBX1M	Peru	R. Naylamp, Lambayeque (nf 1580)
5	OAZ4Z		R. Agricultura del Perú ("La perunísima"), Lima
1	OBX5C		R. Esmeralda, Huanta
0.25	CV159	Uruguay	R. Regional, Constitución
0.25	CW159		R. Regional, Lascano
0.1	CX159A		R. Paso de los Toros, o
10	YV..	Venezuela	R. Deporte, Caracas
1591 kHz			
2.5	CX159	Uruguay	R. Real, Colonia (nf 1590)
1598 kHz			
	O..2.	Peru	R. Luz de la Vida, Cajamarca
1600 kHz			
1.5	LRI2	Argentina	R. Lago Aluminé, Ciudad Aluminé
			R. Esperanza, General Viamonte
0.5	CP153	Bolivia	R. Continental, Punata
	ZYJ...	Brazil	R. Planalto, Ariquemes, RO
0.25	CB160A	Chile	R. Dos Mil, Puente Alto
0.25	CB160B		R. Reloj, Viña del Mar (ti)
1	CC160		R. Regional, Concepción (vf)
1	HJHB	Colombia	Armonías Zipaquireñas, Zipaquirá
5	HJSP		R. Saratama, Pereira
5	HCUP1	Ecuador	R. Unión, Quito (ti)
3	HCPB5		R. Intiñán, Girón
1	OBX2Q	Peru	R. Popular, Chepén
	O..6.		R. Camaná, o
5	CV160	Uruguay	Em. Continental, Pando
0.25	CW160		R. Tacuarembó, o (vf)
0.5	CX160		R. Litoral, Fray Bentos
1602 kHz			
1	OBX9B	Peru	R. Frecuencia Popular, Rioja (nf 1460)
1605 kHz			
1	HCJP2	Ecuador	R. Consular, Playas (nf 1600)

SHORTWAVE STATIONS OF THE WORLD

November 15th 1992

Country Codes: The standard country codes adopted by the International Telecommunication Union (ITU) are used in this Table and throughout the book. These are listed in alphabetical order in the front section of the book, and are also given under each individual country heading in the World Radio section. When space permits, the country name is also mentioned after the station name wheneve this may not be obvious.
Relay Stations are listed under the code for the country where they are actually located.
Transmitter Sites and Powers: Where a station has more than one transmitter site and/or power output, this information is given for each frequency wherever possible. However, due to schedule changes made during the currency of this edition, such information should not be regarded as definitive.
CIS: The frequencies used by Radio Moscow External Services are not included in this Table due to space restrictions. An official site-by-site listing of these frequencies, as well as a complete schedule of Radio Moscow's foreign language broadcasts, may be found in the World Radio section under Russia.
Abbreviations: * = inactive, r = reported, v = variable frequency, irr. = irregular, alt. = alternative frequency, F.Pl. = Future Plans, w = winter frequency.

kHz	kW	Land	Station
2307	5	INS	RRI Jakarta (Prgr. Nasional)
2310	50	AUS	VL8A, Alice Springs, Australia
2325	50		VL8T, Tennant Creek, Australia
2340		CHN	Fujian 1, Fuzhou, China
2350	1	INS	RRI Yogyakarta, Indonesia
		KRE	Sariwon Reg. St, No. Korea
2360	1	GTM	R. Maya, Barillas, Guatemala
2377	1/0.5	INS	RRI Surabaya, Indonesia
2380	0.25	B	R. Educadora, Limeira, Brazil
2389.9	1	GTM	R. La Voz de Atitlán, Guatemala
2390	1	INS	RRI Cirebon, Indonesia
	0.5	MEX	R. Huayacocotla, México
2410.1	5	B	R. Transamazônica, S. Guiomard
	10/2	PNG	R. Enga, Papua New Guinea
2415	1	CHN	Wenzhou 1, China
2420	0.5	B	R. Sao Carlos, Brazil
2433	1	INS	RRI Banda Aceh, Indonesia
	1		RRI Palembang, Indonesia *
2445		CHN	Jiangxi 1, Nanchang, China
2460	1	B	R. Progresso do Acre, Brazil
		CHN	Yunnan G, Kunming, China
2473	1	INS	RRI Purwokerto, Indonesia *
2475	20	CHN	Hangzhou, China
2485	50	AUS	VL8K, Katherine, NT, Australia
2490	0.25	B	R. Oite de Setembro, Descavaldo
v	0.5	INS	RRI Ujung Padang, Indonesia
2495	100	MDG	R. Madagasikara, Madagscar
2500		AUS	VNG, Llandilo (Time Signals)
	5	HWA	WWVH, Hawaii (Time Signals)
	2	J	JJY, Japan (Time Signals)
	2.5	USA	WWV (Time Signals), Ft. Collins
	1	UZB	RCH, Tashkent (Time Signals)
2552	0.175	INS	RRI Wamena, Indonesia,
2560	15	CHN	Urumqi, China (Chinese prgr)
2582	0.3	INS	RPDT2 Timor Tengah Selatan *
2694.9	0.5		RPDK Ende, Indonesia
2850	120	KRE	KCBS, Pyongyang, No. Korea
2857		INS	RPDK Bolaang Mongodow, Indon.
2904	0.5		RPDT2 Ngada, Bajawa, Indonesia
2960	0.3		RKPDT2 Manggarai, Ruteng
3000.0v			RPDT2 Bengkalis, Indonesia
3109			RPDT2 Lombok Timur, Selong *
3142.9	0.3		RPKDT2 Belitung, Indonesia
3159	0.03		R. P. H. I. Tembilahan, Indonesia
3187			RKPDT2 Bima, Indonesia
3200		CHN	Haixia 2, China
	25	SWZ	Trans World Radio, Swaziland
3200.3v	0.5	BOL	R.9 de Abril, Bolivia
3204	10	INS	RRI Bandung, Indonesia
3204.6	1	PRU	R. Paramonga, Barranca, Peru
3205	1	B	R. Riberao Preto, Brazil
	5		R. Vale do Rio Madeira, Humaitá
		DOM	R. Olímpica Int. (F. Pl.)
	50	IND	All India Radio, Itanagar
	50		All India Radio, Lucknow
	10	PNG	R. West Sepik, Vanimo
3210	1	CHN	Xiahe, China * (alt. freq)
3210.5v	100	MOZ	Em. Nacional, Maputo
3215		INS	RRI Manado, Indonesia
3220	50	CHN	CPBS 1, China (wi.)
	10	EQA	HCJB, Quito, Ecuador
v		KRE	Hamhung Reg. St, No. Korea
	10	PNG	R. Morobe, Lae, Papua N. G.
3222.0v	10	TGO	Radio Kara, Togo
3223	2.5	IND	All India Radio, Shimla
	5	INS	RRI Mataram, Indonesia
3225	10	RRI	Tanjung Pinang, Indonesia
3230	100	AFS	R. Oranje, Johannesburg
	10	LBR	ELWA, Monrovia, Liberia *
3230.1	100/5	NPL	R. Nepal, Khumaltar
3230.2	0.4	PRU	R. El Sol de los Andes, Peru
3232	10	INS	RRI Bukittinggi, Indonesia
v	10	MDG	R. TV Malagasy "Frequence 2"
3235	0.5	B	R. Clube de Marília, Brazil
	20	IND	All India Radio, Guwahati B
	2	PNG	R. West New Britain, Kimbe
3240	25	SWZ	Trans World Radio, Swaziland
3240.2	1	EQA	R. Antena Libre, Ecuador
3241	1	INS	RRI Ambon, Indonesia *
3244.9	1	VEN	R. Libertador, Venezeula *
3245	1	B	R. Clube de Varginha, Brazil
	50	IND	All India Radio, Itanagar
	20	INS	RRI Banjarmasin, Indonesia
	10	PNG	R. Gulf, Kerema, Papua N. G
3249.5v	1	HND	R. Luz y Vida, Honduras
3250	120	KRE	Pyongyang BC St, No. Korea
	400/100		R. Pyongyang Ext. Sce, No. Korea
	1	PRU	R. Qollasuyo, Juliaca, Peru
3255	1	B	R. Educadora Cariri, Crato
	5		R. Transamazônica, Brazil (F. Pl.)
	50	IND	All India Radio, Shillong
		LBR	"ELBC", Gbarnga, Liberia
	250	LSO	BBC Lesotho Relay St.
3259.7v	1	INS	RRI Kupang, Indonesia
3260	10	CHN	Guizhou 1, Guiyang, China
	4	NGR	La Voix du Sahel, Niger
	10	PNG	R. Madang, Papua New Guinea
	1	PRU	R. La Voz de Oxapampa, Peru
3265	25	COG	RTV Congolaise, Brazzaville
	10	INS	RRI Bengkulu, Indonesia
	25	MOZ	R. Moçambique Ext. Sce.

3267.0v	10	INS	RRI Gorontalo, Indonesia
3268	50	IND	All India Radio, Kohima A
3270	1	EQA	R. Ecos del Oriente, Ecuador
	100	NMB	Namibian Broadc. Corp. Ch.1
3274.9v	1	VEN	R. Mara, Venezuela (irr.)
3275	10	PNG	R. Southern Highlands, Mendi
3277	50	IND	All India Radio, Srinagar B
	1	INS	RRI Jakarta, Indonesia
3279.8	2.5	EQA	LV del Napo, Tena, Ecuador
3280	1	BOL	R. Em. Chaco, Bolivia *
		CHN	V. O. Pujiang, Shanghai, China
3280.2	1	PRU	Estación Wari, Ayacucho, Peru
3281	100	MOZ	Beira 2, Mozambique
3285	1	B	R. Sentinela de Amazônia
	1	BLZ	Belize Broadc. Network *
3285.3	0.35	EQA	LV del Río Tarqui, Ecuador
3286	1	INS	RRI Madiun, Indonesia
	100	MDG	R. Madagasikara, Madagascar
3289.9	0.5	EQA	R. Centro, Ambato, Ecuador
3290	100	NMB	Namibian Broadc. Corp. Ch.2
	10	PNG	R. Central, Boroko
	0.04	TDC	Tristan (da Cunha) Broadc. Sce.
3295	0.5	INS	RPDT2 Kotamadaya Pematangs. *
3300	10	GTM	R. Cultural, Guatemala
3305	2	IND	All India Radio, Ranchi
		INS	RRI Bukittinggi, Indonesia
	10	PNG	R. Western, Daru
3306	10	INS	RRI Dili, Indonesia
3310	1	PRU	R. Bagua, Peru *
3310.3	1	BOL	R. San Miguel, Bolivia
3315	10	IND	All India Radio, Bhopal A
	2	PNG	R. Manus, Lorengau
3315.1	2	EQA	R. Pastaza, El Puyo, Ecuador *
3316	10	SRL	SLBS, Goderich, Sierra Leone
3320	120	KRE	Pyongyang BC St, No. Korea
3322	1	EQA	R. Dif. Sangay, Macas, Ecuador
3324.8	1	GTM	R. Maya, Barillas, Guatemala
3325	5	B	R. Liberal, Belém, Brazil
	2.5		R. Tupí, Sao Paulo, Brazil
	1.5	EQA	Ondas Quevedeñas, Ecuador
	10	INS	RRI Palangkaraya, Indonesia
	10	PNG	R. North Solomons, Kieta/Rabaul
3326	50	NIG	R. Nigeria Ch.1, Lagos
3330	0.3	BOL	R. Charcas, Bolivia *
	3	CAN	CHU (Time Signals), Canada
	60	COM	R. Comoro, Moroni
v	20	RRW	R. Rwanda, Kigali
3330.3v	5	PRU	Ondas del Huallaga, Peru
3332	3	EQA	R. Cumandá, Coca, Ecuador
3335	5	B	R. Alvorada, Londrina, Brazil
	10	PNG	R. East Sepik, Wewak
	10	TWN	CBS Netw. 5/1, Taipei, Taiwan
3338.2v	10	MOZ	Em. Interprovincial, Maputo
3339.1v	1	PRU	R. Altura, Cerro de Pasco, Peru
3340.2	1	BOL	R. Viloco, Bolivia (n.3350)
3345.0v	10	AGL	EP do Huambo, Angola *
	50	IND	All India Radio, Jaipur A
	2		All India Radio, Jammu A
	10	INS	RRI Ternate, Indonesia
	1	PHL	DZB-2 Mindoro, Philippines
	2/10	PNG	R. Northern, Popondetta
3350		KRE	P'yongsong Reg. St, No. Korea
3350.1	1	BOL	R.27 de Diciembre, Bolivia
3355.0v	10	AGL	R. Nacional, Cazenga, Angola
	10	IND	All India Radio, Kurseong
	1	INS	RRI Jambi, Indonesia
	1		RRI Sumenep, Indonesia
	10	PNG	R. Simbu, Kundiawa
3356	50	BOT	R. Botswana, Gaborone
3360	5	EQA	R. Federación Shuar, Ecuador *
3360.4	1	GTM	LV de Nahualá, Guatemala
3365	1	B	R. Cultura, Araraquara, Brazil
v	0.5	CUB	R. Rebelde, Bauta, Cuba
	10	IND	All India Radio, Delhi
	20		All India Radio, Khampur
	10	PNG	R. Milne Bay, Alotau
	25	SWZ	Trans World Radio, Swaziland
3366	50/10	GHA	GBC-2, Accra, Ghana
3369.9	1	GTM	R. Tezulutlán, Cobán, Guatemala
3370	0.25	EQA	R. Nacional Limón, Ecuador
3370.5	1.2	BOL	R. Florida, Samaipata, Bolivia
3375.0v	10	AGL	R. Nacional, Mulenvos, Angola
v	5	B	R. Clube, Dourados, Brazil
	5		R. Educadora, Guajará Mirim
v	5		R. Nacional, Sao Gabriel
	50	IND	All India Radio, Guwahati A
	20/7.5	INS	RRI Medan, Indonesia
	10	PNG	R. Western Highlands
3376	1	BOL	R. Nuevo Mundo, Sucre, Bolivia*
3377.5	0.3	J	NHK Osaka, Japan (irr.)
3380	1	GTM	R. Chortis, Jocotán, Guatemala
3380.3	1	BOL	R. Cumbre, Tazna, Bolivia *
3380.6v	100	MWI	MBC, Blantyre, Malawi
3381	1	INS	RRI Malang, Indonesia
3384.9	10	PNG	R. East New Britain, Rabaul
3385	5	B	R. Educacao Rural, Tefé, Brazil
	10	INS	RRI Kupang, Indonesia
	10	MLA	RTM Miri, Sarawak, Malaysia
3390.2	1	ZAI	R. CANDIP, Bunia, Zaire
3390.3	1	BOL	R. Emisoras Camargo, Bolivia
3394.8	10	EQA	R. Católica, Ecuador
3395	10	INS	RRI Tanjung Karang, Indonesia
	2	PNG	R. Eastern Highlands, Goroka
3401.0v	2	B	R. Educadora 6 de Agosto
3445	1	INS	RRI Pontianak, Indonesia
3460			RPDT2 Aceh Timur, Indonesia *
3464.0v		PRU	R. Reina de la Selva, Peru
3471.5v	0.5	BOL	R. Dif. Alfonso Padilla, Bolivia
3488	0.04	INS	RPDT2 Tapanuli Selatan, Indon.
3500	5	CLM	LV del Guainía, Colombia
3502.1		PRU	R. Chilía, Chilía, Peru
3568.7v	1.5	B	R. Educadora 3 de Julho, Brazil
3578	0.35	INS	RPDT2 Poso, Indonesia
3607.5	0.9	J	NHK Shobu-Kuki, Tokyo (irr.)
3644.8v	0.5	INS	RRI Fak-Fak, Indonesia
3654	0.2		RPDT2 Buol, Indonesia
3660	1	PAK	Azad Kashmir R, Muzaffarabad
3754.3v		BOL	R.20 de Septiembre, Bolivia
3800.3	1	PRU	R. Oyón, Peru
3805.0rv		CLM	R. Cultural, Colombia
3810	1	EQA	HD2IOA (Time Signals), Ecuador
v		PRU	R. Centenario, Peru *
3815		CHN	CPBS Taiwan 1 Sce, China *
3900			Hailar, China (Chinese prgr)
			Haixia 1, China
3904.9	2	PNG	R. New Ireland, Kavieng
3905	50	IND	All India Radio, Kingsway
		INS	RRI Banda Aceh, Indonesia
	0.5		RRI Merauke, Indonesia
3910		CHN	Yunnan G, Kunming, China *
3915	100	SNG	BBC Far Eastern Relay St.
3920.0v		KRE	Hyesan Reg. St, No. Korea
3925	20	IND	All India Radio, Khampur
	50/10	J	R. Tanpa 1, Tokyo / Sapporo
	10	PNG	NBC Port Moresby, Papua N. G.
3930.0v	10	AFS	Capital Radio, Transkei
	50	CHN	Hohhot, China (Mongolian prgr)
	10	KOR	KBS Prgr.1, Hwasong, So. Korea
3934	5/10	INS	RRI Semarang, Indonesia
3935	1	NZL	ZLXA - Print Disabled R, Levin
3940	10	CHN	Hubei 1, Wuhan, China
	2	HKG	R. TV Hong Kong (irr.)
3945		IND	All India Radio Ext. Sce.
	50		All India Radio, Gorakhpur
	1	INS	RRI Denpasar, Indonesia
	10	J	R. Tanpa 2, Tokyo, Japan
	10	VUT	R. Vanuatu, Malapoa
3946	2.5	INS	RRI Tanjung Karang, Indonesia
3950		CHN	Qinghai 1, Xining, China
3955	250	G	BBC, United Kingdom
	50	RUS	Kazakh R. via Novosibirsk, S
3956	10/1	INS	RRI Palu, Indonesia
3960	50	CHN	Urumqi, China (Nov-Apr only) *
	100	D	RFE/RL, Biblis, Germany
	1	INS	RRI Padang, Indonesia
		KRE	Kanggye Reg. St, No. Korea
	12	MNG	Dalanzadgad, Mongolia
3960	5	MNG	Uliosxai, Mongolia
3970	50	CHN	Hohhot, China (Chinese prgr)
v	4	CME	CRTV, Buea, Cameroon

	100	D	RFE/RL, Biblis, Germany
	0.6	J	NHK Nagoya / Sapporo (irr.)
		KRE	Wonsan Reg. St, No. Korea
r		NIG	R. Nigeria, Enugu (r. testing)
3975	250	G	BBC, United Kingdom
3976	10/2	INS	RRI Surabaya, Indonesia
3980	100	D	Voice of America, Munich
		KRE	Ch'ongjin Reg. St, No. Korea
3981.0r		PRU	R. Caraz, Peru
3985	250	SUI	China Radio Int. via SRI
		250	Swiss Radio International
3986	1	INS	RRI Manokwari, Indonesia
3990	50	CHN	Urumqi, China (Nov-Apr only) *
3995	250	AFS	Channel Africa, Johannesburg
	100	D	Deutsche Welle, Germany
		INS	RRI Pontianak, Indonesia
	50	RUS	Khabarovsk, FE (wi.)
			Kyzyl, S
			R. Ala, Moskva
4000	20	CME	CRTV, Bafoussam, Cameroon
	5	INS	RRI Kendari, Indonesia
4002.2v		PRU	R. San Nicolás, Peru
4003	10	INS	RRI Padang, Indonesia
4010	50	KGZ	Biskek 1, Kyrgyzstan
	100	RUS	Vladivostok, FE (wi.)
4012.1	0.6	PRU	R. Frecuencia Popular, Peru
4020	500/120	CHN	China Radio International
4030		RUS	Anadyr, FE
4035	500/120	CHN	China Radio International
			Lhasa, China (Tibetan prgr)
4039.1	1	PRU	R. Marginal, Tocache, Peru
4040	50	ARM	Yerevan, Armenia
	1.5	RUS	Tura, S
	50		Vladivostok, FE
4050	50	KGZ	Biskek 2, Kyrgyzstan
	50	RUS	Yuzno-Sakhalinsk, FE (SSB)
4055	50		Tver (?), E
4081.0v	50	MNG	Ulaanbaatar 1, Mongolia
4108.5	0.5	PRU	R. Naylamp, Lambayeque, Peru
4118.0v	0.25	B	R. Difusora, Sena Madureira
4141		PRU	R. Gran Patajén, Peru *
4180.0v	0.1		R. Membrillar, Cascas, Peru
4183.2v		BOL	R.11 de Octubre, Bolivia
4190		CHN	CPBS Minority Sce, China
4200	500/120		China Radio International
4211		BOL	R. Perla del Oriente, Bolivia
4220	15	CHN	Urumqi, China (Mongolian prgr)
4232.5		HKG	VPS8 (Time Sigs), Hong Kong
4270.6v	1	EQA	R. Gonzanamá, Loja, Ecuador
4291	10	AFS	ZSC (Time Signals), So. Africa
4298		CHL	CBV (Time Signals), Chile
4299.6	1.3	PRU	R. Inca, Baños del Inca, Peru
4330		CHN	Urumqi, China (Kazakh prgr)
4395		RUS	Yakutsk, S
4409.2	1	BOL	R. Eco, Reyes, Bolivia
4419.0v	0.85	PRU	R. Frec. Líder, Bambamarca, Peru
4420.7	1	BOL	R. Santa Rosa, Bolivia
4435.1			R. CVU, Villa Tunari, Bolivia
4449.9			R. Frontera, Cobija, Bolivia
4460		CHN	CPBS 1, China
4461.8	0.18	PRU	R. Nor Andina, Celendín, Peru
4472.5	1	BOL	R. Movima, Bolivia
4485		PRU	R. Frec. VH, Celendín, Peru
	100	RUS	Petropavlovsk, FE
	50		Ufa, E
4495		PRU	R. San Mateo, Contumazá, Peru
4500		CHN	Urumqi, China (Chinese prgr)
4509.3		PRU	R. Horizonte, Chiclo, Peru
4510		UZB	Fargona, Uzbekistan
4520	50	RUS	Khanty-Mansiysk, S
	5		Palana, FE
4525	50	CHN	Hohhot, China (Mongolian prgr)
4530.1		BOL	R. Hitachi, Guayamerín, Bolivia
4535.0v	1	LAO	Oudomxay Regional R, Laos
4542.5		BOL	R. Horizontes, Cobija, Bolivia
4545		KAZ	Alma-Ata 1, Kazakhstan
4549.3	0.3	BOL	R. Dif. Trópico, Bolivia
4560.0r	1	NCG	R. Miskut, Nicaragua (nom.5970)
4599.7		BOL	R. Emisora Villamontes, Bolivia
4600.1			R. Perla del Acre, Bolivia
4606.5	1	PRU	LV de Melgar, Ayaviri, Peru
4607.4	0.5	INS	RRI Serui, Indonesia
4610	15	RUS	Komsomolsk / R. St. Vostok, FE
4625.7	0.2	BOL	R. Dif. Machpo, S. Ramón, Bolivia
4629.6	1	PRU	R. Tayabamba, Peru
4635	50	TJK	Yangi-Yul, Tadzhikistan
4649.2	1	BOL	R. Santa Ana, Bolivia
4660.0v	1	LAO	Houa Phan Regional R, Laos
4665.0v		VTN	R. Son La, Vietnam
4681.4	5	BOL	R. Paitití, Guayamerín, Bolivia
4690.0r	1	NCG	R. Miskut, Nicaragua (nom.5970)
4697	1.5	INS	RKIP Surabaya, Indonesia
4700.1		PRU	R. Waira, Chota, Peru
4705	0.25		Estación Láser, Rioja, Peru
4712.3	0.5	BOL	R. Abaroa, Riberalta, Bolivia
4719	50	INS	RRI Ujung Padang, Indonesia
4725	50	BRM	VOM Nationalities Sce, Myanmar
4731.5v	1	PRU	R. San Juan, Caraz, Peru
4735		CHN	Urumqi, China (Uighur prgr)
4739.6v	0.5	BOL	R. Mamoré, Guayamerín, Bolivia
4740	100	TJK	Yangi-Yul (t), Tadzhikistan
4750		CHN	Lhasa, China (Chinese prgr)
	20	CME	CRTV, Bertoua, Cameroon
		IRQ	R. Baghdad, Iraq
	12	MNG	Ulgii, Mongolia
4752.0v	0.5	PRU	R. Huanta 2000, Huanta, Peru
4753	20	INS	RRI Ujung Padang, Indonesia
4755	5	B	R. Dif. do Maranhao, Sao Luíz
	10		R. Educ. Rural, Campo Grande
	10	HND	Sani R, Pto. Lempira, Honduras*
4759.0v	0.5	PRU	R. Tingo María, Peru
4760		CHN	Yunnan G, Kunming, China
	5	EQA	Sist. Emis Atalaya, Ecuador *
	10	IND	All India Radio, Leh
	10		All India Radio, Port Blair
	10	LBR	ELWA, Monrovia, Liberia *
	25	SWZ	Trans World Radio, Swaziland
4761	1	VEN	R. Frontera, Venezuela
4765.0v	10	B	R. Integraçao, Cruzeiro do Sul
	10		R. Rural, Santarém, Brazil
	100	COG	RTV Congolaise, Brazzaville
4765.3	0.3	BOL	R. Guanay, Bolivia
4770	50	NIG	R. Nigeria, Kaduna
	1	VEN	R. Mundial Bolívar, Venezuela
4775	5	B	R. Amarela, Rolim de Moura
	1		R. Congonhas, Brazil
4775	1	B	R. Portal da Amazônia, Cuiabá
	20	IND	All India Radio, Guwahati B
4775.2v	3	BOL	R. Los Andes, Tarija, Bolivia
4777	100	GAB	Radiodiffusion Gabonaise
v		INS	RRI Jakarta, Indonesia
4780.0v	5	AGL	EP do Kuando Kubango, Angola
	20	DJI	Radiodiff.-TV de Djibouti
	1	VEN	LV de Carabobo, Venezuela
4783.0v	100	MLI	Radiodiff.-TV du Mali, Bamako
4784.9	1	PRU	R. COSAT, Satipo, Peru
4785	50	AZE	Gyanca, Azerbaidzhan
	1	B	R. Brasil, Campinas, Brazil
	10		R. Caiari, Porto Velho, Brazil
	1	BOL	R. Em. Ballivián, Bolivia
	20	CHN	Hangzhou, China
v	5	CLM	Ecos del Combeima, Colombia
	50	TZA	R. Tanzania National Sce.
4785.4		PRU	R. Chincheros, Peru
4787.0v		VTN	R. and TV Gia Lai, Vietnam
4789.7v	1	INS	RRI Fak-Fak, Indonesia
4790	50	IND	All India Radio, Shillong
	100	PAK	Azad Kashmir R, Trarkhel
v	3	PRU	R. Atlántida, Iquitos, Peru
	100	SWZ	Trans World Radio, Swaziland
4795	1	B	R. Nova Difusora, Aquidauana
	100	CME	CRTV, Douala, Cameroon
		RUS	Ulan-Ude, S
4795.2	5	EQA	LV de los Caras, Ecuador
4795.9	10	BOL	R. Nueva America, La Paz *
4799.8	1	DOM	La N-103, Santiago, Dom. Rep.
	5	EQA	R. Pop. Independiente, Ecuador
	1	GTM	R. Buenas Nuevas, Guatemala
4800		CHN	CPBS 2 / Minority Sce, China
	10	IND	All India Radio, Hyderabad A
		KAZ	Alma-Ata 1, Kazakhstan

4922.0v	5	BOL	R. S. Miguel, Riberalta,Bolivia
4924.1	1	PRU	R. Ondas del Titicaca, Peru
4925	1	B	R. Difusora, Taubaté, Brazil
4925.1	2	CLM	Meridiano 70, Arauca, Colombia
4925.8	0.5	PRU	R. Cobriza 2000, Peru
4926.0v	100	GNE	Batá, Equatorial Guinea
4927	7.5	INS	RRI Jambi, Indonesia
4930		CHN	Gejiu, China (irr.)
	1	DOM	R. Barahona, Barahona, Dom. Rep.
		RUS	Unknown site, S
	50	TKM	Asqabat, Turkmenistan
4932	10	INS	RRI Surakarta, Indonesia
4935	1	B	R. Capixaba, Vitória, Brazil
	2.5		R. Difusora, Jataí, Brazil
	5/100	KEN	KBC General Sce, Nairobi
v	3	PRU	R. Tropical, Tarapoto, Peru
4938	1.5	BOL	R. Norte, Montero, Bolivia
4939.6	1	VEN	R. Continental, Venezuela
4940		CHN	Haixia 1, China
	10	CLN	SLBC English Comm. Sce
	25	CTI	Radiodiff.-TV Ivoirienne
	50	IND	All India Radio, Guwahati A
	0.5	PRU	R. San Juan, Tarma, Peru
	0.5		R. Ucayali, Contamana, Peru
	50	RUS	Yakutsk, S (wi.)
	50	UKR	Kijiv, Ukraine
4942.0v	1	AGL	EP do Malanje, Angola *
4945	1	B	Emiss. Rural, Petrolina, Brazil
	50		R. Progresso, Porto Velho
	10	BOL	R. Illimani, La Paz, Bolivia *
4950		CHN	V. O. Pujiang, Shanghai, China
			Xilinhot, China (Mongolian)
	1	EQA	R. Bahá'i, Otavalo, Ecuador
	2	IND	All India Radio, Jammu A
v	10	MLA	RTM Kuching, Sarawak, Malaysia
4950.1	5	PRU	R. Madre de Dios, Peru
4955.0v	10	AGL	R. Nacional, Mulenvos, Angola
	1	B	R. Cultura, Campos, Brazil
	10		R. Marajoara, Belém, Brazil
	5	PRU	R. Cultural Amauta, Peru
4957.5	50	AZE	Baku, Azerbaidzhan
4960	10	IND	All India Radio, Delhi A
	50		All India Radio, Kingsway
4960.3	1	PRU	R. La Merced, Peru (irr.)
4960.9	5	EQA	R. Federación Shuar, Ecuador
4962	1	DOM	R. Cima, Sto. Domingo, Dom. Rep.
4965	5	B	R. Alvorada, Parintins, Brazil
	3/1	BOL	R. Juan XXIII, Bolivia
	2	CLM	R. Santa Fé, Bogotá, Colombia *
4966.0v	1	PRU	R. San Miguel, Cusco, Peru
4970		CHN	Urumqi, China (Kazakh prgr)
	50	IND	All India Radio, Itanagar
	10	MLA	RTM Kota Kinabalu, Malaysia
v	10	VEN	R. Rumbos, Caracas, Venezuela
4970.4	1	PRU	R. Imagen, Tarapoto, Peru
4974.9	4		R. Del Pacífico, Lima, Peru
4975	1	B	Super R. Tupí, Sao Paulo
	1	BOL	R. María Auxiliadora, Montero
		CHN	Fujian 1, Fuzhou, China
4976	50	UGA	R. Uganda, Kampala
4976.1	2.5	B	R. Timbira do Maranhao, Brazil
4976.2v	1	CLM	Ondas del Orteguaza, Colombia
4980		CHN	Urumqi, China (Mongolian prgr)
4980	10	VEN	Ecos del Torbes, Venezuela
4980.9	1	BOL	R. Dif. Minería, Oruro, Bolivia
4985	10	B	R. Brasil Central, Goiânia
4990	10	CHN	Hunan 1, Changsha, China
		IND	All India Radio Ext. Sce.
v	1	INS	RRI Gorontalo, Indonesia
	50	NIG	R. Nigeria Ch.1, Lagos
4990.9	5	PRU	R. Ancash, Huaraz, Peru
4991.0v	1	LAO	Xieng Khouang, Laos *
4991.2	1	BOL	R. Animas, Chocaya, Bolivia
4995	12	MNG	Choibalsan, Mongolia
4995.8	2	PRU	R. Andina, Huancayo, Peru
4996	5	RUS	RWM (Time Signals), Moscow
5000	2	ARG	LOL (Time Signals), Argentina
	10	AUS	VNG, Llandilo (Time Signals)
	10/20	CHN	BPM (Time Signals), China
	1	EQA	HD2IOA (Time Signals), Ecuador
	10	HWA	WWVH, Hawaii (Time Signals)
	5/1	I	IBF/IAM, Italy (Time Signals)
	8	IND	ATA, New Delhi (Time Sigs.)
	2	J	JJY, Japan (Time Signals)
		KOR	HLA (Time Signals), Korea
	2	TWN	BSF (Time Signals), Taiwan
	10	USA	WWV (Time Signals), Ft. Collins
	1	UZB	RCH, Tashkent (Time Signals)
	1	VEN	YVTO (Time Signals), Venezuela
5004.0v	100	GNE	Batá, Equatorial Guinea
	1	RUS	RID (Time Signals), Irkutsk
5005.0v	10	MLA	RTM Sibu, Sarawak, Malaysia
v	100/5	NPL	R. Nepal, Khumaltar
5005.7	0.35	SUR	R. Apintie, Paramaribo, Surinam
5005.8	5	BOL	R. Libertad, La Paz, Bolivia
5008.7	5	PRU	R. Horizonte, Chachapoyas
5010	0.1	BOL	R. Litoral, Llica, Bolivia *
		CHN	Guangxi 2, Nanning, China
	100	CME	CRTV, Garoua, Cameroon
	50	IND	All India Radio, Thiru'puram
	100	MDG	R. Madagasikara, Madagascar
	10	SNG	SBC Radio One, Singapore
5010.3	1	EQA	Escuelas Radiofon. Pop, Ecuador
5015.0v	1	AGL	EP do Namibe, Angola *
	1	B	R. Brasil Tropical, Cuiabá
v	1		R. Copacabana, RJ, Brazil
	0.5	PRU	R. Juliaca, Peru
	50	RUS	Arkhangelsk, E
	50		Vladivostok, FE
	50	TKM	Asqabat, Turkmenistan
5015.5	0.7	PRU	R. Estación Tarapoto, Peru
5019.8	1	CLM	Ecos del Atrato, Colombia
5020		CHN	Jiangxi 1, Nanchang, China
v	10	CLN	SLBC Tamil Comm. Sce.
	10	EQA	Voz del Upano, Macas, Ecuador
v	20/100	NGR	La Voix du Sahel, Niger
	1	PRU	R. San Cristóbal de Huamanga
	10	SLM	SIBC Honiara, Solomon Islands
v		VTN	Hanoi, Vietnam (Hmong Sce.)
5025	50	AUS	VL8K, Katherine, NT, Australia
	5	B	R. J. da Transamazônica, Brazil
	1		Super R. Borborema, Campina G.
v	10	BEN	ORTB, Parakou, Benin
	50/10	BHU	BBS, Thimpu, Bhutan
	10	CUB	R. Rebelde, Bauta, Cuba
	5	PRU	R. Quillabamba, Peru
5026.0v	20	UGA	R. Uganda, Kampala
5029.9	7.5	VEN	R.(Reloj)Continente, Venezuela
5030	20	CTR	AWR Latin America, Costa Rica
	10	MLA	RTM Kuching, Sarawak, Malaysia
	1	TON	Tonga Broadcasting Commission
5030.1	10	EQA	R. Católica Nal. del Ecuador
5030.6v	5	PRU	R. Los Andes, Huamachuco, Peru
5034.4v	20/100	CAF	R. TV Centrafricaine, Bangui
5035	10	B	R. Aparecida, Brazil
	5		R. Educacao Rural, Coari
		KAZ	Alma-Ata 2, Kazakhstan
5039.3	1	PRU	R. Libertad, "RLJ", Junín, Peru
5040		CHN	Fujian 1, Fuzhou, China
	10	EQA	Voz del Upano, Macas, Ecuador
	100	GEO	Tbilisi 1, Georgia
	1	GTM	LV de Nahualá, Guatemala
	200	RUS	R. Ala, Sankt-Peterburg, E
5040.1	1	VEN	R. Maturín, Venezuela (irr.)
5040.2v	1	AGL	EP de Benguela, Angola
5044.8	5	BOL	R. Altiplano, La Paz, Bolivia
5045	10	B	R. Cultura do Pará, Belém
	1		R. Difusora, Pres. Prudente
	1	INS	RRI Jayapura, Indonesia
5046	20		RRI Yogyakarta, Indonesia *
	3	PRU	R. Rioja, Peru (n.5045)
5047	100	TGO	Radiodiffusion Togolaise, Lomé
5049.8	5	EQA	R. Jesús del Gran Poder, Quito
5050		CHN	Guangxi FBS, Nanning, China
	50	IND	All India Radio, Aizawl
	0.5	PRU	R. Municipal, Cangallo, Peru
5050.1	1	CLM	LV de Yopal, Colombia
	10	TZA	R. Tanzania National Sce.
5052	50	SNG	SBC Radio One, Singapore

5055	1	B	R. Difusora, Cáceres, Brazil
	5	CTR	Faro del Caribe, Costa Rica
	10	GUF	RFO Cayenne, French Guiana
		PRU	R. Onda Imperial, Cusco, Peru
	25	SWZ	Trans World Radio, Swaziland
5055.5	1	INS	RRI Nabire, Indonesia
5059	0.5	PRU	R. Lircay, Peru
5060		CHN	Urumqi, China (Mongolian prgr)
5061.0v		VTN	Radio TV Kontum, Vietnam
5062.0v	2	EQA	Sist. de Em. Progreso, Ecuador
5066.4	1	ZAI	R. CANDIP, Bunia, Zaire
5068.7		PRU	Ondas del Suroriente, Peru
5075	50	CLM	Caracol Colombia, Bogotá
5081	0.5	PRU	R. Huayabamba, Peru
5082.6	0.5		R. Mundo, Cusco, Peru
5085	0.1		R. Celendín, Peru
5090		CHN	CPBS Taiwan 2 Sce, China *
5097.3v	1	PRU	R. Eco, Iquitos, Peru (n.5010)
5125		CHN	CPBS Taiwan 1 Sce, China *
5147.3v		PRU	"RQ", Tingo María, Peru
5156.2	0.2	BOL	R. Galaxia, Guayamerín, Bolivia
5163		CHN	CPBS 2, China
5186.8	0.5	PRU	R. Onda Popular, Peru
5190	2		R. Nuevo Continente, Peru *
5191.2v	5	AGL	EP do Moxico, Angola
5217		BOL	R. Em. Tarabuco, Bolivia *
5235	0.3	PRU	R. San Antonio de Padua, Peru
5240		CHN	Lhasa, China (Chinese prgr)
5250	120		China Radio International
5256	1	INS	RRI Sibolga, Indonesia *
5260		KAZ	Alma-Ata 2, Kazakhstan
5265.0r		PRU	R. Estación "A", Cutervo, Peru
5271.8			R. Nororiental, Peru
5277.6	1.5		R. Grau, Huancabamba, Peru
5286.1v	5	TCD	R. Moundou, Chad
5290	50	RUS	Krasnoyarsk, S
	20		Perm, E
5320		CHN	CPBS 1, China
5323.5	0.5	PRU	R. Origen, Huancavelica, Peru
5340	0.1		R. La Voz de Los Andes, Peru
5344	0.5		R. Pomabamba, Peru (n.5370)
5359.8	0.1		R. Visión, Juanjuí, Peru
5419.1v			R. Sonorama, Saposoa, Peru
5420		CHN	CPBS Minority Sce, China
5430		BLR	Mahileu, Belarus
	10/20	CHN	BPM (Time Signals), China
5440			Urumqi, China (Kazakh prgr)
5445.2		PRU	R. Alto Huallaga, Uchiza, Peru
5500.4		BOL	R. Luz del Oriente, Bolivia
5505.3	0.5		R.2 de Febrero, Bolivia
5525.0v		PRU	R. Comercial, Chota, Peru
5535.5v		CLM	Ecos Celestiales, Colombia
5538.7		PRU	Frecuencia Modular, Peru
5555.0r	1.6	CLM	LV del Valle, Colombia
5560.0r	1	NCG	R. Miskut, Nicaragua (nom.5970)
5569.7	0.2	CLM	R. Nueva Vida, Tibú, Colombia
5580.2v	1	BOL	R. San José, Bolivia
5619.3	0.1	PRU	R. Ilucán, Cutervo, Peru
5645.0v	0.1		R. Bambamarca, Peru
5660.5	0.6		R. La Voz de Cutervo, Peru
5692	0.5	INS	RPDT2 Berau, Indonesia
5700	0.1	PRU	R. San Ignacio, Peru
5780	15	RUS	R. Shark, Ufa, E
5800		CHN	Urumqi, China (Uighur prgr)
		PRU	R. Nuevo Cajamarca, Peru
5816	1		R. LV del Altiplano, Puno, Peru
5825			R. Comercial Cosmos, Bambamarca
	500	USA	WEWN, Birmingham, AL
5833.0v		PRU	R. Uno, Huánuco, Peru
5835.3r			R. Tinajani, Ayaviri, Peru
5850	500	USA	WCSN, Scotts Corners, ME
	500		WSHB, Cypress Creek, SC
5875	500/100	G	BBC, United Kingdom
5880	250	AUS	R. Australia, Carnarvon
		CHN	CPBS 1, China
5882	500/100	CVA	Vatican Radio
5895	250	BUL	R. Sofia, Plovdiv, Bulgaria
	0.1	PRU	R. Hispana, Arequipa, Peru
5900	200/100	BEL	R. Vlaanderen Internationaal
		CHN	Sichuan 2, Chengdu, China
	20	ISR	Kol Israel Arabic Network D
	1	TCD	R. Abéché, Chad
5910	100	F	R. France Internationale
5915	100	BUL	R. Sofia, Kostinbrod, Bulgaria
		CHN	CPBS 1, China
	20/500	ISR	Kol Israel Arabic Network D
	100	KAZ	Alma-Ata Ext. Sce, Kazakhstan
			R. Tashkent, Uzbekistan (relay)
	20	RUS	R. Vedo, Volgograd, E
5920		CHN	Guangxi 3, Nanning, China
5925.0v		BOL	R. Centinela, Tupiza, Bolivia
	100	EST	Tallinn (Laitse), Estonia
	500	F	R. France Internationale
		VTN	Hanoi I, Vietnam
5930	50	RUS	Monchegorsk, E
	250/100	TCH	Czech & Slovak R. International
5935	50	CHN	Lhasa, China (Chinese prgr)
	100	LAT	Riga, Latvia
5935.0v	5	PRG	R. Guairá, Villarica, Paraguay
	100	USA	WWCR, Nashville, TN
5940	50	RUS	Arman, FE
5945	100	AUT	R. Austria International
	100	F	R. France Internationale
	20	KAZ	Alma-Ata 2, Kazakhstan
		UZB	R. Tashkent Ext. Sce, Uzbekistan
5948.7	5	BOL	R. Pio XII, Bolivia
5950	50	CHN	Heilongjiang 2, Harbin, China
	50		Lhasa, China (Tibetan prgr)
	10	GUY	Voice of Guyana, Georgetown
	50	NCG	R. Nicaragua, Managua *
	100	USA	Voice of Free China via WYFR
	100		WYFR, Okeechobee, FL
	300	YEM	Yemen Radio & TV Corp, Sana'a
5951.3	1	PRU	R. Arequipa, Peru
5953.5	1	CTR	R. Casino, Limón, Costa Rica
5955	10	B	R. Gazeta, Sao Paulo, Brazil
	5	CLM	LV de los Centauros, Colombia
	100	D	RFE/RL, Biblis, Germany
	500/100	G	BBC, United Kingdom
	250	GRC	Voice of America, Kavala
	500	HOL	R. Netherlands, Flevoland
	250	PHL	Voice of America, Philippines
	250/120	ROU	R. Romania International
5955.3	1	PRU	R. Huancayo, Peru
5955.5	0.25	GTM	R. Cultural, Guatemala
5960	250	AFS	Channel Africa, Johannesburg
	250	CAN	R. Canada Int, Sackville
	250		R. Japan via Sackville, Canada
	50	CHN	Yunnan E, Kunming, China *
	500/50	D	Deutsche Welle, Germany
	250	G	R. Japan via Skelton, UK
	2	IND	All India Radio, Jammu A
	100	KAZ	Alma-Ata 2, Kazakhstan
	50	MNG	Ulaanbaatar 2, Mongolia
	250/100	TCH	Czech & Slovak R. International
5964.8	1	BOL	R. Nacional de Huanuni, Bolivia
	10	EQA	Voz del Upano, Macas, Ecuador
5965	7.5	B	R. Transamérica, Santa Maria
	250	CAN	BBC via Sackville, NB
	50	CUB	R. Habana, Cuba
	250/100	CYP	BBC Ea. Mediterranean Relay St.
	500/100	D	Deutsche Welle, Germany
	500		Voice of America, Wertachtal
	250	GRC	Voice of America, Kavala
	50		Voice of America, Rhodes
	100	MLA	RTM R1, Kuala Lumpur, Malaysia
	10	NIG	Plateau R. TV Corp, Jos
	500	NOR	R. Norway Int. / R. Denmark
	100	OMA	BBC Eastern Relay Station
	10/2	PNG	NBC Mount Hagen, Papua N. G.
	25	SWZ	Trans World Radio, Swaziland
5969.7	10	PRU	R. El Sol, Lima, Peru (irr.)
5970	1	CHN	Xiahe, China
	50	CTR	AWR Latin America, Costa Rica
	100		R. Exterior de España, Costa R.
	100	D	RFE/RL, Biblis, Germany
	250		RFE/RL, Holzkirchen, Germany
	100		RFE/RL, Lampertheim, Germany
	500		Voice of America, Wertachtal
	500/100	G	BBC, United Kingdom

	250		R. Japan via Skelton, UK
	10	INS	RRI Banjarmasin, Indonesia
	50	KAZ	Alma-Ata 2, Kazakhstan
5970.0v	1	NCG	R. Miskut, Nicaragua
	500	POR	RFE/RL, Maxoqueira, Portugal
	100	YEM	Yemen Radio & TV Corp, Aden
5974.0v	1	BOL	R. Nacional, Cochabamba,Bolivia
5975	50	ALB	R. Tirana, Lushnjë, Albania
	250	ATG	BBC Caribbean Relay St.
	120	CHN	China Radio International
	5	CLM	R. Macarena, Colombia
	500	D	Voice of America, Wertachtal
	500/100	G	BBC, United Kingdom
	100	KOR	R. Korea, Hwasung, So. Korea
	100	OMA	BBC Eastern Relay Station
	250	SNG	BBC Far Eastern Relay St.
5980	10	B	R. Guarujá, Florianópolis
	150	BUL	R. Sofia, Stolnik, Bulgaria
	50	CHN	Hohhot, China (Chinese prgr)
	500/100	D	Deutsche Welle, Germany
	5	EQA	R. Federación Shuar, Ecuador
r		LAO	Lao National R, Vientiane
	0.5	MEX	R. XEUJ, Linares, México
	10	MLA	RTM Kota Kinabalu, Sabah
	5	PRU	R. Programas del Perú, Lima *
	250	TUR	V. of Turkey, Ankara
5981.7v	5	GTM	AWR Latin America, Guatemala
5984		INS	RRI Pekanbaru, Indonesia
5985	50	ALB	R. Tirana, Lushnjë, Albania
	2.5	BOL	R. San José, Oruro, Bolivia *
r	5	CLM	R. Dif. Nac, Tunja, Colombia
	25	COG	RTV Congolaise, Brazzaville
	100	D	RFE/RL, Biblis, Germany
	100		RFE/RL, Lampertheim, Germany
	500		Voice of America, Wertachtal
	250	GRC	Voice of America, Kavala
		KAZ	Unknown site, Kazakhstan
	10	MEX	R. México Internacional
	100	PAK	R. Pakistan Home Sce, Islamabad
	10	PNG	NBC Rabaul, Papua N. G.
	500	POR	RFE/RL, Maxoqueira, Portugal
	250	USA	Voice of America, Delano, CA
	100/50		WYFR, Okeechobee, FL
5988	10	INS	RRI Manado, Indonesia
5990	50	BRM	Voice of Myanmar, Yangon
	100	ETH	V. of Ethiopia, Gedja
	500/100	F	R. France Internationale
	500/100	G	BBC, United Kingdom
	100	I	RAI Ext. Sce, Rome, Italy
		IND	All India Radio Ext. Sce.
	10		All India Radio, Bhopal A
	250/120	ROU	R. Romania International
5995	100	AUS	R. Australia, Shepparton
	0.8	BOL	R. Loyola, Sucre, Bolivia *
	50	CHN	Lhasa, China (Tibetan prgr)
	500	D	Deutsche Welle, Germany
	500		Voice of America, Wertachtal
	300	G	R. Canada Int. via Skelton, UK
	500/100	IRN	IRIB, Tehran, Iran
	50	MLI	Radiodiff.-TV du Mali, Bamako
	20/100	MWI	MBC, Blantyre, Malawi
	2	PRG	LV del Amambay, Paraguay *
	250/120	ROU	R. Romania International
	250	USA	Voice of America, Greenville
	100	UZB	Toskent 1, Uzbekistan
	10	ZAI	La Voix du Zaire, Mbandaka
5995.2	1	PRU	R. Melodía, Arequipa, Peru
5999.3	10	EQA	Voz del Upano, Macas, Ecuador
6000	50	ARS	BSKSA, Riyadh, Saudi Arabia
	5	B	R. Boas Nova, Sao Luíz (C. P)
	10		R. Guaíba, Porto Alegre, Brazil
	50	CHN	Haixia 2, China
	75	CUB	R. Habana, Cuba
	500/100	D	Deutsche Welle, Germany
	10	IND	All India Radio, Leh
	350	S	R. Sweden
	50	SNG	"City Sounds", Singapore
	1.5	URG	Radiomundo, Uruguay *
6005	250	AFS	BBC via Meyerton, South Africa
	250	ASC	BBC Atlantic Relay St.
	0.5	CAN	CFCX Montreal (rel. CFCF)
	10	CLN	SLBC Ext. Sce, Colombo
	100	D	RIAS Berlin, Germany
	500	IRN	IRIB, Mashhad, Iran
	0.6	J	NHK Sapporo, Japan
	50	MRC	Voice of America, Tangier
	20	RUS	Khabarovsk, FE
	250	SEY	BBC Indian Ocean Relay St.
	500	YUG	R. Yugoslavia, Beograd
6006	3	CTR	R. Reloj, San José, Costa Rica
6008.8	0.3	J	NHK Nagoya (SSB) (6008.75)
6009.9	0.25	MEX	R. Mil, México City, México
6010	25	B	R. Inconfidência, Belo H.
	60	BHR	R. Bahrain
	250	CYP	BBC Ea. Mediterranean Relay St.
	500/100	D	Deutsche Welle, Germany
	500/100	G	BBC, United Kingdom
	250		R. Japan via Skelton, UK
	250	GRC	Voice of America, Kavala
	100	I	RAI Ext. Sce, Rome, Italy
	10	IND	All India Radio, Calcutta
	250/50	PAK	R. Pakistan Ext. Sce.
	20	RUS	Lithuanian R. via Balashikha
		UKR	R. Ukraine International
6010.6	1	VEN	R. Los Andes "1040", Venezuela*
6010.9	2.5	PRU	R. América, Lima, Peru (r.6042)
6012	1	USA	AFRTS McMurdo, Antarctica
6014.4v	0.3	PRG	Em. Paraguay, Asunción
6015		BLR	Unknown site, Belarus
	250	CAN	R. Austria Int. via Sackville
	50	CHN	CPBS Taiwan 1 Sce, China *
	500	CTI	Radiodiff.-TV Ivoirienne
	250	CYP	BBC Ea. Mediterranean Relay St.
	500/100	D	Deutsche Welle, Germany
	300	G	Voice of America, Woofferton
	250	GRC	Voice of America, Kavala
	500	HOL	R. Netherlands, Flevoland
	100	KOR	Liberty 1, Hwasong, So. Korea
	50	TZA	R. Tanzania Zanzibar
6015.3	3	CLM	R. Mira, Tumaco, Colombia
6015.8	10	BOL	R. El Mundo, S. Cruz, Bolivia
6020	50	ARS	BSKSA, Jeddah, Saudi Arabia
	250	ATN	R. Netherlands, Bonaire
	100	AUS	R. Australia, Shepparton
	10	B	R. Gaúcha, Porto Alegre, Brazil
	100	E	R. Exterior de España, Madrid
	500	HOL	R. Netherlands, Flevoland
	2.5	IND	All India Radio, Shimla
	300	MDG	R. Netherlands, Madagascar
v	0.25	MEX	LV de Vera Cruz, México
	10	PNG	NBC Kieta, Papua N. G.
	5	PRU	R. Victoria, Lima, Peru *
	25	SWZ	Trans World Radio, Swaziland
	100	UKR	Kijiv, Ukraine
			R. Ukraine International
	10	VEN	R. Anzoátegui, Venezuela *
6025	10	BOL	R. Illimani, La Paz, Bolivia
	15	CHN	Bayanhot, China
	500	D	R. Canada Int. via Wertachtal
v	5	DOM	R. Amanecer Int, Dom. Rep.
	500/100	G	BBC, United Kingdom
	300		R. Canada Int. via Skelton, UK
	250		R. Japan via Skelton, UK
	100	HNG	R. Budapest/Kossuth R, Hungary
	100	MLA	RTM R5, Kuala Lumpur, Malaysia
	250	MLT	Deutsche Welle, Cyclops, Malta
v	10	MOZ	Beira 1, Mozambique
	10	NIG	R. Nigeria, Enugu
	0.7	PRG	R. Nal. del Paraguay, Asunción
6029.7	1	CHL	R. Sta. María, Coyhaique, Chile
6030	10	B	R. Globo, RJ, Brazil
	0.1	CAN	CFVP Calgary (rel. CFCN)
	20	D	Süddeutscher Rundfunk
	500/100	G	BBC, United Kingdom
	250		R. Japan via Skelton, UK
	100	OMA	BBC Eastern Relay Station
	250	PHL	Voice of America, Philippines
	15	RUS	R. St. Onega, Murmansk, E
	500/250	SUI	Swiss Radio International
	500	USA	R. Martí, Greenville, NC
	175		Voice of America, Bethany, OH

6033.1		PRU	R. Landa, Arequipa, Peru
6035	50/10	BHU	BBS, Thimpu, Bhutan
	50	CHN	Yunnan BS, Kunming, China
	12.5	CLN	Trans World Radio, Sri Lanka
	500/100	D	Deutsche Welle, Germany
	50	IND	All India Radio, Itanagar
	100	NZL	R. New Zealand International
	1	URG	R. Nacional, Uruguay *
	500	USA	Voice of America, Greenville
6035.2	5	CLM	LV del Guaviare, Colombia
6040	250	ATG	Deutsche Welle, Antigua
	250	B	Deutsche Welle via R. Bras
	7.5		R. Clube Paranaense, Curitiba
	50	CHN	Jiangxi 1, Nanchang, China
	500	D	Deutsche Welle, Germany
	100	F	R. France Internationale
	500/100	G	BBC, United Kingdom
	250		Voice of America, Woofferton
	250	MLT	Deutsche Welle, Cyclops, Malta
	10	PNG	NBC Alotau, Papua N. G.
	250	POR	Deutsche Welle, Sines
	20	TWN	CBS Netw. 5/1, Taipei, Taiwan
	500	USA	Voice of America, Greenville
6042.0r	2.5	PRU	R. América, Lima, Peru (n.6010)
6045	50	CHN	Hohhot, China (Chinese prgr)
	500	D	Deutsche Welle, Germany
	500/100	F	R. France Internationale
		IND	All India Radio Ext. Sce.
	50		All India Radio, Khampur
	50		All India Radio, Kingsway
	250	KEN	KBC General Sce, Nairobi
	0.25	MEX	R. Univ. de S. L. Potosí, México
	200	RUS	R. Polis, Sankt-Peterburg, E
	500	THA	Voice of America, Udorn Thani
6045.1	1	URG	R. Integr. Americana, Uuruguay
6045.5	1	CLM	R. Melodía, Bogotá, Colombia
	3	PRU	R. Santa Rosa, Lima, Peru
6050	10	B	R. Guarani, Belo Horizonte
	250	CYP	BBC Ea. Mediterranean Relay St.
	100	D	RFE/RL, Lampertheim, Germany
	100	EQA	HCJB, Quito, Ecuador
	500/100	G	BBC, United Kingdom
	300		R. Canada Int. via Skelton, UK
	250		R. Japan via Skelton, UK
	10	MLA	RTM Sibu, Sarawak, Malaysia
	50	NIG	R. Nigeria, Ibadan
6055	500/100	D	Deutsche Welle, Germany
	350	E	R. Exterior de España, Madrid
	50	J	R. Tanpa 1, Tokyo, Japan
	250	KWT	R. Kuwait
	50	RRW	R. Rwanda, Kigali
	250/100	TCH	Czech & Slovak R. International
6055.3	2	PRU	R. Continental, Arequipa, Peru
6060	50	ARG	R. Nal. Buenos Aires, Argentina
	100	AUS	R. Australia, Shepparton
	15	B	R. Universo, Curitiba, Brazil
	50	CHN	Sichuan 1, Xichang, China
	500/50	CUB	R. Habana, Cuba
	500	D	Voice of America, Wertachtal
	300	G	Voice of America, Woofferton
	250	GRC	Voice of America, Kavala
	50	I	RAI Radio 1, Caltanissetta
	20	KAZ	Alma-Ata 2, Kazakhstan
	10	MLA	RTM Miri, Sarawak, Malaysia
	4	NGR	La Voix du Sahel, Niger
	100	NMB	Namibian Broadc. Corp. Ch.1
6061	3	PRU	R. JSV, Huánuco, Peru
6065	1	CLM	R. Super, Bogotá, Colombia
	500	D	Deutsche Welle, Germany
	500	HOL	R. Netherlands, Flevoland
	50	IND	All India Radio, Kohima A
	350	S	R. Sweden
	250/100	SNG	BBC Far Eastern Relay St.
	100	USA	WYFR, Okeechobee, FL
6069	2	BOL	R. El Cóndor, Oruro, Bolivia *
6070	100	AFS	BBC via Meyerton, South Africa
	7.5	B	R. Capital, RJ, Brazil
	250	BUL	R. Sofia, Plovdiv, Bulgaria
	1	CAN	CFRX Toronto (rel. CFRB)
	20	INS	RRI Jayapura, Indonesia
	10	LBR	ELWA, Monrovia, Liberia *
	10	MRL	WSZO, Majuro, Marshall Isl *
	100/10	PAK	R. Pakistan Home Sce, Islamabad
	10	THA	R. Thailand Prgr 1
	500		Voice of America, Udorn Thani
		UKR	R. Ukraine International
6073.3v	2.5	URG	LV de Artigas, Uruguay
6075	50	CLM	Caracol Colombia, Bogotá *
	10	CLN	SLBC Ext. Sce, Colombo
	1	CTR	R. Rumbo, Cartago, Costa Rica *
	500/100	D	Deutsche Welle, Germany
	1	HND	LV del Junco, Honduras (irr.)
	20	KAZ	Alma-Ata 2, Kazakhstan
	250	KEN	KBC National Sce, Nairobi
	250	USA	R. Martí, Bethany, OH
6080	50	ALB	R. Tirana, Lushnjë, Albania
	100	AUS	R. Australia, Shepparton
	5	B	R. Anhanguera, Goiânia, Brazil
	10		R. Novas de Paz, Curitiba
	0.01	CAN	CKFX Vancouver (rel. CKWX)
		CHN	Hailar, China (Mongolian prgr)
	250	CYP	BBC Ea. Mediterranean Relay St.
6080	10	EQA	HCJB, Quito, Ecuador
	10	PNG	NBC Daru, Papua N. G.
	100	SNG	BBC Far Eastern Relay St.
6080.2	1	CHL	R. Patagonia Chilena, Chile
6082	0.85	BOL	R.21 de Diciembre, Bolivia *
6085	5		R. San Gabriel, La Paz, Bolivia
	100	BUL	R. Sofia, Kostinbrod, Bulgaria
	250	CAN	Deutsche Welle via Sackville
v	1	CLM	Ondas del Darién, Colombia
	250	CYP	BBC Ea. Mediterranean Relay St.
	100	D	Bayerischer Rundfunk, Germany
	500		Deutsche Welle, Germany
	250	G	R. Japan via Skelton, UK
	250	GRC	Voice of America, Kavala
	50	IND	All India Radio, Khampur
	250	MLT	Deutsche Welle, Cyclops, Malta
	100	OMA	R. Oman, Seeb & Thumrait
	100	USA	WYFR, Okeechobee, FL
6087	100	TWN	CBS Netw. 4/1, Taipei, Taiwan
6088.7	1	CHL	R. Esperanza, Temuco, Chile
6090	10	B	R. Bandeirantes, Sao Paulo
	100	BOT	Voice of America, Botswana
v		CBG	Phnom Penh, Cambodia
	500	D	Voice of America, Wertachtal
		LBR	"ELBC", Gbarnga, Liberia
	500	LUX	R. Luxembourg, Junglinster
	1	MEX	Música Romántica, México *
v	250	NIG	R. Nigeria, Kaduna
		RUS	R. Dvizhenie, Yekaterinburg, S
	50		R. St. vez Irkurtsk, Irkutsk, S
6093.6	2	PRU	R. Universal, Cusco, Peru
6094.2	1	BOL	R. Cosmos, Cochabamba, Bolivia
6095	50	CHN	CPBS Taiwan 2 Sce, China
	500/100	CVA	Vatican Radio
	100	D	Deutsche Welle, Germany
	250	G	Voice of America, Woofferton
	100	POL	Polish Radio, Warszawa
	500	POR	RFE/RL, Maxoqueira, Portugal
	250		Voice of America, Gloria
6100	100	AFG	R. Afghanistan, Kabul
	0.5	ARG	R. Prov. de S. Cruz, Argentina *
	50	CHN	Urumqi, China (Chinese prgr)
	500/100	D	Deutsche Welle, Germany
	200	KRE	KCBS, Pyongyang, No. Korea
	100	MLA	V. of Malaysia, Kuala Lumpur
	10	NIG	Borno R. TV Corp, Nigeria
	100	POR	R. Renascença, Muge, Portugal
	250/100	YUG	R. Yugoslavia, Beograd
6105	5	B	R. Cançao Nova, C. Paulista
	5		R. Cultura, Foz do Iguaçu
	2	CTR	R. UCR, San José, Costa Rica
	250	D	RFE/RL, Holzkirchen, Germany
	100		RFE/RL, Lampertheim, Germany
v	0.25	MEX	Tus Panteras, Mérida, Mexico
	500	POR	RFE/RL, Maxoqueira, Portugal
	250/120	ROU	R. Romania International
	100	TZA	R. Tanzania Commercial Sce.
	100	USA	WYFR, Okeechobee, FL

SHORTWAVE TABLE

6105.5	7.5	BOL	R. Panamericana, Bolivia
6110	250	ASC	BBC Atlantic Relay St.
	250	ATG	BBC Caribbean Relay St.
	200	AZE	Gyanca, Azerbaidzhan
		CHN	CPBS Minority Sce, China
	500/100	EQA	HCJB, Quito, Ecuador
	500/100	G	BBC, United Kingdom
6110	100	HNG	R. Budapest, Hungary
	50	IND	All India Radio, Srinagar B
	250	PHL	Voice of America, Philippines
6112	10	MOZ	Em. Nacional, Maputo
6114.9	1	MEX	R. Univ. de Sonora, México
6115	50	CHN	Haixia 1, China
	100	D	Deutsche Welle, Germany
	100		RFE/RL, Biblis, Germany
	100	IND	All India Radio, Madras
	50	J	R. Tanpa 2, Tokyo, Japan
	10	PRU	R. Unión, Lima, Peru
	20	RUS	Tatar R, Tolyatti, E
6115.8	10	CLM	LV del Llano, Colombia
6120	100	ALB	R. Tirana, Krujë, Albania
	10	B	R. Globo, Sao Paulo, Brazil
	250	CAN	R. Canada Int, Sackville
	250		R. Japan via Sackville, Canada
	500/100	D	Deutsche Welle, Germany
	500/100	FNL	R. Finland
	20	RUS	Kazan, E
6125	50	CHN	CPBS 1, China
	250	CYP	BBC Ea. Mediterranean Relay St.
	350	E	R. Exterior de España, Madrid
	500/100	G	BBC, United Kingdom
	250		R. Japan via Skelton, UK
	250		Voice of America, Woofferton
	20	KAZ	Alma-Ata 2, Kazakhstan
	400/100	KRE	R. Pyongyang Ext. Sce, No. Korea
	10	ZAI	La Voix du Zaire, Kananga *
6127	1	INS	RRI Nabire, Indonesia
6130	0.5	CAN	CHNX Halifax (rel. CHNS)
	10	CLN	SLBC English Comm. Sce.
	500/50	D	Deutsche Welle, Germany
	250	G	Voice of America, Woofferton
	50/10	GHA	GBC Ext. Sce, Accra, Ghana
	20	KAZ	Akmola 2, Kazakhstan
	25	LAO	Lao National R, Vientiane
	250	PHL	Voice of America, Philippines
	100	POR	R. Portugal, S. Gabriel
	250	USA	BBC via Delano, CA
	250		Voice of America, Delano, CA
	250		Voice of America, Greenville
	100		WYFR, Okeechobee, FL
6133.8	0.6	J	NHK Fukuoka (SSB) (6133.75)
6135	100	AZE	Gyanca, Azerbaidzhan
	25	B	R. Aparecida, Brazil
	10	BOL	R. Santa Cruz, Bolivia
	100	D	RFE/RL, Biblis, Germany
	250		RFE/RL, Holzkirchen, Germany
	100		RFE/RL, Lampertheim, Germany
	1	INS	RRI Samarinda, Indonesia
	100	KAZ	Alma-Ata Ext. Sce, Kazakhstan
	10	KOR	R. Korea / Liberty 2, Hwasong
v	30	MDG	R. Madagasikara, Madagascar
	4	OCE	RFO, Tahiti, French Polynesia
	100	POL	Polish Radio, Warszawa
	500/250	SUI	Swiss Radio International
v		VTN	Hanoi, Vietnam (Hmong. Sce.)
	300	YEM	Yemen Radio & TV Corp, Sana'a
6139.9v	1.5	URG	R. Monte Carlo, Uruguay
6140	10	AUS	VLW6, Wanneroo, Australia
	100	BDI	R. Nat. Burundi, Bujumbura
	500	CHN	China Radio International
	500	D	Deutsche Welle, Germany
	300	G	Voice of America, Woofferton
6140	250	GRC	Voice of America, Kavala
		IND	All India Radio Ext. Sce.
	2		All India Radio, Ranchi A
	10	PNG	NBC Wewak, Papua N. G.
	2.1	PRU	R. Huayllay, Peru
	250	TUR	V. of Turkey, Ankara
6140.2v	1	PRU	R. Concordia, Arequipa, Peru
6140.7	1	BOL	R. Em. Luis de Fuentes, Bolivia
6145	50	ALB	R. Tirana, Lushnjë, Albania
	50	ALG	RTV Algerienne, Ouled Fayet
	250	CAN	R. Korea via Sackville, Canada
	500/100	D	Deutsche Welle, Germany
	500/100	G	BBC, United Kingdom
		KOR	R. Canada Int. via Kimjae, Korea
	50	RUS	R. Pamyat, Yekaterinburg, S
6150	100	ALS	KNLS, Anchor Point, Alaska
	7.5	B	R. Record, Sao Paulo
	250	CAN	R. Canada Int, Sackville
	10	CLM	Caracol Colombia, Bogotá
	10	CLN	SLBC Sinhala Comm. Sce.
	50	CTR	AWR Latin America, Costa Rica
	500/100	CVA	Vatican Radio
	100	CYP	BBC Ea. Mediterranean Relay St.
v	7.5		R. Bayrak II, No. Cyprus
	500	D	Voice of America, Wertachtal
	500/100	F	R. France Internationale
	500/100	G	BBC, United Kingdom
	300		Voice of America, Woofferton
	250	GRC	Voice of America, Kavala
	300	J	R. Canada Int. via Yamata, Japan
	250	KEN	KBC National Sce, Nairobi
	100	MRC	Voice of America, Tangier
	250	POR	R. Canada Int. via Sines
	20	RUS	Belorussian R. via Balashikha
6152.0v	1	AGL	EP de Benguela, Angola
6155	250	ASC	BBC Atlantic Relay St.
	250		Voice of America, Ascension Is
	300	AUT	R. Austria International
	10	BOL	R. Fides, La Paz, Bolivia
	15	CHN	Gansu 1, Lanzhou, China
	100	GUI	Conakry, Guinea Rep. (irr.)
	500	LBY	LJB, Tripoli/Sabrata, Libya
v	0.5	PRU	R. Pucallpa, Peru
	250/120	ROU	R. Romania International
	50	SNG	"Warna", Singapore
	100	SWZ	Swazi Commercial Radio
6160	50	ALG	RTV Algerienne, Ouled Fayet
	250	ATG	Deutsche Welle, Antigua
	10	B	R. Rio Mar, Manaus, Brazil
	0.3	CAN	CKZN St. John's (rel. CBN)
	0.5		CKZU Vancouver (rel. CBU)
	500	D	Voice of America, Wertachtal
	500/100	G	BBC, United Kingdom
	300		Voice of America, Woofferton
	250	GRC	Voice of America, Kavala
	250	POR	Voice of America, Gloria
6160.7	0.5	ARG	R. Malargüe, Argentina
6165	250	ATN	R. Netherlands, Bonaire
		IND	All India Radio Ext. Sce.
	50		All India Radio, Jaipur
	0.5	MEX	LV de América Latina, México
	3	RUS	Kudymkar, E
	250	SUI	China Radio Int. via SRI
	250		Swiss Radio International
	20	ZMB	ZNBC Radio Two, Lusaka, Zambia
6169.2	2.5	CLM	LV de la Selva, Colombia
6170	7.5	B	R. Cultura, Sao Paulo, Brazil
	15	CHN	Haixia 1, China *
	250	CLN	Deutsche Welle, Trincomalee
	100	D	RFE/RL, Biblis, Germany
	100		RFE/RL, Lampertheim, Germany
	4	GUF	RFO Cayenne, French Guiana
6173.8v	5	PRU	R. Tawantinsuyo, Cusco, Peru
6175	250	AFS	BBC via Meyerton, South Africa
	1	BOL	R. Indoamérica, Potosí *
	250	CAN	BBC via Sackville, NB
	2.5	CTR	Faro del Caribe, Costa Rica
	100	F	R. France Internationale
	500/100	IRN	IRIB, Tehran, Iran
	50	MLA	V. of Malaysia, Kuala Lumpur
	100	NMB	Namibian Broadc. Corp. Ch.2
	100	USA	WYFR, Okeechobee, FL
6176	15	CHN	Shaanxi 1, Xi'an, China
6178.8	0.9	J	NHK Tokyo (SSB) (6178.75)
6180.0v	1	ARG	R. Nal. Mendoza, Argentina *
	250	B	R. Nacional da Amazônia, Brazil
	500/50	CUB	R. Habana, Cuba
	250/100	CYP	BBC Ea. Mediterranean Relay St.

	250		CyBC via BBC Relay Station
	100	D	Deutsche Welle, Germany
	250	G	Voice of America, Woofferton
	100	KAZ	Alma-Ata 1, Kazakhstan
	300	TWN	CBS Network 2, Taipei, Taiwan
	500	UAE	R. of the UAE, Abu Dhabi
6184.7	10	CLN	SLBC Tamil Comm. Sce.
6185	3	BOL	R. Batallón Colorados, Bolivia
		CHN	China Huayi BC, Fuzhou, China
	500/100	CVA	Vatican Radio
	100	D	Deutsche Welle, Germany
	100	J	R. Japan, Yamata, Japan
	100	LBY	LJB, Tripoli/Sabrata, Libya
	5	MEX	R. Educación, México City
	350	S	R. Sweden
6187.8	3	PRU	R. Oriente, Yurimaguas, Peru
6189	1	INS	RRI Manokwari, Indonesia
6190	50	D	SFB / Radio Bremen
	10	IND	All India Radio, Delhi A
	50		All India Radio, Kingsway
	0.5	J	NHK Osaka, Japan
	100	LSO	BBC Lesotho Relay St.
	250/120	ROU	R. Romania International
	250	USA	Voice of America, Greenville
6191	10	INS	RRI Padang, Indonesia
6191.7	1	PRU	R. Cusco, Peru (n.6195)
6195	25	ATG	BBC Caribbean Relay St.
	50	AZE	Baku, Azerbaidzhan
v	50	BGD	R. Bangladesh Home Sce.
	50	CHN	Hohhot, China (Mongolian prgr)
	250	CYP	BBC Ea. Mediterranean Relay St.
	250	EGY	R. Cairo, Abis
	500/100	G	BBC, United Kingdom
	250		Voice of America, Woofferton
	100	PAK	R. Pakistan Home Sce, Islamabad
	250	SNG	BBC Far Eastern Relay St.
6200		CHN	Lhasa, China (Tibetan Prgr.)
	20	RUS	R. Dvizhenie, Yekaterinburg, E
	20		R. N, Yekaterinburg, E
6200.6	2	PRU	R. LV de Huamanga, Peru(n.6070)
6205	500/100	EQA	HCJB, Quito, Ecuador
6210	10	HRV	Hrvatski Radio, Zagreb
6210	30	I	European Christian Radio
6220	100	BUL	R. Sofia, Kostinbrod, Bulgaria
6230	250/100	EGY	R. Cairo, Abu Zaabal
	500/100	MCO	Trans World Radio, Monaco
6235	250	BUL	R. Sofia, Plovdiv, Bulgaria
6243.3	0.12	PRU	R. Calca, Peru
6245	80	CVA	Vatican Radio
6250	10	GNE	Malabo, Equatorial Guinea
		KRE	Pyongyang BC St, No. Korea
6260		CHN	Qinghai 1, Xining, China
6261.1		PRU	R. Juanjuí, Yemen
6280	10	LBN	King of Hope, Lebanon
6281.7	1.5	PRU	R. Huancabamba, Peru (n.6290)
6299.4		HND	Sani R, Pto. Lempira, Honduras
6300		TWN	Family Radio via VOFC, Taiwan
6304.7		PRU	R. Acarí, Peru
6325.1r		TUR	Istanbul Polis Radyosu, Turkey
6325.8	0.25	PRU	R. Abancay, Peru
6329.6	0.8		Estación "C", Moyobamba, Peru
6330	1	BOL	R. Dif. Coral, Oruro, Bolivia
6339.5		PRU	R. San Miguel Arcángel, Peru
6343.0v		VTN	R. Son La, Vietnam
6363	0.2	PRU	R. Huallaga, Saposoa, Peru
6370.0v		VTN	R. Yen Bai, Vietnam
6383	50	MNG	Ulaanbaatar 2, Mongolia
6390.9		PRU	R. San Miguel, Peru
6400		KRE	Pyongyang BC St, Kanggye
6429	1	PRU	R. Espinar, Peru (n.6425)
6448		VTN	Hanoi I, Vietnam
6477.0v			R. Cao Bang, Vietnam
6480	100	KOR	R. Korea, Kimjae, So. Korea
6500	50	CHN	Xining, China (Tibetan prgr)
6525.0v		VTN	R. Bac Tai, Vietnam
6540	400/100	KRE	R. Pyongyang Ext. Sce, No. Korea
6550		LBN	R. Voice of Lebanon
6560		IRQ	R. Baghdad, Iraq
	400/100	KRE	R. Pyongyang Ext. Sce, No. Korea
6570	1(?)	BRM	Defence Forces BC, Myanmar
6571.5	0.17	PRU	R. Tacna, Peru
6576	400/100	KRE	R. Pyongyang Ext. Sce, No. Korea
6585.0v		PRU	R. Tradición, Huanta, Peru
6587			R. Santa Fé, Peru
6628.5	0.5		R. LV de S. Antonio, Peru
6647.0v	1	LAO	Pakse Regional R, Laos
6655		PRU	R. Santa Cruz, Peru *
6666.0r	0.01		R. Jepelacio, Peru
6670.2r			R. Santa Mónica, Peru
6690.5v	1	PRU	R. Cutervo, Peru
6724.5	0.15		R. Satélite, Santa Cruz, Peru
6750		CHN	CPBS 1, China
6754.7		PRU	R. La Merced, Tongod, Peru
6781.7			R. Tigre, Sorochuco, Peru
6790		CHN	CPBS Taiwan 2 Sce, China *
6803.2		PRU	Ondas del Mayo, Peru
6815.0v	0.1		R. Universo, Pandalle, Peru
6840		CHN	CPBS 1, China
		E	EBC (Time Signals), Spain
6851.2		PRU	R. Cultural, Cailloma, Peru
6890		CHN	CPBS 2, China
6895.2		PRU	R. Sensación, Huancabamba, Peru
6900	5	TUR	Turkish State Met. Sce.
6910.0rv		GNE	Africa 2000, Malabo, Eq. Guinea
6920	500/120	CHN	China Radio International
6933	500/120		China Radio International
6940		CHN	Yunnan M, Kunming, China
6955	500/120		China Radio International
6974	50		Hohhot, China (Mongolian prgr)
7032.0v		PRU	R. Azángaro, Peru (v. + 20kHz)
7080.0v	7.5	BGD	R. Bangladesh Home Sce.
7090	500/120	CHN	China Radio International
7100		ZAI	La Voix du Zaire, Kinshasa
7105	100	ALB	R. Tirana, Lushnjë, Albania
	250	ASC	BBC Atlantic Relay St.
	50	CHN	Hohhot, China (Chinese prgr)
	25	COG	RTV Congolaise, Brazzaville
	500	D	Deutsche Welle, Germany
	500		Voice of America, Wertachtal
	500/100	G	BBC, United Kingdom
	250	GRC	Voice of America, Kavala
	50	IND	All India Radio, Lucknow B
	100	SNG	BBC Far Eastern Relay St.
	300	TWN	CBS Network 2, Taipei, Taiwan
7110	50	ALB	R. Tirana, Lushnjë, Albania
	150	CHN	China Radio International
	50		Lhasa, China (Tibetan prgr)
	100	D	Deutsche Welle, Germany
	100	ETH	V. of Ethiopia, Gedja
	500/100	G	BBC, United Kingdom
		IND	All India Radio Ext. Sce.
	50		All India Radio, Gorakhpur
	20		All India Radio, Khampur
	250	MLT	IBRA Radio via DW Malta
	10	PAK	R. Pakistan Home Sce, Peshawar
	250	PHL	Voice of America, Philippines
	250	TUR	V. of Turkey, Ankara
	20/10	UGA	R. Uganda, Kampala
7113.0v	10	MOZ	Em. Interprovincial, Maputo
7115.0v	250	BGD	R. Bangladesh Ext. Sce.
	500	BUL	R. Sofia, Plovdiv, Bulgaria
		CHN	V. O. Pujiang, Shanghai, China
	10	CLN	Voice of America, Colombo
	100	D	RFE/RL, Biblis, Germany
	500/100	IRN	IRIB, Tehran, Iran
	20	KAZ	Alma-Ata 2, Kazakhstan
	250	POR	RFE/RL, Gloria, Portugal
	10	THA	R. Thailand Prgr 1
7120	100	BUL	R. Sofia, Kostinbrod, Bulgaria
	100	D	Deutsche Welle, Germany
	500	F	R. France Internationale
	500/100	G	BBC, United Kingdom
	250	KEN	KBC No. Ea.& Coastal Sce, Kenya
	50	PHL	Voice of America, Philippines
	1	SOM	R. Hargeisa, No. Somalia
	100	TCD	Radiodiff. Nat. Tchadienne
7125	10	CLN	Voice of America, Colombo
	500/100	CVA	Vatican Radio

	250	CYP	BBC Ea. Mediterranean Relay St.
	250		CyBC via BBC Relay Station
	500/100	G	BBC, United Kingdom
	100	GUI	Conakry, Guinea Rep. (irr.)
	10	I	IRRS, Milano, Italy
		IND	All India Radio Ext. Sce.
		2	All India Radio, Ranchi A
	20	RUS	Georgian R. via Balashikha
			R. Vedo, Volgograd, E
7130	500/100	D	Deutsche Welle, Germany
	100		RFE/RL, Lampertheim, Germany
	300	G	Voice of America, Woofferton
	250	GRC	Voice of America, Kavala
	500	HOL	R. Netherlands, Flevoland
	10	MLA	RTM Kuching, Sarawak, Malaysia
	250/50	TWN	Voice of Free China, Taiwan
7135	100	CYP	BBC Ea. Mediterranean Relay St.
	500/100	F	R. France Internationale
	250	GAB	R. France Int. via Gabon
7139	10	INS	RRI Ambon, Indonesia *
7140	150	CHN	China Radio International
	100	CYP	BBC Ea. Mediterranean Relay St.
	100	D	Deutsche Welle, Germany
v	0.5	I	R. Italia Int, Spoleto
		IND	All India Radio Ext. Sce.
	10		All India Radio, Hyderabad
	300	J	R. Japan, Yamata, Japan
	100	KEN	KBC National Sce, Nairobi
	50	RUS	Yakutsk, S
7145	100	ALG	RTV Algerienne, Bouchaoui
	500	D	Deutsche Welle, Germany
	100		RFE/RL, Biblis, Germany
	250		RFE/RL, Holzkirchen, Germany
	100	E	R. Exterior de España, Madrid
	500/100	F	R. France Internationale
v	10	LAO	Lao National R. (Ext. Sce.) *
	10	MLA	RTM Kuching, Sarawak, Malaysia
	100	POL	Polish Radio, Warszawa
	250	POR	RFE/RL, Gloria, Portugal
7150.0v	10	AFS	Capital Radio, Transkei
	350	ARS	BSKSA, Riyadh, Saudi Arabia
	250	AUS	R. Australia, Darwin
	500	BUL	R. Sofia, Plovdiv, Bulgaria
		CHN	R. Canada Int. via Xi'an, China
	100	CME	CRTV, Douala, Cameroon
	500/100	D	Deutsche Welle, Germany
	500/100	G	BBC, United Kingdom
	10	IND	All India Radio, Guwahati B
	10	PAK	R. Pakistan Home Sce, Quetta
		RUS	R. Vilnius via European tr. site
	300	TWN	CBS Network 1, Taipei, Taiwan
	1	ZAI	R. CANDIP, Bunia, Zaire
7154.0v	1	AGL	EP do Lobito, Angola
7154.3	20	NGR	La Voix du Sahel, Niger
7155	100	ALB	R. Tirana, Lushnjë, Albania
	100	D	RFE/RL, Biblis, Germany
	100		RFE/RL, Lampertheim, Germany
	250	E	RFE/RL, Playa de Pals, Spain
	300	G	R. Canada Int. via Skelton, UK
	500/100	IRN	IRIB, Tehran, Iran
	250	KOR	R. Canada Int. via Kimjae, Korea
	100	MDG	R. Madagasikara, Madagascar
	250	POR	RFE/RL, Gloria, Portugal
7160	500/100	F	R. France Internationale
	250	GAB	R. France Int. via Gabon
	250	GRC	Voice of America, Kavala
	50	IND	All India Radio, Khampur
	10		All India Radio, Madras A
	100	MCO	Trans World Radio, Monaco
	10	MLA	RTM Kuching, Sarawak, Malaysia
	100	OMA	BBC Eastern Relay Station
	50	PHL	R. Veritas Asia, Philippines
	250	SNG	BBC Far Eastern Relay St.
7164.0v	1	LAO	Luang Prabang Regional R, Laos
7165	100	D	RFE/RL, Biblis, Germany
	100	ETH	V. of Ethiopia, Gedja
	500	NOR	R. Norway Int. / R. Denmark
	250	POR	RFE/RL, Gloria, Portugal
	500		RFE/RL, Maxoqueira, Portugal
	10	SNG	"Olikkalanjiam", Singapore
	500	THA	Voice of America, Udorn Thani
	10	TZA	R. Tanzania National Sce.
	250/100	YUG	R. Yugoslavia, Beograd
7165.2	100/5	NPL	R. Nepal, Khumaltar
7170	50	ALB	R. Tirana, Lushnjë, Albania
	100	BUL	R. Sofia, Kostinbrod, Bulgaria
	500		R. Sofia, Plovdiv, Bulgaria
	50	CHN	Lhasa, China (Chinese prgr)
	500	D	Voice of America, Wertachtal
	300	G	Voice of America, Woofferton
	250	GRC	Voice of America, Kavala
	10	PAK	R. Pakistan Home Sce, Quetta
7170.4	20	NCL	RFO Noumea, New Caledonia
7173		INS	RRI Serui, Indonesia
7175	500	D	Deutsche Welle, Germany
	500	F	R. France Internationale
	500/100	G	BBC, United Kingdom
	50	I	RAI Radio 2, Caltanissetta
	20	RUS	Armenian R. via Balashikha
7180	250	CYP	BBC Ea. Mediterranean Relay St.
	100	D	RFE/RL, Lampertheim, Germany
	500		Voice of America, Wertachtal
	300	G	R. Canada Int. via Skelton, UK
	300		Voice of America, Woofferton
	500	GAB	R. Japan via Moyabi, Gabon
	250	HNG	BBC East Asian Relay St.
	10	IND	All India Radio, Port Blair
	100	SWZ	Trans World Radio, Swaziland
	250	TUR	V. of Turkey, Ankara
		UKR	R. Ukraine International
7184.9	50	MLI	Radiodiff.-TV du Mali, Bamako
7185	50	BRM	Voice of Myanmar, Yangon
	100	D	Deutsche Welle, Germany
	3	RUS	Tatar R, Perm, E
	250	TUR	V. of Turkey, Ankara
7190	20	BEN	ORTB, Parakou, Benin
	10	CLN	SLBC Ext. Sce, Colombo
	100	D	RFE/RL, Biblis, Germany
	100		RFE/RL, Lampertheim, Germany
	250	E	RFE/RL, Playa de Pals, Spain
	250	GRC	Voice of America, Kavala
	50	IND	All India Radio, Shillong
	500/100	IRN	IRIB, Tehran, Iran
	500	POR	RFE/RL, Maxoqueira, Portugal
	100	YEM	Yemen Radio & TV Corp, Aden
7195	15	CHN	Urumqi, China (May-Oct only)
	250/120	ROU	R. Romania International
v	20/10	UGA	R. Uganda, Kampala
		UKR	R. Ukraine International
7200	100/50	AFG	R. Afghanistan, Kabul
	500/100	G	BBC, United Kingdom
v	50	GNE	R. Africa, Malabo, Eq. Guinea
	50	IND	All India Radio, Thiru'puram
	400/100	KRE	R. Pyongyang Ext. Sce, No. Korea
	100	PHL	Voice of America, Philippines
	250	POR	Voice of America, Gloria
	50	RUS	Yakutsk, S
	100	SDN	Omdurman, Sudan
	25	SWZ	Trans World Radio, Swaziland
	50	YUG	R. Beograd 1, Yugoslavia
7205	100	AUS	R. Australia, Shepparton
	250	GRC	Voice of America, Kavala
	50		Voice of America, Rhodes
	10	I	AWR Europe, Forli, Italy
	1	POL	Rozglosina Harcerska, Poland
	10	ZAI	La Voix du Zaire, Lubumbashi
7210	350	ARS	BSKSA, Riyadh, Saudi Arabia
	50	BLR	Miensk, Belarus
	15	CHN	Yunnan G, Kunming, China
	500	D	Voice of America, Wertachtal
	500/100	G	BBC, United Kingdom
	250	GRC	Voice of America, Kavala
		IND	All India Radio Ext. Sce.
	10		All India Radio, Calcutta
	20		All India Radio, Khampur
	100	J	R. Japan, Yamata, Japan
	100	MRC	Voice of America, Tangier
	500	NOR	R. Norway Int. / R. Denmark
	250	POR	R. Canada Int. via Sines
	50	RUS	Khabarovsk / R. St. Vostok, FE
	250	SUI	Red Cross Broadc. Sce, Geneva

7211	1	INS	RRI Biak, Indonesia
7215.0v	10	AGL	R. Nacional, Cazenga, Angola
	50	CHN	V. O. Jinling, Nanjing, China
	20	CTI	Radiodiff.-TV Ivoirienne
	500	NOR	R. Norway Int. / R. Denmark
	250/50	PAK	R. Pakistan Ext. Sce.
	100	SNG	BBC Far Eastern Relay St.
	500	UAE	R. of the UAE, Abu Dhabi
7220	50	ARS	BSKSA, Dirriya, S. Arabia
	20/100	CAF	R. TV Centrafricaine, Bangui
	250	D	RFE/RL, Holzkirchen, Germany
	100		RFE/RL, Lampertheim, Germany
	250	E	RFE/RL, Playa de Pals, Spain
	250/100	HNG	R. Budapest, Hungary
	250/100	YUG	R. Yugoslavia, Beograd
	10	ZMB	ZNBC Radio One, Lusaka, Zambia
7225	100	BUL	R. Sofia, Kostinbrod, Bulgaria
	15	CHN	Sichuan 1, Chengdu, China
	250	CLN	Deutsche Welle, Trincomalee
	100	D	Deutsche Welle, Germany
	300	G	Voice of America, Woofferton
		IND	All India Radio Ext. Sce.
	100		All India Radio, Khampur
	100		All India Radio, Kingsway
	250	MLT	IBRA Radio via DW Malta
	250/120	ROU	R. Romania International
	250	RRW	Deutsche Welle, Kigali, Rwanda
		UZB	R. Tashkent Ext. Sce, Uzbekistan
7230	500	AFS	BBC via Meyerton, South Africa
	250		Channel Africa, Johannesburg
	50	BFA	R. Burkina, Ouagadougou
	500/100	G	BBC, United Kingdom
	300		R. Canada Int. via Skelton, UK
	250		R. Japan via Skelton, UK
	10	I	AWR Europe, Forli, Italy
	20	IND	All India Radio, Kurseong
	0.5	INS	RRI Fak-Fak, Indonesia
	400/100	KRE	R. Pyongyang Ext. Sce, No. Korea
	50	OMA	R. Oman, Seeb
	200	RUS	R. Pamyat, Yekaterinburg, S
7235	250	CYP	BBC Ea. Mediterranean Relay St.
	500/100	D	Deutsche Welle, Germany
	300	G	R. Canada Int. via Skelton, UK
	100	I	RAI Ext. Sce, Rome, Italy
	20	KAZ	Alma-Ata 2, Kazakhstan
	250	MLT	Deutsche Welle, Cyclops, Malta
	100	OMA	BBC Eastern Relay Station
	20	UKR	Unknown site, Ukraine
v	50	ZMB	ZNBC Radio Two, Lusaka, Zambia
7240	10	AUS	R. Australia, Brandon
	100	CME	CRTV, Garoua, Cameroon
	10	IND	All India Radio, Bombay B
		KAZ	Alma-Ata 2, Kazakhstan
	50	PHL	R. Veritas Asia, Philippines
		UKR	R. Ukraine International
7242	100	MOZ	Em. Nacional, Maputo
7245.0v	100/10	AGL	R. Nacional, Mulenvos, Angola
	50	ALG	RTV Algerienne, Ouled Fayet
	100	D	RFE/RL, Biblis, Germany
	250		RFE/RL, Holzkirchen, Germany
	100		RFE/RL, Lampertheim, Germany
	250	GRC	Voice of America, Kavala
	250	POR	RFE/RL, Gloria, Portugal
	100	TJK	Yangi-Yul, Tadzhikistan
	20	UKR	Lugansk, Ukraine
7250	350/50	ARS	BSKSA, Saudi Arabia
	500/100	CVA	Vatican Radio
	400/100	KRE	R. Pyongyang Ext. Sce, No. Korea
	250	MLT	Deutsche Welle, Cyclops, Malta
	50	SNG	"Warna", Singapore
	100	TWN	CBS Network 1, Taipei, Taiwan
	500	YUG	R. Yugoslavia, Beograd
7255	50	BOT	R. Botswana, Gaborone
	500	D	Deutsche Welle, Germany
	100		RFE/RL, Biblis, Germany
	250		RFE/RL, Holzkirchen, Germany
	300	J	R. Canada Int. via Yamata, Japan
	100	KAZ	Alma-Ata Ext. Sce, Kazakhstan
			R. Tashkent, Uzbekistan (relay)
	250	NIG	V. of Nigeria, Lagos
	500	POR	RFE/RL, Maxoqueira, Portugal
	250/120	ROU	R. Romania International
7260.0v		AGL	EP do Kuanza Sul, Angola *
	100	ALB	R. Tirana, Krujë, Albania
	300/100	AUS	R. Australia, Carnarvon
	120	CHN	China Radio International
	60	COM	R. Comoro, Moroni
	250	CYP	BBC Ea. Mediterranean Relay St.
	100	D	Deutsche Welle, Germany
	500/100	G	BBC, United Kingdom
	100	IND	All India Radio, Bombay
	50	MNG	Ulaanbaatar Ext. Sce, Mongolia
	250	PHL	Voice of America, Philippines
	10	VUT	R. Vanuatu, Malapoa
7264	50	MNG	Ulaanbaatar 1, Mongolia
7265	100	BOT	Voice of America, Botswana
	20	D	Südwestfunk, Rohrdorf
		IND	All India Radio Ext. Sce.
	300	J	R. Canada Int. via Yamata, Japan
	250	MLT	Deutsche Welle, Cyclops, Malta
	100	PAK	Azad Kashmir R, Trarkhel
	250	PHL	R. Veritas Asia, Philippines
	50	RUS	Yakutsk, S
v	100	TGO	Radiodiffusion Togolaise, Lomé
7270	250	AFS	Channel Africa, Johannesburg
	100/50	ALB	R. Tirana, Lushnjë / Krujë
	50	CHN	Hohhot, China (Mongolian prgr)
	250/100	CYP	BBC Ea. Mediterranean Relay St.
	500/100	G	BBC, United Kingdom
	100	GAB	Radiodiffusion Gabonaise
	250	GRC	Voice of America, Kavala
	50	INS	RRI Yogyakarta, Indonesia
	250	KEN	KBC General Sce, Nairobi
	100	MCO	Trans World Radio, Monaco
	10	MLA	RTM Kuching, Sarawak, Malaysia
	250	MLT	Deutsche Welle, Cyclops, Malta
	100	OMA	R. Oman, Seeb
	100	POL	Polish Radio, Warszawa
	250	POR	Deutsche Welle, Sines
	250		Voice of America, Gloria
	100	SEY	FEBA Radio, Seychelles
7275	350	ARS	BSKSA, Riyadh, Saudi Arabia
	7.5	CHN	Guizhou 1, Guiyang, China
	100	D	Deutsche Welle, Germany
	250		RFE/RL, Holzkirchen, Germany
	100	E	R. Exterior de España, Madrid
	100	I	RAI Ext. Sce, Rome, Italy
	250	KOR	R. Korea, Kimjae, So. Korea
	10	LBR	"ELBC", Monrovia, Liberia
	100	NIG	R. Nigeria, Kaduna
7280	50	ARS	BSKSA, Jeddah, Saudi Arabia
	100	BOT	Voice of America, Botswana
	50	CHN	Haixia 1, China
	500/100	F	R. France Internationale
	250	GRC	Voice of America, Kavala
	50	IND	All India Radio, Guwahati A
	20	KAZ	Karaganda 2, Kazakhstan
	30	TZA	R. Tanzania National Sce.
7285	250	CYP	BBC Ea. Mediterranean Relay St.
	500	D	Deutsche Welle, Germany
	500/100	G	BBC, United Kingdom
	50	MLI	Radiodiff.-TV du Mali, Bamako
	50	NIG	R. Nigeria Ch.1, Lagos
	100	POL	Polish Radio, Warszawa
	250	POR	Deutsche Welle, Sines
	250		Voice of America, Gloria
	100	TWN	Voice of Asia, Taiwan
7290	100	BUL	R. Sofia, Kostinbrod, Bulgaria
	100	I	RAI Ext. Sce, Rome, Italy
	100	IND	All India Radio, Khampur
	500	MCO	Trans World Radio, Monaco
	0.25	NZL	ZLXA - Print Disabled R, Levin
	250/50	PAK	R. Pakistan Ext. Sce.
7294	0.4	I	R. Europe, Milano, Italy
7295	100	D	RFE/RL, Biblis, Germany
	250		RFE/RL, Holzkirchen, Germany
	100		RFE/RL, Lampertheim, Germany
	500/100	G	BBC, United Kingdom
	300		Voice of America, Woofferton
	100	GHA	GBC-2, Accra, Ghana

SHORTWAVE TABLE

	250	GRC	Voice of America, Kavala
	50	IND	All India Radio, Aizawl
	100	MLA	RTM R4, Kuala Lumpur, Malaysia
		TWN	CBS News Network, Taiwan
	10	ZAI	La Voix du Zaire, Mbujimayi
7300	20	RUS	Azerbaidzhani R. via Balashikha
7305	500/100	CVA	Vatican Radio
	500	F	R. France Internationale
		RUS	China Radio Int. via Russia
7310	500/100	CVA	Vatican Radio
		RUS	AWR Europe via Yekaterinburg
7315	250		Deutsche Welle via Zhigulevsk
	100	USA	WHRI, Noblesville, IN
		UZB	BBC via Tashkent, Uzbekistan
7320	500/100	G	BBC, United Kingdom
	50	RUS	Arman, FE
7325	500/100	G	BBC, United Kingdom
	300		Voice of America, Woofferton
7335	10	CAN	CHU (Time Signals), Canada
7340	1000	RUS	Deutsche Welle via Novosibirsk
7345	50		Yakutsk, S
	250/100	TCH	Czech & Slovak R. International
7350		CHN	Heilongjiang 1, Harbin, China
		IRQ	R. Baghdad, Iraq
7355	100	ALS	KNLS, Anchor Point, Alaska
	100	USA	WHRI, Noblesville, IN
	100		WRNO, New Orleans, LA
	100		WYFR, Okeechobee, FL
7360	500/120	CHN	China Radio International
7365	100	ALS	KNLS, Anchor Point, Alaska
	500/100	CVA	Vatican Radio
	250	USA	KGEI, Redwood City, CA
7370	400	RUS	R. Ala, Yekaterinburg, S
	1	TUR	Turkish Police R, Ankara
7375	500/120	CHN	China Radio International
v	30	CTR	R. For Peace Int, Costa Rica
7380	400	RUS	R. Ala, Sankt-Peterburg, E
7385	500/120	CHN	China Radio International
			Urumqi, China (May-Oct only)
	100	MCO	Trans World Radio, Monaco
7395	500	USA	WEWN, Birmingham, AL
	100		WRNO, New Orleans, LA
	500		WSHB, Cypress Creek, SC
7400	200	RUS	R. Ala, Sankt-Peterburg, E
			R. Vilnius via European tr. site
7405	500/120	CHN	China Radio International
	250	USA	Voice of America, Bethany, OH
	500		Voice of America, Greenville
7412		IND	All India Radio Ext. Sce.
	250		All India Radio, Aligarh
7416		VTN	Voice of Vietnam, Hanoi
7420	500/120	CHN	China Radio International
7435	100	USA	WWCR, Nashville, TN
7440	500	ARM	Yerevan, Armenia (Foreign Sce)
		CHN	CPBS 2, China
7445	100	TWN	Voice of Asia, Taiwan
	250/50		Voice of Free China, Taiwan
7450	250	GRC	Voice of America, Kavala
7465	20/500	ISR	Kol Israel, Jerusalem
	100	USA	WJCR, Millerstown, KY
	100		WRNO, New Orleans, LA
7470	500/120	CHN	China Radio International
7475	100	TUN	ERTT, Tunis
7480	120	CHN	Swiss Radio Int. via Beijing
7490	100	USA	WJCR, Millerstown, KY
7500.0r	500	SYR	R. Damascus, Syria
7504		CHN	CPBS 1, China
7510	100	USA	KTBN, Salt Lake City, UT
	500		WCSN, Scotts Corners, ME
	500		WEWN, Birmingham, AL
7516		CHN	CPBS 2, China
7540.0v	5	I	Voice of Europe, Pordenone
	500	USA	WEWN, Birmingham, AL
7550	100	KOR	R. Korea, Kimjae, So. Korea
7580	400/100	KRE	R. Pyongyang Ext. Sce, No. Korea
7600	1	EQA	HD2IOA (Time Signals), Ecuador
7620		CHN	CPBS Taiwan 1 Sce, China *
7660	500/120		China Radio International
7670	15	BUL	Stolnik, Bulgaria (Home Sce.)
7770		CHN	CPBS 2, China

7780	500/120		China Radio International
7820	500/120		China Radio International
7865	10	ZAI	La Voix du Zaire, Kananga *
7935		CHN	CPBS 1, China
8000	2	J	JJY, Japan (Time Signals)
8065		PRU	R. El Paraíso de los Andes,Peru
8461	10	AFS	ZSC (Time Signals), So. Africa
8473		CLN	4PB (Time Signals), Sri Lanka
8515.0v	0.2	PRU	R. Amistad, Soritor, Peru
8539		HKG	VPS35 (Time Sigs), Hong Kong
8542	3	INS	PKI (Time Signals), Indonesia
8566		CHN	CPBS Minority Sce, China
8638	10	AUS	VNG, Llandilo (Time Signals)
8650		PRU	OBC3 (Time Signals), Peru
8677		CHL	CBV (Time Signals), Chile
8721		B	PPE (Time Signals), Brazil
8930.0v	0.15	PRU	R. Continente, Juanjuí (n.8900)
9022	500/100	IRN	IRIB, Tehran, Iran
9064		CHN	CPBS 2, China
9080			CPBS 1, China
9170			CPBS Taiwan 2 Sce, China
9190	100	SDN	Omdurman, Sudan (alt.9170)
9280	100	TWN	Family Radio via VOFC, Taiwan
	100		Voice of Asia, Taiwan
9290		CHN	CPBS 1, China
9325	400/100	KRE	R. Pyongyang Ext. Sce, No. Korea
9345	400/100		R. Pyongyang Ext. Sce, No. Korea
9350	500	USA	WCSN, Scotts Corners, ME
	500		WEWN, Birmingham, AL
9351	10/20	CHN	BPM (Time Signals), China
9355	100	MRA	KHBI, Saipan, No. Marianas
9365	500/120	CHN	China Radio International
9370	250/50	PAK	R. Pakistan Ext. Sce.
9380		CHN	CPBS Taiwan 1 Sce, China *
9388	20/500	ISR	Kol Israel, Jerusalem
9395	100	GRC	Voice of Greece, Athens
9410	500/100	G	BBC, United Kingdom
	500	USA	WEWN, Birmingham, AL
9420	100	GRC	Voice of Greece, Athens
9425	250		ERT, Thessaloniki via Kavala
	100		Voice of Greece, Athens
9435	20/500	ISR	Kol Israel, Jerusalem
	500/100	MCO	Trans World Radio, Monaco
9440	500/120	CHN	China Radio International
9445	250	TUR	V. of Turkey, Ankara
9450		RUS	Yerevan, Armenia (relay)
9455		CHN	CPBS Taiwan 1 Sce, China
	250	EGY	R. Cairo, Abis
	100	MRA	KHBI, Saipan, No. Marianas
	500	USA	Voice of America, Greenville
	500		WSHB, Cypress Creek, SC
9460	10	PAK	R. Pakistan Home Sce, Islamabad
	500	TUR	V. of Turkey, Ankara
9462	3	PRU	R. Grau, Piura, Peru
9465	100	MRA	KFBS, Saipan, No. Marianas
		TWN	Family Radio via VOFC, Taiwan
	250	USA	Voice of America, Delano, CA
	100		WHRI, Noblesville, IN
	50		WMLK, Bethel, PA
	500		WSHB, Cypress Creek, SC
9475	250	EGY	R. Cairo, Abis
	100	MRA	KFBS, Saipan, No. Marianas
9480	500	ARM	Yerevan, Armenia (Foreign Sce)
	500/120	CHN	China Radio International
	100	MCO	Trans World Radio, Monaco
9485	100		Trans World Radio, Monaco
	250/50	PAK	R. Pakistan Ext. Sce.
9486.0v	0.2	PRU	R. Tacna, Peru
9490	500/120	CHN	China Radio International
			Lhasa, China (alt. freq) *
	500/100	MCO	Trans World Radio, Monaco
9495	100	GUM	KSDA-Adventist World R, Guam
	100	MRA	KFBS, Saipan, No. Marianas
	100		KHBI, Saipan, No. Marianas
	100	PHL	FEBC Manila, Philippines
	100	USA	WHRI, Noblesville, IN
	500		WSHB, Cypress Creek, SC
9500	500/100	CVA	Vatican Radio
	100	SWZ	Trans World Radio, Swaziland

9505		AUT	R. Canada Int. via Moosbrunn
	7.5	B	R. Record, Sao Paulo, Brazil
	50	CHN	Haixia 2, China
	500/50	CUB	R. Habana, Cuba
	100	D	RFE/RL, Biblis, Germany
	250	E	RFE/RL, Playa de Pals, Spain
	20	KAZ	Alma-Ata 2, Kazakhstan
	400/100	KRE	R. Pyongyang Ext. Sce, No. Korea
	100	PHL	Voice of America, Philippines
	250	POR	RFE/RL, Gloria, Portugal
	500		RFE/RL, Maxoqueira, Portugal
	0.25	PRU	R. Tacna, Peru
	250/100	TCH	Czech & Slovak R. International
	500	UAE	R. of the UAE, Abu Dhabi
	100	USA	WYFR, Okeechobee, FL
	500	YUG	R. Yugoslavia, Beograd
9509.0v	50	ALG	RTV Algerienne, Ouled Fayet
9510	250	AUS	R. Australia, Carnarvon
	250	CLN	Deutsche Welle, Trincomalee
	250	MLT	IBRA Radio via DW Malta
	100	NZL	R. New Zealand International
	50	PHL	Voice of America, Philippines
v	5	PRU	R. América, Lima, Peru (irr.)
	250/120	ROU	R. Romania International
	250/50	TWN	Voice of Free China, Taiwan
	50	USA	KJES, Vado, NM
9515	500	AFS	BBC via Meyerton, South Africa
	50	ATN	Trans World Radio, Bonaire
	10	B	R. Novas de Paz, Curitiba
	250	CAN	BBC via Sackville, NB
	100	D	Deutsche Welle, Germany
	50	I	RAI Radio 1, Caltanissetta
	100	KOR	R. Korea, Kimjae, So. Korea
	0.5	MEX	LV de América Latina, México
	250	MLT	Deutsche Welle, Cyclops, Malta
	250/50	PAK	R. Pakistan Ext. Sce.
	100		R. Pakistan Home Sce, Islamabad
	100	SWZ	Trans World Radio, Swaziland
	250	USA	BBC via Delano, CA
	250		Voice of America, Delano, CA
9520	500	BUL	R. Sofia, Plovdiv, Bulgaria
	50	CHN	Hohhot, China (Chinese prgr)
	250	D	RFE/RL, Holzkirchen, Germany
	500/250	E	RFE/RL, Playa de Pals, Spain
	250	PHL	R. Veritas Asia, Philippines
	10	PNG	NBC Port Moresby, Papua N. G.
	5	PRU	R. La Crónica, Lima, Peru *
	100	SWZ	Trans World Radio, Swaziland
9525	250	CLN	Deutsche Welle, Trincomalee
	500	D	Voice of America, Wertachtal
		INS	RRI Jakarta (Prgr. Nasional)
	500/100	IRN	IRIB, Tehran, Iran
	100	KOR	R. Korea, Kimjae, So. Korea
	10	MOZ	Em. Interprovincial, Maputo
	100	PHL	Voice of America, Philippines
	100	POL	Polish Radio, Warszawa
	250	USA	R. Martí, Greenville, NC
	250		Voice of America, Bethany, OH
	250		Voice of America, Greenville
9530	100	ALB	R. Tirana, Lushnjë / Krujë
	500	D	Voice of America, Wertachtal
	350	E	R. Exterior de España, Madrid
	250	GRC	Voice of America, Kavala
	100	GUM	KSDA-Adventist World R, Guam
	100	MRA	KHBI, Saipan, No. Marianas
	100	MRC	Voice of America, Tangier
	250	POR	RFE/RL, Gloria, Portugal
	250/120	ROU	R. Romania International
	50	RUS	Arman, FE
	175	USA	Voice of America, Bethany, OH
9535.0v	100	AGL	R. Nacional, Mulenvos, Angola
	100	ALG	RTV Algerienne, Bouchaoui
	50	ATN	Trans World Radio, Bonaire
	250	CAN	R. Canada Int, Sackville
	250	CLN	Deutsche Welle, Trincomalee
	300		R. Japan, Ekala, Sri Lanka
	500/50	CUB	R. Habana, Cuba
	500/100	D	Deutsche Welle, Germany
	250	GRC	Voice of America, Kavala
	100	IND	All India Radio, Madras
	0.6/0.3	J	NHK Sapporo / Osaka, Japan
	300		R. Japan, Yamata, Japan
		KOR	R. Canada Int. via Kimjae, Korea
	250	SUI	Swiss Radio International
9538.8	0.6	J	NHK Fukuoka (SSB) (9538.75)
9540	250	AUS	R. Australia, Carnarvon
	10	B	R. Educadora da Bahia, Brazil
	250	D	RFE/RL, Holzkirchen, Germany
	100		RFE/RL, Lampertheim, Germany
	100	POL	Polish Radio, Warszawa
		UZB	R. Tashkent Ext. Sce, Uzbekistan
	25	VEN	R. Nacional de Venezuela
9545	250	ATG	Deutsche Welle, Antigua
	250	CAN	Deutsche Welle via Sackville
	500/100	D	Deutsche Welle, Germany
	50	GRC	Voice of America, Rhodes
	250	MLT	Deutsche Welle, Cyclops, Malta
	250	PHL	Voice of America, Philippines
	10	SLM	SIBC Honiara, Solomon Islands
9550	5	B	R. Nova Esperança, Porto Alegre
	500/50	CUB	R. Habana, Cuba
	500/100	F	R. France Internationale
		IND	All India Radio Ext. Sce.
	20	KAZ	Alma-Ata 2, Kazakhstan
	10	LBR	ELWA, Monrovia, Liberia *
	250/120	ROU	R. Romania International
9552	7.5	INS	RRI Ujung Padang, Indonesia
9553.8	0.6	J	NHK Tokyo (SSB) (9553.75)
9555	250	AFS	Channel Africa, Johannesburg
		CHN	R. Canada Int. via Xi'an, China
	100	D	RFE/RL, Biblis, Germany
	300	G	R. Canada Int. via Skelton, UK
	0.5	MEX	La Hora Exacta, México City
	250/50	PHL	R. Veritas Asia, Philippines
	100	POR	R. Portugal, S. Gabriel
	250	USA	Voice of America, Greenville
	100		WYFR, Okeechobee, FL
9560	250	ASC	BBC Atlantic Relay St.
	100	AUS	R. Australia, Carnarvon
	250		R. Australia, Darwin
	100	BUL	R. Sofia, Kostinbrod, Bulgaria
		CHN	Urumqi, China (Chinese prgr)
	250	CYP	BBC Ea. Mediterranean Relay St.
	100	ETH	V. of Ethiopia, Gedja
	500	FNL	R. Finland
	500/100	G	BBC, United Kingdom
	300	J	R. Japan, Yamata, Japan
	500	JOR	R. Jordan, Al Karanah
	250/50	PHL	R. Veritas Asia, Philippines
	20	RUS	FEBC via Khabarovsk, FE
	500/250	SUI	Swiss Radio International
	500	THA	Voice of America, Udorn Thani
9565	350	ARS	BSKSA, Riyadh, Saudi Arabia
	20	B	R. Universo, Curitiba, Brazil
	100	D	Deutsche Welle, Germany
	100		RFE/RL, Biblis, Germany
	250	MLT	Deutsche Welle, Cyclops, Malta
	500/350	NOR	R. Norway Int. / R. Denmark
	250	POR	RFE/RL, Gloria, Portugal
	250	RRW	Deutsche Welle, Kigali, Rwanda
	100	SEY	FEBA Radio, Seychelles
	250	USA	Voice of America, Bethany, OH
	250		V. of the OAS, Bethany, OH
9570	50	ARS	BSKSA, Jeddah, Saudi Arabia
v	250	BGD	R. Bangladesh Ext. Sce.
	300	CHN	China Radio International
	500	D	Deutsche Welle, Germany
	100		RFE/RL, Lampertheim, Germany
	250/100	KOR	R. Korea, Kimjae, So. Korea
	50	NIG	R. Nigeria, Kaduna
	250	PHL	Voice of America, Philippines
	100	POR	R. Portugal, S. Gabriel
	250		Voice of America, Gloria
	250/120	ROU	R. Romania International
	250/100	SNG	BBC Far Eastern Relay St.
9575	100	I	RAI Ext. Sce, Rome, Italy
	250	MRC	R. Mediterranée Int, Nador
	50	PHL	Voice of America, Philippines
	100	POR	R. Renascença, Muge, Portugal
	250		Voice of America, Gloria

SHORTWAVE TABLE

	250	USA	Voice of America, Bethany, OH
9580	100	ALB	R. Tirana, Lushnjë, Albania
	100	AUS	R. Australia, Shepparton
	250	CYP	BBC Ea. Mediterranean Relay St.
	100	E	R. Exterior de España, Madrid
	300	G	Voice of America, Woofferton
	500	GAB	Africa No.1, Moyabi, Gabon
	100	J	R. Japan, Yamata, Japan
	100	OMA	BBC Eastern Relay Station
	1	PRU	R. Concordia, Arequipa, Peru
	100	SNG	BBC Far Eastern Relay St.
	250/100	TCH	Czech & Slovak R. International
	500	YUG	R. Yugoslavia, Beograd
9585	250	AFS	Channel Africa, Johannesburg
	10	B	R. CBN (Globo), Sao Paulo
	100	F	R. France Internationale
	500/100	G	BBC, United Kingdom
	300		Voice of America, Woofferton
	50	GNE	R. Ea. Africa, Malabo, Eq. Guinea
	250	GRC	Voice of America, Kavala
	100	OMA	BBC Eastern Relay Station
	250	POR	Voice of America, Gloria
	250	USA	Voice of America, Greenville
9590	250	ATN	R. Netherlands, Bonaire
	250	CAN	BBC via Sackville, NB
	500/50	CUB	R. Habana, Cuba
	100	CYP	BBC Ea. Mediterranean Relay St.
	100	GUM	KTWR-Trans World Radio, Guam
	300	J	R. Japan, Yamata, Japan
	300	MDG	R. Netherlands, Madagascar
	500/350	NOR	R. Norway Int. / R. Denmark
	100	OMA	BBC Eastern Relay Station
	250	PHL	Voice of America, Philippines
	250/120	ROU	R. Romania International
	250	USA	BBC via Delano, CA
	250		R. Martí, Bethany, OH
	250		Voice of America, Bethany, OH
	250		Voice of America, Delano, CA
	250		Voice of America, Greenville
9595	50	CHN	Urumqi, China (Uighur prgr)
	250	D	RFE/RL, Holzkirchen, Germany
	100		RFE/RL, Lampertheim, Germany
	50	J	R. Tanpa 1, Tokyo, Japan
		KOR	R. Canada Int. via Kimjae, Korea
	300	UAE	UAE Radio & TV, Dubai
	1.5	URG	R. Monte Carlo, Uruguay
9600	250	ASC	BBC Atlantic Relay St.
	50	BOT	R. Botswana, Gaborone
	500/100	CVA	Vatican Radio
	500	EQA	HCJB, Quito, Ecuador
	500/100	G	BBC, United Kingdom
	400/100	KRE	R. Pyongyang Ext. Sce, No. Korea
	100	LBY	LJB, Benghazi, Libya
	1	MEX	R. UNAM, México City
	100	OMA	BBC Eastern Relay Station
	100	POR	R. Portugal, S. Gabriel
	100		R. Renascença, Muge, Portugal
	50	RUS	Arman, FE
	250	SNG	BBC Far Eastern Relay St.
9605	500/100	CVA	Vatican Radio
	500/100	D	Deutsche Welle, Germany
	100		RFE/RL, Lampertheim, Germany
	100	F	R. France Internationale
	500/100	G	BBC, United Kingdom
	300	MDG	R. Netherlands, Madagascar
	250/100	SNG	BBC Far Eastern Relay St.
	250/100	TCH	Czech & Slovak R. International
	500	UAE	R. of the UAE, Abu Dhabi
	100	USA	WYFR, Okeechobee, FL
9610	250	ASC	BBC Atlantic Relay St.
	50/10	AUS	VLW9, Wanneroo, Australia
	50	COG	RTV Congolaise, Brazzaville
	100	D	Deutsche Welle, Germany
		IND	All India Radio Ext. Sce.
	50		All India Radio, Kingsway
v		INS	RRI Samarinda, Indonesia
	500/100	IRN	IRIB, Tehran, Iran
	100	MTN	R. Mauritanie, Nouakchott.
		TWN	CBS News Network, Taiwan
9615	50/10	BHU	BBS, Thimpu, Bhutan
	250	CLN	Deutsche Welle, Trincomalee
	100/50	D	Deutsche Welle, Germany
	500		R. Canada Int. via Wertachtal
	100	IND	All India Radio, Bombay
	50		All India Radio, Kingsway
	100	MRC	Voice of America, Tangier
	500/350	NOR	R. Norway Int. / R. Denmark
	250	PHL	R. Veritas Asia, Philippines
	250	POR	Deutsche Welle, Sines
	250		R. Portugal via Sines
	500	THA	Voice of America, Udorn Thani
	50	USA	KGEI, Redwood City, CA
9615.3	7.5	B	R. Cultura, Sao Paulo, Brazil
9618	25	MOZ	Em. Nacional, Maputo
	120/25		R. Moçambique Ext. Sce.
9620	100	CHN	R. Ext. de España via Beijing
	100	EGY	R. Cairo, Abu Zaabal
	250/50	PHL	Voice of America, Philippines
	350	S	R. Sweden
	250/100	YUG	R. Yugoslavia, Beograd
9620.3	1	URG	S. O. D. R. E, Montevideo, Uruguay
9625	15	BOL	R. Fides, La Paz, Bolivia
	100	CAN	CBC No. Quebec Sce, Canada
	100	D	Deutsche Welle, Germany
	100		RFE/RL, Biblis, Germany
	100		RFE/RL, Lampertheim, Germany
	250	E	RFE/RL, Playa de Pals, Spain
	500	GAB	AWR Africa via Moyabi, Gabon
	250	MLT	Deutsche Welle, Cyclops, Malta
	500	POR	RFE/RL, Maxoqueira, Portugal
	250/120	ROU	R. Romania International
9630	100	AFS	R. Oranje, Johannesburg
	100	ALB	R. Tirana, Lushnjë / Krujë
	250	ATN	R. Netherlands, Bonaire
	10	B	R. Aparecida, Brazil
	10	CHL	R. Agricultura, Chile *
	250	POR	R. Portugal via Sines
	250	SEY	BBC Indian Ocean Relay St.
	300	TWN	CBS Netw. 4/1, Taipei, Taiwan
9635	100	AFG	R. Afghanistan, Kabul
	250	CAN	R. Canada Int, Sackville
	250	CYP	BBC Ea. Mediterranean Relay St.
	500/100	G	BBC, United Kingdom
	250	GRC	Voice of America, Kavala
	50	MLI	Radiodiff.-TV du Mali, Bamako
	100	POR	R. Portugal, S. Gabriel
	50	SNG	"City Sounds", Singapore
9637	100	MOZ	Beira 2, Mozambique
9640	50	ALG	RTV Algerienne, Bouchaoui
	125	ATG	BBC Caribbean Relay St.
	250	B	Deutsche Welle via R. Bras
	500/50	CUB	R. Habana, Cuba
	500/100	D	Deutsche Welle, Germany
	500	IRN	IRIB, Mashhad, Iran
	100	J	R. Japan, Yamata, Japan
	250/100	KOR	R. Korea, Kimjae, So. Korea
	400/100	KRE	R. Pyongyang Ext. Sce, No. Korea
	250	MLT	Deutsche Welle, Cyclops, Malta
	500	NOR	R. Norway Int. / R. Denmark
	100	SWZ	Trans World Radio, Swaziland
	10	VEN	Ecos del Torbes, Venezuela
9645	7.5	B	R. Bandeirantes, Sao Paulo
	35	CLN	Voice of America, Colombo
	0.5	CTR	Faro del Caribe, Costa Rica
	500/100	CVA	Vatican Radio
	500	GAB	R. Japan via Moyabi, Gabon
	500	NOR	R. Norway Int. / R. Denmark
	100	PAK	R. Pakistan Home Sce, Islamabad
	50	PHL	Voice of America, Philippines
9650	250	CAN	R. Canada Int, Sackville
	250		R. Korea via Sackville, Canada
	500/100	CVA	Vatican Radio
	500	D	Deutsche Welle, Germany
	350	E	R. Exterior de España, Madrid
	250	GRC	Voice of America, Kavala
	100	GUI	Conakry, Guinea Rep. (irr.)
	100	GUM	KSDA-Adventist World R, Guam
	500	IRQ	R. Iraq Int, Salah el Deen
	300	J	R. France Int. via Japan

	400/100	KRE	R. Pyongyang Ext. Sce, No. Korea
	100	MCO	Trans World Radio, Monaco
	100	MRC	Voice of America, Tangier
	350	NOR	R. Norway Int. / R. Denmark
	250	POR	Deutsche Welle, Sines
	500/250	SUI	Swiss Radio International
	25	SWZ	Trans World Radio, Swaziland
		ZAI	Kinshasa, Zaire (irr.)
9655	250	B	R. Bras Ext. Sce, Brasilia
	500	CHN	China Radio International
	500/50	CUB	R. Habana, Cuba
	350	NOR	R. Norway Int. / R. Denmark
	1	PRU	R. Nor Peruana, Peru
	20	RUS	R. Vedo, Volgograd, E
	350	S	R. Sweden
	100	SWZ	Trans World Radio, Swaziland
	100	THA	R. Thailand Ext. Sce.
9659.4	10	VEN	R. Rumbos, Caracas, Venezuela
9660	10	AUS	VLQ9, Brisbane, Australia
	250/100	CYP	BBC Ea. Mediterranean Relay St.
	250	D	RFE/RL, Holzkirchen, Germany
	250	E	RFE/RL, Playa de Pals, Spain
	300	G	Voice of America, Woofferton
	300	J	R. Canada Int. via Yamata, Japan
	50	PHL	FEBC Manila, Philippines
	100	SEY	FEBA Radio, Seychelles
9665	100	AFS	R. Suid Afrika, Johannesburg
	10	B	R. Marumby, Florianópolis
	500	CHN	China Radio International
	500/100	D	Deutsche Welle, Germany
v	200	KRE	KCBS, Pyongyang, No. Korea
	250	POR	IBRA Radio via Sines, Portugal
	250	QAT	Qatar Broadc. Sce, Doha
	250/120	ROU	R. Romania International
	250	TUR	V. of Turkey, Ankara
9670	250	ATG	Deutsche Welle, Antigua
	50	CHN	China Radio International
	250	CLN	Deutsche Welle, Trincomalee
	250	CYP	BBC Ea. Mediterranean Relay St.
	250	EGY	R. Cairo, Abis
	100	EQA	HCJB, Quito, Ecuador
	500/100	G	BBC, United Kingdom
	300		R. Canada Int. via Skelton, UK
	250		R. Japan via Skelton, UK
	100	MRA	KFBS, Saipan, No. Marianas
	100	MRC	Voice of America, Tangier
	250	PHL	Voice of America, Philippines
	350	S	R. Sweden
	500/250	USA	Voice of America, Greenville
9675	10	B	R. Cançao Nova, C. Paulista
	100	E	R. Exterior de España, Madrid
	250	EGY	R. Cairo, Abis
	250	GRC	Voice of America, Kavala
	500	GUF	R. Japan via French Guiana
	100	IND	All India Radio, Kingsway
	250	INS	V. of Indonesia, Jakarta
	5	PRU	R. del Pacífico, Lima, Peru
	20	RUS	Lithuanian R. via Balashikha
	100	TUN	ERTT, Tunis
	500/250	TUR	V. of Turkey, Ankara
9680	100	D	Deutsche Welle, Germany
	250	GRC	Voice of America, Kavala
	50	INS	RRI Jakarta (Prgr. Nasional)
	0.5	MEX	La Q Mexicana, México City *
	50	POR	RFE/RL, Gloria, Portugal
	500		RFE/RL, Maxoqueira, Portugal
	100		R. Renascença, Muge, Portugal
	500	THA	Voice of America, Udorn Thani
	50/100	USA	Voice of Free China via WYFR
	100		WYFR, Okeechobee, FL
9685	100	ALG	RTV Algerienne, Ouled Fayet
	7.5	B	R. Gazeta, Sao Paulo, Brazil
	100	E	R. Exterior de España, Madrid
	500	GUF	R. Japan via French Guiana
	2	PRU	R. Continental, Arequipa, Peru
	250	TUR	V. of Turkey, Ankara
9689.2	10	MDG	R. Madagasikara, Madagascar
9690	25	ARG	R. Argentina al Exterior
	250	ATG	Deutsche Welle, Antigua
	500/100	D	Deutsche Welle, Germany

	350	E	China Radio Int. via Spain
	350		R. Exterior de España, Madrid
	20	KAZ	Alma-Ata 2, Kazakhstan
	250	POR	Voice of America, Gloria
	250/120	ROU	R. Romania International
	20	RUS	Kazakh R. via Balashikha
	350	S	R. Sweden
	250	SUI	China Radio Int. via SRI
	10	TWN	CBS Netw. 5/1, Taipei, Taiwan
	250	USA	BBC via Delano, CA
	250		Voice of America, Delano, CA
9695	7.5	B	R. Rio Mar, Manaus, Brazil
	500/100	CVA	Vatican Radio
	100	D	RFE/RL, Lampertheim, Germany
	250	POR	RFE/RL, Gloria, Portugal
	350	S	R. Sweden
	500	UAE	R. of the UAE, Abu Dhabi
9700	250	BUL	R. Sofia, Kostinbrod, Bulgaria
	150		R. Sofia, Stolnik, Bulgaria
	500/100	D	Deutsche Welle, Germany
	100	EGY	R. Cairo, Abu Zaabal
	250	GRC	Voice of America, Kavala
		IND	All India Radio Ext. Sce.
	100	NZL	R. New Zealand International
9705	50	ARS	BSKSA, Jeddah, Saudi Arabia
	7.5	B	R. Nacional, Rio de Janeiro
	100	D	Deutsche Welle, Germany
	100		RFE/RL, Lampertheim, Germany
	250	E	RFE/RL, Playa de Pals, Spain
	250	GRC	Voice of America, Kavala
	20	KAZ	Alma-Ata 2, Kazakhstan
	10	MEX	R. México Internacional
v	100	NGR	La Voix du Sahel, Niger
	250	POR	RFE/RL, Gloria, Portugal
	100		R. Portugal, S. Gabriel
	100/50	USA	WYFR, Okeechobee, FL
9706.4v	100	ETH	V. of Ethiopia, Gedja
9710	100	AUS	R. Australia, Shepparton
	100	I	RAI Ext. Sce, Rome, Italy
	50	LTU	Sitkunai, Lithuania
	50	PHL	R. Veritas Asia, Philippines
	250/120	ROU	R. Romania International
	250	TUR	V. of Turkey, Ankara
		UKR	R. Ukraine International
9715	250	ATN	R. Netherlands, Bonaire
	500/120	CHN	China Radio International
	250	CLN	Deutsche Welle, Trincomalee
	500/100	D	Deutsche Welle, Germany
	500	F	R. France Internationale
	500	GUF	R. France Int, French Guiana
	500	HOL	R. Netherlands, Flevoland
	100	MRC	Voice of America, Tangier
	250	POR	RFE/RL, Gloria, Portugal
	250		Voice of America, Gloria
	50	USA	WYFR, Okeechobee, FL
9717.2	1	BOL	R. La Plata, Sucre, Bolivia
9720.0v	100	AGL	R. Nacional, Mulenvos, Angola
	50	ARS	BSKSA, Saudi Arabia
	250	ATN	R. Netherlands, Bonaire
	100	CLN	SLBC Ext. Sce, Colombo
	500/100	IRN	IRIB, Tehran, Iran
	500	YUG	R. Yugoslavia, Beograd
9725	100	ALB	R. Tirana, Lushnjë, Albania
	7.5	B	R. Clube Paranaense, Curitiba
	50	CHN	China Radio International
	50	CTR	AWR Latin America, Costa Rica
	100	D	RFE/RL, Biblis, Germany
		IRQ	R. Baghdad, Iraq
	500	POR	RFE/RL, Maxoqueira, Portugal
		UZB	BBC via Tashkent, Uzbekistan
9730	100	ALB	R. Tirana, Lushnjë / Krujë
	500	ARS	BSKSA, Riyadh, Saudi Arabia
	50	BRM	Voice of Myanmar, Yangon
	500	D	Deutsche Welle, Germany
	500/250	FNL	R. Finland
	100	NZL	R. New Zealand International
	250/50	TWN	Voice of Free China, Taiwan
9732		VTN	Voice of Vietnam, Hanoi
9735	250	CYP	BBC Ea. Mediterranean Relay St.
	500	D	Deutsche Welle, Germany

	300	G	Voice of America, Woofferton
	250	MLT	Deutsche Welle, Cyclops, Malta
	100	OMA	R. Oman, Thumrait
	250	POR	Voice of America, Gloria
	100	PRG	R. Nal. del Paraguay, Asunción
	250	RRW	Deutsche Welle, Kigali, Rwanda
		UKR	Unknown site, Ukraine
9740	250/100	CYP	BBC Ea. Mediterranean Relay St.
	500	D	Voice of America, Wertachtal
	250	EGY	R. Cairo, Abis
	300	G	R. Canada Int. via Skelton, UK
	250	GRC	Voice of America, Kavala
	35	MRC	Voice of America, Tangier
	100	POR	R. Portugal, S. Gabriel
	250/100	SNG	BBC Far Eastern Relay St.
		UZB	R. Tashkent Ext. Sce, Uzbekistan
9743		INS	RRI Sorong, Indonesia
9745	250	B	R. Bras Ext. Sce, Brasilia
	100	CTR	R. Exterior de España, Costa R.
	100/50	EQA	HCJB, Quito, Ecuador
	100	F	R. France Internationale
	250	GRC	Voice of America, Kavala
	500/100	IRN	IRIB, Tehran, Iran
9746	60	BHR	R. Bahrain
9750	1	CHL	R. Minería, Santiago, Chile *
	50	CHN	Hohhot, China (Mongolian prgr)
	500/50	CUB	R. Habana, Cuba
	250	CYP	BBC Ea. Mediterranean Relay St.
	500	D	Deutsche Welle, Germany
	500/100	G	BBC, United Kingdom
	250	KOR	R. Korea, Kimjae, So. Korea
	500	KWT	R. Kuwait
	100	MLA	V. of Malaysia, Kuala Lumpur
	100	OMA	BBC Eastern Relay Station
	250	POR	RFE/RL, Gloria, Portugal
	250/120	ROU	R. Romania International
9755	250	CAN	R. Canada Int, Sackville
	50	CHN	CPBS 2, China
	500/100	CVA	Vatican Radio
	100	EGY	R. Cairo, Mokattam
	250	PHL	Voice of America, Philippines
	1.2	PRU	R. Sideral, Pucallpa, Peru
9760	100	ALB	R. Tirana, Lushnjë, Albania
	250	CAN	R. Canada Int, Sackville
	500/100	G	BBC, United Kingdom
	50	J	R. Tanpa 2, Tokyo, Japan
	100	MRC	Voice of America, Tangier
	250	PHL	Voice of America, Philippines
	250	POR	Voice of America, Gloria
	500	THA	Voice of America, Udorn Thani
	500	UAE	R. of the UAE, Abu Dhabi
9765	500/100	D	Deutsche Welle, Germany
	50	EQA	HCJB, Quito, Ecuador
	250	MLT	V. of the Mediterranean, Malta
	250/50	TWN	Voice of Free China, Taiwan
9770	250	AUS	R. Australia, Darwin
	100		R. Australia, Shepparton
	250/100	CYP	BBC Ea. Mediterranean Relay St.
	100		CyBC via BBC Relay Station
	500/100	D	Deutsche Welle, Germany
	500		Voice of America, Wertachtal
	250	EGY	R. Cairo, Abis
	500/100	G	BBC, United Kingdom
	250		R. Japan via Skelton, UK
	50	MLI	China Radio Int. via Mali
	250	MLT	Deutsche Welle, Cyclops, Malta
	250	PHL	Voice of America, Philippines
	100	SEY	FEBA Radio, Seychelles
	500	UAE	R. of the UAE, Abu Dhabi
	100	USA	WYFR, Okeechobee, FL
9775	250	ATN	R. Netherlands, Bonaire
		CHN	CPBS 2 / Minority Sce, China
		RUS	AWR Europe via Moscow
	500/250	USA	Voice of America, Greenville
	100		WYFR, Okeechobee, FL
9779.8	50	YEM	Yemen Radio & TV Corp, Sana'a
9780		CHN	Qinghai 1, Xining, China
	50	KAZ	Alma-Ata 1, Kazakhstan
	50	PHL	FEBC Manila, Philippines

9785	100	GUM	KTWR-Trans World Radio, Guam
	50	USA	KVOH, Rancho Simi, CA
9790	500/100	F	R. France Internationale
	250	GAB	R. France Int. via Gabon
	500	GUF	R. France Int, French Guiana
		UZB	R. Tashkent Ext. Sce, Uzbekistan
9795	500	MCO	Trans World Radio, Monaco
	250	TUR	V. of Turkey, Ankara
9800		CHN	CPBS 1, China
	500/100	F	R. France Internationale
	100	GRC	Voice of Greece, Athens
	500	GUF	R. France Int, French Guiana
	50	PHL	FEBC Manila, Philippines
9805	250/100	EGY	R. Cairo, Abu Zaabal
	500/100	F	R. France Internationale
9810	100	SEY	FEBA Radio, Seychelles
	500/250	SUI	Swiss Radio International
	250/100	TCH	Czech & Slovak R. International
9815	20/500	ISR	Kol Israel Arabic Network D
	100	USA	KCBI, Denton, TX
9820	50	PHL	FEBC Manila, Philippines
9825	500/100	G	BBC, United Kingdom
9830	100	F	R. France Internationale
	10	HRV	Hrvatski Radio, Zagreb
	500	JOR	R. Jordan, Al Karanah
	100	PAL	KHBN, Palau
	100	PHL	FEBC Manila, Philippines
9835	500/100	CVA	Vatican Radio
	250	HNG	R. Budapest, Hungary
		RUS	AWR Russia, Novosibirsk
9840	500/120	CHN	China Radio International
	500/250	KWT	R. Kuwait
	250	USA	Voice of America, Greenville
	500		WCSN, Scotts Corners, ME
	100	UZB	Toskent 2 / FS, Uzbekistan
		VTN	Voice of Vietnam, Hanoi
9845	500/120	CHN	China Radio International
	100	HNG	R. France Int. via Hungary
	20/500	ISR	Kol Israel, Jerusalem
	100	PHL	FEBC Manila, Philippines
	250/50	TWN	Voice of Free China, Taiwan
9850	100	BUL	R. Sofia, Kostinbrod, Bulgaria
	250	EGY	R. Cairo, Abis
	500	F	R. France Internationale
	250	GAB	R. France Int. via Gabon
	100	PHL	FEBC Manila, Philippines
	50/100	USA	Voice of Free China via WYFR
	500		WCSN, Scotts Corners, ME
	100		WYFR, Okeechobee, FL
9855	500/120	CHN	China Radio International
	500	HOL	R. Netherlands, Flevoland
9860	300	AUS	R. Australia, Carnarvon
	500	HOL	R. Netherlands, Flevoland
	300	MDG	R. Netherlands, Madagascar
	500/250	SUI	Swiss Radio International
		UKR	R. Ukraine International
9870	50	ARS	BSKSA, Saudi Arabia
	100	GUM	KTWR-Trans World Radio, Guam
	250/100	KOR	R. Korea, Kimjae, So. Korea
	500	USA	WEWN, Birmingham, AL
	500		WSHB, Cypress Creek, SC
9875	500	AUT	R. Austria International
	100	E	R. Exterior de España, Madrid
	100	PHL	FEBC Manila, Philippines
	1000	RUS	Deutsche Welle via Novosibirsk
9877.5v	1	DOM	R. Santiago, Dom. Rep.
9880	500/100	AUT	R. Austria International
	500/250	KWT	R. Kuwait
		RUS	China Radio Int. via Russia
			R Galaxy, Moskva
9885	500	ARS	BSKSA, Riyadh, Saudi Arabia
	100	BOT	Voice of America, Botswana
	500/250	SUI	Swiss Radio International
		UZB	R. Tashkent Ext. Sce, Uzbekistan
9890	50	PHL	Voice of America, Philippines
9895	500	HOL	R. Netherlands, Flevoland
	300	MDG	R. Netherlands, Madagascar
9900	500/120	CHN	China Radio International
	250	EGY	R. Cairo, Abis

9905	200/100	BEL	R. Vlaanderen Internationaal
9910		IND	All India Radio Ext. Sce.
9915	500/100	G	BBC, United Kingdom
9920		CHN	CPBS Minority Sce, China
	500/120		China Radio International
9926.5		EGY	R. Cairo, Abu Zaabal
9930	200/100	BEL	R. Vlaanderen Internationaal
9935	35	GRC	ERT, Thessaloniki, Greece
9940	250	EGY	R. Cairo, Abis
9945	500/120	CHN	China Radio International
9950	250	EGY	R. Cairo, Abis
		IND	All India Radio Ext. Sce.
	50		All India Radio, Kingsway
	500	SYR	R. Damascus, Syria
9955		TWN	Family Radio via VOFC, Taiwan
	250/50		Voice of Free China, Taiwan
	50	USA	WRMI, Miami, FL
9965	500/120	CHN	China Radio International
9977	400/100	KRE	R. Pyongyang Ext. Sce, No. Korea
9985	500	USA	WEWN, Birmingham, AL
	500		WSHB, Cypress Creek, SC
9996	5	RUS	RWM (Time Signals), Moscow
10000	2	ARG	LOL (Time Signals), Argentina
	10/20	CHN	BPM (Time Signals), China
	10	HWA	WWVH, Hawaii (Time Signals)
	8	IND	ATA, New Delhi (Time Sigs.)
	2	J	JJY, Japan (Time Signals)
	5	RUS	RTA (Time Signals), Russia
	10	USA	WWV (Time Signals), Ft. Collins
	1	UZB	RCH, Tashkent (Time Signals)
10004	1	RUS	RID (Time Signals), Irkutsk
10010		VTN	Voice of Vietnam, Hanoi
10059.0v			Hanoi I, Vietnam
10260		CHN	CPBS 2 / Minority Sce, China
10330		IND	All India Radio Ext. Sce.
	50		All India Radio, Kingsway
10440.5		PHL	NPO (Time Sigs), Philippines
11000		CHN	CPBS Taiwan 2 Sce, China
11040			CPBS 2, China
11100			CPBS Taiwan 1 Sce, China
11330			CPBS 1, China
11335	400/100	KRE	R. Pyongyang Ext. Sce, No. Korea
11375		CHN	CPBS Minority Sce, China
11402	10	ISL	Ríkisútvarpid, Reykjavík
11440	3	INS	PLC (Time Signals), Indonesia
11445	500/120	CHN	China Radio International
11455.0r		ZAI	La Voix du Zaire, Kinsangani
11500	500/120	CHN	China Radio International
11515	500/120		China Radio International
11530	25	LBN	Wings of Hope, Lebanon
11550.0v	100	TUN	ERTT, Tunis
		TWN	Family Radio via VOFC, Taiwan
11560	250	EGY	R. Cairo, Abis
11570	250/50	PAK	R. Pakistan Ext. Sce.
11575	500/120	CHN	China Radio International
11580	100	GUM	KTWR-Trans World Radio, Guam
	100	MRA	KHBI, Saipan, No. Marianas
	250	USA	Voice of America, Greenville
	100		WYFR, Okeechobee, FL
11585	20/500	ISR	Kol Israel, Jerusalem
11587	20/500		Kol Israel, Jerusalem
11588	20/500		Kol Israel, Jerusalem
11590		CHN	Haixia 1, China
11595	35	GRC	ERT, Thessaloniki, Greece
11600	500/120	CHN	China Radio International
11605	20/500	ISR	Kol Israel, Jerusalem
11610		CHN	CPBS 2, China
11620	500/100	CVA	Vatican Radio
		IND	All India Radio Ext. Sce.
11625	500/100	CVA	Vatican Radio
11630	100	BUL	R. Sofia, Kostinbrod, Bulgaria
		CHN	CPBS 2 / Minority Sce, China
11635	100	BUL	R. Sofia, Kostinbrod, Bulgaria
11640	500/100	CVA	Vatican Radio
11645	250	GRC	Voice of America, Kavala
	100		Voice of Greece, Athens
11650	500/120	CHN	China Radio International
	120		R. France Int. via X'ian, China
	100	GUM	KTWR-Trans World Radio, Guam
	100	MRA	KFBS, Saipan, No. Marianas
	50	PHL	FEBC Manila, Philippines
	20	RUS	R. Pamyat, Moskva, E
11655	500	HOL	R. Netherlands, Flevoland
	300	MDG	R. Netherlands, Madagascar
		RUS	R. Art, Moskva, E
11660	250	ATN	R. Netherlands, Bonaire
	250	BUL	Plovdiv, Bulgaria (Home Sce.)
	250		R. Sofia, Kostinbrod, Bulgaria
	500/120	CHN	China Radio International
	500/100	F	R. France Internationale
	100	GUM	KTWR-Trans World Radio, Guam
	500	HOL	R. Netherlands, Flevoland
11665	250/100	EGY	R. Cairo, Abu Zaabal
	100	GUM	KTWR-Trans World Radio, Guam
	100	MRA	KFBS, Saipan, No. Marianas
	100	SEY	FEBA Radio, Seychelles
11670	500/100	F	R. France Internationale
	500	GUF	R. France Int, French Guiana
11675	20/500	ISR	Kol Israel, Jerusalem
	1000	RUS	Yerevan, Armenia (relay)
	100	SEY	FEBA Radio, Seychelles
11680	500/100	G	BBC, United Kingdom
	250	GAB	R. France Int. via Gabon
	500	GUF	China Radio Int. via Fr. Guiana
	500		R. France Int, French Guiana
v		KRE	KCBS, Pyongyang, No. Korea
	250	USA	Voice of America, Greenville
11685	500	ARS	BSKSA, Riyadh, Saudi Arabia
	250	GAB	R. France Int. via Gabon
	500	GUF	China R. Int. via French Guiana
	400	RUS	R. Ala, Sankt-Peterburg, E
	250/100	TCH	Czech & Slovak R. International
11690	120	CHN	Swiss Radio Int. via China
	50	PHL	FEBC Manila, Philippines
	120	SUI	Swiss Radio Int. via Beijing
11695	200/100	BEL	R. Vlaanderen Internationaal
	100	F	R. France Internationale
	500	GUF	China Radio Int. via Fr. Guiana
	250	USA	KGEI, Redwood City, CA
	500		Voice of America, Greenville
11700	100	F	R. France Internationale
	100	GUM	KTWR-Trans World Radio, Guam
	400/100	KRE	R. Pyongyang Ext. Sce, No. Korea
	250	POR	IBRA Radio via Sines, Portugal
11705	10	B	R. Transamérica, Santa Maria
v	250	BGD	R. Bangladesh Ext. Sce.
	50	CHN	Hohhot, China (Chinese prgr) *
	35	CLN	Voice of America, Colombo
	500/50	CUB	R. Habana, Cuba
11705	100	D	Deutsche Welle, Germany
	500		Voice of America, Wertachtal
	500/100	F	R. France Internationale
	250	GAB	R. France Int. via Gabon
	300	J	R. Canada Int. via Yamata, Japan
	100	MRA	KFBS, Saipan, No. Marianas
	50	PHL	R. Veritas Asia, Philippines
	350	S	R. Sweden
	500	THA	Voice of America, Udorn Thani
11710	100	ARG	R. Argentina al Exterior
	100	COG	RTV Congolaise, Brazzaville
	250	CYP	BBC Ea. Mediterranean Relay St.
	500	D	Voice of America, Wertachtal
	500	HOL	R. Netherlands, Flevoland
	250/50	PHL	R. Veritas Asia, Philippines
	250	POR	Voice of America, Gloria
	250/120	ROU	R. Romania International
	350	S	R. Sweden
	100	SEY	FEBA Radio, Seychelles
	500	UAE	R. of the UAE, Abu Dhabi
11715.0v	50	ALG	RTV Algerienne, Bouchaoui
	250	CAN	R. Korea via Sackville, Canada
	500/100	CVA	Vatican Radio
	500	D	R. Canada Int. via Wertachtal
	250	EGY	R. Cairo, Abis
	500	HOL	R. Netherlands, Flevoland
		IND	All India Radio Ext. Sce.
	100		All India Radio, Khampur
	500/100	IRN	IRIB, Tehran, Iran
	300	J	R. France Int. via Japan
	50	MLI	China Radio Int. via Mali

	250/50	PAK	R. Pakistan Ext. Sce.
	50	PHL	Voice of America, Philippines
	50	USA	KJES, Vado, NM
	100		WYFR, Okeechobee, FL
11720	250	ATN	R. Netherlands, Bonaire
	10	AUS	R. Australia, Brandon
	250/100	BUL	R. Sofia, Kostinbrod, Bulgaria
	250		R. Sofia, Plovdiv, Bulgaria
	100	CYP	BBC Ea. Mediterranean Relay St.
	100	MRC	Voice of America, Tangier
	100	OMA	BBC Eastern Relay Station
	250/50	PHL	Voice of America, Philippines
	250/120	ROU	R. Romania International
11725	500/50	CUB	R. Habana, Cuba
	250	KOR	R. Korea, Kimjae, So. Korea
	250	POR	RFE/RL, Gloria, Portugal
	500		RFE/RL, Maxoqueira, Portugal
		TWN	CBS News Network, Taiwan
	100	USA	WYFR, Okeechobee, FL
11730	500	ARS	BSKSA, Riyadh, Saudi Arabia
	250	CAN	R. Canada Int, Sackville
	250/100	CYP	BBC Ea. Mediterranean Relay St.
	500/100	EQA	HCJB, Quito, Ecuador
	500	HOL	R. Netherlands, Flevoland
		IND	All India Radio Ext. Sce.
	250	PHL	Voice of America, Philippines
	250	SEY	BBC Indian Ocean Relay St.
11734.2	50	TZA	R. Tanzania Zanzibar
11735	500	BUL	R. Sofia, Plovdiv, Bulgaria
	250	CLN	Deutsche Welle, Trincomalee
	500/100	EQA	HCJB, Quito, Ecuador
	50	GRC	Voice of America, Rhodes
		IND	All India Radio Ext. Sce.
	100		All India Radio, Kingsway
	400/100	KRE	R. Pyongyang Ext. Sce, No. Korea
	500/350	NOR	R. Norway Int. / R. Denmark
	100	NZL	R. New Zealand International
	1	URG	R. Oriental, Uruguay
	500	USA	WEWN, Birmingham, AL
11740	120	CHN	CPBS 2, China
	250	CLN	Deutsche Welle, Trincomalee
	500/100	CVA	Vatican Radio
	100	CYP	BBC Ea. Mediterranean Relay St.
	500	FNL	R. Finland
	500/100	G	BBC, United Kingdom
	250	GRC	Voice of America, Kavala
	100	KOR	R. Korea, Kimjae, So. Korea
	100	OMA	BBC Eastern Relay Station
	50	PHL	R. Veritas Asia, Philippines
	250		Voice of America, Philippines
	100	POR	R. Portugal, S. Gabriel
	250/120	ROU	R. Romania International
	25	SWZ	Trans World Radio, Swaziland
	50/100	USA	Voice of Free China via WYFR
	100		WYFR, Okeechobee, FL
11745	250	AFS	Channel Africa, Johannesburg
	100	ALB	R. Tirana, Lushnjë, Albania
	250	B	R. Bras Ext. Sce, Brasilia
	120	CHN	China Radio International
	500	D	Deutsche Welle, Germany
		IND	All India Radio Ext. Sce.
	250/50	TWN	Voice of Free China, Taiwan
11750	250	ASC	BBC Atlantic Relay St.
	250	CYP	BBC Ea. Mediterranean Relay St.
	500/100	G	BBC, United Kingdom
	100	MRC	Voice of America, Tangier
	100	OMA	BBC Eastern Relay Station
	250	PHL	Voice of America, Philippines
	250/100	SNG	BBC Far Eastern Relay St.
	500	UAE	R. of the UAE, Abu Dhabi
11752.0v	100	INS	V. of Indonesia, Jakarta
11755	240	CHN	China Radio International
	500/250	FNL	R. Finland
11760	0.5	CKH	R. Cook Islands, Rarotonga
	250/100	CUB	R. Habana, Cuba
	500/100	G	BBC, United Kingdom
	300		Voice of America, Woofferton
		IND	All India Radio Ext. Sce.
v	100	INS	RRI Jakarta (Prgr. Nasional) *
	100	MRC	Voice of America, Tangier
	100	OMA	BBC Eastern Relay Station
	250	PHL	Voice of America, Philippines
	100	POR	R. Portugal, S. Gabriel
	5	RUS	R. Vedo, Volgograd, E
11765	250	ASC	BBC Atlantic Relay St.
	100	BUL	R. Sofia, Kostinbrod, Bulgaria
	500	CHN	China Radio International
	250	CLN	Deutsche Welle, Trincomalee
	500/100	D	Deutsche Welle, Germany
	250	HNG	BBC East Asian Relay St.
	250	SNG	BBC Far Eastern Relay St.
11769.7	10	MEX	R. México Internacional
11770	100	IND	All India Radio, Khampur
	250	POR	RFE/RL, Gloria, Portugal
	3	RUS	Kudymkar, E
11775	120	CHN	China Radio International
	300	G	Voice of America, Woofferton
	500	NOR	R. Norway Int. / R. Denmark
	250/120	ROU	R. Romania International
		RUS	AWR Europe via Moscow
	500	TUR	V. of Turkey, Ankara
	300	TWN	CBS Network 2, Taipei. Taiwan
	100	USA	WYFR, Okeechobee, FL
11780	350	ARS	BSKSA, Riyadh, Saudi Arabia
	500	AUT	R. Austria International
	250	B	R. Nacional da Amazônia, Brazil
	50	CHN	China Radio International
	250	D	RFE/RL, Holzkirchen, Germany
	500/100	G	BBC, United Kingdom
	500/100	IRN	IRIB, Tehran, Iran
	400/100	KRE	R. Pyongyang Ext. Sce, No. Korea
	35	MRC	Voice of America, Tangier
11781.2	1	ARG	R. Libertad, Argentina
11785	7.5	B	R. Guaíba, Porto Alegre, Brazil
	250	CLN	Deutsche Welle, Trincomalee
	500/100	D	Deutsche Welle, Germany
	100	EGY	R. Cairo, Abu Zaabal
		IND	All India Radio Ext. Sce.
	250	MLT	Deutsche Welle, Cyclops, Malta
	250	RRW	Deutsche Welle, Kigali, Rwanda
		RUS	AWR Europe via Moscow
	500	THA	Voice of America, Udorn Thani
11790	50	CHN	China Radio International
	100	E	R. Exterior de España, Madrid
	100	F	R. France Internationale
	500/100	IRN	IRIB, Tehran, Iran
	50	MLI	China Radio Int. via Mali
	50	MNG	Ulaanbaatar Ext. Sce, Mongolia
	250/50	PHL	R. Veritas Asia, Philippines
	250/120	ROU	R. Romania International
	240	RUS	Yerevan, Armenia (relay)
	100	SEY	FEBA Radio, Seychelles
	100	USA	WHRI, Noblesville, IN
11795	250	ATG	Deutsche Welle, Antigua
	500/100	D	Deutsche Welle, Germany
	350	NOR	R. Norway Int. / R. Denmark
	250	RRW	Deutsche Welle, Kigali, Rwanda
	1000	RUS	Deutsche Welle via Novosibirsk
	300	UAE	UAE Radio & TV, Dubai
11800	250	AFS	Channel Africa, Johannesburg
	100	AUS	R. Australia, Shepparton
	100	CLN	SLBC Ext. Sce, Colombo
	500	F	R. France Internationale
	250	GAB	R. France Int. via Gabon
	100	I	RAI Ext. Sce, Rome, Italy
	250	POR	R. Portugal via Gloria
11805	10	B	R. Globo, RJ, Brazil
	500	D	Voice of America, Wertachtal
	250	GRC	Voice of America, Kavala
	100	GUM	KTWR-Trans World Radio, Guam
		IND	All India Radio Ext. Sce.
	250	PHL	Voice of America, Philippines
	250	POR	Voice of America, Gloria
	100	SEY	FEBA Radio, Seychelles
	100	SWZ	Trans World Radio, Swaziland
	100/50	USA	Voice of Free China via WYFR
	500	YUG	R. Yugoslavia, Beograd
11810	250	ATG	Deutsche Welle, Antigua
	250	B	Deutsche Welle via R. Bras
	100	D	Deutsche Welle, Germany

	500	FNL	R. Finland
		IND	All India Radio Ext. Sce.
	500	JOR	R. Jordan, Al Karanah
	250	KOR	R. Korea, Kimjae, So. Korea
	250/120	ROU	R. Romania International
	250	RRW	Deutsche Welle, Kigali, Rwanda
	100	SEY	FEBA Radio, Seychelles
11815	250	AFS	Channel Africa, Johannesburg
	100	ATN	Trans World Radio, Bonaire
	7.5	B	R. Brasil Central, Goiânia
	500/120	CHN	China Radio International
	100	D	RFE/RL, Lampertheim, Germany
	500/100	G	BBC, United Kingdom
	300/100	J	R. Japan, Yamata, Japan
	500	LBY	LJB, Tripoli/Sabrata, Libya
	250	POR	RFE/RL, Gloria, Portugal
	500	UAE	R. of the UAE, Abu Dhabi
11818.0v	120	MOZ	Em. Nacional, Maputo
11820	250	ASC	BBC Atlantic Relay St.
	250		Voice of America, Ascension Is
	500/50	CUB	R. Habana, Cuba
	250	HNG	BBC East Asian Relay St.
	500	JOR	R. Jordan, Al Karanah
	50	PHL	R. Veritas Asia, Philippines
	250		Voice of America, Philippines
	100	QAT	Qatar Broadc. Sce, Doha
		RUS	AWR Europe via Moscow
11821.5v	20	CLM	R. Dif. Nal., Bogotá, Colombia
11825	100	ALB	R. Tirana, Lushnjë, Albania
	350	ARS	BSKSA, Riyadh, Saudi Arabia
	500	D	Voice of America, Wertachtal
	500/100	G	BBC, United Kingdom
	250	GRC	Voice of America, Kavala
	20	KAZ	Alma-Ata 2, Kazakhstan
v	20	OCE	RFO, Tahiti, French Polynesia
	250	POR	RFE/RL, Gloria, Portugal
	250		Voice of America, Gloria
	250/50	TWN	Voice of Free China, Taiwan
11830	1	B	R. Anhanguera, Goiânia, Brazil
	500/100	CVA	Vatican Radio
		IND	All India Radio Ext. Sce.
	100		All India Radio, Bombay
	50		All India Radio, Khampur
	10	LBR	ELWA, Monrovia, Liberia *
	250/120	ROU	R. Romania International
	250	USA	Voice of America, Bethany, OH
	250		V. of the OAS, Bethany, OH
	100		WYFR, Okeechobee, FL
11835	100	ALB	R. Tirana, Lushnjë, Albania
	35	CLN	SLBC Ext. Sce, Colombo
	500/50	CUB	R. Habana, Cuba
	50	D	Deutsche Welle, Germany
	500		Voice of America, Wertachtal
	500/100	EQA	HCJB, Quito, Ecuador
	500/100	G	BBC, United Kingdom
	250	GRC	Voice of America, Kavala
	250	PHL	Voice of America, Philippines
	500	USA	Voice of America, Greenville
	500/100	YUG	R. Yugoslavia, Beograd
11835.7	0.5	URG	R. El Espectador, Uruguay
11840	100	ALB	R. Tirana, Lushnjë, Albania
	250	CAN	China Radio Int. via Canada
	300	CLN	R. Japan, Ekala, Sri Lanka
	500	D	Voice of America, Wertachtal
	300/100	J	R. Japan, Yamata, Japan
	250	PHL	Voice of America, Philippines
	100	POL	Polish Radio, Warszawa
	100	POR	R. Portugal, S. Gabriel
	50	RUS	Yuzno-Sakhalinsk, FE
	100	SEY	FEBA Radio, Seychelles
11845	500	AFS	BBC via Meyerton, South Africa
	250	CAN	R. Canada Int, Sackville
	250	CYP	BBC Ea. Mediterranean Relay St.
	500	EQA	HCJB, Quito, Ecuador
	100	F	R. France Internationale
	250	GRC	Voice of America, Kavala
	400/100	KRE	R. Pyongyang Ext. Sce, No. Korea
	100	OMA	BBC Eastern Relay Station
	250	POR	Voice of America, Gloria

11850	250	CAN	R. Canada Int, Sackville
	120	CHN	China Radio International
	500/50	CUB	R. Habana, Cuba
	250	CYP	BBC Ea. Mediterranean Relay St.
	500	D	Deutsche Welle, Germany
	100	HNG	R. France Int. via Hungary
	100	J	R. Japan, Yamata, Japan
	250	MNG	Ulaanbaatar Ext. Sce, Mongolia
	35	MRC	Voice of America, Tangier
	500	NOR	R. Norway Int. / R. Denmark
	100	OMA	BBC Eastern Relay Station
	250	SNG	BBC Far Eastern Relay St.
11855	250	AUS	R. Australia, Carnarvon
	100		R. Australia, Shepparton
	5	B	R. Aparecida, Brazil
	250	CAN	R. Canada Int, Sackville
	500/120	CHN	China Radio International
	100	D	RFE/RL, Lampertheim, Germany
	50	MNG	Ulaanbaatar 1, Mongolia
	250	POR	Voice of America, Gloria
		RUS	AWR Russia, Novosibirsk
	500	THA	Voice of America, Udorn Thani
	100	USA	WYFR, Okeechobee, FL
11860	250	AFS	Channel Africa, Johannesburg
	250	ASC	BBC Atlantic Relay St.
	100	BUL	R. Sofia, Kostinbrod, Bulgaria
		IND	All India Radio Ext. Sce.
	350	NOR	R. Norway Int. / R. Denmark
	250	SEY	BBC Indian Ocean Relay St.
	250/50	TWN	Voice of Free China, Taiwan
11865	50	ATN	Trans World Radio, Bonaire
	50	CHN	Hohhot, China (Mongolian) *
	500/100	D	Deutsche Welle, Germany
	100	J	R. Japan, Yamata, Japan
	250	MLT	Deutsche Welle, Cyclops, Malta
	350	NOR	R. Norway Int. / R. Denmark
	250	POR	Deutsche Welle, Sines
	250		Voice of America, Gloria
	250	SNG	BBC Far Eastern Relay St.
11870	500	BUL	R. Sofia, Plovdiv, Bulgaria
	50	CTR	AWR Latin America, Costa Rica
	50	IND	All India Radio, Khampur
	500	NOR	R. Norway Int. / R. Denmark
	250	PHL	Voice of America, Philippines
11875	250	D	RFE/RL, Holzkirchen, Germany
	100		RFE/RL, Lampertheim, Germany
	250	EGY	R. Cairo, Abis
	250	GRC	Voice of America, Kavala
	300	J	R. Japan, Yamata, Japan
	250	POR	RFE/RL, Gloria, Portugal
	500	TUR	V. of Turkey, Ankara
11880	250	ASC	BBC Atlantic Relay St.
	10	AUS	R. Australia, Brandon
	100		R. Australia, Shepparton
	250	CAN	R. Canada Int, Sackville
	250	CLN	Deutsche Welle, Trincomalee
		IND	All India Radio Ext. Sce.
v	500	IRQ	R. Iraq Int, Salah el Deen
	100	J	R. Japan, Yamata, Japan
11885	100	ATN	Trans World Radio, Bonaire
	250	D	RFE/RL, Holzkirchen, Germany
	100		RFE/RL, Lampertheim, Germany
	250	E	RFE/RL, Playa de Pals, Spain
	500	MLA	V. of Malaysia, Kuala Lumpur
	250	POR	RFE/RL, Gloria, Portugal
		TWN	CBS FM1 Network, Taiwan
	500	USA	WEWN, Birmingham, AL
11890	120	CHN	R. France Int. via Beijing
	350	E	R. Exterior de España, Madrid
	300	MDG	R. Netherlands, Madagascar
	100	OMA	R. Oman, Seeb
	250	USA	Voice of America, Bethany, OH
	500		Voice of America, Greenville
11895	250	ATN	R. Netherlands, Bonaire
	35	CLN	SLBC Ext. Sce, Colombo
	250	D	RFE/RL, Holzkirchen, Germany
	100		Voice of America, Munich
	100	GUM	KTWR-Trans World Radio, Guam
	500/100	IRN	IRIB, Tehran, Iran
	35	MRC	Voice of America, Tangier

SHORTWAVE TABLE

	250	POR	RFE/RL, Gloria, Portugal
	500		RFE/RL, Maxoqueira, Portugal
	250	TUR	V. of Turkey, Ankara
	500	USA	Voice of America, Greenville
11900	500	AFS	Channel Africa, Johannesburg
	500	HOL	R. Netherlands, Flevoland
	100	PAL	KHBN, Palau
		RUS	AWR Europe via Yekaterinburg
11905	10	B	R. Universo, Curitiba, Brazil
	500/100	D	Deutsche Welle, Germany
	250		RFE/RL, Holzkirchen, Germany
	100		Voice of America, Munich
	100	EGY	R. Cairo, Abu Zaabal
	300	G	R. Canada Int. via Skelton, UK
	250	GRC	Voice of America, Kavala
	50		Voice of America, Rhodes
	100	I	RAI Ext. Sce, Rome, Italy
	400/100	KRE	R. Pyongyang Ext. Sce, No. Korea
		RUS	Tatar R, Tolyatti, E
	50	THA	R. Thailand Ext. Sce.
	500		Voice of America, Udorn Thani
	100	TWN	CBS Netw. 1/6, Taipei, Taiwan
11910	100	AUS	R. Australia, Shepparton
	120	CHN	R. Ext. de España via Xi'an
	500/50	CUB	R. Habana, Cuba
	250	EQA	HCJB, Quito, Ecuador
	250	HNG	R. Budapest, Hungary
		IND	All India Radio Ext. Sce.
	500/100	IRN	IRIB, Tehran, Iran
11915	7.5	B	R. Gaúcha, Porto Alegre, Brazil
	250	CLN	Deutsche Welle, Trincomalee
	100	D	Deutsche Welle, Germany
	250	MLT	Deutsche Welle, Cyclops, Malta
	50	PHL	R. Veritas Asia, Philippines
	250	POR	R. Canada Int. via Sines
	250	USA	Voice of America, Greenville
	100/50		Voice of Free China via WYFR
	100		WYFR, Okeechobee, FL
11920	50	IND	All India Radio, Khampur
	50	MRC	RTV Marocaine, Tanger, Morocco
			Voice of America, Tangier
	250	PHL	Voice of America, Philippines
	100	SNG	BBC Far Eastern Relay St.
11925	10	B	R. Bandeirantes, Sao Paulo
	100	EQA	HCJB, Quito, Ecuador
	500	GAB	R. Japan via Moyabi, Gabon
	250	MLT	V. of the Mediterranean, Malta
	250	PHL	Voice of America, Philippines
	250	POR	RFE/RL, Gloria, Portugal
	500	TUR	V. of Turkey, Ankara
	250	USA	Voice of America, Bethany, OH
11930	100	ATN	Trans World Radio, Bonaire
	250/100	EGY	R. Cairo, Abu Zaabal
	500/100	IRN	IRIB, Tehran, Iran
	250	PHL	Voice of America, Philippines
	100	SEY	FEBA Radio, Seychelles
	250	USA	R. Martí, Greenville, NC
	250		Voice of America, Greenville
11935	500	ARS	BSKSA, Riyadh, Saudi Arabia
		CHN	CPBS Taiwan 1 Sce, China
	500/100	CVA	Vatican Radio
	250	E	RFE/RL, Playa de Pals, Spain
	300	G	R. Canada Int. via Skelton, UK
	500	HOL	R. Netherlands, Flevoland
		IND	All India Radio Ext. Sce.
	500	NOR	R. Norway Int. / R. Denmark
	250	POR	RFE/RL, Gloria, Portugal
	500		RFE/RL, Maxoqueira, Portugal
	500	USA	Voice of America, Greenville
11938.0v		CBG	Phnom Penh, Cambodia
11939.5	0.5	PRG	R. Encarnación, Paraguay
11940	500	AFS	BBC via Meyerton, South Africa
	5	B	R. Cançao Nova, C. Paulista
	100	BOT	Voice of America, Botswana
	250	CAN	R. Canada Int, Sackville
	500	CTI	Radiodiff.-TV Ivoirienne
	500	JOR	R. Jordan, Al Karanah
	100	LSO	BBC Lesotho Relay St.
	250/120	ROU	R. Romania International
	50	SNG	SBC Radio One, Singapore
11945	250	CAN	R. Canada Int, Sackville
	150	CHN	China Radio International
	250	CLN	Deutsche Welle, Trincomalee
	350	E	R. Exterior de España, Madrid
	250	GRC	Voice of America, Kavala
	250	HNG	BBC East Asian Relay St.
		IRQ	R. Baghdad, Iraq
	20	RUS	Tatar R. via Balashikha
	300	UAE	UAE Radio & TV, Dubai
	250	USA	Voice of America, Delano, CA
11950	250	ATN	R. Netherlands, Bonaire
	7.5	B	R. MEC, Rio de Janeiro, Brazil
	15	CHN	Lhasa, China (Tibetan prgr) *
	250	CLN	Deutsche Welle, Trincomalee
	250/100	CUB	R. Habana, Cuba
	100	KAZ	Alma-Ata 1, Kazakhstan
	50	PHL	R. Veritas Asia, Philippines
	250	POR	Deutsche Welle, Sines
11955.0v	100	AGL	R. Nacional, Mulenvos, Angola
	250	CYP	BBC Ea. Mediterranean Relay St.
	300	J	R. Canada Int. via Yamata, Japan
	500	JOR	R. Jordan, Al Karanah
	10	LBR	ELWA, Monrovia, Liberia *
	100	OMA	BBC Eastern Relay Station
	250/100	SNG	BBC Far Eastern Relay St.
	500/250	TUR	V. of Turkey, Ankara
11960	250	ATG	Deutsche Welle, Antigua
	100	EQA	HCJB, Quito, Ecuador
	50	GRC	Voice of America, Rhodes
		IND	All India Radio Ext. Sce.
v	50	MLI	Radiodiff.-TV du Mali, Bamako
11965	250	ATG	Deutsche Welle, Antigua
	7.5	B	R. Record, Sao Paulo
	500	D	R. Canada Int. via Wertachtal
	500/100	F	R. France Internationale
	250	PHL	Voice of America, Philippines
	250	RRW	Deutsche Welle, Kigali, Rwanda
	100	RUS	R. Ala, Bolshakovo, E
	250	USA	BBC via Delano, CA
	250		Voice of America, Delano, CA
11970	500/50	CUB	R. Habana, Cuba
	250	D	RFE/RL, Holzkirchen, Germany
	100		RFE/RL, Lampertheim, Germany
	100	IND	All India Radio, Khampur
	250	POR	RFE/RL, Gloria, Portugal
	500		RFE/RL, Maxoqueira, Portugal
	250/120	ROU	R. Romania International
	500	USA	WEWN, Birmingham, AL
11975	250/100	EGY	R. Cairo, Abu Zaabal
	500	F	R. France Internationale
		UZB	R. Tashkent Ext. Sce, Uzbekistan
11980	500	ARM	Yerevan, Armenia (Foreign Sce)
	500/120	CHN	China Radio International
	250	EGY	R. Cairo, Abis
	100	GUM	KSDA-Adventist World R, Guam
11985	250/100	TCH	Czech & Slovak R. International
11990	500/250	KWT	R. Kuwait
	250/100	TCH	Czech & Slovak R. International
11995	100	F	R. France Internationale
	500	GUF	R. France Int, French Guiana
	50	PHL	FEBC Manila, Philippines
	250		Voice of America, Philippines
12000.0r	500/50	CUB	R. Habana, Cuba
12005	100	F	R. France Internationale
	250	GAB	R. France Int. via Gabon
	100	TUN	ERTT, Tunis
12008		E	EBC (Time Signals), Spain
12010	500	MCO	Trans World Radio, Monaco
12015	250	GAB	R. France Int. via Gabon
	50/250	MNG	Ulaanbaatar Ext. Sce, Mongolia
12020	500	MCO	Trans World Radio, Monaco
	100	PHL	FEBC Manila, Philippines
		VTN	Voice of Vietnam, Hanoi
12025	500/100	F	R. France Internationale
	250	GAB	R. France Int. via Gabon
12030.0v	250	BGD	R. Bangladesh Ext. Sce.
	100	RUS	R. Ala, Yekaterinburg, S
	200		R. Pamyat, Jekaterinburg, E
	500/150	SUI	Swiss Radio International

12035	100	E	R. Exterior de España, Madrid
	500	F	R. France Internationale
	500	GAB	Swiss Radio Int. via Gabon
		RUS	China Radio Int. via Russia
	500/250	SUI	Swiss Radio International
		VTN	Voice of Vietnam, Hanoi
12040	500/100	G	BBC, United Kingdom
	250	PHL	Voice of America, Philippines
12045	1000	RUS	Deutsche Welle via Novosibirsk
12050	250	EGY	R. Cairo, Abis
		RUS	Yerevan, Armenia (relay)
12055	500/120	CHN	China Radio International
		RUS	China Radio Int. via Russia
	250		Deutsche Welle via Zhigulevsk
12060	1000	ARM	Yerevan, Armenia (Foreign Sce)
	20	RUS	R. Pamyat, Moskva, E
12080	100	BOT	Voice of America, Botswana
12085	500	SYR	R. Damascus, Syria
12095	500/100	G	BBC, United Kingdom
12105	100	GRC	Voice of Greece, Athens
12110	500/120	CHN	China Radio International
12120			CPBS 1, China
12307		PRU	PBC3 (Time Signals), Peru
12724	10	AFS	ZSC (Time Signals), So. Africa
12745	8	IND	VWC, Calcutta (Time Sigs)
12804		PHL	NPO (Time Sigs), Philippines
12984	3	AUS	VNG, Llandilo (Time Signals)
13020.4		HKG	VPS60 (Time Sigs), Hong Kong
13605	250	AUS	R. Australia, Darwin
	100	BUL	R. Sofia, Kostinbrod, Bulgaria
	500	UAE	R. of the UAE, Abu Dhabi
13610	500/100	D	Deutsche Welle, Germany
13615	500	USA	WEWN, Birmingham, AL
	500		WSHB, Cypress Creek, SC
13620.0v	250	BGD	R. Bangladesh Ext. Sce.
	500/250	KWT	R. Kuwait
	250/50	PAK	R. Pakistan Ext. Sce.
13625	100	MRA	KHBI, Saipan, No. Marianas
13630.0v	2	CTR	R. For Peace Int, Costa Rica
13635	500/250	SUI	Swiss Radio International
13645	250	BUL	R. Sofia, Kostinbrod, Bulgaria
	20	RUS	Belorussian R. via Balashikha
13650	250	CAN	R. Canada Int, Sackville
	400/100	KRE	R. Pyongyang Ext. Sce, No. Korea
13655	200/100	BEL	R. Vlaanderen Internationaal
	500	JOR	R. Jordan, Al Karanah
13660	500/100	G	BBC, United Kingdom
	250/100	HNG	R. France Int. via Hungary
13665	250/50	PAK	R. Pakistan Ext. Sce.
13670	250	BUL	R. Sofia, Plovdiv, Bulgaria
	250	CAN	R. Canada Int, Sackville
	100	KOR	R. Korea, Kimjae, So. Korea
13675	200/100	BEL	R. Vlaanderen Internationaal
	500	UAE	UAE Radio & TV, Dubai
13680	500	IRQ	R. Iraq Int, Salah el Deen
13685	500	GUF	China Radio Int. via Fr. Guiana
	500/250	SUI	Swiss Radio International
13690	100	D	Deutsche Welle, Germany
13695	50	USA	KVOH, Rancho Simi, CA
	100		WYFR, Okeechobee, FL
13700	500	HOL	R. Netherlands, Flevoland
	300	MDG	R. Netherlands, Madagascar
13705	10	I	Voice of Europe (F. Pl.)
13710	200/100	BEL	R. Vlaanderen Internationaal
	100	BOT	Voice of America, Botswana
	20	RUS	R. Vedo, Volgograd, E
13715	250/100	TCH	Czech & Slovak R. International
13720	250	CAN	R. Canada Int, Sackville
	100	GUM	KSDA-Adventist World R, Guam
13730	500/100	AUT	R. Austria International
13740	250	USA	Voice of America, Bethany, OH
	250		Voice of America, Delano, CA
	500		WEWN, Birmingham, AL
13745	500/100	G	BBC, United Kingdom
13750	20/50	CTR	AWR Latin America, Costa Rica
13753	20/500	ISR	Kol Israel, Jerusalem
13755	300/100	AUS	R. Australia, Carnarvon
13760	500	D	Deutsche Welle, Germany
	400/100	KRE	R. Pyongyang Ext. Sce, No. Korea
	100	USA	WHRI, Noblesville, IN
	500		WSHB, Cypress Creek, SC
13770	100	D	Deutsche Welle, Germany
	500	HOL	R. Netherlands, Flevoland
	500	USA	WCSN, Scotts Corners, ME
13775	250		Voice of America, Greenville
13780	500/100	D	Deutsche Welle, Germany
	250	MLT	Deutsche Welle, Cyclops, Malta
	250	MNG	Ulaanbaatar Ext. Sce, Mongolia
13785	400/100	KRE	R. Pyongyang Ext. Sce, No. Korea
13790	500/100	D	Deutsche Welle, Germany
	250	MLT	Deutsche Welle, Cyclops, Malta
13795		UKR	Unknown site, Ukraine
13830	2	HRV	Hrvatski Radio, Zagreb
13835	10	ISL	Ríkisútvarpid, Reykjavík
13840	100	MRA	KHBI, Saipan, No. Marianas
13845	100	USA	WWCR, Nashville, TN
13855	10	ISL	Ríkisútvarpid, Reykjavík
14670	3	CAN	CHU (Time Signals), Canada
14996	8	RUS	RWM (Time Signals), Moscow
15000	2	ARG	LOL (Time Signals), Argentina
	10/20	CHN	BPM (Time Signals), China
	10	HWA	WWVH, Hawaii (Time Signals)
	8	IND	ATA, New Delhi (Time Sigs.)
	2	J	JJY, Japan (Time Signals)
	5	RUS	RTA (Time Signals), Russia
	2	TWN	BSF (Time Signals), Taiwan
	10	USA	WWV (Time Signals), Ft. Collins
15004	1	RUS	RID (Time Signals), Irkutsk
15010		VTN	Voice of Vietnam, Hanoi
15020		IND	All India Radio Ext. Sce.
15030.0v	5	CTR	R. for Peace Int, Costa Rica
15050		IND	All India Radio Ext. Sce.
15070	500/100	G	BBC, United Kingdom
15080		IND	All India Radio Ext. Sce.
	250		All India Radio, Aligarh
15084	500/100	IRN	IRIB, Tehran, Iran
15090	500/100	CVA	Vatican Radio
15095	20/500	ISR	Kol Israel Arabic Network D
	100	PHL	FEBC Manila, Philippines
	500	SYR	R. Damascus, Syria
15100	150	CHN	China Radio International
	100	GUM	KTWR-Trans World Radio, Guam
	50	PHL	FEBC Manila, Philippines
15105	500	AFS	BBC via Meyerton, South Africa
	250	ASC	BBC Atlantic Relay St.
	250	ATG	Deutsche Welle, Antigua
	250	CLN	Deutsche Welle, Trincomalee
	500/100	D	Deutsche Welle, Germany
	250	MLT	Deutsche Welle, Cyclops, Malta
	50	MRC	RTV Marocaine, Tanger, Morocco
	50		Voice of America, Tangier
	100	USA	WHRI, Noblesville, IN
15110	100	CTR	R. Exterior de España, Costa R.
	350	E	R. Exterior de España, Madrid
		IND	All India Radio Ext. Sce.
	250	PHL	R. Veritas Asia, Philippines
15115	100	EGY	R. Cairo, Abu Zaabal
	100	EQA	HCJB, Quito, Ecuador
	500/100	G	BBC, United Kingdom
	400/100	KRE	R. Pyongyang Ext. Sce, No. Korea
	250/50	PAK	R. Pakistan Ext. Sce.
	250	POR	RFE/RL, Gloria, Portugal
	500	USA	Voice of America, Greenville
15120	250	ATN	R. Netherlands, Bonaire
	250	BUL	R. Sofia, Kostinbrod, Bulgaria
	100	CLN	SLBC Ext. Sce, Colombo
	500/100	CVA	Vatican Radio
	500	FNL	R. Finland
		IND	All India Radio Ext. Sce.
	50		All India Radio, Kingsway
	100	NZL	R. New Zealand International
	250	USA	Voice of America, Delano, CA
	250		Voice of America, Greenville
15125	100	IND	All India Radio, Madras
	100	MRC	Voice of America, Tangier
	100	OMA	BBC Eastern Relay Station
	250	PHL	Voice of America, Philippines
		RUS	AWR Europe via Samara
			China Radio Int. via Russia
	100	SEY	FEBA Radio, Seychelles

		TWN	CBS FM1 Network, Pali, Taiwan
15130	500/120	CHN	China Radio International
	100	D	RFE/RL, Lampertheim, Germany
	50	MLI	China Radio Int. via Mali
	50/100	USA	Voice of Free China via WYFR
	100/50		WYFR, Okeechobee, FL
15135	7.5	B	R. Record, Sao Paulo
	120	CHN	China Radio International
	250	CYP	BBC Ea. Mediterranean Relay St.
	500	D	Deutsche Welle, Germany
	100	F	R. France Internationale
	500/100	G	BBC, United Kingdom
		IND	All India Radio Ext. Sce.
	250	MLT	Deutsche Welle, Cyclops, Malta
15139.4	60	CHL	R. Nacional de Chile *
15140	50	ARS	BSKSA, Jeddah, Saudi Arabia
	250	CAN	R. Canada Int, Sackville
	100	EQA	HCJB, Quito, Ecuador
		IND	All India Radio Ext. Sce.
	250	PHL	R. Veritas Asia, Philippines
	500	YUG	R. Yugoslavia, Beograd
15145	250	AFS	Channel Africa, Johannesburg
	500/100	G	BBC, United Kingdom
	250	GRC	Voice of America, Kavala
		IND	All India Radio Ext. Sce.
	50		All India Radio, Khampur
	35	MRC	Voice of America, Tangier
	50	PHL	Voice of America, Philippines
	100	SEY	FEBA Radio, Seychelles
	50	USA	WINB, Red Lion, PA
	100		WYFR, Okeechobee, FL
15150	250	CAN	R. Canada Int, Sackville
	500	IRQ	R. Iraq Int, Salah el Deen
	300	MDG	R. Netherlands, Madagascar
	250	PHL	Voice of America, Philippines
15155	250	ATN	R. Netherlands, Bonaire
	500	D	Voice of America, Wertachtal
	100	EGY	R. Cairo, Abu Zaabal
	100	EQA	HCJB, Quito, Ecuador
	100	F	R. France Internationale
	250	GAB	R. France Int. via Gabon
	250	GRC	Voice of America, Kavala
v	100	INS	RRI Jakarta (Prgr. Nasional)
	20	KAZ	Alma-Ata 2, Kazakhstan
	250	PHL	Voice of America, Philippines
15160	100	ALG	RTV Algerienne, Bouchaoui
	250	ASC	BBC Atlantic Relay St.
	250		Voice of America, Ascension Is
	250	BUL	R. Sofia, Kostinbrod, Bulgaria
	250	GRC	Voice of America, Kavala
	250/100	HNG	R. Budapest, Hungary
	0.5	MEX	LV de América Latina, México
	250	PHL	Voice of America, Philippines
	250	TUR	V. of Turkey, Ankara
	500	USA	Voice of America, Greenville
	250		V. of the OAS, Greenville, NC
15165	100	BOT	Voice of America, Botswana
	500/150	CHN	China Radio International
	250	CYP	BBC Ea. Mediterranean Relay St.
	500/100	G	BBC, United Kingdom
		IND	All India Radio Ext. Sce.
	100		All India Radio, Kingsway
	500	NOR	R. Norway Int. / R. Denmark
	250	PHL	Voice of America, Philippines
	100	UZB	Toskent 2, Uzbekistan
15170	50	ARS	BSKSA, Jeddah, Saudi Arabia
	300	AUS	R. Australia, Carnarvon
	300	J	R. Japan, Yamata, Japan
	250	KOR	R. Korea, Kimjae, So. Korea
	50	MLI	China Radio Int. via Mali
	20	OCE	RFO, Tahiti, French Polynesia
	250	POR	RFE/RL, Gloria, Portugal
	350	S	R. Sweden
	100	USA	WYFR, Okeechobee, FL
15175	250/100	EGY	R. Cairo, Abis / Abu Zaabal
		IND	All India Radio Ext. Sce.
	500	NOR	R. Norway Int. / R. Denmark
	500	YUG	R. Yugoslavia, Beograd
15180	100	F	R. France Internationale
	500/100	G	BBC, United Kingdom
	500	IRQ	R. Iraq Int, Salah el Deen
	400/100	KRE	R. Pyongyang Ext. Sce, No. Korea
	500	NOR	R. Norway Int. / R. Denmark
	50	PHL	Voice of America, Philippines
	500	YUG	R. Yugoslavia, Beograd
15185	250	CLN	Deutsche Welle, Trincomalee
	500/100	D	Deutsche Welle, Germany
	250	G	Voice of America, Woofferton
		IND	All India Radio Ext. Sce.
	250	MLT	Deutsche Welle, Cyclops, Malta
	100/50	PHL	Voice of America, Philippines
	500	USA	Voice of America, Greenville
	50		WINB, Red Lion, PA
15190	250	ASC	BBC Atlantic Relay St.
	5	B	R. Inconfidência, Belo H.
	50	COG	RTV Congolaise, Brazzaville
	250/50	PAK	R. Pakistan Ext. Sce.
	10		R. Pakistan Home Sce, Islamabad
	250	PHL	R. Pilipinas via VOA Tinang
	250		Voice of America, Philippines
15195	250	CLN	Deutsche Welle, Trincomalee
	100	D	Voice of America, Munich
	500		Voice of America, Wertachtal
	100	F	R. France Internationale
	300/100	J	R. Japan, Yamata, Japan
	100	MRC	Voice of America, Tangier
	500	NOR	R. Norway Int. / R. Denmark
	250	PHL	Voice of America, Philippines
15200	250	B	R. Nacional de Amazônia, Brazil
v	250	BGD	R. Bangladesh Ext. Sce.
	500	GUF	R. France Int, French Guiana
	100	GUM	KTWR-Trans World Radio, Guam
	3	RUS	Tatar R, Perm, E
	100	UZB	Toskent 2 / FS, Uzbekistan
15205.0v	100	ALG	RTV Algerienne, Bouchaoui
	250	ATG	Deutsche Welle, Antigua
	500/100	D	Deutsche Welle, Germany
	250	GRC	Voice of America, Kavala
	100/35	MRC	Voice of America, Tangier
	50	PHL	Voice of America, Philippines
	250	USA	Voice of America, Greenville
15210	500/100	CVA	Vatican Radio
	250	EGY	R. Cairo, Abis
	500	IRQ	R. Iraq Int, Salah el Deen
	250	PHL	Voice of America, Philippines
15215.0v	100	ALG	RTV Algerienne, Bouchaoui
	300	J	R. France Int. via Japan
	20	KAZ	Alma-Ata 2, Kazakhstan
	250	PHL	R. Veritas Asia, Philippines
	250		Voice of America, Philippines
	500	POR	RFE/RL, Maxoqueira, Portugal
	50/100	USA	Voice of Free China via WYFR
	100		WYFR, Okeechobee, FL
15220	125	ATG	BBC Caribbean Relay St.
	500/50	CUB	R. Habana, Cuba
	250/100	EGY	R. Cairo, Abis / Abu Zaabal
	100	HNG	R. Budapest, Hungary
	50	IND	All India Radio, Kingsway
	500	NOR	R. Norway Int. / R. Denmark
	500	TUR	V. of Turkey, Ankara
15225	500/120	CHN	China Radio International
	250	GRC	Voice of America, Kavala
	100	GUM	KSDA-Adventist World R, Guam
	50	PHL	R. Veritas Asia, Philippines
	100	POR	R. Portugal, S. Gabriel
	250		Voice of America, Gloria
15230	500/120	CHN	China Radio International
	500/50	CUB	R. Habana, Cuba
	100	J	R. Japan, Yamata, Japan
	400/100	KRE	R. Pyongyang Ext. Sce, No. Korea
	500/350	NOR	R. Norway Int. / R. Denmark
	250	PHL	Voice of America, Philippines
	350	S	R. Sweden
15235	100	CYP	BBC Ea. Mediterranean Relay St.
	500/100	G	BBC, United Kingdom
	300/250		Voice of America, Woofferton
	250	GRC	Voice of America, Kavala
	500	LBY	LJB, Tripoli/Sabrata, Libya
	100	OMA	BBC Eastern Relay Station

15240	250	AUS	R. Australia, Carnarvon
	100		R. Australia, Shepparton
	350	S	R. Sweden
15245	100	CYP	BBC Ea. Mediterranean Relay St.
	500	FNL	R. Finland
	100	I	RAI Ext. Sce, Rome, Italy
	100	MRC	Voice of America, Tangier
	100	OMA	BBC Eastern Relay Station
	250	PHL	R. Veritas Asia, Philippines
	250	POR	Deutsche Welle, Sines
	350	S	R. Sweden
	500	USA	Voice of America, Greenville
15250	35	CLN	Voice of America, Colombo
	250	EQA	HCJB, Quito, Ecuador
	50	IND	All India Radio, Kingsway
	20	KAZ	Alma-Ata 2, Kazakhstan
	50	PHL	Voice of America, Philippines
	100	POR	R. Portugal, S. Gabriel
	250/120	ROU	R. Romania International
15255	250	ASC	Voice of America, Ascension Is
v	250	BGD	R. Bangladesh Ext. Sce.
	100	EGY	R. Cairo, Abu Zaabal
	250/120	ROU	R. Romania International
	200	RUS	R. Ala, Yekaterinburg, S
15260	250	ASC	BBC Atlantic Relay St.
	250	CAN	BBC via Sackville, NB
	250		R. Canada Int, Sackville
	500/100	IRN	IRIB, Tehran, Iran
15265	250	B	R. Bras Ext. Sce, Brasilia
	300	G	Voice of America, Woofferton
		IND	All India Radio Ext. Sce.
	500	UAE	R. of the UAE, Abu Dhabi
	500	USA	Voice of America, Greenville
15270	500	EQA	HCJB, Quito, Ecuador
	300	J	R. Canada Int. via Yamata, Japan
	100		R. Japan, Yamata, Japan
	20	KAZ	Alma-Ata 2, Kazakhstan
	250	RRW	Deutsche Welle, Kigali, Rwanda
	350	S	R. Sweden
	250/50	TWN	Voice of Free China, Taiwan
15275	250	ATG	Deutsche Welle, Antigua
	500/100	D	Deutsche Welle, Germany
15280	250	HNG	BBC East Asian Relay St.
	100	J	R. Japan, Yamata, Japan
		TWN	Voice of June 4th, Taiwan
	50	USA	KGEI, Redwood City, CA
15285	120	CHN	R. France Int. via Beijing
	100	EGY	R. Cairo, Abu Zaabal
	20	KAZ	Alma-Ata 2, Kazakhstan
	250	PHL	Voice of America, Philippines
15290	100	D	RFE/RL, Lampertheim, Germany
	35	MRC	Voice of America, Tangier
	250	PHL	Voice of America, Philippines
	250	POR	RFE/RL, Gloria, Portugal
	500		RFE/RL, Maxoqueira, Portugal
15295	100	EQA	HCJB, Quito, Ecuador
	500	MLA	V. of Malaysia, Kuala Lumpur
v	10	MOZ	Em. Nacional, Maputo
	250/50	PAK	R. Pakistan Ext. Sce.
	100	POR	R. Portugal, S. Gabriel
	50	USA	WINB, Red Lion, PA
15300	500/100	F	R. France Internationale
	250	GAB	R. France Int. via Gabon
	500	GUF	R. France Int, French Guiana
	100	J	R. Japan, Yamata, Japan
	250	PHL	R. Veritas Asia, Philippines
15305	250/50	PAK	R. Pakistan Ext. Sce.
	250	PHL	R. Veritas Asia, Philippines
	35		Voice of America, Philippines
	500	UAE	R. of the UAE, Abu Dhabi
15310	250	CYP	BBC Ea. Mediterranean Relay St.
	100	GUM	KSDA-Adventist World R, Guam
	100	OMA	BBC Eastern Relay Station
15315	250	ATN	R. Netherlands, Bonaire
	250	CAN	R. Canada Int, Sackville
	500/100	F	R. France Internationale
	500/100	IRN	IRIB, Tehran, Iran
	500	UAE	R. of the UAE, Abu Dhabi
	250	USA	BBC via Bethany, OH
	250		Voice of America, Bethany, OH

	100	AUS	R. Australia, Shepparton
	500	IRQ	R. Iraq Int, Salah el Deen
	100	MRA	KFBS, Saipan, No. Marianas
	35	MRC	Voice of America, Tangier
	250	PHL	Voice of America, Philippines
	300	UAE	UAE Radio & TV, Dubai
15325	250	CAN	R. Canada Int, Sackville
	250	CYP	BBC Ea. Mediterranean Relay St.
	500	GUF	R. Japan via French Guiana
		IND	All India Radio Ext. Sce.
	250	POR	R. Canada Int. via Sines
	100	SEY	FEBA Radio, Seychelles
	250	TUR	V. of Turkey, Ankara
15330	250	ATN	R. Netherlands, Bonaire
	1000	FNL	R. Finland (SSB)
	100	I	RAI Ext. Sce, Rome, Italy
	50	MRC	RTV Marocaine, Tanger, Morocco
	500	NOR	R. Norway Int. / R. Denmark
	250	PHL	Voice of America, Philippines
	100	SEY	FEBA Radio, Seychelles
	125	USA	Voice of America, Greenville
15335	500/350	ARS	BSKSA, Riyadh, Saudi Arabia
	250	EGY	R. Cairo, Abis
		IND	All India Radio Ext. Sce.
	100	MRC	RTV Marocaine, Tanger, Morocco
	100		Voice of America, Tangier
	250/120	ROU	R. Romania International
15340	500/50	CUB	R. Habana, Cuba
	100	D	RFE/RL, Lampertheim, Germany
	500/100	G	BBC, United Kingdom
	500	GAB	R. Japan via Moyabi, Gabon
	100	I	RAI Ext. Sce, Rome, Italy
	400/100	KRE	R. Pyongyang Ext. Sce, No. Korea
	250/120	ROU	R. Romania International
	100	SNG	BBC Far Eastern Relay St.
15345	500	ARS	BSKSA, Riyadh, Saudi Arabia
	50	ATN	Trans World Radio, Bonaire
	500/120	CHN	China Radio International
	500	KWT	R. Kuwait
	250/50	TWN	Voice of Free China, Taiwan
15345.2	100	ARG	R. Argentina al Exterior
15350	500	D	Deutsche Welle, Germany
	50	EQA	HCJB, Quito, Ecuador
	100		R. Nal. del Ecuador via HCJB
	500	GUF	R. Japan via French Guiana
	10	LUX	R. Luxembourg, Junglinster
	200	RUS	Deutsche Welle via Novosibirsk
	250	TUR	V. of Turkey, Ankara
15355	100	ATN	Trans World Radio, Bonaire
	500	D	Voice of America, Wertachtal
	500	GAB	R. Japan via Moyabi, Gabon
	250/100	TCH	Czech & Slovak R. International
	50	USA	KGEI, Redwood City, CA
	100		WYFR, Okeechobee, FL
15360	20	KAZ	Alma-Ata 2, Kazakhstan
	50	MRC	RTV Marocaine, Tanger, Morocco
	500/250	NOR	R. Norway Int. / R. Denmark
	100	SNG	BBC Far Eastern Relay St.
15365	100	AUS	R. Australia, Shepparton
	100	CTR	R. Exterior de España, Costa R.
	100	F	R. France Internationale
	500	GUF	R. France Int, French Guiana
		IND	All India Radio Ext. Sce.
	500/100	IRN	IRIB, Tehran, Iran
	250/120	ROU	R. Romania International
	250/50	TWN	Voice of Free China, Taiwan
	500	USA	Voice of America, Greenville
15370	100	BOT	Voice of America, Botswana
	250	D	RFE/RL, Holzkirchen, Germany
	500	POR	RFE/RL, Maxoqueira, Portugal
	250/120	ROU	R. Romania International
	250/50	TWN	Voice of Free China, Taiwan
15375	50	ATN	Trans World Radio, Bonaire
	100	BOT	Voice of America, Botswana
	350	E	R. Exterior de España, Madrid
	100	EGY	R. Cairo, Abu Zaabal
	100	OMA	R. Oman, Thumrait
	100	USA	KCBI, Denton, TX
15380	250	AUS	R. Australia, Darwin
	350	E	R. Exterior de España, Madrid

	100	MRA	KFBS, Saipan, No. Marianas
	250/50	PAK	R. Pakistan Ext. Sce.
	250	POR	RFE/RL, Gloria, Portugal
	500		RFE/RL, Maxoqueira, Portugal
	250/120	ROU	R. Romania International
	250	SNG	BBC Far Eastern Relay St.
	100	SWZ	Trans World Radio, Swaziland
	500	UAE	R. of the UAE, Abu Dhabi
15385	100	I	RAI Ext. Sce, Rome, Italy
	20	KAZ	Alma-Ata 2, Kazakhstan
15390	250	ASC	BBC Atlantic Relay St.
	50	CHN	CPBS 1, China
	250	CLN	Deutsche Welle, Trincomalee
	100	D	Deutsche Welle, Germany
	350	S	R. Sweden
15395	35	CLN	Voice of America, Colombo
	250	PHL	Voice of America, Philippines
15400	250	ASC	BBC Atlantic Relay St.
	100	D	Deutsche Welle, Germany
	500	FNL	R. Finland
	250	GRC	Voice of America, Kavala
	300	UAE	UAE Radio & TV, Dubai
	250	USA	Voice of America, Greenville
	100		WYFR, Okeechobee, FL
15405	500	F	R. France Internationale
	100	MRA	KHBI, Saipan, No. Marianas
	250/50	PHL	R. Veritas Asia, Philippines
	250/120	ROU	R. Romania International
	250	TUR	V. of Turkey, Ankara
15410	250	ATG	Deutsche Welle, Antigua
	100	AUT	R. Austria International
	250	CLN	Deutsche Welle, Trincomalee
	250	PHL	Voice of America, Philippines
	250	RRW	Deutsche Welle, Kigali, Rwanda
	250	USA	Voice of America, Greenville
15415	250	CLN	Deutsche Welle, Trincomalee
v	500	LBY	LJB, Tripoli/Sabrata, Libya
15420	250	AFS	BBC via Meyerton, South Africa
	300/120	CHN	China Radio International
	100	CYP	BBC Ea. Mediterranean Relay St.
	100	EGY	R. Cairo, Abu Zaabal
	100	GUM	KTWR-Trans World Radio, Guam
	50	IND	All India Radio, Khampur
	250	SEY	BBC Indian Ocean Relay St.
	100	USA	WRNO, New Orleans, LA
15425	50	AUS	VLW15, Wanneroo, Australia
	250	CAN	R. Canada Int, Sackville
	35	CLN	SLBC Ext. Sce, Colombo
	100	D	Deutsche Welle, Germany
	100	F	R. France Internationale
	50	PHL	Voice of America, Philippines
15430	500	AFS	Channel Africa, Johannesburg
	500	ARS	BSKSA, Riyadh, Saudi Arabia
	100	AUT	R. Austria International
	100	J	R. Japan, Yamata, Japan
	50	MEX	R. México Internacional
	100	SEY	FEBA Radio, Seychelles
	500/250	SUI	Swiss Radio International
15435	50	ARS	BSKSA, Jeddah, Saudi Arabia
	120	CHN	China Radio International
	500/50	CUB	R. Habana, Cuba
	500/100	D	Deutsche Welle, Germany
	500	GUF	R. France Int, French Guiana
	500	JOR	R. Jordan, Al Karanah
v	500	LBY	LJB, Tripoli/Sabrata, Libya
	500	UAE	UAE Radio & TV, Dubai
15440	250	AFS	Channel Africa, Johannesburg
	120	CHN	China Radio International
	1000	FNL	R. Finland (SSB)
	50	PHL	R. Veritas Asia, Philippines
	50/100	USA	Voice of Free China via WYFR
	100		WYFR, Okeechobee, FL
15445	100	ATN	Trans World Radio, Bonaire
	250	B	China Radio Int. via Brazil
	1		R. Nacional de Amazônia, Brazil
	100	BOT	Voice of America, Botswana
	100	D	RFE/RL, Biblis, Germany
	500		Voice of America, Wertachtal
	250	GRC	Voice of America, Kavala
	250/50	PAK	R. Pakistan Ext. Sce.
	250/120	ROU	R. Romania International
	100	SEY	FEBA Radio, Seychelles
15450	500/120	CHN	China Radio International
	100	TUN	ERTT, Tunis
15460	50	CTR	AWR Latin America, Costa Rica
	100	F	R. France Internationale
	250	HNG	R. France Int. via Hungary
15465	100/50	PHL	FEBC Manila, Philippines
15470	250	POR	Deutsche Welle, Sines
		UZB	R. Tashkent Ext. Sce, Uzbekistan
15474	1	ARG	RN Arc. S. Gabriel, Antarctica *
15475	500	GAB	Africa No.1, Moyabi, Gabon
15480	20/500	ISR	Kol Israel Arabic Network D
15485	100	F	R. France Internationale
	100	GUM	KTWR-Trans World Radio, Guam
15495	100	BOT	Voice of America, Botswana
	500	KWT	R. Kuwait
15500		CHN	CPBS 2, China
15505	500	KWT	R. Kuwait
	500/250	SUI	Swiss Radio International
15515	250/50	PAK	R. Pakistan Ext. Sce.
	50	PHL	FEBC Manila, Philippines
15516	250/50	PAK	R. Pakistan Ext. Sce.
15520.0v	250/50	BGD	R. Bangladesh Home & Ext. Sce.
	250/100	TCH	Czech & Slovak R. International
15530	250	AUS	R. Australia, Darwin
	500/100	F	R. France Internationale
	250/100	HNG	R. France Int. via Hungary
15540	200/100	BEL	R. Vlaanderen Internationaal
		RUS	China Radio Int. via Russia
15545	500/100	D	Deutsche Welle, Germany
15550		CHN	CPBS 1, China
	100	GRC	Voice of Greece, Athens
	250/50	PAK	R. Pakistan Ext. Sce.
	1000	RUS	Deutsche Welle via Novosibirsk
15555	250/50	PAK	R. Pakistan Ext. Sce.
15560	250	ATN	R. Netherlands, Bonaire
	500	HOL	R. Netherlands, Flevoland
15566	100	USA	WYFR, Okeechobee, FL
15570	300	MDG	R. Netherlands, Madagascar
15575	250	AUS	R. Australia, Darwin
	250/100	CYP	BBC Ea. Mediterranean Relay St.
	250	KOR	R. Korea, Kimjae, So. Korea
15580	250	USA	Voice of America, Greenville
15590	250/100	CYP	BBC Ea. Mediterranean Relay St.
	20/500	ISR	Kol Israel, Jerusalem
	100	USA	KTBN, Salt Lake City, UT
15595	500	D	Deutsche Welle, Germany
	250	MLT	Deutsche Welle, Cyclops, Malta
15600	100	BOT	Voice of America, Botswana
	500/120	CHN	China Radio International
15610	100	GUM	KSDA-Adventist World R, Guam
	100		KTWR-Trans World Radio, Guam
15615	20/500	ISR	Kol Israel, Jerusalem
15625	100	BOT	Voice of America, Botswana
15640	20/500	ISR	Kol Israel, Jerusalem
15650	250	GRC	Voice of America, Kavala
	100		Voice of Greece, Athens
15665	100	MRA	KHBI, Saipan, No. Marianas
	500	USA	WSHB, Cypress Creek, SC
15670		CHN	CPBS Minority Sce, China
15685	100	USA	WWCR, Nashville, TN
15710		CHN	CPBS Taiwan 1 Sce, China
15770	10	ISL	Ríkisútvarpid, Reykjavík
15790	10		Ríkisútvarpid, Reykjavík
15880		CHN	CPBS Taiwan 2 Sce, China
16000	5	AUS	VNG, Llandilo (Time Signals)
17018	10	AFS	ZSC (Time Signals), So. Africa
17096		HKG	VPS80 (Time Sigs), Hong Kong
17387		IND	All India Radio Ext. Sce.
17440	0.5	KIR	R. Kiribati, Tarawa
17500	100	TUN	ERTT, Tunis
17510	500	USA	WCSN, Scotts Corners, ME
17525	500/100	CVA	Vatican Radio
	100	GRC	Voice of Greece, Athens
17540	200/100	BEL	R. Vlaanderen Internationaal
17545	200/100		R. Vlaanderen Internationaal
	20/500	ISR	Kol Israel, Jerusalem
17550	200/100	BEL	R. Vlaanderen Internationaal

17555	100	MRA	KHBI, Saipan, No. Marianas
	250/50	PAK	R. Pakistan Ext. Sce.
	500	USA	WSHB, Cypress Creek, SC
17560	100	D	Deutsche Welle, Germany
	500	GUF	R. France Int, French Guiana
17565	500/250	SUI	Swiss Radio International
17575	500	GUF	R. France Int, French Guiana
	20/500	ISR	Kol Israel, Jerusalem
17580	500	HOL	R. Netherlands, Flevoland
	300	MDG	R. Netherlands, Madagascar
17590	20/500	ISR	Kol Israel, Jerusalem
17595	250	EGY	R. Cairo, Abis
	50	MRC	RTV Marocaine, Tanger, Morocco
	50		Voice of America, Tangier
17605	250	ATN	R. Netherlands, Bonaire
		CHN	CPBS 1, China
	20	KAZ	Alma-Ata 2, Kazakhstan
		RUS	AWR Europe via Moscow
			R. Vilnius via Far East tr. site
17610	500	HOL	R. Netherlands, Flevoland
17620	500/100	F	R. France Internationale
	250	GAB	R. France Int. via Gabon
	500	GUF	R. France Int, French Guiana
17625	100	UKR	Yerevan, Armenia (relay)
17630	500/120	CHN	China Radio International
	500	GAB	Africa No.1, Moyabi, Gabon
17635	500/250	SUI	Swiss Radio International
17640	500/100	G	BBC, United Kingdom
	500	USA	Voice of America, Greenville
17650	100	BOT	Voice of America, Botswana
	500/120	CHN	China Radio International
	100	F	R. France Internationale
17670	250	AUS	R. Australia, Darwin
	250/100	EGY	R. Cairo, Abu Zaabal
	500/250	SUI	Swiss Radio International
17675	250	RUS	Deutsche Welle via Zhigulevsk
17680	240	UKR	Yerevan, Armenia (relay)
17690	250	EGY	R. Cairo, Abis
	250	HNG	R. France Int. via Hungary
		RUS	R. Vilnius via Far East tr. site
17695	250	AUS	R. Australia, Darwin
	500/100	G	BBC, United Kingdom
17700	50	CHN	CPBS 2, China
17705	100	BOT	Voice of America, Botswana
	500	CUB	R. Habana, Cuba
	500/100	G	BBC, United Kingdom
		IND	All India Radio Ext. Sce.
	50		All India Radio, Khampur
	50		All India Radio, Kingsway
	300	J	R. France Int. via Japan
	250/50	PAK	R. Pakistan Ext. Sce.
17710	150	CHN	China Radio International
	120		R. France Int. via Beijing
	100	D	Deutsche Welle, Germany
	300	J	R. France Int. via Japan
	250/50	PAK	R. Pakistan Ext. Sce.
	250	USA	Voice of America, Greenville
17715	250	ATG	Deutsche Welle, Antigua
	250	AUS	R. Australia, Darwin
	100		R. Australia, Shepparton
	500/100	D	Deutsche Welle, Germany
	350	E	R. Exterior de España, Madrid
	500/100	G	BBC, United Kingdom
	20	KAZ	Alma-Ata 2, Kazakhstan
	250	USA	BBC via Greenville, NC
	250		Voice of America, Greenville
17720	500	GUF	R. France Int, French Guiana
	100	PHL	Voice of America, Philippines
	250/120	ROU	R. Romania International
	250/50	TWN	Voice of Free China, Taiwan
17725.0r	500	LBY	LJB, Tripoli/Sabrata, Libya
	250/50	PAK	R. Pakistan Ext. Sce.
	250	POR	RFE/RL, Gloria, Portugal
	250/100	TCH	Czech & Slovak R. International
17730	100	AUT	R. Austria International
	250	B	Swiss Radio Int. via Brasilia
	500/100	CVA	Vatican Radio
	20	KAZ	Alma-Ata 2, Kazakhstan
	500	NOR	R. Norway Int. / R. Denmark
	250	USA	Voice of America, Bethany, OH
	500		Voice of America, Greenville
17735	100	D	Deutsche Welle, Germany
		IND	All India Radio Ext. Sce.
	250	PHL	Voice of America, Philippines
	250	POR	RFE/RL, Gloria, Portugal
	200	RUS	Deutsche Welle via Zhigulevsk
17740	500	IRQ	R. Iraq Int, Salah el Deen
	500	NOR	R. Norway Int. / R. Denmark
	250	PHL	Voice of America, Philippines
	300	POR	R. Portugal, S. Gabriel
	250	USA	Voice of America, Greenville
	500	YUG	R. Yugoslavia, Beograd
17745	100	ALG	RTV Algerienne, Bouchaoui
	250	EGY	R. Cairo, Abis
	100	POR	R. Portugal, S. Gabriel
	250/120	ROU	R. Romania International
		UZB	R. Tashkent Ext. Sce, Uzbekistan
17750	300	AUS	R. Australia, Carnarvon
v	250	BGD	R. Bangladesh Ext. Sce.
	250	CYP	BBC Ea. Mediterranean Relay St.
	100	D	RFE/RL, Lampertheim, Germany
	400/100	KRE	R. Pyongyang Ext. Sce, No. Korea
	250/50	PAK	R. Pakistan Ext. Sce.
	100	USA	Voice of Free China via WYFR
	100		WYFR, Okeechobee, FL
17755	250	B	China Radio Int. via Brazil
	250		R. Bras Ext. Sce, Brasilia
17760	500	ARS	BSKSA, Riyadh, Saudi Arabia
	100	D	RFE/RL, Biblis, Germany
	250	PHL	R. Pilipinas via VOA Tinang
	250		Voice of America, Philippines
	100	SEY	FEBA Radio, Seychelles
	500	USA	WEWN, Birmingham, AL
	100		WYFR, Okeechobee, FL
17765	250	CLN	Deutsche Welle, Trincomalee
	500	D	Deutsche Welle, Germany
	300	J	R. Japan, Yamata, Japan
	20	KAZ	Alma-Ata 2, Kazakhstan
	400/100	KRE	R. Pyongyang Ext. Sce, No. Korea
	10	MEX	R. México Internacional
	250	PHL	Voice of America, Philippines
17770	250	EGY	R. Cairo, Abis
	500/100	G	BBC, United Kingdom
	100	NZL	R. New Zealand International
17775	300	CLN	R. Japan, Ekala, Sri Lanka
	100	F	R. France Internationale
	250/120	ROU	R. Romania International
	50	USA	KVOH, Rancho Simi, CA
17780	500	AFS	Channel Africa, Johannesburg
	250	BUL	R. Sofia, Kostinbrod, Bulgaria
	250	CLN	Deutsche Welle, Trincomalee
	500/100	D	Deutsche Welle, Germany
	250	GRC	Voice of America, Kavala
	100	I	RAI Ext. Sce, Rome, Italy
	100	MRA	KHBI, Saipan, No. Marianas
	100	OMA	BBC Eastern Relay Station
	250	PHL	Voice of America, Philippines
17785	250	CYP	BBC Ea. Mediterranean Relay St.
	500	F	R. France Internationale
		IND	All India Radio Ext. Sce.
	100	J	R. Japan, Yamata, Japan
	350	NOR	R. Norway Int. / R. Denmark
	250/50	PAK	R. Pakistan Ext. Sce.
	500/250	USA	Voice of America, Greenville
17790	500	AFS	BBC via Meyerton, South Africa
	250	ASC	BBC Atlantic Relay St.
	500	EQA	HCJB, Quito, Ecuador
	100	OMA	BBC Eastern Relay Station
	100	PHL	Voice of America, Philippines
	250/120	ROU	R. Romania International
	100	SNG	BBC Far Eastern Relay St.
17795	100	AUS	R. Australia, Shepparton
	100	F	R. France Internationale
	100	I	RAI Ext. Sce, Rome, Italy
	50	IND	All India Radio, Khampur
	400/100	KRE	R. Pyongyang Ext. Sce, No. Korea
	500	NOR	R. Norway Int. / R. Denmark
17800	250	CLN	Deutsche Welle, Trincomalee
	500	D	Deutsche Welle, Germany
	250	EGY	R. Cairo, Abis

	100	F	R. France Internationale
	1000	FNL	R. Finland (SSB)
	100	I	RAI Ext. Sce, Rome, Italy
	250/50	PAK	R. Pakistan Ext. Sce.
	50	PHL	R. Veritas Asia, Philippines
	250	QAT	Qatar Broadc. Sce, Doha
	250	RRW	Deutsche Welle, Kigali, Rwanda
	175	USA	Voice of America, Bethany, OH
	250		Voice of America, Greenville
17802	250/50	PAK	R. Pakistan Ext. Sce.
17805		IND	All India Radio Ext. Sce.
	250/120	ROU	R. Romania International
	50/100	USA	Voice of Free China via WYFR
	100		WYFR, Okeechobee, FL
17810	250	ASC	BBC Atlantic Relay St.
	250		Voice of America, Ascension Is
	250	ATG	Deutsche Welle, Antigua
	250	CLN	Deutsche Welle, Trincomalee
	500	D	Deutsche Welle, Germany
	250	GRC	Voice of America, Kavala
	300	J	R. Japan, Yamata, Japan
	20	RUS	Tatar R. via Balashikha
17815	250	AFS	Channel Africa, Johannesburg
	1	B	R. Cultura, Sao Paulo
	250	EGY	R. Cairo, Abis
	500/100	G	BBC, United Kingdom
	50	IND	All India Radio, Khampur
	50	MRC	RTV Marocaine, Tanger, Morocco
	250/120	ROU	R. Romania International
17820	250	CAN	R. Canada Int, Sackville
	250	CLN	Deutsche Welle, Trincomalee
	300		R. Japan, Ekala, Sri Lanka
	100	D	Deutsche Welle, Germany
	100	J	R. Japan, Yamata, Japan
	100	PHL	Voice of America, Philippines
	500	UAE	R. of the UAE, Abu Dhabi
17825.0v	250	BGD	R. Bangladesh Ext. Sce.
	100	BUL	R. Sofia, Kostinbrod, Bulgaria
	250	CLN	Deutsche Welle, Trincomalee
	500	D	Deutsche Welle, Germany
	300/100	J	R. Japan, Yamata, Japan
	250	MLT	Deutsche Welle, Cyclops, Malta
17830	250	ASC	BBC Atlantic Relay St.
	250		Voice of America, Ascension Is
	250	CAN	Deutsche Welle via Sackville
	250	HNG	BBC East Asian Relay St.
		IND	All India Radio Ext. Sce.
	150/100	SNG	BBC Far Eastern Relay St.
	300	UAE	UAE Radio & TV, Dubai
	100	USA	WHRI, Noblesville, IN
17835	100	J	R. Japan, Yamata, Japan
	250	POR	RFE/RL, Gloria, Portugal
17840	125	ATG	BBC Caribbean Relay St.
	250	PHL	R. Pilipinas via VOA Tinang
	250		Voice of America, Philippines
17845	350	ARS	BSKSA, Riyadh, Saudi Arabia
	100	CTR	R. Exterior de España, Costa R.
	500	D	Deutsche Welle, Germany
	350	E	R. Exterior de España, Madrid
	500	F	R. France Internationale
	300	J	R. Japan, Yamata, Japan
	50/100	USA	Voice of Free China via WYFR
	100		WYFR, Okeechobee, FL
17850	35	CLN	SLBC Ext. Sce, Colombo
	500	F	R. France Internationale
		IND	All India Radio Ext. Sce.
	50		All India Radio, Kingsway
	500	KWT	R. Kuwait
17855	500/120	CHN	China Radio International
	300	G	Voice of America, Woofferton
	500	UAE	R. of the UAE, Abu Dhabi
17860	250	ATG	Deutsche Welle, Antigua
	100	D	Deutsche Welle, Germany
	500	GUF	R. France Int, French Guiana
	100	J	R. Japan, Yamata, Japan
	500	NOR	R. Norway Int. / R. Denmark
	250	RRW	Deutsche Welle, Kigali, Rwanda
17865.0v	20	CLM	R. Dif. Nal., Colombia (USB)
	500/100	CVA	Vatican Radio
	100	MRA	KHBI, Saipan, No. Marianas
	100	MRC	Voice of America, Tangier
	500	NOR	R. Norway Int. / R. Denmark
	250	PHL	Voice of America, Philippines
	250	POR	Voice of America, Gloria
17870	100	BOT	Voice of America, Botswana
	350	S	R. Sweden
17875	250	CAN	R. Canada Int, Sackville
	500	D	Deutsche Welle, Germany
	100	EQA	HCJB, Quito, Ecuador
	250	GRC	Voice of America, Kavala
	50	IND	All India Radio, Khampur
	250	PHL	Voice of America, Philippines
17880	250	ASC	BBC Atlantic Relay St.
	250		Voice of America, Ascension Is
	300/250	AUS	R. Australia, Carnarvon
	250		R. Australia, Darwin
17885	250	CYP	BBC Ea. Mediterranean Relay St.
	100	D	Voice of America, Munich
	500	KWT	R. Kuwait
	250	SEY	BBC Indian Ocean Relay St.
17890	350	E	R. Exterior de España, Madrid
	500	GAB	AWR Africa via Moyabi, Gabon
	100	J	R. Japan, Yamata, Japan
	500	USA	WEWN, Birmingham, AL
17895	350	ARS	BSKSA, Riyadh, Saudi Arabia
	250	CYP	BBC Ea. Mediterranean Relay St.
	500	D	Voice of America, Wertachtal
		IND	All India Radio Ext. Sce.
	500	KWT	R. Kuwait
17900	100	I	RAI Ext. Sce, Rome, Italy
	250/50	PAK	R. Pakistan Ext. Sce.
18930	500	USA	WEWN, Birmingham, AL
20000	2.5	WWV	(Time Signals), Ft. Collins
21455	500	D	Voice of America, Wertachtal
	30	EQA	HCJB, Quito, Ecuador
	250	GRC	Voice of America, Kavala
	250	PHL	R. Pilipinas via VOA Tinang
	250		Voice of America, Philippines
21460	100	CTR	R. Exterior de España, Costa R.
21465.0v	2		R. For Peace Int, Costa Rica
	500/100	D	Deutsche Welle, Germany
21470	250	ASC	BBC Atlantic Relay St.
	250	CYP	BBC Ea. Mediterranean Relay St.
21475	250/50	PAK	R. Pakistan Ext. Sce.
	250	PHL	Voice of America, Philippines
21480	250	EQA	HCJB, Quito, Ecuador
	500	HOL	R. Netherlands, Flevoland
	300	MDG	R. Netherlands, Madagascar
21485	250	USA	Voice of America, Greenville
21490	250	ASC	BBC Atlantic Relay St.
	250		Voice of America, Ascension Is
	300/100	AUT	R. Austria International
	20	RUS	Kazakh R. via Balashikha
21495	250/50	PAK	R. Pakistan Ext. Sce.
21500	500	HOL	R. Netherlands, Flevoland
	350	S	R. Sweden
	100	USA	WYFR, Okeechobee, FL
21510	500	ARS	BSKSA, Riyadh, Saudi Arabia
	250	POR	RFE/RL, Gloria, Portugal
21515	250	ATN	R. Netherlands, Bonaire
	500/100	CVA	Vatican Radio
	100	I	RAI Ext. Sce, Rome, Italy
	100	POR	R. Portugal, S. Gabriel
	500	UAE	R. of the UAE, Abu Dhabi
21520	250	GAB	R. France Int. via Gabon
	250	GRC	Voice of America, Kavala
	250/50	PAK	R. Pakistan Ext. Sce.
21525	250	AUS	R. Australia, Darwin
	100	USA	WYFR, Okeechobee, FL
21530	500	F	R. France Internationale
21535	500	D	Voice of America, Wertachtal
	100	I	RAI Ext. Sce, Rome, Italy
	100	TUN	ERTT, Tunis
21540	500	D	Deutsche Welle, Germany
21545	250	CAN	R. Canada Int, Sackville
21550	1000	FNL	R. Finland (SSB)
	50	PHL	Voice of America, Philippines
21555	350	E	R. Exterior de España, Madrid
	250/100	QAT	Qatar Broadc. Sce, Doha

21560	250	CLN	Deutsche Welle, Trincomalee
	100	D	Deutsche Welle, Germany
	100	I	RAI Ext. Sce, Rome, Italy
	250	PHL	R. Veritas Asia, Philippines
21570	500	D	Voice of America, Wertachtal
	250	GRC	Voice of America, Kavala
	250	PHL	Voice of America, Philippines
21575	500	GAB	R. Japan via Moyabi, Gabon
21580	250	CLN	Deutsche Welle, Trincomalee
	500	F	R. France Internationale
	250/50	PAK	R. Pakistan Ext. Sce.
	250	PHL	R. Pilipinas via VOA Tinang
	250		Voice of America, Philippines
	250	USA	Voice of America, Greenville
21585	250	PHL	Voice of America, Philippines
21590	250	ATN	R. Netherlands, Bonaire
	250	AUS	R. Australia, Carnarvon
21595	500	NOR	R. Norway Int. / R. Denmark
	250/50	PAK	R. Pakistan Ext. Sce.
21600	250	CLN	Deutsche Welle, Trincomalee
	500/100	D	Deutsche Welle, Germany
21605	300	UAE	UAE Radio & TV, Dubai
	500	YUG	R. Yugoslavia, Beograd
21610	100	J	R. Japan, Yamata, Japan
	250	USA	Voice of America, Greenville
21615	100		WYFR, Okeechobee, FL
21620	100	F	R. France Internationale
21625	250	POR	Voice of America, Gloria
	350	S	R. Sweden
21630	250	CLN	Deutsche Welle, Trincomalee
	500	F	R. France Internationale
	250	GAB	R. France Int. via Gabon
21635	100	F	R. France Internationale
	500	GAB	R. Japan via Moyabi, Gabon
21640	250	ASC	BBC Atlantic Relay St.
	250	CLN	Deutsche Welle, Trincomalee
	500/100	G	BBC, United Kingdom
	500	GAB	R. Japan via Moyabi, Gabon
	500	HOL	R. Netherlands, Flevoland
	500	USA	WCSN, Scotts Corners, ME
21645	500	F	R. France Internationale
	500	GUF	R. France Int, French Guiana
21650	250	CLN	Deutsche Welle, Trincomalee
	500/100	D	Deutsche Welle, Germany
	250	MLT	Deutsche Welle, Cyclops, Malta
21655	100	POR	R. Portugal, S. Gabriel
21660	250	ASC	BBC Atlantic Relay St.
21665	250/120	ROU	R. Romania International
21670	500	ARS	BSKSA, Riyadh, Saudi Arabia
	500/100	CVA	Vatican Radio
	500	USA	WEWN, Birmingham, AL
21675	500	KWT	R. Kuwait
21680	500/100	D	Deutsche Welle, Germany
	250	POR	Deutsche Welle, Sines
21685	500	F	R. France Internationale
	500	GUF	R. France Int, French Guiana
21690	100	I	RAI Ext. Sce, Rome, Italy
21700	500	GAB	R. Japan via Moyabi, Gabon
	100	POR	R. Portugal, S. Gabriel
	300	UAE	UAE Radio & TV, Dubai
21705	500/350	NOR	R. Norway Int. / R. Denmark
	250/100	TCH	Czech & Slovak R. International
21710	500/100	CVA	Vatican Radio
	350	NOR	R. Norway Int. / R. Denmark
21715	250	HNG	BBC East Asian Relay St.
	100	SNG	BBC Far Eastern Relay St.
21720	250	GRC	Voice of America, Kavala
	100	USA	Voice of Free China via WYFR
	100		WYFR, Okeechobee, FL
21725	250	AUS	R. Australia, Darwin
21730	100	F	R. France Internationale
	250/50	PAK	R. Pakistan Ext. Sce.
21735	500/100	G	BBC, United Kingdom
		IND	All India Radio Ext. Sce.
	500	UAE	R. of the UAE, Abu Dhabi
	500	USA	WEWN, Birmingham, AL
21740	100	AUS	R. Australia, Shepparton
21745	500/100	G	BBC, United Kingdom
	500	HOL	R. Netherlands, Flevoland
	250	POR	RFE/RL, Gloria, Portugal
	250	USA	Voice of America, Bethany, OH
21765	500	GUF	R. France Int, French Guiana
21770	500/250	SUI	Swiss Radio International
21810	200/100	BEL	R. Vlaanderen Internationaal
21815	200/100		R. Vlaanderen Internationaal
21820	500/250	SUI	Swiss Radio International
21840	100	I	RAI Ext. Sce, Rome, Italy
21850	500/100	CVA	Vatican Radio
	100	I	RAI Ext. Sce, Rome, Italy
22455	10	AFS	ZSC (Time Signals), So. Africa
22536		HKG	VPS22 (Time Sigs), Hong Kong
25690	500	UAE	R. of the UAE, Abu Dhabi
25730	500	NOR	R. Norway Int. / R. Denmark
25740	100	D	Deutsche Welle, Germany
25750	250	AUS	R. Australia, Darwin
25820	100	F	R. France Internationale
25970	500	HOL	R. Netherlands, Flevoland

PPL's Typesetting Manager Stanley Blumenthal and WRTH Associate Editor Bart Kuperus discuss technical production procedures.

INTERNATIONAL BROADCASTS IN ENGLISH

Prepared in cooperation with BBC Monitoring & TRS Consultants

Target Areas: Eu = Europe, Af = Africa, ME = Middle East, As = Asia, Pac = Pacific, NAm = No. America, CAm = Ce. America (incl. the Caribbean), SAm = So. America, LAm = Latin America. An upper case N, S, E, W or C preceding these target areas means North, South, East, West or Central respectively. Where a broadcast is targeted to a specific country, the ITU code is used (see list elsewhere in this Handbook).
Station Abbreviations: AWR-CR = Adventist World Radio Costa Rica, AWR-Eu = Adventist World Radio Europe, BBC WS = British Broadcasting Corporation World Service, D. Welle = Deutsche Welle (Germany), FEBA = Far East Broadcasting Association (Seychelles), FEBC = Far East Broadcasting Company (Philippines), IRRS = Italian Radio Relay Service, KoHope = King of Hope (Lebanon), R. Austr. = Radio Australia, RAE = Radio Argentina Al Exterior, R. Aum S = R. Aum Shinrikyo (Japan) via Russian tr's, RCBS = Red Cross Broadcasting Service, RFPI = Radio For Peace International (Costa Rica), RNZI = Radio New Zealand International, SRI = Swiss Radio International, TGNC = R. Cultural (Guatemala), TWR-MC = Trans World Radio Monte Carlo, TWR-NA = Trans World Radio Bonaire (Netherlands Antilles), TWR-SW = Trans World Radio Swaziland, VoAsia = Voice of Asia (Taiwan), VOA = Voice of America, VOA-Eu = VOA Europe, VoEth = Voice of Ethiopia, VoIndon = Voice of Indonesia, VOM = Voice of the Mediterranean (Malta), WoHope = Wings of Hope (Lebanon).
General Abbreviations: * = Frequency starts on the half hour, GOS = General Overseas Service, USB = Upper Side Band, WS = World Service

Times (UTC)	Station	Targets	kHz
0000-0015	Cambodia	As	11938, 1360
0000-0027	Prague	NAm	11990, 9580, 7345
0000-0030	R. Austr.	As/Pac/PNG	17795, 17750, 17715, 15365, 15320, 13605
0000-0030	BBC WS	As	1413E, 5965, 9570, 9580
0000-0030	BBC WS	NAm/CAm	15070, 12095, 9590, 7325, 6175, 5975
0000-0030	Vilnius	NAm/Eu	17690, 17605, 9710, 7150 or 7400
0000-0030SS	Norway	NAm	9645 (Sun/Mon UTC only)
0000-0030	SRI	NAm/CAm	17730, 12035, 9885, 9650, 6135
0000-0050	N. Korea	Am	15115, 13760, 11335
0000-0100	China	NAm	11715, 9770
0000-0100	Moscow	WS	21790, 21770, 21690, 21530, 21480, 17890, 17825*, 17775, 17655, 17620*, 17570, 15425, 15350, 12050, 12015, 11895*, 11775, 11655, 9885, 9870, 9755, 9750, 9735, 9725, 9715, 9695, 9685, 9655, 9625, 9530, 9480, 7390, 7355, 7305*, 7295, 7285, 7225, 7220, 7170, 7135, 7120, 7115, 7110, 6045, 6000, 5985, 4940, 648, 585
0000-0100	Spain	NAm	9530
0000-0100	WYFR	NAm	6085
0000-0100	VOA	As/Pac	17820, 17735, 15290, 15185, 11760, 9770, 7120, 1575 (to 0030)
0000-0100	VOA	CAm	15205, 15120, 11580, 9775, 7405, 5995, 1580 (from 0030), 1530 (from 0030)
0000-0100	VOA	CAm	11695, 9455, 6130, 930, (Caribbean prgr.)
0000-0130MF	Canada	USA	9755, 5960
0000-0200	Cuba	ENAm	6060, 6000USB
0000-0200	KHBI	As	17865SS, 17555
0000-0200	WCSN	CAf/SAf	9850
0000-0200	WSHB/1	SAm	13760
0000-0200	WSHB/2	NAm/Cam	5850
0000-0200SS	Canada	USA	9755, 5960
0000-0230	FEBC	As	15450
0000-0230	VOA	ME/NAf/Eu	1260
0000-0300	BBC WS	Af	6005
0000-0300	VOA	ME/NAf/Eu	792
0000-0330	BBC WS	As	15360
0000-0800	WEWN/1	NAm/HWI	7540
0000-1100	WWCR/1	Eu/ME/NAf	7435
0000-1300	WHRI	ENAm	7315
0000-1600	KTBN	ENAm	7510
0000-2400	CFCX	CAN	6005 (rel. CFCF-600)
0000-2400	CFRX	CAN	6070 (rel. CFRB-1010)
0000-2400	RFPI	LAm/NAm	15030, 13630USB, 7385USB, 7375
0000-2400	WJCR	Am/Eu/ME/As	7490, 7465
0030-0055	Belgium	SAm/NAm	13655, 9930
0030-0100	R. Austr.	As/Pac/PNG	21740, 17880, 17795, 17750, 17715, 15365, 15320, 15240, 13605, 11880
0030-0125	Holland	SAs/ENAm	11835, 11655, 9860, 6165, 6020
0030-0130	Iran	NAm	15315, 15260, 9022
0030-0130	S. Korea	Am	15575
0030-0145	BBC WS	As	5965, 9580, 11955, 11955, 15310
0030-0145	BBC WS	Eu	5965
0030-0230	BBC WS	NAm/CAm	12095, 9915, 9590S, 7325, 6175S, 5975W
0030-0430	HCJB	NAm/SAm	21455USB, 15155, 15115, 9745
0040-0055W	Caracas	SAm	9540 (Tues-Sun UTC)
0100-0120	Italy	NAm	11800, 9575
0100-0130	Beograd	NAm	9580
0100-0130	Laos	As	7145, 1030
0100-0130	Prague	NAm	9580, 7345, 5930
0100-0130	Norway	NAm	9565 (Sun/Mon UTC only)
0100-0130	Sweden	AUS/As	11820, 9695
0100-0130	Tashkent	As	11975, 9740, 7325, 7235
0100-0145	Sofia	NAm	11720, 9700, 7225
0100-0150	D. Welle	NAm	11865, 9770, 9700, 9610, 9565, 9515, 6145, 6085, 6055, 6040
0100-0200MF	RAE	NAm/CAm/SAm	11710 (Tues-Sat UTC)
0100-0200	AWR	IND	11855 via Novosibirsk
0100-0200	VoIndon	As	11752
0100-0200	Japan	As/NAm	17845, 17835, 17810, 15195, 11840S
0100-0200	Moscow	WS	21790, 21770, 21690, 21585*, 21530, 21480, 17890, 17825, 17775, 17655, 17620, 17570, 15425, 15350, 15315, 12050, 12015, 11895, 11775, 11655, 9885, 9870, 9775*, 9755, 9735, 9725, 9715, 9695, 9685, 9655, 9625, 9565, 9530, 9525, 9480, 7390, 7340*, 7305, 7295, 7275, 7170, 7150, 7135, 7115, 7110, 6110, 6045, 6000, 5985, 4975, 4940, 648, 585
0100-0200SS	Canada	LAm	13720, 11940, 11845, 9535 (Sun/Mon UTC)
0100-0200	Spain	NAm	9530
0100-0200	Ukraine	NAm/Eu	9860, 9710, 7240, 7195, 7180, 6070, 6020, 6010, 4825

Times (UTC)	Station	Targets	kHz
0100-0200	WEWN/4	IND/NAs	9825
0100-0200	VOA	CAm	15205, 15120, 11580, 9775, 9455, 7405, 6130, 5995
0100-0300	VOA	SAs	21550, 17740, 15250, 11705, 9740, 7205, 7115
0100-0300SS	BBC WS	As	15280
0100-0330	BBC WS	As	17790
0100-0400	R. Austr.	As/Pac/PNG	21740, 17880, 17795, 17750, 15365, 15320, 15240, 11880, 11720, 11670
0100-1400	WWCR/2	Eu/ME/NAf	5935
0130-0140W	Greece	NAm	11645, 9420, 9395
0130-0155	Albania	NAm	11840, 9580
0130-0200	R. Austr.	NAm/SAm	13730, 9875, 9870
0130-0325	Holland	SAs	11655, 9860
0145-0200	Vatican	As/Pac	15090, 11935
0145-0200SS	BBC WS	As	5965, 9580, 11955, 15310
0145-0200SS	BBC WS	Eu	5965
0200-0230	BBC WS	As	11955, 15380, 1413, 9580
0200-0230SS	Norway	NAm	9565 (Sun/Mon UTC only)
0200-0230	Sweden	NAm	11705, 9695
0200-0230	SRI	NAm	12035, 9885, 9650, 6135
0200-0230MF	VOA	CAm	15205, 15120, 11580, 9775, 7405, 5995, 930 (Tues-Sat UTC)
0200-0230	Beograd	NAm	9580
0200-0250	D. Welle	SAs	12055, 11945, 9690, 9615, 7285, 6035
0200-0255	Romania	NAm	11940, 11830, 9570, 9510, 6155, 5990
0200-0300	BBC WS	As	15310, 6195
0200-0300	IRRS	NAm/AUS/NZ	7125
0200-0300	KSDA	China	13720
0200-0300	Moscow	WS	21790, 21770, 21690, 21585, 21530, 21480, 17890, 17825, 17775, 17655, 17620, 17610*, 17570, 15425, 15350, 15315, 12050, 12015, 11895, 11775, 11655, 9885, 9775*, 9755, 9735, 9725, 9715, 9695, 9685, 9655, 9565, 9530, 9525, 9480, 7390, 7340*, 7305, 7295, 7275, 7170, 7150, 7135, 7115, 7110, 6110, 6045, 6000, 5985, 4975, 4940, 648, 585
0200-0300MF	Canada	LAm	13720, 11940, 11845, 9755, 9535 (Tues-Sat UTC)
0200-0300	S. Africa	EAf/CAf	11745
0200-0300	Taiwan	NAm/As/Pac	15345, 11860, 11740, 9765, 9680, 5950
0200-0330	BBC WS	Af	11730
0200-0330	BBC WS	As	7135
0200-0330	Egypt	NAm	11865, 9475
0200-0400	Cuba	NAm	9655, 6060
0200-0400SS	KHBI	As	17865, 17555
0200-0400	TWR-NA	CAm	800
0200-0400	WCSN	EAf	9350
0200-0400	WSHB/1	SAm	13760
0200-0400	WSHB/2	NAm/MEX	5850
0200-0500	Iraq	NAm	15340 (Arabic/Eng)
0230-0245	Pakistan	As	21730, 17725, 17640, 15115, 9545
0230-0255	Albania	NAm	11840, 9580
0230-0300MF	Lisbon	SAm/NAm	11840, 9705, 9600, 9570 (Tues-Sat UTC)
0230-0330	BBC WS	NAm/CAm	12095, 9915, 7325, 6175S, 5975
0230-0330	Manila	EAs	21580, 17840, 17760
0230-0430	BBC WS	As	11955
0230-1400	KCBI	CAN	9815
0245-0310	Finland	NAm	15400, 11735
0250-0315	Vatican	NAm	9605, 7305
0255-0400	TWR-NA	NAm/CAm	11930, 9535 (from 0300)
0300-0330	BBC WS	Af	3255, 6190, 9600, 9670
0300-0330	BBC WS	As	6195
0300-0330	Prague	NAm	9810, 7345, 5930
0300-0330	Japan	NAm/LAm/As	21610, 17825, 17810, 15325, 15230, 5960
0300-0330MF	BBC WS	As	15380
0300-0350	D. Welle	NAm	9770, 9705, 9640, 9545, 9535, 6120, 6085, 6055, 6045
0300-0400	BBC WS	As	1413
0300-0400	BBC WS	Eu	7120
0300-0400	S. Africa	Af	11745, 7270
0300-0400	China	NAm	11715, 9770, 9690
0300-0400	Hungary	NAm	11910, 9835, 6110
0300-0400	Moscow	WS	21830, 21790, 21770, 21690, 21585, 21530, 17890, 17825, 17775, 17655, 17620, 17610, 17570, 15545*, 15470*, 15425, 15385*, 15375*, 15350, 15315, 15280*, 12050, 12015, 11765*, 9775, 9755, 9725, 9715, 9655, 9530, 9480, 7350, 7340, 7305, 7295, 7270*, 7150, 7135, 7115, 6045, 6000, 5985, 4975, 4940, 648, 585
0300-0400	VOA	Af	15115, 11895, 11835, 9885, 9575, 7405, 7280, 7265, 6090, 6035, 621
0300-0400	Taiwan	NAm/As/Pac	15345, 11745, 9765, 9680, 5950
0300-0400	WRNO	NAm/CAm	7395 (Spanish Tues-Fri)
0300-0425	TGNC	CAm	3300
0300-0445	BBC WS	Eu	1296
0300-0500	WYFR	NAm	9505, 6065
0300-0800	BBC WS	Eu	9410
0300-0815	BBC WS	As	11760, 15310
0300-0900	BBC WS	As	21715
0300-0915	BBC WS	As	15280
0315-0335	Vatican	EAf	11625
0330-0400	R. Austr.	NAm/CAm	13730, 9870
0330-0400	Dubai	NAm	17890, 15400, 13675, 11945
0330-0400	Japan	As/NAm	17810, 11870, 5960
0330-0425	Holland	WNAm	11720, 9590
0330-0430	BBC WS	NAm/CAm	12095, 5975
0330-0500	BBC WS	Eu	6195
0340-0350W	Greece	NAm	11645, 9420, 9395
0340-0355MF	Caracas	SAm	9540 (Tues-Sun UTC)
0345-0400	Armenia	NAm	13645, 11790, 11675
0400-0425	Romania	NAm	11940, 11830, 9570, 9510, 6155, 5990
0400-0430	BBC WS	Af	1197, 3255, 6190, 7105, 9600, 9610, 15420, 17885
0400-0430	BBC WS	Eu	9410
0400-0430	Prague	ME/As/NAm	13715, 11985, 9810, 7345, 5930
0400-0430SS	Norway	WNAm	9650, 9560 (Sun/Mon UTC)
0400-0430	Moscow	WS	21845, 21830, 21790, 21770, 21690, 21615, 21585, 21565, 21530, 17890, 17825, 17775, 17685, 17655, 17625, 17620, 17610, 17590, 17580, 17570, 15545, 15470, 15385, 15375, 15350, 15315, 15280, 15230, 11975, 11765, 9870, 9775, 9665, 9580, 9535, 9530, 9490, 9480, 7340, 7295, 7150, 7135, 7115, 6165, 6110, 6000, 5950, 4975, 4940, 648, 585

BROADCASTS IN ENGLISH

0400-0430	Canada	Eu	9650
0400-0430	Canada	ME	9670, 9505
0400-0430	SRI	WNAm	13635, 12035, 9860, 6135
0400-0430	VOA	Af	15115, 11850, 11835, 9885, 9575, 7405, 7280, 7265, 6155, 6090, 621
0400-0450	D. Welle	Af/ME	13770, 11765, 9765, 9665, 9565, 7275, 7225, 7150, 6130, 6065
0400-0450	N. Korea	SEAs	17765, 15230, 15180
0400-0450	Turkey	NAm	9445
0400-0500	S. Africa	WAf/SAf	15345, 7270
0400-0500	China	WNAm	11680
0400-0500	Cuba	NAm	9655 (to 0430), 6180, 6060
0400-0500	Moscow	WNAm	17605, 12050, 9895, 9825*, 7270
0400-0500	VOA	ME/NAf/Eu	7170, 6140, 6040, 5995, 792 (to 0430)
0400-0500MF	VOA	CAm	1580, 1530 (Tues-Sat UTC)
0400-0530	BBC WS	Eu	6180
0400-0530	Sofia	NAm/Eu	11720, 9700, 7290
0400-0600	R. Austr.	As/Pac/PNG	21740, 17795, 15365, 15320, 11880, 11720
0400-0600	KHBI	As	17780 (+ 17555 Sat/Sun)
0400-0600	WCSN	SAf	9840
0400-0600	WSHB/1	MEX	9870
0400-0600	WSHB/2	NAm	5850
0400-0700	BBC WS	Af	12095
0400-0700	WMLK	Eu/ME	9465 (exc. Sat)
0400-0730	BBC WS	As	15575
0425-0440	Italy	SEu/NAf/ME	9575, 7275
0430-0500	AWR	NEu	15125 via Russia
0430-0500	BBC WS	NAm/CAm	5975
0430-0500	R. Aum S	WS	21845, 21830, 21790, 21770, 21690, 21615, 21585, 21565, 21530, 17890, 17870, 17825, 17775, 17685, 17675, 17655, 17635, 17625, 17620, 17610, 17590, 17580, 17570, 15545, 15470, 15385, 15375, 15350, 15315, 15280, 15230, 11975, 11765, 9870, 9775, 9665, 9580, 9535, 9530, 9490, 9480, 7340, 7295, 7150, 7135, 7115, 6165, 6110, 6000, 5950, 4975, 4940, 648, 585
0430-0500	TWR-Sw	SAf	5055
0430-0500	VOA	Af	15115, 11850, 9575, 7405, 7280, 7265, 6035
0430-0530	TWR-Sw	CAf	9655
0430-0700	TWR-Sw	SAf	5965
0430-0730	BBC WS	Af	21470
0430-0815	BBC WS	Eu	9410
0430-0820	TWR-Sw	SAf	11740
0445-0515	SriLanka	WNAm	15425, 9720
0500-0515	Israel	WEu/NAm	9435
0500-0530	BBC WS	Af	3955
0500-0530	BBC WS	Eu	6195
0500-0530	Vatican	EAf/SAf	17730, 15090, 11625
0500-0530	VOA	ME/NAf/Eu	15205, 11825, 9700, 9530, 7170, 6140, 6040, 5995, 792
0500-0550	D. Welle	NAm	9690, 9670, 9535, 6130, 6120, 6045, 5960
0500-0600MF	Thailand	As	11905, 9655, 4830, 927
0500-0600	S. Africa	WAf	15430
0500-0600	China	WNAm	11840
0500-0600	Japan	As/NAm/Eu	21610, 17860, 17825, 17810, 17765, 15230, 7230, 6085
0500-0600	Moscow	WS	21845, 21830, 21790, 21770, 21690, 21680, 21655, 21615, 21585, 21565, 21550, 21530, 17890, 17870, 17825, 17775, 17755, 17735, 17720, 17685, 17675, 17655, 17635, 17620, 17610, 17590, 17580, 17570, 15550, 15530, 15470, 15465*, 15350, 15280, 15230, 12055, 11975, 11765, 9890, 9870, 9840, 9665, 9535, 9530, 9490, 9450, 7390, 7340, 7330, 7295, 7180, 7165, 7150, 7135, 7130, 6165, 6110, 5950, 5905, 4975, 4940
0500-0600	Moscow	WNAm	12050, 9895, 7270, 7175*, 5905*
0500-0600	Spain	NAm	9530
0500-0600	WYFR	Eu/Af/NAm	13695, 11915, 9880, 5985
0500-0600	VOA	Af	15600, 15115, 11850, 9885, 9575, 7405, 6035
0500-0630	BBC WS	Af	3255, 6190, 9600, 15400, 15420, 17885
0500-0700	BBC WS	Af	1197
0500-0700	Cuba	WNAm	6060
0500-0700	HCJB	NAm	21455USB, 11925
0500-0700	Nigeria	WAf	7255
0500-0730	BBC WS	NAm/CAm	9640, 5975
0500-0800SS	Thailand	As	11905, 9655, 4830, 927
0500-2200	KoHope	ME	6280 (also in other languages)
0500-2200	WoHope	ME	11530 (also in other languages)
0530-0555	Romania	Af	21665, 17790, 17745, 17720, 15380, 11840
0530-0600	Austria	ME/Eu/WNAm	21490, 15410, 13730, 6155, 6015
0530-0600	Dubai	SEAs/AUS	21700, 17830, 15435
0530-0600	VOA	ME/NAf/Eu	15205, 11825, 9700, 9530, 7170, 6140, 6060, 6040, 5995, 3980, 792
0555-0825	Malaysia	AUS/NZ/INS	15295, 9750, 6175
0600-0615	SRI	Eu	9535, 6165, 3985
0600-0620	Vatican	ME/Eu	7250, 6245
0600-0630	R. Austr.	As/Pac/PNG	21740, 17880, 15365, 15320, 11880, 11720, 11670
0600-0630	Laos	As	7145, 1030
0600-0630	SRI	Af	17565, 13635, 9860
0600-0650	D. Welle	WAf	17875, 15435, 15185, 13790, 13610, 11765
0600-0650	N. Korea	SEAs	15230, 15180
0600-0700	AWR-Eu	Eu/NAf/ME	7205
0600-0700	S. Africa	WAf	15430
0600-0700	S. Korea	WNAm	15155, 11945, 7275
0600-0700	Moscow	WS	21845, 21830, 21790, 21770, 21690, 21680, 21655, 21615, 21585, 21565, 21550, 21530, 17890, 17870, 17825, 17775, 17755, 17735, 17720, 17685, 17675, 17655, 17635, 17620, 17610, 17590, 17580, 17570, 15550, 15530, 15470, 15465, 15350, 15280, 15230, 15190*, 13650*, 12055, 11975, 11765, 9890, 9870, 9860, 9840, 9535, 9530, 9490, 9450, 7340, 7330, 7180, 7165, 7150, 7135, 7130, 6165, 6110, 5905, 4940, 1323
0600-0700	Moscow	WNAm	9825, 7345*, 7270, 7260*, 7175, 5905
0600-0700	WYFR	Eu/Af/NAm	13695, 11915, 7355, 5950

0600-0700	VOA	Africa	15600, 15115, 11925, 11850, 9885, 9575, 9530, 7405, 6035
0600-0700	VOA	ME/NAf/Eu	15205, 11825, 11805, 7325, 7170, 6060, 6040, 6005, 5995, 3980, 792
0600-0700	VOM	Eu/NAf/ME	9765, 1557
0600-0730	BBC WS	Af	15575
0600-0730	BBC WS	Eu	15575, 6180
0600-0800	KHBI	EAs	17780, 17555
0600-0800	WCSN	Eu	5850
0600-0800	WSHB/1	MEX	9870
0600-0800	WSHB/2	WNAm	7395
0600-0815	BBC WS	As	17790
0600-0815	BBC WS	Pac	7150
0600-0915	BBC WS	As	15360
0600-0915	BBC WS	Pac	11955
0600-1030	BBC WS	As/Pac	17830
0600-1030	BBC WS	Eu	15070
0615-0700MF	Canada	Af	11935, 9740
0615-0700MF	Canada	Eu	9760, 7155, 6150, 6050
0630-0700	R. Austr.	As/Pac/PNG	21740, 17880, 15365, 15320, 11880, 11720, 11670, 6020
0630-0700	Austria	WNAm	6015
0630-0700	Vatican	Af	17730, 15090, 11625
0630-0730MF	BBC WS	Af	15420, 17885
0630-0900MF	IRRS	Eu/NAF/ME	7125
0630-1000Sat	IRRS	Eu/NAF/ME	7125
0630-1500Sun	IRRS	Eu/NAF/ME	7125
0632-0643	Romania	Eu	11775, 9570, 9510, 7225
0645-0715	Romania	Pac	21665, 17805, 17720, 15335, 11940, 11810
0645-0800	Ghana	WAf	6130
0700-0730	R. Austr.	As/Pac/PNG	21740, 17750, 17695, 15365, 15320, 11880, 11720, 6020
0700-0730	BBC WS	Af	11940, 17790
0700-0730	BBC WS	Eu	1296
0700-0730	Prague	Eu	11990, 9505, 7345, 6055
0700-0735Sat	Vilnius	Eu	9710 (last Sat of month only)
0700-0750	N. Korea	SeAs	17765, 15340
0700-0800	AWR	PAK	11855 via Novosibirsk
0700-0800	Japan	As/Pac/Eu	21575, 17860, 17810, 17765, 15170, 6025U, 5970U
0700-0800	Moscow	WS	21845, 21830, 21790, 21785, 21690, 21680, 21655, 21615, 21585, 21565, 21550, 21450, 17890, 17870, 17825, 17805*, 17775, 17765, 17755, 17735, 17720, 17685, 17675, 17655, 17635, 17620, 17610, 17590, 17580, 17570, 15550, 15540, 15530, 15520*, 15470, 15465, 15350, 15345*, 15280, 15230, 15190, 13705*, 13650, 12055, 12010, 11975, 11765, 9890, 9535, 9530, 9490, 7380*, 7135*, 7130, 6110, 5960*, 5905, 4940, 1323
0700-0800	Moscow	WNAm	9825, 7345, 7270, 7175, 5905
0700-0800	VOA-Eu	Eu	1197
0700-0800	WEWN/3	Af	5825
0700-0800	WYFR	Eu/Af/NAm	13695, 11915, 7355
0700-0800	Taiwan	NAm	5950
0700-0815	BBC WS	Af	6190, 15070, 15400
0700-0815	BBC WS	Eu	7325
0700-0830	HCJB	Eu	21455USB, 11825, 6205
0700-0900SS	BBC WS	Af	15420, 17885
0700-1030	BBC WS	Eu	12095, 9760
0700-1129	BBC WS	Eu	648
0700-1206	RNZI	Pac	9700
0700-2315	BBC WS	Af	12095

0703-0715	Croatia	Eu/NAm	13830, 9830, 6210
0730-0735	Iceland	Eu/NAm	9265
0730-0755	Belgium	Pac/Eu	11695, 9905, 5900, 1512
0730-0800	R. Austr.	As/Pac/PNG	17750, 17695, 15240, 11880, 11720, 6020
0730-0800	Prague	As/Pac	21705, 17725
0730-0800	Vienna	As/Pac/Eu	21490, 15410, 13730, 6155
0730-0815	BBC WS	Af	17790, 21660
0730-0825	Holland	Pac	11895, 9630
0730-0900SS	BBC WS	Af	21470
0730-0900SS	BBC WS	As	15575
0730-0900SS	BBC WS	Eu	9660
0730-1130	HCJB	SPac	21455USB, 11925, 9745
0735-0920MF	TWR-MC	Eu	9480
0735-0935Sat	TWR-MC	Eu	9480
0735-0945Sun	TWR-MC	Eu	9480
0745-0800	Finland	Eu/ME/Af	11755, 9560, 6120, 963, 558, 252
0745-0930	KTWR	As	15200 (Fri/Sat to 0915)
0800-0815	BBC WS	Af	11860, 11940
0800-0820	TWR-Sw	SAf	7200
0800-0830	R. Austr.	As/Pac/PNG	25750, 17750, 17695, 15240, 11720, 9710, 7240, 6080, 6020, 5995
0800-0845	Pakistan	WEu	21520, 17900
0800-0850	N. Korea	SEAs	15230, 15180
0800-0900	BBC WS	Eu	9410
0800-0900	Volndon	As	11752
0800-0900	KNLS	As	7365
0800-0900	S. Korea	Eu	13670, 7550
0800-0900	Moscow	WS	21845, 21830, 21825, 21790, 21785, 21690, 21680, 21655, 21630, 21615, 21550, 21450, 17870, 17860, 17805, 17775, 17765, 17755, 17735, 17720, 17685, 17655, 17635, 17625, 17610, 17580, 17570, 15550, 15540, 15530, 15520, 15475, 15470, 15465, 15440, 15435*, 15350, 15345, 15280, 15230, 15210, 15190, 15150*, 13705, 13650, 12070*, 12055, 12020, 12010, 11975, 11765, 9580, 9535, 7130, 6110, 5960, 4975, 4940, 1323
0800-1000	KHBI	J/AUS/NZ	17555, 15665
0800-1000	WCSN	Eu	9840
0800-1000	WEWN/1	NAm/AUS/NZ	9870
0800-1000	WEWN/3	WEu	5825
0800-1000	WSHB/1	AUS	13615
0800-1000	WSHB/2	SAm	9455
0800-1000	VOA-Eu	Eu/ME	21455, 17770, 15195, 15160, 11735, 1260, 1197W
0800-1100	WHRI	SAm/CAm	7355
0800-1500	BBC WS	Eu/Af	17640
0800-1615	BBC WS	Eu	15070
0815-0900SS	BBC WS	Af	6190, 11940, 15400, 17790, 21660
0815-0915	BBC WS	Eu	7325
0815-2030	BBC WS	Af	15070
0820-0835SS	TWR-Sw	SAf	7200, 11740
0830-0900	R. Austr.	As/Pac/PNG	25750, 17750, 17695, 11720, 9710, 9580, 7240, 6080, 6020, 5995
0830-0900	Vienna	AUS/As	21490, 15450
0830-0925	Holland	Pac	11895
0840-0850	Greece	AUS	17525, 15650
0900-0915	BBC WS	As	11765
0900-0915	BBC WS	Eu	9410
0900-0930	BBC WS	As	15575
0900-0930	Finland	Pac	21550, 17800
0900-0930	SRI	AUS/NZ	21770, 17670, 13685, 9560

BROADCASTS IN ENGLISH

0900-0950	D. Welle	AUS/As/Af	21680, 21650, 21600, 21465, 17820, 17780, 15410, 11915, 9565, 6160
0900-1000	R. Austr.	As/Pac/PNG	21725, 15170, 13605, 9580, 9510, 5995
0900-1000	AWR-Eu	Eu/NAf/ME	7230
0900-1000	China	SPac	17710, 15440, 11755
0900-1000	KTWR	AUS/NZ	11805
0900-1000	Japan	As/Pac/Eu	21610, 17860, 15270, 11840, 11815
0900-1000	Moscow	WS	21845, 21825, 21785, 21755, 21690, 21680, 21655, 21630, 21625*, 21615, 21550, 21450, 17870, 17860, 17805, 17775, 17765, 17755, 17735, 17730, 17695*, 17685, 17655, 17635, 17625, 17610, 17600*, 17565, 15595, 15550, 15540, 15530, 15520, 15495*, 15475*, 15470, 15465, 15440, 15435, 15350, 15345, 15280, 15230, 15210, 15190, 15175, 15150, 13705, 13650, 12070, 12055, 12020, 12010, 11975, 11765, 11710*, 9855*, 9830*, 9780*, 9755*, 9685*, 9535, 7130, 6110, 4975, 4940, 1323
0900-1030	BBC WS	As	9740, 11750
0900-1030	BBC WS	Eu	9410, 9750
0900-1100	BBC WS	As	17790
0900-1100	FEBC	CHN/SEAs	11685, 9800
0900-1130	BBC WS	Af	15400, 15420
0900-1400	BBC WS	Af	17885
0900-1400	BBC WS	As	9740, 11760
0900-1500	BBC WS	Af	17790
0900-1515	BBC WS	As	15310
0900-1515	BBC WS	Eu	9660
0900-1615	BBC WS	Af	17705, 21470
0900-1615	BBC WS	As	6195
0900-1700	BBC WS	Af	6190 11940
0900-1745	BBC WS	Af	21660
0903-0915	Croatia	Eu/NAm	13830, 9830, 6210
0910-0940	Mongolia	Pac	12015, 11850
0930-0955	Finland	EAs	17800, 15245
0930-1025	Holland	Pac	11895, 9720
0930-1500	BBC WS	As	15575
1000-1025W	Belgium	Eu	9905, 5900, 1512
1000-1030	Vietnam	Eu	15010, 12020, 9840
1000-1100	India	As/AUS/NZ	21735, 17895, 15050
1000-1100	R. Austr.	As/Pac/PNG	21725, 9580, 5995
1000-1100	S. Africa	EAf	17780
1000-1100	China	SPac	17710, 15440, 11755
1000-1100	Moscow	WS	21845, 21825, 21785, 21755, 21680, 21630, 21625, 21615, 21550, 17840*, 17805, 17755, 17695, 17685, 17635, 17625, 17600, 17565, 15550, 15530, 15520, 15475, 15470, 15465, 15440, 15435, 15345, 15280, 15230, 15210, 15175, 13650, 12070, 12055, 12020, 12010, 11975, 11765, 11710, 11675, 9855, 9780, 9705*, 9535, 7130, 6110, 4975
1000-1100	WYFR	NAm	5950
1000-1100	VOA	As/Pac	15425, 11720, 5985
1000-1100	VOA	ME/NAf/Eu	21455, 17770, 15195, 15160, 11735
1000-1200	KHBI	CHN/INS	17555, 13625
1000-1200	KHBN	CHN	9820
1000-1200	WCSN	Eu	13770
1000-1200	WSHB/1	ENAm	9495
1000-1200	WSHB/2	SAm	9455
1000-1200	VOA	CAm	15120, 11915, 9590, 930 (to 1100)
1000-1300	WHRI	SAm/CAm	9850
1030-1055	Dubai	Eu/NAf	21605, 15435, 15320, 13675
1030-1100	BBC WS	As	11750
1030-1100	Vienna	AUS/As/Eu	21490, 15450, 13730, 6155
1030-1125	Holland	EAs via Russian relay tr's (freqs n/a)	
1030-1200Sun	BBC WS	Eu	9410, 9750, 9760 12095, 15070
1030-1200W	BBC WS	Eu	9750
1030-1400	BBC WS	As/Pac	9740
1040-1050	Greece	Japan	17525, 15650
1100-1120	Pakistan	WEu	21520, 17595
1100-1130	BBC WS	Pac	9700
1100-1130	Israel	WEu	17545
1100-1130	SRI	FE	21770, 17670, 15505, 13635
1100-1130Sun	RCBS	Eu	7210 (via SRI, last Sun of month)
1100-1145	SRI	Eu	9535, 6165
1100-1150	D. Welle	WAf	21600, 21465, 17860, 17800, 17765, 15410
1100-1150	N. Korea	Am	11335, 9977, 6576
1100-1200	R. Austr.	As/Pac/PNG	21725, 15170, 13605, 9710, 9580, 7240, 6080, 6020, 5995
1100-1200	BBC WS	NAm/CAm	15220, 9515, 6195, 5965
1100-1200	S. Africa	CAf	11900
1100-1200	Japan	As/NAm	11840, 11815, 6120
1100-1200	Jordan	WEu/NAm	13655
1100-1200	S. Korea	ME/Af	15575
1100-1200	Moscow	WS	21825, 21785, 21755, 21680, 21630, 21550, 17775, 17735, 17695, 17675, 17625, 17610, 17600, 17565, 15585*, 15550, 15530, 15520, 15510, 15475, 15470, 15465, 15440, 15435, 15350, 15345, 15280, 15210, 13650, 12070, 12055, 12020, 12010, 11710, 11675, 9780, 9745*, 9705, 7130, 6000, 4940
1100-1200	WYFR	NAm	7355, 5950
1100-1200	VOA	As/Pac	15425, 15155, 11720, 9760, 6110, 5985
1100-1200	VoAsia	SEAs	7445, 585
1100-1300	AWR-CR	CAm/SAm	11870, 9725 (alt.9875)
1100-1400	WHRI	SAm/CAm	11790
1100-1615	BBC WS	As	11750
1100-2400	WWCR/1	Eu/ME/NAf	15685
1120-1130MF	Vatican	ME/Eu	21670, 15210, 11740, 7250, 6245
1130-1157	Prague	Eu	15355, 11990, 9505, 7345, 6055
1130-1159	S. Korea	ENAm	9650
1130-1215W	BBC WS	Af	15420
1130-1225	Iran	EAs/ME	11930, 11910, 11790, 11715, 9525
1130-1230	Thailand	SEAs	11905, 9655, 4830, 927
1130-1300	Sofia	Eu/ME/NAm	17825, 17780, 13670, 11720, 11630
1130-1325	Holland	Eu	5955
1130-1600	HCJB	NAm/SAm	21455USB, 17890, 17490USB, 15115 (to 1430), 11925
1140-1155W	Caracas	SAm	9540
1200-1215	Cambodia	As	11938, 1360
1200-1225	Tashkent	As	17745, 15470, 9540, 5945

1200-1230	R. Austr.	As/Pac/PNG	21725, 9710, 9580, 9510, 7240, 6080, 6020, 5995
1200-1230	Mongolia	FE	12015, 11850 (Mon/Thurs/Sat only)
1200-1230	Norway	IND/As/AUS	21705, 17860
1200-1230	VOA	CAm	930
1200-1300	BBC WS	NAm/CAm	15220S, 9740, 9515S, 6195
1200-1300	China	Pac/As/ENAm	15450, 14440, 11660, 9715, 9655
1200-1300	Moscow	WS	21825, 21785, 21755, 21680, 21630, 21615, 21550, 17860, 17840, 17735, 17625, 17600, 15585, 15540, 15520, 15510, 15480*, 15475, 15465, 15440, 15435, 15345, 15280, 15210, 13705, 13650, 12070, 12025, 12020, 11980, 11850*, 11775, 11765, 11710, 11675, 9895, 9855, 9780, 9745, 9705, 7380*, 7260, 7160, 7130, 6000, 5960, 5950, 5940, 1323, 720, 630
1200-1300	WYFR	NAm	17760, 11830, 7355, 5950
1200-1320	Brazil	NAm/CAm	15445
1200-1330	VOA	As/Pac	15425, 15155, 11715, 9760, 6110, 1143 (to 1230)
1200-1400	BBC WS	Eu	648
1200-1400	China	Pac/SEAs	15440, 11660, 9715
1200-1400	KHBI	AUS/SEAs	15665, 13625
1200-1400	WCSN	Eu	17510
1200-1400	WSHB/1	ENAm	9495
1200-1400	WSHB/2	CAm	13760
1200-1430	KHBN	As/Pac	9830
1200-1615	BBC WS	Eu	9750, 9760
1200-1745	BBC WS	Eu	12095
1200-2030	BBC WS	Eu	9410, 15070
1210-1650	RNZI	Pac	9510 (occ.)
1215-1315	S. Korea	Am	9750
1215-1330	Egypt	SAs	17595
1230-1250MF	Finland	NAm	15400, 11735
1230-1255Sun	Belgium	NAm/SEAs	21810, 17550
1230-1300	R. Austr.	As/Pac/PNG	21725, 9580, 7240, 5995
1230-1300	Austria	As/NAm/Eu	17730, 15450, 13730, 6155
1230-1300	B'desh	Eu	17750, 15200
1230-1300	France	Am/Eu	21645, 15365, 17650, 15195, 15155, 11670, 9805
1230-1300	Vietnam	As	15010, 12020, 9840
1230-1300	Beograd	AUS/NAm	21605, 17740
1235-1245	Greece	NAm	17515, 15650, 15635
1245-1300W	BBC WS	Af	15420
1300-1330	SRI	As/FE	21820, 17670, 15505, 13635, 11690, 7480
1300-1330SS	Norway	WAf/Eu	25730, 9590
1300-1330SS	TWR-NA	CAm	800
1300-1345	BBC WS	Af	15105
1300-1350	N. Korea	SEAs/Am/Eu	15230, 13650, 9640, 9345, 9325
1300-1355	Poland	Eu	11815, 9525, 7145, 6135
1300-1400	R. Austr.	As/Pac/PNG	13755, 11855, 11800, 9580, 7240, 5995
1300-1400	AWR	IND	11855 via Novosibirsk
1300-1400	BBC WS	Af	15420
1300-1400	BBC WS	As	1413, 7180, 11820
1300-1400	BBC WS	NAm/CAm	15220, 11820, 9740, 9515, 6195
1300-1400	KNLS	As	7355
1300-1400	Moscow	WS	21785, 21755, 21680, 21615, 17860, 17840, 17775, 17610, 15550, 15540, 15440, 15345, 15210, 13705, 11980, 11890*, 11840*, 11765, 11715*, 11710, 11675, 9890*, 9855, 9840*, 9780, 9745, 9735*, 9725*, 9705, 7380, 7370*, 7330, 4940, 1323
1300-1400	Romania	Eu	17850, 17720, 15365, 11940
1300-1400	WYFR	NAm	17760, 13695, 11830, 9705, 5950
1300-1400	VoEth	WAf	9560, 7165
1300-1400MF	Canada	CAm	17820
1300-1400MF	Canada	USA	11855
1300-1600	FEBC	SEAs/IND	11995
1300-1600	WEWN/1	NA/AUS/NZ	9870
1300-1600	WEWN/2	EAf	21670
1300-1700	BBC WS	Af	1197
1300-1700	WHRI	ENAm	9465
1302-1502	WYFR	IND	11550 via Taiwan
1303-1315	Croatia	Eu/NAm	13830, 9830, 6210
1320-1630	Jordan	WEu/NAm	9560
1330-1355	Dubai	Eu/NAf	21605, 15435, 15320, 13675
1330-1355	Tashkent	As	17745, 15470, 9540, 5945
1330-1355MF	Finland	NAm	21550, 15400
1330-1357	Canada	SEAs	9535, 6150 via China
1330-1400	Austria	As	17730, 15450
1330-1400	Laos	As	7145, 1030
1330-1400	Sweden	As/Pac	21625, 15240
1330-1400	Turkey	SWAs	9675
1330-1400	Vietnam	EAs	15010, 12020, 9840
1330-1400	VOA	As/Pac	15425, 15155, 9760, 6110
1330-1425	Holland	SAs	17610, 13770
1330-1500	India	SEAs	15120, 11760
1345-1405	Vatican	As/AUS/NZ	21515, 17525, 15090
1400-0100	WWCR/2	Eu/ME/NAf	13845
1400-0230	KCBI	CAN	15375
1400-1405MF	BBC WS	As	1413
1400-1425	Israel	Eu/NAm/As	17590, 17575, 15650, 15640, 11603, 11587
1400-1425W	Belgium	NAm/SEAs	21810, 17550
1400-1430	R. Austr.	As/Pac/PNG	11800, 11855, 9580, 7240, 5995
1400-1430	BBC WS	Eu	648 (exc. Sat)
1400-1430	BBC WS	NAm/CAm	17840, 11820, 9740, 9515, 6195
1400-1500SS	BBC WS	Pac	9740
1400-1500	China	SAs/WNAm	15165, 11815, 7405
400-1500	France	Eu/ME/As	17650, 15405, 11910
1400-1500	Japan	As/Am	11815, 9535
1400-1500	S. Korea	SEAs	9570
1400-1500	Moscow	WS	21785, 21755, 21680, 21615, 17860, 17840, 17730, 15540, 15525, 15520, 15480, 15465, 15440, 15420, 15345, 15210, 12060*, 12050, 12025, 11930, 11840, 11765, 11715, 11675, 11655, 9895, 9890, 9855, 9840, 9755, 9745, 9725, 9705, 7380, 7370, 7335*, 7330, 7260, 7225, 7170, 7135, 7115*, 6055*, 5960, 5905, 4975*, 1323, 630
1400-1500	WYFR	NAm	17760, 11830, 9705
1400-1500	VOA	As/Pac	15425, 15160, 9760, 9645, 6110, 1143
1400-1500	VOA	ME/NAf/Eu	15205
1400-1500	VOA	SAs	15395, 9760, 9645, 7125, 6110
1400-1500	VOM	Eu/NAf/ME	11925
1400-1500MF	Morocco	NAf	17595
1400-1500SS	BBC WS	As	7180, 9740, 11820
1400-1500SS	Finland	NAm	21550, 15400
1400-1600	KHBI	As	13625, 9530
1400-1600	WCSN	Eu	17510
1400-1600	WSHB/1	ENAm/CAm	15665

BROADCASTS IN ENGLISH

1400-1600	WSHB/2	MEX/WNAm	13760
1400-1600Sat	IRRS	Eu/NAf/ME	7125
1400-1700	BBC WS	Af	17880
1400-1700	Iraq	SAs	15240 (Arabic/English)
1400-1700Sun	Canada	CAm	17820
1400-1700Sun	Canada	USA	11955
1400-1800	WHRI	SAm/CAm	15105
1430-1500	BBC WS	NAm/CAm	17840, 11820, 9740, 9515
1430-1500MF	Finland	NAm	21550, 15400
1430-1525	Holland	SAs	17610, 15150, 13770, 9895
1430-1600	R. Austr.	As/Pac/PNG	13755, 11855, 11800, 9580, 9510, 7260, 7240, 5995
1430-1615	BBC WS	Eu	648
1440-1455W	Caracas	SAm	9540
1445-1515	Mongolia	SAs	13780, 7260
1445-1645W	KTWR	SAs/SEAs	15610
1445-1700Sun	KTWR	SAs/SEAs	15610
1458-1600	FEBA	SEAs	15330, 9810 (Tues-Sat)
1500-1515	WYFR	IND	11550 via Taiwan
1500-1515MF	BBC WS	Af	15420, 17790
1500-1515	BBC WS	Pac	9740
1500-1530	Sudan	Regional	9170, 7200
1500-1530	SRI	ME/EAf	21820, 17670, 15505, 13635
1500-1530	Vietnam	Af	15010, 12020, 9840
1500-1549SS	FEBA	SEAs	11710
1500-1550	D. Welle	Af/ME	21600, 17765, 17735, 13610, 11965, 9735
1500-1550	N. Korea	Eu/ME/Af	13785, 9977, 9640, 9325
1500-1555MF	FEBA	SEAs	11710
1500-1600	China	SAs/WNAm	15165, 11815, 7405
1500-1600	KJES	wCAN	11715 (irr.)
1500-1600	Japan	Eu/Af/NAm	15355, 9535
1500-1600	Moscow	WS	21755, 21680, 21615, 17730, 17610, 15550, 15540, 15480, 15465, 15420, 15210, 12065, 12060, 12050, 12045*, 11930, 11890, 11840, 9895, 9890, 9755, 9725, 9590*, 7390, 7370, 7360, 7335, 7330, 7290*, 7260, 7250*, 7185*, 7170*, 7155*, 7115, 6055, 5905, 4975, 4940, 1323
1500-1600	WEWN/4	ME/Eu	18930
1500-1600	WYFR	NAm	17760, 15215, 11830
1500-1615	BBC WS	As	7180, 9740
1500-1615	BBC WS	NAm/CAm	17840, 15260, 15205, 9740, 9515
1500-1630	VOA	ME/NAf/Eu	15205, 9700, 1260 (to 1600)
1500-1700	BBC WS	Af	15400
1500-1700	VOA	SAs	15395, 9760, 9700, 9645, 7125, 6110, 1575 (1530-1600)
1500-2030	BBC WS	Eu	6195
1503-1506	Finland	Eu/ME/Af	21550, 15440, 11755, 9730, 6120, 963, 558, 252
1515-1530	Canada	Eu	21545, 17820, 15325, 15315, 11935, 11915, 9555
1515-1700	BBC WS	As	7215
1515-1830	BBC WS	As	9740, 15310
1530-1540W	Greece	NAm/NEu	17525, 15650, 15630
1530-1555	Albania	Af	9760, 7155
1530-1600	Austria	WAf/Eu/As	21490, 13730, 11780, 6155
1530-1600MF	Lisbon	ME	21515
1530-1625	Holland	sAs	17610, 15150, 13770, 9895
1545-1600	Vatican	As/AUS/NZ	17865, 15090
6080, 5995, 5880			
1600-1615W	BBC WS	Eu	1296
1600-1630	R. Austr.	As/Pac/PNG	13755, 11880, 11855, 9580, 9510, 7260, 7240, 5995
1600-1630	AWR	NEEu	15125 via Russia
1600-1630	Pakistan	EAf/SAf/ME	21495, 17725, 17555, 15555, 13665, 11600
1600-1630	Sweden	NA/ME/EAf	21500, 17870, 15270
1600-1630SS	Norway	Af/ME	15230, 11875
1600-1640	Dubai	Eu/NAf	21605 (alt.15435), 15320, 13675, 11795
1600-1650	D. Welle	SAs	15595, 15105, 11785, 9585, 7305, 7225, 6170
1600-1655	Poland	Eu	11840, 9525, 7285
1600-1700	BBC WS	Af	17860
1600-1700	China	EAf/SAf	15170, 15130, 11575
1600-1700	France	Af/ME/Eu	17850, 17795, 17620, 15530, 12015, 11705, 6175
1600-1700	KSDA	IND	11980
1600-1700	S. Korea	ME/Af	9870, 5975
1600-1700	Moscow	WS	12060, 12045, 11840, 11685, 11655, 9895, 9890, 9830, 9755, 9590, 9540, 7420*, 7370, 7335, 7330, 7265, 7260, 7250, 7205*, 7185, 7170, 7155, (exc. Fri), 7150, 7115, 6130*, 6055, 6000*, 5905, 1494, 1269, 1125
1600-1700	TWR-Sw	EAf	9500
1600-1700	WYFR	Eu/Af/NAm	21615, 21525, 17760, 15355, 15215, 11830
1600-1700	VOA	Africa	15445, 15320, 15255, 13710, 11995, 11920
1600-1745	BBC WS	As	3915
1600-1800	S. Africa	Af	15430, 11900
1600-1800	KHBI	SEAs/IND	13625, 11580
1600-1800	WCSN	EAf	21640
1600-1800	WEWN/2	NAf	21670
1600-1800SS	WSHB/1	ENAm/CAm	15665
1600-2000SS	WSHB/2	MEX/WNAm	17555
1600-2100	Saudi Ar.	Eu	9720, 9705
1600-2100	WINB	Eu/NAf	15295
1600-2300	WRNO	NAm/CAm/Eu	15420
1600-2400	KTBN	ENAm	15590
1600-2400	WEWN/1	NAm/AUS/NZ	13615
1600-2400	WHRI	ENAm/Eu	13760
1615-1700	BBC WS	Af	9630, 15420
1615-1700	BBC WS	As	5975
1615-1715	BBC WS	Af	21470
1615-1745	BBC WS	NAm/CAm	15260S, 9515S
1615-2200	BBC WS	Eu	9410
1630-1657	Canada	IND	9555, 7150 via China
1630-1700	Austria	As	11780
1630-1700	VOA	ME/NAf/Eu	15245, 15205, 11855, 9760, 9700, 6180
1630-1800	R. Austr.	As/Pac/PNG	13755, 11910, 11880, 9580, 9510, 7260, 7240, 6080, 5995
1630-1800	HCJB	ME	21455USB, 21480, 17790, 17490USB
1630-1830	Egypt	CAf/SAf	15255
1645-1850	RNZI	Pac	9675 (exc. Sat)
1700-1730	Canada	Eu	21545, 17820, 15325, 13650, 7235, 5995
1700-1730	Red Cross	Eu	7210 (Mon after last Sun of month)
1700-1730	TWR-Sw	CAf	9520
1700-1730	SRI	Eu	9535, 6165, 3985
1700-1730	SRI	ME/Af	17635, 15430, 13635, 9885
1700-1730SS	Norway	Eu	9655
1700-1730	VOA	CAm	930
1700-1750	N. Korea	Eu/ME/Af	13785, 9977, 9640, 9325
1700-1800	China	EAf/SAf	15345, 11575, 9570
1700-1800	Japan	Af/Am/As	17775, 11815, 9535, 7140
1700-1800	Moscow	WS	12060, 11840, 9895, 9890, 9860*, 9830, 9755, 9685, 9590, 9540, 9515*, 7420, 7340*, 7330, 7260, 7250, 7205, 7170

1700-1800	Pakistan	WEu	15550, 11570
1700-1800	VOA	Af	15445, 15320, 13710, 11995, 11920, 621SS (from 1730)
1700-1800	VOA	SAs	15395, 9700, 9645, 7125, 6110
1700-1800Sat	Morocco	NAf	17815
1700-1800SS	KSDA	IND/EAf	13720
1700-1900	WYFR	Eu/Af	21500
1700-1900	VOA	ME/NAf/Eu	15205, 9760, 9700, 6040, 792 (to 1800)
1700-2000	WMLK	Eu/ME	9465 (exc. Sat)
1700-2030	BBC WS	Eu	6180, 6195
1700-2130	IRRS	Eu/NAf/ME	7125 (Fri/Sat/Sun to 2230)
1715-1730	Vatican	ME/Af/WEu	9645, 7250, 6245
1715-1830	BBC WS	As	5975, 7160
1730-1800	Romania	Af	17745, 17720, 15365, 15340
1730-1800	Vatican	EAf/SAf	17730, 15090, 11625
1730-1925	Holland	Af	21590, 21515, 9605, 6020
1730-2115	TWR-Sw	SAf	1170
1745-1830	BBC WS	Af	1197, 6190, 11940, 15400
1745-1830	BBC WS	As	1413
1745-1945	India	Af/wEu/ME	15080, 11935, 11860, 11620, 9950, 7412
1745-2030	BBC WS	Af	17880
1800-1815	Israel	Af/NAm/Eu	17575, 11675, 11587, 7465
1800-1827	Prague	Eu	9605, 7345, 6055, 5960
1800-1830	BBC WS	As	11750
1800-1830	Vietnam	Eu	15010, 12020, 9840
1800-1830MF	Canada	Af	17820, 15260, 13670
1800-1845	TWR-Sw	EAf	9500
1800-1850	Brazil	Eu	15265
1800-1855	Poland	Eu	9525, 7270
1800-1900	R. Austr.	As/Pac/PNG	11910, 11880, 9580, 7260, 7240, 6080, 5995, 5880
1800-1900	Sudan	Regional	9170, 7200
1800-1900	Moscow	WS	13670, 12050*, 11920, 11840, 11685, 9890, 9875, 9860, 9855, 9830, 9755, 9725, 9625, 9590, 9565, 7420, 7330, 7260, 7250, 7170, 7145, 6130*, 1215
1800-1900	Spain	Eu	11890
1800-1900	VOA	Af	17800, 15580, 15410, 15320, 13710, 11995, 11920, 621
1800-1900MF	RAE	Eu/Af	15345
1800-1900SS	KSDA	EAf	13720
1800-1900SS	Canada	Af	17820, 15260, 13670
1800-2000	Iraq	Eu	15210, 15180 (Arabic/English)
1800-2000	KHBI	Pac/Eu/ME	13840, 9495
1800-2000	WCSN	SAf	21640
1800-2000	WEWN/2	SAf	21670
1800-2000	WEWN/4	Eu	13740
1800-2000	WSHB/1	ENAm/Eu	15665
1800-2100	R. Kuwait	Eu	13620
1800-2200	BBC WS	Pac	11955
1800-2300	WHRI	SAm/CAm	17830
1815-1900	R. B'desh	Eu/ME	12030, 9565
1830-1900	R. Afghan	As	9635
1830-2000	Sofia	NAm/Eu	11720, 9700, 9560, 6235
1830-2030	BBC WS	Af	3255
1830-2030	BBC WS	As	9740
1830-2200	BBC WS	Eu	9410
1840-1850W	Greece	SAf	17525, 15630
1840-1855W	Caracas	SAm	9540
1845-2000	Ghana	WAf	6130
1850-2140	RNZI	Pac	15120 (Sun-Fri)
1900-1925	Belgium	Af/Eu	15540, 5900, 1512
1900-1930	Japan	Pac/Am	11850, 9640, 9535
1900-1930	Canada	Af	17820, 15260, 13670MF
1900-1930	Vietnam	Eu	15010, 12020, 9840
1900-1930MF	BBC WS	Af	1197, 6005, 6190, 15400
1900-1930MF	Portugal	Eu	11740
1900-1930SS	Norway	NZ/Af/SAm	17730, 15220
1900-1950	D. Welle	WAf/ME	17810, 15350, 13790, 11905, 11785, 11765, 9765
1900-2000	Algiers	Eu/ME	9685, 9535, 9509, 891
1900-2000	AWR	SYR	9835 via Novosibirsk
1900-2000	China	WAf/NAf	11515, 9440
1900-2000	HCJB	Eu	21480, 21455USB, 17790, 17490USB
1900-2000	S. Korea	GOS	9650
1900-2000	Moscow	WS	17605*, 15425*, 13670, 12050, 11920, 11840, 11685, 9890, 9875, 9860, 9855, 9785, 9735, 9725, 9685, 9590, 9530, 7390, 7260, 7205, 7195*, 7170, 7145, 5950
1900-2000	Romania	Eu	9690, 7225, 7195, 6105
1900-2000	Spain	Af	15375
1900-2000	KJES	wCAN	9510 (irr.)
1900-2000	WYFR	Eu/Af	21615, 15355
1900-2000	VOA	Africa	17800, 15580, 15495, 15410, 15320, 13710, 11995, 11920, 621
1900-2000	VOA	As/Pac	15180, 11870, 9525
1900-2000Sun	Morocco	NAf	11920
1900-2030	BBC WS	Af	9630
1900-2030	BBC WS	As	1413, 7160
1900-2100	VOA	ME/NAf/Eu	15205, 9760, 9700, 6040, 3980, 1197 (to 2000)
1900-2115	TWR-Sw	SAf	3240
1905-1915W	Greece	Eu	9400, 7450
1915-2000	DLF	Eu	1269
1930-2000	Austria	SAf/ME/Eu	13730, 12010, 6155, 5945
1930-2000	KFBS	As/Pac	9465
1930-2000	Finland	Eu/ME/Af	11755, 9730, 6120, 963, 558, 252
1930-2000	Canada	Eu	21675, 17875, 15325, 13650, 9670, 5995
1930-2000	Beograd	SAf/Eu	15140, 7200, 6100
1930-2025	Holland	WAf	21590, 17605
1930-2025	Poland	Eu	9525, 7145
1930-2030	Iran	Eu/NAm	15260, 9720, 9022
1935-1955	Italy	Eu	11800, 9710, 7275
1940-2010	Mongolia	Eu	11850, 11790
2000-2030	BBC WS	Af	3955
2000-2030	Israel	Af/Eu/Am	17575, 11675, 11603, 11587, 9435, 7465
2000-2030	Moscow	WS	17605, 15425, 13670, 12050, 11920, 11840, 11685, 9890, 9860, 9855, 9795, 9785, 9735, 9725, 9590, 9565, 9515, 9490, 9450, 7390, 7255, 7250, 7245, 7205, 7195, 7170, 7145, 5950, 4795
2000-2030	R. Austr.	As/Pac/PNG	11910, 11880, 11855, 9580, 7260, 7240, 2000-2030
Vatican	Af	15090, 11625, 9645	
2000-2030	Canada	Eu	17875, 15325, 15140, 13650, 11945, 7235U, 5995U
2000-2030	SRI	Af	15505, 13635, 12035, 9885
2000-2030	VOA	Africa	21485, 17800, 15580, 15495, 15410, 15320, 15160, 13710, 11820, 621
2000-2030MF	Portugal	Af	15250
2000-2050	N. Korea	ME/Af/Eu	9977, 9640, 9345, 6576
2000-2100	Angola	Af	9535, 3255, 944
2000-2100	BBC WS	NAm/CAm	5975
2000-2100	China	Af/Eu	15170, 11715, 11500, 9920, 9440
2000-2100	VOA-Eu	Eu	1197

BROADCASTS IN ENGLISH

2000-2100	Volndon	As	9675, 11752
2000-2130	SriLanka	Eu	15120, 9720
2000-2200	BBC WS	Eu	7325
2000-2200	KHBI	AUS/J	13840, 9455
2000-2200	VOA	As/Pac	17735, 15185, 11870
2000-2200	VOA	ME/NAf/Eu	15205, 11960, 9760, 9700, 6040, 1260 (from 2100)
2000-2200	WCSN	Eu/ME	7510
2000-2200	WSHB/1	ENAm/Eu	13770
2000-2200	WSHB/2	SAm	17555
2000-2200	WYFR	Eu/Af	21525, 17750, 15566, 7355
2000-2315	BBC WS	Pac	15340
2005-2105	Syria	Eu/NAm	15095, 12085
2025-2045	Italy	ME	11800, 9575, 7235
2030-2035W	Latvia	Eu	5935
2030-2038	Croatia	UN Forces	13830, 9830, 6210
2030-2100	R. Austr.	As/Pac/PNG	11855, 9580, 7260, 7240, 5995, 5880
2030-2100	R. Aum S	WS	21480, 17655, 17605, 15425, 15130, 13670, 12050, 11920, 11840, 11685, 9890, 9860, 9855, 9795, 9785, 9735, 9725, 9590, 9565, 9530, 9515, 9490, 9450, 7390, 7255, 7250, 7245, 7220, 7205, 7195, 7180, 7170, 7145, 5950, 4795, 648
2030-2100	Canada	Eu	17875, 15325, 15140, 13650, 11945, 7230, 5995
2030-2100	VOA	Africa	21485, 17800, 15580, 15495, 15410, 15320, 13710, 621
2030-2100	Vietnam	Eu	15010, 12020, 9840
2030-2125	Poland	Eu	9525, 7270, 7145, 6135, 6095
2030-2200	BBC WS	Af	15400
2030-2200	Egypt	WAf	15375
2045-2130	Sofia	NAm/Eu	9560, 6235
2045-2230	India	AUS/NZ/WEu	15265, 11715, 11620, 9950, 9910, 7412
2050-2110	Vatican	Eu	7250, 5882
2055-2100	BBC WS	Eu	648 (exc. Sat)
2100-0030	BBC WS	As	6195
2100-0100	KVOH	CAm	17775
2100-0330	BBC WS	Af	1197
2100-0415	BBC WS	Eu	648
2100-2127	Prague	Eu	9605, 7345, 6055, 5960
2100-2200	China	EAf/SAf/Eu	15170, 11715
2100-2130	R. Austr.	As/Pac/PNG	11855, 9760, 9580, 9540, 5995
2100-2130SS	Norway	SAm/AUS/NZ	15180
2100-2150	D. Welle	As/Pac	11785, 9765, 9690, 9670, 6185
2100-2150	Turkey	Eu	9445
2100-2200	Algiers	Eu/ME	9685, 9535, 891
2100-2200	BBC WS	Af	3255, 6005
2100-2200	BBC WS	As	7180
2100-2200	BBC WS	Eu	6180
2100-2200	BBC WS	NAm/CAm	15070, 9590, 5975
2100-2200	China	EAf/SAf/Eu	11500, 9920
2100-2200	Cuba	Eu/Af/ME	17705
2100-2200	Japan	Oc/As/Eu/Af	17890, 17840, 17810, 15430, 11925
2100-2200	Moscow	WS	21480*, 17690*, 17665*, 17655, 17605, 17570*, 15425, 15130, 13670, 12050, 11840, 11655*, 9890, 9885*, 9870*, 9860, 9855, 9795, 9785, 9750*, 9735, 9725, 9695, 9565, 9530, 9490, 9480, 9450, 7390, 7370, 7255, 7245, 7220, 7205, 7195, 7180, 7170, 7150, 7145, 7115, 6055, 6030, 5950, 4795, 1170, 648, 585

2100-2200	Romania	Eu	9690, 7225, 7195, 6105, 5955
2100-2200	VOA	Af	21485, 17800, 15580, 15495, 15410, 15290, 13710, 621
2100-2200	VOA	CAm	930
2100-2245	WINB	Eu/NAf	15185
2100-2315	BBC WS	Af	3955
2100-2315	BBC WS	Eu	6195
2100-2400	KHBN	China	9820
2110-2210	Syria	NAm/AUS/NZ	15095, 12085
2115-2245	Egypt	Eu	9900
2130-2200	R. Austr.	As/Pac/PNG	11855, 9760, 9540
2130-2200	HCJB	Eu	21480, 21455USB, 17790, 17490USB
2130-2200	Sweden	As/Pac/Eu	11955, 9655, 6065, 1179
2130-2200	Canada	Af	17820, 15150, 11880
2140-0700	RNZI	Pacific	17770
2140-2155W	Caracas	SAm	9540
2145-2245	S. Korea	Eu/ME/Af	15575, 7550, 6480
2200-0300	BBC WS	Eu	7325
2200-2225	Albania	Eu	11825, 9760
2200-2225	Belgium	Eu	9905, 5900, 1512
2200-2225	Prague	Eu	9605, 7345, 6055, 5960
2200-2225	Italy	Japan	15330, 11800, 9710
2200-2230	R. Austr.	As/Pac/PNG	17795, 15365, 15320, 11855, 9760, 9540
2200-2230	China	Eu	3985 via SRI
2200-2230	Canada	As	11705 via Japan
2200-2230	SRI	SAm	12035, 9885, 9810, 6030
2200-2230	VOA	ME/NAf/Eu	11960, 11895, 9530, 7195, 7140, 6160
2200-2230	Beograd	Eu	9505, 7200, 6100
2200-2300	BBC WS	As	11955
2200-2300	BBC WS	NAm/CAm	15070, 9590, 5975
2200-2300	China	Eu	7170 via Russia
2200-2300	Cuba	NAm/CAm	6180
2200-2300	Hungary	Eu	11910, 9835, 6110
2200-2300	Moscow	WS	21790*, 21770*, 21480, 17690, 17665, 17655, 17570, 15425, 15130, 12050, 11775*, 11655, 9890, 9885, 9870, 9860, 9750, 9735, 9725, 9695, 9685, 9530, 9520, 9480, 9450, 7390, 7300, 7295, 7285, 7255, 7245, 7220, 7180, 7150, 7145, 7135, 7115, 6060, 6055, 6045, 6030, 5985, 5980, 5920, 4860, 4825, 4795, 1170
2200-2300	Canada	Eu	13650, 11945, 9160, 7180, 5995
2200-2300	Ukraine	Eu	9710, 7240, 7195, 6020
2200-2300	WYFR	Eu/Af	21525, 17750, 7355
2200-2300	Taiwan	Eu	11915, 9850 via WYFR, USA
2200-2315	BBC WS	Af	15400
2200-2315	BBC WS	Eu	9410
2200-2400	Abu Dhabi	NAm	11815, 11710, 9605
2200-2400	BBC WS	As	9570
2200-2400	Iraq	Eu	15210, 15180 (+ 13680 to 2230)
2200-2400	KHBI	EAs/SEAs	15405, 13625
2200-2400	WCSN	SEu/WAf	7510
2200-2400	WSHB/1	ENAm/Cam	9465
2200-2400	WSHB/2	SAm	17555
2200-2400	VOA	As/Pac	17820, 17735, 15305, 15290, 15185, 11760, 9770, 7120
2200-0030	BBC WS	Pac	11955
2203-2215	Croatia	Eu/NAm	13830, 9830, 6210
2215-2240	Finland	As/Am/Eu/Af	11810, 11740, 9730, 963, 558, 252
2230-2300	R. Austr.	As/Pac/PNG	17795, 15365, 15320, 11855, 9760
2230-2300	Israel	LAm/NAm/wEu	17575, 11675, 11603, 11587, 9435, 7465
2230-2300	Vilnius	Eu	9710, 9675, 1557, 666

(Continued on page 451)

LISTEN TO THE WORLD

Contents:

Clandestine Radio Survey

by Mathias Kropf

Introduction

The WRTH is a major reference work for both the international broadcaster and listener. The stations listed in the book are officially registered in at least one country (although another country may not agree with its politics). But that still leaves some stations on the dial which are not listed..the so-called "clandestine stations". This article is a survey of current activities compiled in 1992.

The clandestine broadcasting scene is one of the most interesting and challenging parts of the broadcasting spectrum. Clandestine stations tend to broadcast in rare languages, change their frequencies quite often (sometimes three times within five minutes!) and alter their schedules unexpectedly without any prior announcement. The life of clandestine listeners is complicated further by jamming (yes, it still exists) and the lack of much printed material from and about the stations themselves.

All this may be annoying to the less experienced shortwave listener, but it is a thrill to a few dedicated clandestine monitors. So, what makes the clandestines such a fascinating part of the broadcasting spectrum? When taken seriously, clandestine listening can improve the understanding of world politics in general and of the situation in certain areas of the globe. Clandestine radio listeners know more about the political situation in a country, about rebel groups and about emerging conflicts than your local TV news will ever tell you.

Now that I've mentioned politics, some of you will probably ask yourself "Isn't all this too political, anyway?" Well, yes and no. It is political, but not necessarily biased. It has always been one of my major principles to report about all clandestines, from extreme right-wing to extreme left-wing and whatever you can probably imagine, regardless of my own political opinion. Of course, clandestines often broadcast political propaganda, but it is not necessarily a political statement to report about their existence!

Although the history of clandestine broadcasting is almost as old as the history of broadcasting itself (the first clandestines appeared before World War II), it has only in recent years become more and more accepted by shortwave listeners as a separate part of their hobby that should not be confused with the official, commercial or religious shortwave stations.

The future of clandestine broadcasting on shortwave will be highly dependent on the future of shortwave broadcasting in general. Since 1990 (which was one of the busiest years for clandestine operations mainly due to the Iraq/Kuwait crisis), several new stations have appeared to replace those which went off the air in the 1980's. Although the use of TV, FM and even satellite broadcasts for propaganda transmissions will certainly increase, shortwave will remain the most effective means of transmission in parts of Asia and Africa for many years to come.

Still, there is no generally accepted definition for which station actually is a clandestine and which not. So, I would like to point out that the selection of clandestines on the following pages is my personal opinion. I don't want to go into a long sermon of explanations and definitions here, because I think that the best method to learn more about clandestines is by actually listening to them. So, I hope that the following sections will help you to make the clandestine monitoring easier and more successful. Good hunting!

Clandestine Stations on Shortwave

Freq.	Station, Times in UTC
3250v	V. of Mojahed of Afghanistan: 0100-0200, 1415-1515
3480	V. of the National Salvation: 0255-0700, 0955-1700, 1955-0100 (14)
3557v	V. of Mojahed: 0130-0330, 0900-1000, 1630-1830 (13)
3650v	La Voz Popular: 0200-0310 (Sat/Wed) (8)
3780v	V. of Mojahed: 0130-0330 (13)
3790v	V. of Mojahed: 1630-1830 (13)
3875v	V. of Iranian Kurdistan: 0230-0330, 0800-0930, 1630-1800
3880v	R. Free Bougainville: 0750-1120 (18)
3888v	V. of the Communist Party of Iran: 0315-0445 (Sat-Thurs), 0430-0600 (Fri), 1700-1900 (10)
3888v	V. of the Iranian Revolution: 1430-1530 (10)
3912v	V. of the People: 1400-1530, 1630-1700, 2000-2130
3930v	V. of the People of Kurdistan: 0400-0600, 1500-1720, 1830-1930
3940	V. of the Broad Masses of Eritrea: 0400-0700, 0900-1100, 1400-1800 (7)
3980	V. of the Peninsula, the Arabian Gulf: 1630-1700
3980	Holy Medina R: 1930-2030
3980	V. of Egypt of Arabism: 2030-2100
3985v	R. Echo of Hope: 1300-1700, 2000-2100
4065v	V. of Iranian Kurdistan: 0230-0330, 0800-0930, 1630-1800
4085	V. of Kashmir Freedom: 1630-1730 (9)
4108	V. of the Feda'i: 0230-0315, 0900-0945, 1700-1745 (12)
4120	V. of National Salvation: 0955-1700 (14)
4130	Kurdistan R., the V. of Unification: 0300-0425, 1600-1725
4160v	V. of the Worker: 0300-0530, 1430-1530, 1700-1830
4160v	V. of the Feda'i: 0230-0315, 0900-0945 (12)
4175	V. of Iraqi Kurdistan: 0430-0600, 1600-1730
4400	V. of National Salvation: 0255-0700, 0955-1700, 1955-0100 (14)
4450v	V. of National Salvation: 0255-0700, 0955-1700, 1955-0100 (14)
4470v	V. of Mojahed: 1630-1830 (13)
4475v	V. of the Iranian Revolution: 1430-1530 (10)
4557	V. of National Salvation: 0255-0700, 0955-1700, 1955-0100 (14)
4670v	V. of Mojahed: 0130-0330 (13)
4700v	V. of Mojahed: 0130-0330, 1630-1830 (13)
4750v	V. of Mojahed: 0900-1000 (13)
4880	V. of the Resistance of the Black Cockerel: 1645-2345
5000	V. of the Broad Masses of Eritrea: 0400-0700, 0900-1100, 1400-1800 (7)
5000	V. of Kashmir Freedom: 0230-0330, 1100-1200, 1430-1530, 1630-1730 (9)
5080v	V. of Mojahed: 1630-1830 (13)
5110	V. of the People of Wa State: 1130-1330
5225v	V. of Mojahed: 0130-0330 (13)
5245v	V. of Mojahed 1630-1830 (13)
5408	V. of the Great National Union Front of Cambodia: 0800-1000, 1200-1500, 2330-0200
5590v	V. of Mojahed: 1630-1830 (13)
5600v	V. of Mojahed: 0130-0330 (13)
5630v	V. of Mojahed: 0130-0330 (13)
5740v	V. of Mojahed: 0900-1000, 1630-1830 (13)
5830v	V. of Mojahed: 0900-1000 (13)
5850v	R. Patria Libre: 0030-0110
5910	Al-Quds R: 0600-1100, 1300-1730 (17)
6010	V. of National Salvation: 0955-1400 (14)
6045	V. of the Resistance of the Black Cockerel: 0445-0835
6100	R. Siboney: 1900-2000
6110v	V. of Mojahed: 1630-1830 (13)
6140	R. Mojahedin of Afghanistan: 0200-0330, 1230-1400 (1)
6145	V. of Palestine via RTA Algiers: 1700-1800 (16)
6146	V. of Afghanistan: 0130-0245, 1315-1415 (2)
6160	V. of Palestine via RTA Algiers: 1700-1800 (16)
6160	V. of the People of Kurdistan: 0400-0600, 1500-1720, 1830-1930
6220v	V. of Mojahed: 0130-0330, 0900-1000, 1630-1830 (13)

6300	V. of Kashmir Freedom: 0230-0330, 1100-1200, 1430-1530 and, 1630-1730 (9)
6300v	R. Patria Libre: 1130-1210
6305	La Voz del CID: 0210-0930 (6)
6325	V. of the Khmer: 0400-0700,0900-1400, 2230-0200(Sun 0300) (4)
6340	R. Muhabura: 0415-0515, 1000-1100, 1715-1815/1845 (alt. to 6400)
6348	R. Echo of Hope: 0300-0700, 0800-1200, 2300-0100
6400	R. Muhabura: 0415-0515, 1000-1100, 1715-1815/1845 (alt. to 6340)
6410v	V. of the Communist Party of Iran: 0315-0445 (Sat-Thurs), 0430-0600 (Fri), 1700-1900 (10)
6415v	V. of the Iranian Revolution: 1430-1530 (10)
6510v	V. of Mojahed: 1630-1830 (13)
6600	V. of the People: 0300-0430, 0600-0730, 0830-1000, 2300-0030
6645v	V. of Feda'i: 0330-0415, 1000-1045, 1800-1845 (12)
6760v	V. of the Kurdistan Revolution: 0500-0700, 1400-1545
6720v	V. of Mojahed: 1630-1830 (13)
6770	V. of the Tigre Revolution: 0430-0530, 1345-1445 (Sat), 1500-1600
6940	V. of the Ethiopian People for Peace, Democracy & Freedom: 0330-0430, 1345-1445 (Sun), 1500-1600
6940	V. of the Broad Oromo Masses: 0430-0530, 1600-1700
6958	R. Mogadishu: 0400-0500, 1600-1800
7000v	La Voz Popular: 2315-0025 (Tues/Fri) (8)
7000v	V. of the Iraqi People: 1400-1445
7000v	V. of the Struggle of Iranian Kurdistan: 1600-1700
7070v	V. of Mojahed: 0130-0330, 0900-1000, 1630-1830 (13)
7075v	V. of the People of Kurdistan: 0400-0600, 1500-1720, 1830-1930
7091.5	R. Message of Freedom: 0145-0315, 1400-1500 (3)
7100	R. Voice of Ethiopian Unity: 1400-1600, 1800-2000
7100	V. of the Resistance of the Black Cockerel: 1645-2345
7125	V. of Afghanistan: 1415-1515 (2)
7145	V. of Palestine via RTA Algiers: 1700-1900 (16)
7180	V. of the Movement of Iranian Baluchistan: 0700-0800
7290	V. of the Resistance of the Black Cockerel: 0445-0835, 1045-1435
7300	Vietnamese Resistance R: 0200-0300, 0600-0700, 1000-1100, 1400-1500, 2200-2300
7340	La Voz del CID: 2300-1100 (6)
7380	A Voz de Renamo: 0515-0600
7435	V. of the Struggle of Iranian Kurdistan: 0930-1035, 1430-1530
7450	V. of the Tigre Revolution: 0430-0530, 1345-1445 (Sat only), 1500-1600
7490	V. of the Broad Masses of Eritrea: 0400-0700, 0900-1100, 1400-1800 (7)
7820	V. of the Tigre Revolution: 0430-0530, 1435-1445 (Sat only), 1500-1600
8000	V. of the Ethiopian People for Peace, Democracy & Freedom: 0330-0430, 1345-1445 (Sun), 1500-1600
8000	V. of the Broad Oromo Masses: 0430-0530, 1600-1700
8012v	V. of Rebellious Iraq: 0330-0600, 1130-1400, 1630-1900
9045	Iran's Flag of Freedom R: 0330-0525 (11)
9170v	R. SPLA: 0500-0600, 1100-1200, 1300-1400
9250	Iran's Flag of Freedom R: 0330-0525 (11)
9355	Iran's Flag of Freedom R: 1630-1825 (11)
9400	R. Azadi: 1830-1925
9435v	R. Voice of Ethiopian Unity: 0400-0600, 1800-2000
9475v	R. Mogadishu: 1000-1100, 1400-1600
9545	V. of the Movement of Iranian Baluchistan: 0700-0800
9570	V. of the Iraqi People: 0300-0530, 1100-1700
9630	Holy Medina R: 1930-2030
9640	V. of the Free Sahara via RTA Algiers: 2200-2300 (15)
9650v	V. of Mojahed 0130-0330 (13)
9685	R. Freedom via R Tanzania: 0415-0445 (Tues/Thurs/Sat), 1815-1845 (Mon/Wed/Fri)
9700	V. of the Resistance of the Black Cockerel: 0445-0835
9715	R. Free Afghanistan via RFE/RL: 0230-0330
9770	R. Free Afghanistan via RFE/RL: 0230-0330
9850	V. of the Resistance of the Black Cockerel: 1045-1430
9860	A Voz da Renamo: 0500-0530, 1045-1145, 1600-1700
9941.7	La Voz del CID: 0940-0210 (6)
9950	V. of Iraq via Damascus: 0400-0500, 1700-1900
9965	R. Caiman: 1200-1530, 2300-0400
9980v	V. of the Iraqi People: 1700-2330
10200	R. St. of the Government for the Liberation of the Lao Nation: 0100-0300, 0700-0900
11200	R. SPLA: 0400-0500, 1300-1400
11470	Iran's Flag of Freedom R: 0330-0525, 0645-0730, 1400-1445, 1630-1825 (11)
11635	La Voz del CID: 1100-2300 (6)
11695	V. of Ethiopian Patriotism via R. Moscow: 1530-1600 (Sun/Wed)
11705	V. of the Oromo Liberation: 0900-1045, 1730-1915
11715	V. of Palestine via RTA Algiers: 1700-1900 (16)
11815	R. Free Afghanistan via RFE/RL: 0230-0330
11830	V. of the Resistance of the Black Cockerel: 1045-1435
11860v	Holy Medina R: 0500-0900, 1600-2000
11945	News Centre of Free Iraq: 2200-2355
11970	V. of Mojahadeen of Iranian Baluchistan: 1300-1455
11970	R. Free Afghanistan via RFE/RL: 0230-0330
12085	V. of Palestine in Damascus: 1630-1730 (irr.)
15050v	R. Patria Libre: 0030-0110, 1130-1210
15095	V. of Palestine in Damascus: 1630-1730
15100	Iran's Flag of Freedom R: 0645-0730, 1400-1445 (11)
15160	R. Free Afghanistan via RFE/RL: 1530-1600
15190	V. of the Iraqi People: 0300-0530, 1100-2330
15205	V. of Palestine via RTA Algiers: 1700-1900 (16)
15215	V. of the Free Sahara via RTA Algiers: 2200-2300 (15)
15215	V. of Palestine via RTA Algiers: 1800-1900 (16)
15235	V. of the Popular Revolution in Palestine: 0330-0400, 2130-2200
15280	V. of China: 2100-2200 (5)
15445	R. Free Afghanistan via RFE/RL: 1300-1330
15555	Iran's Flag of Freedom R: 0330-0525 (11)
15605v	V. of the Iraqi People: 0300-0530, 1100-2330
15650	R. Azadi: 0230-0325, 1830-1925
17570	V. of Ethiopian Patriotism via R. Moscow: 1530-1600 (Sun/Wed)
17725	R. Free Afghanistan via RFE/RL: 1300-1330
17725	V. of the Popular Revolution in Palestine: 0330-0400, 2130-2200
17745	V. of Palestine via RTA Algiers: 1800-1900 (16)
17760	R. Free Afghanistan via RFE/RL: 1300-1330
17805	R. Free Afghanistan via RFE/RL: 1530-1600
17835	R. Free Afghanistan via RFE/RL: 1300-1330
17845	Democratic Voice of Burma via R. Norway: 1430-1525
17950v	V. of the Iraqi People: 0300-0530, 1100-2330
21510	R. Free Afghanistan via RFE/RL: 1300-1330, 1530-1600
21785	V. of Ethiopian Patriotism via R. Moscow: 1530-1600 (Sun/Wed)

Long- & Mediumwave Clandestine Stations

153	V. of Palestine via RTA Algiers: 1700-1800
198	V. of Palestine via RTA Algiers: 1700-1800
531	V. of Palestine via RTA Algiers: 1700-1800
549	V. of Palestine via RTA Algiers: 1700-1800
567	V. of Arab Egypt via R Damascus: 1830-1930
567	V. of Palestine in Damascus: 1630-1730
610	V. of Mojahed: 0130-0330
630v	Al-Quds R: 0600-1100, 1300-1730
666	V. of Palestine in Damascus: 1630-1730
684	Holy Medina R: 2000-2030
684	V. of Egypt of Arabism: 2030-2100
684	V. of the Peninsula & the Arabian Gulf: 1630-1700
693	V. of Palestine via RTA Algiers: 1700-1800
702v	Al-Quds R: 0600-1100, 1300-1730
720	V. of the Islamic Revolution of Afghanistan: 1700-1730
747	V. of Palestine in Damascus: 1630-1730
756	Holy Medina R: 2000-2030
756	V. of Egypt of Arabism: 2030-2100
756	V. of the Peninsula & the Arabian Gulf: 1630-1700
783	V. of Arab Egypt via R Damascus: 1830-1930
783	V. of Palestine in Damascus: 1630-1730
828	V. of Iraq via R Damascus: 0400-0500, 1700-1900
828	V. of Palestine in Damascus: 1630-1730
891	V. of Palestine via RTA Algiers: 1700-1800
918	V. of Palestine in Damascus: 1630-1730
945	V. of the Broad Masses of Eritrea: 0400-0700, 0900-1100, 1400-1800
1035	R. Freedom via R. Tanzania: 0415-0445 (Tues/Thurs/Sat), 1815-1845 (Mon/Wed/Fri)
1053	V. of National Salvation: 0255-0700, 0955-1700, 1955-0100

1053	V. of the Iraqi People: 0300-0530, 1100-2330
1089	V. of Palestine via RTA Algiers: 1700-1800
1098	V. of Palestine via RTA Algiers: 1700-1800
1107	Palestine R. via V. of the Arabs: 0600-0800, 1200-1500
1125	V. of Iraq via R Damascus: 0400-0500, 1700-1900
1161	V. of Palestine via RTA Algiers: 1700-1800
1179	V. of the Iraqi People: 0300-0530, 1100-2330
1197	V. of Mojahed: 0130-0330
1224	V. of the Islamic Revolution in Iraq: 1530-1715
1250	V. of the Khmer: 0400-0700, 0900-1400, 2230-0200 (Sun 0300)
1377	V. of the National Alliance for the Liberation of Syria: 1930-2100
1377	V. of Palestine via R Baghdad: 1500-1600
1431	V. of Rebellious Iraq: 0330-0600, 1130-1400, 1630-1900
1431	V. of Palestine via R Baghdad: 1500-1600
1485	V. of Palestine in Damascus: 1630-1730
1545	National R. of the Saharan Democratic Republic: 0600-0800 and 2000-2400 (ex-1355kHz)
1590	V. of Rebellious Iraq: 0330-0600, 1130-1400, 1630-1900
1615	V. of Iraqi Kurdistan: 0430-0600, 1600-1730

Addresses of Clandestine Stations

(1) **R. Mojahedin of Afghanistan,** P.O. Box 204, Peshawar, Pakistan

(2) **V. of Afghanistan,** GPO Box 1207, Peshawar, Pakistan or Postfach 1103, D-7065 Winterbach, Germany

(3) **R. Message of Freedom,** GPO Box 857, University Town, Peshawar,Pakistan

(4) **V. of the Khmer,** c/o KPNLF Office, P.O. Box 22-25, Ramindra Post Office, Bangkok, Thailand or c/o FUNCINPEC Office, P.O. Box 12-1014, Suan Phlue, Bangkok, Thailand

(5) **V. of China,** P.O. Box 11696, Berkeley, CA 94701, USA or P.O. Box 79218, Monkok, Hongkong

(6) **La Voz del CID,** Apartado 8130, San Jose 1000, Costa Rica or Apartado 2600, 08080 Barcelona, Spain or Apartado Postal 76633, el Marquez, Caracas 1070, Venezuela or 10000 SW 37 Terrace, Miami, FL 33165, USA

(7) **V. of the Broad Masses of Eritrea,** EPLF National Guidance Department, Information Branch, Sahel Eritrea, P.O. Box 671, Port Sudan, Sudan or P.O. Box 891, Port Sudan, Sudan or Information Department, R. Branch, Asmara, Eritrea (via Ethiopia)

(8) **La Voz Popular,** Box 19619, Mexico-City 03910, Mexico or Network in Solidarity with the People of Guatemala (NISGUA), 930 F Street NW, Suite 720, Washington, DC 20004, USA

(9) **V. of Kashmir Freedom,** P.O. Box 102, Muzzafarabad, Azad Kashmir, Pakistan

(10) **V. of the Communist Party of Iran, V. of the Iranian Revolution,** O.I.S., Box 50040, S-104 05 Stockholm, Sweden

(11) **Iran's Flag of Freedom R,** Derafsh, 20, rue de Condorcet, F-75009 Paris, France or c/o Reza Farhadi, P.O. Box 19740, Irving, CA 92714, USA or P.O. Box 05559B, 2000 Hamburg, FRG

(12) **V. of the Feda'i,** A.C.A., B.P. 43, F-94210 Fontenay-sous-Bois, France

(13) **V. of Mojahed,** MISS, BM Box 9720, London WC1N 3XX, England

(14) **V. of National Salvation,** The Front for Saving the Nation, Kankoku Minzoku Minshu Tensen, 1-2-1 Hirakawa-machi, Chiyota-ku, Tokyo, Japan

(15) **V. of the Free Sahara,** B.P. 10, El-Mouradia, Algiers, Algeria

(16) **V. of Palestine,** c/o PLO, P.O. Box 3122, Baghdad, Iraq or C.P. 421, Luanda, Angola

(17) **Al-Quds R,** P.O. Box 10412, Tripoli, Libya or P.O. Box 5092, Damascus, Syria or al-Ghubayrah, P.O. Box 25-74, Beirut, Lebanon or P.O. Box 1397, Aden, Yemen or P.O. Box 6170, Aden, Yemen

(18) **R. Free Bougainville,** P.O. Box 1203, Honiara, Solomon Islands

Clandestine without secrets

Afghanistan: At the present time, there are five clandestines broadcasting to Afghanistan. The Voice of Afghanistan, which already started back in 1984, is again active after some breaks. Radio Message of Freedom, which is run by the Hezb-e Islami group, is a relatively new station, as well as Radio Mojahedin of Afghanistan and Voice of Mojahed of Afghanistan are. These three stations are all run by Afghan guerrilla groups and are probably located in Pakistan. Another clandestine programme to Afghanistan, Radio Free Afghanistan, is broadcast via the RFE/RL facilities in Europe.

Angola: Despite recent negotiations on a very high level, the Voice of the Resistance of the Black Cockerel still remains on the air. This UNITA station, which is sometimes also called Vorgan, is thought to be located in Jamba. Radio Siboney, another station hostile to the Angolan government, started back in 1988, but it is not known if it is still active.

Cambodia: Despite the recent peace efforts by the United Nations, at least two clandestines still remain on the air. The Voice of Khmer is run by the KPNLF since 1984. The Voice of the Great National Union Front of Cambodia was actually created in late 1991, when the two former Khmer Rouge stations, the Voice of Democratic Kampuchea and the Voice of the National Army of Democratic Kampuchea, merged.

China, Peoples Republic of: The Voice of China is broadcast through facilities of the BCC in Taiwan, although it is run by an independent organization whose headquarters are probably in the US.

Colombia: Radio Patria Libre is run by the Ejercito de Liberación Nacional (ELN), one of the Colombian rebel groups. Several sources indicate that this station may in fact be located within Colombia.

Cuba: La Voz del CID is one of the few clandestine stations that is actually running a 24 hour schedule and that is heard almost everywhere in the world without any problems. The station started back in 1981 with a single transmitter, but soon extended its network to several stations which were thought to be located in several different countries. Formerly, some of the stations even used their own identifications, like Radio Maximo Gomez, Radio Antonio Guiteras, Radio Ignacio Agramonte, Radio Jose Antonio Echeverria, Radio Antonio Maceo and Radio Frank Pais. Nowadays, the only identification used for all its outlets is "Radio Camilo Cienfuegos - La Voz del CID". One of the most mysterious clandestines in recent years is Radio Caiman, which started back in 1985. The station has never given any hints about who is actually behind it and it has never announced an address or any similar information.

The latest trend in this area is that rather than operating their own transmitters, a large number of Cuban organizations that are hostile to their government have started to hire airtime on religious commercial US shortwave stations and now there is even one station about to come on the air (Radio Miami International) that is entirely devoted to selling airtime to these programmes.

Egypt: During the Iraq/Kuwait crisis in 1990 the Voice of Egypt of Arabism was again started after several years absence. This programme is broadcast by Iraq and was formerly on the air from 1977 to 1982. It is not known if it is still on the air at present.

Ethiopia: After the political developments in Ethiopia in 1991 the status of the former anti-Ethiopian clandestines in under review. At the present time, most of them are thought to operate via government sponsored transmitters. On the other hand, there is a chance that at least the Voice of the Broad Masses of Eritrea may remain a clandestine for some time to come, because the promised independence for Eritrea has so far not been realized. In October 92 a US group which opposes the current government in Ethiopia hired airtime to broadcast short 30 minute segments within Radio Moscow's Amharic transmissions.

Guatemala: The only anti-Guatemalan government clandestine is La Voz Popular, which has been on the air since 1987. This station is operated by the URNG (Unidad Revolucionaria Nacional de Guatemala).

India: Radio Kashmir Freedom, also known as Voice of Independent Kashmir, has been on the air since February 1991 and is hostile to the Indian government. The transmitter is thought to be located in the Pakistani part of Kashmir ("Azad Kashmir").

Iran: Clandestine activity to Iran has been very high for many years. The Voice of the Communist Party of Iran and the Voice of the Iranian Revolution (the latter is run by Komala, the Kurdish wing of the CPI) have been on the air since 1983. These two stations share the same transmitters. The Voice of Iranian Kurdistan (operated by the KDP Kurdish Democratic Party) has been on the air since 1980 and irregularly even since 1973. The Voice of Mojahed (formerly also known as Voice of Crusader and Voice of the Mojahedin-e Khalq) started its broadcasts in 1981. This station is thought to be based in Iraq. At one time back in 1985/1986 it was one of the most extensive clandestine operations with several short- and mediumwave and FM transmitters on the air almost 24 hours per day and plans for a TV station. However, its transmissions were cut back towards the end of the 80s and during the peak of the Iraq/Kuwait crisis it disappeared altogether for about six months. Obviously, its transmitters had to be used for more important purposes. Iran's Flag of Freedom Radio started back in 1983 as Voice of the Salvation of Iran and later it was renamed Voice of the Liberation of Iran, before it started to use the present name a few years ago. The Voice of the Feda'i and the Voice of the Worker are thought to be operated by two different wings of the same organization. The Voice of the Movement of Iranian Baluchistan is aired by the Iraqi government. This programme started back in 1988, but has been on the air only irregularly. Radio Azadi is the replacement station for Radio Iran, which apparently closed down in late 1991.

Iraq: The Voice of Iraqi Kurdistan, which is run by the Kurdish Democratic Party, has been on the air irregularly at least since 1965 and regularly since 1980. The Voice of the Islamic Revolution in Iraq is broadcast via Iranian facilities, but its transmissions are highly erratic. The Voice of the Iraqi People, which is operated by the Communist Party of Iraq, has been on the air since 1982. This station is not related to the "new" Voice of the Iraqi People, which started in early 1991 as the Voice of Free Iraq, was renamed in April 1991 to Voice of the Iraqi Opposition and changed to its present name in October 1991. So, this operation must hold the unofficial record for the most frequent name changes within a rather short period of time. There are still more Kurdish stations that are hostile to the Iraqi government: The Voice of the People of Kurdistan is operated by the Patriotic Union of Kurdistan (PUK). Two stations which started their activities in 1992 are the Voice of the Kurdistan Revolution, run by the People's Democratic Party of Kurdistan, and Kurdistan Radio, the Voice of Unification, which is operated by the Kurdistan Socialist Party (KSP). With the increasing number of Kurdish stations that are hostile to the Iranian and/or Iraqi government, things will certainly get more and more difficult for monitors and listeners. Another programme that is hostile to the Iraqi government and actually originates from Syria is the Voice of Iraq which has been on the air at least since 1986. And finally, the Voice of Rebellious Iraq started back in March 1991 and is apparently run by a group of Shiite fundamentalists led by Muahmmed Baqir al-Hakim.

Israel: All three stations which are hostile to the Israeli government are in support of an independent Palestine. The Voice of Palestine, which has been on the air for at least 20 years, is run by the PLO and broadcast through the facilities of several different Arab countries. From time to time there have also been Voice of Palestine programmes by different groups and factions within and outside the PLO, making things for the listener even more difficult. The Voice of the Popular Revolution in Palestine is aired by the Libyan foreign service, but seems to be on the air only irregularly. Radio Al-Quds is on the air since September 1987 (but its operations on shortwave seem to be highly erratic). Al-Quds is in fact the Arabic name for Jerusalem. The station is run by the PFLP-GC (Popular Front for the Liberation of Palestine - General Command) led by Ahmad Jibril and its transmitters are probably located in Syria. In October 1992 the station issued a statement in Damascus claiming that the Israeli authorities had resumed jamming of the station's mediumwave outlet. Checks by the WRTH in the target area could not confirm this claim.

Korea: Three clandestine stations are still active on the Korean peninsula. The Voice of National Salvation, which started in 1970 as Voice of the Revolutionary Party for Reunification and was renamed in 1985, broadcasts from North to South Korea. The other two stations are active in the opposite direction: Voice of the People and Radio Echo of Hope.

Laos: A clandestine station known as Radio Station of the Government for the Liberation of the Lao Nation has been heard irregularly since 1990.

Mozambique: The Voice of Renamo, which is operated by the Mozambique National Resistance (MNR), has been heard irregularly since late 1991. This is thought to be related to a similar Renamo station which was on the air in the early 80s.

Morocco: The POLISARIO, which is fighting for the independence of the West Sahara, is running two radio stations, one of them on shortwave. The Voice of the Free Sahara has been on the air since 1975 and is relayed by the shortwave transmitters of RTA Alger. The medium-wave station, which has occasionally been heard in Europe, is called National Radio of the Saharan Arab Democratic Republic.

Myanmar: The Voice of the People of Wa State is supposed to be somehow related to the former Voice of the

People of Burma, which had been on the air from 1971 to 1990. The location of the transmitter is thought to be in the People's Republic of China.

Papua New Guinea: Radio Free Bougainville, which is the station of the Bougainville Interim Government, has been on the air since early 1992, after the government of this island declared its independence from the rest of Papua New Guinea. An Australian amateur radio operator played a major role during the setup of this station. The station was active during the summer of 1992, but disappeared in October 92 as PNG forces closed in. At press-time for the WRTH the station was silent.

Saudi-Arabia: A large number of anti-Saudi Arabian government stations became active during the Iraq/Kuwait crisis, but only two of them are still thought to be on the air at present. Holy Medina Radio and the Voice of the Peninsula and the Arabian Gulf both originate from Iraq.

Somalia: A fake Radio Mogadishu came on the air in early 1992 which was trying to imitate the government station. This outlet was set up by a group led by Ali Mahdi Muhammad.

South Africa, Rep. of: Although most of its transmissions via the Foreign Services of several African nations have ceased over the past couple of years, the ANC station "Radio Freedom" is still on the air via Radio Tanzania.

Sudan: Radio SPLA, which is the voice of the Sudanese People's Liberation Army under John Garang, has been on the air since 1984. Its transmitter was formerly thought to be located in Ethiopia, but this may have changed recently.

Vietnam: Although rarely heard outside its target area, Vietnamese Resistance Radio, which is run by the National Unified Front for the Liberation of Vietnam, has been on the air irregularly since 1984.

More Compact Directional SW Antennas

by, O. G. Villard, Jr., G. H. Hagn, J. M. Lomasney
SRI International, 333 Ravenswood, Menlo Park, CA 94025-3493 USA

Interference and fading are the two biggest detractors from the pleasure of listening to shortwave broadcasts. But as a result of recent research, either or both can be significantly improved by means of a reasonably compact directional accessory antenna that is usable indoors and easy to install and operate. Interference can often be reduced by placing a null in the direction of the offending station; the effects of fading in turn can be reduced by a kind of diversity reception employing the same basic antenna design. The following tells how and why these schemes work and gives enough "how to build it" information to permit duplication by a radio amateur or radio serviceman.

Background

This article is a sequel to one appearing in the 1990 WRTH(1) in which interference reduction by use of compact antenna directivity indoors was described. It's important to point out that both the earlier article and this one are made possible by the availability of compact battery-powered shortwave receivers. These greatly simplify the problem of directive reception at HF because there need be no conducting wires to convey RF to the neighborhood of the loop by the antenna-transmission-line effect. A small self-contained receiver with a compact antenna can then place a null on an interferer or measure field direction at a given location without appreciably disturbing the field that is to be measured.

Standard small loops measure ground waves by the null method very well, but they are in general helpless against sky waves, which contain signal components arriving along and near the great circle path at various vertical angles. To null out all these skywave "modes" at once, a directional antenna must have a fan-shaped null in elevation at least in the lower part of the vertical plane. The 1990 article showed several ways to generate such nulls. Although these schemes have desirable unidirectional polar patterns, they are frequency sensitive and require reasonably careful tuning adjustments using more than one knob. We now feel that for general-purpose shortwave listening (SWL) a more satisfactory design is two loops in an end-fire arrangement that we call the Cascaded Loop Array, or CLA. Both loops can be resonated with a single capacitor in series, resulting in desirable one-knob tuning. The receiver can conveniently be located between the loops; for brevity, the combination of loops and such a receiver will be referred to here as a "CLAR."

Description of the Shortwave CLA

A simplified circuit diagram is shown in Figure 1, and approximate far-field azimuthal patterns are shown in Figure 2. The identical loops are wound in opposite directions and are joined by wires that form a short transmission line.

The Figure 1 arrangement gives a null on the array axis that is effective over a wide range of vertical angles, measuring upward from the horizon. The CLAR is therefore more useful than earlier designs for nulling sky-wave signals from nearby transmitters. In addition, the only tuning control (the series capacitor), needs only to be adjusted for greatest sensitivity and does not appreciably affect directivity. With loops vertically disposed, the antenna's sensitivity is greatest for vertical polarization. Response of the loops to the horizontally polarized component of signal energy is minimal for any azimuth.

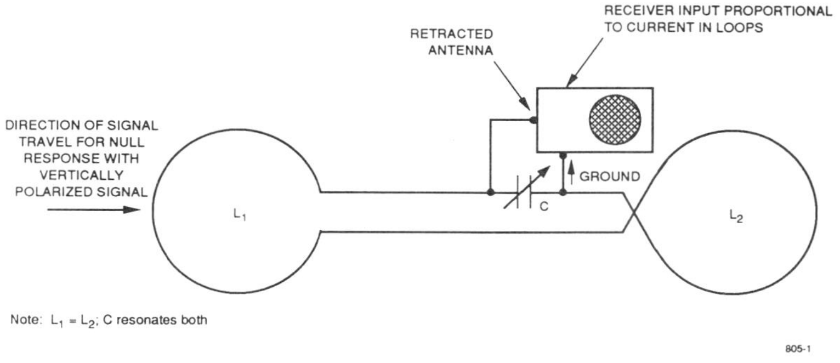

Fig. 1—Simplified Circuit of Shortwave CLAR (loop plane is vertical)

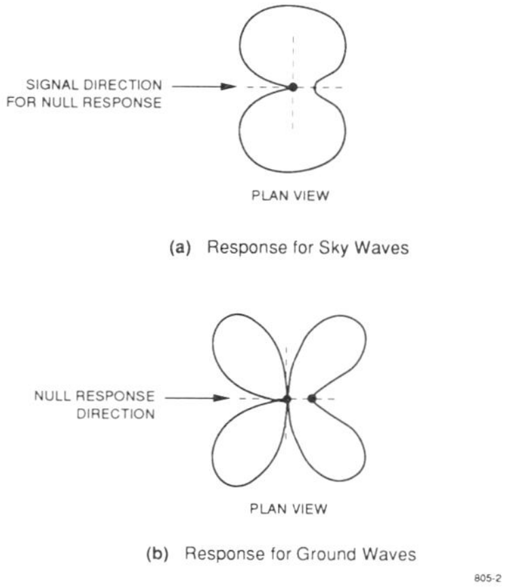

Fig. 2—Rough Far-Field Azimuthal Polar Response Pattern of CLAR. Vertical polarization and vertical loops assumed. Note finite rearward response. Pattern of (a) is essentially independent of vertical angle of arrival for high-angle components.

A great simplification of the CLA scheme results from use of a single tuning capacitor to resonate both loops simultaneously. This idea seems to have occurred independently to the authors and to Mr. F. N. Eddy, who calls a similar device the "Nullerator." He points out its usefulness in tracking down and/or rejecting interference.(2)

The azimuthal response pattern of the CLA is shown in Figure 2. What we can call an "array null" occurs when signal travel is in the plane of the loops. The pattern is not quite bidirectional, having a front-to-back ratio that although small is normally sufficient for azimuth direction-of-arrival ambiguity resolution.

Interestingly, the CLA azimuthal response pattern depends on whether the signal reaches the listener by sky wave or by ground wave. When skywaves are present, there are no stable nulls in the direction perpendicular to that of the station, and the average response in that direction, at least, is as in Figure 2a. But with signals consisting primarily of very low vertical angles, such as ground waves, the response of Figure 2b will be observed. The extra nulls in Figure 2b are the well-known response of the individual loops by themselves, and they can be called element nulls. The depth of the element nulls can, in fact, be used to estimate whether the incoming signal is predominately low or high angle. If high-angle components are present, the loop element nulls become unstable and disappear. In contrast, the extent to which the array nulls stay deep in the presence of such high-angle components is quite impressive.

A single CLA (or a closely spaced CLAR) can be used effectively to combat co-channel interference when the azimuth of arrival of the two signals differs by more than 45 degrees or so.

Directionality of the CLA

The far-field directional pattern null of any device based on loops is degraded by the presence of E-field pickup, which tends to be nondirectional, whereas the H-field or loop pickup is directional. In commercial direction finders, this problem is overcome by shielding. With the CLA, the need for shielding is not as great as it might be because the impedance of the tuned circuit is kept low by using loops with small L and large C, thereby enabling operation over the frequency range 5-17 MHz without bandswitching. Unshielded, the CLA has adequate directionality for interference reduction and other applications. It is low cost and is suitable for home construction. However, for direction finding of the highest accuracy, which would have to be done outdoors in any event, shielding is desirable.

Figure 3 is the complete circuit diagram of a small, demonstration-model shortwave CLA; a photo of this configuration is shown in Figure 4. The radio's whip is retracted and kept as close to the set as possible. Radio and whip function as a voltmeter. In radios that do not have accessible ground terminals, such as the Sony ICFSW1 in the photo, connection can normally be made to the outer ring of the power supply or headphone jacks by means of a small metal sleeve. Use whichever terminal or ring most effectively short-circuits the set's input when the built-in whip, partially extended, is temporarily connected to it.

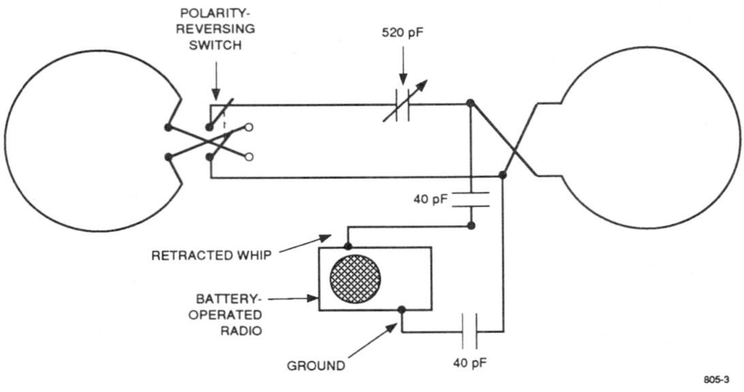

Fig. 3—Complete Circuit Diagram of Typical Shortwave CLAR

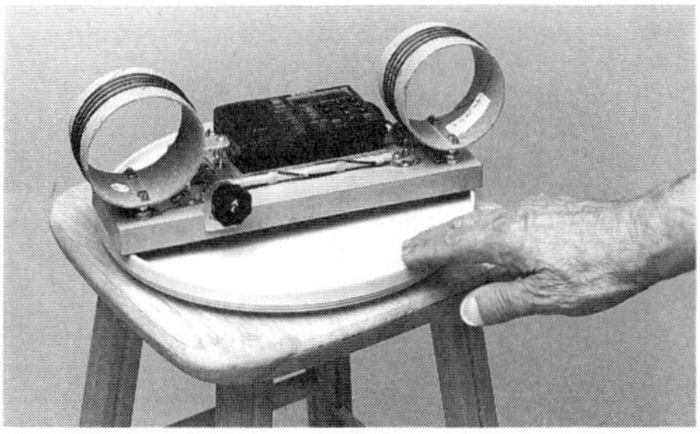

Fig. 4—Demonstration Shortwave CLAR Incorporating Sony ICF-SW1 Receiver. Coil diameter is approximately 10 cm. Sensitivity is roughly equal to that of the set's built-in whip. Tuning range is 5-21 MHz. Note plastic turntable mounted on wooden stool.

Mediumwave CLA

The CLA/CLAR principle is also usable at mediumwaves, but modification is necessary because virtually all radios with this band have inaccessible built-in ferrite loops that cannot easily be disabled by retraction as can a whip. But these same loops serve as a convenient means for transferring energy inductively from a given CLA to a given radio. It would be convenient if radios' ferrite loops would respond only to fields generated by current flow in the associated CLA; however, they also respond directly to the incident fields and these can spoil the CLA null. Fortunately, it is feasible both to couple to the radio and to control distant-field sensitivity by addition of the coil L3 in Figure 5. With the assembly pointing in the null direction, the radio by itself should be oriented for minimum reception of the unwanted signal. Next, with all three coils resonated and in service, a small readjustment of the array direction will balance out stray pickup in the receiver and will ensure a deep, stable null. The size and location of L3 is not critical. A photo of an MW CLA, complete with L3 and ferrite loops for L1 and L2, is shown in Figure 6.

Decorrelated Signal Sources for Fading Reduction

Any antenna receives a combination of signal components (or modes) that at any instant add up to a certain resultant signal at the radio's input. Since the magnitude of this resultant depends on the instantaneous magnitude and phase of its components, it can be appreciated that a change in any of the components, such as might be caused by ionospheric motions, produces a change in the resultant. The resulting variations in received signal level with time are called fading. A useful discussion of this phenomenon will be found in Hagn(3) and the references therein. See also Ames.(4)

The classic way to reduce these amplitude fluctuations has been to provide a second receiver connected to a different antenna that receives a different mode mix. The second receiver's output is then said to be decorrelated from that of the first. The idea, then, is that when the signal is weak in one antenna it may not be weak in the other, thus filling in any break in the program material that would otherwise occur because of loss of signal in the first antenna.

When SW full-size outdoor antennas are used as sources of decorrelated signal energy, they should be spaced 100 m or more to have truly independent fading.(4) Accordingly, it is a surprise to find that when two comparatively small CLAs and their receivers are tuned to the same station and operated indoors side by side with spacing 1 ft or so (Figure 7), their outputs turn out to be decorrelated to a considerable extent.

Independent fading of this sort is much more noticeable when individual indoor antennas are used that, like the CLA, have far-field patterns discriminating against the on-great-circle sky-wave components of a signal. Such antennas understandably receive a larger fraction of energy arriving from off-great-circle directions in azimuth. A typical source is side scattering from reflectors on the ground, which is stronger than most people imagine. Another is backscattering of similar origin. In either case, the received energy travels to one side or the other of the great-circle path. By having two antennas with adjustable-direction nulls, it is possible in one receiver to favor energy arriving from one side, while in the second

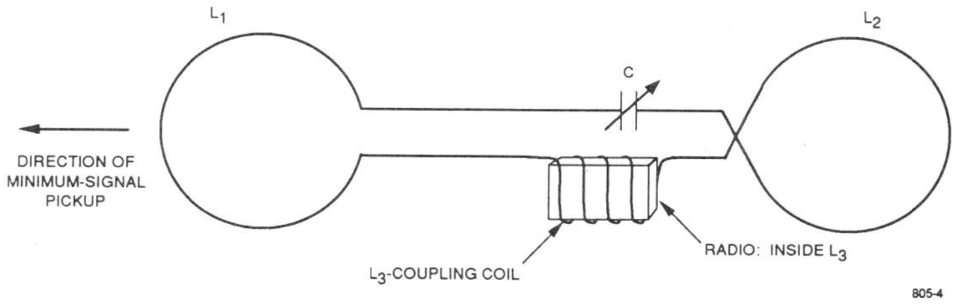

Fig. 5—MW CLAR Circuit Diagram

receiver favoring energy arriving from the other side. The result is a lesser degree of correlation between the receiver outputs.

Complexity of RF Fields Indoors

Another reason for independence of fading in the two closely spaced CLARs is the complexity of RF fields inside a typical house having the usual scatterers or reradiators such as plumbing, electric wiring, and telephone wiring (Figure 8). The near field from these sources is superimposed on—that is, added to—the incident radiated field, and a compact isolated antenna and receiver responds to the mixture at its particular location. Some of the near-field energy making up this mixture is the result of interactions between wires and the incident wave that may have taken place a considerable distance away. The net result is that compact indoor field probes, such as self-contained battery-operated radios, need little change in position to exchange one mixture of mode components for another. Large changes in resultant signal levels are observed for small changes in antenna location.

Buyers of FM auto radios are now offered the option of diversity reception to reduce motion-associated flutter fading in fringe areas.(5) This approach is relatively easy at VHF, where the wider bandwidth enables instantaneous switching of a single receiver input between either of two antennas with independently fading signals on the basis of relative signal strength. Such an economical procedure is not, alas, practical at HF, where bandwidths are too narrow for good suppression of switching transients. Therefore, two HF receivers are required. Since for practical reasons receiver interconnections and modifications are to be avoided, it is reasonable to combine the outputs acoustically, stereo-style. At the same time, it is convenient simultaneously to combine at RF by taking advantage, when possible, of mutual coupling resulting from close CLA spacing.

To handle normal signal-level changes at HF, automatic gain control (AGC) is essentially indispensable. However, whenever the signal received on any kind of an antenna fades deeply—as measured by carrier level—distortion is usually introduced. By increasing amplification at such times, thus making the distorted signal louder, AGC is often counter-productive. Minimizing the audio output of the weaker receiver of a pair is one goal of diversity combining. The present scheme helps in that regard. It uses a full-time combination of acoustic and electrical combining and thereby differs from the traditional diversity approach wherein the strongest channel at any given time is selected in accordance with certain criteria.

Acoustic Diversity Combining

For acoustic combining, the speakers should be separated as in stereo practice. With equal electrical signal levels and identical loudspeakers, each ear should receive the output of the corresponding speaker with approximately equal strength. The listener's head should therefore be approximately halfway between the two loudspeakers whose audio output is very similar. The apparent direction of arrival of a given sound is then determined by its relative strength in the two loudspeakers. The ear can distinguish between two sounds on the basis of perceived direction. This is a facet of the wellknown "cocktail party" effect wherein blocking one ear considerably reduces a person's ability to understand a given speaker in the presence of background noise.(6) In addition to providing directional discrimination, the ear almost automatically concentrates on that which is understandable, rejecting noise or distortion components, such as those accentuated by AGC action.

Apparent Source Motion

It is interesting that a signal continuity improvement with CLAR diversity is accompanied by an impression of motion of the apparent source of the audio signal when the listener is located as previously described. Such motion is a result of uncorrelated fading: When the signal is weaker in one antenna and receiver, it may be stronger in the other. The motion is readily noticeable with CLARs even though they are compact, small compared with the wavelength, and operated indoors.

When the individual-CLA output difference is large, the apparent signal source is perceived to be in the direction of the antenna with the strongest signal. As a result, with typical fading there is motion of the apparent source roughly between the limits set by the CLA positions. At first it may seem unusual that the program modulation appears to originate from a varying area or region instead of from a single point, but the effect is pleasing to most listeners. Even though the signal source is apparently moving, or perhaps because it is apparently moving, the program material is heard as steady and comparatively clear with fewer interruptions, in comparison with the same material as received on one CLAR by itself. It is a curiosity that instability in apparent source direction is associated with improved amplitude stability and better continuity of the received program material.

Electrical Combining

In electrical combining, the CLAs must be close enough together so that inductive coupling can occur. Assume that both CLAs are tuned to the same frequency, and receive equal amounts of signal (i.e., are both heading in the same direction). When spacing between such paral-

Fig. 6—CLAR Designed for Use at MW. L1 and L2 are ferrite-core inductances to save space.

lel CLAs is decreased, thereby increasing coupling, the energy picked up in either one reduces the output of the other. In each case, energy coupled from the other antenna opposes the signal that would otherwise be received, and thus reduces its strength. This effect can readily be demonstrated by manually varying the spacing between two parallel adjacent CLA and receiver assemblies, keeping the CLAs in tune. For normal operation, the aim is to choose a spacing that reduces each CLA's sensitivity by 10 dB or so.

Then, if the output of one antenna decreases, as a consequence of independent fading, the output of the other will tend to increase, assuming there is no change in the signal it receives. Similarly, if one receiver's received signal becomes strong, the other receiver's output is depressed. This action tends to smooth out amplitude fluctuations in the total audio output of the two receivers, and it also tends to suppress distortion associated with the previously described intervals of individual-antenna signal weakness.

Relative Effectiveness of the Two Methods of Combining

It is interesting that two independent methods of improved diversity combining exist. It can also be seen that there is an optimum spacing between the antennas for best fading reduction. If spacing is too close, the acoustical component suffers because the two apparent sound sources nearly coincide, in which case independent fading no longer gives rise to a significant apparent direction change. In addition, the independence of the fading suffers because of the closeness of the antenna spacing. (Decorrelation is not great enough.) On the other hand, if the spacing is too large, the electrical coupling becomes too small to provide a benefit although the independence of the fading is good. For best results, spacing should be such as to maximize all these factors; the configuration shown in Figure 7 appears to be about right. Fortunately, spacing is not particularly critical.

Tests in which CLAR spacing was varied suggest that at the optimum spacing, the electrical and acoustic methods contribute about equally to the effectiveness of the combining action.

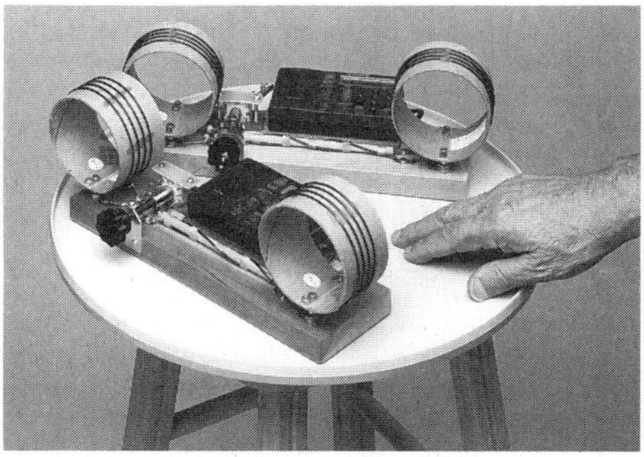

Fig. 7—Two Shortwave CLARs Operated Nearly Side by Side with Nulls Directed to Either Side of the Great Circle Path. For best acoustic diversity reception, there should be an equal distance from the listener's ears to the corresponding loudspeaker.

Benefits of Diversity Reception

The twin-CLAR arrangement does nothing to improve the long, slow strength variations due to ionospheric focusing or polarization rotation. Nor can it help the largescale mode interference effects that cause signal disappearance over a large area.(3) Fortunately such fades are present for a comparatively small fraction of the time.

The twin-CLAR arrangement does help counter the relatively small, frequency-selective interference effects that are present during an appreciable fraction of the time, especially at night. They are more pronounced when paths are long and when multimode propagation is present, and they impart a characteristic wavery sound to the output from a conventional SW antenna and receiver. When the two-CLAR arrangement is used, such signal instability is reduced, and the impression is created in the listener that the audio is solid and continuous most of the time, resembling to some extent a wire circuit. Speech is thus more understandable because momentary intelligibility lapses occur much less frequently. Similarly, music is more natural because sustained notes are reproduced with less of an ionospherically induced waver.

CLAR Circuit Details

The circuit diagram of a practical CLAR (Figure 3) is straightforward. Note the receiver decoupling capacitors, desirable for isolating the receiver, and the polarity reversing switch near one of the loops. This switch is useful for verifying proper operation when the CLAR is oriented in azimuth with the array null toward the distant station (or toward the strongest signal field component indoors). Also, while the phaseopposition loop connection is essential for proper directional performance, it does result in some loss of gain. Therefore, when sensitivity is more important than directivity, it is useful to be able to remove the phase reversal by the switch provided. With the loops in phase, their combined polar pattern is essentially that of one standard loop of twice the size. (The loops must be retuned after switching, of course.)

Operating Considerations

To a first approximation, CLA directionality is independent of tuning, which in practice is always adjusted for maximum response. In the diversity arrangement with the desired coupling and one CLA tuned to resonance, tuning the other to resonance will reduce the output of the first one. This is normal and results from the canceling action previously described. Both CLAs should be simultaneously tuned for maximum output at the common resonant frequency. In normal operation, the two tuned circuits are always undercoupled or critically coupled. If coupling is made excessively tight by too close spacing, confusing interaction of the type well known in the case of overcoupled tuned circuits will occur.

Tuning Procedure for Diversity Reception

1. Tune in the desired signal with one receiver and its CLA.
2. Resonate the CLA and determine the direction to the station by observing the direction of the deepest null. When operating outdoors, the direction can be compared with the expected direction—if this is known.
3. Tune the second receiver and its CLA to the same radio frequency.
4. With both CLAs pointed in the maximum-signal direction (90° from the direction to the station), readjust both sequentially for resonance and set the audio gains to be the same.
5. For best results, the two CLAs should then be pointed slightly to either side of the direction to the station. This configuration ensures that different mixes of signal components are received, thus enhancing the independence of fading. However, the exact direction is not critical. If the listener is oriented as described above with respect to the radio loudspeakers, the diversity action should be readily apparent.

Miscellaneous

- There is admittedly a loss of gain in the diversity arrangement because of the reduction in the on-great-circle components and the additional sensitivity loss due to partial cancellation resulting from the mutual coupling. Fortunately, most broadcasts are far enough above the noise level that these losses can be tolerated. The arrangement of Figure 4 is quite comparable in sensitivity to that obtained using the receiver's whip alone. When a CLA is used with the loops in phase, sensitivity is far superior to that of the whip alone.
- A listener with stereo equipment at his disposal can use it to good effect in this diversity reception technique by placing small microphones close to, but not actually in contact with, the CLA receivers. Each microphone is connected to one stereo input. So long as the interconnecting wires do not actually touch the CLA receivers, there should be little distortion of the ambient fields, and directional performance will not be appreciably affected. And if the microphones are sufficiently close to the corresponding radio speakers, acoustic feedback should not be a problem. By this means, the diversity fading improvement can be enjoyed in a room of some size even when the receivers are small.

REFERENCES

1. O. G. Villard, Jr., "Combatting Interference in Shortwave Reception with Compact Indoor Directive Antennas," World Radio TV Handbook, 1990, pp 547553, Amsterdam, The Netherlands.
2. F. N. Eddy, "Multi-Loop Antenna and Radiometric Processor Techniques for the Rapid Detection, Localization and/or Avoidance of A.C. Power Lines and Other Sources of Man-Made RF Noise," R.F. Design, in press.
3. G. H. Hagn, "Stand on Standards," IEEE Antennas and Propagation Magazine, Vol. 34, No. 4, August 1992, pp. 56-64.
4. John Ames, "Spatial Properties of the Amplitude Fading of Continuous HF Radio Waves," Radio Science, Vol. 68D, No. 12, December 1964, pp 1309 ff.
5. W. R. Rambo, "Switched Diversity for Mobile FM Broadcast Reception," Proceedings of the 38th Annual Broadcast Engineering Conference, published by the National Association of Broadcasters, 1771 "N" St. N.W., Washington, DC 20036, p 508-514, April 1984.
6. See, for example, Spatial Hearing—The Psychophysics of Human Sound Location, by Jens Blavert, the MIT Press (Cambridge, MA, and London, England) 1983.

The authors have been concerned with the design of improved HF indoor receiving antennas for the past ten years. They will be happy to respond to inquiries if accompanied by an International Reply Coupon. Mr. Hagn may also be addressed at SRI Washington, 1611 N. Kent St., Arlington, VA 22209-2111 USA

As this edition of the WRTH goes to press, manufacture of a commercial version of the CLA is under consideration at Palomar Engineers, Box 462222, Escondido, CA, 92046. USA

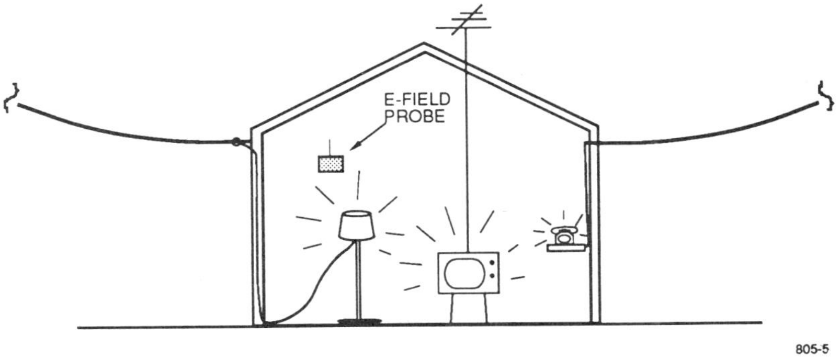

Fig. 8—Reradiators Inside a Typical House. Energy guided in by wires, etc. (electric fields off conductor ends)

The Satellite File

by Rob Ulrich & Bart Kuperus

There was positive reaction to this section in last year's Handbook. With the growth of satellite subcarriers in Europe, there is more and more interest being shown in the radio programmes that are available, as well as the variety of TV channels. This section doesn't pretend to be comprehensive, but it represents thorough practical tests completed by the WRTH team in the course of compiling the WRTH Satellite section. Feedback to Bart Kuperus, Associate Editor, WRTH, P.O. Box 9027, 1006 AA Amsterdam, The Netherlands.

ECHOSTAR SR-50: BACK TO BASICS

Whereas most modern satellite receivers come with a large memory capacity (up to 200 channels) as well as on-screen graphics, Echosphere International surprised many by introducing the SR-50. It has no display, no remote control, no on-screen graphics, no frequency readout, no programmable preselections, no built-in positioner . . . . just some knobs at the front and a few trimmers at the back.

Down-to-Earth technology

In fact the SR-50 looks quite archaic, like a one of the early domestic satellite systems, a variety of knobs and switches on the front panel allowing the user to adjust just about everything. There are controls for variable tuning, variable IF bandwidth, variable left and right audio tuning, variable audio bandwidth, variable LNB switching voltage, variable skew plus a brightly lit VU signal strength meter. In addition, trimmers at the back of the receiver allow you to adjust the video level and the Automatic Gain Control (AGC).

Inside the SR-50 is a well laid-out circuit board with easy access to all components plus lots of empty space. The power supply contains conventional (but reliable) voltage regulators. The IF signal produced by the front end is very conventionally changed to a standard 70 MHz. Nothing earth shattering so far. At the back of the receiver is a 70 MHz IF loop, so it is possible to link up filters to eliminate interference from terrestrial sources. The IF bandwidth is adjustable between 12-26 MHz. A switch at the back of the receiver allows you to invert the video polarity, making the set very suitable for C-Band reception. Furthermore, the SR-50 is fitted with standard phono sockets for baseband, composite video and audio. As far as the sound is concerned, two Phase Locked Loop (PLL) circuits take care of the audio demodulation. The audio bandwidth is variable between 150-350 kHz. However, the audio output level itself is not variable.

In Practice

For testing purposes we connected the SR-50 to a 1.4 metre prime focus dish and an LNB with a maximum noise level of 0.8 dB. We immediately discovered that tuning in to the various frequencies (especially audio) is an acquired skill. Tuning is hampered by the lack of bandspread so that it takes a steady hand to tune accurately. This could have been avoided by installing a multi-turn potentiometer instead of a single-turn version. When a dual polarity LNB is used, the switching voltage (13-18 V) can be adjusted on the front panel. If a regular LNB/servo polarizer combination is used, the polarity is altered by turning the skew knob. It is not possible to use a ferrite/magnetic polarizer with this receiver . . . . this is a serious shortcoming.

Overload Problems

The overall picture quality of the SR-50 is fair. However, when receiving very strong signals (not uncommon with the dish size we used) the receiver clearly overloaded. The receiver does not have a real Automatic Gain Control, but the gain is adjustable with a trimmer at the back of the receiver. Still the channels of the high-power European ASTRA satellite were visible with lots of sparklies caused by overload. Remarkably, most of those sparklies disappeared when we slightly reduced the IF bandwidth. However, the very narrow ASTRA stereo audio subcarriers were difficult to receive. The problems were not solved by connecting a smaller dish. The overloading problem, combined with the coarse tuning makes this receiver unsuitable for ASTRA reception in Europe. When we aimed the dish at other satellites e.g. Eutelsat and Intelsat, the picture and audio quality were noticeably better. However, the sensitivity of the SR-50 when receiving very weak signals was also somewhat disappointing. Another drawback is that the receiver drifts, especially when just switched on. After a few minutes the receiver stabilizes, but this flaw could have been avoided by using Automatic Frequency Control (AFC) circuits. Our measurements also showed that the IF bandwidth control influenced the fine tuning.

Conclusions

The Echostar SR-50 is a fair receiver for the price paid. It is about the cheapest receiver you can get and with its price in Europe of around $200.- it offer good value. The receiver offers plenty of controls to modify the receiver's settings but it will serve best when used as a second receiver to scan the skies. The variable IF bandwidth will especially be appreciated by DX-ers and enthusiasts. Unless modified the Echostar SR-50 is certainly **not** suitable for reception of the ASTRA medium power satellites in Europe!!!

For more information contact:

- **USA**) Echosphere International, P.O. Box 6552, Englewood, Colorado 80155. **Te:** +1 (303) 799 8222 **Fax:** +1 (303) 799-6222
- **Europe**) Echosphere International, Schuilenburglaan 5a, 7604 BJ Almelo, The Netherlands. **Te:** +31 (5490) 15122 **Fax:** +31 (5490) 14691
- **Asia**) Echosphere International, 16A Science Park Drive, Singapore Science Park, 04-03 The Pascal, Singapore. **Te:** +65 779 4160 **Fax:** +65 779 6051

DRAKE ESR 600e: A WOLF IN SHEEP'S CLOTHING

Drake have been in the satellite TV business since the early days Founded by Robert L. Drake in 1943 near the city of Dayton in Ohio (USA), the company has also built a good reputation in the shortwave receiver field too (the famous R7 and R8 models). Over the last ten years they have taken an ever increasing role in the production and development of SMATV and home satellite systems. With the introduction of the ESR 600e, Drake has come up with a remarkably versatile 150 channel multi-standard satellite receiver.

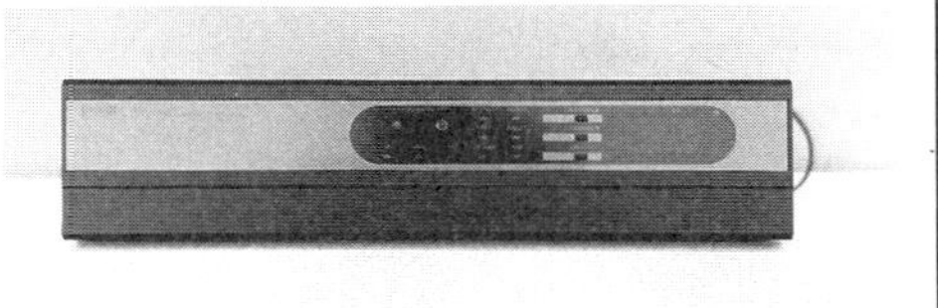

State-of-the-Art Technology

The sober housing of the ESR 600e hides the beautiful insides of this receiver. The clear arrangement of all the boards and components makes it very easy to trace the circuitry. The rather chunky power supply unit occupies the right front section of the receiver. Five voltage regulators are situated at the left hand side of the motherboard and are mounted directly on the chassis. This way the whole chassis serves as a heat sink.

Directly behind the power supply is the Intermediate Frequency (IF) board containing three extremely stable Surface Acoustic Wave (SAW) bandpass filters. The IF range, oddly enough, is fixed at a non-standard 140 MHz, whereas 70 or 480 MHz is more commonly used. The IF loop at the back makes it possible to connect external filters in case of interference problems. The audio detection board, mounted perpendicular to the motherboard, is equipped with high quality ceramic filters.

Many Connection Options

As mentioned, the appearance of the DRAKE ESR 600e is quite plain. The front panel comprises one multi-functional LED-display, three push buttons and a dozen LED indicators. Most of these indicators only light up when one of the 150 preselections is being programmed. Only the mono/stereo LEDs are always visible. The back of the receiver is far more impressive than the front. The DRAKE ESR 600e is fitted with an extensive range of connection possibilities, which interface to almost any decoder or accessory. Even two European SCART sockets are available. Furthermore, the ESR 600e is equipped with regular phono sockets for baseband and audio output. At the right hand side of the back are connections for both magnetic (ferrite) and mechanical polarizers, as well as the possibility to connect two actuators. The built-in 2.5 amp motor control circuit not only drives any conventional motorized system but, together with the optional DRAKE IOA600 inclined orbit adapter, it is also possible to track so called "inclined orbit" satellites. Because some dual- and triple-band LNBs switch at different voltages, the receiver is fitted with a small trimmer to adjust the LNB switching voltage.

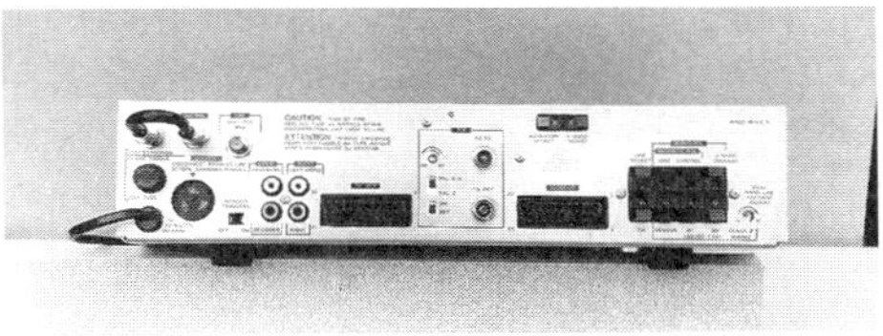

Setting up the DRAKE

The typical DRAKE no-nonsense remote control only has the necessary buttons to operate the receiver. It is possible to change pre-programmed data, or add new data by pressing push buttons on the front of the receiver This means that it is virtually impossible to inadvertently change or delete vital information. Setting up the receiver is not time-consuming. The on-screen graphics immediately indicate what is happening at any given moment. The "Antenna Set-up Menu" even has an option to indicate how many actuator pulses it takes to move the antenna 1 degree. This is vital when you want to find a newly launched satellite. The ESR 600e also offers another handy feature, so called "DATA TRANSFER". It is possible to transfer all data from one ESR 600e to another just by putting them close together. With the memory transfer switch at the back you can put one receiver in "SRC" (source) mode and the other one in "DES" (destination) mode. Within two minutes both receiver memories contain the same information.

Variable IF Bandwidth

Another useful feature of the ESR 600e is the variable bandwidth. It can be set to a fixed 16, 22 or 30 MHz. It can also be altered in 1 MHz steps up to 36 MHz. The option to reduce the bandwidth to 10 MHz comes in very handy and can make very weak signals watchable. The DRAKE ESR 600e has one important shortcoming though..it is not a very sensitive receiver. In this category of high-end satellite receivers, we would have expected the DRAKE's threshold level to be considerably lower than 7dB max. However, this drawback can be overcome by connecting an external threshold extender. The ESR 600e is also a genuine multistandard receiver with separate video filters for PAL, SECAM and NTSC. It is possible to choose between 2 video levels and to invert the video signal. The latter is a must if you want to receive C-Band.

Audio
With the "AUDIO SETTINGS" menu it is possible to tune in sub-carriers between 4.5 - 8.8 MHz. This range is wide enough to cover all satellite radio stations and TV audio frequencies. With the same menu it is possible to alter any of the other audio parameters, such as audio de-emphasis, audio bandwidth and a combination of two subcarrier frequencies to get the stereo. Although the ESR 600e does not feature the famous Wegener-Panda noise reduction system, the audio quality is excellent. Even extremely wide audio subcarriers were no problem.

Decoders
The ESR 600e is fitted with two decoder sockets: one phono and one SCART. The phono socket provides only an unclamped unfiltered baseband signal. The SCART decoder socket has a special "decoder loop-thru" built in. This way the decoded satellite signal is relayed back into the receiver and becomes available through the other SCART socket and the RF modulator. By means of internal jumpers you can change the socket's configuration to:

1) non-de-emphasized/unclamped/unfiltered
2) de-emphasized/unclamped/unfiltered
3) de-emphasized/unclamped/filtered
4) de-emphasized/clamped/filtered.

Only the English users manual explained how to change the various jumper settings. We would have preferred that the decoder parameters were accessible through the software, instead of having to open up the receiver and switching jumpers. Despite this, the ESR 600e's "decoder-friendliness" is excellent.

Conclusions
The DRAKE ESR 600e is a top-of-the-range multi-standard satellite receiver. Picture and sound quality are excellent and the user-friendliness is beyond compare. The DRAKE doesn't come cheap and retails at around US$1,400. The main drawbacks we discovered were: limited number (150) of preselections, disappointing sensitivity and a limited tuner frequency range (1750 MHz). The latter can be overcome by using a semi-triple band LNB for the reception of all three Ku-Band ranges. For more information contact:

- **USA**) R.L. DRAKE Company, 230 Industrial Drive, Franklin, Ohio 45005, USA. **Te:** +1 (513) 746 6990 **Fax:** +1 (513) 746 4683
- **Europe**) R.L. DRAKE Company, Dr. Trueta 1-3, Entresuelo, 08860 Casteldefels, Spain. **Te:** +34 (3) 636 0192 **Fax:** +34 (3) 636 0152

THE WRTH INDUSTRY AWARDS

This is the fifth year of a prestigious award scheme organized by the World Radio TV Handbook. In over 40 years, the WRTH has received many awards itself as the industry reference standard in the field of international broadcasting. We believe in recognising excellence in other parts of the international broadcasting industry. We hope these will act as a stimulus to provide the market place with even better products.

At this moment, thousands of hours of programming are being broadcast from stations round the world. It is NOT our intention to give awards to the programme makers, or individual stations. To be fair we would have to understand a multitude of languages and have an in depth knowledge of the studio facilities available to the producer. The WRTH is a reference book, and we do not feel it is right to pick out a broadcaster as being the "best". In the equipment field, however, we do think that awards are needed.

The main WRTH awards will be given in three categories this year. Here are the results:

- **Best Portable Receiver 1992: Sony ICF-SW55.** This new receiver is not the most sophisticated in the Sony range. But it combines a number of useful tuning features, versatile text based memories and good performance into a handy package. We encourage Sony to develop their receiver technology further to balance recent innovations in the field of tuning convenience.
- **Best Communications Receiver 1992: Yaesu FRG-100.** Although only just launched on the market, Yaesu has succeeded in bringing improved technology and features within the price range of a much wider group of shortwave broadcast listeners. It has been a long time since Yaesu revamped their broadcast receiver range. But for many listeners it will be worth the wait! A good package at an affordable price.
- **Most Innovative Receiver Designer 1992: Lowe Electronics** The English company of Lowe Electronics have already won an award for their HF series of receivers. But we feel they deserve another award, this time for their innovative design and back-to-basics approach to receiver technology, especially the new HF-150. Lowe's chief designer John Thorpe has surprised many observers who proclaimed the European receiver business to be a thing of the past. Lowe's dedication to creating high performance and easy to use receivers is at last getting wider recognition.

Suggestions from readers for next year's awards are always welcomed by the editors. These awards are not restricted to any country, or language. Please note that, due to time constraints, the WRTH editors cannot offer replies to individual equipment problems or trace circuit diagrams.

Associate Editor presents 1992 Award to Panasonic

WARC 1992 and Beyond!

by Richard Dixon

Perhaps the most important World Administrative Radio Conference to affect the future of international radio now seems a distant memory. Held in Spain from 3 February to 3 March 1992, the conference distributed six million pieces of paper weighing some 28 tons and over 16500 international telephone calls were made. Yet the results were not given wide publicity outside technical circles.

Good news for BSS

Towards the end of conference, the plenary approved a 40 MHz worldwide allocation for Broadcast Satellite Sound (BSS) in the range between 1.452 to 1.492 GHz. The new allocation to BSS-sound is for digital sound technology only and despite being subject to all sorts of conditions and limitations, at least on paper, this solution seems what broadcasters wanted. However, it wasn't a worldwide allocation. Ambassador Ian Baron, the head of the United States delegation, said that nobody will see BSS-sound at 1.5 GHz in the USA because of a specific Footnote 722B which allows the USA to allocate that band to the fixed and mobile services on a primary basis. In addition BSS services from bordering countries like Canada and Mexico will be subject to bilateral co-ordination before being brought into use. Another footnote allows China, the Russian Federation and others to develop BSS in the range 2535 to 2655 MHz. In Region 1 the majority of the European countries accept a 40 MHz allocation at 1.452 to 1.492 GHz on primary basis with the exception of Germany, the UK, Spain and Italy which in a footnote have obtained an interim period of 15 years before implementation. An ITU conference will have to be convened not later than 1998 to plan the new BSS band.

On a shorter term point however, just over half of this 40 MHz chunk of space is available after the 12th of October 1993 for digital systems only. That gives the green light to two US private sector groups such as Worldspace and International Radio Satellite Corporation to develop working systems. Worldspace Vice President Wegard Holby announced some details at a Voice of America conference held in November 92 in Washington DC. He said that the development of direct-to-home satellite radio will be driven by the domestic broadcasters in the countries of Africa, the Americas and the Caribbean. Few listeners will want to buy a receiver (around US$100) to pick up just a handful of international broadcasters who can afford to invest in the new technology. In 1992 Radio Netherlands signed an option agreement with Worldspace for a channel on the Afrispace project, currently intended for launch in 1995. Worldspace say they hope to announce three other clients at the start of 1993.

Another set of important decisions affecting international radio listeners were that the shortwave broadcast bands will be extended by 790 kHz in future, although the allocations will not officially come into force until the year 2007. Here are the extra allocations:

Waveband	Extra frequency coverage added at WARC 92 (kHz)
49 m	5900-5950
41m	7300-7350.
31m	9400-9500
25 m	11600-11650 + 12050-12100
22 m	13570-13600 + 13800-13870
19 m	15600-15800
16 m	17480-17550
15 m	18900-19020 (new band!)

You can be sure that many broadcasters won't wait until 2007 before they start to appear on these frequencies. In fact one broadcaster, the new Catholic station WEWN in Alabama USA, started using 18930 kHz in December 92 (13 and 18 hrs UTC beamed to Europe) from a 500 kW transmitter. Of course there have been international broadcasters operating occasionally around 18 MHz before now. Stations like the VOA had Single Sideband links to overseas relay stations on 18782.5 kHz. And for many years an important out of band channel for the BBC's Bengali service was 18080, but remember the new 15 metre broadcast band will be nearer 19 MHz..18900 to 19020 kHz. So bear that in mind when you're in the market for a new shortwave receiver.

The WARC conference also agreed that the next WARC HF meeting should consider the early introduction of Single Sideband as a way of reducing congestion on HF. Currently the target date for the end of standard DSB broadcasting is 31 December 2015. The larger international broadcasters are loathe to make a massive capital investment in upgrading their equipment, especially when SSB does not give a significant improvement in the fidelity.

In September 92 the "USA Digital Radio" group announced at the National Association of Broadcasters convention in New Orleans that digital technology may breath new life into AM. They can squeeze a 15 kHz stereo signal onto mediumwave with amazing results. The group say that they are also considering a solution for shortwave. The secret lies in advanced methods of error correction.

In the meantime some broadcasters are looking at ways of making tuning easier. However, stations are still a long way from issuing their schedules on a credit card. In Europe, FM stations are using the Radio Data System (RDS) which allows specially equipped radios to display the name of the station on the front of the radio. All this without affecting reception on ordinary radios. This useful tuning aid is being adapted for medium and shortwave by the public broadcasters in Germany, including Deutsche Welle.

The whole key to the success of new tuning and transmission systems will depend on whether they can achieve a worldwide standard. Mediumwave stereo failed in the US because the Federal Communications Commission opted to let the marketplace decide. The confusion of standards is hampering the development of wide-screen television in Europe. Few seem to want to invest in an analogue system like MAC if a digital TV system is just around the corner as demonstrated in both US and European trials in mid 1992.

Incredible as it may seem, international radio broadcasters have done little or nothing to stimulate good receiver design. Sangean is one of the world's largest portable radio manufacturers and sells around half a million shortwave portables annually through its European distributor Siemens . Yet Sangean note that they only have regular contact with two radio stations who supply technical feedback on their products. Several manufacturers have developed rather strange receivers which are strong on features but have poor radio performance. That's bad news because people who just discover short-wave usually invest in a budget set first.

It appears that more and more broadcasters are using digital technology to monitor their signals abroad. Philip Collins and Associates Ltd in Sydney Australia has developed a remote control system which allows two receivers to be remotely controlled via any touch-tone phone. The audio can then be fed back down the phone allowing broadcasters to check their reception in the target area. Currently a system costs around US$5000 for the control unit, cable and software. The system announces what it is doing thanks to a built-in voice chip. At the Voice of America an impressive remote monitoring system based on the Macintosh computer has been developed by Bill Whitacre, head of the Monitoring & Frequency Division. This system samples the audio, compresses it digitally, and then feeds it to a analysis centre anywhere in the world.

There are changes in the transmitter business too. In France, the transmitter manufacturer Thomson CSF has developed a whole new concept in short-wave transmitter design . . . namely the modular approach. Imagine a large rotatable multi-configuration antenna on the top of a reinforced building. Inside the building is the transmitter. As a result there is no need for any antenna switching and a whole station of, say, 6 or 8 of these modules is more difficult to silence. Patrick Bureau, marketing manager of the RF department in Thomson CSF says they called the system Alliss after the two shortwave sites in France (Allouis and Issoudun) run by TDF. The changing political climate in Europe means there is a shift towards North-South shortwave broadcast traffic rather than West to East.

You need to have quite a bit of space to set up such a system, but in the long run Thomson says it is cheaper. Fifteen such modules are to be built in France. Thomson say they've patented the antenna design so that it can still operate during wind speeds of up to 160 km per hour. Conventional antennas would buckle and as a result there would be an electrical mismatch with the final stage of the transmitter.

The VOA conference in Washington DC ended the station's 50th anniversary celebrations on November 6th 1992 on a positive note. Providing international broadcasters concentrate on improving their radio programming product, there will always be markets for their material whether it be by shortwave or satellite. Commercial station WRNO reported in 1992 that commercial SW broadcasting has never been successful in the US because the advertisers don't understand the medium. However, it is not all gloom during the worldwide recession. The French government has interests in the popular station Africa Number One based in Gabon. They've announced a 12.8% increase in their profits, making a total of 7.7 million French Francs in 1991. Now they say they has enough cash to invest in a European expansion plan starting with an FM transmitter in Paris which opened on the 4th of September 1992 on 107.5 MHz. That transmitter relays programmes from several public broadcasters in Africa as part of the Africa Number One European package. At last! A positive step to improve the current bias of the northern hemisphere simply shouting a message at the South.

WRTH Equipment Test Bench

with Jonathan Marks and Willem Bos

Updates since 1992 edition

There are a lot of new receivers on the market this year. As usual, the WRTH test bench team has been looking at them critically. Over the years the WRTH has built up a reputation for unbiased reporting. We've been delighted that receiver manufacturers respect our comments and in many cases have taken our notes to heart.

Sony Corporation of Japan made several changes to the ICF-SW77 because of problems reported in WRTH 1992. In August 1992 we tested an off-the-shelf sample of the "revised" ICF-SW77 and found that the synthesizer noise problems had been significantly reduced. Reception on mediumwave is now much better. The SYNC option has also been improved so that the bass response does not cut when the sync is activated. Sony advise us that stock sold in 1993 should only be the new version. Readers reported that most Sony dealers were willing to exchange the first ICF-SW77 versions for revised models.

John Schlipp of the Communications Products Division of the R.L. Drake Company in Miamisburg Ohio USA reports that a number of enhancements have been made to the Drake R8 since we tested that receiver. The AM mode now offers the user an option of displaying down to 10 Hz tuning resolution. The bandwidth filters now revert to the user's last choice when changing modes. The pre-amp disable is now only below 1800 kHz. Finally, all the promised R8 accessories, including MS–DOS PC compatible control software, have been introduced.

WRTH Receiver Survey Results

The 1992 World Radio TV Handbook included a survey questionnaire which readers were asked to return by the end of July 1992. The results were then processed. A total of 4087 valid survey questionnaires were returned and independently analysed.

The survey shows that readers of the World Radio TV Handbook take the international broadcasting business very seriously. On average, they spend some 13 hours a week listening to programmes from outside their own country. The majority of readers have been doing this for between 15 and 20 years. When asked what was the maximum they would be willing to spend on a new radio receiver, the average said it would be up to US$ 1174. This explains the high interest that WRTH readers have in the Equipment Test reviews published annually in the handbook.

Although not part of the direct questioning, it was evident from comments that at least 38% of readers surveyed are in the broadcasting business, either on the engineering side or in management. These people often control equipment budgets amounting to several million dollars. To them the WRTH is a daily reference source.

Results lead to launch of Equipment Buyers Guide

The high response rate to the 92 survey encouraged the editors of the World Radio TV Handbook to examine the market for a standalone book of equipment tests. The survey showed that readers were interested in an updated compilation of previously published tests, since older WRTH's are now out of print. In March 1992 we therefore started the project to revamp and revise. However, quite a lot of tests needed re-writing in the light of new versions or because value for money ratings had changed with the arrival of competing receivers. The result, we hope, is a useful comprehensive market survey, plus a detailed analysis of how the tests are done. The book also includes a revised comparative test of active and passive antennas.

In December 1992 the WRTH therefore launched the WRTH Equipment Buyers Guide to complement to the annual reports in the World Radio TV Handbook. This edition of the WRTH contains some reviews (e.g. Yaesu FRG-100, Sony ICF-SW33) which are not in the WRTH Equipment Buyers Guide because of an earlier deadline. The Buyers Guide includes a new prize survey to judge what material is most desired for any possible future editions. The **WRTH Equipment Buyers Guide** is priced in the USA at US$19.95 and is being distributed worldwide through the same channels as the WRTH.

Digital/Analogue Portable Receivers

DAK DMR3000

In May 1992 DAK Industries launched a more up-market digital short-wave portable which offers better results than the older DAK MR101s. The radio measures 21 x 13 x 5 cm and weighs in at 814 grams including the 4 penlight (AA) batteries. When switched off the liquid crystal display is blank. You have to switch the radio on and select the clock to be able to see the time. The time in a second time zone can also be displayed. Pushing the CLOCK/RADIO button activates the radio and the time disappears. The frequency is shown in MHz. On shortwave the readout is in line with the channel spacing, i.e. 5 kHz. If you are tuned to 11835 and move up by pushing the UP button, the radio jumps to 11840 kHz. No fine tuning within that 5 kHz is possible, and that is a serious disadvantage. FM stereo is activated by plugging in a pair of headphones, otherwise reception is in MONO on the built-in speaker. A switch in the battery compartment allows two types of tuning on AM (530-1710 kHz in 10 kHz steps, or 522 - 1620 kHz when set to 9 kHz).

The radio offers four tuning "ranges". Mediumwave AM, as described above, FM (87.5-108 MHz), and two shortwave ranges, 3580-7735 and 9500-21975 kHz. This means that some important out-of-band frequencies are missing. Note too that coverage of the shortwave bands is NOT continuous. For instance, if you tune up from the 31 metre band you notice the radio jumps from 10135 kHz to 11580 kHz, and you cannot get channels in between. That said, coverage of the broadcast bands is quite generous, only some very out-of-band channels from Beijing or WEWN are not included, plus the BBC's popular 9410 kHz. The new 15 metre broadcast band is also not tuneable. Each of the bands (AM, FM, SW1 and SW2) offers 9 preset memories under the keypad numbers 1-9. You just press the key to recall the memory. Direct entry tuning is more complex. You have to select a direct entry key, tap in the frequency, and then touch a tiny button marked "Enter". People with a physical or visual disability may find that a bit inconvenient. There is also a SEEK function where the radio scans up the dial, stopping when it reaches the next signal on the air. The dial can be illuminated with a light, and a single step attenuator switches in around 25 dB of attenuation for strong local signals. In practice that is useful for mediumwave listening, but its use on shortwave makes the set totally deaf.

The built-in speaker gives fair fidelity, though it is not as rich as more expensive receivers of this size. We found the volume control on the side of the set was very easy to knock and started crackling after a few months' heavy use. The white numbering also came off the keyboard.

In short, at a price of just under US$70 this receiver doesn't offer all that much more than the older DAK MR101s which during December 92 appeared to be on a special offer from DAK for US$29.95. The performance of the DMR 3000 is slightly better, although image rejection is still only fair. For a few dollars more you could get a good analogue portable with far more tuning versatility. The ability for "fine tuning" on the shortwave broadcast bands is important. At presstime (December 1992) DAK Industries was still on the list of companies who have filed for "Chapter 11" bankruptcy protection. However, the company continues to trade normally. More information from DAK Industries Inc, 8200 Remmett Ave, Canoga Park, CA 91304 USA. Tel: (toll-free) 1-800-888-7808, or 1 818 888 8220

Grundig: Not just made in Portugal!

The Grundig line have always been popular in Europe, especially in Germany. Until recently most of the receivers were manufactured at their own factory in Portugal. But rising production costs have forced Grundig to source some of the cheaper radios in the Far East, although most are built to a specification determined by the design team in Fuerth, Southern Germany. Grundig's North American distributor does seem to be putting the Grundig name on radios they import directly from the Far East. They don't seem to be available in Europe and, frankly, it is good that they aren't. The weirdest radio for 1992 cost US$49. We picked it up in the US under the name Grundig Explorer II. It's a combined FM, MW, and SW radio, cassette player, siren and flashlight!

The cassette player is only in mono, doesn't record, nor does it have a button to rewind the tape. The radio offers no bandspread at all, so that the 49 metre band is squeezed into 6mm on the dial! This set might be useful as a last resort on a desert island, but the only redeeming factor is the good quality flashlight. The siren is also loud enough to draw attention in case of an attack. But shortwave performance is dismal.

Grundig Yacht Boy 222

Last year we gave the WRTH award for best analogue receiver to the Grundig Yacht Boy 204/206, the difference between the two sets being a clock-timer option on the more expensive Yacht Boy 206. According to the catalogue next step up from the Yacht Boy 206 is the recently introduced Yacht Boy 222.

Made in Indonesia, the Yacht Boy 222 is a purely analogue receiver which costs around US$35 more than the Yacht Boy 206. In fact for more money you get a little less, with the exception of stereo with FM reception via the headphones. In contrast to the 206, the Yacht Boy 222 offers no coverage of the 120, 90, or 60 metre tropical bands. Control of the volume and band selection is done electrically rather than mechanically as on previous sets. It is true that dust is the main enemy of complicated switches, but we were unable to conclude whether the electronic switching will mean that the Yacht Boy 222 will last longer.

Limited Coverage

The radio doesn't cover all the frequencies currently used by broadcasters, and will miss a number of important channels once the WARC expanded bands take effect. Radio Australia on 5880 kHz is not receivable since 49 metre band coverage starts around 5910 kHz on the sample we purchased. The LED tuning indicator is not much use on shortwave. It stays on permanently when tuning the congested 49 and 41 metre bands. Your ear is a better judge than relying on a little red light for tuning guidance.

Calibration of the analogue dial is only approximate. Radio Sweden on 9670 kHz came in clearly on 9.70 according to the dial. But we also found a mixing product at the end of the dial on 9.45 kHz which was just as strong as the fundamental. BBC on 9410 and 15070 kHz were both impossible to receive because of the lack of out-of-band coverage. WWCR on 15685 kHz was also just outside the coverage range.

The set weighs 398 grams, including the three AA cells needed to power the radio. The radio is more compact than the Yacht Boy 204, but as a consequence the speaker is much smaller and doesn't have anything like the same fidelity on FM. There is no clock, although by pushing a small recessed red button the radio will switch on for a short period as a sleep function.

Power consumption is low, averaging around 35 mA at a comfortable listening volume, even on FM. The limitation of the single-conversion design is evident during night-time listening and during the day on the 21 MHz band. There was a lot of interference from nearby maritime bands as a result of poor image rejection in the receiver.

At a price of around US$85 in Europe this set is a disappointment, especially bearing in mind the performance of the Yacht Boy 206. More information from Grundig, Kurgartenstrasse 37, Fuerth/ Bayern D8510, Germany. Tel: +49 911 7030. In the USA, Grundig has a distributor. This is Lextronix Inc, 3520 Haven Ave. Unit L , Redwood City CA 94063, USA. Tel: 415 361 1611 or in the USA only call (800) 877 2228. In Canada call (800) 637 1648.

Lowe SRX-50

This receiver was launched by Lowe of the UK in July of 1992. The receiver is made for Lowe in the Peoples' Republic of China. That's why Lowe have given it an "SRX" type number to distinguish it from the "HF" series which they make themselves. Coverage is from 153-281 kHz (longwave), 531-1602 kHz (mediumwave) and

5900-15500 kHz (shortwave). FM is offered between 87.5 and 108 MHz, stereo when using headphones. The set measures 18 x 11.5 x 3.5 cm, so is pocket book size.

When switched off, the display shows the time (24 hr format). Turning the radio on, a red LED lights up in the middle of a circle of 5 memory buttons. There is no keypad for direct frequency entry or a rotatable tuning knob. In theory you simply have to use the tear-shaped up and down keys. In practice if you store three channels in the memory at the bottom, middle and top end of the tuning range you can jump quite quickly around the dial.

Audio quality for a portable turned out to be surprisingly good, and the single bandwidth filter is remarkably sharp. A light is included for reading the LCD display in the dark. Tuning on mediumwave is in 9 kHz steps and there appears to be no way to change this to 10 kHz when travelling in North America. Tuning on shortwave is in 5 kHz steps which is adequate for a radio of this type . . . there are only a few shortwave stations using split channels.

The biggest drawback with the receiver is the limited shortwave coverage. A glance at the shortwave broadcast bands table elsewhere in the WRTH shows that even the post-1979 WARC broadcast bands cover a lot more than the range 5900-15500 kHz, not to mention the extensions approved in March 1992. Whilst the bit missing from the 49 metre band is not too serious, the fact that all channels above 15500 kHz are missing is a serious drawback. A few more kHz and this receiver would qualify as the best portable on the market for £40. The synthesizer noise is certainly lower than both DAK receivers. Maybe the original equipment manufacturer can be persuaded to come up with the SRX-55. There is still a market for a low-cost digital portable that doesn't overload as soon as the sun sets. Lowe's address is given under the HF-150 review.

Philips AE 3625

This latest offering from Philips is manufactured in Hong Kong and appears to be their own design (i.e. not a Sangean product carrying a Philips label). This radio was launched in Germany in November 1992. It will be released in other parts of Europe in spring 1993. It has already been released as the Magnavox AE 3625 in North America.

The digital frequency readout is quite large on this receiver, although the display is made up of dots rather than lines. It is easy to read during daylight, but there is no back-light to illuminate the dial at night. Coverage is FM, MW (522-1620 kHz) and 13 SW Bands. These include the tropical bands of 120, 90, 75, 60, metres plus the regions 5950-6200, 7100-7300, 8500-9900, 11650-12050, 13600-13800, 15150-15600, 17550-17900, 21450-21850, 25600-26100 kHz. Although the WARC 79 band extensions are included, no allowance has been made for any of the WARC 92 decisions. So BBC channels of 9410 and 15070 are not receivable, and Radio Vlaanderen Internationaal on 5900 kHz is also just out of range.

Tuning

The set has no keyboard entry possibilities, so moving up and down the dial is achieved with UP/DOWN buttons. You can jump between sections of the shortwave broadcast bands by pressing a separate key. If you want to switch quickly between 13 and 49 metres then the use of one of the 5 memories is possible. Although 20 presets are offered, 5 are for FM, 5 for MW, 5 for the tropical bands, and 5 for the rest of the shortwave dial. The memories are easy to program. Tuning on shortwave is in 1 kHz steps which is perfectly adequate for this type of radio. There is a possibility to switch the MW steps between 9 and 10 kHz by pushing a switch inside the battery compartment.

The receiver uses 40 mA at a comfortable listening volume. The batteries give about 10 hours of useful life. Replacing them is somewhat fiddly since the battery compartment is very tight. Some of the AA rechargeable batteries which are slightly thicker than standard AA alkaline cells will not fit in the radio.

Noisy Synthesizer

The receiver is unable to pick up weak broadcast signals on short-wave because of the very obvious synthesizer noise. Even on strong mediumwave stations the whine of the synthesizer is detectable in the background. This really is something that Philips has to put right. The receiver has only fair image rejection too . . . a channel like 11705 kHz picks up a lot of Morse code utility interference from 900 kHz further up the dial. The limitations of the single conversion design are clearly audible.

The radio is equipped with a dual time-zone display, but this can only be set in steps of a whole hour. So in regions like India where the difference is 5.5 hours you clearly have a problem. Sleep timer and alarm facilities are included.

At a price of around DM250 in Germany and 299 Dutch Guilders in Holland this radio is rather expensive for the performance offered. In the US the price is a more reasonable US$100, but there are still better performers for around the same price (e.g. Sangean ATS800). Note that in Europe the Sony ICF-SW33 is roughly the same price and offers wider coverage as well as FM stereo on headphones. At 527 grams the Philips AE 3625 is also heavier than its rivals. There is some compensation in that the speaker is larger and gives good fidelity on FM.

SONY ICF-SW15: What a shame!

This analogue radio was launched in many markets in the second half of 1992. There is a remarkable similarity both in performance and looks to the Sony ICF-SW20. Although both sets are available at the moment, it looks like the ICF-SW15 will replace the ICF-SW20 in the long term. The price in Germany is DM150, whilst it is on sale in many direct-mail shops in the US for US$99.95.

We tested a pre-production sample in August 1992 and then purchased a sample in November from a shop in Utrecht. The results were so close that we conclude that no last minute changes were made before mass production started.

There's no longwave on this set, so if you rely on that band for listening to stations in Ireland, Britain, France or Germany, for instance, then you need to make another choice. The set has a dual conversion design on shortwave. It offers FM coverage, mediumwave (531-1602 kHz), and seven shortwave bands. On our sample we were able to tune in the following ranges: 5775-6430 kHz, 6880-7525, 9380-10030, 11615-12265, 15000-15650, 17500-18150, and 21325-21980 kHz. On the face of it, the coverage is better than the older ICF-SW20. There is far more out-of-band coverage, e.g. more frequencies outside the 19 metre band. However Sony hasn't done anything about the new 22 metre band (i.e. 13 MHz) so for the holidaymaker in Europe wanting to listen to some of the clearest and sometime the only audible frequencies from Austria, Holland, Russia, and Australia it is simply bad luck. Whilst the bandspread on the radio is clearly better than on sets like the discontinued Sony WA5000, the exclusion of the 22 metre band is going to severely limit the success of this radio, at least in Europe. What a shame!

To conclude, the Sony ICF-7601 is not as compact as the ICF-SW15 but it has much better shortwave coverage, richer audio and better bandspread. At US$120 the Sony ICF-7601 is worth the extra US$20.

Sony ICF-SW33

In November 1992 Sony launched its ICF-SW33 onto the European market, starting in Germany. It will appear in other markets in the course of 1993. Sony has been looking for a budget digital portable to fill the gap between the popular ICF-SW7600 digital receiver and the analogue ICF-7601. Some of the US$100 digital portables on the market these days have such noisy synthesizers that reception of weaker stations is impossible because of the background noise. In the US there is about a US$100 gap between the Sony analogue and digital portables.

The gap has now been closed with the introduction of the ICF-SW33. This set is smaller than the popular ICF-SW7600. It measures 93 by 165 by 29 mm and weighs in at 423 grams with the 3 penlight batteries installed. The radio uses just 35 mA at an average listening volume which is very economical for a set with a PLL synthesizer.

Coverage on most versions is MW, FM and shortwave split into 13 bands. FM is from 87.5-108 MHz except in the Far East where it extends down to 76 MHz. Mediumwave coverage is from 530 - 1710 kHz, thus including the band expansion in North America. Shortwave coverage is not continuous as on the more expensive ICF-SW7600. On the ICF-SW33 you can receive the portions 3700-4200, 4650-5150, 5800-6300, 6950-7450, 9375-10000, 11525-12150, 13375-14000, 14975-15600, 17475-18100, 21320-21950 and 25475-26100 kHz. That means that a few channels and the new 15m band are missing, but most of the band expansions at WARC 92 have been included. The FM coverage on headphones is in stereo. There is no stereo indicator light on the set and if the signal gets too weak the radio switches to mono reception automatically.

Tuning

The radio does not have a keypad. Tuning is made possible with two keys UP/DOWN. On FM the steps are 50 kHz apart, 1 kHz on shortwave and switchable between 9 or 10 kHz on mediumwave. The 1 kHz steps on shortwave mean that the radio is actually easier to fine tune than the ICF-SW7600. There is no Single Sideband capability though. You can jump around the shortwave bands more quickly by pressing two buttons at once. This allows you to find the lower frequency end of each broadcast band very quickly. In addition, there are 5 memory channels for SW, 5 for FM, and 5 for MW. If you want you can use these to jump around the dial even further. A further two channels are programmable so that the receiver can act as an alarm clock waking you with your favorite station (and waking a partner at a different time with a different station). There is a button marked SCAN which starts the radio scanning up the dial until it hits a strong signal. It then pauses for a couple of seconds before resuming its scanning. Pressing the button again cancels the scan function.

The LCD display is clear, showing the metre band as well as the frequency. It has a green backlight which lights for 20 seconds after pushing a small button on the top of the set. The ICF-SW33 also boasts a clock with the time of 38 cities in the memory. Having programmed the time in Amsterdam, you can then read off the time in Bangkok, Jeddah, Chicago etc. Of course the radio doesn't have any knowledge of summer time differences and the use of the radio in areas where there is a half-hour difference with UTC will mean all the city info will be wrong. This would be the case in India or in certain parts of Australia.

Fair-Good Performance

The case of the ICF-SW33 is made of sturdy plastic and the radio has no provision for an external antenna. Thus measurements of sensitivity can only be approximate. The dual conversion circuitry is clearly superior to competing single conversion designs from other manufacturers for around the same price. This is especially noticeable during night-time reception in Europe. There is one selectivity setting . . . the filter being 5 kHz wide (at -6 dB). We found identical performance to the ICF-SW-760C. It is adequate for strong and medium signals. But separating Radio Luxembourg on 6090 kHz and Bavarian Radio on 6085 kHz was difficult from a listening location in The Netherlands.

There is an attenuator on the back of the receiver which cuts signals down by around 20 dB. This is a bit too coarse on some parts of shortwave. But it is fine for mediumwave reception.

Conclusion

The new ICF-SW33 fills an important price gap and it is one of the best performers for its price. The tuning is not as easy as on a set with keypad tuning. But the fine tuning steps on shortwave make more accurate tuning possible. The synthesizer noise is reasonable . . . in fact better than on the early versions of the very expensive ICF-SW77. The audio quality from the small speaker is surprisingly rich and has good fidelity on a pair of hi-fi headphones. In Europe the receiver is priced at DM298 in Germany. It will probably retail for around US$160, but at press-time Sony US had not released a price.

Table-Top Receivers

Grundig Satellit 700

At the time of going to press in December 1992, Grundig is selling off its Satellit 500 receiver in the United States at a discount price; just under 400 dollars depending on the dealer. The replacement receiver, the Satellit 700, has now been introduced worldwide.

In Europe the 500 has already been replaced by production versions of the Satellit 700. Some pre-production models were reviewed at the end of 1991 with mixed results. There were some complaints that the first sets went into oscillation when the volume control was turned up full. Grundig said they were aware of these problems, fixed them, and then gave the OK to the factory in Portugal to start making sets for the shops.

At first glance the Satellit 700 looks similar to its predecessor, the 500. The styling of the case is the same, but the electronics inside are completely new. The case measures 30 by 17 by 7 cm and weighs 1.8 kilos without the 4 Size-D batteries which fit in the back. We think that puts the radio into the transportable category . . . . not for the traveler who likes to keep things light and compact. If you want a radio that will deliver a bit of volume while you're on holiday, let's say in a caravan, this is one of the radios you should consider.

Coverage

Like most portables, you can use this radio for medium, long, and shortwave reception, plus FM stereo. The set has only one speaker, so if you want stereo you have to use headphones or connect an additional matching speaker which is available as an optional extra. Longwave coverage is from 150-353 kHz on most versions, then there's a gap which is normally used by beacon stations. Mediumwave coverage starts at 528 kHz and goes to 1611 kHz where shortwave coverage takes over, and that's continuous until 30 MHz, slightly lower if you buy the set in Italy where there are some legal restrictions. FM is between 87.5 and 108 MHz, and the set claims to be the first portable with the Radio Data System, which means that the set displays the name of the FM station it is tuned to, at least in certain parts of Europe.

Tuning

It is easy to switch on the radio and do some simple tuning. But if you really want to discover the secrets behind the on-board computer you'll have to sit down for a couple of hours and read the instructions. A glance at the instruction book shows where the emphasis lies. There are 32 pages in total. Half a page tells you the frequency bands for broadcast listening, a half page on a few specifications, one and half pages are used to explain the connection of an external antenna and Single Sideband and half a page on fine tuning. The remaining 30 pages are needed to explain how to program the radio to do all kinds of scanning, copy memories, timer functions and so on. The German and Dutch versions of the instructions are much better than the English, which has clearly been translated by someone who hasn't a clue about shortwave terminology. That's a problem because you need a clear set of instructions to get to know this radio. It took us about three days to get the hang of it all. That's not a criticism as such . . . . if you want these programing features, they're all there. The Satellit 700 is clearly aimed at the serious shortwave hobbyist. At press-time it was selling at around US$580 in Europe, whereas US dealers quoted us a price of US$550.

Most of the development costs in the Satellit 700 have apparently gone into the design of the software. But let's look at performance of the radio itself as an instrument for picking up distant radio stations.

External Antenna

If you use this radio at home, chances are you'd try to connect some sort of external antenna to the radio, even if it's a piece of long wire dangling out of the window. The first thing you notice is that the internal ferrite rod antenna is always used for medium and longwave reception. You can't disable it, so you can't get rid of any in-house interference problems on medium and longwave, nor can you connect a directional loop. That we found surprising. When you switch in the external antenna, this becomes active above 1612 kHz.

We've often commented that receiver manufacturers tend to make portable radios that are so sensitive that once you attach an external antenna they can't handle the incoming energy and simply overload. As a result you get all sorts of whistles and stations popping up on parts of the dial where they are not really broadcasting. Measurements show that Grundig have solved this problem when you connect an external antenna to the 700. They've simply made the radio very insensitive! At 1.6 MHz you need 14 microvolts from the antenna to get a readable signal, at 4 MHz the figure is 3.7 microvolts, at 8 MHz sensitivity improves, 1.9 microvolts are needed, at 10 MHz 5 microvolts, and at 20 MHz 2.5 microvolts. These are figures for the AM wide position 20 dB signal to noise ratio with 60% modulation. In the narrow AM filter, the sensitivity improves very slightly. Compare these numbers like 14, 4 and 2 microvolts with a communications receiver where figures of one microvolt are usually the norm, and look at the wide variation in sensitivity the Satellit 700 exhibits across the shortwave dial.

Uneven Sensitivity

The reason for this anomaly in the sensitivity lies in a preselector circuit which has to be peaked for maximum sensitivity. The radio does this peaking for you as you tune around, but clearly it doesn't track with much accuracy. You can peak the preselector manually using the keyboard, which sometimes improves the sensitivity quite a bit, especially around 15 MHz. But having peaked the preselector manually, if you then turn the main tuning knob slightly, the radio switches back to the default preselector position and the extra sensitivity is lost. We couldn't find any way to tune across the bands with maximum sensitivity. Single Sideband results weren't that impressive either.

Sensitivity with the internal telescopic antenna is OK, better than one microvolt. That's because the signals picked up off the telescopic whip are put through a simple antenna preamplifier before being presented to the front end of the radio. Whilst the sensitivity goes up, the amplifier introduces all sorts of unwanted intermodulation products. Listening in Europe to the 20 metre ham band, around 14.2 MHz, you can hear all kinds of broadcast signals, and these are internally mixed signals from the 41 metre shortwave broadcast band. Not what you need for trying to dig weak stations out of the atmospheric noise!

The signal strength meter is simply a tuning guide. It doesn't correspond with real signal strengths because it doesn't start reacting until a very strong signal is heard.

Audio Distortion

Audio distortion is interesting to look at. In AM wide, the figure is an excellent 1.2%, in AM narrow the distortion is then 5%. When the synchronous detection mode is selected the distortion rises to 10%. On FM the distortion is very low. The internal speaker has a characteristic "German" sound when listening to shortwave. There's not much treble, but there's a lot of bass response. In fact the audio amplifier deliberately boosts the bass frequencies to get chest thumping results on FM stations. The bass tone control doesn't do much to remove this if you don't want it. On FM the Grundig Satellit 700 clearly outperforms both the Sony ICF2001D and the Sony ICF-SW77.

Software Bug!

The radio has an intelligent tuning software package, so you can tap in a frequency very quickly. There's one bug though. If you are listening to an SSB station on, let's say, 13630 kHz and you then type in 13640, the receiver assumes you want to switch back to the AM mode. You can avoid this by using the manual tuning knob instead. Fine tuning only works in the Single Sideband mode, allowing you to adjust the frequency plus or minus 150 Hz. The steps in standard AM are 1 kHz on shortwave, and unless you switch to SSB or AM synchronous detection, you can't tune finer than that. That said, if you plan to connect a radio teletype decoder to the receiver, fine tuning in SSB won't be a problem. The automatic gain control is fair, but it seems to start working on signals that are too weak.

Selectivity

Selectivity is the ability of the set to pick out the station you want, and reject the rest. The set gives you two bandwidth choices when listening to broadcast signals. In AM wide the bandwidth is 6.8 kHz at - 6dB, and in AM narrow 4 kHz at - 6dB. The shape factor of the filters is excellent. In practice the maximum selectivity is around 65 dB because the synthesizer in the receiver generates a fair bit of noise.

The synchronous detection system is an improvement over the Satellit 500, but, in our example, for some reason the detector is not symmetrical. The idea is that you tune in the station exactly. In Europe try tuning in the Bayerischer Rundfunk on 6085 kHz. Then press the "sync" option on the keyboard. Now you can turn the tuning knob slightly to listen to either the upper or lower sideband, and in theory avoid strong interference from Radio Luxembourg on nearby 6090 kHz. When the bands are crowded that is often a method to escape interference from a station operating on an adjacent frequency. We could tune down as much as 1 kHz on the lower sideband side before the set lost its lock, but only 700 Hz in upper sideband, which is not enough to be really useful.

Comparison with Lowe HF-150

The synthesiser remains locked on even the weakest signals, but if the signals are really weak then noise and whistles from the synthesiser itself become audible. A comparison with the newly launched Lowe HF-150 show that the sync detector on the Lowe is superior. The Lowe receiver doesn't have a lot of FM features on it though, and none of the memory tricks of the Grundig Satellit 700.

The dynamic range of the Satellit 700 turns out to be only 66 dB. This is OK for a portable, but not good when compared to the performance offered by communications receivers designed for real long distance listening.

Radio Data System Feature

We mentioned that the Satellit 700 has the Radio Data System (RDS) on it. But the full advantages of this new system of tuning haven't been utilised. As with all RDS sets you need an excellent stereo signal before the station name is displayed, so this isn't a new way to identify weak FM stations. You can ask the set to display alternative frequencies for the same network, but you have to store these manually into the memories. And it's a shame they didn't use the RDS clock information to give the radio a super accurate self correcting clock. Of course the RDS mode will only be of use if you live within range of transmitters that use the system..at present that means a select number of countries in Europe.

Pre-programmed Features

You can set a timer to record favorite programs while you're out if you have a suitable cassette recorder. The main frequencies of 15 major international broadcasters are already programmed in the radio, and you can put another 64 stations in yourself, each with up to 8 alternative frequencies. The radio remembers the mode, and bandwidth selection, but not the preselector setting. If that's not enough, you can lift a flap on the front of the set and install extra memory chips to increase the capacity to over 2048 stored frequencies. Some listeners' clubs in Germany are co-operating with Grundig and selling pre-programmed chips for a very reasonable price with the latest frequencies of broadcast and utility stations programmed in to them.

Power Consumption

The set consumes a lot of power. At comfortable listening volume the consumption is around 200 milliamps, so a set of 4 D cells will last 75 hours. You can also put in rechargeable cells and set the radio so that the cells charge up from the mains supply when the set is not in use. A fully charged set of nickel cadmium cells lasts about 20 hours, but it took the radio no less than 4 days before they were fully recharged! That's because the radio trickle-charges the batteries. Not very convenient if you happen to be on holiday and want to do a lot of listening.

Conclusions

It is difficult to summarise the Grundig Satellit 700 in a few words. It is a full feature radio with endless possibilities. Its performance though is only fair, and certainly for a few dollars more the Lowe HF-150 is a much better shortwave communications receiver. But the Lowe doesn't have FM, scanning functions, or RDS. The Satellit 700 is too complicated for the casual listener, and the performance will disappoint some people who expect a portable communications receiver. But if you're looking for extra features, there's nothing on the market at the same price of around US$550. It is up to you to decide whether these features are worth the price. Side by side against the ICF-SW77 from Sony, we conclude that the Grundig gives much better performance on FM, but the Sony gives better results on shortwave above 5 MHz. US readers interested in shortwave reception on a table-top might also want to consider the new Yaesu FRG-100, retailing at US$599. The Yaesu has considerably better shortwave performance, but no FM and far less memory capabilities. More information from Grundig, Kurgartenstrasse 37, Fuerth/Bayern D8510, Germany. Tel: +49 911 7030. In the USA, Grundig has a distributor. This is Lextronix Inc, 3520 Haven Ave. Unit L , Redwood City CA 94063, USA. Tel: 415 361 1611 or in the USA only call (800) 877 2228. In Canada call (800) 637 1648.

Lowe HF-150

Introduction

Holland and Britain seem to be the main marketplace for Lowe receivers so far. This English company started making radios to its own design and specification a couple of years back, with the launch of the HF-125 and HF-225 receivers. The HF-125 has since been discontinued. The approach has been to try and make a simple to operate radio with the best specifications for the price. Compared to Japanese competition the radios look quite plain, but performance wise they score well. At the end of 1991 Lowe announced it was launching a small receiver for the bottom end of the market as an alternative to Japanese push button portables . . . . sets like the Sony ICF2001D. In March 1992 we tested an off-the-shelf example of the HF-150, putting it through a series of laboratory and practical listening tests.

For a price of £329 in Britain (including VAT), you get a table-top communications receiver which at first glance looks surprisingly small. In the United States, Electronic Equipment Bank of Vienna Virginia (Tel: (800) 368 3270) carry the receiver for US$600. The case is made of metal, not plastic, and measures just 185 by 80 by 160 millimetres. It's quite light too, just 1300 grams without the 8 penlight batteries which fit into two special holders at the back of the set.

Simple Controls

From the front there are just 5 controls . . . . a combined on-off switch and volume control, three buttons which have several functions including the selection of the mode and memories, and a large tuning knob. A large 5 digit liquid crystal display shows the frequency you're tuned to within the nearest kiloHertz, If you push a button the display gives you information about the receiver mode and memory number, but normally it shows only the frequency, and there's no light to illuminate it. That's it. Lowe sell a keypad as an optional extra that plugs into the back of the set and you place it in front of the radio as you use it. That's essential if you want to move quickly about the dial . . . otherwise you have to move up and down in frequency by spinning the tuning knob. Getting from 30 kHz right up to 30 MHz, which represents the full coverage of the set, could take some time.

The set has no signal strength meter, you can't add extra filters at a later stage for very narrow bandwidth reception of Morse code, there's no notch filter, no noise blanker, and no tone control. But if these are extras that you can manage without, then what Lowe have put inside the box turns out to be **very** acceptable indeed.

Performance

The dual-conversion superheterodyne design is quite straight forward. Signals come in from an external antenna. You can switch in an antenna amplifier if you're using an indoor whip, although in practice we didn't need that at all in our part of Europe where signals are always strong. Signals go through a 30 MHz low pass filter before they hit the mixing stage of the radio. Of course there's a lot of energy coming off most shortwave antennas, bearing in mind the powers used by broadcast stations. More expensive radios use a series of filters to make sure that if you're listening to 15 MHz shortwave for instance, strong mediumwave signals, or stations in the 41 and 49 metre band are attenuated before they get to the mixing stage of the radio. Too much energy at the front end of the sensitive input circuitry can lead to overloading, and the appearance of signals on the dial which are the result of mixing products inside the radio. Having said that we measured the intercept point as +3 dBm using two signals 30 kHz apart. This gives you a dynamic range of 86 dB which is a fair-to-good value for a radio of this price. We disagree with the instruction book though, that recommends a long wire of up to 30 metres. Our tests in Holland showed that if you connect a wire longer than about 12 metres, you get enormous overloading problems once the sun sets. That will be less of a problem in low signal strength areas such as the Pacific or the American Mid-West.

Attenuation Tricks

It's often assumed that the more signal you pump into a radio, the more distant stations you'll be able to hear. Well that's not the case. We found that late at night, weak and difficult signals were more intelligible if you switch in the 20 dB of attenuation. But that control is on the back of the set which is not easy to get at. If you use the set in Europe you might want to consider a separate aerial attenuator with, say, steps of 6, 12, 20 dB of attenuation, and give it a try on weak signals.

Sensitivity & Modes

We measured sensitivity using a signal modulated at 60% using a 1 kHz tone. We found that our measurements corresponded well with the results given by Lowe in their instruction book. Between 50 and 500 kHz the sensitivity is around 1.8 micro volts, and around 0.8 micro volts for the medium and shortwave part of the dial. There is quite a difference in sensitivity between the wide and narrow filters used in the HF-150.

The radio has various modes: USB, LSB, standard AM, and you can also use what's termed synchronous AM. Unlike other Lowe sets available until now, the HF-150 allows you to listen to either the upper or lower sideband of a broadcast signal whilst in the "sync" mode. That's extremely useful when there's a strong interfering station 5 kHz away from the station you're trying to listen to. You can also use synchronous detection to reduce at least some of the effects of shortwave fading. The use of synchronous detection though in the double sideband mode leads to some slight loss of sensitivity, but that's nothing to be concerned about. The background noise also rises slightly on the example we've tested. The radio takes up to two seconds to lock onto the desired signal, but once it's locked the radio does an excellent job of keeping in lock even when the signal fades to almost nothing.

High Battery Drain

Battery consumption of the receiver is quite high, especially when compared to similar priced competition, anything up to 275 mA at full volume. We put in a set of 8 fresh alkaline batteries and got the radio to work for just 6 hours before they were flat. You can purchase rechargeable nickel cadmium batteries. When the set is switched off they automatically charge up. It takes about 16 hours to get a full charge after which you can use the radio for portable work for about 3 hours before you need to recharge again. The cheapest solution of all is simply to use the supplied external AC adapter which plugs into the back of the radio and gives all the power you need without any hum problems. Be careful if you use alkaline batteries and the mains adapter. Although there is a warning in small print in the instruction booklet, if you use the mains adapter while the alkalines are still installed, the HF-150 will try to charge these up when you switch off the receiver. **Since alkalines are not designed for this they start to heat up and there is a danger that they might explode. We don't think that Lowe explain this clearly enough (there should be a sticker in the battery compartment!).**

The HF-150 has two filters which have a bandwidth of 7 and 2.9 kHz respectively. These ceramic filters have a good shape factor for the price paid. So if the signal you want to listen to is strong you can really sit back and listen to the programming. The design of the automatic gain control is excellent, so there is no unwanted pumping of signals. The signal distortion is very low for a radio of this type, and if you connect the radio to a hi-fi set you'd be surprised what fidelity you can get out of a strong shortwave broadcaster.

Summary

In short, the Lowe HF-150 is an excellent choice as an entry level communications receiver. It gives much better performance than sets like the Trio Kenwood R-1000 which were on the market 10 years ago for the same price, showing that it is still possible to improve on performance and keep the costs reasonable. Lowe is to introduce another receiver in early 1993 . . . the HF-250. Production models of that set were not available when we went to press in December 1992. Further information from: Lowe Electronics, Chesterfield Road, Matlock, Derbyshire, DE4 5LE England. Tel +44 629 580800. Fax: +44 629 580020.

Yaesu FRG-100

Introduction

The FRG-100 is a new neat compact budget-priced receiver, equally suited for use in portable situations (listening expeditions) as well as at home in the shack. It was launched in Japan in August 1992, and started appearing in Europe and North America as this edition of the WRTH went to press. The set measures 238 by 93 by 243 mm and weighs around 3 kg, but note that there is no built-in power supply for the radio. The radio comes complete with a AC mains power adapter, although it can also be operated from a car battery (it takes around 1100 mA at a comfortable listening volume).

Coverage on our version was 130 kHz-30,000 kHz continuously. By holding the FM and SSB mode buttons and then switching on, coverage is extended down to 50 kHz, allowing reception of certain time signal stations and navigation beacons. The radio can receive signals in four modes as standard: AM, LSB, USB, and CW. An option exists for narrow-band FM.

Front panel

At first glance, the front panel of the FRG-100 is deceptively simple. The manual tuning knob dominates the right hand side of the panel. An analogue signal strength meter and LCD frequency/mode/memory/time display are arranged along the top. Two rows of mode and select buttons are placed in the middle of the panel. Volume, squelch and memory channel controls are of the conventional rotary kind. However, by pressing certain buttons before switching the radio on it is possible to change a lot of default settings. Some of these are cosmetic (whether the clock shows 12 or 24 hr display). But other defaults which control the scanning or the sorting of the 52 channel memory are useful additions to the receiver's functionality.

By pressing and holding the SET button and then selecting other buttons on the front panel you can customise tuning steps, vary the beat note on CW between 400 and 700 Hz and select which of the bandwidth filters will be a default when each mode is selected. Fine tuning of the first local oscillator is also possible which is useful if you want to check the receiver's calibration against a know frequency standard such as WWV on 15 MHz. If you find that when you try to listen to AM stations in the SSB mode (so called ECSS) and find that the receiver never exactly beats when the display shows the exact published frequency, this adjustment may also be useful. The radio displays by how much the local oscillator has been adjusted. Having set the offset the receiver stores this information. There is a danger that if you forget about this you may conclude that the receiver is permanently off-frequency.

Direct frequency tuning

The FRG-100 uses a combination of UP-DOWN keys and the manual rotatory tuning knob to tune the radio. There is no direct entry keypad, although the radio does have a connection on the back of the receiver for computer control (Yaesu calls it Computer Aided Tuning or CAT). It seems strange that Yaesu hasn't thought of an optional keypad like the British company Lowe did with their receivers since the UP-DOWN buttons are not as fast as a keypad. That's especially noticeable when you hear an unknown signal, say, on 5955 kHz and want to check whether it is carrying the same programme as a known station on 21455 kHz. A metal stand folds out from underneath the radio to put the front panel off the table and operate the receiver panel at a slight angle. That makes operation much easier.

The radio does not have synchronous detection, so if you are listening to a broadcast station there is no way the receiver will lock on to the carrier frequency. However the receiver is remarkably stable and listening to AM stations in either LSB and USB will allow you to escape annoying heterodynes. The fact that the receiver tunes in 10 Hz steps in the SSB mode means that music is listenable in the SSB mode when properly tuned in. After 10 minutes of listening the pitch of the music was still acceptable. There is an optional crystal which can be installed to improve the receiver's stability. In practice this option isn't needed for broadcast listening.

Memory channels

The radio can store up to 50 stations in memory, including the frequency and the mode of each. Storing frequencies is easy, and there is a useful feature which allows you to re-sort the stored frequencies into ascending order. You can review the contents of each memory channel by turning the memory control knob. If the memories are only partially filled, review of only the active channels is also possible. The memories are fully tuneable. This means that if you have stored 15070 in the memory and select it, you can then start tuning up or down from that frequency. We found this was a useful quick-tuning feature for jumping between broadcast bands by storing a marker frequency in the memory system. An additional two channels in the memory are used for marking purposes.

Scanning of the stored frequencies is versatile. You can use the scan function to keep watch on a particular frequency. The radio checks the chosen channel for activity every 5 seconds. You can be tuning another part of the dial while still keeping an ear out for any activity. The FRG-100 allows you to scan all the 50 channels. Or it is possible to group memories into 5 groups of 10, each memory then being given an alpha-numeric indication. Several types of scans are possible if you decide to divide the memory into 5 groups of 10 channels each.

Yaesu has been particularly keen to provide the user with a number of options to select things like the BFO offset. The manual explains this in detail assuming you are sitting in front of the receiver. Some of the detail may be confusing for the first-time user and Yaesu may have been better off writing a "Short-Cut Guide" for those owners who just want to get started without changing all the defaults on the first day. For some specialist listeners the fact that you can decide to always select the 4 kHz filter instead of the wider 6 kHz filter when AM is selected is a useful feature, especially since the 6 kHz filter is too wide.

The receiver has two built-in clocks, so one can be set to local time the other to UTC. The clocks are independent of each other, so the difference between UTC and local time does not have to be a full hour. A timer function is available although this only switches the radio on and off at a chosen time. It is not as sophisticated as the timer on the Sony ICF-SW77 which allows you to switch the radio on to a chosen frequency in the memory. You can enhance the receiver's flexibility by using computer control.

Sensitivity of FRG-100 using AM below 5 MHz

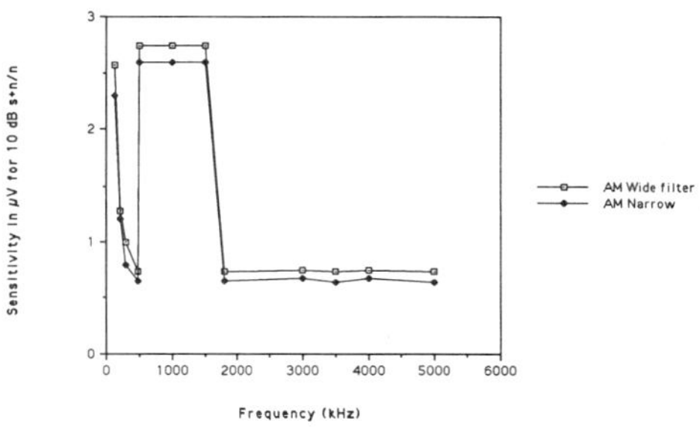

Sensitivity

The sensitivity graph shows that the FRG-100 is rather insensitive at the low end of its coverage range. Sensitivity on mediumwave is deliberately attenuated to prevent a strong local station from overloading the sensitive input circuitry. The figures for shortwave show the set is extremely sensitive . . . in our opinion it is too sensitive in the region 1.8-17 MHz. Above 17 MHz there is not so much atmospheric noise. But connect a reasonable antenna to the FRG-100 and you find the level of atmospheric noise is around 3-4 microvolts. Making the receiver so sensitive brings the overload threshold down as well. For many years we have been trying to destroy the common belief among shortwave listeners that "the more sensitive the set, the more stations it can pick up". If the front end circuitry is so sensitive that intermodulation products become a problem, you will improve the intelligibility of weak signals by switching in a signal attenuator (the FRG-100 has three possible settings, plus the off position) and turning up the volume control!

Sensitivity values for 10 dB s+n/n, measured at 60% modulation depth.

Freq. Range	AM-wide	AM-narrow	SSB/CW wide	CW narrow (MHz)
0.131	2.58	2.30	0.51	0.44
0.200	1.28	1.20	0.23	0.18
0.300	1.00	0.80	0.20	0.17
0.490	0.73	0.65	0.16	0.06
0.5-1.8	2.75	2.60	—	—
1.8-17	0.73	0.65	0.17	0.10
17-30	0.52	0.50	0.17	0.10

A signal is only really intelligible when the (signal+noise) to noise ratio is 20 dB. We examined this using a signal generator modulated at 60% to simulate the sort of signals that shortwave broadcast stations put out (especially with audio processing units like Optimod-HF in use at many transmitter sites). At 10.1 MHz we got the following readings.

Signal needed for 20 dB s+n/n, 10.1 MHz, @ 60% modulation

Filter	Signal mV	dBm
AM wide	3.23	−96.8
AM narrow	3.05	−97.3
SSB/CW wide	0.68	−110.4

Automatic Gain Control

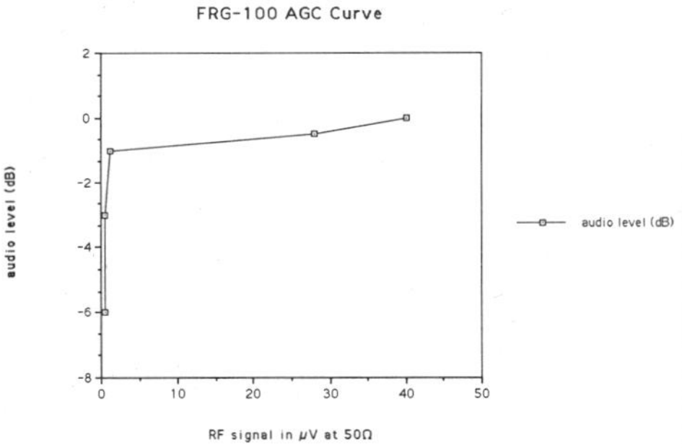

The automatic gain control design on the FRG-100 is excellent, although there doesn't seem a way to disable it altogether. Listening to SSB signals it is often more pleasant to switch off the AGC and wind the RF gain control back to avoid distortion. The gain control on the FRG-100 is such that any intelligible signal is adjusted to give the same audio level output.

RF in @ 50Ω (µV)	Audio level (dB)	s/n ratio (dB)
0.44	–6	6
0.59	–3	9
1.25	–1	15
28	–0.5	—

"S" Meter
The "S" meter is fairly accurate. Under 1 mV (S2) the meter gives an indication that is too low. Above 1.6mV (S4) the meter reads exactly one "S" point too much. The range of the meter is good and for a radio in this price category does a useful job as a tuning indicator. It is linear over a large part of the range.

Squelch
The squelch is useful when you're waiting for a signal to appear and want to silence the background noise until that signal comes on the air. The function is essential on scanners, less so on shortwave radios. The squelch range on the FRG-100 is disappointing. The rotary control moves one complete circle, starting at the "6 o'clock" position. In the range between ½ and ¾ shut the squelch level rises from 2 µV to 9000 µV signal input. The control is so coarse that setting it accurately becomes impossible. The squelch is designed to compensate for fading. It switches in at a signal level of 2 µV when set to the "12 o'clock position" and switches out again when the signal drops to 0.6 µV.

Audio Quality
There is no way to reduce the bass response on the output to either loudspeaker or headphones. In some cases the bass response can overload a pair of hi-fi headphones. The internal speaker is OK since it has a low bass response anyway. The optional Yaesu communications speaker does a better job than the built-in speaker.

The audio amplifier delivers 1.8 watts maximum with 10% distortion. At normal listening levels (100 mW output into 8 at 1 kHz) the distortion on the AM wide position was measured at 5%. That's acceptable, although the distortion on SSB and CW is just 0.3%. That means that distortion on AM is a result of shortcomings in the detector stage rather than the audio amplifier. However this is to be expected in a receiver of this price range. Overall the audio is rated at fair/good. Intelligibility is not affected by the audio distortion.

Dynamic Selectivity/ RF protection ratio.
Manufacturers often quote their selectivity figures based on measurements of the filters before they are installed in the receiver. However it is possible with the compact design of many communications receivers that signals can "leak" round the filters in such a way that the selectivity is reduced. There is therefore an internationally accepted way to measure dynamic selectivity (adopted by the European PTT organisations to measure maritime receivers). We use the so called CEPT method for all the selectivity measurements in the WRTH. This method also ensures that the effects of the synthesizer are also brought into the discussion. The procedure is simple. The FRG-100 is tuned to receive a weak signal which is just understandable, i.e. 20 dB s+n/n. A second modulated signal is applied to the receiver's input at a number of frequencies close to the original signal and a note is made of how strong this interfering signal has to be before the original signal is moderately disturbed (i.e. 14 dB).

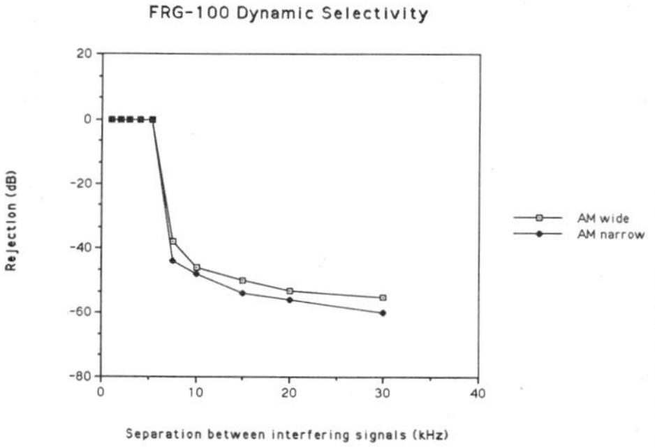

The curves show that if you are listening to a station on the FRG-100 using the AM wide filter, an adjacent station 5 kHz away with the same signal strength will cause annoying interference problems. The narrower AM filter is a bit better, but not by much. In practice the AM narrow filter is the best for listening to weaker broadcast stations. That the maximum rejection of the AM filters is around 50 dB indicates the limitations of the IF amplifier design. More expensive receivers are able to achieve 60-70 dB rejection. But for a receiver in this price class the results are very acceptable.

Attenuator
In Europe we found that the attenuator on the FRG-100 was not a luxury. It has three positions marked 6, 12, and 18 dB. In practice we measured these as 5, 10.4 and 15.4 dB. Nothing wrong there! For listening in Europe we suggest that on the shortwave range between 1.8-17 MHz the 18 dB position is used, whereas the 12 dB is the best above 17 MHz. The 6 dB attenuator is best when listening to mediumwave since the sensitivity is deliberately down on this section of the dial anyway.

Blocking
The blocking level of the FRG-100 is 3.2 mV. That means that an interfering station 100 kHz away from the signal being listened to only has to be 3.2 mV before interference problems arise. This a fair level. We measured the 3rd order intercept point using both AM and SSB filters, and two signals 30 kHz apart. The result was a figure of +3 dBm, a good figure.

Birdies
Every receiver has some internally generated signals. These can interfere with actually received signals when the frequencies co-incide and the level is greater than about 1 μV. We noted a few annoying birdies on the FRG-100, namely at 3070, 4630, 6144, 9218, 12288, 14977, 15280 and 22694 kHz.

Display Buzz
The LCD display in the first production run produced too much noise especially noticeable on the higher shortwave frequencies. Reception on 24859, 24777, 22622, 9415, 8758, 7674 and 4628 kHz was impossible due to the buzz from the frequency display. The level of the interference was reduced when the display was dimmed. Yaesu say they have introduced modifications on subsequent production runs to reduce this problem.

Conclusions
Yaesu say the aim was to produce a better shortwave listeners' receiver for about two thirds the price of the old FRG-8800. They have succeeded very well. The FRG-100 is better than the Lowe HF-150 and offers more features. But it is also more expensive. It does not reach the standard set by the Kenwood R-5000, but then it is cheaper. The price of the FRG-100 varies depending on the country. At press-time we were quoted US$599, UK£475, DM1299 in the USA, UK and Germany respectively. The FRG-100 outperforms older receivers like the Kenwood R-2000 and yet costs much less. Yaesu are therefore to be congratulated for bringing better technology into the mid-range price bracket. This is important as there has been a growing price gap between portable and semi-professional sets.

More information from Yaesu Musen Co.Ltd, C.P.O. Box 1500, Tokyo Japan. Yaesu USA 17210 Edwards Road, Cerritos, California 90701 USA. Yaesu Europe, Snipweg 3, NL-1118 AA Schiphol, The Netherlands. Tel: +31 20 6037299. Fax: +31 20 6480445.

Antenna Update

MLB Scandal: Inferior Copy
In early November 92 we started getting phone calls from people who had seen an advertisement from Palomar Engineers placed in a mail order catalogue. The advert describes a Magnetic Longwire Balun and uses very similar wording to an advertisement for a product with the same name manufactured by RF Systems published in the 1992 WRTH. A spokesman for Doeven Electronics in Holland, which owns the MLB patent, told us that the name appears to have been copied by Palomar without any explanation. The Palomar version is electrically inferior (although it is cheaper) and has a more limited frequency coverage. Palomar have built up an excellent reputation for antenna designs in the past (e.g. their Super Snooper antennas) so this development is rather surprising. Many manufacturers offer the customer similar products but not using the same name, and even the same ordering code. *Caveat Emptor!*

Lowe PR-150

In mid-August 1992 we tested the PR-150 preselector designed to match the HF-150. This will be available in January 1993 at a price of around £160. Overall results were disappointing. Although the unit reduced some of the second order intermodulation products it introduced its own third order problems. The PR-150 was handy when tuning the tropical bands, but offered no better, or even worse performance when listening to broadcast stations around 15 MHz. The set has a one step attenuator (-16 dB) right across 150-30000 kHz. The pre-amp offers an average boost of 7-9 dB depending on the section of the band being tuned.

If you can ensure that the front end of the radio only receives frequencies on or around the part of the dial of interest, then the problems of intermodulation can be reduced. For example, on simple receivers you often hear a lot of extra noise around 14 and 15 MHz which results from mixing products from the 7 MHz broadcast band. The preselector is fine for reducing this. But its own internal circuitry causes mixing products to appear from nearby channels. We measured the third order intercept point as -5dBm using signals at 8 and 10 MHz. If you're using a high quality active antenna such as a DX-7 from RF Systems (third order = 24 dBm) then putting the preselector between the receiver and such an antenna will actually make things worse when listening above around 12 MHz. The PR-150 has a definite advantage if you want to use the HF-150 for monitoring weather fax stations on VLF frequencies.

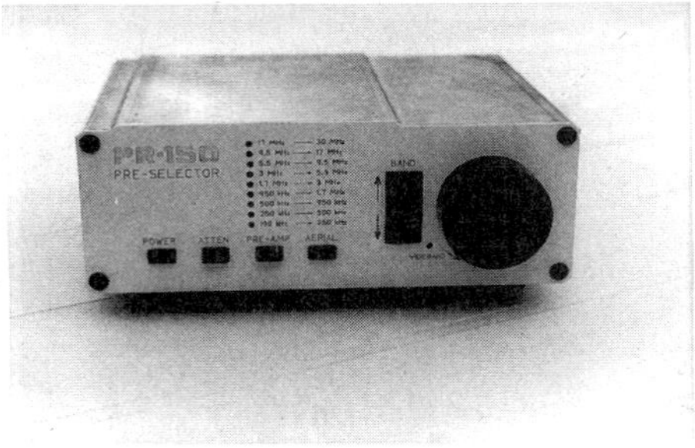

PR-150 Bandwidth

The preselector divides the bands into 9 distinct sections. Lowe are quite conservative in what these cover, thus ensuring that there are no gaps. Our measurements gave the following bands:

Band 1—	13.3-31.7	Band 6—	0.79-1.81
Band 2—	7.4-17.5	Band 7—	0.43-0.98
Band 3—	4.26-10.2	Band 8—	0.2-0.49
Band 4—	2.42-6.0	Band 9—	0.09-0.2
Band 5—	1.44-3.25		

With the exception of the first band we took the middle frequency of each range and measured the bandwidth at that point. Note how the bandwidth gets much wider as the frequency rises.

Bandwidth PR-150

Frequency (kHz)	−6dB point	−20 dB point	−40 dB point
22000	2.5 MHz	6 MHz	—
12500	2.0 MHz	5 MHz	—
7100	1.0 MHz	3 MHz	—
4100	400 kHz	900 kHz	> 2 MHz
2300	300 kHz	650 kHz	2 MHz
1300	125 kHz	300 kHz	1 MHz
700	40 kHz	120 kHz	380 kHz
350	18 kHz	45 kHz	140 kHz
156	10 kHz	20 kHz	55 kHz

RF Systems DX-7 Active Antenna

Launched in August 1992, the DX-7 is a budget version of the famous DX-1 active antenna (tested in WRTH 1988 and the 1993 WRTH Equipment Buyers Guide). It has slightly reduced performance characteristics, but it is also half the price of the DX-1.

The antenna is also far more compact, ideal for someone with limited space for an external antenna. Since the antenna is active it should be mounted on a metal pole outside the interference field of the house (at least 5 metres away from the building). The "coil" design is unique, enabling 2.3 metres of antenna to be wrapped round a sealed polyurethane pipe 50 cm long. The unit requires an indoor power supply. The DX-7 comes with a standard DC connection for a car battery (for portable use). An RF isolated power supply is also available as an optional extra, this operating on any AC power supply between 100-250 volts. Power is fed to the antenna through the coaxial cable (not supplied) and this may be anything up to 50 metres in length. The electronics for the antenna are housed inside the tube which is fully sealed from the outside elements. In order to avoid static build-up during thunderstorms we mounted the DX-7 on a grounded metal pipe (up to 40 mm in diameter is possible), threading the coaxial through the pipe.

The DX-7 produces noticeably less output than the DX-1 or the Dressler ARA-60. However, in comparison to the ARA-60, the intermodulation level is also lower..it is a medium efficiency antenna. The intercept point of our example was +24 dBm, lower than the ARA-60 and DX-1. However it was a quieter antenna than the ARA-60 resulting in less distortion on very weak signals which in turn improves the intelligibility. It is also considerably more compact. The antenna costs .399 in Holland, half the price of the DX-1. More details from: Doeven Elektronics, Schutstraat 58, 7901 EE Hoogeveen, The Netherlands. Tel: +31 5280 69679. Fax: +31 5280 72221. In the US, Universal SW in Reynoldsburg, Ohio carries RF Systems products. Lowe Electronics distributes them in the UK.

Rohde & Schwarz Active Antenna HE011

This German company has an excellent reputation in the professional market place for high-quality receivers. However, the equipment is well beyond the budget of the enthusiast. In August 1992, Rohde & Schwarz announced an active antenna designed for the enthusiast. The HE011 is designed to operate in the region 50 kHz - 30 MHz, although the performance does not degrade too much until 200 MHz. The unit costs in the region DM800 and claims an intercept point of 28 dBm. In September 1992 we tried to order one for testing for the 1993 WRTH. On November 16th there was still no word from Rohde and Schwarz as to a possible delivery date. However, we intend to pursue this active antenna since if it measures up to the specifications sheet it could be an interesting compact solution for many listeners.

THE 1993 WRTH RECEIVER VALUE GUIDE

KEY: The more stars the better, i.e. ***** Excellent **** Good *** Fair ** Poor * Avoid!

Manufacturer & Model Number		Size	Weight	Sensitivity	Selectivity	Dynamic Range	Overall Value
Budget & Travel Portables							
Dak	MR-101s	M	462	****	**	**	***(see text)
Dak	DMR 3000	M	814	***	**	**	**(see text)
Grundig	Explorer II	M	1500	***	*	*	*
Grundig	Yacht Boy 204	M	480	****	***	***	****
Grundig	Yacht Boy 222	S	398	****	**	**	**
Lowe	SRX-50	S	420	****	***	***	***(see text)
Panasonic	RFB-45	M	720	***	***	***	****
Panasonic	RFB-65L	M	775	***	***	***	***
Philips	AE-3625	M	527	****	**	*	** (see text)
Sangean	ATS-808	S	620	****	****	***	*****
Sangean	ATS-803A	L	1588	****	****	***	****
Sony	ICF-7601	M	570	****	***	***	****
Sony	ICF-7600DA (rare)	M	600	****	***	***	*
Sony	SW1-S	S	230	****	****	***	***
Sony	ICF-SW15	S	380	****	***	***	* (see text)
Sony	ICF-SW33	S	423	****	****	****	****
Sony	ICF-SW800	M	660	****	***	***	***
Sony	ICF-SW7600	M	640	***	***	***	****
Serious Shortwave Listener + Semi Professional Receivers							
Drake	R8	L	5897	****	****	****	****(cf WRTH 91)
Grundig	Satellit 700	L	1800	***	***	***	***(see text)
Icom	R-1	S	298	****	*	*	*
Icom	R-100	M	1400	****	**	*	*
Icom	ICR 71E/A/D	L	7500	****	****	****	*****
Icom	R-72E	L	4800	****	***	****	****
Icom	ICR-9000	L	20000	*****	*****	*****	*****
JRC	NRD 535	L	9070	*****	*****	*****	*****
Kenwood	R 2000	L	5488	****	***	***	***
Kenwood	R 5000	L	5600	*****	****	****	*****
Lowe	HF-150	M	1295	****	***	****	*****
Lowe	HF-225	M	1896	****	****	****	*****
Pan Crusader	8000	L	2400	****	**	**	*
Sony	ICF-SW55	M	900	****	****	***	*****
Sony	ICF-SW77	L	1470	****	****	****	****
Sony	ICF-2001D	L	1800	****	****	***	****
Sony	PRO-80/70	M	670	****	***	***	***
Sony	ICF-V21	L	9500	*****	****	***	*
Yaesu-Musen	FRG-100	L	4700	****	***(see text)	****	*****

Notes: Weight is given in grams. The overall value is not just a sum of other columns. We estimate value by comparing performance against average market prices in Europe & North America. Therefore, value ratings may vary from 1991 table. S = small, easily portable, M= Medium size, OK for larger suitcase, L = Large, for table top use only. (rare) means that radio is now getting difficult to find in some markets.

WRTH Receiver Preview 1993

1993 should be an interesting year as far as new portables are concerned. Grundig tell us that the award-winning Yacht Boy 204 is to be replaced with the Yacht Boy 205. To be launched in Germany for the price of DM 99 (i.e. the same price as the Yacht Boy 204) the radio is more compact. It will still be a single-conversion design.

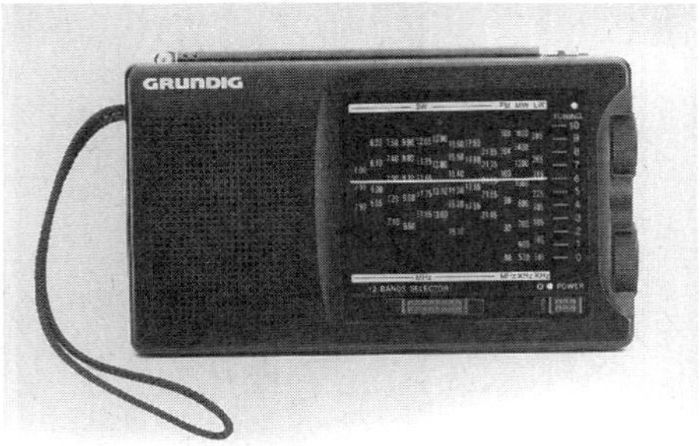

But that will mean the tropical bands coverage will be scrapped, and the internal speaker made smaller. The Yacht Boy 204 had an unusually good speaker so it remains to be seen how the Yacht Boy 205 will perform. More exciting is the extension of the Yacht Boy range to include a PLL synthesized dual conversion receiver. The new Yacht Boy 500 (not to be confused with the discontinued Satellit 500) will offer a compact palm-size radio with a wide range of features. These include USB/LSB and the Radio Data System (only applicable to RDS equipped FM stations in Europe). The power amplifier also has a bass boost system through the 77mm built-in speaker. Tuning is by UP-DOWN buttons or direct keyboard access. The shortwave coverage is complete from 1.6-30 MHz. Some models have longwave. Mediumwave and FM (stereo via headphones) is also included. The LCD display shows the name of a station (up to 8 digits or letters). The set has a capacity of up to 40 stations.

At a price of around DM475 (= US$ 235) in Europe, the radio looks like a very promising addition to the market. Time will tell if the radio performance has to suffer to pay for all the extra features. Launch of the Yacht Boy 500 is expected around April 1993. It will also be sold in North America, although probably at a considerably higher price than in Europe.

Grove SW-100

Originally scheduled for launch in August 1992 this receiver was not available for purchase when we went to press. A message from Bill Grove reported that last minute development of the receiver has delayed the launch until mid-1993 because one of the design engineers had to go into hospital. A hand built example of the Grove SW-100 was on show at the Dayton Hamvention in April of 1992, but this was a non-working mock-up. If the specifications sheet is anything to go by the radio should be an interesting receiver going head to head with the Drake R8 and Lowe HF-250, although more expensive than the Yaesu FRG-100. But all we have at the moment is a photograph.

Lowe Launch Expected for HF-250

The British company of Lowe is expected to ship their new HF-250 in early 1993. This will be based on the successful Lowe HF-150, but contain considerably more features for the serious shortwave listener.
The styling of the radio has also been altered to give it a much more professional look. Ultimately the radio will replace the HF-225.

Compact Trend from Taiwan

Mass production of a new radio from Sangean will start in January 1993. The new ATS-606 is a compacted version of the successful ATS-808. In Europe, the Siemens company will call it the RK-659. It will no doubt also appear through Radio Shack in North America and Roberts Radio in the UK although type numbers had not been announced at press time. The receiver weighs 300 grams without the 3 AA batteries and measures 148 by 89 by 30 mm. Shortwave coverage is continuous. The pre-WARC 1992 bands are easy to access. 45 memories are offered, 9 each for SW, MW, LW and FM (stereo on headphones). Dual clocks have been added to allow independent tracking of local time and a second time zone (e.g. UTC).
What else can we expect in 1993? It is time Kenwood did sometime to replace the R-2000 which is still in their lineup but is now getting long in the tooth. And then, of course, the Grove SW-100 is still to appear. 1993 could be an interesting year!

CREDITS:
"Listen to the World" was compiled from contributions by Richard Dixon, Mathias Kropf, O.G. Villard Jr., Dave Rosenthal, Rob Ulrich, Bart Kuperus, Lou Josephs, and Rocus de Joode. Technical measurements were made by Willem Bos of Universal Micro-Electronics. Additional thanks go to David Ward, Marian Meeuwissen, Fred Haanebeek, Waldemar Kraemer, and Dr. Kim Andrew Elliott for their background research. Section co-ordinator: Jonathan Marks.

WORLD RADIO & TV BY COUNTRY

World Radio & TV By Country *(Cont'd)*

DIRECTORY OF ADVERTISERS

ADVERTISING REPRESENTATIVES

Europe

J.M. Frost, World Radio TV Handbook, Søliljevej 44, DK 2650 Hvidovre, Tel: +45 (31) 78 85 08, Cable: WORATEL

Marian Meeuwissen, World Radio TV Handbook, P.O. Box 9027, 1006 AA Amsterdam, The Netherlands, Tel: +31 (20) 6691911, Fax: +31 (20) 6691941

U.K.

BPI Europe, 23 Ridgmount Street, London WC1E 7AH, Tel: +44 (71) 323-6686, Fax: +44 (71) 323-2314, Telex: 851-262100 (Billb G)

USA and Canada

Glenn Heffernan, Billboard Publications Inc., 1515 Broadway, New York, NY 10036, Tel: +1 (212) 536-5129, Cable: Billboard New York, Telex: 620623 (Billboy New York), Fax: +1 (212) 536-5359

Japan

BPI Communications, Izumi Mansion, Suite 802, 1-44-1 Tomigaya, Shibuya-Ku, Tokyo 150, Japan, Tel: +81 (03) 3498 4641